McGraw-Hill's

NATIONAL ELECTRICAL CODE® 2017 HANDBOOK

Twenty-Ninth Edition

Based on the Current 2017
National Electrical Code®

by

Frederic P. Hartwell

Joseph F. McPartland

and

Brian J. McPartland

Mc Graw Hill Education

New York Chicago San Francisco Athens
London Madrid Mexico City Milan
New Delhi Singapore Sydney Toronto

Cataloging-in-Publication Data is on file with the Library of Congress.

McGraw-Hill books are available at special quantity discounts to use as premiums and sales promotions, or for use in corporate training programs. To contact a representative please visit the Contact Us page at www.mhprofessional.com.

McGraw-Hill's National Electrical Code® 2017 Handbook, Twenty-Ninth Edition

This 29th edition is based on the 2017 NEC®.

5 6 7 8 9 LCR 21 20 19

ISBN 978-1-259-58442-8
MHID 1-259-58442-9

The pages within this book were printed on acid-free paper.

Sponsoring Editor
Michael McCabe

Editorial Supervisor
Donna M. Martone

Acquisitions Coordinator
Lauren Rogers

Project Manager
Anju Joshi,
Cenveo® Publisher Services

Copy Editor
Julie Searls

Indexer
Robert Swanson

Production Supervisor
Pamela A. Pelton

Composition
Cenveo Publisher Services

Art Director, Cover
Jeff Weeks

Contents

Article	Page
Chapter 3	

Article
Chapter 4 *Page*

Chapter 5

About the Authors

Frederic P. Hartwell is a working electrician, is President of Hartwell Electrical Services, Inc., and has been certified by the International Association of Electrical Inspectors as a Certified Master Electrical Inspector. He is the senior member of NEC® CMP 9. He is coauthor of *McGraw-Hill's American Electricians' Handbook, 16th Edition.*

Joseph F. McPartland is an electrical contracting consultant and coauthor of *McGraw-Hill's National Electrical Code® Handbook, 25th Edition.*

Brian J. McPartland is an electrical consultant and educator who teaches the nuts and bolts of the National Electrical Code®. He is coauthor of *McGraw-Hill's National Electrical Code® Handbook, 25th Edition.*

Preface

The 29th edition of *McGraw-Hill's National Electrical Code® Handbook* has been thoroughly revised to reflect the changes given in the 2017 **National Electrical Code**. This is a reference book of commentary, discussion, and analysis on the most commonly encountered rules of the 2017 **National Electrical Code**. Designed to be used in conjunction with the 2017 NE Code book published by the National Fire Protection Association, this Handbook presents thousands of illustrations—diagrams and photos—to supplement the detailed text in explaining and clarifying **NEC** regulations. Description of the background and rationale for specific Code rules is aimed at affording a broader, deeper, and readily developed understanding of the meaning and application of those rules. The style of presentation is conversational and intended to facilitate a quick, practical grasp of the ideas and concepts that are couched in the necessarily terse, stiff, quasi-legal language of the **NEC** document itself.

This Handbook follows the order of "articles" as presented in the NE Code book, starting with "Article 90" and proceeding through the various "Informative Annexes." The Code rules are referenced by "section" numbers (e.g., "250.138. Cord- and Plug-Connected Equipment"). This format ensures quick and easy correlation between **NEC** sections and the discussions and explanations of the rules involved. This companion reference to the **NEC** book expands on the rules and presents common interpretations that have been put on the many difficult and controversial Code requirements. A user of this Handbook should refer to the **NEC** book for the precise wording of a rule and then refer to the corresponding section number in this Handbook for a practical evaluation of the details.

Because many **NEC** rules do not present difficulty in understanding or interpretation, not all sections are referenced. But the vast majority of sections are covered, especially all sections that have proved troublesome or controversial. And particular emphasis is given to changes and additions that have been made in Code rules over recent editions of the **NEC**. Although this new edition, *McGraw-Hill's National Electrical Code® 2017 Handbook*, does not contain the complete wording of the NE Code book, it does contain much greater analysis and interpretation than any other so-called Handbook contains.

Today, the universal importance of the NE Code has been established by the federal government (OSHA and other safety-related departments), by state and local inspection agencies, and by all kinds of private companies and organizations. In addition, national, state, and local licensing or certification as an electrical contractor, master electrician, or electrical inspector will require a firm and confident knowledge of the **NEC**. With requirements for certification or licensing now mandated in nearly every jurisdiction across the country, the need for Code competence is indispensable. To meet the great need for information on the **NEC**, McGraw-Hill has been publishing a handbook on the **National Electrical Code** since 1932. Originally developed by Arthur L. Abbott in that year, the Handbook has been carried on in successive editions for each revision of the **National Electrical Code**.

One final point—words such as "workmanlike" are taken directly from the Code and are intended in a purely generic sense. Their use is in no way meant to deny the role women already play in the electrical industries or their importance to the field.

Frederic P. Hartwell
Joseph F. McPartland
Brian J. McPartland

Introduction to the National Electrical Code®

McGraw-Hill's National Electrical Code® 2017 Handbook is based on the 2017 edition of the National Electrical Code as developed by the National Electrical Code Committee of the American National Standards Institute (ANSI), sponsored by the National Fire Protection Association® (NFPA®). The National Electrical Code is identified by the designation NFPA No. 70-2017. The NFPA adopted the 2017 Code at the NFPA Technical Meeting held in June, 2016.

The National Electrical Code, as its name implies, is a nationally accepted guide to the safe installation of electrical wiring and equipment. The committee sponsoring its development includes all parties of interest having technical competence in the field, working together with the sole objective of safeguarding the public in its utilization of electricity. The procedures under which the Code is prepared provide for the orderly introduction of new developments and improvements in the art, with particular emphasis on safety from the standpoint of its end use. The rules of procedure under which the National Electrical Code Committee operates are published in each official edition of the Code and in separate pamphlet form so that all concerned may have full information and free access to the operating procedures of the sponsoring committee. The Code has been a big factor in the growth and wide acceptance of the use of electrical energy for light and power and for heat, radio, television, signaling, and other purposes from the date of its first appearance (1897) to the present.

The National Electrical Code is primarily designed for use by trained electrical people and is necessarily terse in its wording.

The sponsoring National Electrical Code Committee is composed of a Technical Correlating Committee and 19 Code-making panels, each responsible for one or more Articles in the Code. Each panel is composed of experienced individuals representing balanced interests of all segments of the industry and the public concerned with the subject matter. In an effort to promote clarity and consistency of field interpretations of NEC passages, the National Electrical Code Style Manual was completely rewritten in 1999, with the current version effective in 2015. All Code-making panels have been asked to review their articles for usability and editorial conformity to this publication, and copies are available from the NFPA, Batterymarch Park, Quincy, MA 02269.

The National Fire Protection Association also has organized an Electrical Section to provide the opportunity for NFPA members interested in electrical safety to become better informed and to contribute to the development of NFPA electrical standards. This new Handbook reflects the fact that the National Electrical Code was revised for the 2017 edition, requiring an updating of the previous Handbook, which was based on the 2014 edition of the Code. The established schedule of the National Electrical Code Committee contemplates a new edition of the National Electrical Code every 3 years. Provision is made under the rules of procedure for handling urgent emergency matters through a Tentative Interim Amendment Procedure. The Committee also has established rules for rendering Formal (sometimes called Official) Interpretations. Two general forms of findings for such Interpretations are recognized: (1) those making an interpretation of literal text and (2) those making an interpretation of the intent of the National Electrical Code when a particular rule was adopted. All Tentative Interim Amendments and Formal Interpretations are published by the NFPA as they are issued, and notices are sent to all interested trade papers in the electrical industry.

The National Electrical Code is purely advisory as far as the National Fire Protection Association is concerned but is very widely used as the basis of law and for legal regulatory purposes. The Code is administered by various local inspection agencies, whose decisions govern the actual application of the National Electrical Code to individual installations. Local inspectors are largely members of the International Association of Electrical Inspectors, 901 Waterfall Way, Suite 602, Richardson, TX 75080-7702. This organization, the National Electrical Manufacturers Association, the National Electrical Contractors Association, the Edison Electric Institute, the Underwriters' Laboratories Inc., the International Brotherhood of Electrical Workers, governmental groups, and independent experts all contribute to the development and application of the National Electrical Code.

Brief History of the National Electrical Code®

The **National Electrical Code** was originally drawn in 1897 as a result of the united efforts of various insurance, electrical, architectural, and allied interests. The original Code was prepared by the National Conference on Standard Electrical Rules, composed of delegates from various interested national associations. Prior to this, acting on an 1881 resolution of the National Association of Fire Engineers' meeting in Richmond, Virginia, a basis for the first Code was suggested to cover such items as identification of the white wire, the use of single disconnect devices, and the use of insulated conduit.

In 1911, the National Conference of Standard Electrical Rules was disbanded, and since that year, the National Fire Protection Association (**NFPA**) has acted as sponsor of the **National Electrical Code**. Beginning with the 1920 edition, the **National Electrical Code** has been under the further auspices of the American National Standards Institute (and its predecessor organizations, United States of America Standards Institute, and the American Standards Association), with the **NFPA** continuing in its role as Administrative Sponsor. Since that date, the Committee has been identified as "ANSI Standards Committee C1" (formerly "USAS C1" or "ASA C1").

Major milestones in the continued updating of successive issues of the **National Electrical Code** since 1911 appeared in 1923, when the Code was rearranged and rewritten; in 1937, when it was editorially revised so that all the general rules would appear in the first chapters followed by supplementary rules in the following chapters; and in 1959, when it was editorially revised to incorporate a new numbering system under which each Section of each Article is identified by the Article Number preceding the Section Number. The 1937 edition also included an introduction for the first time, and many of its most important provisions survive, almost verbatim, in Art. 90 today. That article number, together with much of what is familiar about the Code today, began with the 1959 organizational changes.

In addition to an extensive revision, the 1975 **NEC** was the first Code to be dated for the year following its actual release. That is, although it was released in September of 1974, instead of being called the 1974 Code—as was done for the 1971 and all previous editions of the **NEC**—this Code was identified as the 1975 Code. That's the reason there appears to be 4 years, instead of the usual 3, between the 1971 and 1975 editions. The purpose was to have the named code year agree with the effective dates of adoption in, at least, the early adopting jurisdictions.

Due to the proliferation of premises-owned medium-voltage systems, the 1999 Code notably moved those requirements out of the old Art. 710 and into Chapters 1 through 4. The 2005 **NEC** made a notable reorganization of almost all of Chapter 3, resulting in new article numbers for almost every wiring method.

The 2017 edition of the **NEC** continues to modernize in response to developments in the usage of electric power, with five new articles unveiled. Industrial process heating equipment covered in Art. 425 applies to fixed resistance and electrode heating, but only in industrial occupancies. Article 691 covers solar farms with a generating capacity of at least 5000 kW. Article 706 centralizes the requirements applicable to the burgeoning field of on-site electrical storage, including new technologies such as flow batteries. Article 710 will be a central repository for stand-alone system provisions that formerly repeated themselves in other articles, especially Art. 690. Finally, Art. 712 on DC Microgrids will be of use where multiple on-site dc sources such as solar or wind or fuel cells supply functional dc power for utilization.

For many years the **National Electrical Code** was published by the National Board of Fire Underwriters (now American Insurance Association), and this public service of the National Board helped immensely in bringing about the wide public acceptance that the Code now enjoys. It is recognized as the most widely adopted Code of standard practices in the U.S.A. The National Fire Protection Association first printed the document in pamphlet form in 1951 and has, since that year, supplied the Code for distribution to the public through its own office and through the American National Standards Institute. The **National Electrical Code** also appears in the National Fire Codes, issued annually by the National Fire Protection Association.

About the 2017 National Electrical Code®

The trend for very large numbers of proposals for changes and adopted changes in successive editions of the NEC has not reversed itself. The 2017 NEC is based on 4012 public inputs and 1513 public comments that have resulted in literally hundreds of additions, deletions, and other modifications—both minor and major. There are completely new articles covering equipment and applications not previously covered by the Code. There are also new regulations and radical changes in old regulations that affect the widest possible range of everyday electrical design considerations and installation details.

The process of development of the 2017 NEC makes it a watershed edition as NFPA looks to an electronic future in terms of standards development. The process has been transformed into one that is largely based on internet transactions. Instead of the former "proposals," suggestions for NEC changes are now styled as "public inputs" (or, if developed at a panel meeting, as "panel inputs"). Paper submittals are still permitted, but the deadline for submittal is significantly shorter than for the preferred electronic submittals. The panels still meet in an open session to consider the inputs, and a two-thirds vote on a subsequent ballot (now submitted electronically) is still required to advance a change in the NEC. The result (after the customary Correlating Committee review) is a "First Draft Report" instead of the former "Report on Proposals." A customary comment period then opens, and as previously, the panels meet again to consider them prior to issuance of a "Second Draft Report" instead of the former "Report on Comments." This documentation represents the principal source of information relative to the motivation for changes in the Code. This book contains literally thousands of explanations for how Code sections are intended to be applied, and in the vast majority of cases, those explanations were developed from the first and second draft reports, or their predecessors in prior cycles.

The change of title for the reports is of no consequence, but the format is. Both draft reports, as well as all submitted inputs, comments, and related documentation, are only available for review online. NFPA does not make these reports available for download, so even printing them out as a collected report (single inputs and actions can be downloaded) from one's computer is not possible. They can be worked only from an active on-line connection. Another procedural issue relates to the expanded use of task groups prior to the panel meetings under the new process. All public inputs and comments received initial evaluation by appointed task groups comprised of subsets of the panel membership, which met online and through conference telephone calls prior to the meetings. The task groups were not balanced by interest categories, and their reports lacked transparency and frequently had a disproportionate influence on the final results. In general, the task group process was popular because it significantly reduced the workload during the meetings, but has proved problematic in terms of the quality of the final product.

Even the input submittal process is a work in progress. The software is plainly not yet up to the task. The submittals must be formulated using legislative text with changes underlined and deletions struck out, but frequently the software would underline additional text that was unaffected by the input. As a result, many submitters had to resort to explaining in plain English in the substantiation field exactly what they were intending to change in the text field. The software was also problematic for the panel process, presenting such annoyances as not allowing for carriage returns in a ballot comment. This resulted in carefully crafted, properly punctuated voting comments being posted as a single run-on paragraph, and in addition legislative or other formatting options were unavailable as well. The software also imposed restraints on the length of comments, which meant some were relegated to separate Microsoft Word documents almost impossible for the public to find. In many instances the software carried errors that were difficult to detect in the ballot process due to the presentation, and then into print. Because they, technically, were balloted, a TIA will be required to correct them. The new Art. 425 is a case in point, with no fewer than six missed paragraph divisions spread throughout the article, each having substantive impact and in one case making an entire section unusable. This author awaits future improvements in this process.

Everyone involved in the layout, selection, estimation, specification, inspection, as well as installation, maintenance, replacement, etc., of electrical systems and equipment must make every effort to become as thoroughly versed in and completely familiar with the intimate details related to the individual change as is possible. And, this must be done as soon as possible.

Clearly, compliance with the NEC is more important than ever, as evidenced by the skyrocketing numbers of suits filed against *electrical* designers and installers. In addition, inspectors everywhere are more knowledgeable and competent and they are exercising more rigorous enforcement and generally tightening control over the performance of electrical work. Another factor is the Occupational Safety and Health Administration's *Design Safety Standards for Electrical Systems.* That standard, which borrowed heavily from the rules and regulations given in the **NEC**, is federal law and applies to all places of employment in general industry occupancies. Although the OSHA *Design Safety Standards for Electrical Systems* is based heavily on the **NEC**, due to the relatively dynamic nature of the **NEC**, there will eventually be discrepancies. But, for those instances where a more recent edition of the Code permits something that is prohibited by the OSHA standard, OSHA officials have indicated that such an infraction—although still an infraction—will be viewed as what OSHA refers to as a "de minimus violation," which essentially boils down to no fine. Of course that is not always the case. "Listing" and "labeling" of products by third-party testing facilities is *always permitted but frequently not required* by the **NEC**, but it is made *mandatory* in most places of employment by the Occupational Safety and Health Administration (OSHA) *Design Safety Standards for Electrical Systems.* The OSHA requirement for certification may take precedence over the less stringent position of the **NEC** regarding listing of equipment. The impact of the **NEC**—even on OSHA regulations, which are federal law—is a great indicator of the Code's far reaching effect.

The fact that the application of electrical energy for light, power, control, signaling, and voice/data communication, as well as for computer processing and computerized process-control continues to grow at a breakneck pace also demands greater attention to the Code. As the electrical percentage of the construction dollar continues upward, the high-profile and very visible nature of electrical usage demand closer, more penetrating concern for safety in electrical design and installation. In today's sealed buildings, with the entire interior environment dependent on the electrical supply, reliability and continuity of operation has become critical. Those realities demand not only a concern for eliminating shock and fire hazards, but also a concern for continuity of supply, which is essential for the safety of people, and, in today's business and industry, to protect data and processes, as well.

And, of course, one critical factor that, perhaps, emphasizes the importance of Code-expertise more than anything else is the extremely competitive nature of construction and modernization projects, today. The restricted market and the overwhelming pressure to economize have caused some to employ extreme methods to achieve those ends without full attention to safety. The Code represents an effective, commendable, and, in many instances, legally binding standard that *must* be satisfied, which acts as a barrier to any compromises with basic electrical safety. It is a democratically developed consensus standard that the electrical industry has determined to be the essential foundation for safe electrical design and installation; and compliance with the **NEC** will dictate a minimum dollar value for any project.

In this Handbook, the discussion delves into the letter and intent of Code rules. Read and study the material carefully. Talk it over with your associates; engage in as much discussion as possible. In particular, check out any questions or problems with your local inspection authorities. It is true that only time and discussion provide final answers on how some of the rules are to be interpreted. But now is the time to start. Do not delay. Use this Handbook to begin a regular, continuous, and enthusiastic program of updating yourself on this big new Code.

This Handbook's illustrated analysis of the 2017 **NEC** is most effectively used by having your copy of the new Code book at hand and referring to each section as it is discussed. The commentary given here is intended to supplement and clarify the actual wording of the Code rules as given in the Code book itself.

Introduction

ARTICLE 90. INTRODUCTION

90.1. Purpose

(A). Practical Safeguarding. The intent of this section is to establish a clear and definite relationship between the National Electrical Code® and electrical system design as well as field installation. Basically stated, the NEC is intended only to assure that electrical systems installed in commercial, industrial, institutional, and residential occupancies are safe. That is, to provide a system that is "essentially free from hazard." The Code (throughout this manual, the words "Code," "NE Code," and "NEC®" refer to the National Electrical Code) sets forth requirements, recommendations, and suggestions and constitutes a minimum standard for the framework of electrical design. As stated in its own introduction, the Code is concerned with the "practical safeguarding of persons and property from hazards arising from the use of electricity" for light, heat, power, computers, networks, control, signaling, and other purposes.

The NEC is recognized as a legal criterion of safe electrical design and installation. It is used in court litigation and by insurance companies as a basis for insuring buildings. The Code is an important instrument for safe electrical system design and installations. It must be thoroughly understood by all electrical designers and installers. They must be familiar with all sections of the Code and should know the latest accepted interpretations that have been rendered by inspection authorities and how they impact the design and/or installation of electrical systems. They should keep abreast of Formal Interpretations, as well as the issues addressed by Tentative Interim Amendments (TIA) that are issued, periodically, by the NEC committees. They should know the intent of Code requirements (i.e., the spirit as well as the letter of each provision) and be familiar with the safety issue at the heart of the matter. And, most important, they should keep a copy of the NEC and this Code handbook close by for ready reference and repeated study.

The **NEC** is not written for nor intended to be used by untrained individuals. Qualified electricians and engineers spend many years reviewing and thinking about its provisions. It should be noted that the language in Article 90, although considerably embellished over the years, has changed little substantively over the years since this sort of introduction was first presented in the 1937 edition.

(B). Adequacy. It's worth noting that compliance with the provisions of the **National Electrical Code** can effectively minimize fire and accident hazards in any electrical system. A code-compliant installation will be "essentially free from hazard," but not absolutely so. This provision is essentially unchanged since the 1937 edition, and has stood the test of time. Many installations with substantial code violations exist for protracted periods of time without loss experience through good fortune and the fact that protective systems frequently overlap. On the other hand, occasionally a fully compliant installation can fail, usually due to an unforeseen circumstance. Perfect safety is only achievable at infinite cost. Every three years the National Electrical Code Committee wrestles with the concept of "essentially free" as it considers proposed changes to the next edition.

Although the Code assures minimum safety provisions, actual design work must constantly consider safety as required by special types or conditions of electrical application. For example, effective provision of automatic protective devices and selection of control equipment for particular applications involve engineering knowledge above routine adherence to Code requirements. Then, too, designers and installers must know the physical characteristics—application advantages and limitations—of the many materials they use for enclosing, supporting, insulating, isolating, and, in general, protecting electrical equipment. The task of safe application based on skill and experience is particularly important in hazardous locations. Safety is not automatically made a characteristic of a system by simply observing codes. Safety must be designed into a system.

In addition to safety considerations, the need for future expansion and other common sense aspects—such as voltage drop—must be considered and factored into the overall system design. The Code in this section makes it clear that more than Code compliance will be necessary to ensure a system that is not only safe but also functional and capable of providing for future needs, without compromising system-operating continuity or integrity. It is up to the designer and installer, in consultation with the owner, to provide adequate capacity, selectivity, isolation, and protection beyond its minimum requirements in order to achieve the desired system characteristics. Remember, it is always permissible to do more than the Code requires, but *never* permissible to do less than the Code-prescribed minimum.

Addressing voltage drop illustrates these principles. No definite standards have been adopted for the maximum allowable voltage drop in most instances. There is a good reason for this. In most cases, voltage drop is an inefficiency or inconvenience, but it does not rise to the level of a safety hazard. For example, a motor run at 10 percent voltage drop, but with appropriate running overload protection, will have a greatly reduced life span, but not create a shock, fire, or electrocution hazard. The **National Electrical Code** does note, however, in a nonmandatory explanatory note, that if the voltage drop from the point of service entrance to the final outlet does not exceed 5 percent, there will be "reasonable efficiency of operation." The note also explains that not more than 3 percent voltage drop should occur in the feeder system ahead of the branch-circuit supply points, which leaves the other 2 percent for the branch circuit. In the end, the extent to which voltage drop in an electrical system is to be tolerated is the owner's decision, because the NEC does not mandate design flexibility.

There are some instances, however, where voltage drop does directly bear on safety and the NEC contains mandatory rules accordingly. For example, if the conductors to a fire pump are not sized to prevent the voltage drop while starting (i.e., while under locked-rotor conditions) from exceeding 15 percent, measured at the controller terminals, the control contactor for the motor may chatter and not reliably hold in, resulting in a failure to start with disastrous consequences.

(C). Relation to Other International Standards. This section simply states that the National Electrical Code addresses the same safety issues addressed by the International Electrotechnical Commission (IEC) Standard for "Electrical Installations of Buildings." Because the NEC covers the same consideration for safety as related to protection against electrical shock, protection against thermal effects, protection against overcurrent, protection against fault current, protection against overvoltage, faults between circuits, and so forth that are covered by the IEC Standard, it was considered necessary to establish that fact. This statement in this section facilitates the adoption of the Code by foreign countries and is consistent with the ongoing process of harmonizing the NEC and other accepted standards from around the world.

90.2. Scope

(A). Simply stated, the Code applies to all electrical work—indoors and outdoors—other than that work excluded by the rules of part **(B)** in this section. Installation of conductors and equipment, anywhere on the load-side of the point of connection to the serving utility, must comply with the provisions given in the NEC. As of the 2017 NEC, the words "and removal" are added after "installation" to reflect that some NEC rules require wiring to be removed. For example, 590.3(D) requires removal of temporary wiring upon the completion of its function, and numerous limited energy provisions, such as 800.25, require accessible portions of abandoned wiring to be removed.

The scope of the NEC includes the installation of optical fiber cable, part **(A)**. As part of the high-technology revolution in industrial and commercial building operations, the use of light pulses transmitted along optical fiber cables has become an alternative method to electric pulses on metal conductors for data, voice, and video networks, as well as for control and signaling. Although the technology of fiber optics has grown dramatically over recent years, it is still primarily used as a "trunk line" or "backbone" for high-speed networks, while horizontal distribution is generally accomplished via a twisted-pair or coaxial copper medium. Although coaxial cable can handle high rates of data transmission involved in data processing and computer control of machines and processes, optical fiber cables far outperform metallic conductors—even coaxial cable—when it comes to bandwidth as well as cost of materials. (See Fig. 90-1.) NEC Art. 770, "Optical Fiber Cables," covers the installation and use of fiber-optic cables.

Part **(A)(1)** provides a laundry-list of specific indoor installations that must be in compliance with the applicable requirements given in the Code. Note that

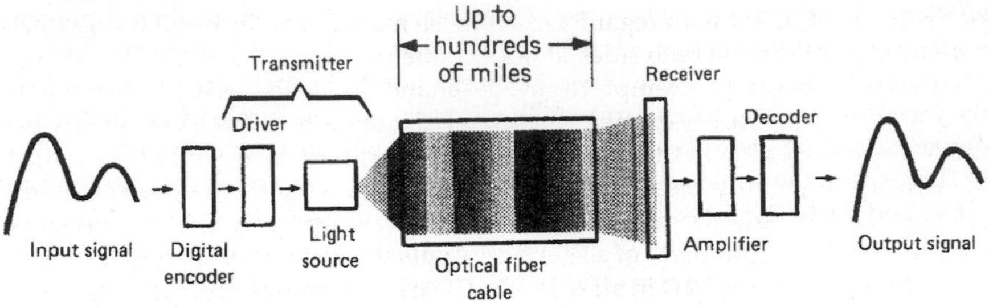
Up to
hundreds
of miles
Transmitter
Driver
Receiver
Decoder
Input signal
Digital
encoder
Light
source
Optical fiber
cable
Amplifier
Output signal

A telephone conversation is first transformed into an electrical signal. This input signal is scanned by a digital encoder, which reduces it to a series of "on's" and "off's." The driver, which activates the laser light source, transmits the digital "on's" as a pulse of light and the "off's" as the absence of a pulse. The light travels through the optical fiber cable to its destination, where it is received, amplified, and fed into a digital decoder. The decoder translates the pulses back into the original electrical signal.

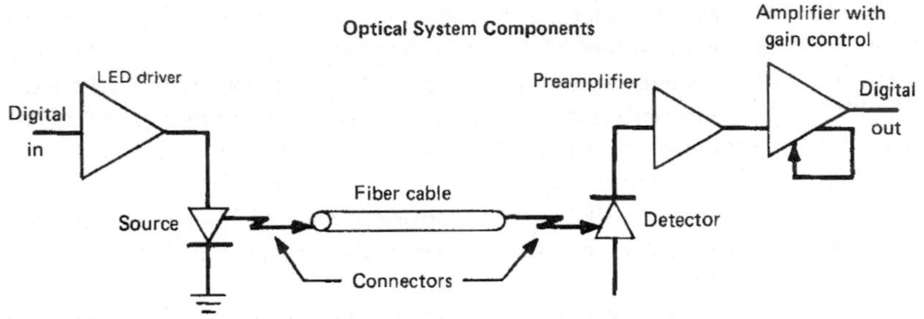
Optical System Components
Amplifier with
gain control
LED driver
Preamplifier
Digital
in
Digital
out
Fiber cable
Source
Detector
Connectors

Fig. 90-1. The NEC covers the technology of fiber optics for communication and data transmission.

this section makes clear that the NEC also applies to "floating buildings" because the safety of Code compliance is required for all places where people are present. Coverage of floating buildings is contained in NEC Art. 553.

Part **(A)(2)** identifies specific outdoor installations, including carnivals and industrial substations, while part **(A)(3)** mandates that supply equipment and conductors—whether supplied from a utility as a service or from on-site generators as a separately derived system—as well as all other outside equipment and conductors must satisfy the rules and regulations of the NEC. Use of the word "equipment" in parts **(A)(2)** and **(A)(3)** makes clear that the NEC applies to electrical circuits, systems, and components in their manner of installation as well as use.

The following discussion and the discussion in 90.2(**B**)(**5**) are very closely related and often hotly debated. Information has been provided from both sides of the discussions as well as the commentary from the Code-making panels (CMPs) where available. The purpose is to allow each designer and installer to make his or her own judgment with regard to how these matters will be resolved based on a full understanding of both sides of the arguments.

Although generally exempt from compliance with the NEC, according to 90.2(**A**)(**4**) certain utility-owned or -operated occupancies must be wired per the NEC. The wording in this section along with the companion rule of 90.2(**B**)(**5**) is intended to identify those utility electrical installations that are subject to the rules of the NEC and those that are not. Basically stated, any utility occupancy that is not an "integral part" of a "generating plant, substation, or control center" must comply with the NEC in all respects. Clearly, any office space, storage area, garage, warehouse, or other nonpower-generating area of a building or structure is not an "integral part" of the generation, transmission, or distribution of electrical energy and therefore is covered by the NEC.

There has been discussion and disagreement over the meaning of the phrase, "*not an integral part of* a generating plant . . . etc." Some feel that the phrase "not an integral part of" applies to the process of generation, and so forth. Others believe that it applies to the building. That is, if an occupancy identified in 90.2(**B**)(**4**) is part of a generating plant, it is exempt from compliance with the NEC. Although that doesn't seem to make sense, past comments made by the CMP indicate that it is the intent of this rule to exempt, say, office spaces within a generating plant. However, this is not completely clear from the wording used. To prevent any problem with this section, one could choose to interpret this rule to require NEC compliance for any occupancy that is "not an integral part of the process" and wire such spaces in accordance with the NEC. Such interpretation cannot be disputed. That is, satisfying the more rigorous NEC requirements cannot be construed as a violation. But, if one does not comply, the potential for legal liability exists.

Some may feel that the term *integral part* should be interpreted to mean "integral part of the process" (i.e., generation, transformation, or distribution of electric energy), according to commentary in the NEC Committee Reports for the 1987 NEC. Others feel that it should be taken to mean an "integral part" of the building or structure. Be aware that the first contention seems more reasonable. That is, just because an office is *in* a generating plant, it shouldn't be exempt from the NEC, especially since these areas will be occupied by the general public. And it seems logical that the same should apply to the cafeteria, bathrooms, and other areas within the plant that are not directly related to the task of generating and delivering electrical energy and will be occupied by other than qualified plant electrical personnel.

With that said, it should be noted that the wording here could be read both ways and it will be up to the local authority having jurisdiction (AHJ) to interpret what is and what is not required to comply with the NEC.

It should be noted that equipment installed by the utility to perform associated functions, such as outdoor lighting at an outdoor substation, is intended to be considered as an "integral" part to the process and is therefore exempt from compliance with the NEC (Figs. 90-2 and 90-3).

(B). Not Covered. The rules of the Code do not apply to the electrical work described in **(1)** through **(5)**. The most common controversy that arises

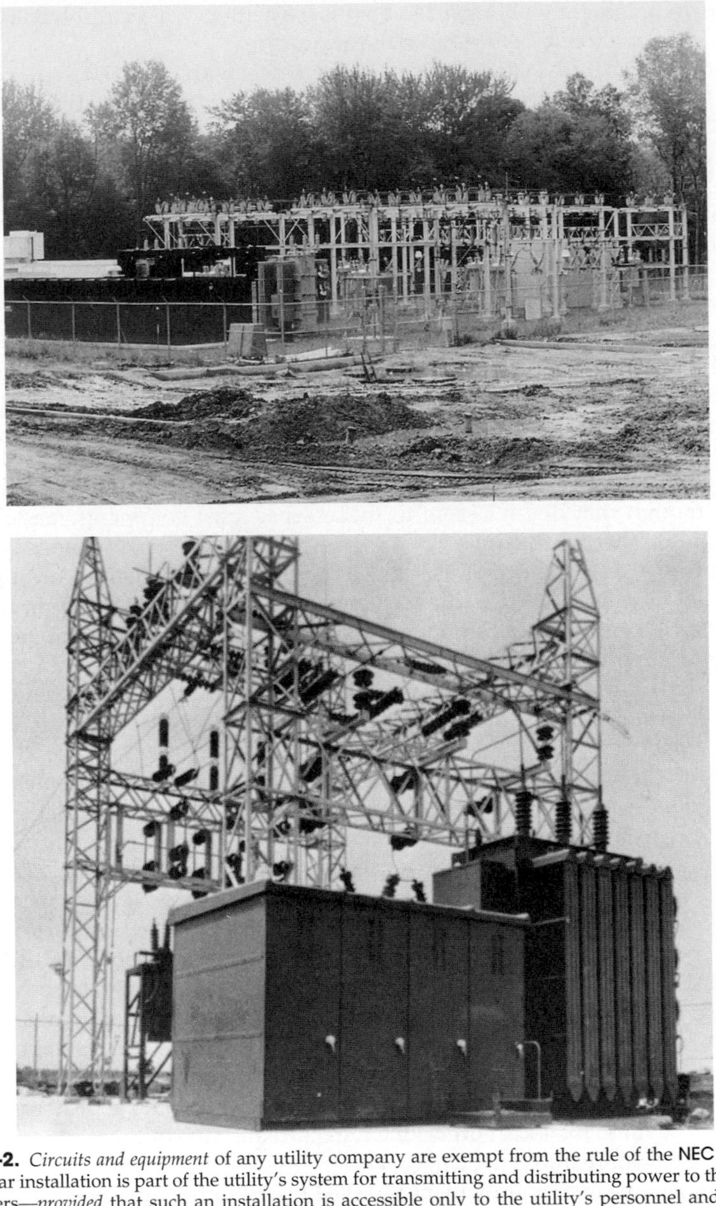

Fig. 90-2. *Circuits and equipment* of any utility company are exempt from the rule of the NEC when the particular installation is part of the utility's system for transmitting and distributing power to the utility's customers—*provided* that such an installation is accessible only to the utility's personnel and access is denied to others. Outdoor, fenced-in utility-controlled substations, transformer mat installations, utility pad-mount enclosures, and equipment isolated by elevation are typical utility areas to which the NEC does not apply. The same is also true of indoor, locked transformer vaults, or electric rooms (Sec. 90.2). But electrical equipment, circuits, and systems that are involved in supplying lighting, heating, motors, signals, communications, and other load devices that serve the needs of personnel in buildings or on premises owned (or leased) and operated by a utility are subject to NEC rules, just like any other commercial or industrial building, provided that the buildings or areas are not integral parts of a generating plant or substation.

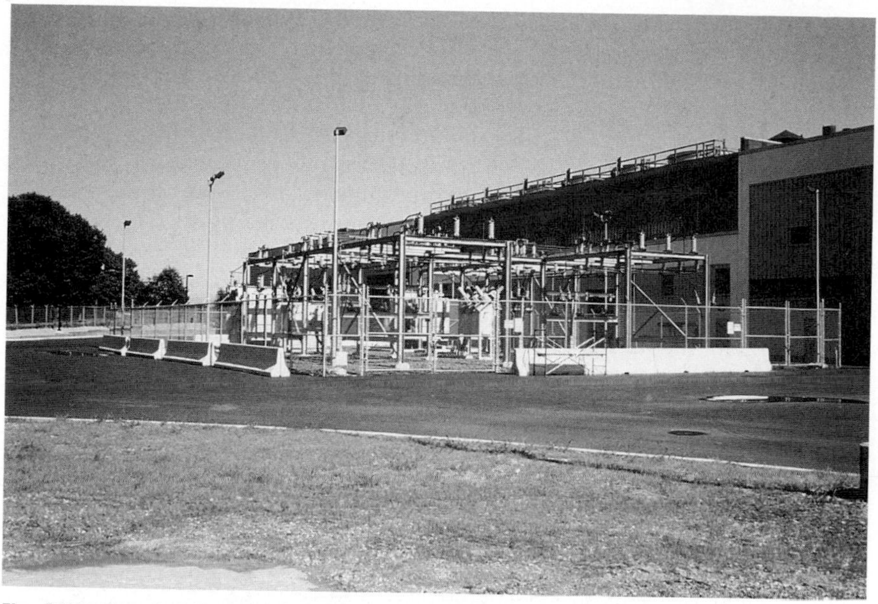

Fig. 90-3. Those buildings and structures that are directly related to the generation, transmission, or distribution of energy are intended to be excluded from compliance with the **NEC**. However, the rules covering this matter also indicate that functionally associated electrical equipment—such as the outdoor lighting for the utility-owned and -operated outdoor substation—is also exempt from the **NEC**.

concerns exclusion of electrical work done by electric utilities (power companies), especially outdoor lighting.

This rule emphatically explains that not all electrical systems and equipment belonging to utilities are exempt from Code compliance. Electrical circuits and equipment in buildings or on premises that are used exclusively for the "generation, control, transformation, transmission, energy storage, or distribution of electric energy" are considered as being safe because of the competence of the utility engineers and electricians who design and install such work. Code rules do not apply to such circuits and equipment—nor to any "communication" or "metering" installations of an electric utility. But, any conventional electrical systems for power, lighting, heating, and other applications within buildings or on structures belonging to utilities *must* comply with Code rules where such places are not "used exclusively by utilities" for the supply of electric power to the utilities' customers.

An example of the kind of utility-owned electrical circuits and equipment *covered* by Code rules would be the electrical installations in, say, an office building of the utility. But, in the Technical Committee Report for the 1987 **NEC**, the Code panel for Art. 90 stated that it is not the intent of this rule to have **NEC** regulations apply to "office buildings, warehouses, and so forth that *are* an *integral* part of a utility-generating plant, substation, or control center." According to comments from the CMP, **NEC** rules would not apply to any wiring or equipment in a utility-generating plant, substation, or control center and would not apply to conventional lighting and power circuits in office areas, warehouses, maintenance shops, or any other areas of utility facilities used for the generation, transmission, or distribution of electric energy for the utility's customers. But **NEC** rules would apply to all electrical work in *other* buildings occupied by utilities—office buildings, warehouses, truck garages, repair shops, etc., that are in separate buildings or structures on the generating facility's premises. And that opinion was reinforced by the statements of the CMP that sat for the 1996 **NEC**. With that said, it should be noted that the actual wording used here in the Code could be read both ways and it will be up to the local AHJ to interpret what areas are and what areas are not required to comply with the **NEC**.

The wording used in 90.2(**B**)(**5**)(**c**) recognizes non-**NEC**-complying utility installations "in legally established easements or rights-of-way." This clearly exempts utility activities on public streets, alleys, and similar areas, even for street and area lighting for adjacent parking lots. However, the 2008 **NEC** deleted the phrase "or by other agreements" from this list. The concern was that this provision opened the door to utility noncompliance throughout a facility, provided an agreement could be struck with the owner and ratified by the governmental authority having jurisdiction over utility practice. Because utilities are governed by the National Electrical Safety Code (NESC), whose provisions are entirely inappropriate for premises wiring, this concern is not inconsequential.

In the 2011 version, an elaborate compromise was struck that addressed numerous areas where formal easements are impractical, including federal lands and military bases, Indian reservations, railroad property, and state agencies, departments, and port authorities. This successfully covered many instances where the removal of the "other agreements" permission caused problems. Its list of governmental entities that are unique to the United States may, unfortunately, create adoption issues in foreign jurisdictions in which the **NEC** is or may in the future be adopted.

However, the change remains extremely controversial because it has the potential to unravel over a century of established precedent regarding site lighting by utilities, where all of the work is on the line side of any service point, or where there is no service point whatsoever, as illustrated in Fig. 90-4. Virtually, every electric utility has permission to supply outdoor lighting according to rates established by the governing authority, and that lighting need not be in a public way or in an easement, provided it is not premises wiring. The key to understanding the problem is the concept of a service point, defined in Art. 100. The NESC applies on the supply side of service points, where they exist. The **NEC** applies on the load side of service points, where they exist. It is instructive to review the premises underlying the 2005 **NEC** language.

The entire premise behind allowing the NESC, substantially different from the **NEC**, to apply to utility work is a simple one: The organizational permanence, engineering supervision, and workforce training in the utility environment are fundamentally different than for premises wiring. Therefore, different standards can be applied to installations under their exclusive control. Whether this also applies to an Energy Service Company (ESCo) doing maintenance under contract with the utility is a regulatory matter that will depend on the degree of command and control exercised by the regulated utility.

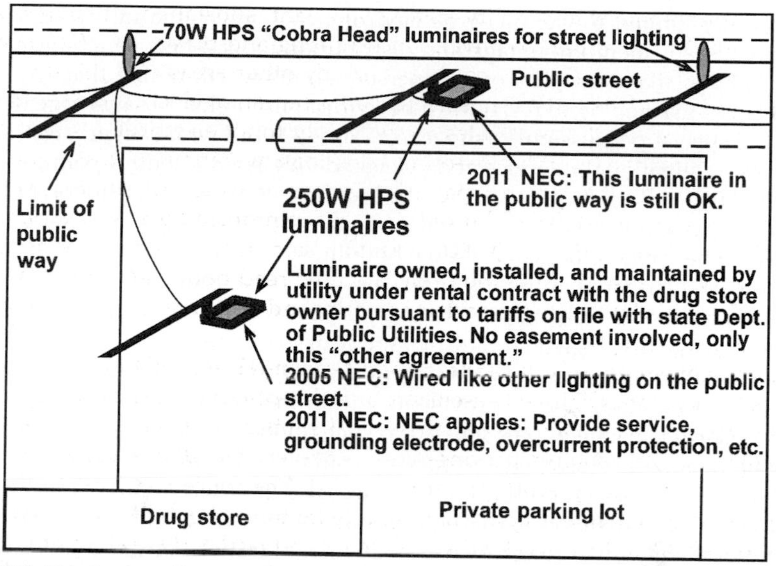
70W HPS "Cobra Head" luminaires for street lighting
Public street
Limit of public way
250W HPS luminaires
2011 NEC: This luminaire in the public way is still OK.
Luminaire owned, installed, and maintained by utility under rental contract with the drug store owner pursuant to tariffs on file with state Dept. of Public Utilities. No easement involved, only this "other agreement."
2005 NEC: Wired like other lighting on the public street.
2011 NEC: NEC applies: Provide service, grounding electrode, overcurrent protection, etc.
Drug store
Private parking lot

Fig. 90-4. This drawing shows an actual example of a practice that is widespread throughout the United States and many other countries. There is no service point, the parking lot luminaire is not premises wiring, and the maintenance will be performed by a utility line crew in the same bucket truck as services the street lighting, at the same time. The drug store is, in effect, buying the 27,500-lm output from each of the two 250-W high-pressure sodium (HPS) lamps. The **NEC** purports to claim jurisdiction over this portion of the parking lot lighting.

Area lighting wired to the NESC will lack local disconnects, specific overcurrent protection, and separate equipment grounding conductors, for just a few examples. Is this a safety issue if a utility line crew does the maintenance? Apparently not, given the ubiquitous presence of street lighting wired this way. Would it be a safety issue if it were premises wiring, maintained by others? Certainly, given that the **NEC** has never allowed such practices over its long history. The fact that these two statements are self-evidently both true leads to this conclusion: You cannot write and apply installation rules without taking the operational context into account. The **NEC** does exactly the same thing over and over again when it creates special exceptions and allowances for work that will be performed under qualified maintenance and supervision.

Part **(C)** gives the AHJ the discretion to permit other than "utilities" the option to install conductors between the utility supply and the service conductors for individual buildings without complying with the **NEC**. Any such waiver is limited to outdoor portions of the electrical installation, or to conductors inside a building but only to the extent necessary to terminate at the nearest readily accessible point to where the conductors enter the building. And as discussed at 230.6(5), as soon as wiring penetrates the outer membrane of a building, whether or not it has fully entered occupied space, it is inside the building for code enforcement purposes. Essentially, it allows the inspector to permit the use of another standard such as in the utilities code, the NESC. Such permission is typically limited to campus-type environments where the utility supply to the premises is medium-voltage and distribution to, and between, buildings is installed and maintained by on-site personnel. It's worth noting that any such permission granted by the AHJ must be *written* permission to satisfy the definition of "special permission," as given in Art. 100. Today, however, such occupancies frequently take service at an elevated voltage at a central point, and all the medium-voltage feeders to serve the buildings are just that, feeders. As soon as the service point becomes a central medium-voltage switch, this provision can no longer be applied to the individual buildings.

There are far more mundane uses for this permission. Many CATV (see Art. 820) companies rely on powered amplifiers mounted near the top of utility poles to keep their signal strength where it needs to be. Those amplifiers will have a small disconnect and overcurrent protective device located adjacent to the amplifiers. There are no provisions within the body of the NEC that allow for a service disconnect to be located at such a location, which is certainly not readily accessible. However, the entire installation is confined to the pole top, and special permission under 90.2(C) is routinely granted in such cases.

90.3. Code Arrangement. This section provides guidance on which rule takes precedence where two rules covering a particular installation are at odds. Basically, the rules in Chaps. 1 through 4 apply at *all* times, except for installations covered by Chap. 8, which stands alone. Installations covered by Chaps. 5, 6, and 7 must always comply with the requirements given in Chaps. 1 through 4, *unless* a specific rule in Chaps. 5 through 7 requires or permits an alternate method. One implication of this principle is that exceptions in Chaps. 1 through 4 that allow for different procedures in Chaps. 5 through 7 are unnecessary. The NEC Style Manual has been rewritten to take this into account, and such exceptions are disappearing from the NEC for this reason. Provisions in Chaps. 1 through 7 of the NEC only apply in Chap. 8 when a Chap. 8 article specifically cites them, and the numbers of such citations in Chap. 8 articles are steadily increasing for this reason. Chapter 9 consists of tables that are mandatory, but only applicable as referenced in earlier articles. The graphic provided in this section facilitates understanding of the relationship between various Code chapters.

As of the 2017 NEC, there is permission granted for Chaps. 5, 6, and 7 articles to modify not just Chapters 1 through 4, but to modify Chapters 5 through 7 as well. This may appear to be of no great effect, but it will prove problematic until inevitable conflicts are sorted out. For example, 517.30(G) relaxes selective coordination rules on essential electrical systems in hospitals. These systems include life-safety systems, which are also within the scope of Article 700, which for its part has extremely strict requirements. Under this rewrite of 90.3, it is arguable whether 517.30(G) amends 700.28, or vice versa. In this particular case, where the Art. 517 provision is there at the direction of the NFPA Standards Council following action in a different NFPA Technical Committee, the ultimately enforceable test is in Art. 517. However, other conflicts will be harder to resolve without clarifying actions in subsequent Code cycles.

90.4. Enforcement. This is one of the most basic and most important of Code rules because it establishes the necessary conditions for use of the Code.

The NEC stipulates that when questions arise about the meaning or intent of any Code rule as it applies to a particular electrical installation, including signaling and communication systems covered by Chap. 8, the electrical inspector having jurisdiction over the installation is the only one authorized by the NEC to make interpretations of the rules. The wording of Sec. 90.4 reserves that power for the local inspection authority along with the authority to approve equipment and material and to grant the special permission for methods and techniques that might be considered alternatives to those Code rules that specifically mention such "special permission."

It should be noted that any deviation from standard Code enforcement *must be* done in accordance with the provisions given in Art. 100 by the definition of "Special Permission." The most salient requirement is the need for documentation. That is, in order to comply with the definition of "special permission," such permission must be in writing. This will serve to provide a written record of the circumstances surrounding the granting of a waiver.

The NEC permits the electrical inspector to "waive specific requirements" or "permit alternate methods" in any type of electrical installation. In residential, commercial, and institutional electrical systems—as well as in industrial—inspectors may accept design and/or installation methods that do not conform to a specific Code rule, provided they are satisfied that the safety objectives of the Code rule are achieved. In other words, there must be a finding of equivalent safety before the permission is granted, and the permission to deviate from them must be provided in writing as required by the first sentence of the second paragraph in this section and stated by the definition of "special permission" given in Art. 100 (Fig. 90-5).

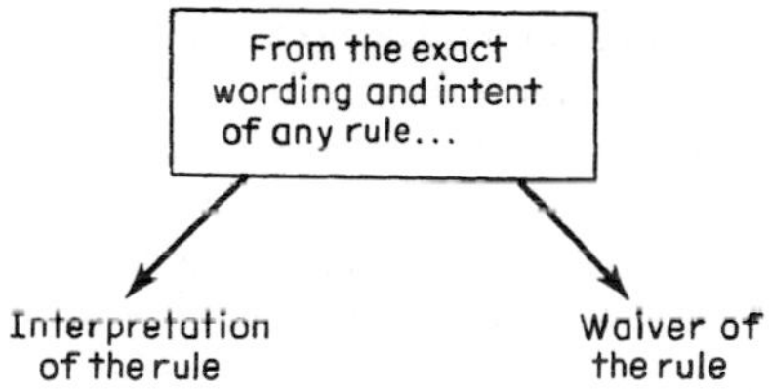
From the exact wording and intent of any rule...
Interpretation of the rule
Waiver of the rule

Fig. 90-5. *Inspector's authority* may be exercised either by enforcement of that individual's interpretation of a Code rule or by waiver of the Code rule when the inspector is satisfied that a specific non-Code-conforming method or technique satisfies the safety intent of the Code. (Sec. 90.4.)

This recognition of practices at variance with the Code is provided only for special conditions and must not be interpreted as a general permission to engage in non-Code methods, techniques, or design procedures. In fact, it is likely that inspectors will exercise this authority only with reluctance and then with great care, because of the great responsibility this places on the inspector. This is especially true because such permission may only be granted in writing. Clearly, this requirement for documentation will give many inspectors pause for reflection and reconsideration. It seems almost certain for the exercise of this prerogative.

This section also covers instances when the NEC has changed and new products are not yet available in the local market to comply with the revised terms. The AHJ may, but is not required to, permit the use of previously compliant products.

90.5. Mandatory Rules and Explanatory Material. This section provides guidance regarding proper application of the NEC. Although the NEC consists essentially of specific regulations on details of electrical design and installation, there is much explanatory material in the form of notes to rules.

Part **(A)** of this section addresses "mandatory" rules, which typically employ the phrases "shall" or "shall not." Compliance with the Code consists of satisfying all requirements and conditions that are stated by use of the word "shall" or "shall not" where used in the body of a Code rule or exception. Those words, anywhere in any rule or exception, designate a mandatory rule. Failure to comply with any mandatory Code rule constitutes a "Code violation."

Part **(B)** of this section indicates the wording that is used in "permissive rules." These rules are typified by the use of phrases such as "shall be permitted" or "shall not be required." Such rules typically provide or accept alternate measures or suspend requirements under certain conditions. It is not necessary to do what these rules permit; it is essentially an optional approach.

Note that under the provisions of the NEC Style Manual, the word "may" is not to be used to set forth a permissive rule. When "may" is being used to indicate permission, it can only be used in the context of a discretionary action of the authority having jurisdiction. For example, NEC 430.26 authorizes, but does not require, an AHJ to permit the application of a demand factor to the loads on a motor feeder being sized under 430.24. This is an excellent example of the appropriate use of the word, as in ". . . the authority having jurisdiction may grant permission for feeder conductors . . . ."

Part **(C)** explains that informational notes [formerly called "fine-print notes" (FPNs), a change that has been applied throughout the NEC as of the 2011 edition] are included, following certain Code rules, to provide additional information regarding related rules or standards. This information is strictly advisory or "explanatory" in nature and presents no rule or additional requirement. The same is true for bracketed information that references other NFPA® standards. The inclusion of the referenced standard is to inform the reader of the origin of "extracted text," where that text is taken from an NFPA standard. However, the reference to another NFPA standard in no way makes the referenced standard part of the Code; nor does such reference oblige compliance with other rules in the referenced standard. Informational notes explain NEC rules, they do not change NEC rules. If, in reading an informational note, it appears to allow or require something different from the rule that precedes it, then you are misreading the rule and you should read the rule again.

Part **(D)**, as of the 2011 edition, covers the annexes at the end of the NEC, changing their titles to include the word "Informative." This paragraph, in concert with the terminology change regarding informational notes, attempts to clarify a distinction between actual rules and merely informative text providing background or interpretive assistance.

90.6. Formal Interpretations. Official interpretations of the National Electrical Code are based on specific sections of specific editions of the Code. In most cases, such official interpretations apply to the stated conditions on given installations. Accordingly, they would not necessarily apply to other situations that vary slightly from the statement on which the official interpretation was issued.

As official interpretations of each edition of the Code are issued, they are published in the *NFPA News*, and press releases are sent to interested trade papers.

All official interpretations issued on a specific Code edition are reviewed by the appropriate CMP. In reviewing a request for a formal interpretation, a Code panel may agree or disagree. The panel will render a simple "yes" or "no" to the question, which places the burden on the questioner to provide a question that can be answered in the affirmative or negative. At some point in future Codes, the CMP might clarify the Code text to avoid further misunderstanding of intent. On the other hand, the Code panel may not recommend any change in the Code text because of the special conditions described in the request for an official interpretation. For these reasons, the NFPA does not catalog official interpretations issued on previous editions of the Code within the Code itself. Such Formal Interpretations can be obtained through the National Fire Protection Association®.

Under NFPA rules, Formal Interpretations require a four-fifths vote, which can easily result in sufficient dissent to preclude their issuance. They are issued on a specific edition of a standard, and are retained until the wording to which they applied changes. In addition, when a Formal Interpretation is issued, the technical committee (in this case a CMP) is encouraged (but not required) to review the disputed text that provoked the request for interpretation when it processes the next edition. A classic example of a Formal Interpretation, on the text of the 1978 NEC, asked whether reinforcing steel in a concrete foundation was "available" for connection after the concrete had hardened. It was common for inspection authorities in Florida at the time to insist that footings be jackhammered and connections be made so as to bring these concrete-encased electrodes into the grounding electrode system. The panel's answer was "No" and that interpretation retained its validity until the 2005 NEC changed the word "available" to "present" in what is now 250.50.

It should be remembered that, according to 90.4, the authority having jurisdiction has the prime responsibility of interpreting Code rules in its area and disagreements on the intent of particular Code rules in its area; and disagreements on the intent of particular Code rules should be resolved at the local level if at all possible. A Formal Interpretation is not really a viable avenue for a couple of reasons. One is the amount of time it will take for the CMP to render its decision, which is generally months. The other is that even if you request a Formal Interpretation and the CMP agrees with your application, there is no guarantee that the authority having jurisdiction will accept the findings of an official interpretation, nor is it required to do so.

Although this section deals with Formal Interpretations, it should be noted that changes in the Code are promulgated in a very similar manner. That is, changes to Code rules are generally precipitated by a request for change from the field. Guidance for submittal of a Code change is provided closely following the Index in the back of the Code.

90.7. Examination of Equipment for Safety. It is not the intent of the National Electrical Code to include the detailed requirements for internal wiring of electrical equipment. Such information is usually contained in individual standards for the equipment concerned. Note that Annex A at the end of the Code book includes the recognized product standards that the testing laboratories use to evaluate the products for which NEC rules require listings.

The last paragraph does not intend to take away the authority of the local inspector to examine and approve equipment, but rather to indicate that the requirements of the National Electrical Code do not generally apply to the internal construction of devices that have been listed by a nationally recognized electrical testing laboratory. Indeed, the last sentence (new as of the 2017 NEC) goes on to require that the testing lab must design its testing protocol in ways that are compatible with the **NEC**, and the third informational note following specifies that Informative Annex A lists only product standards so conceived.

Although the specifics of Code rules on examination of equipment for safety are presented in 110.2 and 110.3, the general Code statement on this matter is made here in 90.7. Although the Code does place emphasis on the need for third-party certification of equipment by independent testing laboratories, it does not make a flat rule to that effect. However, the rules of the U.S. Occupational Safety and Health Administration (OSHA) are very rigid in insisting on product certification.

Codes and standards must be carefully interrelated and followed with care and precision. Modern work that fulfills these demands should be the objective of all electrical construction people.

90.8. Wiring Planning. These two sections address concepts that are essentially design oriented. Part **(A)** alerts the reader to the fact that simply designing to Code-mandated minimums will not provide for any future expansion. Additional capacity in raceways, boxes, enclosures, and so forth, should be considered, but spare capacity is not required. Part **(B)** points out the fact that minimizing the number of circuits within any given enclosure will minimize the consequences of a fault in one or more of those circuits. Additionally, extra room in your raceways (i.e., fewer conductors than the maximum permitted) will also facilitate pulling of the conductors into the raceway. Again, providing extra room in raceways or limiting the number of circuits is only required as indicated elsewhere in the Code (e.g., 314.16 on box fill).

90.9. Metric Units of Measurement. Part **(A)** identifies metric measurements as the preferred measurement, although English units (i.e., inch-pounds-feet) are also provided as indicated by part **(B)**.

In part **(C)**, the Code discusses when one is required to use a "soft" conversion and where a "hard" conversion is permitted. A "soft conversion" is direct mathematical conversion, for example, 1 m = 39.3 in.; a "hard conversion" is more practical, for example, 1 m = 3 ft. It may seem counterintuitive to have a "hard" conversion as the inexact conversion. Another way to express the concept is that a hard conversion is the conversion a hard-core metric user would do, that is, use a round number for his or her metric measurement. The various explanations that follow in the **NEC** at this point regarding "hard" and "soft" conversion are primarily aimed at CMPs. They must make the decisions around which metric unit would unacceptably degrade safety, or cause wholesale changes in industry specifications. For example, CMP 9 used soft conversions in Table 314.16 because hard conversions would result in every steel box being at variance from **NEC** provisions, not by much, but enough to force extensive redesign of manufacturing facilities with no real safety benefit. CMP 1 made the decision that reducing the minimum workspace width in front of a panel from 762 mm (the soft conversion from 30 in.) to 750 mm (the hard conversion) would unacceptably degrade safety, and so that dimension has been retained as a soft conversion.

The rule of part **(D)**, Compliance, addresses the coexistence of the two systems of measurement. There the Code states that use of either SI or English units "shall constitute compliance with this Code." Clearly, designers and installers may use either of the designated values. However, it should be noted that only one, or the other units of measure should be used throughout a given project. Inspectors have raised objection to mixing and matching units of measure.

Chapter One

ARTICLE 100. DEFINITIONS

The NEC reserves Art. 100 to cover the essential definitions required to properly apply its provisions. Not included are general terms that are commonly defined, or technical terms that are used in the same way as in related codes and standards. In addition, if a term is used in only one article, it will be defined within that article and not in Art. 100. Part I of the article applies throughout the NEC; Part II covers definitions that apply only to installations operating over 600 V, nominal. Consult Art. 100 if you are unclear as to how a specialized electrical term is defined that appears in the NEC.

The 2017 NEC began a new practice of calling out the responsible code making panel for each definition. The Correlating Committee, many cycles previous, began assigning actions on specialized definitions to the various panels based on specific expertise. Now these assignments will be transparent to code users.

The Code Making Panel responsible for the requirements in hazardous (classified) locations (CMP-14) is responsible for a group of articles at the beginning of Chap. 5. In the past, relevant definitions were set out in Chap. 5, usually but not exclusively in 500.2. This meant that the same term would commonly appear in at least two articles. Following an instruction from the Correlating Committee for the 2017 cycle, based on the NEC Style Manual but frankly regardless of impact on user friendliness, all such terms are now defined in Article 100. For terminology needing additional explanation, this handbook will continue to address them in their operational locations in Chap. 5. For a summary, refer to Table 100-1 in this handbook located at the end of its coverage of Article 100. Most of the relocated definitions include the following qualifying phrase in their titles: "[as applied to Hazardous (Classified) Locations]" in an attempt to clarify the limited scope of their coverage. However, as noted in Table 100-1, some do not have that qualification.

Accessible (as Applied to Wiring Methods):

Accessible (as Applied to Equipment):

Accessible, Readily (Readily Accessible):

The best way to look at these definitions is to consider all three at the same time because although they are necessarily related, there are important differences. Each of the three terms involves the concept of unimpeded approach. That is, accessible items, whether wiring methods, equipment, or either of these, if readily accessible, must be capable of unimpeded approach as required, but that is about the extent of what these terms have in common.

Wiring methods are accessible if they can be removed or exposed without damaging the building finish or structure. Wiring methods are any of the NEC-recognized techniques for running circuits between equipment, as covered in the articles in Chap. 3 of the Code. Wiring methods are also accessible if they are not permanently closed in by the building structure or finish. Any surface wiring method would obviously qualify if in plain view, but what about above a suspended ceiling? The definition uses the word "exposed," which is also defined in Art. 100 as being on or attached to the surface, or behind panels designed to allow access. Because suspended ceiling panels are clearly designed for that purpose, wiring such as that shown in Fig. 100-1 above a suspended ceiling is exposed, and because it is exposed, it is also accessible.

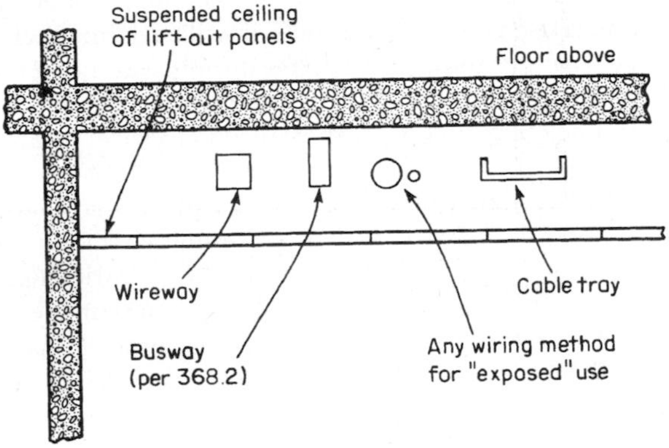
Suspended ceiling
of lift-out panels
Floor above
Wireway
Cable tray
Busway
(per 368.2)
Any wiring method
for "exposed" use

Fig. 100-1.

The same word used to describe equipment does not mean quite the same thing. *Equipment* covers all the products that are connected or hooked up by a recognized wiring method, together with the other components of the wiring system. Equipment *is accessible* if it allows close approach. It is not accessible if it is guarded by a locked door or by height or other barrier that effectively precludes approach by personnel. The word *guarded* is also defined in Art. 100, and it means protected by any of various means to remove the likelihood of "approach or contact by persons or objects to a point of danger."

Consider the busbars in a panelboard located chest high in a corridor, and then think about the panelboard itself, including its enclosing cabinet. Are the busbars themselves "accessible"? No, because they are guarded by the deadfront. Is the panel accessible? Yes, the deadfront makes it safe to approach, and nothing about its location precludes approach. What if the panelboard is for a tenancy, and is located in another tenancy for which access to the supplied tenancy is forbidden? Such a panel would still be accessible, but not to those for whom access is required by the NEC.

This brings us to the final concept, *readily accessible*. This term also applies to equipment, and requires access without climbing over or removing obstacles, or arranging for a ladder or lift to reach the equipment, as covered in Fig. 100-2. Equipment in the open and reachable only by ladder is probably accessible, but could never be considered "readily accessible." Overcurrent (OC) devices are usually required to be readily accessible, but what about a fused switch on an air-conditioning compressor high in the air? This is the reason for the special allowance in 240.24(**A**)(4). It is understood that such equipment is not readily accessible, and a special allowance permits it to be so. Figure 100-3 shows other examples of these special allowances.

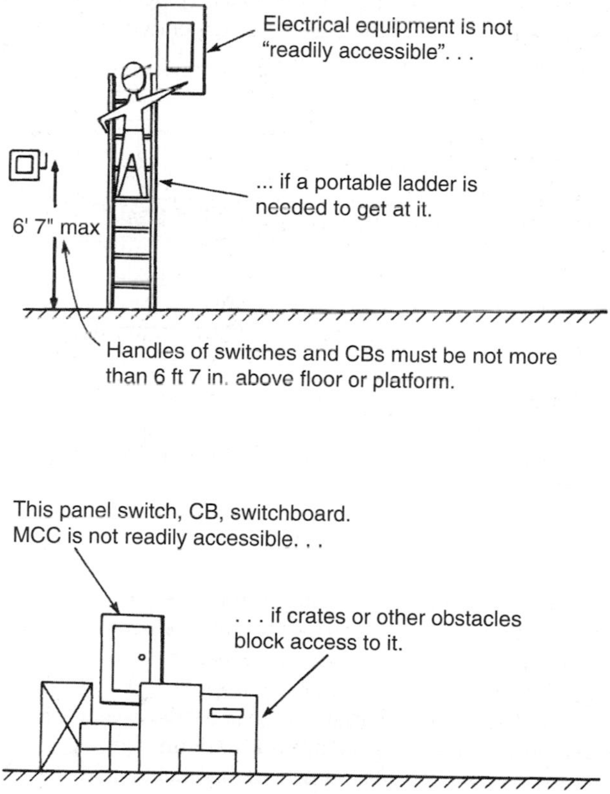
Electrical equipment is not
"readily accessible". . .
... if a portable ladder is
needed to get at it.
6' 7" max
Handles of switches and CBs must be not more
than 6 ft 7 in. above floor or platform.
This panel switch, CB, switchboard.
MCC is not readily accessible. . .
. . . if crates or other obstacles
block access to it.

Fig. 100-2.

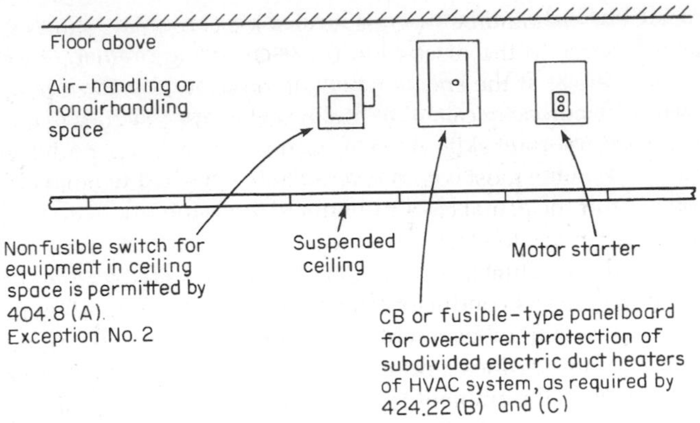
Floor above
Air-handling or nonairhandling space
Nonfusible switch for equipment in ceiling space is permitted by 404.8 (A). Exception No. 2
Suspended ceiling
Motor starter
CB or fusible-type panelboard for overcurrent protection of subdivided electric duct heaters of HVAC system, as required by 424.22 (B) and (C)

Fig. 100-3.

The 2014 NEC first included the concept that the use of tools as a requisite to access disqualified a location as being readily accessible. This was a logical extension of the overall concept of ready access. After all, if local conditions require that those for whom ready access is intended must first find and then operate tools, the inherent delays would likely defeat the purpose. Unfortunately, the sentence structure could have been read to simply require that the location allow for appropriate tools to be used. This has been corrected as of the 2017 NEC. The 2017 NEC also adds "or under" after "to climb over" which may require some discussion with the inspector regarding when an overhead obstruction is low enough to disqualify the access.

There is one other provision in the ready access definition that neatly ties some of the key concepts together. Readily accessible equipment must be reachable quickly by those for *whom ready access is requisite.* This pointedly does not mean everyone. A locked electrical room is a very well-understood concept, and perfectly acceptable as long as those who belong in the room have a key. The 2017 NEC clarified this by adding the parenthetical note "(other than keys)" to the tool restriction, and then adding an informational note that expressly addresses these concepts.

Adjustable Speed Drive (and System):

These two definitions used to be located in Article 430, and were moved to Article 100 in the 2014 cycle in order to apply in both Article 430 and Article 440. The informational note between them properly removes some of the text from the former definition location and reformats it as explanatory material.

Ampacity:

Ampacity is the maximum amount of current in amperes that a conductor may carry continuously under specific conditions of use without exceeding the temperature rating of its insulation. The word "ampacity" is actually a made-up word that does not appear in normal dictionaries and that was first formally defined in the 1965 edition of the NEC. It should be thought of as a logical combination of "amperes" and "capacity." Refer to the discussion on NEC 310.15 in Chap. 3 of this book, together with coverage at the end of this book on Annex D, Example D3(a), for a detailed analysis of ampacity calculations. The calculation of conductor ampacities is one of the most important skills to be learned in the electrical trade, and unfortunately it is also one of the most complicated. There are two key points to raise here, however, in terms of the actual content of the definition.

First, ampacity applies to electrical conductors. Other parts of an electrical system may have current ratings, such as switches, circuit breakers, and motor contactors, but only electrical conductors have an ampacity. Second, ampacity in its true sense cannot be defined by a table in a code book, or even a hundred tables. Every condition of use defines a different ampacity. And every time a condition changes, such as when the ambient temperature changes, the applicable ampacity changes. For example, 12 AWG THHN has an allowable ampacity of 30 A at 30°C with three (or fewer) current-carrying conductors in a raceway. Raise the number of current-carrying conductors in the raceway, or raise the ambient temperature, or both, and the ampacity will decrease by varying degrees, all based on the conditions of actual use.

Approved:

Identified (as Applied to Equipment):

Listed:

These three definitions are covered together in one location, because they cover the three methods of product acceptance recognized by the NEC. They are crucial to the proper application of the Code. Code-making panels (CMPs) have robust discussions every code cycle about which one to apply in a given situation.

The word *approved* means acceptable to the inspectional authority [technically, the authority having jurisdiction (AHJ)], and nothing more or less. It does not mean "identified" unless the inspector chooses to use compliance with the definition of "identified" as the basis for his or her decision. Similarly, it does not mean "listed" unless the inspector chooses that standard as the basis for his or her decision. For this reason, any statements in product literature (and they are common) that something is "approved" by some testing laboratory are necessarily fallacious. A product may be listed by a testing laboratory, but never approved.

The word *identified* is routinely confused with the normal usage in the English language of the word *marked*. It does not mean marked. It means what Art. 100 says it means. It means generally recognizable as suitable for the specific application called out in the NEC requirement. This often comes from product literature generated by manufacturers. This use of the term also correlates with the informational note in 100.3(**A**)(1), where suitability is explained first in terms of "a description marked on or provided with a product to identify the suitability of the product for a specific purpose, environment, or application." The note goes on to indicate that suitability may also be evidenced by listing or labeling, an additional possibility.

For an example of correct usage of this term in a Code rule, the NEC requires two-winding transformers reconnected in the field as autotransformers to be *identified* for use at elevated voltage in 450.4(B). These transformers are frequently listed, but as two-winding transformers. They could not be listed as autotransformers because they do not leave the factory this way, and they have wide application as two-winding transformers. A listing would be excessive because the transformer manufacturers would have to run two production lines with two different labels for the same product. The installer needs to rely on product literature from the manufacturer to verify suitability for reconnection, and fortunately, these manufacturers all provide specific information on how to make the reconnections so the transformers will buck or boost the voltage as desired.

The word *listed* covers the most specific method of product acceptance, because it means that a qualified testing laboratory, usually with testing facilities that an inspector could not possibly duplicate, has performed exhaustive tests to judge the performance of the product under the conditions contemplated in a specific Code rule. The Code note that follows the definition needs some explanation as well. Although the note is written in a general and explanatory manner, in fact, all qualified testing laboratories operating under the current North American electrical safety system do require a label as evidence of the listing. It follows, then, that if a label falls off, the product no longer has the status of being listed. Further, the only way a label can be reapplied is in the presence of an employee of the testing laboratory. Sending labels through the mail is not an option and will result in disciplinary action against the manufacturer by the testing laboratory. The testing laboratories will all send personnel into the field to witness the reapplication of labels.

Be aware that OSHA rules governing workplaces generally require a "listed," "labeled," or otherwise "certified" product to be used in preference to the same "kind" of product that is not recognized by a national testing lab (Fig. 100-4). UL is beginning to use the word "certified" (undefined in the **NEC**) as a synonym for "listed" because the term is more widely recognized in international commerce. See Fig. 100-25(*c*). As of now, the defined term of art in the **NEC** remains "listed."

AN INDEPENDENT LABORATORY • UL ® • TESTING FOR PUBLIC SAFETY •

Fig. 100-4.

Arc-Fault Circuit Interrupter (AFCI):

These are devices designed to protect against arcing failures by recognizing the unique electrical characteristics of the arc and opening the circuit when damaging arcs are present. The definition is not new, but was relocated from 210.12 in the 2011 cycle.

Authority Having Jurisdiction (AHJ):

This definition clarifies the meaning of this term, which is used repeatedly throughout the Code. As indicated by the informational note, the authority having jurisdiction (AHJ) is not necessarily the electrical inspector. In some instances, it may be the head of a fire department or an insurance company representative.

Most jurisdictions have procedures in place that allow for taking an appeal from an adverse decision of an inspectional authority. However, there are inevitable trade-offs in terms of time lost in such a proceeding, so usually only the most compelling instances end up in appellate hearings.

Bathroom:

As defined here and used in the rule of 210.8, a *bathroom* is "an area" (which means it could be a room or a room plus an adjacent space that constitutes a collective area), that contains first a "basin" (usually called a *sink*) and then at least one more plumbing fixture—a toilet, a tub, and/or a shower. A small room with only a basin (a "washroom") is not a bathroom. Neither is a room that contains only a toilet and/or a tub or shower (Fig. 100-5). The 2011 **NEC** added more examples of plumbing fixtures, including showers, urinals, bidets, and then the catch-all

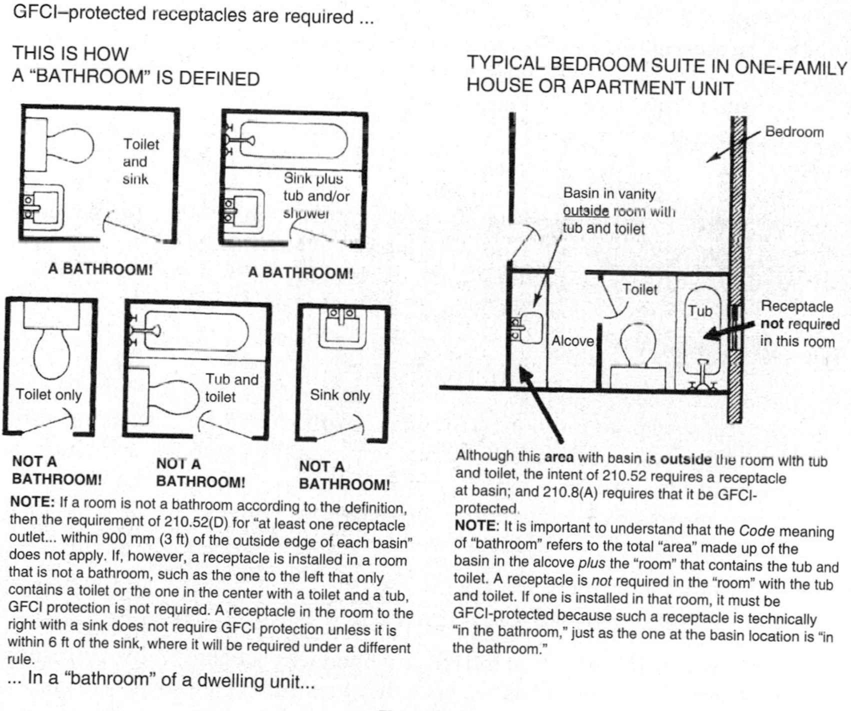
GFCI–protected receptacles are required ...
THIS IS HOW
A "BATHROOM" IS DEFINED
Toilet
and
sink
Sink plus
tub and/or
shower
A BATHROOM!
A BATHROOM!
Toilet only
Tub and
toilet
Sink only
NOT A
BATHROOM!
NOT A
BATHROOM!
NOT A
BATHROOM!
NOTE: If a room is not a bathroom according to the definition, then the requirement of 210.52(D) for "at least one receptacle outlet... within 900 mm (3 ft) of the outside edge of each basin" does not apply. If, however, a receptacle is installed in a room that is not a bathroom, such as the one to the left that only contains a toilet or the one in the center with a toilet and a tub, GFCI protection is not required. A receptacle in the room to the right with a sink does not require GFCI protection unless it is within 6 ft of the sink, where it will be required under a different rule.
... In a "bathroom" of a dwelling unit...
TYPICAL BEDROOM SUITE IN ONE-FAMILY
HOUSE OR APARTMENT UNIT
Bedroom
Basin in vanity
outside room with
tub and toilet
Toilet
Tub
Alcove
Receptacle
not required
in this room
Although this area with basin is outside the room with tub and toilet, the intent of 210.52 requires a receptacle at basin; and 210.8(A) requires that it be GFCI-protected.
NOTE: It is important to understand that the Code meaning of "bathroom" refers to the total "area" made up of the basin in the alcove plus the "room" that contains the tub and toilet. A receptacle is not required in the "room" with the tub and toilet. If one is installed in that room, it must be GFCI-protected because such a receptacle is technically "in the bathroom," just as the one at the basin location is "in the bathroom."

Fig. 100-5.

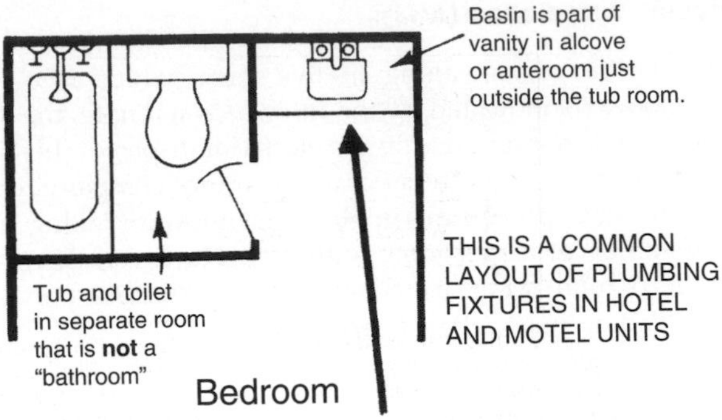
Basin is part of vanity in alcove or anteroom just outside the tub room.
THIS IS A COMMON LAYOUT OF PLUMBING FIXTURES IN HOTEL AND MOTEL UNITS
Tub and toilet in separate room that is **not** a "bathroom"
Bedroom

Guest rooms or suites in hotels and motels are required by 210.60 to have the same receptacle outlets required by 210.52 (A) and (D) for "dwelling units." The requirement for a wall receptacle outlet at the "basin location" applies to bathrooms; and the anteroom area with only a basin is, by definition and intent, part of the "bathroom." Where a guest room or suite has permanent cooking facilities, all receptacle requirements in 210.52 must be satisfied.

... and in a "bathroom" of a hotel/motel guest room.

Fig. 100-6.

wording "similar plumbing fixtures." Figure 100-6 shows applications in hotel and motel bathrooms.

Battery System:

These comprise storage batteries and chargers, at a minimum, and can also include electrical power-processing equipment such as inverters and converters and related control equipment. The definition was relocated from Article 480, and in addition to that article it is also used in Articles 517, 690, 694, 700, and 701.

Bonded (Bonding):

This definition has been simplified and now simply covers the connection of parts in an electrical system to provide continuity and conductivity. This is one of the many definitions and other rules that were impacted by a special task group on grounding and bonding. The definitions have been simplified and the requirements placed in Art. 250, with only special exceptions remaining in other parts of the NEC. The performance criteria for a bonded connection are covered in 250.4.

Bonding Conductor or Jumper:

This is the means of connection between noncurrent-carrying metallic components of the electrical system that are provided to ensure continuity. Examples of bonding jumpers are given in Figs. 100-7 and 100-8. They may be bare, covered,

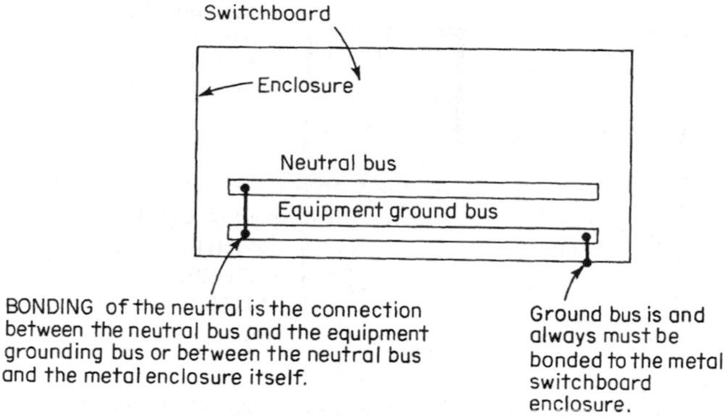
Switchboard
Enclosure
Neutral bus
Equipment ground bus
BONDING of the neutral is the connection between the neutral bus and the equipment grounding bus or between the neutral bus and the metal enclosure itself.
Ground bus is and always must be bonded to the metal switchboard enclosure.

Fig. 100-7.

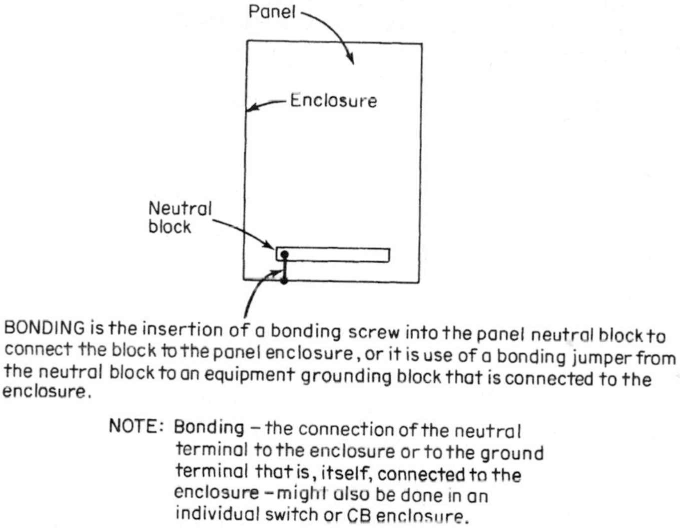
Panel
Enclosure
Neutral block
BONDING is the insertion of a bonding screw into the panel neutral block to connect the block to the panel enclosure, or it is use of a bonding jumper from the neutral block to an equipment grounding block that is connected to the enclosure.
NOTE: Bonding – the connection of the neutral terminal to the enclosure or to the ground terminal that is, itself, connected to the enclosure – might also be done in an individual switch or CB enclosure.

Fig. 100-8.

or insulated conductors, or it may be a mechanical device, such as the 10-32 screws often provided to connect a neutral terminal bar to a service enclosure.

Bonding Jumper, Equipment:

These are bonding connections made between two portions of the equipment grounding system. For example, bonding jumpers are routinely used to ensure an electrically conductive connection between a metal switchboard enclosure and metal conduits entering the open bottom from a concrete floor. If Fig. 100-9 depicted a feeder and not a service, the jumpers from each conduit to the enclosure frame would be equipment bonding jumpers.

Bonding Jumper, Main; System:

A main bonding jumper provides the Code-required connection between the grounded system conductor and the equipment ground bus at the service equipment for a building or structure. The connection between equipment ground and the grounding electrode system in ungrounded services is a "bonding jumper," but not a "main bonding jumper." The connection between the equipment ground bus and the neutral bus in the drawing shown in Fig. 100-9 is an example of a main bonding jumper. The **NEC** maintains a distinction between a main bonding jumper and the same conductor performing the identical function at a separately derived system, which is a *system bonding jumper.* This definition has been relocated from 250.2 to here because it is also used in 708.20(C) Exception, which creates the required second article reference for an Art. 100 definition. Note that the connection may also occur between a grounded circuit conductor and a supply-side bonding jumper, and this is reflected in the current wording.

Branch Circuit:

A branch circuit is that part of a wiring system that (1) extends beyond the final Code-required automatic overcurrent protective device (i.e., fuse or breaker)

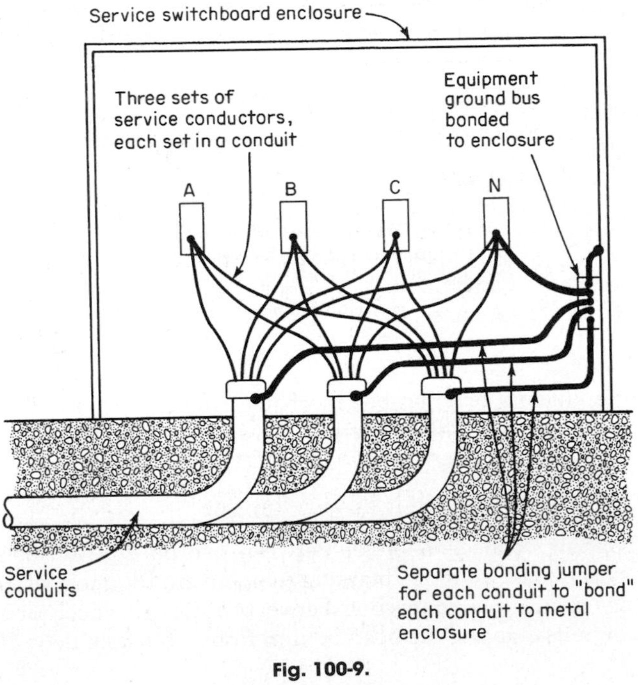

Service switchboard enclosure
Three sets of service conductors, each set in a conduit
Equipment ground bus bonded to enclosure
A
B
C
N
Service conduits
Separate bonding jumper for each conduit to "bond" each conduit to metal enclosure

Fig. 100-9.

which qualifies for use as branch-circuit protection and (2) ends at an outlet, which is another defined term in Art. 100. Thermal cutouts or motor overload devices are not branch-circuit protection. Neither are fuses in luminaires nor in plug connections, which are used for ballast protection or individual fixture protection. Such supplementary overcurrent protection is on the load side of the outlet and is not required by the Code, nor a substitute for the Code-required branch-circuit protection and does not establish the point of origin of a branch circuit. The extent of a branch circuit is illustrated in Fig. 100-10.

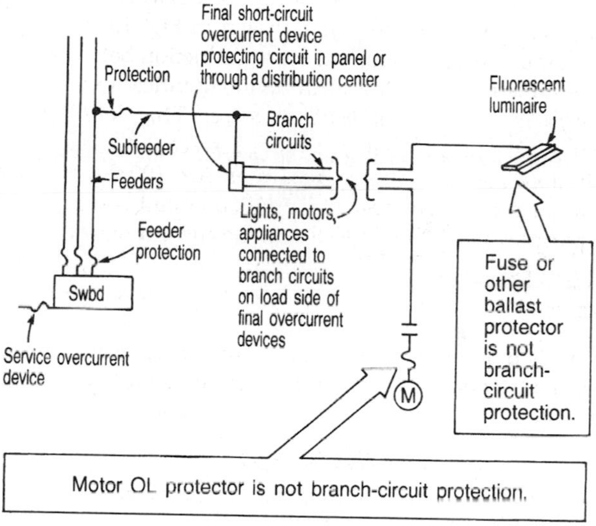
Final short-circuit overcurrent device protecting circuit in panel or through a distribution center
Protection
Branch circuits
Subfeeder
Feeders
Fluorescent luminaire
Lights, motors, appliances connected to branch circuits on load side of final overcurrent devices
Feeder protection
Swbd
Service overcurrent device
Fuse or other ballast protector is not branch-circuit protection.
M
Motor OL protector is not branch-circuit protection.

Fig. 100-10.

Branch Circuit, Appliance:

The point of differentiation between "appliance" branch circuits and "general" branch circuits is related to what is actually connected. For a circuit to be considered an "appliance" branch circuit, it may not supply any lighting, unless that lighting is part of an appliance. Refer to Fig. 100-11.

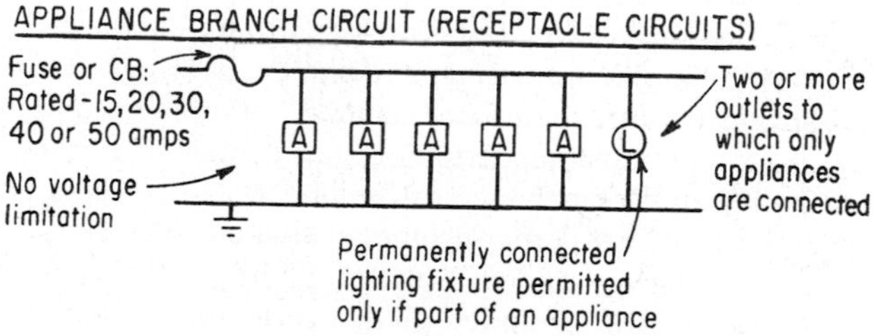
APPLIANCE BRANCH CIRCUIT (RECEPTACLE CIRCUITS)
Fuse or CB: Rated-15,20,30, 40 or 50 amps
A
L
Two or more outlets to which only appliances are connected
No voltage limitation
Permanently connected lighting fixture permitted only if part of an appliance

Fig. 100-11.

Branch Circuit, General Purpose:

Such circuits are identified by the fact that they supply two or more outlets for receptacles, lighting, or appliances. Refer to Fig. 100-12.

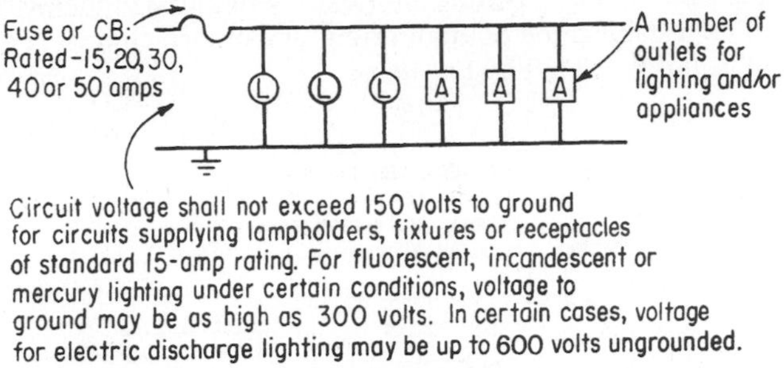
Fuse or CB:
Rated-15,20,30,
40 or 50 amps
L
L
L
A
A
A
A number of
outlets for
lighting and/or
appliances
Circuit voltage shall not exceed 150 volts to ground
for circuits supplying lampholders, fixtures or receptacles
of standard 15-amp rating. For fluorescent, incandescent or
mercury lighting under certain conditions, voltage to
ground may be as high as 300 volts. In certain cases, voltage
for electric discharge lighting may be up to 600 volts ungrounded.

Fig. 100-12.

Branch Circuit, Individual:

As indicated by the term itself, such a branch circuit supplies a single, or "individual" piece of equipment. Refer to Fig. 100-13. A circuit supplying both halves of a duplex receptacle is not an individual branch circuit in most cases, because each half of the duplex is classified as a separate device.

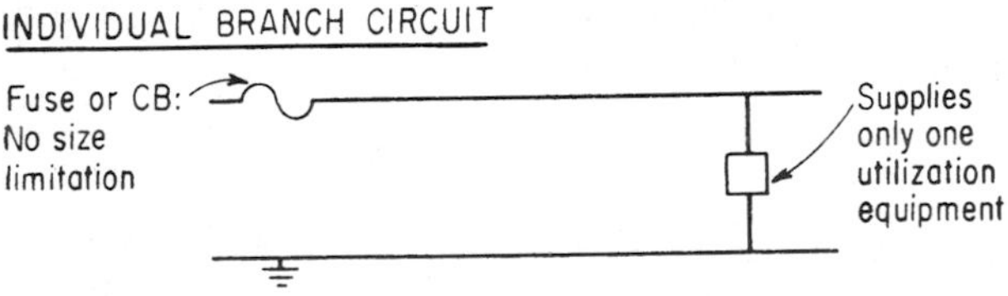
INDIVIDUAL BRANCH CIRCUIT
Fuse or CB:
No size
limitation
Supplies
only one
utilization
equipment

Fig. 100-13.

Branch Circuit, Multiwire:

A multiwire branch circuit must be made up of a neutral or grounded conductor—as in corner-grounded delta systems—and at least two ungrounded or "hot" conductors. The most common multiwire circuits are shown in Fig. 100-14.

A 3-wire, 3-phase circuit (without a neutral or grounded conductor) ungrounded delta system is not a "multiwire branch circuit," even though it does consist of "multi" wires, simply because there is no "neutral" or other grounded conductor. Remember, such a circuit must, by definition, also contain a "grounded" conductor, which may be a neutral, as in the typical 3-phase, 4-wire systems, or a grounded phase conductor, such as in a "corner-grounded" delta system (Fig. 100-15).

Note also that a corner-grounded 3-phase circuit might be interpreted as a multiwire circuit because the grounded phase conductor, while not connected to

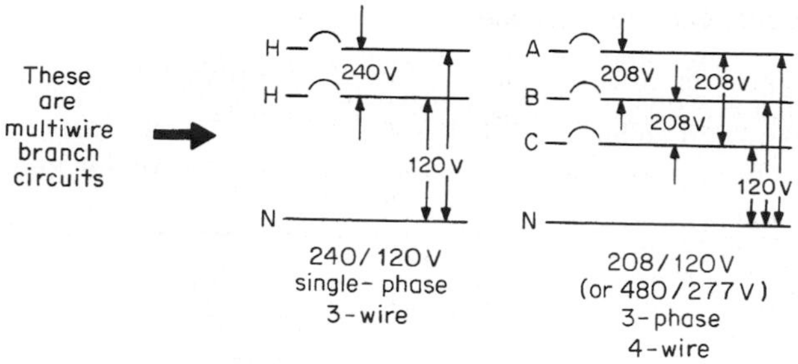
These are multiwire branch circuits
H
H
N
240 V
120 V
240/120 V
single-phase
3-wire
A
B
C
N
208V
208V
208V
120V
208/120V
(or 480/277V)
3-phase
4-wire

Fig. 100-14.

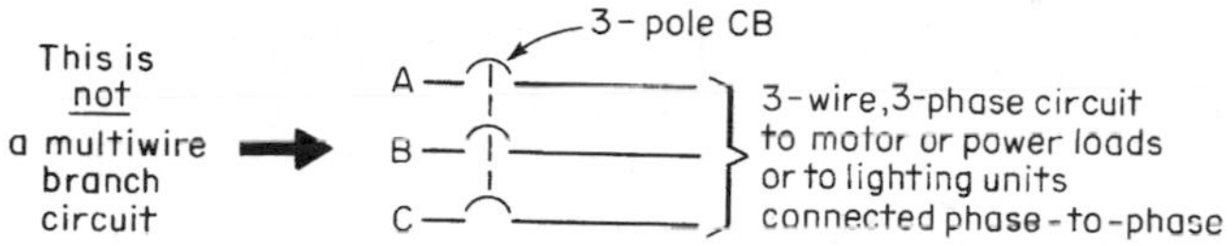
This is not a multiwire branch circuit
3-pole CB
A
B
C
3-wire,3-phase circuit to motor or power loads or to lighting units connected phase-to-phase

Fig. 100-15.

a neutral, is connected to a grounded conductor. The word "or" at this location originated in the 1993 NEC and is only supposed to recognize that neutrals are grounded circuit conductors. It was never intended to describe a 3-phase corner-grounded power circuit. Now that the word "neutral" is formally defined in Art. 100, the simplest way to correct this would be to delete the words "or grounded" from the definition.

Building:

Most areas have building codes to establish the requirements for buildings, and such codes should be used as a basis for deciding the use of the definition given in the National Electrical Code. The use of the term *fire walls* in this definition has resulted in differences of opinion among electrical inspectors and others. Since the definition of a fire wall may differ in each jurisdiction, the processing of an interpretation of a "fire wall" has been studiously avoided in the National Electrical Code because this is a function of building codes and not a responsibility of the National Electrical Code.

In most cases, a "building" is easily recognized by its stand-alone nature. However, one or more "fire walls" also establishes two (or more) buildings in one structure. It is frequently crucial to distinguish between a "fire-separation wall" (or however the local building code describes it) and a "fire wall." As discussed here, a "fire wall" is made of concrete and masonry and will still be standing after a conflagration on one side proceeds to complete destruction. A "fire-separation wall" may consist of several layers of drywall and will have a rating in hours, designed to assure time for the occupants to exit. They are fundamentally different, in kind and not just degree.

Many, many code rules depend on whether a structure comprises multiple buildings, such as whether multiple services will be permitted, which grounding rules will apply at which locations, and whether residential occupancies separated by such construction will be classified as single-family or multiple-family housing. Where in doubt, check with your local electrical inspector for guidance. If the electrical inspector doesn't know, or doesn't have jurisdiction over this particular decision, then the electrical inspector should be able to direct you to the proper authority for a determination. This is a good example of where the AHJ may be the local building commissioner and not the electrical inspector.

Cabinet/Cutout Box:

There are two distinguishing characteristics that differentiate a "cabinet" from a "cutout box." The first is the *physical construction*. The door of a cabinet is (or could be) hinged to a trim covering wiring space, or gutter. The door of a cutout box is hinged (or screwed) directly to the side of the box. The other distinction is *mounting*. Cabinets may be surface- or flush-mounted, while cutout boxes may only be surface-mounted. In terms of use, cabinets usually contain panelboards; cutout boxes contain cutouts, switches, or miscellaneous apparatus.

Cable Routing Assembly:

This is a "channel" and not a raceway that serves as a support and routing mechanism for a variety of power-limited cabling methods, including all forms of communications cabling. It was relocated from Art. 770 because these assemblies apply to wiring covered throughout the limited energy articles in Chap. 7 as well as Chap. 8. Note that when this definition made the trip into Art. 100, the former qualification that it support "high densities" of wiring was dropped. Although such assemblies frequently carry high cable densities, they may start off with few cables. The density aspect is plainly irrelevant to the description of the physical assembly.

Communications Equipment:

Although the definition remains relatively self explanatory, the 2017 NEC added an informational note that will be of increasing importance as the technology of communications increasingly comes to rely on digital processing of information. The distinction between a *signaling circuit* covered in Chap. 7 (Art. 725) and a *communications circuit covered* in Chap. 8 (defined in Art. 800) continues to blur. Pay particular attention to the aspect of the communications utility source as a key factor in the latter definition.

Communications Raceway:

This definition, also relocated from the back end of the Code (in this case Art. 800), describes an "enclosed channel" designed to contain communications cables along with optical fiber cables and data communications wiring for Class-2 and -3 circuits. This is the raceway counterpart to the cable-routing assembly.

Concealed:

Any electrical equipment or conductors that are closed in by structural surfaces are considered to be "concealed," as shown in Fig. 100-16.

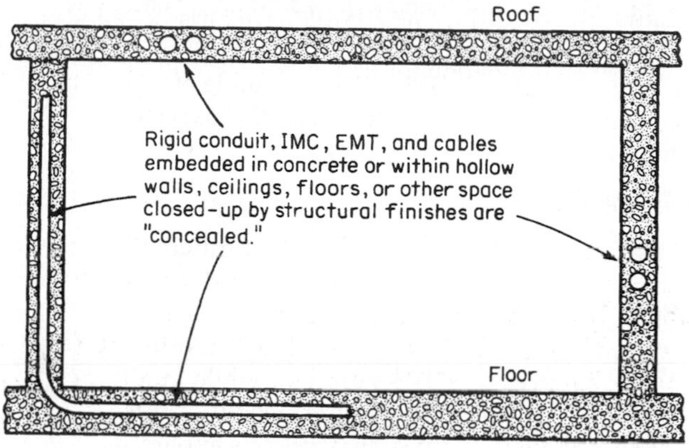
Roof
Rigid conduit, IMC, EMT, and cables embedded in concrete or within hollow walls, ceilings, floors, or other space closed-up by structural finishes are "concealed."
Floor

Fig. 100-16.

Circuits run in an unfinished basement or an accessible attic are not "rendered inaccessible by the structure or finish of the building," and are therefore considered as exposed work rather than a concealed type of wiring. Equipment and wiring in hung-ceiling space behind lift-out panels and underneath raised floors beneath removal tiles are also considered "exposed."

Conduit Body:

The last sentence notes that FS and FD boxes—as well as larger cast or sheet metal boxes—are not considered to be "conduit bodies," as far as the **NEC** is concerned. Although some manufacturers' literature refers to FS and FD boxes as conduit fittings, care must be used to distinguish between "conduit bodies" and "boxes" in specific Code rules. For instance, the first sentence of 314.16(C)(2) limits splicing and use of devices to conduit bodies that are "durably and legibly" marked with their cubic inch capacity by the manufacturer. However, FS and FD boxes are not conduit bodies and may contain splices and/or house devices. Table 314.16(A) lists FS and FD boxes as "boxes." See Fig. 100-17.

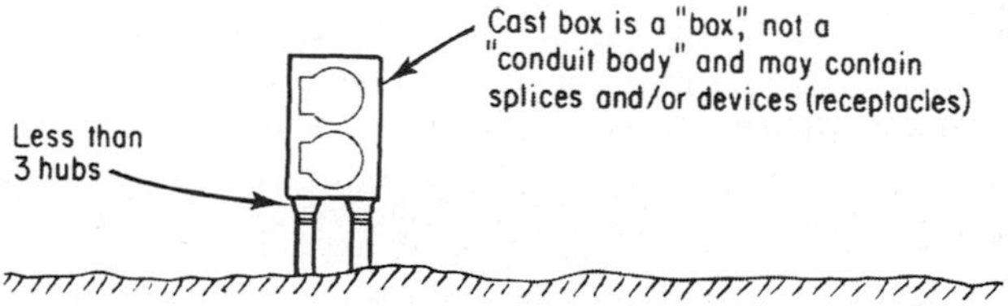
Cast box is a "box," not a "conduit body" and may contain splices and/or devices (receptacles)
Less than 3 hubs

Fig. 100-17.

Continuous Load:

Any condition in which the maximum load current in a circuit flows without interruption for a period of not less than 3 h. Although somewhat arbitrary, the 3-h period establishes whether a given load is continuous. If, for example, a load were energized for 2 h, 59 min, 59 s, then switched off and immediately reenergized, it would technically be a "noncontinuous" load. This is an extreme example, but that is the Code-prescribed evaluation for this important definition.

Control Circuit:

Although this terminology is used in 18 different code articles, it was only defined in Art. 430 prior to the 2014 NEC. It now clearly is to be applied in a consistent way.

Coordination (Selective):

This term refers to the design concept whereby an individual fault will be cleared by the OC protective closest to the faulted circuit or equipment. This design goal is achieved by studying the time–current trip curves of the selected devices and ensuring that the operating characteristics of all selected OC devices are such that the fuse or breaker closest to a fault will blow or open before OC devices upstream (toward the service) operate. This has become mandatory for the main overcurrent protective devices for elevators (620.62), and for protective devices generally for applications covered by 700.32, 701.27, and 708.54, and also for some fire-pump feeders as covered in 695.3(C)(3).

The 2014 NEC removed any doubt that this terminology applies to the effect of any fault, regardless of duration. The intent of those who have made selective coordination mandatory in various code sections is that it be applied in exactly this way. This has far-reaching effects on the design of electrical circuits that are ensnared in this definition. Refer to the discussion of 700.32 in Chap. 7 of this book for an extensive discussion of the significant policy issues raised by code requirements that rely on this definition.

Demand Factor:

The following discussion provides a distinction between two very closely related, but different concepts. For the purposes of NEC application, any design or application of "demand factors" that results in a feeder or service smaller than would be permitted by the applicable rules of the NEC, such as Art. 220, is a violation. From a practical standpoint in new construction, this generally should not be a problem because NEC requirements are essentially bare minimums and provide absolutely no additional capacity. That precludes system expansion and supply of additional loads in the future, which, of course is poor design. Because design goals should, and typically do, include consideration of potential future needs, actual ratings and sizes of selected equipment and conductors should be larger than the Code-required minimum. BUT, if a designer calculates a load that is less than would be permitted by the Code, the larger, Code-mandated load shall be accommodated by selection of equipment and conductors that are adequate to supply the Code-complying load.

Two terms constantly used in electrical design are "demand factor" and "diversity factor." Because there is a very fine difference between the meanings for the words, the terms are often confused.

Demand factor is the ratio of the maximum demand of a system, or part of a system, to the total connected load on the system, or part of the system, under consideration. This factor is always less than unity.

Diversity factor is the ratio of the sum of the individual maximum demands of the various subdivisions of a system, or part of a system, to the maximum demand of the whole system, or part of the system, under consideration. This factor generally varies between 1.00 and 2.00.

Demand factors and diversity factors are used in design. For instance, the sum of the connected loads supplied by a feeder is multiplied by the demand factor to determine the load for which the feed must be sized. This load is termed the maximum demand of the feeder. The sum of the maximum demand loads for a number of subfeeders divided by the diversity factor for the subfeeders will give the maximum demand load to be supplied by the feeder from which the subfeeders are derived.

It is a common and preferred practice in modern design to take unity as the diversity factor in main feeders to loadcenter substations to provide a measure of spare capacity. Main secondary feeders are also commonly sized on the full value of the sum of the demand loads of the subfeeders supplied.

From power distribution practice, however, basic diversity factors have been developed. These provide a general indication of the way in which main feeders can be reduced in capacity below the sum of the demands of the subfeeders they supply. On a radial feeder system, diversity of demands made by a number of transformers reduces the maximum load that the feeder must supply to some value less than the sum of the transformer loads. Typical application of demand and diversity factors for main feeders is shown in Fig. 100-18.

Device:

Switches, fuses, circuit breakers, controllers, receptacles, and lampholders are examples of "devices" that "carry or control" electricity as their principal function. The fact that they may use incidental quantities of power in the process does not affect their principal function.

Dwelling:

Because so many Code rules involve the words "dwelling" and "residential," there have been problems applying Code rules to the various types of "dwellings"—one-family houses, two-family houses, apartment houses, condominium units, dormitories, hotels, motels, etc. The **NEC** includes terminology to eliminate such problems and uses definitions of "dwelling" coordinated with the words used in specific Code rules.

A *dwelling unit* is defined as a single unit that provides "complete and independent living facilities for one or more persons." It must have "permanent provisions for living, sleeping, cooking, and sanitation." A one-family house is a "dwelling unit." So is an apartment in an apartment house or a condominium unit. And a

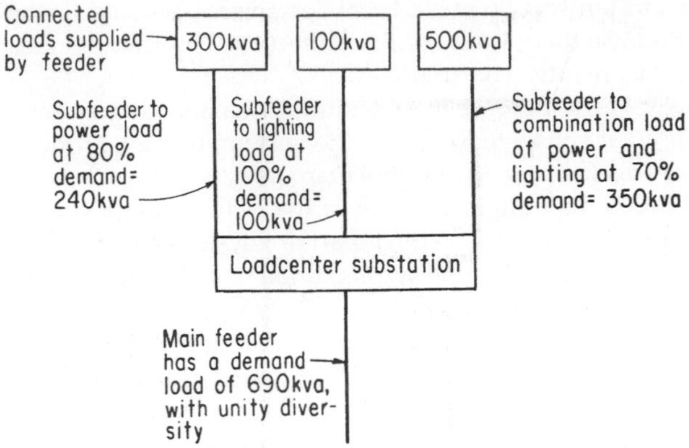
Connected
loads supplied
by feeder
300kva
100kva
500kva
Subfeeder to
power load
at 80%
demand=
240kva
Subfeeder
to lighting
load at
100%
demand=
100kva
Subfeeder to
combination load
of power and
lighting at 70%
demand= 350kva
Loadcenter substation
Main feeder
has a demand
load of 690kva,
with unity diver-
sity

1. Sum of individual demands = 240 + 100 + 350 = 690 kva.
2. Sizing the substation at unity diversity, the required

$$\text{kva} = \frac{690}{1.00} = 690 \text{ kva.}$$

3. To meet this load, use a 750-kva substation.
4. If analysis dictates the use of a diversity factor of 1.4 , the

$$\text{required kva} = \frac{690}{1.40} = 492 \text{ kva.}$$

5. To meet this load, use a 500-kva substation.
6. Primary feeder to unit substation must have capacity to match the substation load.

Fig. 100-18.

guest room in a hotel or motel or a dormitory room or unit is a "dwelling unit" if it contains permanent provisions for "cooking." Refer to the commentary on guest rooms for more information on this point.

Any "dwelling unit" must include all the required elements shown in Fig. 100-19.

Effective Ground-Fault Current Path:

This definition was relocated from 250.2, because it applies in numerous locations in the NEC. It was modified by removing the final phrase "on high-impedance grounded systems" because it may apply in other locations. For example, it is a mandatory feature of most ungrounded systems because a ground fault that develops into simultaneous faults from conductors fed from two different phases must clear as quickly as a faulted phase conductor on a grounded system. Refer to the discussion in 250.4 for more information.

Electrical Circuit Protective System:

This definition was relocated in the 2017 NEC. It is a very important concept to which many provisions in the NEC refer, either implicitly or by express reference.

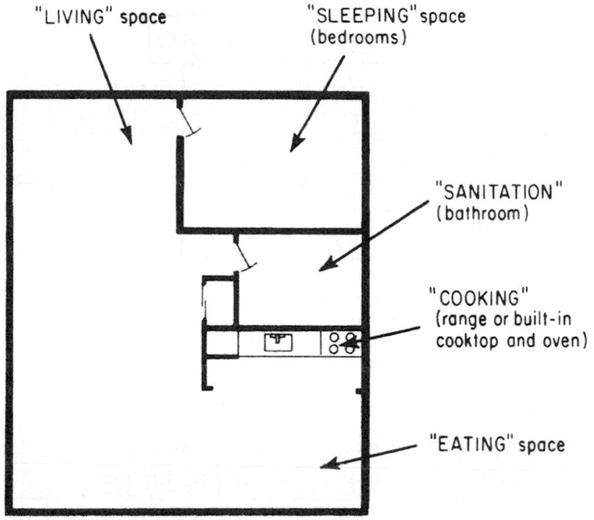
"LIVING" space
"SLEEPING" space
(bedrooms)
"SANITATION"
(bathroom)
"COOKING"
(range or built-in
cooktop and oven)
"EATING" space

NOTE: Eating, living, and sleeping space could be one individual area, as in an efficiency apartment. But the unit must contain a "bathroom," defined in Art. 100 as "an area including a basin with one or more of the following: a toilet, a tub, or a shower." And the unit must contain permanent cooking equipment.

Fig. 100-19.

Now that the NEC has an article devoted to a key aspect involving this topic, this book covers it at 728.4.

Exposed (as Applied to Wiring Methods):

Wiring methods and equipment that are not permanently closed in by building surfaces or finishes are considered to be "exposed." See Fig. 100-20.

Feeder:

A *feeder* is a set of conductors that carry electric power from the service equipment (or from a transformer secondary, a battery bank, or a generator switchboard where power is generated on the premises) to the overcurrent protective devices for branch circuits supplying the various loads. Basically stated, any conductors between the service, separately derived system, or other source of supply and the branch-circuit protective devices are "feeders."

A feeder may originate at a main distribution center and feed one or more subdistribution centers, one or more branch-circuit distribution centers, one or more branch circuits (as in the case of plug-in busway or motor circuit taps to a feeder), or a combination of these. It may be a primary or secondary voltage circuit, but its

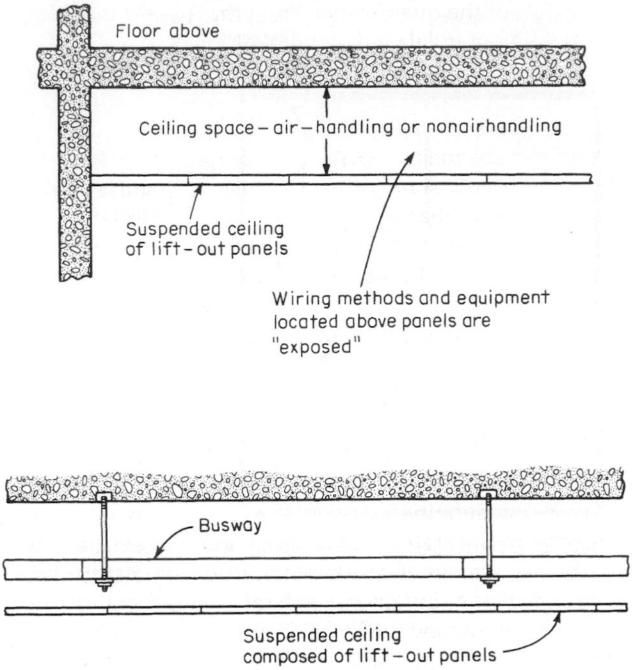
Floor above
Ceiling space – air – handling or nonairhandling
Suspended ceiling
of lift - out panels
Wiring methods and equipment
located above panels are
"exposed"
Busway
Suspended ceiling
composed of lift – out panels

BUSWAY above this suspended ceiling is considered "exposed" as required by 368.10.

Fig. 100-20.

function is always to deliver a block of power from one point to another point at which the power capacity is apportioned among a number of other circuits. In some systems, feeders may be carried from a main distribution switchboard to subdistribution switchboards or panelboards from which subfeeders originate to feed branch-circuit panels or motor branch circuits. In still other systems, either or both of the two foregoing feeder layouts may be incorporated with transformer substations to step the distribution voltage to utilization levels. In any of these described scenarios, the conductors would be considered to be feeders because they interconnect the service and branch circuit.

Field Evaluation Body (FEB):

Field Labeled (as applied to evaluated products):

These definitions (the first being actually extracted text) come from comparatively new NFPA standards addressing the long-standing, and heretofore poorly regulated, process of providing field evaluations of equipment by testing laboratories. NFPA 790 is the Standard for Competency of Third-Party Field Evaluation Bodies, and NFPA 791 is the Recommended Practice and Procedures for Unlabeled Electrical Equipment Evaluation. The inspector now has written standards on which to base an evaluation of the qualifications of a third-party intending to judge the performance suitability of unlabeled equipment.

Ground:

In another example of the major reevaluation of definitions involving grounding concepts, the ground is now simply the planet earth. There is no longer any reference to a conductive body that serves in its place. For example, a little portable generator is no longer classified as being connected to ground just because a connection may have been made to the generator frame. Since the definition no longer refers to connections to the earth, it is no longer correct to refer to insulation failures and the like as grounds; instead, they should be described as the ground faults they really are.

Grounded (Grounding):

Here again, the concept of a conductive body serving in place of the earth has been discontinued. The definition now applies only to connections to the planet earth, either directly or through a conductive body that extends the ground connection. Although the concept of conductive entities serving in place of the earth still survives in such areas as motor vehicles and railroad rolling stock, these areas are generally beyond the scope of the NEC. Recreational vehicles (RVs) are covered, but even there most of the equipment and systems affected by this change are those connected to premises wiring in RV parks, for which a connection to the earth is routine.

Grounded Conductor:

Here the Code distinguishes between a "ground*ing*" conductor and a "ground*ed*" conductor. A ground*ed* conductor is the conductor of an electrical system that is intentionally connected to earth via a grounding electrode conductor and a grounding electrode at the service of a premises, at a transformer secondary, or at a generator or other source of electric power. See Fig. 100-21. It is most commonly a neutral conductor of a single-phase, 3-wire system or 3-phase, 4-wire system but may be one of the phase legs—as in the case of a corner grounded delta system.

Grounding one of the wires of the electrical system is done to limit the voltage upon the circuit that might otherwise occur through exposure to lightning or other voltages higher than that for which the circuit is designed. Another purpose in grounding one of the wires of the system is to limit the maximum voltage to ground under normal operating conditions. Also, a system that operates with one of its conductors intentionally grounded will provide for automatic opening of the circuit if an accidental or fault ground occurs on one of its ungrounded conductors.

Selection of the wiring system conductor to be grounded depends upon the type of system. In 3-wire, single-phase systems, the midpoint of the transformer winding—the point from which the system neutral is derived—is grounded. For grounded 3-phase, 4-wire wiring systems, the neutral point of the wye-connected transformer(s) or generator is usually the point connected to ground. In delta-connected transformer

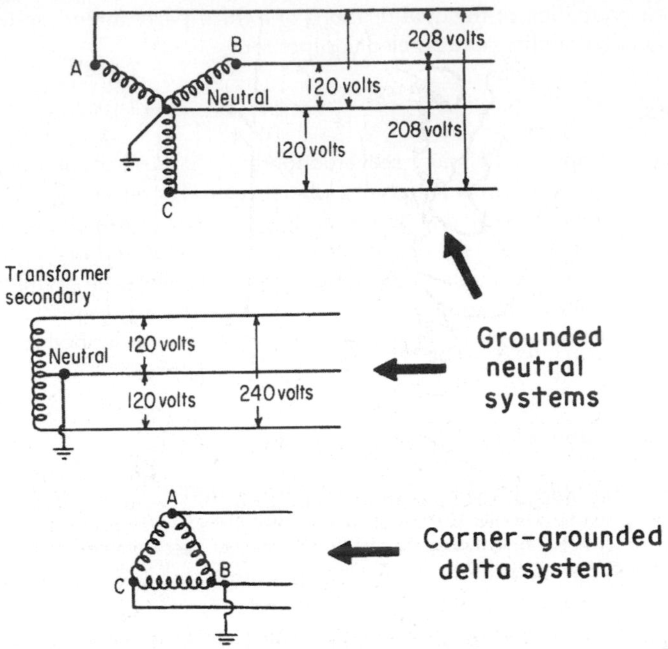

A
B
C
Neutral
208 volts
120 volts
208 volts
120 volts
Transformer secondary
Neutral
120 volts
120 volts
240 volts
Grounded neutral systems
A
B
C
Corner-grounded delta system

Fig. 100-21.

hookups, grounding of the system can be affected by grounding one of the three phase legs, by grounding a center-tap point on one of the transformer windings (as in the 3-phase, 4-wire "red-leg" delta system), or by using a special grounding transformer, which establishes a neutral point of a wye connection that is grounded.

Ground-Fault Circuit Interrupter (GFCI):

This revised definition makes clear that the device described is a GFCI (breaker or receptacle) of the type listed by Underwriters Laboratories Inc. (UL) and intended to eliminate shock hazards to people. "Class A" devices must operate within a definite time from initiation of ground-fault current above the specified trip level (4–6 mA, as specified by UL). See Fig. 100-22. It should be noted that this is *not* the protective device called for by the rule of 210.12. That section calls for the use of a device called an *arc-fault circuit interrupter*, or AFCI, which is required for protection specifically against high-resistance arcing-ground-faults, particularly in residential applications. (See 210.12.)

There are essentially two types of Class A GFCIs: those intended to be permanently installed and those intended for temporary power use. It is important that only those listed as "temporary power" GFCIs be used to satisfy the rules of 590.6 and 525.23(A). That caution is based on the fact that GFCIs listed for temporary power are tested differently than those intended for permanent installation and, as a result, only those listed for temporary power applications may be used for

Fig. 100-22. GFCI protection required for temporary power applications, as covered in 590.6, should be *listed* for temporary power use. Refer to the caption for Fig. 590-13 at the end of Chap. 5 for more information on this topic.

temporary power. There are also "Class B" GFCIs with 20-mA trips; these are only for use with underwater swimming pool luminaires installed before local adoption of the 1965 **NEC** and they are seldom applied today. For all other Code rules requiring GFCIs, those Class A devices listed for permanent installation may be used.

Ground-Fault Current Path:

This definition, like "Effective Ground-Fault Current Path," was also relocated from Art. 250. Unlike its companion, which described the intentionally constructed path, this definition covers what actually happens during a ground fault even if the return is too ineffective to open a circuit-protective device. The informational note bears this out, and the functional differences between these definitions must be kept clearly in mind. This terminology is now included within the definition of equipment grounding conductors.

Ground-Fault Protection of Equipment:

Although any type of ground-fault protection is aimed at protecting personnel using an electrical system, the so-called ground-fault protection required by 215.10, 230.95, and 240.13 for 480Y/277-V disconnects rated 1000 A or more, for example, is identified in 230.95 as "ground-fault protection of equipment (GFPE)." This is essential because a 480/277-V system has an instantaneous peak voltage to ground of 277 V × √3 = 392 V. This voltage is frequently enough to constantly reignite an arc powered by a failed phase leg. The result is an arcing burndown that is extremely destructive. The so-called ground-fault circuit interrupter (GFCI), as described in the previous definition and required by

210.8 for residential receptacles and by other NEC rules, is essentially a "people protector" and is identified in 210.8 as "ground-fault protection for personnel." Because there are Code rules addressing these distinct functions—people protection versus equipment protection—this definition distinguishes between the two types of protection.

Note that there are other protective devices that provide equipment protection and not personnel protection, but that typically operate in the 30-mA range. For example, pipe tracing circuits covered in 427.22 require this protection because a ground fault in a pipe tracing cable can sputter for a very long time without tripping an overcurrent device, given the inherent resistance of this equipment. A GFPE device will de-energize the failed cable promptly.

Grounding Conductor, Equipment (EGC):

The phrase "equipment grounding conductor" is used to describe any of the electrically conductive paths that tie together the noncurrent-carrying metal enclosures of electrical equipment in an electrical system. The term *equipment grounding conductor* includes bare or insulated conductors, metal raceways [rigid metal conduit, intermediate metal conduit, electrical metallic tubing (EMT)], and metal cable jackets where the Code permits such metal raceways and cable enclosures to be used for equipment grounding—which is a basic Code-required concept as follows:

Equipment grounding is the intentional electrical interconnection of all metal enclosures that contain electrical wires or equipment with the grounding electrode conductor (all systems) and with the grounded conductor of the system (grounded systems only). When an insulation failure occurs in such enclosures on ungrounded systems, the result is the system simply becomes corner or otherwise system grounded at the fault, and no hazardous voltage will be present on the enclosures. However, it is still important to correct the insulation failure promptly and the NEC now requires ground detectors on all such systems for this reason. If a second insulation failure happens to occur on a second phase before the first one is fixed, the result will be a line-to-line short circuit flowing through a potentially very long equipment grounding run, perhaps between opposite ends of the factory. A single loose locknut or forgotten setscrew could easily generate a sustained arc in such a case before overcurrent devices operate, with severe consequences and a dangerous voltage on the intervening enclosures while the failure is in progress.

When the insulation failure occurs on a grounded system, equipment grounding serves to ensure adequate current flow to cause the affected circuit's overcurrent protective device to "open," usually in the instantaneous portion of the overcurrent device tripping curve. This prevents the enclosures from remaining energized, which would otherwise constitute a shock or fire hazard. Simply stated, proper connections of all metallic enclosures of electric wires and equipment to each other and to the system grounded conductor (shown in Fig. 100-23) prevent any potential-above-ground on the enclosures.

Workmanship and attention to detail are crucial to the proper implementation of these concepts; a single poor connection can easily reduce the current flowing in a ground fault so it falls into the overload portion of the overcurrent device trip curve. In effect, the fuse or circuit breaker acts as though the arcing fault is a motor

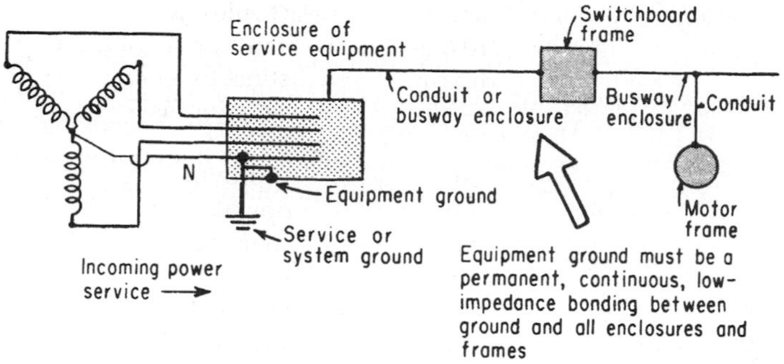
Enclosure of
service equipment
Switchboard
frame
Conduit or
busway enclosure
Busway
enclosure
Conduit
N
Equipment ground
Motor
frame
Service or
system ground
Incoming power
service
Equipment ground must be a
permanent, continuous, low-
impedance bonding between
ground and all enclosures and
frames

Fig. 100-23.

trying to start, and by the time the device finally trips a fire is in progress and the damage to the electrical system can easily involve an outage lasting many weeks.

Grounding Electrode:

The grounding electrode is any one of the building or structural elements recognized in 250.52 that is in actual physical contact with the earth.

Grounding Electrode Conductor:

Basically stated, this is the connection between either the grounded conductor of a grounded electrical system (typically the neutral) and the grounding electrode system, or the connection between the equipment ground bus and the grounding electrode system for ungrounded systems. The conductor that runs from the bonded neutral block or busbar or ground bus at service equipment, separately derived systems, or main building disconnects to the system grounding electrode is clearly and specifically identified as the "grounding electrode conductor." See Fig. 100-24. It should be noted that "main building disconnects" referred to here are those that would be required where one building receives its supply from another as covered in Part II of Art. 225.

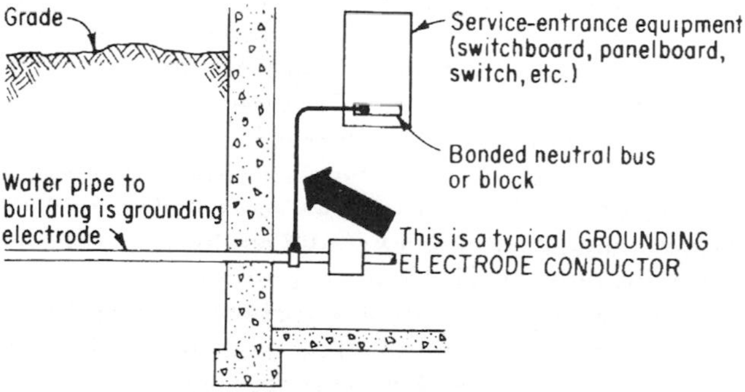
Grade
Service-entrance equipment
(switchboard, panelboard,
switch, etc.)
Bonded neutral bus
or block
Water pipe to
building is grounding
electrode
This is a typical GROUNDING
ELECTRODE CONDUCTOR

Fig. 100-24.

Guest Room:

The only difference between a dwelling unit and a guest room hinges on whether or not provisions for cooking—either permanently installed or cord-and-plug connected—are present. Where microwaves or other types of cooking equipment are not present, then the location is a guest room. If such items are present, then the occupancy is a dwelling unit if the cooking equipment is permanently installed. A loose cord-and-plug-connected microwave oven will not trigger a reclassification, unless it is permanently installed into or below a cabinet.

Handhole Enclosure:

This definition describes any one of a number of small to medium-sized in-ground pull and junction boxes for use in underground distribution, covered in detail in 314.30.

Hermetic Refrigerant Motor-Compressor:

This definition was relocated from Art. 440 and describes a motor operating in its own refrigerant, and which can therefore operate under load conditions that would overload a conventional motor cooled by air.

Hybrid System:

This definition appeared in both Art. 690 and in Art. 705, and has been relocated here where it belongs. The definition clarifies that energy storage mechanisms are not elements of a hybrid power source for code purposes. In addition the 2014 NEC made clear that regenerative power supplied by a descending elevator does not create a hybrid system. The amount of regenerated power is not significant and what is recovered is quickly taken up by adjacent elevators while in ascent.

Identified (as Applied to Equipment):

This term is covered together with "Approved" (and also "Listed") as part of the discussion of "Approved" and related standards for product acceptance near the beginning of this chapter.

In Sight From (Within Sight From, Within Sight):

The phrase "in sight from" or "within sight from" or "within sight" means visible and not more than 50 ft away. These phrases are used in many Code rules to establish installation location of one piece of equipment with respect to another. A typical example is the rule requiring that a motor-circuit disconnect means must be in sight from the controller for the motor [430.102(A)]. This definition in Art. 100 gives a single meaning to the idea expressed by the phrases—not only that any piece of equipment that must be "in sight from" another piece of equipment must be visible, but also that the distance between the two pieces of equipment must not be over 50 ft. If, for example, a motor disconnect is 51 ft away from the motor controller of the same circuit, it is not "within sight from" the controller even though it is actually and readily visible from the controller. In the interests of safety, it is arbitrarily defined that separation of more than 50 ft diminishes visibility to an unacceptable level.

There are places in the NEC where the wording of rules takes these limitations into account. For example, 610.32(2) allows certain crane disconnects to be "within view" (and not "within sight") of certain equipment. This is because on large cranes it may be impossible to meet the 50 ft limitation, and yet the disconnect can still be seen and will be capable of being locked in the open position.

Industrial Control Panel:

This definition was moved from Art. 409 because the term is also used Art. 110. Essentially, an industrial control panel is an enclosure that houses motor starters and/or other control equipment. The definition incorporates all control assemblies of two or more components as being within the scope of these rules. To provide some context, all combination starters, with a disconnect switch and fuse block or a circuit breaker and contactor with running overload protection are included.

Interrupting Rating:

This definition covers both "interrupting ratings" for overcurrent devices (fuses and circuit breakers) and "interrupting ratings" for control devices (switches, relays, contactors, motor starters, etc.).

Labeled:

The label of a nationally recognized testing laboratory on a piece of electrical equipment is a sure and ready way to be assured that the equipment is properly made and will function safely when used in accordance with the application data and limitations established by the testing organization. Each label used on an electrical product gives the exact name of the type of equipment as it appears in the listing book of the testing organization.

Typical labels are shown in Fig. 100-25(*a*).

Underwriters Laboratories Inc., the largest nationally recognized testing laboratory covering the electrical field, describes its "Identification of Listed Products," as shown in Fig. 100-25(*b*).

It should be noted that the definitions for "labeled" and "listed" in the NEC do not require that the testing laboratory be "nationally recognized." But OSHA rules do require such "labeling" or "listing" to be provided by a "nationally recognized" testing lab. Therefore, even though those NEC definitions acknowledge that a local inspector may accept the label or listing of a product by a testing organization that is qualified and capable even though it operates in a small area or section of the country and is not "nationally recognized," OSHA requirements may only be satisfied when "labeling" or "listing" is provided by a "nationally recognized" testing facility.

By universal test lab policies, the *label* is the field evidence of the listing. If the label falls off, the product is no longer presumed to be listed and it can only be

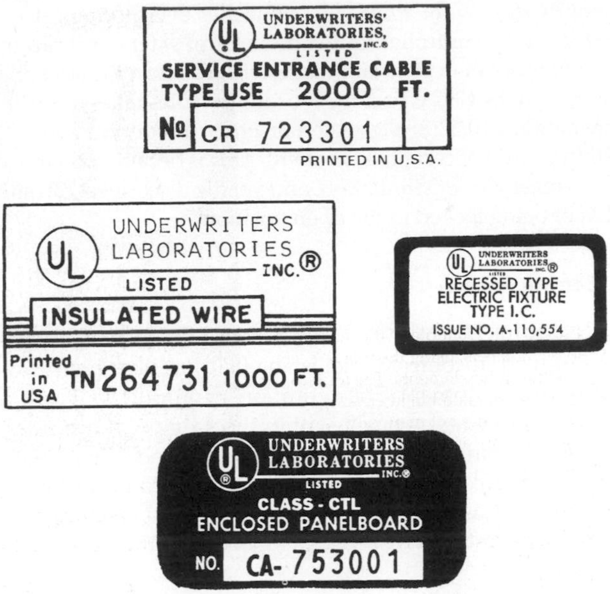
UNDERWRITERS' LABORATORIES, INC.®
LISTED
SERVICE ENTRANCE CABLE
TYPE USE 2000 FT.
№ CR 723301
PRINTED IN U.S.A.
UNDERWRITERS LABORATORIES INC.®
LISTED
INSULATED WIRE
Printed in USA TN 264731 1000 FT.
UNDERWRITERS LABORATORIES INC.®
RECESSED TYPE
ELECTRIC FIXTURE
TYPE I.C.
ISSUE NO. A-110,554
UNDERWRITERS LABORATORIES INC.®
LISTED
CLASS - CTL
ENCLOSED PANELBOARD
NO. CA- 753001

Fig. 100-25(*a*).

The Listing Mark may appear in various forms as authorized by Underwriters Laboratories Inc. Typical forms which may be authorized are shown below:

UNDERWRITERS LABORATORIES INC. ®

UND. LAB. INC. ®

Underwriters Lab. Inc. ®

Underwriters Laboratories Inc. ®

Underwriters Laboratories Inc.®

LISTED

Underwriters Laboratories Inc.®

UNDERWRITERS LABORATORIES INC.®

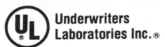

Listing Marks include one of the forms illustrated above, the word "Listed", and a control number assigned by UL. The product name as indicated in this Directory under each of the product categories is generally included as part of the Listing Mark text, but may be omitted when in UL's opinion, the use of the name is superfluous and the Listing Mark is directly and permanently applied to the product by stamping, molding, ink-stamping, silk screening or similar processes.

Separable Listing Marks (not part of a name plate and in the form of decals, stickers or labels) will always include the four elements: UL's name and/or symbol, the word "Listed", the product or category name, and a control number.

The complete four element Listing Mark will appear on the smallest unit container in which the product is packaged when the product is of such a size that only the symbol (UL) can be applied to the product or when the product size, shape, material or surface texture makes it impossible to apply any legible marking to the product.

LOOK FOR THE LISTING MARK

Fig. 100-25(*b*).

UL
CERTIFIED
SAFETY US
E12345

Fig. 100-25(c). Example of the next generation of UL labels, using the word "certified" and modular elements. The term is more easily understood in international commerce. Its use is not mandatory and all existing marks are and will remain valid.

relabeled by or in the presence of a test lab employee; labels cannot simply be sent through the mail. The test labs will send personnel to field locations to witness the application of a label.

Listed:

This term is covered together with "Approved" (and also "Identified") as part of the discussion of "Approved" and related standards for product acceptance near the beginning of this chapter.

Live Parts:

This definition indicates what is meant by that term as it is used throughout the Code. An insulated conductor contains a live part at any time by definition if it is energized (the conductor itself), even if the live part is insulated. For example, 312.2 requires that wiring entries to cabinets, cutout boxes, and meter sockets in wet locations use fittings listed for wet locations if the entry point is above the level of *uninsulated* live parts. The focus of this rule is not insulated conductors that are wet, but only the impact of moisture on uninsulated meter jaws and lugs, etc.

Luminaire:

This definition indicates all elements that are covered by the term *luminaire*. This term was adopted to correlate the **NEC** with other international standards and replace the term "fixture" used in the **NEC** prior to the 2002 Code. There was no intent on the part of the Code-making panels involved to require any change in application; this is simply an editorial revision. In this cycle the definition has been additionally revised to refer to a "light source such as" (but not necessarily) "a lamp or lamps." This allows for light-emitting diode (LED) and other sources that do not involve lamps as technology continues to move ahead.

Motor Control Center:

This definition indicates the necessary elements that would identify a piece of equipment as a motor control center. Such equipment would be subject to all rules aimed at comparable equipment such as 110.26(A) covering required workspace in front of certain types of equipment.

Neutral Conductor:

Neutral Point:

At long last the NEC has actually defined the term *neutral*. It does so by first defining a "neutral point" in a way that is sensible and not controversial, but the definition of "neutral conductor" is more problematic. Refer back to Fig. 100-21. The top two drawings show the most common neutral points, namely the star point of a wye and the center point of a single-phase system. No one would argue that those are neutral points. Since such star or center points must be grounded by rules in Art. 250, any conductor connected to such a point must be a grounded circuit conductor, and must be identified in accordance with 200.6. Therefore, any white wire run in conjunction with a grounded system is now a neutral, whether or not it is neutral between two (or more) associated ungrounded conductors. A 2-wire branch circuit that includes a white wire connected on a neutral bus is now an official neutral all the way to the outlet.

Although this certainly legitimizes common trade slang, it may lead people to believe all white (or gray) wires are neutrals. Not so. Look now at the bottom drawing in Fig. 100-21. That corner-grounded delta system has a white phase conductor, which is not and cannot ever be a neutral because the delta system shown has no neutral point. It remains to be seen whether this effort will add or reduce confusion.

Nonlinear Load:

Those loads that cause distortion of the current waveform are defined as *nonlinear loads*. A typical nonlinear load current and voltage waveform are shown in Fig. 100-26. As can be seen, while the voltage waveform [Fig. 100-26(*b*)] is a sinusoidal, 60-Hz wave, the current waveform [Fig. 100-26(*a*)] is a series of pulses, with rapid rise and fall times, and does not follow the voltage waveform.

The informational note following this definition is not intended to be a complete list, but rather, just a few examples. There are many more such loads. The substantiation for inclusion of this note stated in part:

> It has been known within the entertainment industry for some time that due to the independent single-phase phase-control techniques applied to 3-phase, 4-wire feeder, solid-state dimming can cause neutral currents in excess of the phase currents. This is in addition to the harmonics generated. This situation is dealt with in theaters in 520.27 and 520.51, etc. Dimming is also used in non-theatrical applications such as hotel lobbies, ballrooms, and conference centers. This effect must be taken into account wherever solid-state dimming is employed.

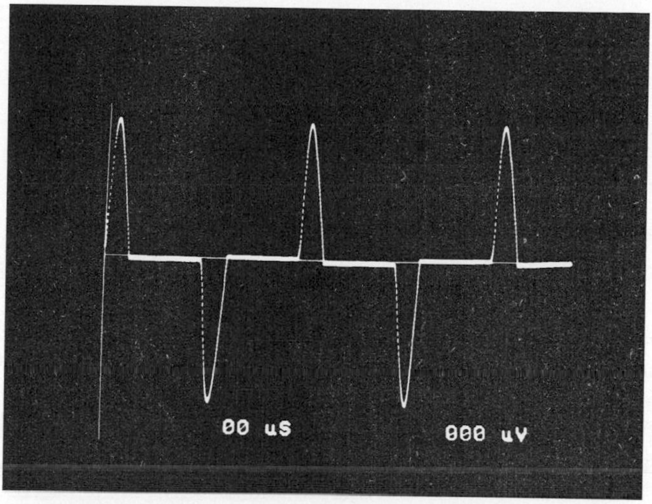
00 uS
000 uV

Fig. 100-26(*a*).

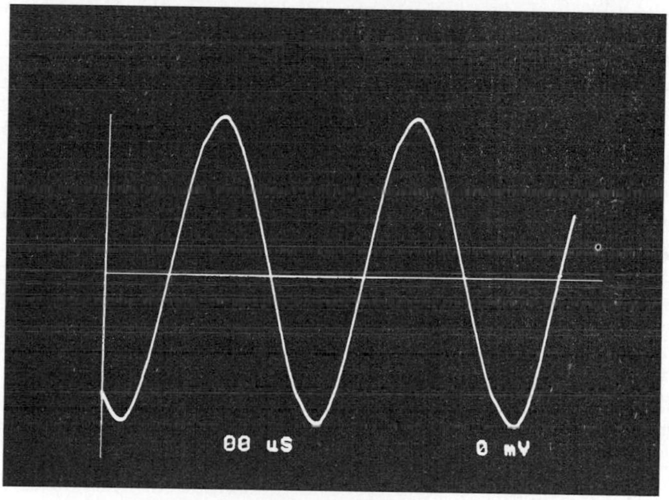
00 uS
0 mV

Fig. 100-26(*b*).

Outlet:

This term refers to a point on a wiring system where current is taken to supply utilization equipment. This is a critical definition because the term is frequently misapplied. For example, a hard-wired fluorescent luminaire set in a suspended ceiling in an office is an outlet and the branch circuit ends at the ballast channel. Article 400 covering flexible cord appears in Chap. 4 (equipment for general use) and not Chap. 3 (wiring methods and materials) because (with limited exceptions) flexible cord is not supposed to substitute for Chap. 3 wiring methods. The terminal housing on a motor, even a motor operating on a 4160-V branch circuit, is the outlet at the end of that medium-voltage branch circuit. Receptacles are outlets, but only a small fraction of the category.

Overcurrent:

Overload:

This is a very important concept. Overcurrent considers current in excess of rated current or ampacity in three different ways. A *short circuit* is a direct line-to-line connection between two circuit conductors, and if it occurs, it can be extremely destructive because of the enormous amounts of energy that will be released unless it is cleared immediately. A *ground fault* is a connection from an ungrounded conductor and an equipment grounding conductor. Although the available energy is somewhat lower, it may be just low enough so that overcurrent devices do not respond immediately. This type of arcing burndown is extremely destructive if not cleared immediately. The third variety of overcurrent is an *overload*. These are sustained currents that are above an equipment full load rating or the ampacity of a conductor, but low enough that it will only cause a problem if it persists for an extended period of time, the period being inversely proportional to the degree of overload.

Overcurrent Device, Branch-Circuit; Supplementary:

Branch-circuit protective devices are capable of providing protection over the full range of overcurrents between the device rating and its interrupting rating, but never less than 5000 A. They are far more robust than the supplementary overcurrent protective devices that offer limited protection for certain applications such as limiting the amount of energy that could enter a luminaire.

Photovoltaic (PV) System:

This term was relocated after the word "Solar" was removed from its title. This refers to the equipment involved in a particular application of solar energy conversion to electric power. This definition correlates to NEC Art. 690 covering design and installation of electrical systems for direct conversion of the sun's light into electric power. While the proliferation of such installations accelerates, remember that any and all installations of photovoltaic equipment at premises covered by the NEC must be performed in accordance with all general requirements given in Chaps. 1 through 4 and the specific requirements given in Art. 690.

Plenum:

The wording of this definition creates an unfortunate inference in the minds of many suggesting that a plenum is something deliberately fabricated to distribute air, in the sense of a piece of duct work to which other ducts are connected. That inference is inconsistent with the way the word is frequently used in NFPA 90A, which is the conditioned air standard. A review of that standard shows four subdefinitions immediately following, which set the appropriate context. Three of the four, namely, "Air Handling Unit Room Plenum," "Ceiling Cavity Plenum," and "Raised Floor Plenum" clearly describe spaces that were not specifically fabricated for air distribution. Since this definition must be used in concert with both NFPA 90A and 300.22(C), as well as other places in the code with similar exposures, it is very important that code users clearly understand that the "compartment or chamber" referred to could be, but is not necessarily, a space specifically fabricated for air transport.

Premises Wiring (System):

Published discussions of the Code panel's meaning of this phrase make clear the panel's intent that premises wiring includes all electrical wiring and equipment on the load side of the "service point," including any electrical work fed from a "source of power"—such as a transformer, generator, computer power distribution center, an uninterruptible power supply (UPS), or a battery bank. Premises wiring includes all electrical work installed on a premises. Specifically, it includes all circuits and equipment fed by the service or fed by a separately derived electrical source (transformer, generator, etc.). This makes clear that all circuiting on the load side of a so-called computer power center or computer distribution center (enclosed assembly of an isolating transformer and panelboard[s]) must satisfy all NEC rules on hookup and grounding, unless the power source in question is listed as "Information Technology Equipment," in which case the rules of Art. 645 would apply. When a "computer power center" is specifically "listed" and supplied with or without factory-wired branch-circuit "whips" (lengths of flexible metal conduit or liquidtight flex—with installed conductors), such equipment may be grounded as indicated by the manufacturer as given in the rules of Art. 645, Information Technology Equipment.

Other sources, such as solar photovoltaic systems or storage batteries, also constitute "separately derived systems." All NEC rules applicable to premises wiring also pertain to the load side wiring of batteries and solar power systems.

Qualified Person:

Here the Code spells out the necessary elements that designate someone as a "qualified person." This rule is used in many sections of the Code and typically compliance with any such rule hinges on the personnel involved being a "qualified person." Notice that it is not simply enough to be knowledgeable about the equipment and/or application, but also, such persons must have "received safety training." Presumably that means attending formal or informal training, or even on-the-job-training, all of which, presumably, must be documented and maintained in a personnel file on "Qualifications" or "Training" or the like. An informational note directs the reader to another NFPA standard (70E) for additional guidance with regard to training requirements.

Raceway:

Whenever this term is used in the Code, it applies to the various enclosed channels that are designed specifically for running conductors between cabinets and housings of electrical distribution components, including busbars, as covered in the relevant wiring method articles in Chap. 3. Prior to the 2014 NEC, the clear implication presented by the choice of wiring methods listed is that raceways are for extended lengths of run, and that more limited enclosed channels such as those within equipment are not to be so classified. This interpretation has been thoroughly tested. If any such enclosed channel were classified as a raceway, then surely an auxiliary gutter would be so classified. In the 1993 **NEC** cycle, the panel initially accepted a proposal to place "auxiliary gutters" into the list, and then unanimously reversed course in the face of negative comments from the current author, NEMA, and others. The issues of auxiliary gutters and panelboard gutter spaces are particularly pressing because 230.7 forbids the sharing of raceways between service conductors and other conductors. If such enclosures are deemed to be raceways, then service panel wiring, as we know it, would be contrary to the **NEC**.

There were other wiring methods omitted from the list as well, and for good reason. A cable tray is a "support system" and not a "raceway." When the Code refers to "conduit," it means only those raceways containing the word "conduit" in their title. Therefore, "EMT" is not conduit. Table 1 of Chap. 9 in the back of the Code book refers to "Conduit and Tubing." The Code, thus, distinguishes between the two. "EMT" is tubing. Notice that cable trays and cablebus are not identified as "raceways." The consequence of their omission is that general rules applying to "raceway" do not apply to cable trays or cablebus.

The 2014 **NEC** deleted the list of raceways from the informational note in favor of a reference to the definitions within the various Chap. 3 articles. However, the intent has not changed; it just requires a little more work to sort out the results. For example, the definitions of a wireway in 376.2 and 378.2 now specifically describe it as a raceway, and the definitions of an auxiliary gutter in 366.2 pointedly do not use that word, relying on the word "enclosure" instead. Similarly, the definition of a cable tray system in 392.2 describes something that fastens or supports cables or raceways, but that is not to be described as a raceway.

Receptacle:

Each place where a plug cap may be inserted is a "receptacle," as shown in Fig. 100-27. Multiple receptacles on one strap are just that, multiple receptacles.

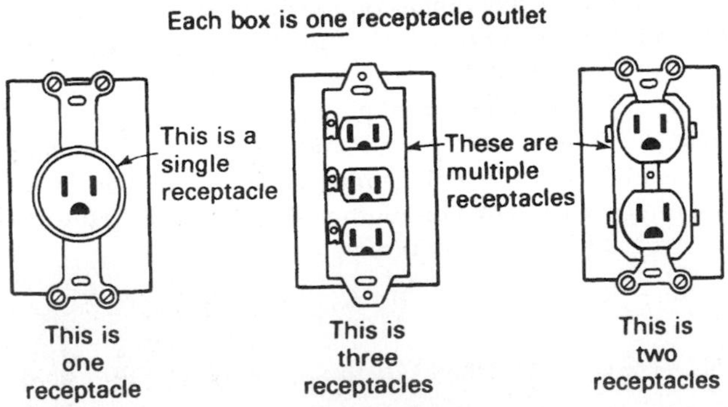
Each box is one receptacle outlet
This is a single receptacle
These are multiple receptacles
This is one receptacle
This is three receptacles
This is two receptacles

Fig. 100-27.

Only a single receptacle can be served by an individual branch circuit. See 210.21(B) and 555.19(A)(3).

The concept of a receptacle was basically unchanged for about a century of usage, but as of the 2017 **NEC**, the concept has been fundamentally expanded. Where formerly the only field attachment to a receptacle was a plug on the end of flexible cord, now the term may also be applied to a female device that receives utilization equipment equipped with a corresponding mating part. Refer to Fig. 100-28 for a detailed photo of the new concept as its originators expect to apply it. The left portion shows the new version of what can now be described as a receptacle.

Fig. 100-28. A new version of a receptacle.

Instead of a plug on a cord, the device will receive the luminaire to the right, which is equipped with a mating contact device. The three contact mechanisms (ground, neutral, hot) are in that order from large to small. The equipment grounding outer ring projects further than the other two and thereby observes the first-make, last-break principle in 406.10(D). The lever projecting from the edge of the luminaire canopy releases the pressure applied to the projecting beads on the contact arm, allowing it to drop freely and thereby disconnecting the luminaire for service.

Note that only the receptacle itself (Fig. 100-28) is listed. The mating component mounted in the luminaire canopy is a recognized component, marked as shown in Fig. 100-29 prior to inclusion in the canopy by the luminaire manufacturer. It is not field installable at present (although in principle, possibly capable of being field labeled) because the testing laboratory needs to test its load-bearing performance and electrical characteristics as part of the luminaire. This mark is issued by UL, but it is not a listing mark. The component recognition program aids efficiency.

Fig. 100-29. An example of a component recognition marking. It is not an **NEC** designation, and its true meaning is often misunderstood.

An inspector stationed in the device factory can review the conformity of the male component to established standards and, if warranted, allow the recognition mark. The component manufacturer can then sell them in quantity to many different luminaire manufacturers. And then the testing lab personnel assigned to the luminaire factory know by the mark that they don't have to investigate the mating part any more than they would have to look at a lampholder with similar recognition. They can simply evaluate the assembled luminaire in accordance with applicable product standards, in this case UL 1598, and then grant the listing if warranted.

Receptacle Outlet:

The *outlet* is the outlet box. But this definition must be carefully related to 220.12(I) for calculating receptacle loads in other than dwelling occupancies. For purposes of calculating load, 220.12(I) requires receptacle outlets to be calculated at not less than 180 for each single or for each multiple receptacle on one yoke. Because a single, duplex, or triplex receptacle is a device on a single mounting strap, the rule requires that 180 VA must be counted for each strap, whether it supports one, two, or three receptacles. On the other hand, the new multi-outlets that feature four or more receptacles permanently molded into a single piece of equipment mounted to an outlet box must be calculated at 90 VA per receptacle.

Remote-Control Circuit:

The circuit that supplies energy to the operating coil of a relay, a magnetic contactor, or a magnetic motor starter is a *remote-control circuit* because that circuit controls the circuit that feeds through the contacts of the relay, contactor, or starter, as shown in Fig. 100-30.

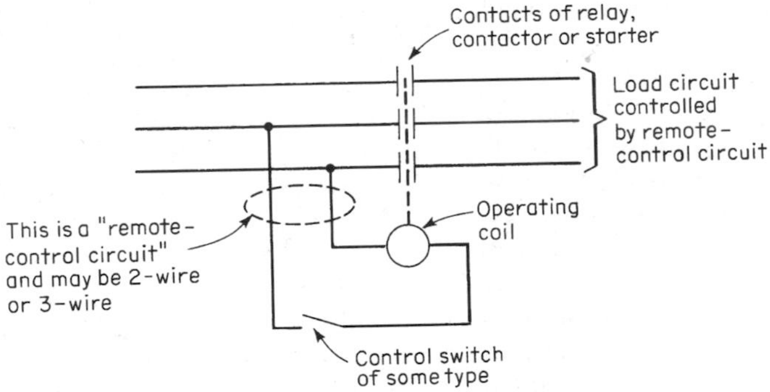
Contacts of relay, contactor or starter
Load circuit controlled by remote-control circuit
This is a "remote-control circuit" and may be 2-wire or 3-wire
Operating coil
Control switch of some type

Fig. 100-30.

Note that the closely related definition of "Control Circuit" was relocated into Art. 100 from Art. 430 because it is used in a variety of NEC articles.

A control circuit, as shown, is any circuit that has as its load device the operating coil of a magnetic motor starter, a magnetic contactor, or a relay. Strictly speaking, it is a circuit that exercises control over one or more other circuits. And these other circuits controlled by the control circuit may themselves be control circuits, or they may be "load" circuits—carrying utilization current to a lighting, heating, power, or signal device. Figure 100-30 clarifies the distinction between control circuits and load circuits.

The elements of a control circuit include all the equipment and devices concerned with the function of the circuit: conductors, raceways, contactor operating coils, source of energy supply to the circuit, overcurrent protective devices, and all switching devices that govern energization of the operating coil.

The **NEC** covers application of remote-control circuits in Art. 725 and in 430.71 through 430.75.

Separately Derived Systems:

This applies to all separate sources of power and includes transformers, generators, battery systems, fuel cells, solar panels, etc., that have no direct connection between circuit conductors of one system to those of another system; however, the definition recognizes that the derived system may share an earth or an equipment grounding connection (whether by wire or by enclosures) with other systems. Virtually, all power transformers are separately derived systems, while a backup generator, for example, may or may not be depending on whether the neutral from the generator is also switched with the phase conductors. Where the grounded (neutral) conductor is switched—such as where a four-pole transfer switch is used on a 3-phase, 4-wire generator output—then the generator is a separately derived system and must comply with the rules of 250.30.

Service:

The word *service* includes all the materials and equipment involved with the transfer of electric power from the utility distribution line to the electrical wiring system of the premises being supplied. Only a utility can supply a service, so if a facility generates its own power, it will have no service, only one or more feeders and building disconnects. The purpose of special rules for actual services is to address the necessary transitional rules that will assure a safe transition from utility work governed by the National Electrical Safety Code (NESC) and premises wiring governed by the **NEC**. Similarly, if a building is supplied by premises wiring in any form, then the disconnect for the entrance of that wiring will be a building disconnect and not a service disconnect.

Although service layouts vary widely, depending upon the voltage and amp rating, the type of premises being served, and the type of equipment selected to do the job, every service generally consists of "service-drop" conductors (for overhead service from a utility pole line) or "service-lateral" conductors (for an underground service from either an overhead or underground utility system)—plus metering equipment, some type of switch or circuit-breaker control, overcurrent protection, and related enclosures and hardware. A typical layout of "service" for a one-family house breaks down as in Fig. 100-31.

The **NEC** does not govern where in a service layout the **NEC** begins and the NESC ends. This is determined by the local public authority that governs public

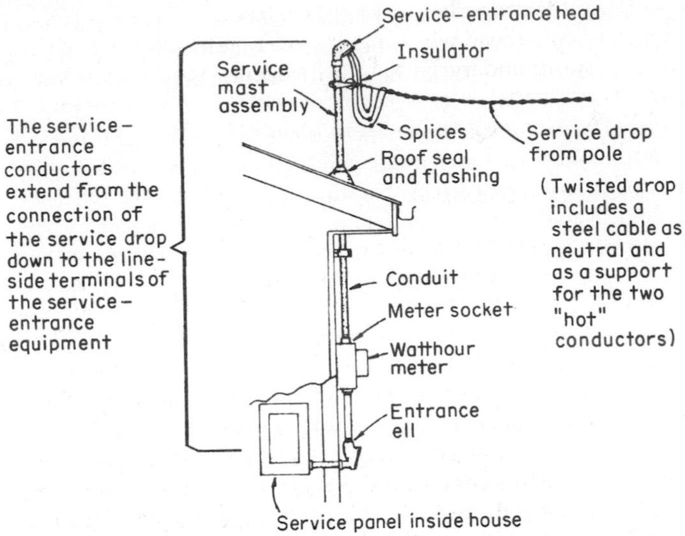
Service-entrance head
Insulator
Service mast assembly
The service-entrance conductors extend from the connection of the service drop down to the line-side terminals of the service-entrance equipment
Splices
Service drop from pole
Roof seal and flashing
(Twisted drop includes a steel cable as neutral and as a support for the two "hot" conductors)
Conduit
Meter socket
Watthour meter
Entrance ell
Service panel inside house

Fig. 100-31. The "service drop" terminology in this drawing is correct if the service point is at the mast (the usual case). If the service point is at the pole, then the supply drop is to be known, as of the 2011 **NEC**, as "overhead service conductors."

utility activities. Although we hear a lot about deregulated utilities, this concept only applies to the generation of electric energy, not its distribution down public streets. Since only one set of line wire can run on any given street, the distribution of electric energy is what economists call a *natural monopoly*; competition is effectively impossible. In such cases, there will be regulation by public authorities. This is the case here. Part of the regulatory process will be determining where the *service points* are allowed to be. If the service point is at the pole, then the **NEC** applies to the overhead service conductors as installed by the electrical contractor, with only the final connections at the street being made by utility personnel. If the service point is at the splices at the bottom of the drip loops, then a utility line crew will install the drop in accordance with the NESC and the **NEC** does not apply.

That part of the electrical system which directly connects to the utility-supply line is referred to as the *service entrance*. Depending upon the type of utility line serving the house, there are two basic types of service entrances—an *overhead* and an *underground* service.

The overhead service has been the most commonly used type of service. In a typical example of this type, the utility supply line is run on wood poles along the street property line or back-lot line of the building, and a cable connection is made high overhead from the utility line to a bracket installed somewhere high up on the building. This wood pole line also carries the telephone lines, and the poles are often called *telephone poles*.

The underground service is one in which the conductors that run from the utility line to the building are carried underground. Such an underground run to a building may be tapped from either an overhead utility pole line or an underground utility distribution system. Although underground utility services tapped from a pole line at the property line have been used for many years to eliminate the unsightliness of overhead wires coming to a building, the use of underground service tapped from an underground utility system has only started to gain widespread usage in residential areas over recent years. This latter technique is called URD—which stands for *underground residential distribution*.

Service Conductors—Overhead, Underground:

These are general terms that cover all the conductors on the load side of the service point used to connect the utility-supply circuit or transformer to the service equipment of the premises served. This term includes both overhead and underground service conductors, as well as "service-entrance" conductors. Although Fig. 100-31 covers an ordinary one-family house, the **NEC** necessarily applies to major industrial occupancies taking power at 69 kV or even higher, and every conceivable size and type of occupancy in between. See also Figs. 100-33 and 100-34.

Where the supply is from an underground distribution system, the underground service conductors may begin at the point of connection to the underground street mains, or at the property line, or at the terminals of the meter socket, or at the terminals of a pad-mounted transformer, again all as governed by state and local rule making around service point locations.

In every case the service conductors terminate at the service equipment, including the service disconnecting means.

Service Drop:

As the name implies, these are the conductors that "drop" from the overhead utility line and connect to the service entrance conductors at their upper end on the building or structure supplied. See Fig. 100-32. As noted in the caption to Fig. 100-31, if the service point is at the utility pole, the "service drop conductors" become "overhead service conductors."

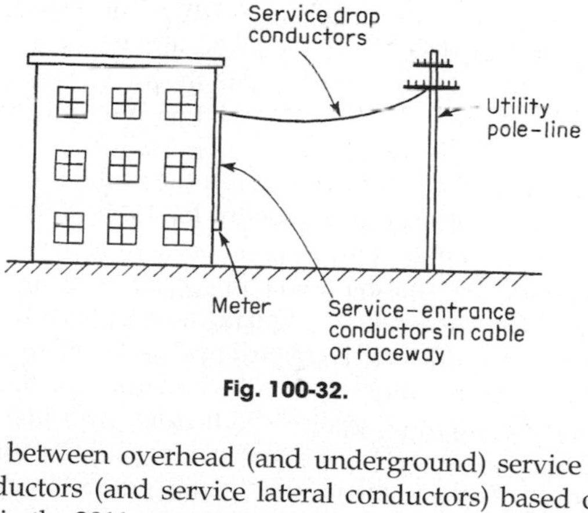
Service drop conductors
Utility pole-line
Meter
Service-entrance conductors in cable or raceway

Fig. 100-32.

The distinction between overhead (and underground) service conductors and service drop conductors (and service lateral conductors) based on service point location was new in the 2011 **NEC**. It upsets established terminology that goes back to the 1933 NEC. Now, only utilities will be responsible for service drops (or laterals), and only property owners will be responsible for overhead (or underground) service conductors. The rearranged terminology offers improved correlation with the "service conductors" definition, which depends on the service point location for its field application.

Service Equipment:

This is the equipment connected to the load end of service entrance conductors for the purpose of providing the principal means to control and disconnect the premises wiring from the source of utility supply. A meter socket is not service equipment in and of itself, but would be part of such equipment in the case of a combination meter-disconnect with the service disconnecting means located at the meter, all in one piece of equipment. The meter and meter socket in Fig. 100-31 is not part of the service equipment. The service disconnecting means will consist of some form of fused switch or circuit breaker because 230.91 requires the overcurrent protective device to be an integral part of the disconnecting means, or located immediately adjacent thereto. Note that any service "overcurrent device" only provides overload protection for service conductors. It cannot possibly respond to an arcing failure in progress on a service conductor located on its line side; such faults must usually burn clear, and this is why the NEC severely limits the exposure of any building to unprotected service conductors.

Service Lateral:

This is the name given to a set of underground service conductors. A service lateral serves a function similar to that of a service drop, as shown in Fig. 100-33. As in the case of a service drop, the terminology only applies on the line side of the service point. For example, if in the bottom drawing of Fig. 100-33 the service point were the transformer bushings, then the underground wiring to the house would be comprised of underground service conductors, and not a service lateral. Refer to the discussion about service drops for additional information on this concept, which was new in the 2011 NEC.

Service Point:

Service point is the "point of connection between the facilities of the serving utility and the premises' wiring." All equipment on the load side of that point is subject to NEC rules. Any equipment on the line side is the concern of the power company and is not regulated by the Code. This definition of "service point" must be construed as establishing that "service conductors" originate at that point. The whole matter of identifying the "service conductors" is covered by this definition.

The definition of "service point" does tell where the NEC becomes applicable, and does pinpoint the origin of service conductors. And that is a critical task, because a corollary of that determination is identification of that equipment which is, technically, "service equipment" subject to all applicable NEC rules on such equipment. Any conductors between the "service point" of a particular

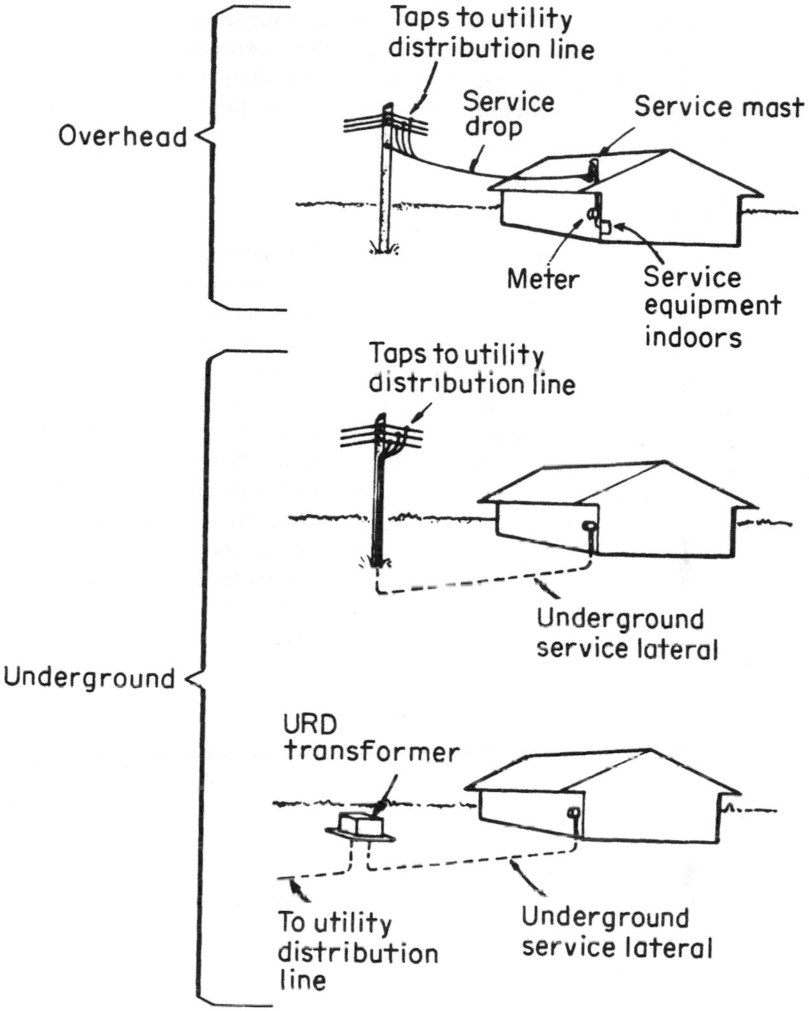
Overhead
Taps to utility distribution line
Service drop
Service mast
Meter
Service equipment indoors
Underground
Taps to utility distribution line
Underground service lateral
URD transformer
To utility distribution line
Underground service lateral

Fig. 100-33.

installation and the service disconnect are identified as service conductors and subject to NEC rules on service conductors (Fig. 100-34).

Special Permission:

It must be carefully noted that any Code reference to *special permission* as a basis for accepting any electrical design or installation technique requires that such "permission" be in written form. Whenever the inspection authority gives "special permission" for an electrical condition that is at variance with Code rules or not covered fully by the rules, the authorization must be "written" and not simply verbal

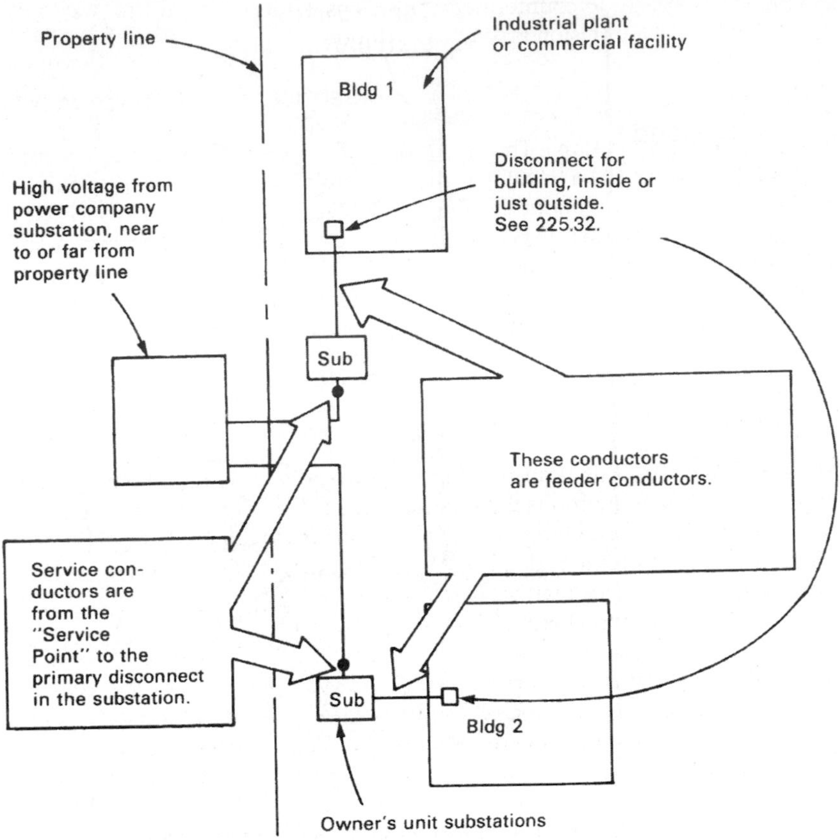
Property line
Industrial plant
or commercial facility
Bldg 1
Disconnect for
building, inside or
just outside.
See 225.32.
High voltage from
power company
substation, near
to or far from
property line
Sub
These conductors
are feeder conductors.
Service con-
ductors are
from the
"Service
Point" to the
primary disconnect
in the substation.
Sub
Bldg 2
Owner's unit substations

Fig. 100-34. NEC rules apply on load side of "service point"—not from property line. (230.200.)

permission. This rule corresponds to the wording used in 90.4, which requires any inspector-authorized deviation from standard Code-prescribed application to be in writing.

Structure:

This definition now disallows the use of the term for equipment. The very simple wording of this definition prior to the 2017 edition, "that which is built or constructed," led to significant controversies. Take a manufactured metering pedestal assembled in the field. That could arguably be included in the earlier definition. And, if so, then it would need a service disconnect, grounding electrode system, and the neutral and equipment grounding conductors would be required to separate from this point out. It is now clear that such equipment does not independently qualify as a structure. Another example involves multiple vehicle charging stations in a parking lot. Sec. 225.30 now has express language allowing such installations in order to countermand rejections based on the former wording of this definition. That revision is now superfluous.

Switches:

Bypass isolation switch This is "a manually operated device" for bypassing the load current around a transfer switch to permit isolating the transfer switch for maintenance or repair without shutting down the system. The second paragraph of 700.5(B) permits a "means" to be provided to bypass and isolate transfer equipment. This definition ties into that rule.

Transfer switch This is a switch for transferring load–conductor connections from one power source to another. Note that such a switch may be automatic or nonautomatic.

Switchgear:

As of the 2014 **NEC** this term replaces the former term "Metal-Enclosed Power Switchgear." The definition is the same, but the terminology tracks current product standards and correlating changes were made throughout the **NEC**. A new informational note provides detailed coverage of current descriptive terminology. The note also extracts the matter of arc-resistant construction from its former location within the definition because it is explanatory in nature.

Questions often arise as to the difference between switchgear and switchboards. They are on the same continuum, but switchgear, which is exclusively metal enclosed, tends to encompass more sophisticated electronics for monitoring and control than current switchboard designs. Although modern switchboards are metal enclosed as well, they still encompass antique designs with open switches on slate backplanes, etc.

Switchgear standards (ANSI C37.20.1 and UL 1558) were developed to focus on the needs of sophisticated industrial enterprises with highly qualified on-site electrical staffing, generally responsible for electrical systems subjected to higher loading on a routine basis. Such facility personnel are also more likely to be capable of routine equipment disassembly and maintenance. For example, switchgear will always exclusively employ draw-out breakers with serviceable arc shields, and contacts, etc. In addition, switchgear will be evaluated for a 30 cycle 3-phase short-circuit test; switchboards (under UL 891) are not subject to this test. This makes for significant differences in the internal characteristics of the support structures for the busbars, the quality of the insulation, and the strength of the enclosure. For example, a switchboard breaker enclosure might be 16 gauge; its counterpart in switchgear will be 11 gauge.

Uninterruptible Power Supply:

This was a new definition in the 2011 **NEC** of a familiar term. Of note is the limitation to alternating current applications and the explanation in the note following that a UPS can (and often does) function as a voltage regulator and power conditioner.

Voltage to Ground:

For a grounded electrical system, voltage to ground is the voltage that exists from any ungrounded circuit conductor to either the grounded circuit conductor (if one is used) or the grounded metal enclosures (conduit, boxes, panelboard cabinets, etc.) or other grounded metal, such as building steel. Examples are given in Fig. 100-35.

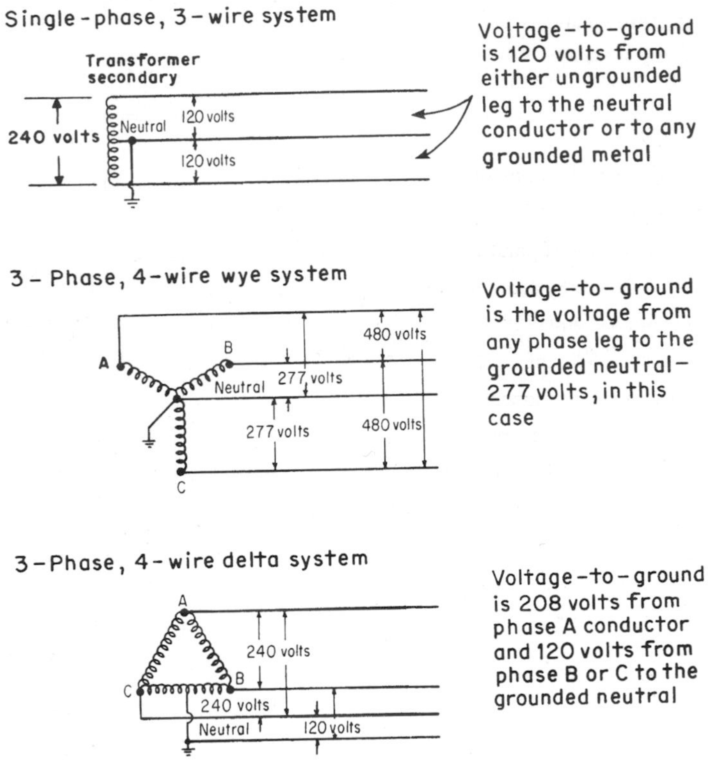
Single-phase, 3-wire system
Transformer secondary
240 volts
Neutral
120 volts
120 volts
Voltage-to-ground is 120 volts from either ungrounded leg to the neutral conductor or to any grounded metal
3-Phase, 4-wire wye system
A
B
C
Neutral
277 volts
277 volts
480 volts
480 volts
Voltage-to-ground is the voltage from any phase leg to the grounded neutral–277 volts, in this case
3-Phase, 4-wire delta system
A
B
C
240 volts
240 volts
Neutral
120 volts
Voltage-to-ground is 208 volts from phase A conductor and 120 volts from phase B or C to the grounded neutral

Fig. 100-35.

For an ungrounded electrical system, voltage-to-ground is taken to be equal to the maximum voltage that exists between any two conductors of the system. This is based on the reality that an accidental ground fault on one of the ungrounded conductors of the system places the other system conductors at a voltage aboveground that is equal to the value of the voltage between conductors. Under such a ground-fault condition, the voltage to ground is the phase-to-phase voltage between the accidentally grounded conductor and any other phase leg of the system. On, say, a 480-V, 3-phase, 3-wire ungrounded delta system, voltage to ground is, therefore, 480 V, as shown in Fig. 100-36.

In many Code rules, it is critically necessary to distinguish between references to "voltage" and to "voltage to ground." The Code also refers to "voltage between conductors," as in 210.6(A) through (E), to make very clear how rules must be observed.

Table 100-1. New and Relocated CMP-14 Definitions in Article 100

Defined Term	Original Location
Associated Apparatus	504.2
Associated Nonincendive Field Wiring Apparatus	506.2
Combustible Dust	500.2; 506.2
Combustible Gas Detection System	500.2; 505.2
Control Drawing	500.2
Cord Connector[1]	new as of 2017 NEC
Dust-Ignitionproof[2]	500.2; 506.2
Dusttight[3,6]	500.2; 506.2; Art. 100
Explosionproof Equipment[6]	Art. 100; not moved
Hermetically Sealed	500.2
Intrinsically Safe Apparatus	504.2
Intrinsically Safe System	504.2
Mobile Equipment[4,6]	513.2
Nonincendive Circuit	500.2; 506.2
Nonincendive Component	500.2
Nonincendive Equipment	500.2; 506.2
Nonincendive Field Wiring	500.2; 506.2
Nonincendive Field Wiring Apparatus	500.2
Oil Immersion	500.2
Portable Equipment[4,6]	513.2
Pressurized	506.2
Process Seal[5]	new as of 2017 NEC
Purged and Pressurized	500.2
Simple Apparatus	504.2
Unclassified Location	500.2; 505.2
Ventilated[6]	Art. 100; not moved

[1]This definition must not be confused with female cord connector bodies as used in 400.10(B) etc. Unlike customary device applications, this definition covers a type of fitting. It assists in the proper selection of fittings designed to attach cabled wiring methods to enclosures in hazardous (classified) locations.

[2]The informational note that follows no longer references NEMA 9 enclosures because they are no longer covered by the NEMA 250 standard. The relevant product requirements will now be the sole province of UL 1202.

[3]See the new Informational Note No. 1, which lists NEMA enclosure types that are considered to be dusttight. Table 110.28 includes an equivalent note, however, the note here conflicts with that note because it also includes Types 4, 4X, 6, and 6P. This will require correlation in a future cycle.

[4]Wording not reviewed by other panels; may or may not be compatible with other NEC rules.

[5]Definition based on applicable ANSI/ISA standard (12.27.01-2011).

[6]Title does not include the hazardous classification limitation.

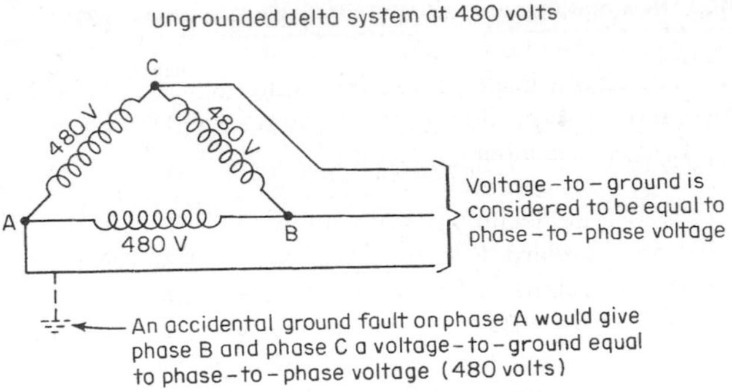
Ungrounded delta system at 480 volts
C
480 V
480 V
A
480 V
B
Voltage-to-ground is considered to be equal to phase-to-phase voltage
An accidental ground fault on phase A would give phase B and phase C a voltage-to-ground equal to phase-to-phase voltage (480 volts)

Fig. 100-36.

Part II. Over 1000 Volts, Nominal

Substation:

This definition was moved from Art. 225 in the 2014 **NEC**. It covers "an assemblage of equipment," which functions to distribute, switch, or modify the characteristics of the incoming electrical energy, or any combination of these functions. As such, they may be field engineered in more complicated cases, or arrive as a listed assembly. They may be located either indoors or outdoors, and operated with or without a medium voltage component. The majority do have a medium voltage supply, and a new Section 490.48 was created at the same time to comprehensively address this work.

ARTICLE 110. REQUIREMENTS FOR ELECTRICAL INSTALLATIONS

110.1. Scope. This article provides a variety of general regulations that govern the installation of equipment and conductors. Part I applies to installations rated 600 V or less and those rated over 600 V, unless specifically modified by another rule in Part III or IV. Part II applies only to systems rated 600 V or less, while Part III provides general rules for systems operating at over 600 V, and Part IV covers electrical systems rated over 600 V used in "tunnel installations." Part V covers the requirements for manholes.

110.2. Approval. As indicated by this section and the companion definition given in Art. 100, all equipment used must be "acceptable to the authority having jurisdiction" (AHJ). That generally means that the local inspector is the final judge of what equipment and conductors may be used in any given application.

Review the discussion in Art. 100 on the three standards for product acceptance (approved, identified, and listed) for more information on this point.

The intent of the **NEC** is to place strong insistence on third-party certification of the essential safety of the equipment and component products used to assemble an electrical installation. And, many Code sections specifically require the equipment to be "listed." But, where any piece of equipment is not required to be listed, then the local inspector is the one who determines if such equipment can be used. It should be noted that many inspectors require equipment to be "listed" if there exists a "listed" version of the type of equipment you're using. Such action helps ensure that the equipment used is inherently safe.

The **NEC** is not the only controlling standard covering electrical installations. The Occupational Safety and Health Administration (OSHA) has a standard, the Electrical Design Safety Standard, that must also be satisfied. Because the **NEC** is the basis for the OSHA standard, and the **NEC** is more dynamic in terms of change, in the vast majority of cases, **NEC** requirements are more stringent than those of the OSHA design standard. And satisfying the **NEC** will ensure compliance with the OSHA regulations. But, while the **NEC** doesn't always mandate the use of listed equipment, the OSHA standard requires that listed equipment be used to the maximum extent possible. That is, as far as OSHA is concerned, if there exists a "listed" piece of equipment of the type you are installing, then you must use the "listed" equipment instead of a nonlisted counterpart. Failure to do so is a direct violation of OSHA's Electrical Design Safety Standard.

In addition, OSHA addresses those instances where there exists no "listed" version of the type of equipment you need. In such cases, the local inspector, plant safety personnel, the manufacturer, or other authority must perform a safety inspection. Although the OSHA standard does not provide any guidance with respect to "what" the safety inspection must entail, it seems reasonable to assume that consideration of the points delineated in 110.3(A)(1) through (8) should serve to satisfy the intent of the OSHA requirement for a safety evaluation.

The use of custom-made equipment is also covered in OSHA rules. Every piece of custom equipment must be evaluated as essentially safe by the local inspector, plant safety personnel, the manufacturer, or other authority and documentary safety-test data of the safety evaluation should be provided to the owner on whose premises the custom equipment is installed. And it seems to be a reasonable conclusion from the whole rule itself that custom-equipment assemblies must make maximum use of "listed," "labeled," or "certified" components, which will serve to mitigate the enormous task of conducting the safety evaluation.

The bottom line is that if an OSHA review is a serious concern, look for listed equipment. However, this usual, overly simplistic, one-size-fits-all approach of a government bureaucracy undermines the integrity of the **NEC** process. In the 2005 **NEC** cycle, NEMA made a serious, official proposal to require all pull boxes to be listed. The panel chair put the question to CMP 9, and out of courtesy, invited the NEMA representative to make the motion, which was to accept the proposal. It was greeted with an extended dead silence, followed by an announcement by the chair that the motion failed because it did not so much as receive a second. This current author then moved to reject the proposal, on the grounds that it was excessive to require a listing, especially on pull boxes that may be made in local sheet metal shops to meet specific dimensional requirements. CMP 9 overwhelmingly followed suit, the second time during this author's tenure on that panel that it has refused to require listings on this equipment.

The **NEC** process is a transparent, open process, fully subject to opportunities for public participation and comment. It is a consensus process requiring a two-thirds vote of a panel for which no interest can have more than a third of the membership, and for which the actual proportion is far lower than that. Every time some bureaucracy tries to require universal listings, it is tantamount to an attempt to make thousands of amendments throughout the NEC, in this case removing "approved" and "identified" and substituting "listed," all without going through the consensus process. The OSHA rules are what they are. This author believes that such agencies would be better off staying out of the way of consensus standards development efforts by agencies that work as well as NFPA does. If OSHA has specific information that a listing is needed where it is not now specified, it should submit a proposal like anyone else. To its credit, the U.S. Consumer Product Safety Commission has been participating in this way for many years.

110.3. Examination, Identification, Installation, Use, and Listing (Product Certification) of Equipment. This section presents general rules for establishing what equipment and conductors may be used. Part **(A)** lists eight factors that must be evaluated in determining acceptability of equipment for Code-recognized use. It's worth noting that these criteria may be used as a basis for the evaluation that is required, but not defined, by OSHA rules for "unlisted" equipment. Remember, as far as OSHA is concerned, use of "unlisted" equipment is only permitted in those cases where no commercially available product of the type to be used is "listed." Where no "listed" version is available, then OSHA would permit the use of the "unlisted" piece of equipment, BUT OSHA requires a safety evaluation be performed.

Part **(A)(1)** states that the "suitability" of the equipment in question must be evaluated with respect to the intended use and installation location. The informational note to 110.3(A)(1) notes that, in addition to "listing" or "labeling" of a product by UL or another test lab to certify the conditions of its use, acceptability may be "identified by a description marked on or provided with a product to identify the suitability of the product for a specific purpose, environment, or application." This is a follow-through on the definition of the word "identified," as given in Art. 100. The requirement for identification of a product as specifically suited to a given use is repeated at many points throughout the Code. As of the 2011 NEC, the note goes on to call attention to the fact that some equipment can only operate safely under limited conditions of use, and such constraints must be observed, all as covered in product markings, instructions, or general guide card limitations in the case of items listed within recognized categories.

As of the 2017 NEC, an informational note expressly recognizes that in addition to new equipment, reconditioned, refurbished, or remanufactured equipment can also be brought into this process. Older electrical equipment is routinely used to extend the useful life of existing equipment. Often the reconditioning process allows for safety upgrades that would not otherwise be possible.

With the exception of items **(3)** and **(8)** in 110.3(A), listing standards will generally cover the concerns listed in items **(2)** and **(4)** through **(7)**. Where "unlisted" equipment is used, these factors must be considered and adequately addressed. Item **(3)**, an important consideration for electrical inspectors to include in their examination to determine suitability of equipment for safe and effective use, is "wire-bending and connection space." See Fig. 110-1. This factor is a function of field-installation and addresses the concern for adequate gutter space to

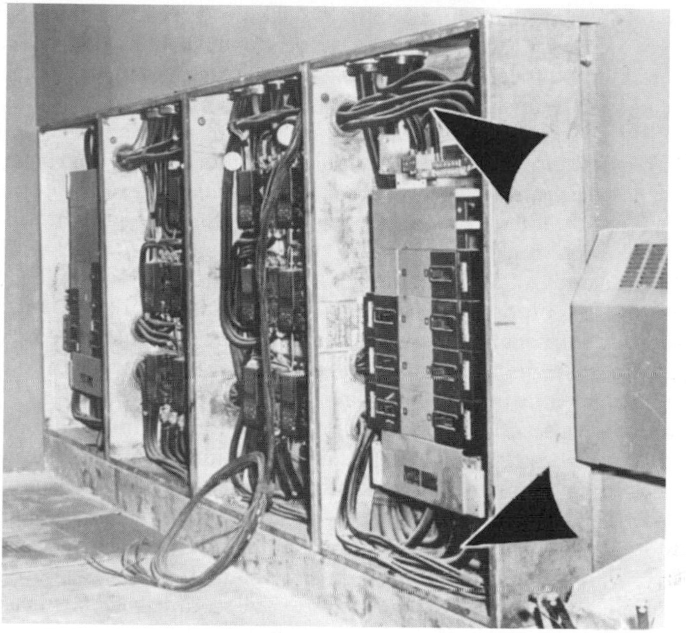

Fig. 110-1. Equipment must be evaluated for adequate gutter space to ensure safe and effective bending of conductors at terminals. [Sec. 110.3(A)(3).]

train conductors for connection at conductor terminal locations in enclosures for switches, CBs, and other control and protection equipment. This general mention of the need for sufficient conductor bending space is aimed at avoiding poor terminations and conductor damage that can result from excessively sharp conductor bends required by tight gutter spaces at terminals. Specific rules that cover this consideration are given in 312.6 on "Deflection of Conductors" at terminals or where entering or leaving cabinets or cutout boxes—covering gutter widths and wire-bending spaces. Item **(8)** is essentially a "catchall" requirement that depends on the designers and installers to use common sense and their knowledge of safe application to identify and correct any condition that may exist or develop relative to the installation they are performing.

Part **(B)** of this section is a critically important Code rule because it incorporates, as part of the **NEC** itself, all the application regulations and limitations published by product-testing organizations, such as UL, Factory Mutual, and ETL.

The most comprehensive collection of this information is in "UL Product Spec," a comprehensive internet tool available to all at http://www.ul.com/productspec. This data is updated on a daily basis, so it is absolutely current. Product Spec offers five different ways to search including by Installation Code/Code Section; Product Type keyword; groupings of Products, Systems and Assemblies; UL Product Category Code; or Master Format number. Short YouTube videos are available to demonstrate how to use UL Product Spec at HYPERLINK "http://www.ul.com/psvideo" www.ul.com/psvideo. It also has some of the disadvantages of an on-line tool, most notably the complete lack of utility without an internet connection. There are other approaches still available to access this material; at least for now.

This material used to be easily available in print, in the form of color-coded reference books updated each year. Specifically, those of most relevance to electrical work were the *Electrical Construction Materials Directory* (Green Book), the *Hazardous Location Equipment Directory* (Red Book), and the *Electrical Appliance and Utilization Equipment Directory* (Orange Book). Each of these directories carried two categories of information. Every single product category carries what UL refers to as the "Guide Card" information for the category. These are based on provisions in the product standard (also identified in the information), and in effect, they are the underlying assumptions UL, and other testing laboratories make about how the product will be used. For ease of access, every product category carries a four-letter code. These product directories also carried lists of the listees in the category as of the effective date of the publication.

Since all users constantly needed the Guide Card information and rarely needed the information about listees, UL extracted the Guide Card information from all the color-coded directories and combined it into a single book, the *Guide Information for Electrical Equipment* (White Book). This book was (and its content remains) essential to correct trade practice. Unfortunately, UL abandoned publishing the White Book with the 2014 edition, for cost reasons. Indeed, it has discontinued all of its printed directories with the sole exception of the Fire Directory. It still makes available the White Book content available electronically as a PDF; the most recent edition is the 2015-2016 version. However, UL recently announced it was seriously considering discontinuing even this effort, in favor of the strictly on-line approach. This would be a major, almost unthinkable inconvenience, and objections are being registered. As of this writing a final decision has not been made regarding the PDF version. To their credit, UL wishes it to be known that they want to hear from users, and concerns with discontinuation of the printed and/or PDF downloadable version of the white book can be directed to UL's Codes and Regulatory Services Department at ULRegulatoryServices@ul.com.

An example will illustrate just how useful, indeed essential, the information in the White Book is to ordinary electrical trade practice. The **NEC** at 430.109(F) says that "a horsepower rated attachment plug and receptacle … shall be permitted to serve as the disconnecting means." Suppose you would like to use this approach on a 7½ hp 460-volt 3-phase motor. Is there a ready source of information regarding such devices, since you would like to routinely connect and disconnect this motor? Yes, indeed. Looking in the White Book index, or after doing a {[Ctrl]F} search on "receptacles" in the electronic version, or on-line using "ProductSpec" after entering "receptacles" as a product type, yields the heading "Receptacles for Plugs and Attachment Plugs (RTRT)." The four-letter code at the end corresponds to the guide card information for this category, and once obtained, greatly simplifies information retrieval.

The guide card information on this product category runs to almost two full pages of small print, and remember, every word of it is effectively incorporated by reference into the **NEC** and is enforceable as such, by 110.3(B). In this case, you will see a table with 50 different NEMA receptacle configurations arranged by amperes, then voltage, then number of phases, then number of poles, then number of wires, then the NEMA configuration designation, and finally, the horsepower rating. In this case, we are looking for a 30 ampere, 3-phase, 3-pole, 4-wire grounding receptacle, and the one listed (NEMA L16-30) has a 10 hp rating. That combination of a locking 30 ampere 3P 4W grounding plug and receptacle combination will meet the 430.109(F) requirement nicely.

Part (C), new as of the 2017 **NEC**, addresses equipment that is not evaluated in accordance with the North America electrical safety system. European procedures allow for self-certification as evidenced by the (CE) mark. This is a one-time self-certification mark. It is not a listing mark. Repeated studies have decisively proven that substantial safety defects are frequently in evidence when equipment so evaluated is submitted for listing to the applicable product standards, that work in concert with the **NEC** and that are included in Annex A in the **NEC**. The informational note that follows addresses the OSHA program to qualify electrical testing laboratories as having suitable testing protocols and follow-up testing to assure suitability for application in accordance with **NEC** requirements.

110.4. Voltages. In all electrical systems there is a normal, predictable spread of voltage values over the impedances of the system equipment. It has been a common practice to assign these basic levels to each nominal system voltage. The highest value of voltage is that at the service entrance or transformer secondary, such as 480Y/277 V. Then considering voltage drop due to impedance in the circuit conductors and equipment, at midsystem the actual voltage would be 460Y/265 V, and finally a "utilization" voltage would be 440Y/254 V. Variations in "nominal" voltages have come about because of (1) differences in utility-supply voltages throughout the country, and (2) varying transformer secondary voltages produced by different and often uncontrolled voltage drops in primary feeders. Note that a system drop from 480 to 440 V is over 8 percent and significantly exceeds the recommendation in the second informational note that follows 215.2(A). This degree of variation inside a facility would be very unusual; however, utility distribution practice does vary, particularly when their distribution systems are under stress such as under summer air-conditioning loads.

110.6. Conductor Sizes. In this country, the American Wire Gauge (AWG) is the standard for copper wire and for aluminum wire used for electrical conductors. The American Wire Gauge is the same as the Brown & Sharpe (B & S) gauge. The largest gauge size is 4/0 AWG; above this size the sizes of wires and cables are stated in thousands of circular mils (kcmil). The circular mil is a unit used for measuring the cross-sectional area of the conductor, or the area of the end of a wire that has been cut square across. One circular mil (commonly abbreviated cmil) is the area of a circle 1/1000 in. in diameter. The area of a circle 1 in. in diameter is 1,000,000 cmil; also, the area of a circle of this size is 0.7854 sq in.

To convert square inches to circular mils, multiply the square inches by 1,273,200.

To convert circular mils to square inches, divide the circular mils by 1,273,200 or multiply the circular mils by 0.7854 and divide by 1,000,000.

In interior wiring the gauge sizes 14, 12, and 10 are frequently solid wire; 8 AWG and larger conductors in raceways are required to be stranded if pulled into a raceway, although in practice this is the usual configuration anyway, even in cable assemblies. [See 310.106(C).]

A cable (if not larger than 1,000,000 cmil) will have one of the following numbers of strands: 7, 19, 37, or 61. In order to make a cable of any standard size, in nearly every case the individual strands cannot be any regular gauge number but must be some special odd size. For example, a Class B stranded 2/0 AWG cable must have a total cross-sectional area of 133,100 cmil and is made up of 19 strands. No. 12 AWG has an area of 6530 cmil and 11 AWG, an area of 8234 cmil; therefore, each strand must be a special size between Nos. 12 and 11.

110.7. Wiring Integrity. This general rule requires that the integrity of the conductor insulation must be maintained. This can be accomplished by observing conduit fill limitations as well as proper pulling techniques. However, basic knowledge of insulation-resistance testing is important.

Measurements of insulation resistance can best be made with a megohmmeter insulation tester. As measured with such an instrument, insulation resistance is the resistance to the flow of direct current (usually at 500 or 1000 V for systems of 600 V or less) through or over the surface of the insulation in electrical equipment. The results are in ohms or megohms, but where the insulation has not been damaged, insulation-resistance readings should be in the megohm range.

110.8. Wiring Methods. All Code-recognized wiring methods are covered in Chap. 3 of the NEC.

110.9. Interrupting Rating. Interrupting rating of electrical equipment is divided into two categories: current at fault levels and current at operating levels.

Equipment intended to clear fault currents must have interrupting rating equal to the maximum fault current that the circuit is capable of delivering at the line (not the load) terminals of the equipment. See Fig. 110-2. The internal impedance of the equipment itself may not be factored in to use the equipment at a point where the available fault current on its line side is greater than the rated, marked interrupting capacity of the equipment.

If overcurrent devices with a specific AIR (ampere interrupting rating) are inserted at a point on a wiring system where the available short-circuit current exceeds the AIR of the device, a resultant downstream solid short circuit between

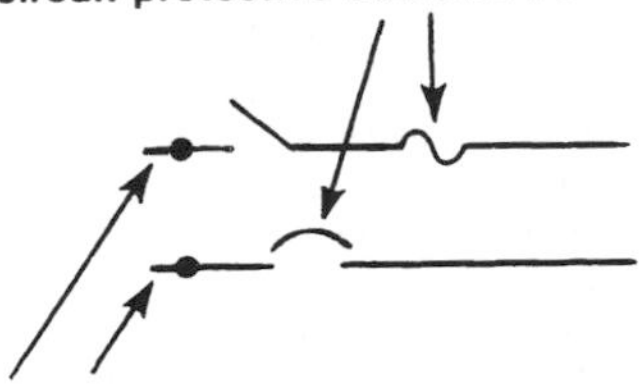
All short-circuit protective devices. . .
. . . must have an interrupting rating at least equal to the maximum fault current that the circuit could deliver into a short circuit on the *line side* of the device.
NOTE: That means that the fault current "available" at the line terminals of all fuses and circuit breakers *must be known* in order to assure that the device has a rating sufficient for the level of fault current.

Fig. 110-2. (Sec. 110.9.)

conductors or between one ungrounded conductor and ground (in grounded systems) could cause serious damage to life and property. Since each electrical installation is different, the selection of overcurrent devices with a proper AIR is not always a simple task. To begin with, the amount of available short-circuit current at the service equipment must be known. Such short-circuit current depends upon the capacity rating of the utility primary supply to the building, transformer impedances, and service conductor impedances. Most utilities will provide this information. But, be aware that 110.9 essentially implies such calculations be performed for all electrical systems, and 110.10 mandates consideration of the available fault current at every point in the system where an overcurrent protective device is applied.

Downstream from the service equipment, AIRs of overcurrent devices generally will be reduced to lower values than those at the service, depending on lengths and sizes of feeders, line impedances, and other factors. However, large motors and capacitors, while in operation, will feed additional current into a fault, and this must be considered when calculating short-circuit currents.

Manufacturers of overcurrent devices have excellent literature on figuring short-circuit currents, including graphs, charts, and one-line-diagram layout sheets to simplify the selection of proper overcurrent devices.

In the last paragraph, the Code recognizes that equipment intended only for control of load or operating currents, such as contactors and unfused switches, must be rated for the current to be interrupted, but does not have to be rated to interrupt available fault current, as shown in Fig. 110-3.

All switches, contactors, starters, relays. . .

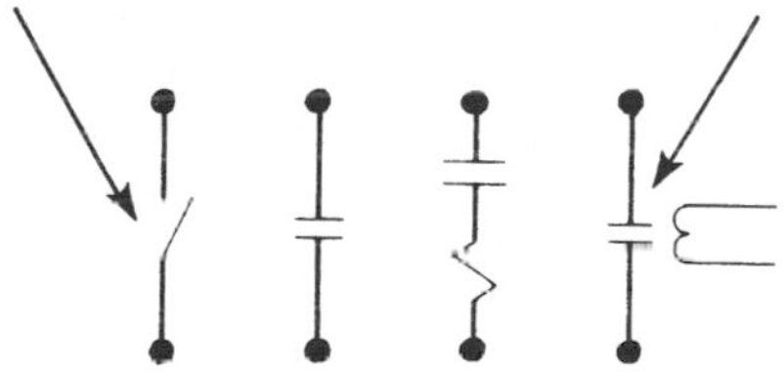

. . . must have an interrupting rating at least equal to "the current that must be interrupted"—which could be full-load current or, in the case of isolating or disconnect switches, some lesser value of operating current (such as transformer magnetizing current).

Fig. 110-3. (Sec. 110.9.)

110.10. Circuit Impedance, Short-Circuit Current Ratings, and Other Characteristics. This section requires that all equipment be rated to withstand the level of fault current that is let through by the circuit protective device in the time it takes to operate—without "extensive damage" to any of the electrical components of the circuit as illustrated in Fig. 110-4.

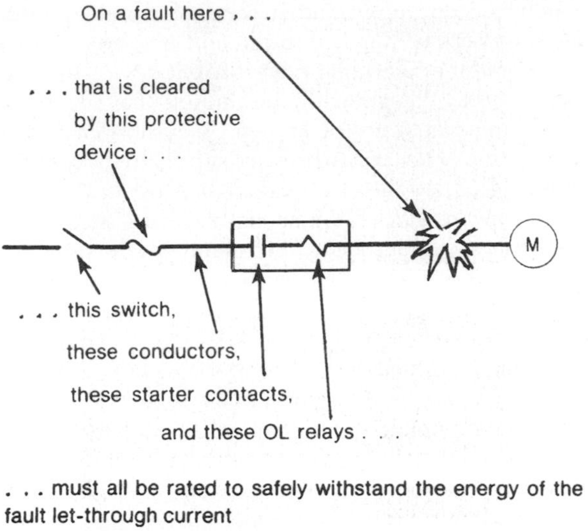
On a fault here . . .
. . . that is cleared by this protective device . . .
M
. . . this switch,
these conductors,
these starter contacts,
and these OL relays . . .
. . . must all be rated to safely withstand the energy of the fault let-through current

Fig. 110-4. (Sec. 110.10.)

The phrase "the component short-circuit current ratings" was added to this rule a few editions back. The intent of this addition is to require all circuit components that are subjected to ground faults or short-circuit faults to be capable of withstanding the thermal and magnetic stresses produced within them from the time a fault occurs until the circuit protective device (fuse or CB) opens to clear the fault, without extensive damage to the components.

The Code-making panel (CMP) responsible for Art. 110 has indicated that this section is not intended to establish a quantifiable amount of damage that is permissible under conditions of short circuit. The general requirement presented here is just that, a general rule. Specifics, regarding what damage is or is not acceptable under fault conditions, are established by the product test standard. For example, as stipulated in UL 508, Industrial Control Equipment, which covers combination motor starters, the permissible damage for a Type E (the so-called self-protected) motor starter is different from the requirements for other types of motor starters. The Type E unit must satisfy a more rigorous performance criterion than the others. Therefore, it is the UL Standard, not 110.10, that requires a more rigorous performance criterion for the Type E starter than is required for the other types of listed motor starters. The last sentence helps clarify that the prevention of "extensive damage" can be achieved by applying listed devices within their listed ratings. That is to say, the NEC does not intend to regulate product safety. Such regulation is the function of the NEMA/UL Product Standards. Any product that satisfies the controlling standard, and is applied in accordance with its ratings, is acceptable to the NEC and will satisfy the intent of this general rule.

110.11. Deteriorating Agents. Equipment must be "identified" for use in the presence of specific deteriorating agents, as shown on the typical nameplate in Fig. 110-5. In addition, equipment not normally suitable for use in wet locations must be protected from permanent damage while exposed to outdoor conditions

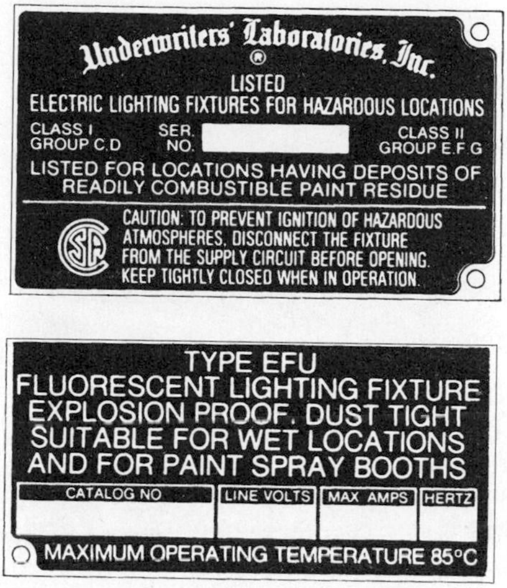
Underwriters' Laboratories, Inc.
LISTED
ELECTRIC LIGHTING FIXTURES FOR HAZARDOUS LOCATIONS
CLASS I GROUP C.D
SER. NO.
CLASS II GROUP E.F.G
LISTED FOR LOCATIONS HAVING DEPOSITS OF READILY COMBUSTIBLE PAINT RESIDUE
CAUTION: TO PREVENT IGNITION OF HAZARDOUS ATMOSPHERES, DISCONNECT THE FIXTURE FROM THE SUPPLY CIRCUIT BEFORE OPENING. KEEP TIGHTLY CLOSED WHEN IN OPERATION.
TYPE EFU
FLUORESCENT LIGHTING FIXTURE
EXPLOSION PROOF, DUST TIGHT
SUITABLE FOR WET LOCATIONS
AND FOR PAINT SPRAY BOOTHS
CATALOG NO
LINE VOLTS
MAX AMPS
HERTZ
MAXIMUM OPERATING TEMPERATURE 85°C

"Identified"—This kind of marking on a product makes it "recognizable as suitable for the specific purpose" as indicated on the labels.

Fig. 110-5.

during construction. The NEC has long stated that a dry location may be temporarily wet during building construction; this provision does not contradict that principle, but requires appropriate care during the construction process.

110.12. Mechanical Execution of Work. This statement has been the source of many conflicts because opinions differ as to what is a "neat and workmanlike manner."

The Code places the responsibility for determining what is acceptable and how it is applied in the particular jurisdiction on the authority having jurisdiction. This basis in most areas is the result of:

1. Competent knowledge and experience of installation methods.
2. What has been the established practice by the qualified journeymen in the particular area.
3. What has been taught in the trade schools having certified electrical training courses for apprentices and journeymen.

Examples that generally would not be considered as "neat and workmanlike" include nonmetallic cables installed with kinks or twists; unsightly exposed runs; wiring improperly trained in enclosures; slack in cables between supports; flattened conduit bends; or improvised fittings, straps, or supports. See Fig. 110-6.

It has long been required in specific Code rules that unused openings in boxes and cabinets be closed by a plug or cap and such rules were presented in what was then Art. 370, now Art. 314, and Art. 373, now Art. 312. The requirement, now given in part **(A)** for such plugging of open holes is also a general rule to provide fire-resistive integrity of all equipment—boxes, raceways, auxiliary gutters,

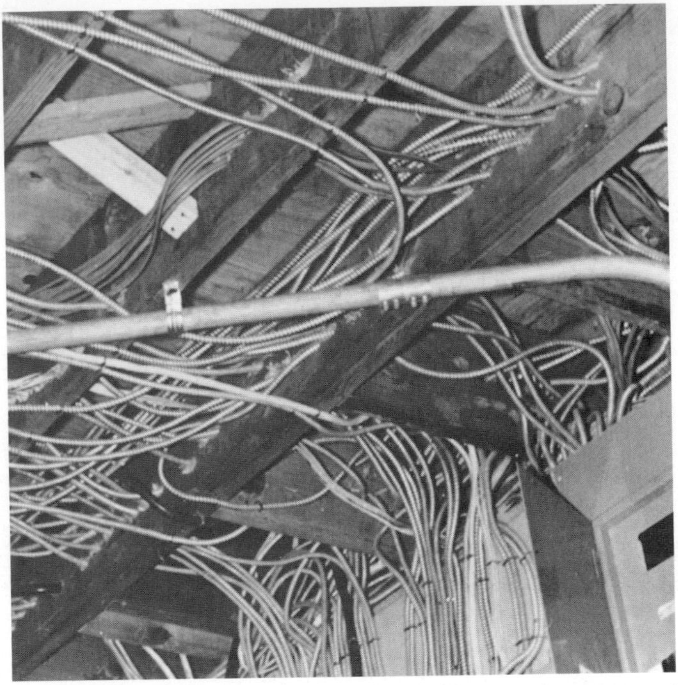

Fig. 110-6. Irregular stapling of BX to bottoms of joists and ragged drilling of joists add up to an unsightly installation that does not appear "workmanlike." (Sec. 110.12.)

cabinets, equipment cases, or housings (Fig. 110-7). This rule does not extend to mounting holes in the back of boxes, etc. Not specifically mentioned, but presumably still permitted, would be weep holes drilled in outdoor enclosures. This is far from clear in the current language. When this rule was part of Arts. 312 and 314, the Code-making panel with jurisdiction (CMP 9) specifically adjusted the wording so weep holes could be drilled in boxes. Now a different panel (CMP 1) has jurisdiction, and it refused to correct the problem in the 2011 cycle. In the 2014 **NEC**, CMP 9 inserted express permission (in 314.15) to add such drainage holes not over 1/4-inch in diameter, provided they were approved. That applies to conduit bodies and boxes, and probably addresses the majority of the problems. Trapped condensation that accumulates in sealed boxes can do great damage over time.

Part **(B)** presents a requirement for current-carrying parts—buswork, terminals, etc.—that is similar to the rule of 250.12. Both rules effectively prohibit conductive surfaces from being rendered nonconductive due to the introduction of paint, lacquer, or other substances. It should be noted that this rule is not intended to prohibit the use of "cleaners." Use of cleaning agents is recognized, but only those agents that do not contaminate conductive surfaces or deteriorate nonmetallic structures within the enclosure, as some spray lubricants are capable of doing. Be certain that any type of cleaner used for maintenance purposes is suitable for the specific application.

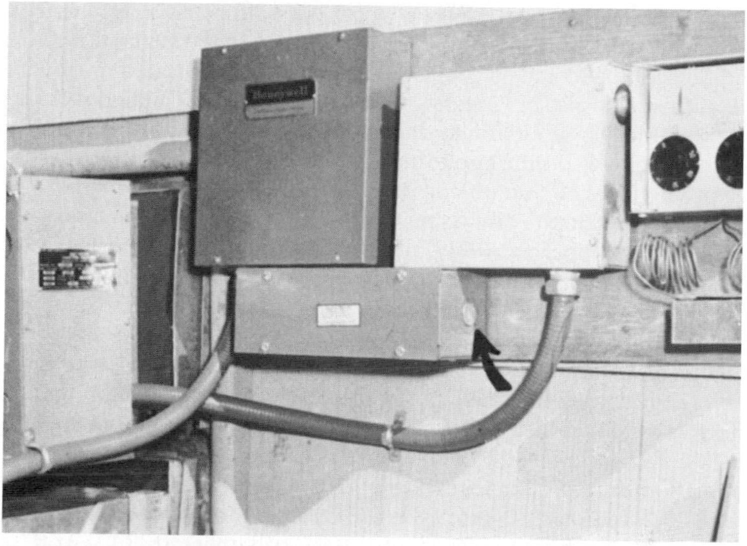

Fig. 110-7. Unused openings in *any* electrical enclosure must be plugged or capped. Any punched knockout that will not be used *must* be closed, as at arrow.

This section also indicates that defective equipment may not be used. Although wording prohibiting the use of damaged or otherwise defective equipment may seem superfluous, apparently many installers were using or reusing damaged equipment. At complete odds with common sense, such practice puts those who use and maintain the system at risk and is expressly forbidden. And although not specifically mentioned, any equipment that is damaged during the construction phase should be considered as covered by the rule of 110.12(B) and should be replaced.

110.14. Electrical Connections. Proper electrical connections at terminals and splices are absolutely essential to ensure a safe installation. Improper connections are the cause of most failures of wiring devices, equipment burndowns, and electrically oriented fires. Remember, field installation of electrical equipment and conductors boils down to the interconnection of manufactured components. The circuit breakers, conductors, cables, raceway systems, switchboards, MCCs, panelboards, locknuts, bushings—everything—only need be connected together. With that in mind it can be easily understood *why* the most critical concern for any designer and installer should be the actual interconnection of the various system components—especially terminations of electrical conductors and bus. Although failure to properly terminate conductors is presently the primary cause of system failure throughout the country, it can easily be overcome by attention to, and compliance with, the rules of this section as well as applicable listing and installation instructions.

Terminals and splicing connectors must be "identified" for the material of the conductor or conductors used with them. Where in previous NEC editions this rule called for conductor terminal and splicing devices to be "suitable" for the material of the conductor (i.e., for aluminum or copper), the wording now requires that terminal and splicing devices must be "identified" for use with the material of the conductor. And devices that combine copper and aluminum conductors in direct contact with each other must also be "identified for the purpose and conditions of use."

The NEC definition of identified does not specifically require that products be marked to designate specific application suitability, as noted in the discussion in Art. 100 on the topic of "Identified." However, the general information from the UL directory quoted in the discussion of 110.3 does say that terminations are generally suitable for copper wire only, and where aluminum is suitable there will be a marking. In addition, the installation instructions furnished with the equipment will clearly indicate whether aluminum terminations are permitted, and how they are to be made and torqued.

In general, pressure-type wire splicing lugs or connectors bear no marking if suitable for only copper wire. If suitable for copper, copper-clad aluminum, and/or aluminum, they are marked "AL-CU"; and if suitable for aluminum only, they are marked "AL." Devices listed by Underwriters Laboratories Inc. indicate the range or combination of wire sizes for which such devices have been listed. Terminals of 15- and 20-A receptacles not marked "CO/ALR" are for use with copper and copper-clad aluminum conductors only. Terminals marked "CO/ALR" are for use with aluminum, copper, and copper-clad aluminum conductors.

The vast majority of distribution equipment has always come from the manufacturer with mechanical set-screw-type lugs for connecting circuit conductors to the equipment terminals. Lugs on such equipment are commonly marked "AL-CU" or "CU-AL," indicating that the set-screw terminal is suitable for use with either copper or aluminum conductors. But, such marking on the lug itself is not sufficient evidence of suitability for use with aluminum conductors. UL requires that equipment with terminals that are found to be suitable for use with either copper or aluminum conductors must be marked to indicate such use on the label or wiring diagram of the equipment—completely independent of a marking like "AL-CU" on the lugs themselves. A typical safety switch, for instance, would have lugs marked "AL-CU," but also must have a notation on the label or nameplate of the switch that reads like this: "Lugs suitable for copper or aluminum conductors."

UL-listed equipment must be used in the condition as supplied by the manufacturer—in accordance with NEC rules and any instructions covered in the UL listing in the Guide Card information for the product category—as required by the NEC's 110.3(B). Unauthorized alteration or modification of equipment in the field is not covered by the UL listing and can lead to very dangerous conditions. For this reason, any arbitrary or unspecified changing of terminal lugs on equipment is not acceptable unless such field modification is recognized by UL and spelled out very carefully in the manufacturers' literature and on the label of the equipment itself.

For instance, UL-listed authorization for field changing of terminals on a safety switch might be described in manufacturers' catalog data and on the switch label itself. It is obvious that field replacement of set-screw lugs with compression-type lugs can be a risky matter if great care is not taken to assure that the size, mounting holes, bolts, and other characteristics of the compression lug line up with and are fully compatible for replacement of the lug that is removed. Careless or makeshift changing of lugs in the field has produced overheating, burning, and failures. To prevent junk-box assembly of replacement lugs, UL requires that any authorized field replacement data must indicate the specific lug to be used and also must indicate the tool to be used in making the crimps. Any crimp connection of a lug should always be done with the tool specified by the lug manufacturer. Otherwise, there is no assurance that the type of crimp produces a sound connection of the lug to the conductor.

The 2011 **NEC** inserted a second paragraph, located as part of the parent text of 110.14 so it applies to both terminals and splices, covering fine-stranded conductors. This equipment must be specifically identified for the class of stranding if the stranding is more fine than that described in a new Table 10 in Chap. 9.

Effective with the 2002 edition of the UL White Book, the Guide Card information on this topic, under the heading "Wire Connectors and Soldering Lugs (ZMVV)" has limited the applicability of conventional splicing and terminating methods to stranding no finer than Class B for aluminum and Class C concentric for copper. The same guide card information specified that "wire connectors additionally rated for use with other Class conductors, such as Class M, are marked with the additional class designation and number of strands." The 2008 **NEC** incorporated this concept in 690.31(F), and it arrived in Chap. 1 for the first time in the next cycle. This concept has been fully enforceable under 110.3(B) since 2002, but has seldom been applied in the field.

One major reason for this has been the relative lack of availability of mechanical solutions for this work. Most available terminations employ some form of crimping with a hydraulically actuated tool. This type of tooling is a significant expense, especially for small shops. Electricians have been making these terminations for a very long time and putting up with occasional failures as a result. Section 410.62(C)(1) recognizes numerous applications of flexible cord (generally Class M) connected to building wire at outlets. Has anyone ever observed a twist-on wire connector marked with the class and number of strands? The overwhelming majority of motor leads are Class H or Class I stranding. Has anyone ever seen a split-bolt connector marked with a stranding classification? For that matter, although plugs and cord bodies are obviously designed to terminate flexible cord, the stranding class and conductor count has not generally been included with these items either.

Nevertheless, this topic will now receive a great deal of attention. Chapter 9, Table 10, in concert with this new paragraph, effectively incorporates the UL limitations in the **NEC**. This process of making an arcane guide card limitation general knowledge by placing it in black-letter Code text has a long pedigree. A relevant example is the concept of limiting the temperature exposure of electrical equipment from terminated conductors, which was fully enforceable under UL guide card limitations for decades, and yet remained largely unknown before 110.14(C) first appeared in the 1993 **NEC**.

The new paragraph does not require a stranding count; however, the guide card limits still ask for this. At this writing it is not clear if UL plans to withdraw this aspect of the requirement. The strand count was specifically stripped out of the rule at the very end of the **NEC** process because of concerns about practicality, so this creates some pressure on UL to follow suit. Regardless, the need for a specific listing evaluation by class of stranding will apply. Fortunately, mechanical

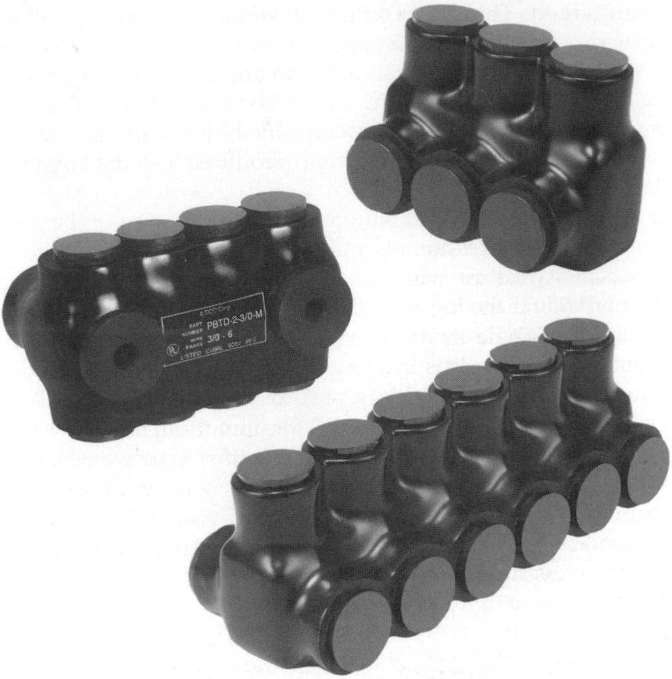

Fig. 110-8. These connectors are listed for a wide variety of finely stranded conductors, including Classes H, I, K, M, and DLO for diesel locomotive use. This vendor has over 50 different configurations with two to six ports and running from 14 AWG up to 350 kcmil, making them suitable for motors up to 200 hp and many other applications. A closely related product line from the same manufacturer covers from 6 AWG up to 750 kcmil. Such terminations routinely involve the use of ferrules to contain the strands, but these do not require them and are therefore reusable without acquiring replacement ferrules. The bearing screw includes a pivoting, nonturning pressure plate that does not cut strands as it tightens. (Courtesy of Ilsco.)

solutions to this are beginning to appear from major vendors with broad market penetration. Figure 110-8 shows such connectors and the caption provides additional background on available configurations.

Part (**A**) prohibits the use of more than one conductor in a terminal (see Fig. 110-9) unless the terminal is identified for the purpose (meaning generally recognizable as suitable for the purpose by appropriate markings or instructions).

Use of the word "identified" in the last sentence of 110.14(A) could be interpreted to require that terminals suited to use with two or more conductors must

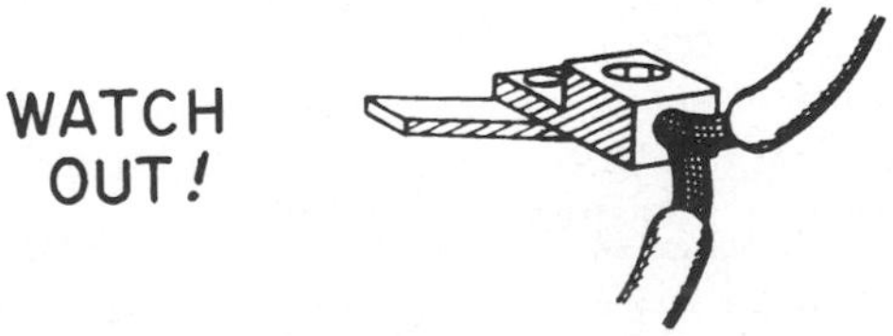
WATCH
OUT!

Fig. 110-9. [Sec. 110.14(A).]

somehow be marked. This is a frequent example of where installers confuse "identified" with "marked" as covered in the discussion of the definition of "identified." For a long time terminals suited to and acceptable for use with aluminum conductors have been marked "AL-CU" or "CU-AL" right on the terminal. Twist-on or crimp-type splicing devices are "identified" both for use with aluminum wires and for the number and sizes of wires permitted in a single terminal—with the identification marked on the box in which the devices are packaged or marked on an enclosed sheet.

For set-screw and compression-type lugs used on equipment or for splicing or tapping-off, suitability for use with two or more conductors in a single barrel of a lug could be marked on the lug in the same way that such lugs are marked with the range of sizes of a single conductor that may be used (e.g., "No. 2 to No. 2/0."). But the intent of the Code rule is that any single-barrel lug used with two or more conductors must be tested for such use (such as in accordance with UL 486B standard), and some indication must be made by the manufacturer that the lug is properly suited and rated for the number and sizes of conductors to be inserted into a single barrel. Again, the best and most effective way to identify a lug for such use is with marking right on the lug, as is done for "AL-CU." But the second sentence of 110.3(A)(1) also allows such identification to be "provided with" a product, such as on the box or on an instruction sheet. See Fig. 110-10.

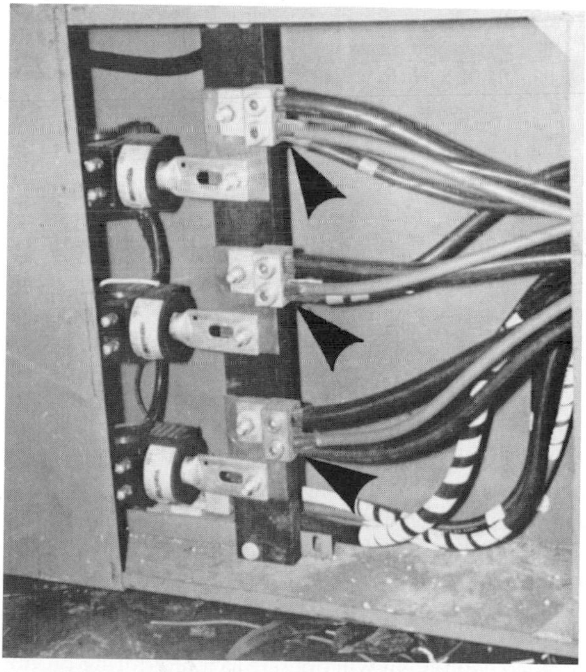

Fig. 110-10. A *terminal* with more than one conductor terminated in a single barrel (hole) of the lug (at arrows) must be "identified" (marked, listed, or otherwise tested and certified as suitable for such use).

Section 110.14(B) covers splice connectors and similar devices used to connect fixture wires to branch-circuit conductors and to splice circuit wires in junction boxes and other enclosures. Much valuable application information on such devices is given in the UL Electrical Construction Materials Directory, under the heading of "Wire Connectors and Soldering Lugs." The new last sentence of 110.14(B) states that connectors or splices used with directly buried conductors must be listed for the application.

This wording makes the use of listed connectors and splice kits mandatory where used directly buried. As indicated by the submitter of this proposal for a change, such equipment *is* listed, *is* commercially available, and *should* be used.

Part **(C)** of 110.14 reiterates the UL rules regarding temperature limitations of terminations, which are made mandatory by 110.3(B). It is worth noting that the information given by UL in the guide card information for "Electrical Equipment for Use in Ordinary Locations" is more detailed. The informational note following this section is intended to indicate that if information in a general or specific UL rule permits or requires different ratings and/or sizes, the UL rule must be followed.

The last sentence indicates acceptability of 90°C-rated wire where applied in accordance with the temperature limitations of the termination.

90°C-insulated conductors may be used in virtually any application that 60°C- or 75°C-rated conductors may be used and in some that the lower-rated conductors cannot. But the ampacity of the 90°C-rated conductor must never be taken to be more than that permitted in the column that corresponds with the temperature rating of the terminations to which the conductor will be connected. And that applies to both ends of the conductor. For example, consider a 6 AWG THHN copper conductor, which has a Table 310.16 ampacity of 55 A in the 60°C column. But, if the 6 AWG is supplied from a CB with, say, a 60/75°C-rating, it may be considered to be a 65-A wire, provided the equipment end is also rated 60/75°C or 75°C. But, if the equipment *at either end of the wire* is rated at 60°C, or unmarked and therefore rated that way by default, the 6 AWG THHN copper conductor may carry no more than the 60°C-ampacity (55 A) shown in Table 310.16 for a 6-AWG copper wire. The wording in parts **(C)(1)(a)(3)** and **(C)(1)(b)(2)** is intended to indicate as much (Fig. 110-11). Note also that most motors go directly to the 75°C ratings regardless of wire size, per **(C)(1)(a)(4)**.

Part **(D)** calls attention to the fact that manufacturers are marking equipment, terminations, packing cartons, and/or catalog sheets with specific values of required tightening torques (pound-inches or pound-feet). Although that puts the installer to the task of finding out appropriate torque values, virtually all manufacturers are presently publishing "recommended" values in their catalogs and spec sheets. In the case of connector and lug manufacturers, such values are even printed on the boxes in which the devices are sold. In 110.3(B), the **NEC** requires that all listed equipment be installed as indicated by the listing instructions that are issued with the product. In virtually all cases, where a mechanical type terminating device is used, the manufacturer will indicate a prescribed torque value. That is the value that was used during product testing. In order for one to be certain that the installed equipment will operate as it did during product certification testing, the equipment must be used in the same manner it was during testing. And that includes torquing the terminating devices to the values prescribed in the

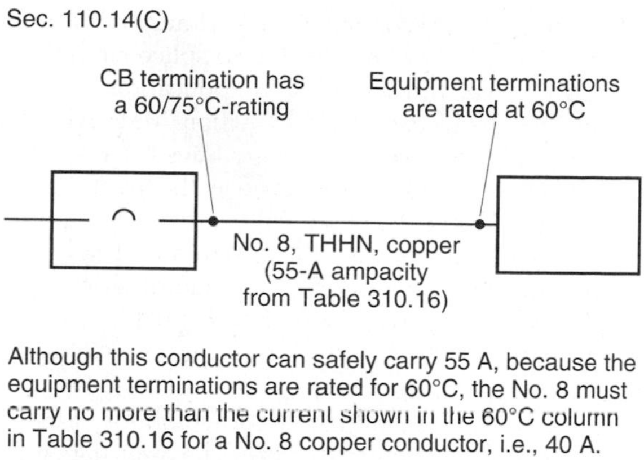
Sec. 110.14(C)
CB termination has a 60/75°C-rating
Equipment terminations are rated at 60°C
No. 8, THHN, copper (55-A ampacity from Table 310.16)
Although this conductor can safely carry 55 A, because the equipment terminations are rated for 60°C, the No. 8 must carry no more than the current shown in the 60°C column in Table 310.16 for a No. 8 copper conductor, i.e., 40 A.

Fig. 110-11. 90°C-insulated conductors may be used even where derating is not required, provided they are taken as having an ampacity not greater than the ampacity shown in the column from Table 310.15(B)(16) that corresponds to the temperature rating of the terminations—at both ends—to which the conductors will be connected.

manufacturer's installation instructions. Failure to torque every terminal to the manufacturer-prescribed value is a clear and direct violation of this Code section.

Torque is what produces the amount of tightness of the screw or bolt in its threaded hole; that is, torque is the measure of the twisting movement that produces rotations around an axis. Such turning tightness is measured in terms of the force applied to the handle of the device that is rotating the screw or bolt and the distance from the axis of rotation to the point where the force is applied to the handle of the wrench or screwdriver:

$$\text{Torque (lb ft)} = \text{force (lb)} \times \text{distance (ft)}$$
$$\text{Torque (lb in.)} - \text{force (lb)} \times \text{distance (in.)}$$

Because there are 12 in. in a foot, a torque of "1 lb-ft" is equal to "12 lb-in." Any value of "pound-feet" is converted to "pound-inches" by multiplying the value of "pound-feet" by 12. To convert from "pound-inches" to "pound-feet," the value of "pound-inches" is divided by 12.

Note: The expressions "pound-feet" and "pound-inches" are preferred to "foot-pounds" or "inch-pounds," although the expressions are used interchangeably. When the unit leads with the distance (such as foot-pounds), it is supposed to be referring to a unit of energy in classical physics, where energy is defined as the product of applied force and the distance over which that force acted.

Torque wrenches and torque screwdrivers are designed, calibrated, and marked to show the torque (or turning force) being exerted at any position of the turning screw or bolt. Fig. 110-12 shows typical torque tools and their application. And they do require periodic calibration. See also the additional coverage in this book at Annex I at the end of the code book. This annex presents default torque specifications based on the size and type of screw supplied with the equipment, and can be used in the event that specifications on or with the equipment are unavailable.

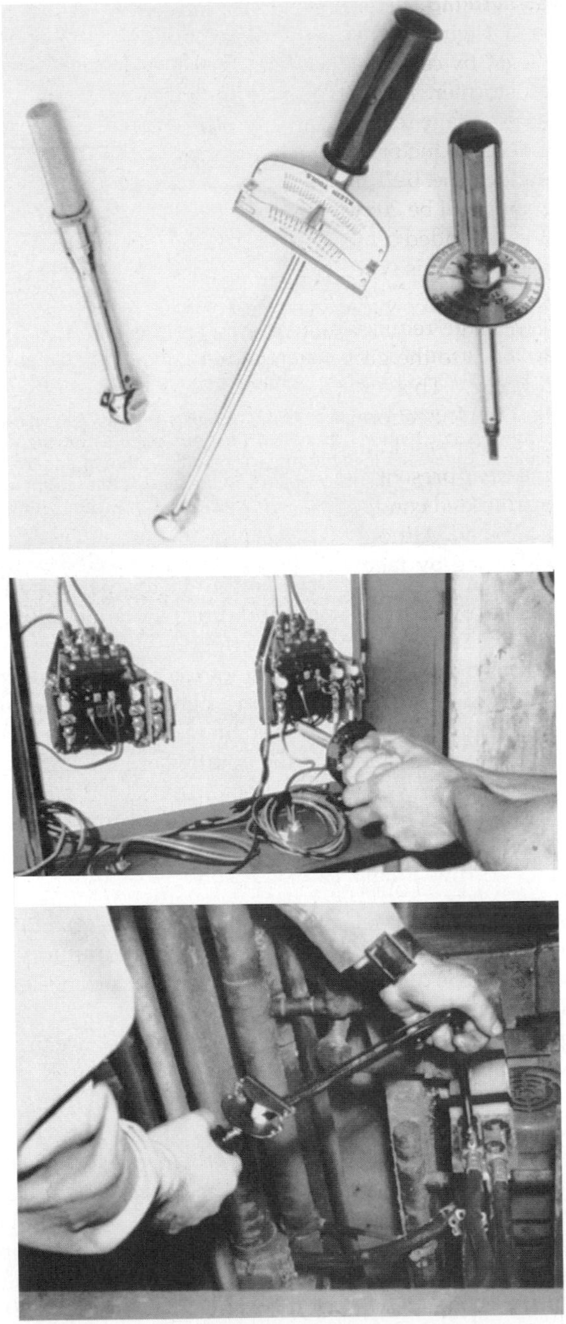

Fig. 110-12. Readily available torque tools are (at top, L–R): torque screwdriver, beam-type torque wrench, and ratchet-type torque wrench. These tools afford ready compliance with the implied requirement of the informational note in the **Code** rule. Sec. 110.14(D).

110.15. High-Leg Marking. Here the Code mandates a specific "color-coding" for the "high leg" in a 4-wire delta system. These systems, used in certain areas, create two 120-V "legs" by center-tapping one of the secondary windings of a 240-V delta-wound transformer. That is, by grounding the center point of one winding, the phase-to-phase voltage remains 240 V, 3-phase, and the voltage between phase A and the grounded conductor, and between phase C and the grounded conductor will be 120 V, single phase. BUT the difference of potential between phase B and the grounded conductor will be 208 V, single-phase. The fact that phase B is at 208 V, with respect to the grounded center-tapped conductor, while phases A and C are at 120 V, is the reason that this type of distribution system is known as a "high-leg" delta system.

As indicated, this rule requires color-coding of the "high leg" (i.e., phase B—the one that's at 208 V to the grounded conductor) or other "effective means" to identify the "high leg." This "identification" must be provided at any point in the system where "a connection is made" and the high leg is run with the other circuit conductors. That would include enclosures where the high leg is itself not connected but merely "present" within the enclosure, and would exclude enclosures where the grounded conductor is not present, such as where the three phase legs supply a motor load. Although the wording used in this section would permit identifying the high leg by tagging with numbers or letters, or other "effective means," where color-coding is used, the high leg must be colored orange. Use of other colors to identify the high leg would seem to be a violation of the wording used here.

110.16. Arc-Flash Hazard Warning. This section calls for a field marking of electrical equipment such as switchboards, panelboards, motor control centers, meter socket enclosures, and industrial control panels—provided by the installer at the time of installation—that indicates that protection against arc-flash hazards is required when maintaining such equipment. Frequently, the hazards encountered will be generic to the equipment and the 2014 **NEC** also allows for, but does not require, factory applied markings to meet this rule. The marking is intended to alert maintenance personnel of the need for protective gear when working on the equipment while it is energized. Such marking must be on the exterior of covers and doors that provide access to energized live parts to satisfy the requirement for the warning to be "clearly visible" "before examination, etc." Refer to **NFPA 70E®** for detailed information on selecting personal protective equipment (PPE) that is appropriate for the degree of arc-flash exposure involved. The rule applies to equipment in "other than dwelling units" so it would not apply to a household service panel or an apartment panel within the apartment. However, it does apply to service equipment and distribution panels for multifamily housing that are in common areas outside any dwelling unit. Typical equipment for single dwelling units is not exposed to the incident energy levels that other equipment may see. Effective with the 2017 **NEC**, in (**B**), large-capacity (1200 amps. or more) non-residential services must include actual numerical values of specific site parameters (voltage, available fault, and clearing time under those conditions) on the label, together with the "date the label was applied." In the case of a pre-applied label by the manufacturer, this should be interpreted as the date the required calculations were made. These numbers allow arc-flash hazard calculations to be done easily. One aspect of the specific wording will require interpretation. The literal text ("in other than dwelling units") raises the question as to whether a large apartment building service is included. Very few individual dwelling services are of such a size. Where the service supplies a building of multiple occupancy, the power is not confined to any dwelling unit, and therefore the rule may apply. It certainly would never be a violation to provide this labeling on an apartment building service. The **NEC** also allows, as covered in Informational Note No. 3, arc-flash labeling as detailed in **NFPA** 70E to substitute for this labeling. This makes sense because the intent of the labeling is to allow for appropriate PPE selections.

Note that the fault current calculation does suffer from the same problem of inevitable inaccuracy based on utility changes over time as the value developed in 110.24. Refer to the discussion in that section for more information.

110.18. Arcing Parts. Complete enclosures are always preferable, but where this is not practicable, all combustible material must be kept well away from the equipment.

110.21. Marking. The marking required in 110.21(A) should be done in a manner that will allow inspectors to examine such marking without removing the equipment from a permanently installed position. It should be noted that the last sentence in 110.21 requires electrical equipment to have a marking durable enough to withstand the environment involved (such as equipment designed for wet or corrosive locations). Effective with the 2017 **NEC**, this section also includes express requirements regarding reconditioned equipment. A marking must identify the responsible entity for the work along with the date of reconditioning. This will establish the beginning of a paper trail through which the inspector can determine the qualifications of the reconditioning party.

Part (B) was new in the 2014 **NEC** and addresses how hazard markings applied in the field should be placed and formatted. This change is even more important than it might first appear, because many sections all across the Code now point to these rules as being applicable to the labels being created in accordance with those diverse sections. The actual code text is quite vague ("adequately warn ... using effective words and/or colors and/or symbols) but the informational note is anything but vague. By pointing to the ANSI Z535.4 standard, an inspectional authority can apply the detailed requirements of that standard. For example, there are three hazard commands:

DANGER (a hazardous condition that if not avoided will result in death or serious injury); label to use red, white, and black

WARNING (a hazardous condition that if not avoided could result in death or serious injury); label to use black, white, and orange

CAUTION (a hazardous condition that if not avoided might result in minor or moderate injury); label to use black, white, and yellow

The new subsection also requires that the label be permanently affixed to its target and have sufficient durability to withstand applicable environmental factors. In addition, the label must not be handwritten, with an exception for a portion of a label that may have variable information or that is otherwise subject to change. In such cases handwriting, which must be legible, is permitted. The information presented here is necessarily incomplete; the Z535.4 standard (developed by NEMA) goes on to address font sizes, colors, placement locations, wording choices, acceptable symbols and appropriate wording, as well as tactics to provide durability under various adverse operating conditions.

110.22. Identification of Disconnecting Means. As shown in Fig. 110-13, it is a mandatory Code rule that all disconnect devices (switches or CBs) for load devices and for circuits be clearly and permanently marked to show the purposes of the disconnects. This is a "must" and, under OSHA, it applies to all existing electrical systems, no matter how old, and also to all new, modernized, expanded, or altered electrical systems. This requirement for marking has been widely neglected in electrical systems in the past. Panelboard circuit directories must be fully and clearly filled out. And all such marking on equipment must be in painted lettering or other substantial identification.

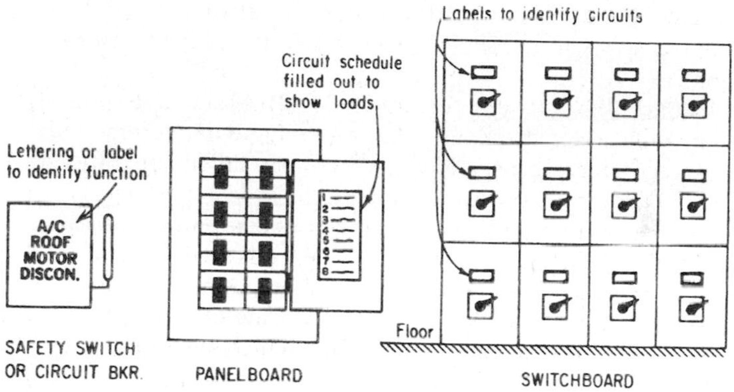
Labels to identify circuits
Circuit schedule filled out to show loads
Lettering or label to identify function
A/C ROOF MOTOR DISCON.
1
2
3
4
5
6
7
8
Floor
SAFETY SWITCH OR CIRCUIT BKR.
PANELBOARD
SWITCHBOARD

Fig. 110-13. All circuits and disconnects must be identified. OSHA regulations make NEC Sec. 110.22 mandatory and retroactive for existing installations and for all new, expanded, or modernized systems—applying to switches as well as circuit breakers. (Sec. 110.22.)

This rule now appears as 110.22(A) because the section has been divided into three lettered paragraphs. Effective identification of all disconnect devices is a critically important safety matter. When a switch or CB has to be opened to de-energize a circuit quickly—as when a threat of injury to personnel dictates—it is absolutely necessary to identify quickly and positively the disconnect for the circuit or equipment that constitutes the hazard to a person or property. Painted labeling or embossed identification plates affixed to enclosures would comply with the requirement that disconnects be "legibly marked" and that the "marking shall be of sufficient durability." Paste-on paper labels or marking with crayon or chalk could be rejected as not complying with the intent of this rule. Ideally, marking should tell exactly what piece of equipment is controlled by a disconnect (switch or CB) and should tell where the controlled equipment is located and how it may be identified. Figure 110-14 shows a case of this kind of identification as used in an industrial facility where all equipment is marked in two languages because personnel speak different languages. And that is an old installation, attesting to the long-standing recognition of this safety feature.

The rule of 110.22 has long required that every disconnect be marked to indicate exactly what it controls. And that marking must be legible and sufficiently durable

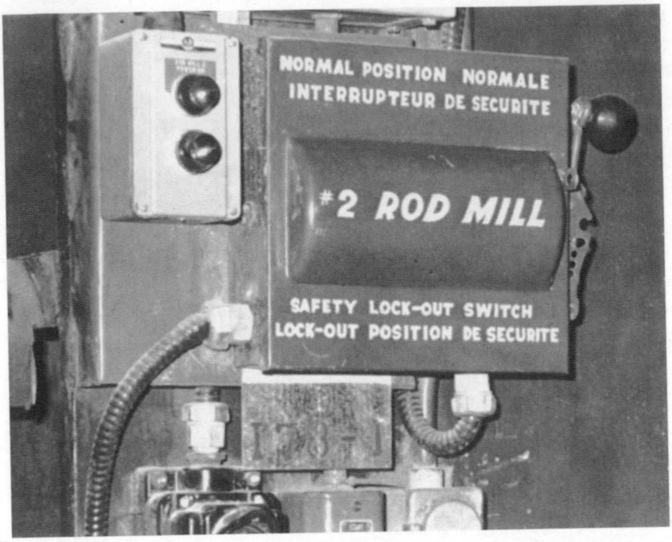
NORMAL POSITION NORMALE
INTERRUPTEUR DE SECURITE
#2 ROD MILL
SAFETY LOCK-OUT SWITCH
LOCK-OUT POSITION DE SECURITE

Fig. 110-14. Identification of disconnect switch and pushbutton stations is "legibly marked" in both English and French—and is of "sufficient durability to withstand the environment"—as required by the **Code** rule. (Sec. 110.22.)

to withstand the environment to which it will be exposed. And, the rule of 408.4 requires that any modifications also be reflected in the circuit directory of panelboards. While most are aware of the requirement in 110.22, it seems as if very few pay any attention to this part of the rule of 408.4.

It should be noted that the marking of disconnects is one of the few requirements that is made retroactive by OSHA. That is, regardless of when the disconnect was installed, or when a modification was performed, the purpose of every disconnect must be marked at the disconnect. If your facility or your customer's facility does not have such markings for each and every disconnect, every effort should be made to ensure that a program to provide such markings is initiated and completed. Failure to do so could result in heavy fines should you be subject to an inspection by OSHA (Fig. 110-14).

The second paragraph, 110.22(B), covers the field marking requirements for series combinations of circuit breakers or fused switches used in an "engineered series combination" with downstream devices that do not have an interrupting rating equal to the available short-circuit current but are dependent for safe operation on upstream protection that is rated for the short-circuit current; enclosure(s) for such "series rated" protective devices must be marked in the field "Caution—Engineered Series Combination System Rated _____Amperes. Identified Replacement Components Required."

Note that this paragraph as well as the one following it specifically cite the new (2014 **NEC**) 110.21(B) for the required field marking.

This provision correlates with the new procedure in 240.86(A) that allows, under strict engineering supervision and only in existing installations, the use of upstream overcurrent protective devices with a let-through current under fault conditions that does not exceed the interrupting rating of a downstream overcurrent device. This allows for adapting existing installations to increases in available fault current resulting from changes in the infrastructure of the serving utility or other factors. The engineer must be able to certify that the lower-rated downstream device will not begin to open or melt during the operating period of the upstream device, or the downstream device may be subjected to the full available fault in excess of its rating, instead of merely the let-through current. This requires a very sophisticated evaluation, and will be completely defeated if the wrong replacement component is selected.

The third paragraph, 110.22(C), says that where circuit breakers or fused switches are used in a "tested series combination" with downstream devices that do not have an interrupting rating equal to the available short-circuit current but are dependent for safe operation on upstream protection that is rated for the short-circuit current, enclosure(s) for such "series rated" protective devices must be marked in the field "Caution—Series Combination System Rated _____Amperes. Identified Replacement Components Required."

Such equipment is typically employed in multimeter distribution equipment for multiple-occupancy buildings—especially residential installations—with equipment containing a main service protective device that has a short-circuit interrupting rating of some value (e.g., 65,000 A) that is connected in series with feeder and branch-circuit protective devices of considerably lower short-circuit interrupting ratings (say 22,000 or 10,000 A). Because all of the protective devices are physically very close together, the feeder and branch-circuit devices do not have to have a rated interrupting capability equal to the available short-circuit current at their points of installation. Although such application is a literal violation of **NEC** 110.9, which calls for all protective devices to be rated for the short circuit current available at their supply terminals, "series rated" equipment takes advantage of the ability of the protective devices to operate in series (or in cascade as it is sometimes called) with a fault current interruption on, say, a branch circuit being shared by the three series protective devices—the main, feeder, and branch circuit. Such operation can enable a properly rated main protective device to protect downstream protective devices that are not rated for the available fault. When manufacturers combine such series protective devices in available distribution equipment, they do so on the basis of careful testing to assure that all of the protective devices can operate without damage to themselves. Then UL tests such equipment to verify its safe and effective operation and will list such equipment as a "Series Rated System."

Because UL listing is based on use of specific models of protective devices to assure safe application, it is critically important that all maintenance on such equipment be based on the specific equipment. For that reason, this Code rule demands that the enclosure(s) for all such equipment be provided with "readily visible" markings to alert all personnel to the critical condition that must always be maintained to ensure safety. Thus, all series-rated equipment enclosure(s) must be marked.

Note that the enhanced requirements for series-rated systems only apply where the series rating is required to satisfy 110.9. For example, on a 22-kA available fault current system at the service disconnect, if it can be shown that there is sufficient static impedance in the length of feeder between the service and a downstream panel so that the available fault current at that panel is only 10 kA, then no series connection listing is involved and this marking requirement does not apply. This is true even though the downstream panel is still in series with the service protective device. Series-connected listings are an economical way to avoid a fully rated system, but they are not required when the conditions in the field are such that a downstream overcurrent device could not be subjected to a fault current in excess of its rating, and therefore comply with 110.9 without any upstream assistance. Since a very little impedance goes a long way in reducing an otherwise very high available fault current, this is a very frequent practical result. Many engineers take this into account when they position downstream equipment.

110.24. Available Fault Current. This rule was new in the 2011 NEC and requires a field marking of the maximum available fault current to be placed on service equipment in other than dwelling units. As in the case of the arc-flash warning in 110.16, the preposition "in" means that the label would be required for a service in a common area for multifamily housing. An exception waives the rule in industrial occupancies with qualified maintenance and supervision.

The rule further requires that this label be updated when modifications occur that change the available fault current. As in the case of the original label, the applicable date when the calculation was performed must be included.

There are serious issues with such a label, beginning with the fact that such a label, if applied, will tend to be believed. Utilities routinely change their distribution patterns without advising property owners of potential changes in fault current availability. This section is in the general part of Art. 110 and therefore also literally applies to medium-voltage installations; routine utility switching operations on medium-voltage networks make the available fault particularly volatile. Frequently, when a utility does provide a fault current number, it does so over a disclaimer that the calculation is only good as of the date on the letter. If the fault current increases, the label is worse than useless if it is relied upon to select additional overcurrent devices.

Although utilities will always make fault current assumptions known to customers on request, they do not inform them prospectively and cannot be expected to know where such labels have been applied and update them. And by far the single most likely modification that would affect a 110.9 calculation would be utility activities on their system. We are left with a label that likely will not be updated, and that would be hazardous if erroneous and acted upon. The 2017 NEC additionally requires that the fault current calculation be documented and made available to interested parties; however that does not address the concerns raised here.

Section 110.9 addresses bolted short-circuit available current, and it must be properly applied. A failure, however, normally only occurs on start-up of a new installation. For example, a parallel makeup on a large feeder, if the wires are crossed, will produce such a fault. This is not a failure that will occur after the commencement of occupancy, where the normal failure mode is about ground faults. This is why documented bolted short-circuit failures are vanishingly rare in occupied buildings, and why utilities see little gain in publishing changes to their required fault duty levels.

There is an additional issue with this label, involving incident energy calculations for personal protective equipment (PPE) to address arc-flash hazards. The number on this label should not be used for this work, but preliminary field discussions indicate that is how it will frequently be applied.

For just this reason the 2014 NEC placed an informational note at this location describing its intended application and differentiating it from *NFPA 70E* that addresses workplace safety requirements.

110.25 Lockable Disconnecting Means. For some time, section after section in the NEC has been devoting considerable real estate to what constitutes an acceptable disconnecting means, particularly when the disconnect is supposed to be lockable. Now a universal rule appears in Art. 110, which will apply throughout the Code, and in almost 50 locations this verbiage has been withdrawn in favor of a reference to this rule. These disconnecting means must lock open, and the lockable portion must be permanently installed so as to remain in place whether or not a padlock is in place. Other arrangements are too easily defeated, because they can usually be removed with the padlock still closed. Note that special conditions do arise. For example, 490.46 covers medium-voltage circuit breakers, and in addition to referencing this section, that section also integrates instances where a drawout mechanism is relied upon in this context.

II 1000 Volts, Nominal, or Less

110.26. Spaces about Electric Equipment. Although the upper voltage parameter throughout most of the code has been raised from 600 to 1000 V largely to accommodate higher dc voltages from PV applications, the 600-V limit did not change in this article. The basic rule of 110.26 calls for "sufficient access and working space" to be provided in all cases to permit ready and safe operation and maintenance of electrical equipment. This rule applies to receptacles and all electrical equipment. However, the specific work space dimensions and other rules that are in the lettered paragraphs following only apply under the conditions set forth in those paragraphs. For example, a hydromassage bathtub motor and receptacle that has no access door provided is in violation of this section (as well as some prescriptive criteria in Art. 680). However, the full panoply of required work space widths and depths does not apply because such equipment does not need to be worked hot.

The wording of 110.26(A) calls for compliance with parts **(A)(1)**, **(2)**, and **(3)**. Those three subparts define the work space zone needed at electrical equipment. These rules are slightly modified and expanded upon by parts **(B)**, **(C)**, **(D)**, and **(E)**. Part **(F)** has nothing whatever to do with work space and covers the dedicated wiring space above (and below) certain pieces of equipment.

As indicated by 110.26(A)(1), the dimensions [shown in Table 110.26(A)] of working space in the direction of access to live electrical parts for equipment operating at 1000 V or less—where live parts are exposed—or to the equipment enclosure—in the usual case where the live parts are enclosed—must be carefully observed. The minimum clearance is 3 ft. The minimum of 3 ft was adopted for Code Table 110.26(A)(1) to make all electrical equipment—panelboards, switches, breakers, starters, etc.—subject to the same 3-ft (914-mm) minimum to increase the level of safety and assure consistent, uniform spacing where anyone might be exposed to the hazard of working on any kind of live equipment. Application of Code Table 110.26(A)(1) to the three "conditions" described in the note below Table 110.26(A)(1) is shown in the sketches making up this handbook's Table 110-1. The 2017 **NEC** added a line to the table for 601–1000 volts. This made no change in **NEC** requirements, because the required spacings agree with Table 110.34(A), which formerly applied. This rearrangement is part of the gradual inclusion of this subject matter in rules that formerly went only up to 600 volts. Figure 110-15 shows a typical example of Condition 3.

Table 110-1. Clearance Needed in Direction of Access to Live Parts in Enclosures for Switchboards, Panelboards, Switches, CBs, or Other Electrical Equipment—Plan Views [110.26(A)]

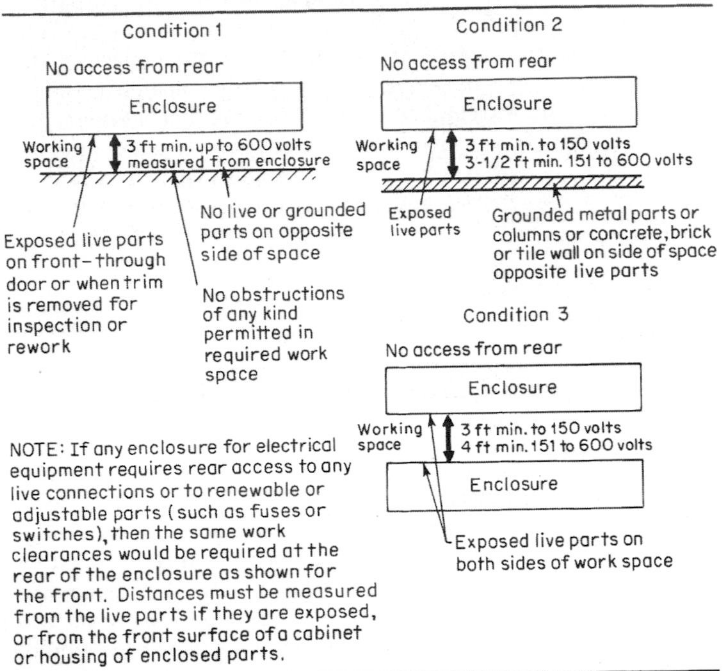
Condition 1
No access from rear
Enclosure
Working space
3 ft min. up to 600 volts measured from enclosure
No live or grounded parts on opposite side of space
Exposed live parts on front–through door or when trim is removed for inspection or rework
No obstructions of any kind permitted in required work space
Condition 2
No access from rear
Enclosure
Working space
3 ft min. to 150 volts
3-1/2 ft min. 151 to 600 volts
Exposed live parts
Grounded metal parts or columns or concrete, brick or tile wall on side of space opposite live parts
Condition 3
No access from rear
Enclosure
Working space
3 ft min. to 150 volts
4 ft min. 151 to 600 volts
Enclosure
Exposed live parts on both sides of work space
NOTE: If any enclosure for electrical equipment requires rear access to any live connections or to renewable or adjustable parts (such as fuses or switches), then the same work clearances would be required at the rear of the enclosure as shown for the front. Distances must be measured from the live parts if they are exposed, or from the front surface of a cabinet or housing of enclosed parts.

For applications above 600 volts but not exceeding 1000 volts, Condition 1 clearances (3 ft) are unchanged, but Condition 2 clearances increase to 4 ft and Condition 3 clearances increase to 5 ft.

According to 110.26(A)(1)(a), a "minimum" depth of work space behind equipment rated 1000 V, and less, must be provided where access is needed when deenergized. The past few editions of the **NEC** have required a minimum depth of work space behind equipment rated over 600 V where access was required only when the equipment was deenergized. The 2002 **NEC** strengthened these requirements by changing the references from "deenergized parts" to "nonelectrical parts," concluding that merely shutting off power to such equipment was not sufficient. An example of such access would be periodic replacement of an air filter (Fig. 110-16).

Fig. 110-15. Condition 3 in Code Table 110.26(A) for the rule covered by Sec. 110.26(A) applies to the case of face-to-face enclosures, as shown here where two switchboards face each other. The distance indicated must be at least 3 or 4 ft depending on the voltage of enclosed parts. [Sec. 110.26(A).]

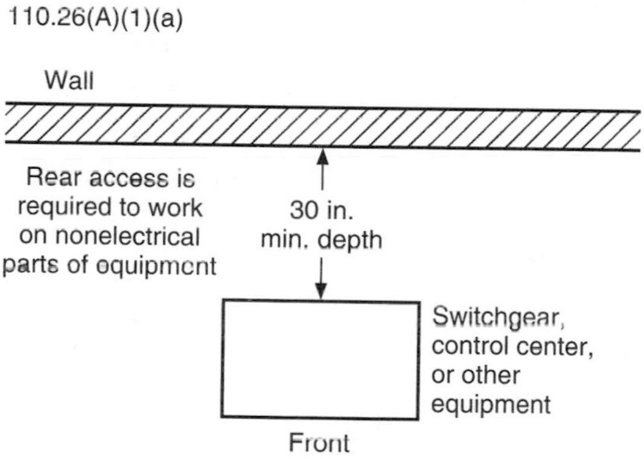
110.26(A)(1)(a)
Wall
Rear access is required to work on nonelectrical parts of equipment
30 in. min. depth
Switchgear, control center, or other equipment
Front

Fig. 110-16. Where access is needed, but only when the equipment is deenergized, the work space need only be 30 in. deep. This addition to part **(a)** in Sec. 110.26(A) applies *only* to those cases where access is needed only to service *nonelectrical parts.* As always, if access is needed at the rear for "examination, adjustment, servicing, or maintenance" when the equipment is energized, the depth, as well as the other aspects of work space must satisfy the basic rule. And, if there is never a need to gain access to the rear of the equipment, there is no minimum depth required by the Code, but careful attention should be paid to any clearances required by the equipment manufacturer.

Part (1)(b) to 110.26(A) allows working clearance of less than the distances given in Table 110.26(A) for live parts that are operating at not over 30 V RMS, 42 V peak, or 60 V dc. The last phrase recognizes the inherent safety of low-voltage circuits like the Class 2 and Class 3 control and power-limited circuits covered by Art. 725, as well as certain other low-voltage systems recognized in Chaps. 7 and 8. This exception allows less than the 3-ft minimum spacing of Table 110.26(A) for live parts of low-voltage communication, control, or power-limited circuits. BUT, only where "special permission" is granted. Remember, any such "special permission" must be in writing.

Part **(1)(c)** to 110.26(A) permits smaller work space when replacing equipment at existing facilities, provided procedures are established to ensure safety. This allowance responds to widespread misinterpretation of these rules at the time 480- and 600-V motor control centers were being installed in industrial occupancies with Condition 2 dimensioned aisles. Many engineers considered this to be a 3½-ft clearance because when staff would be working one side hot the other side would be closed, and therefore present a grounded and not energized surface. This was never the intent of the rule, because often both sides are worked hot at the same time. The **NEC** has been reworded to preclude this misinterpretation from continuing.

However, when it came time to upgrade such existing nonconforming installations, facilities were in a quandary because now the room and the conduits were in place. The solution was to provide limited relief for existing applications that allows Condition 2, but only if qualified personnel are involved and there are written procedures in place that preclude working both sides of the aisle hot.

In part **(A)(2)** of 110.26, the Code mandates a minimum width for the required clear work space. For all equipment, the work space must be 762 mm (30 in.) or the width of the equipment, whichever is greater. Note that there is no requirement to center the equipment in the clear space, only the requirement to provide the space, which may even begin right at one edge of the equipment and then extend beyond the equipment on the other side.

And, as required by the second sentence of Sec. 110.26(A)(2), clear work space in front of any enclosure for electrical equipment must be deep enough to allow doors, hinged panels, or covers on the enclosure to be opened to an angle of at least 90°. Any door or cover on a panelboard or cabinet that is obstructed from opening to at least a 90° position makes it difficult for any personnel to install, maintain, or inspect the equipment in the enclosure safely. Full opening provides safer access to the enclosure and minimizes potential hazards (Fig. 110-15). Although this rule seems to be related to "depth" of work space, it is covered in 110.26(A)(2), "Width of Working Space." (See Fig. 110-17.)

Part **(A)(3)** covers the minimum headroom, which must extend from the floor or work platform to 6½ ft (2 m) or (if higher) the height of the equipment. The rule

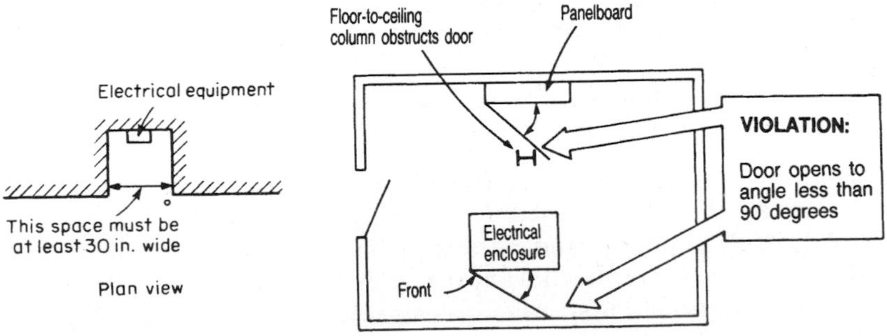
Electrical equipment
This space must be at least 30 in. wide
Plan view
Floor-to-ceiling column obstructs door
Panelboard
VIOLATION:
Door opens to angle less than 90 degrees
Electrical enclosure
Front

Fig. 110-17. Working space required in front of electrical equipment for side-to-side clearance and door opening. [Sec. 110.26(A).]

applies to any equipment covered in 110.26(A), and generally includes service equipment, switchboards, panelboards, and motor control centers. The first exception permits "service equipment or panelboards, in existing dwelling units, that do not exceed 200 A" to be installed with less than 6½ ft of headroom—such as in crawl spaces under single-family houses. But this exception applies only in existing "dwelling units" that meet the definition of that phrase. In any space other than an existing dwelling unit, all indoor service equipment, switchboards, panelboards, or control centers must have headroom at the equipment that is at least 6½ ft high, but never less than the height of the equipment.

Some details on headroom are shown in the bottom portion of Fig. 110-20. But, in that sketch, it should be noted that the 2.0-m (6½-ft) headroom must be available for the entire length of the work space. There must be a 2.0-m (6½-ft) clearance from the floor up to the bottom of the light fixture or to any other overhead obstruction—and not simply to the ceiling or bottom of the joists. Note that this rule now incorporates the provisions in former 110.26(E), which has been deleted.

The wording here does permit a limited intrusion into the required work space. Equipment associated with the electrical installation, such as wireways, and pull boxes, may protrude into the work space, but not more than 150 mm (6 in.) beyond the front of the electrical equipment that requires the dedicated space. This intrusion is permitted either below or above the equipment in question. It is also permitted for even the items specified in 110.26(E) (service equipment, panelboards, distribution boards, and motor control centers) because all 110.26(E) does is to establish the extent of the vertical dimension. The rule in this location, and no other, determines the extent to which the required work space may be intruded upon. Note that this allowance does not permit large transformers or other equipment that extend further into the work space. The second exception, covering electric meters, allows a further intrusion over and above the 6-in. allowance for general electrical equipment. The meter socket that supports the meter, however, is subject to the normal 6-in. intrusion limitation.

The rules in 424.66(B) regarding work space above suspended ceilings were moved to 110.26(A) in the 2017 **NEC** as a new (4). In the process, the rules now apply to equipment generally, not just duct heaters, and parallel rules were inaugurated for crawl spaces as well. Fig. 110-18 shows a typical example above a

Fig. 110-18. Example of typical cramped working space above suspended ceilings.

suspended ceiling. Note that the metric dimensions are unrounded soft conversions. This is obviously an inappropriate use of soft conversions. Clearly, a couple millimeters is of no consequence here and, ironically, soft conversions were being inserted while the Code-making panel, elsewhere in Article 110, was busily and correctly changing soft conversions to hard conversions.

The minimum opening to the space has been chosen to match a typical square lay-in tile clear opening between the T-bar edges. The minimum width follows the usual rules, being that of the equipment and not less than 30 in., and also such that the enclosure doors open not less than 90°. The front space must also meet the usual depth requirement as given in Table 110.26(A)(1). The final set of rules, in (d), will undoubtedly require some discussion with the inspector, for several reasons. First, the "maximum" height is specified, beginning with the required height to install the equipment. It would seem the appropriate word would be "minimum." Obviously, there should be no code objection to a ceiling cavity (or, for that matter, a crawl space height) of any height that happens to be greater than that necessary for the equipment to function. In effect, this provision should be read, as common sense would indicate, as a simple recognition that the maximum space available in such installations is an acceptable working space height.

Another issue that will likely initiate discussion relates to the allowance for "a horizontal ceiling structural member" in this space. The original text in 2014 NEC 424.66(B)(4) specifically mentioned T-bars; the meaning here is less clear because crawl spaces are also covered. Crawl spaces generally have ceilings comprised of horizontal ceiling structural members that support the floor immediately above. The fact that the sentence also includes "access panel," in wording joined by the coordinating conjunction "or," suggests the provision only applies to suspended ceilings but the inspector will need to decide.

There is at least one more issue. This wording remained in Art. 424 through the input period, only coming over completely in the comment period. The Code-making panel responsible for Art. 424 (remember, applied only to ceilings) also accepted additional wording in the input period mandating that the working space be unobstructed to the floor. This is crucial to a safe work environment. Often suspended ceiling cavities extend over fixed partition walls that only extend up to the suspended ceiling. If the equipment access panel is located above one of those partitions, it will be impossible to position a stepladder below the working space. The electrician working on the equipment will then be forced to lean, often to a hazardous extent, to the side of the ladder, defeating the objective of the requirement. The wording to address this issue was inexplicably lost when the requirement was relocated.

Part **(A)** also includes, as of the 2017 NEC, item (5) which is essentially a mirror image of 110.34(B) from the standpoint of lower voltage equipment. The exception was not duplicated because the paragraph, as written in this location, ends up with the same requirements as the original section.

Part **(B)** in 110.26 presents a very important requirement regarding the use of work space. The three-dimensional area identified by parts **(A)(1)**, **(A)(2)**, and **(A)(3)** of 110.26 must not only be provided but must be maintained! That is, such space must be viewed and treated as an "exclusion zone." There may be no other things in the work space—not even on a "temporary" basis.

The second sentence of 110.26(B) addresses maintenance situations on equipment located in "passageways or general open space." The concern here is for unqualified personnel, coming into the proximity of and potentially in contact with live electrical parts. Where equipment located in areas accessible to the general population of a building or facility is opened to perform maintenance or repair, the work space area must be cordoned off to keep unqualified persons from approaching the live parts. Failure to do so clearly constitutes a violation of the NEC.

In 110.26(C), "Entrance to Working Space," the Code regulates the necessary entrance/exit to the work space. In part **(A)**, a general statement calls for at least one entrance, of sufficient size, be provided to allow the work space to be entered/exited. Although the wording used here—"of sufficient area"—does not clearly define the dimensions needed to ensure compliance, it seems safe to assume that compliance with the dimensions spelled out in 110.26(C)(2) [i.e., 610 mm (24 in.) wide × 2.0 m (6½ ft) high] would be acceptable to the vast majority of electrical inspectors, if not all electrical inspectors.

As an added safety measure, to prevent personnel from being trapped in the working space around burning or arcing electrical equipment, the rule of 110.26(C)(2) requires *two* "entrances" or directions of access to the working space around any equipment enclosure that contains "overcurrent devices, switching devices, or control devices," where such equipment is rated 1200 A or more, and which is also larger than 1.8 m (6 ft) wide; both conditions must be met before the enhanced access rules in this provision apply. This rule covers all types of equipment. That is, 110.26(C)(2) requires *two* "entrances" or directions of access to the working space around switchboards, motor-control centers, distribution centers, panelboard lineups, UPS cubicles, rectifier modules, substations, power conditioners, and any other equipment that is rated 1200 A or more.

At each end of the working space at such equipment, an entranceway or access route at least 610 mm (24 in.) wide and at least 2.0 m (6½ ft) high must be provided. Because personnel have been trapped in work spaces by fire between them and the only route of exit from the space, rigid enforcement of this rule is likely. Certainly, design engineers should make two paths of exit a standard requirement in their drawings and specs. Although the rule does not require two doors into an electrical equipment room, it may be necessary to use two doors in order to obtain the required two entrances to the required work space—especially where the switchboard or control panel is in tight quarters and does not afford a 24-in.-wide path of exit at each end of the work space.

In Fig. 110-19, sketch "A" shows compliance with the Code rule—providing two areas for entering or leaving the defined dimensions of the work space. In that sketch, placing the switchboard with its front to the larger area of the room and/or other layouts would also satisfy the intent of the rule. It is only necessary to have an assured means of exit from the defined work space. If the space in front of the equipment is deeper than the required depth of work space, then a person could simply move back out of the work space at any point along the length of the equipment. That is confirmed by the wording in 110.26(C)(2)(a). A similar idea is behind the objective of part (b) to 110.26(C)(2), which recognizes that where the space in front of equipment is twice the minimum depth of working space required by Table 110.26(A) for the voltage of the equipment and the conditions described,

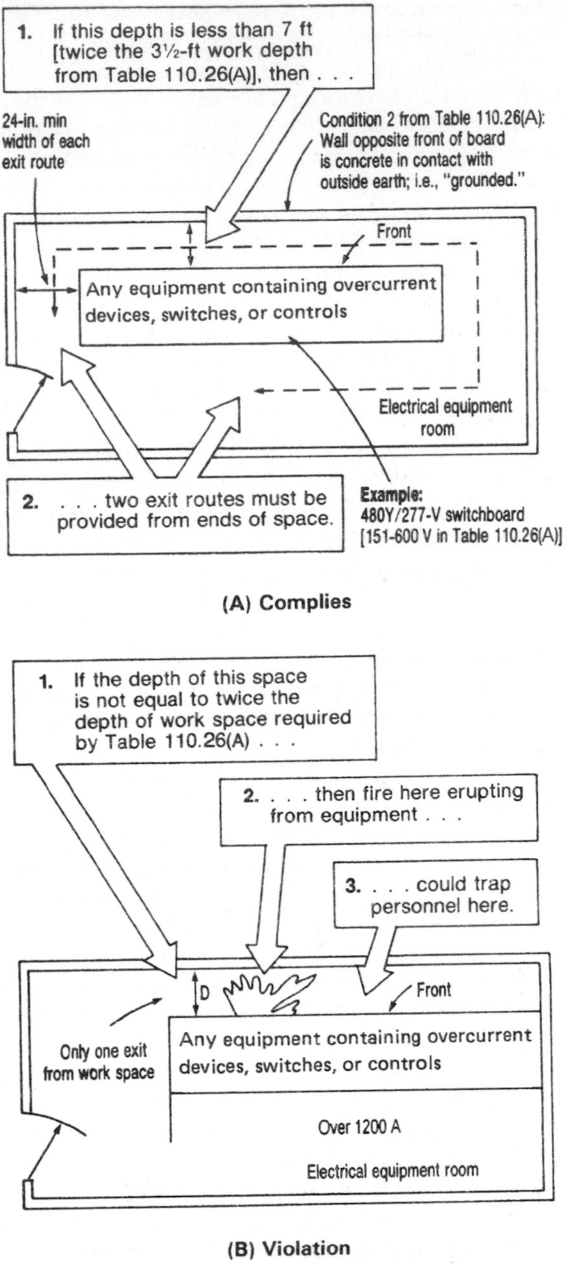
1. If this depth is less than 7 ft [twice the 3½-ft work depth from Table 110.26(A)], then . . .
24-in. min width of each exit route
Condition 2 from Table 110.26(A): Wall opposite front of board is concrete in contact with outside earth; i.e., "grounded."
Front
Any equipment containing overcurrent devices, switches, or controls
Electrical equipment room
2. . . . two exit routes must be provided from ends of space.
Example: 480Y/277-V switchboard [151-600 V in Table 110.26(A)]
(A) Complies
1. If the depth of this space is not equal to twice the depth of work space required by Table 110.26(A) . . .
2. . . . then fire here erupting from equipment . . .
3. . . . could trap personnel here.
D
Front
Only one exit from work space
Any equipment containing overcurrent devices, switches, or controls
Over 1200 A
Electrical equipment room
(B) Violation

Fig. 110-19. There must be two paths out of the work space required in front of any equipment containing fuses, circuit breakers, motor starters, switches, and/or any other control or protective devices, where the equipment is rated 1200 A or more and is more than 1.8 m (6 ft) wide. [Sec. 110.26(C).]

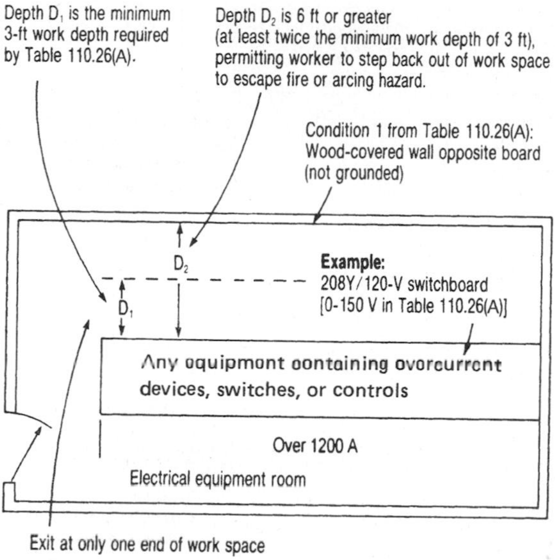
Depth D1 is the minimum 3-ft work depth required by Table 110.26(A).
Depth D2 is 6 ft or greater (at least twice the minimum work depth of 3 ft), permitting worker to step back out of work space to escape fire or arcing hazard.
Condition 1 from Table 110.26(A): Wood-covered wall opposite board (not grounded)
D2
D1
Example: 208Y/120-V switchboard [0-150 V in Table 110.26(A)]
Any equipment containing overcurrent devices, switches, or controls
Over 1200 A
Electrical equipment room
Exit at only one end of work space

Fig. 110-20. This satisfies Sec. 110.26(C)(2)(b).

it is not necessary to have an entrance at each end of the space (Fig. 110-20). In such cases, a worker can move directly back out of the working space to avoid fire. For any case where the depth of space is not twice the depth value given in Table 110.26(A) for working space, an entranceway or access route at least 24 in. wide must be provided at each end of the working space in front of the equipment.

Sketch "B" in Fig. 110-19 shows the layout that must be avoided. With sufficient space available in the room, layout of any equipment rated 1200 A or more and over 1.8 m (6 ft) wide with only one exit route from the required work space would be a clear violation of the rule. As shown in sketch "B," a door at the right end of the working space would eliminate the violation. *But,* if the depth *D* in sketch "B" is equal to or greater than twice the minimum required depth of working space from Table 110.26(A) for the voltage and "conditions" of installation, then a door at the right is not needed and the layout would not be a violation.

The last sentence in 110.26(C)(2)(b) states that when the defined work space in front of an electrical switchboard or other equipment has an entranceway at only one end of the space, the edge of the entrance nearest the equipment must be at least 3, 3½, or 4 ft (900 mm, 1.0 m, or 1.2 m) away from the equipment—as designated in Table 110.26(A) for the voltage and conditions of installation of the particular equipment. This Code requirement requires careful determination in satisfying the precise wording of the rule. Figure 110-21 shows a few of the many possible applications that would be subject to the rule.

The third numbered paragraph in 110.26(C) puts forth an additional requirement for installations where doors are used as a means of access. Here the NEC mandates that any such door be provided with special hardware to facilitate exit where a maintenance person has lost the use of his or her hands, as could be the case in a

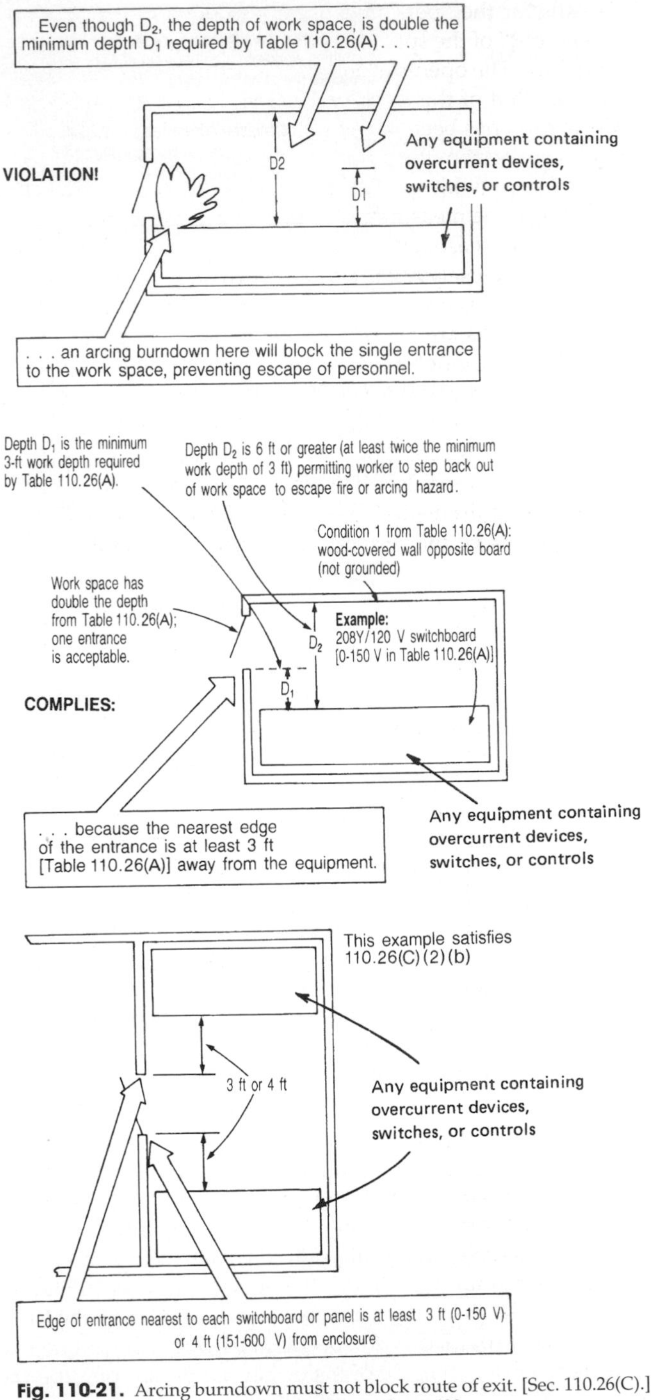

Even though D_2, the depth of work space, is double the minimum depth D_1 required by Table 110.26(A) . . .
Any equipment containing overcurrent devices, switches, or controls
VIOLATION!
D2
D1
. . . an arcing burndown here will block the single entrance to the work space, preventing escape of personnel.
Depth D_1 is the minimum 3-ft work depth required by Table 110.26(A).
Depth D_2 is 6 ft or greater (at least twice the minimum work depth of 3 ft) permitting worker to step back out of work space to escape fire or arcing hazard.
Condition 1 from Table 110.26(A): wood-covered wall opposite board (not grounded)
Work space has double the depth from Table 110.26(A); one entrance is acceptable.
Example: 208Y/120 V switchboard [0-150 V in Table 110.26(A)]
D_2
D_1
COMPLIES:
. . . because the nearest edge of the entrance is at least 3 ft [Table 110.26(A)] away from the equipment.
Any equipment containing overcurrent devices, switches, or controls
This example satisfies 110.26(C)(2)(b)
3 ft or 4 ft
Any equipment containing overcurrent devices, switches, or controls
Edge of entrance nearest to each switchboard or panel is at least 3 ft (0-150 V) or 4 ft (151-600 V) from enclosure

Fig. 110-21. Arcing burndown must not block route of exit. [Sec. 110.26(C).] Note that the spacings increase for voltages above 600.

fire. The Code calls for the "personnel" doors to open in the direction of egress, which of course is "out" of the spaces. Additionally, such doors must be fitted with listed panic hardware. The open-out and the panic-hardware rules now apply to all egress doors within 25 ft of the work space area. The issue is making sure that an injured electrician who has been burned and cannot use his hands to turn a knob can get far enough away from the burndown to seek help, without unduly burdening the rest of the facility with special hardware requirements. The NEC is unclear as to how the 25 ft is to be measured, whether in a straight line on a plan view or by proceeding through a reasonably assumed route of travel, but the route of travel would seem to be more closely related to the motivation for the requirement.

This requirement applies only to large equipment rated 800 A or more, where it is assumed that the risks are greater just as it is assumed that enhanced egress rules are appropriate. Although not required for other than "Large Equipment" work space access, recommending and providing such a means of egress is not a Code violation and could be viewed as an added safety feature.

Note that, as in the past, this parameter is an equal or exceed number, which differs from the enhanced egress rule that uses a measurement that must be exceeded in order to trigger the requirement. However, the 2014 NEC is the first time that the trigger numbers differ, with the larger number applicable to the enhanced egress requirements only.

110.26(D) requires lighting of work space at "service equipment, switchboards, panelboards, or motor control centers installed indoors." The basic rule is shown in Fig. 110-22.

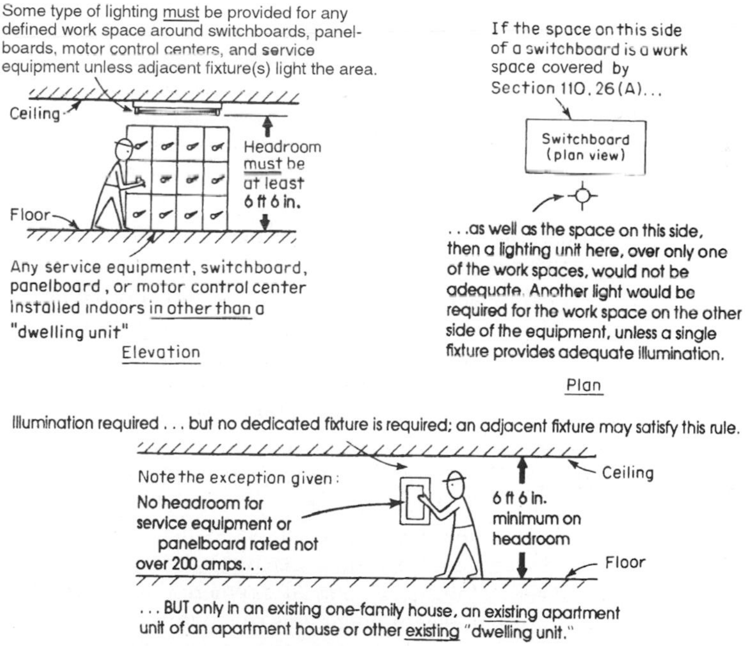
Some type of lighting must be provided for any defined work space around switchboards, panelboards, motor control centers, and service equipment unless adjacent fixture(s) light the area.
Ceiling
Headroom must be at least 6 ft 6 in.
Floor
Any service equipment, switchboard, panelboard, or motor control center installed indoors in other than a "dwelling unit"
Elevation
If the space on this side of a switchboard is a work space covered by Section 110.26(A)...
Switchboard (plan view)
...as well as the space on this side, then a lighting unit here, over only one of the work spaces, would not be adequate. Another light would be required for the work space on the other side of the equipment, unless a single fixture provides adequate illumination.
Plan
Illumination required ... but no dedicated fixture is required; an adjacent fixture may satisfy this rule.
Note the exception given:
No headroom for service equipment or panelboard rated not over 200 amps...
6 ft 6 in. minimum on headroom
Ceiling
Floor
... BUT only in an existing one-family house, an existing apartment unit of an apartment house or other existing "dwelling unit."

Fig. 110-22. Electrical equipment requires lighting and 6½-ft headroom at *all* work spaces around certain equipment. [Secs. 110.26(D) and 110.26(A)(3).]

The second to last sentence in 110.26(D) points out the Code-making panel's intent. That is, if an adjacent fixture provides adequate illumination, another fixture is not required. In dwelling units where the identified equipment is located in a habitable room, a switched receptacle outlet, as permitted by 210.70(A)(1) Exception No. 1, would also satisfy the requirement for illumination given here. And, lastly, control of the required lighting outlet at electrical equipment by automatic means, only, is prohibited. But, the use of automatic control along with a manual override would meet the spirit and the letter of this rule.

It should be noted that although lighting is required for safety of personnel in work spaces, nothing specific is said about the kind of lighting (incandescent, fluorescent, mercury-vapor), no minimum footcandle level is set, and such details as the position and mounting of lighting equipment are omitted. All that is left to the designer and/or installer, with the inspector the final judge of acceptability.

110.26(E). Dedicated Equipment Space. Pipes, ducts, etc., must be kept out of the way of circuits from panelboards and switchboards. This rule is aimed at ensuring clean, unobstructed space for proper installation of switchboards and panelboards, along with the connecting wiring methods used with such equipment.

The wording of this rule has created much confusion among electrical people as to its intent and correct application in everyday electrical work. And, it seems that one can develop a complete understanding of this rule only by repeated readings. On first reading, there are certain observations about the rule that can be made clearly and without question:

1. Although the rule is aimed at eliminating the undesirable effects of water or other liquids running down onto electrical equipment and entering and contacting live parts—which should always be avoided both indoors and outdoors—the wording of the first sentence limits the requirement to switchboards, switchgear, panelboards, and to motor control centers. Individual switches and CBs and all other equipment are not subject to the rule—although the same concern for protection against liquid penetrations ought to be applied to all such other equipment. The reason for this is rooted in the history of the requirement and not technical merit. The rule originated in what is now Art. 408 under the control of a different Code-making panel, and in that location the other types of equipment were not within the scope of that article. The material was relocated by order of the Correlating Committee, but the wording here still is focused on the old applications.
2. The designated electrical equipment covered by the rule (switchboards, panelboards, etc.) does not have to be installed in rooms dedicated exclusively to such equipment, although it may be. This rule applies only to the area above the equipment, for the width and depth of the equipment.

Part **(E)(1)(a)** (for indoor installations) of this rule very clearly defines the "zones" for electrical equipment to include any open space above the equipment up to 1.8 m (6 ft) above the top of the gear, or the structural ceiling, whichever is lower. In any case, the dedicated clear space above switchboards, panelboards, distribution panels, and motor control centers extends to the structural ceiling if it is less than 1.8 m (6 ft) above the equipment. And, where the structural ceiling is higher than 1.8 m (6 ft) above the equipment, this rule permits water piping, sanitary drain lines, and similar piping for liquids to be located above switchboards, etc., if such piping is at least 1.8 m (6 ft) above the equipment. The permission for switchboards and panelboards to be installed below liquid piping that is located more than 6 ft above the equipment must be carefully considered, even though containment systems, etc., are required to prevent damage from dripping or leaking liquids. The object is to keep foreign piping (chilled-water pipes, steam pipes, cold-water pipes, and other piping) from passing directly over electrical equipment and thereby eliminate even the possibility of water leaking from the piping and overflowing the drain onto the equipment (Fig. 110-23). This rule completely prohibits any intrusion on the dedicated area, up to 6 ft above the equipment. And, where piping, etc., is to be run over the 6-ft minimum dedicated area, it must be provided with some means to prevent any discharge or condensation from coming into contact with the equipment below.

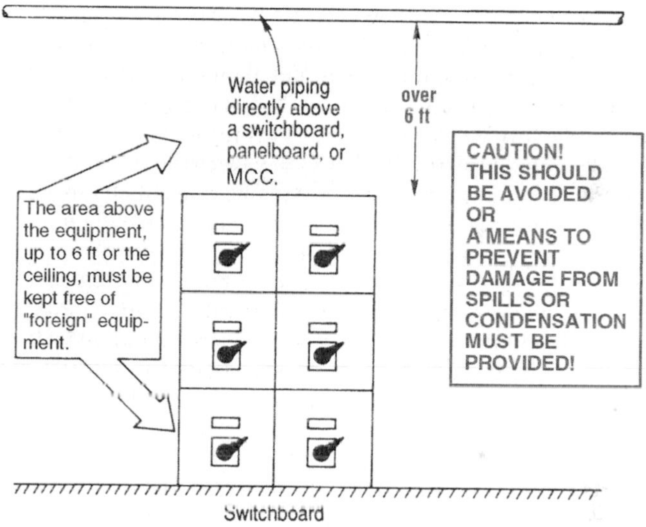
Water piping directly above a switchboard, panelboard, or MCC.
over 6 ft
CAUTION! THIS SHOULD BE AVOIDED OR A MEANS TO PREVENT DAMAGE FROM SPILLS OR CONDENSATION MUST BE PROVIDED!
The area above the equipment, up to 6 ft or the ceiling, must be kept free of "foreign" equipment.
Switchboard

Fig. 110-23. Water pipes and other "foreign" piping must not be located less than 6 ft above switchboard. [Sec. 110.26(E).]

The exception recognizes that suspended ceiling systems with removable tiles may occupy the dedicated space above switchboards, etc.

110.26(E)(1)(b) identifies the area where "foreign systems" are permitted to be installed. As one would imagine, this zone begins at a distance of 6 ft above the top of the electrical equipment and extends to the structural ceiling. As indicated, protection against damage due to leaking of the foreign piping systems must be provided.

Note carefully: It is not a requirement of this rule that "foreign" piping, ducts, etc., must always be excluded from the entire area above electrical equipment. Although the rules require that the "foreign" piping, ducts, etc., must be kept out of the "space" dedicated to the electrical equipment, the rule, literally, permits such "foreign" piping, ducts, etc., to be installed *above* the dedicated space, above the equipment. BUT, protection must be provided in the form of drain gutters or containment systems of some sort to prevent damage to the electrical equipment, below. However, it is much wiser to eliminate any foreign piping—even sprinkler piping used for fire suppression—from the area above electrical equipment. Where such installation is not possible, then take great care to ensure an adequate system of protection for the electrical equipment. (See Fig. 110-23.)

As covered in part **(E)(1)(c),** sprinkler piping, which is intended to provide fire suppression in the event of electrical ignition or arcing fault, would not be foreign to the electrical equipment and would not be objectionable to the Code rule. Another confirmation of Code acceptance of sprinkler protection for electrical equipment (which means sprinkler piping within electrical equipment and even directly over electrical equipment) is very specifically verified by 450.47, which states, "Any pipe or duct system foreign to the electrical installation shall not enter or pass through a transformer vault. Piping or other facilities provided for vault fire protection or for transformer cooling shall not be considered foreign to the electrical installation." BUT, the wording here only permits installation of sprinkler piping above the dedicated space "where the piping complies with this section." That means the sprinkler piping would have to be at least 6 ft above the equipment to comply with 110.26(E)(1)(a) *and* be provided with the "protection"—a gutter or containment system—required by 110.26(E)(1)(b). As long as the containment system only falls below the sprinkler pipe and not underneath the sprinkler head, the sprinkler will still be able to perform its function. In other words, route the sprinkler piping in such a manner that it is offset and not directly over the equipment, or at the least, make sure that the suppression system is arranged so the sprinkler head is not directly over the electrical equipment, allowing it to discharge on a fire in the gear from either the front or sides or both. Layouts of piping can be made to assure effective fire suppression by water from the sprinkler heads when needed, without exposing equipment to shorts and ground faults that can be caused by accidental water leaks from the piping. That will prevent any conflict with the rule given here and provide for the desired fire suppression.

Part **(2)** of this rule covers outdoor installations. The first portion covers the required enclosure types, reiterates the normal work space requirement, and requires protection against accidental vehicular contact. This portion continues unchanged from the previous five Code cycles. The second portion is new as of the 2014 **NEC**. It takes the dedicated electrical space concept for indoor installations and makes it applicable, essentially unchanged, to outdoor work. The idea is to reserve a 6-ft zone above the equipment (and down to grade) that is not to be invaded by foreign systems. As of the 2017 **NEC**, a misplaced exception [to (b) instead of (c)] permits structural overhangs and roof extensions into this area.

The final lettered paragraph, 110.26(F), used to be the last sentence of 110.26. It makes clear that the use of a locked door or enclosure is acceptable, where the key or combination is available to "qualified personnel" (e.g., the house electrician or serving contractor's journeyman). Under such conditions, a lock does not inhibit the ready accessibility contemplated in the definition of that term.

110.27. Guarding of Live Parts. Part **(A)** of this rule generally requires that "live parts" (i.e., energized parts of equipment) be "guarded" to prevent accidental contact. It applies to all systems operating at 50 V or more. This is typically accomplished through the use of manufacturer-provided enclosures. However, where live parts are not enclosed with a suitable enclosure, the alternate methods described in parts **(A)(1)** through **(A)(4)** can be employed to satisfy this requirement.

Part **(B)** of 110.27 addresses an additional concern for protection of electrical equipment. After the 1968 **NEC**, old Sec. 110.17(a)(3), accepting guardrails as suitable for guarding live parts, was deleted. It was felt that a guardrail is not proper or adequate protection in areas accessible to other than qualified persons. However, where electrical equipment is exposed to physical damage—such as where installed alongside a driveway, or a loading dock, or other locations subjected to vehicular traffic—the use of guardrails is clearly acceptable and required by this rule. Failure to protect equipment against contact by vehicles is a violation of this section.

Live parts of equipment should in general be protected from accidental contact by complete enclosure (i.e., the equipment should be "dead-front"). Such construction is not practicable in some large control panels, and in such cases the apparatus should be isolated or guarded as required by these rules.

The **NEC** also recognizes guarding through elevation at a height of at least 8 ft above the floor. The 2014 **NEC** has fine-tuned this requirement in order to correlate with NESC requirements, and increased this height requirement for voltages over 300 and up to 600 by 6 in. (to 8½ ft). The 2017 **NEC** went a slight step further, adding a third height limit of 8 ft 7 in. (1 in. higher than the next lower height) above the floor for 600-1000 volts. In addition, all the voltages covered in (4) are now clarified as voltages between ungrounded conductors.

110.28. Enclosure Types. This section and table cover selection criteria on types of enclosures. The second paragraph of this section makes the selection criteria set forth in the table mandatory. The material has been relocated from 430.91 because it does not just cover motor control centers but has general applicability for all installations. This section gives selection data, with characteristics tabulated, for application of the various NEMA types of motor controller enclosures for use in specific nonhazardous locations operating at 600 V and below. Note that this table is relocated into Part II (1000 Volts Nominal, or Less) from Part I (General) because it does not apply to medium-voltage equipment.

III. Over 1000 V, Nominal

110.30. General. Figure 110-24 notes that high-voltage switches and circuit breakers must be marked to indicate the circuit or equipment controlled. This requirement arises because 110.30 says that high-voltage equipment must comply with preceding sections in Part I of Art. 110. Therefore, the rule of 110.22 calling for marking of all disconnecting means must be observed for high-voltage equipment as well as for equipment rated up to 1000 V.

The second sentence states clearly, and emphatically, that the rules given in Part II of Art. 110 apply only to equipment on the load side of the service. That is, only high-voltage equipment installed on the load side of the "service point" is covered. In no case shall these rules be applied to high-voltage equipment that is owned and operated by the utility.

110.31. Enclosure for Electrical Installations. The last sentence of the first paragraph indicates that there may be instances where additional precautions or special design would be necessary, due to the specifics related to the application.

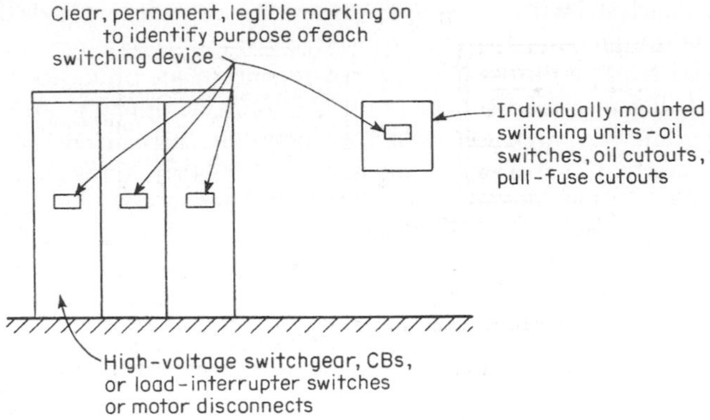
Clear, permanent, legible marking on to identify purpose of each switching device
Individually mounted switching units - oil switches, oil cutouts, pull - fuse cutouts
High-voltage switchgear, CBs, or load - interrupter switches or motor disconnects

Fig. 110-24. High-voltage switches and breakers must be properly marked to indicate their function. (Sec. 110.30.)

Always check the manufacturer's installation instructions and appropriate UL data to ensure that the enclosure meets the specific hazard encountered.

Table 110.31 provides minimum clearance requirements between the required fencing to live parts.

Section 110.31(A) repeats key construction requirements from the basic rules in Part III of Art. 450 for transformer vaults and makes them applicable to comparable rooms without transformers, but with other medium-voltage equipment. It is presently unclear, however, when such a vault requirement would be triggered by a medium-voltage installation. The opening text simply applies the requirement when the vault is "required or specified." For the 2011 NEC, additional text was imported from Art. 450, including the allowance for a fire-rating reduction to 1 h if an automatic fire suppression system is installed.

Section 110.31(B) covers enclosed areas or rooms in interior locations. Figure 110-25 illustrates the rules that cover installation of high-voltage equipment indoors in places accessible to unqualified persons in part **(B)(1)**. Installation must be in a locked vault or locked area, or equipment must be metal-enclosed and locked. The Code is quite clear that a lock and key is the only acceptable means to provide positive control [110.34(C)]. The basic concern is related to unqualified persons coming into proximity or contact with high voltage. This rule states what is considered as adequate to provide the desired exclusion of other than "qualified personnel."

For equipment that is not enclosed, as described in 110.31(C), another enclosure—or, more accurately, a barrier—must be constructed around the entire area where unenclosed high-voltage equipment is "accessible" to other than qualified persons. Such fencing must be no lower than 7 ft in total height. This may be 7 ft of fencing, or a 6-ft fence supplemented by at least three strands of barbed wire, or the "equivalent."

Part **(B)(1)** also calls for "appropriate caution signs" for all enclosures, boxes, or "similar associated equipment." This is a field marking that must be provided by the installer. For any equipment, rooms, or enclosures where the voltage

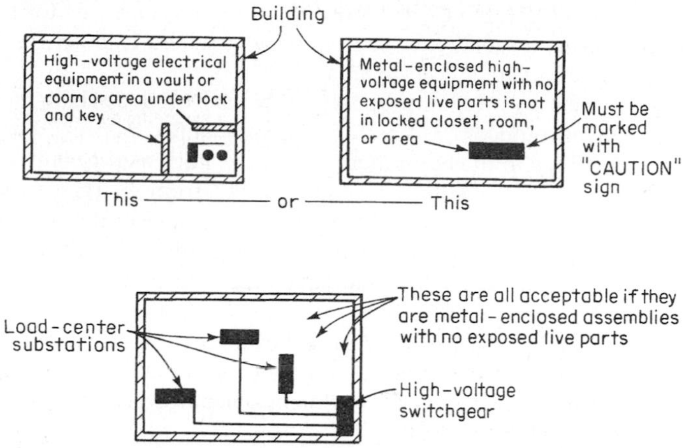
Building
High-voltage electrical equipment in a vault or room or area under lock and key
Metal-enclosed high-voltage equipment with no exposed live parts is not in locked closet, room, or area
Must be marked with "CAUTION" sign
This
or
This
Load-center substations
These are all acceptable if they are metal-enclosed assemblies with no exposed live parts
High-voltage switchgear

Fig. 110-25. NEC rules on high-voltage equipment installations in buildings accessible to electrically unqualified persons. [Sec. 110.31(A).]

exceeds 600 V, permanent and conspicuous warning signs reading "DANGER—HIGH VOLTAGE, KEEP OUT" must always be provided. It is a safety measure that alerts unfamiliar or unqualified persons who may, for some reason, attempt to gain access to a locked, high-voltage area. Note that Sec. 110.31(B)(1) does not require locking indoor metal-enclosed equipment that is accessible to unqualified persons, but such equipment is required to be marked with "WARNING" signs [see 110.34(C)].

And, the last sentence of part **(A)(1)** requires that manufacturers design their equipment to ensure that unqualified persons can't come into contact with live parts of the high-voltage equipment.

Part **(B)(2)** applies to areas accessible only to qualified persons. In such areas, no guarding or enclosing of the live high-voltage parts is called for. The rule simply requires compliance with the rules given in 110.34, 110.36, and 490.24. In 110.34, the Code covers clear "working space" and the methods of "guarding" for systems rated over 600 V. Section 110.36 describes the acceptable wiring methods and 490.24 covers internal spacings in medium-voltage equipment that are field wired, and fabricated. Note that the spacings in Table 490.24 do not apply to internal spacings on equipment "designed, manufactured, and tested in accordance with accepted national standards."

In 110.31(C)(1), the Code requires compliance with the rules for equipment rated over 1000 V, given in all parts of Art. 225. And, **(C)(2)** requires compliance with 110.34, 110.36, and 490.24 where outdoor high-voltage equipment is accessible only to qualified persons.

110.31(D) essentially specifies that outdoor installations with exposed live parts must not provide access to unqualified persons.

For equipment rated over 600 V, nominal, 110.31 requires that access be limited to qualified persons only, by installing such equipment within a "vault, room, closet, or in an area that is surrounded" by a fence, etc., with locks on the doors.

In part **(D)** of Sec. 110.31, the Code identifies additional methods for preventing unauthorized access to metal-enclosed equipment where it is *not* installed in a locked room or in a locked, fenced-in area.

110.31(D) provides a variety of precautions that are needed where high-voltage equipment, installed outdoors, is accessible to unqualified persons. For example, the design of openings in the equipment enclosure, such as for ventilation, must be such that they deflect inserted "foreign objects" from energized parts. Also, "guards" must be provided where the equipment is subject to damage by cars, trucks, and so forth. Enclosed equipment must be equipped with nuts and bolts that are not "readily removed." In addition, elevation may be used to prevent access [110.34(E)], or equipment may be enclosed by a wall, screen, or fence under lock and key, as shown in Fig. 110-26. Where the bottom of high-voltage equipment is *not* mounted at least 8 ft above the floor or grade, the equipment enclosures must be kept locked. And covers on junction boxes, pull boxes, and so forth, must be secured using a lock, bolt, or nut.

DANGER
HIGH VOLTAGE
KEEP OUT

Fig. 110-26. High-voltage equipment enclosed by a wall, screen, or fence at least 7 ft (2.13 m) high with a lockable door or gate is considered as accessible only to qualified persons. [Sec. 110.31(B).]

That sentence in 110.31(D) recognizes a difference in safety concern between high-voltage equipment accessible to "unqualified persons"—who may not be qualified as electrical personnel but are adults who have the ability to recognize warning signs and the good sense to stay out of electrical enclosures—and "the general public," which includes children who cannot read and/or are not wary enough to stay out of unlocked enclosures (Fig. 110-27).

The rationale submitted with the proposal that led to this change in the Code rule noted:

> Where metal-enclosed equipment rated above 600 V is accessible to the general public and located at an elevation less than 8 ft, the doors should be kept locked to prevent children and others who may be unaware of the contents of such enclosures from opening the doors.

Fig. 110-27. *Metal-enclosed* high-voltage equipment accessible to the general public—such as pad-mount transformers or switchgear units installed outdoors or in indoor areas where the general public is not excluded—must have doors or hinged covers locked (arrow) if the bottom of the enclosure is less than 8 ft (2.5 m) above the ground or above the floor.

However, in a controlled environment where the equipment is marked with appropriate caution signs as required elsewhere in the NEC, and only knowledgeable people have access to the equipment, the requirement to lock the doors on all metal-enclosed equipment rated above 600 V and located less than 8 ft (2.5 m) above the floor does not contribute to safety and may place a burden on the safe operation of systems by delaying access to the equipment.

For equipment rated over 600 V, 110.31(D) has required that the equipment cover or door be locked unless the enclosure is mounted with its bottom at least 8 ft off the ground. In that way, access to the general public is restricted and controlled. In addition, the bolts or screws used to secure a cover or door may serve to satisfy the rule of Sec. 110.31(D), provided the enclosure is used only as a pull, splice, or junction box. Where accessible to the general public, an enclosure used for any other purpose must have its cover locked unless it is mounted with its bottom at least 8 ft (2.5 m) above the floor (Fig. 110-28).

In the last two sentences of 110.31(D), "bolted or screwed-on" covers, as well as in-ground box covers over 100 lb (45.4 kg) are recognized as preventing access to the general public. The last sentence recognizes that there is no need to secure the cover on an in-ground box that weighs at least 100 lb. This correlates with the rule of 314.72(E), which states that covers weighing 100 lb (45.4 kg) satisfy the basic requirement for securing covers given in this section.

Fig. 110-28. Access by the general public to any metal enclosure containing circuits or equipment rated over 600 V, nominal, must be prevented. The wording recognizes the bolts or screws on the covers of boxes used as pull, splice, and junction boxes as satisfying the requirement for preventing access. And, additional wording in this rule recognizes that covers weighing over 100 lb are inherently secured and do not require bolts or screws for the cover or door. *Remember*, the permission given in the basic rule is for pull, splice, and junction boxes, *only*.

110.32. Work Space about Equipment. Figures 110-29 and 110-30 point out the basic Code rule of 110.32 relating to working space around electrical equipment.

Figure 110-31 shows required side-to-side working space for adequate elbow room in front of high-voltage equipment.

110.33. Entrance to Enclosures and Access to Working Space. Entrances and access to working space around high-voltage equipment must comply with the rules shown in Fig. 110-32. Section 110.33(A)(1) says that if the depth of space in

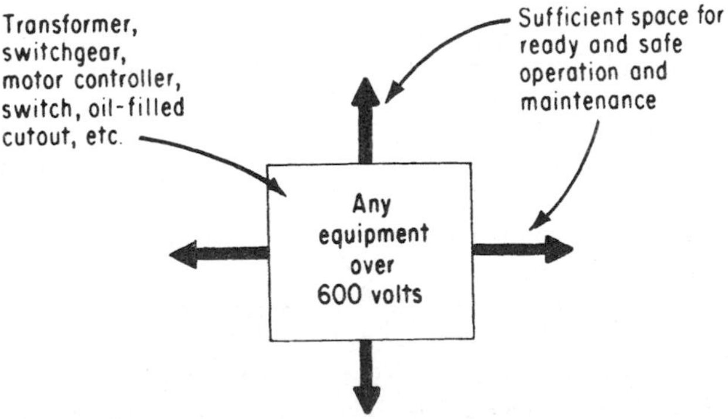
Transformer, switchgear, motor controller, switch, oil-filled cutout, etc.
Sufficient space for ready and safe operation and maintenance
Any equipment over 600 volts

Fig. 110-29. This is the general rule for work space around any high-voltage equipment. (Sec. 110.32.)

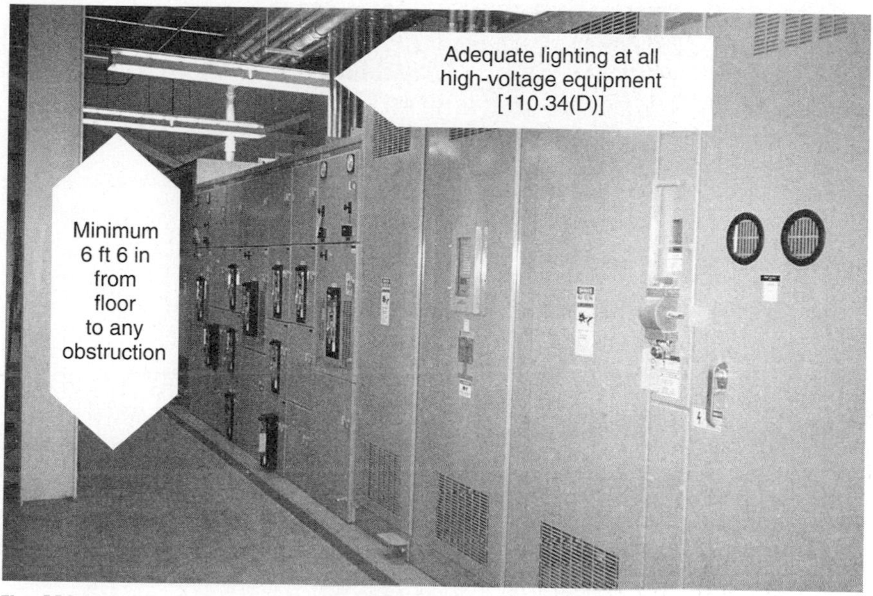
Adequate lighting at all
high-voltage equipment
[110.34(D)]
Minimum
6 ft 6 in
from
floor
to any
obstruction

Fig. 110-30. Sufficient headroom and adequate lighting are essential to safe operation, maintenance, and repair of high voltage equipment. (Sec. 110.32.)

front of a switchboard is at least twice the minimum required depth of working space from Table 110.34(A), any person in the working space would be capable of moving back out of the working space to escape any fire, arcing, or other hazardous condition. In such cases there is no need for a path of exit at either end or at both ends of the working space. But where the depth of space is not equal to twice the minimum required depth of working space, there must be an exit path at each end of the working space in front of switchgear or control equipment enclosures that are wider than 6 ft (1.8 m). And what applies to the front of a switchboard also applies to working space at the rear of the board if rear access is required to work on energized parts.

The wording of 110.33(A)(1)(b) specifies minimum clearance distance between high-voltage equipment and edge of entranceway to the defined work space in front of the equipment, where only one access route is provided. Based on Table 110.34(A)—which gives minimum depths of clear working space in front of equipment operating at over 600 V—the rule in this section presents the same type of requirement described by 110.26(C)(2). Based on the particular voltage and the conditions of installation of the high-voltage switchgear, control panel, or other equipment enclosure, the nearest edge of an entranceway must be a prescribed distance from the equipment enclosure. Refer to the sketches given for 110.26(C).

Section 110.33(A) concludes with paragraph (3) on personnel doors, how they must open and what hardware is required for them. This rule is identical to the comparable rule for large equipment in 110.26(C)(3), and raises the same issues. Refer to that discussion for more information. Section 110.33(B) requires permanent provisions for

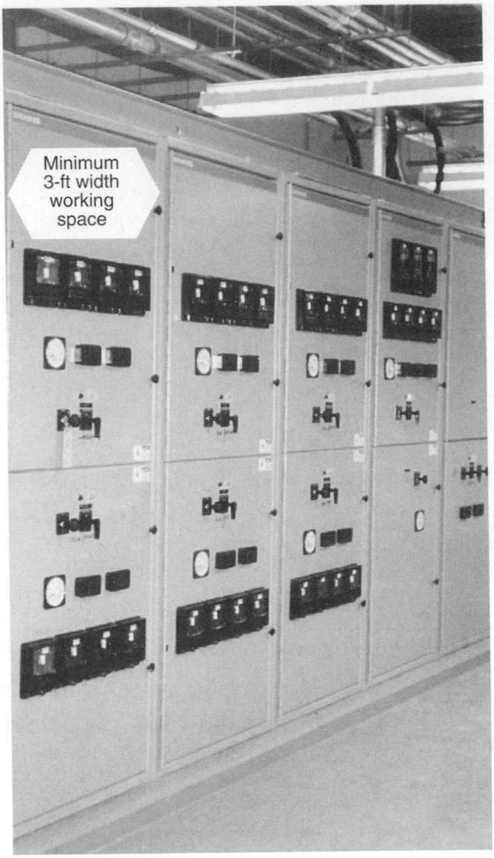
Minimum
3-ft width
working
space

Fig. 110-31. Working space in front of equipment must be at least 3 ft (900 mm) wide, measured parallel to front surface of the enclosure. (Sec. 110.32.)

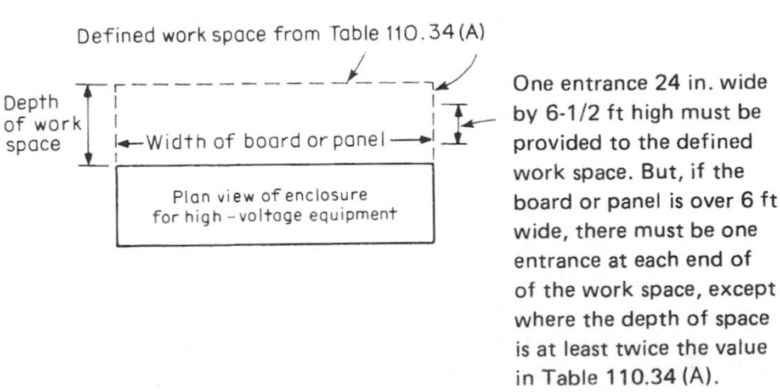
Defined work space from Table 110.34(A)
Depth of work space
Width of board or panel
Plan view of enclosure for high-voltage equipment
One entrance 24 in. wide by 6-1/2 ft high must be provided to the defined work space. But, if the board or panel is over 6 ft wide, there must be one entrance at each end of of the work space, except where the depth of space is at least twice the value in Table 110.34 (A).

Fig. 110-32. Access to required work space around high-voltage equipment must be ensured. (Sec. 110.33.)

access in the form of ladders or stairways to the required work space about medium-voltage equipment on balconies, rooftops, attics, platforms, etc.

110.34. Work Space and Guarding. Application of Code Table 110.34(A) to working space around high-voltage equipment is made in the same way, as shown for Code Table 110.26(A)—except that the depths are greater to provide more room because of the higher voltages.

As shown in Fig. 110-33, a 30-in. (762-mm)-deep work space is required behind enclosed high-voltage equipment that requires rear access to "deenergized" parts. Section 110.34(A)(1) notes that working space is not required behind dead-front equipment when there are no fuses, switches, other parts, or connections requiring rear access. But the rule adds that if rear access is necessary to permit work on "nonelectrical" parts of the enclosed assembly, the work space must be at least 30 in. (762 mm) deep. This is intended to prohibit cases where switchgear requiring rear access is installed too close to a wall behind it, and personnel have to work in cramped quarters to reach taps, splices, and terminations. However, it must be noted that this applies only where "nonelectrical" parts are accessible from the back of the equipment. If energized parts are accessible, then Condition 2 of 110.34(A) would exist, and the depth of working space would have to be anywhere from 4 to 10 ft (1.2 to 3.0 m) depending upon the voltage [see Table 110.34(A)].

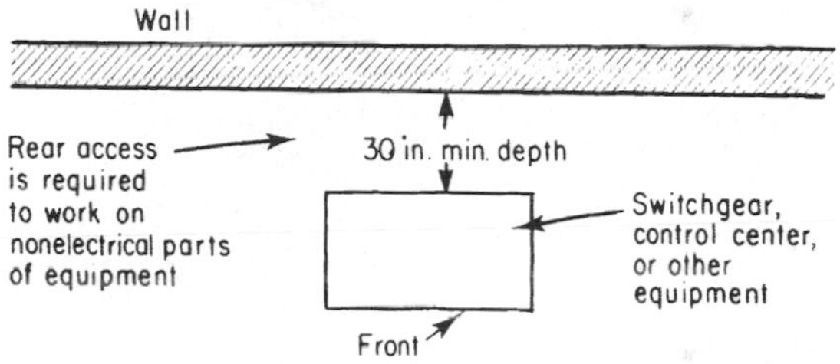
Wall
Rear access is required to work on nonelectrical parts of equipment
30 in. min. depth
Switchgear, control center, or other equipment
Front

Fig. 110-33. Space for safe work on nonelectrical parts. [Sec. 110.34(A).] An example would be an intake air filter.

Section 110.34(B) covers the common occurrence of medium-voltage equipment or transformer rooms with exposed live parts, and what that means for 600 V and lower equipment that may be in the same room. In such cases the medium-voltage equipment must be separated by a screen, fence, or other partition. However, this separation rule does not apply to lower voltage equipment that is only serving the room where the medium-voltage equipment is located. For example, a snap switch and a luminaire in the room would not provoke the separation rule.

Section 110.34(C) requires that the entrances to all buildings, rooms, or enclosures containing live parts or exposed conductors operating in excess of 600 V be kept locked, except where such entrances are under the observation of a qualified attendant at all times. The last paragraph in this section requires use of warning signs to deter unauthorized personnel. The rule of 110.34(D) on lighting of high-voltage work space is shown in Fig. 110-30. Note that the rule calls for "adequate illumination," but does not specify a footcandle level or any other characteristics.

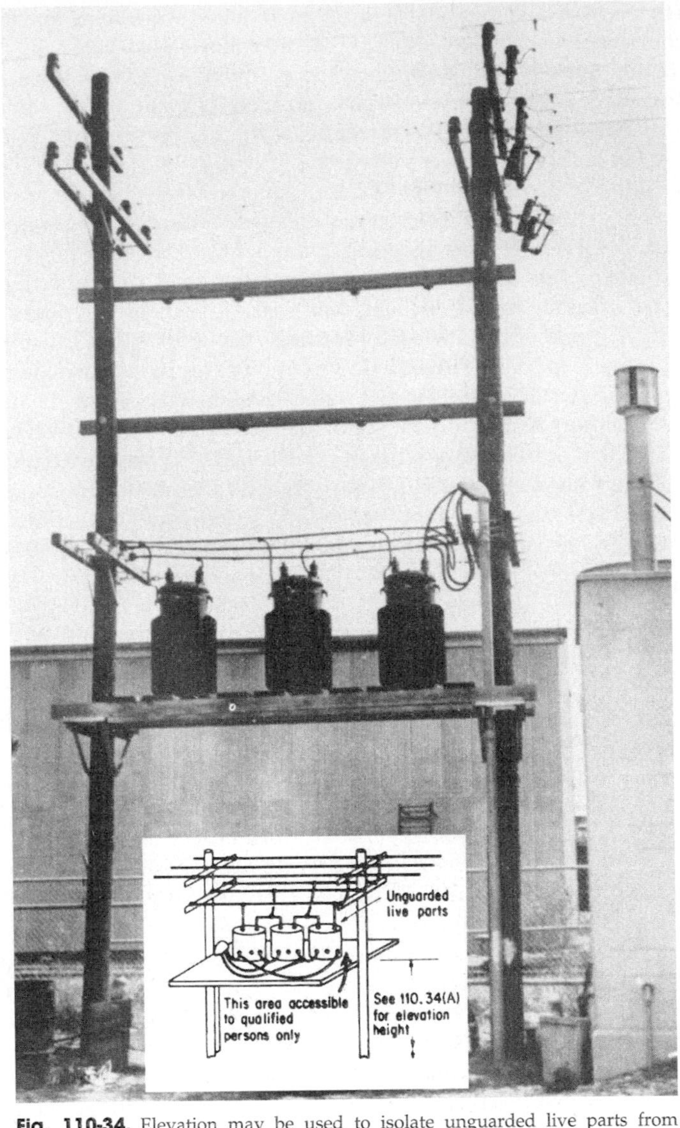
Unguarded
live parts
This area accessible
to qualified
persons only
See 110.34(A)
for elevation
height

Fig. 110-34. Elevation may be used to isolate unguarded live parts from unqualified persons. [Sec. 110.34(E).]

Figure 110-34 shows how "elevation" may be used to protect high-voltage live parts from unauthorized persons.

110.34(F). Protection of Service Equipment, Metal-Enclosed Power Switchgear, and Industrial Control Assemblies. The basic rule of the first sentence in this section excludes "pipes or ducts foreign to the electrical installation" from the "vicinity of the service equipment, metal-enclosed power switchgear, or industrial control assemblies." Then, addressing the case where foreign piping is unavoidably close to the designated electrical equipment, the next sentence calls for "protection" (such as a hood or shield above such equipment) to prevent damage to the equipment by "leaks or breaks in such foreign systems."

Piping for supplying a fire protection medium for the electrical equipment is not considered to be "foreign" and may be installed at the high-voltage gear. The reason given for that sentence was to prevent the first sentence from being "interpreted to mean that no sprinklers should be installed." Fire suppression at such locations may use water sprinklers or protection systems of dry chemicals and/or gases specifically designed to extinguish fires in the equipment without jeopardizing the equipment. Water is sometimes found to be objectionable; leaks in piping or malfunction of a sprinkler head could reduce the switchgear integrity by exposing it to a flashover and thereby initiate a fire.

110.40. Temperature Limitations at Terminations. Terminations for equipment supplied by conductors rated over 2000 V must carry not more than the 90°C ampacity values given in Tables 310.60(C)(67) through 310.60(C)(86), unless the conductors and equipment to which the conductors are connected are "identified" for the 105°C ampacity.

The proposal to include this section pointed out that the ampacity values given in tables for conductors rated above 2000 V were all 90°C-rated values. And, with the increased attention that has been focused on the coordination between conductor ampacity and temperature limitations of the equipment, some question had been raised regarding the use of the 90°C ampacity values in Tables 310.60(C)(67) through 310.60(C)(86) with equipment intended to be supplied by conductors rated over 2000 V. The rule of 110.40 allows the conductors covered in Tables 310.60(C)(67) through 310.60(C)(86) to carry the full 90°C ampacity and be connected virtually without concern for the equipment terminations, unless otherwise marked to indicate that such application is *not* permitted.

This rule was accepted based, in part, on information provided in the proposal regarding American National Standards Institute (ANSI) acceptance of the use of such conductors at their full 90°C ampacity, where tested for such operation. The Code-making panel (CMP) added the qualifying statement "unless otherwise identified" to indicate that such application is permitted where equipment has been so tested. In fact, the wording accepted actually assumes that equipment intended to be supplied by conductors rated over 2000 V—i.e., the conductors covered by Tables 310.60(C)(67) through 310.60(C)(86)—is tested at the full 90°C ampacity. But, if the equipment is otherwise "identified," it must be used as indicated by the manufacturer. It should be noted that although the rule is contained in Part III of Art. 110, which covers equipment rated over 600 V, because the tables mentioned cover conductors rated over 2000 V, it only applies to the terminations on equipment intended to be supplied by conductors rated over 2000 V. The terminations on all other equipment supplied by conductors rated from 601 to 2000 V must be coordinated with the ampacity value corresponding to the temperature rating of the terminations (Fig. 110-35).

110.41. Inspections and Tests. Acceptance testing is of equal, perhaps even greater importance for indoor applications as for outdoor installations. This section, new as of the 2017 NEC, creates a basis for the inspector to request and then

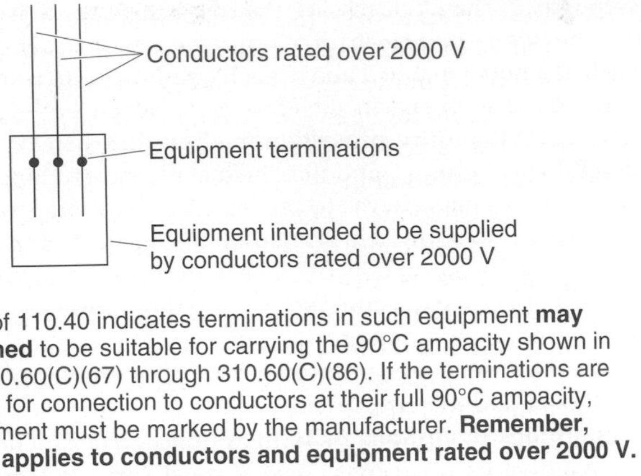
Conductors rated over 2000 V
Equipment terminations
Equipment intended to be supplied by conductors rated over 2000 V
Wording of 110.40 indicates terminations in such equipment **may be assumed** to be suitable for carrying the 90°C ampacity shown in Tables 310.60(C)(67) through 310.60(C)(86). If the terminations are not suited for connection to conductors at their full 90°C ampacity, the equipment must be marked by the manufacturer. **Remember, this only applies to conductors and equipment rated over 2000 V.**

Fig. 110-35. When derating with conductors rated over 2000 V, the temperature rating of the terminations may be assumed to be 90°C unless otherwise marked.

review such testing. The inspiration for this section came from Part III (medium voltage) of Article 225, primarily 225.56. Its location here makes it a likely point of reference for future NEC requirements.

Part IV of Art. 110 covers high-voltage—600 V or more—installations in tunnels. Given that mines and surface mining equipment are not regulated by the NEC, the types of tunnels covered here are those used for trains, cars, irrigation, or whatever—but NOT for mines. Installation of all high-voltage power and distribution equipment, as well as the tunneling equipment identified here, must be protected and installed in accordance with 110.51 through 110.59.

V. Manholes and Other Electric Enclosures Intended for Personnel Entry, All Voltages

110.70. General. The rules in this part apply unless an industrial occupancy demonstrates appropriate engineering supervision that supports design differences; those differences being subject to documentation and review by the inspector. The NEC requires that manholes must be designed under engineering supervision and that they must withstand the loading likely to be imposed.

110.72. Cabling Work Space. A clear work space must be provided not less than 3 ft (900 mm) wide where cables run on both sides, and 2½ ft (750 mm) wide if the cables are on only one side, with vertical headroom not less than 6 ft (1.8 m) unless the opening is within 1 ft (300 mm) of the adjacent interior side wall of the manhole. If the only wiring in the manhole is power-limited fire alarm or signaling circuits, or optical-fiber cabling, then one of the work space dimensions can drop to 2 ft (600 mm) if the other horizontal clear work space is increased so the sum of both dimensions is not less than 6 ft (1.8 m).

110.73. Equipment Work Space. If the manhole includes equipment with live parts that will require work while energized, then the normal NEC work space rules apply. The headroom rises to 6½ ft (2.0 m), and there must be clear work space at least 3 ft deep and 30 in. (762 mm) wide (wider if the equipment is wider). The depth rises to 3½ ft or 1.1 m over 150 V to ground, and then goes up to 4 ft (1.2 m) for up to 2.5 kV, 5 ft (1.5 m) for up to 9 kV, 6 ft for up to 25 kV, and deeper for higher voltages, as covered in Table 110.34(A).

110.74. Conductor Installation. Essentially, manholes are pull boxes that are large enough for personnel to enter, and therefore they need to be sized to accommodate the installed conductors without violating the rules that normally apply to sizing pull boxes, as covered for comparable applications in Art. 314. For medium-voltage applications, where the same conductor passes straight through the manhole (e.g., entering the south wall and leaving the north wall) the minimum distance is 48 times the largest shielded cable diameter (32 times the largest unshielded cable diameter). If a conductor enters one side of a manhole and then makes a right-angle turn, the dimension drops to 36 times (24 times for unshielded conductors), but it is measured in both directions, and it is increased by the sum of the outside diameters of all other cable entries on the same wall. If there are multiple rows or columns of duct openings in any direction, use the row or column that gives you the largest sizing calculation and ignore all other cable entries.

110.75. Access to Manholes. The NEC requires access to be at least 26 × 22 in. (650 × 550 mm) if rectangular, otherwise at least 26 in. (650 mm) in diameter. This allows for a ladder to rest against the edge of the opening; if the manhole has a fixed ladder permanently mounted, then the diameter can be reduced to 24 in. (600 mm). A similar reduction is permitted if the only wiring in the manhole is power-limited fire alarm or signaling circuits, or optical-fiber cabling. The access opening must not be directly over electrical equipment or conductors, but if this is not practicable, the manhole must be fitted with a protective barrier or a fixed ladder. The cover, if rectangular, must be restrained so it cannot fall into the manhole. Covers must "prominently" identify the manhole's function by wording (e.g., "ELECTRIC") or a logo, and they must weigh at least 100 lb (45 kg); if not, then they must be secure so a tool will be required for access.

Chapter Two

ARTICLE 200. USE AND IDENTIFICATION OF GROUNDED CONDUCTORS

200.2. General. Generally, all grounded conductors used in premises wiring systems must be identified as described in 200.6. This section used to require grounded conductors unless exempted, and a laundry list of such exemptions was included. It turned out that that list was not all inclusive, and furthermore 250.24 and 250.30(A) adequately cover instances where a grounded conductor should be required. Therefore, the entire content of this paragraph was deleted in favor of a parent sentence introducing lettered paragraphs (A) and (B), which remain in effect.

Grounded conductors must have the same insulation voltage rating as the ungrounded conductors in all circuits rated up to 1000 V—which means in all the commonly used 240/120-, 208/120-, and 480/277-V circuits. To correlate with 250.184 on minimum voltage rating of insulation on grounded neutrals of high-voltage systems, 250.184 and 200.2 state that where an insulated, solidly grounded neutral conductor is used with any circuit rated over 1000 V—such as in 4160/2400- or 13,200/7600-V solidly grounded neutral circuits—the neutral conductor does not have to have insulation rated for either phase-to-phase or phase-to-neutral voltage, but must have insulation rated for at least 600 V. See 250.184. (Of course, a bare, solidly grounded neutral conductor may be used in such circuits that constitute service-entrance conductors, are direct-buried portions of feeders, or are installed overhead, outdoors—as specified in Sec. 250.184. But when an insulated neutral is used, the previously noted rule on 600-V rating applies.) Both 250.184 and 200.2(A) represent exceptions to 310.2(A) requiring conductors to be insulated.

Part (**B**) of this section is new as of the 2008 **NEC**. The continuity of a grounded circuit conductor must not depend on connections to enclosures, raceways, or cable armor. This problem frequently arises in service panelboards with multiple busbars. Figure 200-1 shows an example of the problem, and how to correct it. The NEC Committee has spent considerable effort in recent years, trying to assure that normal circuit current is confined to recognized conductors, and does not pass over raceways and enclosures that were never designed to be current-carrying conductors.

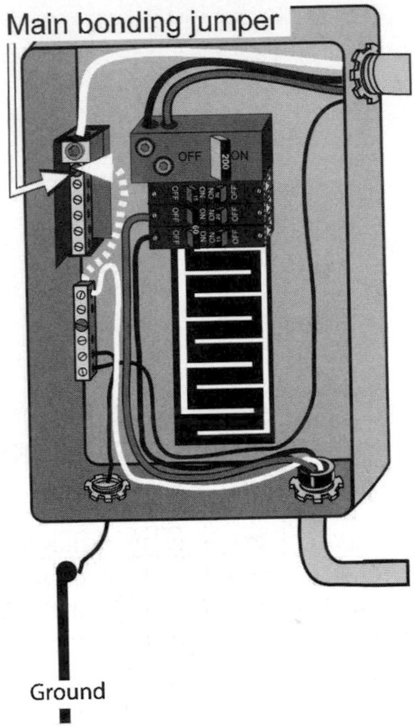
Main bonding jumper
OFF
200
ON
Ground

Fig. 200-1. *Violation!* The feeder neutral has been terminated on the equipment grounding bus in this service panelboard. The neutral current must flow over the enclosure in order to reach the service neutral, thereby making the continuity of the grounded conductor depend on the enclosure. The feeder neutral must be reterminated on the neutral busbar above.

200.3. Connection to Grounded System. Here the Code prohibits "connection" of a grounded conductor in a premises wiring system to any supply system—the utility feed or generator—that does not also have a grounded conductor. The second sentence clarifies that the "connection" referred to here is a direct connection. Supply of grounded conductor through a transformer is acceptable, even if the supply system does not contain a grounded conductor.

200.4. Neutral Conductor. Part **(A)** of this section, new as of the 2011 NEC, prohibits using an oversize neutral conductor in association with plural ungrounded conductors on the same line or phase leg, or in association with plural multiwire branch circuits, unless the Code specifically allows the practice. For these purposes a multiwire branch circuit is considered a single circuit and is not addressed here. Section 215.4 allows the practice for up to two sets of 4-wire feeders and up to three sets of 3-wire feeders, and Sec. 225.7 allows the practice for lighting branch circuits installed outdoors.

Part **(B)** of this section is new as of the 2014 NEC and significantly broadened the requirement formerly of 210.4(D) for grouping multiwire circuit conductors with their associated neutrals in the originating panelboards. Now all circuits with grounded conductors, whether two-wire or more, must have the conductors grouped within any enclosure. Wire markers are also allowed on the grounded conductors that specify the corresponding circuit(s) in lieu of a cable tie or other grouping mechanism. There are two exceptions. The first waives the rule if the grouping is obvious due to a wiring entry unique to the circuit, such as a raceway entry with a single circuit or a cable entry through a single cable connector. The second, in effect, waives the requirement in instances where the conductors are in a box or conduit body and are so short that it would violate 300.14 if someone attempted to make a splice or termination.

200.6. Means of Identifying Grounded Conductors. The basic rule in part **(A)**, covering 6 AWG and smaller wires, generally requires, as covered in the first four list items, that any grounded neutral conductor or other circuit conductor that is operated intentionally grounded must have a white (or gray) outer finish for the entire length of the conductor, or a conductor with three white stripes encircling other than green insulation is also permitted, or colored threads in white or gray insulation, if the conductor is 6 AWG size or smaller. See Fig. 200-2.

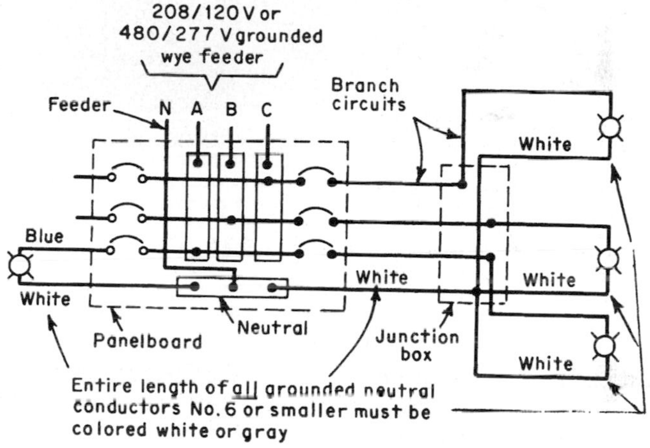
208/120 V or
480/277 V grounded
wye feeder
Feeder
N A B C
Branch
circuits
White
Blue
White
White
White
Panelboard
Neutral
Junction
box
White
Entire length of all grounded neutral
conductors No. 6 or smaller must be
colored white or gray

Fig. 200-2. Generally any grounded circuit conductor that is No. 6 size or smaller must have a continuous white or gray outer finish. [Sec. 200.6(A).]

Note that there is no language in Art. 200 to exempt dc circuits from its provisions. One of the global changes throughout the NEC commencing with the 2014 edition is the introduction of a color code for dc circuits, with red for positive and black for negative. However, if either a positive or negative dc conductor is operated as a grounded circuit conductor, then its color must become either white or gray, or meet one of the other identification provisions of this section.

Exempted by parts **(A)(5)** through **(A)(8)** from the requirement of 200.6 for a white, gray, or three white (or gray) striped neutral are mineral-insulated, metal-sheathed cable; single conductors used as the grounded conductor in photovoltaic systems provided the conductor's insulation is rated for outdoor use and is "sunlight resistant"; fixture wires as covered by 402.8; and neutrals of aerial cable—which may have a raised ridge on the exterior of the neutral to identify them.

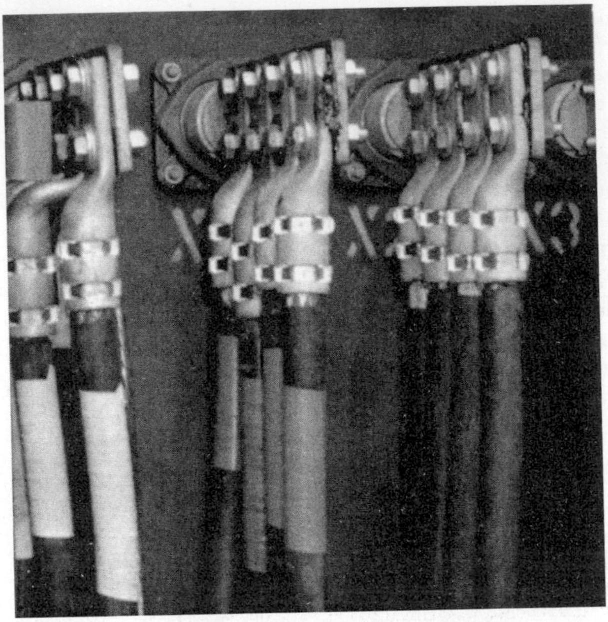

Fig. 200-3. Conductors of colors other than white or gray—in sizes larger than No. 6—may be used as grounded neutrals or grounded phase legs if marked white at all terminations—such as by white tape on the grounded feeder neutrals, at left. [Other color tapes are used on other circuit conductors to identify the three phases as A, B, and C—as required by 210.5(C) for branch circuits and 215.12(C) for feeders.] [Sec. 200.6(B).]

The rule of 200.6(B) requires any grounded conductor larger than No. 6 to either comply with the usual identification rules, or to be marked with white or gray identification (such as white tape) encircling the conductor at all terminations at the time of installation. This is the usual approach in the field, since colored insulation is seldom available as a stock item on larger conductors. See Fig. 200-3.

In the rule of part **(D)**, color coding must distinguish between grounded circuit conductors where branch circuits and/or feeders of different systems are in the same raceway or enclosure. This rule ensures that differentiation between grounded circuit conductors of different wiring systems in the same raceway or other enclosure is provided for feeder circuits as well as branch circuits. (See Fig. 200-4.) Because gray is now permitted as a color choice for grounded conductors, identifying two systems in an enclosure is easily done with white wire for one system and gray for the other. You can also use white or gray wire with a stripe, which would become a requirement if there are three or more systems in a common enclosure, although such wires are usually only available on special order and with a very large minimum length. Whatever method of identification is chosen, it must be posted on the panelboard or other circuit source, or otherwise documented in a readily available manner. Refer to the discussion at 210.5(C) for more information on this point.

For approximately 75 years (beginning with 1923), the **NEC** described the customary identification rule in terms of "white or natural gray" coloring. This originally

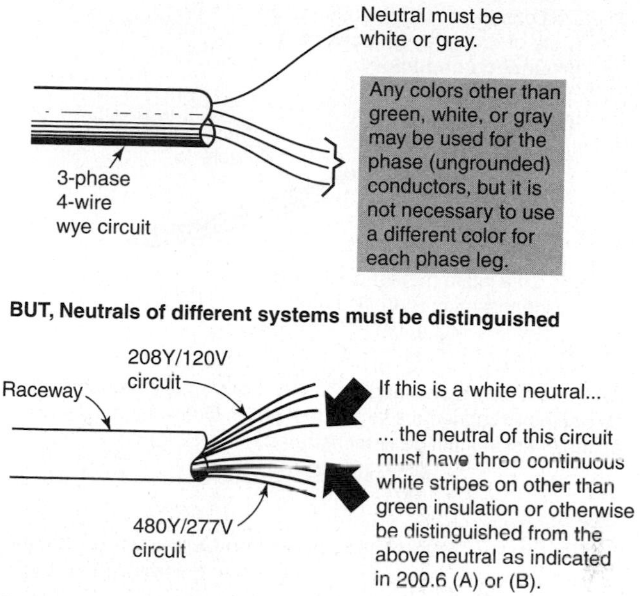
Neutral must be white or gray.
Any colors other than green, white, or gray may be used for the phase (ungrounded) conductors, but it is not necessary to use a different color for each phase leg.
3-phase
4-wire
wye circuit
BUT, Neutrals of different systems must be distinguished
208Y/120V circuit
Raceway
If this is a white neutral...
... the neutral of this circuit must have three continuous white stripes on other than green insulation or otherwise be distinguished from the above neutral as indicated in 200.6 (A) or (B).
480Y/277V circuit

Fig. 200-4. Grounded circuit conductors must have color identification and must be distinguishable by system wherever they enter a cable assembly, common raceway, or other common enclosure. [Sec. 200.6(D).]

referred to the color of latex insulation and the unbleached muslin put over it. It wasn't exactly either white or gray, but installers knew what it was. It was never intended to be the controlled color gray, and conductors manufactured in this way have not been produced for many decades. In fact, the controlled color gray could always have been used, and occasionally was used as an ungrounded conductor. However, with the advent of 480Y/277-V systems, the controlled color gray was increasingly used as an identified conductor based on an improper interpretation of the old terminology "natural gray." The 2002 **NEC** ratified what had become the convention, dropped the term "natural gray" completely, and recognized the controlled color gray as a permitted color for identified conductors for the first time. However, since gray wires were permitted, at least theoretically, for use as ungrounded conductors, the **NEC** advises caution when working with gray wires on existing systems.

The basic rules of 200.6(A) and (B) require the use of continuous white or gray or three continuous white (or gray) stripes running the entire length of an insulated grounded conductor (such as grounded neutral). But the Code permits the use of a conductor of other colors (black, purple, yellow, etc.) for a grounded conductor in a multiconductor cable under certain conditions (see Fig. 200-5):

1. That such a conductor is used only where qualified persons supervise and do service or maintenance on the cable—such as in industrial and mining applications.
2. That every grounded conductor of color other than white or gray will be effectively and permanently identified at all terminations by distinctive white marking or other effective means applied at the time of installation.

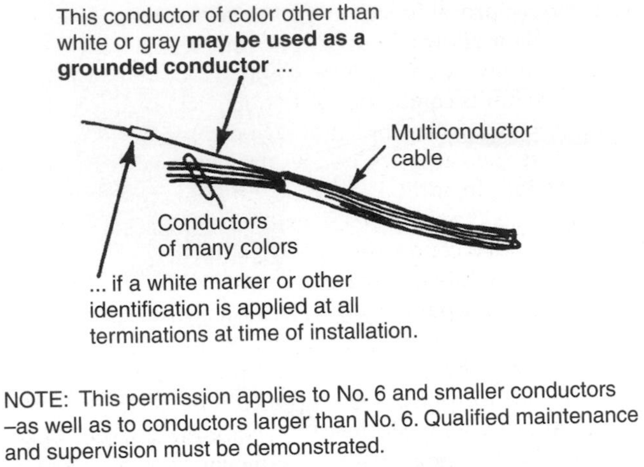
This conductor of color other than white or gray may be used as a grounded conductor ...
Multiconductor cable
Conductors of many colors
... if a white marker or other identification is applied at all terminations at time of installation.
NOTE: This permission applies to No. 6 and smaller conductors –as well as to conductors larger than No. 6. Qualified maintenance and supervision must be demonstrated.

Fig. 200-5. [Sec. 200.6(E).]

This permission for such use of grounded conductors in multiconductor cable allows the practice in those industrial facilities where multiconductor cables are commonly used—although the rule does not limit the use to industrial occupancies. Be aware that this permission *does not apply to conductors in a raceway,* regardless of the degree of supervision. In a raceway, it is assumed that there is no good reason why a conductor with the wrong color insulation cannot be replaced with one having the appropriate color insulation if its function changes. See also Sec. 200.7 and Fig. 200-6.

200.7. Use of Insulation of a White or Gray Color or Three Continuous White (or Gray) Stripes. The previous section covered how to identify grounded conductors, the usual, but not the only approach being white or gray color coding.

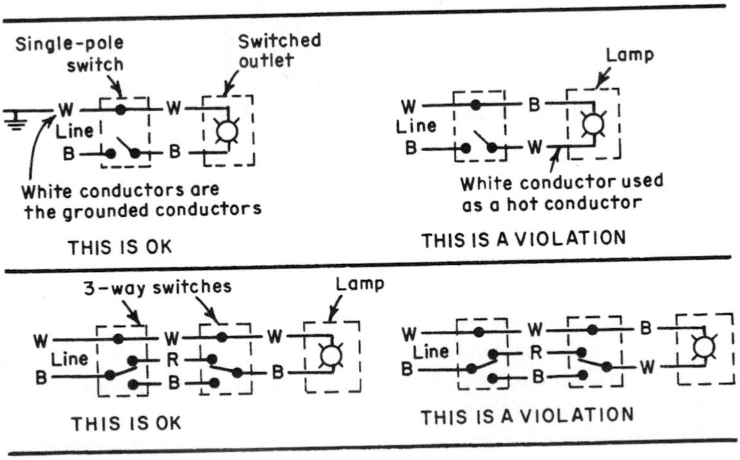
Single-pole switch
Switched outlet
W
Line
B
W
B
White conductors are the grounded conductors
THIS IS OK
Lamp
W
Line
B
B
W
White conductor used as a hot conductor
THIS IS A VIOLATION
3-way switches
Lamp
W
Line
B
W
R
B
W
B
THIS IS OK
W
Line
B
W
R
B
B
W
THIS IS A VIOLATION

Fig. 200-6. A white- or gray-colored conductor must normally be used only as a grounded conductor (the grounded circuit neutral or grounded phase leg of a delta system). (Sec. 200.7.)

This section has the reciprocal function of covering how the colors white and gray are to be limited in their allowable uses. It is a subtle difference, but taking these sections together definitively covers white/gray usage in the **NEC**.

The basic rule here limits conductors with outer covering colored white or gray or with three continuous white (or gray) stripes on other colors to use only as grounded conductors (i.e., as grounded neutral or grounded phase or line conductors [see Fig. 200-6]). In addition, those conductors reidentified at the time of installation as "grounded" conductors (usually the neutral of a grounded system) must actually be grounded conductors. [200.7(A).]

Figure 200-7 shows a white-colored conductor used for an ungrounded phase conductor of a feeder to a panelboard. As shown in the left side of the panel bottom gutter, the white conductor has black tape wrapped around its end for a length of a few inches. The Code used to permit a white conductor to be used for an ungrounded (a hot phase leg) conductor if the white is "permanently reidentified"—such as by wrapping with black or other color tape—to indicate clearly and effectively that the conductor is ungrounded. However, the permission given for such application of white or gray, or even the three white stripes on conductors of other colors, has been eliminated for other than cable assemblies, multiconductor flexible cord, and for circuits "of less than 50 V."

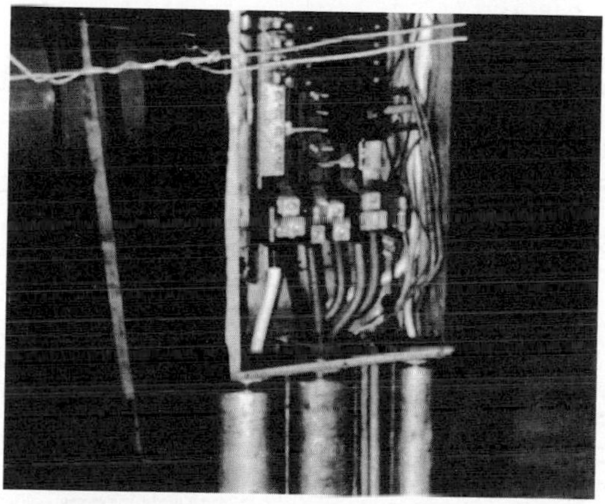

Fig. 200-7. *Violation!* White conductor in lower left of panel gutter is used as an ungrounded phase conductor of a feeder, with black tape wrapped around the conductor end to "reidentify" the conductor as *not* a grounded conductor. Although such practice was previously permitted, the **NEC** no longer recognizes it. (Sec. 200.7.)

Part (B) of 200.7 covers the use of conductors whose insulation is white, gray, or has three continuous white (or gray) stripes for circuits operating at 50 V, or less. Circuit conductors in such systems that have an insulation coloring or configuration reserved for "grounded" conductors are not required to be grounded unless required by 250.20, which identifies those systems that must be operated with a grounded conductor. If the low-voltage system in question is supplied from a transformer whose primary supply voltage is over 150 V to ground; or if the supply transformer's primary conductors are not grounded; or where the low-voltage system is run overhead outdoors, 250.20(A) would mandate grounding of one of the circuit conductors. And therefore, reidentifying a conductor with an overall outer covering or insulation that is one of the colors or configurations reserved for grounded conductors, as an ungrounded conductor, is prohibited.

Part **(C)(1)** indicates conditions under which a white conductor in a cable (such as BX or nonmetallic-sheathed cable) may be used for an ungrounded (hot-leg) conductor. When used as described, the white conductor is acceptable even though it is not a grounded conductor, provided it is reidentified (such as by painting or taping). Figure 200-8 shows examples of correct and incorrect hookups of switch loops where the hot supply is run first to the switched outlet, then to switches.

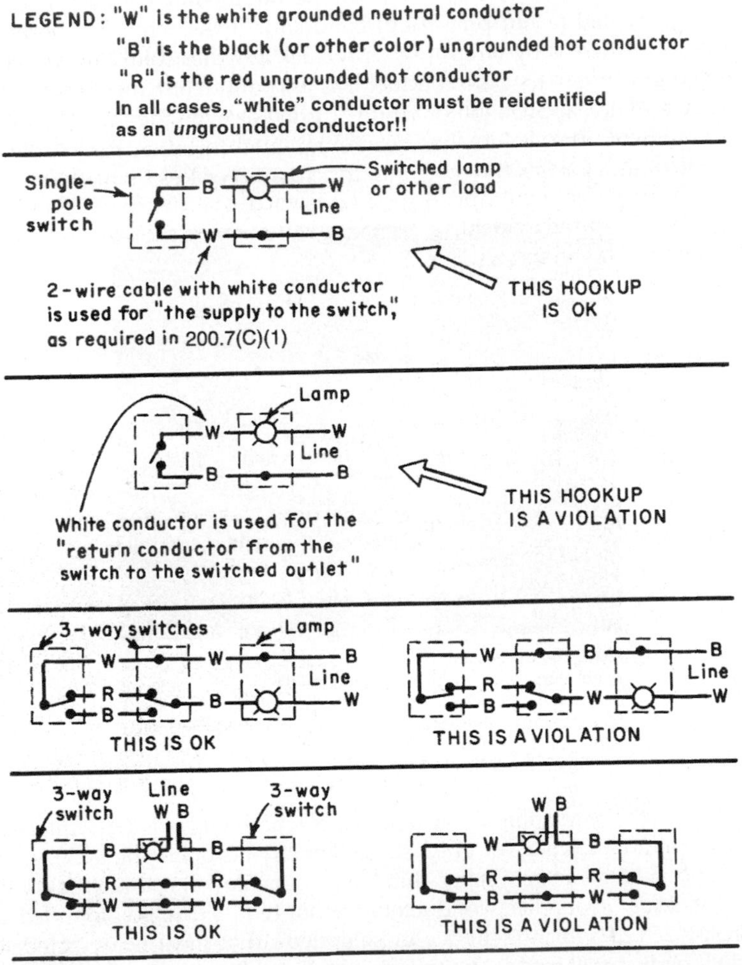
LEGEND: "W" is the white grounded neutral conductor
"B" is the black (or other color) ungrounded hot conductor
"R" is the red ungrounded hot conductor
In all cases, "white" conductor must be reidentified as an ungrounded conductor!!
Single-pole switch
Switched lamp or other load
Line
2-wire cable with white conductor is used for "the supply to the switch," as required in 200.7(C)(1)
THIS HOOKUP IS OK
Lamp
White conductor is used for the "return conductor from the switch to the switched outlet"
THIS HOOKUP IS A VIOLATION
3-way switches
THIS IS OK
THIS IS A VIOLATION
3-way switch
3-way switch
THIS IS OK
THIS IS A VIOLATION

Fig. 200-8. For switch loops from load outlets with hot supply to the load outlet, white conductor in cable must be the "supply to the switch." Also, the white conductors must be reidentified at the time of installation. [Sec. 200.7(C)(1).]

The former unrestricted allowance to use the white wire in a cable assembly as the supply side of a switch leg, something every apprentice learns in the first year, is still in the Code but now the white wire must be reidentified at terminations and other places where it is "visible and accessible." Although the original substantiation for this change was questionable, the requirement has taken on increased importance given the likely steady increase in actual grounded white wire connections to switches with electronic components, as anticipated by 404.2(C).

200.7(C)(2) covers flexible cords for connecting any equipment recognized by 400.10 for cord-and-plug connection to a receptacle outlet.

200.10. Identification of Terminals. Part **(B)** permits a grounded terminal on a receptacle to be identified by the word "white" or the letter "W" marked on the receptacle as an alternative to the use of terminal parts (screw, etc.) that are "substantially white in color."

Marking of the word "white" or the letter "W" provides the required identification of the neutral terminal on receptacles that require white-colored plating on all terminals of a receptacle for purposes of corrosion resistance or for connection of aluminum conductors. Obviously, if all terminals are white colored, color no longer serves to identify or distinguish the neutral as it does if the hot-conductor terminals are brass colored. And as the rule is worded, the marking "white" or the letter "W" may be used to identify the neutral terminal on receptacles that have all brass-colored terminal screws. See Fig. 200-9.

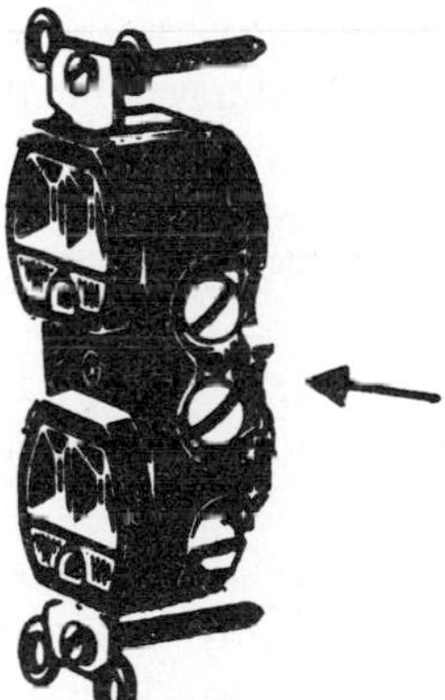

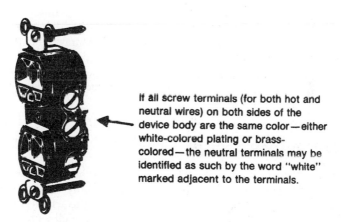

If all screw terminals (for both hot and neutral wires) on both sides of the device body are the same color—either white-colored plating or brass-colored—the neutral terminals may be identified as such by the word "white" marked adjacent to the terminals.

Fig. 200-9.

Subpart **(2)** of part **(B)** permits a push-in-type wire terminal to be identified as the neutral (grounded) conductor terminal either by marking the word "white" or the letter "W" on the receptacle body adjacent to the conductor entrance hole or by coloring the entrance hole white—as with a white-painted ring around the edge of the hole.

The rule of part **(C)** is shown in Fig. 200-10.

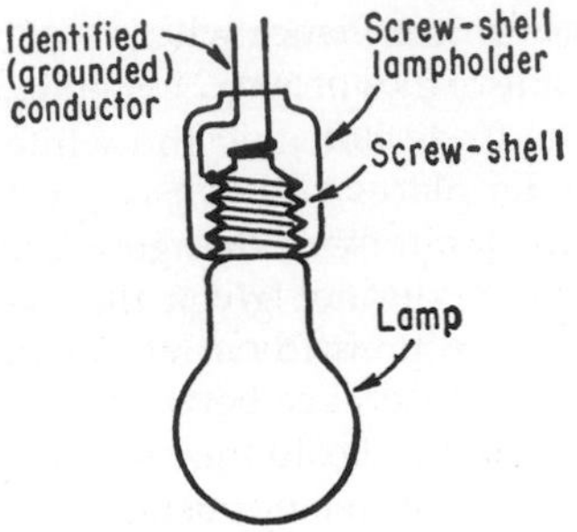
Identified (grounded) conductor
Screw-shell lampholder
Screw-shell
Lamp

SCREW-SHELL LAMPHOLDERS are wired so that ungrounded conductor is connected to center terminal to reduce shock hazard. The identified (grounded) conductor is connected to the screw-shell.

Fig. 200-10. Screw-shell sockets must have the grounded wire (the neutral) connected to the screw-shell part. [Sec. 200.10(C).]

Part **(E)** of Sec. 200.10 requires that the grounded conductor terminal of appliances be identified—to provide proper connection of field-installed wiring (either fixed wiring connection or attachment of a cord set).

The rule applies to "appliances that have a single-pole switch or a single-pole overcurrent device in the line or any line-connected screw-shell lampholder" and requires simply that some "means" (instead of "marking") be provided to identify the neutral. As a result, use of white color instead of marking is clearly recognized for such neutral terminals of appliances.

200.11. Polarity of Connections. This rule makes failure to observe the proper polarity when terminating conductors a Code violation. Installers are required to ensure that each and every grounded conductor is connected to the termination specifically identified as the neutral point of connection. Any connection either of grounded conductors to "other" termination points, or the connection of an ungrounded conductor to an identified "grounded" conductor connection point, is clearly and specifically prohibited.

ARTICLE 210. BRANCH CIRCUITS

210.1. Scope. Article 210 covers all branch circuits other than those "specific-purpose branch circuits" such as those that supply only motor loads, which are covered in Art. 430. This section makes clear that the article covers branch circuits supplying lighting and/or appliance loads as well as branch circuits supplying any combination of those loads plus motor loads or motor-operated appliances, unless the branch circuit is one identified in Table 210.2, "Specific-Purpose Branch Circuits." Where motors or motor-operated appliances are connected to branch circuits supplying lighting and/or appliance loads, the rules of both Arts. 210 and 430 apply. Article 430 alone applies to branch circuits that supply only motor loads.

210.3. Other Articles for Specific-Purpose Branch Circuits. This rule provides correlation with specific branch-circuiting requirements in other articles. There are a number of "specific-purpose" circuits identified in this rule that must be laid-out and installed in compliance with the specific requirements of those rules shown. However, all the rules of Art. 210 continue to apply, except to the extent modified by the other provisions.

210.4. Multiwire Branch Circuits. A "branch circuit," as covered by Art. 210, may be a 2-wire circuit or may be a "multiwire" branch circuit. A "multiwire" branch circuit consists of two or more ungrounded conductors having a potential difference between them and an identified grounded conductor having equal potential difference between it and each of the ungrounded conductors and which is connected to the neutral conductor of the system. Thus, a 3-wire circuit consisting of two opposite-polarity ungrounded conductors and a neutral derived from a 3-wire, single-phase system or a 4-wire circuit consisting of three different phase conductors and a neutral of a 3-phase, 4-wire system is a single multiwire branch circuit. This is only one circuit, even though it involves two or three single-pole protective devices in the panelboard (Fig. 210-1). This is important, because other sections of the Code refer to conditions involving "one branch circuit" or "the single branch circuit." (See 250.32 Exception and 110.64(C).)

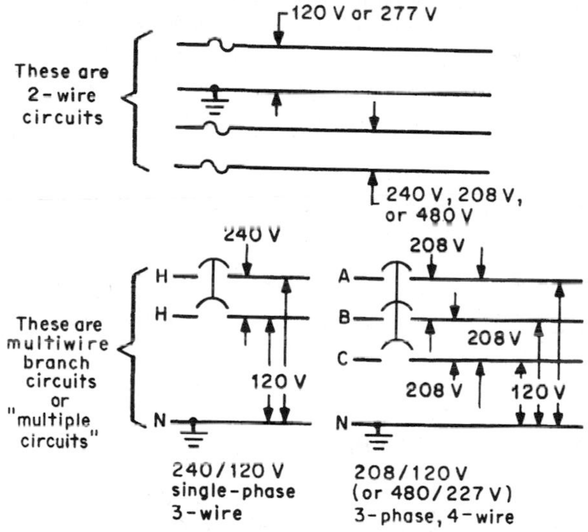
120 V or 277 V
These are 2-wire circuits
240 V, 208 V, or 480 V
240 V
208 V
H
H
N
A
B
C
N
These are multiwire branch circuits or "multiple circuits"
208 V
120 V
208 V
120 V
240/120 V single-phase 3-wire
208/120 V (or 480/227 V) 3-phase, 4-wire

Fig. 210-1. Branch circuits may be 2-wire or multiwire type. (Sec. 210.4.)

The wording of part **(A)** of this section makes clear that a multiwire branch circuit may be considered to be either "a single circuit" or "multiple circuits." This coordinates with other Code rules that refer to multiwire circuits as well as rules that call for two or more circuits. For instance, 210.11(C)(1) requires that at least two 20-A small appliance branch circuits be provided for receptacle outlets in those areas specified in 210.52(B)—that is, the kitchen, dining room, pantry, and breakfast room of a dwelling unit. The wording of this rule recognizes that a single 3-wire, single-phase 240/120-V circuit run to the receptacles in those rooms is equivalent to two 120-V circuits and satisfies the rule of 210.11(C)(1).

In addition, a "multiwire" branch circuit is considered to be a single circuit of multiple-wire makeup. That will satisfy the rule in 410.64(C), which recognizes that a multiwire circuit is a single circuit when run through end-to-end connected lighting fixtures that are used as a raceway for the circuit conductors. Only one principal circuit—either a 2-wire circuit or a multiwire (3- or 4-wire) circuit—may be run through fixtures connected in a line.

The first informational note following part **(A)** of 210.4 warns of the potential for "neutral overload" where line-to-neutral nonlinear loads are supplied. This results from the additive harmonics that will be carried by the neutral in multiwire branch circuits. In some cases, where the load to be supplied consists of, or is expected to consist of, so-called nonlinear loads that are connected line-to-neutral, it may be necessary to use an oversized neutral (up to two sizes larger), or each phase conductor could be run with an individual full-size neutral. Either way, a derating of 80 percent would be required for the number of conductors [see 310.15(B)(5)(c)].

The second informational note calls attention to 300.13(B), which requires maintaining the continuity of the grounded neutral wire in a multiwire branch circuit by pigtailing the neutral to the neutral terminal of a receptacle. Exception No. 2 of 210.4(C) and 300.13(B) are both aimed at the same safety objective—to prevent damage to electrical equipment that can result when two loads of unequal impedances are series-connected from hot leg to hot leg as a result of opening the neutral of an energized multiwire branch circuit or are series-connected from hot leg to neutral. 300.13(B) prohibits dependency upon device terminals (such as internally connected screw terminals of duplex receptacles) for the splicing of neutral conductors in multiwire (3- or 4-wire) circuits. Grounded neutral wires must not depend on device connection (such as the break-off tab between duplex-receptacle screw terminals) for continuity. White wires can be spliced together, with a pigtail to the neutral terminal on the receptacle. If the receptacle is removed, the neutral will not be opened.

This rule is intended to prevent the establishment of unbalanced voltages should a neutral conductor be opened first when a receptacle or similar device is replaced on energized circuits. In such cases, the line-to-neutral connections downstream from this point (farther from the point of supply) could result in a considerably higher-than-normal voltage on one part of a multiwire circuit and damage equipment, because of the "open" neutral, if the downstream line-to-neutral loads are appreciably unbalanced. Refer to the description given in 300.13 of this book.

Part **(B)** of this section requires a "means" to simultaneously disconnect all ungrounded conductors of a multiwire branch circuit "at the point where the branch circuit originates." Although at one time this was a dwelling-unit provision for split-wired receptacles, and then it applied in all occupancies to multiple devices on one yoke, *it now applies to all multiwire circuits serving any loads in all occupancies.* There is a long and unfortunate history of unqualified persons creating havoc when working on multiwire circuits without protecting against the consequences of open neutrals and of voltage backfeeding into an outlet from a different leg than the one thought to be at issue. Now a common disconnecting means will be in an obvious and prominent location when the branch circuit is being disconnected.

A multipole circuit breaker (CB) certainly complies with this rule, as would a multipole fused switch. Single-pole CBs connected together with identified handle ties definitely qualify due to clarifying language in 240.15(B), as now cross-referenced in an informational note. Remember that handle ties are for operation by hand; they are not rated to automatically open the companion breaker if only one leg trips. Even less clear is a multipole switch located immediately adjacent to the panel where the circuit originates. This would be the only practical option on an existing fusible panelboard with no internal switching.

The objective is to assure that when someone goes to deenergize an ungrounded conductor of some equipment being maintained or replaced, that person will open all the conductors and thereby preclude line voltage from appearing on the load-side neutral conductor through loads connected on another leg of the circuit. In other words, this rule serves a maintenance function. If the purpose were electrical, even fuses in a multipole fused switch, would have been disallowed because they are inherently single-pole devices and if one opens, the others still provide power to the other legs. In this regard, note that the wording here differs from the requirement in 210.4(C) Exception No. 2, which serves an electrical function and clearly does require a multipole CB for other reasons. On this basis a good case can be made for the multipole switch adjacent to the panel, but this is certainly subject to local interpretation.

The basic rule of part **(C)** addresses the need for personnel safety. To help minimize the possibility of shock or electrocution during maintenance or repair, this section states that multiwire branch circuits (such as 240/120-V, 3-wire, single-phase and 3-phase, 4-wire circuits at 208/120 or 480/277 V) may be used only with loads connected from a hot or phase leg to the neutral conductor (Fig. 210-2). However, while generally prohibited, where additional measures are taken to

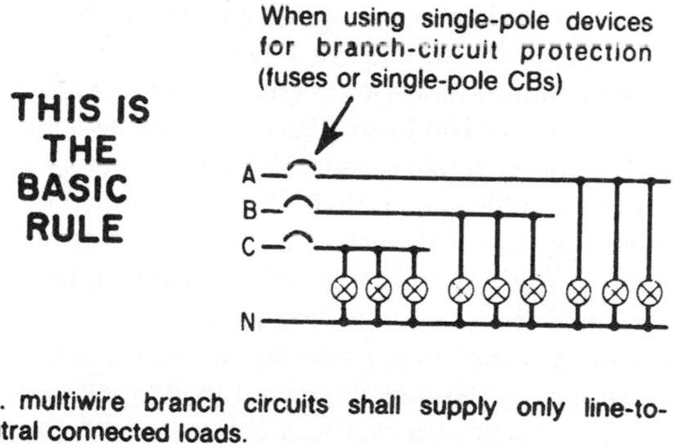
When using single-pole devices for branch-circuit protection (fuses or single-pole CBs)
THIS IS THE BASIC RULE
A
B
C
N
. . . multiwire branch circuits shall supply only line-to-neutral connected loads.

Fig. 210-2. With single-pole protection only line-to-neutral loads may be fed. (Sec. 210.4.)

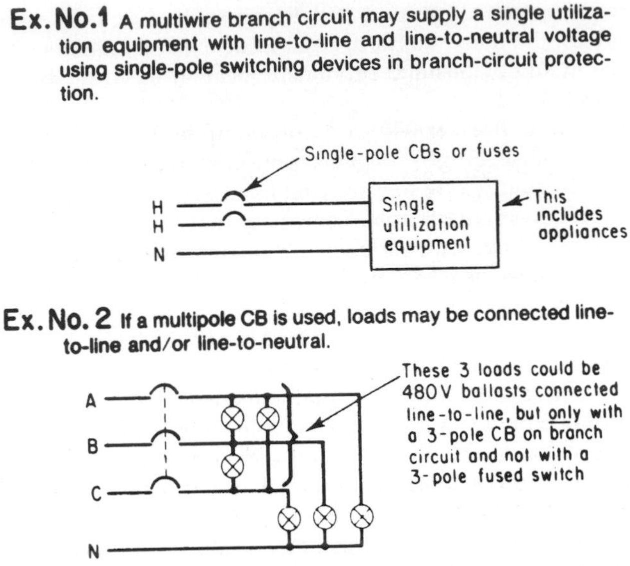
Ex. No.1 A multiwire branch circuit may supply a single utilization equipment with line-to-line and line-to-neutral voltage using single-pole switching devices in branch-circuit protection.
Single-pole CBs or fuses
H
H
N
Single utilization equipment
This includes appliances
Ex. No. 2 If a multipole CB is used, loads may be connected line-to-line and/or line-to-neutral.
These 3 loads could be 480V ballasts connected line-to-line, but only with a 3-pole CB on branch circuit and not with a 3-pole fused switch
A
B
C
N

Fig. 210-3. Line-to-line loads may only be connected on multiwire circuits that conform to the exceptions given. Note that the single-pole breakers in the upper drawing would require a handle tie to comply with the common disconnecting means requirements. (Sec. 210.4.)

protect personnel, the two exceptions to this rule permit supplying "other than line-to-neutral loads" from multiwire branch circuits. The two exceptions to that rule are shown in Fig. 210-3.

Exception No. 1 permits use of single-pole protective devices for an individual circuit to "only one utilization equipment"—in which the load may be connected line-to-line as well as line-to-neutral. "Utilization equipment," as defined in Art. 100, is "equipment which utilizes electric energy for electronic, electromechanical, chemical, heating, lighting, or similar purposes." The definition of "appliance," in Art. 100, notes that an *appliance* is "utilization equipment, generally other than industrial, that is normally built in standardized sizes or types and is installed or connected as a unit to perform one or more functions such as washing clothes, air conditioning, food mixing, deep frying, and so forth." Because of those definitions, the wording of Exception No. 1 opens its application to commercial and industrial equipment as well as residential. It should be noted that 210.4(B) applies in these cases, and therefore means must still be provided, such as handle ties, to provide for simultaneous opening of a set of single-pole breakers installed for this equipment.

Exception No. 2 permits a multiwire branch-circuit to supply line-to-line connected loads, but only when it is protected by a multipole CB. The intent of Exception No. 2 is that line-to-line connected loads may be used (other than in Exception No. 1) only where the poles of the circuit protective device operate together, or simultaneously. A multipole CB satisfies the rule, but a fused multipole switch would not comply because the hot circuit conductors are not "opened

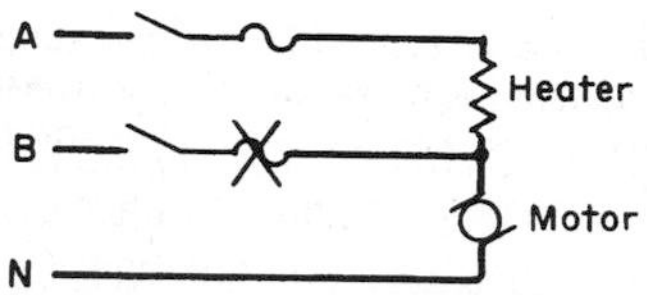
A
B
N
Heater
Motor

Should fuse B open, the heater and motor would be in series on 115 volts, and the motor could burn out if not properly protected.

Fig. 210-4. Single-pole protection can expose equipment to damage. (Sec. 210.4.)

simultaneously by the branch-circuit overcurrent device." This rule requiring a multipole CB for any circuit that supplies line-to-line connected loads as well as line-to-neutral loads was put in the Code to prevent equipment loss under the conditions shown in Fig. 210-4. Use of a 2-pole CB in the sketch would cause opening of both hot legs on any fault and prevent the condition shown.

The UL Guide Card information addressing this point can easily be misinterpreted. The relevant text under the category "DIVQ" reads as follows: "Single-pole or multipole independent trip CBs with handle ties rated 120/240 V ac, are suitable for use on multiwire circuits with line-to-line or line-to-neutral connected loads." This is not an open-ended permission to use single-pole breakers on applications such as Fig. 210-4. The only time such breakers can be used is when they are supplying a single appliance that requires both hot legs and a neutral.

Figure 210-5 shows that a 2-pole CB, two single-pole CBs with a handle tie that enables them to be used as a 2-pole disconnect, or a 2-pole switch ahead of branch-circuit fuse protection will satisfy the requirement that both hot legs must be interrupted when the disconnect means is opened to deenergize a multiwire circuit to a split-wired receptacle. This Code rule provides the greater safety of disconnecting both hot conductors simultaneously to prevent shock hazard in replacing or maintaining any piece of electrical equipment where only one of two hot supply conductors has been opened.

It should also be noted that although a 2-pole switch ahead of fuses may satisfy as the simultaneous disconnect required ahead of split-wired receptacles, such a switch does not satisfy as the simultaneous multipole "branch-circuit protective device" that is required by Exception No. 2 of 210.4 when a multiwire circuit supplies any loads connected phase-to-phase (other than the single appliance covered in Exception No. 1). In such a case, a 2-pole CB must be used because fuses are single-pole devices and do not ensure simultaneous opening of all hot legs on overcurrent or ground fault.

It should be noted that the threat of motor burnout, shown in the diagram of Fig. 210-4, may exist just as readily where the 230-V resistance device and the 115-V motor are fed from a dual-voltage (240-V, 120-V) duplex receptacle as where loads are fixed wired. As shown in Fig. 210-6, the rule of 210.4 does clearly call for a 2-pole CB (and not single-pole CBs or fuses) for a circuit supplying a dual-voltage receptacle. In such a case, a line-to-line load and a line-to-neutral load could be connected and subjected to the condition shown in Fig. 210-4.

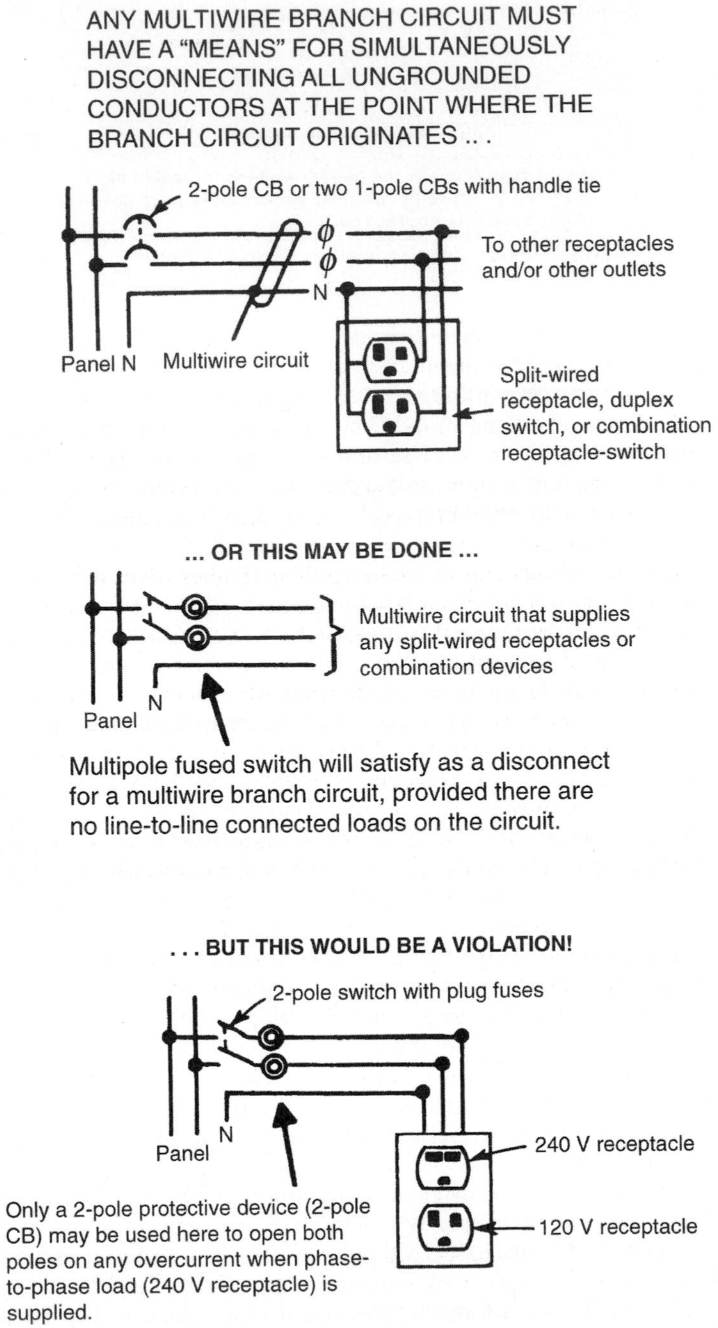
ANY MULTIWIRE BRANCH CIRCUIT MUST HAVE A "MEANS" FOR SIMULTANEOUSLY DISCONNECTING ALL UNGROUNDED CONDUCTORS AT THE POINT WHERE THE BRANCH CIRCUIT ORIGINATES . . .
2-pole CB or two 1-pole CBs with handle tie
φ
φ
N
To other receptacles and/or other outlets
Panel N
Multiwire circuit
Split-wired receptacle, duplex switch, or combination receptacle-switch
... OR THIS MAY BE DONE ...
Multiwire circuit that supplies any split-wired receptacles or combination devices
Panel
N
Multipole fused switch will satisfy as a disconnect for a multiwire branch circuit, provided there are no line-to-line connected loads on the circuit.
. . . BUT THIS WOULD BE A VIOLATION!
2-pole switch with plug fuses
Panel
N
240 V receptacle
120 V receptacle
Only a 2-pole protective device (2-pole CB) may be used here to open both poles on any overcurrent when phase-to-phase load (240 V receptacle) is supplied.

Fig. 210-5.

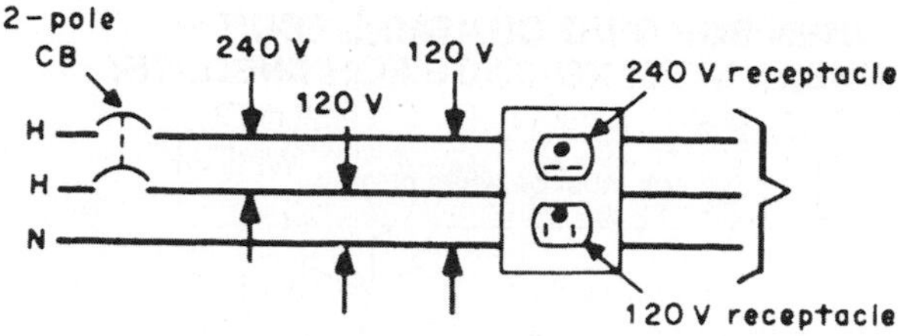
2-pole
CB
240 V
120 V
120 V
240 V receptacle
H
H
N
120 V receptacle

Fig. 210-6. A dual-voltage receptacle requires a 2-pole CB on its circuit. (Sec. 210.4.)

210.5. Identification for Branch Circuits. For grounding and grounded conductors this section simply directs the reader to comply with other Code rules that cover conductor color-coding or color-identification schemes. It directs that "grounded" and "grounding" conductors in branch circuits utilize the specific color identification given in 200.6 and 250.119. Those rules generally reserve the color green for equipment grounding conductors and white, gray, or three continuous white stripes on other than green-colored insulation for the grounded conductors in branch circuits.

It should be noted that rules on color coding of conductors given in Art. 210 apply only to branch-circuit conductors and do not directly require color coding of feeder conductors. But the rules given in 200.6 and 250.119 must generally be observed, and would apply to feeder and service conductors. 215.12 also requires identification of phase legs of feeders to panelboards, switchboards, and so forth—and that requires some technique for marking the phase legs; those provisions are now harmonized with the ones here for branch circuits. Note that many design engineers have insisted on color coding of feeder conductors all along to afford effective balancing of loads on the different phase legs.

Color identification for branch-circuit conductors is divided into three categories:

Grounded conductor As indicated, grounded conductors must satisfy 200.6. That rule generally requires that the grounded conductor of a branch circuit (the neutral of a wye system or a grounded phase of a delta) must be identified by a continuous white or gray color for the entire length of the conductor, or have three continuous white (or gray) stripes for its entire length on other than green insulation. Where wires of different systems (such as 208/120 and 480/277) are installed in the same raceway, box, or other enclosure, the neutral or grounded wire of one system must be white or gray or have the three continuous white (or gray) stripes on other than green insulation; and the neutral of the other system must be white with a color stripe, or be gray if the first one is white, etc., or it must be otherwise distinguished—such as by painting or taping. The point is that neutrals of different systems must be distinguished from each other when they are in the same enclosure [200.6(D) and Fig. 210-7]. For more, see 200.6.

Hot conductor The NEC requires that individual hot conductors be identified where a building has more than one nominal voltage system. In contrast to the rule for grounded circuit conductors, the coding rules for these wires apply anytime multiple voltage systems exist in a building, whether or not they happen to share an enclosure. Another difference is that the grounded conductor identification

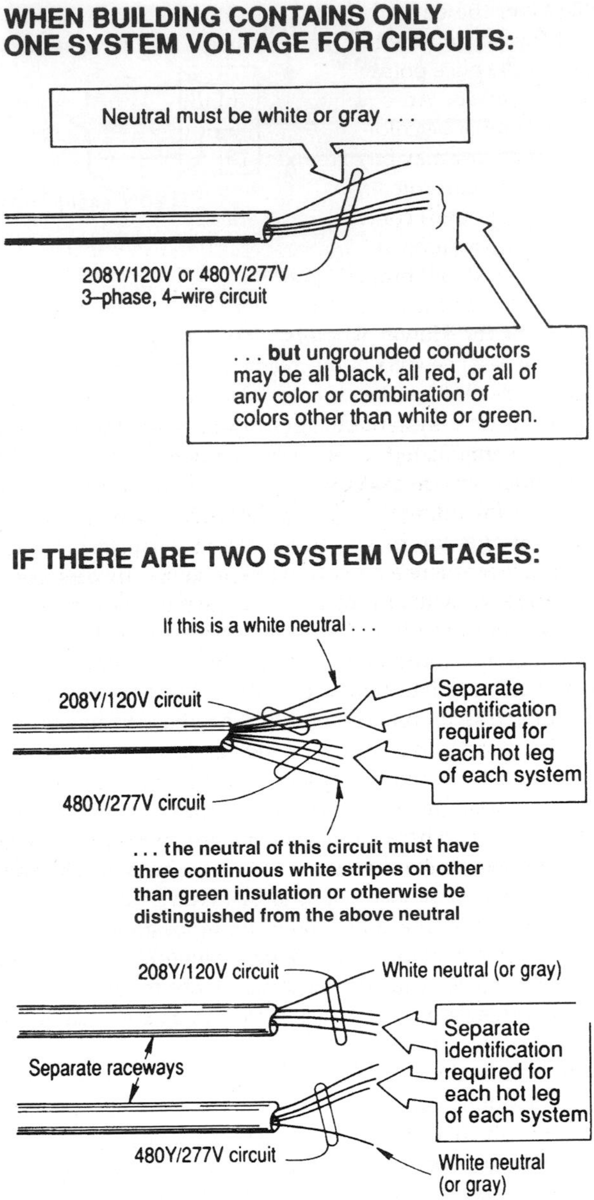
WHEN BUILDING CONTAINS ONLY ONE SYSTEM VOLTAGE FOR CIRCUITS:
Neutral must be white or gray . . .
208Y/120V or 480Y/277V
3-phase, 4-wire circuit
. . . but ungrounded conductors may be all black, all red, or all of any color or combination of colors other than white or green.
IF THERE ARE TWO SYSTEM VOLTAGES:
If this is a white neutral . . .
208Y/120V circuit
Separate identification required for each hot leg of each system
480Y/277V circuit
. . . the neutral of this circuit must have three continuous white stripes on other than green insulation or otherwise be distinguished from the above neutral
208Y/120V circuit
White neutral (or gray)
Separate raceways
Separate identification required for each hot leg of each system
480Y/277V circuit
White neutral (or gray)

Fig. 210-7. Separate identification of ungrounded conductors is required only if a building utilizes more than one nominal voltage system. Neutrals must be color-distinguished if circuits of two voltage systems are used in the same raceway, but not if different voltage systems are run in separate raceways. [Sec. 210.5(C).]

scheme applies over the entire length of the conductor for 6 AWG and smaller conductors, but the ungrounded conductors need only be identified at "termination, connection, and splice points."

Grounding conductor An equipment grounding conductor of a branch circuit (if one is used) must be color-coded green or green with one or more yellow stripes—or the conductor may be bare [250.119].

In part **(C)**, an important rule for branch circuits requires some means of identification of hot (ungrounded) conductors of branch circuits in a building that contains wiring systems operating at two or more different voltage levels. That means that one needs to identify all branch circuits including individual branch circuits, as well as single-phase and three-phase power circuits whether or not a neutral is part of the branch circuit. However, every branch-circuit panelboard—in both the 208Y/120-V system and the 408Y/277-V system—must have the means of identification marked on it—but in a key clarification for 2008, the panel identification label need only specify the system in use for circuits originating within it. It is not necessary to create complicated, fully reciprocal labels that describe every color code for every voltage system in the building. Such identification is also required in 215.12 for feeders, including the marking of feeder panels.

As of the 2017 **NEC**, the burden of circuit wiring identification has been significantly reduced in a case where a new voltage system is being added to an existing, already wired occupancy with one or more voltage systems in place. The exception allows the new system to be identified using whatever system has been designed for it, without triggering a requirement that the existing wiring be changed to accommodate and differentiate from the new system. The exception is contingent on markings being present on all distribution equipment that originates branch circuits on the new system, in addition to the posting required in (b), which specifically includes the phrase "other unidentified systems exist on the premises."

This Code rule and that given in 215.12 restore the need to identify phase legs of branch and feeder circuits where more than one voltage system is used in a building. For instance, a building that utilizes both 208Y/120-V circuits and 480Y/277-V circuits must have separate and distinct color coding of the hot legs of the two voltage systems—or must have some means other than color coding such as tagging, marking tape (color or numbers), or some other identification that will satisfy the inspecting agency. And this new rule further states that the "means of identification must be permanently posted at each branch-circuit panelboard or similar branch-circuit distribution equipment"—to tell how the individual phases in each of the different voltage systems are identified (Fig. 210-7).

The wording of the new rule requires that the "means of identification" must distinguish between all conductors, including the system. But, if a building uses only one voltage system—such as 208Y/120 V or 240/120 V single phase, no identification is required for the circuit phase (the "hot" or ungrounded) legs. And where a building utilizes two or more voltage systems, the separate, individual identification of ungrounded conductors must be done whether the circuits of the different voltages are run in the same or separate raceways.

Color coding of circuit conductors (or some other method of identifying them), as required by 210.5(C), is a wiring consideration that deserves the close, careful, complete attention of all electrical people. Of all the means available to provide for the ready identification of the two- or three-phase legs and neutrals in wiring systems, color coding is the easiest and surest way of balancing loads among the phase legs, thereby providing full, safe, effective use of total circuit capacities. In circuits where color coding is not used, loads or phases get unbalanced, many conductors are either badly underloaded or excessively loaded, and breakers or fuses sometimes are increased in size to eliminate tripping due to overload on only one phase leg. Modern electrical usage—for reasons of safety and energy conservation, as well as full, economic application of system equipment and materials—demands the many real benefits that color coding can provide.

For the greater period of its existence, the **NEC** required a very clear, rigid color coding of branch circuits for good and obvious safety reasons. Color coding of hot legs to provide load balancing is a safety matter. 210.11(B) requires balancing of loads from branch-circuit hot legs to neutral. The rule of 220.61 bases sizing of feeder neutrals on clear knowledge of load balance in order to determine "maximum unbalance." And mandatory differentiation of voltage levels is in the safety interests of electricians and others maintaining or working on electrical circuits, to warn of different levels of hazard.

Because the vast majority of electrical systems involve no more than two voltage configurations for circuits up to 1000 V, and because there has been great standardization in circuit voltage levels, there should be industry-wide standardization on circuit conductor identifications. A clear, simple set of rules could cover the preponderant majority of installations, with exceptions made for the relatively small number of cases where unusual conditions exist and the local inspector may authorize other techniques. Color coding should follow some basic pattern—such as the following:

- 120-V, 2-wire circuit: grounded neutral—white; ungrounded leg—black
- 240/120-V, 3-wire, single-phase circuit: grounded neutral—white; one hot leg—black; the other hot leg—red
- 208Y/120-V, 3-phase, 4-wire: grounded neutral—white; one hot leg—black; one hot leg—red; one hot leg—blue
- 240-V, delta, 3-phase, 3-wire: one hot leg—black; one hot leg—red; one hot leg—blue
- 240/120-V, 3-phase, 4-wire, high-leg delta: grounded neutral—white; high leg (208-V to neutral)—orange; one hot leg—black; one hot leg—blue
- 480Y/277-V, 3-phase, 4-wire: grounded neutral—gray; one hot leg—brown; one hot leg—orange; one hot leg—yellow
- 480-V, delta, 3-phase, 3-wire: one hot leg—brown; one hot leg—orange; one hot leg—yellow

By making color coding a set of simple, specific color designations, standardization will ensure all the safety and operating advantages of color coding to all electrical systems. Particularly today, with all electrical systems being subjected to an unprecedented amount of alterations and additions because of continuing development and expansion in electrical usage, conductor identification is a regular safety need over the entire life of the system. (Fig. 210-8.)

Of course, there are alternatives to "color" identification throughout the length of conductors. Color differentiation is almost worthless for color-blind electricians. And it can be argued that color identification of conductors poses problems

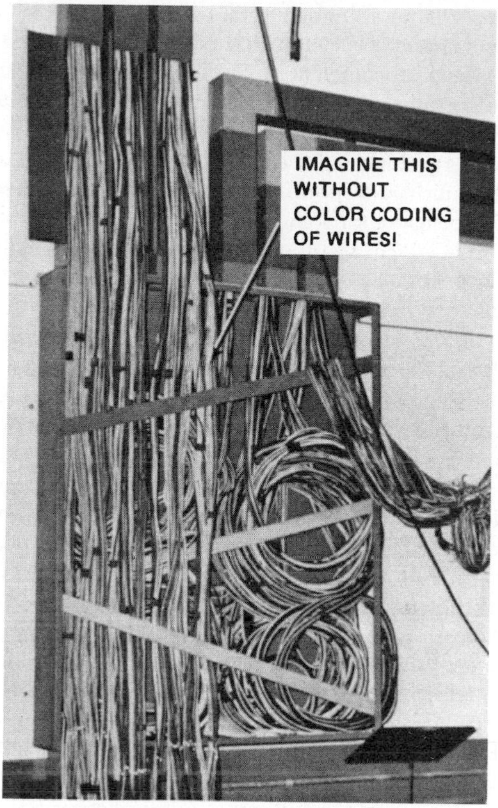
IMAGINE THIS
WITHOUT
COLOR CODING
OF WIRES!

Fig. 210-8. Although only required for branch circuits in buildings with more than one nominal voltage, color identification of branch-circuit phase legs is needed for safe and effective work on grouped circuits. [Sec. 210.5(C).]

because electrical work is commonly done in darkened areas where color perception is reduced even for those with good eyesight. The NEC already recognizes white tape or paint over the conductor insulation end at terminals to identify neutrals (200.6). Number markings spaced along the length of a conductor on the insulation (1, 2, 3, etc.)—particularly, say, white numerals on black insulation—might prove very effective for identifying and differentiating conductors. Or the letters "A," "B," and "C" could be used to designate specific phases. Or a combination of color and markings could be used. But some kind of conductor identification is essential to safe, effective hookup of the ever-expanding array of conductors used throughout buildings and systems today. And the method used for identifying ungrounded circuit conductors must be posted at each branch-circuit panelboard to comply with requirements of 210.5(C). Note that 200.6(D) imposes the same requirement on grounded conductors if different systems share a common enclosure at any point. For usability, the marking protocol for both types of conductors should be grouped on the same legend plate or within the same documentation.

The NEC addresses this by recognizing that some occupancies with very sophisticated operations maintain the circuit identification protocols in documentation at central points. If such documentation is "readily available," Sec. 210.5(C) and 200.6(D) now make the same allowance for grounded conductors and allow such on-site records or manuals to substitute for panelboard markings. This degree of sophistication becomes important when, for example, multiple branch circuits running at the same voltage but derived from differing separately derived systems happen to arrive in a common enclosure for some reason. In such cases, it may be very useful to know which wire is which, and the simple use of color would probably not be adequate for this purpose.

Part (C)(2) of this section is new as of the 2014 NEC and covers color coding for the ungrounded conductors of branch circuits that are derived from dc systems operating above 60 V. The requirement follows the customary industry practice of using red for positive and black for negative, either as the entire insulation color or in the form of a stripe along the entire length. For obvious reasons, if a stripe is used, the contrasting color cannot be that of the opposite polarity, or white, gray, or green. Imprinted positive (+) and negative (–) signs on the conductor insulation at least every 2 ft in accordance with 310.120(B) is also permitted. Another option, new as of the 2017 NEC, is using approved permanent marking methods such as shrink tubing with the same identifying markings. Just as in the case of 200.6 for coding grounded conductors, the rule for 6 AWG and smaller applies over the length of the conductor. For 4 AWG and larger, the rule only applies to "termination, connection, and splice points" and can be met by taping or tagging or other approved means. Note that in the case of approved permanent markings (shrink tubing, etc.), the same limitation applies equally to smaller conductors. It would clearly be prohibitive to require such methods to be used over the entire length of the conductor.

210.6. Branch-Circuit Voltage Limitations. Voltage limitations for branch circuits are presented here in 210.6. In general, branch circuits serving lampholders, fixtures, cord-and-plug-connected loads up to 12 A, or motor loads rated ¼ hp or less are limited to operation at a maximum voltage rating of 120 V. It should be noted that these rules, for the most part, are aimed at the manufacturers. But designers and installers should be aware of these limitations so that they do not unwittingly apply a given piece of equipment in an other than acceptable manner.

Part **(A)**, occupancy limitation, applies specifically to dwelling units—one-family houses, apartment units in multifamily dwellings, and condominium and co-op units—and to guest rooms and suites in hotels and motels and similar residential occupancies, including college dormitories. In such occupancies, any luminaire or any receptacle for plug-connected loads rated up to 1440 VA or for motor loads of less than ¼ hp must be supplied at not over 120 V between conductors.

Note: The 120-V supply to these types of loads may be derived from (1) a 120-V, 2-wire branch circuit; (2) a 240/120-V, 3-wire branch circuit; or (3) a 208/120-V, 3-phase, 4-wire branch circuit. Appliances rated more than 1440 VA (e.g., ranges, dryers, water heaters) may be supplied by 240/120-V or 208/120-V circuits in accordance with 210.6(C)(6).

Caution: The concept of maximum voltage not over "120 V . . . between conductors," as stated in 210.6(A), has caused considerable discussion and controversy in the past when applied to split-wired receptacles and duplex receptacles of two voltage levels. It can be argued that split-wired general-purpose duplex receptacles are not acceptable in dwelling units and in hotel and motel guest rooms because they are supplied by conductors with more than 120 V between them—that is, 240 V on the 3-wire, single-phase, 120/240-V circuit so commonly used in residences. The two hot legs connect to the brass-colored terminals on the receptacle, with the shorting tab broken off, and the voltage between those conductors does exceed 120 V. The same condition applies when a 120/240-V duplex receptacle is used—the 240-V receptacle is fed by conductors with more than 120 V between them.

That interpretation is not supported by the definition of a receptacle, by which a duplex receptacle is actually two receptacles on a single yoke, and each of those receptacles is considered as a separate device. In addition, the rule limits loads over 1440 VA, not devices, and until the load is plugged in, there is no issue. This rule is primarily of interest to manufacturers, who are obliged not to manufacture appliances in violation of these limits. All of that said, there is a legitimate concern with respect to the voltage on the yoke when maintenance is being performed, but the current requirements for a common disconnecting means in 210.4(B) and 210.7 fully address those issues. See Fig. 210-9.

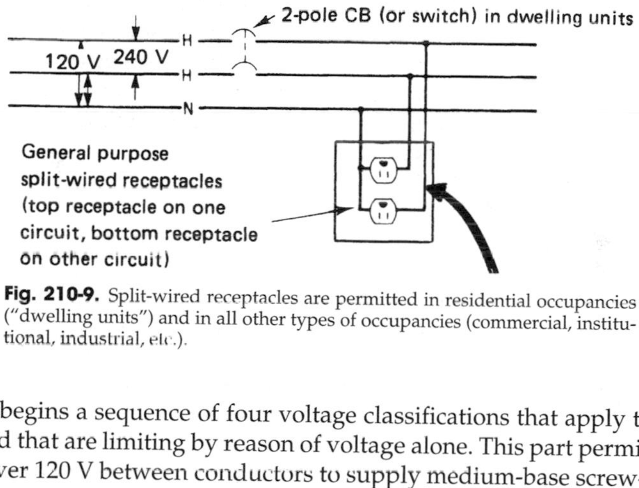
2-pole CB (or switch) in dwelling units
H
H
N
120 V
240 V
General purpose
split-wired receptacles
(top receptacle on one
circuit, bottom receptacle
on other circuit)

Fig. 210-9. Split-wired receptacles are permitted in residential occupancies ("dwelling units") and in all other types of occupancies (commercial, institutional, industrial, etc.).

Part **(B)** begins a sequence of four voltage classifications that apply to all occupancies and that are limiting by reason of voltage alone. This part permits a circuit with not over 120 V between conductors to supply medium-base screw-shell lampholders, ballasts for fluorescent or HID lighting fixtures, and plug-connected or hard-wired appliances—in any type of building or on any premises (Fig. 210-10).

Part **(C)** applies to circuits with over 120 V between conductors (208, 240, 277, or 480 V) but not over 277 V (nominal) to ground. This is shown in Fig. 210-11, where all of the circuits are "circuits exceeding 120 V, nominal, between conductors and not exceeding 277 V, nominal, to ground." Circuits of any of those voltages are permitted to supply incandescent luminaires with mogul-base screw-shell lampholders, electric-discharge and LED luminaires, or plug-connected or hard-wired appliances, or other utilization equipment.

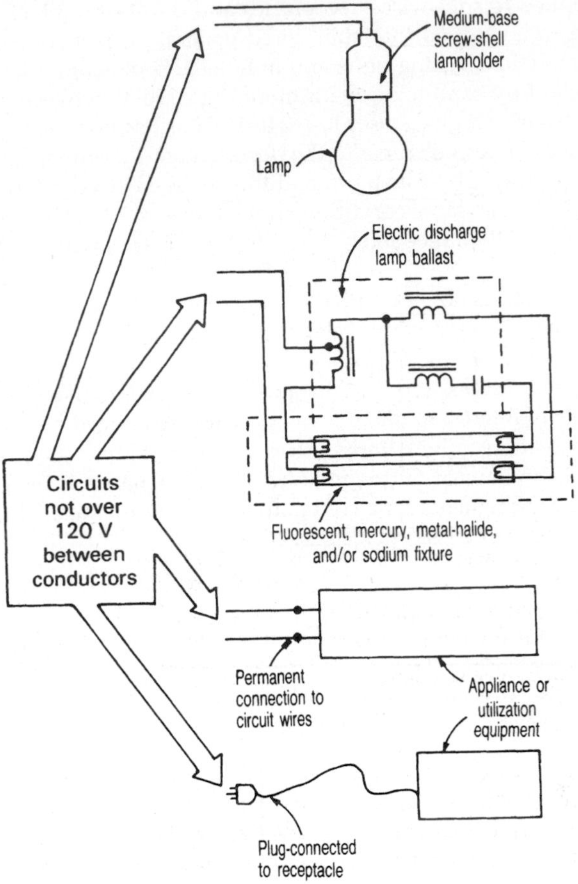
Medium-base
screw-shell
lampholder
Lamp
Electric discharge
lamp ballast
Circuits
not over
120 V
between
conductors
Fluorescent, mercury, metal-halide,
and/or sodium fixture
Permanent
connection to
circuit wires
Appliance or
utilization
equipment
Plug-connected
to receptacle

Fig. 210-10. In any occupancy, 120-V circuits may supply these loads. [Sec. 210.6(B).]

It is important to note that this section no longer contains the requirement for a minimum 8-ft (2.5-m) mounting height for incandescent or electric-discharge fixtures with mogul-base screw-shell lampholders used on 480/277-V systems. Previously, this still had to be correlated with 225.7(C), which formerly required that luminaires connected to circuits over 120 V to ground up to 277 V not be located within 3 ft of "windows, platforms, fire escapes, and the like." So, you could walk up and hug a 277-V bollard-style luminaire on the edge of a sidewalk, but a comparable luminaire on the side of a building had to have been out of reach. Fortunately, all these limitations in Art. 225 were removed in the 2011 **NEC**, so these positioning limitations no longer apply.

A UL-listed electric-discharge luminaire rated at 277 V nominal may be equipped with a medium-base screw-shell lampholder and does not require a mogul-base screw-shell. The use of the medium-base lampholder, however, is limited to

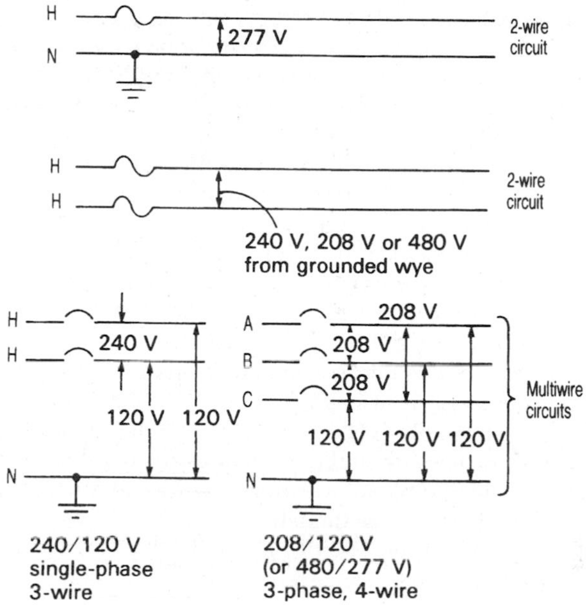
H
N
277 V
2-wire circuit
H
H
2-wire circuit
240 V, 208 V or 480 V from grounded wye
H
H
240 V
120 V 120 V
N
A
B
C
208 V
208 V
208 V
120 V 120 V 120 V
N
Multiwire circuits
240/120 V single-phase 3-wire
208/120 V (or 480/277 V) 3-phase, 4-wire

Fig. 210-11. These circuits may supply incandescent lighting with mogul-base screw-shell lampholders for over 120 V between conductors, electric-discharge ballasts, and cord-connected or permanently wired appliances or utilization equipment. [Sec. 210.6(C).]

"listed electric-discharge" or "listed light-emitting diode-type luminaires." The NEC also permits listed incandescent luminaires with the usual medium bases to be supplied, provided the luminaire contains an integral autotransformer that reduces the voltage at the lamp to 120. For 277-V incandescent fixtures, 210.6(C)(3) continues the requirement that such fixtures be equipped with "mogul-based screw-shell lampholders."

Fluorescent, mercury-vapor, metal-halide, high-pressure sodium, low-pressure sodium, and/or incandescent fixtures may be supplied by 480/277-V, grounded-wye circuits—with loads connected phase-to-neutral and/or phase-to-phase. Such circuits operate at 277 V to ground even, say, when 480-V ballasts are connected phase-to-phase on such circuits. Or lighting could be supplied by 240-V delta systems—either ungrounded or with one of the phase legs grounded, because such systems operate at not more than 277 V to ground.

On a neutral-grounded 480/277-V system, incandescent, fluorescent, mercury-vapor, metal-halide, high-pressure sodium, and low-pressure sodium equipment can be connected from phase-to-neutral on the 277-V circuits. If fluorescent or mercury-vapor fixtures are to be connected phase-to-phase, some Code authorities contend that autotransformer-type ballasts cannot be used when these ballasts raise the voltage to more than 300 V, because, they contend, the NEC calls for connection to a circuit made up of a grounded wire and a hot wire. (See 410.138.) On phase-to-phase connection these ballasts would require use of 2-winding, electrically isolating ballast transformers according to this interpretation. However, the actual wording in 410.138 states the restriction in terms not of whether the supply conductors are grounded, but rather that the supply system be a grounded one, and a 480-V luminaire connected to a 480Y/277-V system is connected to a grounded system.

210.6(C)(6) clearly permits either "cord-and-plug-connected or permanently connected utilization equipment" to be supplied by a circuit with voltage between conductors in excess of 120 V, and permission is intended for the use of 277-V heaters in dwelling units, as used in high-rise apartment buildings and similar large buildings that may be served at 480/277 V. This is OK in such locations as long as such equipment, if cord-and-plug-connected, is larger than the 1440 VA threshold set in the occupancy limitation in 210.6(A).

In 210.6(D), the **NEC** permits fluorescent and/or high-intensity discharge units to be installed on circuits rated over 277 V (nominal) to ground and up to 600 V between conductors—but only where the lamps are mounted in permanently installed luminaires on poles or similar structures for the illumination of areas such as highways, bridges, athletic fields, parking lots, at a height not less than 22 ft, or on other structures such as tunnels at a height not less than 18 ft (5.5 m). (See Fig. 210-12.) Part **(D)** covers use of lighting fixtures on 480-V ungrounded circuits—such as fed from a 480-V delta-connected or wye-connected ungrounded transformer secondary.

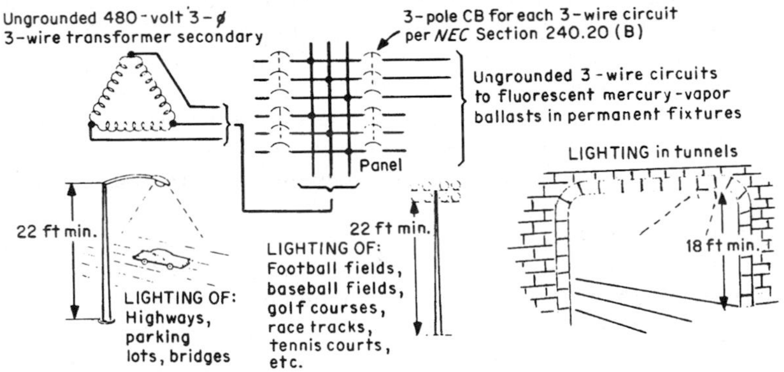
Ungrounded 480-volt 3-ø
3-wire transformer secondary
3-pole CB for each 3-wire circuit
per NEC Section 240.20 (B)
Ungrounded 3-wire circuits
to fluorescent mercury-vapor
ballasts in permanent fixtures
Panel
LIGHTING in tunnels
22 ft min.
22 ft min.
18 ft min.
LIGHTING OF:
Highways,
parking
lots, bridges
LIGHTING OF:
Football fields,
baseball fields,
golf courses,
race tracks,
tennis courts,
etc.

Fig. 210-12. Ungrounded circuits, at up to 600 V between conductors, may supply lighting only as shown. [Sec. 210.6(D).]

This permission for use of fluorescent and mercury units under the conditions described is based on phase-to-phase voltage rather than on phase-to-ground voltage. This rule has the effect of permitting the use of 240- or 480-V ungrounded circuits for the lighting applications described. But as described previously, autotransformer-type ballasts may not be permitted on an ungrounded system if they raise the voltage to more than 300 V (410.138). In such cases, ballasts with 2-winding transformation would have to be used.

Certain electric railway applications utilize higher circuit voltages. Infrared lamp industrial heating applications may be used on higher circuit voltages as allowed in 422.14 of the Code. 210.6(D)(2) allows utilization equipment other than luminaires to be connected at these voltages, whether hard-wired or cord-and-plug-connected. 210.6(D)(3) allows dc luminaires operating at these voltages, provided they are listed with an isolating ballast that only allows conventional voltages on the lamp circuit and where there would otherwise be a shock hazard while changing lamps. Additionally, luminaires with no provisions for changing lamps are permitted. This provision addresses luminaires that can run directly off photovoltaic circuits that easily run over 300 V dc; such luminaires can now be connected directly instead of relying on the inverter.

Part **(E)** covers medium-voltage circuits, limited to locations with qualified maintenance and supervision. Such circuits generally supply motors running at 2300, 4160, or even 13,800 V.

210.7. Multiple Branch Circuits. This section is very important because it extends the common-disconnect principle for multiwire branch circuits [210.4(B)] to all devices on a single strap or yoke. If a multiwire branch circuit arrives at a split receptacle, 210.4(B) will require that a common disconnect be installed because that is now a requirement for all multiwire branch circuits in all occupancies. However, what if two 2-wire branch circuits arrive at the same location? This provision assures that both these circuits will have a common disconnect as well, also for maintenance purposes.

This rule is functionally identical to the rule in 210.4(B) in terms of how the disconnect is defined. It is reasonably clear that handle ties could be used, or even a multipole fused switch. The rule is pointedly not written like 210.4(C) Exception No. 2, which requires an actual multipole CB to meet the electrical requirements that lie behind that provision. And, just as covered in the earlier discussion on this point under the 210.4(B) heading, a multipole switch immediately adjacent to the panel would be the only option for a fusible panel. It is also the only option when the two branch circuits leave the same panel from nonadjacent locations.

For example, suppose you wanted to use a snap-switch controlled receptacle for the lighting outlet in a dining room. The **NEC** specifically permits this arrangement in 210.52(B)(1) Exception No. 1; however, the required receptacle placements must still be observed, and this switched receptacle must not be on a small-appliance branch circuit (covered later). One way to do this is to split both sides of the receptacle, with the switch-controlled receptacle on the lighting circuit and the always-on receptacle connected to the appliance circuit. There are three options at this point. First, you can rearrange the panel so the lighting and the appliance circuits come off adjacent breakers and handle-tie those breakers together. That would definitely meet code. You could use a 2-gang opening, with one receptacle (either single or duplex) entirely controlled by the snap switch, and the receptacles on the adjacent yoke being connected to the appliance circuit. That would definitely meet Code.

Of course, the snap-switch-controlled receptacle(s) could even be in their own wall openings, as long as the switch-controlled receptacle(s) were not relied upon to meet the receptacle spacing rules in Sec. 210.52 generally. Remember that any receptacle outlet not controlled by a wall switch in a dining room must be on the small appliance circuit. Finally, you could use a two-pole snap switch immediately adjacent to the panel, and have it disconnect both circuits. This last option requires a local interpretation of whether immediately adjacent to the panel satisfies the "at the point at which the branch circuits originate" wording in this section. As was discussed under 210.4(B), the case for allowing this practice is strong but not conclusive.

210.8. Ground-Fault Circuit-Interrupter Protection for Personnel. Since GFCIs need periodic testing, the parent language of this section requires all such devices, wherever located, to be readily accessible. The 2017 NEC settled a major, recurring area of controversy by adding unequivocal parent language at the beginning of 210.8 addressing how distances to receptacles potentially requiring GFCI protection are to be measured. They are measured along the path a cord could take without "piercing" a wall or doorway, etc., similar to the long established rule in 680.22(A)(5). The two receptacles shown in Fig. 210-13 will require GFCI protection; one on the other side of the door, regardless of proximity, would not. Part **(A)** is headed "Dwelling Units." The very specific definition of those words, as given in Art. 100 of the NEC, indicates that all the GFCI rules apply to:

- All one-family houses
- Each dwelling unit in a two-family house
- Each apartment in an apartment house
- Each dwelling unit in a condominium

GFCI protection is required by 210.8 for all 125-V, single-phase, 15- and 20-A receptacles installed in bathrooms of dwelling units [part **(A)(1)**] and all other occupancies [part **(B)(1)**] and in garages of dwelling units (Fig. 210-14). The requirement

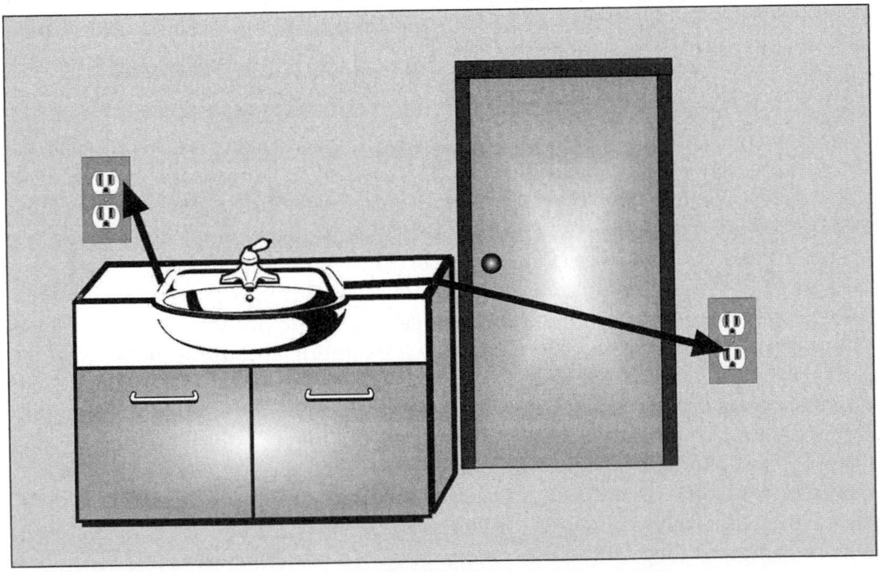

Fig. 210-13. This drawing shows receptacle placements near a kitchen sink.

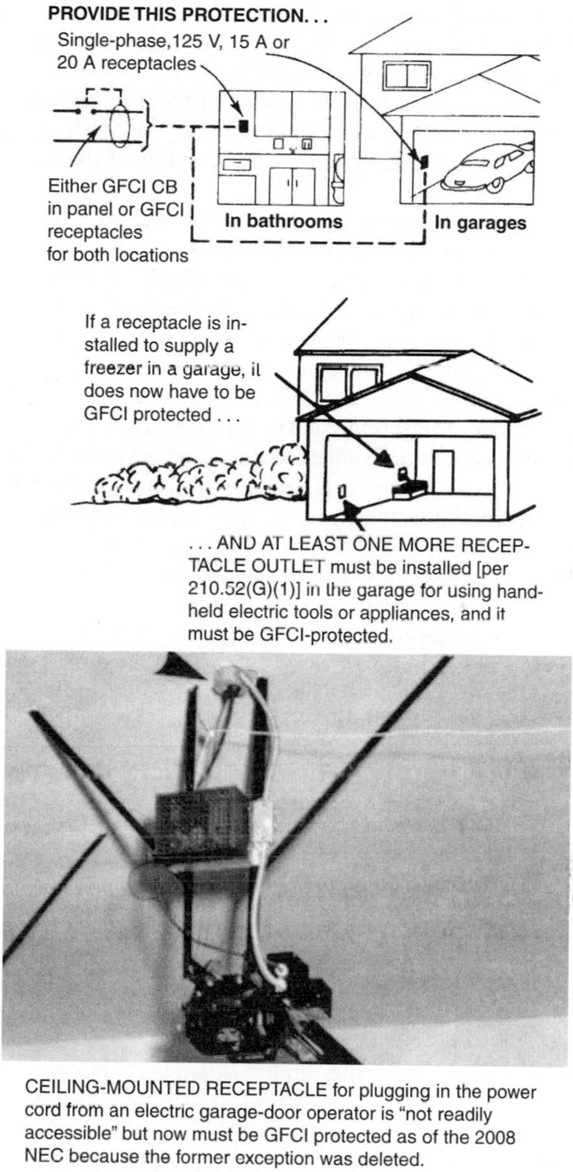
PROVIDE THIS PROTECTION. . .
Single-phase,125 V, 15 A or 20 A receptacles
Either GFCI CB in panel or GFCI receptacles for both locations
In bathrooms
In garages
If a receptacle is installed to supply a freezer in a garage, it does now have to be GFCI protected . . .
. . . AND AT LEAST ONE MORE RECEPTACLE OUTLET must be installed [per 210.52(G)(1)] in the garage for using hand-held electric tools or appliances, and it must be GFCI-protected.
CEILING-MOUNTED RECEPTACLE for plugging in the power cord from an electric garage-door operator is "not readily accessible" but now must be GFCI protected as of the 2008 NEC because the former exception was deleted.

Fig. 210-14. GFCI protection is required for receptacles in garages as well as in bathrooms. [Sec. 210.8(A)(2).]

for GFCI protection in "garages" is included because home owners do use outdoor appliances (lawn mowers, hedge trimmers, etc.) plugged into garage receptacles. Such receptacles require GFCI protection for the same reason as "outdoor" receptacles. In either place, GFCI protection may be provided by a GFCI CB that protects the whole circuit and any receptacles connected to it, or the receptacle may be a GFCI type that incorporates the components that give it the necessary tripping capability on low-level ground faults.

As just noted, GFCI protection is required by 210.8(B)(1) in bathrooms of all occupancies. This includes commercial office buildings, industrial facilities, schools, dormitories, theaters—bathrooms in ALL nondwelling occupancies. The rule here extends the same protection of GFCI breakers and receptacles to bathrooms in all nondwelling-type occupancies as for receptacles in bathrooms of dwelling units. It should be noted that there is no requirement to install a receptacle in bathrooms of other than dwelling units. But, if a 15- or 20-A, 125-V receptacle *is* installed in the bathroom of, say, an office building, then GFCI protection is required.

It is important to remember that a bathroom is defined in Art. 100 as an area that includes a basin and no fewer than one additional plumbing fixture. The 2014 **NEC** addressed a new residential architectural concept, where a master bedroom (or any other room) is configured with a shower (or tub or both) on one side of the room, and a basin with additional plumbing fixtures on the other side of the room. That makes one side of the room a bathroom, but not the other side because it plainly is not in the same area as the basin. To close this loophole, **(A)(9)** took effect as of the 2014 **NEC** and requires receptacles within 6 ft of the outside edge of a tub or shower stall in a dwelling to have GFCI protection.

The rule of 210.8(A)(2) requiring GFCI protection in garages applies to both attached garages and detached (or separate) garages associated with "dwelling units"—such as one-family houses or multifamily houses where each unit has its own garage. In 210.52 the Code requires at least one receptacle in an attached garage and in a detached garage if electric power is run out to the garage.

Part **(A)(2)** of Sec. 210.8 says that 15- and 20-A receptacles in tool huts, workshops, storage sheds, and other "accessory buildings" with a "floor located at or below grade level" at dwellings must be GFCI-protected. In addition to requiring GFCI protection for receptacles installed in a garage at a dwelling unit, other outbuildings, such as tool sheds and the like, must have GFCI protection for all 15- and 20-A, 125-V receptacles. In the 1996 **NEC**, the rule only applied to receptacles installed at "grade level portions." The rewording as of the 1999 **NEC** requires GFCI protection for all 15- and 20-A, 125-V receptacles installed in an accessory building where the *building* has a floor that is "at or below grade level." Obviously, that wording would eliminate the need for GFCI protection if the building's floor is raised above "grade level," such as by use of cinder blocks or stilts. Note, however, that the garage requirement applies wherever it is located in relation to grade level, even if you have to drive up a ramp.

It should be noted that this rule in no way requires a receptacle to be installed in such a building. But, where a 15- or 20-A, 125-V receptacle is installed in such a location and if the area is "not intended as (one or more) habitable rooms" but instead "limited to storage areas, work areas, and areas of similar use," it must be GFCI protected.

Note that the former exceptions for receptacles that were not readily accessible, such as for garage door openers, and single receptacles for dedicated uses such as freezers, have been entirely eliminated as of the 2008 **NEC** edition. Any receptacle of the specified amperage and voltage and phasing as described must have GFCI protection. The panel made the decision that the reliability of these devices has reached the point where special allowances need not be given.

Part **(A)(3)** of Sec. 210.8, on outdoor receptacles, requires GFCI protection of all 125-V, single-phase, 15- and 20-A receptacles installed "outdoors" at dwelling units. Because hotels, motels, and dormitories are not "dwelling units" in the meaning of the Code definition, outdoor receptacles at such buildings are not covered here; see 210.8(B)(4) for those rules. The rule specifies that such protection of outdoor receptacles is required for all receptacles outdoors at dwellings (Fig. 210-15). The phrase "direct grade level access" was deleted from part **(A)(3)** a number of Code editions ago. Because the qualifier "grade-level access" was deleted, apartment units constructed above ground level would need GFCI protection of receptacles installed outdoors on balconies. Likewise, GFCI protection would be required for any outdoor receptacle installed on a porch or other raised part of even a one-family house even though there is no "grade-level access" to the receptacle, as in the examples of Fig. 210-15.

The only exception to the rule of 210.8(A)(3) is for 15- and 20-A, 125-V receptacles that are installed to supply snow-melting and deicing equipment in accordance with Art. 426, and also pipe tracing as covered in Art. 427. Such a receptacle does not require GFCI protection as called for by Sec. 210.8(A)(3), but must have GFPE applied to the equipment as described in 426.28 (or 427.22), provided it is installed on a dedicated circuit and in a not-readily accessible location. Under those circumstances to supply deicing and snow-melting equipment only, GFCI protection called for by this Code section may be omitted.

According to the rule of 210.8(A)(4) and (5), all 125-V, single-phase, 15- and 20-A receptacles installed in crawl spaces at or below grade and/or in unfinished basements must be GFCI-protected. This is intended to apply only to those basements or portions thereof that are unfinished (not habitable), and limited to "storage areas, work areas, and the like." The rule of 210.52(G) requires that at least one receptacle outlet must be installed in the basement of a one-family dwelling, in addition to any installed for laundry equipment. The requirement that a receptacle be installed applies to basements of all one-family houses but not to apartment houses, hotels, motels, dormitories, and the like.

As in the case of garage locations, the former exceptions for dedicated use and for receptacles that were not readily accessible have been deleted, and for the same reasons. And here again this is very controversial, with particular concern registered around freezers and sump pumps. Note that it is at least theoretically possible to hard-wire critical equipment and avoid the issue; the requirement runs to the receptacle and not to the equipment. An exception here specifically exempts receptacles supplying "fire alarm and burglar alarm systems" from the need for GFCI protection. However, such a receptacle must be a single receptacle. This is not a conventional line-voltage smoke detector setup; the exception refers to a full fire alarm control panel instead. The receptacle is powering the internal power supply and stand-by battery charger in the unit.

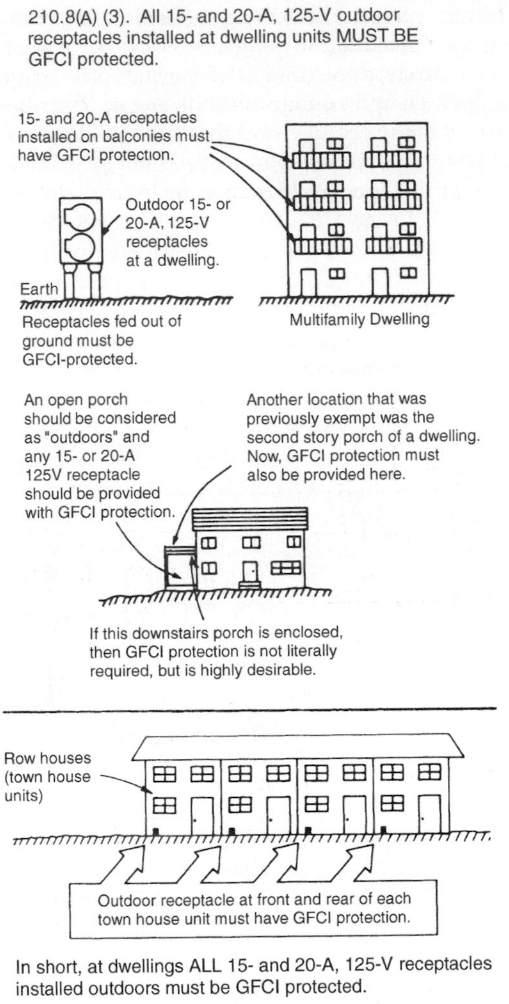
210.8(A) (3). All 15- and 20-A, 125-V outdoor receptacles installed at dwelling units MUST BE GFCI protected.
15- and 20-A receptacles installed on balconies must have GFCI protection.
Outdoor 15- or 20-A, 125-V receptacles at a dwelling.
Earth
Receptacles fed out of ground must be GFCI-protected.
Multifamily Dwelling
An open porch should be considered as "outdoors" and any 15- or 20-A 125V receptacle should be provided with GFCI protection.
Another location that was previously exempt was the second story porch of a dwelling. Now, GFCI protection must also be provided here.
If this downstairs porch is enclosed, then GFCI protection is not literally required, but is highly desirable.
Row houses (town house units)
Outdoor receptacle at front and rear of each town house unit must have GFCI protection.
In short, at dwellings ALL 15- and 20-A, 125-V receptacles installed outdoors must be GFCI protected.

Fig. 210-15. For dwelling units, all outdoor receptacles require GFCI protection. [Sec. 210.8(A)(3).]

According to part **(A)(6),** GFCI protection is required for all 125-V, single-phase, 15- or 20-A receptacles installed in any kitchen of a dwelling unit where such receptacles are serving the countertop area. This will provide GFCI-protected receptacles for appliances used on countertops in kitchens in dwelling units. This would include any receptacles installed in the vertical surfaces of a kitchen "island" that includes countertop surfaces with or without additional hardware such as a range, grill, or even a sink. Because so many kitchen appliances are equipped with only 2-wire cords (toasters, coffee makers, electric fry pans, etc.), their metal frames are not grounded and are subject to being energized by internal insulation failure, making them shock and electrocution hazards. Use of such appliances close to any grounded metal—the range, a cooktop, a sink—creates the strong possibility that a person might touch the energized frame of such an appliance and at the same time make contact with a faucet or other grounded part—thereby exposing the person to shock hazard. Use of GFCI receptacles within the kitchen will protect personnel by opening the circuit under conditions of dangerous fault current flow through the person's body (Fig. 210-16).

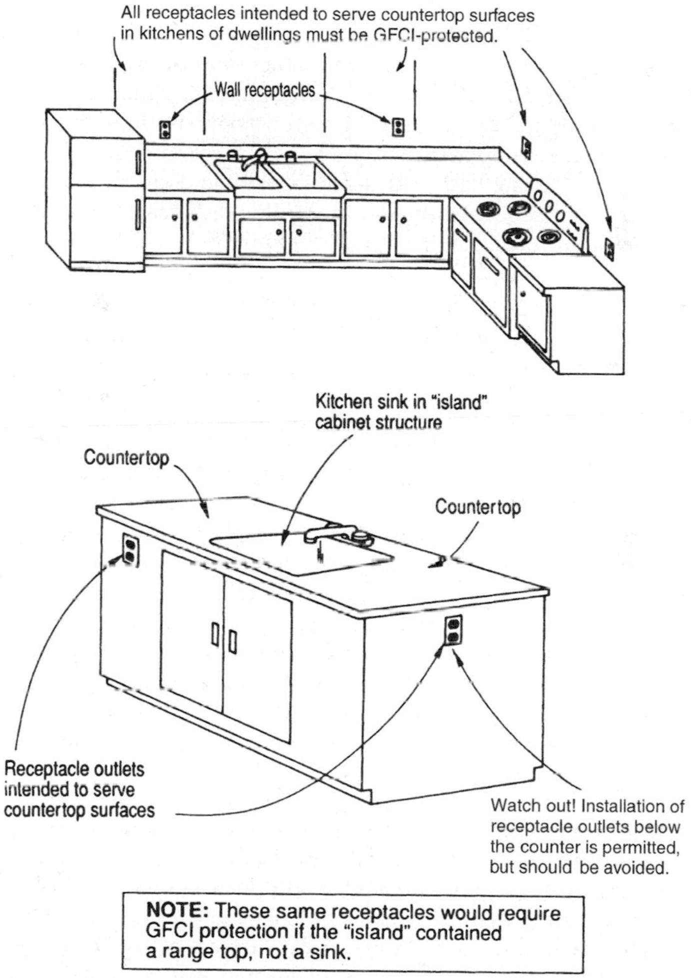
All receptacles intended to serve countertop surfaces in kitchens of dwellings must be GFCI-protected.
Wall receptacles
Kitchen sink in "island" cabinet structure
Countertop
Countertop
Receptacle outlets intended to serve countertop surfaces
Watch out! Installation of receptacle outlets below the counter is permitted, but should be avoided.
NOTE: These same receptacles would require GFCI protection if the "island" contained a range top, not a sink.

Fig. 210-16. GFCI protection must be provided for receptacles in kitchen. Receptacles in face of island cabinet structure in kitchen, if permitted, must be GFCI protected. [Sec. 210.8(A)(6).]

Part **(A)(7)** requires that 15- and 20-A, 125-V countertop receptacles installed within 6 ft (1.8 m) of a sink, such as a laundry, utility, or wet bar sink, be GFCI protected. Note that such receptacles may *not* be installed in the face-up position of the wet bar countertop, as covered in 406.5(G).

Although the requirement for GFCI protection of kitchen countertop receptacles is no longer based on their distance from the kitchen sink, the 6-ft (1.8 m) limitation is still the determining factor with wet bar countertop receptacles, or any receptacle located within 6 ft of a sink. Any 15- or 20-A receptacles installed within 6 ft (1.8 m) from the outside edge of a sink must be provided with GFCI protection. Wording in the 2014 NEC, which removed the qualifying phrase "to serve countertop surfaces," had the effect of requiring GFCI protection for under-sink receptacles for garbage disposers and other appliances. This was problematic, since these receptacles were generally not readily accessible locations. The 2017 NEC may have resolved this by specifying that the distance measurement be taken from the top inside edge of the sink. This must be read in concert with the clarification that distances are to be taken as the path of a cord, and that transit through a doorway doesn't count. Putting these concepts together, it appears that once again these receptacles are exempted provided they are in a sink base with a fully closeable door, the usual condition. Note, however, the subsequent discussion regarding dishwasher circuits [210.8(D)].

210.8(A)(8) calls for GFCI protection of 15- and 20-A, 125-V-rated receptacles installed at dwelling unit boathouses. Boathouses are comparable to garages but with a greater exposure to water, and some dwellings come with boathouses that are sited with direct aquatic access to a body of water.

As of the 2014 NEC, all dwelling receptacles in "laundry areas" as covered in 210.8(A)(10) require GFCI protection. The term laundry area is undefined, but definitely includes the receptacle mandated by 210.52(F). A laundry hookup in an unfinished basement or one in a bathroom is already covered, but this rule covers other locations. It apparently was motivated by end-of-life reliability concerns of certain appliance manufacturers.

210.8(B). Other than Dwelling Units. These rules cover GFCI requirements for receptacles installed at commercial, industrial, and institutional occupancies. As of the 2017 NEC, all receptacles rated not over 150 V to ground (or less) and installed in accordance with the ten numbered applications that follow must have GFCI protection unless they are rated over 50 A single-phase, or over 100 A three-phase. The panel decided that the shock hazard was comparable on the larger circuits, and they were assured the technology was suitably evolved. All prior language about 15- and 20-A has been removed in favor of the new parameters. This is obviously a quantum leap from all prior code editions that only included 15- or 20-A, 125-V configurations. There is no delayed effective date on this provision, but in the case of three-phase receptacle circuits there don't seem to be products available. Under 90.4 (third paragraph) the inspector may (is not required to do so) permit equipment compliant with the prior edition of the NEC. As given in (B)(1), all receptacles installed in bathrooms of such occupancies must be GFCI protected. There is no requirement for the installation of receptacles in bathrooms of these occupancies, but if a receptacle is installed, this rule calls for GFCI protection of that receptacle.

Part **(B)(2)** requires GFCI protection for receptacles installed in "kitchens"—regardless of accessibility or equipment supplied. The definition has been moved to Art. 100, and includes the phrase "with a sink and permanent facilities for preparing and cooking," which excludes receptacles from the requirement for GFCI protection where installed in other areas of a commercial or institutional food service facility, such as a serving line or cafeteria area.

Part (**B**)(**3**) requires all rooftop receptacles to be GFCI protected, and 210.8(B)(4) mandates GFCI protection for similar receptacles installed outdoors, now also in all locations regardless of accessibility. The first exception to parts (3) and (4) eliminates the need for GFCI protection of receptacles installed to supply snow-melting or deicing equipment, provided the receptacles are "not readily accessible." The same allowance now extends to pipe tracing cables and similar installations covered by Art. 427. A second exception covering industrial occupancies with qualified maintenance and supervision correlates with 590.6(A) exception in waiving GFCI protection for equipment that would pose a greater hazard under a nonorderly shutdown, or for equipment designs that are inherently incompatible with GFCI protective devices. These instances are extremely unusual.

The exception following (3) addresses rooftop receptacles and makes the normal readily accessible requirement for GFCI receptacles applicable from the roof location, but not from grade or inside the building. It would be absurd and it is not necessary to construct a permanent stair to a rooftop location just so a GFCI receptacle could be accessed. Note that the rule under exception is in the parent wording of 210.8, and that is where this exception actually belongs.

The requirement [210.8(B)(5)] is to protect any receptacle within 1.8 m (6 ft) of a sink, similar to the rule in 210.8(A)(7). This rule applies to all sinks of any description, not just laundry, utility, and wet bar sinks, however, it comes with an exception for receptacles adjacent to sinks in industrial laboratories where the removal of power could create a greater hazard. An example would be a receptacle adjacent to a lab hood sink for which a showing can be made that power to a mixer or other process is essential to the orderly, perhaps even nonexplosive, completion of reactions carried out in those locations.

A second exception exempts GFCI protections for receptacles near sinks in the patient care areas of hospitals, although the GFCI receptacle requirements in hospital bathrooms continue in effect. This allowance recognizes that in some areas, particularly in critical care areas, there will often be sinks within 6 ft of the "minimum of 14 receptacles" required by 517.19(B)(1). These receptacles require the very highest standard of reliability, and for that reason must be connected to two different supply sources (normal and emergency) from different transfer switches. An outage here could literally kill a critically ill patient reliant on life-support equipment of some sort that is plugged into one of these receptacles.

Three new areas were added to the list effective with the 2011 **NEC**. Item (6) requires the protection for any indoor wet location and item (7) requires the protection for locker rooms if they have showering areas associated with them. Item (8) takes the long-standing requirement in 511.12 for GFCI protection of service area receptacles in commercial repair garages and extends it to all comparable receptacles in any nonresidential occupancy. The broadened requirement does not apply to vehicle exhibition halls and showrooms.

The 2017 **NEC** added two new applications that mirror customary residential applications. Now provisions in (B)(9) and (B)(10) match (A)(4) and (A)(5) in requiring protection for crawl spaces and for unfinished basements or unfinished areas of basements that do not constitute habitable room.

210.8(C) requires GFCI protection for all boat hoist "outlets" on all circuits operating not over 240 V, regardless of amperage, that are provided in "dwelling-unit locations." As a standalone provision, it covers hard-wired hoists as distinct from receptacle connections covered by **(A)**.

210.8(D), new as of the 2014 **NEC**, requires GFCI protection for any outlet supplying a dwelling unit dishwasher. The receptacle protection rules in (A) do not guarantee GFCI coverage because a hard-wired appliance would be otherwise uncovered. As in the case of laundry equipment, some appliance manufacturers raised concerns about end-of-life appliance failures. As in the case of appliances, rules addressing receptacles in crawl spaces do not protect luminaires. As of the 2017 **NEC**, crawl space lighting outlets now require GFCI protection [Part (E)] on the circuit level, and in all occupancies.

210.9. Circuits Derived from Autotransformers. The top of Fig. 210-17 shows how a 110-V system for lighting may be derived from a 220-V system by means of an autotransformer. The 220-V system either may be single phase or may be one leg of a 3-phase system. That hookup complies with the basic rule. In the case illustrated, the "supplied" system has a grounded wire solidly connected to a grounded wire of the "supplying" system: 220-V single-phase system with one conductor grounded.

Autotransformers are commonly used to supply reduced voltage for starting induction motors.

Exception No. 1 permits the use of an autotransformer without a grounded conductor connection for modest 208- to 240-V transformations or vice versa (see Fig. 210-17). Typical applications include cooking equipment, heaters, motors, and air-conditioning equipment. This has been allowed since the 1971 **NEC**.

Buck or boost transformers are designed for use on single- or 3-phase circuits to supply 12/24 or 16/32-V secondaries with a 120/240-V primary. When connected as autotransformers the kVA load they will handle is large in comparison with their physical size and relative cost.

Exception No. 2 permits 480- to 600-V or 600- to 480-V autotransformers without connection to grounded conductor—but only for industrial occupancies with qualified maintenance and supervision. The reason for this basic rule requiring continuity of a grounded circuit conductor has to do with predictability of voltage to ground. If the circuit in Fig. 210-17 is fed right to left (600 V ungrounded in, 480 V ungrounded out), and if the top conductor becomes grounded due to an insulation failure, the bottom conductor (common to both sides) will now be running 600 V to ground. This means that the 480-V derived system on the left will now run 480 V line-to-line, but 600 V to ground. The result is OK with appropriate supervision, and it has a very long track record of successful applications, but it must be taken into consideration at all times.

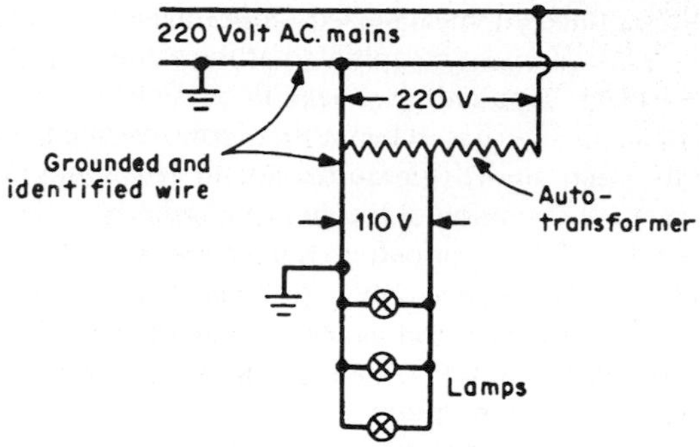
220 Volt A.C. mains
220 V
Grounded and identified wire
Auto-transformer
110 V
Lamps

Autotransformer used to derive a 2-wire 110-V system for lighting from a 220-V power system. (210.9.)

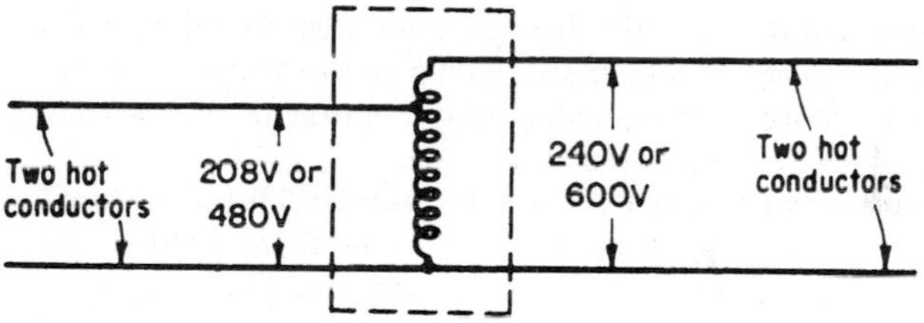
Two hot conductors
208V or 480V
240V or 600V
Two hot conductors

Fig. 210-17. Autotransformers with and without grounded conductors are recognized. (Sec. 210.9, Exceptions Nos. 1 and 2.)

210.10. Ungrounded Conductors Tapped from Grounded Systems. This section permits use of 2-wire branch circuits tapped from the outside conductors of systems, where the neutral is grounded on 3-wire dc or single phase, 4-wire, 3-phase, and 5-wire, 2-phase systems.

Figure 210-18 illustrates the use of unidentified 2-wire branch circuits to supply small motors, the circuits being tapped from the outside conductors of a 3-wire dc or single-phase system and a 4-wire, 3-phase wye system.

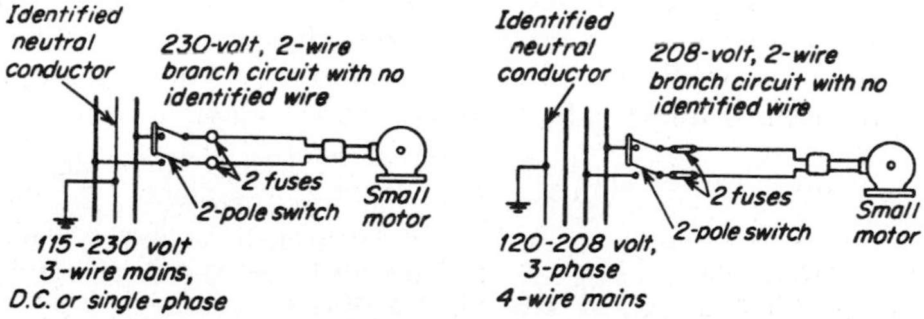
Identified neutral conductor
230-volt, 2-wire branch circuit with no identified wire
2 fuses
2-pole switch
Small motor
115-230 volt 3-wire mains, D.C. or single-phase
Identified neutral conductor
208-volt, 2-wire branch circuit with no identified wire
2 fuses
2-pole switch
Small motor
120-208 volt, 3-phase 4-wire mains

Fig. 210-18. Tapping circuits of ungrounded conductors from the hot legs of grounded systems. (Sec. 210.10.)

All poles of the disconnecting means used for branch circuits supplying permanently connected appliances must be operated at the same time. This requirement applies where the circuit is supplied by either CBs or switches.

In the case of fuses and switches, when a fuse blows in one pole, the other pole may not necessarily open, and the requirement to "manually switch together" involves only the manual operation of the switch. Similarly, when a pair of CBs is connected with handle ties, an overload on one of the conductors with the return circuit through the neutral may open only one of the CBs; but the manual operation of the pair when used as a disconnecting means will open both poles. The words "manually switch together" should be considered as "operating at the same time," that is, during the same operating interval, and apply to the equipment used as a disconnecting means and not as an overcurrent protective device.

CBs with handle ties are, therefore, considered as providing the disconnection required by this section. The requirement to "manually switch together" can be achieved by a "master handle" or "handle tie" since the operation is intended to be effected by manual operation. The intent was not to require a common trip for the switching device but to require that it have the ability to disconnect ungrounded conductors by one movement of the hand. For service disconnecting means, see Sec. 230-71.

210.11. Branch Circuits Required. After following the rules of 220.10 to ensure that adequate branch-circuit capacity is available for the various types of load that might be connected to such circuits, the rule in 210.11(A) requires that the minimum required number of branch circuits be determined from the total computed load, as covered in 220.10, and from the load rating of the branch circuits used.

For example, a 15-A, 120-V, 2-wire branch circuit has a load rating of 15 A times 120 V, or 1800 VA. If the load is resistive, like incandescent lighting or electric heaters, that capacity is 1800 W. If the total load of lighting that was computed from 220.12 were, say, 3600 VA, then exactly two 15-A, 120-V, 2-wire branch circuits would be adequate to handle the load, provided that the load on the circuit is not a "continuous" load (one that operates steadily for 3 h or more). Because 210.19(A) requires that branch circuits supplying a continuous load be loaded to not more than 80 percent of the branch-circuit rating, if the above load of 3600 VA was a continuous load, it could *not* be supplied by *two* 15-A, 120-V circuits loaded to full capacity. A continuous load of 3600 VA could be fed by *three* 15-A, 120-V circuits—divided among the three circuits in such a way that no circuit has a load of over 15 A times 120 V times 80 percent, or 2880 VA. If 20-A, 120-V circuits are used, because each such circuit has a continuous load rating of 20 times 120 times 80 percent, or 1920 VA, the total load of 3450 VA can be divided between two 20-A, 120-V circuits. The examples here use 120 V and not 115 or 110 V because 120 V is the standard voltage required to be used for load calculations in 220.5(A).

example Given the required unit load of 3 VA/sq ft for dwelling units (Table 220.12), the Code-minimum number of 20-A, 120-V branch circuits required to supply general lighting and general-purpose receptacles (not small appliance receptacles in kitchen, dining room, etc.) in a 2200-sq-ft one-family house is three circuits. Each such 20-A circuit has a capacity of 2400 VA. The required total circuit capacity is 2200 times 3 VA/sq ft, or 6600 VA. The next step is to divide 6600 by 2400, which equals 2.75. Thus, at least three such circuits would be needed.

example In 220.12, the NEC requires a minimum unit load of 3 VA/sq ft for general lighting in a school, as shown in Table 220.12. For a small school of 1500 ft^2, *minimum capacity for general lighting* would be

$$1500 \text{ ft}^2 \times 3 \text{ VA/ft}^2 \quad \text{or} \quad 4500 \text{ VA}$$

By using 120-V circuits, when the total load capacity of branch circuits for general lighting is known, it is a simple matter to determine how many lighting circuits are needed. By dividing the total load by 120 V, the total current capacity of circuits is determined:

$$\frac{4500 \text{ VA}}{120 \text{ V}} = 37.5 \text{ A}$$

But, because the circuits will be supplying continuous lighting loads (over 3 h), it is necessary to multiply that value by 1.25 in order to keep the load on any circuit to not more than 80 percent of the circuit rating. 37.5 × 1.25 = 46.9. Then, using either 15- or 20-A, 2-wire, 115-V circuits gives

$$\frac{1.25 \times 37.5 \text{ A}}{15 \text{ A}} = 3.1$$

which means four 15-A circuits, or

$$\frac{1.25 \times 37.5 \text{ A}}{20 \text{ A}} = 2.3$$

which means three 20-A circuits. And then each circuit must be loaded without exceeding the 80 percent maximum on any circuit.

Part **(B)** of 210.11 makes clear that a feeder to a branch-circuit panelboard and the main busbars in the panelboard must have a minimum ampacity to serve the *calculated* total load of lighting, appliances, motors, and other loads supplied. And the amount of feeder and panel ampacity required for the general lighting load must not be less than the amp value determined from the circuit voltage and the total volt-amperes resulting from the minimum unit load from Table 220.12 (volt-amperes per square foot) times the area of the occupancy supplied by the feeder—even if the actual connected load is less than the calculated load determined on the volt-amperes-per-square-foot basis. (Of course, if the connected load is greater than that calculated on the volt-amperes-per-square-foot basis, the greater value of load must be used in determining the number of branch circuits, the panelboard capacity, and the feeder capacity.) Then, because this is actually a feeder calculation, the lighting loads determined by Table 220.12 then can be made subject to the demand factors in Table 220.42 as applicable for the specific occupancy.

It should be carefully noted that the first sentence of 210.11(B) states, "Where the load is computed on a volt-amperes-per-square-foot basis, the *wiring system* up to and including the branch-circuit panelboard(s) shall be provided to serve not less than the calculated load." Use of the phrase "wiring system up to and including" requires that a feeder must have capacity for the total minimum branch-circuit load determined from square-foot area times the minimum unit load (volt-amperes per square foot from Table 220.12). And the phrase clearly requires that amount of capacity to be allowed in every part of the distribution system supplying the load.

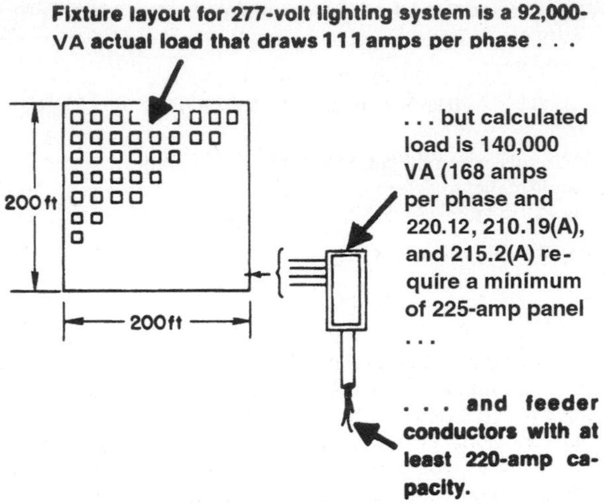
Fixture layout for 277-volt lighting system is a 92,000-VA actual load that draws 111 amps per phase . . .
200 ft
200 ft
. . . but calculated load is 140,000 VA (168 amps per phase and 220.12, 210.19(A), and 215.2(A) require a minimum of 225-amp panel . . .
. . . and feeder conductors with at least 220-amp capacity.

Fig. 210-19. Capacity must be provided in service and feeder conductors, as well as a branch-circuit panelboard that is adequate for the calculated load.

The required capacity would, for instance, be required in a subfeeder to the panel, in the main feeder from which the subfeeder is tapped, and in the service conductors supplying the whole system (Fig. 210-19).

Actually, reference to "wiring system" in the wording of 210.11(B) presents a requirement that goes beyond the heading, "Branch Circuits Required," of 210.11 and, in fact, constitutes a requirement on *feeder* capacity that supplements the rule of the second sentence of 215.2(A). This requires a feeder to be sized to have enough capacity for the load—as determined by part **(A)** of this article (which means, as computed in accordance with 220.10). However, as previously noted, the required feeder capacity can be reduced to the extent the NEC permits for the feeder in question.

The second part of 210.11(B) affects the required minimum number of branch circuits. Although the feeder and panelboard must have a minimum capacity to serve the *calculated* load, it is only necessary to install the number of branch-circuit overcurrent devices and circuits required to handle the actual connected load in those cases where it is less than the calculated load. The last sentence of 210.11(B) is clearly an exception to the basic rule of the first sentence of 210.11(A), which says that "The minimum number of branch circuits *shall* be determined from the *total computed* load. …" Instead of having to supply *that minimum* number of branch circuits, it is necessary to have only the number of branch circuits required for the actual total "connected load." However, the branch-circuit panelboard would also need to have sufficient space to install the numbers of circuits calculated, because that panel is part of the wiring system at the feeder level.

example For an office area of 200 × 200 ft, a 3-phase, 4-wire, 480/277-V feeder and branch-circuit panelboard must be selected to supply 277-V HID lighting that will operate continuously (3 h or more). The actual continuous connected load of all the luminaires is 92 kVA.

What is the minimum size of feeder conductors and panelboard rating that must be used to satisfy Sec. 210.11(B)?

$$200 \text{ ft} \times 200 \text{ ft} = 40{,}000 \text{ sq ft}$$
$$40{,}000 \text{ sq ft @ minimum of } 3.5 \text{ VA/sq ft} = 140{,}000 \text{ VA}$$

The minimum computed load for the feeder for the lighting is

$$140{,}000 \text{ VA} \div [(480 \text{ V})(1.732)] = 168 \text{ A per phase}$$

The actual connected lighting load for the area, calculated from the lighting design, is

$$92{,}000 \text{ VA} \div [(480 \text{ V})(1.732)] = 111 \text{ A per phase}$$

Sizing of the feeder and panelboard must be based on 168 A, *not* 111 A, to satisfy 210.11(B). The next step is to correlate the rules of Sec. 210.11(A) and (B) with those of 215.2(A)(1). The rule of 215.2(A)(1) requires a feeder to be sized for the "computed load" as determined by parts II, III, and IV in Art. 220. Because the entire lighting load is assumed to operate continuously in this type of occupancy, the feeder to supply the continuous calculated load of 168 A must have an ampacity at least equal to that load times 1.25. This is not for the sake of the wire, whose ampacity is by definition a *continuous* current-carrying capacity expressed in amperes. This is for the sake of the internal calibration of a conventional CB, which requires the heat sink effect of a cooler wire bolted to it. Therefore, 215.2(A)(1) assures that in any load calculation under conditions of continuous loading, a phantom capacity will be built into the feeder size. Further, 110.14(C)(1)(b) requires that the terminations on the circuit breaker be made based on wire sizes evaluated under the 75°C column of Table 310.15(B)(16). This is true whenever using a CB or fused switch that is not UL-listed for continuous operation at 100 percent of rating, as required in 215.3. Finally, since this 3-phase, 4-wire feeder will be feeding predominantly electric-discharge lighting with a strong triplen harmonic content, 310.15(B)(5)(c) will require that the neutral be counted as current carrying, and with four wires carrying current in the same raceway, 310.15(B)(3)(a) will then impose an 80 percent derating factor on the feeder conductors for mutual conductor heating.

$$168 \text{ A} \times 1.25 = 210 \text{ A} \quad [215.2(A)(1)]$$

1. Assuming use of a non-100 percent rated protective device, the overcurrent device must be rated not less than 1.25 × 168 A, or 210 A—which calls for a standard 225-A circuit breaker or fuses (the standard rating above 210 A).
2. Although feeder conductors with an ampacity of 210 A would satisfy the rule of 215.2(A) and be adequate for the load, they might not be properly protected (240.4) by a 225-A device after derating. The feeder must have a 75°C table ampacity that is not less than 210 A (168 A × 1.25) before derating *but* must also be properly protected by the 225-A rated device *after* derating. Additional calculations are required to make a final determination.
3. Using Table 310.15(B)(16), the smallest size of feeder conductor that would be protected by 225-A protection after 80 percent derating for a number of conductors is a No. 4/0 THHN or XHHW copper, with a 230 A value in the 75°C column and a 90°C ampacity of 260 A before derating (260 A × 0.8 = 208 A). Remember: To satisfy 215.2(A)(1), the 75°C column is used. And, where 90°C-insulated conductors are used, any "deratings" needed may be applied against the ampacity value shown in the 90°C column in Table 310.15(B)(16).
4. Because the UL and 110.14(C)(1)(b) requires that conductors larger than No. 1 AWG must be used at no more than their 75°C ampacities to limit heat rise in equipment terminals, the selected No. 4/0 THHN or XHHW copper conductor must not operate at more than 230 A—which is the table value of ampacity for a 75°C No. 4/0 copper

conductor. And the load current of 168 A is well within that 230 A maximum. Further, the 225-A CB will protect the 4/0 feeder conductors under the conditions of use, because the final ampacity of these conductors is 208 A and a 225-A overcurrent device is the next higher standard size, allowable in these size ranges by 240.4(B).

Thus, all requirements of 215.2(A) and UL are satisfied.

5. Note that the minimum feeder size came out 210 A, and the ampacity of the feeder conductors chosen came out 208 A in item 3 above. To some, this may look like the feeder size needs to be further increased, but not so. Comparing these two numbers is comparing apples and oranges. The required feeder size to overcome mutual conductor heating in distant parts of the feeder raceway is one calculation, involving the middle of the wire. And it turns out that a 4/0 feeder will carry this load safely because under the conditions of use its ampacity (continuous load capacity) is 208 A, although actually loaded to 168 A. Further, the same size wire will fulfill its heat sink responsibilities at the CB terminals because, using the 75°C column of Table 310.15(B)(16), the ampacity of this wire is 230 A, and all it needs to be, inclusive of the required 25 percent phantom loading, is 210 A. These two calculations are entirely independent and should be done on separate pieces of paper.
6. The calculation of ampacities for conductors carrying continuous loads under nonstandard conditions of use and correlating those calculations with other rules on allowable terminating sizing involves some of the most complex analysis of any Code application. Any such analysis involves integrating key general rules in the first three chapters of the **NEC** together with specific application provisions in the remainder of the **NEC**. For this reason, and because the end of the Code, in Annex D Example 3(a), contains a fully developed example of one of these calculations fully worked out, line-by-line and with all applicable **NEC** provisions specifically cited, a full explanation of this calculation process will be reserved for the end of the book. Refer to Chap. 9 for this information.

The rule in 408.30 requires the panelboard here to have a rating not less than the loads as calculated in Art. 220—which, in this case, means the panel must have a busbar rating not less than 168 A. Since the bus assembly must, in this case, be assumed to be distributing this load on a continuous basis, the 125 percent rule in 215.2(A)(1) will apply here as well, resulting in a minimum bus size of 210 A. A 225-A panelboard (i.e., the next standard rating of panelboard above the minimum calculated value of load current—210 A) is therefore required, even though it might seem that a 125-A panel would be adequate for the actual load current of 111 A.

The number of branch-circuit protective devices required in the panel (the number of branch circuits) is based on the size of branch circuits used and their capacity related to connected load. If, say, all circuits are to be 20-A, 277-V phase-to-neutral, each pole may be loaded not more than 16 A because 210.19(A)(1) requires the load to be limited to 80 percent of the 20-A protection rating. With the 111 A of connected load per phase, a single-circuit load of 16 A calls for a minimum of 111 ÷ 16, or 8 poles per phase leg after rounding up. Thus, a 225-A panelboard with 24 breaker poles, assuming perfectly balanced loading across all of the connected branch circuits.

However, the minimum number of positions available in the panel must accommodate the calculated load of 168 A per phase. The result would probably be a 42-circuit panel (the usual next larger panel size above the calculated result of 11 per phase or 33 circuits), but with some number between 33 and 42 left as spare or vacant positions. Whether those positions could be reassigned to other loads depends on how the inspector interprets 210.11(B). The prudent option is to combine a modest increase in the number of connected circuits so each is not fully loaded, and then leave the balance of the calculated vacant positions for future developments.

Part **(C)(1)** of 210.11 requires that two or more 20-A branch circuits be provided to supply all the receptacle outlets required by 210.52(B) in the kitchen, pantry, dining room, breakfast room, and any similar area of any dwelling unit—one-family houses, apartments, and motel and hotel suites with cooking facilities or serving pantries. That means that at least one 3-wire, 20-A, 240/120- or 208/120-V circuit shall be provided to serve only receptacles for the small-appliance load in the kitchen, pantry, dining room, and breakfast room of any dwelling unit. Of course, two 2-wire, 20-A, 120-V circuits are equivalent to the 3-wire circuit and could be used. If a 3-wire, 240/120-V circuit is used to provide the required two-circuit capacity for small appliances, the 3-wire circuit can be split-wired to receptacle outlets in these areas, provided a common disconnecting means is installed to meet 210.4(B) and 210.7.

Part **(C)(2)** of 210.11 requires that at least one 120-V 20-A branch circuit be provided for the one or more laundry receptacles installed, as required by 210.52(F), at the laundry location in a dwelling unit. Further, the last sentence of part **(C)(2)**, in conjunction with 210.52(F), prohibits use of the laundry circuit for supplying outlets that are not for laundry equipment. Receptacle outlets for the laundry must be located at any anticipated laundry equipment locations because 210.50(C) requires them to be within 6 ft (1.8 m) of the intended appliance location (Fig. 210-20).

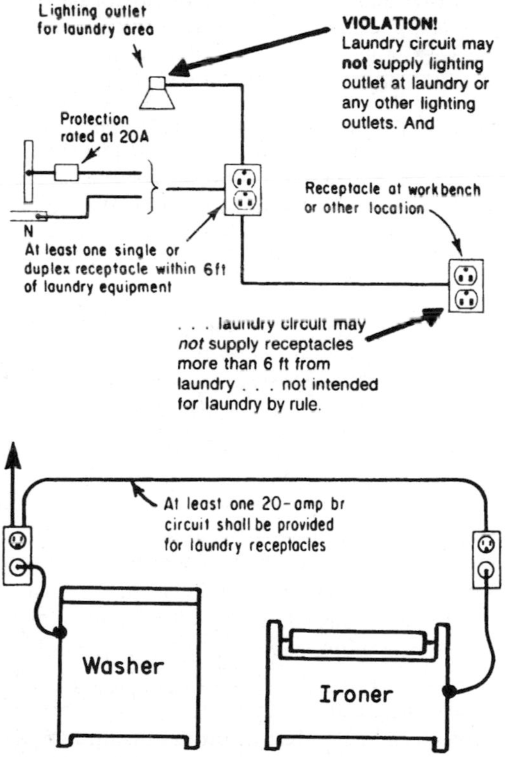
Lighting outlet for laundry area
VIOLATION! Laundry circuit may not supply lighting outlet at laundry or any other lighting outlets. And
Protection rated at 20A
N
Receptacle at workbench or other location
At least one single or duplex receptacle within 6ft of laundry equipment
. . . laundry circuit may not supply receptacles more than 6 ft from laundry . . . not intended for laundry by rule.
At least one 20-amp br circuit shall be provided for laundry receptacles
Washer
Ironer

Fig. 210-20. *No* "other outlets" are permitted on 20-A circuit required for laundry receptacles. [Sec. 210.11(C).]

Part **(C)(3)** of 210.11 requires a dedicated branch circuit to supply receptacle outlets within a dwelling unit's bathrooms. This must be a 120-V 20-A circuit, and it may supply receptacles in bathrooms only! The wording recognizes supplying more than one bathroom from a single 20-A circuit. And the exception allows limited installation of "other" outlets on this circuit where the circuit supplies only a single bathroom. So the basic rule puts all bathroom receptacles on a single circuit, or more, provided such circuits serve only bathroom receptacles. Then, as a trade-off, a 20-A circuit can supply all the loads in a bathroom, but as soon as it serves some load other than a receptacle, it must serve only a single bathroom. This exception allows for a simple and practical method of complying with listing instructions that often apply to exhaust fans over a tub, where they will generally require GFCI protection. Use the exception, and make sure that the fan is on the load side of the receptacle. Of course at this point all the wiring in a bath wired this way would need to be 12 AWG to match the 20-A circuit configuration.

As printed through 2016, the 2014 **NEC** included the indefinite article "a" ahead of "bathroom receptacle outlet(s)." This literally meant that once you supplied any bathroom receptacle, no matter how unlikely to be used, you had met the minimum requirement. This has been corrected in the 2017 **NEC**, which says "supply the bathroom receptacle outlet(s)" [emphasis supplied]. There was no documentation to support the printed version, and it was recently formally declared an erratum by NFPA. That resolves the issue in terms of 2014 **NEC** enforcement.

New as of the 2017 **NEC** is a requirement (**D**) [partially relocated from 210.52(G)] to add a 20-A, 120-V dedicated circuit for dwelling unit garage receptacle outlets. Only one circuit is required regardless of the number of vehicle spaces under cover, and whether or not multiple garages are present. The one or more receptacles in a detached garage are also included, if the detached garage is powered. As in previous Codes, this is not a requirement to provide power to detached garages, only how to circuit them if they are powered. There is an exception to the dedicated circuit rule; "readily accessible outdoor receptacle outlets" may also be connected to this circuit.

210.12. Arc-Fault Circuit Interrupter Protection. The arc-fault circuit interrupter (AFCI) devices are similar to, but different from, the more commonly recognized GFCIs. But while the GFCI operates on the basis that any current difference between the hot and the neutral (or the hot and the hot for 250-V single-phase devices) greater than 5 mA is "unauthorized" current flow to ground, causing it to open the circuit under such conditions, the AFCI operates to open the circuit either on a low-level current imbalance exceeding about 30 mA or when it senses a specific waveform anomaly that is indicative of an arcing fault.

Advances in electronics have made it possible for the internal chip to recognize the specific waveform characteristics of an arcing-type fault and to operate a mechanical ratchet to open the circuit, thereby providing a greater level of protection against the potential for shock, electrocution, and property damage that these typically high-impedance, low-current malfunctions can present.

There are two broad classifications of arcing failures that can be addressed by AFCI technology, namely, a line-to-line or line-to-ground failure that can occur in parallel with a connected load or even with no energized load in operation, or a failure between two severed ends of the same conductor, or at a poor connection point for such a conductor, either one in series with a connected load. The first AFCI devices in wide usage, configured as an additional tripping provision in certain CBs (and designated as "Branch/Feeder AFCIs" by UL), addressed the more common parallel events only, and the **NEC** permitted the use of this more limited protection until January 1, 2008. Meanwhile, the only design prior to that time that addressed the series failure was configured as part of a duplex receptacle akin to a GFCI receptacle. After that time, the default AFCI protective device had to be a "combination type" device that could respond to both parallel and series failures. These are designated by UL as an "Outlet/Branch Circuit AFCI," and if located as the first outlet on a branch circuit, provide series protection for the entire branch circuit and parallel protection for all downstream portions of the circuit. The history and terminology presented in this paragraph and those following is essential to fully understand the expanded content as of the 2014 **NEC**, which includes no fewer than six protective options.

AFCI protective schemes must combine the best protective features of both parallel and series protective devices, preferably in the form of what the **NEC** refers to as a "combination-type" device, meaning one that responds to both parallel and series arcing failures. *This designation has nothing to do with a device that provides both AFCI and GFCI protection*, although the technology is mutually compatible. All AFCI protective devices, whether in the form of a receptacle or a CB, must be mounted in a readily accessible location so the test and reset functions can be easily performed.

Much of the activity over the prior five code cycles in this section has involved successive attempts to integrate the different forms of AFCI devices into a practical framework that involved as broad a spectrum of manufacturing involvement as possible. There are untold millions of dwellings with overcurrent systems that will not accept modern CB designs, and for which the owners will likely resist the expense of a new service.

The "outlet/branch-circuit" devices are just now becoming routinely available on the market. The reason for this has been the severely restrictive nature of the **NEC** conditions for which such a device had been permitted to qualify as the required AFCI protection for a branch circuit. Under the 2005 **NEC** the device had to be located not over 6 ft from the point of branch circuit origination with the distance to be measured along the conductors, and metallic wiring methods employed between the two locations. Under the 2008 **NEC**, the distance could be of any length, but the wiring methods had to be of steel, either as a cable assembly (e.g., steel Type AC, but not the usual Type MC with aluminum armoring) or as one of three specified steel tubular raceways (IMC, RMC, or EMT) but not, technically, wireways or other wiring methods even if made of steel. The 2011 **NEC** added conventional Type MC cable (whether steel or aluminum) to the list, but Type AC cable must be steel only. The panel substantiated this disparity because Type AC cable lacks a crush resistance test.

The 2011 **NEC** also added a second exception recognizing nonmetallic conduit encased in concrete as a permitted wiring method to the first outlet. This is feasible in high-rise construction where poured concrete is often used to embed wiring systems.

AFCIs are now required to protect all circuits that supply outlets (receptacle and lighting) in dwelling-unit kitchens (as of the 2014 **NEC**), family rooms, dining rooms, living rooms, parlors, libraries, dens, bedrooms, sunrooms, recreation rooms, closets, hallways, laundry areas (as of the 2014 **NEC**), or similar rooms or areas. Most receptacle outlets in bathrooms, basements, garages, and outdoors require GFCI protection. The 2014 **NEC** also expanded this requirement to include devices, which results in AFCI coverage to an indoor snap switch controlling an outdoor light, which would not have otherwise been covered. Due to a lack of consensus in the voting at the comment stage for the 2017 **NEC**, no changes were made in 210.12(A) because the relevant text under the rule reverted to previous edition text. The result is that virtually all 120-V wiring on 15-A and 20-A branch circuits in a dwelling unit must now have some form of residual current detection, and in most areas a failure in the branch-circuit wiring itself will also be detected and opened. Although it would be theoretically possible to omit protection of a lighting circuit that serves only a bathroom or a garage, as a practical matter all general-purpose lighting and receptacle outlets throughout the entire dwelling must be protected on installations governed as of the 2014 **NEC** and thereafter.

An exception allows wiring to a fire alarm system to omit AFCI protection entirely on an individual branch circuit run in any of the common steel wiring methods, although steel wireways were omitted due to a panel mistake in the 2014 cycle. An effort to correct this in the 2017 edition failed when, as previously noted, the entire text reverted to 2014 **NEC** language. Remember that this reference is to a full red-box fire alarm control panel governed by 760.41(B) or 760.121(B), and not the usual 120 V reciprocally alarming residential smoke detector installation.

There are still problems in the overall picture, but unmistakable progress as well. Not all manufacturers are currently making two-pole AFCI CBs, although most now are. This is important because, just as in the case of GFCI protection, a two-pole device is required to be used with a multiwire branch circuit. Some, but not all manufacturers make AFCIs in bolt-on configurations frequently specified for commercial and multifamily residential applications. Of course, the bolt-on configuration can be used anywhere but it is more prevalent in the larger quasi-commercial applications. Until the supply chain becomes completely up to speed on current applications, this will be an issue, particularly on retrofits. Note that at least one manufacturer of two-pole AFCIs also makes classified AFCI CBs that are rated for a number of competitors' panels, so that could be another way out for now. Be careful because such classification ratings are limited to some extent, particularly for applications involving available fault currents over 10 kA.

Perhaps the most encouraging development has been in the area of receptacle style AFCIs. The steady expansion of mandated AFCI applications, together with the January 1, 2014, passing of the effective date for AFCI protection on replaced receptacles in 406.4(D)(4), and the advent of additional options for installing outlet branch-circuit AFCI devices (see below) has combined to result in the first wide-scale market entry of these protective devices.

Sec. 210.12(A) has been reformatted into a list of six acceptable AFCI installation procedures, three of which break new ground.

(1) A listed combination-type device ahead of the entire circuit (Fig. 210.21). This has been the usual method and has been acceptable as long as these devices have been on the market. If GFCI protection is also required, add a GFCI receptacle or

TEST
AFCI

Fig. 210-21. A combination AFCI CB meets all requirements [210.12(A)(1)].

receptacles as needed. At least one manufacturer offered a "dual function" device (this is the term of art, because this is fundamentally different from a "combination AFCI" that protects against both parallel and series arcing events), but the AFCI portion was a branch/feeder device that is no longer recognized as providing full protection. Now that kitchens are included in the required coverage there will probably be renewed manufacturing interest in true dual function devices.

(2) A listed branch/feeder AFCI device (Fig. 210.22) ahead of the entire circuit in concert with a listed outlet branch-circuit device (Fig. 210.23) at the first

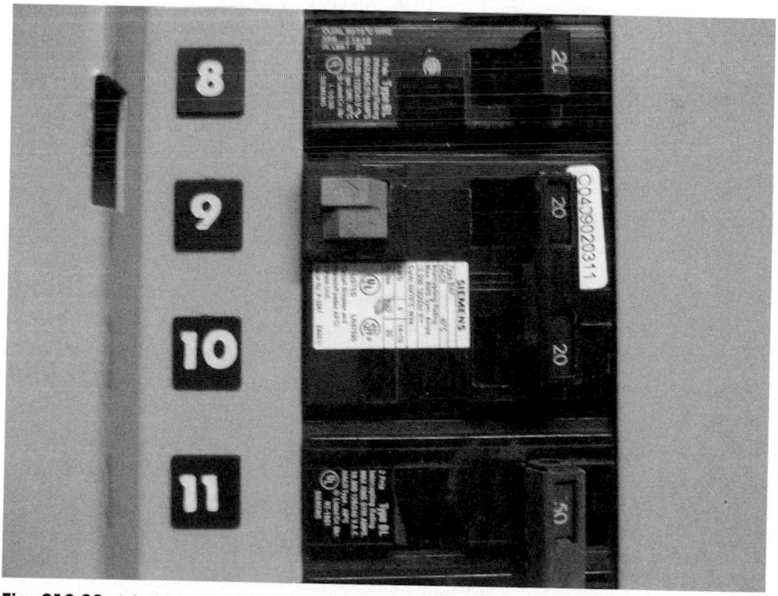
8
9
10
11
20
20
20
50
C0409020311
SIEMENS

Fig. 210-22. A bolt-on two-pole branch/feeder AFCI installed in 2003 [210.12(A)(2)].

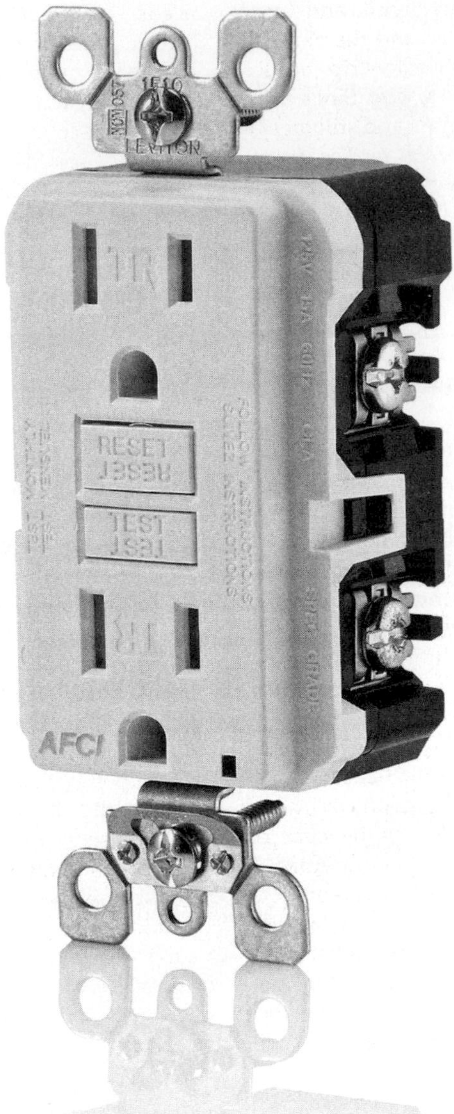
RESET
TEST
AFCI

Fig. 210-23. An outlet branch-circuit receptacle AFCI [210.12(A)(2-6)]. (Courtesy, Leviton) "outlet box" on the circuit, and this box needs to be marked as such. If GFCI protection is required and the panel manufacturer offers a dual function breaker use that, or place a master-trip (faceless) GFCI ahead of any receptacles (or other loads) requiring protection. Since a two-gang box is a single outlet, this combination could be so located and still meet all requirements. *Note that there are no wiring method limitations on this method,* in contrast with (5) and (6). This was new as of the 2014 NEC, and there are some issues with the literal text. Since any box under 100 cu. in. and not equipped with device yoke support drillings is classified by product standards as an outlet box, that puts the downstream AFCI protection at the first outlet box after the panel, perhaps a 4-in. sq. box on a basement ceiling, and in addition the text does not accommodate multiple downstream AFCI devices protecting multiple branches of a circuit. These questions will need to be addressed with the inspector, because the likely intent was to address the first box that is at the location of an outlet, and there should be no objection if the circuit splits as long as all outlets are protected. Note that the box in the wall and not the faceplate outside the wall gets the first outlet marking. The concept is for the inspector to be able to see the wiring arrangement during the rough inspection.

(3) A listed supplemental arc protection CB ahead of the entire circuit with an outlet branch-circuit protective device as the first outlet. This also was new as of the 2014 NEC, and as of this writing this type of CB is still not available on the market. The rules, and the problems with the wording of those rules, are the same as in (2). Additional limitations include a requirement that the branch-circuit wiring from breaker to first device be "continuous," which is not further defined but presumably means without joint or splice. Finally, the wiring cannot extend over 50 ft for 14 AWG or 70 ft for 12 AWG wiring.

(4) A listed outlet branch-circuit AFCI in combination with a listed branch-circuit protective device. This option was also new as of the 2014 NEC and follows the same requirements (and also has the same editorial problems) as in (3), with one additional and very important requirement. The exact combination must be "identified" as meeting applicable requirements for a "system combination-type AFCI" and be listed as such. This is a misuse of the word "identified" since any listed combination would automatically be generally recognizable as suitable for the purpose. Presumably, the intent was to require a marking by the manufacturer. This option is particularly intriguing because it opens the door to the possibility of fuses qualifying as the overcurrent device. The current UL outline of investigation allows for reductions in the permitted circuit length to the first device, and also calls for reciprocal marking on both devices warning against replacing a device at either end without replacing both at the same time.

The reciprocal marking requirement is the issue that caused further progress on AFCIs in this cycle to founder. The CB manufacturers decided that they would not mark their usual breakers as being suitable for the front end of a system combination AFCI regardless of markings that might occur on particular outlet branch-circuit devices. The AFCI receptacle manufacturers accused the breaker manufacturers of, in effect rent seeking, using the regulatory process to maintain their market share. They requested the panel to remove the limitation of a reciprocal marking on the breaker, and the panel supported that position, but in the end not by the required two-thirds majority. This contentious issue will be an emotional, major focus of the panel's attention in the 2020 NEC cycle.

(5) A listed outlet branch-circuit AFCI placed at the first outlet of a circuit wired using steel wiring methods or MC cable of any composition. This method is carried over from the 2011 NEC.

(6) A listed outlet branch-circuit AFCI placed at the first outlet of circuit wired using MC cable or either metallic or nonmetallic raceways embedded in concrete (at least 2 in. under the surface) for the portion from the panel to the first outlet. This method is also carried over from the 2011 NEC.

Part **(B)** is new as of the 2014 NEC and applies the AFCI requirements for dwelling units to student rooms in dormitories. The 2017 NEC added "and devices," which aligns this with 210.12(A) for conventional dwelling units. Note that the paragraph title is "Dormitory Units," which suggests that the rules apply with designated units and not to collective lounge spaces and hallways, but this is far from clear and subject to interpretation.

Part **(C)** brings the guest rooms and suites of hotels into the same set of AFCI protective options that govern dwelling units. This requirement is a new addition to the 2017 NEC.

Part **(D)**, originally from the 2011 code edition, requires any wiring in areas subject to 210.12 that is modified, extended, or replaced, be brought into compliance with AFCI provisions. This can be accomplished in the usual way of providing a conventional branch/feeder AFCI in the panel. It can also be accomplished, with an outlet branch-circuit AFCI, located at the first receptacle outlet of the existing branch circuit. The 2017 NEC extended these provisions to dormitory units.

Note that the wiring method limitations for new installations do not apply in these cases. The first receptacle outlet could have been wired in Type NM cable and the outlet/branch-circuit AFCI could still be installed. This language addresses the reality that many older dwelling units are supplied through obsolete panels that cannot accommodate modern branch/feeder AFCIs. However, the presence of an obsolete panel is not required to invoke this wiring extension/modification permission. The 2014 NEC did grant relief from the AFCI upgrade requirement for incidental circuit extensions not over 6 ft in length and which do not involve any additional devices or outlets. The exception allows for such common activities as the collection of existing branch circuit wiring into a pull box that then connects to a nearby panel as part of a service upgrade.

210.13. Ground-Fault Protection of Equipment. This rule, new in the 2014 NEC, creates a GFPE requirement for large branch circuits that parallels the one in 215.10 for feeders. Although harmless, this rule will have little practical application. Most cases cited for it, under close examination, turn out to be feeders and not branch circuits.

210.17. Guest Rooms and Guest Suites. A guest room in a hotel or motel, if it contains permanent provisions for cooking, must meet all the rules for outlet circuiting and receptacle placements that a dwelling unit must meet. A plug-in microwave oven wouldn't, by itself, trigger this classification, but a permanent cooktop certainly would. If so, the guest room or suite would be subject to AFCI coverage, no fewer than two small-appliance branch circuits, etc.

210.18. Rating. A branch circuit is rated according to the rating of the overcurrent device used to protect the circuit. A branch circuit with more than one outlet must normally be rated at 15, 20, 30, 40, or 50 A (see Fig. 210-24). That is, the protective device must generally have one of those ratings for multioutlet circuits, and the conductors must meet the other size requirements of Art. 210.

BASIC RULE —

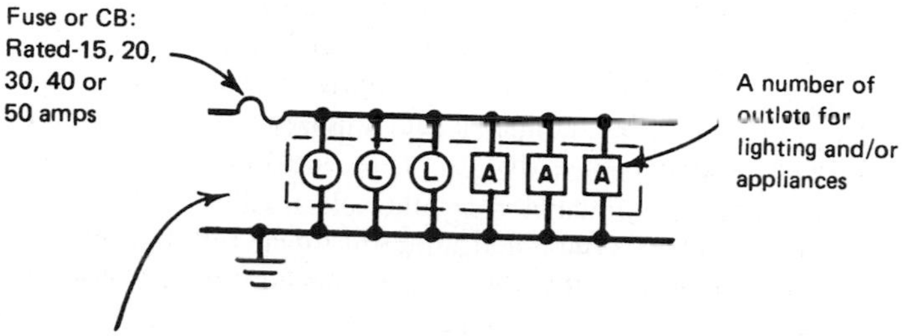
Fuse or CB:
Rated-15, 20,
30, 40 or
50 amps
A number of
outlets for
lighting and/or
appliances
L
A

Circuit voltage shall not exceed 150 volts to ground for circuits supplying lampholders, fixtures or receptacles of standard 15-amp rating. For incandescent or electric-discharge lighting under certain conditions, voltage to ground may be as high as 300 volts. In certain cases, voltage for electric discharge lighting may be up to 600 volts "between conductors" and may be an ungrounded circuit.

EXCEPTION —

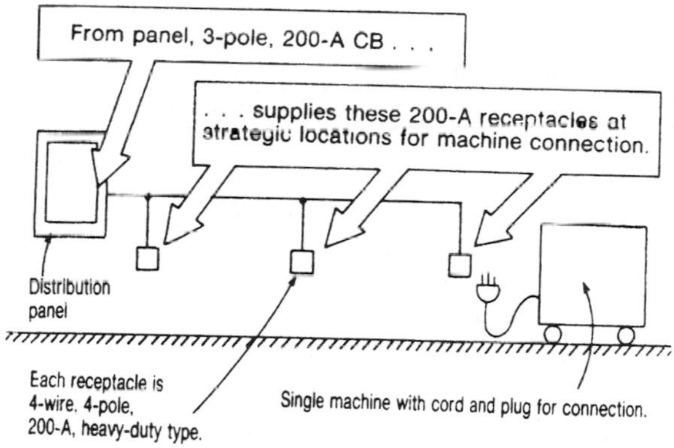
From panel, 3-pole, 200-A CB . . .
. . . supplies these 200-A receptacles at strategic locations for machine connection.
Distribution panel
Each receptacle is 4-wire, 4-pole, 200-A, heavy-duty type.
Single machine with cord and plug for connection.

NOTE: Typical receptacles supplied by such layout could be rated 60 A, 100 A, 200 A or 400 A—with their supply circuit of the same rating. Such hookups are common in jet airplane hangars for supplying cord-connected equipment used for servicing individual planes in their hangar bays.

Fig. 210-24. A multioutlet branch circuit must usually have a rating (of its overcurrent protective device) at one of the five values set by 210.18. (Sec. 210.18.)

Under the definition for "receptacle" in NEC Art. 100, it clearly provides that a duplex receptacle is two receptacles and not one—even though there is only one box and therefore one outlet. However, a circuit that supplies only one duplex receptacle is still usually not an "individual branch circuit" because it normally will be likely to supply more than one utilization equipment through its separate receptacles, and therefore flunk the definition of "individual branch circuit" in Art. 100. If an individual branch circuit is required for any reason, and the purpose is to supply cord-and-plug connected utilization equipment, a single receptacle must be installed. One example is the individual branch circuit required in 422.16(B)(4)(5) for a cord-and-plug connected range hood.

The exception to the rule of 210.18 gives limited permission to use multioutlet branch circuits rated over 50 A—but only to supply nonlighting loads and only in industrial places where maintenance and supervision ensure that only qualified persons will service the installation. This exception recognizes a real need in industrial plants where a machine or other electrically operated equipment is going to be provided with its own dedicated branch circuit of adequate capacity—in effect, an individual branch circuit—but where such machine or equipment is required to be moved around and used at more than one location, requiring multiple points of outlet from the individual branch circuit to provide for connection of the machine or equipment at any one of its intended locations (see Fig. 210-25). For instance, there could be a 200-A branch circuit to a special receptacle outlet or a 300-A branch circuit to a single machine. In fact, the wording used here actually recognizes the use of such a circuit to supply more than one machine at a time, but other realities of application make such an approach impractical.

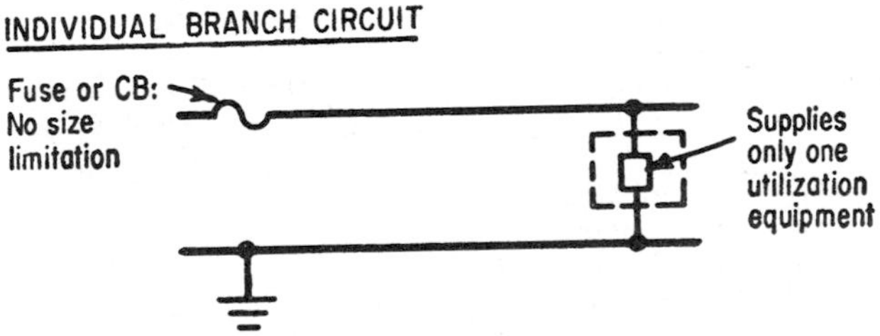
INDIVIDUAL BRANCH CIRCUIT
Fuse or CB: No size limitation
Supplies only one utilization equipment

Fig. 210-25. A circuit to a single load device or equipment may have any rating. (Sec. 210.18.)

It is important to note that it is the size of the overcurrent device that actually determines the rating of any circuit covered by Art. 210, even when the conductors used for the branch circuit have an ampere rating higher than that of the protective device. In a typical case, for example, a 20-A CB in a panelboard might be used to protect a branch circuit in which 10 AWG conductors are used as the circuit wires. Although the load on the circuit does not exceed 20 A, and 12 AWG conductors would have sufficient current-carrying capacity to be used in the circuit, the 10 AWG conductors with their rating of 30 A were selected to reduce the voltage drop in a long homerun. The rating of the circuit is 20 A, because that is the size of the overcurrent device. The current rating of the wire does not enter into the ampere classification of the circuit.

210.19. Conductors—Minimum Ampacity and Size. In past NEC editions, the basic rule of this section has said—and still does say—that the conductors of a branch circuit must have an ampacity that is not less than the maximum current load that the circuit will supply. Obviously, that is a simple and straightforward rule to ensure that the conductors are not operated under overload conditions. But, where the load to be served is "continuous," other concerns must be addressed.

210.19(A)(1). General. Part (A)(1) of this section says that branch circuit wiring supplying a continuous current load must have an ampacity (the continuous current-carrying capacity expressed in amperes as evaluated under the conditions of use) not less than 125 percent of the continuous-current load portion of the circuit, plus any noncontinuous load. The idea of the rule is that 125 percent of a total continuous-load current portion of the circuit plus the noncontinuous load gives a circuit rating such that the continuous-load current does "not exceed 80 percent of the rating of the branch circuit evaluated after subtracting the noncontinuous load." One is the flip side of the other. (See Fig. 210-26 for a simple application.)

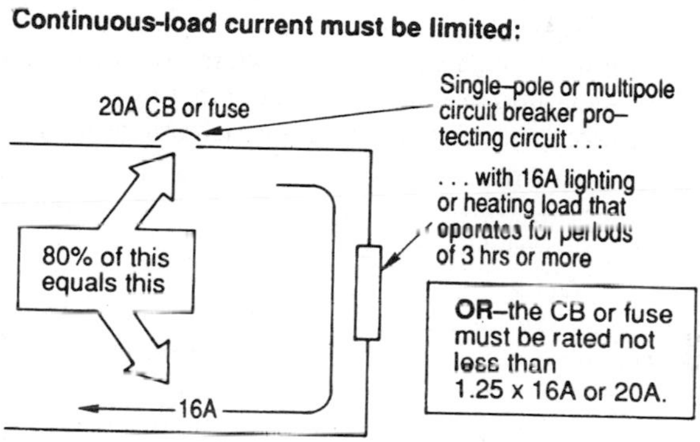
Continuous-load current must be limited:
20A CB or fuse
Single–pole or multipole circuit breaker pro–tecting circuit . . .
. . . with 16A lighting or heating load that operates for periods of 3 hrs or more
80% of this equals this
16A
OR–the CB or fuse must be rated not less than 1.25 x 16A or 20A.

Fig. 210-26. Branch-circuit protective device must be rated not less than 125 percent of the continuous-load current. [Sec. 210.19(A).]

This portion of the rule covers how the wire and the device or load interrelate. The second half of the rule, new for the 2014 NEC, covers how the wire relates to its environment. After any applicable adjustment factors for ambient heat or mutual conductor heating have been applied, the resulting ampacity must be at least equal to the load whether or not that load is continuous.

The exception covers an overcurrent device including the assembly in which it is installed that has been listed for operation at 100 percent of its rating for continuous duty. In such cases, there is no derating for the continuous portion of the load. Remember that the ampacity of a conductor reflects its ability to carry current on a continuous basis. Continuous loads do not bother conductors, but they do cause problems with overcurrent devices and the NEC builds in additional capacity in the circuit wiring so, where it is connected to the overcurrent device, it will be cool and capable of providing a heat sink for the device to which it is connected. A 100 percent-rated device does not require this feature.

Note, however, that as a practical matter this allowance will never be used for branch circuits. The smallest CB with this capability has a 400-A frame size and tripping elements set for 100 or 125 A, depending on the manufacturer. Although there are industrial applications for this provision, the far more common application will be on feeder circuits. This book includes extensive coverage of this topic in its coverage of Art. 215.

The entire process of correlating **NEC** rules for continuously loaded conductors with the requirements for derating and with the restrictions on conductor sizing at terminations is very possibly the most complicated calculation process in the **NEC**, and also one of the most essential to learn correctly. After all, what do electricians do but select and install wires? Because this process involves rules from many different locations in the code, and because the **NEC** now includes a comprehensive example written by this author that correlates this information, please refer to the discussion at the end of this book on Annex D, Example D3(a) for a systematic walk-through of how to apply the rules. They are not simple, but there are some basic principles to keep you on the right track, such as every wire having a middle and two ends and not confusing the rules that apply to one part of a circuit with rules that apply only to another part.

This part of 210.19(A) concludes with an informational note addressing voltage drop. This topic is fully addressed in this book in its coverage of Art. 215.

The wording of the rule in 210.19(A)(2) requires the circuit conductors to have an ampacity not less than "the rating of the branch circuit" only for a multioutlet branch circuit that supplies receptacles for cord- and plug-connected loads. The concept here is that receptacles provide for random, indeterminate loading of the circuit; and, by matching conductor ampacity to the amp rating of the circuit fuse or CB, overloading of the conductors can be avoided. But for multioutlet branch circuits that supply fixed outlets—such as lighting fixture outlets or hard-wired connections to electric heaters or other appliances—it is acceptable to have a condition where the conductor ampacity is adequate for the load current. But where there is no standard rating of protective device that corresponds to the conductor ampacity, the circuit fuse or CB rating is the next higher standard rating of protective device above the ampacity value of the conductor (Fig. 210-27).

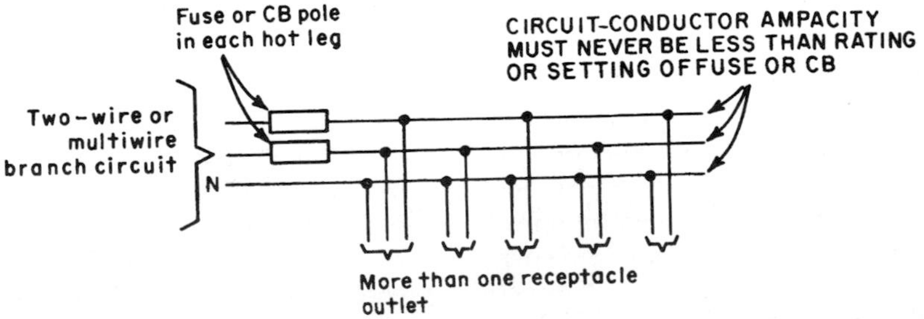
Fuse or CB pole in each hot leg
CIRCUIT-CONDUCTOR AMPACITY MUST NEVER BE LESS THAN RATING OR SETTING OF FUSE OR CB
Two-wire or multiwire branch circuit
N
More than one receptacle outlet

Fig. 210-27. This is the basic rule for any multioutlet branch circuit supplying one or more receptacles. [Sec. 210.19(B).]

The receptacle limitation now correlates with 240.4(B)(1), which disallows the use of the familiar next-higher-standard-size-device permission for circuits that supply multiple receptacles. This rule [210.19(A)(2)] requires fully sized conductors if more than one receptacle is supplied by the branch circuit; the rule in 240.4(B)(1) now also applies when more than one receptacle is supplied. The title of this section has also been adjusted by adding the words "with more than one receptacle." A review of the definition of the word "receptacle" in Art. 100 shows that if you consider a circuit with just one duplex receptacle, the rule in 210.19(A)(2) requires fully sized branch circuit conductors, and the conflict that used to exist with 240.4(B)(1) has been removed.

For multioutlet branch circuits (rated at 15, 20, 30, 40, or 50 A), the ampacities of conductors usually correspond to standard ratings of protective devices when there is only one circuit in a cable or conduit. But when circuits are combined in a single conduit so that more than three current-carrying conductors are involved, the ampacity derating factors of Table 310.15(B)(3)(a) often result in reduced ampacity values that do not correspond to standard fuse or CB ratings. It is to such cases that the rule of 210.19(A)(2) may be applied.

For instance, assume that two 3-phase, 4-wire multioutlet circuits are run in a single conduit. Two questions arise: (1) How much load current may be put on the conductors? and (2) What is the maximum rating of overcurrent protection that may be used for each of the six hot legs? Evaluate this problem assuming that the outlets supply receptacle outlets, and then evaluate it again assuming that the circuit supplies fluorescent lighting.

The eight wires in the single conduit (six phases and two neutrals) must be taken as eight conductors when applying 310.15(B)(3)(a) because the neutrals to electric-discharge lighting carry harmonic currents and must be counted as current-carrying conductors [310.15(B)(5)(c)]. Table 310.15(B)(3)(a) then shows that the No. 14 wires must have their ampacity reduced to 70 percent (for 7 to 9 wires) of the 20-A ampacity given in Table 310.15(B)(16) for 14 AWG THW, assuming that is the insulation type. With the eight No. 14 wires in the one conduit, then, each has an ampacity of 0.7×20, or 14 A. Because 210.19(A)(2) requires circuit wires to have an ampacity at least equal to the rating of the circuit fuse or CB if the circuit is supplying receptacles, use of a 15-A fuse or 15-A CB would not be acceptable in such a case because the 14-A ampacity of each wire is less than "the rating of the branch circuit" (15 A) if more than one receptacle is supplied. On the other hand, if the circuits here are supplying fixed lighting outlets, 210.19(A)(2) would not apply and 210.19(A)(1) would accept the 15-A protection on wires with 14-A ampacity. In such a case, it is not only necessary that the design load current on each phase must not exceed 14 A, but if the lighting load is continuous (operating steadily for 3 h or more), the load on each 15-A CB or fuse must not exceed 0.8×15, or 12 A [as required by 210.19(A)(1)].

In part **(A)(3)**, the rule also calls for the same approach to sizing conductors for branch circuits to household electric ranges, wall-mounted ovens, counter-mounted cooking units, and other household cooking appliances (Fig. 210-28).

The maximum demand for a range of 12-kW rating or less is sized from NEC Table 220.55 as a load of 8 kW. And 8000 W divided by 240 V is approximately 33 A. Therefore, No. 8 conductors with an ampacity of 40 A may be used for the range branch circuit.

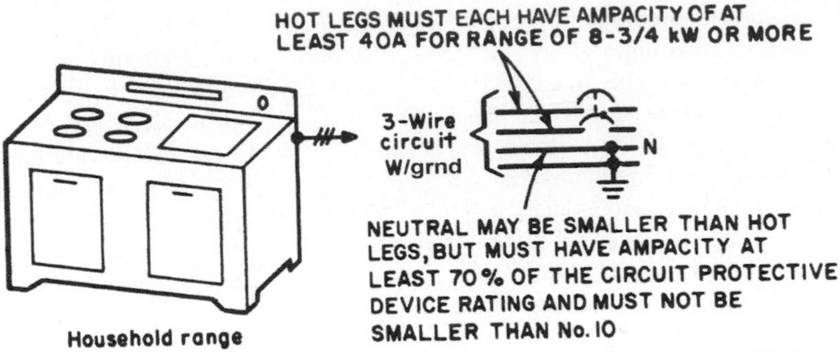
HOT LEGS MUST EACH HAVE AMPACITY OF AT LEAST 40A FOR RANGE OF 8-3/4 kW OR MORE
3-Wire circuit W/grnd
N
NEUTRAL MAY BE SMALLER THAN HOT LEGS, BUT MUST HAVE AMPACITY AT LEAST 70% OF THE CIRCUIT PROTECTIVE DEVICE RATING AND MUST NOT BE SMALLER THAN No. 10
Household range

Fig. 210-28. Sizing circuit conductors for household electric range. [Sec. 210.19(A)(3).]

On modern ranges the heating elements of surface units are controlled by five-heat unit switches. The surface-unit heating elements will not draw current from the neutral unless the unit switch is in one of the low-heating positions. This is also true to a greater degree as far as the oven-heating elements are concerned, so the maximum current in the neutral of the range circuit seldom exceeds 20 A. Because of that condition, Exception No. 2 permits a smaller size neutral than the ungrounded conductors, but not smaller than No. 10.

A reduced-size neutral for a branch circuit to a range, wall-mounted oven, or cooktop must have an ampacity of not less than 70 percent of the circuit rating, which is determined by the current rating or setting of the branch-circuit protective device. This is a change from previous wording that required a reduced neutral to have an ampacity of at least 70 percent of "the ampacity of the ungrounded conductors." Under that wording, a 40-A circuit (rating of protective device) made up of No. 8 TW wires for the hot legs could use a No. 10 TW neutral—because its 30-A ampacity is at least 70 percent of the 40-A ampacity of a No. 8 TW hot leg (0.7×40 A = 28 A). But if No. 8 THHN (55-A ampacity) is used for the hot legs with the same 40-A protected circuit, the neutral ampacity would have to be at least 70 percent of 55 A (0.7×55 A = 38.5 A) and a No. 10 TW (30 A) or a No. 10 THW (35 A) could not have been used. The newer wording bases neutral size at 70 percent of the protective-device rating (0.7×40 A = 28 A), thereby permitting any of the No. 10 wires to be used, and does not penalize use of higher temperature wires (THHN) for the hot legs.

Exception No. 1 permits taps from electric cooking circuits (Fig. 210-29). Because Exception No. 1 says that taps on a 50-A circuit must have an ampacity of at least 20 A, No. 14 conductors—which have an ampacity of 20 A in Table 310.15(B)(16)—may be used.

Exception No. 1 applies to a 50-A branch circuit run to a counter-mounted electric cooking unit and wall-mounted electric oven. The tap to each unit must be as short as possible and should be made in a junction box immediately adjacent to each unit. The words "not longer than necessary for servicing the appliance" mean that it should be necessary only to move the unit to one side in order that the splices in the junction box become accessible.

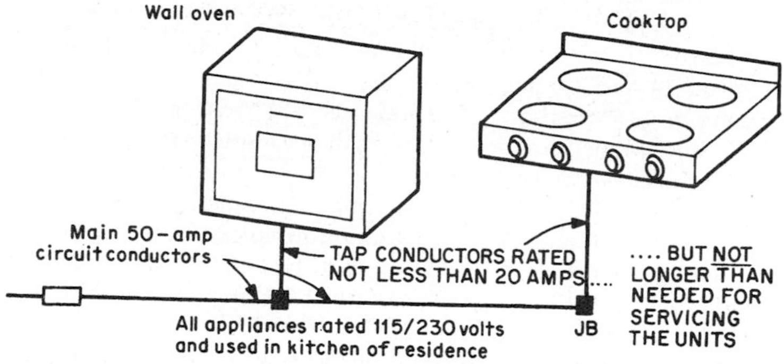
Wall oven
Cooktop
Main 50-amp circuit conductors
TAP CONDUCTORS RATED NOT LESS THAN 20 AMPS....
.... BUT NOT LONGER THAN NEEDED FOR SERVICING THE UNITS
All appliances rated 115/230 volts and used in kitchen of residence
JB

Fig. 210-29. Tap conductors may be smaller than wires of cooking circuit. [Sec. 210.19(A)(3), Exception No. 1.]

210.19(A)(4) sets No. 14 as the smallest size of general-purpose circuit conductors. But tap conductors of smaller sizes are permitted as explained in Exceptions Nos. 1 and 2 (Fig. 210-30). No. 14 wire, not longer than 18 in. (450 mm), may be used to supply an outlet unless the circuit is a 40- or 50-A branch circuit, in which event the minimum size of the tap conductor must be No. 12.

The wording of 210.19(A)(4), Exception No. 1, specifically excludes receptacles from being installed as indicated here because they are not tested for such use. That is, when tested for listing, receptacles are not evaluated using 18-in. (450-mm) taps of the size specified in Table 210.24 and protected as indicated by 210.19(A)(4),

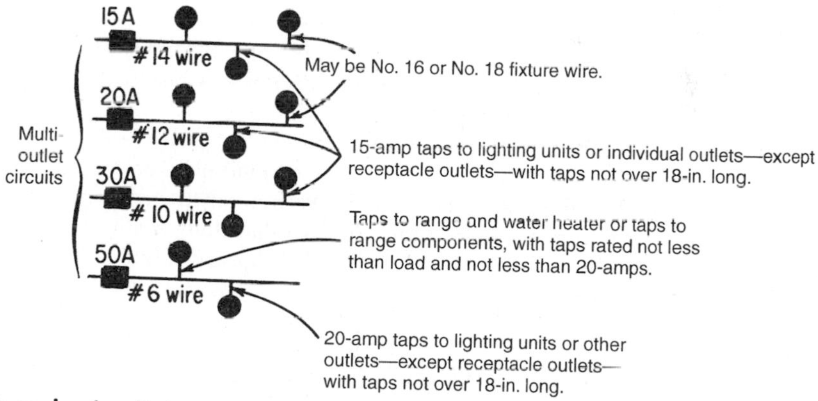
15A
14 wire
May be No. 16 or No. 18 fixture wire.
20A
12 wire
Multi-outlet circuits
15-amp taps to lighting units or individual outlets—except receptacle outlets—with taps not over 18-in. long.
30A
10 wire
Taps to range and water heater or taps to range components, with taps rated not less than load and not less than 20-amps.
50A
6 wire
20-amp taps to lighting units or other outlets—except receptacle outlets—with taps not over 18-in. long.

Branch circuit taps—as covered in 210.19 and 210.20—are considered protected by the branch circuit overcurrent devices.

Fig. 210-30. Tap conductors may be smaller than circuit wires. [Sec. 210.19(A)(3), Exception Nos. 1 and 2.]

Exception No. 1. As a result, receptacles have been prohibited from being supplied by tap conductors, as is permitted by this exception for other loads. It is not permitted to install 14 AWG pigtails on receptacles connected to 20-A branch circuits.

210.20. Overcurrent Protection. The previous section covered the minimum size of a wire used in a branch circuit; this covers the permitted size of a branch-circuit overcurrent protective device. And here again, the overcurrent device must be rated at not less than 125 percent of any continuous loading plus 100 percent of any noncontinuous loading. This is another example of a code rule that requires correlation while making calculations so as to be certain that all requirements are satisfied. Refer to the detailed discussion at the end of this book [Annex D, Example D3(a)] where all the concepts are integrated.

210.21. Outlet Devices. Specific limitations are placed on outlet devices for branch circuits: Lampholders must not have a rating lower than the load to be served; and lampholders connected to circuits rated over 20 A must be heavy-duty type (i.e., rated at least 660 W if it is an "admedium" type and at least 750 W for other types). Because fluorescent lampholders are not of the heavy-duty type, this excludes the use of fluorescent luminaires on 30-, 40-, and 50-A circuits. The intent is to limit the rating of lighting branch circuits supplying fluorescent fixtures to 20 A. The ballast is connected to the branch circuit rather than the lamp, but by controlling the lampholder rating, a 20-A limit is established for the ballast circuit. Most lampholders manufactured and intended for use with electric-discharge lighting for illumination purposes are rated less than 750 W and are not classified as heavy-duty lampholders. If the luminaires are individually protected, such as by a fuse in the cord plug of a luminaire cord connected to, say, a 50-A trolley or plug-in busway, some inspectors have permitted use of fluorescent luminaires on 30-, 40-, and 50-A circuits. But such protection in the cord plug or in the luminaire is supplementary (240.10), and branch-circuit protection of 30-, 40-, or 50-A rating would still exclude use of fluorescent fixtures according to 210.21(A). High-intensity discharge lighting such as metal-halide luminaires frequently incorporates heavy-duty mogul-base lampholders and would not be limited by this rule.

210.21(B) contains four paragraphs of importance. Part **(B)(1)** reads: "A single receptacle installed on an individual branch circuit shall have an ampere rating of not less than that of the branch circuit." Since the branch-circuit overcurrent device determines the branch-circuit rating (or classification), a single receptacle (not a duplex receptacle) supplied by an individual branch circuit cannot have a rating less than the branch-circuit overcurrent device, as shown in Fig. 210-31.

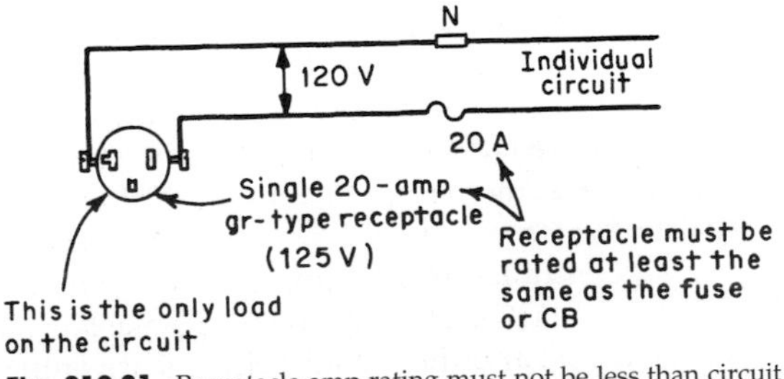
N
120 V
Individual circuit
20 A
Single 20-amp gr-type receptacle (125 V)
Receptacle must be rated at least the same as the fuse or CB
This is the only load on the circuit

Fig. 210-31. Receptacle amp rating must not be less than circuit protection rating for an individual circuit. [Sec. 210.21(B).]

Exceptions apply for specialized applications where the single receptacle configuration must correlate with motor rules in one case and with the inherently special short-time usages associated with welders in the other.

Part (B)(2) requires that receptacles installed in multiple on a branch circuit, *including just one duplex receptacle installed on a branch circuit with only the one outlet,* must not have a cord-and-plug-connected load in excess of 80 percent of the receptacle rating. Per Part (B)(3), on circuits having two or more receptacles or outlets, receptacles shall be rated as follows:

- On 15-A circuits—not over 15-A rating
- On 20-A circuits—15- or 20-A rating
- On 30-A circuits—30-A rating
- On 40-A circuits—40- or 50-A rating
- On 50-A circuits—50-A rating

Note that on 15-A circuits, a 20-A configured receptacle is not permitted, even though a 15-A receptacle is permitted on a 20-A circuit (unless it is a single receptacle on an individual branch circuit). The Code entitles any user to believe that if a 20-A plug will fit into the receptacle, the circuit will have the capability of safely supplying that load. Exceptions apply in instances where a multioutlet branch circuit is used for multiple cord-and-plug-connected welders (to correlate with Art. 630) and for electric-discharge lighting applications where a receptacle rating of not less than 125 percent of the load is sufficient. For multioutlet branch circuits rated over 50 A, as permitted under the limited conditions described in the discussion on the Exception to 210.18, every receptacle must have a rating not less than the branch-circuit rating. Part (D)(4) allows range receptacle configurations to use the same Table 220.55 loading calculations as other elements of the circuit.

210.22 and 210.23. Permissible Loads. As of the 2014 **NEC**, what was just Sec. 210.23 has been divided for editorial purposes. Sec. 210.22 covers individual branch circuits and provides that a single branch circuit to one outlet or load may serve any load and is unrestricted as to amp rating. Sec. 210.23 covers circuits with more than one outlet, which are subject to **NEC** limitations on use as follows (the word "appliance" stands for any type of utilization equipment):

1. Branch circuits rated 15 and 20 A may serve lighting units and/or appliances. The rating of any one cord- and plug-connected appliance shall not exceed 80 percent of the branch-circuit rating. Appliances fastened in place may be connected to a circuit serving lighting units and/or plug-connected appliances, provided the total rating of the fixed appliances fastened in place does not exceed 50 percent of the circuit rating (Fig. 210-32). Example: 50 percent of a 15-A branch circuit = 7.5 A. A permanently connected ventilating fan/light combination installed in a bathroom ceiling and drawing, say, 2.5 A, is permitted to be connected to a lighting circuit. However, the same appliance configured with a heating element drawing an additional 9 A could not be connected to the aforementioned lighting circuit. However, no hard-wired loads, such as range hoods or other appliances, regardless of current draw, are permitted to be connected to the specialized appliance circuits covered in 210.11(C). The bathroom receptacle circuits, the small-appliance branch circuits, and the laundry circuits are entirely reserved for cord-and-plug-connected loads in the designated areas.

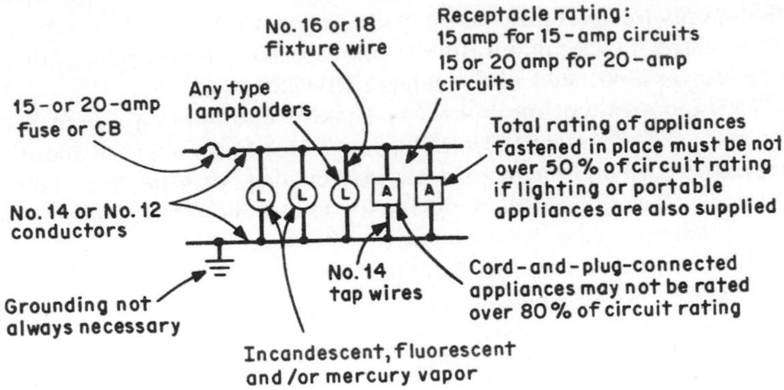
No. 16 or 18
fixture wire
Receptacle rating:
15 amp for 15-amp circuits
15 or 20 amp for 20-amp
circuits
15-or 20-amp
fuse or CB
Any type
lampholders
Total rating of appliances
fastened in place must be not
over 50 % of circuit rating
if lighting or portable
appliances are also supplied
No. 14 or No. 12
conductors
L
A
No. 14
tap wires
Cord-and-plug-connected
appliances may not be rated
over 80 % of circuit rating
Grounding not
always necessary
Incandescent, fluorescent
and /or mercury vapor

Fig. 210-32. General-purpose branch circuits—15 or 20 A. [Sec. 210.23(A).]

However, modern design usually provides separate circuits for individual fixed appliances of any significant load. In commercial and industrial buildings, separate circuits should be provided for lighting and separate circuits for receptacles.

2. Branch circuits rated 30 A may serve fixed lighting units (with heavy-duty-type lampholders) in other than dwelling units or appliances in any occupancy. Any individual cord-and-plug-connected appliance that draws more than 24 A may not be connected to this type of circuit (Fig. 210-33).

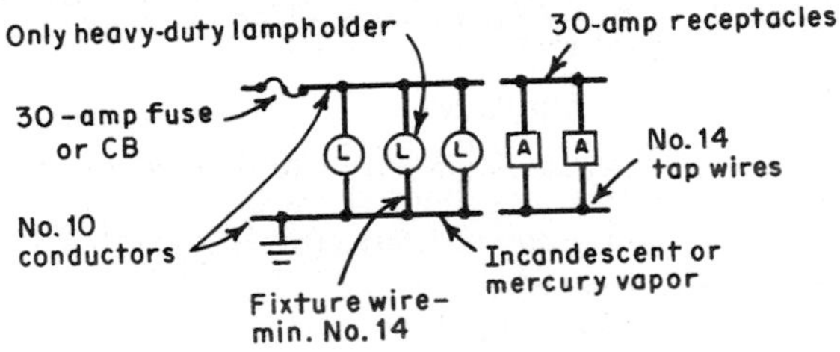
Only heavy-duty lampholder
30-amp receptacles
30-amp fuse
or CB
L
A
No. 14
tap wires
No. 10
conductors
Incandescent or
mercury vapor
Fixture wire-
min. No. 14

Fig. 210-33. Multioutlet 30-A circuits. [Sec. 210.23(B).]

Because an individual branch circuit—that is, a branch circuit supplying a single outlet or load—may be rated at any ampere value, it is important to note that the omission of recognition of a 25-A multioutlet branch circuit does not affect the full acceptability of a 25-A individual branch circuit supplying a single outlet. A typical application of such a circuit would be use of No. 10 TW aluminum conductors [rated at 25 A in Table 310.15(B)(16)], protected by 25-A fuses or CB, supplying, say, a 4500-W water heater at 240 V. The water heater is a load of 4500 ÷ 240, or 18.75 A, which is taken as 19 A per 220.5(A). Then, because 422.13 designates water heaters as continuous loads (in tank capacities up to 120 gal), the 19-A load current multiplied by 125 percent equals 24 A and satisfies 422.10(A) on the required minimum branch-circuit rating. The 25-A rating of the circuit overcurrent device also satisfies 422.11(E)(3), which says that the overcurrent protection must not exceed

150 percent of the ampere rating of the water heater. Note that although the 25-A circuit is permitted in this case, a 30-A circuit is also permitted, and far more common. The 19-A load applied at 150 percent is just under 30 A, and the next-higher standard-sized protective device is permitted.

3. Branch circuits rated 40 and 50 A may serve fixed lighting units (with heavy-duty lampholders) or infrared heating units in other than dwelling units or cooking appliances in any occupancy (Fig. 210-34). It should be noted that a

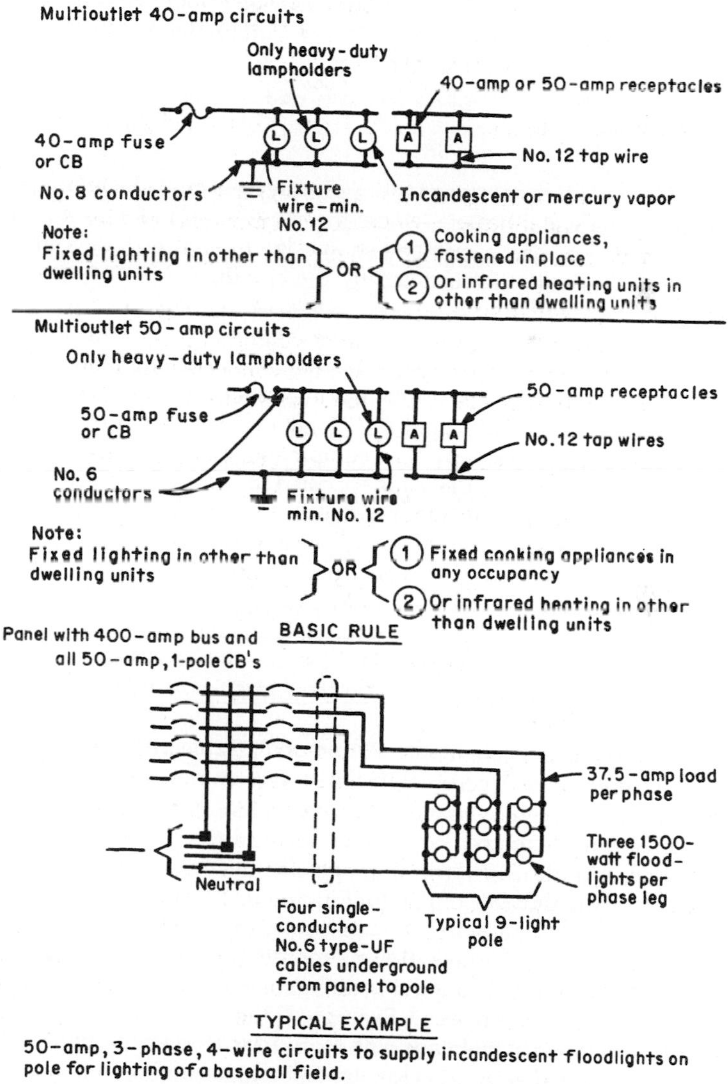
Multioutlet 40-amp circuits
Only heavy-duty lampholders
40-amp or 50-amp receptacles
40-amp fuse or CB
No. 12 tap wire
No. 8 conductors
Fixture wire-min. No. 12
Incandescent or mercury vapor
Note:
Fixed lighting in other than dwelling units
OR
1 Cooking appliances, fastened in place
2 Or infrared heating units in other than dwelling units
Multioutlet 50-amp circuits
Only heavy-duty lampholders
50-amp receptacles
50-amp fuse or CB
No.12 tap wires
No. 6 conductors
Fixture wire min. No. 12
Note:
Fixed lighting in other than dwelling units
OR
1 Fixed cooking appliances in any occupancy
2 Or infrared heating in other than dwelling units
BASIC RULE
Panel with 400-amp bus and all 50-amp, 1-pole CB's
37.5-amp load per phase
Three 1500-watt flood-lights per phase leg
Neutral
Four single-conductor No.6 type-UF cables underground from panel to pole
Typical 9-light pole
TYPICAL EXAMPLE
50-amp, 3-phase, 4-wire circuits to supply incandescent floodlights on pole for lighting of a baseball field.

Fig. 210-34. Larger circuits. [Sec. 210.23(C).]

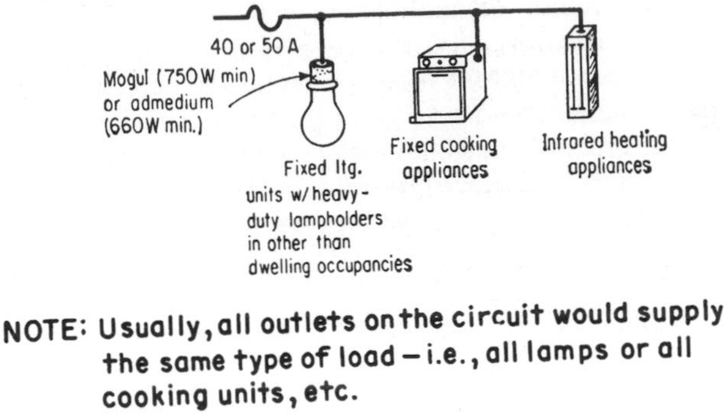
40 or 50 A
Mogul (750W min)
or admedium
(660W min.)
Fixed ltg.
units w/heavy-
duty lampholders
in other than
dwelling occupancies
Fixed cooking
appliances
Infrared heating
appliances
NOTE: Usually, all outlets on the circuit would supply
the same type of load – i.e., all lamps or all
cooking units, etc.

Fig. 210-35. Only specified loads may be used for multioutlet circuit. [Sec. 210.23(C).]

40- or 50-A circuit may be used to supply any kind of load equipment—such as a dryer or a water heater—where the circuit is an individual circuit to a single appliance. The conditions shown in that figure apply only where more than one outlet is supplied by the circuit. Figure 210-35 shows the combination of loads.

4. A multioutlet branch circuit rated over 50 A—as permitted by 210.18—is limited to use only for supplying industrial utilization equipment (machines, welders, etc.) and may not supply lighting outlets.

Except as permitted in 660.4 (and 517.71 for medical purposes) for portable, mobile, and transportable medical x-ray equipment, branch circuits having two or more outlets may supply only the loads specified in each of the preceding categories. It should be noted that any other circuit is not permitted to have more than one outlet and would be an individual branch circuit.

It should be noted that the requirement calling for heavy-duty-type lampholders for lighting units on 30-, 40-, and 50-A multioutlet branch circuits excludes the use of fluorescent lighting on these circuits because lampholders are not rated "heavy-duty" in accordance with 210.21(A) (Fig. 210-36). High-intensity discharge units with mogul lampholders may be used on these circuits provided tap conductor requirements are satisfied.

210.24. Branch-Circuit Requirements—Summary. Table 210.24 summarizes the requirements for the size of conductors where two or more outlets are supplied. The asterisk note also indicates that these ampacities are for copper conductors where derating is not required. Where more than three conductors are contained in a raceway or a cable, 310.15(B)(3)(a) specifies the load-current derating factors to apply for the number of conductors involved. A 20-A branch circuit is required to have conductors that have an ampacity of 20 A and also must have the overcurrent protection rated 20 A where the branch circuit supplies two or more outlets. Refer to the detailed discussion of conductor ampacity and load-current limits covered at Annex D, Example D3(a).

210.25. Branch Circuits in Buildings with More Than One Occupancy. The first part of this rule states that branch circuits within a dwelling unit may not supply loads in any other dwelling or its associated loads. This is a basic safety concern.

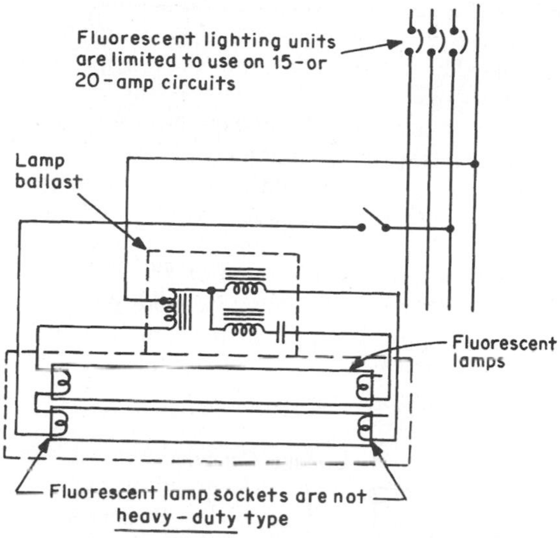
Fluorescent lighting units are limited to use on 15-or 20-amp circuits
Lamp ballast
Fluorescent lamps
Fluorescent lamp sockets are not heavy-duty type

Fig. 210-36. Watch out for this limitation on fluorescent equipment. (Sec. 210.23.)

In the past, there have been cases where the supply of loads in adjacent dwellings has resulted in injury and death where people mistakenly thought everything was electrically isolated when it was not. As a result, supply of any loads other than those "within that dwelling unit or loads associated only with that dwelling unit" has long been prohibited.

It should be noted that a common area panel is required in virtually every two-family and multifamily dwelling, and now, in multioccupant commercial buildings as well. The explosion of local ordinances regarding interconnected smoke detectors in such occupancies, as well as the growth of the so-called common area and the vast array of equipment that may be supplied in such an area, today, has assured us that a common area panel must be provided. Indeed, in some of the more expensive complexes, the common load may be equal to, or greater than, the combined load of all of the dwellings. Remember that loads such as lighting for the parking lot, landscape, hallways, stairways, walkways, and entrance ways, as well as fountain pumps, sprinkler systems and so forth—in short, any common area load—must be supplied from this common area panel. At one time, and until the 2008 NEC for commercial occupancies, the landlord could reach an agreement with a tenant to trade-off rent for coverage of common-area electrical charges. Those days are over.

The second part [210.25(B)] addresses installation of the common area panel at two-family and multifamily dwellings, and multioccupancy commercial buildings. Basically stated, a separate panel to supply common area loads must be provided and it must be supplied directly from the service conductors, have its own meter, be suitable for use as service equipment, or be supplied from a disconnect that is, and so forth. That statement is based on the change in wording that now prohibits supplying the common area panel from "equipment that supplies an individual dwelling unit or tenant space." Clearly, if a meter supplying any individual unit was also used to monitor usage on the common area panel, the literal wording of 210.25 would be violated because that "equipment" (the meter) supplies "an individual dwelling unit."

The literal wording would also be satisfied if the whole building were on a single meter. In such a case, the common area panel would be supplied from "equipment" that supplies many dwellings or tenant spaces, not "an individual" unit. But, even then, the common area panel would have to be supplied directly from the single meter and satisfy other rules (e.g., be suitable as service equipment) as necessary. In no case may the common area panel be supplied from a panel in another dwelling or, as it now states, from any equipment that supplies a single unit (Fig. 210-37). Although supplying the entire building at the expense of

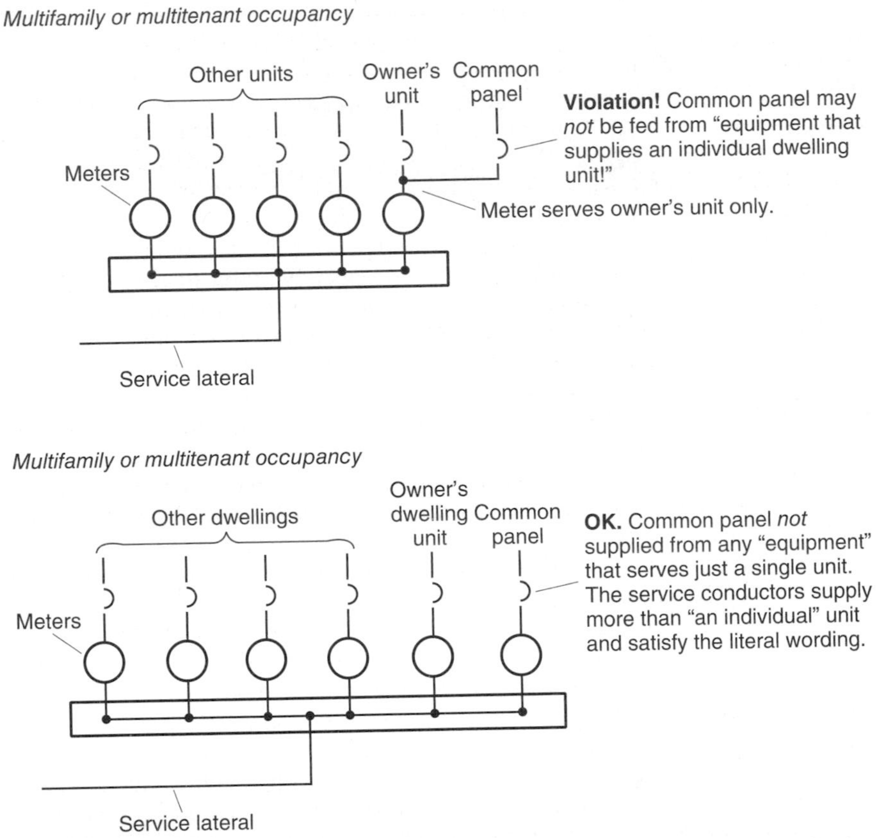
Multifamily or multitenant occupancy
Other units
Owner's unit
Common panel
Violation! Common panel may not be fed from "equipment that supplies an individual dwelling unit!"
Meters
Meter serves owner's unit only.
Service lateral
Multifamily or multitenant occupancy
Other dwellings
Owner's dwelling unit
Common panel
OK. Common panel not supplied from any "equipment" that serves just a single unit. The service conductors supply more than "an individual" unit and satisfy the literal wording.
Meters
Service lateral

Fig. 210-37. Rewording of this rule has answered a number of questions regarding its application. The rule clearly prohibits the supply of the common area panel from any individual unit's "equipment." The term "equipment" is defined in Art. 100 and includes virtually every part of an electrical installation. As indicated by the literal wording, the common area panel must be supplied from a point in the system that serves more than a single unit. In the diagram, that would be the service lateral because beginning with the taps to the individual meters, the "equipment" is serving one unit. And the common area panel may not be supplied from such equipment.

the owner is still an option, with the escalating cost of energy such arrangements are almost unheard of in new construction.

210.50. General. Part **(B)** simply requires that wherever it is known that cord-and-plug-connected equipment is going to be used, receptacle outlets must be installed. That is a general rule that applies to any electrical system in any type of occupancy or premises. This rule is of critical importance in commercial occupancies, because there are no prescriptive requirements regarding receptacle placements in such locations, and abuses are very common. For example, a receptionist's station in an insurance office, located in the middle of the floor plan with at least 5 ft separating the nearest part of the desk to any wall, and the usual desktop electrical equipment at least 7 ft from any wall, was wired from day one on extension cords with no floor receptacles at the station. The inspector could, and did use this rule to require power to be brought out to the location. Another example came up in a renovated college office building. On the rough inspection the inspector noted that in a 4.5-m (15-ft) square room there were only two receptacle outlets. The inspector pointed out that although there were no specific rules regarding receptacle placements in the room, if at any time he came back and found electrical equipment in a seemingly semipermanent location connected by extension cords, he would fail the work under this section. Perhaps, he suggested, with the walls still open the college might consider additional receptacles, and then duly documented the conversation. Additional receptacles were provided.

Part **(C)** applies to dwelling units and requires receptacles for specific appliances, such as a receptacle for a washing machine, be within 1.8 m (6 ft) of the appliance location. If possible, good design would result in a far closer placement.

210.52. Dwelling-Unit Receptacle Outlets. This section sets forth a whole list of rules requiring specific installations of receptacle outlets in all "dwelling units"—that is, one-family houses, apartments in apartment houses, and other places that conform to the definition of "dwelling unit." As indicated, receptacle outlets on fixed spacing must be installed in every room of a dwelling unit except the bathroom. The Code rule lists the specific rooms that are covered by the rule requiring receptacles spaced no greater than 12 ft (3.7 m) apart in any continuous length of "wall space."

What immediately follows is a list of locations that automatically disqualify a receptacle from being counted as satisfying one of the mandatory placement requirements that follow. Any receptacle that is an integral part of a lighting fixture or an appliance or in a cabinet may not be used to satisfy a placement requirement. For instance, a receptacle in a medicine cabinet or lighting fixture may not serve as the required bathroom receptacle. And a receptacle in a post light may not serve as the required outdoor receptacle for a one-family dwelling. Any receptacle located over 1.7 m (5½ ft) above the floor does not qualify, and any receptacle that is controlled by a wall switch in accordance with 210.70(A)(1) Exception No. 1 does not qualify either.

This last provision was new in the 2008 **NEC**. It is very common, and still permitted to use the allowance in 210.70(A) and leave a floor or table lamp plugged in to a receptacle controlled by a wall switch as the light in the room. Now, however, the receptacle used for this purpose does not qualify for the perimeter placement rules in 210.52(A). The simple way to address this is to split the hot side of a duplex receptacle so one half can be on all the time and the other half controlled

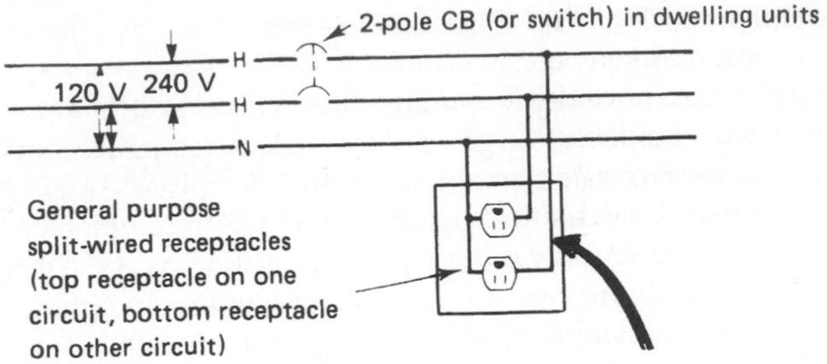
2-pole CB (or switch) in dwelling units
H
H
N
120 V
240 V
General purpose
split-wired receptacles
(top receptacle on one
circuit, bottom receptacle
on other circuit)

Fig. 210-38. Split-wiring of receptacles to control one of the receptacles may be done from the same hot leg of a 2-wire circuit or with separate hot legs of a 3-wire, 240/120-V circuit.

by the switch (Fig. 210-38). This meets all **NEC** rules because 210.52(A) does not require two receptacles at each location, only one, although duplex receptacles are far more commonly used for obvious reasons. Be careful, however, about using this procedure if multiple circuits are involved, because the common disconnecting means requirements in 210.7 will apply. It is always permitted to install a quadruplex (double duplex) receptacle outlet, with the different circuits supplying different device yokes without any common-disconnect limitations.

In part **(A)**, the required receptacles must be spaced around the designated rooms and any "similar room or area of dwelling units." The wording of this section ensures that receptacles are provided—the correct number with the indicated spacing—in those unidentified areas so commonly used today in residential architectural design, such as greatrooms and other big areas that combine living, dining, and/or recreation areas.

As shown in Fig. 210-39, general-purpose convenience receptacles, usually of the duplex type, must be laid out around the perimeters of living room, bedrooms, and

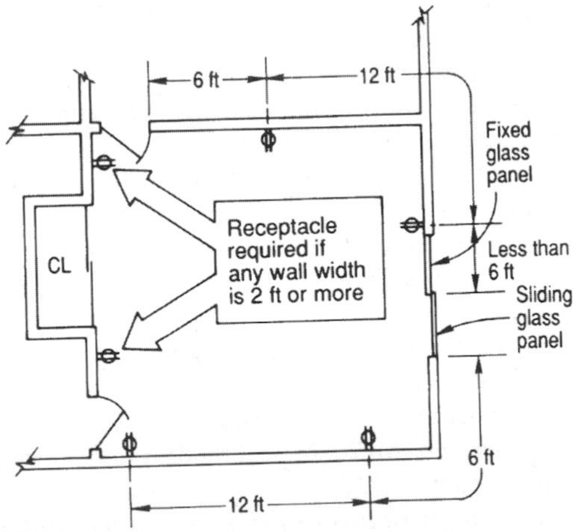
6 ft
12 ft
Fixed
glass
panel
Receptacle
required if
any wall width
is 2 ft or more
CL
Less than
6 ft
Sliding
glass
panel
6 ft
12 ft

Fig. 210-39. From any point along wall, at floor line, a receptacle must be not more than 6 ft away. Required receptacle spacing considers a fixed glass panel as wall space and a sliding panel as a doorway. [Sec. 210.52(A)(1) and (2).]

all the other rooms. Spacing of receptacle outlets should be such that no point along the floor line of an unbroken wall is more than 6 ft (1.8 m) from a receptacle outlet. Care should be taken to provide receptacle outlets in smaller sections of wall space segregated by doors, fireplaces, bookcases, or windows.

The 2011 **NEC** made two additions to the list in 210.52(A)(2)(1). First, "and similar openings" went in as catch-all wording. The other addition is more interesting and perhaps more far reaching than the panel anticipated. The list of spaces not considered as candidates for the perimeter receptacle spacing requirements now includes "fixed cabinets." There are many rooms with extensive fixed cabinets that fully occupy multiple walls of the room. By the literal text of this new exclusion, it might be possible to supply such a room with only one or two receptacles, even though there will be obvious furniture placements in front of such cabinets. The 2017 **NEC** made a stab at fixing this, by modifying the cabinet exclusion so it only applied to cabinets without "countertops or similar work surfaces." However, by not addressing the fundamental issue, the result can still be a library styled room with no requirements for receptacle outlets.

This provision came into the **NEC** in the context of the new 210.52(A)(4) that excludes receptacles serving the countertops (and similar work surfaces) specified in 210.52(C) from being counted as perimeter receptacles in 210.52(A). Without the fixed cabinet exclusion, it would be literally required to install, for an extended run of wall-mounted counters, both receptacles to serve the countertops, and additional receptacles, perhaps in the toe spaces, to meet the 6-ft rule in 210.52(A)(1). This would have been preposterous.

However, as written the "fixed cabinet" exclusion applies to any habitable room, not just the rooms covered by 210.52(C). As previously noted some rooms could lose most of their receptacles and end up with a plethora of extension cords, exactly opposite to the intent of this section generally. There are other problems as well. A wall cutout for a refrigerator now gets a receptacle because it now would be considered isolated wall space over 2 ft wide. It is no longer permissible to place the outlet for the appliance above the counter just to the side of the cutout, even though many owners appreciate the convenience of not pulling out a heavy appliance to unplug it. Even more absurd is the literal requirement to position a receptacle outlet in the cutout space for an electric range (in addition to the range receptacle if provided) because that too is now isolated wall space over 2 ft wide.

At one time, there were no prescriptive rules for countertop receptacle placements and the only rules that applied were the usual 6-ft and 2-ft provisions. When the 2-ft and 1-ft countertop rules entered the **NEC**, the counters became much more heavily populated with receptacles (and deservedly so), but the usual perimeter spacing rules for contiguous walls never failed to apply, until the 2011 **NEC**. Previously kitchens, etc., in the process of complying with 210.52(C), simply had many more receptacles than the minimum number required by 210.52(A), and that was fine. This will require considerable effort to correct in a subsequent code cycle. Until it is, expect considerable controversy regarding the apparently unintended consequences of these changes.

In determining the location of a receptacle outlet, the measurement is to be made along the floor line of the wall and is to continue around corners of the room, but is not to extend across doorways, archways, fireplaces, passageways, or other space unsuitable for having a flexible cord extended across it. The location of outlets for special appliances within 6 ft (1.8 m) of the appliance [Sec. 210.50(C)] does not affect the spacing of general-use convenience outlets but merely adds a requirement for special-use outlets.

Figure 210-40 shows two wall sections 9 ft and 3 ft wide extending from the same corner of the room. The receptacle shown located in the wider section of the wall will permit the plugging in of a lamp or appliance located within 6 ft (1.8 m) of either side of the receptacle.

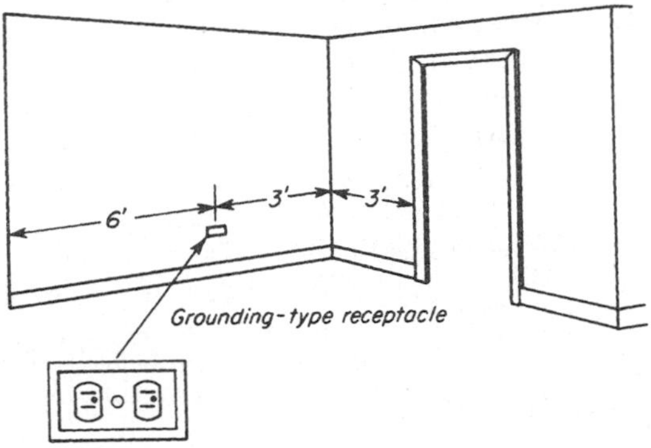
6'
3'
3'
Grounding-type receptacle

Fig. 210-40. Location of the receptacle as shown will permit the plugging in of a lamp or appliance located 6 ft on either side of the receptacle. [Sec. 210.52(A).]

Receptacle outlets shall be provided for all wall space within the room except individual isolated sections that are less than 600 mm (2 ft) in width. For example, a wall space 23 in. wide and located between two doors would not need a receptacle outlet.

The Code-making panel receives proposals almost every code cycle to not count spaces behind a door swing, or in wider spaces than the 600 mm (2 ft) considered here, etc., and consistently rejects them. The panel is aware that often the rules will end up with receptacles in locations for which permanent furniture placements are unlikely, and still intends the rules to apply as written. The reason is to assure that at least some receptacle outlets will not be obstructed by furniture placements, and thereby be available for vacuum cleaner plugs and other transient uses.

In measuring receptacle spacing for exterior walls of rooms, the fixed section of a sliding glass door assembly is considered to be "wall space" and the sliding glass panel is considered to be a doorway. Many years ago the entire width of a sliding glass door assembly—both the fixed and movable panels—was required to be treated as wall space in laying out receptacles "so that no point along the floor line in any wall space is more than 1.8 m (6 ft)" from a receptacle outlet. The wording takes any fixed glass panel to be a continuation of the wall space adjoining it, but the sliding glass panel is taken to be the same as any other doorway (such as with hinged doors) (Fig. 210-39). Although this change was generally viewed as reducing the number of receptacles, this is not necessarily the case. If two sliding glass units are mulled together, the frequent result is an isolated glass panel roughly

900 mm (3 ft) wide. Since this glass panel is wall space, and since it is more than 600 mm (2 ft) wide, a receptacle outlet must be provided in this space. And since it is obviously impracticable to put a receptacle in a glass window, the only solution is a floor receptacle (see further).

Part **(A)(2)(3)** requires fixed room dividers and railings to be considered in spacing receptacles. This is illustrated by the sketch of Fig. 210-41. In effect, the two side faces of the room divider provide additional wall space, and a table lamp placed as shown would be more than 6 ft (1.8 m) from both receptacles "A" and "B." Also, even though no place on the wall is more than 6 ft (1.8 m) from either "A" or "B," a lamp or other appliance placed at a point such as "C" would be more than 6 ft (1.8 m) from "B" and out of reach from "A" because of the divider. This rule would ensure placement of a receptacle in the wall on both sides of the divider or in the divider itself if its construction so permitted.

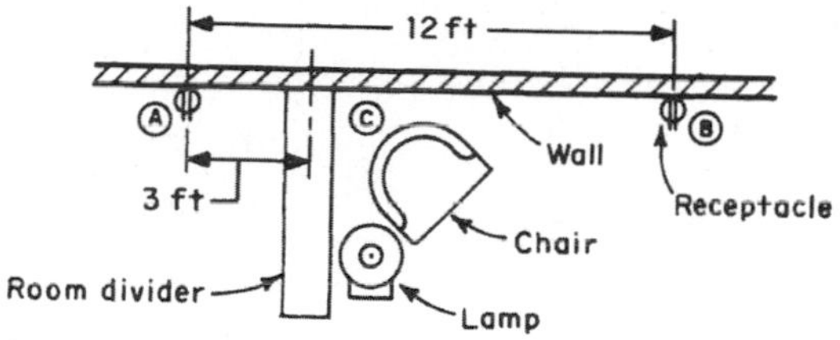
12 ft
A
C
B
Wall
3 ft
Receptacle
Chair
Room divider
Lamp

Fig. 210-41. Fixed room dividers must be counted as wall space requiring receptacles. [Sec. 210.52(A).]

Note that nothing limits the usual 1.8-m (6-ft) rule from extending around the base of a fixed wall divider, as long as a cord doesn't traverse a walkway. For example, if "A" were 600 mm (2 ft) from the divider, and if the divider only projected 600 mm (2 ft), then "C" would still be considered to be covered by "A" because a 1.8-m (6-ft) cord run out and back along the divider and then over to "A" would not have to rely on an extension cord. Although the usual design preference is to put the receptacle in the room, there are construction difficulties that arise from time to time that make this approach worth considering.

Recessed or surface-mounted floor receptacles must be within 18 in. (450 mm) of the wall to qualify as one of the "required" receptacle outlets in a dwelling. The previous wording used in Sec. 210.52(A)(3) indicated floor-mounted receptacles were not considered to fulfill the requirement of 210.52(A) unless they were "located close to the wall." The use of a specific dimension, regardless of its arbitrary nature, is much more desirable than the relative term "close." The use of nonspecific, relative, and subjectively interpreted terms—such as "close" or "large"—opens the door for conflict and makes applying or enforcing a given rule much more difficult.

The use of either surface-mounted or recessed receptacle outlets has grown since "railings" were required to be counted as "wall space" by the 1993 Code. Now, where floor-mounted receptacle outlets are provided—either surface-mounted or recessed—to serve as a required receptacle outlet in a dwelling for any so-called wall space, such an outlet must be no more than 18 in. (450 mm) from the wall (Fig. 210-42).

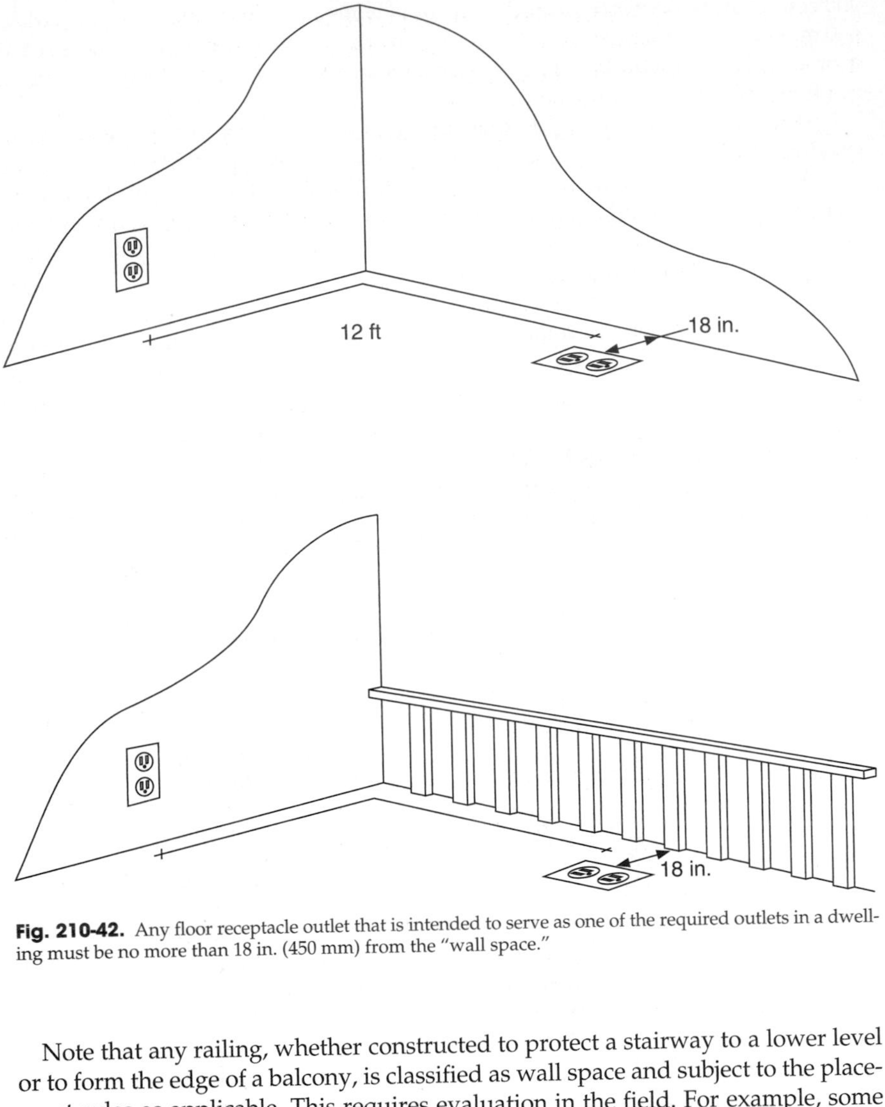
12 ft
18 in.
18 in.

Fig. 210-42. Any floor receptacle outlet that is intended to serve as one of the required outlets in a dwelling must be no more than 18 in. (450 mm) from the "wall space."

Note that any railing, whether constructed to protect a stairway to a lower level or to form the edge of a balcony, is classified as wall space and subject to the placement rules as applicable. This requires evaluation in the field. For example, some balcony spaces are long and narrow with doorways opening along their long dimension. Generally such spaces are classified as hallways and need only one receptacle, assuming they are over 3 m (10 ft) long; and such a receptacle could be put in conventional wall space. In other cases balconies are, to all intents and purposes, the wall of a habitable room. In such cases, they will likely be used for furniture placements and the usual spacing rules will apply.

In spacing receptacle outlets so that no floor point along the wall space of the rooms designated by 210.52(A) is more than 6 ft (1.8 m) from a receptacle, a receptacle that is part of an appliance must not generally be counted as one of the required spaced receptacles. However, the second paragraph of 210.52 states that a receptacle that is "factory installed" in a "permanently installed electric baseboard heater" (not a portable heater) may be counted as one of the required spaced receptacles for the wall space occupied by the heater. Or a receptacle "provided as a separate assembly by the manufacturer" may also be counted as a required spaced receptacle. But, such receptacles must not be connected to the circuit that supplies the electric heater. Such a receptacle must be connected to another circuit.

Because of the increasing popularity of low-density electric baseboard heaters, their lengths are frequently so long (up to 14 ft) that required maximum spacing of receptacles places receptacles above heaters and produces the undesirable and dangerous condition where cord sets to lamps, radios, TVs, and the like will droop over the heater and might droop into the heated-air outlet. And UL rules prohibit use of receptacles above almost all electric baseboard heaters for that reason. Receptacles in heaters can afford the required spaced receptacle units without mounting any above heater units. They satisfy the UL concern and also the preceding note near the end of 210.52(A) that calls for the need to minimize the use of cords across doorways, fireplaces, and similar openings—and the heated-air outlet along a baseboard heater is such an opening that must be guarded (Fig. 210-43).

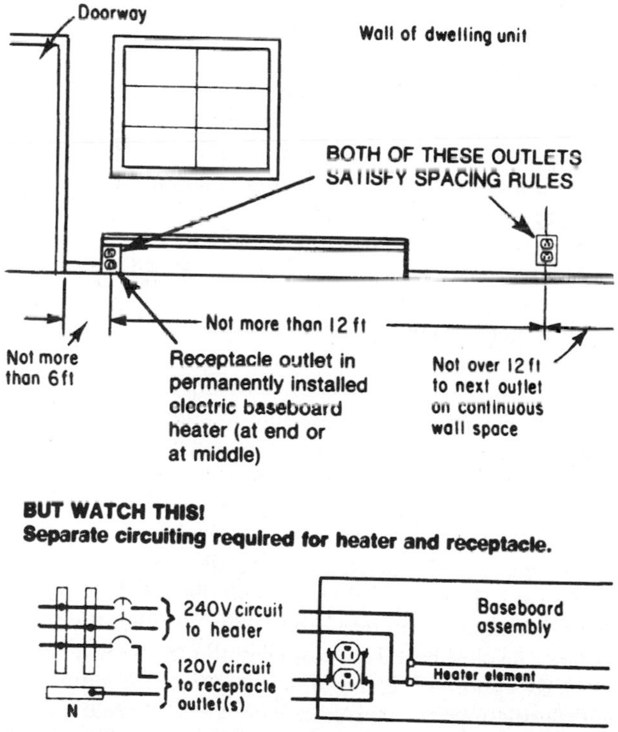
Doorway
Wall of dwelling unit
BOTH OF THESE OUTLETS SATISFY SPACING RULES
Not more than 12 ft
Not more than 6 ft
Receptacle outlet in permanently installed electric baseboard heater (at end or at middle)
Not over 12 ft to next outlet on continuous wall space
BUT WATCH THIS!
Separate circuiting required for heater and receptacle.
240V circuit to heater
120V circuit to receptacle outlet(s)
N
Baseboard assembly
Heater element

Fig. 210-43. Receptacles in baseboard heaters may serve as "required" receptacles. [Sec. 210.52(A).]

An informational note at the end of 210.52 points out that the UL instructions for baseboard heaters (marked on the heater) may prohibit the use of receptacles above the heater because cords plugged into the receptacle are exposed to heat damage if they drape into the convection channel of the heater and contact the energized heating element. The insulation can melt, causing the cord to fail.

A rewrite of 210.52(B) serves to clarify application and prohibits one longtime practice. Part **(B)(1)** of this section requires two, or more, 20-A branch circuits to supply all receptacle outlets required by 210.52(A) and (C) in the kitchen, and so forth. And part **(B)(2)** states that no other outlets may be supplied from those small appliance branch circuits. Those two requirements had both been contained in part **(B)(1)** of the 1993 and previous Codes. However, because the two rules were combined in a single paragraph, it was not always easy to determine to which part a given exception applied.

The basic rule of 210.52(B)(1) states that those receptacles required every 12 ft [Sec. 210.52(A)], those that serve countertop space [210.52(C)], and the refrigerator receptacle in the kitchen, dining room, pantry, and so forth, must be supplied by one of the two, or more, 20-A small appliance branch circuits.

The wording used here must be carefully examined. Because the wording only specifically permits the refrigerator receptacle and those receptacles required by 210.52(A) and (C) on the small appliance branch circuits, the installation of any other receptacles on the small appliance branch circuits is effectively prohibited. Any receptacle installed for specific equipment, such as dishwashers, garbage disposals, and trash compactors—which are *not* required by part **(A)** or **(C)**—must be supplied from a different 15- or 20-A branch circuit, which could be a multioutlet general-purpose branch circuit if the load meets the 50 percent test in 210.23(A).

The exceptions to part **(B)(1)** are exceptions to the rule that *all required* receptacle outlets must be supplied from the two, or more, 20-A branch circuits. The first exception recognizes the use of a switched receptacle supplied from a general-purpose branch circuit where such a receptacle is provided instead of a lighting outlet in accordance with Exception No. 1 to 210.70(A). That rule specifically excludes kitchens from employing a switched receptacle instead of a lighting outlet, but, in those other rooms and areas identified in 210.52(B)(1), particularly dining rooms, a wall-switched receptacle outlet supplied from a general-purpose branch circuit is permitted and should count as a required receptacle (Fig. 210-44).

Be very careful, however, about attempting to split a duplex receptacle to do this. Since the entire yoke must be disconnectable in a single motion at the panel, the lighting circuit and the small-appliance branch circuit must have a common disconnect. This may or may not be feasible. If not, and the owner insists on wall switch control of floor or table lighting in the room, a two-gang opening must be provided, with one yoke connected to a small-appliance branch circuit, and the other to the local lighting circuit. Note also that no "always on" receptacle can be connected to a lighting circuit in this room, so either the switched side is a single receptacle, or it could be a duplex with both halves switched.

In 210.52(B)(1), Exception No. 2, the Code recognizes the supply of the required receptacle for a specific appliance (broadened in the 2017 cycle from the prior limitation to refrigerators) from an individual 15-A branch circuit. Many refrigerators and other specific appliances installed in dwellings are rated at 12 A or less, often

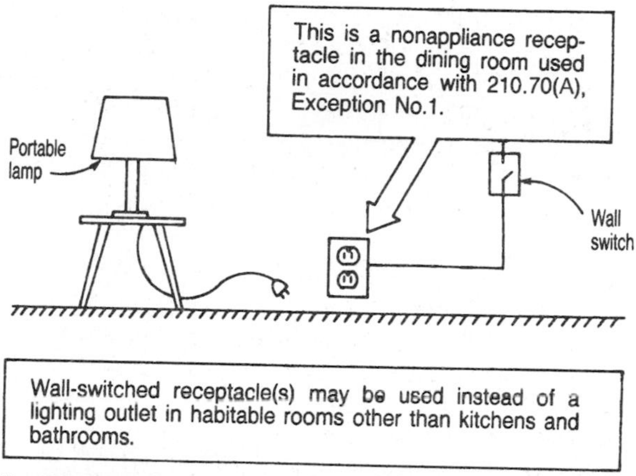
This is a nonappliance receptacle in the dining room used in accordance with 210.70(A), Exception No.1.
Portable lamp
Wall switch
Wall-switched receptacle(s) may be used instead of a lighting outlet in habitable rooms other than kitchens and bathrooms.

Fig. 210-44. For those rooms and areas identified by Sec. 210.52.(B)(1), other than the kitchen, a wall-switch-controlled receptacle may be supplied from a general-purpose branch circuit and serve as one of the required receptacles. In the drawing, both halves must be switched, and as depicted, the receptacle does not qualify as meeting the normal 210.52(A) placement rules for this location.

far less, and could be supplied from a 15-A circuit. Rather than mandate the use of a 20-A-rated circuit for those cases where a 15-A circuit is adequate, it is permissible to use a 15-A-rated circuit, provided the supply to the refrigerator (or comparable) receptacle is a dedicated branch circuit—that is, *no* other outlets supplied. Remember that in this case the 15 A receptacle must be a single receptacle.

It should be noted that it is no longer permissible to supply an outdoor receptacle from the small appliance branch circuit. This was recognized years ago in 210.52(B)(1), Exception No. 2, and served to limit the number of GFCIs needed at a dwelling. That is, because grade-level-accessible outdoor receptacles were required to have GFCI protection, the Code permitted supplying the outdoor receptacle using the feed-through capability of the GFCIs installed in the kitchen rather than require an additional GFCI device, which provided for economy. Now, however, supplying an outdoor receptacle from the small appliance branch circuit is prohibited (Fig. 210-45).

210.52(B)(2) states that only those outlets identified in part **(B)(1)**—and no other outlets—may be installed on the two, or more, small appliance branch circuits. Outlets for lighting and hard-wired appliances, as well as "unrequired" receptacles for equipment, must be installed on 15- or 20-A general-purpose circuits.

The first exception to 210.52(B)(2) allows a clock hanger receptacle to be installed on a small appliance branch circuit, or it may be supplied from a general-purpose circuit. The second recognizes a receptacle provided for control power or clock, fan, or light in a gas-fired cooking unit. Note that only a *receptacle* outlet is permitted. Any hard-wired connection for such auxiliary functions on a gas-fired unit must be supplied from a general-purpose branch circuit and *not* from the small appliance branch circuits (Fig. 210-46).

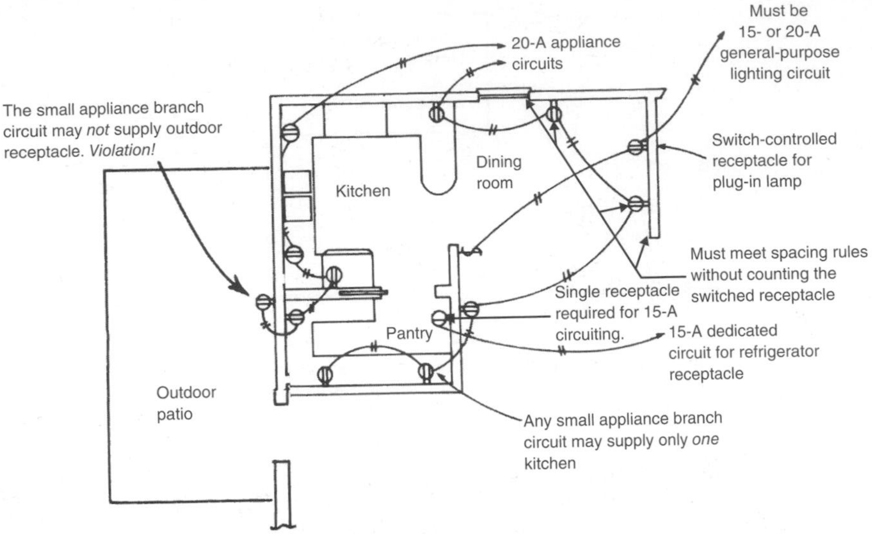

20-A appliance
circuits
Must be
15- or 20-A
general-purpose
lighting circuit
The small appliance branch
circuit may *not* supply outdoor
receptacle. *Violation!*
Switch-controlled
receptacle for
plug-in lamp
Dining
room
Kitchen
Must meet spacing rules
without counting the
switched receptacle
Single receptacle
required for 15-A
circuiting.
15-A dedicated
circuit for refrigerator
receptacle
Pantry
Outdoor
patio
Any small appliance branch
circuit may supply only *one*
kitchen

Fig. 210-45. Summary of Sec. 210.52(B)(1) and its two exceptions. As indicated, supply of an outdoor receptacle from any of the two, or more, 20-A small appliance branch circuits is prohibited. The switched receptacle cannot be used to meet spacing requirements, and a 15-A refrigerator circuit, if used (and not shown in this drawing), must be an individual branch circuit.

The rule of 210.52(B)(3) places a limit on the number of "kitchens" that a small appliance branch circuit may serve. Any given 20-A small appliance branch circuit may supply only a single kitchen. Given the reality that some dwellings are equipped with more than one kitchen, this rule will ensure that adequate capacity is available for countertop receptacles in *both* kitchens. Note that it is still permissible to supply a kitchen and, say, a dining room, or any other rooms or "similar areas," from a given 20-A small appliance branch circuit. Although this is not desirable, *all* rooms identified in part **(B)(1)** may be supplied from such a circuit.

Section 210.52(C) presents requirements and restrictions regarding installation of countertop receptacles in kitchens and dining rooms. The 2017 **NEC** has extended this concept to add "and Work Surfaces." This includes the other permanent surfaces now commonly added in kitchens, such as built-in desks for computer work, paying bills.

This section is broken into five subparts—**(C)(1)** through **(5)**. The first four subparts identify those counter spaces in the kitchen and dining room that must be provided with receptacle outlets and indicate the number required, while the last subpart, **(C)(5),** indicates where the receptacle outlet must be installed.

In part **(C)(1)**, the **NEC** puts forth the spacing requirements for receptacle outlets installed at counter spaces along the wall. Basically stated, each wall counter space that is 12 in. (300 mm) or wider must have at least one receptacle outlet to supply cord-and-plug-connected loads. The receptacles must be placed so that no point along the wall line is more than 24 in. (600 mm) from an outlet. That translates into one outlet every 4 ft. It should be noted that the term "measured horizontally" is intended to recognize application, as shown in Fig. 210-47. The wording is

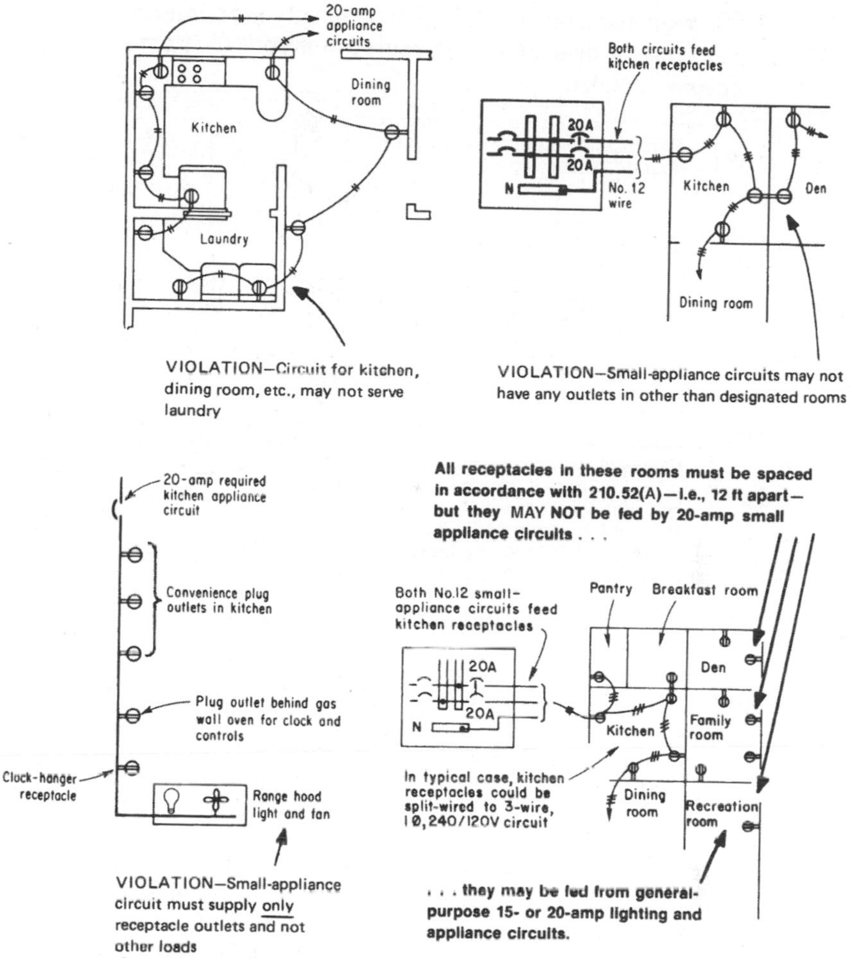
20-amp appliance circuits
Dining room
Kitchen
Laundry
VIOLATION—Circuit for kitchen, dining room, etc., may not serve laundry
Both circuits feed kitchen receptacles
20A
20A
N
No. 12 wire
Kitchen
Den
Dining room
VIOLATION—Small-appliance circuits may not have any outlets in other than designated rooms
20-amp required kitchen appliance circuit
Convenience plug outlets in kitchen
Plug outlet behind gas wall oven for clock and controls
Clock-hanger receptacle
Range hood light and fan
VIOLATION—Small-appliance circuit must supply only receptacle outlets and not other loads
All receptacles in these rooms must be spaced in accordance with 210.52(A)—i.e., 12 ft apart—but they MAY NOT be fed by 20-amp small appliance circuits . . .
Both No.12 small-appliance circuits feed kitchen receptacles
Pantry
Breakfast room
Den
20A
20A
N
Family room
Kitchen
In typical case, kitchen receptacles could be split-wired to 3-wire, 1Ø, 240/120V circuit
Dining room
Recreation room
. . . they may be fed from general-purpose 15- or 20-amp lighting and appliance circuits.

Fig. 210-46. A summary of Sec. 210-52(B)(2) and its two exceptions. A clock-hanger receptacle and/or a receptacle for the supply of auxiliary equipment on a gas-fired range, oven, or cooktop may also be supplied from the two, or more, small appliance branch circuits.

supposed to indicate that there is no need to measure "around the corner" in that case. But, watch out! Some inspectors believe that such application is in violation. In doing so, they are applying the same logic that measures the dimensions of two walls meeting at a corner per 210.52(A)(1) in this manner, and the wording of the two provisions is parallel ("along the floor line" and "along the wall line"). Check with your local electrical inspector to verify acceptability. In spite of numerous attempts to get this language clarified, the Code-making panel has yet to do so.

The exception to 210.52(C)(1) eliminates the need for receptacles where the countertop wall space is behind a range or sink. Where the dimensions are equal

Any point along the wall line of each length of countertop must *not* be over 24 in., measured horizontally, from a receptacle outlet.

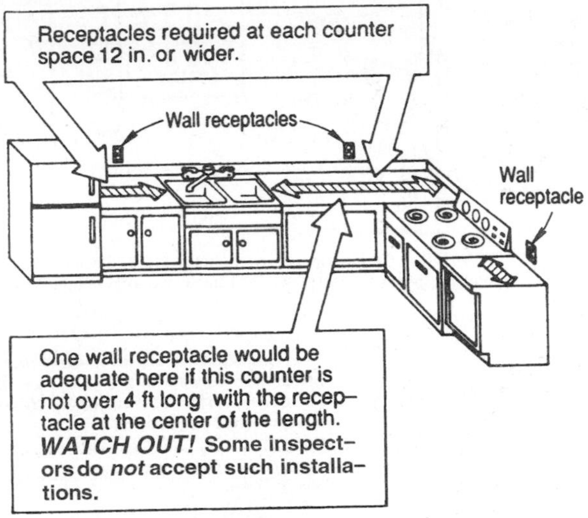
Receptacles required at each counter space 12 in. or wider.
Wall receptacles
Wall receptacle
One wall receptacle would be adequate here if this counter is not over 4 ft long with the receptacle at the center of the length. *WATCH OUT!* Some inspectors do *not* accept such installations.

Fig. 210-47. The term "measured horizontally" can essentially be translated as "when you are facing the counter." There is no need to measure around the corner here because that would effectively measure the area twice. However, this is far from clear. Many inspectors apply the same principles that apply to wall placements, and therefore measure the entire unbroken wall line. Repeated efforts to clarify this have failed.

to or less than those shown in Fig. 210-48, a receptacle outlet is not required. Note, however, that for the majority of corner applications where the sink or range is on the diagonal, the distance to the corner will exceed 450 mm (18 in.) and qualify for a receptacle in that space. The 450-mm (18-in.) dimension is the altitude of an isosceles triangle whose base is 900 mm (36 in.) and many sinks or ranges will at least equal that width. If this is the case, great care and foresight may be required to avoid major construction problems. Such locations often involve window placements or other difficulties that must be carefully anticipated. Further, the decision to count such spaces in the placement rules is appropriate. The space behind such a sink, for example, works well for a number of appliances, including electric teakettles, that must be routinely refilled with water.

As given in subpart **(C)(2)** and **(C)(3)**, each freestanding (island) countertop that measures 24 in. (600 mm) or more by 12 in. (300 mm) or more must be provided with one receptacle outlet. The same dimensions apply to peninsular countertops, which are countertops that extend from another counter or a wall.

The dimensions are to be measured, as of the 2017 NEC, from the "connected perpendicular wall." Prior editions back to the 1993 edition measured from the "connecting edge." Because the orientation of the long dimension was

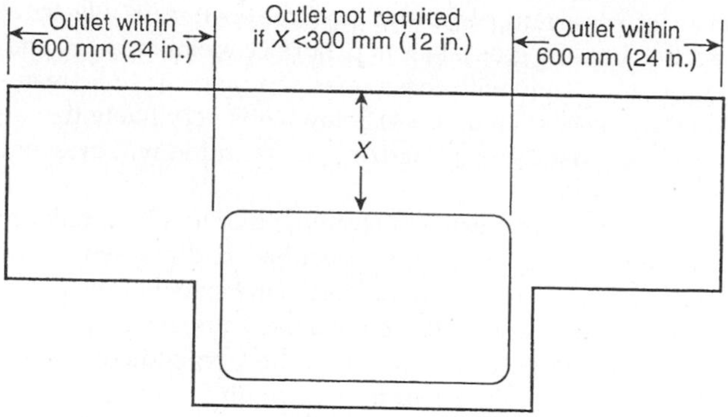
Outlet within
600 mm (24 in.)
Outlet not required
if X <300 mm (12 in.)
Outlet within
600 mm (24 in.)
X

Sink or range extending from face of counter

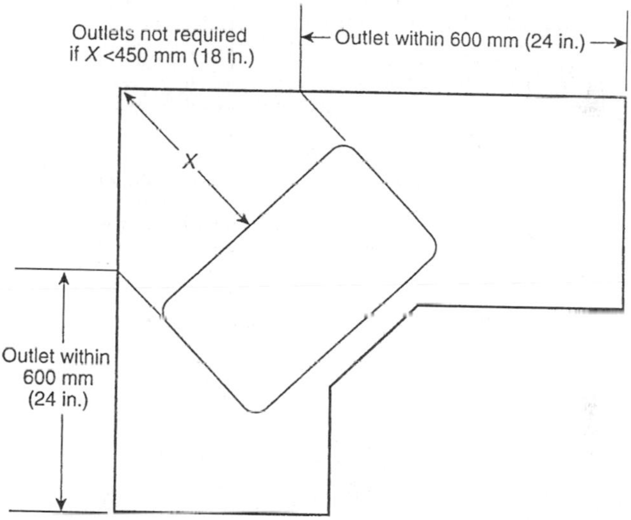
Outlets not required
if X <450 mm (18 in.)
Outlet within 600 mm (24 in.)
X
Outlet within
600 mm
(24 in.)

Sink or range mounted in corner

Fig. 210-48. This diagram—Fig. 210.52 in the NEC—provides guidance for countertop receptacles located behind sinks and range tops.

unspecified, that wording meant that even a 1-ft peninsula (assuming standard 25-inch counter stock) needed its own receptacle. This often resulted in ferocious resistance on the part of owners who did not want to see their decorative peninsula ends molested by receptacles.

Island and peninsula receptacles often involve significant construction issues. Not all peninsulas have base cabinets underneath them. Many are in effect a permanently attached kitchen table, with not much more than a table leg to support the open end and bar stools arranged around the three open sides. Electrical contractors have been forced to literally glue a length of surface metal raceway and box to their underside in order to comply with these rules. If a range or sink creates two peninsulas out of one (see discussion at (C)(4) below), it is very likely that at least one of the two peninsulas or islands resulting from the partition will present construction difficulties.

As shown in Fig. 210-49, a creative alternative that involved redefining the peninsula often avoided these construction difficulties and pleased the owners. If the wall-mounted receptacle were within the horizontal extent of the peninsula dimension, then that receptacle became the peninsular countertop receptacle. The 2017 NEC, by defining the measurement point as the perpendicular connecting wall, has effectively provided that the heretofore alternative approach is now the correct approach. If a rule says to measure the size of something, and it also says where to begin your measurement, then it follows that where you start your measurement is necessarily where the measured object starts.

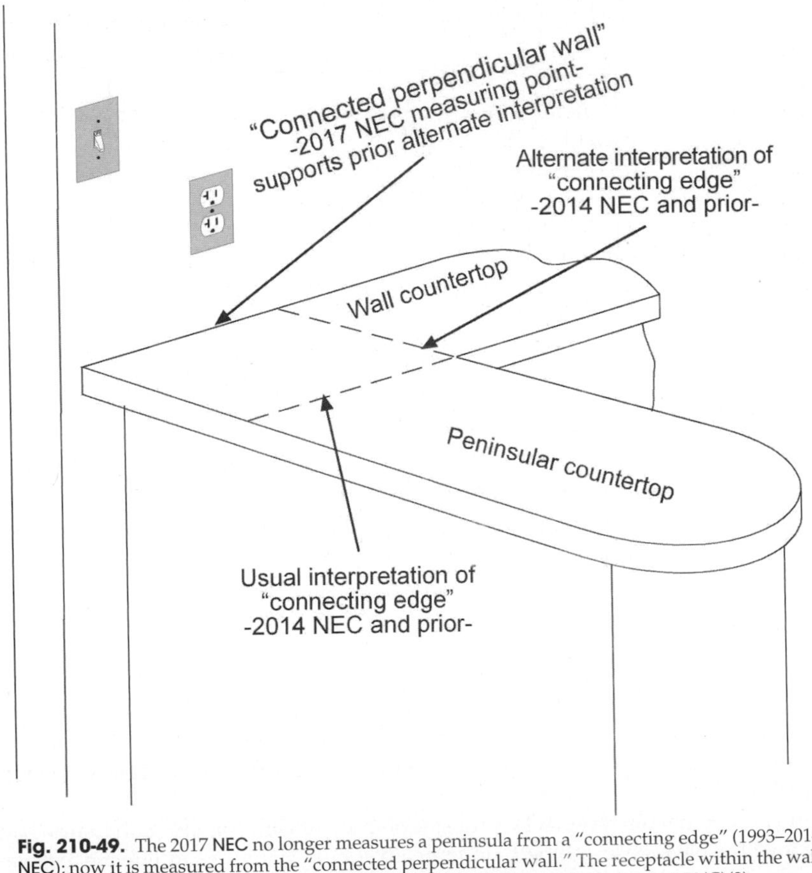
"Connected perpendicular wall"
-2017 NEC measuring point-
supports prior alternate interpretation
Alternate interpretation of
"connecting edge"
-2014 NEC and prior-
Wall countertop
Peninsular countertop
Usual interpretation of
"connecting edge"
-2014 NEC and prior-

Fig. 210-49. The 2017 NEC no longer measures a peninsula from a "connecting edge" (1993–2014 NEC); now it is measured from the "connected perpendicular wall." The receptacle within the wall space of its extension to the wall therefore serves the peninsula and satisfies 210.52(C)(3).

The **NEC** did not formerly cover, with any clarity, another related problem, namely, how long can a peninsula or island get before more than one receptacle is required. There was an answer to this, but it also involved interpretation, in this case as to whether 210.52(C) supersedes 210.52(A), or merely augments 210.52(A) and both rules continue to apply. The plethora of receptacles required on a wall-mounted counter clearly meets the required 210.52(A) placements many times over, but what about peninsulas and islands?

A kitchen island or peninsula clearly and permanently divides the room, and as we have seen, a fixed room divider in the form of "a free-standing bar-type counter" invokes conventional placement requirements, as per 210.52(A)(2)(3). On this basis, it was not unreasonable to require an additional receptacle on an island or peninsula that is over 1.8 m (6 ft long). And, in conjunction with the interpretation offered as to how to place the connecting edge of a peninsula and thereby measure its length, that would be the cut point on how far out a peninsula could extend (or an island could be long) before an additional receptacle would be required. However, the new text of 210.52(A)(4) makes this approach obsolete. In perhaps another unintended consequence of the new rule, an island or peninsula can now be of indefinite length; since 210.52(A) cannot be applied to them, the only remaining rule is 210.52(C)(2) or (3) as applicable, namely, no fewer than one receptacle for either.

Subpart **(C)(4)** covers the longtime rule regarding pieces of countertops that are separated by cooktops, sinks, and so forth. As indicated, each such piece must be treated as an individual counter. And, if the dimensions are as described in parts **(C)(1)**, **(C)(2)**, or **(C)(3)**, as applicable, of this section, at least one receptacle outlet must be provided. The concluding language makes it clear that if a range, counter-mounted cooktop or sink has less than 300 mm (12 in.) of space behind it, the counter is considered continuous and the rules will apply to each side independently. For example, if a range is located in a peninsula, the outer end of the peninsula, for code purposes, is now an island. The inner end of the peninsula is now, for code purposes, a short peninsula. In both instances, the applicable rules must be applied independently to each segment in order to determine whether one or more receptacles are required in that segment.

The requirements given in 210.52(C)(5) mandate where the required receptacle outlets may be installed. That is, a given receptacle outlet may not be counted as one of the required outlets unless it is installed on top of, but not more than 500 mm (20 in.) above, the counter it is intended to serve, and note that no receptacle may be installed face-up in a countertop per 406.5(G). That is a "make sense" proposition inasmuch as a receptacle installed face-up would eventually become a "drain" for soup, milk, water, or whatever else is eventually spilled on the counter. Only the so-called tombstone or doghouse enclosures would be acceptable for surface mounting. However, there are now pop-up receptacles designed for countertop use that withdraw into the countertop when not in use, and when withdrawn, only show a small, almost flat metal circle. This is not a face-up mounting and as of the 2011 edition, the **NEC** now expressly recognizes these constructions. Although not entirely clear, it is assumed that the 20 in. must be measured from the counter surface, *not* the top of the backsplash. In addition, and in correlation with 210.52(3), a receptacle mounted inside an appliance garage, or otherwise not readily accessible due to the placement of a sink, range top, or other appliance fastened in place, does not qualify as providing the required countertop coverage.

Note that the basic rule generally requires the outlet to be mounted above, or on top of, the counter. The basic rule does *not* recognize installation of an outlet *below* the counter space. However, where the counter does not extend more than 6 in. (150 mm) beyond "its support base," the exception to 210.52(C)(5) permits installation of a receptacle outlet below, but not more than 300 mm (12 in.) below the counter. And where the outer edge of the countertop *does* extend more than 6 in. beyond its support base, any "below-the-countertop" receptacles must be installed so that the receptacle, itself, is not more than 6 in. from the outer edge. That would necessitate the use of either a surface-mounted receptacle or plug-mold type of receptacle to ensure that when measuring from the face of the receptacle or edge of the plug mold, the distance to the outer edge of the countertop overhang is not more than 6 in.

Note that the allowance for a receptacle mounted below the countertop only applies to island and peninsular counters, and even on those counters the exception will not apply unless the counter is flat. If there are two levels to the counter for any reason, then the receptacle has to be placed in the vertical rise between the two levels. Further, if there are suspended upper cabinets over the island or peninsula such that a receptacle could be mounted so as to be not more than 500 mm (20 in.) above the counter, then that opportunity must be used, and a placement on the side of a base cabinet is not allowed.

The **NEC** is trying to minimize instances where the exception will be used in order to lessen the opportunities for toddlers to pull over dangerous appliances that could severely injure them by spilling hot liquid on their heads or otherwise. On the other hand, the **NEC** also recognizes that a flat island or peninsula is like someone's kitchen table, and a mandatory tombstone outlet is a nonstarter in the mind of the public. The result is a reasonable compromise that minimizes the instances of base-mounted receptacles; ultimately the parents are responsible for policing how they connect kitchen appliances when toddlers are in the house. The other exception allows receptacles below a counter, whether for an island or otherwise, in construction for the physically disabled. The usual procedure here is to mount surface metal raceway with receptacles under the counter lip where someone in a wheel chair can reach it easily (Fig. 210-50).

Part **(D)** requires the installation of at least one receptacle outlet adjacent to and within 3 ft of each washbasin location in bathrooms of dwelling units—and 210.60 requires the same receptacle in bathrooms of hotel and motel guest rooms and suites. This receptacle may be mounted in an adjacent wall, or partition, or within the side or face of the vanity not more than 12 in. below the vanity's countertop. The Code requires a dedicated circuit for bathroom receptacles installed in dwellings. In every bathroom, at least one receptacle outlet must be installed at each basin and any such outlet(s) must be supplied from the dedicated 20-A branch circuit (Fig. 210-51), as required by 210.11(C)(3). If a bathroom has two basins, two receptacles will usually be required unless the basins are small and very close together, or unless a receptacle is mounted between the basins, perhaps horizontally in a backsplash. And yes, receptacle outlets have been successfully installed in, yes in, bathroom mirrors, but a receptacle on each side is usually more cost effective.

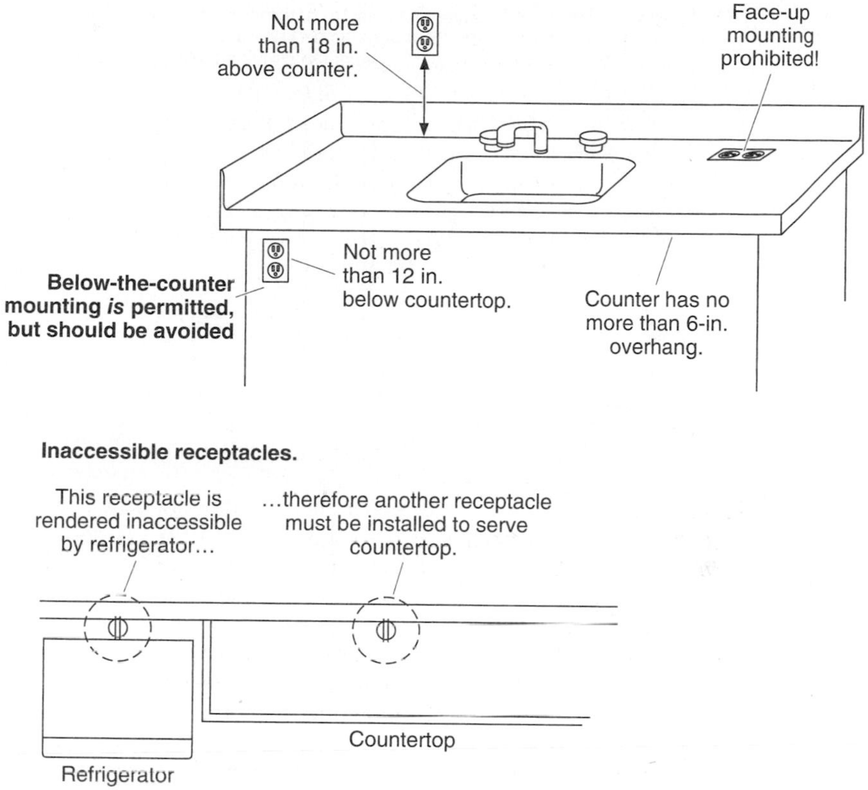
Not more than 18 in. above counter.
Face-up mounting prohibited!
Below-the-counter mounting *is* permitted, but should be avoided
Not more than 12 in. below countertop.
Counter has no more than 6-in. overhang.
Inaccessible receptacles.
This receptacle is rendered inaccessible by refrigerator...
...therefore another receptacle must be installed to serve countertop.
Countertop
Refrigerator

Receptacle located behind an appliance, making the receptacle inaccessible, does not count as one of the required "countertop" receptacles.

Fig. 210-50. This section indicates where required outlets intended to serve counter space in dwellings must be installed. Remember that below-the-counter mounting is only permitted on a flat island or peninsula, and where the overhang does not exceed 150 mm (6 in.).

The NEC also recognizes the same pop-up style receptacles for mounting in bathroom countertops to comply with these requirements as it does for kitchen countertops in 210.52(C)(5).

Part **(E)** requires that at least one outdoor receptacle "readily accessible from grade" and not more than 6 ft 6 in. (2.0 m) above grade" must be installed at the front and back for every one-family house ("a one-family dwelling") and grade-level accessible unit in a two-family dwelling. Receptacles on decks and porches now are eligible to count among the required outdoor receptacles provided they are within the prescribed height limitation and can be reached easily from grade, such as through permanent stairs. And note that there will be a receptacle outlet on that deck or porch. Why? Because the NEC now requires that every deck, balcony, and porch that is accessible from within any dwelling unit (whether one,

210.52(D). Receptacle outlets installed in bathrooms in dwellings must be supplied from dedicated—no other outlets—20-A branch circuit . . . unless the circuit supplies only one bathroom.

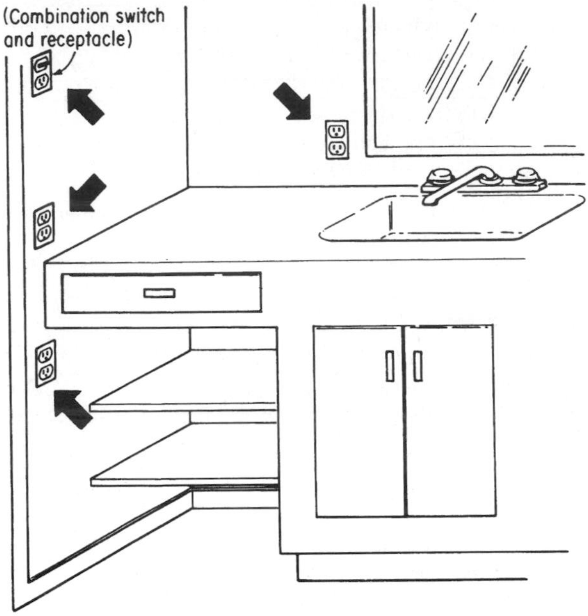
(Combination switch
and receptacle)

LOCATION of receptacle will vary, depending upon available wall space. Arrows show several possibilities. A receptacle in a medicine cabinet or in the bathroom lighting fixture does not satisfy this rule. The receptacle must also be located not more than 36 in. from the edge of the sink. It should be noted that mounting on an adjacent wall or partition is clearly permitted.

Fig. 210-51. A dedicated 20-A branch circuit must be provided to supply required receptacle outlets installed "adjacent" to bathroom sinks, as well as any other receptacle outlets installed in the bathroom. The outlet installed below the counter space here is now clearly acceptable, provided it is not more than 300 mm (12 in.) below the basin countertop. The 2014 NEC clarified that this vertical distance limitation applied whether the receptacle was wired in the basin cabinet or on an adjacent wall.

two, or multifamily construction) (and attached thereto) have a receptacle outlet for the outdoor space, located not over 2.0 m (6½ ft) above the walking surface, and accessible from that outdoor space. This wording is less restrictive in the 2014 NEC, having formerly been required to be within the actual perimeter of the space. The exception for small porches and balconies in the 2008 Code has not returned. Any balcony or porch, no matter how small, is viewed as a likely candidate for seasonal lighting if nothing else, and therefore will require a receptacle outlet.

Part **(E)** also requires that townhouse-type multifamily dwellings be provided with at least one GFCI-protected outdoor receptacle outlet not over 2.0 m (6½ ft) above grade. The front-and-back provision does not apply to multifamily construction, but the outdoor porch, deck, and balcony provisions do apply. Thus, for multifamily housing only one outdoor receptacle is required, and located at either the front or the back, and anywhere accessible from grade, such as up some porch steps. The distinction between multifamily housing and single- or two-family construction is not obvious because it relies on construction details that must be reviewed, preferably with the local building official. Specifically, it must be determined whether the separations between occupancies are fire separation walls of specified hourly ratings, or true fire walls that are made of masonry or concrete, run from grade to roof line or above, and that will survive a conflagration on one side.

Fire walls define buildings per Art. 100, so much depends on this evaluation. Due to their expense, most fire separations are not fire walls but rather other fire-resistant construction that has enough of an hourly rating to allow the occupants to escape in an orderly way, as defined by the local building code. In some cases, a building will have both for other reasons. For example, there are many examples of structures with eight dwelling units as depicted in Fig. 210-52 that have a fire

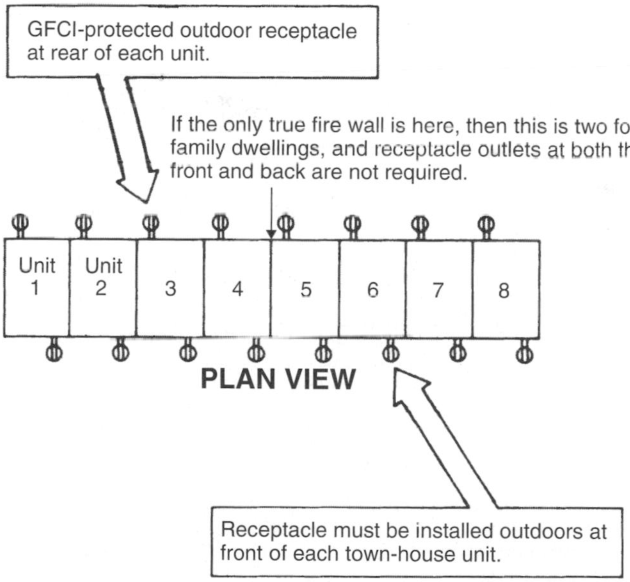
GFCI-protected outdoor receptacle at rear of each unit.
If the only true fire wall is here, then this is two four-family dwellings, and receptacle outlets at both the front and back are not required.
Unit 1
Unit 2
3
4
5
6
7
8
PLAN VIEW
Receptacle must be installed outdoors at front of each town-house unit.

Fig. 210-52. Front and rear-receptacle outlets are required outdoors for town-house-type structures as shown, but only if true fire walls divide the structure into multiple single- or two-family dwellings. With mere fire separation walls and not true fire walls, this is multifamily housing and only one outdoor receptacle per unit is required. [Sec. 210.52(E).]

wall down the middle, creating two four-family dwellings. This has been done at various times to avoid expensive fire alarm systems and sprinklers that would otherwise have been required at the time if the structure had qualified as a single building. Most electrical inspectors will defer to the judgment of the local building official as to how the local building code classifies the nature of an occupancy separation.

Part **(F)** requires that at least one receptacle—single or duplex or triplex—must be installed for the laundry area of a dwelling unit. Such a receptacle and any other receptacles for special appliances must be placed within 6 ft (1.8 m) of the intended location of the appliance. Exceptions apply for multifamily housing where central laundry facilities are provided on site for the occupants, or in other than one-family dwellings in instances where on-site laundry facilities are not to be installed. And part **(G)** requires a receptacle outlet in a basement in addition to any receptacle outlet(s) that may be provided as the required receptacle(s) to serve a laundry area or other designated equipment such as a whole-house vacuum system in the basement. One receptacle in the basement at the laundry area located there may *not* serve as *both* the required "laundry" receptacle and the required "basement" receptacle. A separate receptacle has to be provided for each requirement to satisfy the Code rules.

210.52**(G)** requires that at least one receptacle outlet be provided in each portion of the basement that is "unfinished" in a one- or a two-family house, in addition to any required for a basement laundry or other dedicated use (Fig. 210-53). In other words, if a finished section of a basement results in two noncontiguous unfinished sections in different parts of a basement, then both of those unfinished

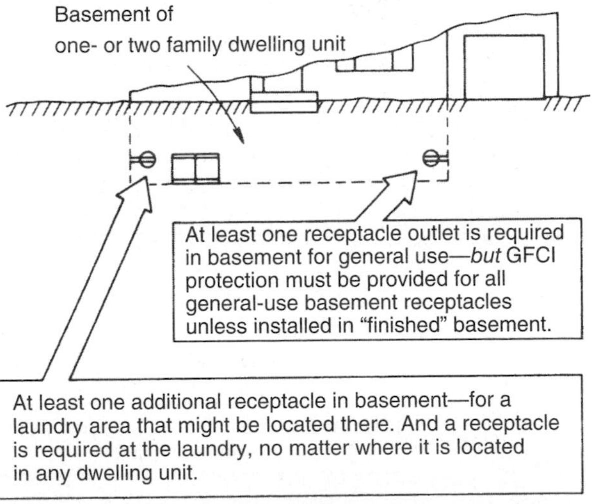
Basement of
one- or two family dwelling unit
At least one receptacle outlet is required in basement for general use—*but* GFCI protection must be provided for all general-use basement receptacles unless installed in "finished" basement.
At least one additional receptacle in basement—for a laundry area that might be located there. And a receptacle is required at the laundry, no matter where it is located in any dwelling unit.

Fig. 210-53. Only one basement receptacle is required (in addition to any for the laundry), but *all* general-purpose receptacles in *unfinished* basements must be GFCI protected. As of the 2017 **NEC**, this rule also applies to two-family houses. [Sec. 210.52(G).]

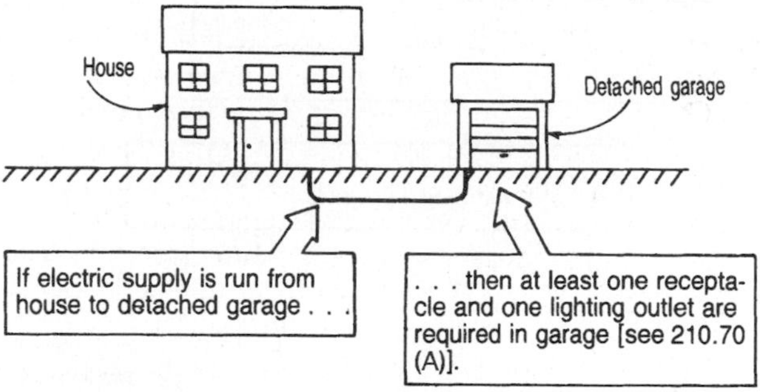

House
Detached garage
If electric supply is run from house to detached garage . . .
. . . then at least one receptacle and one lighting outlet are required in garage [see 210.70 (A)].

Fig. 210-54. Detached garage may be required to have a receptacle and lighting outlet. [See. 210.52(G).]

sections must have a receptacle installed on a general-purpose branch circuit. It calls for at least one receptacle in an attached garage of a one- or two-family house. But for a detached garage, the rule simply requires that one receptacle outlet must be installed in the detached garage *if*—for some reason other than the **NEC**—electric power is run to the garage, such as where the owner might desire it or some local code might require it (Fig. 210-54). The rule itself does not require that electric power be run to a detached garage to supply a receptacle there. Effective with the 2011 **NEC**, the same rule also applies to detached accessory buildings.

The 2014 **NEC** required a receptacle outlet for each vehicle space, but did not specify a location. The 2017 **NEC** has closed this loophole by requiring placement "in each vehicle bay" and not over 5½ ft above the floor, matching the height limit in 210.52(4). As now covered in 210.11(C)(4), at least one circuit must be dedicated to these outlets (with an outdoor receptacle also permitted). Figure 210-55 shows some of the required receptacles.

In part **(H)**, a receptacle outlet is required in any dwelling-unit hallway that is 10 ft (3.0 m) or more in length. This provides for connection of plug-in appliances that are commonly used in halls—lamps, vacuum cleaners, and so forth. The length of a hall is measured along its centerline. Although 210.52(H) calls for one receptacle outlet for each dwelling-unit hallway that is 10 ft (3.0 m) or more in length, part **(H)** does not specify location or require more than a single receptacle outlet. However, good design practice would dictate that a convenience receptacle should be provided for each 10 ft (3.0 m) of hall length. And they should be located as close as possible to the middle of the hall. Note that the rule applies within dwelling units only, and therefore does not apply to a common hallway of a multifamily building or a hotel, although it would be obviously good design to provide receptacles in those other buildings.

Part **(I)**, as of the 2011 **NEC**, requires receptacle coverage in foyers. Foyers, or entrance hallways, have been previously considered as hallways and only a single receptacle outlet has been enforceable, and then only if the total length was 10 ft (3.0 m) or longer. This new paragraph only applies to foyers that are not "part" of a hallway qualifying as such under 210.52(H), and which exceed 60 sq ft (5.6 sq m)

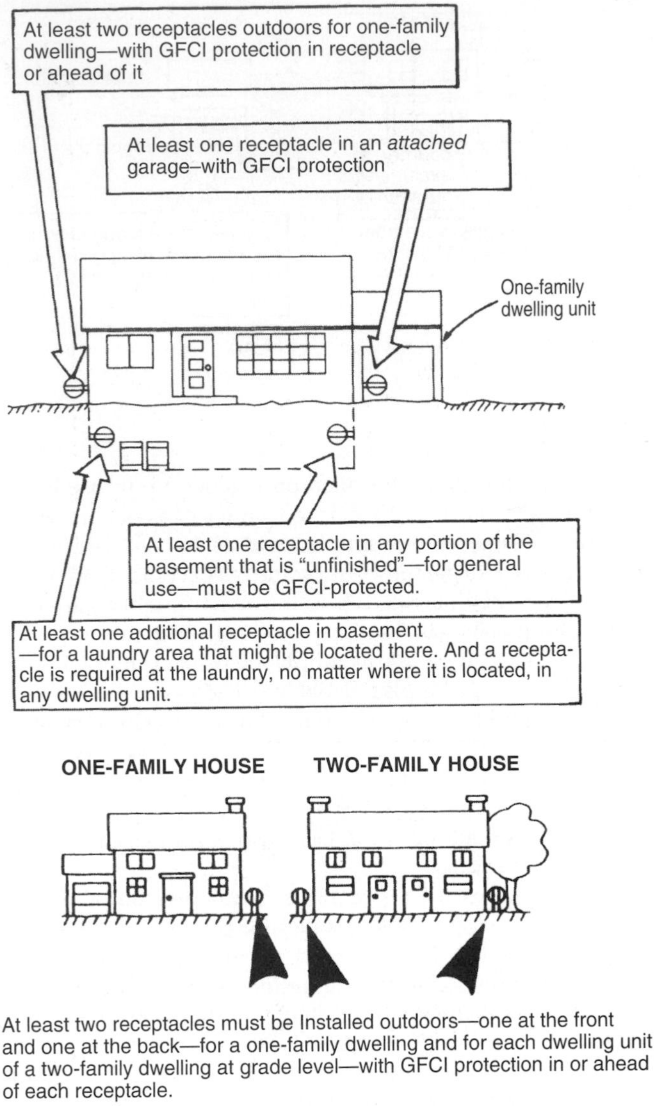
At least two receptacles outdoors for one-family dwelling—with GFCI protection in receptacle or ahead of it
At least one receptacle in an *attached* garage–with GFCI protection
One-family dwelling unit
At least one receptacle in any portion of the basement that is "unfinished"—for general use—must be GFCI-protected.
At least one additional receptacle in basement —for a laundry area that might be located there. And a receptacle is required at the laundry, no matter where it is located, in any dwelling unit.
ONE-FAMILY HOUSE
TWO-FAMILY HOUSE
At least two receptacles must be Installed outdoors—one at the front and one at the back—for a one-family dwelling and for each dwelling unit of a two-family dwelling at grade level—with GFCI protection in or ahead of each receptacle.

Fig. 210-55. These specific receptacles are required for dwelling occupancies. [Sec. 210.52(E), (F), and (G).]

in area. It also does not apply to entrance areas that are part of a room otherwise covered in 210.52(A). Figure 210-56 shows a foyer that qualifies under this new provision.

If a foyer qualifies under this rule, at least one receptacle outlet must be installed in each wall space 3 ft (900 mm) or wider. This measurement includes contiguous

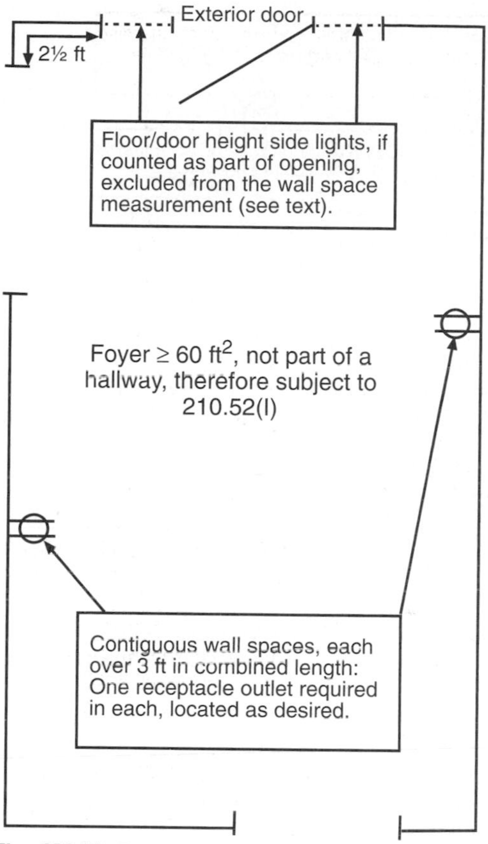

Exterior door
2½ ft
Floor/door height side lights, if counted as part of opening, excluded from the wall space measurement (see text).
Foyer ≥ 60 ft², not part of a hallway, therefore subject to 210.52(I)
Contiguous wall spaces, each over 3 ft in combined length: One receptacle outlet required in each, located as desired.

Fig. 210-56. Receptacles are required for large foyers. [Sec. 210.52(I).]

wall space unbroken at the floor line by doorways, door-side windows that extend to the floor, and similar openings.

210.60. Guest Rooms, Guest Suites, Dormitories, and Similar Occupancies. The number of receptacles in a guest room of a hotel or motel, or an equivalent sleeping room of a dormitory or similar location, must be determined by the every-12-ft rule of 210.52(A) but *may* be located where convenient for the furniture layout. In other words, first lay out the room on paper as though 210.52(A) applied as written, and count the receptacles that result. Then, shift the receptacles if desired to accommodate permanent furniture arrangements, but do not decrease the total number. In addition, make sure that no fewer than two of the receptacle outlets are readily accessible. If a receptacle outlet falls behind the bed, install some means of assuring that when the bed is moved against the wall, any cords plugged into the receptacle are not damaged (a wet-location "in-use" cover might suffice), or locate the receptacle low enough so as to be out of the way. Where "permanent provisions

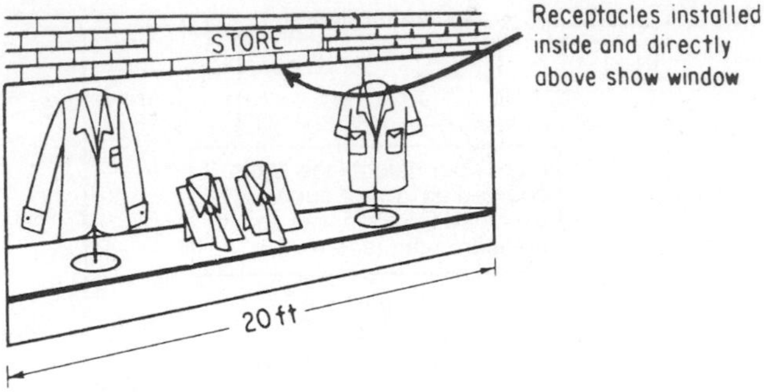

Receptacles installed inside and directly above show window
STORE
20 ft
For a 20-ft-long store show window,
a minimum of two receptacles must be installed, one for each 12 linear ft or major fraction thereof of show window length.

Fig. 210-57. Receptacles are required for show windows in stores or other buildings. (Sec. 210.62.)

for cooking" are installed in guest rooms or suites, the installation must satisfy the rule of 210.17 and be wired just as if it were an individual dwelling unit within a multifamily dwelling.

210.62. Show Windows. The rule here calls for one receptacle in a show window for each 3.7 m (12 ft), or major fraction thereof, of length (measured horizontally) to accommodate portable window signs and other electrified displays (Fig. 210-57). The receptacles (125-V, 15- or 20-A, single phase) must be installed within 450 mm (18 in.) of the top of the window to count.

210.63. Heating, Air-Conditioning, and Refrigeration Equipment Outlet. A general-purpose 125-V receptacle outlet must be installed within 25 ft (7.5 m) of heating, air-conditioning, and refrigeration equipment. Although the 1999 NEC limited application of this rule to equipment located on rooftops *and* in attics and crawl spaces (Fig. 210-58), the maintenance receptacle required by this rule must be provided wherever the equipment is installed, and now the rule applies wherever the equipment is located. The receptacles must be on the same level as the equipment, which is occasionally an issue on discontinuous rooftops where a "nearby" receptacle within 7.5 m (25 ft) as the crow flies in actuality requires a ladder to reach. Common sense would dictate that an additional receptacle need not be installed where there is another properly rated receptacle within 25 ft and on the same level as the equipment. For example, if one of the outdoor receptacles installed at a dwelling unit falls within 7.5 m (25 ft) of this equipment, an additional receptacle need not be installed. The service receptacle must not be connected to the load side of the disconnect switch for the equipment for obvious reasons. Note that the rule does not apply to pure ventilating equipment (the "V" in the familiar acronym "HVAC") and it also does not apply to evaporative coolers (the so-called "swamp coolers") at single- and two-family dwellings.

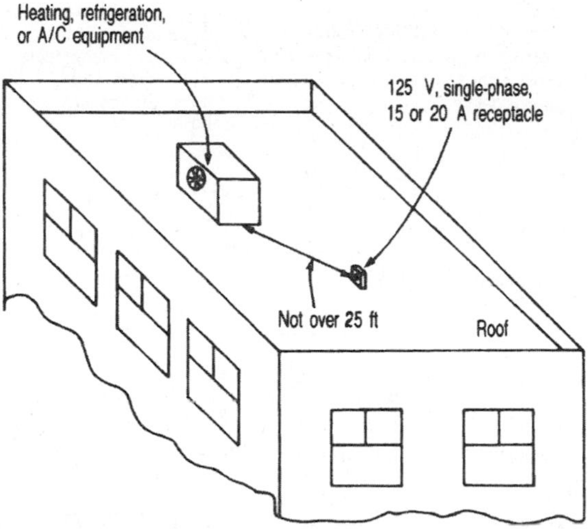
Heating, refrigeration, or A/C equipment
125 V, single-phase, 15 or 20 A receptacle
Not over 25 ft
Roof

Fig. 210-58. Maintenance receptacle outlet required for rooftop mechanical equipment as well as for such equipment in all other locations—indoors and outdoors. (Sec. 210.63.)

210.64. Electrical Service Areas. In other than one- and two-family dwellings, at least one general-purpose receptacle outlet must be provided in an accessible location within 25 ft and located in the same room as the service equipment. This will allow for orderly connections of monitoring equipment for data analysis and other purposes in such areas. As of the 2017 **NEC**, this requirement is waived for service equipment operating above 120 volts to ground at electric irrigation machinery and at natural and artificial bodies of water.

210.70. Lighting Outlets Required. The basic rule of part **(A)(1)** requires at least one wall-switch controlled lighting outlet in habitable rooms and in bathrooms of dwelling units. The 2017 **NEC** added language to clarify that kitchens are also included. The rule of part **(A)(2)** calls for wall-switch-controlled lighting outlets in halls, stairways, attached garages, "detached garages with electric power," and at outdoor entrances and exits in dwelling units.

The rule of 210.70(A)(1) basically requires that a "wall-switched-controlled" lighting outlet be provided in every room that would be required to be provided with a receptacle outlet as defined in 210.52(A). While that requirement constitutes the basic rule, a wall-switch-controlled receptacle will satisfy Exception No. 1 to 210.70(A)(1) where used in every "habitable room" except the kitchen—which, like the bathrooms in dwelling units, must *always* have a wall-switch-controlled lighting outlet. Other areas of dwelling units are covered in part 210.70(A)(2), while storage and equipment space are governed by part **(A)(3)**.

The word *bathrooms* is in the basic rule because various building codes do not include bathrooms under their definition of "habitable rooms." So the word "bathroom" was needed to ensure that the rule covered bathrooms. The rule does not stipulate that the required "lighting outlets" must be ceiling lighting outlets; they also may be wall-mounted lighting outlets (Fig. 210-59).

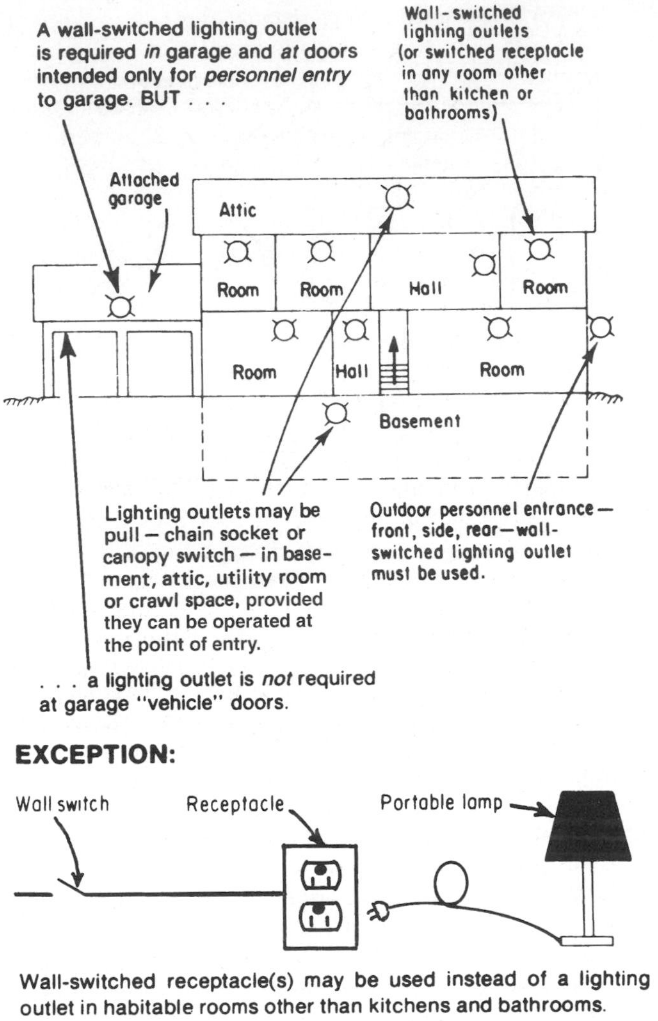
A wall-switched lighting outlet is required *in* garage and *at* doors intended only for *personnel entry* to garage. BUT . . .
Wall-switched lighting outlets (or switched receptacle in any room other than kitchen or bathrooms)
Attached garage
Attic
Room
Room
Hall
Room
Room
Hall
Room
Basement
Lighting outlets may be pull – chain socket or canopy switch – in basement, attic, utility room or crawl space, provided they can be operated at the point of entry.
Outdoor personnel entrance – front, side, rear – wall-switched lighting outlet must be used.
. . . a lighting outlet is *not* required at garage "vehicle" doors.
EXCEPTION:
Wall switch
Receptacle
Portable lamp
Wall-switched receptacle(s) may be used instead of a lighting outlet in habitable rooms other than kitchens and bathrooms.

Fig. 210-59. Lighting outlets required in dwelling units. (Sec. 210.70.)

Two exceptions are given to the basic requirements in **(A)(1)**. Exception No. 1 notes that in rooms other than kitchens and bathrooms, a wall-switch-controlled receptacle outlet may be used instead of a wall-switch-controlled lighting outlet. The receptacle outlet can serve to supply a portable lamp, which would give the necessary lighting for the room.

Part **(A)(1)**, Exception No. 2, indicates two conditions under which the use of an occupancy sensor is permitted to control any of the required lighting outlets designated in the basic rule: (1) where used in addition to the required wall switch and (2) where the sensor is equipped with manual override and is mounted at the "customary" switch location. Notice that the literal wording only permits such control for "lighting outlets." Therefore, even though it is not entirely clear, it must be assumed that occupancy sensor control of a receptacle outlet—installed in accordance with the first exception to this section—is *not* permitted. In addition, occupancy sensors are generally rated not to exceed specified loads, and a receptacle load is not controlled, which would likely result in a listing violation [110.3(B)]. Note that 404.14(E) imposes a similar reservation on dimmer switches.

In 210.70(A)(2)(a), the Code calls for wall-switched lighting outlets for all those "other" areas in dwellings. Specifically, hallways and stairways must be provided with at least one wall-switch-controlled lighting outlet.

Part **(A)(2)(1)** also requires a wall-switch-controlled lighting outlet in every *attached* garage of a dwelling unit (such as a one-family house). But, for a *detached* garage of a dwelling unit, a switch-controlled lighting outlet is required *only* if the garage is provided with electric power—whether the provision of power is done as an optional choice or is required by a local code. Note that the NEC rule here does not itself require running power to the detached garage for the lighting outlet, but simply says that the lighting outlet must be provided *if* power is run to the garage.

Part **(A)(2)(2)** requires that outdoor entrances for personnel that afford grade-level access at dwelling unit garages be provided with an exterior lighting outlet. The second sentence clarifies that "a vehicle door in an attached garage is *not* considered as an outdoor entrance." This makes it clear that the Code does not require such a light outlet at any garage door that is provided as a vehicle entrance because the lights of the car provide adequate illumination when such a door is being used during darkness. But the wording of this sentence does suggest that a rear or side door that is provided for personnel entry to an attached garage would be "considered as an outdoor entrance" because the note excludes only "vehicle" doors. Such personnel entrances from outdoors to the garage would seem to require a wall-switched lighting outlet.

At least one lighting outlet must be installed in every attic, underfloor space, utility room, and basement if it is used for storage or if it contains equipment requiring servicing. In such cases, the lighting outlet must frequently be controlled by a wall switch, but not always. The rule specifically allows a pull-chain lampholder or luminaire, provided it is reachable ("at least one point of control") at the usual point of entry. However, for equipment requiring servicing, the lighting outlet must also be located at or near the equipment. If the equipment is at the point of entry, or if there is not such equipment and the space is only used for storage, then the pull chain at the entry will do. However, if there is such equipment and it is remote from the entry, then a wall switch is the only practical method because the luminaire or lampholder will be over at the equipment.

Part **(A)(2)(3)** requires that stairways between levels that have six or more risers have wall switch control from both ends of the stairway, with additional control required if the stairway includes a landing with an "entryway" between floor levels, as is common with split-level houses. The "entryway" provision closes a loophole in the top-and-bottom rule that would otherwise have no wall switch control over the stairway lighting available to an occupant returning through the front door. Note that this rule applies even to a stairway connecting to an unfinished attic, and therefore a three-way switch loop or some other control must be arranged at both ends of the attic stair in this case. The 2017 **NEC** added the limitation that a conventional dimmer switch cannot be used at one end of a stair. If left in the fully dimmed position, the switch at the other end of the stair becomes non-functional. However, dimming is permitted if it can be controlled from both locations.

Where using any occupancy sensor for control of lighting, the use of a sensor that fails in the "on" position would be preferable to one that fails "off" or one that fails "as is." Such a fail-safe feature on any sensor is not required but is preferable because control of the lighting outlet can be provided from the conventional wall switch or manual override until such time as the sensor can be replaced (Fig. 210-60). But, in the exception to **(A)(2)(1, 2, and 3)**, the Code states that "in hallways, stairways, and at outdoor entrances remote, central, or automatic control of lighting shall be permitted." This latter recognition appears to accept remote, central, or automatic control as an alternative to the wall switch control mentioned in the basic rules.

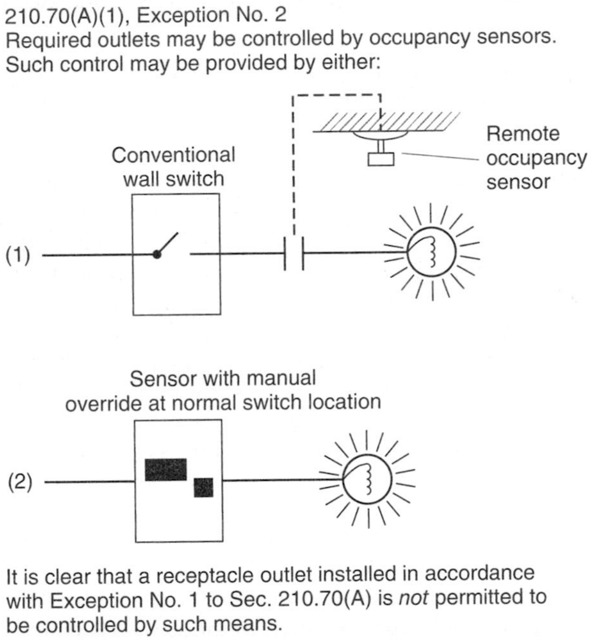
210.70(A)(1), Exception No. 2
Required outlets may be controlled by occupancy sensors.
Such control may be provided by either:
Conventional wall switch
Remote occupancy sensor
(1)
Sensor with manual override at normal switch location
(2)
It is clear that a receptacle outlet installed in accordance with Exception No. 1 to Sec. 210.70(A) is *not* permitted to be controlled by such means.

Fig. 210-60. Occupancy sensor control for lighting outlets in dwellings must be as shown here. Either a sensor and a conventional wall switch or a sensor with a manual override installed at the "customary" wall-switch location must be provided. Remember, the rule here only applies to dwellings.

210.70(B) notes that at least one wall-switch-controlled lighting outlet is to be provided for guest rooms in hotels, motels, or similar occupancies. As in the case of conventional dwelling units, exceptions provide for wall switch control of receptacles in other than bathrooms, and in kitchens where provided (Fig. 210-61). In addition, the exceptions also allow for occupancy sensors on the same basis as for dwelling units generally.

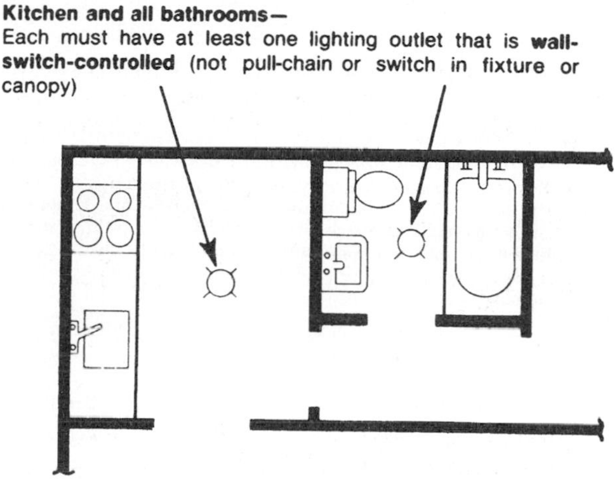
Kitchen and all bathrooms—
Each must have at least one lighting outlet that is wall-switch-controlled (not pull-chain or switch in fixture or canopy)

Fig. 210-61. Switch-controlled lighting outlet in kitchen and bathroom of guest rooms and suites. [Sec. 210.70(A).]

Part **(C)** of 210.70 requires that either a lighting outlet containing a switch—such as the familiar pull-chain porcelain lampholder—or a wall-switch-controlled lighting outlet must be provided in attics or underfloor spaces, utility rooms, and basements—*in all occupancies* (including dwelling units). The lighting outlet must be located at or near the equipment to provide effective illumination. And the control wall switch must be installed at the point of entry to the space.

210.71. Meeting Rooms. As of the 2017 NEC, meeting rooms equal in size or smaller than 1000 sq ft must comply with a mandatory convenience receptacle placement provision. Larger rooms need not comply, but the evaluation must be carried out with any relocatable partitions so arranged as to create the smallest possible room size. The informational notes provide some helpful context, by explaining the intended coverage involves seated gatherings of conferees, typically using portable electronic equipment along with projectors and the like. Panel meetings of the National Electrical Code Committee are excellent examples of such activities. Rooms styled as auditoriums, schoolrooms, and coffee shops are specifically cited as not the intended application of this new section. Figure 210-62 shows the process at work in a 4000 sq ft conference wing of a hotel.

The process considers both fixed walls and the open floor as candidates for receptacles. For the fixed walls, in a similar procedure to hotel rooms, this section requires

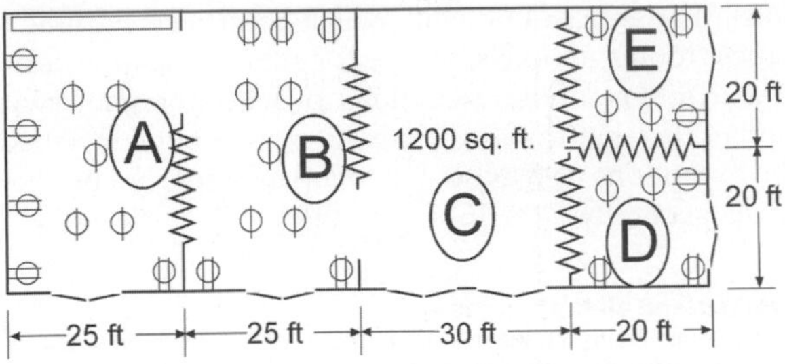
A
B
1200 sq. ft.
C
E
D
20 ft
20 ft
25 ft
25 ft
30 ft
20 ft

Fig. 210-62. A 4000 sq ft conference area. The receptacle count is based on the relocatable partitions being closed to minimize floor areas. The left room "A" has fixed cabinets in the back without work surfaces, and therefore has its receptacle census reduced [210.52(A)(4)]. Room "B" has the receptacles on the rear fixed wall repositioned to accommodate a design decision by the owner. Both Room "A" and that room are 1000 sq ft, and have the full complement of floor receptacles (shown as single receptacles for the purposes of this illustration; four plus a major fraction means five required). Room "C" is too large to be subject to the requirement. Smaller rooms "D" and "E" follow the customary procedures (210.71).

that a total number of receptacles be determined as though the normal residential rules in 210.52(A) were being applied, and then a number of receptacle outlets at least equal to that be installed. And, also comparable to hotel rooms, the specific locations of those receptacles can be located according to the design objectives of the owner. Any floor of a conference room must also be populated if it is at least 12 ft wide, and at least 215 sq ft in extent, in this case with floor receptacles. Here, the total number is determined by dividing the floor area in square feet by 215, with any major fractions rounded up. The positioning of floor receptacles is left as a design choice, but they must always be at least 6 ft from a fixed wall to be counted as satisfying the requirements.

ARTICLE 215. FEEDERS

215.1. Scope. *Feeders* are the conductors that carry electric power from the service equipment (or generator switchboard, where power is generated on the premises) to the overcurrent protective devices for branch circuits supplying the various loads. *Subfeeders* originate at a distribution center other than the service equipment or generator switchboard and supply one or more other distribution panelboards, branch-circuit panelboards, or branch circuits. Code rules on feeders apply also to all subfeeders for the simple reason that all subfeeders meet the definition of a feeder, and for this reason the **NEC** does not recognize the term (Fig. 215-1). Its use is being gradually phased out of this book.

As a matter of good design, for the given circuit voltage, feeders and subfeeders must be capable of carrying the amount of current required by the load,

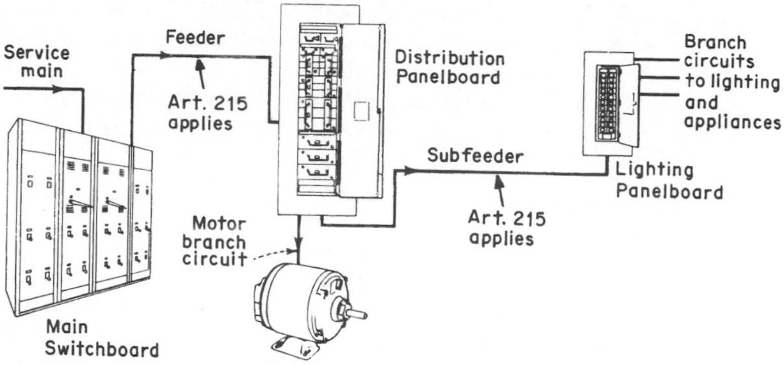
Service main
Feeder
Art. 215 applies
Distribution Panelboard
Branch circuits to lighting and appliances
Subfeeder
Lighting Panelboard
Art. 215 applies
Motor branch circuit
Main Switchboard

Fig. 215-1. Article 215 applies only to those circuits that conform to the NEC definition of "feeder." (Sec. 215.1.)

plus any current that may be required in the future. Selection of the size of a feeder depends on the size and nature of the known load computed from branch-circuit data, the anticipated future load requirements, and voltage drop. However, the NEC does not require owners and installers to be wise about the future, and a feeder that will carry the load connected to it as determined by Art. 220 is NEC compliant. Section 90.1(B) clearly states that an NEC-compliant installation will be "essentially free from hazard, but not necessarily efficient, convenient, or adequate for good service or future expansion of electrical use."

Article 215 deals with the determination of the minimum sizes of feeder conductors necessary for safety. Overloading of conductors may result in insulation breakdowns due to overheating, overheating of switches, busbars, and terminals; the blowing of fuses and consequent overfusing; excessive voltage drop; and excessive copper losses. Thus, the overloading will in many cases create a fire risk and is sure to result in very unsatisfactory service.

215.2. Minimum Rating and Size. The actual maximum load on a feeder depends upon the total load connected to the feeder and the demand factor(s) as established by the rules in parts III, IV, and V of Art. 220.

From an engineering viewpoint, there are two steps in the process of predetermining the maximum load that a feeder will be required to carry: first, a reasonable estimate must be made of the probable connected load; and, second, a reasonable value for the demand factor must be assumed. From a survey of a large number of buildings, the average connected loads and demand factors have been ascertained for lighting and small appliance loads in buildings of the more common classes of occupancy, and these data are presented in parts III, IV, and V of Art. 220 as minimum requirements. That is, it is not permissible to assume any demand factor that would result in a calculated load that is less than the Code-prescribed minimum. However, given that calculations in accordance with the NEC provide for no future growth or additional loading, providing capacity for less than the Code-prescribed minimum represents poor design and is a violation of the NEC.

The load is specified in terms of volt-amperes per square foot for certain occupancies. These loads are here referred to as standard loads, because they are minimum standards established by the Code in order to ensure that the service, feeder, and branch-circuit conductors will have sufficient carrying capacity for safety.

Calculating Feeder Load

The key to accurate determination of required feeder conductor capacity in amperes is effective calculation of the total load to be supplied by the feeder. Feeders and subfeeders are sized to provide sufficient power to the circuits they supply. For the given circuit voltage, they must be capable of carrying the amount of current required by the load, plus any current which may be required in the future. The size of a feeder depends upon known load, future load, and voltage drop.

The minimum load capacity which must be provided in any feeder or subfeeder can be determined by considering **NEC** requirements on feeder load. As presented in Sec. 215.2, these rules establish the minimum load capacity to be provided for all types of loads.

The first sentence of 215.2(A)(1) requires feeder conductors to have ampacity at least equal to the sum of loads on the feeder, as determined in accordance with Art. 220. And 215.3 gives rules on the rating of any feeder protective device.

If an overcurrent protective device for feeder conductors is not UL-listed for continuous operation at 100 percent of its rating, the load on the device must not exceed the noncontinuous load plus 125 percent of the continuous load. 215.3 applies to feeder overcurrent devices—CBs and fuses in switch assemblies—and requires that the rating of any such protective device must generally never be less than the amount of noncontinuous load of the circuit (that amount of current that will not be flowing for 3 h or longer) plus 125 percent of the amount of load current that will be continuous (flowing steadily for 3 h or longer) (Fig. 215-2).

For any given load to be supplied by a feeder, after the minimum rating of the overcurrent device is determined from the preceding calculation (noncontinuous plus 125 percent of continuous), then a suitable size of feeder conductor must be selected. For each ungrounded leg of the feeder (the so-called phase or line legs of the circuit), the conductor must have a table ampacity in the 75°C column that is at least equal to the amount of noncontinuous current plus the amount of continuous current, from the **NEC** tables of ampacity [Tables 310.15(B)(16) through 310.15(B)(21)].

Although the rules of 210.19 and 215.2 are aimed at limiting the load on the circuit protective device, the conductor's ampacity also must be based on the nature of the load. Just as is required for the overcurrent device, the conductor's ampacity must not be less than the noncontinuous load plus 125 percent of the continuous load, *except* where derating—either for the number of conductors, 310.15(B)(3)(a), or elevated ambient temperature, which must be derated by the factor shown in the Ambient Temperature Correction Factors in Tables 310.15(B)(2)(a) or 310.15(B)(2)(b) as applicable—is needed. In those cases, the conductor's table ampacity in the 75°C column must be not less than the sum of noncontinuous plus 125 percent of the continuous load *before* any derating is applied. *And,* after derating *is* applied,

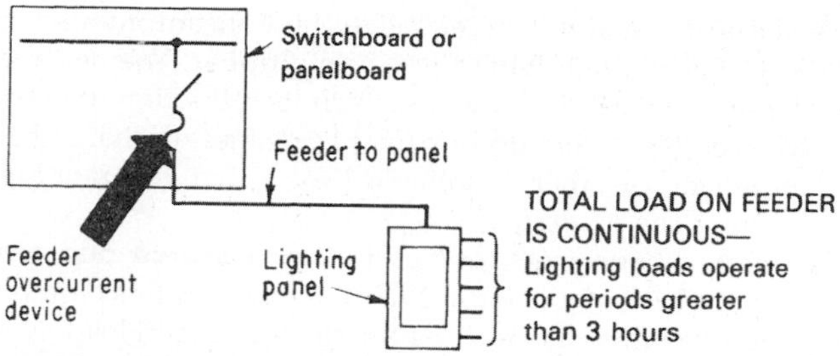
Switchboard or
panelboard
Feeder to panel
TOTAL LOAD ON FEEDER
IS CONTINUOUS—
Lighting loads operate
for periods greater
than 3 hours
Feeder
overcurrent
device
Lighting
panel

FEEDER OVERCURRENT DEVICES AND CONDUCTORS must be rated not less than 125% of the continuous load *and* the feeder conductors must be sized so they have an ampacity such that they are properly protected by the rating of the feeder CB or fuses, as required by 240.4. Another way of saying that is "the continuous load must not exceed 80% of the rating of the protection."

EXAMPLE: For this feeder, with conductors rated at 420 A, (smallest size at least 125% of the continuous load), the maximum continuous load permitted for a conventional fused switch is 80% of the 400-A fuse rating [400 × 0.8 = 320A].

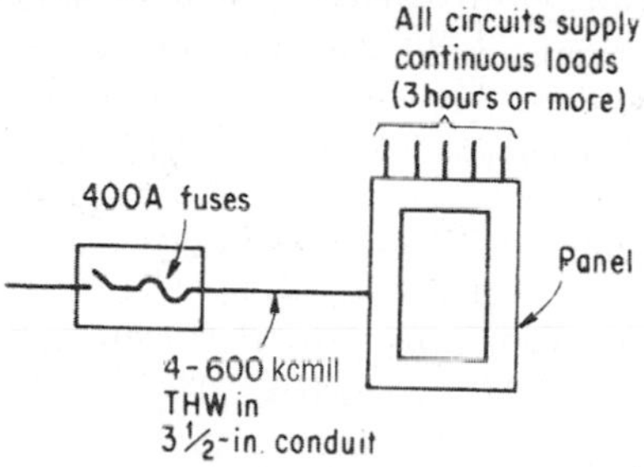
All circuits supply
continuous loads
(3 hours or more)
400A fuses
Panel
4-600 kcmil
THW in
3½-in. conduit

Rating of fuses must be at least equal to 125% times the continuous load and the 400-A rating is proper protection for conductors with an ampacity of 420 A. But, the load must be limited to 320 A to prevent over heating of OC device.

Fig. 215-2. Feeders must generally be loaded to no more than 80 percent for a continuous load. [Sec. 215.2(A)(1)]

the conductor's ampacity must be such that the overcurrent device protects the conductor as required or permitted by 240.4.

Note that the conductor size increase previously described applies only to the ungrounded or phase conductors because they are the ones that must be properly protected by the rating of the protective device. A neutral or grounded conductor of a feeder does not have to be increased; its size must simply have ampacity sufficient for the neutral load as determined from 220.61.

The exceptions for 215.2(A) and 215.3 note that a CB or fused switch that is UL-listed for continuous operation at 100 percent of its rating may be loaded right up to a current equal to the device rating. Feeder ungrounded conductors must be selected to have ampacity equal to the noncontinuous load plus the continuous load—without applying the 1.25 multiplier. The neutral conductor is sized in accordance with 220.61, which permits reduction of neutral size for feeders loaded over 200 A that do not supply electric-discharge lighting, data processing equipment, or other "nonlinear loads" that generate high levels of harmonic currents in the neutral.

Feeders run with differing ampacities between source and load A new Exception No. 2 to 215.2(A)(1)(a) in the 2017 **NEC** modifies this requirement in the event the feeder conductors change size between source and load. Normally, this is never done; the process of installing a pull box and splicing and re-splicing conductors between line and load ends is generally not cost-effective. However, there are instances where it is extremely advantageous, and even required under some conditions. The remainder of the coverage in this book on Article 215 assumes that this exception is not being used because it usually isn't; however, it is now a live option.

One example is when Type MI cable, with its inherent fire resistance characteristics, is used between source and load. Sec. 700.10(D(1)(2) requires a 2-hour fire resistance rating for emergency circuit feeders under the occupancy conditions specified in 700.10(D), and Type MI cable is often used for this purpose. Now, the ampacity of MI cable is essentially limited only by the melting point of copper because its insulation is magnesium oxide, which is a refractory compound with a melting point far higher than copper. This is why 332.80 essentially allows free-air ampacities or better for this wiring, and the MI cable industry has step-up connectors that allow for the direct termination of small MI conductors into connectors that allow for integral connections to larger conductor tails. Another example is cablebus, which is also allowed free-air ampacities by 370.80 and is essentially designed for large-capacity industrial feeders. There have also been a number of field installations done with custom Type MC cable constructions where the extreme length and units costs of a large-capacity industrial feeder supplying continuous loads justified the cost of running the greater length of the circuit with conductors sized to ampacity, and then extensions at each end provided to accommodate **NEC** rules for actual device terminations in 110.14(C).

The underlying principle is that wires have middles and ends, and the rules for one differ from the other. It is indeed possible to separate the two concepts in the process of wiring a feeder through the medium of 110.14(C)(2). Unlike conventional device terminations that must be utilized in concert with the special rules governing the wire sizes permitted to land on them, connectors applied under this procedure are rated only by their own chemistry and design. Properly positioned and wired, these connectors allow an installer to marry one size wire that might be required at a termination with another size wire that works for the length of the run. Power distribution blocks, covered in 314.28(E) and 376.56(B), are extremely well suited to these applications. The terms of the exception assure that undersized conductors never enter the enclosure housing a terminating device. For complete details, please refer to the exhaustive discussion of ampacity calculation and wire selection that is in Chap. 9 of this book, located at the coverage of Example D3(A).

Fuses for feeder protection The rating of a fuse is taken as 100 percent of rated nameplate current when enclosed by a switch or panel housing. But, because of the heat generated by many fuses, the maximum continuous load permitted on a fused switch is restricted by a number of NEMA, UL, and **NEC** rules to 80 percent of the rating of the fuses. Limitation of circuit-load current to *no more than* 80 percent of the current rating of fuses in equipment is done to protect the switch or other piece of equipment from the heat produced in the fuse element—and also to protect attached circuit wires from excessive heating close to the terminals. The fuse itself can actually carry 100 percent of its current rating continuously without damage to itself, but its heat is conducted into the adjacent wiring and switch components.

NEMA standards require that a fused, enclosed switch be marked, as part of the electrical rating, "Continuous Load Current Not to Exceed 80 Percent of the Rating of Fuses Employed in Other Than Motor Circuits" (Fig. 215-3). That derating

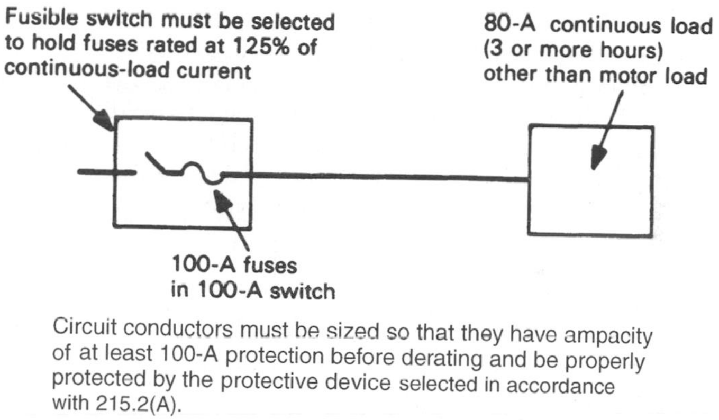
Fusible switch must be selected to hold fuses rated at 125% of continuous-load current
80-A continuous load (3 or more hours) other than motor load
100-A fuses in 100-A switch
Circuit conductors must be sized so that they have ampacity of at least 100-A protection before derating and be properly protected by the protective device selected in accordance with 215.2(A).

Fig. 215-3. For branch circuit or feeder, fuses in enclosed switch must be limited for continuous duty. [Sec. 215.2(A)(1).]

compensates for the extra heat produced by continuous operation. Motor circuits are excluded from that rule, but a motor circuit is required by the NEC to have conductors rated at least 125 percent of the motor full-load current—which, in effect, limits the load current to 80 percent of the conductor ampacity and limits the load on the fuses rated to protect those conductors. But, the UL *Electrical Construction Materials Directory* does recognize fused bolted-pressure switches and high-pressure butt-contact switches for use at 100 percent of their rating on circuits with available fault currents of 100,000, 150,000, or 200,000 rms symmetrical A—as marked (Fig. 215-4). (See "Fused Power Circuit Devices" in that UL publication.)

Manual and electrically operated switches designed to be used with Class L current-limiting fuses rated 601 to 4000 A, 600 V ac are listed by UL as "Fused Power Circuit Devices." This category covers bolted-pressure-contact switches and high-pressure, butt-type-contact switches suitable for use as feeder devices or service switches if marked "Suitable for Use As Service Equipment." Such devices "have been investigated for use at *100 percent of their rating* on circuits having available fault currents of 100,000, 150,000, or 200,000 rms symmetrical amperes" as marked.

CB for feeder protection The nominal or theoretical continuous-current rating of a CB generally is taken to be the same as its trip setting—the value of current at which the breaker will open, either instantaneously or after some intentional

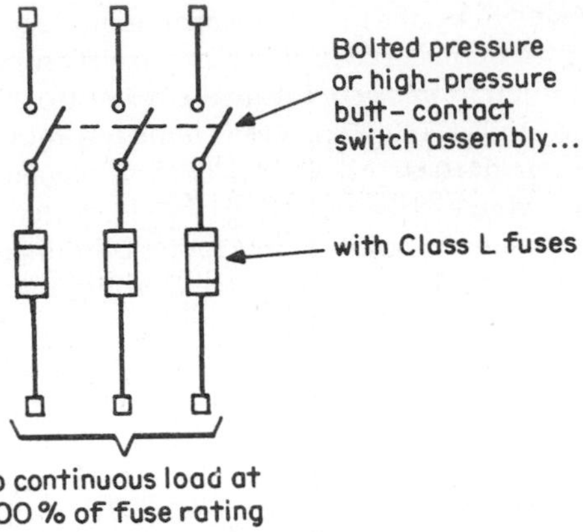
Bolted pressure
or high-pressure
butt-contact
switch assembly...
with Class L fuses
To continuous load at
100% of fuse rating

Fig. 215-4. Some fused switches and CBs may be used at 100 percent rating for continuous load. [Sec. 215.2(A)(1) Exception No. 1.]

time delay. But, as described previously for fuses, the real continuous-current rating of a CB—the value of current that it can safely and properly carry for periods of 3 h or more—frequently is reduced to 80 percent of the nameplate value by codes and standards rules.

The UL *Electrical Construction Materials Directory* contains a clear, simple rule in the instructions under "Circuit Breakers, Molded-Case." It says:

> Unless otherwise marked, circuit breakers should not be loaded to exceed 80 percent of their current rating, where in normal operation the load will continue for three or more hours.

A load that continues for 3 h or more is a *continuous* load. If a breaker is marked for *continuous* operation, it may be loaded to 100 percent of its rating and operate continuously.

There are some CBs available for continuous operation at 100 percent of their current rating, but they must be used in the mounting and enclosure arrangements established by UL for 100 percent rating. Molded-case CBs of the 100 percent continuous type are made in ratings from 225 A up. Information on use of 100-percent–rated breakers is given on their nameplates.

Figure 215-5 shows two examples of CB nameplate data for two types of UL-listed 2000-A, molded-case CBs that are specifically tested and listed for continuous operation at 100 percent of their 2000-A rating—*but* only under the conditions described on the nameplate. These two typical nameplates clearly indicate that ventilation may or may not be required. Because most switchboards have fairly large interior volumes, the "minimum enclosure" dimensions shown on these nameplates (45 by 38 by 20 in.) usually are readily achieved. *But*, special UL tests must be performed if these dimensions are not met. Where busbar extensions and lugs are connected to the CB within the switchboard, the caution about copper conductors does not apply, and aluminum conductors may be used.

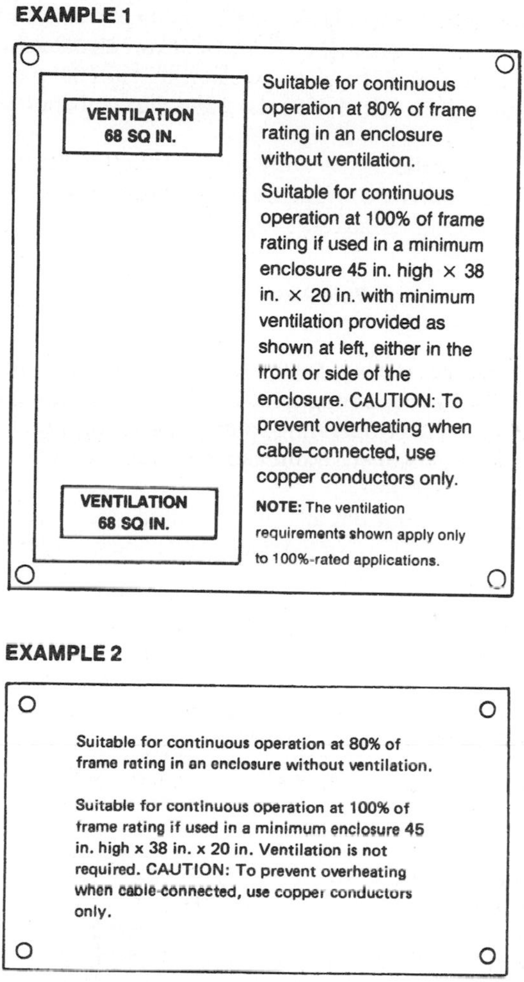
EXAMPLE 1
VENTILATION
68 SQ IN.
Suitable for continuous operation at 80% of frame rating in an enclosure without ventilation.
Suitable for continuous operation at 100% of frame rating if used in a minimum enclosure 45 in. high × 38 in. × 20 in. with minimum ventilation provided as shown at left, either in the front or side of the enclosure. CAUTION: To prevent overheating when cable-connected, use copper conductors only.
NOTE: The ventilation requirements shown apply only to 100%-rated applications.
VENTILATION
68 SQ IN.
EXAMPLE 2
Suitable for continuous operation at 80% of frame rating in an enclosure without ventilation.
Suitable for continuous operation at 100% of frame rating if used in a minimum enclosure 45 in. high x 38 in. x 20 in. Ventilation is not required. CAUTION: To prevent overheating when cable-connected, use copper conductors only.

Fig. 215-5. Nameplates from CBs rated for 100 percent continuous loading. [Sec. 215.2(A)(1) Exception No. 1.]

If the ventilation pattern of a switchboard does not meet the ventilation pattern and the required enclosure size specified on the nameplate, the CB must be applied at 80 percent rating. Switchboard manufacturers have UL tests conducted with a CB installed in a specific enclosure, and the enclosure may receive a listing for 100-percent-rated operation even though the ventilation pattern or overall enclosure size may not meet the specifications. In cases where the breaker nameplate specifications are not met by the switchboard, the customer would have to request a letter from the manufacturer certifying that a 100-percent-rated listing has been received. Otherwise, the breaker must be applied at 80 percent.

To realize savings with devices listed by UL at 100 percent of their continuous-current rating, use must be made of a CB manufacturer's data sheet to determine the types and ampere ratings of breakers available that are 100-percent-rated, along with the frame sizes, approved enclosure sizes, and the ventilation patterns required by UL, if any.

It is essential to check the instructions given in the UL listing to determine *if* and under what conditions a CB (or a fuse in a switch) is rated for continuous operation at 100 percent of its current rating.

A Comparison: 100-Percent-Rated versus Non-100-Percent-Rated OC Devices

OCPD (overcurrent protective device) rating For the purpose of comparison, let's consider a feeder supplying an 800-A fluorescent lighting load connected line-to-neutral, assumed to be operating continuously, and supplied from a circuit made up of three sets of parallel conductors, with each set run in a separate raceway (Fig. 215-6).

Non-100-percent-protected circuit As indicated, the rules of 215.3 call for a non-100-percent-rated OCPD to be rated not less than the sum of noncontinuous load (in this case 0 A) plus 125 percent of the continuous load, or 1000 A (800 A × 1.25), which means a minimum rating of 1000 A (0 A + 1000 A).

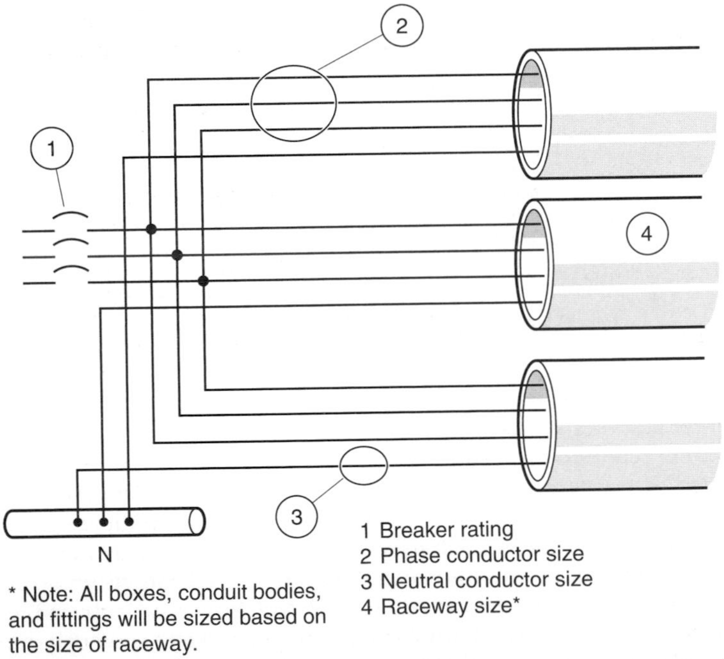
1
2
3
4
N
1 Breaker rating
2 Phase conductor size
3 Neutral conductor size
4 Raceway size*
* Note: All boxes, conduit bodies, and fittings will be sized based on the size of raceway.

Fig. 215-6. The decision to use or not use a 100-percent-rated device will affect the size/rating of these circuit components.

100-percent-protected circuit The rating required for a 100-percent-rated breaker to protect this lighting feeder need be no greater than the actual connected lighting load. Therefore, an 800-A, 100-percent-rated breaker would fully satisfy all applicable rules.

Phase Conductor Sizing

Non-100-percent-protected circuit Another disadvantage to using the non-100-percent-rated breaker is the Code-mandated procedure that must be used to establish the minimum acceptable conductor size for such circuits. As previously indicated, the conductors used with the 1000-A non-100-percent-rated breaker to supply the 800-A lighting load must also have additional capacity. And that capacity must be determined by the method prescribed in 215.2(A)(1)(a). The 2014 NEC has made this procedure somewhat more straightforward at the expense of some technical accuracy in certain cases. There are two independent calculations, which reflect a core principle, namely, that any wire has a middle and two ends. The rules that apply to the end of a wire are influenced by how connected devices and equipment tend to react to conductor operating temperatures and to the effect of continuous loads. Working from the default temperature assumption in 110.14(C)(1)(b) of 75°C and applying 215.2(A)(1)(a), we have a minimum size based on 100% of the noncontinuous load plus 125% of the continuous load, evaluated under the 75°C column in Table 310.15(B)(16).

The rules that apply to the middle of a wire reflect how ambient temperatures and mutual conductor-heating change conductor ampacity, because they change the conditions of use. Whether the load is continuous or not, and how the conductor size affects termination performance and heating is irrelevant to this calculation as this is the middle of the wire and the very definition of ampacity already assumes continuous load conditions. This is why 215.2(A)(1)(b) is an independent calculation. Take the connected load and then upsize to counteract any effects of mutual heating and increased ambient temperatures. Then, as covered in the parent language in 215.2(A)(1), size the wire to accommodate the worst case.

The 2014 NEC language works very well when the same size and type of wire lands on the line and load-side equipment and also runs between both ends without changing size or other ratings. This covers the overwhelming majority of installations. However, for very large and long feeders, at a certain price point it becomes cost-effective to place a pull box near both ends and decrease the feeder size between the pull boxes. Sec. 110.14(C)(2) expressly recognizes pressure connectors that can make such splices, limited only by the temperature rating of the connector, and in this regard full 90°C connectors are readily available. This makes it technically feasible, and entirely justified in terms of engineering principles, to leave a fully sized conductor heat sink in effect at each end, and then reduce the size of the middle of the wire (between the pull boxes) and rely on the increased ampacity resulting from a permitted higher operating temperature. The simplified wording in the 2014 NEC has inadvertently complicated this procedure to the point where an inspector would need to apply Sec. 90.4. There are a few other considerations in this regard.

- Where deratings are required, it is permissible that the derated ampacity be less than the sum of noncontinuous load plus 125 percent of the continuous load. But the conductors must always be properly protected by the OCPD, as required by Sec. 240.4.
- If the derated ampacity is *not* properly protected in accordance with 240.4 by the rating of the overcurrent device selected, then the next larger size of conductor must be evaluated. However, that evaluation can be limited to just the conductor's ampacity because the concern for satisfying 215.2(A)(1) was addressed when the previously considered smaller conductor was evaluated.
- The final point cannot be repeated too much. Remember which temperature column in Table 310.15(B)(16) must be used in satisfying the rule of Sec. 215.2(A)(1). As most are aware, Table 310.15(B)(16) has three different columns identified as 60°C, 75°C, and 90°C, under which are listed different insulating materials and the corresponding ampacity for each conductor size. Because current directly equates with heat, the higher-temperature-rated insulations are permitted to carry more current. Even though it is permissible to use the 90°C values for derating purposes, it must be remembered that for the requirement given in Sec. 215.2(A)(1) the ampacity value shown in the 75°C column of Table 310.15(B)(16) must be used! Where 90°C insulation (e.g., THHN) is used, it is still permissible to use the 90°C ampacity for the purposes of derating, as has long been recognized, but such practice is *not* permitted in selecting the minimum conductor size to supply continuous loads. Do not confuse the two rules.

To repeat, every conductor has a middle and two ends. The rules that apply to determining whether a wire will overheat in use involve an entirely separate set of calculations from those that determine how a termination will behave, and whether the wire as connected will have a sufficient heat sink capability so the connected OCPD will function as its manufacturer and UL tested it. At the end of this book, in conjunction with the discussion of Annex D, Example D3(a), we will tie all of these rules together in one place. The procedures that follow here will be consistent with that discussion, but focus on this concrete example of the cost comparison between 100 percent and conventional OCPDs.

For our example, we must first select a conductor that has a 75°C ampacity in Table 310.15(B)(16) which when multiplied by 3 (the number of conductors that will be paralleled to make up each phase) equals at least 1000 A.

In Table 310.15(B)(16), we see that at 75°C, three 400-kcmil copper conductors (335 A each, for a total of 1005 A) are adequate to satisfy the rule of 215.2(A)(1). However, because the load is fluorescent lighting, the rule of 310.15(B)(5)(c) would require us to count the neutral conductor as a current-carrying conductor. And 310.15(B)(3)(a) would call for an 80 percent derating of the Table 310.15(B)(16) ampacity because there are now more than three current-carrying conductors in the raceway. As indicated, if THHN (90°C-rated) insulation is used, then the derating may be applied against the 90°C ampacity for 400-kcmil copper. Multiplying the 90°C ampacity of 400 kcmil (380 A) times 3 gives us 1140 A. But, when that value is derated by 80 percent to satisfy 310.15(B)(3)(a) for more than three current-carrying conductors, we end up with an ampacity of only 912 A. According to the rule of 240.4(C), where the OCPD is rated at more than 800 A, the conductor's ampacity may *not* be less than the rating of the OC device.

The conductor's ampacity must be equal to, or greater than, the rating of the OC device used. Therefore, because the derated ampacity of the 400-kcmil THHN copper conductors is less than 1000 A, it is *not* properly protected by the 1000-A-rated device. To supply this load, the next larger size of conductor (500 kcmil) must be evaluated—but *only* for ampacity. That's because we have already established that the 400-kcmil copper conductor satisfies 215.2(A). And, if the smaller 400-kcmil conductors were adequately sized, then so should the larger 500s.

To determine the ampacity of the THHN-insulated, 500-kcmil copper conductors under these conditions of use, because we're using THHN conductors, we can apply the 80 percent derating against the 90°C ampacity value for 500-kcmil copper. Multiply the 430-A table value times the number of conductors (3) and derate by 80 percent. Or,

$$\begin{aligned}\text{Derated ampacity} &= (430\ \text{A/wire} \times 3\ \text{wires/phase}) \times 0.80\\ &= 1290\ \text{A/phase} \times 0.80\\ &= 1032\ \text{A}\end{aligned}$$

Therefore, the minimum size permitted for the phase conductors protected by the non-100-percent-rated device is THHN-insulated 500-kcmil copper.

100-percent-protected circuit As indicated by the exception to 215.2(A), the overcurrent device and the conductor size need only be adequate for the load to be served. Therefore, just as the CB need only be rated for 800 A, the circuit conductors need only be rated for 800 A. Where a 100-percent-rated OCPD is used, the basic rule in 215.2(A)(1) does *not* apply. Therefore, all we're really concerned with is conductor ampacity. That is, we must select a conductor with a table ampacity that when multiplied by 3 (number of conductors per phase) and derated by 80 percent (because there are still more than three current-carrying conductors in each raceway) is still equal to or greater than 800 A.

From Table 310.15(B)(16), we select a THHN-insulated 350-kcmil copper conductor with a 90°C table ampacity of 350 A. (*Remember:* The table ampacity shown in the 90°C column may be used in applying deratings to 90°C-insulated conductors.) As before, multiply the table value by 3 and derate by 80 percent, or:

$$\begin{aligned}\text{Derated ampacity} &= (350\ \text{A/wire} \times 3\ \text{wires/phase}) \times 0.80\\ &= 1050\ \text{A/phase} \times 0.80\\ &= 840\ \text{A}\end{aligned}$$

Therefore, the use of three THHN-insulated 350-kcmil copper conductors per phase would satisfy all rules regarding the minimum acceptable conductor size permitted for circuiting a 100-percent-rated 800-A device supplying this fluorescent lighting load.

Neutral Conductor Sizing

Non-100-percent-protected circuit It is worth noting that the rules given in 215.2(A) do *not* affect the sizing of the neutral. That is, because the neutral conductor does not generally connect to the CB, there is no need to be concerned with the nature of the load (i.e., continuous or a combination of continuous and noncontinuous). Remember, neutral sizing must satisfy the rules of 220.61. And, in that section, there is no requirement for additional capacity in the neutral where the load to be supplied is a continuous load. However, it certainly seems as if common sense should be applied to the sizing of the neutral for the non-100-percent-protected circuit. Furthermore, effective with the 2008 NEC, there is now express language [215.2(A)(1) Exception No. 2] that specifically exempts a feeder neutral from the upsizing rules at terminations, provided it runs from busbar to busbar and does not land on an OCPD.

The 2005 NEC introduced an additional wrinkle in sizing neutrals on these systems. The neutral must have sufficient ampacity to safely carry a line-to-neutral short circuit without damaging itself or blowing open. This rule is now in its own numbered paragraph, at 215.2(A)(2). In this case, the neutral must have a size that is not smaller than the required size of an equipment grounding conductor for the system, as determined by Table 250.122. As we have seen when we looked at conduit fill for nonmetallic conduit, the required size is 2/0 AWG for this load if a 125 percent OCPD were used (1000 A) and 1/0 if the 100 percent OCPD (800 A) is used. Since the total neutral capacity will be 3 times a 350-kcmil conductor, this will not be a factor in this case. This issue has significance in instances where the overwhelming majority of the load is line-to-line, resulting in calculated neutral sizes that may be very small, to the point of not being able to handle a short circuit.

Note also that although 250.122(F) normally requires equipment grounding conductors in each parallel raceway to qualify independently as fully rated and sized conductors based on the full rating of the OCPD, this special provision in 215.2(A)(2) waives this rule and allows the three neutrals to be considered in terms of their collective ampacity as a group. In addition, 310.10(H)(1) sets the minimum size for paralleled conductors at 1/0 AWG, so in installations such as this that sets yet another floor under the minimum sizing on grounded circuit conductors generally. Again, for this installation with three sets of 350-kcmil neutral conductors, compliance with both these rules is not in doubt. However, because this is an NEC Handbook, the topic must be considered.

100-percent-protected circuit As was just indicated, whether a 100-percent-rated CB or a non-100-percent-rated CB is used, the Code permits the neutral conductor for both circuits to be the same size. For the example at hand, a neutral conductor of 350-kcmil, THHN-insulated copper conductor would be acceptable regardless of the type of CB used.

Raceway size As has long been the rule, the minimum acceptable size of raceway must be based on the amount of space occupied by the circuit conductors. And in no case may the cross-sectional area of the enclosed conductors exceed 40 percent of the raceway's cross-sectional area where three or more conductors are run within the raceway, as indicated in Chap. 9 of the NEC.

Non-100-percent-protected circuit In accordance with Note 6 to the tables in Chap. 9, where a mix of conductor sizes is to be run, conduit fill must be determined by using the specific dimensions given for conductors and raceway fill in Tables 5 and 5A and Table 4, respectively.

The phase conductors are 500 kcmil and the neutral is 350 kcmil. From Table 5, we take the square-inch area of a 500-kcmil THHN, which is 0.7073 sq in. That value is multiplied by 3, which is the number of phase conductors in each raceway.

The product of that multiplication is then added to the Table 5 value given for the 350-kcmil THHN neutral conductor (0.5242 sq in.) as follows:

$$\begin{aligned}\text{Total area} &= (0.7073 \text{ sq in.} \times 3) + 0.5242 \text{ sq in.}\\ &= 2.1219 \text{ sq in.} + 0.5242 \text{ sq in.}\\ &= 2.6461 \text{ sq in.}\end{aligned}$$

Next, using the data given for the individual raceways in Table 4, we can determine the minimum acceptable size of raceway. Moving down the 40 percent-fill column of each raceway's table we find the size that is equal to, or greater than, 2.6461 sq in. The minimum size permitted for this combination of conductors in the common metal raceways—rigid metal conduit, intermediate metal conduit, or electrical metallic tubing—is 3 in. in each case.

To determine the minimum size of nonmetallic conduit, add the square-inch area of a No. 2/0 grounding conductor—from Table 5, if insulated, or from Table 8, if bare—to the 2.6461 sq in. total just determined. Then go to Table 4 and find the minimum size raceway that has a square-inch value in the 40 percent-fill column that is equal to, or greater than, the total of this combination.

100-percent-protected circuit As covered in Note 1 to the tables of Chap. 9, where all the conductors are of the same size and have the same insulation, the data given in the tables of App. C are permitted to be used.

For rigid metal conduit, Table C8 shows that the minimum-size pipe permitted to contain four 350-kcmil THHN insulated conductors is metric designator 78 (trade size 3), which may contain five such 350s. However, Table C4 covering IMC permits four 350-kcmil THHN conductors in a single metric designator 63 (trade size 2½) raceway. And Table C1 also recognizes four 350-kcmil THHN conductors in metric designator 63 (trade size 2½) EMT.

Summary The choice we have to supply this 800 A of fluorescent lighting is between:

1. 1000-A CB/500-kcmil phase conductors/metric designator 78 (trade size 3) conduit
2. 800-A CB/350-kcmil phase conductors/metric designator 63 (trade size 2½) conduit (where IMC or EMT are used)

If larger raceway is used, then larger fittings, boxes, and so forth will also be required. If nonmetallic raceway is used, the equipment grounding conductor would be required to be larger in the 1000-A circuit (2/0 versus 1/0). And in either case, the labor costs will rise.

In addition to those economic realities, a check of one manufacturer's pricing indicates that the 100-percent-rated 800-A CB is actually *less expensive* than the non-100-percent-rated 1000-A device, primarily because of the jump in frame size. That is, the 1000-A CB has a 1200-A frame size while the 800-A breaker has an 800-A frame. The larger frame sizes are generally more robust. That is, they are designed to be capable of withstanding greater electrical stresses than the smaller frame sizes. As a result, the larger frame size will require more material, engineering, production costs, and so forth; therefore, a greater price is charged. There is an immediate savings to be realized simply by selecting the 800-A, 100-percent-rated device.

The final benefit that should be realized is the elimination of the need for equipment ground fault protection (GFPE). Remember Sec. 215.10 would require any feeder disconnect rated 1000 A or more on a 480Y/277-V system to be provided with equipment ground fault protection of the type required at services. Of course, this protection is *not* required where such equipment protection is provided ahead of the feeder disconnect, as long as that GFP has not been desensitized by a connection between ground and neutral on the line side of the disconnect, such as at the output of a separately derived system (e.g., a transformer). A 2008 **NEC** revision to 215.10 Exception No. 2 now expressly covers this point, and disallows any GFPE on the line side of a transformer from being considered as protecting any load side feeders.

One last thing to remember is that the exception to 215.2(A)(1) requires that the conductors used with 100-percent-rated OC devices must have an ampacity at least equal to the sum of continuous and noncontinuous loads. In the preceding example, the conductor size selected satisfies this requirement because of its 840-A ampacity, which includes all required deratings. If the conductor had a final ampacity of, say, 760 A, it would still satisfy the rules of 240.4 because the circuit is not rated over 800 A. *But*, the exception to this rule requires a minimum ampacity—as covered in 310.15 and the accompanying tables—that is *not less than* the total load supplied.

It certainly seems as if 100-percent-rated devices are the way to go in this particular case. Such an approach will allow maximum utilization and will do so at lower cost! (See Fig. 215-7.)

Load: 800-A fluorescent lighting

Makeup: Three parallel sets of conductors in three separate conduits

Non–100-percent–rated circuit		100-percent–rated circuit	
• Breaker:	1000 A	• Breaker:	800 A
• Phase conductors:	500 kcmil (3/phase)	• Phase conductors:	350 kcmil (3/phase)
• Neutral conductors:	350 kcmil (1/pipe)	• Neutral conductors:	350 kcmil (1/pipe)
• Raceway size* (IMC):	3 in.	• Raceway size* (IMC):	2-1/2 in.

* All pullboxes, junction boxes, conduit bodies, etc., must be based on the raceway size.

Fig. 215-7. Non-100-percent-rated circuit versus 100-percent-rated circuit.

As shown in Fig. 215-8, the rule of part **(A)(3)** of this section requires that the ampacity of feeder conductors must be at least equal to that of the service conductors where the total service current is carried by the feeder conductors. In the case shown, No. 4 TW aluminum is taken as equivalent to No. 6 TW copper and has the same ampacity (55 A).

In this section, part **(A)(4)** formerly noted that it was never necessary for feeder conductors at mobile homes or "individual dwelling units" to be larger than the

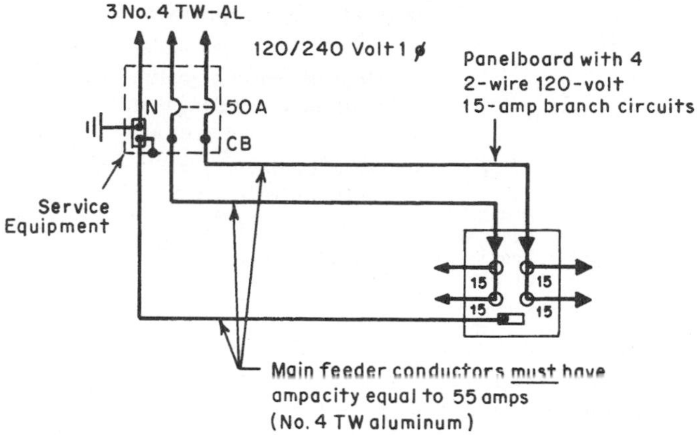
3 No. 4 TW-AL
120/240 Volt 1 ø
Panelboard with 4
2-wire 120-volt
15-amp branch circuits
N
50A
CB
Service
Equipment
15
15
15
15
Main feeder conductors must have
ampacity equal to 55 amps
(No. 4 TW aluminum)

Fig. 215-8. Feeder conductors must not have ampacity less than service conductors. [Sec. 215.2(A)(3).]

service-entrance conductors (assuming use of the same conductor material and the same insulation). In particular, this was aimed at those cases where the size of service-entrance conductors for a dwelling unit is selected in accordance with the higher-than-normal ampacities permitted by 310.15(B)(7) for services to residential occupancies. If a set of service conductors for an individual dwelling unit is brought in to a single service disconnect (a single fused switch or CB) and load and the service conductors are sized for the increased ampacity value permitted by 310.15(B)(7), diversity on the load-side feeder conductors gives them the same reduced heat-loading that enables the service conductors to be assigned the higher ampacity. This rule simply extended the permission of 310.15(B)(7) to those feeders and was applicable for any such feeder for a dwelling unit (a one-family house or an apartment in a two-family or multifamily dwelling) or for a mobile-home feeder (Fig. 215-9). Since 310.15(B)(7) now clearly covers such feeders, the 2014 **NEC** deleted this rule.

Informational note No. 2 in 215.2(A)(1) comments on voltage drop in feeders. It should be carefully noted that with extremely few and very specific exceptions the **NEC** does not establish any mandatory rules on voltage drop for either branch circuits or feeders. The references to 3 and 5 percent voltage drops are purely advisory—that is, recommended maximum values of voltage drop. The Code normally does not consider excessive voltage drop to be unsafe.

The voltage-drop note suggests not more than 3 percent for feeders supplying power, heating, or lighting loads. It also provides for a maximum drop of 5 percent for the conductors between the service-entrance equipment and the connected load. If the feeders have an actual voltage drop of 3 percent, then only 2 percent is left for the branch circuits. If a lower voltage drop is obtained in the feeder, then the branch circuit has more voltage drop available, provided that the total drop does not exceed 5 percent. For any one load, the total voltage drop is made up of the voltage drop in the one or more feeders plus the voltage drop in the branch circuit supplying that load.

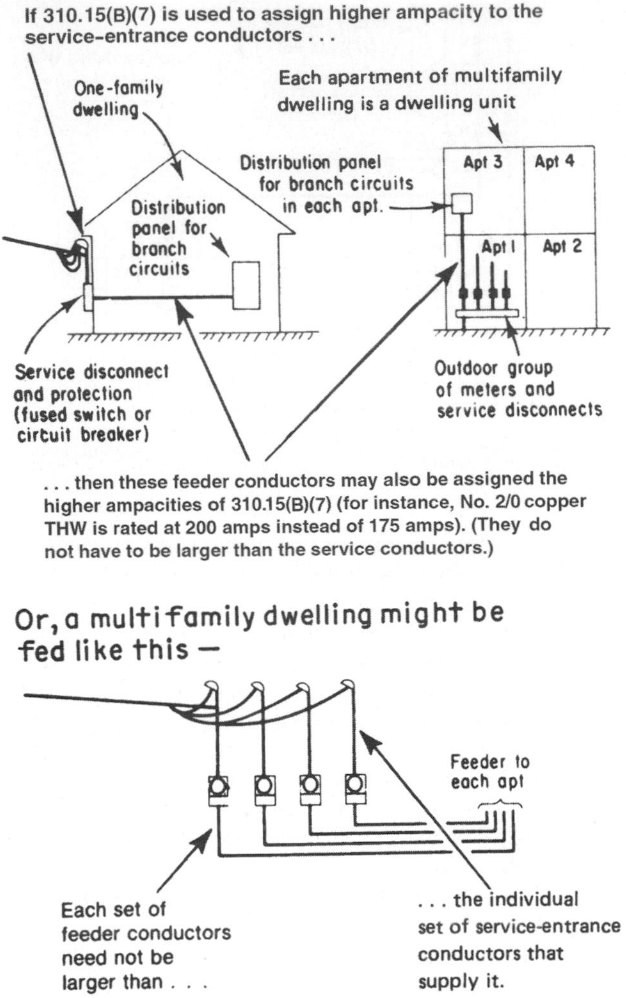
If 310.15(B)(7) is used to assign higher ampacity to the service-entrance conductors . . .
One-family dwelling
Each apartment of multifamily dwelling is a dwelling unit
Distribution panel for branch circuits in each apt.
Apt 3
Apt 4
Distribution panel for branch circuits
Apt 1
Apt 2
Service disconnect and protection (fused switch or circuit breaker)
Outdoor group of meters and service disconnects
. . . then these feeder conductors may also be assigned the higher ampacities of 310.15(B)(7) (for instance, No. 2/0 copper THW is rated at 200 amps instead of 175 amps). (They do not have to be larger than the service conductors.)
Or, a multifamily dwelling might be fed like this —
Feeder to each apt
Each set of feeder conductors need not be larger than . . .
. . . the individual set of service-entrance conductors that supply it.

Fig. 215-9. Feeder conductors need not be larger than service-entrance conductors when higher ampacity is used. [Sec. 215.2(A)(3).]

Again, however, values stated in the informational note on voltage drop are recommended values and are not intended to be enforced as a requirement.

Voltage drop should always be carefully considered in sizing feeder conductors, and calculations should be made for peak load conditions. For maximum efficiency, the size of feeder conductors should be such that voltage drop up to the branch-circuit panelboards or point of branch-circuit origin is not more than 1 percent for lighting loads or combined lighting, heating, and power loads and not more than 2 percent for power or heating loads. Voltage drop in most cases is a design concern only, and the applicable design specification may impose lower

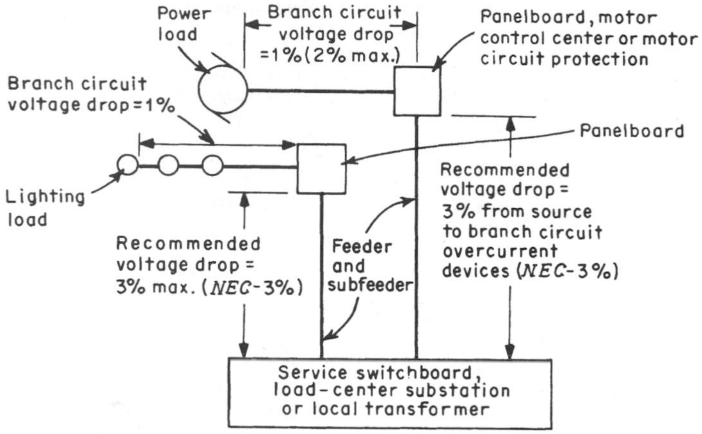
Power load
Branch circuit voltage drop =1%(2% max.)
Panelboard, motor control center or motor circuit protection
Branch circuit voltage drop=1%
Panelboard
Lighting load
Recommended voltage drop = 3% from source to branch circuit overcurrent devices (NEC-3%)
Recommended voltage drop = 3% max. (NEC-3%)
Feeder and subfeeder
Service switchboard, load-center substation or local transformer

Fig. 215-10. Recommended basic limitations on voltage drop. (Sec. 215.2, informational note No. 2.)

limits of voltage drop. Voltage-drop limitations are shown in Fig. 215-10 for NEC levels and better levels of drop, as follows:

1. For combinations of lighting and power loads on feeders and branch circuits, use the voltage-drop percentages for lighting load (at left in Fig. 215-10).
2. The word *feeder* here refers to the overall run of conductors carrying power from the source to the point of final branch-circuit distribution, including feeders, subfeeders, sub-subfeeders, and so forth. As previously noted, the prefix "sub" is no longer correct code terminology.
3. The voltage-drop percentages are based on nominal circuit voltage at the source of each voltage level. Indicated limitations should be observed for each voltage level in the distribution system.

There are many cases in which the previously mentioned limits of voltage drop (1 percent for lighting feeders, etc.) should be relaxed in the interests of reducing the prohibitive costs of conductors and conduits required by such low drops. In many installations a 5 percent drop in feeders is not critical or unsafe—such as in apartment houses.

Voltage-drop tables and calculators are available from a good number of electrical equipment manufacturers. Voltage-drop calculations vary according to the actual circuit parameters (e.g., ac or dc, single- or multiphase, power factor, circuit impedance, line reactance, types of enclosures [nonmetallic or metallic], length and size of conductors, and conductor material [copper, copper-clad aluminum, or aluminum]).

Calculations of voltage drop in any set of feeders can be made in accordance with the formulas given in the electrical design literature, such as those shown in Fig. 215-11. From this calculation, it can be determined if the conductor size initially selected to handle the load will be adequate to maintain voltage drop within given limits. If it is not, the size of the conductors must be increased (or other steps taken where conductor reactance is not negligible) until the voltage

Two-wire, single-phase circuits (inductance negligible)

$$V = \frac{2k \times L \times I}{d^2} = 2R \times L \times I$$

$$d^2 = \frac{2k \times L \times I}{V}$$

V = drop in circuit voltage (volts)
R = resistance per ft of conductor (ohms/ft)
I = current in conductor (amperes)

Three-wire, single-phase circuits (inductance negligible)

$$V = \frac{2k \times L \times I}{d^2}$$

V = drop between outside conductors (volts)
I = current in more-heavily loaded outside conductor (amps)

Three-wire, three-phase circuits (inductance negligible)

$$V = \frac{2k \times L \times I}{d^2} \times 0.866$$

V = voltage drop of 3-phase circuit

Four-wire, three-phase balanced circuits (inductance negligible)

Lighting loads
Voltage drop between one outside conductor and neutral equals one-half of drop calculated by formula above for 2-wire circuits.

Motor loads
Voltage drop between any two outside conductors equals 0.866 times the drop determined by formula above for two-wire circuits.

In the above formulas:

L = one-way length of circuit (ft)
d^2 = cross-section area of conductor (circular mils)
k = resistivity of conductor metal (cir mil-ohms/ft)
= 12 for circuits loaded to more than 50% of allowable circuit capacity
= 11 for circuits loaded less than 50%
= 18 for aluminum or copper-clad aluminum conductors

Example: **230-V two-wire heating circuit. Load is 24 A. Circuit size is No. 10 AWG copper, and the one-way circuit length is 200 ft.**

$$VD = \frac{24 \times 200 \times 24}{10{,}380} = \frac{115{,}200}{10{,}380} = 11$$

An 11-V drop on a 230-V circuit is about a 5 percent drop (11/230 = 0.0478). No. 8 AWG copper conductors would be needed to reduce the voltage drop to 3 percent on the branch circuit and allow 2 percent more on the feeder.

Fig. 215-11. Calculating voltage drop in feeder circuits. (Sec. 215.2, informational note No. 2.)

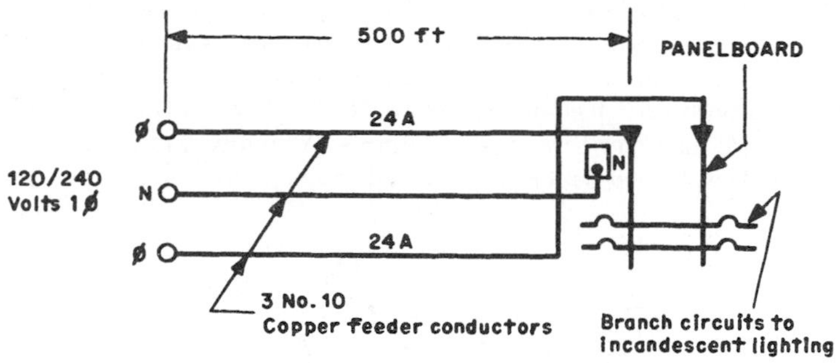
500 ft
PANELBOARD
24 A
120/240
Volts 1ø
ø
N
ø
N
24 A
3 No. 10
Copper feeder conductors
Branch circuits to
incandescent lighting

1. No. 10 copper conductor has a resistance of 1.018 ohms per 1000 ft (Table 8, Chapter 9).
2. The two 500-ft lengths of circuit conductors total 1000 ft and have a resistance of 1.018 ohms.
3. Voltage Drop = load current x conductor resistance
   = 24 amps x 1.018 ohms = 24.43 volts
4. $\frac{24.43}{240}$ = 10.2% VOLTAGE DROP–*NEC* SUGGESTS MAX. 3%

Fig. 215-12. Feeder voltage drop should be checked. (Sec. 215.2, informational note No. 2.)

drop is within prescribed limits. Many such graphs and tabulated data on voltage drop are available in handbooks and from manufacturers. Figure 215-12 shows an example of excessive voltage drop—over 10 percent in the feeder.

215.2(B). Feeders Over 600 V. This part of the section covers medium-voltage feeders, for which the ampacity rules are far different. The basic rule is that the ampacity of a medium-voltage feeder supplying transformers must match the sum of the primary ratings of the transformers supplied. If such a feeder also supplies utilization equipment, then the minimum ampacity is the sum of any transformer primaries supplied plus the utilization load taken at 125 percent of its maximum design loading based on the maximum current that would be drawn at any one time, thereby allowing for noncoincident loads. If, however, a facility has engineering staff with documented training and experience working with medium-voltage power systems, and they exercise supervision over the monitoring, maintenance, and service required for the system, then the engineering staff may alter the sizing of the feeder conductors that are to be installed. Because the majority of the NEC is adjusting upwards the threshold for applying high-voltage rules from 600 V to 1000V, 215.3 Exception No. 2 has been modified so only circuits at or over 1000V move into Part X of Art. 240.

215.4. Feeders with Common Neutral Conductor. A frequently discussed Code requirement is that of 215.4, covering the use of a common neutral with more than one set of feeders. This section says that a common neutral feeder may be used for two or three sets of 3-wire feeders or two sets of 4-wire feeders. It further requires that all conductors of feeder circuits employing a common neutral feeder must be within the same enclosure when the enclosure or raceway containing them is metal.

A common neutral is a single neutral conductor used as the neutral for more than one set of feeder conductors. It must have current-carrying capacity equal to the sum of the neutral conductor capacities if an individual neutral conductor were used with each feeder set. Figure 215-13 shows a typical example of a common neutral, used for three-feeder circuits. A common neutral may be used only with feeders. It may never be used with branch circuits, except outdoors as

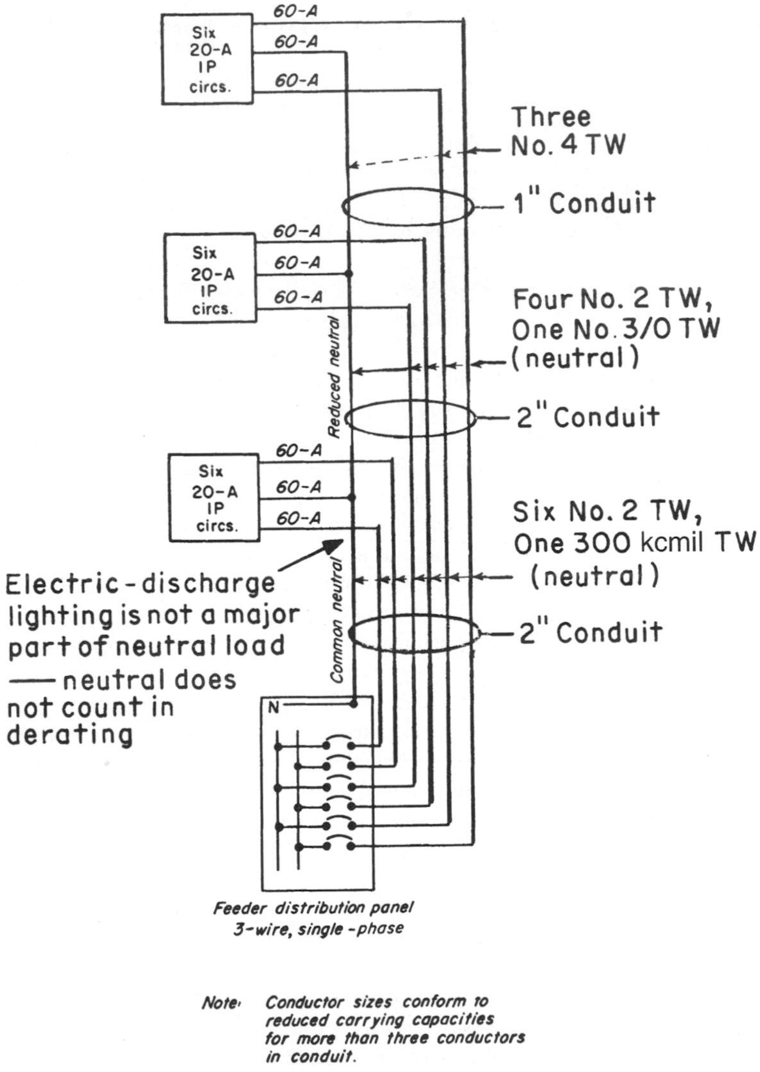

60-A
Six 20-A IP circs.
60-A
60-A
Three No. 4 TW
1" Conduit
60-A
Six 20-A IP circs.
60-A
60-A
Reduced neutral
Four No. 2 TW, One No. 3/0 TW (neutral)
2" Conduit
60-A
Six 20-A IP circs.
60-A
60-A
Six No. 2 TW, One 300 kcmil TW (neutral)
Electric-discharge lighting is not a major part of neutral load — neutral does not count in derating
Common neutral
2" Conduit
N
Feeder distribution panel 3-wire, single-phase
Note: Conductor sizes conform to reduced carrying capacities for more than three conductors in conduit.

Fig. 215-13. Example of three feeder circuits using a single, "common neutral"—with neutral size reduced as permitted. (Sec. 215.4.)

covered in 225.7. A single neutral of a multiwire branch circuit is not a "common neutral." It is the neutral of only a single circuit, even though the circuit may consist of 3 or 4 wires. A feeder common neutral is used with more than one feeder circuit.

215.5. Diagrams of Feeders. This is the code section that authorizes the inspection community to request, and to insist on if necessary, feeder diagrams and load calculations, stipulations of demand factors applied, and recitations of wire sizes and insulation types, etc.

215.7. Ungrounded Conductors Tapped from Grounded Systems. Refer to 210.19 for a discussion that applies as well to feeder circuits as to branch circuits.

215.9. Ground-Fault Circuit-Interrupter Protection for Personnel. A ground-fault circuit interrupter may be located in the feeder and protect all branch circuits connected to that feeder. In such cases, the provisions of 210.8 and Art. 590 on temporary wiring will be satisfied and additional *downstream* ground-fault protection on the individual branch circuits would not be required. It should be mentioned, however, that downstream ground-fault protection is more desirable than ground-fault protection in the feeder because less equipment will be deenergized when the ground-fault circuit interrupter opens the supply in response to a line-to-ground fault.

As shown in Fig. 215-14, if a ground-fault protector is installed in the feeder to a panel for branch circuits to outdoor residential receptacles, this protector will satisfy the NEC as the ground-fault protection required by 210.8 for such outdoor receptacles.

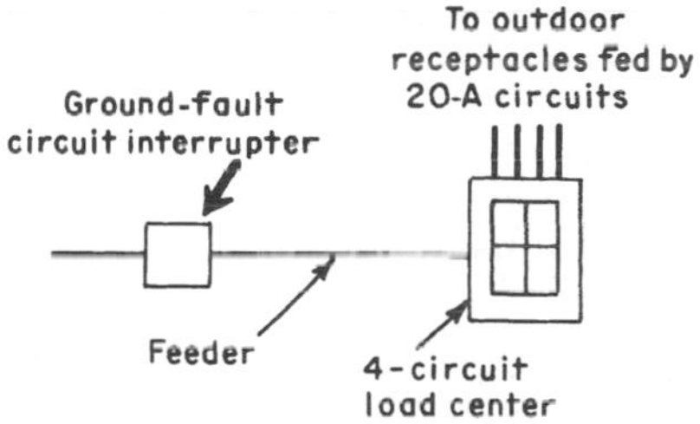
To outdoor receptacles fed by 20-A circuits
Ground-fault circuit interrupter
Feeder
4-circuit load center

Fig. 215-14. GFCI in feeder does satisfy as protection for branch circuits. (Sec. 215.9.)

215.10. Ground-Fault Protection of Equipment. This section mandates equipment ground-fault protection for every feeder disconnect switch or circuit used on a 480Y/277-V, 3-phase, 4-wire feeder where the disconnect is rated 1000 A or more, as shown in Fig. 215-15. This is a very significant Code requirement for ground-fault protection of the same type that has long been required by 230.95 for every *service* disconnect rated 1000 A or more on a 480Y/277-V service.

As indicated by Exception No. 1, GFPE of equipment may be omitted for "continuous industrial" processes, but only if "additional" or "increased" hazards will result where a process is shut down in a nonorderly manner. An example of this principle at work is 695.6(G), which forbids the application of GFPE to a fire pump circuit. Since Chap. 6 provisions automatically vary the requirements

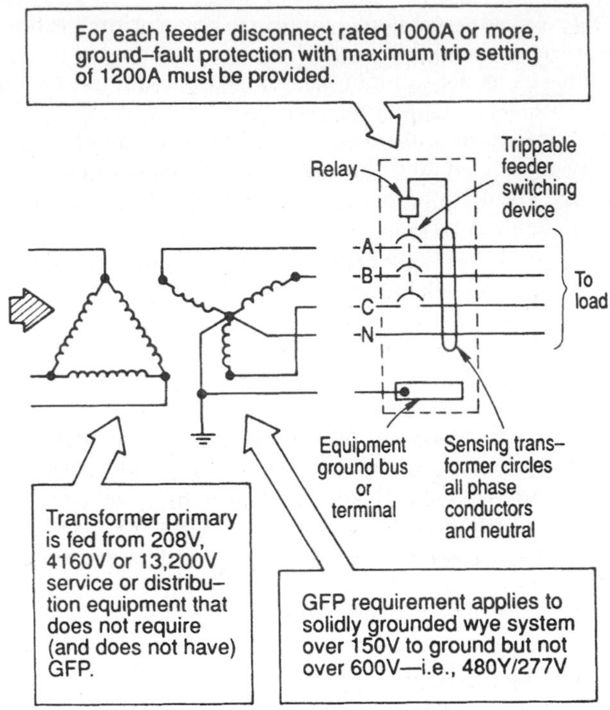
For each feeder disconnect rated 1000A or more, ground–fault protection with maximum trip setting of 1200A must be provided.
Relay
Trippable feeder switching device
-A
-B
-C
-N
To load
Equipment ground bus or terminal
Sensing trans– former circles all phase conductors and neutral
Transformer primary is fed from 208V, 4160V or 13,200V service or distribu– tion equipment that does not require (and does not have) GFP.
GFP requirement applies to solidly grounded wye system over 150V to ground but not over 600V—i.e., 480Y/277V

Fig. 215-15. A 480Y/277-V feeder disconnect rated 1000 A or more must have ground-fault protection (GFP) if there is not GFP on its supply side. (Sec. 215.10.)

of Chap. 1 through 4 per 90.3, the former Exception No. 2 in this location that excluded fire pump disconnects from the need for equipment GFPE has been deleted as unnecessary.

Exception No. 2 notes that feeder ground-fault protection is not required on a feeder disconnect if equipment ground-fault protection is provided on the supply (line) side of the feeder disconnect and, as noted previously, this exception does not apply when a transformer is interposed between the GFPE and the large feeder to be protected.

The substantiation submitted as the basis for the addition of this new rule stated as follows:

> Substantiation: The need for ground-fault equipment protection for 1000 amp or larger 277/480 grounded system is recognized and required when the service equipment is 277/480 volts. This proposal will require the same needed protection when the service equipment is not 277/480 volts. Past proposals attempted to require these feeders be treated as services in order to achieve this protection, but treating a feeder like a service created many other concerns. This proposal only addresses the feeder equipment ground-fault protection needs when it is not provided in the service equipment.

As noted, this rule calls for this type of feeder ground-fault protection when ground-fault protection is not provided on the supply side of the feeder disconnect, such as where a building has a high-voltage service (say, 13,200 V) or has, say, a 208Y/120-V service with a load-side transformer stepping-up the voltage to 480Y/277 V—because a service at either one of these voltages (e.g., 13.2 kV and 208Y/120 V) is not required by 230.95 to have GFP. The wording of this exception also prohibits GFPE on the supply side of a transformer from qualifying as protecting conductors on its load side for any other reason, such as instances where a one-to-one isolating transformer is used to supply feeder conductors where the applied voltage and current require GFPE under this section.

215.11. Circuits Derived from Autotransformers. This section recognizes application of autotransformers for supplying a feeder to a panelboard or group of overcurrent devices. This is the same permission given for branch circuits (see 210.9), and with comparable exceptions.

215.12. Identification for Feeders. The rules in parts **(A)**, **(B)**, and **(C)** of this section present the same requirements for feeder conductors that are given for branch circuits in 210.5. See 210.5 for a discussion of these rules. Note that this section has been correlated with the changes in 210.5 made in the 2014 NEC to systematically cover dc distributions.

ARTICLE 220. BRANCH-CIRCUIT, FEEDER, AND SERVICE LOAD CALCULATIONS

220.1. Scope. The revision of this article in the 2005 edition of the Code provides a more logical and coherent approach for establishing the minimum rating of conductors used as branch circuits, feeders, and service conductors. Better arrangement and segregation of rules applying to branch circuits from those that apply to feeders and/or service conductors and vice versa has reduced confusion and made the intent of the Code more understandable. The same Code-making panel (CMP 2) has jurisdiction over Arts. 210, 215, and 220. The article reorganization in 2005 followed an earlier reassignment of provisions among these three articles with the objective that Art. 220 should only address how to calculate a load, whether expressed in amperes of current on a wire or in terms of volt-amperes of power required for some portion of an electrical system. For this reason, the discussion that follows will only address load calculations. How to translate those results into an appropriate wire selection is a task left to other locations in the NEC and in this book, particularly at the end where Annex D, Example D3(a) is covered. The load calculation examples in Annex D, including D3(a), are also under the jurisdiction of CMP 2, so the examples should correlate with procedures in Art. 220.

The general rules related to calculating branch-circuit feeder and service conductors are presented in Part I and these sections apply across the board. The rules for calculating the minimum ratings and sizes for branch circuits are given in Part II, while those for feeders and service conductors are covered in Part III. Part IV covers the alternate methods of calculating feeder loads in certain cases.

Part V provides procedures for calculating minimum sizes of feeders and service conductors supplying a farm.

All the calculations and design procedures covered by Art. 220 involve mathematical manipulation of units of voltage, current, resistance, and other measures of electrical conditions or characteristics.

220.5. Calculations. NEC references to voltages vary considerably. The Code contains references to 120, 125, 115/230, 120/240, and 120/208 V. Standard voltages to be used for the calculations that have to be made to observe the rules of Art. 220 are 120, 120/240, 208Y/120, 240, 480Y/277, 480, 600Y/347, and 600 V. But use of lower voltage values (115, 230, 440, etc.) as denominators in calculations would not be a Code violation because the higher current values that result would ensure Code compliance because of greater capacity in circuit wires and other equipment. Nevertheless, the approach in this book will be to always use the standard values. If a circuit is to be oversized for design reasons, which is often an excellent idea, it is preferable to add the allowance for future growth openly, instead of hiding that decision by using a bogus voltage. Remember that for the majority of loads a decreased voltage results in decreased load current, which is why electric utilities, when faced with demand they cannot cope with, reduce voltage (brownout).

In all electrical systems, there is a normal, predictable spread of voltage values over the impedances of the system equipment. It has been common practice to assign these basic levels to each nominal system voltage. The highest value of voltage is that at the service entrance or transformer secondary, such as 480Y/277 V. Then considering a voltage drop due to impedance in the circuit conductors and equipment, a "nominal" mid-system voltage designation would be 460Y/265. Variations in "nominal" voltages have come about because of (1) differences in utility-supply voltages throughout the country, (2) varying transformer secondary voltages produced by different and often uncontrolled voltage drops in primary feeders, and (3) preferences of different engineers and other design authorities. Although an uncommon voltage system within the United States, the reference to 600Y/347V is one of many additions that were made to the 1996 NEC that attempt to harmonize the NEC with the Canadian Electrical Code.

Because the NEC is produced by contributors from all over the nation and of varying technical experiences, it is understandable that diversity of designations would creep in. As with many other things, we just have to live with problems until we solve them.

To standardize calculations, Annex. D covering the examples at the end of the code book also specifies that nominal voltages of 120, 120/240, 240, and 208/120 V are to be used in computing the ampere load on a conductor. The reason for this is the plain wording of 220.5(A).

In some places, the NEC adopts 115 V as the basic operating voltage of equipment designed for operation at 110–125 V. That is indicated in Tables 430.248 to 430.251. References are made to "rated motor voltages" of 115, 230, 460, 575, and 2300 V—all values over 115 are integral multiples of 115. The last note in Tables 430.249 and 430.250 indicates that motors of those voltage ratings are applicable on systems rated 110–120, 220–240, 440–480, and 550–600 V. Although the motors can operate satisfactorily within those ranges, it is better to design circuits to deliver rated voltage. These Code voltage designations for motors are consistent with the trend over recent years for manufacturers to rate equipment for corresponding values of voltage. These voltages are used in the motor rules for historical reasons and nowhere else in the Code.

In this context, the issue of significant figures needs to be addressed. Many calculations involve dividing volt-amperes by volts to get amperes, or something comparable. Usually, such a calculation will result in some form of infinitely repeating decimal. Given today's calculators, work with these machines is simple and potentially misleading due to the illusion of precision. Load calculations are not an exact science. The inherent nature of large-scale electrical power systems is such that three significant figures is probably more than can actually be relied on as having any meaning whatsoever. The newest example in Annex D, Example D3(a), squarely addresses this issue for the first time, stating "For reasonable precision, volt-ampere calculations are carried to three significant figures only." The example goes on to state that "Where loads are converted to amperes, the results are rounded to the nearest whole ampere" and due reference is made to 220.5(B).

Rounding to the nearest whole ampere is fully in accordance with these concepts and fully validated statistical procedures. Some numbers go up a little and others go down; there is no real decrease in safety or ultimate results. The majority of tax calculations are now being done as whole dollar calculations for the same reasons. Sec. 220.5(B) specifically authorizes the process for load calculations, with results of 0.5 and up resulting in the number to the left of the decimal increasing by one, and results of less than 0.5 being discarded.

220.12. Lighting Load for Specified Occupancies. Article 220 gives the basic rules on calculation of loads for branch circuits and feeders. The task of calculating a branch-circuit load and then determining the size of circuit conductors required to feed that load is common to all electrical system calculations. Although it may seem to be a simple matter (and it usually is), there are many conditions which make the problem confusing (and sometimes controversial) because of the **NEC** rules that must be observed.

Code Table 220.12 lists certain occupancies (types of buildings) for which a minimum general lighting load is specified in volt-amperes per square meter (square foot). In each type of building, there must be adequate branch-circuit capacity to handle the total load that is represented by the product of volt-amperes per square foot times the square-foot area of the building. For instance, if one floor of an office building is 40,000 sq ft in area, that floor must have a total branch-circuit capacity of 3700 m^2 (40,000 ft^2) times 39 VA/m^2 (3½ VA/ft^2) (Code Table 220.12) for general lighting. Note that the total load to be used in calculating required circuit capacity must never be taken at less than the indicated volt-amperes per unit area times the area for those occupancies listed. Of course, if the branch-circuit load for lighting is determined from a lighting layout of specific fixtures of known volt-ampere rating, the load value must meet the previous volt-amperes-per-unit area minimum; and if the load from a known lighting layout is greater, then the greater volt-ampere value must be taken as the required branch-circuit capacity.

Note that the bottom of Table 220.12 requires a minimum general lighting load of 6 VA/m^2 (½ VA/ft^2) to cover branch-circuit and feeder capacity for halls, corridors, closets, and all stairways. Likewise, an additional 3 VA/m^2 (¼ VA/ft^2) must be provided for storage areas.

As indicated in 220.12, when the load is determined on a volt-amperes-per-unit area basis, open porches, garages, unfinished basements, and unused areas are not counted as part of the area for dwelling unit calculations. Also note that area calculations must be made using the *outside* dimensions of the "building, apartment, or other area involved."

When fluorescent or HID lighting is used on branch circuits, the presence of the inductive effect of the ballast or transformer creates a power factor consideration. Determination of the load in such cases must be based on the total of the volt-ampere rating of the units and not on the wattage of the lamps.

Based on extensive analysis of load densities for general lighting in office buildings, Table 220.12 requires a minimum unit load of only 3½ VA/sq ft—rather than the previous unit value of 5—for "office buildings" and for "banks."

A footnote at the bottom of the table requires compliance with 220.14(K) for banks and office buildings. That rule establishes the minimum load capacity required for receptacles in such occupancies. The rule in 220.14(K) calls for capacity to be provided based on the larger of the following: 11 VA/m^2 (1 VA/ft^2) or the actual connected number of receptacles at 1.5 A apiece. In those cases where the actual number of receptacles is not known at the time feeder and branch-circuit capacities are being calculated, it seems that a unit load of 4½ VA/sq ft must be used, and the calculation based on that figure will yield minimum feeder and branch-circuit capacity for both general lighting and all general-purpose receptacles that may later be installed.

Of course, where the actual number of general-purpose receptacles is known, the general lighting load is taken at 3½ VA/sq ft for branch-circuit and feeder capacity, and each strap or yoke containing a single, duplex, or triplex receptacle is taken as a load of 180 VA to get the total required branch-circuit capacity, with the demand factors of Table 220.44 applied to get the minimum required feeder capacity for receptacle loads. Again, where the actual connected known receptacle load is less than 1 VA/sq ft, then a value of 1 VA/sq ft must be added to the total.

The 2014 **NEC** added an exception to this rule, which may be the beginning of more extensive changes. For the first time an Art. 220 load, in this case for lighting, can be based on something external to the **NEC**. In this case the local energy code results can be used if the building was designed and constructed to meet those requirements. This process is conditioned on a power monitoring system being in place that watches the entire "general lighting load of the building" and generates an alarm if that load exceeds the value in the energy code. In addition, if you decide to use this exception, the demand factors in Table 220.42 are off limits.

Until this exception was added, the system load had to be figured strictly by Art. 220 procedures. This has far-reaching effects, because 210.11(B) requires the electrical system upstream of branch-circuit panelboards to fully accommodate Art. 220 results; only the number of branch circuits actually installed can be reduced to match the connected load. Facility engineers have complained for generations that they had to provide feeder and service capacity far in excess of realistic demand, and the electric utilities routinely scoff at **NEC** generated demand results. The mandated alarm systems should be easily capable of being programmed to record actual demand. Assuming the sky does not fall, look for public input going forward aimed at further adjustments of Art. 220 results based on this experience.

The 2017 NEC, as predicted, added to this direction through a second exception here. Now, if a bank or office building is designed and constructed in accordance with a local building code that limits the lighting density to not over 1.2 VA/ft^2, a reduction from table lighting loads in those categories by 1 VA/ft^2 can be applied.

220.14. Other Loads—All Occupancies. This section covers rules on providing branch-circuit capacity for loads other than general lighting and designates specific amounts of load that must be allowed for each outlet. This rule establishes the minimum loads that must be allowed in computing the minimum required branch-circuit capacity for general-use receptacles and "outlets not used for general illumination." 220.14(D) requires that the actual volt-ampere rating of a recessed lighting fixture be taken as the amount of load that must be included in branch-circuit capacity. This permits local and/or decorative lighting fixtures to be taken at their actual load value rather than having them be taken as "other outlets," which would require a load allowance of "180 volt-amperes per outlet"—even if each such fixture were lamped at, say, 25 W. Or, in the case where a recessed fixture contained a 300-W lamp, allowance of only 180 VA would be inadequate. Note that these loads are not the general lighting loads addressed in 220.12 and included in volt-ampere per unit area calculations, but rather specialized lighting that may apply in certain applications as described.

Similarly, sign and outline lighting must also be considered separately. Such lighting is *not* part of the general lighting load and therefore must be accounted for as indicated in the specific sections that cover those types of equipment. In this case Art. 600 has a mandatory minimum circuiting allowance that must be built in to most commercial load calculations. Of course, if the actual load is known to be larger, then the larger number enters the load calculation.

Receptacle Outlets

The last sentence of 220.14(I) calls for "each single or each multiple receptacle *on one yoke*" to be taken as a load of "not less than 180 volt-amperes"—in commercial, institutional, and industrial occupancies. The rule requires that every general-purpose, single or duplex or triplex convenience receptacle outlet in nonresidential occupancies be taken as a load of 180 VA, and that amount of circuit capacity must be provided for each such outlet (Fig. 220-1). Code intent is that each individual device yoke—whether it holds one, two, or three receptacles—is a load of 180 VA. This rule makes clear that branch-circuit and feeder capacity must be provided for receptacles in nonresidential occupancies in accordance with loads calculated at 180 VA per receptacle yoke.

If a 15-A, 120-V circuit is used to supply *only* receptacle outlets, then the maximum number of general-purpose receptacle outlets that may be fed by that circuit is

$$15\text{ A} \times 120\text{ V} + 180\text{ VA or } 10 \text{ receptacle outlets}$$

For a 20-A, 115-V circuit, the maximum number of general-purpose receptacle outlets is

$$20\text{ A} \times 120\text{ V} + 180\text{ VA or } 13 \text{ receptacle outlets}$$

See Fig. 220-2.

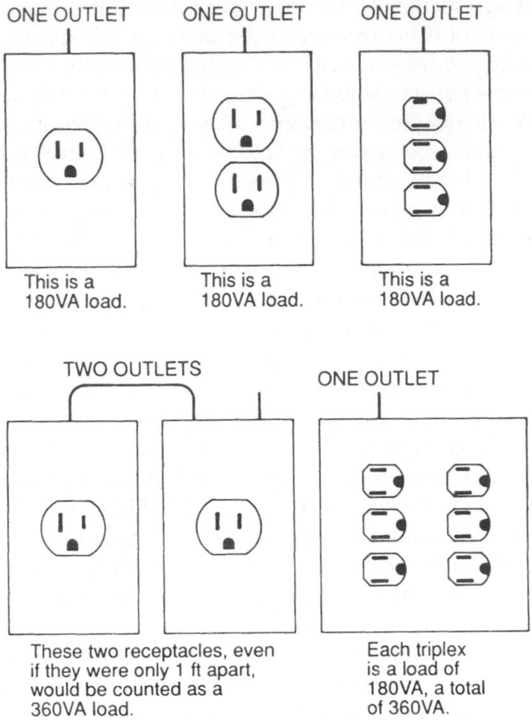
ONE OUTLET
ONE OUTLET
ONE OUTLET
This is a 180VA load.
This is a 180VA load.
This is a 180VA load.
TWO OUTLETS
ONE OUTLET
These two receptacles, even if they were only 1 ft apart, would be counted as a 360VA load.
Each triplex is a load of 180VA, a total of 360VA.

Fig. 220-1. Classification of single, duplex, and triplex receptacles. [Sec. 220.14(I).]

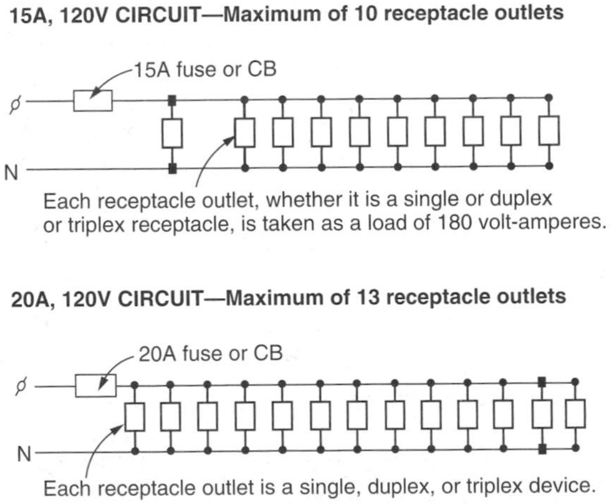
15A, 120V CIRCUIT—Maximum of 10 receptacle outlets
15A fuse or CB
ϕ
N
Each receptacle outlet, whether it is a single or duplex or triplex receptacle, is taken as a load of 180 volt-amperes.
20A, 120V CIRCUIT—Maximum of 13 receptacle outlets
20A fuse or CB
ϕ
N
Each receptacle outlet is a single, duplex, or triplex device.

Fig. 220-2. Number of receptacles per circuit, nonresidential occupancy. As shown in the drawing, there is one yoke per outlet. Where multiple yokes are installed at an outlet, the 180-VA requirement applies per yoke. [Sec. 220.14(I).]

Although the Code gives the previously described data on maximum permitted number of receptacle outlets in commercial, industrial, institutional, and other nonresidential installations, there are no such limitations on the number of receptacle outlets on residential branch circuits. There are reasons for this approach.

In 210.52, the Code specifies where and when receptacle outlets are required on branch circuits. Note that there are no specific requirements for receptacle outlets in commercial, industrial, and institutional installations other than for store show windows in 210.62 and equipment for conditioned air and refrigeration in 210.63. There is the general rule that receptacles do have to be installed where flexible cords are used. In nonresidential buildings, if flexible cords are not used, there is no *requirement* for receptacle outlets. They have to be installed only where they are needed, and the number and spacing of receptacles are completely up to the designer. But because the Code takes the position that receptacles in nonresidential buildings only have to be installed where needed for connection of specific flexible cords and caps, it demands that where such receptacles are installed, each must be taken as a load of 180 VA.

A different approach is used for receptacles in dwelling-type occupancies. The Code simply assumes that cord-connected appliances will always be used in all residential buildings and requires general-purpose receptacle outlets of the number and spacing indicated in 210.52 and 210.60. These rules cover one family houses, apartments in multifamily houses, guest rooms in hotels and motels, living quarters in dormitories, and so forth. But because so many receptacle outlets are required in such occupancies and because use of plug-connected loads is intermittent and has great diversity of load values and operating cycles, the Code notes at the bottom of Table 220.12 that the loads connected to such receptacles are adequately served by the branch-circuit capacity required by 210.11, and no additional load calculations are required for such outlets.

In dwelling occupancies, it is necessary to first calculate the total "general lighting load" from 220.12 and Table 220.12 (at 33 VA/m^2 [3 VA/ft^2] for dwellings or 22 VA/m^2 [2 VA/ft^2] for hotels and motels, including apartment houses without provisions for cooking by tenants) and then provide the minimum required number and rating of 15-A and/or 20-A general-purpose branch circuits to handle that load as covered in 210.11(A). As long as that basic circuit capacity is provided, any number of lighting outlets may be connected to any general-purpose branch circuit, up to the rating of the branch circuit if loads are known. The lighting outlets should be evenly distributed among all the circuits. Although residential lamp wattages cannot be anticipated, the Code method covers fairly heavy loading.

When the preceding Code rules on circuits and outlets for general lighting in dwelling units, guest rooms of hotels and motels, and similar occupancies are satisfied, general-purpose convenience receptacle outlets may be connected on circuits supplying lighting outlets; or receptacles only may be connected on one or more of the required branch circuits; or additional circuits (over and above those required by Code) may be used to supply the receptacles. But no matter how general-purpose receptacle outlets are circuited, *any number* of general-purpose receptacle outlets may be connected on a residential branch circuit—with or without lighting outlets on the same circuit.

And when small-appliance branch circuits are provided in accordance with the requirements of 210.11(C)(1), *any number* of small-appliance receptacle outlets may be connected on the 20-A small-appliance circuits—*but only* receptacle outlets may be connected to these circuits and only in the specified rooms.

210.52(A) applies to spacing of receptacles connected on the 20-A small-appliance circuits, as well as spacing of general-purpose receptacle outlets. That section, therefore, establishes the *minimum* number of receptacles that must be installed for greater convenience of use.

220.14(H) requires branch-circuit capacity to be calculated for multioutlet assemblies (prewired surface metal raceway with plug outlets spaced along its length). Part **(H)(2)** says that each 300-mm (1-ft) length of such strip must be taken as a 180-VA load when the strip is used where a number of appliances are likely to be used simultaneously. For instance, in the case of industrial applications on assembly lines involving frequent, simultaneous use of plugged-in tools, the loading of 180 VA/ft must be used. Part **(H)(1)** allows loading of 180 VA for each 1.5-m (5-ft) section in commercial or institutional applications of multioutlet assemblies when use of plug-in tools or appliances is not heavy. Figure 220-3 shows an example of the more intensive load calculation.

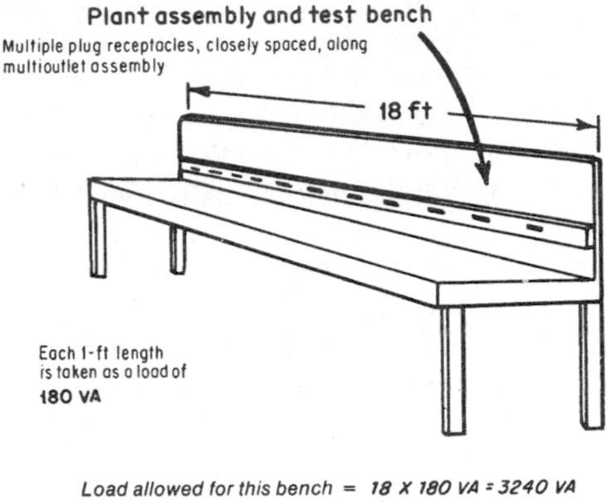
Plant assembly and test bench
Multiple plug receptacles, closely spaced, along multioutlet assembly
18 ft
Each 1-ft length is taken as a load of 180 VA
Load allowed for this bench = 18 X 180 VA = 3240 VA

Fig. 220-3. Calculating required branch-circuit capacity for multioutlet assembly. [Sec. 220.14(H).]

Part **(G)** permits branch-circuit capacity for the outlets required by 210.62 for show windows to be calculated, as shown in Fig. 220-4—instead of using the load-per-outlet value (180 VA) from part **(I)** of 220.14.

As noted by 220.14(B), 220.54 is permitted to be used in calculating the size of branch-circuit conductors for household clothes dryers; this results in a load of 5000 VA being used for a household electric dryer when the actual dryer rating is not known. This is essentially an exception to 220.14(A), which specifies that the "ampere rating of appliance or load served" shall be taken as the branch-circuit load for an outlet for a specific appliance. A comparable calculation applies to a household cooking appliance; here again the number from Part III if the article on feeders is allowed to be used for a branch circuit. For household ranges, this correlates with 210.21(B)(4)

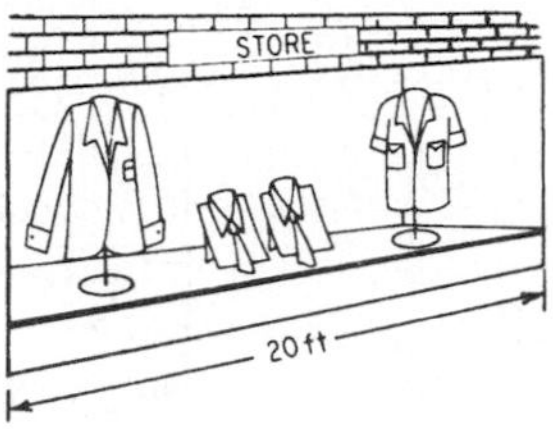
STORE
20 ft
Required branch-circuit capacity for show window
= 200 VA X 20 linear ft
= 4000 VA

Fig. 220-4. Alternate method for calculating show-window circuit capacity. [Sec. 220.14(G).]

that bases the range receptacle ampere rating on a single range demand load as given in Table 220.55, and that configuration also correlates with the rating of the branch-circuit overcurrent device through 210.20(D).

Part III. Feeder and Service Load Calculations

220.42. General Lighting. For general illumination, a feeder must have capacity to carry the total load of lighting branch circuits determined as part of the lighting design and not less than a minimum branch-circuit load determined on a volt-amperes-per-unit-area basis from the table in 220.12.

Demand factor permits sizing of a feeder according to the amount of load that operates simultaneously.

Demand factor is the ratio of the maximum amount of load that will be operating at any one time on a feeder to the total connected load on the feeder under consideration. This factor is frequently less than 1. The sum of the connected loads supplied by a feeder is multiplied by the demand factor to determine the load which the feeder must be sized to serve. This load is termed the *maximum demand* of the feeder:

$$\text{Maximum demand load} = \text{connected load} \times \text{demand factor}$$

Tables of demand and diversity factors have been developed from experience with various types of load concentrations and various layouts of feeders and subfeeders supplying such loads. Table 220.42 of the NEC presents common demand factors for feeders to general lighting loads in various types of buildings (Fig. 220-5).

The demand factors given in Table 220.42 may be applied to the total branch-circuit load to get required feeder capacity for lighting (but must not be used in calculating branch-circuit capacity). Note that a feeder may have a capacity of less than 100 percent of the total branch-circuit load for only the types of buildings designated in Table 220.42, that is, for dwelling units, hospitals, hotels, motels, and storage warehouses. In all other types of occupancies, it is assumed that *all* general lighting will be operating at the same time, and each feeder in those occupancies must have capacity (ampacity) for 100 percent of the volt-amperes of branch-circuit load of general lighting that the feeder supplies.

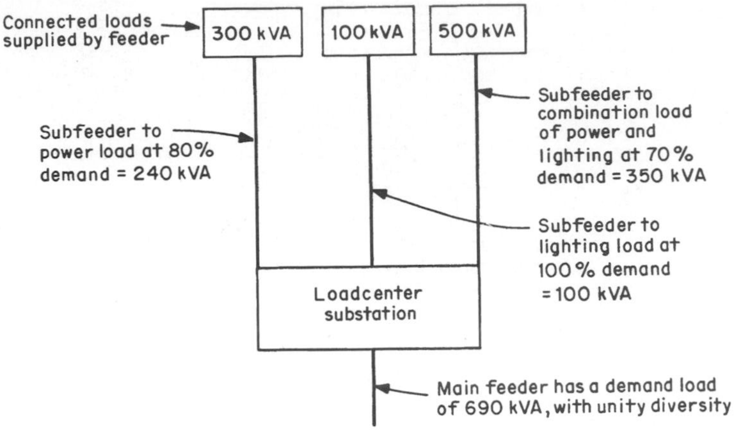
Connected loads supplied by feeder
300 kVA
100 kVA
500 kVA
Subfeeder to combination load of power and lighting at 70% demand = 350 kVA
Subfeeder to power load at 80% demand = 240 kVA
Subfeeder to lighting load at 100% demand = 100 kVA
Loadcenter substation
Main feeder has a demand load of 690 kVA, with unity diversity

Fig. 220-5. How demand factors are applied to connected loads. (Sec. 220.42.)

example If a warehouse feeder fed a total branch-circuit load of 20,000 VA of general lighting, the minimum capacity in that feeder to supply that load must be equal to 12,500 VA plus 50 percent times (20,000 – 12,500) VA. That works out to be 12,500 plus 0.5 × 7500 or 16,250 VA.

But, the note to Table 220.42 warns against using any value less than 100 percent of branch-circuit load for sizing any feeder that supplies loads that will all be energized at the same time.

220.43. Show-Window and Track Lighting. In providing minimum required capacity in feeders, a load of 150 VA must be allowed for each 2 ft (600 mm) or fraction thereof of lighting track. That amount of load capacity must be provided in feeders and service conductors (see Fig. 220-6) in nonresidential installations and would have to be added in addition to the general lighting load in volt-amperes per square foot from **NEC** Table 220.12. For residential applications, track lighting is considered to be included in the 33 VA/m^2 (3 VA/ft^2) calculations and no additional load need be added.

A new exception as of the 2011 **NEC** grants relief from this loading requirement if the total load that will be carried by the track is limited by a device designed for this purpose, the reduced load can be used to design the feeder capacity. The substantiation for this proposal documented an instance where the track load calculated under the usual provisions of 220.43(B) resulted in a feeder load of 11,250 VA. However, the applicable energy code set an upper limit of 4500 W. Differences like this produce significant changes in transformer and feeder sizing. For this reason, the original proposal requested a blanket exception allowing energy code limits to replace those in the **NEC**. This was another in a very long and continuing line of proposed changes involving complaints about excessive load capabilities arising in building electrical designs from Art. 220 results. Refer to the discussion of the new (in the 2014 **NEC**) 220.12 Exception for the latest chapter in this process.

The panel compromised and agreed to allow for a reduction in the calculated load, but only if an actually installed device permanently restricted the load that could be connected to the track. This final load may or may not have any relation

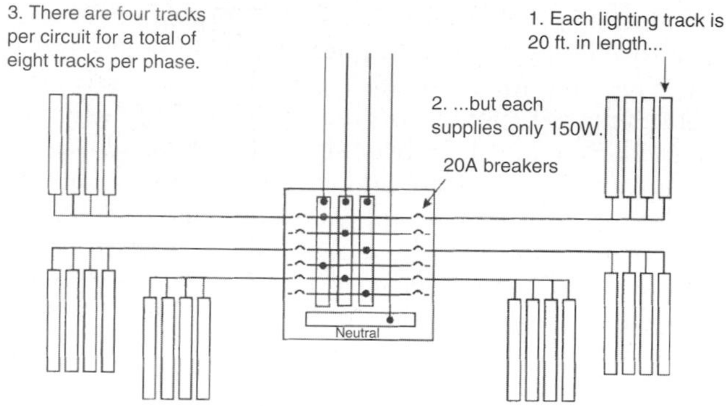
3. There are four tracks per circuit for a total of eight tracks per phase.
1. Each lighting track is 20 ft. in length...
2. ...but each supplies only 150W.
20A breakers
Neutral

1. According to 220.43(B), the feeder must have capacity equal to 75 VA/300 mm (/1 ft) [150 VA/600 mm (/2 ft)]. Therefore, each phase conductor must have capacity for:
8 tracks × 6.0 m (20 ft)/ track = minimum VA capacity
12,000 VA = minimum VA capacity
This is all line-to-neutral load; to find amperes, divide by 120 V
12,000 VA ÷ 120 V = 100 A

2. Due to the continuous nature of the load, the minimum rating of the overcurrent protective device must be sized to accept 125% of this load: 100 A × 1.25 = 125 A.

3. The conductors must have a similar rating where terminating on an overcurrent device, evaluated under the 75°C column of Table 310.15(B)(16). No. 1AWG THHN (130 A) will work.

4. If the luminaires are HIDs, the neutral will be current carrying per 310.15(B)(5)(c) and mutual conductor heating in the raceway must be evaluated, based on the 90°C column of Table 310.15(B)(16), taken at 80% per 310.15(B)(3)(a): 150 A × 0.8 = 120 A; this is permitted because the 125A OCPD protects the conductor per 240.4(B), and the load (100 A) does not exceed the actual ampacity (120 A).

Fig. 220-6. A single-circuit lighting track must be taken as a load of 150 VA for each 2 ft (610 mm) or fraction thereof, divided among the number of circuits. For a 2-circuit lighting track, each 2-ft (610-mm) length is a 75-VA load for each circuit. For a 3-circuit track, each 2-ft (610-mm) length is a 50 VA load for each circuit. (Sec. 220.43.)

to the applicable energy code, and could result from a simple design preference on the part of the owner. When such a device is installed, its upper let-through loading is now permitted to be the track lighting contribution to Art. 220 feeder load for the system. The limiting device can be as simple as the branch-circuit overcurrent device actually installed, or a supplementary overcurrent device installed for this purpose.

If show-window lighting is supplied by a feeder, capacity must be included in the feeder to handle 200 VA per linear 300 mm (1 ft) of show-window length. Note that on the feeder level, the 200 VA per 300 mm (1 ft) is the only allowable calculation, unlike the branch circuit calculations that can be done either this way or on the basis of the receptacles actually in use taken at 180 VA each.

220.44. Receptacle Loads—Other Than Dwelling Units. This rule permits two possible approaches in determining the required feeder ampacity to supply receptacle loads in "other than dwelling units," where a load of 180 VA of feeder capacity must be provided for all general-purpose 15- and 20-A receptacle outlets. (In dwelling units and in guest rooms of hotels and motels, no feeder capacity is required for 15- or 20-A general-purpose receptacle outlets. Such load is considered sufficiently covered by the load capacity provided for general lighting.) But in other than dwelling units, where a load of 180 VA of feeder capacity must be provided for all general-purpose 15- and 20-A receptacle outlets, a *demand factor* may be applied to the total calculated receptacle load as follows. Wording of this rule makes clear that *either* Table 220.42 or Table 220.44 may be used to apply demand factors to the total load of 180-VA receptacle loads when calculating required ampacity of a feeder supplying receptacle loads connected on branch circuits.

In other than dwelling units, the branch-circuit load for receptacle outlets, for which 180 VA was allowed per outlet, may be added to the general lighting load and may be reduced by the demand factors in Table 220.42. That is the basic rule of 220.44 and, in effect, requires any feeder to have capacity for the total number of receptacles it feeds and requires that capacity to be equal to 180 VA (per single or multiple receptacle) times the total number of receptacles (yokes)—with a reduction from 100 percent of that value permitted only for the occupancies listed in Table 220.42.

Because the demand factor of Table 220.42 is shown as 100 percent for "All Other" types of occupancies, the basic rule of 220.44 as it appeared prior to the 1978 **NEC** required a feeder to have ampacity for a load equal to 180 VA times the number of general-purpose receptacle outlets that the feeder supplied. That is no longer required. Recognizing that there is great diversity in use of receptacles in office buildings, stores, schools, and all the other occupancies that come under "All Others" in Tables 220.42, 220.44 contains a table to permit reduction of feeder capacity for receptacle loads on feeders. Those demand factors apply to *any* "non-dwelling" occupancy.

The amount of feeder capacity for a typical case where a feeder, say, supplies panelboards that serve a total of 500 receptacles is shown in Fig. 220-7.

Although the calculation of Fig. 220-7 cannot always be taken as realistically related to usage of receptacles, it is realistic relief from the 100-percent-demand factor, which presumed that all receptacles were supplying 180-VA loads simultaneously.

220.50. Motors. Any feeder that supplies a motor load or a combination load (motors plus lighting and/or other electrical loads) must satisfy the indicated **NEC** sections of Art. 430. Feeder capacity for motor loads is usually taken at 125 percent of the full-load current rating of the largest motor supplied, plus the sum of the full-load currents of the other motors supplied. However, 430.26 allows the application of demand factors in certain cases as determined by the inspector.

Specifically, 430.26 operates by "special permission," which is the written permission from the authority enforcing the code. Some authorities recommend that no demand factor be used in determining the size of circuit to install so that the additional current capacity, thus allowed in the circuit, will give some spare capacity for growth. On the other hand, as noted in the discussion at 220.12 Exception, one of the major and repeated areas of discussion by the Code-making panel responsible for load calculations involves the repeated and well-documented

Take the total number of general-purpose receptacle outlets fed by a given feeder. . .

Example:
500 receptacles

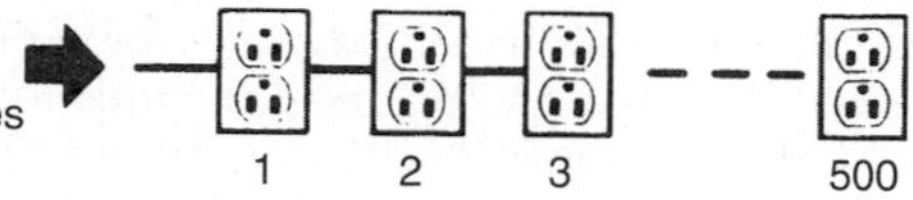
1
2
3
500

. . . multiply the total by 180 volt-amperes [required load of 220.3(B) (9) for each receptacle]. . .

500 × 180 VA = 90,000 VA

Then apply the demand factors from Table 220.13:

First 10 k VA or less @ 100% demand = 10,000 VA
Remainder over 10 kVA @ 50% demand
= (90,000 − 10,000) × 50%
= 80,000 × 0.5 = 40,000 VA

Minimum demand-load total = 50,000 VA

Therefore, the feeder must have a capacity of 50 kVA for the total receptacle load. Required minimum ampacity for that load is then determined from the voltage and phase-makeup (single- or 3-phase) of the feeder.

Fig. 220-7. Table 220.13 permits demand factor in calculating feeder demand load for general-purpose receptacles. (Sec. 220.44.)

oversizing of industrial feeders in this area. Electric utilities know how much power they provide to their customers, and it too often does not compare well with load calculations run without appropriate demand factors. The express allowance for the judicious use of demand factors in the NEC for these loads is something that well deserves careful consideration. The factors given in Div. 12 of our sister book, the *American Electricians' Handbook*, are an excellent place to start.

220.51. Fixed Electric Space Heating. Capacity required in a feeder to supply fixed electrical space-heating equipment is determined on the basis of a load equal to the total connected load of heaters on all branch circuits served from the feeder. Under conditions of intermittent operation or where all units cannot operate at the same time, permission may be granted for use of less than a 100 percent demand factor in sizing the feeder. 220.82, 220.83, and 220.84 permit alternate calculations of electric heat load for feeders or service-entrance conductors (which constitute a service feeder) in dwelling units. But reduction of the feeder capacity to less than 100 percent of connected load must be authorized by the local electrical inspector. Feeder load current for heating must not be less than the rating of the largest heating branch circuit supplied.

220.52. Small-Appliance and Laundry Loads—Dwelling Unit. For a feeder or service conductors in a single-family dwelling, in an individual apartment of a multifamily dwelling with provisions for cooking by tenants, or in a hotel or motel suite with cooking facilities or a serving pantry, at least 1500 VA of load must be provided for each 2-wire, 20-A small-appliance circuit (to handle the small appliance load in kitchen, pantry, and dining areas) that is actually installed. The total small-appliance load determined in this way may be added to the general lighting load, and the resulting total load may be reduced by the demand factors given in Table 220.42.

Note that in a major clarification, the 2008 **NEC** changed the verb in this rule from "required by 210.11(C)(1)" to "covered by 210.11(C)(1)." A key point of contention for a very long time has been whether 3000 VA based on two small appliance branch circuits was all that was needed to put into a load calculation, even if many more were actually installed in a given dwelling unit. The theory behind this was that 210.11(C)(1) requires only two such circuits. The opposing viewpoint noted that 210.11(C)(1) actually mandated "two or more" such circuits, so all that were provided should be counted in the load calculation. By changing the word, it is now clear that although only two such circuits are required, if you choose to install more, the load calculation must include every one that is in place, taken at 1500 VA each. The same change was made in (B) following, so if multiple laundry circuits are provided, each will enter the load calculation at 1500 VA.

A feeder load of at least 1500 VA must be added for each 2-wire, 20-A laundry circuit installed as covered by 210.1(C)(2). And that load may also be added to the general lighting load and subjected to the demand factors in Table 220.42.

220.53. Appliance Load—Dwelling Unit(s). For fixed appliances (fastened in place) other than ranges, clothes dryers, air-conditioning equipment, and space-heating equipment, feeder capacity in dwelling occupancies must be provided for the sum of these loads; but, if there are at least four such fixed appliances, the total load of four or more such appliances may be reduced by a demand factor of 75 percent (**NEC** 220.53). Wording of this rule makes clear that a "fixed appliance" is one that is "fastened in place."

As an example of application of this Code provision, consider the following calculation of feeder capacity for fixed appliances in a single-family house. The calculation is made to determine how much capacity must be provided in the service-entrance conductors (the service feeder):

Water heater	2500 W	240 V =	10.4
Kitchen disposal	½ hp	120 V = 9.8 A + 25% =	12.3
Furnace motor	¼ hp	120 V =	5.8 A
Attic fan	¼ hp	120 V = 5.8 A	0.0 A
Water pump	½ hp	240 V =	4.9 A

Load in amperes on each ungrounded leg of feeder = 33.4 A

To comply with 430.24, 25 percent is added to the full-load current of the ½-hp, 120-V appliance motor because it is the highest-rated motor in the group. Since it is assumed that the load on the 120/240-V feeder will be balanced and each of the ¼-hp motors will be connected to different ungrounded conductors, only one is counted in the preceding calculation. Except for the 120-V motors, all the other appliance loads are connected to both ungrounded conductors and are automatically balanced. Since there are four or more fixed appliances in addition to a range, clothes dryer, etc., a demand factor of 75 percent may be applied to the total load of these appliances. Seventy-five percent of 33.4 = 25 A, which is the current to be added to that computed for the lighting and other loads to determine the total current to be carried by the ungrounded (outside) service-entrance conductors.

The preceding demand factor may be applied to similar loads in two-family or multifamily dwellings.

220.54. Electric Clothes Dryers—Dwelling Unit(s). This rule prescribes a minimum demand of 5 kVA for 120/240-V electric clothes dryers in determining branch-circuit and feeder sizes. Note that this rule applies only to "household" electric clothes dryers, and not to commercial applications. This rule is helpful because the ratings of electric clothes dryers are not usually known in the planning stages when feeder calculations must be determined (Fig. 220-8).

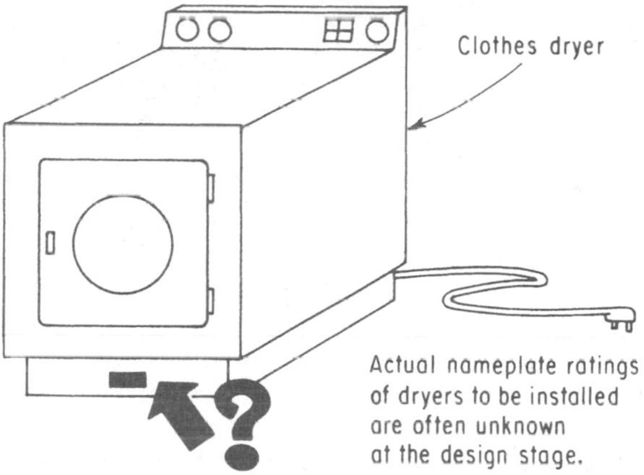
Clothes dryer
Actual nameplate ratings
of dryers to be installed
are often unknown
at the design stage.

Fig. 220-8. Feeder load of 5 kVA per dryer must be provided if actual load is not known. (Sec. 220.54.)

When sizing a feeder for one or more electric clothes dryers, a load of 5000 VA or the nameplate rating, whichever is larger, shall be included for each dryer—subject to the demand factors of Table 220.54 when the feeder supplies a number of clothes dryers, as in an apartment house. At one time this table periodically generated paradoxical load calculations; for some load brackets, adding additional clothes dryers actually decreased the calculated load for the feeder. This has been corrected, and now adding a clothes dryer always results in at least some additional load capacity required in the feeder.

220.55. Electric Cooking Appliances in Dwelling Units and Household Cooking Appliances Used in Instructional Programs. Feeder capacity must be allowed for household electric cooking appliances rated over 1¾ kW, in accordance with Table 220.55 of the Code. Feeder demand loads for a number of cooking appliances on a feeder may be obtained from Table 220.55.

Note 4 to Table 220.55 permits sizing of a branch circuit to supply a single electric range, a wall-mounted oven, or a counter-mounted cooking unit in accordance with that table. That table is also used in sizing a feeder (or service conductors) that supplies one or more electric ranges or cooking units. Note that 220.55 and Table 220.55 apply only to such cooking appliances in a "dwelling unit" and do not cover commercial or institutional applications, although ranges in vocational school kitchens are covered.

Figure 220-9 shows a typical **NEC** calculation of the minimum demand load to be used in sizing the branch circuit to the range. The same value of demand load is also used in sizing a feeder (or service conductors) from which the range circuit is fed. Calculation is as follows:

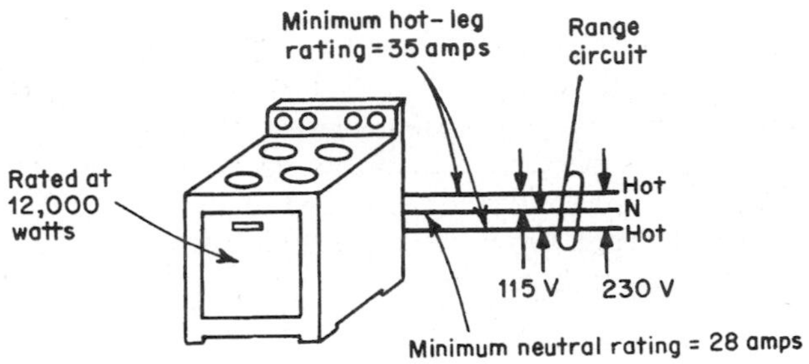
Minimum hot- leg rating = 35 amps
Range circuit
Rated at 12,000 watts
Hot
N
Hot
115 V
230 V
Minimum neutral rating = 28 amps

Fig. 220-9. Minimum amp rating of branch-circuit conductors for a 12-kW range. (Sec. 220.55.)

A branch circuit for the 12-kW range is selected in accordance with Note 4 of Table 220.55, which says that the branch-circuit load for a range may be selected from the table itself. Under the heading "Number of Appliances," read across from "1." The maximum demand to be used in sizing the range circuit for a 12-kW range is shown under the heading "Maximum Demand" to be not less than 8 kW. The minimum rating of the range-circuit ungrounded conductors will be

$$\frac{8000\text{ W}}{240\text{ V}} = 33.3\text{ A or }33\text{ A}$$

NEC Table 310.15(B)(16) shows that the minimum size of copper conductors that may be used is 8 AWG (TW—40 A, THW—45 A, XHHW or THHN—50 A). No. 8 AWG is also designated in 210.19(A)(3) as the minimum size of conductor for any range rated 8¾ kW or more because the circuit must be at least rated 40 A.

The overload protection for this circuit of No. 8 TW conductors would be *40-A fuses or a 40-A circuit breaker*. If THW, THHN, or XHHW wires are used for the circuit, they must be taken as having an ampacity of not more than 40 A and protected at that value. That requirement follows from the UL rule that conductors up to No. 1 AWG size must be used at the 60°C ampacity for the size of conductor, regardless of the actual temperature rating of the insulation—which may be 75°C or 90°C. Similarly, 110.14(C)(1)(a) brings the same listing limitations into the **NEC** itself. The ampacity used must be that of TW wire of the given size unless both the overcurrent device at the beginning of the circuit and the range receptacle at the end of the circuit, if used, are marked to allow for 75°C connections.

Although the two hot legs of the 120/240-V, 3-wire circuit must be not smaller than No. 8, Exception No. 2 to Sec. 210.19(A)(3) permits the neutral conductor to be smaller, but it specifies that it must have an ampacity not less than 70 percent of the rating of the branch-circuit CB or fuse and may never be smaller than No. 10.

For the range circuit in this example, the neutral may be rated

$$70\% \times 40 \text{ A (rating of branch-circuit protection)} = 28 \text{ A}$$

This calls for a No. 10 neutral.

Figure 220-10 shows a more involved calculation for a range rated over 12 kW. Figure 220-11 shows two units that total 12 kW and are taken at a demand load of 8 kW, as if they were a single range. Figure 220-12 shows another calculation for separate cooking units on one circuit. And a feeder that would be used to supply any of the cooking installations shown in Figs. 220-9 through 220-12 would have to include capacity equal to the demand load used in sizing the branch circuit.

A feeder supplying more than one range (rated not over 12 kW) must have ampacity sufficient for the maximum demand load given in Table 220.55 for the number of ranges fed. For instance, a feeder to 10 such ranges would have to have ampacity for a load of 25 kW.

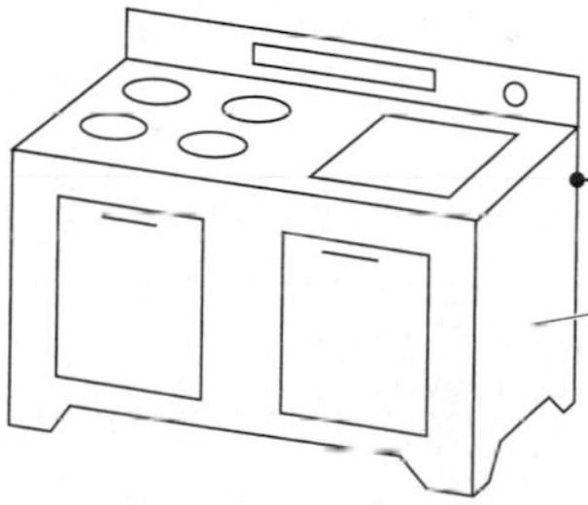
4-wire cable
16.6-kW household range,
115/230 volts

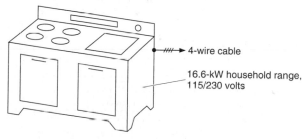

Refer to *NEC* Table 220.55.

1. Column C applies to ranges rated not over 12 kW, but this range is rated 16.6 kW.
2. Note 1, below the Table, tells how to use the Table for ranges over 12 kW and up to 27 kW. For such ranges, the maximum demand in Column C must be increased by 5% for each additional kW of rating (or major fraction) above 12 kW.
3. This 16.6-kW range exceeds 12 kW by 4.6 kW.
4. 5% of the demand in Column C for a single range is 400 watts (8000 watts × 0.05).
5. The maximum demand for this 16.6-kW range must be increased above 8 kW by 2000 watts:
   400 watts (5% of Column C) × 5 (4 kW + 1 for the remaining 0.6 kW)
6. The required branch circuit must be sized, therefore, for a total demand load of 8000 watts + 2000 watts = 10,000 watts
7. Required size of branch circuit—

$$\text{Amp rating} = \frac{10{,}000 \text{ W}}{240 \text{ V}} = 42 \text{ A}$$

USING 60C CONDUCTORS, AS REQUIRED BY 110.14(C)(1)(a), THE UNGROUNDED BRANCH CIRCUIT CONDUCTORS WOULD CONSIST OF 6 AWG CONDUCTORS. THIS IS THE DEFAULT; 75C CONNECTIONS ARE PERMITTED AS COVERED IN 110.14(C)(1)(a)(3).

Fig. 220-10. Sizing a branch circuit for a household range over 12 kW. (Sec. 220.55.)

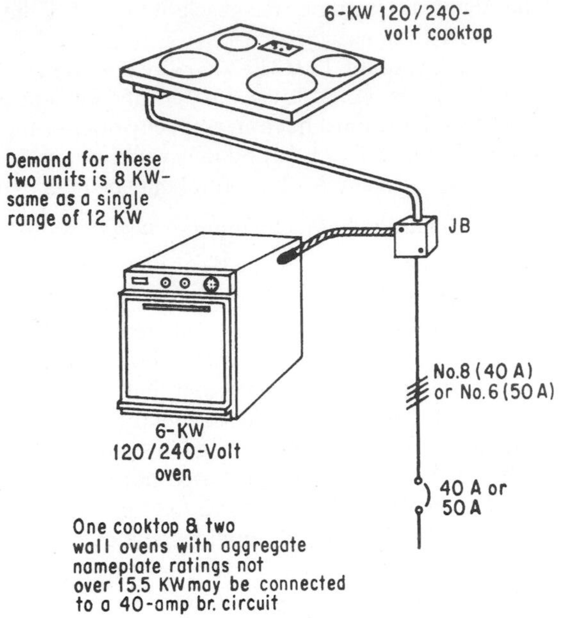
6-KW 120/240-
volt cooktop
Demand for these
two units is 8 KW-
same as a single
range of 12 KW
JB
No.8 (40 A)
or No.6 (50 A)
6-KW
120/240-Volt
oven
40 A or
50 A
One cooktop & two
wall ovens with aggregate
nameplate ratings not
over 15.5 KW may be connected
to a 40-amp br. circuit

Fig. 220-11. Two units treated as a single-range load. (Sec. 220.55.)

Other Calculations on Electric Cooking Appliances

The following "roundup" points out step-by-step methods of wiring the various types of household electric cooking equipment (ranges, counter-mounted cooking units, and wall-mounted ovens) according to the NEC.

Tap Conductors

210.19(A)(3), Exception No. 2, gives permission to reduce the size of the neutral conductor of a 3-wire range branch circuit to 70 percent of the rating of the CB or fuses protecting the branch circuit. However, this rule does not apply to smaller taps connected to a 50-A circuit—where the smaller taps (none less than 20-A ratings) must all be the same size. Further, it does not apply when individual branch circuits supply each wall- or counter-mounted cooking unit and all circuit conductors are of the same size and less than No. 10.

210.19(A)(3), Exception No. 1, permits tap conductors, rated not less than 20 A, to be connected to 50-A branch circuits that supply ranges, wall-mounted ovens, and counter-mounted cooking units. These taps cannot be any longer than necessary for servicing. Figure 220-13 illustrates the application of this rule.

In 210.19(A)(3), Exception No. 1, the wording "no longer than necessary for servicing" encourages the location of circuit junction boxes as close as possible to each cooking and oven unit connected to 50-A circuits. A number of counter-mounted

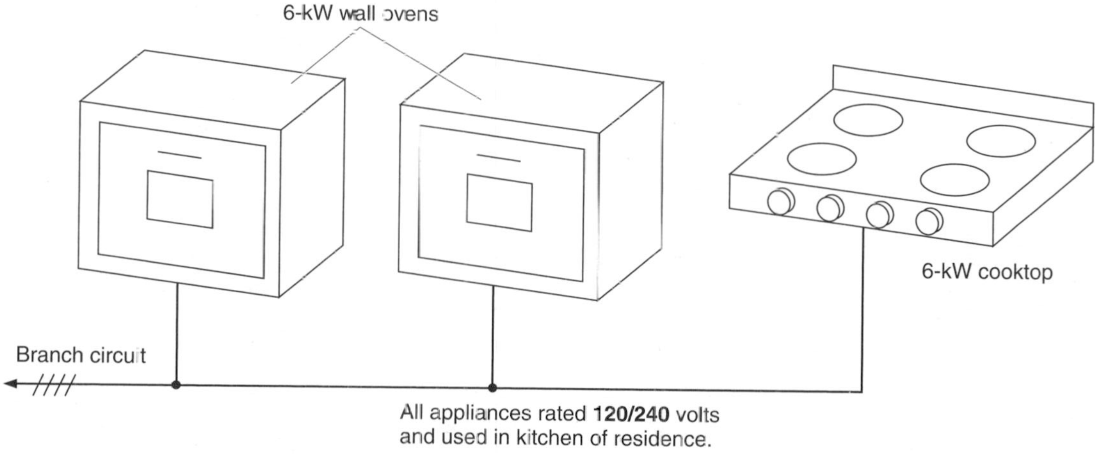
6-kW wall ovens
6-kW cooktop
Branch circuit
All appliances rated **120/240** volts
and used in kitchen of residence.

1. Note 4 of Table 220.55 says that the branch-circuit load for a counter-mounted cooking unit and not more than two wall-mounted ovens, all supplied from a single branch circuit and located in the same room, shall be computed by adding the nameplate ratings of the individual appliances and treating this total as a single range.

2. Therefore, the three appliances shown may be considered to be a single range of 18-kW rating (6 kW + 6 kW + 6 kW).

3. From Note 1 of Table 220.55, such a range exceeds 12 kW by 6 kW and the 8-kW demand of Column C must be increased by 400 watts (5% of 8000 watts) for each of the 6 additional kilowatts above 12 kW.

4. Thus, the branch-circuit demand load is—

8000 WATTS + (6 × 400 WATTS) = 10,400 WATTS

A 50-AMP CIRCUIT IS REQUIRED.

Fig. 220-12. Determining branch-circuit load for separate cooking appliances on a single circuit. (Sec. 220.55.)

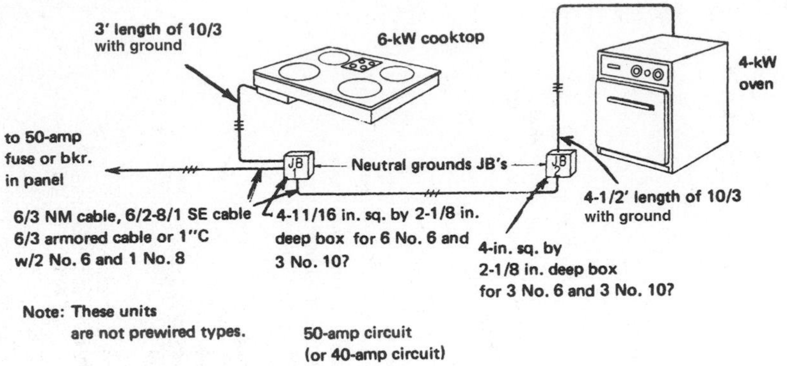
3' length of 10/3
with ground
6-kW cooktop
4-kW
oven
to 50-amp
fuse or bkr.
in panel
JB
Neutral grounds JB's
JB
4-1/2' length of 10/3
with ground
6/3 NM cable, 6/2-8/1 SE cable
6/3 armored cable or 1"C
w/2 No. 6 and 1 No. 8
4-11/16 in. sq. by 2-1/8 in.
deep box for 6 No. 6 and
3 No. 10?
4-in. sq. by
2-1/8 in. deep box
for 3 No. 6 and 3 No. 10?
Note: These units
are not prewired types.
50-amp circuit
(or 40-amp circuit)

NEC rules permit a 50-amp circuit to supply cooktops and ovens. Typical arrangement shows such a circuit. Junction box sizes are computed from Table 314.16(A) and Table 314.16(B) for No. 6 conductor combinations. Taps to each unit are No. 10.

Fig. 220-13. One branch circuit to cooking units. (Sec. 220.55.)

cooking units have integral supply leads about 36 in. (914 mm) long, and some ovens come with supply conduit and wire in lengths of 48 to 54 in. Therefore, a box should be installed close enough to connect these leads.

Feeder and Circuit Calculations

220.55 permits the use of Table 220.55 for calculating the feeder load for ranges and other cooking appliances that are individually rated more than 1¾ kW.

Note 4 of the table reads: "The branch-circuit load for one wall-mounted oven or one counter-mounted cooking unit shall be the nameplate rating of the appliance." Figure 220-14 shows a separate branch circuit to each cooking unit, as permitted.

Common sense dictates that there is no difference in demand factor between a single range of 12 kW and a wall-mounted oven and surface-mounted cooking unit totaling 12 kW. This is explained in the last sentence of Note 4 of Table 220.55. The mere division of a complete range into two or more units does not change the demand factor. Therefore, the most direct and accurate method of computing the branch-circuit and feeder calculations for wall-mounted ovens and surface-mounted cooking units within each occupancy is to total the kilowatt ratings of these appliances and treat this total kilowatt rating as a single range of the same rating. For example, a particular dwelling has an 8-kW, 4-burner, surface-mounted cooking unit and a 4-kW wall-mounted oven. This is a total of 12 kW, and the maximum permissible demand given in Column C of Table 220.55 for a single 12-kW range is 8 kW.

Similarly, it follows that if the ratings of a 2-burner, counter-mounted cooking unit and a wall-mounted oven are each 3.5 kW, the total of the two would be 7 kW—the same total as a small 7-kW range. Because the 7-kW load is less than 8¾ kW, Note 3 of Table 220.55 permits Column B of Table 220.55 to be used in lieu of Column C. The demand load is 5.6 kW (7 kW times 0.80). Range or total cooking

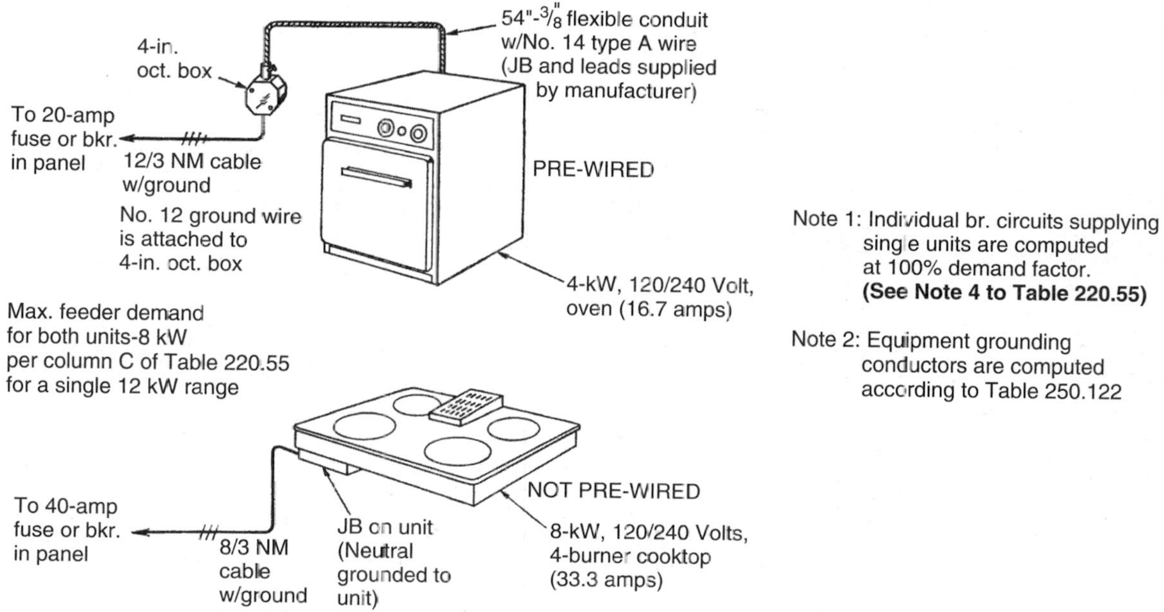
4-in.
oct. box
54"-3/8" flexible conduit
w/No. 14 type A wire
(JB and leads supplied
by manufacturer)
To 20-amp
fuse or bkr.
in panel
12/3 NM cable
w/ground
PRE-WIRED
No. 12 ground wire
is attached to
4-in. oct. box
Note 1: Individual br. circuits supplying
single units are computed
at 100% demand factor.
(See Note 4 to Table 220.55)
4-kW, 120/240 Volt,
oven (16.7 amps)
Max. feeder demand
for both units-8 kW
per column C of Table 220.55
for a single 12 kW range
Note 2: Equipment grounding
conductors are computed
according to Table 250.122
NOT PRE-WIRED
To 40-amp
fuse or bkr.
in panel
8/3 NM
cable
w/ground
JB on unit
(Neutral
grounded to
unit)
8-kW, 120/240 Volts,
4-burner cooktop
(33.3 amps)

An 8-kW cooktop is supplied by an individual No. 8 (40-amp) branch circuit, and a No. 12 (20-amp) branch circuit supplies a 4-kW oven. Such circuits are calculated on the basis of the nameplate rating of the appliance. In most instances individual branch circuits cost less than 50-amp, multi-outlet circuits for cooking and oven units.

Fig. 220-14. Separate branch circuit to cooking units. (Sec. 220.55.)

and oven unit ratings less than 8¾ kW are more likely to be found in small apartment units of multifamily dwellings than in single-family dwellings.

Because the demand loads in Column C of Table 220.55 apply to ranges not exceeding 12 kW, they also apply to wall-mounted ovens and counter-mounted cooking units within each individual occupancy by totaling their aggregate nameplate kilowatt ratings. Then if the total rating exceeds 12 kW, Note 1 to the table should be used as if the units were a single range of equal rating. For example, assume that the total rating of a counter-mounted cooking unit and two wall-mounted ovens is 16 kW in a dwelling unit. The maximum demand for a single 12-kW range is given as 8 kW in Column C. Note 1 requires that the maximum demand in Column C be increased 5 percent for each additional kilowatt or major fraction thereof that exceeds 12 kW. In this case 16 kW exceeds 12 kW by 4 kW. Therefore, 5 percent times 4 equals 20 percent, and 20 percent of 8 kW is 1.6 kW. The maximum feeder and branch-circuit demand is then 9.6 kW (8 kW plus 1.6 kW). A 9600-W load would draw exactly 40 A at 240 V, thereby just fitting on a circuit rated 40 A.

For the range or cooking unit demand factors in a multifamily dwelling, say a 12-unit apartment building, a specific calculation must be made, as follows:

1. Each apartment has a 6-kW counter-mounted cooking unit and a 4-kW wall-mounted oven. And each apartment is served by a separate feeder from a main switchboard. The maximum cooking demand in each apartment feeder should be computed in the same manner as previously described for single-family dwellings. Since the total rating of cooking and oven units in each apartment is 10 kW (6 kW plus 4 kW), Column C of Table 220.55 for one appliance should apply. Thus, the maximum cooking demand load on each feeder is 8 kW.
2. In figuring the size of the main service feeder, Column C should be used for 12 appliances. Thus, the demand would be 27 kW.

As an alternate calculation, assume that each of the 12 apartments has a 4-kW counter-mounted cooking unit and a 4-kW wall-mounted oven. This would total 8 kW per apartment. In this case, Column B of Table 220.55 can be used to determine the cooking load in each separate feeder. By applying Column B on the basis of a single 8-kW range, the maximum demand is 6.4 kW (8 kW times 0.80). Therefore, 6.4 kW is the cooking load to be included in the calculation of each feeder. Notice that this is 1.6 kW less than the previous example where cooking and oven units, totaling 10 kW, had a demand load of 8 kW. And this is logical, because smaller units should produce a smaller total kilowatt demand.

On the other hand, it is advantageous to use Column C instead of Column B for computing the main service feeder capacity for twelve 8-kW cooking loads. The reason for this is that Column B gives a higher result where more than five 8-kW ranges (or combinations) and more than twelve 7-kW ranges (or combinations) are to be used. In these instances, calculations made on the basis of Column B result in a demand load greater than that of Column C for the same number of ranges. As an example, twelve 8-kW ranges have a demand load of 30.72 kW (12 times 8 kW times 0.32) in applying Column B, but only a demand load of 27 kW in Column C. And in Column C the 27 kW is based on twelve 12-kW ranges. This discrepancy dictates use of Column B only on the limited basis previously outlined.

The reason for the higher demand factor for a smaller range is that the smaller appliances, while in use, have more of their current consuming functions in operation at the same time. This is the reason that the demand factors in Column A are never lower and usually higher for the same number of appliances listed in Column B. When the cooking function is subdivided over several appliances, as in this example, the effect is to create one large range, and collectively the appliances will behave as a larger range with its attendant effectively smaller demand factor in Column C. Note 3 gives the option of using either Column C or Columns A and/or B to make this calculation for this reason.

Branch-Circuit Wiring

Where individual branch circuits supply each counter-mounted cooking unit and wall-mounted oven, there appears to be no particular problem. Figure 220-14 gives the details for wiring units on individual branch circuits.

Figure 220-13 shows an example of how typical counter-mounted cooking units and wall-mounted ovens are connected to a 50-A branch circuit.

Several manufacturers of cooking units provide an attached flexible metal conduit with supply leads and a floating 4-in. octagon box as a part of each unit. These units are commonly called "prewired types." With this arrangement, an electrician does not have to make any supply connections in the appliance. Where such units are connected to a 50-A circuit, the 4-in. octagon box is removed, and the flexible conduit is connected to a larger circuit junction box, which contains the 6 AWG circuit conductors.

On the other hand, some manufacturers do not furnish supply leads with their cooking units. As a result, the electrical contractor must supply the tap conductors to these units from the 50-A circuit junction box (see Fig. 220-13). In this case, connections must be made in the appliance as well as in the junction box. Figure 220-15 shows a single branch circuit supplying the same units, as shown in Fig. 220-13.

40-A Circuits

The **NEC** does recognize a 40-A circuit for two or more outlets, as noted in 210.23(C). Because an 8 AWG (40-A) circuit can supply a single range rated not over 16.4 kW, it can also supply counter- and wall-mounted units not exceeding the same total of 16.4 kW. The rating of 16.4 kW is determined as the maximum rating of equipment that may be supplied by a 40-A branch circuit, which has a capacity of 9600 W (40 A × 240 V). From Note 1 to Table 220.55, a 15.4-kW load would require a demand capacity equal to 8000 W plus [(16.4 × 12) × 0.05 × 8000] = 8000 W plus 4 × 0.05 × 8000 = 8000 plus 1600 = 9600 W.

Figure 220-16 shows an arrangement of a 8-AWG (40-A) branch circuit supplying one 7.5-kW cooking unit and one 4-kW oven. Or individual branch circuits may be run to the units.

220.56. Kitchen Equipment—Other than Dwelling Unit(s). Commercial electric cooking loads must comply with 220.56 and its table of feeder demand factors for commercial electric cooking equipment—including dishwasher booster

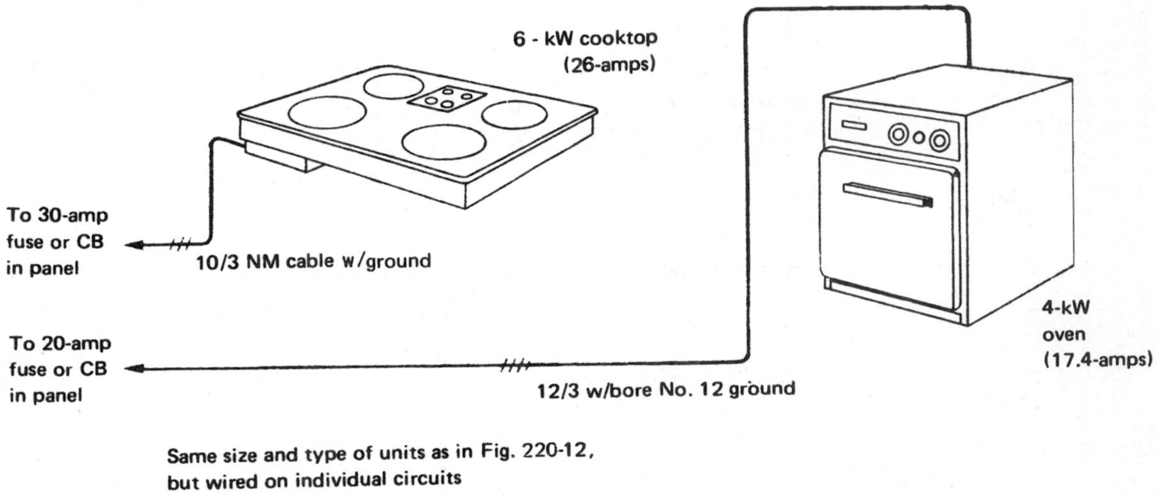
6 - kW cooktop
(26-amps)
To 30-amp
fuse or CB
in panel
10/3 NM cable w/ground
4-kW
oven
(17.4-amps)
To 20-amp
fuse or CB
in panel
12/3 w/bore No. 12 ground
Same size and type of units as in Fig. 220-12,
but wired on individual circuits

Individual branch circuits supply the same units that appear in Fig. 220-12. With this arrangement, smaller branch circuits supply each unit with no JBs required. Although two additional fuse or CB poles are required in a panelboard, overall labor/material costs are less than the 50-amp circuit shown in Fig. 220-12. However, one disadvantage to individual circuits is that smaller size circuits will not handle larger units, which may be installed at a later date.

Fig. 220-15. Separate circuits have advantages. (Sec. 220.55.)

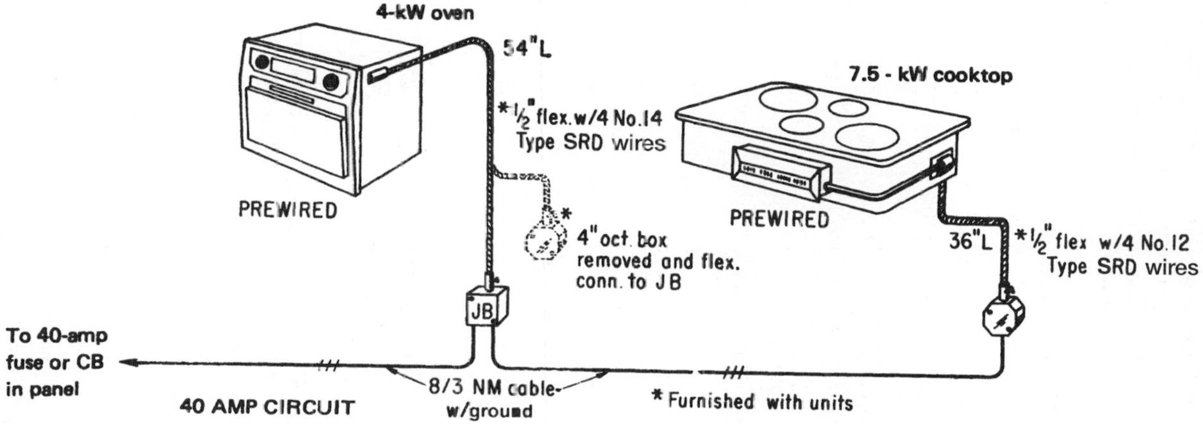
4-kW oven
54"L
* 1/2" flex. w/4 No. 14
Type SRD wires
PREWIRED
7.5 - kW cooktop
PREWIRED
36"L
* 1/2" flex w/4 No. 12
Type SRD wires
4" oct. box
removed and flex.
conn. to JB
JB
To 40-amp
fuse or CB
in panel
40 AMP CIRCUIT
8/3 NM cable
w/ground
* Furnished with units

The *NEC* permits 40-amp circuits in lieu of 50-amp circuits where the aggregate nameplate rating of cooktops and ovens is less than 15.5 kW. Most ranges are less than 15.5 kW and so are most combinations of cooktops and ovens.

Fig. 220-16. A single 40-A circuit may supply units. (Sec. 220.55.)

heaters, water heaters, and other kitchen equipment. Space-heating, ventilating, and/or air-conditioning equipment is excluded from the phrase "other kitchen equipment."

At one time, the Code did not recognize demand factors for such equipment. Code Table 220.56 is the result of extensive research on the part of electric utilities. The demand factors given in Table 220.56 may be applied to *all* equipment (except the excluded heating, ventilating, and air-conditioning loads) that is *either* thermostatically controlled *or* is used only on an intermittent basis. Continuously operating loads, such as infrared heat lamps used for food warming, would be taken at 100 percent demand and not counted in the "Number of Units" that are subject to the demand factors of Table 220.56.

The rule says that the minimum load to be used in sizing a feeder to commercial kitchen equipment must not be less than the sum of the largest two kitchen equipment loads. If the feeder load determined by using Table 220.56 on the total number of appliances that are controlled or intermittent and then adding the sum of load ratings of continuous loads like heat lamps is less than the sum of load ratings of the two largest load units—then the minimum feeder load must be taken as the sum of the two largest load units.

example Find the minimum demand load to be used in sizing a feeder supplying a 20-kW quick-recovery water heater, a 5-kW fryer, a dough mixer with a 3-phase 1½-hp, 208-V motor, and four continuously operating 250-W food-warmer infrared lamps—with a 208Y/120V, 3-phase, 4-wire supply.

Although the water heater, the fryer, and the four lamps are a total of 1 + 1 + 4, or 6, unit loads, the 250-W lamps may not be counted in using Table 220.56 because they are continuous loads. For the water heater and the fryer, Table 220-56 indicates that a 90 percent demand must be used where the "Number of Units of Equipment" is 3. The motor must be taken at 125 percent per 430.24, and based on Table 430.250 per 430.6(A)(1). The table current is 6.6 A, and 125 percent of that is 8.3 A.

To convert the motor load to volt-amperes, do the following multiplication:

8.3 A × 208 V × $\sqrt{3}$. Although this should be familiar to you, and it is the usual 1.732 term used in three-phase work all the time, remember that 208 V is simply the line-to-neutral voltage, also multiplied by the same term, thus the same multiplication can be written:

8.3 A × 120 V × $\sqrt{3} \times \sqrt{3}$. Since $\sqrt{3} \times \sqrt{3} = 3$, the multiplication is simply 8.3 A × 360 V. Working with 360 in 208Y/120V systems (and 831 in 480Y/277V systems) is much simpler and faster than using the square root of three and the line-to-line voltage.

Therefore, 8.3 A × 360 V = 3 kVA, and feeder minimum load (kW = kVA) must then be taken as

Water heater @ 90%	18.0 kVA
Fryer @ 90%	4.5 kVA
Dough mixer @ 90%	+2.7 kVA
	25.2 kVA
Four 250-W lamps @ 100%	+1.0 kVA = 26.2 kVA

Then, the feeder must be sized for a minimum current load of

$$\frac{26.2 \times 1000}{360} = 73 \text{ A}$$

The two largest equipment loads are the water heater and the dryer:

$$20 \text{ kVA} + 5 \text{ kVA} = 25 \text{ kVA}$$

and they draw

$$\frac{25 \times 1000}{360} = 69 \text{ A}$$

Therefore, the 73-A demand load calculated from Table 220.56 satisfies the last sentence of the rule because that value is "not less than" the sum of the largest two kitchen equipment loads. The feeder must be sized to have at least 73 A of capacity for this part of the total building load.

Figure 220-17 shows another example of reduced sizing for a feeder to kitchen appliances.

220.60. Noncoincident Loads. When dissimilar loads (such as space heating and air cooling in a building) are supplied by the same feeder, the smaller of the two loads may be omitted from the total capacity required for the feeder if it is unlikely that the two loads will operate at the same time.

220.61. Feeder or Service Neutral Load. This section covers requirements for sizing the neutral conductor in a feeder, that is, determining the required ampere rating of the neutral conductor. The basic rule of this section says that the minimum required ampacity of a neutral conductor must be at least equal to the "feeder neutral load"—which is the "maximum unbalance" of the feeder load.

"The maximum unbalanced load shall be the maximum net computed load between the neutral and any one ungrounded conductor. . . ." In a 3-wire, 120/240-V, single-phase feeder, the neutral must have a current-carrying capacity at least equal to the current drawn by the total 120-V load connected between the more heavily loaded hot leg and the neutral. As shown in Fig. 220-18, under unbalanced conditions, with one hot leg fully loaded to 60 A and the other leg open, the neutral would carry 60 A and must have the same rating as the loaded hot leg. Thus No. 6 THW hot legs would require No. 6 THW neutral (copper).

It should be noted that straight 240-V loads, connected between the two hot legs, do not place any load on the neutral. As a result, the neutral conductor of such a feeder must be sized to make up a 2-wire, 120-V circuit with the more heavily loaded hot leg. Actually, the 120-V circuit loads on such a feeder would be considered as balanced on both sides of the neutral. The neutral, then, would be the same size as each of the hot legs if only 120-V loads were supplied by the feeder. If 240-V loads also were supplied, the hot legs would be sized for the total load; but the neutral would be sized for only the total 120-V load connected between one hot leg and the neutral, as shown in Fig. 220-19.

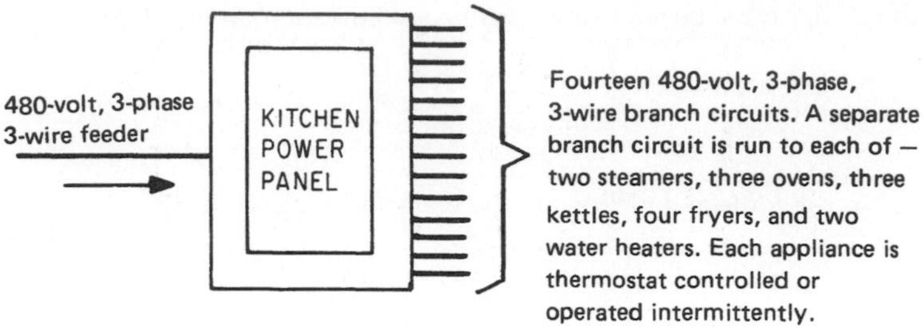
480-volt, 3-phase
3-wire feeder
KITCHEN
POWER
PANEL
Fourteen 480-volt, 3-phase, 3-wire branch circuits. A separate branch circuit is run to each of – two steamers, three ovens, three kettles, four fryers, and two water heaters. Each appliance is thermostat controlled or operated intermittently.

Kitchen panel supplies fourteen 480-volt, 3-phase, 3-wire branch circuits.
A separate branch circuit is run to each of–
Two steamers,
Three ovens,
Three kettles,
Four fryers, and
Two water heaters
The 14 appliances make up a total connected load of 303.3 kVA

QUESTION:
Is a full-capacity feeder (303.3 kVA/480 X 1.73 = 366 amps) required here? Or can a demand factor be applied?

ANSWER:
Although it is possible that all of the appliances might operate simultaneously, it is not expected that they will all be operating at full connected load. Table 220.56 of the *NE Code* does permit use of a demand factor on a feeder for commercial electric cooking equipment (including dishwasher, booster heaters, water heaters and other kitchen equipment). As shown in the Table, for six or more units, a demand factor of 65% can be applied to the feeder sizing:

366 amps × 0.65 = 238 amps

The feeder must have a least that much capacity, and that amp rating must be at least equal to or greater than the sum of the amp ratings of the two largest load appliances served. Capacity must be included in the building service entrance conductors for this load.

Fig. 220-17. Demand factor for commercial-kitchen feeder. (Sec. 220.56.)

But, there are qualifications on the basic rule of 220.61, as follows:

1. When a feeder supplies household electric ranges, wall-mounted ovens, counter-mounted cooking units, and/or electric dryers, the neutral conductor may be smaller than the hot conductors but must have a carrying capacity at least equal to 70 percent of the current capacity required in the

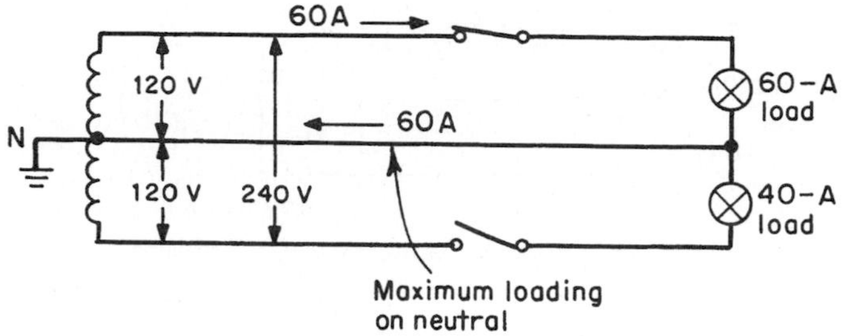
60A
120 V
60 A
N
120 V
240 V
60-A load
40-A load
Maximum loading on neutral

Fig. 220-18. Neutral must be sized the same as hot leg with heavier load. (Sec. 220.61.)

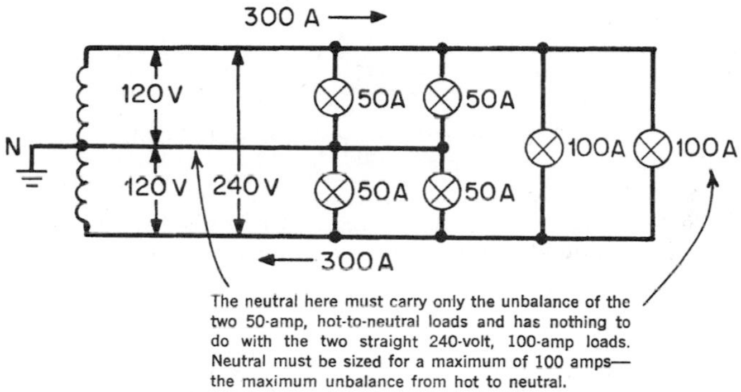
300 A
120 V
50A
50A
N
100A
100A
120 V
240 V
50A
50A
300A
The neutral here must carry only the unbalance of the two 50-amp, hot-to-neutral loads and has nothing to do with the two straight 240-volt, 100-amp loads. Neutral must be sized for a maximum of 100 amps—the maximum unbalance from hot to neutral.

Fig. 220-19. Neutral sizing is not related to phase-to-phase loads. (Sec. 220.61.)

ungrounded conductors to handle the load (i.e., 70 percent of the load on the ungrounded conductors). Table 220.56 gives the demand loads to be used in sizing feeders that supply electric ranges and other cooking appliances. Table 220.55 gives demand factors for sizing the ungrounded circuit conductors for feeders to electric dryers. The 70 percent demand factor may be applied to the minimum required size of a feeder phase (or hot) leg in order to determine the minimum permitted size of neutral, as shown in Fig. 220-20.

2. For feeders of three or more conductors—3-wire, dc; 3-wire, single-phase; and 4-wire, 3-phase—a further demand factor of 70 percent may be applied to that portion of the unbalanced load in excess of 200 A. That is, in a feeder supplying only 120-V loads evenly divided between each ungrounded conductor and the neutral, the neutral conductor must be the same size as each ungrounded conductor up to 200-A capacity, but may be reduced from the size of the ungrounded conductors for loads above 200 A by adding to the 200 A only 70 percent of the amount of load current above 200 A in computing the size of the neutral. It should be noted that this 70 percent demand factor is applicable to

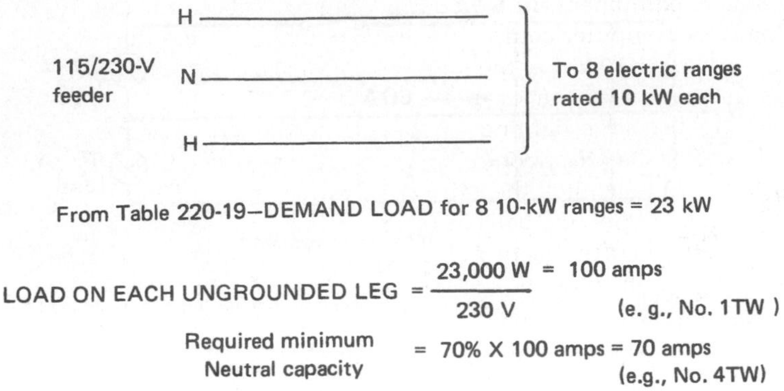
H
115/230-V feeder
N
H
To 8 electric ranges rated 10 kW each
From Table 220-19—DEMAND LOAD for 8 10-kW ranges = 23 kW
LOAD ON EACH UNGROUNDED LEG = 23,000 W / 230 V = 100 amps
(e. g., No. 1TW)
Required minimum Neutral capacity = 70% X 100 amps = 70 amps
(e.g., No. 4TW)

Fig. 220-20. Sizing the neutral of a feeder to electric ranges. (Sec. 220.61.) This calculation uses 230 V because it generates a round number result; the more exact approach would use 240 V resulting in a load of 96 A resulting a neutral ampacity of 67 A.

the unbalanced load in excess of 200 A and not simply to the total load, which in many cases may include 240-V loads on 120/240-V, 3-wire, single-phase feeders or 3-phase loads or phase-to-phase connected loads on 3-phase feeders. Figure 220-21 shows an example of neutral reduction as permitted by 220.61.

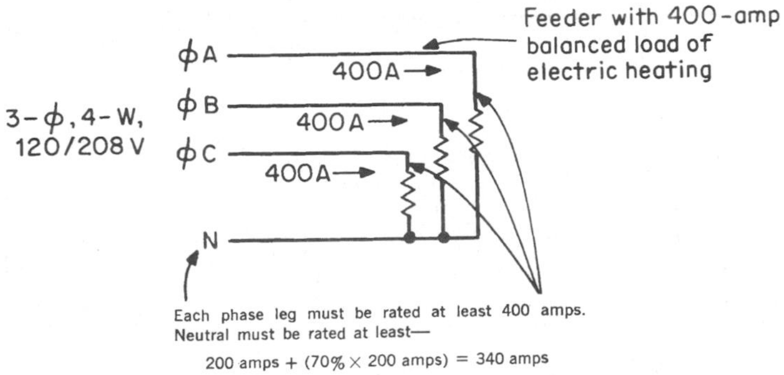
Feeder with 400-amp balanced load of electric heating
ϕA
ϕB
ϕC
N
400A
400A
400A
3-ϕ, 4-W, 120/208 V
Each phase leg must be rated at least 400 amps.
Neutral must be rated at least—
200 amps + (70% × 200 amps) = 340 amps

Fig. 220-21. Neutral may be smaller than hot-leg conductors on feeders over 200 A. (Sec. 220.61.)

WATCH OUT!

The size of a feeder neutral conductor may *not* be based on less than the current load on the feeder phase legs when the load consists of electric-discharge lighting, data-processing equipment, or similar equipment. The foregoing reduction of the neutral to 200 A plus 70 percent of the current over 200 A does not apply when all or most of the load on the feeder consists of electric-discharge lighting, electronic data-processing equipment, and similar electromagnetic or solid-state equipment. In a feeder supplying ballasts for electric-discharge lamps and/or computer equipment, there must not be a reduction of the neutral capacity for that part of the load which consists of discharge light sources, such as fluorescent mercury-vapor or other HID lamps. For feeders supplying only electric-discharge lighting or computers, the neutral conductor must be the same size as the phase conductors no matter how big the total load may be (Fig. 220-22). Full-sizing of the neutral of such feeders is required because, in a balanced circuit supplying ballasts or computer loads, neutral current approximating the phase current is produced by third (and other odd-order) harmonics developed by the ballasts. For large electric-discharge lighting or computer loads, this factor affects sizing of neutrals all the way back to the service. It also affects rating of conductors in conduit because such a feeder circuit consists of *four* current-carrying wires, which requires application of an 80 percent reduction factor. [See 310.15(B)(3)(a) and 310.15(B)(5)(c).]

In the case of a feeder supplying, say, 200 A of fluorescent lighting and 200 A of incandescent, there can be no reduction of the neutral below the required 400-A capacity of the phase legs, because the 200 A of fluorescent lighting load cannot be

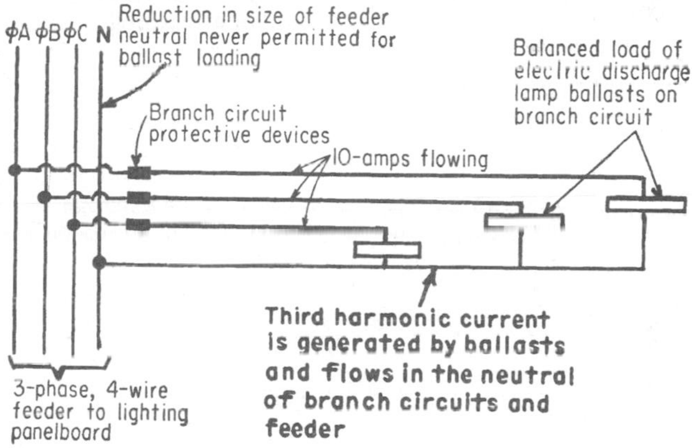
ϕA ϕB ϕC N
Reduction in size of feeder neutral never permitted for ballast loading
Balanced load of electric discharge lamp ballasts on branch circuit
Branch circuit protective devices
10-amps flowing
Third harmonic current is generated by ballasts and flows in the neutral of branch circuits and feeder
3-phase, 4-wire feeder to lighting panelboard

EXAMPLE:

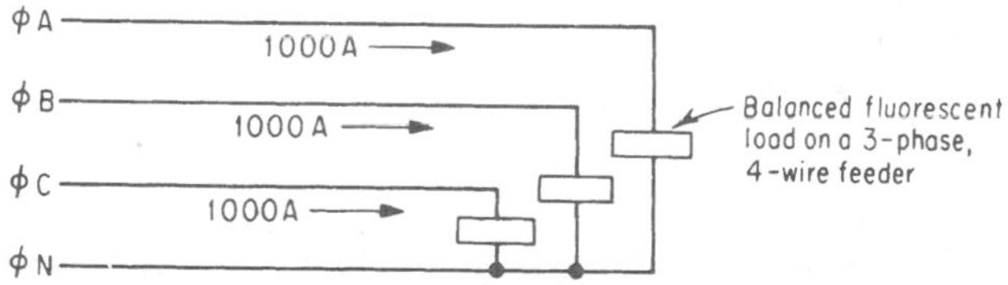
ϕA
1000A
ϕB
1000 A
ϕC
1000A
ϕN
Balanced fluorescent load on a 3-phase, 4-wire feeder

There must be no reduction in amp rating of this neutral.
It must have 1000-amp rating like the phase conductors.

Fig. 220-22. Full-size neutral for feeders to ballast loads or computers. (Sec. 220.61.)

used in any way to take advantage of the 70 percent demand factor on that part of the load in excess of 200 A.

It should be noted that the Code wording in 220.61 permits reduction in the size of the neutral when electric-discharge lighting and/or computers are used, if the feeder supplying the electric-discharge lighting load over 200 A happens to be a 120/240-V, 3-wire, single-phase feeder. In such a feeder, the harmonic currents in the hot legs are 180° out of phase with each other and, therefore, would not be additive in the neutral as they are in a 3-phase, 4-wire circuit. In the 3-phase, 4-wire circuit, the third harmonic components of the phase currents are in phase with each other and add together in the neutral instead of canceling out. Figure 220-23 shows a 120/240-V circuit.

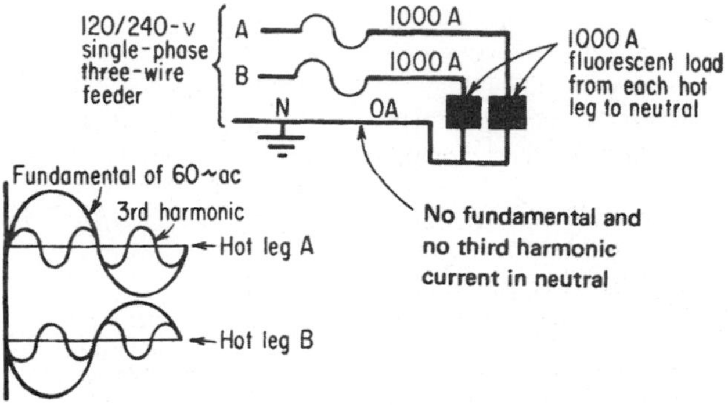
120/240-v single-phase three-wire feeder
A
B
N
1000 A
1000 A
0A
1000 A fluorescent load from each hot leg to neutral
Fundamental of 60~ac
3rd harmonic
Hot leg A
Hot leg B
No fundamental and no third harmonic current in neutral

As shown, both the fundamental and harmonic currents are 180° out of phase and both cancel in the neutral. Under balanced conditions, the neutral current is zero.

Fig. 220-23. Harmonic loading on true single-phase distributions does not interfere with permitted size reductions over 200 A (220.61).

Figure 220-24 shows a number of circuit conditions involving the rules on sizing a feeder neutral.

Part IV

This part of Art. 220 offers a number of alternative methods for establishing the minimum required current-carrying capacity of service or feeder conductors. Remember that each of the requirements specified within each individual optional method must be satisfied.

220.82. Dwelling Unit. This section sets forth an optional method of calculating service demand load for a residence. This method may be used instead of the standard method as follows:

1. Only for a one-family residence or an apartment in a multifamily dwelling, or other "dwelling unit"
2. Served by a 120/240-V or 120/208-V 3-wire, 100-A or larger service or feeder
3. Where the total load of the dwelling unit is supplied by one set of service-entrance or feeder conductors that have an ampacity of 100 A or greater

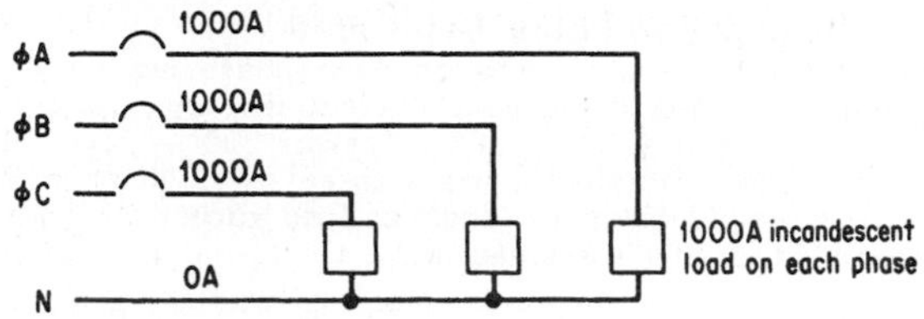
1000A
1000A
1000A
ϕA
ϕB
ϕC
N
0A
1000A incandescent load on each phase

1. Incandescent lighting only

Serving an incandescent load, each phase conductor must be rated for 1000 amps. But neutral only has to be rated for 200 amps plus (70% x 800 amps) or 200 + 560 = 760 amps.

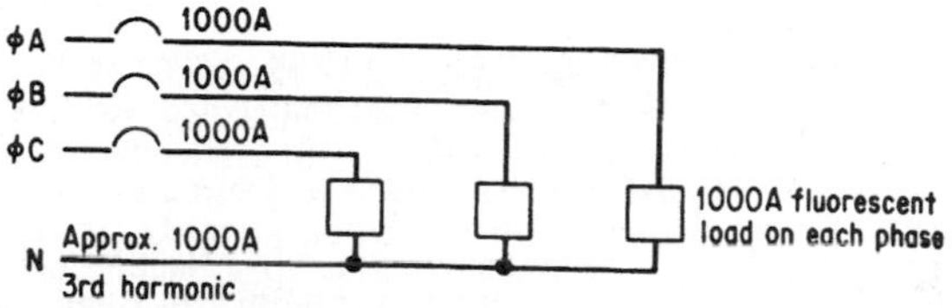
ϕA
ϕB
ϕC
1000A
1000A
1000A
N
Approx. 1000A
3rd harmonic
1000A fluorescent load on each phase

2. Electric discharge lighting only

Because load is electric discharge lighting, there can be no reduction in the size of the neutral. Neutral must be rated for 1000 amps, the same as the phase conductors, because the third harmonic currents of the phase legs add together in the neutral. This applies also when the load is mercury-vapor or other metallic-vapor lighting.

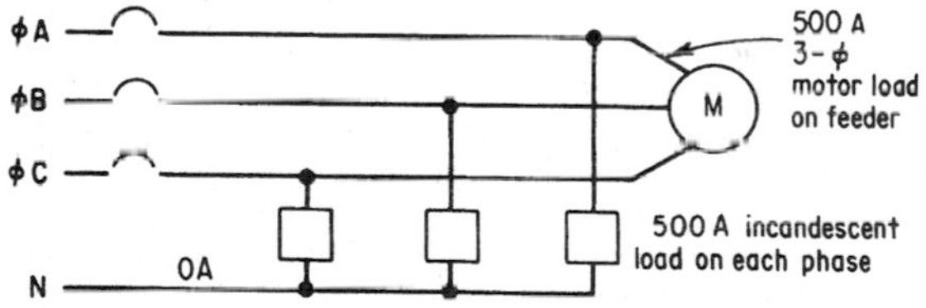
ϕA
ϕB
ϕC
N
0A
M
500 A 3-ϕ motor load on feeder
500 A incandescent load on each phase

Fig. 220-24. Sizing the feeder neutral for different conditions of loading. (Sec. 220.61.)

This method recognizes the greater diversity attainable in large-capacity installations. It therefore permits a smaller size of service-entrance conductors for such installations than would be permitted by using the load calculations of 220.40 through 220.61.

In making this calculation, as described by 220.82(C), the heating load or the air-conditioning load may be disregarded as a "noncoincident load," where it is unlikely that two dissimilar loads (such as heating and air conditioning) will be operated simultaneously. In the present **NEC**, 100 percent of the air-conditioning load is compared with only *40 percent* of the total connected load of four or more

3. Incandescent plus motor load

Although 1000 amps flow on each phase leg, only 500 amps is related to the neutral. Neutral, then, is sized for 200 amps plus (70% x 300 amps) or 200 + 210 = 410 amps. The amount of current taken for 3-phase motors cannot be "unbalanced load" and no capacity has to be provided for this in the neutral.

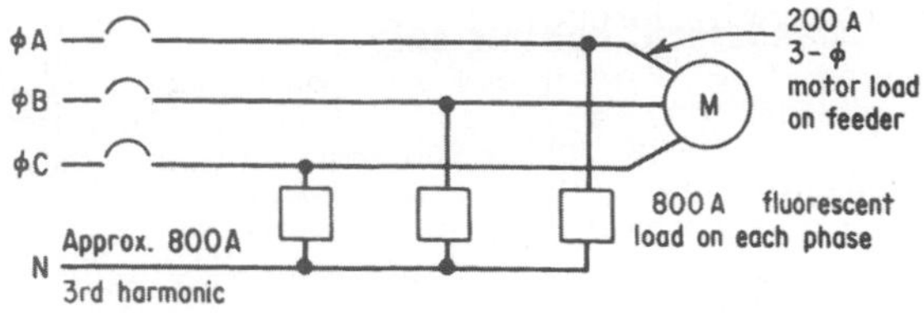

φA
φB
φC
N
Approx. 800A
3rd harmonic
M
200 A
3-φ
motor load
on feeder
800 A fluorescent
load on each phase

4. Electric discharge lighting plus motor load

Here, again, the only possible load that could flow on the neutral is the 800 amps flowing over each phase to the fluorescent lighting. But because it is fluorescent lighting there can be no reduction of neutral capacity below the 800-amp value on each phase. The 70% factor for that current above 200 amps DOES NOT APPLY in such cases.

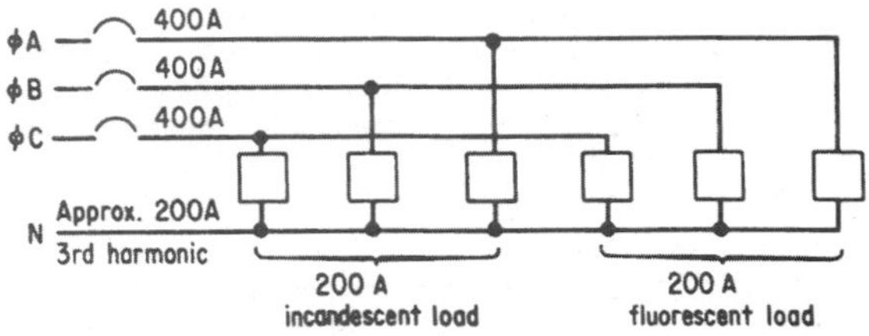

φA
φB
φC
400A
400A
400A
N
Approx. 200A
3rd harmonic
200 A
incandescent load
200 A
fluorescent load

5. Incandescent plus electric discharge lighting

Each phase leg carries a total of 400 amps to supply the incandescent load plus the fluorescent load. But because there can be no reduction of neutral capacity for the fluorescent and because the incandescent load is not over 200 amps, the neutral must be sized for the maximum possible unbalance, which is 400 amps.

Fig. 220-24. *(Continued)*

electric space heaters [220.82(C)(6)], and the lower value is omitted from the calculation. Or, where there are less than four separately controlled electric heating units, the 100 percent value of the air-conditioning load is compared with 65 percent of the load where the electric heating system has less than four separately controlled units.

example A typical application of the data and table of 220.82, in calculating the minimum required size of service conductors, is as follows:

A 1500-sq-ft (139.5-m^2) house (excluding unoccupied basement, unfinished attic, and open porches) contains the following specific electric appliances:

12-kW range
2.5-kW water heater
1.2-kW dishwasher
9 kW of electric heat (in five rooms)
5-kW clothes dryer
6-A, 230-V air-conditioning unit

When using the optional method, if a house has air conditioning as well as electric heating, there is recognition in 220.60 that if "it is unlikely that two dissimilar loads will be in use simultaneously," it is permissible to omit the smaller of the two in calculating required capacity in feeder or in service-entrance conductors. In 220.82, that concept is spelled out in the subparts of 220.82(C) to require adding only the largest of the loads described in this rule. Where the dwelling in question has air conditioning and four separately controlled electric heating units, we add capacity equal to either the total air-conditioning load or 40 percent of the connected load of *four or more* separately controlled electric space-heating units. For the residence considered here, these loads would be as follows:

$$\text{Air conditioning} = 6\ \text{A} \times 230\ \text{V} = 1.38\ \text{kVA}$$

Note: The air conditioner voltage and current ratings are from the equipment nameplate, and therefore, when converting to kilovolt-amperes, must be taken as is. The overall calculation at the end, which determines the service amperage on a 120/240 system, uses the rated system voltages from 220.5(A).

$$40\%\ \text{of heating (five separate units)} = 9\ \text{kW} \times 0.4 = 3.6\ \text{kW (3600 VA)}$$

Because 3.6 kW is greater than 1.38 kVA, it is permissible to omit the air-conditioning load and provide a capacity of 3.6 kW in the service or feeder conductors to cover *both* the heating and air-conditioning loads.

The "other loads" must be totaled up in accordance with 220.82:

	Volt-amperes
1. 1500 VA for each of two small-appliance circuits (2-wire, 20-A)	3,000
Laundry branch circuit (3-wire, 20-A)	1,500
2. 3 VA/sq ft of floor area for general lighting and general-use receptacles (3 × 1500 sq ft)	4,500
3. Nameplate rating of fixed appliances:	
Range	12,000
Water heater	2,500
Dishwasher	1,200
Clothes dryer	5,000
Total	29,700

In reference to 220.82(C), load categories 1, 2, 3, 4, and 5 are not applicable here: "Air conditioning" has already been excluded as a load because 40 percent of the heating load is greater. The dwelling does not have a heat pump without a controller that prevents simultaneous operation of the compressor and supplemental heating. There is no thermal heating unit. There is no "central" electric space heating; and there are *not* "less than four" separately controlled electric space-heating units.

The total load of 29,700 VA, as previously summed up, includes "all other load," as referred to in 220.82. *Then:*

1. Take 40% of the 9000-W heating load	3,600
2. Take 10 kVA of "all other load" at 100% demand	10,000
3. Take the "remainder of other load" at 40% demand factor: (29,700 – 10,000) × 40% = 19,700 × 0.4	7,880
Total demand	21,480

Using 240- and 120-V values, ampacities may then be calculated. At 240 V, single phase, the *ampacity of each service hot leg would then have to be*

$$\frac{21{,}480\ \text{W}}{240\ \text{V}} = 89.5 \text{ or } 90\ \text{A}$$

Minimum service conductor required 100 A

Then the neutral service-entrance conductor is calculated in accordance with 220.61. All 240-V loads have no relation to required neutral capacity. The water heater and electric space-heating units operate at 240 V, 2-wire and have no neutrals. By considering only those loads served by a circuit with a neutral conductor and determining their maximum unbalance, the minimum required size of neutral conductor can be determined.

When a 3-wire, 240/120-V circuit serves a total load that is balanced from each hot leg to neutral—that is, half the total load is connected from one hot leg to neutral and the other half of total load from the other hot leg to neutral—the condition of maximum unbalance occurs when all the load fed by one hot leg is operating and all the load fed by the other hot leg is off. Under that condition, the neutral current and hot-leg current are equal to half the total load watts divided by 120 V (half the volts between hot legs). But that current is exactly the same as the current that results from dividing the *total* load (connected hot leg to hot leg) by 240 V (which is twice the voltage from hot leg to neutral). Because of this relationship, it is easy to determine neutral-current load by simply calculating hot-leg current load—total load from hot leg to hot leg divided by 240 V.

In the example here, the neutral-current load is determined from the following steps that sum up the components of the neutral load:

	Volt-amperes
1. Take 1500 sq ft at 3 VA/sq ft (Table 220.12)	4,500
2. Add three small-appliance circuits (two kitchen, one laundry) at 1500 VA each (220.52)	4,500
Total lighting and small-appliance load	9,000
3. Take 3000 VA of that value at 100% demand factor (220.42 & 220.52; Table 220.42)	3,000
4. Take the balance of the load (9000 – 3000) at 35% demand factor: 6000 VA × 0.35	2,100
Total of 3 and 4	5,100

Assuming an even balance of this load on the two hot legs, the neutral load under maximum unbalance will be the same as the total load (5100 VA) divided by 240 V (Fig. 220-25) (all results are carried to three significant figures):

$$\frac{5100\ \text{VA}}{240\ \text{V}} = 21.3\ \text{A}$$

And the neutral unbalanced current for the range load can be taken as equal to the 8000-W range demand load multiplied by the 70 percent demand factor permitted by 220.61 and then divided by 240 V (Fig. 220-26):

$$\frac{8000 \times 0.7}{240} = \frac{5600}{240} = 23.3\ \text{A}$$

The clothes dryer contributes neutral load due to the 115-V motor, its controls, and a light. As allowed in 220.61, the neutral load of the dryer may be taken at 70 percent of the load on the ungrounded hot legs. Therefore, the neutral capacity required to accommodate the dryer contribution is (5000 W × 0.7) + 240 V = 14.6 A.

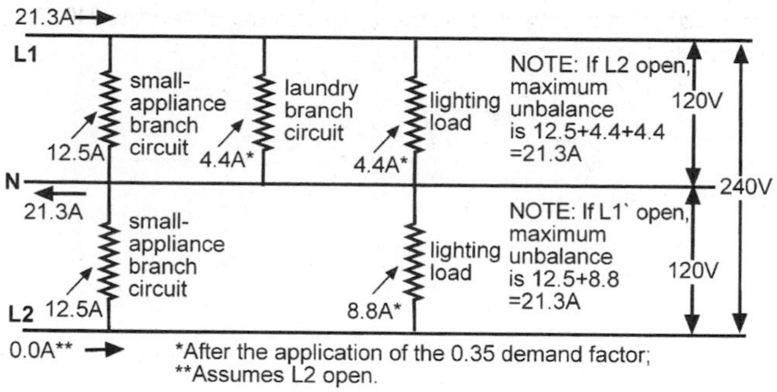
21.3A
L1
small-appliance branch circuit
12.5A
laundry branch circuit
4.4A*
lighting load
4.4A*
NOTE: If L2 open, maximum unbalance is 12.5+4.4+4.4 =21.3A
120V
N
21.3A
240V
small-appliance branch circuit
lighting load
8.8A*
NOTE: If L1` open, maximum unbalance is 12.5+8.8 =21.3A
120V
L2 12.5A
0.0A**
*After the application of the 0.35 demand factor;
**Assumes L2 open.

Fig. 220-25. Neutral current for lighting and receptacles. The 4.4-A net load equivalence between the laundry circuit and one portion of the lighting circuit is a coincidence. (Sec. 220.82.)

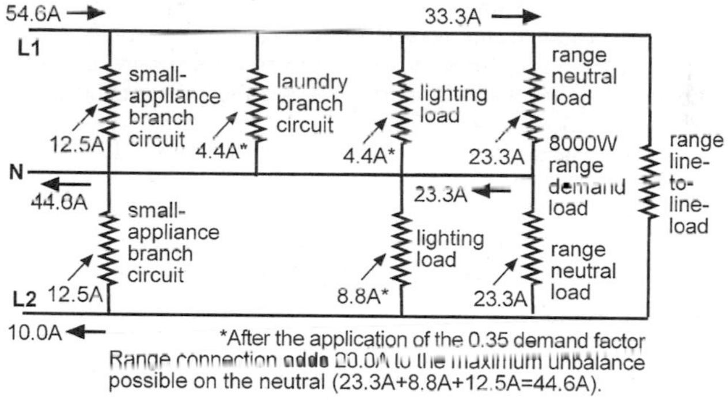
54.6A
33.3A
L1
small-appliance branch circuit
12.5A
laundry branch circuit
4.4A*
lighting load
4.4A*
range neutral load
23.3A
8000W range demand load
range line-to-line-load
N
44.6A
23.3A
small-appliance branch circuit
lighting load
8.8A*
range neutral load
23.3A
L2 12.5A
10.0A
*After the application of the 0.35 demand factor
Range connection adds 20.0A to the maximum unbalance possible on the neutral (23.3A+8.8A+12.5A=44.6A).

Fig. 220-26. Neutral for lighting, receptacles, and range. (Sec. 220.82.)

Then, the neutral-current load that is added by the 120-V, 1200-W dishwasher must be added (Fig. 220-27): Although it could be argued that since this load is entirely on a single leg of the system, it should be added to the neutral directly, but that is not the case, as evidenced by the result in Annex D, Example D2(a) from which these numbers are taken. Making the calculation in this way only artificially inflates the size of the neutral. In the real world, the dishwasher and the small-appliance and the laundry and the general lighting circuits all originate from the same panel. There are, in this example and typically, three appliance circuits (laundry and small appliance) all taken at 1500 VA each. Obviously two will be on one line and one on the other. The number of lighting circuits is unknown but there will be some imbalance there as well. The 1200-W imbalance represented by the dishwasher will be totally lost in the distribution of loads in this panel. Simply placing it against one of the appliance circuits erases its contribution. Further, the 1200-W is not a steady load but one that cycles, depending on whether the booster element is in operation and whether the motor is running. This is why load calculations are usually done on a volt-ampere basis throughout, changing over to amperes only at the point where a conductor size must be determined.

The load calculations for neutral conductors assume reasonable balance for the branch circuits connected to them, as is generally required by 210.11(B). The most meticulously balanced distribution on paper will be defeated by poor panel work. For example, this

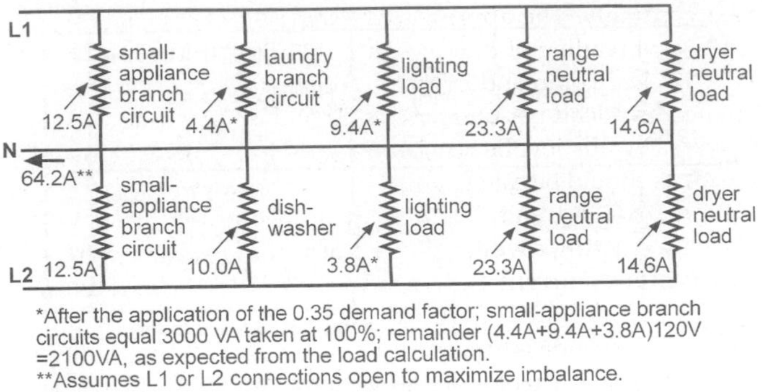
L1
small-
appliance
branch
circuit
12.5A
laundry
branch
circuit
4.4A*
lighting
load
9.4A*
range
neutral
load
23.3A
dryer
neutral
load
14.6A
N
64.2A**
small-
appliance
branch
circuit
L2 12.5A
dish-
washer
10.0A
lighting
load
3.8A*
range
neutral
load
23.3A
dryer
neutral
load
14.6A
*After the application of the 0.35 demand factor; small-appliance branch circuits equal 3000 VA taken at 100%; remainder (4.4A+9.4A+3.8A)120V =2100VA, as expected from the load calculation.
**Assumes L1 or L2 connections open to maximize imbalance.

Fig. 220-27. Neutral current for all loads in the example. The lighting load division between line connections is to equalize the load, thereby minimizing the maximum imbalance and the need for a larger feeder neutral. (Sec. 220.82.)

calculation starts with 21.3 A of line-to-neutral load due to small-appliance and general lighting circuits. If these were arranged to connect to the same line bus, the result would be 42.6 A of current from these sources routinely. And all this could occur before the dishwasher is even connected.

The minimum required neutral capacity is, therefore,

	21.3 A
	23.3 A
	14.6 A
	5.0 A
Total:	64.2 A

From Code Table 310.15(B)(16), the neutral minimum for 64 A would be:

4 AWG copper TW or 3 AWG aluminum
6 AWG THW copper* or 4 AWG aluminum*

*If the terminations are evaluated for 75°C connections; and 90°C insulation is permitted provided the conductor size remains as described. Note that if this panel were located remote from the service disconnecting means through a feeder that carried the entire load current, all of these conductors would satisfy 215.2(A)(2), requiring a reduced neutral to have enough size to carry a line-to-neutral short circuit. The minimum size in this case, per 250.122, is 8 AWG copper or 6 AWG aluminum.

This book is a handbook based on the 2017 **NEC**, and the above discussion is the best case the author can make to support the conclusion that is now in the **NEC** for this example. The method in this example has been unchanged since the 1984 **NEC** introduced the present 120/240-V nominal voltages, resulting in the recalculation of all the examples. It is also consistent with other neutral calculations throughout the examples, with one notable exception. It is glaringly inconsistent, however, with Example D(1)(b), which has been unchanged for the same amount of time. That example tracks the line contributions of specified 120-V appliances, and the end result shifts accordingly. The approach in Example D(1)(b) can also be supported based on a different reading of the rules in 210.11(B) than the one alluded to above, because the only clear command in that section is to balance loads on circuits that were determined on a load per unit area basis, and a dishwasher load was certainly not part of an area evaluation. Be advised that no clear conclusion can be reached as to how these loads should be calculated at this time.

220.83. Existing Dwelling Unit. This covers an optional calculation for additional loads in an existing dwelling unit that contains a 120/240- or 208/120-V, 3-wire service of any current rating. The method of calculation is similar to that in 220.82.

The purpose of this section is to permit the *maximum load* to be applied to an *existing* service without the necessity of increasing the size of the service. The calculations are based on numerous load surveys and tests made by local utilities throughout the country. This optional method would seem to be particularly advantageous when smaller loads such as window air conditioners or bathroom heaters are to be installed in a dwelling with, say, an existing 60-A service, as follows:

If there is an existing electric range, say, 12 kW (and no electric water heater), it would not be possible to add any load of substantial rating. The first 8000 VA is taken at 100 percent, leaving the remainder of permissible load to be calculated at 40 percent. Use the formula 14,400 VA (240 V × 60 A) = 8000 VA + 0.4(X VA), where the quantity (X) is the amount of other load to be evaluated. Rearranging terms gives 6400 VA = 0.4X, so X = 16,000 VA, and therefore, the total "gross load" that can be connected to an existing 120/240-V, 60-A service would be 16,000 VA + 8000 VA = 24,000 VA.

example Thus, an existing 1000-sq-ft dwelling with a 12-kW electric range, two 20-A appliance circuits, a 750-W furnace circuit, and a 60-A service would have a gross load of:

	Volt-amperes
1000 sq ft × 3 VA/sq ft	3,000
Two 20-A appliance circuits @ 1500 VA each	3,000
One electric range @	12,000
Furnace circuit @	750
Gross volt-amperes	18,750

Since the *maximum* permitted gross load is 24,000 VA, an appliance not exceeding 5250 VA could be added to this existing 60-A service. However, the tabulation at the end of this section lists air-conditioning equipment, central space heating, and less than four separately controlled space-heating units at 100 percent demands; and if the appliance to be added is one of these, then it would be limited even more:

From the 18,750-VA gross load we already have 8000 VA @ 100 percent demand and (10,750 VA [18,750 − 8000] × 0.40) or 4300 VA. The total for the 100 percent and the 40 percent calculation brackets is the sum of 8000 VA and 4300 VA, or 12,300 VA. Then, 14,400 VA (60 × 240 V) − 12,300 VA = 2100 VA for an appliance listed at *100 percent demand.*

Although this procedure is limited with respect to saving 60-A services, it can also be applied, with considerably more headroom, to existing 100-A services.

220.84. Multifamily Dwelling. This section provides an optional method of calculating the load in a multifamily dwelling with a fairly high connected load, by reason of electric cooking equipment in all units as well as electric space heating or air conditioning or both. Any house loads are over and above the calculation results from this section, and are to be figured using the standard method in Part III. The connected load list for each dwelling unit is formatted the same as the calculation in 220.82, with two differences. There is no 40 percent bracket; instead, all connected loads are simply totaled. In addition, the heating/air-conditioning line is quite simple; just pick the larger number whether the one for heat or the one for air conditioning. Multiply the total per/unit calculation by the number of units, and then by demand factor based on the number of units in Table 220.84.

If the load for a multifamily housing project without electric cooking (and therefore does not initially qualify to use this procedure), as determined by the traditional procedures in Part III, turns out to exceed the numbers that come from Table 220.84, the smaller load is permitted to be used.

220.85. Two Dwelling Units. This section provides an optional calculation for sizing a feeder to "two dwelling units." It notes that if calculation of such a feeder according to the basic long method of calculating given in Part III of Art. 220 exceeds the minimum load ampacity permitted by 220.84 for three identical dwelling units, then the *lesser* of the two loads may be used. This rule was added to eliminate the obvious illogic of requiring a greater feeder ampacity for two dwelling units than for the three units of the same load makeup. Now optional calculations provide for a feeder to one dwelling unit, two dwelling units, or three or more dwelling units.

220.86. Schools. The optional calculation for feeders and service-entrance conductors for a school makes clear that feeders "within the building or structure" must be calculated in accordance with the standard long calculation procedure established by Part III of Art. 220. *But* the ampacity of any individual feeder does not have to be greater than the minimum required ampacity for the whole building, regardless of the calculation result from Part III. Note that these calculations differ from most in that they are based on actual load density. The entire connected load is added together, and then divided by the area of the school to generate a load per unit area, whether per square foot or meter. Then the unit load is reduced according to the table, in progressive steps. Finally, the applicable number of volt-amperes per unit area from each step is multiplied by the area of the building to get the final load.

The last sentence in this section excludes portable classroom buildings from the optional calculation method to prevent the possibility that the demand factors of Table 220.86 would result in a feeder or service of lower ampacity than the connected load. Such portable classrooms have air-conditioning loads that are not adequately covered by using a watts-per-square-foot calculation with the small area of such classrooms.

220.87. Determining Existing Loads. Because of the universal practice of adding more loads to feeders and services in all kinds of existing premises, this calculation procedure is given in the Code. To determine how much more load may be added to a feeder or set of service-entrance conductors, at least one year's accumulation of measured maximum-demand data must be available. Then, the required spare capacity may be calculated as follows:

$$\text{Additional load capacity} = \text{ampacity of feeder or service conductors} - ([1.25 \times \text{existing demand kVA} \times 1000] \div \text{circuit voltage})$$

where "circuit voltage" is the line-to-line value for single-phase circuits and $\sqrt{3}$ (1.732) times the phase-to-phase value for 3-phase circuits.

A third required condition is that the feeder or service conductors be protected against overcurrent, in accordance with applicable Code rules on such protection.

If the full-year demand data is not available, an exception allows for a month of monitoring by a continuously recording ammeter, based on maximum demand for the period as defined by the highest average kilowatts reached and maintained for a 15-minute interval. This phrasing was clarified by a simple but important bit of panel work during the 2017 **NEC** cycle. The building must be occupied so the readings will be realistic. In addition, periodic or seasonal loading must be accounted for, either by direct measurements or by calculation, so that the larger of the heating and cooling loads will be included. Although utility demand meter data is commonly available for a full year for service loads, this alternative method is extremely important when the feeder in question is not subject to utility metering, such as a feeder to one part of a building or a feeder connected to a separately derived system.

220.88. New Restaurants. This calculation is available to new restaurants, and produces two different results based on whether or not the restaurant has gas-fired cooking equipment. The numbers for the all-electric restaurants are, of course, significantly higher. The demand table looks somewhat different from comparable tables elsewhere in Art. 220, with one entry providing a 10-percent increment for additional loading over a base number, and others providing far different increments ranging from 20 to 50 percent, and in different relative orders based on the type of restaurant. The table entries are correct.

When this table first came into the **NEC**, it looked like a conventional table, but it turned out to generate paradoxical results. In one instance adding a few kVA to a load took about 80 A off the ending service calculation, and this was not in just one location. The only way to be sure that additional connected load actually resulted in additional service or feeder capacity was to go back and carefully copyfit demand curves to the utility data that provided the substantiation for the change. That data is well documented, but the resulting curves have some interesting shapes, and the current values in Table 220.88 accurately predict the electrical demand.

Part V. Farm Load Calculations This part of the article stands alone because farms usually have both a dwelling and a commercial operation connected to a single distribution point. Therefore, some of the loads are eligible for optional treatments in Part IV and others are not, requiring a direct transition from the branch circuit calculations in Part II to the calculations here. In general, the farmhouse is a dwelling unit and qualifies for the optional dwelling unit procedures in Part IV. However, if the dwelling has electric heat and the farm operation uses electric grain drying systems, the dwelling must be calculated under the conventional procedures in Part III if it and the barn have a common service.

The electrical equipment for farm operations is taken through Table 220.102, using a master compilation of farm loads in terms of those that will operate continuously and otherwise. Beginning with the continuous loads take the first 60 A at 100 percent and then add the next 60 A of load at 50 percent, and then all other load at 25 percent. Then, for the total farm load, use Table 220.103, which is organized by load. If different buildings or loads have the same function, then those loads can be combined into a single load for these purposes. After the load analysis is complete, add the farm dwelling load calculated as noted above.

ARTICLE 225. OUTSIDE BRANCH CIRCUITS AND FEEDERS

225.1 Scope. This article covers all outdoor installations of conductors "on or between buildings, structures, or poles on the premises" and utilization equipment mounted on the outside of buildings, or outdoors on other structures or poles. This rule is followed by the rule of Sec. 225.2, which indicates other Code rules that bear upon the installation of equipment and conductors outdoors on buildings, structures, or poles. Like the large majority, but not all NEC articles commencing with the 2014 edition, the elevated voltage cut point has been raised from 600 to 1000 V throughout.

225.4. Conductor Covering. The wiring method known as "open wiring" is recognized in Art. 225 as suitable for overhead use outdoors—"run on or between buildings, structures, or poles" (Fig. 225-1). This is derived from Secs. 225.1, 225.4, 225.14, 225.18, and 225.19. In Sec. 225.4 the Code requires open wiring to be insulated for the nominal voltage if it comes within 10 ft (3.0 m) of any building or other structure, which it must do if it attaches to the building or structure. Insulated conductors have a dielectric covering that prevents conductive contact with the conductor when it is energized. This is new as of the 2014 NEC; this section used to permit such wires to be "insulated or covered." Covered conductors, now no longer recognized—such as braided, weatherproof conductors—have a certain mechanical protection for the conductor but are not rated as having insulation, and thus there is no protection against conductive contact with the energized conductor.

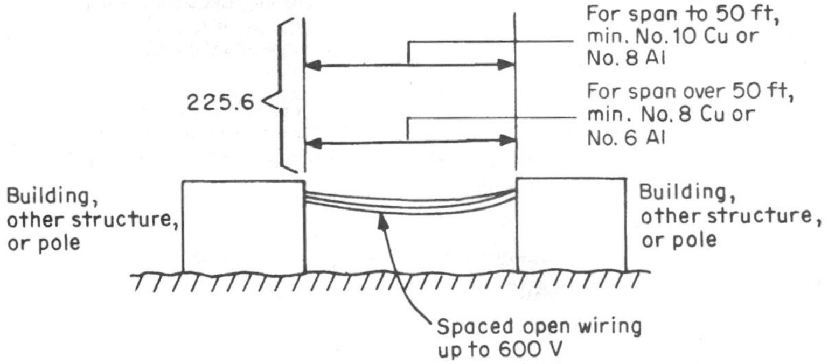
For span to 50 ft, min. No.10 Cu or No. 8 Al
225.6
For span over 50 ft, min. No. 8 Cu or No. 6 Al
Building, other structure, or pole
Building, other structure, or pole
Spaced open wiring up to 600 V

Fig. 225-1. Open wiring is OK for overhead circuits. (Sec. 225.4.)

Because 225.4 says that conductors in "*cables*" (except Type MI) must be of the rubber or thermoplastic type, a number of questions arise.

1. What kind of "*cable*" does the Code recognize for overhead spans between buildings, structures, and/or poles?
2. May an overhead circuit from one building to another or from lighting fixture to lighting fixture on poles use service-entrance cable, UF cable, or Type NM or NMC nonmetallic-sheathed cable?

The Code covers specific types of cables in turn (Chap. 3), but only in Art. 396 does the Code refer to use for outdoor overhead applications. Effective with the 2008 **NEC**, 396.30(B) and (C) for the first time cover the use of messenger cable assemblies with the messenger performing an electrical function, thereby closing a long-standing gap in **NEC** coverage. In addition, the use of service-entrance cable between buildings, structures, and/or poles is supported by Art. 338 in 338.12(A)(3), which points to Part II of Art. 396, thereby including the express reference in 396.10(A).

Use of Type MI, MC, or UF cable for outdoor, overhead circuits is supported by Sec. 396.10. There are no exceptions given to the support requirements in Sec. 334.30 that would let NM or NMC be used aerially, and such cables are not recognized by Sec. 396.10(A) for use as "messenger supported wiring."

Service-Drop Cable

The **NEC** has Art. 396, "Messenger Supported Wiring," which covers use of "service-drop" cable, but the UL has no listing for or reference to such cable. There is a listing for a suitable medium-voltage cable, but traditionally the principal customer for service-drop cables has been the electric utilities. And, since they are not usually subject to the **NEC** because most of their applications are on the line side of the service point, there has not been a large market for a listed product. Article 396 does not require a listing for this product. The **NEC** does make reference to it; and its use for aerial circuits between buildings, structures, and/or poles is particularly dictated (Fig. 225-2). Experience with this cable is very extensive and highly satisfactory. It is an engineered product specifically designed and used for outdoor, overhead circuiting.

NEC rules in 230.21 through 230.29 cover use of service-drop cable for overhead service conductors. Because the general rules of Art. 225 on outside branch circuits and feeders do make frequent references to other sections of Art. 230, it is logical to equate cables for overhead branch circuits and feeders to cables for overhead services. Although the rules of Art. 396 refer to a variety of messenger-supported cable assemblies, for outdoor circuits, use of service-drop cable is the best choice—because such cable is covered by the application rules of 230.21 through 230.29. Other types of available aerial cable assemblies, although not listed by UL, might satisfy some inspection agencies. But, in these times of OSHA emphasis on codes and standards, use of service-drop cable has the strongest sanction.

One important consideration in the use of service-drop cable as a branch circuit or feeder is the general Code prohibition against use of bare circuit conductors. 310.106(D) generally requires conductors to be insulated, although bare or covered conductors may be used where "specifically permitted." Bare *equipment grounding conductors* are permitted in 250.118. A bare conductor for SE cable is permitted in 338.100. Bare neutrals are permitted for service-entrance conductors in the exception to 230.41, for underground service-entrance conductors in the exception to 230.30, and for *overhead service conductors* in 230.22, exception *when used as service conductors.* When service-drop cable is used as a feeder or branch circuit, however, there is no permission for use of a bare circuit conductor—although it may be acceptable to use the bare conductor of the service-drop cable as

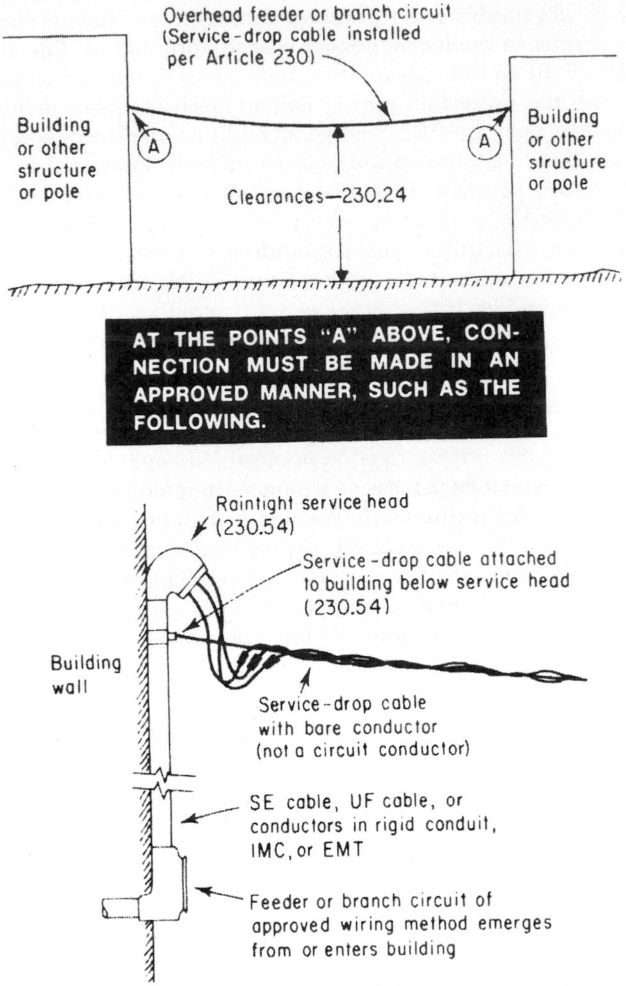
Overhead feeder or branch circuit
(Service-drop cable installed
per Article 230)
Building
or other
structure
or pole
A
A
Building
or other
structure
or pole
Clearances—230.24
AT THE POINTS "A" ABOVE, CONNECTION MUST BE MADE IN AN APPROVED MANNER, SUCH AS THE FOLLOWING.
Raintight service head
(230.54)
Service-drop cable attached
to building below service head
(230.54)
Building
wall
Service-drop cable
with bare conductor
(not a circuit conductor)
SE cable, UF cable, or
conductors in rigid conduit,
IMC, or EMT
Feeder or branch circuit of
approved wiring method emerges
from or enters building

Fig. 225-2. Aerial cable for overhead circuits. (Sec. 225.4 and 225.11.)

an equipment grounding conductor. And where service-drop cable is used as a feeder from one building to another, it would seem that a bare neutral could be acceptable as a grounded neutral conductor—as permitted in the last sentence of 338.3(B), first paragraph. This is where the new language in 396.30 is useful, because it expressly recognizes this use when it complies with 225.4, and 225.4 Exception allows the bare neutral where recognized elsewhere, such as where a regrounded neutral is permitted as covered in 250.32(B)(1) Exception No. 1. Note that such regrounded neutrals are only permitted for existing premises wiring systems, so this use will gradually disappear.

When service-drop cable is used between buildings, the method for leaving one building and entering another *must* satisfy 230.52 and 230.54. This is required in Sec. 225.11.

The exception to 225.4 excludes equipment grounding conductors and grounded circuit conductors from the rules on conductor covering. This exception permits equipment-grounding conductor *and* grounded circuit conductors (neutrals) to be bare or simply covered (but not insulated) as permitted by other Code rules.

Because the matter of outdoor, overhead circuiting is complex, check with local inspection agencies on required methods. As NEC 90.4 says, the local inspector has the responsibility for making interpretations of the rules.

225.5. Size of Conductors 1000 V Nominal or Less. This rule calls for conductor ampacity to be determined in accordance with 310.15 and the rules of Art. 220. But remember, where the load to be supplied is continuous or a combination of continuous and noncontinuous loads, then the rules of 210.19(A) and 210.20(A) or 215.2(A) and 215.3(A), covering conductor sizing and OC protection for branch circuits and feeders supplying continuous loads, must be observed, as well.

225.6. Conductor Size and Support. Open wiring must be of the minimum sizes indicated in 225.6 for the various lengths of spans indicated.

Article 100 gives a definition of *festoon lighting* as "a string of outdoor lights suspended between two points" (Fig. 225-3). Such lighting is used at carnivals, displays, used-car lots, etc. Such application of lighting is limited because it has a generally poor appearance and does not enhance commercial activities.

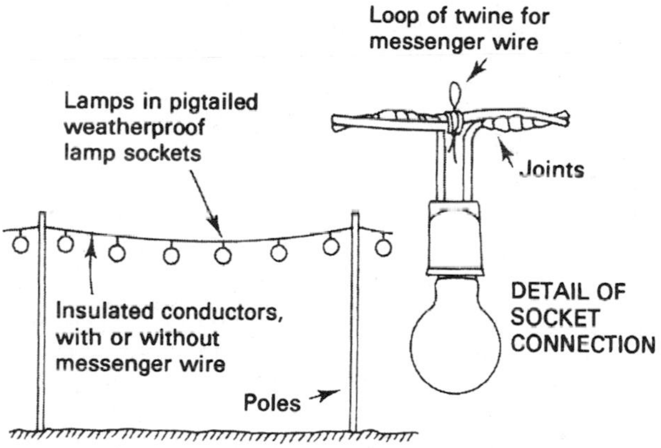
Loop of twine for messenger wire
Lamps in pigtailed weatherproof lamp sockets
Joints
DETAIL OF SOCKET CONNECTION
Insulated conductors, with or without messenger wire
Poles

Fig. 225-3. Festoon lighting is permitted outdoors. (Sec. 225.6.) Note that 590.4(F) requires a guard on lampholder such as the one shown here if the use qualifies as temporary wiring.

As covered in 225.6(B), overhead conductors for festoon lighting must not be smaller than No. 12; and where any span is over 40 ft (12.0 m), the conductors must be supported by a messenger wire, which itself must be properly secured to strain insulators. But the rules on festoon lighting do not apply to overhead circuits between buildings, structures, and/or poles.

225.7. Lighting Equipment Installed Outdoors. Part **(B)** permits a common neutral for both outdoor branch circuits and feeders—something not permitted for indoor branch circuits (a neutral of a 3-phase, 4-wire circuit is *not* a common neutral), although 215.4 grants limited permission for feeders with common neutrals. For two 208Y/120-V multiwire circuits consisting of six ungrounded conductors (two from each phase) and a single neutral (serving both circuits) feeding a bank of floodlights on a pole, if the maximum calculated load on any one circuit is 12 A and the maximum calculated load on any one phase is 24 A, the ungrounded circuit conductors may be No. 14, but the neutral must be at least No. 10. This rule clearly states the need to size a common neutral for the *maximum* (most heavily loaded) phase leg made up by multiple conductors connected to any one phase and supplying loads connected phase-to-neutral.

Part **(C)** covers use of 480/277-V systems for supplying incandescent and electric-discharge lighting fixtures. The former requirement that outdoor fixtures installed for lighting "outdoor areas" at commercial or public buildings must be not less than 3 ft (900 mm) from "windows, platforms, fire escapes, and the like" was deleted for the 2011 edition, thereby finally eliminating the obvious inconsistency between a 277-V bollard luminaire that a three-year-old could walk up to and hug because there was no objection in 210.6(C), and yet the same luminaire covered here had to be out of reach.

225.8. Calculation of Loads 1000 Volts, Nominal, or Less. Part **(A)** of this section calls for branch circuits to be sized in accordance with the rules of 220.10. And, part **(B)** of this rule calls for compliance with Part III of Art. 220 when sizing outdoor feeder conductors run "on or attached to" buildings, and so forth.

225.10. Wiring on Buildings (or Other Structures). This section identifies those wiring methods, rated up to 1000 V, that are permitted to be mounted on the exterior of buildings. Note that rigid nonmetallic conduit (PVC or RTRC) may be used for outside wiring on buildings, as well as the other raceway and cable methods covered in this section. For a long time, rigid PVC was not permitted for such application. Installation of conductors rated over 1000 V must comply with the provisions of 300.37, and electric signs and outline lighting must be installed as dictated by the rules of Art. 600.

225.14. Open-Conductor Spacings. Open wiring runs must have a minimum spacing between individual conductors (as noted in 225.14) in accordance with Table 230.51(C), which gives the spacing of the insulator supports on a building surface and the clearance between individual conductors on the building or where run in spans (Fig. 225-4).

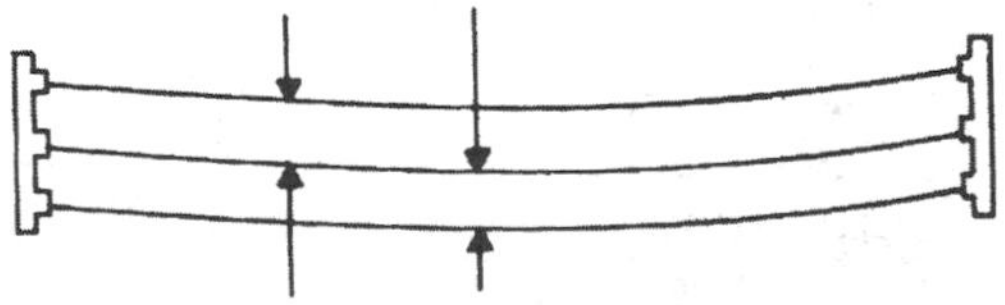
Minimum clearance distance between
conductors is given in Table 230.51 (C)

Fig. 225-4. Spacing of open-wiring conductors. (Sec. 225.14.)

It should be noted that 225.14 and Table 230.51(C) require that the *minimum spacing* between individual conductors in spans run overhead be 3 in. (75 mm) for circuits up to 300 V (such as 120, 120/240, 120/208, and 240 V). For circuits up to 1000 V, such as 480 Δ and 480/277 V, the *minimum spacing* between individual conductors must be at least 6 in. (150 mm).

225.17. Masts as Supports. Masts must have sufficient rigidity to handle the strain, or they must be guyed accordingly. If a raceway mast is used, any fittings must be identified for use with masts including the hub at the base, and only the feeder and branch-circuit conductors within the scope of Art. 225 can be attached to the mast. For example, a telephone drop is not permitted to be attached to a power circuit mast, regardless of the strength of the raceway or the amount of guy wire support provided.

As of the 2014 NEC, there is now a complete prohibition against attaching the drop to any mast above the highest coupling unless there is a "point of securement" above it. This is not controversial because an additional attachment to the side of a building is easily accomplished. However, the new rule goes on to prohibit, in all cases, the coupling above the roof. This is, of course, the preferred design, because the process of threading a steel conduit inevitably weakens it substantially particularly in reference to a shearing load. It is of interest that an attempt to allow for an exposed coupling in the event the strain was offset by a guy wire failed because the panel was concerned about the guy wire being released during roofing activities.

225.18. Clearance for Overhead Conductors and Cables. Overhead spans of open conductors and open multiconductor cables must be protected from contact by persons by keeping them high enough above ground or above other positions where people might be standing. And they must not present an obstruction to vehicle passage or other activities below the lines (Fig. 225-5).

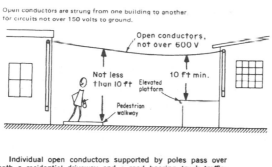

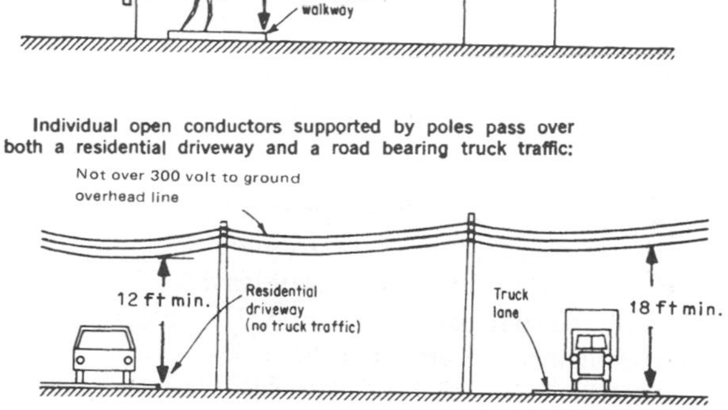
Open conductors are strung from one building to another for circuits not over 150 volts to ground.
Open conductors, not over 600 V
Not less than 10 ft
10 ft min.
Elevated platform
Pedestrian walkway
Individual open conductors supported by poles pass over both a residential driveway and a road bearing truck traffic:
Not over 300 volt to ground overhead line
12 ft min.
Residential driveway (no truck traffic)
Truck lane
18 ft min.

Fig. 225-5. Conductor clearance from ground. (Sec. 225.18.)

The rule of 225.18 applies to "open conductors and open multiconductor cables" and gives the conditions under which clearances must be 10, 12, 15, 18, or 24.5 ft (3.0, 3.7, 4.5, 5.5, or 7.5 m)—for conductors that make up either a branch circuit or a feeder [not service-drop conductors, which are subject to 230.24(B)]. The 24.5 ft (7.5 m) span height is new as of the 2011 edition and covers overhead crossings over railroad tracks. The dimension comes from the NESC. Although the wording used here is the same as that referring to corresponding clearances in 230.24(B), 225.18 covers those "open conductors and open multiconductor cables"—such as triplex and quadruplex cables—that do not meet the definition of service conductors, which would be regulated by Art. 230. Article 225 gives minimum clearances for triplex or quadruplex cables, as well as open individual conductors, commonly used for outdoor overhead branch circuits and feeders.

As 225.18 stands, "open conductors and multiconductor cables" for an overhead *branch circuit* or *feeder* require only a 10-ft (3.05-m) clearance from ground for circuits up to 150 V to ground; just as service-drop conductors up to 150 V must have a clearance of not less than 10 ft (3.05 m) from ground.

The rules of this section agree with the clearances and conditions set forth in the *NESC* (*National Electrical Safety Code*) for open conductors outdoors. The distances given for clearance from ground must conform to maximum voltage at which certain heights are permitted.

225.19. Clearances from Buildings for Conductors of Not Over 1000 Volts, Nominal. The basic minimum required clearance for outdoor conductors running above a roof is 8-ft (2.5-m) vertical clearance from the roof surface.

The basic ideas behind the rules are as follows:

1. Any branch-circuit or feeder conductors—whether insulated, simply covered, or bare—must have a clearance of at least 8 ft (2.5 m) vertically from a roof surface over which they pass. And that clearance must be maintained not less than 3 ft (900 mm) from the edge of the roof in all directions.
2. A roof that is subject to "pedestrian or vehicular traffic" must have conductor clearances "in accordance with the clearance requirements of 225.18." That reference essentially requires a clearance of 12 ft above a roof that serves as driveway or parking area, not subject to "truck traffic," and where the voltage to ground does not exceed 300 V to ground. Where the voltage to ground exceeds 300 V to ground, then a minimum clearance of 15 ft must be provided. And, if the area is subject to truck traffic and the conductors are operated at more than 300 V to ground, then a minimum of clearance above the roof of 18 ft must be provided.

In parts **(B)** and **(C)**, overhead conductor clearance from signs, chimneys, antennas, and other nonbuilding or nonbridge structures must be at least 3 ft (900 mm)—vertically, horizontally, or diagonally.

Part **(D)** addresses installation details regarding clearance of "final spans." In part **(D)(1)**, the Code requires that the connection point of overhead branch circuit and feeder conductors to the building be kept at least 3 ft from any of the building openings identified by the first part of this rule. The rule exempts windows that do not open from compliance.

But in part **(D)(2)** the Code addresses those final spans that run above areas that people may occupy. Where the final span's connection point runs above, or is within 3 ft horizontally from, "platforms, projections, or surfaces" where a person could come into contact with the conductors or cable, the clearances given in 225.18 must be observed.

Part **(D)(3)** prohibits installation of outside branch-circuit and feeder conductors beneath, or where they obstruct the entrance to, building openings through which material or equipment is intended to be moved. Barns provide a good example of the type of building opening this rule is intended to cover. Although only "farm and commercial buildings" are mentioned, they are only held up as examples. The wording used extends this requirement to any such opening, at any occupancy.

As indicated in Fig. 225-6, Exception No. 2 to 225.19(A) may apply to circuits that are operated at 300 V or less.

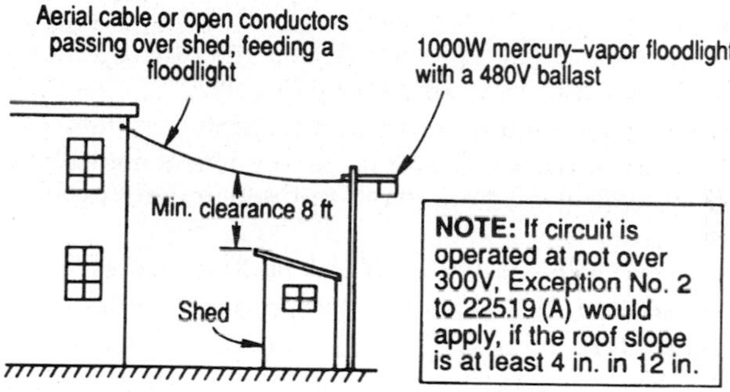
Aerial cable or open conductors passing over shed, feeding a floodlight
1000W mercury–vapor floodlight with a 480V ballast
Min. clearance 8 ft
Shed
NOTE: If circuit is operated at not over 300V, Exception No. 2 to 225.19 (A) would apply, if the roof slope is at least 4 in. in 12 in.

Fig. 225-6. Conductors—whether or not they are fully insulated for the circuit voltage—must have at least 8-ft (2.44-m) vertical clearance above a roof over which they pass. (Sec. 225.19.)

Part **(E)** covers a preferred exclusion zone (required only if practicable) in which overhead lines should not be run adjacent to high-rise buildings in order that fire ladders can be set up.

225.22. Raceways on the Exterior Surfaces of Buildings or Other Structures. Condensation of moisture is very likely to take place in conduit or tubing located outdoors. The conduit or tubing should be considered suitably drained when it is installed so that any moisture condensing inside the raceway or entering from the outside cannot accumulate in the raceway or fittings. This requires that the raceway shall be installed without "pockets," that long runs shall not be truly horizontal but shall always be pitched, and that fittings at low points be provided with drainage openings.

An excellent way to accomplish this is to drill a weep hole in an appropriate location in a conduit body, such as in the heel of an LB at the bottom of a vertical run or in the bottom side of a "C" fitting in a horizontal run. This practice is now expressly allowed in 314.15.

In order to be raintight, all conduit fittings must be made up wrench-tight. Couplings and connectors used with electrical metallic tubing shall be listed as "raintight."

225.24. Outdoor Lampholders. This section applies particularly to lampholders used in festoons. Where "pigtail" lampholders are used, the splices should be staggered (made a distance apart) in order to avoid the possibility of short circuits, in case the taping for any reason should become ineffective.

According to the UL Standard for Edison-Base Lampholders, "pin-type" terminals shall be employed only in lampholders for temporary lighting or decorations, signs, or specifically approved applications. The NEC requires that such lampholders only be used on stranded wire.

225.25. Location of Outdoor Lamps. In some types of outdoor lighting it would be difficult to keep all electrical equipment above the lamps, and hence a disconnecting means may be required. A disconnecting means should be provided for the equipment on each individual pole, tower, or other structure if the conditions are such that lamp replacements may be necessary while the lighting system is in use. It may be assumed that grounded metal conduit or tubing extending below the lamps would not constitute a condition requiring that a disconnecting means must be provided.

225.26. Vegetation As Support. Trees or any other "vegetation" must not be used "for support of overhead conductor spans." Note that the wording used here does not include electric equipment, but, rather, only prohibits "overhead conductor spans" from being supported by "vegetation." The effect is to permit outdoor lighting fixtures to be mounted on trees and to be supplied by an approved wiring method—conductors in a raceway or Type UF cable—attached to the surface of the tree (Fig. 225-7).

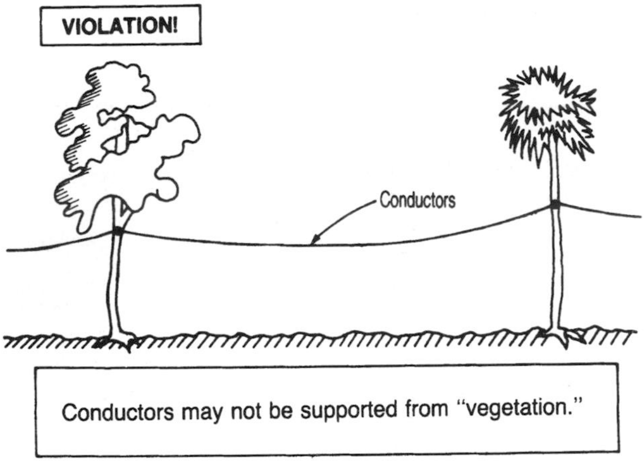
VIOLATION!
Conductors
Conductors may not be supported from "vegetation."

Fig. 225-7. The rule of 225.26 prohibits overhead wiring installed on trees.

225.27. Raceway Seal. This is simply the existing requirement for underground services (230.8) copied into Art. 225, so the concept will apply to additional buildings suppled by feeders under similar conditions.

Part II. Buildings or Other Structures Supplied by a Feeder(s) or Branch Circuit(s)

225.30. Number of Supplies. These rules cover those installations where several buildings are supplied from a single service. Although technically "feeder" or "branch-circuit" conductors, given that the supply conductors to the other buildings are the effective equivalent of service conductors, the rules given in Part II of Art. 225 are very nearly identical to those given for service conductors by Art. 230. In fact, many of the rules here were simply lifted from Art. 230 and modified as needed for use in Art. 225.

In 225.30, the Code stipulates the number of sources of supply to one building from another. The basic rule is similar to 230.2 for services, which calls for no more than one source of supply. Of course, as with 230.2, the rules in 225.30(A) through (E) present a number of circumstances where it would be permissible to supply one building or structure from another with more than one source. The wording used here is identical to that used in Sec. 230.2, which regulates the number of services permitted, simply because the "feeder" from the main building (i.e., the one where the service is installed) is essentially or effectively the "service" to the second building. For a group of buildings under single management, disconnect means must be provided for each building, as in Fig. 225-8.

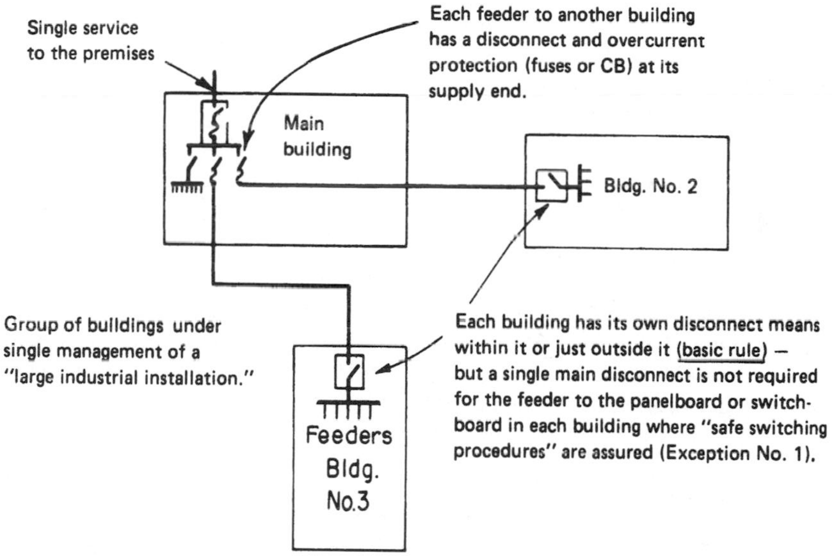
Single service
to the premises
Each feeder to another building
has a disconnect and overcurrent
protection (fuses or CB) at its
supply end.
Main
building
Bldg. No. 2
Group of buildings under
single management of a
"large industrial installation."
Each building has its own disconnect means
within it or just outside it (basic rule) –
but a single main disconnect is not required
for the feeder to the panelboard or switchboard in each building where "safe switching
procedures" are assured (Exception No. 1).
Feeders
Bldg.
No.3

Fig. 225-8. Each building must have its own disconnect means. (Sec. 225.32.)

An important 2011 NEC additional paragraph to this section addresses instances where feeders leave building No. 1, arrive at building No. 2 to greet a generator in building No. 2 which has been designed to carry loads in both buildings. The literal text does not say this, but that is what was described in the documentation. When this happens, no more than one feeder or branch circuit is permitted to go back to building No. 1 unless specifically permitted under paragraphs (A) through (E) following.

The rules for services will be exhaustively covered in Art. 230, so this discussion will only focus on the differences between parts II and III of this article and the service article. The principal general difference is that wiring from one building to another, although superficially like service wiring, is not service wiring. A *service entrance* is the interface between premises wiring and the facilities of the serving utility. It is also the interface between wiring governed by the National Electrical Safety Code (NESC) and premises wiring governed by the **National Electrical Code**. These codes are very different because the NESC presumes that the maintenance of those systems will be performed by a highly regulated utility workforce operating under a unique workplace culture and environment, and that this will continue for the foreseeable future, given the organizational permanence of utility enterprises. If there is no utility interface, then there is no service. For this reason, it is incorrect to label the disconnecting means for one building fed from another as a "service" disconnect when it is actually a "building" disconnect. The real significance of the creation of this part of Art. 225 is the clarity it brings to what is encompassed by a service, and more importantly, what is not. Specific differences are as follows:

As previously noted, only one source of supply is permitted as a general rule. This is the same general rule as for services, but there are subtle differences. The permission to run multiple service laterals from a common connection to disconnected loads does not apply. The allowance for multiple supplies where the capacity requirement exceeds that which the local utility supplies through one service does not apply for obvious reasons, and the allowance for an exemption on capacity grounds by special permission does not apply either. The final difference is the permission, granted here and not for services, for additional supplies where there are "documented switching procedures" in place for safe disconnection. This normally arises on large, campus-style industrial distributions.

The 2017 **NEC** added a seventh allowance, for multiple supplies to electrical vehicle charging stations, since they frequently are installed in multiple, and must be circuited as individual branch circuits for each unit. Some inspectors had been rejecting these arrangements based on the previous definition of "structure" in Art. 100 that could have been interpreted as describing the resulting installations as structures capable of receiving only one supply circuit. This allowance is unnecessary because the definition was changed, and it can presumably be removed in the next revision cycle. Refer to the discussion in Art. 100 for more information.

225.31. Disconnecting Means. This rule mandates the installation of a disconnecting means to permit the feeder or branch-circuit conductors that supply "or run through" a building to be deenergized.

The location and other details are presented by 225.32.

As could be anticipated, the required disconnecting means for the building supply conductors must be installed at a "readily accessible location" as defined in Art. 100 of Chap. 1.

225.32 Location. This rule calls for the disconnecting means to be located at the point where the supply conductors enter the building if the disconnecting means are located inside or outside the building or structure served. This rule has been questioned by some experts. The wording here is essentially the same as it is for service conductors, although the supply conductors are fully protected against overcurrent. Where installed inside an auxiliary building or structure, supplied from a service in another building, the disconnecting means required by this rule must be at the point of conductor entry.

The reason is to back up the essential nature of the disconnect rules in this part of the article, namely, if there is a problem with the wiring system in a remote building (or structure), it should never be necessary to go to another location to disconnect the wiring, with only rare and very specific exceptions. This even applies to a powered garage at a detached single-family dwelling. Note also that 700.12(B)(6), 701.12(B)(5), and 702.12 create Chap. 7 amendments to this Chap. 2 requirement by allowing remote disconnects on outdoor generators provided they are "within sight" of the occupancy served. In addition, 700.12(B)(6) Exception allows out-of-sight disconnecting means under conditions that correlate with 225.32 Exception No. 1.

The last sentence of 225.32 states that the remedies provided in 230.6 apply to the feeder and branch-circuit conductors supplying an outbuilding. This allowance can be used, just as for service conductors, to artificially extend the point of entry into a building to an interior disconnecting means.

Exception No. 1 applies to commercial, industrial, and institutional occupancies where a full-time staff provides maintenance. If such a facility's maintenance staff have established—in writing—"safe switching procedures," then the disconnecting means required by the basic rule may be located remotely from the building supplied.

For any "integrated electrical system" as defined and regulated by Art. 685, Exception No. 2 suspends the basic rule calling for a feeder disconnect at each building.

Exception No. 3 eliminates the need for individual disconnects for individual lighting standards. The literal wording calls for a disconnect at each "structure." The addition of this exception indicates the CMP's intent, which is to permit one disconnect for a number of lighting poles. And Exception No. 4 extends similar recognition to "poles or similar structures" that support signs (Art. 600).

225.33. Maximum Number of Disconnects. The six-disconnect rule is almost the same as for services, but in this application, disconnects for surge-protective equipment and power monitoring equipment are not exempted from the allowable total of six. The access requirements, the grouping requirement, the requirement to segregate disconnects for certain critical systems, and the specification of minimum ratings for certain applications are the same as for services.

225.36. Type of Disconnecting Means. This section has been completely revised as of the 2014 NEC, and in the process two crucial changes to prior practice resulted. First, unless the supply to the second building has a regrounded neutral as covered in 250.32(B)(1) Exception No. 1 and therefore only for existing installations wired many years ago when such connections were routinely permitted, a disconnecting means that is suitable for use as service equipment is no longer required. Such disconnects have increased spacings and provisions to reground the neutral, and have no relevance to most of today's applications.

Any disconnecting means that complies with 225.38, including a snap switch, is permitted if it has the required number of poles and the applicable voltage and current ratings. Note that the former exception in 225.38 allowing three-way and four-way switches for residential outbuildings, even though they were not indicating as normally required in 225.38(D), has been deleted. Therefore, although snap switches are permitted, three-way and four-way switches cannot be used as the required disconnecting means.

Each disconnecting means must be grouped and marked to indicate its function and the load served as covered in 225.34(A). A residential compromise in 225.37 Exception No. 2 involves a waiver of the reciprocal labeling rule when multiple circuits supply a dwelling outbuilding. The Code-making panel decided that reciprocal signage in such a building to the effect of "This is disconnect 1 of 2, controlling the overhead light; disconnect 2 of 2 located on the west side of the garage door controls the GFCI receptacle" would be excessive. The reciprocal labeling waiver does not, however, waive the identification rule on each switch; nor does it waive any other disconnecting requirements covered here. Specifically:

1. Ungrounded conductors supplying a load intended to stay energized, such as a receptacle, must pass through a disconnecting means located at a readily accessible point nearest the point of entrance.
2. A snap switch is a permissible disconnecting means as long as it is indicating.
3. The switches associated with a single source of supply, such as a single branch circuit, must be grouped, although they needn't be as close as adjacent snap switches in a two-gang box.
4. Each switch must be marked with its function. If that function is obvious, such as the overhead light, **NEC** 110.22 allows some basis for omitting this marking. However, by providing the marking you will avoid challenges.

Suppose you install a receptacle that will supply a freezer, and the owner wants to be assured that it won't be turned off inadvertently. Assume there will also be a light controlled from the house and the garage using three-way switches. Run the three-way switch travelers through a two-pole snap switch in another box near the point of entry, perhaps at an odd height, say 3 ft above the floor, and mark it **LIGHT**. Run the receptacle feed through a single-pole snap switch near the disconnect for the three-way switch and mark it **RECEP DISC**. or similar. It would be permitted, but not required, to install some sort of weatherproof cover over these disconnect switches to preclude inadvertent operation, as long as the disconnect functions could be read with the cover closed.

Part III, Over 1000 Volts

This part covers medium-voltage supply wiring (over 1000 V).

225.50. Sizing of Conductors. This incorporates the usual medium-voltage circuit sizing provisions from 215.2(B) (for feeders) and 210.19(B) for branch circuits.

225.51. Isolating Switches. This carries over the provisions of 230.204 so they will apply to applications within the scope of Art. 225.

225.52. Disconnecting Means. Part **(A)** and **(B)** (principal text) are unchanged from the 2008 **NEC** and essentially follow 230.205(A) and (C) with one major difference. The grade-operable pole-top disconnect now found in 230.205(A) has now, as of the 2014 **NEC**, been included in these rules.

The exception to the simultaneous disconnecting means requirement for the ungrounded conductors is new as of the 2011 **NEC** and covers fused cutouts, which can be safely opened as single-pole devices provided the load has been removed by some other means. There must be a sign installed adjacent to the cutouts stating this requirement. Therefore, the exception now mandates the wording on the sign to read: "DISCONNECT LOAD BEFORE OPENING CUTOUTS."

Parts **(C), (D),** and **(E),** although new as of 2011, cover some of the usual features of disconnects, including a lock-open function, a clear indication of their state, (off) or (on), and that when oriented vertically up is to be the on position unless a double-throw switch makes that irrelevant. Part **(F)** is also new and restates 225.37 [and thereby 230.2(E)], but without either the dwelling unit exception (obvious) or the safe switching procedures exception for multibuilding industrial installations (not a single word of substantiation having been offered for effectively removing the more usual application of that exception).

225.56. Inspections and Tests. This section sets forth the parameters for acceptance testing generally prior to placing the installation into service, although the tests in 225.56(A)(7) necessarily occur after operation is underway. See also 110.41, which provides a more universal platform for these procedures.

225.60. Clearances over Roadways, Walkways, Rail, Water, and Open Land. Medium-voltage feeders run overhead need to meet enhanced clearances reflecting the increased hazard involved for circuits operating over 1000 V.

225.61. Clearances over Buildings and Other Structures. These conductors must be installed so that they are at least 2.3 m (7.5 ft) away (horizontally separated) from building walls, projections, and windows. They must observe the same horizontal spacing from balconies, catwalks, and similar areas that people would have access to, and the same distance applies to other structures. The same conductors must be at least 3.8 m (12.5 ft) either above or below a roof (or other projection) where run at that level. For roofs accessible to vehicles (but not truck traffic) such as parking garages, the vertical clearance rises to 4.1 m (13.5 ft), 5.6 m (18.5 ft) if truck traffic uses the roof. Clearances over open ground vary by the type of traffic as well, beginning at 4.1 m (12.5 ft) for walkways and then rising a foot to 4.4 m (14.5 ft) for pedestrian ways and restricted traffic. Water areas not suitable for boating come in at 5.2 m (17 ft) and then open land suitable for grazing, cultivation, or vehicles, along with ways subject to vehicular access generally, including roads, driveways, alleys, and parking lots, all at 5.6 m (18.5 ft). The highest prescribed clearance, 8.1 m (26.5 ft), applies to runs over railways. These clearances all apply to medium voltages up to and including 22 kV *measured to ground*. Higher voltages add 10 mm (0.4 in) per kilovolt above 22 kV, and special cases, including clearances over navigable waters, and areas with large vehicles such as mining operations may require special engineering and review by the authority having jurisdiction.

ARTICLE 230. SERVICES

230.2. Number of Services. For any building, the service consists of the conductors and equipment used to deliver electric energy from the utility supply lines to the interior distribution system. Service may be made to a building either overhead or underground, from a utility pole line or from an underground transformer vault.

The first sentence of this rule requires that a building or structure be supplied by "only one service." Because "service" is defined in Art. 100 as "The *conductors* and equipment for delivering energy from the serving utility to the wiring system of the premises served," use of one "service" corresponds to use of one "service drop" or one "service lateral," or one set of "overhead service conductors" or set of "underground service conductors," depending on the location of the service point. For more information, refer to the discussion in Chap. 1 of this book regarding crucial changes in definitions of service terms in Art. 100.

These changes have forced interesting terminology changes throughout Art. 230. Now that service drops and service laterals are exclusively under utility control on the line side of the service point, many rules in the NEC no longer mention these terms. Other rules in the article (and explanations in this book) mention both drops and laterals along with overhead and underground service conductors. It depends on whether or not a construction rule applies to how a drop or lateral interfaces with the rule. For example, the rules in 230.40 on the permitted number of service-entrance conductor sets apply whether they connect to a service drop or overhead service conductors, or to a service lateral or underground service conductors, and the rule is written accordingly. On the other hand, the majority of the rules in the article only refer to components on the load side of the service point, and the old "service drop" and "service lateral" terminology has been discarded. As an example of this, the ampacity requirement in 230.31(A) only applies to underground service conductors. A utility is free to use the NESC to size a service lateral because conductors qualifying as such are now necessarily on the utility side of the service point and under their exclusive control.

The basic rule of this section requires that a building or other structure be fed by only one service drop (overhead service) or by only one set of service lateral conductors (underground service). As shown in Fig. 230-1, a building with only one service drop to it satisfies the basic rule even when more than one set of service-entrance conductors are tapped from the single drop (or from a single lateral circuit). Also note that only a utility may supply a service. A power source consisting of a generator, or even an on-site electric plant, is a separately derived system and the applicable rules for disconnects, etc., will be found in Part II of Art. 225 and not this article. And when such energized conductors reach the premises in question, they will pass through a "building disconnect" and not a "service disconnect." Review the coverage at the beginning of Part II of Art. 225 for more information on this crucial topic.

230.2 adds an important qualification of that rule as it applies only to 230.40, Exception No. 2, covering service-entrance layouts where two to six service disconnects are to be fed from one drop, lateral, or set of service conductors and are installed in separate individual enclosures at one location, with each disconnect supplying a

Fig. 230-1. One set of service-drop conductors supplies this building from utility line (coming from upper left) and two sets of SE conductors are tapped through separate metering CTs. In this case, the incoming wires are probably service drops because of the location of the metering CTs, but not necessarily. If the service point is at the utility pole, then they are overhead service conductors instead. (Sec. 230.2.)

separate load. As described in 230.40, Exception No. 2, such a service equipment layout may have a separate set of service-entrance conductors run to *"each or several"* of the two to six enclosures. The second sentence in 230.2 notes that where a separate set of underground conductors of size 1/0 or larger is run to each or several of the two to six service disconnects, the several sets of underground conductors are considered to be one service even though they are run as separate circuits, that is, connected together at their supply end (at the transformer on the pole or in the pad-mount enclosure or vault) *but not* connected together at their load ends. The several sets of conductors are taken to be "one service" in the meaning of 230.2, although they actually function as separate circuits (Fig. 230-2).

Although 230.40, Exception No. 2, applies to "service-entrance conductors" and service equipment layouts fed by *either* an overhead or underground service, the second sentence in 230.2 is addressed specifically and only to underground service conductors because of the need for clarification based on the Code definitions of "service conductors" (both overhead and underground), "service drop," "service lateral," "service-entrance conductors, overhead system," and "service-entrance conductors, underground system." (Refer to these definitions in the Code book to clearly understand the intent of this part of 230.2 and its relation to 230.40, Exception No. 2.)

The matter involves these separate but related considerations:

1. Because a "service lateral" or set of underground service conductors may (and usually does) run directly from a transformer on a pole or in a pad-mount enclosure to gutter taps where short tap conductors feed the terminals of the service disconnects, most layouts of that type literally do not have any "service-entrance conductors" that would be subject to the application

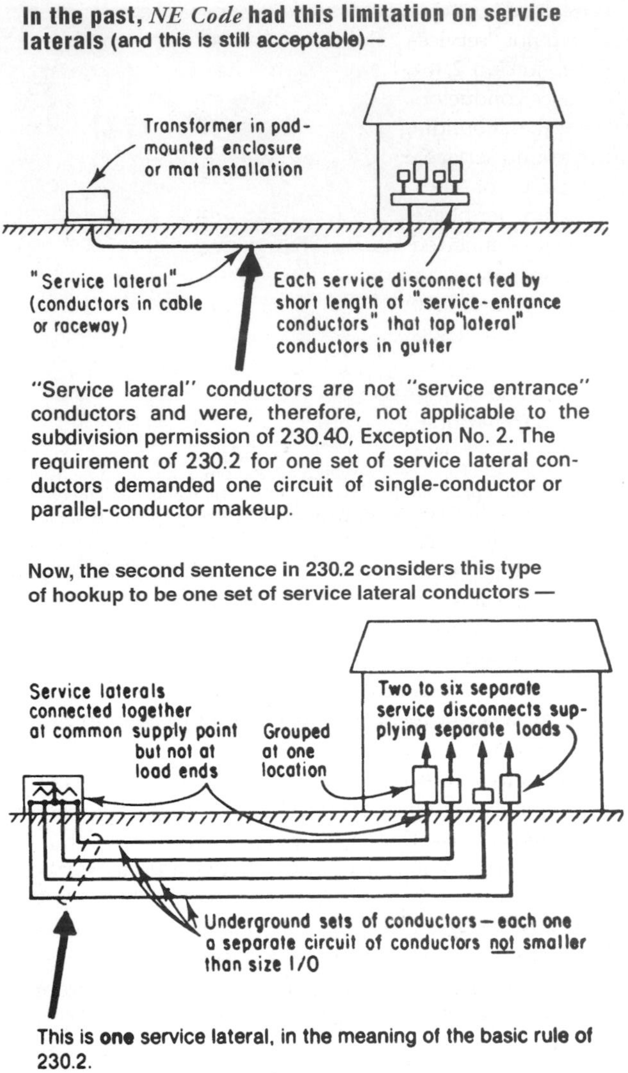
In the past, NE Code had this limitation on service laterals (and this is still acceptable)—
Transformer in pad-mounted enclosure or mat installation
"Service lateral" (conductors in cable or raceway)
Each service disconnect fed by short length of "service-entrance conductors" that tap "lateral" conductors in gutter
"Service lateral" conductors are not "service entrance" conductors and were, therefore, not applicable to the subdivision permission of 230.40, Exception No. 2. The requirement of 230.2 for one set of service lateral conductors demanded one circuit of single-conductor or parallel-conductor makeup.
Now, the second sentence in 230.2 considers this type of hookup to be one set of service lateral conductors —
Service laterals connected together at common supply point but not at load ends
Grouped at one location
Two to six separate service disconnects supplying separate loads
Underground sets of conductors — each one a separate circuit of conductors not smaller than size 1/0
This is one service lateral, in the meaning of the basic rule of 230.2.

Fig. 230-2. "One" service lateral may be made up of several circuits. Depending on the service point location, these may be sets of underground service conductors instead of service laterals. See Art. 100. (Sec. 230.2.)

permitted by 230.40, Exception No. 2—other than the short lengths of tap conductors in the gutter or box where splices are made to the underground conductors.

2. Because 230.40, Exception No. 2, refers only to sets of "service-entrance conductors" as being acceptable for individual supply circuits tapped from *one* drop or lateral to feed the separate service disconnects, that rule clearly does not apply

to "service lateral" or "underground service conductors" conductors which by definition are not "service-entrance conductors." So there is no permission in 230.40, Exception No. 2, to split up the capacity of either laterals or sets of underground service conductors. And the first sentence of 230.2 has the clear, direct requirement that a building or structure be supplied through only *one* run for any underground service. That is, either the underground service wiring must be a single circuit of one set of conductors, or if circuit capacity requires multiple conductors per phase leg, the wiring must be made up of sets of conductors in parallel—connected together at *both* the supply and load ends—in order to constitute a single circuit.

3. 230.2 permits "underground sets of conductors" to be subdivided into separate, nonparallel sets of conductors in the way that 230.40, Exception No. 2, permits such use for "service-entrance conductors"—*but only* for conductors of 1/0 and larger and *only* where each separate set of conductors (each separate underground circuit) supplies *one* or *several* of the two to six service disconnects.

230.2 recognizes the importance of subdividing the total service capacity among a number of sets of smaller conductors rather than a single parallel circuit (i.e., a number of sets of conductors connected together at *both* their *supply and load* ends). The single parallel circuit would have much lower impedance and would, therefore, require a higher short-circuit interrupting rating in the service equipment. The higher impedance of each separate set of lateral conductors (not connected together at their load ends) would limit short-circuit current and reduce short-circuit duty at the service equipment, permitting lower AIR (ampere interrupting rating)-rated equipment and reducing the destructive capability of any faults at the service equipment.

Subparts **(A)** through **(F)** cover cases where two or more service drops or laterals, or sets of overhead or underground service conductors may supply a single building or structure.

230.2(A) permits a separate supply to a fire pump and/or to emergency electrical systems, such as emergency lighting or exit lights and/or standby systems.

Part **(A)**, which is essentially an exception to the basic rule that a building "shall be supplied by only one service," also recognizes use of an additional power supply to a building from any "parallel power production systems." This would permit a building to be fed by a solar photovoltaic, wind, or other electric power source—in addition to a utility service—just as an emergency or standby power source is also permitted (Fig. 230-3). Fire pumps may be supplied from separate service conductors, and service from optional standby systems, legally required standby systems, and emergency systems is permitted in addition to the "one" service required by the basic rule in 230.2, "Number of Services." The final condition covered in this part is the "system designed for connection to multiple sources of supply for the purpose of enhanced reliability." This is the widely used double-ended switchboard with services provided to each end, and some form of throw-over in the middle so if one end goes down, the other end can supply the entire occupancy.

In 230.2(B) the Code recognizes other situations in which more than one service (i.e., more than one service drop or lateral, or set of overhead or underground

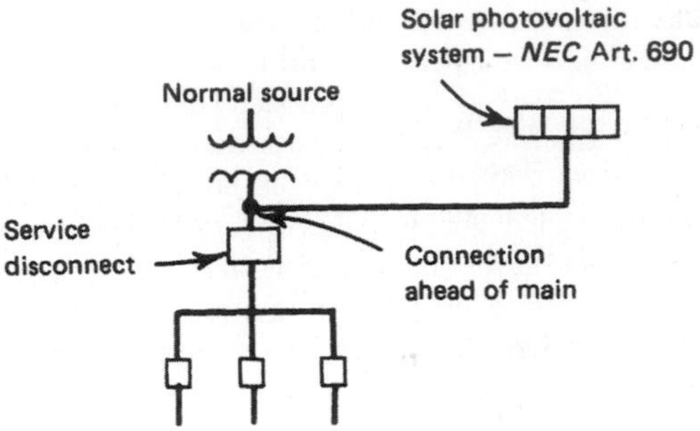
Solar photovoltaic
system – NEC Art. 690
Normal source
Service
disconnect
Connection
ahead of main

Fig. 230-3. Electric power generated by a solar voltaic assembly or by a wind-driven generator may be used as a source of power in "parallel" with the normal service. (Sec. 230.2.)

service conductors depending on the service point location) may be used. By "special permission" of the inspection authority, more than one service may be used for a multitenant building when there is no single space that would make service equipment available to all tenants. Do not confuse this allowance with the similar one in 230.40 Exception No. 1 where a single service supplies an unlimited number of occupancies using multiple sets of service conductors originating from the service location. Refer to the discussion at that point for extensive coverage on this topic.

Part **(B)(2)** requires special permission to install more than one service to buildings of large area. Examples of large-area buildings are high-rise buildings, shopping centers, and major industrial plants. In granting special permission the authority having jurisdiction must examine the availability of utility supplies for a given building, load concentrations within the building, and the ability of the utility to supply more than one service. Any of the special-permission clauses in 230.2 require close cooperation and consultation between the authority having jurisdiction and the serving utility. And, as always, such "special permission" must be provided in writing to satisfy the wording of the definition for "special permission" given in Art. 100.

Two or more services to one building are also permitted by part **(C)** when the total demand load of all the feeders is more than 2000 A, up to 1000 V, where a single-phase service needs more than one supply as determined by the policies of the serving utility, or by special permission (Fig. 230-4). 230.2(C) relates capacity to permitted services. Where requirements exceed 2000 A, two or more sets of service conductors may be installed. Below this value, special permission is required to install more than one set. The term "capacity requirements" appears to apply to the total calculated load for sizing service-entrance conductors and service equipment for a given installation, which would mean that the load calculated in accordance with Art. 220 must exceed 2000 A before one can assume permission for more than one set of service conductors.

Cases of separate light and power services to a single building and separate services to water heaters for purposes of different rate schedules are also permitted.

230.2 (C)

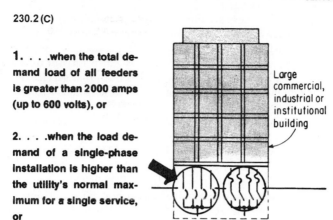

1. . . .when the total demand load of all feeders is greater than 2000 amps (up to 600 volts), or

2. . . .when the load demand of a single-phase installation is higher than the utility's normal maximum for a single service, or

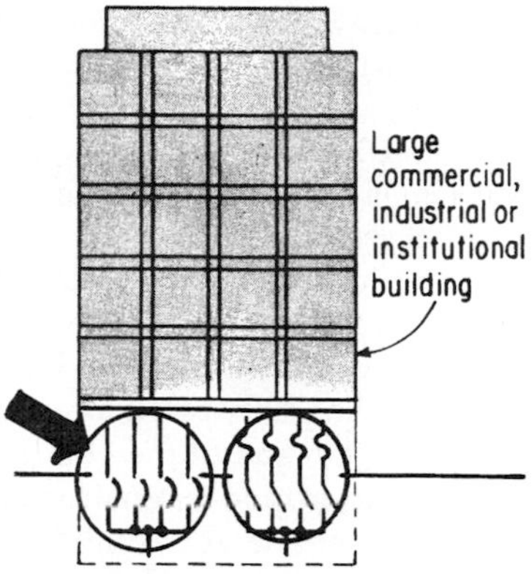
Large
commercial,
industrial or
institutional
building

3. . . .when special permission is obtained from the inspection authority.

NOTE: "Two or more services" means two or more service drops or service laterals—not sets of service-entrance conductors tapped from one drop or lateral.

230.2 (D)

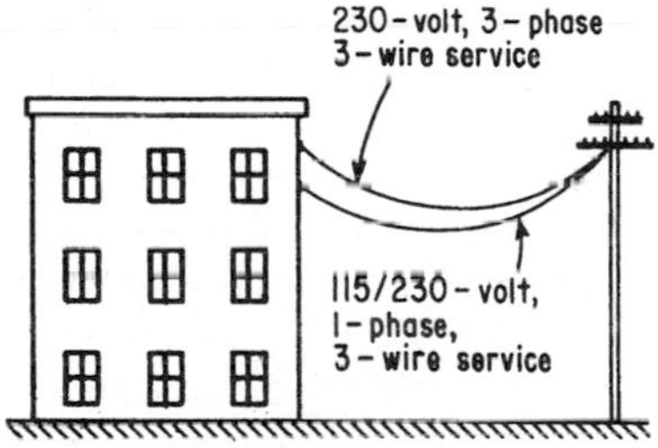
230-volt, 3-phase
3-wire service
115/230-volt,
1-phase,
3-wire service

Fig. 230-4. Exceptions to 230.2 permit two or more services under certain conditions. Depending on the location of the service point, the reference in the note to the upper drawing may refer to sets of either overhead or underground service conductors, as the case may be. (Sec. 230.2.)

And if a single building is so large that one service cannot handle the load, special permission can be given for additional services.

230.2(D) is illustrated at the bottom of Fig. 230-4.

The last part of the rules in 230.2, part **(E),** introduces a requirement that applies to any installation where more than one service is permitted by the Code to supply one building. It requires a "permanent plaque or directory" to be mounted or placed "at each service drop or lateral or at each service-equipment location" to advise personnel that there are other services to the premises and to tell where such other services are and what building parts they supply.

This directory must be placed at *each* service. So, if there are two services, there should be two plaques at *each* service location. The directory (or directories) must identify all feeders and branch circuits supplied from that service. Further, such directories must be *fully reciprocal* both as to load descriptions and locations. That is, if plaque number one says: "This is Service #1 of 2, for the north half of the building. Service #2, located in the middle of the south wall, is for the south end of the building" then there must be another such plaque in the middle of the south end of the building, reading something like this: "This is Service #2 of 2, for the south half of the building. Service #1, located in the middle of the north end of the building, is for the north end of the building." Labeling that provides this type of fully reciprocal information is required at every service equipment location if more than one service arrives at the building, for whatever reason. In addition, in some cases a building will be fed directly by a service and also by a feeder from another building or perhaps from a separately derived system. The reciprocal labeling rules apply to those supply systems just as if they were services.

230.3. One Building or Other Structure Not to Be Supplied Through Another. For the most part, the service conductors supplying each building or structure shall not *pass through the inside* of another building. The concern here is related to the fact that service-entrance conductors have no overcurrent protection at their line end. They are simply connected to the utility's supply without any type of OC device. Although the utility may have fuses in its lines, the fuses probably won't open the circuit unless there is a bolted-fault, which represents the smallest percentage of all faults. Effectively speaking, this means that any other fault in the service conductor is expected to "burn" itself clear. That being the case, the Code here prohibits running unprotected service conductors through one building to another, unless they are in a raceway encased by 2 in. (50 mm) of concrete or masonry (Fig. 230-5). 230.6 points out that conductors in a raceway enclosed within 2 in. (50 mm) of concrete or masonry are considered to be "outside" the building even when they are run within the building.

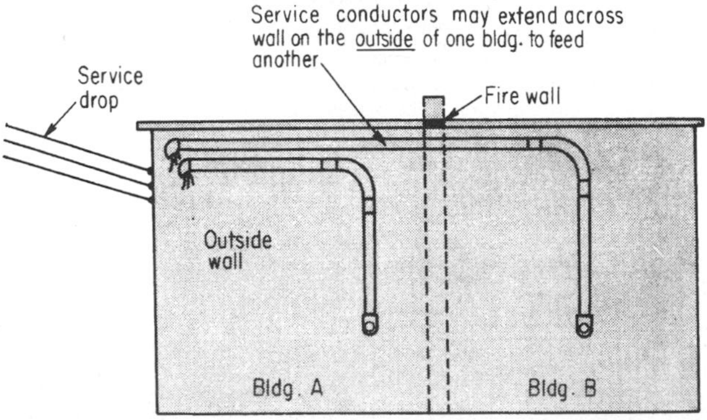
Service conductors may extend across wall on the outside of one bldg. to feed another
Service drop
Fire wall
Outside wall
Bldg. A
Bldg. B

Fig. 230-5. This is not a violation of the basic rule of 230.3. (Sec. 230.3.)

A building as defined in Art. 100 is a "structure which stands alone or which is cut off from adjoining structures by fire walls with all openings therein protected by approved fire doors." A building divided into four units by such fire walls may be supplied by four separate sets of service conductors, but a similar building without the fire walls may be supplied by only one set of service conductors, except as permitted in 230.2.

A commercial building may be a single building but may be occupied by two or more tenants whose quarters are separate, in which case it might be undesirable to supply the building through one set of service conductors. Under these conditions special permission may be given to install more than one service.

230.6. Conductors Considered Outside the Building. A complement to the requirement in 230.3, this section presents certain criteria that, when satisfied, render the service equipment and/or conductors "outside" the building. For example, in part **(1)**, the Code states that service conductors are considered "outside" the building when run in conduit under a building, and the conduit is covered with either brick or concrete at least 2 in. thick.

In part **(2)** the Code recognizes that conductors in conduit or duct *encased* by concrete or brick not less than 2 in. (50 mm) thick are considered to be outside the building, even though they are actually run within the building. The word "encased" replaced the word "enclosed" in the 1990 **NEC** to indicate actual embedment of the wiring in mortar and masonry (or concrete) and not just a masonry or concrete chase. Figure 230-6 shows how a service conduit was encased within

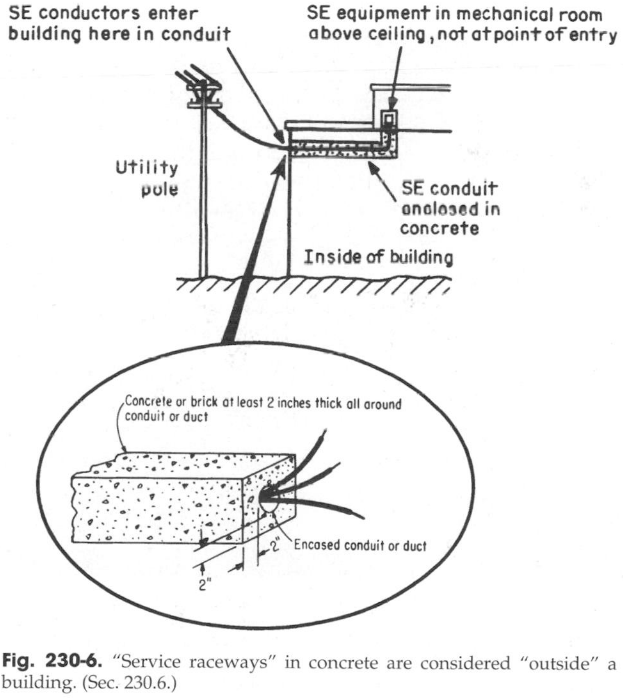
SE conductors enter
building here in conduit
SE equipment in mechanical room
above ceiling, not at point of entry
Utility
pole
SE conduit
enclosed in
concrete
Inside of building
Concrete or brick at least 2 inches thick all around
conduit or duct
Encased conduit or duct
2"
2"

Fig. 230-6. "Service raceways" in concrete are considered "outside" a building. (Sec. 230.6.)

a building so that the conductors are considered as entering the building right at the service protection and disconnect where the conductors emerge from the concrete, to satisfy the rule of 230.70(A), which requires the service disconnect to be as close as reasonably possible to the point where the SE conductors enter the building. Figure 230-7 shows an actual case of this application, where forms were

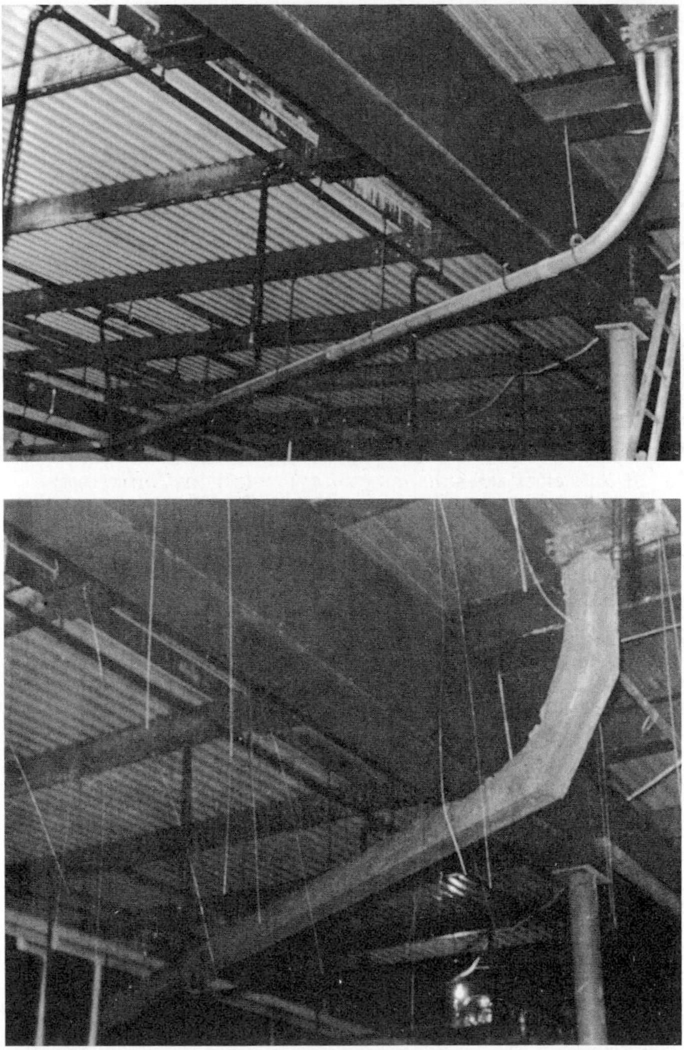

Fig. 230-7. Top photo shows service conduit carried above suspended ceiling, without the SE disconnect located at the point of entry. When conduit was concrete-encased, the service conductors then "enter" the building at the SE disconnect—where they emerge from the concrete. Service conduit enters building at lower left and turns up into SE disconnect (right) in roof electrical room.

hung around the service conduit and then filled with concrete to form the required concrete encasement.

Part **(3)** considers service equipment and conductors, installed within a fire-rated vault that conforms with the Code rules for transformer vaults given in part **III** of Art. 450, to be outside the building. Part **(4)** presents recognition similar to that given in part **(1)**, but this covers conductors buried in raceways under at least 18 in. of earth and running below a building or other structure.

Part **(5)** was new in the 2011 **NEC** and covers overhead service masts that typically penetrate at least the outer membrane of a building envelope in order to rise above the roof to establish a required clearance in 230.24. It would be unheard of to require concrete encasement for a typical heavy-wall steel or aluminum conduit mast that passes directly through an eave cavity, but that is what the previous literal text theoretically required.

Some eave cavities, particularly in old-fashioned Victorian balloon frame construction, are completely open to the adjoining attic and such a mast is well within the interior of the building. More modern construction typically has the eaves as part of a soffit that is separated from the interior of the building. In either case, however, once the outer membrane of the building is penetrated the disconnect location rule in 230.70(A) applies, making this allowance necessary.

As of the 2014 **NEC**, this provision has been severely limited to prevent abuse. It is now limited to heavy wall metal conduits passing directly through the eave cavity. The now prohibited procedure had actually been attempted using PVC conduit running through a roof, passing horizontally through 15 ft of the eave cavity, and then descending to a meter on the side of a building below the soffit.

230.7. Other Conductors in Raceway or Cable. Although the basic rule permits only service-entrance conductors to be used in a service raceway or service cable, exceptions do recognize the use of grounding conductors in a service raceway or cable and also permit conductors for a time switch if overcurrent protection is provided for the conductors, as shown in Fig. 230-8. Refer to the

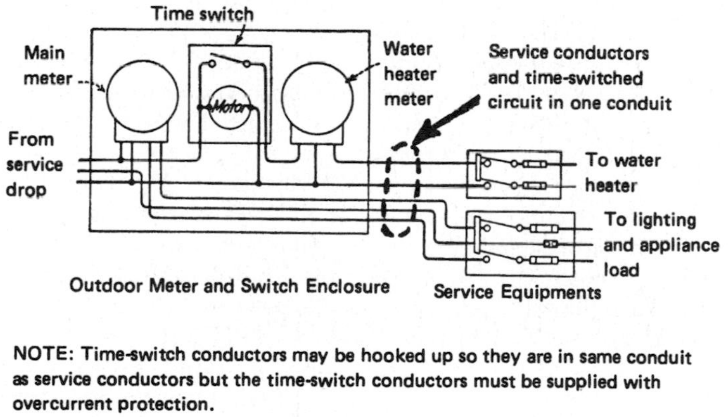

Time switch
Main meter
Motor
Water heater meter
Service conductors and time-switched circuit in one conduit
From service drop
To water heater
To lighting and appliance load
Outdoor Meter and Switch Enclosure
Service Equipments
NOTE: Time-switch conductors may be hooked up so they are in same conduit as service conductors but the time-switch conductors must be supplied with overcurrent protection.

Fig. 230-8. A time switch with its control circuit connected on the supply side of the service equipment. (Sec. 230.7.)

discussion of the definition of "raceway" in Art. 100 (Chap. 1 in this book) for more information regarding the status of auxiliary gutters with respect to this rule and why, for this reason, they are not classified as raceways.

230.8. Raceway Seal. Figure 230-9 indicates that Sec. 300.5(G) applies to underground service conduits. Where service raceways are required to be sealed—as where they enter a building from underground—the sealing compound used must be marked on its container or elsewhere as suitable for safe and effective use with the particular cable insulation, with the type of shielding used, and with any other components it contacts. Some sealants attack certain insulations, semiconducting shielding layers, and so forth.

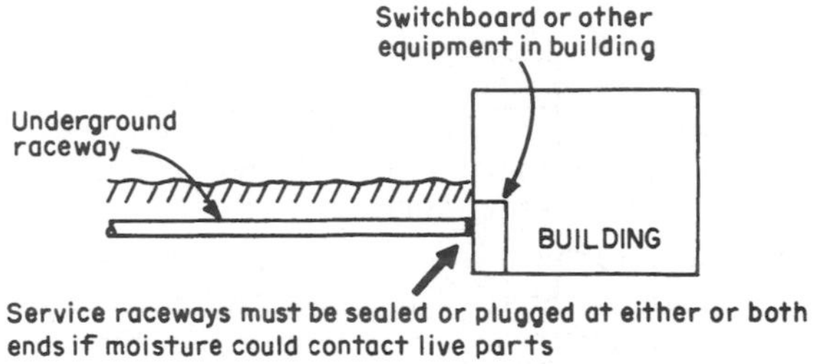
Switchboard or other equipment in building
Underground raceway
BUILDING
Service raceways must be sealed or plugged at either or both ends if moisture could contact live parts

Fig. 230-9. Service raceways may have to be sealed. (Sec. 230.8.)

230.9. Clearances on Buildings. Parts **(A)**, **(B)**, and **(C)** cover the clearance requirements for service conductors, including the final portion of overhead spans and their point of connection to the building or structure.

In part **(A)**, the Code makes clear that any overhead service conductors—open wiring or multiplex drop cable—must have the 3-ft (900-mm) clearance from windows, doors, porches, and so forth, to prevent mechanical damage to and accidental contact with service conductors (Figs. 230-10 and 230-11). The clearances required in 230.24, 230.26, and 230.29 are based on safety-to-life considerations in that wires are required to be kept a reasonable distance from people who stand, reach, walk, or drive under overhead service conductors. As the exception notes, conductors that run above the top level of a window do not have to be 3 ft (900 mm) away from the window.

Note that if the window in the upper drawing of Fig. 230-10 is not "designed to be opened" the clearance rules do not apply and there would be no violation. This topic was a frequent source of discussion. Was the intent of the rule to prevent a reaching exposure, or was it to address window washing hazards? CMP 4 decided that it should only address the reaching exposure and added the clarifying language in the 1993 NEC, thereby also correlating with the NESC coverage on this topic.

The rule of 230.9(B) recognizes clearances of less than 3 ft horizontally from porches, balconies, and so on, provided the minimum vertical clearance, measured from the floor of the porch or balcony, is in accordance with 230.24(B).

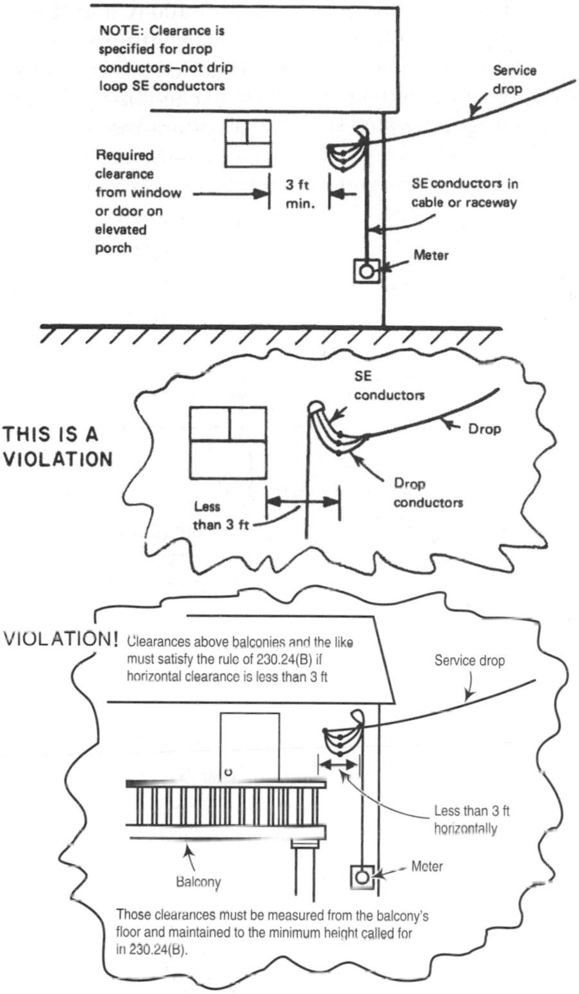
NOTE: Clearance is specified for drop conductors—not drip loop SE conductors
Service drop
Required clearance from window or door on elevated porch
3 ft min.
SE conductors in cable or raceway
Meter
THIS IS A VIOLATION
SE conductors
Drop
Drop conductors
Less than 3 ft
VIOLATION! Clearances above balconies and the like must satisfy the rule of 230.24(B) if horizontal clearance is less than 3 ft
Service drop
Less than 3 ft horizontally
Meter
Balcony
Those clearances must be measured from the balcony's floor and maintained to the minimum height called for in 230.24(B).

Fig. 230-10. Drop conductors must have clearance from building openings. If the service point is remote from the building, these are overhead service conductors, but the same clearance rules apply. (Sec. 230.9.)

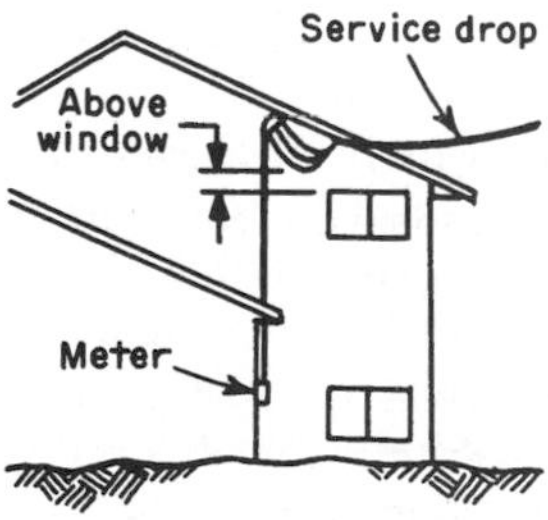
Service drop
Above window
Meter

Fig. 230-11. Drop conductors above top level of a window or door do not require 3-ft horizontal clearance. (Sec. 230.9, Exception.)

The service-drop conductors shown in the drawing at the bottom of Fig. 230-10 would have to be either 10 or 12 ft above the balcony's floor surface.

In part **(C)** of this section the Code says that service-drop or service-entrance conductors must not be mounted on or secured to a building wall directly beneath an elevated opening through which supplies or materials are moved into and out of the building. Such installations of conductors—say, beneath a high door to a barn loft—would obstruct access to the opening and present a hazard to personnel (Fig. 230-12).

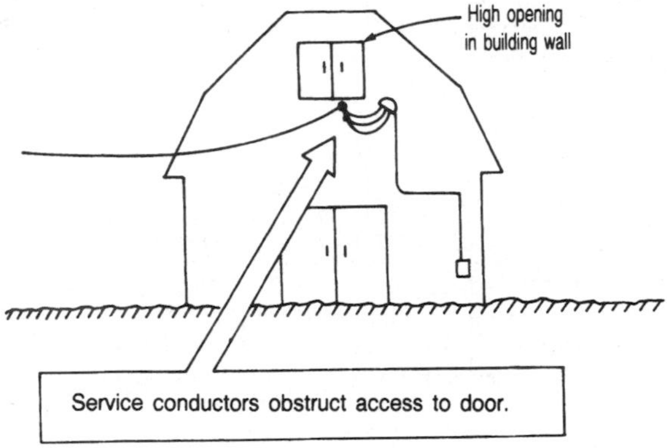
High opening
in building wall
Service conductors obstruct access to door.

Fig. 230-12. This violates the rule of the last paragraph of 230.9.

230.22. Insulation or Covering. In the past, the use of "covered"—not "insulated"—wire, such as triple-braid weatherproof wire, resulted in quite a few tragic accidents, including a number of electrocutions. As a result, for many years now, only the use of insulated wire for ungrounded conductors was permitted with service conductors. For the 2002 **NEC**, however, Code-making panel (CMP) 4 has, with very little substantiation, gone ahead and again recognized the use of "covered" instead of insulated conductors for overhead service conductors. The panel did specifically mention the legitimacy of this application for medium-voltage applications. There were no comments received during the comment period, either positive or negative, on this change.

The exception recognizes the use of a bare grounded (neutral) conductor of a multiconductor cable. The exception only covers multiconductor cables, and therefore grounded neutral of open wiring must be insulated or covered just as the ungrounded conductors.

230.24. Clearances. There are five exceptions to the basic rule of part **(A)** that service-drop conductors must have at least an 8-ft (2.5-m) vertical clearance from the highest point of roofs over which they pass.

Exception No. 1 to the basic rule calling for 8-ft (2.5-m) clearance of overhead service conductors above a roof requires that clearance above a flat roof subject to pedestrian traffic or used for auto and/or truck traffic must observe the heights for clearance of these conductors from the ground as given in part **(B)** of 230.24.

The intent of Exception No. 2 is that where the roof has a slope greater than 4 in. (100 mm) in 12 in. (300 mm), it is considered difficult to walk upon, and the height of conductors could then be less than 8 ft (2.5 m) from the highest point over which they pass but in no case less than 3 ft (900 mm) except as permitted in Exception No. 3. Figure 230-13 shows the rule. Exception No. 4 eliminates the need for maintaining the 8-ft minimum for 3 ft vertically in all directions where the final span attaches to the side of the building. This exception is particularly useful for a service drop hitting a building on a front corner above a porch roof below it. Without this provision, the drop would have to attach at a great height or else a second pole would be required to redirect the drop so it missed the projected footprint of the porch roof. Figures 230-14 and 230-15 show the conditions permitted by Exception Nos. 3 and 4.

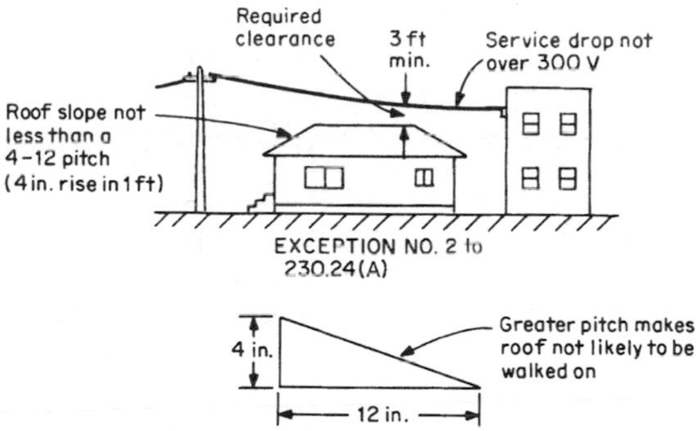
Required clearance
3 ft min.
Service drop not over 300 V
Roof slope not less than a 4-12 pitch (4 in. rise in 1 ft)
EXCEPTION NO. 2 to 230.24(A)
4 in.
12 in.
Greater pitch makes roof not likely to be walked on

Fig. 230-13. Service-drop conductors may have less than 8-ft (2.5-m) roof clearance. (Sec. 230.24.)

Exception No. 5 was new for the 2011 **NEC** and covers roof areas that are guarded or isolated. If the voltage between conductors does not exceed 300, the clearance can be reduced to 900 mm (3 ft). This provision brings the **NEC** in closer agreement with the NESC, although the voltage limitation in the latter standard is 750 V.

Part **(B)** covers the clearance to grade of overhead service conductor spans, as shown in Fig. 230-16. The five dimensions of clearance from ground—10, 12, 15, 18, and 24½ ft (3.0, 3.7, 4.5, 5.5, and 7.5 m)—are qualified by voltage levels and, for the 10-ft (3.0-m) mounting height, by the phrase "only for service-drop cables." Note that in this case the terminology "service drop cables" is correct, even if the span meets the definition for overhead service conductors. The terminology is referring to a type of wiring construction defined in product standards as service drop cable, and consisting of insulated conductors spiraled around a bare messenger wire. These **NEC** rules are generally in agreement with the *National Electrical Safety Code*. Where mast-type service risers are provided, the clearances in 230.24(B) will have to be considered by the installer.

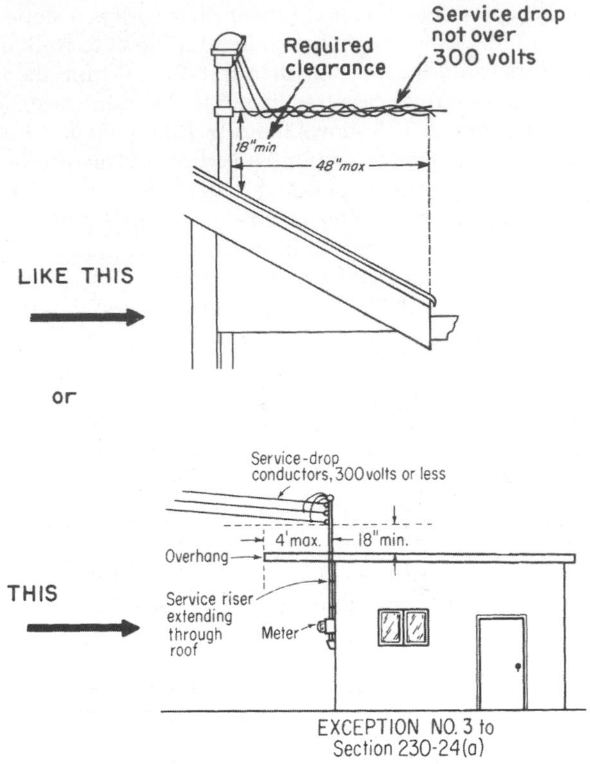
Service drop
not over
300 volts
Required
clearance
18"min
48"max
LIKE THIS
or
Service-drop
conductors, 300 volts or less
4' max.
18" min.
Overhang
THIS
Service riser
extending
through
roof
Meter
EXCEPTION NO. 3 to
Section 230-24(a)

Fig. 230-14. Reduced clearance for overhead service conductors. (Sec. 230.24.)

230.28. Service Masts as Supports. Figure 230-17 illustrates this rule. The last sentence of the rule is both important and controversial. It disallows all drops except the service drop from colocating on a service mast. No telephone drops, coaxial cable drops, or any other drop. This is an absolute prohibition and it applies no matter how stout the mast, no matter how well the mast is guyed, no matter how long the mast, and no matter what spacing would be provided between the service drop and other prospective drops from other utilities.

As of the 2014 **NEC**, there is now a complete prohibition against attaching the drop to any mast above the highest coupling unless there is a "point of securement" above it. This is not controversial because an additional attachment to the side of a building is easily accomplished. However, the new rule goes on to prohibit, in all cases, the coupling above the roof. This is, of course, the preferred design, because the process of threading a steel conduit inevitably weakens it substantially particularly in reference to a shearing load. As in the case of 225.17, it is of interest that an attempt to allow for an exposed coupling in the event the strain was offset by a guy wire failed because the panel was concerned about the guy wire being released during roofing activities.

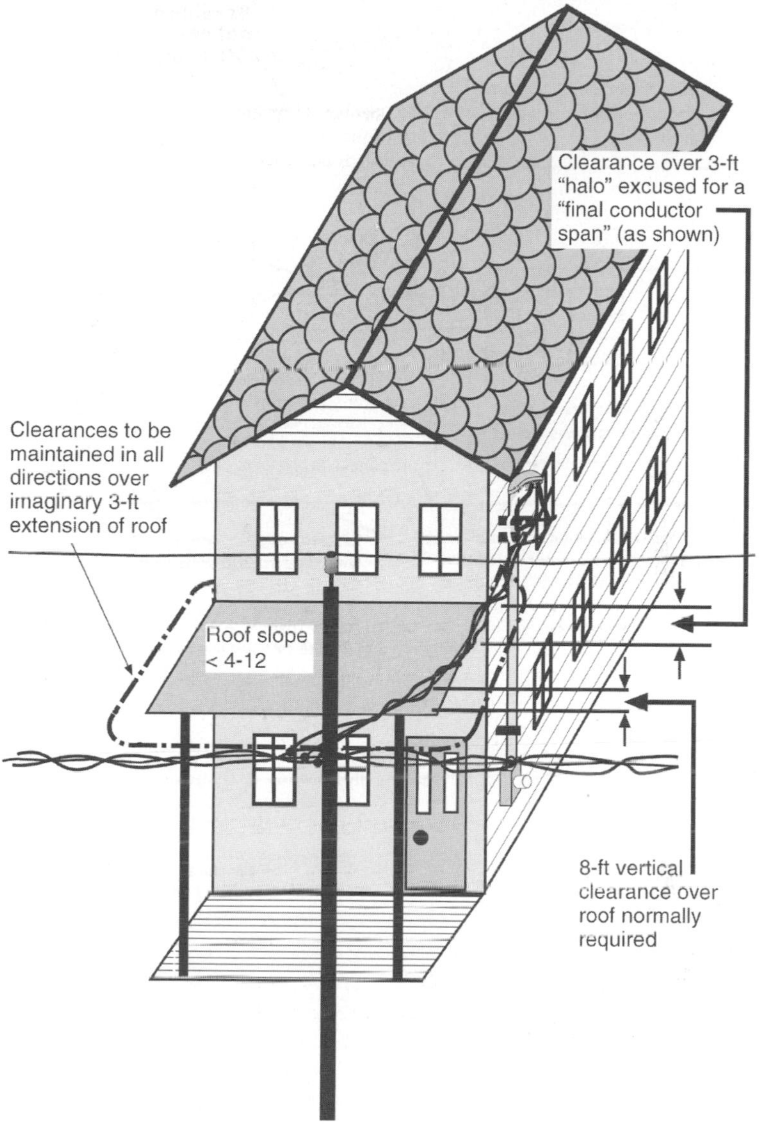
Clearance over 3-ft
"halo" excused for a
"final conductor
span" (as shown)
Clearances to be
maintained in all
directions over
imaginary 3-ft
extension of roof
Roof slope
< 4-12
8-ft vertical
clearance over
roof normally
required

Fig. 230-15. Reduced clearances permitted for final spans, such as adjacent to this porch roof with a gradual pitch. (Sec. 230.24 Exception No. 4.) (From *Practical Electrical Wiring*, 22nd edition, © Park Publishing, 2014, all rights reserved.)

230.29. Supports over Buildings. The 2017 NEC added an important requirement with respect to the use of metal support structures for overhead services using a grounded circuit conductor. This section now expressly requires that the metal support structure be bonded to the grounded conductor.

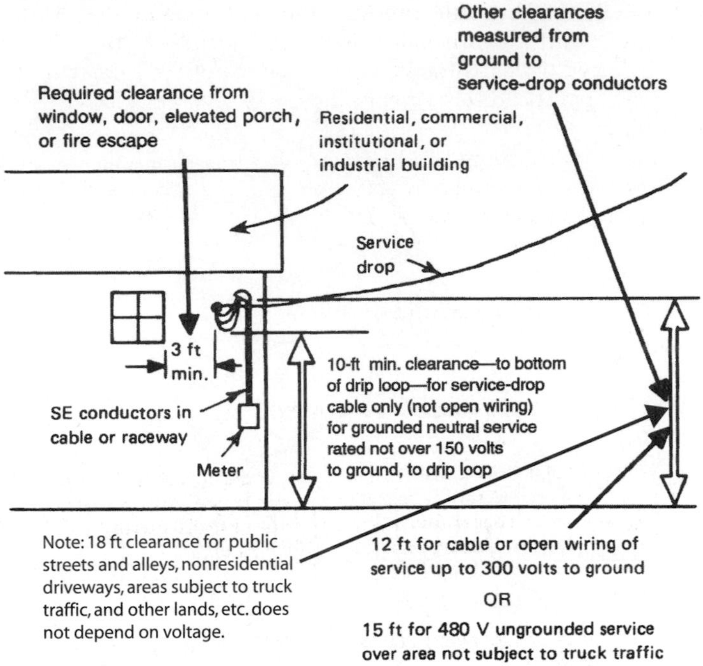
Other clearances measured from ground to service-drop conductors
Required clearance from window, door, elevated porch, or fire escape
Residential, commercial, institutional, or industrial building
Service drop
3 ft min.
SE conductors in cable or raceway
Meter
10-ft min. clearance—to bottom of drip loop—for service-drop cable only (not open wiring) for grounded neutral service rated not over 150 volts to ground, to drip loop
Note: 18 ft clearance for public streets and alleys, nonresidential driveways, areas subject to truck traffic, and other lands, etc. does not depend on voltage.
12 ft for cable or open wiring of service up to 300 volts to ground
OR
15 ft for 480 V ungrounded service over area not subject to truck traffic

Fig. 230-16. Service-drop clearance to "final grade." (Sec. 230.24.)

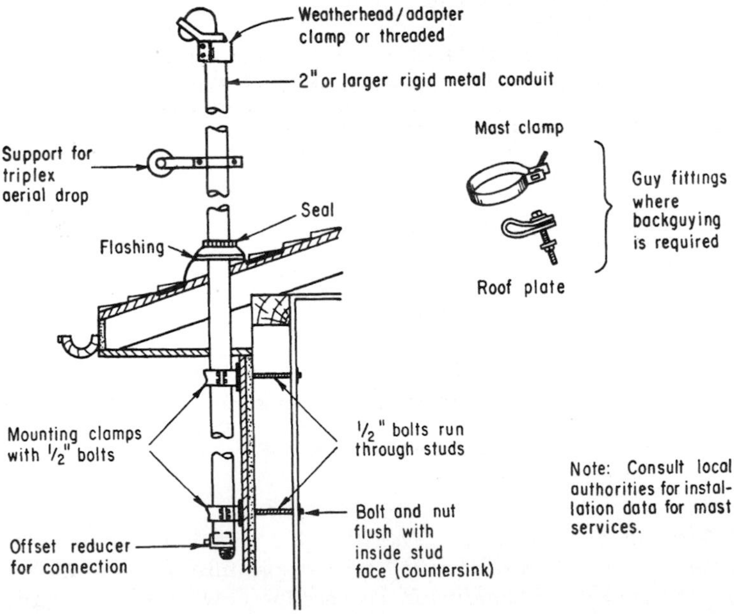
Weatherhead / adapter clamp or threaded
2" or larger rigid metal conduit
Mast clamp
Support for triplex aerial drop
Guy fittings where backguying is required
Seal
Flashing
Roof plate
Mounting clamps with 1/2" bolts
1/2" bolts run through studs
Note: Consult local authorities for installation data for mast services.
Bolt and nut flush with inside stud face (countersink)
Offset reducer for connection

Fig. 230-17. Service mast must provide adequate support for connecting drop conductors. The hub at the base of the mast must be identified for use with service equipment. (Sec. 230.28.)

230.30. Installation. This rule presents the requirement that underground service conductors must be insulated. Although service-drop conductors were previously required to be insulated, CMP 4 has seen fit to reinstate permission to use "covered" overhead service conductors, as given in 230.22. No such permission is granted here.

The exceptions to 230.30 and 230.41 clarify the use of aluminum, copper-clad aluminum, and bare copper conductors used as grounded conductors in underground service and service-entrance conductors (Fig. 230-18).

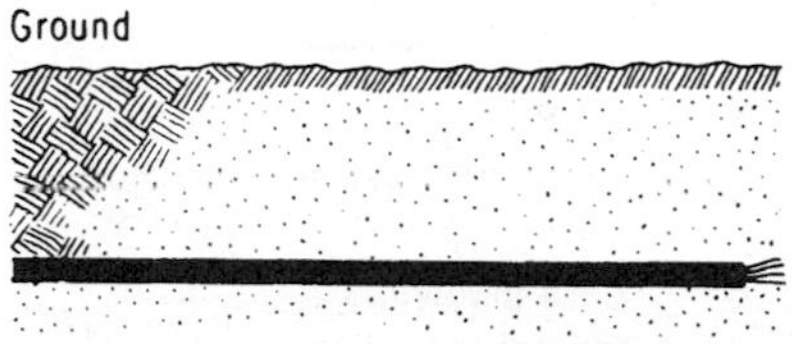
Ground

INSULATED PHASE CONDUCTORS and a bare copper neutral for an underground service lateral in buried raceway. Note: A bare aluminum or copper-clad aluminum neutral could be used here when part of a moisture- and fungus-resistant cable.

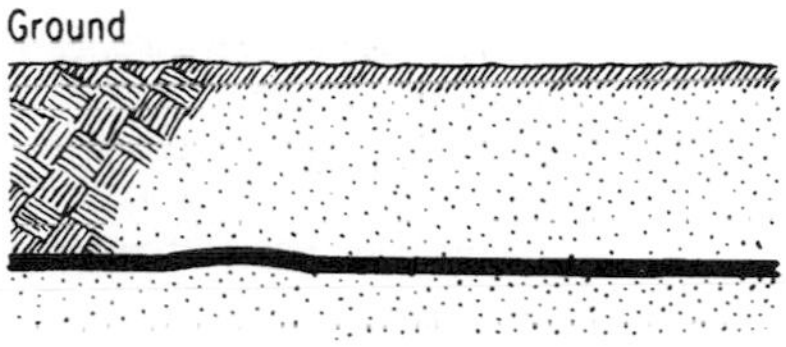
Ground

BARE COPPER NEUTRAL in a direct-buried cable assembly with moisture- and fungus-resistant outer covering. Note: A bare aluminum or copper-clad aluminum could be used like this, but it must be within the same type of cable assembly

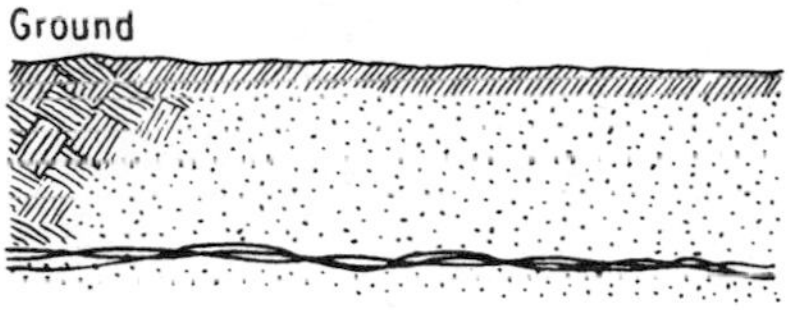
Ground

TYPE USE PHASE CONDUCTORS and a bare copper neutral directly buried where soil conditions are suitable for the bare copper.

Fig. 230-18. 230.30 and 230.41 permit bare neutrals for service conductors. (Secs. 230.30 and 230.41.)

For underground service conductors, an individual grounded conductor (such as a grounded neutral) of *aluminum* or *copper-clad aluminum* without insulation or covering may *not* be used in a raceway underground. A bare *copper* neutral may be used—in a raceway, in a cable assembly, or even directly buried in soil where local experience establishes that soil conditions do not attack copper.

The wording of part **(4)** of the exception permits an aluminum grounded conductor of an underground service lateral to be without individual insulation or covering "when part of a cable assembly identified for underground use" where the cable is directly buried or run in a raceway. Of course, a lateral made up of individual insulated phase legs and an *insulated* neutral is acceptable in underground conduit or raceway (Fig. 230-19).

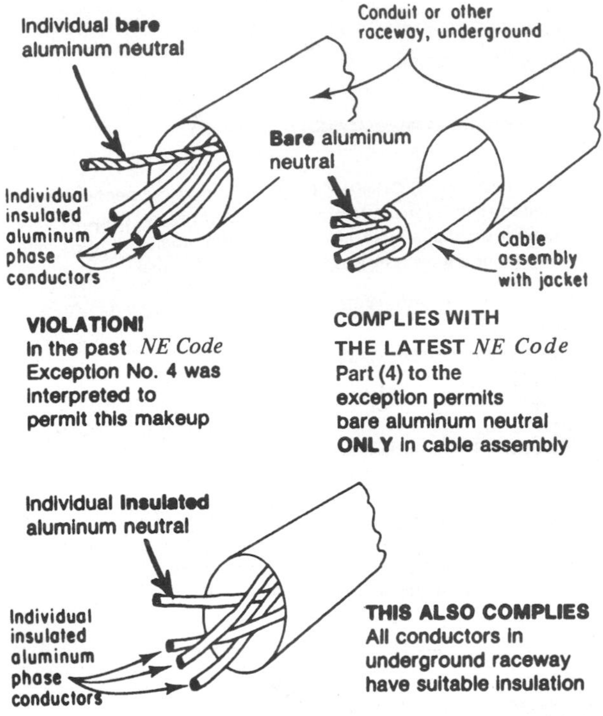
Individual bare aluminum neutral
Conduit or other raceway, underground
Bare aluminum neutral
Individual insulated aluminum phase conductors
Cable assembly with jacket
VIOLATION!
In the past NE Code Exception No. 4 was interpreted to permit this makeup
COMPLIES WITH THE LATEST NE Code
Part (4) to the exception permits bare aluminum neutral ONLY in cable assembly
Individual insulated aluminum neutral
Individual insulated aluminum phase conductors
THIS ALSO COMPLIES
All conductors in underground raceway have suitable insulation

Fig. 230-19. Underground bare aluminum grounded leg must always be in a cable assembly. (Sec. 230.30.)

Part **(B)**, which is new as of the 2014 NEC, presents in list form the acceptable wiring methods for underground service conductors. Item (9) groups MC and MV cables together with a requirement for suitability for direct burial. This will raise a question as to whether this rating is still required in the event the wiring is pulled into a buried raceway also appearing in the list. For example, it is very common to pull MV cable into PVC raceways either directly buried or run with concrete encasement and it is difficult to understand why the direct burial rating requirement should apply. It would seem that in such cases the method should simply be interpreted as that of the enclosing raceway.

230.32. Protection Against Damage. Underground service conductors—whether directly buried cables, conductors in metal conduit, or conductors in nonmetallic conduit—must comply with 300.5 for protection against physical damage. But WATCH OUT! Where conductors are buried at depths of 450 mm (18 in.) or more below grade, compliance with a special rule in Sec. 300.5 for service conductors is mandatory. As called for by Sec. 300.5(D)(3), the local inspector will always require that a warning ribbon be buried in the trench not less than a certain distance (i.e., 300 mm [12 in.]) *above* the buried service lateral or buried service entrance conductors (Fig. 230-20).

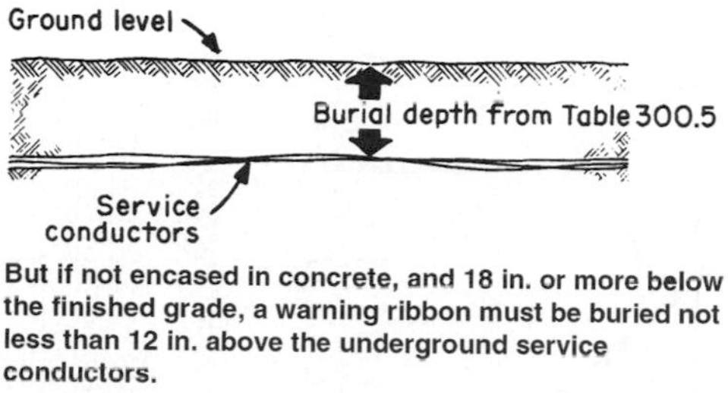
Ground level
Burial depth from Table 300.5
Service conductors
But if not encased in concrete, and 18 in. or more below the finished grade, a warning ribbon must be buried not less than 12 in. above the underground service conductors.

Fig. 230-20. Protecting underground service conductors. (Sec. 230.32.)

230.33. Spliced Conductors. Service conductors in the form of underground service laterals and all service entrance conductors are permitted to be spliced as long as the splicing method complies with the usual rules in the NEC for general wiring of comparable size and location. The NEC does not expressly cover splices in overhead service runs, but given the other rules as long as the splice meets industry standards for strain tolerance and workmanship, it would normally be permitted subject to the judgment of the authority having jurisdiction.

230.40. Number of Service-Entrance Conductor Sets. As a logical follow-up to the basic rule of 230.2, which requires that a single building or structure must be supplied "by only one service," this rule calls for only one set of service-entrance (SE) conductors to be supplied by each service drop or lateral, or by each set of overhead or underground service conductors as the case may be, that is permitted for a building. Exception No. 1 covers a multiple-occupancy building (a two-family or multifamily building, a multitenant office building, or a store building, etc.). In such cases, a set of SE conductors for each occupancy or for groups of occupancies is permitted to be tapped from a single drop or lateral (Fig. 230-21).

When a multiple-occupancy building has a separate set of SE conductors run to each occupancy, in order to comply with 230.70(A), the conductors should either be run on the outside of the building to each occupancy or, if run inside the building, be encased in 2 in. (50.8 mm) of concrete or masonry in accordance with 230.6. In either case, the service equipment should be located "nearest to the entrance of the conductors inside the building," and each occupant would have to have "access to his disconnecting means."

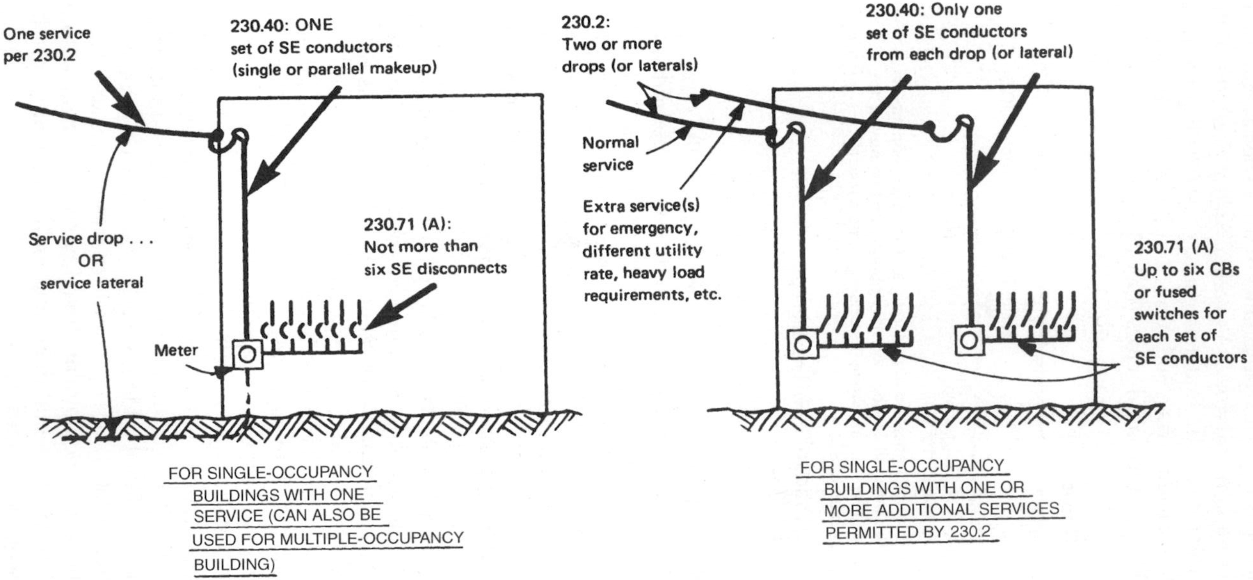
One service
per 230.2
230.40: ONE
set of SE conductors
(single or parallel makeup)
Service drop . . .
OR
service lateral
230.71 (A):
Not more than
six SE disconnects
Meter
FOR SINGLE-OCCUPANCY
BUILDINGS WITH ONE
SERVICE (CAN ALSO BE
USED FOR MULTIPLE-OCCUPANCY
BUILDING)
230.2:
Two or more
drops (or laterals)
Normal
service
Extra service(s)
for emergency,
different utility
rate, heavy load
requirements, etc.
230.40: Only one
set of SE conductors
from each drop (or lateral)
230.71 (A)
Up to six CBs
or fused
switches for
each set of
SE conductors
FOR SINGLE-OCCUPANCY
BUILDINGS WITH ONE OR
MORE ADDITIONAL SERVICES
PERMITTED BY 230.2

Fig. 230-21. Service layouts must simultaneously satisfy 230.2, 230.40, 230.71, and all other **NEC** rules that are applicable. (Sec. 230.40.)

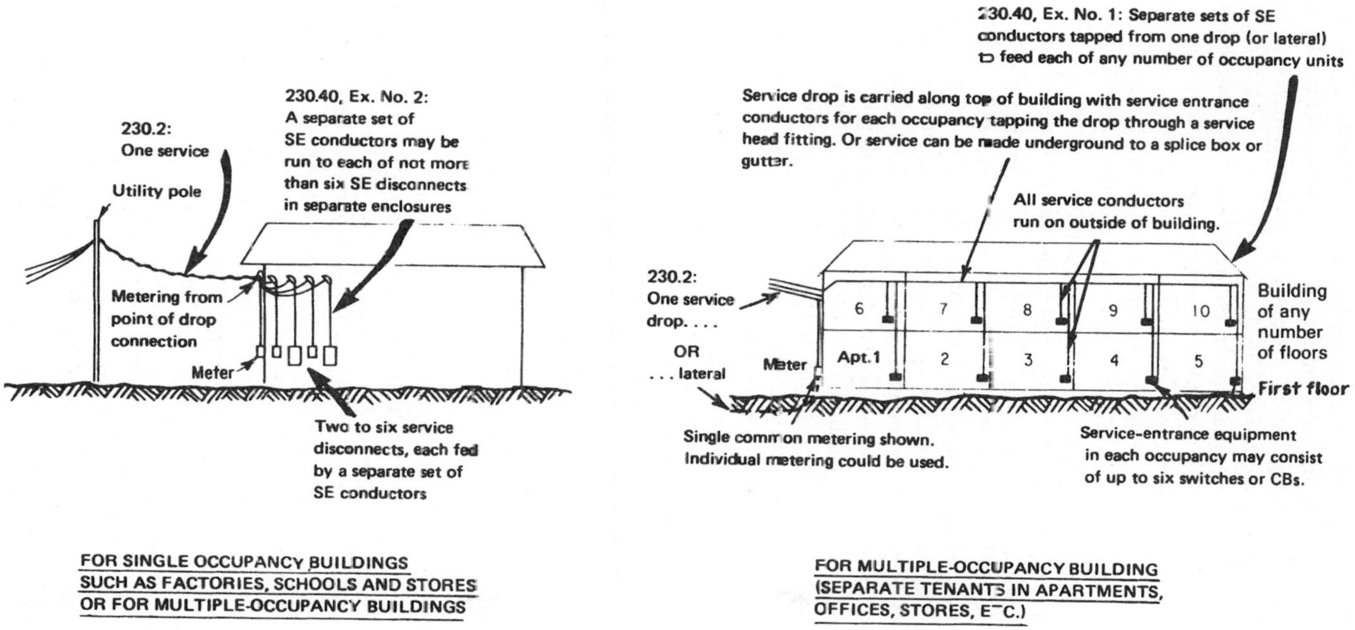
230.2:
One service
Utility pole
230.40, Ex. No. 2:
A separate set of
SE conductors may be
run to each of not more
than six SE disconnects
in separate enclosures
Metering from
point of drop
connection
Meter
Two to six service
disconnects, each fed
by a separate set of
SE conductors
FOR SINGLE OCCUPANCY BUILDINGS
SUCH AS FACTORIES, SCHOOLS AND STORES
OR FOR MULTIPLE-OCCUPANCY BUILDINGS
230.40, Ex. No. 1: Separate sets of SE
conductors tapped from one drop (or lateral)
to feed each of any number of occupancy units
Service drop is carried along top of building with service entrance
conductors for each occupancy tapping the drop through a service
head fitting. Or service can be made underground to a splice box or
gutter.
All service conductors
run on outside of building.
230.2:
One service
drop. . . .
OR
. . . lateral
Meter
6
7
8
9
10
Apt.1
2
3
4
5
Building
of any
number
of floors
First floor
Single common metering shown.
Individual metering could be used.
Service-entrance equipment
in each occupancy may consist
of up to six switches or CBs.
FOR MULTIPLE-OCCUPANCY BUILDING
(SEPARATE TENANTS IN APARTMENTS,
OFFICES, STORES, ETC.)

Fig. 230-21. *(Continued)*

Any desired number of sets of service-entrance conductors may be tapped from the service drop or lateral, or two or more subsets of service-entrance conductors may be tapped from a single set of main service conductors, as shown for the multiple-occupancy building in Fig. 230-21.

As written, there are no limitations on this permission comparable to the parallel allowance in 230.2(B)(1) that allows multiple services to supply individual occupancies where there is no common location available for a conventional service. That allowance only operates by special permission. Part of the special permission process can and usually does allow a review of reciprocal labeling. In the case of 230.40 Exception No. 1, since there is only one service, 230.2(E) does not apply. For example, suppose a multitenant building has seven occupancies. This allowance can result in a group of six disconnects (that being the limit in any one location) in the vicinity of the service drop or lateral, and then a seventh set of service entrance conductors extended around the building (or through concrete) to the seventh occupancy. Since 230.2(E) does not apply, there was absolutely no requirement to post a sign or directory at the principal service location advising emergency personnel that opening the six disconnects at that location does not, in fact, disconnect the entire building. Given the cost of providing a master disconnect for a group of services, this was far from a purely academic concern. Some jurisdictions placed limits on this allowance for that reason.

The 2011 NEC made a major step forward in addressing this problem through extensive additional text included at the end of this exception. First, if the number of disconnects for any classification of service does not exceed six, then the reciprocal labeling rules of 230.2(E) are incorporated by reference. Second, if the number of disconnects for any classification of supply exceeds six, then full reciprocal labeling is deemed too cumbersome and confusing, and a complete description of disconnect locations, using graphics, text, or both must be created on one or more plaques. This information must be mounted in an approved, readily accessible location on the building and as near as practicable to the point of arrival of the supply conductors. This will put emergency service personnel on notice as to how to disconnect the entire location when needed.

Exception No. 2 permits two to six disconnecting means to be supplied from a single set of service conductors where each disconnect supplies a separate load (Fig. 230-22). Exception No. 2 recognizes the use of, say, six 400-A sets of service-entrance conductors to a single-occupancy or multiple-occupancy building in lieu of a single main 2500-A service. It recognizes the use of up to six subdivided loads extending from a single set of conductors in a *single-occupancy* as well as multiple-occupancy building. Where single metering is required, doughnut-type CTs could be installed at the service drop.

The real importance of this rule is to eliminate the need for "paralleling" conductors of large-capacity services, as widely required by inspection authorities to satisfy previous editions of the NEC (Fig. 230-22). This same approach could be used in subdividing services into smaller load blocks to avoid the use of the equipment ground-fault circuit protection required by 230.95.

This rule can also facilitate expansion of an existing service. Where less than six sets of service-entrance conductors were used initially, one or more

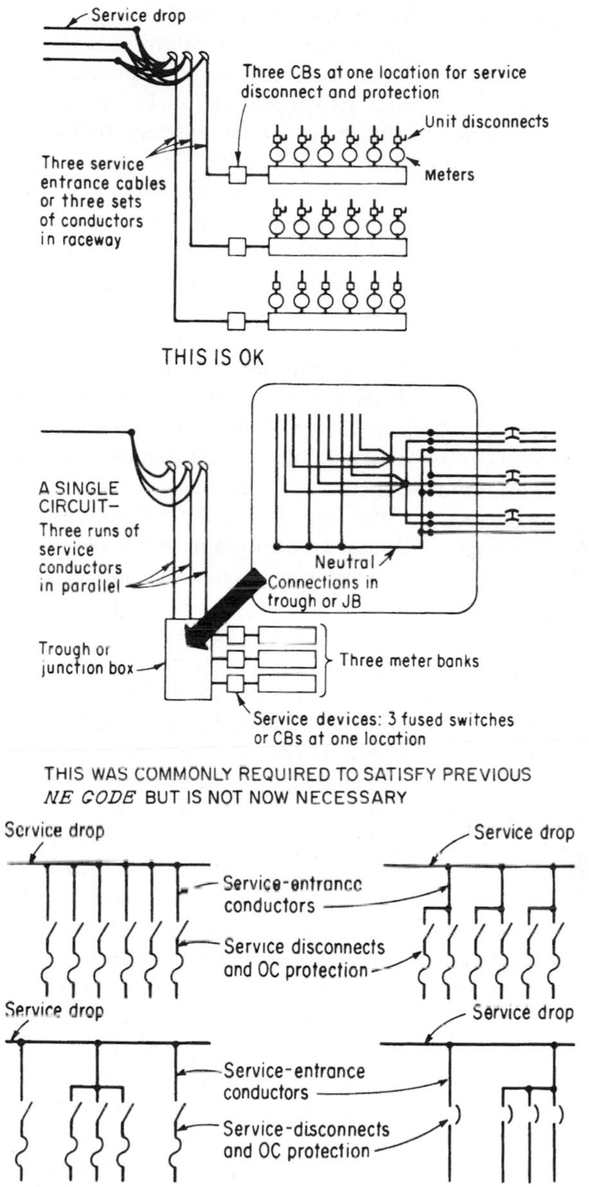
Service drop
Three CBs at one location for service disconnect and protection
Unit disconnects
Three service entrance cables or three sets of conductors in raceway
Meters
THIS IS OK
A SINGLE CIRCUIT–
Three runs of service conductors in parallel
Neutral
Connections in trough or JB
Trough or junction box
Three meter banks
Service devices: 3 fused switches or CBs at one location
THIS WAS COMMONLY REQUIRED TO SATISFY PREVIOUS *NE CODE* BUT IS NOT NOW NECESSARY
Service drop
Service drop
Service-entrance conductors
Service disconnects and OC protection
Service drop
Service drop
Service-entrance conductors
Service-disconnects and OC protection

From two to six separate sets of service-entrance conductors may be supplied by a single service drop for either single- or multiple-occupancy buildings. Disconnects can be of same or different ratings, and each set of service-entrance conductors can be installed using any approved wiring method.

Fig. 230-22. Tapping sets of service-entrance conductors from one drop (or lateral). (Sec. 230.40.)

additional sets can be installed subsequently without completely replacing the original service. Of course, metering considerations will affect the layout.

But, the two to six disconnects (CBs or fused switches) must be installed close together at one location and not spread out in a building. Since under this exception the disconnects are still grouped, the objection formerly raised under Exception No. 1 does not apply.

Exception No. 3 recognizes tapping two sets of service conductors from a single drop or lateral at a dwelling unit to supply the dwelling *and* one other building. Exception No. 4 recognizes an additional set of service entrance conductors to supply the "common area" panel required by Sec. 210.25. And Exception No. 5 specifically recognizes supplying "other" equipment as indicated in 230.82(5) and 230.82(6).

230.41. Insulation of Service-Entrance Conductors. Except for use of a bare neutral, as permitted, all service-entrance conductors must be insulated and may not simply be "covered"—as discussed under 230.22. The wording used in part **(3)** of the exception in 230.41 is slightly different from that described previously for 230.30. In this section, the reference is to "service-entrance conductors" instead of "service lateral conductors." But here, a *bare individual* aluminum or copper-clad aluminum grounded conductor (grounded neutral or grounded phase leg) may be used in a raceway or a cable assembly or for direct burial where "identified" for direct burial.

Aluminum SE cable with a bare neutral may be used aboveground as an SE conductor. *But,* an aluminum SE cable with a bare neutral may be used underground *only* if it is "identified" for underground use in a raceway or directly buried. Conventional-style SE-U aluminum SE cable with a bare neutral is not "identified" for use underground but may be used, as the first sentence of 230.40 describes, for "service-entrance conductors entering or on the exterior of buildings or other structures." In "SE-U," the "U" stands for "unarmored" not "underground."

230.42. Minimum Size and Rating. Sizing of service-entrance conductors involves the same type of step-by-step procedure as set forth for sizing feeders covered in Art. 220. A set of service-entrance conductors is sized just as if it were a feeder. In general, the service-entrance conductors must have a minimum ampacity—current-carrying capacity—selected in accordance with the ampacity tables and rules of 310.15, as well as the rules for "continuous loading" following part **(A),** here, that is sufficient to handle the total lighting and power load as calculated in accordance with Art. 220. Where the Code gives demand factors to use or allows the use of acceptable demand factors based on sound engineering determination of less than 100-percent-demand requirement, the lighting and power loads may be modified.

According to the last sentence of 230.42(A), the "maximum allowable current" of busways used as service-entrance conductors must be taken to be the ampere value for which the busway has been listed or labeled. This is an "exception" to the basic rule that requires the ampacity of service-entrance conductors to be "determined from 310.15"—which does not give ampacities of busways. Parts **(A)(1)** and **(A)(2)** repeat the rule of 215.2(A). This now includes the exception for sizing continuously loaded feeder neutrals at 100 percent and not 125 percent if they arrive at a busbar and not an overcurrent device. (See 215.2.)

From the analysis and calculations given in the feeder-circuit section, a total power and lighting load can be developed to use in sizing service-entrance conductors. Of course, where separate power and lighting services are used, the sizing procedure should be divided into two separate procedures.

When a total load has been established for the service-entrance conductors, the required current-carrying capacity is easily determined by dividing the total load in kilovolt-amperes (or kilowatts with proper correction for power factor or the load) by the voltage of the service.

From the required current rating of conductors, the required size of conductors is determined. Sizing of the service neutral is the same as for feeders, and the 2014 **NEC** added 230.42(A)(2) to correlate with 215.2(A)(1) Exception No. 2 and thereby avoid upsizing for continuous loads. Although suitably insulated conductors must be used for the phase conductors of service-entrance feeders—except as permitted for overhead conductors as described in 230.22—the **NEC** does permit use of bare grounded conductors (such as neutrals) under the conditions covered in 230.30 and 230.41.

An extremely important element of service design is that of fault consideration. Service busway and other service conductor arrangements must be sized and designed to ensure safe application with the service disconnect and protection. That is, service conductors must be capable of withstanding the let-through thermal and magnetic stresses on a fault.

After calculating the required circuits for all the loads in the electrical system, the next step is to determine the minimum required size of service-entrance conductors to supply the entire connected load. The **NEC** procedure for sizing SE conductors is the same as for sizing feeder conductors for the entire load—as set forth in 215.2(A).

Basically, the service "feeder" capacity must be not less than the sum of the loads on the branch circuits for the different applications.

The *general lighting load* is subject to demand factors from Table 220.12, which takes into account the fact that simultaneous operation of all branch-circuit loads, or even a large part of them, is highly unlikely. Thus, service or feeder capacity does not have to equal the connected load. The other provisions of Art. 220 are then factored in.

Reference to 230.79 in part **(B)** of 230.42 makes a 100-A service conductor ampacity a mandatory minimum if the system supplied is a one-family dwelling. And for all other occupancies where more than two 2-wire circuits are supplied, the minimum rating of the service conductors may not be less than 60 A. This reference is not intended to require the service conductors to *always* have an ampacity equal to the rating of the service disconnect(s). That is, for those installations described in 230.79, the service conductors must have the minimum ampacity required by 230.79. But for all other installations, the ampacity as established in accordance with 310.15 must be not less than the calculated load as determined in accordance with Art. 220, and the service conductors must be sized as required by 230.42(A), for continuous loading.

Another point of confusion is the wording here, "less than the rating of the service disconnecting means specified...." This does not mean that if you install a 400-A fused service switch for a 250-A load and with 250-A fuses, the service needs to be cabled for 400 A. It simply means that in the specified applications, the conductors must meet the minimums set in 230.79. The following table correlates the requirements for service lateral, service drop, and service entrance conductors in a single location.

Conductors	*Minimum Allowable Copper Size*
Service entrance for all installations, except ones specified below	6 AWG (to supply a 60 A minimum disconnect size using 75°C terminations)
Service entrance supplying a single branch circuit	14 AWG (to supply a 15 A minimum disconnect size)
Service entrance supplying a two-circuit installation	10 AWG (to supply a 30 A minimum disconnect size)
Service entrance supplying a single-family house	4 AWG [to supply a 100 A minimum disconnect size, using 310.15(B)(7)]
Overhead or underground service	12 AWG (overhead: hard-drawn) for a single circuit installation
Overhead or underground service	8 AWG, for other installations
Service of over 1000 V	6 AWG, except that in multiconductor cables 8 AWG may be used

230.43. Wiring Methods for 1000 Volts, Nominal, or Less. The list of acceptable wiring methods for running service-entrance conductors does include flexible metal conduit (Greenfield) and liquidtight flexible metal conduit, but limits use of such raceways to a maximum length of 6 ft (1.8 m), and an equipment bonding conductor must be run with it. Although such raceways were prohibited at one time, effectively bonded flexible metal conduit and liquidtight flexible metal conduit in a length not over 6 ft (1.8 m) may be used as a raceway for service-entrance conductors (Fig. 230-23). A length of flex or liquidtight flex not longer than 6 ft (1.8 m)—in total—may be used as a service raceway, provided an equipment bonding conductor sized from Table 250.66 (and with a cross-sectional area at least 12½ percent of the csa of the largest service phase conductor for conductors larger than 1100 MCM copper or 1750 MCM aluminum) is used. This rule recognizes that the flexibility of such raceway is often needed or desirable in routing service-entrance conductors around obstructions in the path of connections between metering equipment and service-entrance switchboards, panelboards, or similar enclosures. The required equipment grounding conductor may be installed either inside or outside the flex, using acceptable fittings and termination techniques for the grounding conductor.

It should be noted that *liquidtight* flexible metal conduit is recognized as an acceptable service raceway, provided the bonding requirements given in 250.102 are satisfied. And, liquidtight flexible *nonmetallic* conduit—of any length—may be used as a service raceway containing service-entrance conductors.

230.44. Cable Trays. This section recognizes the use of a cable tray for the support of service-entrance conductors, provided the cable tray contains only service-entrance conductors. Cable trays are now limited as to the service wiring they are allowed to contain, namely Types SE, MC, MI, and IGS cables, along with single conductors not smaller than 1/0 AWG provided they are marked for cable tray use (a "CT" rating). They must also be field labeled with the

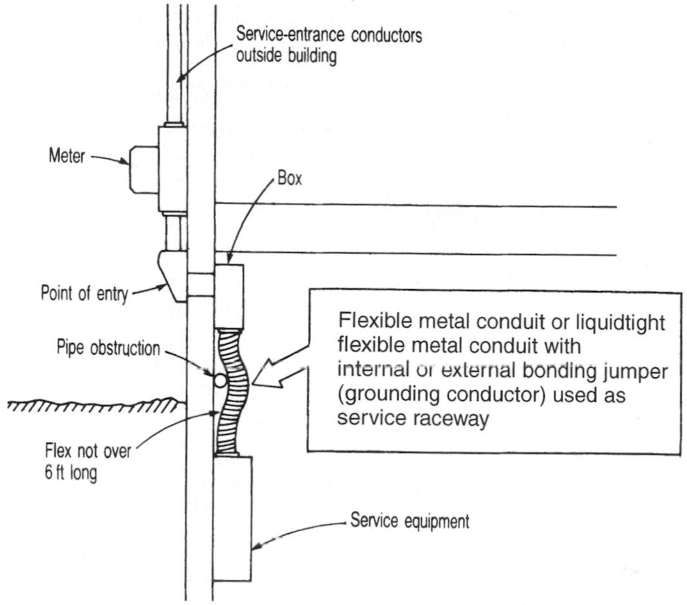
Service-entrance conductors outside building
Meter
Box
Point of entry
Pipe obstruction
Flex not over 6 ft long
Flexible metal conduit or liquidtight flexible metal conduit with internal or external bonding jumper (grounding conductor) used as service raceway
Service equipment

Fig. 230-23. These two flexible conduits may be used for service raceway. (Sec. 230.43.)

designation "Service-Entrance Conductors" in a manner that is visible after the tray is installed and at intervals and in such positions that the service conductors can be easily tracked over the full length of the tray installation, but no interval between labels can exceed 10 ft. Note that the relief granted for medium-voltage warning labeling for cable trays in 392.18(H) Exception for qualified industrial occupancies does not govern the labels required here. Formerly, this marking rule only applied when the tray was divided into two sections, one with service conductors and one for other purposes. As in prior Codes, if the tray is to be shared in this way, the requirement is for a solid barrier compatible with the tray to be installed. Note that 392.10(B) will additionally limit the use of single conductors in cable tray to industrial occupancies with qualified maintenance and supervision, using certain types of tray and certain rung intervals if ladder tray is used for conductors smaller than 250 kcmil.

230.46. Spliced Conductors. The wording in this section used to prohibit splicing of SE conductors, and that prohibition was followed by a number of exceptions. However, that long-standing rule was eliminated in the 1999 NEC. The Code now makes specific references to recognized methods, that is, "clamped or bolted connections." Now splicing of SE conductors may be accomplished using the methods described by the applicable rules in 110.14, 300.13, and 300.15, which recognize use of the splicing methods previously described. (See Fig. 230-24.)

230.50. Protection Against Physical Damage. The wording in part **(A)** of this section no longer contains the "laundry list" of instances where physical

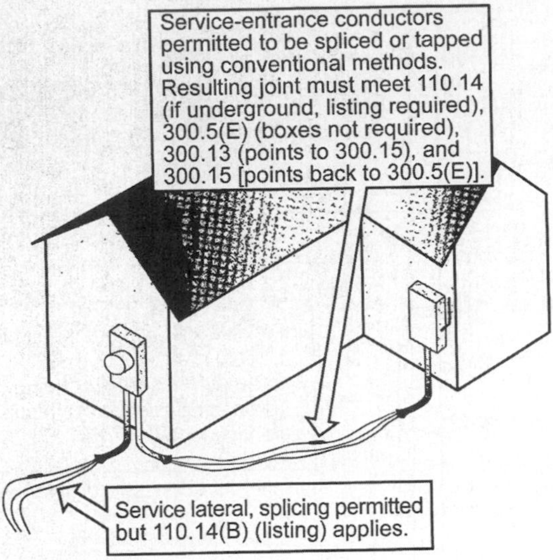
Service-entrance conductors permitted to be spliced or tapped using conventional methods. Resulting joint must meet 110.14 (if underground, listing required), 300.5(E) (boxes not required), 300.13 (points to 300.15), and 300.15 [points back to 300.5(E)].
Service lateral, splicing permitted but 110.14(B) (listing) applies.

Fig. 230-24. Permitted splices in service-entrance conductors. The comment on the "service lateral" now applies to underground service entrance conductors only. Service laterals are now utility wiring only on the line side of the service point, and the **NEC** has no jurisdiction over how a utility may choose to repair one of its underground conductors. (Sec. 230.46.)

protection would be required. Instead, it is now strictly up to the local inspector to determine where such protection must be provided. It seems reasonable to assume that those cases previously identified would be covered.

In parts **(A)(1)** through **(6)**, the Code indicates specific methods that may be utilized to achieve the desired and required physical protection. In keeping with the broad discretion given by the wording used in this rule, part **(A)(6)** recognizes "other approved means." Given the definition of the term "approved" in Art. 100, that wording essentially means "whatever the inspector will accept." Since it is up to the inspector to decide when such protection is needed, it seems reasonable to grant the inspector the latitude to establish another method to ensure the required physical protection needed for service cables (Fig. 230-25).

Part **(B)** exception allows use of type MI and MC cables for service-entrance or service lateral applications, without need for mounting at least 10 ft (3.0 m) above grade—provided they are not exposed to damage or are protected.

230.51. Mounting Supports. Service-entrance cable must be clamped to building surface by straps at intervals not over 30 in. (750 mm). And the cable must still be clamped within 12 in. (300 mm) of the service weatherhead and within 12 in. (300 mm) of cable connection to a raceway or enclosure.

Where the cable assembly is not listed for attachment to building surfaces, in part **(B)** the Code calls for such cables to be mounted on insulators that provide at least 2 in. of clearance from the building surface, and the insulators must be mounted no more than 15 ft apart. Similar requirements are given in part **(C)** for

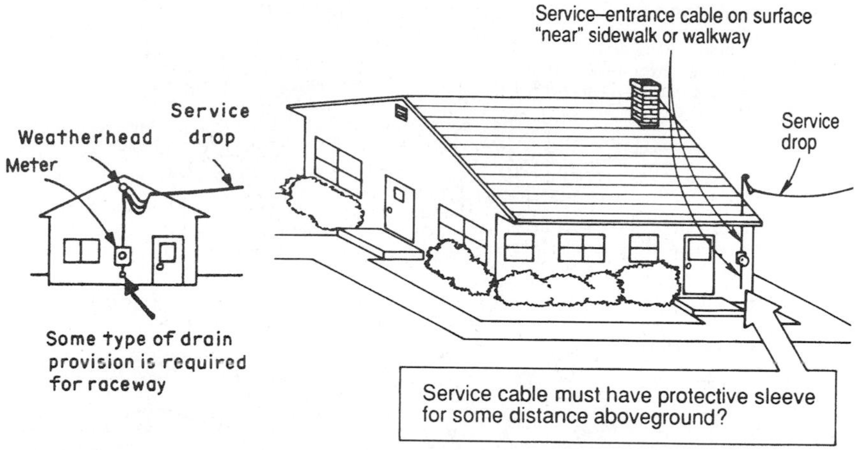
Service drop
Weatherhead
Meter
Some type of drain provision is required for raceway
Service–entrance cable on surface "near" sidewalk or walkway
Service drop
Service cable must have protective sleeve for some distance aboveground?

Fig. 230-25. Outdoor service raceway must be raintight *and* drained and SE cable must be protected if subject to physical damage. (Secs. 230.50 and 230.53.)

service conductors run as open individual conductors. Table 230.51(C) spells out the maximum spacing for the insulators and the minimum clearance from the building surface.

230.53. Raceways to Drain. Service-entrance conductors in conduit must be made raintight, using raintight raceway fittings, and must be equipped with a drain hole in the service ell at the bottom of the run or must be otherwise provided with a means of draining off condensation (Fig. 230-25). Use of a "pitch" equivalent to 1/8 in. per foot, or 1 in. per 8 ft (10.4 mm per meter) will serve to satisfy the rule given by the last sentence for arranging conduit to drain where embedded in masonry. The same rule applies in 225.22. Refer to that discussion for more information about potential conflicts with 110.12(A).

230.54. Overhead Service Locations. When conduit or tubing is used for a service, the raceway must be provided with a service head (or weatherhead). Figure 230-26 shows details of a service-head installation. As covered in the exception to part **(B)**, service cable may be installed without a service head, provided it is bent to form a "gooseneck"; then tape the end with a "self-sealing water-resistant thermoplastic": that is, where no service head is used at the upper end of a service cable, the cable should be bent over so that the individual conductors leaving the cable will extend in a downward direction, and the end of the cable should be carefully taped and painted or sealed with water-resistant tape to exclude moisture.

Part **(C)** of this section requires that service heads be located above the service-drop attachment. Although this arrangement alone will not always prevent water from entering service raceways and equipment, such an arrangement will solve most of the water-entrance problems. An exception to this rule permits a service head to be located not more than 24 in. (600 mm) from the service-drop termination where it is found that it is impractical for the service head to be located above the service-drop termination. In such cases, a *mechanical connector* is advisable at

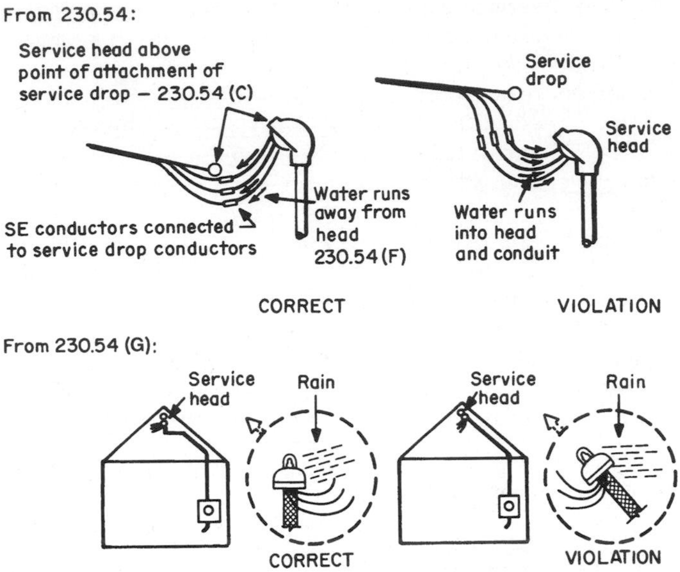
From 230.54:
Service head above point of attachment of service drop – 230.54 (C)
Service drop
Service head
Water runs away from head 230.54 (F)
Water runs into head and conduit
SE conductors connected to service drop conductors
CORRECT
VIOLATION
From 230.54 (G):
Service head
Rain
Service head
Rain
CORRECT
VIOLATION

Fig. 230-26. Location of service head minimizes entrance of rain. (Sec. 230.54.) The violation at the lower right can often be cured by twisting the service head so it points straight down, thereby avoiding the arrangement to the left that many consider unsightly, particularly if run in a large raceway for which the required bends will occupy considerable area on the side of the building.

the lowest point in the drip loop to prevent siphoning. This exception will permit the Code-enforcing authority to handle hardship cases that may occur.

As covered by the wording in part **(D)**, service cables shall be "held securely in place." And each phase and neutral must be routed through an individual bushed opening in the service head to satisfy the basic rule in 230.54(E). But, the exception following this rule permits deviation from the one-phase one-bushed opening where the service conductors are in a jacketed multiconductor cable, as would be required in the case of a gooseneck.

The intent of parts **(F)** and **(G)** is to require use of connections or conductor arrangements, both at the pole and at the service, so that water will not enter connections and siphon under head pressure into service raceways or equipment.

230.56. Service Conductor with the Higher Voltage to Ground. This Code rule presents the requirement that the "high" leg (the 208-V-to-ground leg) of a 240/120-V 3-phase, 4-wire delta system must be identified by marking to distinguish it from the other hot legs, which are only 120 V to ground. One method permitted is color-coding the so-called high leg orange. The rule recognizes "other means," but any such "identification" must be provided "at each termination or junction point." Clearly, the use of an overall orange-colored insulation will most easily satisfy this rule.

230.66. Marking. Service equipment must be marked as being suitable for the use. Since individual meter socket enclosures do not function as the main cutoff and control of the supply, they are not service equipment as defined in Art. 100 and therefore not subject to this rule. All service equipment must, as of the 2011 **NEC**, be listed. Although as noted meter sockets are not service equipment, the 2017 **NEC** instituted a listing requirement for service meter sockets, unless they were supplied by and remained under the exclusive control of the serving utility.

230.70. General. Parts **(A)(1), (2),** and **(3)** cover the place of installation of a service disconnect. The disconnecting means required for every set of service-entrance conductors must be located at a readily accessible point outside the building; or, where installed inside, nearest to the point at which the service conductors enter the building (Fig. 230-27). The service disconnect switch (or CB) is generally placed on the inside of the building as near as possible to the point at which the conductors come in.

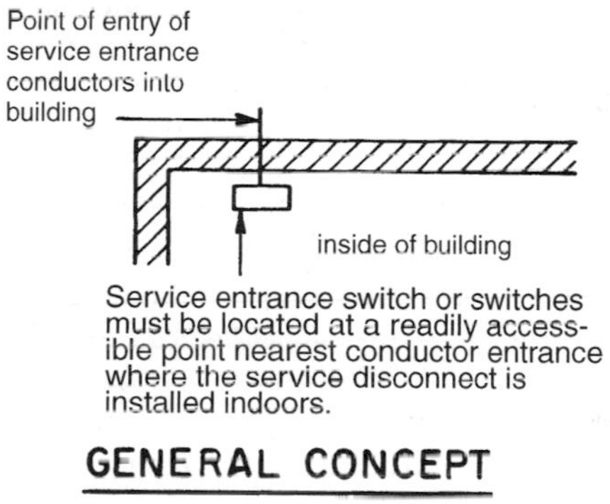
Point of entry of service entrance conductors into building
inside of building
Service entrance switch or switches must be located at a readily accessible point nearest conductor entrance where the service disconnect is installed indoors.
GENERAL CONCEPT

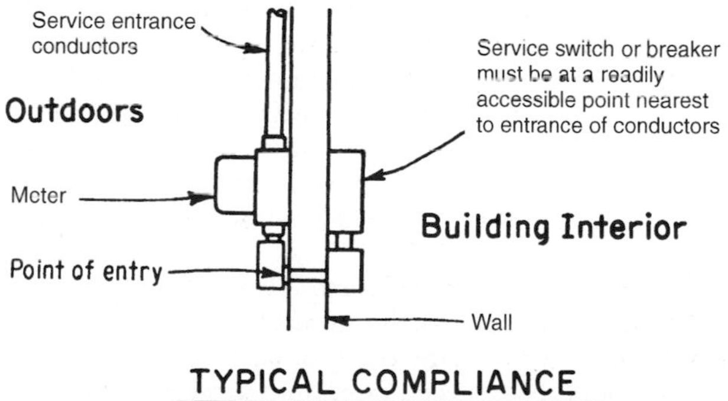
Service entrance conductors
Outdoors
Service switch or breaker must be at a readily accessible point nearest to entrance of conductors
Meter
Building Interior
Point of entry
Wall
TYPICAL COMPLIANCE

Fig. 230-27. Service disconnect must open current for any conductors within building. (Sec. 230.70.)

And part **(B)** requires lettering or a sign on the disconnect(s) to identify it (them) as "Service Disconnect." There are no exceptions to this marking requirement. Most panels with an integral main breaker come with an embossed or stamped marking "MAIN" next to that breaker, along with "SERVICE DISCONNECT" labels to be applied in the event the panel will be used as service equipment. These labels must be applied as appropriate, and nowhere else. A main breaker and a service disconnect are two different things. A building disconnect and a service disconnect are two different things. In many cases, the function is obvious, but not all. This author recalls having been called by fire officials to a convenience store with a line up of six identically sized panels, all with "Main" breakers and no other designation. Two of those breakers were service disconnects and four were not, as definitively determined only after spending 20 min actually removing the dead fronts and tracing conductors. Fortunately there was no emergency at the time; the department was engaged in a valuable exercise in advance preparation. While part **(C)** calls for the equipment to be "suitable" for use as service equipment, in all practicality this means that such disconnects must be listed and marked as being suitable for use as service equipment.

Although the Code does not set any maximum distance from the point of conductor entry to the service disconnect, various inspection agencies set maximum limits on this distance. For instance, service cable may not run within the building more than 18 in. (450 mm) from its point of entry to the point at which it enters the disconnect. Or, service conductors in conduit must enter the disconnect within 10 ft (3.0 m) of the point of entry. Or, as one agency requires, the disconnect must be within 10 ft (3.0 m) of the point of entry, but overcurrent protection must be provided for the conductors right at the point at which they emerge from the wall into the building. The concern is to minimize the very real and proven potential hazard of having unprotected service conductors within the building. Faults in such unprotected service conductors must burn themselves clear and such application has caused fires and fatalities. Check with your local inspection agency to find out what it intends to enforce. Such action should serve to prevent any surprises on the job.

Often shifting the point of entry, with somewhat longer conductors run outdoors, will solve a problem. In extreme cases a combination meter and overcurrent protection/disconnect can be installed, which then allows the conductors to run anywhere in the building. Many jobs have issues with obstructions such as oil tanks that need to be discussed with the inspector. Every cycle the Code-making panel receives, and rejects, proposals to set a specific allowable distance on the indoor length of service conductors. The panel intends that this remain a topic of negotiation and discussion between installers and inspectors with respect to the specific problems that arise in the field.

230.71. Maximum Number of Disconnects. Service-entrance conductors must be equipped with a readily accessible means of disconnecting the conductors from their source of supply. As stated in part **(A)**, the disconnect means for each service and each set of SE conductors permitted by Sec. 230.2 and Sec. 230.40 Exception Nos. 1 and 3, respectively, may consist of not more than six switches or six CBs, in a common enclosure or grouped individual enclosures, located either "within sight of" and outside the building wall, or inside, as close as possible to the point at which the conductors enter the building. Figure 230-28 shows the basic application of that rule to a single set of SE conductors.

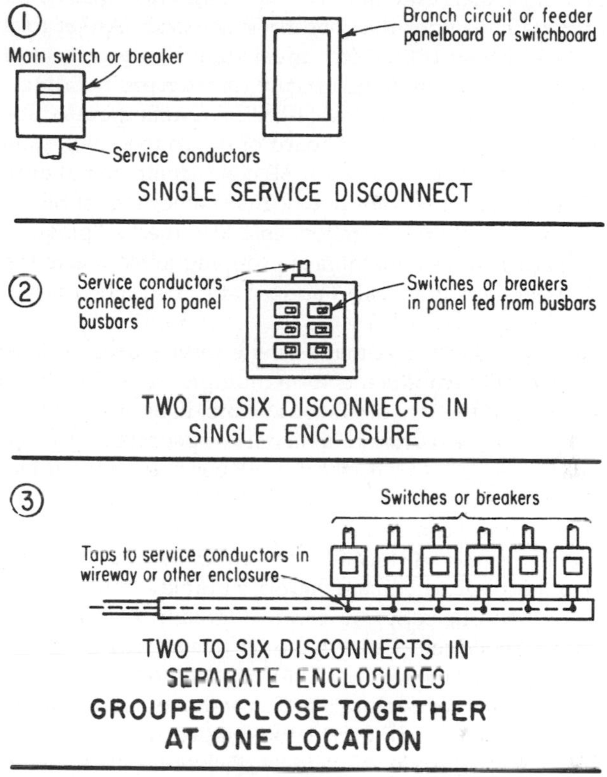
1
Main switch or breaker
Branch circuit or feeder panelboard or switchboard
Service conductors
SINGLE SERVICE DISCONNECT
2
Service conductors connected to panel busbars
Switches or breakers in panel fed from busbars
TWO TO SIX DISCONNECTS IN SINGLE ENCLOSURE
3
Switches or breakers
Taps to service conductors in wireway or other enclosure
TWO TO SIX DISCONNECTS IN SEPARATE ENCLOSURES
GROUPED CLOSE TOGETHER AT ONE LOCATION

Fig. 230-28. The three basic ways to provide service disconnect means. (Sec. 230.71.)

The last sentence in part **(A)** identifies a number of specific applications where the disconnecting means is not to be counted as "service disconnecting means" and applied against the maximum number of six disconnects permitted in the first two sentences. That is, when control power for a ground-fault protection system is tapped from the line side of the service disconnect means, the disconnect for the control power circuit is not counted as one of the six permitted disconnects for a service. A ground-fault-protected switch or CB supplying power to the building electrical system *counts* as one of the six permitted disconnects. But a disconnect supplying only the control-circuit power for a ground-fault protection system, installed as part of the listed equipment, does not count as one of the six service disconnects. The same idea applies to all other specifically identified equipment disconnects that are generally located at services, including surge protective devices and power monitoring equipment.

The rule of this section correlates "number of disconnects" with 230.2 and 230.40, which permit a separate set of SE conductors to be run to each occupancy (or group of occupancies) in a multiple-occupancy building, as follows:

230.2 permits more than one "service" to a building—that is, more than one service drop or lateral—under the conditions set forth. As set forth in the first sentence of 230.40 each such "service" must supply *only* one set of SE conductors in a building that is a single-occupancy (one-tenant) building, and each set of SE conductors may supply up to six SE disconnects grouped together at one location—in the same panel or switchboard or in grouped individual enclosures. If the grouped disconnects for one set of SE conductors are not at the same location as the grouped disconnects for one or more other sets of SE conductors, for those situations described and permitted in 230.2, then a "plaque or directory" must be placed at each service-disconnect grouping to tell where the other group (or groups) of disconnects are located and what loads each group of disconnects serves.

Exception No. 1 to 230.40 says that a single service drop or lateral may supply *more than one set* of SE conductors for a multiple-occupancy building. Then at the load end of each of the sets of SE conductors, in an individual occupancy or adjacent to a group of occupancy units (apartments, office, stores), up to six SE disconnects may be supplied by each set of SE conductors. And Exception No. 3 recognizes two sets of SE conductors at a dwelling unit to supply the dwelling and one other separate "structure."

The first sentence of part **(A)** to 230.71 ties directly into 230.40, Exception No. 1. It is the intent of this basic rule that, where a multiple-occupancy building is provided with more than one set of SE conductors tapped from a drop or lateral, each set of those SE conductors may have up to six switches or CBs to serve as the service disconnect means for that set of SE conductors. The rule does recognize that six disconnects for each set of SE conductors at a multiple-occupancy building with, say, 10 sets of SE conductors tapped from a drop or lateral does result in a total of 6 × 10, or 60, disconnect devices for completely isolating the building's electrical system from the utility supply. 230.72(B) also recognizes use of up to six disconnects for each of the "separate" services for fire pumps, emergency lighting, and so on, which are recognized in 230.2 as being separate services for specific purposes. And if service is provided for different classes of service, six disconnects could be provided for *each* class of service to *each* occupancy, resulting in 120 disconnects! This is why 230.40 Exception No. 1 was modified in the 2011 **NEC** to address the required labeling and signage in a coherent way.

Although the basic rule of 230.40 specifies that only one set of SE conductors may be tapped from a single drop for a building with single occupancy, Exception No. 2 to 230.40 recognizes that a separate set of SE conductors may be run from a single service drop or lateral to each of up to six service disconnects mounted in separate enclosures at one location, constituting the disconnect means for a single service to a single-occupancy building.

For any type of occupancy, a panel containing up to six switches or CBs may be used as service equipment with the enclosed six or fewer breakers comprising the service disconnecting means under a special exception (408.36 Exception No. 1). A panel used as service equipment for renovation of an existing service in an individual residential occupancy (but not for new installations) may have up to six main breakers or fused switches under 408.36 Exception No. 3. A panel meeting the old 42-circuit limitation for the former lighting and appliance branch-circuit panelboards may have up to two main breakers (or sets of fuses), per 408.36 Exception No. 2. However, a panel used without these limitations and used as service equipment for new buildings of any type must have not more than a single main device—with its rating not greater than the panel bus rating. See 408.36.

The first sentence of 230.71(A) and that of 230.72(A) note that from one to six switches (or CBs) may serve as the service disconnecting means for each class of service for a building. For example, if a *single-occupancy* building has a 3-phase service and a separate single-phase service, each such service may have up to six disconnects (Fig. 230-29). Where the two sets of service equipment are not located adjacent to each other, a plaque or directory must be installed at each service-equipment location indicating where the other service equipment is—as required by 230.2(E).

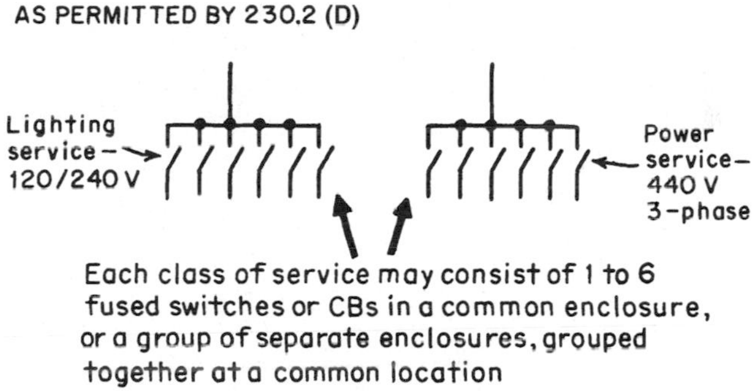
AS PERMITTED BY 230.2 (D)
Lighting service—120/240 V
Power service—440 V 3-phase
Each class of service may consist of 1 to 6 fused switches or CBs in a common enclosure, or a group of separate enclosures, grouped together at a common location

Fig. 230-29. Each separate service may have up to six disconnect devices. (Sec. 230.71.)

230.71(B) notes that single-pole switches or circuit breakers equipped with handle ties may be used in groups as single disconnects for multiwire circuits, simultaneously providing overcurrent protection for the service (Fig. 230-30). Multipole switches and circuit breakers may also be used as single disconnects. The requirements of the Code are satisfied if all the service-entrance conductors can be disconnected with no more than six operations of the hand—regardless of whether each hand motion operates a single-pole unit, a multipole unit, or a group of single-pole units with "handle ties" or a "master handle" controlled by a single hand motion. Of course, a single main device for service disconnect and overcurrent protection—such as a main CB or fused switch—gives better protection to the service conductors.

The informational note to this section refers to 408.36 Exceptions 1 and 3, which vary from the individual protection requirements that now apply to panelboards generally, and as discussed previously. The reference to 430.95 is intended to point out the limitations associated with installations where the service equipment is within a motor control center. For such installations, the rule of 430.95 mandates the use of a single main disconnect.

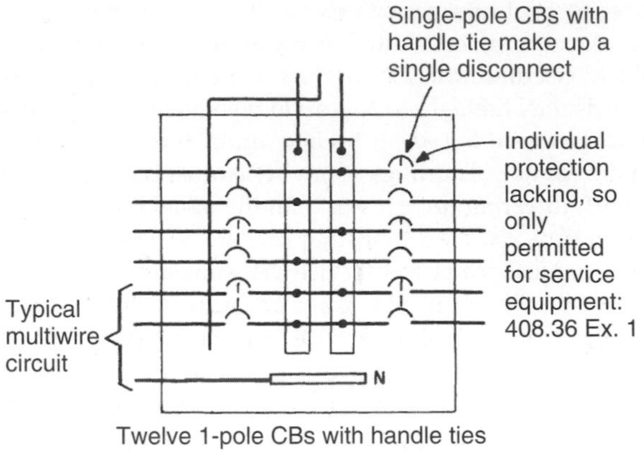
Single-pole CBs with handle tie make up a single disconnect
Individual protection lacking, so only permitted for service equipment: 408.36 Ex. 1
Typical multiwire circuit
N
Twelve 1-pole CBs with handle ties

Fig. 230-30. This arrangement constitutes six disconnects. (Sec. 230.71.)

230.72. Grouping of Disconnects. The basic rule of part **(A)** requires that for a service disconnect arrangement of more than one disconnect—such as where two to six disconnect switches or CBs are used, as permitted by 230.71(A)—all the disconnects making up the service equipment "for each service" must be grouped and not spread out at different locations. The basic idea is that anyone operating the two to six disconnects must be able to do it while standing at one location. Service conductors must be able to be readily disconnected from all loads at one place. And each of the individual disconnects must have lettering or a sign to tell what load it supplies (Fig. 230-31).

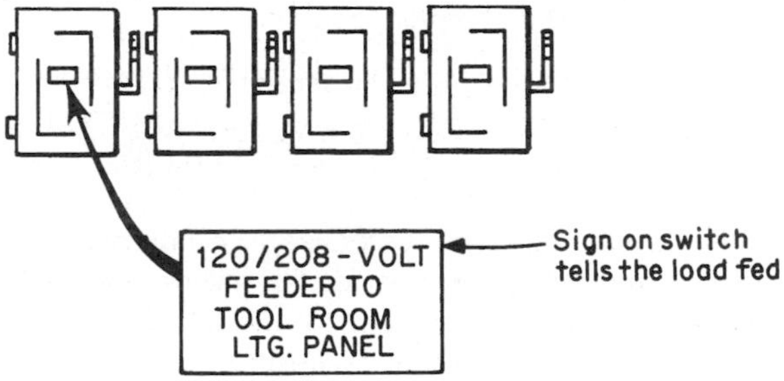
120/208-VOLT FEEDER TO TOOL ROOM LTG. PANEL
Sign on switch tells the load fed

Fig. 230-31. Two to six disconnect switches or CBs must be grouped and *identified.* (Sec. 230.72.)

This rule makes clear that the two to six service disconnects that are permitted by 230.71(A) for *each* "service" or for "*each* set of SE conductors" at a multiple-occupancy building must be grouped. But, where permitted by 230.2, the individual groups of two to six breakers or switches do not have to be together, and if they are not together, a sign at each location must tell where the other service disconnects are. (See 230.2.) Each grouping of two to six disconnects may be within a unit occupancy—such as an apartment—of the building.

The exception to part **(A)** *permits* (*Note:* It permits, it does not *require*, but read the next paragraphs) one of the two to six service disconnects to be located remote from the other disconnecting means that are grouped in accordance with the basic rule—PROVIDED THAT *the remote disconnect is used only to supply a water pump that is intended to provide fire protection.* In a residence or other building that gets its water supply from a well, a spring, or a lake, the use of a remote disconnect for the water pump will afford improved reliability of the water supply for fire suppression in the event that fire or other faults disable the normal service equipment. And it will distinguish the water-pump disconnect from the other normal service disconnects, minimizing the chance that firefighters will unknowingly open the pump circuit when they routinely open service disconnects during a fire. This exception ties into the rule of 230.72(B), which *requires* (not simply permits) remote installation of a true, Art. 695-compliant fire-pump disconnect switch that is required to be tapped ahead of the one to six switches or CBs that constitute the normal service disconnecting means [see 230.82(5)]. The exception provides remote installation of a *normal service disconnect* when it is used for the comparable purpose (water pump used for fire fighting) as an authentic fire-pump disconnecting means, which joins various standby disconnecting means that are specifically recognized as meriting remote placement from the rest of the service disconnecting means for the purposes of increasing reliability covered in 230.72(B). In both cases, remote installation of the pump disconnect isolates the critically important pump circuit from interruption or shutdown due to fire, arcing-fault burndown, or any other fault that might knock out the main (normal) service disconnects.

The 2011 **NEC** added a very important qualification to the remote location exception; when it applies, there must be a plaque at the principal service disconnect location advising as to the location of this remote disconnecting means.

A wide variety of layouts can be made to satisfy the Code *permission* for remote installation of a disconnect switch or CB service as a *normal* service disconnect (one of a maximum of six) supplying a water pump. Figure 230-32 shows three typical arrangements that would basically provide the isolated fire-pump disconnect.

Part **(B)**, as noted above, makes it mandatory to install emergency disconnect devices where they would not be disabled or affected by any fault or violent electrical failure in the normal service equipment (Fig. 230-33). Figure 230-34 shows a service disconnect for emergency and exit lighting installed very close to the normal service switchboard. An equipment burndown or fire near the main switchboard might knock out the emergency circuit. And the tap for the switch, which is made in the switchboard ahead of the service main, is particularly susceptible to being opened by *an arcing failure in the board.* The switch should be 10 or 15 ft (3.0 or 4.5 m) away from the board. And because the switchboard is fed from an outdoor transformer-mat layout directly outside the building, the tap to the safety switch would have greater reliability if it was made from the transformer secondary terminals rather than from the switchboard service terminals. Although the rule sets no specific distance of separation, remote locating of emergency disconnects is a mandatory Code rule.

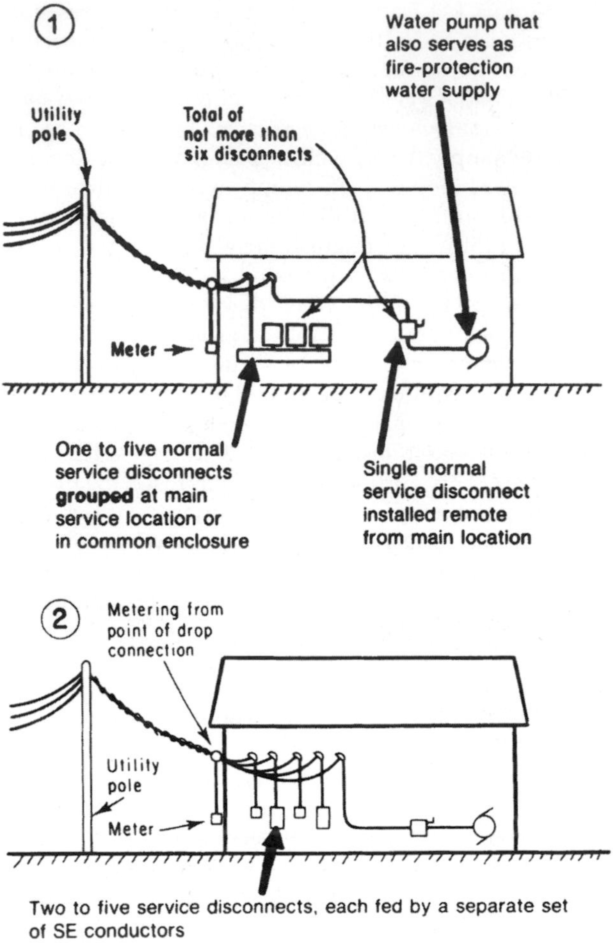
1
Water pump that also serves as fire-protection water supply
Utility pole
Total of not more than six disconnects
Meter
One to five normal service disconnects grouped at main service location or in common enclosure
Single normal service disconnect installed remote from main location
2
Metering from point of drop connection
Utility pole
Meter
Two to five service disconnects, each fed by a separate set of SE conductors

Fig. 230-32. Rule *permits* remote installation of one of the two to six service disconnects to protect fire-pump circuits (typical layouts). (Sec. 230.72.)

In part **(B)**, the phrase "permitted by 230.2" makes clear that each separate service permitted for fire pumps or for either legally required or optional standby service may be equipped with up to six disconnects in the same way as the normal service—or any service—may have up to six SE disconnects. And the disconnect or disconnects for a fire-pump or standby services must be remote from the normal service disconnects, as shown in Fig. 230-33.

Part **(C)** applies to applications of service disconnect for multiple-occupancy buildings—such as apartment houses, condominiums, town houses, office buildings, and shopping centers. Part **(C)** requires that the disconnect means for each occupant in a multiple-occupancy building be accessible to each occupant.

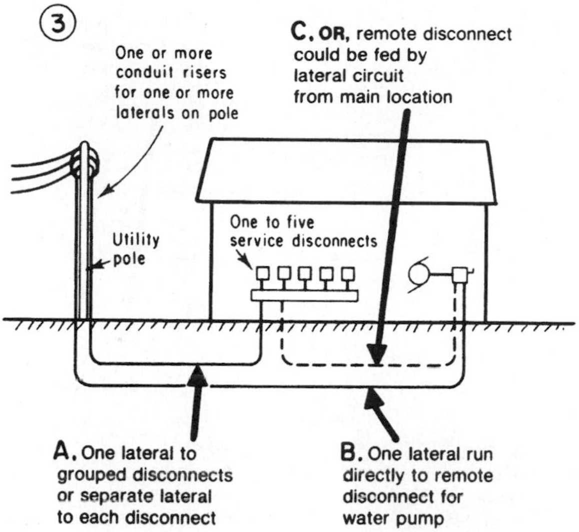
3
One or more
conduit risers
for one or more
laterals on pole
C. OR, remote disconnect
could be fed by
lateral circuit
from main location
Utility
pole
One to five
service disconnects
A. One lateral to
grouped disconnects
or separate lateral
to each disconnect
B. One lateral run
directly to remote
disconnect for
water pump

Fig. 230-32. *(Continued)*

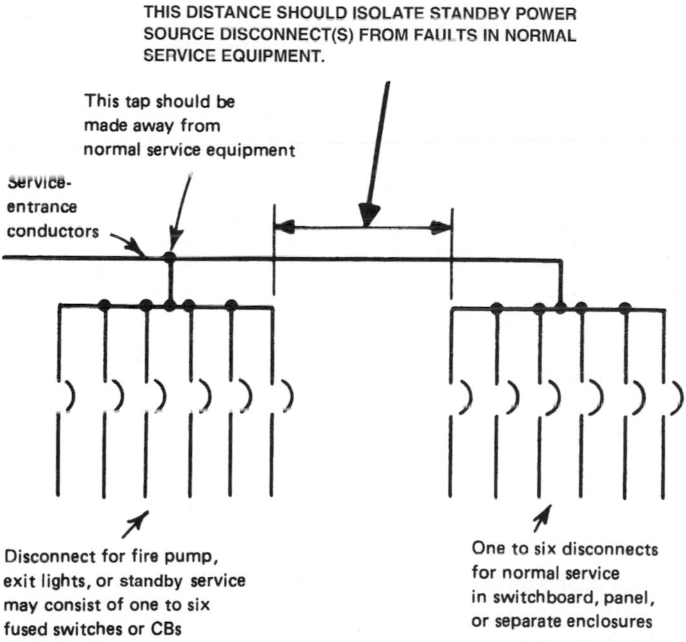
THIS DISTANCE SHOULD ISOLATE STANDBY POWER
SOURCE DISCONNECT(S) FROM FAULTS IN NORMAL
SERVICE EQUIPMENT.
This tap should be
made away from
normal service equipment
Service-
entrance
conductors
Disconnect for fire pump,
exit lights, or standby service
may consist of one to six
fused switches or CBs
One to six disconnects
for normal service
in switchboard, panel,
or separate enclosures

Fig. 230-33. Emergency service disconnects must be isolated from faults in normal SE equipment. Note that taps ahead of the service main are no longer recognized in 700.12 as an acceptable source of emergency circuits, but that a separate service is permitted if allowed for this purpose by the AHJ. Therefore, this drawing must be understood as applicable to installations made under the 1993 or prior editions of the NEC that recognized such connections, or as covering installations supplied by two entirely separate services that comply with the separation requirements in 700.12(D). (Sec. 230.72.)

INTERNA
ELECTRI

Fig. 230-34. Emergency disconnect close to service switchboard and fed by tap from it could readily be disabled by fault in board. (Sec. 230.72.)

For instance, for the occupant of an apartment in an apartment house, the disconnect means for deenergizing the circuits in the apartment must be in the apartment (such as a panel), in an accessible place in the hall, or in a place in the basement or outdoors where it can be reached.

As covered by the exception to part **(C)**, the access for each occupant as required by paragraph **(C)** would be modified where the building was under the management of a building superintendent or the equivalent and where electrical service and maintenance were furnished. In such a case, the disconnect means for more than one occupancy may be accessible only to authorized personnel. Figure 230-35 summarizes the way the grouping requirements are typically applied to multiple occupancy applications.

230.75. Disconnection of Grounded Conductor. In this section, the other means for disconnecting the grounded conductor from the interior wiring may be a screw or bolted lug on the neutral terminal block. The grounded conductor must not be run straight through the service equipment enclosure with no means of disconnection.

230.76. Manually or Power Operable. Any switch or CB used for service disconnect must be manually operable. In addition to manual operation, the switch may have provision for electrical operation—such as for remote control of the switch, provided it can be manually operated to the open or OFF position.

Code wording clearly indicates that an electrically operated breaker with a mechanical trip button that will open the breaker even if the supply power is dead is suitable for use as a service disconnect. The manually operated trip button

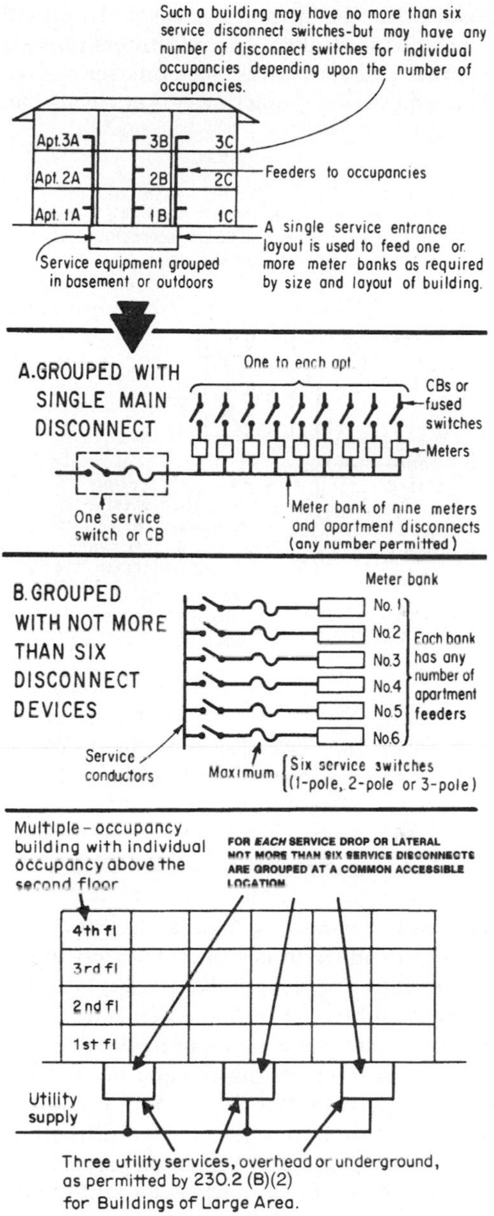
Such a building may have no more than six service disconnect switches-but may have any number of disconnect switches for individual occupancies depending upon the number of occupancies.
Apt. 3A
3B
3C
Apt. 2A
2B
2C
Apt. 1A
1B
1C
Feeders to occupancies
A single service entrance layout is used to feed one or more meter banks as required by size and layout of building.
Service equipment grouped in basement or outdoors
A. GROUPED WITH SINGLE MAIN DISCONNECT
One to each apt.
CBs or fused switches
Meters
One service switch or CB
Meter bank of nine meters and apartment disconnects (any number permitted)
B. GROUPED WITH NOT MORE THAN SIX DISCONNECT DEVICES
Meter bank
No. 1
No. 2
No. 3
No. 4
No. 5
No. 6
Each bank has any number of apartment feeders
Service conductors
Maximum
Six service switches (1-pole, 2-pole or 3-pole)
Multiple-occupancy building with individual occupancy above the second floor
FOR EACH SERVICE DROP OR LATERAL NOT MORE THAN SIX SERVICE DISCONNECTS ARE GROUPED AT A COMMON ACCESSIBLE LOCATION
4th fl
3rd fl
2nd fl
1st fl
Utility supply
Three utility services, overhead or underground, as permitted by 230.2 (B)(2) for Buildings of Large Area.

Fig. 230-35. These groupings of service disconnects represent good and acceptable practice that has been followed widely. (Sec. 230.72.)

ensures that the breaker "can be opened by hand." To provide manual closing of electrically operated circuit breakers, manufacturers provide emergency manual handles as standard accessories. Thus, such breaker mechanisms can be both closed and opened manually if operating power is not available, which fully satisfies this rule (Fig. 230-36).

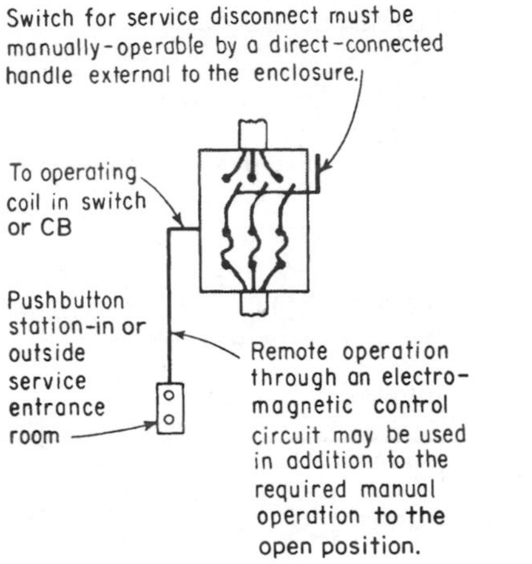
Switch for service disconnect must be manually-operable by a direct-connected handle external to the enclosure.
To operating coil in switch or CB
Pushbutton station-in or outside service entrance room
Remote operation through an electromagnetic control circuit may be used in addition to the required manual operation to the open position.
Note: Some local codes require both manual and electrical means of operation.

Fig. 230-36. Manual operation of any service switch is required. (Sec. 230.76.)

Local requirements on the use of electrically operated service disconnects should be considered in selecting such devices.

230.79. Rating of Service Disconnecting Means. Aside from the limited conditions covered in parts **(A)** and **(B)**, this section requires that service equipment (in general) shall have a rating not less than 60 A, applicable to both fusible and CB equipment. Part **(C)** requires 100-A minimum rating of a single switch or CB used in the service disconnect for *any* "one-family dwelling." It should be noted that the rule applies to one-family houses only, because of the definition of "one-family dwelling" as given in Art. 100. It does not apply to apartments or similar dwelling units that are in two-family or multifamily dwellings. These rules and the requirements in 230.42(B) must be carefully correlated. Review the discussion at 230.42(B) in this chapter for more information on this topic.

Even if the demand on a total connected load, as calculated from 220.40 through 220.61 or any of the applicable optional calculations permitted by part **(C)** of Art. 220, is less than 10 kVA, a 100-A service disconnect, as well as 100-A rated service-entrance conductors [230.42(B)], must be used.

If a 100-A service is used, the demand load may be as high as 24 kVA. By using the optional service calculations of Table 220.82, a 24-kVA demand load is obtained from a connected load of as much as 45 kVA, depending on how the load is configured. This shows the effect of diversity on large-capacity installations.

230.80. Combined Rating of Disconnects. Figure 230-37 shows an application of this rule, based on determining what rating of a single disconnect would be required *if* a single disconnect were used instead of multiple ones. It should be noted that the sum of ratings above 400 A does comply with the rule of this section and with Exception No. 3 of 230.90(A), even though the 400-A service-entrance conductors could be heavily overloaded. Exception No. 3 exempts this type of layout from the need to protect the conductors at their rated ampacity, as required in the basic rule of 230.90. The Code assumes that the 400-A rating of the service-entrance conductors was carefully calculated from Art. 220 to be adequate for the maximum sum of the demand loads fed by the five disconnects shown in the layout.

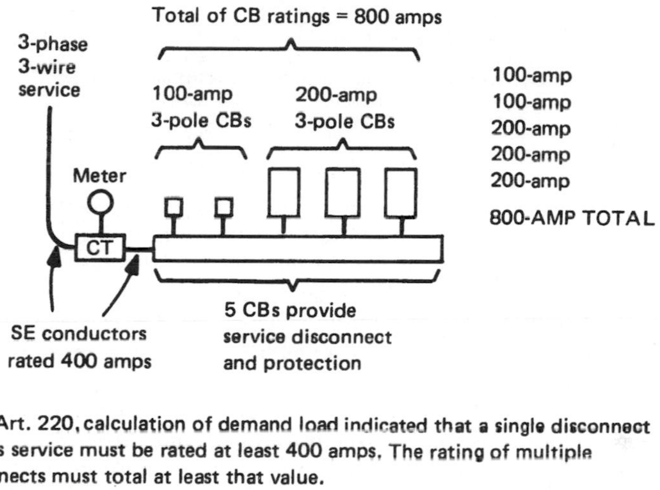
Total of CB ratings = 800 amps
3-phase
3-wire
service
100-amp
3-pole CBs
200-amp
3-pole CBs
Meter
CT
100-amp
100-amp
200-amp
200-amp
200-amp
800-AMP TOTAL
SE conductors
rated 400 amps
5 CBs provide
service disconnect
and protection
From Art. 220, calculation of demand load indicated that a single disconnect for this service must be rated at least 400 amps. The rating of multiple disconnects must total at least that value.

Fig. 230-37. Multiple disconnects must have their sum of ratings at least equal to the minimum rating of a single disconnect. (Sec. 230.80.)

230.82. Equipment Connected to the Supply Side of Service Disconnect. Cable limiters, fuses or CBs away from the building, high-impedance shunt circuits (such as potential coils of meters), supply conductors for time switches, surge-protective capacitors, instrument transformers, lightning arresters and circuits for emergency systems, fire-pump equipment, and fire and sprinkler alarms may be connected on the supply side of the disconnecting means. Emergency-lighting circuits, surge-protective capacitors, and fire-alarm and other protective signaling circuits, when placed ahead of the regular service disconnecting means, must have separate disconnects and overcurrent protection.

Part **(1)** of the rule prohibiting equipment connections on the line side of the service disconnect permits "cable limiters or other current-limiting devices" to be so connected.

Cable limiters are used to provide protection for individual conductors that are used in parallel (in multiple) to make up one phase leg of a high-capacity circuit, such as service conductors. A cable limiter is a cable connection device that contains a fusible element rated to protect the conductor to which it is connected.

As indicated in Part **(2)**, meters, and meter sockets can be connected on the supply side of the service disconnecting means and OCPDs if the meters are connected to service not in excess of 1000 V where the grounded conductor bonds the meter equipment cases and enclosures to the grounding electrode. (See Fig. 230-38.)

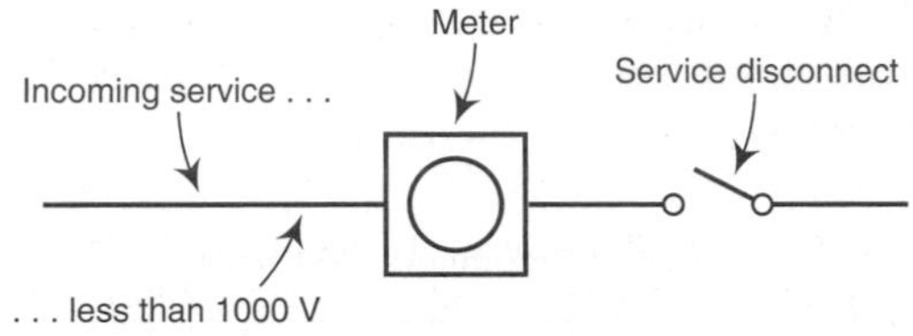
Meter
Service disconnect
Incoming service . . .
. . . less than 1000 V

Fig. 230-38. A "meter" may be connected on the supply side of the service disconnect. [Sec. 230.82(2).]

Part **(3)** allows meter disconnect switches ahead of a service disconnect, provided they have a suitable load rating and fault-current interrupting capability. WARNING: Many bypass meter sockets contain a switch that is used to maintain continuity across the meter jaws when the meter is removed. Utility personnel will close this switch, remove the meter for service, replace the meter, and then open the switch. It is of paramount importance that *this switch never be opened or closed under load*. The switches are unsuitable for load switching, only to maintain continuity.

Electrical utilities are requiring meter disconnect switches with ever-increasing frequency for the safety of their metering departments, so they can work the metering equipment cold. For this reason they are an important safety enhancement. From a code enforcement perspective, they present some challenges and opportunities for conflict in the field, especially since they are now fully load-break rated. If such a fuse were fused to contribute to the fault duty of the switch, the switch would arguably contain all the elements of service equipment as defined in the NEC.

This immediately raises the question, did the location of the service disconnect just move? The orderly application of a plethora of NEC rules depends on common agreement on exactly what device constitutes the service disconnecting means for any premises wiring application. If the service disconnect, in effect, relocates, then the number of conductors to the building changes because the neutral cannot be used for equipment grounding, the permitted point of connecting the grounding electrode conductor changes, and on and on. In fact, there has never been any intent to change the intended service disconnect locations, and this switch is only for the use of utility employees, who may well lock the switch in the ON position except while it is under active use. However, since it may also meet the Art. 100 definition, it would be wise to review the plans and the utility site policies with the inspector so there will not be surprises at the time of the final service inspection. The 2014 **NEC** took a giant step toward making sure that the function would not be misinterpreted. A meter disconnect must now be field labeled as such, and with the additional words "NOT SERVICE EQUIPMENT."

It must also be noted that this rule does not mandate the electrical location of this disconnect. Different utilities have different policies as to whether the switch is ahead (cold sequence) or after (hot sequence) the meter, policies that can change based on the voltage and currents involved. Most self-contained meters operating on 480 V distributions use cold sequence switching, because a fault created in the process of removing a meter will tend to be self-sustaining, particularly in the likely event of a transfer to the grounded meter socket. If that happens the damage will be extensive. On the other hand, some utilities prefer hot sequence metering on 208Y/120 V systems because they reduce the likelihood of power theft. These provisions simply allow for certain equipment ahead of service disconnects; the exact location is left to the serving utility that would be the specifier for the equipment.

Part **(4)** recognizes the connection of current and voltage transformers, high-impedance shunts, and surge protection on the line side of the service disconnecting means, *but* only where such devices are listed for such application. Where supply conductors are installed as service conductors, load management devices, fire alarm and suppression equipment, and standby power systems are permitted to be connected on the line side of the service disconnecting means by part **(5).** This part allows transfer switches ahead of the service disconnecting means in accordance with the limitations spelled out in 700.5, 701.5, and 702.5. Transfer switches are also available with ratings for service disconnecting means, in which case they will be marked accordingly.

As permitted by part **(6)**, an electric power production source that is auxiliary or supplemental to the normal utility service to a premises may be connected to the supply (incoming) side of the normal service disconnecting means. This part of the rule permits connection of a solar photovoltaic system, fuel cells, or interconnected power sources into the electrical supply for a building or other premises, to operate as a parallel power supply. The 2017 **NEC** added wind electric systems and energy storage systems to this list.

Where properly protected and provided with suitable disconnects, control circuits for power-operated service disconnects may be connected ahead of the service disconnect as recognized by subpart **(7)**. And part **(8)** recognizes that control power for a ground-fault protection system may be tapped from the supply side of the service disconnecting means. Where a control circuit for a ground-fault system is tapped ahead of the service main and "installed as part of listed equipment," suitable overcurrent protection and a disconnect must be provided for the control-power circuit.

Part **(9)** was new in the 2011 edition and covers communications equipment provided by the utility and under its exclusive control. This equipment must be listed and equipped with both a disconnecting means and overcurrent protection. The disconnect need not be installed if the equipment is enclosed within a metering compartment so access is impossible unless the meter is removed. This sort of allowance will be critical to enable increasingly sophisticated methods of sophisticated metering and other smart grid technologies to be put in place.

230.90. Where Required. The intent in paragraph **(A)** is to ensure that the overcurrent protection required in the service-entrance equipment protects the service-entrance conductors from "overload." It is obvious that these overcurrent devices cannot provide "fault" protection for the service-entrance conductors if the fault occurs in the service-entrance conductors (which are on the line side of the service overcurrent devices), but they can protect the conductors from overload where so selected as to have proper rating. Conductors on the load side of the service equipment are considered as feeders or branch circuits and are required by the Code to be protected as described in Arts. 210, 215, and 240.

Part **(A)** states that the term "set" of fuses means all the fuses required to protect all the ungrounded service-entrance conductors in a given circuit.

Each ungrounded service-entrance conductor must be protected by an overcurrent device in series with the conductor (Fig. 230-39). The overcurrent device must have a rating or setting not higher than the allowable current capacity of the conductor, with the exceptions noted.

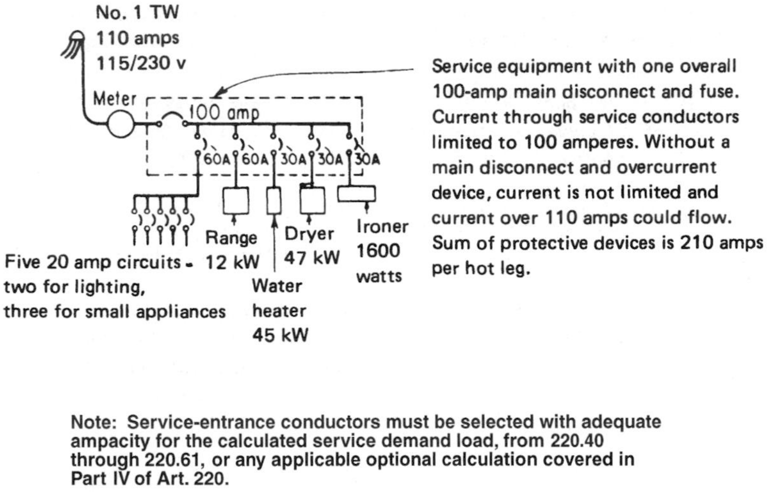
No. 1 TW
110 amps
115/230 v
Meter
100 amp
60A
60A
30A
30A
30A
Range
12 kW
Dryer
47 kW
Ironer
1600
watts
Five 20 amp circuits -
two for lighting,
three for small appliances
Water
heater
45 kW
Service equipment with one overall 100-amp main disconnect and fuse. Current through service conductors limited to 100 amperes. Without a main disconnect and overcurrent device, current is not limited and current over 110 amps could flow. Sum of protective devices is 210 amps per hot leg.
Note: Service-entrance conductors must be selected with adequate ampacity for the calculated service demand load, from 220.40 through 220.61, or any applicable optional calculation covered in Part IV of Art. 220.

Fig. 230-39. Single main service protection must not exceed conductor ampacity (or may be next higher rated device above conductor ampacity). (Sec. 230.90.)

The rule of Exception No. 1 says that if the service supplies one motor in addition to other load (such as lighting and heating), the overcurrent device may be rated or set in accordance with the required protection for a branch circuit supplying the one motor (430.52) plus the other load, as shown in Fig. 230-40. Use of 175-A fuses where the calculation calls for 170-A conforms to Exception No. 2 of 230.90—next higher standard rating of fuse (240.6). For motor branch circuits and feeders, Arts. 220 and 430 permit the use of overcurrent devices having ratings or settings higher than the capacities of the conductors. Article 230 makes similar provisions for services where the service supplies a motor load or a combination load of both motors and other loads.

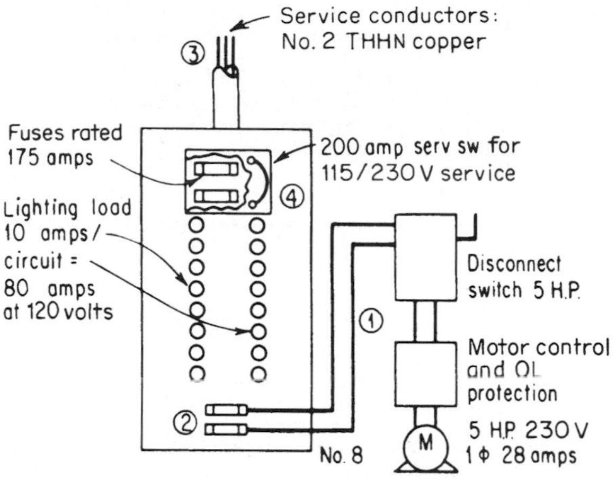
Service conductors:
No. 2 THHN copper
3
Fuses rated
175 amps
200 amp serv sw for
115/230 V service
4
Lighting load
10 amps/
circuit =
80 amps
at 120 volts
Disconnect
switch 5 H.P.
1
Motor control
and OL
protection
2
No. 8
M
5 H.P. 230 V
1 φ 28 amps

1. Size of motor branch circuit conductors: 125% x 28 amps equals 35 amps. This requires No. 8's.

2. Size of motor branch circuit fuses: 300% x 28 amps equals 84 amps. This requires maximum fuse size of 90 amps. Smaller fuses, such as time-delay type, may be used.

3. Size of service entrance conductors must be adequate for a load of 125% x 28 amps plus 100 amps (continuous lighting load of 80 amps x 1.25) or 135 amps.

4. Size of main fuses: 90 amps (from 2 above) plus 100 amps (80 amps x 1.25) equals 190 amps. This requires maximum fuse size of 200 amps. Again, smaller fuses may and should be used where possible to improve the overload protection on the circuit conductors.

Fig. 230-40. Service protection for lighting plus motor load. (Sec. 230.90.)

If the service supplies two or more motors as well as other load, then the overcurrent protection must be rated in accordance with the required protection for a feeder supplying several motors plus the other load (430.63). Or if the service supplies only a multimotor load (with no other load fed), then 430.62 sets the maximum permitted rating of overcurrent protection.

Exception No. 3. Not more than six CBs or six sets of fuses may serve as overcurrent protection for the service-entrance conductors even though the sum of the ratings of the overcurrent devices is in excess of the ampacity of the service conductors supplying the devices—as illustrated in Fig. 230-41. The grouping of single-pole CBs as multipole devices, as permitted for disconnect means, may also apply to overcurrent protection.

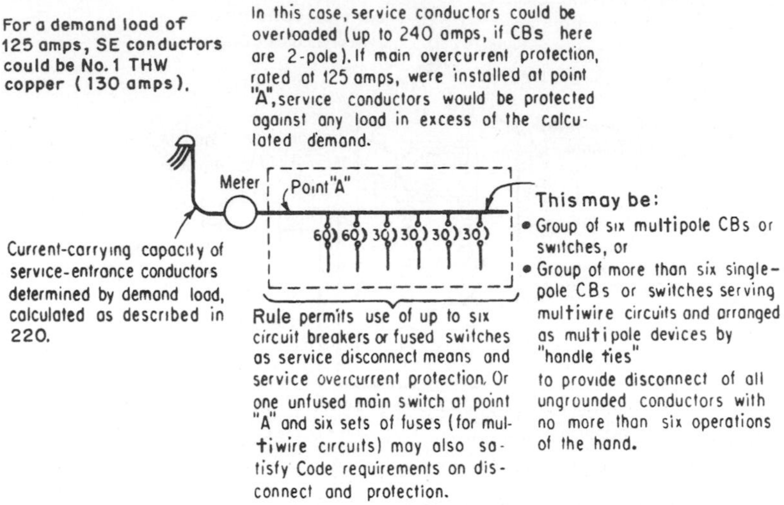
For a demand load of 125 amps, SE conductors could be No. 1 THW copper (130 amps).
In this case, service conductors could be overloaded (up to 240 amps, if CBs here are 2-pole). If main overcurrent protection, rated at 125 amps, were installed at point "A", service conductors would be protected against any load in excess of the calculated demand.
Meter
Point "A"
60 60 30 30 30 30
This may be:
• Group of six multipole CBs or switches, or
• Group of more than six single-pole CBs or switches serving multiwire circuits and arranged as multipole devices by "handle ties"
to provide disconnect of all ungrounded conductors with no more than six operations of the hand.
Current-carrying capacity of service-entrance conductors determined by demand load, calculated as described in 220.
Rule permits use of up to six circuit breakers or fused switches as service disconnect means and service overcurrent protection. Or one unfused main switch at point "A" and six sets of fuses (for multiwire circuits) may also satisfy Code requirements on disconnect and protection.

Fig. 230-41. With six subdivisions of protection, conductors could be theoretically overloaded. In effect, the service conductors are being protected from overload by the accuracy of the Art. 220 load calculation that sized them. (Sec. 230.90.)

Exception No. 3 ties into 230.80. Service conductors are sized for the *total* maximum demand load—applying permitted demand factors from Art. 220. Then each of the two to six feeders fed by the SE conductors is also sized from Art. 220 based on the load fed by each feeder. When those feeders are given overcurrent protection in accordance with their ampacities, it is frequently found that the sum of those overcurrent devices is greater than the ampacity of the SE conductors, which were sized by applying the applicable demand factors to the total connected load of all the feeders. Exception No. 3 recognizes that possibility as acceptable even though it departs from the rule in the first sentence of 230.90(A). The assumption is that if calculation of demand load for the SE conductors is correctly made, there will be no overloading of those conductors because the diversity of feeder loads (some loads "on," some "off") will be adequate to limit load on the SE conductors.

Assume that the load of a building computed in accordance with Art. 220 is 255 A. Under 240.4(B), 300-A fuses or a 300-A CB may be considered as the proper-size overcurrent protection for service conductors rated between 255 and 300 A if a single service disconnect is used.

If the load is separated in such a manner that six 70-A CBs could be used instead of a single service disconnecting means, total rating of the CBs would be greater than the ampacity of the service-entrance conductors. And that would be acceptable.

Exception No. 4 to 230.90(A) is shown in Fig. 230-42 and is intended to prevent opening of the fire-pump circuit on any overload up to and including stalling or even seizing of the pump motor. Because the conductors are "outside the building," operating overload is no hazard; and, under fire conditions, the pump must

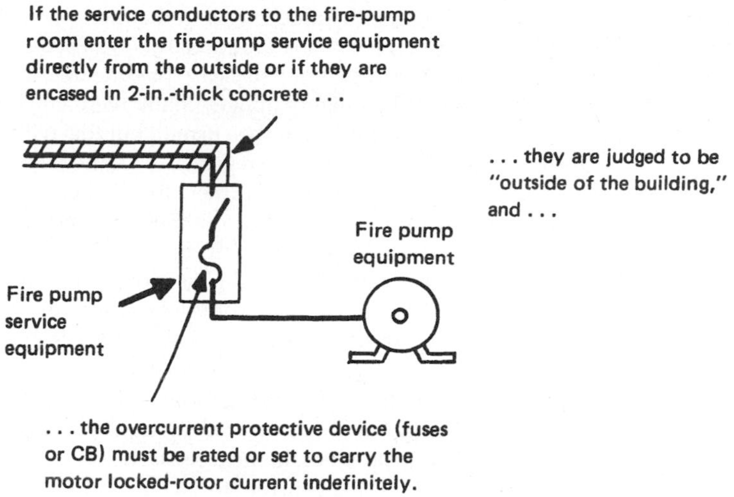
If the service conductors to the fire-pump room enter the fire-pump service equipment directly from the outside or if they are encased in 2-in.-thick concrete . . .
. . . they are judged to be "outside of the building," and . . .
Fire pump equipment
Fire pump service equipment
. . . the overcurrent protective device (fuses or CB) must be rated or set to carry the motor locked-rotor current indefinitely.

Fig. 230-42. (Sec. 230.90.)

have no prohibition on its operation. It is better to lose the motor than attempt to protect it against overload when it is needed.

Exception No. 5 specifically recognizes the use of conductors in accordance with 310.15(B)(7). There the Code considers the conductor sizes in Table 310.15(B)(7) to be adequately protected by the value of OC protection indicated.

230.91. Location. The 2017 NEC added a requirement that if fuses are used as the overcurrent protection, they must be on the load side of the disconnect. This, in general, is common sense, but it was ill-advised here because the information is incomplete. For one thing a fusible limiter, which is expressly allowed elsewhere by 230.82(1), may get caught up in this. The entire, surprisingly complicated topic is fully covered in 240.40, and the extra sentence here should be disregarded.

230.95. Ground-Fault Protection of Equipment. Fuses and CBs, applied as described in the previous section on "Overcurrent Protection," are sized to protect conductors in accordance with their current-carrying capacities. The function of a fuse or CB is to open the circuit if current exceeds the rating of the protective device. This excessive current might be caused by operating overload, by a ground fault, or by a short circuit. Thus, a 1000-A fuse will blow if current in excess of that value flows over the circuit. It will blow early on heavy overcurrent and later on low overcurrents. But it will blow, and the circuit and equipment will be protected against the damage of the overcurrent. But, there is another type of fault condition which is very common in grounded systems and will not be cleared by conventional overcurrent devices. That is the phase-to-ground fault (usually arcing) which has a current value less than the rating of the overcurrent device.

On any high-capacity feeder, a line-to-ground fault (i.e., a fault from a phase conductor to a conduit, to a junction box, or to some other metallic equipment enclosure) can, and frequently does, draw current of a value less than the rating or setting of the circuit protective device. For instance, a 500-A ground fault on a 2000-A protective device which has only a 1200-A load will not be cleared by the device. If such a fault is a "bolted" line-to-ground fault, a highly unlikely fault, there will be a certain amount of heat generated by the I^2R effect of the current; but this will usually not be dangerous, and such fault current will merely register as additional operating load, with wasted energy (wattage) in the system. Further, such bolted faults usually draw large values of current, particularly if the equipment grounding system has been installed correctly, and the result will be a trip in very short order. But, bolted phase-to-ground faults are very rare. The usual phase-to-ground fault exists as an intermittent or arcing fault, and an arcing fault of the same current rating as the essentially harmless bolted fault can be fantastically destructive because of the intense heat of the arc.

Of course, any ground-fault current (bolted or arcing) above the rating or setting of the circuit protective device will normally be cleared by the device. But, even where the protective device eventually operates, in the case of a heavy ground-fault current which adds to the normal circuit load current to produce a total current in excess of the rating of the normal circuit protective device (fuse or CB), the time delay of the device may be minutes or even hours—more than enough time for the arcing-fault current to burn out conduit and enclosures, acting just like a torch, and even propagating flame to create a fire hazard.

In the interests of safety, definitive engineering design must account for protection against high-impedance ground faults, as is required by this Code rule. Phase OCPDs are normally limited in their effectiveness because (1) they must have a time delay and a setting somewhat higher than full load to ride through normal inrushes and (2) they are unable to distinguish between normal currents and low-magnitude fault currents, which, when combined, may be less than the trip rating of the OCPD.

Dangerous temperatures and magnetic forces are proportional to current for overloads and short circuits; therefore, OCPDs usually are adequate to protect against such faults. However, the temperatures of arcing faults are, generally, independent of current magnitude; and arcs of great and extensive destructive capability can be sustained by currents not exceeding the overcurrent device settings. Other means of protection are therefore necessary. A ground-detection device, which "sees" only ground-fault current, is coupled to an automatic switching device to open all three phases when a line-to-ground fault exists on the circuit. Such protective systems are readily available in listed configurations from electrical equipment manufacturers, which eases compliance with this rule. Careful attention to manufacturer's installation instructions is mandatory to ensure proper operation and the desired level of protection.

230.95 requires ground-fault protection of equipment (GFPE) to be provided for each service *disconnecting means* rated 1000 A or more in a solidly grounded-wye electrical service that operates with its ungrounded legs at more than 150 V to ground. Note that this applies to the rating of the disconnect, not to the rating of the overcurrent devices or to the capacity of the service-entrance conductors.

The wording of the first sentence of this section makes clear that service GFPE (ground-fault protection of equipment) is required under specific conditions: only for grounded-wye systems that have voltage over 150 V to ground and less than 1000 V phase-to-phase. In effect, that means the rule applies only to 480/277-V grounded-wye and *not* to 120/208-V systems or any other commonly used systems (Fig. 230-43). Recent recognition of the 600Y/347-V distribution systems—used in Canada—would subject any system so rated to the rule of Sec. 230.95. And, each disconnect rated 1000 A, or more, must be provided with equipment

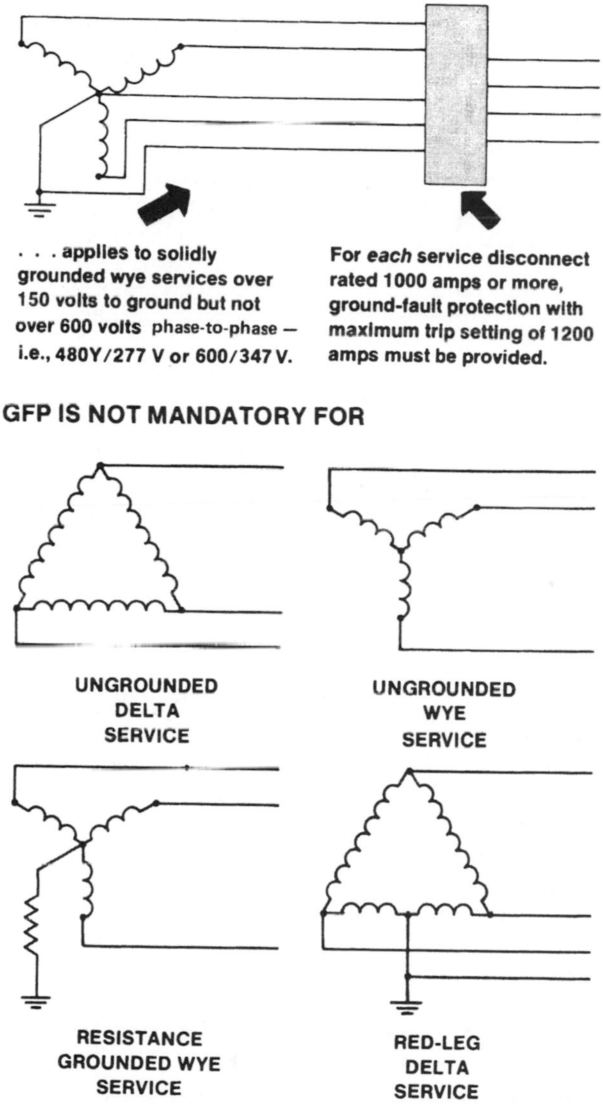
. . . applies to solidly grounded wye services over 150 volts to ground but not over 600 volts phase-to-phase – i.e., 480Y/277 V or 600/347 V.
For *each* service disconnect rated 1000 amps or more, ground-fault protection with maximum trip setting of 1200 amps must be provided.
GFP IS NOT MANDATORY FOR
UNGROUNDED DELTA SERVICE
UNGROUNDED WYE SERVICE
RESISTANCE GROUNDED WYE SERVICE
RED-LEG DELTA SERVICE

Fig. 230-43. Service ground-fault protection is mandatory. (Sec. 230.95.)

GFPE. GFPE is *not* required on any systems operating over 1000 V phase-to-phase. The reason for this voltage parameter is that on ac systems an arc naturally self-extinguishes at every current zero, 120 times per second. The likelihood of damage is directly related to the likelihood of the arc restriking, and that is related to the peak voltage in the system. Testing has shown this is a major problem above about 375 V. Since the peak voltage of a 120-V-to-ground system is 170 V ($120 \times \sqrt{2}$), these systems aren't so much of a problem. However, on 480Y/277-V systems, the peak voltage is about 390 V, easily high enough to keep the arc in business.

The second paragraph clearly indicates that the "rating" to be considered is the rating of the largest fuse a switch can accommodate, or the longtime trip rating of a nonadjustable CB, or the maximum "setting" for adjustable-trip CBs. If the fusible switch, nonadjustable CB, or adjustable-trip CB used as the service disconnect is rated at, or can be set at, 1000 A or more, then ground-fault protection is required.

In a typical GFPE hookup, as shown in Fig. 230-44, part **(A)** of the section specifies that a ground-fault current of 1200 A or more must cause the disconnect to open all ungrounded conductors. Thus, the maximum GF pick-up setting permitted is 1200 A, although it may be set lower.

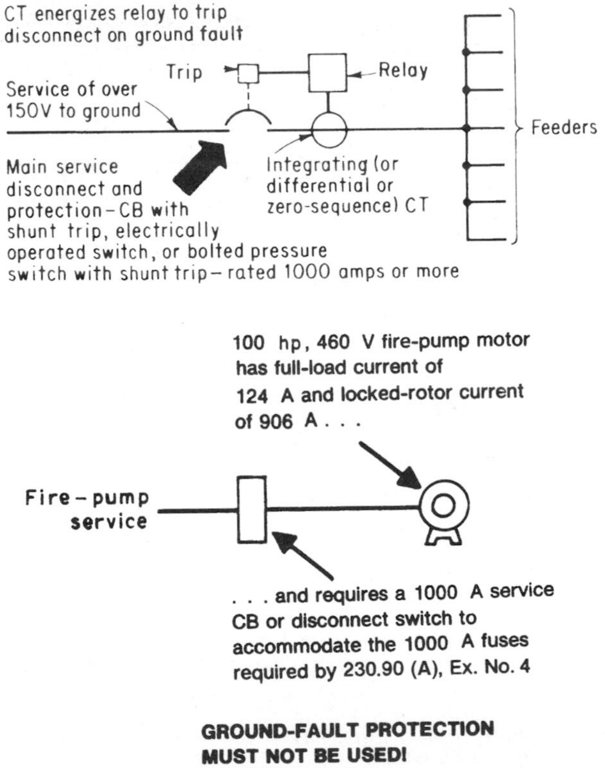
CT energizes relay to trip
disconnect on ground fault
Trip
Relay
Service of over
150V to ground
Feeders
Main service
disconnect and
protection-CB with
shunt trip, electrically
operated switch, or bolted pressure
switch with shunt trip-rated 1000 amps or more
Integrating (or
differential or
zero-sequence) CT
100 hp, 460 V fire-pump motor
has full-load current of
124 A and locked-rotor current
of 906 A . . .
Fire-pump
service
. . . and requires a 1000 A service
CB or disconnect switch to
accommodate the 1000 A fuses
required by 230.90 (A), Ex. No. 4
GROUND-FAULT PROTECTION
MUST NOT BE USED!

Fig. 230-44. GFPE is required for each disconnect rated 1000 A or more, but not for a fire-pump disconnect. (Sec. 230.95.)

With a GFPE system, at the service entrance a ground fault anywhere in the system is immediately sensed in the ground-relay system, but its action to open the circuit usually is delayed to allow some normal overcurrent device near the point of fault to open if it can. As a practical procedure, such time delay is designed to be only a few cycles or seconds, depending on the voltage of the circuit, the time–current characteristics of the overcurrent devices in the system, and the location of the ground-fault relay in the distribution system. Should any of the conventional short-circuit OCPDs fail to operate in the time predetermined to clear the circuit, and if the fault continues, the ground-fault protective relays will open the circuit. This provides added overcurrent protection not available by any other means.

The rule requiring GFPE for any service disconnect rated 1000 A or more (on 480/277-V or 600/347-V services) specifies a maximum *time delay of 1 s for ground-fault currents of 3000 A or more* (Fig. 230-45).

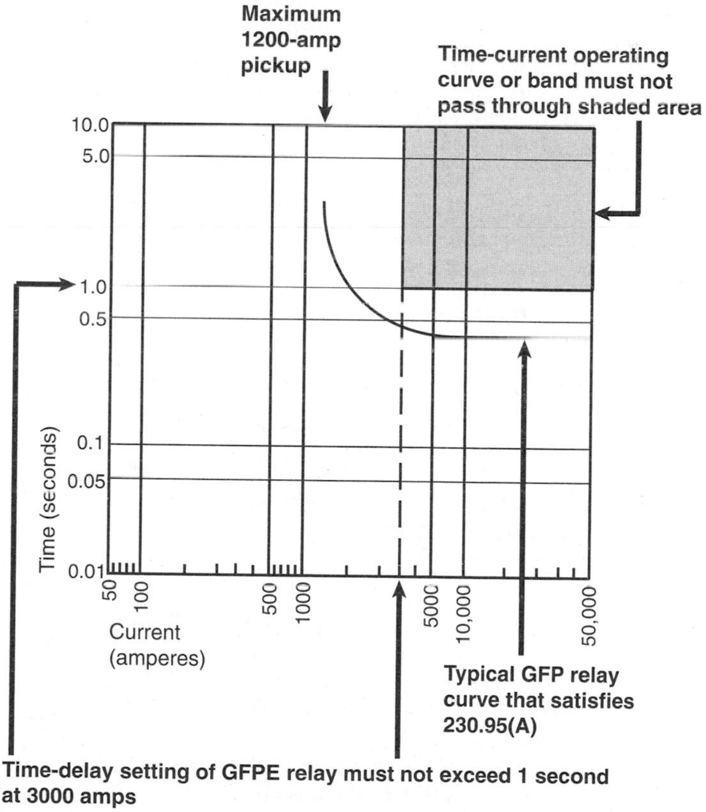
Maximum 1200-amp pickup
Time-current operating curve or band must not pass through shaded area
10.0
5.0
1.0
0.5
0.1
0.05
0.01
Time (seconds)
50
100
500
1000
5000
10,000
50,000
Current (amperes)
Typical GFP relay curve that satisfies 230.95(A)
Time-delay setting of GFPE relay must not exceed 1 second at 3000 amps

Fig. 230-45. The rule specifies maximum energy let-through for GFPE operation. [Sec. 230.95(A).]

The maximum permitted setting of a service GFPE hookup is 1200 A, but the time-current trip characteristic of the relay must ensure opening of the disconnect in not more than 1 s for any ground-fault current of 3000 A or more. This change in the Code was made to establish a specific level of protection under GFPE by setting a maximum limit on I^2t of fault energy.

The reasoning behind this change was explained as follows:

The amount of damage done by an arcing fault is directly proportional to the time it is allowed to burn. Commercially available GFPE systems can easily meet the 1-s limit. Some users are requesting time delays up to 60 s so all downstream overcurrent devices can have plenty of time to trip thermally before the GFP on the main disconnect trips. However, an arcing fault lasting 60 s can virtually destroy a service equipment installation. Coordination with downstream overcurrent devices can and should be achieved by adding GFPE on feeder circuits where needed. The Code should require a reasonable time limit for GFP. Now, 3000 A is 250 percent of 1200 A, and 250 percent of setting is a calibrating point specified in ANSI 37.17. Specifying a maximum time delay starting at this current value will allow either flat or inverse time-delay characteristics for ground-fault relays with approximately the same level of protection.

Selective coordination between GFPE and conventional protective devices (fuses and CBs) on service and feeder circuits is now a very clear and specific task as a result of rewording of 230.95(A) that calls for a maximum time delay of 1 s at any ground-fault current value of 3000 A or more.

For applying the rule of 230.95, the rating of any service disconnect means shall be determined, as shown in Fig. 230-46.

Because the rule on required service GFPE applies to the rating of each service disconnect, there are many instances where GFPE would be required if a single service main disconnect is used but *not* if the service subdivision option of 230.71(A) is taken, as shown in Fig. 230-47.

By the exception to 230.95, continuous industrial process operations are exempted from the GFPE rules of parts **(A), (B),** and **(C)** where the electrical system is under the supervision of qualified persons who will effect orderly shutdown of the system and thereby avoid hazards, greater than ground fault itself, that would result from the nonorderly, automatic interruption that GFPE would produce in the supply to such critical continuous operations. The exception excludes GFPE requirements where a nonorderly shutdown will introduce additional or increased hazards. The idea behind that is to provide maximum protection against service outage of such industrial processes. With highly trained personnel at such locations, design and maintenance of the electrical system can often accomplish safety objectives more readily without GFPE on the service. Electrical design can account for any danger to personnel resulting from loss of process power versus damage to electrical equipment.

The former Exception No. 2 at this location excluded fire-pump service disconnects from the basic rule that requires ground-fault protection on any service disconnect rated 1000 A or more on a grounded-wye 600/347-V or 480/277-V system. This exception has been deleted, not because it is a bad idea, but because the fire pump article now has a clear statement [at 695.6(G)] that forbids the use of GFPE on a fire pump circuit. Since per 90.3 a provision in Chap. 6 (or 5 or 7)

FUSED SWITCH (bolted pressure switch, service protector, etc.)

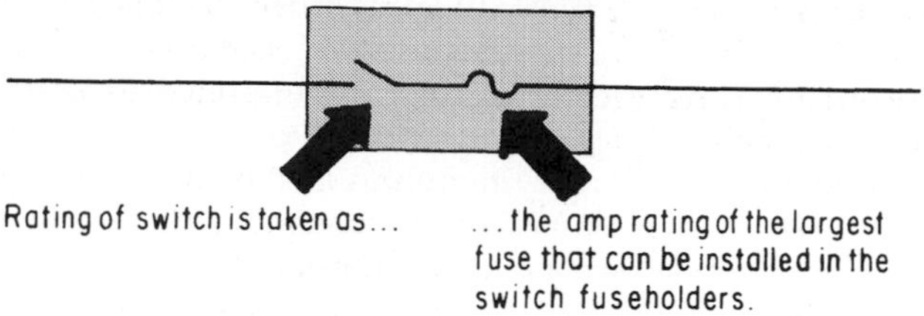
Rating of switch is taken as...
...the amp rating of the largest fuse that can be installed in the switch fuseholders.

EXAMPLE

If 900-amp fuses are used in this service switch, ground-fault protection would be required, because the switch can take fuses rated 1200 amps—which is above the 1000-amp level at which GFPE becomes mandatory.

CIRCUIT BREAKER

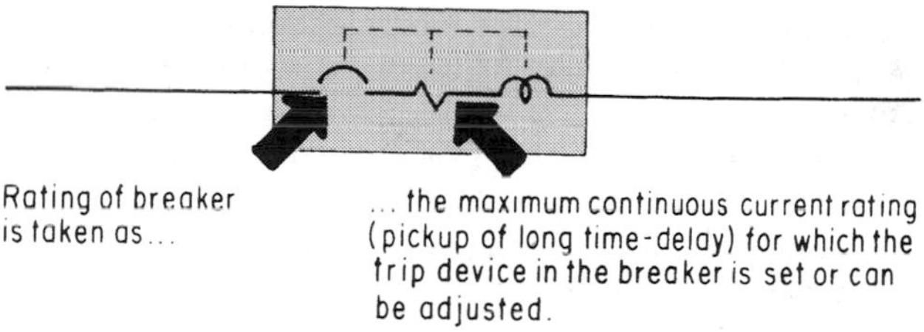
Rating of breaker is taken as...
...the maximum continuous current rating (pickup of long time-delay) for which the trip device in the breaker is set or can be adjusted.

Example: GFPE would be required for a service CB with, say, an 000-amp trip setting if the CB had a trip device that can be adjusted to 1000 amps or more.

Fig. 230-46. Determining rating of service disconnect for GFPE rule. (Sec. 230.95.)

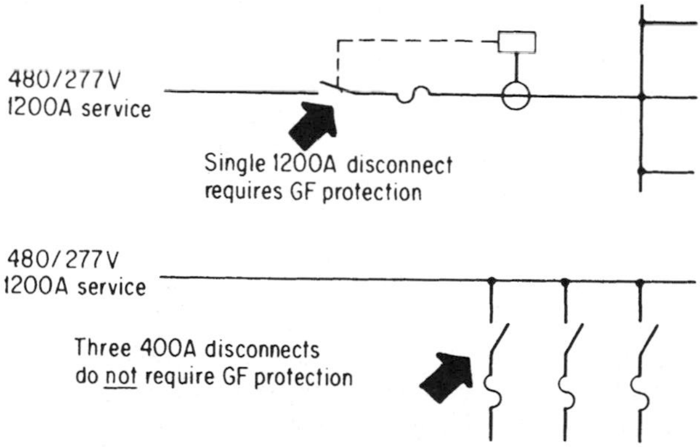
480/277V
1200A service
Single 1200A disconnect requires GF protection
480/277V
1200A service
Three 400A disconnects do not require GF protection

Fig. 230-47. Subdivision option on disconnects affects GFPE rule. (Sec. 230.95.)

automatically supersedes a contrary provision in Chaps. 1 through 4, an exception to the same effect here is a waste of space. There are good reasons for the prohibition.

Because fire pumps are required by 230.90, Exception No. 4, to have overcurrent protection devices large enough to permit locked-rotor current of the pump motor to flow without interruption, larger fire pumps (100 hp and more) would have disconnects rated 1000 A or more. Without Exception No. 2, those fire-pump disconnects would be subject to the basic rule and would have to be equipped with ground-fault protection. But GFP on any fire pump is objectionable on the same basis that 230.90, Exception No. 4, wants nothing less than protection rated for locked rotor. The intent is to give the pump motor every chance to operate when it functions during a fire, to prevent opening of the motor circuit or any overload up to and including stalling or seizing of the shaft or bearings. For the same reason, 430.31 exempts fire pumps from the need for overload protection, and 430.72(C) exception requires overcurrent protection to be omitted from the control circuit of a starter for a fire pump.

Important considerations are given in the informational notes in this section. Obviously, the selection of ground-fault equipment for a given installation merits a detailed study. The option of subdividing services discussed under *six service entrances from one lateral* [230.2(A)(1)] should be evaluated. A 4000-A service, for example, could be divided using five 800-A disconnecting means, and in such cases GFPE would not be required.

One very important note in 230.95 warns about potential desensitizing of ground-fault sensing hookups when an emergency generator and transfer switch are provided in conjunction with the normal service to a building. The note applies to those cases where a solid neutral connection from the normal service is made to the neutral of the generator through a 3-pole transfer switch. With the neutral grounded at the normal service and the neutral bonded to the generator frame, ground-fault current on the load side of the transfer switch can return over two paths, one of which will escape detection by the GFPE sensor, as shown in Fig. 230-48. Such a hookup can also cause nuisance tripping of the GFPE due to normal neutral current. Under normal (nonfaulted) conditions, neutral current due to normal load unbalance on the phase legs can divide at common neutral connection in transfer switch, with some current flowing toward the generator and returning to the service main on the conduit—indicating falsely that a ground fault exists and causing nuisance tripping of GFPE. The note points out that "means or devices" (such as a 4-pole, neutral-switched transfer switch) "may be needed" to ensure proper, effective operation of the GFPE hookup (Fig. 230-49).

Very Important!

Because of so many reports of improper and/or unsafe operation (or failure to operate) of ground-fault protective hookups, part **(C)** of 230.95 *requires* (a mandatory rule) that *every* GFPE hookup be "performance tested when first installed." And the testing MUST *be done on the job site!* Factory testing of a GFPE system does not satisfy this Code rule. This rule requires that such testing be done according to

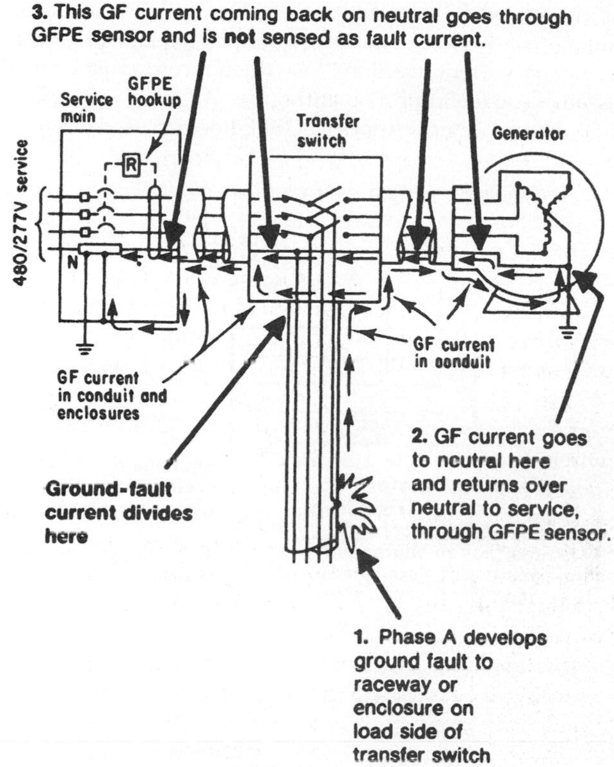
3. This GF current coming back on neutral goes through GFPE sensor and is not sensed as fault current.
Service main
GFPE hookup
Transfer switch
Generator
480/277V service
N
R
GF current in conduit and enclosures
GF current in conduit
Ground-fault current divides here
2. GF current goes to neutral here and returns over neutral to service, through GFPE sensor.
1. Phase A develops ground fault to raceway or enclosure on load side of transfer switch

Fig. 230-48. Improper operation of GFPE can result from emergency system transfer switch. (Sec. 230.95.)

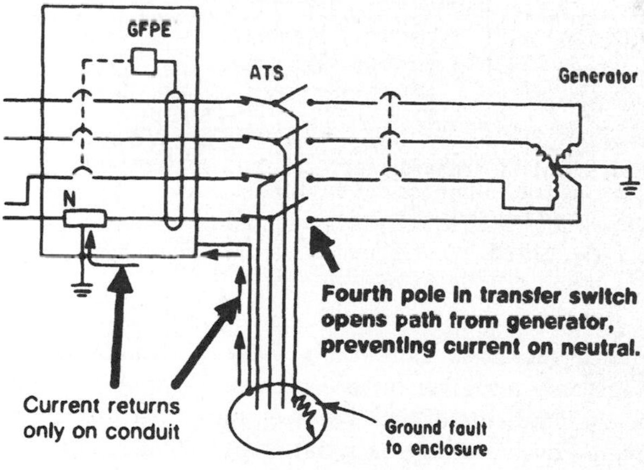
GFPE
ATS
Generator
N
Fourth pole in transfer switch opens path from generator, preventing current on neutral.
Current returns only on conduit
Ground fault to enclosure

Refer to Fig. 230–48

Fig. 230-49. Four-pole transfer switch is one way to avoid desensitizing GFPE. (Sec. 230.95.)

"instructions . . . provided with the equipment." The 2017 NEC took all this a step further by mandating that the on-site testing be done by a qualified person and include a "process of primary current injection." A written record must be made of the test and must be available to the inspection authority.

Figure 230-50 shows two basic types of GFPE hookup used at service entrances.

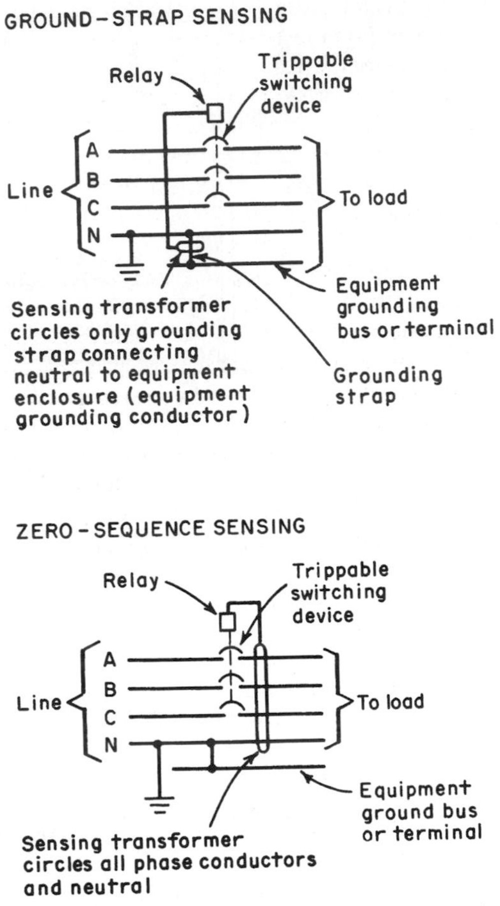
GROUND-STRAP SENSING
Relay
Trippable switching device
A
B
C
N
Line
To load
Equipment grounding bus or terminal
Sensing transformer circles only grounding strap connecting neutral to equipment enclosure (equipment grounding conductor)
Grounding strap
ZERO-SEQUENCE SENSING
Relay
Trippable switching device
A
B
C
N
Line
To load
Equipment ground bus or terminal
Sensing transformer circles all phase conductors and neutral

Fig. 230-50. Types of ground-fault detection that may be selected for use at services. (Sec. 230.95.)

230.200. General (Services Exceeding 1000 V, Nominal). The rules on medium-voltage services given in the provisions of Art. 230 apply only to equipment on the load side of the "service-point." Because there has been so much controversy over identifying what is and what is not "service" equipment in the many complicated layouts of outdoor high-voltage circuits and transformers, the definition in Art. 100 provides clarification. In any particular installation, identification of that point can be made by the utility company and design personnel. The definition clarifies that the property line is not the determinant as to where **NEC** rules must begin to be applied. This is particularly important in cases of multibuilding industrial complexes where the utility has distribution circuits on the property. See "Service Point" in Art. 100.

230.200 says that "service conductors and equipment used on circuits exceeding 1000 V" must comply with *all* the rules in Art. 230 (including any "applicable provisions" that cover services up to 1000 V). And Art. 100 says that for services up to 1000 V, the "service conductors" are those conductors—whether on the primary or secondary of a step-down transformer or transformers—that carry current from the "service point" (where the utility connects to the customer's wiring) to the service disconnecting means for a building or structure. See "Service Point" in Art. 100. All conductors between the defined points—"service point" and "service disconnecting means"—must comply with all requirements for service conductors, whether above or below 1000 V.

Design and layout of any "service" are critically related to safety, adequacy, economics, and effective use of the whole system. It is absolutely essential that we know clearly and surely what circuits and equipment of any electrical system constitute the "service" and what parts of the system are not involved in the "service." For instance, in a system with utility feed at 13.2 kV and step down to 480/277 V, the mandatory application of 230.95 requiring GFPE hinges on establishing whether the "service" is on the primary or secondary side of the transformers. If the secondary is the service, where the step-down transformer belongs to the utility and the "service point" is on its secondary, we have a mandatory need for GFPE and none of the Code rules on service would apply to any of the 13.2-kV circuits—regardless of their length or location. If the transformer belongs to the customer and the "service point" is on the primary side, the primary is the service, 230.95 does not require GFPE on services over 1000 V phase-to-phase, all the primary circuit and equipment must comply with all of Art. 230, and the secondary circuits are feeders and do not have to comply with any of the service regulations. This potential loophole has been closed by the addition of comparable GFPE rules to feeder circuits in 215.10.

The whole problem involved here is complex and requires careful, individual study to see clearly the many interrelated considerations. Let us look at a few important things to note about Code definitions as given in Art. 100:

1. "Service conductors" run to the *service disconnect* of the premises supplied. Note that they run to "premises" and are not required to run to a "building." The Code does not define the word *premises*, but a typical dictionary definition is "a tract of land, including its buildings" (Fig. 230-51).
2. "Service equipment" *usually* consists of "a circuit breaker or switch and fuses, and their accessories, located near the point of entrance of *supply* conductors to a building or other structure, or an otherwise defined area." Note that the service equipment is the means of cutoff of the supply, and the service conductors may enter "a building" or "other structure" or a "defined area." But, again, a service does not necessarily have to be to "a building." It could be to such a "structure" as an outdoor switchgear or unit substation enclosure.

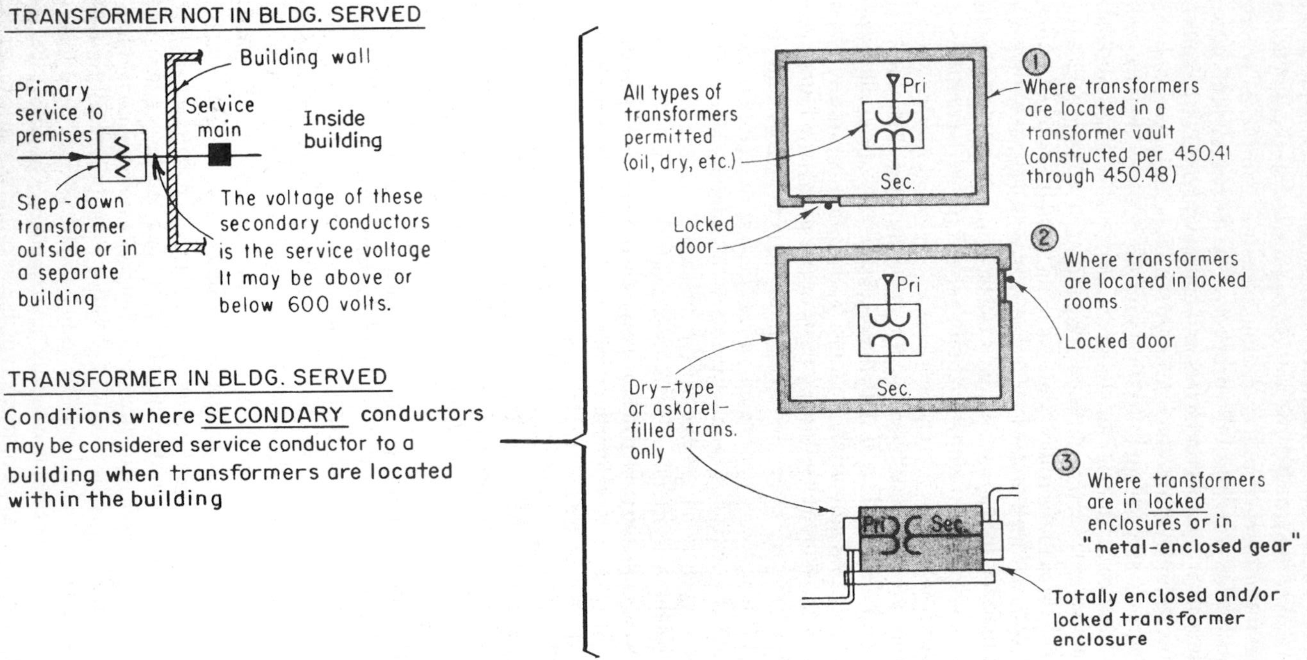
TRANSFORMER NOT IN BLDG. SERVED
Building wall
Primary service to premises
Service main
Inside building
Step-down transformer outside or in a separate building
The voltage of these secondary conductors is the service voltage It may be above or below 600 volts.
TRANSFORMER IN BLDG. SERVED
Conditions where SECONDARY conductors may be considered service conductor to a building when transformers are located within the building
All types of transformers permitted (oil, dry, etc.)
Pri
Sec.
Locked door
1 Where transformers are located in a transformer vault (constructed per 450.41 through 450.48)
2 Where transformers are located in locked rooms
Locked door
Dry-type or askarel-filled trans. only
3 Where transformers are in locked enclosures or in "metal-enclosed gear"
Totally enclosed and/or locked transformer enclosure

Fig. 230-51. Where the transformer belongs to the utility, the "service point" is on its secondary and the secondary conductors are the service conductors to the building or structure. (Sec. 230.202 and Art. 100.)

The wording in Art. 100 bases identification of "service conductors" as extending from the "service point." Because of the definition of "service point," it is essential to determine whether the transformers belong to the power company or the property owner.

If a utility-owned transformer that handles the electrical load for a building is in a locked room or locked enclosure (accessible only to qualified persons) in the building and is fed, say, by an underground medium-voltage (over 1000 V) utility line from outdoors, the secondary conductors from the transformer would be the "service conductors" to the building. *And* the switching and control devices (up to six CBs on fused switches) on the secondary would constitute the "service equipment" for the building. Under such a condition, if any of the secondary section "service disconnects" were rated 1000 A or more, at 480/277-V grounded wye, they would have to comply with 230.95, requiring GFP for the service disconnects.

However, if the utility made primary feed to a transformer or unit substation belonging to the owner, then the primary conductors would be the service conductors and the primary switch or CB would be the "service disconnect." In that case, no GFPE would be needed on the "service disconnect" because 230.95 applies only up to 1000 V, and there is no requirement for GFPE on medium-voltage services (Fig. 230-52). But in that case, there would be a need for GFPE on the secondary

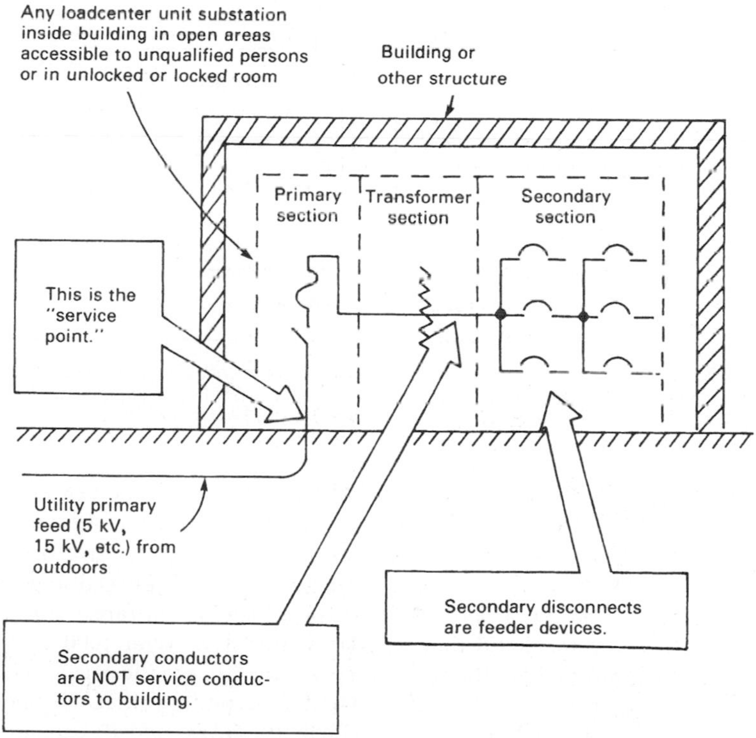
Any loadcenter unit substation inside building in open areas accessible to unqualified persons or in unlocked or locked room
Building or other structure
Primary section
Transformer section
Secondary section
This is the "service point."
Utility primary feed (5 kV, 15 kV, etc.) from outdoors
Secondary disconnects are feeder devices.
Secondary conductors are NOT service conductors to building.

Fig. 230-52. The primary is the "service" for any indoor transformer belonging to the owner and fed by utility line. (Sec. 230.200.)

section disconnects, even though they would not be "service disconnects"—and those are the same disconnects that might be subject to 230.95 if the transformer belonged to the utility. However, 215.10 or 240.13 may require such protection for these secondary section disconnects. (See also Fig. 230-53.)

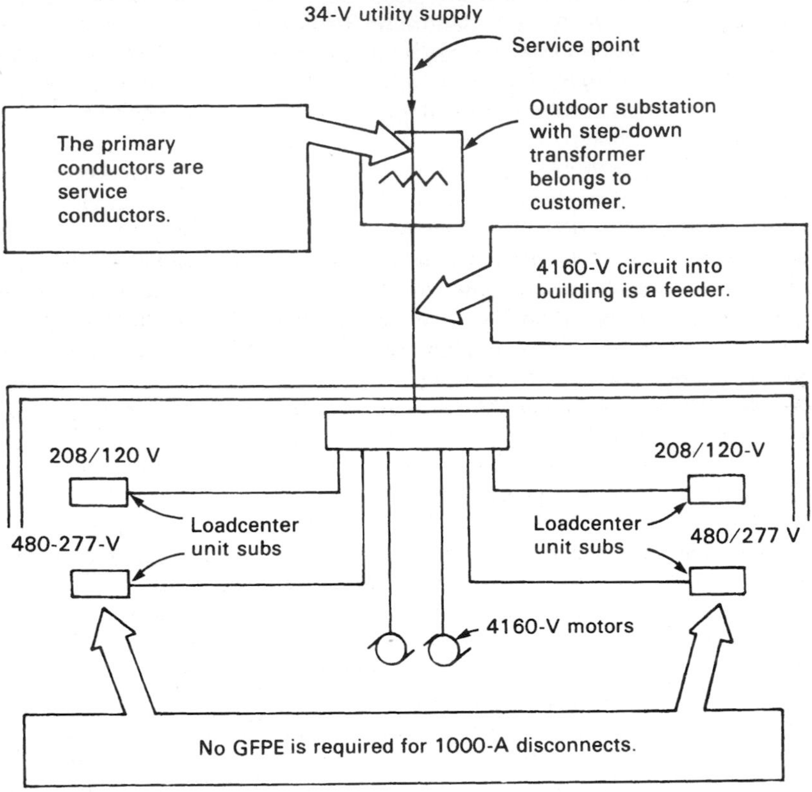
34-V utility supply
Service point
The primary conductors are service conductors.
Outdoor substation with step-down transformer belongs to customer.
4160-V circuit into building is a feeder.
208/120 V
208/120-V
Loadcenter unit subs
Loadcenter unit subs
480-277-V
480/277 V
4160-V motors
No GFPE is required for 1000-A disconnects.

Fig. 230-53. The primary circuit must be taken as the "service conductors" where the "service point" is on the primary side of an outdoor transformer. Although GFPE is not required by 230.95 for the 480Y/277-V disconnects shown, it still is required because 215.10 applies instead.

230.202. Service-Entrance Conductors. This section specifies the minimum conductor size, that is, No. 6 in a raceway and No. 8 in a multiconductor cable. In addition, it indicates that only those wiring methods given in Secs. 300.37 (aboveground) and 300.50 (underground) may be used. That section gives the wiring methods that are acceptable for use as service-entrance conductors where it has been established that primary conductors (over 1000 V) are the service conductors or where the secondary conductors are the service conductors and operate at more than 1000 V. The basic conduits that may be used are rigid metal conduit, intermediate metal conduit, RTRC and PVC nonmetallic conduits, and electrical metallic tubing. In addition, cable tray, cable bus, or "other identified" raceways or even type MC may be used. Note, too, that bare conductors, bare bus-work, or open runs of type MV are permitted as indicated. And the **NEC** no longer requires concrete encasement of the nonmetallic conduit.

Section 300.37 points out that *cable tray* systems are also acceptable for high-voltage services. However, any such application for service work would still require application of the cable tray rules in 230.44 regarding segregation and marking if used for dual voltages. Medium-voltage (over 1000 V) service-entrance cables may be used if they meet the requirements for the cables in Art. 426 and the rules for cable trays with medium-voltage wiring in Art. 392. Details of this section are shown in Fig. 230-54.

230.204. Isolating Switches. An air-break isolating switch capable of visible verification of the blade position must be used ahead of an oil switch or an air, oil, vacuum, or sulfur hexafluoride CB used as a service disconnecting means, unless removable truck panels or metal-enclosed units are used providing disconnection of all live parts in the removed position. In addition, such removable equipment must not be openable unless the circuit is disconnected. This line-side disconnect ensures safety to personnel in maintenance (Fig. 230-55). Part **(D)** requires a grounding connection for an isolating switch, as in Fig. 230-56.

230.205. Disconnecting Means. In part **(A)**, the basic rule requires a high-voltage service disconnecting means to be located "outside and within sight of, or inside nearest the point of entrance of, the service conductors" into the building or structure being supplied—as for 1000-V equipment in 230.70. A new provision in 2008 allows this disconnecting means to be located where it is not readily accessible if part of an "overhead or underground primary distribution system." The intent was to recognize the customary load break switches at the top of utility poles. Now it is true that the switch mechanism itself is not readily accessible. However, it is operable through a mechanical linkage at the pole base, and although mechanical, this meets all the provisions of 230.205(C) in that the switch is at a separate structure and operated remotely.

The real problem with this wording under the 2008 **NEC**, however, was that it avoided compliance with the rest of 230.205(C), which is that any such remote disconnecting provision be located in a readily accessible location. In effect, however unconscionable, it could have been considered acceptable to install a pole-top switch with no linkage to the pole base, thereby relying on personnel working with a hot stick out of a bucket truck to open the switch. This has all been corrected as of the 2011 edition. The operating handle must be readily accessible, or electronic control per 230.205(C) is another option.

Part **(B)**, covering the electrical fault characteristics, requires that the service disconnect be *capable of closing*, safely and effectively, on a fault equal to or greater than the maximum short-circuit current that is available at the line terminals of the disconnect. The last sentence notes that where fuses are used within the disconnect or in conjunction with it, the fuse characteristics may contribute to fault-closing rating of the disconnect. This provision recognizes that some medium-voltage fuses have current-limiting characteristics, and that having them in place will make it possible to close the switch safely. This might be seen as a modification of 110.9, which normally requires equipment that will

300.37 HIGH-VOLTAGE SERVICE CONDUCTORS FOR LOCATIONS ACCESSIBLE TO OTHER THAN QUALIFIED PERSONS

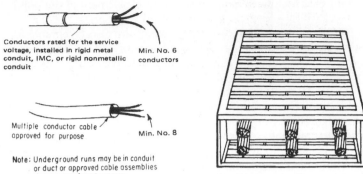

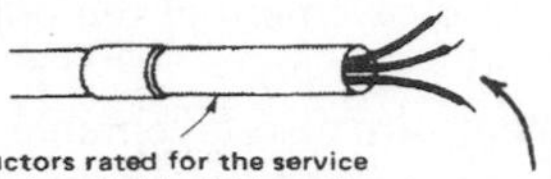
Conductors rated for the service voltage, installed in rigid metal conduit, IMC, or rigid nonmetallic conduit
Min. No. 6 conductors

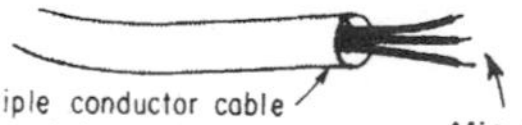
Multiple conductor cable approved for purpose
Min. No. 8

Note: Underground runs may be in conduit or duct or approved cable assemblies and must conform to 300.50

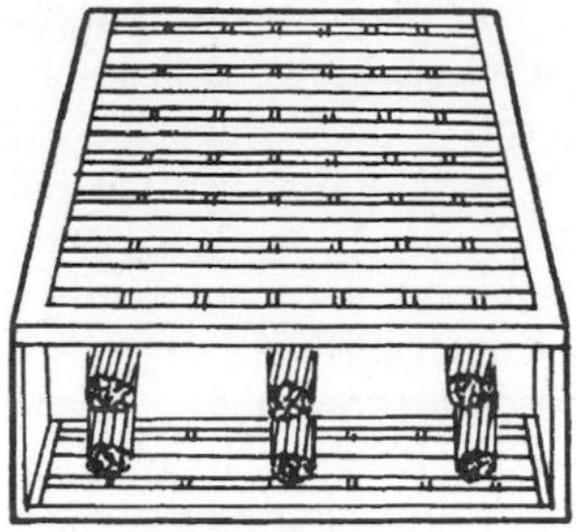

In cablebus–5 kV to 35 kV
[Article 370]

230.212 SERVICE CONDUCTORS OPERATING AT MORE THAN 35 kV

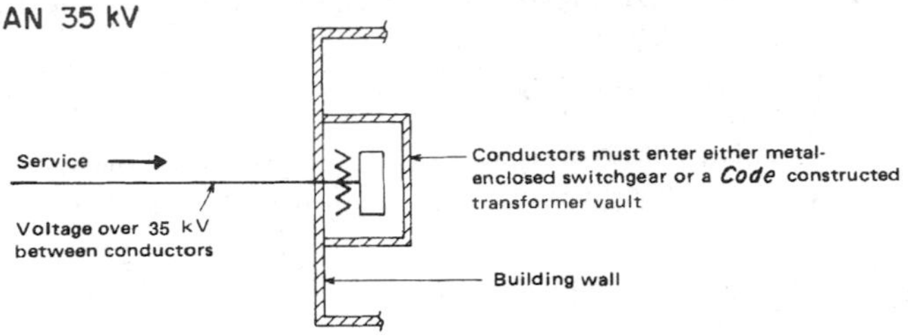
Service
Voltage over 35 kV between conductors
Conductors must enter either metal-enclosed switchgear or a *Code* constructed transformer vault
Building wall

300.42 POTHEAD ON SERVICE CONDUCTORS

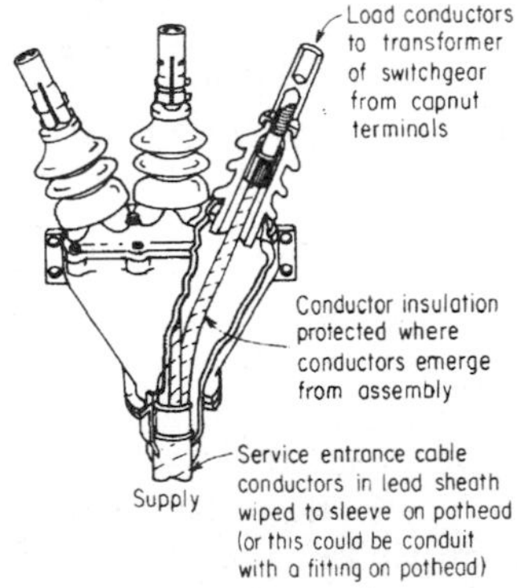
Load conductors to transformer of switchgear from capnut terminals
Conductor insulation protected where conductors emerge from assembly
Service entrance cable conductors in lead sheath wiped to sleeve on pothead (or this could be conduit with a fitting on pothead)
Supply

REMEMBER, 230.6 ALSO APPLIES TO HIGH-VOLTAGE CONDUCTORS (OVER 600 V NOMINAL) AS GIVEN IN 230.200, THEREFORE CONDUCTORS ENCLOSED IN MASONRY ARE CONSIDERED AS INSTALLED OUTSIDE THE BUILDING

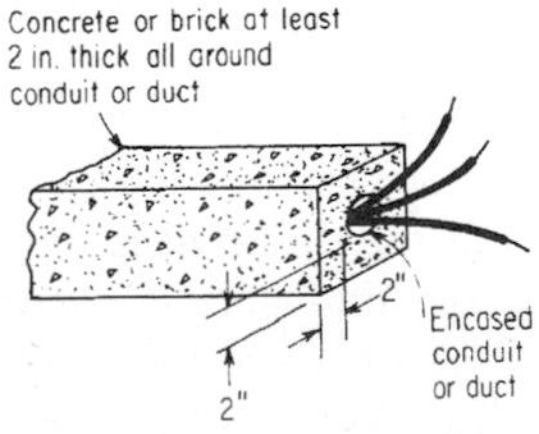
Concrete or brick at least 2 in. thick all around conduit or duct
2"
2"
Encased conduit or duct

Fig. 230-54. Provisions for service conductors rated over 1000 V (refer to subpart letter identification of rules). (Sec. 230.202.)

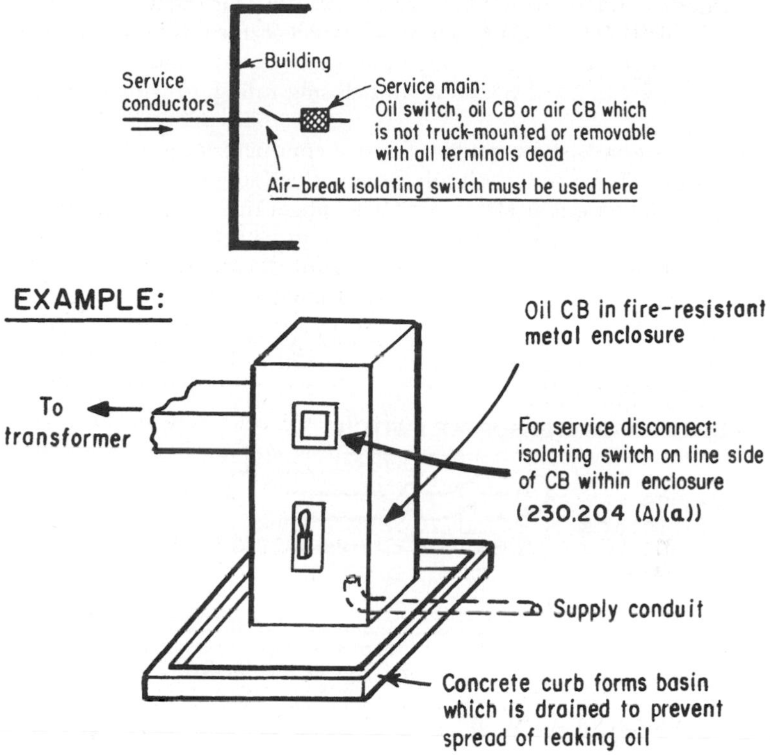
Building
Service conductors
Service main:
Oil switch, oil CB or air CB which is not truck-mounted or removable with all terminals dead
Air-break isolating switch must be used here
EXAMPLE:
Oil CB in fire-resistant metal enclosure
To transformer
For service disconnect: isolating switch on line side of CB within enclosure (230.204 (A)(a))
Supply conduit
Concrete curb forms basin which is drained to prevent spread of leaking oil

Fig. 230-55. Isolating switch may be needed to kill line terminals of service disconnect. (Sec. 230.204.)

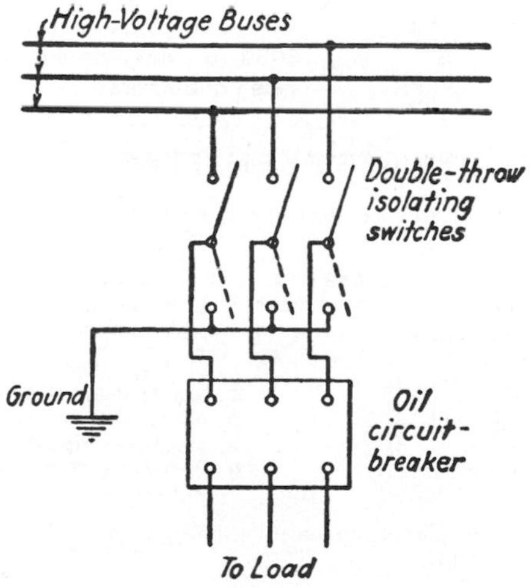
High-Voltage Buses
Double-throw isolating switches
Ground
Oil circuit-breaker
To Load

Fig. 230-56. One method for grounding the load side of an open isolating switch. (Sec. 230.204.)

interrupt circuits under fault conditions to be rated for the available fault current as their supply terminals. In this case, however, there is a distinction drawn between a fault clearing rating, which the fuse will have and is normally quite high, and covered in 230.208, and a fault closing rating, to which the fuse may safely contribute.

230.208. Protection Requirements. Service conductors operating at voltages over 1000 V must have a short-circuit (not overload) device in each ungrounded conductor, installed either (1) on load side of service disconnect, or (2) as an integral part of the service disconnect.

All devices must be able to detect and interrupt all values of current in excess of their rating or trip setting, which must be as shown in Fig. 230-57.

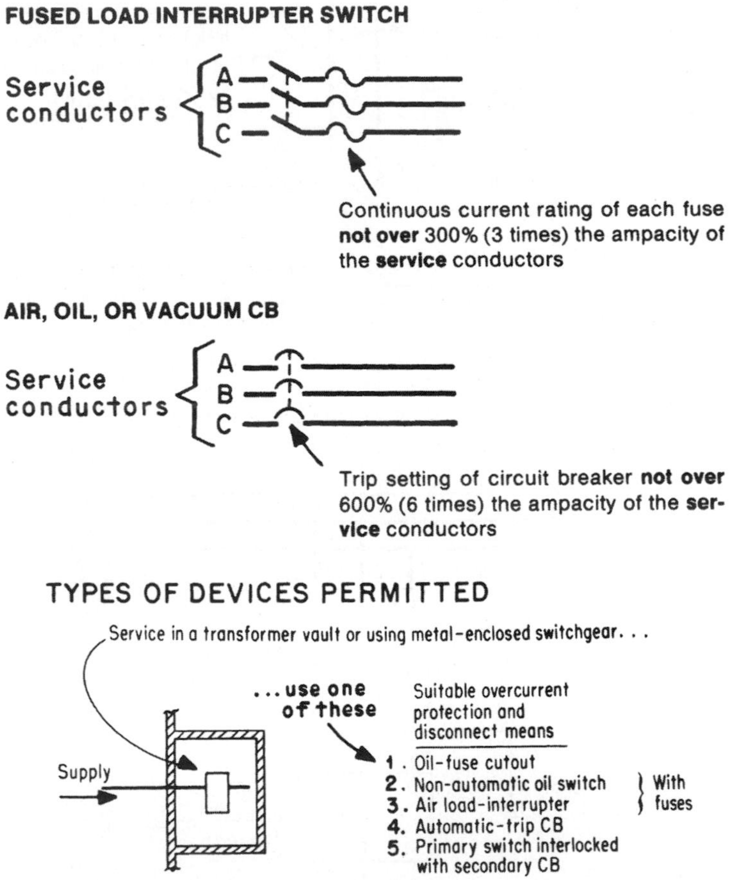
FUSED LOAD INTERRUPTER SWITCH
Service conductors
A
B
C
Continuous current rating of each fuse not over 300% (3 times) the ampacity of the service conductors
AIR, OIL, OR VACUUM CB
Service conductors
A
B
C
Trip setting of circuit breaker not over 600% (6 times) the ampacity of the service conductors
TYPES OF DEVICES PERMITTED
Service in a transformer vault or using metal-enclosed switchgear. . .
. . . use one of these
Supply
Suitable overcurrent protection and disconnect means
1. Oil-fuse cutout
2. Non-automatic oil switch
3. Air load-interrupter
With fuses
4. Automatic-trip CB
5. Primary switch interlocked with secondary CB

Fig. 230-57. Maximum permitted rating or setting of high-voltage overcurrent protection for service. (Sec. 230.208.)

The difference between 300 percent for fuses and 600 percent for CBs is explained as follows:

The American National Standards Institute (ANSI) publishes standards for power fuses. The continuous-current ratings of power fuses are given with the letter "E" following the number of continuous amps—for instance, 65E or 200E or 400E. The letter "E" indicates that the fuse has a melting time-current characteristic in accordance with the standard for E-rated fuses:

> The melting time–current characteristics of fuse units, refill units, and links for power fuses shall be as follows:
>
> (1) The current-responsive element with ratings 100 amperes or below shall melt in 300 seconds at an rms current within the range of 200 or 240 percent of the continuous current rating of the fuse unit, refill unit, or fuse link.
>
> (2) The current-responsive element with ratings above 100 amperes shall melt in 600 seconds at an rms current within the range of 220–264 percent of the continuous current rating of the fuse unit, refill unit, or fuse link.
>
> (3) The melting time–current characteristic of a power fuse at any current higher than the 200 to 240 or 264 percent specified in (1) or (2) above shall be shown by each manufacturer's published time current curves, since the current-responsive element is a distinctive feature of each manufacturer.
>
> (4) For any given melting time, the maximum steady-state rms current shall not exceed the minimum by more than 20 percent.

The fact that E-rated fuses are given melting times at 200 percent or more of their continuous-current rating explains why **NEC** 230.208 and 240.100 set 300 percent of conductor ampacity as the maximum fuse rating but permit CBs up to 600 percent. In effect, the 300 percent for fuses times 2 (200 percent) becomes 600 percent—the same as for CBs.

Part **(B)** of this section permits overcurrent protection for services over 1000 V to be loaded up to 100 percent of its rating even on continuous loads (operating for periods of 3 h or more). The greater spacings in medium-voltage equipment permit this latitude safely; see also 110.40 for allowances to use 90°C ratings on medium-voltage terminations for the same reason.

ARTICLE 240. OVERCURRENT PROTECTION

240.1. Scope. For any electrical system, required current-carrying capacities are determined for the various circuits—feeders, subfeeders, and branch circuits. Then these required capacities are converted into standard circuit conductors that have sufficient current-carrying capacities based on the size of the conductors, the type of insulation on the conductors, the ambient temperature at the place of installation, the number of conductors in each conduit, the type and continuity of load, and judicious determination of spare capacity to meet future load growth. Or if busway, armored cable, or other cable assemblies are to be used, similar considerations go into selection of conductors with required current-carrying capacities. In any case, the next step is to provide overcurrent protection for each and every circuit.

The overcurrent device for conductors or equipment must automatically open the circuit it protects if the current flowing in that circuit reaches a value that will cause an excessive or dangerous temperature in the conductor or conductor installation.

Overcurrent protection for conductors must also be rated for safe operation at the level of fault current obtainable at the point of their application. Every fuse and circuit breaker for short-circuit protection must be applied in such a way that the fault current produced by a bolted short circuit on its lead terminals will not damage or destroy the device. Specifically, this requires that a short-circuit overcurrent device have a proven interrupting capacity at least equal to the current that the electrical system can deliver into a short on its line terminals. That is, the calculation for the short-circuit interrupting rating must *not* include the impedance of the device itself. That impedance may only be applied to the calculation for the next device downstream.

But safe application of a protective device does not stop with adequate interrupting capacity for its own use at the point of installation in the system. The speed of operation of the device must then be analyzed in relation to the thermal and magnetic energy which the device permits to flow in the faulted circuit. A very important consideration is the provision of conductor size to meet the potential heating load of short-circuit currents in cables. With expanded use of circuit-breaker overcurrent protection, coordination of protection from loads back to the source has introduced time delays in operation of overcurrent devices. Cables in such systems must be able to withstand any impressed short-circuit currents for the durations of overcurrent delay. For example, a motor circuit to a 100-hp motor might be required to carry as much as 15,000 A for a number of seconds. To limit damage to the cable due to heating effect, a much larger size conductor than necessary for the load current alone may be required.

A device may be able to break a given short-circuit current without damaging itself in the operation; but in the time it takes to open the faulted circuit, enough energy may get through to damage or destroy other equipment in series with the fault. This other equipment might be a cable or busway or a switch or motor controller—any circuit component which simply cannot withstand the few cycles of short-circuit current that flows in the period of time between initiation of the fault and interruption of the current flow.

The informational note following Sec. 240.1 often raises questions about the approved use of conductors and overcurrent protection to withstand faults.

example Assume a panelboard with 20-A breakers rated 10,000 AIR (ampere interrupting rating) and No. 12 copper branch-circuit wiring. Available fault current at the point of breaker application is 8000 A. The short-circuit withstand capability of a No. 12 copper conductor with plastic or polyethylene insulation rated 60°C would be approximately 3000 A of fault or short-circuit current for one cycle.

question: Assuming that the CB (circuit breaker) will take at least one cycle to operate, would use of the conductor where exposed to 8000 A violate Sec. 110.9 or 110.10? These sections state that overcurrent protection for conductors and equipment is provided for the purpose of opening the electrical circuit if the current reaches a value that will cause an excessive or dangerous temperature in the conductor or conductor insulation. The 8000-A available fault current would seem to call for use of conductors with that rating of short-circuit withstand.

This could mean that branch-circuit wiring from all 20-A CBs in this panelboard must be *No. 6 copper* (the next larger size suitable for an 8000-A fault current).

answer: As noted in UL Standard 489, a CB is required to operate safely in a circuit where the available fault current is up to the short-circuit current value for which the breaker is rated. The CB must clear the fault without damage to the insulation of conductors of proper size for the rating of the CB. A UL-listed, 20-A breaker is, therefore, tested and rated to be used with 20-A-rated wire (say, No. 12 THW) and will protect the wire in accordance with 240.4 when applied at a point in a circuit where the short-circuit current available does not exceed the value for which the breaker is rated. This is also true of a 15-A breaker on No. 14 (15-A) wire, for a 30-A breaker on No. 10 (30-A) wire, and all wire sizes.

UL 489 states:

> A circuit breaker shall perform successfully when operated under conditions as described in paragraphs 21.2 and 21.3. There shall be no electrical or mechanical breakdown of the device, and the fuse that is indicated in paragraph 12.16 shall not have cleared. Cotton indicators as described in paragraphs 21.4 and 21.6 shall not be ignited. There shall be no damage to the insulation on conductors used to wire the device. After the final operation, the circuit breaker shall have continuity in the closed position at rated voltage.

240.2. Definitions. Here, the Code provides a number of additional definitions that apply to this article on overcurrent protection. These definitions must be considered when interpreting the requirements given in this Code article. They will be referred to in the context of subsequent discussion of the relevant topics.

240.3. Other Articles. Here, the Code reminds us that the rules given in Art. 240 are essentially general requirements for conductor protection. The individual articles for the equipment specifically indicated by this section have overcurrent protection requirements that are different from the "general rules" given in Art. 240 for protection of conductors and flexible cords. Where installing overcurrent protection for circuits for the equipment, and in the locations, identified here, the rules for overcurrent protection given in the indicated articles supersede the requirements given in Art. 240. In effect, compliance with those rules satisfies the rule of 240.3.

240.4. Protection of Conductors. Aside from flexible cords and fixture wires, conductors for all other circuits must conform to the rules of 240.4.

Clearly, the rule wants overcurrent devices to prevent conductors from being subjected to currents in excess of the ampacity values for which the conductors are rated by 310.15 and Tables 310.15(B)(16) through 310.15(B)(21).

The wording mentions 310.15, which includes Tables 310.15(B)(16) through 310.15(B)(21). That is important because it points out that when conductors have their ampacities derated because of conduit fill [310.15(B)(3)(a)] or because of elevated ambient temperature, the conductors must be protected at the *derated* ampacities and *not* at the values given in the tables.

As of the 2011 **NEC** there is a new informational note at this point referencing an ICEA paper (P-32-382-2007) providing information on allowable short-circuit currents for insulated conductors. The inclusion of this note was very controversial, with the NEMA, UL, and IAEI representatives all voting against its inclusion. The concerns in the voting involved lack of confidence in the validity of the data in the ICEA document relative to actual field conditions, and the fact that overcurrent devices are all tested with actual conductor samples attached to them; if the conductor is not protected under the worst-case fault condition for which the device is being rated, then the device fails the test. Finally, the note could be misapplied to require wire sizes above those given in Table 310.15(B)(16) although no substantiation was presented to invalidate the ampacity table, and Table 310.15(B)(16) is much easier to enforce than the calculations in the ICEA paper. Conductors do not have withstand ratings.

This information has been previously proposed on several occasions for general application in the NEC, including the 1987 cycle when it made it all the way through the normal process but was turned aside on the floor of the NFPA Annual Meeting. One reason was that the ICEA indicated that its paper was not appropriate for a reference in the NEC. A comparable effort on equipment grounding conductors failed in the 1996 NEC cycle. It does appear in Part **VIII** of Art. 240 (Supervised Industrial Installations), but this is a more appropriate context because it is not generally applicable and primarily addresses taps, which by definition have overcurrent protection that is either beyond the conductor rating or absent.

The basic rule of Sec. 240.4(A) represents a basic concept in Code application. When conductors supply a load to which loss of power would create a hazard, this rule states it is not necessary to provide "overload protection" for such conductors, *but* "short-circuit protection" *must* be provided. By "overload protection," this means "protection at the conductors' ampacity"—that is, protection that would prevent overload (Fig. 240-1).

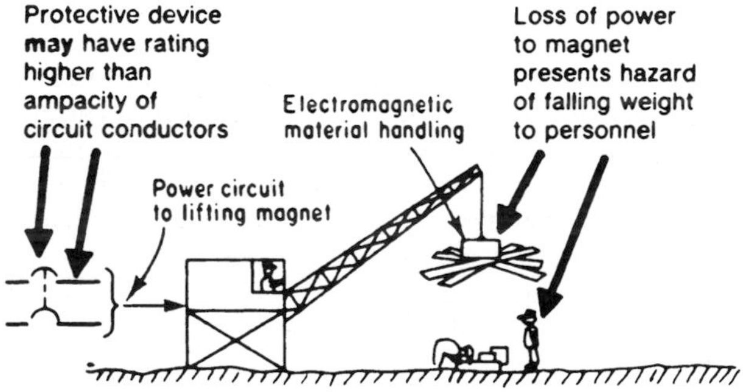
Protective device may have rating higher than ampacity of circuit conductors
Electromagnetic material handling
Loss of power to magnet presents hazard of falling weight to personnel
Power circuit to lifting magnet

Fig. 240-1. If "overload protection" creates a hazard, it *may* be eliminated. (Sec. 240.4.)

Several points should be noted about this rule.

1. This requirement is reserved only for applications where circuit opening on "overload" would be more objectionable than the overload itself, "such as in a material handling magnet circuit." In that example mentioned in the rule, loss of power to such a magnet while it is lifting a heavy load of steel would cause the steel to fall and would certainly be a serious hazard to personnel working below or near the lifting magnet. To minimize the

hazard created by such power loss, the circuit to it *need not* be protected at the conductor ampacity. A higher value of protection may be used—letting the circuit sustain an overload rather than opening on it and dropping the steel. Because such lifting operations are usually short-time, intermittent tasks, occasional overload is far less a safety concern than the dropping of the magnet's load.

2. The rule to eliminate *only* "overload protection" is not limited to a lifting-magnet circuit, which is mentioned simply as an example. Other electrical applications that present a similar concern for "hazard" would be equally open to use of this rule. Fire pump circuits are required to implement this principle.
3. Although 240.4(A) *allows* elimination of overload protection and requires short-circuit protection, it gives no guidance on selecting the actual rating of protection that must be used. For such circuits, fuses or a CB rated, say, 200 to 400 percent of the full-load operating current would give freedom from overload opening. Of course, the protective device ought to be selected with as low a rating as would be compatible with the operating characteristics of the electrical load. And it must have sufficient interrupting capacity for the circuit's available short-circuit current.
4. Finally, it should be noted that this is *not* a mandatory rule but a *permissible* application. It says ". . . overload protection *shall not be required* . . . "; it does not say that overload protection "shall *not be used.*" Overload protection *may be used,* or it *may be eliminated.* Obviously, careful study should always go into application of this requirement.

Specifically, the general rule is that the device must be rated to protect conductors in accordance with their safe allowable current-carrying capacities. Of course, there will be cases where standard ampere ratings and settings of overcurrent devices will not correspond with conductor capacities. In such cases, part **(B)** permits the next larger standard size of overcurrent device to be used where the rating of the protective device is 800 A or less, unless the circuit in question is a multioutlet receptacle circuit for cord- and plug-connected portable loads, in which case the next smaller standard size overcurrent device *must* be used. Therefore, a basic guide to effective selection of the amp rating of overcurrent devices for any feeder or service application is given in various subsections [**(A)** through **(G)**].

For example, if a circuit conductor of, say, 500-kcmil THW copper (not more than three in a conduit at not over 86°F [30°C] ambient) satisfies design requirements and **NEC** rules for a particular load current not in excess of the conductor's table ampacity of 380 A, then the conductor *may* be protected by a 400-A rated fuse or CB.

240.6, which gives the "Standard Ampere Ratings" of protective devices to correspond to the word "standard" in part **(B)**, shows devices rated at 350 and 400 A, but none at 380 A. In such a case, the **NEC** accepts a 400-A-rated device as "the next higher standard device rating" above the conductor ampacity of 380 A.

But, such a 400-A device would permit load increase above the 380 A that is the safe maximum limit for the conductor. Better conductor protection could be achieved by using a 350-A-rated device, which will prevent such overload.

For application of fuses and CBs, parts **(B)** and **(C)** have this effect:

1. If the ampacity of a conductor does not correspond to the rating of a standard-size fuse, the next *larger* rating of fuse may be used only where that rating is 800 A or less. Over 800 A, the next *smaller* fuse must be used, as covered in part **(C)**. For any circuit over 800 A, 240.4(C) prohibits the use of "the next higher standard" rating of protective device (fuse or CB) when the ampacity of the circuit conductors does not correspond with a standard ampere rating of fuse or CB. The rating of the protection may not exceed the conductor ampacity. Although it would be acceptable to use a protective device of the next lower standard rating (from 240.6) below the conductor ampacity, there are many times when greater use of the conductor ampacity may be made by using a fuse or CB of rating lower than the conductor ampacity but not as low as the next lower standard rating. Listed fuses and CBs are made with ratings between the standard values shown in 240.6.

   For example, if the ampacity of conductors for a feeder circuit is calculated to be 1540 A, 240.4(C) does not permit protecting such a conductor by using the next higher standard rating above 1540–1600 A. The next lower standard rating of fuse or CB shown in 240.6 is 1200 A. Such protection could be used, but that would sacrifice 340 A (1540 – 1200) of conductor ampacity. Because listed 1500-A protective devices are available and would provide for effective use of almost all the conductor's 1540-A capacity, this rule specifically recognizes such an application as safe and sound practice. Such application is specifically recognized by the second sentence of 240.6.

   In general, 240.6 is not intended to require that all fuses or CBs be of the standard ratings shown. Intermediate values of protective device ratings may be used, provided all Code rules on protection—especially, the basic first sentence of 240.4, which requires conductors to be protected at their ampacities—are satisfied (Fig. 240-2).

2. A nonadjustable-trip breaker (one without overload trip adjustment above its rating—although it may have adjustable short-circuit trip) must be rated in accordance with the current-carrying capacity of the conductors it protects—except that the next higher standard rating of CB may be used if the ampacity of the conductor does not correspond to a standard unit rating. In such a case, the next higher standard setting may be used only where the rating is 800 A or less. An example of such application is shown in Fig. 240-1, where a nonadjustable CB with a rating of 1200 A is used to protect the conductors of a feeder circuit, which are rated at 1140 A. As shown there, use of that size CB to protect a circuit rated at 1140 A (3 × 380 A = 1140 A) clearly violates 240.4(C) because the CB is the next higher rating above the ampacity of the conductors—on a circuit rated over 800 A. With a feeder circuit as shown (three 500-kcmil THW, each rated at 380 A), the CB must *not* be rated over 1140 A. A standard 1000-A CB would satisfy the Code rule—being the *next lower* rated protective device from 240.6. Or a 1100-A fuse could be used. Of course, if 500-kcmil THHN or XHHW conductors are used instead of THW conductors, then each 500 is rated at 430 A, three per phase would give the circuit an ampacity of 1290 A (3 × 430), and the 1200-A CB would satisfy the

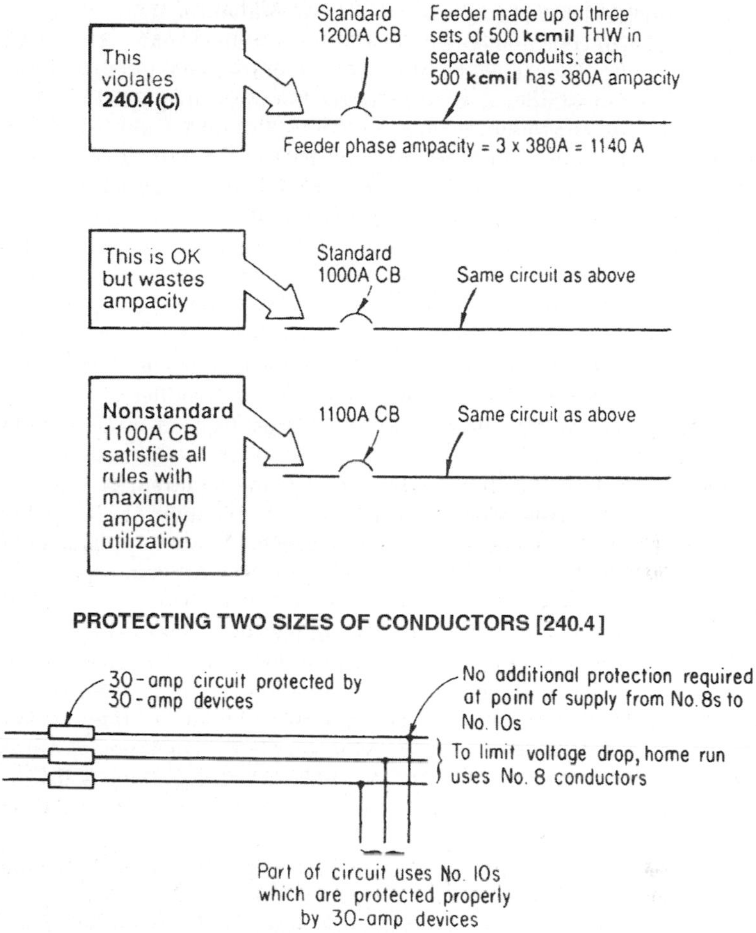

This violates 240.4(C)
Standard 1200A CB
Feeder made up of three sets of 500 kcmil THW in separate conduits; each 500 kcmil has 380A ampacity
Feeder phase ampacity = 3 x 380A = 1140 A
This is OK but wastes ampacity
Standard 1000A CB
Same circuit as above
Nonstandard 1100A CB satisfies all rules with maximum ampacity utilization
1100A CB
Same circuit as above
PROTECTING TWO SIZES OF CONDUCTORS [240.4]
30-amp circuit protected by 30-amp devices
No additional protection required at point of supply from No. 8s to No. 10s
To limit voltage drop, home run uses No. 8 conductors
Part of circuit uses No. 10s which are protected properly by 30-amp devices

Fig. 240-2. Protection in accordance with 240.4(C) may use standard or nonstandard rated fuses or circuit breakers. And, smaller conductors are considered protected as covered in 240.4(D). (Secs. 240.4 and 240.6.)

basic rule in 240.4(C). *But,* given that 500-kcmil conductors would be operating at 90°C when carrying the 430 A of current, such conductors could never be loaded to that value as there is no equipment rated for use with conductors operating at 90°C. To satisfy the termination temperature limitations of 110.14(C), the load would be prohibited from exceeding the 75°C value, or 3 × 380 A = 1140 A. Alternatively, a circuit breaker listed for terminations operating at 90°C could be used. These are only available in very large frame sizes, such as those in this example.

It should be noted, however, that 240.4(B) requires that the rating of overcurrent protection must *never* exceed the ampacity of circuit conductors supplying one or more receptacle outlets on a branch circuit with more than one outlet. This wording in 240.4(B) coordinates with the rules described under 210.19(A)(2) on conductor ampacity. The effect of that rule is to require that the rating of the overcurrent protection must not exceed the Code-table ampacity [**NEC** Table 310.15(B)(16)] or the derated ampacity dictated by 310.15(B)(3)(a) for any conductor of a multioutlet branch circuit supplying any receptacles for cord-and-plug-connected portable loads. If a standard rating of fuse or CB does not match the ampacity (or derated ampacity) of such a circuit, the next lower standard rating of protective device must be used. *But,* where branch-circuit conductors of an individual circuit to a single load or a multioutlet circuit supply *only* fixed connected (hardwired) loads—such as lighting outlets or permanently connected appliances—the next larger standard rating of protective device *may* be used in those cases where the ampacity (or derated ampacity) of the conductor does not correspond to a standard rating of protective device—but, again, that is permitted only up to 800 A, above which the next lower rating of fuse or CB must be used, as described under 210.19(A)(2).

The rules of 240.4 must also be correlated with the requirement for minimum conductor size where continuous loading or a combination of continuous and noncontinuous loading is supplied. Where such loads are supplied, 210.19(A), 210.20(A), 215.2(A), 215.3(A), and 230.42 require that additional capacity be provided where the branch-circuit, feeder, or service conductors and overcurrent devices supply continuous loads. After that minimum size has been established, the OCPD must be rated such that it either protects the conductors in accordance with their ampacity, or is the next larger rated overcurrent device—up to 800 A. Above 800 A, 240.4(C) would mandate use of the next smaller rated overcurrent device, which may not be adequately rated to supply the continuous load. Careful correlation of the rules here and in 210.19, 210.20, 215.2, 215.3, and 230.42 is especially important to ensure that the selected conductors and OC protection are properly rated.

At the end of this book, as part of the detailed coverage of ampacity calculations, all of these code requirements are integrated in one location. The coverage focuses on Annex D, Example D3(a), which is the new example devoted to ampacity calculations as distinguished from load calculations.

Part **(D)** of 240.4 covers a long-standing requirement for protection of the smaller sizes of conductors, that is, No. 14, No. 12, and No. 10. Although such conductors have greater ampacities, as shown in Table 310.15(B)(16), this rule requires that the maximum rating of overcurrent protection be 15, 20, and 30 A, respectively. This limitation on the rating of overcurrent devices is related to the fact that listed overcurrent devices cannot protect against conductor damage under short-circuit testing where, say, a No. 12 copper THW conductor is protected by a 25-A CB. Although the No. 12 has an ampacity of 25 A and can carry that current when used under the conditions described in the heading of Table 310.15(B)(16), a 25-A CB will not operate fast enough to prevent the conductor from burning open during the short-circuit test. It was established that lower-rated breakers, such as a 20-A CB, can protect the No. 12 conductor from damage under short-circuit conditions. Because this is not a problem with CBs rated for protection of conductor sizes No. 8 and larger, only No. 14, No. 12, and No. 10, copper, as well as No. 12 and

No. 10 aluminum and copper-clad aluminum conductors, are specially limited with regard to the maximum rating of their overcurrent protection, where they are required to be protected in accordance with their ampacity. The 2011 **NEC** changed the 60°C column ampacities for these conductors so they now agree with the limits in these paragraphs, and is true that 110.14(C)(1)(a) uses this temperature as the default termination temperature starting points for these wire sizes. However, the limitations here need to stay in place because other columns can be used in some cases, and unlike the ampacity tables, these limits do not vary with insulation temperature ratings.

In the 2008 **NEC**, this part of 240.4 was expanded to cover even smaller conductors (18 and 16 AWG) and the allowable overcurrent devices that can be used to protect them when they are not considered to be protected by ordinary branch-circuit protective devices due to special provisions in various Code rules. The principal application of these ampacities involves industrial machinery covered by *NFPA 79* where international competition dictates smaller wires for small motors that are components of complicated equipment. See 430.22(G).

240.4(G) gives a list of "specific conductor applications" that are exempt from the basic rules of 240.4. For example, this rule refers the matter of protecting motor-control circuits to Art. 430 on motors.

Table 240.4(G) also applies to the protection of the remote-control circuit that energizes the operating coil of a magnetic contactor, as distinguished from a magnetic motor starter (Fig. 240-3).

725.45(C) covers control wires for magnetic contactors used for control of lighting or heating loads, but not motor loads. 430.72 covers that requirement for

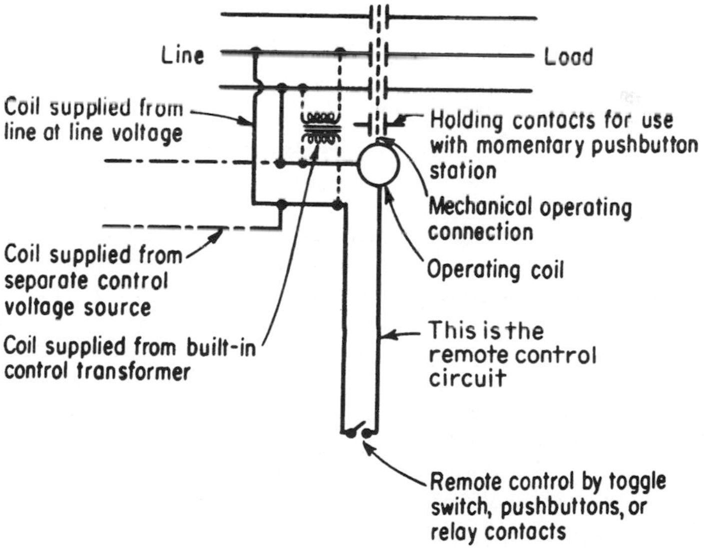
Line
Load
Coil supplied from line at line voltage
Holding contacts for use with momentary pushbutton station
Mechanical operating connection
Operating coil
Coil supplied from separate control voltage source
Coil supplied from built-in control transformer
This is the remote control circuit
Remote control by toggle switch, pushbuttons, or relay contacts

Fig. 240-3. Coil-circuit wires of magnetic contactor must be protected as required by 725.43. (Sec. 240.4.)

motor-control circuits. In Fig. 240-4, the remote-control conductors may be considered properly protected by the branch-circuit overcurrent devices (A) if these devices are rated or set at not more than 300 percent of (3 times) the current rating of the control conductors. If the branch-circuit overcurrent devices were rated or set at more than 300 percent of the rating of the control conductors, the control conductors would have to be protected by a separate protective device located at the point (B) where the conductor to be protected receives its supply. [See 725.45(C).]

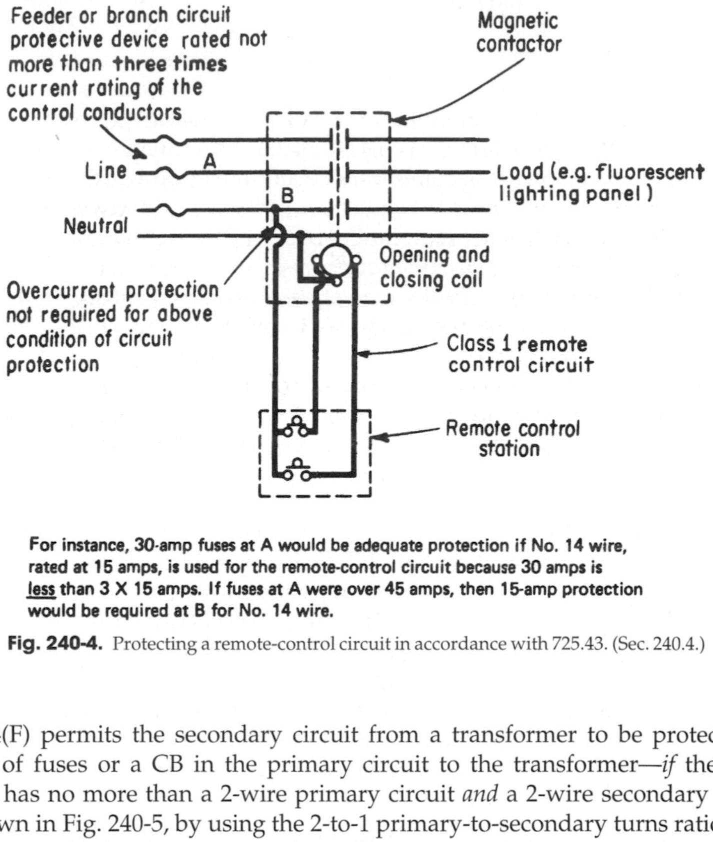
Feeder or branch circuit protective device rated not more than three times current rating of the control conductors
Magnetic contactor
Line
A
B
Neutral
Load (e.g. fluorescent lighting panel)
Opening and closing coil
Overcurrent protection not required for above condition of circuit protection
Class 1 remote control circuit
Remote control station
For instance, 30-amp fuses at A would be adequate protection if No. 14 wire, rated at 15 amps, is used for the remote-control circuit because 30 amps is less than 3 X 15 amps. If fuses at A were over 45 amps, then 15-amp protection would be required at B for No. 14 wire.

Fig. 240-4. Protecting a remote-control circuit in accordance with 725.43. (Sec. 240.4.)

240.4(F) permits the secondary circuit from a transformer to be protected by means of fuses or a CB in the primary circuit to the transformer—*if* the transformer has no more than a 2-wire primary circuit *and* a 2-wire secondary circuit. As shown in Fig. 240-5, by using the 2-to-1 primary-to-secondary turns ratio of the transformer, 20-A primary protection will protect against any secondary current in excess of 40 A—thereby protecting, say, secondary No. 8 TW wires rated at 40 A. As the wording of the rule states, the protection on the primary (20 A) must not exceed the value of the secondary conductor ampacity (40 A) multiplied by the secondary-to-primary transformer voltage ratio ($120 \div 240 = 0.5$). Thus, $40 \text{ A} \times 0.5 = 20 \text{ A}$. But it should be carefully noted that the rating of the primary protection must comply with the rules of 450.3(B).

The rule of part **(F)** also recognizes protection of the secondary conductors by the primary OCPD for delta-delta-wound transformers. This permission recognizes

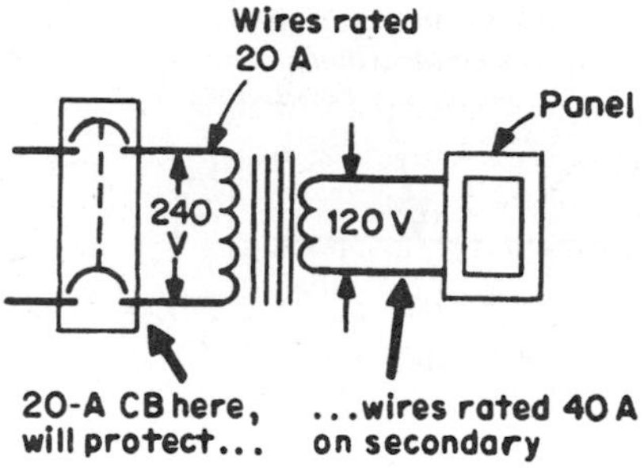
Wires rated
20 A
Panel
240
V
120 V
20-A CB here,
will protect...
...wires rated 40 A
on secondary

Fig. 240-5. Primary fuses or CB may protect secondary circuit for 2-wire to 2-wire transformer. (Sec. 240.4.)

that the "per-unit" current value on the secondary side will be equal to or less than the per-unit current value on the primary conductors. And, because a directly proportional current will be carried by both conductors, the overcurrent device on the primary side can protect both sets of conductors, the primary and secondary. For 3- and 4-wire delta-wye-wound transformers, separate overcurrent protection is required for the primary conductors and secondary conductors.

To put this another way, no conductors connected to a dual-voltage transformer secondary can be protected on the primary side by relying on a turns ratio. Consider a 480-V to 120/240-V transformer of the type commonly used to create separately derived single-phase systems for local lighting and receptacles. Suppose the panel on the secondary side is rated 100 A, the secondary conductors are 3 AWG, and the primary-side circuit breaker is rated 50 A. The winding ratio from 480 to 240 V is 2:1, so the maximum current that could flow over the secondary conductors is 100 A, right?

Wrong. If the load in the panel is perfectly balanced, then when the load on the panel exceeds 100 A, the primary side protection will open, true enough. But now suppose the panel load is not balanced. In fact, suppose the worst case happens, and 100 percent of the line-to-neutral load is on only one of the line legs. Now the transformer is, in effect, operating as a 4:1 (480:120 V) transformer. At this point, 100 A of load on the secondary, at 120 V, will cause only 25 A of current to flow in the primary. The transformer will be quite happy, and the primary side protection will be nowhere close to opening. Meanwhile, up to 200 percent of rated current (in this case 200 A) could be drawn on the secondary side before the primary side would open. The so-called protection on the primary side does protect the transformer, but it is absolutely useless in terms of reliably protecting the conductors and other equipment on the secondary side. 725.45(D) clearly makes this point regarding Class 1 control circuit conductors, and 240.21(C)(1) reiterates the point made here in 240.4(F) for power circuits. Figure 240-6 previews the rules in 240.21(C) and Fig. 240-7 gives another example of the problems with potential imbalances on a multiwire transformer secondary.

240.5. Protection of Flexible Cords, Flexible Cables, and Fixture Wires. The basic rules of part **(A)** are that

1. *All flexible cords and extension cords* must be protected at the ampacity given for each size and type of cord or cable in **NEC** Tables 400.5(A)(1) and

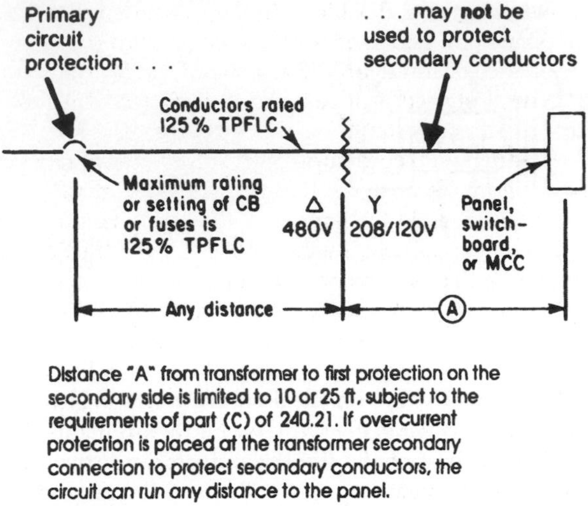
Primary
circuit
protection . . .
. . . may not be
used to protect
secondary conductors
Conductors rated
125% TPFLC
Maximum rating
or setting of CB
or fuses is
125% TPFLC
Δ
480V
Y
208/120V
Panel,
switch-
board,
or MCC
Any distance
A
Distance "A" from transformer to first protection on the
secondary side is limited to 10 or 25 ft, subject to the
requirements of part (C) of 240.21. If overcurrent
protection is placed at the transformer secondary
connection to protect secondary conductors, the
circuit can run any distance to the panel.

Fig. 240-6. Part **(C)** *clearly* resolves long-standing controversy. (Sec. 240.4.)

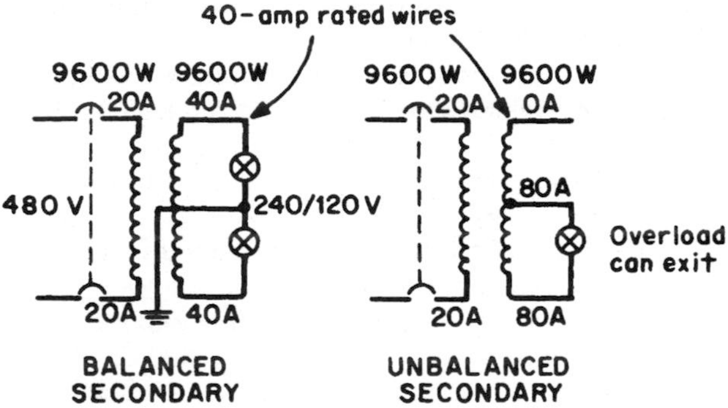
40-amp rated wires
9600W 9600W
20A 40A
480 V
240/120 V
20A 40A
9600W 9600W
20A 0A
80A
Overload
can exit
20A 80A
BALANCED
SECONDARY
UNBALANCED
SECONDARY

Fig. 240-7. Why primary protection may not do the job for 3-wire or 4-wire secondary 40-A-rated wires. (Sec. 240.4.)

400.5(A)(2). "Flexible cords" includes "tinsel cord"—No. 27 AWG wires in a cord that is attached directly or by a special plug to a portable appliance rated not over 50 W.

2. *All fixture wires* must be protected in accordance with their ampacities, as given in Table 402.5.
3. The required protection may be provided by use of supplementary OCPD (usually fuses), instead of having branch-circuit protection rated at the low values involved.

Then, the basic rules are modified by the rules in parts **(B)(1)** and **(B)(2)** applying to each of the preceding rules:

Part **(B)(1)** applies only to a flexible cord or a tinsel cord (not an "extension cord") that is "approved for and used with a specific *listed* (by UL or other recognized test lab) appliance or luminaire." Such a cord, under the conditions stated, is not required to be protected at its ampacity from **NEC** Table 400.5. The 2008 **NEC** removed the qualifier "portable" as a descriptive term for the light, thereby removing a direct conflict with cord-supplied luminaires that rely on flexible cord dropping out of a canopy because the luminaire is supported with aircraft cable that can be adjusted in the field to change the mounting height. Such luminaires are not portable and they are necessarily connected with flexible cord, but they need not be provided with overcurrent protection.

Note that "extension cords" are *not* covered by part **(B)(1)** because they are *not* "approved for and used with a specific listed appliance." They are covered in part **(B)(3)** and **(B)(4)**, depending on whether they are a listed extension cord set or field assembled. If they are listed, then there are no longer any prescriptive rules and they are only limited by the listing requirements. If they are field assembled from listed components, then they are only limited by the rules in 400.5, but only where constructed from 14 AWG and larger cord. If they employ 16 AWG cord they can be connected to up to a 20-A (and no larger) branch circuit, and if they are 18 AWG they revert to the default limits of 7 or 10 A (from Table 400.5). Refer to the bottom half of Fig. 240-8. Some cords are now available with 18 AWG cord, but such cords have supplementary overcurrent protection in the form of fuses in their plugs, in deference to these rules and in accordance with the requirements of 240.4(D)(1).

Part **(B)(2)** gives the conditions under which fixture wire does not have to be protected at the ampacity value given in Table 402.5 for its particular size *if* the fixture wire is any one of the following:

- No. 18 wire, not over 15 m (50 ft) long, connected to a branch circuit rated not over 20 A
- No. 16 wire, not over 30 m (100 ft) long, connected to a branch circuit rated not over 20 A
- No. 14 or larger wire, of any length, connected to a branch circuit rated not over 30 A
- No. 12 or larger wire, of any length, connected to a branch circuit rated not over 50 A

From those rules, No. 16 or No. 18 fixture wire may be connected on any 20-A branch circuit, provided the "run length" (the length of any one of the wires used in the raceway) is not more than 15 m (50 ft)—such as for 450 mm to 1.8 m (1½ to 6-ft) fixture whips [410.117(C)], as illustrated in the top part of Fig. 240-8. *But,* for remote-control circuits run in a raceway from a magnetic motor starter or contactor to a remote pushbutton station or other pilot-control device, 430.72(B) and 725.43 require that a No. 18 wire be protected at not over 7 A and a No. 16 wire at not over 10 A—where fixture wires are used for remote-control circuit wiring, as permitted by Sec. 725.49(A) and (B).

240.6. Standard Ampere Ratings. This is a listing of the "standard ampere ratings" of fuses and CBs for purposes of Code application, now presented in a tabular form. However, an important qualification is made by the second sentence of this section. Although this **NEC** section designates "STANDARD ampere ratings"

240.5 (B)(2)

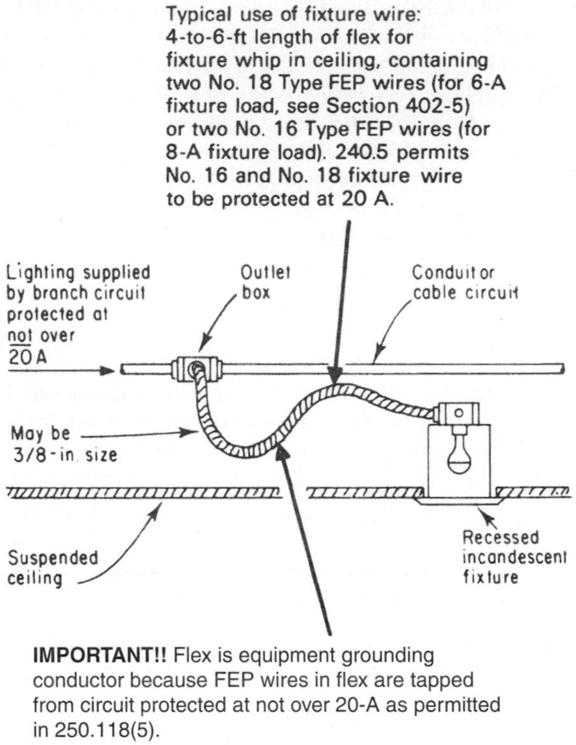
Typical use of fixture wire: 4-to-6-ft length of flex for fixture whip in ceiling, containing two No. 18 Type FEP wires (for 6-A fixture load, see Section 402-5) or two No. 16 Type FEP wires (for 8-A fixture load). 240.5 permits No. 16 and No. 18 fixture wire to be protected at 20 A.
Lighting supplied by branch circuit protected at not over 20 A
Outlet box
Conduit or cable circuit
May be 3/8-in. size
Suspended ceiling
Recessed incandescent fixture
IMPORTANT!! Flex is equipment grounding conductor because FEP wires in flex are tapped from circuit protected at not over 20-A as permitted in 250.118(5).

240.5 (B)(3)

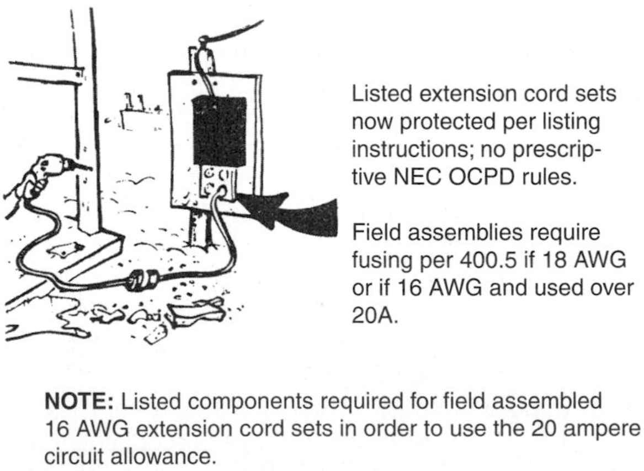
Listed extension cord sets now protected per listing instructions; no prescriptive NEC OCPD rules.
Field assemblies require fusing per 400.5 if 18 AWG or if 16 AWG and used over 20A.
NOTE: Listed components required for field assembled 16 AWG extension cord sets in order to use the 20 ampere circuit allowance.

Fig. 240-8. Separate rules cover fixture wires and extension cords. (Sec. 240.5.)

for fuses and circuit breakers, UL-listed fuses and circuit breakers of other intermediate ratings are available and may be used if their ratings satisfy Code rules on protection. For instance, 240.6 shows standard rated fuses at 1200 A, then 1600 A. But if a circuit was found to have an ampacity of, say, 1530 A and, because 240.4(C) says such a circuit may not be protected by 1600-A fuses, it is not necessary to drop down to 1200-A fuses (the next lower standard size). This final sentence fully intends to recognize use of 1500-A fuses—which would satisfy the basic rule of 240.4(C) for protection rated over 800 A. (Fig. 240-1.)

The last sentence in part **(A)** of 240.6 designates specific "additional standard ratings" of *fuses* at 1, 3, 6, 10, and 601 A. These values apply *only* to fuses and *not* to CBs. The 601-A rating gives Code recognition to use of Class L fuses rated less than 700 A. The reasoning of the Code panel was:

> An examination of fuse manufacturers' catalogs will show that 601 amperes is a commonly listed current rating for the Class L nontime-delay fuse. Section [430.52(C)(1) (Exception No. 2d)] also lists this current rating as a break point in application rules.
>
> Without a 601 ampere rating, the smallest standard fuse which can be used in Class L fuse clips is rated 700 amperes. Since the intent of Table 430.152 and 430.52 is to encourage closer short-circuit protection, it seems prudent to encourage availability and use of 601-ampere fuses in combination motor controllers having Class L fuse clips.
>
> Because ratings of inverse time circuit breakers are not related to fuse clip size, a distinction between 600 and 601 amperes in circuit breakers would serve no useful purpose. Hence, inverse-time circuit breaker ratings are listed separately. Such separation also facilitates recognition of other fuse ratings as standard.

The smaller sizes of fuses (1, 3, 6, and 10 A) listed as "standard ratings" provide more effective short-circuit and ground-fault protection for motor circuits—in accordance with 430.52, 430.40, and UL requirements for protecting the overload relays in controllers for very small motors. The Code panel reasoning was as follows:

> Fuses rated less than 15 amperes are often required to provide short circuit and ground-fault protection for motor branch circuits in accordance with 430.52.
>
> Tests indicate that fuses rated 1, 3, 6 and 10 amperes can provide the intended protection in motor branch circuits for motors having full load currents less than 3.75 amperes ($3.75 \times 400\% = 15$). These ratings are also those most commonly shown on control manufacturers' overload relay tables. Overload relay elements for very small full load motor currents have such a high resistance that a bolted fault at the controller load terminals produces a short-circuit current of less than 15 amperes, regardless of the available current at the line terminals. An overcurrent protective device rated or set for 15 amperes is unable to offer the short circuit or ground fault protection required by 110.10 in such circuits.
>
> An examination of fuse manufacturers' catalogs will show that fuses with these ratings are commercially available. Having these ampere ratings established as standard should improve product availability at the user level and result in better overcurrent protection.
>
> Since inverse time circuit breakers are not readily available in the sizes added, it seems appropriate to list them separately.

Listing of those smaller fuse ratings has a significant effect on use of several small motors (fractional and small-integral-horsepower sizes) on a single branch circuit as described under 430.53(B).

240.6(B) states that if a circuit breaker has external means for changing its continuous-current rating (the value of current above which the inverse-time overload—or longtime delay—trip mechanism would be activated), the breaker must be considered to be a protective device of the maximum continuous current (or overload trip rating) for which it might be set. This type of CB adjustment is available on molded-case, insulated-case, and air power circuit breakers. As a result of that rule and 240.4, the circuit conductors connected to the load terminals of such a circuit breaker must be of sufficient ampacity as to be properly protected by the maximum current value to which the adjustable trip might be set. That means that the CB rating must not exceed the ampacity of the circuit conductors, except that where the ampacity of the conductor does not correspond to a standard rating of CB, the next higher standard rating of CB may be used, up to 800 A (Fig. 240-9).

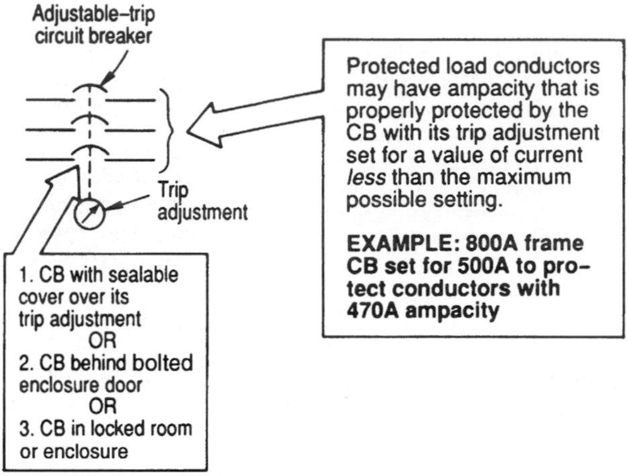
Adjustable–trip circuit breaker
Trip adjustment
1. CB with sealable cover over its trip adjustment
OR
2. CB behind bolted enclosure door
OR
3. CB in locked room or enclosure
Protected load conductors may have ampacity that is properly protected by the CB with its trip adjustment set for a value of current *less* than the maximum possible setting.
EXAMPLE: 800A frame CB set for 500A to protect conductors with 470A ampacity

Fig. 240-9. An adjustable-trip circuit breaker that has access to its trip adjustment limited only to qualified persons may be taken to have a rating less than the maximum value to which the continuous rating (the longtime or overload adjustment) might be set. (Sec. 240.6.)

Prior to the 1987 edition, the NEC did not require that a circuit breaker with adjustable or changeable trip rating must have load-circuit conductors of an ampacity at least equal to the highest trip rating at which the breaker might be used. Conductors of an ampacity less than the highest possible trip rating could be used, provided that the actual trip setting being used did protect the conductor in accordance with its ampacity, as required in NEC 240.4. Since the 1990 edition, such application may be made only in accordance with the rule in part **(C)**, which says that an adjustable-trip circuit breaker may be used as a protective device of a rating lower than its maximum setting and used to protect conductors of a corresponding ampacity in accordance with 240.4(B) *if* the trip-adjustment is

1. Located behind a removable and sealable cover, or
2. Part of a circuit breaker which is itself located behind bolted equipment enclosure doors accessible only to qualified persons, or
3. Part of a circuit breaker that is locked behind doors (such as in a room) accessible only to qualified persons

Although this rule permits use of conductors with ampacity lower than the maximum possible trip setting of a CB under the conditions given, this does not apply to fusible switches, and it is never necessary for a fusible switch to have its connected load-circuit conductors of ampacity equal to the maximum rating of a fuse that might be installed in the switch—provided that the actual rating of the fuse used in the switch does protect the conductor at its ampacity.

240.8. Fuses or Circuit Breakers in Parallel. This rule allows fuses or CBs in parallel where "factory assembled" and "listed as a unit." Such units are used to increase the rating of overcurrent protection in marine, over-the-road, off-road, commercial, and industrial installations. Use of other than listed units, including any composed of field assembled devices, is a clear and direct violation.

240.10. Supplementary Overcurrent Protection. Supplementary overcurrent protection is commonly used in lighting fixtures, heating circuits, appliances, or other utilization equipment to provide individual protection for specific components within the equipment itself. Such protection is not branch-circuit protection and *the* NEC *does not require supplemental overcurrent protective devices to be readily accessible.* Typical applications of supplemental overcurrent protection are fuses installed in fluorescent fixtures and cooking or heating equipment where the devices are sized to provide lower overcurrent protection than that of the branch circuit supplying such equipment. This is discussed under 424.19 and 424.22 on electric space-heating equipment.

Years ago there was no allowance for conventional OCPDs to be in locations that were not readily accessible, and so they were classified, essentially at the convenience of the engineer, as supplementary. One common example is the combination plug fuse and snap switch assemblies that come premounted in box covers or handy box covers, particularly where mounted in not-readily-accessible locations such as adjacent to ceiling-mounted equipment and/or fractional-horsepower motors. Since the plug fuse is actually rated for branch-circuit protection, this wasn't really correct until the rule in 240.24(A)(4) caught up with the very long-standing allowance in 404.8(A) Exception No. 2. Now that 240.24(A)(4) allows this openly (although only adjacent to the equipment supplied), the need to artificially classify branch-circuit rated protective devices as supplementary devices has largely gone away.

240.12. Electrical System Coordination. This rule applies to any electrical installation where hazard to personnel would result from disorderly shutdown of electrical equipment under fault conditions. The purpose of this rule is to permit elimination of "overload" protection—that is, protection of conductors at their ampacities—and to eliminate unknown or random relation between operating time of overcurrent devices connected in series.

The section recognizes two requirements, both of which must be fulfilled to perform the task of "orderly shutdown."

One is selective coordination of the time-current characteristics of the short-circuit protective devices in series from the service to any load—so that, automatically, any fault will actuate only the short-circuit protective device closest to the fault on the line side of the fault, thereby minimizing the extent of electrical outage due to a fault.

The other technique that must also be included if *overload* protection is eliminated is "overload indication based on monitoring systems or devices." A note to this section gives brief descriptions of both requirements and establishes only a generalized understanding of "overload indication." Effective application of this rule depends on careful design and coordination with inspection authorities.

It should be noted, however, that it says that the technique of eliminating overload protection to afford orderly shutdown "shall be permitted"—but does *not require* such application. Although it could be argued that the wording implies a mandatory rule, consultation with electrical inspection authorities on this matter is advisable because of the safety implications in nonorderly shutdown due to overload. Emergency systems (700.32), legally required standby systems (701.27), critical operations power systems (708.54), fire pump protective devices connected to campus-style multibuilding distributions [695.3(C)(3)], and main elevator feeders, and multiple elevator driving machines on a single feeder (620.62) now require selective coordination, a specifically defined term in Art. 100, within their scope.

240.13. Ground-Fault Protection of Equipment (GFPE). Equipment ground-fault protection—of the type required for 480Y/227-V service disconnects—is now required for each disconnect rated 1000 A or more that serves as a main disconnect for a building or structure. Like 215.10, this section expands the application of protection against destructive arcing burndowns of electrical equipment. The intent is to equip a main building disconnect with GFPE whether the disconnect is technically a service disconnect or a building disconnect on the load side of service equipment located elsewhere. This was specifically devised to cover those cases where a building or structure is supplied by a 480Y/277-V feeder from another building or from outdoor service equipment. Because the main disconnect (or disconnects) for such a building serves essentially the same function as a service disconnect, this requirement makes such disconnects subject to all of the rules of 230.95, covering GFPE for services (Fig. 240-10).

The last part of this section is intended to clarify that the rule applies to the rating of *individual* disconnects and *not* to the sum of disconnects. Where an individual disconnect is rated 1000 A, or more, GFPE must be provided.

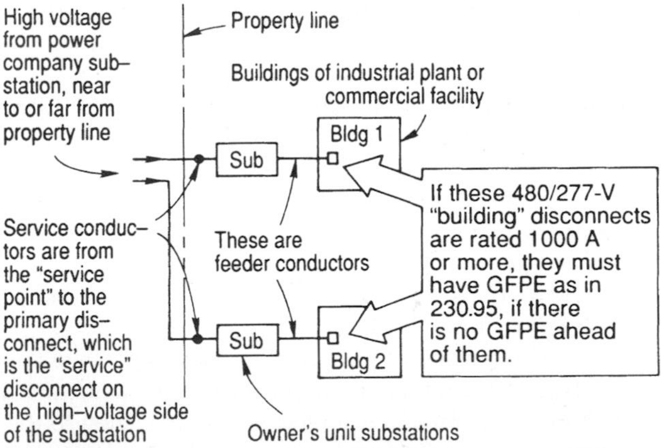
High voltage from power company sub-station, near to or far from property line
Property line
Buildings of industrial plant or commercial facility
Bldg 1
Sub
If these 480/277-V "building" disconnects are rated 1000 A or more, they must have GFPE as in 230.95, if there is no GFPE ahead of them.
Service conductors are from the "service point" to the primary dis-connect, which is the "service" disconnect on the high-voltage side of the substation
These are feeder conductors
Sub
Bldg 2
Owner's unit substations

Fig. 240-10. Ground-fault protection is required for the feeder disconnect for each building—either at the building or at the substation secondary. (Sec. 240.13.)

There are three conditions under which GFPE may be omitted. The first condition here excluded from the need for such GFPE disconnects for critical processes where automatic shutdown would introduce additional or different hazards. And as with service GFPE, the requirement does not apply to fire-pump disconnects.

As covered in 240.13(2), the need for GFPE on a building or structure disconnect is suspended *if* such protection is provided on the upstream (line) side—either service or feeder disconnect GFPE—of the feeder disconnect. The rule (eliminated in the 1996 **NEC**) used to stipulate that there must not be any desensitizing of the ground-fault protection because of downstream neutral regrounding, that is, bonding to the equipment grounding conductor and grounding electrode conductor in the downstream building disconnect. If this were done, and it is now prohibited in these cases by 250.32(B)(1) Exception No. 1, any ground-fault current in the downstream building that develops will pass over the bonding connection and return to the upstream GFPE not as unbalanced and detectable fault current, but rather as perfectly balanced and undetectable neutral load current.

The problem with this was that the rule recognizing the upstream protection was in the form of an exception. The "requirement" to avoid desensitization was added to the exception. However, since it was part of an exception, it was unenforceable. If someone desensitized the upstream GFPE, what rule was broken? True, the exception became inoperable and therefore GFPE was now required at the building disconnect. However, if there were additional downstream cross-connections, then neither GFPE device would work properly. The real solution was to address the problem in the second-building regrounding rules in Art. 250. This was successfully done in the 1999 **NEC**, eliminating the problem.

240.15. Ungrounded Conductors. A fuse or circuit breaker must be connected in series with each ungrounded circuit conductor—usually at the supply end of the conductor. A current transformer and relay that actuates contacts of a CB is considered to be an overcurrent trip unit, like a fuse or a direct-acting CB (Fig. 240-11).

Although part **(B)** basically requires a CB to open all ungrounded conductors of a circuit simultaneously, that is, to be a multipole circuit breaker, parts **(1)**, **(2)**, **(3)**, and **(4)** cover acceptable uses of a number of single-pole CBs instead of multipole CBs.

The permission in (1) for use of single-pole breakers on multiwire branch circuits requires handle-ties on all such applications, thereby correlating with 210.4(B). Note that for 277-V multiwire branch circuits originating from a 480-V distribution, handle ties are not an option because the UL Guide Card entry does not recognize voltages above 120/240 for this application. Therefore, actual two- or three-pole breakers (or fused switches) are required for these circuits. In addition, a multipole circuit breaker that simultaneously opens all ungrounded conductors is required for multiwire branch circuits that supply both line-to-line loads and line-to-neutral loads as permitted by 210.4(C) Exception No. 2.

The next three paragraphs allow handle-tied breakers for exclusively line-to-line loads such as baseboard electric heaters on grounded single-phase ac systems **(2)** and three-phase ac systems **(3)** (which also includes 5-wire 2-phase systems). In both cases the circuit must be derived from a grounded distribution and use a slash voltage rated circuit breaker, as covered in the second paragraph of 240.85.

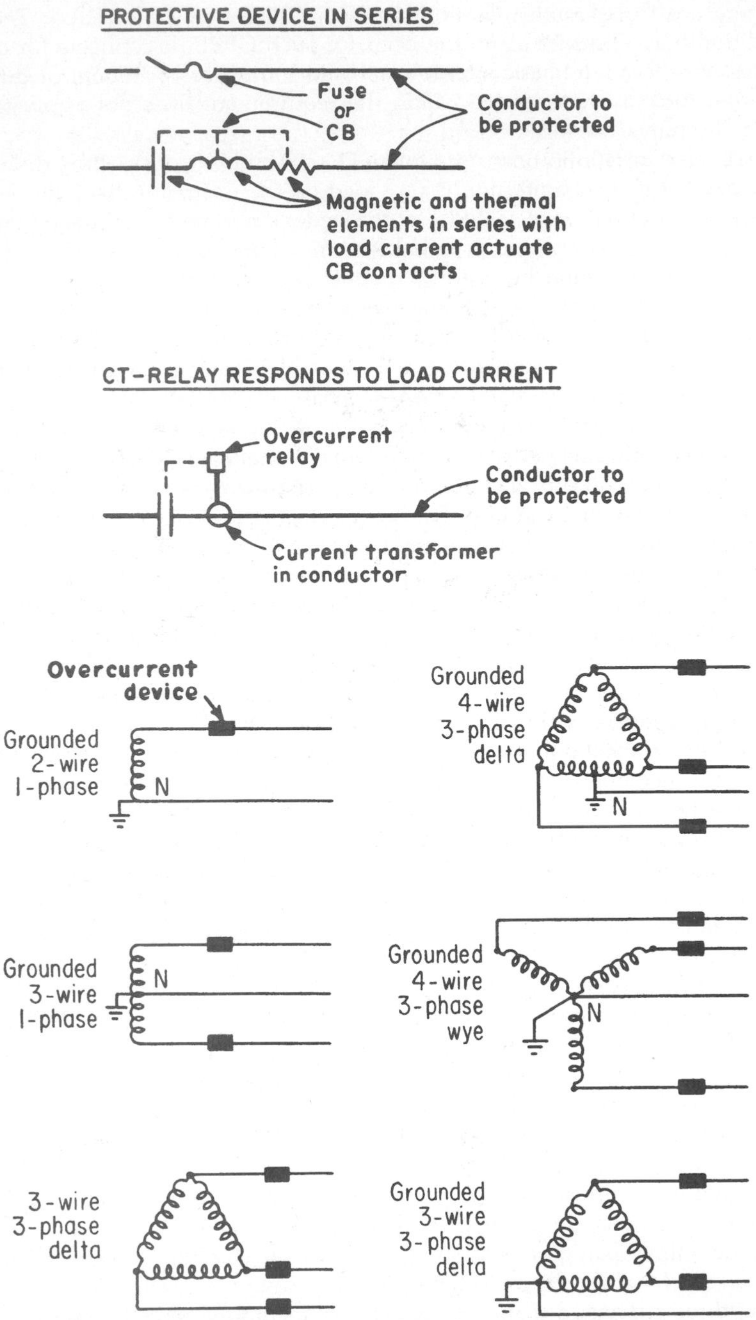
PROTECTIVE DEVICE IN SERIES
Fuse
or
CB
Conductor to
be protected
Magnetic and thermal
elements in series with
load current actuate
CB contacts
CT-RELAY RESPONDS TO LOAD CURRENT
Overcurrent
relay
Conductor to
be protected
Current transformer
in conductor
Overcurrent
device
Grounded
2-wire
1-phase
N
Grounded
4-wire
3-phase
delta
N
Grounded
3-wire
1-phase
N
Grounded
4-wire
3-phase
wye
N
3-wire
3-phase
delta
Grounded
3-wire
3-phase
delta

Fig. 240-11. A fuse or overcurrent trip unit must be connected in series with each ungrounded conductor. (Sec. 240.20.)

Paragraph **(4)** imposes a similar limitation on dc circuits, namely, that they be supplied from a system with a grounded neutral.

In all three cases [**(2)**, **(3)**, and **(4)**] the systems must be grounded, but it is not necessary for the neutral to be running to the load supplied. In addition, as of the 2011 edition, there is an express rather than implied voltage system limit in the **NEC** for these applications. Line-to-line loads that employ single-pole breakers with handle ties are limited to 120/240-V applications, or 125/250-V dc applications, or both if dual rated. Prior codes (going back to the 1959 edition) referred to the voltage limitations of 210.6. Now these three paragraphs line up with the UL Guide Card information on circuit breakers.

Note: Two single-pole circuit breakers may not be used on "ungrounded 2-wire circuits"—such as 240-, 480-, or 600-V single-phase, 2-wire circuits. A 2-pole CB must be used if protection is provided by CBs. Use of single-pole CBs with handle ties but not common-trip is not allowed. These circuits become corner grounded on the first ground fault, and must then clear on a ground fault elsewhere in a different phase connection; only the straight two-pole breaker will have been tested without the assistance of a reduced voltage to ground. However, use of fuses for protection of such a circuit is permitted even though it will present the same chance of a fault condition as shown in Fig. 240-12.

Although 1-pole CBs may be used, as noted, it is better practice to use multipole CBs for circuits to individual load devices that are supplied by two or more ungrounded conductors. It is never wrong to use a multipole CB; but, based on the rules given here and in 210.4, it may be a violation to use two single-pole CB units (see Fig. 240-13). A 3-pole CB must always be used for a 3-phase, 3-wire circuit supplying phase-to-phase loads fed from an ungrounded delta system, such as 480-V outdoor lighting for a parking lot, as permitted by 210.6(B). In addition, there is a significant problem with availability of handle ties for three single-pole breakers used on three-phase wye multiwire branch circuits.

Just because a two-pole breaker ships as a single unit from a manufacturer does not necessarily mean that it is actually a common-trip device; to be so it must have an actual internal common trip, and an external handle tie, no matter how robust, does not qualify the breaker as a common-trip device. UL requires such breakers to be marked as "independent trip" or "no common trip" where directly visible when the dead-front is removed.

240.21. Location in Circuit. The basic rule of this section is shown in Fig. 240-14. A very important qualification that applies to all tap conductors is this: A tap cannot be tapped. Any conductor that originates under one of the provisions of 240.21(**A** through **H**) cannot supply any other conductor unless the next conductor has protection at its supply end with a conventional overcurrent device meeting all the rules in 240.4.

Although basic Code requirements dictate the use of an overcurrent device at the point at which a conductor received its supply, subparts **(A)** through **(H)** effectively present exceptions to this rule in the case of taps to feeders. That is, to meet the practical demands of field application, certain lengths of unprotected conductors may be used to tap energy from protected feeder conductors.

These "exceptions" to the rule for protecting conductors at their points of supply are made in the case of 10-, 25-, and 100-ft (3.0-, 7.5-, and 30.0-m) taps from

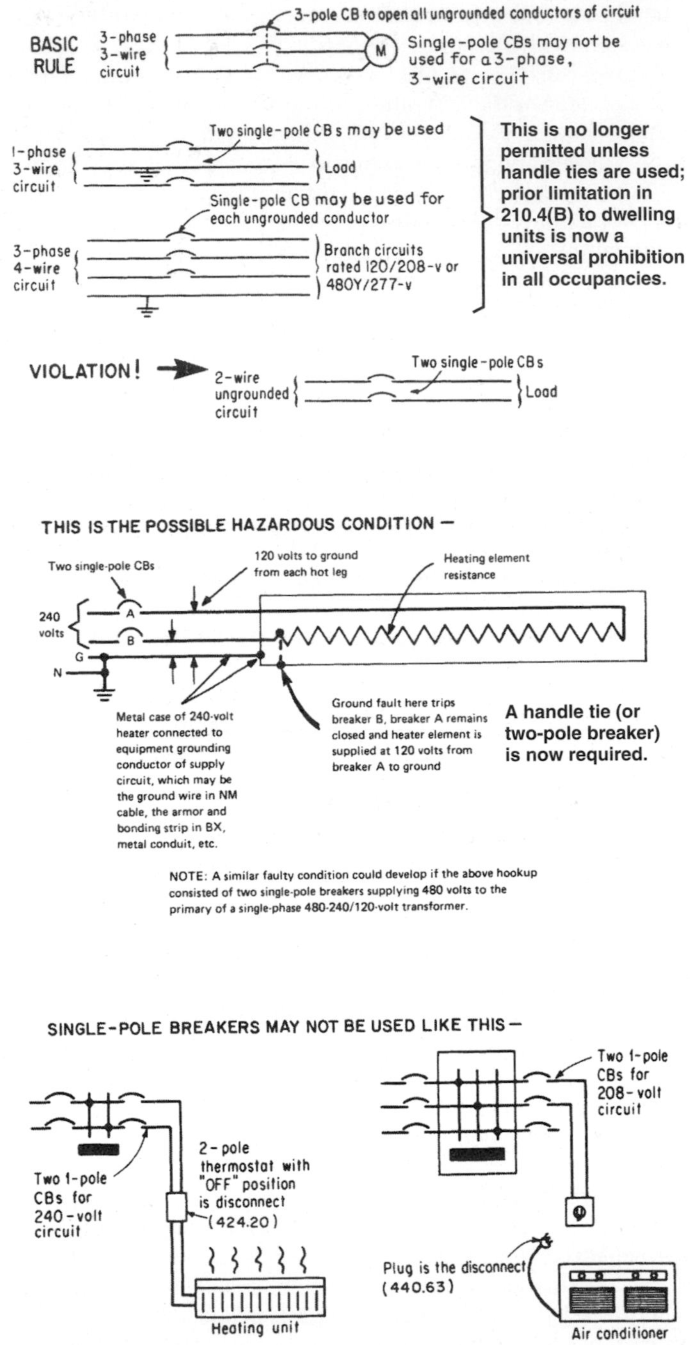
3-pole CB to open all ungrounded conductors of circuit
BASIC RULE
3-phase 3-wire circuit
M
Single-pole CBs may not be used for a 3-phase, 3-wire circuit
Two single-pole CBs may be used
1-phase 3-wire circuit
Load
Single-pole CB may be used for each ungrounded conductor
3-phase 4-wire circuit
Branch circuits rated 120/208-v or 480Y/277-v
This is no longer permitted unless handle ties are used; prior limitation in 210.4(B) to dwelling units is now a universal prohibition in all occupancies.
VIOLATION!
2-wire ungrounded circuit
Two single-pole CBs
Load
THIS IS THE POSSIBLE HAZARDOUS CONDITION —
Two single-pole CBs
120 volts to ground from each hot leg
Heating element resistance
240 volts
A
B
G
N
Metal case of 240-volt heater connected to equipment grounding conductor of supply circuit, which may be the ground wire in NM cable, the armor and bonding strip in BX, metal conduit, etc.
Ground fault here trips breaker B, breaker A remains closed and heater element is supplied at 120 volts from breaker A to ground
A handle tie (or two-pole breaker) is now required.
NOTE: A similar faulty condition could develop if the above hookup consisted of two single-pole breakers supplying 480 volts to the primary of a single-phase 480-240/120-volt transformer.
SINGLE-POLE BREAKERS MAY NOT BE USED LIKE THIS —
Two 1-pole CBs for 240-volt circuit
2-pole thermostat with "OFF" position is disconnect (424.20)
Heating unit
Two 1-pole CBs for 208-volt circuit
Plug is the disconnect (440.63)
Air conditioner

Fig. 240-12. Single-pole versus multipole breakers. (Sec. 240.15.)

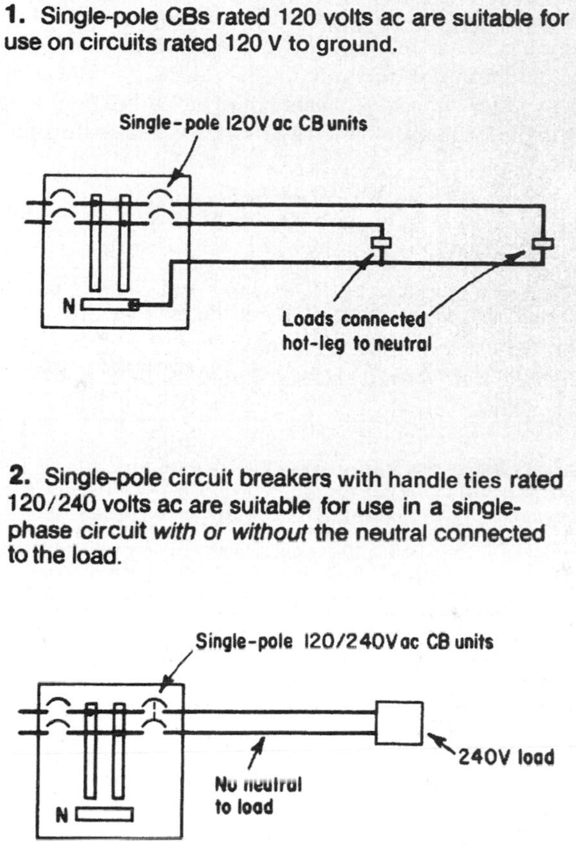
1. Single-pole CBs rated 120 volts ac are suitable for use on circuits rated 120 V to ground.
Single-pole 120V ac CB units
N
Loads connected hot-leg to neutral
2. Single-pole circuit breakers with handle ties rated 120/240 volts ac are suitable for use in a single-phase circuit *with or without* the neutral connected to the load.
Single-pole 120/240V ac CB units
N
240V load
No neutral to load

Fig. 240-13. NEC rules must be correlated with these UL requirements. (Sec. 240.20.)

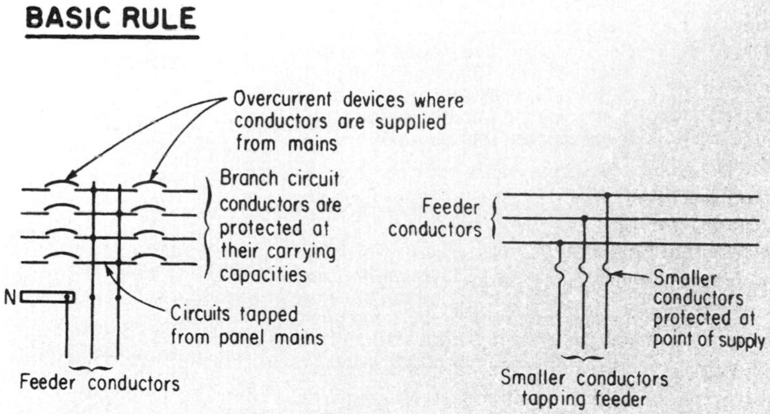
BASIC RULE
Overcurrent devices where conductors are supplied from mains
Branch circuit conductors are protected at their carrying capacities
N
Circuits tapped from panel mains
Feeder conductors
Feeder conductors
Smaller conductors protected at point of supply
Smaller conductors tapping feeder

Fig. 240-14. Conductors must be protected at their supply ends. (Sec. 240.21.)

a feeder, as described in 240.21, parts **(B)(1)**, **(B)(2)**, and **(B)(4)**. Application of the tap rules should be made carefully to effectively minimize any sacrifice in safety. The taps are permitted without OCPDs at the point of supply.

240.21**(B)(1)** says that unprotected taps not over 10 ft (3.0 m) long (Fig. 240-15) may be made from feeders, and 240.21**(C)(2)** for transformer secondaries, provided:

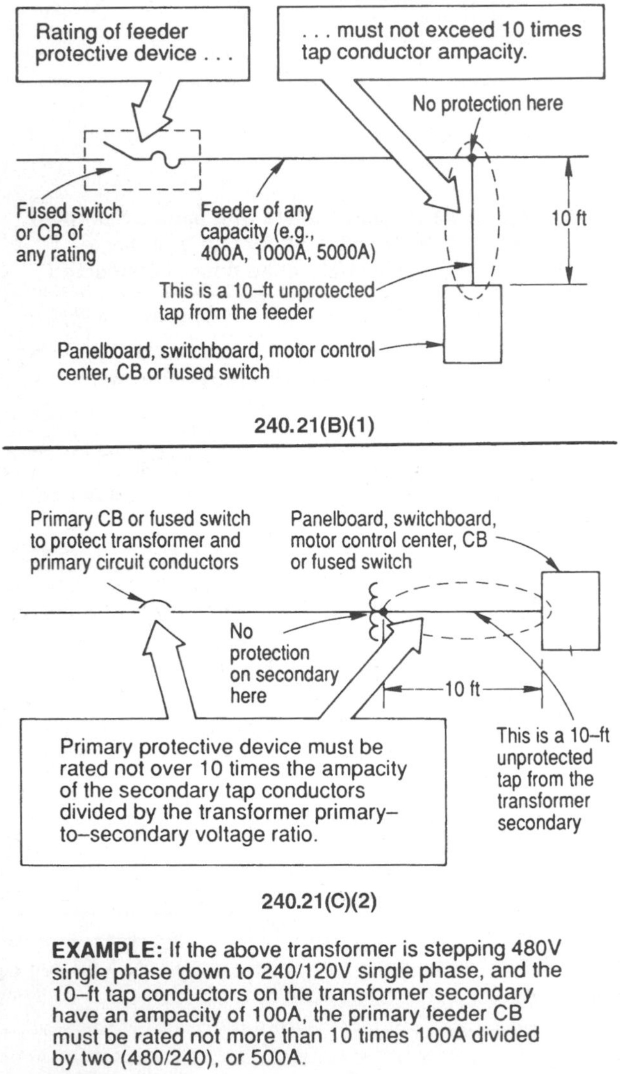
Rating of feeder protective device . . .
. . . must not exceed 10 times tap conductor ampacity.
No protection here
Fused switch or CB of any rating
Feeder of any capacity (e.g., 400A, 1000A, 5000A)
10 ft
This is a 10-ft unprotected tap from the feeder
Panelboard, switchboard, motor control center, CB or fused switch
240.21(B)(1)
Primary CB or fused switch to protect transformer and primary circuit conductors
Panelboard, switchboard, motor control center, CB or fused switch
No protection on secondary here
10 ft
Primary protective device must be rated not over 10 times the ampacity of the secondary tap conductors divided by the transformer primary–to–secondary voltage ratio.
This is a 10–ft unprotected tap from the transformer secondary
240.21(C)(2)
EXAMPLE: If the above transformer is stepping 480V single phase down to 240/120V single phase, and the 10–ft tap conductors on the transformer secondary have an ampacity of 100A, the primary feeder CB must be rated not more than 10 times 100A divided by two (480/240), or 500A.

Fig. 240-15. Ten-foot taps may be made from a feeder or a transformer secondary. (Sec. 240.21.)

1. The smaller conductors have a current rating that is not less than the combined computed loads of the circuits supplied by the tap conductors and must have ampacity of—

   Not less than the rating of the equipment containing an overcurrent device(s) supplied by the tap conductors, or (which formerly included the bus structure of a main lug only panelboard but given changes in 408.36, an overcurrent device is now generally required) and

   Not less than the rating of the overcurrent device (fuses or CB) that is installed at the termination of the tap conductors. This provision does not apply to listed equipment such as some surge protective devices that include specific instructions covering minimum conductor sizing; in such cases, the manufacturer's instructions apply.

   Important Limitation: For any 10 ft (3-m) unprotected feeder tap installed in the field, the rule limits its connection to a feeder that has protection rated *not* more than 1000 percent of (10 times) the ampacity of the tap conductor where the tap conductors do not remain within the enclosure or vault in which the tap is made. This provision recognizes that taps present little threat while they remain within the confines of a transformer vault. It also recognizes the practical issues of sensor wiring within enclosures. For example, if a voltmeter is installed in the enclosure door of a 2000 A switchboard, 10 percent of 2000 A would otherwise require 3/0 conductors to run to the meter. Under the rule, unprotected No. 14 tap conductors are not permitted to tap a feeder any larger than 1000 percent of the 20-A ampacity of No. 14 copper conductors—which would limit such a tap for use with a maximum feeder protective device of not over 10 × 20 A, or 200 A.
2. The tap does not extend beyond the switchboard, panelboard, disconnect, or control device which it supplies.
3. The tap conductors are enclosed in conduit, EMT, metal gutter, or other approved raceway when not a part of the switchboard or panelboard.

240.21(C)(2) specifically recognizes that a 10-ft (3-m) tap may be made from a transformer secondary in the same way it has always been permitted from a feeder. In either case, the tap conductors must not be over 10 ft (3 m) long and must have ampacity not less than the amp rating of the switchboard, panelboard, disconnect, or control device—or the tap conductors may be terminated in an overcurrent protective device rated not more than the ampacity of the tap conductors. As in the case of the equivalent feeder tap, this provision does not apply to listed equipment, such as some surge protective devices, that includes specific instructions covering minimum conductor sizing; in such cases the manufacturer's instructions apply.

In the case of an unprotected tap from a transformer secondary, the ampacity of the 10-ft (3-m) tap conductors would have to be related through the transformer voltage ratio to the size of the transformer primary protective device—which in such a case would be "the device on the line side of the tap conductors." Just as in the case of the feeder tap, there is a 1000 percent ratio limitation (in this case multiplied by the applicable transformer winding ratio) except once again where the secondary conductors don't leave the vault or the enclosure where they originate the 1000 percent (10 times) factor does not apply.

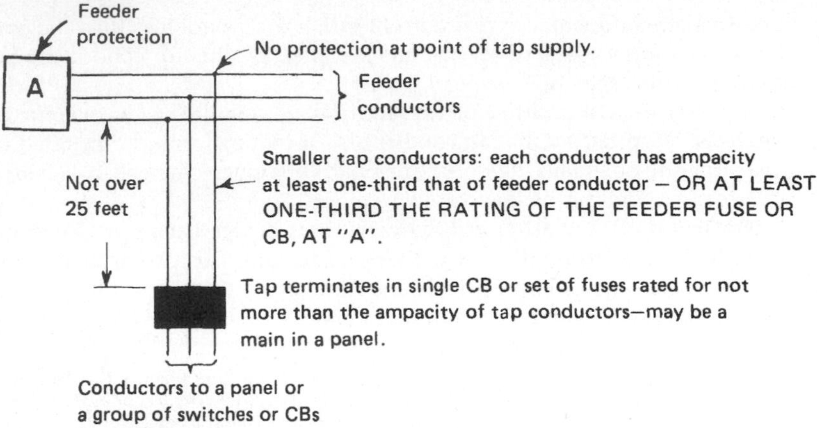
Feeder protection
No protection at point of tap supply.
A
Feeder conductors
Not over 25 feet
Smaller tap conductors: each conductor has ampacity at least one-third that of feeder conductor – OR AT LEAST ONE-THIRD THE RATING OF THE FEEDER FUSE OR CB, AT "A".
Tap terminates in single CB or set of fuses rated for not more than the ampacity of tap conductors—may be a main in a panel.
Conductors to a panel or a group of switches or CBs

Fig. 240-16. Sizing feeder taps not over 25 ft (7.5 m) long. (Sec. 240.21.)

Taps not over 25 ft (7.5 m) long (Fig. 240-16) may be made from feeders, as noted in part **(B)(2)** of 240.21, provided:

1. The smaller conductors have a current rating at least one-third that of the feeder overcurrent device rating or of the conductors from which they are tapped.
2. The tap conductors are suitably protected from mechanical damage. In previous Code editions, the 25-ft (7.5-m) feeder tap without overcurrent protection at its supply end simply had to be "suitably protected from physical damage"—which could accept use of cable for such a tap. Now, the rule requires such tap conductors to be "enclosed in an approved raceway or by other approved means"—strongly suggesting, but not quite mandating a raceway as has always been required for 10-ft (3-m) tap conductors.
3. The tap is terminated in a single CB or set of fuses, which will limit the load on the tap to the ampacity of the tap conductors.

Examples of Taps

Figure 240-17 shows use of a 10-ft (3-m) feeder tap to supply a single motor branch circuit. The conduit feeder may be a horizontal run or a vertical run, such as a riser. If the tap conductors are of such size that they have a current rating at least one-third that of the feeder conductors (or protection rating) from which they are tapped, they could be run a distance of 25 ft (7.5 m) without protection at the point of tap-off from the feeder because they would comply with the rules of 240.21(B)(2), which permit a 25-ft (7.5-m) feeder tap if the conductors terminate in a single protective device rated not more than the conductor ampacity. 368.17(C) generally requires that any busway used as a feeder must have overcurrent protection on the busway for any subfeeder or branch circuit tapped from the busway. The use of a cable-tap box on a busway without overcurrent protection (as shown in the conduit installation of Fig. 240-17) would usually be a violation. But, Exception No. 1 to 368.17(C) clearly eliminates such protection where making taps. Refer to 240.24 and 368.17.

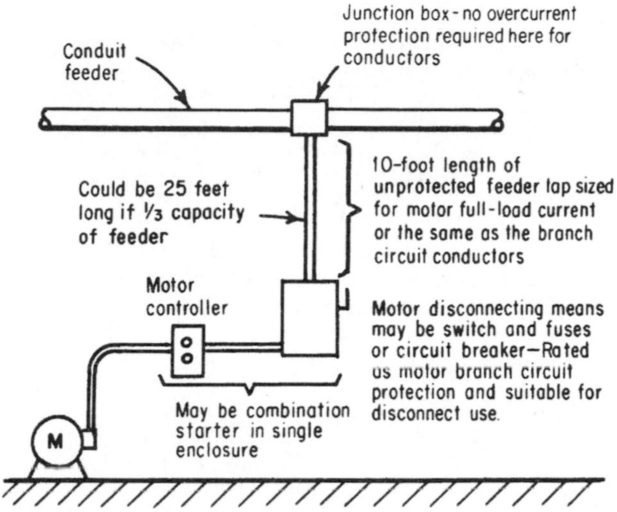
Junction box - no overcurrent protection required here for conductors
Conduit feeder
Could be 25 feet long if 1/3 capacity of feeder
10-foot length of unprotected feeder tap sized for motor full-load current or the same as the branch circuit conductors
Motor controller
Motor disconnecting means may be switch and fuses or circuit breaker—Rated as motor branch circuit protection and suitable for disconnect use.
May be combination starter in single enclosure
M

Fig. 240-17. A 10-ft (3-m) tap for a single motor circuit. (Sec. 240.21.)

A common application of the 10-ft (3.0-m) tap is the supply of panelboards from conduit feeders or busways, as shown in Fig. 240-18. The case shows an interesting requirement that arises from 408.36, which requires that all panelboards be protected on their supply side by overcurrent protection rated not more than the rating of the panelboard busbars. If the feeder is a busway, the protection must be placed [a requirement of 368.17(C)] at the point of tap on the busway. In that

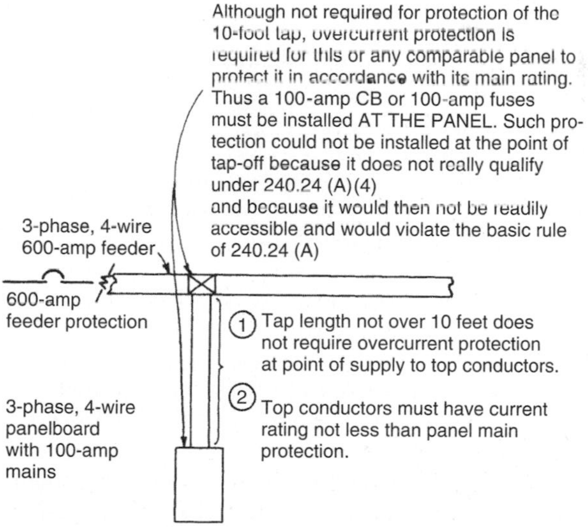
Although not required for protection of the 10-foot tap, overcurrent protection is required for this or any comparable panel to protect it in accordance with its main rating. Thus a 100-amp CB or 100-amp fuses must be installed AT THE PANEL. Such protection could not be installed at the point of tap-off because it does not really qualify under 240.24 (A)(4) and because it would then not be readily accessible and would violate the basic rule of 240.24 (A)
3-phase, 4-wire 600-amp feeder
600-amp feeder protection
1 Tap length not over 10 feet does not require overcurrent protection at point of supply to tap conductors.
2 Tap conductors must have current rating not less than panel main protection.
3-phase, 4-wire panelboard with 100-amp mains

Fig. 240-18. A 10-ft (3.0 m) tap to lighting panel with unprotected conductors. (Sec. 240.21.)

case a 100-A CB or fused switch on the busway would provide the required protection of the panel, and the panel would not require a main in it. But, if the feeder circuit is in conduit, the 100-A panel protection would have to be in the panel or just ahead of it.

For transformer applications, typical 10- and 25-ft (3.0- and 7.5-m) tap considerations are shown in Fig. 240-19.

The bottom half of Fig. 240-19 illustrates an important concept that was just clarified in the 2008 NEC. A transformer (assuming appropriate capacity and primary-side protection) can supply any number of sets of secondary conductors, each of which is considered independently when applying the various rules for transformer secondary conductors covered in 240.21(C). If five sets of secondary conductors were supplied from a common secondary, in raceway and feeding a suitable overcurrent device at their load end, each could be 7.5 m (25 ft) long. It would not be necessary to keep them all 1.5 m (5 ft) long or other lengths such that the total did not exceed the 7.5 m (25 ft) limit overall.

Figure 240-20 shows application of part **(B)(3)** of 240.21 in conjunction with the rule of 450.3(B), covering transformer protection. As shown in Example 1, the 100-A main protection in the panel is sufficient protection for the transformer and the primary and secondary conductors when these conditions are met:

1. Tap conductors have ampacity at least one-third that of the 125-A feeder conductors.
2. Secondary conductors are rated at least one-third the ampacity of the 125-A feeder conductors, based on the primary-to-secondary transformer ratio.
3. Total tap is not over 25 ft (7.5 m), primary plus secondary.
4. All conductors are in "approved raceway or other approved means."
5. Secondary conductors terminate in the 100-A main protection that limits secondary load to the ampacity of the secondary conductors and simultaneously provides the protection required by the lighting panel.
6. Primary feeder protection is not over 250 percent of transformer rated primary current, as recognized by 450.3(B), and the 100-A main breaker in the panel satisfies as the required "overcurrent device on the secondary side rated or set at not more than 125 percent of the rated secondary current of the transformer." Alternatively, if the primary protection meets the 125 percent rule in 450.3(B), the secondary protection would not be required for the transformer, and would therefore be limited only by the requirements of protecting the secondary conductors and of protecting the panelboard.

Frequently the wiring under this rule uses conductors on the line side of the transformer that are not reduced in any way from the size of the conductors of the feeder to which they are connected. In this case, the length of wire on the primary size that has to be figured in to the 7.5 m (25 ft) limitation under this rule is zero, and the secondary conductors can take the full 7.5 m (25 ft) if necessary.

Example 2 of Fig. 240-20 shows multiple sets of tap conductors from the primary feeder to a group of transformers. In such cases the primary taps are frequently reduced because the primary feeder must have the capacity for several load groups. In such cases the length of the primary side is not zero, and must be subtracted from the permitted overall total. The allowable protection for that

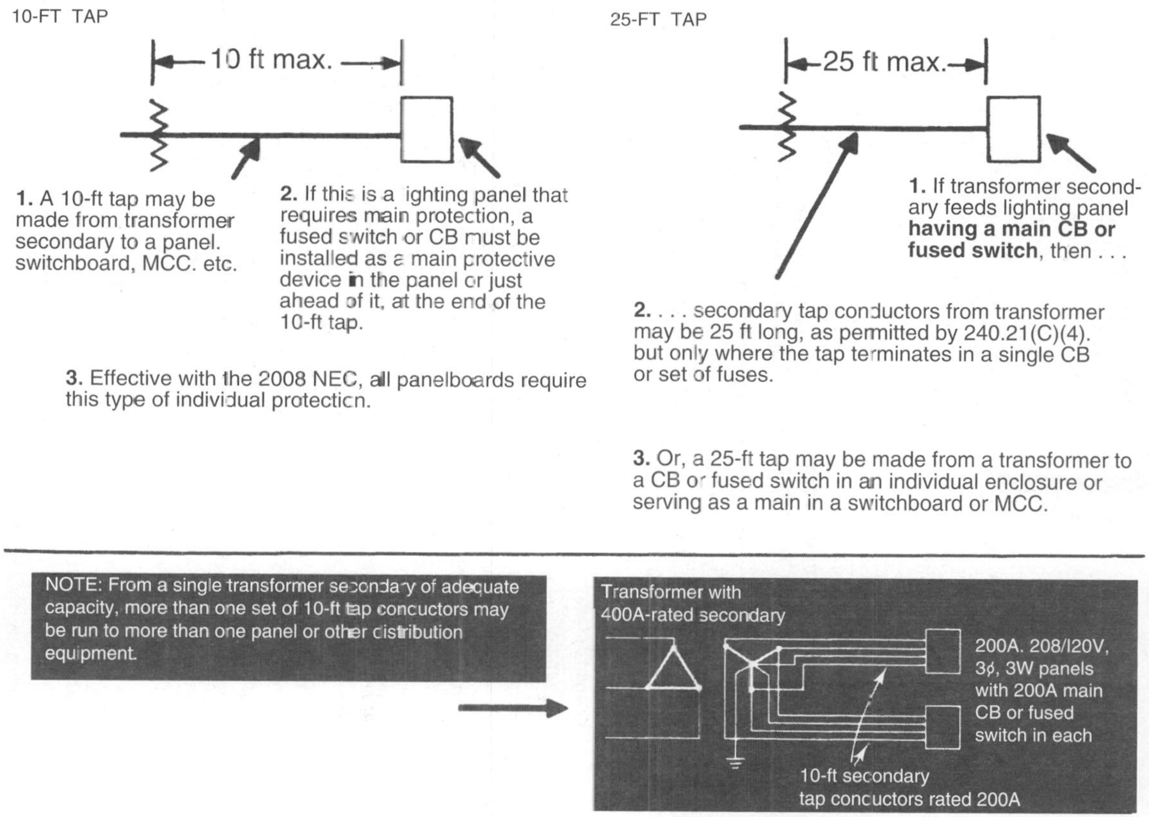

10-FT TAP
10 ft max.
1. A 10-ft tap may be made from transformer secondary to a panel. switchboard, MCC. etc.
2. If this is a lighting panel that requires main protection, a fused switch or CB must be installed as a main protective device in the panel or just ahead of it, at the end of the 10-ft tap.
3. Effective with the 2008 NEC, all panelboards require this type of individual protection.
25-FT TAP
25 ft max.
1. If transformer secondary feeds lighting panel having a main CB or fused switch, then . . .
2. . . . secondary tap conductors from transformer may be 25 ft long, as permitted by 240.21(C)(4). but only where the tap terminates in a single CB or set of fuses.
3. Or, a 25-ft tap may be made from a transformer to a CB or fused switch in an individual enclosure or serving as a main in a switchboard or MCC.
NOTE: From a single transformer secondary of adequate capacity, more than one set of 10-ft tap conductors may be run to more than one panel or other distribution equipment.
Transformer with 400A-rated secondary
200A. 208/120V, 3ϕ, 3W panels with 200A main CB or fused switch in each
10-ft secondary tap conductors rated 200A

Fig. 240-19. Taps from transformer secondaries. (Sec. 240.21.)

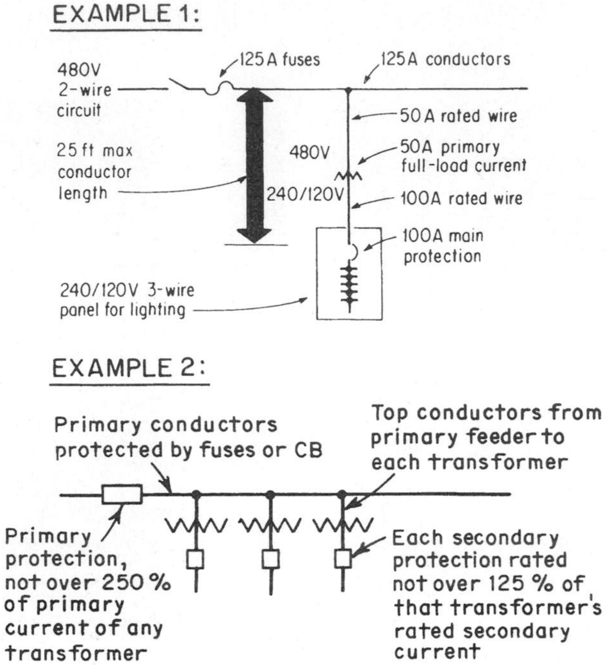
EXAMPLE 1:
480V
2-wire
circuit
125 A fuses
125A conductors
50 A rated wire
25 ft max
conductor
length
480V
50A primary
full-load current
240/120V
100A rated wire
100A main
protection
240/120V 3-wire
panel for lighting
EXAMPLE 2:
Primary conductors
protected by fuses or CB
Top conductors from
primary feeder to
each transformer
Primary
protection,
not over 250 %
of primary
current of any
transformer
Each secondary
protection rated
not over 125 % of
that transformer's
rated secondary
current

Fig. 240-20. Feeder tap of primary-plus-secondary not over 25 ft (7.5 m) long. (Sec. 240.21.)

parent feeder must meet both 240.4 for the feeder conductors employed, and also provide protection for *each* of the transformers supplied, at a value therefore based on 250 percent of the primary rating of the *smallest* transformer served. Figure 240-21 shows this process at work, although in this example the primary conductors were not reduced in size, allowing a full-length secondary.

This is as good an illustration as any of a crucial principle that we will discuss again in 450.3, namely, the rules in Art. 240 for conductor protection stand alone from the rules in Art. 450 for transformer protection. However, if it is intended that a single protective device perform both functions, then both sets of rules must be applied. Make separate calculations, and select for the worst case. If the result is one you don't want to live with, add additional devices until you do meet all the rules.

Figure 240-22 compares the two different 25-ft (7.5-m) tap techniques covered by part **(B)(2)** and the equivalent distance with a transformer secondary interposed, 240.21(C)(5), as just covered in 240.21(B)(3). This rule in part **(C)** simply provides correlation with 240.21(B)(3) because that other rule also covers a transformer secondary.

Part **(B)(4)** is another departure from the rule that conductors must be provided with overcurrent protection at their supply ends, where they receive current from larger feeder conductors. 240.21(B)(4) permits a longer length than the 10-ft unprotected tap of part **(B)(1)** and the 25-ft (7.5-m) tap of part **(B)(2).** Under specified

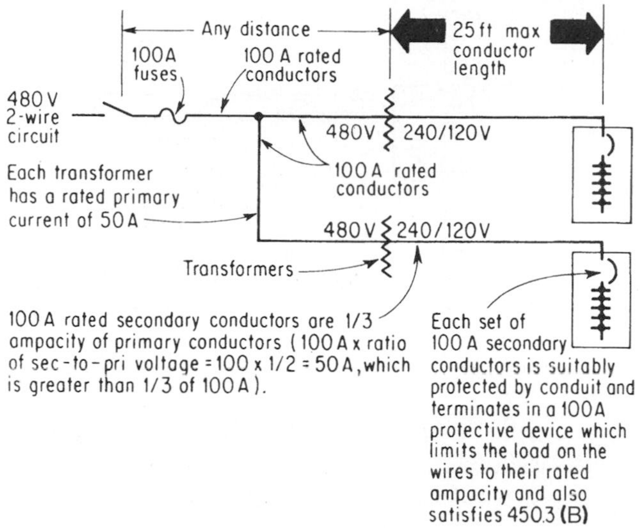
Any distance
25 ft max conductor length
100A fuses
100 A rated conductors
480 V 2-wire circuit
480V 240/120V
Each transformer has a rated primary current of 50 A
100 A rated conductors
480 V 240/120V
Transformers
100 A rated secondary conductors are 1/3 ampacity of primary conductors (100 A x ratio of sec-to-pri voltage = 100 x 1/2 = 50 A, which is greater than 1/3 of 100 A).
Each set of 100 A secondary conductors is suitably protected by conduit and terminates in a 100A protective device which limits the load on the wires to their rated ampacity and also satisfies 450.3 (B)

Fig. 240-21. Sizing a 25-ft (7.5-m) tap and transformer protection. (Sec. 240.21.)

conditions that are similar to the requirements of the 25-ft-tap exception, an unprotected tap up to 100 ft (30.0 m) in length may be used in "high-bay manufacturing buildings" that are over 35 ft (11.0 m) high *at the walls*—but only "where conditions of maintenance and supervision assure that only qualified persons will service the system." Obviously, that last phrase can lead to some very subjective and individualistic determinations by the authorities enforcing the Code. And the phrase "35 ft (11.0 m) high at the walls" means that this rule cannot be applied where the height is over 35 ft (11.0 m) at the peak of a triangular or curved roof section but less than 35 ft (11.0 m) at the walls.

The 100-ft (30.0-m) tap exception must meet specific conditions:

1. "Qualified" persons must maintain the system.
2. From the point at which the tap is made to a larger feeder, the tap run must not have more than 25 ft (7.5 m) of its length run horizontally, and the sum of horizontal run and vertical run must not exceed 100 ft (30.0 m). Figure 240-23 shows some of the almost limitless configurations of tap layout that would fall within the dimension limitations.
3. The tap conductors must have an ampacity equal to at least one-third of the rating of the overcurrent device protecting the larger feeder conductors from which the tap is made.
4. The tap conductors must terminate in a circuit breaker or fused switch, where the rating of overcurrent protection is not greater than the tap-conductor ampacity.
5. The tap conductors must be protected from physical damage and must be installed in "an approved raceway or other approved means."
6. There must be no splices in the total length of each of the conductors of the tap.

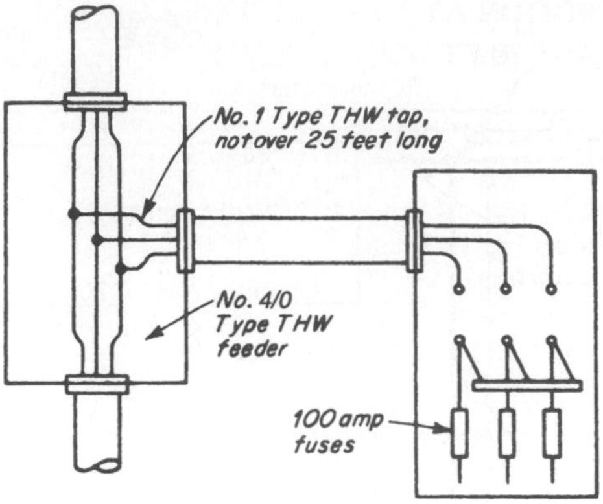

No. 1 Type THW tap,
not over 25 feet long
No. 4/0
Type THW
feeder
100 amp
fuses

25-ft tap – 240.21(B)(2)

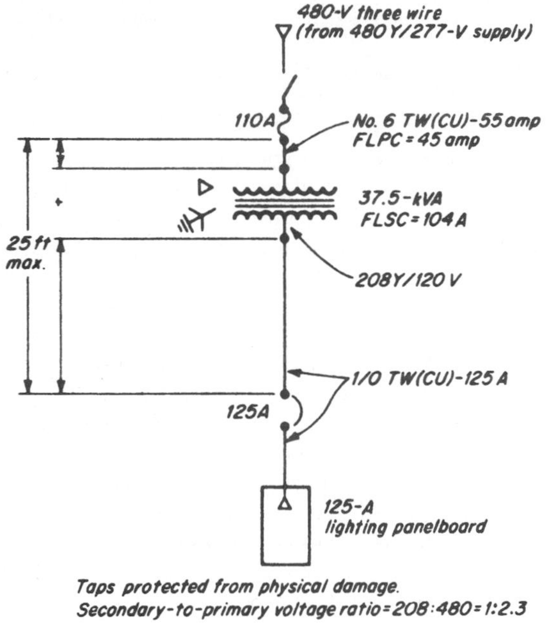

480-V three wire
(from 480Y/277-V supply)
110A
No. 6 TW(CU)-55 amp
FLPC = 45 amp
37.5-kVA
FLSC = 104 A
25 ft
max.
208Y/120 V
1/0 TW(CU)-125 A
125A
125-A
lighting panelboard
Taps protected from physical damage.
Secondary-to-primary voltage ratio = 208:480 = 1:2.3

25-ft tap – 240.21(C)(5)

Fig. 240-22. Examples show difference between the two types of 25-ft (7.5-m) taps covered by parts **(C)** and **(C)(5).** (Sec. 240.21.)

7. The tap conductors must not be smaller than 6 AWG copper or 4 AWG aluminum.
8. The tap conductors must not pass through walls, floors, or ceilings.
9. The point at which the tap conductors connect to the feeder conductors must be at least 30 ft (9.0 m) above the floor of the building.

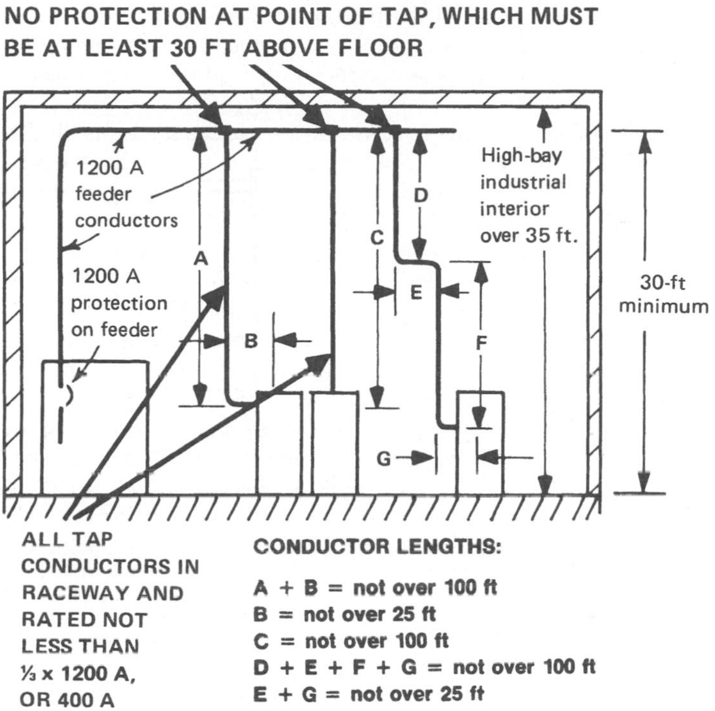
NO PROTECTION AT POINT OF TAP, WHICH MUST BE AT LEAST 30 FT ABOVE FLOOR
1200 A feeder conductors
1200 A protection on feeder
High-bay industrial interior over 35 ft.
30-ft minimum
A
B
C
D
E
F
G
ALL TAP CONDUCTORS IN RACEWAY AND RATED NOT LESS THAN ⅓ x 1200 A, OR 400 A
CONDUCTOR LENGTHS:
A + B = not over 100 ft
B = not over 25 ft
C = not over 100 ft
D + E + F + G = not over 100 ft
E + G = not over 25 ft

Fig. 240-23. Unprotected taps up to 100 ft long may be used in "high-bay manufacturing buildings."

As shown in Fig. 240-23, the tap conductors from a feeder protected at 1200 A are rated at not less than one-third the protection rating, or 400 A. Although 500-kcmil THW copper is rated at 380 A, that value does not satisfy the minimum requirement for 400 A. But if 500-kcmil THHN or XHHW copper, with an ampacity of 430 A, were used for the tap conductors, the rule would be satisfied. However, in such a case, those conductors would have to be used as if their ampacity were 380 A for the purpose of load calculation because of the general UL rule of 75°C conductor terminations for connecting to equipment rated over 100 A—such as the panelboard, switch, motor-control center, or other equipment fed by the taps. And the conductors for the main feeder being tapped could be rated less than the 1200 A shown in the sketch if the 1200-A protection on the feeder was selected in accordance with 430.62 or 430.63 for supplying a motor load or motor and lighting load. In such cases, the overcurrent protection may be rated considerably higher than the feeder conductor ampacity. But the tap conductors must have ampacity at least equal to one-third the *feeder protection rating.*

The 1200-A feeder that was tapped in this example raises another point of discussion. That feeder, unless from a busway, almost certainly was run with multiple conductors in parallel. For the sake of argument, suppose the feeder consists of three sets of 600-kcmil conductors. The 400-A tap, as noted, could be 500 kcmil THHN.

The question constantly arises in the field, is it necessary to connect each phase of the tap to all of the corresponding phase conductors in the overhead feeder? Certainly tapping only one of those conductors would be a far simpler task. The answer is no.

The feeder as connected to its overcurrent protective device is all three runs. Separating one of the sets of the supplied conductors means that the tap is being applied to only one-third of the feeder. In effect the tap is being made to another tap, namely, one that begins at the 1200 A breaker. That tap would not comply with any known allowance in the **NEC** given its length, location, etc. Further, the actual field tap covered here would then be made from this undefined tap, in violation of the clear prohibition of making taps from other taps.

240.21(C)(3) applies exclusively to industrial electrical systems. Conductors up to 25 ft (7.5 m) long may be tapped from a transformer secondary without overcurrent protection at their supply end and without need for a single-circuit breaker or set of fuses at their load end. Normally, a transformer secondary tap over 10 ft (3 m) long and up to 25 ft (7.5 m) long must comply with the rules of 240.21(C)(5) or (C)(6)—which call for such a transformer secondary tap to be made with conductors that require no overcurrent protection at their supply end but are required to terminate at their load end in a single CB or single set of fuses with a setting or rating not over the conductor ampacity. However, 240.21(C)(3) permits a 10- to 25-ft (3- to 7.5-m) tap from a transformer secondary without termination in a single main overcurrent device—*but* it limits the application to "industrial installations." The tap conductor ampacity must be at least equal to the transformer's secondary current rating and must be at least equal to the sum of the ratings of overcurrent devices supplied by the tap conductors.

As clarified in the 2014 **NEC**, this provision is limited to tap conductors arriving at the main lugs of a switchboard, as in Fig. 240-24. A motor control center could never have qualified, because overcurrent protection in the form of a singular device is required in accordance with the rating of the common power bus, as covered in 430.94. Power panels no longer comply because all panelboards now require individual overcurrent protection, with exceptions that would not apply here (see 408.36). If the tap arrived at a wireway or auxiliary gutter over the collection of loads intended to be supplied, as shown at the bottom of Fig. 240-24, the individual taps to each of the loads would almost certainly have violated the prohibition against tapping taps, certainly so if they were reduced in size to meet the likely termination limitations of the smaller equipment.

The rule of parts **(B)(5)** and **(C)(4)** allows outdoor feeder taps and unprotected secondary conductors from outdoor transformers to run for any distance *outdoors*. Physical protection for the conductors must be provided and they must terminate in a single CB or set of fuses. The CB or set of fuses must be part of, or adjacent to, the disconnect, which may be installed anywhere outdoors or indoors as close as possible to the point of conductor entry. Both sections emphasize that such unprotected conductors must not be run within any building or structure. As is the case with service conductors, these tap conductors must be terminated at an OC device as soon as they enter. Also, as in the case of services, the rules of 230.6 (concrete encasement, etc.) can be used to artificially extend the point of entrance if necessary.

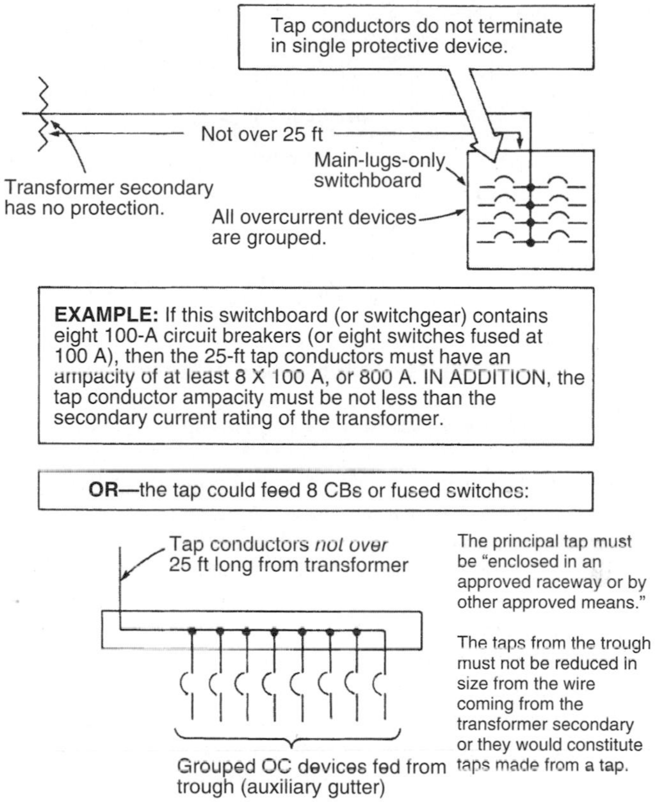
Tap conductors do not terminate in single protective device.
Not over 25 ft
Main-lugs-only switchboard
Transformer secondary has no protection.
All overcurrent devices are grouped.
EXAMPLE: If this switchboard (or switchgear) contains eight 100-A circuit breakers (or eight switches fused at 100 A), then the 25-ft tap conductors must have an ampacity of at least 8 X 100 A, or 800 A. IN ADDITION, the tap conductor ampacity must be not less than the secondary current rating of the transformer.
OR—the tap could feed 8 CBs or fused switches:
Tap conductors not over 25 ft long from transformer
The principal tap must be "enclosed in an approved raceway or by other approved means."
The taps from the trough must not be reduced in size from the wire coming from the transformer secondary or they would constitute taps made from a tap.
Grouped OC devices fed from trough (auxiliary gutter)

Fig. 240-24. These tap applications are permitted for transformer secondaries only in "industrial" electrical systems. The setup at the bottom of the figure would always have been a virtual impossibility to wire given the terminations limitations of the supplied devices, and as of the 2014 **NEC** the wiring must be as shown in the upper drawing.

As shown in Fig. 240-25, 240.21(G) gives permission for unprotected taps to be made from generator terminals to the first overcurrent device it supplies—such as in the fusible switch or circuit breakers used for control and protection of the circuit that the generator supplies. No maximum length is specified for the generator tap conductors, although various limits have been proposed over the years. Note also that 445.13, which is referenced, requires the tap conductors to have an ampacity of at least 115 percent of the generator nameplate current rating.

Section 240.21(H), new in the 2008 **NEC**, allows the location of overcurrent protection for battery output conductors to be as close as practicable to the battery room and still be out of range of the hazardous location boundary, if such a classification has been established. Note that 480.5 requires the disconnecting means for conductors supplied from a stationary battery system operating over 30 V to be readily accessible and within sight of the battery system. While batteries are charging, the current flowing over the conductors is controlled by the charging

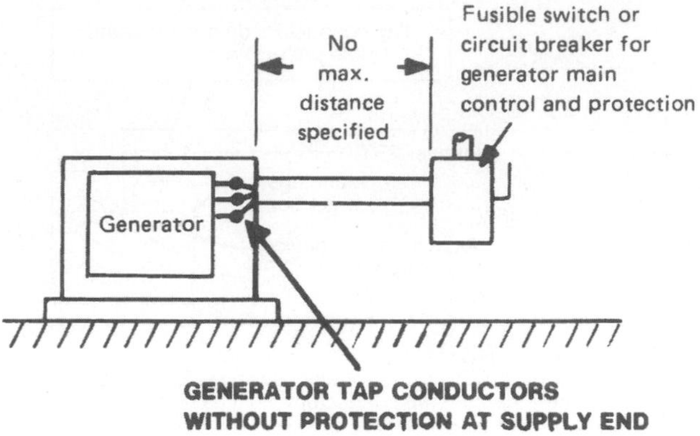
Fusible switch or
circuit breaker for
generator main
control and protection
No
max.
distance
specified
Generator
GENERATOR TAP CONDUCTORS
WITHOUT PROTECTION AT SUPPLY END

Fig. 240-25. Unprotected tap may be made from a generator's output terminals to the first overcurrent device. [Sec. 240.21(G).]

system, but when the batteries are actually supplying power overcurrent protection is necessary.

240.22. Grounded Conductor. The basic rule prohibits use of a fuse or CB in any conductor that is intentionally grounded—such as a grounded neutral or a grounded phase leg of a delta system. Figure 240-26 shows the two "exceptions" to that rule and a clear violation of the basic rule.

240.23. Change in Size of Grounded Conductor. In effect, this recognizes the fact that if the neutral is the same size as the ungrounded conductor, it will be protected wherever the ungrounded conductor is protected. One of the most obvious places where this is encountered is in a distribution center where a small grounded conductor may be connected directly to a large grounded feeder conductor.

240.24. Location in or on Premises. According to part **(A),** overcurrent devices must be readily accessible. And in accordance with the definition of "readily accessible" in Art. 100, they must be "capable of being reached quickly for operation, renewal, or inspections, without requiring those to whom ready access is requisite to climb over or remove obstacles or to resort to portable ladders, chairs, etc." (Fig. 240-27).

Although the Code gives no maximum heights at which OCPDs are considered readily accessible, some guidance can be obtained from 404.8, which provides detailed requirements for location of switches and CBs. This section states that switches and CBs shall be so installed that the center of the grip of the operating handle, when in its highest position, will not be more than 6 ft 7 in. (2.0 m) above the floor or working platform.

There are certain applications where the rules for ready accessibility are waived.

Part **(A)(1)** covers any case where an overcurrent device is used in a busway plug-in unit to tap a branch circuit from the busway. 368.12 requires that such devices consist of an externally operable CB or an externally operable fusible switch. These devices must be capable of being operated from the floor by means of ropes, chains, or sticks. Part **(A)(2)** refers to 240.10, which states that where supplementary overcurrent protection is used, such as for lighting fixtures, appliances, or internal circuits or components of equipment, this supplementary protection is not required to be readily accessible. An example of this would be

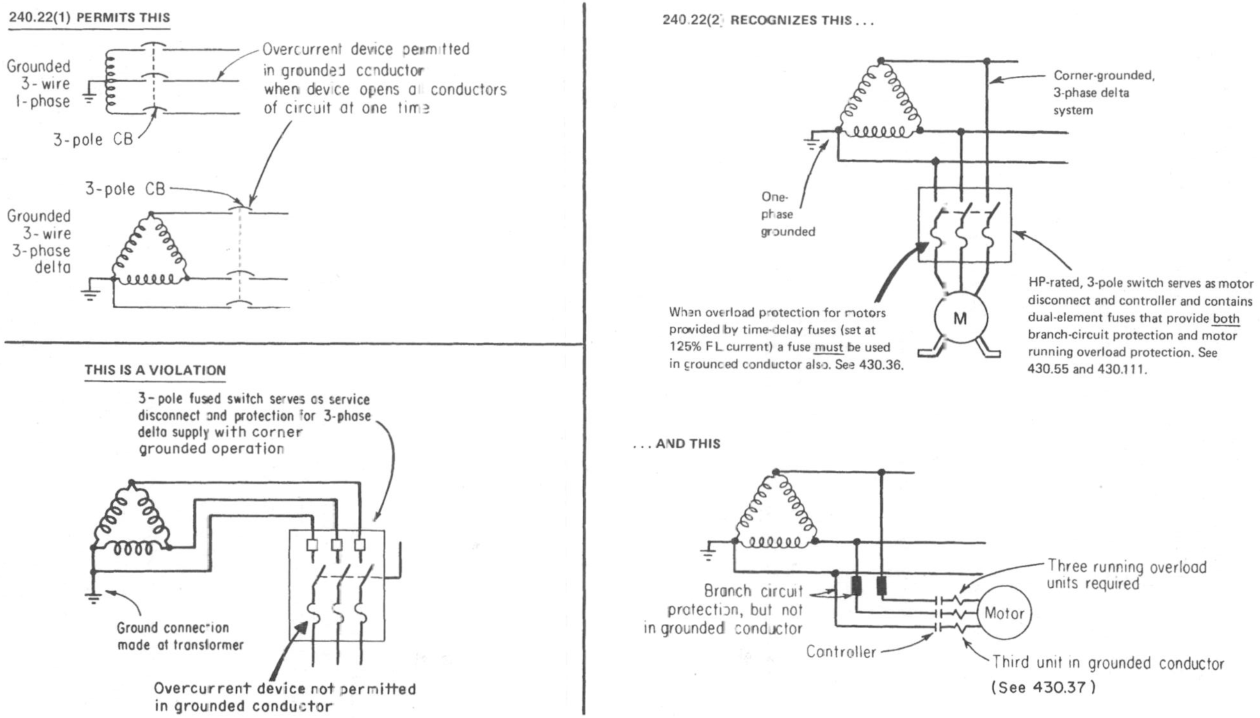
240.22(1) PERMITS THIS
Grounded 3-wire 1-phase
3-pole CB
Overcurrent device permitted in grounded conductor when device opens all conductors of circuit at one time
3-pole CB
Grounded 3-wire 3-phase delta
THIS IS A VIOLATION
3-pole fused switch serves as service disconnect and protection for 3-phase delta supply with corner grounded operation
Ground connection made at transformer
Overcurrent device not permitted in grounded conductor
240.22(2) RECOGNIZES THIS . . .
Corner-grounded, 3-phase delta system
One-phase grounded
M
When overload protection for motors provided by time-delay fuses (set at 125% FL current) a fuse must be used in grounded conductor also. See 430.36.
HP-rated, 3-pole switch serves as motor disconnect and controller and contains dual-element fuses that provide both branch-circuit protection and motor running overload protection. See 430.55 and 430.111.
. . . AND THIS
Three running overload units required
Branch circuit protection, but not in grounded conductor
Motor
Controller
Third unit in grounded conductor (See 430.37)

Fig. 240-26. Overcurrent protection in grounded conductor. (Sec. 240.22.)

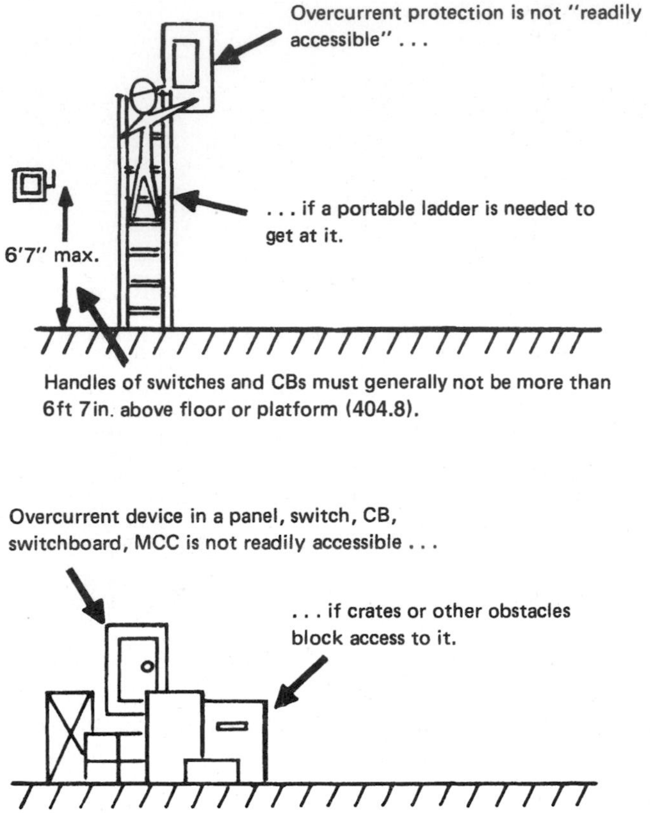
Overcurrent protection is not "readily accessible" . . .
. . . if a portable ladder is needed to get at it.
6′7″ max.
Handles of switches and CBs must generally not be more than 6 ft 7 in. above floor or platform (404.8).
Overcurrent device in a panel, switch, CB, switchboard, MCC is not readily accessible . . .
. . . if crates or other obstacles block access to it.

Fig. 240-27. Overcurrent devices must be "readily accessible." (Sec. 240.24.)

an overcurrent device mounted in the cord plug of a fixed or semifixed luminaire supplied from a trolley busway or mounted on a luminaire that is plugged directly into a busway. Part **(A)(3)** acknowledges that 230.92 permits service overcurrent protection to be sealed, locked, or otherwise made not readily accessible. Figure 240-28 shows these details.

240.24 clarifies the use of plug-in OCPDs on busways for protection of circuits tapped from the busway. After making the general rule that OCPDs must be readily accessible (capable of being reached without stepping on a chair or table or resorting to a portable ladder), part **(A)(1)** notes that it is not only *permissible* to use busway protective devices up on the busway—it is *required* by 368.17(C). Such devices on high-mounted busways are not "readily accessible" (not within reach of a person standing on the floor). The wording of 368.17(C) makes clear that this requirement for overcurrent protection in the device on the busway applies to subfeeders tapped from the busway as well as branch circuits tapped from the busway.

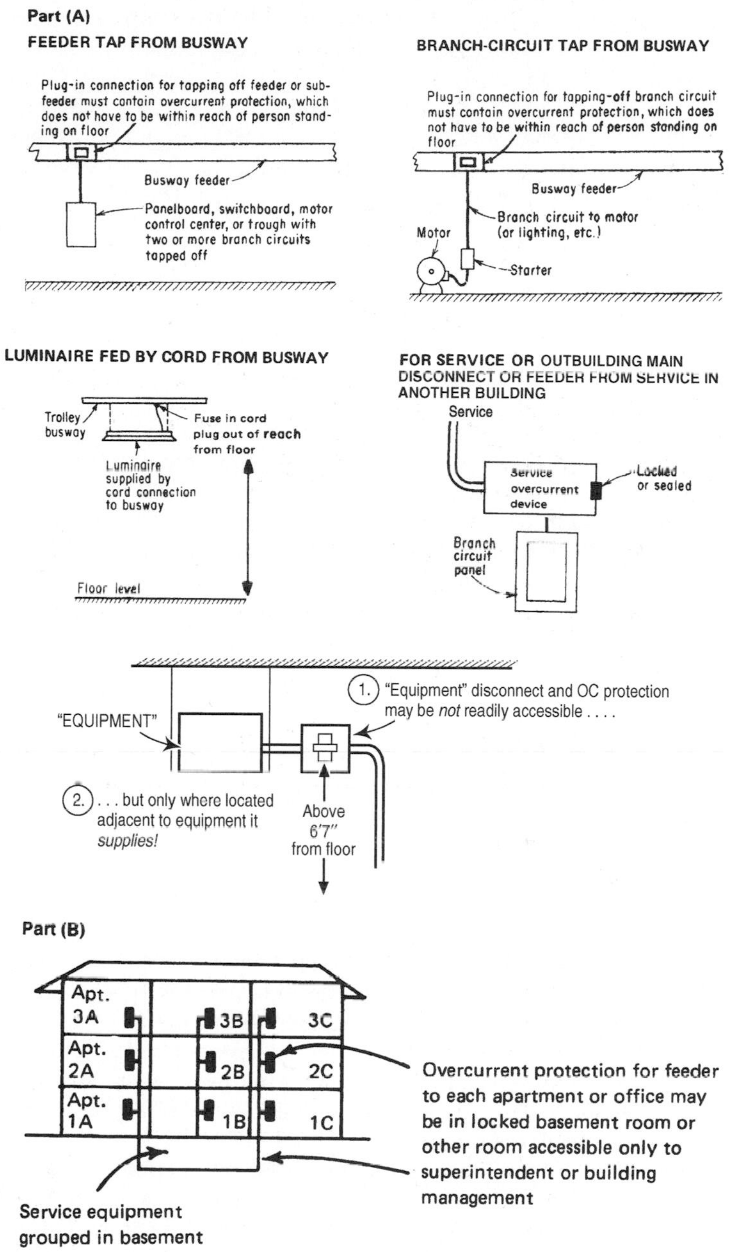

Part (A)
FEEDER TAP FROM BUSWAY
Plug-in connection for tapping off feeder or sub-feeder must contain overcurrent protection, which does not have to be within reach of person standing on floor
Busway feeder
Panelboard, switchboard, motor control center, or trough with two or more branch circuits tapped off
BRANCH-CIRCUIT TAP FROM BUSWAY
Plug-in connection for tapping-off branch circuit must contain overcurrent protection, which does not have to be within reach of person standing on floor
Busway feeder
Branch circuit to motor (or lighting, etc.)
Motor
Starter
LUMINAIRE FED BY CORD FROM BUSWAY
Trolley busway
Fuse in cord plug out of reach from floor
Luminaire supplied by cord connection to busway
Floor level
FOR SERVICE OR OUTBUILDING MAIN DISCONNECT OR FEEDER FROM SERVICE IN ANOTHER BUILDING
Service
Service overcurrent device
Locked or sealed
Branch circuit panel
"EQUIPMENT"
1. "Equipment" disconnect and OC protection may be *not* readily accessible . . . .
2. . . . but only where located adjacent to equipment it *supplies!*
Above 6′7″ from floor
Part (B)
Apt. 3A
3B
3C
Apt. 2A
2B
2C
Apt. 1A
1B
1C
Overcurrent protection for feeder to each apartment or office may be in locked basement room or other room accessible only to superintendent or building management
Service equipment grouped in basement

Fig. 240-28. Fuses or CBs that are permitted to be *not* readily accessible. (Sec. 240.24.)

The rule of **(A)(4)** recognizes the installation of an OC device in an inaccessible location where mounted adjacent to "utilization equipment they supply." The term "equipment" is defined in Art. 100. That definition seems to give broad permission for application of this rule. It seems that locating OC devices for conductor protection in other than a readily accessible location would not be permitted. Clearly, for motors, appliances, and transformers, the OC device that *supplies* such "equipment" may be mounted in an inaccessible location. The rules of NEC 240.24, 368.17(C), and 404.8 must be correlated with each other to assure effective Code compliance. And indeed, the customary application of this provision is to the local disconnecting means that is required to be adjacent to the utilization equipment of whatever kind in accordance with an applicable rule in the code, where that disconnecting means is also an overcurrent device in the form of a fused switch or a circuit breaker. This provision does not apply to the more usual application where overcurrent devices are provided to protect branch-circuit or feeder conductors. The revised definition of "readily accessible" in Art. 100, which prohibited the use of tools as a means of reaching readily accessible equipment, created significant mischief in terms of how this section can be applied. The 2017 NEC took a small step in the right direction by permitting the use of tools on industrial control panels and similar equipment. However, much more needs to be done. Until that happens, inspectors should be encouraged to take an expansive approach to the words "similar enclosures." For just one example, NEMA 4X stainless steel panelboards are routinely environmentally protected by stainless steel rim bolts placed around the perimeter of a solid cover. A panelboard is not normally something that would be judged as similar to an industrial control panel, but for now it must be because we clearly are not going to simply stop installing this equipment. Note also, that just because such a panelboard (or other similar equipment) may qualify under this reading, when the doors are open the contents must still be held to traditional provisions, such as the maximum permissible height of an operating handle.

Part **(B)** applies to apartment houses and other multiple-occupancy buildings—such as hotel guest rooms and suites, as described in Fig. 240-28.

In addition, it is important to note that parts **(C)** and **(D)** of 240.24 require that overcurrent devices be located where they will not be exposed to physical damage or in the vicinity of easily ignitable material. Panelboards, fused switches, and circuit breakers may *not* be installed in *clothes closets* in any type of occupancy—residential, commercial, institutional, or industrial. But they may be installed in other closets that do not have easily ignitable materials within them—provided that the working clearances of 110.26 (30-in. [752 mm] wide work space in front of the equipment, 6 ft 6 in. [2.0-m] headroom, illumination, etc.) are observed and the work space is "not used for storage," as required by 110.26(B).

240.24(E) flatly prohibits what was a somewhat common practice for dwellings, as well as guest rooms and suites in hotels and motels. In certain areas of the nation, OCPDs were located in areas such as kitchens and bathrooms. Although it is still permissible to locate the OCPDs in the kitchen, the rule of part **(E)** now forbids location of the overcurrent devices within the bathroom of a dwelling, dormitory, or hotel guest room or suite.

Part **(F)**, new for the 2008 NEC, flatly prohibits locating overcurrent devices over the inclined portion of a stairway. The literal text prohibits the location over "steps" which is presumably different from a "landing." There is no dimension given as to when a step becomes wide enough to be a landing, but that should be relatively obvious and interpreted consistently. Presumably, the required working space width would be a good starting point.

240.33. Vertical Position. Figure 240-29 shows the basic requirements of 240.30, 240.32, and 240.33. The rule in 240.33 is frequently misunderstood as favoring vertical mounting in the sense of having the operator move up and down, as distinguished from moving from side to side. That is the topic of 240.81 but is incorrect here. This section addresses the plane in which the overcurrent device is mounted, and favors a vertical plane as in mounting on a wall, and discourages mounting in a horizontal plane as in face up or face down.

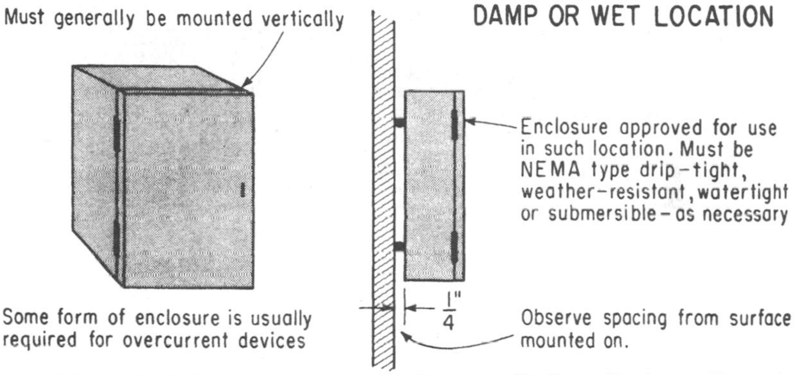
Must generally be mounted vertically
DAMP OR WET LOCATION
Enclosure approved for use in such location. Must be NEMA type drip-tight, weather-resistant, watertight or submersible - as necessary
Some form of enclosure is usually required for overcurrent devices
1"/4
Observe spacing from surface mounted on.

Fig. 240-29. Enclosures for overcurrent protection. (Sec. 240.30.)

This rule has been in the NEC for over 90 years, having first appeared in the 1926 edition. The commentary in the 8th edition of this *Handbook*, on the 1953 NEC, is instructive as to the intent of this rule:

> Installing cabinets or cutout boxes on ceilings is a practice that should be avoided wherever possible. Section 2435 [corresponds to 240.24 in the 2017 NEC] calls for cutouts and circuit breakers to be readily accessible, and a box on a ceiling is seldom readily accessible. In a box so installed, one end of a cartridge fuse may fall out of the terminals and make contact with the door of the box, thus grounding the circuit.

In addition to ceiling mounting issues, there have been some occasions for horizontal mounting in other circumstances. Some small panels, with perhaps four to six circuits, have been horizontally mounted, face-up with a door, in the top section of a short but deep wall housing special equipment. The circuit breakers were readily accessible, there was no good alternative, and the inspector agreed with the result. That said, wall mounting is almost always preferable. The rule also makes allowances for listed busway plug-in units that may have been designed for a horizontal orientation when the busway is in certain positions.

240.40. Disconnecting Means for Fuses. The basic rules are shown in Fig. 240-30. The second sentence covers cable limiters, and as covered in 230.82(1) they can be

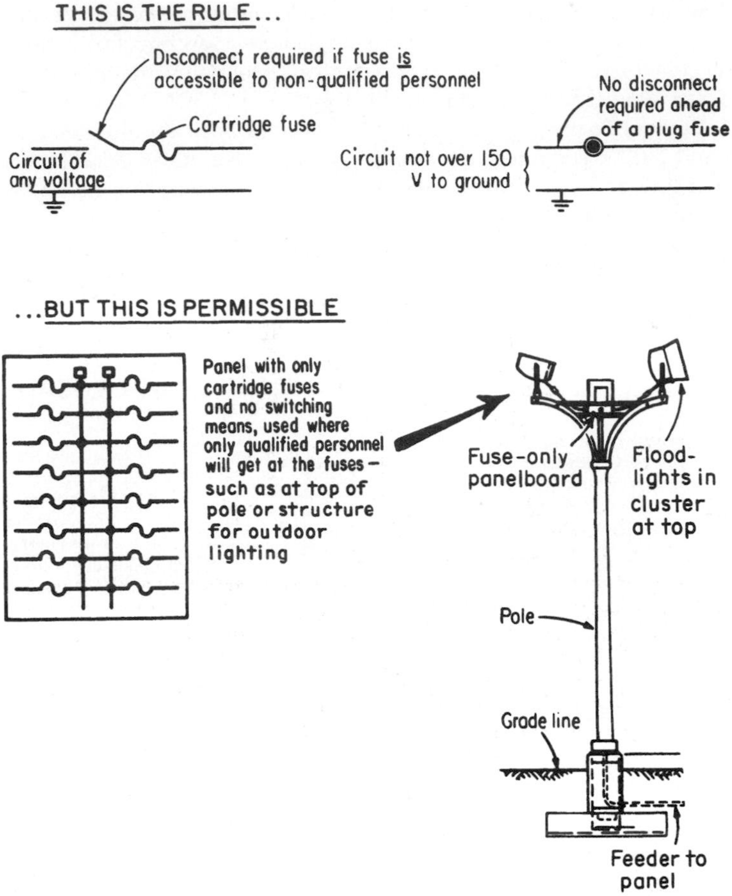
THIS IS THE RULE...
Disconnect required if fuse is accessible to non-qualified personnel
Cartridge fuse
Circuit of any voltage
No disconnect required ahead of a plug fuse
Circuit not over 150 V to ground
...BUT THIS IS PERMISSIBLE
Panel with only cartridge fuses and no switching means, used where only qualified personnel will get at the fuses – such as at top of pole or structure for outdoor lighting
Fuse-only panelboard
Flood-lights in cluster at top
Pole
Grade line
Feeder to panel

Fig. 240-30. Disconnect means for fuses. (Sec. 240.40.)

located ahead of the service disconnect, where no switch is required. The rule presented by the last sentence is illustrated in Fig. 240-31.

240.50. General (Plug Fuses). Plug fuses must not be used in circuits of more than 125 V between conductors, but they may be used in grounded-neutral systems where the circuits have more than 125 V between ungrounded conductors but not more than 150 V between any ungrounded conductor and ground (Fig. 240-32). And the screw-shell of plug fuseholders must be connected to the load side of the circuit.

240.51. Edison-Base Fuses. 240-52. Edison-Base Fuseholders. 240-53. Type S Fuses. 240.54. Type S Fuses, Adapters, and Fuseholders. Rated up to 30 A, plug fuses are Edison-base or Type S. 240.51(B) limits the use of Edison-base fuses to replacements of existing fuses of this type, and even then, they must be replaced if there is evidence of tampering or overfusing. Type S plug fuses are required by 240.53 for

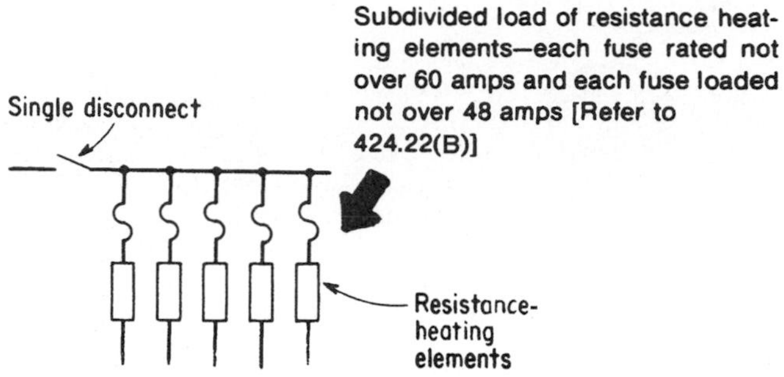
Subdivided load of resistance heating elements—each fuse rated not over 60 amps and each fuse loaded not over 48 amps [Refer to 424.22(B)]
Single disconnect
Resistance-heating elements

Fig. 240-31. Single disconnect for one set of fuses is permitted for electric space heating with subdivided resistance-type heating elements. (Sec. 240.40.)

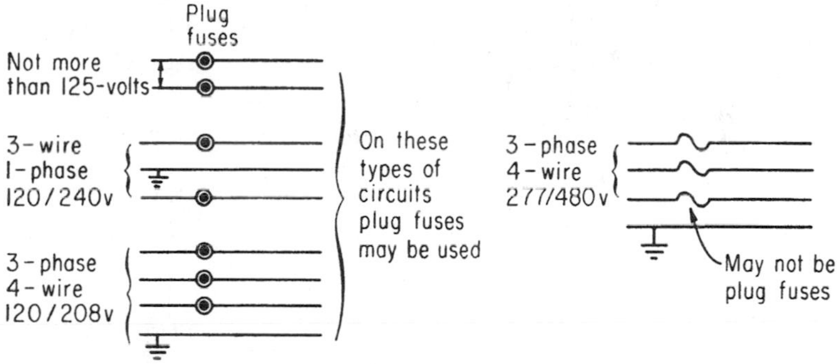
Plug fuses
Not more than 125-volts
3-wire 1-phase 120/240v
3-phase 4-wire 120/208v
On these types of circuits plug fuses may be used
3-phase 4-wire 277/480v
May not be plug fuses

Fig. 240-32. Using plug fuses. (Sec. 240.50.)

all new plug-fuse installations, and 240.52 requires new Edison-base fuseholders to be converted to Type S. These adapters are designed to go in but not come out. Once converted to Type S, an Edison-base fuseholder cannot be unconverted without the use of a special tool that destroys the adapter in the process. An unqualified person is unlikely to successfully attempt this process. Type S plug fuses must be used in Type S fuseholders or in Edison-base fuseholders with a Type S adapter inserted, so that a Type S fuse of one ampere classification cannot be replaced with a higher-amp rated fuse (Fig. 240-33). Type S fuses, fuseholders, and adapters are rated for three classifications based on amp rating and are noninterchangeable from one classification to another. The classifications are 0 to 15, 16 to 20, and 21 to 30 A. The 0- to 15-A fuseholders or adapters must not be able to take any fuse rated over 15 A, etc. The purpose of this rule is to prevent overfusing of 15- and 20-A circuits.

240.60. General (Cartridge Fuses). The last sentence of part **(B)** must always be carefully observed. It is concerned with an extremely important matter:

The installation of current-limiting fuses demands extreme care in the selection of the fuse clips to be used. Because current-limiting fuses have an additional

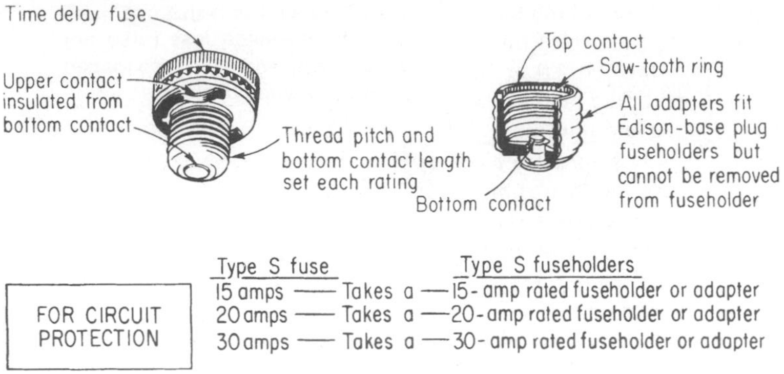
Time delay fuse
Upper contact insulated from bottom contact
Thread pitch and bottom contact length set each rating
Top contact
Saw-tooth ring
All adapters fit Edison-base plug fuseholders but cannot be removed from fuseholder
Bottom contact
FOR CIRCUIT PROTECTION
Type S fuse
Type S fuseholders
15 amps — Takes a — 15-amp rated fuseholder or adapter
20 amps — Takes a — 20-amp rated fuseholder or adapter
30 amps — Takes a — 30-amp rated fuseholder or adapter

Fig. 240-33. Type S plug fuse. (Sec. 240.53.)

protective feature (that of current limitation, that is, extremely fast operation to prevent the flow of the extremely high currents which many modern circuits can produce into a ground fault or short circuit) as compared to noncurrent-limiting fuses, some condition of the mounting arrangement for current-limiting fuses must prevent replacement of the current-limiting fuses by noncurrent-limiting. This is necessary to maintain safety in applications where, for example, the busbars of a switchboard or motor control center are braced in accordance with the maximum let-through current of current-limiting fuses which protect the busbars, but would be exposed to a much higher potential value of fault let-through current if noncurrent-limiting fuses were used to replace the current-limiting fuses. The possibility of higher current flow than that for which the busbars are braced is created by the lack of current limitation in the noncurrent-limiting fuses.

240.60(B) takes the above matter into consideration when it rules that "fuseholders for current-limiting fuses shall not permit insertion of fuses that are not current limiting." To afford compliance with the Code and to obtain the necessary safety of installation, fuse manufacturers provide current-limiting fuses with special ferrules or knife blades for insertion only in special fuse clips. Such special ferrules and blades do permit the insertion of current-limiting fuses into standard **NEC** fuse clips, to cover those cases where current-limiting fuses (with their higher type of protection) might be used to replace noncurrent-limiting fuses. But the special rejection-type fuseholders will not accept noncurrent-limiting fuses—thereby ensuring replacement only with current-limiting fuses.

The very real problem of Code compliance and safety is created by the fact that many fuses with standard ferrules and knife-blade terminals are of the current-limiting type and are made in the same construction and dimensions as corresponding sizes of noncurrent-limiting fuses, for use in standard fuseholders. Such current-limiting fuses are not marked "current limiting" but may be used to obtain limitation of energy let-through. Replacement of them by standard nonlimiting fuses could be hazardous. Note that 240.60(C) covers the required markings on fuses, and in this regard pay close attention to the interrupting rating, which must always be marked if other than the default value of 10,000 A.

Class J and L fuses Both the Class J (0–600 A, 600 V ac) and Class L (601–6000 A, 600 V ac) fuses are current-limiting, high-interrupting-capacity types. The interrupting ratings are 100,000 or 200,000 rms symmetrical amperes, and the designated rating is marked on the label of each Class J or L fuse. Class J and L fuses are also marked "current limiting," as required in part **(C)** of 240.60.

Class J fuse dimensions are different from those for standard Class H cartridge fuses of the same voltage rating and ampere classification. As such, they will require special fuseholders that will not accept noncurrent-limiting fuses. This arrangement complies with the last sentence of **NEC** 240.60(B).

Class K fuses These are subdivided into Classes K-1, K-5, and K-9. Class K fuses have the same dimensions as Class H (standard **NEC**) fuses and are interchangeable with them. Classes K-1, K-5, and K-9 fuses have different degrees of current limitation but are not permitted to be labeled "current limiting" because physical characteristics permit these fuses to be interchanged with noncurrent-limiting types. Use of these fuses, for instance, to protect equipment busbars that are braced to withstand 40,000 A of fault current at a point where, say, 60,000 A of current would be available if noncurrent-limiting fuses were used is a clear violation of the last sentence of part **(B)**. As shown in Fig. 240-34, because such fuses can be replaced with nonlimiting fuses, the equipment bus structure would be exposed to dangerous failure. Classes R and T have been developed to provide current limitation and prevent interchangeability with noncurrent-limiting types.

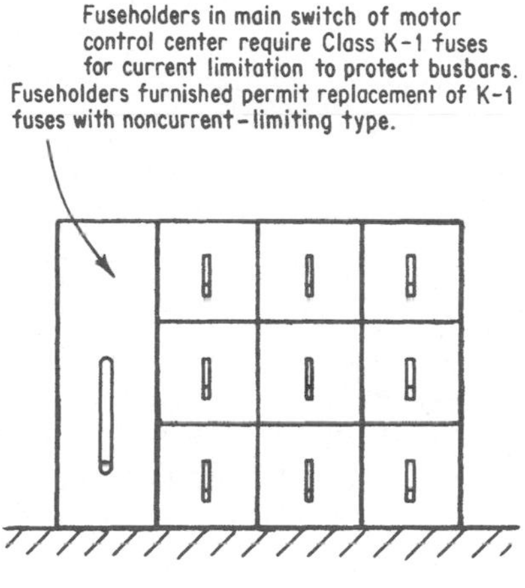
Fuseholders in main switch of motor control center require Class K-1 fuses for current limitation to protect busbars.
Fuseholders furnished permit replacement of K-1 fuses with noncurrent-limiting type.

Fig. 240-34. Current-limiting fuseholders must be rejection type. (Sec. 240.60.)

Class R fuses These fuses are made in two designations: RK1 and RK5. UL data are as follows:

Fuses marked "Class RK1" or "Class RK5" are high-interrupting-capacity types and *are* marked "current limiting." Although these fuses will fit into standard fuseholders that take Class H and K fuses, special rejection-type fuseholders designed for Class RK1 and RK5 fuses will not accept Class H and K fuses. In that way, circuits and equipment protected in accordance with the characteristics of RK1 or RK5 fuses cannot have that protection reduced by the insertion of other fuses of a lower protective level.

Other UL application data that affect selection of various types of fuses are as follows:

Fuses designated as Class CC (0 to 20 A, 600 V ac) are high-interrupting-capacity types and are marked "current limiting." They are not interchangeable with fuses of higher voltage or interrupting rating or lower current rating.

Class G fuses (0 to 60 A, 300 V ac) are high-interrupting-capacity types and are marked "current limiting." They are not interchangeable with other fuses mentioned preceding and following.

Fuses designated as Class T (0 to 600 A, 250 and 600 V ac) are high-interrupting-capacity types and are marked "current limiting." They are not interchangeable with other fuses mentioned previously.

Part **(C)** requires use of fuses to conform to the marking on them. Fuses that are intended to be used for current limitation must be marked "current limiting."

Class K-1, K-5, and K-9 fuses are marked, in addition to their regular voltage and current ratings, with an interrupting rating of 200,000, 100,000, or 50,000 A (rms symmetrical). (See Fig. 240-35.)

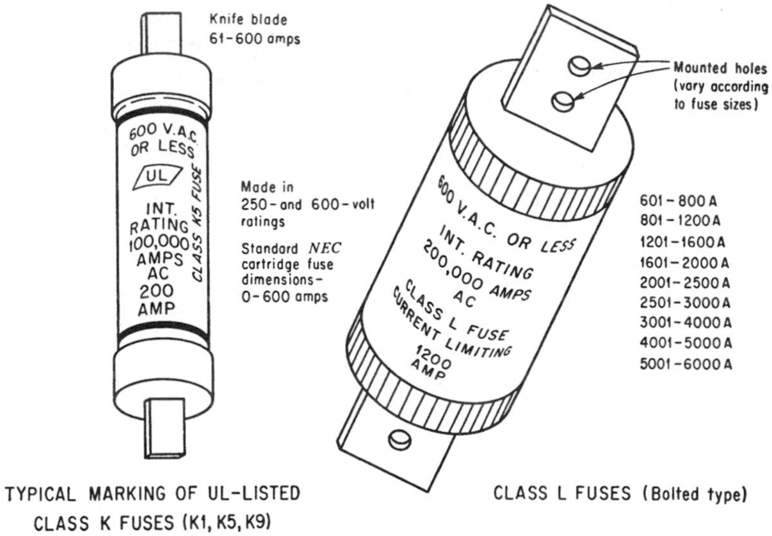
Knife blade
61-600 amps
600 V.A.C. OR LESS
UL
INT. RATING 100,000 AMPS AC
200 AMP
CLASS K5 FUSE
Made in 250- and 600-volt ratings
Standard NEC cartridge fuse dimensions-0-600 amps
Mounted holes (vary according to fuse sizes)
600 V.A.C. OR LESS
INT. RATING 200,000 AMPS AC
CLASS L FUSE
CURRENT LIMITING
1200 AMP
601-800 A
801-1200 A
1201-1600 A
1601-2000 A
2001-2500 A
2501-3000 A
3001-4000 A
4001-5000 A
5001-6000 A
TYPICAL MARKING OF UL-LISTED CLASS K FUSES (K1, K5, K9)
CLASS L FUSES (Bolted type)

Fig. 240-35. Fuses must be applied in accordance with marked ratings. (Sec. 240.60.)

Class CC, RK1, RK5, J, L, and T fuses are marked, in addition to their regular voltage and current ratings, with an interrupting rating of 200,000 A (rms symmetrical).

Although it is not required by the Code, manufacturers are in a position to provide fuses that are advertised and marked indicating they have "time-delay" characteristics. In the case of Class CC, G, H, K, and RK fuses, time-delay characteristics of fuses (minimum blowing time) have been investigated. Class G or CC fuses, which can carry 200 percent of rated current for 12 s or more, and Class H, K, or RK fuses, which can carry 500 percent of rated current for 10 s or more, may be marked with "D," "time delay," or some equivalent designation. Class L fuses are permitted to be marked "time delay" but have not been evaluated for such performance. Class J and T fuses are not permitted to be marked "time delay."

240.61. Classification. This section notes that any fuse may be used at its voltage rating or at any voltage below its voltage rating.

240.67. Arc Energy Reduction. In a major step towards evening the playing field between fuses and circuit breakers, the 2017 NEC began applying parallel language taken largely from 240.87 aimed at establishing mandatory requirements to reduce the incident energy of arc flash events, this time in systems relying on fusible switches for overcurrent protection. This section becomes effective at the beginning of 2020. The basic requirement stipulates that a fuse must clear in 0.07 s, which is just over 4 cycles. The value was chosen because typical zone selective interlocking arrangements [240.87(B)(1)] take about the same amount of time to function, so this seemed a fair comparison.

240.80. Method of Operation (Circuit Breakers). This rule requiring trip-free manual operation of circuit breakers ties in with that in 230.76, although this rule requires manual operation to *both* the closed and the open positions of the CB. According to 230.76, a power-operated circuit breaker used as a service disconnecting means must be capable of being opened by hand but does not have to be capable of being closed by hand. The general rule of 240.80 requires circuit breakers to be "capable of being *closed and opened* by manual operation." That rule also says that if a CB is electrically or pneumatically operated, it must also provide for manual operation (Fig. 240-36).

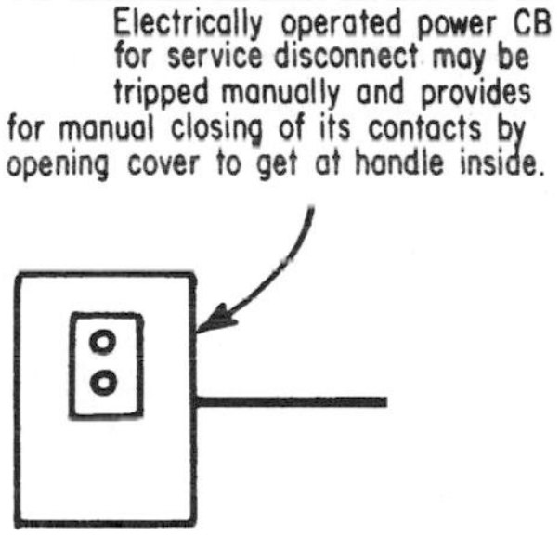
Electrically operated power CB for service disconnect may be tripped manually and provides for manual closing of its contacts by opening cover to get at handle inside.

Fig. 240-36. Every CB must be manually operable. (Sec. 240.80.)

240.81. Indicating. This rule requires the up position to be the ON position for any CB. *All* circuit breakers—not just those "on switchboards or in panelboards"—must be ON in the up position and OFF in the down position if their handles operate vertically rather than rotationally or horizontally. This is an expansion of the rule that previously applied only to circuit breakers on switchboards or in panelboards. This brings the rule into agreement with that of the second paragraph of 404.7—which makes the identical requirement for *all* circuit breakers *and* switches in individual enclosures. Switches and circuit breakers in individual enclosures must be marked to clearly show ON and OFF positions and vertically operated switches and CBs must be ON when in the up position (Fig. 240-37).

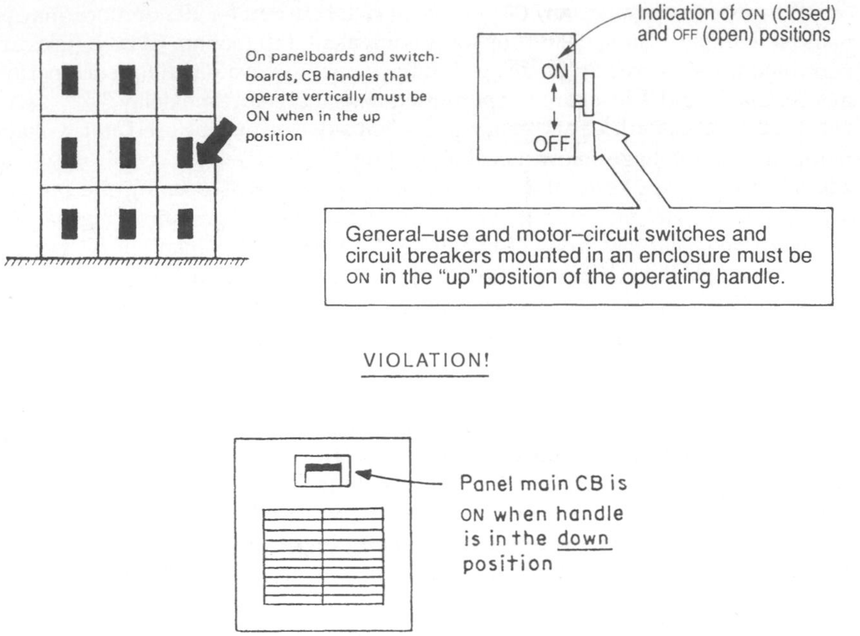
Indication of ON (closed) and OFF (open) positions
On panelboards and switchboards, CB handles that operate vertically must be ON when in the up position
ON
OFF
General–use and motor–circuit switches and circuit breakers mounted in an enclosure must be ON in the "up" position of the operating handle.
VIOLATION!
Panel main CB is ON when handle is in the down position

Fig. 240-37. Handle position of CB in any kind of enclosure must be ON in the up position. (Sec. 240.81.)

240.83. Marking. Part **(A)** requires that the marking of a CB's ampere rating must be durable and visible after installation. That marking is permitted to be made visible by removing the trim or cover of the CB.

In part **(B),** the Code mandates that the ampere rating be marked on the CB's handle (or escutcheon area) when it is rated 100 A or less. Part **(C)** presents the same requirement that UL does with regard to the marking of the OC device's ampere interrupting rating (AIR). Where an OC device has more than a 5000 AIR, the AIR must be marked on the CB by the manufacturer.

Part **(D)** of this section requires that any CB used to switch 120- or 277-V fluorescent lighting be listed for the purpose and be marked "SWD" or "HID." Note that the "HID" rating is somewhat more robust, and therefore such a breaker can be used for fluorescent lighting, but the reverse is not the case and an "SWD" breaker is only good for fluorescent lighting (Fig. 240-38). In commercial and industrial electrical systems, ON–OFF control of lighting is commonly done by the breakers in the lighting panel, eliminating any local wiring-device switches. Be careful to integrate the requirements in 210.4(B) with this process on new installations. If the lighting circuits are configured as multiwire branch circuits, multipole breakers will generally be in order, and a much larger area will go off and on when the breaker operates. However, with the recent focus on energy conservation, large numbers of these lighting zones are being provided with occupancy sensors or other automated methods to run the lights only where needed, so this is probably not the concern it was years ago.

The rule of part **(E)** requires specific voltage markings on circuit breakers.

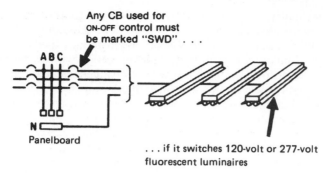

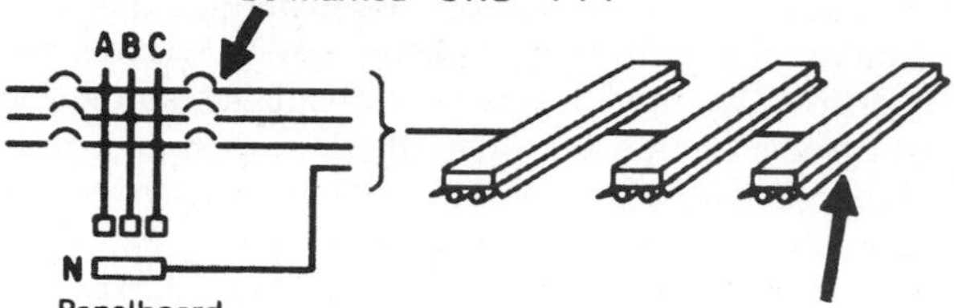
Any CB used for
ON-OFF control must
be marked "SWD" . . .
A B C
N
Panelboard
. . . if it switches 120-volt or 277-volt
fluorescent luminaires

Fig. 240-38. Circuit breakers used for switching lights must be SWD type. [Sec. 240.83(D).]

240.85. Applications. This section repeats UL data regarding interpretation of voltage markings. The wording explains circuit-breaker voltage markings in terms of the device's suitability for grounded and ungrounded systems. Designation of only a phase-to-phase rating—such as "480 V"—indicates suitability for grounded or ungrounded systems. But voltage designations showing a phase-to-neutral voltage by "slash" markings—like 480Y/277V or 120/240V—indicate that such circuit breakers are limited exclusively to use in grounded neutral electrical systems. Specifically, a slash-rated breaker must only be used where all ungrounded conductors to which it will be connected operate at the lower voltage to ground. This makes a real difference in a center-tapped delta system (capable of traditional three-phase 240-V connections and 120/240-V connections across one pair of phases). The other phase, the so-called high leg, will be at 208 V to ground on such systems. Any two-pole circuit breaker connected to the high leg will be (1) operating correctly in terms of line-to-line voltage, but (2) operating beyond its ratings in terms of line-to-ground voltage. A line-to-ground fault will require the breaker to clear a fault that is in progress using only one of its poles at a significantly higher voltage than it was tested.

Breakers without the slash markings are internally braced to withstand and clear full line-to-line voltage faults that can easily flow through only one pole of the breaker, particularly on corner-grounded systems. This requires a far more robust construction than the usual grounded neutral system, where any ground fault that involves only one pole will be at only the line-to-neutral voltage, and for any line-to-line short circuit the interrupting effort will be shared between two poles of the breaker. For this reason track this rule carefully when laying out jobs. Three-pole breakers are generally available without relying on a slash marking, but two-pole breakers without the slash markings are frequently only available by special order and sell at a substantial cost premium.

The last sentence in the first part of Sec. 240.85 calls attention to the marking that identifies a two-pole breaker's suitability for use on corner-grounded systems. Two-pole devices marked 240 or 480 V must be further identified by a marking "1φ-3φ" to be used on corner-grounded delta systems. These breakers undergo special testing, including some consideration of the "individual pole interrupting capability" discussed in the informational note at the end of the section.

240.86. Series Ratings. This section recognizes the use of the "series-rated" OC devices to ensure adequate fault-current protection. These devices, when operated in series with each other, allow the fault-interrupting capability of the main breaker, under fault conditions, to assist feeder or branch breakers that are applied at a point in the distribution system where the available fault current is greater than the AIR of the feeder or branch breaker. By sharing the arc, and operating in series, the circuit components will be provided the protection required by 110.10, even though a downstream protective device in the series may not have an adequate AIR for the point in the system where it is installed. Application of such OCPDs must satisfy the requirements given here.

When considering the concept of series rated circuit breakers, a key controversy quickly arises, and most of the changes in this part of the **NEC** over the last several code cycles have involved attempts to address this concern. This is the question of how to deal with dynamic impedance. When a circuit breaker trips and begins the process of clearing a fault, its contacts begin to separate and as they do, they draw an arc. Electrical arcs have significant impedance, and that impedance changes rapidly as the internal contacts separate. This is an oversimplification; however, the contacts of a smaller breaker, having less inertia, may open more quickly than those of a larger breaker. In the worst case, the smaller breaker can introduce just enough impedance into the circuit that the larger breaker may not unlatch, and ride out the fault. Assuming that the fault was well beyond the interrupting rating of the smaller breaker, the consequence of the upstream large breaker riding out the fault can easily result in the complete destruction of the downstream breaker.

It turns out that this process is very difficult to accurately predict by engineering modeling, even with second-order differential equations. Therefore, the circuit breaker manufacturers have resigned themselves to bench testing every conceivable combination of breakers in their product lines. The result is the "tested combinations" of 240.86(B), and the mandated marking required in 110.22(C). (See Fig. 240-39 for an example.) Every combination marked on a panelboard label has been bench tested to verify that the combination of this large breaker ahead

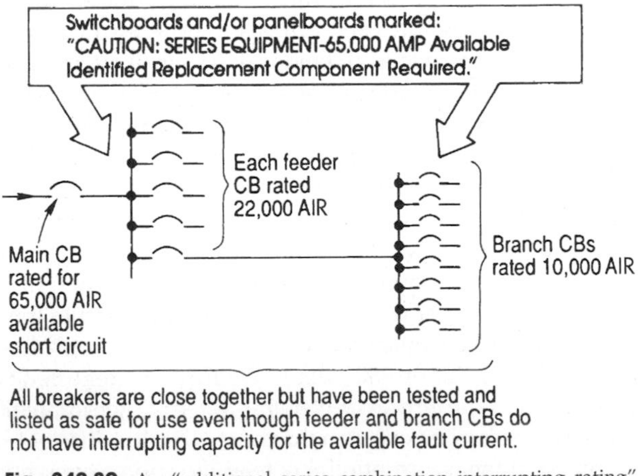
Switchboards and/or panelboards marked:
"CAUTION: SERIES EQUIPMENT-65,000 AMP Available
Identified Replacement Component Required."
Each feeder
CB rated
22,000 AIR
Branch CBs
rated 10,000 AIR
Main CB
rated for
65,000 AIR
available
short circuit
All breakers are close together but have been tested and
listed as safe for use even though feeder and branch CBs do
not have interrupting capacity for the available fault current.

Fig. 240-39. An "additional series combination interrupting rating" must be "marked" on equipment. (Sec. 240.86.)

of that small breaker, ranging from comparatively low fault to the specified maximum available fault current under prescribed test conditions, will clear and both upstream and downstream devices will live to protect again after the interruption is complete. These combinations undergo intermediate testing as well as testing under maximum fault current exposures to ensure that the combination will function in accordance with applicable standards under any overcurrent applied, not just bolted fault conditions. If a combination fails, the manufacturer has two choices: either leave that combination off the label, or make subtle changes in its breakers so the combination will pass reliably.

However, not all combinations, particularly combinations involving obsolete breakers, can be tested, and available fault currents steadily increase as the utility infrastructure stiffens in response to population increases and demands for increased reliability. If that upgrade crosses the previously designed available fault current line at a major industrial facility, the result is that large sections of the facility distribution system may drop dangerously below the interrupting ratings of the existing protective devices. To preserve safety, the facility must now consider buying and installing completely new gear with available fault current ratings that ensure appropriate performance under all overcurrent conditions. This process could involve, quite clearly, an astronomical expense.

That being the case, facilities that have been confronted by this exposure have tried to find a way to address it in some other way, and the fuse industry would love to be part of the solution. For decades the fuse industry has published "let through" calculations and data on its products. The customary approach is what is called an "up-over-and-down" analysis using published current-limiting graphs for the style of fuse considered. Beginning with the available rms fault current on the horizontal axis, read straight up to the diagonal index line for the proposed fuse size, then straight over to a line at a 45° slope, and then read straight down to the horizontal axis once again. The number there is the worst-case let-through rms current for the fuse in question when it is applied in the system being analyzed. If that number is less than interrupting rating of the old circuit breakers, can the problem be solved by adding a fuse?

Not necessarily because the dynamic impedance problem can defeat this design. If, and only if the circuit breaker can be guaranteed to not unlatch for several cycles, then yes, problem solved. And there are some old air-frame power breakers that won't unlatch for three cycles or so, giving the fuse time to clear the fault. But modern molded case circuit breakers have mechanisms, even those that aren't officially current limiting, that have internal current paths for which the magnetic forces on large faults tend to oppose each other and blow the contacts apart. The fault is often not a bolted fault but an arcing fault. If that happens and the arc adds enough impedance to take the current below the current-limiting range of the fuse, then the fuse will delay its response and the breaker will take the hit. It is not always impossible to design for this, but frequently very difficult to impossible. This brings us to current **NEC** requirements.

In addition to bench-tested combinations in 240.86(B), there now exists a procedure to field engineer a series-connected rating, as given in 240.86(A). There are significant restrictions on this approach. First, the procedure can only be used in an existing facility. It must be designed by a licensed professional engineer with appropriate training. He or she must document the selection and stamp the design, which must be made available to the local inspector and all others who will be working with the system. The rating, including the identity of the downstream device, must be field marked on the end use equipment in the manner specified in 110.22(B).

This entire issue continues as one of the most difficult to address in the history of the **NEC** for a particularly compelling reason. Both sides are right. The circuit breaker manufacturers are right to object to oversimplifications by some who market fuses. And the fuse manufacturers are right to point to the astronomical expenses involved in reworking existing plant, and the existence of at least some applications that seem amenable to field engineering. Remember that 90.1(B) doesn't promise electrical installations will be free from hazard, only that they will be essentially free from hazard. So the limited engineering approach has merit, but just when this author was getting comfortable with the 2005 **NEC** provisions that ushered this approach into the **NEC**, along comes documented adverse experience where a facility applied the field engineering process. The methodology appeared to this author to have been competent. Fortunately, the owner was willing to pay to have the engineered combination tested. All five tests failed and the project was redesigned with a separate transformer vault that subdivided the load through smaller gear and avoided the problem. Had the actual bench testing not been done, the engineer would have stamped the plans and this system would be in service today.

To this end, the **NEC** now incorporates an additional paragraph that requires the engineer to *ensure* that the downstream breaker(s) that are part of the series design remain passive while the upstream current-limiting device is interrupting the fault. It remains to be seen how many engineers will put their professional status on the line to offer such assurances. It is significant that both the UL and NEMA representatives on the panel remain opposed to this procedure. Of equal significance is the fact that the allowance remains in the **NEC**, having retained the necessary consensus of the panel. Remember, both sides are right. This controversy will continue.

Part **(C)** in this section addresses the concern related to applications where motor contribution to downstream faults may render the "lower-rated" device incapable of safely clearing the faulted circuit. Remember that at the instant of an outage a rotating motor is a generator, fully capable of adding current. Any short-circuit current study necessarily considers motor contributions to the fault current available. In any application of "series-rated" OC devices, if the "sum" of motor full-load currents that may be contributed to the lower-rated device—without passing through the higher-rated device—exceeds the lower-rated device's rating by 1 percent (e.g., 100 A of contributed motor current to a 10,000-AIR device), then series-rated devices may *not* be used. This disqualification applies to both bench-tested applications in (B) and to field engineered combinations in (A) because the parent language in the section requires compliance with (A) or (B) as applicable, and (C) in all cases.

240.87. Arc Energy Reduction. Some sophisticated power circuit breakers include features that allow the breaker's normal instantaneous pickup to be retarded or shut off. For a breaker in an upstream position, this might be desired in order to allow it to selectively coordinate with downstream levels of the distribution, as is now required by 700.32 and a few other articles. There are drawbacks to doing this centering on the increased amount of damage that could be done during an arcing failure, and the increased amount of incident energy capable of injuring a worker if the equipment must be worked hot for any reason.

The 2011 **NEC** recognized these breakers for the first time, and to counteract this drawback mandates the use of additional features in an attempt to counteract them. The 2014 **NEC** added a significant limitation in that this section is now limited to circuit breakers that have a maximum trip setting greater than or equal to 1200 A, and the 2017 **NEC** added more options. One of the six specified strategies (or an approved alternative) must be in place, and there must be documentation available for those who work on or inspect the system showing where such breakers are located.

The first strategy is zone-selective interlocking. This uses a process of communication between an upstream and downstream breaker. When the downstream breaker detects a current rise it signals the upstream breaker not to open and the upstream breaker restrains its operation accordingly. If the fault is between the two breakers, there will be no signal from the downstream breaker and the upstream breaker will exercise its instantaneous function and trip.

The second approach uses relaying to compare the current on the line side of the upstream breaker with the current leaving the downstream breakers collectively. If there is no difference, then the fault cannot be in the intervening feeder and the upstream breaker restrains its operation accordingly. If there is a difference in current, the instantaneous trip function in the upstream breaker activates. Note that this concept and an explanatory note currently reside at 240.92(C)(1)(2).

The third approach involves a manual activation of a control function on the upstream breaker to restore its instantaneous trip capability during a maintenance activity. When the instantaneous function is active, the amount of incident energy possible during a failure is reduced, possibly allowing a reduced hazard risk category exposure for energized work if it is required. If this is done, the status of the control must be clearly indicated. During the maintenance period, selective coordination will have been lost.

The fourth approach was new in the 2014 **NEC** and relies on an "energy-reducing active arc flash mitigation system." This is only somewhat explained in its associated informational note. These systems typically involve the retrofit of some form of rapid-acting interrupting mechanism, perhaps using a vacuum circuit breaker on the medium voltage side of the local unit substation, which has been engineered to act in conjunction with current transformers on the secondary side and a microprocessor supported relay package. The systems are designed to mitigate the amount of energy released in an arc flash event. By way of example, one major manufacturer's literature provides an example of a reduction in arc flash potential from 2-second duration and 62 Cal/sq · cm. incident energy to 0.18-second duration and 5.7 Cal/sq · cm. incident energy.

The fifth and sixth options are new as of the 2017 **NEC** and involve instantaneous trip settings or overrides that are less than the available arcing current. They focus on coordinating the instantaneous trip settings of the breaker such that it is less than the "available arcing current." If this can be arranged, then the amount of fault energy available to an arc will be reduced, which is the focus of this section. Instantaneous trip settings typically run between a half cycle for molded case breakers to three cycles for the older "air-frame" breakers. New Informational Note No. 3

explains the role of instantaneous trip settings, and new Note No. 4 cites an IEEE publication that describes how to perform arc flash hazard calculations.

All of these approaches are controversial. There is considerable debate in the engineering community about the effectiveness and practicability of these schemes during a high-level fault, when these coordinating functions must successfully proceed in a time interval that is on the order of milliseconds. They also burden a distribution system based on circuit breakers when a comparable fusible distribution, typically based on large Class L fuses, will often have even higher amounts of let-through energy. In fairness, the development of 240.67 in the 2017 cycle does address this question.

Part VIII. Supervised Industrial Installations A supervised industrial installation is limited to the industrial portions of a facility that meet the following three criteria:

- There is qualified maintenance and engineering supervision such that only qualified personnel are running the system.
- The premises electrical system supporting the industrial processes or manufacturing activities or both (and not including any office or other indirect support loads) has a calculated load per Art. 220 that is not less than 2500 kVA.
- The premises electrical system is comprised of not less than one service or feeder that runs over 150 V to ground and over 300 V phase-to-phase.

These installations must comply with all requirements in Art. 240, except as modified in this part. And any such modifications that are applied in the facilities must not extend beyond the manufacturing or process control environment.

240.91. Protection of Conductors. This provision, new as of the 2011 NEC, amends 240.4(C) in these supervised environments by allowing limited protection of large conductors above their Table 310.15(B)(16) ampacity. Specifically covering overcurrent devices rated over 800 A, if the conductor ampacity in the proposed application is at least 95 percent of the next higher size overcurrent device [or setting, as covered in 240.6(C)], that next higher rating or setting is allowed.

This allowance only applies in the context of an overcurrent coordination study that establishes short-circuit time/current exposures for the proposed conductors and demonstrates that they would be safely protected. In addition, the overcurrent devices must be listed and marked for the application.

This section is nothing but an elaborate attempt to use 500 kcmil conductors in applications that would otherwise require 600 kcmil conductors. The Table 310.15(B)(16) 75°C ampacity for 500 kcmil wires is 380 A, which just happens to be 95 percent of 400. If you are wiring a large feeder (1200, 1600, 2000 A or higher), this section would allow the use of multiple runs of 500 kcmil instead of the current practice of using 600 kcmil conductors.

The first condition appropriately addresses how the conductor can be expected to perform while it is in use. The second condition implicitly addresses the other side of the equation, namely, how the overcurrent protective device will perform with diminished conductor sizes connected to it. Remember that the reason for 110.14(C) limits has to do with the ability of connected conductors to function as heat sinks for the devices connected to them. A reduction from 600 to 500 kcmil is very significant in this regard. This is why the equipment *at both ends of the wiring* must be "listed and marked for the application."

The costs involved in product development, testing, and distribution of this sort of these product lines are enormous and the number of potential applications in these supervised environments is small. It is extremely unlikely that such equipment will come to market. And it bears restating, that *both* the originating overcurrent device *and* the terminating device or other equipment must have been so evaluated, listed, and marked.

The next section, 240.92, covers the principal impact of Part VIII. This section makes significant modifications to the tap rules in 240.21, as follows:

240.92(B) provides that a short-circuit analysis can be performed based on the short-circuit current rating of the conductors to be protected, using a table that is long familiar to the electrical engineering community but that was new to the NEC as of the 2008 cycle. If sensors are arranged to monitor the variables that make up the table formulas, then there will be no adverse outcome. The three eligible taps for this treatment are the 25-ft taps in 240.21(B)(2), the combined primary and secondary 25-ft taps in 240.21(B)(3), and the high-bay manufacturing facility taps in 240.21(B)(4).

240.92(C) allows transformer secondary conductors to be protected using an approach that divorces the short-circuit and ground-fault protection from the overload protection function. The short-circuit and ground-fault protection can be arranged in one of three ways by 240.92(C)(1). The first option liberalizes the winding-ratio limitation of 240.21(C)(1) by allowing secondary conductors, even those extended from a multiwire secondary, to run up to 30 m (100 ft) with primary side protection only, set at not more than 150 percent of tap ampacity after adjusting for the winding ratio. The second option recognizes a differential current relay arranged to operate a shunt-trip mechanism on the upstream overcurrent device. The third option is to verify under engineering supervision that the system as configured will protect the conductors under short-circuit and ground-fault conditions; the new Table 240.92(B) would be one tool in this analysis.

Of course, there is another half of this puzzle, involving overload protection. There are four options per 240.92(C)(2) to provide this protection, the simplest being to terminate in a single overcurrent device sized to the conductor ampacity. Almost as simple is to terminate at a group of protective devices selected so the sum of all their ratings doesn't exceed the conductor ampacity. Although based on 240.21(C)(3), there is no limit on the secondary conductor length. The devices must be grouped, and not exceed six, which also happens to be the limit of the sum-of-the-ratings rule for transformer secondary protection in Notes #2 to the 450.3 protection tables. Remember, nothing in this article can amend the transformer protection rules in Art. 450. If taps to the individual devices are needed, the fact that that limitation also occurs in 240.21 suggests that these smaller taps are also permitted here. The other two approaches, using overcurrent relaying or engineering supervision, directly parallel the comparable provisions for short-circuit and ground-fault protection.

The third issue, covered in 240.92(C)(3) is to provide physical protection for the conductors by enclosing them in a raceway or "by other approved means." This rule directly tracks comparable rules in numerous places in 240.21.

The next major modification involves rules for outside feeder taps as covered in 240.92(D). This rule largely parallels comparable coverage in 240.21(B)(5) and

240.21(C)(4). There is one major departure, that being the normal requirement for a single device at the building termination does not apply. Instead up to six devices can be grouped, with the required protection to comprise the sum of the ratings of the terminating devices.

The final major change, 240.92(E), completely removes 240.21(C)(1) from consideration because in this case the primary protection for the transformer, after being reproportioned by the winding ratio, is allowed to protect secondary conductors whether or not there is a multiwire secondary.

240.100. Feeders and Branch Circuits (Over 1000 V, Nominal). This section and 240.101 present rules on overcurrent protection for medium-voltage (over 1000 V) feeder conductors. It requires overcurrent protection located at the point of supply, or elsewhere if the alternate location has been designed under engineering supervision based on fault current analysis, conductor damage curves, and coordination analysis as required. The overcurrent protection can be in the form of fuses or using CTs and relays. Although the rule calls for "short-circuit" protection, it does *not* require that conductors be protected in accordance with their rated ampacities (Fig. 240-40). Remember that the ampacity rules for medium-voltage feeders as given in 215.3(B) and in 210.19(B) for branch circuits pretty much assure that overloads are unlikely. By long history, the overcurrent protection rules here focus on short circuits and ground faults.

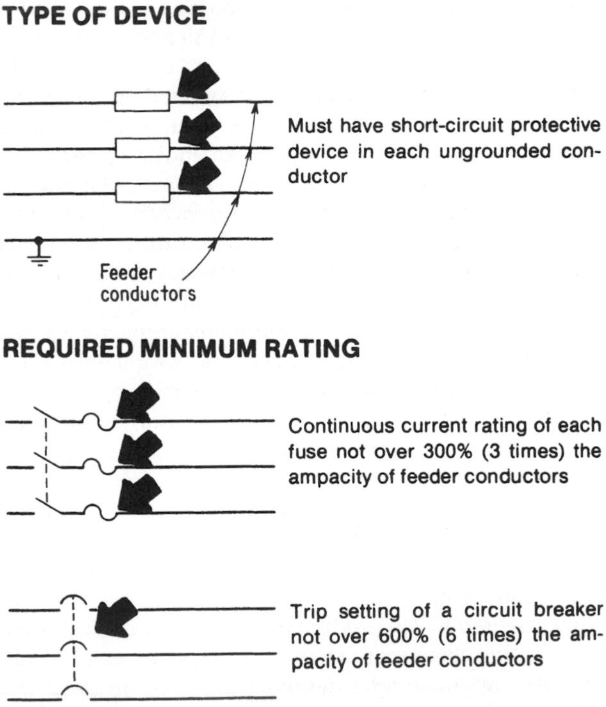
TYPE OF DEVICE
Must have short-circuit protective device in each ungrounded conductor
Feeder conductors
REQUIRED MINIMUM RATING
Continuous current rating of each fuse not over 300% (3 times) the ampacity of feeder conductors
Trip setting of a circuit breaker not over 600% (6 times) the ampacity of feeder conductors

Fig. 240-40. Overcurrent protection of medium-voltage (over 600 V) branch-circuit and feeder conductors. (Secs. 240.100 and 240.101.)

ARTICLE 250. GROUNDING AND BONDING

250.2. Definitions—Bonding Jumper, Supply-Side. This new definition for the 2011 NEC describes a conductor on the supply side of a power source, whether a service or a separately derived system, or within service equipment, whose function is to provide the required low-impedance pathway between metallic elements of the system requiring grounding continuity.

These conductors were previously described as "equipment bonding jumpers" and were covered in 250.30(A)(2) and 250.102(C). The problem was that when installed on the load side of associated overcurrent devices they were part of the equipment grounding system and sized under 250.122 [see 250.102(D)]. When installed on the supply side of the main overcurrent device (whether for a service or separately derived system) their only function was bonding and they were sized per 250.102(C). To avoid confusion, these separate applications (before and after the main disconnect and overcurrent protection) are now uniquely defined, and the old term "equipment bonding jumper" only applies on the load side of the main. See Fig. 250-21 and the discussion at 250.30(A)(2) for more information.

250.4. General Requirements for Grounding and Bonding. This section creates an overall context for everything that follows in the article, because it sets the performance requirements for grounding and bonding. That is, it sets out what grounding and bonding are supposed to achieve in an electrical system. The prescriptive requirements that comprise the remainder of the article constitute the methods which, if followed, will result in the electrical system achieving the objectives stated here.

One of the most important, but least understood, considerations in design of electrical systems is that of grounding. The use of the word *grounding* comes from the fact that part of the technique involves making a low-resistance connection to the earth. Remember that "ground" has been redefined to simply mean the earth, as in the planet. The term *grounding* also refers to the "safety ground" that facilitates sensing of faults and provides for automatic operation of the circuit OCPDs by ensuring a low-impedance return path in the event of a fault, but this is only true for grounded systems, and not all systems are grounded. *Bonding* is the process of interconnecting parts such that electrical continuity and conductivity are assured. Specific rules then require bonding noncurrent-carrying metallic components of the distribution system to each other and, in some instances, to noncurrent-carrying components of other systems, such as metal ladders and diving boards at swimming pools, to ensure all noncurrent-carrying metal pieces are at a common potential with respect to ground.

These are examples of an "effective ground-fault current path" which is an intentionally constructed low-impedance path that has been designed to carry ground-fault current safely from the fault location to the electrical supply source. It will facilitate the prompt operation of OCPDs on a grounded system. It will also cause the operation of ground-fault detectors on high-impedance grounded systems and also on ungrounded systems, which, in general, must now be incorporated. It will also provide a safe path for current between two phases of an ungrounded or high-impedance grounded system in the event of two ground faults from different phases in different locations.

For any given piece of equipment or circuit, the connection to earth may be a direct wire connection to the grounding electrode that is buried in the earth; or it may be a connection to some other conductive metallic element (such as conduit or switchboard enclosure) that, through bonding as required in this article, is electrically connected to a grounding electrode.

The combined purpose of grounding and bonding is to provide protection of personnel, equipment, and circuits by largely eliminating the possibility of continuing dangerous or excessive voltages that could pose a shock hazard, and that could damage equipment in the event of overvoltage imposed on the conductors supplying such equipment.

There are two distinct considerations in grounding for grounded electrical systems, covered in Part **(A)** of this section: grounding of one of the conductors of the wiring system, and grounding of all metal enclosures containing electrical wires or equipment, where an insulation failure in such enclosures might place a potential on the enclosures and constitute a shock or fire hazard. The types of grounding are:

1. *Wiring system ground.* This is covered in (A)(1) and consists of grounding one of the wires of the electrical system, such as the neutral, to limit the voltage upon the circuit that might otherwise occur through exposure to lightning or other voltages higher than that for which the circuit is designed. Another purpose in grounding one of the wires of the system is to limit the maximum voltage to ground under normal operating conditions. Also, a system that operates with one of its conductors intentionally grounded will provide for automatic opening of the circuit if an accidental or fault ground occurs on one of its ungrounded conductors (Fig. 250-1).

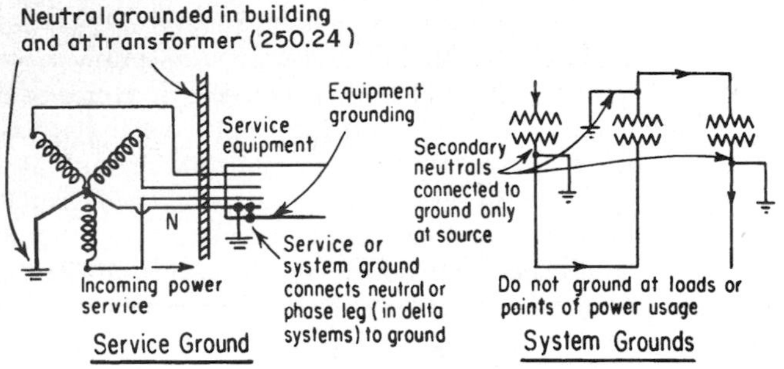
Neutral grounded in building
and at transformer (250.24)
Service
equipment
Equipment
grounding
Secondary
neutrals
connected to
ground only
at source
N
Service or
system ground
connects neutral or
phase leg (in delta
systems) to ground
Incoming power
service
Service Ground
Do not ground at loads or
points of power usage
System Grounds

Fig. 250-1. Operating a system with one circuit conductor grounded. (Sec. 250.1.)

A new informational note at this location calls attention to the fact that an important aspect of limiting the voltage to ground includes keeping grounding electrode conductors as short as possible consistent with making the required connection, and in particular avoiding loops and bends as much as possible. This avoids high-frequency reactance issues that are not problems at 60 Hz, but are very significant on lightning transients and the like. Although lightning is inherently a dc event, it is a rapidly interrupted dc event capable of Fourier analysis that approximates a 1000-Hz signal. Loops and long-circuit lengths have progressively higher impact on increasing impedance as the frequency increases. The wording about disturbing the permanent parts of the installation suggests that heroic measures such as drilling partitions and block walls are unnecessary, but the straighter and shorter the path, the better, all things being equal. NFPA 780, the *Standard for the Installation of Lightning Protection Systems,* also addresses this topic for the same reason.

2. *Equipment ground or "safety" ground.* This is covered in (A)(2), (A)(3), (A)(4), and (A)(5) on grounded systems. The first topic is the grounding objective, by which noncurrent-carrying metal parts that enclose electrical equipment or conductors, or that comprise such equipment, are connected to earth to limit the voltage to ground on such materials. In conjunction with this process is the bonding objective, which results in the same materials connected together to establish both conductivity and continuity across the entire system, and in the process establishes an effective ground-fault current path that will allow the current to flow such that the operation of the automatic operation of the overcurrent protective device is facilitated (Fig. 250-2). This path must be capable of safely carrying such currents wherever they are imposed and running back to the source of the supply system.

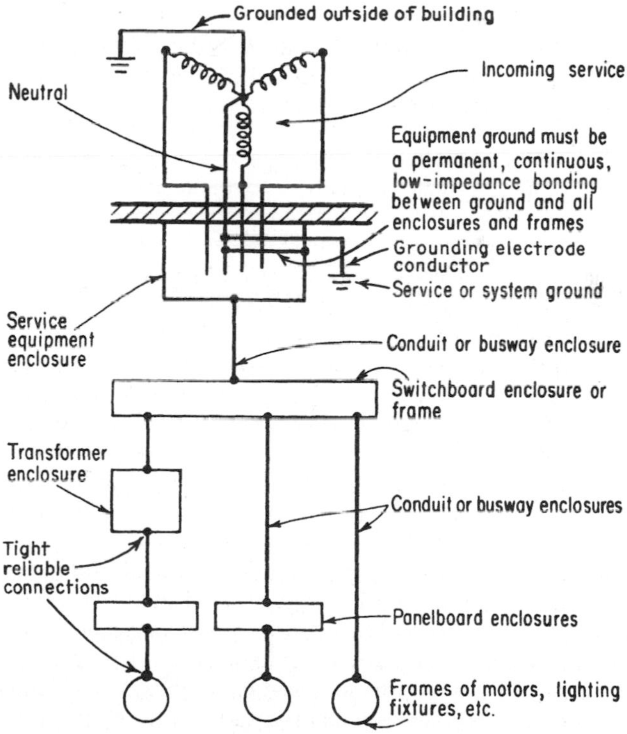
Grounded outside of building
Incoming service
Neutral
Equipment ground must be a permanent, continuous, low-impedance bonding between ground and all enclosures and frames
Grounding electrode conductor
Service or system ground
Service equipment enclosure
Conduit or busway enclosure
Switchboard enclosure or frame
Transformer enclosure
Conduit or busway enclosures
Tight reliable connections
Panelboard enclosures
Frames of motors, lighting fixtures, etc.

Fig. 250-2. Equipment grounding is interconnection of metal enclosures of equipment and their connection to ground. [Sec. 250.4(A)(3).]

In a grounded electrical system with a ground-fault current path that has excessive impedance due to installation or maintenance issues, if one of the phase conductors of the system (i.e., one of the ungrounded conductors of the wiring system) should accidentally come in contact with one of the metal enclosures in which the wires are run, it might produce a condition where not enough fault current would flow to operate the overcurrent devices. In such a case, the faulted circuit would not automatically open, and a dangerous voltage would be present on the conduit and other metal enclosures. This voltage presents a shock hazard and a fire hazard due to possible arcing or sparking from the energized conduit to some grounded pipe or other piece of grounded metal. Section 250.4(A)(5) places three requirements on these connections so the system will operate as intended.

1. That every effective ground-fault current path be installed by proper mounting, coupling, and terminating of the conductor or raceway intended to serve as the grounding conductor. Also, the condition can be visually checked by the electrical inspector, the design engineer, and/or any other authority concerned.
2. That every grounding conductor be "capable of safely carrying the maximum ground-fault current likely to be imposed on it" can be established by falling back on those other Code rules [Secs. 250.24(B), 250.28, 250.30, 250.66, 250.122, 250.166, 680.25(A)(D)(E), etc.] that specifically establish a minimum required size of grounding conductor. Although it is reasonable to conclude that adequate sizing of grounding conductors in accordance with those rules provides adequate capacity, such may not always be the case. Where high levels of fault current are available, use of the Code-recommended "minimum" may be inadequate. There are available a number of recognized methods promulgated by such organizations as the International Electrical and Electronic Engineers (IEEE) that can be consulted to determine if the Code-prescribed minimum size of grounding conductor actually is adequate and capable of "safely carrying the maximum fault." If Code-prescribed minimums cannot safely carry the available fault current, it certainly seems as if it would be a violation of this rule to use a grounding conductor of the Code-prescribed size.
3. When we come to a condition put forth by part **(A)(5)** of Sec. 250.4, "creates a low-impedance circuit facilitating the operation of the overcurrent device" questions arise as to the intent of the rule; and whether specific testing is required to evaluate the result. Here again the answer is in the parent text, namely that if the prescriptive requirements in the article are met, then compliance is usually assumed for enforcement purposes. However, that is not always the case. For example, if the equipment grounding conductor is a wire sized to Table 250.122 limits, and if it could be demonstrated that given the length of run or for any other reason even a solid ground fault would not draw enough current to put the circuit breaker into its instantaneous tripping range, then the note at the bottom of that table would support increasing the size of the equipment grounding conductor or taking other steps to decrease the impedance so the overcurrent device would act promptly. There are a number of software programs that will make these calculations for a variety of grounding conductors including the various steel tubular raceways (Fig. 250-3). To *know* for sure that impedance of any and every grounding conductor is "sufficiently low to limit . . . etc." requires that the actual value

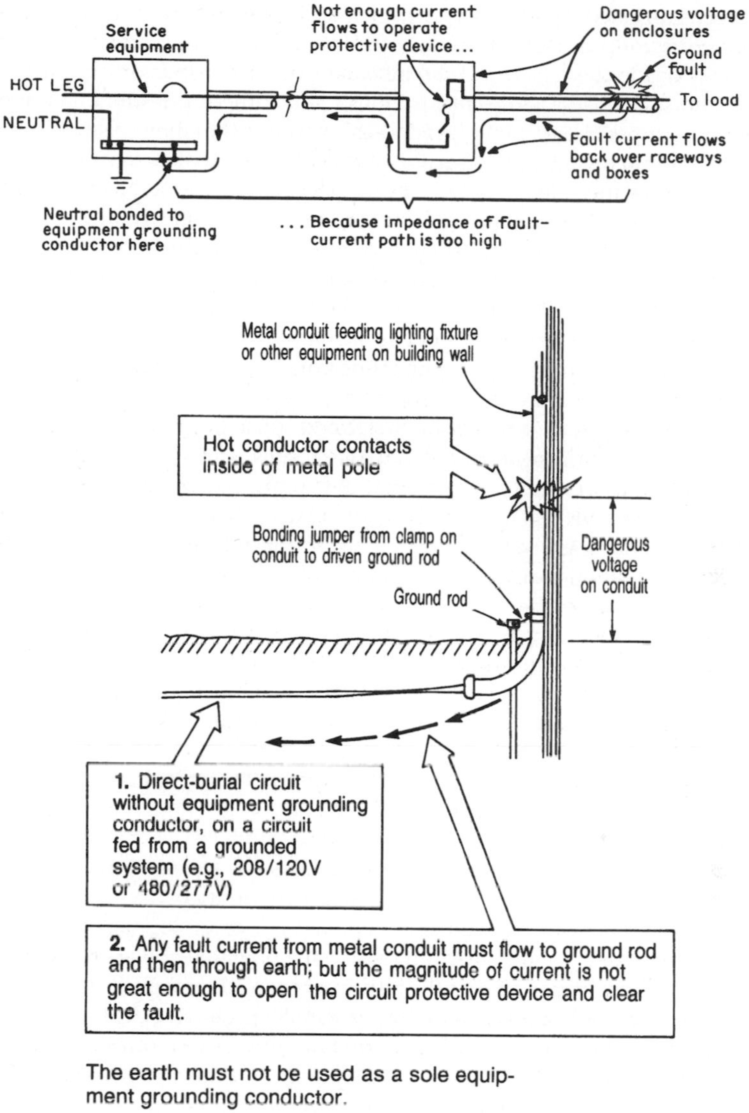
Service equipment
Not enough current flows to operate protective device...
Dangerous voltage on enclosures
Ground fault
HOT LEG
NEUTRAL
To load
Fault current flows back over raceways and boxes
Neutral bonded to equipment grounding conductor here
... Because impedance of fault-current path is too high
Metal conduit feeding lighting fixture or other equipment on building wall
Hot conductor contacts inside of metal pole
Bonding jumper from clamp on conduit to driven ground rod
Ground rod
Dangerous voltage on conduit
1. Direct-burial circuit without equipment grounding conductor, on a circuit fed from a grounded system (e.g., 208/120V or 480/277V)
2. Any fault current from metal conduit must flow to ground rod and then through earth; but the magnitude of current is not great enough to open the circuit protective device and clear the fault.
The earth must not be used as a sole equipment grounding conductor.

Fig. 250-3. These are violations of the basic concept of effective grounding. (Sec. 250.4.)

of impedance be measured; and such measurement not only involves use of testing equipment but also demands a broad and deep knowledge of the often sophisticated technology of testing in circuits operating on alternating current where inductance and capacitance are operative factors. In short, what is "a low-impedance circuit" and what does "facilitating the operation of the overcurrent device" mean? And if testing *is* done, is it necessary to test every equipment grounding conductor? In the end, comply with the prescriptive rules in Art. 250 unless there is compelling reason to go beyond them, and in the real world the installers and the inspectors will usually be on the same page. However, thinking about these questions is important because it positions you to respond to questions that may arise.

The last sentence in this section prohibits the use of current flow through the earth as the sole equipment grounding conductor because earth impedance is too high and restricts fault-current flow, as shown at the bottom of Fig. 250-3. Inspectors as well as computer, telecommunications, data systems, and CATV installers have often overlooked this very important Code rule. The one thing to remember with current flow through earth is that it can do no appreciable "work"—it won't light a 40-W bulb—but, it can and will kill!

4. *Bonding.* This term refers to connecting components together in such a manner as to ensure conductivity and continuity. Once this is done, such components may have other functions as defined subsequently in the article.

   Simply stated, grounding of all metal enclosures of electric wires and equipment minimizes any potential above ground on the enclosures. Such bonding together and grounding of all metal enclosures are required for both grounded electrical systems (those systems in which one of the circuit conductors is intentionally grounded) and ungrounded electrical systems (systems with none of the circuit wires intentionally grounded).

Effective equipment grounding is extremely important for grounded electrical systems to provide the automatic fault clearing that is one of the important advantages of grounded electrical systems. A low-impedance path for fault current is necessary to permit enough current to flow to operate the fuses or CB protecting the circuit.

Note that 250.4(A)(4) addresses the "bonding" of other metallic building components and systems that are "likely" to become energized. These connections are not truly ground-return paths because the impedance of such connections is unknown. The return path through building steel or a metal water piping system will be quite high because each item is not necessarily in close proximity to the phase conductors, which of course will result in a higher impedance. However, it is better to make such connections than to leave the steel and piping at a potential above ground should a fault energize them. Depending on the rating of the OC device protecting the faulted circuit, there may be enough current flow to trip the protective device.

Part (B) of this section covers ungrounded systems. These rules omit any counterpart to 250.4(A)(1) because there is no system grounding by definition. However, the bonding and grounding rules are comparable. Although these systems are not set up to facilitate the operation of an overcurrent device in the event of a ground fault, they absolutely must provide a low-impedance path for fault current. If the insulation on one phase conductor fails at one end of the plant, and a similar failure occurs at the other end of the plant on a different phase before the first failure is cleared, the result is a line-to-line short circuit over the intervening equipment grounding system. If a very high standard of workmanship was not adhered to, such an event will produce elevated voltages on metal raceways, etc., and dangerous showers of sparks at every random locknut or other joint not made wrench tight in accordance with 250.120(A).

250.6. Objectionable Current. Although parts **(A)** and **(B)** of this section permit "arrangement" and "alterations" of electrical systems to prevent and/or eliminate objectionable flow of currents over "grounding conductors or grounding paths," part **(D)** specifically prohibits any exemptions from **NEC** rules on grounding for "electronic equipment" and states that "currents that introduce noise or data errors" in electronic data-processing and computer equipment are not "objectionable" currents that allow modification of grounding rules.

This paragraph emphasizes the Code's intent that electronic data-processing equipment has its input and output circuits in full compliance with all **NEC** rules on neutral grounding, equipment grounding, and bonding and grounding of neutral and ground terminal buses. 250.6(B) does offer alternative methods for correcting "objectionable current over grounding conductors," but part **(D)** specifically states that such modifications or alternative methods are not applicable to the on-site wiring for electronic or data processing equipment if the only purpose is to eliminate "noise or data errors" in the electronic equipment. This paragraph amplifies the wording of "Premises Wiring" as given in Art. 100.

250.8. Connection of Grounding and Bonding Equipment. This rule has been reformatted and significantly expanded to better cover the topic. There are now eight possibilities for making these attachments, and in a rather obvious 2014 **NEC** clarification, one or more of them can be used. Note that 250.12 still requires clean surfaces unless the attachment means digs through the paint. This is more likely for a locknut than a small screw.

1. *A listed pressure connector.* This includes conventional twist-on-wire connectors, settling a long-standing controversy. One side contended that, with the exception of the green-style connectors with the hole in the end of the connector that are specifically listed for grounding, other such connectors must not be used. The opposite side said that since any fault current passed through conventional connectors on the way to the fault, they should be acceptable for use in the path returning from the fault. Now we know that the second side won the argument.
2. *A terminal bar.* This is common in panels, switchboards, and motor control centers, and also in boxes where other **NEC** rules such as 680.23(F)(2) forbid conventional splicing devices.
3. *A pressure connector listed as grounding and bonding equipment.* These include the green twist-on connectors referred to in item 1 above. Ground-rod clamps and water pipe clamps would also fit in this category.
4. *A connection made by the thermite ("exothermic welding") process.*
5. *A machine screw that engages no fewer than two threads.* For example, conventional steel boxes are 1.59 mm (1⁄16 in.) thick per 314.40(B). A 10-32 screw with 32 threads per inch will have a thread every 1/32nd of an inch, and thereby engage two threads in the box. If the metal wall of the enclosure is less than this, as many panels are, there are several options, the best being to only use the screws provided by the manufacturer, who will have anticipated this problem. If that doesn't work, the **NEC** allows you to substitute a nut. If getting behind the enclosure is a problem, try making a small, recessed, steeply angled dimple in the enclosure wall with a prick punch. Then drill the center of the now conical indent with the No. 21 tap drill. When you tap the hole,

the tap will engage the bottom sides of the cone as well as the enclosure wall itself. As long as there is at least one good thread in the bottom of the cone formed by the prick punch, a 32-pitch screw will meet this rule, since for sure there will be one more thread in the drilled hole in the enclosure wall making the required total of two.

6. *A thread-forming machine screw under the same requirements as discussed in item 5 above.* Note that this is not a "teck" screw with sheet metal threads. This is a thread forming screw, but with machine threads. They are self-tapping and some (not all) come with drill points that avoid the need for a separate drill bit. Sheet-metal screws must not be allowed and are not recognized. Their thread pitch is far too coarse to produce enough force to reliably hold the connector in place, since any screw is an inclined plane wrapped around a shaft and thus the mechanical advantage decreases as the pitch coarsens.
7. *A connection that is part of a listed assembly.* Many assemblies come equipped with grounding terminals already part of the equipment. These can be used in accordance with the listing, and the test lab will have evaluated the likely connections at this point as part of the listing process.
8. *Other listed means.* This opens the door for other approaches, provided they are listed.

250.10. Protection of Ground Clamps and Fittings. This rule addresses the need to ensure that grounding connections are protected from physical damage. Obviously, no protection is required where the connection is not subject to damage, such as where it is made within an enclosure. However, where a physical damage potential exists, a fabricated enclosure of wood, metal, or the "equivalent" can be used to protect those connections that may be vulnerable to physical damage. Exactly what constitutes the equivalent will be up to the authority having jurisdiction, usually the local electrical inspector, as will any determination of what is subject to potential damage.

250.20. Alternating-Current Systems to Be Grounded. Part **(A)** does recognize use of ungrounded circuits or systems when operating at less than 50 V. But system grounding of circuits under 50 V is sometimes required, as shown in Fig. 250-4. Note that if (and only if) system grounding is required, 250.112(I) will then require that equipment grounding be implemented on the low-voltage circuit.

According to part **(B)(1)** of this rule, all alternating-current wiring systems from 50 to 1000 V *must* be grounded if they can be so grounded that the maximum voltage to ground does not exceed 150 V. This rule makes it *mandatory* that the following systems or circuits operate with one conductor grounded:

1. 120-V, 2-wire systems or circuits must have one of their wires grounded.
2. 120/240-V, 3-wire, single-phase systems or circuits must have their neutral conductor grounded.
3. 208Y/120-V, 3-phase, 4-wire, wye-connected systems or circuits must be operated with the neutral conductor grounded.

In all the foregoing systems or circuits, the neutrals must be grounded because *the maximum voltage to ground does not exceed 150 V* from any other conductor of the system when the neutral conductor is grounded.

In parts **(2)** and **(3)** of this section, all systems of any voltage up to 1000 V must operate with the neutral conductor solidly grounded whenever any loads are

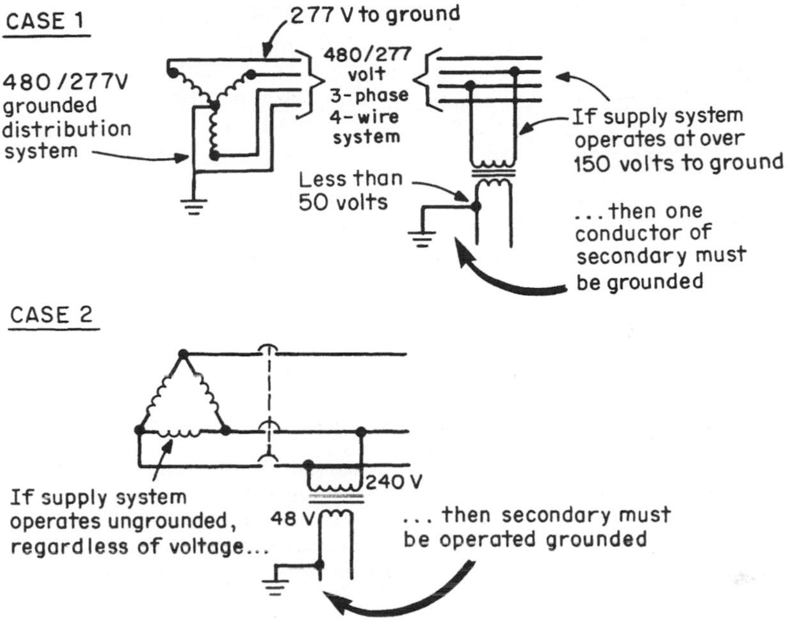
CASE 1
277 V to ground
480/277V grounded distribution system
480/277 volt 3-phase 4-wire system
Less than 50 volts
If supply system operates at over 150 volts to ground
...then one conductor of secondary must be grounded
CASE 2
240 V
48 V
If supply system operates ungrounded, regardless of voltage...
... then secondary must be operated grounded

Fig. 250-4. Circuits under 50 V may have to be grounded. (Sec. 250.20.) In addition to the two cases illustrated above, system grounding is also required if the transformer primary conductors came in from outdoors as overhead conductors.

connected phase-to-neutral, so that the neutral carries load current. All 3-phase, 4-wire wye-connected systems and all 3-phase, 4-wire delta systems (the so-called *red-leg systems*) must operate with the neutral conductor solidly grounded if they are used as a circuit conductor. That means:

1. The neutral conductor of a 240/120-V, 3-phase, 4-wire system (with the neutral taken from the midpoint of one phase) must be grounded.
2. It is also mandatory that 480Y/277-V, 3-phase, 4-wire interior wiring systems have the neutral grounded if the neutral is to be used as a circuit conductor—such as for 277-V lighting.
3. Also, if 480-V autotransformer-type fluorescent or mercury-vapor ballasts are to be supplied from 480/277-V systems, then the neutral conductor will have to be grounded at the voltage source to conform to 410.138, even though the neutral is not used as a circuit conductor. Of course, it should be noted that 480/277-V systems are usually operated with the neutral grounded to obtain automatic fault clearing of a grounded system (Fig. 250-5).

As covered by the rule of part **(C)**, any ac system of 1000 V or more must be grounded if it supplies portable equipment. Otherwise, such systems do not have to be grounded, although they *may* be grounded. There are several forms of system grounding that can be applied to medium-voltage systems, as covered in Part **X** of Art. 250. Because of the higher voltages involved, substantial

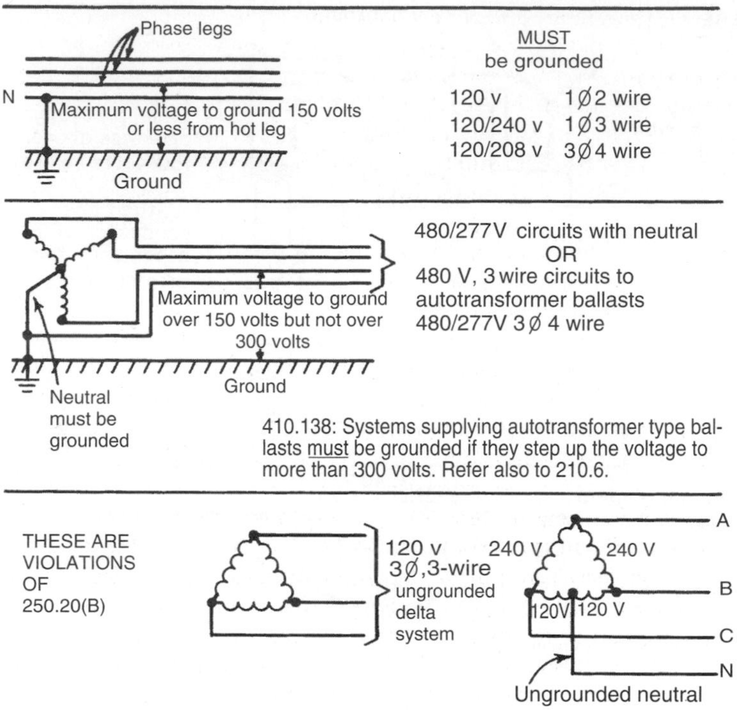

Phase legs
N
Maximum voltage to ground 150 volts or less from hot leg
Ground
MUST be grounded
120 v 1Ø2 wire
120/240 v 1Ø3 wire
120/208 v 3Ø4 wire
Maximum voltage to ground over 150 volts but not over 300 volts
Ground
Neutral must be grounded
480/277V circuits with neutral
OR
480 V, 3 wire circuits to autotransformer ballasts
480/277V 3Ø 4 wire
410.138: Systems supplying autotransformer type ballasts must be grounded if they step up the voltage to more than 300 volts. Refer also to 210.6.
THESE ARE VIOLATIONS OF 250.20(B)
120 v 3Ø,3-wire ungrounded delta system
240 V
240 V
120V
120 V
A
B
C
N
Ungrounded neutral

Fig. 250-5. Some systems or circuits must be grounded. (Sec. 250.20.)

amounts of current will flow into the earth in the event of a failure on a grounded system, whether solidly grounded using either single-point grounded neutrals or multigrounded neutrals, or impedance grounded. Engineers take this into account to design protective designs that address lack of customer tolerance to outages through high-impedance grounding, or lack of tolerance for fault damage through low-impedance grounding. The amount of ground-fault current flow provides enough headroom to design relaying that allows for orderly shutdowns, or shuts down automatically but at current levels that will not cause extensive system damage.

A very important permission is given in 250.20**(D)**. This rule, by recognizing the requirements in 250.36 (or 250.186 for medium voltage), correlates this section with the high-impedance grounding provisions elsewhere in the article.

250.21. AC Systems of 50 Volts to 1000 Volts Not Required to Be Grounded. Although the NEC does not require grounding of electrical systems in which the voltage to ground would exceed 150 V, it now requires that ground-fault detectors be used with ungrounded systems that operate at more than 150 and less than 1000 V. Such detectors indicate when an accidental ground fault develops on one of the phase legs of ungrounded systems. Then the indicated ground fault can be removed during downtime of the industrial operation—that is, when the production machinery is not running. Although not required in previous editions of the Code before the 2005 edition—prior to then the use of ground-fault detectors was mentioned in a fine-print note—such equipment is mandated by Part **(B)** of this section. Effective with the 2011 **NEC**, an additional rule mandates that the ground detection equipment be placed as close as practicable to the point of system supply. In addition, 250.21(C) requires ungrounded systems to be marked "ungrounded system" at the system source or first disconnecting means.

Many industrial plants prefer to use an ungrounded system with ground-fault detectors instead of a grounded system. With a grounded system, the occurrence of a ground fault is supposed to draw enough current to operate the overcurrent device protecting the circuit. But such fault clearing opens the circuit—which may be a branch circuit supplying a motor or other power load or may be a feeder that supplies a number of power loads—and many industrial plants object to the loss of production caused by downtime. They would rather use the ungrounded system and have the system kept operative with a single ground fault and clear the fault when the production machinery is not in use. In some plants, the cost of downtime of production machines can run to thousands of dollars per minute. In other plants, interruption of critical processes is extremely costly.

The difference between a grounded and an ungrounded system is that in a grounded system a single ground fault will automatically cause opening of the circuit, which may shut down operations. In an ungrounded system the first ground fault will register at the ground detectors but will not interrupt operations. However, there is the very important negative aspect that the presence of a single ground fault in an ungrounded system exposes the system to the very destructive possibilities of a phase-to-phase short if another ground fault should simultaneously develop on a different phase leg of the system (Fig. 250-6).

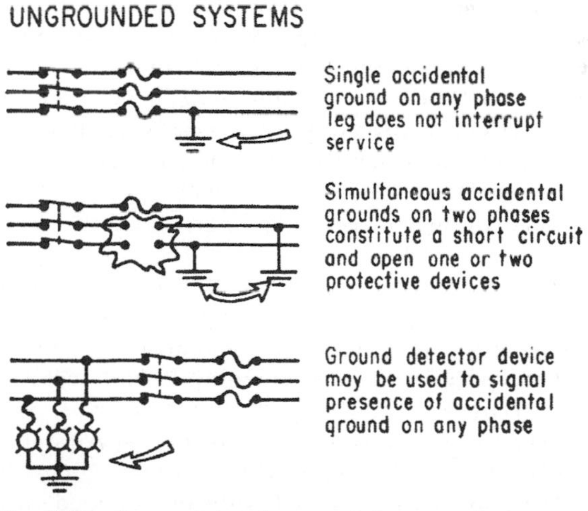
UNGROUNDED SYSTEMS
Single accidental ground on any phase leg does not interrupt service
Simultaneous accidental grounds on two phases constitute a short circuit and open one or two protective devices
Ground detector device may be used to signal presence of accidental ground on any phase

Fig. 250-6. Characteristics of ungrounded systems. (Sec. 250.21.) Ground detectors are now mandatory for most ungrounded system applications.

Grounded neutral systems are generally recommended for high-voltage (over 600 V) distribution. Although ungrounded systems do not undergo a power outage with only one-phase ground faults, the time and money spent in tracing faults indicated by ground detectors and other disadvantages of ungrounded systems have favored use of grounded neutral systems. Another design issue is that transient overvoltages have no way out of an ungrounded system, but they can be removed easily on a grounded system. This has encouraged many engineers to look into high-impedance grounded systems. These systems have a way to bleed out transients, but as in the case of ungrounded systems the first fault will not disrupt power. Grounded systems are more economical in operation and maintenance if a process outage can be tolerated. In such a system, if a fault occurs, it is isolated immediately and automatically.

Grounded neutral systems have many other advantages. The elimination of multiple faults caused by undetected restriking grounds greatly increases service reliability. The lower voltage to ground that results from grounding the neutral offers greater safety for personnel and requires lower equipment voltage ratings. And on high-voltage (above 600 V) systems, residual relays can be used to detect ground faults before they become phase-to-phase faults that have substantial destructive ability.

Part **(3)** of the basic rule recognizes use of ungrounded control circuits derived from transformers. According to the rules of 250.20(B), any 120-V, 2-wire circuit must normally have one of its conductors grounded; the neutral conductor of any 240/120-V, 3-wire, single-phase circuit must be grounded; and the neutral of a 208/120-V, 3-phase, 4-wire circuit must be grounded. Those requirements have often caused difficulty when applied to control circuits derived from the secondary of a control transformer that supplies power to the operating coils of motor starters, contactors, and relays. For instance, there are cases where a ground fault on the hot leg of a grounded control circuit can cause a hazard to personnel by actuating the control circuit fuse or CB and shutting down an industrial process in a sudden, unexpected, nonorderly way. A metal-casting facility is an example of an installation where sudden shutdown due to a ground fault in the hot leg of a grounded control circuit could be objectionable. Because designers often wish to operate such 120-V control circuits ungrounded, 250.21(3) permits ungrounded control circuits under certain specified conditions.

A 120-V control circuit may be operated ungrounded when all the following exist:

1. The circuit is derived from a transformer that has a primary rating less than 1000 V.
2. Whether in a commercial, institutional, or industrial facility, supervision will assure that only persons qualified in electrical work will maintain and service the control circuits.
3. There is a need for preventing circuit opening on a ground fault—that is, continuity of power is required for safety or for operating reliability.
4. Some type of ground detector is used on the ungrounded system to alert personnel to the presence of any ground fault, enabling them to clear the ground fault in normal downtime of the system (Fig. 250-7).

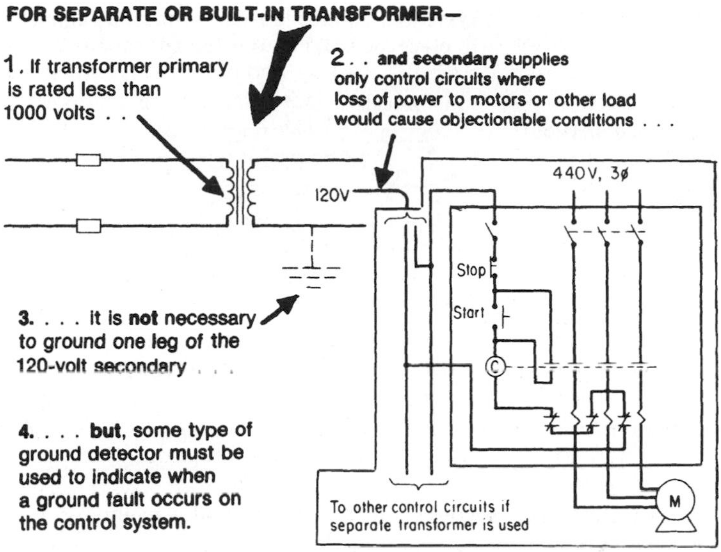
FOR SEPARATE OR BUILT-IN TRANSFORMER—
1. If transformer primary is rated less than 1000 volts . .
2. . and secondary supplies only control circuits where loss of power to motors or other load would cause objectionable conditions . . .
120V
440V, 3ø
Stop
Start
C
M
3. . . . it is not necessary to ground one leg of the 120-volt secondary . . .
4. . . . but, some type of ground detector must be used to indicate when a ground fault occurs on the control system.
To other control circuits if separate transformer is used

Fig. 250-7. Ungrounded 120-V circuits may be used for controls. (Sec. 250.21.)

Although no mention is made of secondary voltage in this Code rule, this rule permitting ungrounded control circuits is primarily significant only for 120-V control circuits. The NEC has long permitted 240- and 480- and even 600-V control circuits to be operated ungrounded. Application of this rule can be made for any 120-V control circuit derived from a control transformer in an individual motor starter or for a separate control transformer that supplies control power for a number of motor starters or magnetic contactors. Of course, the rule could also be used to permit ungrounded 277-V control circuits under the same conditions.

250.24. Grounding Service-Supplied Alternating-Current Systems. As noted in parts **(A)** and **(A)(1),** when a premises is supplied by an electrical system that has to be operated with one conductor grounded—either because it is required by the Code (e.g., 240/120-V, single phase) or because it is desired by the system designer (e.g., 240 V, 3 phase, corner grounded)—a connection to the grounding electrode must be made at the service entrance (Fig. 250-8). That is, the neutral conductor or other conductor to be grounded must be connected at the service equipment to a conductor that runs to a grounding electrode. The conductor that runs to the grounding electrode is called the "grounding electrode conductor"—an official definition in the NEC.

The Code rule of 250.24(A)(1) says that the connection of the grounding electrode conductor to the system conductor that is to be grounded must be made "at any accessible point from the *load end* of the overhead service conductors, service drop, underground service conductors, or service lateral" to the service disconnecting means. This means that the grounding electrode conductor (which runs to building steel and/or water pipe or driven ground rod) must be connected to the system neutral or other system wire to be grounded either in the enclosure for the

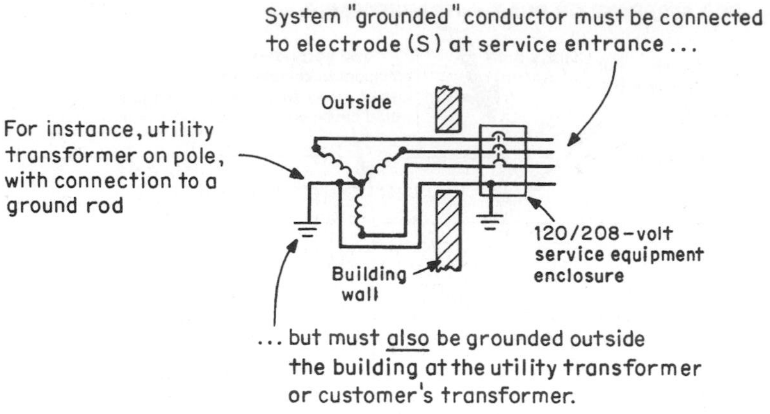
System "grounded" conductor must be connected to electrode (S) at service entrance...
Outside
For instance, utility transformer on pole, with connection to a ground rod
120/208-volt service equipment enclosure
Building wall
... but must also be grounded outside the building at the utility transformer or customer's transformer.

Fig. 250-8. Grounded interior systems must have *two* grounding points. (Sec. 250.24.)

service disconnect or in some enclosure on the supply side of the service disconnect. Such connection may be made, for instance, in the main service switch or CB or in a service panelboard or switchboard. Or, the grounding electrode conductor may be connected to the system grounded conductor in a gutter, CT cabinet, or meter housing on the supply side of the service disconnect (Fig. 250-9). The utility company should be checked on grounding connections in meter sockets or other metering equipment. In some areas, the connection really is literally at the load end of the service drop (or lateral), in the form of connections made right below the weatherhead in the case of a service drop.

As a result of this requirement, if a service is fed to a building from a meter enclosure on a pole or other structure some distance away, as is commonly done on farm properties, and an overhead or underground run of service conductors is made to the service disconnect in the building, the grounding electrode conductor will not satisfy the Code if it is connected to the neutral in the meter enclosure but must be connected at the load end of the underground or overhead service conductors. The connection should preferably be made within the service-disconnect enclosure.

This rule on grounding connections is shown in Fig. 250-10. If, instead of an underground lateral, an overhead run were made to the building from the pole, the overhead line would be a "service drop." The rule of 250.24(A)(1) would likewise require the grounding connection at the load end of the service drop. If a fused switch or CB is installed as service disconnect and protection at the load side of the meter on the pole, then that would establish the service at that point, and the grounding electrode connection to the bonded neutral terminal would be required at that point. The circuit from that point to the building would be a feeder and not service conductors. But electrical safety and effective operation would require that an equipment grounding conductor be run with the feeder circuit conductors for grounding the interconnected system of conduits and metal equipment enclosures along with metal piping systems and building steel within the building. Or, if an equipment grounding conductor is not in the circuit from the pole to the building,

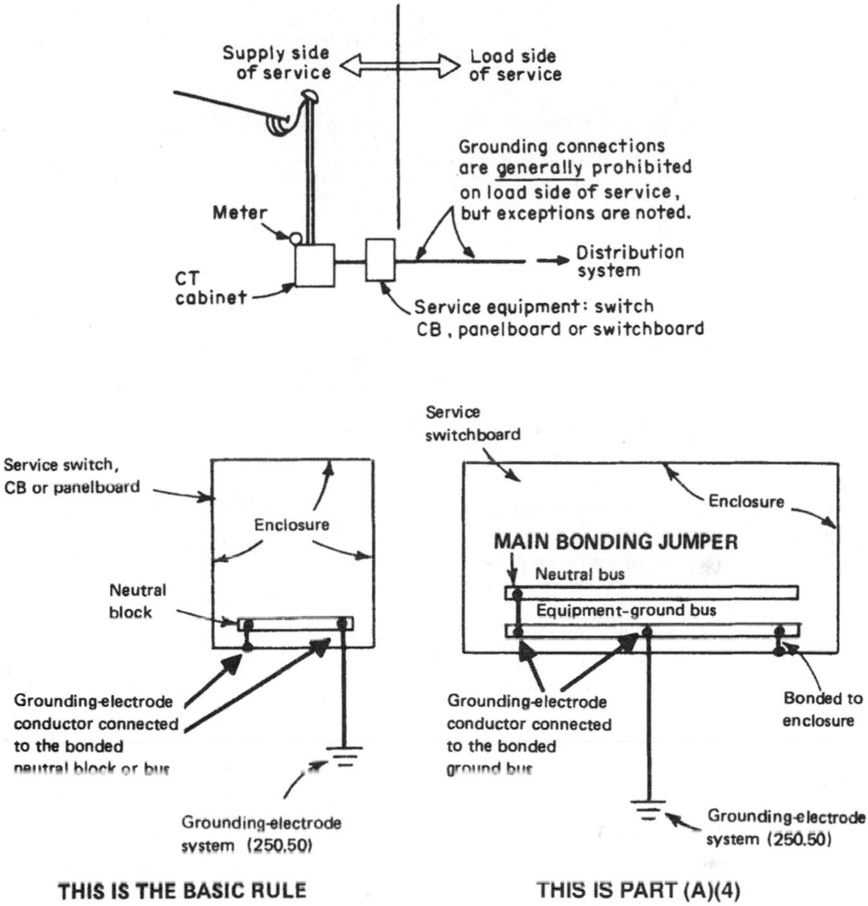
Supply side of service
Load side of service
Grounding connections are generally prohibited on load side of service, but exceptions are noted.
Meter
CT cabinet
Distribution system
Service equipment: switch CB, panelboard or switchboard
Service switch, CB or panelboard
Enclosure
Neutral block
Grounding-electrode conductor connected to the bonded neutral block or bus
Grounding-electrode system (250.50)
Service switchboard
Enclosure
MAIN BONDING JUMPER
Neutral bus
Equipment-ground bus
Grounding-electrode conductor connected to the bonded ground bus
Bonded to enclosure
Grounding-electrode system (250.50)
THIS IS THE BASIC RULE
THIS IS PART (A)(4)

Fig. 250-9. Grounding connection must be made in SE equipment or on its line side. (Sec. 250.24.)

the neutral could be bonded to the main disconnect enclosure in the building and a grounding electrode connection made at that point also. However, that alternative is now restricted to existing applications only.

In addition to the grounding connection for the grounded system conductor at the point of service entrance to the premises, according to 250.24**(A)(2)**, it is further required that another grounding connection be made to the same grounded conductor at the transformer that supplies the system. This means, for example, that a grounded service to a building must have the grounded neutral connected to another grounding electrode at the utility transformer on the pole, away from the building, as well as having the neutral grounded to a water pipe and/or other suitable electrode at the building, as shown in Fig. 250-8. And in the case of a building served from an outdoor transformer pad or mat installation, the conductor that is grounded in the building must also be grounded at the transformer pad

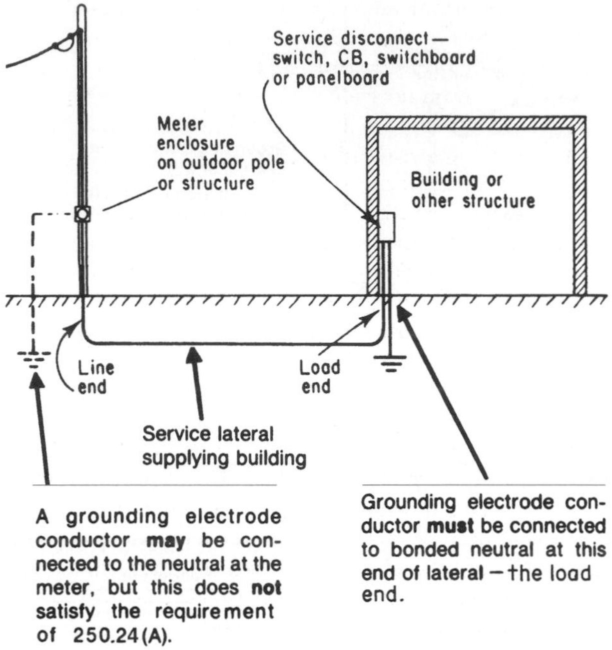
Service disconnect—
switch, CB, switchboard
or panelboard
Meter
enclosure
on outdoor pole
or structure
Building or
other structure
Line
end
Load
end
Service lateral
supplying building
A grounding electrode conductor **may** be connected to the neutral at the meter, but this does **not** satisfy the requirement of 250.24(A).
Grounding electrode conductor **must** be connected to bonded neutral at this end of lateral — the load end.

Fig. 250-10. Connection to grounded conductor at load end of lateral or drop. (Sec. 250.24.)

or mat, per 250.24(A)(2). However, this connection must not be made if the facility will be using a high-impedance grounded neutral system. On these systems, the second electrode at the transformer would allow return current to leave the building system through the earth, bypassing the monitoring equipment.

250.24**(A)(4)** permits the grounding electrode conductor to be connected to the equipment grounding bus in the service-disconnect enclosure—instead of the neutral block or bus—for instance, where such connection is considered necessary to prevent desensitizing of a service GFPE hookup that senses fault current by a CT-type sensor on the ground strap between the neutral bus and the ground bus. (See Fig. 250-9.) However, in any particular installation, the choice between connecting to the neutral bus or to the ground bus will depend on the number and types of grounding electrodes, the presence or absence of grounded building structural steel, bonding between electrical raceways and other metal piping on the load side of the service equipment, and the number and locations of bonding connections. The grounding electrode conductor may be connected to either the neutral bus or terminal lug or the ground bus or block in any system that has a conductor or a busbar bonding the neutral bus or terminal to the equipment grounding block or bus. Where the neutral is bonded to the enclosure simply by a bonding screw, the grounding electrode conductor must be connected to the neutral in all cases, because screw bonding is not suited to passing high lightning currents to earth.

One of the most important and widely discussed regulations of the entire Code revolves around this matter of making a grounding connection to the system grounded neutral or grounded phase wire. The Code says in part **(A)(5)**, "A grounded conductor shall not be connected to normally noncurrent-carrying metal parts of equipment, to equipment-grounding conductor(s), or be reconnected to ground on the load side of the service disconnecting means." Once a neutral or other circuit conductor is connected to a grounding electrode at the service equipment, the general rule is that the neutral or other grounded leg must be insulated from all equipment enclosures or any other grounded parts on the load side of the service. That is, bonding of equipment-grounding and neutral buses within subpanels (or any other connection between the neutral or other grounded conductor and equipment enclosures) is prohibited by the NEC.

There are some situations that are essentially "exceptions" to that rule, but they are few and are very specific:

1. In a system, even though it is on the load side of the service, when voltage is stepped down by a transformer, a grounding connection must be made to the secondary neutral to satisfy Secs: 250.20(B) and 250.30. Since this is a separately derived system, it isn't really an exception, because separately derived systems have no conductor that is common to the service-supplied system, and therefore the service neutral is not being regrounded in this case. Further, 250.30(A) now expressly forbids regrounding any grounded conductor of a separately derived system without specific code authorization.
2. When a circuit is run from one building to another, it may be necessary or prohibited to connect the system "grounded" conductor to a grounding electrode at the other building—as covered by 250.32. This is now generally prohibited in new installations as well, although it remains an option for "installations made in compliance with previous editions of this Code that permitted such connection" only.
3. In grounding of ranges and dryers, where supplied by an existing circuit as covered by 250.140, and Exception No. 1 to 250.142(B).

The Code makes it a violation to bond the neutral block in a panelboard to the panel enclosure in other than a service panel, or in a comparable location at the beginning of a separately derived system where no such connection has been placed at the energy source for that system. In a panelboard used as service equipment, the neutral block (terminal block) is bonded to the panel cabinet by the bonding screw provided. And such bonding is required to tie the grounded conductor to the interconnected system of metal enclosures for the system (e.g., service-equipment enclosures, conduits, busway, boxes, panel cabinets). It is this connection that provides for flow of fault current and operation of the overcurrent device (fuse or breaker) when a ground fault occurs. However, there must not be any connection between the grounded system conductor and the grounded metal enclosure system at any point on the load side of the service equipment, because such connection would constitute connection of the grounded system conductor to a grounding electrode (through the enclosure and raceway system to the water pipe or driven ground rod). Such connections, like bonding of subpanels, can be dangerous, as shown in Fig. 250-11.

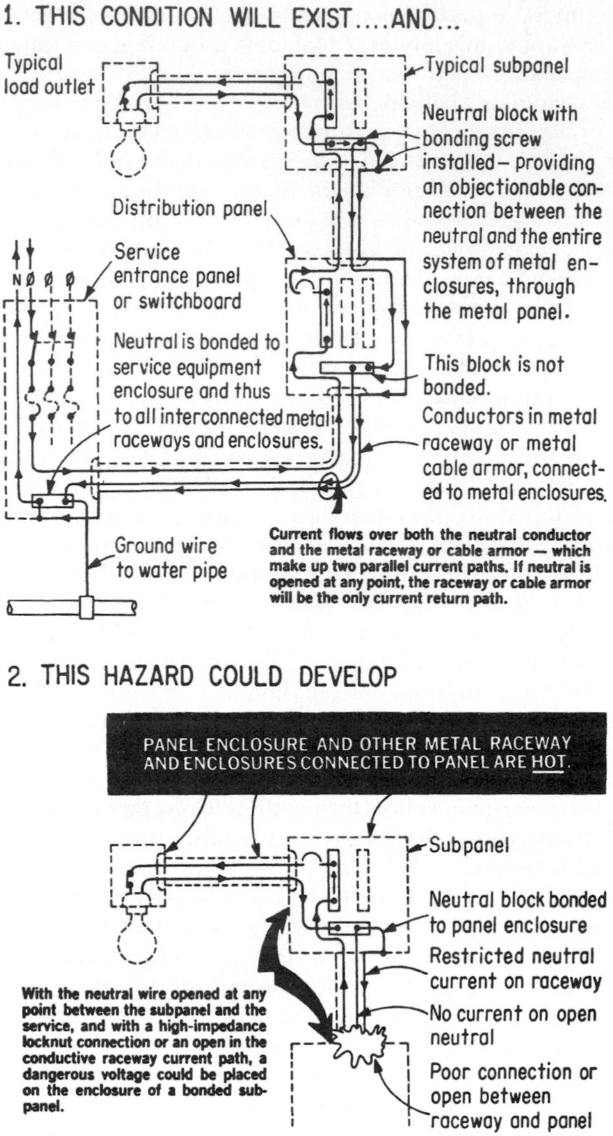
1. THIS CONDITION WILL EXIST....AND...
Typical load outlet
Typical subpanel
Neutral block with bonding screw installed– providing an objectionable connection between the neutral and the entire system of metal enclosures, through the metal panel.
Distribution panel
Service entrance panel or switchboard
N Ø Ø Ø
Neutral is bonded to service equipment enclosure and thus to all interconnected metal raceways and enclosures.
This block is not bonded.
Conductors in metal raceway or metal cable armor, connected to metal enclosures.
Ground wire to water pipe
Current flows over both the neutral conductor and the metal raceway or cable armor — which make up two parallel current paths. If neutral is opened at any point, the raceway or cable armor will be the only current return path.
2. THIS HAZARD COULD DEVELOP
PANEL ENCLOSURE AND OTHER METAL RACEWAY AND ENCLOSURES CONNECTED TO PANEL ARE HOT.
Subpanel
Neutral block bonded to panel enclosure
Restricted neutral current on raceway
No current on open neutral
Poor connection or open between raceway and panel
With the neutral wire opened at any point between the subpanel and the service, and with a high-impedance locknut connection or an open in the conductive raceway current path, a dangerous voltage could be placed on the enclosure of a bonded subpanel.

Fig. 250-11. NEC prohibits bonding of subpanels because of these reasons. (Sec. 250.24.)

This rule on not connecting the grounded system wire to a grounding electrode on the load side of the service disconnect must not be confused with the rule of 250.140, which permits the grounded system conductor to be used for grounding the frames of electric ranges, wall ovens, counter-mounted cooking units, and electric clothes dryers, but only from existing branch circuits. The connection referred to in 250.140 is that of an ungrounded metal enclosure to the grounded conductor for the purpose of grounding the enclosure. If a new circuit is run, it must be provided with an equipment grounding conductor and a 4-wire plug and receptacle must be used.

There is an important "exception" to the rule that each and every service for a grounded ac system have a grounding electrode conductor connected to the grounded system conductor anywhere on the supply side of the service-disconnecting means (preferably within the service-equipment enclosure) and that the grounding electrode conductor be run to a grounding electrode at the service. Because controversy has arisen in the past about how many grounding electrode conductors have to be run for a dual-feed (double-ended) service with a secondary tie, part **(A)(3)** recognizes the use of a single grounding electrode conductor connection for such dual services. It says that the single grounding electrode connection may be made to the "tie point of the grounded circuit conductors from each power source." The explanation on this Code permission was made by NEMA, the sponsor of the rule, as follows:

> Unless center neutral point grounding and the omission of all other secondary grounding is permitted, the selective ground-fault protection schemes now available for dual power source systems with secondary ties will not work. Dual power source systems are utilized for maximum service continuity. Without selectivity, both sources would be shut down by any ground fault. This proposal permits selectivity so that one source can remain operative for half the load, after a ground fault on the other half of the system.

Figure 250-12 shows two cases involving the concept of single grounding point on a dual-fed service:

- In case 1, if the double-ended unit substation is in a locked room in a building it serves or consists of metal-enclosed gear or a locked enclosure for each transformer, the secondary circuit from each transformer is a "service" to the building. The question then arises, "Does there have to be a separate grounding electrode conductor run from each secondary service to a grounding electrode?"
- In case 2, if each of the two transformers is located outdoors, in a separate building from the one they serve, in a transformer vault in the building they serve, or in a locked room or enclosure and accessible to qualified persons only or in metal-enclosed gear, then the secondary circuit from each transformer constitutes a service to the building. Again, is a separate grounding connection required for each service?
- In both cases, a single grounding electrode connection may serve both services, as shown at the bottom of Fig. 250-12.

The Code rule in part **(A)(3)** refers to "services that are dual fed (double ended) in a common enclosure or grouped together." The phrase *common enclosure* can readily cover use of a double-ended loadcenter unit substation in a single, common enclosure. But the phrase *grouped together* can lend itself to many interpretations and has caused difficulties. For instance, if each of two separate services is a single-ended unit substation, do both the unit substations have to be in the same room or within the same fenced area outdoors? How far apart may they be and still be considered grouped together? As shown in case 2 of Fig. 250-12, if separate transformers and switchboards are used instead of unit subsubstations, may one of the transformers and its switchboard be installed at the opposite end

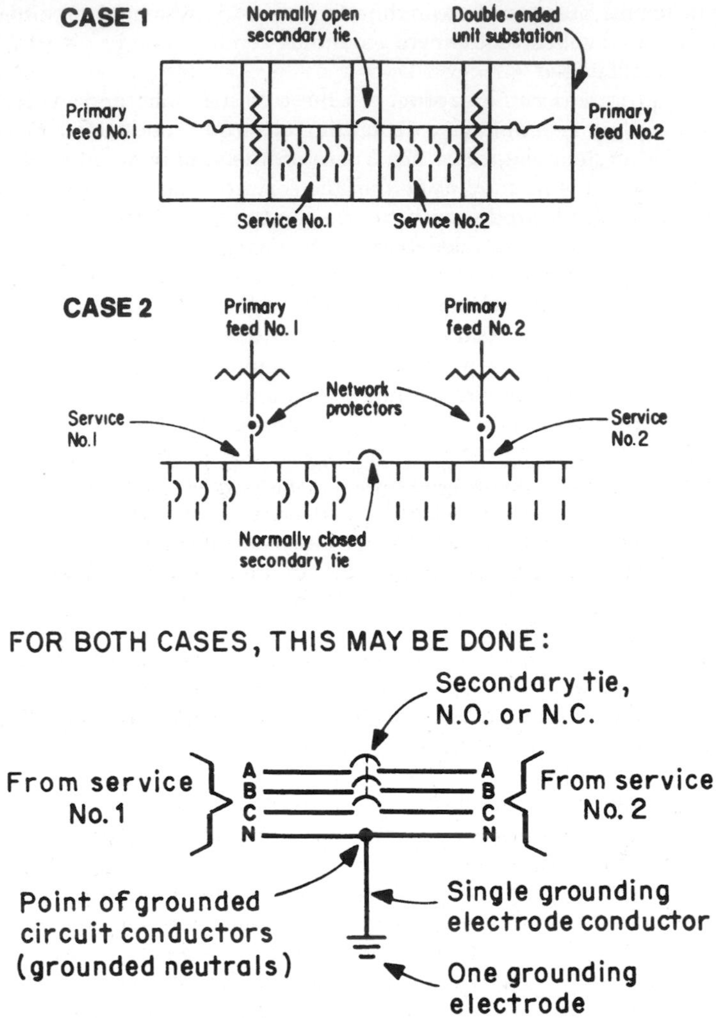
CASE 1
Normally open secondary tie
Double-ended unit substation
Primary feed No.1
Primary feed No.2
Service No.1
Service No.2
CASE 2
Primary feed No. 1
Primary feed No.2
Network protectors
Service No.1
Service No.2
Normally closed secondary tie
FOR BOTH CASES, THIS MAY BE DONE:
Secondary tie, N.O. or N.C.
From service No. 1
A
B
C
N
From service No. 2
Point of grounded circuit conductors (grounded neutrals)
Single grounding electrode conductor
One grounding electrode

Fig. 250-12. *One* grounding connection permitted for a double-end service. (Sec. 250.24.)

of the building from the other one? The Code does not answer those questions, but it seems clear that the wording does suggest that both the services must be physically close and at least in the same room or vault or fenced area. That understanding has always been applied to other Code rules calling for grouping—such as for switches and CBs in 404.8 and for service disconnects in 230.72(A).

Part **(B)** of 250.24 mandates a connection between the grounded conductor—usually a neutral conductor—and the equipment grounding conductor for all systems required or desired to be grounded. Such installation must satisfy the requirements of 250.28. This connection uses the *main bonding jumper* and it is the most important single connection in the entire electrical system because without it overcurrent devices will not clear ground faults.

Part **(C)** requires that whenever a service is derived from a grounded neutral system, the grounded neutral conductor must be brought into the service entrance equipment, even if the grounded conductor is not needed for the load supplied by the service. A service of less than 1000 V that is grounded outdoors at the service transformer (pad mount, mat, or unit substation) must have the grounded conductor routed with the ungrounded service conductors and run to "each service disconnecting means" and bonded to the separate enclosure for "each" service disconnect. If two to six normal service disconnects [as permitted by 230.71(A)] are installed in separate enclosures (or even additional disconnect switches or circuit breakers for emergency, fire pump, etc.), the grounded circuit conductor must be run to a bonded neutral terminal in each of the separate disconnect enclosures fed from the service conductors. The exception to this rule clarifies that if multiple service disconnect switches or circuit breakers are installed within "an assembly listed for use as service equipment"—such as in a service panelboard, switchboard, or multimeter distribution assembly—only a single grounded (neutral) conductor has to be run to the single, common assembly enclosure and bonded to it.

Running the grounded conductor to each individual service disconnect enclosure is required to provide a low-impedance ground-fault current return path to the neutral to ensure operation of the overcurrent device for safety to personnel and property. (See Fig. 250-13.) In such cases, the neutral functions strictly as an equipment grounding conductor, to provide a closed circuit back to the transformer for automatic circuit opening in the event of a phase-to-ground fault anywhere on the load side of the service equipment. Only one phase leg is shown in these diagrams to simplify the concept. The other two phase legs have the same relation to the neutral.

The same requirements apply to installation of separate power and light services derived from a common 3-phase, 4-wire, grounded "red-leg" delta system. The neutral from the center-tapped transformer winding must be brought into the 3-phase power service equipment as well as into the lighting service, even though the neutral will not be used for power loads. This is shown in Fig. 250-14 and is also required by 250.24(C)(1). This grounded conductor must be sized as covered in Table 250.102(C)(1). If the installation is large enough to use paralleled conductors in two or more raceways, then the grounded conductors are to be no smaller than 1/0 AWG and run in each of the paralleled raceways, sized by bringing the size of the largest ungrounded resulting phase connection into the new Table 250.102(C)(1). Remember, for example, that a four-conductor paralleled hookup might consist of two raceways containing two sets of conductors each, and in such a case the area used in Table 250.102(C)(1) would be twice the area of the largest phase conductor.

In any system where the neutral is required on the load side of the service—such as where 208Y/120-V or 480Y/277-V, 3-phase, 4-wire distribution is to be made on the premises—the neutral from the supply transformer to the service equipment is needed to provide for neutral current flow under conditions of load unbalance on the phase legs of the premises distribution system. But, even in a premises where all distribution on the load side of the service is to be solely 3-phase,

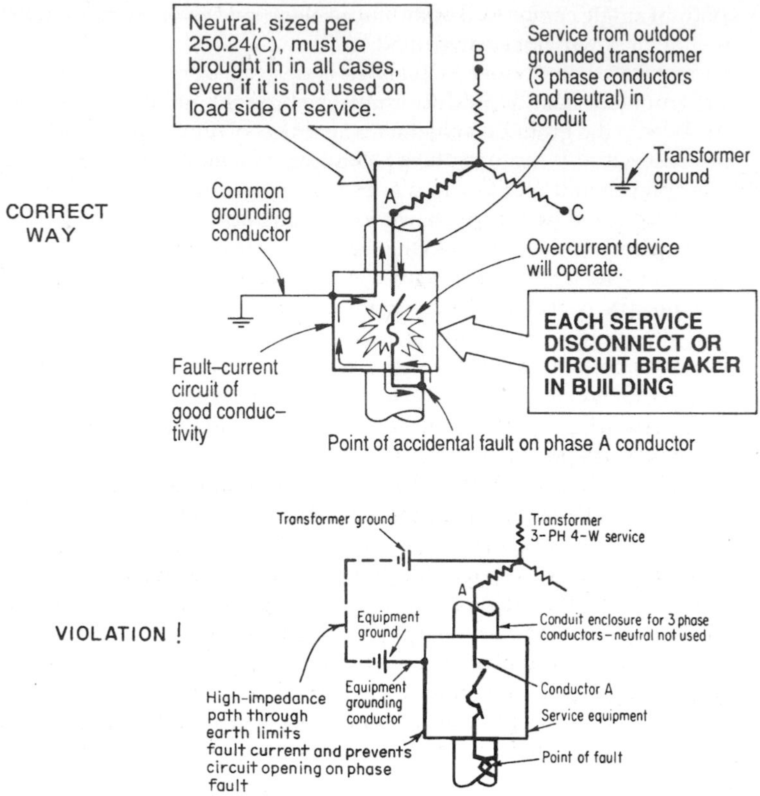
CORRECT WAY
Neutral, sized per 250.24(C), must be brought in in all cases, even if it is not used on load side of service.
B
Service from outdoor grounded transformer (3 phase conductors and neutral) in conduit
Transformer ground
Common grounding conductor
A
C
Overcurrent device will operate.
EACH SERVICE DISCONNECT OR CIRCUIT BREAKER IN BUILDING
Fault–current circuit of good conduc–tivity
Point of accidental fault on phase A conductor
VIOLATION !
Transformer ground
Transformer 3-PH 4-W service
A
Equipment ground
Conduit enclosure for 3 phase conductors – neutral not used
Equipment grounding conductor
Conductor A
High-impedance path through earth limits fault current and prevents circuit opening on phase fault
Service equipment
Point of fault

Fig. 250-13. Clearing of ground faults on the load side of any service disconnect depends on fault-current return over a grounded circuit conductor (usually a neutral) brought into each and every enclosure for service disconnect switch or CB. (Sec. 250.24.)

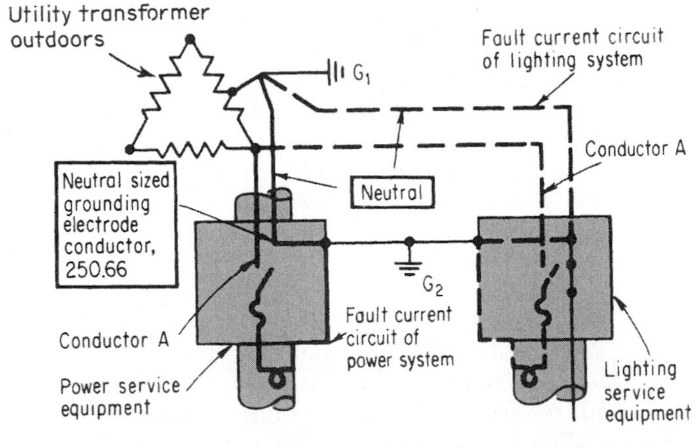
Utility transformer outdoors
G_1
Fault current circuit of lighting system
Conductor A
Neutral sized grounding electrode conductor, 250.66
Neutral
G_2
Fault current circuit of power system
Conductor A
Power service equipment
Lighting service equipment

Fig. 250-14. Neutral must be brought in to each service equipment and bonded to enclosure. (Sec. 250.24.)

3-wire (such as 480-V, 3-phase, 3-wire distribution) and the neutral conductor is not required in the premises system, this Code rule says that the neutral must still run from the supply transformer to the service equipment.

It should be noted that 250.24(C)(3) calls for the grounded conductor to be no smaller than the ungrounded conductors where the supply is a corner-grounded delta system. That makes sense because the grounded conductor, in such an application, is a phase conductor. Given that the load to be supplied will be carried by all three supply conductors—including the grounded phase conductor—the grounded conductor must be sized as an ungrounded conductor to satisfy this rule.

Part **(C)(2)** of the rule covers cases where the service phase conductors are paralleled, with two or more conductors in parallel per phase leg and neutral, and requires that the size of the grounded neutral be calculated on the equivalent area for parallel conductors. If the calculated size of the neutral (at least 12½ percent of the phase leg cross section) is to be divided among two or more conduits, and if dividing the calculated size by the number of conduits being used calls for a neutral conductor smaller than 1/0 in each conduit, the informational note calls attention to 310.10(H), which gives No. 1/0 as the minimum size of conductor that may be used in parallel in multiple conduits. For that reason, each neutral would have to be at least a No. 1/0, even though the calculated size might be, say, No. 1 or No. 2 or some other size smaller than No. 1/0. But the Code rule does permit subdividing the required minimum 12½ percent grounded (neutral) conductor size by the total number of conduits used in a parallel run, thereby permitting a multiple makeup using a smaller neutral in each pipe.

As shown in Fig. 250-15, the minimum required size for the grounded neutral conductor run from the supply transformer to the service is based on the size of the service phase conductors. In this case, the overall size of the service phase

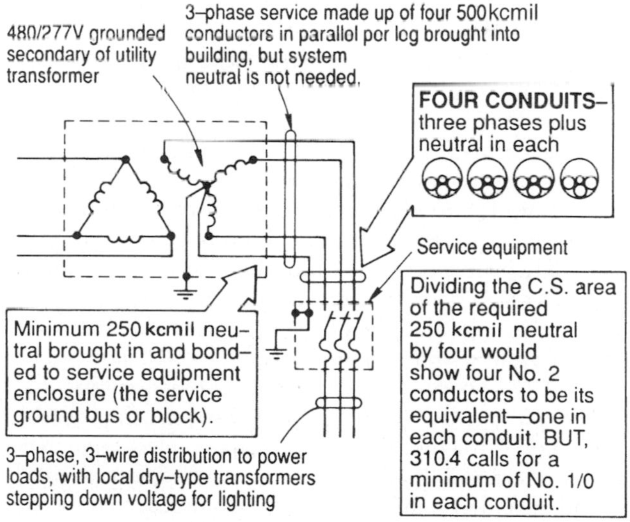
3-phase service made up of four 500kcmil conductors in parallel per leg brought into building, but system neutral is not needed.
480/277V grounded secondary of utility transformer
FOUR CONDUITS– three phases plus neutral in each
Service equipment
Dividing the C.S. area of the required 250 kcmil neutral by four would show four No. 2 conductors to be its equivalent—one in each conduit. BUT, 310.4 calls for a minimum of No. 1/0 in each conduit.
Minimum 250 kcmil neutral brought in and bonded to service equipment enclosure (the service ground bus or block).
3-phase, 3-wire distribution to power loads, with local dry-type transformers stepping down voltage for lighting

Fig. 250-15. Grounded service conductor must *always* be brought in. (Sec. 250.24.)

conductors is 4 × 500 kcmil per phase leg, or 2000 kcmil. Because that is larger than 1100 kcmil, it is not permitted to simply use Table 250.66 in sizing the neutral. Instead, 2000 kcmil must be multiplied by 12½ percent. Then 2000 kcmil × 0.125 equals 250 kcmil—the minimum permitted size of the neutral conductor run from the transformer to the service equipment. It is the Code's intent to permit the required 250-kcmil-sized neutral to be divided by the number of conduits. From NEC Table 8 in Chap. 9, it can be seen that four No. 2 conductors, each with a cross-sectional area of 66,360 circular mils, would approximate the area of one 250 kcmil (250,000 circular mils divided by 4 = 62,500 circular mils). But, because No. 1/0 is the smallest conductor that is permitted by 310.4 to be used in parallel for a circuit of this type, it would be necessary to use a No. 1/0 copper conductor in each of the four conduits, along with the phase legs. Remember, however, that corner-grounded delta services must not have the grounded conductor reduced in size from that of the other 2-phase legs.

250.24(D) requires all the bonded components—the service-equipment enclosure, the grounded neutral or grounded phase leg, and any equipment grounding conductors that come into the service enclosure—to be connected to a common grounding electrode (250.58) by the single grounding electrode conductor. A common grounding electrode conductor shall be run from the common point so obtained to the grounding electrode as required by Codes 250.24 and 250.58 (Fig. 250-16). Connection of the system neutral to the switchboard frame or ground bus within the switchboard provides the lowest impedance for the equipment ground return to the neutral.

250.24(E) covers ungrounded systems. They too require grounding electrode conductors, also connected at any accessible point from the load end of the service lateral or drop to the enclosure for the disconnecting means. This connection accomplishes the ground reference objective as set forth in 250.4(B)(1). The system equipment grounding conductors and the service enclosure are connected here, but there is, of course, no connection to a phase conductor.

250.26. Conductor to Be Grounded—Alternating-Current Systems. Selection of the wiring system conductor to be grounded depends upon the type of system. In 3-wire, single-phase systems, the midpoint of the transformer winding—the point from which the system neutral is derived—is grounded. For grounded 3-phase wiring systems, the neutral point of the wye-connected transformer(s) or generator is the point connected to ground. In delta-connected transformer hookups, grounding of the system can be effected by grounding one of the three phase legs, by grounding a center tap point on one of the transformer windings (as in the 3-phase, 4-wire "red-leg" delta system), or by using a special grounding transformer that establishes a neutral point of a wye connection that is grounded.

250.28. Main and System Bonding Jumpers. The NEC now makes a semantic distinction between two conductors with identical functions and essentially identical installation requirements. For any grounded system, the arguably single most important connection is the (usually) single-point connection between the equipment grounding system and the grounded circuit conductor. If this connection is compromised, no meaningful fault current can complete a connection to the system source, and in that process thereby remove voltage from enclosures and create the high-current return that will cause OCPDs to open immediately. In the case

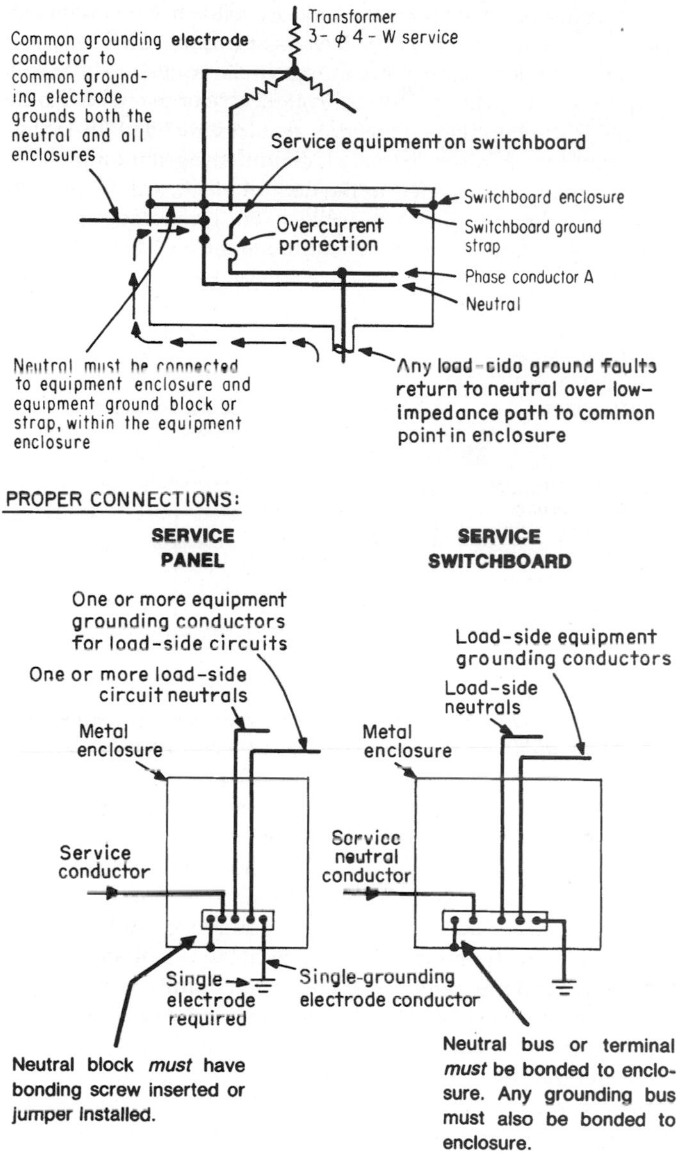
Transformer
3- ϕ 4 - W service
Common grounding electrode conductor to common grounding electrode grounds both the neutral and all enclosures
Service equipment on switchboard
Switchboard enclosure
Switchboard ground strap
Overcurrent protection
Phase conductor A
Neutral
Neutral must be connected to equipment enclosure and equipment ground block or strap, within the equipment enclosure
Any load-side ground faults return to neutral over low-impedance path to common point in enclosure
PROPER CONNECTIONS:
SERVICE PANEL
SERVICE SWITCHBOARD
One or more equipment grounding conductors for load-side circuits
One or more load-side circuit neutrals
Metal enclosure
Service conductor
Single electrode required
Single-grounding electrode conductor
Neutral block *must* have bonding screw inserted or jumper installed.
Load-side equipment grounding conductors
Load-side neutrals
Metal enclosure
Service neutral conductor
Neutral bus or terminal *must* be bonded to enclosure. Any grounding bus must also be bonded to enclosure.

Fig. 250-16. Common grounding electrode conductor for service and equipment ground. [Sec. 250.24(D).]

of a system supplied by a service, this conductor is the "main bonding jumper." In the case of a separately derived system, this conductor is the "system bonding jumper."

As required by the wording of 250.24(B), a "main bonding jumper" must be installed between the grounded and grounding conductors at or before the service disconnect.

The main bonding jumper that bonds the service enclosure and equipment grounding conductors (which may be either conductors or conduit, EMT, etc., as permitted by 250.118) to the grounded conductor of the system is required to be installed within the service equipment or within a service conductor enclosure on the line side of the service. This is the bonding connection required by 250.130(A) (Fig. 250-17). It should be noted that in a service panel, equipment grounding conductors for load-side circuits may be connected to the neutral block, and there is no need for an equipment grounding terminal bar or block. For grounded separately derived systems, 250.30(A)(1) imposes the same requirement for system bonding jumpers.

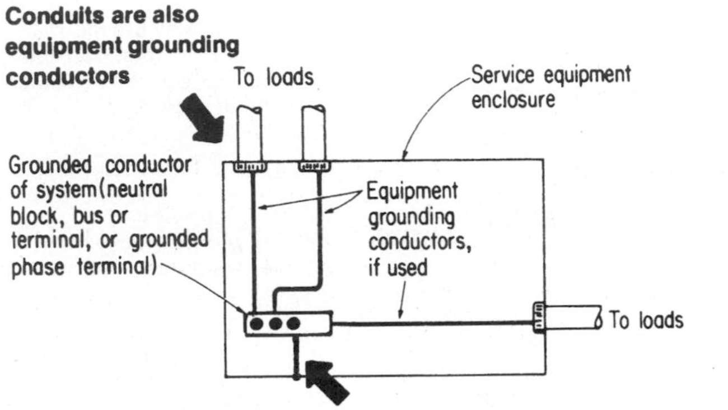
Conduits are also equipment grounding conductors
To loads
Service equipment enclosure
Grounded conductor of system (neutral block, bus or terminal, or grounded phase terminal)
Equipment grounding conductors, if used
To loads
Main bonding jumper (wire, bus, screw or similar conductor) must be *within* service equipment or service conductor enclosure

Fig. 250-17. Main bonding jumper must be within SE enclosure. (Sec. 250.28.)

If a grounding conductor were used to ground the neutral to the water pipe or other grounding electrode and a separate grounding conductor were used to ground the switchboard frame and housing to the water pipe or other electrode, without the neutral and the frame being connected in the switchboard, the length and impedance of the ground path would be increased. The proven hazard is that the impedance of the fault-current path can limit fault current to a level too low to operate the overcurrent devices "protecting" the faulted circuit.

Note that a number of grounding electrodes that are bonded together, as required by 250.50, are considered to be one grounding electrode.

Part **(A)** calls for use of copper or other "corrosion-resistant" conductor material—which does include aluminum and copper-clad aluminum. Part **(B)** notes that if the bonding jumper is in the form of a screw, the screw head must be finished with a green coloring. This allows the inspector to zero in on the required connection, and not confuse the bonding screw with other screws that may not be making the required connections. Part **(C)** demands use of connectors, lugs, and other fittings that have been designed, tested, and listed for the particular application, as covered in 250.8.

Part **(D)** covers sizing of any bonding jumper within the service equipment enclosure or on the line or supply side of that enclosure. Refer to the definitions of "Bonding Jumper, Main" and "Bonding Jumper, System" in Art. 100.

The minimum required size of this jumper for this installation is determined by calculating the size of one service phase leg. For example, with three 500 kcmil per phase, that works out to 1500-kcmil copper per phase. Because that value is in excess of 1100-kcmil copper, as noted in Table 250.102(C)(1), the minimum size of the main bonding jumper must equal at least 12½ percent of the phase leg cross-sectional area. Then,

$$12\frac{1}{2}\% \times 1500 \text{ kcmil} = 0.125 \times 1500 = 187.5 \text{ kcmil}$$

Referring to Table 8 in Chap. 9 in the back of the Code book, the smallest conductor with at least that cross-sectional area (csa) is No. 4/0, with a csa of 211,600 cmil or 211.6 kcmil. Note that No. 3/0 has a csa of only 167.8 kcmil. Thus No. 4/0 copper with any type of insulation would satisfy the Code.

If there are multiple service enclosures, the main bonding jumper in each is sized on the basis of the size of the largest ungrounded service-entrance line or phase conductor supplying that enclosure. If the multiple enclosures are fed from a common wireway with large service conductors tapped with smaller conductors feeding each actual service disconnect enclosure, the grounding in the parent wireway requires additional consideration. Any bonding connections necessarily take place on the supply side of the service disconnects, and therefore the bonding conductors within the wireway will be "supply-side bonding jumpers" (refer to the discussion at the definition in 250.2). These jumpers will be sized like grounding electrode conductors, increased if necessary in the event the ungrounded conductors exceed the tabulated values in Table 250.102(C)(1), as covered in the example in the previous paragraphs. If the grounding electrode conductor departs from this common wireway, it is sized in accordance with the normal rules based on the size of the entering common service conductors, as covered in 250.64(D)(3).

For separately derived systems with multiple enclosures, the same procedure can be used, or a system bonding jumper can be located at the derived system source. In this case, the bonding jumper will be sized on the basis of the largest ungrounded phase or line conductor cross-section area figured collectively across comparable conductors as represented in all the feeders supplied by the system.

250.30. Grounding Separately Derived Alternating-Current Systems. A separately derived ac wiring system is a source derived from an on-site generator (emergency or standby), a battery-inverter, or the secondary winding(s) of a transformer, or from any of the many other sources that do not share a direct connection with other sources, including PV systems. The 2014 **NEC** also inserted language at this point covering large separately derived systems connected in parallel, such as with multiple transformers supplying double-ended switchgear. These systems must comply with 250.20, 250.21, 250.22, and 250.26. Therefore:

1. Any system that operates at over 50 V but not more than 150 V to ground must be grounded [250.20(B)].
2. This requires the grounding of generator windings and secondaries of transformers serving 208/120-V, 3-phase or 240/120-V, single-phase circuits for

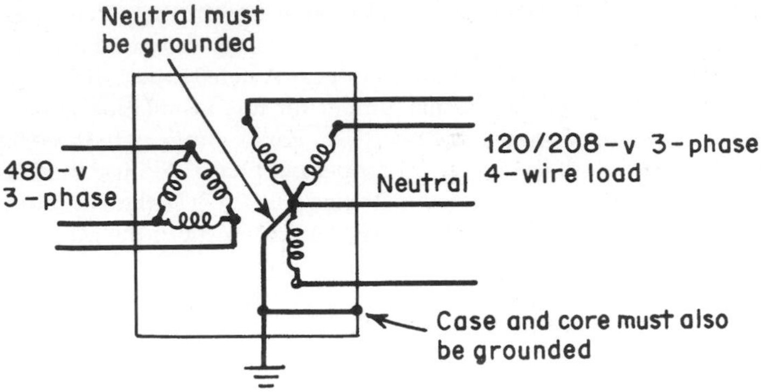
Neutral must
be grounded
480-v
3-phase
Neutral
120/208-v 3-phase
4-wire load
Case and core must also
be grounded

Fig. 250-18. Grounding is required for "separately derived" systems. (Sec. 250.30.)

lighting and appliance outlets and receptacles, at loadcenters throughout a building, as shown for the very common application of dry-type transformers in Fig. 250-18.

3. All Code rules applying to both system and equipment grounding must be satisfied in such installations.

The first informational note in this section specifically identifies an on-site generator (emergency or standby) as "not a separately derived system" if the neutral conductor from the generator is connected solidly through a terminal lug in a transfer switch to the neutral conductor from the normal (usually, the power company) service to the premises (Fig. 250-19). Therefore, the generator neutral point does not have to be bonded to the frame and connected to a grounding electrode. In fact, such a bonding point is prohibited in this case. It squarely violates 250.24(A)(5) because the downstream (from the service) connection from neutral to equipment ground in the generator remains active regardless of the position of the transfer switch. But, a grounding conductor selected from Table 250.122, based on the size of the generator's OC protection, must be connected to the generator frame/enclosure and the equipment ground bus within the transfer switch.

The second informational note essentially cautions that the neutral conductor from a generator to a transfer switch must be sized at least equal to 12½ percent of the cross-sectional area of the largest associated phase conductor (445.13) to ensure adequate conductivity (low impedance) for fault current that might return over that neutral when the generator is supplying the premises load, the neutral of both the generator and the normal service are connected solidly through the transfer switch (making the generator not a separately derived system), and the generator neutral is not bonded to the generator case and grounded at the generator. Under such a set of conditions, fault current from a ground fault in the premises wiring system would have to return to the point at the normal service equipment where the equipment grounding conductor (service equipment enclosure, metal conduits, etc.) is bonded to the service neutral. Only from that point can the fault current return over the neutral conductors, through the transfer switch to the neutral point of the generator winding. 445.13 effectively requires that such a generator neutral must satisfy 250.30(A)(3), which says that a neutral that might function

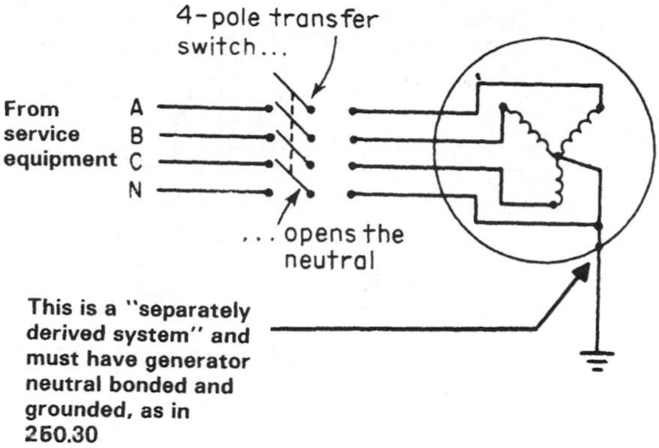
4-pole transfer switch...
From service equipment
A
B
C
N
...opens the neutral
This is a "separately derived system" and must have generator neutral bonded and grounded, as in 250.30

Generator neutral *must* be bonded when neutral is opened.

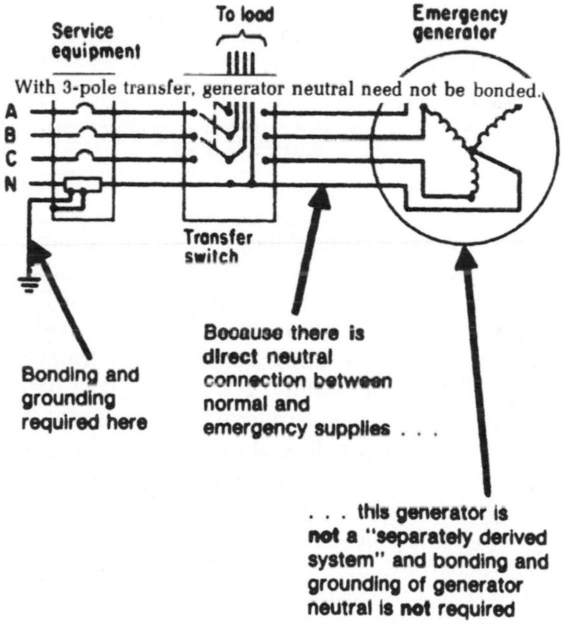
Service equipment
To load
Emergency generator
With 3-pole transfer, generator neutral need not be bonded.
A
B
C
N
Transfer switch
Bonding and grounding required here
Because there is direct neutral connection between normal and emergency supplies . . .
. . . this generator is not a "separately derived system" and bonding and grounding of generator neutral is not required

With 3-pole transfer, generator neutral need not be bonded.

Fig. 250-19. Not all generators are wired in a way that creates a separately derived system.

as an equipment grounding conductor must have a cross-sectional area at least equal to 12½ percent of the cross-sectional area of the largest phase conductor of the generator circuit to the transfer switch (Fig. 250-20). The actual reference is to 250.30(A), which in turn now references Table 250.102(C)(1) for sizing, and is over

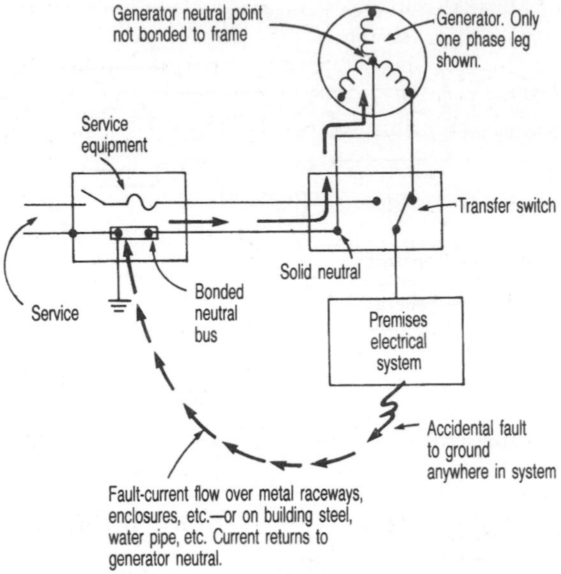
Generator neutral point
not bonded to frame
Generator. Only
one phase leg
shown.
Service
equipment
Transfer switch
Solid neutral
Bonded
neutral
bus
Service
Premises
electrical
system
Accidental fault
to ground
anywhere in system
Fault-current flow over metal raceways,
enclosures, etc.—or on building steel,
water pipe, etc. Current returns to
generator neutral.

Fig. 250-20. Neutral conductor from service equipment to generator neutral point must be sized at least equal to 12½ percent of the cross-sectional area of the generator phase leg. (Sec. 445.13).

two pages long; as a service to readers the relevant provisions have been identified here.

The effect of the rules on transfer switches is as follows:

- *3-pole transfer switch.* If a solid neutral connection is made from the service neutral, through the transfer switch, to the generator neutral, then bonding and grounding of the neutral at the generator are not required because the neutral is already bonded and grounded at the service equipment. And if bonding and grounding were done at the generator, it could be considered a violation of 250.6(A) and would have to be corrected by 250.6(B) (Fig. 250-6). It also squarely violates 250.24(A)(5) because the downstream (from the service) connection in the generator remains active regardless of the position of the transfer switch.
- *4-pole transfer switch.* Because there is no direct electrical connection of either the hot legs or the neutral between the service and the generator, the generator in such a hookup is a separately derived system and must be grounded and bonded to the generator case at the generator (Fig. 250-19).

It should be noted that the 4-pole transfer switch and other neutral-switching techniques came into use to eliminate problems of GFPE desensitizing that were caused by use of a 3-pole transfer switch when the neutral of the generator was bonded to the generator housing. As a result, the use of a 4-pole transfer switch or some other technique that opens the neutral is the only effective way to avoid GFPE desensitizing. Ground-fault protection is not compatible with a solid neutral tie between the service and an emergency generator—with or without its neutral bonded.

However, for smaller systems that are not using GFPE, 3-pole transfer switches with a solid neutral are still very common.

Refer to the discussion under 250.24(A)(4) on the relationship between GFPE desensitizing and the point of connection of the grounding electrode conductor.

Referring to Fig. 250-21, the steps involved in satisfying the Code rules are as follows:

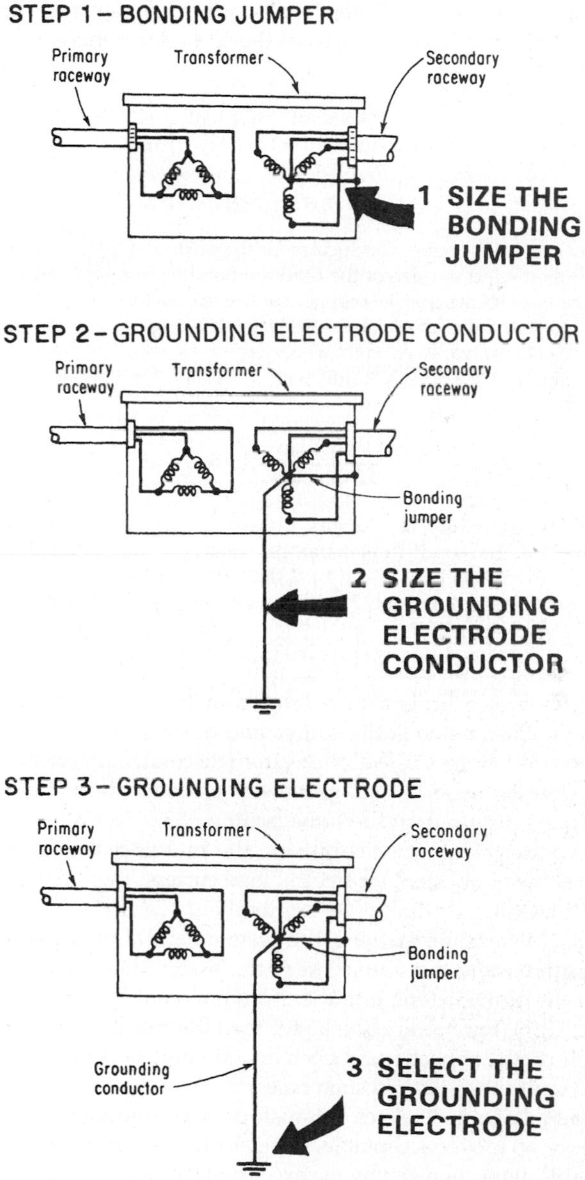
STEP 1 – BONDING JUMPER
Primary raceway
Transformer
Secondary raceway
1 SIZE THE BONDING JUMPER
STEP 2 – GROUNDING ELECTRODE CONDUCTOR
Primary raceway
Transformer
Secondary raceway
Bonding jumper
2 SIZE THE GROUNDING ELECTRODE CONDUCTOR
STEP 3 – GROUNDING ELECTRODE
Primary raceway
Transformer
Secondary raceway
Bonding jumper
Grounding conductor
3 SELECT THE GROUNDING ELECTRODE

Fig. 250-21. Grounding a transformer secondary. (Sec. 250.30.)

Step 1—250.30(A)(1)

A system bonding jumper must be installed between the transformer secondary neutral terminal and the metal case of the transformer. The size of this system bonding conductor is based on 250.28(D) and is selected from Table 250.102(C)(1) of the Code, based on the size of the transformer secondary phase conductors. For cases where the transformer secondary circuit is larger than 1100-kcmil copper or 1750-kcmil aluminum per phase leg, the bonding jumper must be not less than 12½ percent of the cross-sectional area of the secondary phase leg as described in Note 1 to that table.

example Assume this is a 75-kVA transformer with a 120/208-V, 3-phase, 4-wire secondary. Such a unit would have a full-load secondary current of

$$75{,}000 \div (208 \times 1.732) \text{ or } 209 \text{ A}$$

If we use No. 4/0 THW copper conductors for the secondary phase legs (with a 230-A rating), we would then select the size of the required bonding jumper from Table 250.102(C)(1) accordingly. The table shows that 4/0 copper service conductors require a minimum of No. 2 copper or No. 1/0 aluminum for a grounding electrode conductor. The bonding jumper would have to be either of those two sizes.

If the transformer were a 500-kVA unit with a 120/208-V secondary, its rated secondary current would be

$$\frac{500 \times 1000}{1.732 \times 208} = 1388 \text{ A}$$

Using, say, THW aluminum conductors, the size of each secondary phase leg would be four 700-kcmil aluminum conductors in parallel (each 700-kcmil THW aluminum is rated at 375 A; four are 4 × 375 or 1500 A, which suits the 1388-A load).

Then, because 4 × 700 kcmil equals 2800 kcmil per phase leg and is in excess of 1750 kcmil, Note 1 to Table 250.102(C)(1) requires the bonding jumper from the case to the neutral terminal to be at least equal to 12½ percent × 2800 kcmil (0.125 × 2800) or 350-kcmil aluminum.

The system bonding jumper must remain within the enclosure. Figure 250-22 shows the connection made at the source and dashed lines showing the alternate possibility of connections made at the system disconnecting means. However, the location of the system bonding jumper and the connection to the grounding electrode conductor must occur at the same point.

In the event the transformer is outside the building, with no likelihood of a parallel path being established for neutral load current, then **(A)(1)** Exception No. 2 together with **(A)(2)** exception team up to allow system bonding connections at both locations. In such applications, the system neutral functions as the fault current return path back to the source. As such, in accordance with **(A)(3)** it must be able to carry the fault current, and therefore must have a minimum size as set in Table 250.102(C)(1) (but need not be larger than the largest ungrounded conductor). The inevitable earth connections at each end are not considered to establish a parallel fault current return path in such cases.

Figure 250-23 shows techniques of transformer grounding that have been used in the past but are no longer acceptable, along with an example of "case grounding," which is specifically recognized by the exceptions to 250.30(A)(1), (A)(5), and (A)(6).

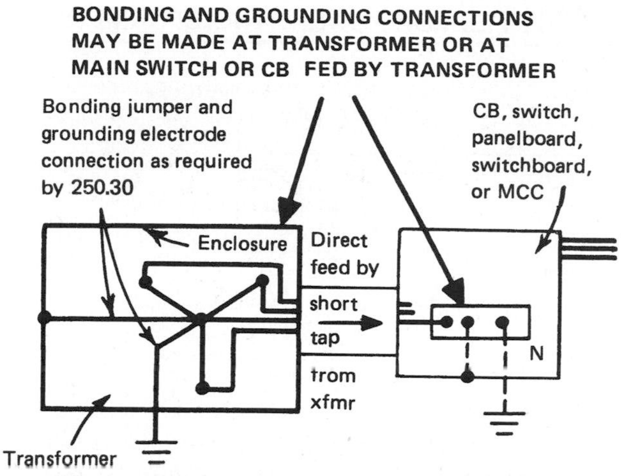
BONDING AND GROUNDING CONNECTIONS
MAY BE MADE AT TRANSFORMER OR AT
MAIN SWITCH OR CB FED BY TRANSFORMER
Bonding jumper and
grounding electrode
connection as required
by 250.30
CB, switch,
panelboard,
switchboard,
or MCC
Enclosure
Direct
feed by
short
tap
from
xfmr
N
Transformer

Fig. 250-22. Transformer secondary bonding and grounding must be "at the source" or at a secondary disconnect or protective device. Either the connections at the left, or the ones at the right are permitted, but never both. And in either case, the transformer enclosure and the disconnect enclosure must be bonded by a supply-side bonding conductor in accordance with 250.30(A)(2), unless the transformer is outside the building in accordance with 250.30(A)(1) Exception No. 2. (Sec. 250.30.)

Exception No. 3 to Sec. 250.30**(A)(1)** exempts small transformers for control, signal, or power-limited circuits from the basic requirement for a grounding electrode conductor run from the bonded secondary grounded conductor (such as a neutral) to an extension of a grounding electrode (nearby building steel or a water pipe). This exception, along with Exception No. 3 to 250.30(A)(5) and Exception No. 2 to (A)(6), applies to transformers used to derive control circuits, signal circuits, or power-limited circuits, such as circuits to damper motors in air conditioning systems. A Class 1, Class 2, or Class 3 remote-control or signaling transformer that is rated not over 1000 VA simply has to have a grounded secondary conductor bonded to the metal case of the transformer. No grounding electrode conductor is needed, provided that the metal transformer case itself is properly grounded by grounded metal raceway that supplies its primary or by means of a suitable (Sec. 250.118) equipment grounding conductor that ties the case back to the grounding electrode for the primary system, as indicated at the bottom of Fig. 250-24. Exception No. 3 to 250.30(A)(1) permits use of a No. 14 copper conductor to bond the grounded leg of the transformer secondary to the transformer frame, leaving the supply conduit to the transformer to provide the path to ground back to the main service ground, but depending on the connection between neutral and frame to provide effective return for clearing faults, as shown. Grounding of transformer housings must be made by connection to grounded cable armor or metal raceway or by use of a grounding conductor run with circuit conductor (either a bare conductor or a conductor with green covering).

Because the rule on bonding jumpers for the secondary neutral point of a transformer refers to 250.28, and therefore ties into Table 250.102(C)(1), the

NOT PERMITTED FOR POWER TRANSFORMERS

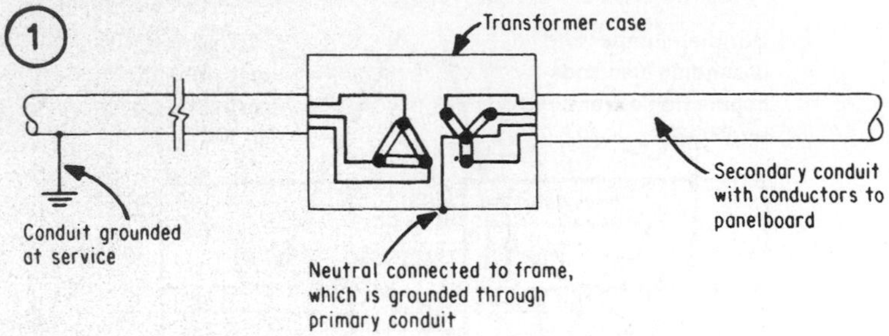
1
Transformer case
Secondary conduit with conductors to panelboard
Conduit grounded at service
Neutral connected to frame, which is grounded through primary conduit

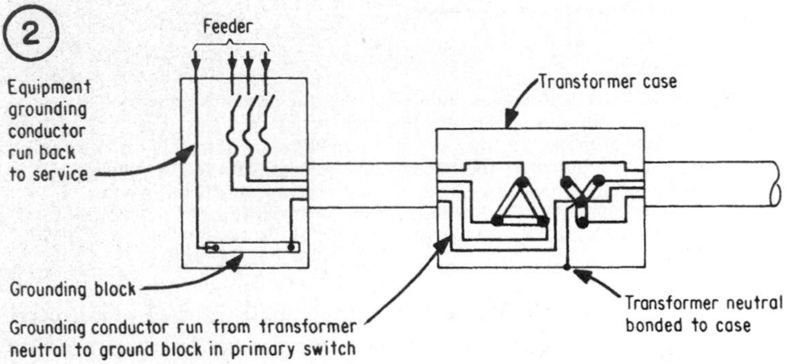
2
Feeder
Equipment grounding conductor run back to service
Transformer case
Grounding block
Grounding conductor run from transformer neutral to ground block in primary switch
Transformer neutral bonded to case

THIS IS PERMITTED

TRANSFORMER NOT OVER 1000 VA — FOR
Class 1, Class 2, or Class 3 circuits

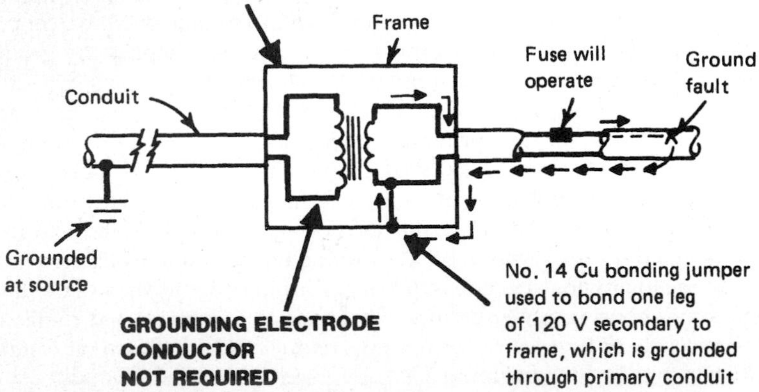
Frame
Fuse will operate
Ground fault
Conduit
Grounded at source
GROUNDING ELECTRODE CONDUCTOR NOT REQUIRED
No. 14 Cu bonding jumper used to bond one leg of 120 V secondary to frame, which is grounded through primary conduit

Fig. 250-23. Code rules regulate specific hookups for grounding transformer secondaries. (Sec. 250.30.)

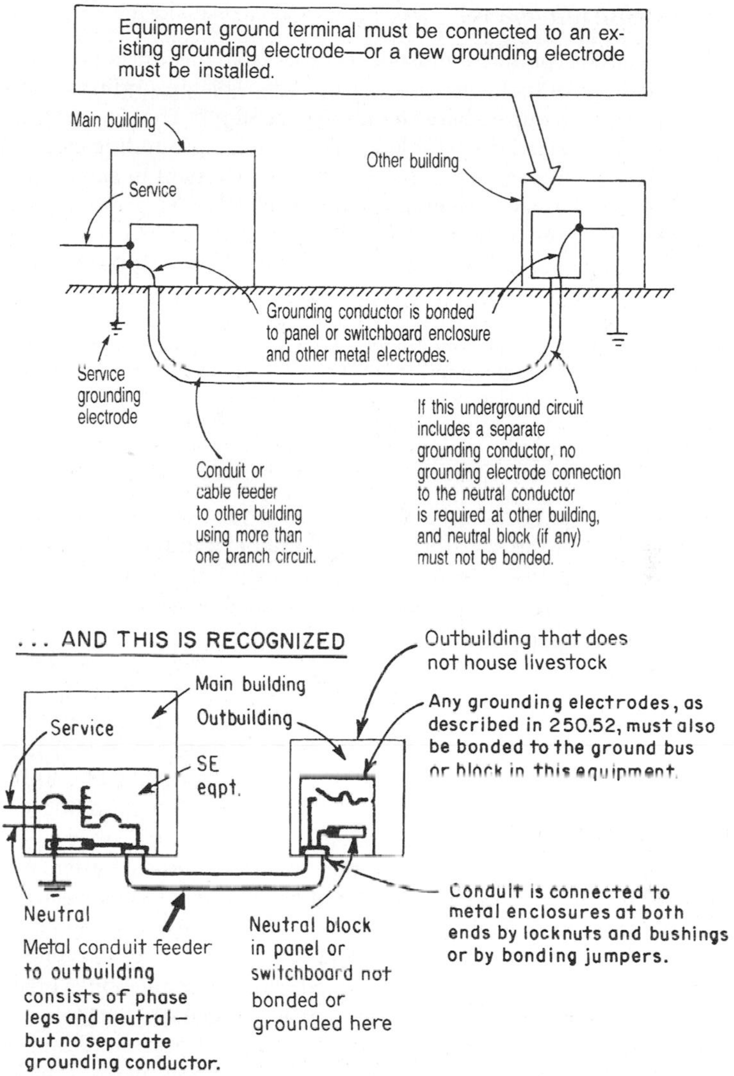
Equipment ground terminal must be connected to an existing grounding electrode—or a new grounding electrode must be installed.
Main building
Other building
Service
Grounding conductor is bonded to panel or switchboard enclosure and other metal electrodes.
Service grounding electrode
If this underground circuit includes a separate grounding conductor, no grounding electrode connection to the neutral conductor is required at other building, and neutral block (if any) must not be bonded.
Conduit or cable feeder to other building using more than one branch circuit.
. . . AND THIS IS RECOGNIZED
Outbuilding that does not house livestock
Main building
Outbuilding
Service
Any grounding electrodes, as described in 250.52, must also be bonded to the ground bus or block in this equipment.
SE eqpt.
Conduit is connected to metal enclosures at both ends by locknuts and bushings or by bonding jumpers.
Neutral
Metal conduit feeder to outbuilding consists of phase legs and neutral—but no separate grounding conductor.
Neutral block in panel or switchboard not bonded or grounded here

Fig. 250-24. Equipment grounding rules for outbuildings now parallel comparable rules for wiring within a building. (Sec. 250.32.)

smallest size that may be used is No. 8 copper, as shown in that table. But for small transformers—such as those used for Class 1, Class 2, or Class 3 remote-control or signaling circuits—that large a bonding jumper is not necessary and is not suited to termination provisions. For that reason, Exceptions No. 3 to 250.30(A)(1) and 250.30(A)(5), and also Exception No. 2 to 250.30(A)(6), permit the bonding jumper for such transformers rated not over 1000 VA to be smaller than No. 8. The jumper simply has to be at least the same size as the secondary phase legs of the transformer and in no case smaller than No. 14 copper or No. 12 aluminum.

Step 2—250.30(A)(5) and (6)

A grounding electrode conductor must be installed from the transformer secondary neutral terminal to a suitable grounding electrode. This grounding conductor is sized the same as the required bonding jumper in Step 1. That is, this grounding electrode conductor is sized from Table 250.66 as if it is a grounding electrode conductor for a service with service-entrance conductors equal in size to the phase conductors used on the transformer secondary side. But this grounding electrode conductor does not have to be larger than 3/0 copper or 250-kcmil aluminum when the transformer secondary circuit is over 1100-kcmil copper or 1750-kcmil aluminum.

example For the 75-kVA transformer in Step 1, the grounding electrode conductor must be not smaller than the required minimum size shown in Table 250.102(C)(1) for 4/0 phase legs, which makes it the same size as the bonding jumper—that is, No. 2 copper or No. 1/0 aluminum. But, for the 500-kVA transformer, the grounding electrode conductor is sized directly from Table 250.66—which requires 3/0 copper or 250-kcmil aluminum where the phase legs are over 1100-kcmil copper or 1750-kcmil aluminum.

The rule of 250.30(A)(1) permits the bonding and grounding connections to be made either right at the transformer or generator or at the first disconnect or overcurrent device fed from the transformer or generator, as in Fig. 250-22, but wherever the bonding jumper is connected, the grounding electrode conductor must be attached at the same point.

Part **(A)(6)(a)** recognizes the use of a "common grounding electrode conductor" as the electrode for a separately derived system. This represents relief from the requirement for connection of a separate grounding electrode conductor from each separately derived system to the water piping system within 5 ft from its point of entry to the building when building steel is not available. In high-rise construction where no building steel is available, a single grounding electrode conductor, connected to, say, the service enclosure, could be run in a shaft and be used as the "grounding electrode" for the separately derived systems.

Basically stated, this rule permits what amounts to a "grounding electrode" bus. First, as would be expected, all connections must be made at accessible locations. Next, there must be a positive means of connection employing irreversible pressure connectors, exothermic welding, or listed connectors to copper busbars not less than ¼ in. × 2 in. In addition, the minimum acceptable size for this continuous grounding electrode conductor is 3/0 copper, and 250 kcmil for aluminum, and the installation must also satisfy the requirements given in 250.64, which covers the installation requirements for grounding electrode conductors.

It should be noted that although high-rise construction without structural steel is perhaps the most obvious use for this permission, the wording used seems to permit horizontal distribution as well. If a situation presents itself where a continuous grounding electrode conductor run horizontally would be a benefit, then such application would seem to be acceptable.

In addition to a 3/0 conductor, the 2017 **NEC** added two more possibilities for the common grounding electrode conductor. First, a metal water pipe can be used if it complies with 250.68(C)(1). Normally this is trivial, since the total length permitted is 5 ft from where the water pipe electrode emerges from ground and changes function into a sort of grounding electrode conductor. However, the nonresidential exception is anything but trivial. Unchanged from prior editions, it allows a water pipe extension of indefinite length as long as it is not concealed. This may be a very plausible possibility in many occupancies to accomplish this function.

The other addition is the metal structural framework of a building as covered in 250.68(C)(2). The rules in Sec. 250.68 underwent considerable refinement in the 2017 NEC; refer to the additional discussion in this book at that location for the details. Obviously a metal building frame offers extensive opportunities for remote connection, so this will be a very practical alternative in many situations.

The exception following part **(A)** of 250.30 exempts high-impedance grounded transformer secondaries or generator outputs from the need to provide direct (solid) bonding and grounding electrode connections of the neutral, as required in parts **(A)(1)**, **(A)(3)**, and **(A)(4)**. This simply states an exception to each part that is necessary to operate a high-impedance grounded system.

Step 3—250.30(A)(4)

The 2017 NEC drastically simplified the rules for the grounding electrode system on separately derived systems. Make the connection to the building or structure grounding electrode system. Simple, end of story, and in the same vein if the separately derived system originates in listed equipment suitable for use as service equipment, the grounding electrode for that equipment can be used for the separately derived system. This latter provision was Exception No. 2 in prior Code and now is the only exception here. If the separately derived system is outdoors, then the rules in 250.30(C), specifically developed for this condition, are the ones to follow.

Step 4—250.30(A)(2) and (A)(3)

A supply-side bonding jumper must connect the source and the separately derived system disconnecting means. In the case of the systems illustrated in Figs. 250-21 and 250-22, that is simply the steel conduit, as recognized in 250.30(A)(2). However, in cases where there is no "nonflexible metal raceway," a wire or busbar will need to be installed. Because this occurs on the supply side of the first overcurrent protective device, it cannot be sized as an equipment grounding conductor in accordance with 250.122. Instead, 250.102(C) sets the size, which is the same calculation as in 250.28(D) for the main and system bonding jumpers: Use Table 250.102(C)(1) and for large systems where the ungrounded conductors exceed the range of the table, use a wire no smaller than ⅛ (12½%) the cross-sectional area as the phase legs. If a busbar is to be used, its cross-sectional area must be calculated (or obtained from the manufacturer) and it must be no smaller than a wire used for the same purpose.

A closely related case concerns the grounded circuit conductor, covered in 250.30(A)(3). If the system bonding jumper is installed at the main disconnect, then the grounded circuit conductor becomes the low-impedance path between the separately derived system loads and the source. As such, just as in the case of 250.24(C), the grounded circuit conductor is sized for fault return duty using the same procedures. Review the discussion on that topic and also Fig. 250-15 for complete information. Corner grounded delta systems also follow the same procedures for services because the provisions of 250.30(A)(3)(c) and 250.24(C)(3) are the same.

250.30(B). Ungrounded Systems. These systems must also have a grounding electrode conductor connected to a grounding electrode. The conductor is sized in accordance with Table 250.66 just as in the case of grounded systems, and the electrode must match up with the one specified in 250.30(A)(4). In addition, a supply-side bonding conductor must connect the source and the disconnecting means, and it must comply with 250.30(A)(2) just as in the case of grounded systems.

250.30(C). Outdoor Sources. Just as additional grounding rules apply to service sources originating off site [See 250.24(A)(2)], separately derived systems must have a local grounding electrode conductor connected between the source and a qualified grounding electrode.

The exception following this part of 250.30 exempts high-impedance grounded transformer secondaries or generator outputs from the need to provide direct (solid) bonding and grounding electrode connections of the neutral, as normally required in parts **(A)(1)**, **(A)(3)**, and **(A)(4)**. This simply states an exception to each part that is necessary to operate a high-impedance grounded system. The repetition of this exception allows this procedure to be used when the transformer source is, for example, in an outdoor unit substation and it is desired to make all connections at that point.

250.32. Buildings or Structures Supplied by a Feeder(s) or Branch Circuit(s). In 250.24(A), bonding of a panel neutral block (or the neutral bus or terminal in a switchboard, switch, or circuit breaker) to the enclosure is required in service equipment. The informational note following part **(A)(5)** in that section calls attention to the fact that 250.32 covers grounding connections in those cases where a panelboard (or switchboard, switch, etc.) is used to supply circuits in a building and the panel is fed from another building. Where two or more buildings are supplied from a common service to a main building, and therefore by feeders or branch circuits or both, and not by a service, a grounding electrode at each other building shall be connected to the ac system equipment grounding conductor. There shall be no such connection to a grounded conductor under the normal rules. In other words, the wiring in the second building is now treated exactly the same as any wiring within the originating building that originates in a subpanel. The previous allowance for bonding equipment grounding and grounded circuit conductors at the building disconnect for the second building has been largely revoked. It now lives on, but only as an exception covered later. That is, there will be a system grounding electrode system that must satisfy the basic rules covered in parts **(B)** or **(C)** of this section, but the only connection will be to the local equipment grounding system at the building disconnect. (See Fig. 250-24.)

There is an exception to part **(A)** that provides that for a separate building supplied by only one branch circuit where the branch circuit has an equipment grounding conductor run with it, a grounding electrode is not required. A multiwire branch circuit qualifies a single circuit under the wording of the exception. An example would be a small residential garage with a single lighting outlet and a receptacle. As long as an equipment grounding conductor is run with the circuit conductors, then no grounding electrode system need be provided.

Note that if two or more two-wire or multiwire branch circuits supply the outbuilding, then the grounding electrode must be provided and connected. This may not be straightforward. A grounding electrode conductor cannot be smaller than 8 AWG, and then only if run in raceway; 6 AWG is required otherwise. Terminating a 6 AWG conductor in a small device box, or daisy-chaining it through in multiple device boxes for the several circuits involved all of which require disconnecting means in accordance with 225.31, may be a challenge. If a feeder supplies the second building at a small panel the task is, of course, a simple one.

It follows that the supply to any outbuilding, whether a large feeder or a single branch circuit, must be run with an equipment grounding conductor of any type recognized by 250.118 along with the circuit conductors. (See Fig. 250-24.) As shown at the bottom of that illustration, a grounding electrode connection to the grounded neutral conductor at the outbuilding is prohibited. If the separate building has an approved grounding electrode and/or interior metallic piping system, the equipment grounding conductor shall be bonded to the electrode and/or piping system and the neutral conductor is connected to the neutral bus without a bonding jumper between the neutral and ground busses. However, if the separate building does not have a grounding electrode—that is, does not have 10 ft (3.0 m) or more of underground metal water pipe, does not have grounded structural steel, and does not have any of the other electrodes recognized by 250.52(A)(1) through (4)—then at least one of the other recognized grounding electrodes given in 250-52(A)(5) through (7) [or (8) if desired and available] must be installed unless the supply is a single two-wire or multiwire branch circuit as just covered above. That would most likely be a rod, pipe, or plate electrode—such as a driven ground rod—and it must be bonded to the equipment ground terminal or equipment grounding bus in the enclosure of the panel, switchboard, circuit breaker, or switch in which the feeder terminates (Fig. 250-24).

For "installations made in compliance with previous editions of this Code that permitted such connection," a special exception does allow a system grounding connection to the local grounding electrode conductor and the equipment grounding conductors, just as if the building were supplied by a service. The qualifying wording for this exception is a significant improvement over prior wording (existing premises wiring systems) because it settles whether the wiring system had to have been in existence for the premises generally (which would include all wiring on the parcel), or whether the second building had to have been previously wired. For example, if building No. 1 was wired in 1965, there was an existing premises wiring system that predated the code change. The new wording clarifies that building No. 2 actually had to have been wired prior to the 2008 edition of the **NEC** (when this grounding method was very severely limited) in order to have the wiring scheme legitimized. This practice was (and still is) used in uncounted millions of locations because it was the default procedure for the first 100 years of **NEC** editions. It is now headed down the road to extinction. There are additional conditions on its use. There must be no parallel metallic return paths that would allow current that should flow over the grounded circuit conductor to instead return to the service through other paths. Examples include an equipment grounding conductor, including a wiring method to the second building that is itself an equipment grounding conductor, such as rigid metal conduit. In such a case, a system grounding connection in the second building would send normal circuit current through the conduit in parallel with the enclosed grounded circuit conductor (usually a neutral). Another example would be a metallic water piping system common to both buildings; since such systems must be bonded to the grounding systems in each building the water pipes would become parallel conductors for the same reason.

In addition, as covered in 230.95, there must be no GFPE installed in any parent location because any line-to-ground fault in the second building will return over the neutral and look like ordinary load to the GFPE sensor. Finally, the neutral must have sufficient ampacity to perform both as a neutral (220.61) and as an equipment grounding conductor (250.122). New as of the 2014 **NEC** is a second exception to correlate with the fact that under the conditions governing 250.30(A)(1) Exception No. 2 (outdoor transformer supply) a bonded neutral condition will exist. This exception is essentially trivial because no load beyond the feeder it supplies will be connected at this remote location. It is also questionable because it is neither a building nor a structure and seemingly out of the scope of 250.32.

Part **(2)** is new as of the 2011 edition and covers instances where the remote building or structure is supplied by a separately derived system. For example, a large facility might distribute power at 480 V and set pad-mount transformers outside detached process buildings. If there is overcurrent protection where these feeders originate, the wiring doesn't differ from conventional feeder (or branch circuit) wiring as covered in 250.32(B)(1). If, however, the conductors arrive with no overcurrent protection, as could be done under the outdoor transformer secondary tap rule in 240.21(C)(4), then the wiring reverts to that for separately derived systems and must comply with 250.30(A). In the usual case where a supply-side bonding jumper would be necessary, it must be connected to the building disconnect and to the local grounding electrode conductor. Refer to the discussion at 250.30(A)(2) for more information.

Figure 250-25 shows another condition in which a grounding electrode connection must be made at the other building, as specified in the basic rule of 250.32(C).

If the 3-phase, 3-wire, ungrounded feeder circuit to the outbuilding had been run with a separate equipment grounding conductor that effectively connected the metal enclosure of the disconnect in the outbuilding to the grounding electrode conductor in the SE equipment of the main building, a connection to a grounding electrode would still be required. There is no distinction regarding the presence of an equipment grounding conductor or ground path. Under all situations where an ungrounded system supplies another building, a grounding electrode connection would be required at the outbuilding, and then the equipment grounding conductor run to the outbuilding would have to be bonded to any grounding electrodes that were "existing" at that building—such as an underground metal water service pipe and/or a grounded metal frame of the building. All grounding electrodes that exist at the outbuilding must be bonded to the ground bus or terminal in the disconnect at the outbuilding, whether or not an equipment grounding conductor is run with the circuit conductors from the main building.

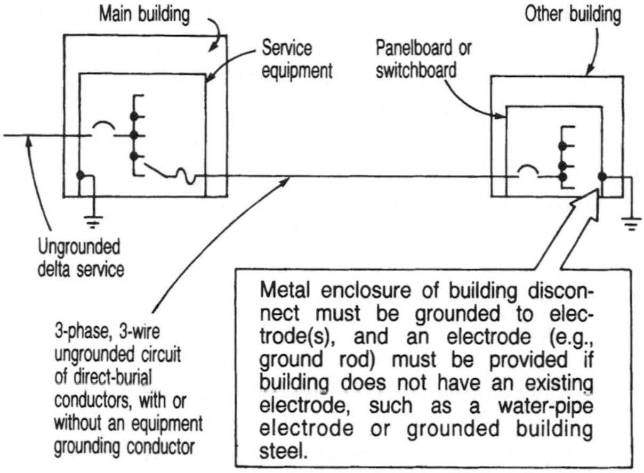
Main building
Service equipment
Panelboard or switchboard
Other building
Ungrounded delta service
3-phase, 3-wire ungrounded circuit of direct-burial conductors, with or without an equipment grounding conductor
Metal enclosure of building disconnect must be grounded to electrode(s), and an electrode (e.g., ground rod) must be provided if building does not have an existing electrode, such as a water-pipe electrode or grounded building steel.

Fig. 250-25. Grounding connection for an ungrounded supply to outbuilding. (Sec. 250.32.) Under current requirements, the equipment grounding conductor must be provided with the supply conductors. [Sec. 250.32(C)(1)].

Just as in the case of grounded systems, this topic now has two subheadings, one for a conventional supply and one for a supply through a separately derived system. For a conventional supply, or a separately derived supply with overcurrent protection established at the source, an equipment grounding conductor must be provided and be connected to the building disconnect and to the local grounding electrode. If the overcurrent protection is at the building, then the arrangement follows 250.30(B) and any separate supply side bonding jumper must make the same connections.

Part **(D)** covers design of the grounding arrangement for a feeder from one building to another building when the main disconnect for the feeder is at a remote location from the building being supplied—such as in the other building where the feeder originates. The rule prohibits grounding and bonding of a feeder to a building from another building if the disconnect for the building being fed is located in the building where the feeder originates. Part **(D)** correlates the grounding concepts of 250.32 with the disconnect requirements of 225.32, Exceptions Nos. 1 and 2. The rule also incorporates consideration of a standby generator as a source of supply where the generator is located remote from the building supplied, as covered in 700.12(B)(6), 701.12(B)(5), and 702.12. The exceptions in 225.32 address industrial situations where buildings may have no local disconnects, and instead rely on "documented safe switching procedures" and the behavior and knowledge of highly trained staff to accomplish the discontinuation of electric power in an emergency.

In all of these cases, there must still be a grounding electrode conductor, but special provisions must be made to address how the associated grounding electrode conductor will be connected to the local electrical system. There are three requirements. First, regrounding the neutral at the building supplied is prohibited. Second, the feeder must include an equipment grounding conductor, which must connect to the on-premises equipment grounding system and to an on-site grounding electrode unless only one branch circuit is supplied. Third, the equipment grounding and grounding electrode interconnection must occur in a junction box to be located immediately inside or outside the building supplied.

250.32(E) clarifies that the sizing rules for grounding electrode conductors located in buildings supplied by branch circuits or feeders or both follow the sizing rules for such conductors generally. They are based on the size of the ungrounded conductors that are the source of supply, with the usual Table 250.66 upper limit of 3/0 AWG as the maximum required size.

250.34. Portable and Vehicle-Mounted Generators. Part **(A),** which covers portable generators, rules that the frame of a portable generator does not have to be grounded if the generator supplies only equipment mounted on the generator and/or plug-connected equipment through receptacles mounted on the generator, provided that the noncurrent-carrying metal parts of equipment and the equipment grounding conductor terminals are bonded to the generator frame. (See Fig. 250-26.)

A clarification in part **(A)** points out that, where a portable generator is used with its frame *not* grounded, the frame is permitted to act as the grounding point for any cord-connected tools or appliances plugged into the generator's receptacles. This ensures that tools and appliances that are required by 250.114 to be grounded do satisfy the Code when plugged into a receptacle on the ungrounded frame of a portable generator.

Part **(B)** notes that the frame of a vehicle-mounted generator may be bonded to the vehicle frame, which then serves as a grounding point—it is not a grounding electrode, and since "ground" is defined as the earth, actually both the portable and vehicle-mounted generators covered here are operating ungrounded. This is only permitted when the generator supplies only equipment mounted on the vehicle and/or cord- and plug-connected equipment through receptacles on the vehicle or generator. When the frame of a vehicle is used as the grounding point for a generator mounted on the vehicle, grounding terminals of receptacles on the generator must be bonded to the generator frame, which must be bonded to the vehicle frame.

If either a portable or vehicle-mounted generator supplies a fixed wiring system external to the generator assembly, it must then be grounded as required for any separately derived system (as, for instance, a transformer secondary), as covered in 250.30.

The wording of part **(C)** brings application of portable and vehicle-mounted generators into compliance with the concept previously described in 250.30 on grounding and bonding of the generator neutral conductor. A generator neutral *must be* bonded to the generator frame when the generator is a truly separately derived source, such as the sole source of power to the loads it feeds. If the neutral is solidly connected to the building's utility service neutral, then such a supply would not be considered separately derived, and would not be subject to the bonding and grounding requirements given in 250.30. And if the generator neutral is *not* tied into the neutral conductor of the building's normal supply, such as where connected through a 4-pole transfer switch as part of a *normal/emergency* hookup—for feeding the load normally from the utility service and from the generator on an emergency or standby basis—then the generator would have to comply with the

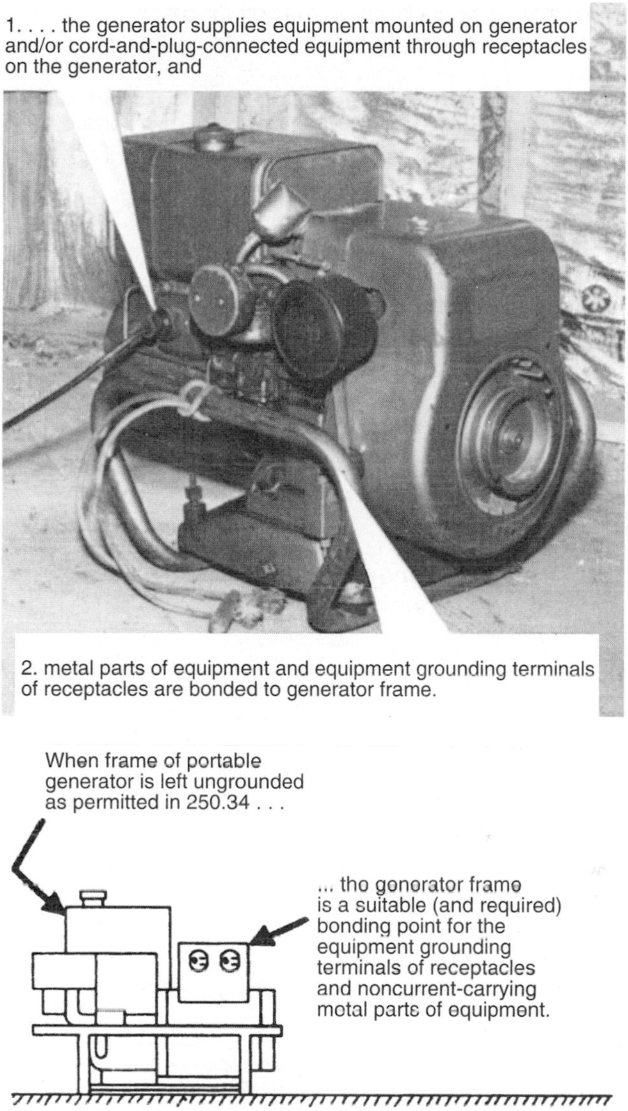
1. . . . the generator supplies equipment mounted on generator and/or cord-and-plug-connected equipment through receptacles on the generator, and
2. metal parts of equipment and equipment grounding terminals of receptacles are bonded to generator frame.
When frame of portable generator is left ungrounded as permitted in 250.34 . . .
... tho gonorator framo is a suitable (and required) bonding point for the equipment grounding terminals of receptacles and noncurrent-carrying motal parts of equipment.

Fig. 250-26. Grounding details for a portable generator.

rule of 250.30, which covers grounding of separately derived systems (Fig. 250-27). A note to this section refers to 250.30, and that rule is applicable to grounding and bonding of portable generators that supply a fixed wiring system on a premises. In such a case, bonding of the neutral to the generator frame is not required if there is a solid neutral connection from the utility service, through a transfer switch to the generator, as shown in the bottom sketch of Fig. 250-27.

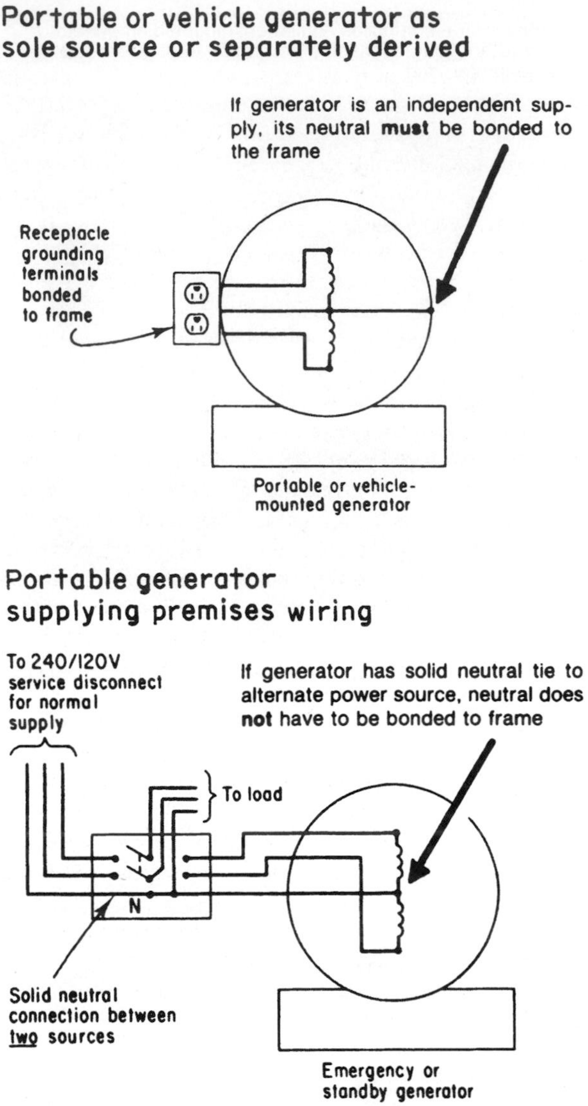
Portable or vehicle generator as sole source or separately derived
If generator is an independent supply, its neutral must be bonded to the frame
Receptacle grounding terminals bonded to frame
Portable or vehicle-mounted generator
Portable generator supplying premises wiring
To 240/120V service disconnect for normal supply
If generator has solid neutral tie to alternate power source, neutral does not have to be bonded to frame
To load
N
Solid neutral connection between two sources
Emergency or standby generator

Fig. 250-27. Generator neutral may be required to be grounded. (Sec. 250.34.)

Watch Out! The recently released UL standard on Portable Engine Generator Assemblies (UL 2201) requires all 15-kW or smaller portable generators to have bonded neutral connections. However, the bonded neutral also means that their output wiring must be wired as a separately derived system if it is used to power a premises wiring system. This, in turn, makes them only suitable for transfer switches that transfer the grounded circuit conductor along with the ungrounded conductors. This is incompatible with the overwhelming majority of transfer arrangements now in place and available in the market for residential and light commercial applications. This outcome is proving to be very controversial; many generators are not listed and do not have bonded neutrals. Many generator companies believe that operating a small generator with a floating neutral is actually safer, and a perfectly valid alternative to a bonded neutral. One company now has its small generators with a bonding connection that is easily reconfigured in the field, so they can be run either way.

However, although UL released the new generator standard in March of 2009, not a single portable generator has ever been listed to its provisions. The generator industry adamantly opposes its provisions, and as of the 2014 NEC 445.20 creates an opening for a generator with an on-board twist-lock receptacle that, when in use, disables any on-board 15- and 20-A receptacles. The concept is that powering a building and utilizing its main bonding jumper is harmless in such cases. Refer to the discussion at 445.20 for more information.

250.35. Permanently Installed Generators. This is a new section as of the 2008 NEC covering permanently installed generators. There must be an appropriately designed fault current path so wiring faults will be cleared properly. If the generator neutral is not connected to any other neutral source in the supplied building, in consequence of the generator qualifying as a separately derived system and having its neutral controlled in the transfer switch, then the grounded circuit conductor of the transfer switch simply complies with all the rules in 250.30. In other words, if the generator is the energy source for a separately derived system, then it is wired like any other separately derived system.

On the other hand, if the generator neutral is permanently connected to the premises neutral through a transfer switch with a solid neutral, there are two possibilities for sizing that neutral depending on where the overcurrent device for the generator output is located. If it is on the generator, then the fault-current path will be over an *equipment grounding conductor* sized in accordance with Table 250.122 [technically, 250.102(D) but that section immediately points to 250.122] based on the size of the OCPD. Example: A standby generator rated 50 kVA, 208Y/120-V has a 150-A circuit breaker mounted on the unit. An equipment grounding conductor must be run with the supply conductors, not smaller than 6 AWG.

If the OCPD is at the transfer switch, then the fault current path will be over a *supply-side bonding conductor* sized in accordance with 250.102(C), which means it will follow Table 250.102(C) with upward sizing, if necessary, in instances where the associated current-carrying load conductors exceed 1100 kcmil. Example: Same generator as before, output conductors sized per 445.13 at 115 percent of FLC. Therefore, 1.15(50,000 VA ÷ 360 V) = 160 A; 2/0 AWG conductors selected for the supply. From Table 250.102(C), the associated supply-side bonding jumper is 4 AWG copper.

250.36. High-Impedance Grounded Neutral Systems. (Adapted from Practical Electrical Wiring, 22nd ed., © Park Publishing, 2014, all rights reserved.) These systems combine the best features of the ungrounded systems in terms of reliability, and the best features of the grounded systems in terms of their ability to dissipate energy surges due to their grounding connection. They are permitted for 3-phase ac systems running from 480 to 1000 V, provided no line-to-neutral loads are connected, there is qualified maintenance and supervision, and ground detectors are installed.

These systems behave like ungrounded systems in that the first ground fault will not cause an overcurrent device to operate. Instead, detectors required by NEC 250.36(2) will alert qualified supervisory personnel. Remember, a capacitor is two conductive plates separated by a dielectric. A plant wiring system consists of miles and miles of wires, all of which are separated by their insulation. This means that a plant wiring system is a giant though very inefficient capacitor, and it will charge and discharge 120 times each second. The resistance is set such that the current under fault conditions is only slightly higher than the capacitive charging current of the system. Since a fault will often continue until an orderly shutdown can be arranged, the resistor must be continuously rated to handle this duty safely.

As shown in Fig. 250-28, the grounding impedance must be installed between the system neutral [250.36(A)] and the grounding electrode conductor. Where a system neutral is not available, the grounding impedance must be installed between the neutral derived from a grounding transformer [see 450.5(B)] and the grounding electrode conductor, as shown in Fig. 250-29. The neutral conductor between the neutral point and the grounding impedance must be fully insulated. Size it at 8 AWG minimum. This size is for mechanical concerns; the actual current is on the order of 10 A or less [250.36(B)].

Contrary to the normal procedure of terminating a neutral at a service disconnecting means enclosure, when the system is high-impedance grounded, the grounded conductor is prohibited from being connected to ground except through the grounding impedance [250.36(C)]. In addition, the neutral conductor connecting the transformer neutral point to the grounding impedance is not required to be installed with the phase conductors. It can be installed in a separate raceway to the grounding impedance [250.36(D)]. The normal procedure (usually performed by the utility) of adding a grounding electrode outside the building at the source of a grounded system (should one be used as the energy source for an impedance-grounded system) must *not* be observed [250.24(A)(2) Exception], because any grounding currents returning through the earth to the outdoor electrode will bypass and therefore desensitize the monitor.

An equipment bonding jumper [250.36(E)] must be installed unspliced from the first system disconnecting means or overcurrent device to the grounded side of the grounding impedance. The grounding electrode conductor can be attached at any point from the grounded side of the grounding impedance to the equipment grounding connection at the service equipment of the first system disconnecting means [250.36(F)]. Note that the size of the equipment bonding jumper depends on the end to which the grounding electrode conductor is connected, as shown in Fig. 250-29 [250.36(G)]. A connection at the impedance makes the bonding jumper a functional extension of the grounding electrode conductor, and it must be sized accordingly. A connection at the load end of the equipment grounding bus makes the bonding jumper a functional extension of the neutral, normally sized at 8 AWG [250.36(B)].

250.50. Grounding Electrode System. The rule in this section covers the grounding electrode arrangement required at the service entrance of a premise or in a

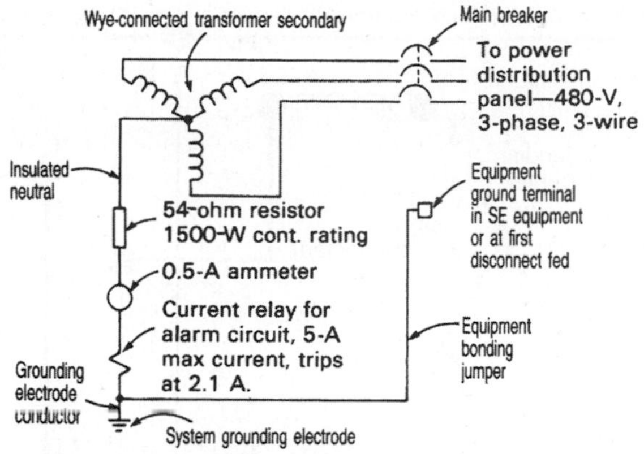
Wye-connected transformer secondary
Main breaker
To power distribution panel—480-V, 3-phase, 3-wire
Insulated neutral
54-ohm resistor 1500-W cont. rating
Equipment ground terminal in SE equipment or at first disconnect fed
0.5-A ammeter
Current relay for alarm circuit, 5-A max current, trips at 2.1 A.
Equipment bonding jumper
Grounding electrode conductor
System grounding electrode

1—Diagram of resistance-grounded system shows how neutral of main transformer(s) is grounded through a resistor, ammeter and current relay. Ground fault on any phase causes current to flow from fault through ground at transformer, through current relay, ammeter and resistor back to transformer neutral. Resistor limits fault current; current relay, which trips at any current level above 2.1 amps, initiates alarm.

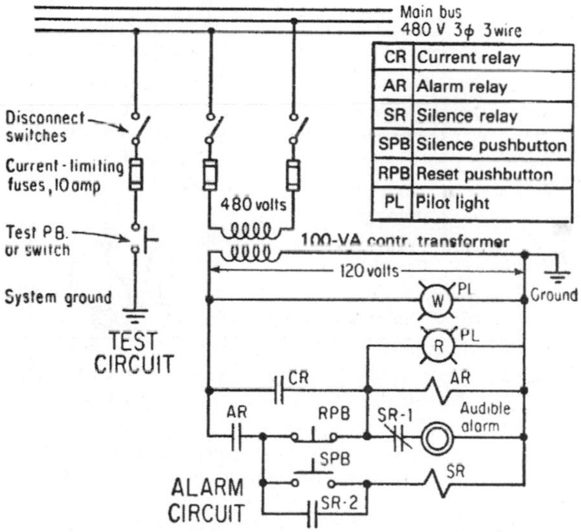
Main bus 480 V 3ϕ 3wire
CR Current relay
AR Alarm relay
SR Silence relay
SPB Silence pushbutton
RPB Reset pushbutton
PL Pilot light
Disconnect switches
Current-limiting fuses, 10 amp
480 volts
100-VA contr. transformer
120 volts
Test P.B. or switch
System ground
Ground
W
PL
R
TEST CIRCUIT
CR
AR
AR
RPB
SR-1
Audible alarm
SPB
SR
ALARM CIRCUIT
SR-2

2—Test circuit (on left of diagram) allows application of temporary ground to system to test alarm function. In alarm circuit, contacts CR close when current relay, which is connected in neutral-conductor ground-circuit at main transformer, senses a ground fault (see Fig. 1). Operation of contacts CR initiates audible and visual alarm.

Fig. 250-28. This is a typical application of resistance-grounded system operation on a wye-connected supply secondary. (Sec. 250.36.)

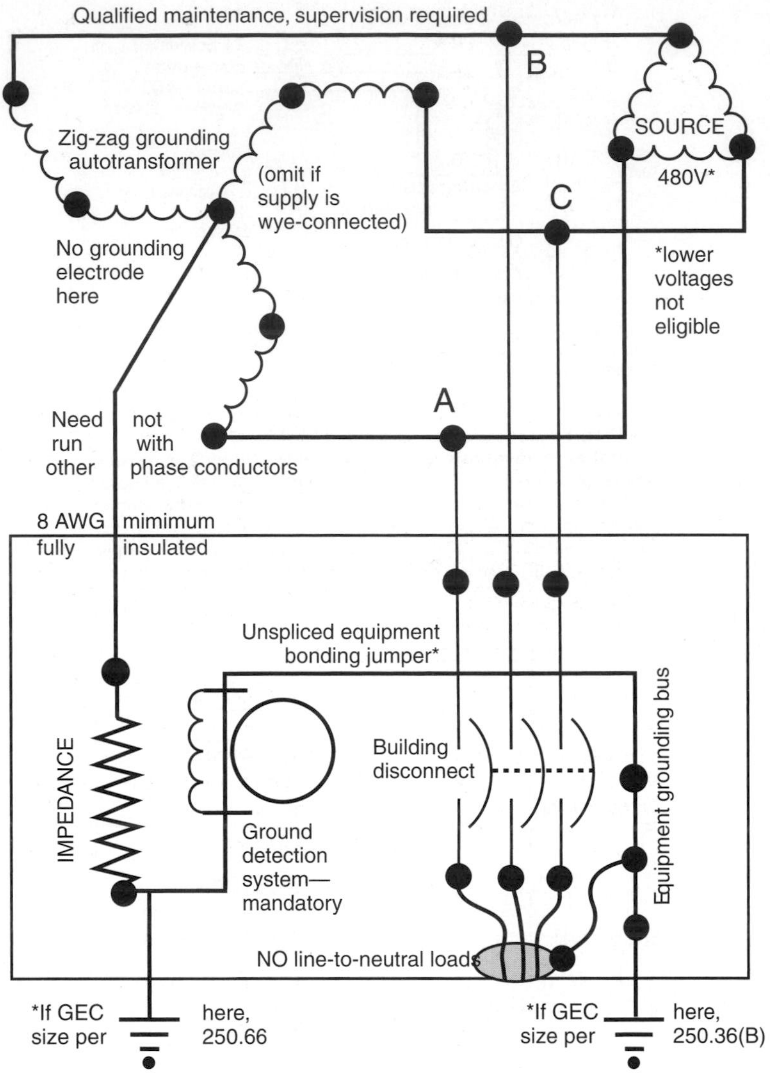
Qualified maintenance, supervision required
B
SOURCE
480V*
Zig-zag grounding autotransformer
(omit if supply is wye-connected)
C
No grounding electrode here
*lower voltages not eligible
A
Need not run with other phase conductors
8 AWG mimimum fully insulated
Unspliced equipment bonding jumper*
IMPEDANCE
Building disconnect
Ground detection system—mandatory
Equipment grounding bus
NO line-to-neutral loads
*If GEC here, size per 250.66
*If GEC here, size per 250.36(B)

Fig. 250-29. High-impedance grounded neutral systems can be retrofitted on systems supplied by ungrounded delta distributions. This drawing illustrates most of the rules in 250.36.

building or other structure fed from a service in another building or other structure, as covered in 250.32. This section mandates interconnection of the grounding electrodes specified in 250.52(A)(1) through (A)(7), which describe certain building components and other recognized electrodes that must be hooked together to form a "grounding electrode system." Figure 250-30 shows a number of potential elements in a grounding electrode system as envisioned by the **NEC**. Where the grounding electrodes described in 250.52(A)(1) through (A)(4) are not *present* at the

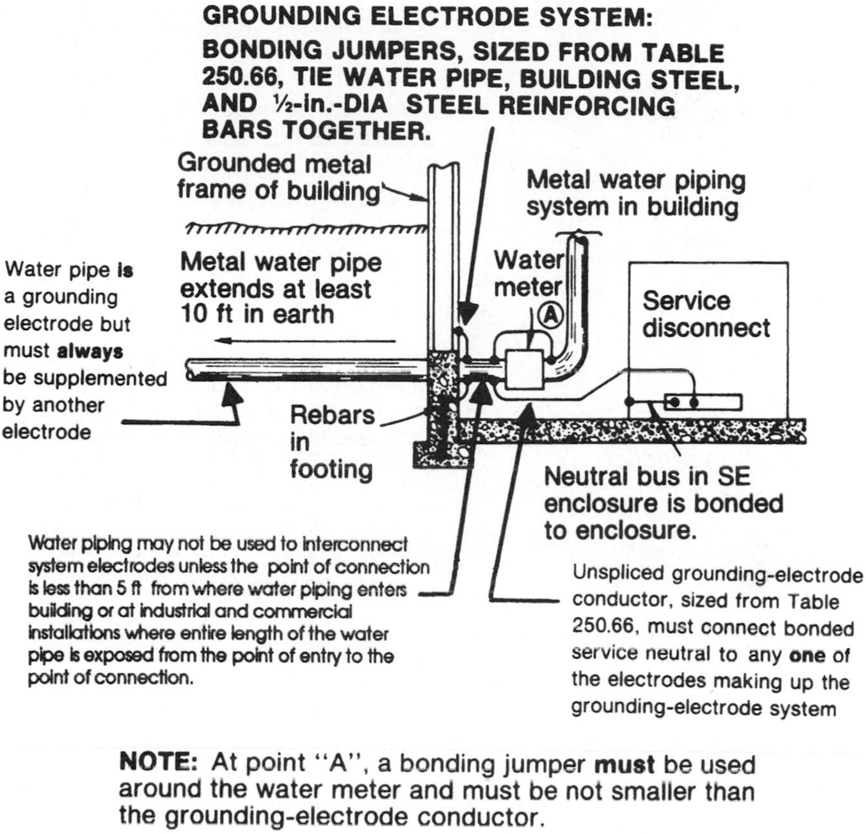
GROUNDING ELECTRODE SYSTEM:
BONDING JUMPERS, SIZED FROM TABLE 250.66, TIE WATER PIPE, BUILDING STEEL, AND ½-in.-DIA STEEL REINFORCING BARS TOGETHER.
Grounded metal frame of building
Metal water piping system in building
Water pipe is a grounding electrode but must always be supplemented by another electrode
Metal water pipe extends at least 10 ft in earth
Water meter
A
Service disconnect
Rebars in footing
Neutral bus in SE enclosure is bonded to enclosure.
Water piping may not be used to interconnect system electrodes unless the point of connection is less than 5 ft from where water piping enters building or at industrial and commercial installations where entire length of the water pipe is exposed from the point of entry to the point of connection.
Unspliced grounding-electrode conductor, sized from Table 250.66, must connect bonded service neutral to any one of the electrodes making up the grounding-electrode system
NOTE: At point "A", a bonding jumper must be used around the water meter and must be not smaller than the grounding-electrode conductor.

Fig. 250-30. Metal in-ground supporting structures and reinforcing bars must be used as an electrode if present. (Sec. 250.50.)

building or structure served—either by its own service or supplied from another building—then at least one of the grounding electrodes identified in parts **(A)(4)** through **(A)(8)** must be installed to form a grounding electrode system.

The use of the word "present," which in the 2005 edition replaced "available" in the 2002 and many previous NEC editions, was probably the most far-reaching change in the NEC achieved by changing a single word. The effect of the change was to bring qualifying concrete-encased electrodes (illustrated in Fig. 250-31) into grounding electrode system if they are present, not just if they are available. Therefore, the order of construction on building projects frequently had to change because as soon as the building steel in the footings was set and tied, an electrical connection had to be made and an electrical inspection performed. Another approach involves bringing a segment of reinforcing steel out of the pour that is tightly tied to the segment(s) making up the qualified electrode; however, most electrical inspectors will want to know that some disinterested and qualified third party witnessed the other end of such steel before the concrete truck arrived. And

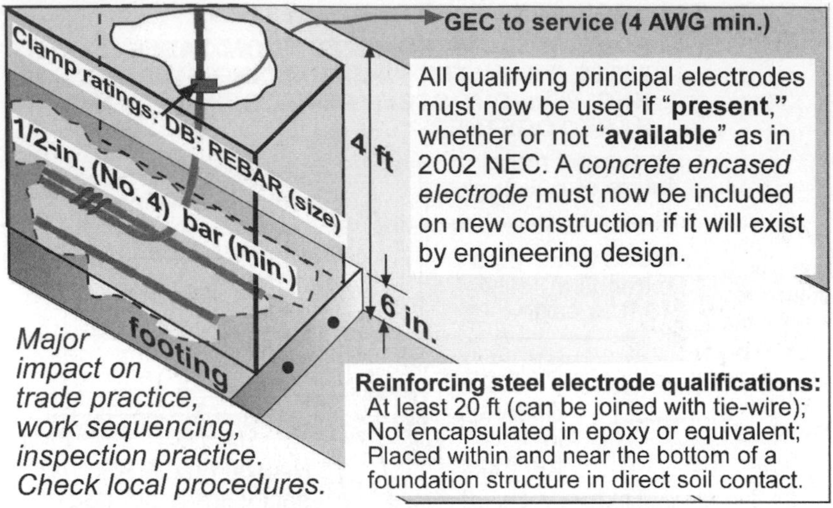
GEC to service (4 AWG min.)
All qualifying principal electrodes must now be used if "present," whether or not "available" as in 2002 NEC. A concrete encased electrode must now be included on new construction if it will exist by engineering design.
Clamp ratings: DB; REBAR (size)
1/2-in. (No. 4) bar (min.)
4 ft
6 in.
footing
Major impact on trade practice, work sequencing, inspection practice. Check local procedures.
Reinforcing steel electrode qualifications:
At least 20 ft (can be joined with tie-wire);
Not encapsulated in epoxy or equivalent;
Placed within and near the bottom of a foundation structure in direct soil contact.

Fig. 250-31. Concrete-encased electrode, connections to reinforcing steel. This rule can have a significant impact on trade sequencing. It is a superior electrode, and can be easily created in any foundation with 20 ft of bare copper arranged to be encased in the pour. The **NEC** also recognizes placements in other portions of the foundation as long as the concrete is in direct contact with the soil. (Sec. 250.50.) As of the 2011 **NEC**, the wording changed and now allows a horizontal run above the footing provided it is below the grade line. However, the deeper the better, because there is more permanent moisture, and a placement below frost counteracts the enormous increases in resistance that occur when the ground freezes.

if that inspection does not take place, the general contractor risks having to dismantle a foundation and start over.

The problem of connecting to these electrodes after the concrete has dried is an obvious one, which is why there is an exception to 250.50 that waives the rule for concrete-encased electrodes in existing buildings when the concrete would have to be disturbed in order to complete a connection.

The overlapping of this rule—mentioning part **(A)(4)** twice—is a little strange, but the wording of the rule here would recognize a ground ring, if present, as the grounding electrode system. And where a water pipe, building steel, or rebars in the footing or foundation—as covered in 250.52(A)(1) through (A)(3)—*and* a ground ring [250.52(A)(4)] are *not* available at the building or structure, then the Code would accept a ground ring, as described in 250.52(A)(4), that is installed specifically to serve as the grounding electrode system for the building or structure. The concept here is that (A)(1) through (A)(3) electrodes are extremely unlikely to be capable of being added in the field, but the others, including ground rings, are capable of field installation after the building is in place.

The electrodes identified in 250.52(A)(8) are never required to be connected, unless one desires to do so. That is, by excluding part **(A)(8)** from the first sentence, it is never mandatory to hook up such equipment wherever it exists. Rather, the way this rule is worded, such underground metal piping systems or metal structures *may be* used as a grounding electrode, but are *not required* to always be connected to the grounding electrode system.

It should be noted that the requirements for service or building grounding electrode systems given here now apply to the grounding of a separately derived system, such as a local step-down transformer, which is covered by part **(A)(4)** of 250.30 (Fig. 250-32). The simplification of electrode choices in that section means that any qualified building electrode can be used for such systems.

250.50 calls for a "grounding electrode system" instead of simply a "grounding electrode" as required at one time. Up to the 1978 **NEC**, the "water-pipe" electrode was the premier electrode for service grounding, and "other electrodes" or "made

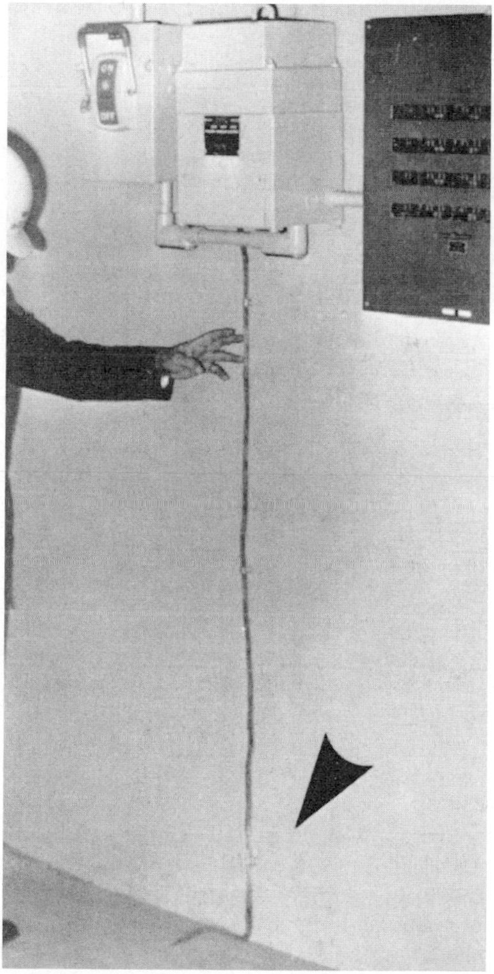

Fig. 250-32. Grounding electrode conductor from the bonded secondary neutral of this local transformer was connected to grounded building steel before concrete floor was poured. This installation is not covered by the rules of 250.50, but is covered by 250.30 and complies with those rules. (Sec. 250.50.)

electrodes" were acceptable *only* "where a water system (electrode) . . . is not available." If a metal water pipe to a building had at least 10 ft (3.0 m) of its length buried in the ground, that *had* to be used as the grounding electrode and no other electrode was required. The underground water pipe was the preferred electrode, the *best* electrode.

For many years now, and in the present NEC, of all the electrodes previously and still recognized by the NEC, the water pipe is the least acceptable electrode and is the only one that may never be used by itself as the *sole* electrode. It must always be supplemented by at least one "additional" grounding electrode (Fig. 250-33). Any one of the other grounding electrodes recognized by the NEC *is* acceptable as the *sole* grounding electrode, by itself.

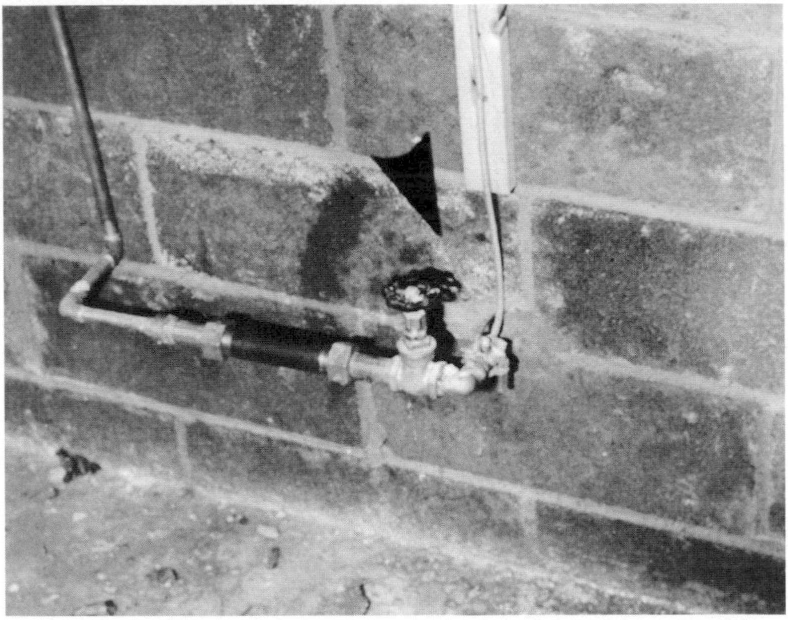

Fig. 250-33. Connection to an underground metal water-supply pipe is never adequate grounding for electric service equipment. (Sec. 250.53.)

Take a typical water supply of 12-in. (305-mm)-diameter metal pipe running, say, 400 ft (122 m) underground to a building with a 4000-A service. In 250.53 the Code requires that the water pipe, connected by a 3/0 copper conductor to the bonded service-equipment neutral may *not* serve as the only grounding electrode. It must be supplemented by one of the other electrodes from 250.52. So the installation can be made acceptable by, say, running a No. 6 copper grounding electrode conductor from the bonded service neutral to an 8-ft (2.44-m), ⅝-in. (15.87-mm)-diameter ground rod. Although that seems like using a mouse to help an elephant pull a load, it is the literal requirement of 250.53. And if the same building did not have 10 ft (3.0 m) of metal water pipe in the ground, the 8-ft (2.44-m) ground rod would be entirely acceptable as the *only* electrode, provided it has a "ground-resistance measurement" of 25 Ω or less, as required by 250.53(A)(2) Exception. And even if that resistance can't be met, then one more such electrode cures the code compliance problem regardless of resistance.

In fairness, the panel certainly recognizes the mouse/elephant issue. The problem has never been the suitability of a water pipe, which will always be a good electrode. The problem is that the **NEC** cannot predict when water supply companies will, often without warning, remove metal water pipes and substitute polyethylene or some other plastic in its place. This is an ongoing and very prevalent problem. Think of the ground rod therefore not as a "supplemental" electrode, but as a reserve electrode instead. This is why 250.53(D)(2) requires these supplemental electrodes to meet the same 25-Ω rule as where the electrodes are the sole electrodes present. A water pipe does not actually need supplementation in terms of its electrode function, as long as it is in the ground. But if it is removed, something has to remain in its place.

250.52. Grounding Electrodes. This section identifies those building components and other equipment that are recognized as "grounding electrodes." The basic rule of 250.50 requires that all or any of the electrodes specified in 250.52(A)(1) through (A)(7), if they are present on the premises, must be bonded together to form a "grounding electrode system." It should be understood and remembered that the grounding electrodes described in parts **(A)(1)** through **(A)(4)** are not required to be provided. But, if such building components or a ground ring are present at the building or structure, then it is required that they be interconnected. The electrodes described in parts **(A)(5),** rod and pipe electrodes, or **(A)(6),** plate electrodes, or **(A)(7),** other listed electrodes such as chemically enhanced designs, would also be required to be interconnected, if they already existed, and at least one would have to be installed if none of the grounding electrodes given in **(A)(1)** through **(A)(4)** were available. Alternately, the grounding electrode—"Other Local Metal Underground Systems and Structures"—recognized by part **(A)(8)** would be acceptable as the sole grounding electrode, where available, and its presence would eliminate the need to drive a rod, pipe, or plate electrode, or install a ground ring.

Note: The other underground metal systems and structures are not mandated to always be interconnected; rather, such systems and structures *may be* used as, or interconnected with the "grounding electrode system."

If present at the building or premises supplied, the following shall be interconnected:

(A)(1) If there is at least a 10-ft (3.0-m) length of underground metal water pipe, connection of a grounding electrode conductor must, generally, be made to the water pipe at a point less than 5 ft (1.52 m) from where the water piping enters the building. As a general principle, grounding electrodes are conductive objects in actual contact with earth; even if an unbroken length of water pipe extends out of the earth, only the portion in the earth is the electrode, and the portion extending out of the ground is a conductive object that may or may not be permitted to be utilized to extend the electrode for the purposes of making a connection.

(A)(2) The 2017 **NEC** no longer describes this as a metal building frame. It is now a "metal in-ground support structure." This represents an additional step in an ongoing effort to confine the term "grounding electrode" to conductive objects that can actually transfer electric current into the earth (see Fig. 250-34.) A new informational note drives that home by listing typical examples. The paragraph also contains the permission in the prior **NEC** to only bond one of multiple in-ground structures into the grounding electrode system in the event there are multiple instances. The former rules in this location about the hold-down bolts and their connection to concrete-encased electrodes have been moved to 250.68(C) as part of the larger editorial plan.

(A)(3) If there is at least a total of 20 ft (6.0 m) of one or more ½-in. (13-mm)-diameter (No. 4 or larger) steel reinforcing bars or rods embedded in the concrete footing or foundation low enough so the enclosing concrete is in direct contact with earth, or at least 6.0 m (20 ft) of 4 AWG copper wire likewise embedded in concrete, a bonding connection must be made to the bare wire or to one of the rebars—and obviously that has to be done before concrete is poured for the footing or foundation. The electrode length, in whole or in part, can also be measured vertically as long as the concrete surrounding the portion encased is in direct contact with earth. Second, if the reinforcing steel is discontinuous so that multiple qualified concrete-encased electrodes exist on any given building, it is sufficient (although not required or even advisable) to bond just one of them into the grounding electrode system. In fact, since the rule simply refers to multiple concrete-encased electrodes, if 6.0 m (20 ft) of 4 AWG bare copper, which independently qualifies, were added, the requirement would be met without making a connection to the steel. Refer to the beginning of coverage on 250.53 for a more extensive discussion of how these electrodes must be installed.

(A)(4) If present, a "ground ring" consisting of a buried, bare copper conductor, 2 AWG or larger, that is at least 6.0 m (20 ft) long and in direct contact with the earth is supplied, a bonding connection to it must be made. No minimum depth is given because the installation of ground rings is covered by part **(E)** of 250.53, which calls for a minimum cover of 750 mm (2½ ft).

If a building has all or some of the electrodes described, the preceding applications are mandatory to the extent they are present. If it has none, then any one of the electrodes described in 250.52(A)(4) through (A)(8) must be installed and/or used as the grounding electrode system for service grounding or outbuilding grounding.

An electrode by definition is an object in physical, conductive contact with earth. However, under some circumstances the **NEC** has allowed them to extend out of the earth for connection purposes. The 2011 **NEC** placed the rules for such electrode extensions together in one location, 250.68(C). Paragraph (1) at that location addresses water pipes, covered in (A)(1) above, and Paragraph (2) at that location addresses metal building frames, described in (A)(2) above. The 2014 **NEC** added concrete-encased electrodes, covered in (A)(3) above, as Paragraph (3). The principle is that when an electrode extends beyond its contact with earth, it has become a grounding electrode conductor of some form and the exposed portion is no longer an electrode. To correlate with this principle, 250.62 as of the 2014 **NEC** now effectively describes these extensions as covered in 250.68(C) under a heading

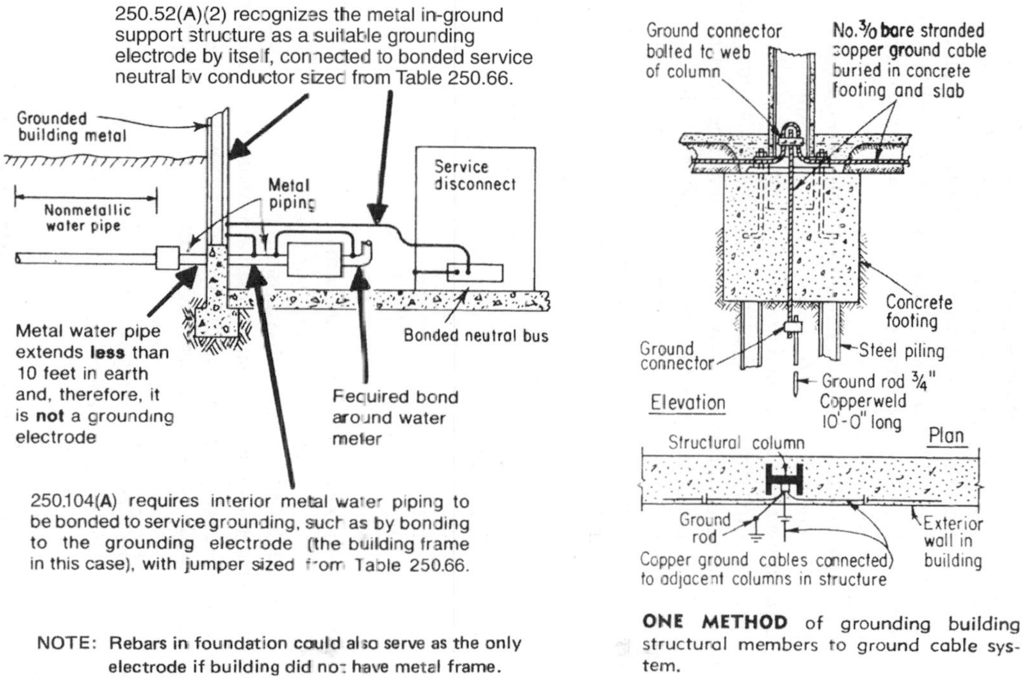
250.52(A)(2) recognizes the metal in-ground support structure as a suitable grounding electrode by itself, connected to bonded service neutral by conductor sized from Table 250.66.
Grounded building metal
Nonmetallic water pipe
Metal piping
Service disconnect
Metal water pipe extends less than 10 feet in earth and, therefore, it is not a grounding electrode
Bonded neutral bus
Required bond around water meter
250.104(A) requires interior metal water piping to be bonded to service grounding, such as by bonding to the grounding electrode (the building frame in this case), with jumper sized from Table 250.66.
NOTE: Rebars in foundation could also serve as the only electrode if building did not have metal frame.
Ground connector bolted to web of column
No. 3/0 bare stranded copper ground cable buried in concrete footing and slab
Concrete footing
Ground connector
Steel piling
Ground rod 3/4" Copperweld 10'-0" long
Elevation
Structural column
Plan
Ground rod
Exterior wall in building
Copper ground cables connected to adjacent columns in structure
ONE METHOD of grounding building structural members to ground cable system.

Fig. 250-34. Metal in-ground support structures may be the sole grounding electrode. (Sec. 250.52.)

"grounding electrode conductor material." Please refer to coverage at 250.68(C) for more information on this important topic.

(A)(5) describes rods and pipes that would be recognized as grounding electrodes. Whether a pipe or a rod, the minimum is 2.44 m (8 ft). Pipe not smaller than metric designator 21 (trade size ¾) is permitted, but generally requires corrosion protection where made of iron or steel. Rods must not be smaller than 15.87 mm (⅝ in.) where made of iron or steel. Stainless-steel rods that are less than 15.87 mm (⅝ in.) and nonferrous rods, such as brass, copper, or "their equivalent," must be listed for use as a grounding electrode and be not smaller than 12.70 mm (½ in.).

(A)(6) covers "other listed grounding electrodes" and includes specially designed and listed products such as those made of punched copper pipe prefilled with chemical additives to enhance effectiveness.

(A)(7) covers plate electrodes, which must have a surface area such that not less than 0.186 m^2 (2 ft^2) is exposed to the soil when buried. Note that as worded, a plate with soil exposure on two sides need only have a footprint of 0.093 m^2 (1 ft^2). The minimum thickness for steel or iron plate electrodes is 6.4 mm (¼ in.), but where nonferrous plate electrodes are used, the minimum thickness required is 1.5 mm (0.06 in.). Clearly, use of listed electrodes, exclusively, will go a long way toward ensuring a safe and acceptable installation.

These last three electrodes are listed in the UL *Electrical Construction Materials Directory* under the heading "Grounding and Bonding Equipment"—which also covers bonding devices, ground clamps, grounding and bonding bushings, ground rods, armored grounding wire, protector grounding wire, grounding wedges, ground clips for securing the ground wire to an outlet box, water-meter shunts, and similar equipment. Only listed devices are acceptable for use. And listed equipment is suitable only for use with copper, unless it is marked "AL" and "CU."

The last grounding electrode recognized by 250.52(A) is **(A)(8),** which covers "Other Metal Underground Systems and Structures." The basic thrust of the rule in 250.50 is that these underground piping systems or tanks, if metallic, may be used as the grounding electrode in lieu of the other electrodes described in parts **(A)(4)** through **(A)(6).** It is never *required* that a connection be made to such underground systems or structures; but, if desired, such a connection would constitute compliance with the rule in 250.50 calling for a grounding electrode system. Note that underground metal well casings have been specifically added to the list of examples of such electrodes. One reason for making these electrodes optional is that they may be at some distance from the system connection. As discussed in the informational note to 250.4(A)(1), grounding electrode conductors should be as short as practicable. This is due to the effects of distance on increasing the impedance of high-frequency transients.

In this context, it is necessary to revisit the language in 250.52(A)(1) (the water pipe electrode description) about bonding to a metal well casing. This is not a requirement to use metal well casings as electrodes generally. However, if the metal portion of a water pipe goes all the way out to the side of a well casing, and then continues down into the depths of the well, the casing has to be bonded to the pipe at the upper end. This arrangement is uncommon but not unknown, particularly for large or deep wells for which a steel riser is considered necessary.

The bonding requirement in this case is no different from, and serves the same function as, the familiar requirement to bond both ends of a steel conduit to an enclosed grounding electrode conductor. In this case, the metal water pipe is the conductor, the well casing is the ferrous enclosure, and the bond at the upper end in conjunction with both pipes being in contact with a common destination at the bottom addresses the impedance problem.

Part **(B)(1)** warns that a metal underground gas piping system must *never* be used as a grounding electrode. A metal underground gas piping system has been flatly disallowed as an acceptable grounding electrode because gas utility companies reject such practice and such use is in conflict with other industry standards.

As a general rule, if a water piping system or other approved electrode is not available, a driven rod or pipe is used as the grounding electrode system (Fig. 250-35). A rod or pipe driven into the ground does not always provide as low a ground resistance as is desirable, particularly where the soil becomes very dry, or where soil chemistry reduces conductivity. In the granitic soils of the Northeast, for example, the resistance on a conventional 8-ft copper-plated ground frequently approaches 1000 Ω in moist soil, significantly higher than test results demonstrated for comparable electrodes in the desert outside Las Vegas. A famous demonstration carried out by a ground resistance tester vendor in moist soil in central Massachusetts in the 1980s forever settled this point in the minds of the many inspectors from around New England who witnessed it. Ten fully driven ground rods, widely spaced so as to not have their cones of influence overlap, were connected in parallel. The resistance tester was connected to

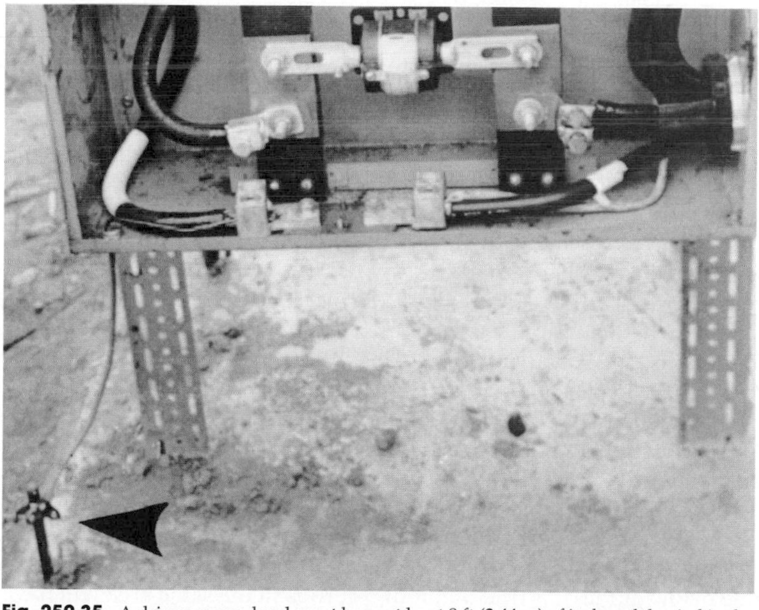

Fig. 250-35. A driven ground rod must have at least 8 ft (2.44 m) of its length buried in the ground, and if the end of the rod is above ground (arrow), both the rod and its grounding-electrode-conductor attachment must be protected against physical damage.

the field and the reference electrode for the tester was spaced out a great distance so as to get a reliable test result from so large a field. The overall result was 320 Ω! A ground rod is certainly better than no electrode, but not always by very much. Part **(B)(2)** of 250.52 prohibits use of an aluminum grounding electrode. The requirement to use concrete-encased electrodes wherever present in new construction is providing a welcome improvement in this area.

Part **(B)(3)**, new in the 2017 **NEC**, forbids the use of swimming pool bonding structures as grounding electrodes. The intent of Art. 680 is that these structures must retain their focus on equipotential bonding to mitigate voltage gradients in the pool area. Directly connecting them (through a main or system bonding jumper) to a system neutral is asking for the injection of voltage gradients they are intended to eliminate.

250.53. Grounding Electrode System Installation. When two or more grounding electrodes of the types described in 250.52 are to be combined into a "grounding electrode system," as required by 250.50, the rules of this section govern such installation and provide additional conditions, restrictions, and requirements.

Although not called out for special installation rules in this section, the greatly increased importance of concrete-encased electrodes has raised some areas of discussion regarding installation details. These electrodes are known as the "Ufer system," and they have particular merit in new construction where the bare copper conductor can be readily installed in a foundation or footing form before concrete is poured, even if no reinforcing steel is scheduled to be installed and thereby become a mandatory electrode. Installations of this type using a bare copper conductor have been installed as far back as 1940, and tests have proved this system to be highly effective.

These electrodes must be completely encased within the concrete, which means simply laying the electrode on the dirt at the bottom of a form does not comply. The electrode must be elevated at least 50 mm (2 in.) into the pour either by positioning on supports or by lifting after the pour and while the concrete is still wet. The latter approach is effective, but creates the logistical problem of needing the inspector present at that exact moment to witness the encasement. The footing itself must be in direct contact with the earth, which means that dry gravel or polyethylene sheets between the footing and the earth are not permitted (Fig. 250-36).

The **NEC** requires these electrodes to be "continuous" and also says that the encased electrode can be made of "multiple pieces connected together." The panel intends that where multiple segments are used, they still extend in the same sort of line the 20-ft long solid bar would make. And yes, this would mean for a vertical application the pier or other concrete structure would need to extend at least 20 ft down, or perhaps somewhat less if the electrode were oriented on a slant. However, the segment allowance is not intended to vary from the geometrical concept of a 20-ft long solid reinforcing bar. The reason for this is that Herbert Ufer's original testing used a horizontal, straight-line configuration. There apparently is no subsequent testing that has been done on other geometries, such as four 5-ft lengths bonded together and installed in four piers with at least 5 ft of concrete below grade. It is true that four 5-ft rods in the same area will have their areas of influence overlap and their paralleling efficiency diminished in comparison to a continuous run, hence the long-standing spacing limitation in 250.53(B)

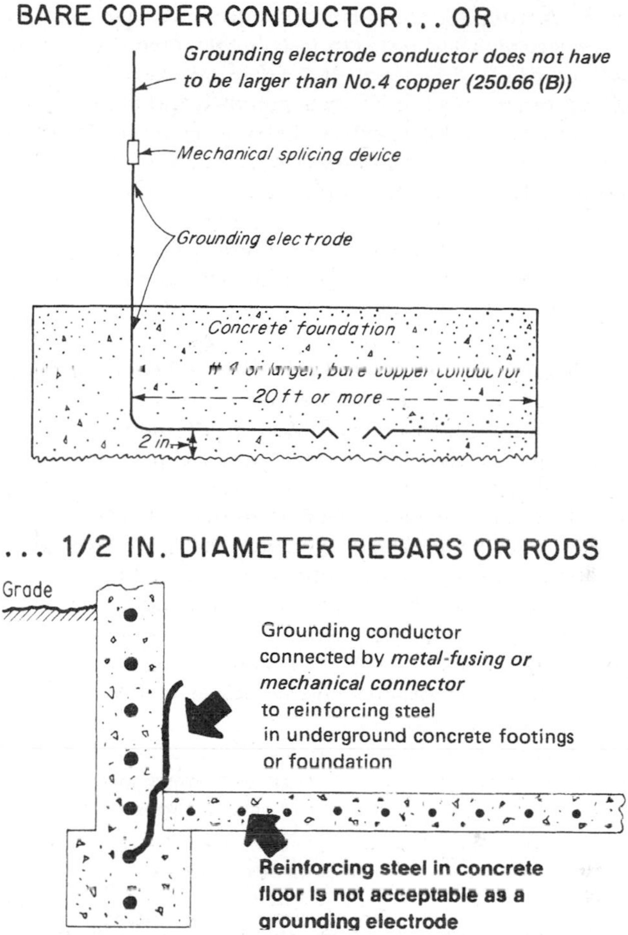
BARE COPPER CONDUCTOR... OR
Grounding electrode conductor does not have to be larger than No.4 copper (250.66 (B))
Mechanical splicing device
Grounding electrode
Concrete foundation
20 ft or more
2 in.
... 1/2 IN. DIAMETER REBARS OR RODS
Grade
Grounding conductor connected by metal-fusing or mechanical connector to reinforcing steel in underground concrete footings or foundation
Reinforcing steel in concrete floor is not acceptable as a grounding electrode

Fig. 250-36. The "Ufer" grounding electrode is concrete-encased, and the grounding electrode conductor does not have to be larger than 4 AWG copper in either case. [Sec. 250.52(A)(3).]

for paralleled ground rods. This subject was not revisited in either the 2014 NEC or the 2017 NEC.

It is generally advisable, depending on the additives that may be in the concrete, to provide additional corrosion protection in the form of plastic tubing or sheath at the point where the grounding electrode leaves the concrete foundation. Do not use ferrous raceways for this, however, or the result will be a magnetic choke unless bonding at both ends has been arranged to the enclosed conductor, which is seldom practical.

For concrete-encased steel reinforcing bar or rod systems used as a grounding electrode in underground footings or foundations, welded-type connections (metal-fusing methods) may be used for connections that are encased in concrete.

Compression or other types of mechanical connectors may also be used. *Conventional "acorn" ground rod clamps are not suitable for this purpose.* The connectors listed for use on rebar are made of a special alloy formulated to break through the oxide coating on the reinforcing steel, and they are marked with the size of bar for which they have been designed, along with a "DB" designation (direct burial) that is required for any grounding or bonding product that will be used below grade or embedded in concrete. These instructions, along with any torque specifications, must be followed exactly per 110.3(B).

In 250.53**(A)** the Code calls for the upper end of the rod to be buried below "permanent moisture level," where "practicable." That wording clearly shows that the Code intends the ground rod or plate to be completely buried, unless something prevents such installation. However, the rule of part **(G)** in this section does provide remedies for problem installations, and includes a requirement for "protection" of the ground rod and clamp where the rod is not "flush with or below ground." Suffice it to say that to the maximum extent possible, always ensure that rod, pipe, and plate be buried below the permanent moisture level. [See 250.53(G).]

The second part of 250.53(A) includes the specific concepts in former 250.56 on ground rod (or pipe or plate) resistances, which was deleted. Because of marginal resistance readings in so many cases for these electrodes, the default instruction is now to supplement any one of these electrodes with another, even another of the same type (and thereby resulting in the usual two ground rods in such cases). If a single electrode does qualify for the 25 Ω criterion, then it may continue to be used without supplementation, but now that is found in a new exception to the general rule. (See. Fig. 250-37.)

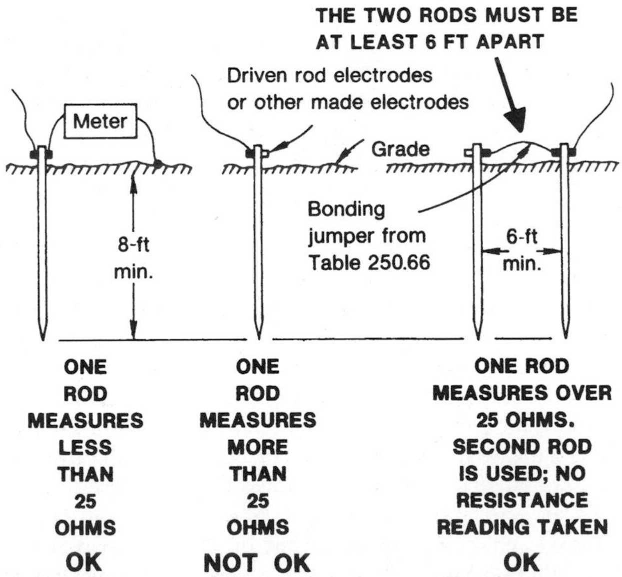
THE TWO RODS MUST BE AT LEAST 6 FT APART
Driven rod electrodes or other made electrodes
Meter
Grade
Bonding jumper from Table 250.66
8-ft min.
6-ft min.
ONE ROD MEASURES LESS THAN 25 OHMS
OK
ONE ROD MEASURES MORE THAN 25 OHMS
NOT OK
ONE ROD MEASURES OVER 25 OHMS. SECOND ROD IS USED; NO RESISTANCE READING TAKEN
OK

Fig. 250-37. Earth resistance of ground rod must be considered. [Sec. 250.53(A)(2) and exception.]

Insofar as rod, pipe, or plate electrodes are concerned, there is a wide variation of resistance to be expected, and the present requirements of the National Electrical Code concerning the use of such electrodes do not provide for a system that is in any way comparable to that which can be expected where a good underground metallic piping can be utilized.

It is recognized that some types of soil may create a high rate of corrosion and will result in a need for periodic replacement of grounding electrodes. It should also be noted that the intimate contact of two dissimilar metals, such as iron and copper, when subjected to wet conditions can result in electrolytic corrosion.

Under abnormal conditions, when a cross occurs between a high-tension conductor and one of the conductors of the low-tension secondaries, the electrode may be called upon to conduct a heavy current into the earth. The voltage drop in the ground connection, including the conductor leading to the electrode and the earth immediately around the electrode, will be equal to the current multiplied by the resistance. This results in a difference of potential between the grounded conductor of the wiring system and the ground. It is therefore important that the resistance be as low as practicable.

Where rod, pipe, or plate electrodes are used for grounding interior wiring systems, resistance tests should be conducted on a sufficient number of electrodes to determine the conditions prevailing in each locality. The tests should be repeated several times a year to determine whether the conditions have changed because of corrosion of the electrodes or drying out of the soil.

Figure 250-38 shows a ground tester being used for measuring the ground resistance of a driven electrode. Two auxiliary rod or pipe electrodes are driven to a depth of 1 or 2 ft (300 to 600 mm), the distances A and B in the figure being 50 ft (15 m) or more. Connections are made as shown between the tester and the electrodes; then the crank is turned to generate the necessary current, and the pointer on the instrument indicates the resistance to earth of the electrode being tested. In place of the two driven electrodes, a water piping system, if available, may be used as the reference ground, in which case terminals P and C are to be connected to the water pipe.

But, as previously noted, where two rods, pipes, or plate electrodes are used, it is not necessary to take a resistance reading, which is required in the case of fulfilling the requirement of 25 Ω to ground for one such electrode.

The 25-Ω rule has a very long history in the **NEC**, beginning with the 1918 edition. In what was then Class C (Inside Work), Rule 15A (Method of Grounding, when Protective Grounding is Required), subheading (t), the following text appeared: *"The combined resistances of the ground wires and connections of any grounded circuit, equipment, or lightning arrester should not exceed 3 ohms for water pipe connections nor 25 ohms for artificial grounds where these must be used. Where, because of dry or other high resistance soils it is impracticable to obtain artificial ground resistance as low as 25 ohms, two such grounds six feet apart if practicable must be installed, and no requirement will be made as to resistance."*

The water pipe specification was changed to a fine print note in the 1933 **NEC**, which stated that in general the resistance to ground of water pipes didn't exceed 3 Ω. The note was quite extensive, with commentary and recommendations on other electrodes. This arrangement continued until the 1975 **NEC** removed the note

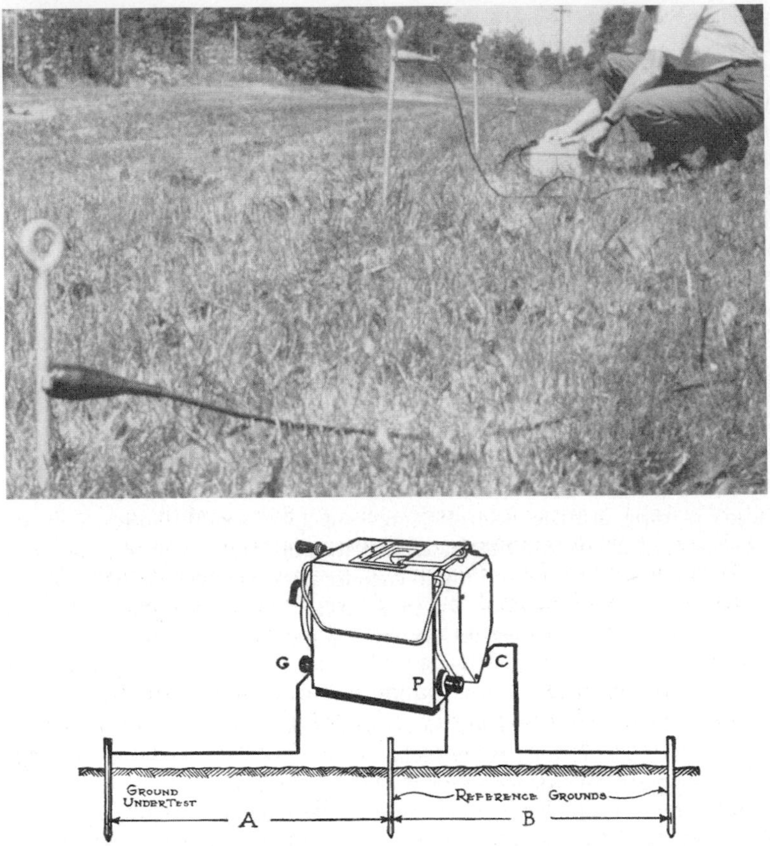
G
C
P
GROUND UNDER TEST
REFERENCE GROUNDS
A
B

Fig. 250-38. Ground-resistance testing must be done with the proper instrument and in strict accordance with the manufacturer's instructions. [Sec. 250.53(A)(2) and exception.]

entirely, as unnecessary. However, the 25-Ω parameter for rod, pipe, and plate electrodes still survives.

250.53**(B)** says that where it is necessary to bury more than one pipe or rod or plate in order to lower the resistance to ground, they should be placed at least 6 ft (1.8 m) apart. If they were placed closer together, there would be little improvement.

Where two driven or buried electrodes are used for grounding two different systems that should be kept entirely separate from one another, such as a grounding electrode of a wiring system for light and power and a grounding electrode for a lightning rod, care must be taken to guard against the conditions of low resistance between the two electrodes and high resistance from each electrode to ground. If two driven rods or pipes are located 6 ft (1.83 m) apart, the resistance between the two is sufficiently high and cannot be greatly increased by increasing the spacing. The rule of this section requires at least 6 ft (1.83 m) of spacing between electrodes serving different systems.

As covered by 250.53**(C)**, the size of the bonding jumper between pairs of electrodes must not be smaller than the size of grounding electrode conductor indicated in 250.66, both the table of sizes based on the largest associated ungrounded conductor(s) and the individual provisions that are based on particular electrodes. The installation must satisfy the indicated rules of 250.64, and must be connected as required by 250.64 and 250.70.

Part **(D)** of 250.53 presents additional criteria for the hookup of the grounding electrode system.

A very important sentence of 250.53(D)(1) says that "continuity of the grounding path or the bonding connection to interior piping shall not rely on water meters or filtering devices and similar equipment." The intent of that rule is that a bonding jumper always *must be used* around a water meter. This rule is included because of the chance of loss of grounding if the water meter is removed or replaced with a nonmetallic water meter. The bonding jumper around a meter must be sized in accordance with Table 250.66. Although the Code rule does not specify that the bonding jumper around a meter be sized from that table, the reference to "bonding jumper . . . sized in accordance with 250.66," as stated in 250.53(C), would logically apply to the water-meter bond. The reference to filters and the like is even more crucial, because they are commonly nonmetallic, and therefore continuity would be lost permanently and not just when the equipment is out for servicing.

It usually saves both material costs and labor expense to integrate compliance with 250.104(A)(1) at the same time as making any required grounding electrode connection to a water pipe electrode. The other rule requires an interior metal water piping system to be bonded to the service or to the grounding electrode conductor using the same size bonding jumper as is required for the grounding electrode conductor run to a water pipe. Since a water meter or filter raises continuity issues, a connection on the street side does not satisfy 250.104(A)(1). The very simple solution is to leave the grounding electrode conductor long enough to attach to both sides of a meter or filter as necessary. For example, the installation in Fig. 250-34 almost certainly will have a water meter installed subsequently, and the electrician would have saved money and time if he had connected to both sides with a slightly longer ground wire.

In the last sentence of 250.53(D)(2), an electrode (such as a driven ground rod) that supplements an underground water-pipe electrode must be "bonded" to any one of several points in the service arrangement. It may be "bonded" to (1) the grounding electrode conductor or (2) the grounded service conductor (grounded neutral), such as by connection to the neutral block or bus in the service panel or switchboard or in a CT cabinet, meter socket, or other enclosure on the supply side of the service disconnect or (3) grounded metal service raceway or (4) any grounded metal enclosure that is part of the service. (See Fig. 250-39.) It may also be bonded to an interior part of a metal water system in those occupancies where the interior part of the water pipe is allowable for grounding electrode connections. Item (5) is new as of 2011 and reminds users that if they are working with an electrode in a second building fed from another, 250.32(B)(1) Exception No. 1 will limit the permissibility of connecting a grounding electrode conductor to the grounded circuit conductor in those cases. This item is somewhat incongruous in a list of permitted supplemental electrodes.

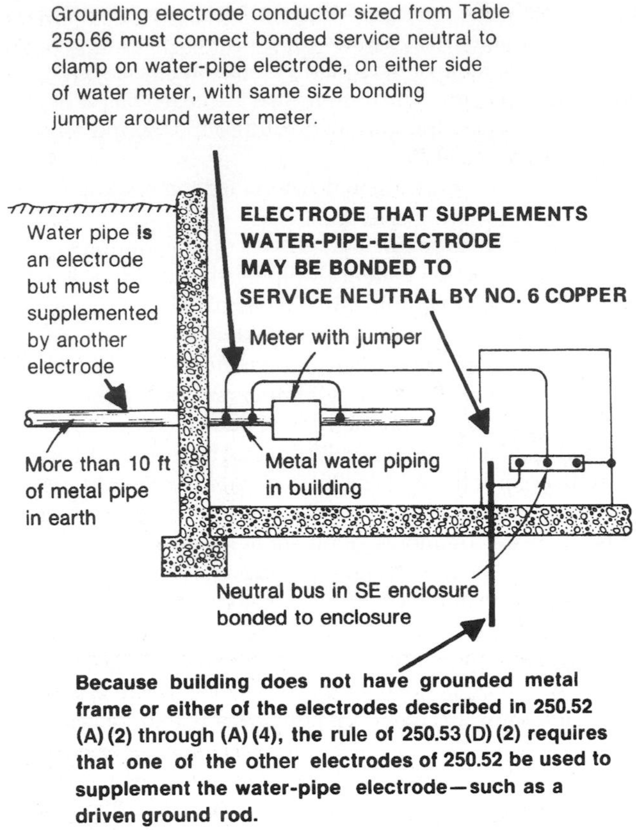
Grounding electrode conductor sized from Table 250.66 must connect bonded service neutral to clamp on water-pipe electrode, on either side of water meter, with same size bonding jumper around water meter.
ELECTRODE THAT SUPPLEMENTS WATER-PIPE-ELECTRODE MAY BE BONDED TO SERVICE NEUTRAL BY NO. 6 COPPER
Water pipe is an electrode but must be supplemented by another electrode
Meter with jumper
More than 10 ft of metal pipe in earth
Metal water piping in building
Neutral bus in SE enclosure bonded to enclosure
Because building does not have grounded metal frame or either of the electrodes described in 250.52 (A) (2) through (A) (4), the rule of 250.53 (D) (2) requires that one of the other electrodes of 250.52 be used to supplement the water-pipe electrode—such as a driven ground rod.

Fig. 250-39. Supplementing water-pipe electrode in building without metal frame. The 25 Ω requirement applies to rods used as supplemental electrodes to the same extent as where they are the only electrodes. As in those cases, a second rod or plate is always a cure for the resistance problem in terms of meeting the minimum requirement. (Sec. 250.53.)

The rule of 250.53(E) makes very clear that a ground rod, pipe, or plate electrode that is used to supplement a water-pipe electrode does not require any larger than a No. 6 copper (or No. 4 aluminum) conductor for a bonding jumper that is the only connection from the ground rod to the grounding-electrode conductor, to the bonded neutral block or bus in the service equipment, to any grounded service enclosure or raceway, or to interior metal water piping.

The basic rule of 250.53(G) calls for a ground rod to be driven straight down into the earth, with at least 8 ft (2.44 m) of its length in the ground (in contact with soil). *This means that if you can see the end of a 2.44-m (8-ft) rod above the ground surface, even a little bit, it cannot possibly have been driven far enough to meet the requirement.* If rock bottom is hit before the rod is 8 ft (2.44 m) into the earth, it is permissible to drive it into the ground at an angle—not over 45° from the vertical—to have at least 8 ft (2.44 m) of its length in the ground. However, if rock bottom is so shallow that it is not possible to get 8 ft (2.44 m) of the rod in the earth at a 45° angle, then it is necessary to dig a 2½-ft (750-mm)-deep trench and lay the rod horizontally in the trench. Figure 250-40 shows these techniques. Note that for any of these installations the ground rod clamp must be suitable for direct burial, and that means there will be a marking to that effect.

A second requirement calls for the upper end of the rod to be flush with or below

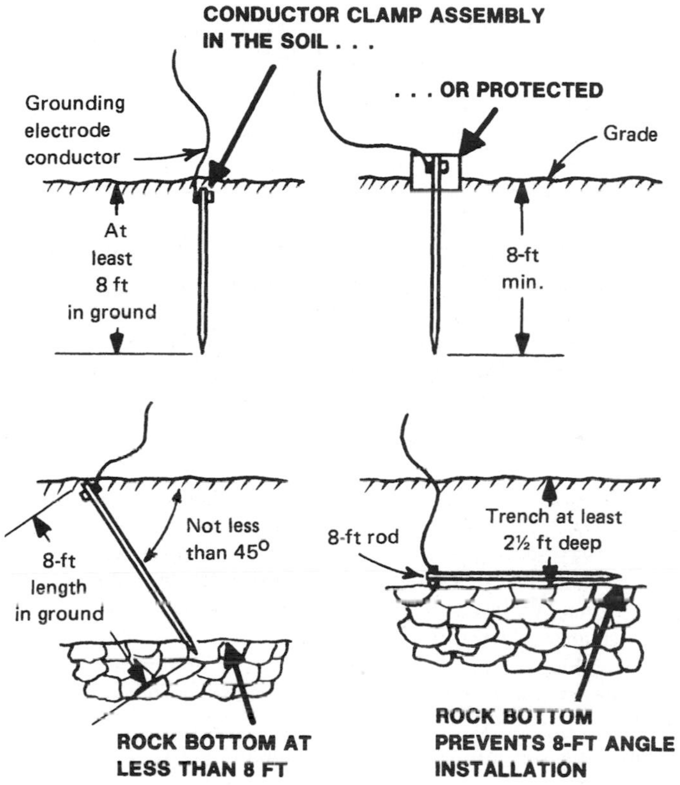
CONDUCTOR CLAMP ASSEMBLY
IN THE SOIL . . .
. . . OR PROTECTED
Grounding
electrode
conductor
Grade
At
least
8 ft
in ground
8-ft
min.
Not less
than 45°
8-ft
length
in ground
8-ft rod
Trench at least
2½ ft deep
ROCK BOTTOM AT
LESS THAN 8 FT
ROCK BOTTOM
PREVENTS 8-FT ANGLE
INSTALLATION

Fig. 250-40. In all cases, a ground rod must have at least 8 ft (2.44 m) of its length in contact with the soil.

ground level—*unless* the aboveground end and the conductor clamp are protected either by locating it in a place where damage is unlikely or by using some kind of metal, wood, or plastic box or enclosure over the end (Sec. 250.10). In the case of an 8-ft rod this is not an issue because as noted, if you can see the end, the installation does not meet the NEC. However, there are 3.0-m (10-ft) ground rods, and if they are not fully driven this provision may come into play.

This two-part rule was added to the Code because it had become a common practice to use an 8-ft (2.44-m) ground rod driven, say, 6½ ft (1.98 m) into the ground with the grounding electrode conductor clamped to the top of the rod and run over to the building. Not only is the connection subject to damage or disconnection by lawnmowers or vehicles, but also the length of unprotected, unsupported conductor from the rod to the building is a tripping hazard. The rule says—bury everything or protect it! Of course, the buried conductor-clamp assembly that is flush with or below grade must be resistant to rusting or corrosion that might affect its integrity, as required by 250.70.

250.54. Auxiliary Grounding Electrodes. This is an extremely important rule that has particular impact on the use of electrical equipment outdoors. The first part of the rule accepts the use of "auxiliary grounding electrodes"—such as a ground rod—to "augment" the equipment grounding conductor; BUT an equipment grounding conductor must always be used where needed and the connection of outdoor metal electrical enclosures to a ground rod is never a satisfactory alternative to the use of an equipment grounding conductor. The use of just ground-rod grounding would have the earth as "the sole equipment grounding conductor," and that is expressly prohibited by the last clause of this rule.

Such an earth return path has impedance that is too high, limiting the current to such a low value that the circuit protective device does not operate. In that case, a conductor that has faulted (made conductive contact) to a metal standard, pole, or conduit will put a dangerous voltage on the metal—exposing persons to shock or electrocution hazard as long as the fault exists. The basic concept of this problem—and Code violation—is revealed in Fig. 250-41.

This violation results from a fundamental confusion around the distinction between bonding requirements that create an effective ground-fault current path, and grounding requirements that create a local ground reference for reasons that have nothing to do with clearing faults. The 480-V panel in Fig. 250-42 is an extreme electrocution hazard. The sketch in Fig. 250-43 shows the correct procedure. Note that any acceptable equipment grounding conductor, including one of the metal raceways listed in 250.118, would produce a safe installation. The drawing in Fig. 250-44 highlights some practical issues on terminating branch circuits at lighting equipment on poles. The caption focuses on a very common problem of how to deal with a metal conduit sweep inserted because of its resistance to damage from heavy pulling forces, and that is stranded in a nonmetallic conduit run.

250.58. Common Grounding Electrode. The same electrode(s) that is used to ground the neutral or other grounded conductor of an ac system must also be used for grounding the entire system of interconnected raceways, boxes, and enclosures. The single, common grounding electrode conductor required by 250.24 connects to the single grounding electrode and thereby grounds the bonded point of the system and equipment grounds. See 250.50.

250.60. Use of Strike Termination Devices. This rule requires an individual "grounding electrode system" for grounding of the grounded circuit conductor (e.g., the neutral) and the equipment enclosures of electrical systems, and it prohibits use of the lightning ground electrode system for grounding the electrical system. Although this rule does *not* generally prohibit or require bonding between different grounding electrode systems (such as for lightning and for electric systems), it does note that the prohibition against using a lightning protection grounding system for power system grounding must not be read as prohibiting the required bonding of the two grounding systems, as covered in 250.106. And the note calls attention to the advantage of such bonding. There have been cases where fires and shocks have been caused by a potential difference between separate ground electrodes and the neutral of ac electrical circuits.

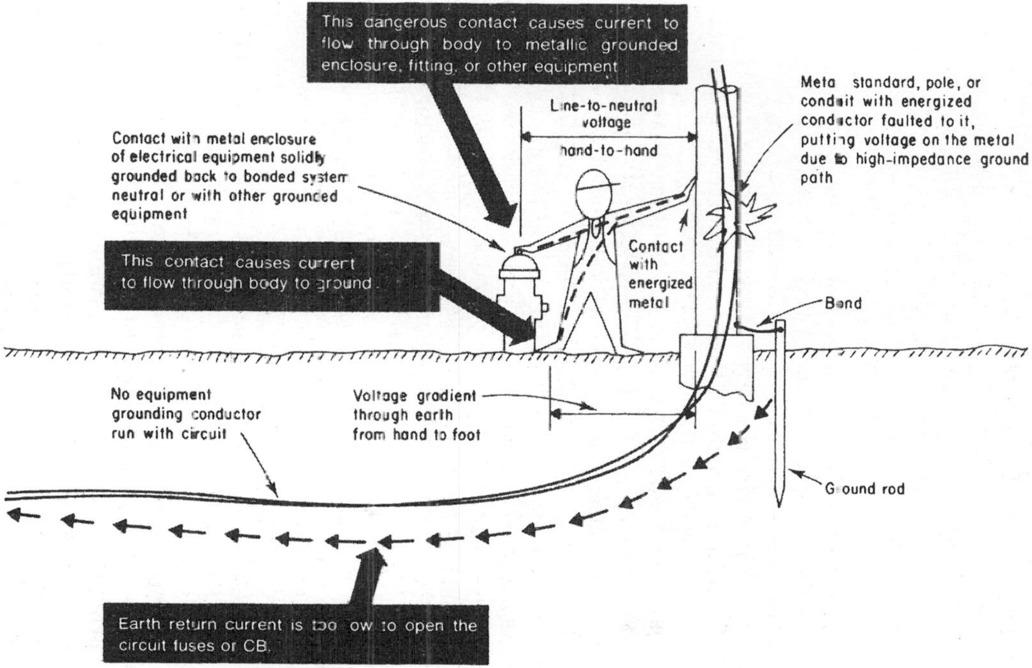
This dangerous contact causes current to flow through body to metallic grounded enclosure, fitting, or other equipment
Line-to-neutral voltage
hand-to-hand
Contact with metal enclosure of electrical equipment solidly grounded back to bonded system neutral or with other grounded equipment
This contact causes current to flow through body to ground
Contact with energized metal
Meta standard, pole, or conduit with energized conductor faulted to it, putting voltage on the metal due to high-impedance ground path
Bond
No equipment grounding conductor run with circuit
Voltage gradient through earth from hand to foot
Ground rod
Earth return current is too low to open the circuit fuses or CB.

Fig. 250-41. Ineffective grounding creates shock hazards. (Sec. 250.54.)

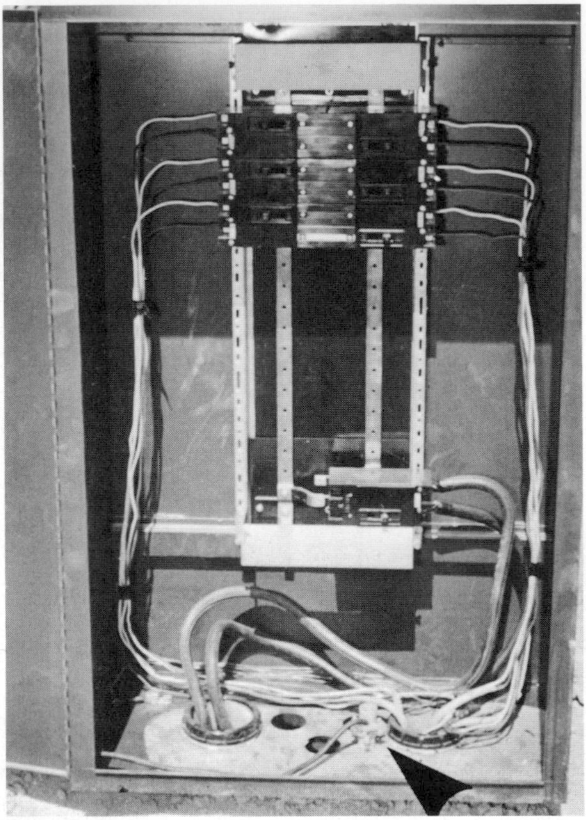

Fig. 250-42. Driven ground rod (arrow) has conductor run to it from a large lug at the left rear of the enclosure. All of the equipment grounding conductors from UF 480-V circuits to pole lights are connected to that lug. But the ground rod and earth path are the sole return paths for fault currents. The two larger conductors make up a 480-V underground USE circuit, without the neutral or an equipment grounding conductor brought to the panel, leaving "earth" as the sole return path. (Sec. 250.54.)

250.62. Grounding Electrode Conductor Material. Figure 250-45 shows the typical use of copper, aluminum, or copper-clad aluminum conductor to connect the bonded neutral and equipment ground terminal of service equipment to each of the one or more grounding electrodes used at a service. Controversy has been common on the permitted color of an insulated (or covered) grounding electrode conductor. 200.7(A) generally prohibits use of white or gray color for any conductor other than a "grounded conductor"—such as the grounded neutral or phase leg, as described in the definition of "grounded conductor." Grounding conductors must usually be green if insulated, but there is no reciprocal limitation on the use of the color for other than ungrounded conductors. This means that although equipment grounding conductors must be green or bare, there is no Code rule clearly prohibiting a green grounding electrode conductor. Refer to 250.119. The

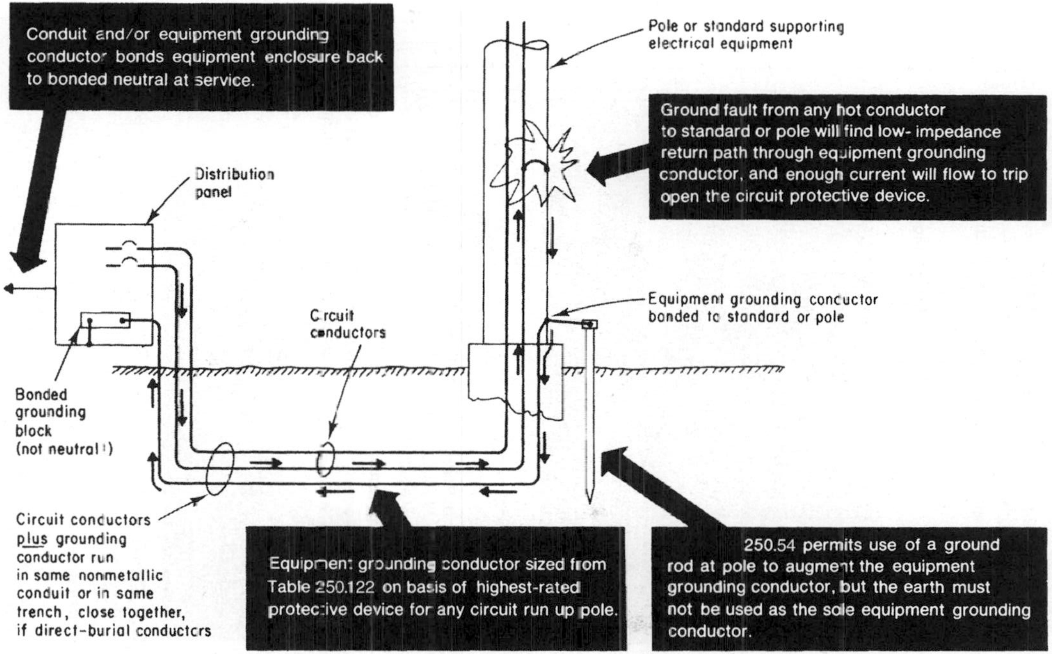
Conduit and/or equipment grounding conductor bonds equipment enclosure back to bonded neutral at service.
Pole or standard supporting electrical equipment
Ground fault from any hot conductor to standard or pole will find low-impedance return path through equipment grounding conductor, and enough current will flow to trip open the circuit protective device.
Distribution panel
Equipment grounding conductor bonded to standard or pole
Circuit conductors
Bonded grounding block (not neutral!)
Circuit conductors plus grounding conductor run in same nonmetallic conduit or in same trench, close together, if direct-burial conductors
Equipment grounding conductor sized from Table 250.122 on basis of highest-rated protective device for any circuit run up pole.
250.54 permits use of a ground rod at pole to augment the equipment grounding conductor, but the earth must not be used as the sole equipment grounding conductor.

Fig. 250-43. Equipment grounding conductor ensures effective fault clearing. (Sec. 250.54.)

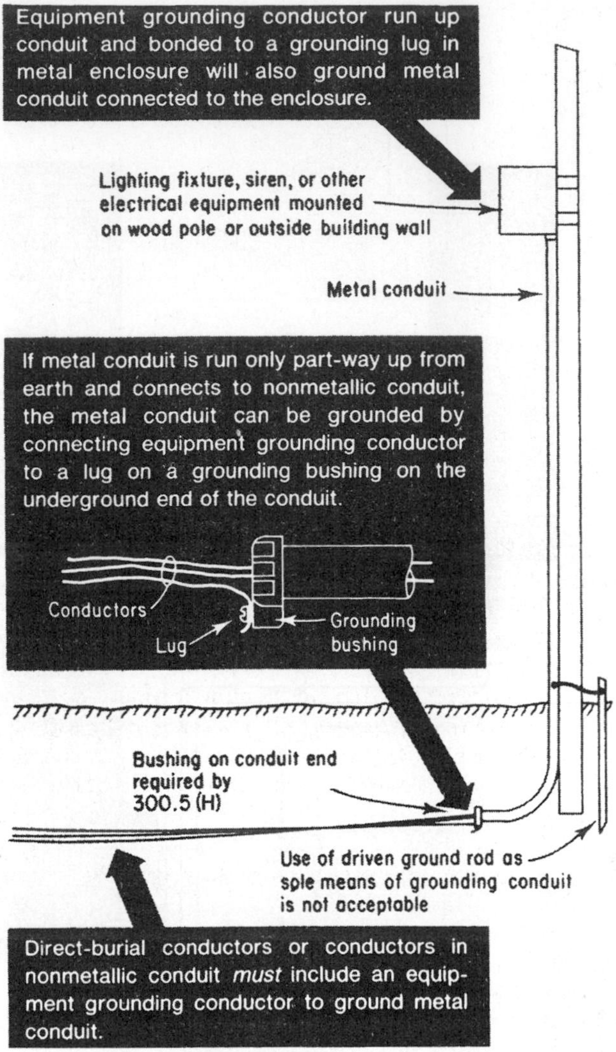
Equipment grounding conductor run up conduit and bonded to a grounding lug in metal enclosure will also ground metal conduit connected to the enclosure.
Lighting fixture, siren, or other electrical equipment mounted on wood pole or outside building wall
Metal conduit
If metal conduit is run only part-way up from earth and connects to nonmetallic conduit, the metal conduit can be grounded by connecting equipment grounding conductor to a lug on a grounding bushing on the underground end of the conduit.
Conductors
Lug
Grounding bushing
Bushing on conduit end required by 300.5 (H)
Use of driven ground rod as sole means of grounding conduit is not acceptable
Direct-burial conductors or conductors in nonmetallic conduit *must* include an equipment grounding conductor to ground metal conduit.

Fig. 250-44. Watch out for grounding details like these! (Sec. 250.54.) With respect to the center call-out on a transition to nonmetallic conduit, there are important issues to consider. The bonding bushing at the bottom of the sweep almost certainly does not have a direct burial listing and would require modifications and inspection approval to be used in this way. A better approach is to put the bushing on the aboveground end of the conduit which, in an all-conduit run would only work with a box, which may be objectionable. There are other options. One is to use the exception in 250.102**(E)** to route a bonding jumper from the sweep up the pole to a location where it can be connected to the equipment grounding conductor. There are "U-bolt" style ground clamps that are listed for direct burial. Another is to bury the sweep low enough so its upper end is still 450 mm (18 in.) below grade level, in which case bonding is not required. (250.86 Exception No. 3).

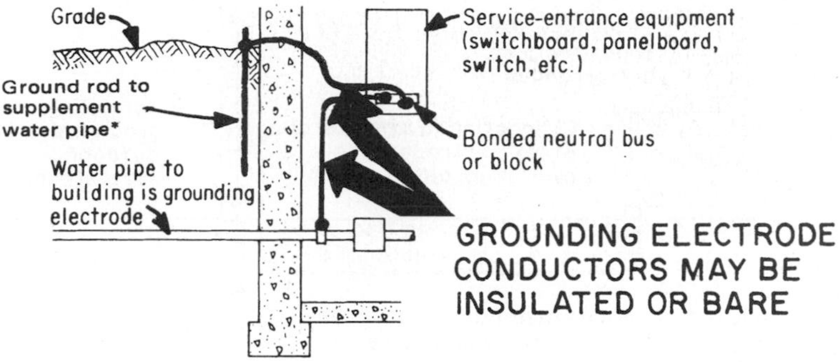
Grade
Ground rod to supplement water pipe*
Water pipe to building is grounding electrode
Service-entrance equipment (switchboard, panelboard, switch, etc.)
Bonded neutral bus or block
GROUNDING ELECTRODE CONDUCTORS MAY BE INSULATED OR BARE

Fig. 250-45. A grounding electrode conductor may be bare, covered, or insulated, and any color other than white or gray, which is reserved for grounded circuit conductors by 200.7(A). (Sec. 250.62.) The color green is permitted because 250.119 only excludes it for grounded or ungrounded circuit conductors, and a grounding electrode conductor is neither.

2014 NEC added the electrode extensions covered in 250.68(C) to this section as well; refer to that section for complete information.

250.64. Grounding Electrode Conductor Installation. This section covers all grounding electrode conductor installations, whether for services, or for buildings or other structures supplied with a feeder or branch circuit when the requirements in 250.32 require such conductors, or for separately derived systems when provisions in 250.30 produce the same result. Part **(A)** limits the use of aluminum conductors, but only in part. First, bare aluminum or copper-clad aluminum conductors cannot be used in direct contact with masonry or the earth or where subject to corrosive conditions. (See Fig. 250-46.) Aluminum is a chemically reactive metal that relies on its oxide coating to retain its integrity. There are compounds

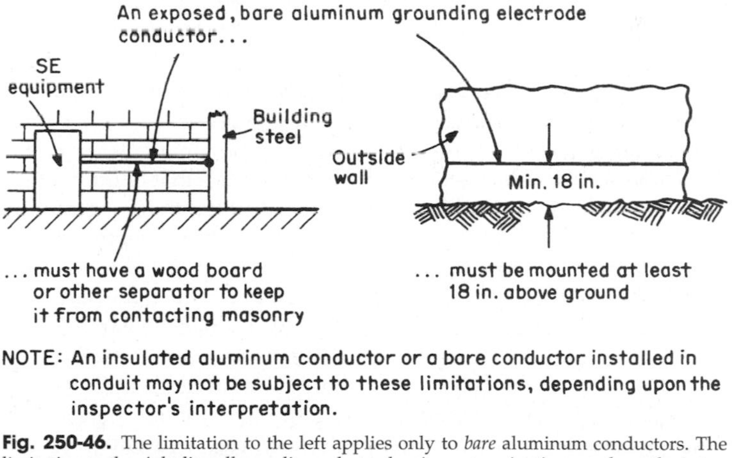
An exposed, bare aluminum grounding electrode conductor...
SE equipment
Building steel
Outside wall
Min. 18 in.
... must have a wood board or other separator to keep it from contacting masonry
... must be mounted at least 18 in. above ground
NOTE: An insulated aluminum conductor or a bare conductor installed in conduit may not be subject to these limitations, depending upon the inspector's interpretation.

Fig. 250-46. The limitation to the left applies only to *bare* aluminum conductors. The limitation to the right literally applies only to aluminum terminations, and not the intervening route. However, it cannot be run in contact with masonry or earth if it is bare. [Sec. 250.64(A).]

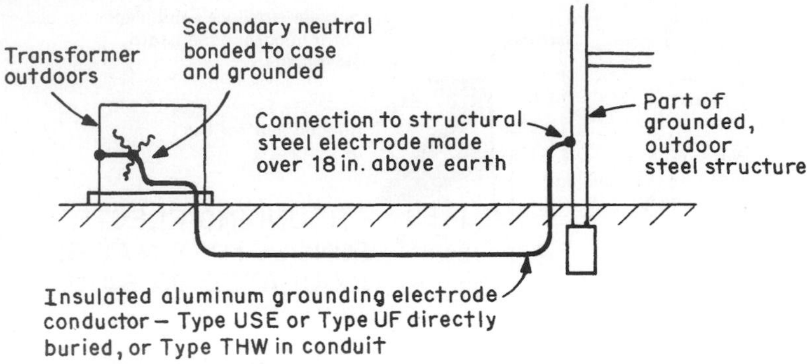
Secondary neutral
bonded to case
and grounded
Transformer
outdoors
Connection to structural
steel electrode made
over 18 in. above earth
Part of
grounded,
outdoor
steel structure
Insulated aluminum grounding electrode
conductor – Type USE or Type UF directly
buried, or Type THW in conduit

Fig. 250-47. This has been accepted but does violate literal Code wording in an outdoor pad-mounted transformer if the X_0 termination is less than 450 mm (18 in.) above grade level. [Sec. 250.64(A).]

in masonry and soils that will attack the oxide coating and the metal will corrode because of it. Insulated aluminum conductors are more forgiving, but where used outdoors, they must not be terminated within 450 mm (18 in.) above grade. (See Fig. 250-47.)

Part **(B)** covers the rules that limit exposure to physical damage, based on the size of the grounding electrode conductor. In all cases and sizes, the conductor must be securely fastened to any surface on which it runs. It may also pass through framing members. A 4 AWG or larger conductor can run without other limitation unless it is "exposed to physical damage." Note that this provision used to say "severe physical damage." The substantiation for the change (2005 **NEC**) was editorial, to the effect that any physical damage was unacceptable. However, whether it will be consistently applied that way in the field is uncertain. A 6 AWG conductor that is free of exposure to physical damage can additionally run "along the surface of the building construction" without additional protection; this is generally understood to include the sides of floor joists but not from joist to joist. An 8 AWG conductor, the smallest size permitted by 250.66, must run in a raceway or cable armor. (See Fig. 250-48.) The larger conductors, where threatened with damage because of local conditions, must be protected with a raceway or cable armor as well.

The 2014 **NEC** added a final sentence to this paragraph that definitively answers the question as to whether there is a mandated burial depth for a grounding electrode conductor run outdoors. The answer is no, because no depth is given, and a very clear statement is made that the burial depths specified for underground wiring in 300.5 do not apply to these conductors. They must, however, be buried at least to some extent or else otherwise protected if subject to damage.

Part **(C)** is the continuous length rule. This rule was reorganized for the 2008 **NEC**. Now the parts of the requirement chiefly involved with where the conductors originate, at service equipment and the like, remain here. Specifically, grounding electrode conductors are preferably run without joint or splice, but there are major exceptions to this. First, busbar segments must be bolted together in the field. Second, splices in a wire-type conductor must be made

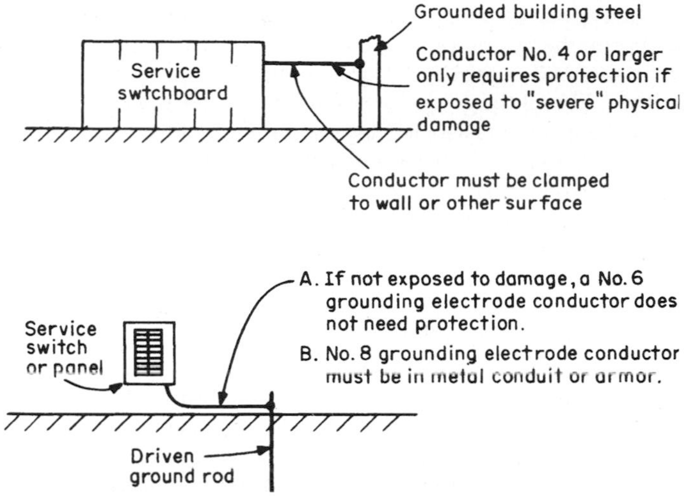
Grounded building steel
Service
swtchboard
Conductor No. 4 or larger
only requires protection if
exposed to "severe" physical
damage
Conductor must be clamped
to wall or other surface
A. If not exposed to damage, a No. 6
grounding electrode conductor does
not need protection.
B. No. 8 grounding electrode conductor
must be in metal conduit or armor.
Service
switch
or panel
Driven
ground rod

Fig. 250-48. Protection for grounding electrode conductor. (Sec. 250.64.)

with a high degree of permanence, defined as having been made using the thermite ("exothermic welding") process or a compression connector applied with a tool that makes it irreversible. Because grounding electrodes such as building steel and water pipes are now only those metallic elements in direct with earth, and because it is frequently necessary to create extensions of those electrodes, they are now to be described as some sort of grounding electrode conductor instead of a grounding electrode. Therefore, this section had to make allowances, because these extensions [see 250.68(C)] obviously cannot be installed without joints. Therefore, two additional allowances had to be made in this section for these construction materials, first for building steel elements, and second for water piping, and they follow the traditional ones for busbars and wires. Note that no mention is included for concrete-encased electrode extension in this context. This is because the only permitted form for these extensions is the very end of an otherwise embedded rod, and there will be no splice. Refer to 250.68(C)(3) for the details.

The other half of this requirement, covering where and how these conductors end up arriving at the electrodes is now covered in 250.64**(F)**. Specifically, the unspliced grounding electrode conductor can run to any convenient electrode (assuming more than one exists) in the grounding electrode system, provided these individual electrodes are bonded together per 250.53(C), and the size of any grounding electrode conductor employed is no smaller than the largest size conductor required among them by 250.66.

If, for instance, a grounding electrode system consists of a metal underground water-pipe electrode supplemented by a driven ground rod, the grounding electrode conductor to the water pipe would have to be sized from Table 250.66; and on, say, a 2000-A service, it would have to be a 3/0 AWG copper or 250-kcmil aluminum, connected to the water-pipe electrode, which would require that larger size of grounding electrode conductor. A bonding jumper from the bonded grounding terminal or bus in the SE equipment to the driven ground rod would not have to be larger than a 6 AWG copper or 4 AWG aluminum grounding electrode conductor, just as it would be if the ground rod is used by itself as a grounding electrode, provided it had a ground resistance, as established by testing, of 25 Ω, or less, to ground. A bonding jumper between the water-pipe electrode and the ground rod would also have to be that size. There is negligible benefit in running larger than a 6 AWG copper or 4 AWG aluminum to a rod, pipe, or plate electrode, because the rod itself is the limiting resistance to earth.

The other option is a ground bus spotted in an accessible location. The language makes it very clear that the unspliced grounding electrode conductor can terminate on the busbar, which must be made of copper or aluminum (only where over 450 mm [18 in.] above grade) and measure not less than 6 mm × 50 mm (¼ in. × 2 in.) in cross section. Then bonding jumpers to individual or groups of grounding electrode conductors can leave the busbar as is convenient for the specific installation. Further, the termination rules allow for both exothermic welding terminations and "listed connectors" on this busbar. That includes most mechanical lugs without the requirement of irreversible crimping tools.

Part **(D)** covers services with multiple enclosures, as covered in 230.71(A). This presents a very large number of possible applications, since 230.2 allows multiple services for a variety of good reasons, only some of which imply that the service enclosures will be remote from each other. For example, if a facility had a 480Y/277-V and a 208Y/120-V service, the two sets of service equipment could be (but need not be) next to each other. Other rules allow two-to-six disconnecting means per set of service entrance conductors, such as 230.40 Exception No. 4 that allows an owner's meter and service equipment in addition to the dwelling provisions all on a single set of service entrance conductors. In addition, absent from 230.71(A), is 230.40, Exception No. 2 where multiple disconnects next to each other are fed from multiple sets of service entrance conductors originating at one tap or lateral. In the 2008 **NEC**, (D)(1) opened with a reference to 230.40 Exception No, 2 but this was deleted in the 2011 **NEC** as an unnecessary reference due to the parent text of (D). The panel apparently failed to realize that 230.71(A) as referenced in (D) does not include 230.40 Exception No. 2 applications within its scope. This issue has been successfully addressed in the 2014 **NEC** through the removal of the reference to 230.71(A) in favor of a generic reference to plural enclosures fed by feeders or services.

Paragraph (D)(1) describes the tap method as shown in Fig. 250-49, lower left, with taps extending into each enclosure. The unspliced grounding electrode conductor, is sized by 250-66 based on the largest sum of the cross-sectional areas of each phase or line leg, calculated by each combined phase (for polyphase applications) or each combined line (for single-phase applications). Then taps run into each enclosure, sized by 250.66 based on the largest phase or line feeding each service enclosure. Figure 250-50 shows a detail of this process. The taps must be made without cutting the common grounding electrode conductor and the joints must be made either by exothermic welding or by splicing methods that are listed as grounding and bonding equipment under UL 467. There are mechanical split bolt connectors available that have UL 467 listings, so this need

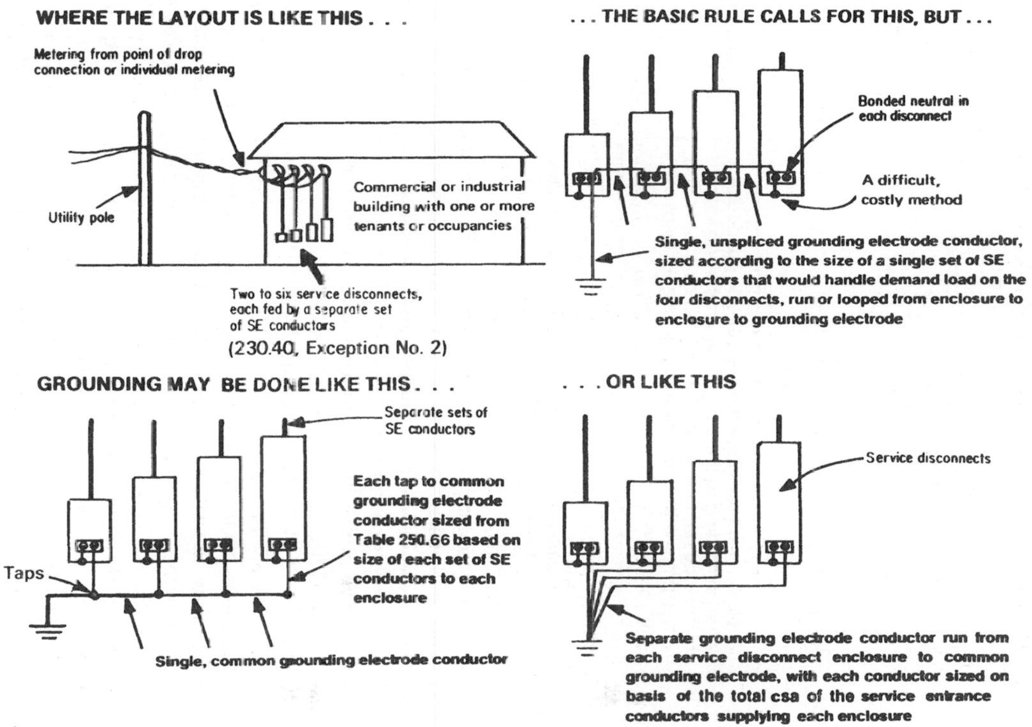
WHERE THE LAYOUT IS LIKE THIS . . .
Metering from point of drop connection or individual metering
Utility pole
Commercial or industrial building with one or more tenants or occupancies
Two to six service disconnects, each fed by a separate set of SE conductors
(230.40, Exception No. 2)
. . . THE BASIC RULE CALLS FOR THIS, BUT . . .
Bonded neutral in each disconnect
A difficult, costly method
Single, unspliced grounding electrode conductor, sized according to the size of a single set of SE conductors that would handle demand load on the four disconnects, run or looped from enclosure to enclosure to grounding electrode
GROUNDING MAY BE DONE LIKE THIS . . .
Separate sets of SE conductors
Each tap to common grounding electrode conductor sized from Table 250.66 based on size of each set of SE conductors to each enclosure
Taps
Single, common grounding electrode conductor
. . . OR LIKE THIS
Service disconnects
Separate grounding electrode conductor run from each service disconnect enclosure to common grounding electrode, with each conductor sized on basis of the total csa of the service entrance conductors supplying each enclosure

Fig. 250-49. Grounding electrode conductor may be tapped for multiple service disconnects. The procedure at the upper right will likely turn out to be impracticable in almost all cases due to termination limitations. (Sec. 250.64.) The text has generally been modified throughout 250.64(D) to provide the same coverage for installations for buildings and structures supplied by feeders as has always been the case for services, thereby filling an important gap.

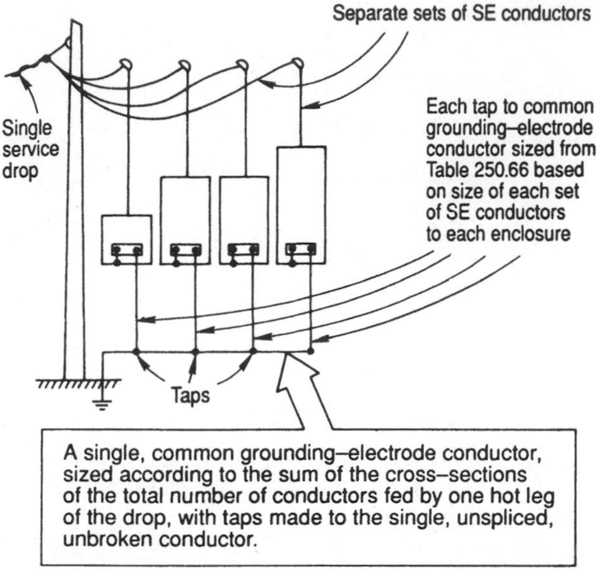
Separate sets of SE conductors
Single service drop
Each tap to common grounding–electrode conductor sized from Table 250.66 based on size of each set of SE conductors to each enclosure
Taps
A single, common grounding–electrode conductor, sized according to the sum of the cross–sections of the total number of conductors fed by one hot leg of the drop, with taps made to the single, unspliced, unbroken conductor.

Fig. 250-50. Rule covers sizing main and taps of grounding electrode conductor at multiple-disconnect services. A single common grounding electrode conductor must be "without splice or joint," with taps made to the grounding electrode conductor. (Sec. 250.64.)

not involve compression tooling. In addition, a grounding busbar of the same type and dimensions, and using the same connectors as described in the earlier discussion of 250.64(F), can be used for this purpose, effective with the 2011 NEC. The 2014 NEC added a requirement that this busbar be long enough to accommodate the necessary terminations; this is obvious, but as of now uncorrelated with its counterpart in 250.64(F).

Paragraph (D)(2) covers the case of running individual grounding electrode conductors, enclosure by enclosure, sized by 250.66 on the size of the supply conductors for that enclosure, as depicted in Fig. 250-49, the lower right drawing.

Paragraph (D)(3) covers the case where a wireway or auxiliary gutter is installed adjacent to and on the supply side of the service line-up, or where manufactured equipment is preconfigured in this way such as on a multimetering setup with a tap enclosure connected to the common buswork. A common grounding electrode conductor is connected to the grounded service conductor that is common to the adjacent service equipment, using a connector that is listed as grounding and bonding equipment under UL 467. The common grounding electrode conductor is sized per 250.66 based on the largest ungrounded phase or line conductor supplying the common location.

Part **(E)** covers the critical importance of maintaining the electrical continuity of all ferrous metal enclosures from the point a grounding electrode conductor begins at a system enclosure, or the point any bonding jumper is attached to a grounding electrode conductor, all the way to the point where the grounding electrode conductors and bonding jumpers terminate on an electrode. If this path is

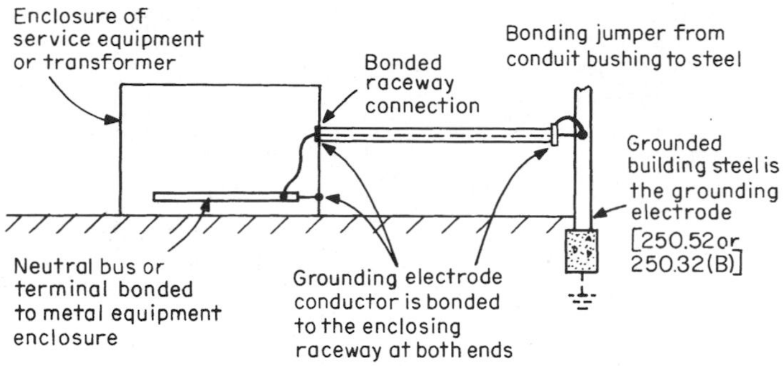
Enclosure of
service equipment
or transformer
Bonded
raceway
connection
Bonding jumper from
conduit bushing to steel
Grounded
building steel is
the grounding
electrode
[250.52 or
250.32(B)]
Neutral bus or
terminal bonded
to metal equipment
enclosure
Grounding electrode
conductor is bonded
to the enclosing
raceway at both ends

Fig. 250-51. Grounding electrode conductor must be electrically in parallel with enclosing ferrous raceway and other enclosures. (Sec. 250.64.)

interrupted or left incomplete for any reason, the reactance will seriously degrade the performance of the grounding electrode, especially under high-fault conditions. Figure 250-51 gives an overview of this procedure, and Fig. 250-52 provides some construction detail on the process.

On alternating current circuits, when a steel conduit is properly bonded to an enclosed grounding electrode conductor at both ends and a fault develops, the current will not flow where you might expect. Figure 250-53 shows the test setup to measure the results. Actual testing with 30 m (100 ft) of metric designator 21 (¾ trade size) rigid conduit enclosing 6 AWG copper wire showed that with 100 A of current entering the circuit, 97 A flows over the conduit and 3 A flows over the copper wire. Another test showed that with 2/0 AWG wire in the same length of metric designator 35 (trade size 1¼ in.) rigid conduit, 300 A of current pushed through resulted in 295 A over the conduit and 5A on the wire. If you break the continuity, the full current will flow through the copper wire, but at approximately double the impedance. Lightning and other electric discharges to earth through the grounding conductor will find a high-impedance path. Figure 250-54 shows the correct procedure from a transformer enclosure for a separately derived system, and Fig. 250-55 shows blatant, but distressingly common, violations of this rule.

The question has come up as to whether the required bonding jumper to a ferrous raceway might need to be larger than the enclosed grounding electrode conductor. This came up because bonding jumpers generally don't stop increasing in size at the 250.66 cut-off point; they keep increasing on the basis of one-eighth (12.5 percent) of the cross-sectional area of the largest phase or line conductor. At one time the literal text of the **NEC** did impose that requirement, even though it made no sense. Now 250.64(E) squarely ends that discussion by setting the size of the grounding electrode conductor as the size reference. Figure 250-56 shows two more examples of the wrong way to terminate ferrous metal enclosures at grounding electrodes.

PVC conduit may be used to protect grounding electrode conductors of any size used in accordance with this section. Use of nonmetallic raceways for enclosing grounding electrode conductors will reduce the impedance below that of the same

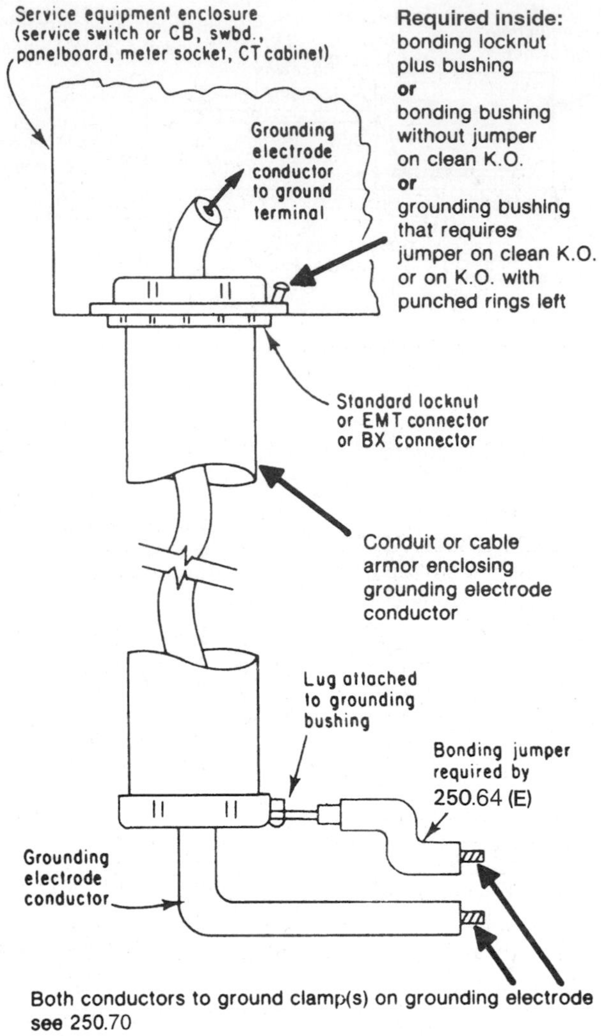
Service equipment enclosure
(service switch or CB, swbd.,
panelboard, meter socket, CT cabinet)
Required inside:
bonding locknut
plus bushing
or
bonding bushing
without jumper
on clean K.O.
or
grounding bushing
that requires
jumper on clean K.O.
or on K.O. with
punched rings left
Grounding
electrode
conductor
to ground
terminal
Standard locknut
or EMT connector
or BX connector
Conduit or cable
armor enclosing
grounding electrode
conductor
Lug attached
to grounding
bushing
Bonding jumper
required by
250.64 (E)
Grounding
electrode
conductor
Both conductors to ground clamp(s) on grounding electrode
see 250.70

Fig. 250-52. Grounding-conductor enclosure must be "bonded" at both ends of ferrous metal enclosures. [Sec. 250.64(E)]

conductor in a steel raceway. The grounding electrode conductor will perform its function whether enclosed or not, the principal function of the enclosure being to protect the conductor from physical damage. Rigid nonmetallic conduit will satisfy this function.

250.66. Size of Alternating-Current Grounding Electrode Conductor. For copper wire, a minimum size of No. 8 is specified in order to provide sufficient carrying capacity to ensure an effective ground and sufficient mechanical strength to be permanent. Where one of the service conductors is a grounded conductor, the

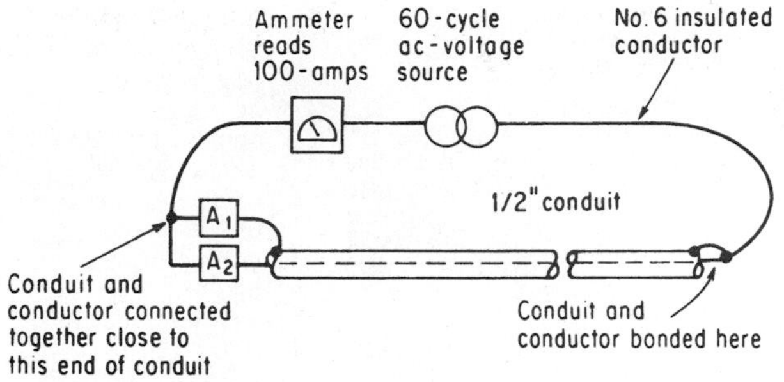
Ammeter
reads
100-amps
60-cycle
ac-voltage
source
No. 6 insulated
conductor
1/2" conduit
A1
A2
Conduit and
conductor connected
together close to
this end of conduit
Conduit and
conductor bonded here

Ammeter A_1 (indicates amount of current in conduit) = 97 amps

Ammeter A_2 (indicates amount of current in conductor) = 3 amps

Fig. 250-53. Enclosing conduit is more important than the enclosed grounding electrode conductor. (Sec. 250.64.)

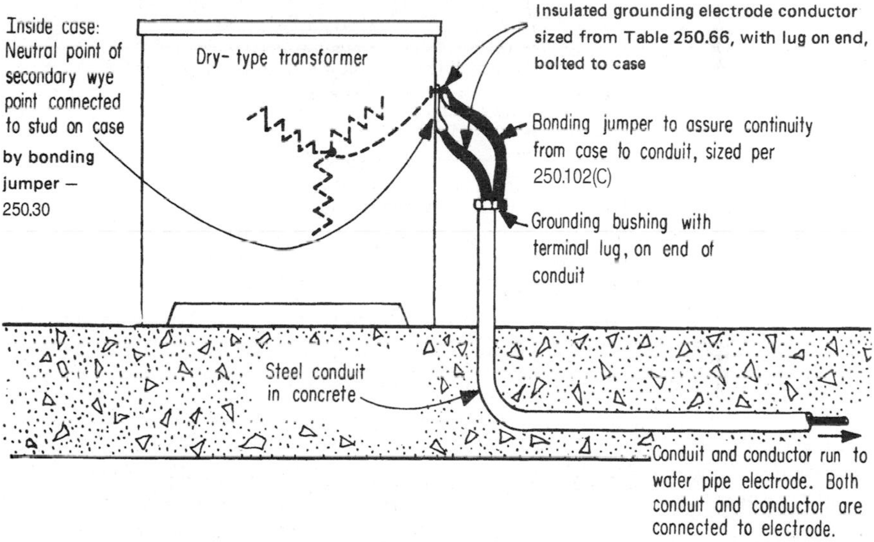
Inside case:
Neutral point of
secondary wye
point connected
to stud on case
by bonding
jumper –
250.30
Dry- type transformer
Insulated grounding electrode conductor
sized from Table 250.66, with lug on end,
bolted to case
Bonding jumper to assure continuity
from case to conduit, sized per
250.102(C)
Grounding bushing with
terminal lug, on end of
conduit
Steel conduit
in concrete
Conduit and conductor run to
water pipe electrode. Both
conduit and conductor are
connected to electrode.

Fig. 250-54. Protective metal conduit on grounding conductor must always be electrically in parallel with conductor. (Sec. 250.64.)

same grounding electrode conductor is used for grounding both the system and the equipment. Where the service is from an ungrounded 3-phase power system, a grounding electrode conductor of the size given in Table 250.66 is required at the service.

If the sizes of service-entrance conductors for an ac system are known, the minimum acceptable size of grounding electrode conductor can be determined from **NEC** Table 250.66. Where the service consists of only one conductor for each hot leg or phase, selection of the minimum permitted size of grounding electrode

Fig. 250-55. Grounding electrode conductors are run in conduit from their connections to an equipment grounding bus in an electrical room to the point where they connect to the grounding electrodes. Without a bonding jumper from each conduit to the ground bus, this is a clear VIOLATION of the second sentence rule of 250.64(C). (Sec. 250.64.)

conductor is a relatively simple, straightforward task. If the largest phase leg is, say, a 500-kcmil copper THW, Table 250.66 shows 1/0 AWG copper or No. 3/0 aluminum (reading across from "Over 350 kcmil thru 600 kcmil") as the minimum size of a grounding electrode conductor.

But, use of the table for services with multiple conductors per phase leg (e.g., four 500 kcmil for each of three phase legs of a service) is more involved.

The heading over the left-hand columns of this table is "Size of Largest Service-Entrance Conductor or Equivalent for Parallel Conductors." To make proper use of this table, the meaning of the word "equivalent" must be clearly understood. "Equivalent" means that parallel conductors per phase are to be converted to a single conductor per phase that has a cross-section area of its conductor material at least equal to the sum of the cross-section areas of the conductor materials of the two or more parallel conductors per phase. (The cross-sectional area of the insulation must be excluded.)

For instance, two parallel 500-kcmil copper RHH conductors in separate conduits would be equivalent to a single conductor with a cross-section area of 500 + 500, or 1000 kcmil. From Table 250.66, the minimum size of grounding electrode conductor required is shown to be 2/0 AWG copper or 4/0 AWG aluminum—opposite the left column entry, "Over 600 kcmil thru 1,100 kcmil." Note that use of this table is based solely on the size of the conductor material itself, regardless of the type of insulation. No reference is made at all to the kind of insulation.

Figure 250-57 shows a typical case where a grounding electrode conductor must be sized for a multiple-conductor service. A 208/120-V, 3-phase, 4-wire service is

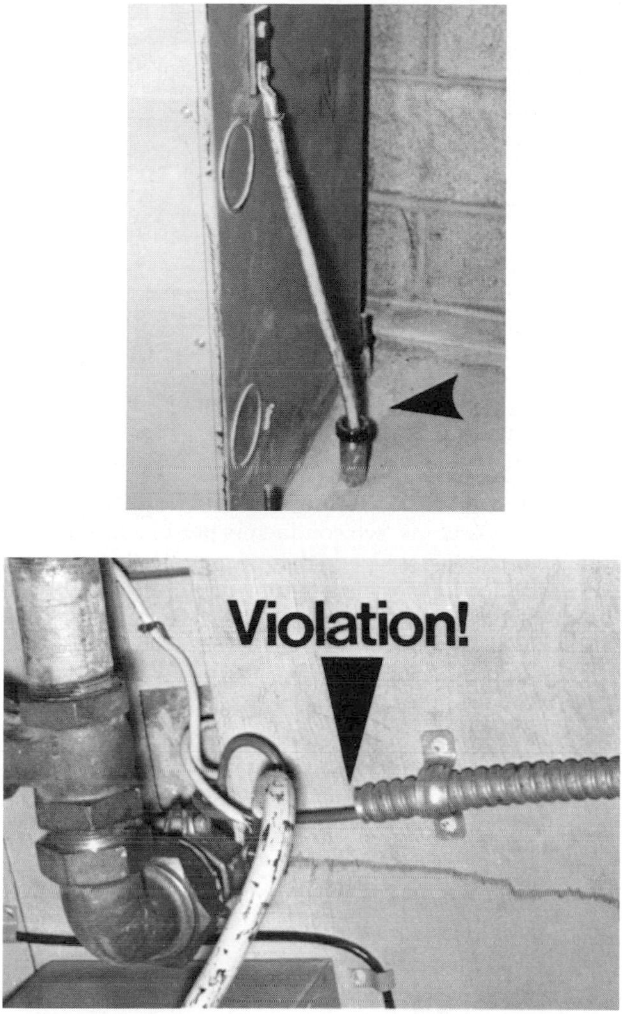
Violation!

Fig. 250-56. Two examples of very clear violations of the rule that requires enclosing metal raceways (rigid metal conduit at top and flex at right) to be bonded at both ends to a grounding electrode conductor within the raceway. [Secs. 250.92 and 250.64(C).]

made up of two sets of parallel copper conductors of the sizes shown in the sketch. The minimum size of grounding electrode conductor which may be used with these service-entrance conductors is determined by first adding together the physical size of the two 2/0 AWG conductors which make up each phase leg of the service:

1. From **NEC** Table 8 in Chap. 9 in the back of the Code book, which gives physical dimensions of the conductor material itself (excluding insulation cross-sectional area), each of the phase conductors has a cross-sectional area (csa) of 133,100 kcmil. Two such conductors per phase have a total csa of 266,200 kcmil.

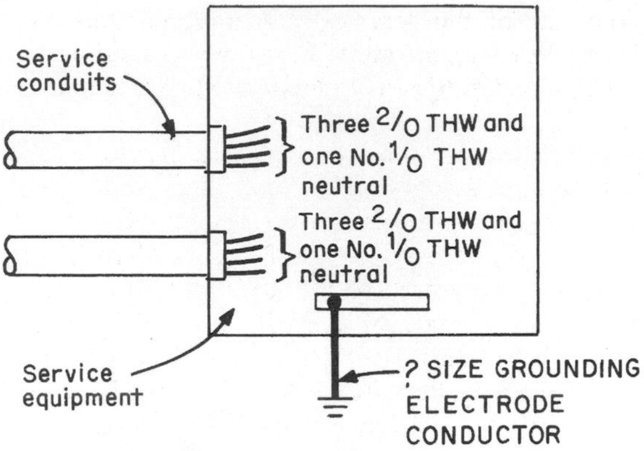
Service
conduits
Three 2/0 THW and
one No. 1/0 THW
neutral
Three 2/0 THW and
one No. 1/0 THW
neutral
Service
equipment
? SIZE GROUNDING
ELECTRODE
CONDUCTOR

Fig. 250-57. Typical task of sizing the conductor to the grounding electrode. (Sec. 250.66.)

2. The same table shows that the single conductor which has a csa at least equal to the total csa of the two conductors per phase is a 300-kcmil size of conductor. That conductor size is then located in the left-hand column of Table 250.66 to determine the minimum size of grounding electrode conductor, which turns out to be 2 AWG copper or 1/0 AWG aluminum or copper-clad aluminum.

Figure 250-58 shows another example of conductor sizing, as follows:

1. The grounding electrode conductor A connects to the street side of the water meter of a metallic water supply to a building. The metallic pipe extends 30 ft (9.14 m) underground outside the building.
2. Because the underground metallic water piping is at least 10 ft (3.0 m) long, the underground piping system is a grounding electrode and must be used as such.

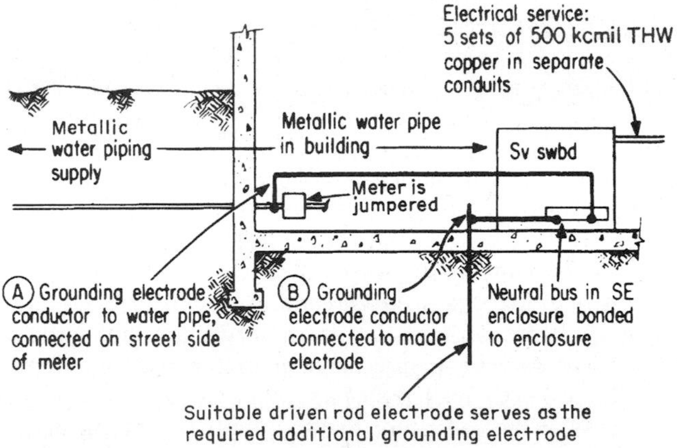
Electrical service:
5 sets of 500 kcmil THW
copper in separate
conduits
Metallic
water piping
supply
Metallic water pipe
in building
Sv swbd
Meter is
jumpered
(A) Grounding electrode
conductor to water pipe,
connected on street side
of meter
(B) Grounding
electrode conductor
connected to made
electrode
Neutral bus in SE
enclosure bonded
to enclosure
Suitable driven rod electrode serves as the
required additional grounding electrode

Fig. 250-58. Two different sizes of grounding electrode conductors are required for installations like this. (Sec. 250.66.)

3. Based on the size of the service-entrance conductors (5 × 500 kcmil = 2500 kcmil per phase leg), the minimum size of grounding electrode conductor to the water pipe is 3/0 AWG copper or 250-kcmil aluminum or copper-clad aluminum.
4. The connection to the ground rod at B satisfies the rule of 250.53(D)(2), requiring a water-pipe electrode to be supplemented by another electrode.
5. But, the size of grounding electrode conductor B required between the neutral bus and the rod, pipe, or plate electrode is 6 AWG copper or 4 AWG aluminum, as covered by part **(A)** of 250.66. And the Code does not require the conductor to a rod, pipe, or plate electrode to be larger than 6 AWG, regardless of the size of the service phases. As discussed under Sec. 250.53(C) and shown in Fig. 250-40, the conductor at B in Fig. 250-58 can be considered to be a bonding jumper, as covered by the sizing rule in 250.64(F) which also says in effect that the conductor to the ground rod need not be larger than 6 AWG copper or 4 AWG aluminum due to the reference to 250.66(A). And the 2014 **NEC** has clarified that even if multiple rod, pipe, or plate electrodes, or any combination thereof are used with a grounding conductor, the size still need not increase above 6 AWG copper. Closing a minor loophole, the 2017 **NEC** placed clarifying language in all three of the special sizing allowances in (A), (B), and (C), to the effect that they could not be used if the electrode(s) in question were wired en route to another electrode that required a larger sized grounding electrode conductor.

Parts **(B)** and **(C)** of the rule make clear that a grounding electrode conductor does not have to be larger than a conductor-type electrode to which it connects. 250.52(A)(3) recognizes a "concrete-encased" electrode—which must be at least 20 ft (6.0 m) of one or more ½-in. (13-mm)-diameter steel reinforcing bars or rods in the concrete or at least 20 ft (6.0 m) of bare 4 AWG copper conductor (or a larger conductor), concrete-encased in the footing or foundation of a building or structure. Section 250.52(A)(4) recognizes a "ground-ring" electrode made up of at least 20 ft (6.0 m) of 2 AWG bare copper conductor (or larger), buried directly in the earth at a depth of at least 2½ ft (750 mm). Those electrodes described under 250.52(A)(1) through (4) are *not* "rod, pipe, or plate electrodes." As electrodes from 250.52(A)(1) through (4), which must be bonded where present, such electrodes would normally be subject to the basic rule of 250.66, which calls for connection to any such electrode by a grounding electrode conductor sized from Table 250.66—requiring up to No. 3/0 copper for use on high-capacity services. *But,* that is not required, as explained in these paragraphs of the two parts.

Parts **(B)** and **(C)** recognize that there is no reason to use a grounding electrode conductor that is larger than a conductor-electrode to which it connects. The grounding electrode conductor need not be larger than 4 AWG copper for a 4 AWG concrete-encased electrode and need not be larger than 2 AWG copper if it connects to a ground-ring electrode—as in part **(C)** of 250.66. Where Table 250.66 would permit a grounding electrode conductor smaller than 4 AWG or 2 AWG (based on size of service conductors), the smaller conductor may be used—but the electrode itself must not be smaller than 4 AWG or 2 AWG. See Fig. 250-36.

The first note under Table 250.66 correlates to 230.40, Exception No. 2, and 250.64(D), as follows:

When two to six service disconnects in separate enclosures are used at a service, with a separate set of SE conductors run to each disconnect, the size of a single common grounding electrode conductor must be based on the largest sum of the cross sections of the same phase leg of each of the several sets of SE conductors. When using multiple service disconnects in separate enclosures, with a set of SE conductors run to each from the drop or lateral (230.40, Exception No. 2) and using a single, common grounding electrode conductor, either run continuous and unspliced from one disconnect to another and then to the grounding electrode, as in the upper right drawing of Fig. 250-49 (which will be impracticable in almost all cases due to termination restrictions in the enclosures), or with taps from each disconnect to a common grounding electrode conductor run to the electrode—as in 250.64(D)(1), this note is used to determine the size of the common grounding electrode conductor from Table 250.66. The "equivalent area" of the size of SE conductors is the largest sum of the cross-sectional areas of one ungrounded leg of each of the several sets of SE conductors.

250.68. Grounding Electrode Conductor Connection and Bonding Jumper Connections to Grounding Electrodes. The rule requires that the connection of a grounding electrode conductor to the grounding electrode "shall be accessible" (Fig. 250-59). [250.104(A) also requires that any clamp for a bonding jumper to interior metal water piping must be accessible.] Inspectors want to be able to see and/or be able to get at any connection to a grounding electrode. But because there are electrodes permitted in 250.52 that would require underground or concrete-encased connections, an exception was added to the basic rule to permit inaccessible connections in such cases (Fig. 250-60). Electrode connections that are *not* encased or buried—such as where they are made to exposed parts of electrodes that are encased, driven, or buried—*must* be accessible. This section now places the burden on the installer to make such connections accessible wherever possible.

The second exception to 250.68(A) covers accessibility when connections are made to steel framing members that will be subsequently encased in fireproofing compounds, rendering such connections inaccessible. In such cases, there are often two connection issues in the same location. The first issue is the connection of the grounding electrode connection to a lug, which must be an exothermic weld or an irreversible crimp in order to be acceptable without remaining accessible. The second issue is how the lug is attached to the metal framing. In this case any mechanical connection is acceptable, even if reversible.

The second part of this section requires an effective grounding path, and to make this to be the case on a metal piping system electrode, use the proper clamp shown in Fig. 250-59. If the connection is to be remote from the point of entry for the water pipe into the building, as shown in Fig. 250-60, make sure the interior home run for the piping qualifies in terms of access to the piping and the qualifications of those supervising the installation, as covered in 250.68(C)(1) Exception. In addition, add bonding jumpers around any insulated joints or around equipment likely to be removed for repair, as also covered in 250.53(D)(1). In a typical case of grounding for a local transformer within a building, 250.30(A)(4)(1) and 250.68(C)(1) note that grounding of the secondary neutral, where there is no building steel available, must be made to the water pipe within 1.5 m (5 ft)

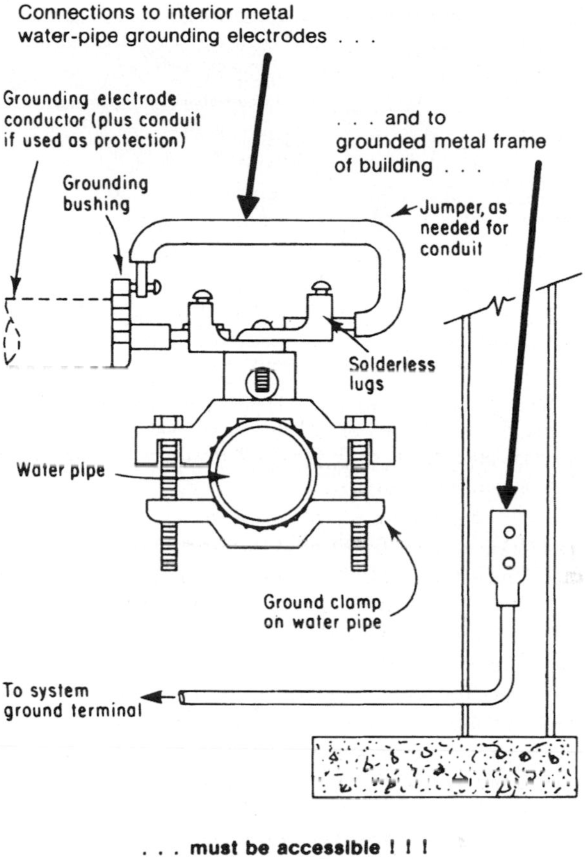
Connections to interior metal
water-pipe grounding electrodes . . .
Grounding electrode
conductor (plus conduit
if used as protection)
. . . and to
grounded metal frame
of building . . .
Grounding
bushing
Jumper, as
needed for
conduit
Solderless
lugs
Water pipe
Ground clamp
on water pipe
To system
ground terminal
. . . must be accessible ! ! !

Fig. 250-59. Whenever possible, connections to grounding electrodes must be "accessible." Note that the water-pipe clamp, if used on copper water tubing as opposed to heavy-wall red brass or galvanized steel pipe, must be marked for this service, as required by the UL Grounding and Bonding Standard, #467. [Secs. 250.68, and 110.3(B).]

from its point of entry in the building unless the piping qualifies for the exception in 250.68(C). And 250.68(B) requires that bonding jumpers be used to ensure continuity of the ground path back to the underground pipe for that portion permitted to serve as a grounding electrode by 250.52(A)(1) wherever the piping may contain insulating sections or is liable to become disconnected. Bonding jumpers around unions, valves, water meters, and other points where a water piping system electrode might be opened must have enough slack to permit removal of the part. Hazard is created when bonding jumpers are so short that they have to be removed to remove the equipment they jumper. Dangerous conditions have been reported about this matter. Bonding jumpers must be long enough to ensure grounding integrity along piping systems under any conditions of maintenance or repair.

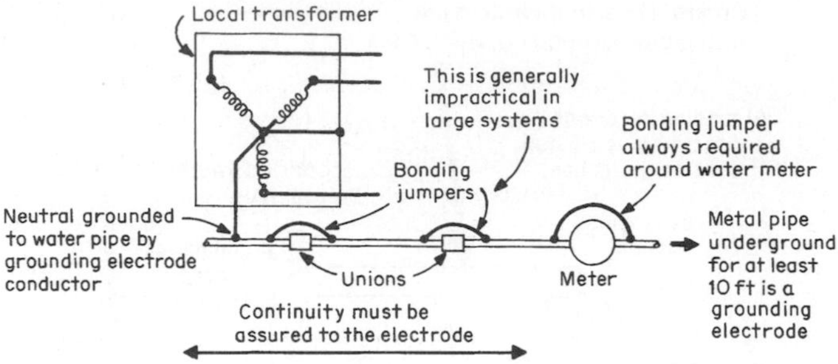
Local transformer
This is generally impractical in large systems
Bonding jumper always required around water meter
Bonding jumpers
Neutral grounded to water pipe by grounding electrode conductor
Metal pipe underground for at least 10 ft is a grounding electrode
Unions
Meter
Continuity must be assured to the electrode

Fig. 250-60. Although required, bonding of metal piping can pose problems. And even though the connection to the pipe may be made without regard to the pipe's point of entry in some occupancies, that permission only applies to water pipes that are supervisable. Unless this is an "industrial" or "commercial" installation with qualified maintenance people and the water pipe is completely exposed and visible for the entire distance from its point of entry to the grounding electrode conductor connection, such connection from the grounded conductor must be made no further than 5 ft (1.52 m) from the pipe's entry point. (Sec. 250.68(C)(1) Exception.)

Part **(C)** of 250.68 is the logical culmination of numerous discussions over many code cycles regarding an interesting question: What is the boundary, the physical extent, of an electrode? For example, a standard plumbing trade length of rigid pipe, whether ABS sewer pipe or copper water tubing, is 20 ft. If 17 ft of a standard trade length of copper pipe is in the earth, and the remaining 3 ft sticks out and enters a water meter, how long is the electrode, 17 ft or 20 ft? The answer now is, 17 ft. Only the portion of any conductive object used as an electrode that is actually in contact with the earth counts as an electrode.

From this result, which has good internal logic, it is apparent that the exposed portion of a metal object being used as an electrode is something else, requiring adjustments in the **NEC** to correlate with this outcome. Those adjustments are in this subsection. The parent language describes, for a water pipe, a "conductor" and for building steel a "bonding conductor." It should also be noted that all of the water pipe content describing connections to the electrode have been relocated here from their former home in 250.52(A)(1), for the same reason. If what you are connecting to is not the electrode, then don't describe it in terms of being a part of the electrode. These issues chiefly concern building steel and water pipes, so they are the ones addressed at this point.

Paragraph (1) covers water pipes, and the default allowance over many cycles continues: You can make connection within the first 5 ft (1.5 m), but note the literal text: you are connecting to "water piping that is electrically continuous with a metal underground water pipe . . ." For dwelling occupancies, including large multifamily buildings, this is all that is available.

The exception covers the nondwelling occupancy allowance to use a longer conductor, one which is a "bonding conductor to interconnect electrodes" as long as it is exposed over its entire length. Since the spaces above suspended ceilings with lift-out panels qualify under the definition of "exposed," those locations are acceptable as well. Obviously, a wall or floor penetration would make the pipe run concealed for the short passage through the partition, but such short passages directly through the partition are acceptable as well. However, a pipe which is concealed in the long dimension of a fixed partition loses its eligibility for a remote grounding electrode conductor attachment at that point. In probably the great majority of cases, however, a main water service pipe will make it far into the building before it loses this qualification. The other qualification that applies to this method is that there must be a showing that those servicing the installation are qualified persons.

Paragraph (2) covers a structural metal building frame. As of the 2014 edition, it no longer matters whether connected structural steel building elements are themselves extended into physical contact with soil. It only matters that they are elements of a structural building framework presumed to be inherently conductive. And if they are, then quite properly they are capable of providing a bonding path between electrodes, even as 250.30(A)(4) allows this by right for separately derived system connections. If one end of a building has a water or a sprinkler entrance or both, then those pipe electrodes may be connected to local building steel. And that steel may be relied upon to then bond those electrodes to whatever other qualified electrodes are available at the service location, or even to a connection to the service itself. Additional text in this paragraph arrived in the 2017 edition, after being relocated from 250.52(A)(2). It describes how the hold-down bolts that secure the upper building structure are to be connected to a concrete-encased electrode to provide continuity to earth.

Paragraph (3) explicitly recognizes a useful method of establishing a connection to a concrete-encased electrode that avoids the necessity for making the electrical connection ahead of the actual concrete pour. If a conductive element is extended up and out so it has an end that will be accessible afterwards, then the connection out of the pour is permitted.

250.70. Methods of Grounding and Bonding Conductor Connection to Electrodes. Because 250.53(G) requires *buried* or protected connections of grounding electrode conductors to ground rods, the third sentence of this section requires that a buried ground clamp be of such material and construction that it has been designed, tested, and marked for use directly in the earth. And any clamp that is used with two or more conductors must be designed, tested, and marked for the number and types of conductors that may be used with it. This is shown in the bottom drawing of Fig. 250-61. Connections depending on solder cannot be used due to the likelihood that the solder will melt during a sustained fault. Figure 250-59 gives some listing information about water-pipe clamps.

250.86. Other Conductor Enclosures and Raceways. The basic rule requires all enclosures and raceways to be grounded. But, Exception No. 1 permits the installation of short runs as extensions from existing open wiring, knob-and-tube work, or nonmetallic-sheathed cable without grounding where there is little likelihood of an accidental connection to ground or of a person touching both the conduit, raceway, or armor and any grounded metal or other grounded surface at the same time. Additionally, the exception permits "short sections" of enclosures and raceways to be used as sleeving or to otherwise support cables. The exact length of such ungrounded metallic enclosures and raceways is *not* given. That determination will ultimately be made by the local electrical inspector.

The last exception covers instances where metal conduit sweeps are inserted into nonmetallic conduit runs to prevent the pulling rope, when under high tension,

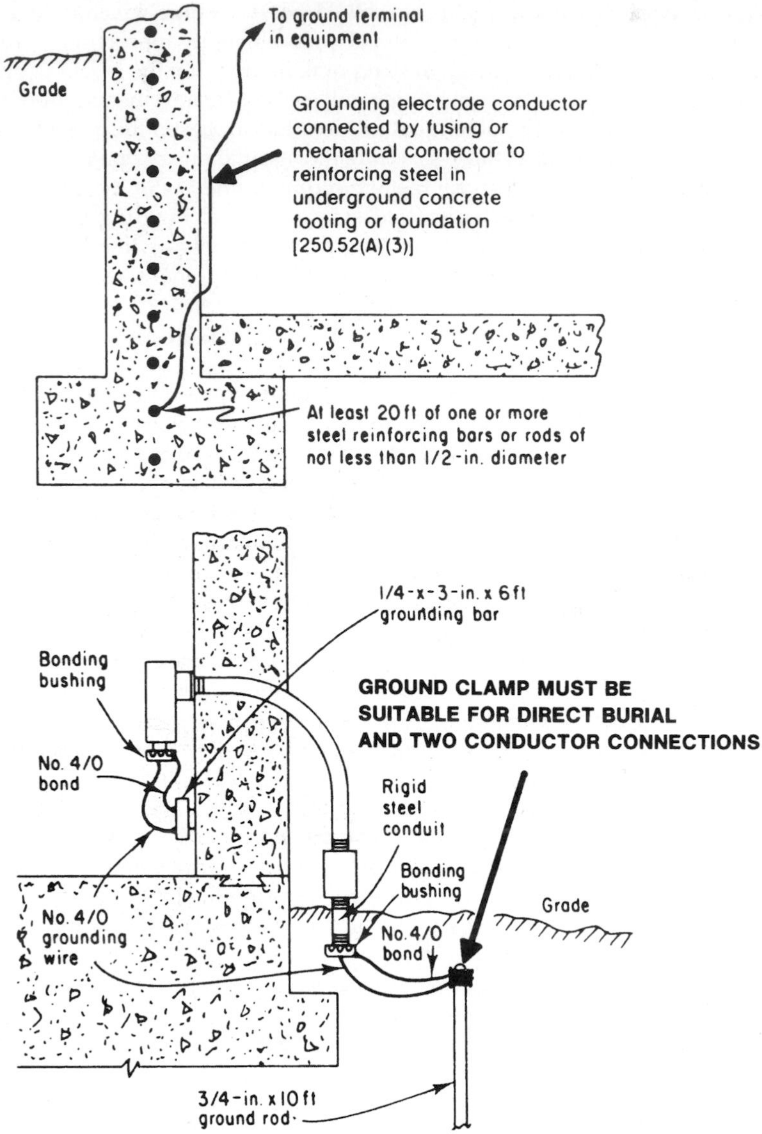
To ground terminal in equipment
Grade
Grounding electrode conductor connected by fusing or mechanical connector to reinforcing steel in underground concrete footing or foundation [250.52(A)(3)]
At least 20 ft of one or more steel reinforcing bars or rods of not less than 1/2-in. diameter
1/4-x-3-in. x 6 ft grounding bar
Bonding bushing
GROUND CLAMP MUST BE SUITABLE FOR DIRECT BURIAL AND TWO CONDUCTOR CONNECTIONS
No. 4/0 bond
Rigid steel conduit
Bonding bushing
Grade
No. 4/0 grounding wire
No. 4/0 bond
3/4-in. x 10 ft ground rod

Fig. 250-61. Encased and buried electrode connections are permitted by exception to the basic rule. The bonding bushing at the bottom of the sweep almost certainly does not have a direct burial listing and would require modifications and inspection approval to be used in this way. A better option in this case would be a "U-bolt" style ground clamp around the conduit at the end of the sweep. They are available in constructions that are listed for direct burial. Another option would be to use PVC conduit and skip the bonding issue altogether. (Sec. 250.68.)

from sawing through the inner radius of the bend and destroying the integrity of the conduit system. If the sweep is buried low enough so its upper end is still 450 mm (18 in.) below grade level, or if it is entirely encased including the ends in 50 mm (2 in.) of concrete (i.e., the concrete has to encase the steel/PVC raceway system so it overlaps the ends of the steel sweep by at least 50 mm [2 in.]), bonding is not required. The 2017 **NEC** extended this concept of isolated metal conduit elbows to metal components generally. The same extension of concept was made for service applications in 250.80.

Part V. Bonding

250.90. General. One of the most interesting and controversial phases of electrical work involves the grounding and bonding of secondary-voltage service-entrance equipment. Modern practice in such work varies according to local interpretations of Code requirements and specifications of design engineers. In all cases, however, the basic intent is to provide an installation which is essentially in compliance with **National Electrical Code** rules on the subject, using practical methods for achieving objectives.

In order to ensure electrical continuity of the grounding circuit, bonding (special precautions to ensure a permanent, low-resistance connection) is required at all conduit connections in the service equipment and where any nonconductive coating exists which might impair such continuity. This includes bonding at connections between service raceways, service cable armor, all service-equipment enclosures containing service-entrance conductors, including meter fittings, boxes, and the like.

The need for effective grounding and bonding of service equipment arises from the electrical characteristics of utility-supply circuits. In the common arrangement, service conductors are run to a building and the service overcurrent protection is placed near the point of entry of the conductors into the building, at the load end of the conductors. With such a layout, the service conductors are not properly protected against ground faults or shorts occurring on the supply side of the service overcurrent protection. Generally, the only protection for the service conductors is on the primary side of the utility's distribution transformer. By providing "bonded" connections (connecting with special care to reliable conductivity), any short circuit in the service-drop or service-entrance conductors is given the greatest chance of burning itself clear—because there is no effective overcurrent protection ahead of those conductors to provide opening of the circuit on such heavy fault currents. And for any contact between an energized service conductor and a grounded service raceway, fittings, or enclosures, bonding provides discharge of the fault current to the system grounding electrode—and again burning the fault clear. This condition of services is shown in Fig. 250-62.

250.92. Services. Because of the requirement set forth in 250.92, all enclosures for service conductors must be grounded to prevent a potential above ground on the enclosures as a result of fault—which would be a very definite hazard—and to facilitate operation of overcurrent devices anywhere on the supply side of the service conductors. However, because of the distant location of the protection and the normal impedance of supply cables, it is important that any fault to an enclosure of a hot service conductor of a grounded electrical system find a

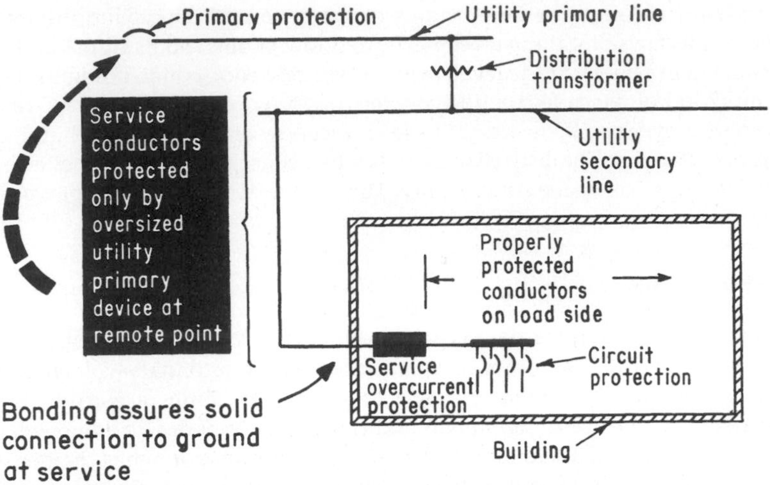
Primary protection
Utility primary line
Distribution transformer
Service conductors protected only by oversized utility primary device at remote point
Utility secondary line
Properly protected conductors on load side
Service overcurrent protection
Circuit protection
Bonding assures solid connection to ground at service
Building

Fig. 250-62. Service bonding must ensure burn-clear on shorts and grounds in service conductors. (Sec. 250.90.)

firm, continuous, low-impedance path to ground to ensure sufficient current flow to operate the primary protective device or to burn the fault clear quickly. This means that all enclosures containing the service conductors—service raceway, cable armor, boxes, fittings, cabinets—must be effectively bonded together; that is, they must have low impedance through themselves and must be securely connected to each other to ensure a continuous path of sufficient conductivity to the conductor which makes the connection to ground (Fig. 250-63).

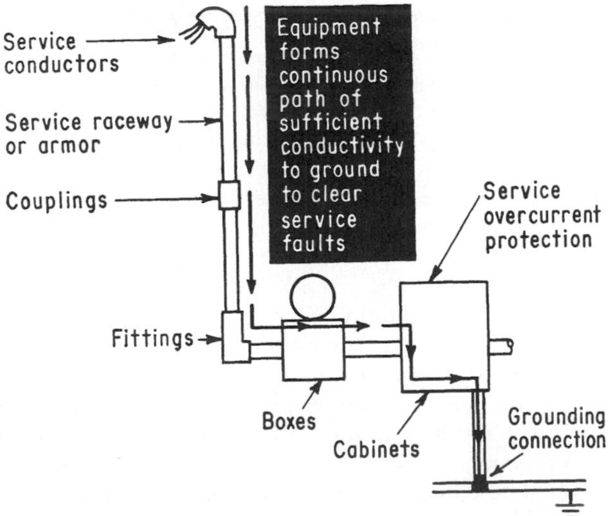
Service conductors
Service raceway or armor
Couplings
Equipment forms continuous path of sufficient conductivity to ground to clear service faults
Service overcurrent protection
Fittings
Boxes
Cabinets
Grounding connection

Fig. 250-63. Bonding ensures low-impedance path through all service conductor enclosures. (Sec. 250.92.)

The spirit of the Code and good engineering practice have long recognized that the conductivity of any equipment ground path should be at least equivalent to 25 percent of the conductivity of any phase conductor with which the ground path will act as a circuit conductor on a ground fault. Or, to put it another way, making the relationship without reference to insulation or temperature rise, the impedance of the ground path must not be greater than four times the impedance of any phase conductor with which it is associated.

In ungrounded electrical systems, the same careful attention should be paid to the matter of bonding together the noncurrent-carrying metal parts of all enclosures containing service conductors. Such a low-impedance ground path will quickly and surely corner ground any hot conductor which might accidentally become common with the enclosure system, and if a second fault occurs from a different phase, the low-impedance path will pass enough current so the fault will burn free.

Specific **NEC** requirements on grounding and bonding are as follows:

1. 250.80 requires that metal enclosures and raceways for service conductors as well as service equipment be grounded, either to the grounded circuit conductor (grounded systems) or to the grounding electrode conductor (ungrounded systems). And, the Code requires that flexible metal conduit or liquidtight flexible metal conduit used in a run of service raceway must be bonded around (Fig. 250-64). 230.43 states that rule on flex and lists the *only* types of raceway that may enclose service-entrance conductors.

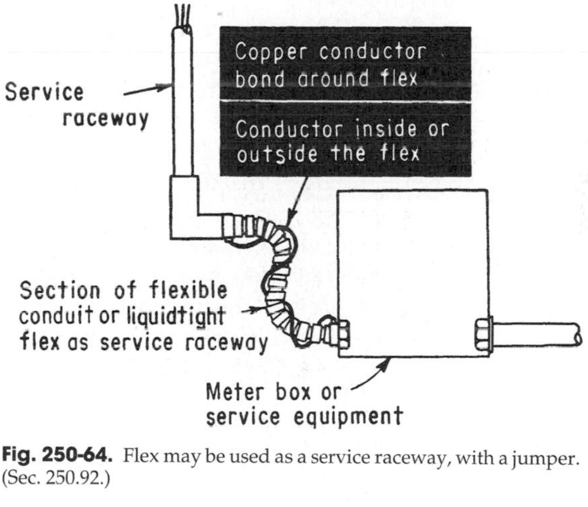
Service raceway
Copper conductor bond around flex
Conductor inside or outside the flex
Section of flexible conduit or liquidtight flex as service raceway
Meter box or service equipment

Fig. 250-64. Flex may be used as a service raceway, with a jumper. (Sec. 250.92.)

2. 250.92 sets forth the service equipment which must be bonded—that is, the equipment for which the continuity of the grounding path must be specifically ensured by using specific connecting devices or techniques. As indicated in Fig. 250-65, this equipment includes (1) service raceway, cable trays, cable sheath, and cable armor; and (2) all service-equipment enclosures containing service-entrance conductors, including meter fittings, boxes, etc., interposed

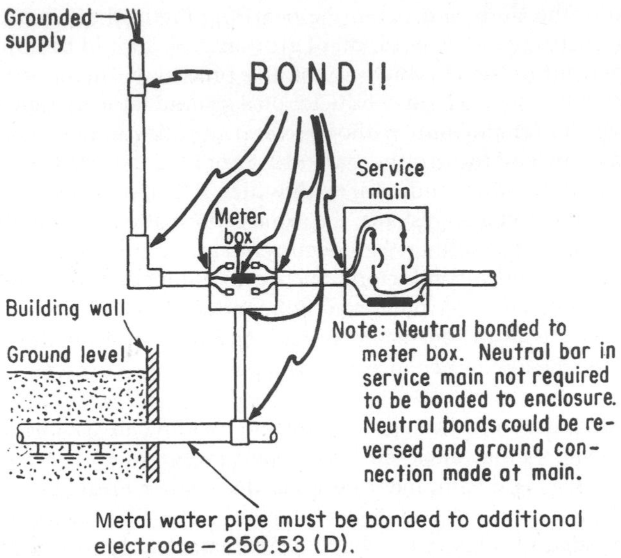

Grounded supply
BOND!!
Service main
Meter box
Building wall
Ground level
Note: Neutral bonded to meter box. Neutral bar in service main not required to be bonded to enclosure. Neutral bonds could be reversed and ground connection made at main.
Metal water pipe must be bonded to additional electrode – 250.53 (D).

Fig. 250-65. "Bonding" consists of using prescribed fittings and/or methods for connecting components enclosing SE conductors. (Sec. 250.92.)1

in the service raceway or armor. There is what is effectively an exception for 250.80. Presumably this is to correlate with the optional unbonded metal elbow in the exception, and that would be correct. However, before the 2011 **NEC**, this language pointed to what is now 250.84, which addresses lead-sheathed cables. Removing this reference was incorrect, what should have happened was to refer both sections. However, since for environmental reasons lead-sheathed cables are no longer used, there will be little practical effect.

250.92**(B)** covers "Method of Bonding at the Service." 250.92(A) is very specific in listing the many types of equipment that require bonding connections, but the actual "how to" is often hazy. For virtually every individual situation where a bonding connection must be made, there is a variety of products available in the market which present the installer with a choice of different methods.

This section sets forth the specific means which may be used to connect service-conductor enclosures together to satisfy the bonding requirements of 250.92(A). These means include:

1. Bonding equipment to the grounded service conductor by means of suitable lugs, pressure connectors, clamps, or other approved means—except that soldered connections must not be used. 250.142 permits grounding of meter housings and service equipment to the grounded service conductor on the supply side of the service disconnecting means.
2. Threaded couplings in rigid metal conduit or IMC (intermediate metal conduit) runs and threaded bosses on enclosures to which rigid metal conduit or IMC connects.
3. Threadless couplings and connectors made up tight for rigid metal conduit, IMC, or electrical metallic tubing, or for metal-clad cables.

4. Other devices (not standard locknuts and bushings) listed for the purpose, such as bonding locknuts, wedges, and bushings.

In general, bonding jumpers must be used around concentric or eccentric knockouts which are punched or otherwise formed in such a manner that would impair the electrical current flow through the reduced cross section of metal that bridges between the enclosure wall and punched ring of the KO (knockout). The 2011 NEC included reducing washers at this point; they are not listed for use in service bonding applications for good reason because they are notoriously poor at maintaining continuity, especially on enameled enclosures that are commonplace in service work. Very few inspectors ever accepted them without bonding jumpers anyway. And the bonding jumpers must be sized from 250.102(C).

Based on those briefly worded Code requirements, modern practice follows more or less standard methods.

Where a rigid conduit is the service raceway, threaded or threadless couplings are used to couple sections of a conduit together. A conduit connection to a meter socket may be made by connecting a threaded conduit end to a threaded hub or boss on the socket housing, where the housing is so constructed; by a locknut and bonding bushing; by a locknut outside with a bonding wedge or bonding locknut and a standard metal or completely insulating bushing inside; or sometimes by a locknut and standard bushing where the socket enclosure is bonded to the grounded service conductor. Conduit connections to KOs in sheet metal enclosures can be made with a bonding locknut (Fig. 250-66), a bonding wedge, or a bonding bushing where no KO rings remain around the opening through which the conduit enters and where the box is listed for such use, even where the KOs have not all been removed. But, generally, where a KO ring does remain around the conduit entry hole, a bonding bushing or wedge with a jumper wire must be used to assure a path of continuity from the conduit to the enclosure. Figure 250-67 summarizes the various acceptable techniques. It should be noted that the use of the common locknut and bushing type of connection is not allowed. Neither is the use of double locknuts—one inside, one outside—and a bushing, although that is permitted on the load side of the service equipment. The special methods set forth in 250.92(B) are designed to prevent poor connections or loosening of connections due to vibration. This minimizes the possibility of arcing and consequent damage which might result when a service conductor faults to the grounded equipment.

Similar provisions are used to ensure continuity of the ground path when EMT is the service raceway or when armored cable is used. EMT is coupled or connected by threadless devices—compression-type, indenter-type, or set-screw type, using raintight type outdoors. Although a threadless box connector is suitable to provide a bonded connection of the connector to the metal raceway (rigid metal, IMC, EMT), it is also necessary to provide a "bonded" connection between the connector and the metal enclosure. A threadless box connector on the end of EMT used as a service raceway provides satisfactory bonding of the EMT to the connector, but the last sentence of this rule says that a standard locknut or a standard bushing connected to the threaded end of the connector does not provide the required bonding of the connector to the metal service equipment to which the connector is connected. On the end of the connector, a bonding locknut or bonding bushings with or without jumpers must be used if the knockout is clean (all rings

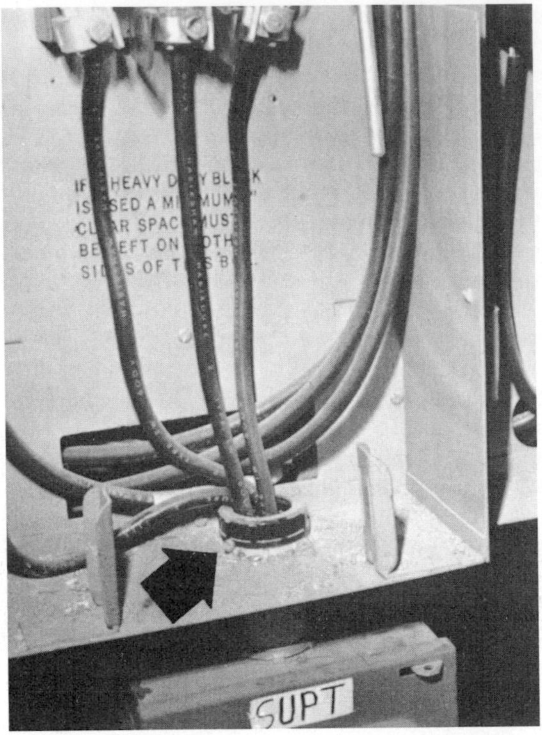
SUPT

Fig. 250-66. Bonding locknut is a recognized method for bonding a service conduit nipple to a meter socket, when the KO is clean (no rings left in enclosure wall) or is cut on the job. With plastic bushing permitted, this is the most economical of the several methods for making a bonded conduit termination. (Sec. 250.92.)

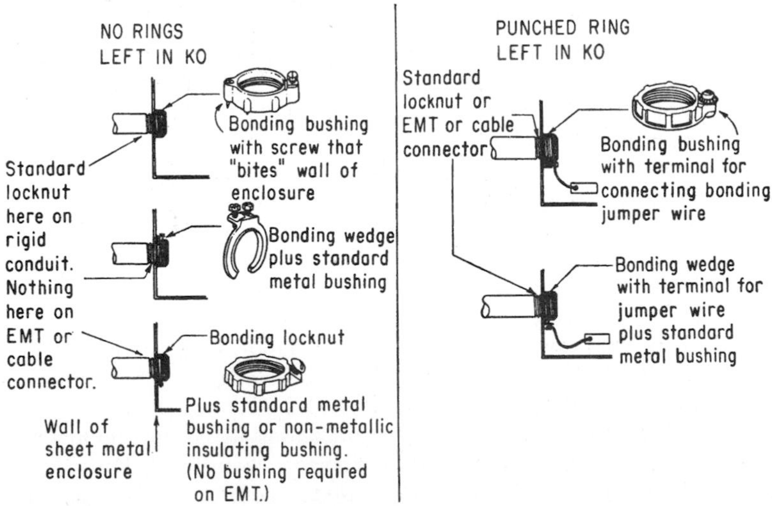
NO RINGS
LEFT IN KO
Bonding bushing
with screw that
"bites" wall of
enclosure
Standard
locknut
here on
rigid
conduit.
Nothing
here on
EMT or
cable
connector.
Bonding wedge
plus standard
metal bushing
Bonding locknut
Plus standard metal
bushing or non-metallic
insulating bushing.
(No bushing required
on EMT.)
Wall of
sheet metal
enclosure
PUNCHED RING
LEFT IN KO
Standard
locknut or
EMT or cable
connector
Bonding bushing
with terminal for
connecting bonding
jumper wire
Bonding wedge
with terminal for
jumper wire
plus standard
metal bushing

Fig. 250-67. Methods for "bonding" wiring methods to sheet metal enclosures. (Sec. 250.92.)

punched out or clean knockout punched on the job). If concentric or eccentric rings are left, a grounding locknut with a jumper, a grounding bushing with a jumper, or a grounding wedge with a jumper must be used to provide bonded connection around the perforated knockout. And fittings used with service cable armor must assure the same degree of continuity of the ground path.

The use of bonding bushings, bonding wedges, and bonding locknuts is recognized without reference to types of raceways or types of connectors used with the raceways or cable armor. As a result, common sense and experience have molded modern field practice in making raceway and armored service cable connection to service cabinets. The top of Fig. 250-68 shows how a bonding wedge is used on existing connections at services or for raceway connections on the load side of the

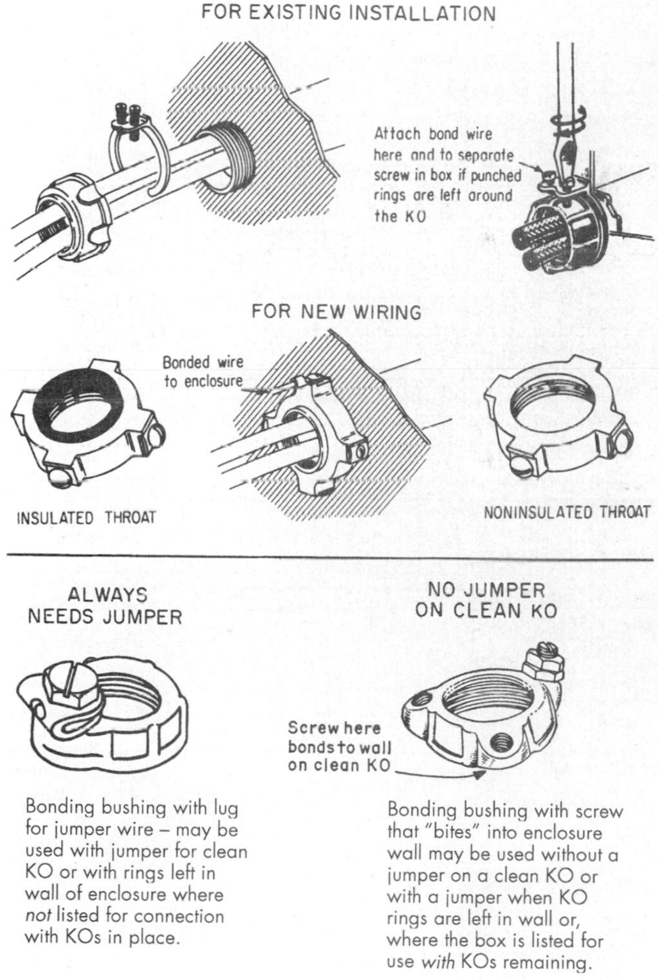
FOR EXISTING INSTALLATION
Attach bond wire here and to separate screw in box if punched rings are left around the KO
FOR NEW WIRING
Bonded wire to enclosure
INSULATED THROAT
NONINSULATED THROAT
ALWAYS NEEDS JUMPER
NO JUMPER ON CLEAN KO
Screw here bonds to wall on clean KO
Bonding bushing with lug for jumper wire – may be used with jumper for clean KO or with rings left in wall of enclosure where *not* listed for connection with KOs in place.
Bonding bushing with screw that "bites" into enclosure wall may be used without a jumper on a clean KO or with a jumper when KO rings are left in wall or, where the box is listed for use *with* KOs remaining.

Fig. 250-68. Bonding bushings and similar fittings must be used in their intended manners. (Sec. 250.92.)

service—such as required by 501.30(B) for Class I hazardous locations. A bonding bushing with provision for connecting a bonding jumper, is the common method for new service installations where some of the concentric or eccentric "doughnuts" (knockout rings) are left in the wall of the enclosure, therefore requiring a bonding jumper. Great care must be taken to ensure that each and every type of bushing, locknut, or other fitting is used in the way for which it is intended to best perform the bonding function.

Figure 250-69 shows detailed application of the preceding rules to typical meter-socket installations. Meter-enclosure bonding techniques are shown in Fig. 250-70. Bonding details for current-transformer installations are shown in Fig. 250-71. Those illustrations are intended to portray typical field practice aimed at satisfying the various Code rules.

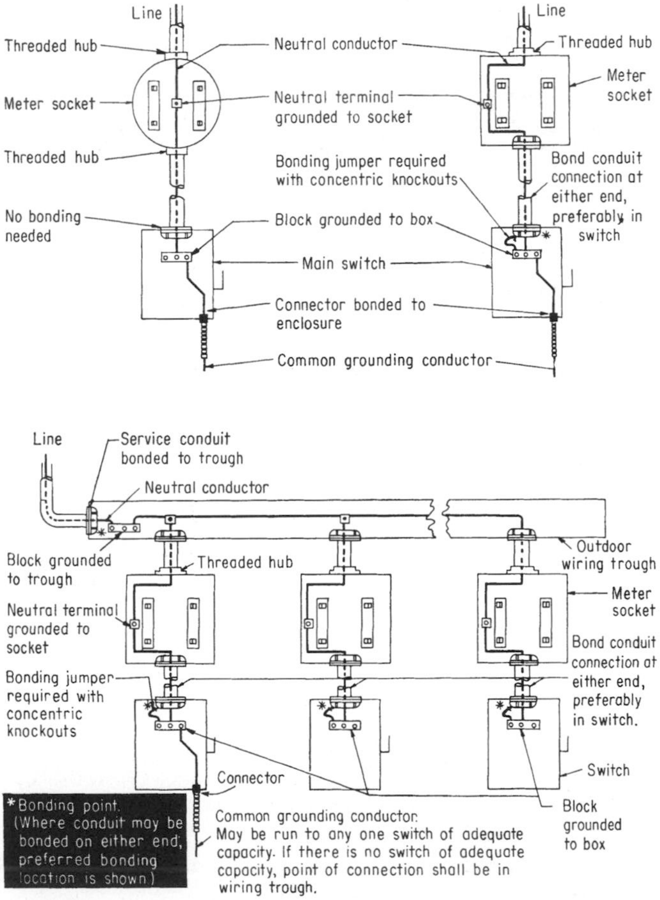
Line
Line
Threaded hub
Neutral conductor
Threaded hub
Meter socket
Meter socket
Neutral terminal grounded to socket
Threaded hub
Bonding jumper required with concentric knockouts
Bond conduit connection at either end, preferably in switch
No bonding needed
Block grounded to box
Main switch
Connector bonded to enclosure
Common grounding conductor
Line
Service conduit bonded to trough
Neutral conductor
Block grounded to trough
Threaded hub
Outdoor wiring trough
Neutral terminal grounded to socket
Meter socket
Bond conduit connection at either end, preferably in switch.
Bonding jumper required with concentric knockouts
Switch
Connector
Block grounded to box
*Bonding point. (Where conduit may be bonded on either end; preferred bonding location is shown.)
Common grounding conductor. May be run to any one switch of adequate capacity. If there is no switch of adequate capacity, point of connection shall be in wiring trough.

Fig. 250-69. Typical meter socket applications. (Sec. 250.92.)

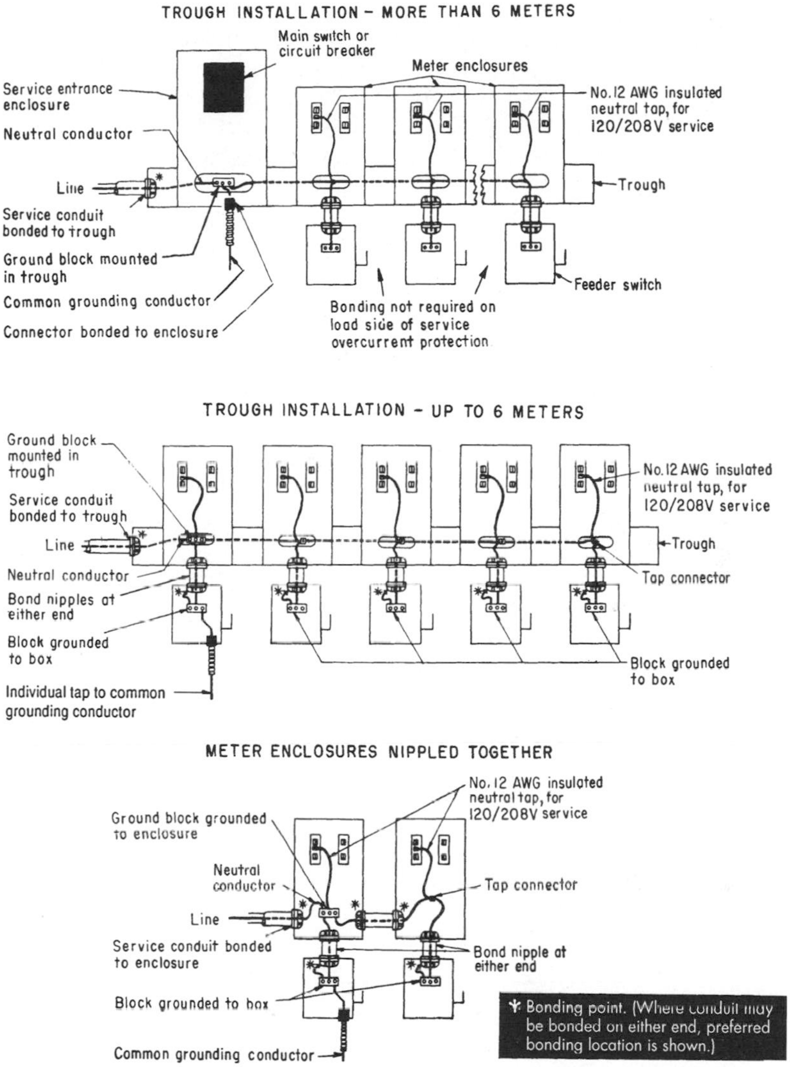
TROUGH INSTALLATION – MORE THAN 6 METERS
Main switch or circuit breaker
Meter enclosures
Service entrance enclosure
Neutral conductor
No.12 AWG insulated neutral tap, for 120/208V service
Line
Trough
Service conduit bonded to trough
Ground block mounted in trough
Common grounding conductor
Connector bonded to enclosure
Bonding not required on load side of service overcurrent protection
Feeder switch
TROUGH INSTALLATION – UP TO 6 METERS
Ground block mounted in trough
Service conduit bonded to trough
Line
Neutral conductor
Bond nipples at either end
Block grounded to box
Individual tap to common grounding conductor
No.12 AWG insulated neutral tap, for 120/208V service
Trough
Tap connector
Block grounded to box
METER ENCLOSURES NIPPLED TOGETHER
No. 12 AWG insulated neutral tap, for 120/208V service
Ground block grounded to enclosure
Neutral conductor
Line
Tap connector
Service conduit bonded to enclosure
Bond nipple at either end
Block grounded to box
Common grounding conductor
* Bonding point. (Where conduit may be bonded on either end, preferred bonding location is shown.)

Fig. 250-70. Typical meter-enclosure installations (120/208- or 120/240-V services). (Sec. 250.92.)

250.94. Bonding for Communications Systems. This section calls for a ready, effective "intersystem" bonding and grounding of communications systems, such as telephone, and community antenna television (CATV) systems at the service equipment for *all* buildings, not just dwellings, and at both services and at building disconnects. The rule requires that there be an "intersystem bonding termination . . . accessible for connection and inspection . . . [with the] capacity for connection of not less than three intersystem bonding conductors." This makes provision

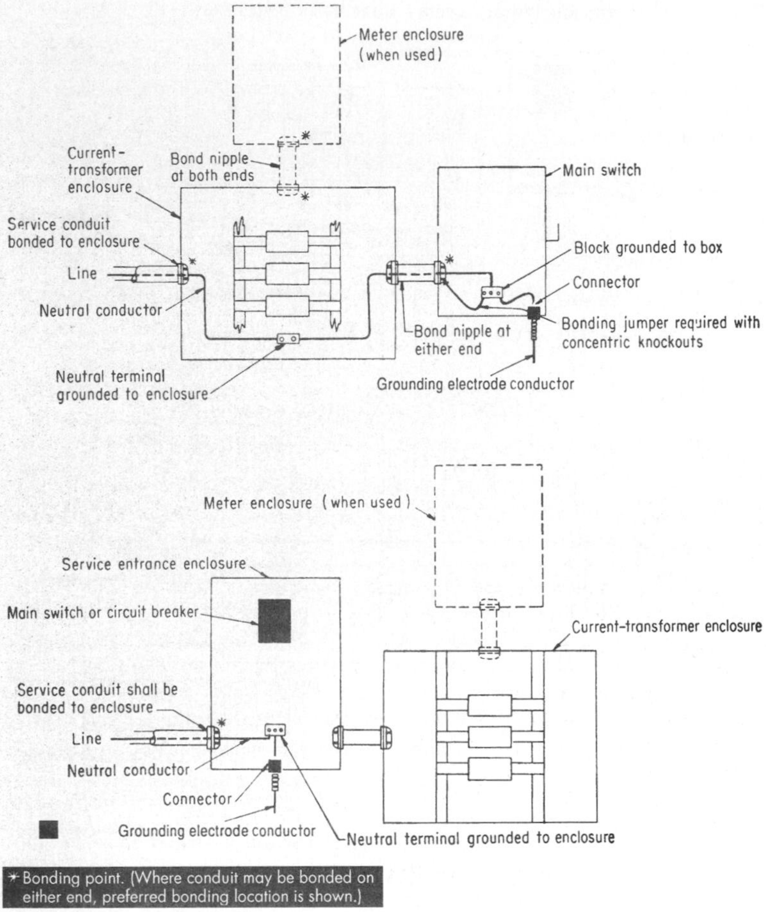
Meter enclosure
(when used)
Current-
transformer
enclosure
Bond nipple
at both ends
Main switch
Service conduit
bonded to enclosure
Block grounded to box
Line
Connector
Neutral conductor
Bonding jumper required with
concentric knockouts
Bond nipple at
either end
Neutral terminal
grounded to enclosure
Grounding electrode conductor
Meter enclosure (when used)
Service entrance enclosure
Main switch or circuit breaker
Current-transformer enclosure
Service conduit shall be
bonded to enclosure
Line
Neutral conductor
Connector
Grounding electrode conductor
Neutral terminal grounded to enclosure
* Bonding point. (Where conduit may be bonded on
either end, preferred bonding location is shown.)

Fig. 250-71. Bonding at CT cabinets. (Sec. 250.94.)

for bonding metal enclosures of, say, telephone equipment to metal enclosures of electrical system components to reduce voltage differences between such metal enclosures as a result of lightning or power contacts. This rule was placed in the NEC because 800.100(D) and 800.100(B)(1) require bonding interconnection between a building's power grounding electrode system and the "protector ground" (grounding electrode conductor) of telephone and other communications systems, and because making that bonded interconnection has become more difficult. Both 810.21(J) and 810.21(F)(1) and 820.100(D) and 820.100(B)(1) require such grounding interconnections, as does 830.100(D) and 830.100(B)(1), which cover "Network-Powered Broadband Communications Systems." This section in the 2014 and prior editions referred to "other" rather than "communications" systems. Several attempts to make the Code hospitable to landing other systems, such as bonding wires from CSST (corrugated stainless steel tubing) for gas lines were rejected, and this section was retitled as a result. Two panel statements are worthy of note:

> "Intersystem bonding terminations are required to be listed per **NEC** 250.94. These products are listed in accordance with UL467 Grounding and Bonding Equipment. Intersystem bonding terminations were created by the telecommunications industry to provide a bonding location for their systems. Expanding this application could reduce the product capacity for telecom circuits and add the possibility of the introduction of noise into the communications system."
>
> "The NFPA research report indicates that the bonding that is needed for CSST systems has many more restrictions than what is permitted for the connections to the grounding electrode system of a structure for communication systems which utilize an IBT. For induced and indirect lightning bonding at the entrance of the installation will help, however the IBT mandated by the **NEC** is not required to be at the entrance of the CSST. Bonding should also be limited to short lengths with minimum number of bends, but tests have not determined the recommended lengths yet nor maximum number of bends or permitted bending radii. The IBT is defined in both UL 467 and the **NEC** as an intersystem bonding device for communication systems which does not include the bonding of other systems that may need to be bonded."

The proposal for this Code addition in its original form for the 1981 **NEC** included the following commentary:

> In the past, the bond between communications and power systems was usually achieved by connecting the communications protector grounds to an interior water pipe. Where the power was grounded to a ground rod, the bond was connected to the power grounding-electrode conductor or to metallic service conduit, which were usually accessible. With growing use of plastic water pipe, the tendency for service equipment to be installed in finished areas where the grounding electrode conductor is often concealed, and the use of plastic entrance conduit, communications installers no longer have an easily identifiable point for connecting bonds or grounds.
>
> Where lightning or external power fault currents flow in protective grounding systems, there can be dangerous potential differences between the equipment of those systems. Even with the required common or bonded electrodes, lightning currents flowing in noncommon portions of the grounding system result in significant potential differences as a result of inductive voltage drop in the noncommon conductor. If a current flows through a noncommon grounding conductor 10 ft (3.05 m) long, there can be an inductive voltage drop as high as 4000 volts. If that noncommon conductor is either the power grounding-electrode conductor or the communications-protector grounding conductor, the voltage will appear between communication-equipment and power-equipment enclosures. The best technical solution to minimizing that voltage is with a short bond between the service equipment and the communications-protector ground terminal. The conductor to the grounding electrode is then common, and the voltage drop in it does not result in a potential difference between systems.
>
> An externally accessible point for intersystem bonding should be provided at the electrical service if accessible metallic service-entrance conduit is not present or if the grounding-electrode conductor is not accessible. This point could be in the form of a connector, tapped hole, external stud, a combination connector-SE cable clamp, or some other approved means located at the meter base or service equipment enclosure.

Prior to the 2008 **NEC** intersystem bonding terminations were done using exposed portions of service equipment, or grounding electrode conductors, and if they were not accessible for external connections, as in finished basements, then a short length of 6 AWG was to be left or other approved means. Where this has been done in existing construction, it can remain by virtue of an exception that, in effect, was the prior rule.

A definite shock hazard can arise if a *common* grounding electrode conductor is *not* used to ground *both* the bonded service neutral *and* the communications protector. The problem can be solved by simply bonding the ground terminal of the protector to a grounded enclosure of the service equipment (the service panelboard enclosure or the meter socket) and not using the separate telephone grounding electrode conductor. The basic concept has been in the NEC for a very long time. The changes in 2008 mandated specific design components and performance criteria, including the requirement that the terminating device not obstruct the opening of an enclosure. Note that many manufacturers have come to market with intersystem-bonding equipment, however, any terminal bar listed as grounding and bonding equipment should be acceptable (inspector's decision) as long as it is applied within its ratings and installed in accordance with the rules for these connections. Examples for dry locations include those readily available as equipment grounding bars from panelboard manufacturers.

The 2017 NEC has added the usage of a customary grounding electrode termination plate [250.64(D)(1)(3)] for this purpose in the new 250.94(B) as another option. Note that, perhaps at cross purposes to the panel objections (quoted at length just previously) regarding electrical noise injection into communications systems, the 250.94(B) busbar terminations are not literally limited to communications systems as long as the three-terminal minimum is met. Arguably, one of these busbars could even be used for CSST bonding. This will undoubtedly receive more attention in the 2020 NEC.

250.96. Bonding Other Enclosures. Metal raceways, cable sheaths, equipment frames and enclosures, and all other metal noncurrent-carrying parts must be carefully interconnected with Code-recognized fittings and methods to ensure a low-impedance equipment grounding path for fault current—whether or not an equipment grounding conductor (a ground wire) is run within the raceway and connected enclosures. The interconnected system of metal raceways and enclosures must itself form a Code-conforming equipment grounding path—even if a "supplementary equipment grounding conductor" is used within the metal-enclosure grounding system (Fig. 250-72).

The rule of 250.96(B) recognizes that to help reduce electromagnetic noise or interference on a grounding circuit, an insulating "spacer or fitting" may be used to interrupt the electrical continuity of a metallic raceway system used to enclose the branch-circuit conductors at the point of connection to the metal enclosure of a single piece of equipment.

This rule permits interrupting the current path between a metal equipment enclosure and the metal conduit that supplies the enclosure—*but only* if the metal conduit is grounded at its supply end *and* an equipment grounding conductor is run through the conduit into the metal enclosure and is connected to an equipment grounding terminal of the enclosure, to provide safety grounding of the metal enclosure. Provisions for an equipment ground reference separate from the metallic raceway system are covered by 250.146(D) for electronic equipment that is cord-and-plug-connected. This rule covers a separate equipment ground reference for hard-wired-sensitive electronic equipment.

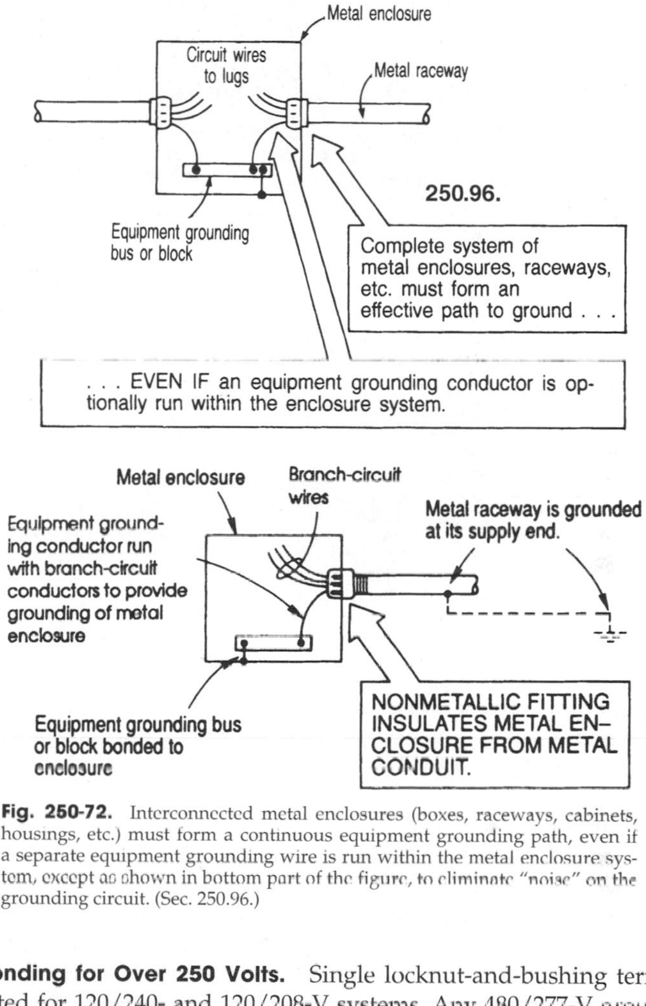
Metal enclosure
Circuit wires to lugs
Metal raceway
250.96.
Equipment grounding bus or block
Complete system of metal enclosures, raceways, etc. must form an effective path to ground . . .
. . . EVEN IF an equipment grounding conductor is optionally run within the enclosure system.
Metal enclosure
Branch-circuit wires
Metal raceway is grounded at its supply end.
Equipment grounding conductor run with branch-circuit conductors to provide grounding of metal enclosure
Equipment grounding bus or block bonded to enclosure
NONMETALLIC FITTING INSULATES METAL ENCLOSURE FROM METAL CONDUIT.

Fig. 250-72. Interconnected metal enclosures (boxes, raceways, cabinets, housings, etc.) must form a continuous equipment grounding path, even if a separate equipment grounding wire is run within the metal enclosure system, except as shown in bottom part of the figure, to eliminate "noise" on the grounding circuit. (Sec. 250.96.)

250.97. Bonding for Over 250 Volts. Single locknut-and-bushing terminations are permitted for 120/240- and 120/208-V systems. Any 480/277-V grounded system, 480-V ungrounded system, or higher must generally use double locknut-and-bushing terminals on clean knockouts of sheet metal enclosures (no concentric rings in wall) for rigid metal conduit and IMC (Fig. 250-73).

Where good electrical continuity is desired on installations of rigid metal conduit or IMC, two locknuts should always be provided on clean knockouts (no rings left) of sheet metal enclosures so that the metal of the box can be solidly clamped between the locknuts, one being on the outside and one on the inside. The reason for not relying on the bushing in place of the inside locknut is that both the conduit and the box may be secured in place and if the conduit is placed so that it extends into the box to a greater distance than the thickness of the bushing, the bushing will not make contact with the inside surface of the box. But that possible weakness in the single-locknut termination does not exclude it from use on systems up to 250 V to ground.

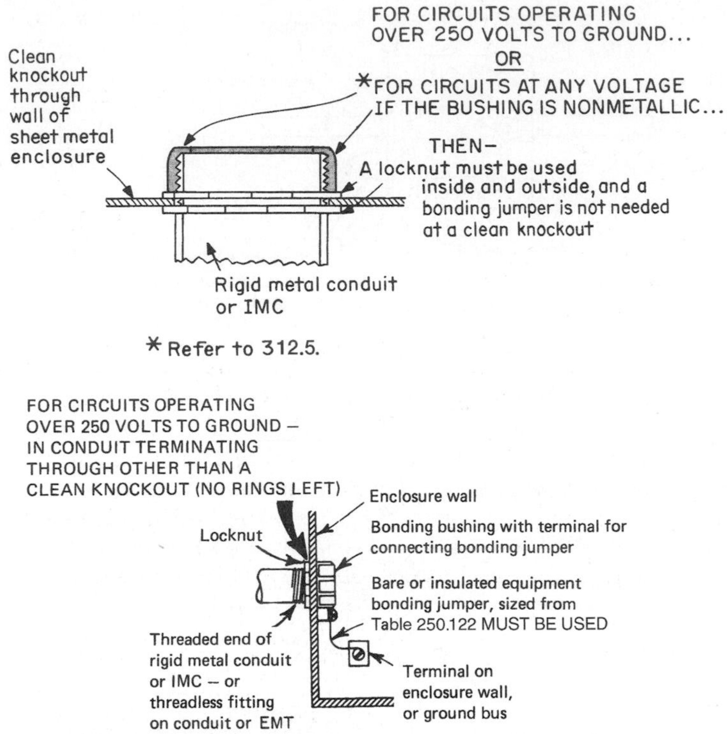
FOR CIRCUITS OPERATING
OVER 250 VOLTS TO GROUND...
OR
*FOR CIRCUITS AT ANY VOLTAGE
IF THE BUSHING IS NONMETALLIC...
THEN–
A locknut must be used
inside and outside, and a
bonding jumper is not needed
at a clean knockout
Clean
knockout
through
wall of
sheet metal
enclosure
Rigid metal conduit
or IMC
* Refer to 312.5.
FOR CIRCUITS OPERATING
OVER 250 VOLTS TO GROUND –
IN CONDUIT TERMINATING
THROUGH OTHER THAN A
CLEAN KNOCKOUT (NO RINGS LEFT)
Enclosure wall
Locknut
Bonding bushing with terminal for
connecting bonding jumper
Bare or insulated equipment
bonding jumper, sized from
Table 250.122 MUST BE USED
Threaded end of
rigid metal conduit
or IMC – or
threadless fitting
on conduit or EMT
Terminal on
enclosure wall,
or ground bus

Fig. 250-73. For circuits over 250 V to ground, a bonding jumper may be needed at conduit termination. (Sec. 250.97.)

The Exception to the main rule here has the effect of requiring that a bonding jumper must be used around any "oversized, concentric, or eccentric knockouts" in enclosures for circuits over 250 V to ground that are run in a metal raceway or cable unless the enclosure or fittings have been investigated and listed for use without a bonding jumper. Many such enclosures and fittings are so listed and are readily available, in particular, most conventional steel outlet boxes with concentric knockouts. Clearly, the use of that equipment will serve to reduce labor costs. But, generally for such circuits, a bonding jumper must be used at any conduit or cable termination in other than a clean, unimpaired opening in an enclosure (Fig. 250-73). For example, the majority of larger enclosures with concentric knockouts, including most panelboards and wireways, are not listed as providing the required continuity and a bonding jumper will be required.

In any case where all the punched rings (the "doughnuts") are not removed and the box is *not* listed for use without a bonding jumper, or, where all the rings are removed but a reducing washer is used to accept a smaller size of conduit, a bonding jumper must be installed from a suitable ground terminal in the enclosure to a lug on the bushing or locknut of the termination of any conduit or cable containing conductors operating at over 250 V to ground. Such circuits include 480/277-V circuits (grounded or ungrounded); 480-, 550-, 600/347-, and 600-V circuits; and higher-voltage circuits. A bonding jumper is not needed for terminations of conduit that carry such circuits through KOs that are punched on the job to accept the corresponding size of conduit. *But,* double locknuts (one inside and one outside the enclosure) must be used on threaded conduit ends, or suitable threadless connectors or other fittings must be used on rigid or flexible conduit, EMT, or cable.

250.98. Bonding Loosely Jointed Metal Raceways. Provision must be made for possible expansion and contraction in concrete slabs due to temperature changes by installing expansion joints in long runs of raceways run through slabs. See 300.7(B). Because such expansion joints are loosely jointed to permit back-and-forth movement to handle changes in gap between butting slabs, bonding jumpers must be used for equipment grounding continuity (Fig. 250-74). Expansion fittings may be selected as vibration dampers and deflection mediums as well as to provide for movement between building sections or for expansion and contraction due to temperature changes in long conduit runs. The fitting diagrammed in Fig. 250-74 provides for movement from the normal in all directions plus 30° deflection, is available up to 4-in. (101.5 mm) in diameter, and may be installed in concrete. Expansion fittings are available from some manufacturers that have special internal designs and which have enabled them to be listed as providing the required continuity without the use of supplemental bonding jumpers.

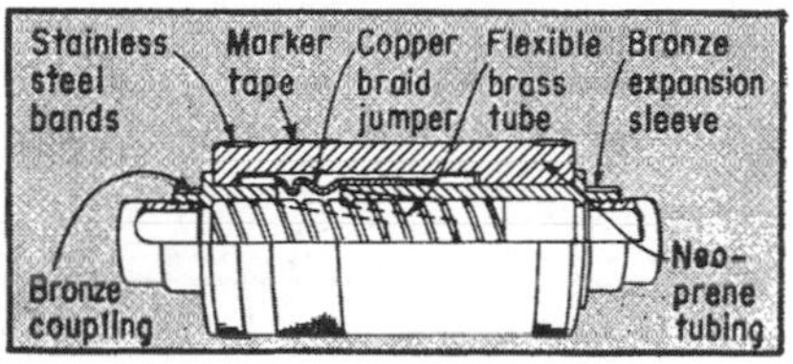
Stainless steel bands
Marker tape
Copper braid jumper
Flexible brass tube
Bronze expansion sleeve
Bronze coupling
Neoprene tubing

Fig. 250-74. Conduit expansion fitting includes bonding jumper for ground continuity. (Sec. 250.98.)

250.100. Bonding in Hazardous (Classified) Locations. All raceway terminations in hazardous locations must be made by one of the techniques shown in Fig. 250-67 for service raceways. And as required by 501.30(A) such bonding techniques must be used in "all intervening raceways, fittings, boxes, enclosures, etc., between hazardous areas and the point of grounding for service equipment." Refer to 501.30(A), 502.30(A), 503.30(A), 505.25, and 506.25 (Fig. 250-75).

250.102. Grounded Conductors, Bonding Conductors, and Jumpers. The jumper shown in Fig. 250-76 running from one conduit bushing to the other and then to the equipment ground bus is defined by the NEC as an "equipment bonding jumper." Equipment bonding jumpers must be made of copper or other corrosion-resistant material, and they can be in the form of wires, busbars, screws, or any other suitable conductor. They must be connected using one of the methods in 250.8 (or 250.70 if connecting to a grounding electrode). The rest of this section covers sizing calculations on the supply side of the service (Part C), sizing calculations on the load side of the service (Part D), and how to install these jumpers in the event they run with their associated conductors (Part E) instead of simply making bonding connections at terminations, as shown in Fig. 250-76.

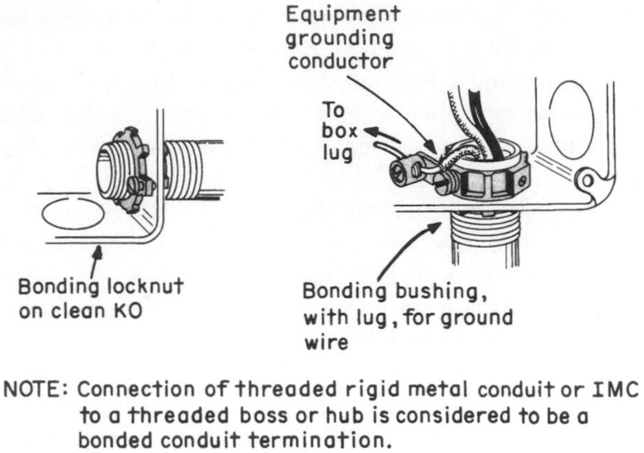
Equipment
grounding
conductor
To
box
lug
Bonding locknut
on clean KO
Bonding bushing,
with lug, for ground
wire
NOTE: Connection of threaded rigid metal conduit or IMC
to a threaded boss or hub is considered to be a
bonded conduit termination.

Fig. 250-75. Bonded raceway terminations must be used at sheet metal KOs in hazardous areas. (Sec. 250.100.)

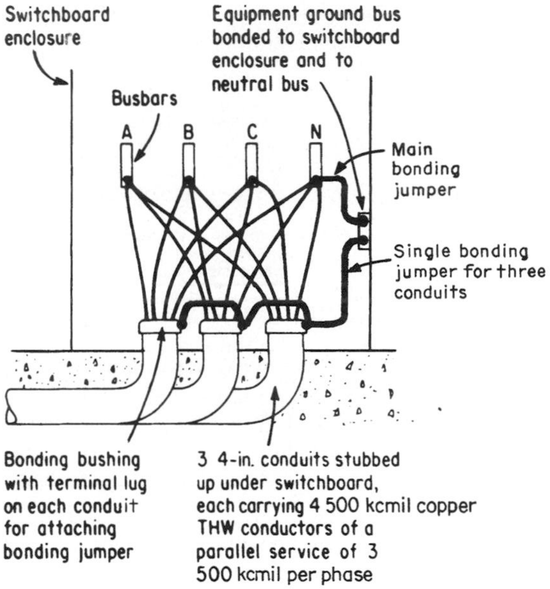
Switchboard
enclosure
Equipment ground bus
bonded to switchboard
enclosure and to
neutral bus
Busbars
A
B
C
N
Main
bonding
jumper
Single bonding
jumper for three
conduits
Bonding bushing
with terminal lug
on each conduit
for attaching
bonding jumper
3 4-in. conduits stubbed
up under switchboard,
each carrying 4 500 kcmil copper
THW conductors of a
parallel service of 3
500 kcmil per phase

Fig. 250-76. Sizing main bonding jumper and other jumpers at service equipment. (Sec. 250.102.)

Part (C) applies to the supply side of services. As of the 2014 **NEC**, the content of Table 250.66, with additional notes and requirements to reflect that resulting conductor sizes increase beyond those of grounding electrode conductors, has been placed here in a new Table 250.102(C)(1).

Part (C)(2) covers paralleled installations. It actually, incorporates many different calculations in one paragraph. The first involves one jumper to all the raceways of the circuit (Fig. 250-76). Since, in the process of making its connection to the bus, it sees all the raceways, it is sized on the basis of the entire service, for which the largest phase conductor (in this as in most cases all are equal) is 3 × 500 kcmil = 1500 kcmil. Since this is larger than the coverage of Table 250.102(C)(1), its size must be calculated on the basis of ⅛ (12.5 percent) of the phase area just determined. 1,500,000 cm ÷ 8 = 187,500 cm. Consulting Table 8 in Chap. 9 of the **NEC**, the next larger wire size is 4/0 copper, and that is the size required for this wire. This procedure is identical to the one in 250.28(D)(1) for a main bonding jumper.

In the sketch of Fig. 250-76, if each of the three 4-in. (101.5-mm) conduits has a separate bonding jumper connecting each one individually to the equipment ground bus, then part **(C)(2)** may be applied to an individual bonding jumper for each separate conduit (Fig. 250-77). The size of a separate bonding jumper for each

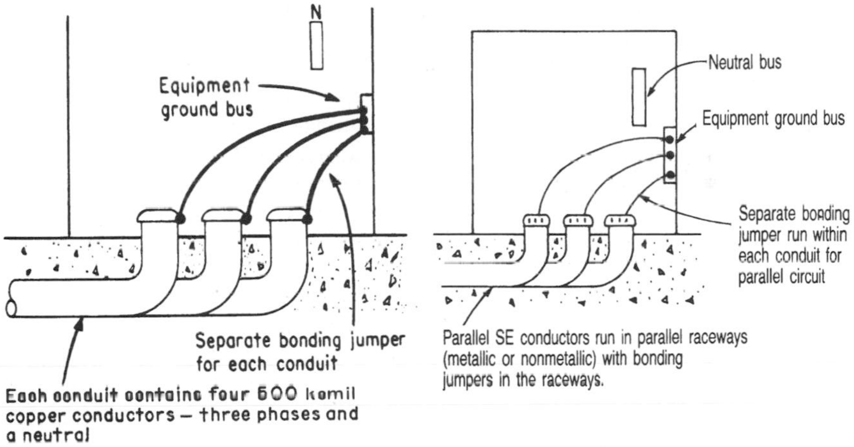
N
Equipment ground bus
Separate bonding jumper for each conduit
Each conduit contains four 500 kcmil copper conductors – three phases and a neutral
Neutral bus
Equipment ground bus
Separate bonding jumper run within each conduit for parallel circuit
Parallel SE conductors run in parallel raceways (metallic or nonmetallic) with bonding jumpers in the raceways.

Fig. 250-77. An individual bonding jumper may be used for each conduit (left) and *must* be used as shown at right. [Sec. 250.102(C).]

conduit in a parallel service must be not less than the size of the grounding electrode conductor for a service of the size of the phase conductor used in each conduit. Referring again to Table 250.102(C)(1), a 500-kcmil copper service calls for at least a No. 1/0 grounding electrode conductor. Therefore, the bonding jumper run from the bushing lug on each conduit to the ground bus must be at least a No. 1/0 copper (or 3/0 aluminum). Since these are bonding jumpers and not grounding electrode conductors, on a very large service or if multiple parallel sets of conductors ran in single conduits, the area of the largest phase conductor might exceed the reach of Table 250.102(C)(1), and in such an instance the ⅛ (12.5 percent) procedure would apply to determine the area of the bonding conductor.

Part **(C)(2)** *requires* separate bonding jumpers when the service is made up of multiple conduits and the equipment bonding jumper is run within each raceway (such as plastic pipe) for grounding service enclosures. According to part **(C),** when service-entrance conductors are paralleled in two or more raceways, a supply-side bonding jumper that is routed within the raceways must also be run in parallel, one in each raceway, as at the right in Fig. 250-77. This clarifies application of nonmetallic service raceway where parallel conduits are used for parallel service-entrance conductors. As worded, the rule applies to both nonmetallic and metallic conduits where the bonding jumper is run within the raceways rather than from lugs on bonding bushings on the conduit ends. But for metallic conduits stubbed up under service equipment, if the conduit ends are to be bonded to the service equipment enclosure by jumpers from lugs on the conduit bushings, either a single large common bonding jumper may be used—from one lug, to another lug, to another, and so on, and then to the ground bus—or an individual bonding jumper (of smaller size from Table 250.66, based on the size of conductors in each conduit) may be run from each bushing lug to the ground bus.

Do not confuse a supply-side bonding jumper run in a raceway with a grounded circuit conductor run in the same raceway. For grounded systems, that will be the usual application, and the only equipment bonding will be to the metallic service raceways in the event a metallic wiring method is in use. Remember that the grounded circuit conductor is permitted to, and usually does provide, the supply-side bonding in upstream enclosures such as metering cabinets, as covered in 250.142(A)(1). This rule generally applies to ungrounded or impedance-grounded services where there are no grounded circuit conductors running through the raceways, and the necessity is to maintain bonding of conductive surfaces upstream of the service disconnect.

One other point in this regard. These wires are bonding jumpers, not circuit conductors. The normal 1/0 AWG lower threshold for paralleled "phase, polarity, neutral, or grounded circuit conductors" [310.10(H)(1)] does not apply in this case. For example, if the conduit fill consisted of sets of 350 kcmil, the Table 250.102(C)(1) result for the bonding conductors would be 2 AWG, and they could be run in the conduit without being increased to 1/0 AWG.

Note 2 to Table 250.102(C)(1) sets minimum sizes of copper *and* aluminum service-entrance conductors above which a service bonding jumper must have a cross-sectional area "not less than 12½ percent of the area of the largest phase conductor." And the rule states that if the service conductors and the bonding jumper are of different material (i.e., service conductors are copper, say, and the jumper is aluminum), the minimum size of the jumper shall be based on the assumed use of phase conductors of the same material as the jumper and with an ampacity equivalent to that of the installed phase conductors (Fig. 250-78). In this case, three sets of 750 kcmil aluminum at 75°C have a combined ampacity of 1155 A. A copper service equal to or above this number would be three sets of 600 kcmil, or 1,800,000 cm. 1,800,000 ÷ 8 (or, × 0.125) = 225,000 cm. Reviewing Table 8 in Chap. 9, the next large size wire would be a 250 kcmil copper bonding jumper. The 2017 **NEC** further clarified this note by conditioning it on the supply conductors being above 1100 kcmil copper or 1750 kcmil aluminum because below these sizes, whether or not the materials are mixed the table should still be used.

Part **(D)** requires a bonding jumper on the load side of the service to be sized as if it were an equipment grounding conductor for the largest circuit with which it is used. And sizing would have to be done from Table 250.122, as follows.

Figure 250-79 shows a floor trench in the switchboard room of a large hotel. The conductors are feeder conductors carried from circuit breakers in the main switchboard (just visible in upper right corner of photo) to feeder conduits going out at left, through the concrete wall of the trench, and under the slab floor to the

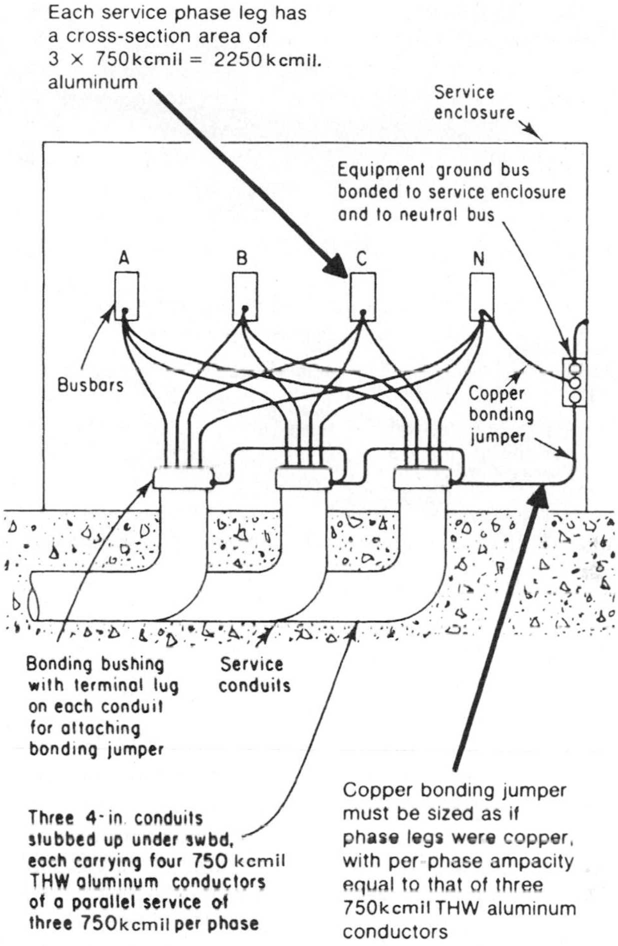
Each service phase leg has
a cross-section area of
3 × 750kcmil = 2250kcmil.
aluminum
Service
enclosure
Equipment ground bus
bonded to service enclosure
and to neutral bus
A
B
C
N
Busbars
Copper
bonding
jumper
Bonding bushing
with terminal lug
on each conduit
for attaching
bonding jumper
Service
conduits
Three 4-in. conduits
stubbed up under swbd,
each carrying four 750 kcmil
THW aluminum conductors
of a parallel service of
three 750kcmil per phase
Copper bonding jumper
must be sized as if
phase legs were copper,
with per-phase ampacity
equal to that of three
750kcmil THW aluminum
conductors

Fig. 250-78. Sizing a copper bonding jumper for aluminum service conductors. [Sec. 250.102(C).]

various distribution panels throughout the building. Because the conduits themselves are not metallically connected to the metal switchboard enclosure, bonding must be provided from the conduits to the switchboard ground bus to ensure electrical continuity and conductivity as required by **NEC** Secs. 250.4(A)(5), 250.86, 250.110(5), 250.120, and 250.134.

1. The single, common, continuous bonding conductor that bonds all the conduits to the switchboard must be sized in accordance with **NEC** Table 250.122, based on the highest rating of CB or fuses protecting any one of the total number of circuits run in all the conduits.
2. Sizing of the single, common bonding jumper would be based on the highest rating of overcurrent protection for any one of the circuits run in the group of

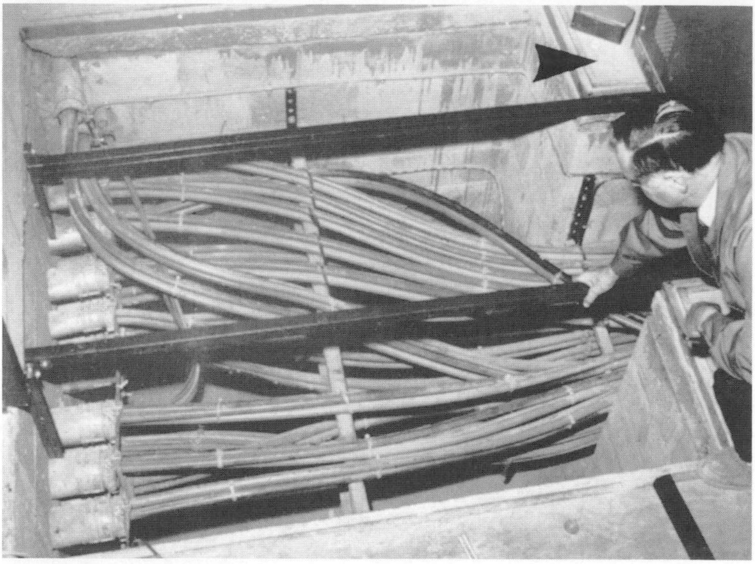

Fig. 250-79. Conduits in trench carry feeder conductors from switchboard at right (arrow) out to various panels and control centers. A single, common bonding jumper—run continuously from bushing to bushing—may be used to bond all conduits to the switchboard ground bus. [Sec. 250.102(D).]

conduits. For instance, some of the circuits could be 400-A circuits made up of 500 kcmils in individual 3-in. (76-mm) conduits, and others could be parallel-circuit makeups in multiple conduits—such as 800-A circuits, with two conduits per circuit, and 1200-A circuits, with three conduits. If, for instance, the highest-rated feeder in the group was protected by a 2000-A circuit breaker, then the single, common bonding jumper for all the conduits would have to be 250-kcmil copper or 400-kcmil aluminum—determined readily from Table 250.122, by simply going down the left column to the value of "2000" and then reading across. The single conductor is run through a lug on each of the conduit bushings and then to the switchboard ground bus.

In the case shown in Fig. 250-79, however, because the bonding jumper from the conduit ends to the switchboard is much longer than a jumper would be if the conduits stubbed up under the switchboard, better engineering design might dictate that a separate equipment grounding *conductor* (rather than a "jumper") be used for each individual circuit in the group. If one of the conduits is a metric designator 78 (trade size 3) conduit carrying three 500-kcmil conductors from a 400-A CB in the switchboard, the minimum acceptable size of bonding jumper (or equipment grounding conductor) from a grounding bushing on the conduit end to the switchboard ground bus would be a 3 AWG copper or 1 AWG aluminum or copper-clad aluminum, as shown opposite the value of 400 A in the left column of Table 250.122. If another two of the metric designator 78 (trade size 3) conduits are used for a feeder consisting of two parallel sets of three 500-kcmil conductors (each set of three 500-kcmil conductors in a separate conduit) for a circuit protected at 800 A, a single bonding jumper could be used, run from one grounding bushing to the other grounding bushing and then to the switchboard ground bus. This single bonding jumper would have to be a minimum 1/0 AWG copper, from **NEC** Table 250.122 on the basis of the 800-A rating of the feeder overcurrent protective device.

With such a long run for a jumper, as shown in Fig. 250-79, Code rules could be interpreted to require that the bonding jumper be subject to the rules of 250.122; that is, use of a bonding jumper must conform to the requirements for equipment grounding conductors. As a result, bonding of conduits for a parallel circuit makeup would have to comply with part **(F)** in 250.122, which requires equipment grounding conductors to be run in parallel "where conductors are run in parallel in multiple raceways. . . ." That would then be taken to require that bonding jumpers *also* must be run in parallel for multiple-conduit circuits.

That interpretation is incorrect. The rules in 250.122(F) assure an appropriate fault current path based on the possibility of backfeed into a fault from a downstream parallel bus connection, resulting in the maximum current flow from the overcurrent protective device. Therefore the equipment grounding conductor in each parallel raceway will be paralleled and will be based on the full trip setting of the breaker. When the parallel raceways reach the switchboard the equipment grounding conductors no longer need to be paralleled; they only need to remain sized based on Table 250.122 for the overcurrent device. This is entirely consistent with 250.122(C), which expressly allows a single equipment grounding conductor to protect any number of circuits as long as it is sized based on the highest rating of overcurrent protection on any of the associated conductors. However, it is certainly permissible to run individual conductors. Figure 250-80 shows the two possible arrangements.

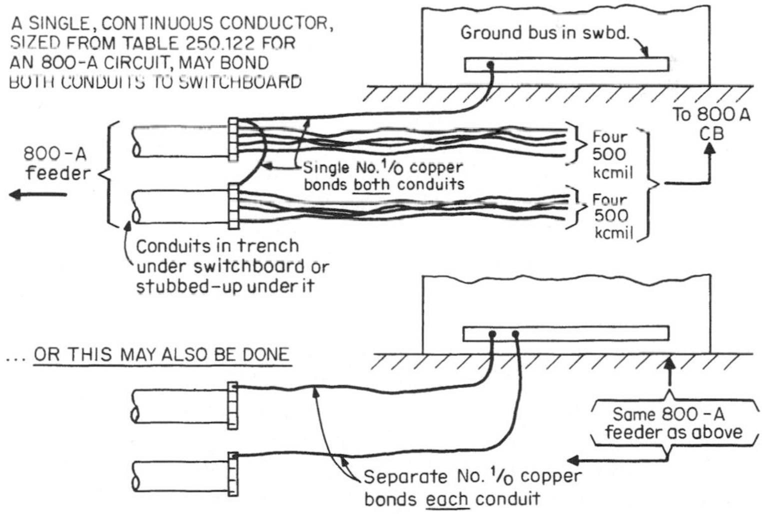
A SINGLE, CONTINUOUS CONDUCTOR, SIZED FROM TABLE 250.122 FOR AN 800-A CIRCUIT, MAY BOND BOTH CONDUITS TO SWITCHBOARD
Ground bus in swbd.
To 800 A CB
800-A feeder
Single No. 1/0 copper bonds both conduits
Four 500 kcmil
Four 500 kcmil
Conduits in trench under switchboard or stubbed-up under it
... OR THIS MAY ALSO BE DONE
Same 800-A feeder as above
Separate No. 1/0 copper bonds each conduit

Fig. 250-80. One jumper may be used to bond two or more conduits on the load side of the service. [Sec. 250.102(D).]

Part **(E)** of 250.102 follows the thinking that is described in 250.136 for external grounding of equipment attached to a properly grounded metal rack or structure. A short length of flexible metal conduit, liquidtight flex, or any other raceway may, if the raceway itself is not acceptable as a grounding conductor, be provided with grounding by a "bonding jumper" (*note: not* an "equipment grounding conductor") run *either* inside or *outside* the raceway or enclosure PROVIDED THAT the *length* of the *equipment bonding jumper* is *not more* than 6 ft (1.8 m) and the jumper is routed with the raceway or enclosure.

Where an equipment bonding conductor is installed within a raceway, it must comply with all the Code rules on identification of equipment grounding conductors. A bonding jumper installed in flexible metal conduit or liquidtight flex serves essentially the same function as an equipment grounding conductor. For that reason, a bonding jumper should comply with the identification rules of 250.119—on the use of bare, green-insulated, or green-taped conductors for equipment grounding.

Note that this application has limited use for the conditions specified and is a special variation from the concept of 250.134(B), which requires grounding conductors run inside raceways. Its big application is for external bonding of short lengths of liquidtight or standard flex, under those conditions where the particular type of flex itself is not suitable for providing the grounding continuity required by 348.60 and 350.60. Refer also to 250.64.

The top of Fig. 250-81 shows how an external bonding jumper may be used with standard flexible metallic conduit (so-called Greenfield). If the length of the flex is not over 6 ft (1.8 m), but the conductors run within the flex are protected at more than 20 A, a bonding jumper *must* be used either inside or outside the flex. An outside jumper must comply as shown. For a length of listed flex not over 6 ft (1.8 m), containing conductors that are protected at not more than 20 A and used with conduit termination fittings that are approved for grounding, a bonding jumper is *not* required.

The bottom of Fig. 250-81 shows use of an external bonding jumper with liquidtight flexible metallic conduit. If liquidtight flex is not over 6 ft (1.8 m) long *but* is larger than 1¼-in. trade size, a bonding jumper must be used, installed *either* inside or outside the liquidtight. An outside jumper must comply as shown. If a length of liquidtight flex larger than 1¼ in. is short enough to permit an external bonding jumper that is not more than 6 ft (1.8 m) long between external grounding-type connectors at the ends of the flex, an external bonding jumper may be used. BUT WATCH OUT! The rule says the *jumper,* not the flex, must not exceed 6 ft (1.8 m) in length *AND* the jumper "shall be routed with the raceway"—that is, run along the flex surface and not separated from the flex.

The exception at the end of this section provides some practical assistance to isolated steel sweeps that are frequently installed in runs of nonmetallic conduit to avoid problems associated with pulling ropes sawing through the inner radius of a nonmetallic sweep during a heavy pull. Since these sweeps are seldom located where a box could be conveniently located, making a connection to the enclosed equipment grounding conductor is problematic. In addition to the isolation provisions of 250.80 Exception and 250.86 Exception No. 3, the exception at this location provides help on a steel sweep positioned at the base of a pole. The exception also

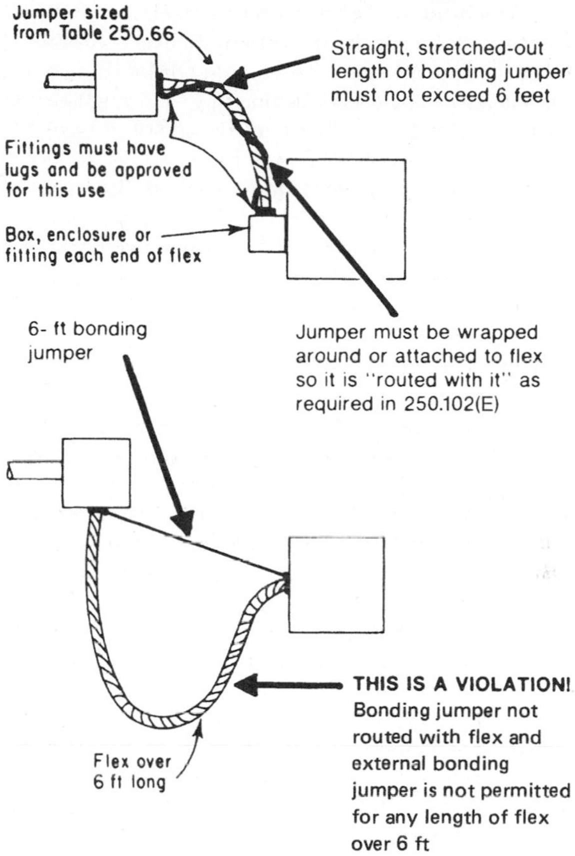
Jumper sized from Table 250.66
Straight, stretched-out length of bonding jumper must not exceed 6 feet
Fittings must have lugs and be approved for this use
Box, enclosure or fitting each end of flex
6- ft bonding jumper
Jumper must be wrapped around or attached to flex so it is "routed with it" as required in 250.102(E)
THIS IS A VIOLATION!
Bonding jumper not routed with flex and external bonding jumper is not permitted for any length of flex over 6 ft
Flex over 6 ft long

Bonding jumper required for flex may be outside the flex.

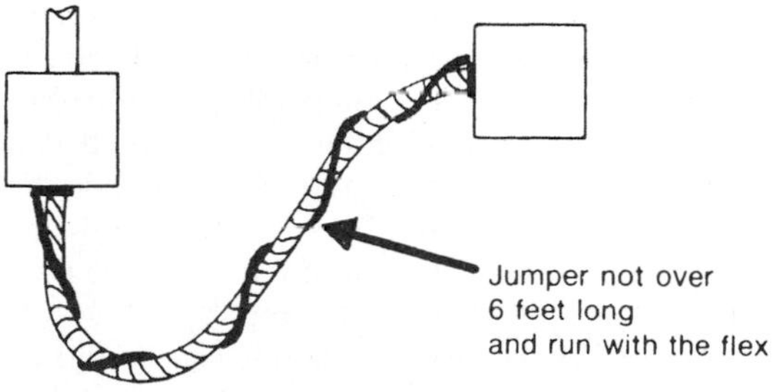
Jumper not over 6 feet long and run with the flex

Liquidtight flex larger than 1¼ in. size must have bonding jumper — inside or outside — in any length up to 6 ft.

Fig. 250-81. Bonding jumper rules for standard flex and liquidtight flex. [Sec. 250.102(E).]

covers instances where not only the elbow is steel, but the first pipe section on the pole is steel as well, for mechanical strength. The exception allows a bonding jumper to be attached to the steel at a convenient location, and then run up the pole to a point where the enclosed equipment grounding conductor (load side of service) or grounded circuit conductor (supply side of service) is available for connection, even if the length exceeds 1.8 m (6 ft).

250.104. Bonding of Piping Systems and Exposed Structural Metal. This section on bonding of piping systems in buildings begins with two parts—metal *water* piping and *other* metal piping. This section is a rather elaborate sequence of phrases that may be understood in several ways. Of course, the basic concept is to ground any metal pipes that would present a hazard if energized by an electrical circuit. In general, bear in mind that these bonding requirements apply to "piping systems." With the increasing use of nonmetallic piping in many occupancies, controversies are arising over how short a section of piping has to be before it no longer qualifies as a "piping system." This is a decision for the inspector, perhaps based in part on the proximity to other electrical equipment or wiring. However, it is quite clear that an isolated chrome faucet does not create a metal piping system, but metal piping covering an entire floor presumably is a system. The third part of the section covers interconnected structural metal framing that supports a building, and the section concludes by correlating the grounding rules for separately derived systems in 250.30 with the bonding rules in this part of the article.

Part **(A)** requires any "metal water piping system(s) installed in or attached to a building or structure" to be bonded to the service-equipment enclosure, the grounded conductor (usually, a neutral) at the service, the grounding electrode conductor, *or* the one or more grounding electrodes used (with the obvious proviso that GEC to the electrode be of the appropriate size). All points of attachment of bonding jumpers for metal water piping systems must be accessible. Only the connections (and not the entire length) of water-pipe bonding jumpers are required to be accessible for inspection. This rule applies where the metal water piping system does not have 10 ft (3.0 m) of metal pipe buried in the earth and is, therefore, not a grounding electrode. In such cases, though, this rule makes clear that the water piping system must be bonded to the service grounding arrangement. And the bonding jumper used to connect the interior water piping to, say, the grounded neutral bus or terminal (or to the ground bus or terminal) must be sized from Table 250.102(C)(1) based on the size of the service conductors. The jumper is sized from that table because that is the table that would have been used *if* the water piping had 10 ft (3.0 m) buried under the ground, making it suitable as a grounding electrode. Note that the "bonding jumper" is sized from Table 250.102(C)(1), which means it never has to be larger than 3/0 AWG copper or 250-kcmil aluminum. Refer to the illustrations for 250.52 and 250.53, which also cover bonding of water piping.

Note that the status of the water piping system as a bonding conductor used to extend the connection to the actual water pipe electrode, particularly where compliance with 250.68(C)(1) Exception has been achieved, frequently extends very long distances into a building. Other sections of the **NEC** allow users to count on this when grounding other systems, including separately derived systems as covered in 250.30(A)(4)(1). If the appropriate bonding has not been performed and another system is connected, the result will be a hazard.

Part **(A)(2)** permits "isolated" metal water piping to be bonded to the main electrical enclosure (panelboard or switchboard) in each unit of a multitenant building—such as in each apartment of an apartment house, each store of a shopping center, or each office unit of a multitenant office building. See the top of Fig. 250-82. This rule is intended to provide a realistic and effective way to bond interior metal water piping to the electrical grounding system in multitenant buildings where the metal water piping in each tenant's unit is fed from a main water distribution system of nonmetallic piping and is isolated from the metal water piping in other units. In apartment houses, multistory buildings, and the like, it would be difficult, costly, and ineffective to use long bonding jumpers to tie the isolated piping in all the units back to the equipment grounding point of the building's service equipment—as required

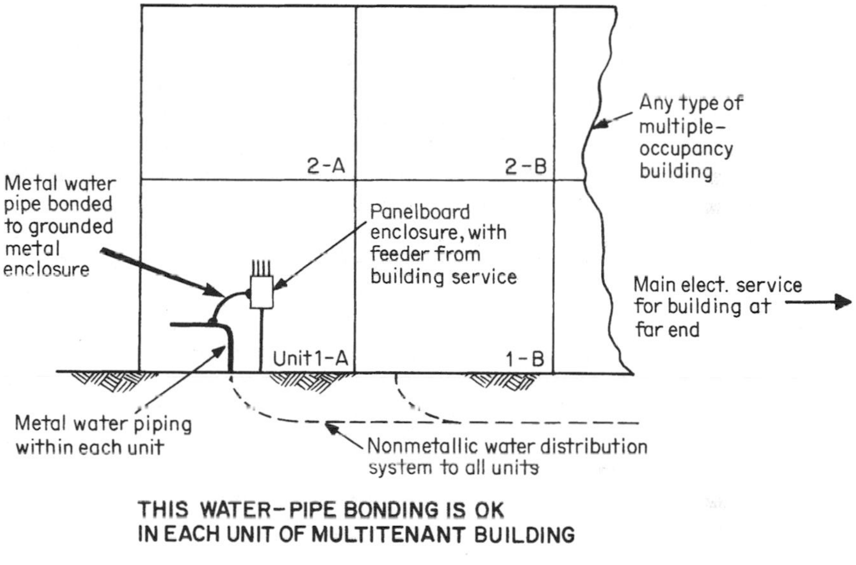
Any type of multiple-occupancy building
2-A
2-B
Metal water pipe bonded to grounded metal enclosure
Panelboard enclosure, with feeder from building service
Main elect. service for building at far end
Unit 1-A
1-B
Metal water piping within each unit
Nonmetallic water distribution system to all units
THIS WATER-PIPE BONDING IS OK IN EACH UNIT OF MULTITENANT BUILDING

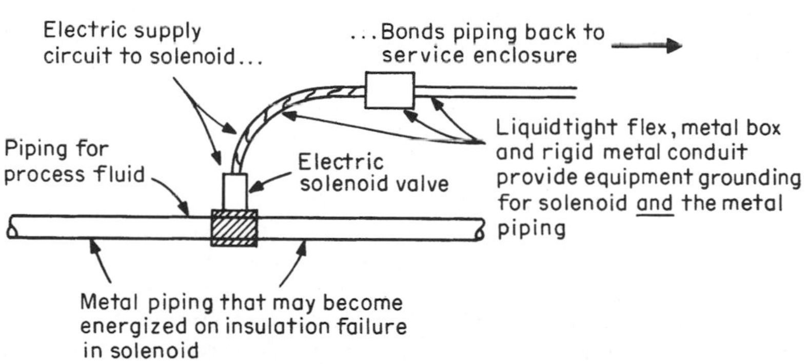
Electric supply circuit to solenoid...
...Bonds piping back to service enclosure
Liquidtight flex, metal box and rigid metal conduit provide equipment grounding for solenoid and the metal piping
Piping for process fluid
Electric solenoid valve
Metal piping that may become energized on insulation failure in solenoid

Fig. 250-82. Certain techniques are permitted as alternatives to the basic rules on grounding of metal piping systems. (Sec. 250.104.)

by the basic rule of Sec. 250.104(A). The objective of the basic rule is better achieved in such cases by simply bonding the isolated water piping in each occupancy to the equipment ground bus of the panelboard or switchboard serving the occupancy. The bonding jumper must be sized from Table 250.122 (*not* 250.66)—based on the rating of the protective device for the circuit supplying the occupancy.

Part **(A)(3)** covers instances where multiple buildings or structures under common ownership have water piping systems. In such cases the bonding follows the rules for isolated tenancies in one building, as just covered, except that the reference point in this case is the building disconnect, and the sizing applies Table 250.102(C)(1) to the size of the feeder or branch-circuit conductors that supply the building, and not larger than the supply conductors. If there is no building disconnect the connection can be made to the equipment grounding conductor of the supply circuit, or to the grounding electrode conductor for the building served.

Part **(B)** requires a bonding connection from "other" (than water) metal piping systems—such as process liquids or fluids—that "is likely to become energized" to the grounded neutral, the service ground terminal, the grounding electrode conductor, or the grounding electrodes. *But,* for these *other* piping systems the bonding jumper is sized from Table 250.122, using the rating of the overcurrent device of the circuit that *may* energize the piping. Since as a practical matter it is almost impossible to wire anything without an equipment grounding conductor in the circuit, and since equipment grounding conductors must be sized in accordance with 250.122, if there is any conductive connection between the equipment supplied by the branch circuit and the piping system, the equipment grounding connection will automatically bond the piping with no further effort required. Most solenoids and other control devices for piping qualify as bonded under this principle (see bottom of Fig. 250-82).

The phrase "likely to become energized" is used frequently in electrical standards. It basically means that it is capable of being energized if a single insulation failure occurs. For example, a motor supplied by a conventional motor circuit is "likely to become energized" because a single insulation failure on an ungrounded conductor in the usual metal termination box will energize the motor frame, subject to the operation of OCPDs. Double-insulated appliances are not "likely to become energized" because the supply conductors enter a nonmetallic housing, and two points of insulation failure (the conductor insulation and the nonmetallic connection housing) would have to fail before a voltage could be present on the surface of the equipment.

Part **(C)** of 250.104 requires exposed building steel that is *not* effectively grounded to be bonded to the grounding electrode system. The connection may be made at the service equipment, to the grounded conductor, to the grounding electrode conductor where the grounding electrode conductor is "of sufficient size," or to any of the electrodes used. If the building or structure is supplied from another, as covered in 250.32, then the bonding may be made to the local building disconnect. The reference to "sufficient size" for connection to the grounding electrode conductor is intended to indicate that such a conductor must be sized per 250.102(C)(1) based on the size of the largest service phase conductor. But, where, say, 6 AWG copper is used for connection to a driven ground rod—as permitted in part **(A)** to 250.102(C)(1)—the bonding jumper from the nongrounded building must *not* be connected to that 6 AWG conductor, unless it also satisfies the basic rule in 250.102(C)(1). The same concept applies to connections to concrete-encased electrodes and ground rings, which have permission for less than full-sized conductors and therefore may or may not satisfy the "of sufficient size" criterion.

In part **(D)**, this section puts forth the bonding requirements for separately derived systems in three parts. Paragraph **(1)** provides that the grounded conductor must be bonded to the interior metal water piping system within the area served by the separately derived system. This is in addition to the bonding connection required between the grounded conductor and the building steel, or where no building steel is available, the connection to the water pipe, not more than 5 ft from its point of entry. This local connection ensures that the metal water piping within the area supplied by the separately derived system is at the same potential with respect to ground as the metal enclosures and raceways associated with the separately derived system. Further, the electrical connections to both the bonding conductor and the grounding electrode conductor must occur at the same point. For example, if the grounding electrode conductor is connected in the secondary compartment of a transformer, then the bonding conductor must be connected in the same transformer compartment. The bonding conductor size is governed by Table 250.102(C)(1) based on the largest ungrounded conductor of the derived system.

If the provisions of 250.68(C)(1) Exception successfully apply as far as the separately derived system location, then this bonding connection will also qualify as the termination point for the grounding electrode conductor. Since the wiring and sizing rules are the same for both, no adjustments are necessary. On the other hand, if 250.68(C)(1) Exception does not apply, then the grounding electrode connection and the bonding connection will usually be in two entirely different areas. In addition, if the grounding electrode for the derived system is local building steel, and the steel is bonded to the local water piping system, there is no point in making a second bonding connection to water piping in the area served.

Paragraph **(2)** covers structural metal framing that is exposed in an area supplied by a separately derived system. Here again the system grounded conductor and the conductive framing must be bonded in the area served, and both the bonding connection and the grounding electrode connection to this conductor must occur at the same point. For example, if in this case the grounding electrode conductor connection takes place in the main disconnect enclosure for the derived system, then the bonding connection must connect there and not in the transformer housing. If the local metal framing is the grounding electrode for the derived system, then an additional bonding connection is not required for obvious reasons. Likewise, if the local water piping system is the grounding electrode for the derived system, and the water system and the metal framing are bonded in the area served by the derived system, then no further bonding is required. Here again, the bonding conductor size is governed by Table 250.102(C)(1) based on the largest ungrounded conductor of the derived system.

Paragraph **(3)** covers bonding in areas where a common grounding electrode conductor is available, such as might be run vertically in a high-rise building, as discussed in 250.30(A)(6). In this case, the derived system will be using the common grounding electrode conductor as its ground reference point. Both the local water piping (if any in the area served) and the local exposed structural metal framing (if any in the area served) must be bonded to the derived system. Presumably (although the NEC does not specifically say this) the connection should be made at the same location in the derived system as where the grounding electrode connection is made, in order to correlate this requirement with the other two. If these connections are separated, there is a possibility of a potential difference and circulating currents that would be objectionable under 250.6(A).

250.106. Lightning Protection Systems. Lightning discharges, with their steep wave fronts, build up tremendous voltages to metal near the strike termination devices and the down conductors. Although the grounding electrode systems will be bonded, the actual electrodes must be separated, and NFPA 780, Standard for Lightning Protection Systems has detailed information on grounding, bonding, sideflash distances, and required spacings for the safe operation of these systems. Former specific references to such dimensions as "1.8 m (6 ft)" have been removed from the associated notes in this section, because they are far too simplistic to be of any real value,

250.110. Equipment Fastened in Place (Fixed) or Connected by Permanent Wiring Methods. The word "fixed" as applied to equipment requiring grounding now applies to "equipment supplied by or enclosing conductors or components that are likely to become energized" as covered in the section title and shown in Fig. 250-83. This is another example of the "likely to become energized" phrasing; refer to the discussion at 250.104(B) for an explanation of this concept. As noted in Exception No. 3, enclosures for listed equipment—such as information technology equipment and listed office equipment—operating at over 150 V to ground do not have to be grounded if protected by a system of double insulation or its equivalent. Since equipment grounding conductors travel with circuit conductors almost everywhere in modern wiring, item (5) alone on the list mandates an equipment grounding connection, even if the equipment does not meet the vertical or horizontal parameters in item (1).

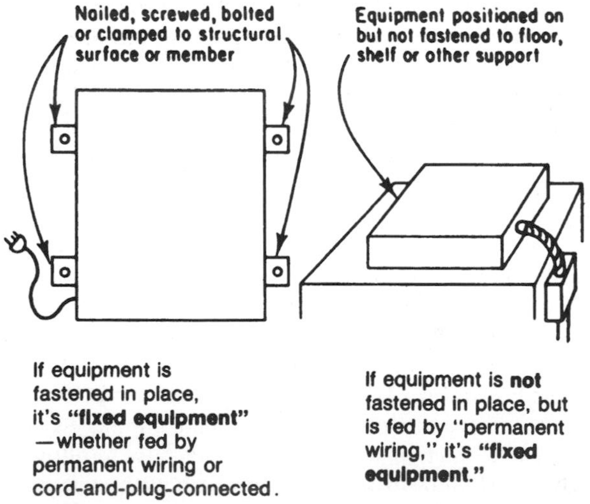
Nailed, screwed, bolted or clamped to structural surface or member
Equipment positioned on but not fastened to floor, shelf or other support
If equipment is fastened in place, it's "fixed equipment" —whether fed by permanent wiring or cord-and-plug-connected.
If equipment is not fastened in place, but is fed by "permanent wiring," it's "fixed equipment."

Fig. 250-83. "Fixed" equipment is now clearly and readily identified for grounding rules. (Sec. 250.110.)

250.112. Specific Equipment Fastened in Place (Fixed) or Connected by Permanent Wiring Methods. This section presents what amounts to an incomplete, yet binding, roster of equipment that must be connected to an equipment grounding conductor. The various subparts identify those items that are *specifically* required to be grounded.

Part **(I)** requires special commentary because of its close and often poorly understood connection with 250.20(A) and 250.162. If system grounding is required by either of these rules, then the remote control, signaling, or fire alarm circuit supplied must incorporate an equipment grounding conductor. For example, a 277-V duct heater with an integral Class 2 transformer to run the thermostatic control circuit, or some Class 2 or Class 3 lighting controls connected to 277-V branch circuits will qualify for mandatory system grounding, with the associated identified grounded circuit conductor.

Should such an application arise, there are additional considerations. Such a grounded system must be connected to a grounding electrode, although this is seldom a problem because the systems normally qualify as separately derived, and **NEC** 250.30(A)(5) Exception No. 3 generally allows the use of the equipment grounding conductor connection for the supply transformer (limited in the exception to 1 kVA) to function as a grounding electrode conductor in such cases. Another wrinkle is the requirement here that these systems incorporate an equipment grounding conductor. Such a conductor must be colored green unless it is uninsulated because 250.119 Exception correlates with 250.112(I) and preserves the color code in these applications. Note also that the size of this wire can be the same as the others in the circuit because 250.122(A) never requires an equipment grounding conductor to be larger than the associated circuit conductors. In addition, the color white (or gray) will be reserved for the grounded circuit conductor, because 200.7(B) reserves that color when, you guessed it, system grounding is required by 250.20(A). The likely result is two wires in the power-limited cable assembly (white and green) will have code required functions and connections at odds with the traditional color code for HVAC wiring diagrams. Take care to plan accordingly.

As required by part **(L)**, motor-operated water pumps, including the submersible type, must have their metal frames grounded. This Code rule clarifies an issue that was a subject of controversy. It means that a circuit down to a submersible pump in a well or cistern must include a conductor to ground the pump's metal frame, even though the frame is not accessible or exposed to contact by persons. This is why the parent sentence of the rule refers to exposed conductive parts on items (A) through (K), and all such parts of items (L) and (M). The water pump and the well casing must be bonded even if not exposed to direct contact.

Notice, too, that 250.112(M) makes grounding of metal well-casings mandatory for ALL types of occupancies. The grounding of metal well-casings was formerly specified for "Agricultural Buildings" in 547.8(D) until the issuance of the 1996 **NEC**. In recognition of the fact that the shock hazard exists where a metal well-casing is used—regardless of the type of occupancy or installation—this section requires grounding ALL metal well-casings used with "submersible pumps." However, that rule in Art. 547 was eliminated because it was viewed as redundant. Remember, according to the rule of 90.3, any requirements put forth in Chaps. 1

to 4 apply to those special occupancies covered in Chap. 5, unless otherwise modified by the rules in Chap. 5. Therefore, extension of this requirement to other occupancies by inclusion in 250.112 made the wording in 547.8(D) superfluous and that section was deleted.

250.114. Equipment Connected by Cord and Plug. Figure 250-84 shows cord-connected loads that must either be operated grounded or be double insulated. Except when supplied through an isolating transformer as permitted by the exception following **(4)(g)** of this section, the frames of portable tools should be grounded by means of an equipment grounding conductor in the cord or cable through which the motor is supplied. Portable hand lamps used inside boilers or metal tanks should preferably be supplied through isolating transformers having a secondary voltage of 50 V or less, with the secondary ungrounded. Code-recognized double-insulated tools and appliances may be used in all types of occupancies other than hazardous locations, in lieu of required grounding. Note that cord-and-plug-connected equipment operating over 150 V to ground in instances where it consists of motors that are guarded, or if it consists of metal-framed electrically heated appliances exempted by special permission, and where the frames are permanently insulated from ground, is exempted from the usual equipment grounding requirement.

OSHA regulations have made NEC 250.114 retroactive, requiring grounded operation of cord- and plug-connected appliances in all existing as well as new installations. Check on local rulings on that matter.

250.118. Types of Equipment Grounding Conductors. This section describes the various types of conductors and metallic cables or raceways that are considered suitable for use as equipment grounding conductors. And the Code recognizes cable tray as an equipment grounding conductor as permitted by Art. 392.

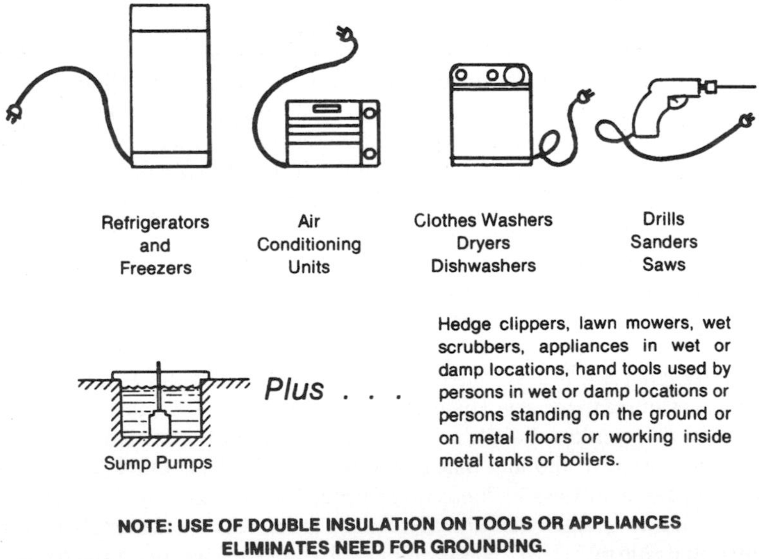
Refrigerators and Freezers
Air Conditioning Units
Clothes Washers Dryers Dishwashers
Drills Sanders Saws
Sump Pumps
Plus . . .
Hedge clippers, lawn mowers, wet scrubbers, appliances in wet or damp locations, hand tools used by persons in wet or damp locations or persons standing on the ground or on metal floors or working inside metal tanks or boilers.
NOTE: USE OF DOUBLE INSULATION ON TOOLS OR APPLIANCES ELIMINATES NEED FOR GROUNDING.

Fig. 250-84. Grounding cord and plug cap are required for shock protection. (Sec. 250.114.)

250.118(5), (6), and (7) recognize listed flexible metal conduit, listed liquidtight flexible metal conduit, and flexible metallic tubing, with termination fittings UL-listed for use as a grounding means (without a separate equipment grounding wire) if the total length of flexible methods is not over 6 ft (1.8 m) and the contained circuit conductors are protected by overcurrent devices rated at 20 A or less.

Standard flexible metal conduit (also known as "Greenfield") must be listed by UL but the former allowance for flex listed as a grounding means has been deleted because none has, shall we say, a credible listing. However, part **(5)** permits flex to be used without any supplemental grounding conductor when any length of flex in a ground return path is not over 6 ft (1.8 m) and the conductors contained in the flex are protected by overcurrent devices rated not over 20 A and the *fittings* are listed as suitable for grounding (Fig. 250-85). The **NEC** rules follow:

1. When conductors within a length of flex up to 6 ft (1.83 m) are protected at more than 20 A, equipment grounding may not be provided by the flex, but a separate conductor must be used for grounding. If a length of flex is short enough to permit a bonding jumper not over 6 ft (1.83 m) long to be run between external grounding-type connectors at the flex ends, while keeping the jumper *along* the flex, such an external jumper may be used where equipment grounding is required—as for a short length of flex with circuit conductors in it protected at more than 20 A. Of course, such short lengths of flex may also be "bonded" by a bonding jumper inside the flex, instead of external. Refer to 250.102(E).

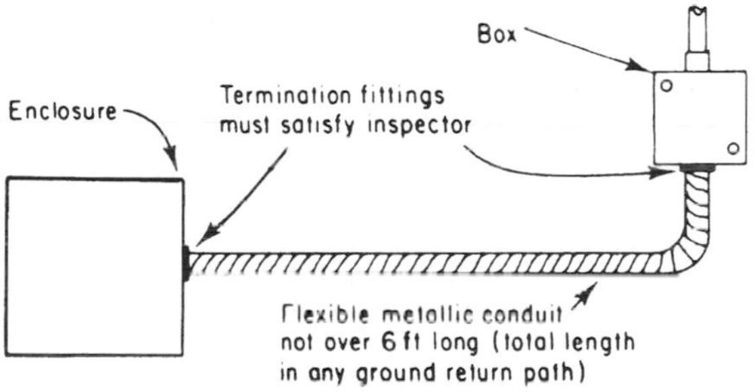
Box
Enclosure
Termination fittings
must satisfy inspector
Flexible metallic conduit
not over 6 ft long (total length
in any ground return path)

Flex not over 6 ft long is suitable as a grounding means (without a separate ground wire) if the conductors in it are protected by OC devices rated not more than 20 amps.

Fig. 250-85. Standard flex is limited in use without an equipment ground wire. (Sec. 250.118.)

2. Any length of standard flex that would require a bonding jumper longer than 6 ft (1.8 m) may not use an external jumper. In the Code sense, when the length of such a grounding conductor exceeds 6 ft (1.8 m), it is *not* a BONDING JUMPER BUT *IS* AN EQUIPMENT GROUNDING CONDUCTOR AND MUST BE RUN ONLY *INSIDE* THE FLEX, AS REQUIRED BY 250.134(B). Combining UL data with the rule of Sec. 250.102, *every* length of flex that is over 6 ft (1.8 m) must contain an equipment grounding conductor run *only* inside the flex (Fig. 250-86).

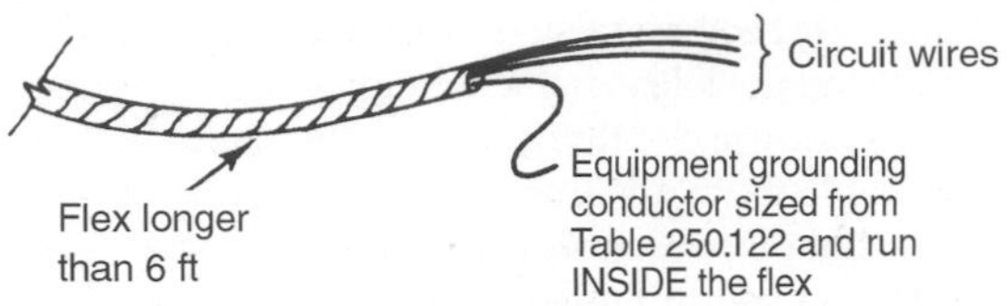
Circuit wires
Equipment grounding conductor sized from Table 250.122 and run INSIDE the flex
Flex longer than 6 ft

Fig. 250-86. Internal equipment grounding is required for any flex over 6 ft (1.8 m) long. (Sec. 250.118.)

In part **(5)(c)**, the 2017 **NEC** has now ratified an earlier posting in the UL guide card limitations for flex that additionally condition the equipment grounding suitability on not being more than a maximum raceway size, in this case 1¼ trade size. Even if the circuit protection is held at 20 amps as been acceptable for generations, now larger trade sizes must have a separate equipment grounding conductor installed.

In part **(5)(d)**, it should be noted that exemption from the need for an equipment grounding conductor applies only to flex where there is not over 6 ft (1.8 m) of "total length in the same ground return path." That means that from any branch-circuit load device—lighting fixture, motor, and so forth—all the way back to the service ground, the total permitted length of flex without a ground wire is 6 ft (1.8 m). In the total circuit run from the service to any outlet, there could be one 6-ft (1.8-m) length of flex or two 3-ft (900-mm) lengths or three 2-ft (600-mm) lengths or a 4-ft (1.2-m) and a 2-ft (600-mm) length—where the flex lengths are in series as equipment ground return paths. In any circuit run—feeder to subfeeder to branch circuit—any length of flex that would make the total series length over 6 ft (1.8 m) would have to use an internal or external bonding jumper, regardless of any other factors. Further, this principle extends to all varieties of flex covered here, taken in their collective lengths. For example, 600 mm (2 ft) of flexible metal conduit plus the same length of liquidtight flexible metal conduit plus the same length of flexible metallic tubing just equals the longest permitted length (1.8 m [6 ft]) without a separate equipment grounding conductor. If any one of the three wiring methods in this example were even slightly longer in the same fault current path, then a separate grounding conductor would be required.

In all cases, sizing of bonding jumpers for all flex applications is made according to Sec. 250.102, which requires the same minimum size for bonding jumpers as is required for equipment grounding conductors, or grounding electrode conductors for service applications as allowed in 230.43(15). In such cases, the size of the conductor is selected from Table 250.122, based on the maximum rating of the overcurrent devices protecting the circuit conductors that are within the flex. For service work, substitute Table 250.66 for obvious reasons.

In part **(5)(e)**, the flex will require a separate equipment grounding conductor, regardless of whether it would otherwise qualify in terms of length or circuit protection, if "flexibility is necessary . . . after installation." A run to a swinging sign is a clear application of this principle. The 2011 **NEC** has clarified that the phrase "flexibility is necessary" includes a run of flex used as a vibration isolator for a transformer. A short run at a motor to allow flexibility so a belt can be tightened would also void the exemption from running a separate equipment grounding conductor.

The "after installation" provision is in the **NEC** in order to protect a run of flex that has been installed as part of a fixed raceway layout to get around an obstruction. Flexibility was clearly required to install the flex; that is why it was installed in the first place, but just as clearly after the installation is complete it will never move again, and there will be no motion stress applied to its connecting fittings.

Part **(6)** presents conditions under which *liquidtight* flexible metal conduit may be used without need for a separate equipment grounding conductor:

1. Both part **(6)** and the UL's Guide Card information [see discussion at 110.3(B) for details] note that any listed liquidtight flex in metric designator 35 (trade size 1¼) and smaller, in a length not over 6 ft (1.8 m), may be satisfactorily used as a grounding means through the metal core of the flex, without need of a bonding jumper (or equipment grounding conductor) either internal or external (Fig. 250-87), depending on the rating of the overcurrent device ahead of it (next paragraph). The permitted length is exactly the same, and involves the other flexible metal raceways in exactly the same way, as flexible metal conduit, analyzed previously. In addition, there is an identical limitation for uses where flexibility is required after installation.

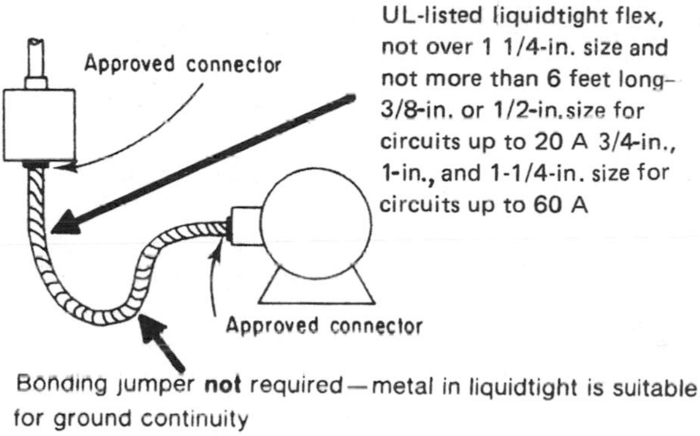
UL-listed liquidtight flex, not over 1 1/4-in. size and not more than 6 feet long–3/8-in. or 1/2-in. size for circuits up to 20 A 3/4-in., 1-in., and 1-1/4-in. size for circuits up to 60 A
Approved connector
Approved connector
Bonding jumper **not** required—metal in liquidtight is suitable for ground continuity

Fig. 250-87. Liquidtight flex may be used with a separate ground wire. (Sec. 250.118.)

Where terminated in fittings investigated for grounding and where installed with not more than 6 ft (1.8 m) (total length) in any ground return path, liquidtight flexible metal conduit in the metric designators 12 and 16 (trade sizes ⅜ and ½) is suitable for grounding where used on circuits rated 20 A or less, and the metric designators 21, 27, and 35 (trade sizes ¾, 1, and 1¼) are suitable for grounding where used on circuits rated 60 A or less. See the category "Conduit Fittings" (DWTT) with respect to fittings suitable as a grounding means.

The following are not considered to be suitable as a grounding means:

a. The metric designator 41 (trade size 1½) and larger sizes
b. The metric designators 12 and 16 (trade sizes ⅜ and ½) where used on circuits rated higher than 20 A, or where the total length in the ground return path is greater than 6 ft (1.8 m)

c. The metric designators 21, 27, and 35 (trade sizes ¾, 1, and 1¼) where used on circuits rated higher than 60 A, or where the total length in the ground return path is greater than 6 ft (1.8 m)

2. For liquidtight flex over metric designator 35 (trade size 1¼), UL does not list any as suitable for equipment grounding, thereby requiring use of a separate equipment grounding conductor installed in *any* length of the flex, as required by Code. If a length of liquidtight flex larger than this is short enough to permit an external bonding jumper not more than 6 ft (1.8 m) long between external grounding-type connectors at the ends of the flex, an external bonding jumper may be used. BUT WATCH OUT! The rule says the *jumper*, not the flex, must not exceed 6 ft (1.8 m) in length AND the jumper "shall be routed with the raceway"—that is, run along the flex surface and not separated from the flex.
3. If any length of flex is *over* 6 ft (1.8 m), then the flex is not a suitable grounding conductor, regardless of the trade size of the flex, whether it is larger or smaller than metric designator 35 (trade size 1¼). In such cases, an *equipment grounding conductor* (not a "bonding jumper"—the phrase reserved for short lengths) must be used to provide grounding continuity and IT MUST BE RUN *INSIDE* THE FLEX, NOT EXTERNAL TO IT, IN ACCORDANCE WITH 250.134(B).
4. The rule for combining lengths of the three flexible wiring method raceways in the 6-ft limitation, and the clarification that the use for vibration isolation requires a separate equipment grounding conductor, apply to this wiring method as well. Refer to the discussion on flexible metal conduit for more information.

Part **(10)** covers metal-clad cable. After years of discussion, this part of 250.118 finally gives clear, unambiguous recognition (in "b.") to the new style of MC cable that includes a bare equipment grounding conductor immediately beneath its interlocking armor. As such, it is somewhat analogous to Type ac cable, except that the bonding tape is replaced by a full-size, usually aluminum conductor.

At conventional terminations, this aluminum is simply cut off and does not need to enter the enclosure. However, at panelboard enclosures and other large cabinets and pull boxes where the inner portions of concentric knockouts are not recognized for equipment grounding continuity (see the discussion at 250.97), the aluminum conductor is not cut off at the connector, but allowed to enter the enclosure and thereby bond around the impaired connection point. The armor on this cable is also available for health-care facility use because of its independent qualification as an equipment grounding conductor [see 517.13(A)]; such make-ups have a conventional green-insulated separate equipment grounding conductor within them to meet 517.13(B).

The first paragraph of this metal-clad cable provision ("a.") primarily covers conventional interlocking-armor metal-clad cable that contains a normally sized equipment grounding conductor, which always enters the wired enclosures for termination. The third paragraph ("c.") covers smooth and corrugated styles of metal-clad cable that in many sizes and configurations have no separate equipment grounding conductors at all, because their armor has enough metal in its cross section to fully qualify as an equipment grounding conductor. However, some larger sizes of this cable (for larger feeders) still require an internal grounding conductor and the wording allows for this.

250.119. Identification of Equipment Grounding Conductors. Although the default identification for equipment grounding conductors is green (next paragraph), what if equipment grounding is not required? This is indeed the case on many Class 2 and Class 3 control circuit applications supplied by the secondary of a 120-V transformer. Many air-conditioning and other applications use standard color codes for thermostat and related wiring that include the color green as an ungrounded control conductor, which may violate both 250.112(I) and 250.119 here. A new exception allows this use on power limited circuits. The 2011 **NEC** correlated the wording of the exception with 250.112(I) so these two provisions are synchronized. Review the discussion at 250.112(I) for a detailed analysis of the code rules that intersect in this area.

The 2014 **NEC** added two other exceptions here. The first, relatively trivial example, involves flexible cord with no equipment ground and a monolithic green color to the insulation; such cords as the 2-conductor parallel cords frequently used as extension cords, especially during the Christmas holidays. The second is very important because it resolves a conflict that has caused periodic inspectional problems for many decades. The standard default color code for the ungrounded conductors run from traffic signal control cabinets and the the actual signals follows the color of the displayed signal. So, a red light is usually powered with a red wire along with the white wire, and a yellow light is usually connected to a yellow wire, and, you guessed it, a green light is usually powered with a green ungrounded conductor. Virtually every traffic signal is wired this way, and now the **NEC** squarely addresses the topic. Because the color green is used in this way, the **NEC** goes on to require that yes, there will be an equipment grounding connection meeting 250.110, and if it is of the wire type (the usual case) it will be either bare or green with one or more yellow stripes. Usually, the equipment grounding wires bond the metal pole bases and that is sufficient given the robust mechanical connections between powered components and the poles. Sec. 396.30(C) would recognize a span wire as an equipment ground if need be, but the elevation would likely excuse a connection in such cases given 250.110(1).

Part **(A)** recognizes conductors of colors other than green for use as equipment grounding conductors if the conductor is stripped for its exposed length within an enclosure, so it appears bare, or if green coloring, green tape, or a green label is used on the conductor at the termination. As shown in Fig. 250-88, the phase legs may or may not be required to be "identified by phase and system" [see 210.5(C)]. If color coding is used, the phase legs may be any color other than white, gray, or green. The neutral may be white or gray or any other color than green if it is larger than No. 6 and if white tape, marking, or paint is applied to the neutral near its terminations. The grounding conductor may be green or may be any insulated conductor of any color if all insulation is stripped off for the exposed length. Alternatives to stripping the black insulated conductor used for equipment ground include (1) coloring the exposed insulation green or (2) marking the exposed insulation with green tape or green adhesive labels; however, any marking must "encircle" the grounding conductor. Relief is granted to the marking rule on conductors larger than 6 AWG where they pass through conduit bodies that have no unused hubs and no splices.

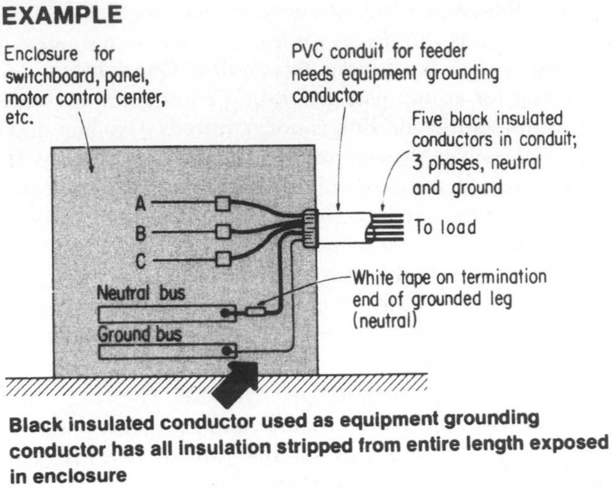
EXAMPLE
Enclosure for switchboard, panel, motor control center, etc.
PVC conduit for feeder needs equipment grounding conductor
Five black insulated conductors in conduit; 3 phases, neutral and ground
A
B
C
To load
Neutral bus
Ground bus
White tape on termination end of grounded leg (neutral)
Black insulated conductor used as equipment grounding conductor has all insulation stripped from entire length exposed in enclosure

Fig. 250-88. Equipment grounding conductor larger than No. 6 may be a stripped conductor of any color covering. (Sec. 250.119.)

Part **(B)** permits specific on-the-job identification of an insulated conductor used as an equipment grounding conductor in a multiconductor cable. Such a conductor, regardless of size, may be identified in the same manner permitted by 250.119(A) for conductors larger than No. 6 used in raceway. The conductor may be stripped bare or colored green to indicate that it is a grounding conductor. But such usage of multiconductor cables is recognized only for commercial, institutional-, and industrial-type systems under conditions of qualified maintenance and supervision. Such field-applied re-identification must encircle the conductor.

250.120. Equipment Grounding Conductor Installation. This rule contains a number of crucial requirements. Part **(A)** requires equipment grounding conductors in the form of raceways or cable assemblies to be installed with their code requirements. The last sentence, however, is a favorite with the inspection community: "All connections, joints, and fittings shall be made tight using suitable tools." This is the requirement that directly reaches every untightened set-screw and every loose locknut. Attention to detail is absolutely critical in constructing an effective ground-fault current path required [250.4(A)(5)] on every part of every grounded electrical system, and the low-impedance path for fault current required [250.4(B)(4)] on every ungrounded system. A single poor connection can be the one that delays the operation of a circuit protective device such that a fire or even worse results.

This part also includes an informational note pointing to "Electrical Circuit Protective Systems." If one of these systems is being used, it can affect, in surprising ways, numerous aspects of the process of providing an equipment grounding return path. For example, it may affect the choice of a raceway wiring method, the methods by which segments of such a wiring method are joined together, and even the choice of insulation type used on an equipment grounding conductor run within the raceway. Refer to more complete coverage of this topic at 728.4 in this book.

Part **(B)** correlates the equipment grounding rules for aluminum with the comparable rules for aluminum used for grounding electrode conductors, in 250.64(A). The analysis at that location also applies here.

Part **(C)** states the requirement that equipment grounding conductors must be protected within cable armor or raceway unless run within hollow walls or other locations where not subject to damage. The hollow-wall allowance primarily applies to where an old branch circuit with no equipment grounding conductor is being extended and a grounding conductor needs to be fished in to make it comply, as covered in 250.130(C).

250.121. Use of Equipment Grounding Conductors. This section, new as of the 2011 NEC, prohibits the use of an equipment grounding conductor as a grounding electrode conductor. Occasionally particularly creative electricians have figured out how to take a wire sized for grounding electrode conductor purposes (and generally larger than that required for the equipment grounding purpose at hand) and install it between the required equipment grounding termination points and then continue it without splice to a grounding electrode, in full conformity with the rules for both conductors.

This was difficult, but occasionally easier than installing an additional raceway or creating a passage for a required grounding electrode conductor, perhaps from a transformer used to originate a separately derived system. No rule expressly prohibited this. In the 2014 NEC, by virtue of a new exception, this practice is once again allowable, and now very clearly so, provided there are no objectionable currents introduced [hence the reference to 250.6(A)] and the arrangements meet all requirements for both functions.

250.122. Size of Equipment Grounding Conductors. When an individual equipment grounding conductor is used in a raceway—either in a nonmetallic raceway or in a metal raceway where such a conductor is used for grounding reliability even though 250.118 usually accepts metal raceways as a suitable grounding conductor—the grounding conductor must have a minimum size as shown in Table 250.122.

The basic rules are covered in 250.122**(A)**, and one of the more important is this: In no case is an equipment grounding conductor required to be larger than its associated circuit conductors. This can come up on motor circuits, where the short-circuit and ground-fault protective device settings may be high enough to cause this problem. For example, a motor circuit using nontime-delay fuses for 12 AWG THHN wire [75°C ampacity, per 110.14(C)(1)(4) = 25 A] could be protected with a 80 A fuse per 430.52(C)(1) Exception No. 1. This circuit, by 250.122(A), should normally have an 8 AWG equipment grounding conductor run with the 12 AWG motor circuit conductors; this rule avoids that and would allow a 12 AWG grounding conductor for this purpose.

Part **(G)** covers another application of this concept, where wire-type equipment grounding conductors run with feeder taps are sized in accordance with the rating of the protective device next back on the line. Here again, the size of an equipment grounding conductor selected for this purpose need not exceed the size of the ungrounded tap conductors with which it is associated. Raceway-type equipment grounding conductors recognized in 250.118 may always be used without consideration as to their size (unless so required in 250.118) and are not subject to this rule.

Part **(A)** also contains the allowance for equipment grounding conductors made up in multiconductor cable to be sectioned, and therefore run in parallel with individual wires smaller than the Table 250.122 specifications. This is often done in larger metal-clad cables to assist with the internal geometry and efficiency of cable manufacturing. The allowance is not new to the **NEC**, but was relocated from Art. 310 because it only applies to equipment grounding conductors.

The minimum acceptable size of an equipment grounding conductor is based on the rating of the overcurrent device (fuse or CB) protecting the circuit, run in the same raceway, for which the equipment grounding conductor is intended to provide an effective ground-fault current path (Fig. 250-89). Each size of grounding conductor in the table is adequate to carry enough current to blow the fuse or trip the CB of the rating indicated beside it in the left-hand column. In Fig. 250-89, if the fuses are rated at 60 A, Table 250.122 shows that the equipment grounding conductor used with that circuit must be at least a 10 AWG copper or a 8 AWG aluminum or copper-clad aluminum.

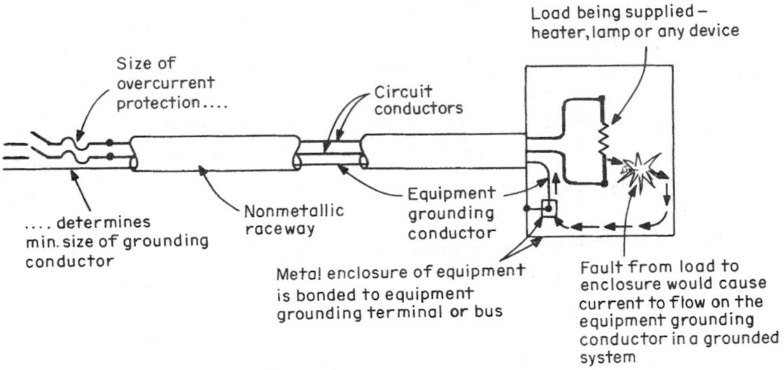
Load being supplied – heater, lamp or any device
Size of overcurrent protection....
Circuit conductors
.... determines min. size of grounding conductor
Nonmetallic raceway
Equipment grounding conductor
Metal enclosure of equipment is bonded to equipment grounding terminal or bus
Fault from load to enclosure would cause current to flow on the equipment grounding conductor in a grounded system

Fig. 250-89. Size of grounding conductor must carry enough current to operate circuit overcurrent device. (Sec. 250.122.)

Whenever an equipment grounding conductor is used for a circuit that consists of only one conductor for each hot leg (or phase leg), the grounding conductor is sized simply and directly from Table 250.122, as described. When a circuit is made up of parallel conductors per phase, say an 800-A circuit with two conductors per phase, an equipment grounding conductor is also sized in the same way and would, in that case, have to be at least a 1/0 AWG copper or 3/0 AWG aluminum. *And* if all the paralleled conductors run together in one raceway or cable assembly, only one equipment grounding conductor of the prescribed size is required. *But,* if such a circuit is made up using two conduits—that is, three phase legs and a neutral in each conduit—250.122**(F)** requires that an individual grounding conductor be run in each of the conduits *and* each of the two grounding conductors must be at least 1/0 AWG copper or 3/0 AWG aluminum (Fig. 250-90). Another example is shown in Fig. 250-91, where a 1200-A protective device on a parallel circuit calls for a 3/0 AWG copper or 250-kcmil aluminum grounding conductor.

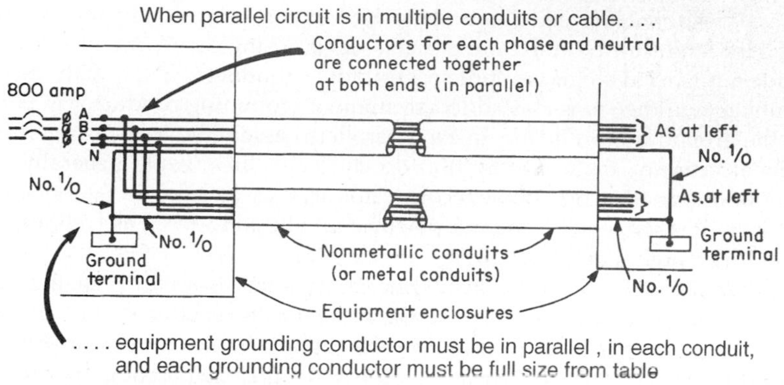
When parallel circuit is in multiple conduits or cable....
Conductors for each phase and neutral are connected together at both ends (in parallel)
800 amp
ø A
ø B
ø C
N
No. 1/0
No. 1/0
Ground terminal
As at left
No. 1/0
As at left
Ground terminal
Nonmetallic conduits (or metal conduits)
No. 1/0
Equipment enclosures
.... equipment grounding conductor must be in parallel, in each conduit, and each grounding conductor must be full size from table

Fig. 250-90. Grounding conductor must be used in each conduit for parallel conductor circuits. (Sec. 250.122.)

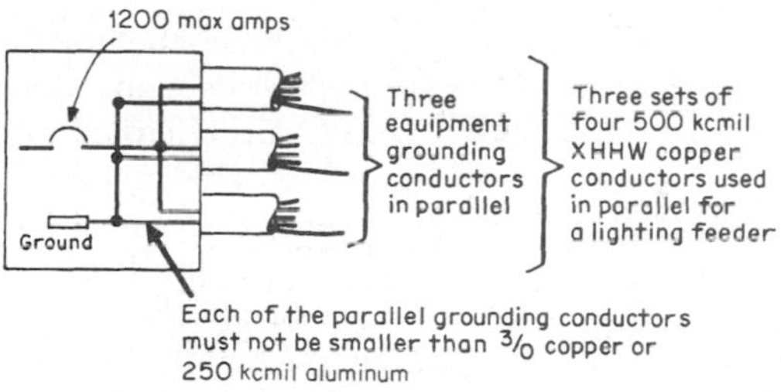
1200 max amps
Ground
Three equipment grounding conductors in parallel
Three sets of four 500 kcmil XHHW copper conductors used in parallel for a lighting feeder
Each of the parallel grounding conductors must not be smaller than 3/0 copper or 250 kcmil aluminum

Fig. 250-91. Using equipment grounding conductors in parallel. (Sec. 250.122.)

Part **(F)** covering parallel setups, as of the 2017 NEC, separates rules for equipment grounding conductors serving conventional raceway installations [in what is now Part **(1)**] and as previously covered here, and those involving large circuits using paralleled cable assemblies, now covered in a new **(2)** with four paragraphs. Taking the provisions one at a time, but not in order, **(d)** sets what is in effect the default arrangement: Each cable assembly must be arranged with an equipment grounding conductor in each cable sized in accordance with Table 250.122 based on the full ampere rating circuit overcurrent protective device. This is identical to the rules in (1) for raceway installations using single conductors. Two of the other three paragraphs do not break new ground either: **(c)** effectively enforces the 4 AWG minimum size threshold for single equipment grounding conductors run in cable tray [392.10(B)(1)(c)] and **(a)** reinforces the general rule in 310.10(H)(1) that parallel runs have all the originating wire ends tied together and likewise for the terminating ends.

Paragraph **(b)** does break new ground, however, because it provides a path to utilize multiple cable assemblies with conventional (not special order) make-ups. These are generally designed for one cable on one circuit, with the incorporated equipment grounding conductor sized out of Table 250.122 for that typical circuit size, and not a much larger circuit comprised of paralleled conductors. The new procedure is to run the full parallel make-up in a single raceway or cable tray, provide a full-sized single equipment grounding conductor along with the cable assemblies, and then wire this added equipment grounding conductor in parallel with the grounding conductors in each paralleled assembly. In the case of a conventional raceway, the 310.15(B)(3)(a) derating penalties would generally make this option a non-starter. However, in cable tray as covered in 392.80(A)(1)(a) there would be no penalty, and a wireway could be considered as well up to the 30-conductor limitation [376.22(B)].

Part **(D)** of 250.122 covers another concern for unnecessarily oversizing equipment grounding conductors. Because the minimum acceptable size of an equipment grounding conductor is based on the rating of the overcurrent protective device (fuse or CB) protecting the circuit for which the equipment grounding conductor is intended to provide a path of ground-fault return, a problem arises when a motor circuit is protected by a magnetic-only (a so-called instantaneous) circuit breaker. Because 430.52 and Table 430.52 permit an instantaneous-trip CB with a setting of 800 percent of (8 times) the motor full-load running current—and even up to 1700 percent for an instantaneous CB or MSCP (motor short-circuit protector), if needed to handle motor inrush current—use of those high values of current rating permitted in Table 430.52 would result in excessively large equipment grounding conductors. Because such large sizing is unreasonable and not necessary, the rule says when sizing an equipment grounding conductor from Table 250.122 for a circuit protected by an instantaneous-only circuit breaker or by an MSCP, there is a special calculation procedure which has significantly changed in the 2008 **NEC**.

Begin with the ampacity for the applicable size of the motor circuit conductors. Multiply the ampacity [using the 75°C column for all sizes, per 110.14(C)(1)(4)] by the percentage stated for a dual-element (time-delay) fuse in Table 430.52 (usually 175 percent). In accordance with 430.52(C)(1) Exception No. 1, round the resulting number up to the next higher standard sized fuse, as listed in 240.6(A), if it does not correspond to a standard size. This result, and not the much lower setting of the running overload protection as had been allowed in the **NEC** for the six previous cycles, is the one that you use to enter Table 250.122 (Fig. 250-92).

The last sentence of 250.122(A) points out that metal raceways and cable armor are recognized as equipment grounding conductors; Table 250.122 does not apply to them. However, they must still provide an effective fault current path, as covered in the note at the bottom of Table 250.122.

In 250.122**(B)**, the Code places a mandatory requirement for equipment grounding conductors to be "increased in size" where the phase conductors are "upsized" for any reason, but traditionally to overcome voltage drop on long runs. Where any upsizing is provided to ensure adequate voltage at the point of equipment installation, the equipment grounding conductor must also be upsized to ensure adequate current flow under fault conditions. That is, if voltage drop presents a problem for the phase conductors, then it also presents a problem to the equipment grounding conductors, because the grounding conductors will be run for the same distance as the phase conductors.

The current rule to increase an equipment grounding conductor if the other circuit conductors, quite simply, are "increased in size" was unclear and required

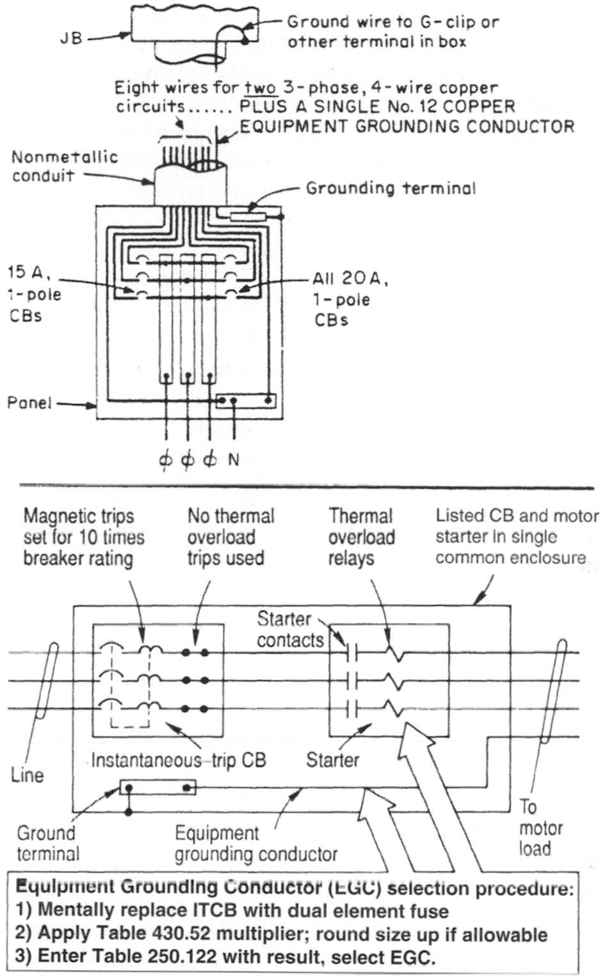
JB
Ground wire to G-clip or other terminal in box
Eight wires for two 3-phase, 4-wire copper circuits...... PLUS A SINGLE No. 12 COPPER EQUIPMENT GROUNDING CONDUCTOR
Nonmetallic conduit
Grounding terminal
15 A, 1-pole CBs
All 20 A, 1-pole CBs
Panel
ϕ ϕ ϕ N
Magnetic trips set for 10 times breaker rating
No thermal overload trips used
Thermal overload relays
Listed CB and motor starter in single common enclosure
Starter contacts
Instantaneous-trip CB
Starter
Line
Ground terminal
Equipment grounding conductor
To motor load
Equipment Grounding Conductor (EGC) selection procedure:
1) Mentally replace ITCB with dual element fuse
2) Apply Table 430.52 multiplier; round size up if allowable
3) Enter Table 250.122 with result, select EGC.

Fig. 250-92. These applications are covered by parts **(C)** and **(D)** of 250.122. (Sec. 250.122.)

field interpretation, particularly in the event the increase was simply to satisfy the minimum size requirements in the NEC. The 2014 NEC, by adding text to the effect that the floor for beginning an up-size analysis is the minimum NEC required ampacity for the intended installation, has clarified that size increases required to overcome ampacity deratings for high ambient temperatures and mutual conductor heating do not provoke an increase in equipment grounding conductor sizing. For example, suppose a 3-phase, 4-wire wye-connected feeder runs with significant harmonic loading. The ampacity of the conductors must be derated in accordance with 310.15(B)(5)(c) and 310.15(B)(3)(a). The conductor size must be increased to overcome the mandatory derating factor applied. It is now clear that this "increase in size" does not invoke a larger equipment grounding conductor.

It can be shown that when conductors are increased to counteract issues of ambient temperature and mutual conductor heating, the result tends to be a net decrease in the total impedance in the fault current path (i.e., from the fuse or circuit breaker out to a fault on a power conductor and back on an equipment grounding conductor), even when that conductor is taken straight from Table 250.122 with no adjustments. In Annex D, Example D3(a) there are comprehensive calculations that address problems of continuous loading, termination issues, etc. This issue is more fully examined there.

Calculation example: Suppose you have a feeder that would, in its most basic application, use 1 AWG conductors, and they are being increased to 2/0 conductors to decrease voltage drop. The Table 250.122 equipment grounding result based on the circuit protection is 6 AWG. Here is how to figure proportional increases to an equipment grounding conductor size if necessary. Perform the following calculation, using the numbers in Table 8 of **NEC** Chap. 9, where "CSA" means "cross-sectional area" and "EGC" means "equipment grounding conductor":

$$\text{New EGC CSA} = \frac{\text{ungrounded conductor CSA} \times \text{base EGC CSA}}{\text{CSA of base ungrounded conductors from table}}$$

$$\text{New EGC} = \frac{(2/0\ \text{AWG} = 133100\ \text{cmil}) \times (6\ \text{AWG} = 26240\ \text{cmil})}{(1\ \text{AWG} = 83690\ \text{cmil})}$$

$$= (41740\ \text{cmil} = 4\ \text{AWG})$$

Figure 250-92 shows details of a controversy that often arises about 250.122**(C)** and 250.134. When two or more circuits are used in the same conduit, it is logical to conclude that a single equipment grounding conductor within the conduit may serve as the required grounding conductor for each circuit if it satisfies Table 250.122 for the circuit with the highest rated overcurrent protection. The common contention is that if a single metal conduit is adequate as the equipment grounding conductor for all the contained circuits, a single grounding conductor can serve the same purpose when installed in a nonmetallic conduit that connects two metal enclosures (such as a panel and a home-run junction box) where both circuits are within both enclosures. As shown, a 12 AWG copper conductor satisfies Table 250.122 as an equipment grounding conductor for the circuit protected at 20 A. The same 12 AWG also may serve for the circuit protected at 15 A, for which a grounding conductor must not be smaller than 14 AWG copper. Such an application is specifically permitted by part **(C)** of 250.122. Although this will have primary application with PVC conduit where an equipment grounding conductor is required, it may also apply to circuits in EMT, IMC, or rigid metal conduit when an equipment grounding conductor is run with the circuit conductors to supplement the metal raceway as an equipment grounding return path.

250.130. Equipment Grounding Conductor Connections. Part **(A)** requires that the equipment grounding conductor at a service—such as the ground bus or terminal in the service-equipment enclosure, or the enclosure itself—must be connected to the system *grounded* conductor (the neutral or grounded phase leg). The equipment ground and the neutral or other grounded leg must be bonded together and it

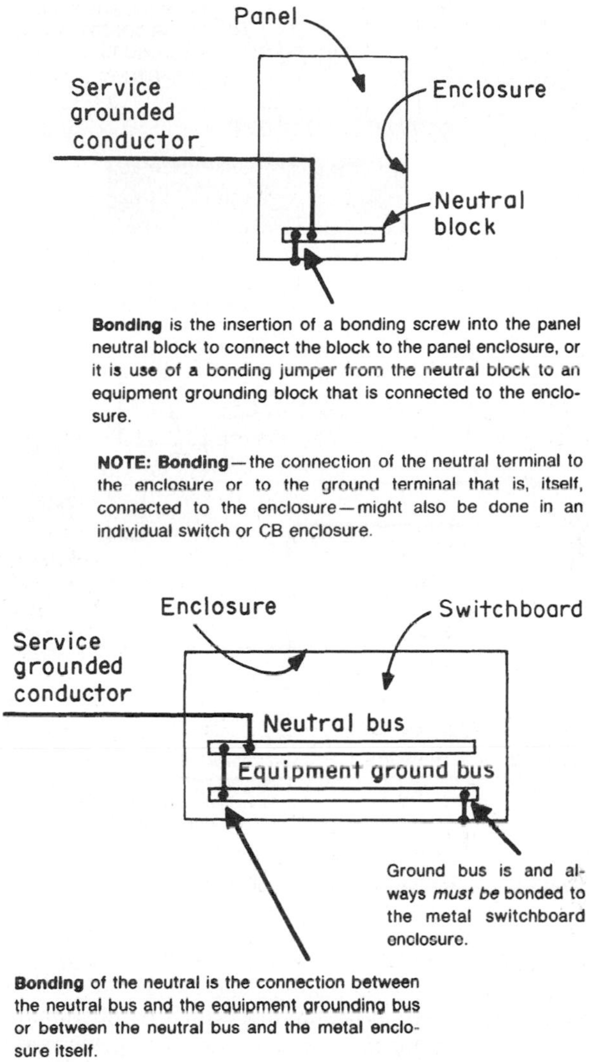
Panel
Service grounded conductor
Enclosure
Neutral block
Bonding is the insertion of a bonding screw into the panel neutral block to connect the block to the panel enclosure, or it is use of a bonding jumper from the neutral block to an equipment grounding block that is connected to the enclosure.
NOTE: Bonding—the connection of the neutral terminal to the enclosure or to the ground terminal that is, itself, connected to the enclosure—might also be done in an individual switch or CB enclosure.
Enclosure
Switchboard
Service grounded conductor
Neutral bus
Equipment ground bus
Ground bus is and always *must be* bonded to the metal switchboard enclosure.
Bonding of the neutral is the connection between the neutral bus and the equipment grounding bus or between the neutral bus and the metal enclosure itself.

Fig. 250-93. Equipment ground must be "bonded" to grounded conductor at the service equipment. (Sec. 250.130.)

must be done on the supply side of the service disconnecting means—which means either *within* or *ahead* of the enclosure for the service equipment (Fig. 250-93).

Part **(B)** requires the ground bus or the enclosure to be simply bonded to the grounding electrode conductor within or ahead of the service disconnect for an ungrounded system.

As shown at the top of Fig. 250-94, some switchboard sections or interiors include neutral busbars factory-bonded to the switchboard enclosure and are

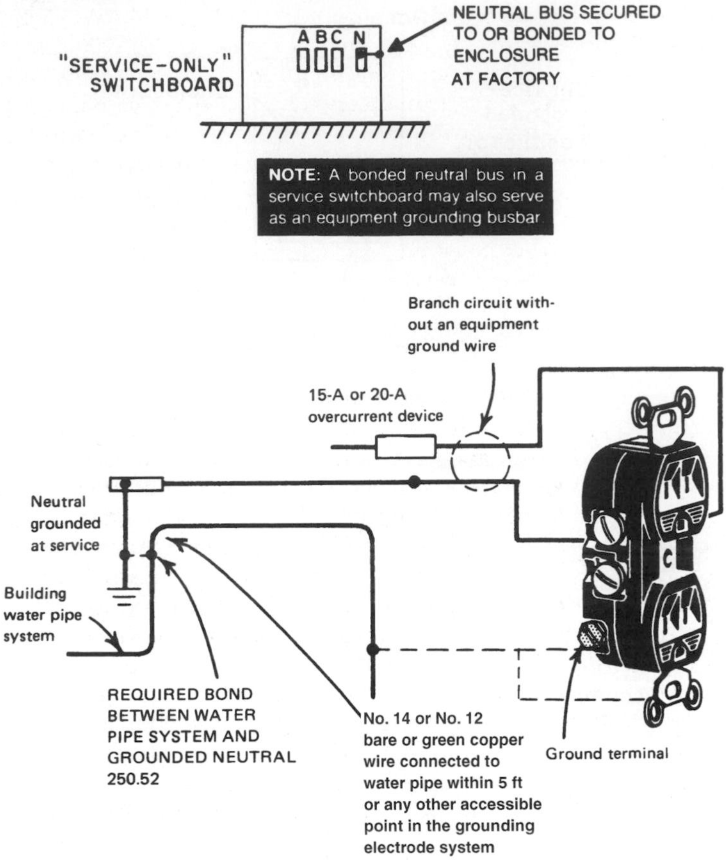
"SERVICE-ONLY" SWITCHBOARD
A B C N
NEUTRAL BUS SECURED TO OR BONDED TO ENCLOSURE AT FACTORY
NOTE: A bonded neutral bus in a service switchboard may also serve as an equipment grounding busbar.
Branch circuit without an equipment ground wire
15-A or 20-A overcurrent device
Neutral grounded at service
Building water pipe system
REQUIRED BOND BETWEEN WATER PIPE SYSTEM AND GROUNDED NEUTRAL 250.52
No. 14 or No. 12 bare or green copper wire connected to water pipe within 5 ft or any other accessible point in the grounding electrode system
Ground terminal

Fig. 250-94. These are bonding and grounding details covered by 250.130**(C)**.

marked "suitable for use only as service equipment." They may not be used as subdistribution switchboards—that is, they may not be used on the load side of the service except where used, with the inspector's permission, as the first disconnecting means fed by a transformer secondary or a generator and where the bonded neutral satisfies 250.30 for a separately derived system.

Part **(C)** covers the retrofit of equipment grounding connections for receptacles connected to, or for extensions of, *existing* branch circuits that do not include any recognized equipment grounding conductor (Fig. 250-94, bottom), such as concealed knob-and-tube wiring (Art. 394). An equipment grounding conductor can be brought to these locations without rewiring the circuit, and used to make a grounding connection on a receptacle grounding terminal, or to the equipment grounding provisions of a wiring method being used to extend an existing branch circuit. For example, if a concealed knob-and-tube circuit were being extended with Type NM cable, which includes a bare copper wire for equipment grounding, an equipment grounding conductor could be fished to the location of the branch circuit being extended.

The equipment grounding conductor, presumably a 12 or 14 AWG wire, would be run to the location, probably using the provisions of 250.120(C) that give a waiver on raceway or cable armor protection in such cases for runs within hollow walls or joist cavities, or where otherwise protected from damage. It must originate from one of the following locations: (1) the grounding electrode system at an accessible point, or on the grounding electrode conductor; (2) the equipment grounding terminal bar within the enclosure where the circuit originates; (3) the service equipment grounding terminal bar (ungrounded systems) or the grounded service conductor within the service equipment, or the equivalent grounding terminal within the service equipment for ungrounded systems. The 2014 NEC added a connection to an equipment grounding conductor of any other branch circuit that originates within the same enclosure as where the branch circuit being extended (or supporting the receptacle being grounded) originates.

At one time, this rule allowed connections to nearby water pipes, but not since the 1993 NEC. Now all the likely connection points are in a basement or the first level. You are unlikely to be searching for a method of grounding concealed knob-and-tube wiring in a steel-frame building. Rather you will be attempting this in old wood-frame buildings, probably residential. In such occupancies, even if the water supply lateral is metallic, the water piping system ceases to be considered as an electrode beyond 5 ft from the point of entry. This means fishing into the basement. If you can fish a ground wire down into the basement, you can fish a modern circuit up in the reverse direction and avoid all the problems. Meanwhile, the note at the end of this part points to the procedure in 406.4(D) that allows grounding-configured GFCI receptacles to be installed with no equipment ground provided. This entire procedure is of marginal significance today. The 2014 NEC modification does have the virtue of possibly allowing a connection point potentially closer to the work being undertaken.

250.134. Equipment Fastened in Place or Connected by Permanent Wiring Methods (Fixed)—Grounding. This section requires that metal equipment enclosures, boxes, and cabinets to be grounded must be grounded by metal cable armor or by the metal raceway that supplies such enclosures (rigid metal conduit, intermediate metal conduit, EMT, flex or liquidtight flex), *or* by an equipment grounding conductor, such as where the equipment is fed by rigid nonmetallic conduit. Refer to 250.118. As illustrated at the bottom of Fig. 250-94, Exception No. 1 for parts **(A)** and **(B)** recognizes the accepted technique given in 250.130—for using grounding-type receptacles for replacement of existing nongrounding devices or for circuit extensions—on wiring systems that do not include an equipment grounding conductor.

In Sec. 250.134(B), the rule explicitly requires that *when* a separate equipment grounding conductor (i.e., other than the metal raceway or metal cable armor) is used for alternating-current circuits, it *must* be contained *within* the same raceway, cable, or cord or otherwise run with the circuit conductors (Fig. 250-95). External grounding of equipment enclosures or frames or housings is a violation for ac equipment. It is not acceptable, for instance, to feed an ac motor with a nonmetallic conduit or cable, without a grounding conductor in the conduit or cable, and then provide grounding of the metal frame by a grounding conductor connected to the metal frame and run to building steel or to a grounding-grid conductor. An equipment grounding conductor *must always* be run with the circuit conductors.

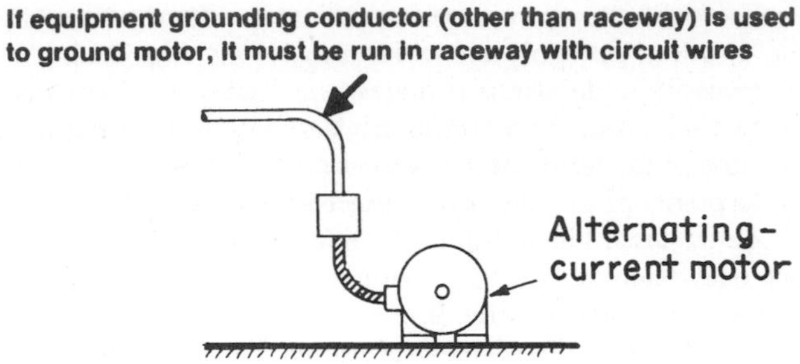
If equipment grounding conductor (other than raceway) is used to ground motor, it must be run in raceway with circuit wires
Alternating-current motor

Fig. 250-95. Equipment grounding conductor must be in raceway or cable with circuit conductors for ac equipment. (Sec. 250.134.)

The rule in Sec. 250.134(B) which insists on keeping an equipment grounding conductor physically close to ac circuit supply conductors is a logical follow-up to the rules of 250.4 which call for minimum impedance in grounding current paths to provide most effective clearing of ground faults. When an equipment grounding conductor is kept physically close to any circuit conductor that would be supplying the fault current (i.e., the grounding conductor is in the "same raceway, cable, or cord or otherwise run with the circuit conductors"), the impedance of the fault circuit has minimum inductive reactance and minimum ac resistance because of mutual cancellation of the magnetic fields around the conductors and the reduced skin effect. Under such condition of a "low-impedance circuit" the meaning of 250.4(A)(5) is best fulfilled—voltage to ground is limited to the greatest extent, the fault current is higher because of minimized impedance, the circuit overcurrent device will operate at a faster point in its time-current characteristic to ensure maximum fault-clearing speed, and the entire effect will be "facilitating the operation of the overcurrent device. . . ."

The arrangement shown in Fig. 250-96 violates the basic rule of 250.134(B) because the lighting fixture, which must be grounded to satisfy 250.112, is not grounded in accordance with 250.134 and 250.118 or by an equipment grounding conductor contained within the cord, as noted in 250.138.

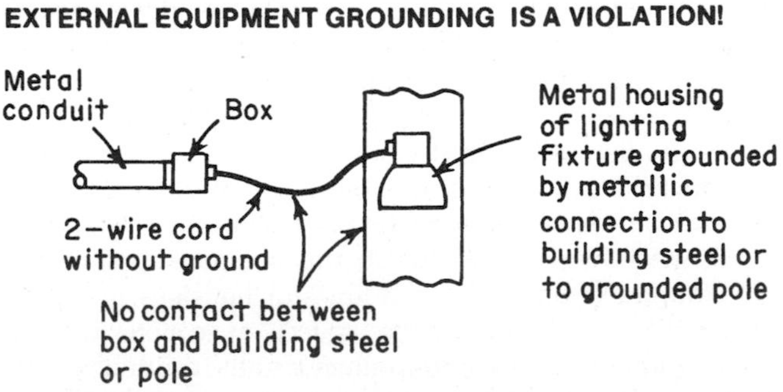
EXTERNAL EQUIPMENT GROUNDING IS A VIOLATION!
Metal conduit
Box
Metal housing of lighting fixture grounded by metallic connection to building steel or to grounded pole
2-wire cord without ground
No contact between box and building steel or pole

Fig. 250-96. Supply to ac equipment must include equipment grounding conductor. (Sec. 250.134.)

Note that 250.134(B) refers very clearly to an "equipment grounding conductor contained within the same raceway, cable, or otherwise run with the circuit conductors." Except for dc circuits, replacement receptacles [250.130(C)], and isolated, ungrounded power sources [517.19(G)], an equipment grounding conductor of any type must not be run separately from the circuit conductors. The engineering reason for keeping the ground return path and the phase legs in close proximity (i.e., in the same raceway) is to minimize the impedance of the fault circuit by placing conductors so their magnetic fields mutually cancel each other, keeping inductive reactance down, and allowing sufficient current to flow to "facilitating the operation of the overcurrent device...." as required by 250.4(A)(5).

The hookup in Fig. 250-96 also violates the rule of the last sentence in 250.136(A), which prohibits use of building steel as the equipment grounding conductor for ac equipment. And the rules of 250.136 often have to be considered in relation to the rules of 250.134.

Note: CARE MUST BE TAKEN TO DISTINGUISH BETWEEN AN "EQUIPMENT *GROUNDING CONDUCTOR*" AS COVERED BY 250.134 AND AN "EQUIPMENT *BONDING JUMPER*" AS COVERED BY 250.102(E). A "*BONDING JUMPER*" MAY BE USED EXTERNAL TO EQUIPMENT BUT IT MUST NOT BE OVER 6 ft (1.8 m) LONG.

250.136. Equipment Considered Grounded. This rule clarifies the way in which structural metal may be used as an equipment grounding conductor, consistent with the rule of 250.134(B) requiring a grounding conductor to be kept physically close to the conductors of any ac circuit for which the grounding conductor provides the fault return path.

Part **(A)** notes that if a piece of electrical equipment is attached and electrically conductive to a metal rack or structure supporting the equipment, the metal enclosure of the equipment is considered suitably grounded by connection to the metal rack, PROVIDED THAT the metal rack itself is effectively grounded by metal raceway enclosing the circuit conductors supplying the equipment or by an equipment grounding conductor run with the circuit supplying the equipment. An example of such application is shown in Fig. 250-97. Although this example shows grounding of lighting fixtures to a rack, the Code rule recognizes any "electric equipment" when this basic grounding concept is observed. It is important to note that if a ground fault developed in equipment so grounded (as at point A), the fault current would take the path indicated by the small arrows. In such case, although the fault-current path through the steel rack is not close to the hot conductor in the flexible cord that is feeding the fault—as normally required by 250.134(B)—the distance of the external ground path is not great, from the fixture to the panel enclosure or box. Because such a short external ground path produces only a relatively slight increase in ground-path impedance, 250.136(A) permits it. The permission for external bonding of flexible metal conduit and liquidtight flex in 250.102(E) is based on the same acceptance of only slight increase of overall impedance of the ground path.

The second sentence of 250.136(A) clearly prohibits using structural building steel as an equipment grounding conductor for equipment mounted on or fastened to the building steel—IF THE SUPPLY CIRCUIT TO THE EQUIPMENT OPERATES ON ALTERNATING CURRENT. BUT, structural building steel that is effectively

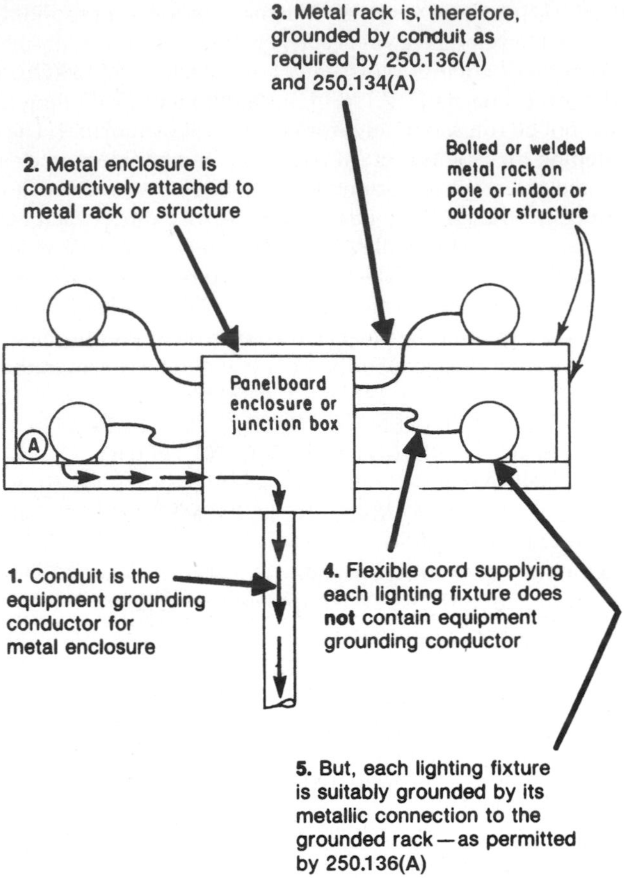
3. Metal rack is, therefore, grounded by conduit as required by 250.136(A) and 250.134(A)
2. Metal enclosure is conductively attached to metal rack or structure
Bolted or welded metal rack on pole or indoor or outdoor structure
Panelboard enclosure or junction box
A
1. Conduit is the equipment grounding conductor for metal enclosure
4. Flexible cord supplying each lighting fixture does not contain equipment grounding conductor
5. But, each lighting fixture is suitably grounded by its metallic connection to the grounded rack—as permitted by 250.136(A)

Fig. 250-97. This use of metal rack as equipment ground is permitted. [Sec. 250.136(**A**).]

grounded and bonded to the grounded circuit conductor of a dc supply system may be used as the equipment grounding conductor for the metal enclosure of dc-operated equipment that is conductively attached to the building steel.

It is important to understand the basis for the Code rules of 250.134(B) and 250.118 and their relation to the concept of 250.136(A):

Note that 250.134(B) refers very clearly to an "equipment grounding conductor contained within the same raceway or cable or otherwise run with the circuit conductors." Except for dc circuits (250.168) and for isolated, ungrounded power sources [517.19(F) and (G)], an equipment grounding conductor of any type must not be run separately from the circuit conductors. Keeping the ground return path and the phase legs in close proximity (i.e., in the same raceway) minimizes the impedance of the fault circuit by placing conductors so their magnetic fields mutually cancel each other, keeping inductive reactance down and allowing sufficient current to flow to "facilitating the operation of the overcurrent device..." as required by 250.4(A)(5).

The second sentence of 250.136(A) applies the concept of ground-fault impedance to the metal frame of a building and prohibits its use as an equipment grounding conductor for ac equipment enclosures. As shown in Fig. 250-98, use of building steel as a grounding conductor provides a long fault return path of very high impedance because the path is separated from the feeder circuit hot legs—thereby violating 250.4. Ground-fault current returning over building steel to the point where the building steel is bonded to the ac system neutral (or other grounded) conductor is separated from the circuit conductor that is providing the fault current. Impedance is, therefore, elevated and the optimum conditions required by 250.4 are not present, so that the grounding cannot be counted on for "facilitating the operation" of the fuse or CB protecting the faulted circuit. The current may not be high enough to provide fast and certain clearing of the fault.

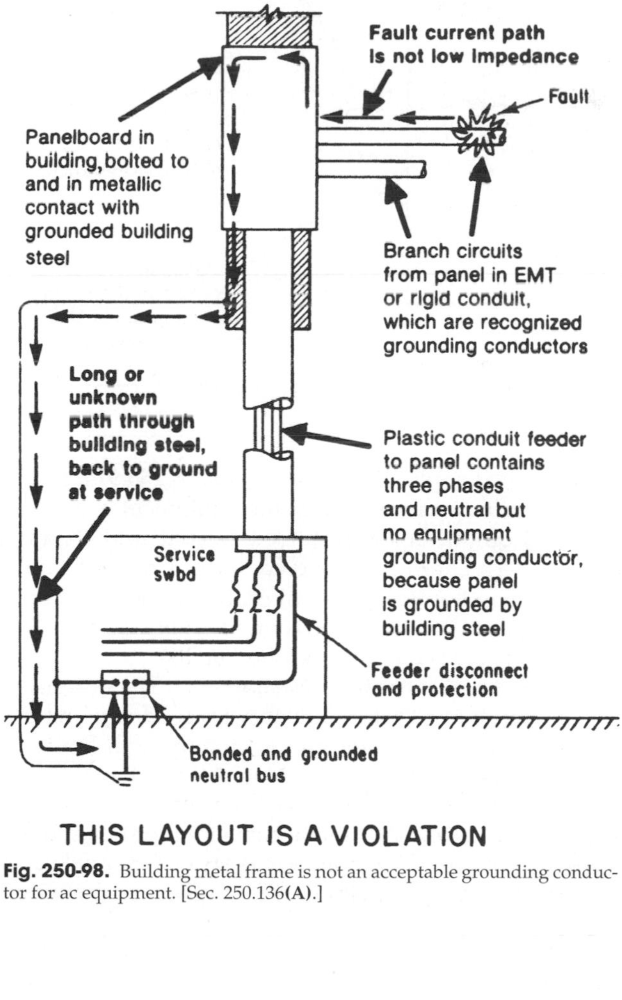
Fault current path
is not low impedance
Fault
Panelboard in building, bolted to and in metallic contact with grounded building steel
Branch circuits from panel in EMT or rigid conduit, which are recognized grounding conductors
Long or unknown path through building steel, back to ground at service
Plastic conduit feeder to panel contains three phases and neutral but no equipment grounding conductor, because panel is grounded by building steel
Service swbd
Feeder disconnect and protection
Bonded and grounded neutral bus
THIS LAYOUT IS A VIOLATION

Fig. 250-98. Building metal frame is not an acceptable grounding conductor for ac equipment. [Sec. 250.136**(A)**.]

The first sentence of 250.136(A) accepts a limited variation from the basic concept of keeping circuit hot legs and equipment grounding conductors physically close to each other. When equipment is grounded by connection to a "metal rack or structure" that is specifically provided to support the equipment and *is* grounded, the separation between the circuit hot legs and the rack, which serves as the equipment grounding conductor, exists only for a very short length that will not significantly raise the overall impedance of the ground-fault path. Figure 250-99 shows another application of that type, similar to the one shown in Fig. 250-97. Although this shows a 2-wire cord as being acceptable, use of a 3-wire cord (two circuit wires and an equipment grounding wire) is better practice, at very slight cost increase.

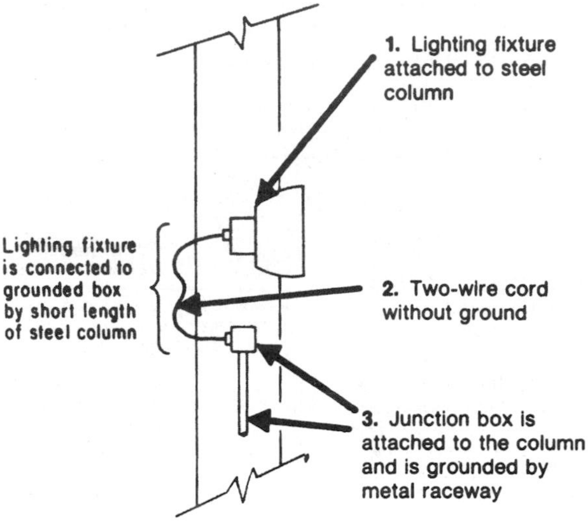
1. Lighting fixture attached to steel column
Lighting fixture is connected to grounded box by short length of steel column
2. Two-wire cord without ground
3. Junction box is attached to the column and is grounded by metal raceway

Fig. 250-99. This satisfies basic rule of 250.136**(A)**. [Sec. 250.136**(A)**.]

Aside from the limited applications shown in Figs. 250-97 and 250-99, *required* equipment grounding must always keep the equipment grounding conductor alongside the circuit conductor for grounded ac systems. Of course, as long as required grounding techniques are observed, there is no objection to additional connection of equipment frames and housings to building steel or to grounding grids to provide equalized potentials to ground. But the external grounding path is not suitable for clearing ac equipment ground faults.

250.138. Cord-and-Plug-Connected Equipment. The proper method of grounding portable equipment is through an extra conductor in the supply cord. Then if the attachment plug and receptacle comply with the requirements of 250.138, the grounding connection will be completed when the plug is inserted in the receptacle.

A grounding-type receptacle and an attachment plug should be used where it is desired to provide for grounding the frames of small portable appliances. The receptacle will receive standard 2-pole attachment plugs, so grounding is optional with the user. The grounding contacts in the receptacle are electrically connected to the supporting yoke so that when the box is surface-mounted, the connection to ground is provided by a direct metal-to-metal contact between the device yoke and the box. For a recessed box a grounding jumper must be used on the receptacle or a self-grounding receptacle must be used. See 250.146 and 250.148.

Figure 250-100 shows a grounding-type attachment plug with a movable, self-restoring grounding member—as previously covered in an exception to this section. Although previously recognized, availability of such a device is questionable, and use of such a product is no longer permitted.

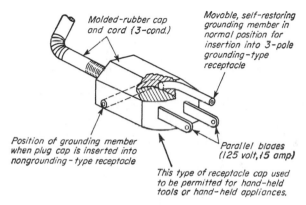

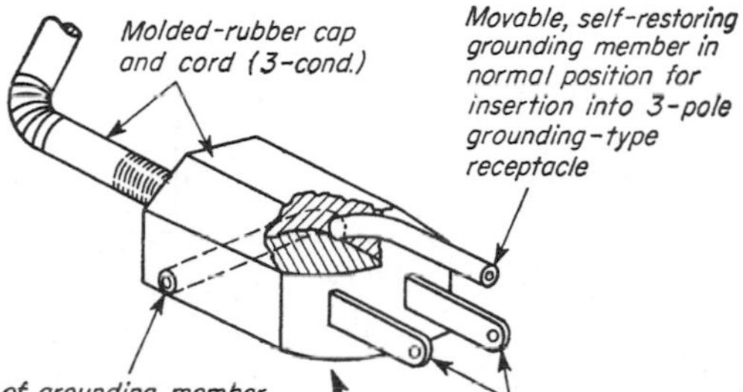
Molded-rubber cap
and cord (3-cond.)
Movable, self-restoring
grounding member in
normal position for
insertion into 3-pole
grounding-type
receptacle
Position of grounding member
when plug cap is inserted into
nongrounding-type receptacle
Parallel blades
(125 volt, 15 amp)
This type of receptacle cap used
to be permitted for hand-held
tools or hand-held appliances.

Fig. 250-100. This type of plug cap is no longer recognized on cords for tools and appliances. However, the exception in 250.138**(A)** *does* recognize "moveable" ground pins on "grounding-type, plug-in" GFCIs where the voltage is not greater than 150 V. [See 250.138(A) Exception.]

250.140. Frames of Ranges and Clothes Dryers. Prior to the 1996 edition of the NEC, the frame of an electric range, wall-mounted oven, or counter-mounted cooking unit could be grounded by direct connection to the grounded circuit conductor (the grounded neutral) and thus could be supplied by a 3-wire cord set and range receptacle irrespective of whether the conductor to the receptacle contains a separate grounding conductor.

The NEC prohibits such applications except on "existing branch circuits." That wording doesn't permit grounding of ranges or dryers with the neutral, unless the circuit itself—not the occupancy—is an existing circuit. For *all* new circuits and new construction, the neutral may *not* be used as a grounding conductor.

Where permitted to be so grounded parts **(1)** and **(2)** clarify the use of a No. 10 or larger grounded neutral conductor of a *120/208-* or *120/240-V* circuit for grounding the frames of electric ranges, wall-mounted ovens, counter-mounted units, or clothes dryers. This method is acceptable whether the 3-wire supply is 120/208 or 120/240 V. Normally, the grounded conductor must be insulated. However, a provision, applicable to both 3-wire supply voltages, does require that when using service-entrance cable having an uninsulated neutral conductor, the branch circuit must originate at the service-entrance equipment. The purpose of this provision is to prevent the uninsulated neutral from coming in contact with a panelboard supplied by a feeder and a separate grounding conductor (in the case of nonmetallic-sheathed cable). This would place the neutral in parallel with the grounding conductor, or with feeder *metal* raceways or cables if they were used. Insulated neutrals in such situations will prevent this (Fig. 250-101).

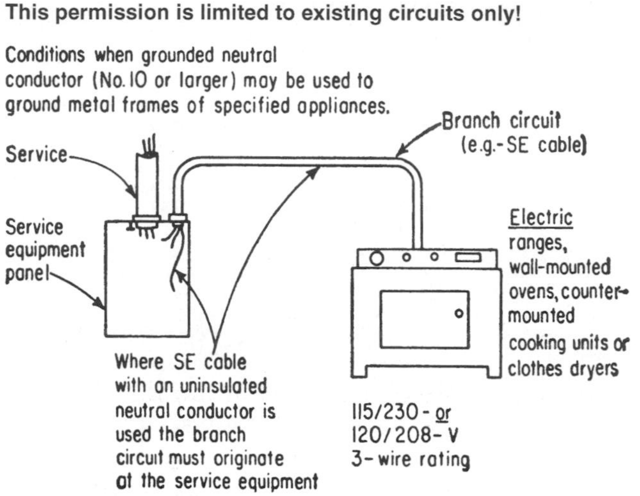
This permission is limited to existing circuits only!
Conditions when grounded neutral conductor (No. 10 or larger) may be used to ground metal frames of specified appliances.
Branch circuit (e.g.-SE cable)
Service
Service equipment panel
Electric ranges, wall-mounted ovens, counter-mounted cooking units or clothes dryers
Where SE cable with an uninsulated neutral conductor is used the branch circuit must originate at the service equipment
115/230 - or 120/208-V 3-wire rating

Fig. 250-101. Ranges and dryers may be grounded to the circuit neutral, but only on existing circuits. All new installations must provide an equipment grounding conductor with the 3-wire supply to the range or dryer. (Sec. 250.140.)

Wording of the rule that permits frames of ranges and clothes dryers to be grounded by connection to the grounded neutral conductor of their supply circuits also permits the same method of grounding of "outlet or junction boxes" serving such appliances. The rule permits grounding of an outlet or junction box, as well as cooking unit or dryer, by the circuit grounded neutral (Fig. 250-102). That practice has been common for many years but has raised questions about the suitability of the neutral for such grounding. However, the wording of this rule makes clear that such grounding of the box is acceptable. Figure 250-103 shows other details of such application. Without this permission to ground the metal box to the grounded neutral, it would be necessary to run a 4-wire supply cable to the box, with one of the wires serving as an equipment grounding conductor sized from Table 250.122, which is always required for other than existing branch circuits that supply dryers and ranges.

Important: As shown in the asterisk note under Fig. 250-102, if a nonmetallic-sheathed cable was used, say, to supply a wall oven or cooktop, such cable is required by 250.140 to have an *insulated* neutral. It would be a violation, for instance, to use a 10/2 NM cable with a bare 10 AWG grounding conductor to supply a cooking appliance—connecting the two insulated 10 AWG wires to the hot

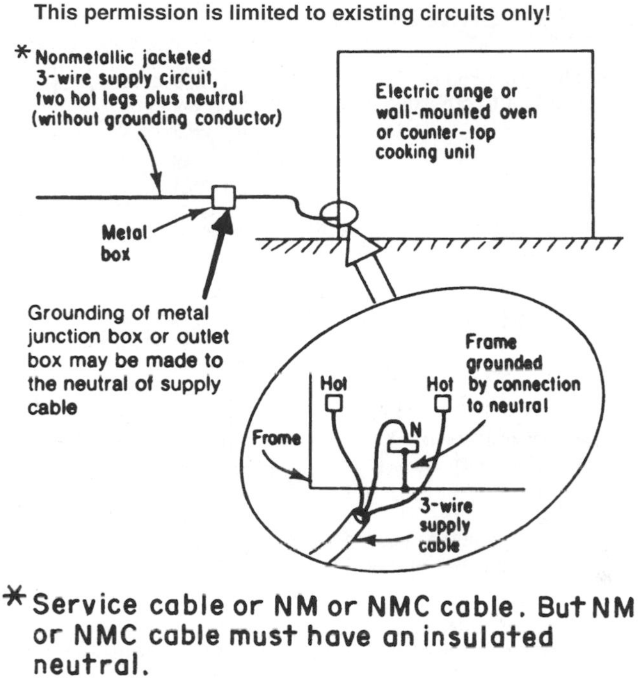
This permission is limited to existing circuits only!
* Nonmetallic jacketed 3-wire supply circuit, two hot legs plus neutral (without grounding conductor)
Electric range or wall-mounted oven or counter-top cooking unit
Metal box
Grounding of metal junction box or outlet box may be made to the neutral of supply cable
Hot
Hot
Frame grounded by connection to neutral
N
Frame
3-wire supply cable
* Service cable or NM or NMC cable. But NM or NMC cable must have an insulated neutral.

Fig. 250-102. Neutral may be used to ground boxes as well as appliances. (Sec. 250.140.)

terminals and using the bare 10 AWG as a neutral conductor to ground the appliance. An uninsulated grounded neutral may be used only when part of a service-entrance cable. Where an existing branch circuit is made up with 10/2 Type NM cable, a new branch circuit with an equipment ground must be installed to supply the dryer or range.

250.142. Use of Grounded Circuit Conductor for Grounding Equipment. Part **(A)** permits connection between a grounded neutral (or grounded phase leg) and equipment enclosures, for the purpose of grounding the enclosures to the grounded circuit conductor. The grounded conductor (usually the neutral) of a circuit may be used to ground metal equipment enclosures and raceways on the supply side of the service disconnect or the supply side of the first disconnect fed from a separately derived transformer secondary or generator output or on the supply side of a main disconnect for a separate building. The wording here includes the supply side of a separately derived system as a place where metal equipment parts or enclosures may be grounded by connection to the grounded circuit conductor (usually a neutral). It is important to note that, in the meaning of the code (as covered in 250.30 and in 250.24), the phrase "on the supply side of the disconnecting means" includes connection within the enclosure of the disconnecting means. Note that the supply side of the separately derived system language correlates with the permission in 250.30(A)(1) to locate the system bonding jumper

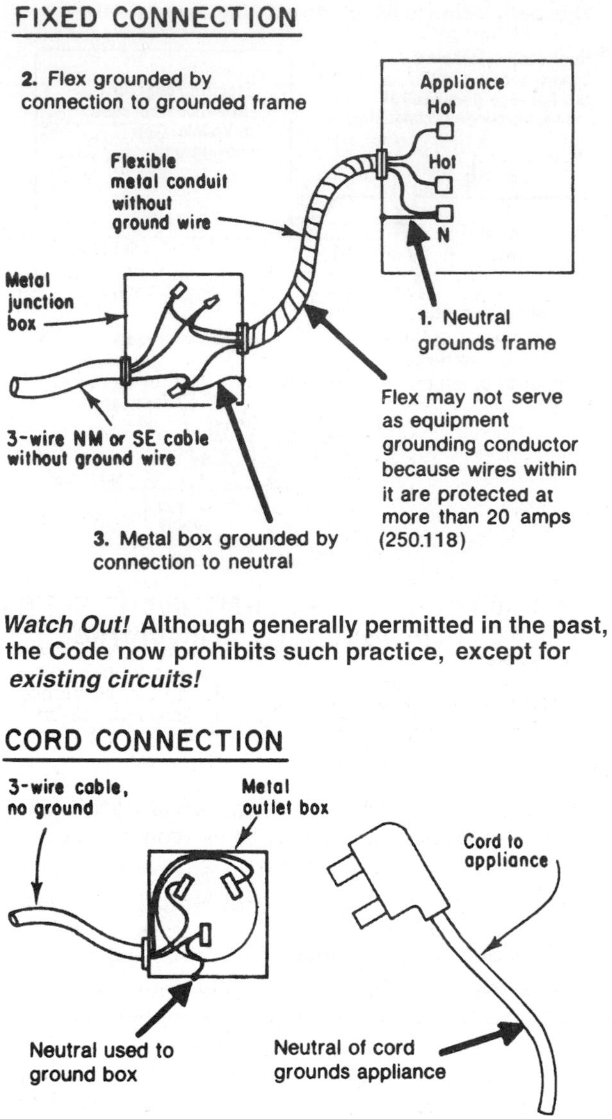
FIXED CONNECTION
2. Flex grounded by connection to grounded frame
Appliance
Hot
Hot
N
Flexible metal conduit without ground wire
Metal junction box
1. Neutral grounds frame
Flex may not serve as equipment grounding conductor because wires within it are protected at more than 20 amps (250.118)
3-wire NM or SE cable without ground wire
3. Metal box grounded by connection to neutral
Watch Out! Although generally permitted in the past, the Code now prohibits such practice, except for existing circuits!
CORD CONNECTION
3-wire cable, no ground
Metal outlet box
Cord to appliance
Neutral used to ground box
Neutral of cord grounds appliance

Fig. 250-103. These techniques may be used to ground boxes on existing circuits only! (Sec. 250.140.)

at a point from the source to the first system disconnecting means. Also the permission on outbuildings reflects a continuing, but now extremely limited, permission in those areas to reground a neutral on existing premises wiring systems only.

Figure 250-104 shows such applications. At A, the grounded service neutral is bonded to the meter housing by means of the bonded neutral terminal lug in the

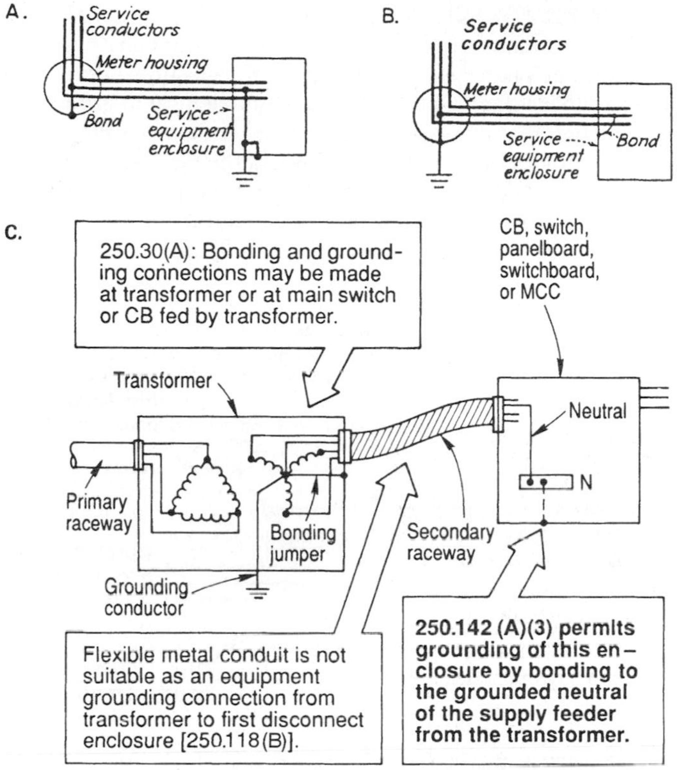
A.
Service conductors
Meter housing
Bond
Service equipment enclosure
B.
Service conductors
Meter housing
Service equipment enclosure
Bond
C.
250.30(A): Bonding and grounding connections may be made at transformer or at main switch or CB fed by transformer.
CB, switch, panelboard, switchboard, or MCC
Transformer
Neutral
N
Primary raceway
Bonding jumper
Secondary raceway
Grounding conductor
Flexible metal conduit is not suitable as an equipment grounding connection from transformer to first disconnect enclosure [250.118(B)].
250.142 (A)(3) permits grounding of this enclosure by bonding to the grounded neutral of the supply feeder from the transformer.

Fig. 250-104. Using grounded circuit conductor to ground equipment housings on line side of service or separately derived system. (Sec. 250.142.)

socket—and the housing is thereby grounded by this connection to the grounded neutral, which itself is grounded at the service equipment as well as at the utility transformer secondary supplying the service. At B, the service equipment enclosure is grounded by connection (bonding) to the grounded neutral—which itself is grounded at the meter socket and at the supply transformer. These same types of grounding connections may be made for CT cabinets, auxiliary gutters, and other enclosures on the line side of the service-entrance disconnect means, including the enclosure for the service disconnect. In some areas, the utilities and inspection departments will not permit the arrangement shown in Fig. 250-104 because the connecting lug in the meter housing is not always accessible for inspection and testing purposes. At C, equipment is grounded to the neutral on the line (supply) side of the first disconnect fed from a step-down transformer (a separately derived system).

Aside from the permission given in the three exceptions to the rule of part **(B)**, and separately derived systems or main building disconnects (250.30 and 250.32), the wording of part **(B)** prohibits connection between a grounded neutral and equipment enclosures on the load side of the service. The wording supports the prohibition in 250.24(A)(5) and 250.30(A) of grounding connections. So aside from the few specific exceptions mentioned, bonding between any system grounded conductor, neutral or phase leg, and equipment enclosures is prohibited on the load side of the service (Fig. 250-105). The use of a neutral to ground panelboard or other equipment (other than specified in the exceptions) on the load side of service equipment would be extremely hazardous if the neutral became loosened or disconnected. In such cases any line-to-neutral load would energize all metal components connected to the neutral, creating a dangerous potential above ground. Hence, the prohibition of such a practice. This is fully described in Fig. 250-11.

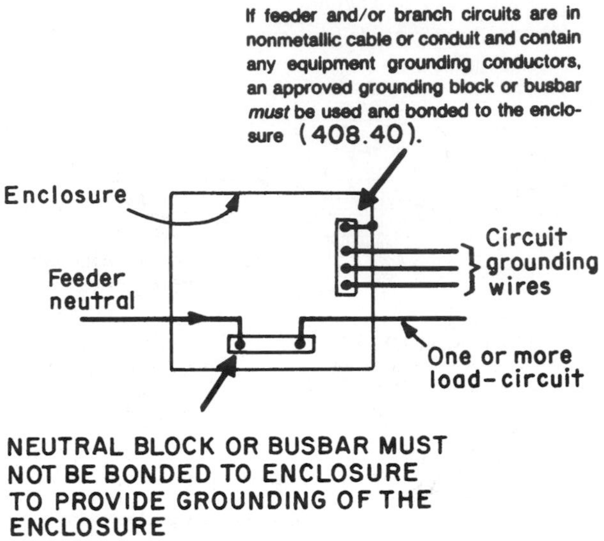
If feeder and/or branch circuits are in nonmetallic cable or conduit and contain any equipment grounding conductors, an approved grounding block or busbar *must* be used and bonded to the enclosure (408.40).
Enclosure
Circuit grounding wires
Feeder neutral
One or more load-circuit
NEUTRAL BLOCK OR BUSBAR MUST NOT BE BONDED TO ENCLOSURE TO PROVIDE GROUNDING OF THE ENCLOSURE

Fig. 250-105. Panel, switchboard, CB, and switch on load side of service within a single building. (Sec. 250.142.)

When a circuit is run from one building to another, it may be necessary, simply permissible, or expressly prohibited to connect the system "grounded" conductor to a grounding electrode at the other building—as covered by 250.32. Separately derived systems—as covered in 250.30—are also exempted from the basic requirement by the first sentence of 250.142(B).

Although this rule of the Code prohibits neutral bonding on the load side of the service, 250.130(A) and 250.24(A) clearly require such bonding at the service entrance. And the exceptions to prohibiting load-side neutral bonding to enclosures are few and very specific:

- Exception No. 1 of 250.142(B) permits frames of ranges, wall ovens, countertop cook units, and clothes dryers to be "grounded" by connection to the grounded neutral of their supply circuit, but only for *existing* circuits (250.140).
- Exception No. 2 to 250.142(B) permits grounding of meter enclosures to the grounded circuit conductor (generally, the grounded neutral) on the *load side*

of the service disconnect if the meter enclosures are located immediately adjacent to the service disconnect, the service is not equipped with ground-fault protection, and the neutral is not less than the minimum required by 250.122, based on the rating of the service overcurrent device. This rule applies, of course, to multioccupancy buildings (apartments, office buildings, etc.) with individual tenant metering (Fig. 250-106).

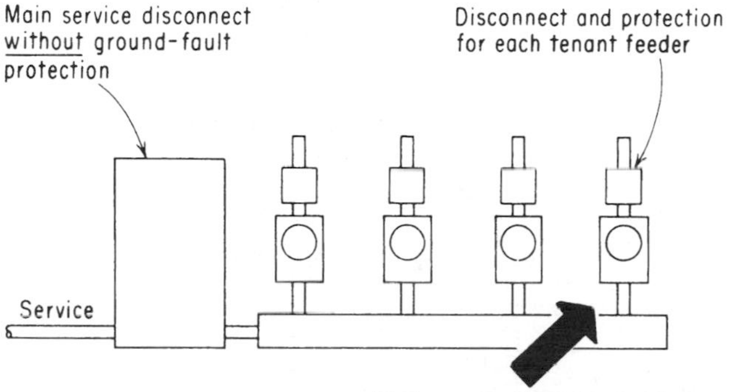
Main service disconnect
without ground-fault
protection
Disconnect and protection
for each tenant feeder
Service
Meter sockets on load side may
be grounded to grounded
neutral

Fig. 250-106. Grounding meter enclosures to grounded conductor on *load side* of service disconnect, with meters located "immediately adjacent" to the service disconnect. (Sec. 250.142.)

- Exception No. 3 permits dc systems to be grounded on the load side of the service disconnect as described in 250.164.
- Exception No. 4 permits medium-voltage electrode type boilers to operate with their neutral conductors bonded to the pressure vessel and with ancillary electrical equipment bonded to the vessel or to the equipment, as covered in Part V of Art. 490.

If a meter bank is on the upper floor of a building, as in a high-rise apartment house, or otherwise away from service disconnect, such meter enclosures would not meet the rule that they must be "immediately adjacent to" the service disconnect. In such cases, the enclosures must not be grounded to the neutral. And if the service has ground-fault protection, meter enclosures on the load side must not be connected to the neutral, even if they are "immediately adjacent to" the service disconnect.

250.146. Connecting Receptacle Grounding Terminal to Box. The first paragraph requires that a jumper be used when the outlet box is installed in the wall (Fig. 250-107). Because boxes installed in walls are very seldom found to be perfectly flush with the wall, direct contact between device screws and yokes and boxes is seldom achieved. Screws and yokes currently in use were designed solely for the support of devices rather than as part of the grounding circuit. Although the general rule states that a flush-type box, installed in a wall for a receptacle outlet, does require a bonding jumper from a grounded box to the

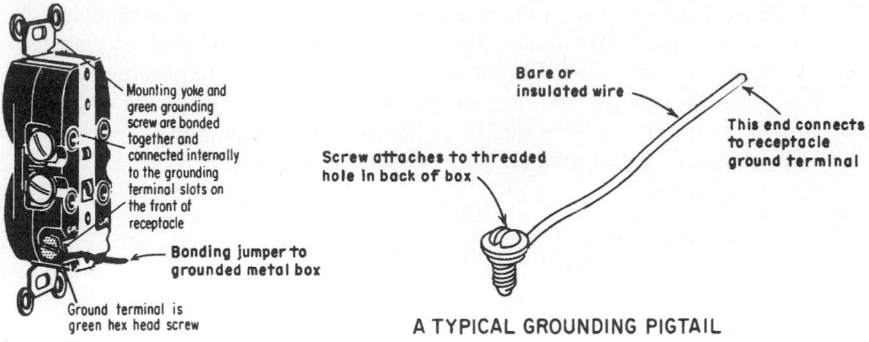
Mounting yoke and green grounding screw are bonded together and connected internally to the grounding terminal slots on the front of receptacle
Bonding jumper to grounded metal box
Ground terminal is green hex head screw
Bare or insulated wire
This end connects to receptacle ground terminal
Screw attaches to threaded hole in back of box
A TYPICAL GROUNDING PIGTAIL

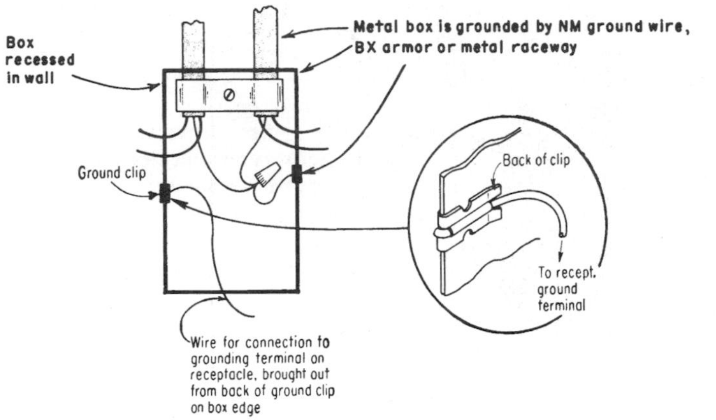
Box recessed in wall
Metal box is grounded by NM ground wire, BX armor or metal raceway
Ground clip
Back of clip
To recept. ground terminal
Wire for connection to grounding terminal on receptacle, brought out from back of ground clip on box edge

Fig. 250-107. Bonding jumper connects receptacle ground to grounded box. (Sec. 250.146.)

receptacle grounding terminal, part **(A)** pertains to surface-mounted boxes and eliminates the need for a separate bonding jumper between a surface-mounted box and the receptacle grounding terminal under the conditions described. But, where such grounding is selected, at least one of the insulating washers must be removed from the device's securing screws.

Although part **(A)** generally exempts surface-mounted boxes from the need for a bonding jumper from the box to the ground terminal of a receptacle installed in the box—because there is solid contact between the receptacle's grounded mounting yoke and the ears on the box when installed—that is not applicable to a receptacle mounted in a raised box cover. (See Fig. 250-108.) There are several issues to consider with respect to mounting receptacles in raised covers. The first is 406.5(C), which requires (generally) no fewer than two screws to hold the receptacle in place. Contrary to common belief, this rule has nothing to do with grounding; it has to do with documented loss experience where a single screw holding the center of a duplex receptacle loosened to the point of allowing the receptacle to fall

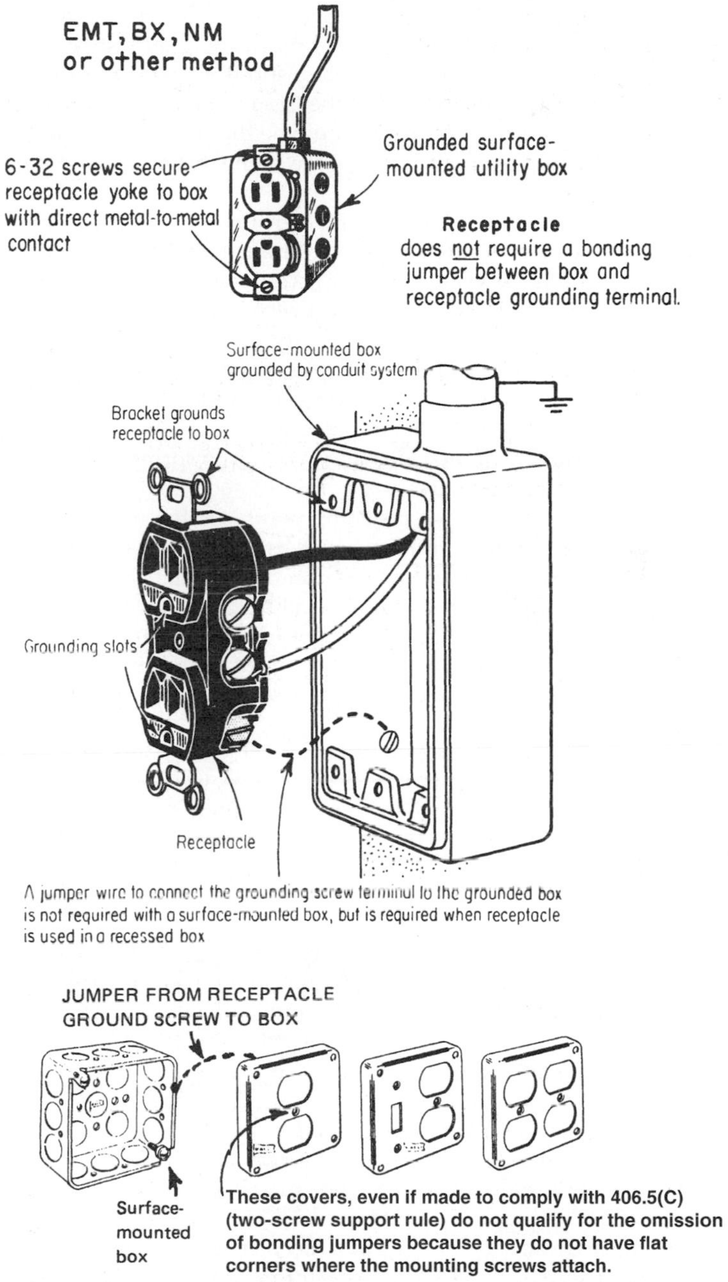
EMT, BX, NM
or other method
6-32 screws secure receptacle yoke to box with direct metal-to-metal contact
Grounded surface-mounted utility box
Receptacle
does not require a bonding jumper between box and receptacle grounding terminal.
Surface-mounted box grounded by conduit system
Bracket grounds receptacle to box
Grounding slots
Receptacle
A jumper wire to connect the grounding screw terminal to the grounded box is not required with a surface-mounted box, but is required when receptacle is used in a recessed box
JUMPER FROM RECEPTACLE
GROUND SCREW TO BOX
Surface-mounted box
These covers, even if made to comply with 406.5(C) (two-screw support rule) do not qualify for the omission of bonding jumpers because they do not have flat corners where the mounting screws attach.

Listed, raised covers with flat corners and "screw locking" means for receptacle attachments, no fewer than two points of support, now recognized for receptacle mounting without bonding jumpers.

Fig. 250-108. Typical applications where a surface box does and does not need a receptacle bonding jumper. (Sec. 250.146.)

far enough behind the cover that it would begin to twist. If a plug were inserted at the time (and this happened a number of times), and the twist occurred in the direction of the ungrounded blade on the plug, and if the plug were even slightly loose, the energized blade would contact the grounded cover at the edge of the duplex punch-out. The result, of course, would be the compete destruction of the plug, the receptacle, the cover, and quite a few items nearby. The double-screw rule has ended this problem.

The reluctance to recognize the cover as the grounding contact has more to do with the permanence of the entire setup. The language in Part A about a box and cover combination being listed as providing acceptable continuity has never been of much help; it went into the **NEC** to protect certain explosionproof receptacles from needing bonding jumpers to their heavily bolted supports. This problem has been addressed in the 2008 **NEC**, however, through additional language in Part A. Now, for the first time a raised cover is permitted to support receptacles without bonding jumpers to the box, provided it is of "crushed corner" construction. If the cover has totally flat corners, there will be no spring in the sheet metal, and the 8-32 corner screws on the box will hold securely. The cover must be listed and with no fewer than two fastening points for the receptacle. The support hardware must either be permanent, such as a rivet, or use "thread locking" or "screw locking" means. Thread locking means presumably a jam nut of some type, and it would be somewhat difficult to get hold of between the edge of the receptacle and the raised part of the cover. Screw locking was clarified by adding the words "or nut" and thereby includes the current raised cover designs, which come with knurled 6-32 nuts that grab the underside of the yoke when the screw is tightened.

Figure 250-109 illustrates a grounding device which is intended to provide the electrical grounding continuity between the receptacle yoke and the box on which it is mounted and serves the dual purpose of both a mounting screw and a means of providing electrical grounding continuity in lieu of the required bonding jumper. As shown in the sketch, special wire springs and four-lobed machine

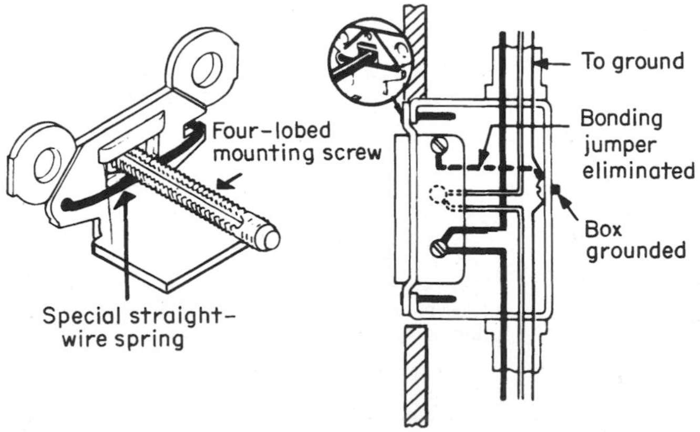
Four-lobed
mounting screw
Special straight-
wire spring
To ground
Bonding
jumper
eliminated
Box
grounded

Fig. 250-109. Self-grounding screws ground receptacle in recessed box without bonding jumper. (Sec. 250.146.)

screws are part of a receptacle design for use without a bonding jumper to box. This complies with 250.146(B).

250.146(C) permits non–self-grounding receptacles without an equipment grounding jumper to be used in floor boxes which are designed for and listed as providing proper continuity between the box and the receptacle mounting yoke.

Part **(D)** of 250.146 allows the use of a receptacle with an isolated grounding terminal (no connection between the receptacle grounding terminal and the yoke). Sensitive electronic equipment that is grounded normally through the building ground is often adversely affected by pickup of transient signals, which cause an imbalance in the delicate circuits. This is particularly true with highly intricate medical and communications equipment, which often picks up unwanted currents, even of very low magnitude.

The use of an isolated grounding receptacle allows a "pure" path to be established back to the system grounding terminal, in the service disconnecting means, without terminating in any other intervening panelboard. In Fig. 250-110, a

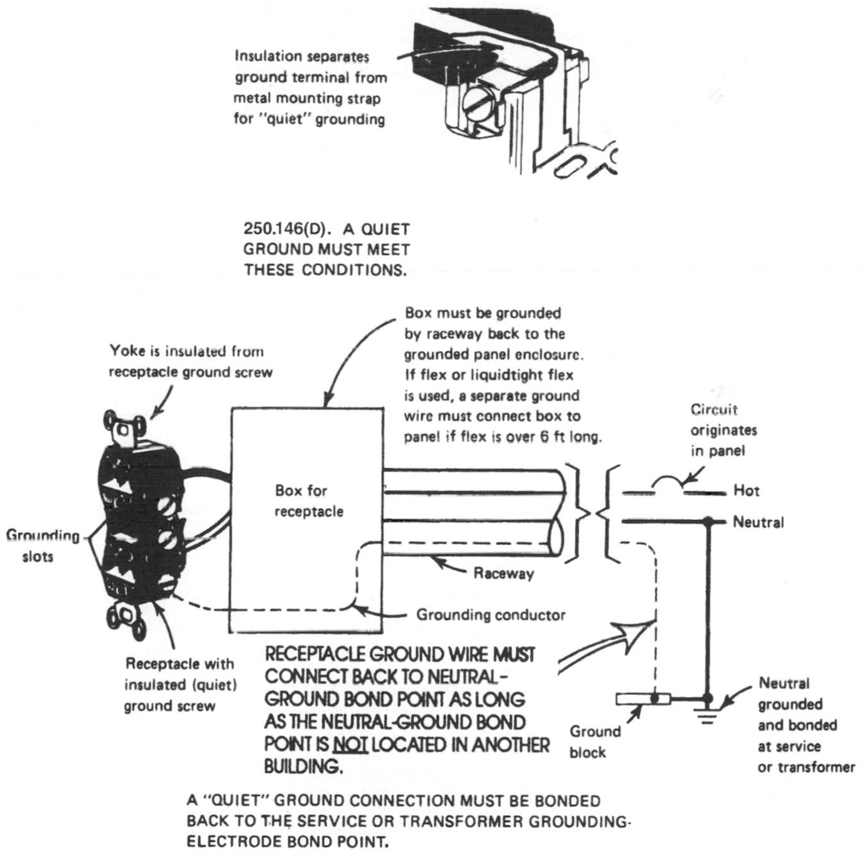
Insulation separates ground terminal from metal mounting strap for "quiet" grounding
250.146(D). A QUIET GROUND MUST MEET THESE CONDITIONS.
Yoke is insulated from receptacle ground screw
Box must be grounded by raceway back to the grounded panel enclosure. If flex or liquidtight flex is used, a separate ground wire must connect box to panel if flex is over 6 ft long.
Circuit originates in panel
Hot
Neutral
Box for receptacle
Grounding slots
Raceway
Grounding conductor
Receptacle with insulated (quiet) ground screw
RECEPTACLE GROUND WIRE MUST CONNECT BACK TO NEUTRAL-GROUND BOND POINT AS LONG AS THE NEUTRAL-GROUND BOND POINT IS NOT LOCATED IN ANOTHER BUILDING.
Ground block
Neutral grounded and bonded at service or transformer
A "QUIET" GROUND CONNECTION MUST BE BONDED BACK TO THE SERVICE OR TRANSFORMER GROUNDING-ELECTRODE BOND POINT.

Fig. 250-110. Receptacles with isolated ground terminal are used with "clean" or "quiet" ground. (Sec. 250.146.)

cutaway of an isolated grounding receptacle shows the insulation between the grounding screw and the yoke (top), and the hookup of the insulated grounding conductor to the common neutral-equipment-ground point of the electrical system (bottom).

The last sentence of part **(D)** permits an equipment grounding conductor from the insulated (quiet) ground terminal of a receptacle to be run, unbroken, all the way back to the ground terminal bus that is bonded to the neutral at the service equipment or at the secondary of a step-down transformer—but in no case may the isolated ground extend beyond the building in which it is used. That is, it must be bonded to a ground bus within the building it is run in, even if the "service equipment" is located in another building. Or the equipment grounding conductor may be connected to any ground bus in an intermediate panelboard fed from the service or transformer. *But,* the important point is to be sure the insulated ground terminal of the receptacle does tie into the equipment ground system that is bonded to the neutral.

This rule must be observed very carefully to avoid violations that have been commonly encountered in the application of branch circuits to computer equipment—where manufacturers of computers and so-called computer power centers specified connection of "quiet" receptacle ground terminals to a grounding electrode that is independent of (not bonded to) the neutral and bonded equipment ground bus of the electrical system. This practice developed to eliminate computer operating problems that were attributed to "electrical noise." Such isolation of the receptacle ground terminal does not provide an effective return path for fault-current flow and, therefore, constitutes a hazard.

Any receptacle grounding terminal (the green hex-head screw)—whether it is the common type with the mounting yoke or the type insulated from the yoke—must be connected back to the point at which the system neutral is bonded to the equipment grounding terminal and to the grounding electrode, thereby providing the "effective ground-fault current path." That common (bonded) point may be at the service equipment (where there is no voltage step-down from the service to the receptacle), or the common neutral-equipment-ground point may be at a panelboard fed from a step-down transformer (as used in computer power centers).

When an isolated ground connection is made for the receptacle ground terminal, the box containing the receptacle must be grounded by the raceway supplying it and/or by another equipment grounding conductor run with the circuit wires. And those grounding conductors must tie into the same neutral-equipment-ground point to which the receptacle isolated ground terminal is connected. The equipment grounding conductor that actually lands on the grounding terminal of the receptacle, in addition to passing through intervening panels without joining with other equipment grounding conductors, is also permitted (2008 **NEC** clarification) to pass through intervening boxes and other enclosures, also without bonding at those intermediate points. This was always implied in the rule but it is now expressly stated.

See comments that follow 408.40, Exception.

250.148. Continuity and Attachment of Equipment Grounding Conductors to Boxes. The basic rule requires that if circuit conductors enter a box and get spliced or terminated on equipment in or supported by the box, all equipment grounding conductors *associated with those conductors* must be connected as well, in accordance with five specific provisions (below). This wording means that when wires pass directly through a box, there is no requirement to break the continuity of any of the conductors. Of course, that does not relieve the requirement to connect the box to an equipment grounding conductor as required by 314.4. However, if a box is connected to a metal raceway, there is no reason to interrupt the continuity of the unbroken conductors. In addition, there is no reason to break the continuity of any equipment grounding conductors associated with circuit conductors that are unbroken in a box, even if conductors on or associated with a different circuit are spliced or terminated with the box and therefore subject to the five rules that follow:

Part **(A)** requires that all connections and splices meet 110.14**(B)**, except that insulation is not required.

Part **(B)** requires that the grounding connections in a box be arranged so removing a device or other piece of equipment will not interrupt the continuity of the equipment grounding conductor to any other loads on the branch circuit. As a practical matter this means using pigtail connections so the grounding connection to any item can be released with the splice still intact.

Part **(C)** requires that for a metal box, a connection must be made between the one or more equipment grounding conductors that enter in association with conductors that are spliced or terminated (and that are subject to these five rules) and also a connection must be made between them and the box itself, using a grounding screw that is used for no other purpose, or other equipment or device listed for grounding.

Part **(D)** requires that equipment grounding conductors entering a nonmetallic box be arranged so a grounding connection can be made to any item in the box that requires the use of a grounding connection.

Part **(E)** forbids the use of connections that depend solely on solder.

Many issues follow from Part **(C)** above, beginning with the fact that grounding conductors in any metal box (for which the associated ungrounded conductors are spliced or terminated) must be connected to each other and to the box itself. Figure 250-111 shows a method of connecting ground wires in a box to satisfy the letter of 250.148(C). Note that the two ground wires are solidly connected to each other by means of a crimped-on spade-tongue terminal, with one of the ground wires (arrow) cut long enough so that it is bent back out of the crimp lug to provide connection to the green hex-head screw on a receptacle outlet (if required by 250.146). The spade lug is secured firmly under a screw head, bonding the lug to the box. Of course, the specific connections could be made in other ways. For instance, the ground wires could be connected to each other by twist-on splicing devices; and connection of the ground wires to the box could be made by simply wrapping a single wire under the screw head or by connecting a wire from the splice connector to an approved grounding clip on the edge of the box (Fig. 250-112).

In all the drawings here, connection to the box is made either by use of a screw in a threaded hole in the side or back of the box or by an approved ground clip device which tightly wedges a ground wire to the edge of the box wall, as shown

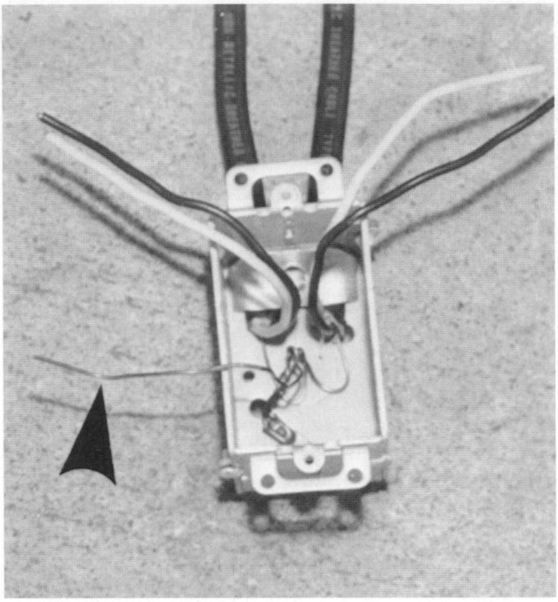

Fig. 250-111. Both ground wires are solidly bonded together in the crimped barrel of the spade lug, which is screwed to the back of the metal box. (Sec. 250.148.) Note that length of the branch-circuit equipment grounding conductors in this photo are probably less than 6 in. long from cable sheath to the lug. This would violate 300.14, which, contrary to widespread belief, applies to all conductors entering a box including equipment grounds.

in Fig. 250-113. Preassembled pigtail wires with attached screws are available for connecting either a receptacle or the system ground wire to the box.

Figure 250-114 shows connection of two cable ground wires by means of two grounding clips on the box edges (arrow). In the past, such use has been disallowed by some inspection authorities because the ground wires are not actually connected to each other but are connected only through the box. The text ". . . between the one or more... and a metal box by means of... equipment listed for grounding . . .," however, permits such practice as long as the clips are listed for grounding.

Figure 250-115 shows another method that has been objected to as clear violation of NEC 250.148(C) which requires that a screw used for connection of grounding conductors to a box "shall be used for no other purpose." Use of this screw, simultaneously, to hold the clamp is for "other purpose" than grounding. Objection is not generally made to use of the clamp screw for ground connection when, in cases where the clamp is not in use, the clamp is removed and the screw serves only the one purpose—to ground the grounding wires.

The exception to this rule eliminates the need for connecting an isolated grounding conductor to all other grounding conductors.

250.164. Point of Connection for Direct-Current Systems. On a 2- or a 3-wire dc distribution system, a neutral that is required to be grounded must be grounded at the supply station only.

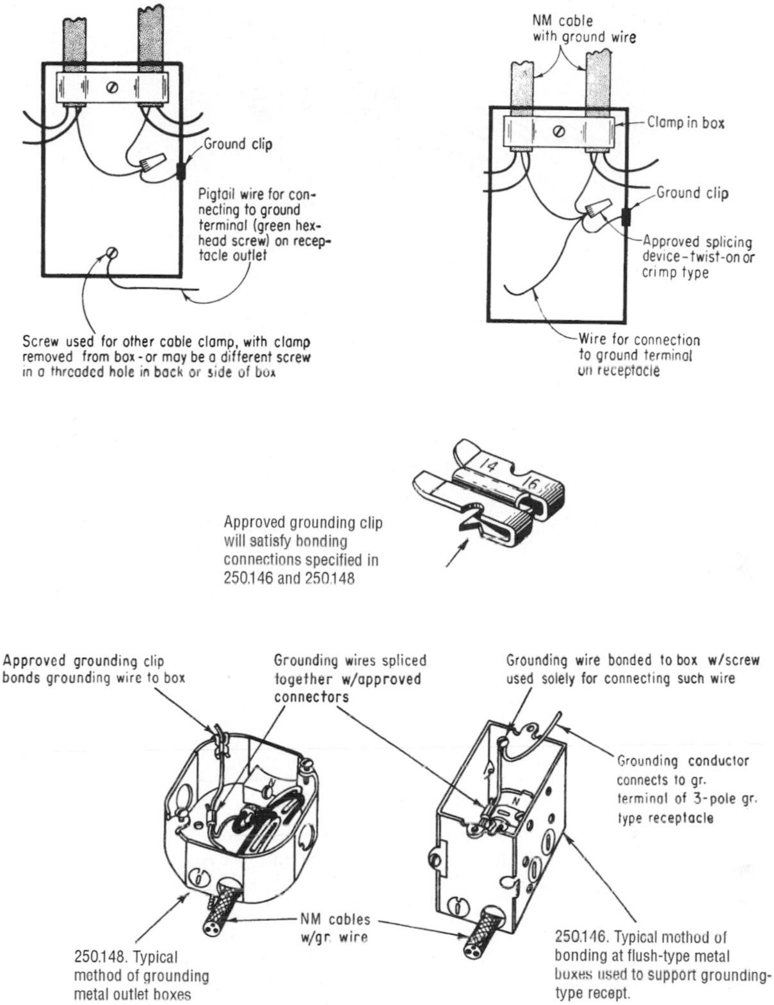
Ground clip
Pigtail wire for connecting to ground terminal (green hex-head screw) on receptacle outlet
Screw used for other cable clamp, with clamp removed from box - or may be a different screw in a threaded hole in back or side of box
NM cable with ground wire
Clamp in box
Ground clip
Approved splicing device-twist-on or crimp type
Wire for connection to ground terminal on receptacle
Approved grounding clip will satisfy bonding connections specified in 250.146 and 250.148
14
16
Approved grounding clip bonds grounding wire to box
Grounding wires spliced together w/approved connectors
Grounding wire bonded to box w/screw used solely for connecting such wire
Grounding conductor connects to gr. terminal of 3-pole gr. type receptacle
NM cables w/gr. wire
250.148. Typical method of grounding metal outlet boxes
250.146. Typical method of bonding at flush-type metal boxes used to support grounding-type recept.

Fig. 250-112. All these techniques bond the ground wires together and to the box. (Sec. 250.148.)

As noted in part **(B)**, an on-site supply for a dc system must have a required grounding connection made at either the source of the dc supply or at the first disconnect or overcurrent device supplied. Because the basic rule says a dc source (from outside a premises) must have a required grounding connection made at "one or more supply stations" and *not* at "any point on premises wiring," an on-site dc source would be prohibited from having a grounding connection that might be required. This rule resolves that basic problem by referring to a "dc system source . . . located on the premises."

250.166. Size of the Direct-Current Grounding Electrode Conductor. Figure 250-116 is a diagram of a balancer set used with a 2-wire 230-V generator to

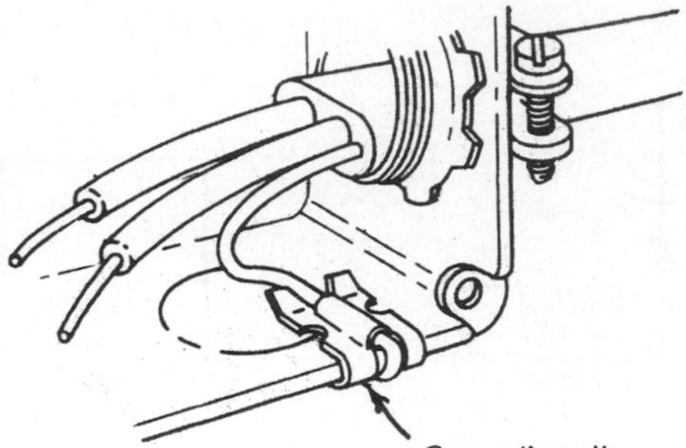

Grounding clip
for No. 14 or No. 12
ground wire

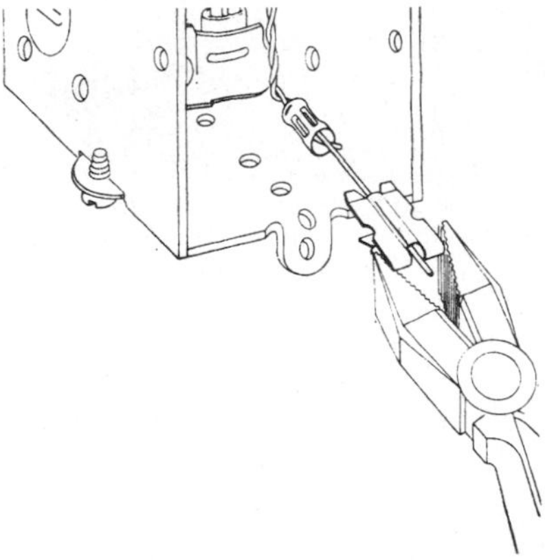

Installation of grounding clip.

Fig. 250-113. Ground clip is "identified" for use as called for by 110.14**(B)**. (Sec. 250.148.) Note that the length of free equipment grounding conductor shown here is likely less than 150 mm (6 in.) and therefore in violation of 300.14. Contrary to widely held opinion, nothing in the wording of 300.14 limits it to current-carrying conductors; it applies equally to equipment grounding conductors.

supply a 3-wire system as referred to in 445.12(D) and covered by part **(A)** of this section. These rules require the grounding electrode conductor to be no smaller than the neutral if there is a balancer set or balancer winding and not smaller than 8 AWG copper or 6 AWG aluminum. Otherwise the grounding conductor must be as large as the largest conductor supplied by the system. This rule is applicable to solar photovoltaic sources among other places. The rules also incorporate the special electrode sizing applicable to the same specific electrodes called out in 250.66, and with the same wire sizes specified. The 2014 **NEC** further correlated these requirements and Table 250.66 by setting 3/0 copper or 250 kcmil aluminum as the largest grounding electrode conductors required for dc systems.

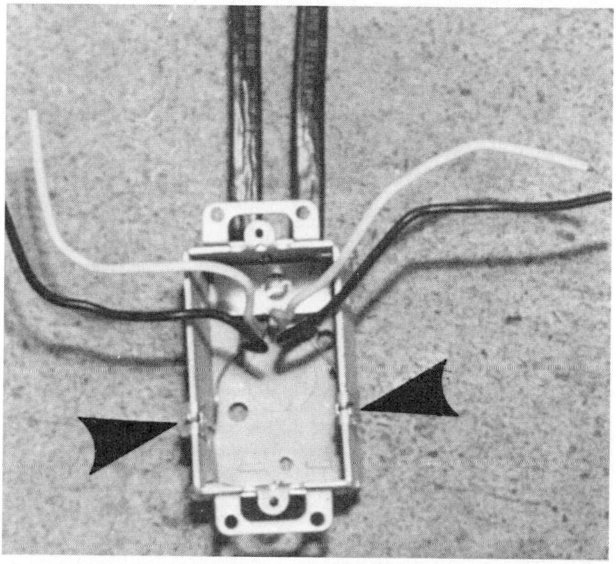

Fig. 250-114. Each ground wire is connected to the metal box by a ground clip (one on each side at arrows). The first sentence of 110.14(B) permits ground wires to be "spliced or joined" by use of ground clips or ground screw terminals in the box. (Sec. 250.148.)

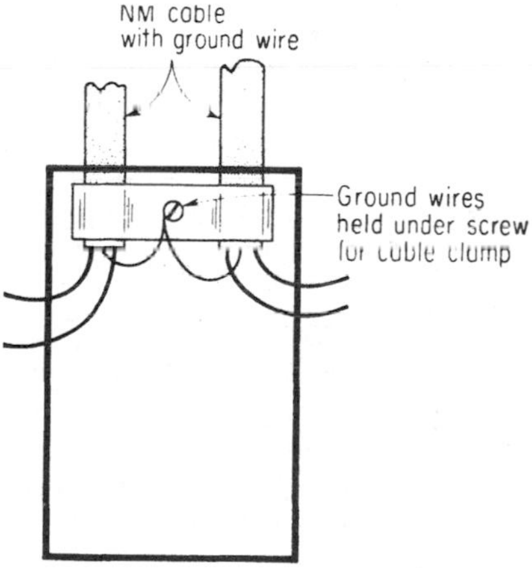
NM cable
with ground wire
Ground wires
held under screw
for cable clamp

Fig. 250-115. This clearly violates 250.148(C) because the screw is also used to anchor the cable clamp. (Sec. 250.148.)

250.167. Direct-Current Ground-Fault Detection. This section is new as of the 2014 NEC. DC systems that are ungrounded must be equipped with ground-fault detection, and those that are grounded are permitted to use it. They must be marked at their source of first disconnecting means as to the type of grounding system that is applicable. An informational note points to the 2012 edition of NFPA 70E as a source of information about the four applicable grounding types.

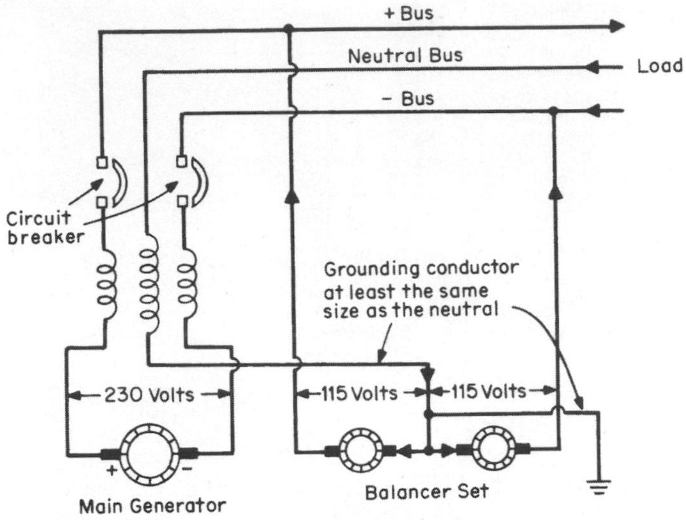
+ Bus
Neutral Bus
- Bus
Load
Circuit breaker
Grounding conductor at least the same size as the neutral
230 Volts
115 Volts
115 Volts
Main Generator
Balancer Set

Fig. 250-116. Sizing a dc system grounding conductor. (Sec. 250.166.)

The four grounding types listed in the Informational Note attached to Sec. 320.3(C)(2) in the 2012 edition of that standard are:

(1) Type 1. Ungrounded dc usually equipped [and now **NEC** mandated] with ground detection alarms.
(2) Type 2. Solidly grounded dc with either the most positive or most negative pole grounded; ground detection not typically deployed.
(3) Type 3. Resistance grounded dc; a fault enables activation of an alarm and a second fault may result in a short circuit.
(4) Type 4. Solidly grounded dc at a center point; a ground fault on either polarity can result in a short circuit, and ground-fault detection is not typically deployed.

250.184. Solidly Grounded Neutral Systems. Figure 250-117 shows the details of this set of rules. This section does permit a neutral conductor of a solidly grounded "Y" system to have insulation rated at only 600 V, instead of requiring insulation rated for the high voltage (over 1000 V). It also points out that a bare copper neutral may be used in such systems for service-entrance conductors or for direct buried feeders, and bare copper or copper-clad aluminum may be used for overhead sections of outdoor circuits.

250.186. Grounding Service-Supplied Alternating-Current Systems. This section is new as of the 2014 **NEC**. This section is a medium-voltage counterpart to Sec. 250.24(C) and is intended to assure a low impedance path, wherever this would be reasonably possible, for returning fault current by requiring that a grounded circuit conductor of some description [refer to the two possibilities covered in (A) and (B) below] is solidly connected at the service equipment. Part **(A)** covers distributions with a grounded conductor at the service point. In this case 250.184 will still apply, and in accordance with 250.184(A)(2) the minimum neutral ampacity

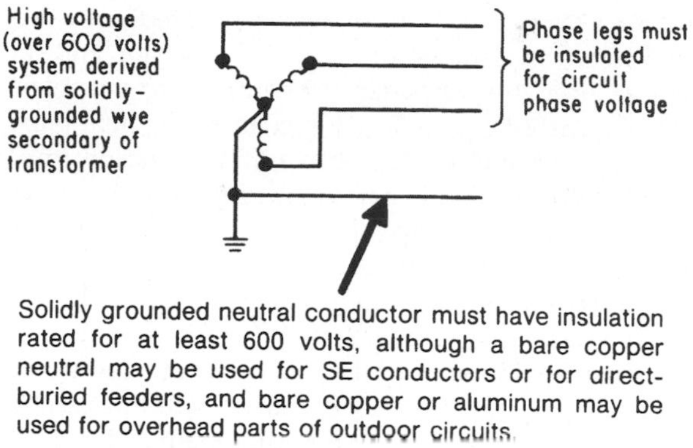
High voltage
(over 600 volts)
system derived
from solidly-
grounded wye
secondary of
transformer
Phase legs must
be insulated
for circuit
phase voltage
Solidly grounded neutral conductor must have insulation rated for at least 600 volts, although a bare copper neutral may be used for SE conductors or for direct-buried feeders, and bare copper or aluminum may be used for overhead parts of outdoor circuits.

Fig. 250-117. Neutral of high-voltage system generally must be insulated for 600 V. (Sec. 250.184.)

will depend on the load profile but never less than one-third or one-fifth of the ampacity of the phase conductors, depending on engineering supervision. This new section sets a different floor based on fault return performance, one that can be higher but never less than the minimum size required to service the load. Both calculations will need to be completed, because the worst case will vary and is not obvious. For a single raceway or overhead run, this minimum sizing depends on whether the service is configured as a single run or in parallel. A single run follows the usual rules in 250.102(C)(1). If the service entrance wiring is made up in parallel, whether overhead or underground, the grounded conductor shall also run in parallel, with the size in each run also figured on the usual 250.102(C)(1) process, but based on associated ungrounded conductor sizes in that run, and never smaller than 1/0. As in the case of 250.24(C), an identically worded exception allows for multiple disconnecting means within an appropriately listed assembly. Rounding out the picture, a corner-grounded delta connected service mirrors 250.24(C)(3), with the grounded conductor at least equal in size to its ungrounded partners, and an impedance-grounded system neutral is sized, for medium voltage, in accordance with 250.187.

Part **(B)**, covering distributions without a grounded conductor at the service point, does not have a direct 250.24(C) counterpart, and this makes it more interesting. Note, however, that even this part contains the wording "is grounded at any point" in its first sentence. Even the NEC Committee cannot make a completely ungrounded system grounded. The substantiation on this point concludes with the following sentence, that puts this in a reasonable perspective: "Where a utility does not provide a neutral conductor there is generally a static line or other ground fault return path where the supply side bonding jumper can be connected to, completing the return circuit." That statement is undoubtedly correct. The sizing and routing requirements for the grounding conductor installed under this part are completely identical to those in (A), the only difference being the identity of the wire. This conductor is not a grounded conductor, it is a supply-side bonding jumper. Left entirely unanswered in this new section is the insulation level to be applied to this wiring. Presumably the rules in 250.184(A)(1) would be the place to start.

250.187. Impedance Grounded Neutral Systems. Medium-voltage systems are often grounded using the neutral point of a wye connection and connecting it to a grounding electrode through an impedance. There are two general types of systems, generally described as either high or low impedance. The high impedance systems operate in a similar manner as those described at 250.36. The low impedance systems use a much lower impedance that is not intended to keep the system running, but instead allows for orderly and prompt operation of circuit protective devices while the destructive effects of the fault are greatly diminished, allowing for less extensive repairs. The neutrals on such systems are insulated based on the maximum neutral voltage that would be encountered, which is $\frac{1}{2}\sqrt{3}$ or 58% of the line-to-line voltage.

250.190. Grounding of Equipment. All noncurrent-carrying metal parts of equipment, housings, and enclosures must be grounded, including associated fences and supporting structures. The exception covers equipment isolated from ground and placed so it would be impossible for anyone in contact with the ground to touch such metal parts when the equipment is energized. This correlates with 250.110 Exception No. 2 that exempts wooden-pole-mounted distribution equipment that is over 2.5 m (8 ft) high from the usual equipment grounding requirements.

Part **(B)** requires that if a grounding electrode conductor is installed and connects noncurrent-carrying metal parts to ground, it must be sized in accordance with Table 250.66 as applies to such conductors generally. Enter the table using the size of the largest associated ungrounded conductor at that location, with a minimum size of 6 AWG (or 4 AWG for aluminum).

Part **(C)** covers equipment grounding conductors. The first paragraph requires that equipment grounding conductors that are not an integral part of a cable assembly must be no smaller than 6 AWG copper or 4 AWG aluminum. This requirement establishes the usual size of conductor that will be installed routinely in the field for transformer vaults, switchboards, and at other medium-voltage cable termination locations to make the required grounding connections to medium-voltage cable shielding.

The second paragraph covers shielded cables. Shielding is normally used to confine the voltage stresses to the insulation, thereby avoiding damage from corona discharge. However, limited permission is granted for the use of shielding to function as an equipment grounding conductor. A concentric neutral type of arrangement with the collective size of the concentric conductors, if collectively equalling or exceeding the sizing rule in the third paragraph, is generally acceptable. Only on systems that are grounded through an impedance, or ungrounded, metallic tape or drain wire insulation shields are permitted for equipment grounding, provided the shield is rated for the clearing time of the ground-fault current protective device.

The third paragraph incorporates the equipment grounding conductor sizing table in 250.122, but applies it to the current rating of the upstream fuse or the setting of the protective relay, as applicable. An informational note points out that medium-voltage circuit breakers use current transformers and a protective relay to actuate; combining the current pickup setting with the CT winding ratio determines the actual breaker current setting.

250.194. Grounding and Bonding of Fences and Other Metal Structures. This section is new as of the 2014 NEC. It provides prescriptive requirements that will be extremely useful with respect to addressing a very common construction feature of outdoor medium-voltage substations, particularly those with "exposed electrical conductors and equipment." These fences and other metal structures must be arranged so step, touch, and transfer voltages are limited. Refer to the coverage in this book of these voltages at Sec. 682.33. Transfer voltage is, in effect, a special case of touch voltage; it refers to instances where a metallic conductor entering or leaving a substation and connected to its grounding system sees the entire ground potential rise during a fault condition.

In all cases, the requirements reference, directly or indirectly, connects to "the grounding electrode system." These systems depend on careful engineering because they must go far beyond any conventional prescriptive requirements in the **NEC** in order to address applicable voltages, soil resistivity, and potential influences of other conductive elements of the electrical distribution system. However, when that system is established, the requirements in this new section are comparatively easy to apply and inspect.

ARTICLE 280. SURGE ARRESTERS, OVER 1 KV

280.1. Scope. This article provides rules and regulations covering the application of "surge arresters" or, as they are more commonly known, "lightning arresters." Here, as stated the Code provides general, installation, and connection requirements for lightning equipment and systems (Fig. 280-1). The major change as of the 2008 **NEC** is that this article now only applies to over 1 kV applications. All other applications at conventional utilization voltages have been transferred to Art. 285.

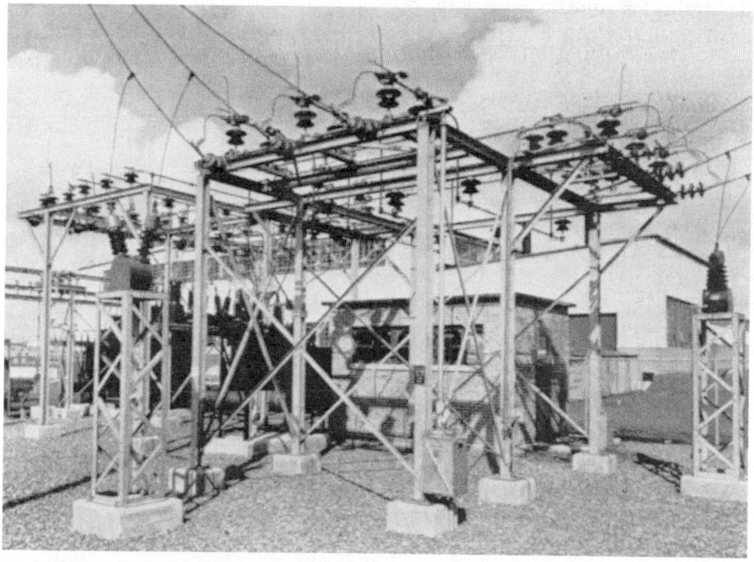

Fig. 280-1. Surge (lightning) arresters in an electric substation serving an industrial plant are commonly used in areas where lightning is a problem. (Sec. 280.1.)

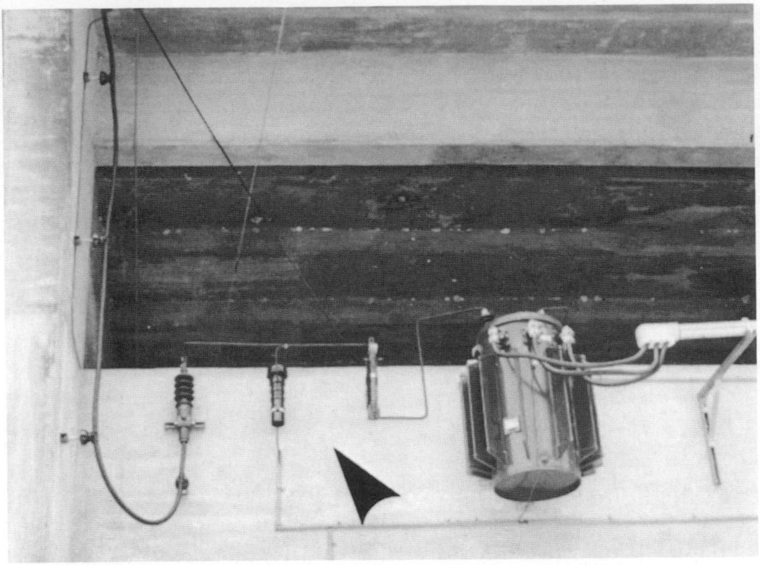

Fig. 280-2. Lightning arrester (arrow) is a typical "surge arrester" and, where used, one arrester must be connected to each ungrounded circuit conductor—such as shown here for a 2400-V grounded circuit supplying a transformer for stepping voltage down to supply lighting at this athletic stadium. (Sec. 280.3.)ì

Figure 280-2 shows a lightning arrester used on one of several high-voltage circuits serving the heavy electrical needs of a modern sports stadium.

280.3. Number Required. A double-throw switch which disconnects the outside circuits from the station generator and connects these circuits to ground would satisfy the condition for a single set of arresters for a station bus, as covered in the second sentence of this section.

280.4. Surge Arrester Selection. Figure 280-3 shows the position of a choke coil where it is used as a lightning-protection accessory to an arrester.

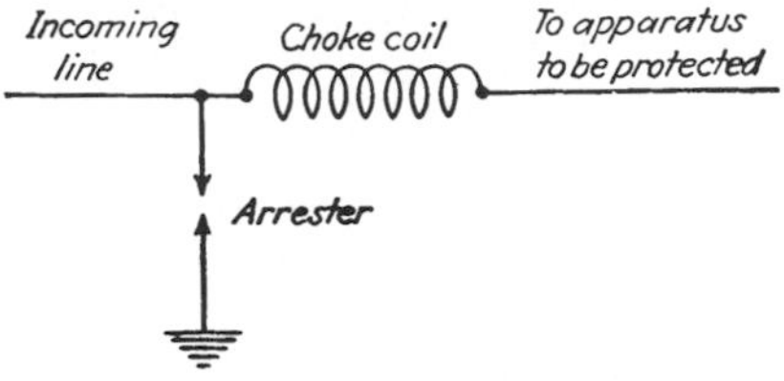
Incoming line
Choke coil
To apparatus to be protected
Arrester

Fig. 280-3. Using a choke coil as an accessory to an arrester. (Sec. 280.4.)

In 280.4(B), ratings of surge arresters are covered by the basic rule that applies to silicon-carbide-type surge arresters with an informational note pointing up the difference in voltage rating of metal-oxide-varistor-type arresters. This addresses the high-technology operating nature of the metal-oxide surge arrester as applied to premises wiring systems. The concern is to make an effective distinction between gapped silicon-carbide arresters, widely used in the past, and the newer metal-oxide block arresters. Manufacturers' application data on rating and other characteristics and the minimum duty-cycle voltage rating of an arrester for a particular method of system grounding must be observed carefully.

280.12. Uses Not Permitted. A surge arrester must only be installed in a location where the voltage rating of the arrester is at least equal to the maximum available continuous voltage to ground at the available frequency.

280.14. Routing of Surge Arrester Grounding Conductors. This rule is particularly important because bends and turns enormously increase the impedance to lightning discharges and therefore tend to nullify the effectiveness of a grounding conductor.

280.24. Interconnections. These rules are aimed at ensuring more effective lightning protection of transformers. Lightning protection of a transformer cannot be provided by a primary arrester that is connected only to a separate electrode. Common grounding of gaps or other devices must be used to limit voltage stresses between windings and from windings to case.

280.25. Grounding Electrode Conductor Connections and Enclosures. This section refers to Art. 250, which also covers connection of lightning arresters. The second sentence covers the need to keep grounding conductors electrically in parallel with their enclosing metal raceway. For instance, assume that a lightning arrester is installed at the service head on a conduit service riser, with the grounding conductor run inside the service conduit, bonded to the meter socket at the grounding lug, then run through a hole in the meter socket to the grounding electrode without a metal enclosure from the drilled hole to the electrode. In such a hookup, this rule requires the grounding conductor to be bonded to the conduit at the service head (Fig. 280-4). Ordinarily, the meter enclosure has a threaded hub,

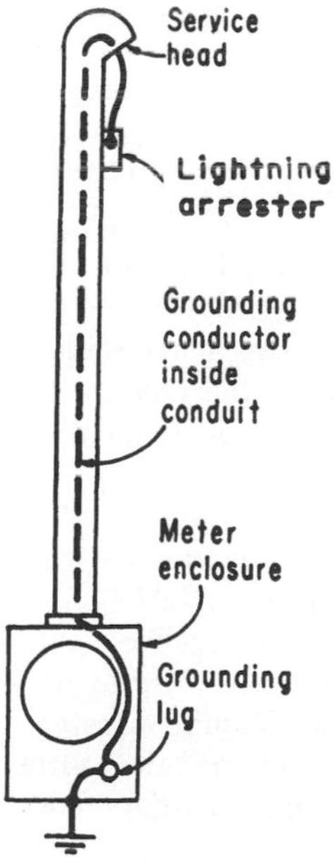
Service head
Lightning arrester
Grounding conductor inside conduit
Meter enclosure
Grounding lug

Fig. 280-4. Arrester grounding conductor must be bonded to both ends of enclosing metal raceway (or other enclosure). (Sec. 280.25.) Since surge arresters now only apply to installations over 1 kV; this drawing probably shows a surge protective device covered in Art. 285, however, it correctly illustrates the concept of the rule.

which would mean the conduit would be in good electrical contact with the meter enclosure and would be bonded at the meter socket end. However, 280.25 requires that the grounding conductor, if in a metallic enclosure, be bonded at both ends. Therefore, bonding at the service head is necessary.

The reason given for putting that rule in the NEC was explained as follows:

> When conducting lightning currents, the impedance of a lightning arrester grounding conductor is materially increased if run through a metallic enclosure, especially if of magnetic material. The voltage drop in this impedance may be sufficient to cause arcing to the enclosure, and in any event it reduces the effectiveness of the lightning arrester. Bonding of the conductor to both ends of the enclosure is necessary to eliminate this detrimental effect where metallic enclosures are used.

ARTICLE 285. SURGE-PROTECTIVE DEVICES, 1 KV OR LESS

285.1. Scope. The term "surge-protective device" (SPDs) is now, as of the 2008 NEC, the term of art for devices that shunt voltage spikes to ground. At one time the term was "lightning arrester" but, effective with the 1981 NEC, the terminology changed to "surge arrester" in recognition that voltage surges had many origins in addition to lightning. The next step occurred in the 2002 NEC, when a new article recognized "transient voltage surge suppressors" in addition to surge arresters. These devices were normally located on the load side of the service equipment, but could be on the line side if special arrangements were made. This has now changed, and the terminology "surge arrester" is now reserved for medium-voltage (over 1 kV) applications exclusively. The term "surge protective device" covers the entire installation spectrum for 600 V and lower applications.

285.3. Uses Not Permitted. SPDs may not be used on circuits exceeding 600 V or ungrounded electrical systems unless specifically listed for these connections. That is, if the system in question does not have one of its circuit conductors intentionally connected to earth, such as with a 480-V, ungrounded delta system, then use of surge suppressors is prohibited unless specifically listed for ungrounded systems. The same limitation applies to the use of these devices on high-impedance grounded systems, or on corner-grounded delta systems. In addition, an SPD must never be applied at a point where its rating is less than the maximum continuous phase-to-ground voltage using the available power frequency at the point of application.

285.6. Listing. Surge protective devices must be listed products for NEC applications.

285.7. Short-Circuit Current Rating. Care must be taken when installing SPDs to ensure that the device, which is required to be listed by 285.5, has a listed fault-current rating that is at least equal to the fault current available at the point in the distribution system where it is installed. This rule doesn't apply to receptacles.

285.12. Routing of Connections. As in the case of surge arresters running over 1 kV, this rule is particularly important because bends and turns enormously increase the impedance to lightning discharges and therefore tend to nullify the effectiveness of a grounding conductor.

285.13. Type 4 and Other Component Type SPDs. This section, new as of the 2014 NEC, clarifies that Type 4 surge protective devices along with other component-type protection are not appropriate for field installation; they are to be installed by a manufacturer only as a component of some other product.

285.21. Connection. As given in Secs. 285.23, 285.24, and 285.25, which cover permitted application locations of specific SPDs, such devices may be installed on the load side of the service OC protection, the load side of the OC protection at a main building disconnect, or on the load side of the first OC device fed from a separately derived system. Take care to find and use appropriate termination methods for the conductors from these devices. Particular attention must be paid to the requirement in 110.14(A) that terminals for more than one conductor must be so identified. The frequent practice of shoving the leads from an SPD into the lugs of a main circuit breaker along with the principal circuit conductors will not comply with any known listing recognition for such a combination of conductors. Installation of such devices at other locations would constitute a violation of this rule.

The NEC recognizes three types of surge protective devices as suitable for field installations. Type 1 devices correspond to the old surge arresters and are suitable for installation on the line side of the service equipment. Type 2 devices can be installed at any point on the load side of a service disconnect. They are also permitted on the load side of the first overcurrent device in a building or other structure supplied by a feeder. When special arrangements are made for overcurrent protection, these devices can also go on the line side of the service. Type 3 devices, the least robust, are permitted only on the load side of branch circuit protective devices and with the further restriction that they are at least 30 ft, measured along the conductors, from the service or local feeder disconnect for the building. The 2008 NEC imposed this limitation generally. However, some manufacturers are testing them to Type 2 tolerances and the distance restriction need not apply. For the 2011 and subsequent editions of the NEC, if and only if the restriction does apply, then the installation instructions will contain this limitation.

Chapter Three

ARTICLE 300. GENERAL REQUIREMENTS FOR WIRING METHODS AND MATERIALS

300.1. Scope. Part **(A)** of this section indicates that the rules of Art. 300 apply to all installations of the wiring methods covered by this article and the remainder of Chap. 3. As is generally the case, clearly not all the general requirements in Art. 300 apply to specialized installations, such as remote-control circuits, to signal circuits, to low-energy circuits, to fire-protective signaling circuits, and to communications systems. The wording "unless modified by other articles" is intended to convey that idea.

Part **(B)** establishes what amounts to the boundary line between the NEC and UL or other equipment testing labs, at least with respect to this article, which has no "construction" part. In fact, the NEC contains requirements throughout that are effectively aimed at the equipment manufacturers, in the process of setting policy around how products need to perform in order to be used safely, based on a consensus vote of a Code-making panel (CMP). For only one of countless examples, the reason there is foam-core ("nonhomogeneous") PVC conduit available is that a Code-making panel voted it in [352.10(G)], but then (for one cycle) also limited its use to underground applications. The reason that UL, with well-earned pride, points to its representation on all the Code-making panels is that it is then in a position to make sure that its product standards correctly implement the policy decisions reached through the NFPA consensus process. This rule clearly states that there is no intent, or permission, to apply Code rules within Art. 300 to the interior wiring of equipment to be connected. That is the province of the testing labs. If a product is "listed," it has been tested and found to be essentially free from hazards. Generally, the equipment is tested for use in accordance with the Code, which includes any Code-prescribed construction or performance requirements.

In part **(C)** of this section, a table is provided to allow proper selection of standard trade size raceways and tubing whether applying the Code in the United States or abroad. Simply determine the needed raceway size, in either system, and use Table 300.1**(C)** to establish the correct metric or English system size equivalent. The table also establishes that the dimensions are not real dimensions, but only to be used for identification purposes. Generally throughout this book, we are following the NEC style, and therefore if we are referring to what is known on the street as "3-in. conduit" we refer to it here, in the customary NEC practice of listing the metric size first and the English unit following in parentheses, as "metric designator 78 (trade size 3) conduit."

300.3. Conductors. Part **(A)** requires that single conductors described in Table 310.104(A) must be used only as part of one of the wiring methods covered in Chap. 3. This basically means that the various insulated conductors recognized in Table 310.104(A) may not simply be strung overhead, without benefit of being incorporated into a cable assembly or otherwise protected and supported by conduit or tubing. The exception to part **(A)** recognizes the permission given in 225.6, where individual conductors *are* permitted to be run as "open conductors" in overhead feeders and branch circuits installed outdoors, as well as in festoon lighting.

Part **(B)** requires that all conductors of the same circuit—including the neutral and all equipment grounding conductors—must be run in the same raceway, cable tray, trench, cable, or cord. Part **(B)(1)** recognizes the use of separate raceways and cables, where circuits are made up of multiple (two or more) sets of conductors or cables in parallel. The exception that follows correlates with 300.5(I) Exception No. 2 for isolated phase installations. Part **(B)(2)** in that rule notes those very specific and unusual sections of the NEC, where an equipment grounding conductor may be run separately from the other conductors of the circuit such as for dc circuits and for retrofits under the provisions of 250.130(C).

Part **(B)(3)** requires taking steps to prevent induced currents where nonmetallic or nonmagnetic sheathed conductors are run through metallic enclosure walls with magnetic properties as covered in 300.20(B). The second sentence gives a similar warning for Type MI cables and references 332.31, which addresses the concern for induced currents. Both of those other rules discuss methods for preventing induced currents.

Part **(4)** of 300.3(B) permits limited use of a pull box equipped with a terminal block for the connection of the system neutral, as the point of origin for branch-circuit neutral conductors. That is, a properly sized neutral is run to the pull box—which is connected by an auxiliary gutter to a column-width panelboard—from the panelboard, and the individual branch-circuit neutrals may be run from the pull box and need not go back to the panelboard where the hot conductors originate. This saves space within the panelboard, but is only permitted for column-width panelboards connected by an auxiliary gutter to a pull box that is manufacturer-equipped with a neutral terminal block. Note that there is no inductive heating in this arrangement and it is arguably permitted even without this permission. Every ampere that comes up the auxiliary gutter from the panel on a branch-circuit conductor is equaled by an ampere of current moving down the same gutter space to the panel over one of the feeder conductors.

Part **(C)** covers system separations under ordinary conditions. The first topic is conductors operating on different voltages, but not over 1000 V. This rule clearly allows conductors of different voltage systems, and whether dc or ac, to occupy the same raceway, cable assembly, or enclosure. The only limitation here is that all the conductors must be insulated for the maximum voltage that will be present in the common location. For example, nonpower-limited 6-V circuit conductors running to a valve motor can run in the same raceway as 277-V circuit wiring to a duct heater, if both sets of conductors have 600-V (technically 277-V) insulation.

This brings up Fig. 300-1. Whoever wired this furnace must have read this section, because when he wanted to get his 24-V thermostat circuit neatly down to the burner control, he carefully ran 600-V THHN down the EMT riser along with the 120-V circuit for the burner motor, making the transition to thermostat wire where the THHN poked out of the upper end of the riser. A nice neat job, and a serious Code violation. The thermostat circuit is a Class 2 power-limited circuit, covered under the enhanced system separation rule found in 725.136(A). This comes up so often that the **NEC** now has a note at this location. A Class 2 circuit is a *special condition*, covered in Chap. 7 of the **NEC**. As such, the rules in Chap. 7 automatically supersede or modify information in Chaps. 1 through 4 of the **NEC**, as per 90.3. And for Class 2 wiring, it is forbidden to ever rely on insulation alone to define a circuit separation. These systems are presumed to be incapable of fire and electrocution hazard by virtue of the limitations built into their power supplies and the only way to be really sure of this is to keep them out of common raceways and enclosures, except under extremely well-controlled situations. There is even a special exception to the rule forbidding

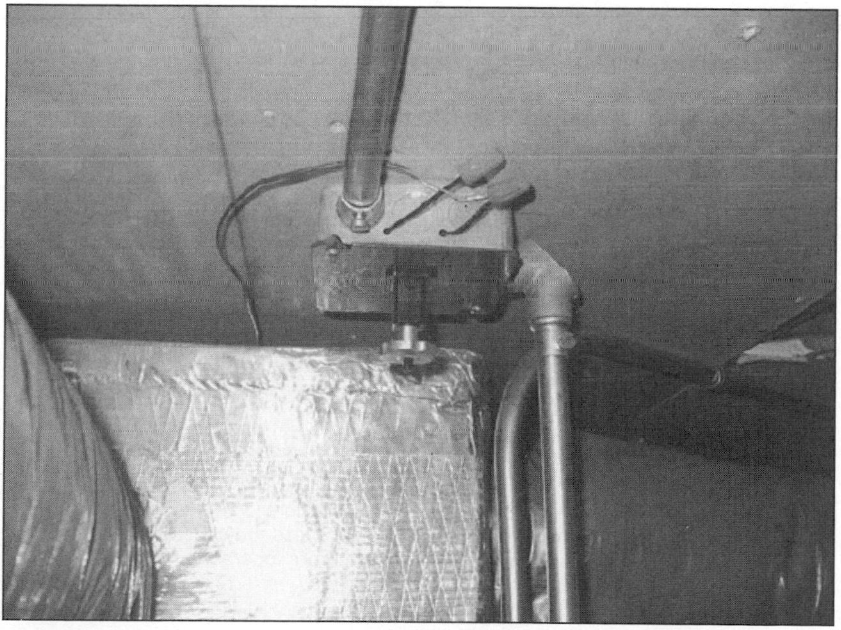

Fig. 300-1. The 14 AWG THHN leaving the box from the EMT riser and connected to 18-2 thermostat wire for a 24-V Class 2 circuit does comply with 300.3(C)(1), but violates 725.136(A).

cables to be attached to electrical raceways [300.11(B)(2)] that allows the Class 2 wiring to be attached to the outside of the raceway, and that would have been the way to wire the oil burner in Fig. 300-1.

The second informational note here reminds users that photovoltaic source and output circuits are prohibited by 690.4(B) from sharing enclosures or raceways unless they are contained within a separately partitioned area. There are other NEC rules that require circuit separation, notably including 700.10(B) for emergency circuits and 695.6(A)(2)(a) for fire pump supply conductors, so these two notes do not begin to cover the subject.

Part **(C)(2)** of 300.3 states that conductors operating at more than 1000 V must not occupy the same equipment wiring enclosure, cable, or raceway with conductors of 1000 V or less. There are four conditions where this can be relaxed; however, note that none of these involve a common raceway. Raceways remain fully segregated, but there are limited applications where having both systems in a common enclosure is essential. Part **(C)(2)(b)** is intended to apply to enclosures, not raceways, such as used for high-voltage motor starters, permitting the high-voltage conductors operating at over 600 V to occupy the same controller housing as the control conductors operating at less than 600 V (Fig. 300-2).

Note that the former 600-V parameter has been routinely changed to 1000 V as of the 2014 NEC. In large part, this change has been driven by increased voltages in photovoltaic systems, and the changes here follow the global revisions throughout Chap. 3 (and the large majority of other NEC locations), as covered in 300.2(A).

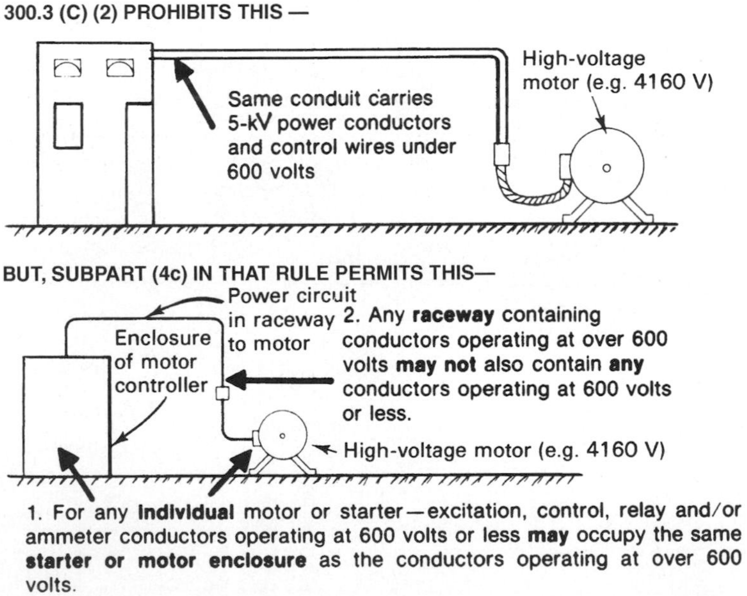
300.3 (C) (2) PROHIBITS THIS —
Same conduit carries 5-kV power conductors and control wires under 600 volts
High-voltage motor (e.g. 4160 V)
BUT, SUBPART (4c) IN THAT RULE PERMITS THIS—
Power circuit in raceway to motor
Enclosure of motor controller
2. Any **raceway** containing conductors operating at over 600 volts **may not** also contain **any** conductors operating at 600 volts or less.
High-voltage motor (e.g. 4160 V)
1. For any **Individual** motor or starter—excitation, control, relay and/or ammeter conductors operating at 600 volts or less **may** occupy the same **starter or motor enclosure** as the conductors operating at over 600 volts.

Fig. 300-2. Control wires for high-voltage starters may be used in the starter enclosure, but not in *raceway* with power conductors. (Sec. 300.3.)

This paragraph [(C)(2)] is a case in point; it used to have five enumerated conditions, the first of which addressed 1000-V secondary wiring to electric discharge lamps. Since the voltage went up generally to 1000 V, that item became superfluous in this location and it was relocated as a second paragraph to (C)(1).

The last sentence in part **(C)(2)** is intended to prohibit unshielded conductors (now limited by 310.10(E) to 5 kV) from occupying the same enclosure, raceway, or cable, unless the actual voltage carried is the same. Now that the allowable voltage for unshielded conductors has been reduced from 8 kV, the application of this in new construction is limited, but there are a lot of higher voltage unshielded conductors still in use. Some feel that the normal voltage discharge and leakage current will be increased where a difference of potential exists between two unshielded conductors. It is theorized that the result of the increased discharge and leakage will be premature insulation failure. Although the assertion of premature failure has not been determined empirically, there is no data to suggest otherwise. Any problem can be avoided by simply complying with this requirement and using separate raceway systems.

300.4. Protection Against Physical Damage. This rule presents a general requirement for ensuring that conductors, raceways, and cables are properly protected where "subject to physical damage." Common sense must be applied and the normal operation of the facility in question must be considered. If it appears that, during normal operations, a given conductor or cable is likely to be damaged, then physical protection in the form of a raceway sleeve, an enclosure, kickplate, etc., must be provided. This requirement is a catchall. That is, the specific situations spelled out in the subsequent parts of 300.4 must be provided with the physical protection prescribed, under the circumstances described. But, in any other location where conductors or cables are exposed to damage, suitable remedies must be employed to satisfy this most basic requirement. The 2017 **NEC** added an informational note here advising that minor damage to cable sheaths is not necessarily a problem for the conductors within.

Part **(A)** gives the rules on protection required for cables and raceways run through wood framing members, as shown in Fig. 300-3. Where the edge of a hole in a wood member is less than 1¼ in. (32 mm) from the nearest edge of the member, a $^1/_{16}$ in. (1.6 mm) thick steel plate must be used to protect any cable or flexible conduit against driven nails or screws. The same protection is required for any cable or flexible conduit laid in a notch in the wood. But, as given in Exception No. 1, rigid metal conduit, electrical metallic tubing (EMT), intermediate metal conduit (IMC), and PVC conduits do not require such protection. And, Exception No. 2 allows for the use of steel plates that are less than $^1/_{16}$-in. thick where the plate is "listed and marked" as providing "equal or better protection" from nails, etc.

Clearance must be provided from the edge of a hole in a wood member to the edge of the wood member. The **NEC** requires only 1¼ in. (32 mm). This permits realistic compliance when drilling holes in studs that are 3½ in. (89 mm) deep. It also was taken into consideration that the nails commonly used to attach wall surfaces to studs are of such length that the 1¼-in. (32-mm) clearance to the edge of the cable hole affords entirely adequate protection against possible penetration of the cable by the nail.

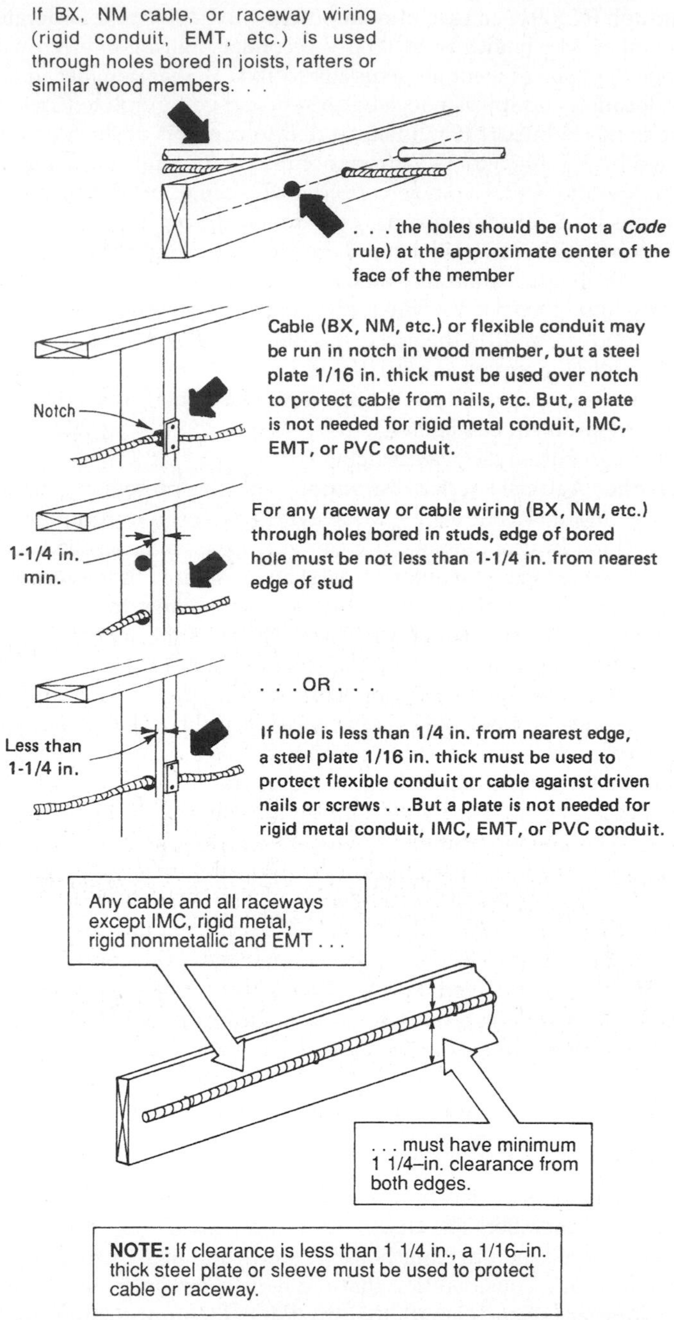
If BX, NM cable, or raceway wiring (rigid conduit, EMT, etc.) is used through holes bored in joists, rafters or similar wood members. . .
. . . the holes should be (not a *Code* rule) at the approximate center of the face of the member
Cable (BX, NM, etc.) or flexible conduit may be run in notch in wood member, but a steel plate 1/16 in. thick must be used over notch to protect cable from nails, etc. But, a plate is not needed for rigid metal conduit, IMC, EMT, or PVC conduit.
Notch
For any raceway or cable wiring (BX, NM, etc.) through holes bored in studs, edge of bored hole must be not less than 1-1/4 in. from nearest edge of stud
1-1/4 in. min.
. . . OR . . .
If hole is less than 1/4 in. from nearest edge, a steel plate 1/16 in. thick must be used to protect flexible conduit or cable against driven nails or screws . . .But a plate is not needed for rigid metal conduit, IMC, EMT, or PVC conduit.
Less than 1-1/4 in.
Any cable and all raceways except IMC, rigid metal, rigid nonmetallic and EMT . . .
. . . must have minimum 1 1/4–in. clearance from both edges.
NOTE: If clearance is less than 1 1/4 in., a 1/16–in. thick steel plate or sleeve must be used to protect cable or raceway.

Wiring on structural members must have clearance for protection against nails.

Fig. 300-3. Holes in wood framing must not weaken structure or expose cable to nail puncture. (Sec. 300.4.)

Drilled holes at the center of the face of a joist do not reduce the structural strength of the joist.

Signal and alarm wiring is run through the same stud holes as the NM cables. The NEC does not prohibit use of more than one cable through a single hole.

Fig. 300-4. Holes or notches in joists and studs must not weaken the structure of a building. (Sec. 300.4.)

Figure 300-4 shows typical application of cable through drilled studs, with holes at centers and adequate clearance to the edge of the stud. Figure 300-5 shows an objectionable example of a drilled hole, violating the rule of part **(2)** of this section, which warns against "weakening the building structure." Figure 300-6 shows an acceptable way of protecting cables run through holes in wood members.

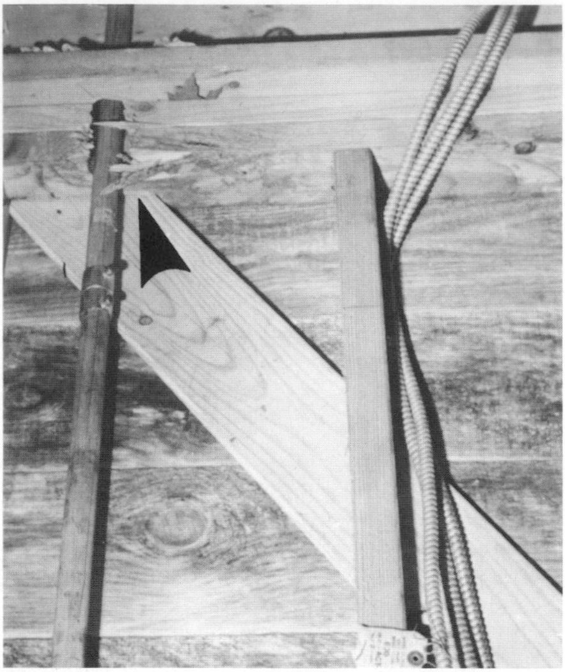

Fig. 300-5. Excessive drilling of structural wood members can result in dangerous notching (arrow) that weakens the structure, violating Sec. 300.4(A)(2). (Sec. 300.4.)

Fig. 300-6. Steel plates are attached to wood structure member to protect cable from penetration by nail or screw driven into finished wall, where the edge of the cable hole is less than ¼ in. (31.8 mm) from the edge of the wooden member. (Sec. 300.4.)

In part **(B)**, the rules on installations through metal framing members apply to nonmetallic-sheathed cable and to electrical nonmetallic tubing (ENT). Part **(1)** of 300.4(B) applies to NM cable run through slots or holes in metal framing members and requires that such holes must always be provided with bushings or grommets installed in the openings before the cable is pulled. But that requirement on protection by bushings or grommets in the holes does not apply to ENT where run through holes in metal framing members. The grommet must encircle the entire hole; "V"-shaped grommets that are open at the top are not permitted for this purpose. The reason is that experience pulling cable into such holes has shown repeated instances where the cable rose up when under tension and the outer jacket was severely damaged.

Part **(2)** applies to both NM cable and ENT and requires that the cable or tubing be protected by a steel sleeve, a steel plate, or a clip when run through metal framing members in any case where nails or screws might be driven into the cable or tubing.

Part **(C)** requires cables and raceways above lift-out ceiling panels to be supported as they are required to be when installed in the open. They may not be treated as if they were being run through closed-in building spaces or fished through hollow spaces of masonry block.

Part **(D)** requires cables and raceways run along (parallel with) framing members (studs, joists, rafters), as well as furring strips, to have at least a 1¼-in. (32-mm) clearance from the nearest edge of the member; otherwise, the cable or raceway must be protected against nail or screw penetrations by a steel plate or sleeve at least $^1/_{16}$ in. (1.6 mm) thick. The application of the 32 mm (1¼-in.) spacing rule to furring strips means that cables can be stapled across the lower edges of ceiling joists on a furred ceiling, provided the cables are secured at least that distance away from the edges of the furring strips. In that way, an errant drywall screw won't bother the cable, as the rule intends. For conventional framing members, the distance applies from both the edges of the member, which makes a conventional 2 by 3 incapable of supporting a cable because at 2½ in. deep the prescribed distance from each face leaves no room for wiring. There are some products designed that stack cables at some distance from the framing, which may solve this problem.

Objections have been raised to this rule from the manufacturers of metal-clad cables (Types MC, AC, etc.) because actual tests have shown that these cables tend to roll out of the way of a screw, and are actually far more difficult to penetrate than rigid raceways. For similar reasons, proposals have also been made to relocate this rule into Art. 334 on nonmetallic sheathed cable, which is the wiring method most likely to be damaged. However, none of these attempts has succeeded over the years.

The rule applies in both exposed and concealed locations to most wiring methods, with two exceptions, with a third being the customary allowance for listed shielding plates to be of thinner steel than the default 1.6 mm ($^1/_{16}$ in.) thickness.

Exception No. 1 excludes IMC, rigid metal conduit, rigid nonmetallic conduit, and EMT from the rule. The rule will apply to Romex (Type NM), BX (Type AC), flexible metal conduit, ENT, Type MC cable, and all other cables and raceways, except those excluded by Exception No. 1. (see Fig. 300-3, bottom).

Exception No. 2 excludes from this rule concealed work in finished buildings and finished panels in prefab buildings, where cables may be fished.

Part **(E)**, on wiring under roof decking, is new as of the 2008 **NEC** but was expanded and clarified in the 2011 **NEC**. It requires all wiring, whether cables or raceways (subject to an exception for rigid and for intermediate metal conduits) as well as boxes to be mounted far enough below "metal-corrugated sheet roof decking" so subsequent reroofing and repairs don't end up impaling the wiring with the long screws commonly used to hold down the new surface, as explained in the note. This change follows actual loss experience.

The text calls for 38 mm (1½ in.) spacing between the top of the wiring method (meaning nearest to the roof decking) and the "lowest" edge of the roofing (meaning nearest to an observer on the floor). This puts the wiring well below the lowest part of the corrugations, which is appropriate because subsequent roofing activities proceed blindly, with screws sized to penetrate to a bottom of a corrugation as easily as an upper portion. Figure 300-7 shows how the clearances are to be measured. The 2011 **NEC** extended the coverage of this rule to include boxes, in addition to general wiring methods, and has also applied the prohibition to the concealed cavity between the decking and the roof membrane, as shown in Fig. 300-7.

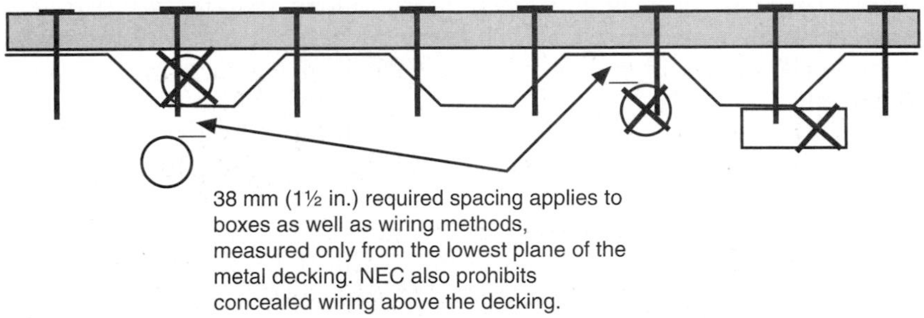
38 mm (1½ in.) required spacing applies to
boxes as well as wiring methods,
measured only from the lowest plane of the
metal decking. NEC also prohibits
concealed wiring above the decking.

Fig. 300-7. This new (2008 **NEC**) rule to avoid damage from roof repair has been clarified and the clearance measurement is to be applied from the plane of the lowest surface of the roof decking. It also has been expanded to cover boxes and the concealed spaces above the decking. [Sec. 300.4(E).]

Although the exception waives the rule for installations using heavy-wall steel (rigid or intermediate) conduit, take care in using this exception. As previously noted, tests run by UL on a variety of wiring methods to determine their susceptibility to penetration under conditions covered by the related rules in 300.4(D) showed that even heavy wall conduits were no match for drill-point screws. Interestingly, only unrestrained steel-armored cables that could roll out of the way typically survived the testing, a result not reflected in the Code.

Part **(F)** presents protection requirements where cables are run in "shallow grooves." This rule calls for the same $^1/_{16}$ in. (1.6 mm) thick steel kick plate, sleeving, or the equivalent. Alternately, one may locate the cable with at least 1¼ in. (32 mm) of "free space" for the cable's entire concealed length. The "free space" is to be measured from the top of the groove to the top of the cable, not to the bottom of the groove. This rule is aimed at applications with exposed beam construction where grooves or channels are cut in the beams for supply conductors to lighting fixtures or ceiling fans.

Part **(G)** applies to all conductors of size No. 4 or larger entering a cabinet or box from rigid metal conduit, flexible metal conduit, electrical metallic tubing, and so forth. As indicated in 300.4(G), to protect the conductors from cutting or abrasion, a smoothly rounded insulating surface is required. While many fittings are provided with insulated sleeves or linings, it is also possible to use a separate insulating lining or sleeve to meet the requirements of the Code.

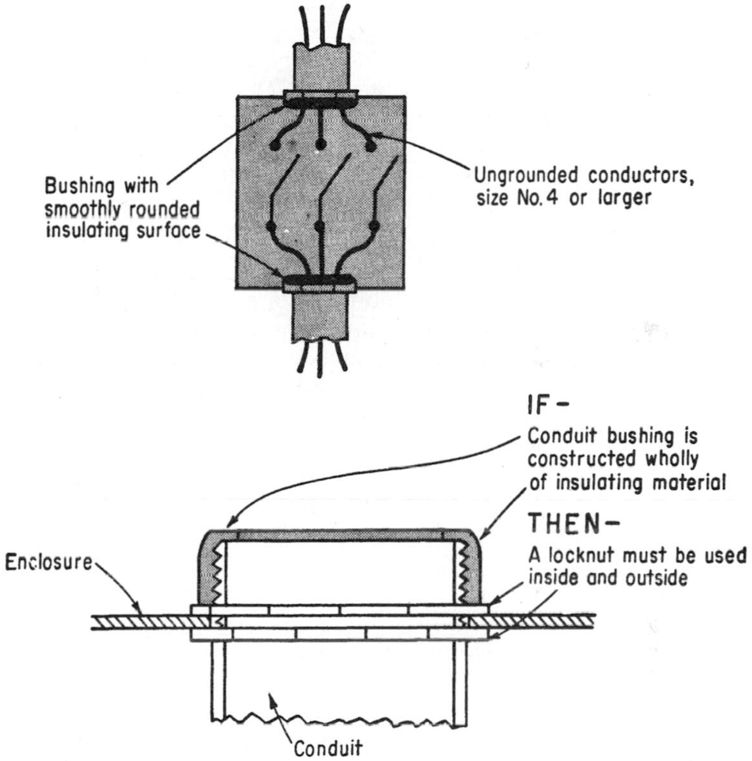
Bushing with
smoothly rounded
insulating surface
Ungrounded conductors,
size No. 4 or larger
IF -
Conduit bushing is
constructed wholly
of insulating material
THEN -
A locknut must be used
inside and outside
Enclosure
Conduit

Fig. 300-8. An insulated-throat bushing or other protection must be used at enclosure openings. [Sec. 300.4(G).]

Figure 300-8 shows use of a bushing with an insulated edge or a completely nonmetallic bushing to satisfy this rule. Figure 300-9 shows an approved sleeve which may be used to separate the conductors from the raceway fitting, which may be installed after the conductors are already installed and connected.

In the exception to 300.4(G), an insulated throat is not required for conductor protection on enclosure threaded hubs or bosses that have a rounded or flared entry surface. This is recognition of a long-standing reality—that there is no need for protective insulating material around the interior opening of integral hubs and bosses on equipment enclosures. Insulated-throat bushings and connectors are needed only for entries through KOs in sheet-metal enclosures.

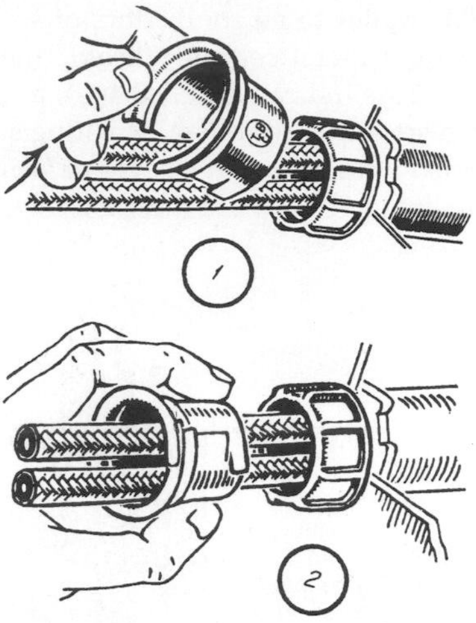
1
2

Fig. 300-9. Slip-over nonmetallic sleeve may be used to cover metal bushing throat. [Sec. 300.4(G).]

The last paragraph of 300.4(G) prohibits use of a plastic or phenolic bushing ("wholly of insulating material") as a device for securing conduit to an enclosure wall. On a KO, there must be a metal locknut outside and a metal locknut inside to provide tight clamping to the enclosure wall, with the nonmetallic bushing put on after the inside locknut. An EMT or conduit connector must also be secured in position by a metal locknut and not by a nonmetallic bushing. This paragraph also requires that any insulating bushing or insulating material used to protect conductors from abrasion must have a temperature rating at least equal to the temperature rating of the conductors.

It is important to remember that this is a rule that applies to wiring being pulled into raceways, and therefore there is no requirement to put an insulating bushing on the end of an SE cable connector, for example. If an enclosure has a molded hub, it will be sufficiently rounded on the inside and no additional protection is required.

Part **(H)**, new in the 2011 edition, requires the use of listed expansion or deflection fittings, or other approved methods, as required for the particular application when a raceway crosses a structural joint in a building or other structure that is intended to shift while in use. These joints are common in bridges and parking structures, where wide areas of concrete must be allowed to expand and contract due to temperature changes, and they are also applied in some designs to counteract the effects of earthquakes.

300.5. Underground Installations. This section is a comprehensive set of rules on installation of underground circuits for circuits up to 600 V. (Higher-voltage circuits must satisfy 300.50.) Table 300.5 in the Code book establishes minimum earth cover needed for specific conditions of use. Figure 300-10 shows cover requirements for rigid metal conduit and IMC. Figure 300-11 shows the basic depth requirements for the various wiring methods. In all instances, "cover" is *not the depth of the trench.* It is the depth of the trench plus the diameter of the wiring method, which *equals the amount of dirt on top of the wiring method* when the job is complete.

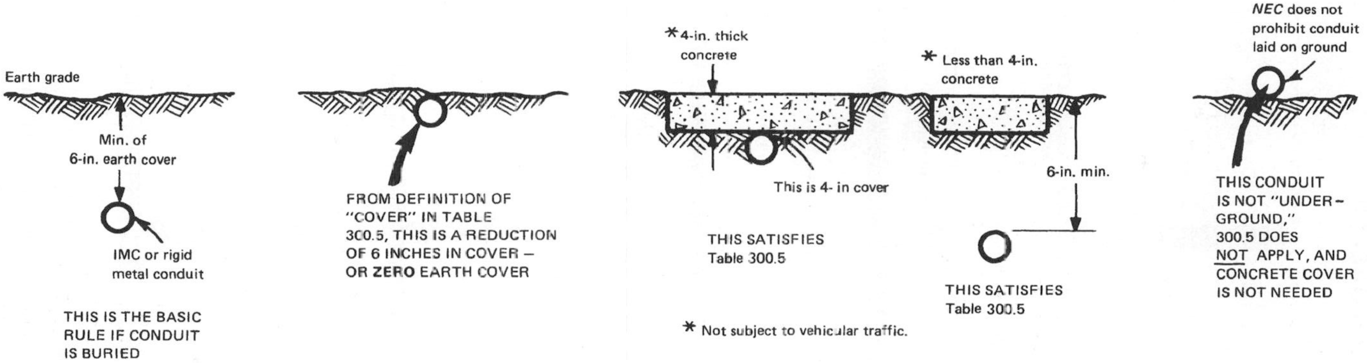
Earth grade
Min. of
6-in. earth cover
IMC or rigid
metal conduit
THIS IS THE BASIC
RULE IF CONDUIT
IS BURIED
FROM DEFINITION OF
"COVER" IN TABLE
300.5, THIS IS A REDUCTION
OF 6 INCHES IN COVER –
OR ZERO EARTH COVER
*4-in. thick
concrete
This is 4- in cover
THIS SATISFIES
Table 300.5
* Less than 4-in.
concrete
6-in. min.
THIS SATISFIES
Table 300.5
* Not subject to vehicular traffic.
NEC does not
prohibit conduit
laid on ground
THIS CONDUIT
IS NOT "UNDER-
GROUND,"
300.5 DOES
NOT APPLY, AND
CONCRETE COVER
IS NOT NEEDED

Fig. 300-10. These are the details involved with the use of rigid metal conduit and IMC underground.

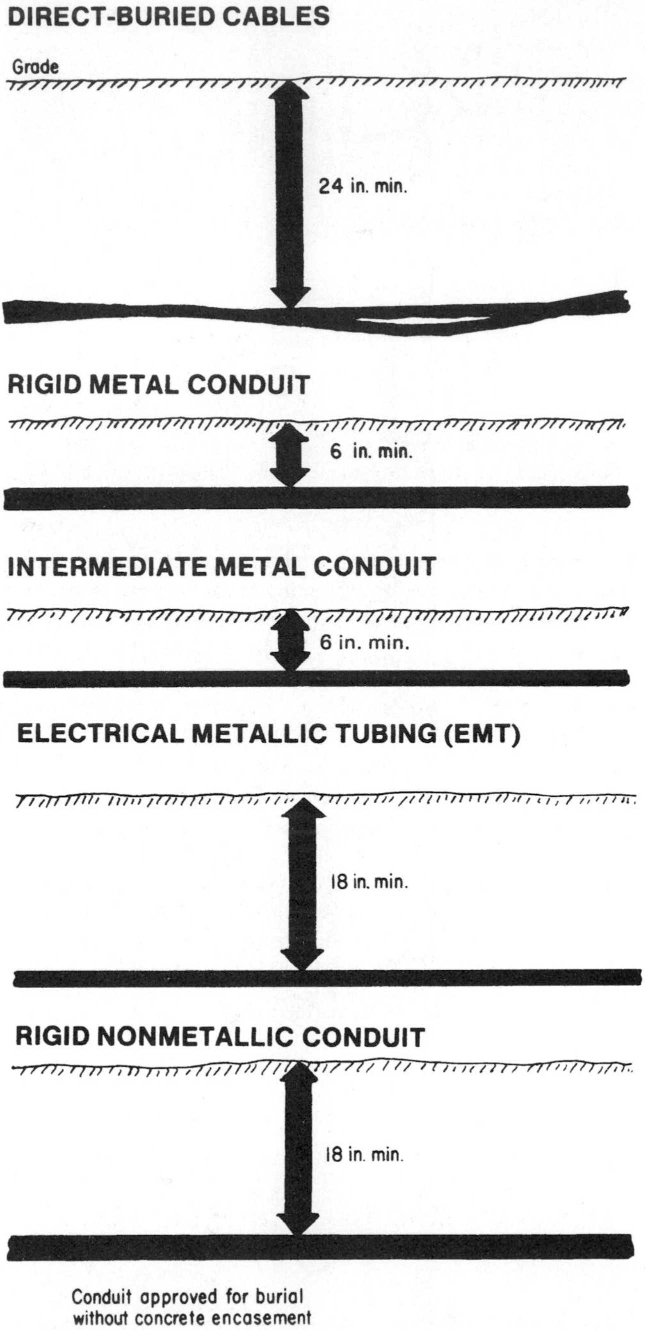
DIRECT-BURIED CABLES
Grade
24 in. min.
RIGID METAL CONDUIT
6 in. min.
INTERMEDIATE METAL CONDUIT
6 in. min.
ELECTRICAL METALLIC TUBING (EMT)
18 in. min.
RIGID NONMETALLIC CONDUIT
18 in. min.
Conduit approved for burial
without concrete encasement

Fig. 300-11. These are the *basic* burial depths, but variations are recognized in Table 300.5 for certain conditions. (Sec. 300.5.)

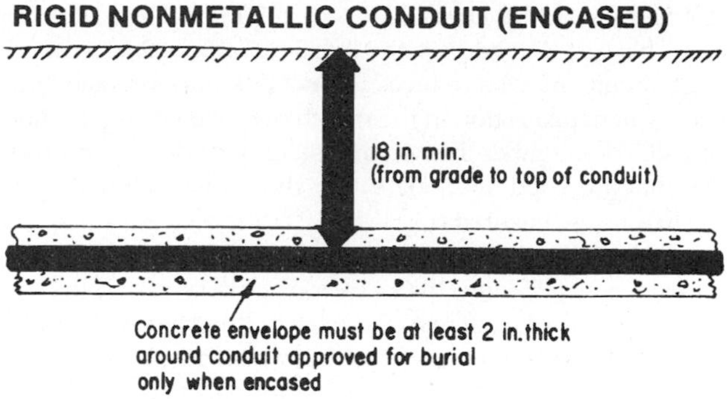
RIGID NONMETALLIC CONDUIT (ENCASED)
18 in. min.
(from grade to top of conduit)
Concrete envelope must be at least 2 in. thick
around conduit approved for burial
only when encased

Fig. 300-11. (*Continued*)

Because Table 300.5 does not specifically mention EMT, it could be taken to indicate that the NEC does not recognize EMT for underground use. But 358.10(B) does recognize EMT for direct earth burial if "protected by corrosion protection and judged suitable for the condition," and so does UL, with this stipulation: "In general, electrical metallic tubing in contact with soil requires supplementary corrosion protection." Note that such protection is not literally mandatory in all instances, but it is close. The UL note means to indicate that EMT might be buried without a protective coating (like asphalt paint) where local experience verifies that soil conditions do not attack and corrode the EMT. On the other hand, UL does not evaluate supplementary corrosion protection on this product with respect to the corrosive influences of soils, so the inspector is on his own in choosing to permit this. Supplementary corrosion protection must always be applied unless there is a solid local record of positive experience, which is very unusual. On balance, EMT should not be used for direct burial. Refer to the more extensive discussion at 358.10 in this book for additional information on this point.

Figure 300-12 shows modifications of basic cover requirements. If a 2 in. (50 mm) thick or thicker concrete pad is used in the trench over an underground circuit other than rigid metal conduit or IMC, the basic burial depth in Table 300.5 may be reduced by 6 in. (150 mm). It should be noted that the intent here is for the concrete pad to be in the trench, right over the cable or raceway. The wording must be taken to mean that it may not be a walk or other concrete at grade level. And the burial

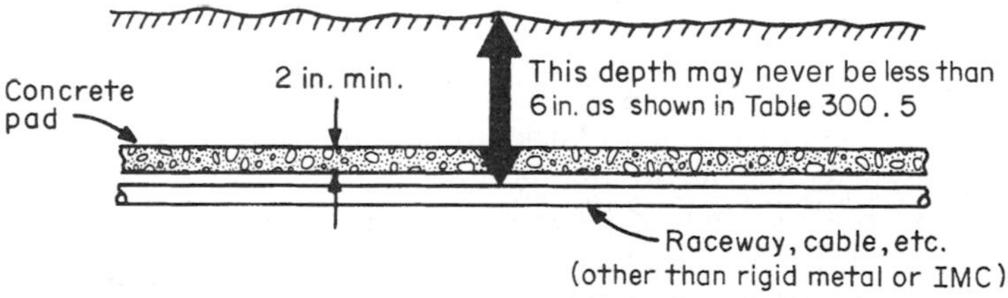
Concrete pad
2 in. min.
This depth may never be less than
6 in. as shown in Table 300.5
Raceway, cable, etc.
(other than rigid metal or IMC)

Fig. 300-12. Concrete pad "in trench" permits what amounts to a 6-in. (150-mm) reduction of burial depth for circuits up to 600 V for other than RMC, IMC, and applications covered in column 5. (Sec. 300.5.)

depth may not be reduced by more than 6 in. (150 mm) no matter how thick the concrete pad is. This rule is at odds with Note 3 to Table 300.50, where burial depth for high-voltage circuits may be reduced "6 in. (150 mm) for each 2 in. (50 mm) of concrete" or equivalent protection in the trench over the wiring method (other than rigid metal or IMC). Remember this when installing underground runs of conductors and cables that are rated over 600 V. For those installations, compliance with 300.50, not 300.5, must be ensured.

Note that in Table 300.5, rigid metal conduit or IMC that is buried in the ground must have at least a 6 in. (150 mm) thick cover of earth or earth plus concrete—even if it has a 2 in. (50 mm) thick concrete pad over it. But rigid metal conduit, IMC, or other raceways may be installed directly under a 4 in. or thicker exterior slab that is not subject to vehicle traffic, without any need for earth cover. Given the fact that rigid metal conduit or IMC may be laid directly on the ground (which supports it for its entire length) and would not necessarily require any concrete cover, there is no reason why it cannot be laid on the ground or flush with the ground and covered with at least 4 in. (102 mm) of concrete. (See Fig. 300-10.)

Section 300.5 only applies to "underground installations" and is not applicable if the conduit is laid directly on the ground. No Code rule prohibits conduit laid on the ground, provided the conduit is "securely fastened in place" (300.13) and is not exposed to physical damage such as vehicular traffic, and many such installations have been made for years. But, when conduit is installed in the ground, there is serious concern about damage due to digging in the ground, which Sec. 300.5 addresses.

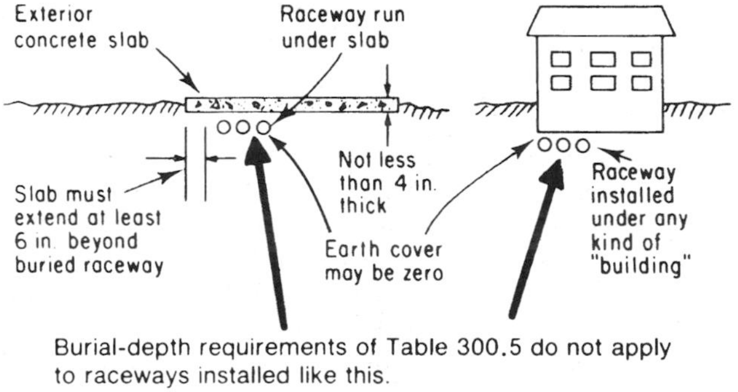
Exterior concrete slab
Raceway run under slab
Not less than 4 in. thick
Slab must extend at least 6 in. beyond buried raceway
Earth cover may be zero
Raceway installed under any kind of "building"
Burial-depth requirements of Table 300.5 do not apply to raceways installed like this.

Fig. 300-13. Table 300.5 eliminates burial-depth requirements for direct buried raceways under specified conditions. (Sec. 300.5.)

As shown in Fig. 300-13, Table 300.5 recognizes that raceways run under concrete slabs at least 4 in. (102 mm) thick or under buildings have sufficient protection against digging and are not required to be subject to the burial-depth requirements given in the top line of Table 300.5. Where raceways are so installed, the rule requires that the slab extend at least 6 in. (150 mm) beyond the underground raceway, as follows:

1. Any direct burial cable run under a building must be installed in a raceway, as required by 300.5(C), and the raceway may be installed in the earth,

immediately under the bottom of the building, without any earth cover. Some cables, as covered in exceptions to 300.5(C), are now permitted for direct burial under buildings.

2. Any direct buried cable under a slab at least 4 in. (102 mm) thick and not subject to vehicles is subject to the 18-in. (450-mm) minimum burial-depth requirement of Table 300-5. The reason for the equivalence in cover between the 2-in. (50-mm) concrete-in-trench rule and the 4-in. (102-mm) concrete slab rule is that when the concrete is buried in the trench an excavator will recognize it as a protective structure. The surface slab may not be easily recognized as performing an additional protective function.

Figure 300-14 shows the mandatory 24 in. (600 mm) of earth cover given in Table 300.5 for any wiring methods buried under public or private roads, alleys, driveways, parking lots, or other areas subject to car and truck traffic. A minimum earth cover of 24 in. (600 mm) is required for any underground cable or raceway wiring that is installed under vehicle traffic, regardless of concrete encasement or any other protective measure. This requires the minimum 2 ft (600 mm) earth cover for wiring under the designated areas, including driveways and parking areas of private multifamily residences. The minimum earth cover for cables and raceways under driveways and parking areas for one- and two-family dwellings is only 18 in. (450 mm).

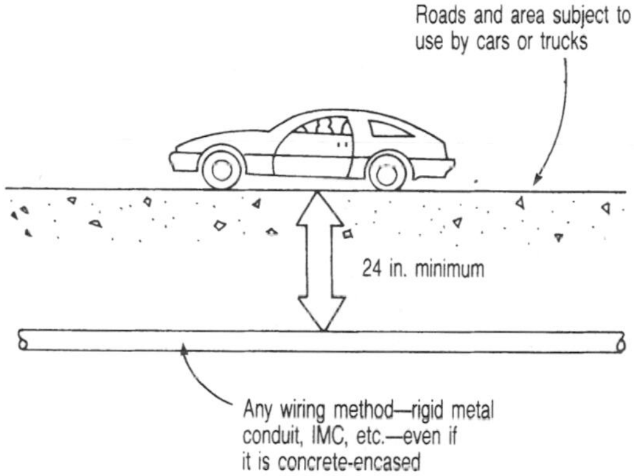
Roads and area subject to use by cars or trucks
24 in. minimum
Any wiring method—rigid metal conduit, IMC, etc.—even if it is concrete-encased

Fig. 300-14. All wiring methods must be at least 2 ft (600 mm) under vehicular traffic. (Sec. 300.5.)

Table 300.5 (second vertical column from right) gives limited use of lesser burial depth for the residential circuits described, as shown in Fig. 300-15. Any GFCI-protected residential "branch circuit" not over 120 V and protected at 20 A or less may be buried only 12 in. (300 mm) below grade, instead of, say, 24 in. (600 mm), as required for Type UF cable for any nonresidential use or for a residential "feeder."

Figure 300-16 shows three other special conditions for burial depth. Table 300.5 (vertical column at right) recognizes reduced burial depth for low-voltage landscape lighting circuits and supply circuits to lawn sprinkler and irrigation

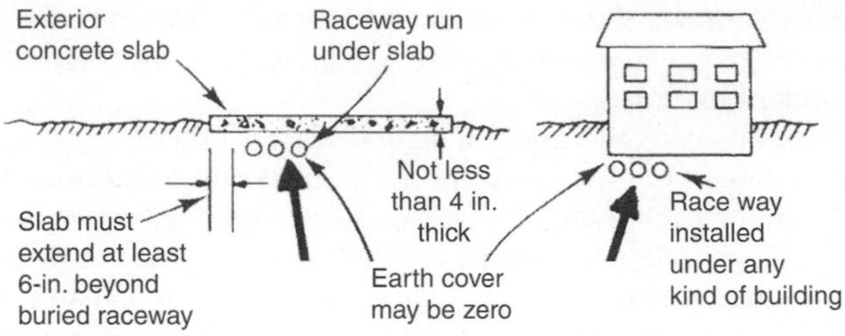
Exterior
concrete slab
Raceway run
under slab
Not less
than 4 in.
thick
Slab must
extend at least
6-in. beyond
buried raceway
Earth cover
may be zero
Race way
installed
under any
kind of building

Fig. 300-15. This is OK only for a residential branch circuit rated not over 20 A and protected by a GFCI circuit breaker. (Sec. 300.5.)

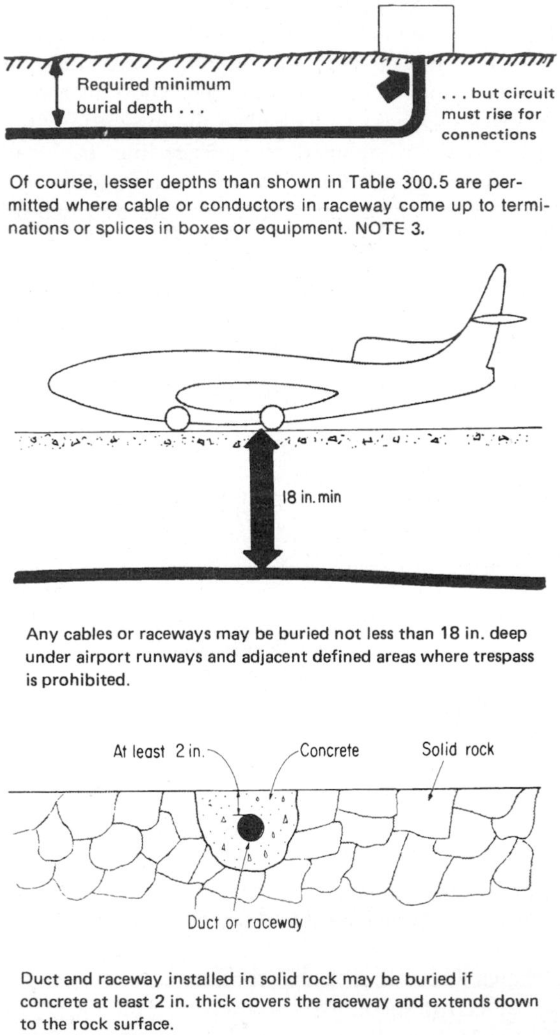
Required minimum
burial depth . . .
. . . but circuit
must rise for
connections
Of course, lesser depths than shown in Table 300.5 are permitted where cable or conductors in raceway come up to terminations or splices in boxes or equipment. NOTE 3.
18 in. min
Any cables or raceways may be buried not less than 18 in. deep under airport runways and adjacent defined areas where trespass is prohibited.
At least 2 in.
Concrete
Solid rock
Duct or raceway
Duct and raceway installed in solid rock may be buried if concrete at least 2 in. thick covers the raceway and extends down to the rock surface.

Fig. 300-16. These applications are also covered in the table burial depths. (Sec. 300.5.)

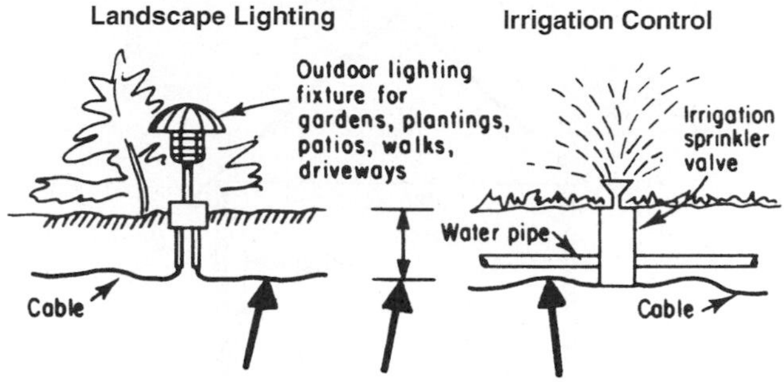
Landscape Lighting
Irrigation Control
Outdoor lighting fixture for gardens, plantings, patios, walks, driveways
Irrigation sprinkler valve
Water pipe
Cable
Cable
Minimum burial depth of 6 inches if cable is Type UF and circuit operates at not more than 30 volts

Fig. 300-17. Reduced burial depth for low-voltage landscape lighting and lawn-sprinkler controls. (Sec. 300.5.)

valves, as shown in Fig. 300-17. This recognizes the reduced hazards and safety considerations for circuits operating at not more than 30 V. However, such systems must be supplied by Type UF cable or other "identified" cable or raceway. Such identification is generally required to be marked by the manufacturer. On the cables and conductor, the marking should include a "W" indicating suitability for use in a "wet location" to ensure compliance with part **(B)**. The 2017 NEC added recognition, in terms of two table footnotes, of listed low-voltage landscape lighting systems. In general, such wiring can use even lesser cover in accordance with the product instructions. In addition, for use around swimming pools and if run in nonmetallic raceway and not operating over 30 volts, such systems can be used with a 6-in. burial depth.

Part **(B)** of 300.5 requires all electrical cables and conductors to be listed for use in "wet locations" where installed underground, even if they are contained within a raceway or other enclosure (Fig. 300-18). This effectively applies to all raceways, cables, boxes, enclosures, etc., covered by Chap. 3.

Figure 300-19 shows the rule of part **(C)**. Note that this rule has been clarified to avoid the former implication that the raceway always had to exit beyond

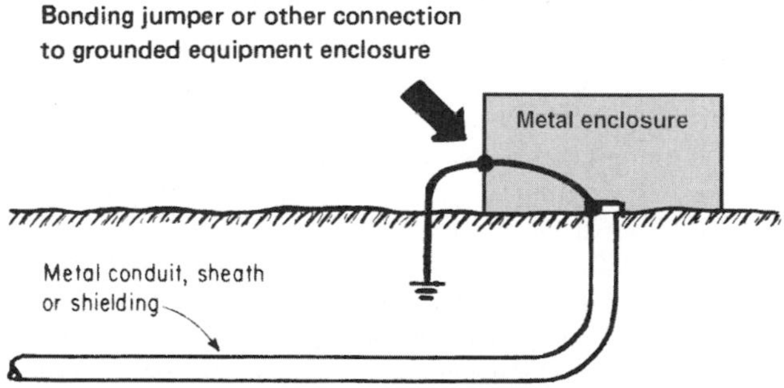
Bonding jumper or other connection to grounded equipment enclosure
Metal enclosure
Metal conduit, sheath or shielding

Fig. 300-18. Conductors run underground must be listed for "wet locations." (Sec. 300.5.)

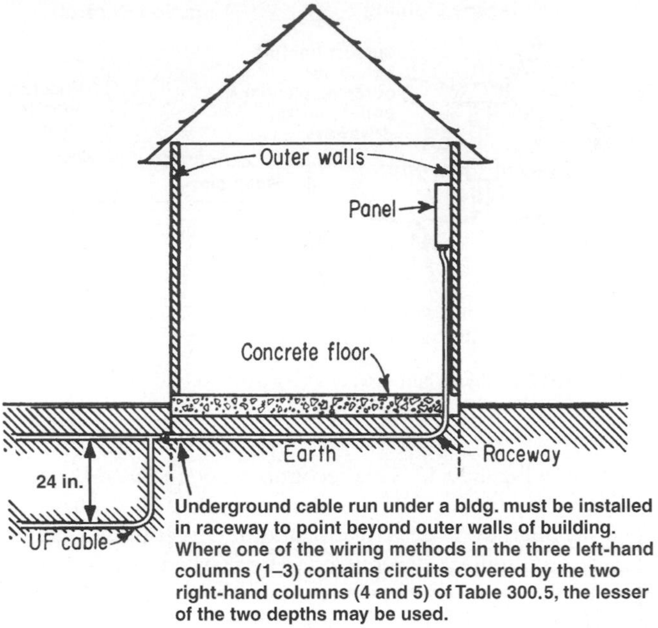
Outer walls
Panel
Concrete floor
Earth
Raceway
24 in.
UF cable
Underground cable run under a bldg. must be installed in raceway to point beyond outer walls of building. Where one of the wiring methods in the three left-hand columns (1–3) contains circuits covered by the two right-hand columns (4 and 5) of Table 300.5, the lesser of the two depths may be used.

Fig. 300-19. Burial of most cables in earth is not permitted under a building. Exceptions include some MI and MC cables. (Sec. 300.5.)

the building perimeter. It is now clear that a raceway can go from one point to another below the floor grade of a building without needing to pass beyond the wall line. However, the requirement for all cables under a building to be in raceway still applies.

Two exceptions to this general prohibition against running cables under buildings without raceway protection were added in the 2011 **NEC**. Type MI cable is permitted if suitably protected against damage and corrosion, and Type MC cable is permitted if identified for direct burial and it meets the construction requirements for wet locations. Note that the first, fourth, and fifth columns in Table 300.5 have been changed to reflect these new exceptions. These applications presumably meet minimum safety requirements, but the design implications of putting any direct burial wiring method under a building need careful consideration. If the wiring fails it may not be repairable as a practical matter.

As shown at the top of Fig. 300-20, direct buried conductors or cables coming up a pole or on a building from underground installation must be protected from the minimum required burial distance below grade (from Table 300.5, but never required to be more than 18 in. [450 mm] into the ground) to at least 8 ft (2.5 m) above grade, as required by part **(D)** of this section. Where exposed to physical damage, raceways on buildings and raceways on poles must be rigid conduit, IMC, RTRC-XW or PVC Schedule 80, or equivalent, and the raceway or other enclosure for underground conductors must extend from below the ground line up to 8 ft (2.5 m) above the finished grade. The 2017 **NEC** added EMT to this list; note, however, that 358.12(1) would

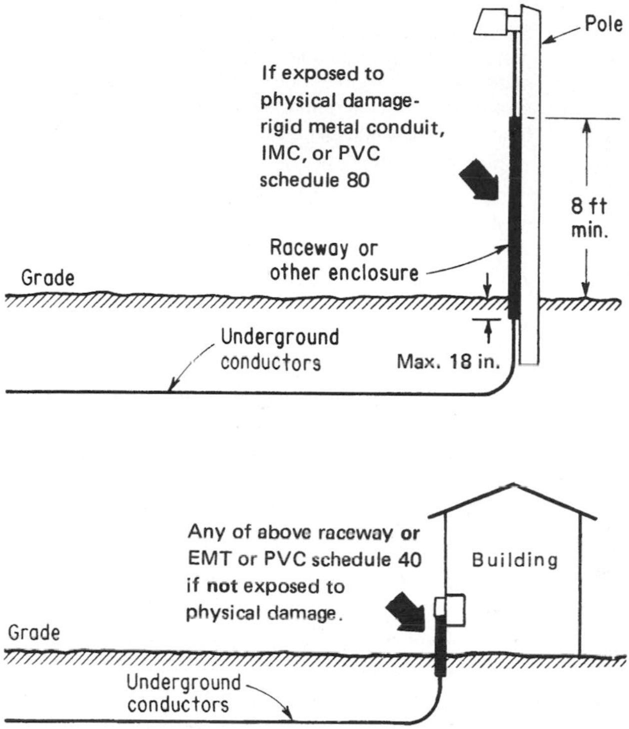

Pole
If exposed to physical damage-rigid metal conduit, IMC, or PVC schedule 80
8 ft min.
Raceway or other enclosure
Grade
Underground conductors
Max. 18 in.
Any of above raceway or EMT or PVC schedule 40 if not exposed to physical damage.
Building
Grade
Underground conductors

Fig. 300-20. Conductors from underground must be protected. (Sec. 300.5.)

prohibit it if exposed to severe physical damage. If a raceway on a building or on a pole is not subject to physical damage, EMT or Schedule 40 PVC may be used instead of other raceways.

Figure 300-21 shows the service lateral protection ribbon rule from 300.5(D)(3), along with other requirements including the ground movement accommodation rule in 300.5(J) and a warning on an important limitation on straight underground USE cable not colisted with a building wire designation.

Although always intended to apply to buried PVC (or RTRC) raceways containing service conductors, the literal text of the parent language may limit the reach of this requirement to direct burial conductors and cables. The 18-in. parameter in this rule (as opposed to the 24-in. specification for direct-burial cables) was chosen to reflect the usual minimum burial depth for nonmetallic raceways. Attempts to fix this problem in the 2014 and 2017 **NEC** cycles failed, so it is still addressed incorrectly. Another attempt will be made in the 2020 **NEC** cycle.

Figures 300-22 and 300-23 show other rules of 300.5. Part **(E)** covers direct burial splices, which do not require boxes but do need to comply with 110.14(B), which requires the splicing method to be listed for this application. There are twist-on wire connectors listed for direct burial, and they are an appropriate solution for single-conductor type USE; however, do not use these connectors

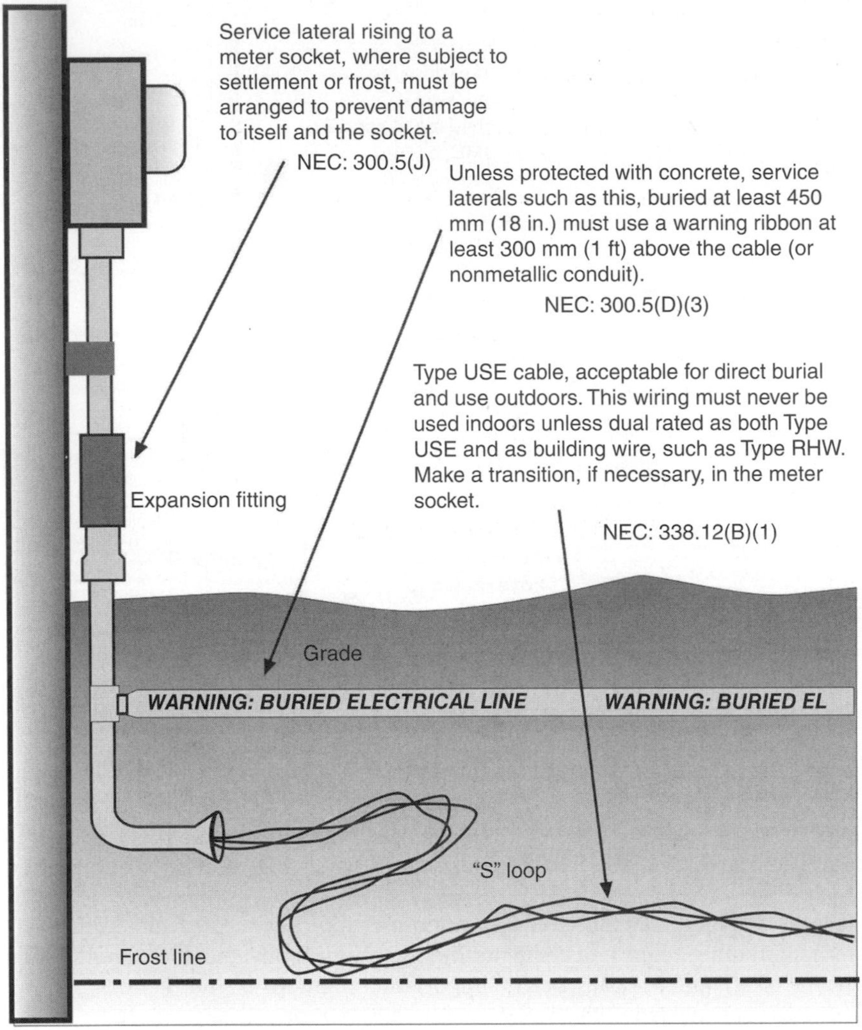
Service lateral rising to a meter socket, where subject to settlement or frost, must be arranged to prevent damage to itself and the socket.
NEC: 300.5(J)
Unless protected with concrete, service laterals such as this, buried at least 450 mm (18 in.) must use a warning ribbon at least 300 mm (1 ft) above the cable (or nonmetallic conduit).
NEC: 300.5(D)(3)
Type USE cable, acceptable for direct burial and use outdoors. This wiring must never be used indoors unless dual rated as both Type USE and as building wire, such as Type RHW. Make a transition, if necessary, in the meter socket.
NEC: 338.12(B)(1)
Expansion fitting
Grade
WARNING: BURIED ELECTRICAL LINE
WARNING: BURIED EL
"S" loop
Frost line

Fig. 300-21. Underground installations must accommodate ground movement. [Sec. 300.5(J).] Note that the bell end on the bottom end of the PVC ell (could also have been a bushing) meets the provisions of part **(H)**.

for conventional Type UF cable. The direct-burial listing on Type UF cable applies to the entire cable assembly and not to the individual conductors within the jacket. Therefore, if a splice is made with the jacket stripped back and these connectors made up, the resulting exposure of individual conductors within the UF cable assembly to direct soil burial violates the UF cable listing. The splicing method must provide an outer surface that is continuous from cable jacket to cable jacket.

Note that part **(F)** specifically requires that backfilled trenches must contain any necessary protection for raceways or cables buried in the trench. It specifies that

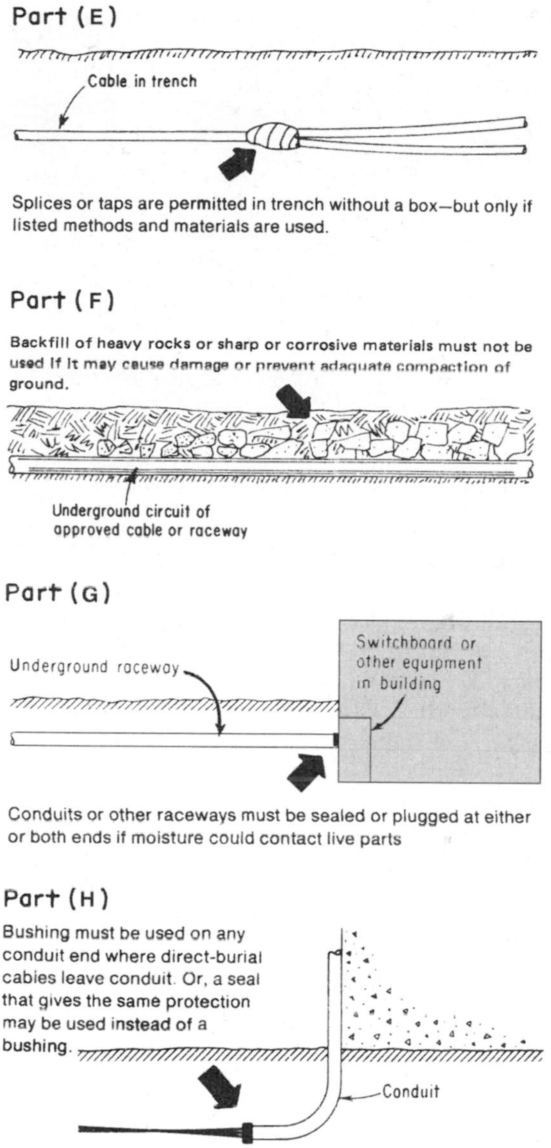
Part (E)
Cable in trench
Splices or taps are permitted in trench without a box—but only if listed methods and materials are used.
Part (F)
Backfill of heavy rocks or sharp or corrosive materials must not be used if it may cause damage or prevent adaquate compaction of ground.
Underground circuit of approved cable or raceway
Part (G)
Underground raceway
Switchboard or other equipment in building
Conduits or other raceways must be sealed or plugged at either or both ends if moisture could contact live parts
Part (H)
Bushing must be used on any conduit end where direct-burial cables leave conduit. Or, a seal that gives the same protection may be used instead of a bushing.
Conduit

Fig. 300-22. Underground wiring must satisfy these requirements. Part G requires the sealing material to be identified as suitable for the wiring in the raceway. The sealing rule also applies to spare raceways. (Sec. 300.5.)

sand or suitable running boards of wood or concrete or other protection must be afforded in those cases where backfill consists of heavy stones or sharp objects that otherwise would present the possibility of damage to the cable or raceway. Refer to the discussion at 310.10(F) for additional information on this topic.

Fig. 300-23. For direct burial underground conductors, a box must be used at splice points, with conductors brought up in sweep ells and the box properly grounded—*unless* listed materials are used to make directly buried splices in the conductors in accordance with Sec. 110.14(B). (Sec. 300.5.)

Part **(I)** of this section requires that an underground circuit made up of single-conductor cables for direct burial must have all conductors of the circuit run in the same trench. That rule raises the question: When an underground direct burial circuit is made up of conductors in multiple, must all the conductors be installed in the same trench? And if they are, is derating required for more than three current-carrying conductors in a trench, just as it would be for more than three conductors in a single raceway? The answer to both questions is yes.

The wording of the rule in part **(I)** clearly indicates that all the conductors making up a direct burial circuit of single conductors in parallel must be run in the same trench and must be "in close proximity." The wording of 300.5(I) also requires that all conductors of a circuit be run in the same raceway if a raceway is used (with building wire suitable for wet locations, such as THW or THWN). But Exception No. 1 permits parallel conductor makeup in multiple raceways, with each raceway containing all hot, grounded, and (if used) grounding conductors of the circuit. And Exception No. 2 to 300.5(I) recognizes the use of "isolated phase" installations, provided the rules for paralleling conductors—given in 310.4—and the rules for reducing the effects of inductive heating in adjacent metallic materials—as given in 300.20—are satisfied.

When multiple-conductor makeup of a circuit is installed with all the parallel-circuit conductors in the same trench, it is necessary to observe the rule of 310.15(B)(3) of the NEC, which states:

> Where single conductors . . . are installed without maintaining spacing for a continuous length longer than 24 in. (600 mm) and are not installed in raceways, the allowable ampacity of each conductor *shall* be reduced as shown . . .

This means that the same deratings must be made as when more than three conductors are used in a single conduit, as explained in 310.15(B)(3)(a).

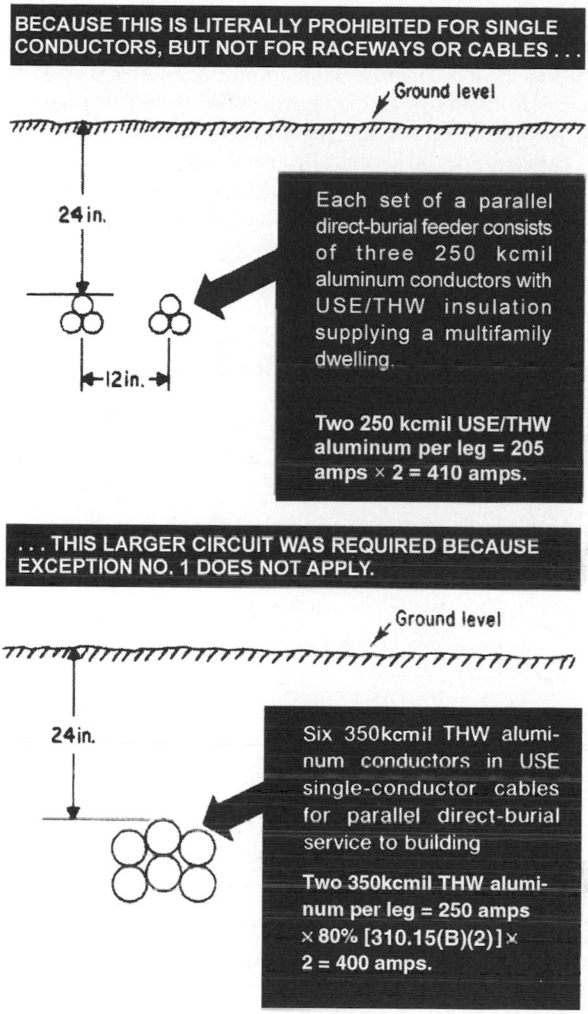
BECAUSE THIS IS LITERALLY PROHIBITED FOR SINGLE CONDUCTORS, BUT NOT FOR RACEWAYS OR CABLES . . .
Ground level
24 in.
12 in.
Each set of a parallel direct-burial feeder consists of three 250 kcmil aluminum conductors with USE/THW insulation supplying a multifamily dwelling.
Two 250 kcmil USE/THW aluminum per leg = 205 amps × 2 = 410 amps.
. . . THIS LARGER CIRCUIT WAS REQUIRED BECAUSE EXCEPTION NO. 1 DOES NOT APPLY.
Ground level
24 in.
Six 350kcmil THW aluminum conductors in USE single-conductor cables for parallel direct-burial service to building
Two 350kcmil THW aluminum per leg = 250 amps × 80% [310.15(B)(2)] × 2 = 400 amps.

Fig. 300-24. Literal application of Code rules often imposes stiff requirements. (Sec. 300.5.) The rewording of Exception No. 1 for the 2011 NEC made it possible to run two trenches and avoid this outcome.

Certainly, direct burial single conductors are covered by that requirement because Table 310.15(B)(16) specifically covers direct burial conductors.

These Code rules often make for tricky and troublesome applications. For instance, as shown in Fig. 300-24, an underground circuit of Type USE insulated aluminum conductors might be used for a 3-wire, single-phase service to a multifamily dwelling. Because that is a residential service, 310.15(B)(7) might be considered to gain a higher-than-normal ampacity for the conductors; however, the rule has been rewritten to exclude multifamily applications and it only applies to applications where the conductors only supply a one-family dwelling unit. Assuming that a 400-A conductor ampacity is indicated by the calculated demand load from Art. 220, each phase leg of the service feeder must have an ampacity of 400 A. Refer to the 75°C column in Table 310.15(B)(16): A 250 kcmil THW aluminum has an ampacity of 205 A. Two such conductors per hot leg and two for the neutral would give the required 400-A capacity for the service. But how should the parallel circuit be run? All the circuit conductors must be run in close proximity in the same trench, as required by 300.5(I). At first blush that would mean that all six USE conductors are in the same trench; and, because the neutrals do not count as current-carrying conductors, the derating of these conductors run without spacing must be to 80 percent of the 200-A ampacity, as required for four conductors in the table of Table 310.15(B)(3). With each 250 kcmil THW aluminum now derated to 161 A (0.8 × 205), the ampacity of each hot leg would only be 322 A (2 × 161).

In referring to Table 310.15(B)(16), it would therefore seem necessary to pick a larger size of THW aluminum, such that, when derated to 80 percent of their Table 310.15(B)(16) value, the two of them will still provide the required 400-A rating. A 350-kcmil THW aluminum has a normal rating of 250 A. Derated to 80 percent (250 × 0.8), it has the needed ampacity of 200 A, so that two in parallel per hot leg and neutral will have the ampacity of 400 A. Fortunately, Exception No. 1 was rewritten in the 2011 cycle to avoid this result. In addition to directly buried multiconductor cables and raceways, the exception now also applies to sets of directly buried conductors. As a result, two parallel trenches can be run, in this case with a set of 250 kcmil THW aluminum conductors in each, each running at their normal ampacity value of 205 A. If the two parallel sets of conductors could have been run in separate trenches, the 250 kcmil THW aluminum conductors would have met the need in terms of ampacity. Obviously, the conductors selected for actual direct burial must be Type USE, or combine that rating with THW (or RHW).

Exception No. 1 of part **(I)** permits an underground circuit to be made up in parallel in two or more raceways, without need for derating. But, in such cases, each raceway must contain one of each of the phase legs, a neutral (if used), and an equipment grounding conductor (if used). With "A-B-C-N" in each raceway of a multiple group, that would be the same type of multiple-conduit-parallel-conductor makeup as required and commonly used for aboveground circuits.

One question that often arises when designing this work is how much distance to allow between parallel trenches before heat generated in one trench begins to affect the other. The **NEC** does not answer this directly for 600 V and below applications. However, running each set of cables in parallel trenches would generally line up with Detail 8 in Fig. 310.60(C)(3), which shows a spacing of 24 in. in order to still allow the application of the related medium-voltage ampacity tables. This distance is probably somewhat conservative for utilization voltages, so it should be more than sufficient. Closer spacings could be applied if justified by engineering support.

Exception No. 2 permits "isolated-phase" makeup of underground circuits in multiple conduits—all phase A conductors in one conduit, all phase B conductors in a second conduit, all phase C conductors in a third conduit, all neutrals in a fourth conduit—with an equipment grounding conductor (or conductors, if needed) installed in a fifth conduit or installed in each of the three conduits carrying the phase conductors. However, that makeup is permitted only where the conduits are nonmetallic and are "in close proximity" to each other. (See 300.20.) Remember that the raceways must terminate in a manner that avoids inductive heating using methods in 300.20(B), such as entering below an open-bottom switchboard, or through a nonferrous window.

Isolated phase installations on ac systems inherently increase the circuit impedance because the magnetic fields surrounding each phase conductor do not cancel. This means the impedance (and therefore voltage-drop) increase with the distance between conductor groups; on the other hand, some engineers employ this technique to reduce the available fault current at the end of the run.

Warning! Because of difficulties that can be encountered with pulling lines sawing through the inner radii of nonmetallic sweeps, some wiring designs show PVC everywhere except at both ends of an underground installation, where steel sweeps are specified. In Art. 250, 250.102(E)(2) Exception, 250.80 exception, and 250.86 Exception No. 3 all facilitate this practice. Never use this wiring procedure on an isolated phase installation covered here. The failure to comply with 300.20(A) in such instances will result, especially for the large current installations for which this procedure was designed, in significant inductive heating within the steel sweeps. A leading cable engineer described to the author actual field instances where the conductor insulation within the steel sweeps had completely melted. Aluminum sweeps with supplemental corrosion protection would be one option, and another is staying with PVC and switching to a polyester pulling line, as covered in the wire pulling analysis offered in our sister publication, the *American Electricians' Handbook*.

Part **(J)** makes it mandatory that frost heave and settling be accommodated. In those areas of the country that experience cyclical freezes and thaws, concern about frost damage to buried raceways and cables has generally been addressed either by installation below the frost line or by the use of expansion fittings and direct buried cables with slack. This is done to ensure that the raceway remains intact and its contained circuits remain operational. The Code now mandates the implementation of those materials and methods that will ensure the underground raceways and cables are not damaged by frost heave. The informational note immediately following part **(J)** identifies methods to accomplish the desired effect.

Figure 300-21 shows the rules in action. Note that this rule applies regardless of whether the run from grade to termination is enough to raise a concern about thermal expansion and contraction [300.7(B)]. It also applies regardless of the thermal expansion coefficient of the wiring method, because it is the ground that is moving, not the raceway changing length, and therefore it applies equally to metallic and nonmetallic raceways alike. There is considerable loss experience associated with steel raceways being forced vertically upward through concentric knockouts until the raceway shorted against a line bus.

In addition, do not make the mistake of assuming that this rule only applies in areas subject to frost. It applies to "earth movement." This includes settling conditions that tend to drag raceways downward to the point of exposing the service conductors within to touch, or pulling downward on intact locknut connections to the point where enclosures are removed from a wall or severely deformed to the point that the covers no longer work properly. The usual way to address this rule is through the use of expansion fittings, which are certainly acceptable, but expensive in the case of metallic models. In some cases other approaches have worked, such as an elliptical hole where a conduit body entered a building on a small branch circuit.

This author once reviewed a medium-voltage application where the feeder emerged from grade and went up the side of a new industrial occupancy to the roof. The engineers were so concerned about the likelihood of settlement that they specified a special pull box with telescoping sides that would allow for the conductors to gently sweep up out of the ground, crest, and go downward almost to grade, and then rise again and enter the vertical raceways up the side of the building. The arrangement was designed to allow the grade to settle several feet without compromising the required bend radius on the Type MV cables.

In part **(K)** the Code mandates that underground cables and raceways installed using "direct boring equipment" must be "approved" for such installation. "Approved" means acceptable to the inspector. In this case, wording in the manufacturer's instructions indicating that such installation is permitted will generally satisfy the local inspector.

300.6. Protection Against Corrosion and Deterioration. This section provides some specific wording with regard to what must be provided to achieve compliance with the rule here. Ferrous-metal equipment is covered in part **(A)**. An exception exempts stainless steel from the necessity to have protective coatings.

The 2014 NEC has added a new informational note explaining what a field-cut thread is. This was necessary because it was suggested that a thread applied in a mechanic's shop or anywhere else beside the field location where the conduit would be installed was not field cut. It is now clear than any thread not applied by the manufacturer is a field-cut thread for these purposes, and if corrosion protection is necessary, the threads must be coated with a compound that is both corrosion resistant and conductive, as approved by the inspector.

Part **(A)(1)** limits the use of ferrous-metal equipment protected solely by enamel to damp and dry indoor locations. Part **(A)(2)** allows organic coatings to be applied to metallic boxes or cabinets to prevent corrosion when used outdoors, in lieu of the standard "4-dip" zinc galvanizing method. And part **(A)(3)** addresses the use of ferrous metal equipment in concrete, earth, or other hostile environments, provided either the material or the corrosion protection is approved for the location.

Part **(B)** covers aluminum raceways, enclosures, etc. All such equipment must be provided with supplemental where in concrete or the earth. In a previous cycle, this part referred to nonferrous metal, which was an error because red brass is not subject to this limitation. Part **(C)** covers nonmetallic equipment, which must be suitable for use in direct sunlight and impervious to other environmental hazards based on its chemical makeup.

Figure 300-25 shows the right and wrong ways of installing equipment in indoor wet locations, as covered in part **(D)** of this section. Note that the exception that allows nonmetallic products to avoid the 6-mm (¼-in.) air space rule is limited to surfaces that are impervious to moisture, such as tile or concrete.

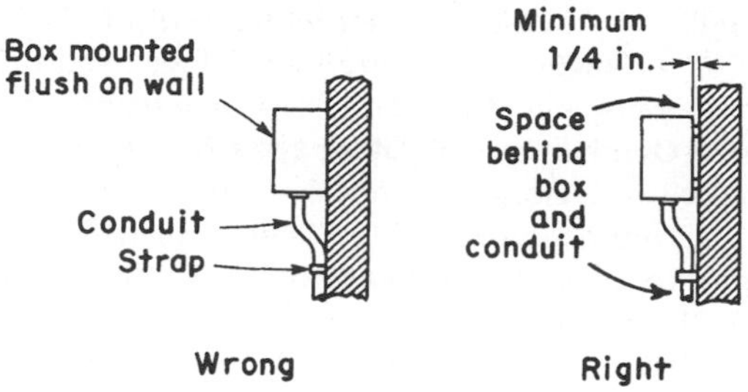
Box mounted
flush on wall
Conduit
Strap
Minimum
1/4 in.
Space
behind
box
and
conduit
Wrong
Right

Fig. 300-25. Water or moisture must not be trapped in contact with metal. (Sec. 300.6.)

300.7. Raceways Exposed to Different Temperatures. Part **(A)** requires protection against moisture accumulation. If air is allowed to circulate from the warmer to the colder section of the raceway, moisture in the warm air will condense in the cold section of the raceway. This can usually be eliminated by sealing the raceway just outside the cold rooms so as to prevent the circulation of air. Sealing may be accomplished by stuffing a suitable compound in the end of the pipe (Fig. 300-26). Note that this rule has been expanded in recent years, and now expressly includes a passage

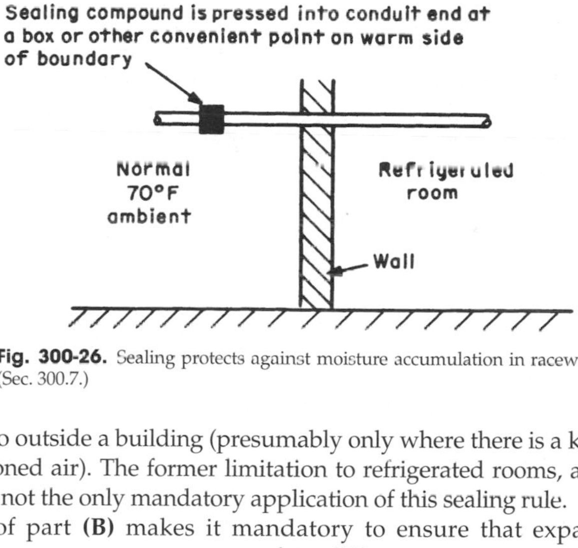
Sealing compound is pressed into conduit end at
a box or other convenient point on warm side
of boundary
Normal
70°F
ambient
Refrigerated
room
Wall

Fig. 300-26. Sealing protects against moisture accumulation in raceway. (Sec. 300.7.)

from inside to outside a building (presumably only where there is a known issue, as with conditioned air). The former limitation to refrigerated rooms, as suggested in the sketch, is not the only mandatory application of this sealing rule.

The rule of part **(B)** makes it mandatory to ensure that expansion fittings are used where raceways are exposed to different temperatures to ensure that the raceways, fittings, and enclosures are not damaged and compromised by the inherent expansion and contraction that results from thermal cycling. The informational note directs the reader to data regarding the use of expansion fittings for PVC conduit and provides guidance and data that applies to the expansion characteristics of IMC, EMT, and rigid steel conduit. Do not confuse this with ground movement as covered in 300.5(J). They are two different rules and different conditions mandate their application, although the usual solution for both problems is an expansion fitting. In some cases thermal action results in deflection rather than simple expansion or contraction. There are also fittings to address this issue, and some are evaluated for both deflection and expansion/contraction.

300.8. Installation of Conductors with Other Systems. Any raceway or cable tray that contains electric conductors must not contain "any pipe, tube, or equal for steam, water, air, gas, drainage, or any service other than electrical."

300.9. Raceways in Wet Locations Above Grade. This new (2008 **NEC**) rule clarifies once and for all that the inside of a raceway in a wet location is presumed to be wet, and the enclosed wiring must be suitable for wet locations.

300.10. Electrical Continuity of Metal Raceways and Enclosures. This is the basic rule requiring a permanent and continuous bonding together (i.e., connecting together) of all noncurrent-carrying metal parts of equipment enclosures—conduit, boxes, cabinets, enclosures, housings, frames of motors, and lighting fixtures—and connection of this interconnected system of enclosures to the system grounding electrode at the service or transformer (Fig. 300-27). The interconnection of all metal enclosures must be made to connect all metal to the grounding electrode and to provide a low-impedance path for fault-current flow along the enclosures to ensure operation of overcurrent devices that will open a circuit in the event of

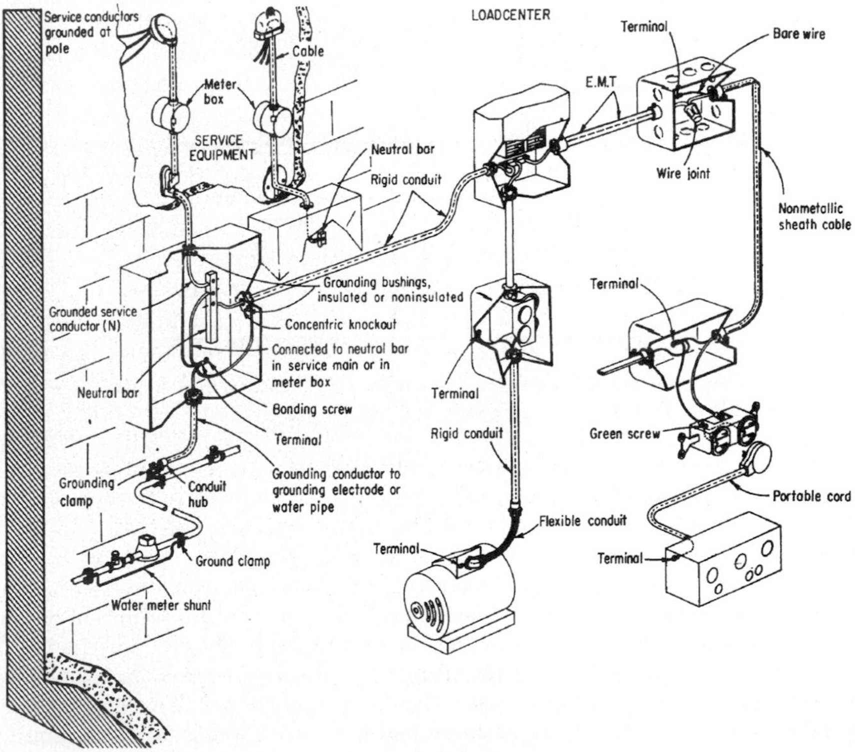
Service conductors grounded at pole
Cable
Meter box
SERVICE EQUIPMENT
Neutral bar
Rigid conduit
LOADCENTER
Terminal
Bare wire
E.M.T
Wire joint
Nonmetallic sheath cable
Grounding bushings, insulated or noninsulated
Terminal
Grounded service conductor (N)
Concentric knockout
Connected to neutral bar in service main or in meter box
Neutral bar
Terminal
Bonding screw
Terminal
Rigid conduit
Green screw
Grounding clamp
Conduit hub
Grounding conductor to grounding electrode or water pipe
Portable cord
Flexible conduit
Terminal
Terminal
Ground clamp
Water meter shunt

Fig. 300-27. All metal enclosures must be interconnected to form "a continuous electric conductor." (Sec. 300.10.)

a fault. By opening a faulted circuit, the system prevents dangerous voltages from being present on equipment enclosures that could be touched by personnel, with consequent electric shock to such personnel.

Simply stated, this interconnection of all metal enclosures of electric wires and equipment prevents any potential aboveground on the enclosures. Such bonding together and grounding of all metal enclosures are required for both grounded electrical systems (those systems in which one of the circuit conductors is intentionally grounded) and ungrounded electrical systems (systems with none of the circuit wires intentionally grounded).

Exception No. 1 to this rule recognizes the specific permission given in Art. 250 to depart from the requirement for "electrical continuity," where an isolated metallic elbow is used underground. As indicated in 250.86, Exception No. 2, short sections of metallic piping used for providing physical protection for cables are not subject to the basic requirement that such metallic raceways, cable jackets, etc., all be electrically continuous. But effective equipment interconnection and grounding are extremely important for grounded electrical systems to provide the automatic fault-clearing that is one of the important advantages of grounded electrical systems. A low-impedance path for fault current is necessary to permit enough current to flow to operate the fuses or CB protecting the circuit.

Exception No. 2 recognizes the permission given in 250.96(B) to interrupt the electrical continuity of a raceway containing a so-called isolated ground circuit. But, such installation must fully comply with all requirements given in 250.96(B) to ensure a modicum of safety.

300.11. Securing and Supporting. The basic rule in part **(A)** requires all raceways, cables, boxes, and so on to be "securely fastened in place." That wording effectively prohibits simply laying raceways, cables, boxes, and so on on the floor, on piping, on ceiling tiles, or anywhere. Although such installation may provide "support," the basic rule calls for fastening, not just support. Regardless of where raceways, cables, boxes, and the like are installed, they must be fastened in place using a recognized method. Generally speaking, the rule in part **(B)** permits the use of "ceiling support" wires to secure wiring methods and equipment; however, they must be "in addition to the ceiling grid support wires." That is, the ceiling will have a designed pattern of support wires, and tie wires for electrical equipment are to be in addition to the ones required to hold up the ceiling. This is accomplished by using appropriate commercially available fasteners. Note that where ceiling support wires are used, they must be secured at both ends. The last sentence simply reiterates a long-held interpretation that asserts that unfastened cables and raceways are prohibited, and that if tied to a support grid the adjacent ceiling panel will be very difficult or impossible to remove. Note that 300.23 specifically requires that any wiring above such panels be arranged so the panels can be removed and equipment accessed above the ceiling.

In part **(B)(1)**, the Code calls for an independent means of support for wiring methods within the cavity above a suspended ceiling that is an integral part of fire-rated floor-ceiling or roof-ceiling assembly. This rule covers those installations where the fire rating of the hung ceiling is being used to establish the fire separation rating—typically specified by local building codes—for the space above and beneath the ceiling. Such installations are covered by **(B)(1).** Nonfire-rated hung-ceiling assemblies are covered under part **(B)(2).**

A true **(B)(1)** application is somewhat unusual, and it is important to know exactly what it is and why the rules make the distinctions they do. Building codes typically establish required fire separation ratings between floors of major buildings. Usually, the building is simply designed with enough concrete so the structural ceiling possesses the required fire separation by itself. In these cases the integrity of the suspended ceiling has no bearing on how much time it would take for a fire on one floor to compromise the next higher floor. These ceilings are covered in **(B)(2)**.

However, there are other ways to achieve a particular floor/ceiling separation. Some ceilings are designed with a thinner slab than one that would establish the required fire separation on its own. In this case, the suspended ceiling is an integral part of the overall fire rating. Qualified testing laboratories have evaluated many such designs, using specified ceiling panel materials. These tests involve actually constructing such a room and ceiling inside a giant furnace, so they are extremely expensive to run. Invariably, the designs specify, in precise detail, exactly how many support wires must be used, how thick they must be, how and where they must be attached, how fixtures are to be supported, how air ducts must be constructed and run, etc.

In making these tests, the testing laboratories do not assume any additional weight loads on these support wires, nor do they (or could they) evaluate how such a ceiling might deform under fire conditions with such loading. In these cases, the wiring methods must be supported securely and independently of the support wires. The only exception allows for the remote possibility that someone might rerun a ceiling assembly test with specified wiring attached to specified supports. The fact that the rated designs specify the support wire locations makes it essential that both electrical and building inspectors be able to know which support wires were placed in addition to the essential elements of the ceiling construction.

Therefore, no raceways or cables may be supported by the wires installed to support the ceiling. But, if additional wires are installed, such wires may be used to "securely support" the raceways or cables within the ceiling. And because of the critical nature and precise construction detail that goes into such a ceiling design, where additional ceiling support wires are used as the support for wiring methods within a hung ceiling, such wires must be identified via marking, tagging, painting, or virtually any method that will permit the ceiling support wires to be readily distinguished from the electrical system support wires.

In part **(B)(2)** the Code addresses the same consideration for wiring method support within nonfire-rated hung ceilings. In the case of nonfire-rated applications, you have to add wires to those required for ceiling support, and due to a change in the 2011 **NEC** they must be identifiable in the same way as those used for fire-rated assemblies in 300.11(B)(1). The language in this paragraph talks of an "independent means of secure support" which has led some to conclude that all support wires were off limits. That has never been the case and the issue was clarified in the 2008 **NEC** which added the phrase "and shall be permitted to be attached to the assembly." Thus, as in the case of the **(B)(1)** applications, you can't use the ceiling design wires but you can add to them, provided they are attached at both ends so as to "provide secure support" as covered in the parent language in 300.11(A). The extension of the identification rule makes this easy to police.

Both **(B)(1)** and **(B)(2)** have rather impractical exceptions that recognize the use of the ceiling design wires if the design includes the wiring. The **(B)(1)** exception would require the entire ceiling system to go back in for testing, which is so costly as to be unimaginable. Remember, the exact location of the wiring would need to be specified, and that is something that changes with every installation. The **(B)(2)** exception is predicated on the ceiling manufacturer choosing to recognize wiring on his specified ceiling support protocol. Frankly the likelihood of that happening is only slightly better than infinitesimal, but the exception is probably harmless as long as no one expects to use it.

In part **(C)**, raceways are prohibited from being used as a means of support for cables or nonelectrical equipment. Telephone or other communication, signal, or control cables must not be fastened to electrical conduits, such as by plastic straps or any other means. Although raceways must not be used as a means of support for other raceways, cables, or nonelectric equipment, part **(C)(1)** might permit large conduits with hanger bars or fittings intended to support smaller raceways, but this application would have to be called out on some manufacturer's literature, as provided in the definition of "identified." Part **(C)(2)** permits such applications as tying Class 2 thermostat cable to a conduit carrying power-supply conductors for electrically controlled heating and air-conditioning equipment that is controlled by the Class 2 wires. Part **(C)(3)** applies to those instances where raceways are permitted to support boxes or conduit bodies (314.23) or luminaires [410.36(E)].

In part **(D)**, the Code also prohibits the use of cable assemblies as a means of support for any equipment, such as other power cables, raceways, boxes, cable TV, or phone lines. In short, nothing may be supported by any of the cable assemblies recognized as wiring methods by the NEC. Two cables tie-wrapped together are a support for neither. The reason for the cable-on-other-wiring prohibitions is that the ampacity calculation procedures in the NEC presume that wiring methods are free to radiate heat. Adding a blanket of cables can lead to overheated wiring.

300.13. Mechanical and Electrical Continuity—Conductors. The general rule of part **(A)** is: Where conductors are spliced, a box is needed. The second sentence then identifies those limited number of cases where the Code does allow a splice without a box.

Part **(B)** prohibits dependency upon device terminals (such as internally connected screw terminals of duplex receptacles) for the splicing of neutral conductors in multiwire (3-wire or 4-wire) circuits. Grounded neutral wires must not depend on device connection (such as the break-off tab between duplex receptacle screw terminals) for continuity. White wires can be spliced together with a pigtail to the neutral terminal on the receptacle. If the receptacle is removed, the neutral will not be opened (Fig. 300-28).

This rule is intended to prevent the establishment of unbalanced voltages should a neutral conductor be opened first when a receptacle or similar device is replaced on energized circuits. In such cases, the line-to-neutral connections downstream from this point (farther from the point of supply) could result in

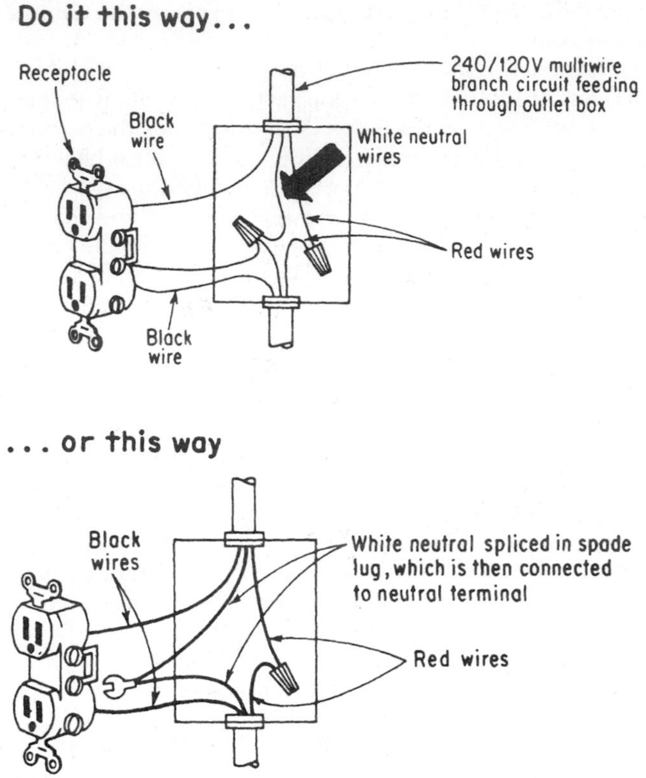
Do it this way...
Receptacle
Black wire
240/120V multiwire branch circuit feeding through outlet box
White neutral wires
Red wires
Black wire
... or this way
Black wires
White neutral spliced in spade lug, which is then connected to neutral terminal
Red wires

Fig. 300-28. Neutrals of multiwire circuit must *not* be spliced at receptacle terminals. (Sec. 300.13.)

a considerably higher-than-normal voltage on one part of a multiwire circuit and damage to equipment, because of the "open" neutral, if the downstream line-to-neutral loads are appreciably unbalanced.

Note that this paragraph does not apply to 2-wire circuits or circuits that do not have a grounded conductor. This rule applies only where multiwire circuits feed receptacles or lampholders. This would most commonly be a 3-wire 240/120-V or a 3- or 4-wire 208/120-V or even a 480/277-V branch circuit.

The reason for the pigtailing requirement is to prevent the neutral conductor from being broken and creating downstream hazards. The problem lies in the inclination of electricians to work on hot circuits. Assume that a duplex receptacle on a 240/120-V 3-wire circuit becomes defective, and the first thing the electrician does, working hot, is to disconnect the neutral wires from the receptacle. Downstream, 2.4- and 12-A loads have been operating (plugged into additional receptacles on the multiwire circuit), each connected to a different hot leg. When the neutral is broken by the electrician upstream, normal operation of the loads reverts to the condition shown in Fig. 300-29. The two loads are now

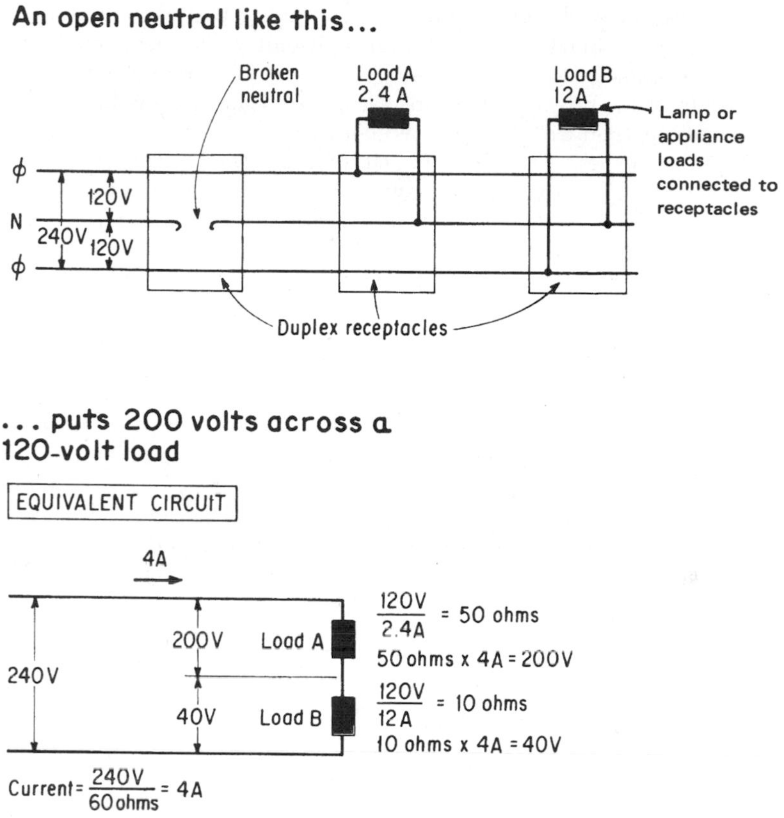
An open neutral like this...
Broken neutral
Load A 2.4 A
Load B 12A
Lamp or appliance loads connected to receptacles
120V
N
240V
120V
Duplex receptacles
... puts 200 volts across a 120-volt load
EQUIVALENT CIRCUIT
4A
200V
Load A
240V
40V
Load B
120V / 2.4A = 50 ohms
50 ohms x 4A = 200V
120V / 12 A = 10 ohms
10 ohms x 4A = 40V
Current = 240V / 60 ohms = 4A

Fig. 300-29. Splicing neutrals on receptacle screws causes "open" in neutral if receptacle is removed. (Sec. 300.13.)

in series across 240 V. As shown, load A now has 200 V impressed across it. It could run extremely hot and burn out. Load B now has only 40 V across it; if it is a motor-operated device, the low voltage could cause the motor to burn up. Both could cause injuries. Also, in disconnecting the neutral, the electrician could get a 120-V shock if both the disconnected neutral conductor going downstream and the box were touched—not unlikely, since the neutral is usually considered to be dead—that is, at ground potential.

300.14. Length of Free Conductors at Outlets, Junctions, and Switch Points. The rule here applies only to the length of the conductor at its end. The exception covers wires running through the box. Wires looping through the box and intended for connection to outlets at the box need have only sufficient slack that any connections can be made easily. Generally, at least 150 mm (6 in.) of slack must be provided. This dimension is measured from the point the conductor emerges from its raceway or cable sheath. In addition, for smaller boxes (i.e., those smaller than 200 mm [8 in.] in any dimension), the minimum length of slack is based on the amount of conductor that extends outside the box—presumably when held at a 90° angle to its entry. For these smaller boxes, the conductor must be long enough so that at least 75 mm (3 in.) of conductor can extend from the box opening. This recognizes that boxes with small openings prevent a person from getting two hands into the box for the purpose of connecting devices, etc. Therefore, the Code mandates the conductors extend beyond the face of the box for not less than 75 mm (3 in.) where the box has an opening that is less than 200 mm (8 in.) in width or height.

Note the "emerges from its . . . cable sheath" provision. Those who don't remove cable jackets prior to the walls being closed, and strip the jackets only at the finish wiring stage at the point where the cable assembly comes out of the wall may want to reconsider this practice. The 150-mm (6-in.) dimension would then start at the wall, resulting in excessive box fill. Although 314.16 has no prescriptive volume rule to cover this, it does require "free space for all enclosed conductors," One of the reasons for specifying the point of emergence on cable assemblies as being the end of the jacket was to discourage this practice.

300.15. Boxes, Conduit Bodies, or Fittings—Where Required. The basic rule requires either a "box" or a "conduit body" to be used at splice points or connection points in cable and raceway systems. Figure 300-30 covers the basic

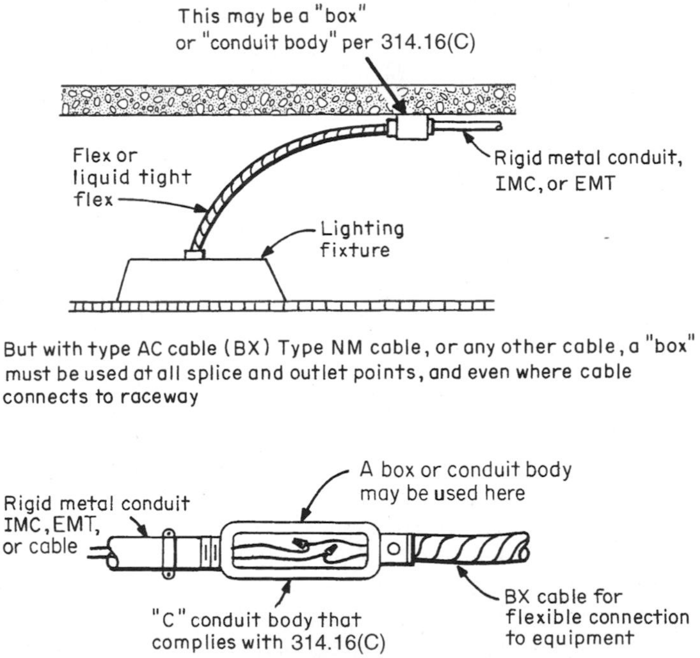
This may be a "box"
or "conduit body" per 314.16(C)
Flex or
liquid tight
flex
Rigid metal conduit,
IMC, or EMT
Lighting
fixture
But with type AC cable (BX) Type NM cable, or any other cable, a "box" must be used at all splice and outlet points, and even where cable connects to raceway
A box or conduit body
may be used here
Rigid metal conduit
IMC, EMT,
or cable
"C" conduit body that
complies with 314.16(C)
BX cable for
flexible connection
to equipment

Fig. 300-30. Raceways and cables may use boxes or conduit bodies at conductor splice points. (Sec. 300.15.) Note that based on current typical conduit body volumes, splices such as those shown are problematic because they usually will not meet the volume requirements in 314.16.

principles. In addition, concealed knob-and-tube wiring must have a box at outlet and switch points. This sets the default stance of the **NEC**. The lettered paragraphs that follow, (A) through (L), set forth the only circumstances under which wiring methods can have any form of termination, pull, or outlet point without a box or conduit body.

The second sentence in 300.15, also part of the default stance, was included to address the almost universal misuse of Type NM (Romex) connectors with other cables and even cords. The wording as it appears in the Code, however, requires that *any* fitting or connector be "designed and listed" for the wiring method used. Be aware that, while it will be up to the manufacturers of this equipment to obtain the listing for these products, it is the designer-installer's responsibility to specify or use only those fittings and connectors specifically listed for the application. Connectors for Type MC cable are another source of abuse; many listed connectors for Type AC cable are misapplied on aluminum armored MC cable. The permitted cable types, including the allowable cable diameters and whether the cable is listed for smooth or corrugated or interlocking-armor cable or some combination of the three, will be listed on the smallest shipping unit carton.

In part **(A)** the Code eliminates the need for a box or conduit body where certain enclosures that have removable covers are used. And part **(B)** recognizes equipment that is provided with adequately sized junction boxes or wiring compartments, such as motor terminal housings, or wiring compartments in HID and other luminaires that are arranged for a direct wiring entry.

Part **(C)** calls for abrasion protection for cables where they enter/exit a conduit or tubing. This ensures that rough or sharp edges on the conduit or tubing will not damage the cable. This applies to kick pipes and also to short risers from outlet and switch points on unfinished basement walls. Part **(D)** exempts a specific Type MI splice, but only where the actual splice is accessible.

Part **(E)** of this section recognizes the use of wiring devices that have "integral enclosures." These are the so-called boxless devices made and acceptable for use in nonmetallic sheathed cable systems (Type NM). Such listed devices [see 334.30(C) for the rules on these devices] do not require a separate box at each outlet because the construction of the device forms an integral box in itself. These devices are flush, modern versions of the old surface-mounted devices covered in **(H)**. The devices in this latter paragraph are primarily used for surface installations of Type NM cable as covered in 334.40(B), although fished-in applications are recognized as well.

Part **(F)** recognizes transition from Type AC or MC cable to raceway without the need for a box, provided that no splice or termination is made in the conductors. This permits the common practice of changing from, say, AC to EMT for a run down a wall, with the armor stripped from a long length of the AC and the exposed wires run in the EMT. A suitable fitting made for connecting AC to EMT must be used (Fig. 300-31).

Part **(G)** correlates with the rules of 300.5(E) and 110.14(B), which permit splicing of underground direct burial cables and conductors using listed splice kits without a box.

Part **(I)** covers terminations and pull points in large accessible enclosures much larger than the wiring methods, where a box would serve no purpose.

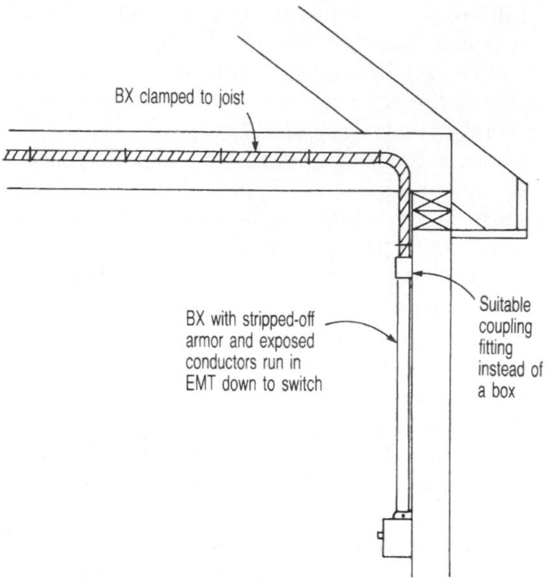
BX clamped to joist
BX with stripped-off armor and exposed conductors run in EMT down to switch
Suitable coupling fitting instead of a box

Fig. 300-31. This type of no-box connection for cable-to-raceway change is permitted by part **(F).** [Sec. 300.15(F).]

Examples include switchboards, motor control centers, and cabinets and cutout boxes generally.

Part **(J)** permits splices to be made within luminaire wiring compartments where the branch-circuit wires are spliced to fixture or ballast wires and the wiring compartment qualifies for consideration as a raceway under one of the three conditions provided in 410.64.

Part **(K)** correlates with provisions for radiant electric heating cable connections that are done with the connections embedded with the cables.

Part **(L)** covers handholes and manholes unless there is an actual connection to electrical equipment contained in the handhole or manhole, as opposed to a simple conductor or cable splice.

300.16. Raceway or Cable to Open or Concealed Wiring. Where the wires are run in conduit, tubing, metal raceway, or armored cable and are brought out for connection to open wiring or concealed knob-and-tube work, a fitting such as that shown in Fig. 300-32 may be used. Where the terminal fitting is an accessible outlet box, the installation may be made as shown in Fig. 300-32.

300.17. Number and Size of Conductors in Raceway. This section contains the general principle that raceway installations must permit the ready installation and withdrawal of the enclosed conductors. The various raceway articles have rules that generally limit the number of bends in tubular raceways between pull points, all as part of carrying out the general objectives of this rule. Note, however, that no individual raceway article can impose a bend limit on two different raceways joined together with a changeover fitting and no pull point

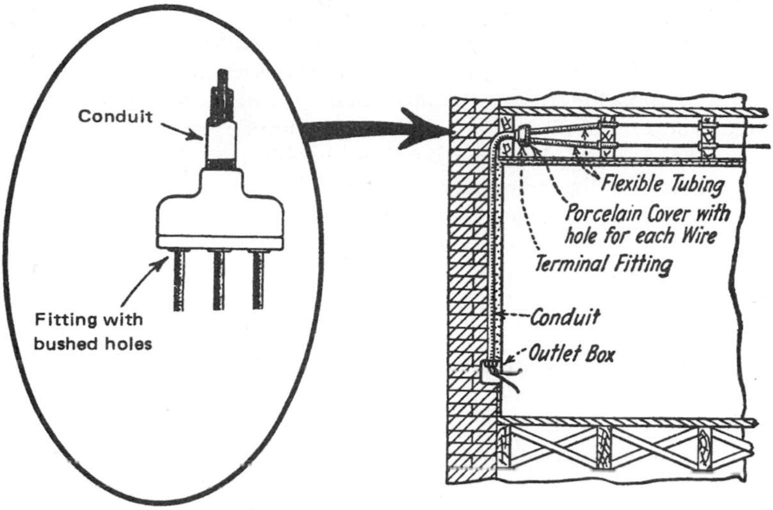
Conduit
Fitting with
bushed holes
Flexible Tubing
Porcelain Cover with
hole for each Wire
Terminal Fitting
Conduit
Outlet Box

A terminal fitting satisfies the rule on transition from raceway to knob and tube.

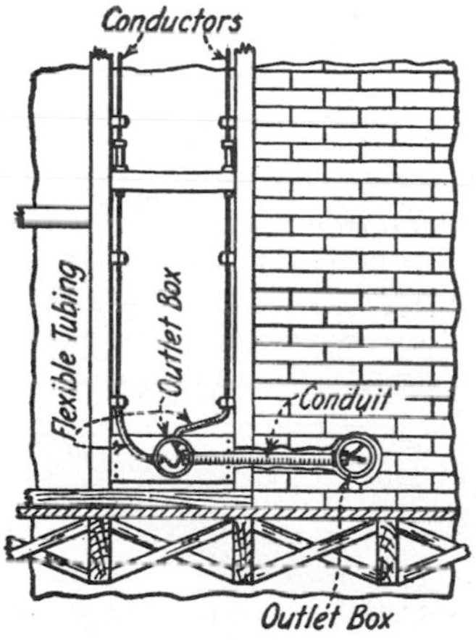
Conductors
Flexible Tubing
Outlet Box
Conduit
Outlet Box

An outlet box may be used where raceway connects to open wiring or concealed knob and tube.

Fig. 300-32. These are techniques for connecting conduit to open wiring. (Sec. 300.16.)

at the transition. On the literal text of those articles, it would be possible to have 360° in bends in a run of intermediate metal conduit connected to a run of EMT, also with 360° of bends. This section can and should be applied to challenge that practice. It frequently arises with respect to transitions to flexible wiring methods at the end of a run.

300.18. Raceway Installations. This section presents a general Code concept requiring raceways to be installed as a complete system, with associated boxes and other enclosures, before pulling conductors into the raceway and box system. This rule is in the NEC because of reports of damage to conductors being pulled into incomplete raceway systems. There is another, more fundamental reason for this rule. Any raceway system is supposed to be capable of having its enclosed conductors withdrawn and replaced without damage (see 300.17). On very long runs, the frictional forces can overcome the tensile strength of the conductors and the ability of the insulation to stand up to the sidewall bearing pressures that will develop on difficult pulls. To avoid these problems, some contractors have pulled wire in segments, adding raceway lengths as they go forward, eventually resulting in a raceway that might as well be classified as a cable because there will be no possibility that the wires could ever be withdrawn. This section provides a partial defense against this by assuring that, at least, someone actually pulled the wires in through all the bends, from access point to access point.

The last sentence of (A) does recognize prewired assemblies, but only where the raceway article specifically permits this to occur. An example is 362.10(8) for ENT. The rule also intends to permit wiring of motors and fixture whips after the basic raceway system has been wired, as well as covering prewired assemblies. Figure 300-33 shows wires pulled into an incomplete raceway system. That used to be a Code violation, but the rule prohibiting it was removed as long as the raceways are complete between pull points. The raceway to the right is still OK because it has no wires in it. The exception recognizes the permission given in 250.86 Exception No. 2. Part **(B)** prohibits welds made to raceways unless they were specifically designed for this purpose, or some Code provision specifically allows it.

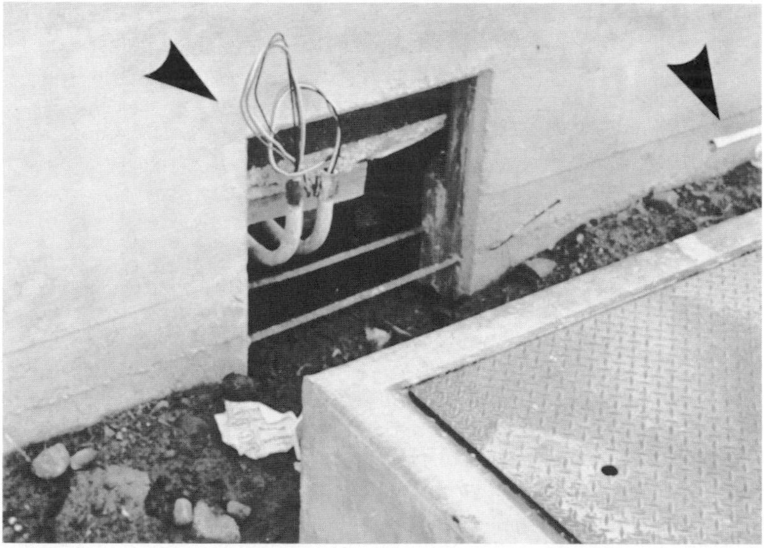

Fig. 300-33. Conductors shown here have been pulled into the conduit before boxes and continuation of the raceway system were installed to supply underground circuits to outdoor building lighting. Under previous Code editions, this was a violation of Sec. 300.18. That section and its rules were removed after the 1981 **NEC**, but restored six years later.

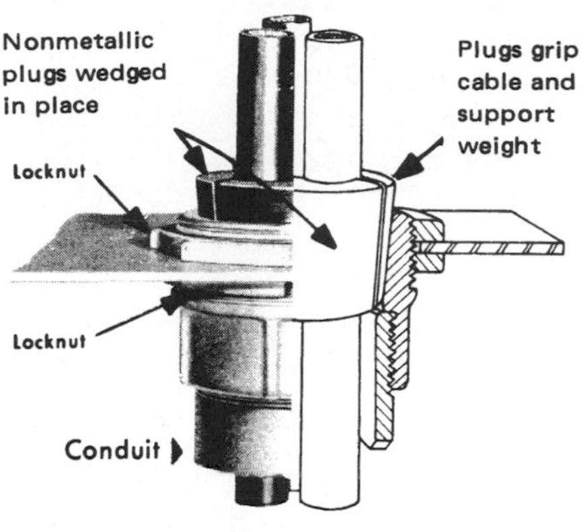
Nonmetallic plugs wedged in place
Plugs grip cable and support weight
Locknut
Locknut
Conduit

Conductor-support bushing screwed on end of conduit at a cabinet, pull box, or conductor-support box. (*Russell & Stoll.*)

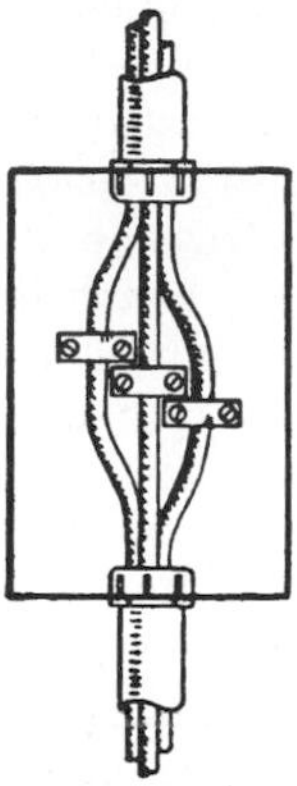

Conductor-support box with single-wire cleats to clamp conductors.

Fig. 300-34. Some type of support must carry the weight of conductors in long risers. (Sec. 300.19.)

300.19. Supporting Conductors in Vertical Raceways. Long vertical runs of conductors should not be supported by the terminal to which they are connected. Supports, as shown in Fig. 300-34, may be used to comply with 300.19(A), as may the other support methods recognized by part **(C)**. (See also Figs. 300-35 and 300-36.) Note that support intervals for fire-rated applications, as covered generically in part **(B)**, may and frequently do differ substantially from the normal requirements for the wiring method. For example, the normal interval for Type MI cable is 1.8 m (6 ft), but if the cable is employed as a component of an electrical circuit protective system the support interval decreases significantly. The exact spacing is specified for each system employed, along with the support method (generally involving steel components, etc.) that will withstand both the heating tests for the time duration associated with the rating, as well as a hose stream test designed to simulate fire suppression activity afterwards.

300.20. Induced Currents in Ferrous Metal Enclosures or Ferrous Metal Raceways. When all conductors of an ac circuit are kept close together—in a raceway or a box or other enclosure—the magnetic fields around the conductors tend to oppose or cancel each other, thereby minimizing the inductive reactance of the circuit and also minimizing the amount of magnetic flux that can cause heat due to hysteresis loss (magnetic friction) in steel or iron and due to the I^2R losses of currents that are induced in adjacent metal. The rule of this section calls for always running a neutral conductor with the phase legs of an ac circuit to minimize such induction heating. The equipment grounding conductor must also be run close to the circuit conductors to achieve the reduction in inductive reactance and minimize the impedance of the fault-current return path when a fault does occur, thereby assuring the fastest possible operation of the protective device (fuse or CB) in the circuit. The reference to

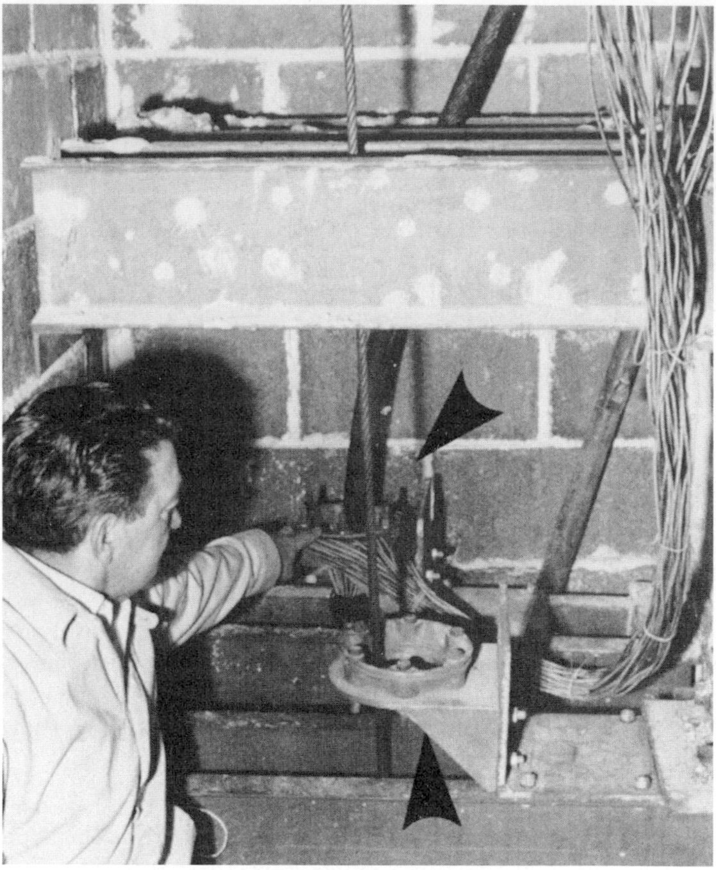

Fig. 300-35. Bore-hole cable, with steel wire armor, is permitted by the exception to be supported only at the top of very high risers because the steel armor supports the length of the cable when the steel wires are properly clamped in the support ring of the type of fitting shown here. (Sec. 300.19.)

"all" grounding conductors in the basic rule is aimed at the so-called isolated grounding conductors, which must also be run with the circuit conductors from the outlet to the panel where that circuit originates. Where the isolated ground continues through that panel, it should be run with the feeder that supplies that panel, although that is not entirely clear (Fig. 300-37).

When an ac circuit is arranged in such a way that the individual conductors are not physically close for mutual cancellation of their field flux, it is particularly important to take precautions where a single conductor passes through a hole in any magnetic material, such as a steel enclosure surface. The presence of the magnetic material forms a closed (circular) magnetic core that raises the flux density of the magnetic field around the conductor (i.e., it greatly strengthens the magnetic field). Under such conditions, there can be substantial heating

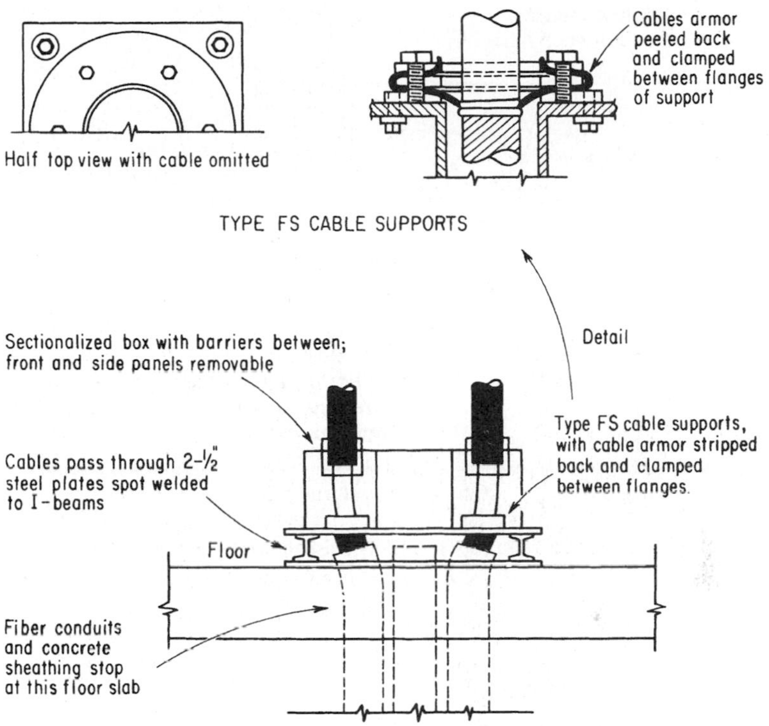
Half top view with cable omitted
Cables armor peeled back and clamped between flanges of support
TYPE FS CABLE SUPPORTS
Detail
Sectionalized box with barriers between; front and side panels removable
Type FS cable supports, with cable armor stripped back and clamped between flanges.
Cables pass through 2-½" steel plates spot welded to I-beams
Floor
Fiber conduits and concrete sheathing stop at this floor slab

Fig. 300-36. Separate strands of cable armor are snubbed between flanges of support fitting at top of run. Partitioned enclosure protects unarmored sections of cable. (Sec. 300.19.)

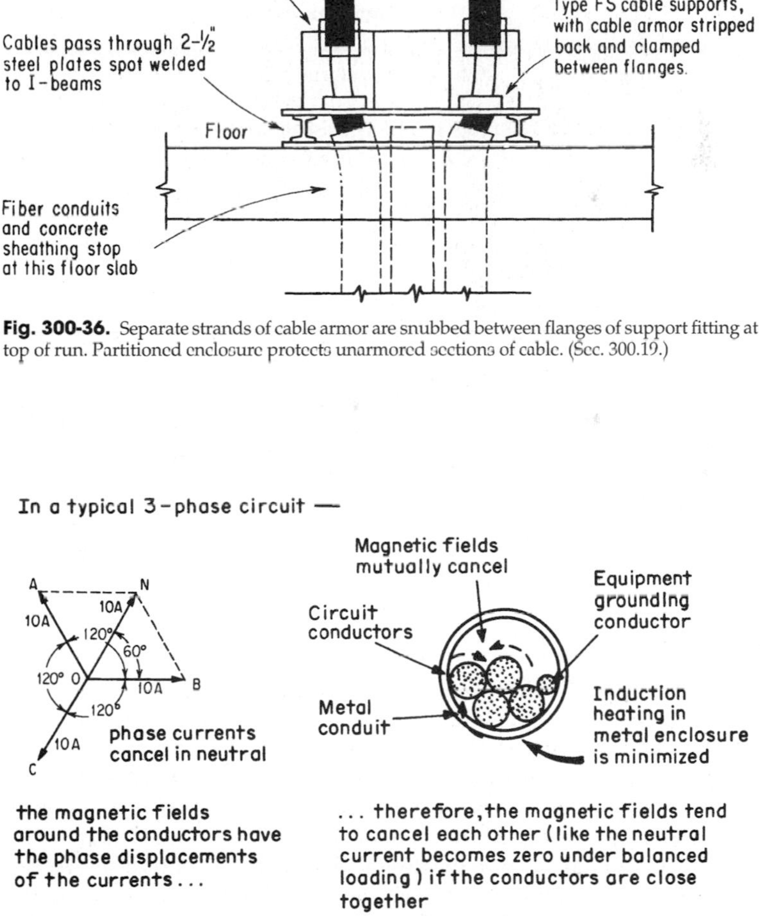
In a typical 3-phase circuit —
A
N
10A
10A
120°
60°
120°
O
10A
B
120°
10A
C
phase currents cancel in neutral
Magnetic fields mutually cancel
Circuit conductors
Equipment grounding conductor
Metal conduit
Induction heating in metal enclosure is minimized
the magnetic fields around the conductors have the phase displacements of the currents . . .
. . . therefore, the magnetic fields tend to cancel each other (like the neutral current becomes zero under balanced loading) if the conductors are close together

Fig. 300-37. Close placement of ac conductors minimizes magnetic fields and induction. (Sec. 300.20.)

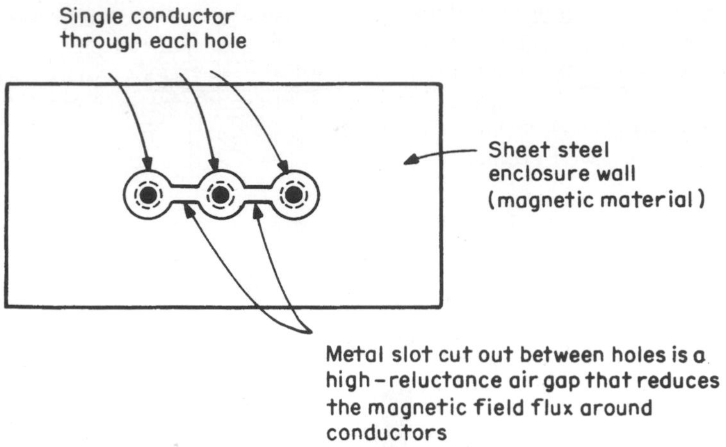
Single conductor through each hole
Sheet steel enclosure wall (magnetic material)
Metal slot cut out between holes is a high-reluctance air gap that reduces the magnetic field flux around conductors

Fig. 300-38. Induction heating is reduced by opening the magnetic core. (Sec. 300.20.)

in the enclosure due to hysteresis (friction produced by the alternating reversals of the magnetic domains in the steel) and due to currents induced in the steel by the strong magnetic field. To minimize those effects, the second paragraph of Sec. 300.20 requires special treatment, such as that shown in Fig. 300-38. Or a rigid, nonmetallic board (fiberglass, plastic, etc.) should be used for the enclosure wall that the conductors pass through.

300.21. Spread of Fire or Products of Combustion. Application of this section to all kinds of building constructions is a very broad and expanding controversy in modern electrical work, in particular because of the phrase *substantially increased.* The rule here requires that electrical installations shall be made to substantially protect the integrity of rated fire walls, fire-resistant or fire-stopped walls, partitions, ceilings, and floors. Electrical installations must be so made that the possible spread of fire through hollow spaces, vertical shafts, and ventilating or air-handling ducts will be reduced to a minimum. These rules require close cooperation with building officials to avoid destruction of fire ratings when electrical installations extend through such areas.

Floor Penetrations

Certainly, poke-through wiring—that technique in which floor outlets in commercial buildings are wired through holes in concrete slab floors—is an acceptable wiring method if use is made of UL-listed poke-through fittings that have been tested and found to preserve the fire rating of the concrete floor. Throughout the country, poke-through wiring continues to be a popular and very effective method of wiring floor outlets in office areas and other commercial and industrial locations. Holes are cut or drilled in concrete floors at the desired locations of floor outlets, and floor box assemblies are installed and wired from the ceiling space of the floor below. The method permits installation of every floor box at the precise location that best serves the layout of desks and other office equipment.

The wiring of each floor outlet at a poke-through location may be done basically in either of two ways—by some job-fabricated assembly of pipe nipples and boxes or by means of a manufactured through-floor assembly (Fig. 300-39) made expressly for the purpose and tested and listed by a nationally recognized testing lab, such as UL.

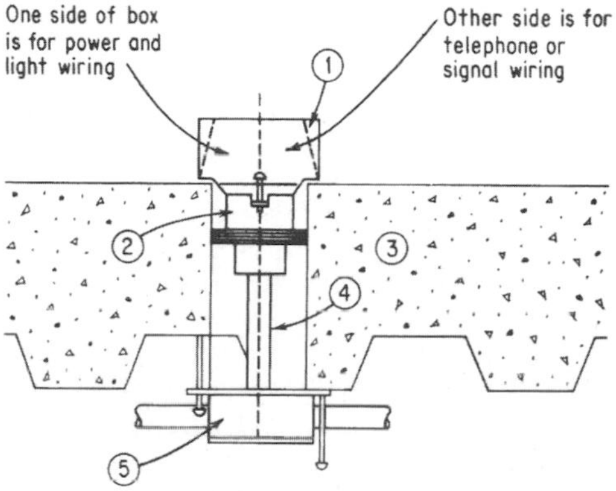
One side of box is for power and light wiring
Other side is for telephone or signal wiring
1
2
3
4
5

Numbered components include (1) combination floor service box, (2) fire-rated center coupling, (3) concrete slab, (4) barriered extension, and (5) barriered junction box.

UL-listed assembly is fire-rated for thickest concrete slab.

Bottom end of UL-listed poke-through fitting consists of a partitioned box for 120-V circuit and telephone circuit run through vertical channels of fitting into dual-service floor outlet box on top of slab.

Fig. 300-39. Several manufacturers make UL-listed poke-through assemblies. (Sec. 300.21.)

May either of the methods be used? A clear regulation of the Occupational Safety and Health Administration (OSHA) appears to rule decisively on this question. OSHA clearly and flatly demands that an installation or equipment determined to be safe by a nationally recognized testing lab must always be used in preference to any equipment not certified by a testing lab. Thus, if a UL-listed poke-through fitting is available, then the use of any nonlisted, homemade assembly that has not been determined to be safe appears to be clearly not acceptable to OSHA.

300.21 also applies to cable and/or conduit penetrations of fire-rated walls, floors, or ceilings without altering the fire rating of the structural surface (Figs. 300-40 and 300-41).

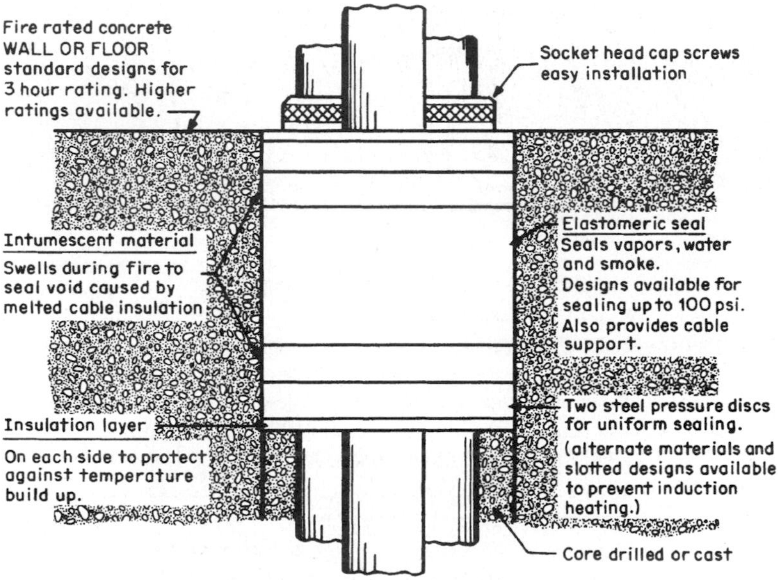
Fire rated concrete
WALL OR FLOOR
standard designs for
3 hour rating. Higher
ratings available.
Socket head cap screws
easy installation
Intumescent material
Swells during fire to
seal void caused by
melted cable insulation
Elastomeric seal
Seals vapors, water
and smoke.
Designs available for
sealing up to 100 psi.
Also provides cable
support.
Insulation layer
On each side to protect
against temperature
build up.
Two steel pressure discs
for uniform sealing.
(alternate materials and
slotted designs available
to prevent induction
heating.)
Core drilled or cast

Fig. 300-40. Fire-stop fitting for passing cables or conduit through a fire-rated wall, floor slab, or similar concrete surface, without altering the fire rating of the surface. (Sec. 300.21.)

Ceiling Penetrations

Another similar concern covered by this section is the installation of lay-in lighting fixtures in a fire-rated suspended ceiling. Suspended ceilings are usually evaluated only for their esthetic and acoustic value, but they also serve as fire-protective membranes for the floor above. Although concrete floor structures have various fire-resistance ratings by themselves (depending on the concrete thickness and aggregate used), some assemblies require some type of protective cover. When this is the case, the ceiling is tested in combination with the floor-slab structure for which the rating is desired. Such a ceiling properly serves its function of fire protection until an installer cuts holes in it, such as for recessed lighting fixtures

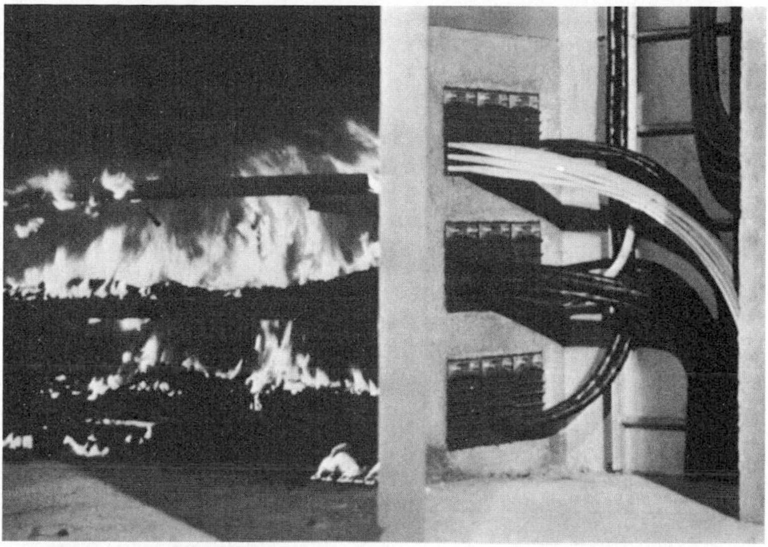

Fig. 300-41. Another type of device to provide for passing cable and/or conduit through fire-rated building surfaces without altering the conditions of fire resistance. (Sec. 300.21.)

or for air diffusers or grilles. Because of that, the acceptability of the overall ceiling system must be carefully determined. Refer to the discussion at 300.11(A)(1) for more information on this part of the process.

The informational note following the basic rule provides some guidance, but the following presents a comprehensive step-by-step procedure for using the publications vaguely referenced in the note. First, check the Underwriters Laboratories' Guide Card information [see discussion at 110.3(B) for details], which notes that recessed fixtures that have been shown to provide a degree of fire resistance with the floor, roof, or ceiling assemblies with which they have been tested are labeled as follows: "Recessed-type electric fixture classified for fire resistance; fire-resistance classification floor and ceiling Design No. ——." Next, find the design referred to in the UL *Fire Resistance Index.* This booklet follows the format to that used with UL Guide Card information. Refer to a design of the required fire rating and be sure that the fixtures are listed for use with that design.

Designers must specify the particular UL design that suits their requirement, note this in the specifications, and be certain that the lighting fixtures are fire-rated in accordance. But it is advisable for the electrical contractor to investigate the ceiling design for possible fire rating in all cases and to receive from the designer written confirmation of the exact nature and value of the rating, if one exists.

In the UL *Building Materials Directory,* various fire-rated assemblies are listed by design number and by rating time. A companion publication, the *Fire Resistance Index,* contains detailed cross-sectional drawings of the assemblies, with all critical dimensions shown. Each pertinent element is usually flagged with an identifying number. Keyed to the number are clarifying statements listing additional critical limitations (such as the size and number of penetrations in the ceiling).

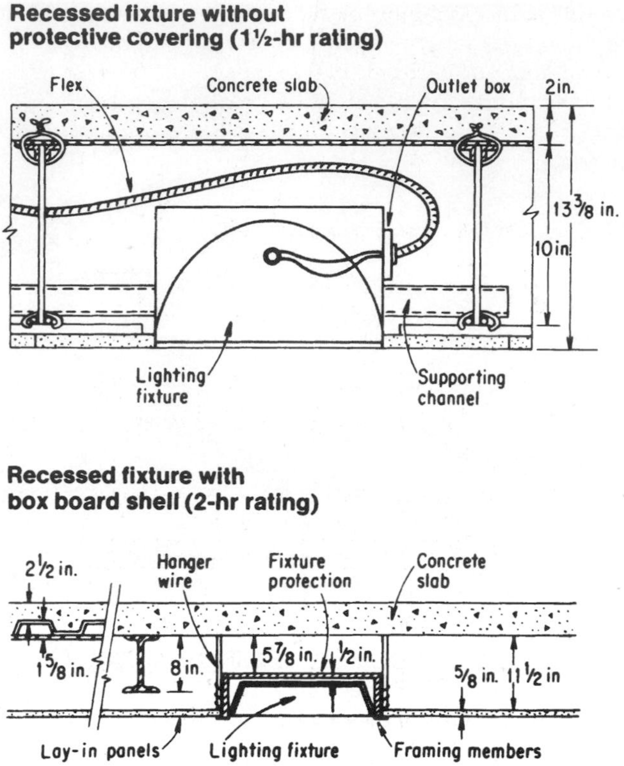
Recessed fixture without
protective covering (1½-hr rating)
Flex
Concrete slab
Outlet box
2in.
13⅜ in.
10in.
Lighting
fixture
Supporting
channel
Recessed fixture with
box board shell (2-hr rating)
2½ in.
Hanger
wire
Fixture
protection
Concrete
slab
1⅝ in.
8 in.
5⅞ in.
½ in.
⅝ in.
11½ in
Lay-in panels
Lighting fixture
Framing members

Fig. 300-42. The complete assembly of concrete slab plus fixture and ceiling gets a fire rating. (Sec. 300.21.)

The top installation in Fig. 300-42 was tested and given a 1½-h rating. No protective material was used between the fixture and the floor slab above. A somewhat better rating could have been obtained had protection been provided over the fixture.

At the bottom of Fig. 300-42 is a fixture with protection. When this construction was tested, failure occurred after 2 h 48 min, and it received a 2-h time rating. Even with this type of protection, the UL listing will limit the area occupied by fixtures to 25 percent of the total ceiling area. (But a coffered ceiling may contain 100 percent lighted vaulted modules.)

Other Penetrations

Plasterboard (gypsum board) panels used so commonly for interior wall construction in modern buildings are fire-rated. UL and other labs make tests and assign fire ratings (in hours) to wall assemblies or constructions that make use of plasterboard. For instance, a wall made up of wood or metal studs with a single course of $^5/_8$-in. (15.9-mm) plasterboard on each side of the studs would be assigned a 1-h fire rating. A wall with two courses of ½- or $^5/_8$-in. (12.7- or 15.9-mm) plasterboard on each side of the studs would be a 2-h wall (Fig. 300-43). The assigned fire ratings are based on the thickness and number of courses of plasterboard. And the fire rating is for the wall assembly *without any penetrations into the wall.*

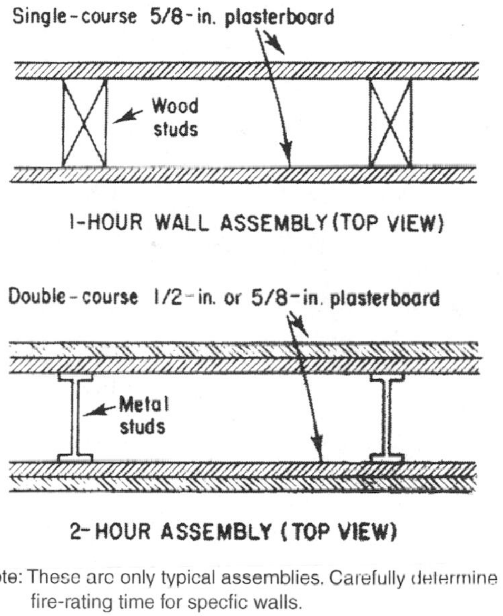
Single-course 5/8-in. plasterboard
Wood studs
1-HOUR WALL ASSEMBLY (TOP VIEW)
Double-course 1/2-in. or 5/8-in. plasterboard
Metal studs
2-HOUR ASSEMBLY (TOP VIEW)
Note: These are only typical assemblies. Carefully determine fire-rating time for specfic walls.

Fig. 300-43. Wall assemblies using plasterboard are fire-rated by UL and others. (Sec. 300.21.)

Because of the fire rating assigned to the assembly, any wall so constructed is fire-rated. The wall may be between rooms or between a room and a corridor or stairwell. No distinction is made between an interior wall of an apartment and a wall that separates one apartment from another. All wall assemblies using plasterboard are fire-rated and immediately raise concern over violation of 300.21 if any electrical equipment is recessed in the wall. However, this is one area where an intelligent use of the phrase "substantially increased" may be useful. If the building code considers both sides of a given wall, such as the one between a dwelling unit bedroom and the hallway leading to it, as being in the same smoke compartment and not deserving of any particular fire rating by code, then it is difficult to imagine how using fire-rated electrical penetrations accomplishes any improvement in safety.

Building inspectors and electrical inspectors have generally permitted installation of wall switches, thermostats, dimmers, and receptacles in boxes recessed in plasterboard walls. In single-family houses, the entire interior is not considered to be compartmented. It is assumed that individual rooms or areas are not normally closed off from each other and that fire or smoke spread would not be affected at all by those penetrations. The consensus has been that such small openings cut in the plasterboard do not violate the letter or intent of 300.21.

In apartment houses, office buildings, and other multioccupancy buildings, however, inspectors could logically question use of wiring devices installed in common walls between apartments or between an apartment and a corridor or stairwell. Such walls are assumed to be between interior spaces that are normally closed off from each other by the main doors to the individual apartments. Fire and/or smoke spread, which are normally restricted by the closed doors, might be considered substantially increased by any penetrations of those fire-rated walls (Fig. 300-44). Although switch and receptacle boxes are usually accepted, use of a panelboard in a common wall has been rejected.

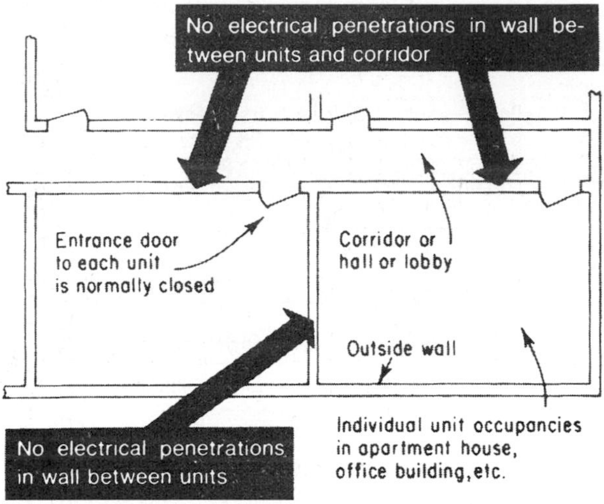
No electrical penetrations in wall between units and corridor
Entrance door to each unit is normally closed
Corridor or hall or lobby
Outside wall
No electrical penetrations in wall between units
Individual unit occupancies in apartment house, office building, etc.

Fig. 300-44. Walls separating closed-off spaces must have maintained fire rating. (Sec. 300.21.)

Another issue frequently arises in mirror image construction in apartment houses, where one kitchen will back up to the kitchen in the next unit to save on plumbing costs. The default separation rule on boxes is 600 mm (24 in.) of horizontal separation, and that dimension applies even if framing members intervene. With boxes required every few feet in kitchen counters, these rules need careful attention. There are now listed nonmetallic boxes that do not require the full horizontal separation. In addition, there are now listed "putty pads" that can be used to wrap around metal boxes that also dramatically reduce the required separation.

For larger electrical equipment, such as panelboards, the same general analysis would apply. For interior walls of private houses or individual unit occupancies in apartment houses, hotels, dormitories, office buildings, and the like, a panel installed in a wall between two rooms or spaces that are normally not closed off from each other cannot "substantially" contribute to greater fire and/or smoke spread. However, panelboards and similar large equipment should normally not be installed in fire-rated walls between spaces that are closed off from each other by doors that are normally closed.

When it is necessary to install a panelboard or other large equipment in a wall between areas that are normally closed off from each other, a boxed recess in the wall should be constructed of the fire-rated plasterboard to maintain the fire rating of the wall (Fig. 300-45). This is also a common practice for installing recessed enclosures for fire extinguishers in corridor walls and medicine cabinets mounted in walls between apartment units.

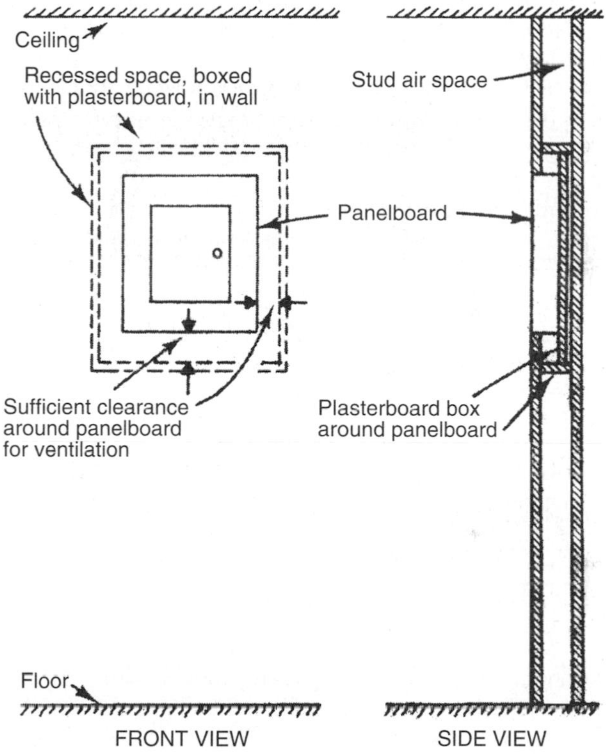
Ceiling
Recessed space, boxed with plasterboard, in wall
Stud air space
Panelboard
Panelboard
Sufficient clearance around panelboard for ventilation
Plasterboard box around panelboard
Floor
FRONT VIEW
SIDE VIEW

Fig. 300-45. Boxing of large-area penetrations has been required in fire walls. (Sec. 300.21.)

Another technique that has been used to maintain fire rating where a panelboard is installed in a wall between individual apartments is to glue pieces of plasterboard to the top, bottom, sides, and back of the recessed panel. In one particular job, this was done as a corrective measure where panelboards had first been installed in such walls without attention to maintaining the fire rating of the wall. But the use of plasterboard directly affixed to the panelboard surfaces could be considered an unauthorized modification of the panel that voids UL listing because of improper application.

There are now fire protection materials that are designed to be applied to panelboards that allow the preservation of up to a 2-hour fire separation, and they have the listings to support those claims. Typically they come with specifications as to the numbers and sizes of steel raceways that can enter which walls of the enclosure. If it is necessary to vary from the anticipated limits, the manufacturer will accept requests for variances and will quickly provide engineered analyses to back up such variances if appropriate. The author recently used this process to retrofit a panel into a 2-hour wall.

300.22. Wiring in Ducts Not Used for Air Handling, Fabricated Ducts for Environmental Air, and Other Spaces for Environmental Air (Plenums). Part **(A)** of this section applies only to wiring in the types of ducts described, which can contain no wiring systems of any type. Part **(B)** covers use of wiring methods and equipment within "ducts or plenums," which are channels or chambers intended, specifically fabricated, and used only for supply or return of conditioned air. Such ducts or plenums are sheet metal or other types of enclosures that are provided expressly for air handling and must be distinguished from "Other Space Used for Environmental Air"—such as the space between a suspended ceiling and the floor slab above it. Space of that type is covered by part **(C)** of this section. The space between a raised floor (Fig. 300-46) and the slab below the raised floor is also covered by part **(C)**, unless the air-handling raised floor is within a computer room. If the raised floor is in a computer room, as defined in Art. 645, part **(D)** states that such an air-handling raised floor used for data-processing circuits must comply with Art. 645.

Part **(B)** addresses actual duct systems with air movement and excludes all wiring with any nonmetallic coverings and with pervious enclosures. Thus, Type MC

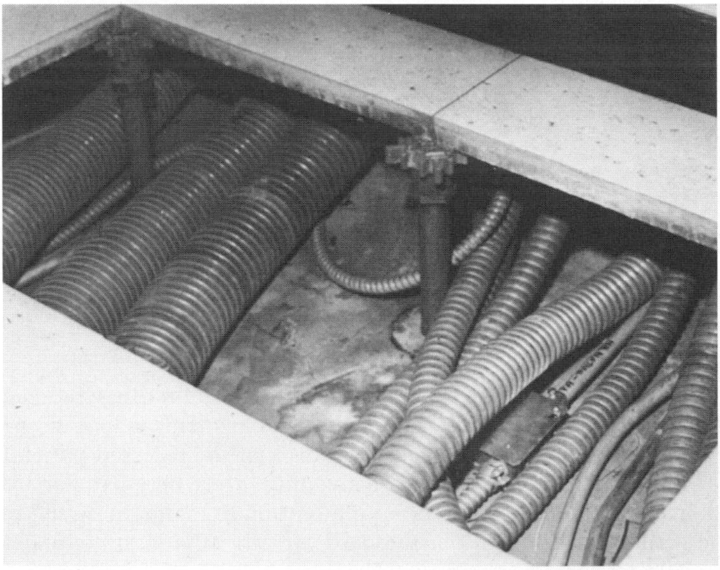

Fig. 300-46. Space under a raised floor, which is commonly used for circuits to data-processing equipment and provides for passage of conditioned air to the room and to the equipment is covered in part **(D)**. (Sec. 300.22.)

cable is acceptable, but only in its corrugated or smooth forms, and never with a nonmetallic jacket. Type MI cable is acceptable as well. Flexible metallic tubing, rigid and intermediate metal conduit, and electrical metallic tubing are permitted for the same reasons. Greenfield is permitted, but only where absolutely needed to repositionable equipment required to be there and never in lengths over 1.2 m (4 ft). A new 2017 **NEC** exception based on correlation with the NFPA 90A standard effectively permits comparable lengths (up to 4 ft) of other wiring suitable for Part (C) (plenum) applications to be used under similar conditions as the principal rule here (direct action on the air stream). Enclosed and gasketed luminaires are also permitted in association with other necessary equipment.

Part **(C)** covers other spaces used for environmental air movement, particularly plenum cavity ceiling spaces, where the space above a hung ceiling point is used as a central collection point for return air to a central air handler. These spaces are now formally considered to be, in accordance with the subsection title, actual plenums. In effect, now the word "plenums" freely substitutes for the former phrasing "other air handling spaces." Prior to the 2011 **NEC** a plenum was specifically fabricated for air transport; its use now stands that concept on its head, and plenum cavity ceilings are now plenums. This new usage is consistent with mechanical codes and also the way the **NFPA** air conditioning and ventilation standard uses the word. Refer also to the commentary in this book on the Art. 100 definition.

These spaces present a less intensive, but comparable concern as actual ductwork carrying the same air, and the **NEC** rules are also comparable, but slightly less intense as well. There is an important exception at this point, generally used for Type NM cable, that excludes joist or stud cavities used for return air movement, as shown in Fig. 300-47. This exception, however, has two very specific limitations. First, it only applies to dwelling units. Second, it only

Fig. 300-47. A joist space through which Type NM cable passes "perpendicular to the long dimension" of the space may be closed in to form a duct-like space for the cold-air return of a hot-air heating system—but only in a "dwelling unit."

applies if the wiring involved passes straight across the short dimension. That means no turns, no device boxes, etc.; just straight across.

This part permits use of totally enclosed, nonventilated, insulated busway in an air-handling ceiling space, provided it is a non–plug-in-type busway that cannot accommodate plug-in switches or breakers. This one specific busway wiring method was added for hung-ceiling space used for environmental air. The methods in part B can be used, except that interlocking-armor Type MC cable can be used as well (but still no nonmetallic covering is allowed). Surface metal raceway or wireway with metal covers or solid-bottom metal cable tray with solid metal covers may be used in air-handling ceiling space, provided that the raceway is accessible, such as above lift-out panels. Other factory-assembled cable assemblies (without nonmetallic jackets) for use with Art. 604 manufactured wiring systems are allowable as well. (See Fig. 300-48.) The 2014 NEC has

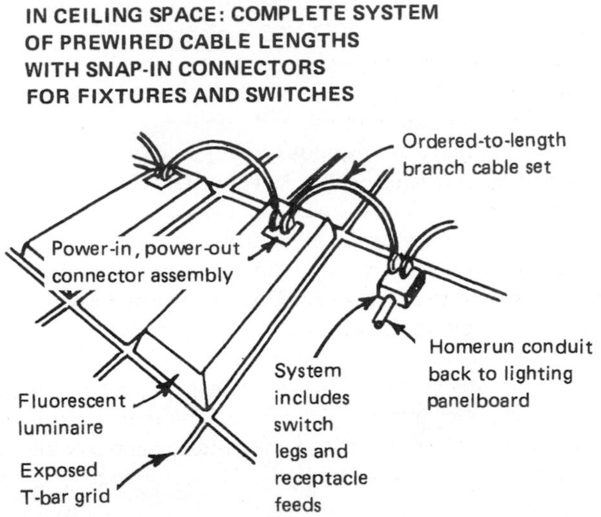
IN CEILING SPACE: COMPLETE SYSTEM OF PREWIRED CABLE LENGTHS WITH SNAP-IN CONNECTORS FOR FIXTURES AND SWITCHES
Ordered-to-length branch cable set
Power-in, power-out connector assembly
Homerun conduit back to lighting panelboard
System includes switch legs and receptacle feeds
Fluorescent luminaire
Exposed T-bar grid

Fig. 300-48. Modular wiring systems, as recognized by Art. 604, are permitted to be used in air-handling ceiling spaces.

added a further limitation at this point, disallowing any nonmetallic cable ties or "nonmetallic cable accessories" unless they are listed as having "low smoke and heat release properties." This requirement also correlates with work done in NFPA 90A. The air-handling space under a raised floor in a data-processing location is covered by Art. 645 on information technology equipment. Figure 300-49 shows wiring methods for use in an air-handling ceiling space. Note that acceptable cable tray applications now appear in their own numbered paragraph (2). Cable trays must be constructed of metal, and if they contain wiring methods other than those normally permitted in these spaces, they must be the functional equivalent of metal wireways, with solid metal bottoms, sides, and covers.

The Code panel has made it clear that they generally oppose nonmetallic wiring methods in ducts and plenums and in air-handling ceilings, except for nonmetallic

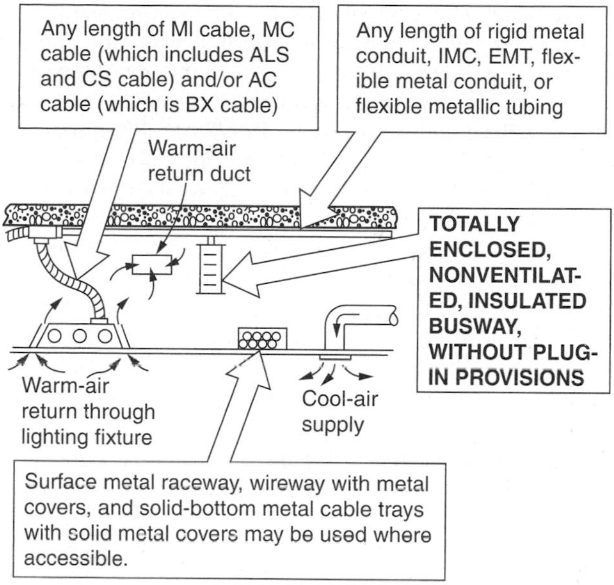
Any length of MI cable, MC cable (which includes ALS and CS cable) and/or AC cable (which is BX cable)
Any length of rigid metal conduit, IMC, EMT, flexible metal conduit, or flexible metallic tubing
Warm-air return duct
TOTALLY ENCLOSED, NONVENTILATED, INSULATED BUSWAY, WITHOUT PLUG-IN PROVISIONS
Warm-air return through lighting fixture
Cool-air supply
Surface metal raceway, wireway with metal covers, and solid-bottom metal cable trays with solid metal covers may be used where accessible.

Fig. 300-49. Any wiring method other than these is a violation in air-handling space. (Sec. 300.22.)

cable assemblies that are specifically listed for such use. It is also the intent of the Code that cables with an outer overall nonmetallic jacket should not be permitted in ducts or plenums. Although the jacket material—usually PVC—would not propagate a fire, it would contribute to the smoke and provide additional flammable material in the air duct. The last sentence in the opening language of part **(C)** is intended to exclude from the requirements those areas that may be occupied by people. Hallways and habitable rooms are being used today as portions of air-return systems, and while they have air of a heating or cooling system passing through them, the prime purpose of these spaces is obviously not air handling.

Part **(C)(2)** splits off cable tray applications from the general wiring method coverage in **(C)(1)**. Metal cable trays can always be used provided they only support cabled wiring methods that are allowable in such spaces anyway, such as Type MC cables without plastic coverings. In addition, other wiring methods are permitted in cable tray types that are the functional equivalent of a metal wireway, with solid bottoms, sides, and covers.

Part **(C)(3)** permits use of nonmetallic equipment enclosures and wiring that are specifically UL-listed or classified for use in air-handling ceiling spaces. The basic condition that must be satisfied is that the wiring materials and other construction of the equipment must be suitable for the expected ambient temperature to which they will be subjected, and constructed of materials tested in terms of minimal degrees of both heat release and smoke production.

In effect, the rules in parts **(B)** and **(C)** exclude from use in all air-handling spaces any wiring that is not metal-jacketed or metal-enclosed, to minimize the creation of toxic fumes due to burning plastic under fire conditions. 800.113(C), and similar rules in Art. 810, 820, and 830 covering radio, cable TV, and power broadband networks, basically require telephone, intercom, and other communications circuits to be wired with Type CMP cable or other identified types installed in compliance with 300.22 when such circuits are used in ducts or plenums or air-handling ceilings. Wiring in air-handling space under raised floors in computer centers must use the wiring methods described in 645.5(E). Ventilation in the raised-floor space must be used only for the data-processing area and the data-processing equipment.

Application of that Code permission on use of equipment calls for substantial interpretation. The designer and/or installer must check carefully with equipment manufacturers and with inspection agencies to determine what is acceptable in the air-handling space above a suspended ceiling. Practice in the field varies widely on this rule, and Code interpretation has proved difficult.

The exception to part **(C)(3)** recognizes the installation of integral fan systems where such equipment has been specifically approved for the purpose. Equipment of this type is UL-listed and may be found in the *Guide Card information* under the heading "Heating and Ventilating Equipment."

300.34. Conductor Bending Radius. This section covers the acceptable final bending radius of medium-voltage conductors in enclosures. This section is important because it is referenced in other locations in the NEC. For example, if a medium-voltage cable enters the back of a pull box opposite to a removable cover, this section sets the minimum depth allowed for the pull box. The dimensions are based on the diameters of the cables being installed. Great care must be used at medium-voltage cable terminations in enclosures, which may only be wide enough for straight-in terminations. Many failures have occurred after cables were bent into loops prior to landing in enclosures not wide enough to allow such loops to be formed without overstressing the shielding.

300.35. Protection Against Inductive Heating. This rule echoes 300.20, but it is worth noting that the higher power and voltages involved in these applications have produced dramatic failures in comparatively short order, as in within days of becoming energized.

300.37. Aboveground Wiring Methods. Conductors aboveground must be in rigid metal conduit, in intermediate metal conduit (IMC), in electrical metallic tubing, in auxiliary gutters, in cable trays, in a cable bus, in busways, in other identified raceways, or as open runs of metal-clad cable suitable for the use and purpose (Fig. 300-50). Rigid nonmetallic conduit has been added to the list of raceways acceptable for running medium-voltage circuits aboveground—and the PVC (and RTRC) conduit does not have to be encased in concrete. The phrase "other identified raceways" includes wireways; auxiliary gutters are specifically added because they are not raceways but should be permissible in this context. Refer to the discussion in Art. 100 on the raceway definition on this point.

In locations accessible to qualified persons only, open runs of Type MV cable, bare conductors, and bare busbars may be used. The 2017 NEC additionally recognized medium-voltage airfield lighting series circuits for exposed applications in restricted access lighting vaults.

300.38. Raceways in Wet Locations Above Grade. This was a new section as of the 2014 NEC and it carries the existing rule in 300.9 for work below 1000 V to Part II for higher voltages.

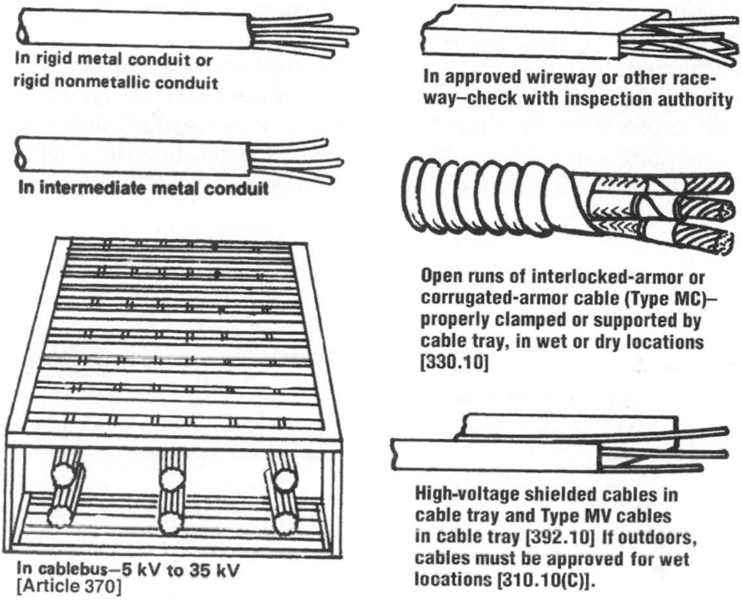
In rigid metal conduit or
rigid nonmetallic conduit
In approved wireway or other raceway–check with inspection authority
In intermediate metal conduit
Open runs of interlocked-armor or
corrugated-armor cable (Type MC)–
properly clamped or supported by
cable tray, in wet or dry locations
[330.10]
High-voltage shielded cables in
cable tray and Type MV cables
in cable tray [392.10] If outdoors,
cables must be approved for wet
locations [310.10(C)].
In cablebus–5 kV to 35 kV
[Article 370]

Fig. 300-50. A variety of wiring methods may be used for aboveground high-voltage circuits. (Sec. 300.37.)

300.40. Insulation Shielding. One of the basic decisions to make in selecting medium-voltage conductors is whether or not electrostatic insulation shielding is required on the cable. The basic requirements on electrostatic shielding of medium-voltage conductors are presented later in this chapter in the coverage of 310.10(E).

This section sets forth very general rules on terminating shielded medium-voltage conductors. The metallic shielding or any other conducting or semiconducting static shielding components on shielded cable must be stripped back to a safe distance according to the circuit voltage, at all terminations of the shielding. At such points, stress reduction must be provided by such methods as the use of potheads, terminators, stress cones, or similar devices. The wording of this regulation makes it clear that the need for shield termination using stress cones or similar terminating devices applies to semiconducting insulation shielding as well as to metallic-wire insulation shielding systems.

A stress cone is a field-installed device or a field-assembled buildup of insulating tape and shielding braid that must be made at a terminal of medium-voltage shielded cable, whether a pothead is used or not. A stress-relief cone is required to relieve the electrical stress concentration in cable insulation directly under the end of cable shielding. Some cable constructions contain stress-control components that afford the cable sufficient stress relief without the need for stress-relief cones. If a cable contains inherent stress-relief components in its construction, that would satisfy 300.40 as doing the work of a stress cone. As a result, separate stress cones would not have to be installed at the ends of such cable. Or, heat-shrinkable tubing terminations may be used with stress-control material that provides the needed relief of electrostatic stress.

At a cable terminal, the shielding must be cut back some distance from the end of the conductor to prevent any arcing over from the hot conductor to the grounded shield. When the shield is cut back, a stress is produced in the insulation. Providing a flare-out of the shield, that is, extending the shield a short distance in the shape of a cone, relieves the stress, as shown in Fig. 300-51.

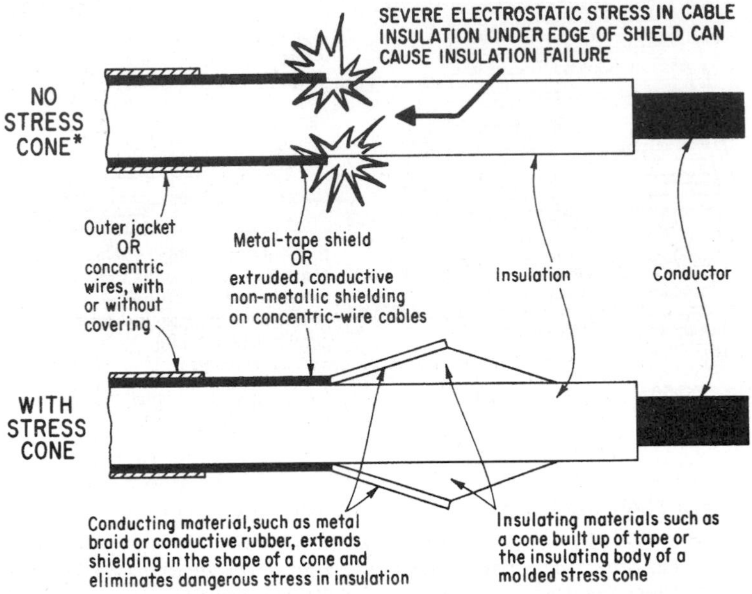
SEVERE ELECTROSTATIC STRESS IN CABLE INSULATION UNDER EDGE OF SHIELD CAN CAUSE INSULATION FAILURE
NO STRESS CONE*
Outer jacket OR concentric wires, with or without covering
Metal-tape shield OR extruded, conductive non-metallic shielding on concentric-wire cables
Insulation
Conductor
WITH STRESS CONE
Conducting material, such as metal braid or conductive rubber, extends shielding in the shape of a cone and eliminates dangerous stress in insulation
Insulating materials such as a cone built up of tape or the insulating body of a molded stress cone

Fig. 300-51. Here is how a stress cone protects insulation at cable ends. (Sec. 300.40.)

Stress cones provide that protection against insulation failure at the terminals of shielded medium-voltage cables. Manufacturers provide special preformed stress cones (Fig. 300-52) and kits for preparing cable terminals with stress cones for cables operating at specified levels of medium voltage (Fig. 300-53). A wide assortment of stress-relief terminators are made for all the medium-voltage cable assemblies used today.

Metallic shielding tape must be grounded, as required by 300.40 and 300.5(B), which refer to "metallic sheathing," as in Fig. 300-54. The shield on shielded cables must be grounded at one end at least. It is better to ground the shield at two or more points. Grounding of the shield at all terminals and splices will keep the entire length of the shield at about ground potential for the safest most effective operation of the cable. Cable with improperly or ineffectively grounded shielding can present more hazards than unshielded cable.

In addition to the requirements at terminations that apply to shielding, there is a related topic, that of the actual connector at the end, such as a lug. It should be designed to be compatible with medium-voltage applications. Its external

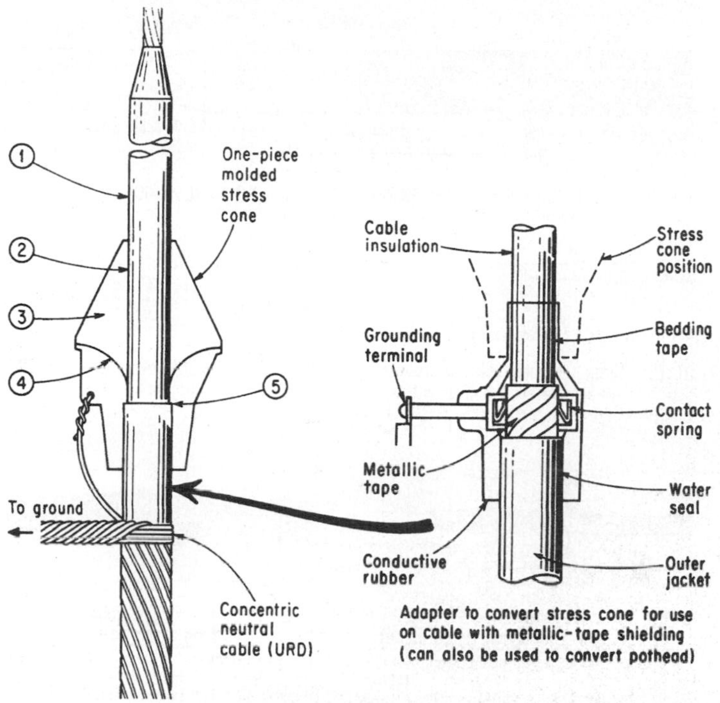
One-piece
molded
stress
cone
1
2
3
4
5
To ground
Concentric
neutral
cable (URD)
Cable
insulation
Stress
cone
position
Bedding
tape
Grounding
terminal
Contact
spring
Metallic
tape
Water
seal
Conductive
rubber
Outer
jacket
Adapter to convert stress cone for use
on cable with metallic-tape shielding
(can also be used to convert pothead)

APPLICATION:
Cable shield is cut back about 12 in. Then, using silicone lubricant, the cable insulation surface and the inner bore of the stress cone are lubricated. The stress cone is simply pushed down over the cable end until it bottoms on the cable shield. After cable is prepared, termination takes about 30 seconds.

REFER TO DIAGRAM:
1. Cable insulation with shielding cut back
2. Tight fit between insulation and bore of stress cone
3. Insulating rubber
4. Stress relief provided by conductive rubber flaring away from insulation along bond between insulating rubber and conductive rubber
5. Conductive rubber of cone tightly fit to conductive cable shield

Fig. 300-52. Typical preformed stress cone is readily applied on cables up to 35 kV indoors. (Sec. 300.40.)

surfaces should be well rounded to minimize the production of corona discharge. These lugs are available as listed items.

300.42. Moisture or Mechanical Protection for Metal-Sheathed Cables. A pothead is one specific form of stress-reduction means referred to in 300.40 and has long been a common means of protecting insulation against moisture or mechanical injury where conductors emerge from a metal sheath (Fig. 300-55).

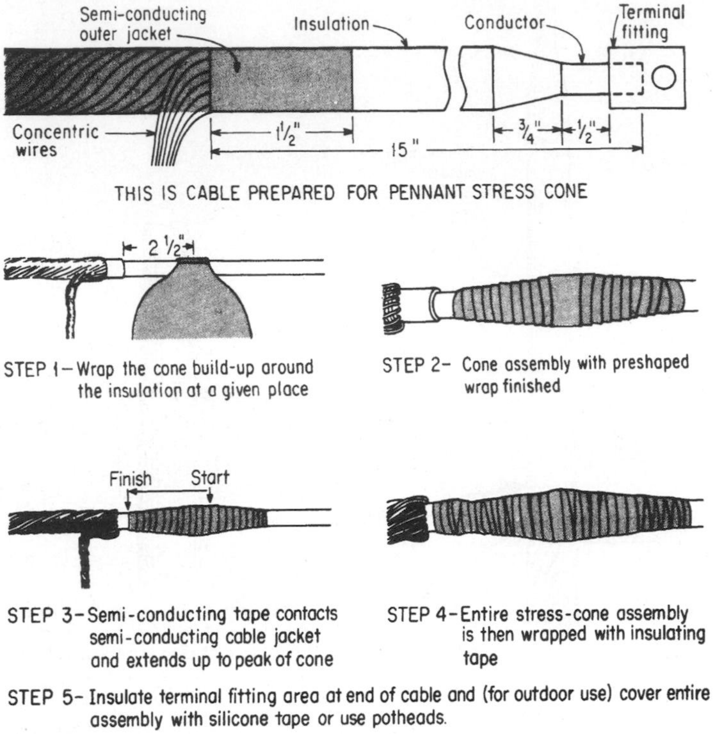
Semi-conducting outer jacket
Insulation
Conductor
Terminal fitting
Concentric wires
1½"
15"
¾"
½"
THIS IS CABLE PREPARED FOR PENNANT STRESS CONE
2½"
STEP 1 – Wrap the cone build-up around the insulation at a given place
STEP 2 – Cone assembly with preshaped wrap finished
Finish
Start
STEP 3 – Semi-conducting tape contacts semi-conducting cable jacket and extends up to peak of cone
STEP 4 – Entire stress-cone assembly is then wrapped with insulating tape
STEP 5 – Insulate terminal fitting area at end of cable and (for outdoor use) cover entire assembly with silicone tape or use potheads.

Fig. 300-53. "Pennant" method is one of a variety of job-site termination buildups. (Sec. 300.40.)

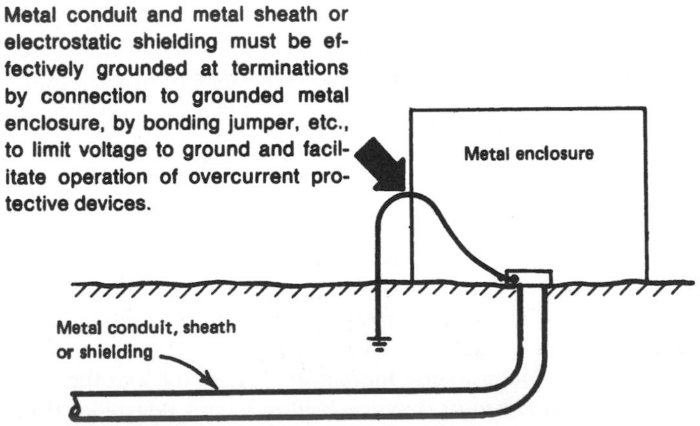
Metal conduit and metal sheath or electrostatic shielding must be effectively grounded at terminations by connection to grounded metal enclosure, by bonding jumper, etc., to limit voltage to ground and facilitate operation of overcurrent protective devices.
Metal enclosure
Metal conduit, sheath or shielding

Fig. 300-54. Metallic shielding must be grounded for all high-voltage conductors—under- or aboveground. (Sec. 300.40.)

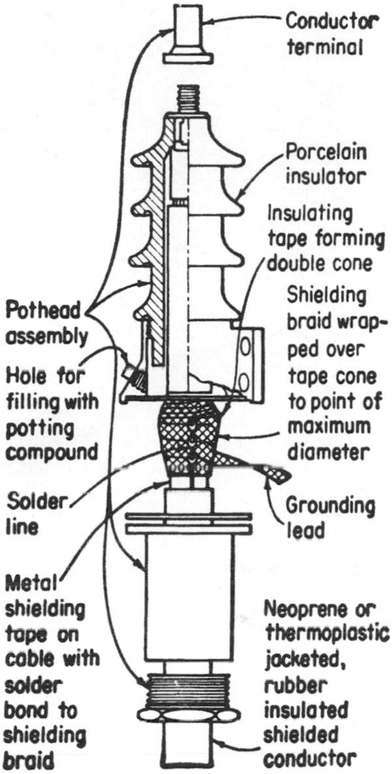
Conductor terminal
Porcelain insulator
Insulating tape forming double cone
Shielding braid wrapped over tape cone to point of maximum diameter
Grounding lead
Pothead assembly
Hole for filling with potting compound
Solder line
Metal shielding tape on cable with solder bond to shielding braid
Neoprene or thermoplastic jacketed, rubber insulated shielded conductor

Fig. 300-55. Typical single-conductor pothead protects metallic- or nonmetallic-jacketed cable. (Sec. 300.42.)

Such protection for metal-sheathed cables (such as lead-covered, paper-insulated cables) is required by this section. A *pothead* is a cable terminal that provides sealing to the sheath of the cable for making a moisture-proof connection between the wires within the cable and those outside.

When metal-jacketed medium-voltage cables are terminated outdoors exposed to the weather, a pothead is commonly used to protect the insulation of conductors against moisture or mechanical injury where conductors emerge from a metal sheath, as shown in Fig. 300-56.

On use of potheads:

1. Paper-insulated cables must be terminated in potheads. This requirement also extends to such cables operated at under 600 V.
2. Varnished-cambric-insulated cables should be terminated in potheads but may be terminated with taped connections in dry locations.
3. Rubber-insulated cables are commonly terminated in potheads in locations where moisture protection is critical but may be terminated without potheads in accordance with manufacturer's instructions.
4. Although many modern medium-voltage cables can be terminated without potheads, many engineers consider potheads the best terminations for medium-voltage cable.

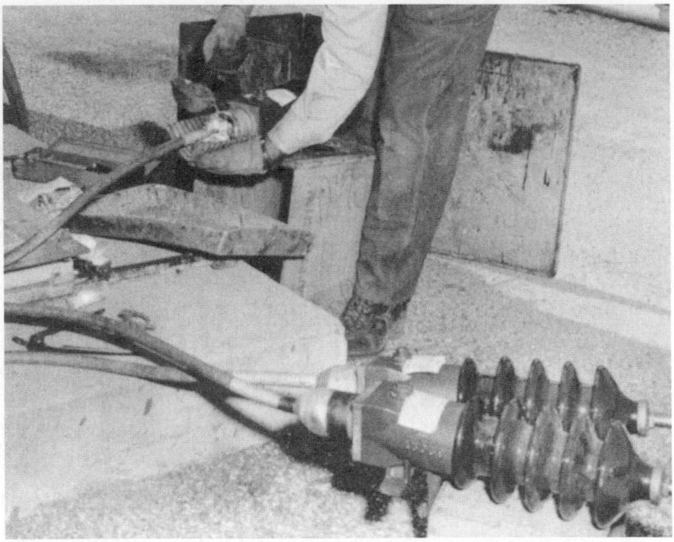

Fig. 300-56. A pothead is used on the end of each paper-insulated, lead-covered cable to protect against entry of moisture, with a wiped lead joint at the terminal. (Sec. 300.42.)

5. The use of potheads offers a number of advantages:
   a. Seals cable ends against moisture that would damage the insulation
   b. Provides a compartment for surrounding the termination with insulating compound to increase the strength of electrical insulation
   c. Seals cable ends against loss of insulating oils
   d. Provides engineered support of connections

In the past, medium-voltage circuits used for commercial and industrial feeders, both outdoors and indoors, were covered in Art. 710. The 1999 Code eliminated Art. 710 and dispersed the rules formerly given in that article throughout the Code. Rules on wiring method are now contained in Part II of Art. 300. Although voltages up to 15,000 V (15 kV) are most commonly used for these circuits today, higher voltages (26 and 35 kV) are often used because they can offer economy for extremely large installations. Typical circuits today operate at 4160/2400 and 13,800/8000 V—both 3-phase, 4-wire wye hookups.

Modern medium-voltage circuits for buildings include overhead bare or covered conductors installed with space between the conductors, which are supported by insulators at the top of wood poles or metal tower structures, overhead aerial cable assemblies of insulated conductors entwined together, supported on building walls or on poles or metal structures; insulated conductors installed in metal or nonmetallic conduits or ducts run underground, either directly buried in the earth or encased in a concrete envelope under the ground insulated conductors in conduit run within buildings; and multiple-conductor cable assemblies (such as nonmetallic jacketed cables, lead-sheathed cable, or interlocked armor cable) installed in conduit or on cable racks or trays or othe types of supports. Another wiring method gaining wide acceptance consists of plastic conduit containing factory-installed conductors, affording a readily used direct earth burial cable assembly for underground circuits but still permitting removal of the cable for repair or replacement.

300.45. Warning Signs. This is a general duty to warn relative to elevated voltages at points of medium-voltage conductor access within conduit and cable wiring systems. Although 314.72(E) for boxes and 490.53 for mobile equipment and 392.18(H) for cable trays have this requirement, there are many other comparable locations such as wireways that do not. This rule was revised in the 2014 NEC and relocated from Art. 225.

300.50. Underground Installations. Directly buried nonmetallic conduit-carrying medium-voltage conductors do not have to be concrete-encased they are is a type approved for use without concrete encasement. If concrete encasement is required, it will be indicated on the UL label and in the listing. 300.50 permits direct burial rigid nonmetallic conduit (without concrete encasement) for medium-voltage circuits. Part **(B)** extends to medium-voltage installations, the long-existing concept (see 300.9 for applications at 600 V and below) that the insides of raceways placed in wet locations are to be considered wet locations.

Part **(C)** requires protection of conductors where they emerge from underground (Fig. 300-57).

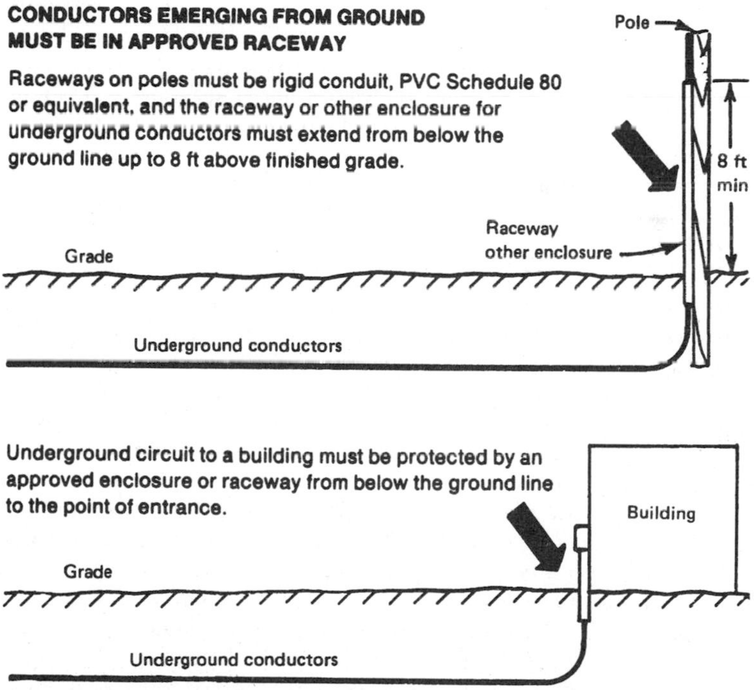

CONDUCTORS EMERGING FROM GROUND
MUST BE IN APPROVED RACEWAY
Raceways on poles must be rigid conduit, PVC Schedule 80 or equivalent, and the raceway or other enclosure for underground conductors must extend from below the ground line up to 8 ft above finished grade.
Pole
8 ft min
Raceway other enclosure
Grade
Underground conductors
Underground circuit to a building must be protected by an approved enclosure or raceway from below the ground line to the point of entrance.
Building
Grade
Underground conductors

Fig. 300-57. Direct burial cables must be protected aboveground. (Sec. 300.50.)

300.50 makes it clear that underground circuits may be installed in rigid metal conduit, in intermediate metal conduit, or in rigid nonmetallic conduit. Rigid metal conduit or IMC does not have to be concrete-encased, but it may be, of course. Direct burial nonmetallic conduit must be an approved (UL-listed and labeled) type specifically recognized for use without concrete encasement. If rigid nonmetallic conduit is approved for use only with concrete encasement, at least 2 in. (50 mm) of concrete must enclose the conduit. All applications of the various types of nonmetallic conduit must conform to the data made available by UL in the applicable Guide Card information.

Specific footnote c to column 4 of Table 300.50 eliminates the need for any burial depth in earth for conduits or other raceways that are run under a building or exterior concrete slab not less than 4 in. (100 mm) thick and extending at least 6 in. (150 mm) "beyond the underground installation"—that is, overlapping the raceway by 6 in. (150 mm) on each side. Note that the 4 in. (100 mm) thick concrete may be up at grade level in the form of a slab or patio or similar concrete area not subject to vehicular traffic. Figure 300-58 covers burial of underground medium-voltage circuits.

Part **(A)(2)** is new as of the 2014 **NEC** and covers the use of unshielded 2000-V conductors in qualified industrial locations where the conductors are directly buried, with the appropriate listing. The principal use of this provision is likely to be for PV farms on industrial property.

Part **(A)(3)** in 300.50 notes that other unshielded cable (i.e., cable without electrostatic shielding on the insulation)—other than within a metallic-sheathed cable assembly—must be installed in rigid metal conduit, in IMC, or in rigid nonmetallic conduit encased in not less than 3 in. (75 mm) of concrete. The effect of this rule is that unshielded, or nonshielded, cables may not be used directly buried in the earth. By reference to 310.10(F), nonshielded cables may be directly buried up to a rating of 2000 V, except that metal-encased, nonshielded conductors (as in Type MC or lead-jacketed cables) may be used in ratings up to 2400 V (other unshielded cables allowed up to 5000 V for series lighting at airfields). But all direct burial cables must be identified for such use. As indicated above, when nonshielded cable (of the nonmetallic-jacket type) is used underground in rigid nonmetallic conduit, the conduit must have a 3 in. (75 mm) thick concrete encasement. But if the same nonshielded cable is used in rigid nonmetallic conduit aboveground, concrete encasement is not required.

Figure 300-58 demonstrates uses of direct burial medium-voltage cables in relation to the Code rules. Figure 300-59 covers underground medium-voltage circuits in raceways. Figure 300-60 covers the basic rules of Table 300.50, subject to the considerations noted in the rules and exceptions, as follows:

1. Where run under a building or slab that is at least 4 in. (102 mm) thick and that extends 6 in. (150 mm) in each direction beyond the cable or raceway.
2. Areas of heavy traffic (public roads, commercial parking areas, etc.) must have minimum burial depth of 24 in. (600 mm) for any wiring method.
3. In industrial locations only, with qualified maintenance and supervision the minimum cover requirements for other than steel conduits can be reduced by 150 mm (6 in.) for each 50 mm (2 in.) of concrete or equivalent placed entirely within the trench over the underground installation. This

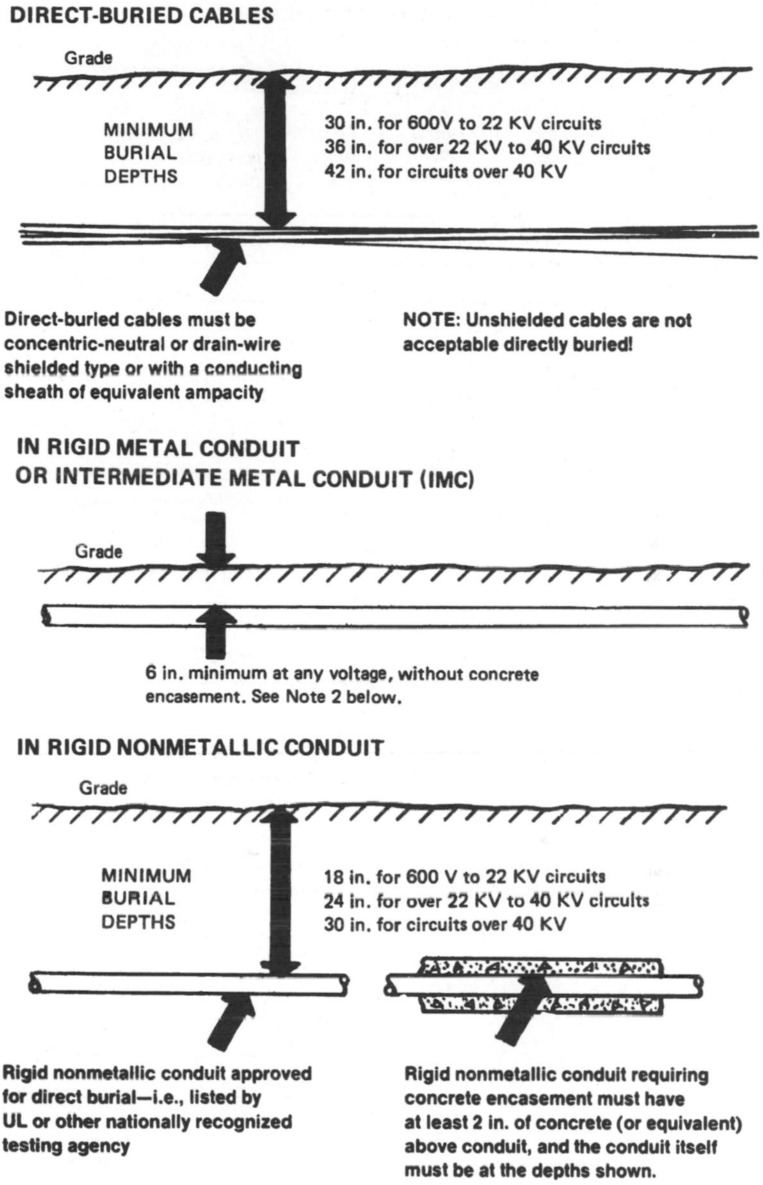
DIRECT-BURIED CABLES
Grade
MINIMUM
BURIAL
DEPTHS
30 in. for 600V to 22 KV circuits
36 in. for over 22 KV to 40 KV circuits
42 in. for circuits over 40 KV
Direct-buried cables must be concentric-neutral or drain-wire shielded type or with a conducting sheath of equivalent ampacity
NOTE: Unshielded cables are not acceptable directly buried!
IN RIGID METAL CONDUIT
OR INTERMEDIATE METAL CONDUIT (IMC)
Grade
6 in. minimum at any voltage, without concrete encasement. See Note 2 below.
IN RIGID NONMETALLIC CONDUIT
Grade
MINIMUM
BURIAL
DEPTHS
18 in. for 600 V to 22 KV circuits
24 in. for over 22 KV to 40 KV circuits
30 in. for circuits over 40 KV
Rigid nonmetallic conduit approved for direct burial—i.e., listed by UL or other nationally recognized testing agency
Rigid nonmetallic conduit requiring concrete encasement must have at least 2 in. of concrete (or equivalent) above conduit, and the conduit itself must be at the depths shown.

Fig. 300-58. Direct burial high-voltage cables must be of correct type, at specified depth. (Sec. 300.50.)

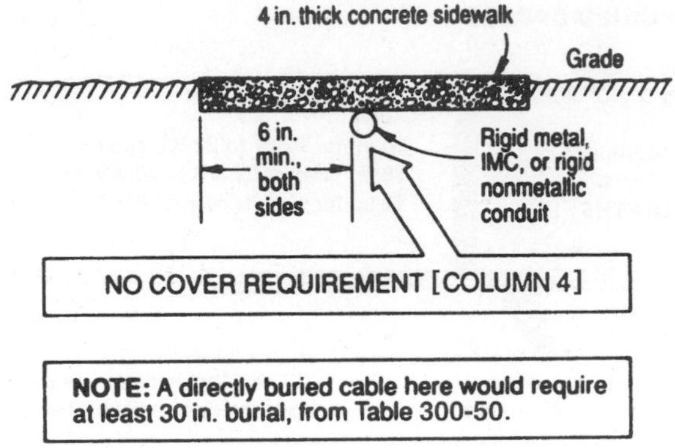

4 in. thick concrete sidewalk
Grade
6 in. min., both sides
Rigid metal, IMC, or rigid nonmetallic conduit
NO COVER REQUIREMENT [COLUMN 4]
NOTE: A directly buried cable here would require at least 30 in. burial, from Table 300-50.

Fig. 300-58. (*Continued*)

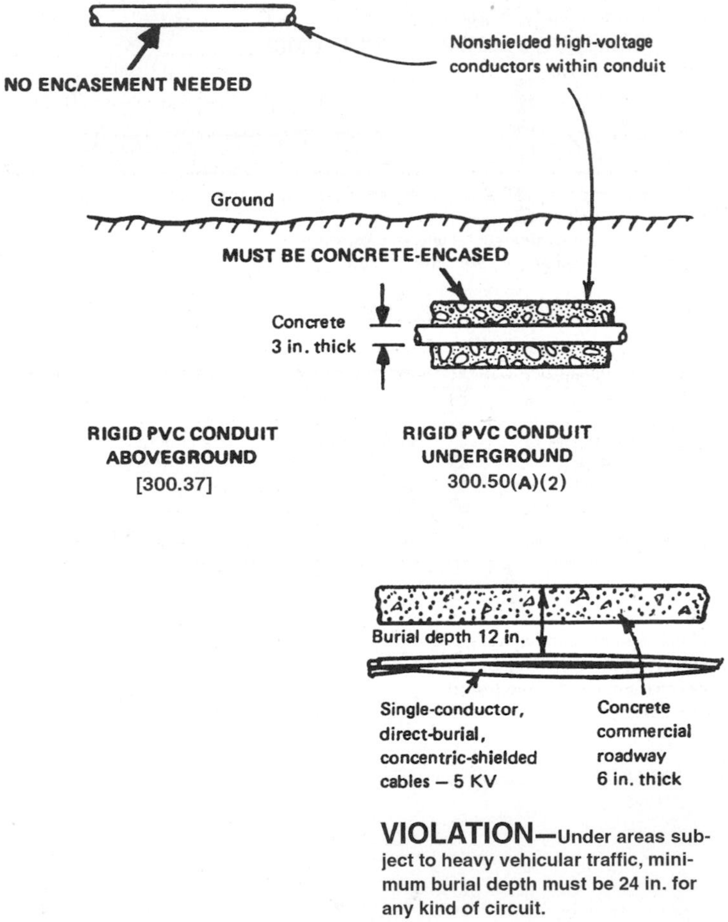

NO ENCASEMENT NEEDED
Nonshielded high-voltage conductors within conduit
Ground
MUST BE CONCRETE-ENCASED
Concrete 3 in. thick
RIGID PVC CONDUIT ABOVEGROUND [300.37]
RIGID PVC CONDUIT UNDERGROUND 300.50(A)(2)
Burial depth 12 in.
Single-conductor, direct-burial, concentric-shielded cables – 5 KV
Concrete commercial roadway 6 in. thick
VIOLATION—Under areas subject to heavy vehicular traffic, minimum burial depth must be 24 in. for any kind of circuit.

Fig. 300-59. Underground raceway circuits may vary widely in acceptable conditions of use. (Sec. 300.50.)

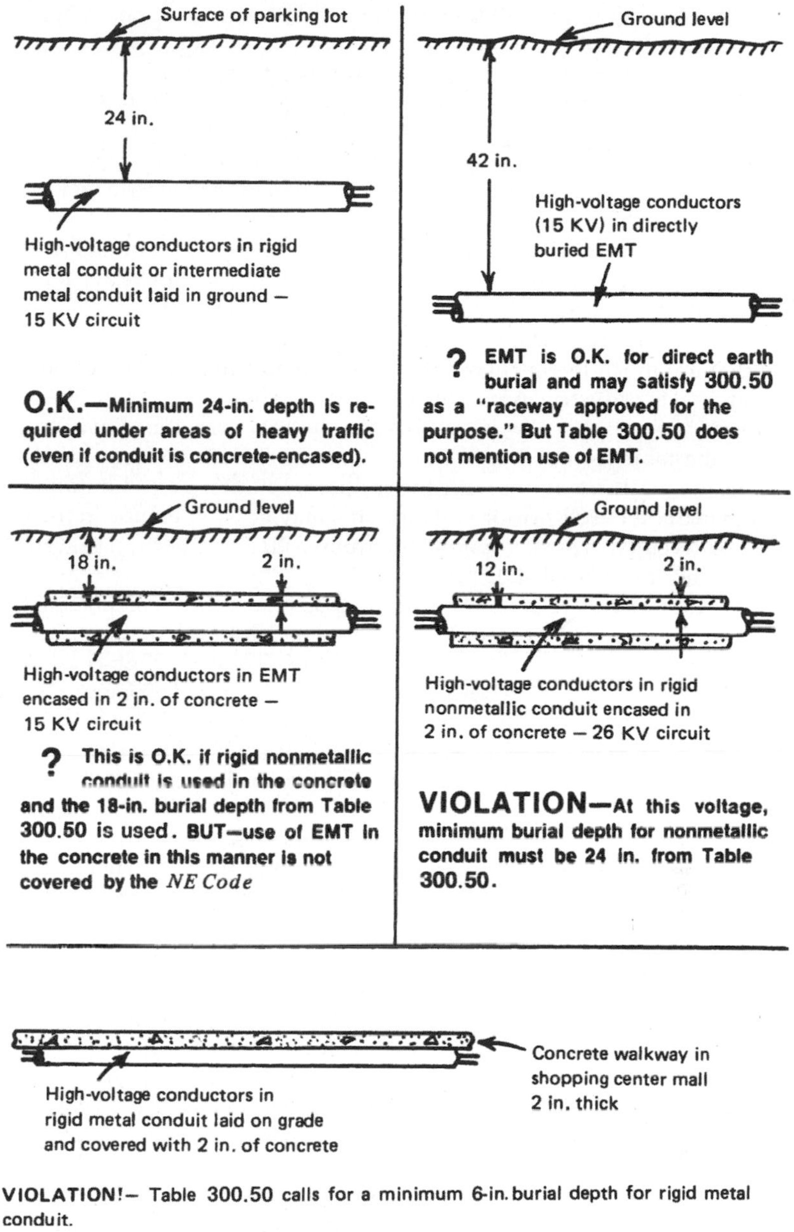
Surface of parking lot
24 in.
High-voltage conductors in rigid metal conduit or intermediate metal conduit laid in ground — 15 KV circuit
O.K.—Minimum 24-in. depth is required under areas of heavy traffic (even if conduit is concrete-encased).
Ground level
42 in.
High-voltage conductors (15 KV) in directly buried EMT
? EMT is O.K. for direct earth burial and may satisfy 300.50 as a "raceway approved for the purpose." But Table 300.50 does not mention use of EMT.
Ground level
18 in.
2 in.
High-voltage conductors in EMT encased in 2 in. of concrete — 15 KV circuit
? This is O.K. if rigid nonmetallic conduit is used in the concrete and the 18-in. burial depth from Table 300.50 is used. BUT—use of EMT in the concrete in this manner is not covered by the NE Code
Ground level
12 in.
2 in.
High-voltage conductors in rigid nonmetallic conduit encased in 2 in. of concrete — 26 KV circuit
VIOLATION—At this voltage, minimum burial depth for nonmetallic conduit must be 24 in. from Table 300.50.
Concrete walkway in shopping center mall 2 in. thick
High-voltage conductors in rigid metal conduit laid on grade and covered with 2 in. of concrete
VIOLATION!— Table 300.50 calls for a minimum 6-in. burial depth for rigid metal conduit.

Fig. 300-60. Underground high-voltage circuits must observe burial depths. (Sec. 300.50.) EMT, for the reasons given in the discussion at 300.5, is not a good choice for direct burial.

applies to the general condition columns (1), (2), and (3). It does not modify the specific condition columns, which apply if applicable.

4. Lesser depths are permitted where wiring rises for termination.
5. Airport runways may have cables buried not less than 18 in. (450 mm) deep, without raceway or concrete encasement.
6. Conduits installed in solid rock may be buried at lesser depths than shown in diagrams when covered by at least 2 in. (50 mm) of concrete that extends to the rock surface.

Item 3 directly corresponds to Table 300.5, row 2 (for 600 V and below), with two differences. First, Table 300.50 allows the 50 mm (2 in.) provision to be duplicated, with successive reductions in burial for each thickness of concrete applied. This is appropriate because Table 300.50 has some required burial depths that are much lower than Table 300.5. In addition, Table 300.50 limits the permission to industrial occupancies, which is not done in Table 300.5. This permission (for medium voltage) had been in the NEC since 1968 but it dropped out when the table format was created in the 2005 NEC, without substantiation. It has now returned, but with the industrial limitation, a limitation that was never included in any prior version. This may be problematic in the many institutional occupancies that operate medium-voltage systems in campus layouts with similar qualified staffing.

As of the 2008 NEC there is now a warning tape requirement comparable to the one for underground service conductors in 300.5(D)(3), but with one significant difference. This rule appears as a footnote "d" to the first column of Table 300.50, and therefore only applies to direct-burial cable work, and not underground raceways whether steel or nonmetallic. The other provisions ["in the trench," at least 300 mm (12 in.) above the underground installation, and applies only if no concrete protection or encasement] match up to the service protection rules generally. Refer to the discussion at 300.5(D)(3) for more information.

As noted in part **(D)**, splices or taps are permitted in a trench without a box, but only if approved methods and materials are used—that is, listed splice kits. Taps and splices must be watertight and protected from mechanical injury. For shielded cables, the shielding must be continuous across the splice or tap. There is an exception to this continuity rule for engineered systems using direct-buried single-conductor cables with maintained spacing between phases. Under these conditions, the shielding is permitted to be made discontinuous, provided the severed ends of the shielding are overlapped at the point of discontinuity. Each shielded section is then separately grounded at a single point for each end. This interrupts the circulating currents from capacitive charging that become significant on long runs under these conditions. The currents can be high enough such that the shield heating seriously degrades the cable ampacity.

Figure 300-61 shows a permanent straight splice for joining one end of a cable off a reel to the start of a cable off another reel, or for repairing a cable that is cut through accidentally by a backhoe or other tool digging into the ground. T and Y splices are made with similar techniques. Disconnectable splice devices provide watertight plug-and-receptacle assembly for all types of shielded cables and are fully submersible.

Figure 300-62 covers the rules of parts **(D)** and **(E)**.

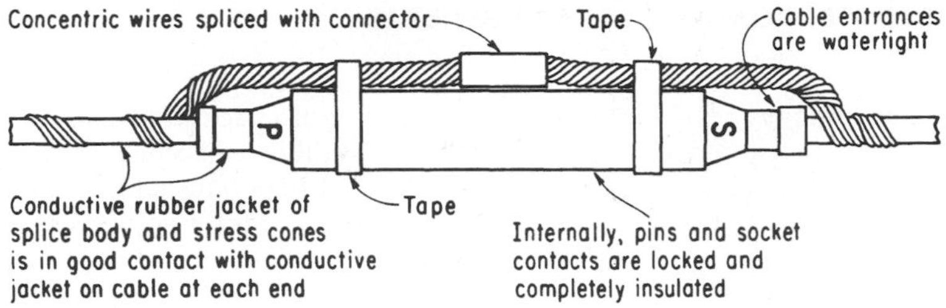
Concentric wires spliced with connector
Tape
Cable entrances are watertight
P
S
Conductive rubber jacket of splice body and stress cones is in good contact with conductive jacket on cable at each end
Tape
Internally, pins and socket contacts are locked and completely insulated

Fig. 300-61. Splice may be made in direct burial cable if suitable materials are used. (Sec. 300.50.)

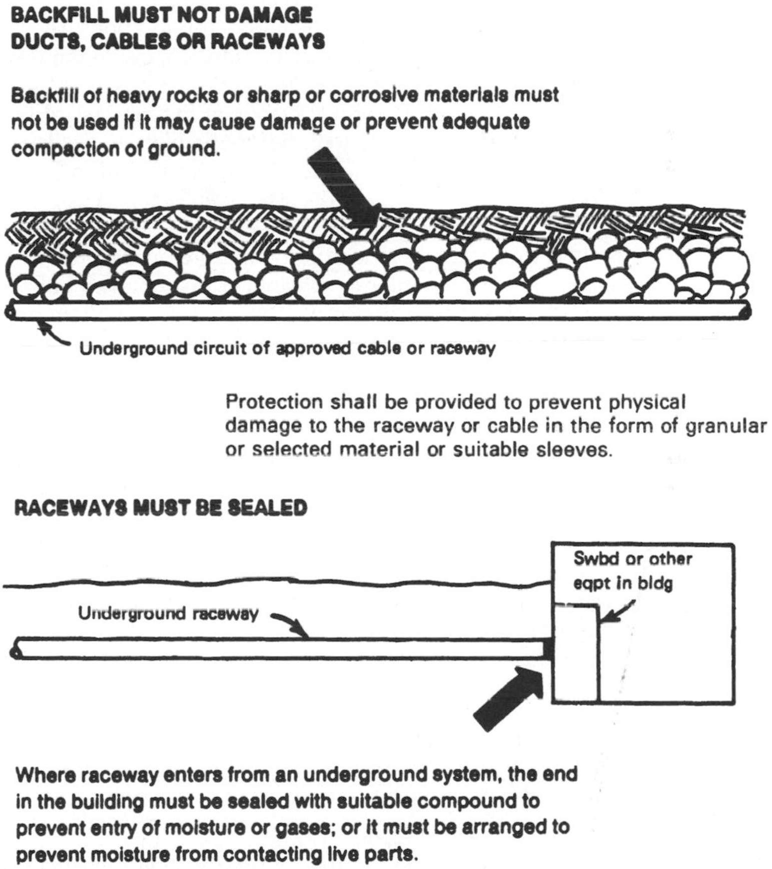
BACKFILL MUST NOT DAMAGE DUCTS, CABLES OR RACEWAYS
Backfill of heavy rocks or sharp or corrosive materials must not be used if it may cause damage or prevent adequate compaction of ground.
Underground circuit of approved cable or raceway
Protection shall be provided to prevent physical damage to the raceway or cable in the form of granular or selected material or suitable sleeves.
RACEWAYS MUST BE SEALED
Swbd or other eqpt in bldg
Underground raceway
Where raceway enters from an underground system, the end in the building must be sealed with suitable compound to prevent entry of moisture or gases; or it must be arranged to prevent moisture from contacting live parts.

Fig. 300-62. Circuits must be protected and sealed where they enter equipment. (Sec. 300.50.)

ARTICLE 310. CONDUCTORS FOR GENERAL WIRING

310.2. Definitions. The definitions are those for electrical ducts and thermal resistivity that appeared in Codes prior to the 2011 edition adjacent to the medium-voltage ampacity tables. They have been relocated here (and edited to be pure definitions that do not state requirements) because Sec. 2 is the section designated in the NEC *Style Manual* for this material.

310.10. Uses Permitted. In order to make this section look like wiring method sections (which is not required since this is not a wiring method article, but that is another story), as of the 2011 NEC former secs. 4, 6, 7, 8, and 9 have been combined here. There is no "Uses Not Permitted" section, so each of the relocated sections are pretty much the same, stating both what is and what is not permitted, except they have been demoted to lettered subsections.

310.10(A), (B), (C), and (D). Locations. Part **(A)** on dry locations simply says to use wires as specified in the Code. Part **(B)** on dry and damp locations lists insulation types that generally, but not always, have a "W" in their insulation designation. Part **(C)** covers wet locations. Any conductor used in a "wet location" (refer to the definition under "location" in Art. 100) must be one of the designated types—each of which has the letter W in its marking to indicate suitability to wet locations. Any conduit run underground is assumed to be subject to water infiltration and is, therefore, a wet location, requiring use of only the listed conductor types within the raceway. In addition, "moisture-impervious metal-sheathed" cables and conductors can be used; until the 1999 NEC this was a permission to use lead-covered wire; the reason for making the change in terminology was never documented although it presumably had to do with the market unacceptability of lead for environmental reasons. Finally, any type listed for use in wet locations, even if not on the specific insulation list, can be used as well.

Part **(D)** covers locations exposed to direct sunlight, and requires that wiring directly exposed to the sun to have been evaluated for that exposure, or be protected by tape or sleeving that has been so evaluated. UL does those tests as part of its listing evaluations for Type SE cable, so there isn't a problem there. Incidentally, Type SE cable is an example of wiring that is listed sunlight resistant but the individual wires are not marked accordingly. If they were, then by Code usage of the cable would be "listed and marked"; instead it qualifies as "listed."

Either one is OK. However, on conduit services this whole requirement has been controversial, since almost every drip loop ever installed before there even was an NEC (1897) has used whatever building wire was on hand, provided it was good for wet locations. There have been a number of proposals to exempt drip loops, which emerge from separately bushed holes in the weatherhead, to no avail. Listed coverings are beginning to find their way to market, as are conductors. One major manufacturer is now selling all of its RHH/RHW-2/USE-2 as sunlight resistant, plus its XHHW-2 and its THHN/THWN-2 in all sizes of 2 AWG and larger as having been listed and marked as sunlight resistant.

310.10(E). Shielding. The default effect of this Code rule is to require all open conductors operating over 2 kV to be shielded, unless the conductor is UL-listed for operation unshielded at voltages above 2 kV. However, metal-clad cables, Type MC, using an unshielded cable makeup, are now permitted for use up to 5 kV in industrial establishments with qualified maintenance and supervision. This expands the permitted voltage range of the unshielded product from the

2.4 kV limit in the 2008 **NEC**, and will accommodate the many 4160 distributions still in use. The **NEC** permitted even higher voltages for approximately 60 years prior to the 2005 edition. At that time the limit dropped to 2.4 from 8 kV where it had stood since 1975. It stood from 3 to 6 kV, depending on distribution characteristics (2 kV, rising to 3 kV in 1959, for wet or damp locations), but without a listing requirement since 1947. There is a very large installed base of industrial equipment that lacks the terminating enclosure sizing to accommodate the stress cones associated with shielded cables. This is one of two provisions that provides some relief in this cycle, trading off the higher voltage for the requirement to use a metal-clad construction.

Exception No. 2 is the other provision in this cycle that addresses the installed base of medium-voltage equipment. If the connected equipment is *existing*, listed nonshielded cable can be used for replacement purposes provided it carries the appropriate listing and there is qualified maintenance and supervision. The exception provisions are essentially identical to the ones that applied to comparable voltage cables prior to the 2005 **NEC**. An informational note following this exception cautions that replaced or relocated equipment may not qualify as "existing" as it is being used in the exception. The provisions that formerly allowed unshielded cables up to 8 kV have been deleted, replaced only by the 2400-V limit. Because 2300-V delta (which is over 2 kV) is the lowest general-purpose, medium-voltage circuit in use today, unlisted conductors must be shielded for such circuits. *But note this:* UL does list 2.4-kV unshielded conductors for use in accordance with Sec. 310.15(10)(E), Table 310.104(D), and other Code rules (Fig. 310-1). Exception No. 1 allows those cables to be generally used.

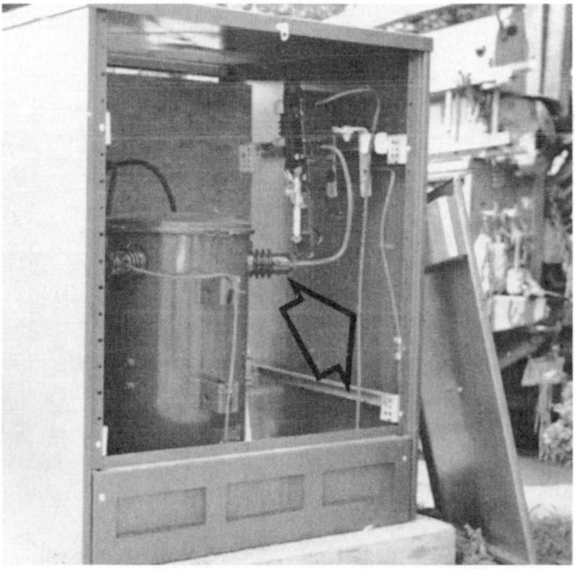

Fig. 310-1. A nonshielded conductor (arrow) is permitted for use on a 2300-V circuit (phase-to-neutral), as shown here, *only* if the conductor is listed by UL or another national test lab and approved for use without electrostatic shielding. [Sec. 310.10(E).]

UL also lists shielded conductors up to 35 kV. And, in accordance with NEC Table 310.104(B), UL has been listing Type RHH insulated conductors (rubber or cross-linked polyethylene insulation) with electrostatic shielding for operation up to 2 kV.

In addition to applicable NEC, Insulated Power Cable Engineers Association (IPCEA), and UL data on use of cable shielding, manufacturers' data should be consulted to determine the need for shielding on the various types and constructions of available cables.

Shielding of medium-voltage cables protects the conductor assembly against surface discharge or burning (due to corona discharge in ionized air), which can be destructive to the insulation and jacketing. It does this by confining and distributing stress in the insulation and eliminating charging current drain to intermittent grounds. It also prevents ionization of any tiny air spaces at the surface of the insulation by confining electrical stress to the insulation. Shielding, which is required by this Code rule to be effectively grounded, increases safety by eliminating the shock hazard presented by the external surface of unshielded cables. By preventing electrical discharges from cable surfaces to ground, shielding also reduces fire or explosive hazards and minimizes any radio interference medium-voltage circuits might cause.

Electrostatic shielding of cables makes use of both nonmetallic and metallic materials. As shown in accompanying sketches of typical cable assemblies, semiconductive tapes or extruded coverings of semiconductive materials are combined with metal shielding to perform the shielding function. Metallic shielding may be done with:

1. A copper shielding tape wrapped over a semiconducting shielding of nonmetallic tape that is applied over the conductor insulation (Fig. 310-2)

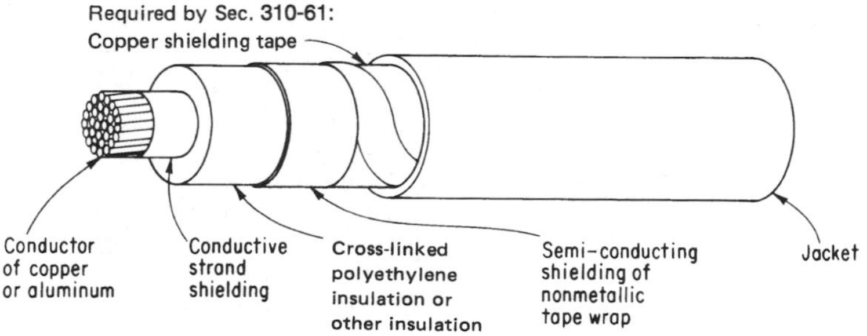
Required by Sec. 310-61:
Copper shielding tape
Conductor of copper or aluminum
Conductive strand shielding
Cross-linked polyethylene insulation or other insulation
Semi-conducting shielding of nonmetallic tape wrap
Jacket

Fig. 310-2. A flat copper tape spiraled over the insulation is an electrostatic shield. [Sec. 310.10(E).]

2. A concentric wrapping of bare wires over a semiconducting, nonmetallic jacket over the conductor insulation (Fig. 310-3)
3. Bare wires embedded in the semiconducting nonmetallic jacket that is applied over the insulation (Fig. 310-4)
4. A metal sheath over the conductor insulation, as with lead-jacketed cable

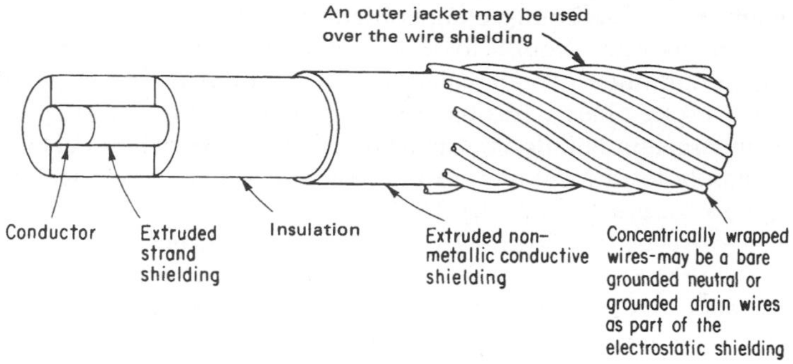
An outer jacket may be used over the wire shielding
Conductor
Extruded strand shielding
Insulation
Extruded non-metallic conductive shielding
Concentrically wrapped wires-may be a bare grounded neutral or grounded drain wires as part of the electrostatic shielding

Fig. 310-3. Wires, instead of metal tape, are also used for electrostatic shielding (URD and UD type). [Sec. 310.10(E).]

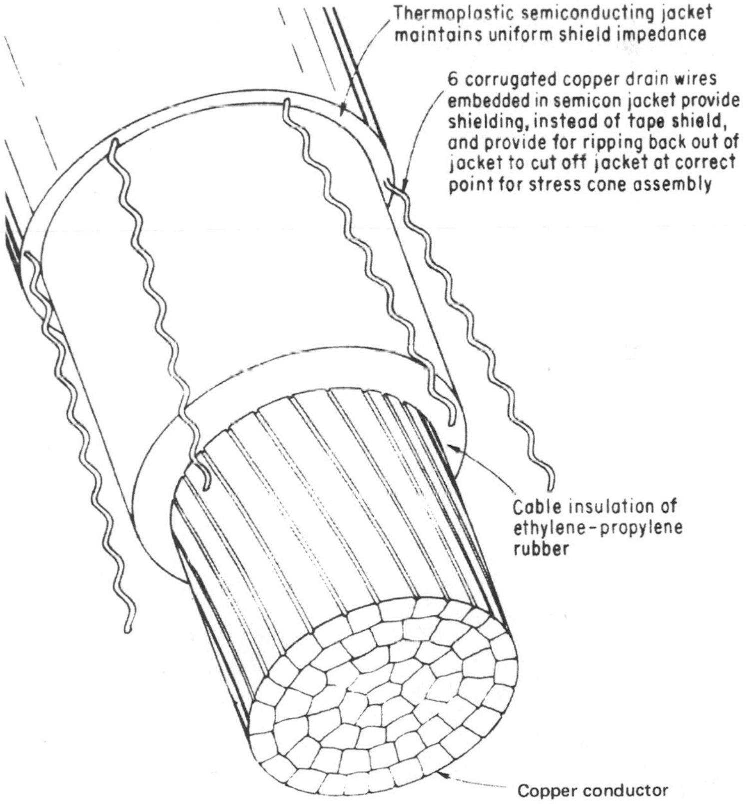
Thermoplastic semiconducting jacket maintains uniform shield impedance
6 corrugated copper drain wires embedded in semicon jacket provide shielding, instead of tape shield, and provide for ripping back out of jacket to cut off jacket at correct point for stress cone assembly
Cable insulation of ethylene-propylene rubber
Copper conductor

Fig. 310-4. Wires embedded in semiconducting jacket form another type of shielding. [Sec. 310.10(E).]

For many years, medium-voltage shielded power cables for indoor distribution circuits rated from 5 to 15 kV were of the type using copper tape shielding and an outer overall jacket. But in recent years, cables shielded by concentric-wrapped bare wires have also come into widespread use, particularly for underground outdoor systems up to 15 kV. These latter cables are the ones commonly used for underground residential distribution (called URD). Such a conductor is shown in Fig. 310-3.

In addition to use for URD (directly buried with the concentric-wire shield serving as the neutral or second conductor of the circuit), concentric-wire-shielded cables are also available for indoor power circuits, such as in conduit, with a nonmetallic outer jacket over the concentric wires. Such cable assemblies are commonly called *drain-wire-shielded* cable rather than *concentric-neutral* cable because the bare wires are used only as part of the electrostatic shielding and not also as a neutral. Smaller-gauge wires are used where they serve only for shielding and not as a neutral.

Figure 310-4 shows drain-wire-shielded medium-voltage cable with electrostatic shielding by means of drain wires embedded in a semiconducting jacket over the conductor insulation. This type of drain-wire-shielded conductor is designed to be used for medium-voltage circuits in conduit or duct for commercial and industrial distribution as an alternative to tape-shielded cables. For the same conductor size, this type of embedded drain-wire-shielded cable has a smaller outside diameter and lighter weight than a conventional tape-shielded cable. For the drain-wire cable the assembly difference reduces installation labor, permits reduced bending radius for tight conditions and easier pulling in conduit, and affords faster terminations (with stress cones) and splices. An extremely important result of the smaller overall cross-sectional area (csa) of the drain-wire-shielded cable is the chance to use smaller conduits—with lower material and labor costs—when conduit is filled to 40 percent of its csa based on the actual cable csa, as covered by Note 5 to the tables in Chap. 9 of the **NEC**.

Another consideration in conductor assemblies is that of strand shielding. As shown in Fig. 310-5, a semiconducting material is tape-wrapped or extruded onto the conductor strands and prevents voids between the insulation and the strands, thereby reducing possibilities of corona cutting on the inside of the insulation.

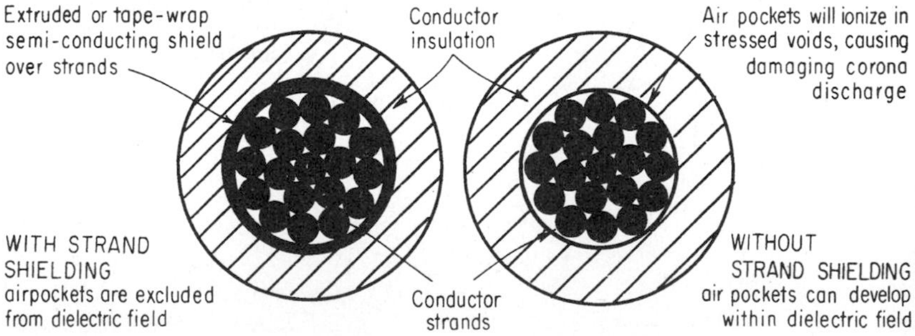
Extruded or tape-wrap semi-conducting shield over strands
Conductor insulation
Air pockets will ionize in stressed voids, causing damaging corona discharge
WITH STRAND SHIELDING
airpockets are excluded from dielectric field
Conductor strands
WITHOUT STRAND SHIELDING
air pockets can develop within dielectric field

Fig. 310-5. Strand shielding is part of the overall electrostatic shielding system on the conductor. [Sec. 310.10(E).]

Refer to 300.40 on terminating and grounding shielded conductors. Refer to 250.190(C)(2) for equipment grounding considerations on shielded cables.

310.10(F). Direct Burial Conductors. The first sentence of this section applies to conductors rated for 600 V and lower systems as well as medium-voltage systems, and as noted, it requires conductors used for direct burial applications to be identified for this purpose. The remainder of the Code rules on underground use of conductors rated up to 600 V are covered in the following paragraphs.

Figure 310-6 shows a clear violation of 310.10(F). In the photo, conductors marked RHW are run from the junction box below the magnetic contactor, directly buried in the ground. Although Type RHW is suitable for wet locations, it is not approved for direct burial. If, however, the conductors were of the type that is marked RHW-USE—that is, it is listed and recognized as both a single-conductor RHW and a single conductor Type USE (underground service entrance) cable—then such conductors would satisfy this section.

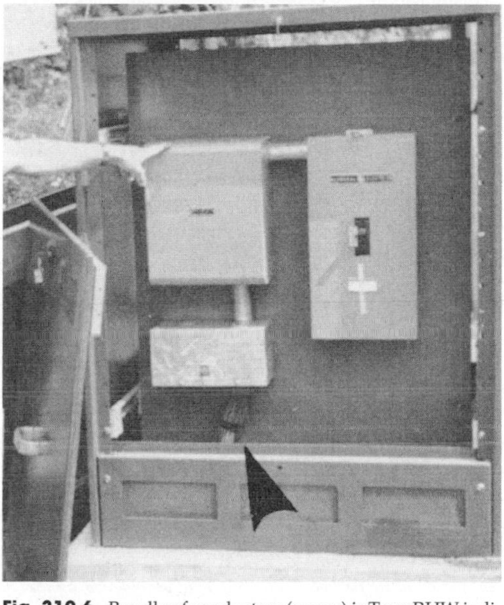

Fig. 310-6. Bundle of conductors (arrow) is Type RHW individual building conductors that would be suitable for installation in conduit underground but are not marked "USE"; and their use here, run directly buried to outdoor lighting poles, constitutes a violation of the last sentence of Sec. 310.10(F), first paragraph.

Type UF cable is acceptable for direct earth burial. Section 338.10(B)(4)(b) says USE cable used for branch circuits and feeders is to be installed in accordance with Part II of Art. 340 on Type UF cable, and 340.10(A) specifically says the cable is acceptable for direct burial, so that concludes the issue. In addition, the UL *Guide Card information* [see discussion at 110.3(B) for details] notes that listed USE cable is recognized for burial directly in the earth (Fig. 310-7).

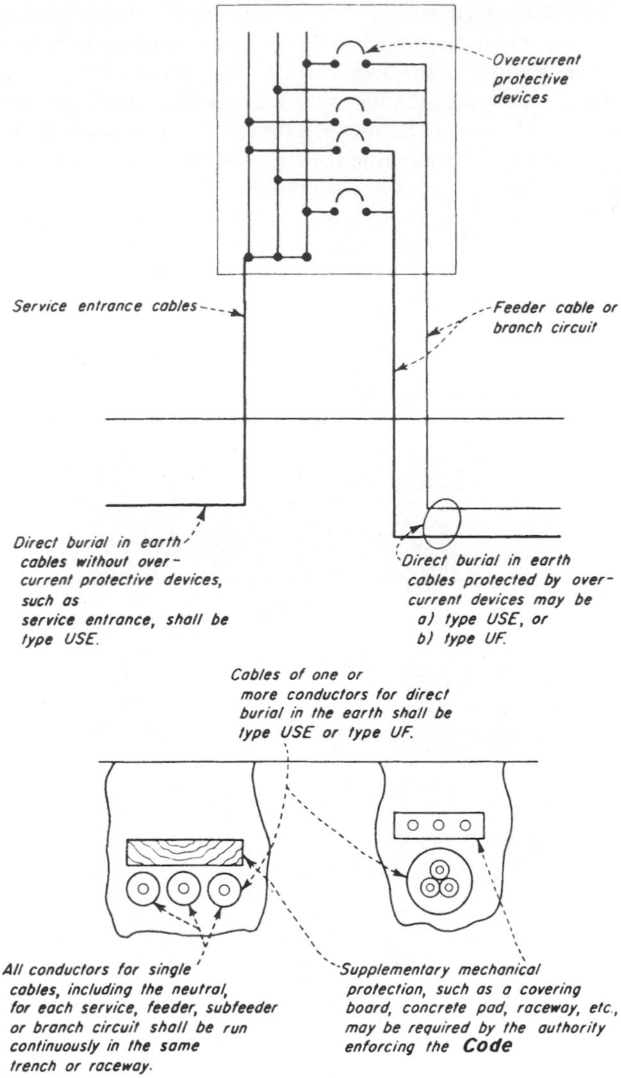

Overcurrent protective devices
Service entrance cables
Feeder cable or branch circuit
Direct burial in earth cables without over-current protective devices, such as service entrance, shall be type USE.
Direct burial in earth cables protected by over-current devices may be a) type USE, or b) type UF.
Cables of one or more conductors for direct burial in the earth shall be type USE or type UF.
All conductors for single cables, including the neutral, for each service, feeder, subfeeder or branch circuit shall be run continuously in the same trench or raceway.
Supplementary mechanical protection, such as a covering board, concrete pad, raceway, etc., may be required by the authority enforcing the Code

Fig. 310-7. Types USE and UF cables are designated by the letter U for underground use. [Sec. 310.10(F) first paragraph.]

Besides UF and USE, then, what other cables can be directly buried? 330.10(A)(5) does recognize MC cable for direct burial "where identified for such use." 332.10(10) recognizes MI cable for direct burial. Note that 332.10(6) permits MI in "fill" below grade, and 332.10(10) permits it for underground runs with suitable damage and corrosion protection.

For burial-depth requirements on directly buried cables, refer to 300.5 and Table 300.5. Cables approved for direct earth burial must be installed a minimum of 24 in. (600 mm) below grade, as given in Table 300.5, or at least 30 in. (750 mm) below grade for medium-voltage cables as covered in Table 300.50 with its footnotes.

Direct burial conductors should be trench-laid without crossovers; should be slightly "snaked" to allow for possible earth settlement, movement, or heaving due to frost action; and should have cushions and covers of sand or screened fill to protect conductors against sharp objects in trenches or backfill. Figure 310-8 shows some recommended details on installing direct burial cables. Moreover, when conductors are routed beneath roadways or railroads, they should be additionally protected by conduits. To guard against damage that might occur during future digging, conductors in soft fill should be covered by concrete slabs or treated planks.

Where prewired cable-in-conduit (Art. 354) is being buried, it also should be slightly snaked, although it is unnecessary to provide sand beds or screen the backfill. Inasmuch as these complete conductor-raceway assemblies can be delivered on reels in specified factory-cut lengths, installation is simplified and expedited.

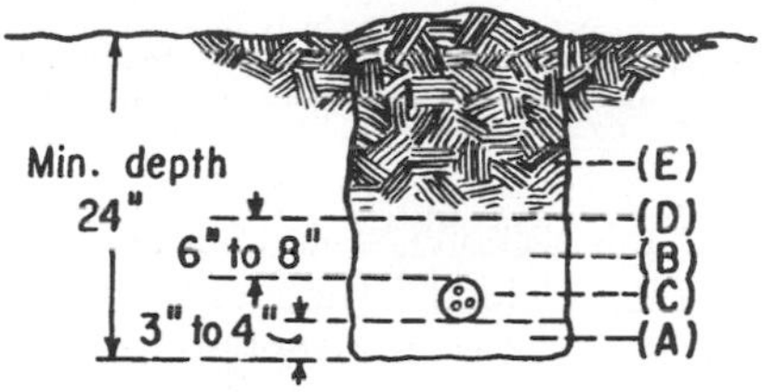
Min. depth
24"
6" to 8"
3" to 4"
(E)
(D)
(B)
(C)
(A)

A—Soft bed of sand or screened fill.

B—Blanket of sand or screened fill 6 in. to 8 in. above top of cable.

C—Cable "snaked" slightly in trench for slack when earth settles. Keep single-conductor cables uniformly apart about 6 in. in trench. Avoid cable crossovers. Keep cable below frost line.

D—Add protective slab (creosoted plank, etc.) on sand fill in areas where future digging might occur. Enclose cable in pipe or conduit under highways or rail tracks.

E—Normal backfill.

Fig. 310-8. This satisfies the intent of Sec. 300.5(F); note that cover requirements, where given in the Code for minimum burial depths, are taken from the top of the electrical conductor, cable, or raceway. The minimum trench depth is, therefore, the cover dimension plus the diameter of the wiring method. If the wiring is not below the frost line, or in ground subject to settlement, requirements in 300.5(J) must be met as well. [Sec. 310.10(F).]

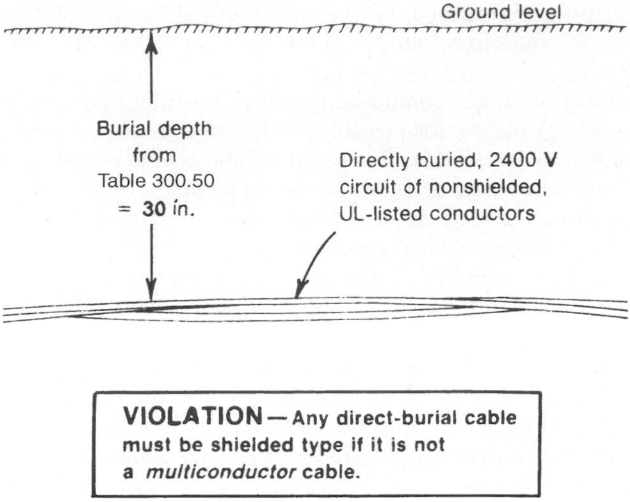
Ground level
Burial depth
from
Table 300.50
= 30 in.
Directly buried, 2400 V
circuit of nonshielded,
UL-listed conductors
VIOLATION — Any direct-burial cable
must be shielded type if it is not
a *multiconductor* cable.

Fig. 310-9. This application is covered by both Secs. 310.10(F) second paragraph and 300.50. The 2.4 kV cable must be identified for the purpose and has an overall metallic sheath or armor. [Sec. 310.10(F) second paragraph.]

The second sentence in this section is a standalone paragraph reading "Cables rated above 2000 V shall be shielded." Therefore direct burial cables over 2 kV must be shielded (Fig. 310-9), with an exception for multiconductor cables up to 2.4 kV if the cable has an overall metallic sheath or armor. There is no NEC listing requirement for this cable, which is different from the aboveground requirement in 310.10(E). However, the section has generic language in its first sentence requiring any direct burial cables to be "identified for such use." Review the discussion in this book at the definition of "identified" in Art. 100 for more information on this point. The metallic shielding must be connected to a grounding busbar or a grounding electrode. Sec. 250.190 still applies and addresses the issues comprehensively.

There is also an exception to retain the former allowance for unshielded 5-kV cabling for direct burial applications of series-connected airport runway lighting.

Series lighting is still used in some locations by utilities to provide street lighting, although it is disappearing. It is widely used, with FAA sanction, for airport runways, in applications that are squarely premises wiring and not protected by the utility exemption.

310.10(G). Corrosive Conditions. Figure 310-10 shows how conductors such as Type THHN-THWN are marked to indicate that they are gasoline- and oil-resistant, for use in gasoline stations and similar places.

TYPE THHN 600 V OIL AND GASOLINE RESISTANT

Fig. 310-10. Typical marking indicates suitability of conductors for use under unusual environmental conditions. [Sec. 310.10(G).]

310.10(H). Conductors in Parallel. The requirements of part **(H)** for conductors in parallel recognize copper, copper-clad aluminum, and aluminum conductors in sizes 1/0 AWG and larger. Also, this section makes it clear that the rules for paralleling conductors apply to grounding conductors (except for sizing, which is accomplished in accordance with 250.122) when they are used with conductors in multiple. Conductors that are permitted to be used in parallel (in multiple) include "phase" conductors, "polarity" conductors, "neutral" conductors, and "grounded circuit" conductors. In the places where this section describes parallel makeup of circuits, a grounded circuit conductor is identified along with phase, polarity, and neutral conductors to extend the same permission for paralleling to grounded legs of corner-grounded delta systems and also dc circuits.

This section recognizes the use of conductors in sizes 1/0 and larger for use in parallel under the conditions stated, to allow a practical means of installing large-capacity feeders and services. Paralleling of conductors relies on a number of factors to ensure equal division of current, and thus all these factors must be satisfied in order to ensure that none of the individual conductors will become overloaded.

When conductors are used in parallel, all the conductors making up each phase, polarity, neutral, or grounded circuit conductor must satisfy the five conditions of part **(2)** in this section. Those characteristics—same length, same conductor material (copper or aluminum), same size, same insulation, and same terminating device—apply only to the paralleled conductors making up each phase, polarity, or neutral of a parallel-makeup circuit. All the conductors of any phase, polarity, or the neutral must satisfy the rule, but phase A conductors (all of which must be the same length, same size, etc.) may be different in length, material, size, etc., from the conductors making up phase B or phase C or the neutral. All phase B conductors must be the same length, same size, and so on; phase C conductors must all be the same; and neutral conductors must all be alike (Fig. 310-11). It is not the intent of the Code rule to require that conductors of one phase be the same as those of another phase or of the neutral. Note that the word "polarity" expressly supports the application of these provisions to dc circuits, if applicable. The only concern for safe operation of a parallel-makeup circuit is that all the conductors in parallel per phase leg (or neutral or grounded conductor) will evenly divide the load current and thereby prevent overloading of any one of the conductors. Of course, the realities of material purchase and application and good design practice will dictate that all the conductors of all phases and neutral will use the same conductor material, will have the same insulation, will have as nearly the same length as possible to prevent voltage drop from causing objectionable voltage unbalance on the phases, and will be terminated in the same way. The size of conductors may vary from phase to phase or in the neutral, depending upon load currents.

Figure 310-12 shows two examples of parallel-conductor circuit makeup. The photo at bottom shows six conductors used per phase and neutral to obtain 2000-A capacity per phase, which simply could not be done without parallel conductors per phase leg. Note that a fusible limiter lug is used to terminate each individual conductor. Although limiter lugs are required by the **NEC** only as used in 450.6(A)(3), they may be used to protect each conductor of any parallel circuit against current

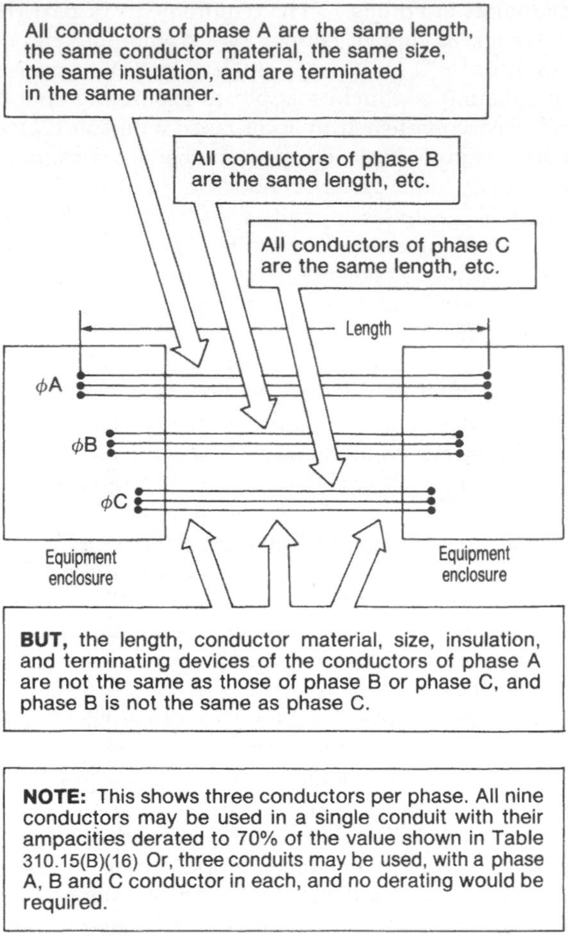

All conductors of phase A are the same length, the same conductor material, the same size, the same insulation, and are terminated in the same manner.
All conductors of phase B are the same length, etc.
All conductors of phase C are the same length, etc.
Length
ϕA
ϕB
ϕC
Equipment enclosure
Equipment enclosure
BUT, the length, conductor material, size, insulation, and terminating devices of the conductors of phase A are not the same as those of phase B or phase C, and phase B is not the same as phase C.
NOTE: This shows three conductors per phase. All nine conductors may be used in a single conduit with their ampacities derated to 70% of the value shown in Table 310.15(B)(16) Or, three conduits may be used, with a phase A, B and C conductor in each, and no derating would be required.

Fig. 310-11. This is the basic rule on conductors used for parallel circuit makeup. [Sec. 310.10(H).]

in excess of the ampacity of the particular size of conductor. In addition to protecting against unequal division of current, limiters placed at both ends also increase overall reliability, because in the event of a ground fault on one of the conductors, the limiters will immediately deenergize only the faulted cable, often allowing the circuit to remain energized at reduced load until a complete shutdown can be arranged. The CB or fuses on such circuits are rated much higher than the ampacity of each individual conductor.

Where large currents are involved, it is particularly important that the separate phase conductors be located close together to avoid excessive voltage drop and ensure equal division of current. It is also essential that each phase and the

Multiple conductors (two in parallel for each phase leg) are used for normal and emergency feeder through this automatic transfer switch.

Six conductors in parallel make up each phase leg and the neutral of this feeder. Fusible limiter lug on each conductor, although not required by **Code** on other than transformer tie circuits, is sized for the conductor to protect against division of current among the six conductors that would put excessive current on any conductor.

Fig. 310-12. These are examples of circuit makeup using conductors in parallel. [Sec. 310.10(H).]

neutral, and grounding wires, if any, be run in each conduit even where the conduit is of nonmetallic material.

Part **(3)** of the section requires the use of the same type of raceway or enclosure for conductors in parallel. The impedance of the circuit in a nonferrous raceway will be different from the same circuit in a ferrous raceway or enclosure. (See 300.20.) In addition, each raceway must have the same number of conductors per phase. This is a recent addition to the **NEC**, after it was substantiated that someone actually ran three sets of parallel conductors through two raceways—two sets in one and one set in the other. Since the heat developed in the raceway with two sets of conductors will significantly differ from the other, and since the resistance of a conductor differs with temperature, the current will divide unequally.

From the Code tables of current-carrying capacities of various sizes of conductors, it can be seen that small conductor sizes carry more current per circular mil of cross section than do large conductors. This results from rating conductor capacity according to temperature rise. The larger a cable, the less the radiating surface per circular mil of cross section. Loss due to "skin effect" (apparent higher resistance of conductors to alternating current than to direct current) is also higher in the larger conductor sizes. And larger conductors cost more per ampere than smaller conductors.

All the foregoing factors point to the advisability of using a number of smaller conductors in multiple to get a particular carrying capacity, rather than using a single conductor of that capacity. In many cases, multiple conductors for feeders provide distinct operating advantages and are more economical than the equivalent-capacity single-conductor makeup of a feeder. However, it should be noted that the reduced overall cross section of conductor resulting from multiple conductors instead of a single conductor per leg produces higher resistance and greater voltage drop than the same length as a single conductor per leg. Voltage drop may be a limitation.

Figure 310-13 shows a typical application of copper conductors in multiple, with the advantages of such use. Where more than three conductors are installed in a single conduit, the ampacity of each conductor must be derated from the ampacity value shown in **NEC** Table 310.15(B)(16). The four circuit makeups show:

1. Without ampacity derating because there are more than three conductors in the conduit, circuit 2 would be equivalent to circuit 1.
2. A circuit of six 400 kcmil can be made equivalent in ampacity to a circuit of three 2000 kcmil by dividing the 400s between two conduits [3 conductors/3-in. (76-mm) rigid metal conduit]. If three different phases are used in each of two 3-in. (76-mm) conduits for this circuit, the multiple circuit would not require ampacity derating to 80 percent, and its 670-A rating would exceed the 665-A rating of circuit 1.
3. Circuit 2 is almost equivalent to circuit 3 in ampacity.
4. Circuit 4 is equivalent to circuit 1 in ampacity, but uses less conductor copper and a smaller conduit. The advantages are obtained even with the ampacity derating for conduit fill.

Except where the conductor size is governed by conditions of voltage drop it is seldom economical to use conductors of sizes larger than 1000 kcmil

1. 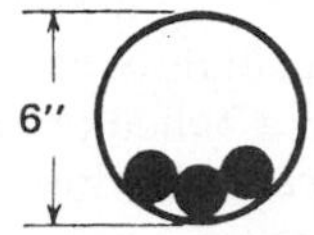
6"

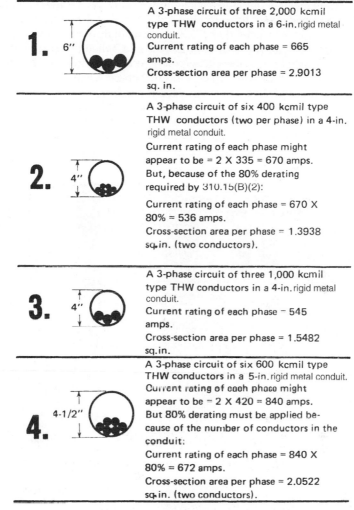

A 3-phase circuit of three 2,000 kcmil type THW conductors in a 6-in. rigid metal conduit.
Current rating of each phase = 665 amps.
Cross-section area per phase = 2.9013 sq. in.

2. 
4"

A 3-phase circuit of six 400 kcmil type THW conductors (two per phase) in a 4-in. rigid metal conduit.
Current rating of each phase might appear to be = 2 X 335 = 670 amps.
But, because of the 80% derating required by 310.15(B)(2):

Current rating of each phase = 670 X 80% = 536 amps.
Cross-section area per phase = 1.3938 sq.in. (two conductors).

3. 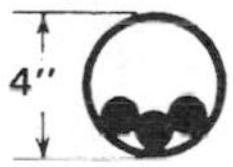
4"

A 3-phase circuit of three 1,000 kcmil type THW conductors in a 4-in. rigid metal conduit.
Current rating of each phase - 545 amps.
Cross-section area per phase = 1.5482 sq.in.

4. 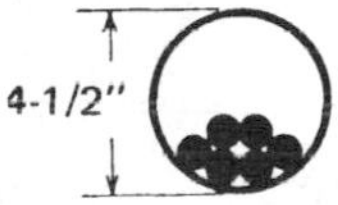
4-1/2"

A 3-phase circuit of six 600 kcmil type THW conductors in a 5-in. rigid metal conduit.
Current rating of each phase might appear to be = 2 X 420 = 840 amps.
But 80% derating must be applied because of the number of conductors in the conduit:
Current rating of each phase = 840 X 80% = 672 amps.
Cross-section area per phase = 2.0522 sq.in. (two conductors).

Fig. 310-13. These circuit makeups represent typical considerations in the application of multiple-conductor circuits. [Sec. 310.10(H).]

because above this size the increase in ampacity is very small in proportion to the increase in the size of the conductor. Thus, for a 50 percent increase in the conductor size, that is, from 1,000,000 to 1,500,000 cmil, the ampacity of a Type THW conductor increases only 80 A, or less than 15 percent, and for an increase in size from 1,000,000 to 2,000,000 cmil, a 100 percent increase, the ampacity increases only 120 A, or about 20 percent. In any case where single conductors larger than 500,000 cmil would be required, it is worthwhile to compute the total installation cost using single conductors and the cost using two (or more) conductors in parallel.

Figure 310-14 shows an interesting application of parallel conductors. A 1200-A riser is made up of three conduits, each carrying three phases and a neutral. At the basement switchboard, the 1200-A circuit of three conductors per phase plus three conductors for the neutral originates in a bolted-pressure switch with a 1200-A fuse in each of the three-phase poles. Because the total of 12 conductors make up a single 3-phase, 4-wire circuit, a 400-A, 3-phase, 4-wire tap-off must tap all the conductors in the junction box at top. That is, the three-phase A legs (one from each conduit) must be skinned and bugged together and then the phase A tap made from that common point to one of the lugs on the 400-A CB. Phases B and C must be treated the same way—as well as the neutral. The method shown in the photo was selected by the installer on the basis that the conductors in the right-hand conduit

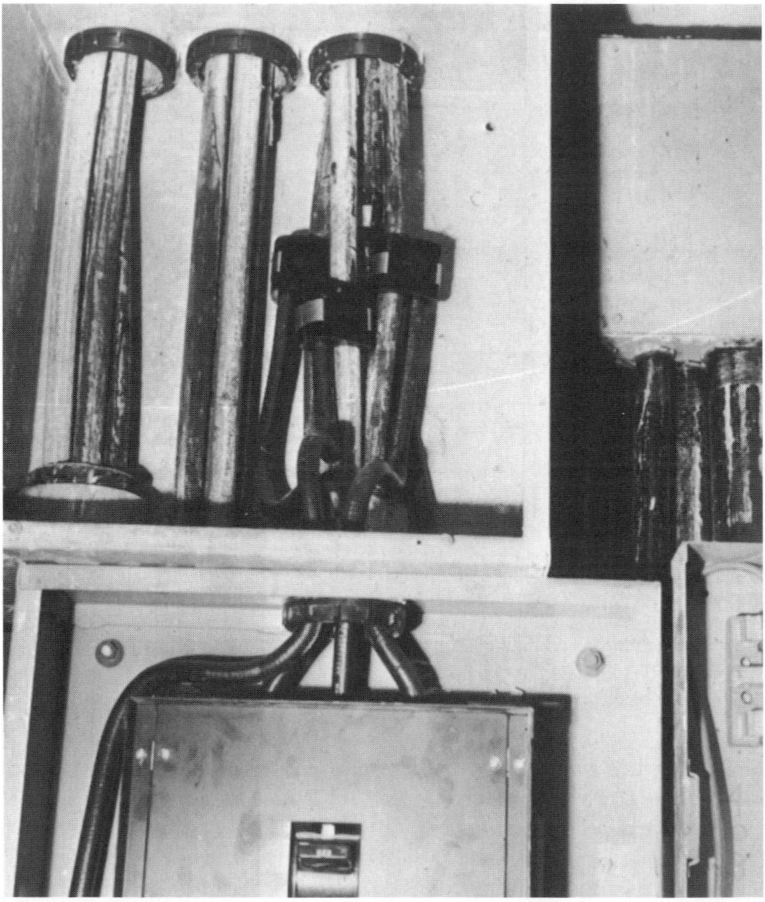

Fig. 310-14. A 1200-A circuit of three sets of four 500-kcmil conductors (top) is tapped by a single set of 500-kcmil to a 400-A CB (bottom) that feeds an adjacent meter center in an apartment house. This was ruled a violation because the tap must be made from all the conductors of the 1200-A circuit. (*Note:* The conduits feeding the splice box at top are behind the CB enclosure at bottom.) [Sec. 310.10(H).]

are tapped on this floor, the center-conduit conductors are tapped to a 400-A CB on the floor above, and the left-conduit conductors are tapped to a 400-A CB on the floor above that. But such a hookup can produce excessive current on some of the 500 kcmil. Because it does not have the parallel conductors of equal length at points of load-tap, the currents will not divide equally, and this is a violation of the second paragraph of 310.4, which calls for parallel conductors to "be the same length."

This arrangement is also an unauthorized tap, in violation of 240.21 because it amounts to creating a (in this case) one-third tap, and then (depending on conductor sizing) either using that tap at an excessive length, or tapping that tap, both of which violate the **NEC**. Review the discussion associated with Fig. 240-23 in Chap. 2 for more discussion of this concept.

Exception No. 1 of 310.(H)(1) permits parallel-circuit makeup using conductors smaller than 1/0 AWG—but all the conditions given must be observed.

This exception permits use of smaller conductors in parallel for circuit applications where it is necessary to reduce conductor capacitance effect or to reduce voltage drop over long-circuit runs. It also recognizes the benefits in terms of lower reactance for high-frequency circuits (360 Hz and higher) as are commonly used in the aerospace industry. As it was argued in the proposal for this exception:

> If a 14 AWG conductor, for example, is adequate to carry some load of not more than the 15-A rating of the wire, there can be no reduction in safety by using two 14 AWG wires per circuit leg to reduce voltage drop to acceptable limits—with a 15-A fuse or CB pole protecting each pair of 14 AWGs making up each leg of the circuit.

Where conductors are used in parallel in accordance with this exception, the rule requires that all the conductors be installed in the same raceway or cable.

And that will dictate application of the last sentence of 310.4: "Conductors installed in parallel shall comply with the provisions of 310.15(B)(3)a" which applies to wire of 0 to 2000 V. Thus a single-phase, 2-wire control circuit made up of two 14 AWGs for each of the two legs of the circuit would have to be considered as four conductors in a conduit, and the "ampacity" of each 14 AWG would be reduced to 80 percent of the value shown in Table 310.15(B)(16). If TW wires are used for the circuit described, the ampacity of each is no longer the value of 15 A, as shown in Table 310.15(B)(16). With four of them in a conduit, each would have an ampacity of 0.8 × 15, or 12 A. Then using a 15-A fuse or CB pole for each pair of 14 AWGs would properly protect the conductors and would also comply with the 240.4(D)(3) which says that No. 14 must not have overcurrent protection greater than 15 A. (See Fig. 310-15.)

Exception No. 2 now recognizes the use of parallel conductors in sizes down to No. 2 where used as a neutral, but only in an existing installation. This is a good idea where necessary to accommodate additive harmonics on the neutral of multiwire circuits. The use of two No. 2s provides about 25 percent less cross-sectional area while at the same time providing 25 percent more surface area than a 3/0. This serves to reduce the heating caused by skin effect because the "skin" area has been increased. This exception is limited to existing installations under engineering supervision.

Equipment grounding conductors are to be sized in accordance with 250.122, which means they are sized to the overcurrent device and need not comply with the 1/0 AWG minimum size limitation that applies to load-bearing circuit

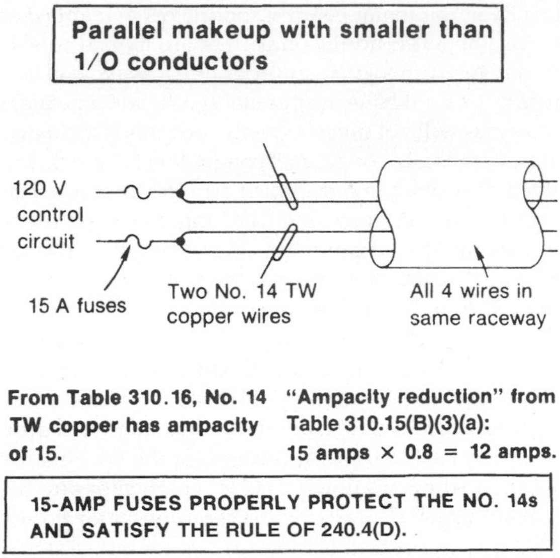
Parallel makeup with smaller than 1/O conductors
120 V control circuit
15 A fuses
Two No. 14 TW copper wires
All 4 wires in same raceway
From Table 310.16, No. 14 TW copper has ampacity of 15.
"Ampacity reduction" from Table 310.15(B)(3)(a): 15 amps × 0.8 = 12 amps.
15-AMP FUSES PROPERLY PROTECT THE NO. 14s AND SATISFY THE RULE OF 240.4(D).

Fig. 310-15. Overcurrent protection must be rated not in excess of the ampacity of one conductor when conductors smaller than No. 1/0 are used in parallel. Fifteen amperes is the rating of the next higher standard size of overcurrent protection, and therefore does protect the 14 AWG conductors in accordance with 240.4(B). The permitted loads under this exception do not include multiple receptacle outlets. [Sec. 310.10(H)(1) Exception No. 1.]

conductors as itemized in 310.10(H)(1). Equipment (and, by virtue of the 250.102 reference, supply-side) bonding jumpers are also governed by Art. 250 rules and not the 1/0 limit.

310.15. Ampacities for Conductors Rated 0 to 2000 Volts. 310.15 states that ampacities of conductors may be determined by either of the two methods. The first method is described in part **(B)** and is the old, tested, and familiar method of the NEC, based on Tables 310.16 through 310.21, as they were numbered prior to the 2011 NEC edition. The second permitted method is covered in part **(C)** of the NEC and is the complex, confusing, incomplete, and problematic procedure that was presented in the 1987 NEC as the basic method, based on an elaborate formula to be applied under very sophisticated engineering supervision, as stated there.

The NEC ampacity determination procedure using the formula is permitted as an optional alternative "under engineering supervision." The formula is generally known as the "Neher-McGrath" formula after the authors of a seminal 1957 paper describing a more precise way to measure ampacities, particularly for underground wiring. It takes into account all the factors that impede heat loss from a loaded conductor, including the ability of a raceway wall and the surrounding soil to conduct heat away from the conductors. These topics are considerably complicated by the effect of airspaces within the wiring, hysteresis losses in magnetic conduits (if applicable), and skin effects (depending in part on conductor size). These and other effects become the subject of elaborate side calculations that eventually generate the terms that are included in the overall formula in 310.15(C). Therefore the elaborate formula in the Code is actually deceivingly simple. A proper application of the Neher-McGrath formula requires an extremely experienced electrical engineer who has made these calculations a major part of his or her professional specialty. Such engineers are hard to find and generally employed by major high-voltage cable vendors.

All of the ampacity tables based on the formula are in Annex B in the back of the Code book, where information on the formula method and its related ampacity tables is introduced with the sentence, "This informative annex is **not** part of the requirements of this NFPA document but is included for informational purposes only." Thus, the 1987 NEC ampacity method is given as a nonmandatory, optional alternative to the old standby method. The tables do, however, allow for a more accessible format because they present results instead of requiring use of the formula in the field. Also included in Annex B is the old Note 8 ampacity reduction factors [Table B-310.15(B)(3)(11)] for multiple conductors in a raceway.

The second paragraph of **(A)** conveys the basic "weak-link-in-the-chain" principle that says when multiple ampacities apply to a conductor over the course of its run, then the lowest calculated ampacity must be used to determine the characteristics of the circuit. The exception allows a waiver in instances where adjacent portions of the conductor are capable of functioning as a heat sink, allowing the short section with a low ampacity to be ignored.

The exception has been in the NEC since the 1990 edition, and until the 2017 edition only applied to adjacent portions of a circuit. For example, suppose a circuit went through a boiler room, with the ampacity diminished by ambient temperatures in that space, on its way to its ultimate destination through areas with normal temperatures. The exception recognized that, to a degree, the heating from the boiler room could be offset by the process of thermal conduction. The adjacent parts of the circuit, to a limited degree, could be relied on as a heat sink for the boiler room exposure. The limits placed in the exception, 10 ft or 10 percent of the adjacent part of the circuit, whichever was less, were the result of a conversation with a renowned electrical engineer then serving on the code panel, based on his personal experience in Texas. The 2017 edition has retained the parameters, but removed the requirement of proximity. The substantiation for this completely ignored the essential concept of thermal conduction that was the basis for the exception in the first place. Until this change is reversed, the inspection community should consider challenging installations making use of it by using the next numbered paragraph.

The third paragraph of **(A)** is titled: **Temperature Limitation of Conductors.** This used to be 310.10 until the 2011 cycle. This rule says that no matter how your theoretical calculations seem to come out, no matter what number an ampacity table might seem to say is justified, if at the end of the day a wire will be consistently operating at a temperature above the ultimate temperature rating of its insulation, its operating conditions must change accordingly. The factors that influence this are set out in the detailed note following the rule. This rule is seldom cited by electrical inspectors. Instead, it is usually cited indirectly, such as through enforcement of 310.15(B)(3)(a) around a failure to allow for mutual conductor heating (thereby implicitly applying list item 4 from the note), or the failure to account for an elevated ambient temperature derating factor when using the pertinent ampacity table through enforcement of 240.4 (thereby implicitly applying list item 1 from the note), or a failure to account for heating in a neutral in response to triplen harmonic loading through enforcement of 310.15(B)(5)(c) (thereby implicitly applying list item 2 from the note), etc.

However, there are circumstances where this rule must be cited directly, because no other rule in the **NEC** can be cited. What follows is a real-world example, together with the underlying research that made it come together. This author was called on to inspect a single-family whole-house renovation that included a feeder from the basement to a panel in a second-floor storage area. The feeder was protected with a 100 A circuit breaker, and was wired with 2/3 aluminum SER cable with a 4 AWG equipment grounding conductor. The feeder ran through an outside wall of the building, where it went up two stories before swinging over through the attic floor joists and down to the panel. The house had been completely gutted and entirely new insulation consisting of polyurethane foam had been sprayed into all outside wall and ceiling joist cavities, completely embedding this cable.

Now to begin with, since the feeder was not carrying the full load of this single-family house, the special procedure in 310.15(B)(7) does not apply, and therefore the normal values in Table 310.15(B)(16) apply, which in this case would be 90 A, because the 75°C column must be used per 110.14(C)(1)(b). There were no continuous loads on this feeder. However, reducing the feeder protection to 90 A, although fixing the preceding violations, would still have completely failed to cure a blatant violation of 310.15(A)(3). The reasons for this make good reading.

When wiring is embedded in thermal insulation, the ampacity is significantly degraded, and foamed-in-place polyurethane foam has about the best "R" value of any commercially available thermal insulation. The results tend to get progressively worse for larger cable sizes, because larger cables are generally installed to meet an expectation of higher-current draw. The heat generated by a conductor is I^2R. Larger cables carrying larger currents reduce the heat by the first power due to larger radiating surface areas, and increase the heat by the square of the current increase. This means that large cables tend to be worse than smaller ones, because the latter increase in heat ($+\Delta I^2$) usually overwhelms the former decrease ($-\Delta R$).

In the 1987 **NEC** cycle, this was confirmed by actual NEMA testing, with dramatic results. The test used 2 AWG aluminum Type SE cable of the "SEU" style (two insulated conductors and the grounded conductor configured as a spirally wrapped conductor around the ungrounded conductors). This cable is used routinely for 100 A residential services. It was run through thermal insulation under controlled test conditions. Specifically, the cable was run embedded in cellulose insulation with 7 in. above it. Thermocouples were placed on the cable, and the various loads under test were maintained for long periods. This cable has a Table 310.15(B)(16) ampacity (terminations not considered for this purpose) of 100 A. When the cable was loaded to 100 A, the cable jacket was "completely charred" as well as adjacent "charred wood members," all while the cable was operating within its table ampacity limitations. In fact, the testing showed (65 A caused 96°C operation) the cable exceeded its rated operating temperature at any time the continuous current exceeded about two-thirds of the table ampacity.

These results strongly demonstrate the fact that the ampacity of a conductor is not necessarily what a table may predict. The ampacity of a conductor is its ability to carry current continuously under the conditions of use. This is determined by thermodynamics, not Code tables. The ampacity of 2 AWG aluminum

XHHW configured as 2/3 Al SEU cable and embedded in cellulose insulation is approximately 60 A. Actually, the ampacity corresponding to Table 310.15(B)(16) would be even less than that, because that table is structured around three current-carrying conductors and this testing only used two of the three conductors in the test circuit.

This testing focused on Type SE cable, but a test run using 6 AWG copper in metal conduit embedded in foam insulation overheated those conductors as well, although not as badly. Typical branch circuits do OK by starting in the usually mandated 60°C ampacity column. There is no ampacity table for thermal insulation, and it would be impossible to create one, because variations in R factors significantly change end results. However, it is fair to say, for example, that the ampacity of 4/0 Al SE cable (table ampacity of 205 A) embedded in thermal insulation is far less than even the 60°C table result of 150 A would predict, perhaps as low as 100 A. **NEC** 310.15(A)(3), now focusing on item 3 of its application note, is the only **NEC** rule that covers this point, and it is referenced in notes to a number of cable articles for this reason. To the extent practicable, good practice is to route larger circuits in such a way as to avoid contact with thermal insulation.

And so the citation was 310.15(A)(3), cured with a 60 A circuit breaker, whose terminals were fortunately large enough to accept the 2 AWG feeder conductors, that ran to a panel where, fortunately again, the load per Art. 220 was probably not over 35 A.

In part **(B)**, an informational note points out that "Tables 310.15(B)(16) through 310.15(B)(19) are application tables that are for use in determining conductor size on loads calculated in accordance with Art. 220." Inasmuch as the **NEC** itself requires that Art. 220 be used at all times in calculating loads, the ampacity-determination method of part **(B)** is completely adequate for all conductor sizing in accordance with all Code rules.

Correctly integrating load calculation results with ampacity table values, ambient temperature adjustment factors, mutual conductor heating derating factors, termination temperature limitations, etc., is quite possibly the most challenging task confronting electrical personnel today, and also one of the most necessary. After all, if we do nothing else in this trade, we select and install wires. Since there is now a comprehensive example that focuses on these issues at the end of the Code in Annex D [Example D3(a)], the place to integrate these rules is there. Refer to the comprehensive discussion of these issues in Chap. 9 of this book. In addition, there is an analysis with step-by-step calculations on Table B-310.15(B)(3)(11) so you can see how to bring the old Note 8 back to life.

Subparts to 310.15(B) (not in order)

(B)(2) Instead of repeating the same temperature correction factors under all the ampacity tables, the information is now only given once, and the ampacity tables all have a footnote that points here. Note that there are two tables, the first (a) applying to a default temperature of 30°C and that therefore applies to the familiar Table 310.15(B)(16). The second (b) applies to a default temperature of 40°C and applies to all the rest except 310.15(B)(17). The material concludes with the mathematical formula that allows an ampacity adjustment for any temperature to be calculated exactly if necessary.

(B)(7) This rule used to feature a special ampacity table for residential services and main feeders. The 2014 **NEC** removed the table and completely reformatted the rule, but in a way that makes the conductors selected under its provisions identical to those selected under previous **NEC** editions. The 2017 **NEC** has resurrected the old table, and inserted it into the calculation example of this procedure that is in Annex D as Example D7. For qualifying applications, the procedure now is, as before, to begin with a load calculation rounded up to the next higher standard size overcurrent device that will protect the relevant service or principal feeder. Then, instead of taking that number into a special ampacity table, multiply that number by 0.83 instead. If adjustments are required for mutual conductor heating [310.15(B)(3)(a)] or ambient temperature [310.15(B)(2)(a or b)], apply them. Then take the resulting number into Table 310.15(B)(16) and select a conductor with an ampacity, usually in the 75°C column as required by 110.14(C)(1)(b), that equals or exceeds the resulting number. There is a new example (D7) at the end of the Code book that shows this at work for a 175-A service. The calculation gives a result of 145 A, which translates to a 1/0 copper (or 3/0 aluminum) conductor, exactly the same as many previous decades. Note that although this is now a calculation instead of a table, it is not open ended as to size. Both services and feeders are still subject to the 400-A limit of the previous table approach.

This rule has been in the **NEC** in some form for a long time and permits use of certain conductors at ampacity values higher than those shown for the conductors in Table 310.15(B)(16). For instance, a 2/0 AWG THW copper conductor may be used protected at 200 A instead of at 175 A, as shown in Table 310.15(B)(16). This permission has been given by the **NEC** in recognition of the reality that residential service conductors are supplying loads of great diversity and of short operating periods or cycles, so that the conductors almost never see full demand load approaching their ampacity and certainly not for continuous operation (3 h or more). There are crucial conditions that underlie the acceptability of this rule; if any of these conditions is not met, then the rule is invalid and cannot be used.

1. The conductors are for an individual dwelling unit. They must not be for a collection of dwelling units such as an apartment dwelling, even though it might be argued that the entire load is residential in nature. The collective load of a multifamily dwelling will be calculated differently and may not have the diversity, after the applications of relevant Art. 220 demand factors, of an individual dwelling unit load. The rule has always been based on those individual applications and no technical substantiation has been presented over the years to support broadening the application.
2. The conductors must see the entire load of the dwelling unit, including loads such as outdoor air conditioning units and swimming pools, together with any outbuildings if provided with electric power. For example, removing an air conditioner from a load profile obviously reduces the load. It also increases the intensity of the load that remains on the feeder, and thereby works to undermine the validity of the ampacity values allowed in this section. The load that remains on those conductors will be the subject of a new load calculation reflecting the actual connected load that remains. This process of load shedding could continue until there was only a single load left which might even run continuously.

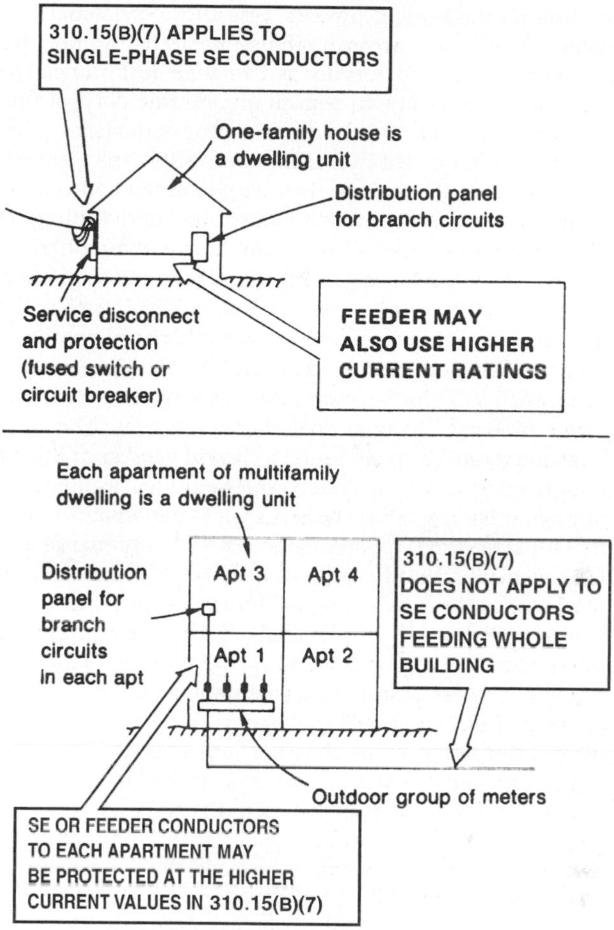
310.15(B)(7) APPLIES TO
SINGLE-PHASE SE CONDUCTORS
One-family house is
a dwelling unit
Distribution panel
for branch circuits
Service disconnect
and protection
(fused switch or
circuit breaker)
FEEDER MAY
ALSO USE HIGHER
CURRENT RATINGS
Each apartment of multifamily
dwelling is a dwelling unit
Distribution
panel for
branch
circuits
in each apt
Apt 3
Apt 4
Apt 1
Apt 2
310.15(B)(7)
DOES NOT APPLY TO
SE CONDUCTORS
FEEDING WHOLE
BUILDING
Outdoor group of meters
SE OR FEEDER CONDUCTORS
TO EACH APARTMENT MAY
BE PROTECTED AT THE HIGHER
CURRENT VALUES IN 310.15(B)(7)

Fig. 310-16. Higher ratings of protective devices may be used for service conductors to "dwelling units." [Sec. 310.15(B)(7).]

The provisions of (4) of 310.15(B)(7) permit the neutral conductor of these 3-wire services and 3-wire feeders to be smaller than the hot conductors because the neutral carries only the unbalanced current of the hot legs and is not at all involved with 2-wire, 240-V loads. However, it must not be sized smaller than 215.2, 220.61, and 230.42 requirements.

Note, however, that this fourth paragraph allowance for reduced neutral sizing does not apply to circuits originating from two phase legs of a 208Y/120-V three-phase distribution.

This limitation on neutral size reductions is all that remains, as of the 2017 NEC, of the former blanket prohibition regarding the use of 310.15(B)(7) for dwelling units, typically in large apartment buildings, that were supplied by such distributions. The action to remove this prohibition is extremely controversial, because, quite frankly, it defies the laws of physics. One of the principal reasons the traditional allowance works is that on a single-phase distribution the worst-case heating occurs with two conductors loaded to their full ampacity. This occurs when the entire load is either line to neutral on one side only, or line to line. The heating declines in proportion to the square of any reductions in current as the load becomes more evenly distributed. Since ampacity tables are based on three conductors operating fully loaded, the result is a justified bonus allowance under conditions that guarantee only two-thirds of the calculated heating will be present.

The 2017 **NEC** applies this deserved bonus for single-phase distributions equally to a distribution where all three conductors will see essentially full-loading assuming complete line-to-neutral loading, the worst case. The lowest loading will occur when one line conductor is at 100 percent and the other at 50 percent, in which case the neutral current is reduced to 86 percent ($\frac{1}{2}\sqrt{3}$) of the higher loaded phase conductor. This is why 310.15(B)(5)(b) requires such neutrals to be taken at 100 percent, and, indeed, the **NEC** is not changing that rule in any way. The evident technical validity of that requirement disqualifies the technical validity of this **NEC** change.

The substantiation for this change made essentially two arguments, based on the experience in Canada. The Canadian Code (CEC) allows what is now allowed here, and has done so for some time, apparently with no loss experience to point to as an objection. What this more properly suggests is that the CEC, as is the case with the **NEC**, overstates the actual load on residential feeders. The electric utilities have been telling us this for many decades. If you routinely overbuild a class of feeders, then you can tolerate a bogus allowance on how you figure ampacity. However, as Art. 220 becomes more genuine in this area, and it is gradually doing so, we will begin to have problems. Neither the CEC nor the **NEC** gets to ignore the laws of physics.

Part (B)(4) 310.15(B)(4) provides that if an uninsulated conductor is used with insulated conductors in a raceway or cable, its size shall be the size that would be required based on it being a conductor having the same insulation as the lowest insulation temperature rating of the adjacent insulated conductors. The point is to assure that its surface temperature will not exceed the temperature rating of any adjacent conductor insulation (Fig. 310-17).

Bare conductor has the capacity of a conductor of its size that has...
... the same insulation as used on the insulated conductors run with the bare conductor.

Fig. 310-17. How to figure ampacity of a bare conductor, where permitted. (Sec. 310.15.)

example Two 6 AWG Type THW conductors and one bare 6 AWG conductor in a raceway or cable. The ampacity of the bare conductor would be 65 A.

If the insulated conductor were Type TW, the ampacity of the bare conductor would be 55 A.

Part (B)(3) Where more than three current-carrying conductors are used in a raceway or cable, their current-carrying capacities must be reduced to compensate for the proximity heating effect and reduced heat dissipation due to reduced ventilation of the individual conductors that are bunched or form an enclosed group of closely placed conductors. Where the number of conductors in a raceway or cable exceeds three, the ampacity of each conductor shall be reduced as indicated in the table of part **(B)(3).**

If, for instance, four 8 AWG THW copper conductors are used in a conduit, the ampacity of each 8 AWG is reduced from the 50-A value shown in the table to 80 percent of that value. In such a case, each 8 AWG then has a new reduced ampacity of 0.8×50 A, or 40 A. And, from 240.4, "Conductors shall be protected in accordance with their ampacities." Thus, 40-A-rated fuses or CB poles would be required for overcurrent protection as the general rule.

The application of those part **(B)(3)** conductors and their protection rating is based on the general concept behind the NEC tables of maximum allowable current-carrying capacities (called "ampacities"). The NEC tables of ampacities of insulated conductors installed in a raceway or cable have always set the maximum continuous current that a given size of conductor can carry continuously (for 3 h or longer) without exceeding the temperature limitation of the insulation on the conductor, that is, the current above which the insulation would be damaged. But, because the overcurrent devices were tested with conductors sized at 125 percent of the continuous current plus the noncontinuous, the conductor's prederated ampacity must be increased where supplying a continuous load [see 210.19(A)].

This concept has always been verified in the informational note to 240.1, where the wording has been virtually identical for over 30 years and says, "Overcurrent protection for conductors and equipment is provided to open the circuit if the current reaches a value that will cause an excessive or dangerous temperature in conductors or conductor insulation." To correspond with that objective, 240.4 says, "Conductors, other than flexible cords and fixture wires, shall be protected against overcurrent in accordance with their ampacities as specified in 310.15."

Table 310.15(B)(16), for instance, gives ampacities under the conditions that the raceway or cable containing the conductors is operating in an ambient not over 30°C (86°F) and that there are not more than three current-carrying conductors in the raceway or cable. Under those conditions, the ampacities shown correspond to the thermal limit of the particular insulations. But if either of the two conditions is exceeded, allowable ampacities have to be reduced to keep heat from exceeding the temperature limits of the insulation:

1. If ambient is above 30°C, the ampacity must be reduced in accordance with the correction factors given in Table 310.15(B)(2)(a) [or (b) depending on the ampacity table under consideration].
2. If more than three current-carrying conductors are used in a single cable or raceway, the conductors tend to be bundled in such a way that their heat-dissipating capability is reduced and excessive heating would occur at the ampacities shown in the table. As a result, part **(B)(3)** requires reduction of ampacity, and conductors must be protected at the reduced ampacity.

(*Note:* It should be clearly understood that any reduced ampacity—required for higher ambient and/or conductor bundling—has the same meaning as the value shown in a table: Each represents a current value above which excessive heating would occur under the particular conditions. And if there are two conditions that reduce heat dissipation, then more reduction of current is required than for one condition of reduced dissipation.)

Part (B)(3) Requires Derating of Ampacity

Part **(B)(3)(a)** says, "Where the number of current-carrying conductors in a raceway or cable exceeds three . . . the allowable ampacity of each conductor shall be reduced as shown in Table 310.15(B)(3)(a)." This table has a heading on the right to require that any ampacity adjustment for elevated ambient temperature must be made in addition to the one for number of conductors. If, for instance, four No. 8 THHN current-carrying copper conductors are used in a conduit, the ampacity of each No. 8 is reduced from the 55-A value shown in Table 310.15(B)(16) to 80 percent of that value. Each No. 8 then has a new (reduced) ampacity of 0.8 × 55 A, or 44 A. Then, if a derating factor must be applied because the conductors are in a conduit where the ambient temperature is, say, 40°C instead of 30°C, the factor of 0.91 (36°C to 40°C) from Table 310.15(B)(2)(a) must be applied to the 44-A current value to determine the final value of ampacity for the conductors (44 × 0.91 = 40 A). Moreover, 240.4 of the **NEC** states, "Conductors, other than flexible cords and fixture wires, shall be protected against overcurrent in accordance with their ampacities as specified in 310.15." Thus, fuses or CB poles rated at 40 A would be required. The ampacity of the conductors is changed and the conductors must be protected in accordance with the derated ampacity value and not in accordance with the tabulated value.

Because conductor ampacity is reduced when more than three conductors are used in a conduit, the overcurrent protection for each phase leg of a parallel makeup in a single conduit would generally have to be rated at not more than the sum of the derated ampacities of the number of conductors used per phase leg. That would satisfy 240.4, which requires conductors to be protected at their ampacities. Because ampacity is reduced in accordance with the percentage factors given in part **(B)(3)** for more than three conductors in a single conduit, that derating dictates the use of multiple conduits for parallel-makeup circuits to avoid the penalty of loss of ampacity.

Figure 310-18 shows examples of circuit makeups based on the unsafe concept of load limitation instead of ampacity derating, as applied to overcurrent rating and conductor ampacity—which is a Code violation, because the conductors are not protected in accordance with their ampacities.

Although part **(B)(5)** exempts only neutral conductors from those conductors that must be counted in determining load-limiting factors for more than three conductors in a raceway or cable [per part **(B)(3)**], similar exemption is, as of the 2014 **NEC**, allowed for one of the "travelers" in a three-way (or three- and four-way) switch circuit. As shown in Fig. 310-19, only one of the two conductors is a current-carrying conductor at any one time; therefore, the other should not be counted for load limitation purposes where such switch legs are run in conduit or EMT along with other circuit conductors. No change in Code language is needed to accomplish this; when one of two wires in a conduit necessarily carries zero amperes at all times, even if the wire at zero switches its identity from time to time, the number of current-carrying conductors countable in that group of two is—exactly one.

The reason the 2011 **NEC** dropped the term "current-carrying" from the table heading was because of concerns that spare conductors could be too easily brought into service without benefit of inspection. Therefore, spare conductors are now counted for ampacity adjustment purposes. The "current-carrying" limitation originated in the 1993 **NEC**, and the 2011 **NEC** position stands the 1993

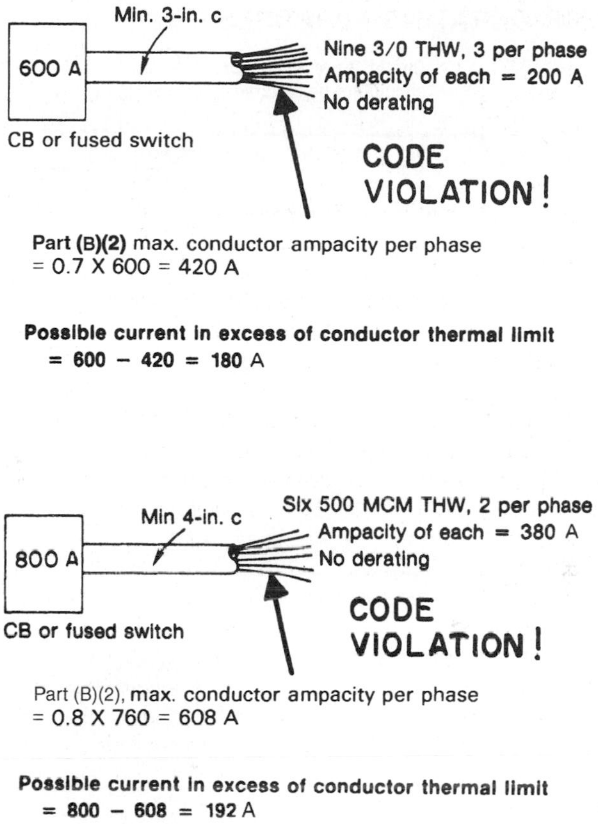
Min. 3-in. c
600 A
CB or fused switch
Nine 3/0 THW, 3 per phase
Ampacity of each = 200 A
No derating
CODE
VIOLATION!
Part (B)(2) max. conductor ampacity per phase
= 0.7 X 600 = 420 A
Possible current in excess of conductor thermal limit
= 600 − 420 = 180 A
Min 4-in. c
800 A
CB or fused switch
Six 500 MCM THW, 2 per phase
Ampacity of each = 380 A
No derating
CODE
VIOLATION!
Part (B)(2), max. conductor ampacity per phase
= 0.8 X 760 = 608 A
Possible current in excess of conductor thermal limit
= 800 − 608 = 192 A

Fig. 310-18. Parallel-conductor makeup must not be used in single conduits without ampacity reduction, even if load current is limited to the conductor ampacity as shown here.

substantiation on its head, which noted that spare conductors are a heat sink and improve the thermal performance of the assembly.

The 2014 NEC has split the difference in the process of revising the table note. Spare wires are now, unambiguously, counted in the wire fill, but wires subject to noncoincident loading are not. Although the three-way switch travelers are viewed by some as trivial, they can make a difference that is technically unjustified. There are other cases, particularly in industrial applications, where this issue is extremely serious. One instance wired by the author concerned two large, identical 3-phase motors fed through a common raceway that were wired, to save money, to a single-variable frequency drive through an interlocking contactor assembly that precluded simultaneous operation. The nature of the process involved made this practical because there were two sewage lines for which the plumbing arrangements precluded simultaneous operation. Literal enforcement of the prior table note would have caused an increase in wire size that in turn would have required an increase in raceway size.

Figure 310-20 shows a condition of bunched or bundled (now clarified as "installed without maintaining spacing for a continuous length") Type NM cables where they

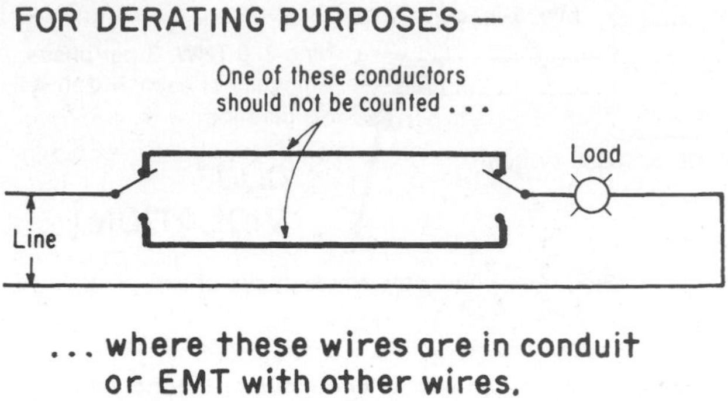
FOR DERATING PURPOSES —
One of these conductors should not be counted . . .
Load
Line
. . . where these wires are in conduit or EMT with other wires.

Fig. 310-19. The three-wire run in conduit between three-way switches contains only two current-carrying conductors. Whether both travelers (resulting in a total of three wires) must be counted depends on how the inspector applies the wording in the parent section ("current-carrying conductors") to the table column heading ("conductors"). The same issue applies to any instance where multiple sets of conductors serving noncoincident loads run together. (Sec. 310.15.)

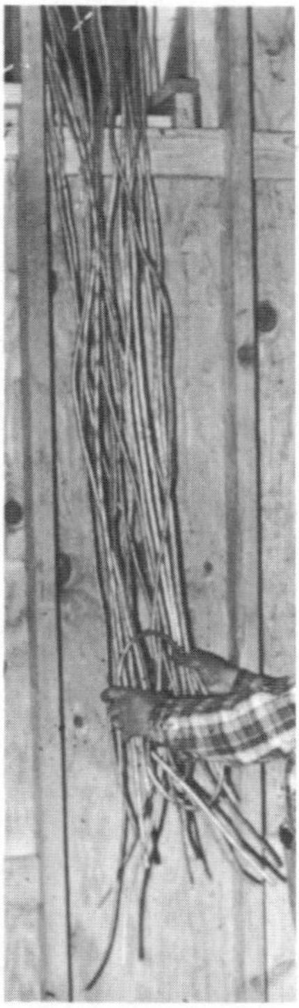

Fig. 310-20. If a large number of multiconductor cables are bundled together in a stud space, capacity derating in accordance with part **(B)(3)** would be required. If the individual cables are spaced apart and stapled, then the conductors in the cables may be loaded up to their rated ampacity values from Table 310.15(B)(16). (Sec. 310.15.)

come together at a panelboard location. A better example would be the common practice of running an entire group of cables across a basement ceiling through a single set of holes through adjacent joists. It looks very workmanlike but it will compromise ampacity, especially if there are enough conductors represented to drop the ampacity to a lower standard size overcurrent device rating. This is a particular problem for receptacle circuits, because 240.4(B)(1) will not allow the next higher standard size device to protect such conductors. The rule of 310.15(B)(3)(a) requires conductors in bundled cables to have their load currents reduced from the ampacity values shown in Table 310.15(B)(16). The ampacity derating is required for conductor stacks or bundles that are longer than 2 ft (24 in. or 600 mm). For shorter bundles, derating is not required in general; however, note that Type NM cable has an effective length of not much more than zero if multiple cables run through draft- or fire-stopped holes in framing members. Item No. (2) excludes the need for derating groups of four or more conductors installed in nipples not over 24-in. (600 mm) long.

Item No. (3) in 310.15(B)(3)(a) says that underground conductors that are brought up aboveground in a protective raceway [300.5(D)(1)] do not require derating if not more than four conductors are used and if the protective conduit has a length not over 10 ft (3.0 m) "above grade." The total length of raceway may exceed 10 ft (3.0 m). The phrase *above grade* clearly limits the length of protective conduit that may contain conductors without derating in accordance with Table 310.15(B)(3)(a). The 10-ft (3.0-m) length covers the length of 8 ft (2.5 m) above grade but not the 1½ ft (450 mm) into the earth given by 300.5(D)(1) on conductors emerging from underground.

310.15(B)(3) does not apply to conductors in metal wireways and auxiliary gutters, as covered in 366.23 and 376.22. Metal wireways or auxiliary gutters may contain up to 30 current-carrying conductors at any cross section [excluding signal circuits and control conductors used for starting duty only between a motor and its starter in auxiliary gutters, 366.22(A)] before limitations for mutual conductor heating take effect. Load-limiting factors for more than three conductors do not apply to a metal wireway, the way they do to wires in conduit. However, if the derating factors from Table 310.15(B)(3)(a) are used, there is no limit to the number of wires permitted in a wireway or an auxiliary gutter. However, the sum of the cross-sectional areas of all contained conductors at any cross section of the wireway must never exceed 20 percent of the cross-sectional area of the wireway or auxiliary gutter. More than 30 conductors may be used under those conditions.

Part **(B)(3)(b)** says that spacing between "conduits, tubing, or raceways shall be maintained." The most significant thing about this provision is not what it says, but what it very nearly said, and was stopped from saying at the last minute.

When the **NEC** panel began to seriously question the adequacy of derating factors for mutual conductor heating in the 1987 Code edition, it didn't stop with the new values (which are still in the **NEC**) in what is now (a). It also was going to add a table to what is now (b). This table would have mandated an *additional* set of derating factors for conduit and tubing run on trapeze hangers where the spacing was less than a full raceway diameter apart! The table was multidimensional in that the additional penalties would have applied not just side-to-side, but also vertically in the event more than one trapeze was stacked up the same set of threaded rods. For example, three trapeze hangers with six raceway runs each, and with less than a raceway diameter separation in both dimensions, would have produced an additional derating factor of 0.74. Even two conduits on a trapeze hanger, if the conduits were separated by less than a full raceway diameter, would have generated an additional factor of 0.94. There was also language that probably would have been applied as a complete prohibition against any separation less than a quarter of a raceway diameter.

UL, as it has so many times, saved the industry by running tests in the comment period showing that with any type of conventional conduit clamps on adjacent conduits in the trapeze hanger, there was enough separation and air flow so the mutual heating was negligible. Anyone with an ounce of common sense and field experience already knew that. Since no one is throwing bushel baskets of conduits on trapeze hangers without securing them with some kind of hardware, we can pretty much ignore (b). No loss experience was ever documented to support the inclusion of either the note or the derating factors in the original proposal. However, be advised that it is a Code violation to throw bushel baskets of conduits on trapeze hangers without securing them individually with some kind of hardware that provides a modicum of spacing.

Part **(B)(3)(c)** was new in the 2008 NEC, and is based on substantial research surrounding the effects of heat re-radiating from rooftops on raceways and cables exposed to direct sun. The rule, substantially decreased in severity in the 2017 NEC, now only applies rooftop adders to wiring mounted within 23 mm ($^7/_8$ in.) of the roof. In such cases, a 33°C (60°F) temperature correction is applied to the rated outdoor ambient temperature. The dimension was chosen as an approximation of the spacing afforded by shallow profile strut. The resulting temperature to be taken through the ampacity correction factors in Table 310.15(B)(16) becomes the sum of the number in the table based on raceway position, and the normal ambient temperature for the geographic location. There is a note that suggests using the ASHRAE Handbook to look up ambient temperature data. An exception in the 2017 NEC exempts conductors with XHHW-2 from these requirements due to the robust thermal performance of the insulation. Note that the 2017 NEC added a new thermoset conductor insulation style for wet locations (XHWN-2) but that is not as yet correlated here.

The ASHRAE publication is based on 30 years of data, and looks at three different percentiles for the 3-month period from June through August. The 0.4 percent number is the temperature that will be exceeded for 0.4 percent of the total time in a 1-month period. Since a month is 720 h, this temperature will be exceeded for 3 h a month. The tables also calculate 1.0 and 2.0 percent numbers the same way. The number usually employed for these purposes is the 2.0 percent number, corresponding to 14 h a month. In the example that follows, we will use Hartford, Connecticut, for which the 2.0 percent number is 32°C (90°F). The Copper Development Association has an extensive extract of this data from many U.S. cities available online.

example An IMC conduit will supply a rooftop air-conditioning unit, routed 4.5 m (15 ft) across the roof running on Minerallac style conduit supports that hold the conduit ½ in. off the roof. The unit has a nameplate FLA of 120 A and it will be connected to a 480Y/277-V system using THHN/THWN-2 copper wire, with the conduit as the equipment grounding conductor. What size wire and conduit is required?

answer: The wire size must be initially selected on the basis of the 75°C column in Table 310.15(B)(16) based on 110.14(C)(1)(b), but using 125 percent of the FLA per 440.32.

120 A × 1.25 = 150 A. Note that this is to be used as the operating current for Code purposes. The wire cannot be smaller than 1/0 AWG in order to not overheat the terminations. However, now we need to look at conductor ampacity. The ampacity of this wire is 170 A from the table before adjustments. As previously noted, the conduit supports hold the raceway less than the minimum limit of 23 mm ($^7/_8$ in.) away from the roof, so the prescribed temperature adder must be applied of 33°C (60°F), resulting in an applicable temperature of 150°F (66°C). According to the correction factors in Table 310.16(B)(2)(a), this results in a correction factor of 0.58. Since the table ampacity is 170 A, the ampacity on the roof will be 170 A × 0.58 = 99 A. The wire is far too small, but would generally have been allowed prior to the 2008 **NEC**. Without the rooftop adjustment the ampacity correction factor is 0.96, for 163 A. The 33°C (60°F) adder lowered the resulting ampacity by 64 A.

Three sizes up, 4/0 AWG (table ampacity = 260 A), just works: 260 A × 0.58 = 151 A.

Part (B)(5) In the determination of conduit size, neutral conductors must be included in the total number of conductors because they occupy space as well as phase conductors. A completely separate consideration, however, is the relation of neutral conductors to the number of conductors, which determines whether ampacity derating must be applied to conductors in a conduit, as follows.

Neutral conductors that carry only unbalanced current from phase conductors (as in the case of normally balanced 3-wire, single-phase or 4-wire, 3-phase circuits supplying resistive loads) are not counted in determining ampacity derating of conductors on the basis of the number in a conduit, as described. A neutral conductor used with two phase legs of a 4-wire, 3-phase system to make up a 3-wire feeder is not a true neutral in the sense of carrying only current unbalance. Such a neutral carries the same current as the other two conductors under balanced load conditions and must be counted as a phase conductor when more than three conductors in conduit are derated.

The technical basis for the full-current loading on the 3-wire feeder neutral from a wye system is based on phasor analysis that shows that if you have a circuit consisting of two-phase conductors and a neutral from a wye distribution with 100 A of line-to-neutral current on each phase conductor, perfectly balanced, the neutral will carry 100 A as well. Note that the identical load profile connected to a single-phase distribution would result in a zero ampere loading on the neutral. Further, as the load on one of the phase conductors in this example decreases, the load on the neutral does decline slightly, reaching a minimum of about 87 A ($\frac{1}{2}\sqrt{3}$ × highest load) and then rising again to 100 A when the load on the more lightly loaded phasor reaches zero and we are left with 100 A in what has become effectively a two-wire circuit. Therefore, these neutrals are effectively fully loaded at all times and must be counted accordingly. By the way, the minimum loading in these cases occurs when one phasor carries 100 A and the other is carrying one-half that amount or 50 A.

Because the neutral of a 3-phase, 4-wire wye branch circuit or feeder to a load of fluorescent, metal-halide, mercury, or sodium lamp lighting or to electronic data processing equipment—the so-called information technology equipment—or any other nonlinear load will carry harmonic current even under balanced loading on the phases (refer to 220.61), such a neutral is not a true noncurrent-carrying conductor and must be counted as a phase wire when the number of conductors to arrive at an ampacity derating factor is determined for more than three conductors in a conduit. As a result, all the conductors of a 3-phase, 4-wire branch circuit or feeder to a fluorescent load would have an ampacity of only 80 percent of their nominal ampacity from Table 310.15(B)(16) or other ampacity table. Because the 80 percent is a derating of ampacity, the conductors must be protected at the derated ampacity value.

Figure 310-21 shows four basic conditions of neutral loading and the need for counting the neutral conductor in loading a circuit to fluorescent or mercury ballasts, as follows:

Case 1—With balanced loads of equal power factor, there is no neutral current, and consequently no heating contributed by the neutral conductor. For purposes of heat derating according to the Code, this circuit produces the heating effect of only three conductors.

Case 2—With two phases loaded and the third unloaded, the neutral carries the same as the phases, but there is still the heating effect of only three conductors.

Case 3—With two phases fully loaded and the third phase partially loaded, the neutral carries the difference in current between the full phase value and the partial phase value, so that again there is the heating effect of only three full-load phases.

Case 4—With a balanced load of fluorescent ballasts, third-harmonic current generation causes a neutral current approximating phase current, and there will be the heating effect of four conductors. Such a neutral conductor must be counted with the phase conductors when the load-current limitation due to conduit occupancy is determined, as required in part **(C)** of part **(B)(5).**

Part (B)(6) This note makes it clear that an equipment grounding conductor or bonding conductor, which under normal conditions is carrying no current, does not have to be counted in determining ampacity derating of conductors when more than three conductors are used in a raceway or cable. As a result, equipment grounding and bonding conductors do not have to be factored into the calculation of required ampacity derating specified in part **(B)(3).** Part **(C)** is the Neher-McGrath method of ampacity calculations. Refer to the extensive discussion in 310.15(A)(1) for more information on this approach and how it contrasts with the ampacity table approach.

310.60. Conductors Rated 2001 to 35,000 Volts. Since the 1975 edition, the NEC has added a vast amount of information and data for conductors rated over 600 V, up to 35 kV. For instance, Tables 310.60(C)(67) through (86) give maximum continuous ampacities for copper and aluminum solid dielectric insulated conductors rated from 2001 to 35,000 V. Note that there are two ampacity columns in the tables, one for 90°C and one for 105°C. This sets up a comparable situation to the one at 600 V and below where there are termination rules and ampacity rules. In this case, however, the terminations take on the 90°C column per 110.40 regardless of conductor choice, but the 105°C numbers can be used as the starting point for ampacity calculations based on environmental factors. This section also includes, in 310.60(C)(4), an ambient temperature ampacity correction table and corresponding formula to the one at 310.15(B)(2) for 600 V and lower applications.

310.104. Conductor Constructions and Applications. Table 310.104(A) presents application and construction data on the wide range of 600 V insulated, individual conductors recognized by the NEC, with the appropriate letter designation used to identify each type of insulated conductor. The table title states that its

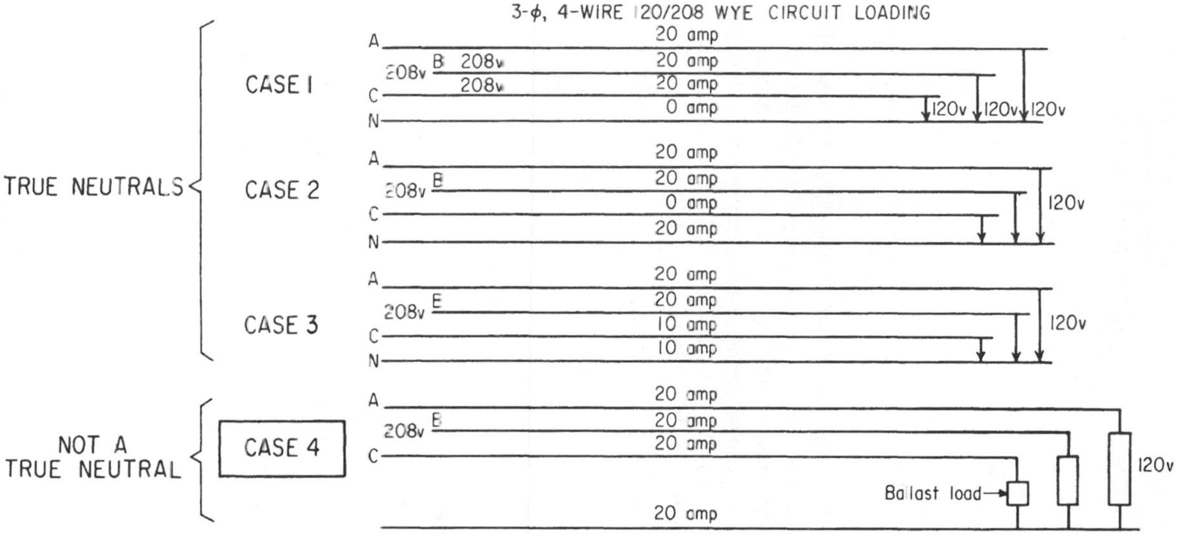
3-ϕ, 4-WIRE 120/208 WYE CIRCUIT LOADING
TRUE NEUTRALS
CASE 1
CASE 2
CASE 3
NOT A TRUE NEUTRAL
CASE 4
A
B
C
N
208v
20 amp
0 amp
10 amp
120v
Ballast load

Fig. 310-21. All neutrals count for conduit fill but only "true neutrals" do not count in determining ampacity derating for number of conductors in a raceway or cable [Part (B)(5)(b)]. (Sec. 310.15.)

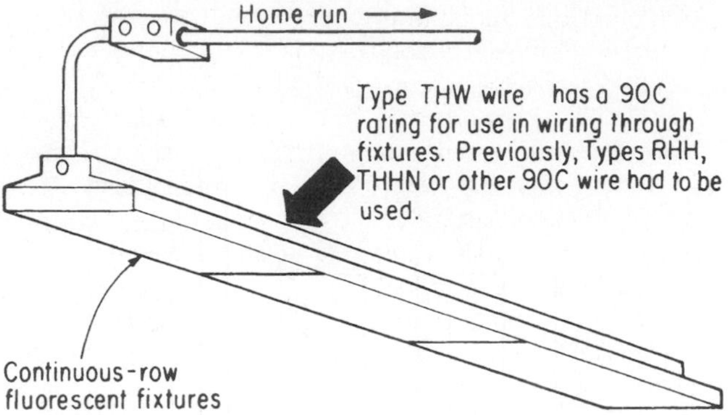
Home run
Type THW wire has a 90C rating for use in wiring through fixtures. Previously, Types RHH, THHN or other 90C wire had to be used.
Continuous-row fluorescent fixtures

Fig. 310-22. THW wire has the 90°C rating required of conductors within 3 in. (76 mm) of a ballast. (Secs. 410.68 and 310.104.)

contents apply to conductors "rated 600 V." The 2014 NEC has added an interesting footnote to the title and therefore applicable to all conductors throughout, stating that "Conductors can be rated up to 1000 V if listed and marked." Figure 310-22 shows a typical detail on applications, as covered for Type THW conductors in NEC Table 310.104(A). Type THW wire has a special application provision for electric-discharge lighting, which makes THW an acceptable answer for installers needing a 90°C conductor for wiring end-to-end fixtures in compliance with Sec. 410.68.

Important data that should be noted in Table 310.104(A) are as follows:

1. The designation for "thousand circular mils" is "kcmil," which has been substituted for the long-time designation "MCM" in this table and throughout the NEC.
2. Type MI (mineral insulated) cable may have either a copper or an alloy steel sheath.
3. Type RHW-2 is a conductor insulation that is moisture- and heat-resistant rubber with a 90°C rating, for use in dry and wet locations.
4. Type XHHW-2 is a moisture- and heat-resistant cross-linked synthetic polymer with a 90°C rating, for use in dry and wet locations.
5. The suffix "LS" designates a conductor insulation to be "low smoke" producing and flame retardant. For example, Type THHN/LS is a THHN conductor with a limited smoke-producing characteristic.
6. Type THHW is a moisture- and heat-resistant insulation, rated at 75°C for wet locations and 90°C for dry locations. This is similar to THWN and THHN without the outer nylon covering but with thicker insulation. These insulation styles are also available with a "–2" suffix, enabling use in wet applications at the full 90°C rating.
7. All insulations using asbestos—A, AA, AI, AIA, AVA, etc.—have been deleted from Table 310.104(A) because they are no longer made.

Conductors intended for 600 V (and up to 2000 V) general wiring under the requirements of the National Electrical Code are required to be one of the recognized types listed in Code Table 310.104(A) and not smaller than No. 14 AWG. The National Electrical Code does not contain detailed requirements for insulated conductors because these are covered in separate standards such as those of Underwriters Laboratories Inc.

Table 310.104(A) permits maximum operating temperatures of 90°C (194°F) in dry and damp (but not "wet") locations for Types FEP, FEPB, RHH, XHHW, and THHN wire; but the load-current ratings for 14, 12, and 10 AWG copper conductors and 12 and 10 AWG aluminum conductors are limited to those permitted by the maximum overcurrent protection ratings given in 240.4(D). One reason is the inability of 15-, 20-, and 30-A CBs to protect these sized conductors against damage under short-circuit conditions. The other reason is that the wiring devices that are commonly connected by these sizes of conductors are not suitable for conditions encountered at higher current loadings.

Conductors for medium-voltage circuits (over 2000 V) must satisfy Tables 310.104(B) through (E) as applicable, and the ampacity values given in Tables 310.60(C)(67) through (86) subject to the conditions in 310.60.

310.106. Conductors. Part **(C)** generally requires that 8 AWG and larger conductors must be stranded when they are installed in conduit, EMT, or any other "raceway." The use of an insulated or stranded 8 AWG copper conductor is permitted as the equipment bonding conductor required by 680.23(B)(2)(b). But only a solid 8-AWG copper conductor is permitted by 680.26(B) at swimming pools for bonding together noncurrent-carrying metal parts of pool equipment—metal ladders, diving board stands, pump motor frames, lighting fixtures in wet niches, and so on. The reason is that in the first instance the wire will be pulled into a raceway, and in the second the conductor will be used for multiple bonding connections in exposed or corrosive locations where stranded wire does not survive.

Although conductors are generally required by part **(D)** to be insulated for the phase-to-phase voltage between any pair of conductors, bare conductors may be used for equipment grounding conductors, for bonding jumpers, for grounding electrode conductors generally, and for grounded neutral conductors in certain locations (230.22, 230.30, 230.41, 230.62, 250.140, and 338.3).

The application shown in Fig. 310-23 is a commonly encountered violation of 310.106(D) because it involves an unauthorized use of a bare conductor. 250.140

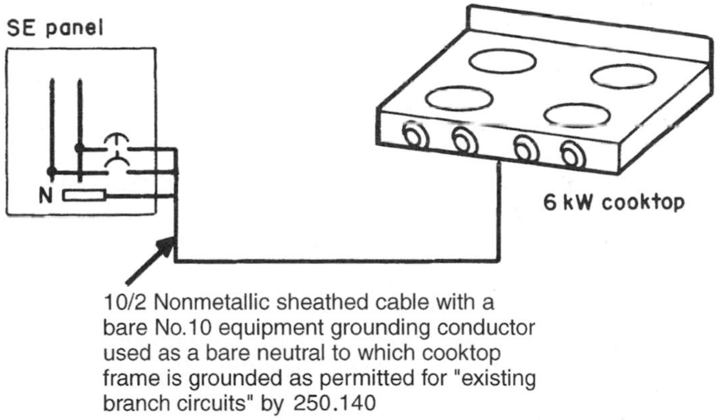
SE panel
N
6 kW cooktop
10/2 Nonmetallic sheathed cable with a bare No.10 equipment grounding conductor used as a bare neutral to which cooktop frame is grounded as permitted for "existing branch circuits" by 250.140

Fig. 310-23. This is a controversial application because many believe it to be authorized by 250.140. Actually, however, it is one that violates Secs. 310.106(D) and 250.140, because it does not conform to Condition (3) of the exception thereto, and therefore is not "specifically permitted elsewhere in this Code to be covered or bare." [Sec. 310.106(D).]

exception permits grounding of existing ranges, cooktops, and ovens to the neutral conductor, and then only where "the grounded conductor (the neutral) is insulated" or is a bare neutral of an SE cable that originates at the service equipment. Where supplied from a new circuit, a separate equipment ground must be provided.

310.106(D) states that "conductors shall be insulated," except where covered or bare conductors (see definition in Art. 100) are specifically permitted elsewhere in the Code. As noted above, the applicable rule in Art. 250 does not allow this arrangement on new installation, or on uninsulated neutrals under any other circumstance than Type SE cable is used, and only if the branch circuit originates at the service panel. The same objection would apply to any other wiring method.

Although the basic rule of this section requires conductors to be insulated, a note refers to 250.184 on the use of solidly grounded neutral conductors in medium-voltage systems. As an exception to the general rule that conductors must be insulated, 250.184 does permit a neutral conductor of a solidly grounded wye system to have insulation rated at only 600 V (Fig. 310-24). It also points out that a bare copper neutral may be used for service-entrance conductors or for direct buried feeders, and bare copper or copper-clad aluminum may be used for overhead sections of outdoor circuits.

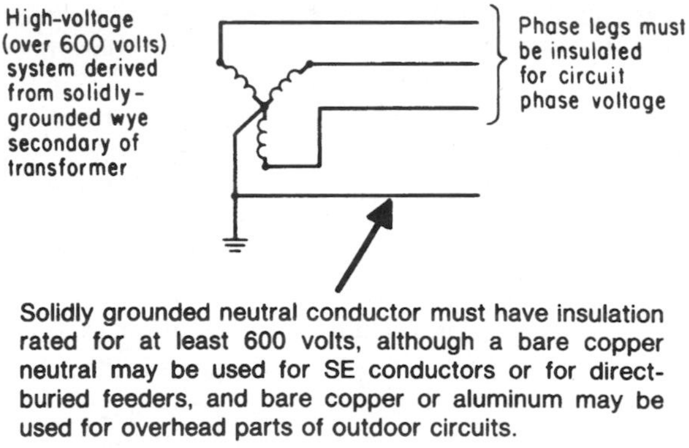
High-voltage (over 600 volts) system derived from solidly-grounded wye secondary of transformer
Phase legs must be insulated for circuit phase voltage
Solidly grounded neutral conductor must have insulation rated for at least 600 volts, although a bare copper neutral may be used for SE conductors or for direct-buried feeders, and bare copper or aluminum may be used for overhead parts of outdoor circuits.

Fig. 310-24. A note refers to neutral conductors of solidly grounded high-voltage systems. [Secs. 250.184 and 310.106(D).]

310.110. Conductor Identification. For part **(A),** refer to the discussion given for 200.6 and 200.7. For part **(B),** refer to 250.119(A) and (B). 310.12(A) and 200.6(E) recognize the use in multiconductor cables of a grounded conductor that is not white throughout its entire length, provided that only qualified persons will service the installation. The rule requires that such grounded conductors be identified by white marking at their termination at the time of installation.

Similarly, a grounding conductor in a multiconductor cable may be identified at each end and at every point where the conductor is accessible by stripping the insulation from the entire exposed length or by coloring the exposed insulation green or by marking with green tape or green adhesive labels.

Part (C) for ungrounded conductors now simply defers to 210.5(C) for branch circuit and 215.12 for feeders on color (or other method of) coding. The method used cannot render the required conductor markings in 310.120(B)(1) unreadable

ARTICLE 312. CABINETS, CUTOUT BOXES, AND METER SOCKET ENCLOSURES

312.1. Scope. Cabinets and cutout boxes, according to the definitions in Art. 100, must have doors (cutout boxes can also have telescoping covers), and are thus distinguished from large boxes with covers consisting of plates attached with screws or bolts. Article 312 applies to all boxes used to enclose operating apparatus—that is, apparatus having moving parts or requiring inspection or attention, such as panelboards, cutouts, switches, circuit breakers, control apparatus, and meter socket enclosures. The article does not apply to medium-voltage applications unless "specifically referenced" in another provision. One example of such a reference is 490.3(B).

312.2. Damp and Wet Locations. This section addresses "damp and wet locations." The Code requires all equipment covered by this article to be "placed or equipped" so as to ensure water does not enter or collect within the enclosure. Additionally, a minimum clearance from the mounting surface of ¼ in. must be maintained to prevent corrosion of the enclosure. And, fittings at raceway or cable entries in metallic enclosures must be listed for use in "wet locations" where such entry is above the level of any uninsulated live parts. This qualification reflects the fact that any energized part, whether insulated or not, is a live part per the Art. 100 definition. The clearance to the mounting surface is not required for nonmetallic enclosures on nonabsorbent surfaces, as noted by the exception. The rule about wet location fittings being required for entries above a certain level answers a long-standing question as to whether a conventional SE cable connector suitable for dry locations can be used with SE cable exiting from the bottom of an outdoor meter socket. The answer is that it can be used, but the fitting for the cable entry in the top of the socket does have to be listed for wet locations.

312.3. Position in Wall. Figure 312-1 shows how the ¼-in. (6.35-mm) setback relates to cabinets installed in noncombustible walls.

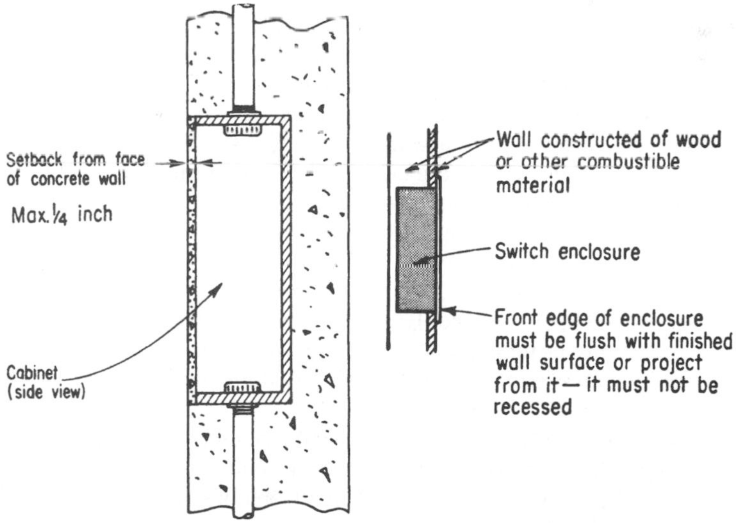
Setback from face of concrete wall
Max. ¼ inch
Cabinet (side view)
Wall constructed of wood or other combustible material
Switch enclosure
Front edge of enclosure must be flush with finished wall surface or project from it— it must not be recessed

Fig. 312-1. In masonry wall, cabinet does not have to be flush with wall surface—as it does in wood wall. (Sec. 312.3.)

312.4. Repairing Noncombustible Surfaces. The gap between the enclosure that uses a flush-mounted cover and adjacent finished surfaces—where the surface is plaster, drywall, or plasterboard—shall not be greater than 3 mm ($^1/_8$ in.). This makes the rule identical to the comparable rule in 314.21. Both 312.3 and 312.4, and their counterparts in Art. 314, address issues related to when ordinary building construction will be allowed to complete an electrical enclosure. Review the commentary following 314.21 for more information regarding the intent behind these requirements.

312.5. Cabinets, Cutout Boxes, and Meter Socket Enclosures. Part (C) makes clear that all cables used with cabinets or cutout boxes must be attached to the enclosure. NM cable, for instance, does not have to be connected by clamp or connector device to a single-gang nonmetallic outlet box as in 314.17(C), but must *always* be connected to KOs in panelboard enclosures and other cabinets (Fig. 312-2).

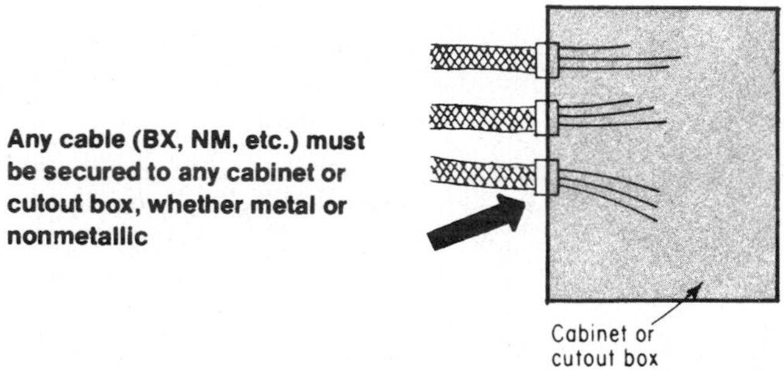
Any cable (BX, NM, etc.) must be secured to any cabinet or cutout box, whether metal or nonmetallic
Cabinet or cutout box

Fig. 312-2. All cables must be secured to all cabinets or cutout boxes. (Sec. 312.5.)

If installations require concealed wiring, spare conduits should be stubbed to accessible areas, such as above the lift-out ceiling-panel construction in common use today. In such areas this section includes an allowance for running multiple nonmetallic cable assemblies through a vertical riser. This facilitates the use of Type NM cables for light commercial construction; a surface panel can go on the wall at a convenient point with a single large riser conduit stub poked through the suspended ceiling to receive the cables. There are numerous limitations to be observed, but the neat and workmanlike result often makes this approach worthwhile. Specifically, the installation must meet the following conditions:

- The cables must have entirely nonmetallic sheaths.
- The raceway must enter the top of the cabinet.
- The raceway must be nonflexible, such as EMT.
- The raceway must be no shorter than 450 mm (18 in.), nor longer than 3.0 m (10 ft).
- Every cable entering the raceway end must be secured not over 300 mm (12 in.) from the entry point to the raceway.
- The raceway must extend directly above the cabinet.
- The raceway must not penetrate a structural ceiling.
- The raceway must be equipped with fittings on both ends to protect the cables from abrasion.

- The outer raceway end must be sealed or plugged by an approved means to prevent debris from falling into the cabinet. Duct seal is often used for this purpose.
- The cables that run to the enclosure must have their sheaths intact through the entire raceway section.
- The raceway must be supported in accordance with the rules that apply to the wiring method employed.
- The cable-fill limitations in Chap. 9 of the **NEC** must be observed. This means the cables cannot occupy more than 40 percent of the raceway internal cross section, and for elliptical cables, the cable cross section is taken as the area of a circle whose diameter equals the major axis of the ellipse. The raceway sleeve does not constitute a complete wiring system; however, the terms of this exception still reference the fill limits and, for correlation, the exception disqualifies the complete system limitation in Note 2 to the Chap. 9 tables.
- The mutual conductor heating rules, as covered in 310.15(B)(3)(a), must be applied. This is perhaps the most problematic constraint; however, if the raceway riser does not exceed 600 mm (24 in.) in length, those rules can be ignored because the raceway now qualifies as a nipple.

312.6. Deflection of Conductors. Parts **(A)** and **(B)** cover a basic Code rule that is referenced in a number of Code articles to ensure safety and effective conductor application by providing enough space to bend conductors within enclosures.

This section addresses a number of major considerations in terms of evaluating the minimum amount of space that must be allowed in enclosures to allow for bending and landing conductors safely. There are two basic starting points for making this analysis, based on whether the conductor enters the enclosure opposite to the terminal, or at right angles to the terminal. In addition, the conductor material must be considered, because the current aluminum alloy for conductors [AA-8000, in accordance with 310.106(B)] has less spring-back and is easier to bend than the same size in copper. As of the 2017 **NEC**, the allowable bending space dimensions now differ in both of the size/space tables presented in this section for that reason.

One example of how this particular change turned out to be unusually consequential concerned a common installation technique for residential 200A services. If the service cable enters the rear of the panelboard, the rules in 408.55(C), instituted in the 2014 **NEC** and illustrated in Fig. 408-23, will apply. If using aluminum SE cable, the usual approach is to use a 4/0 cable. The enclosure depth must at least equal the required bend radius, and on the prior version of Table 312.6(A) that required a 4 in. cabinet depth. It turns out that many manufacturers standardize on a 3¾ or 3½ in. depth. Adding the revised bending spacing for aluminum wiring (3½ in.) avoided a costly redesign of this equipment.

The rules for the two basic orientation of entry approaches follow:

1. The conductor does not enter (or leave) the enclosure through the wall opposite its terminals. This would be any case where the conductor passes through a wall of the enclosure at right angles to the wall opposite the terminal lugs to which the conductor is connected or at the opposite end of the enclosure. In all such cases, the bend at the terminals is a single-angle bend (90° bend), and the conductor then passes out of the bending space. It is also called an *L bend,* as shown at the top left of Fig. 312-3. For bends of

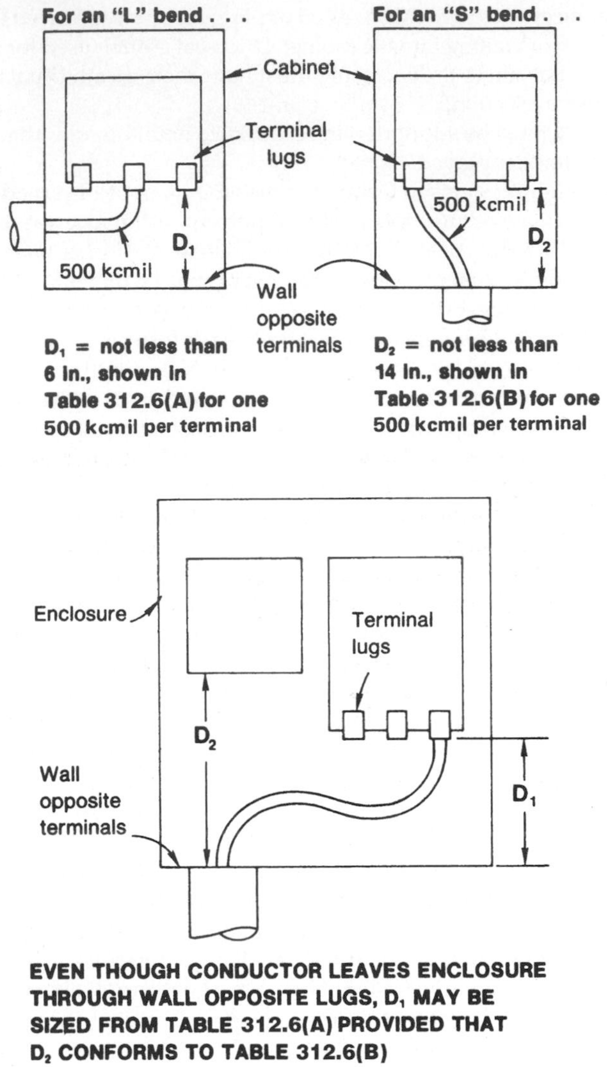

For an "L" bend . . .
For an "S" bend . . .
Cabinet
Terminal lugs
500 kcmil
D_1
500 kcmil
D_2
Wall opposite terminals
D_1 = not less than 6 in., shown in Table 312.6(A) for one 500 kcmil per terminal
D_2 = not less than 14 in., shown in Table 312.6(B) for one 500 kcmil per terminal
Enclosure
Terminal lugs
D_2
D_1
Wall opposite terminals
EVEN THOUGH CONDUCTOR LEAVES ENCLOSURE THROUGH WALL OPPOSITE LUGS, D_1 MAY BE SIZED FROM TABLE 312.6(A) PROVIDED THAT D_2 CONFORMS TO TABLE 312.6(B)

Fig. 312-3. These clearances are minimums that must be observed. (Sec. 312.6.)

that type, the distance from the terminal lugs to the wall opposite the lugs must conform to Table 312.6(A), which requires lesser distances than those of Table 312.6(B) because single bends are more easily made in conductors, and they are easier to land on a terminal.

2. The conductor enters (or leaves) the enclosure through the wall opposite its terminals. This is a more difficult condition because the conductor must make an offset or double bend to go from the terminal and then align with the raceway or cable entrance. This is also called an *S* or a *Z bend* because

of its configuration, as shown at the top right of Fig. 312-3. For such bends, Table 312.6(B) specifies a greater distance from the end of the lug to the opposite wall to accommodate the two 45° bends, which are made difficult by the short lateral space between lugs and the stiffness of conductors (especially with the plastic insulations in cold weather).

Table 312.6(B) provides increased bending space to accommodate use of factory-installed connectors that are not of the lay-in or removable type and to allow use of field-installed terminals that are not designated by the manufacturer as part of the equipment marking. Exception No. 1 to part **(B)** is shown in the bottom drawing of Fig. 312-3. This setup is often used in CT enclosures.

Note: For providing Code-required bending space at terminals for enclosed switches or individually enclosed circuit breakers, refer to 404.18. For conductor bending space at panelboard terminals, refer to 408.55. In Fig. 312-3, the clearances shown are determined from Table 312.6(A) or Table 312.6(B), under the column for one wire per terminal. For multiple-conductor circuit makeups, the clearance at terminals and inside gutters has to be greater, as shown under two, three, four, etc., wires per terminal.

Exception No. 2 of part **(B)** covers application of conductors entering or leaving a meter-socket enclosure, and was based on a study of 100- and 200-A meter sockets with lay-in terminals.

312.8. Switch and Overcurrent Device Enclosures. The basic rule here is a follow-up to the rule of 312.7. This section has been reorganized to present the traditional applications to splices, tapping, and feeding through in **(A)**, and then a separate topic of power monitoring equipment in **(B)**.

Most enclosures for switches and/or overcurrent devices have been designed to accommodate only those conductors intended to be connected to terminals within such enclosures. And in designing such equipment it would be virtually impossible for manufacturers to anticipate various types of "foreign" circuits, feed-through circuits, or numerous splices or taps.

The rule here states that enclosures for switches, CBs, panelboards, or other operating equipment must not be used as junction boxes, troughs, or raceways for conductors feeding through or tapping off, unless designs suitable for the purpose are employed to provide adequate space. This rule affects installations in which a number of branch circuits or subfeeder circuits are to be tapped from feeder conductors in an auxiliary gutter, using fused switches to provide disconnect and overcurrent protection for the branch or subfeeder circuits. It also applies to feeder taps in panelboard cabinets.

In general, the most satisfactory way to connect various enclosures together is through the use of properly sized wireways or auxiliary gutters (Fig. 312-4) or junction boxes. Figure 312-5 shows a hookup of three motor disconnects, using a junction box to make the feeder taps. Following this concept, enclosures for switches and/or overcurrent devices will not be overcrowded.

There are cases where large enclosures for switches and/or overcurrent devices will accommodate additional conductors, and this is generally where the 40 percent (conductor space) and 75 percent (splices or taps) at one cross section would apply. An example would be control circuits tapped off or extending through 200-A or larger fusible switches or CB enclosures. The csa within such enclosures is the *free gutter wiring space* intended for conductors.

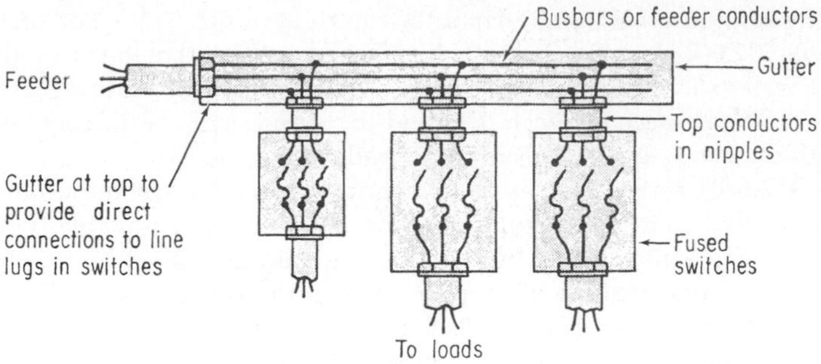
Busbars or feeder conductors
Feeder
Gutter
Top conductors
in nipples
Gutter at top to
provide direct
connections to line
lugs in switches
Fused
switches
To loads

Fig. 312-4. Feeder taps in auxiliary gutter keep feeder cables and tap connectors out of switch enclosures. (Sec. 312.8.)

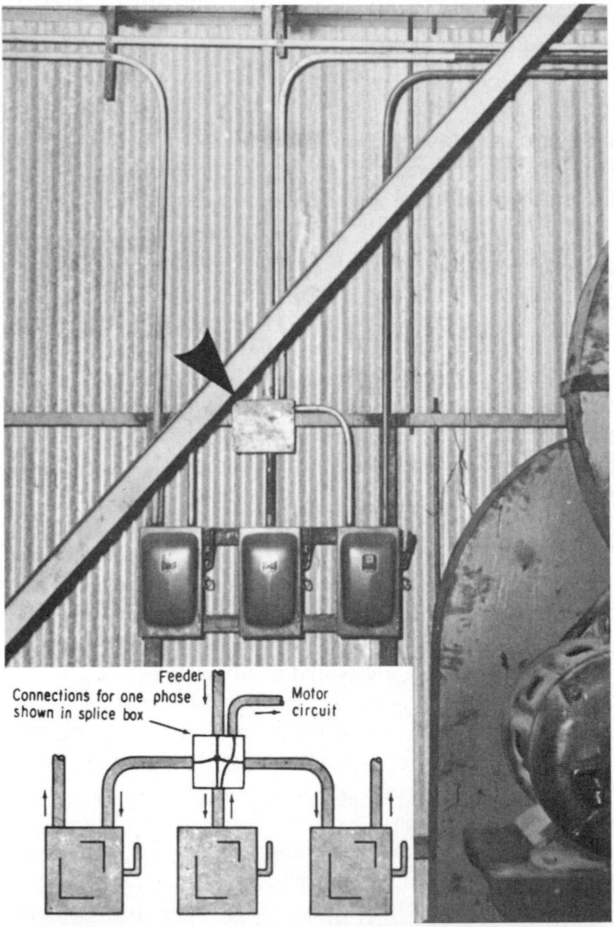
Feeder
Connections for one phase
shown in splice box
Motor
circuit

Fig. 312-5. Junction box (arrow) is used for tapping feeder conductors to supply individual motor branch circuits—as shown in inset diagram. (Sec. 312.8.)

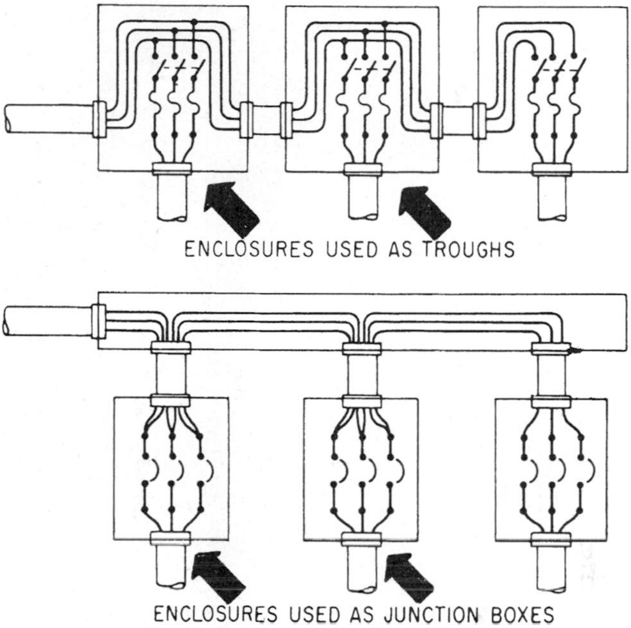
ENCLOSURES USED AS TROUGHS
ENCLOSURES USED AS JUNCTION BOXES

Fig. 312-6. These hookups are permitted where space in enclosure gutters satisfies the exception to the basic rule. (Sec. 312.8.)

Parts **(1)** and **(2)** of this rule are shown in Fig. 312-6 and are applied as follows:

example If an enclosure has a gutter space of 3 by 3 in., the csa would be 9 sq in. Thus, the total conductor fill (use Table 5, Chap. 9) at any cross section (including conductors and splices) could not exceed 6.75 sq in. (9 × 0.75). The actual conductor fill at any cross section (just the wires) could not exceed 3.6 sq in. (9 × 0.4).

In the case of large conductors, a splice other than a wire-to-wire "C" or "tube" splice would not be acceptable if the conductors at the cross section are near a 40 percent fill, because this would leave only a 35 percent space for the splice. Most splices for larger conductors with split-bolt connectors or similar types are usually twice the size of the conductors being spliced. Accordingly, where larger conductors are to be spliced within enclosures, the total conductor fill should not exceed *20 percent* to allow for any bulky splice at a cross section.

Figure 312-7 shows an example of feeder taps made in panelboard side gutter where the cabinet is provided with adequate space for the large feeder conductors and for the bulk of the tap devices with their insulating tape wrap. In accordance with condition (3) of the rule, which was new in the 2011 **NEC**, if any of these conductors feed through, a warning label must be applied to the cover of the enclosure that puts personnel on notice as to the location of the nearest disconnecting means that is ahead of those conductors. In the case of a panelboard, the requirement for the circuit directory to identify the source of its supply feeder in Sec. 408.4. would be sufficient, unless there are additional conductors passing through that do not energize this particular panel. The intent is to ensure that personnel working in such an enclosure are

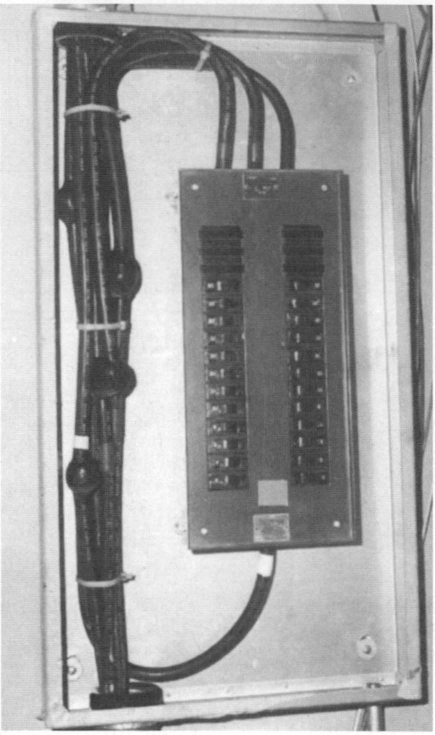

Fig. 312-7. In 312.11(C), the Code permits feeding through and tapping off in cabinets for panelboards on feeder risers, where the side gutter is specially oversized for the application. (Secs. 312.8 and 312.11)

aware of all sources of energized conductors within it, especially if some of them are not obvious because they do not energize loads within the enclosure.

Power monitoring equipment is covered in **(B)**, which only applies to enclosures for switches and overcurrent devices, such as panelboard cabinets. This topic has been extensively researched and a UL outline of investigation released that forms an orderly basis for the manufacture and installation of products. The equipment will be allowed in two related forms. The first is a field-installable accessory that is available as an add-on to listed equipment, not unlike surge protectors. The other option will be a listed kit that is field installable in different enclosures. Figure 312-8 shows an example of the former.

The panel actively considered a much broader requirement, one that would have required listings on any equipment placed in a switch or panelboard enclosure. After receiving extensive negative comments, and reflecting on a wide variety of innocuous intrusions, the panel decided to limit the reach to power monitoring equipment. One example consisted of the tail end of a 50-amp 3-pole 4-wire locking receptacle mounted to the side of a panelboard cabinet. The location was carefully chosen to not create potential conflict with future branch-circuit connections in the wiring gutter. The particular installation was well policed by the inspector. The panel realized that there was no possible way the device manufacturer would ever have been able to obtain a listing that could have anticipated that application, and there was certainly nothing wrong with it. The panel instead took care to leave the section structure in a way that could easily include rules on other topics if a compelling need were demonstrated in a future cycle.

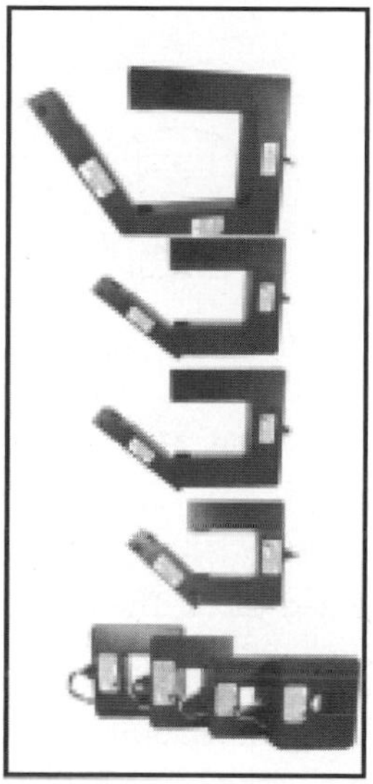

Fig. 312-8. Field-installable current transformers, listed to UL 2808 and suitable for placement in panelboard wiring gutter spaces. [312.8(B)(1).] (Courtesy, Siemens)

ARTICLE 314. OUTLET, DEVICE, PULL, AND JUNCTION BOXES; CONDUIT BODIES; FITTINGS; AND HANDHOLE ENCLOSURES

314.1. Scope. This rule makes clear that Art. 314 regulates use of conduit bodies when they are used for splicing, tapping, or pulling conductors. And this article does refer specifically to conduit bodies, to more effectively distinguish rules covering *boxes, conduit bodies,* and *fittings.* The rules of Art. 314 must be evaluated in accordance with the definitions given in Art. 100 for "conduit body" and "fitting." Capped elbows and SE elbows are fittings, not conduit bodies, and must not contain splices, taps, or devices. The pieces of equipment described in Art. 314 tie into 300.15, "Boxes or fittings—where required."

314.2. Round Boxes. The purpose of this rule is to require the use of rectangular or octagonal metal boxes having, at each knockout or opening, a flat bearing surface for the locknut or bushing or connector device to seat against a flat surface. But round outlet boxes may be used with nonmetallic-sheathed cable because the cable is brought into the box through a knockout, without the use of a box connector to secure the cable to the box. However, the exception to 314.17(C) permits only boxes of "nominal size 2¼ in. by 4 in.," the so-called single gang boxes, to be used without securing the NM or NMC cable to the box itself—as long as it is stapled to the stud or joist within 8 in. (203 mm) of the box. Because round boxes are *not* single gang boxes, it appears that all such round outlet boxes must be equipped with cable clamps to satisfy the exception to 314.17(C) (Fig. 314-1). Shallow metal boxes with internal clamps for NM cable are acceptable as round boxes.

314.3. Nonmetallic Boxes. Growth in the application of nonmetallic boxes over past years is the basis for the two exceptions to this section, which regulate the conditions under which nonmetallic boxes may be used with metal raceways

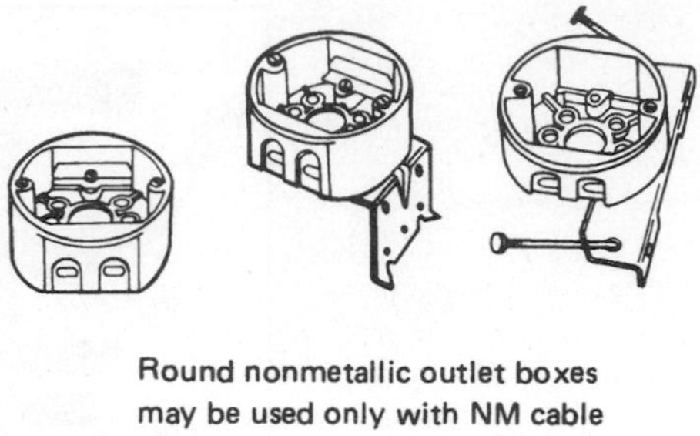
Round nonmetallic outlet boxes
may be used only with NM cable

Fig. 314-1. Round boxes may be used only for connecting cables with internal clamps—such as NM or BX cable. (Sec. 314.2.)

or metal-sheathed cable. The need for and popularity of these boxes developed out of industrial applications where corrosive environments dictated their use to resist the ravages of various punishing atmospheres. In many applications it is desirable to use nonmetallic boxes along with plastic-coated metal conduits for a total corrosion-resistant system. Such application is recognized by the Code in the exceptions of this section, although a limitation is placed requiring "internal" or "integral" bonding means in such boxes (Fig. 314-2).

According to the basic rule in the first sentence of this rule, nonmetallic boxes are permitted to be used only with open wiring on insulators, concealed knob-and-tube wiring, nonmetallic sheathed cable, electrical nonmetallic tubing, and

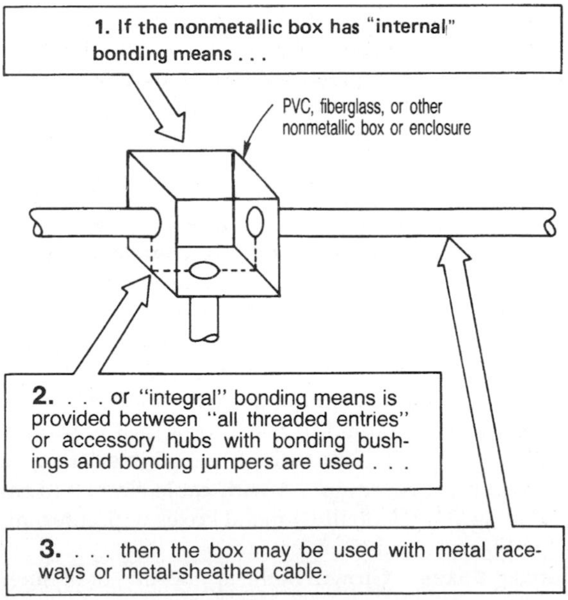
1. If the nonmetallic box has "internal" bonding means . . .
PVC, fiberglass, or other nonmetallic box or enclosure
2. . . . or "integral" bonding means is provided between "all threaded entries" or accessory hubs with bonding bushings and bonding jumpers are used . . .
3. . . . then the box may be used with metal raceways or metal-sheathed cable.

Fig. 314-2. Nonmetallic boxes are recognized for use with metal raceways and metal-sheathed cable. (Sec. 314.3.)

rigid nonmetallic conduit (any "nonmetallic raceways"). Exception No. 1 requires internal bonding means in such boxes used with metal cable or raceways. The permission used to apply only to nonmetallic boxes sufficiently large—that is, over 100 cu in. Now, *any* size of box—PVC boxes, fiberglass boxes, or other nonmetallic boxes or enclosures—may be used with metal raceways or metal-sheathed cable. However, for each entry, a bonding bushing would have to be applied in order to provide continuity in the equipment grounding conductor.

Exception No. 2 requires that "integral" bonding means between all "threaded" raceway and cable entries must be provided in the box for all metal conduits or metal-jacketed cables. This exception addresses an entirely different box construction, where a metallic web is positioned within and throughout a nonmetallic (usually fiberglass) enclosure. Many of these boxes are used for hazardous locations that are also extremely corrosive. Some of them even meet the requirements for Class I Division 1. When hubs are added to these boxes using the methods prescribed in the installation instructions, continuity is achieved through the integral web, almost as if the box were all metal. The only constraint imposed in this case is that an equipment grounding terminal must be provided within the box by its manufacturer, which would reach the bonding web. That way if there is equipment in the box that requires a bonding connection, it will be available.

314.4. Metal Boxes. This is now a very simple rule. Ground all metal boxes. Period. The only exception is when 250.112(I) relieves the need for equipment grounding.

314.15. Damp or Wet Locations. In damp or wet locations, equipment within the scope of Art. 314 must be placed or equipped to prevent moisture from entering or accumulating within the equipment. Boxes, conduit bodies, and fittings used in wet locations must be listed for use in those areas. The left part of Fig. 314-3 shows a fitting which is generally available as a listed item and it has been designed with the openings for the conductors so placed that rain or snow cannot enter the fitting. On the right, it shows a fitting also suitable for a wet location listing; it has a metal cover that slides under flanges on the face of the fitting, and, as required by 230.53,

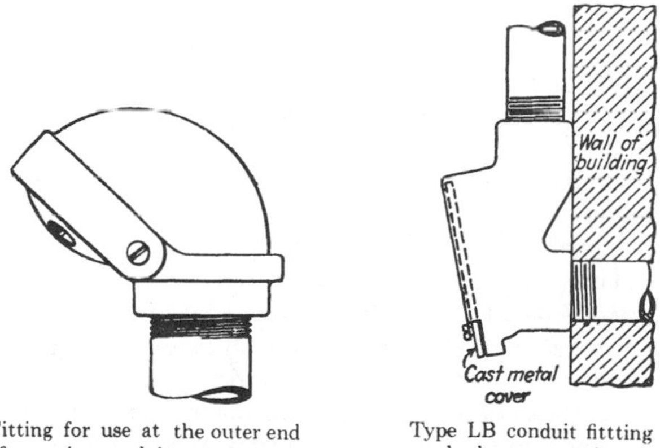
Wall of building
Cast metal cover

Fitting for use at the outer end of a service conduit.

Type LB conduit fittting used where a service conduit passes through a building wall.

Fig. 314-3. Fittings must be suited to use in wet locations. (Sec. 314.15.)

an opening is provided through which any moisture condensing in the conduit can drain out. Note that this is a "service entrance elbow." It is a short-radius conduit body covered in 314.16(C)(3). It is functionally obsolete for service applications because it does not have the required bending radius for any conductors 4 AWG or larger. It does retain theoretical acceptability for a service to a single small load such as a telephone booth, as covered in 230.42(B) and 230.79(A).

The 2014 **NEC** has added "outlet box hoods" (the so-called "bubble covers" for in-use receptacle outlets) to the list of enclosures in the first sentence that are required to be placed or arranged to avoid water entry or accumulation in damp or wet locations. In addition, official recognition was granted to the widespread and common practice of drilling weep holes in the underside of boxes to allow condensation to escape instead of slowly degrading the box and its contents from the inside out. The **NEC** now specifies the permitted size and geometry of these holes ($^1/_8$ to ¼ in., diameter specified, only applies to a circle) to assure they stay open but still internal access. Larger sizes are allowable to accommodate a listed drain fitting.

314.16. Number of Conductors in Outlet, Device, and Junction Boxes, and Conduit Bodies. Note that motor terminal housings are excluded from the rules on box conductor fill. And where any box or conduit body contains 4 AWG or larger conductors, all the requirements of 314.28 on pull boxes must be satisfied. Refer to 314.28 for applications of conduit bodies as pull boxes.

Selection of any outlet or junction box for use in any electrical circuit work must take into consideration the maximum number of wires permitted in the box by 314.16. Safe electrical practice demands that wires *not* be jammed into boxes because of the possibility of nicks or other damage to insulation—posing the threat of grounds or shorts.

This section is broken down into three subparts. Part **(A)** establishes the volume of a box. Part **(B)** describes the method for determining how much volume is used by the various conductors, devices, and the like. Part **(C)** applies to conduit bodies, only.

As of the 2017 **NEC**, the parent language in **(A)** also includes a volume rule that applies to both of its numbered paragraphs following. If a box contains a barrier, either for system separation between power and class 2, as required in 725.136(A), 800.133(A)(1)(d), etc., or for voltage separation as required in 404.8(B) and 406.5(J), or any other purpose, that barrier must be accounted for in the volume calculation. Figure 314-4 shows a typical example. If the barrier is marked with its volume, then that volume must be apportioned to each side. If it is not marked, and most will not be at present, then metal barriers are assumed to have a volume of ½ in^3 and for a nonmetallic barrier a volume of 1.0 in^3. The number for steel barriers was developed by taking a length of 4 in., a height of 2 in. and a thickness of 0.0625 in. as required by 314.40(B). A typical nonmetallic barrier was assumed to be about double the thickness as for steel. As noted, any manufacturer who objects is always free to mark his barriers if they differ.

As stated in part **(A)(1)** of this section, Table 314.16(A) shows the maximum number of wires permitted in the *standard* metal boxes listed in that table. But that table applies only where all wires in a given box are the same size; that is, all 14 AWG or all 12 AWG Table 314.16(B) is provided for sizing a box where all the wires in the box are not the same size, by using so much cubic-inch space for each size of wire.

Table 314.16(A) includes the maximum number of 18 AWG and 16 AWG conductors that may be used in various sizes of boxes, and Table 314.16(B) gives the required box space for those sizes of conductors. Because of the

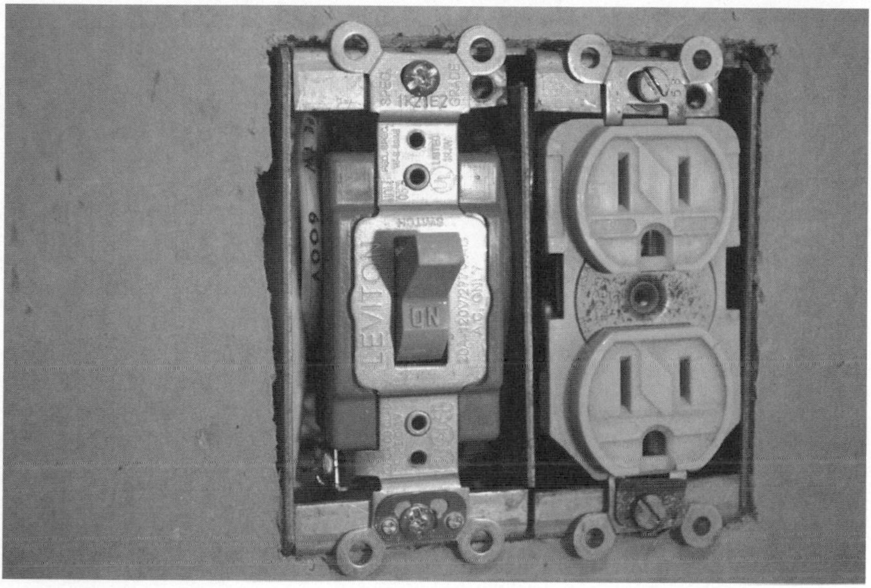

Fig. 314-4. The steel barrier, installed for voltage separation between the two gangs of this gangable masonry box, is assumed to have a volume of 1.0 in^3, which reduces the volume on each side by half that number, or 0.5 in^3. [314.16(A).]

extensive use of 18 AWG and 16 AWG wires for fixture wires and for control, signal, and communications circuits, these data are needed to assure safe box fill for modern electrical systems.

As stated in part **(A)(2),** all other boxes—nonstandard, nonmetallic, or those metal boxes covered by part **(A)(1)** that are stamped with their cubic-inch capacity by the manufacturer—must consider their volume to be that which is stamped on them. And, as stated in the first paragraph of 314.16, the value of volume [part **(A)**] must never be less than the fill [part **(B)**].

Part **(A)(2)** of 314.16 covers boxes—metal and nonmetallic—that are not listed in Table 314.16(A). And the basic way of determining correct wire fill is to count wires in accordance with the intent of 314.16(B) and then calculate the required volume of the box or conduit body by totaling up the volumes for the various wires from Table 314.16(B).

Part **(A)(2)** covers wire fill for metal boxes, up to 1650 cm^3 (100 in.^{3}) volume, that are not listed in Table 314.16(A) and for nonmetallic outlet and junction boxes. Although Code rules have long regulated the maximum number of conductors permitted in metal wiring boxes [such as given in Table 314.16(A)], there was no regulation on the use of conductors in nonmetallic device boxes up to the 1978 **NEC**. Since that time, 314.16(A)(2) requires that *both* metal boxes not listed in Table 314.16(A) and nonmetallic boxes be durably and legibly marked by their manufacturer with their cubic inch capacities to permit calculation of the maximum number of wires that the Code will permit in the box. Calculation of the conductor fill for these boxes will be based on the marked box volume and the method of counting conductors set forth in 314.16(B). The conductor volume will be taken at the values given in Table 314.16(B), and allocations of space as required for wiring devices or for clamps must be made

Fig. 314-5. Every nonmetallic box must be "durably and legibly marked by the manufacturer" with its cubic-inch capacity to permit calculation of number of wires permitted in the box—using Table 314.16(B) and the additions of wire space required to satisfy 314.16(A)(1). (Sec. 314.16.)

in accordance with the rules of 314.16(B). This requirement for marking of both metal and nonmetallic boxes arises from the wording of 314.16(A)(2), which refers to boxes other than those described in Table 314.16(A) and to nonmetallic boxes.

As shown in Fig. 314-5, a nonmetallic box for a switch has two 14/2 NM cables, each with a 14 AWG ground. The wire count is four 14 AWG insulated wires, plus two for the switch to be installed, and one for the two ground wires. That is a total of seven 14 AWG wires. From Table 314.16(B), at least 32.8 cm^3 (2 $in.^3$) of box volume must be allowed for each 14 AWG. This box must, therefore, be marked to show that it has a capacity of at least 7×2, or 14, cu in. As shown, the ground wires are connected by a twist-on connector, with one end of the wire brought out to connect to a ground screw on the switch mounting yoke. Such a technique is required to provide grounding of a metal switchplate. Refer to 404.9(B).

Part **(B)** of 314.16 describes the detailed way of counting wires in a box and subtracting from the permitted number of wires shown in Table 314.16(A) where cable clamps, fittings, or devices like switches or receptacles take up box space. The first sentence requires all the allowances applicable to a box assembled from separate components, such as a ganged device box, be added together to determine the volume of box required from Part (A). The second sentence addresses the reverse issue of barriers, and requires that the allowances applicable to either side be calculated separately in terms of determining the total volume required within each side.

Important details of the wire-counting procedure of part **(B)** are as follows:

1. From the wording, it is clear that no matter how many ground wires come into a box, whether they are ground wires in NM cable or ground wires run in metal or nonmetallic raceways, a fill allowance of only one conductor must be made from the number of wires shown in Table 314.16(A) (Fig. 314-6). Or, as will be

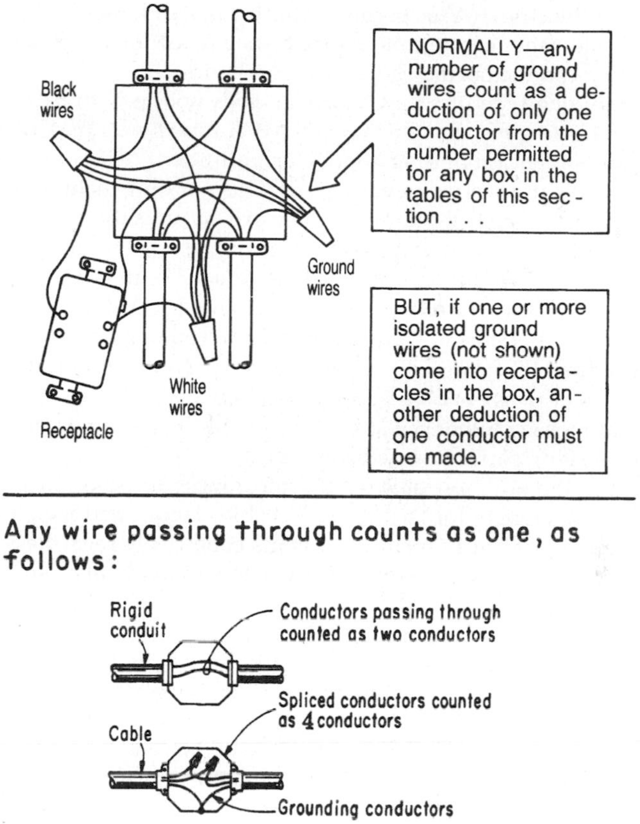
Black wires
NORMALLY—any number of ground wires count as a deduction of only one conductor from the number permitted for any box in the tables of this section . . .
Ground wires
BUT, if one or more isolated ground wires (not shown) come into receptacles in the box, another deduction of one conductor must be made.
White wires
Receptacle
Any wire passing through counts as one, as follows:
Rigid conduit
Conductors passing through counted as two conductors
Spliced conductors counted as 4 conductors
Cable
Grounding conductors

Fig. 314-6. Count all ground wires as *one* wire (or two wires if isolated-ground wires are also used) of the largest size of ground wire in the box. (Sec. 314.16.)

shown in later examples, one or more ground wires in a box must be counted as a single wire of the size of the largest ground wire in the box. Any wire running unbroken through a box counts as one wire, provided it is not long enough to be cut in the center of the loop and used as two wires, as defined in 300.14; if it is long enough to count as two wires, then it must be counted as two wires, and the count goes up further on multiple coils. A small hand loop that is only there to assist routing the wire is still only counted once. Each wire coming into a splice device (crimp or twist-on type) is counted as one wire. And each wire coming into the box and connecting to a wiring device terminal is *one* wire.

When a number of "isolated-ground" equipment grounding conductors for receptacles come into a box along with conventional equipment grounding wires, each type of equipment ground wires must be counted as one conductor for purposes of wire count when determining the maximum number of wires permitted in a box. When a number of isolated-ground receptacles are used in a box (as for computer wiring), all the isolated-ground conductors count as one fill allowance counted against the number of wires given in Table 314.16(A) as permitted for the particular size of box. And *then*, another allowance of one conductor must be made for any other equipment grounding conductors (*not* isolated-ground wires).

There is one type of wire entry that does not need to be counted, and that is fixture wires smaller than 14 AWG (maximum of 4 with an additional equipment grounding conductor permitted) that enter a box from a domed luminaire canopy or equivalent. This wording includes the canopies on paddle fan hanger assemblies. The word *domed* is there to assure at least some volume contribution from the canopy; there are luminaires that incorporate what for all intents and purposes is a 100-mm (4-in.) octagon blank cover; in such cases the fixture wires are counted. The domed canopy allowance makes it possible to hang luminaires on ceiling pans that usually have no more than 98.4 cm^3 (6.0 $in.^3$) for volume. Until comparatively recently wires smaller than 14 AWG had no mandatory wire fill assigned; however, now that those volumes are assigned, something had to be done to be certain that ceiling pans remained workable.

2. Regarding the addition of wire fill allowances to other such allowances under consideration, for fixture studs, cable clamps, and hickeys, does this apply to the previously mentioned items collectively regardless of number and combination, or does it apply to each item individually—such as clamps, plus one; studs, plus one; and so forth?

Answer: It is the intent of parts **(B)(2)** and **(3)** to clarify that a fill allowance of one must be made from the number in the table for each *type* of device used in a box, and this must use the largest conductor in the box as the reference point for the calculation. A fill allowance of one must be made if the box contains cable clamps—whether one clamp or two clamps, a count of only one has to be made. A fill allowance of one must be made if the box contains a fixture stud. A fill allowance of one must be made if the box contains a hickey. Thus, a box containing two clamps but no fixture studs or hickeys would have a fill allowance of one from the table number of wires for the clamps. If a box contained one clamp and one fixture stud, a fill allowance of two would be made because there are two *types* of devices in the box.

Then, as given in part **(B)(4)**, in addition to the allowances for clamps, hickeys, and/or studs, an allowance of two conductors must be made for each mounting strap that supports a receptacle, switch, or combination device. Note that for large devices that cannot fit in single-gang metal device boxes listed in Table 314.16(A), such as NEMA 14-50R (3-pole 4-wire 50-A 125/250-V) receptacles for ranges, allowances must be taken on the basis of the number of such ganged device boxes required for the device. In this case that would be two such boxes, for a total of four allowances, and if 6 AWG conductors are used, this rule adds another 163.8 cm^3 (10 in^3.) to the volume.

It should be noted that not all 14-50R devices are alike in today's market; some device manufacturers have shrunk the body size just enough that, even with the actual configuration remaining constant for obvious reasons, the devices will fit in a single gang opening and accept a single-gang faceplate similar to the traditional 3-pole, 3-wire 125/250-V (NEMA 10-50R) receptacle.

used prior to the 1996 **NEC**. In the example given, such a receptacle, even if configured 3-pole 4-wire grounding, would only add half the volume.

3. Must unused cable clamps be removed from a box? And if clamps are not used at all in a box, must they be removed to avoid the one-wire fill allowance?

   Answer: Unused cable clamps may be removed to gain space or fill in the box, or they may be left in the box if adequate space is available without the removal of the clamp or clamps. If one clamp is left, the one-wire allowance must be made. If no clamps are used at all in a box, such as where the cable is attached to the box by box connectors, the one-wire allowance does not take effect.

4. Is the short jumper installed between the grounding screw on a grounding-type receptacle and the box in which the receptacle is contained officially classified as a *bonding jumper?* And is this conductor counted when the box wire count is taken?

   Answer: The jumper is classed as a *bonding jumper.* 250.146 uses the wording "bonding jumper" in the section pertaining to this subject. However, whatever it is called, this conductor is not counted because it does not leave the box. The last sentence of 314.16(B)(1) covers that point.

An example of wire counting and correct wire fill for ganged boxes is included in the following examples. *Note:* In the examples given here, the same rules apply to wires in boxes for any wiring method—conduit, EMT, BX, NM.

Examples of Box Wire Fill

The top example in Fig. 314-7 shows how wire fill allowances must be added to account for a box containing cable clamps and a fixture stud. Note this is an example of a domed canopy and the two fixture wires from the luminaire are not counted. The example at the bottom shows a nonmetallic-sheathed cable with three 14 AWG copper conductors supplying a 15-A duplex receptacle (one ungrounded conductor, one grounded conductor, and one "bare" grounding conductor).

After supplying the receptacle, these conductors are extended to other outlets and the conductor count would be as follows:

Circuit conductors	4
Grounding conductors	1
For internal cable clamps	1
For receptacle	2
Total	8

The 14 AWG conductor column of Table 314.16(A) indicates that a device box not less than 3 by 2 by 3½ in. (75 by 50 by 90 mm) is required. Where a square box with plaster ring is used, a minimum of 4- by 1¼-in. (100- by 32-mm) size is required.

Table 314.16(A) includes the most popular types of metal "trade-size" boxes used with wires 14 AWG to 6 AWG. Cubic-inch capacities are listed for each box shown in the table. According to paragraph **(A)(2),** boxes other than those shown in Table 314.16(A) are required to be marked with the cubic-inch content so wire combinations can be readily computed.

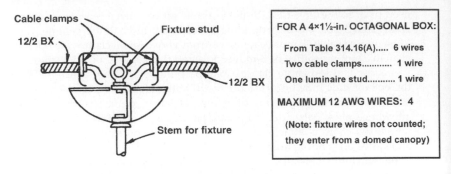

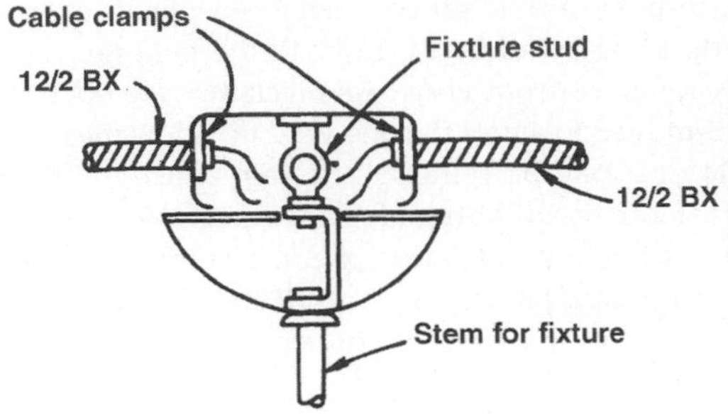
Cable clamps
Fixture stud
12/2 BX
12/2 BX
Stem for fixture

FOR A 4×1½-in. OCTAGONAL BOX:

From Table 314.16(A).....	6 wires
Two cable clamps............	1 wire
One luminaire stud...........	1 wire

MAXIMUM 12 AWG WIRES: 4

(Note: fixture wires not counted; they enter from a domed canopy)

NOTE: If NM cable were used, another deduction of one for the two ground wires would make use of this box a violation.

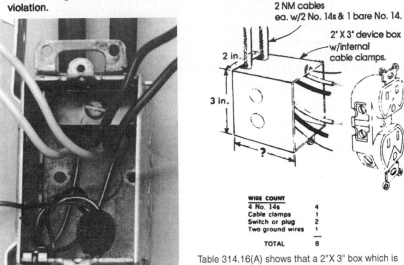

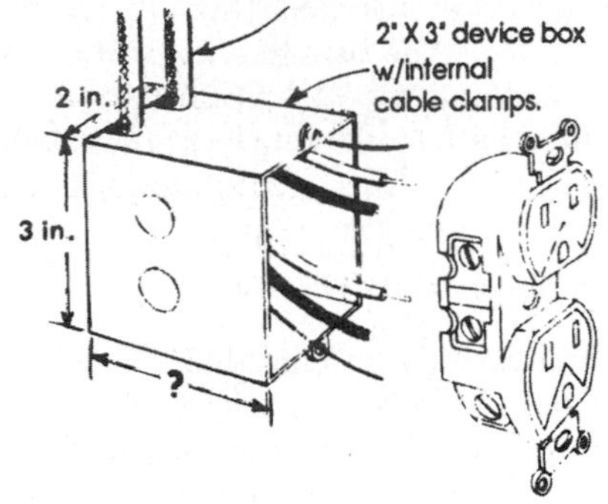
2 NM cables
ea. w/2 No. 14s & 1 bare No. 14.
2" X 3" device box
w/internal
cable clamps.
2 in.
3 in.
?

WIRE COUNT	
4 No. 14s	4
Cable clamps	1
Switch or plug	2
Two ground wires	1
TOTAL	8

Table 314.16(A) shows that a 2"X 3" box which is suitable for use with 8 No. 14 wires must be 3 1/2" deep.

Fig. 314-7. Correct wire count determines proper minimum size of outlet box. (Sec. 314.16.)

Figure 314-8 shows another example with the counting data in the caption. The wire fill in this case may violate the limit set by 314.16(A).

Figure 314-9 shows an example of wire-fill calculation for a number of ganged sections of sectional boxes. The photo shows a four-gang assembly of 3- by 2- by 3½-in. (75- by 50- by 90-mm) box sections with six 14/2 NM cables, each with a 14 AWG ground wire and one 14/3 NM cable with a 14 AWG ground. The feed to the box is 14/3 cable (at right side), with its black wire supplying the receptacle which will be installed in the right-hand section. The red wire serves as feed to three combination devices—one in each of the other sections—each device

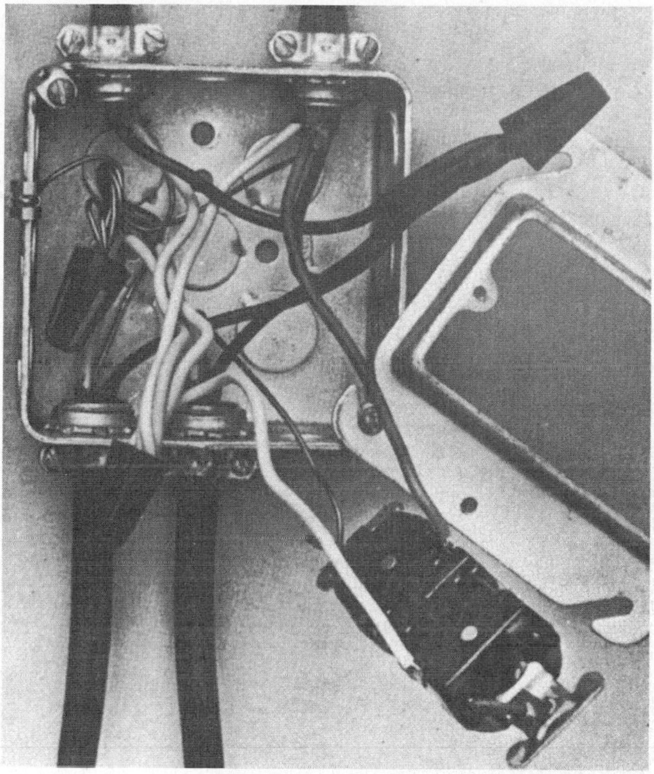

Fig. 314-8 THIS MAY BE A CODE VIOLATION! A 4- × 4- × 1½-in. (102- × 102- × 38-mm) square metal box, generally referred to as a *1900 box*, has four NM cables coming into it. At upper right is a 14/3 cable with 14 AWG ground. The other three cables are 14/2 NM, each with a 14 AWG ground. The red wire of the 14/3 cable feeds the receptacle to be installed in the one-gang plaster ring. The black wire of the 14/3 feeds the black wires of the three 14/2 cables. All the whites are spliced together, with one brought out to the receptacle, as required by 300.13(B). All the ground wires are spliced together, with one brought out to the grounding terminal on the receptacle and one brought out to the ground clip on the left side of the box. The wire count is as follows: nine 14 AWG insulated wires, plus one for all of the ground wires and two for the receptacle. That is a total of 12 14 AWGs. Note that box connectors are used instead of clamps and there is, therefore, no addition of one conductor for clamps. But Table 314.16(A) shows that a 4- × 4- × 1½-in. (102- × 102- × 38-mm) square box may contain only 10 14 AWG wires. Some think that this is *not* a violation. They say that because the area provided by the cover has not been considered. But unless the cubic-inch capacity is marked on the cover, it may *not* be considered. The box is only 49.2 cm^3 (3 in.^{3}) short of the required volume. Practical experience in working with this equipment strongly suggests that if this plaster ring (with at least a 13-mm (½-in.) rise is marked with a volume, that volume would be sufficient. Therefore, this comes down to the existence of a marking, which may well be on the other side. (Sec. 314.16.)

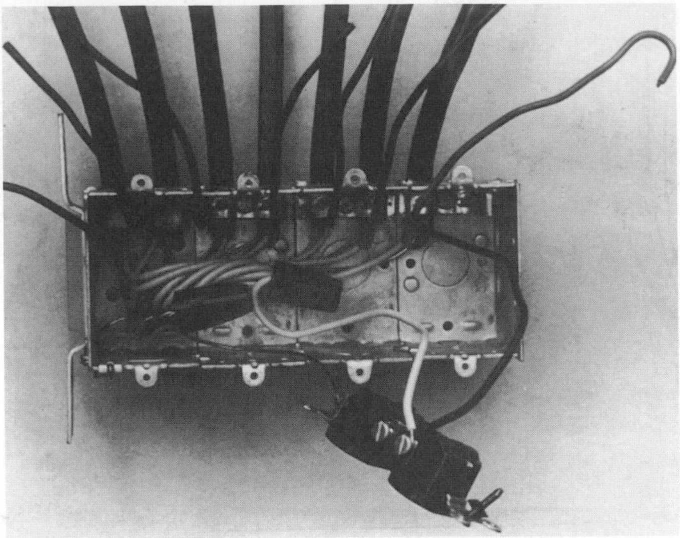

Fig. 314-9. Calculation of the proper minimum box size for the number of conductors used in ganged boxes must follow 314.16(A)(1), taking the assembly as a single box of the sum of the volumes of the ganged sections and filling it to the sum of the conductor count. (Sec. 314.16.)

consisting of two switches on a single strap. When finished, the four-gang box will contain a total of six switches and one duplex receptacle. Each of the 14/2 cables will feed a switched load. All the white neutrals are spliced together and the seven bare 14 AWG ground wires are spliced together, with one bare wire brought out to the receptacle ground terminal and one to the ground clip on the bottom of the left-hand section. The four-gang assembly is taken as a box of volume equal to 4 times the volume of one 3- by 2- by 3½-in. (75- by 50- by 90-mm) box. From Table 314.16(A), that volume is 18 cu in. for each sectional box. Then, for the four-gang assembly, the volume of the resultant box is 4 × 18, or 72, cu in. This is the front half of the calculation, set it aside.

Remember that the box is to be considered as one box, indivisible. Do not waste time trying to apportion volumes or fill allowances by gang. Do not make wire counts by Table 314.16(A) numbers per gang, because they are truncated. For example, the square box in the previous example holds 10 14 AWG wires, but this number is rounded down because the Code limit is a not to exceed number. Its capacity is actually 344 cm^3 (21.0 $in.^3$); if these were gangable, and several were put together, and you figured on the basis of 10 wire each, you would lose one wire for every two boxes put together.

Now sum the fill allowances. There are 15 current-carrying 14 AWG wires i this box, for a fill allowance of: 15 14 AWG conductors. The equipment groundin conductors count collectively as 1 wire, so add 1 14 AWG conductor. The interna cable clamps count collectively as 1 wire, so add 1 14 AWG conductor. The fou devices (counted by the strap, not the individual devices) add 2 each for a total c 8 14 AWG conductors. The grand total is 25, and at 32.8 cm^3 (2.00 $in.^3$) for each, th total required volume is 820 cm^3 (50 $in.^3$). This is the back half of the calculation; compare it with the front half to determine compliance. The total required is far less than the total provided, so the layout is acceptable.

When different sizes of wires are used in a box, Table 314.16(B) must be used in establishing adequate box size. Using the same method of counting conductors as described in Sec. 314.16(B), the volume of cubic inches shown in Table 314.16(B) must be allowed for each wire, depending on its size. Where two or more ground wires of different sizes come into a box, they must all be counted as a single wire of the largest size used.

When allowances are made from the number of wires permitted in a box [Table 314.16(A)], as when devices, fixture studs, and the like are in the box, those allowances must "be based on the largest conductors entering the box" in any case where the conductors are of different sizes. This principle is true in such cases because the volume allowances equally affect all the conductors, regardless of size. However, when it comes to counting strap allowances, you only need to count based on the basis of the largest conductor arriving on a given strap, not the largest in the box.

Figure 314-10 shows a calculation with different wire sizes in a box. When conduit or EMT is used, there are no internal box clamps and, therefore, no addition for clamps. In this example, the metal raceway is the equipment grounding

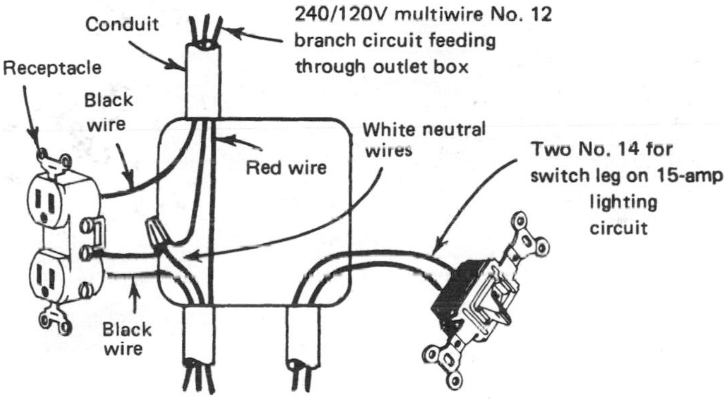
Conduit
240/120V multiwire No. 12
branch circuit feeding
through outlet box
Receptacle
Black
wire
White neutral
wires
Red wire
Two No. 14 for
switch leg on 15-amp
lighting
circuit
Black
wire

Two white No. 12s in splice	2 wires
Pigtail to neutral terminal does not count	
Red wire running directly through box	1 wire
One receptacle strap	2 wires
One switch strap	2 wires
Total No. 12s =	7 wires
Two black No. 12s to receptacle	2 wires
Two black No. 14s to switch	2 wires
Total No. 14s =	4 wires

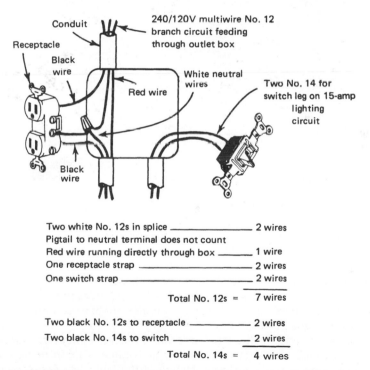

Fig. 314-10. When wires are of different sizes, volumes from Table 314.16(B) must be used. (Sec. 314.16.)

conductor—so no addition has to be made for one or more ground wires. And the red wire is counted as one wire because it is run through the box without splice or tap. As shown in the wire count under the sketch, the way to account for the space taken up by the wiring devices is to take each one as two wires of the same size as the largest wire attached to the device—that is, 12 AWG—as required in the end of the first sentence of part **(A)(2).** Note that the neutral pigtail required by 300.3(B) is excluded from the wire count as it would be under 314.16(A)(1).

From Table 314.16(B), each 12 AWG must be provided with 36.9 cm^3 (2.25 $in.^3$)—a total of 7 × 36.9 cm^3 (2.25 $in.^3$), or 258.3 cm^3 (15.75 $in.^3$) for the No. 12s. Then each No. 14 is taken at 32.8 cm^3 (2.0 $in.^3$)—a total of 4 × 32.8 cm^3 (2.0 $in.^3$), or 131.2 cm^3 (8.0 $in.^3$) for both. Adding the two resultant volumes—258.3 + 131.2 (15.75 + 8)—gives a minimum required box volume of 389.5 cm^3 (23.75 $in.^3$). From Table 314.16(A), a 100 × 100 × 54 mm 4- by 4-in. square box $2^1/_8$ in. deep, with 497 cm^3 (30.3 $in.^3$) interior volume, would satisfy this application.

For the many kinds of tricky control and power wire hookups so commonly encountered today—such as shown in Fig. 314-11—care must be taken to count all sizes of wires and make the proper volume provisions of Table 314.16(B).

Fig. 314-11. Many boxes contain several sizes of wires—some running through, some spliced, and some connected to wiring devices. Calculation of minimum acceptable box size must be carefully made. The combination switch and receptacle here is on a single mounting strap, which is taken as two wires of the size of wires connected to it. (Sec. 314.16.)

Table 314.16(A) gives the maximum number of wires permitted in boxes. But the last sentence of the first paragraph of 314.16(A)(2) does indicate that boxes may contain more wires if their internal volumes are marked and are greater than shown in Table 314.16(A).

Because the volumes in the table are minimums, most manufacturers continue to mark their products with the actual volume. This in many cases is somewhat greater than the volumes shown in the table. The last sentence of 314.16(A)(2) says that boxes that are marked to show a cubic-inch capacity greater than the minimums in the table may have conductor fill calculated in accordance with their actual volume, using the volume per conductor given in Table 314.16(A)(2).

The 2014 **NEC** has added a new paragraph to 314.16(B)(2) to recognize an invention (Fig. 314-12) designed to speed up the wiring process, especially in residential applications. The NM cable wires do not count in box fill, but those furnished with the clamp assembly do count. The clamp assembly does not require a fill allowance, but the box volume must be measured such that the volume of any internal portions of the clamp is excluded from its marked capacity. The clamp only accepts comparatively flat cables, and the manufacturer has arranged for the availability of 14/3 and 12/3 NM cable in a flat configuration to work with the assembly.

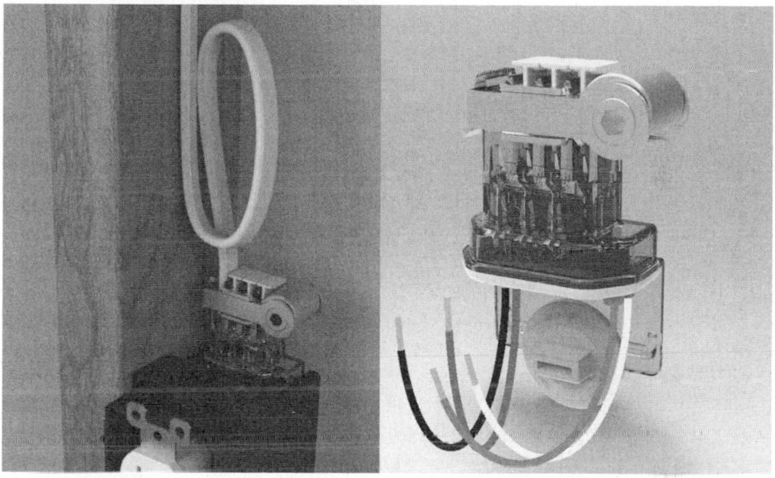

Fig. 314-12. This combination of a termination fitting and nonmetallic box must be part of a listed combination. [Sec. 314.16(B)(2).]

Conduit bodies must be marked with their cubic-inch capacity, and conductor fill is determined on the basis of Table 314.16(C). Such conduit bodies may contain splices or taps. An example of such application is shown in Fig. 314-13. Each of the eight 12 AWG wires that are "counted" as shown at bottom must be provided with at least 36.9 cm^3 (2.25 $in.^3$), from Table 314.16(B). The conduit body must, therefore, be marked to show a capacity of not less than 8 × 36.9 cm^3 (2.25 $in.^3$), or 295.2 cm^3 (18 $in.^3$).

The point of the conduit body rule is simple. If you use a conduit body to add an access point to a raceway or change its direction, but do nothing to the enclosed wires, then its function is to be part of the raceway system, and its size is whatever Chap. 9 requires for the raceways. On the other hand if you use a conduit body to make a splice or support a device or a lampholder, then you are using it as a box and you will size it as if it were a box.

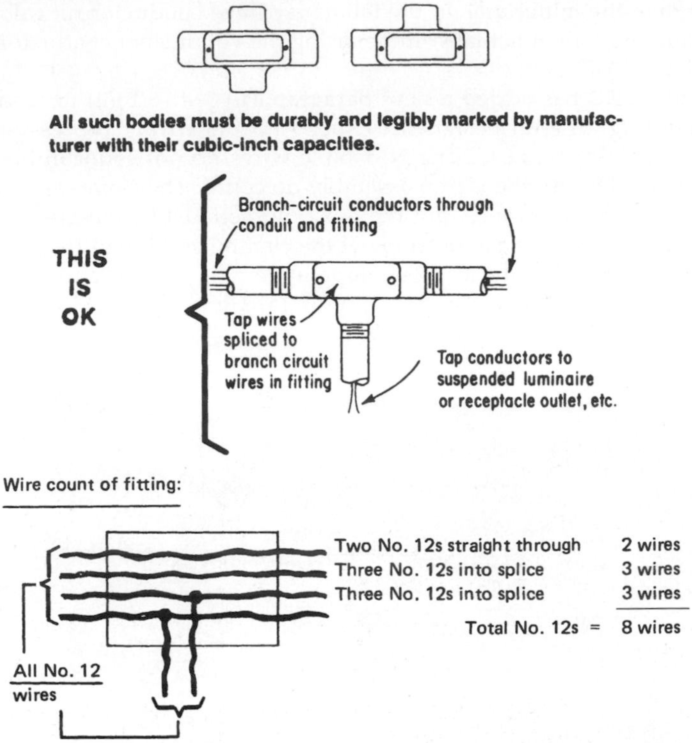
All such bodies must be durably and legibly marked by manufacturer with their cubic-inch capacities.
Branch-circuit conductors through conduit and fitting
THIS IS OK
Tap wires spliced to branch circuit wires in fitting
Tap conductors to suspended luminaire or receptacle outlet, etc.
Wire count of fitting:
Two No. 12s straight through 2 wires
Three No. 12s into splice 3 wires
Three No. 12s into splice 3 wires
Total No. 12s = 8 wires
All No. 12 wires

Fig. 314-13. Conduit bodies no longer must have "more than two entries" for conduit bodies to contain splices or taps. (Sec. 314.16.)

This leads to a practical consideration, and that is that it is almost inconceivable that a conduit body will actually work as a box in cases like this. This example, a very modest run of 4 12 AWG wires with two of them branching out at the tee fitting, comes out requiring 295.2 cm^3 (18 in.^{3}) in volume. The wires will fit easily in a metric designator 16 (trade size ½) raceway. Casual inspection of manufacturer's data for conduit bodies strongly suggests that the smallest tee conduit body with this much volume is a metric designator 35 (trade size 1¼) conduit body. As a practical matter, who in their right mind would be so enamored of conduit bodies that they would redesign the raceway upward by three trade sizes just to make this splice? And be careful about suggesting bushing the larger conduit body down to the size of the raceway. That can be done on rigid metal conduit and intermediate metal conduit, but on no other wiring methods, and only when they are actually threaded into the enclosure, and only if the volume is below 1650 cm^3 (100 in.^{3}), which is true in this case. If those conditions are not met, such as the wiring method being EMT or PVC, then 314.23(E) will have the tee fitting independently supported to structure, instead of being able to rely on the entering raceways. This job would be done with a box in a heartbeat.

Part **(C)** of 314.16 contains a number of provisions that must be carefully evaluated. Figure 314-14 shows the first rule. For instance, in that drawing, if a conduit body is connected to metric designator 16 (trade size ½) conduit, the conduit and the conduit body may contain seven 12 AWG TW wires—as indicated in Table C8, App. C, for rigid metal conduit—and the conduit body must have a csa at least equal to 2 × 204 mm^2 (0.314 $in.^2$) (the csa of metric designator 16 [trade size ½] conduit), or 408 mm^2 (0.628 $in.^3$). That is really a matter for the fitting manufacturers to observe.

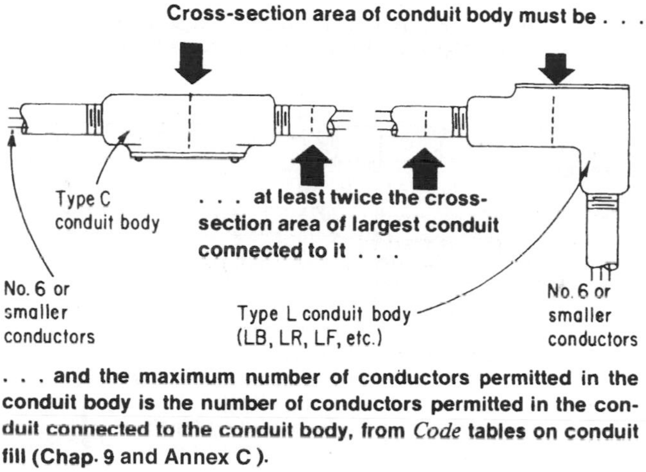
Cross-section area of conduit body must be . . .
Type C
conduit body
. . . at least twice the cross-section area of largest conduit connected to it . . .
No. 6 or
smaller
conductors
Type L conduit body
(LB, LR, LF, etc.)
No. 6 or
smaller
conductors
. . . and the maximum number of conductors permitted in the conduit body is the number of conductors permitted in the conduit connected to the conduit body, from Code tables on conduit fill (Chap. 9 and Annex C).

Fig. 314-14. For No. 6 and smaller conductors, conduit body must have a csa twice that of largest conduit. (Sec. 314.16.)

The second paragraph of part **(C)** covers the details shown in Fig. 314-15. The rule requires that where fittings are used as shown in the drawing, they must be supported in a rigid and secure manner. Because 314.23 establishes the correct methods for supporting of boxes and fittings, it must be observed, and that section refers to support by "conduits"—which seems to exclude such use on EMT because the **NEC** distinguishes between "conduit" and "tubing" (EMT), as in the headline for Table 1 of Chap. 9 in the back of the Code book. Figure 314-15 shows typical applications of those conduit bodies for splicing.

Refer to the discussion in the text associated with Fig. 314-13 for a full discussion of how unrealistic it is to expect to actually make use of the splicing allowances for conduit bodies. The only possible way out would be if manufacturers made oversize conduit bodies with much larger internal volumes, and that hasn't happened in the 18 years since these same rules have been in the **NEC**. The applications in Fig. 314-16 are indeed common, and they speak to a failure in the industry to enforce the rules. The floodlight needs to go on a box, and the box needs support arranged for it. The connection at the motor should be

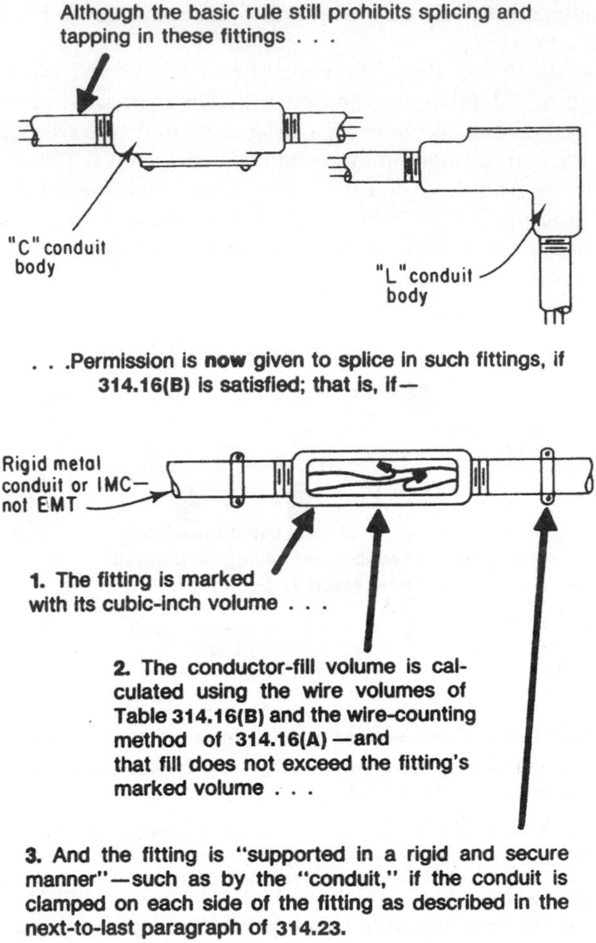
Although the basic rule still prohibits splicing and tapping in these fittings . . .
"C" conduit body
"L" conduit body
. . .Permission is now given to splice in such fittings, if 314.16(B) is satisfied; that is, if—
Rigid metal conduit or IMC—not EMT
1. The fitting is marked with its cubic-inch volume . . .
2. The conductor-fill volume is calculated using the wire volumes of Table 314.16(B) and the wire-counting method of 314.16(A)—and that fill does not exceed the fitting's marked volume . . .
3. And the fitting is "supported in a rigid and secure manner"—such as by the "conduit," if the conduit is clamped on each side of the fitting as described in the next-to-last paragraph of 314.23.

Fig. 314-15. Splices may be made in C and L conduit bodies—if the conditions shown in this illustration are satisfied. (Sec. 314.16.)

in the terminal housing, and yes, the LB needs more support; one way is to use a conduit nipple and threaded coupling out of the LB to a straight Greenfield connector. That would allow the support that should have been provided. That said, there are many small devices out there, such as solenoids, which come with short leads, with no splicing enclosure, and located far from any support point. It is unfortunate indeed when conscientious installers are driven to 90.4 in order to provide workmanlike installations in this area of the trade.

The third paragraph covers "short-radius conduit bodies." For generations the conduit bodies listed here and shown in Fig. 314-17 have been in a sort of "no-man's land." They meet the Art. 100 definition of a conduit body. They fail the bending space rules in 314.28(A)(2). They are obviously unsuited to enclose splices or devices. So now they have their own special classification, along with some

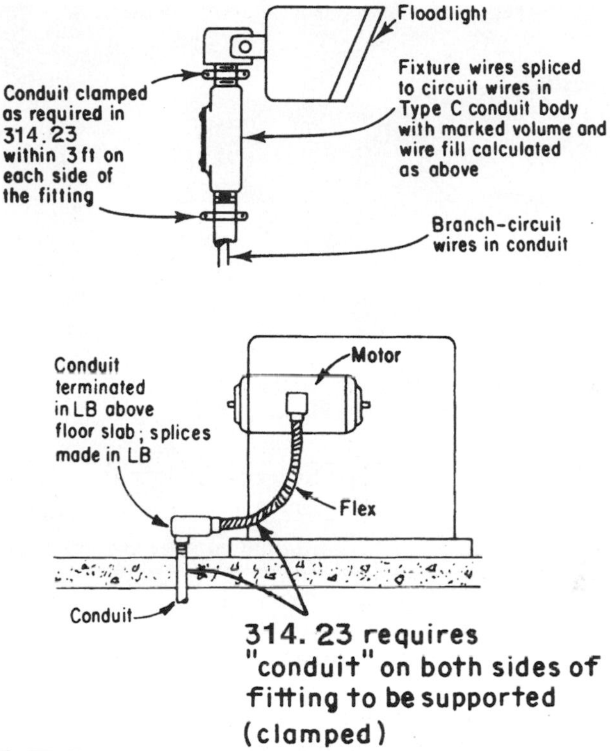
Floodlight
Fixture wires spliced to circuit wires in Type C conduit body with marked volume and wire fill calculated as above
Conduit clamped as required in 314.23 within 3 ft on each side of the fitting
Branch-circuit wires in conduit
Conduit terminated in LB above floor slab; splices made in LB
Motor
Flex
Conduit
314. 23 requires "conduit" on both sides of fitting to be supported (clamped)

Fig. 314-16. Splicing in C or L conduit bodies is common practice. (Sec. 314.16.)

rules to make sure they are applied properly. They must not be used with any conductors 4 AWG and larger, for which bending space rules apply, and they must only serve to change a raceway direction, nothing more. In addition to the items named in the rule, the classification includes flat 90° fittings with the removable covers that are coplanar with the axis of the raceway entries, and also the little "handy ells" often used at the edge of a box. This requirement used to be a standalone Sec. 314.5, but was moved here because it fits with the other material on conduit bodies.

314.17. Conductors Entering Boxes, Conduit Bodies, or Fittings. Part **(A)** requires "openings" to be adequately closed. Compliance with this rule can be achieved using a properly sized and listed enclosure, as is generally required by 300.16. Part **(B)** requires cables or raceways to be secured to *all metal* outlet boxes, conduit bodies, or fittings—such as by threaded connections, connector devices, or internal box clamps. It also addresses the rules for bringing loom into a box safely so the cable clamps don't damage the wire. It also requires that when a nonmetallic cable assembly enters a metal box, the sheath must clear the box and any cable clamping mechanism by at least ¼ in. This will avoid a metal clamping edge from abrading the insulation on one of the enclosed conductors. This 2017 **NEC** change directly

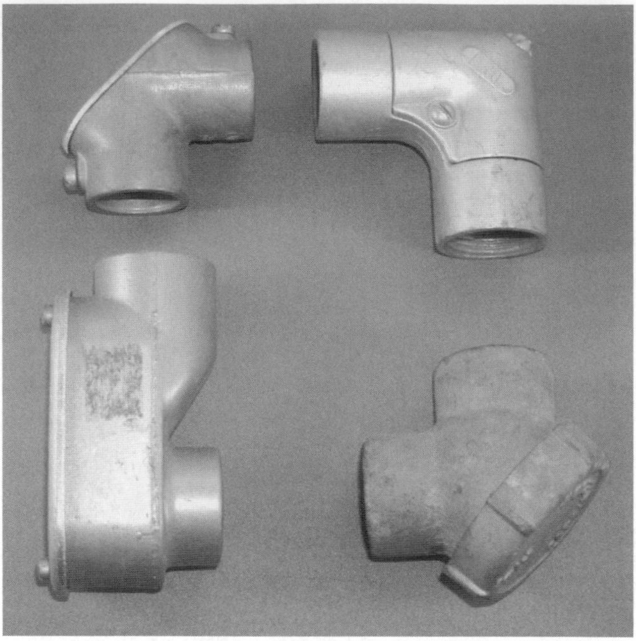

Fig. 314-17. Four examples of short-radius conduit bodies. [314.16(C)(3).]

correlates with the longstanding rule in (C) for nonmetallic boxes. The first sentence of part **(C)** requires that a nonmetallic box must have a temperature rating at least equal to the lowest-temperature-rated conductor entering the box. This rule assumes that the lowest-temperature-rated conductor in a box must be suited to the temperature in the box. The box would, therefore, be properly applied if it has a temperature rating at least equal to that of the lowest-rated conductor. At least ¼ in. of the cable sheath must be brought inside the box. Here too, it also addresses the rules for bringing loom into a box safely so the cable clamps don't damage the wire.

Another very important limitation in this Code section applies to the need for clamping nonmetallic-sheathed cable at a KO where the cable enters anything other than a single gang box. The Code has always accepted the use of nonmetallic-sheathed cable without box clamps or any type of connector where the cable is stapled within 8 in. (200 mm) of the box. The cable is then brought into the box through an NM cable KO on the box, without any kind of a connector at the KO or any clamps in the box (Fig. 314-18). But the intention of the exception to part **(C)** is that boxes or enclosures other than single gang boxes must be provided with a clamp or connector to secure nonmetallic-sheathed cable to such boxes (Fig. 314-19). *Only single gang nonmetallic boxes may be used without a cable clamp at the box KOs.* Where the Code permits elimination of a cable clamp if the cable is clamped to the stud within 8 in. (200 mm) of the box, the rule specifies that the 8-in. (200-mm) length be measured *along the cable* and not simply from the point of the cable strap to the box edge itself. The idea is to reduce the likelihood that pushing the device back into the box after wiring will be able to force the cable assembly back out of the box.

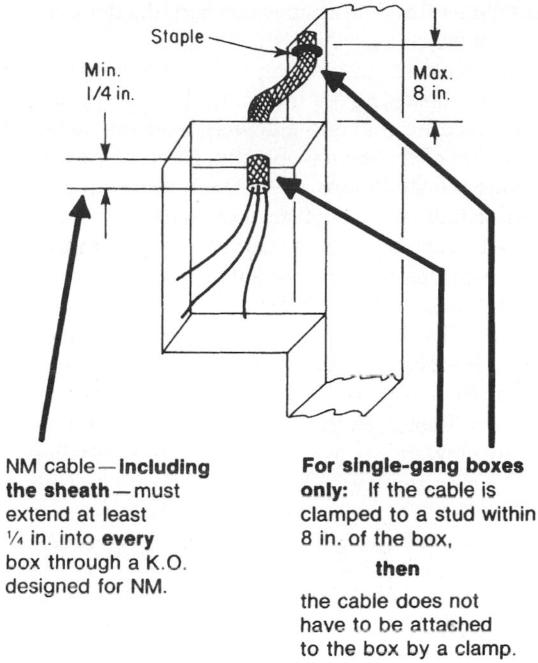
Staple
Min.
1/4 in.
Max.
8 in.
NM cable—**including the sheath**—must extend at least ¼ in. into **every** box through a K.O. designed for NM.
For single-gang boxes only: If the cable is clamped to a stud within 8 in. of the box,
then
the cable does not have to be attached to the box by a clamp.

Fig. 314-18. NM cable does not have to be clamped to single gang boxes. (Sec. 314.17.)

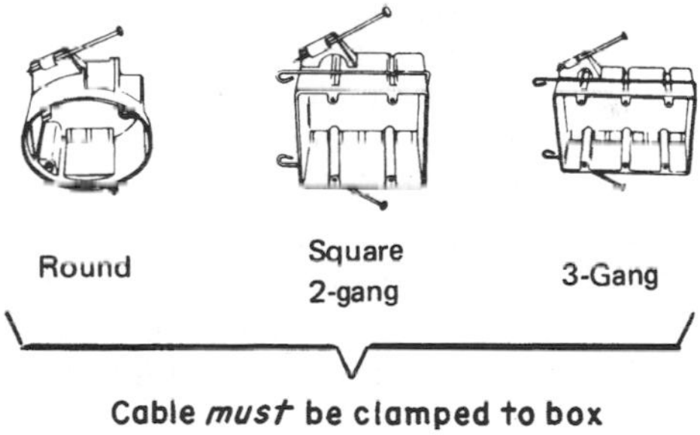
Round
Square
2-gang
3-Gang
Cable *must* be clamped to box

Fig. 314-19. NM cable must be clamped to all nonmetallic boxes that are *not* single gang boxes. (Sec. 314.17.)

When used with open wiring on insulators, knob-and-tube work, or nonmetallic-sheathed cable, nonmetallic boxes have the advantage that an accidental contact between a "hot" wire and the box will not create a hazard.

314.19. Boxes Enclosing Flush Devices. A through-the-wall box is a box which is manufactured to be installed in a partition wall so that a receptacle or switch may be attached to either side; therefore, it is not necessary to use two standard boxes, one facing each side, connected by a jumper. After the devices have been installed on both sides, and the required faceplates secured, the devices are enclosed on all sides and the intent of the rule is satisfied.

If the screws used for attaching the receptacles and switches to boxes were used also for the mounting of boxes, a poor mechanical job would result, since the boxes would be insecurely held whenever the devices were not installed and the screws loosened for adjustment of the device position. Hence the prohibition.

314.20. Flush-Mounted Installations. For flush-mounting applications, a box must be flush with a combustible surface or even project forward; if the surface is not combustible the box can be recessed but not more than 6 mm (¼ in.). As used here, the term *box* includes the components that make up the final enclosure, including extension boxes and plaster rings. If the box is recessed more than is allowed by this rule, a "listed extender" is permitted to make up the difference.

314.21. Repairing Noncombustible Surfaces. For boxes using flush covers and recessed in drywall or plaster surfaces, the plaster or drywall must be repaired so that the gap between the box and the surface does not exceed 3 mm ($^1/_8$ in.). The spacing allowance provides some relief to installers, because for several generations the requirement was a complete repair. The dimension comes from the fact that when UL evaluates electrical box penetrations in fire-rated assemblies, they assume this gap will exist around the perimeter of the box being tested.

The purpose of 314.20 and 314.21 is to prevent openings around the edge of the box through which fire could be readily communicated to combustible material in the wall or ceiling. Both 314.20 and 314.21, and their counterparts in Art. 312, address issues related to when ordinary building construction will be allowed to complete an electrical enclosure. 314.20 is looking at the cut surface that is perpendicular to the building surface and extending into the opening and concluding that it is acceptable as long as it isn't combustible and doesn't extend more than 6 mm (¼ in.). 314.21 is looking at the plane of the building surface itself and asking, if the enclosure is behind the plane of the finished wall to some extent, how well does the cut-out need to be repaired to prevent a fault in the box from getting past the flush cover and into the wall along the side of the enclosure, and concluding the answer is 3 mm ($^1/_8$ in.). 314.21 also concludes (implicitly) that it isn't necessary to repair the wall edge if the wall is combustible, because in such cases the enclosure must come all the way out, and at that point the flush cover will pretty well seal to the enclosure walls, as extended if necessary. Therefore, it is not required to bring plastic wood filler to jobs in wood paneling.

None of this discussion is relevant to a surface-mounted style enclosure cover that telescopes over the enclosure walls or seats flat on all four sides, such as a raised cover on an outlet box designed for surface mounting. The box being recessed in the wall to a greater or lesser degree doesn't change the fact that a surface cover will seal to the box and complete the enclosure. If the enclosure is half in the wall and half out, for example, and if the building surface isn't repaired at all and looks ugly, so be it. If it is a penetration in a fire-rated partition, the repair will be required by 300.21, but many applications are not and it is not appropriate to write this rule as though a repair should always be required. The electrical enclosure is complete, and if the owner doesn't like how the wall looks, it can be repaired later. This is why these rules have been modified in recent code cycles to limit their application to flush-mounting situations where, as noted, the building finish is being called upon to complete an electrical enclosure.

314.22. Surface Extensions. The extension should be made as illustrated in Fig. 314-20. The extension ring is secured to the original box by two screws passing through ears attached to the box.

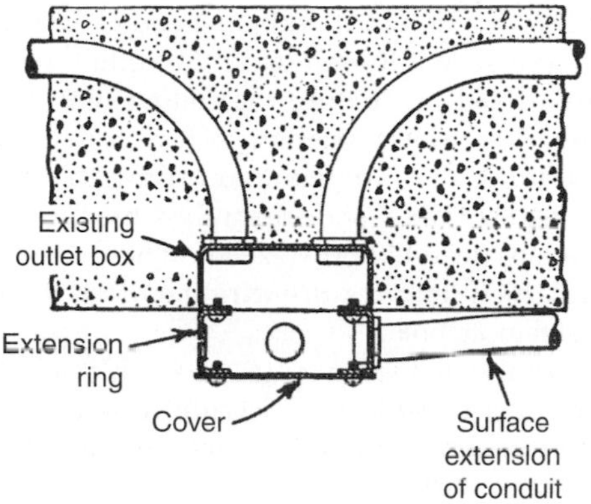
Existing
outlet box
Extension
ring
Cover
Surface
extension
of conduit

Fig. 314-20. Extension ring must be secured to box for surface extension. (Sec. 314-22.)

As noted in the exception, a surface extension may be made from the *cover* of a concealed box, *if* the cover mounting design is secure, the extension wiring method is flexible, and grounding does not depend upon connection between the box and the cover. This exception to the basic rule, which requires use of a box or extension ring over the concealed box, permits a method that provides a high degree of reliability.

The wording about the cover not falling off if the securing means loosens may need some explanation. An extension from a blank handy box cover or the like is not a problem because the 6-32 screws would have to come completely out of the box to release the cover. The rule is aimed at extensions from concrete rings or octagon boxes (or round plaster rings) on ceilings that have a usual keyhole-shaped slot for at least one of the 8-32 mounting screws. If those screws loosen, even a little, the cover can turn the required few degrees until the wide end of the slot lines up with the screw head and the cover falls free. To counteract this, some cover designs have a raised detent on the edge of the keyhole slot. The cover cannot turn past this point unless the screw is loosened at least another (roughly) three complete turns, and then the cover is pushed up against gravity and turned the rest of the way. The wording requires this or some other design feature that will keep the cover in place until it is really supposed to come down.

314.23. Supports. The Code rule strictly requires all boxes, conduit bodies, and fittings to be fastened in their installed position—and the various paragraphs of this section cover different conditions of box support for commonly encountered enclosure applications. The one widely accepted exception to that rule—although actually not recognized by the Code—is the so-called throw-away or floating box, which is a junction box used to connect flexible metal conduit from a recessed fixture to flex or BX branch-circuit wiring, in accordance with Sec. 410.117(C) (Fig. 314-21). In such cases, the connection of the fixture "whip" (the 18 in. to 6 ft length of flex with high-temperature wires, for example, 150°C Type AF) is made to the branch-circuit junction box which hangs down through the ceiling opening, and then the junction box is pushed back out of the way in the ceiling space, and the fixture is raised into position. But with suspended ceilings of lift-out panels, there is no need to leave such a loose box in the ceiling space, because connection can be made to a fixed box before the ceiling tiles are laid in place. It is important to remember that this procedure is functionally obsolete for another reason. Most recessed luminaires are required to have thermal protection. This is arranged through a self-contained connection box mounted as part of the luminaire because the test labs need to know exactly where the device will be located so they can test it. Any high-temperature wiring ends in this box, and is not provided in the field to a remote box.

Figure 314-22 shows the rule of part **(B)(1)** of this section. Note that since everyone has screw guns these days, the rule now addresses screws where nails would have been used previously. Screws have sharp edges all along their shafts, which can damage wiring pushed blindly into the box. So, if screws are used with exposed threads in the box, some kind of sleeving has to be applied over the raw threads. Periodically, it is necessary to drill a mounting hole in a more convenient location than where the manufacturer pre-made hole; now when this is done the inspector gets asked for approval.

An outlet box built into a concrete ceiling, as shown in Fig. 314-23, seldom needs any special support, per 314.23**(G)**. At such an outlet, if it is intended for a fixture of great weight to be safely hung on an ordinary ⅜-in. fixture stud, a special fixture support consisting of a threaded pipe or rod is required, such as is shown in Fig. 314-24.

In a tile arch floor (Fig. 314-24), a large opening must be cut through the tile to receive the conduit and outlet box.

The requirement of metal or wood supports for boxes applies to concealed work in walls and floors of wood-frame construction and other types of construction having open spaces in which the wiring is installed. In walls or floors of concrete, brick, or tile where conduit and boxes are solidly built into the wall or floor material, special box supports are not usually necessary.

As covered in part **(C)**, in an existing building, boxes may be flush mounted on plaster or any other ceiling or wall finish. Where no structural members are available for support, boxes may be affixed with approved anchors or clamps. Figure 314-25 illustrates that. For cutting metal boxes into existing walls, "Madison Holdits" are used to clamp the box tightly in position in the opening. Actually, the local inspector can determine acceptable methods of securing "cut-in" device boxes because this provision provides appreciable latitude for such decisions.

And according to part **(D)**, framing members of suspended ceiling systems may be used to support boxes if the framing members are rigidly supported and securely fastened to each other and to the building structure.

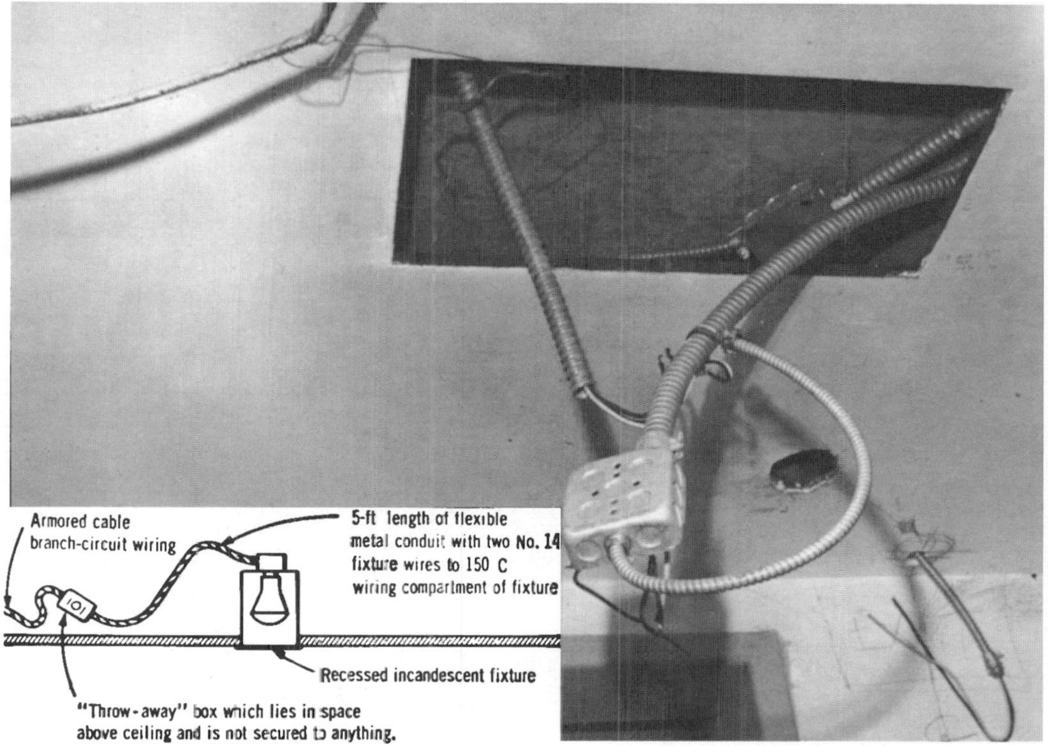
Armored cable
branch-circuit wiring
5-ft length of flexible
metal conduit with two No. 14
fixture wires to 150 C
wiring compartment of fixture
Recessed incandescent fixture
"Throw-away" box which lies in space
above ceiling and is not secured to anything.

Fig. 314-21. Fixture supply flex is tapped out of junction box fed by flex or BX branch-circuit wiring in ceiling space. Box is later "thrown away," unattached, into ceiling space. This is not permitted. (Sec. 314.23.)

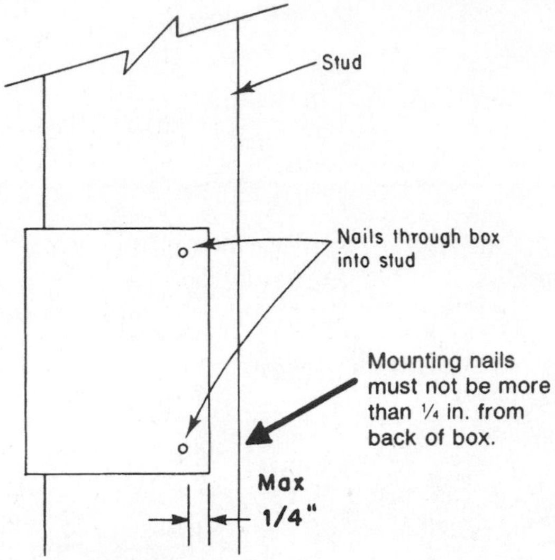
Stud
Nails through box
into stud
Mounting nails
must not be more
than ¼ in. from
back of box.
Max
1/4"

Fig. 314-22. Box-mounting nails must not obstruct box interior space. (Sec. 314.23.)

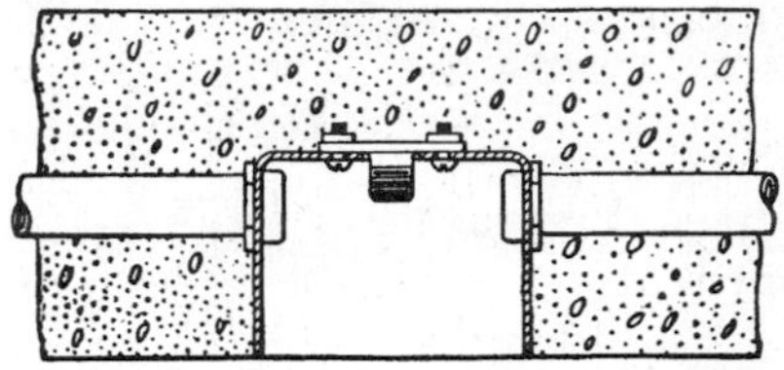

Fig. 314-23. Box in concrete is securely supported. (Sec. 314.23.)

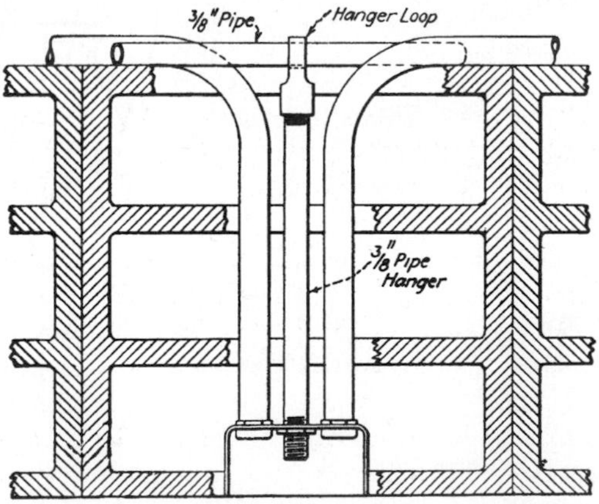
3/8" Pipe
Hanger Loop
3/8" Pipe
Hanger

Fig. 314-24. Box in tile arch ceiling requires pipe-hanger support if very heavy lighting fixture is to be attached to the box stud. (314.23.)

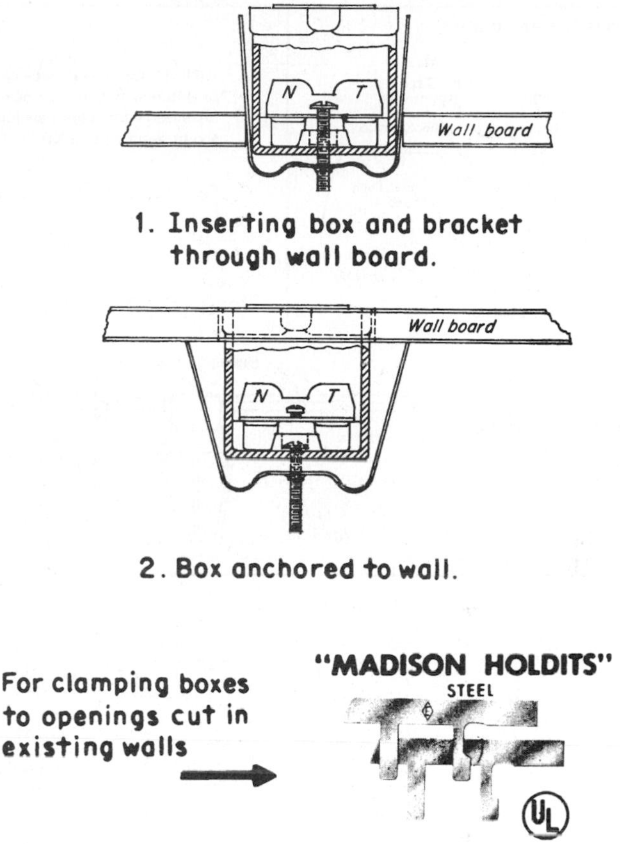
N
T
Wall board
1. Inserting box and bracket through wall board.
Wall board
N
T
2. Box anchored to wall.
For clamping boxes to openings cut in existing walls
"MADISON HOLDITS"
STEEL

Fig. 314-25. Part **(C)** of 314.23 refers to these types of clamping devices. (Sec. 314.23.)

Figure 314-26 shows box-support methods that are covered by part **(E)** of 314.23 for boxes not over 100 in.^{3} (1650 cm^3) in volume that do not contain devices or support luminaires. Pay close attention because a great number of rules are packed into this paragraph of Code text. The rule there applies to "conduit" (rigid metal conduit and IMC only; see exception following for others) used to support boxes—as for overhead conduit runs. A box may be supported by two properly clamped conduit runs (rigid metal conduit or IMC) that are *threaded into* entries on the box or into field-installed hubs attached to the box. The words "threaded wrenchtight into" purposely exclude a set-screw or compression connector applied to the unthreaded end of a run of conduit. The conduit must thread into the box. If that can't be done because of problems turning the conduit, then a short nipple and a union can be used, but the conduit system will thread into the box. These words, along with 352.10(H), also exclude PVC conduit as a support. The associated permission to use separate, field-installed hubs offers a relaxation of the

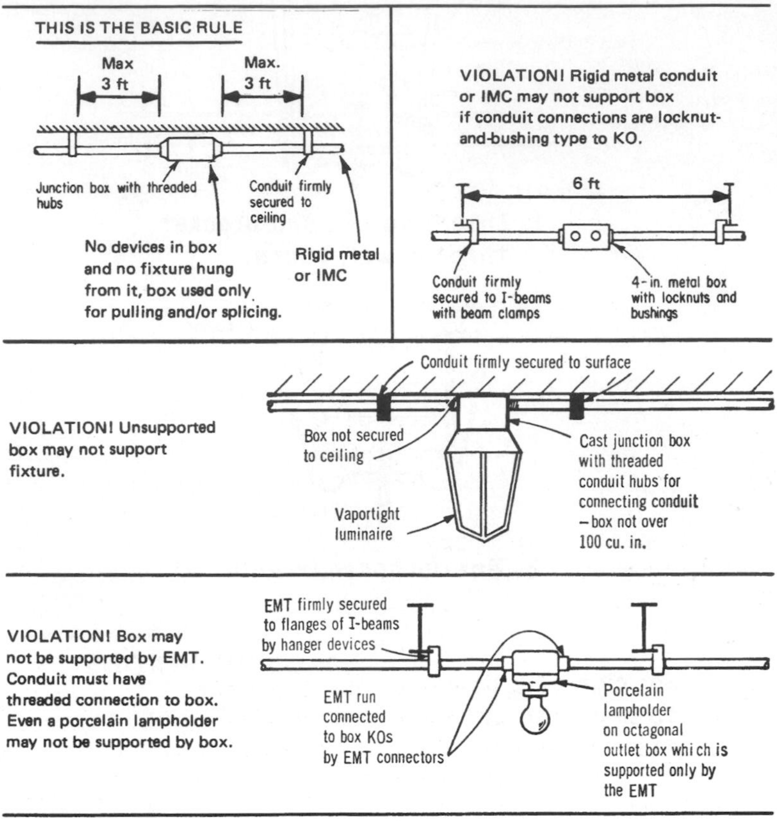
THIS IS THE BASIC RULE
Max 3 ft
Max. 3 ft
Junction box with threaded hubs
Conduit firmly secured to ceiling
No devices in box and no fixture hung from it, box used only for pulling and/or splicing.
Rigid metal or IMC
VIOLATION! Rigid metal conduit or IMC may not support box if conduit connections are locknut-and-bushing type to KO.
6 ft
Conduit firmly secured to I-beams with beam clamps
4-in. metal box with locknuts and bushings
Conduit firmly secured to surface
VIOLATION! Unsupported box may not support fixture.
Box not secured to ceiling
Cast junction box with threaded conduit hubs for connecting conduit – box not over 100 cu. in.
Vaportight luminaire
EMT firmly secured to flanges of I-beams by hanger devices
VIOLATION! Box may not be supported by EMT. Conduit must have threaded connection to box. Even a porcelain lampholder may not be supported by box.
EMT run connected to box KOs by EMT connectors
Porcelain lampholder on octagonal outlet box which is supported only by the EMT

Fig. 314-26. Box may be supported by "conduit" that is clamped, but box must not contain or support anything. (Sec. 314.23.) Note that the vapor-tight luminaire installation, as shown in the center panel, probably would comply with 314.23(F) if the wiring method were rigid or intermediate metal conduit.

previous demand for hubs manufactured with the box, such as FS boxes. This is the only permissible way to support a sheet metal box on the entering raceways: add hubs (the so-called "Myers Hubs") to the box, and then thread RMC or IMC into the hubs, as before. Where locknut and bushing connections are used instead of threaded conduit connections to a threaded hub or similar connection, the box must be independently fastened in place. The orientation of the conduit entries also matters. They must enter on two or more sides because of the substantial distance allowed to the support point, 900 mm (3 ft). If that support distance is reduced to 450 mm (18 in.), then the conduit entries can be on the same side.

The rule as worded would not allow even a conduit body on EMT or PVC to be supported by its raceway entries, and calls for direct support to structure instead. This would be an absurd result and the exception is written to take care of these problems. The exception and the rule are very tightly integrated, and must be read together. The exception begins with a list of wiring methods which include RMC and IMC. Why? Because the rule does not allow for raceway supported enclosures above 1650 cm^3 (100 $in.^3$), and many conduit bodies for the larger conduit sizes well exceed this size threshold. However, a conduit body installed under this exception must be no larger than its largest entering raceway. The exception also recognizes the use of "E" fittings that have only a single conduit entry, for obvious reasons.

Note that this does vary from the principle that enclosures must never be supported by a single entry because an enclosure so mounted can easily untwist. "E" fittings are not much larger than the supporting conduit and so there is not much mechanical advantage presented by the fitting in terms of torque in comparison to a FS box or other comparable enclosure.

The question now arises, can a (for example) metric designator 27 (trade size 1) tee fitting be bushed down for all metric designator 21 (trade size ¾) entries? This brings up an important rule of reading Code text, namely, if you comply with a rule, it does not matter what a permissive exception following the rule happens to say. (Mandatory exceptions are extensions of the rule, must be followed if applicable, and in any list of exceptions must be listed first for this reason.) In this case, if the wiring is RMC or IMC and all the rules regarding support are followed, the installation is perfectly OK because the exception following is permissive. However, if you depart from the rule for any reason, then you have to abide by the terms of the exception.

The rules of part **(F)** of 314.23 are shown in Fig. 314-27. The basic rules and first exception closely follow the rules in (E), except that the support distance drops to 450 mm (18 in.) which is allowed to be all on one side. The exception does not list EMT or PVC because now there is considerable load on the conduit body. Figure 314-28 shows several installations that are in violation of these rules.

Part **(F)** has a second exception to cover cantilevered applications where luminaires are on the ends of conduit stems, such as some commercial lighting over sidewalks and billboard lighting, as shown in the upper left of Fig. 314-29. This exception and the work in Part **(H)(2)** below, was developed in conjunction with a task group with representation from the steel conduit makers, and focused in part on how much force typical threaded joints could be expected to withstand. Note that these are rules for field installations of conduit and boxes. If the conduit stem is provided with the luminaire and is covered by the luminaire listing, then these rules do not apply and the luminaire installation directions take precedence.

Part **(H)** permits a pendant box (such as one containing a START-STOP button) to be supported from a multiconductor cable, using, say, a strain-relief connector threaded into the hub on the box, or some other satisfactory protection for the conductors. The second part, **(H)(2)**, covers the common practice, especially in industrial occupancies, of hanging HID luminaires from conduit stems, as illustrated in the lower right of Fig. 314-29. These rules also focus on limiting the breaking forces that could be imposed on the threads. Note that the maximum dimension for luminaire horizontal extension from the conduit entry, 300 mm (12 in.), precludes hanging 900 mm (3 ft) or 1.2 m (4 ft) fluorescent luminaires from a single conduit entry support. This is a significant issue with T5 fluorescent luminaire retrofits for previous HID applications.

314.24. Depth of Boxes. Sufficient space should be provided inside the box so that the wires do not have to be jammed together or against the box, and the box should

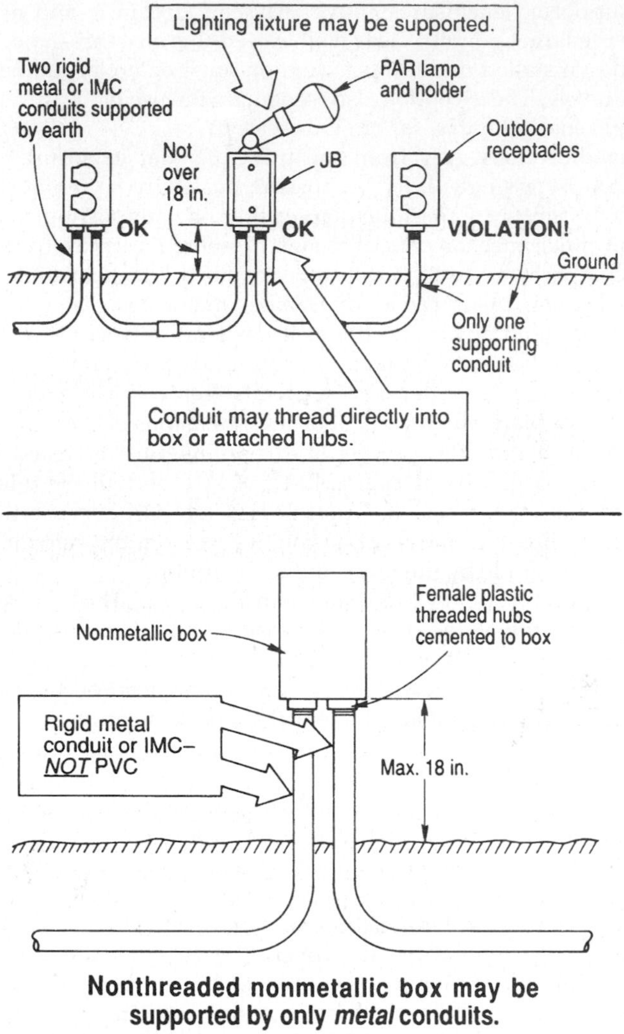
Lighting fixture may be supported.
Two rigid metal or IMC conduits supported by earth
PAR lamp and holder
Outdoor receptacles
Not over 18 in.
JB
OK
OK
VIOLATION!
Ground
Only one supporting conduit
Conduit may thread directly into box or attached hubs.
Nonmetallic box
Female plastic threaded hubs cemented to box
Rigid metal conduit or IMC– NOT PVC
Max. 18 in.
Nonthreaded nonmetallic box may be supported by only metal conduits.

Fig. 314-27. Boxes fed out of the ground or a concrete floor, patio, or walk must observe these rules. (Sec. 314.23.)

provide enough of an enclosure so that in case of trouble, burning insulation cannot readily ignite flammable material outside the box. The 2008 NEC expanded the scope of this section considerably in response to proposals that substantiated fire alarm equipment in particular that was causing installation problems in even large boxes. The proposals wanted to recalculate the volume rules in 314.16. That would not have solved the problem because an item with small volume can still have a long reach backward. The solution was to add language here regulating box depth.

Parts **(A)** and **(B)(5)** were in the Code previously and cover ceiling pans and the 25.4 mm (1-in.) deep boxes used in some special applications such as for doorjamb

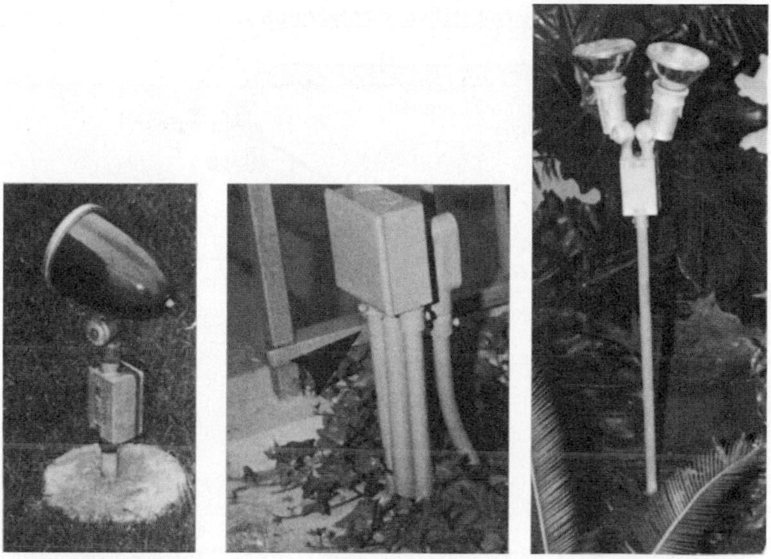

Fig. 314-28. A *single* rigid metal conduit may not support a box, even with concrete fill in the ground (left). Box may not be supported on EMT, even with several connections used (center). Method at right is a violation on three counts: EMT, not "conduit," supports the box; only one hub on box is connected; box is more than 18 in. above ground. (Sec. 314.23.)

switches. The fire alarm equipment problem is addressed in **(C)(1)**. Equipment that has been listed for use in certain boxes may, of course, use those boxes.

This section also covers devices, and sets depth requirements based on wire sizes. The dimensions were chosen based on task group research into actual device and conductor dimensions and comparing that with standard box dimensions. The objective is to prevent damage to conductors resulting from devices pinning wires tightly against the back wall of a box in the process of installation. Remember that the inner back planes of many outlet boxes have inwardly punched knockouts as well as the tops and shoulders of mounting hardware, any of which could easily damage a conductor if clearance is not provided.

Item **(2)** covers large power outlets that usually mount in cast boxes and are designed accordingly. Item **(3)** covers the larger devices, such as 3-pole 4-wire range receptacles. The dimension chosen correlates with the standard sized $2^1/_8$-in. box depth; as an internal measurement it is 1/16-in. smaller than the standard nominal box dimension. Item **(4)** correlates typical device sizes with standard minimum device box depths, after allowing for the thickness of the steel. This paragraph also contains a backup requirement for a minimum enforceable clearance of ¼ in. because some devices in this category approach 1¾ in. in depth, and a deeper box will be required. Item **(5)** picks up the requirement that existed in the **NEC** since the 1975 edition. It was originally designed to accommodate a device box mounted on a masonry wall with ¾-in. furring plus paneling. Although largely obsolete because of changes over the years in volume allowances, it can still be used, particularly with some sort of box extension. It is limited to 14 AWG wiring accordingly.

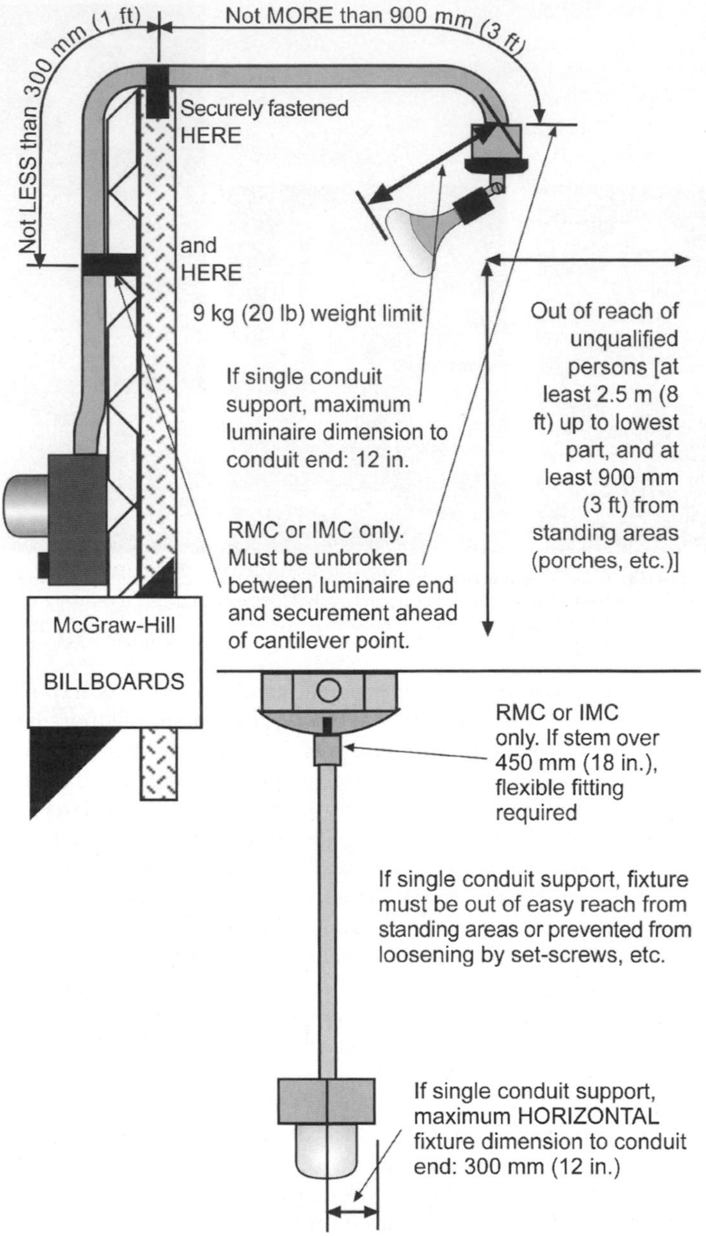
Not LESS than 300 mm (1 ft)
Not MORE than 900 mm (3 ft)
Securely fastened
HERE
and
HERE
9 kg (20 lb) weight limit
If single conduit support, maximum luminaire dimension to conduit end: 12 in.
RMC or IMC only. Must be unbroken between luminaire end and securement ahead of cantilever point.
Out of reach of unqualified persons [at least 2.5 m (8 ft) up to lowest part, and at least 900 mm (3 ft) from standing areas (porches, etc.)]
McGraw-Hill
BILLBOARDS
RMC or IMC only. If stem over 450 mm (18 in.), flexible fitting required
If single conduit support, fixture must be out of easy reach from standing areas or prevented from loosening by set-screws, etc.
If single conduit support, maximum HORIZONTAL fixture dimension to conduit end: 300 mm (12 in.)

Fig. 314-29. Luminaires on conduit stems. [314.23(F) Exception No. 2 and 314.23(H)(2).]

314.25. Covers and Canopies. This rule requires every outlet box to be covered up—by a cover plate, a fixture canopy, or a faceplate, which has the openings for a receptacle, snap switch, or other device installed in a box. An electric discharge luminaire mounted over an outlet box is permitted to substitute for the cover normally required, but it must comply with 410.24(B) in the process, which means it will be punched to allow access to the interior of the box from the luminaire ballast channel without removing the luminaire from the wall surface. The 2014 **NEC** has added a rule requiring cover mounting screws to either match the thread tapping of a box or line up with the manufacturer's instructions; the point is to eliminate the use of random drywall screws for this purpose.

Part **(A)** requires all metal faceplates to be grounded as required by 250.110. As noted in the commentary under that topic, the fact that any wiring method with an equipment ground must connect exposed metal surfaces to the equipment grounding conductor has transformed the practical implementation of the overall requirement. Instead of measuring to grounded surfaces, just look for the equipment ground, and make the connection. For faceplates (covered in Fig. 314-30) use a device with a grounding terminal. This rule is frequently violated for no good reason. For a cover a ground clip (Fig. 250-112) will easily attach to the side of a metal cover, creating a tail that is easily connected. It should be further noted that for a snap switch faceplate, such as the one that would be installed on the box in Fig. 314-30, the grounding termination on the snap switch is generally required whether or not a nonmetallic faceplate is first installed, as covered in 404.9(B).

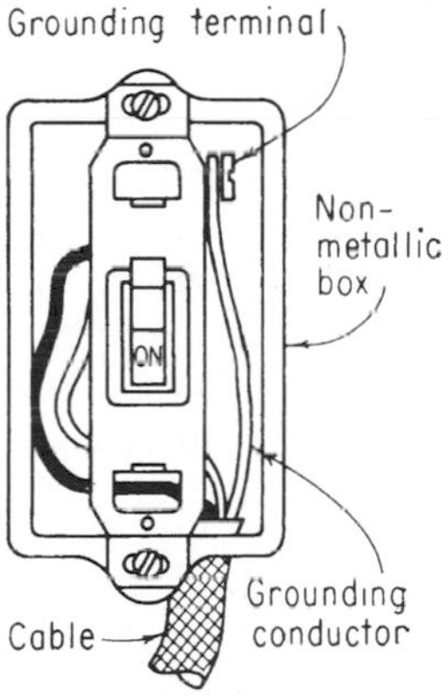
Grounding terminal
Non-metallic box
ON
Grounding conductor
Cable

Fig. 314-30. Grounding switch must be used for metal faceplate on any nonmetallic box. (Sec. 314.25.)

In part **(B)** of 314.25, if the ceiling or wall finish is of combustible material, the canopy and box must form a complete enclosure. The chief purpose of this rule is to require that no open space be left between the canopy and the edge of the box where the finish is wood or other combustible material; however, a 2014 **NEC** change in 410.23 has virtually eliminated this requirement in any practical sense. Where the wall or ceiling finish is plaster, the requirement does not apply, because plaster is not classed as a combustible material; however, the plaster must be continuous up to the box, leaving no opening around the box beyond the 3 mm ($^1/_8$ in.) allowed in 314.21. Refer to the coverage of the sister requirement in 410.23 for more information on this topic and the impact of the major 2014 **NEC** change at that location.

314.27. Outlet Boxes. Part **(A)** requires that any box used at luminaire (or lamp holder) outlets in a ceiling be designed for this purpose and be capable of supporting a luminaire weighing not less than 23 kg (50 lb). Boxes used for the same purpose in a wall must also be designed for the purpose, but if they are not capable of supporting a 23 kg (50 lb) luminaire, they can be used, and must be marked on the inside of the box with the weight they can support if it's anything other than 23 kg (50 lb) (some are more). This requirement means that any boxes with the marking "FOR FIXTURE SUPPORT" on the unit carton will have the required construction and be suitable for supporting a 23 kg (50 lb) luminaire. There have been some nonmetallic boxes that had been listed for intermediate support weights, such as 6.9 kg (15 lb). Although they were marked, most installers weren't looking for the marking, believing that if it had the usual 8-32 screws in one of the usual mounting configurations, it was suitable for a 23 kg (50 lb) load. Boxes suitable for higher weights are marked accordingly. Note that this section formerly did not address box support limits where they were mounted on open framing instead of wall surfaces. The 2014 **NEC** has closed this loophole by changing the subject heading to "vertical surface" (instead of "wall") outlets. The intent for both numbered paragraphs is the same: Outlet boxes designed for luminaire support are marked to that effect and must be rated for the ability to support at least 50 lb. This is the default and need not be marked. Then, if the manufacturer produces an outlet box that will carry a greater weight and has a listing to prove it, it can be used for up to that weight. The box must be marked, on its interior so the inspector can see it after the large luminaire is in evidence, with the designed capacity.

The exception allows other boxes to support luminaires weighing up to 3 kg (6 lb) to be supported on other boxes not designed for luminaire support, such as device boxes or single-gang plaster rings with 6-32 mounting screw provisions. At least two 6-32 screws must be used to support the luminaire, and this exception only applies when the screws are in shear and not in tension, that is, for a wall- and never a ceiling-mounted luminaire. This exception does not apply to smoke detectors, etc.; refer to **(D)** for utilization equipment coverage.

Part **(B)** requires floor boxes to be completely suitable for the particular way in which they are used. Adjustable floor boxes and associated service receptacles can be installed in every type of floor construction. A metal cap keeps the assembly clean during pouring of concrete slabs. After the concrete has cured, this cap can then be removed and discarded and floor plates and service fittings added.

There are two types of floor receptacles, the usual ones and ones for the elevated floors of show windows, etc., where they will not receive the same floor traffic, as per the exception. These are classified by UL as "display receptacles." Although they have a superficial similarity to floor receptacles, floor receptacles are sold with a box and the combination has passed additional tests, including a scrub-water test. Do not confuse the two.

As noted in part **(C)**, a ceiling paddle fan must not be supported from a ceiling outlet box—unless the box is UL-listed as suitable as the sole support means for a fan (Fig. 314-31). The vibration of ceiling fans places severe dynamic loads on the screw attachment points of boxes. But boxes designed and listed for this application pose no safety problems. There are two weight limits in this section. No fan box may serve as the direct support of a paddle fan weighing over 32 kg (70 lb). All fan boxes need

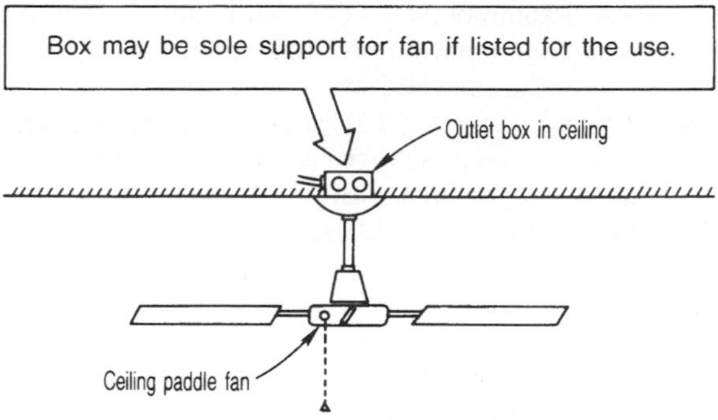
Box may be sole support for fan if listed for the use.
Outlet box in ceiling
Ceiling paddle fan

Fig. 314-31. Support for ceiling fans must be suited to the dynamic loading of the vibrating action. [Sec. 314.27(D).]

to be listed and marked that they are suitable for fan support, and the default weight limit for suitability is a 16 kg (35 lb) fan. Any manufacturer who chooses to design a box for a heavier fan may do so, provided (1) it will not be expected to hold more than 32 kg (70 lb) and (2) the designed fan weight, whatever it is between 16 kg (35 lb) and 32 kg (70 lb), must be marked on the box. A fan box capable of supporting a 16 kg (35 lb) fan does not require a weight marking, because that is the default suitability.

The second paragraph addresses widespread reports of attempts at deceit on the part of some unscrupulous builders, who have bedrooms roughed in without the expense of fan boxes, but wired to tacitly encourage their customers to hang fans anyway. If a spare, separately switched conductor is run to a ceiling box in a location that will accommodate a paddle fan (e.g., ceiling boxes 18 in. from a wall, perhaps for track lighting with wall washer luminaires, would not be caught in this requirement) then a listed fan box must be installed. There have been many attempts to address this, all failing on the legitimate uses of multiple switched conductors run to a ceiling box, including chandeliers with split loads.

This wording received a unanimous vote due to the magical word "spare." If the owner or builder doesn't want the fan box, then the owner doesn't run a spare wire. And there are legitimate reasons to avoid a fan box requirement. Not all ceiling boxes are simple, particularly in locations with raceway wiring; in such areas it is common to branch the perimeter receptacle outlets from a ceiling overhead luminaire outlet box. This may involve large steel boxes, even with extension boxes, and plaster rings. Combinations such as these will never be available as listed fan boxes because there is insufficient market to support their development.

In addition, it is very important to remember that it never has been and still is not a requirement to hang a paddle fan on a listed fan box. It is only a requirement to use the fan box *if it will be the sole support of the fan*. Often ceilings are deliberately framed so as to be capable of receiving ¼-in. lag bolts run through the fan bracket, through the crow-foot pattern mounting holes in the back of the box, and into the framing member. Such an arrangement does not require a listed fan box. At least one major paddle fan manufacturer sells its fans with such lag bolts, and includes instructions (these are listed fans so the instructions are enforceable) to run the bolts into framing, and not use a fan box. The wording as of the 2011 **NEC**, by focusing on the word "spare," neatly accommodates these legitimate concerns; again, if one wants to avoid the fan box requirement, simply don't run the spare wire.

Part **(D)** was new for the 2008 **NEC** and covers utilization equipment mounted on outlet boxes. The rules in 314.27(A) carry over, but the 3 kg- (6-lb) exception differs in one key respect because there is no orientation limitation; the box can be mounted on the ceiling or on a wall.

Part **(E)**, new for the 2017 **NEC**, addresses the new locking support receptacles now covered in the revised definition of "receptacle" in Article 100. Refer to Fig. 100-28 for a photo of one of these receptacles, together with a luminaire to the side outfitted with compatible hardware. The locking support and mounting receptacle must be listed for the support of compatible attachment fittings, and the combination must be identified for the allowable weight and orientation. In other words, if it can be used on a vertical surface it must say so, etc. If the receptacle is mounted within a box, as in the case of the one shown in Fig. 100-28, then a double volume allowance must be applied, in accordance with 314.16(B)(4). Note that 422.18(2) contains correlating language that recognizes the support of paddle fans in accordance with these rules.

314.28. Pull and Junction Boxes and Conduit Bodies. As noted in 314.16, conduit bodies must be sized the same as pull boxes when they contain 4 AWG or larger conductors, if the conductors are required to be insulated. Grounding electrode conductors are routinely installed in conduit for protection, in raceways sized to carry the one conductor. Conduit bodies for the smaller raceways do not meet the dimensional requirements of this section. Because the size rules have primarily to do with protecting the long-term integrity of insulated conductors, the rules need not be applied in these cases.

For raceways containing conductors of 4 AWG or larger size, the **NEC** specifies certain minimum dimensions for a pull or junction box installed in a raceway run. These rules also apply to pull and junction boxes in cable runs—but instead of using the cable diameter, the minimum trade size raceway required for the number and size of conductors in the cable must be used in the calculations. Basically, there are two types of pulls—straight pulls and angle pulls. Figure 314-32 covers straight pulls. Figure 314-33 covers angle pulls. In all the cases shown in those illustrations, the depth of the box only has to be sufficient to permit installation of the locknuts and bushings on the largest conduit. And the spacing between adjacent conduit entries is also determined by the diameters of locknuts and bushings—to provide proper installation. Depth is the dimension not shown in the sketches.

According to the rule of part (A)(2), in sizing a pull or junction box for an angle or U pull, if a box wall has more than one row of conduits, "each row shall be calculated separately and the single row that provides the maximum distance shall be used." Consider the following:

A pull box has two rows of conduits entering one side (or wall) of the box for a right-angle pull. What is the minimum required inside distance from the wall with the two rows of conduit entries to the opposite wall of the box?

Row 1: One metric designator 63 (trade size 2½) and one metric designator 27 (trade size 1) conduit

Row 2: One metric designator 16 (trade size ½), two metric designator 35 (trade size 1¼), one metric designator 41 (trade size 1½), and two metric designator 21 (trade size ¾) conduits

EXAMPLES:

1.

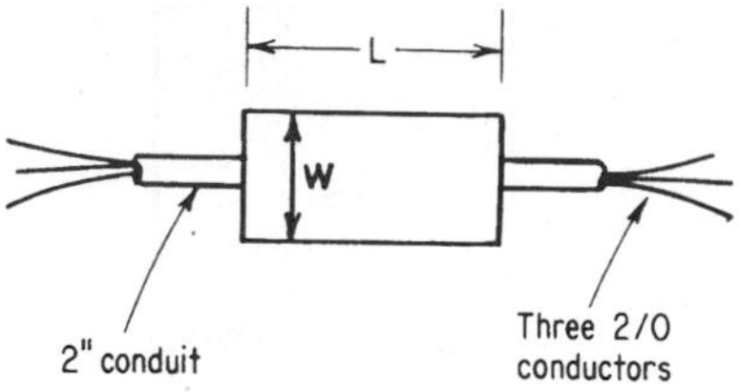
L
W
2" conduit
Three 2/0
conductors

L = 8 × 2 in. = 16 in. minimum
W = Whatever width is necessary to provide proper installation of the conduit locknuts and bushings within the enclosure.

2.

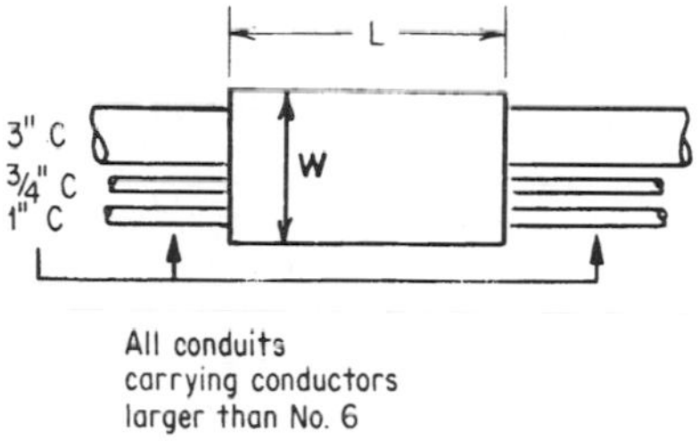
L
W
3" C
3/4" C
1" C
All conduits
carrying conductors
larger than No. 6

The 3-in. conduit is the largest.
Therefore—
L = 8 × 3 in. = 24 in. minimum
W = Width necessary for conduit locknuts and bushings.

Fig. 314-32. In straight pulls, the length of the box must be not less than 8 times the trade diameter of the largest raceway. (Sec. 314.28.)

Calculate as prescribed by this rule. Note that in accordance with the metrication rule in 314.28(A), these calculations must be done expressing the raceway size using the metric designator (trade size) units as actual dimensions in the system of measurement being employed. Based on the expected market for this book, the calculations going forward will be in the English system of units.

Calculating each row separately and taking the box dimension from the row that gives the maximum distance:

Row 1: 6 × largest raceway (2½ in.) + other entries (1 in.) = 16 in.
Row 2: 6 × the largest raceway (1½ in.) + other entries [½ in. + (2 × 1¼ in.) + (2 × ¾ in.)] = 9 in. + ½ in. + 4 in. = 13½ in.
Result: The minimum box dimension must be the 16-in. dimension from run No. 1, which "provides the maximum distance" calculated.

EXAMPLE:

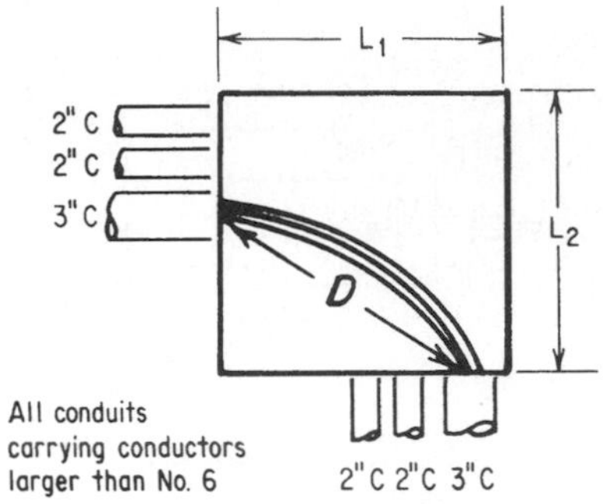
L1
2" C
2" C
3" C
L2
D
All conduits carrying conductors larger than No. 6
2" C 2" C 3" C

The 3-in. conduit is the largest.
Therefore—
$L_1 = 6 \times 3$ in. + (2 in. + 2 in.) = 22 in. min.
$L_2 = 6 \times 3$ in. + (2 in. + 2 in.) = 22 in. min.
$D = 6 \times 3$ in. = 18 in., minimum distance between raceway entries enclosing the same conductors

Fig. 314-33. Box size must be calculated for angle pulls. For boxes in which the conductors are pulled at an angle or in a U, the distance between each raceway entry inside the box and the opposite wall of the box must not be less than 6 times the trade diameter of the largest raceway in a row. And the distance must be increased for additional raceway entries by the amount of the maximum sum of the diameters of all other raceway entries in the same row on the same wall of the box. The distance between raceway entries enclosing the same conductors must not be less than 6 times the trade diameter of the larger raceway. (Sec. 314.28.)

Some of the rules for angle pull spacings also apply to splices in pull boxes. As shown in Fig. 314-34, when a pull box is used for a splice enclosure, the distance between conduit entries and the opposite wall must meet the six times rule, increased, if applicable, by other entries on the same side. However, because the conductors are spliced together, they are necessarily not "the same conductor" and therefore there is no rule governing the distance between the conduits of the same circuit. The principles are illustrated in Fig. 314-34.

Figure 314-35 shows a more complicated conduit and pull box arrangement, which requires more extensive calculation of the minimum permitted size. In the particular layout shown, the upper metric designator 78 (trade size 3) conduits running straight through the box represent a problem separate from the metric designator 53 (trade size 2) conduit angle pulls. In this case the metric designator 78 (trade size 3) conduit establishes the box length in excess of that required for the metric designator 53 (trade size 2) conduit. After computing the metric designator 78 (trade size 3) requirements, the box size was calculated for the angle pull involving the metric designator 53 (trade size 2) conduit.

Figure 314-36 shows another consideration in sizing a pull box for angle condui layouts. A pull box is to be installed to make a right-angle turn in a group of conduit consisting of two metric designator 78 (trade size 3), two metric designator 63 (trad size 21/2), and four metric designator 53 (trade size 2) conduits. Subparagraph **(2**

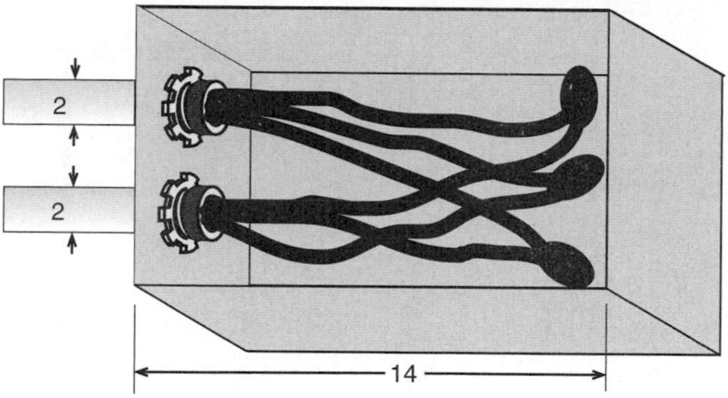
2
2
14

Fig. 314-34. The opposite wall spacing rule applies to splices; however, the spacing rule between the same conduits of a circuit does not. [Sec. 314.28(A)(2).]

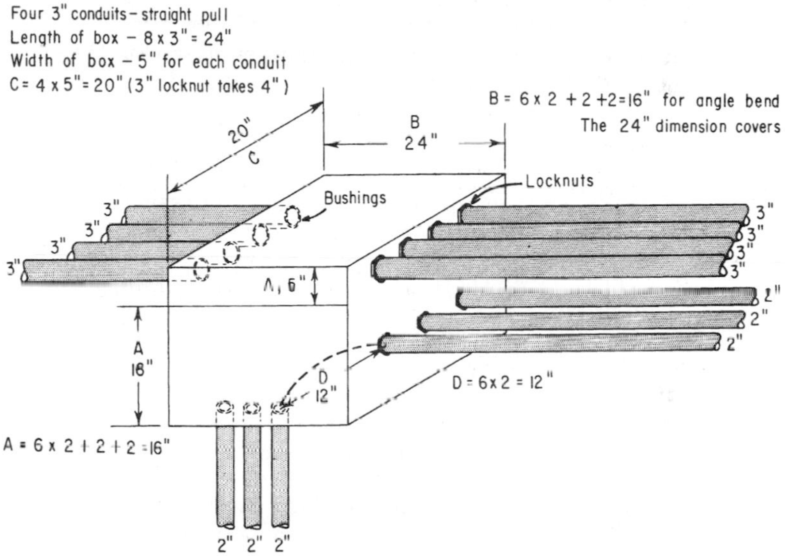
Four 3" conduits – straight pull
Length of box – 8 x 3" = 24"
Width of box – 5" for each conduit
C = 4 x 5" = 20" (3" locknut takes 4")
B = 6 x 2 + 2 + 2 = 16" for angle bend
The 24" dimension covers
20"
C
B
24"
Locknuts
Bushings
3"
3"
3"
3"
A, 6"
A
16"
D
12"
D = 6 x 2 = 12"
A = 6 x 2 + 2 + 2 = 16"
2" 2" 2"

Fig. 314-35. A number of calculations are involved when angle and straight pulls are made in different directions and different planes. (Sec. 314.28.)

of 314.28(A) gives two methods for computing the box dimensions, and both must be met; here again for calculations we will use English units.

First method:

$$
\begin{aligned}
6 \times 3 \text{ in.} &= 18 \text{ in.} \\
1 \times 3 \text{ in.} &= 3 \\
2 \times 2\tfrac{1}{2} \text{ in.} &= 5 \\
4 \times 2 \text{ in.} &= 8 \\
\hline
\text{Total} &= 34 \text{ in.}
\end{aligned}
$$

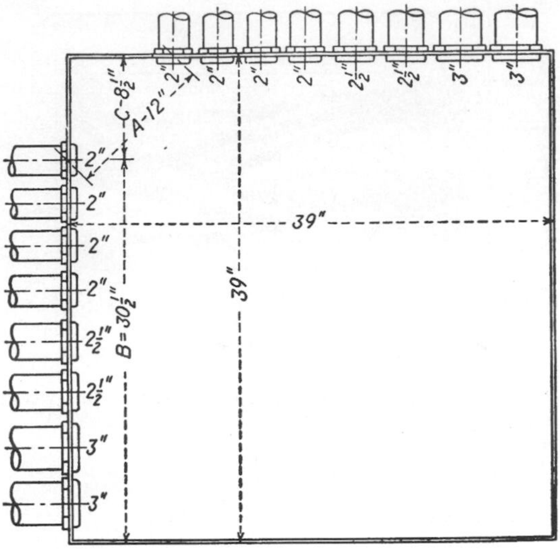
C-8½"
A-12"
2"
2"
2"
2"
2½"
2½"
3"
3"
39"
39"
B=30½"
2"
2"
2"
2"
2½"
2½"
3"
3"

Fig. 314-36. Distance between conduits carrying same cables has great impact on box size. (Sec. 314.28.)

Second method:

Assuming that the conduits are to leave the box in the same order in which they enter, the arrangement is shown in Fig. 314-37, and the distance A between the ends of the two conduits must be not less than 6 × 2 in. = 12 in. It can be assumed that this measurement is to be made between the centers of the two conduits. By calculation, or by laying out the corner of the box, it is found that the distance C should be about 8½ in.

The distance *B* should be not less than 30½ in. (774.7 mm) approximately, as determined by applying practical data for the spacing between centers of conduits,

30½ in. + 8½ in. = 39 in.

In this case the box dimensions are governed by the second method. The largest dimension computed by either of the two methods is of course the one to be used. Of course, if conduit positions for conduits carrying the same cables are transposed—as in Fig. 314-32—then box size can be minimized.

The most practical method of determining the proper size of a pull box is to sketch the box layout with its contained conductors on a paper.

Figure 314-37 shows how the rules of 314.28(A) apply to conduit bodies.

Important: The exception given in 314.28(A)(2) establishes the minimum dimension of L2 for angle runs, but this exception applies only to conduit bodies which have the removable cover opposite one of the entries, such as a Type LB body. Types LR, LL, and LF do not qualify under that exception, and for such conduit bodies the dimension L2 would have to be at least equal to the dimension L1 (i.e., 6 times raceway diameter). Be very careful in shopping for type "LB" conduit bodies that will enclose large conductors in big raceways

STRAIGHT RUN

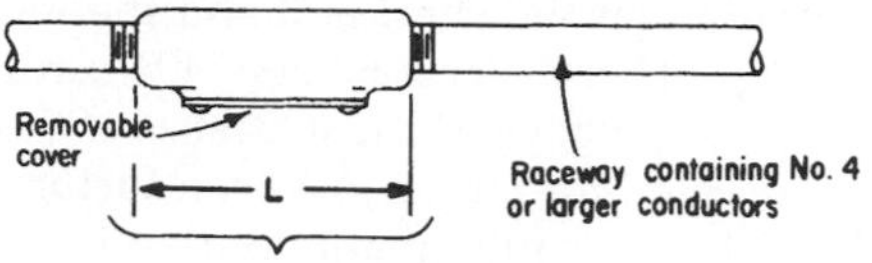
Removable cover
L
Raceway containing No. 4 or larger conductors

Type C conduit body must have length L equal to 8 times diameter of the raceway

Examples

If four No. 4 THHN are used in 1-in. conduit, conduit body must be at least 8 in. long.

If four 500 kcmil XHHW are used in 3-in. conduit, conduit body must be at least 24 in. long.

ANGLE RUN

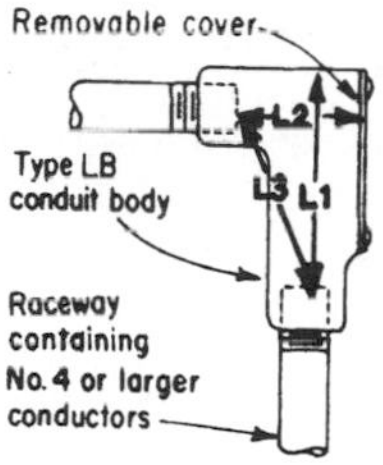
Removable cover
L2
Type LB conduit body
L3
L1
Raceway containing No. 4 or larger conductors

From 314.28(A)(2):

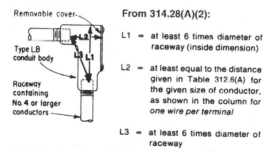

L1 = at least 6 times diameter of raceway (inside dimension)

L2 = at least equal to the distance given in Table 312.6(A) for the given size of conductor, as shown in the column for *one wire per terminal*

L3 = at least 6 times diameter of raceway

Examples

If four No. 4 THW conductors are used in 1¼-in. conduit, min imum dimensions would be calculated as follows:

L1 = 6 × 1¼ in. = 7.5 in.
L2 = 2 in., from Table 312.6(A) for one No. 4 conductor per terminal
L3 = 6 × 1¼ in. = 7.5 in.

If four 500 kcmil THW conductors are used in 3½-in. conduit, minimum dimensions would be:

L1 = 6 × 3½ in. = 21 in.
L2 = 6 in., from Table 312.6(A)
L3 = 6 × 3½ in. = 21 in.

Fig. 314-37. Conduit bodies must be sized as pull boxes under these conditions. (Sec. 314.28.)

Many manufacturers have read 314.28(A)(2) and make so-called mogul conduit bodies that meet the 6 times rule. However not many of them have read the entire rule all the way through, and made their large conduit mogul fittings with extended noses that accommodate Table 312.6(A) dimensions for large conductors. For example, a metric designator 78 (trade size 3) "LB" conduit body expected to carry 3 600 kcmil THW conductors must have 8 in. of distance between the conduit stop on the short end and the inside of the cover, dimension "L2" in Fig. 314-37. There is at least one "LB" fitting on the market designed exactly this way, but very few. Most, at least as of this writing, violate these rules and should be rejected for larger sized conductors.

Subparagraph **(3)** of 314.28(A) permits smaller pull or junction boxes where such boxes have been listed for and marked with the maximum number and size of conductors and the conduit fills are *less* than the maximum permitted in Table 1, Chap. 9. This rule provides guidelines for boxes that have been widely used for years, but which have been smaller than the sizes normally required in subparagraphs **(1)** and **(2)**. These smaller pull boxes must be listed by UL under this rule. The usual application of this provision is not to pull boxes but to conduit bodies marked for a fill that is smaller than the entering raceways could normally carry.

For example, a conduit body might be marked "3 4/0 XHHW MAX." If you are using three XHHW conductors, then the marking means exactly what it says. Of course that doesn't happen often. For any other insulation style or number of conductors, take the marking and translate it to a total cross-sectional area of wire fill, using the numbers in Table 5 of Chap. 9 at the end of the Code book. That is your maximum fill, regardless of what Table 4 might say the raceway can hold. Compare that maximum with what you need to install, and decide whether you need to look for a full mogul conduit body or whether you can use the smaller one. In the case of PVC, you will usually be stuck with a reduced size, and the only alternative will be to either increase the raceway size or use a pull box sized to the rules given here. Note that the Canadian Electrical Code does not contain a comparable rule. During times of market instability, PVC conduit bodies manufactured only for use in Canada often show up in the USA without the required markings and should be rejected.

There is another issue regarding the markings for reduced fill. At one time, the markings customarily stated not only the size of the conductor, but also the style of insulation that was used during the pull test. This is critical information because as described in the previous paragraph, if you are using different insulation or perhaps a combination of conductor sizes, you need the insulation type in order to establish the upper limit of cross-sectional area that can be used for the allowable fill of the conduit body at hand. However, the current edition of the applicable product standard (UL 514B) does not require the insulation type to be included in the marking. For this reason the insulation type has been slowly disappearing However Sec. 8.4 of UL 514B effectively specifies XHHW as the insulation type to be used in the pull test in most cases. The 2017 **NEC** has now fully addressed this issue, beginning with an amendment to the main rule to expressly validate the procedure just outlined. Then, an informational note was added to explain why the default insulation type to take into a calculation is XHHW.

Actual field example:
Sewage lift station requiring 100A 3-phase 4-wire feeder

Note: minimal neutral load sized by 215.2(A)(2) = 8 AWG
Conduit Body: PVC 1¼ trade size LB, marked "3 WIRE MAX FILL SIZE AWG 2

Step 1, Maximum allowable fill expressed as cross-sectional area:

3(2 AWG XHHW-2) = 3(0.1146 in^2) = 0.3438 in^2

Step 2, Check proposed fill using smallest cross-sectional area conductors:

3(3 AWG XHHW-2) = 3(0.0962 in^2) = 0.2886 in^2

1(8 AWG THWN-2) = 1(0.0366 in^2) = 0.0366 in^2

1(8 AWG EGC bare) = 0.0170 in^2 = 0.0170 in^2

Total Fill = 0.3422 in^2

Calculation succeeds by 0.0016 in^2

Note: 3(3 AWG THWN-2) = 3(0.0973 in^2) = 0.2919 in^2

Calculation fails by 0.0017 in^2

Note: 1(8 AWG XHHW-2) = 1(0.0437 in^2) = 0.0437 in^2

Calculation fails by 0.0055 in^2

Therefore, any other wire selection would increase the raceway size.

The 2014 **NEC** formally recognizes for the first time, in a new paragraph in 314.28(A)(3), a conduit body that changes the direction of a pipe run by the usual 90°, but for which, unlike a conventional LB, the body of the fitting and the access cover are set at a 45° angle to both entering raceways. They are not widely used

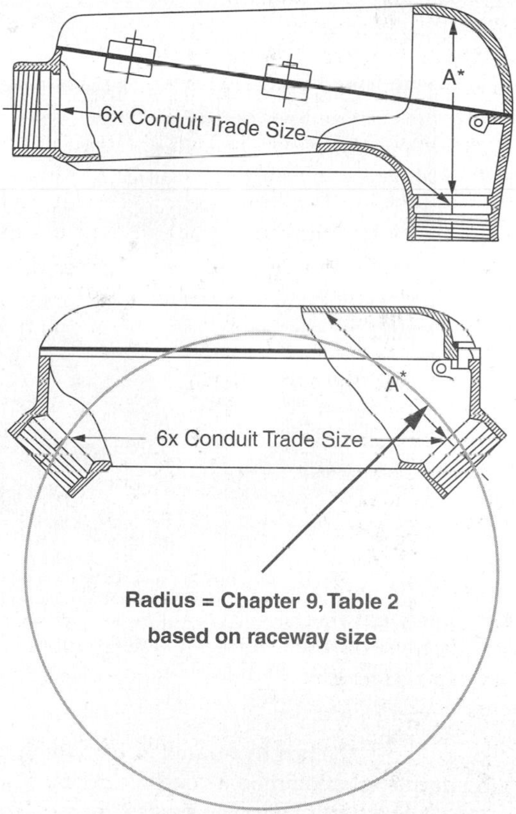
A*
6x Conduit Trade Size
A*
6x Conduit Trade Size
Radius = Chapter 9, Table 2
based on raceway size

Fig. 314-38. Comparison of special design right-angle conduit body recognized for relaxed bend radius and conventional mogul style LB. [Sec. 314.28(A)(3).]

because they will not lay up tight to a building corner. However, where they can be used they are much easier to pull through because the resulting conductor loop need not be forced into a full right angle at either end.

These conduit bodies generally, on close examination, result in an actual bend radius imposed on the conductors that is no different or even greater than the radius permitted by Table 2 of Chap. 9 for the entering raceway if the conduit body were replaced by a conduit sweep. Therefore, such a conduit body results in no reduction of the permitted fill beyond the normal limits imposed by Table 1 of Chap. 9 for the entering raceway. The conduit body must be listed and marked in such a way as to demonstrate that they have been evaluated in accordance with this concept.

314.28 applies particularly to the pull boxes commonly placed above distribution switchboards and which are often, and with good reason, termed *tangle boxes.* In such boxes, all conductors of each circuit should be cabled together by securing them with tie-wraps so as to form a self-supporting assembly that can be formed into shape, or the conductors should be supported in an orderly manner on racks, as required by part **(B)** of 314.28. The conductors should not rest directly on any metalwork inside the box, and insulating bushings should be provided wherever required by 300.4(G). Figure 314-39 shows an example of this process.

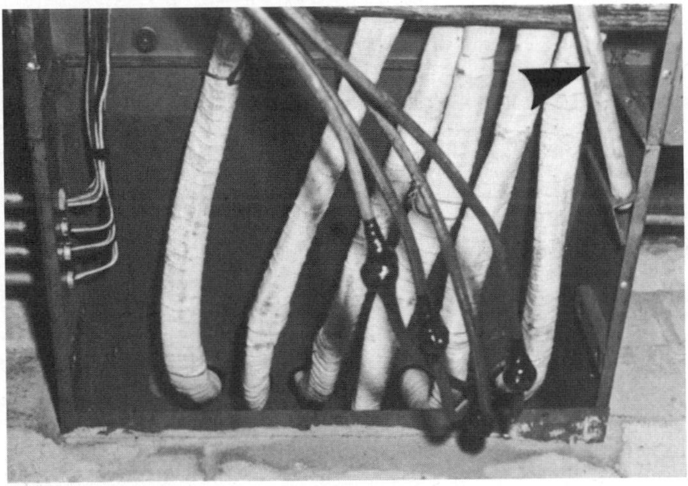

Fig. 314-39. If a pull box has *any* dimension over 6 ft (1.8 m), the conductors within it must be supported by suitable racking (arrow) or cabling, as shown here for arcproofed bundles of feeder conductors, to keep the weight of the many conductors off the sheet metal cover that attaches to the bottom of the box. (Sec. 314.28.)

For example, the box illustrated in Fig. 314-36 could be approximately 125 m (5 in.) deep and accommodate one horizontal row of conduits. By making it twic as deep, two horizontal rows or twice the number of conduits could be installed

Insulating racks are occasionally placed between conductor layers, and space must be allowed for them when they are used.

The rule in **(C)** requires covers to be provided on boxes that are compatible with the box construction and suitable for the conditions of use. The rule concludes by reiterating the grounding requirement of 314.25(A).

To line up with the recognition in 376.56 of power distribution blocks being added to wireways in the field, part **(E)** now covers these devices field added into pull and junction boxes. Refer to Fig. 376-3 for a photograph of this equipment in use. The rules in Art. 314 permit this equipment in large boxes (over 100 cu. in.) with an exception permitting equipment grounding terminal bars in smaller boxes. This allows for correlation, for example, with the requirement in 680.23(F)(2)(b) that equipment grounding conductors, if broken, must be spliced on terminal bars and not rely on a twist-on wire connector for continuity.

The requirements for listing and for a cover over the uninsulated live parts line up with the wireway rules. The bending space rule points to 312.6 instead of specifying 312.6(B) because unlike wireways, the conductors terminating in a pull box could be approaching by way of a right angle turn or straight into its terminals. As in the case of wireways, these blocks cannot be applied on the line side of service equipment unless they are listed and marked "suitable for use on the line side of service equipment" or equal. In the short term there are some issues with this, refer to the discussion at 376.56(B) for more information.

314.29. Boxes, Conduit Bodies, and Handhole Enclosures to Be Accessible. This is the rule that prevents boxes from buries in walls or otherwise behind building finish without suitable access to their enclosed conductors. For outdoor applications, buried boxes are permitted under the terms of the exception; these terms are both specific and intentionally restrictive. First the box must be below soil that can be easily shoveled, and second, the location of the box must be "effectively identified." This might involve a map attached to the panelboard supplying the circuit(s) in the box, or some other way of assuring that those who come afterward don't need to contend with a junction point that is entirely invisible.

314.30. Handhole Enclosures. A handhole enclosure is defined in Art. 100 as an enclosure for use in underground systems that is sized to allow personnel to reach into, but not enter, for the purpose of installing wiring and maintaining it afterward. They may contain electrical equipment, and they can be made with either an open or closed bottom. Most handholes are made with open bottoms, and are used as pull boxes for underground systems.

They must be identified for use on underground systems, and they must be designed and installed to withstand all loads likely to be imposed. The **NEC** dentifies a standard (ANSI/SCTE 77-2002) that can be used to evaluate loadng on these enclosures. This standard identifies a series of tiers that can be pecified based on the expectation of vehicular loading. Tier 5 is for pedestrian use with a safety factor for occasional nondeliberate vehicle traffic; it carries a 000 lb vertical design load rating. Tier 8 is for sidewalk use with a safety factor for nondeliberate vehicular traffic, with a comparable design load rating of 8000 lb. Tier 15 is for parking lots, driveways, and off roadway uses that are subject to ccasional nondeliberate heavy vehicular traffic; here the vertical load rating is

15,000 lb. They are now commonly available in designs evaluated to an even more robust performance level of Tier 22.

Handhole enclosures have the same sizing rules related to wire bending space as for manholes. There is an instance where a different set of rules applies, however. If a raceway enters through the bottom plane of a bottomless enclosure, the measurements are taken from the end of the conduit or cable assembly. When such an entry is opposite a removable cover, as it would be in this case, the spacing rules change. For wires operating at 600 V and below, the minimum distance becomes the one-wire-per-terminal distance in Table 312.6(A) for the conductor sizes involved, as previously discussed. For medium-voltage wiring the distance is 12 times the shielded cable diameter and 8 times the diameter for unshielded conductors.

Handhole enclosures are renowned for filling up with sand and water, particularly for the ones without bottoms, which is usually the case. All wiring and splicing provisions must be listed for wet locations for this reason. In addition, if a handhole enclosure has a metal cover, it must be bonded to an equipment grounding conductor (usual case) or to a grounded circuit conductor if the application is on the line side of service equipment. This is a critical safety issue. There have been many fatalities, of both people and their pets, from energized handhole covers.

314.40. Metal Boxes, Conduit Bodies, and Fittings. This section through 314.44 covers construction of boxes. UL data on application of boxes supplement this Code data as follows:

1. Cable clamps in outlet boxes are marked to indicate the one or more types of cables that are suitable for use with that clamp.
2. Box clamps have been tested for securing only one cable per clamp, except that multiple-section clamps may secure one cable under each section of the clamp, with each cable entering the box through a separate KO.

Part **(B)** covers the thickness of metal for boxes in the usual outlet and device box sizes, and is primarily of interest to manufacturers and testing laboratories. However, it is important to recognize that boxes are not required to be listed, although almost all of these boxes are. Note that permission is granted in Exception No. 1 for boxes to be made of different alloys or with special construction details (such as ribbing), and if listed as having equivalent strength in comparison to the plain vanilla product, they can be used.

Part **(C)** of this section covers the pull boxes regulated by 314.28. UL data on such boxes are important and must be related to the Code rules. Listed pull and junction boxes may be sheet metal, cast metal, or nonmetallic, and all of these have a volume greater than 100 cu in. (1650 mm^3). These boxes are frequently made in sheet metal shops to accommodate particular field conditions and the decision not to require listings on these boxes is a very deliberate one in the **NEC**. Therefore, these construction requirements, together with the applicable UL standard (UL 50), have frequent field applications. Boxes marked "raintight" or "rainproof" are tested under a condition simulating exposure to beating rain. *Raintight* means water will not enter the box. *Rainproof* means that exposure to beating rain will not interfere with proper operation of the apparatus within the enclosure. Use of a box with either designation must satisfy 314.15, which notes that boxes in wet locations (such as outdoors wher exposed to rain or indoors where exposed to water spray) must prevent moisture from entering *or* accumulating within the box. That is, water *may* enter the box if it does not accumulate in the box, where the box is drained. A box that is raintight or rainproof may satisfy that rule. Be sure, though, that any equipment installed in a box labeled "rainproof" is mounted within the location restrictions marked in the box. Boxes in wet locations must be fully listed for wet locations.

In part **(D)** of 314.40, *connection provisions for a grounding conductor is required in metal box.* This rule is intended to ensure a suitable means within a metal box to connect the equipment grounding conductor that is required to be used with such wiring methods to provide equipment grounding.

Part IV. Pull and Junction Boxes for Use on Systems Over 1000 V, Nominal.

314.70. General. This part of the article covers pull and junction boxes that are components of systems operating over 600 V. The dimensional requirements in this part squarely address conduit bodies and handhole enclosures for the first time in the 2011 **NEC**. It would generally be difficult, but not necessarily impossible on smaller cables, particularly unshielded ones at lower voltages, to meet the 8 times (unshielded) or 12 times (shielded) the cable diameter provision in 300.34 for bending radii opposite removable covers (think in terms of "LB" conduit bodies). Note that 314.70(B)(2) was amended in the 2017 **NEC** to reference 314.28(A)(3). This is an attempt by the code-panel to call attention to the right angle pull conduit bodies manufactured with two 45° hubs in terms of being potentially applicable to medium-voltage applications due to the lengthened pull radius. See Fig. 314-38 and associated commentary.

Handhole enclosures also make their debut for medium-voltage applications. This application is very common; the dimensional requirements were already in 314.30(A) by reference, so nothing changes in this regard.

314.71. Size of Pull and Junction Boxes, Conduit Bodies, and Handhole Enclosures. Figure 314-40 shows the rules on sizing of pull boxes for high-voltage circuits. In addition, this part of Art. 314 includes spacing rules for a conduit entry opposite a removable cover that correlates with a similar provision in 314.28(A)(2) Exception. Here the requirement uses the minimum bend radius as given in 300.34.

314.72. Construction and Installation Requirements. Part **(E)** requires that covers of pull and junction boxes for systems operating at over 600 V must be marked with readily visible lettering at least ½ in. (12.7 mm) high, warning "DANGER HIGH VOLTAGE KEEP OUT."

All required warning signs must be properly worded to include the command "KEEP OUT." While certain sections of the Code, such as this one, as well as 110.34(C), clearly require the inclusion of the command "KEEP OUT," be aware that courts have held that warning signs that fail to include some sort of instruction or command with respect to an appropriate action that must be taken are inadequate and constitute negligence on the part of the individual posting the sign. Always include some phrase that will tell the individual what to do about the condition or hazard that exists.

STRAIGHT PULLS

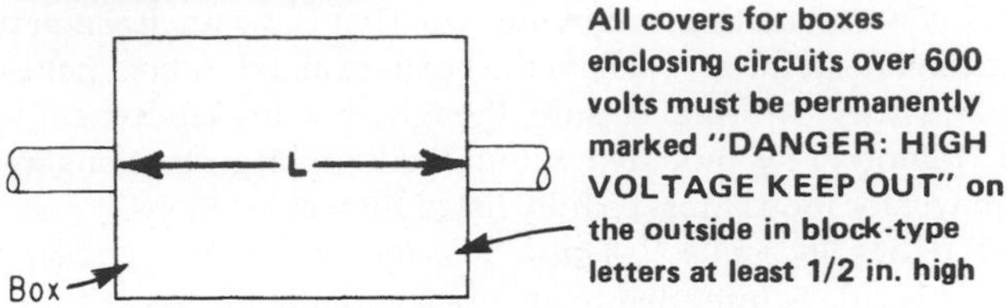
All covers for boxes enclosing circuits over 600 volts must be permanently marked "DANGER: HIGH VOLTAGE KEEP OUT" on the outside in block-type letters at least 1/2 in. high
L
Box

L - not less than 48 times the outside diameter, over sheath, of the largest shielded or lead-covered *conductor* or *cable* entering the box, *OR* not less than 32 times the outside diameter of the largest nonshielded conductor or cable.

NOTE: The box length must be 48 times the conductor or cable diameter, *not the conduit* diameter.

ANGLE PULLS

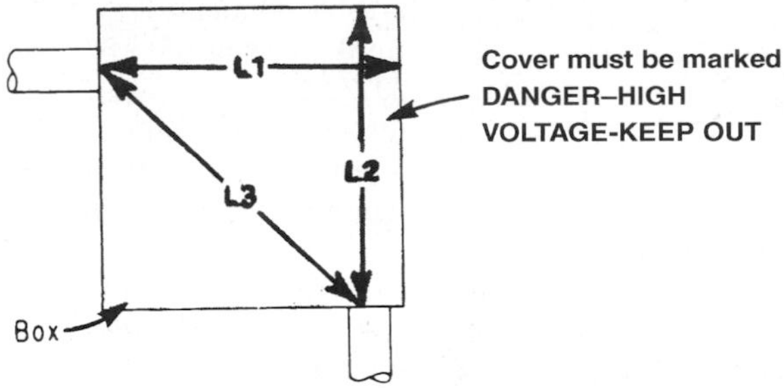
Cover must be marked DANGER–HIGH VOLTAGE-KEEP OUT
L1
L2
L3
Box

L1, L2, L3—not less than 36 times the outside diameter, over sheath, of the largest *conductor* or *cable*

Fig. 314-40. Minimum dimensions are set for high-voltage pull and junction boxes. For multiple angle pulls, increase the dimension by the size of the additional entries on the same wall. [Sec. 314.71.]

ARTICLE 320. ARMORED CABLE: TYPE AC

320.2. Definition. Armored cable (Type AC) contains insulated conductors of a type accepted for general wiring applications in the **NEC**. The conductors are enclosed in an armor comprised of steel or aluminum interlocking tape. The armor is arranged with an internal bonding strip of aluminum or copper "in intimate contact with the armor for its entire length." Type AC cable, which is commonly called BX, has largely been supplanted in the market by the interlocking-armor style of metal-clad cable. However, certain applications, especially patient care areas of health care facilities, require a wiring method where the outer margin of the wiring method, whether raceway or cable assembly, qualifies as an equipment grounding return path, and Type AC cable inherently qualifies for this use. A recent addition to this list is 210.12(A)(5), which specifically enumerates the steel version of this cable as an acceptable cabled wiring method acceptable under those provisions (Fig. 320-1).

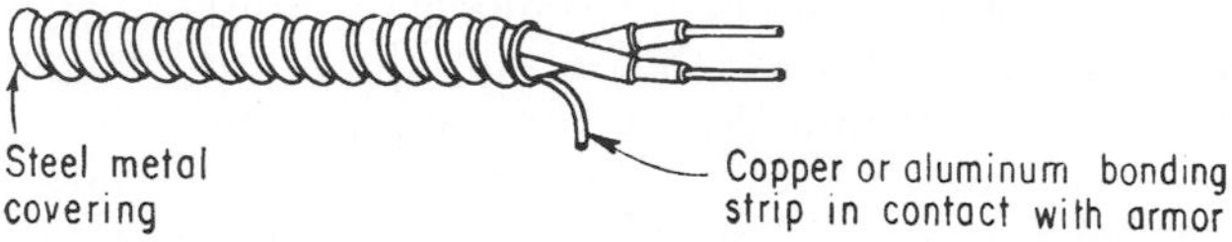
Steel metal covering
Copper or aluminum bonding strip in contact with armor

Fig. 320-1. Type AC cable contains insulated conductors plus bonding conductor under the armor. (Sec. 320.2.)

Armored cable assemblies of 2, 3, 4, or more conductors in sizes No. 14 AWG to No. 1 AWG—such conductors may even incorporate an optical-fiber cable—conform to the standards of the Underwriters Laboratories. These standards cover multiple-conductor armored cables for use in accordance with the **National Electrical Code**, in wiring systems of 600 V or less, with conductors having insulation rated for temperatures of 90°C, as called for by 320.80 where the cable is embedded in thermal insulation. As of the 2017 **NEC**, Type AC cable must be listed (320.6).

320.10. Uses Permitted. Type AC armored cable can be used in all types of electrical systems for power and light branch circuits and feeders. Figure 320-2 shows

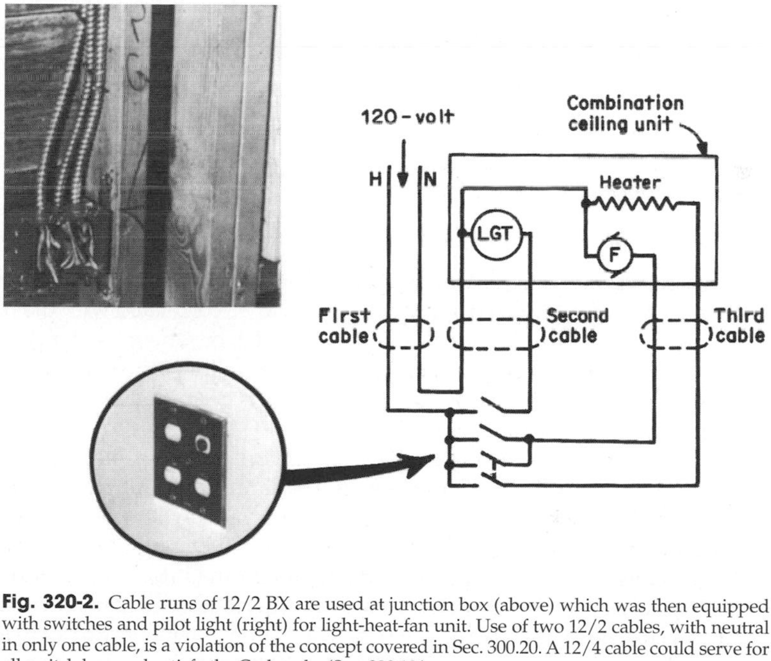
120-volt
Combination ceiling unit
H
N
Heater
LGT
F
First cable
Second cable
Third cable

Fig. 320-2. Cable runs of 12/2 BX are used at junction box (above) which was then equipped with switches and pilot light (right) for light-heat-fan unit. Use of two 12/2 cables, with neutral in only one cable, is a violation of the concept covered in Sec. 300.20. A 12/4 cable could serve for all switch legs and satisfy the Code rule. (Sec. 320.10.)

use of three runs of 12/2 BX for the supply and two switch legs to a combination light-heat-fan unit in a bathroom. One 12/2 is the supply and the other cables control the appliance as shown in the wiring diagram. But the use of two 12/2 cables for the switch legs violates 300.20 because the neutral is not kept with all the conductors it serves. As a result, induction heating could be produced. A single run of 12/4 cable to the appliance would be necessary.

As stated in the note following this section, the list of items in this "uses permitted" section is not intended to completely enumerate all permitted uses. These sections must be read in conjunction with "uses not permitted" section that follows to zero in on where particular issues lie that the NEC wants to address. For example, it is clear that this cable can be used embedded in a plaster finish on masonry construction, but only where those locations are classified as dry. It can also run in hollow block voids, but only if not subject to excessive moisture or dampness, which could well be all wet locations but probably not all damp locations; the local inspector would need to determine that in the field. Figure 320-3 shows an example of a permitted hazardous location use of the cable, which is also not (and need not be) itemized on the list.

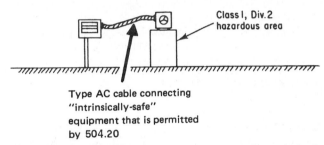

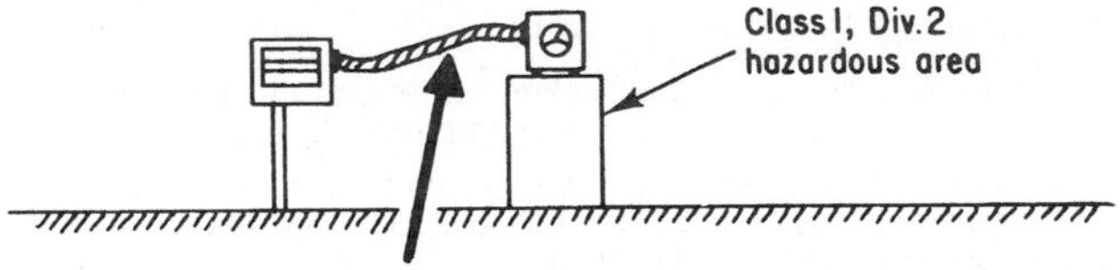
Class I, Div. 2
hazardous area
Type AC cable connecting
"intrinsically-safe"
equipment that is permitted
by 504.20

Fig. 320-3. The informational note following 320.10 indicates the list is not all inclusive and 504.20 recognizes BX cable for limited use in hazardous locations. (Sec. 320.10.)

320.15. Exposed Work. This section requires cables to "closely follow the surface of the building" where unexposed. Of course, this is to be correlated with the rule in 320.30. That is, where needed for flexibility [part **(2)**], or where used as fixture whips [part **(3)**] of 320.30, the requirement to "closely follow" the contour of the wall space is waived. In addition, use of joist bottoms is permitted where not subject to physical damage.

320.23. In Accessible Attics. These rules also apply to Type NM cable installation because 334.23 specifically incorporates them by reference. Part (A) applies to all attic and roof spaces accessible by a permanent ladder or stair, and to other accessible attic locations within 1.8 m (6 ft) of the outside edges of the scuttle hole. Any cables run on top of floor joists, or across rafters or studs within 2.1 m (7 ft) of the floor or the tops of the floor joists must be protected by guard strips at least as high as the cable diameter. Cables run on and parallel to the sides of framing members do not require guard strips but must meet the spacing rules in 300.4(D).

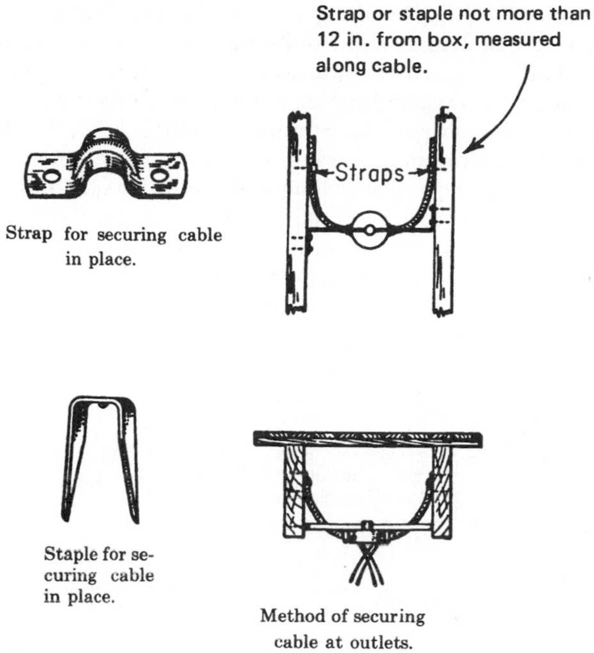
Strap or staple not more than
12 in. from box, measured
along cable.
Straps
Strap for securing cable
in place.
Staple for se-
curing cable
in place.
Method of securing
cable at outlets.

Fig. 320-4. BX must be clamped every 4½ ft (1.4 m) and within 12 in. (300 mm) of terminations. Cable ties, if listed for "securement and support," are also permitted. (Sec. 320.30.)

320.30. Securing and Supporting. Armored cable must be secured and supported by approved staples, straps, or similar fittings, as shown in Fig. 320-4. Note that a requirement to secure a cable is more restrictive than a requirement to support the cable. For example, Type AC cable must be supported at intervals, a condition that can be met by routing it through successive holes in framing member, and it must be secured at terminations, a condition that requires preventing all movement such as by using a staple.

In exposed work, both as a precaution against physical damage and to ensure a workmanlike appearance, fastenings should be spaced not more than 600 to 750 mm (24 to 30 in.) apart. In concealed work in new buildings, the cable must be supported at intervals of not over 1.4 m (4½ ft) for Type AC to keep it out of the way of possible injury by mechanics of other trades. In either exposed work or concealed work, the cable should be securely fastened in place within 300 mm (1 ft) of each outlet box or fitting so that there will be no tendency for the cable to pull away from the box connector. Take care to observe the minimum bend radius for this cable, 5 times its diameter measured to the inner edge of the bend, as given in 320.24. Sharper bends can and often will break the convolutions, creating a hazardous condition.

Part **(2)** of 320.30**(D)** limits Type AC cable to not over a 600-mm (2-ft) unclamped length for flexibility where such a cable feeds motorized equipment (such as a fan or unit heater) or connects to any enclosure or equipment where the flexibility of the ac length will isolate and suppress vibrations. Similarly, part **(3)** of this section permits lengths of ac up to 1.8 m (6 ft) long to be used without any staples, clamps, or other support where used in a hung ceiling as a lighting fixture whip or similar whip to other equipment (Fig. 320-5). This permits use of ac in the same manner as permitted for flexible metal conduit (Greenfield) or liquidtight flexible metal conduit in lengths from 0.45 to 1.8 m (1½ to 6 ft) as a connection from a circuit outlet box to a recessed lighting fixture [410.117(C)]. This use of unclamped BX is an exception to the basic rule that it be clamped every 1.4 m (4½ ft) and within 300 mm (12 in.) of any outlet box or fitting. Part **(1)** recognizes the acceptability of fishing in Type AC cables, for which additional securing and supporting is not required.

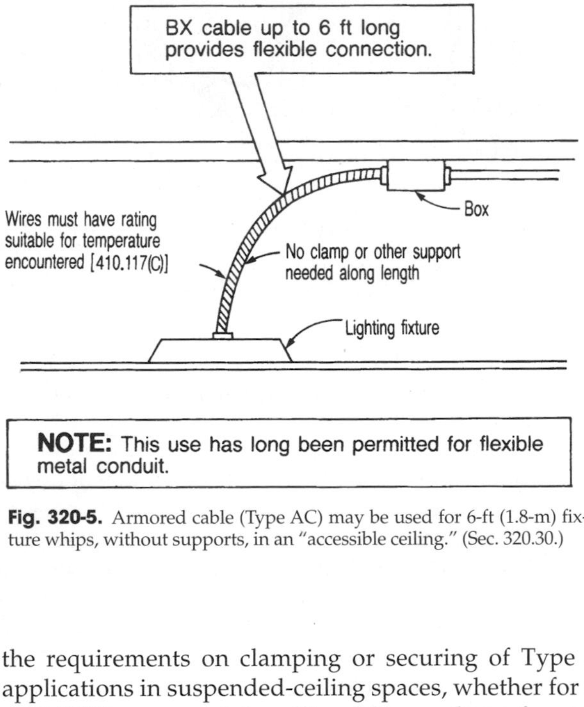
BX cable up to 6 ft long provides flexible connection.
Box
Wires must have rating suitable for temperature encountered [410.117(C)]
No clamp or other support needed along length
Lighting fixture
NOTE: This use has long been permitted for flexible metal conduit.

Fig. 320-5. Armored cable (Type AC) may be used for 6-ft (1.8-m) fixture whips, without supports, in an "accessible ceiling." (Sec. 320.30.)

Note that the requirements on clamping or securing of Type AC must be observed for applications in suspended-ceiling spaces, whether for air handling, as covered in 300.22(C), or non-air handling. The wording of part **(C)** in 320.30 recognizes that in horizontal runs the hole in the framing member is to be considered as satisfying the support requirements given in this section.

320.40. Boxes and Fittings. Note that a termination fitting—that is, a box connector—must be used at every end of Type AC cable entering an enclosure or a box (Fig. 320-6) unless the box has an approved built-in clamp to hold the cable armor, provide for the bonding of the armor to the metal box, and protect the wires in the cable from abrasion.

Fig. 320-6. Connectors for BX entering a panelboard cabinet or other enclosure must use approved fittings—some type of single-connector or duplex type (as shown, with two cables terminated at each connector through a single KO). (Sec. 320.40.)

A standard type of box connector for securing the cable to knockouts or other openings in outlet boxes and cabinets is shown in Fig. 320-7. A plastic bushing, as shown, must be inserted between the armor and the conductors. The plastic bushing, which can be seen through slots in the connector after installation, prevents the sharp edges of the armor from cutting into the insulation on the conductors and so grounding the copper wire.

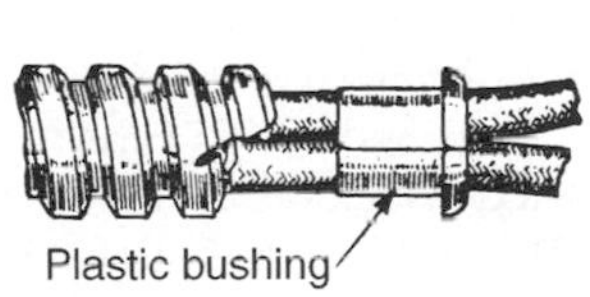
Plastic bushing

Plastic bushing to protect the conductors in Type AC cable from the sharp edges of the armor.

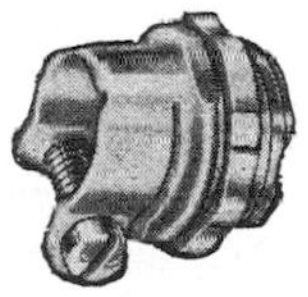

Box connector for Type AC cable

Fig. 320-7. Every BX termination must be equipped with a protective bushing and a box connector or clamp built into the box. (Sec. 320.40.)

The box shown in Fig. 320-8 is equipped with clamps to secure Type AC cables, making it unnecessary to use separate box connectors. The other box shown is similar but has the cable clamps outside, thus permitting one more conductor in the box. See 314.16(A)(1).

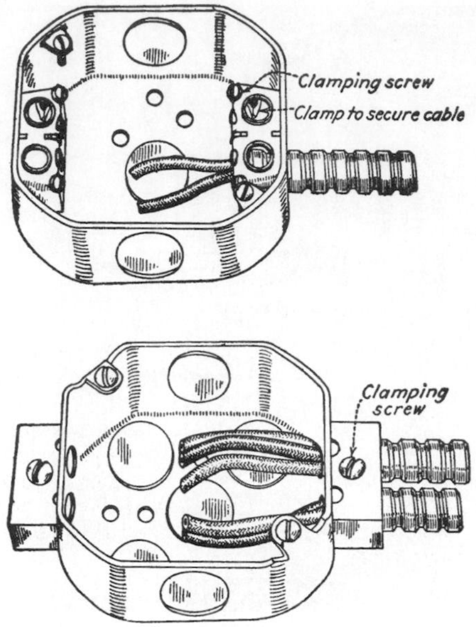
Clamping screw
Clamp to secure cable
Clamping screw

Fig. 320-8. A box connector fitting is not required if box includes cable clamps for Type AC cable. (Sec. 320.40.)

As covered in the last sentence of 320.40, "a box, fitting, or conduit body"—such as a C conduit body—must be used where Type AC cable is connected to another wiring method. Figure 320-9 shows a typical application of this technique, in accordance with 300.15(F).

320.80. Ampacity. This section makes it clear that the current-carrying capacity of Type AC cable must be established in accordance with the rules of 310.15. The last part of this rule places an additional limitation on the use of Type AC run in thermal insulation. As is the case with Types NM, NMB, and NMC, Type AC cable must be provided by the manufacturer with 90°C-rated insulation, *but* it must be loaded to no more than the 60°C value shown in Table 310.15(B)(16). Although it is permissible to use the 90°C current value shown in Table 310.15(B)(16) for the purposes of derating, the actual load carried must be no greater than the current value shown for the particular size and conductor material.

The wording of this section actually understates the true problem for cables run in thermal insulation. By allowing derating to occur from the 90°C column for such cables, the NEC allows users to ignore factors that increase the retention of heat in the cable until they reach the point of reducing the ampacity below the 60°C value, which in most cases is to allow them to be ignored altogether. Meanwhile, particularly for large, heavily loaded cables, even the 60°C numbers are too high. Review the discussion associated with 310.15(A)(3) for detailed information on this topic.

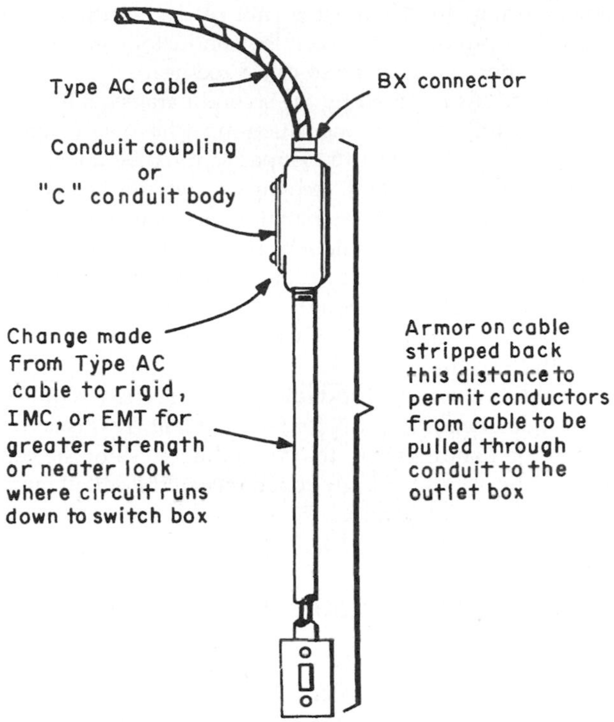
Type AC cable
BX connector
Conduit coupling
or
"C" conduit body
Change made
from Type AC
cable to rigid,
IMC, or EMT for
greater strength
or neater look
where circuit runs
down to switch box
Armor on cable
stripped back
this distance to
permit conductors
from cable to be
pulled through
conduit to the
outlet box

Fig. 320-9. This connection of BX to conduit or other cable is specifically recognized. Where there is no splice, a "from-to" connector may be used. (Sec. 320.40; see also 300.15(F) with the conductors unspliced at the transition.)

320.100. Construction. Note that Type AC cable is recognized for branch circuits and feeders, but *not* for service-entrance conductors, which *must* be one of the cables or wiring methods specified in 230.43. Type MC (metal-clad) cable, such as interlocked armor cable or the other cables covered in Art. 334, is recognized by 230.43 for use as service-entrance conductors.

Because the armor of Type AC cable is recognized as an equipment grounding conductor by 250.118(9), its effectiveness must be ensured by using an "internal bonding strip," or conductor, under the armor and shorting the turns of the steel jacket. The ohmic resistance of finished armor, including the bonding conductor that is required to be furnished as a part of all, except lead-covered armored cable, must be within values specified by UL and checked during manufacturing. The bonding conductor run within the armor of the cable assembly is required by the UL standard.

Because the function of the bonding conductor in Type AC cable is simply to short adjacent turns of the spiral-wrapped armor, there is no need to make any connection of the bonding conductor at cable ends in enclosures or equipment. The conductor may simply be cut off at the armor end.

Construction of armored cable must permit ready insertion of an insulating bushing or equivalent protection between the conductors and the armor at each termination of the armor—such as the so-called red head.

320.104. Conductors. As required by the second paragraph in 320.80, armored cable (BX) installed within thermal insulation must have 90°C-rated conductors (Types THHN, RHH, XHHW), but the ampacity must be taken as that of 60°C rated conductors. This requirement recognizes that the heat rise on conductors operating with reduced heat-dissipating ability (such as those surrounded by fiberglass or similar thermal insulation) requires that the conductors have a 90°C-rated insulation. That temperature might be reached even with the wires carrying only 60°C ampacities. Although the wires must have 90°C insulation, they must not be loaded over those ampacity values permitted for TW (60°C), as shown in Table 310.15(B)(16).

320.108. Equipment Grounding Conductor. This section requires that Type AC cable provide "an adequate path" for fault current. This is accomplished by the No. 16 aluminum bonding strip that runs the length of the cable's sheath. This bonding strip shorts out the high impedance of the coiled metal jacket and provides a UL-listed ground path.

ARTICLE 322. FLAT CABLE ASSEMBLIES: TYPE FC

322.2. Definition. Type FC cable is a flat assembly with three or four parallel 10 AWG special stranded copper conductors. The assembly is installed in an approved U-channel surface metal raceway with one side open. Then tap devices can be inserted anywhere along the run. Connections from tap devices to the flat cable assembly are made by *pin-type* contacts when the tap devices are fastened in place. The pin-type contacts penetrate the insulation of the cable assembly and contact the multistranded conductors in a matched phase sequence (phase 1 to neutral, phase 2 to neutral, and phase 3 to neutral).

Covers are required when the installation is less than 8 ft from the floor. The maximum branch-circuit rating is 30 A.

Figure 322-1 shows the basic components of this wiring method.

322.10. Uses Permitted. Figure 322-2 shows a Type FC installation supplying lighting fixtures. As shown in the details, one tap device provides for circuit tap-off to splice to cord wires in the junction box; the other device is simply a fitting to support the fixture from the lips of the channel. Flat cable assemblies must be listed (322.6).

ARTICLE 324. FLAT CONDUCTOR CABLE: TYPE FCC

324.1. Scope. This article covers design and installation regulations on a branch-circuit wiring system that supplies floor outlets in office areas and other commercial and institutional interiors. (See Fig. 324-1.) The method may be used for new buildings or for modernization or expansion in existing interiors. FCC wiring

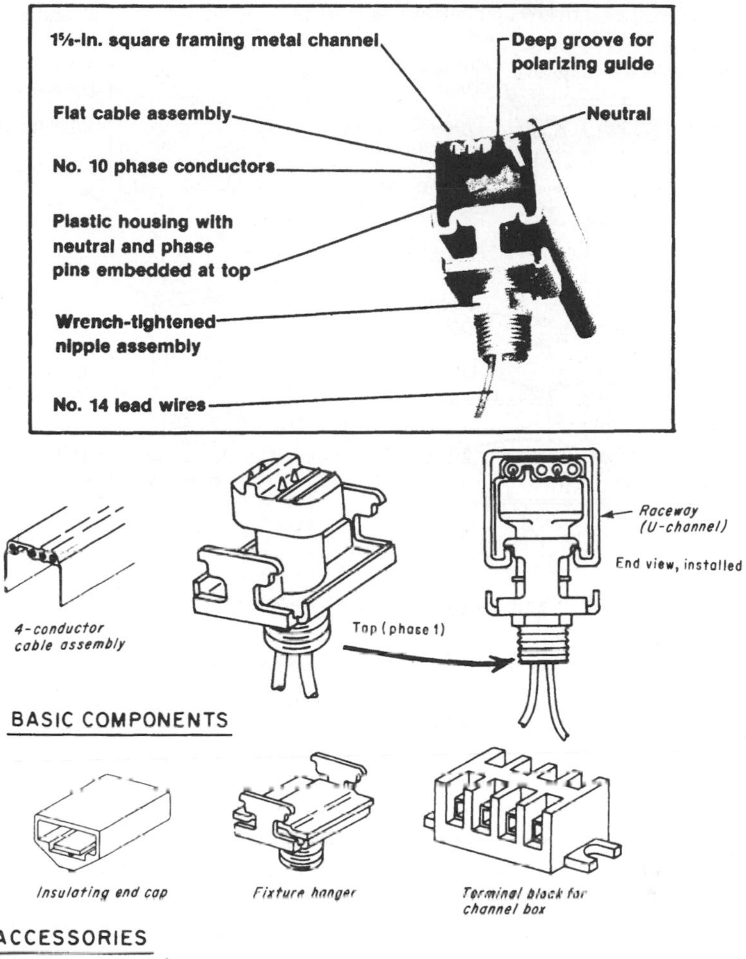
1⅝-In. square framing metal channel
Deep groove for polarizing guide
Flat cable assembly
Neutral
No. 10 phase conductors
Plastic housing with neutral and phase pins embedded at top
Wrench-tightened nipple assembly
No. 14 lead wires
Raceway (U-channel)
End view, installed
4-conductor cable assembly
Tap (phase 1)
BASIC COMPONENTS
Insulating end cap
Fixture hanger
Terminal block for channel box
ACCESSORIES

Fig. 322-1. Type FC wiring system uses cable in channel, with tap devices to loads. (Sec. 322.2.)

may be used on any hard, sound, smooth floor surface—concrete, wood, ceramic, and so forth. The great flexibility and ease of installation of this surface-mounted flat-cable wiring system meet the need that arises from the fact that the average floor power outlet in an office area is relocated every 2 years.

Undercarpet wiring to floor outlets eliminates any need for core drilling of concrete floors—avoiding noise, water dripping, falling debris, and disruption of normal activities in an office area. Alterations or additions to Type FCC circuit runs are neat, clean, and simple and may be done during office working hours—not requiring the overtime labor rates incurred by floor drilling, which must be done at night or on weekends. The FCC method eliminates use of conduit or cable, along with the need to fish conductors.

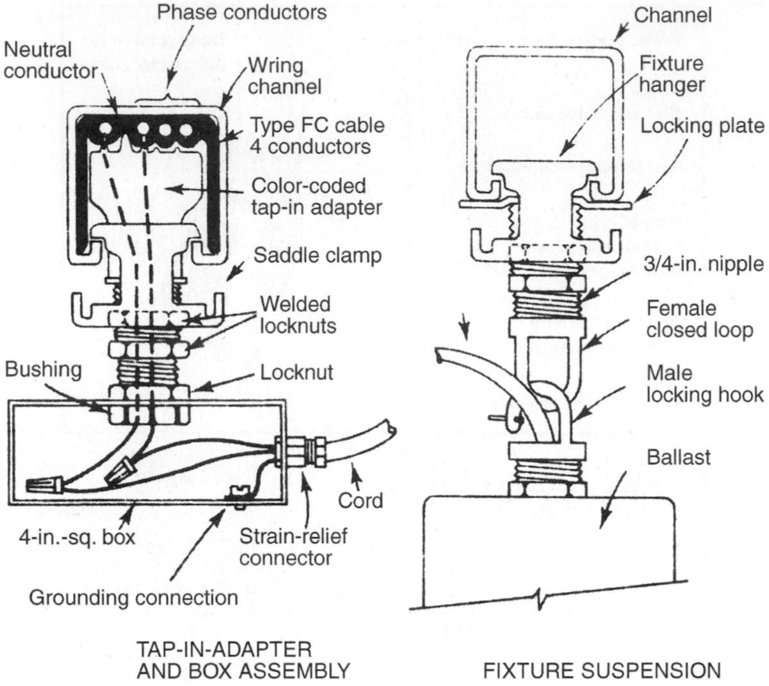
Phase conductors
Neutral conductor
Wring channel
Type FC cable 4 conductors
Color-coded tap-in adapter
Saddle clamp
Welded locknuts
Bushing
Locknut
Cord
4-in.-sq. box
Strain-relief connector
Grounding connection
Channel
Fixture hanger
Locking plate
3/4-in. nipple
Female closed loop
Male locking hook
Ballast
TAP-IN-ADAPTER AND BOX ASSEMBLY
FIXTURE SUSPENSION

Fig. 322-2. Limited application of Type FC cable system includes use as branch-circuit wiring method to supply luminaries. (Sec. 322.10.)

Type FCC wiring offers versatile supply to floor outlets for power and communication—at any location on the floor. The flat cable is inconspicuous under the carpet squares. Elimination of floor penetrations maintains the fire integrity of the floor, as required by 300.21.

A typical system might use separate flat-cable circuit layouts for 120-V power to floor-pedestal receptacles, telephone circuits, and data communications lines for CRT displays and computer units. For 120-V power, the flat cable contains three flat, color-coded (black, white, and green), 12 AWG copper conductors for 20-A circuits—one hot conductor, one neutral, and one equipment grounding conductor. Telephone circuits use flat, 3-pair, 26 AWG gauge conductors. And data connection circuits use flat RG62A/U coaxial cable that is only 0.09 in. (2.25 mm) high.

324.2. Definitions. The various components of a Type FCC system are described here. Figure 324-2 shows typical components of an FCC system. Sec. 324.6 requires the cable and fittings used with it to be listed.

324.10. Uses Permitted. This section describes the acceptable uses for Type FCC. Part **(A)** recognizes the use of Type FCC for both general-use branch circuits and individual branch circuits, while **(B)(1)** and **(2)** regulate the maximum voltage rating (not more than 300 V between conductors) and current rating (not more than 20 A for general-purpose branch circuits, or 30 A for individual branch circuits) permitted for Type FCC.

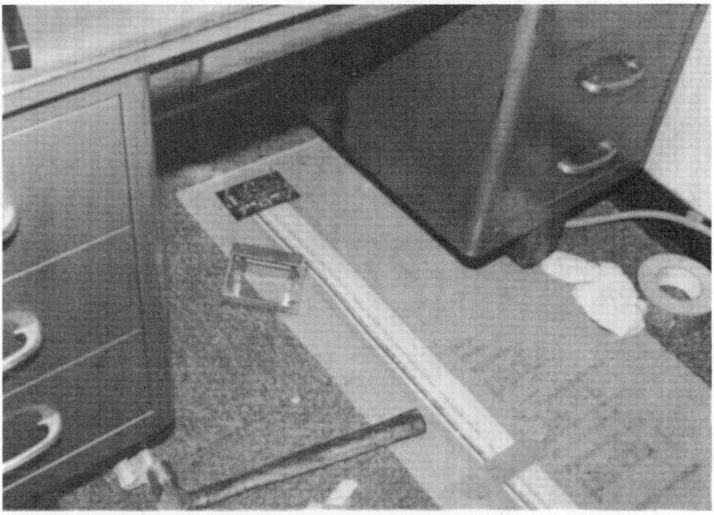

Fig. 324-1. Flat conductor cable (FCC) supplies terminal base for floor-outlet pedestal at exact location required for desk in office area. FCC is taped in position over an insulating bedding tape and then covered with a flat steel tape (not yet installed here) to protect the three conductors (hot leg, neutral, and equipment grounding conductor) in the flat cable. Carpet squares are used to cover the finished cable runs.

324.12. Uses Not Permitted. This wiring method, prior to the 2017 NEC, was prohibited from residential, school, and hospital buildings, even in administrative areas. The fear was that such areas would be converted to other purposes and unauthorized persons would have routine access to the wiring. Although this prior concern was never really addressed, the panel went ahead and the NEC now allows the wiring method in the administrative areas of school and hospital buildings.

324.18. Crossings. Not more than two FCC cable runs may be crossed over each other at any one point. To prevent lumping under the floor carpets, this rule permits no more than two Type FCC cables to be crossed over each other at a single point. This applies to FCC power cable and FCC communications and data cables.

In 324.41, "Floor Coverings," the Code restricts the maximum size of carpet squares used to cover the Type FCC. Carpet squares used to cover Type FCC wiring must not be larger than 1 m (39.37 in.) square. This rule eliminates questions that arose about the possibility of using single "squares" of broadloom carpet large enough to cover a floor from wall to wall. In addition, the carpet squares must be put down with a release-type adhesive.

In making an undercarpet installation, usual thinking would dictate installation of the cable layout first and then placement of the floor covering of carpet squares over the entire area. But some installers have found it easier and less expensive to first cover the entire floor area with the self-adhesive carpet squares and then plan the circuit layouts to keep the cable runs along the centerlines of carpet squares and away from the edges of the squares. After the layout is determined, it is a simple matter to lift only those carpet squares

(a)

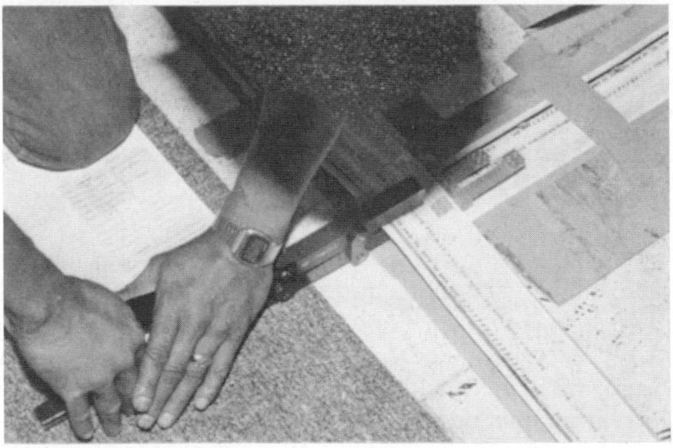

(b)

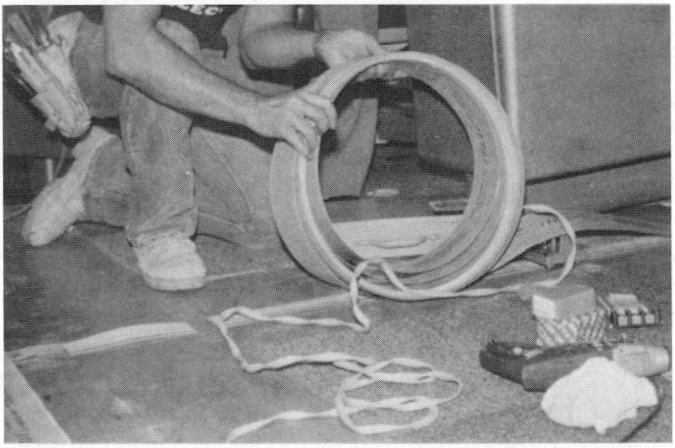

(c)

Fig. 324-2. Typical components of a Type FCC system: (*a*) Bottom shield in place. (*b*) Connecting the conductor. (*c*) Coil of top shield.

along the route of each run, install the cable and pedestal bases, and replace the self-stick carpet squares to restore the overall floor covering. That approach has proved effective and keeps carpet cutting to the middle of any square.

In 324.60, the Code calls for grounding of metallic shields—commonly employed in Type FCC assemblies to prevent interference on communication circuits. This rule further stipulates that those connectors be specifically designed for connecting the metallic shield to ground.

ARTICLE 326. INTEGRATED GAS SPACER CABLE: TYPE IGS

326.2. Definition. Type IGS cable is a "factory assembly of one or more conductors, each individually insulated and enclosed in a loose-fit nonmetallic flexible conduit." The cable is for underground use—including direct earth burial—for service conductors, feeders, or branch circuits. It must not, however, be used as interior wiring or exposed in contact with any building. Sizes available for this product run are metric designators 53, 78, and 103 (trade sizes 2, 3, and 4). The introduction of this cable to the NEC was recommended on the following basis:

> Underground cable costs are increasing at a high rate. A need exists for lower material costs and reduced cost for installation. Failures on underground cables are increasing, particularly direct-burial types.

The new cable system overcomes all the above problems. The new cable system has the advantage of low first cost for materials and low installation cost. It eliminates the need for field pulling of cables into conduits and eliminates the cost of assembly of conduit in the field. The new system may be directly buried, plowed in, or bored in for further savings. It is a cable and conduit system.

A tough natural-gas-approved pipe is used as the conduit. When it is pressurized, it will withstand much abuse. The gas pressure keeps out moisture and serves to monitor the cable for damage by insects or mechanical damage that can lead to future failure. The gas pressure can even be attached to an alarm to sound a loss of pressure or to trip a CB for hazardous locations. However, a loss of pressure in the cable will not cause it to fail. Even on dig-ins, the gas serves to warn the digger. The gas prevents combustion and burning on cable failure. The SF_6 gas is nontoxic, odorless, tasteless, and will not support combustion. It acts to put out a fire.

UL has tested a 3/C 250 MCM [kcmil] Type IGS-EC 600-V cable in 2-in. conduit. The UL test at zero gauge pressure shows a breakdown voltage between conductors of 14,000 V after numerous short-circuit, breakdown, and humidity tests. When the cable is single conductor, the breakdown voltage is even higher on loss of pressure, as the polyethylene pipe or conduit provides additional insulating value.

An award-winning installation was made in 1979 at 5 kV. Three installations of 3/C 250 MCM [kcmil] Type IGS-EC cable have been made for residential underground service entrances in Oakland, California. The first was made in May of 1979 and all have been successful.

326.80. Ampacity. This section indicates that Type IGS cable must have its ampacity determined in accordance with Table 326.80 for either single or multiconductor cable.

ARTICLE 328. MEDIUM-VOLTAGE CABLE: TYPE MV

328.2. Definition. This is a very limited definition of a Code designation—Type MV. This cable must be listed (328.6). This cable type is constructed in accordance with provisions published by Underwriters Laboratories as part of the applicable Guide Card information, as follows:

Medium-Voltage Cable (PITY)

Medium-voltage cables are rated 2400 to 35,000 volts.

They are single or multiconductor, aluminum or copper, with solid extruded dielectric insulation and may have an extruded jacket, metallic covering, or combination of both over the single conductors or over the assembled conductors in a multiconductor power cable.

All insulated conductors rated higher than 2400 volts have electrostatic shielding. Cable rated 2400 volts is nonshielded.

Nonshielded cables are intended for use where conditions of maintenance and supervision ensure that only competent individuals service and have access to the installation.

Shielded cable is marked either "MV-90" or "MV-105" and is suitable for use in wet or dry locations at 90° or 105°C.

Nonshielded cable is marked either "MV-90" indicating suitability for use in wet or dry locations at 90°C maximum or "MV-90 Dry Locations Only" indicating suitability for use only in dry locations at 90°C maximum.

Cables marked "oil resistant I" or "oil resistant II" are suitable for exposure to mineral oil at 60° or 75°C, respectively.

Cables marked "sunlight resistant" may be exposed to the direct rays of the sun.

Cables intended for installation in cable trays in accordance with Art. 392 of the National Electrical Code are marked "For Use in Cable Trays" (or "For CT Use").

Cables with aluminum conductors are marked with the word "aluminum" or the letters "AL."

Cables are marked with their conductor size, voltage rating, and insulation level (100% or 133%).

The basic standard used to investigate products in this category is UL1072, "Medium-Voltage Power Cables."

The Listing Mark of Underwriters Laboratories Inc. on the product is the only method provided by UL to identify products manufactured under its Listing and Follow-Up Service. The Listing Mark for these products includes the name and/or symbol of Underwriters Laboratories Inc. (as illustrated in the Introduction of this Directory) together with the word "LISTED," a control number and the following product name: "Medium-Voltage Cable."

Medium-Voltage Cable, Classified in Accordance with UL 1072, with Metric Conductor Sizes (PIVW)

This category covers medium-voltage cables rated 2001 to 35,000 volts and in conductor sizes 10 through 500 sq mm.

The cable complies with all requirements specified in UL 1072 "Medium-Voltage Power Cables," except that metric conductor sizes are used instead of AWG sizes. The cable is for use in jurisdictions where metric conductor sizes are required or permitted.

The cable is single or multiconductor, aluminum or copper, with solid extruded dielectric insulation. An extruded jacket, metallic covering, or combination of both may be provided over single conductors or over the assembled conductors in a multiconductor power cable.

All insulated conductors rated 2400 V and higher have electrostatic shielding. Cables rated 2400 V are nonshielded.

Nonshielded cables are intended for use where conditions of maintenance and supervision ensure that only competent individuals service and have access to the installation.

Shielded cable is marked "MV-90" or "MV-105" and is suitable for use in wet or dry locations at 90° or 105°C.

Nonshielded cable is marked either "MV-90" indicating suitability for use in wet or dry locations at 90°C maximum, or "MV-90 Dry Locations Only."

Cable marked "oil resistant I" or "oil resistant II" is suitable for exposure to mineral oil at 60° or 75°C, respectively.

Cable marked "sunlight resistant" may be exposed to the direct rays of the sun.

Cable intended for installation in cable trays is marked "For CT Use" or "For Use In Cable Trays."

Cables with aluminum conductors are marked with the word "Aluminum" or the letters "AL."

Cables are marked with conductor size in sq mm, voltage rating and insulation level (100% or 133%).

The basic standard used to investigate products in this category is UL1072, "Medium-Voltage Power Cables."

The Certification Mark of Underwriters Laboratories Inc. on the product, the attached tag, the reel, or the smallest unit container in which the product is packaged is the only method provided by UL to identify these products manufactured under its Classification and Follow-Up Service.

328.10. Uses Permitted. Because the Code has an article and cable designation (Type MV) for cables operating above 2000 V up to 35,000 V, it may be expected that electrical inspection authorities will insist that all cables in that voltage range must be Type MV to satisfy the NEC.

Great care should be exercised in determining the attitude of local inspection authorities toward the meaning of this article. In particular, the relationship of 110.8 to Art. 328 should be determined. 110.8 states that "only wiring methods recognized as suitable are included in this Code." The question to be answered is: Will electrical inspection agencies require all medium-voltage conductors to be Type MV? Or will inspection agencies accept medium-voltage conductors not specifically designated Type MV? In other words, because the Code has an accepted type of high-voltage cable, will it be permissible to use high-voltage cables that are not of this accepted type? The answer seems to be no. This is especially true when the OSHA insistence on listed equipment is considered. In addition, for circuits in common use up to 600 V, 110.8 has consistently been interpreted to require that *any* conductor or cable must be one of the types specifically designated in the Code—Table 310.104(A) or elsewhere in Arts. 300 to 398. That is, conductors must be Type TW, THW, or one of the other designated types, and cable must be Type AC, NM, MI, MC, or other designated cable. It would be a Code violation to use any non–Code-designated wire or cable for systems up to 2000 V. It would, therefore, seem to be similarly contrary to Code to use a non–Code-designated cable for higher-voltage circuits inasmuch as there is a Code-designated type (Type MV) for such applications. In addition, the 2011 NEC added a new Sec. 328.14 requiring these cables to only be installed, terminated, and tested by qualified persons.

Refer to Table 310.104(B) on Type MV conductors and to 392.22(C) and 392.80(B) for use of Type MV cables in tray.

ARTICLE 330. METAL-CLAD CABLE: TYPE MC

330.2. Definition. This article covers "Metal-Clad Cable," as listed by UL under that heading in the *Guide Card information* [see discussion at 110.3(B) for details]. This section defines the type of cable assemblies covered by this article (Fig. 330-1). The definition for metal-clad cable—"a factory assembly of one or more insulated circuit conductors with or without optical fiber members enclosed in an armor of interlocking metal tape, or a smooth or corrugated tube"—makes this category the successor to former Type ALS and Type CS cables.

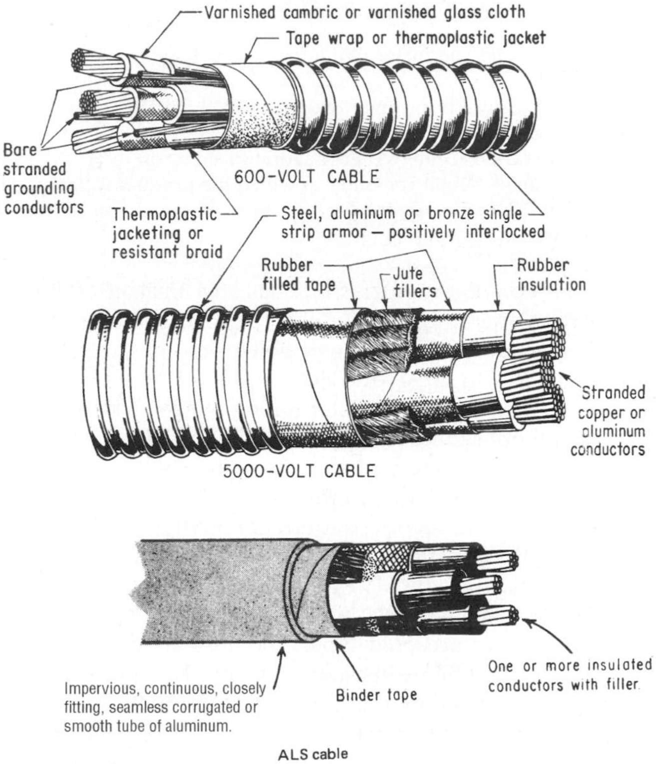
Varnished cambric or varnished glass cloth
Tape wrap or thermoplastic jacket
Bare stranded grounding conductors
600-VOLT CABLE
Thermoplastic jacketing or resistant braid
Steel, aluminum or bronze single strip armor — positively interlocked
Rubber filled tape
Jute fillers
Rubber insulation
Stranded copper or aluminum conductors
5000-VOLT CABLE
Impervious, continuous, closely fitting, seamless corrugated or smooth tube of aluminum.
Binder tape
One or more insulated conductors with filler.
ALS cable

Fig. 330-1. These are some of the constructions in which Type MC cable is available. (Sec. 330.2.) Note that the conductors comprising the center cable (5000 V) would require shielding to meet the current requirements in 310.10(E).

Aluminum-sheathed (ALS) cable had insulated conductors with color-coded coverings, cable fillers, and overall wrap of Mylar tape—all in an impervious, continuous, closely fitting, seamless tube of aluminum. It was used for both exposed and concealed work in dry or wet locations, with approved fittings. CS cable was very similar, with a copper exterior sheath instead of aluminum.

Because the rules of these three cable types have been compiled into a single article, use of any one of the Type MC cables must be evaluated against the specific rules that now generally apply to all such cables. The Code no longer contains the designations Type ALS and Type CS. They are included now as either smooth or corrugated styles as applicable along with interlocked armored cable as Type MC cables.

Type MC is rated by UL for use up to 2000 V. Cable rated 2400 to 35,000 V is listed as Type MV. Type MC cable is recognized in three basic armor designs: (1) interlocked metal tape, (2) corrugated, or (3) smooth metallic sheath.

330.6. Listing Requirements. This cable, as of the 2017 **NEC**, is required to be listed, along with its connecting fittings. The application rules for these connectors require careful study. Only approved, UL-listed connectors and fittings are permitted to be used with any Type MC cable. Such fittings are listed in the UL Guide Card listings as "Metal-Clad Cable Connectors (PJOX)." Figure 330-2 shows

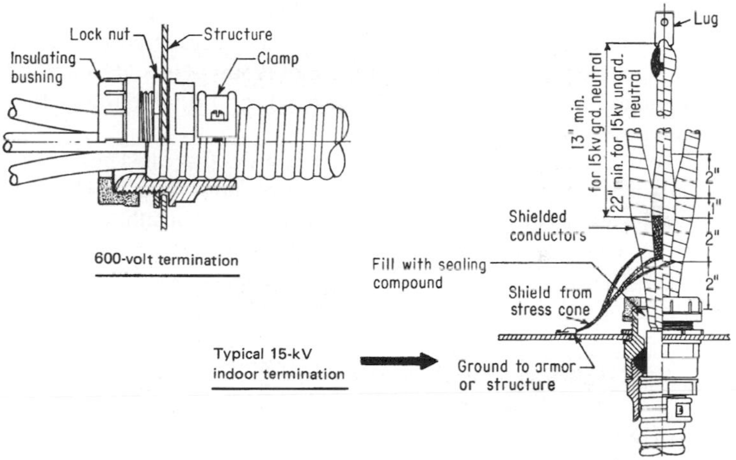
Lock nut
Structure
Insulating bushing
Clamp
600-volt termination
Typical 15-kV indoor termination
Lug
13" min. for 15kv grd. neutral
22" min. for 15kv ungrd. neutral
Shielded conductors
2"
1"
2"
2"
Fill with sealing compound
Shield from stress cone
Ground to armor or structure

Fig. 330-2. Terminations for interlocked-armor cables must be approved devices, correctly installed. (Sec. 330.6.)

typical approved connectors for interlocked-armor Type MC cable. As shown at left, 600-V terminations for interlocked-armor cable to switchgear or other enclosures in dry locations can be made with connectors, a locknut, and a bushing in the typical basic assembly shown. The bushing is generally optional because Type MC cable is not a raceway and 300.4(G) does not apply. In damp locations, compound-filled or other protective terminations may be desired. Medium-voltage connectors (5 through 35 kV) are generally filled with sealing compound, and individual conductors are terminated in a suitable manner, depending upon whether the conductors are shielded. Or the IA cable may terminate in a pothead for positively sealed and insulated terminations indoors or outdoors.

Type MC cable must terminate in connectors designed for the particular cable involved. Pay particular attention to differences between MC and AC cable terminations, for example. The **NEC** doesn't require anti-short bushings for Type MC cable, although one leading manufacturer does make them available for that purpose, and many contractors choose to use them anyway. They are not required because the throat designs of listed connectors keep the conductors away from the cut edges of the armor. In addition, since these connectors may have to handle ground-fault currents, they are tested with their designed cable types for this duty. Those tests have no validity beyond the cable types actually tested.

Before you run MC (or any other cable), look closely at the box or shipping carton for the connectors. Particularly in the case of Type MC cable, you'll find very specific size ranges given. For example, the box might indicate suitability for smooth Type MC in a particular range of cable diameters, corrugated in a range of diameters, and interlocking with a range of conductor configurations. If you don't find your particular application, select a different connector. The same principle applies to internal box connectors. Look at the shipping carton for the box, because the label (for a listed box) will always tell you what cables the internal clamps are designed for.

330.10. Uses Permitted. Although this section clearly lists all the permitted applications of any of the various forms of Type MC cable, care must be taken to distinguish between the different constructions, based on the Code rules (Fig. 330-3). For a long time, the interlocked-armor Type MC and the corrugated-sheath Type MC have been designated by UL as "intended for aboveground use." But part **(5)** of this section recognizes Type MC cable as suitable for direct burial in the earth, when it is identified for such use.

Note that item (11) includes the specific characteristics that must be met in order to qualify a particular cable style as suitable for wet locations. The 2014 NEC modified this rule by requiring a corrosion-resistant jacket over the metallic covering regardless of the inner construction detail. Item (12) recognizes single-conductor Type MC cable; care must be taken in such cases to track inductive heating issues 300.21(B), special ampacity calculations [330.80(B)], and termination issues [110.14(C)].

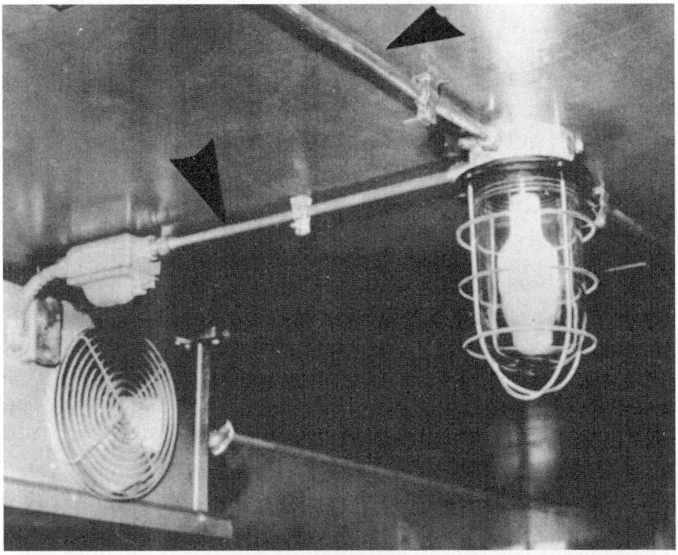

Fig. 330-3. ALS (aluminum-sheathed) Type MC cable was used for extensive power and light wiring in refrigerated rooms and storage areas of a store. The ALS was surface mounted (exposed) on clamps in this damp location. Because the cable assembly is a tight grouping of conductors within the sheath, there would be no passage of warm air from adjacent nonrefrigerated areas through the cable which crosses the boundaries between the areas. It was therefore not necessary to seal the cables to satisfy Sec. 300.7(A). (Sec. 330.10.)

330.12. Uses Not Permitted. Type MC cables are permitted by 330.10 to be used exposed or concealed in dry or wet locations. But such cable must not be subjected to destructive, corrosive conditions—such as direct burial in the earth, in concrete, or exposed to cinder fills, strong chlorides, caustic alkalis, or vapors of chlorine or of hydrochloric acids, unless protected by materials suitable for the condition.

330.15. Exposed Work. This wiring method, especially in the smaller branch circuit sizes, has largely supplanted type AC cable in the market. Because it is increasingly used for comparable applications, this section, new in the 2017 NEC, has been cut and pasted from 320.15. Refer to further discussion at that location for more information.

330.24. Bending Radius. Figure 330-4 shows the bending-radius rules for the "smooth-sheath" Type MC cables. Cable with interlocked or corrugated armor must have a bending radius not less than 7 times the outside diameter of the cable armor. To conform to ICEA rules on bending radius for shielded conductors in MC cable, the minimum value must be either 12 times the diameter of one of the conductors within the cable or 7 times the diameter of the MC cable itself, whichever is greater. For medium-voltage applications with shielded cables, the minimum bending radius is 12 times the cable diameter for a single-conductor makeup, and 7 times the overall diameter of a multiconductor makeup, as applicable.

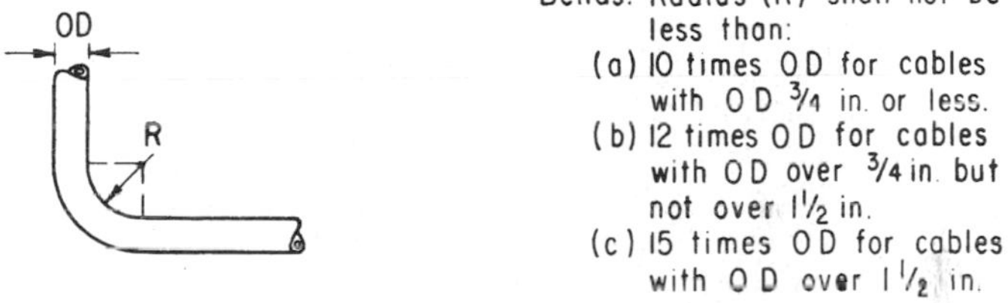
OD
R
Bends: Radius (R) shall not be less than:
(a) 10 times OD for cables with OD 3/4 in. or less.
(b) 12 times OD for cables with OD over 3/4 in. but not over 1 1/2 in.
(c) 15 times OD for cables with OD over 1 1/2 in.

Fig. 330-4. Minimum radius values prevent excessively sharp, destructive bending of Type MC cable, or its predecessors, Type ALS or CS cable. (Sec. 330.24.)

330.30. Securing and Supporting. Figure 330-5 shows the maximum permitted spacing of supports for the larger-sized Type MC cable. The interlocked-armor Type MC has commonly been used on cable tray, as permitted in part **(A)(6)** of 330.10 if it is identified for this use (Fig. 330-6). In addition to the 1.8 m (6 ft) interval, smaller cables—those with four or fewer conductors smaller than No. 8—must also be secured within 300 mm (12 in.) of "each box, cabinet, fitting, or other cable termination." In addition, part **(C)** recognizes the common practice of entraining cables through successive holes in framing members.

The 2014 NEC has made two changes that extend support intervals in this section. Small cables whose normal first point of support is 12 in. can, in the case of the interlocking armor type of cable, have that distance go out to 36 in. in instances where flexibility is required after the installation is complete. It also added permission to extend the support interval for vertical runs of very large

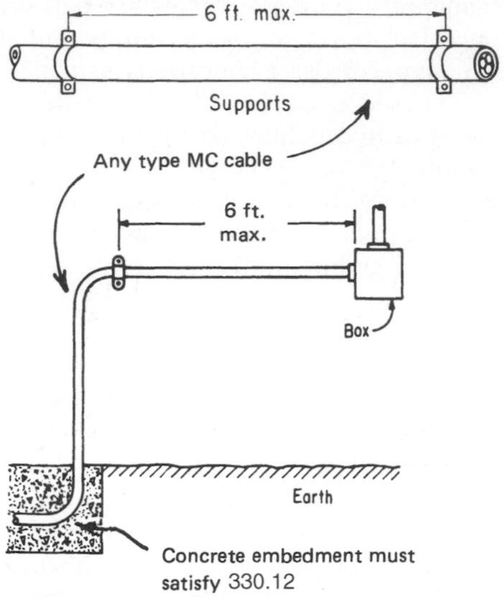
6 ft. max.
Supports
Any type MC cable
6 ft.
max.
Box
Earth
Concrete embedment must
satisfy 330.12

Fig. 330-5. Surface mounting of Type MC cable must be secured. Smaller sizes are additionally required to be secured within 300 mm (12 in.) of terminations. (Sec. 330.30.)

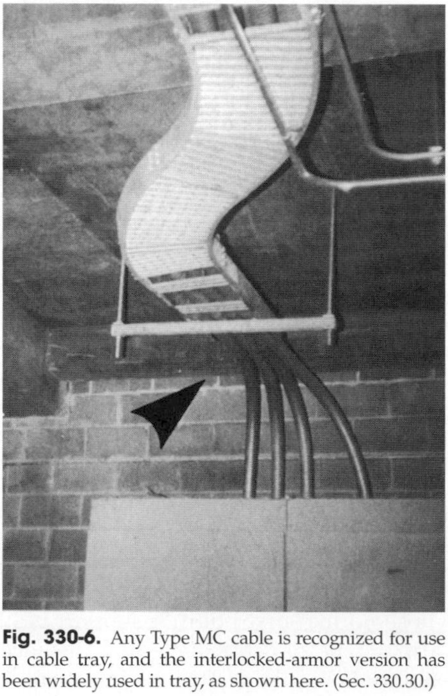

Fig. 330-6. Any Type MC cable is recognized for use in cable tray, and the interlocked-armor version has been widely used in tray, as shown here. (Sec. 330.30.)

MC cables, those with ungrounded conductors no smaller than 250 kcmil, from the customary 6 to 10 ft.

Part **(D)** recognizes that Type MC cable can be fished without concern for periodic supports. It also recognizes that a cable connector can, in effect, be considered a point of support, at least in suspended ceilings. As in the case of Type AC cable, these cable ties must be listed for this function. This is an extremely practical allowance that avoids the necessity for supports near the cable terminations. Note that this also supersedes the normal requirement for securing within 300 mm (12 in.) of terminations for smaller cables. See also the discussion at 320.30 for Type AC cable for more information on this topic.

330.80. Ampacity. In addition to the normal ampacity considerations, it should be noted that Type MC cable is available in single-conductor configurations, with special ampacity considerations. The usual application for these configurations is cable tray, and the allowance to use messenger cable ampacities correlates with 392.80(A)(2)(d) and 392.80(B)(2)(c). Note that these enhanced ampacities apply to the middle of a cable run and do not excuse compliance with 110.14(C) [or 110.40 for medium-voltage applications] at wiring terminations. Step-up terminators with tails that comply with applicable Table 310.15(B)(16) ampacities may be available, or the run can be brought through a pull box with the wire sizes adjusted prior to arrival at equipment terminals.

330.108. Equipment Grounding Conductor. Generally the smooth and corrugated types of MC cable have sufficient cross-sectional area of metal in their armors such that the armor itself qualifies as the equipment grounding conductor, although not always; if additional grounding conductors are incorporated, they must be used as part of the equipment grounding system, right along with the cable armor. Interlocking metal tape, without modification, is never suitable for equipment grounding because the spiral path adds too much impedance. Type AC cable solves this problem with the bonding tape and the paper filler, because the tape is held in firm contact with the armor, and as such it shorts the convolutions. Traditionally, Type MC with interlocking armor always carried an additional equipment grounding conductor, usually with green insulation and sized per 250.122. This conductor had to terminate in boxes and it counted for a required fill allowance in 314.16.

Recently one manufacturer solved the grounding continuity problem across interlocking MC cable armor by using a bare, fully sized aluminum equipment grounding conductor against the armor and on the outside of the plastic wrap that goes over the circuit conductors. This product can be used similarly to Type AC cable, without a separate grounding conductor entering the box. In addition, for instances where the connector and locknut cannot be relied upon for grounding continuity, such as for 480Y/277 system circuits running into cabinets with concentric knockouts, the bare wire can be left as long as the circuit conductors and brought to an equipment grounding terminal in the enclosure. Even Type AC cable isn't capable of that trick because the little aluminum bonding strip does not qualify as an equipment grounding conductor. Here again, make certain to correlate the listing information for the connectors with the type of MC cable being used. Refer also to the commentary on this point contained in the coverage at 250.118(10), where this method of equipment grounding is recognized.

ARTICLE 332. MINERAL-INSULATED, METAL-SHEATHED CABLE: TYPE MI

332.2. Definition. The data from the UL Guide Card information [see discussion at 110.3(B) for details] expand on the definition (Fig. 332-1) and cover application notes as follows.

Mineral-insulated metal-sheathed cable is labeled in a single-conductor construction from 16 AWG through 500 kcmil single conductor, two- and three-conductor from 16 AWG through No. 4 AWG, four-conductor from No. 16 AWG through No. 6 AWG, and seven-conductor Nos. 16, 14, 12, and 10 AWG. The exterior sheath may be of copper or alloy steel. There is also a signaling circuit style available with a 300 V rating and configured for 16 and 18 AWG in 2-, 3-, 4-, and 7-conductor put-ups. It is available with a copper sheath (the usual configuration) in which case the sheath is qualified as an equipment grounding conductor, and also with an alloy steel sheath that does not so qualify.

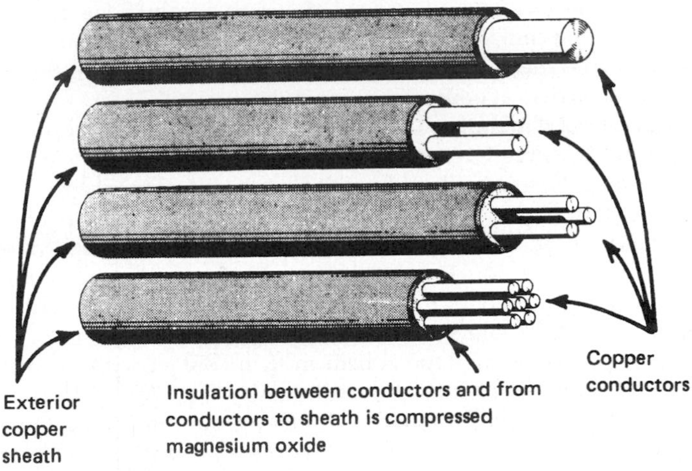
Exterior copper sheath
Insulation between conductors and from conductors to sheath is compressed magnesium oxide
Copper conductors

Fig. 332-1. Type MI is a single- or multiconductor cable that requires special termination. (Sec. 332.2.)

The standard length in which any size is furnished depends on the final diameter of the cable. The smallest cable, 1/C No. 16 AWG, has a diameter of 0.216 in. and can be furnished in lengths of approximately 1900 ft. Cables of larger diameter have proportionally shorter lengths. The cable is shipped in paper-wrapped coils ranging in diameter from 3 to 5 ft. As of the 2017 NEC, this cable must be listed.

The original intent behind development of this cable was to provide a wiring material that would be completely noncombustible, thus eliminating the fire hazards resulting from faults or excessive overloads on electrical circuits. To accomplish this, it is constructed entirely of inorganic materials. The conductors, sheath, and protective armor are of metal. The insulation is highly compressed magnesium oxide, which is extremely stable at high temperatures (fusion temperature of 2800°C).

332.10. Uses Permitted. This section describes the general use of mineral-insulated metal-sheathed cable, designated Type MI. Briefly, it basically includes general use as services, feeders, and branch circuits in exposed and concealed work, in dry and wet locations, for underplaster extensions and embedded in plaster, masonry, concrete, or fill, for underground runs, or where exposed to weather, continuous moisture, oil, or other conditions not having a deteriorating effect on the metallic sheath (Fig. 332-2). The temperature rating of the cable for conventional power applications is limited by the rating of the end-seal fitting, which necessarily includes organic compounds for sealing purposes. UL sets the maximum rating of current end-seal designs as covered in the Guide Card information that is current as of this writing at 90°C in dry locations and 60°C in wet locations. The cable itself, however, is recognized for 250°C in special applications. Permissible current ratings will be those given in Table 310.15(B)(16) (or, under engineering supervision, in Table B.310.3 in Annex B in the back of the NEC book). Type MI cable in its many sizes and constructions is suitable for all power circuits up to 600 V.

There is no question that MI cable can be used "in underground runs" as indicated in 332.10(10). But there is the matter of the additional qualification

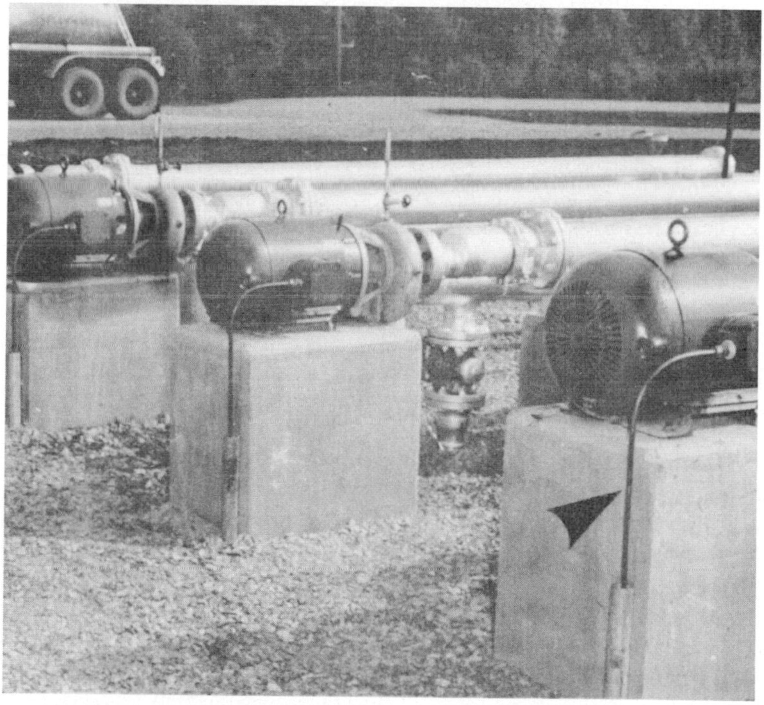

Fig. 332-2. Type MI is recognized for an extremely broad range of applications—for any kind of circuit, indoors or outdoors, wet or dry, and even in hazardous locations, as where MI motor branch circuits supply pumps in areas subject to flammable gases or vapors. (Sec. 332.10.)

that it be "suitably protected against physical damage and corrosive conditions." Although the copper sheath of MI cable has good resistance to corrosion, acid soils may be harmful to the copper sheath. Direct earth burial in alkaline and neutral soils would generally be expected to create no problems, but in any direct burial application MI cable with an outer plastic or neoprene jacket would ensure effective application and provide compliance with the additional language in part **(10)** of 332.10. Such jacketed MI cables are available, and have been successfully used in direct burial applications. As in the case of Type MC cable, the permitted minimum bending radius is limited by Sec. 332.24, to amounts that are related to the outer diameter of the cable.

332.40. Boxes and Fittings. Part **(A)** calls for connections of Type MI cable to be carefully made in accordance with UL and manufacturers' application data to ensure effective operation (Fig. 332-3).

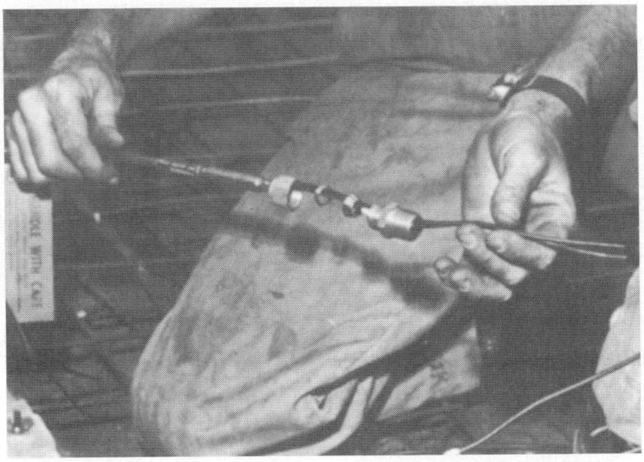

Fig. 332-3. Termination fitting for Type MI cable must be an approved connector, with its component parts assembled in proper sequence. [Sec. 332.40(A).]

Part **(B)** of 332.40, "Terminal Seals," presents a rule that is applied in conjunction with that of 332.40(A) to ensure *both* sealing of the cable end and means for connecting to enclosures (Fig. 332-4).

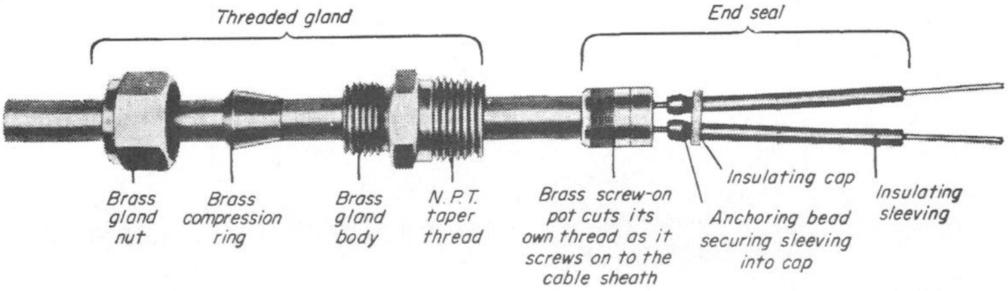
Threaded gland
End seal
Brass gland nut
Brass compression ring
Brass gland body
N.P.T. taper thread
Brass screw-on pot cuts its own thread as it screws on to the cable sheath
Insulating cap
Anchoring bead securing sleeving into cap
Insulating sleeving
This typical fitting is approved for MI termination in hazardous locations, in accordance with 501.10.

Fig. 332-4. MI cable termination must provide end sealing and connection means. [Sec. 332.40(B).]

332.80. Ampacity. The amount of heat that a run of MI cable can tolerate is governed by the temperature rating of the end-seal connector, which is 90°C in dry locations. For multiconductor configurations, look in Table 310.15(B)(16) in that column and read out the ampacity. If the conductor size does not change prior to landing on a circuit breaker, then 110.14(C) applies and a lower-temperature ampacity column may need to be substituted. This is really no different from other multiconductor cables with 90°C insulation, such as Type MC cable.

However, this cable can be and very frequently is run as single-conductor cable, and 332.80(B) allows its ampacity to be calculated from Table 310.15(B)(17). This permission is conditioned on a maintained spacing between adjacent cable groupings equal to not less than 2.15 times the outside diameter of the single largest cable in a group. For single conductor 4/0, that is 17.4 mm (0.684 in.) in diameter, the spacing would need to be about 37.4 mm (1.5 in.) between cable bundles. Even after the spacing is settled, however, there are other concerns that have to be addressed, as follows:

When it is run in this form, 332.21 requires that the cables be grouped to minimize the induced voltages on the cable sheaths. When this happens, the individual cables are not free to radiate their heat to the extent normally presumed for a conductor using free-air ampacities from Table 310.17. Consider a 3-phase, 4-wire makeup using the 4/0 AWG Type MI cable already mentioned, which comes out of the 90°C column of Table 310.17 at 405 A. Is this a way to avoid running 500 kcmil as pipe and wire? It might be, but read on.

When the cable is grouped, each cable does not radiate its heat as well. And in fact, loading 4/0 AWG MI cable to 405 A after it has been grouped or bundled results in a steady-state temperature within the bundle significantly higher than 90°C. This is a known fact; however, the MI cable manufacturers correctly pointed out that there are no end seals in the middle of one of these runs. The only thing there is copper and magnesium oxide. These items can withstand temperatures far higher than even temperatures over 90°C indefinitely without damage. Therefore, it has been conclusively demonstrated that yes, the cable bundles will run hot, but Table 310.15(B)(17) ampacities are not unreasonable for this type of cable applied in just this way.

The next issue is to bring this cable, in this heated state, to the panel, and then into the panel for terminations. This means getting the hot cables in the bundle to cool off enough so the end seal temperature operates within its temperature rating (90°C). This means spreading the cables out to a certain amount for a certain length, neither of which is specified in the **NEC**. That means requesting information from the cable manufacturer's engineering department, and then following their recommendations, which should be documented. Strictly by way of illustration, informal conversations have resulted in informal suggestions in the 300 to 450 mm (1 to 1½ ft) range, but that will need to be determined properly.

The final step is to make certain that the terminations stay within the usual 75°C limits prescribed in 110.14(C)(1)(b). Here the MI cable manufacturers seem to be quite aware of this issue. They have a range of step-up sizing terminations that incorporate upsizing terminating tails that match typical ampacity combinations. For example, a step-up connector rated for 4/0 MI cable and providing a 500 kcmil tail for termination purposes is a stock item with the leading manufacturer of Type MI cable.

ARTICLE 334. NONMETALLIC-SHEATHED CABLE: TYPES NM, NMC, AND NMS

334.1. Scope. This section makes clear that nonmetallic-sheathed cable must be installed *and* manufactured as required here.

334.2. Definition. Nonmetallic-sheathed cable is one of the most widely used cables for branch circuits and feeders in residential and light commercial systems (Fig. 334-1). Such cable is commonly and generally called "Romex" by electrical construction people, even though the word *Romex* is a registered trade name of the General Cable Corp. Industry usage has made the trade name a generic title so that nonmetallic-sheathed cable made by any manufacturer might be called Romex. This generic usage of a trade name also applies to the term *BX*, which is commonly used to describe any standard armored cable, made by any manufacturer—even though the term *BX* is a registered trade name of General Electric Co. Type NM cable has an overall covering of fibrous or plastic material that is flame-retardant and moisture-resistant. Type NMC is similar, but the overall covering is also fungus-resistant and corrosion-resistant. The letter C indicates that it is corrosion-resistant Type NM cable must be listed (334.6), along with its associated fittings.

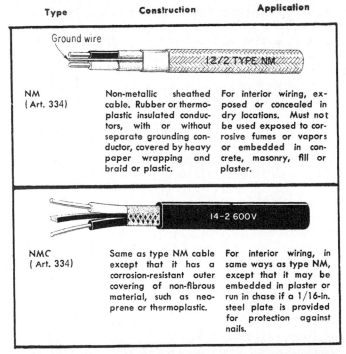

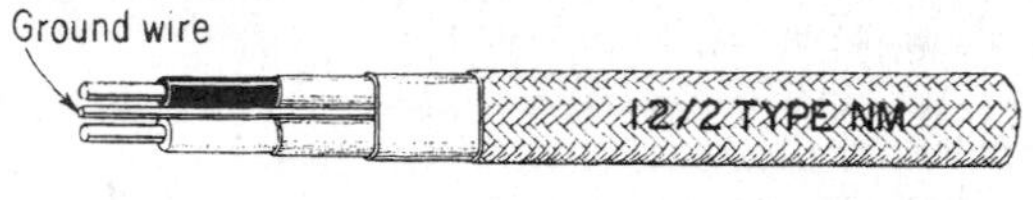
Ground wire
12/2 TYPE NM

Type	Construction	Application
NM (Art. 334)	Non-metallic sheathed cable. Rubber or thermoplastic insulated conductors, with or without separate grounding conductor, covered by heavy paper wrapping and braid or plastic.	For interior wiring, exposed or concealed in dry locations. Must not be used exposed to corrosive fumes or vapors or embedded in concrete, masonry, fill or plaster.

14-2 600V

Type	Construction	Application
NMC (Art. 334)	Same as type NM cable except that it has a corrosion-resistant outer covering of non-fibrous material, such as neoprene or thermoplastic.	For interior wiring, in same ways as type NM, except that it may be embedded in plaster or run in chase if a 1/16-in. steel plate is provided for protection against nails.

Fig. 334-1. Two of the three separate types of nonmetallic-sheathed cable are shown here. (Sec. 334.100.) Since these drawings were made, changes in the product standard have made conventional Type NM cable much smaller in overall size because the paper wrap has been eliminated (subject to a cold-weather pull-through-the-joists test) and now only the bare equipment grounding conductor has paper around it. Type NMC has largely been supplanted in the market by Type UF cable.

334.10. Uses Permitted. This type of wiring may be used either for exposed or for concealed wiring (Fig. 334-2) in any kind of building or structure.

1. NM cable may be used in one-family dwellings, together with their attached or detached garages and their storage buildings.
2. NM cable may be used in two-family dwellings, together with the same list of accessory buildings that applies to single-family houses.
3. NM cable may be used in multifamily dwellings.
4. NM cable may be used in other structures.

But for multifamily dwellings or other buildings or structures, the use of NM is permitted *only* if these locations are *permitted to be* of Type III, Type IV, or Type V construction as given in NFPA *5000*® (see Code Annex E). This is a critical point. The acceptability of NM cable depends not on how the building is actually

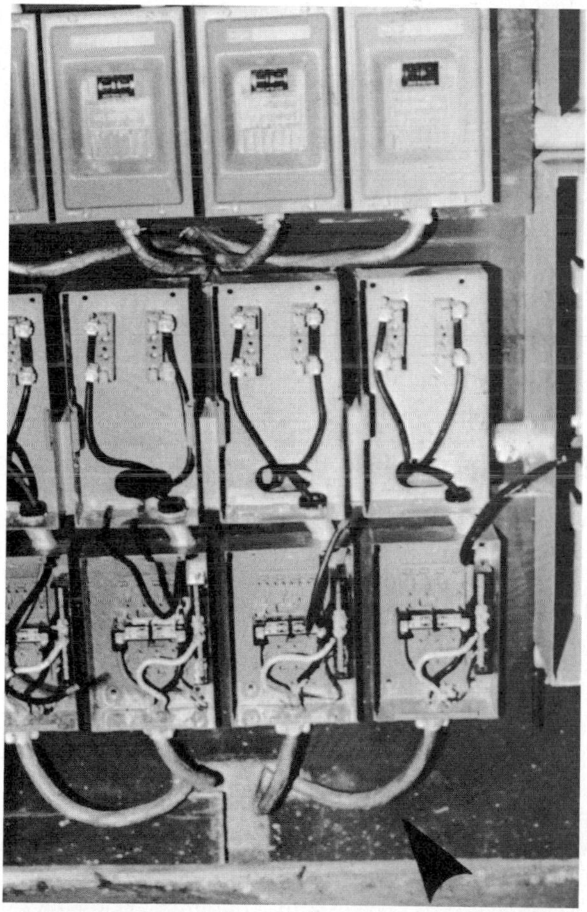

Fig. 334-2. Although NM cable is most widely used for branch circuits, the larger sizes (No. 8 and up) are commonly used for feeders, as run here from apartment disconnects to tenant panelboards. (Sec. 334.10.)

constructed, but rather on what the applicable building code would permit the construction to be based on the occupancy. A small multifamily dwelling that meets the size and area tables of the local building code for Type V construction, and yet was constructed to Type II standards, may still use Type NM cable. For nonresidential applications, however, Type NM cable must be concealed within stud or joist cavities (or other protected areas) that have a finish construction that provides a minimum of a 15-min finish rating as defined in listings of fire-rated assemblies. This generally includes 12.7 mm (½ in.) drywall.

Note that there is also permission [334.10(5)] for NM cables to be pulled in to Chap. 3 raceways that would be allowed in the construction. This has a certain internal logic since if EMT is safe with THHN inside, why shouldn't it be safe with NM cable built around THHN in the same location? Why anyone would buy and install NM cable where straight THHN would do is another question, but the exception appears to be harmless.

Type NM or NMC cable must be "identified" for use in cable trays. This requirement essentially calls for UL listing and marking on the cable to make it "recognized" as suitable for installation in cable trays. Type NM or NMC cable can also be fished into hollow voids of masonry block or tile construction.

Although NM cable is limited to use in "normally dry locations," NMC—the corrosion-resistant type—is permitted in "dry, damp, moist, or corrosive locations." Because it has been widely used in barns and other animals' quarters where the atmosphere is damp and corrosive (due to animal vapors), NMC cable is sometimes referred to as "barn wiring."

Part **(B)** says that Type NMC that is run in a shallow chase in masonry, concrete, or adobe and covered over must be protected against nail or screw penetration by a 1.59-mm (1/16-in.) (minimum) steel plate. This allowance does not extend to NM or NMS cables.

Part **(C)** specifies those permissible uses for Type NMS, which is a hybrid cable containing power, signaling, communication—voice/data/video—and even optical-fiber cable. This cable type was originally put in the Code to correlate with the former Art. 780 on closed-loop circuiting, where all outlets remained de-energized until a suitable plug was connected. This action would wake up a central controller and announce that a stipulated load was now connected, so a power draw could be provided that corresponded to the load. If the appli- ance malfunctioned and started to draw too much power, the circuit would be de-energized by the controller. The system was essentially foolproof, but never managed to gain any market share, and as of the 2008 Code cycle, Art. 780 has been deleted. Type NMS cable now relates to intelligent energy managemen systems that rely on signaling between an outlet and a central control system, anc it can carry other systems as well, such as communications. The constructior requirements in 334.116(C) assure that the system separation rules in Chaps. and 8 of the **NEC** are respected in these cable constructions.

334.12. Uses Not Permitted. The first subpart of this rule eliminates the us of any nonmetallic-sheathed cable where it is not already permitted by 334.1 regardless of occupancy type (Fig. 334-3). Guidance regarding building typ can be found on the building permit, and Table E.1 in Annex E of the Code, pre sented here as Tables 334-1A and 334-1B, can also be of assistance. Perhaps th best guidance can be provided by the local building department. Remembe

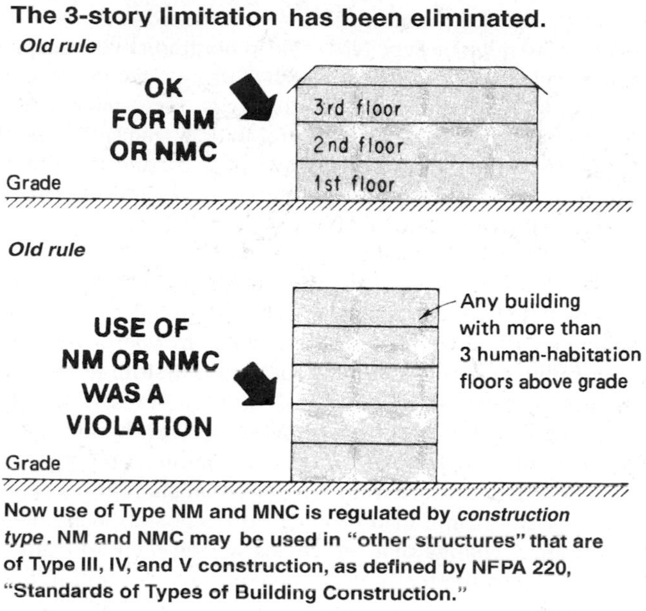
The 3-story limitation has been eliminated.
Old rule
OK
FOR NM
OR NMC
3rd floor
2nd floor
1st floor
Grade
Old rule
USE OF
NM OR NMC
WAS A
VIOLATION
Any building
with more than
3 human-habitation
floors above grade
Grade
Now use of Type NM and MNC is regulated by construction type. NM and NMC may be used in "other structures" that are of Type III, IV, and V construction, as defined by NFPA 220, "Standards of Types of Building Construction."

Fig. 334-3. Nonmetallic-sheathed cable is limited in application. [Sec. 334.12(A).]

the rule is based not on what the building is, but what it could be, and that is entirely within the purview of the building department.

The whole approach of using building construction types to define the permitted scope of a wiring method is unique in the history of the NEC, although some rules of transformer installations do include these references. The wording of the change originated in a special task group that met in the summer of 1999 to try to sort out where the real issues were in NM cable usage. The reversal of the three-story limitation was also implemented in a very unusual way because it was imposed by the NFPA Standards Council in the 2002 NEC after it had been rejected at all steps in the usual Code amendment process. The Council action was based on the entire record before it, including NFPA fire statistics that showed no association between Type NM cable and fire prevalence, and the report of the NFPA Toxicity Advisory Committee to the effect that in a fire the contribution of Type NM cable jacketing materials is a negligible fraction of the total smoke load.

It was also very clear at the time that major housing interests were prepared to expend significant resources litigating any continuing three-story limitation on grounds that would challenge the technical validity of the rule, and quite possibly the technical objectivity of the NEC panel. Furthermore, the rule had attracted the attention of housing activists, who identified it as a source of artificially inflated housing costs, and who were advocating state legislative attacks on the rule that would have undermined the traditional political independence of the local NEC adoption process in some major jurisdictions. The Standards Council decision seems to have been successful; those attacks are now moot, and the political challenges in this area have lessened.

The response from the **NEC** Committee after the Standards Council action has been a steady erosion of the reach of Type NM cable in nontraditional occupancies, while carefully leaving unchanged the basic approach of using construction types to define the applicability of the wiring method. No change is more emblematic of this process than 334.12(A)(2), which is a backdoor prohibition against the use of NM cable above any suspended ceiling in smaller commercial occupancies, a practice that had been done under the old one- to three-stories rule for decades. Because all of these spaces are accessible through removal of ceiling tiles, the areas meet the definition of "exposed" in Art. 100, and the prohibition is complete.

This 2005 **NEC** change replaced wording from the task group, implemented in the 2001 Standards Council action, which prohibited running NM cable in nonresidential suspended ceilings where it was "in open runs." The concept was that these ceilings were often used for random storage, and if NM cable ran from bar joist to bar joist on 1.2-m (4-ft) centers, the regular support rule would still be met but the cable would be subject to damage. The "in open runs" prohibition still allowed the cable in the ceiling, as long as it was protected with a running board or equal. The only substantiation behind making the change (2005 **NEC**) from "in the open" to "exposed" was that "exposed" was a defined term. A change that removes a wiring method from a good portion of a market, and in the teeth of public comments detailing the gravity of the change, does not turn on an editorial desire to use a defined term. This and comparable issues will undoubtedly fester for some time to come.

As covered in part **(12)(B)(2),** Types NM and NMS *are* effectively prohibited from embedding in plaster and so forth (Fig. 334-4). Given the robust nature of NMC, it is specifically permitted for such application, but it is also necessary to protect the cable against the possibility of being damaged by driven nails—such as nails used to hang pictures or add construction elements on the wall. Sufficient protection against nail puncture of the cable is provided by a cover of corrosion-resistant coated steel of at least 1.59 mm ($^1/_{16}$ in.) thickness (or equivalent protection from a listed nail plate). Such metal protection must be

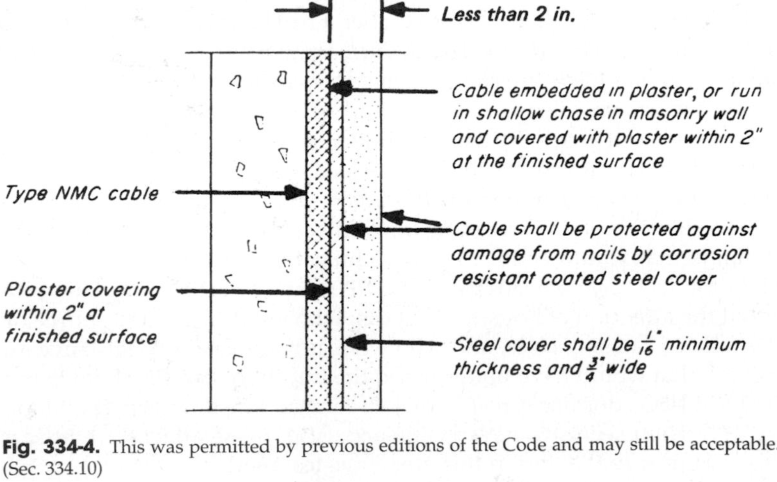
Less than 2 in.
Cable embedded in plaster, or run in shallow chase in masonry wall and covered with plaster within 2" at the finished surface
Type NMC cable
Cable shall be protected against damage from nails by corrosion resistant coated steel cover
Plaster covering within 2" at finished surface
Steel cover shall be 1/16" minimum thickness and 3/4" wide

Fig. 334-4. This was permitted by previous editions of the Code and may still be acceptable. (Sec. 334.10)

Table 334-1A Fire Resistance Ratings for Type I Through Type V Construction (h)

	Type I		Type II			Type III		Type IV	Type V	
	443	332	222	111	000	211	200	2HH	111	000
Exterior bearing walls[a]										
Supporting more than one floor, columns, or other bearing walls	4	3	2	1	0[b]	2	2	2	1	0[b]
Supporting one floor only	4	3	2	1	0[b]	2	2	2	1	0[b]
Supporting a roof only	4	3	1	1	0[b]	2	2	2	1	0[b]
Interior bearing walls										
Supporting more than one floor, columns, or other bearing walls	4	3	2	1	0	1	0	2	1	0
Supporting one floor only	3	2	2	1	0	1	0	1	1	0
Supporting roofs only	3	2	1	1	0	1	0	1	1	0
Columns										
Supporting more than one floor, columns, or other bearing walls	4	3	2	1	0	1	0	H	1	0
Supporting one floor only	3	2	2	1	0	1	0	H	1	0
Supporting roofs only	3	2	1	1	0	1	0	H	1	0
Beams, girders, trusses, & arches										
Supporting more than one floor, columns, or other bearing walls	4	3	2	1	0	1	0	H	1	0
Supporting one floor only	2	2	2	1	0	1	0	H	1	0
Supporting roofs only	2	2	1	1	0	1	0	H	1	0
Floor-ceiling assemblies	2	2	2	1	0	1	0	H	1	0
Roof-ceiling assemblies	2	1½	1	1	0	1	0	H	1	0
Interior nonbearing walls	0	0	0	0	0	0	0	0	0	0
Exterior nonbearing walls[c]	0[b]	0[b]	0[b]	0[b]	0[b]	0[b]	0[b]	0[b]	0[b]	0[b]

H: Heavy timber members

[a]See 7.3.2.1 in NFPA 5000.

[b]See Sec. 7.3 in NFPA 5000.

[c]See 7.2.3.2.12, 7.2.4.2.3, and 7.2.5.6.8 in NFPA 5000.

Source: Table 7.2.1.1 from NFPA *5000®*, *Building Construction and Safety Code®*, 2009 ed.

run for the entire length of the cable where it is run "in shallow chase." The metal strip protection must be run in the chase and then covered with the plaster or adobe (or similar) finish. But it must be carefully noted that NM, NMC, and NMS are prohibited by 334.12(A)(9) from embedment in cement, concrete, or aggregate—which is distinguished from plaster.

334.15. Exposed Work. Figure 334-5 shows the details described in parts **(A)** and **(B)** of this section. The rules of this section tie into the rules of part **(C)**, covering use in unfinished basements, which are really places of "exposed work."

As covered in part **(C)** to 334.15, cables containing Nos. 14, 12, or 10 conductors must be run through holes drilled through joists, or installed on running boards. When running parallel to joists, any cable must be stapled to the wide, vertical face of a joist and never to the bottom edge. But, as shown in Fig. 334-6,

Table 334-1B Maximum Number of Stories for Types V, IV, and III Construction

Construction Type	Maximum Number of Stories Permitted
V Rated	2
V Rated, sprinklered	3
V One-hour rated	3
V One-hour rated, sprinklered	4
IV Heavy timber	4
IV Heavy timber, sprinklered	5
III Nonrated	2
III Nonrated, sprinklered	3
III One-hour rated	4
III One-hour rated, sprinklered	5

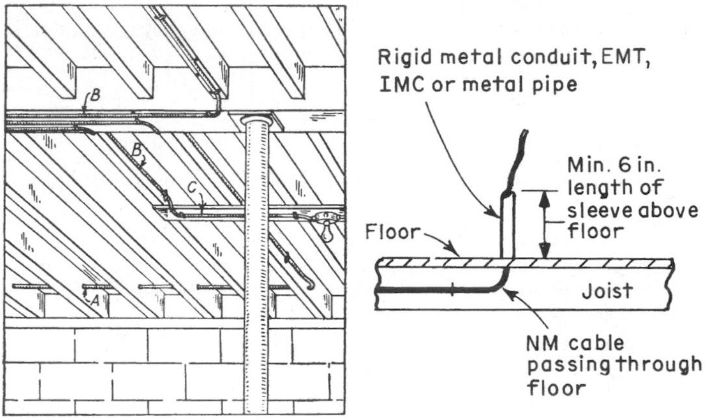
B
B
C
A
Rigid metal conduit, EMT, IMC or metal pipe
Min. 6 in. length of sleeve above floor
Floor
Joist
NM cable passing through floor

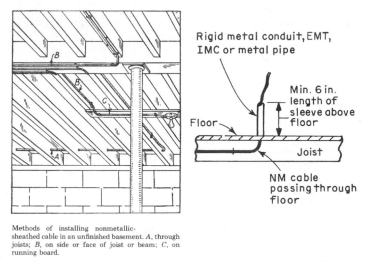

Methods of installing nonmetallic-sheathed cable in an unfinished basement. *A*, through joists; *B*, on side or face of joist or beam; *C*, on running board.

Fig. 334-5. This applies to unfinished basements and other exposed applications. (Sec. 334.15.)

larger cables may be attached to the bottom of joists when run at an angle to the joists. For mounting on the unfinished wall of a basement, there is now language in this location that allows NM cable to be sleeved in a raceway with a protective fitting at the upper end. Of course, this permission has already been in the **NEC** [300.15(C)] for over 35 years.

334.17. Through or Parallel to Framing Members. This rule reiterates requirements on 300.4, and emphasizes that the grommets required for the use of NM cable in steel-stud construction must be designed to remain in place and be listed for the purpose of cable protection. There is one additional requirement in 300.4(B)(1)

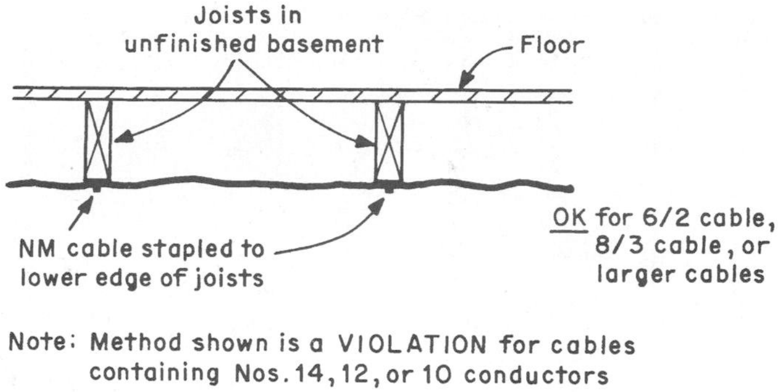
Joists in
unfinished basement
Floor
NM cable stapled to
lower edge of joists
OK for 6/2 cable,
8/3 cable, or
larger cables
Note: Method shown is a VIOLATION for cables
containing Nos. 14, 12, or 10 conductors

Fig. 334-6. Only large cables may be stapled to bottom edge of floor joists. [Sec. 334.15(C).]

not mentioned here that needs to be emphasized in this regard: for NM cable, the grommet must encompass the entire cut opening. There is a V-shaped grommet design that is open at the top; these are not acceptable for NM cable. In the process of pulling through, especially if the holes don't line up perfectly, the unprotected upper edge has been known to severely damage cables.

334.30. Securing and Supporting. Figure 334-7 shows support requirements for NM or NMC cable. Figure 334-8 shows a violation, both of this section and also 300.11(D) because cables tied together as shown amount to cables supporting cables. In concealed work the cable should, if possible, be so installed that it will be out of reach of nails. Care should be taken to avoid wherever possible the parts of a wall where the trim will be nailed in place—for example, door

Fig. 334-7. NM or NMC cables must be stapled every 4½ ft (1.4 m) where attached to the surfaces of studs, joints, and other wood structural members. It is not necessary to use staples or straps on runs that are supported by the drilled holes through which the cable is pulled. But there must be a staple within 12 in. (300 mm) of every box or enclosure in which the cable terminates. This illustrates the distinction between support intervals, which the stud drillings provide, and securement, which only the staples provide. Cable ties, if listed for "securement and support," are also permitted; their usual application is above suspended ceilings. (Sec. 334.30.)

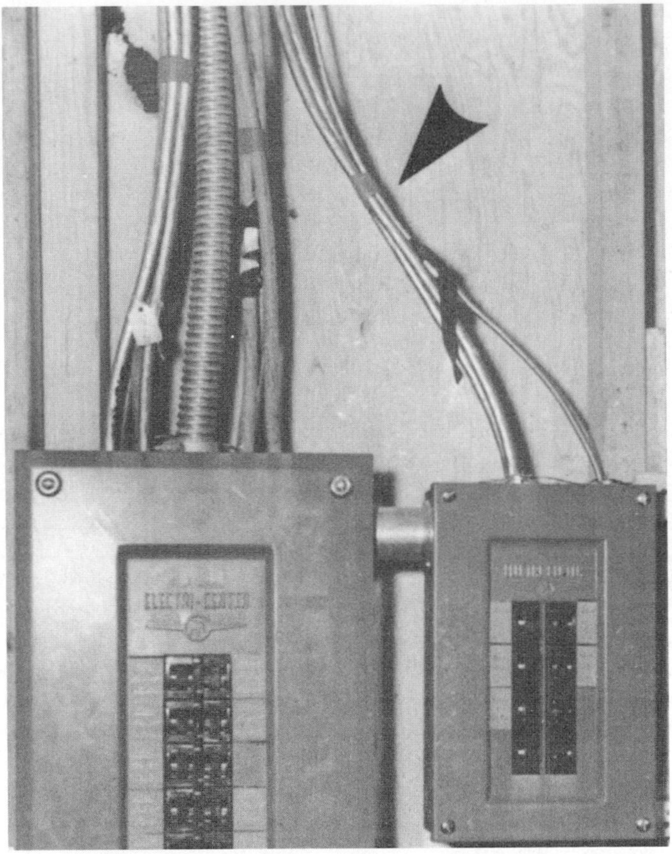

Fig. 334-8. Absence of stapling of the NM cables within 12 in. (300 mm) of entry into the panelboards is a clear violation of Sec. 334.15. (Sec. 334.30.) Support of one cable from another also violates 300.11(D).

and window casings, baseboards, and picture moldings. See 300.4. NM cable is also permitted to be fished, and run in suspended ceilings [residential only per 334.12(A)(2)] with up to 1.4 m (4½ ft) of unsupported cable to a luminaire.

Connectors listed for use with Type NM, NMS, or NMC cable (nonmetallic-sheathed cable) are also suitable for use with flexible cord or service-entrance cable *if* such additional use is indicated on the device or carton. Connectors listed under the classifications "Armored Cable Connectors" and "Conduit Fittings" may be used with nonmetallic-sheathed cable when that is specifically indicated on the device or carton. Connectors for NM, NMS, or NMC cable are also suitable for use on Type UF cable (underground feeder and branch-circuit cable—**NEC** Art. 340) in dry locations, unless otherwise indicated on the carton. Each connector covered in the listing is recognized for connecting only one cable or cord—unless it is a duplex connector for connecting two cables or if the carton is marked to indicate use with more than one cable or cord.

Part **(C)** covers a wiring device configured with a self-contained enclosure that does not require a separate outlet or device box. There must be not less than 300 mm (1 ft) of an unbroken cable loop (or 150 mm [6 in.] of free cable ends) left at the opening so the device can be serviced. This application is also covered in 300.15(E), and in the majority of cases the result is a flush installation in a building wall. They are also noted in 334.40(C).

334.40. Boxes and Fittings. In part **(A)**, the Code presents requirements for "Boxes of Insulating Material." By using nonmetallic outlet and switch boxes, a completely nonmetallic wiring system is provided. Such a system has economic and other advantages in locations where corrosive vapors are present. See 314.3.

In 334.40(B), "Devices of Insulating Material," note that use of switch and outlet devices without boxes is limited to exposed cable systems and for repair wiring in existing buildings. These are primarily surface-mounted devices also covered in 300.15(H). They were far more common about 50 years ago, and were commonly used to provide snap switches and receptacles in summer cottages with NM cable surface mounted on unfinished framing. Unlike the surface-mounted switches and receptacles of many decades ago, the "interconnector" devices are routinely used today. The terminology and related applications were revisited in the 2014 **NEC** cycle in order to address the permitted uses of these concealable NM cable splicing devices that have been recognized in 545.13 for decades. They are recognized here without boxes for concealed "repair wiring" only. This reference must not be confused with that of subpart **(C)** as given in 334.30, which refers to approved wiring devices that incorporate their own wiring boxes, so they are devices "without a *separate* outlet box" and not devices "without boxes."

334.80. Ampacity. The second sentence requires that NM, NMS, and NMC cables always have their conductors applied to the ampacity of Type TW wire—that is, the 60°C ampacity from Table 310.15(B)(16). However, the insulation on the conductors must be rated at 90°C. This provides an additional margin for error and helps prevent overheating of Type NM where installed, say, in the attic of a residential occupancy located in the New Mexico desert. The ambient temperature would soar in the summer, unless the attic was air conditioned.

The last paragraph of 334.80 correlates the use of NM cable to the rule of 310.15(B)(3), which says:

> ...where single conductors or multiconductor cables are stacked or bundled longer than 24 in. without maintaining spacing and are not installed in raceways, the allowable ampacity of each conductor shall be reduced as shown...

Bundled NM, NMS, or NMC cables will require ampacity derating in accordance with 310.15(B)(3) when the whole bundle is tightly packed, thereby losing the ability of the inside cables to dissipate the heat generated in them. An example of this is shown in Fig. 310-20. This is true of NM cables as well as any other cables. And the derating percentage from the table in 310.15(B)(3)(a) must be based on the total number of insulated conductors in the group. For instance, fourteen 3-wire 14 AWG cables would have to be ampacity derated to 35 percent of the conductor ampacity [$14 \times 3 = 42$ conductors at 35 percent, from Table 310.15(B)(3)(a)], which, at 9 A ($25\text{ A} \times 0.35 = 9\text{ A}$) would immediately disqualify them from use, particularly for receptacle circuits. Also note that since this rule now goes directly to Table 310.15(B)(3)(a) and bypasses the heat-sink exception, there is no way around this except for drilling more holes. The same issue applies to the draft-stop limitation following, where if ever more than two NM cables are draft- (or fire-) stopped in the same framing hole, regardless of the length of penetration, the table derating factors apply.

The wording of this section understates the problem for installations that are embedded in thermal insulation. By allowing derating to occur from the 90°C column for such cables, the NEC allows users to ignore factors that increase the retention of heat in the cable until they reach the point of reducing the ampacity below the 60°C value, which in most cases is to allow them to be ignored altogether. Meanwhile, particularly for large, heavily loaded cables, even the 60°C numbers are too high. Review the discussion associated with 310.15(A)(3) for detailed information on this topic.

ARTICLE 336. POWER AND CONTROL TRAY CABLE: TYPE TC

336.2. Definition. This article covers the use of a nonmetallic-sheathed power and control cable, designated Type TC cable (TC is the abbreviation for *tray cable*) (Fig. 336-1). The cable is an assembly of two or more insulated conductors, with or without associated bare or covered grounding conductors, under a nonmetallic jacket. Type TC cable must be listed (336.6). It is a cable that can be used for power and lighting applications, as well as control and signaling functions, including Class 1 control circuits as well as non-power-limited fire alarm wiring.

TRAY CABLE (UL)

Fig. 336-1. This typical 3-conductor tray cable contains bare equipment grounding conductors. (Sec. 336.2.)

336.10. Use Permitted; 336.12. Uses Not Permitted. It is not permitted to be made with a metallic cable armor either under or over the nonmetallic jacket; however, metal shielding is permitted, and where employed, the minimum bending radius of the cable must not be less than 12 times the cable diameter. Otherwise, the minimum allowable bend radius is based on the cable diameter, with 4 times the diameter allowed for cables 25 mm (1 in.) or thinner, 5 times for cables up to 50 mm (2 in.), and 6 times for larger cables. In addition to cable tray usage, Type TC cable is permitted to run within raceways and on messenger wires, but for outdoor applications in direct sun it must be identified for that use. It may be directly buried, but also only if so identified, and it may also be used in wet locations if made resistant to moisture and corrosive agents.

Type TC cable is available with a more robust configuration [covered in (10)(7)] that will meet the crush and impact tests that apply to metal-clad cable, Type MC. This form is identified as Type TC-ER. For industrial occupancies with qualified maintenance and supervision, Type TC-ER is permitted to exit a cable tray and run to utilization equipment or devices, provided also that it has continuous mechanical support, such as strut, angles, or channels. In addition, if not threatened with physical damage, Type TC-ER is permitted to run unsupported between cable tray transitions, or from cable trays to utilization equipment or devices as long as the unsupported distance does not exceed 1.8 m (6 ft). Where the cable exits the tray, mechanical support must be provided so the required bend radius is maintained.

The 2017 NEC, responding to a input from the standby generator industry, now recognizes a new style of TC-ER tray cable, permitted in one- and two-family dwelling units. It is made up of both power and control conductors, and must carry an additional designation "JP" (which stands for joist pull, and means the cable assembly has undergone the UL cold environment pull through joists test). It must generally meet the installation requirements for Type NM cable; however, where used to run between a permanent standby generator and the transfer equipment, it avoids the ampacity limits in 334.80. Some varieties of this cable carry the control wiring conductors cabled with power conductors, but within their own nonmetallic jacket that meets the construction requirements of Type NM cable. This effectively makes the arrangement acceptable in terms of system separation rules. Refer to the discussion at 725.136(I) for details.

UL data on "Power and Control Tray Cable" include the following:

Power and Control Tray Cable (QPOR)

This category covers Type TC power and control tray cable intended for use in accordance with Art. 336 of ANSI/NFPA 70, "National Electrical Code" (NEC). The cable consists of one or more pairs of thermocouple extension wires or two or more insulated conductors, with or without one or more grounding conductors, with or without one or more optical fiber members and covered with a nonmetallic jacket. A single grounding conductor may be insulated or bare and may be sectioned. Any additional grounding conductor is fully insulated and has a distinctive surface marking. The cable is rated 600 or 2000 V.

The cable is certified in conductor sizes 18 AWG to 1000 kcmil copper or 12 AWG to 1000 kcmil aluminum or copper-clad aluminum. Conductor sizes within a cable may be mixed. Thermocouple extension conductors are certified in sizes 24 to 12 AWG.

PRODUCT MARKINGS

Cable with copper-clad aluminum conductors is surfaced printed "AL (CU-CLAD)" or "Cu-clad Al."

Cable with aluminum conductors is surface printed "AL."

Cable employing compact-stranded copper conductors is so identified directly following the conductor size, wherever it appears (surface, tag, carton, or reel), by "compact copper." The abbreviations "CMPCT" and "CU" may be used for compact and copper, respectively.

Tags, reels, and cartons for products employing compact-stranded copper conductors have the marking: "Terminate with connectors identified for use with compact-stranded copper conductors." For termination information, see Electrical Equipment for Use in Ordinary Locations (AALZ).

If the type designation of the conductors is marked on the outside surface of the cable, the temperature rating of the cable corresponds to the rating of the individual conductors. When this marking does not appear, the temperature rating of the cable is 60°C unless otherwise marked on the surface of the cable.

Cable investigated for use where exposed to direct rays of the sun is marked "sunlight resistant."

Cable investigated for direct burial in the earth is so identified.

Cable suitable for use between cable trays and utilization equipment in accordance with NEC 336.10(7) is surface marked with the suffix "-ER."

Cable consisting of thermocouple extension wires is surface marked "THCPL EXTN," "For thermocouple extension use only" or "Thermocouple extension wire only."

Cable surface marked "Oil Resistant I" or "Oil Res I" is suitable for exposure to mineral oil at 60°C. Cable suitable for exposure to mineral oil at 75°C is surface marked "Oil Resistant II" or "Oil Res II."

Cable that complies with the Limited Smoke Test requirements specified in UL 1685, "Vertical-Tray Fire-Propagation and Smoke-Release Test for Electrical and Optical-Fiber Cables," is surface marked with the suffix "-LS."

Cable containing optical fiber members is identified with the suffix "-OF."

Regarding cable seals outlined in Art. 501 of the **NEC**, Type TC cable has a sheath that is considered to be gas/vapor tight but the cable has not been investigated for transmission of gases or vapors through its core.

RELATED PRODUCTS

Connectors and fittings for use with this cable are covered under Power and Control Tray Cable Connectors (QPOZ).

Some connectors and fittings covered under Outlet Bushings and Fittings (QCRV), Nonmetallic-sheathed Cable Connectors (PXJV), and Service Entrance Cable Fittings (TYZX) are also suitable for use with this cable when specifically marked on the device or carton.

ADDITIONAL INFORMATION

For additional information, see Electrical Equipment for Use in Ordinary Locations (AALZ).

REQUIREMENTS

The basic standard used to investigate products in this category is ANSI/UL 1277, "Electrical Power and Control Tray Cables with Optional Optical-Fiber Members."

UL MARK

The Certification Mark of UL on the attached tag, reel, or on the smallest unit package in which the product is packaged is the only method provided by UL to identify products manufactured under its Certification and Follow-Up Service. The Certification Mark for these products includes the UL symbol, the words "CERTIFIED" and "SAFETY," the geographic identifiers, and s file number.

ALTERNATE UL MARK

The Listing Mark of Underwriters Laboratories Inc. on the product is the only method provided by UL to identify products manufactured under its Listing and Follow-Up Service. The Listing Mark for these products includes the UL symbol (as illustrated in the Introduction of this Directory) together with the word "LISTED," a control number, and the product name as appropriate: Power and control tray cable that contains copper or copper-clad aluminum conductors has the product name "Power and Control Tray Cable Type TC"; power and control tray cable that contains aluminum conductors has the product name "Aluminum Power and Control Tray Cable Type TC."

Note that this cable appears to be for cable tray only, but Type TC is recognized by 336.2 for use in raceway or with messenger support, in addition to use in tray.

Although item **(4)** of 336.12 has the effect of prohibiting the use of Type TC tray cable directly buried in the earth, the rule is modified by the phrase "unless identified for such use." The result of this wording is to permit Type TC cable to be directly buried in the earth where the cable is marked or otherwise generally recognizable as suitable for this use. The product listing information (above) directly recognizes this possibility. This permission for direct burial was added

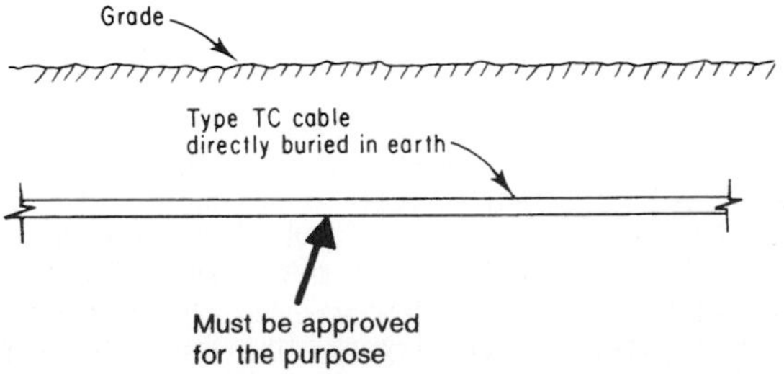
Grade
Type TC cable
directly buried in earth
Must be approved
for the purpose

Fig. 336-2. Type TC (power and control tray cable) is recognized for direct earth burial. (Sec. 336.12.)

because the cable assembly was designed to withstand such application and because Type TC cable has been successfully and effectively used directly for years in many installations (Fig. 336-2 with burial conforming to 300.5). Such cable is listed for direct earth burial by UL, and the performance record has been excellent.

ARTICLE 338. SERVICE-ENTRANCE CABLE: TYPES SE AND USE

338.2. Definition. The Code contains few specifications for the construction of this cable; it is required to be listed (338.6) and beyond a few general specifications in Part III of the article, it is left to Underwriters Laboratories Inc. to determine what types of cable should be approved for this purpose. The types listed by the Laboratories conform to the following data:

> This category covers service-entrance cable (TYLZ) designated Type SE and Type USE for use in accordance with Art. 338 of ANSI/NFPA 70, "National Electrical Code" (NEC).
>
> Service-entrance cable, rated 600 V, is Listed in sizes 14 AWG and larger for copper, and 12 AWG and larger for aluminum or copper-clad aluminum.
>
> The cable is designated as follows:
>
> Type SE—Indicates cable for aboveground installation. Both the individual insulated conductors and the outer jacket or finish of Type SE are suitable for use where exposed to sun. Type SE cable contains Type RHW, RHW-2, XHHW, XHHW-2, THWN, or THWN-2 conductors. Maximum size is 4/0 AWG copper or 300 kcmil aluminum or copper-clad aluminum.
>
> Types USE and USE-2—Indicates cable for underground installation including direct burial in the earth. Maximum size is 2000 kcmil. Cable in sizes 4/0 AWG copper, aluminum, or copper-clad aluminum, and smaller and having all conductors insulated is suitable for all of the underground uses for which Type UF cable is permitted by the NEC. Multiconductor Type USE cable contains conductors with insulation equivalent to RHW or XHHW. Multiconductor Type USE-2 contains insulation equivalent to RHW-2 or XHHW-2 and is rated 90°C wet or dry. Single- and multiconductor Types USE and USE-2 are not suitable for use in premises. Single- and multiconductor Types USE and USE-2 are not suitable aboveground except to terminate at the service equipment or metering equipment. Both the insulation and the outer covering, when used, on single- and multiconductor Types USE and USE-2, are suitable for use where exposed to sun.
>
> Submersible Water Pump Cable—Indicates a multiconductor cable in which 2, 3, or 4 single-conductor Types USE or USE-2 cables are provided in a flat or twisted assembly. The cable is certified in sizes 14 AWG to 4/0 AWG inclusive, copper, and 12 AWG to 4/0 AWG inclusive, aluminum or copper-clad aluminum. The cable is tag marked "For use within the well casing for wiring deep-well water pumps where the cable is not subject to repetitive handling caused by frequent servicing of the pump units." The insulation may also be surface marked "Pump Cable." The cable may be directly buried in the earth in conjunction with this use.
>
> For termination information, see Electrical Equipment for Use in Ordinary Locations (AALZ).
>
> Based upon tests which have been made involving the maximum heating that can be produced, an uninsulated conductor employed in a service cable assembly is considered to have the same current-carrying capacity as the insulated conductors even though it may be smaller in size.

Figure 338-1 shows two basic styles of service-entrance cable for aboveground use. The one without an armor over the conductors is referred to as Type SE Style U—the letter U standing for "unarmored." That cable is sometimes designated as Type SEU. The cable assembly with the armor is designated Type ASE cable, with the A standing for "armored."

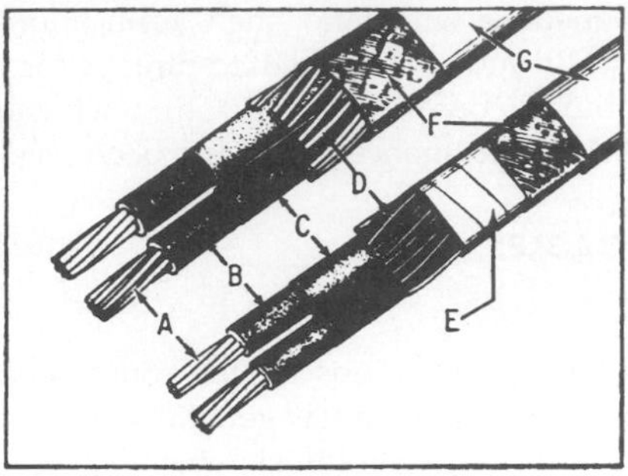
G
F
D
C
B
A
E

SERVICE ENTRANCE cables may consist of either copper or aluminum phase conductors (A) covered by heat resistant insulation (B) and moisture-resistant braid or tape (C) color coded for circuit identification, while basic assembly is enclosed by concentric neutral (D). Unarmored Type SE Style U is covered by variety of tapes (F) and outer braid (G) such as glass and cotton impregnated with moisture resistant and flame retardant finish labelled with pertinent data. Armored Style A additionally contains flat steel armor (E) as protection against physical abuse.

Fig. 338-1. Two types of aboveground SE cable. (Sec. 338.2.)

Each phase leg is an insulated conductor

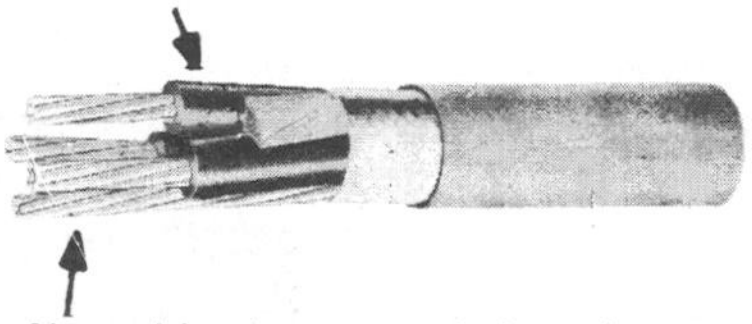

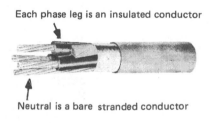

Neutral is a bare stranded conductor

Fig. 338-2. Style SER cable contains individual conductors and no concentric neutral. (Sec. 338.2.)

Figure 338-2 shows another type of SE cable, known as Style SER—the letter R standing for "round." In a typical assembly of that cable, three conductors insulated with Type XHHW cross-link polyethylene are cabled together with fillers and one bare ground conductor with a tape over them and gray PVC overall jacket. For use aboveground in buildings, it is suitable for operation at 90°C in dry locations or 75°C in wet locations unless the insulation designation has the suffix "-2" in which case it is suitable for 90°C for both locations.

For 3-phase, 4-wire grounded services, the three insulated conductors—a black, a red, and a blue—are used as the phase legs of the service, and the bare conductor is used as the neutral. For other applications that cable configuration can be used as a feeder or branch circuit to a power load with a bare equipment grounding conductor and this form of cable is also available with black, red, and white insulated conductors for single-phase applications, including the supply of downstream panels, and black, red, blue, and white 5-wire configurations for 3-phase equivalent loads.

Fig. 338-3. Type USE cable may be multiconductor or single-conductor cable. (Sec. 338.2.)

Figure 338-3 shows multiconductor Type USE cable for underground (including direct earth burial) applications of service or other circuits. Type USE may consist of one, two, or three conductors. If it has no building wire designation, its insulation will burn, and it is absolutely excluded from any interior wiring whatsoever by 338.12(B)(1). Where used aboveground it must terminate where it emerges from the ground, such as at a meter socket, as covered in 338.12(B)(2). It is recognized for use as aerial cable, provided it is in the multiconductor form, identified as suitable for aboveground use, and run on a messenger. Some cables have both a "USE" designation and also a building wire designation, such as "RHW." This cable can continue into a building. The basic temperature rating for this cable is 75°C and that limit applies unless a different number is marked on the cable.

Depending on whether USE cable is used for service entrance, for a feeder, or for a branch circuit, burial depth must conform to 300.5 and its many specific rules on direct burial cable.

338.10. Uses Permitted. As would be expected, where used as service conductors, Type SE must comply with Art. 230, "Services."

Part **(B)** recognizes use of service-entrance cable for branch circuits and feeders within buildings or structures, provided that all circuit conductors, including the neutral of the circuit, are insulated. Such use must conform to Art. 334 on installation methods—the same as those for Type NM cable.

Part **(B)(2)** covers, in the exception, the permitted uses of service-entrance cable that contains a bare conductor for the neutral, of which there are only two. This exception permits the use of SE cable for circuits supplying existing ranges, wall-mounted ovens, and countermounted cooking units (Fig. 338-4). And in such cases, the bare conductor may be used as the neutral of an existing branch circuit as well as the equipment grounding conductor (see 250.140).

SE cable is also permitted to be used as a feeder from one building to another building, with the bare conductor used as a grounded neutral, but now only for an existing premises wiring system in accordance with 250.32(B)(1) Exception No. 1. Or an SE cable with a bare neutral may be used as a feeder within a building, if the bare neutral is used *only* as the equipment grounding conductor and one of the insulated conductors within the cable is used as the neutral of the feeder. (See Fig. 338-5.)

Part **(B)(3)** requires that SE cable used to supply appliances not be subject to conductor temperatures in excess of the temperature specified for the insulation involved. The insulated conductors of SE cables are 75° or 90°C, and if they are rated at 90°C, such marking will appear on the outer sheath. A cooking unit or oven that requires 90°C supply conductors would be an application for the use of SE cables, rated at 90°C. However, a review of UL listings for cooking units and ovens indicates that most such units do not require supply conductor ratings to exceed 60°C. The details in Fig. 338-4 show a method of connecting cooking units where the supply conductors are required to be 75° or 90°C.

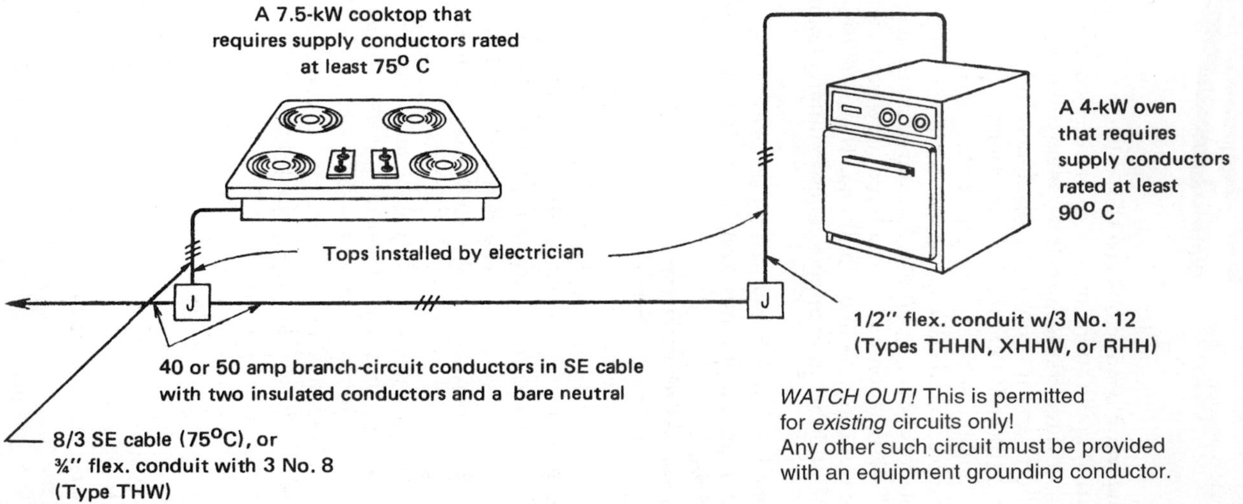
A 7.5-kW cooktop that requires supply conductors rated at least 75° C
Tops installed by electrician
A 4-kW oven that requires supply conductors rated at least 90° C
J
J
1/2'' flex. conduit w/3 No. 12 (Types THHN, XHHW, or RHH)
40 or 50 amp branch-circuit conductors in SE cable with two insulated conductors and a bare neutral
8/3 SE cable (75°C), or ¾'' flex. conduit with 3 No. 8 (Type THW)
WATCH OUT! This is permitted for existing circuits only!
Any other such circuit must be provided with an equipment grounding conductor.

Fig. 338-4. Although previously permitted, SE cable with bare neutral may be used for branch circuits to a range or other cooking units, but *only* on "existing installations." (Sec. 338.10.)

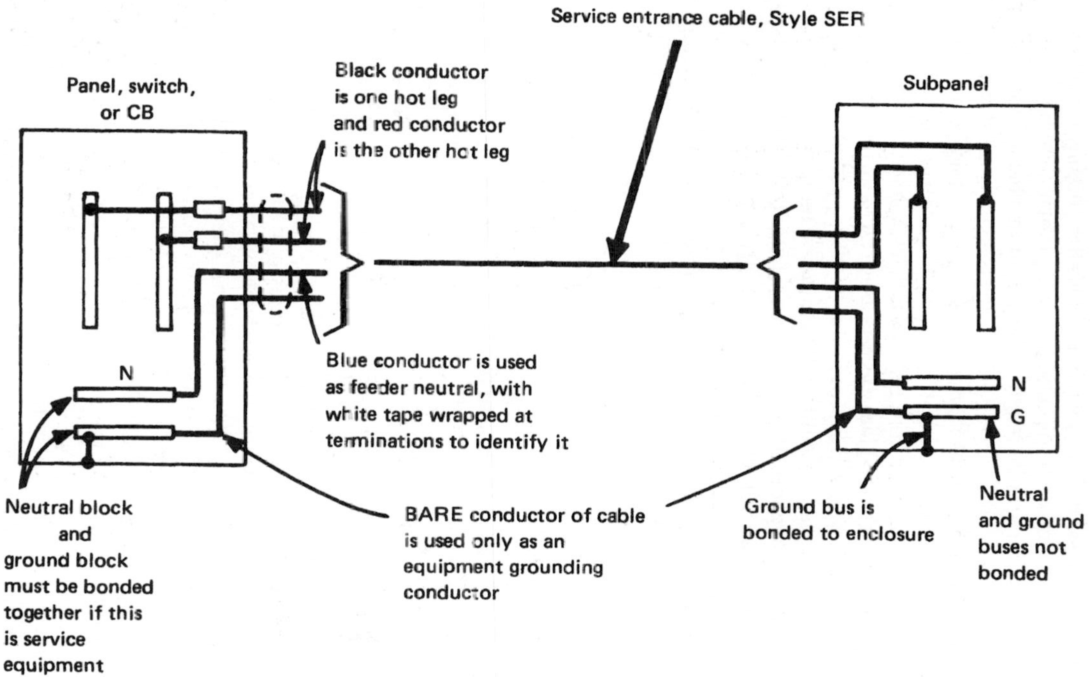
Service entrance cable, Style SER
Panel, switch, or CB
Black conductor is one hot leg and red conductor is the other hot leg
Subpanel
N
Blue conductor is used as feeder neutral, with white tape wrapped at terminations to identify it
N
G
Neutral block and ground block must be bonded together if this is service equipment
BARE conductor of cable is used only as an equipment grounding conductor
Ground bus is bonded to enclosure
Neutral and ground buses not bonded

Fig. 338-5. Typical application of SE cable with a bare neutral for use as a feeder within a building. (Sec. 338.10.)

In part **(B)(4)(a),** "Interior Installation," Type SE cable is permitted for use as interior branch circuits or feeders, but where so used, the installation must satisfy all of the general wiring rules of Art. 300. This section requires the installation of unarmored SE cable (which is the usual type of SE cable) to satisfy Art. 334 on nonmetallic-sheathed cable (Type NM). All the rules of Art. 334 that cover *how* cable is installed must be satisfied. This notably includes the building construction limitations in 334.10, and remember the discussion at that point in this book; these are limits based on what the building code requires the minimum characteristics of the buildings to be, and not necessarily on how the building is actually constructed. As a result, the use of SE cable as a feeder, as shown in Fig. 338-5, would be a violation in any building required to be constructed of Type I or II construction. There is, however, one notable point of separation from the two articles, namely, the incorporation of the ampacity provisions in 334.80 into this article by reference has been discontinued.

The wording that replaces it restores the effective permission for this cable to be applied at the rating of its internal conductor temperature ratings (either 75°C or 90°C depending on the specific construction). This means that, for example, 4/0 AWG Type SER cable used as a feeder indoors is once again a 180 A conductor (after applying termination limitations). Since very few feeders run in the window between 180 and 200 A, this cable goes back to being the customary feeder it had been for decades, protected at the next higher standard size, or 200 A. And in instances where it carried the entire load of a dwelling occupancy, 310.15(B)(7), makes it a 200 A feeder outright.

This allowance is tempered by a crucial limitation, however, one that is squarely based on actual experimental results. Review the discussion in this book regarding 310.15(A)(3) and the performance of Type SE cable embedded in thermal insulation. To partially respond to that experience, under the 2014 **NEC**, if this cable were embedded in thermal insulation, its final ampacity would have been that of a 60°C conductor. In the case of 4/0 AWG cables embedded in insulation, under these conditions this was a 150 A cable, and no higher. However, the 2017 **NEC** has gutted that limitation, now confining it to 10 AWG and smaller, sizes almost never used.

In fact, even the previous wording actually somewhat understated the true problem for cables run in thermal insulation. By allowing derating to occur from the 90°C column for such cables, the **NEC** allows users to ignore factors that increase the retention of heat in the cable until they reach the point of reducing the ampacity below the 60°C value, which in most cases is to allow them to be ignored altogether. Meanwhile, particularly for large, heavily loaded cables, even the 60°C numbers are too high. Again, review the discussion associated with 310.15(A)(3) for detailed information on this topic. For example, 1/0 AWG aluminum SE cable has a 75° limit for terminations, which is 120 A in Table 310.15(B)(16). Since very few feeders run in the window between 120 and 125 A for actual calculated load, this cable has been sold by the mile for 125 A services and, in SER configurations, as feeder cable to apartments and commercial 3-phase, 4-wire feeder cable to stores and in restaurants, etc. If run embedded in thermal insulation, this cable is now 100 A cable, period. However, the true ampacity of this cable where actually embedded in thermal insulation is probably less than 70 A and only a diligent application of 310.15(A)(3) stands in the way.

The ampacity issues regarding Type SE cables generally do not apply to underground applications, and the blanket application of ampacity limitations on Type UF cables in 340.80 need not be applied to Type USE cables. An exception, new as of the 2014 **NEC**, limits the reach of the final sentence in 338.10(B)(4)(b) by waiving those limitations. Note that the wording includes the term "multi-rated USE conductors." This term is undefined, but based on the documentation connected to the change, and the reference is intended to describe Type USE cable that also has a building wire rating, such as USE-2/RHW-2 or the like.

ARTICLE 340. UNDERGROUND FEEDER AND BRANCH-CIRCUIT CABLE: TYPE UF

340.2. Definition. Figure 340-1 shows a violation of the Code rule that a bare conductor in a UF cable is for grounding purposes only. Type UF cable must be listed.

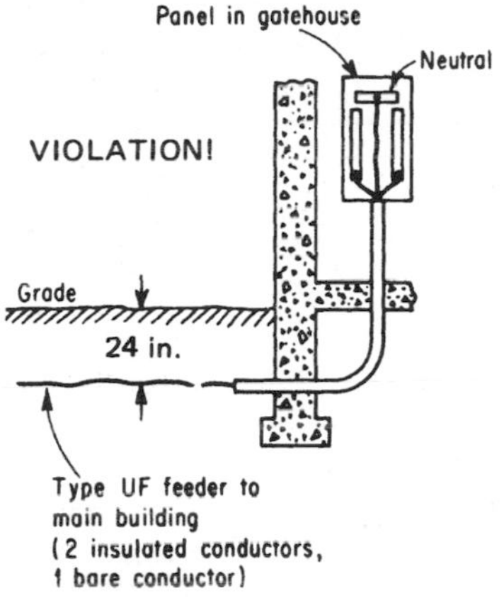
Panel in gatehouse
Neutral
VIOLATION!
Grade
24 in.
Type UF feeder to main building (2 insulated conductors, 1 bare conductor)

Fig. 340-1. Bare conductor in UF cable may not be used as a neutral. (Sec. 340.2.)

340.10. Uses Permitted. As called for in part **(1)**, Figs. 340-2 and 340-3 show details on compliance of UF cable with 300.5. Where UF comes up out of the ground, it must be protected for 8 ft up on a pole and as described in 300.5(D).

The rules of part **(1)** are shown in Fig. 340-4 and must be correlated to the rules of 300.5 on direct burial cables. The rule of **(2)** corresponds to that 300.5(I). If multiple conductors are used per phase and neutral to make up a high-current circuit, this rule requires all conductors to be run in the same trench or raceway and therefore subject to the derating factors of 310.15(B)(3). Refer to the paragraph covering 310.15(B)(3). Also see discussion under 300.5(I).

UF cable may be used underground, including direct burial in the earth, as feeder or branch-circuit cable when provided with overcurrent protection not in excess of the rated ampacity of the individual conductors. If single-conductor cables are installed, all cables of the feeder circuit, subfeeder, or branch circuit,

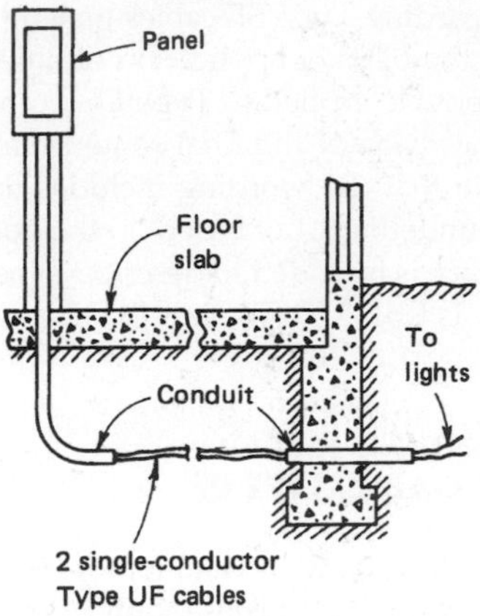
Panel
Floor
slab
To
lights
Conduit
2 single-conductor
Type UF cables

VIOLATION for cable to run under any building if not totally in raceway [300.5]

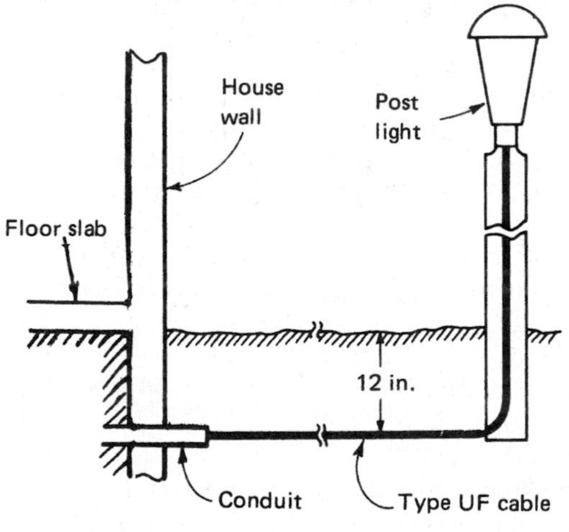
House
wall
Post
light
Floor slab
12 in.
Conduit
Type UF cable

ACCEPTABLE for burial at only 12-in. depth when used for residential branch-circuit rated not more than 20 A at 120 V or less where GFCI protection is also provided [Table 300.5]

Fig. 340-2. UF cable must conform to Sec. 300.5 on direct burial cables. This includes the requirement in 300.5(C), illustrated here, of sleeving the cable in a raceway where it runs under a building. (Sec. 340.10.)

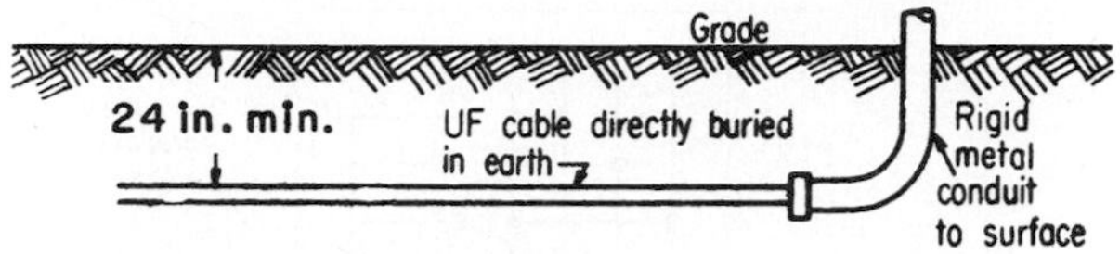
Grade
24 in. min.
UF cable directly buried in earth
Rigid metal conduit to surface
① UF CABLE WITHOUT SUPPLEMENTAL PROTECTION

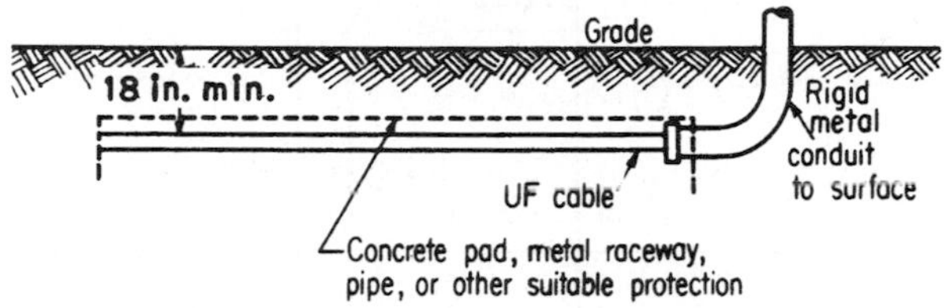
Grade
18 in. min.
Rigid metal conduit to surface
UF cable
Concrete pad, metal raceway, pipe, or other suitable protection
② UF CABLE WITH SUPPLEMENTAL PROTECTIVE COVERINGS

Fig. 340-3. The second qualifier under "Location of Wiring Method or Circuit" in Table 300.5 permits a 6-in. reduction of UF burial depth. The concrete must be above the cable, without embedding it. (Sec. 340.10.)

including the neutral cable, must be run together in close proximity in the same trench or raceway. It may be necessary in some installations to provide additional mechanical protection, such as a covering board, concrete pad, raceway, or the like, when required by the authority enforcing the Code. Multiple-conductor Type UF cable (but not single-conductor Type UF cables) may also be used for interior wiring when used in the same way as Type NM cable, complying with the provisions of Art. 334 of the Code. And UF may be used in wet locations.

The effect of the wording in part **(4)** where UF cable is used for interior wiring, is to require that its conductors must be rated at 90°C, with loading based on 60°C ampacity. This rule is a follow-up to the requirement that UF for interior wiring must satisfy the rules of Art. 334 on nonmetallic-sheathed cable (see 334.80). It is also subject to the construction type limitations in 334.10.

As noted in **340.12(8)**, single-conductor Type UF cable embedded in poured cement, concrete, or aggregate may be used for nonheating leads of fixed electric space heating cables, as covered in 424.43.

Application data (Category YDUX) of the UL are as follows:

> This category covers underground feeder and branch circuit cable, rated 600 V, in sizes 14 to 4/0 AWG inclusive, copper, and 12 to 4/0 AWG inclusive, aluminum or copper-clad aluminum, for single and multiple conductor cables. It is designated as Type UF cable and is intended for use in accordance with Art. 340 of ANSI/NFPA 70, "National Electrical Code" (NEC).
>
> Some multiconductor cable is surface marked with the suffix "B" immediately following the type letters to indicate the usage of conductors employing 90°C rated insulation.
>
> Such cable may also be installed as nonmetallic-sheathed cable, per Section 340.10(4) of the NEC. The ampacities of Type UF cable, with or without the suffix "B," are those of 60°C rated conductors as specified in the latest edition of the NEC.

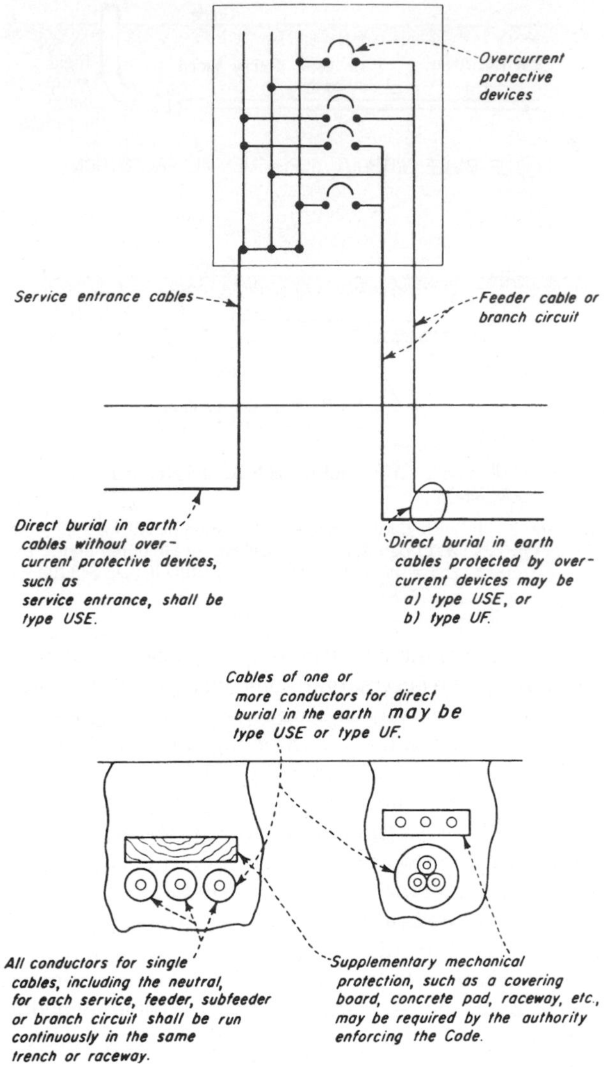
Overcurrent protective devices
Service entrance cables
Feeder cable or branch circuit
Direct burial in earth cables without over-current protective devices, such as service entrance, shall be type USE.
Direct burial in earth cables protected by over-current devices may be a) type USE, or b) type UF.
Cables of one or more conductors for direct burial in the earth may be type USE or type UF.
All conductors for single cables, including the neutral, for each service, feeder, subfeeder or branch circuit shall be run continuously in the same trench or raceway.
Supplementary mechanical protection, such as a covering board, concrete pad, raceway, etc., may be required by the authority enforcing the Code.

Fig. 340-4. UF cable may be used only as feeders or branch circuits. (Sec. 336.10.)

Submersible Water Pump Cable—Indicates multiconductor cable in which 2, 3, or 4 single-conductor Type UF cables are provided in a flat or twisted assembly. The cable is listed in sizes from 14 AWG to 4/0 AWG inclusive, copper, and from 12 AWG to 4/0 AWG inclusive, aluminum or copper-clad aluminum. The cable is tag marked "For use within the well casing for wiring deep well water pumps where the cable is not subject to repetitive handling caused by frequent servicing of the pump units." The insulation may also be surface marked "Pump Cable." The cable may be directly buried in the earth in conjunction with this use.

This cable may employ copper, aluminum, or copper-clad aluminum conductors. Cable with copper-clad aluminum conductors is surface printed "AL (CU-CLAD)" or "Cu-Clad Al." Cable with aluminum conductors is surface printed "AL."

Cable employing compact-stranded copper conductors is so identified directly following the conductor size wherever it appears (surface, tag, carton, or reel) by "compact copper." The abbreviations "CMPCT" and "CU" may be used for compact and copper, respectively.

Tags, reels, and cartons for products employing compact-stranded copper conductors have the marking: "Terminate with connectors identified for use with compact-stranded copper conductors." For conductor termination information, see Electrical Equipment for Use in Ordinary Locations (AALZ).

This cable may be terminated at boxes and other enclosures by using nonmetallic-sheathed cable connectors (see Nonmetallic-Sheathed Cable Connectors [PXJV]).

Cable suitable for exposure to direct rays of the sun is indicated by tag marking and marking on the surface of the cable with the designation "Sunlight Resistant."

Only multiconductor Type UF cable may be used in cable tray, in accordance with 340.10(7).

ARTICLE 342. INTERMEDIATE METAL CONDUIT: TYPE IMC

This article covers a conduit with wall thickness less than that of rigid metal conduit but greater than that of EMT. Called "IMC," this intermediate metal conduit uses the same threading method and standard fittings for rigid metal conduit and has the same general application rules as rigid metal conduit. Intermediate metal conduit actually is a lightweight rigid steel conduit which requires about 25 percent less steel than heavy-wall rigid conduit. Acceptance into the Code was based on a UL fact-finding report which showed through research and comparative tests that IMC performs as well as rigid steel conduit in many cases and surpasses rigid aluminum and EMT in most cases. The wiring method must be listed (Sec. 342.6).

IMC may be used in any application for which rigid metal conduit is recognized by the **NEC**, including use in all classes and divisions of hazardous locations as covered in 501.4, 502.4, and 503.3. Its thinner wall makes it lighter and less expensive than standard rigid metal conduit, but it has physical properties from differences in the alloy that give it outstanding strength. The lighter weight facilitates handling and installation at lower labor units than rigid metal conduit. Because it has the same outside diameter as rigid metal conduit of the same trade size, it has greater interior cross-sectional area (Fig. 342-1). In the past this extra space was not recognized by the **NEC** to permit the use of more conductors than can be used in the same size of rigid metal conduit. However, with the elimination of Tables 2, 3A, 3B, and 3C, as well as the revisions of Tables 4 and 5 to more correctly reflect the interior area of raceways and the dimensions of conductors, the Code does permit greater fill in IMC (see Tables C4 and C4A in Annex C).

342.10. Uses Permitted. The data of the UL supplement (Category DYBY) the requirements of part **(A)** on use of IMC, as follows:

3/4" TRADE SIZES

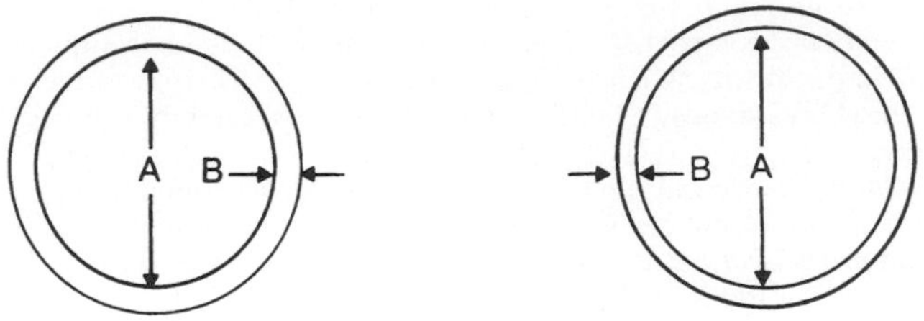
A B
B A

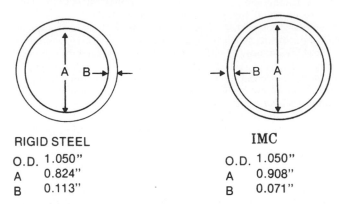

RIGID STEEL		IMC	
O.D.	1.050"	O.D.	1.050"
A	0.824"	A	0.908"
B	0.113"	B	0.071"

Fig. 342-1. Typical comparison between rigid conduit and IMC shows interior space difference. This is now recognized by Tables C4 and C4A in Annex C. (Sec. 342.1.)

Listing of Intermediate Ferrous Metal Conduit includes standard 10 ft lengths of straight conduit, with a coupling, special length either shorter or longer, with or without a coupling for specific applications or uses, elbows, bends, and nipples in trade sizes ½ to 4 in. incl. for installation in accordance with Art. 342 of the **National Electrical Code.**

Galvanized intermediate steel conduit installed in concrete does not require supplementary corrosion protection.

Galvanized intermediate steel conduit installed in contact with soil does not generally require supplementary corrosion protection.

In the absence of specific local experience, soils producing severe corrosive effects are generally characterized by low resistivity less than 2000 ohm-centimeters.

Wherever ferrous metal conduit runs directly from concrete encasement to soil burial, severe corrosive effects are likely to occur on the metal in contact with the soil.

Fittings for use with unthreaded intermediate ferrous metal conduit are listed under conduit fittings (Guide DWTT) and are suitable only for the type of conduit indicated by the marking on the carton.

Although literature on IMC at one time referred to Type I and Type II IMC because of slight differences in dimensions due to manufacturing methods, the **NEC** considers IMC to be a single type of product and the rules of Art. 342 apply to all IMC. Sec. 342.10, in **(A)**, generally allows IMC above ground in all occupancies and conditions.

Note that the wording in the UL data includes the word "generally" in stating that IMC does not need additional protective material applied to the conduit when used in soil. That is intended to indicate that local soil conditions (acid versus alkaline) may require protection of the conduit against corrosion. And the UL note about corrosion of conduit running from concrete to soil must be observed. Refer to comments under 344.2 covering these conditions.

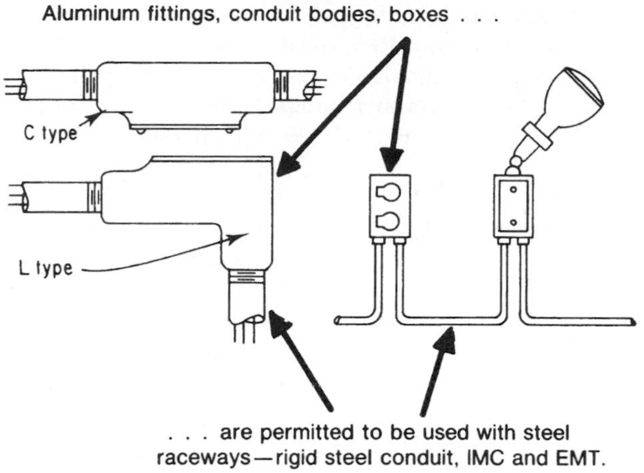
Aluminum fittings, conduit bodies, boxes . . .
C type
L type
. . . are permitted to be used with steel raceways—rigid steel conduit, IMC and EMT.

Fig. 342-2. NEC warning against use of dissimilar metals does not apply to this. (Sec. 342.14.)

In part **(B)**, wording of the rule intends to make clear that the galvanizing or zinc coating on the IMC does give it the measure of protection required when used in concrete or when directly buried in the earth. The last phrase, "judged suitable for the condition," refers to the need to comply with UL regulations as included in the Guide Card information applicable to each specific raceway, advising how and when steel raceways and other metal raceways may be used in concrete or directly buried in earth.

The UL data point out that there are soils where some difficulties may be encountered, and there are other soil conditions that present no problem to the use of steel or other metal raceways. The phrase "judged suitable for the condition" implies that a correlation was made between the soil conditions or the concrete conditions at the place of installation and the particular raceway to be used. This means that it is up to the designers and/or installers to satisfy themselves as to the suitability of any raceway for use in concrete or for use in particular soil conditions at a given geographic location. Of course, all such determinations would have to be cleared with the electrical inspection authority to be consistent with the meaning of Code enforcement.

For use of IMC in or under cinder fill, part **(C)** gives the limiting conditions. See 344.10.

342.14. Dissimilar Metals. This section has been significantly revised based on recent testing by NEMA as part of the 2017 NEC revision cycle. Interconnections between galvanized IMC and aluminum fittings, conduit elbows, etc. are still mostly permitted but not if the conditions are deemed extremely corrosive. Perhaps surprisingly, stainless steel IMC has been ruled out for contact with aluminum as well as other types of steel, and strictly limited to other stainless steel fittings, boxes, and other accessories. Similar language is now in place in other related raceway articles. The actual NEMA supporting argument for this change is reprinted in this handbook at 344.14.

342.20. Size. This is the only difference between this product and the ferrous versions of rigid metal conduit; RMC is available up to metric designator 155 (6 trade size) and IMC is only available up to metric designator 103 (trade size 4).

342.22. Number of Conductors in Conduit. The rules on conduit fill are the same for IMC, rigid metal conduit, EMT, flexible metal conduit, flexible metallic tubing, and liquidtight flexible metallic tubing—for conduits ½ in. size and larger although different tables are used. Refer to 344.20.

342.30. Securing and Supporting. The basic rule on clamping IMC is simple and straightforward (Fig. 342-3). Spacing may be increased to a maximum of 5 ft (1.5 m) where necessary because no structural member is available. But the distance must not be extended, except as permitted by the subsections. The subsections allowing wider spacing of supports are the same as those covered in 344.30 for rigid metal conduit.

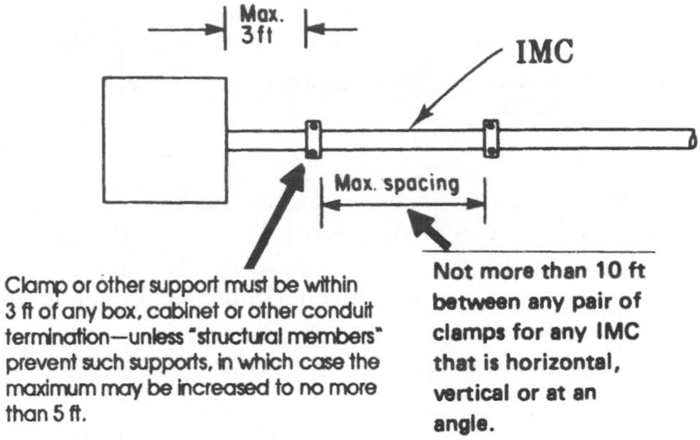
Max.
3ft
IMC
Max. spacing
Clamp or other support must be within 3 ft of any box, cabinet or other conduit termination—unless "structural members" prevent such supports, in which case the maximum may be increased to no more than 5 ft.
Not more than 10 ft between any pair of clamps for any IMC that is horizontal, vertical or at an angle.

Fig. 342-3. All runs of IMC must be clamped in this way. (Sec. 342.30.)

Spacing between supports for IMC (greater than every 10 ft [3.0 m]) is the same as the spacing allowed for rigid metal conduit. The subparts recognize the essential equality between the strengths of IMC and rigid metal conduit.

The special support limitations for conduit nipples imposed in the 2008 **NEC** have been revoked and the text returned to that of the 2005 edition. Any nipple not over the general limit on the first point of support in length, namely 900 mm (3 ft), is considered supported by its terminating hardware and no additional support is generally required.

For the rest of the run, as before, the conduit must be "supported" (as distinguished from "secured") at not over 3 m (10 ft) intervals, with greater support distances allowed if threaded couplings are used and the geometry of the run is such that the weight of the conduit is not imposed on a terminating enclosure. A second provision makes a very practical allowance for conduit drops from a high-bay ceiling. The support interval, if necessary due to a lack of intervening supports, can be extended to 6 m (20 ft) if threaded couplings are used and the top and bottom of the conduit run are securely fastened. This is often arranged with a tee conduit body at the equipment level, with the vertical run continuing using an empty conduit nipple ending at a threaded floor flange. The side branch of the tee is usually extended to the equipment termination with a flexible wiring method. Note, however, that an additional support must be arranged on the riser above the tee in order to comply with the provisions of both this section and also 314.23(E) with respect to the conduit body.

A somewhat related change is to recognize approved IMC mast risers with no support within 900 mm (3 ft) of the service head, for obvious reasons.

ARTICLE 344. RIGID METAL CONDUIT: TYPE RMC

344.10. Uses Permitted. UL data on rigid metal conduit are similar to those on IMC and supplement the rules of this section, as follows:

> Galvanized rigid steel conduit installed in concrete does not require supplementary corrosion protection.
>
> Galvanized rigid steel conduit installed in contact with soil does not generally require supplementary corrosion protection.
>
> In the absence of specific local experience, soils producing severe corrosive effects are generally characterized by low resistivity (less than 2000 ohm-centimeters).
>
> Wherever ferrous metal conduit runs directly from concrete encasement to soil burial, severe corrosive effects are likely to occur on the metal in contact with the soil.
>
> Conduit that is provided with a metallic or nonmetallic coating, or a combination of both, has been investigated for resistance to atmospheric corrosion. Nonmetallic outer coatings that are part of the required resistance to corrosion have been additionally investigated for resistance to the effects of sunlight.
>
> Nonmetallic outer coatings of greater than 0.010-in. thickness are investigated with respect to flame propagation detrimental effects to any underlying corrosion protection, the fit of fittings, and electrical continuity of the connection of conduit to fittings.
>
> Conduit with nonmetallic coatings has not been investigated for use in ducts, plenums, or other environmental air spaces in accordance with the NEC.
>
> Rigid metal conduit with or without a nonmetallic coating has not been investigated for severely corrosive conditions.

For nonferrous rigid metal conduits, the UL application notes state:

> Aluminum conduit used in concrete, in contact with soil, or in severely corrosive conditions, requires supplementary corrosion protection.
>
> The Listing Mark for these products includes the UL symbol together with the word "LISTED," a control number, and the product name "Electrical Rigid Metal Conduit-Aluminum" (or "ERMC-A"), Electrical Rigid Metal Conduit-Red Brass" (or "ERMC-RB"), or "Electrical Rigid Metal Conduit-Stainless Steel" ("ERMC-SS").

For direct earth burial of rigid conduit and IMC, the UL notes must be carefully studied and observed:

1. Galvanized rigid steel conduit and galvanized intermediate steel conduit directly buried in soil do not *generally* require supplementary corrosion protection. The use of the word "generally" in the UL instructions indicated that it is still the responsibility of the designer and/or installer to use supplementary protection where certain soils are known to produce corrosion of such conduits. Where corrosion of underground galvanized conduit is known to be a problem, a protective jacketing or a field-applied coating of asphalt paint or equivalent material must be used on the conduit. *But,* UL notes must be observed for resistance to corrosion under "severely corrosive conditions" because when steel conduits pass from concrete to direct earth burial, the juncture is classified as "severely corrosive" and nonmetallic coatings

have not been investigated for these conditions. An agreement will need to be worked out with the inspector if severely corrosive effects are anticipated in any location. In the case of underground/concrete interfaces, supplementary protective coating on conduit at the crossing line can eliminate the conditions shown in Fig. 344-1.

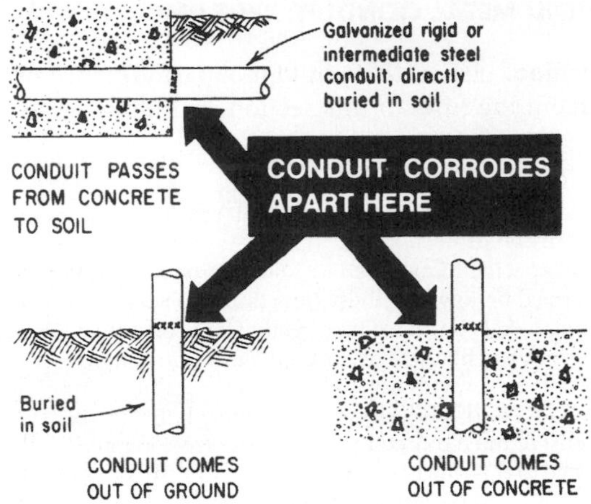
Galvanized rigid or intermediate steel conduit, directly buried in soil
CONDUIT PASSES FROM CONCRETE TO SOIL
CONDUIT CORRODES APART HERE
Buried in soil
CONDUIT COMES OUT OF GROUND
CONDUIT COMES OUT OF CONCRETE

Fig. 344-1. Protective coating on section of conduit can prevent this corrosion problem. (Sec. 342.2.)

2. Aluminum conduit used directly buried in soil requires supplementary corrosion protection. Exactly what that could be in this case will require some research and the approval of the inspector. At one time UL declared that such coatings "presently used" have not been recognized for resistance to corrosion, but that statement no longer appears in the Guide Card information for this category.
3. Red brass conduit is permitted by right for underground and swimming pool applications. Note that all rigid metal conduit is required to be listed per 344.6. This may be a difficult issue on this product. An admittedly nonscientific survey of all current listees in the nonferrous rigid metal conduit category found a number of producers of aluminum conduit, a few producers of stainless steel conduit, and no producers of red brass conduit. For swimming pool applications run to a forming shell this has been a known issue for some time, which is why 680.23(B)(2)(a) specifically allows brass conduit to be approved and not listed, thereby constituting a deliberate Chap. 6 amendment of the Chap. 3 rule in 344.6. Some plumbing supply houses carry heavy wall red brass pipe, often in 12-ft lengths, that takes a conventional pipe thread extremely well, and is a very robust product with an extremely smooth interior that, if anything, is somewhat more difficult to bend than IMC or RMC. Lack of heavy foot pressure with excessive force on the handle won't kink the product, but will bend the bender handle. Approval is at the discretion of the local inspector, but this product should certainly be considered unless a listed alternative becomes more available.

As indicated by the rule of part **(C)**, care must be taken where cinder fill is used. Cinders usually contain sulfur, and if there is much moisture, sulfuric acid is formed, which attacks steel conduit. A cinder fill outdoors should be considered as "subject to permanent moisture." In such a place conduit runs should be provided corrosion protection as described, encased in 2 in. (50 mm) of concrete, or buried in the ground at least 18 in. (450 mm) below the fill. This would not apply if cinders were not present.

344.14. Dissimilar Metals. This section has been significantly revised based on recent testing by NEMA as part of the 2017 NEC revision cycle. Interconnections between galvanized heavy wall conduit and aluminum fittings, conduit elbows, etc. are still mostly permitted but not if the conditions are deemed extremely corrosive. Perhaps surprisingly, stainless steel IMC has been ruled out for contact with aluminum as well as other types of steel, and strictly limited to other stainless steel fittings, boxes, and other accessories. Similar language is now in place in other related raceway articles.

The NEMA commentary on this topic is both surprising and informative: "Stainless steel is considerably more noble (or cathodic) than aluminum and is also considerably more noble than steel and zinc (galvanized steel) and would be subject to more aggressive galvanic attack in the presence of an electrolyte. Stainless steel conduit used with aluminum or galvanized fittings, accessories, outlet boxes and enclosures may result in a galvanic action, leading to corrosion."

344.22. Number of Conductors. The basic NEC rule on the maximum number of conductors that may be pulled into rigid metal conduit, rigid nonmetallic conduit, intermediate metal conduit, electrical metallic tubing, flexible metal conduit, and liquidtight flexible metal conduit is contained in the single sentence of this section.

The number of conductors permitted in a particular size of conduit or tubing is covered in Chap. 9 of the Code, and in Tables C1 through C12A in Annex C for conductors all of the same size used for either new work or rewiring. Tables 4 and 5 of Chap. 9 cover combinations of conductors of different sizes when used for new work or rewiring. For nonlead-covered conductors, three or more to a conduit, the sum of the cross-sectional areas of the individual conductors must not exceed 40 percent of the interior cross-sectional area (csa) of the conduit or tubing for new work or for rewiring existing conduit or tubing (Fig. 344-2). Note 4 preceding all the tables in Chap. 9, in the back of the Code book, permits a 60 percent fill of conduit nipples not over 24 in. (600 mm) long and no derating of ampacities is needed.

When all conductors in a rigid metal conduit are the same size, Tables C8 and C8A in App. C give the maximum allowable fill for conductors depending on conductor type up to 2000 kcmil, for metric designator 16 to 155 (trade size ½ to 6) rigid metal conduit.

question What is the minimum size of rigid metal conduit required for six 10 THHN AWG wires?

answer Table C.8, Annex C, shows that six 10 THHN AWG wires may be pulled into a metric designator 16 (trade size ½) rigid metal conduit.

question What size conduit is the minimum for use with four 6 RHH AWG conductors with outer covering?

answer Table C.8, Annex C, shows that a metric designator 35 (trade size 1¼) minimum conduit size must be used for from four to six RHH AWG conductors with outer covering. If they lacked the outer covering, the answer would be a metric designator 27 (trade size 1) conduit.

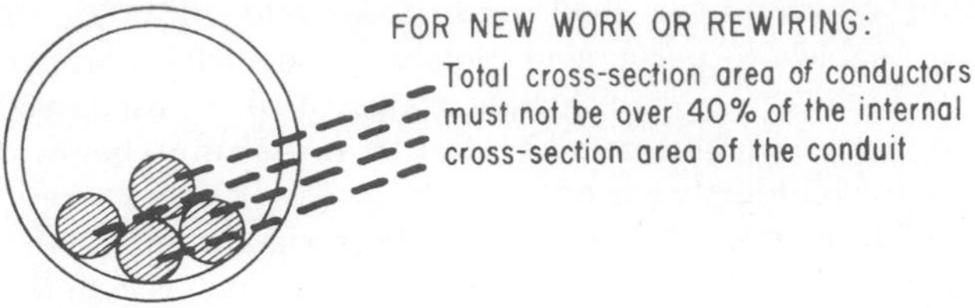
FOR NEW WORK OR REWIRING:
Total cross-section area of conductors must not be over 40% of the internal cross-section area of the conduit

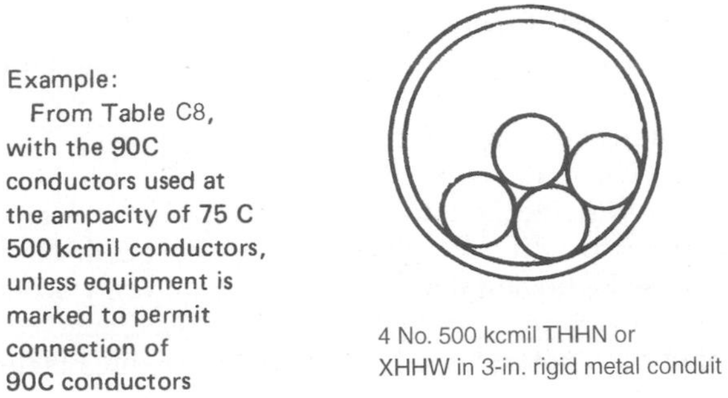
Example:
From Table C8, with the 90C conductors used at the ampacity of 75 C 500 kcmil conductors, unless equipment is marked to permit connection of 90C conductors
4 No. 500 kcmil THHN or XHHW in 3-in. rigid metal conduit

Fig. 344-2. For three or more conductors the sum of their areas must not exceed 40 percent of the conduit area. (Sec. 344.20.)

question What is the minimum size conduit required for four 500-kcmil XHHW conductors?

answer Table C.8 shows that metric designator 78 (trade size 3) conduit may contain four 500-kcmil XHHW (or THHN) conductors.

When all the conductors in a conduit or tubing are not the same size, the minimum required size of conduit or tubing must be calculated. Table 1, Chap. 9, says that conduit containing three or more conductors of any type except lead-covered, for new work or rewiring, may be filled to 40 percent of the conduit csa. Note 6 to this table refers to Tables 4 through 8, Chap. 9, for dimensions of conductors, conduit, and tubing to be used in calculating conduit fill for combinations of conductors of different sizes.

example What size rigid metal conduit is the minimum required for enclosing six 10 AWG THHN, three 4 AWG RHH (with outer covering), and two 12 AWG TW conductors (Fig. 344-3)?

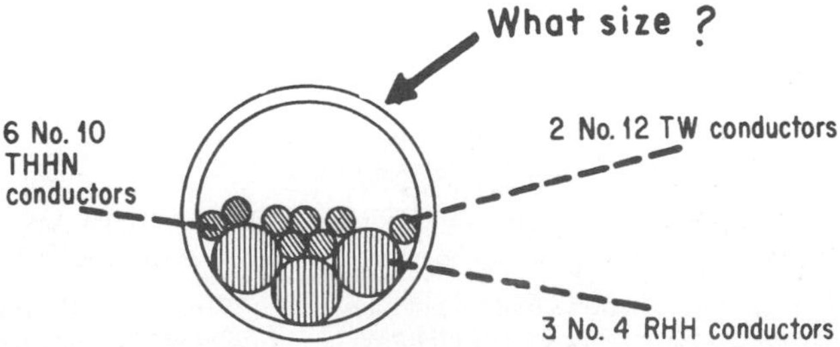
What size ?
6 No. 10 THHN conductors
2 No. 12 TW conductors
3 No. 4 RHH conductors

Fig. 344-3. Minimum permitted conduit size must be calculated when conductors are not all the same size. (Sec. 344.22.)

Cross-sectional areas of conductors:
From Table 5, Chap. 9:

10 AWG THHN	0.0211 sq in.
4 AWG RHH	0.1333 sq in.
12 AWG TW	0.0181 sq in.

Total area occupied by conductors:

6 10 AWG THHN	6 × 0.0211 = 0.1266 sq in.
3 4 AWG RHH	3 × 0.1333 = 0.3999 sq in.
2 12 AWG TW	2 × 0.0181 = 0.0362 sq in.
Total area occupied by conductors	0.5627 sq in.

Referring to Table 4, Chap. 9:

The fourth column from the left gives the amount of square-inch area that is 40 percent of the csa of the sizes of conduit given in the farthest column from the left. The 40 percent column in the table on rigid metal conduits shows that 0.355 sq in. is 40 percent fill of a 1-in. conduit, and 0.610 sq in. is 40 percent fill of a 1¼-in. conduit. Therefore, a 1-in. conduit would be too small and—

A metric designator 35 (trade size 1¼) rigid metal conduit is the smallest for these 11 conductors.

example What is the minimum size of conduit for four No. 4/0 TW and four No. 4/0 XHHW conductors?

From Table 5, a No. 4/0 TW has a csa of 0.3718 sq in. Four of these come to 4 × 0.3718 or 1.4872 sq in.

From Table 5 we find that four No. 4/0 XHHW have a csa of 1.2788 sq in.

$$1.4872 + 1.2788 = 2.766 \text{ sq in.}$$

From Table 4, 40 percent of the csa of 3-in. rigid metal conduit is 3.000 sq in. A 2½-in. conduit would be too small. Therefore—

A metric designator 78 (trade size 3) conduit must be used.

Figure 344-4 shows how a conduit nipple is excluded from the normal 40 percent limitation on conduit fill. In this typical example, the nipple between a panelboard and a wireway contains 12 10 AWG TW wires, 6 14 AWG THHN wires, 3 8 AWG THW wires, and 2 2 AWG RHH wires (without outer covering). The minimum trade size of nipple that can be used in this case is metric designator 35 (trade size 1¼). (Nipple may be filled to 60 percent of its csa if it is not over 24 in. [610 mm] long. Area of conductors = 12 × 0.0243 sq in. [csa of each 10 AWG TW] plus 6 × 0.0097 sq in.

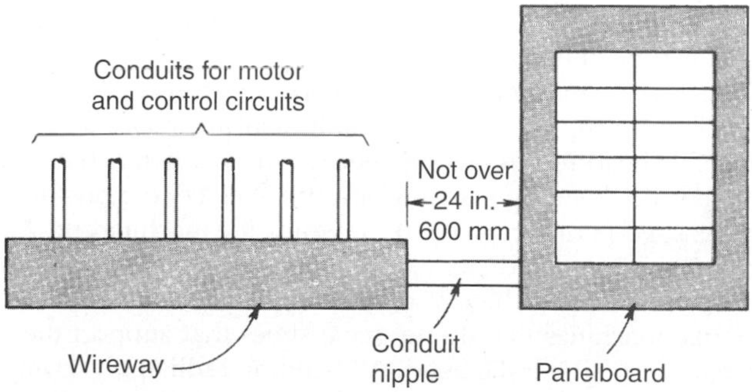
Conduits for motor
and control circuits
Not over
24 in.
600 mm
Wireway
Conduit
nipple
Panelboard

Fig. 344-4. Conduit nipples may be filled to 60 percent of csa and no derating is required. (Sec. 344.22.)

[each 14 AWG THHN] plus 3 × 0.0437 sq in. [each 8 AWG THW] plus 2 × 0.1333, or a total of 0.7475 sq in. If we divide this number by 0.6, the result will be the minimum area that the nipple can be sized at; 0.7475 ÷ 0.6 = 1.246 sq in. The next higher sized rigid conduit raceway from Table 4, Chap. 9 is a metric designator 35 [trade size 1¼] conduit, so that is the size to select. Note that the Table 4 raceway sizes now include 60 percent columns done out, and entering the 0.7475 sq in. conductor summation in the 60 percent column shows it to be too large for the metric designator 27 [trade size 1] conduit but well within the next larger size.) And the conductors do *not* have to be derated in accordance with 310.15(B)(3)(a). If the nipple had been 25 in. long, calculation at 40 percent fill would have called for a metric designator 41 (trade size 1½) size, and all conductors would have had to be derated per 310.15(B)(3).

THWN and THHN are the smallest-diameter building wires in many common sizes. The greatly reduced insulation wall on Type THWN or THHN gives these thin-insulated conductors greater conduit fill than TW, THW, or RHH. And the nylon jacket on THWN and THHN has an extremely low coefficient of friction. THWN is a 75°C-rated wire for general circuit use in dry or wet locations; however, it is routinely being supplied in its THWN-2 variety, which has the same 90°C rating as THHN. THHN is a 90°C rated wire for dry locations only.

Although the same procedure applies, the tables in Annex C and the various parts of Table 4 must be correlated with the type of raceway to be used. This change of approach in the 1996 **NEC** was a major departure from the editions of the **NEC** that preceded it, but provides for more realistic fill. Remember, that spare fill capacity may be desirable in certain applications—such as long underground runs to outbuildings. The Code permits fill to 40 percent but no more. If a raceway is filled to the 40 percent maximum permitted in Chap. 9, a new raceway will be required if additional circuits are desired at a later date.

To fill conduit to the Code, maximum allowance is frequently difficult or impossible from the mechanical standpoint of pulling the conductors into the conduit, because of twisting and bending of the conductors within the conduit. Bigger-than-minimum conduit should generally be used to provide some measure of spare capacity for load growth; and, in many cases, the conduit to be used should be upsized considerably to allow future installation of some larger anticipated size of conductors.

344.24. Bends—How Made. The basic rule here provides general, common sense requirements regarding the bending of rigid metal conduits. The next part of the rule refers to Table 2 in Chap. 9. That table provides the same data that appeared in Table 344.24 in the 2002 **NEC**. Table 2 in Chap. 9 now shows the acceptable radii, depending on the type of bend that is made.

Table 2, Chap. 9 gives minimum bending radii for bends in rigid metal conduit, IMC, or EMT using any approved bending equipment and methods. (See Fig. 344-5.) However, the table headings in this rule permit sharper bends (i.e., smaller bending radii) if a one-shot bending machine or a full-shoe bender, including full-shoe hand benders, is used in making a bend for which the machine and its accessories are designed. The minimum radii for one-shot bends are given in the center column of Table 2, Chap. 9. The "other bends" column applies to bends made with hickeys or with hydraulic machines that do not have shoes that support the walls of the conduit or tubing throughout the bend. All bending radii apply to any amount of bend—that is, 45°, 90°, and so on.

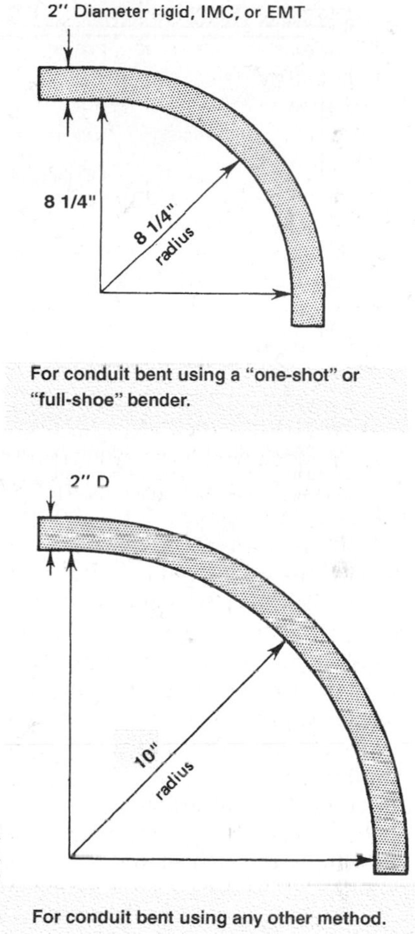
2" Diameter rigid, IMC, or EMT
8 1/4"
8 1/4" radius
For conduit bent using a "one-shot" or "full-shoe" bender.
2" D
10" radius
For conduit bent using any other method.

Fig. 344-5. Minimum bending radii are specified to protect conductors from damage during pull-in. Note that the actual minimum code radius is applied to the centerline of the conduit, and not its inside edge. (Sec. 344.24.)

344.26. Bends—Number in One Run. There must be not more than the equivalent of four quarter bends (360°) between any two "pull points"—conduit bodies and boxes, as shown in Fig. 344-6. In previous Codes, the 360° of bends was permitted between boxes and "fittings" and even between "fitting and fitting." Because the word *fitting* is defined in Art. 100 and the term does include conduit couplings, bushings, and so forth, there could be very many bends in an overall run, totaling far more than 360° if the equivalent of four quarter bends could be made between each pair of conduit couplings. The present wording limits the 360° of bends to conduit runs between "pull points"—such as between switchboards and panelboards, between housings, boxes, and conduit bodies—all of which are "pull points."

Note that the bends to be included in the summation include all deflections from a straight line, including a box kick at the end (about 20°), and bends of long radius, such as when conduits are joined outside of a trench and then

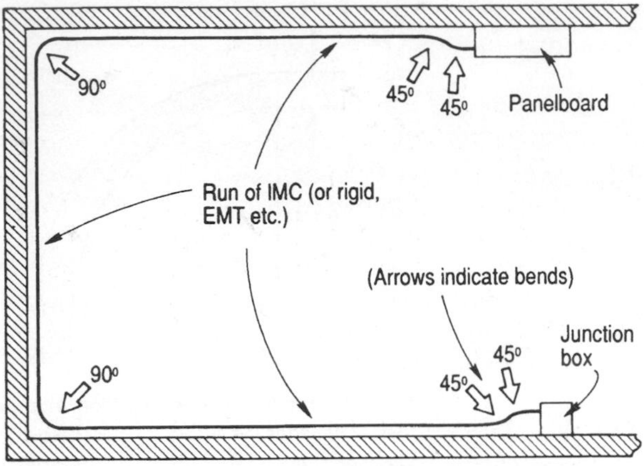
90°
45°
45°
Panelboard
Run of IMC (or rigid, EMT etc.)
(Arrows indicate bends)
Junction box
45°
45°
90°

Fig. 344-6. Rigid metal conduit—like all other types of conduits—is limited to not over 360° of bends between "pull points," such as the panelboard and junction box shown here. (Sec. 344.26.)

forced in. There is no language that waives any bend of any radius, because pull force calculations show that the force to overcome a change in direction is independent of the radius; a longer radius results in more friction over the longer length involved. The sidewall bearing pressure is reduced on a longer radius bend, but that is only one of the forces to be considered.

The same concept of the number of bends permitted is given in all of the NEC articles on raceways—ENT, EMT, IMC, rigid metal conduit, rigid nonmetallic conduit, and so on.

344.28. Reaming and Threading. As with IMC, rigid metal conduit always requires a bushing on the conduit end using locknuts and bushing for connection to knockouts in sheet metal enclosures (Fig. 344-7). But simply because

Fig. 344-7. Conduit terminations, other than threaded connections to threaded fittings or enclosure hubs, must be provided with bushings for protection of the conductors. (Sec. 344.28.)

a conduit can be secured to a sheet metal KO with two locknuts (one inside and one outside—as required by 250.97), it does not mean the bushing may be eliminated. Of course, no bushing is needed where the conduit threads into a hub or boss on a fitting or an enclosure.

344.30. Securing and Supporting. As with Art. 342, the support distances to the first support can be increased to up to 1.5 m (5 ft) on a showing that there are no readily available support points at the default point of 900 mm (3 ft). In addition, where threaded couplings are used and the supports are arranged to prevent stress in the run and as may result from bends in the run from being visited on terminations, the greater support distances in the associated table can be used. The special support limitations for conduit nipples imposed in the 2008 NEC have been revoked and the text returned to that of the 2005 edition. Any nipple not over the general limit on the first point of support in length, namely 900 mm (3 ft), is considered supported by its terminating hardware and no additional support is generally required. The permissions for longer distances between supports for certain run geometries, service masts, and for high-bay ceiling drops are the same as for IMC; refer to this topic in 342.30 for more information on these topics.

344.42. Couplings and Connectors. Figure 344-8 shows a threadless connection of rigid metal conduit to the hub on a fitting. It is effective both mechanically and electrically if any nonconducting coating is removed from the conduit.

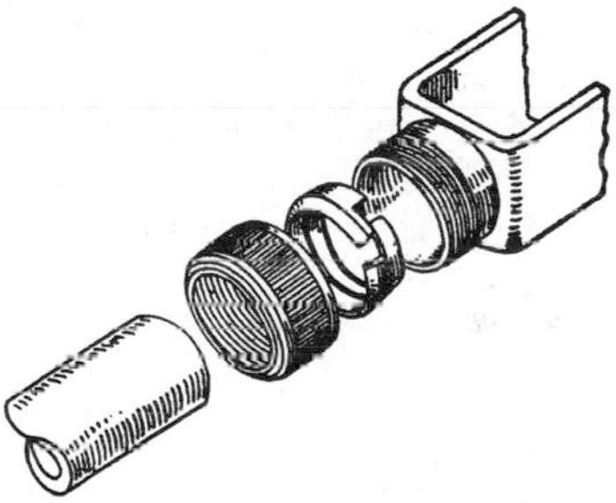

Fig. 344-8. Threadless connectors may be used on unthreaded end of conduit. (Sec. 344.42.)

A running thread is considered mechanically weak and has poor electrical conductivity.

Where two lengths of conduit must be coupled together but it is impossible to screw both lengths into an ordinary coupling, the Erickson coupling or a swivel-coupling may be used (Fig. 344-9). They make a rigid joint which is both mechanically and electrically effective. Also, bolted split couplings are available.

It is not intended that conduit threads be treated with paint or other materials in order to ensure watertightness. It is assumed that the conductors are approved for the locations and that the prime purpose of the conduit is for protection from physical damage and easy withdrawal of conductors for replacement. There are

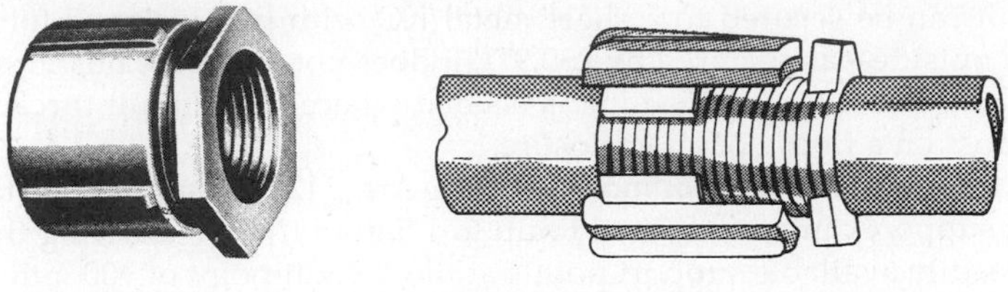

Erickson coupling

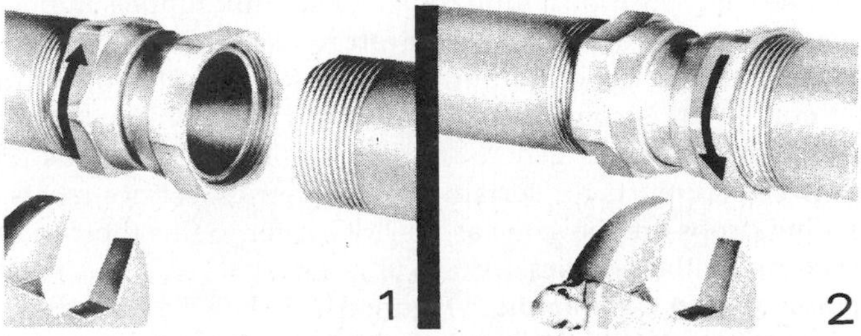
1
2

Swivel coupling

Fig. 344-9. Fittings provide for coupling conduits where conduits cannot be rotated (turned). (Sec. 344.42.)

available pipe-joint compounds that seal against water without interrupting electr cal conductivity; see 300.6 for more information on this topic. Note also that threac less connectors and couplings, as covered in 344.42(A), must not be applied to threaded conduit end unless listed for that application because they are tested o conduit ends that are a full diameter in size; without testing the performance of or of these items it cannot be ensured.

ARTICLE 348. FLEXIBLE METAL CONDUIT: TYPE FMC

348.12. Uses Not Permitted. UL data (Code DXUZ) supplement the Code dat on use of standard flexible metal conduit—known also as "Greenfield" or simpl "flex." The UL data note:

> This category covers flexible aluminum and steel conduit in trade sizes ⅜ to 4 (metr designators 12 to 103) inclusive, flexible aluminum and steel conduit Type RW (reduce wall), flexible aluminum and steel conduit Type XRW (extra reduced wall) in trade siz from ⅜ to 3 (16 to 78) inclusive, for installation in accordance with Art. 348 of ANSI NFPA 70, "National Electrical Code" (NEC), for conductors in circuits of 600 V, nomina or less. This product may also be used for installation of conductors in motor circuit electric signs, and outline lighting in accordance with the NEC.
>
> Flexible metal conduit (steel or aluminum) should not be used underground (direct buried or in duct which is buried) or embedded in poured concrete or aggregate, or

direct contact with earth or where subjected to corrosive conditions. In addition, flexible aluminum conduit should not be installed in direct contact with masonry in damp locations.

For flexible metal conduit in 1-¼ (35) trade size and smaller, where terminated in fittings investigated for grounding and where installed with not more than 6 ft (total length) in any ground-return path, flexible metal conduit is suitable for grounding where used on circuits rated 20 A or less. See Conduit Fittings (DWTT) with respect to fittings suitable as a grounding means.

The following are not considered suitable as a grounding means:

1. The 1½ (41) and larger trade sizes.
2. The 1¼ (35) trade size and smaller where used on circuits rated higher than 20 A, or where the total length in the ground-return path is greater than 6 ft.

To prevent possible damage to flexible aluminum conduit, flexible aluminum, and steel conduit Types RW and XRW, care must be exercised when installing connectors employing direct bearing set screws.

PRODUCT MARKINGS

Flexible aluminum conduit is marked at intervals of not more than one ft with the letters "AL."

Flexible aluminum conduit Type RW is marked at intervals of not more than one ft with the letters "AL" and "RW."

Flexible steel conduit Type RW is marked at intervals of not more than one ft with the letters "RW."

Flexible aluminum conduit Type XRW is marked at intervals of not more than one ft with the letters "AL" and "XRW."

Flexible steel conduit Type XRW is marked at intervals of not more than one ft with the letters "XRW."

Note that 348.12 was used to permit this wiring method in wet locations under several conditions; that permission has been revoked and now this wiring method is simply not permitted in a wet location.

348.20. Size. Part **(A)(2)** to this rule permits metric designator 12 (trade size ⅜) flexible metal conduit to be used in lengths up to 6 ft (1.83 m) for connections to lighting fixtures. This provides correlation with 410.117(C), which includes 18 in. to 6 ft (0.45 to 1.8 m) of metal raceway for connecting recessed luminaires (generally the nonwired types). Figure 348-1 shows such application, and it is permissible to use 16 AWG or 18 AWG 150°C fixture wire, as shown in Fig. 360-1, for flex tubing.

The usual field application of trade designator 12 (trade size ⅜) flexible metal conduit is covered in (A)(2)a, where the 1.8 m (6 ft) length (or shorter) can be used for any utilization equipment.

Part **(A)(5)** permits trade designator 12 (trade size ⅜) flex if it is "part of a listed assembly," which assumes it is supplied as part of UL-listed equipment, in lengths up to 25 ft (7.5 m) as specified in 410.137(C) to connect "wired luminaire sections."

Part **(A)(3)** permits flex in trade designator 12 (trade size ⅜) to be used for the cable assemblies of modular wiring systems, typically in hung ceilings [so-called manufactured wiring systems covered by 604.100(A)]. This is directed specifically to ceiling modular wiring. And the equipment grounding conductor run in such flex wiring assemblies may be either bare or insulated [see 604.100(A)(2)] Fig. 348-1).

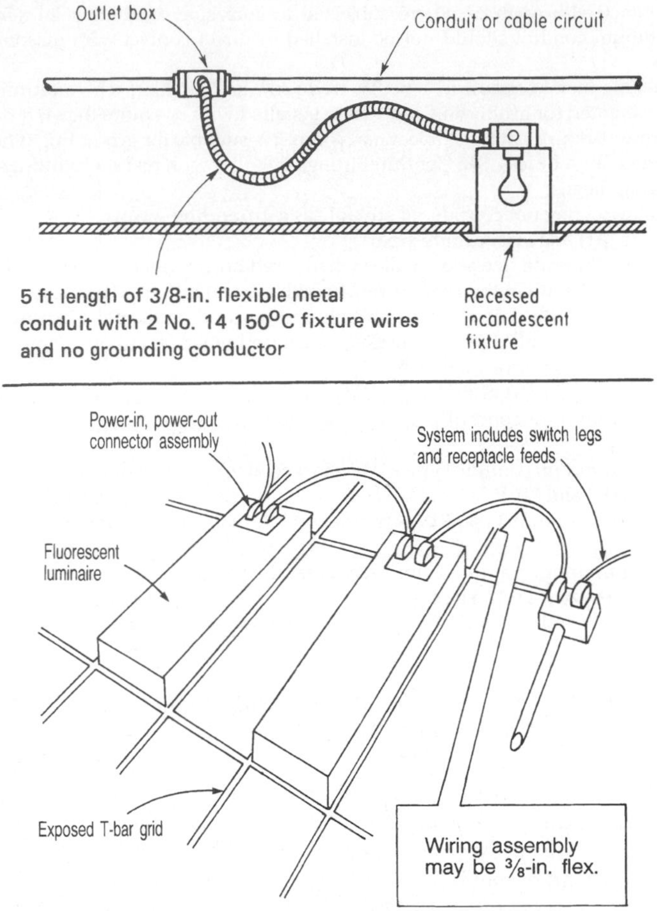
Outlet box
Conduit or cable circuit
5 ft length of 3/8-in. flexible metal conduit with 2 No. 14 150°C fixture wires and no grounding conductor
Recessed incandescent fixture
Power-in, power-out connector assembly
System includes switch legs and receptacle feeds
Fluorescent luminaire
Exposed T-bar grid
Wiring assembly may be ⅜-in. flex.

complete system in a ceiling space consisting of prewired cable lengths with snap-in connectors for fixtures and switches, may use ⅜-in. flexible metal conduit.

Fig. 348-1. Flex of ⅜-in. size may be used for fixture "whip." (Sec. 348.20.)

348.22. Number of Conductors. This section specifies that Table 1 of NEC Chap. 9 must be used in determining the maximum permitted number of conductors in ½ through 4-in. flex. Flexible metal conduit is permitted the same conductor fill procedure, although customized to the actual inner diameter as other types of conduit and tubing. The number of conductors permitted in metric designator 12 (trade size ⅜) flex is given in Table 348.22.

348.26. Bends—Number in One Run. Figure 348-2 shows the details of this section.

In this section, the limitation to no more than a total of 360° of bends between outlets applies to both exposed and concealed applications of standard metal flex and liquidtight metal flex.

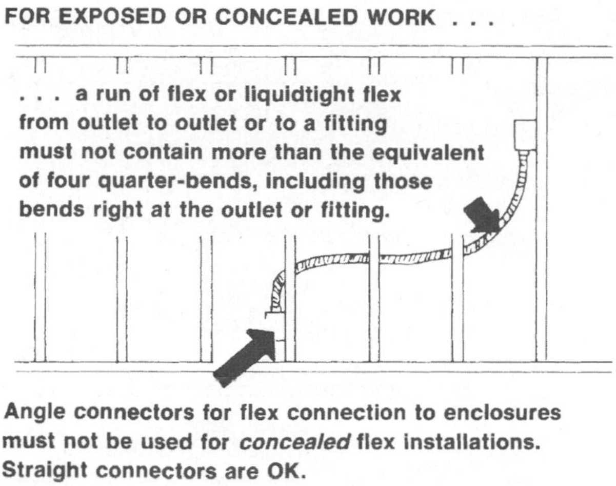
FOR EXPOSED OR CONCEALED WORK . . .
. . . a run of flex or liquidtight flex from outlet to outlet or to a fitting must not contain more than the equivalent of four quarter-bends, including those bends right at the outlet or fitting.
Angle connectors for flex connection to enclosures must not be used for *concealed* flex installations. Straight connectors are OK.

Fig. 348-2. Concealed or exposed flex must not have too many bends that could damage wires on pull-in. (Sec. 348.24.)

Without restriction on the maximum number of bends in exposed and concealed work, bends could result in damage to conductors in a run with an excessive number of bends or could encourage installation of conductors prior to conduit installation, with conduit then installed as a cable system. A limit on number of bends for exposed and concealed work conforms to the requirements for other raceway systems.

348.30. Securing and Supporting. Straps or other means of securing the conduit in place should be spaced much closer together (every 4½ ft [1.4 m] and within 12 in. [300 mm] of each end) for flexible conduit than is necessary for rigid conduit. Every bend should be rigidly secured so that it will not be deformed when the wires are being pulled in, thus causing the wires to bind. Note that there is some relief on the larger sizes, with an additional 300 mm (foot) to the first support for metric designator 41 and 53 (trade sizes 1½ and 2) and yet another 300 mm (foot) for larger sizes (Exception No. 2) but only where flexibility is required after installation.

Figure 348-3 shows use of unclamped lengths of flex that are short enough to comply with the main rule. The rigid raceways to which they are connected must be secured within the distances of their ends as required by their articles. The application shown does not require flexibility after installation and therefore Exception No. 2 does not apply. The raceway in front may exceed the basic 300 mm (12 in.) distance, but that is unclear from the photo. Figure 348-4 shows another example. Exception No. 3 is illustrated in Fig. 348-1. There is also a 1.8 m (6 ft) allowance for luminaire connections above hung ceilings, and the first exception allows for it to be fished between access points in finished construction. As with other flexible wiring methods, listed cable ties evaluated for support and securement are permitted, and frequently used above suspended ceilings.

348.60. Grounding and Bonding. As shown in the UL data under 348.12, flex in any length over 6 ft (1.83 m) is not suitable as an equipment grounding

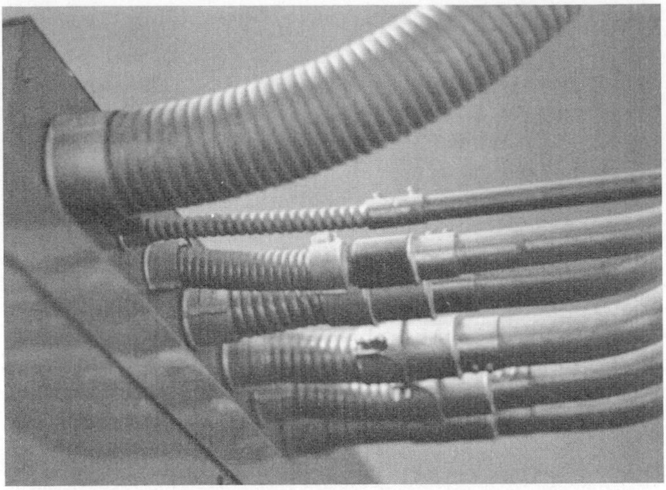

Fig. 348-3. Lengths of flex not over 3 ft (900 mm) long may be used without clamps or straps where the flex is used at terminals to provide flexibility for vibration isolation or for alignment of connections to knockouts. (Sec. 348.30.)

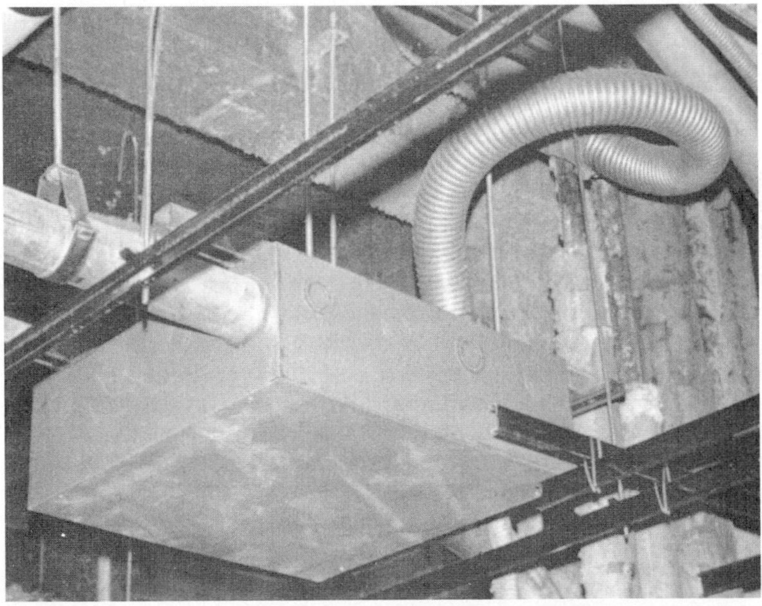

Fig. 348-4. A length of flex not over 3 ft (900 mm) long connects conduit to pull box in modernization job, providing the flexibility to feed from fixed conduit to box. (Sec. 348.30.) Note that this run of flex already has almost 270° of bend before any bends in the rigid conduit are counted, and 300.18(A) requires raceway to be complete between pull points before wires are pulled in. Although there is no express language in the NEC that requires bends in succeeding raceway types joined end-to-end to adhere to the customary 360° rule, if there are several bends in the rigid conduit, the pull is likely to violate 300.17.

conductor, and an equipment grounding conductor must be used within the flex to ground metal enclosures fed by the flex. 250.118 permits flex as an equipment grounding conductor *only* under the given conditions—which would be the same as shown in Fig. 360-1 for flex tubing. Refer to 250.118 and to the discussion of grounding and bonding in 250.102.

The fourth paragraph of 348.60 essentially recognizes that an equipment bonding jumper used with flexible metal conduit may be installed inside the conduit or outside the conduit when installed in accordance with the limitations of 250.102. 250.118 and this rule make clear that use of flexible metal conduit as an equipment grounding conductor in itself is permitted only where a length of not over 1.8 m (6 ft) is inserted in any ground return path. The wording indicates that the total length of flex in any ground return path must not exceed 1.8 m (6 ft). That is, it may be a single 1.8 m (6 ft) length. Or, it may be two 900 mm (3 ft) lengths, three 600 mm (2 ft) lengths, or any total equivalent of 1.8 m (6 ft). If the total length of flex in any ground return path exceeds 1.8 m (6 ft), the rule requires an equipment grounding conductor to be run within or outside any length of flex beyond the permitted 1.8 m (6 ft) that is acceptable as a ground return path in itself. This rule also includes other flexible wiring methods in the ground return path; review 250.118(5) in this book for the complete story.

It should be noted that 250.118 is not applicable to the use of flex in a hazardous location. The rules in 501.30(B) and 502.30(B) simply require bonding for flex, with only a very narrow exception given for instrumentation (Fig. 348-5).

Effective with the 2012 edition of the UL Guide Card information [see discussion at 110.3(B) for details], there was a direct conflict between a key UL limitation on the use of a flexible metal conduit and the expressly permitted uses in 250.118(5)

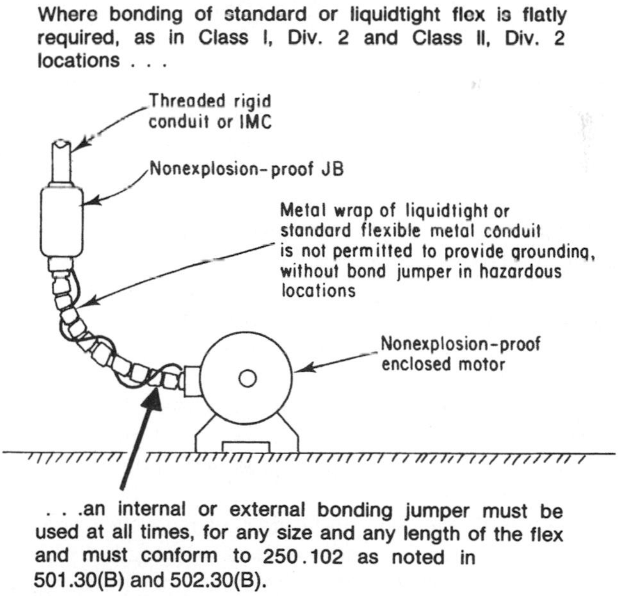
Where bonding of standard or liquidtight flex is flatly required, as in Class I, Div. 2 and Class II, Div. 2 locations . . .
Threaded rigid conduit or IMC
Nonexplosion-proof JB
Metal wrap of liquidtight or standard flexible metal conduit is not permitted to provide grounding, without bond jumper in hazardous locations
Nonexplosion-proof enclosed motor
. . .an internal or external bonding jumper must be used at all times, for any size and any length of the flex and must conform to 250.102 as noted in 501.30(B) and 502.30(B).

Fig. 348-5. Flex must always be bonded in Class I and Class II hazardous locations. (Sec. 348.60.)

for this wiring method going back 42 years. There was never an upper size limitation on the use of this product as an equipment grounding conductor. The length limit of 6 ft continues to this day, along with the severe limitation on the associated circuit protection of 20 A, which also continues. The conflict has been eliminated as of the 2017 edition of the NEC because it has been incorporated into 250.118(5).

250.118(5)d says that an equipment grounding conductor (or jumper) must *always* be installed for a length of metal flex that is used to supply equipment "where flexibility is necessary after installation," such as equipment that is not fixed in place, or equipment that tends to vibrate, such as motors or transformers. That wording actually modifies the conditions under which a 6-ft or shorter length of metal flex (Greenfield) may be used for grounding through the metal of its own assembly, without need for a bonding wire. Because experience has indicated many instances of loss of ground connection through the flex metal due to repeated movement of a flex whip connected to equipment that vibrates, or flex supplying movable equipment, the last sentence requires use of an equipment bonding jumper, either inside or outside the flex in all cases where vibrating or movable equipment is supplied—for ensured safety of grounding continuity. The caption for Fig. 348-3 lists an example of each possibility. Lining up connections to a pull box does not involve flexibility after installation, but vibration isolation does. The 2011 NEC clarified that when this wiring method is used for vibration isolation, it is employing its flexibility after the initial installation and the supplemental bonding conductor is required.

ARTICLE 350. LIQUIDTIGHT FLEXIBLE METAL CONDUIT: TYPE LFMC

350.1. Scope. This article covers *metallic* liquidtight flex. Liquidtight metal flex (often called "Sealtite" as a generic term in industry usage, although that word is the registered trade name of the liquidtight flex made by Anaconda Metal Hose Division) is similar in construction to the common type of flexible metal conduit, but is covered with an outer sheath of thermoplastic material (Fig. 350-1).

350.6. Listing Requirements. Most raceways are required to be listed, as is this one, but in this case it should be noted that unlike other raceways, enormous quantities of unlisted liquidtight flexible metal conduit were sold for many decades. Frequently supply houses would never stock the listed product and it was only available by special order. And the differences in this case are substantial. Only the listed product has copper wound into its convolutions that improve its ground-return effectiveness. In addition there are chemical differences in the stability of the nonmetallic jacket that bear on how well the product holds up in outdoor environments over time. In looking at existing installations, keep in mind that lengths of this product already in use may very well be substandard.

350.10. Uses Permitted. UL data on liquidtight metal flex present a variety of restrictions on the use of LFMC. And the basic wording of 350.10(A) essentially refers the reader to the limitations placed on the use of a particular type of LFMC by its specific listing.

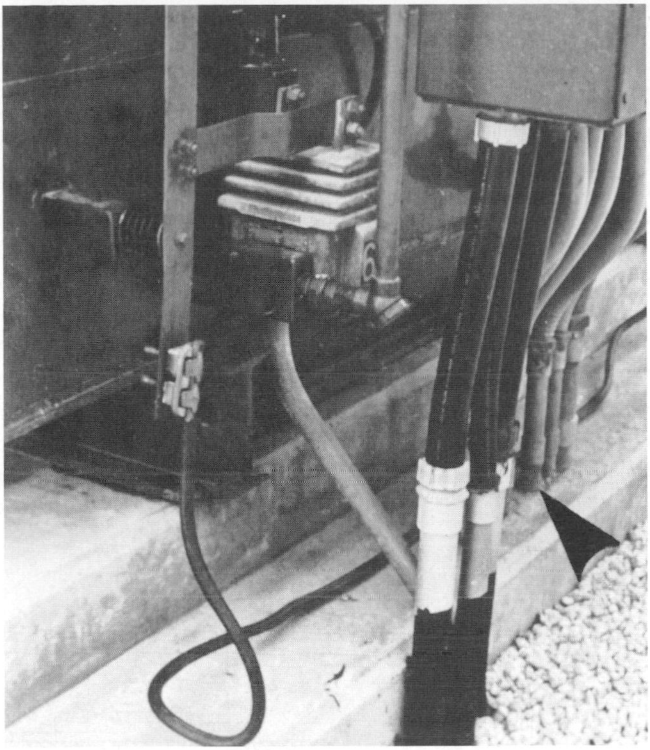

Fig. 350-1. Plastic jacket on liquidtight flex suits it to outdoor use exposed to rain or indoor locations where water or other liquids or vapors must be excluded from the raceway and associated enclosures. In lengths under 6 ft (1.8 m), UL-listed metal liquidtight flex may not require a bonding jumper depending on size and overcurrent protection. See text. (Sec. 350.60.)

As noted in the UL data, liquidtight flexible metal conduit is permitted for use directly buried in the earth if it is "so marked on the product." The rule in 350.10 extends Code recognition to direct burial of liquidtight flexible metal conduit if it is "listed and marked" for such use. Based on many years of such application, liquidtight metal flex is recognized for direct burial, but any such use is permitted only for liquidtight flex that is "listed" by UL, or some other test lab, and is "marked" to indicate suitability for direct burial, to assure the installer and inspector of Code compliance. In the past, successful applications have been made in the earth and in concrete. Standard flexible metal conduit is prohibited from being used "underground or embedded in poured concrete or aggregate." But that prohibition is not placed on liquidtight metal flex.

350.12**(2)** covers issues with operating temperatures that must be kept in mind. The UL Guide Card information (DXHR) provides the following points with respect to how to apply markings that may be found on the product:

> Liquidtight flexible metal conduit suitable for direct burial and in poured concrete is marked "Direct Burial," "Burial," "Dir Burial," or "Dir Bur."

Liquidtight flexible metal conduit not marked with a temperature designation or marked "60 C" is intended for use at temperatures not in excess of 60°C (140°F).

Conduit intended for use in dry or oily locations at a temperature higher than 60°C (140°F) is marked "____ C dry, 60 C wet, 70 C oil res" or "____ C dry, 60 C wet, 70 C oil resistant" with "80" or "105" inserted as the dry-locations temperature.

Conduit marked "80 C dry, 60 C wet, 60 C oil res" or "80 C dry, 60 C oil resistant" is intended for use at 80°C (176°F) and lower temperatures in air, and at 60°C (140°F) and lower temperatures where exposed to water, oil or coolants.

Conduit that has not been investigated for use where exposed to oil is marked "OIL-FREE ENVIRONMENTS ONLY."

350.20. Size. Refer to 348.20, which has the same rules. Figure 350-2 satisfies this subpart if the No. 12 wires are stranded, as required in 430.245(B). Table 348.22 accepts three No. 12 THHN plus an equipment grounding conductor of the same size in metric designator 12 (trade size ⅜) Greenfield or liquidtight.

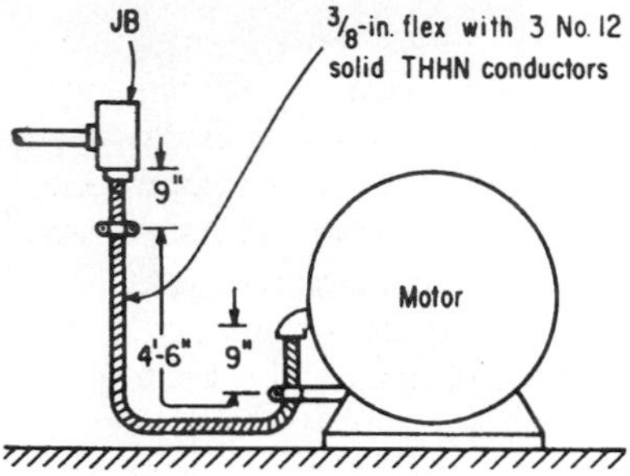
JB
3/8-in. flex with 3 No. 12 solid THHN conductors
9"
Motor
4'-6"
9"
NOTE: The 6 ft length of liquidtight is suitable as an equipment grounding conductor, without bonding.

Fig. 350-2. Both standard flexible metal conduit and liquidtight may be used here. If the connection to the motor is viewed as a vibration isolator, or the motor can be repositioned to adjust belt tension, etc., then the flexible connection will be classified as necessary after installation and an equipment grounding conductor will need to be pulled in with the circuit conductors. (Sec. 350.20.)

350.22. Number of Conductors. The maximum number of conductors must satisfy the rules of Table 1, Chap. 9. Refer to Tables C.5 and C.5(A) in Annex C for metric designators 16 to 103 (trade sizes ½ to 4), and to Table 348.22 for metric designator 12 (trade size ⅜) data.

350.26. Bends—Number in One Run. Figure 348-2 shows this rule.

350.28. Trimming. This is a new section in the 2017 edition, requiring the removal of rough edges after cutting. The panel made an interesting point, namely, that often the end ferrule furnished with the connector won't seat properly if it can't be twisted all the way in because of rough edges. And, if that happens, grounding continuity will be impaired.

350.30. Securing and Supporting. As shown in Fig. 350-3, the rule permits a length of liquidtight flexible metal conduit not over 900 mm (3 ft) long to be used at terminals where flexibility is required without any need for clamping or strapping. Obviously, the use of flex requires this permission for short

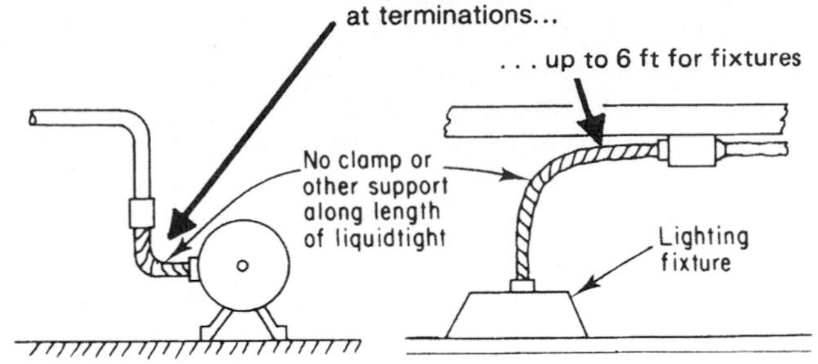
A length not over 3 ft
may be used for flexibility
at terminations...
. . . up to 6 ft for fixtures
No clamp or
other support
along length
of liquidtight
Lighting
fixture

Fig. 350-3. Unsupported length of liquidtight flex is okay at terminations. (Sec. 350.30.)

lengths without support. As in the case of flexible metal conduit, relief is now available for the distance to the first support point for the larger sizes.

In addition, liquidtight metal flex is specifically recognized in lengths up to 1.8 m (6 ft) for fixture "whip," without clamping of the flex. This covers a practice that has long been common. Either standard or liquidtight metal flex may be used to carry supply conductors to lighting fixtures—such as required by 410.117(C), where high-temperature wires must be run to a fixture terminal box. As in the case of other flexible raceways, cable ties are permitted if listed for support and securement. They are frequently used above suspended ceilings.

350.42. Couplings and Connectors. As in the case of 348.42, it is a Code violation to conceal an angle connector, because the eventual outcome will be someone deciding he or she absolutely has to pull through it, and damaging insulation in the process. The 2014 **NEC** has added a clarification here that straight fittings are permitted for direct burial where marked accordingly. Angle connectors belong at the very end of a run, and where they can be disassembled and then reconnected in the field.

350.60. Grounding and Bonding. According to the rule of 250.118, where flexible metal conduit and fittings have not been specifically listed as a grounding means, a separate grounding conductor (insulated or bare) shall be run inside the conduit (or outside, for lengths not over 6 ft [1.8 m]) and bonded at each box or similar equipment to which the conduit is connected. Refer to the discussion at 250.118(5 and 6) for full details of the circumstances under which this wiring method qualifies as an equipment grounding conductor. In addition, UL has some Guide Card information because it presents some of the same information in a mirror image from that in 250.118(6), because it clearly states the types of liquidtight flexible metal conduit that are *not* to be considered as a qualified equipment grounding conductor, as follows:

1. The metric designator 41 (trade size 1½) and larger sizes.
2. The metric designators 12 and 16 (trade sizes ⅜ and ½) sizes where used on circuits rated higher than 20 A, or where the total length in the ground return path is greater than 6 ft.

3. The metric designators 21, 27, and 35 (trade sizes ¾, 1, and 1¼) sizes where used on circuits rated higher than 60 A, or where the total length in the ground return path is greater than 6 ft.

As in the case of flexible metal conduit, where liquidtight flexible metal conduit requires flexibility as part of its normal usage pattern after the installation is complete, a separate equipment grounding conductor must be installed regardless of the size or length or protective ampere rating ahead of the flexible wiring method. As noted previously, the 2011 **NEC** clarified that when this wiring method is used for vibration isolation, it is employing its flexibility after the initial installation and the supplemental bonding conductor is required.

ARTICLE 352. RIGID POLYVINYL CHLORIDE CONDUIT: TYPE PVC

352.2. Definition. Rigid polyvinyl chloride conduit wiring systems include a wide assortment of products (Fig. 352-1). This article was formerly titled "Rigid Nonmetallic Conduit" but in the 2008 **NEC**, it was reserved to the PVC variety, with the fiberglass version (Type RTRC) given its own article (Art. 355). Note that since there is now no **NEC** article to cover nonmetallic conduit in the generic sense, and only three **NEC** articles cover rigid nonmetallic conduits (352 on PVC, 353 on HDPE, and 355 on RTRC), and since 110.8 spells out that only wiring methods recognized in the **NEC** as suitable are the ones enumerated

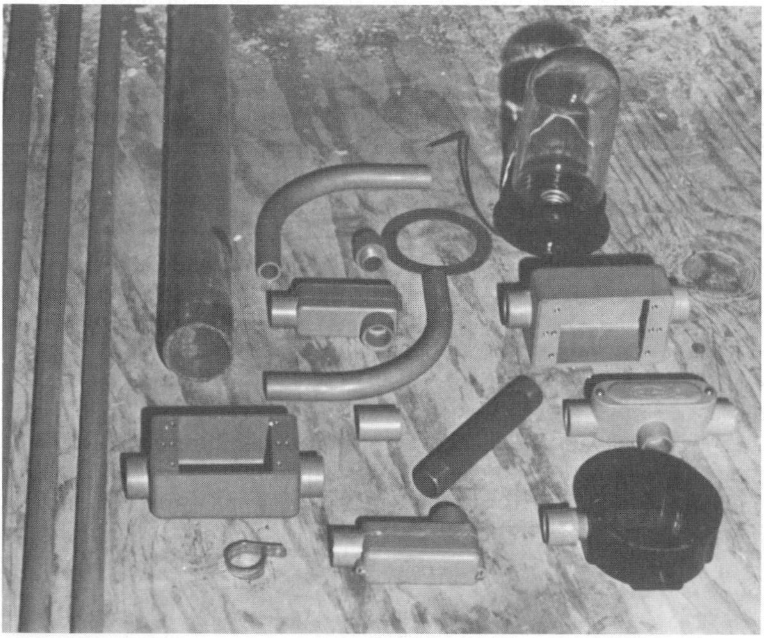

Fig. 352-1. Rigid nonmetallic conduit systems are made up of a wide variety of components—conduit, fittings, elbows, nipples, couplings, boxes, straps. (Sec. 352.2.)

therein, all other forms of rigid nonmetallic conduit, such as styrene, fiber, tile, asbestos cement, soapstone, and others that have been used in the past for underground use are no longer recognized as suitable under the NEC.

UL application data are detailed and divide rigid polyvinyl chloride conduit into two categories. Note that only the first, covering Schedule 40 and 80 PVC, involve wiring methods that can be used above grade (Fig. 352-2). Schedule 80

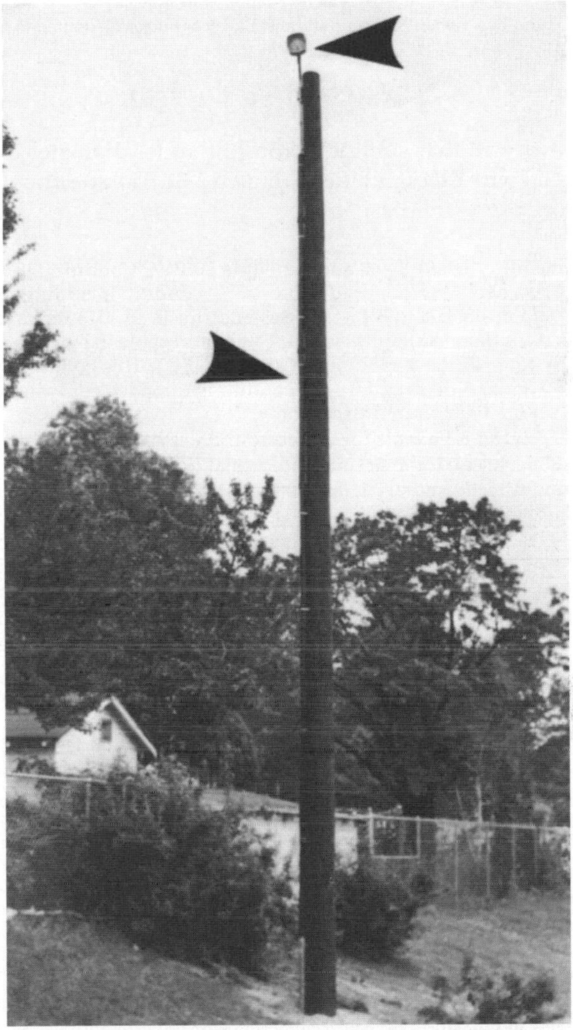

Fig. 352-2. PVC and RTRC conduits are the only rigid nonmetallic conduits that may be used aboveground. And when enclosing conductors run up a pole (shown here feeding a floodlight at top), the PVC conduit must be Schedule 80 PVC conduit if it is exposed to physical damage, such as possible impact by trucks or cars. If the conduit is not so exposed, it may be Schedule 40 PVC conduit. See Sec. 300.5(D). (Sec. 352.2.)

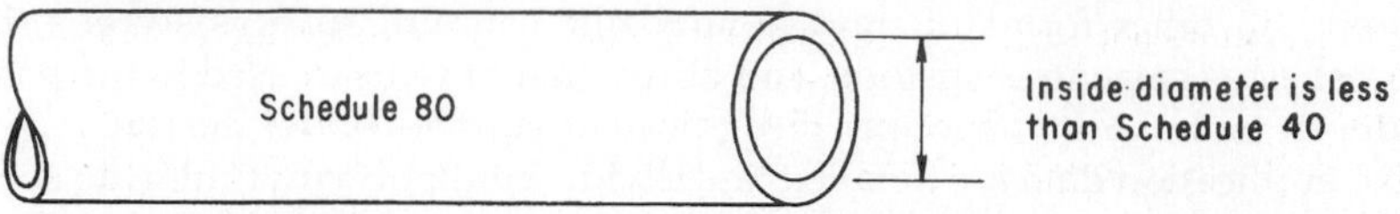
Schedule 80
Inside diameter is less than Schedule 40

For conductor fill to 40% of the cross-section area, refer to data on wire-fill capacity marked on conduit surface.

Fig. 352-3. Extra-heavy-wall PVC conduit must have conductor fill limited to its reduced csa. (Sec. 352.2.)

has a very heavy wall that subtracts from the inner diameter of the raceway, leading to special wire fill calculations (Fig. 352-3). The specific UL instructions on each category are as follows:

Rigid Nonmetallic, Schedule 40 and Schedule 80 PVC Conduit (DZYR)

This category covers rigid nonmetallic PVC conduit including straight conduit and elbows in trade sizes ½ to 6 (metric designators 16 to 155) inclusive, intended for installation as rigid nonmetallic raceway for wire and cable in accordance with Art. 352 of ANSI/NFPA 70, "National Electrical Code" (NEC). This conduit may be Schedule 40, Schedule 80, Type A or Type EB. The conduit is intended for installation and use in accordance with the following information.

Schedule 40 conduit is suitable for underground use by direct burial or encasement in concrete. Schedule 40 conduit marked "Directional Boring" (or "Dir. Boring") is suitable for underground directional boring applications. Schedule 40 conduit is also suitable for aboveground use indoors or outdoors exposed to sunlight and weather where not subject to physical damage. Schedule 40 conduit marked "Underground Use Only" is only suitable for underground applications.

Schedule 80 conduit has a reduced cross-sectional area available for wiring space and is suitable for use wherever Schedule 40 conduit may be used. The marking "Schedule 80 PVC" identifies conduit suitable for use where exposed to physical damage and for installation on poles in accordance with the NEC.

Type A, Type EB, and Schedule 40 are intended for underground use under the following conditions, as indicated in the Certification Mark:

Type A—Installed with its entire length in concrete in any underground location.

Type EB—Installed with its entire length in concrete in trenches outside of buildings.

Schedule 40—Direct burial with or without being encased in concrete.

Where conduit emerges from underground installation, the wiring method is intended to be of a type recognized by the NEC for the purpose.

Unless marked for higher temperature, rigid nonmetallic conduit PVC is intended for use with wire rated 75°C or less including where it is encased in concrete within buildings and where ambient temperature is 50°C or less. Where encased in concrete in trenches outside of buildings it is suitable for use with wires rated 90°C or less.

Certified PVC conduit is inherently resistant to atmosphere containing common industrial corrosive agents and will also withstand vapors or mist of caustic, pickling acids, plating bath, and hydrofluoric and chromic acids.

PVC conduit and elbows (including couplings) that have been investigated for direct exposure to other reagents may be identified by the designation "Reagent Resistant" printed on the surface of the product. Such special uses are described as follows: Where exposed to the following reagents at 60°C or less: Acetic, Nitric (25°C only) acids in concentrations not exceeding ½ normal; hydrochloric acid in concentrations not exceeding 30 percent; sulfuric acid in concentrations not exceeding 10 normal; sulfuric acid in concentrations not exceeding 80 percent (25°C only); concentrated or dilute ammonium hydroxide; sodium hydroxide solutions in concentrations not exceeding 50 percent; saturated or dilute sodium chloride solution; cottonseed oil, or ASTM 3 petroleum oil.

Schedule 40, Schedule 80, Type EB, and Type A PVC conduit are designed for connection to all PVC couplings, fittings, and boxes by the use of a suitable solvent-type cement. Instructions supplied by the solvent-type cement manufacturer describe the method of assembly and precautions to be followed.

Note: As a result of the wording and intent of NEC 110.3(B), all the preceding application data constitute mandatory rules of the NEC itself—subject to the same enforcement as any other NEC rules.

When equipment grounding is required for metal enclosures of equipment used with rigid nonmetallic conduit, an equipment grounding conductor must be provided. Such a conductor must be installed in the conduit along with the circuit conductors (Fig. 352-4).

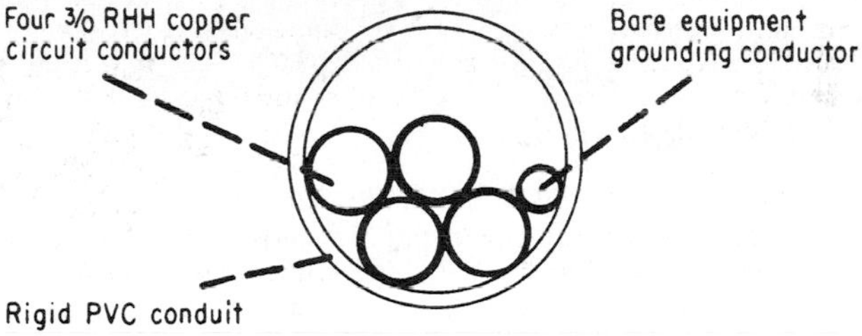
Four 3/0 RHH copper circuit conductors
Bare equipment grounding conductor
Rigid PVC conduit

Equipment grounding conductor is sized from Table 250.122, based on the rating of the fuses or CB protecting the circuit conductor in the conduit. With 200-amp protection for the 3/0 conductors here, the equipment grounding conductor must be at least a No. 6 copper or No. 4 aluminum.

Fig. 352-4. Equipment grounding conductor must be used "within" the rigid nonmetallic conduit. (Sec. 352.60).

352.10. Uses Permitted. This section applies to use of the conduit for circuits operating at any voltage (up to 1000 V and at higher voltages). The rules make rigid nonmetallic conduit a general-purpose raceway for interior and exterior wiring, concealed or exposed in wood or masonry construction—under the conditions stated. The Schedule 80 variety is acceptable where there would be a strong likelihood of damage to a less robust conduit.

Rigid nonmetallic conduit may be used aboveground to carry high-voltage circuits without need for encasing the conduit in concrete. That permission is also given in 300.37. Aboveground use is permitted indoors and outdoors.

Part **(G)** covers underground applications of all the types of rigid nonmetallic conduit—for circuits up to 1000 V, as regulated by 300.5; and for circuits over 1000 V, as covered by 300.50 (Fig. 352-5). Directly buried nonmetallic conduit carrying high-voltage conductors does not have to be concrete-encased if

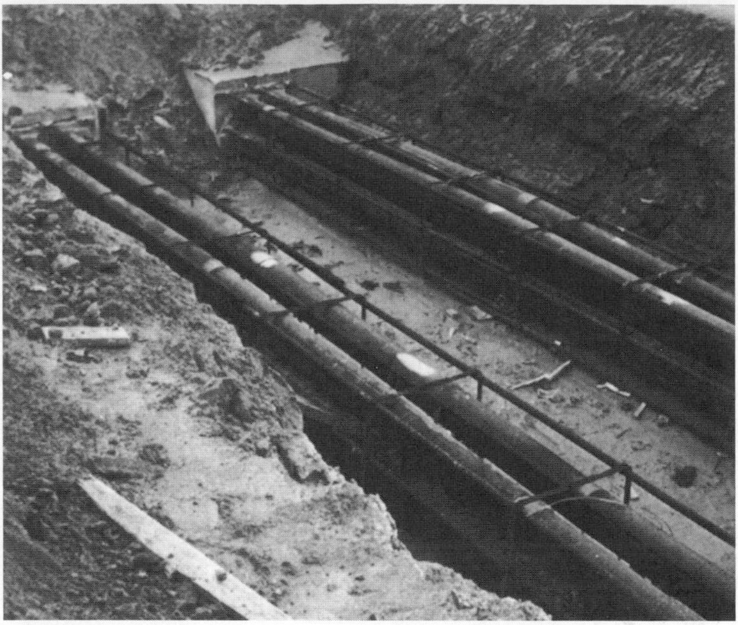

Fig. 352-5. All UL-listed rigid nonmetallic conduits are acceptable for use underground. PVC Schedule 40 and Schedule 80 do not require concrete encasement. Other types must observe UL and **NEC** rules on concrete encasement. (Sec. 352.10.)

it is a type listed for use without concrete encasement. If concrete encasemen is required, it will be indicated on the UL label and in the listing.

Figure 352-6 shows both underground and aboveground application. Referring to the circled numbers: (1) The burial depth must be at least 18 in. (450 mm) for any circuit up to 1000 V. The buried conduit may be Schedule 40 or Schedule 80 (either without concrete encasement) or Type A or Type EB (both require concrete encasement). Refer to the opening coverage here and 300.5. (2) The concrete encasement where the conduit comes up from its 18-in. (450-mm) depth was required at one time by the **NEC**, but is no longer required. (See 300.5.) (3) The radius of the bend must comply with Table 2, Chap. 9 (minimum 18 in. [457.2 mm] if done in the field with a bending box; 330.2 mm [13 in.] if done in the factory with some form of one-shot bender). The conduit aboveground, on a pole or on a building wall, must be Schedule 80 if the conduit is exposed to impact by cars or trucks or to other physical damage. If the conduit is not exposed to damage it may be Schedule 40.

In many cases where nonmetallic conduit is used to enclose conductors suitable for direct burial in the earth, inspectors and engineering authorities have accepted use of any type of conduit—PVC, polyethylene, styrene, and so on—without concrete encasement and without considering application of Code rules to the conduit. The reasoning is that because the cables are suitable for direct burial in the earth, the conduit itself is not required at all and its use

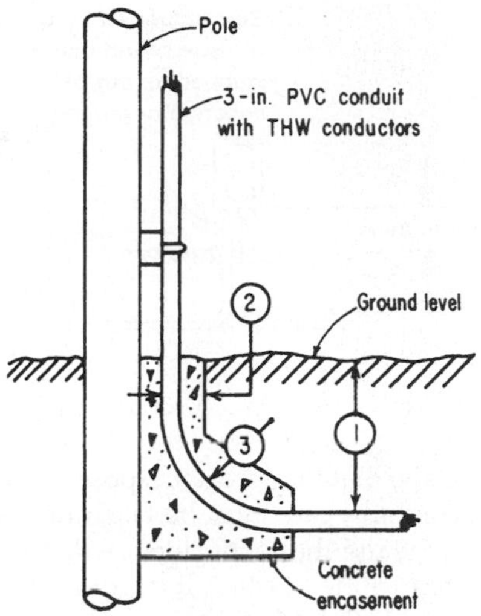
Pole
3-in. PVC conduit with THW conductors
2
Ground level
3
1
Concrete encasement

Numbers in circles refer to text.

Fig. 352-6. Schedule 80 PVC conduit may run up pole from earth to aboveground use. (Sec. 352.10.)

is above and beyond Code rules. But temperature considerations are real and related to effective, long-time operation of an installation. Temperature effects must not be disregarded in any conduit-conductor application.

New as of the 2008 **NEC** was the recognition of a foam-core PVC product. This conduit is significantly lighter than the customary solid product. It was referred to in the 2008 **NEC** as "nonhomogeneous" PVC and it was implicity permitted underground as both a direct-burial wiring method and where encased in concrete. As of the 2011 edition, it is no longer mentioned. The intent is to fully recognize it for all applications where conventional PVC is used. It is tested to the exact same product standard as the conventional product. The lack of mention now simply reflects that it is PVC conduit and can be used in accordance with this article.

Although it is generally prohibited from supporting "equipment," part **(H)** correlates with 314.23(E) Exception and recognizes the use of rigid nonmetallic conduit to support nonmetallic conduit bodies, provided the conduit body is no larger than the largest conduit that is providing the support.

Probably the most important use of part **(I)** will be the use of PVC to enclose MV-105 conductors that have a 105°C temperature rating but that are being operated so as to not exceed 90°C. This is often done to increase reliability and to extend the expected project life of a job.

352.12. Uses Not Permitted. It should be noted that nonmetallic conduit is not permitted in ducts, plenums, and other air-handling spaces. See 300.21 and the comments following 300.22.

Figure 352-7 shows a difference in application rules between rigid nonmetallic conduit and metal conduit with respect to supporting equipment. Part **(B)** allows limited use of rigid nonmetallic conduit for support of nonmetallic conduit

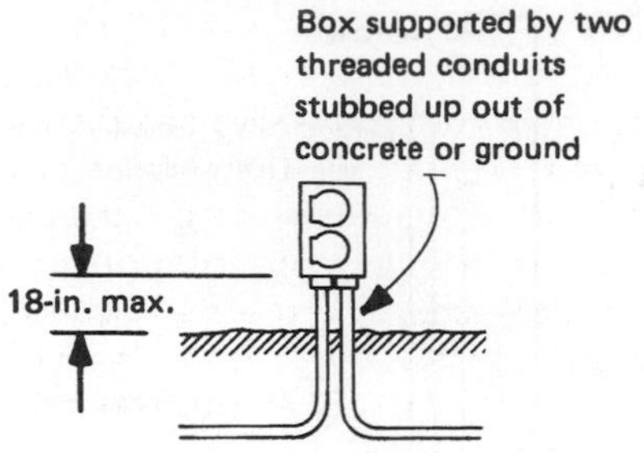
Box supported by two threaded conduits stubbed up out of concrete or ground
18-in. max.

Fig. 352-7. This is okay for rigid metal conduit but not for rigid nonmetallic conduit. (Sec. 352.12.)

bodies that do *not* contain devices or fixtures, as described in 352.10(H). This provision correlates with 314.23(F).

Part **(D)** requires care in use of the conduits so that they are not exposed to damaging temperatures. In using nonmetallic conduits, care must be taken to ensure temperature compatibility between the conduit and the conductors used in it. For instance, a conduit that has a 75°C temperature rating at which it might melt and/or deform must not be used with conductors which have a 90°C temperature rating and which will be loaded so they are operating at their top temperature limit. There is available PVC rigid conduit listed by UL and marked to indicate its suitability for use with all 90°C-rated conductors, thereby suiting the conduit to use with 90°C-rated conductors. The UL data described in 352.2 give the acceptable ambient temperatures and conductor temperature ratings that correlate to these NEC rules. Conductors with 90°C insulation may be used at the higher ampacities of that temperature rating only when the conduit is concrete encased, or marked with evidence of suitability for this temperature exposure (Fig. 352-8).

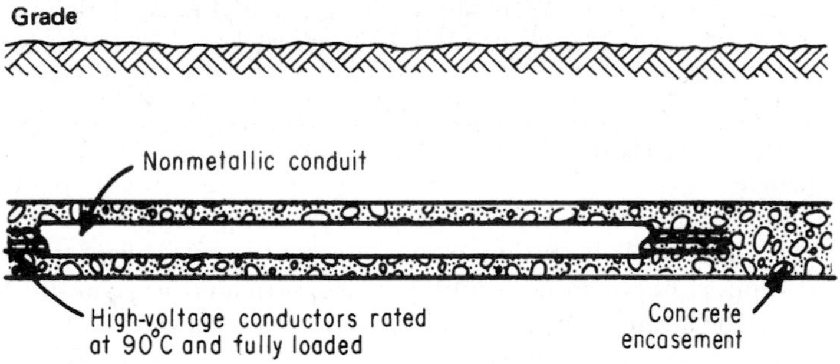
Grade
Nonmetallic conduit
High-voltage conductors rated at 90°C and fully loaded
Concrete encasement

Fig. 352-8. UL data indicate that this violates Sec. 352.12(E) unless it is marked as suitable for such applications. (Sec. 352.12.)

352.22. Number of Conductors. Refer to 344.22.

352.24. Bends—How Made. Refer to 344.24 and the discussion in this chapter in 352.10. The radius of field bends should be based on the "One Shot and Full Shoe" column in Chap. 9, Table 2 provided bending equipment identified for the purpose is used properly. Usually the conduit is heated with a special electric blanket or placed in a bending box designed for this purpose. A blow torch is most certainly not appropriate and most inspectors, seeing the tell-tale scorch marks, will fail a job where this has been done. Larger conduits must have closure plugs placed in the ends prior to heating, which results in heated compressed air providing interior support to the conduit until the bend has been cooled and set. Smaller size conduits can also be bent cold using special removable interior springs that support the walls while the conduit is bent, using a full-shoe rigid conduit bender if necessary.

352.26. Bends—Number in One Run. Refer to 344.26.

352.28. Trimming. See Fig. 352-9.

Fig. 352-9. PVC conduit is designed for connection to couplings and enclosures by an approved cement, but leaving rough edges in the conduit end is a clear violation of Sec. 352.28. (Sec. 352.28.)

352.30. Securing and Supporting. In this section, Table 352.30(B), giving the maximum distance between supports for rigid nonmetallic conduit, permits greater spacing than some previous NEC editions. For each size of rigid nonmetallic conduit, a single maximum spacing between supports, in feet, is given for all temperature ratings of conductors used in rigid nonmetallic conduit raceways (Fig. 352-10).

The wording in the paragraph of part (B) here is similar to subparts in Code articles covering other raceways and cables. This wording specifically recognizes holes in framing members as providing support for rigid nonmetallic conduit.

The special support limitations for conduit nipples imposed in the 2008 NEC have been revoked and the text returned to that of the 2005 edition. Any nipple not over the general limit on the first point of support in length, namely 900 mm

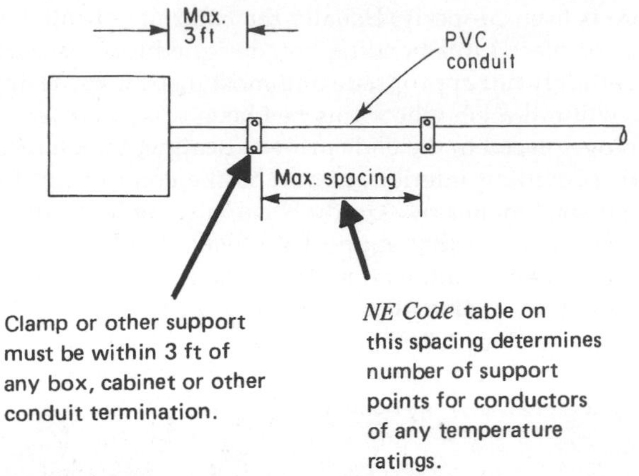
Max. 3 ft
PVC conduit
Max. spacing
Clamp or other support must be within 3 ft of any box, cabinet or other conduit termination.
NE Code table on this spacing determines number of support points for conductors of any temperature ratings.

Fig. 352-10. Support rules on nonmetallic conduit are simple and direct. (Sec. 352.30.)

(3 ft), is considered supported by its terminating hardware and no additional support is generally required.

352.44. Expansion Fittings. In applications in which the conduit installation will be subject to constantly changing temperatures and the runs are long, consideration must be given to expansion and contraction of PVC conduit. In such instances an expansion coupling should be installed near the fixed end of the run to take up any expansion or contraction that may occur. The normal expansion range of these fittings is about 6 in. (150 mm). The coefficient of linear expansion of PVC conduit is given in Table 352.44 of the NEC, and exceeds the expansion coefficient of steel by a factor of five. Without properly applied expansion fittings an aboveground installation subject to the usual range of outdoor temperatures will fail. If the job is installed on a cold winter day, and the conduit cannot freely move when the hot weather arrives, the result will look like an accordion. If the same job is installed on a hot day, come the winter months the conduit will contract to the point of pulling out of glue joints and exposing the conductors within.

In addition, when PVC conduit is exposed to direct sun, it absorbs even more heat than the surface wired over. A leading maker of PVC conduit recommends that at least 140°F be used as the upper temperature design parameter for this reason. If the minimum temperature were 0°F, that change in temperature would cause a 100 ft (30 m) length of conduit to expand and contract through a range of almost 6 in. (150 mm). The wiring system must be arranged with this in mind. Of course, some parts of the system will handle this movement without difficulty. For example, a service riser running straight up from a meter socket and ending at a weatherhead, if properly supported with clamps designed to allow movement, can expand and contract at will.

Caution is in order when one end of a straight length of PVC conduit is securely held, such as when it emerges from concrete embedment, and connects to an enclosure. This section does not require an expansion fitting if the expected expansion and contraction does not exceed 6 mm (¼ in.). If that dimension is distributed over both ends of the conduit run, the result is reasonable. However, if the ¼ in. occurs entirely at one end, that degree of movement is easily enough to break anchor points and crack nonmetallic enclosures on a cold day.

Expansion couplings are normally used where conduits are exposed. In underground or slab applications such couplings are seldom used because expansion and contraction can be controlled by *bowing* the conduit slightly. However, the rule of Sec. 300.5(J) now mandates that ground movement be addressed in underground installations. The informational note indicates methods—expansion joints in vertical risers—that may be used to satisfy the rule. Conduits left exposed for an extended period of time without expansion fittings during widely variable temperature conditions should be examined to see if contraction has occurred and compromised the glue joints.

ARTICLE 353. HIGH-DENSITY POLYETHYLENE CONDUIT: TYPE HDPE CONDUIT

353.1. Scope. This article covers the requirements for the installation and use of Type HDPE conduit. It is a tubular raceway of circular cross section, and is available in discrete lengths, or in continuous lengths on a reel. It is available from metric designator 16 (trade size ½) up to and including metric designator 155 (trade size 6). This type of nonmetallic raceway has been used by utilities in various jurisdictions across the country for years. Its high durability and flexibility of application regardless of soil conditions makes it a very good choice for underground installations of electrical conductors. 353.2 provides a definition and 353.6 mandates the use of listed HDPE only.

353.6. Listing. As with the case for all raceway wiring methods, this wiring method must be listed, and with this comes important information from the UL Guide Card information:

This category covers plastic types of rigid, nonmetallic high-density polyethylene (HDPE) conduit, including straight conduit, elbows, and other bends, in sizes ½ to 6 (metric designators 16 through 155) inclusive, intended for installation underground as raceway for wire and cable in accordance with Art. 353 of ANSI/NFPA 70, "National Electrical Code" (NEC). This conduit may be HDPE Schedule 40, Schedule 80, EPEC A, or EPEC B. This conduit is intended for installation and use in accordance with the following information.

The conduit is intended for underground use under the following condition, as indicated in the Certification Mark: Direct burial with or without being encased in concrete (HDPE Schedule 40, Schedule 80, EPEC A, EPEC B). The conduit is intended for use in ambient temperatures of 50°C or less.

Unless marked otherwise, HDPE conduit is intended for use with wire rated 75°C or less, or when directly buried or encased in concrete in trenches outside of buildings, it may be used with wire rated 90°C or less.

Where conduit emerges from underground installation, the wiring method is intended to be of a type recognized by the NEC for the purpose.

HDPE conduit is designed for joining by threaded couplings, drive-on couplings, or a butt-fusing process. Instructions supplied by the solvent-type cement manufacturer describe the method of assembly and precautions to be followed.

353.10. Use Permitted. Part **(1)** indicates that Type HDPE is available in both individual cut-lengths or it may be supplied on a reel. It is suited for use in severely corrosive environments, cinder fill, and underground in direct contact with earth or concrete.

Part **(6)** includes similar recognition for conductors with insulation ratings that exceed those of the HDPE conduit, under the condition that they are not operated at such a temperature. Refer to the discussion at 352.10(I) for more information on this topic.

353.12. Uses Not Permitted. It must not be used above 50°C, either by reason of a high ambient temperature, or high operating temperatures of the enclosed conductors, or both. Polyethylene is flammable, and for that reason it is generally limited to direct burial applications. If not specifically prohibited, it is permitted above grade if encased in a concrete envelope not less than 50 mm (2 in.) thick. It is not permitted to be exposed, and it must not be used inside buildings.

353.22. Number of Conductors. Refer to 344.22.

353.24. Bends—How Made. NEC Table 354.24 specifies the minimum bending radius for this material, which is more restrictive than most tubular raceways. The table begins with a 250 mm (10 in.) radius for metric designator 16 (trade size ½) conduit, and rises to a 1.5 m (5 ft) minimum radius for the trade size 4 product. There is no table entry as of the 2014 NEC for metric designator 129 or 155 (trade size 5 or 6) conduit, so the manufacturer's directions need to be consulted for these sizes.

353.26. Bends—Number in One Run. Refer to 344.26.

353.48. Joints. All joints must be made by an approved method. The 2008 NEC added a note at this location suggesting the three methods of successfully splicing this wiring method, including heat fusion or electrofusion along with mechanical fittings.

ARTICLE 354. NONMETALLIC UNDERGROUND CONDUIT WITH CONDUCTORS: TYPE NUCC

This is the same product as Type HDPE conduit, and it follows the same installation rules, but conductors are preinstalled and shipped with the product by the manufacturer. Note that even though conductors arrive with the product preinstalled, it is still classified as a raceway, and the 360° maximum bends-in-the-run rule continues to apply. In this form, the upper size limit is metric designator 103 (trade size 4) conduit, also using the special bend radius Table 354.24.

ARTICLE 355. REINFORCED THERMOSETTING RESIN CONDUIT: TYPE RTRC

355.2. Definition. This is the fiberglass entry in the nonmetallic conduit market. At one time it was only permitted for below-grade applications, but advances in chemistry have resulted in materials that meet above-grade fire resistance tests and it is now permitted for use in buildings where concealed in walls, floors, and ceilings, and also where exposed if identified for the application. It is stiffer than PVC and has a much lower coefficient of thermal expansion (about 45 percent of the value for PVC). Although the NEC uses the same support distance table as for PVC conduit, it does allow for longer support intervals if the product is listed for larger distances. It is more difficult to bend in the field, although an extensive range of different bend angle sweeps is available to accommodate field installation issues. It must not be used above 50°C unless listed for a higher temperature. The article is, in effect, a carbon copy of Art. 352 with only one significant difference: it has its own thermal expansion table, with a coefficient of expansion appreciably less (about 45 percent) of that for rigid PVC conduit. The "XW" variety of this conduit (see below) is its counterpart to Schedule 80 PVC. The UL Guide Card information (Guide Card DYJC) follows:

This category covers reinforced thermosetting resin conduit and fittings intended for installation in accordance with Art. 355 of ANSI/NFPA 70, "National Electrical Code" (NEC).

Reinforced thermosetting resin conduit is certified in trade sizes ½ through 6 (metric designators 16 through 155) inclusive, in IPS and ID dimensions, and in trade sizes ¾ through 6 (metric designators 21 through 155) inclusive, in XW dimensions as marked on the product. Certification includes straight conduit, elbows, and other fittings, unless otherwise noted.

Reinforced thermosetting resin conduit has been investigated for use at –40°C (–40°F) to 110°C (230°F).

Reinforced thermosetting resin conduit is designed for connection to couplings, fittings, and boxes by use of a suitable epoxy type cement or drive-on bell and spigot. Instructions supplied by the epoxy-type-cement manufacturer describe the method of assembly and precautions to be followed.

The conduit is designated "EB" (Encased Burial) or "DB" (Direct Burial), which refers to specific wall thicknesses. EB conduit is suitable for encasement in concrete. DB conduit is suitable for encasement in concrete and direct burial. Conduit marked "Below Ground" (or "BG") has been investigated for underground use only for direct burial, with or without being encased in concrete.

Conduit marked "Above Ground" (or "AG") has been investigated for use aboveground, underground, and for direct burial with or without encasement in concrete. This conduit has been investigated for concealed or exposed work where not subject to physical damage. The conduit is designated "SW" (Standard Wall) or "HW" (Heavy Wall), which refers to specific wall thicknesses.

XW-type reinforced thermosetting resin conduit, which refers to specific wall thicknesses, is certified as suitable for use where exposed to physical damage in accordance with the NEC and is suitable for use wherever IPS and ID conduit may be used. The marking "AG, XW, RTRC" identifies conduit suitable for aboveground use and where exposed to physical damage in accordance with the NEC.

Reinforced thermosetting resin conduit, elbows, and other fittings investigated for direct exposure to reagents are identified by the designation "Reagent Resistant" and are marked to indicate the specific reagents.

ARTICLE 356. LIQUIDTIGHT FLEXIBLE NONMETALLIC CONDUIT: TYPE LFNC

356.2. Definition. This is a flexible nonmetallic raceway of circular cross section, available in three forms. Type LFNC-A has a smooth, seamless inner core and cover bonded together, with reinforcement between the core and cover layers. Type LFNC-B has a smooth inner surface together with reinforcement within the conduit wall. This is its most usual form. Type LFNC-C has a corrugated inner and outer surface, with no reinforcement in the wall. It is generally limited to 6 ft (1.8 m) lengths unless a longer length is required for the amount of flexibility called for at the point of use. However, the "B" style does not carry this length limitation.

356.10. Uses Permitted. Liquidtight flexible nonmetallic conduit may be used exposed or concealed and also may be used for direct burial in earth if "listed and marked for the purpose." This extends similar permission to liquidtight flexible *nonmetallic* conduit that was given for liquidtight flexible *metallic* conduit in the 1987 NEC. And 356.10 recognizes this nonmetallic flex for "concealed" as well as exposed locations. This product [356.10(6)] is also authorized to be produced as a listed prewired assembly in the metric designator 16 through 27 (trade size ½ through 1) sizes. The 2014 NEC removed the over 600-V prohibition from 356.12, thereby making a formerly specialized application for neon sign work in 600.32(A) completely unrestricted in this article. Note that this application is not, as yet, correlated with the permissible wiring methods for over 1000-V work in 300.37 or for motors in 430.223. Note that the former limitation for uses longer than 6 ft to the "B" style product [in (5)] was lifted in the 2017 NEC, and so any of the three styles can now be used for long runs as long as the support rules in 356.30 are met. Supposedly the relevant product testing for all three styles is the same.

356.20. Size. Although metric designator 16 (trade size ½) trade size is the smallest recognized size of liquidtight flexible nonmetallic conduit for general use, Part **(A)(1)** notes that metric designator 12 (trade size ⅜) liquidtight flexible metal conduit may be used for motor leads. This was added to coordinate with 430.245(B) for motors with detached junction boxes. The other part allows 1.8 m (6 ft) lengths for utilization equipment connections and where "part of a listed assembly for tap connections to luminaires as required in 410.117(C)." The upper size limit is metric designator 103 (trade size 4).

ARTICLE 358. ELECTRICAL METALLIC TUBING: TYPE EMT

358.10. Uses Permitted. As is the case with most other raceways, the NEC only recognizes the use of *listed* EMT. EMT is a general-purpose raceway of the same nature as rigid metal conduit and IMC. Although rigid metal conduit and IMC afford maximum protection for conductors under all installation conditions, in many instances it is permissible, feasible, and more economical to use EMT to enclose circuit wiring. Because EMT is lighter than conduit however, and is less rugged in construction and connection details, the NEC restricts its use (Art. 358) to locations (either exposed or concealed) where it will not be subjected to severe physical damage or (unless suitably protected) to corrosive agents.

EMT distribution systems are constructed by combining wide assortments of related fittings and boxes. Connection is simplified by employing threadless components that include compression, indentation, and set-screw types.

Some questions have been raised about the acceptability of EMT directly buried in soil. These questions are even more difficult following substantial changes in content and organization of this material in the 2017 NEC. At first glance, the answer appears now to be an unqualified "yes" for galvanized steel EMT. Sec. 358.10(A) now gives permission for use "in concrete, in direct contact with the earth, or in areas subject to severe corrosive influences." The sentence then concludes with a reference to 358.10(B); the question immediately arises, does this phrase modify the entire sentence or just the last of the three items? The way the sentence is structured, including the use of the coordinating conjunction "or" suggests all three topics are indeed being referred to 358.10(B)(1), since that rule, after the permissive opening, identifies the three topics in exactly the same way as (A). The bottom line appears to be that it can be buried without additional corrosion protection beyond the galvanizing, provided the inspector approves it as suitable for the local soil conditions. However, the UL Guide Card (FJMX) information for this product is as yet unchanged. In comparing At this point it is useful to compare the UL listing information on EMT and galvanized steel rigid conduit. The relative acceptability of the different steel raceways will come down to the limitations placed on these listed products by the testing laboratories and local experience with soil conditions. In the UL listing on "Electrical Metallic Tubing," a note says that "galvanized steel electrical metallic tubing in a concrete slab below grade level *may* require supplementary corrosion protection." (That word *may* leaves the decision up to the designer and/or installer, subject to final review by the inspector.)

The next note says, "In general, galvanized steel electrical metallic tubing in contact with soil requires supplementary corrosion protection." That sentence virtually requires that direct burial use include supplementary corrosion protection. Now compare the equivalent sentence from the Guide Card information on rigid ferrous metal conduit, "Galvanized rigid ferrous metal conduit installed in contact with soil does not generally require supplementary corrosion protection." The identical wording goes with IMC, and both sentences create a presumption of acceptability for the heavy wall products that is not there for EMT. Further, UL does not evaluate supplementary corrosion protection on EMT for this use. Although EMT may be available with supplementary protection, the Guide Card information on this topic reads: "Galvanized electrical metallic tubing that is provided with a metallic or nonmetallic coating, or a combination of both, has been investigated for *resistance to atmospheric corrosion*" (emphasis supplied). This means that inspectors are entirely on their own if they recognize any form of supplementary protection with respect to the corrosive influences of soil. And supplementary corrosion protection must always be applied unless there is a solid local record of positive experience, which is very unusual. On balance, EMT should not be used for direct burial absent a solid local track record of benign soil conditions.

The first paragraph of Part B also addresses stainless steel EMT. Intriguingly, the UL information specifically indicates that this product, along with its galvanized steel counterpart, does not need supplementary protection in concrete either on grade or above grade. However, there is no mention one way or another as to how the stainless steel product would perform below grade. That, then would be the inspector's call, but in most cases a stainless product would probably survive soil burial.

Part **(B)**, in the second paragraph, addresses aluminum EMT, and here there is no question; supplementary protection must be provided for both within concrete and also for direct burial. And this does line up perfectly with the UL Guide Card information on that topic.

358.14. Dissimilar Metals. This section was added in the 2017 cycle based on recent testing by NEMA as part of the 2017 **NEC** revision cycle, and the fact that stainless steel EMT is now on the market. Interconnections between galvanized heavy wall conduit and aluminum fittings, conduit elbows, etc. are still mostly permitted but not if the conditions are deemed extremely corrosive. Perhaps surprisingly, and similar to other comparable raceways, stainless steel IMC has been ruled out for contact with aluminum as well as other types of steel, and strictly limited to other stainless steel fittings, boxes, and other accessories. The actual NEMA supporting argument for this change is reprinted in this handbook at 344.14.

358.20. Size. The whole concern and discussion regarding the differences of actual cross-sectional area between the various raceways has been rendered moot. That is, in recognition of the differences between actual csa from one conduit or tubing, the table in Chap. 9 covering csa—Table 4—and the tables covering maximum number of conductors of all the same size and insulation within a given size of raceway—now Tables C1 and C1A—have been completely rearranged and revised. The procedure remains the same, but the permitted fill is raceway-specific. And conductor dimensions have been corrected.

358.22. Number of Conductors. Conductor fill for EMT is the same as described under 344.22 for rigid metal conduit.

358.24. Bends—How Made. Refer to 344.24.

358.26. Bends—Number in One Run. Figure 358-1 shows EMT run from a panelboard to a junction box (JB) along the wall—with a total of exactly 360° of bend (from the panel: 45°, 45°, 90°, 90°, 45°, 45°). Note that assigning two opposing 45° angles to ordinary box kicks is unduly restrictive. Opposing 10° angles, so that two such kicks equal about a 45° bend, is more realistic and reasonable.

358.28. Reaming and Threading. Here, the rules clarify Code intent. Threading of electrical metallic tubing is prohibited, but integral couplings used on EMT shall be permitted to be factory threaded. Such equipment has been used successfully in the past and has been found satisfactory. The revised Code rule recognizes such use. But it should be noted that this applies to EMT using *integral threaded fittings,* that is, fittings which are part of the EMT itself as part of

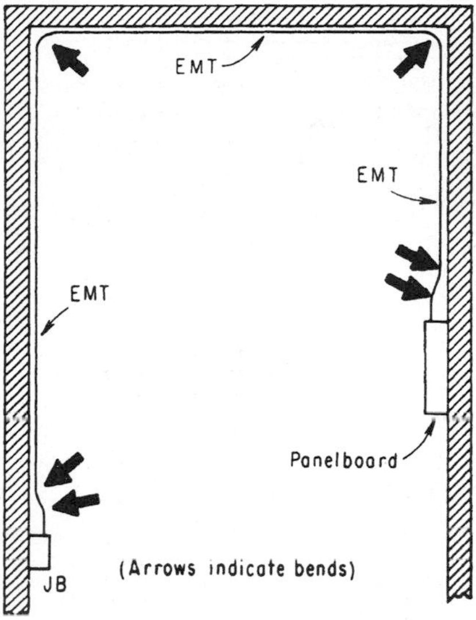
EMT
EMT
EMT
Panelboard
(Arrows indicate bends)
JB

Fig. 358-1. EMT, like other conduit runs, is limited to not over 360° of bends between raceway ends. (Sec. 358.26.)

the manufacturing process. A listing still exists on this product, but apparently it has not been in actual production for many years.

358.30. Securing and Supporting. Figure 358-2 shows this rule applied to an EMT layout. As stated in the basic rule of this section, EMT must be supported every 10 ft (3.0 m) and within 3 ft (900 mm) of each "outlet box, junction box, device box, cabinet, conduit body, or other tubing terminations." Prior to the 1993 **NEC**, this section referred to "each outlet box, junction box, cabinet, and fitting." If the word *fitting* is taken to include couplings, then a strap must be used within 3 ft (900 mm) of each coupling. The definition of *fitting*, given in Art. 100, includes locknuts and bushings. That wording was changed to provide a "laundry list" of enclosures that are covered by this rule. The intent was to clarify that supports are not required within 3 ft (900 mm) of EMT couplings.

As permitted by Exception No. 1, clamps on unbroken lengths of EMT may be placed up to 5 ft (1.5 m) from each termination at an outlet box or fitting where structural support members do not *readily* permit support within 3 ft (900 mm). This exception allows the first clamp to be up to 5 ft (1.5 m) from a termination of EMT at an outlet box. This is like the comparable permission for heavy-wall steel conduits but with an important difference. The EMT between the support 5 ft back and the termination must be unbroken, without coupling. Exception No. 2 allows EMT to be fished; although this may seem odd, it has been done successfully where there is room to stage the unbroken length required.

Part **(B)**, following the two exceptions, makes clear that no additional means of support or securing are needed where framing provides support for horizontal runs at least every 10 ft (3.0 m); however, in addition, the EMT must be secured within

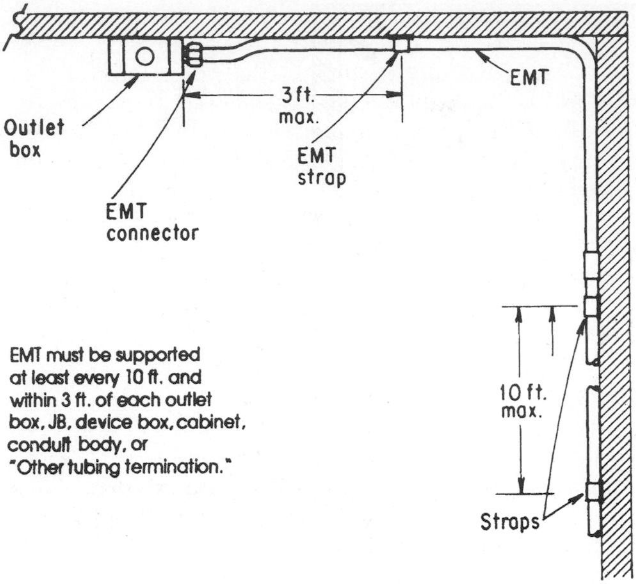
Outlet box
EMT connector
3 ft. max.
EMT strap
EMT
EMT must be supported at least every 10 ft. and within 3 ft. of each outlet box, JB, device box, cabinet, conduit body, or "Other tubing termination."
10 ft. max.
Straps

Fig. 358-2. EMT must be clamped within 3 ft (900 mm) of every enclosure or "fitting." (Sec. 358.30.)

3 ft (900 mm) of every termination. The special support limitations for conduit nipples imposed in the 2008 NEC have been revoked and the text returned to that of the 2005 edition. Any nipple not over the general limit on the first point of support in length, namely 900 mm (3 ft), is considered supported by its terminating hardware and no additional support is generally required.

358.42. Couplings and Connectors. Couplings of the raintight type are required wherever electrical metallic tubing is used on the exteriors of buildings. (See 225.22 and 230.53.) Note that the ability of conventional compression connectors and couplings to be actually raintight has been questioned and UL has revised the product standard, greatly toughening the rain tests on fittings that claim wet location suitability. The result was that every standard compression connector failed. Do not assume that because you are holding a compression connector, you have a fitting that meets the current test requirements for wet locations; look for specific labeling on the product carton. Wet-location suitable fittings now have special designs, usually including internal nonmetallic glands, to pass the new tests. One manufacturer developed a connector with a ferrule similar to what plumbers use to attach brass supply tubes to angle stops under sinks and toilets. This ferrule must engage steel EMT instead of soft brass, however, and the torque specification that comes with the fitting is impressive. It requires a crow-foot open-end wrench extended from a torque wrench, with the wrench reading adjusted for the extra reach of the crow foot, in order to properly seat this fitting.

314.17 requires that conductors entering a box, cabinet, or fitting be protected from abrasion. The end of an EMT connector projecting inside a box, cabinet, or fitting must have smooth, well-rounded edges so that the covering of the wire will not be abraded while the wire is being pulled in. Where ungrounded conductors of size 4 AWG or larger enter a raceway in a cabinet, the EMT connector must have an insulated throat (insulation set around the edge of the connector opening) to protect the conductors. See 300.4(G). For conductors smaller than 4 AWG, an EMT connector does *not* have to be the insulated-throat type. Using THW conductors, a circuit of 4 AWG conductors (a 2- or 3-wire circuit) requires a metric designator 27 (trade size 1) EMT (Table C.1, Annex C, NEC). Therefore, for THW or TW wire, there is no requirement for insulated-throat EMT connectors in the metric designator 16 and 21 (trade size ½ and ¾) sizes. A circuit, say, of three 1 AWG THW wires would call for metric designator 35 (trade size 1¼) EMT, which would require use of insulated-throat connectors—or noninsulated-throat connector with a nonmetallic bushing on the connector end. In the larger sizes, the economics on the makeups can be significantly different. A metric designator 103 (trade size 4) insulated-throat EMT connector might cost $18, whereas a noninsulated-throat connector in that size might cost $10 and $2 for a plastic bushing (Fig. 358-3).

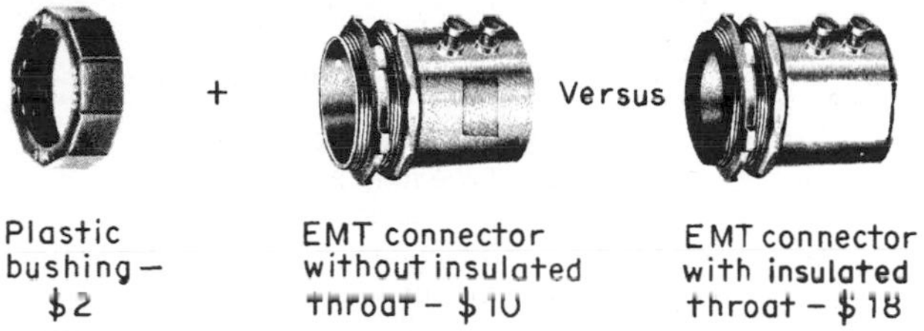
+
Versus
Plastic bushing— $2
EMT connector without insulated throat - $10
EMT connector with insulated throat - $18

Fig. 358-3. Different-cost makeups for 4-in. (102-mm) EMT satisfy Code rules on EMT termination. (Sec. 358.42.)

When an EMT connector is used—either with or without an insulated throat to satisfy 300.4(G)—there is no requirement in Art. 358 that a bushing be used on the connector end. Note, however, that a bushing is required for rigid metal conduit and for IMC as covered in 344.46 and 342.46.

ARTICLE 360. FLEXIBLE METALLIC TUBING: TYPE FMT

360.2. Definition. This section defines this NEC raceway. The rule indicates that flexible metallic tubing is a *raceway*. Use of that term makes clear that all rules applying to "raceways" within the Code apply to Type FMT as well. The rule further indicates that flexible metallic tubing is intended for use where "not subject to physical damage" and gives use above suspended ceilings as an example. Although this wording does not limit its use to air-handling ceilings, it does raise some questions for electrical inspectors with respect to accepting flexible metallic tubing as a general-purpose raceway.

360.6. Listing Requirements. This rule makes it a violation of the Code to use any flexible metallic tubing that is not specifically listed for use with electrical conductors. Ensure that any flexible metallic tubing is listed and marked as Type FMT, which indicates its suitability for use with electrical conductors.

360.10. Uses Permitted. This section limits the use of flexible metallic tubing to branch circuits. In addition, branch-circuit conductors can only be installed in "dry locations"—either concealed or accessible—with systems rated no more than 1000 V. This product has particular utility for making connections in an air-handling ceiling because it completely excludes the transmission of air and it has no nonmetallic elements that could be a source of smoke or fumes. For these reasons it is permitted not just in other spaces for environmental air [300.22(B)], but in actual ductwork for connections, as covered in 300.22(B).

360.12. Uses Not Permitted. Here the Code states those applications for which Type FMT is not permitted. When the proposal was made to add flexible metallic tubing to the Code as a suitable raceway, it was indicated that it had been designed for certain specific applications and not for general use. It was specifically intended for use as the fixture whip on recessed fixtures where high-temperature wire is run from the branch-circuit junction box to the hot wiring compartment in lighting fixtures, an application long filled by flexible metallic conduit (Fig. 360-1). Today the cold-lead applications have largely disappeared with integrally wired thermally-protected luminaires, but this wiring method works

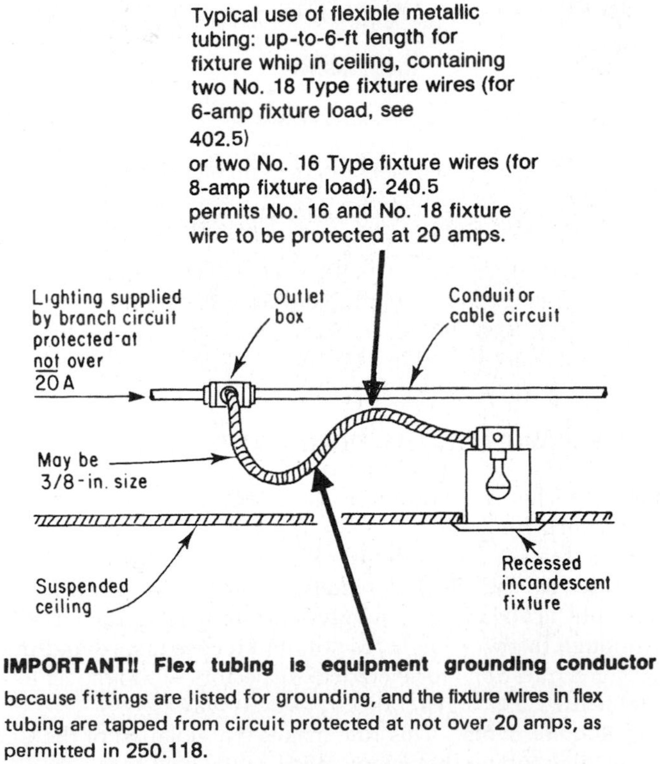
Typical use of flexible metallic tubing: up-to-6-ft length for fixture whip in ceiling, containing two No. 18 Type fixture wires (for 6-amp fixture load, see 402.5) or two No. 16 Type fixture wires (for 8-amp fixture load). 240.5 permits No. 16 and No. 18 fixture wire to be protected at 20 amps.
Lighting supplied by branch circuit protected at not over 20 A
Outlet box
Conduit or cable circuit
May be 3/8-in. size
Suspended ceiling
Recessed incandescent fixture
IMPORTANT!! Flex tubing is equipment grounding conductor because fittings are listed for grounding, and the fixture wires in flex tubing are tapped from circuit protected at not over 20 amps, as permitted in 250.118.

Fig. 360-1. Flexible metallic tubing has limited application. (Sec. 360.12.)

equally well for conventional connections in hung ceilings. It should be noted that the limitation given in part **(6)** of this section limits the use of this raceway to lengths not exceeding 6 ft (1.8 m), which has the effect of effectively limiting this use of this product to the application for which it was originally intended—fixture whips.

360.20. Size. FMT is only available for general purposes in the metric designator 16 and 21 (trade size ½ and ¾) sizes, with the metric designator 12 (trade size ⅜) available for ductwork connections [300.22(B)], connections in other spaces for environmental air [300.22(C)], for luminaire connections, especially in accessible ceilings, and as part of listed assemblies.

360.24. Bends. The allowable bend radius for this product varies significantly based on whether it will be flexed, which should only be infrequently, after the initial installation. There are two tables that provide the required data.

ARTICLE 362. ELECTRICAL NONMETALLIC TUBING: TYPE ENT

362.2. Definition. One type of plastic raceway defined in the Code (Fig. 362-1) is *ENT* (electrical nonmetallic tubing), which is "a pliable corrugated raceway of circular cross section with integral or associated couplings, connectors, and fittings listed for the installation of electrical conductors. It is composed of a material that is resistant to moisture [and] chemical atmospheres and is flame retardant." ENT can be bent by hand, when being installed, to establish direction and lengths of runs.

Fig. 362-1. ENT is a pliable, bendable plastic raceway for general-purpose use for feeders and branch circuits.

362.10. Uses Permitted. Electrical nonmetallic tubing is permitted to be used as a general-purpose, flexible-type conduit in any type of occupancy (Fig. 362-2). ENT is not limited to use in buildings up to three stories high. But, where the building does *not* exceed three stories above grade, ENT may be used in "exposed" locations. Where concealed throughout (and not just above the first three stories), ENT may be used in a building of any height—subject to conditions given in 362.10 and 362.12. ENT may be used:

1. Concealed in walls, floors, and ceilings that provide a thermal barrier with at least a 15-min fire rating from listings of fire-rated assemblies. In the case of walls, this is fairly easy to arrange, since most ½-in. drywall used in commercial construction carries this rating. The same holds true above a drywall ceiling. However, if there is a suspended ceiling (common in commercial occupancies), check with the building inspector. The support grid and the ceiling panels need to be identified as a combination for this duty. For example, having 15-min panels would do no good if the T-bars dumped those panels onto the floor after 11 min of fire exposure. Note that although this is an "exposed" use not normally permitted in high-rise construction, there is specific permission to use this procedure in 362.10(5).

   As previously indicated, ENT may be used *exposed* without these limitations in a building that is not over three floors above grade. The first floor is defined as the one with at least half its exterior wall area at or above grade level; one additional floor level at the base is allowed for vehicle parking or storage, provided it is not designed for human habitation. This is limited permission that recognizes ENT for exposed use under the same limitations that were placed on use of Romex in the past, provided the building finish has a 15-min fire

Fig. 362-2. ENT may be used in residential and nonresidential buildings. (Sec. 362.10.)

rating. In addition, where sprinklers are provided on all floors so as to provide complete occupancy protection, not just the areas in which the ENT is proposed, ENT may be used even if exposed.

2. In severe corrosive locations where suited to resist the particular atmosphere (but not "exposed").
3. In concealed, dry, and damp locations not prohibited by 362.12.
4. Above suspended ceilings with at least a 15-min fire rating (see commentary above).
5. Embedded in poured concrete with fittings that are listed or otherwise identified for that use.
6. In indoor wet locations, or in slab on or below grade, with fittings listed for this application.
7. Metric designator 16 through 27 (trade sizes ½ through 1) sizes are authorized to be prewired as a listed manufactured assembly. Even if installed prewired, it is still a raceway and not a cable and must observe the four quarter bend rule, etc.
8. High-temperature wiring must not be used unless it is certain to be operating, due to limited loading, at temperatures below the rating of the ENT.

362.12. Uses Not Permitted. ENT may not be used in *exposed* locations, except above suspended ceilings of 15-min fire-rated material in buildings of any height above grade. This section excludes ENT from hazardous locations—except for intrinsically safe circuits per Art. 504—from supporting fixtures or equipment, from use where the ambient temperature exceeds that for which the ENT is rated, from direct burial, and from exposed use, with exceptions as noted.

362.20. Size. ENT is Code-recognized in metric designator 16 to 53 (trade size ½ to 2) sizes. A full line of plastic couplings, box connectors, and fittings is available, which are attached to the ENT by mechanical method or cement adhesive (Fig. 362-3).

362.26. Bends—Number in One Run. ENT runs between "pull points"—boxes, enclosures, and conduit bodies—must not contain more than the equivalent of four quarter-bends (360°).

362.30. Securing and Supporting. ENT has enhanced support requirements, with the basic requirement of being securely fastened every 900 mm (3 ft). There is the usual 1.8 m (6 ft) allowance for luminaire and equipment whips above suspended ceilings, and it can also be fished in finished buildings as necessary. Horizontal runs through framing members are also recognized as sufficient provided, as in the case of many other wiring methods, the ends of a run are actually secured and not just supported. Appropriately listed cable ties also make an appearance as being eligible for use with this wiring method.

ARTICLE 366. AUXILIARY GUTTERS

366.1. Scope. Auxiliary gutters differ from wireways only by the way they are applied in the field; they are usually listed for both purposes as they leave a manufacturer. They are available in both sheet metal and nonmetallic forms. They have hinged or removable covers that allow for conductors to be laid

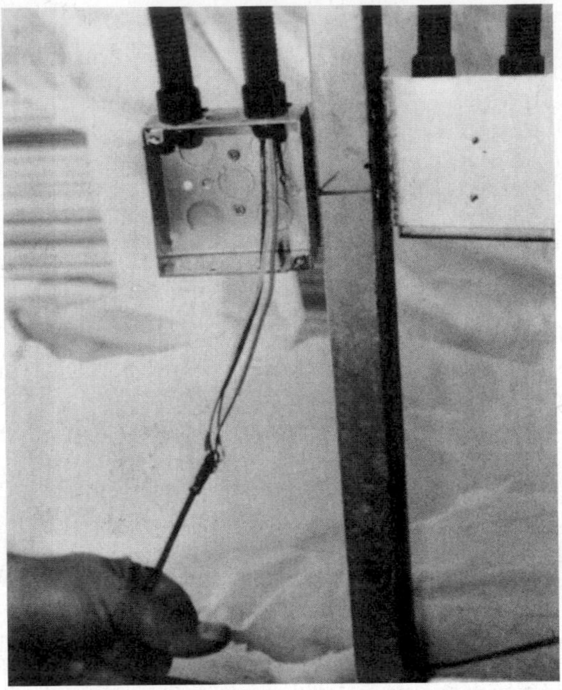

Fig. 362-3. Available in ½-, ¾-, and 1-in. sizes, ENT has a full line of couplings and box connectors. (Sec. 362.2.)

in place after the system is complete. Their function is to supplement wiring spaces at meter centers, distribution centers, switchboards, and similar locations in a wiring system (Fig. 366-1). Note that a careful review of the definitions in Sec. 2 shows that auxiliary gutters are not classified as raceways for the reasons discussed at length in Chap. 1 as part of the analysis of the changes associated with the revision to the raceway definition, and as covered in this commentary.

They are not wireways, which are unlimited in length and intended as a circuit wiring method that connects a line and a load. An auxiliary gutter with a rectangular opening cut to match a similar opening cut in a panelboard, and used to contribute to the wire bending space in the panelboard would be an excellent example This concept is why the auxiliary gutter article has current-carrying limitations for busbars placed in the enclosure, but the wireway article does not address the topic. If a conductor needs to be pulled from the gutter through a nipple to a panel or switchboard, the use may be crossing over into the wireway article. Tha said, the basic field installation rules for conductor fill, derating thresholds, use a pull boxes, and distinctions between metallic and nonmetallic versions are simila Refer to wireway topics in Arts. 376 and 378 for more information.

Auxiliary gutters are available in both metal and nonmetallic forms, but onl the nonmetallic form requires listing, with specific listing requirements give that differ based on whether the use will be outdoors or not.

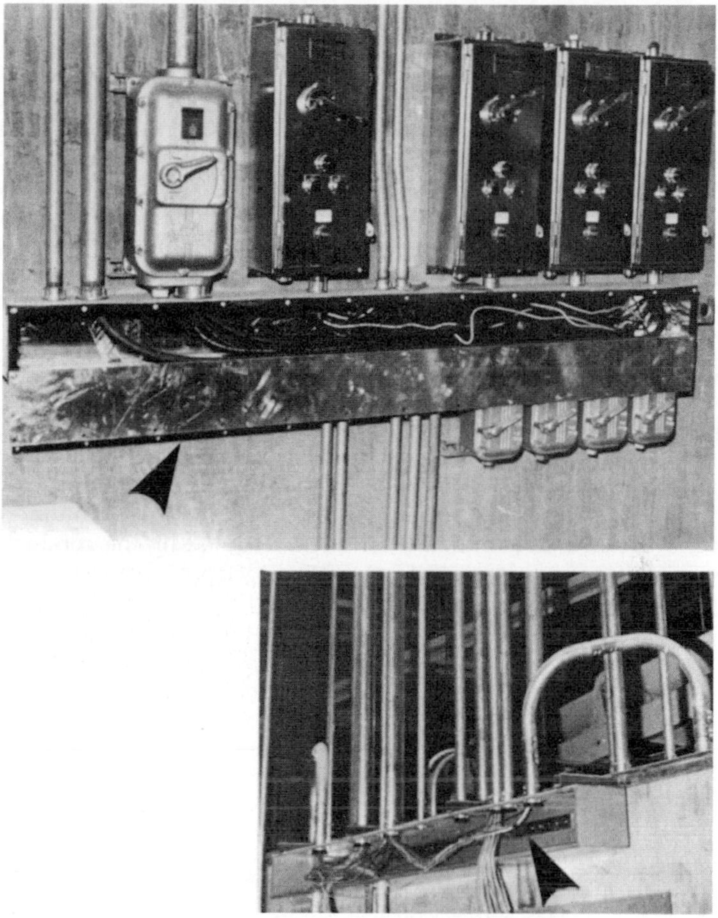

Fig. 366-1. Typical applications of auxiliary gutters provide the necessary space to make taps, splices, and other conductor connections involved where a number of switches or CBs are fed by a feeder (top) or for multiple-circuit routing, as at top of a motor control center (right) shown with a ground bus in gutter (arrow). (Sec. 366.1.)

366.12. Uses Not Permitted. Auxiliary gutters are not intended to be a type of general raceway and are not permitted to extend more than 30 ft (9.0 m) beyond the equipment which they supplement, except in elevator work. Where an extension beyond 30 ft (9.0 m) is necessary, Arts. 376 or 378 for wireways must be complied with. The label of Underwriters Laboratories Inc. on each length of trough bears the legend "Wireways or Auxiliary Gutters," which indicates that they may be identical troughs but are distinguished one from the other by their use. See comments following 376.2 in this handbook.

366.20. Conductors Connected in Parallel. Refer to the commentary at 376.20. The same new provision was made in both articles, along with Art. 378.

366.22. Number of Conductors. The rules on permitted conductor fill for auxiliary gutters are basically the same as those for wireways. Refer to 376.22. Note that part rule permits more than 30 current-carrying conductors—including neutrals in some cases, as described under 310.15(B)(5); but where over 30 such wires are installed, the correction factors specified in 310.15(B)(3)(a) must be applied to all the wires. One of the key differences between metallic and nonmetallic gutters is the fact that the 30-conductor allowance before derating is imposed does not apply to nonmetallic gutters. For nonmetallic auxiliary gutters the derating factors apply after the first three current-carrying conductors, just like any tubular raceway. Metal gutters are much better at providing a heat sink and a surface that easily radiates heat away from itself.

No limit is placed on the size of conductors that may be installed in an auxiliary gutter.

Figure 366-2 shows a typical gutter application where the conductor sizes and fill must be calculated to determine the acceptable csa of the gutter. There are several factors involved in sizing auxiliary gutters that often lead to selecting the wrong size. The two main factors are how conductors enter the gutter and the contained conductors at any cross section. The minimum required width of a gutter is determined by the csa occupied by the conductors and splices and the space necessary for bending conductors entering or leaving the gutter. The total csa occupied by the conductors at any cross section of the gutter must not be greater than 20 percent of the gutter interior csa at that point (366.22). The total csa occupied by the mass of conductors and splices at any cross section of the gutter must not be greater than 75 percent of the gutter interior csa at that point [366.56(A)].

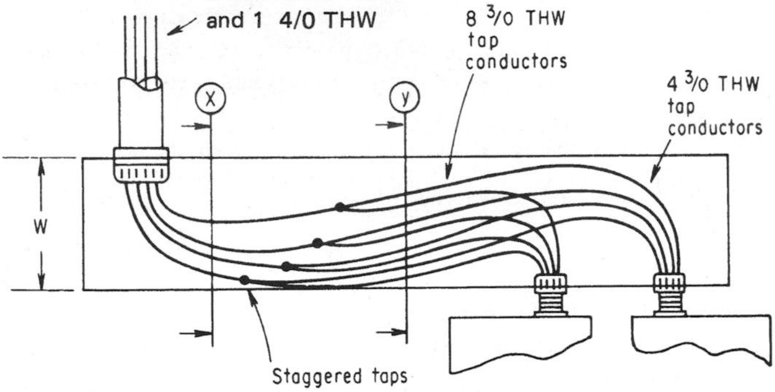
Supply conductors
3-300 kcmil THW
and 1 4/0 THW
8 3/0 THW
tap
conductors
4 3/0 THW
tap
conductors
X
y
W
Staggered taps

Fig. 366-2. Minimum acceptable gutter cross section and depth must be calculated. (Sec. 366.22.)

In the gutter installation shown in Fig. 366-2, assume that staggering of the splice has been done to minimize the area taken up at any cross section—to keep the mas of splices from all adding up at the same cross section. The greatest conductor cor centration is therefore either at section *x*, where there are three 300-kcmil and on

4/0 THW conductors, or at section y, where there are eight 3/0 THW conductors. To determine at which of these two cross sections the fill is greater, apply the appropriate csas of THW conductors as given in Table 5, Chap. 9:

1. The total conductor csa at section x is 3×0.5281 sq in. plus 1×0.3718, or 1.9561 sq in.
2. The total conductor csa at section y is 8×0.3117 sq in. or 2.4936 sq in.

Section y is, therefore, the determining consideration. Because that fill of 2.4936 sq in. can at most be 20 percent of the gutter csa, the total gutter area must be at least 5 times this conductor fill area, or 12.468 sq in.

Assuming the gutter has a square cross section (all sides of equal width) and the sides have an integral number of inches, the nearest square value would be 16 sq in., indicating a 4- by 4-in. gutter, and that would be suitable if the 300-kcmil conductors entered the end of the gutter instead of the top. But because those conductors are deflected entering and leaving the gutter, the first two columns of Table 312.6(A) must also be applied to determine whether the width of 4 in. affords sufficient space for bending the conductors. That consideration is required by 366.58. The worst condition (largest conductors) is where the supply conductors enter; therefore the 300-kcmil cable will determine the required space.

Table 312.6(A) shows that a circuit of one 300 kcmil per phase leg (or wire per terminal) requires a bending space at least 5 in. deep (in the direction of the entry of the 300-kcmil conductors), calling for a standard 6- by 6-in. gutter for this application.

In Fig. 366-2, if the 300-kcmil conductors entered at the left-hand end of the gutter instead of at the top, 366.58 would require Table 312.6(A) to be applied only to the deflection of the No. 3/0 conductors. The table shows, under one wire per terminal, a minimum depth of 4 in. is required. In that case, a 4- by 4-in. (102- by 102-mm) gutter would satisfy.

366.56. Splices and Taps. Part **(A)** is discussed under 366.22.

Part **(B)** covers cases where bare busbar conductors are used in gutters. The insulation might be cut by resting on the sharp edge of the bar or the bar might become hot enough to damage the insulation. When taps are made to bare conductors in a gutter, care should be taken so as to place and form the wires in such a manner that they will remain permanently separated from the bare bars.

Part **(C)** requires that identification be provided wherever it is not clearly evident what apparatus is supplied by the tap. Thus if a single set of tap conductors is carried through a short length of conduit from a gutter to a switch and the conduit is in plain view, the tap is fully identified and needs no special marking; but if two or more sets of taps are carried in a single conduit to two or more different pieces of apparatus, each tap should be identified by some marking such as a small tag secured to each wire.

ARTICLE 368. BUSWAYS

368.2. Definition. Busways consist of metal enclosures containing insulator-supported busbars. Varieties are so extensive that possibilities for 600-V distribution purposes are practically unlimited. Busways are available for either indoor or outdoor use as point-to-point feeders or as plug-in takeoff routes for power. Progressive improvements in busway designs have enhanced their electrical and mechanical characteristics, reduced their physical size, and simplified the methods used to connect and support them. These developments have in turn reduced installation labor to the extent that busways are most favorably considered when it is required to move large blocks of power to loadcenters (via low-impedance feeder busway), to distribute current to closely spaced power utilization points (via plug-in busways), or to energize rows of lighting fixtures or power tools (via trolley busways).

Busways classed as indoor low-reactance assemblies can be obtained in small incremental steps up to 6000 A for copper busbars and 5000 A for aluminum. Enclosed outdoor busways are similarly rated. In the plug-in category, special assemblies are available up to 5000 A, although normal 600-V ac requirements generally are satisfied by standard busways in the 225- to 1000-A range. Where power requirements are limited, small compact busways are available with ratings from 250 down to 20 A.

Plug-in and clamp-on devices include fused and nonfusible switches and plug-in circuit breakers (CBs) rated up to about 800 A. Other plug-in devices include ground detectors, temperature indicators, capacitors, and transformers designed to mount directly on the busway.

Busways are listed by UL with the following general information:

> This category covers busways and associated fittings, rated 600 V or less, 6000 A or less. Busways are grounded metal enclosures containing factory-mounted bare or insulated conductors, which are usually copper or aluminum bars, rods, or tubes. These enclosures and, in some cases an additional ground bus, are intended for use as equipment grounding conductors.
>
> Some busways are not intended for use ahead of service equipment and are marked with the maximum rating of overcurrent protection to be used on the supply side of the busway.

Busways may be of one of the following designs:

> *Lighting Busway*—Busway intended to supply and support industrial and commercial luminaires. Lighting busway is limited to a maximum current rating of 50 A.
>
> *Trolley Busway*—Busway having provision for continuous contact with a trolley by means of a slot in the enclosure. Trolley busway may be additionally marked "Lighting Busway" if intended to supply and support industrial and commercial luminaires.
>
> *Continuous Plug-In Busway*—Busway provided with provision for the insertion of plug-in devices at any point along the length of the busway. Continuous plug-in busway is intended for general use and may be installed within reach of persons. Busways of this design are limited to a maximum current rating of 225 A.
>
> *Short-Run Busway*—Unventilated busway intended for a maximum run of 30 ft horizontally, 10 ft vertically and are primarily used to supply switchboards. Except for transformer stubs, short-run busway is not intended to have intermediate taps.

368.10. Uses Permitted. Figure 368-1 shows the most common way in which busways are installed—in the open. Note that the former use of the term "concealed" in (B), as was the case for many Code cycles and as can be seen from context, squarely violated the definition of this term in Art. 100. This faulty terminology was corrected in the 2011 **NEC**, now correctly reading "Behind Access Panels."

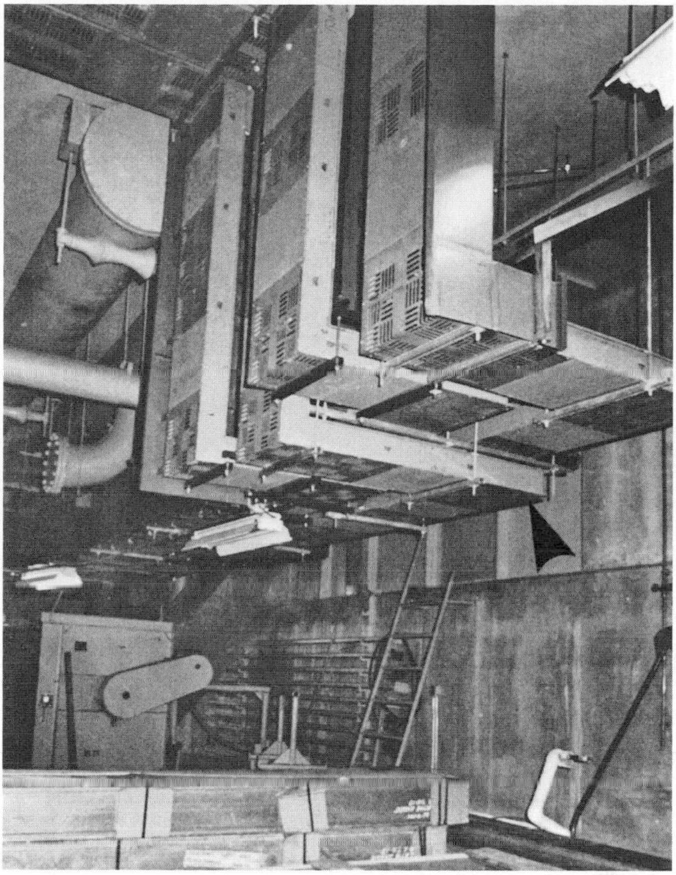

Fig. 368-1. Ventilated-type (with open grills for ventilation) busways may be used only "in the open" and must be "visible." Only the totally enclosed, nonventilating type may be used above a suspended ceiling. (Sec. 368.10.)

Wiring methods above lift-out ceiling panels are considered to be "exposed"—because the definition of that word includes reference to "behind panels designed to allow access." This section calls for busways to be "located in the open and visible" and therefore does not allow them above suspended ceilings, except with the limitations given in (B). In such locations, the busway is permitted provided means of access are provided, the joints and fittings can be reached for maintenance, and the ceiling is either not air-handling, or if air-handling, the busway conductors are insulated and there are no provisions for plug-in connections. Figure 368-2 shows how other Code rules tie into this section. Note that the wording of 368.10(B)(2) directly correlates to the permission for such busway use in spaces used for environmental air—as stated in 300.22(C).

Special rules govern the routing of busways through floor slabs. Figure 368-3 illustrates the prohibition against the use of ventilated busways on the pass-through

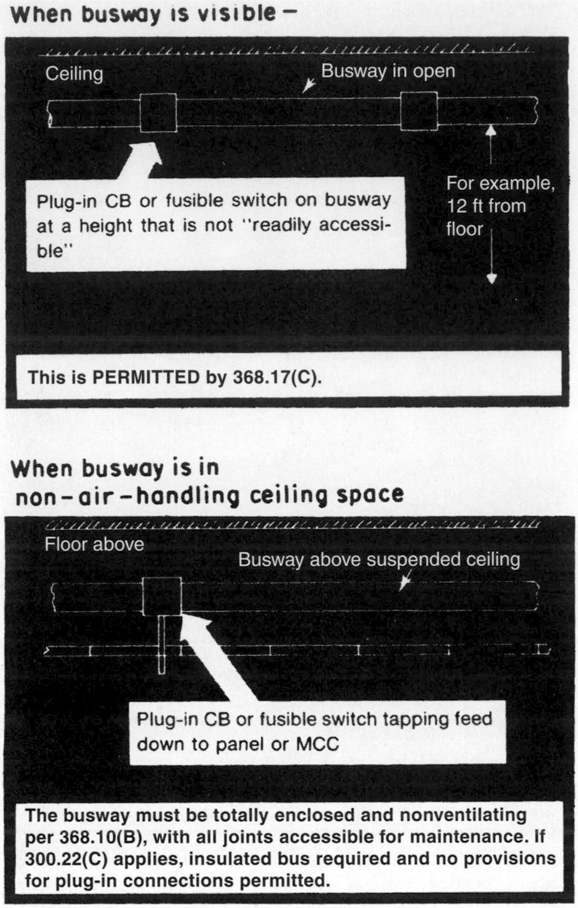
When busway is visible –
Ceiling
Busway in open
Plug-in CB or fusible switch on busway at a height that is not "readily accessible"
For example, 12 ft from floor
This is PERMITTED by 368.17(C).
When busway is in non-air-handling ceiling space
Floor above
Busway above suspended ceiling
Plug-in CB or fusible switch tapping feed down to panel or MCC
The busway must be totally enclosed and nonventilating per 368.10(B), with all joints accessible for maintenance. If 300.22(C) applies, insulated bus required and no provisions for plug-in connections permitted.

Fig. 368-2. Use of busways involves NEC rules on accessibility of overcurrent devices. (Sec. 368.10.)

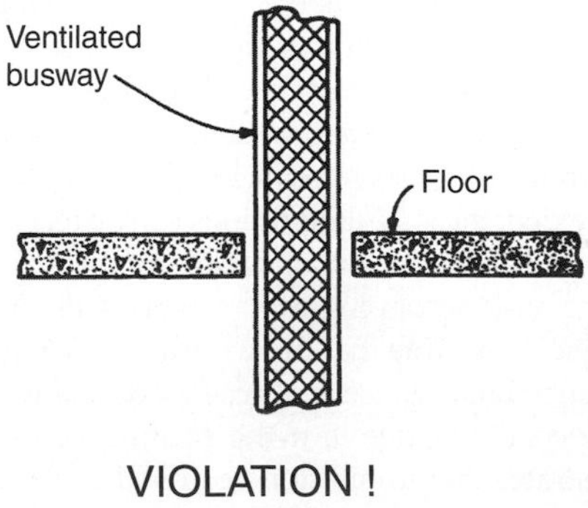
Ventilated busway
Floor
VIOLATION !

Fig. 368-3. Ventilated busways may not be used through a floor slab and for 6 ft (1.8 m) above the floor. [Sec. 368.10(C)(2)(a).]

Fig. 368-4. Opening for a busway riser through a slab must be curbed in most cases, and fire-stopped if the floor is part of a required fire separation, as required by 300.21. [Sec. 368.10(C).]

and up at least 1.8 m (6 ft) above the slab, as covered in 368.10(C)(2)(a). Figure 368-4 shows a close-up of a floor slab penetration. Depending on the occupancy, curbing is frequently required by 368.10(C)(2)(b).

Other data limiting applications of busways are contained in the UL regulations on listed busways—all of which information becomes mandatory Code rules because of **NEC** 110.3(B). Such UL data are as follows:

> Busways are intended for installation in accordance with Art. 368 of ANSI/**NFPA 70, "National Electrical Code" (NEC)**, and the manufacturer's installation instructions.
>
> Busways investigated to determine their suitability for
>
> - installation in a specified position,
> - for use in a vertical run, or for support at intervals greater than 5 ft,
> - for outdoor use
>
> are so marked. This marking is on or contiguous with the nameplate incorporating the manufacturer's name and electrical rating.
>
> A busway or fitting containing a vapor seal is so marked, but unless marked otherwise, the busway or fitting has not been investigated for passage through a fire-rated wall.
>
> Busway marked "Lighting Busway" and protected by overcurrent devices rated in excess of 20 A is intended for use only with luminaires employing heavy-duty lampholders unless additional overcurrent protection is provided for the luminaire in accordance with the **NEC**. [See Fig. 368-5 for examples; refer in this handbook to the discussion at 210.21(A) and 210.23 with which this provision correlates for more information.]
>
> Trolley busway should be installed out of the reach of persons or be otherwise installed to prevent accidental contact with exposed conductors.
>
> Some busways have a number of short stubs and are marked for use with certain compatible equipment.
>
> Busways and fittings covered under this category are intended for use with copper conductors unless marked to indicate which terminals are suitable for use with aluminum conductors. Such marking is independent of any marking on the terminal connectors and is on a wiring diagram or other readily visible location.
>
> Unless the equipment is marked to indicate otherwise, the termination provisions are based on the use of 60°C ampacities for wire sizes 14 to 1 AWG, and 75°C ampacities for wire sizes 1/0 AWG and larger as specified in Table 310.15(B)(16) of the **NEC**. Termination provisions are determined based on values provided in Table 310.15(B)(16) or Section 310.15(B)(7), with no adjustment made for correction factors.
>
> Some fittings are suitable for use as service equipment and are so marked.

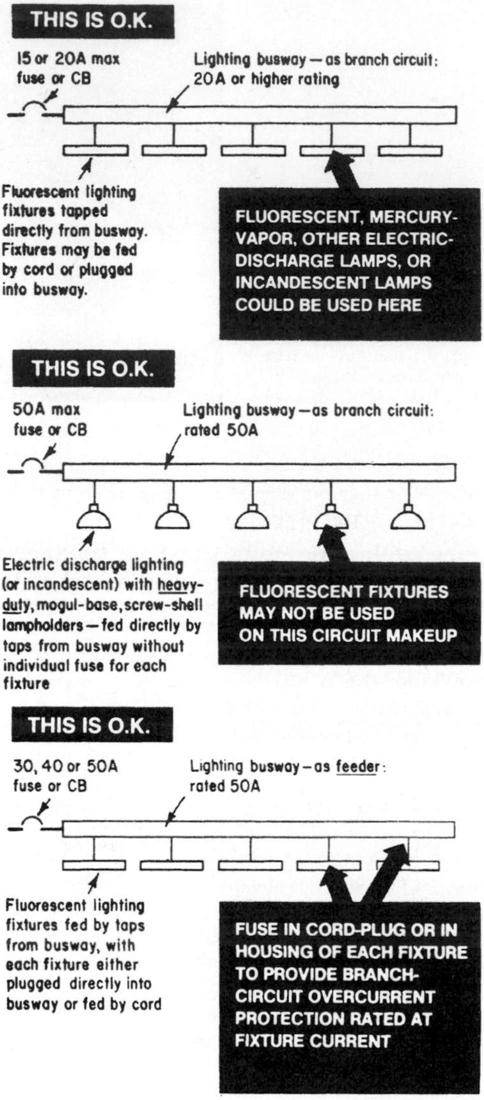
THIS IS O.K.
15 or 20A max fuse or CB
Lighting busway—as branch circuit: 20A or higher rating
Fluorescent lighting fixtures tapped directly from busway. Fixtures may be fed by cord or plugged into busway.
FLUORESCENT, MERCURY-VAPOR, OTHER ELECTRIC-DISCHARGE LAMPS, OR INCANDESCENT LAMPS COULD BE USED HERE
THIS IS O.K.
50A max fuse or CB
Lighting busway—as branch circuit: rated 50A
Electric discharge lighting (or incandescent) with heavy-duty, mogul-base, screw-shell lampholders—fed directly by taps from busway without individual fuse for each fixture
FLUORESCENT FIXTURES MAY NOT BE USED ON THIS CIRCUIT MAKEUP
THIS IS O.K.
30, 40 or 50A fuse or CB
Lighting busway—as feeder: rated 50A
Fluorescent lighting fixtures fed by taps from busway, with each fixture either plugged directly into busway or fed by cord
FUSE IN CORD-PLUG OR IN HOUSING OF EACH FIXTURE TO PROVIDE BRANCH-CIRCUIT OVERCURRENT PROTECTION RATED AT FIXTURE CURRENT

Fig. 368-5. These applications involve UL data and several Code sections. [Sec. 368.17(A).]

Note that fluorescent lighting fixtures can be fed by a 50-A lighting busway with each fixture individually fused at a few amps to protect its nonheavy-duty lampholders using the fuse in each fixture or in its attachment plug, as is permitted in the UL application information as well as by **NEC** 368.17(C), Exception Nos. 2 or 3. In that case, the lighting busway for Code purposes is halfway between a feeder and a branch circuit. Each fixture tap might be understood to be a branch circuit, and the wording "overcurrent device" is intended to require protection that will qualify as a "branch circuit overcurrent device" as defined in Art. 100. If it so qualifies, then the arrangement fully complies with 210.23(C).

368.17. Overcurrent Protection. As given in part **(A),** Rating of Overcurrent Protection—Feeders, the rated ampacity of a busway is fixed by the allowable temperature rise of the conductors. The ampacity can be determined in the field only by reference to the nameplate.

The rule of part **(B)** covers Reduction in Ampacity Size of Busway. Overcurrent protection—either a fused-switch or CB—is usually required in each busway subfeeder tapping power from a busway feeder of higher ampacity, protected at the higher ampacity. This is necessary to protect the lower current-carrying capacity of the subfeeder and should be placed at the point at which the subfeeder connects into the feeder. However, the exception to this section provides that overcurrent protection may be omitted where busways are reduced in size, if the smaller busway does not extend more than 50 ft and has a current rating at least equal to one-third the rating or setting of the overcurrent device protecting the main busway feeder (Figs. 368-6 and 368-7), but only at an "industrial establishment." For all other installations, the basic rule for overcurrent protection at the point where the busway is reduced in size must be satisfied.

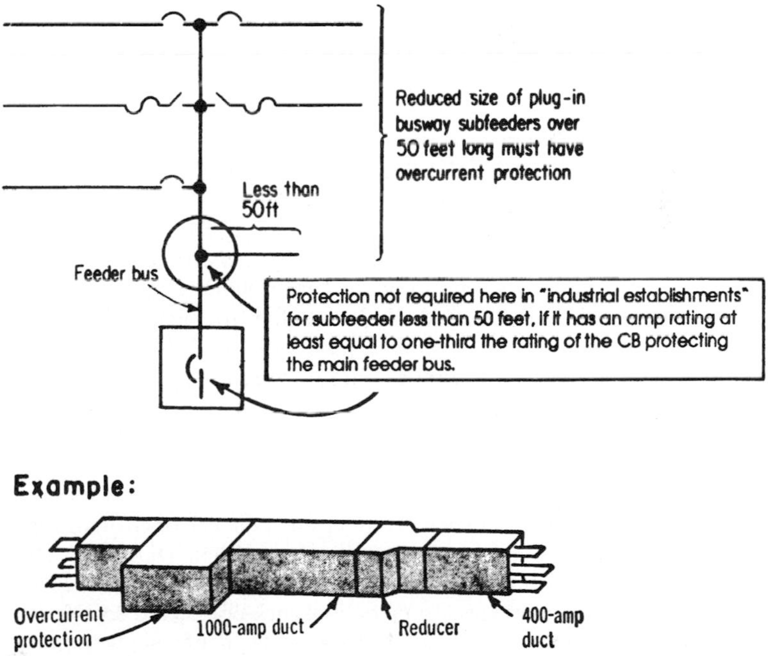
Reduced size of plug-in busway subfeeders over 50 feet long must have overcurrent protection
Less than 50ft
Feeder bus
Protection not required here in "industrial establishments" for subfeeder less than 50 feet, if it has an amp rating at least equal to one-third the rating of the CB protecting the main feeder bus.
Example:
Overcurrent protection
1000-amp duct
Reducer
400-amp duct

Fig. 368-6. A busway subfeeder may sometimes be used without protection. [Sec. 368.17(B).]

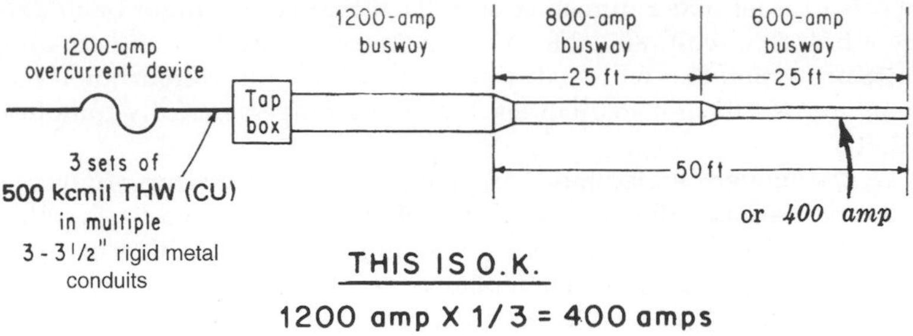
1200-amp overcurrent device
1200-amp busway
800-amp busway
25 ft
600-amp busway
25 ft
Tap box
50 ft
3 sets of 500 kcmil THW (CU) in multiple 3 - 3 1/2" rigid metal conduits
or 400 amp
THIS IS O.K.
1200 amp X 1/3 = 400 amps

Fig. 368-7. Total length of a reduced busway is not over 50 ft (15.24 m). [Sec. 368.17(B).]

Where the smaller busway is kept within the limits specified, the hazards involved at industrial installations are very slight and the additional cost of providing overcurrent protection at the point where the size is changed is not considered as being warranted.

The rules of part **(C)** are interrelated with those of 240.24 and 404.8. The basic rule of this section makes it clear that branch circuits or subfeeders tapped from a busway must have overcurrent protection on the busway at the point of tap. And if they are out of reach from the floor, all fused switches and CBs must be provided with some means for a person to operate the handle of the device from the floor (hookstick, chain operator, rope-pull operator, etc.).

Although no definition is given for "out of reach" from the floor, the wording of 404.8(A) can logically be taken to indicate that a switch or CB *is* "out of reach" if the center of its operating handle, when in its highest position, is more than 2.0 m (6 ft 7 in.) above the floor or platform on which the operator would be standing. Thus, a busway over 2.0 m (6 ft 7 in.) above the floor would require some means (hookstick, etc.) to operate the handles of any switches or CBs on the busway.

Figure 368-8 relates the rules of 240.24 and Exception No. 1 to 368.17(C)—with the rule of 240.24 *permitting* overcurrent devices to be "not readily accessible" when used up on a high-mounted busway and 368.17 *requiring* such protection to be mounted on the busway. To get at overcurrent protection in either case, personnel might have to use a portable ladder or chair or some other climbing technique. Again, 6 ft 7 in. (2.0 m) could be taken as the height above which the overcurrent protection is not "readily accessible"—or the height above which the Code considers that some type of climbing technique (ladder, chair, etc.) may be needed by some persons to reach the protective device.

Then, where the plug-in switch or CB on the busway is "out of reach" from the floor (i.e., over 6 ft 7 in. [2.0 m] above the floor), provision must be made for operating such switches or CBs from the floor, as shown in Figs. 368-9 and 368-10. The plug-in switch or CB unit must be able to be operated by a hookstick or chain or rope operator if the unit is mounted out of reach up on a busway. Section 404.8 says all busway switches and CBs must be operable from the floor. Refer to 404.8(A) Exception No. 1 Figure 368-10 shows a typical application of hookstick-operated disconnects.

Figure 368-11 shows an application that has caused controversy because 368.17 says that any busway used as a feeder must have overcurrent protection on the busway for any subfeeder or branch circuit tapped from the busway. Therefore

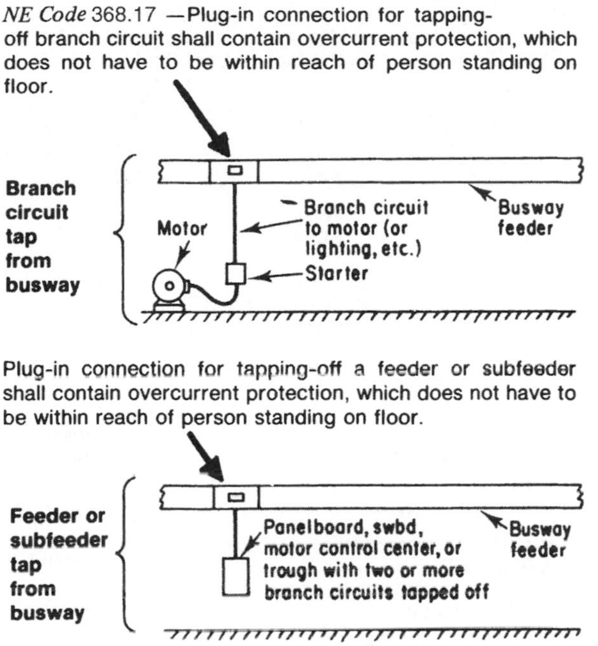
NE Code 368.17 —Plug-in connection for tapping-off branch circuit shall contain overcurrent protection, which does not have to be within reach of person standing on floor.
Branch circuit tap from busway
Motor
Branch circuit to motor (or lighting, etc.)
Starter
Busway feeder
Plug-in connection for tapping-off a feeder or subfeeder shall contain overcurrent protection, which does not have to be within reach of person standing on floor.
Feeder or subfeeder tap from busway
Panelboard, swbd, motor control center, or trough with two or more branch circuits tapped off
Busway feeder

Fig. 368-8. Protection must always be used on busway for these taps—regardless of busway mounting height. (Sec. 368.17.)

use of a cable-tap box on busway without overcurrent protection could be ruled a Code violation. It can be argued that the installation shown—a 10- or 25-ft (3.0- or 7.5-m) tap without overcurrent protection on the busway—is covered by Exception No. 1 of that section, which recognizes taps as permitted in 240.21—including 10- and 25-ft (3.0- or 7.5-m) taps. But, as is now clearly spelled out in Exception No. 1 to 368.17(C), busways may be tapped as would any feeder. It is to be viewed simply as a conductor. Then, the rules given in the referenced parts of 240.21 must be satisfied. Therefore, the application shown in Fig. 368-11 does satisfy the Code.

The 2017 **NEC** added a fourth exception to (C) providing still further relief. If the branch-circuit overcurrent device directly supplies a readily accessible disconnect, it is not necessary to provide for operation of that branch-circuit device from the floor. Note that Exception No. 1 already provides relief where the installation meets a tap rule in 240.21. This exception must not become a vehicle for violating the tap rules when Exception No. 1 calls for them to be applied. On the contrary, this is simply an allowance to mount the disconnect remote from the busway. The overcurrent device will stay on the busway, but a hookstick need not be provided.

368.17(D). Rating of Overcurrent Protection—Branch Circuits. Refer to data on busways on lighting branch circuits in 368.10 and Fig. 368-5.

368.30. Support. As shown in Fig. 368-12, busway risers may be supported by a variety of spring-loaded hangers, wall brackets, or channel arrangements where busways pierce floor slabs or are supported on masonry walls or columns. Fig. 368-4 shows an example of spring mounts for vertical busways which may be

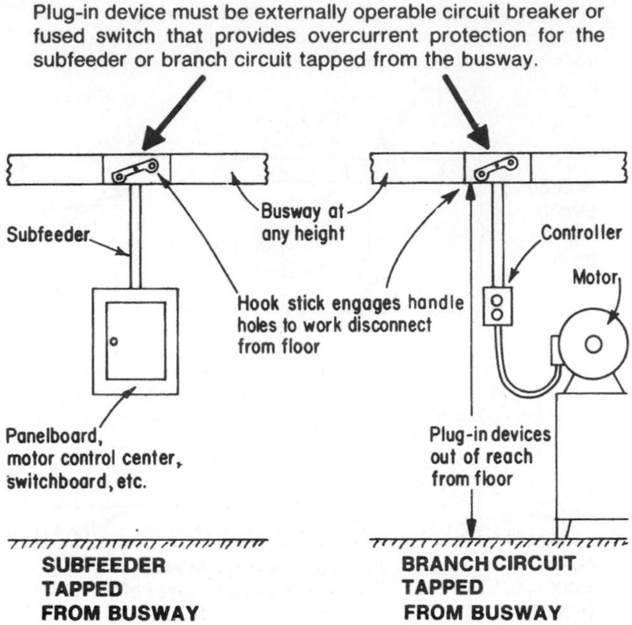
Plug-in device must be externally operable circuit breaker or fused switch that provides overcurrent protection for the subfeeder or branch circuit tapped from the busway.
Busway at any height
Subfeeder
Controller
Motor
Hook stick engages handle holes to work disconnect from floor
Panelboard, motor control center, switchboard, etc.
Plug-in devices out of reach from floor
SUBFEEDER TAPPED FROM BUSWAY
BRANCH CIRCUIT TAPPED FROM BUSWAY

Fig. 368-9. Busway plug-in devices must be operable from the floor or platform where operator stands. (Sec. 368.17.)

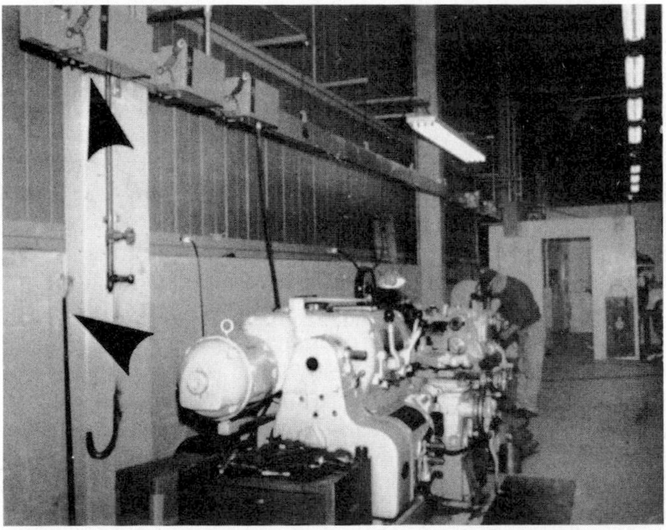

Fig. 368-10. Disconnects mounted up on the busway (top arrow) are out of reach from the floor but do have hook-eye lever operators to provide operation by person standing in front of machines. Although the **NEC** does not literally require ready availability of a hookstick, it is certainly the intent of the Code that one be handy (lower arrow). (Sec. 368.17.)

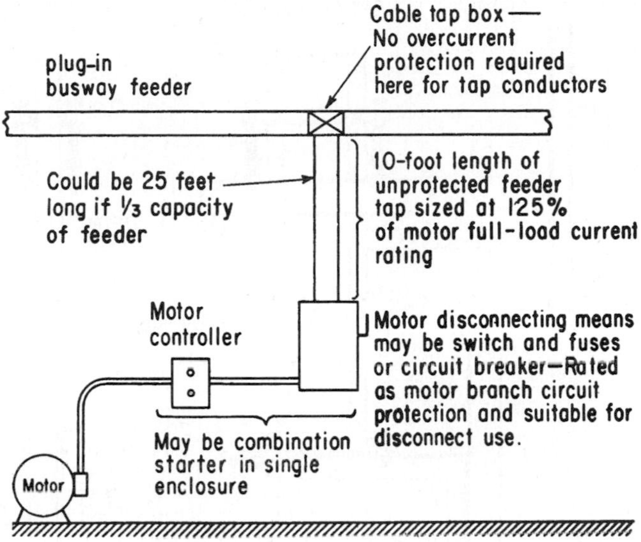
Cable tap box—
No overcurrent
protection required
here for tap conductors
plug-in
busway feeder
Could be 25 feet
long if 1/3 capacity
of feeder
10-foot length of
unprotected feeder
tap sized at 125%
of motor full-load current
rating
Motor
controller
Motor disconnecting means
may be switch and fuses
or circuit breaker—Rated
as motor branch circuit
protection and suitable for
disconnect use.
May be combination
starter in single
enclosure
Motor

Fig. 368-11. This use of unprotected tap from busway does not conflict with 368.17 as long as the tap meets the applicable requirements in 240.21(B) and thereby satisfies 368.17(C) Exception No. 1. (Sec. 368.17.)

located at successive floor-slab levels or, as indicated in Fig. 368-12, supported by wall brackets located at intermediate elevations. Springs provide floating cradles for absorbing transient vibrations or physical shocks. Fire-resistant material is packed into space between the busway and the edges of slab-piercing throat.

368.56. Branches from Busways. Busway branches can be made into a wide variety of Chap. 3 wiring methods. For cord connections, the rule here requires that a cord connecting to a plug-in switch or CB on a busway must be supported by a "tension take-up support device" with the swag not longer than 1.8 m (6 ft). Industrial occupancies can extend that limit indefinitely, provided the cord is supported at 2.5 m (8 ft) or shorter intervals. "Bus Drop Cable" has its own coverage in UL (ZIMX) apart from most flexible cords. It is not covered in Art. 400, because it is comprised of either three or four finely stranded building wires of one of seven insulation types (TW, THW, THHN, THWN, XHHW, RHW, or RHH) together with a grounding conductor, all cabled within an overall jacket. Fig. 368-13 shows a number of details regarding branches from buways, and Fig. 368-14 shows a busway lineup, using bus-drop cable connections.

368.320. Marking. Busway is also available for medium-voltage distribution, as covered in Part IV of Art. 368. The UL data that apply to this category follow:

> This category covers metal-enclosed busways of the nonsegregated phase type, for use in accordance with Art. 368 of ANSI/NFPA 70, "National Electrical Code." Nonsegregated phase busway is one in which all phase conductors are in a common metal enclosure without barriers between the phases.
>
> These are assemblies of metal-enclosed conductors, together with associated interconnections, enclosures, and supporting structures.

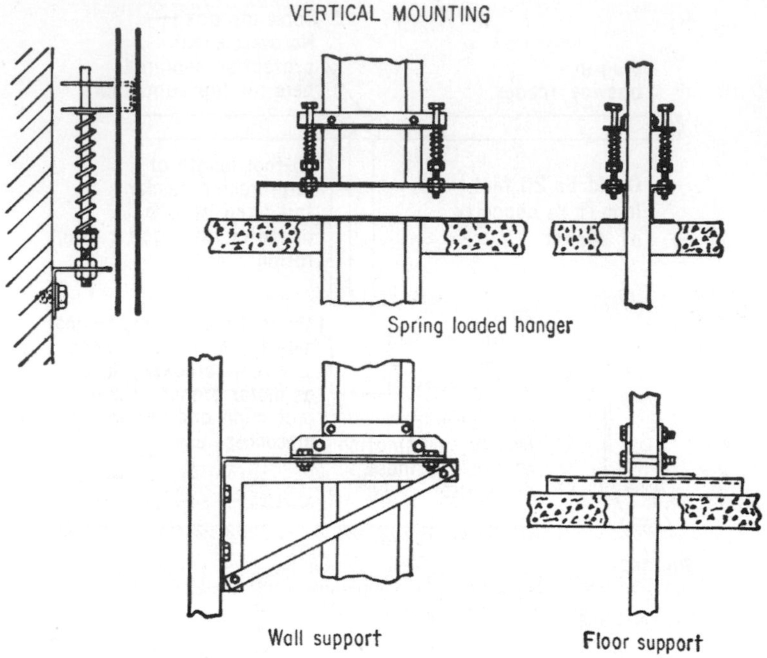
VERTICAL MOUNTING
Spring loaded hanger
Wall support
Floor support

Fig. 368-12. Vertical busway runs should be supported at least every 5 ft, unless designed and marked for another support interval. (Sec. 368.30.)

These assemblies are intended for use on systems with nominal rated voltages from 601 V to 38 kV ac. Current ratings are from 600 to 10,000 A.

These assemblies may be intended for either indoor or outdoor applications. An assembly that has been investigated to determine that it is rainproof is marked "Rainproof," "Outdoor," or "3R."

Enclosures are of the ventilated or nonventilated type. A ventilated enclosure is provided with means to permit circulation of sufficient air to remove excess heat.

A nonventilated enclosure is constructed to provide no intentional circulation of external air through the enclosure.

These products are marked with the following electrical ratings: rated voltage, rated continuous current, insulation (BIL) level, frequency, rated frequency withstand voltage (dry), and rated short-circuit withstand current (momentary current). When shipped in sections, each section is marked.

ARTICLE 370. CABLEBUS

Cablebus is an approved assembly of insulated conductors mounted in "spaced" relationship in a *ventilated* metal-protective supporting structure, including fittings and conductor terminations. In general, cablebus is assembled at the point of installation from components furnished by the manufacturer.

The definition was revised in the 2014 **NEC** to increase the emphasis on the role of the cablebus structure as a means to "securely fasten or support conductors" and as such to separate it from most raceways.

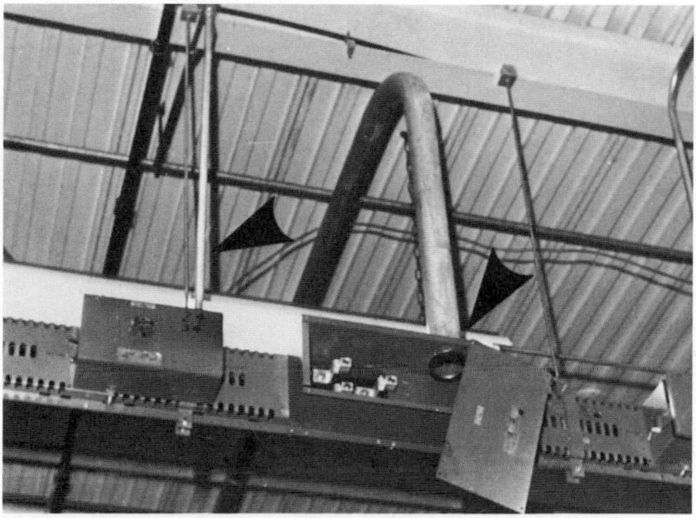

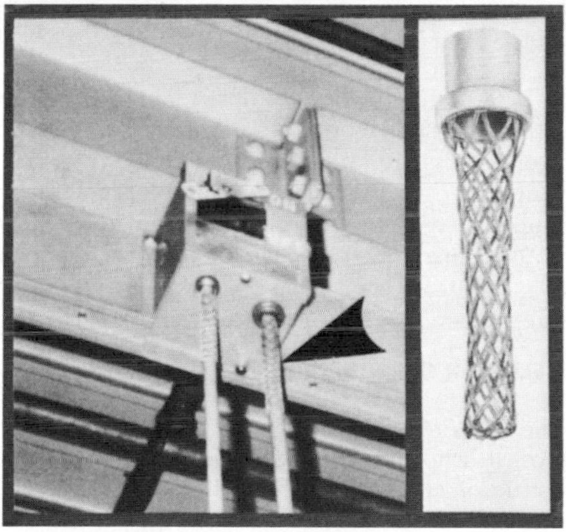

Fig. 368-13. Wiring details at busway connections must be carefully observed. (Sec. 368.56.) Circuits fed from busways may be run in any conventional wiring method—such as EMT or rigid conduit (left) or as "suitable cord," such as "bus-drop" cable down to machines. And cable-tap boxes may be used (arrow at right) to connect feeder conductors that supply power to busway. SOME TYPE of tension-relief device must be used on bus-drop cable or other suitable cord where it connects to a plug-in switch or CB on busway, per 368.56(B)(4). Photo shows strain relief connector with mesh grip (arrow) on cord to bus-tap CB, which is equipped with hook-eye lever mechanism to provide operation of the CB by a hookstick from floor level—as required by 404.8(A), Ex. No. 1 and 368.17(C).

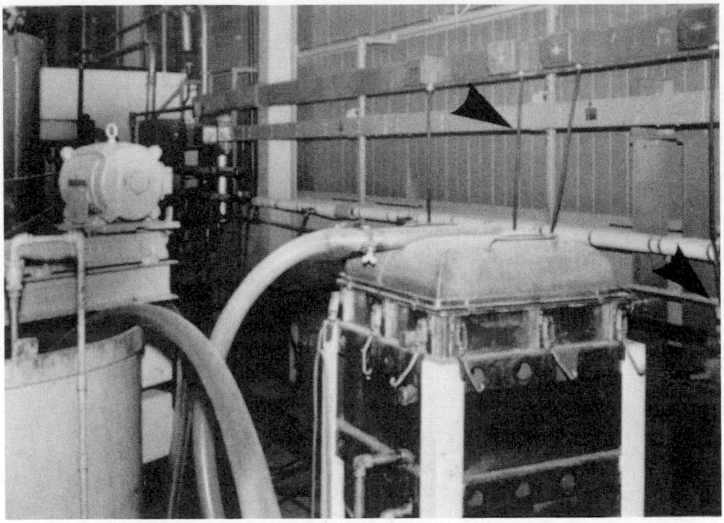

Fig. 368-14. Bus-drop power cables are flexible cables listed by UL for feeding power down from busway to supply machines. Cables here have connector bodies on their ends for machine cord caps to plug into. [Sec. 368.56(B).]

Field-assembly details are shown in Fig. 370-1. First, the cablebus framework is installed in a manner similar to continuous rigid cable support systems. Next, insulated conductors are pulled into the cablebus framework. Then the conductors are supported on special insulating blocks at specified intervals. And finally, a removable (ventilated) top is installed. Refer to the discussion at 215.2(A)(1)(a) Ex. 2 for an approach to meeting NEC termination rules.

ARTICLE 372. CELLULAR CONCRETE FLOOR RACEWAYS

372.1. Scope. The term *Precast cellular concrete floor,* as mentioned in 372.2, refers to a type of floor construction designed for use in steel frame, concrete frame, and wall-bearing construction, in which the monolithically precast reinforced concrete floor members form the structural floor and are supported by beams or bearing walls. The floor members are precast with hollow voids that form smooth round cells. The cells are of various sizes depending on the size of floor member used.

The cells form raceways which by means of suitable fittings can be adapted for use as underfloor raceways. A precast cellular concrete floor is fire-resistant and requires no additional fireproofing.

372.18. Cellular Concrete Floor Raceways Installation. The 2017 NEC reorganized this article almost completely, without substantive change. Five former sections became lettered subsections under this title.

(A) Header. Connections to the cells are made by means of *headers* secured to the precast concrete floor, extending from cabinets and across the cells. A header

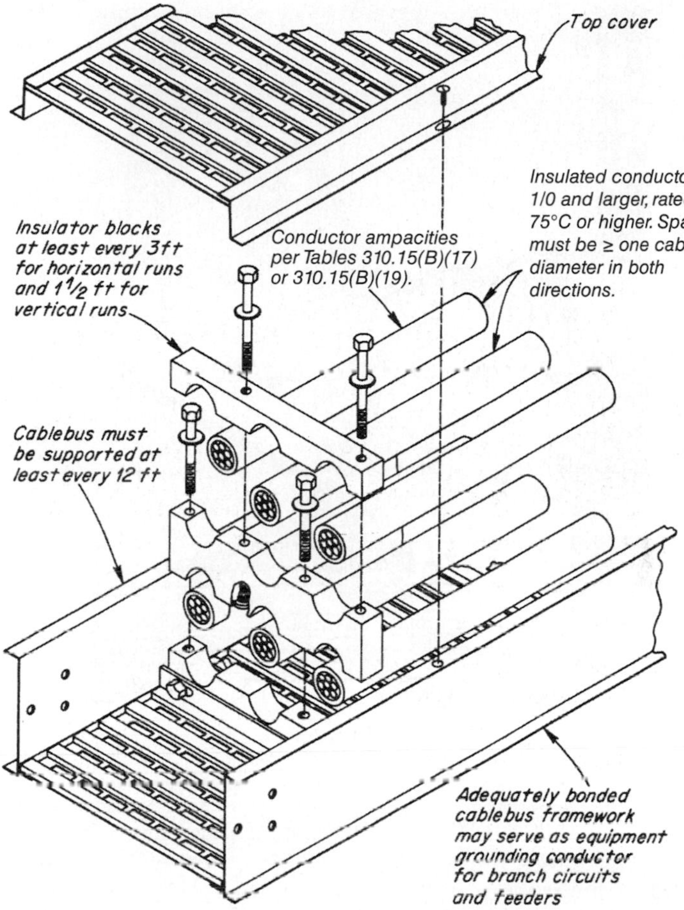
Top cover
Insulated conductors 1/0 and larger, rated 75°C or higher. Spacing must be ≥ one cable diameter in both directions.
Insulator blocks at least every 3ft for horizontal runs and 1½ ft for vertical runs
Conductor ampacities per Tables 310.15(B)(17) or 310.15(B)(19).
Cablebus must be supported at least every 12 ft
Adequately bonded cablebus framework may serve as equipment grounding conductor for branch circuits and feeders

Fig. 370-1. Cablebus systems are field assembled from manufactured components. (Sec. 370.2.)

connects only those cells that are used as raceways for conductors. Two or three separate headers, connected to different sets of cells, may be used for different systems (e.g., for light and power, signaling, and telephones).

Figure 372-1 shows three headers installed, each header connecting a cabinet with separate groups of cells. Special elbows extend the header to the cabinet.

(B) Connection to Cabinets and Other Enclosures. The require listed raceways and fittings.

(C) Junction Boxes. Figure 372-2 shows how a JB must be arranged where a header connects to a cell.

(D) Inserts. A $1\frac{7}{8}$-in. diameter hole is cut through the floor and into the center of a cell with a concrete drill bit. A plug is driven into the hole and a nipple is

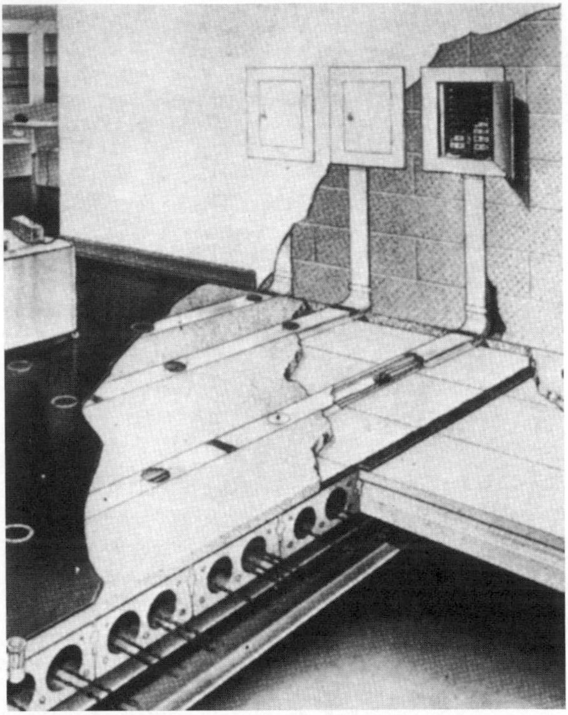

Fig. 372-1. Headers, flush with finished concrete pour, carry wiring to cells. (Sec. 372.5.)

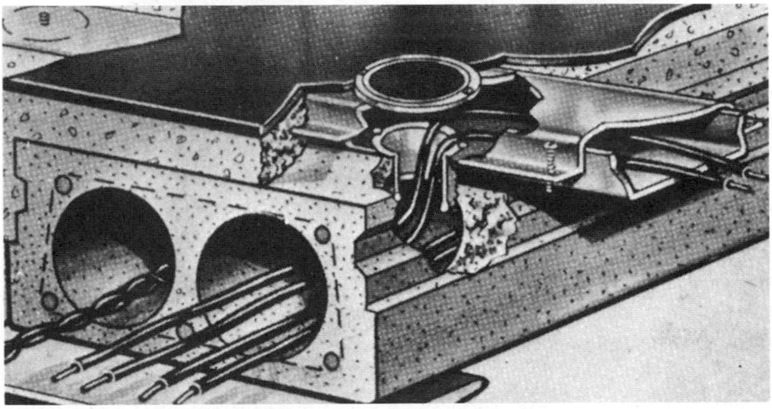

Fig. 372-2. Junction box is used to provide conductor installation from header to cell. (Sec. 372.7.)

screwed into the plug. The nipple is designed to receive an outlet with a duplex electrical receptacle or an outlet designed for a telephone or signal system.

(E) Markers. Markers used with this system are special flat-head brass screws which are installed level with the finished floor. One type of marker marks the location of an access point between a header and a spare cell reserved for, but not connected to, the header. A junction box can be installed at the point located by the marker if the spare cell is needed in the future. The screw for this type of marker is installed in the center of a special knockout provided in the top of the header at the access point. The second type of marker is installed over the center of cells at various points on the floor to locate and identify the cells below. Screws with specially designed heads identify the type of service in the cell.

372.23. Ampacity of Conductors. This section makes clear that the rules given in 310.15(B)(3) requiring derating of conductors where there are more than three current-carrying conductors within a raceway, cable, or trench apply to conductors installed within cellar concrete floor raceways. Note the comments on discontinued outlets, Sec. 58.

372.58. Discontinued Outlets. When an outlet is discontinued, the conductors supplying the outlet shall be removed from the raceway. The general practice is to loop wire intermediate receptacles between the header and the end of the run. This requirement assures that reinsulated conductors will not be resting in the raceway below an abandoned outlet. This in turn prevents a fish wire inserted afterward from a downstream location from getting caught on a reinsulated conductor, with very destructive consequences.

It is often advisable to wire each outlet on its own pair of conductors back to the header or other junction point. Then, when an outlet is abandoned, the associated pair of conductors can be withdrawn without disrupting other outlets on the run. Take care, however, to keep track of the total number of conductors at all portions of the duct, because the ampacity derating factors for mutual conductor heating (see 372.23) will apply. Some jurisdictions provide limited waivers to these derating factors in underfloor applications, however, in order to encourage this practice and thereby discourage the reinsulation of conductors. When the conductors are withdrawn, leave a pull string in their place so the outlet can be easily reactivated in the future as necessary.

ARTICLE 374. CELLULAR METAL FLOOR RACEWAYS

374.2. Definitions. This is a type of floor construction designed for use in steel frame buildings in which the members supporting the floor between the beams consist of sheet steel rolled into shapes that are so combined as to form cells, or closed passageways, extending across the building. The cells are of various shapes and sizes, depending upon the structural strength required.

The cellular members of this type of floor construction form raceways. A cross-sectional view of one type of cellular metal floor is shown in Fig. 374-1.

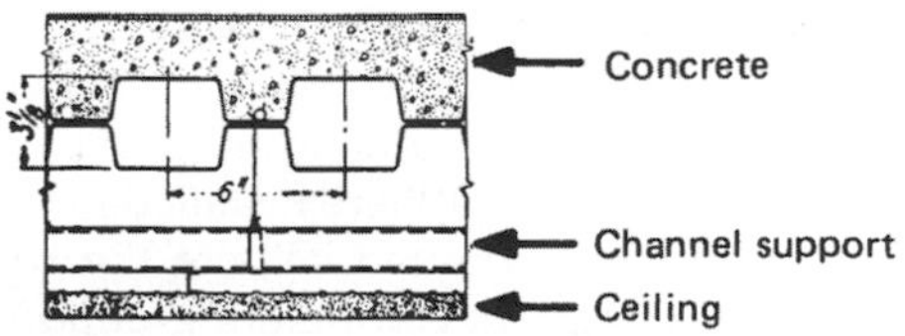
Concrete
Channel support
Ceiling

Fig. 374-1. Cross section of one type of cellular-method floor construction. (Sec. 374.2.)

374.18. Cellular Metal Floor Raceways Installation. Connections to the ducts are made by means of *headers* extending across the cells. A header connects only to those cells that are to be used as raceways for conductors. Two or three separate headers, connecting to different sets of cells, may be used for different systems (e.g., for light and power, signaling systems, and public telephones).

Figure 374-2 shows the cells, or ducts, with header ducts in place. By means of a special elbow fitting the header is extended up to a cabinet or distribution center on a wall or column. A junction box or access fitting is provided at each point where the header crosses a cell to which it connects.

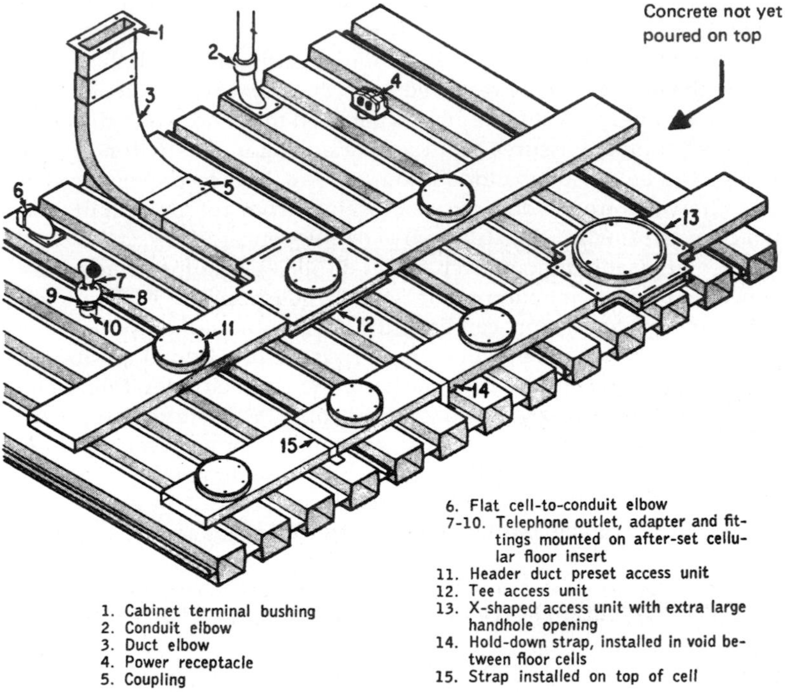
Concrete not yet poured on top
1
2
3
4
5
6
7
8
9
10
11
12
13
14
15
1. Cabinet terminal bushing
2. Conduit elbow
3. Duct elbow
4. Power receptacle
5. Coupling
6. Flat cell-to-conduit elbow
7-10. Telephone outlet, adapter and fittings mounted on after-set cellular floor insert
11. Header duct preset access unit
12. Tee access unit
13. X-shaped access unit with extra large handhole opening
14. Hold-down strap, installed in void between floor cells
15. Strap installed on top of cell

CELLULAR STEEL FLOOR contains unlimited number of channels for enclosing and isolating various electrical services. Wiring is routed from distribution panels to floor outlets through header ducts as shown.

Fig. 374-2. Components for electrical usage in cellular metal floor must be properly applied. (Sec. 374.3.)

(A) Connection to Cabinets and Extensions from Cells. This section establishes the acceptable methods for connection to equipment and enclosures supplied from a cellular metal floor raceway. Flex and liquidtight flex may be used, provided they are not installed in concrete. Where installed in concrete, RMC, IMC EMT, or "approved fittings" are permitted. Where the equipment or enclosure has provisions for connecting and equipment grounding conductor, then nonmetallic conduit, ENT, and LFNMC are permitted. As stated by the last sentence, where listed for installation in concrete, LFMC and LFNMC are recognized to supply equipment and enclosures from cellular metal floor raceways.

(B) Junction Boxes. The fittings with round covers shown in Fig. 374-2 are termed access *fittings* by the manufacturer but actually serve as junction boxes. Where additional junction boxes are needed, a similar fitting of larger size is provided, which may be attached to a cell at any point.

(C) Inserts. The construction of an insert is shown in Fig. 374-3. A $1\frac{5}{8}$-in. diameter hole is cut in the top of the cell with a special tool. The lower end of the insert is provided with coarse threads of such form that the insert can be screwed into the hole in the cell, thus forming a substantial mechanical and electrical connection.

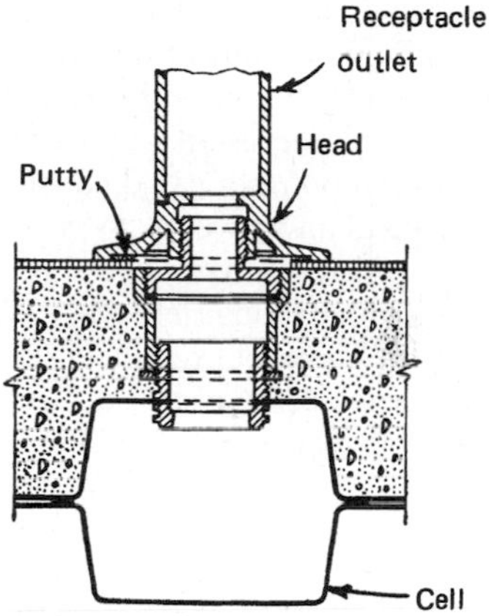
Receptacle outlet
Head
Putty
Cell

Fig. 374-3. Typical insert for connecting from cell-to-floor outlet assembly. (Sec. 374.10.)

(D) Markers. The markers used with this system consist of special flat-head brass screws, screwed into the upper side of the cells with their heads flush with the floor finish.

374.23. Ampacity of Conductors. This section makes clear that the rules given in 310.15(B)(3) requiring derating of conductors where there are more than three current-carrying conductors within a raceway, cable, or trench apply to conductors installed within cellular metal floor raceways. See the extensive comments on the topic of discontinued outlets at 374.58.

374.58. Discontinued Outlets. When an outlet is discontinued, the conductors supplying the outlet shall be removed from the raceway. The general practice is to loop wire intermediate receptacles between the header and the end of the run. This requirement assures that reinsulated conductors will not be resting in the raceway below an abandoned outlet. This in turn prevents a fish wire inserted afterward from a downstream location from getting caught on a reinsulated conductor, with very destructive consequences.

It is often advisable to wire each outlet on its own pair of conductors back to the header or other junction point. Then, when an outlet is abandoned, the associated pair of conductors can be withdrawn without disrupting other outlets on the run. Take care, however, to keep track of the total number of conductors at all portions of the duct, because the ampacity derating factors for mutual conductor heating (see 374.23) will apply. Some jurisdictions provide limited waivers to these derating factors in underfloor applications, however, in order to encourage this practice and thereby discourage the reinsulation of conductors. When the conductors are withdrawn, leave a pull string in their place so the outlet can be easily reactivated in the future as necessary.

ARTICLE 376. METAL WIREWAYS

376.2. Definition. Metal wireways are sheet metal troughs in which conductors are laid in place after the wireway has been installed as a complete system. Wireway is available in standard lengths of 1, 2, 3, 4, 5, and 10 ft (0.30, 0.61, 0.91, 1.22, 1.52, and 3.05 m), so runs of any exact number of feet can be made up without cutting the duct. The cover may be a hinged or removable type. Unlike auxiliary gutters, wireways represent a type of wiring, because they are used to carry conductors between points located considerable distances apart.

The purpose of a wireway is to provide a flexible system of wiring in which the circuits can be changed to meet changing conditions, and one of its principal uses is for exposed work in industrial plants. Wireways are also used to carry control wires from the control board to remotely controlled stage switchboard equipment. The definition here explicitly uses the term "raceway" in contradistinction to 366.2 for auxiliary gutters which does not. An installation of wireway is shown in Fig. 376-1.

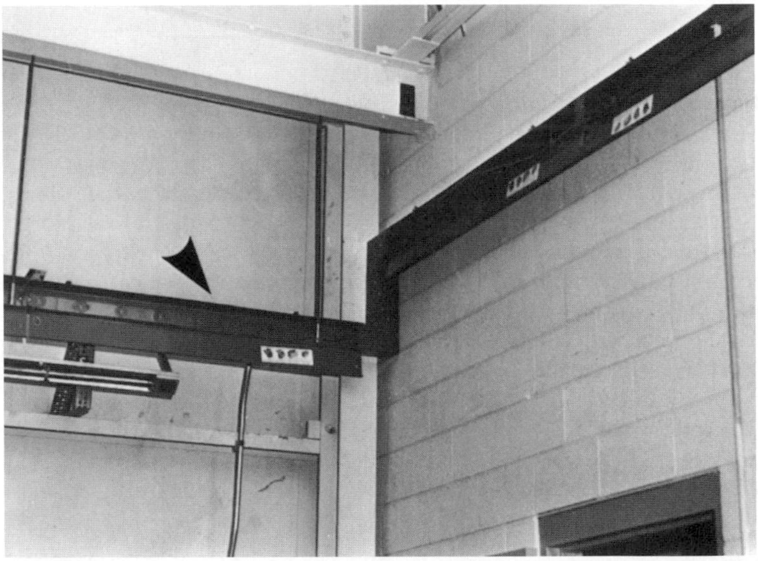

Fig. 376-1. Wireway in industrial plant—installed exposed, as required by 376.10 provides highly flexible wiring system that provides easy changes in the number, sizes, and routing of circuit conductors for machines and controls. Hinged covers swing down for ready access. Sec. 376.56 permits splicing and tapping in wireway. And 376.70 covers use of conduit for taking circuits out of wireway. (Sec. 376.10.)

376.10. Uses Permitted. Figure 376-1 shows a typical Code-approved application of 4- by 4-in. wireway in an exposed location.

376.20. Conductors Connected in Parallel. Where single conductor cables comprising each phase, neutral, or grounded conductor of an alternating-current circuit are connected in parallel, the conductors must be installed in groups consisting of not more than one conductor per phase, neutral, or grounded conductor. This is being required to prevent current imbalance in the paralleled conductors due to inductive reactance. The requirement here parallels equivalent text for cable trays in 392.20. A failure to apply this procedure resulted in documented uneven current distribution in a very large feeder run with parallel conductors in a wireway, which was then fully corrected by rearranging the conductors into phase groups. The same rule applies to auxiliary gutters and to nonmetallic wireway.

376.22. Number of Conductors and Ampacity. Wireways have two general conductor limitations covered in this section. The first addresses the physical realities of dealing with large conductor fills with a multitude of potential destinations. The second addresses mutual conductor heating in a context where there will be greater air circulation than in a tubular raceway, and where the conductors are enclosed in a steel enclosure that will function as a heat sink and which affords a ready heat transfer ability to the ambient air.

Part **(A)** limits the combined cross sectional area of all contained conductors at any given cross section to 20 percent of the cross-sectional area of the wireway. This rule covers both power and control conductors taken together as conductors occupying any given cross section of a wireway.

Part **(B)** has two sentences. The second correlates with 725.51(B)(2), assuming the control wiring carries its usual load profile of less than 10 percent of its ampacity being carried on any continuous basis. In such cases, based on the I^2R relationship, a 10 percent load will equal 0.1^2 or 1 percent of its potential heat contribution in comparison to its heat when loaded to its ampacity. Such a miniscule heat contribution can be disregarded, which is why the allowance exists in Art. 725 and this sentence makes it so for wireways. With these conductors disregarded, the first sentence tackles power circuits. If the number of current-carrying conductors does not exceed 30, including only those neutrals that must be counted as required by 310.15(B)(5), *mutual conductor heating is reflected in the adjustment factors that are normally applied in accordance with 310.15(B)(3)(a) is ignored and the adjustment factors are not applied.*

Note that spares are not counted here—refer to the discussion regarding Table 310.15(B)(3)(a) and the change in its left-hand column heading for more information on the significance of this consideration. If more than 30 conductors at any cross section are current-carrying, then the adjustment factors for mutual conductor heating will apply to every conductor passing through that cross section. Referring to Table 310.15(B)(3)(a), a 31st current-carrying conductor at a cross section dooms the entire fill at that cross section to have its allowable ampacities reduced by 60 percent, which in most cases will be unacceptable and sufficient to redesign the application. This permission, as far as it goes, is a crucial characteristic of metal wireways.

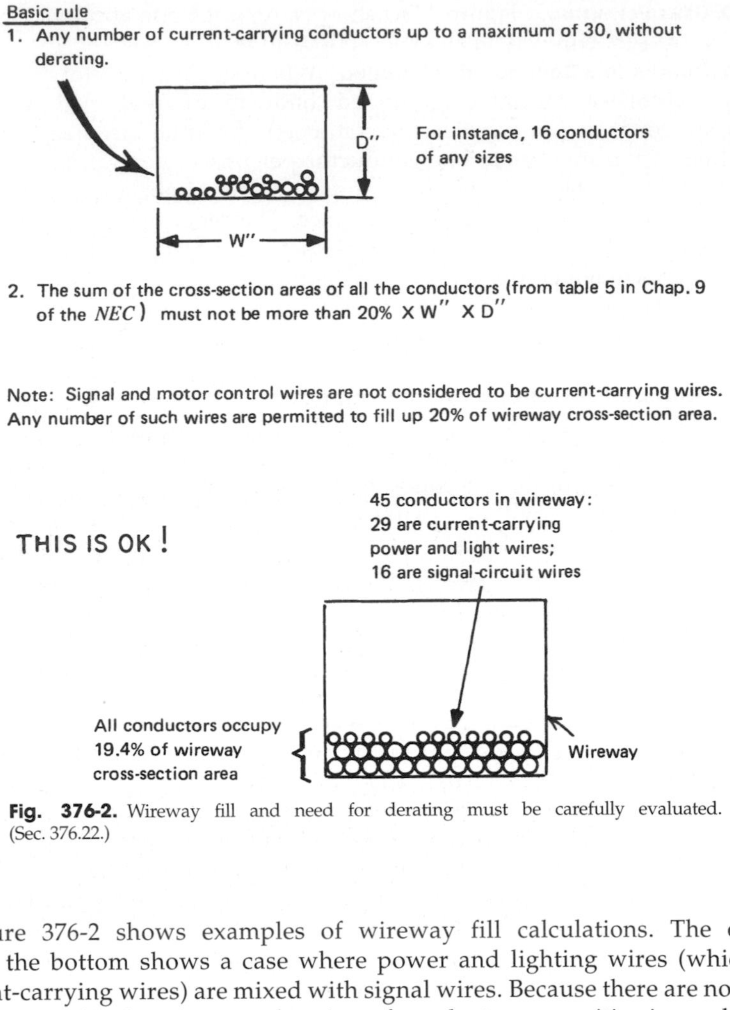
Basic rule
1. Any number of current-carrying conductors up to a maximum of 30, without derating.
D''
For instance, 16 conductors of any sizes
W''
2. The sum of the cross-section areas of all the conductors (from table 5 in Chap. 9 of the NEC) must not be more than 20% X W'' X D''
Note: Signal and motor control wires are not considered to be current-carrying wires. Any number of such wires are permitted to fill up 20% of wireway cross-section area.
THIS IS OK!
45 conductors in wireway:
29 are current-carrying
power and light wires;
16 are signal-circuit wires
All conductors occupy
19.4% of wireway
cross-section area
Wireway

Fig. 376-2. Wireway fill and need for derating must be carefully evaluated. (Sec. 376.22.)

Figure 376-2 shows examples of wireway fill calculations. The example at the bottom shows a case where power and lighting wires (which *are* current-carrying wires) are mixed with signal wires. Because there are not over 30 power and light wires, no derating of conductor ampacities is needed. If, say, 31 power and light wires were in the wireway, then the power and light conductors would be subject to derating. If all 49 conductors were signal and/or control wires, no derating would be required. But, in all cases, wireway fill must not be over 20 percent.

376.23. Insulated Conductors. Deflected conductors in wireways must observe the rules on adequate enclosure space given in 312.6. This section is based on the following substantiation for its submittal for the 1990 NEC (the internal Code references have been updated to agree with current numbering within Chap. 3):

> Although wireways don't contain terminals or supplement spaces with terminals, pull boxes and conduit bodies don't either. This rule borrows language from both 366.58(A) and 314.28(A)(2), Exception, in an attempt to produce a consistent approach in the Code. Although in some cases the deflected conductors travel long distances in the wireway

and are therefore easily inserted, in other cases the conductors are deflected again within inches of the first entry. The result is even more stress on the insulation than if they were entering a conduit body.

The next logical step, as in part **(B)** of this section, is to squarely address short wireway sections that are used as pull boxes. This part now assures that the usual requirements in 314.28 will be applied.

376.30. Securing and Supporting. Wireway must be supported every 5 ft (1.5 m). Wireway lengths over 5 ft (1.5 m) must be supported at *each end* or *joint, unless* listed for other support. In no case should the distance between supports for wireway exceed 10 ft (3.0 m). Vertical runs can use a 4.5 m (15 ft) interval, provided there is no more than a single joint between supports.

376.56. Splices, Taps, and Power Distribution Blocks. The conductors should be reasonably accessible so that any circuit can be replaced with conductors of a different size if necessary and so that taps can readily be made to supply motors or other equipment. Accessibility is ensured by limiting the number of conductors and the space they occupy as provided in 376.22 and 376.23. For power distribution blocks, the bending space requirements of 312.6(B) (the bigger distance table) apply even if a particular entry may happen to be arranged at a right angle. In addition, the block design must include an insulating cover so the uninsulated live parts are not exposed after the block installation is complete, even with the wireway cover open. Figure 376-3 shows an example of the devices in use. The 2014 NEC has added another qualification, lifted from 314.28(E)(5), that the wiring end up arranged so the terminals on the blocks remain unobstructed after the completion of work.

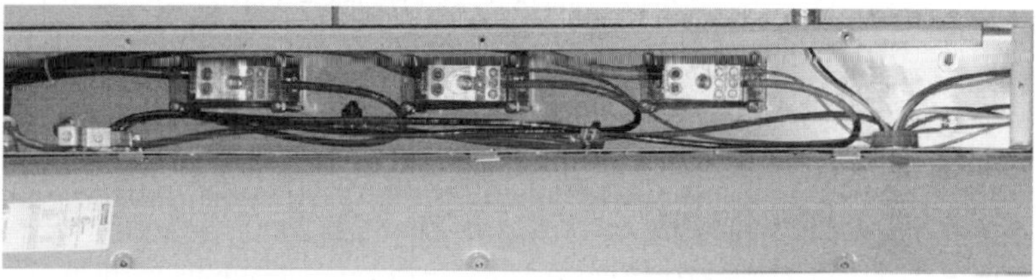

Fig. 376-3. Power distribution blocks in a wireway; although difficult to see in this photo, the blocks are covered with clear polycarbonate. (Sec. 376.56.)

The 2014 NEC has placed a listing restriction from the UL Guide Card information (category QPQS) on these blocks, one that most installers have been completely unaware of, by adding it to 376.56(B)(1) as black letter code. They are not permitted on the line side of service equipment. This cancels, for the time being, one of the more elegant applications of this equipment, namely, the subdivision of large service conductors within a wireway section into 2 to 6 m and switches permitted at one location. It remains to be seen whether the industry will take the necessary steps to accommodate this practice, and at best it is likely to be a lengthy process.

376.70. Extensions from Metal Wireways. Knockouts are provided (or they can be field-punched) in wireways so that circuits can be run to motors or other apparatus at any point. Conduits connect to such knockouts, as shown in Fig. 376-1.

Sections of wireways are joined to one another by means of flanges which are bolted together, thus providing rigid mechanical connection and electrical continuity. Fittings with bolted flanges are provided for elbows, tees, and crosses and for connections to cabinets. See 250.118.

ARTICLE 378. NONMETALLIC WIREWAYS

This article covers nonmetallic wireways. As given in 378.22, the maximum conductor fill is determined in the same manner as for metal wireways, but support requirements are slightly different. 378.30 requires support at 3-ft (0.9-m) intervals, instead of 5-ft (1.5-m) spacing. The most significant difference is found in 378.22 with respect to mutual conductor heating derating factors. Because nonmetallic wireways are not efficient heat sinks and transfer heat comparatively poorly, the 30-conductor allowance prior to the imposition of derating factors that pertains to metal wireways does not apply to nonmetallic wireways. The derating factors begin to apply with the fourth current-carrying conductor, just like ordinary tubular raceways.

ARTICLE 380. MULTIOUTLET ASSEMBLY

380.1. Scope. UL data (Guide Card PVGT) are as follows:

> This category covers multioutlet assemblies, accessories for use with multioutlet assemblies, and factory-assembled wiring kits intended for installation into multioutlet assemblies.
>
> Multioutlet assemblies consist of an enclosure or raceway and outlet wiring devices that provide power for connection of utilization equipment. Multioutlet assemblies are intended for use in dry locations, other than hazardous (classified) locations, in accordance with ANSI/NFPA 70, "National Electrical Code" (NEC). Multioutlet assemblies are intended to be connected to permanently installed branch circuits operating at frequencies between 50 and 400 Hz and dc (direct current) circuits.
>
> A multioutlet assembly may be provided with channels for additional power circuits, control circuits, power-limited circuits and communication-circuit wiring for audio, video and data.
>
> Accessories are parts that may be added to a multioutlet assembly either by the manufacturer or by the installer to add functionality, e.g., hangers, retainers, luminaires, remote-control modules, signs.
>
> Wiring kits are assemblies of conductors and devices, such as receptacle outlets and switches, which are supplied as a wiring system for use in specific multioutlet assemblies.
>
> A part used to connect, change direction, or terminate a multioutlet assembly (e.g., a transition coupler, an end cap, a corner, a tee, an adapter, a box) or a specific wiring device that completes the system is covered under Multioutlet Assembly Fittings (PVUR).

380.2. Uses Permitted. These assemblies are intended for surface mounting, except that the metal type may be surrounded by the building finish or recessed so long as the front is not covered. The nonmetallic type may be recessed in baseboards. Section 380.76 allows them, if metal, to extend through a dry partition provided no outlet is left within the partition and provided the removable cover can be removed on each side from all exposed portions. In calculating the load for branch circuits supplying multioutlet assembly, see 220.14(H).

380.23. Insulated Conductors. This new section incorporates the wireway conductor bending space provisions applicable to 4 AWG and larger conductors in multioutlet assemblies. Although this may seem a stretch, the proposal substantiation demonstrated the existence of multioutlet assemblies as large as 10 in. wide and 5 in. deep, containing very large devices. The rule only applies to field-assembled equipment.

ARTICLE 382. NONMETALLIC EXTENSIONS

A *nonmetallic extension* is an assembly of two insulated conductors within a nonmetallic jacket or an extruded thermoplastic covering. The assembly is mounted directly on the surface of walls or ceilings. Nonmetallic extensions are permitted only if (1) the extension is from an existing outlet on a 15- or 20-A branch circuit and (2) the extension is run exposed and in a dry location. Nonmetallic extensions are limited to residential or office buildings that do not exceed three stories above grade. This category also, as a new 2008 **NEC** category in this article, includes *concealable nonmetallic extensions*, which comprise two, three, or four insulated circuit conductors that mount on wall and ceiling surfaces in a flat configuration that is capable of concealment behind paint, joint compound, wallpaper, etc. In this form, if identified for the purpose, concealable extensions are permitted in buildings of more than three stories. As of this writing, no products have been listed by UL in this category, but there are manufacturers who are strongly interested in the concept.

ARTICLE 384. STRUT-TYPE CHANNEL RACEWAY

384.1. Scope. This article covers the installation and use of channel raceways. It provides requirements for the channel manufacturer as well as for the designer and installer. As given in 384.2, such raceways must be resistant to moisture or protected against corrosion. The next sentence specifically recognizes galvanized steel, stainless steel, and enameled or PVC-coated aluminum and steel as satisfying this requirement for moisture and corrosion resistance.

384.10. Uses Permitted. Channel raceways are permitted for a variety of applications. Analysis of the locations and conditions of permitted use spelled out here indicates that channel raceway may be used within buildings where exposed in dry locations—and for limited cases in "damp or corrosive" locations—for power poles to feed receptacles, electrified partitions, under-carpet or under-floor applications, and the like. Such raceways may be used only for systems rated up to 600 V, and in hazardous locations as described in 501.4(B).

384.12. Uses Not Permitted. Here the Code prohibits the use of channel raceway systems where concealed and permits their use only indoors when constructed with ferrous metal (e.g., steel or iron) and protected only by enamel coating.

384.22. Number of Conductors. To determine the number of conductors in a strut channel layout, first note whether you are using couplings that mount inside the channel or outside it. Then read out the area available for conductor fill from Table 384.22. For example, a standard $1^5/_8 \times 1^5/_8$ strut with an inside coupling ("joiner") has 327 mm² (0.507 in.²) available for wire fill. If all the wires you want to insert are the same size, divide this number by the csa of the wire from Chap. 9 Table 5. Otherwise, add the proposed fill and compare it with what you just determined was available.

example If 3 8 AWG THHN wires will be used in the above strut, how many 10 AWG THHN could be added?

Step 1: As determined above, the maximum possible fill area is 0.507 in.².

Step 2: Subtract the known fill, 3 8 AWG = 3 × 0.0366 in.² = 0.1098 in.². Therefore, remaining fill area is 0.507 – 0.110 (three significant figures) = 0.397 in.².

Step 3: Divide the remaining fill area by the csa of a 10 AWG wire based on Table 5; 10 AWG THHN csa is: 0.0211 in.². 0.397 ÷ 0.0211 = 18.8 conductors; therefore, the answer is 18 10 AWG THHNs can be installed with the 3 8 AWG THHNs.

Note that there is a similar calculation to be done as part of a three-part procedure to determine if the installation qualifies for a waiver from the ampacity derating rules. This allowance depends on three factors being true, and then allows a similar waiver as applies to metal wireways. The second and third conditions are not difficult to meet. The first condition, however, apparently based on 386.22 for surface metal raceways, is impossible to meet. None of the raceways recognized by this article meet the 2500 mm² (4 in.²) threshold in total cross-sectional area. The largest one, the 1½ × 3 size just misses at 2487 mm² (3.854 in.²), and the others are all smaller. Therefore, this allowance cannot be used, all the wires are subject to the penalties in 310.15(B)(3)(a), and the exemption procedure should be ignored.

ARTICLE 386. SURFACE METAL RACEWAYS

386.1. Use. At one time, this article was titled "Surface Metal Raceways and Surface Nonmetallic Raceways." The article now covers only metallic surface raceways (Fig. 386-1).

386.21. Size of Conductors. Manufacturers of metal surface raceways provide illustrations and details on wire sizes and conductor fill for their various types of raceway. It is important to refer to their specification and application data.

386.22. Number of Conductors in Raceways. The rules of conductor fill may now be applied to surface metal raceway in very much the same way as standard wireway (Fig. 386-2). This rule applies wireway conductor fill and ampacity determination to any surface metal raceway that is over 4 sq in. in cross section. As with wireway, if there are not more than 30 conductors in the raceway and they do not fill the cross-sectional area to more than 20 percent of its value, the conductors may be used without any conductor ampacity derating

Right: Typical use of small metal surface raceway for extensions from existing receptacle outlets.
Below: Shallow switch or receptacle box for surface raceway.

Fig. 386-1. Surface raceway has become popular for new works as well as for modernization. (Sec. 386.10.)

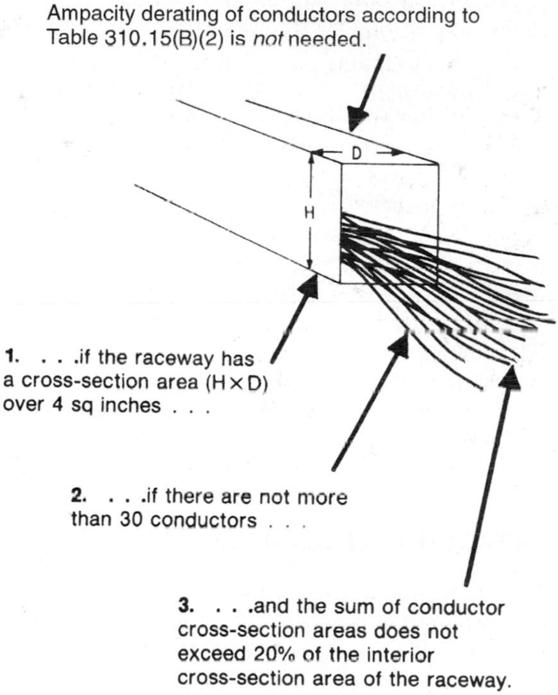
Ampacity derating of conductors according to Table 310.15(B)(2) is *not* needed.
D
H
1. . . .if the raceway has a cross-section area (H × D) over 4 sq inches . . .
2. . . .if there are not more than 30 conductors . . .
3. . . .and the sum of conductor cross-section areas does not exceed 20% of the interior cross-section area of the raceway.

Fig. 386-2. NEC rule permits conductor fill of metal surface raceway *without* ampacity derating of wires. (Sec. 386.22.)

accordance with 310.15(B)(3). This allowance only applies to raceways ith over 2500 mm^2 (4 in.^{2}) in cross-sectional area. Only as an example, in the 'iremold product line, that parameter applies to only two raceways, the 4000 ndivided only, 7.2 in.^{2}) and the 6000 (16 in.^{2}).

386.60. Grounding. In every type of wiring having a metal enclosure around the conductors, it is important that the metal be mechanically continuous in order to provide protection for the conductors and that the metal form a continuous electrical conductor of low impedance from the last outlet on the run to the cabinet or cutout box. A path to ground is thus provided through the box or cabinet, in case any conductor comes in contact with the metal enclosure, an outlet box, or any other fitting. See 250.118. This rule also requires transition boxes and fittings that are used to interface with other wiring methods, such as surface device-box extensions, to have an equipment grounding provision for attaching a grounding conductor from the other wiring method.

386.70. Combination Raceways. Metal surface raceways may contain separated systems as shown in Fig. 386-3. The separate compartments must be consistently distinguishable through the use of stamping, imprinting, or color coding of the interior finish. This is why the divider strips that are used to split these raceways in the field are usually painted with two different colors. It is very important to apply that color code consistently. The **NEC** used to insist that the relative orientation of the divided segments be maintained throughout. That requirement has been deleted, and now only the color code identifies which divided segment is to be which.

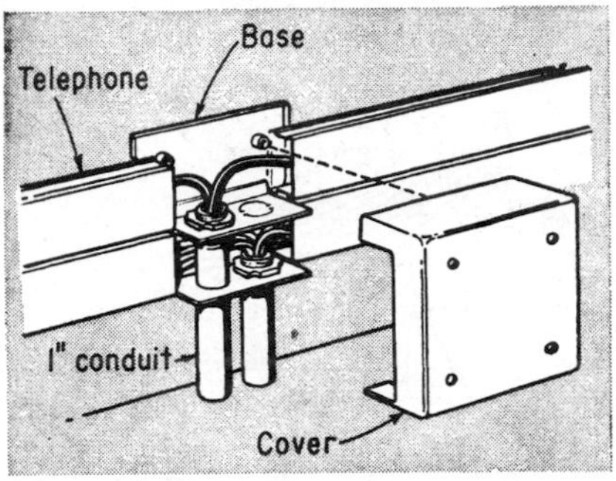
Base
Telephone
1" conduit
Cover

Fig. 386-3. For separating high and low potentials, combination raceway or tiered separate raceways may be used with barriered box assembly. (Sec. 386.70.)

ARTICLE 388. SURFACE NONMETALLIC RACEWAYS

388.1. Scope. Although covered under the same article as surface metal raceway systems—which includes the associated fittings—this wiring method is now covered separately in its own article.

388.2. Definition. The Code describes the various characteristics that surface nonmetallic raceways must possess. Obviously, for any nonmetallic raceway to have the necessary moisture and corrosion resistance, mechanical strength, flame retardance, and low-smoke-producing characteristics, etc., such characteristics must be designed into the raceway system, which means that the manufacturer is responsible for ensuring compliance with this rule. Designers and installers are required to select a manufacturer whose product meets the requirements given here; that is, a *listed* product must be used. Remember to look for the "LS" (low-smoke producing) marking on the selected nonmetallic raceway system.

ARTICLE 390. UNDERFLOOR RACEWAYS

390.3. Use. Underfloor raceway (now defined in 390.2) was developed to provide a practical means of bringing conductors for lighting, power, and signaling systems to office desks and tables (Fig. 390-1). It is also used in large retail stores, making it possible to secure connections for display-case lighting at any desired location.

Fig. 390-1. Underfloor raceway system, with spaced grouping of three ducts (one for power, one for telephone, one for signals), is covered with concrete after installation on first slab pour. (Sec. 390.3.)

This wiring method makes it possible to place a desk or table in any location and it will always be over, or very near to, a duct line. The wiring method for lighting and power between cabinets and the raceway junction boxes may be conduit, underfloor raceway, wall elbows, and cabinet connectors.

390.4. Covering. The intent in paragraphs **(A)** and **(B)** is to provide a sufficient amount of concrete over the ducts to prevent cracks in a cement, tile, or similar floor finish. Figure 390-2 shows a violation. Two 1½- by 4½-in. underfloor

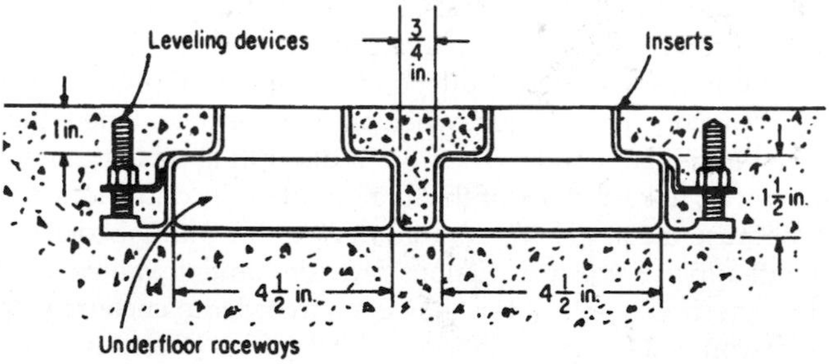
Leveling devices
3/4 in.
Inserts
1 in.
1 1/2 in.
4 1/2 in.
4 1/2 in.
Underfloor raceways

Fig. 390-2. The 1-in. cover is inadequate for raceways less than an inch apart. (Sec. 390.4.)

raceways with 1 in. high inserts are spaced ¾ in. apart by adjustable-height supports resting directly on a base floor slab, as shown. After raceways are aligned, leveled, and secured, concrete fill is poured level with insert tops. But spacing between raceways must be at least 1 in.; otherwise the concrete cover must be 1½ in. deep.

390.7. Splices and Taps. This section has a second paragraph that recognizes "loop wiring" where "unbroken" wires extend from underfloor raceways to terminals of attached receptacles, and then back into the raceway to other outlets. For purposes of this Code rule *only,* the loop connection method is not considered a splice or tap (Fig. 390-3).

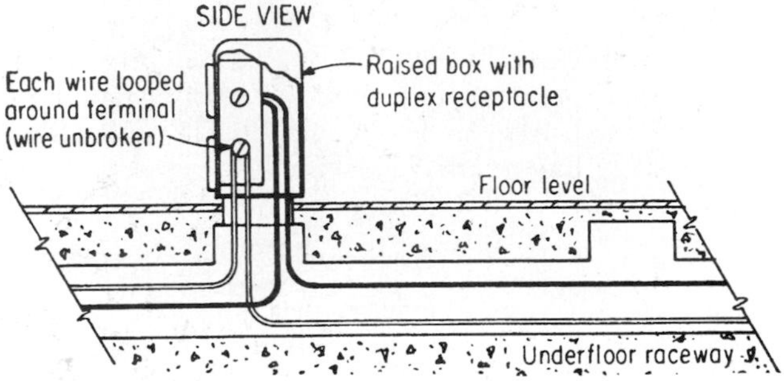
SIDE VIEW
Each wire looped around terminal (wire unbroken)
Raised box with duplex receptacle
Floor level
Underfloor raceway

Fig. 390-3. "Loop" method permitted at outlets supplied from underfloor raceways. (Sec. 390.7.)

Note: As noted in the exception, splices and taps may be made in trench-type flush raceway with an accessible removable cover. The removable cover of the trench duct must be accessible after installation, and the splices and taps must not fill the raceway to more than 75 percent of its cross-sectional area.

390.8. Discontinued Outlets. When an outlet is discontinued, the conductors supplying the outlet shall be removed from the raceway. The general practice is to loop wire intermediate receptacles between the header and the end of the run. This requirement assures that reinsulated conductors will not be resting in the raceway below an abandoned outlet. This in turn prevents a fish wire inserted afterward from a downstream location from getting caught on a reinsulated conductor, with very destructive consequences.

It is often advisable to wire each outlet on its own pair of conductors back to the header or other junction point. Then, when an outlet is abandoned, the associated pair of conductors can be withdrawn without disrupting other outlets on the run. Take care, however, to keep track of the total number of conductors at all portions of the duct, because the ampacity derating factors for mutual conductor heating (see 390.17) will apply. Some jurisdictions provide limited waivers to these derating factors in underfloor applications, however, in order to encourage this practice and thereby discourage the reinsulation of conductors. When the conductors are withdrawn, leave a pull string in their place so the outlet can be easily reactivated in the future as necessary.

390.17. Ampacity of Conductors. This section makes clear that the rules given in 310.15(B)(3) requiring derating of conductors where there are more than three current-carrying conductors within a raceway, cable, or trench apply to conductors installed within underfloor raceways. See the above comment.

ARTICLE 392. CABLE TRAYS

392.2. Definition. Cable trays are open, raceway-like support assemblies made of metal or suitable nonmetallic material and are widely used for supporting and routing circuits in many types of buildings. Troughs of metal mesh construction provide a sturdy, flexible system for supporting feeder cables, particularly where routing of the runs is devious or where provision for change or modification in circuiting is important. Ladder-type cable trays are used for supporting interlocked-armor cable feeders in many installations (Fig. 392-1). This article also covers cable tray with solid bottoms in numerous places, and wire mesh tray has been added to the fill tables [Tables 392.22(A) and 392.22(B)(1)]

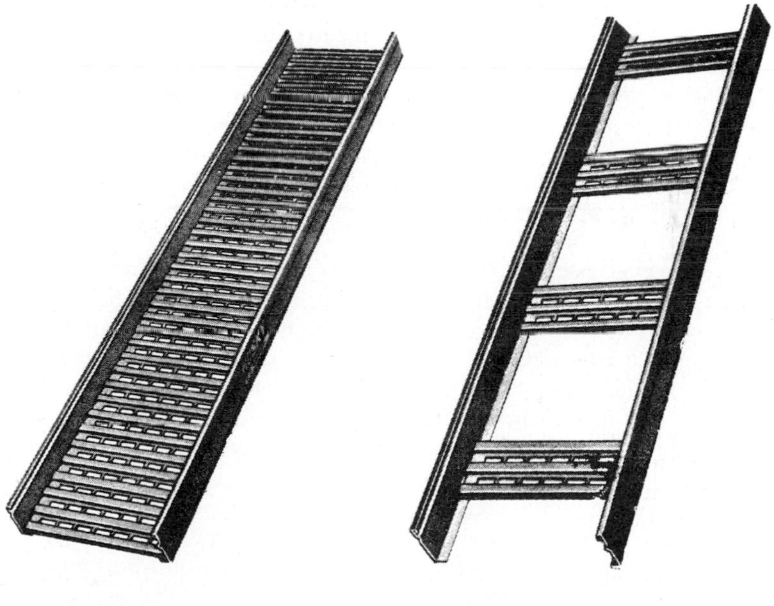
Trough-type (or expanded-metal-type) tray
Ladder-type tray

Fig. 392-1. Two basic types of cable tray. (Sec. 392.2.)

as of the 2011 Code edition. Where past Code editions treated a cable tray simply as a support system for cables, in the same category as a clamp or hanger, the Code today recognizes a cable tray as a conductor support method, somewhat like a raceway, under prescribed and extremely limited conditions, and an integral part of a Code-approved wiring method. However, the definition carefully avoids any mention of the term *raceway* within its provisions. Refer to the extensive commentary in this book at the definition of the term "raceway" in Art. 100. Any "raceway" must be an "enclosed" channel for conductors. A cable tray is a *support system,* not a raceway.

Cable trays are not generally available as a listed product. UL does classify it, but only for its suitability for use as an equipment grounding conductor. The relevant text in the Guide Card information reads as follows:

> This category covers cable trays intended for assembly in the field and for use in accordance with Art. 392 of ANSI/**NFPA 70**, "National Electrical Code" (**NEC**). They have been Classified as to their suitability for use as equipment grounding conductors in accordance with Secs. 392.60(A) and 392.60(B) of the **NEC**. The cable trays are marked on the outer surface of the sidewall of the tray indicating the cross-sectional area of the grounding metal.

Many have misunderstood the ramifications of the limited classification, supposing that if extensive modifications are made in the field, a listing would be compromised. There is no listing to compromise. However, if after all bonding connections have been made, any field changes affect the effective cross-sectional area of metal in a potential ground-fault return path, then the classification is necessarily void at that point for any disturbed sections of tray. Of course, if the tray will not be used as a fault current return path, (the usual case) then the absence of a classification means nothing in this context. See the discussion at 392.60 for more information.

392.10. Uses Permitted. The **NEC** recognizes a cable tray as a support for wiring methods that may be used without a tray (metal-clad cable, conductors in EMT, IMC, or rigid conduit, etc.), and a cable tray may be used in either commercial, industrial, or institutional buildings or premises (Fig. 392-2). Where cables are available in both single-conductor and multiconductor types—such as SE (service entrance) and UF cable—only the multiconductor type may be used in a tray. However, Sec. 392.10(B)(1) permits use of single-conductor building wires in a tray. Single-conductor cables for use in a tray must be 1/0 AWG or larger, listed for use in a tray, and "marked on the surface" as suitable for tray applications. In earlier **NEC** editions, single-conductor cable in a tray had to be 250 kcmil or larger. Sizes 1/0 through 4/0 AWG single-conductor cables may now be used but if used in a ladder-type tray the rung spacing must not be over 225 mm (9 in.) apart. Sizes 250 kcmil and larger may be used in any kind of tray. This rule states that such use of building wire is permitted in industrial establishments only, where conditions of maintenance and supervision ensure that only competent individuals will service the installed cable tray system. This applies to ladder-type trays, ventilated troughs, solid bottom, or ventilated channel-type cable trays.

Single-conductor cables used in a cable tray must be a type specifically "listed for use in cable trays." This is a qualification on the rule that was in previous codes permitting use of single-conductor building wire (RHH, USE, THW, MV

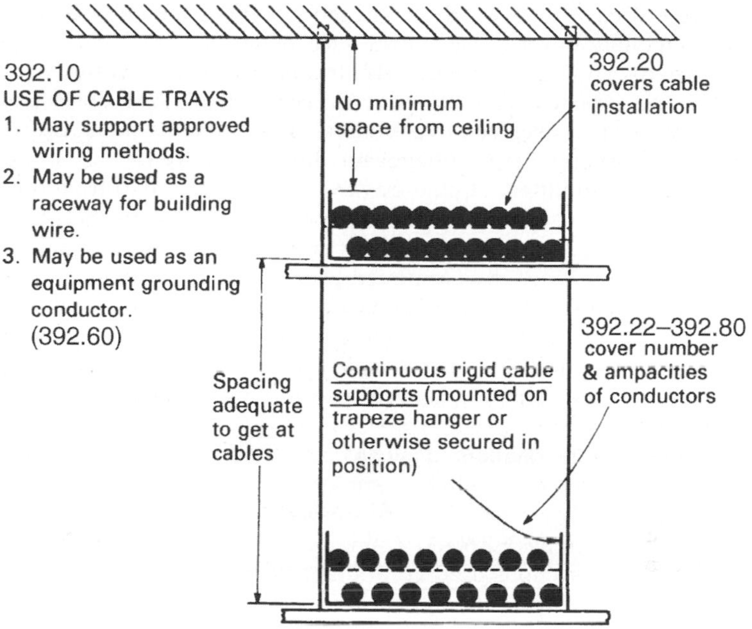
392.10
USE OF CABLE TRAYS
1. May support approved wiring methods.
2. May be used as a raceway for building wire.
3. May be used as an equipment grounding conductor.
(392.60)
No minimum space from ceiling
392.20
covers cable installation
392.22–392.80
cover number & ampacities of conductors
Spacing adequate to get at cables
Continuous rigid cable supports (mounted on trapeze hanger or otherwise secured in position)

Fig. 392-2. Cable-tray use is subject to many specific rules in Art. 392. (Sec. 392.10.)

in a cable tray. The wording permits any choice of conductor types that may be used, simply requiring that any type must be listed. It adds thin-wall-insulated cables, like THHN or XHHW, to the other types mentioned. Present UL standards make reference to cables designated "for CT (cable tray) use" or "for use in cable trays"—which is marked on the outside of the cable jacket. Such cables are subjected to a "vertical tray flame test," as used for Type TC tray cable and other cables. Only cables so tested and marked "VW-1" would be recognized for use in a cable tray.

A cable tray is not a wireway without a cover. This is one of the most abused provisions in the entire **NEC**. Single-conductor cable tray has been found in exhibition halls and hospitals, and just about every other nonindustrial occupancy, except dwellings. None of those occupancies qualifies for single-conductor tray. Single-conductor tray has also been seen over industrial motor control centers, a qualified occupancy to be sure, but filled with motor branch circuit and control conductors running from 12 AWG up to 3/0 AWG. Again, a cable tray is not a wireway without a cover. Single conductors in cable trays are only permitted in the very restricted circumstances itemized here. Specifically, single conductors are only permitted in industrial occupancies with qualified maintenance and supervision, in sizes 1/0 AWG and larger. Medium-voltage conductors run as Type MV cable are permitted to run as single conductors, provided they follow the same rules.

Section 392.10**(C)** specifically uses the word "only" when referring to cable types that are permitted to be used in cable trays in hazardous locations.

In previous Code editions, wording was more open-ended and permitted specific cables without limiting use to only such cables.

As covered in part **(D),** nonmetallic cable tray may be used in corrosive locations. This permits use of nonmetallic tray—such as fiberglass tray—in industrial or other areas where severe corrosive atmospheres would attack a metal tray. Such a tray is also permitted where "voltage isolation" is required.

392.12. Uses Not Permitted. Cable tray systems are excluded from hoistways and areas subject to severe physical damage. The language that used to be in this location limiting cable trays in other spaces for environment air, such as plenum cavity ceilings, was removed because 300.22(C) does permit metal cable trays in such spaces provided they only support wiring methods, such as metal-clad cables without nonmetallic coverings, that are allowed in such spaces generally. In addition, metal cable trays with solid bottoms and sides, and a solid cover, are permitted in such spaces to contain other wiring methods. See 300.22(C)(2).

392.18. Cable Tray Installation. Part **(A)** makes clear that cable trays *must* be used as a complete system—that is, straight sections, angle sections, offsets, saddles, and so forth—to form a cable support system that is continuous and grounded as required by 392.60(A). Cable trays must not be installed with separate, unconnected sections used at randomly spaced positions to support the cable, but must provide electrical continuity and the support required for the cabled wiring methods they contain. Manufactured fittings or field-bent sections of tray may be used for changes in direction or elevation. However, the degree to which this requires absolute end-to-end continuity has significantly evolved.

In the 1971 NEC, part **(c)** of Sec. 318-4 on Installation [now 392.18(A)] read as follows:

> **(c)** Continuous rigid cable supports shall be mechanically connected to any enclosure or raceway into which the cables contained in the continuous rigid cable support extend or terminate.

That wording clearly made a violation of the kind of hookup shown in Fig. 392-3, where the tray does not connect to the transformer enclosures—and is not bonded by jumpers to those enclosures. However, an informational note added in the 1993 Code—which was incorporated in the basic rule in 392.6(A) of the 1996 Code [now 392.18(A)]—indicates that use of "discontinuous segments" is intended to be acceptable—*but,* ground-continuity between the enclosure and the cable tray must be ensured. Refer to 392.30(B)(3) for allowable spacing on discontinuous tray segments containing single conductors, and 392.60(C) for bonding requirements between such sections for more information.

Figure 392-4 shows the rule of part **(F).** The first paragraph of part **(G)** covers the termination rules for cables and raceways arriving at the tray. In general although raceways must always be secured to a tray they service, the tray is not to be used as a qualified point of support, and therefore the next support is the final one before the termination, and therefore must be, typically, within 900 mm (3 ft) of the tray. However, with qualified maintenance and supervision in industrial facilities, and with the knowledge that the tray system ha

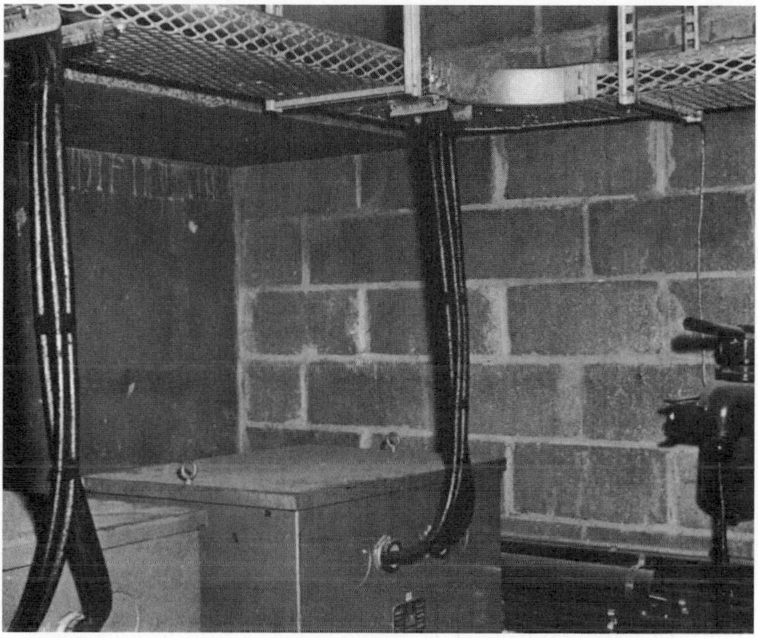

Fig. 392-3. This was clearly a violation of 318-4(c) in the 1971 **NEC** because the tray does not connect to the transformer enclosures. The tray continuity required by 392.18(A) of the present **NEC** is not absolute, provided the installed cables have the support the **NEC** requires. For single conductors as shown here, 392.30(B)(3) permits this discontinuity provided the distance does not exceed 1.8 m (6 ft) and this installation clearly complies in terms of distance, although bonding in accordance with 392.60(C) is not clearly evident. [Sec. 392.18.]

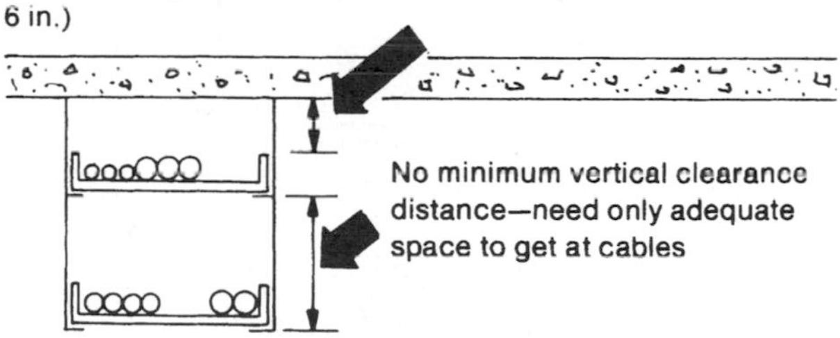
No minimum vertical clearance distance from tray top to ceiling, beam or other obstruction (used to be 6 in.)
No minimum vertical clearance distance—need only adequate space to get at cables

Fig. 392-4. Tray spacing must simply be adequate for cable installation and maintenance. (Sec. 392.18.)

been designed and installed to support the load, then the cable tray qualifies as being "permitted to support raceways and cables and boxes and conduit bodies." In this case, the tray clamp or adapter for the raceway, which must be listed for this duty, becomes the support at a termination [within 900 mm (3 ft)] of the actual raceway bushing) and therefore the next point of support could be 3.0 m (10 ft) away, or at whatever the support interval is in the article for the arriving wiring method. This is a much more flexible and realistic procedure for most industrial occupancies. The other material addresses how to route wiring methods and boxes on the underside of a tray. Part **(H)** is new as of the 2011 **NEC** and requires cable trays supporting medium-voltage wiring to have a readily visible warning label advising of this fact and posted at not over 3 m (10 ft) intervals.

The 2014 edition granted limited relief to qualified industrial occupancies. For the portions of the tray that are not accessible as Art. 100 applies the term to equipment (admitting close approach, not guarded by ... elevation or other effective means), the warning labels can be omitted except as necessary to ensure safety.

392.20. Cable and Conductor Installation. Part **(A)** notes that any multiconductor cables rated 600 V or less may be used in the same cable tray, and Part **(B)** points out that high-voltage cables and low-voltage cables may be used in the same tray if a solid, fixed barrier is installed in the tray to separate high-voltage cables from low-voltage cables. And where the high-voltage (over 600 V) cables are Type MC, it is not necessary to have a barrier in the cable tray, and MC cables operating above 600 V may be used in the same tray with MC or other cables operating less than 600 V. But for high-voltage cables other than Type MC, a barrier must be used in the tray to separate high-voltage from low-voltage cables (Fig. 392-5), or another tray must be used.

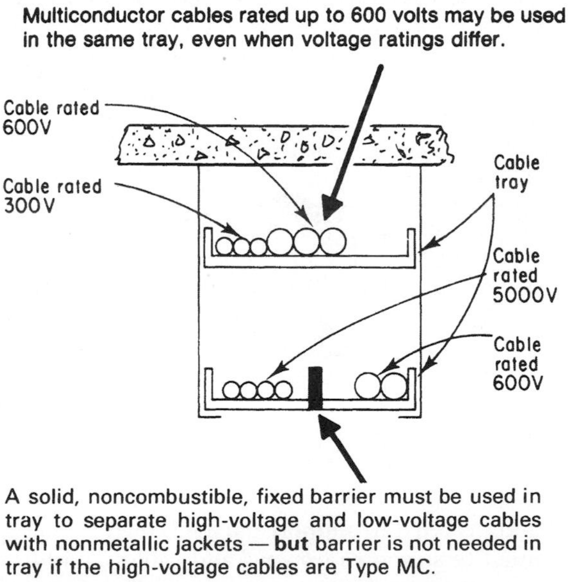
Multiconductor cables rated up to 600 volts may be used in the same tray, even when voltage ratings differ.
Cable rated 600V
Cable rated 300V
Cable tray
Cable rated 5000V
Cable rated 600V
A solid, noncombustible, fixed barrier must be used in tray to separate high-voltage and low-voltage cables with nonmetallic jackets — **but** barrier is not needed in tray if the high-voltage cables are Type MC.

Fig. 392-5. Cables of different voltage ratings may be used in the same tray. (Sec. 392.20.) The 2014 **NEC** has clarified that this rule turns on the actual cable operating voltage and not its rated voltage.

Figure 392-6 shows how single-conductor cables must be grouped to satisfy part **(C)** of this section for a 1200-A circuit made up of three 500-kcmil copper XHHW conductors per phase and three for the neutral. By distributing the phases and neutral among three groups of four and alternating positions, more effective cancellation of magnetic fluxes results from the more symmetrical placement—thereby tending to balance current by balancing inductive reactance of the overall 1200-A circuit.

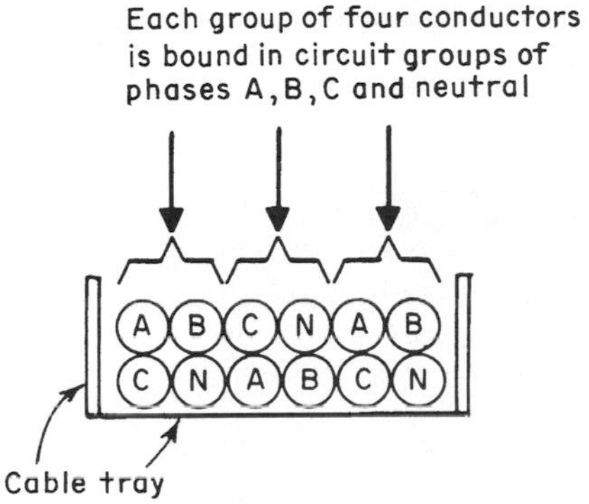
Each group of four conductors
is bound in circuit groups of
phases A, B, C and neutral
A B C N A B
C N A B C N
Cable tray

Fig. 392-6. A parallel 1200-A circuit must have conductors grouped for reduced reactance and effective current balance. (Sec. 392.20.)

Part **(D)** prohibits stacking of single conductors 1/0 to 4/0 AWG in ladder or ventilated trough cable trays, unless bound together as a circuit group.

392.22(A). Number of Multiconductor Cables, Rated 2000 V or Less, in Cable Trays. These rules apply to multiconductor cables rated 2000 V or less. For cables rated 2001 V or higher, the number permitted in a cable tray is now covered in 392.22(C). Note that if a tray is used with a divider, each resulting section is calculated separately in terms of allowable fill.

Section 392.22(A) is broken down into parts **(1)**, **(2)**, **(3)**, **(4)**, **(5)**, and **(6)**, each part covering a different condition of use. Section 392.22(A)(1) applies to ladder or ventilated trough cable trays containing multiconductor power or lighting cables or any mixture of multiconductor power, lighting, control, and signal cables.

Section 392.22(A)(1) has three subdivisions:

(a) Where all the multiconductor cables are made up of conductors 4/0 AWG or larger, the sum of the outside diameters of all the multiconductor cables in the tray must not be greater than the cable tray width, and the cables *must* be placed side by side in the tray in a single layer, as shown in Fig. 392-7. If this applies, and 392.80(A)(1)(c) is used to determine free-air ampacities for the multiconductor cables using the Neher-McGrath engineering calculation, then the cable tray must be wide enough to show that the cables can be spread out as required.

(b) Where all the multiconductor cables in the tray are made up of conductors smaller than 4/0 AWG, the sum of the cross-sectional areas of all cables *must not* exceed the maximum allowable cable fill area in column 1 of

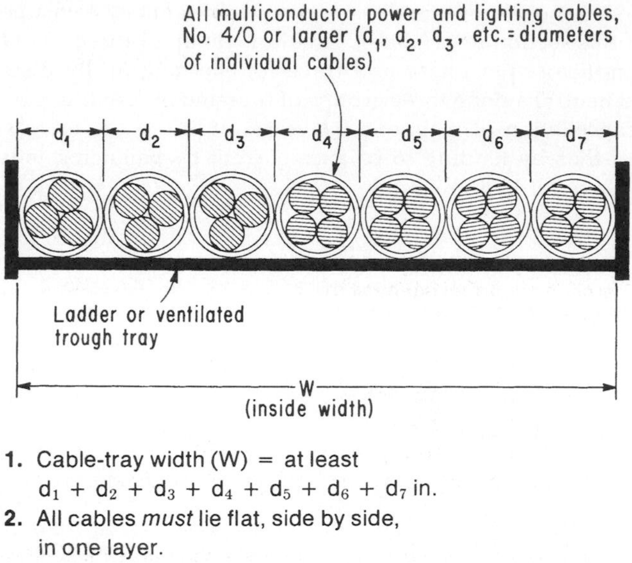
All multiconductor power and lighting cables, No. 4/0 or larger (d_1, d_2, d_3, etc. = diameters of individual cables)
d_1 d_2 d_3 d_4 d_5 d_6 d_7
Ladder or ventilated trough tray
W (inside width)
1. Cable-tray width (W) = at least $d_1 + d_2 + d_3 + d_4 + d_5 + d_6 + d_7$ in.
2. All cables *must* lie flat, side by side, in one layer.

Fig. 392-7. No. 4/0 and larger multiconductor cables *must* be in a *single* layer. [Sec.392.22(A).]

Table 392.22(A) for the particular width of cable tray being used. The tabl shows, for instance, that if a 450 mm (18 in.) wide ladder or ventilate trough cable tray is used with multiconductor cables smaller than 4/ AWG, column 1 sets 13,500 sq mm (21 sq in.) as the maximum value for th sum of the overall cross-sectional areas of all the cables permitted in tha tray, as in Fig. 392-8.

(c) Where a tray contains one or more multiconductor cables 4/0 AWG or large along with one or more multiconductor cables smaller than 4/0 AWG, ther are two steps in determining the maximum fill of the tray.

First, the sum of the outside cross-sectional areas of all the cables smalle than 4/0 AWG must not be greater than the maximum permitted fill are resulting from the computation in column 2 of Table 392.22(A) for the partic ular cable tray width. Then, the multiconductor cables that are 4/0 AWG c larger must be installed in a single layer, and no other cables may be place on top of them (Fig. 392-9). Note that the available cross-sectional area of tray that can properly accommodate cables smaller than 4/0 AWG installe in a tray along with 4/0 AWG or larger cables is, in effect, equal to the allow able fill area from column 1 for each width of tray minus 1.2 times the sum c the outside diameters of the 4/0 AWG or larger cables.

Another way to look at this is to consider that, for any cable tray, the sur of the cross-sectional areas of cables smaller than 4/0 AWG, when added t 1.2 times the sum of the diameters of cables 4/0 AWG or larger, must nc exceed the value given in column 1 of Table 392.22(A) for a particular cabl tray width.

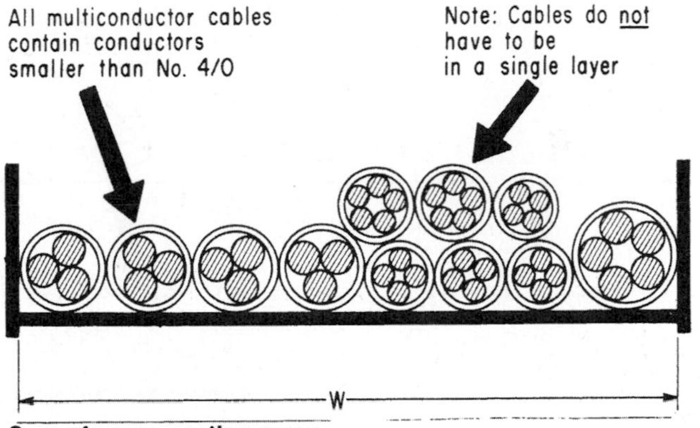
All multiconductor cables
contain conductors
smaller than No. 4/0
Note: Cables do not
have to be
in a single layer
W

Sum of cross-section areas of all cables in cable tray
= not less than cable fill in sq in. given in column 1 of Table 392.22(A) for the particular cable-tray width involved.

Note: Cross-section area (in sq in.) of each cable can be obtained from cable manufacturers' catalogs or spec sheets. If, in the case shown here, the sum of the cross-section areas of the 10 cables in the tray came to, say, 26 sq in., the smallest permissible width of cable tray would be 24 in., as shown in column 1 of Table 392.22(A). An 18-in.-wide cable tray would be good only for a sum of cable areas up to 21 sq in.

Fig. 392-8. Smaller than No. 4/0 cables may be stacked in tray. [Sec. 392.22(A).]

For the installation shown in Fig. 392-9, assume that the sum of the cross-sectional areas of the seven cables smaller than 4/0 AWG is 16 sq in. and assume that the diameters of the four 4/0 AWG or larger cables are 3 in., 3.5 in., 4 in., and 4 in. The abbreviation "Sd" in column 2 of Table 392.22(A) represents "sum of the diameters" of 4/0 AWG and larger cables installed in the same tray with cables smaller than 4/0 AWG. In the example here, then, Sd is equal to $3 + 3.5 + 4 + 4 = 14.5$, and $1.2 \times 14.5 = 17.4$. Then we add the 16-sq-in. total of the cables smaller than 4/0 AWG to the 17.4 and get $17.4 + 16 = 33.4$. Note that this sum is over the limit of 28 sq in., which is the maximum permitted fill given in column 1 for a 24-in. wide cable tray. And column 1 shows that a 30-in. wide tray (with 35-sq-in. fill capacity) would be required for the 33.4 sq in. determined from the calculation of column 2, Table 392.22(A).

Section 392.22(A)**(2)** covers use of multiconductor control and/or signal cables (not power and/or lighting cables) in ladder or ventilated trough with a usable inside depth of 6 in. or less. For such cables in ladder or ventilated trough cable tray, the sum of the cross-sectional areas of all cables at any cross section of the tray *must not* exceed 50 percent of the interior cross-sectional area of the cable tray. And it is important to note that a depth of 150 mm (6 in.)

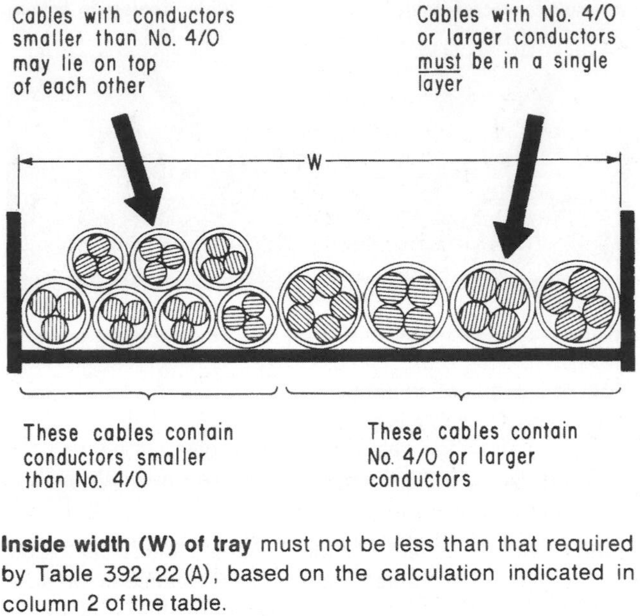
Cables with conductors smaller than No. 4/0 may lie on top of each other
Cables with No. 4/0 or larger conductors must be in a single layer
W
These cables contain conductors smaller than No. 4/0
These cables contain No. 4/0 or larger conductors
Inside width (W) of tray must not be less than that required by Table 392.22 (A), based on the calculation indicated in column 2 of the table.

Fig. 392-9. Large and small cables have a more complex tray-fill formula. [Sec. 392.22(A).]

must be used in computing the allowable interior cross-sectional area of any tray that has a usable inside depth of more than 150 mm (6 in.) (Fig. 392-10).

Section 392.22(A)**(3)** applies to solid-bottom cable trays with multiconductor power or lighting cables or mixtures of power, lighting, control, and signal

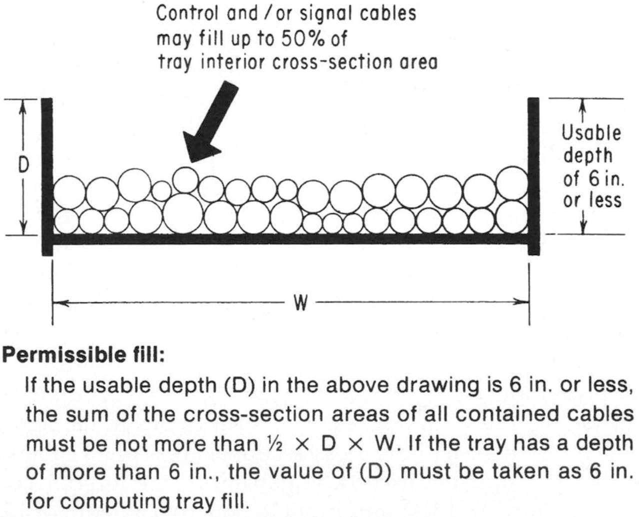
Control and / or signal cables may fill up to 50% of tray interior cross-section area
D
Usable depth of 6 in. or less
W
Permissible fill:
If the usable depth (D) in the above drawing is 6 in. or less, the sum of the cross-section areas of all contained cables must be not more than ½ × D × W. If the tray has a depth of more than 6 in., the value of (D) must be taken as 6 in. for computing tray fill.

Fig. 392-10. Tray fill for multiconductor control and/or signal cables is readily determined. [Sec. 392.22(A).]

cables. The maximum number of cables is covered in three paragraphs that depend on whether or not 4/0 AWG and larger cables comprise the fill, are included in the fill, or are absent from the fill. Section 392.22(A)**(4)** also covers fill for solid-bottom tray, but for control or signaling cables only.

Section 392.22(A)**(5)** covers the fill allowances for ventilated channel tray, with the requirements reflected in a side table that is based on whether or not the tray supports a single cable or multiple cables. Section 392.22(A)**(6)** covers the same topic in the same way for solid channel tray.

392.22(B). Number of Single Conductor Cables, Rated 2000 V or Less, in Cable Trays. This section covers the maximum permitted number of single-conductor cables in a cable tray and stipulates that the conductors must be evenly distributed on the cable tray. This section differentiates between **(1)** ladder or ventilated trough tray and **(2)** ventilated channel-type cable trays.

In Ladder or Ventilated Trough Tray

(a) Where all cables are 1000 kcmil or larger, the sum of the diameters of all single-conductor cables must not be greater than the cable tray width, as shown in Fig. 392-11. That means the cable tray width must be at least equal to the sum of the diameters of the individual cables.

(b) Where all cables are from 250 up to and including 900 kcmil, the sum of the cross-sectional areas of all cables must not be greater than the maximum allowable cable fill areas in square inches, as shown in column 1 of Table 392.22(B)(1) for the particular cable tray width.

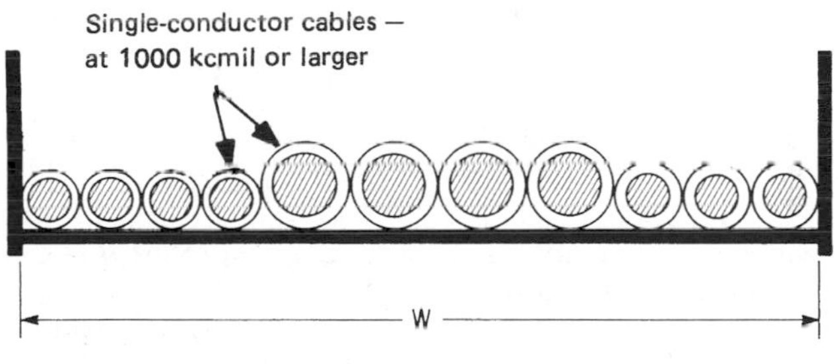
Single-conductor cables —
at 1000 kcmil or larger
W
Tray must be wide enough to hold all cables side-by-side, as shown.

Fig. 392-11. For large cables, tray width must at least equal sum of cable diameters. [Sec. 392.22(B).]

example 1 Assume a number of cables, all smaller than 1000 kcmil, have a total csa of 11 sq in. (7095 mm^2). Column 1 of Table 392.22(B)(1) shows that a fill of 11 sq in. (7095 mm^2) is greater than that allowed for 6 in. wide tray (6.5 sq in.) but less than the maximum fill of 13 sq in. permitted for 12 in. wide tray. Thus, 12 in. wide tray would be acceptable.

example 2 Assume four 4-wire sets of single-conductor, 600-kcmil RHH cables are used as power feeder conductors in a cable tray. Table 5 in Chap. 9 of the **NEC** shows that the overall csa of each 600-kcmil RHH conductor (without outer covering) is 0.9729 sq in. The total area of 16 such conductors would be 16 × 0.9729, or 15.6 sq in. From Table 392.22(B)(1), column 1, 15.6 sq in. is over the maximum permissible fill for 12 in. wide tray, but it is well below the maximum fill of 17.5 sq in. permitted for 16 in. wide tray. Thus, 16 in. wide tray is acceptable.

(c) Where 1000-kcmil or larger single-conductor cables are installed in the same tray with single-conductor cables smaller than 1000 kcmil, the fill must not exceed the maximum fill determined by the calculation indicated in column 2 of Table 392.22(B)(1)—in a manner similar to the calculations indicated above for multiconductor cables. Note that (d) following requires that if 1/0 through 4/0 AWG single-conductor cables are included in the fill, then the sum of all single-conductor diameters must not exceed the tray width.

example If nine 800-kcmil THW conductors are in a tray with six 1000-kcmil THW conductors, the required minimum width (*W*) of the tray would be determined as follows:

1. The sum of the csa of the nine 800-kcmil conductors (those smaller than 1000 kcmil) is equal to 9 × 1.2272 sq in. (from column 5, Table 5, Chap. 9, **NEC**), or 11.04 sq in.
2. Each 1000-kcmil THW conductor has an outside diameter of 1.372 in. The sum of the diameters of the 1000-kcmil conductors is, then, 6 × 1.372, or 8.232 in.
3. Column 2 of Table 392.22(B)(1) says, in effect, that to determine the minimum required width of cable tray it is necessary to add 11.04 sq in. (from 1 above) to 1.1 × 8.232 (from 2 above) and use the total to check against column 1 of Table 392.22(B)(1) to get the tray width:

$$11.04 + (1.1 \times 8.232) = 11.04 + 9.05 = 20.1 \text{ sq in.}$$

From column 1, Table 392.22(B)(1), the fill of 20.1 sq in. is greater than the 19.5 sq in. permitted for 18 in. wide tray. But, this fill is within the permitted fill of 21.5 sq in. for 20 in. wide tray. The 20 in. wide tray is, therefore, the minimum size tray that is acceptable.

Where *any* cables in the tray are sizes 1/0 through 4/0, then all cables *must* be installed in a single layer. And the sum of the single-conductor cable diameters must not exceed the cable tray width as required for "Multiconductor Cables Rated 2000 V, Nominal, or Less" as covered in 392.22(A)(1)(a).

Section 392.22**(B)(2)** governs cable fill in ventilated channel cable trays. Where single-conductor cables are installed in 50 mm (2 in.), 75 mm (3 in.), 100 mm (4 in.), or 150 mm (6 in.) wide, ventilated channel-type trays, the sum of the diameters of all single conductors must not exceed the inside width of the channel.

392.22(C). Number of Type MV and Type MC Cables (2001 V or Over) in Cable Trays. This section applies only to high-voltage circuits in a tray. Type MV cable is a high-voltage cable now covered by Art. 328. Type MC cable is the metal-clad cable operating above 2000 V—a cable assembly long known as interlocked armor cable. [Type MC or other armored cable [e.g., ALS or CS] operating at voltages up to 2000 V must conform to 392.22(A) and 392.80(A) on number and ampacities of cables when used in tray.] Type MV and Type MC medium-voltage cables must conform to the tray fill shown in Fig. 392-12.

392.30. Securing and Supporting. The NEC does not have prescriptive language regarding cable tray support intervals, instead leaving it to the various manufacturers to make specifications based on their product designs. If tray is used in a sloping or vertical orientation, the contained cabling must be fastened to the transverse portions of the tray system at points of crossing. Support is also required within the tray to relieve stress on cables as they enter raceways from the tray.

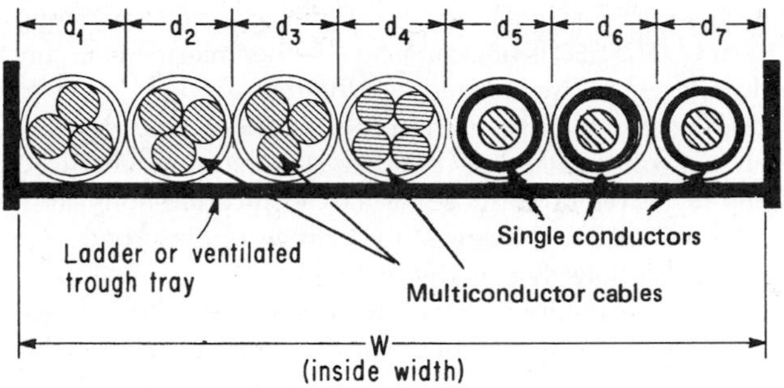
d1
d2
d3
d4
d5
d6
d7
Single conductors
Ladder or ventilated
trough tray
Multiconductor cables
W
(inside width)

1. Cable-tray width (W) = at least $d_1 + d_2 + d_3 + d_4 + d_5 + d_6 + d_7$ in.
2. All cables *must* lie flat, side by side, in one layer.

Fig. 392-12. Tray must be wide enough for all medium-voltage cables in a single layer. [Sec. 392.22(C).]

For discontinuous portions of a cable tray system, as anticipated in 392.18(A), the discontinuity cannot exceed the support distance limitations for the wiring methods supported. For single conductors the maximum spacing is 1.8 m (6 ft), which correlates with the support interval for Type MC cable. The conductors must be secured to the tray (or to the terminating location) at both ends. Refer also to 392.60(C) for required bonding at these transitions.

392.46. Bushed Conduit and Tubing. Cables or conductors can drop out of a tray in conduits or tubing provided they have protective bushings and are clamped to the tray side rail by cable tray conduit clamps to provide the bonded connections required by 392.60(B)(4). The clamps must also comply with 392.18(G), which requires listing for this equipment.

392.56. Cable Splices. Although splices are generally limited to use in conductor enclosures with covers and are prohibited in the various conduits, this section permits splicing of conductors in cable trays. Splices are now permitted to project above the side rails, if not subject to damage.

392.60. Grounding and Bonding. Part (A) specifically recognizes the optional use of the metal length of a cable tray as an equipment grounding conductor for the circuit(s) in the tray—in both commercial, industrial, and institutional premises where qualified maintenance personnel are available to ensure the integrity of the grounding path. Tray is not required to be, and frequently is not, used as an equipment grounding conductor. If it is, then it must meet the bonding requirements in 250.96 (which, in the terms of that section, only applies to enclosures used for equipment grounding).

However, any cable tray used to support electrical conductors must also meet the bonding requirements in Part IV or Art. 250, of which 250.86 applies in these cases generally, and requires a connection to the relevant equipment grounding conductor. Parallel rules apply to service conductor applications, as covered in 250.80.

These connections are easily made at the originating and terminating enclosures at each end of the tray. The NEC is unclear as to whether intermediate connections must be made, and under what circumstances. If the tray supports making frequent connections to building steel that has been properly bonded, and to which the enclosed grounding conductors are ultimately bonded as required in other NEC articles, then additional bonding serves no purpose. Where such connections are not made, one simple approach is to extend the largest relevant grounding conductor from end-to-end, picking up any sections that are discontinuous.

This procedure is sufficient where the tray supports cable assemblies with their enclosed equipment grounding conductors, because such tray applications are not considered to be likely to become energized. This is because a combination of at least two failures (cable jacket and conductor insulation) must both occur in order to energize the tray. However, if the tray supports individual conductors, as only permitted in accordance with 392.10(B), the tray should be considered likely to become energized, and the periodic bonding must comply with the note to Table 250.122 because energized tray that does not promptly open an overcurrent device under the conditions cited could lead to catastrophic results. This would mean making an engineering judgment relative to the impedance of the steel members of the tray, which is frequently higher than that of the applicable equipment grounding conductor.

A new provision at this point is a requirement for a bonding jumper (size unspecified) for all discontinuous metal cable tray segments for cable tray applications that support only "nonpower conductors." An informational note explains that such conductors include, but presumably are not limited to, nonconductive optical fiber cables (which are not conductors at all) and Class 2 and Class 3 power limited circuits. This rather odd rule says nothing about what these trays are supposed to be bonded to; only that their segments are to be bonded together. The substantiation pointed to a NECA/NEMA installation standard, #105, paragraph 4.7.3.2, which states that "Cable tray systems containing conductors outside the scope of Art. 250 (such as communications, data, signal cables, etc.) still require bonding and grounding for system operation and performance." This does not describe a safety justification for the rule. When challenged, the panel cited 250.122 (or 250.66) "tied to the grounding system" but didn't include this in the rule, and 250.112(I) waives equipment grounding for these circuits in most cases.

Part **(B)** of 392.60 permits (does not require and for very large circuits prohibits) steel or aluminum cable tray to serve as an equipment grounding conductor for the circuits in the tray. Even though metallic cable trays are permitted to act as an equipment grounding conductor in the same way that a metallic conduit, tubing, or cable sheath might be, it should be noted that a cable tray is not a "raceway" as defined in Art. 100. Therefore, other Code rules that apply to "raceways" (e.g., 200.7 on grounded conductor identification) do not apply to cable trays.

Note that paragraph **(4)** under part **(B)** requires all tray system components, as well as "connected raceways," to be bonded together—either by the bolting means provided with the tray sections or fittings or by bonding jumpers, as shown in Fig. 392-13.

Part **(C)** is the bonding rule for discontinuous cable tray segments, and requires the isolated segments to be bonded with jumpers sized in accordance

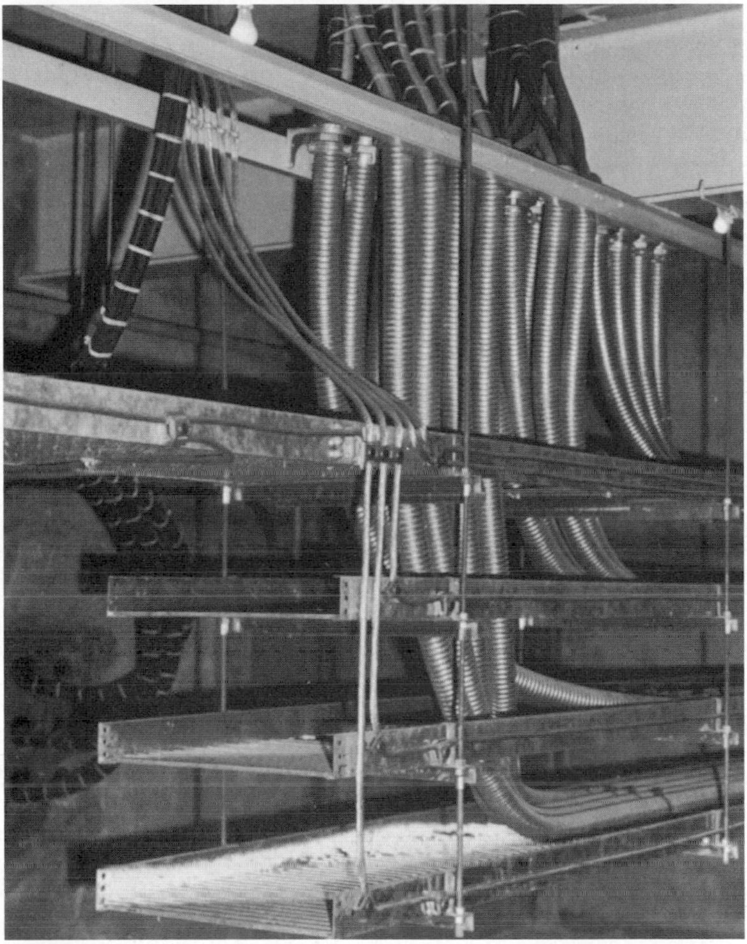

Fig. 392-13. Bare equipment bonding jumpers tie together all tray runs, with jumpers carried up to the equipment grounding bus in the switchboard above. (Sec. 392.60.)

with 250.102 and installed in accordance with 250.96. The jumpers in Fig. 392-13 clearly comply with this requirement. As previously noted, 250.96 does not actually apply to the majority of applications that do not employ the tray as an equipment grounding return path. The reference in (A) to Part IV of Art. 250 is the correct one in these cases, and it will apply throughout the tray system, including any discontinuous segments.

392.80(A). Ampacity of Cables, Rated 2000 Volts or Less, in Cable Trays

(1) Multiconductor Cables

When cable assemblies of more than one conductor are installed as required by 392.22(A), each conductor in any of the cables will have an ampacity as given in Table 310.15(B)(16) or 310.15(B)(18). Those are the standard tables of ampacities for cables with not more than three current-carrying conductors within the cable (excluding neutral conductors that carry current only during load unbalance on the phases). The ampacity of any conductor in a cable is based on the size of the conductor and the type of insulation on the conductor, as shown in those tables. For cables not installed in a cable tray, if a cable contains more than three current-carrying conductors, derating of the conductor ampacities must be made in accordance with 310.15(B)(3)(a). But the last sentence of 392.80(A)(1)(a) flatly exempts cables in a tray from these penalties unless the cable itself includes more than three current-carrying conductors, and even then the penalties only apply based on the cable content itself and not on the basis of the total cable population in the tray.

Section 392.11(A)(1)(b) reduces the above determination of conductor ampacities in the case of any cable tray with more than 6 ft (1.83 m) of continuous, solid, unventilated covers. In such cases, the conductors in the cable have an ampacity of not more than 95 percent of the ampacities given in Tables 310.15(B)(16) and 310.15(B)(18).

Section 392.80(A)(1)(c) applies to a single layer of multiconductor cables with maintained spacing (one cable width) between cables, installed in uncovered tray. In such a case the ampacity can be calculated based on the Neher-McGrath method in 310.15(C). An informational note points to a table in Annex B [Table B.310.15(B)(2)(3)] that has ampacities (for use under engineering supervision) calculated under what is effectively a free-air condition.

(2) Single-Conductor Cables

The ampacity of any single-conductor cable or single conductors twisted together is determined as follows:

(a) 600 kcmil and larger—Where installed in accordance with 392.22(B), the ampacity of any 600-kcmil or larger single-conductor cable in an uncovered tray is *not more* than 75 percent of the ampacity given for the size and insulation of conductors in Tables 310.15(B)(17) and 310.15(B)(19). Note that this means 75 percent of the free-air ampacity of the conductor. And if more than 6 ft (1.8 m) of the tray is continuously covered with a solid, unventilated cover, the ampacities for 600-kcmil and larger conductors must not exceed 70 percent of the ampacity value in the same tables.

(b) No. 1/0 through 500 kcmil—For any single-conductor cable in this range, installed in accordance with 392.10 in uncovered tray, its ampacity is not more than 65 percent of the ampacity value shown in Tables 310.15(B)(17) and 310.15(B)(19). And if any such cables in this range are used in a tray that is continuously covered for more than 6 ft (1.8 m) with a solid, unventilated cover, the ampacities must not exceed 60 percent of the ampacity values in the same tables.

(c) Where No. 1/0 and larger single-conductor cables are installed in a single layer in an uncovered cable tray with a maintained spacing of not less than one cable diameter between individual conductors, the ampacities of such conductors are equal to the free-air ampacities given in Tables 310.15(B)(17) and 310.15(B)(19), as shown in Fig. 392-14. However, if the tray has a solid bottom, this procedure

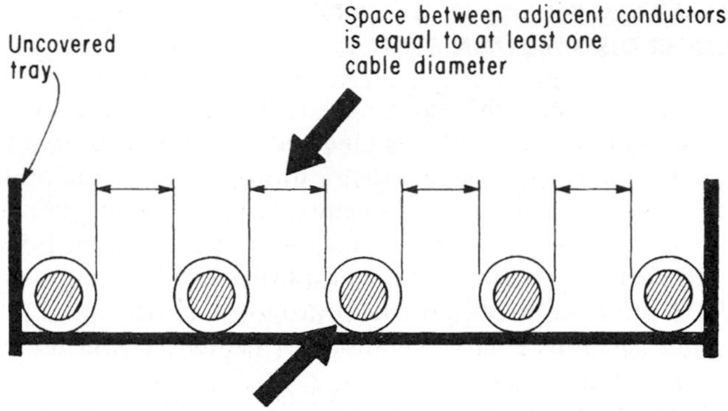
Space between adjacent conductors
is equal to at least one
cable diameter
Uncovered
tray
500 kcmil THW single-conductor copper cables may be used at
their 620 amp rating, from Table 310.17

Fig. 392-14. With spacing, cables in tray may operate at free-air ampacity. [Sec. 392.80(A).]

is not allowed because the cables will overheat, and the calculation must be done using the Neher-McGrath method of 310.15(C).

(d) If the single conductors are arranged in a triangle or diamond configuration, with a maintained free space not less than 2.15 times a conductor diameter between bundles, the ampacities of 1/0 AWG and larger conductors can be calculated on the basis of two- or three-conductors in a bundle of conductors run on a messenger, using 310.15(B). This in effect allows the use of (and an informational note points to) the messenger cable ampacity table, Table 310.15(B)(20) for these purposes.

(3) Combinations of Multiconductor and Single-Conductor Cables—When both types are mixed, the ampacities for the single cables and the ampacities for the multiconductor cables are calculated independently using the applicable procedures above. However, this only applies if (1) the multiconductor fill per 392.22(A) and then expressed as a percentage of the total allowable fill in the cable tray, (2) the single conductor fill per 392.22(B) and then expressed as a percentage of the total allowable fill in the cable tray, and (3) total together not over 100 percent. In addition, the multiconductor cables must be installed in accordance with 392.22(A) and the single-conductor cables must be installed in accordance with 392.22(B) and 392.20(C) and (D) with respect to binding parallel makeups and balanced, crossed phase groupings of conductors (A + B + C + N) + (A + B + C + N), etc.

392.80(B). Ampacity of Type MV and Type MC Cables (2001 V or Over) in Cable Trays. This section covers the ampacities of MV and MC cables operating above 2000 V in cable trays—both single-conductor and multiconductor. 392.80(B)(1)(b) recognizes the improved heat dissipation afforded by spacing of the cables and allows use of the free-air ampacity tables in loading multiconductor cables. The spacing of "one cable diameter" is also recognized for single conductors in 392.80(B)(2)(b), with a separate provision in (c) for triangular and square configurations.

ARTICLE 393. LOW-VOLTAGE SUSPENDED CEILING POWER DISTRIBUTION SYSTEMS

393.1 Scope. The heart of this system is a specially constructed suspended ceiling grid work system. The grid is engineered to simultaneously carry ceiling panels and support luminaires, while distributing a low-voltage power system to power luminaires and other utilization equipment capable of operating within Class 2 parameters. Current designs include T-bars that look much like conventional hung ceiling supports but with the power bus available on the very top of the inverted T (Fig. 393-1) to support LED drop-in luminaires that can operate on Class 2 limits, and split T-bars that allow for pendant power takeoffs to suspended luminaires (Fig. 393-2).

The basic electrical approach to this system is very similar to the approach taken in Art. 411, in this case with the voltage set not over 30 V ac or 60 V dc and with the Class 2 circuits mandated as ungrounded in 393.60(B). Figure 393-3 shows a power supply for one of these systems arranged to perch above a ceiling tile. Sec. 393.45 implements the general rule in 725.121(B) prohibiting interconnections between Class 2 sources unless listed for that possibility, because if not managed properly, such an interconnection could result in a circuit operating at a higher power level than permitted for a Class 2 system, compromising safety.

As in the case of Art. 411, there is an option to install the entire system completely as a listed entity [393-6(A)], or to assemble the system from components, all listed by function as covered in 393.6(B). As noted, the rails themselves are listed elements of the whole, and 393.30 includes support requirements for the

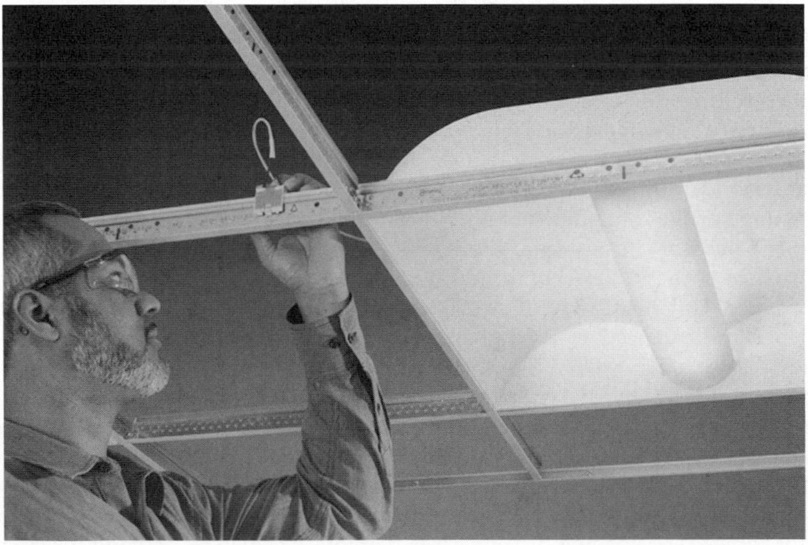

Fig. 393-1. The top of this T-bar has a conductor running on both sides of it carrying a Class 2 circuit accessible through the fitting being installed in the photo to supply the luminaire. [Sec. 393.40(A)(1).] (Courtesy of EMerge Alliance®.)

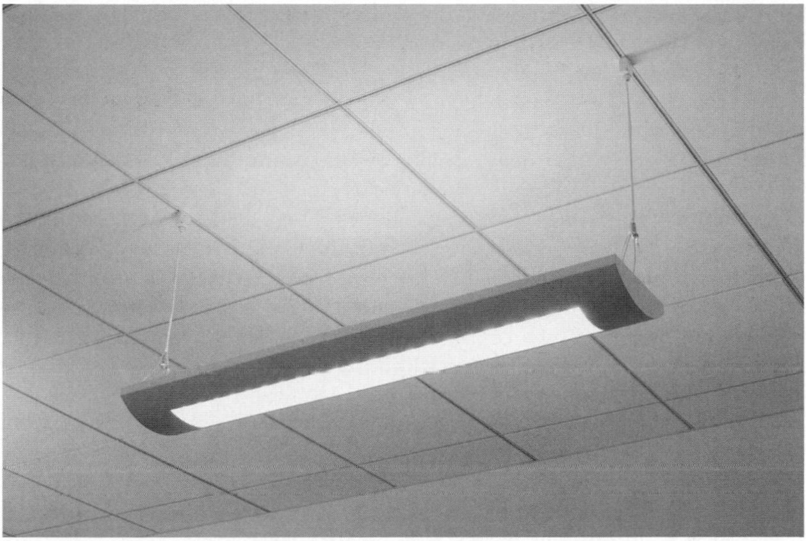

Fig. 393-2. This luminaire is fed from a pendant fitting that attaches to the inside of a specially constructed grid. [Sec. 393.40(A)(2)]. (Courtesy of EMerge Alliance®.)

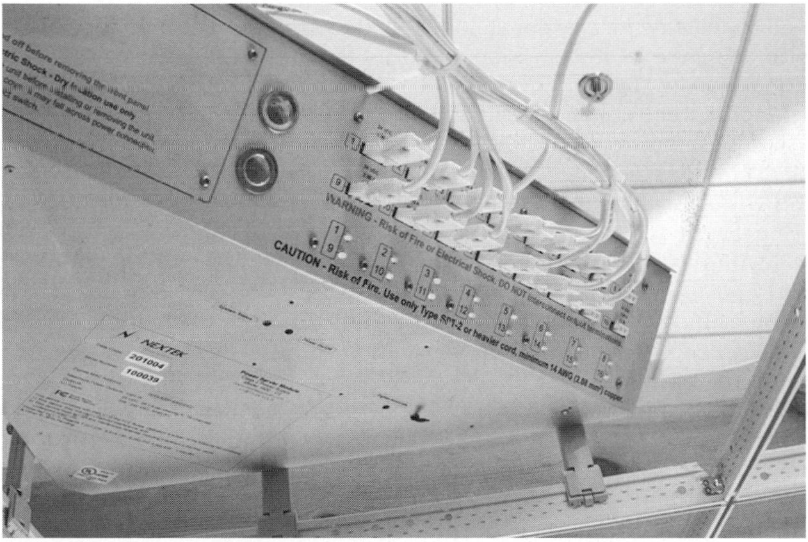
WARNING - Risk of Fire or Electrical Shock. DO NOT
CAUTION - Risk of Fire. Use only Type SPT-2 or heavier cord, minimum 14 AWG
NEXTEK
201004
100039

Fig. 393-3. This power supply will supply 16 separate grid lines, which the manufacturer estimates will cover between 800 and 1000 sq. ft of ceiling. The listing mark is just visible at the opposite corner. [Sec. 393.6(B)(2).] (Courtesy of EMerge Alliance®.)

grid structure. The hardware needs to come with the system or be specified in detail in the installation instructions.

These systems are extremely interesting and the article is well drawn and belongs in the Code. However it seems a poor fit in Chap. 3 because it describes an equipment assembly and not a wiring method. This article probably belongs in either Chap. 4 or 6, either near Art. 604 or 605 (Manufactured Wiring Systems or Office Furnishings) or 411 given the obvious close technical connections to that article on lighting operating on similar voltages. These systems will undoubtedly raise administrative issues out of the scope of the Code with respect to the identity of the personnel permitted to install the system. The manufacturer has taken a stab at this by suggesting the grid go up as specified by an interior designer or architect and installed by a ceiling installer along with lay-in panels, and the remainder of the work be specified by a lighting designer or an electrical engineer and installed by an electrical contractor.

ARTICLE 394. CONCEALED KNOB-AND-TUBE WIRING

394.10. Uses Permitted. Note that this wiring method is restricted to use only for extensions of existing installations and is not Code-acceptable as a general-purpose wiring method for new electrical work. Under the conditions specified in **(1)** and **(2)**, concealed knob-and-tube wiring may be used to extend an existing installation, and in all other cases only if special permission is granted by the local inspection authority having jurisdiction as noted in the second sentence of 90.4.

394.12. Uses Not Permitted. Of particular interest here is item (5), which forbids this wiring method to be used in joist or stud cavities that are insulated in a manner that envelops the conductors. This, for all practical purposes prohibits the use of this wiring method in exterior walls. Fiberglass, cellulose, blown foam will all envelop the conductors, in violation of this rule. Probably the only way to insulate a wall with this wiring method in it is to use rigid foam board, and that would require opening the wall. With the wall open, no one would be likely to try to save a wiring method that has no equipment grounding conductor, and therefore this wiring method is almost incapable of extension without tripping over a number of other serious code issues.

394.23. In Accessible Attics. Where the wiring is installed at any time after the building is completed, in a roof space having less than 3-ft (900-mm) headroom at any point, the wires may be run on knobs across the faces of the joists, studs, or rafters or through or on the sides of the joists, studs, or rafters. Such a space would not be used for storage purposes, and the wiring installed may be considered as concealed knob-and-tube work.

An attic or roof space is considered accessible if it can be reached by means of a stairway or a permanent ladder. In any such attic or roof space wires run through the floor joists where there is no floor must be protected by a running board and wires run through the studs or rafters must be protected by a running board if within 7 ft (2.1 m) from the floor or floor joists.

394.104. Conductors. Conductors for concealed knob-and-tube work may be any of the general-use types listed in Table 310.104(A) for "dry" locations and "dry and wet" locations such as TW, THW, XHHW, and RHH.

ARTICLE 396. MESSENGER-SUPPORTED WIRING

396.2. Definition. This article covers a wiring system that has long been manufactured and widely used in industrial installations. The basic construction of the wiring method has been used for many years as service-drop cable for utility supply to all kinds of commercial and residential properties.

From long-time application, messenger-supported wiring is actually an old standard method, even though it has not been covered by the **NEC** up until recent years. In 225.6(A)(1), the rule refers to "supported by messenger wire," and that phrase has been in the Code for over 60 years. Figure 396-1 shows an example of triplex service-drop cable used for supplying floodlights at an outdoor athletic field. So coverage of this type of wiring is important—especially for the vast amounts of outdoor use where messenger-cable wiring offers so many advantages over open wiring. But messenger-supported wiring is recognized for both indoor and outdoor branch circuits and/or feeders. Refer to the discussion under 225.6 for outdoor use of messenger supported wiring.

396.10. Uses Permitted. Messenger support is permitted for a wide range of cables and conductors—for use in commercial and industrial applications. Part (B) covers ordinary building wires supported on a messenger and recognizes use of messenger supported wiring in "industrial establishments *only*" with "any of the conductor types given in Table 310.104(A) or Table 310.104(B)." These tables include single-conductor Types MV, RHH, RHW, and THW and also accept all the other single-conductor types, such as THHN, XHHW, TW. All such application is recognized either indoors or outdoors—provided that any conductors exposed to the weather are "listed for use in wet locations" and are "sunlight-resistant" if exposed to direct rays of the sun. The ampacity of the wiring method is easy to determine now that Table 310.15(B)(20) has come out of Annex B and is now available for general use. Any of the cabled wiring methods in Table 396.10(A) can now, by right, be supported by a messenger, including tray cable, multiconductor service entrance cable, UF cable, etc., along with Type MV where consistent with other rules governing medium-voltage wiring.

396.30. Messenger. Part (B) of this section, through a somewhat circular reference to 225.4, now recognizes the use of this wiring method under the first exception to 250.32(B) where the neutral is also the equipment grounding conductor and therefore both a current-carrying conductor and a grounding conductor. Its normal function is as a grounding conductor that does not routinely carry current.

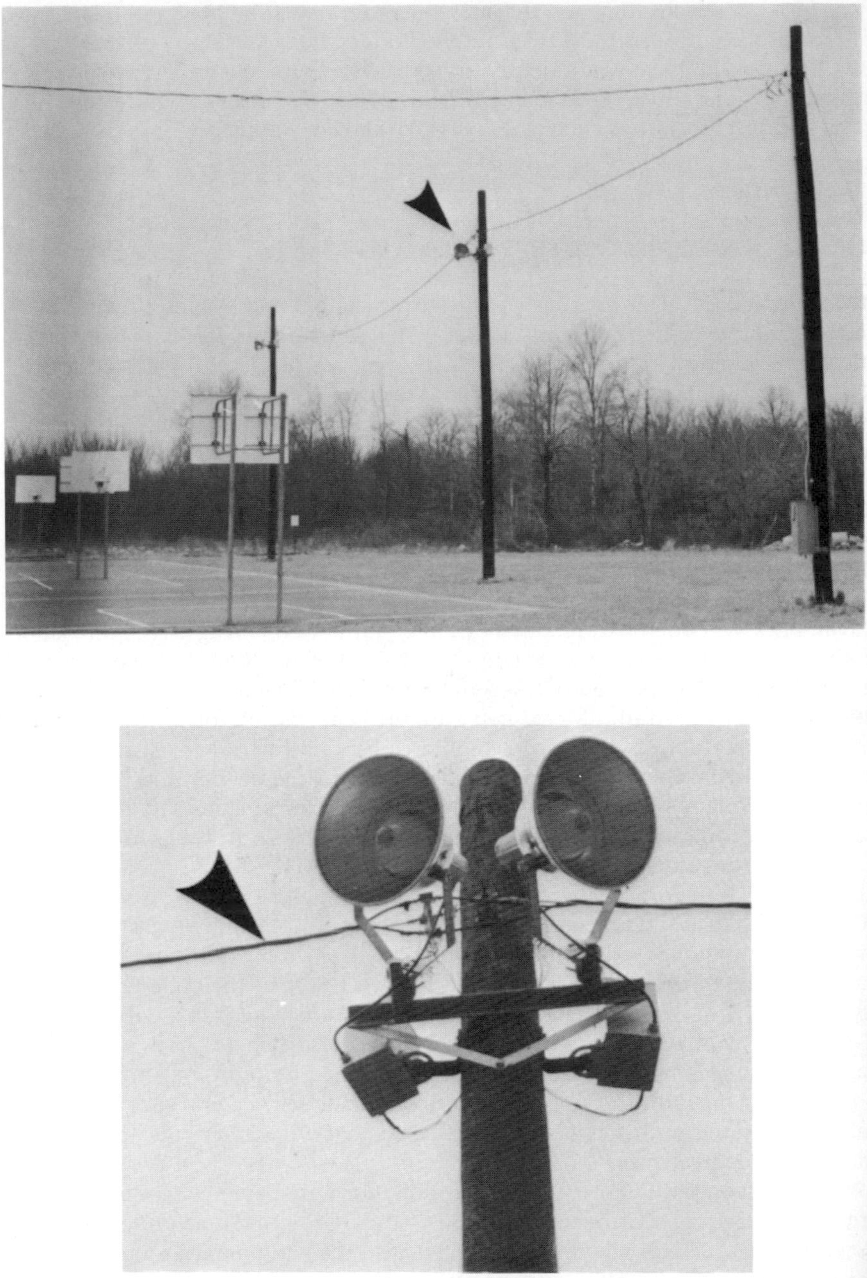

Fig. 396-1. Messenger-supported cable, used here to supply pole-mounted floodlights, may be constructed in a number of different assemblies, such as this service-drop cable with an ACSR messenger cabled with insulated conductors.

ARTICLE 398. OPEN WIRING ON INSULATORS

398.2. Definition. Conductors for open wiring may be any of the general-use types listed in Table 310.104(A) for "dry" locations and "dry and wet" locations such as THW, XHHW, THHN, and so on.

The conductors are secured to and supported by insulators of porcelain, glass, or other composition materials. In modern wiring practice open wiring is used for high-tension work in transformer vaults and substations. It is very commonly used for temporary work and is used for runs of heavy conductors for feeders and power circuits, as in manholes and trenches under or adjacent to switchboards, to facilitate the routing of large numbers of circuits fed into conduits.

398.10. Uses Permitted. This section limits open wiring on insulators to industrial or agricultural establishments, up to 600 V.

398.19. Clearances. The additional insulation on the wire referred to in this rule, is to prevent the wire from coming in contact with the adjacent pipe or other metals.

398.23 spells out such installations in unfinished attics and roof spaces.

398.30. Securing and Supporting. Methods of dead-ending open cable runs are shown in Fig. 398-1.

Method of dead-ending heavy conductors used in open wiring.

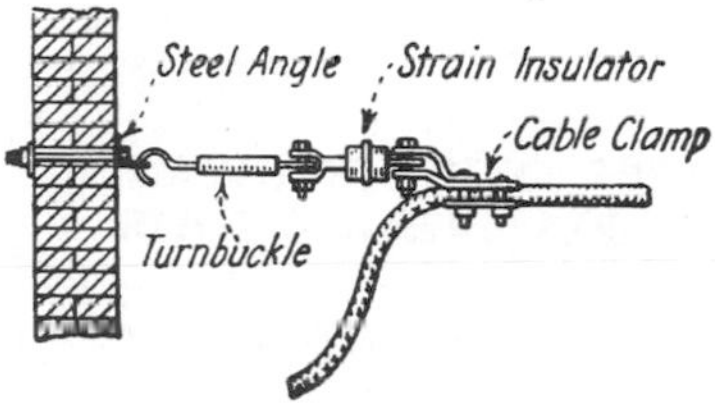
Steel Angle
Strain Insulator
Cable Clamp
Turnbuckle

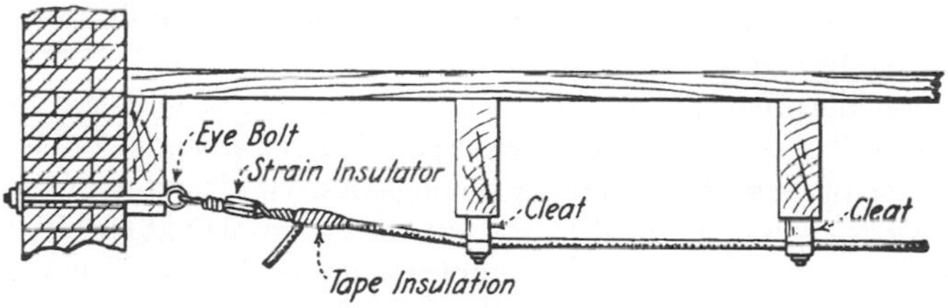
Eye Bolt
Strain Insulator
Cleat
Cleat
Tape Insulation

Method of dead-ending heavy conductors used in open wiring.

Fig. 398-1. Proven methods must be used for dead-ending open wiring. (Sec. 398.30.)

Where heavy ac feeders are run as open wiring, the reactance of the circuit s reduced and hence the voltage drop is reduced by using a small spacing etween the conductors. Up to a distance of 15 ft between supports the 2½-in. pacing may be used if spacers are clamped to the conductors at intervals not xceeding 4½ ft. A spacer consists of the three porcelain pieces of the same orm as used in the support, with a metal clamping ring.

In the rule of part **(B)**, reference to "mill construction" is generally understood to mean the type of building in which the floors are supported on wooden beams spaced about 14 to 16 ft apart. Wires not smaller than No. 8 may safely span such a distance where the ceilings are high and the space is free from obstructions.

Figure 398-2 illustrates the rules of subpart **(D)** on mounting of knobs and cleats for the support of No. 14, No. 12, and No. 10 conductors. For conductors of larger size, solid knobs with tie wires or single-wire cleats should be used.

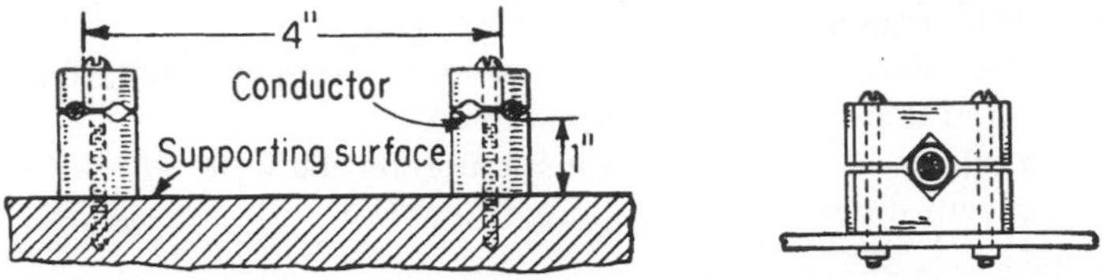
4"
Conductor
Supporting surface
1"

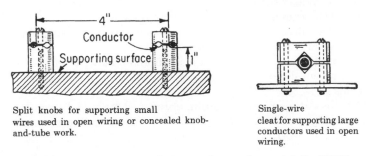

Split knobs for supporting small wires used in open wiring or concealed knob-and-tube work.

Single-wire cleat for supporting large conductors used in open wiring.

Fig. 398-2. Proper wiring support devices must be correctly mounted. (Sec. 398.30.)

ARTICLE 399. OUTDOOR OVERHEAD CONDUCTORS OVER 600 VOLTS

399.2. Definition. These conductors are single conductors, whether insulated, covered, or bare, that are run outdoors on supporting structures.

This article, new as of the 2011 **NEC**, is quite odd. At slightly less than 12-column inches it is shorter than all but a handful of articles in the Code, and it contains no prescriptive information.

The components are essentially all to be as evaluated and specified by a professional engineer. It describes neither a wiring method nor material, but instead covers wiring design. As such, it probably belongs in Chap. 2 and not Chap. 3 of the Code, and not as a standalone article. It would be a natural fit in Part III of Art. 225, which already addresses the topic and which is under the jurisdiction of a Code-making panel with direct experience in this subject matter.

These wiring designs are frequently employed by major industrial occupancies to supply campus style layouts when they are accepting service at medium voltage or even high voltage (over 69,000 V by IEEE definition). The real significance of the article is that when such distributions are premises wiring the authority having jurisdiction now has a basis to review the engineering documentation on the distribution layout, and enforce changes if required. An informational note points to the National Electrical Safety Code, which does have comprehensive requirements for these systems because electrical utilities have been performing this work for over a century. The topic certainly deserve to be covered in the Code.

On balance it is probably correct to leave the prescriptive design information in the NESC. There is precedent for this approach. When manholes came into the NEC through a proposal by this author, there was also a great deal of prescriptive information in the NESC that could have entered the NEC, but for very little benefit. The significance of the informational note now at 110.71, under a section that simply requires a design under engineering supervision, is that there are cases where too much information doesn't really help. Manholes have been in the NEC now for many Code cycles, and no proposal has ever been offered to expand this section. That said, there is definitely room to question the article status and chapter location.

Chapter Four

ARTICLE 400. FLEXIBLE CORDS AND FLEXIBLE CABLES

400.1. Scope. Flexible cords are equipment, not wiring methods, which is why the rules covering them reside in Chap. 4. In addition the rules in the article apply to flexible cords even after they have been manufactured into power supply cords and extension cords (cord sets). A new informational note as of the 2017 **NEC** notes the relationship and points to the rules on permissible use in 400.10 and 400.12. The text in those sections has been carefully adjusted to resolve a major topic of recent discussion relative to whether those rules still applied after the cord had been manufactured into a cord set. The terminology throughout the article, including its title, was adjusted at the same time to include the word "flexible" ahead of "cables" to additionally clarify that the cables covered here are not Chap. 3 wiring methods.

400.3. Suitability. This rule requires that any application of flexible cord or cable may require use of "hard usage" cord (such as SJ cord) or "extra hard usage" cord (such as S or SO cord) if the cord is used where it is exposed to abrasion or dragging or repetitive flexing and/or pulling, depending on severity of use. As noted in Table 400.4, cords for portable heaters must be one of those types when used in damp places. Determination of the need for a particular cable on the basis of use severity is subjective. Table 400.4 also indicates the types of portable cable—that is, for data processing and elevator circuits—and conditions under which each type is suitable, as for hazardous or nonhazardous locations. Flexible cord is not a wiring method, which is why it appears in Chap. 4 as equipment, and not in Chap. 3.

Table 400.4 and its accompanying endnotes contain a great deal of important information that is necessary to proper field application of flexible cord. For example, many are aware that the letter "W" in a cord type suffix indicates wet location suitability, as in, for example SOOW (hard service, 600-V rating, oil-resistant conductor insulation and oil-resistant outer jacket, wet location suitability).

However, the "W" carries an additional meaning clarified in the 2011 NEC. It also denotes sunlight resistance, as covered in Footnote 9 to the table. This has obvious implications for usage where it could be left unattended for long periods, such as for an outdoor swimming pool filter motor field connected with a cord.

400.5. Ampacities for Flexible Cords and Flexible Cables. Unlike other current-carrying conductors and cables recognized by the NEC, the permissible value of current permitted for those assemblies listed as "flexible cords and flexible cables" is determined in accordance with this section, instead of the rules given in 310.15. Generally speaking, compliance with these requirements can easily be accomplished by selecting listed cords and cables and using them within their rating.

A three-conductor cord set is permitted by 250.140 to be used with one conductor serving as *both* the neutral conductor *and* the equipment grounding conductor, with the frame of the range or dryer grounded by connection to the neutral, *but* only for an existing branch circuit. It should be noted that the common neutral-grounding conductor does not count as a current-carrying conductor, thereby making the 3-wire cord suitable for use at the higher ampacity shown under column B in Table 400.5A—which is for cord with not more than two wires.

This section now addresses the effects of elevated temperature by specifically incorporating the ampacity correction factors for ambient temperatures other than 30°C that are included in Table 310.15(B)(2)(a). In addition, cords that are rated at 105°C now have their ampacities referred for elevated ambient temperatures to the 90°C column of the table for the simple reason that no 105°C column exists for that table. Note, however, that the parent language in 310.15(B)(2) permits the use of an adjustment formula in lieu of table values should this be problematic.

example The design load from an Art. 220 load calculation for a floating commercial enterprise moored in a tidal basin on the southeastern coast of Texas is 105 A from a 208Y/120V system, of which 60 A is continuous and comprised of primarily nonlinear loads. The feeder conductors will be connected directly to the terminals of a 125-A molded-case circuit breaker. The building will be subjected to constant tidal motion of the base water level, in addition to periodic wave effects. The design temperature based on the ASHRAE 2 percent temperature database is 95°F.

Flexible cord solution Section 553.7(B) allows portable power cable rated for extra hard usage, sunlight resistance, and wet locations to be used as a feeder to a floating building.

Conductor calculation The load current is 105 A, but there will be mutual conductor heating because over half the load is nonlinear and will result in harmonic currents flowing on the neutral, per 400.5(A). This will result in an 80 percent factor from Table 400.5(A)(3), together with a 0.96 percent correction factor based on a 90°C conductor rating. Therefore, the minimum size of the cord to address the conditions of use is: 105 A ÷ 0.8 ÷ 0.96 = 137 A. For this, the ampacity can be taken from Table 400.5(A)(2) using the 90°C column. A 2 AWG Type W power cable could be used based on this calculation, ampacity 152 A.

Termination calculation The size of the conductor on the terminals of th circuit breaker must accommodate the continuous portion of the load taken a 125 percent plus the noncontinuous load, or 1.25(60 A) + 45 A = 120 A. The wir size, per 110.14(C)(1)(b), must be referenced to the 75°C column of Table 310.15((16) and not Table 400.5(A)(1) or (A)(2) per 110.14(C)(1). The smallest size coppe wire that will allow the overcurrent device to function as intended on this load profile is 1 AWG copper. Because the design constraint is that the cord conductors terminate directly at the circuit breaker, this calculation produces the worst-case, highest conductor size, and therefore this is the correct answer. Because this is flexible cord, undoubtedly more finely stranded than Class C concentric, the terminal on the circuit breaker will most likely not accommodate the stranding. Listed ferrules rated for the class of stranding will need to be crimped on to the cord conductors, and then landed on the breaker. Refer to the discussion at 110.14 for more information.

400.10. Uses Permitted. Figure 400-1 shows accepted uses for flexible cord. Flexible cord may be used for luminaires under **(A)(2).** Refer to 410.24(A) for limitations on use with electric-discharge luminaires and 410.62(B) and (C) for luminaires that require aiming or adjusting after installation and conditional use with other electric discharge lighting. The 2014 **NEC** added an eleventh expressly permitted use, and the context will likely be extremely controversial. The original proposal responded to kits being offered at home stores that allow for homeowners to wire flush receptacle outlets for large-screen, wall-mounted televisions without needing to cope with line-voltage wiring. In fact, the installation instructions expressly forbid any direct connection between the contents of the kit and the wiring system in the building. These kits are being marketed as compliant with the **NEC**; however, the actual contents and installation instructions for these kits appear to raise major **NEC** compliance issues.

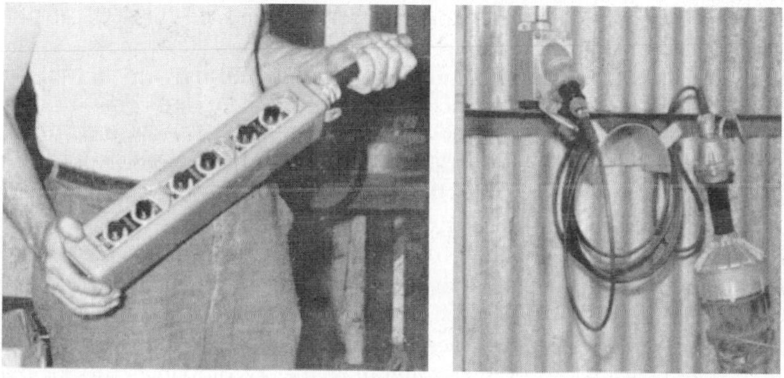

Fig. 400-1. Permitted uses for flexible cable and cord include pendant pushbutton station for crane and hoist controls (left), and connection of portable lamps. (Sec. 400.10.)

The kits generally come with two integral device boxes within an overall self-imming enclosure flange so that once the assembly is secured to the wall the evice opening and mounting is complete. The self-trimming enclosures are also vailable with an additional slot next to the enclosure through which power lim-ed cables related to the audio/video system can enter and exit the wall, resulting no cables being visible around the screen. The two devices are in one case a cessed flanged inlet, installed at normal receptacle height, and in the other case ther a single or a duplex receptacle, also recessed, installed behind the flat-screen appliance. A short run of NM cable that comes with the kit and is fished in the wall connects the two device openings. The homeowner then takes what amounts to an extension cord, which comes with the kit, and plugs the inlet into an adjacent perimeter receptacle.

The new permitted use comes in the context of likely concern that the semi-permanent extension cord from the receptacle to inlet amounts to the substitution of flexible cord for permanent wiring in the building. The new permitted use would make that issue disappear. However, the actual requirement states "the inlet, receptacle outlet, and Chap. 3 wiring method and fittings shall be a *listed assembly specific for this application*" [emphasis supplied]. This lines up with the substantiation that came with the wording, which expressed concern over users providing a smaller AWG size in the cord with resulting overheating. This accompanying substantiation, says in relevant part "The use of *non-standard configuration inlets* or the use of overcurrent protection at the inlet are two possible methods that could be provided in the listing of the assembly to address the hazard" [emphasis supplied].

The cord set and devices supplied with the kit, particularly including the inlet and the female cord body supplied with the extension cord, have conventional NEMA 5-15 configurations that cannot, particularly in this context, be considered as "specific for this installation." Actually, the cord set appears to be a conventional 6-ft extension cord. Therefore, compliance with the 2014 **NEC** is unlikely at best. In addition, the installation instructions and optional equipment raise many other issues. One installation option is to connect an NM cable lead to both enclosures and then connect them using a splicing device that is buried in the wall. The picture of the splicing device shows something that looks very much like the devices covered in 334.40(B). These devices, revisited in the 2014 **NEC**, can only be used for "repair wiring" in this context, and clearly a new flat-screen television is not an example of repair wiring. Another instruction calls for 4 in. of cable in the outlet box (on this point the instructions are inconsistent and some models call for 6 in.) with 3 in. of sheath removed. This plainly fails to comply with 300.14, which requires 6 in. of free conductor "measured from the point in the box where it emerges from its raceway or cable sheath." Although the default assembly anticipates not over 5 ft separating the two boxes, the instructions include an option to purchase enough cable to allow for a 75-ft separation between the two boxes. This raises a host of issues relative to how such a length of cable installed in most applications would be secured and supported, especially by untrained installers.

All of this together raises the question as to how such a kit might have been certified as a "listed assembly." The literature that accompanies this kit plainly implies that it has been so listed. It provides the logo of a particular testing laboratory (not UL) followed by the word "LISTED" and then the word "conforms to UL Std. 514C." UL 514C is the standard that covers nonmetallic outlet boxes, flush device boxes, and covers. Although it is very possible, in fact probable, that the device boxes built into the kit components comply with that standard, compliance therewith clearly cannot justify compliance with the many other standards that come into play, including those covering the flexible cord, the devices, the in-wall cable assembly, etc.

Part **(B)** states that *if* flexible cord is used to connect portable lamps or appliances, stationary equipment to facilitate frequent interchange, or fixed or stationary appliances to facilitate removal or disconnection for maintenance or repair, the cord "shall be equipped with an *attachment plug* and shall be energized from a *receptacle outlet or cord connector body (Fig. 400-2).*"

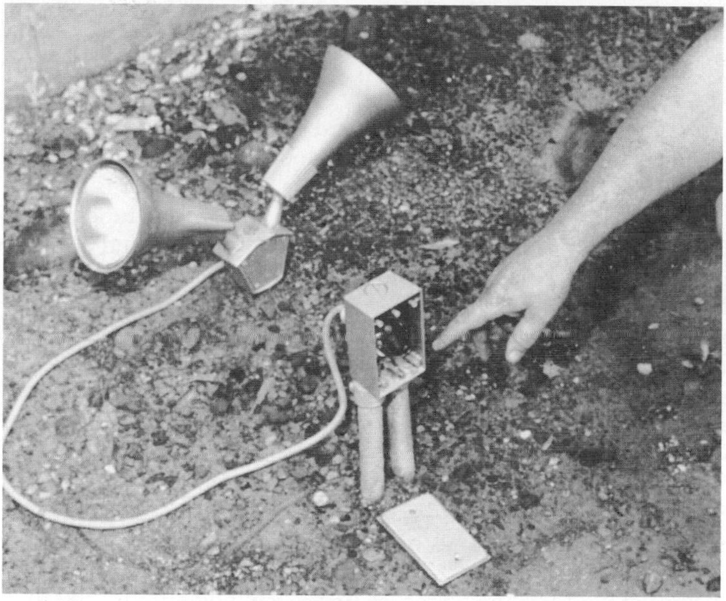

Fig. 400-2. This use of flexible cord to supply an outdoor lampholder assembly can readily be described as a "substitute for fixed wiring"—which is a prohibited use of cord. Here, the lampholders could have been attached to one or more threaded openings on an outlet box. (Sec. 400.12.)

It should be noted that the cords referred to under this section are the cords attached to the appliance and not extension cords supplementing or extending the regular supply cords. The use of an extension cord would represent a conflict with the requirements of the Code in that it would serve as a substitute for a receptacle to be located near the appliance, thereby violating 210.50(B).

Extension cords are intended for temporary use with portable appliances, tools, and similar equipment which are not normally used at one specific location.

Be careful not to confuse flexible cord, regulated within the scope of Art. 400, with bus-drop cable that is not classified as flexible cord and that is not therefore covered in Art. 400. Refer to the coverage in this book of 368.56(B) for more information. This cable may be used to feed down to machines in factories.

400.12. Uses Not Permitted. Although 400.10 says that flexible cord may be used for "wiring of luminaires," that must be applied in the context of 400.12(1) that prohibits the use of cord as a substitute for fixed wiring. That rule could be strictly enforced to require all luminaires to be supplied by fixed wiring methods—approved, Code-recognized cables like NM or BX or by a standard raceway method (EMT, rigid, flex, etc.). This sort of narrow reading raises the question, "Is there ever a case where a luminaire could *not* be fed by a fixed wiring method?" Certainly, any luminaire that might be supplied by a cord connection from a junction box to the luminaire could just as easily be fed by conductors in flexible metal conduit or in liquidtight flexible metal conduit—both of which conduit-and-wire connections are considered "fixed wiring" methods. If there are no cases where a luminaire could not be fed by such a fixed wiring connection, then every use of flexible cord to supply a luminaire is "a substitute for fixed wiring." The relationship between 400.10(A)(2), 400.12(1), 410.24(A), 410.62(B), and 410.62(C) must be carefully evaluated to assure ready compliance with Code rules—particularly because cord connection of luminaires has been used so long and so successfully for both indoor and outdoor applications.

The best way to approach wording issues like this in the **NEC** is to apply an old principle of statutory interpretation that goes back many centuries in the courts: Read both rules that could be viewed as being in conflict, and apply an interpretation that gives the maximum effect to both of the rules. In this case, 400.10 and many other rules in the **NEC** that provide specific instances where cord connections can be used should be honored; however, any expansion of cord usage beyond that point should be closely questioned under the substitution for fixed wiring methods prohibition. In other words, in those cases where the **NEC** specifically permits the use of cord, it is presumed not to be a substitute for fixed wiring methods.

Figure 400-2 shows one of a number of twin floodlight units that were installed outdoors for lighting of the facade of a building. The use of cord from a junction box to a stab-in-the-ground twin lampholder assembly does not comply with "(2) wiring of luminaires" in 400.10 because it does not satisfy 410.62, which regulates use of cord for luminaires, as previously noted under 400.10. And the application does not comply with the other permitted uses in 400.10. Because floodlights could have been installed in lampholders that thread into hubs on a weatherproof box, use of the cord is an evasion of a fixed or permanent connection technique that would totally avoid the potential shock hazard of cord pull-out or breakage. Mounting the floodlights on the box would still allow adjustment. This use of cord is a substitute for fixed wiring and is a violation.

Flexible cord is not permitted to be attached to a building surface **(4)**, unless it is part of a busway branch as covered in 368.56(B). This rule clearly limits the once common practice of running wireway overhead in industrial occupancies in a manner very similar to busway, and feeding cord drops to machine tools and other equipment. Although permitted by 376.70, the wireway must be carefully located because the cord must drop very nearly straight down to the equipment. Figure 400-3 gives another example of this permission, because the strain relief at the box cover is not a connection to a building surface. A swag out to a strain relief over the equipment is a connection to a building surface, and as such is disallowed for all cord connections, except the busway applications.

The prohibition in **(5)** against concealed usage behind structural surfaces together with the ban on usage above suspended ceilings has received extensive attention recently, particularly regarding suspended ceilings. Many types of small scale equipment are routinely furnished with flexible cords and plugs, such as condensate pumps for just one example. Generally, this equipment is also available with an option for hard wiring. Take it. The 2017 **NEC** now includes an exception her

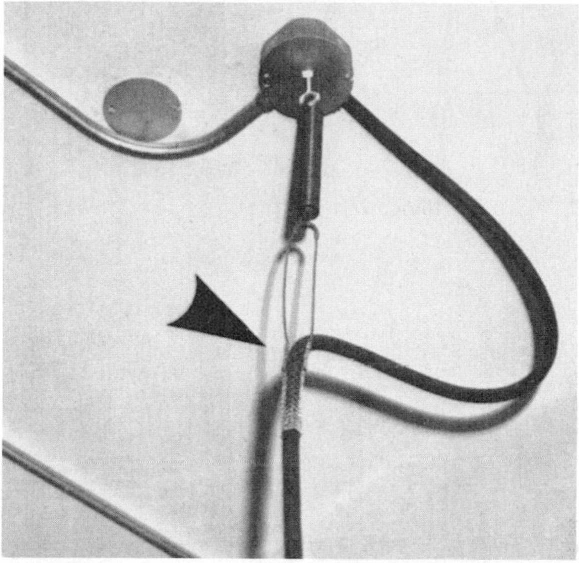

Fig. 400-3. Strain-relief for flexible cord must protect cable jacketing from damage at box connectors and protect wire terminations from pull-out. Spring-loaded come-along support supports cable against weight on bottom end of pendant and also provides up-and-down movement of cable end. (Sec. 400.14.)

that, although phrased correctly in general terms, illustrates extremely well just how tightly this rule is being policed. A manufacturer of AV equipment created an open metal housing for its equipment, designed for ceiling grid applications, that was open only in the downward direction, and that had been evaluated for use in an air-handling ceiling. The equipment required flexible cord connections because some of it was capable of reposition for obvious reasons. Because this equipment resided above the plane of the suspended ceiling, the manufacturer had encountered extensive resistance from inspectors. The exception that resulted was narrowly crafted to apply to that sort of application only.

Flexible cords and cables may not be installed in raceways, except where the NEC specifically recognizes such uses. Part **(6)** clarifies such use of flexible cords and cables, limiting their use in raceways to applications described or inferred in 400.10 and other Code rules—such as 550.10(G) for sleeving of a mobile home power-supply cord, 551.46(A)(2) on the same technique for recreational vehicles, 645.5(B)(3) for computer-room connecting cables, and 680.23(B) on the flexible cord run in conduit for a wet-niche luminaire—and similar limited applications. One very important allowance occurs in 400.14. The reason for this limitation is that flexible cord ampacities, as expressed in Table 400.5(A) and (B), are evaluated on the ability of the cord to freely radiate its heat.

400.14. Pull at Joints and Terminals. Figure 400-4 shows methods of strain relief for cords. The "Underwriters' knot" has been used for many years and is a good method for taking the strain from the socket terminals where lamp cord

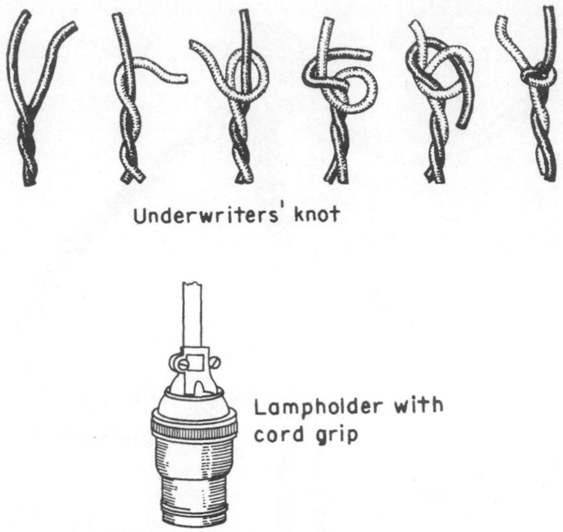
Underwriters' knot
Lampholder with
cord grip

Fig. 400-4. Strain-relief must be provided at cord connections to devices. (Sec. 400.14.)

is used for the pendant, through the hole in the lampholder or switch device. For reinforced cords and junior hard-service cords, sockets with cord grips such as shown in Fig. 400-3 provide an effective means of relieving the terminals of all strain. Figure 400-3 shows a support technique that comes under "other approved means."

400.15. In Show Windows and Showcases. Because of the flammable material nearly always present in show windows, great care should be taken to ensure that only approved types of cords are used and that they are maintained in good condition.

400.17. Protection from Damage. Flexible cord must be protected from damage where it passes over sharp edges through the use of bushings, etc. The second paragraph recognizes a very common practice on machine tools in industrial occupancies, where cord is used to wire sensors or solenoids on flexible parts or other instances where the flexibility of cord is crucial to orderly equipment function. Cord is permitted for these uses generally, but as soon as the need for flexibility stops then a wiring method transition must be arranged. This in turn usually means a box with splices inside, the very location where problems crop up most often in electrica systems. The language in the second paragraph allows a raceway to extend from the industrial control panel (Art. 409) (or other point of origin) out to the desired location, with a raceway-to-cord transition fitting at the outer end. This entirely avoids a box needing support plus a set of splices out on the equipment. If mutual conductor heating is an issue, the derating factors apply to all conductors in the raceway, and not cord by cord. The raceway cannot exceed 15 m (50 ft) in length, which correlate with rules in NFPA 79, *Electrical Standard for Industrial Machinery.*

400.22. Grounded-Conductor Identification. These are the rules for identify ing the grounded circuit conductor in cord sets, which itemize many method in addition to white and gray insulation. One of the more important is item (F which includes a ridge on the outer surface of the cord. This is commonly used for floor and table lamps wired with SPT 2 ("zip cord") wire, and also for pendant applications where the cord weaves through chain links holding a chandelier. It is absolutely essential to know how to recognize and verify proper connections to this wiring, because otherwise Edison-base screw-shells of lampholders will be energized any time the luminaire is lit. This is a hazard during relamping because the screw-shell of the lamp will be hot as soon as it comes in contact with the lampholder, even while someone's finger is touching it.

ARTICLE 402. FIXTURE WIRES

402.5. Allowable Ampacities for Fixture Wires. Note that Table 402.5 gives the ampacity for each size of fixture wire *regardless of the type of insulation used on the wire.* For instance, an 18 AWG fixture wire is rated for 6 A whether it is Type TFN or PF or any other type.

402.7. Number of Conductors in Conduit or Tubing. The maximum number of any size and type of fixture wire permitted in a given size and type of conduit is selected from the same tables as the ones used for determining conduit fill for building wire (THW, THHN, etc.). This must be carefully observed, especially when using fixture wires for Class 1 remote-control, signaling, or power-limited circuits, as permitted and regulated by 725.49 and 725.51.

402.8. Grounded Conductor Identification. This is where the rules in 400.22 become incorporated as rules for identifying fixture wire as well. For example, 400.22(B) allows two fixture wires with yellow braids to be used for a luminaire, with the one finished with a contrasting tracer color, such as red, being the grounded circuit conductor. Many people are fooled by this, and think the red tracer is the hot conductor because its color does not suggest white or gray.

402.10. Uses Permitted. Fixture wires may be used for internal wiring of luminaires and other utilization devices. They may also be used for connecting luminaires to the junction box of the branch circuit—such as by a flex whip to satisfy 410.117(C) (Fig. 402-1).

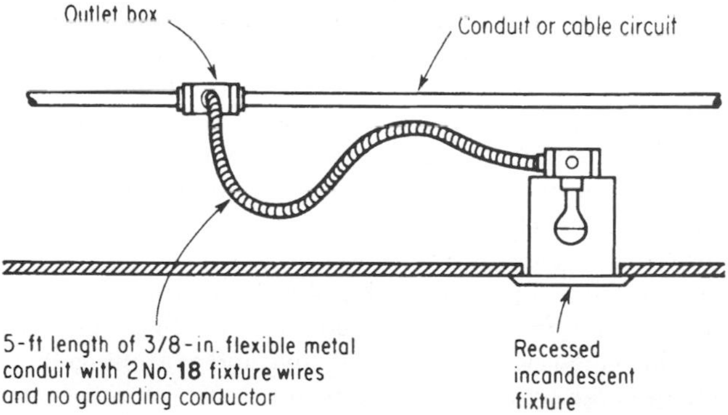
Outlet box
Conduit or cable circuit
5-ft length of 3/8-in. flexible metal conduit with 2 No. 18 fixture wires and no grounding conductor
Recessed incandescent fixture

Fig. 402-1. Fixture wires may connect luminaires to branch-circuit wires. (Sec. 402.10.)

402.12. Uses Not Permitted. With the exception of their use for remote-control, signaling, or power-limited circuits, fixture wires are not to be used as general-purpose branch-circuit wires. An example of the use permitted by 725.49 would be, say, 18 AWG fixture wires run as remote-control wires in a raceway from a motor starter to a remote pushbutton station, where the 6-A rating of the wire is adequate for the operating current of the coil in the starter. Another issue that often comes up is lighting installed for commercial cooking hoods using the special luminaires required by 410.10(C). These luminaires usually have 110°C minimum supply temperature conductor ratings. Because fixture wires of the required temperature rating are quite common, and because branch conductors rated above 90°C are difficult to find on a supply house shelf, and because each luminaire is an outlet on a branch circuit, this section makes it mandatory to order the branch-circuit conductors with the required temperature ratings.

402.14. Overcurrent Protection. This rule refers to 240.5, which covers overcurrent (OC) protection requirements for fixture wire as a separate matter from other building wire and cable, and permits No. 18 and No. 16 fixture wire of any type to be protected by a 15- or 20-A fuse or CB, provided the distance limitations in 240.5(B)(2) are met. That covers use of 18 AWG or 16 AWG in luminaire "whips" on 15- or 20-A branch circuits. Coverage for remote-control, signaling, or power-limited circuits is addressed in 725.43 and as frequently modified for taps in 725.45(C).

ARTICLE 404. SWITCHES

404.1. Scope. Note that all the provisions of this article that cover switches *also* apply to circuit breakers, which are operated exactly as a switch whenever they are manually moved to the ON or OFF position. The scope also now makes clear that the article only covers equipment operating at 600 V and below. Medium-voltage switches are covered in Art. 490.

404.2. Switch Connections. The rule of part **(A)** is shown in Fig. 404-1. Keeping both the supply and return conductors in the same raceway or cable minimizes inductive heating, as described under 300.20(A). The routing of both the supply and return conductors within the same raceway provides for mutual cancellation of both conductors' magnetic fields and thereby reduces inductive heating, as required under 300.20(A). This is also an issue with multiple 2-conductor (as opposed to a single 3- or 4-conductor) Type NM cable run to steel boxes and entering through separate knockouts for multiple loads on a single circuit (such as a bathroom fan-light unit) or a three-way switch loop run with the travelers in one cable. In either case there will be inductive heating unless both cables enter a single knockout. Fortunately there are numerous cable connectors now listed for two cables that will solve this problem.

The rule of part **(B)** is illustrated in Fig. 404-2. The three-pole switch satisfies the exception. Opening only the grounded wire of a 2-wire circuit would leave all devices that are connected to the circuit alive and at a voltage to ground equal to the voltage between wires on the mains. In case of an accidental ground on the grounded wire, the circuit would not be controlled by the single-pole switch

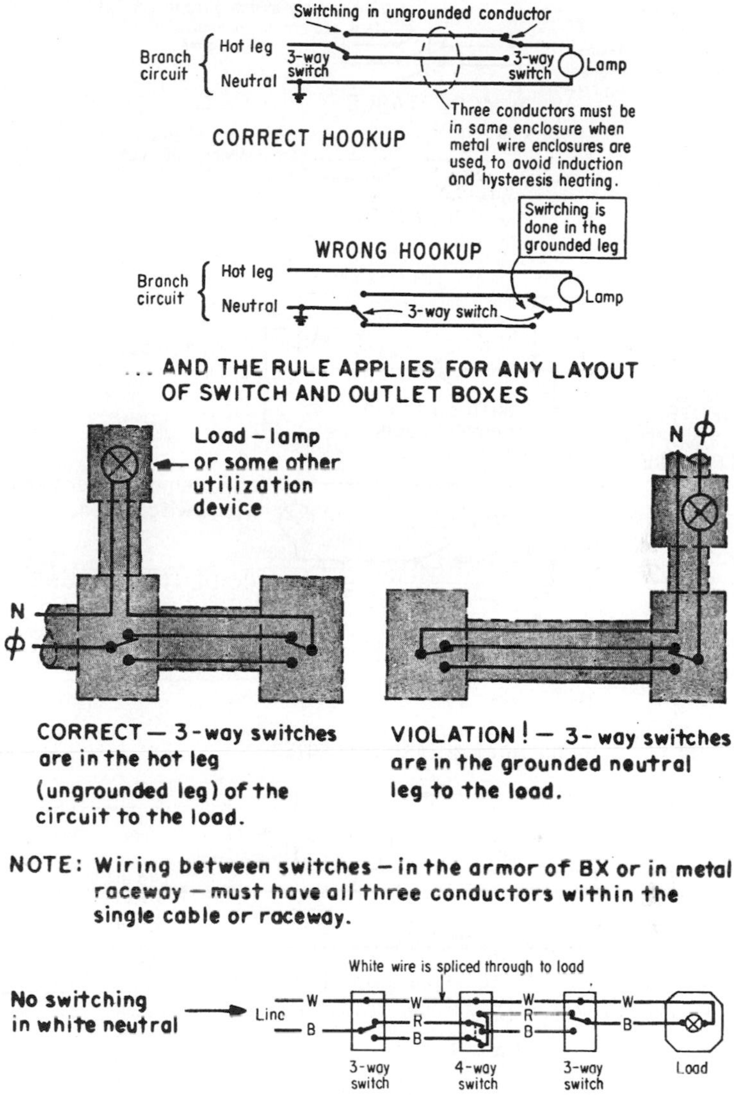
Switching in ungrounded conductor
Branch circuit
Hot leg
Neutral
3-way switch
3-way switch
Lamp
Three conductors must be in same enclosure when metal wire enclosures are used, to avoid induction and hysteresis heating.
CORRECT HOOKUP
Switching is done in the grounded leg
WRONG HOOKUP
Branch circuit
Hot leg
Neutral
3-way switch
Lamp
. . . AND THE RULE APPLIES FOR ANY LAYOUT OF SWITCH AND OUTLET BOXES
Load – lamp or some other utilization device
N
ϕ
N ϕ
CORRECT — 3-way switches are in the hot leg (ungrounded leg) of the circuit to the load.
VIOLATION! — 3-way switches are in the grounded neutral leg to the load.
NOTE: Wiring between switches – in the armor of BX or in metal raceway – must have all three conductors within the single cable or raceway.
White wire is spliced through to load
No switching in white neutral
Line
W
B
W
R
B
W
R
B
W
B
3-way switch
4-way switch
3-way switch
Load

Fig. 404-1. All three-way and four-way switches must be placed in the hot conductor to the load. (Sec. 404.2.)

In Fig. 404-3, the load consists of lamps connected between the neutral and the two outer wires and is not balanced. Opening the neutral while the other wires are connected would cause the voltages to become unbalanced and might burn out all lamps on the more lightly loaded side.

Except for 514.11(A), which requires a switch in a grounded neutral for a circuit to a pump at a gas station, the neutral does not need to be switched. But where

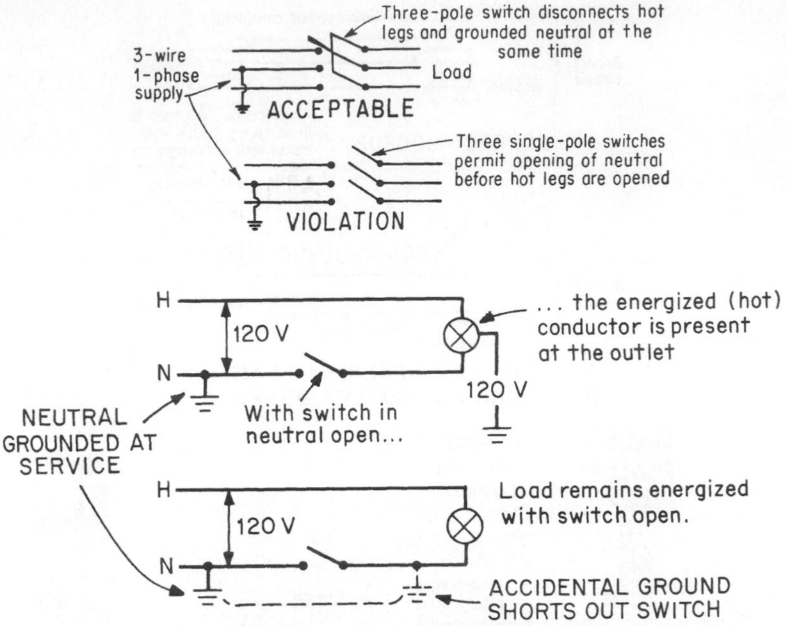
Three-pole switch disconnects hot legs and grounded neutral at the same time
3-wire 1-phase supply
Load
ACCEPTABLE
Three single-pole switches permit opening of neutral before hot legs are opened
VIOLATION
H
120 V
... the energized (hot) conductor is present at the outlet
N
120 V
NEUTRAL GROUNDED AT SERVICE
With switch in neutral open...
H
120 V
Load remains energized with switch open.
N
ACCIDENTAL GROUND SHORTS OUT SWITCH

Fig. 404-2. A single-pole switch must not be used in a grounded circuit conductor. (Sec. 404.2.)

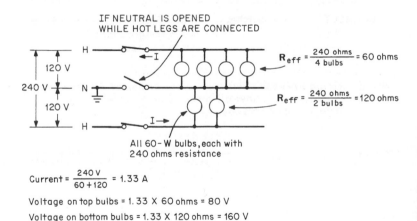

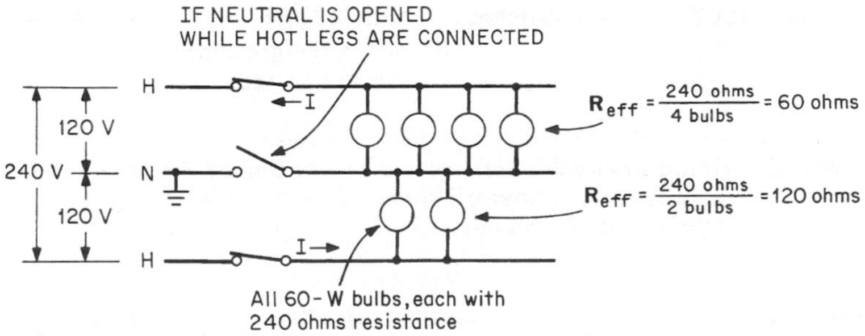
IF NEUTRAL IS OPENED WHILE HOT LEGS ARE CONNECTED
H
120 V
240 V
N
120 V
H
I
I
$R_{eff} = \frac{240 \text{ ohms}}{4 \text{ bulbs}} = 60$ ohms
$R_{eff} = \frac{240 \text{ ohms}}{2 \text{ bulbs}} = 120$ ohms
All 60-W bulbs, each with 240 ohms resistance

$$\text{Current} = \frac{240 \text{ V}}{60 + 120} = 1.33 \text{ A}$$

Voltage on top bulbs = 1.33 X 60 ohms = 80 V

Voltage on bottom bulbs = 1.33 X 120 ohms = 160 V

Fig. 404-3. A single-pole switch in neutral can cause damaging load unbalance if opened. (Sec. 404.2.)

a grounded neutral or grounded phase leg is switched, it must never be by a single-pole switch or single-pole CB, even if the CBs are provided with handle ties.

A switch may be arranged to open the grounded conductor if the switch simultaneously opens all the other conductors of the circuit.

Part **(C)**, new for the 2011 **NEC**, generally requires accommodating switch locations to receive electronic switching devices, such as occupancy sensors, for lighting loads. The rule applies in all occupancies with limitations that apply to wiring configurations that allow the white wire to be installed subsequently with relative ease. With increasing national concerns around energy conservation, building codes are increasingly mandating the installation of occupancy-sensing switching, so lights will not be inadvertently left on. An example of such a switch is shown in Fig. 404-4. This switch is capable of replacing either a conventional single-pole or three-way switch; it has three leads for the switch, and if used as a single-pole switch, one of the three leads is simply capped off. If used as a three-way switch, however, it must be wired to replace the three-way switch that is electrically closest to the load. It is capable of a 180° field of vision over a floor area of approximately 20 ft wide by 20 ft deep. In accordance with 210.70(A)(1) Exception No. 2, the switch is equipped with a manual bypass, so it can be turned on or off at its location regardless of the actual state of occupancy.

Cat. No

Fig. 404-4. An occupancy-sensor type of switch, designed to replace conventional switches, the four leads are color-coded, and allow for the switch to function as either a single-pole or three-way switch. The fourth lead is green and grounds the switch yoke as well as allowing the electronic controls in the switch to function with the lights off. [Sec. 404.2(C).]

The particular switch illustrated here shows some of the limitations involved with this technology at the current time. The switch, purchased new at the end of 2010, is rated for 500 VA incandescent lighting, or 400 VA rapid-start fluorescent lighting, but the fluorescent lighting ballasts must be the old-fashioned magnetic ballasts and not the modern electronic ballasts. Some sensor switches are more robust than the one in Fig. 404-4, even with small horsepower ratings, but none can control a receptacle. This is due to the unknown loading that can result. However, the **NEC** clearly allows such uses generally including the express permission for residential lighting, and such switches are exempted accordingly from the need to run standby neutrals. The **NEC** waives the requirement in other instances where the application is such that an occupancy switch would not or could not be installed. The neutral is not required where the switch does not serve a habitable room other than a bathroom; a frequently cited example of this is a door-jamb switch controlling a closet light. A neutral does not need to be brought to a switch with an integral enclosure [covered in 300.15(E) and 334.30(C)] because it couldn't be landed once it arrived anyway. If there are multiple switch locations that "see" the entire lit space, then the neutral need be brought to only one of those locations. The term "habitable" in some building codes is limited to residential uses; the 2017 **NEC** adds "or occupancy" to address this issue. It was always intended that these rules would be applied generally and not just in dwellings. Some locations use snap switches in series with occupancy sensors so the lights are sure to go off when the room is unoccupied, and also forced off by a switch in order to guarantee darkness, such as for a projected presentation. Such switch locations clearly do not require a neutral. Note that a neutral is required, however, for a switch location outside but that still controls the lit area. Although an occupancy sensor would never replace such a switch, some high-end dimmers require a neutral and therefore the Code-making panel elected to leave the requirement in place for these locations.

The switch in Fig. 404-4 does not require a neutral connection to operate. In order for the switch to function with the lights off, it must be connected to a complete circuit so its electronic sensor will function. Current product standards allow such switches to "leak" up to 0.5 milliamperes to the equipment grounding system for this purpose. This is a very small amount of current, but if many such switches are installed, there can be appreciable current imposed on equipment grounding conductors, which are only supposed to carry current in the event of a ground fault. The Code change must be understood in this context. Over time the **NEC** Committee always intended that the product standard for switches such as these would require them to have terminals for the connection to a grounded circuit conductor, and that the intentional leakage of routine current into the equipment grounding system would be discontinued.

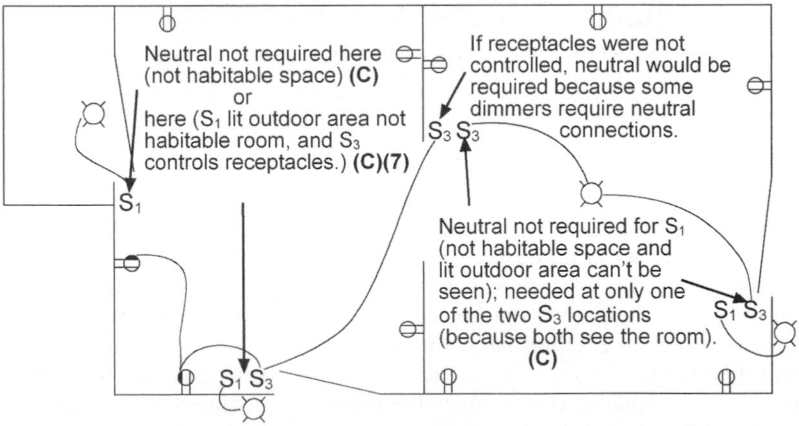
Neutral not required here
(not habitable space) (C)
or
here (S1 lit outdoor area not
habitable room, and S3
controls receptacles.) (C)(7)
If receptacles were not
controlled, neutral would be
required because some
dimmers require neutral
connections.
S3 S3
S1
S1 S3
Neutral not required for S1
(not habitable space and
lit outdoor area can't be
seen); needed at only one
of the two S3 locations
(because both see the room).
(C)
S1 S3

Fig. 404-5. The 2014 **NEC** has added important exemptions to the neutral at every switch rule to accommodate applications where a neutral would not be useful. [Sec. 404.2(C)].

That time is now firmly in view. The 2017 **NEC** added a second paragraph to the rule, requiring the neutral to be extended to the switch location if it weren't already there (such as through the allowance to run it later if the raceway wiring method had sufficient room to add it), and then to actually land it on any switch requiring line-to-neutral voltage to operate. The paragraph also includes a reference to 404.22 in Part II of the article (Construction Specifications). This part of an article chiefly addresses the product standards for the equipment governed by the article. Sec. 404.22, also new for 2017, largely puts the old electronic switch designs on the road to extinction. As of Jan. 1, 2020, the old-style designs will no longer be recognized, with one important exception, and this is correlated with the exception that follows the new paragraph in 404.2(C).

The two exceptions cover an old-style design that will remain acceptable for an indefinite time period, but limited to replacement/retrofit applications only. By the parent wording in 404.22, all electronic switches must be listed (effective immediately); those for old-style applications must be specifically listed and marked by their manufacturer for use in replacement or retrofit applications only. As would be expected, with respect to field wiring as covered in 404.2(C) Exception, the same effective date applies with respect to the requirement to land a neutral on an electronic switch. Obviously installers can't be expected to wire switches that aren't available. However, the exception has several additional features that require careful attention.

First, the exception applies only to switch locations wired prior to local adoption of 404.2(C). In general, that would be prior to local adoption of the 2011 (or later) **NEC**. In addition, the switch location in question must be difficult to rewire. Specifically, creating a neutral connection must involve the removal of surface materials. For example, if the switch is wired with a raceway method to which the neutral can be added, then it must be added. Another example would involve a switch in an office with open-walled partitions extending to open suspended ceilings. With very little effort a new cable that included the neutral could be sent down a 4-ft open cavity to the switch box. The inspector will make the final decision, but the clear intent is to minimize the necessity to apply the exception.

Second, the degree of leakage current is restricted through numerical limits (5 for a branch circuit and 25 for a feeder) on the number of electronic devices connected. At 0.5 mA each, the leakage current on any branch circuit will be limited to 2.5 mA and a feeder to 12.5 mA. The numbers are conservative but high enough to provide a necessary degree of flexibility for existing occupancies. For example, the branch-circuit number is half the customary trip setting for a GFCI.

Finally, the exception makes it very clear that a neutral bus in service equipment equipped with a functioning main bonding jumper carries no limits. That is the location where neutral and equipment ground join, and any leakage currents arriving become ordinary neutral currents and need not be limited. The reference to 200.2(B) is potentially important in this regard. Review Fig. 200-1 in this book. If the violation shown in that drawing exists, then the leakage currents returning over the neutral will only exacerbate the problem that led to 200.2(B) in the first place. That violation is comparatively simple to correct, and must be completed if one of these retrofits is to go forward.

404.3. Enclosure. Figure 404-6 shows the basic rule of this section and 404.4. This rule also requires adequate wire bending space at terminals and inside gutters of switch enclosures. In this section and in other sections applying to wiring space around other types of equipment, it is a mandatory Code requirement that wire bending space and side gutter wiring space conform to the requirements of Table 312.6(A) for side gutters and to Table 312.6(B) for wire bending space at the line and load terminals, as described under 404.18. Those tables establish the minimum distance from wire terminals to enclosure surface or from the sides of equipment to enclosure side based on the size of conductors being used, as shown in Fig. 404-7.

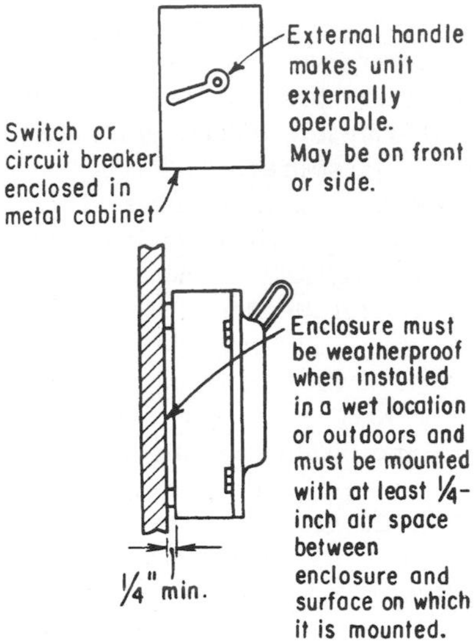
Switch or
circuit breaker
enclosed in
metal cabinet
External handle
makes unit
externally
operable.
May be on front
or side.
Enclosure must
be weatherproof
when installed
in a wet location
or outdoors and
must be mounted
with at least ¼-
inch air space
between
enclosure and
surface on which
it is mounted.
¼" min.

Fig. 404-6. Switch and CB enclosures must be suitable. (Sec. 404.3.)

This whole concern for adequate wiring space in all kinds of equipment enclosures reflects a repeated theme in many Code sections as well as in Art. 110 on general installation methods. One of the most commonly heard complaints from constructors and installers in the field concerns the inadequacy of wiring space at equipment terminals. 404.3 is designed to ensure sufficient space for the necessary conductors run into and through switch enclosures.

404.4. Damp or Wet Locations. Refer to Fig. 404.6 and discussion under 312.2. In addition, this rule prohibits switches from being installed in the wet location of a tub or shower space. Although the yoke would be grounded by other rules, the concern is how long the gasket will hold up before water, particularly from a shower, gets behind it and starts rotting out the connections, including the grounding connections. An exception exempts a switch in a "listed tub or shower assembly" under the understanding that some hydromassage tubs or spas may have air or low-voltage switching to control the water or airflow; such controls are carefully controlled by the listing process in terms of potential hazards.

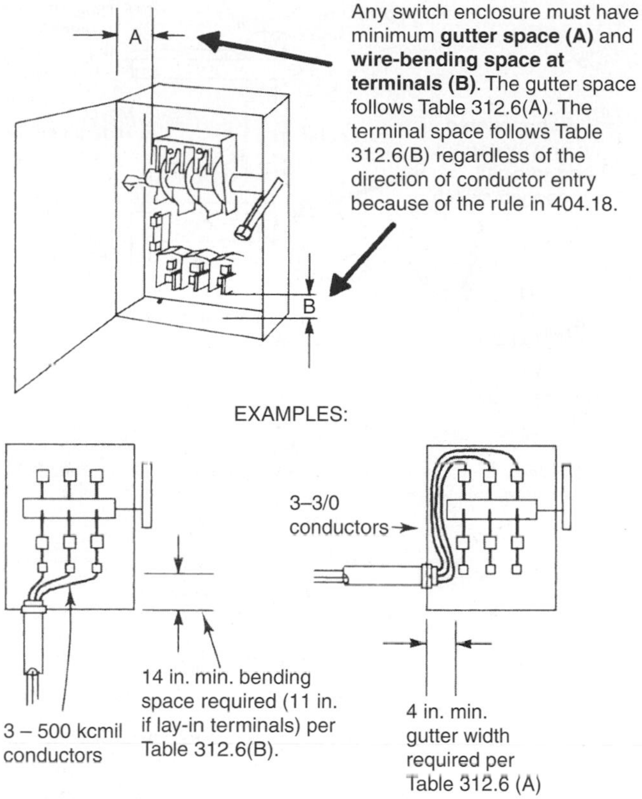
A
Any switch enclosure must have minimum **gutter space (A)** and **wire-bending space at terminals (B)**. The gutter space follows Table 312.6(A). The terminal space follows Table 312.6(B) regardless of the direction of conductor entry because of the rule in 404.18.
B
EXAMPLES:
3–3/0 conductors
14 in. min. bending space required (11 in. if lay-in terminals) per Table 312.6(B).
3 – 500 kcmil conductors
4 in. min. gutter width required per Table 312.6 (A)

Fig. 404-7. Terminating and gutter space in switch enclosures must be measured. (Sec. 404.3.)

404.5. Time Switches, Flashers, and Similar Devices. Any automatic switching device should be enclosed in a metal box unless it is a part of a switchboard or control panel which is located as required for live-front switchboards. Such devices must not present exposed energized parts, except under very limited conditions where they are accessible only to qualified persons.

404.6. Position and Connection of Knife Switches. The NEC requires that knife switches be so mounted that gravity will tend to open them rather than close them (Fig. 404-8). But the Code recognizes use of an upside-down or reverse-mounted knife switch where provision is made on the switch to prevent gravity from actually closing the switch contacts. This permission is given in recognition of the much broader use of underground distribution, with the intent of providing a switch with its line terminals fed from the bottom and its load terminals connected at the top (Fig. 404-9). With such a configuration, an upside-down knife switch provides the necessary locations of such terminals, that is, "line" at bottom and "load" at top. However, use of any knife switch in the reverse or upside-down position is contingent upon the switch being approved for such use, which virtually means

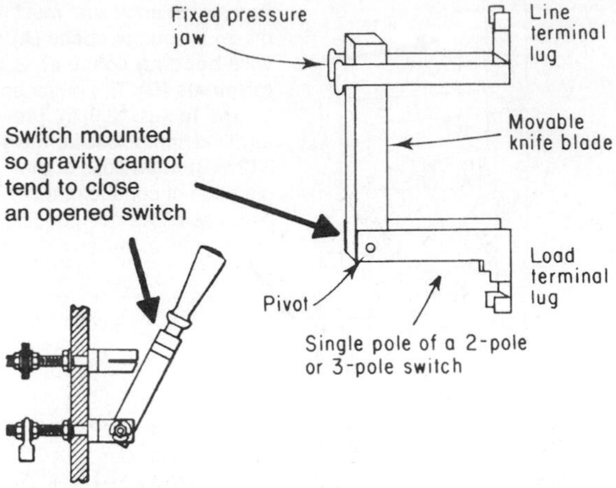
Fixed pressure jaw
Line terminal lug
Movable knife blade
Switch mounted so gravity cannot tend to close an opened switch
Load terminal lug
Pivot
Single pole of a 2-pole or 3-pole switch

Fig. 404-8. Movable knife blade of a knife switch must be pivoted at its bottom. (Sec. 404.6.)

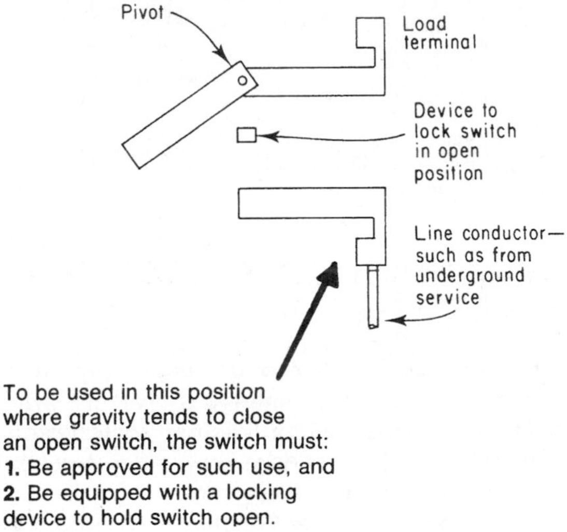
Pivot
Load terminal
Device to lock switch in open position
Line conductor—such as from underground service
To be used in this position where gravity tends to close an open switch, the switch must:
1. Be approved for such use, and
2. Be equipped with a locking device to hold switch open.

Fig. 404-9. This type of knife-switch operation is permitted. (Sec. 404.6.)

UL-listed for that application, and also upon the switch being equipped with a locking device that will prevent gravity from closing the switch. The same type of operation is permitted for double-throw knife switches.

As required by part **(C)**, knife-switch blades must be "dead" in the open position, except where a warning sign is used (Fig. 404-10). In a number of electrical system hookups—UPS systems, transformer secondary ties, and emergency generator layouts—electrical backfeed can be set up in such a way as to make the load terminals, blades, and fuses of a switch energized when the switch is

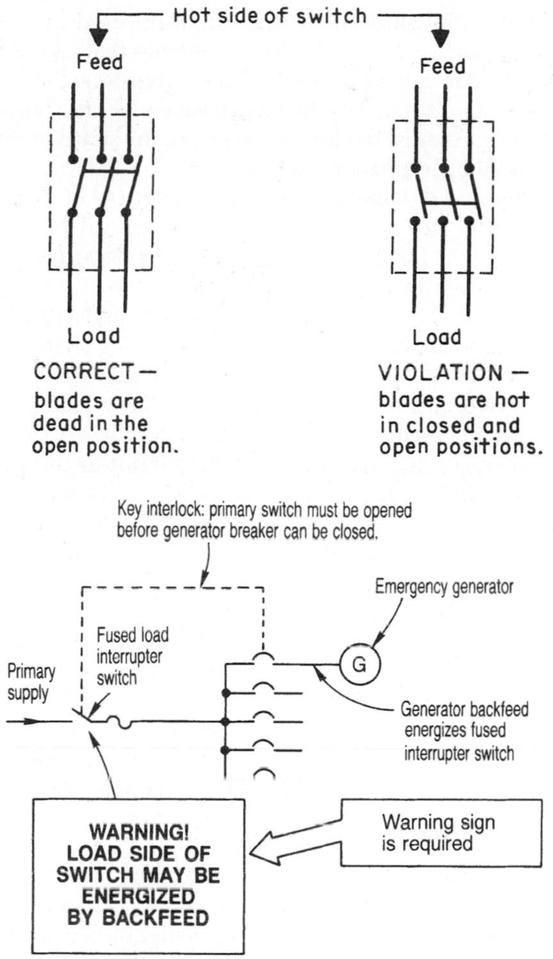
Hot side of switch
Feed
Feed
Load
Load
CORRECT—
blades are
dead in the
open position.
VIOLATION —
blades are hot
in closed and
open positions.
Key interlock: primary switch must be opened
before generator breaker can be closed.
Emergency generator
Fused load
interrupter
switch
Primary
supply
G
Generator backfeed
energizes fused
interrupter switch
WARNING!
LOAD SIDE OF
SWITCH MAY BE
ENERGIZED
BY BACKFEED
Warning sign
is required

Fig. 404-10. Supply conductors must connect to "LINE" terminals of switch, but backfeed is permitted if carefully marked. (Sec. 404.6.)

in the OFF or open position. Where that might happen, the exception to this section says a permanent sign must be prominently placed at or near the switch to warn of the danger. The sign must read, "Warning—switch may be energized by backfeed."

This potential hazard has long been recognized for high-voltage systems (over 600 V), and 490.21(E) covers the matter. This rule that the load side of the switch be de-energized when the switch is open applies the same concept to systems operating up to 600 V.

404.7. Indicating. Switches and circuit breakers in individual enclosures must be marked to clearly show ON and OFF positions, and vertically operated switches and CBs must be ON when in the up position. This is basically a repetition of the rule that has been in 240.81 for circuit breakers used in switchboards and panelboards. Exception No. 2 covers busway switches that have a center-pivoting switch, so with one side pulled down the switch is ON and with the other side down the switch is OFF. This makes the switch easy to work from the floor with a hook stick. The positions must be clearly marked so the status of the switch will be visible from the usual operating location.

404.8. Accessibility and Grouping. The rule of part **(A)** of this section presents certain requirements regarding the *accessibility* (as defined in Art. 100 for equipment) generally required for disconnects and OC devices. This rule, along with the exceptions, is shown in Fig. 404-11. Exception No. 1 cross-references with 368.17(C). Exception No. 2 correlates with 240.24(A)(4), and Exception No. 3 makes the hook stick principle apply generally to other isolating switches.

Part **(B)** of this section applies where 277-V switches, mounted in a common box (such as two- or three-ganged), control 277-V loads, with the voltage between *adjacent* switches in the common box being 480 V. Anyone changing one of the switches without disconnecting the circuit at the panel could contact 480 V, as shown in Fig. 404-12. The rule of this section requires permanent barriers between adjacent switches located in the same box where the voltage between such switches exceeds 300 V. This rule applies anytime adjacent devices operate over 300 V between them, whether or not the terminals are exposed or use pigtail connections. Of course, the hookup shown would be acceptable if a separate single-gang box and plate are used for each switch, or a common wire from only one phase (A, B, or C) supplies all the three switches in the three-gang box.

This rule has been broadened to include snap switches mounted with other devices, such as receptacles. Although the actual voltage between a 277-V snap switch and a 125-V receptacle is not necessarily over 300 V (depending on which phases of the two systems are involved and the amount of phase shift between them), the intent is to classify this as impermissible without the grounded barrier between the devices. This rule has been reinforced by similar language appearing in 406.5(G), and means that the old practice of putting a snap switch for a 277-V lighting system together with a 125-V general-purpose receptacle as the only opening in small rooms with specialized equipment is no longer possible the way it used to be done. However, square outlet boxes with 2-gang plaster rings are generally manufactured with notches and drillings that easily accommodate these barriers, and nonmetallic box manufacturers now have similar products available.

Part **(C)** covers multipole snap switches. It effectively ratifies the following statement in the UL Guide Card requirements: "Multi-pole, general-use snap switches have not been investigated for more than single-circuit operation unless marked '2-circuit' or '3-circuit.'" Since no such switches have been in production with that marking except by special order, almost any practical use would be a violation of this limitation, which is citable under 110.3(B). A 2008 **NEC** change attempted to trump this limitation by allowing it by right, provided the line-to-line voltage didn't exceed the snap switch rating.

Facing concerted industry opposition, the Code-making panel backed away from the challenge. Enforcement of this particular listing limitation has been sporadic at best, but may increase in view of the recent activity.

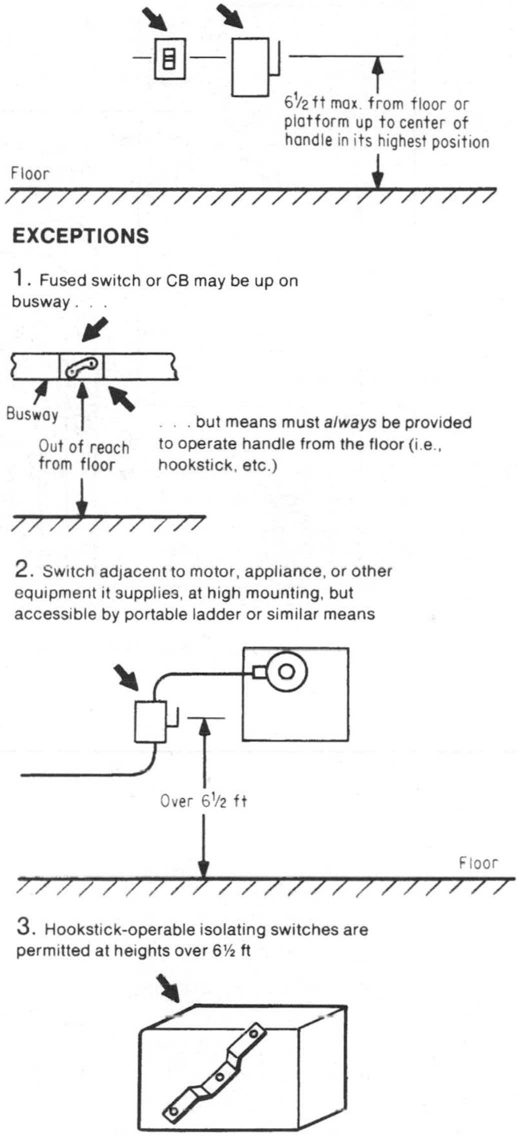
6½ ft max. from floor or platform up to center of handle in its highest position
Floor
EXCEPTIONS
1. Fused switch or CB may be up on busway . . .
Busway
Out of reach from floor
. . . but means must *always* be provided to operate handle from the floor (i.e., hookstick, etc.)
2. Switch adjacent to motor, appliance, or other equipment it supplies, at high mounting, but accessible by portable ladder or similar means
Over 6½ ft
Floor
3. Hookstick-operable isolating switches are permitted at heights over 6½ ft

Fig. 404-11. All switches and circuit breakers used as switches must be capable of being operated by a person from a readily accessible place. (Sec. 404.8.) Note that the maximum height dimension has been slightly changed to 2.0 m (6 ft 7 in.) from 6½ ft.

404.9. Provisions for General-Use Snap Switches. The first sentence of this section [Part **(A)**] requires that faceplates be installed to cover the wall opening completely to ensure that the box behind the faceplate is properly covered and to prevent any openings that could afford penetration to energized parts.

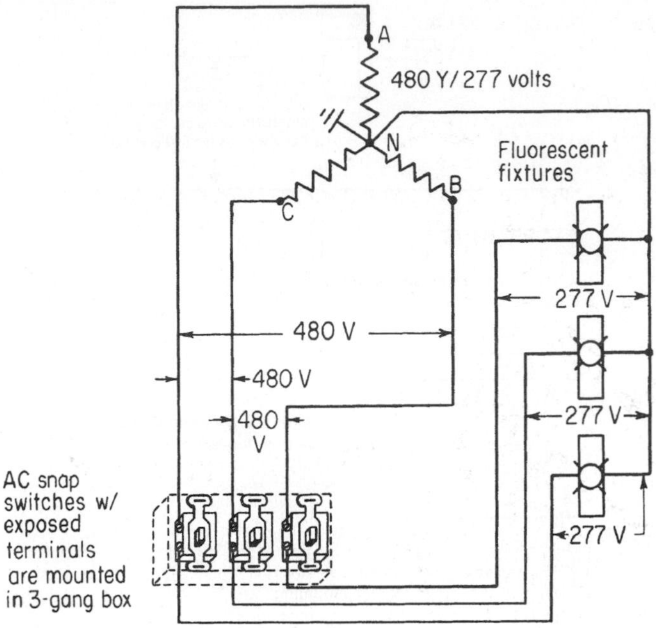
A
480 Y/277 volts
N
Fluorescent
fixtures
B
C
277 V
480 V
480 V
480
V
277 V
AC snap
switches w/
exposed
terminals
are mounted
in 3-gang box
277 V

Fig. 404-12. This is a violation if barriers are not used between switches in the box. (Sec. 404.8.)

Part **(B)** requires that metal faceplates be grounded, even if they are installed in the future. This is done by writing the rule to require that snap switches, including dimmer and similar control switches, be connected to an equipment grounding conductor, and that in the process the snap switch provide a means to connect the faceplate to the grounding conductor, "whether or not a metal faceplate is installed." Although "what if" code writing is generally a poor practice, in this case it is necessary because unqualified persons change faceplates all the time without benefit of permits and inspections. This way, whenever a metal faceplate is installed, it will be safely grounded in the process.

There are two methods of accomplishing this objective. The first is to use an actual bonding jumper between an equipment grounding terminal on the device yoke and the equipment grounding system for the box. The second is to rely on a conductive connection between the device yoke and a metal box or metal cover that is connected to an equipment grounding conductor for the supply circuit, or to a nonmetallic box with an integral equipment grounding conductor. Put simply, this could be two 6-32 screws from yoke to metal device box, or from the yoke to a raised cover.

At this point, many ask how this compares with 250.146, where receptacles are only permitted to rely on the 6-32 screws when the box is surface mounted, or in a raised cover with crushed corners and locking means for the cover screws, or where a special yoke modification holds the 6-32 screw securely. The answer is simple: this wording is more lenient, and very intentionally so. A receptacle is a portal to a quasi-branch-circuit extension whenever it supplies a cord-and-plug-connected load. This is not the case for a snap switch, which has to protect only itself. The panel is not aware of any credible loss experience that would support revisiting the quality of this connection.

Figure 404-13 shows the basic operation of the exception to this part of the section. Note that the recessed metal box is not grounded and that is not acceptable by 314.4; however, the **NEC** is not retroactive and at one time this was acceptable because the box is not exposed to contact, and this exception is for replacement duty only. Note also that, as an alternative, and in correlation with 410.42(B) Exception No. 2 for ungrounded luminaires with exposed conductive parts, if the switch circuit has GFCI protection arranged ahead of the switch, it can have a metal faceplate.

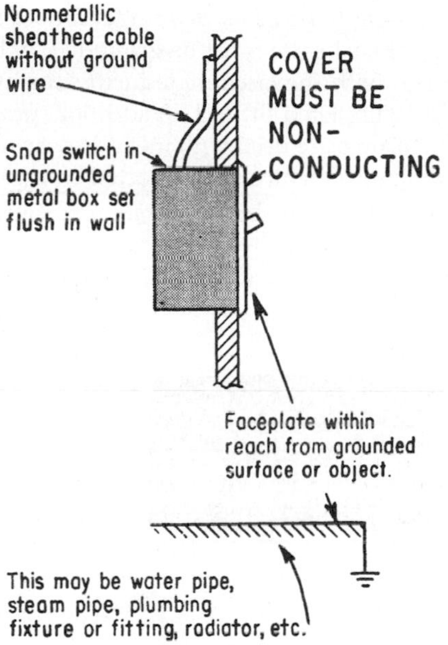
Nonmetallic sheathed cable without ground wire
COVER MUST BE NON-CONDUCTING
Snap switch in ungrounded metal box set flush in wall
Faceplate within reach from grounded surface or object.
This may be water pipe, steam pipe, plumbing fixture or fitting, radiator, etc.

Fig. 404-13. Nonmetallic faceplate eliminates shock hazard. (Sec. 404.9.) The faceplate screws must also be nonmetallic.

Two additional exceptions apply to special circumstances. The first correlates with 406.5(D) Exception and addresses a product line of devices that come with proprietary nonmetallic faceplates that cannot be mounted on any other device line, and that cannot accept a metal faceplate. A grounding connection in such an instance is impossible, as is the case with the second exception that correlates with 300.15(E) and 334.30(C). These devices, designed for nonmetallic-sheathed cable connected without a box, have an integral nonconductive faceplate and cannot be retrofitted with a metal faceplate.

404.10. Mounting of Snap Switches. The purpose of paragraph **(B)** is to prevent "loose switches" where openings around *recessed* boxes provide no means of seating the switch mounting yoke against the box "ears" properly. It also permits the maximum projection of switch handles through the installed switch plate. The cooperation of other crafts, such as drywall installers, will be required to satisfy this rule. The 2014 NEC added language effectively outlawing drywall or other unrecognized screws from being used to support snap switch yokes.

404.11. Circuit Breakers as Switches. Molded-case CBs are intended to be mounted on a vertical surface in an upright position or on their side. Use in any other position requires evaluation for such use. ON and OFF legends on CBs and switches are not intended to be mounted upside down. This general permission does not specifically state that any CB used as a switch must be listed for such use and be marked "SWD," which indicates only that the CB has been evaluated for a particular type of switching duty during product testing for listing, namely, fluorescent lighting banks. [See 240.83(D).]

404.12. Grounding of Enclosures. This section calls for the metallic enclosures and metal switch boxes used to house switches and circuit breakers to be connected to an equipment grounding conductor. 250.110 requires *all* exposed metal parts (including enclosures) of fixed equipment to be grounded under any of the conditions described. And any switch or CB enclosure must be grounded, either by an equipment grounding conductor run with the circuit conductors, or by a metallic conduit or a metal-sheathed cable with listed fittings. In addition, provisions must be made when nonmetallic enclosures are used with metallic raceways and cables to ensure grounding continuity between all interconnected raceways, cables, and any equipment within the enclosure.

404.13. Knife Switches. UL data on ratings and application correspond to the Code data (Fig. 404-14).

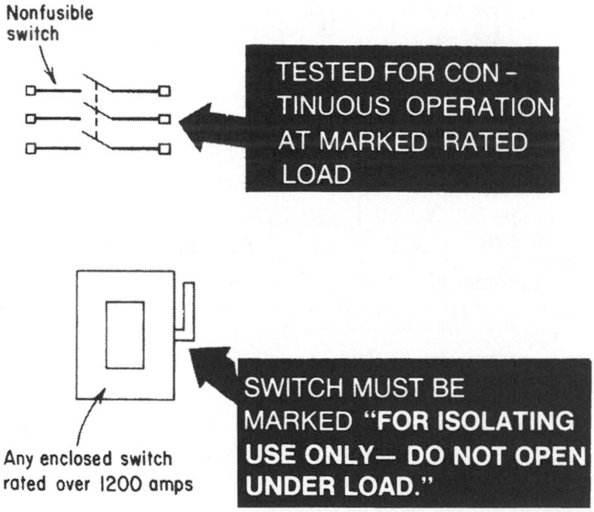
Nonfusible switch
TESTED FOR CON - TINUOUS OPERATION AT MARKED RATED LOAD
Any enclosed switch rated over 1200 amps
SWITCH MUST BE MARKED "FOR ISOLATING USE ONLY— DO NOT OPEN UNDER LOAD."

Fig. 404-14. UL data must be correlated to the Code rules. (Sec. 404.13.)

404.14. Rating and Use of Switches. The usual snap switch is covered in **(A)**, which is the snap switch rated for ac only. These switches will control inductive and resistive loads up to the switch rating; motor loads up to 80 percent of the switch rating, and tungsten filament lamp loads up to the switch rating on a 120-V circuit.

The ac and dc snap switch will control dc loads, although that is seldom a concern on today's branch circuits and these switches are comparatively unusual. Because they have a dc rating, which is a more difficult switch function because there is no current zero, they are built differently and have a much louder "click" when actuated. Some people actually specified them for their children's bedrooms because they wanted to hear if their kids turned the lights back on after bedtime. They will handle the full rating of the switch for resistive loads, the full rating for tungsten filament lamp loads (but only if "T" rated) and inductive loads up to 50 percent of the switch rating.

Part **(C)** covers the special devices with an enhanced design on their terminal screws rated for direct aluminum terminations, known by the mandatory labeling "CO/ALR." For snap switches rated 20 A or less, these are the only permissible switches unless the aluminum wires are pigtailed to copper for the final device terminations.

Part **(D)** covers the 347-V snap switches primarily used in the Canadian market. Instead of using 480Y/277V systems, they more commonly use 600Y/347V systems, and these switches are engineered accordingly. They will handle most applicable inductive and resistive loads up to the switch rating; however, any limitations imposed as part of a listing process must be observed. They must not be rated less than 15 A, and they must be physically configured so that flush applications in device boxes will not be readily interchangeable with switches identified for 120/240, or 208Y/120, or 480Y/277V systems.

Part **(E)** prohibits the use of dimmer switches to control receptacles; they are only permitted to control fixed incandescent lighting, unless specifically listed otherwise.

Part **(F)**, new as of the 2011 **NEC**, and extended to "control devices" in the 2017 edition to capture electronic control equipment, addresses a previously unforeseen consequence of the long-standing permission in 210.21(B)(3) for 20-A branch circuits with multiple receptacles to allow for either 15 or 20 A-configured plugs. The usual 20-A NEMA 5-20R tee-slot receptacle will accept either a 5-20P 20-A plug (uncommon in practice but they are used) or a 5-15P 15-A parallel blade plug (common). If the 20-A receptacle is controlled by a snap switch and receives the 5-20P plug of an appliance drawing more than 15 A, a 15-A snap switch will be overloaded. The new rule requires that such a snap switch be rated at not less than the maximum branch-circuit rating that correlates with 210.21(B) and the receptacle configuration. In this case, a 20-A snap switch must be used ahead of the 20-A receptacle. The exception covers the unusual condition where only a single 15-A receptacle is installed on a multireceptacle 20-A branch circuit. This is permitted by 210.21(B)(3) for 20-A branch circuits. In this case the snap switch rating can follow that of the receptacle.

404.20. Marking. Part **(B)** of this section requires any switch with a marked "OFF" position to completely disconnect all ungrounded conductors to the load. This is comparable to the rule for electric heating thermostats with "OFF" positions, as covered in 424.24(A). The reason for this was incidents where electronic switches, such as occupancy sensors, were installed on circuits with no grounded conductor present, and the switch leaked a small amount of current through the load in order for its electronics to work. Workers in the room replacing ballasts assumed that since the lights were out, there was no voltage at the luminaires. The amount of current is regulated by the product standards and is not a shock hazard, but it was enough to startle the workers to the point of falling off the ladder. Now these switches must use some other indication, such as "standby." Be aware that a switch with some marking other than "OFF" very likely will pass voltage to the luminaires even when they appear to be off. A good example is the lit-handle snap switches that have the toggle handle illuminated with the load off (pilot-handle switches, lit with the load on, use a grounded circuit conductor connection and are OK). These devices also pass current through the load in the OFF position, and for that reason the normally OFF position of the toggle no longer reads "OFF."

404.22. Electronic Lighting Control Switches. This is the product standard counterpart to the wiring requirement in 404.2(C). Refer to that location, particularly the new second paragraph, for the complete discussion.

404.28. Wire Bending Space. At terminals of individually enclosed switches and circuit breakers, the spacing from lugs to the opposite wall *must* be at least equal to that of Table 312.6(B) for the given size and number of conductors per lug. The larger spacing of that table, rather than the smaller spacing of Table 312.6(A), must be used regardless of how conductors enter or leave the enclosure—on the sides or opposite terminals.

Figure 404-15 shows the rule on wire bending space in switch or CB enclosures.

Fig. 404-15. Top wire-bending space contains offset (S) bends and must have dimensions from Table 312.6(B). Bottom wiring space has single (L) bends—but must conform to Table 312.6(B), not Table 312.6(A). That varies from 312.6(B)(4), which permits terminal space from Table 312.6(A) when wires go out the side of the enclosure.

ARTICLE 406. RECEPTACLES, CORD CONNECTORS, AND ATTACHMENT PLUGS (CAPS)

406.1. Scope. Previously part of Art. 410 in the 1999 NEC, this article gives a number of requirements for the proper use of receptacles, cord connectors, and attachment plugs.

406.3 Receptacle Rating and Type. In part **(A),** the Code makes a general statement regarding any receptacle that is used. *All* receptacles must be listed by a third-party testing agency, such as Underwriters Laboratories, for the specific application. This wording effectively prohibits the use of any receptacle that is not listed, as well as the use of listed receptacles that are not specifically listed for that use. The only way to ensure compliance with this requirement is to read and understand the product literature supplied with the receptacle. Then one can evaluate the acceptability of any receptacle for any application, based on the listing of the receptacle in question.

Part **(B)** requires that receptacles be rated no less than 15 A at either 125 or 250 V. This establishes the minimum acceptable ampere rating for receptacles recognized by the Code for general wiring, in any occupancy.

Part **(C)** requires that receptacles rated 20 A or less for direct aluminum connections be listed as and marked CO/ALR type.

Part **(D)** covers the so-called isolated-ground (IG) receptacles that are intended for use in accordance with 250.146(D). Such receptacles may *only* be used as indicated by 250.146(D). That is, if used, the ground *must* be isolated. This prevents anyone coming along later from being fooled into believing the receptacle is an isolated-ground receptacle when it's not. All IG receptacles *must* be isolated; however, there is no **NEC** requirement as to how far upstream the isolation must be maintained because that is a design issue. The isolation might stop at the local branch-circuit panelboard, or it might continue as far as the main disconnect for the building, system, or service except as limited in 250.146(D). If the receptacle is installed in a conventional nonmetallic box with no integral mechanism to bond the device yoke, then only nonmetallic faceplates are permitted.

It should be noted that isolated-ground receptacles must be identified by an orange triangle on the face. Although receptacles that are orange overall are the most common form of these receptacles, even with the overall color if you look carefully, there will be an orange triangle, and technically the only orange on that receptacle face that indicates the isolating function is the orange within the triangle. It is entirely legal to make an orange receptacle with no triangle, and sell it as a conventional receptacle for Halloween celebrations, or for whatever reason orange is desired. Conversely, the receptacle face can be any other color, but with an orange triangle, and it will be one that employs an isolated equipment grounding construction.

Part (E) is new as of the 2014 **NEC** and requires all customary straight-blade 15- and 20-A, 125-V receptacles that are automatically controlled by any form of building automation including energy management to be marked with the symbol as shown in Fig. 406-1. The final language in the 2017 edition requires explanation on some key points. The first paragraph is very close to the 2014 text, and refers to two different control modalities. The first, "controlled by an automatic control device" refers to some upstream relay control, not incorporated into the receptacle, that switches the power that flows over the wires connected to the receptacle. The second, "incorporate control features that remove power from the receptacle" refers to a receptacle that includes ("incorporates") within it some sort of relay function responsive to an external signal that can disable the receptacle.

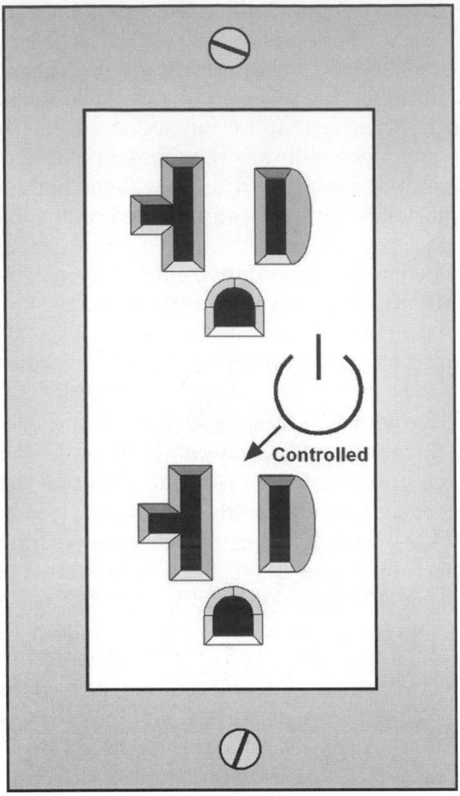
Controlled

Fig. 406-1. A compliant marking of a controlled receptacle. The common connection tab(s) has (have) been broken open as required so only the lower half of this duplex receptacle is controlled automatically. The upper half is always available. [406.3(E)].

The first paragraph then, at its end, requires the device to be "permanently marked (new for 2017) with the specified symbol, so a faceplate marking does nc suffice. That symbol is like the one for 2014, but includes the word "Controlled underneath it. The paramount concern was that without the word being addec far too many people would be completely unaware of the true significance of th symbol alone. The only way to realistically mark the device itself is to have it don by the manufacturer. Indeed receptacles are now available from several manufac turers that have both the symbol and the word. In general, the market is showin a very strong preference for a split receptacle, so one side will always be availabl To that end, most factory-marked receptacles will need to installed as split recep tacles; receptacles that show both ends as controlled will likely only be availab on special order.

The new second paragraph further clarifies that if option one above (the auto matic control device) is used, the receptacle must be marked on its face (again, th face of the receptacle, not the faceplate) and in a way that is visible after installa tion. And, again, even without a dedicated paragraph, an option two receptac still also needs a permanent marking and the word "controlled" on the manage receptacle(s). The final paragraph, quite properly working from the establishe principle that a duplex receptacle is not a single device but consists of two receptacles, clarifies that the function of each of the two receptacles of a duplex is to be apparent upon looking at them as installed, and visible regardless of whatever faceplate is being used.

Although it is difficult to imagine how a manually operated snap switch in a dwelling unit and similar locations (such as those addressed in exceptions within 210.70) could be a constituent of such a control system, a harmless exception nevertheless exempts such receptacles from the new marking requirement.

Part (F), new as of the 2017 **NEC**, covers receptacles that incorporate USB charging slots for cell phone and comparable charging equipment. Such equipment must be made so the USB circuitry is integral with the receptacle. In this way the USB power circuit wiring, which operated under Class 2 parameters, is precluded from being in contact with the branch-circuit wiring. There have been separate, add-on modules available without the system separation required by 725.136(A). However, the integral design permitted here complies with 725.136(B), and in addition, the entire assembly must be listed in its entirety.

406.4. General Installation Requirements. Parts **(A)** through **(F)** of this section provide the general requirements for installing receptacles in any occupancy. Part **(A)** of this rule calls for the use of 15- and 20-A receptacles that are factory-equipped with a grounding terminal. The next sentence prohibits the use of receptacles at currents and voltages for which they are not rated, except as indicated in Tables 210.21(B)(2) and (B)(3), which give the maximum current rating of the circuits to which the standard-rated receptacles may be safely connected. And the first sentence of this rule recognizes the use of nongrounding-type receptacles for replacement as described in 406.4(D).

Collectively these requirements provide a general framework that is an essential element of electrical safety, namely, receptacle configurations "advertise" circuit characteristics, and electrical equipment users are generally entitled to believe what they see. If they have a 20-A 125-cord cap (with the grounded conductor blade at right angles to the ungrounded conductor blade), they are entitled to believe that there is a 20-A overcurrent device and 20-A capable wiring ahead of the receptacle. A receptacle with a grounding terminal can be counted on as providing a connection to an equipment grounding conductor unless marked otherwise, as covered in (D)(2)(b or c).

Part **(B)** calls for the receptacle's "grounding contacts," usually the U-shaped connector, to be grounded, and part **(C)**, which generally requires receptacles to be grounded by connection to the circuit equipment grounding conductor, indicates how this must be accomplished. The exceptions to part **(B)** exempt receptacles on vehicle-mounted generators and replacement receptacles from the need for connection to an equipment grounding conductor, as permitted by 250.34 and 406.4(D).

The basic requirement of 406.4**(D)** mandates that when receptacles are replaced for any reason, the replacements must be in accordance with the requirements indicated in parts **(D)(1)**, **(2)**, **(3)**, and **(4)**. New language in the 2014 **NEC** mandated that both GFCI and AFCI receptacles be installed in readily accessible locations. This will facilitate the monthly testing they are supposed to undergo. Note that accessibility requirements for receptacles will be of very questionable enforceability due to the inherent lack of inspection following furniture rearrangements. Although the location of this rule only applies to replacement devices, the requirement applies generally because it backs up the general circuit requirements to the same effect in 210.8 and 210.12.

As covered in 406.4(D)(1), the Code generally requires any two-slot, nongrounding receptacle that becomes defective to be replaced with a grounding-type receptacle if there is a "grounding means" in the outlet box. The grounding means may be the equipment grounding conductor generally required to be run with the circuit conductors by the rule of 406.4(C) to provide for grounding of receptacles. Or, it may be an equipment ground installed in accordance with 250.130(C). In either case, a grounding-type receptacle must be used as a replacement for the defective nongrounding type. And the green ground screw on the receptacle must be connected to the available grounding means. It should be noted that it is not necessary to install a ground in accordance with 250.130(C), but if one chooses to do so, the use of a grounding-type receptacle becomes mandatory. And in either case, the grounding conductor installed in accordance with 250.130(C) must be connected to the green equipment ground screw on the receptacle.

Part **(2)** of 406.4(D) covers those instances where a replacement must be made at a receptacle enclosure that does not contain a grounding means. In that situation the Code offers three options for replacement. It should be noted that there is no evidence in any available documentation to suggest that this section is presenting these options in an order of preference. Any of the three indicated replacement alternatives is equally acceptable.

The first option, in part **(a)**, is to replace the existing two-slot (nongrounding) receptacle with another nongrounding-type receptacle. Although acceptable, such replacement may not be the most desirable where the receptacle has to supply grounded equipment. This has the virtue of strict accordance with the principle of truth in advertising discussed in **(A)**, that is quite simply, no ground equals no ground holes. At this point extensive debates begin. "They'll cut off the ground pins" or "They'll avoid the ungrounded connection" or "If they cut off the ground pin they know they are doing something bad" or "They'll plug into the grounding-type (but not functionally grounded) receptacle with no understanding of a possible problem" or "GFCI is no substitute for grounding" or "Surge protective devices will appear to be installed properly and fail," etc. All of these arguments have merit, but the issue has been settled in the **NEC**. Figure 406-2 summarizes the options.

The rule of part **(D)(2)(b)** recognizes a GFCI receptacle as replacement of a nongrounding-type receptacle where there is no ground in the box. Such a receptacle will provide automatic opening of the supply under fault conditions without connection to an equipment grounding conductor. This is accomplished through the internal electronic circuitry of the GFCI device. Given the much lower threshold (5 ± 1 mA) at which the GFCI receptacle operates compared to the instantaneous trip for a 15-A CB, the GFCI offers a much better level of shock protection.

The next sentence requires that a GFCI receptacle installed as a replacement for a nongrounding receptacle, in a box where a ground is not available, be marked "No Equipment Ground." This requirement is intended to alert anyone using that outlet that, even though the GFCI receptacle has a U-shaped ground slot

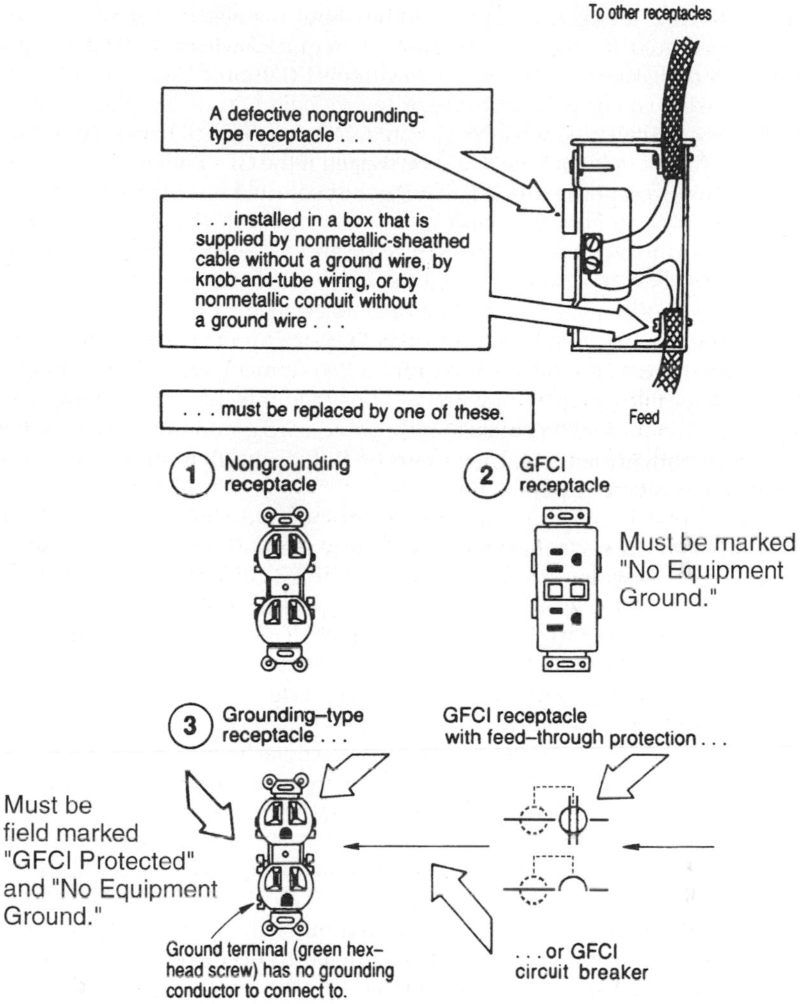
To other receptacles
A defective nongrounding-type receptacle . . .
. . . installed in a box that is supplied by nonmetallic-sheathed cable without a ground wire, by knob-and-tube wiring, or by nonmetallic conduit without a ground wire . . .
. . . must be replaced by one of these.
Feed
1 Nongrounding receptacle
2 GFCI receptacle
Must be marked "No Equipment Ground."
3 Grounding–type receptacle . . .
GFCI receptacle with feed–through protection . . .
Must be field marked "GFCI Protected" and "No Equipment Ground."
Ground terminal (green hex-head screw) has no grounding conductor to connect to.
. . . or GFCI circuit breaker

Fig. 406-2. A nongrounding-type receptacle, a GFCI-type receptacle, or GFCI protection, must always be used when replacing a nongrounding receptacle in any case where the box does not contain an equipment grounding conductor. [Sec. 406.4(D).]

here is no connection to the grounding system. This is important where supply-ng electronic equipment, such as computers, electronic cash registers, and the ike, because an equipment grounding conductor is critical for proper operation f such equipment. The field marking "No Equipment Ground" will serve to revent a user from being fooled into thinking that there is a ground connection. he required marking may be placed on the device or on the cover plate.

If there is no ground in the box, 406.4(D)(2)(c) permits replacement of a ongrounding-type receptacle with a GFCI-protected grounding-type receptacle. nd, in addition to the marking "No Equipment Ground," the receptacle must be marked "GFCI Protected." The warning about the absence of an equipment ground is required for the same reason it is required where a GFCI receptacle is used as a replacement in accordance with part **(D)(2)(b).** And the "GFCI Protected" marking can be on the device or (more likely) on its faceplate, and must be visible after installation wherever located. It is provided to help service electricians in the future. A lot of time could be wasted if the GFCI device protecting the grounding-type receptacle tripped, leaving the protected receptacle de-energized, and the service electrician was unaware of the upstream GFCI device. By providing such a field marking, any service electrician will be immediately advised of the possibility that the upstream GFCI device—receptacle or breaker—may have de-energized the circuit. Two informational notes address the fact that a GFCI is not a substitute for a ground-fault return path. Some appliances come with installation instructions that effectively require an equipment grounding connection, and 250.114 specifies equipment with the same requirement. The marking rules in this paragraph put users on notice that the three-wire grounding configuration to the contrary notwithstanding, the nongrounding connection offered in this context does not meet the requirements.

The only possible replacement combination not addressed here, because it is covered in other ways, is where a two-slot (nongrounding) receptacle is used as a replacement at a location that is now required to be GFCI-protected. There is no intent to prohibit the use of a nongrounding-type receptacle and a GFCI breaker would satisfy the need for replacement and GFCI protection. Such an application fully complies with all the requirements. The GFCI protection in the circuit breaker meets 406.4(D)(3) because the new receptacle is "GFCI protected" and the nongrounding receptacle meets 406.4(D)(2)(a).

Next, 406.4(D)(3) requires that where any receptacle is to be replaced, if the Code now requires a receptacle in that location to be GFCI protected, a GFCI-protected receptacle must be provided. That requirement has the effect of making all GFCI rules retroactive where a replacement is made. That is, normally there is no requirement to apply a new Code to an existing building and bring that building into compliance with the new Code. But, in an existing installation—say, in a bathroom of a commercial office building—if a receptacle should need replacement, 210.8(B)(1) in conjunction with 406.4(D)(3) would require that the replacement receptacle be GFCI protected.

The 2014 **NEC** added an exception that still remains and that allows upstream GFCI protection of a replacement receptacle in the event a GFCI receptacle in the subject location is impracticable, such as in cases where the existing outlet box is too small. This exception creates the erroneous impression that 406.4(D)(3) requires the use of a GFCI receptacle. It does not. It requires the use of a "ground-fault circuit interrupter protected receptacle." Any conventional receptacle on the load side of a GFCI protective device complies with this rule regardless of motivation. The **NEC** exception is completely unnecessary.

The exception is also technically incorrect because it requires the use of labeling indicating "GFCI protected" and "no equipment ground." The latter label should not be used if an equipment ground is present and yet would be required. There are many applications of small outlet boxes containing old wiring that nevertheless do provide an equipment grounding return path. This exception, which is permissive and therefore does not vary the rule it follows, can technically be ignored. However, it is very likely to cause confusion.

Three numbered paragraphs extend the rules for receptacle replacements. Item **(4)** is the most intriguing and controversial. If a dwelling unit receptacle, or any other receptacle should later editions of the **NEC** require AFCI protection for it, fails at an outlet that, if newly wired would require AFCI protection, then AFCI protection must be arranged for the replacement receptacle. This requirement took effect on January 1, 2014, in jurisdictions enforcing the 2011 **NEC** whether or not the 2014 **NEC** was in effect on that date.

If the panelboard and circuit protection are fairly new, an AFCI circuit breaker could be wired ahead of the replacement receptacle, assuming AFCI breakers are available for the wiring configuration; for example, even after all this time certain major manufacturers of overcurrent protective devices still do not have two-pole models, and if the damaged receptacle is on a multiwire branch circuit that will not be an option. It also will not be an option on any of the untold numbers of obsolete panelboard designs still in service. However, AFCI receptacles (the outlet branch circuit style) will be able to be applied either ahead of or even at the damaged receptacle location.

By placing the 3-year delay in the 2011 **NEC**, the Code-making panel placed a bet that the AFCI receptacles would be available in the market. The panel won that bet because the devices are indeed available. Refer to the coverage at 210.12 for further information. Two exceptions follow as of the 2017 **NEC**, the first allowing the use of GFCI protection [406.4(D)(2)(b)] in lieu of AFCI protection. This exception is extremely limited, with three conditions that clearly indicate the context. The first limits its use to where there is no equipment grounding conductor in the box, *and* it is impracticable to install one using a separate grounding conductor as covered in 250.130(C). The second limits its use to applications where the branch circuit originates in archaic equipment for which no combination-style AFCI device is available. The third limits its use to the time period prior to commercial availability of a combination GFCI/AFCI receptacle. If any of the three limits are not true when the receptacle is replaced, then conventional GFCI substitution does not apply. In fact, well before the beginning of 2017, combination AFCI/GFCI receptacles have become commonplace in supply houses and even home stores, making the first exception largely moot.

The second exception addresses a misapplication of 210.12(D) Exception, which is intended to apply to panelboard replacements on the supply side of branch circuits where the conductors need minor extensions in order to complete the work. It was never intended that one could add a few inches of conductor within the outlet box, count that as an extension of less than 6 ft, and then escape from arc fault protection requirements. That foolishness now ends. Note that under the **NEC** Style Manual this exception should have been listed first because it is a mandatory exception next to a permissive exception.

Items **(5)** and **(6)** require that replacement receptacles in locations requiring tamper-resistant receptacles or weather-resistant receptacles, respectively, under the applicable **NEC** provisions for new wiring, be replaced with new receptacles that comply with those provisions. This is just as has been the case for many years for GFCI receptacles. Intriguingly, as of the 2017 **NEC**, a nongrounding receptacle in an outlet normally required to be tamper resistant, can be replaced by another of the same style, because these receptacles are not manufactured in tamper-resistant configurations. Note that the new (2017) language here is unnecessary because 406.12 Exception already waives the tamper-resistant requirement in such cases. On the other hand, 406.4(D)(6) offers no such relief and 406.9(B)(1) does not either.

406.5 Receptacle Mounting. As required by this part, receptacles must be securely mounted in outlet boxes or "assemblies" (e.g., power poles) that are "identified" (meaning, as per Art. 100, generally recognizable as being suitable). This wording should be interpreted to mean that the box or assembly must be listed for use in the particular location (e.g., wet, dry, damp, indoors, outdoors, hazardous location). The last part of this sentence [in **(A)**] requires a receptacle to be "held rigidly at the finished surface" in any instance where the box is set back, as allowed for some building surfaces in 314.20. And the screws need to be actual 6-32 machine screws or otherwise as the enclosure manufacturer intended, and not drywall screws or other random fastenings.

This is an important requirement, because receptacles must accept considerable force in while cord caps are plugged in and pulled out. Some wall finishes are installed with close enough spacing (see also 314.21) so the plaster ears on the device yokes have enough bearing surface to comply with this rule, but many do not. In such cases spacers (#6 washers, ¼ in. nuts, whatever is handy) need to be mounted between the box ears and the underside of the yoke so when the 6-32 yoke screws are turned down, the yoke is securely held at the plane of the finished surface, and the receptacle yoke has enough bearing surface to not rock or twist in use, even with no faceplate installed. Then, and only then, add the faceplate. For boxes not set back, covered in **(B)**, secure the receptacle tightly to the box, and for boxes that project from the wall or that are surface mounted, remove at least one of the fiber retaining washers if the yoke connections are to be relied upon for grounding continuity, per 250.146(A).

406.5**(C)** requires receptacles mounted in a raised cover use more than a single screw for this purpose. This is not done for grounding continuity. There were numerous instances where a single screw in a duplex receptacle loosened to the point where the receptacle fell back far enough into the box to actually pivot on the remaining screw threads. If a cord cap was at all loose, and the pivot moved toward the ungrounded blade, the result was a shower of sparks and a destroyed cord cap. Raised cover designs now all address this requirement and include the required hardware. See the discussion at 250.146(A) for more information on the use of this hardware for grounding.

406.5**(D)** is intended to prevent short circuits when attachment plugs (caps) are inserted in receptacles mounted with metal faceplates—in which case, the metal of the plate could short (or bridge) the blades of the plug cap if the faceplate is not set back from the receptacle face. The rule requires that the "faces" of receptacles project at least 0.015 in. (0.4 mm) through the faceplate opening when the faceplate is metallic. And it is necessary to assure a solid backing for receptacles so that attachment plugs can be inserted without difficulty. The requirement for receptacle face to project at least 0.015 in. (0.4 mm) from installed metal faceplates will also prevent faults caused by countless existing attachment plugs with exposed bare terminal screws. With receptacle faces and faceplates installed according to 406.5(D), attachment plugs can be fully inserted into receptacles and will provide a better contact. The cooperation of other crafts, such as plasterers or drywall applicators, will be needed to satisfy the requirements. Note that compliance with this requiremen is best assured by rigorous enforcement of the rule in (A), because the product standards require receptacle faces to project forward from their yokes by enough margin to comply with these rules. Of course, this only works if the yokes are at the building surface the way they are supposed to be.

406.5**(E)** and **(F)**, new as of the 2017 NEC, address "countertops" and "work surfaces" in separate paragraphs. In so doing they respond to differences in the testing protocols between receptacles seemingly designed to be applied in comparable locations. It turns out that liquid spillage on a "work surface" is assumed comparable to a cup of liquid spilled, and similar spillage on a "countertop" (assumed to be in a kitchen or bathroom) assumes a much larger volume of water (actually ½ gallon of saline solution, such as a pot going over). For that reason, a countertop exposure requires a countertop listing, and a work surface exposure requires a work surface listing, but a countertop listing is also permitted because the requirements are either similar to or exceed those for work surfaces. Another key difference between the paragraphs is that the product standard connected to work surfaces is essentially about furniture capable of relocation. The standard connected to the countertop rules is focused on locations such as kitchen counters that are part of the building.

406.5**(G)** as of the 2014 cycle applies to countertops in all occupancies and not just for residential applications. It prohibits the placement of a receptacle in a face-up orientation. This rule does not prohibit pop-up receptacle assemblies that fully recess when not in use, whether supporting conventional or GFCI receptacles. These are also discussed in this book as part of the requirements for kitchen counters in 210.52(C)(5).

406.5**(H)** was entirely new as of the 2014 NEC and covers receptacle mounting in "seating areas and other similar surfaces." The focus of this rule is on receptacles now increasingly popular in airport gate seating areas that support travelers with laptop computers. Many of these devices also directly support USB charging equipment for smaller electronic devices. The pop-up receptacles discussed in (E) above are the third of four options for these applications, with the traditional floor box as the fourth. The first two options address listed assemblies that are part of a "furniture power distribution unit" if cord-and-plug-connected, and part of listed household or commercial furnishings if permanently located.

406.6. Receptacle Faceplates (Cover Plates). The 2017 NEC added a fascinating new subsection (D) covering faceplate assemblies that are constructed with insulated probes that are designed to make contact with the terminals on each side of a duplex receptacle (see Fig. 406-3). A small amount of power is taken and used to provide either night-lighting or a USB charging point or both. As in the case of 406.3(F), to assure safety this equipment must now be listed and designed so any Class 2 circuitry is integral with the cover plate.

406.7. Attachment Plugs, Cord Connectors, and Flanged Surface Devices. These rules contain the requirement for dead-front construction of cord caps, so the simple loss of a fiber or paper cover over the terminal screws will not result in energized parts near metal faceplates. Part **(B)** prohibits wiring schemes that result in circuits being energized through a receptacle connection of some sort to an energized plug. This has been a problem for years when untrained individuals attempt to connect fixed loads to the output of a portable generator, etc. The proper method is to have an energized connector body connect to a flanged inlet, as covered in (D) and shown in Fig. 406-4.

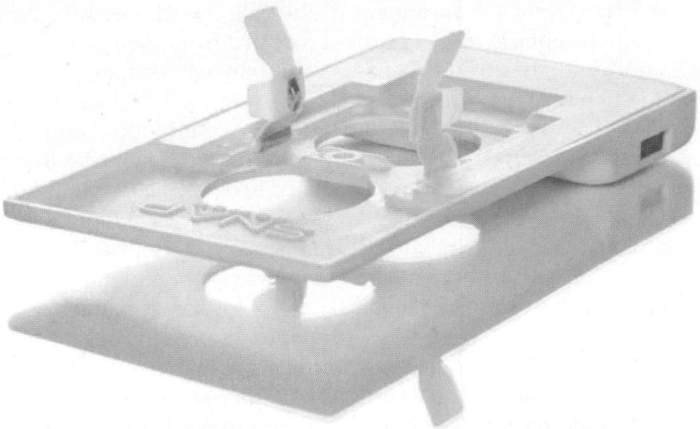

Fig. 406-3. This faceplate is also a device, in this case a USB charging point powered by the conductive pick-ups that engage the energized terminal screws on a duplex receptacle. [Sec. 406.6(D).]

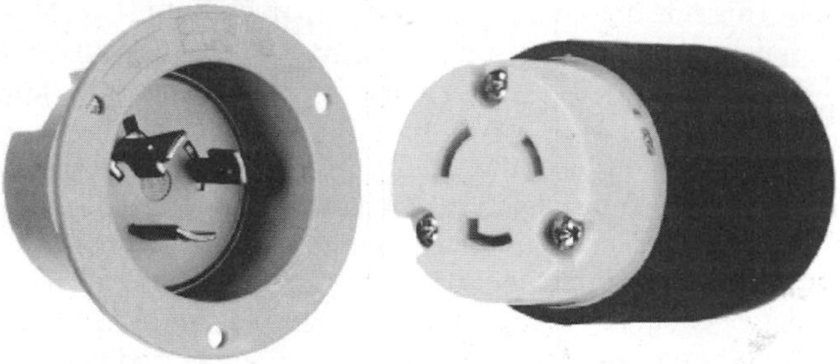

Fig. 406-4. This flanged inlet is energized by the mating connector body, and therefore no exposed and energized plug blades are possible. [Sec. 406.7(B)].

406.9. Receptacles in Damp or Wet Locations. The definition of *location* ir Art. 100 describes places that would be considered wet or damp locations. Any receptacle used in a damp location—such as an open or screened-in porch witl a roof or overhang above it—may not be equipped with a conventional recep tacle cover plate. It must be provided with a cover that will make the receptacle(s weatherproof when the cover or covers are in place—no plug inserted. The type o cover plate that has a thread-on, gasketed metal cap held captive by a short meta chain or a closeable cover would be acceptable for damp locations (Fig. 406-5). Th type of receptacle cover that has horizontally opening hinged flaps (doors) to cove the receptacles may be used in damp but not wet locations as described in part **(B)** even if the flaps are self-closing (i.e., if the flaps cannot stay open; Fig. 406-6 Of course, any cover plate that is listed for weatherproof use may also be used i damp locations.

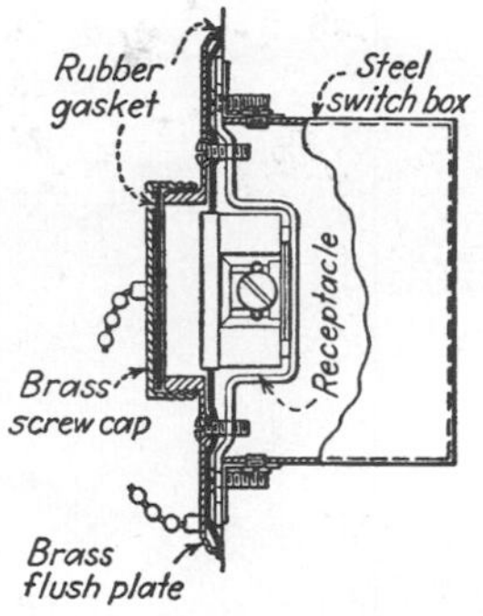
Rubber gasket
Steel switch box
Receptacle
Brass screw cap
Brass flush plate

Fig. 406-5. Chain-held screw-cap cover is suitable for damp, but not wet, locations. (Sec. 406.9.)

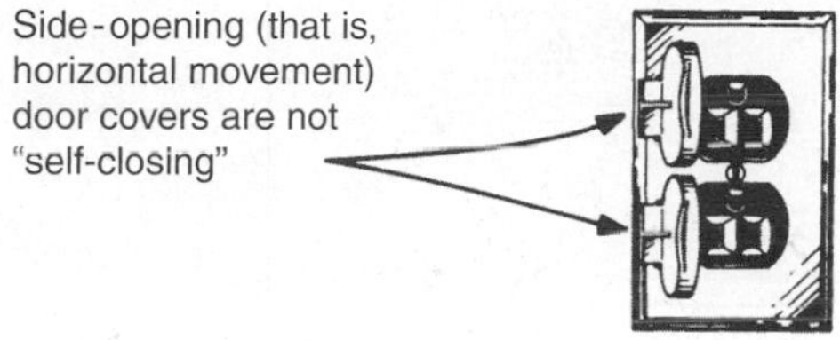
Side-opening (that is, horizontal movement) door covers are not "self-closing"

Note : Whether or not the doors are self-closing, and whether or not attended, this cover cannot be used in a wet location because the cover is not weatherproof with a cord cap inserted.

Fig. 406-6. Cover assembly with stay-open doors may be used in damp but not in wet locations. (Sec. 406.9.)

A new requirement as of the 2008 **NEC**, and that covers both damp and wet locations equally, in all occupancies, is that 15- and 20-A, 125- and 250-V non-locking receptacles be listed as "weather resistant." This makes the receptacles themselves a second line of defense in case the wet-location cover required to be functional with a plug inserted (required in wet locations) is left open or falls off. These receptacles have additional corrosion-resistant features on their metal parts, as well as UV and cold-weather impact resistance on their nonmetallic parts. They are identified in the field with a "WR" marking.

In part **(B)(1),** in addition to the requirement for weather resistance, the **NEC** clearly requires that *all* 15- and 20-A, 125- and 250-V receptacles installed in wet locations outdoors must be provided with covers that are weatherproof when the plug cap is inserted (a so-called "in-use" cover, shown in Fig. 406-7). This applies to all occupancies, and unlike the weather-resistance rule, to both straight-blade and locking-configured receptacles. There is no distinction within the wording used that bases compliance on occupancy. However, in-use covers are not NEMA 4 rated and will not stand up to a hose stream, as many wet locations, particularly wet locations indoors, are characterized by. For this reason, an exception allows covers that will be subject to "routine high-pressure spray washing" to have an enclosure that is weatherproof only when the plug is removed. Figure 406-5 is one classic style of a receptacle cover that is suitable for such exposures.

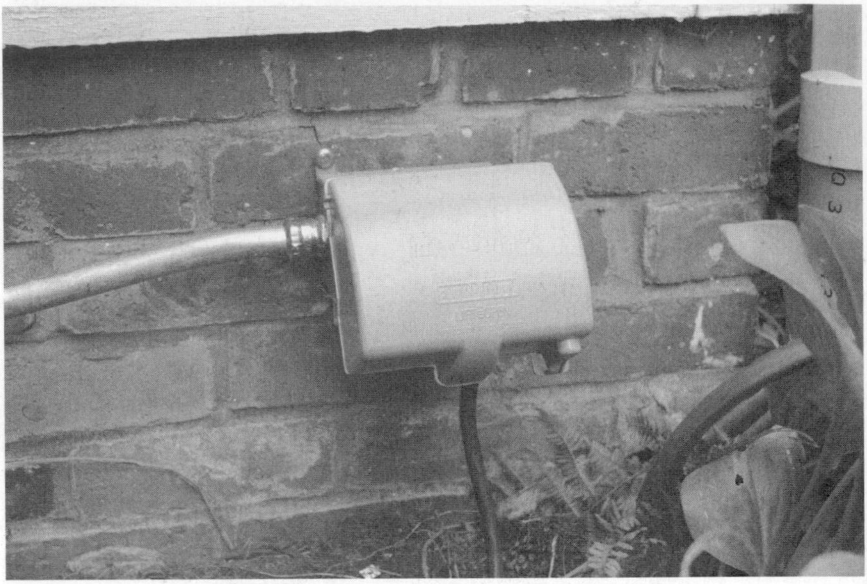

Fig. 406-7. A listed "outlet box hood" additionally identified as "extra duty." This cover will protect the receptacle even when a cord-and plug-connected load is in use. (Sec. 406.9.)

A new concept accepted in the 2011 NEC involves an "extra-duty outlet box hood." These hoods (Fig. 406-7) are now defined in 406.2, and are more robust versions of the "in-use" or "bubble" covers required before that time, and which tended to fail with disturbing regularity. As of the 2014 NEC, these are now required for all wet location exposures, whether or not residential, and however mounted, whether or not from grade.

As of the 2017 NEC, useful alternatives involving installations that use other listed assemblies or enclosures that do not employ outlet box hoods are relieved of the necessity of obtaining the "extra duty" marking. The NEC also added very useful information to Informational Note No. 1 that describes important limitations to the "extra duty" classification and expressly recognizes outdoor enclosure types covered in Table 110.28 that do not use an outlet box hood. Figure 406-8 shows an installation in an outdoor location that dates from when the only requirement was that the enclosure be weatherproof whether or not the attachment plug was inserted. Nevertheless, it anticipated the 2017 allowances very nicely. There is room through a removable plate in the bottom to admit as necessary up to 24 hard service extension cords along with a 6/4 hard service cord with a right-angle cord cap, and when the cover is closed and latched every plug and receptacle combination is immune to weather.

There are other more common applications of this concept, such as the traditional mobile home or RV receptacle in a common enclosure with a duplex GFCI convenience receptacle, all within an overall enclosure door that assures that the elements do not bother the plugged-in connections while they are in use.

Wet-location receptacles other than those covered by part **(B)(1)** are covered in **(B)(2)**, such as receptacles for higher voltages and current ratings. As given here

Fig. 406-8. A grouping of 12 duplex 125 V 20A duplex receptacles (each connected to a 20A GFCI circuit breaker) along with a 125/250-V 50A 3-pole 4-wire grounding receptacle, all within a fully gasketed enclosure with a hinged door.

all other wet-location receptacles must be used with either of two types of cover assemblies: If the receptacle will be used only while attended, such as for portable tools, a receptacle such as the one in Fig. 406-5 can be used, that is, one that is only weatherproof with the plug removed and the cover in place. If the receptacle will be kept in use, then some arrangement must be used that will maintain its integrity with the plug inserted. Some wet-location covers have a flat area around the receptacle that will seal to a wet location boot applied over a plug. Another approach is to use a pendant receptacle body on a short length of flexible cord with an expanding or flanged rubber boot, which will mate to a similar boot applied to the plug.

Part **(C)** of 406.9 prohibits the installation of receptacle outlets "within or directly over a bathtub or shower stall." The minimum exclusion area is that area, measured from the outside edge of the tub or shower enclosure, from the floor to the ceiling. Although luminaires are permitted when installed at least 8 ft (2.5 m) above the maximum water level, installation of a receptacle would be a violation of the wording given here. Note the differences between this wording and its counterpart in 404.4 for switches. Part **(D)** requires standpipes to floor-mounted receptacles to "allow floor cleaning" without damaging the floor-mounted device. Part **(E)** pertains to flush-mounted boxes in which receptacles are installed.

406.10. Grounding-Type Receptacles, Adapters, Cord Connectors, and Attachment Plugs. Paragraph **(3)** of part **(B)** requires a "rigid" terminal for equipment grounding connection in grounding adapters for insertion into nongrounding receptacles (Fig. 406-9). Adapters with pigtail leads are not acceptable to the rule.

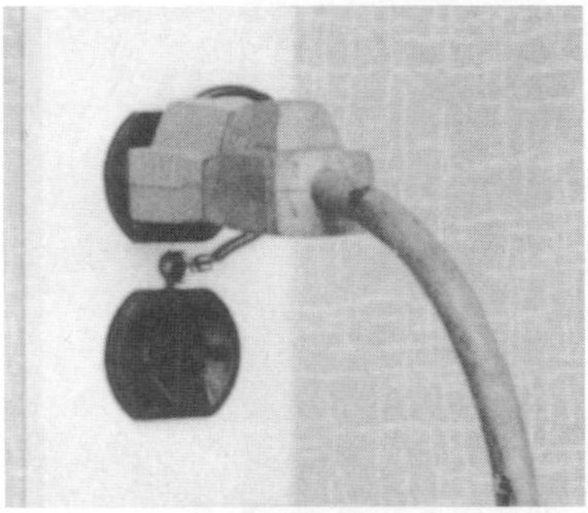

NE Code **AND UL VIOLATION! The** *NE Code* **prohibits use of the pigtail grounding adapters that have been available and widely used for many years. Such devices also do not satisfy UL construction standards.**

THIS IS UL AND *NE Code* **RECOGNIZED. Grounding adapter has rigid tab with spade end for connecting grounding terminal of the adapter to metal screw contacting the grounded metal yoke that mounts receptacle to the grounded metal box of the outlet (or to an equipment grounding conductor in NM cable used with a nonmetallic outlet box). Different-width blades on adapter polarize it for insertion in only one way into the polarized blade-openings on the receptacle.**

Fig. 406-9. This is the NEC and UL position on grounding adapters. (Sec. 406.10.)

This rule prevents the forked grounding terminal at the end of a pigtail from finding the energized slot of a receptacle (Fig. 406-10), and also eliminates the possibility of using two such adapters at a duplex receptacle, or one on a single receptacle either as a conventional device or as is common in pull-chain lampholders. Remember that these adapters must be polarized, and therefore only one of them can be used at a duplex receptacle where it will fit under the cover screw.

Figure 406-11 shows a very important rule from part **(E)** of 406.10.

406.12. Tamper-Resistant Receptacles. Another major change as of the 2008 NEC concerns the requirement for tamper-resistant receptacles at virtually all receptacle outlets in dwelling units. The same requirements were also extended to guest rooms and guest suites of hotels and motels, and to child care facilities (defined in 406.2 as for more than four children 7 years of age and under). These devices have shutter mechanisms that will move out of the way when two blades of a plug, of equal length, hit both shutters at the same time, which then allow the plug to be inserted (see Fig. 406-12). A toddler experimenting with a bobby pin

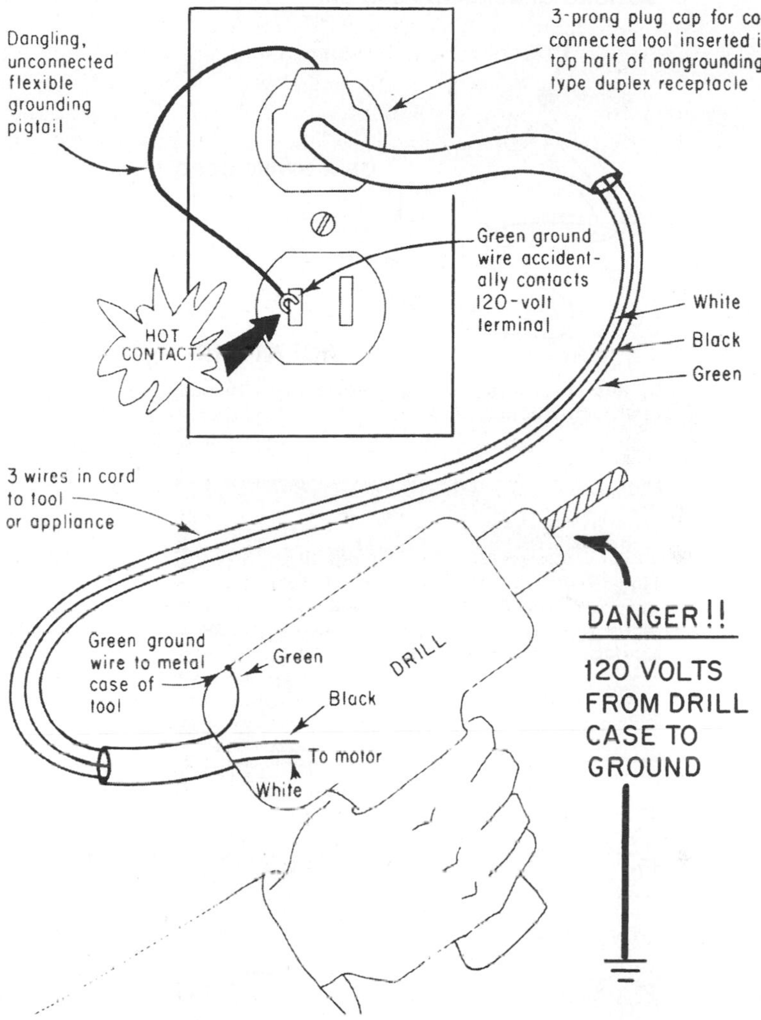

Dangling, unconnected flexible grounding pigtail
3-prong plug cap for cord-connected tool inserted into top half of nongrounding-type duplex receptacle
Green ground wire accidentally contacts 120-volt terminal
HOT CONTACT
White
Black
Green
3 wires in cord to tool or appliance
Green ground wire to metal case of tool
Green
DRILL
Black
To motor
White
DANGER!!
120 VOLTS FROM DRILL CASE TO GROUND

Fig. 406-10. Pigtail adapter can present shock hazard to personnel. (Sec. 406.10.)

or other similar object will see only an immovable shutter. These receptacles have been required for a very long time in pediatric wards of hospitals. This requirement, extended to 250-volt receptacles in the 2017 NEC, applies to "all 15- and 20-ampere non-locking-type receptacles in areas specified in 210.52 and 550.13."

This last reference, also new in the 2017 NEC, addresses comparable exposures in mobile/manufactured homes, and as a Chap. 4 amendment of Chap. 5 rules sets an interesting precedent. Generally, Art. 550 is at least a cycle behind the rest of residential requirements and often more, because these units move

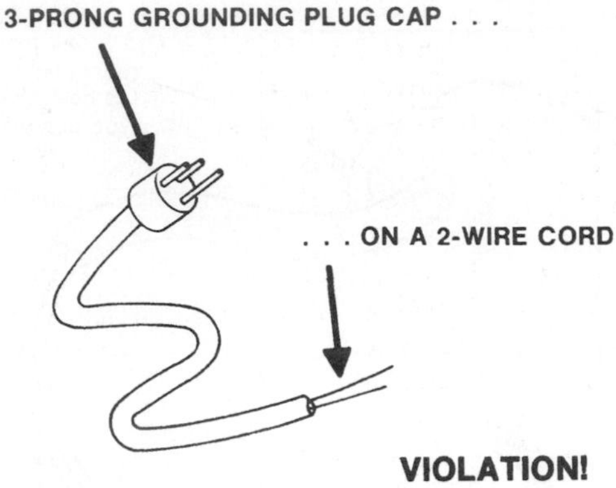
3-PRONG GROUNDING PLUG CAP . . .
. . . ON A 2-WIRE CORD
VIOLATION!

Fig. 406-11. A 3-wire cord must be used when connecting a grounding-type plug-cap. (Sec. 406.10.)

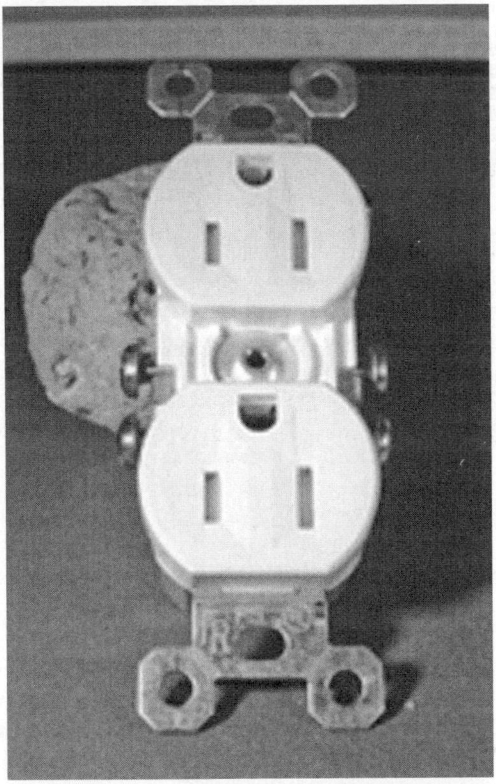

Fig. 406-12. A residential grade tamper-resistant receptacle. Note the shutters over the slot contact points, and the "TR" mark on the lower left corner of the yoke. With the faceplate off, this marking must be fully visible with the receptacle otherwise completely installed. (Sec. 406.12.)

in interstate commerce and there are issues at the federal regulatory levels. Art. 550, being in Chap. 5, has generally been considered as effectively requiring at least some degree of regulatory consent before imposing the latest requirements from the rest of the **NEC**. The regulations governing mobile/manufactured home construction were authorized under a congressional enactment that specifically invoked the supremacy clause in the U.S. Constitution. The **NEC** is not a federal document. It is a set of rules adopted by the several states under the police powers implicitly reserved to the states under the 10th Amendment. Any attempt at electrical inspection and enforcement of the **NEC** without at least the acquiescence of federal authorities is unlikely to succeed, and place local inspectors in a very difficult position if they attempt to enforce these provisions when one of these homes arrives on site.

The reference to 210.52 has been clarified through the placement of an exception that correctly correlates this requirement in a sensible way with 210.52(A) and with 210.50(B). For example, 210.52(A) does not count receptacles in luminaires and appliances, or over 1.7 m (5½ ft) above the floor; now 406.12 exempts those areas as well. The high receptacles are Out of toddler reach and the others would never be available in a tamper-resistant form. Receptacles located behind what used to be called stationary appliances to meet 210.50(B) are clearly no threat to a toddler either. The last exemption in the exception is for nongrounding replacement receptacles allowed by 406.4(D)(2)(a), which will never be available in a tamper-resistant form.

Note that this requirement also applies to GFCI receptacles, and they are available with the tamper-resistant features. The requirement also applies to outdoor receptacles installed to meet 210.52(E), so those receptacles will also need to meet the weather-resistance rules (TR + WR). The 2017 **NEC** drastically expanded the reach of this section, by effectively adding all locations where children are likely to congregate. List item (4) adds preschool and elementary schools. The literal text uses the word "facilities" so this would include areas devoted to schooling children of those age levels within other occupancies. Item (5) reaches business offices, waiting rooms and corridors of most medical occupancies. Item (6) reaches four of the assembly occupancies included in 518.2(A), and item (7) adds dormitories.

ARTICLE 408. SWITCHBOARDS AND PANELBOARDS

408.1. Scope. As in the case of Art. 404, the scope now clarifies that the provisions within Art. 408 do not apply to medium-voltage equipment, which is exclusively addressed in Art. 490. In addition, the Code-making panel removed, as of the 2011 cycle, the reference to "distribution boards." This terminology had been a source of endless speculation as to what a distribution board might actually be, and the panel concluded that it was an anachronism that should be discarded. This terminology was literally a century old, originating in Rule 70 of the 1911 **NEC** as part of the phrasing (under a heading covering manufacturing requirements for

"Cabinets") "For panel and distributing boards, cutouts and switches." From this and subsequent **NEC** language, it seems quite clear that the terminology simply reflected parallel usage long ago that described the same equipment, and there was no technical issue or problem raised in deleting the term.

The term "switchgear" was folded in throughout this article (in fact, throughout the 2014 **NEC**) with parallel coverage to switchboards. Modern versions of both are quite similar, being exclusively metal enclosed, and follow whatever product standard their manufacturers submitted them under when seeking a listing. Switchgear tends to encompass more sophisticated electronics for monitoring and control than current switchboard designs. The coverage of the term "switchgear" in this book in Art. 100 includes additional information relative to differences between the two types of equipment. Although the product standards differ, the Code rules for both are identical. The term "switchboard" still encompasses antique designs with open switches on slate backplanes, etc.

Because the differences primarily reflect design preferences on the part of the owner or specifying engineer, and do not have an impact on the minimum applicable **NEC** requirements, in this book the usual term will continue to be "switchboard" for the sake of brevity. Unless stated otherwise, the minimum requirements apply equally to switchgear.

408.3. Support and Arrangement of Busbars and Conductors. Part **(A)(2)** requires that all service switchboards have a barrier installed within the switchboard to isolate the service busbars and the service terminals from the remainder of the switchboard as shown in Fig. 408-1. Because it is usually impossible to kill the circuit feeding a service switchboard, it has become very common practice for mechanics to work on switchboards with the service bus energized. The hazard associated with this has caused concern and is the reason for this addition to the Code.

Switchboard manufacturers in many parts of the country have been supplying switchboards with these barriers in place; this Code rule aims to make such protection for personnel a standard requirement. With a barrier of this type installed in a service switchboard, mechanics working on feeder devices for other sections of the switchboard will not be exposed to accidental or surprise contact with the energized parts of the service equipment itself.

As of the 2017 **NEC**, comparable requirements are now in place for panelboards incorporating service disconnects. See Fig. 408-2 for one example of how this rule is expected to be implemented. The change has been proposed for many cycles and is now in place. This is not directly comparable to the Canadian rules, and still allows branch circuit entries in the service protection end of the panel.

Part **(A)(3)** notes that only those conductors intended for termination in a vertical section of a switchboard may be run within that section, other than required inner connections and/or control wiring. This rule was intended to prevent repetition of the many cases on record of damage to switchboards having been caused by termination failures in one section being transmitted to other parts of the switchboard. In order to comply with this requirement, it will be necessary in some cases to provide auxiliary gutters. The basic concept behind the rule is that any load conductors originating at the load terminals of switches or breakers in a switchboard must be carried vertically up or down, so that they leave the switchboard from that vertical section. Such conductors may not be carried horizontally to or through any other vertical section of the switchboard, except as indicated in the sketch at the top of Fig. 408-1. Because of field

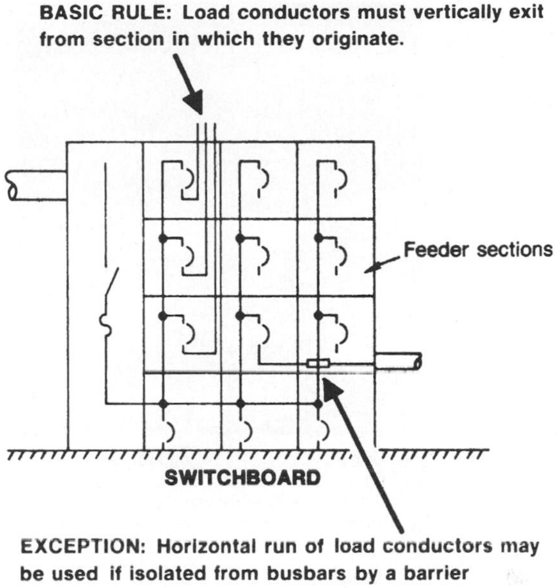
BASIC RULE: Load conductors must vertically exit from section in which they originate.
Feeder sections
SWITCHBOARD
EXCEPTION: Horizontal run of load conductors may be used if isolated from busbars by a barrier

All conductor terminals within a switchboard must be used only for connection to conductors that leave the switchboard vertically from the same switchboard section in which the terminals are located—with the conductors run out the top or out of the bottom of the switchboard.

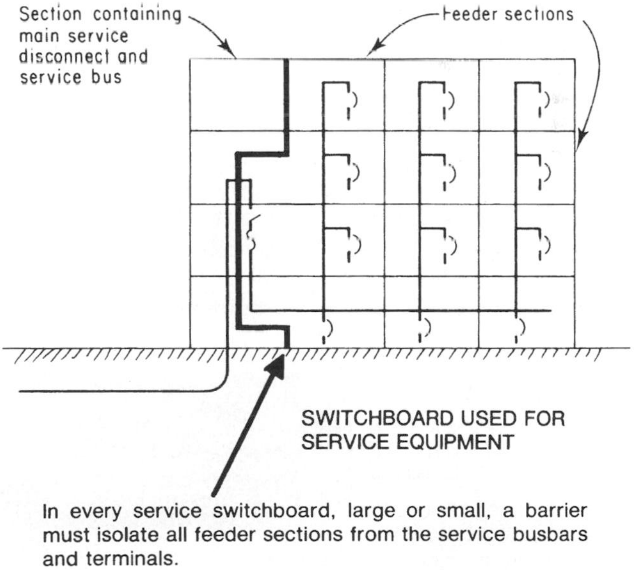
Section containing main service disconnect and service bus
Feeder sections
SWITCHBOARD USED FOR SERVICE EQUIPMENT
In every service switchboard, large or small, a barrier must isolate all feeder sections from the service busbars and terminals.

Fig. 408-1. These are basic rules on switchboard wiring. (Sec. 408.3.)

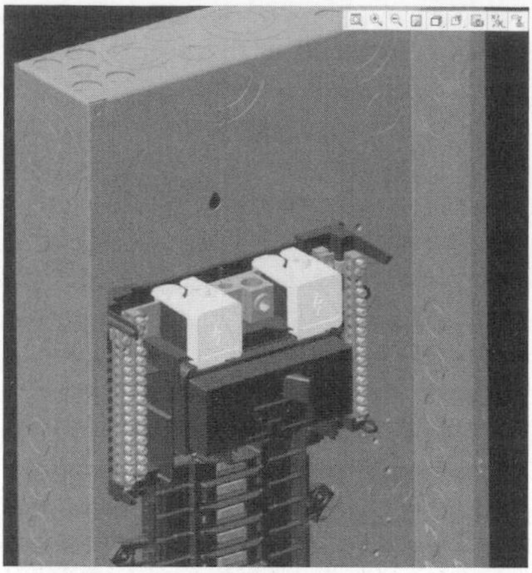

Fig. 408-2. An example of making ungrounded and uninsulated service terminals unexposed to inadvertent contact while energized. [408.3(A)(2).] (Courtesy, Square D/Schneider)

conditions, in which the installer does not know the location of protective devices in a switchboard at the time of conduit installations, the exception modifies the requirement that conductors within a vertical section of a switchboard terminate in that section. Conductors may pass horizontally from one vertical section to another, *provided* that the conductors are isolated from switchboard busbars by some kind of barrier.

Part **(C)** requires a bonding jumper in a switchboard or panelboard used for service equipment to connect the grounded neutral or grounded phase leg to the equipment grounding conductor (the metal frame or enclosure of the equipment). UL data apply to this rule:

1. Switchboard sections or interiors are optionally intended for use either as a feeder distribution switchboard or as a service switchboard. For service use, a switchboard must be marked "Suitable for use as service equipment."
2. Some switchboard sections or interiors include neutral busbars factory-bonded to the switchboard enclosure. Such switchboards are marked "Suitable *only* for use as service equipment" and may *not* be used as sub-distribution switchboards (Fig. 408-3). A bonded neutral bus in a service switchboard may also serve as an equipment grounding busbar.

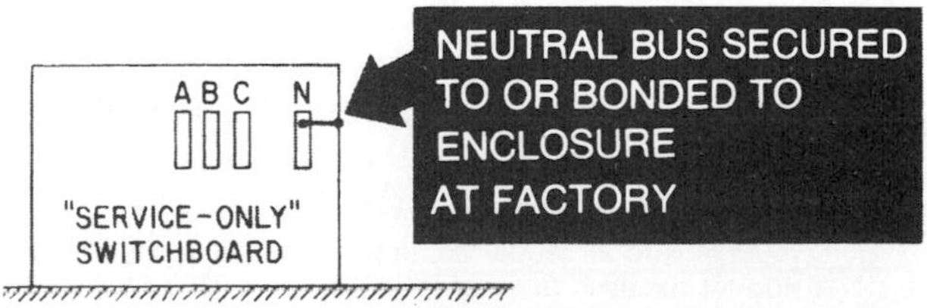
A B C N
"SERVICE-ONLY"
SWITCHBOARD
NEUTRAL BUS SECURED TO OR BONDED TO ENCLOSURE AT FACTORY

Fig. 408-3. Bonded neutral bus limits switchboard to service applications. (Sec. 408.3.)

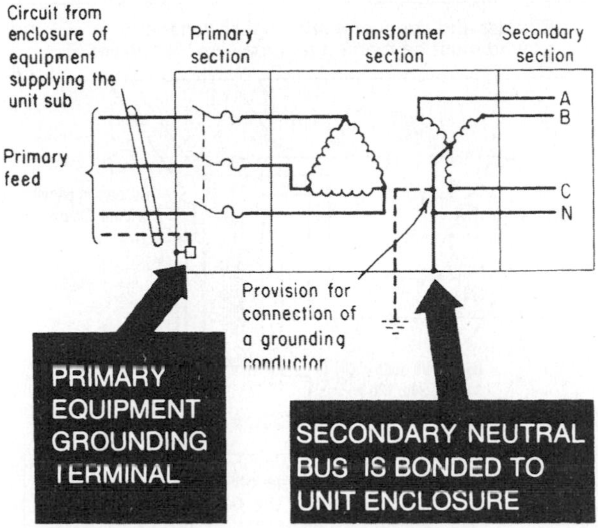
Circuit from enclosure of equipment supplying the unit sub
Primary section
Transformer section
Secondary section
Primary feed
A
B
C
N
Provision for connection of a grounding conductor
PRIMARY EQUIPMENT GROUNDING TERMINAL
SECONDARY NEUTRAL BUS IS BONDED TO UNIT ENCLOSURE

Fig. 408-4. Unit subs have bonded neutral in secondary switchboards. (Sec. 408.3.)

3. UL-listed unit substations have the secondary neutral bonded to the enclosure and have provision on the neutral for connection of a grounding conductor, as shown in Fig. 408-4. A terminal is also provided on the enclosure near the line terminals for use with an equipment grounding conductor run from the enclosure of primary equipment feeding the unit sub to the enclosure of the unit sub. Connection of such an equipment grounding conductor provides proper bonding together of equipment enclosures where the primary feed to the unit sub is directly buried underground or is run in nonmetallic conduit without a metal conduit connection in the primary feed.
4. Unless marked otherwise (with both the size and temperature rating of wire to be used), the termination provisions on switchboards are for 60°C wire from No. 14 to No. 1 and 75°C for No. 1/0 and larger wires.

The rule of 110.15 is shown in Fig. 408-5 and correlated to the rule of part **(E).** On a 3-phase, 4-wire delta-connected system (the so-called red-leg delta, with the midpoint of one phase grounded), the phase busbar or conductor having the higher voltage to ground must be marked, and the higher leg to ground must be phase B, as required by 110.15. Without identification of the higher voltage leg, an installer connecting 120-V loads (lamps, motor starter coils, appliances) to the panelboard shown in the diagram might accidentally connect the loads from the high leg to neutral, exposing the loads to burnout with 208 V across such loads.

408.3(E) requires a fixed arrangement (or phase sequence) of busbars in panels or switchboards. The installer must observe this sequence in hooking up such equipment and must therefore know the phase sequence (or rotation) of the feeder or service conductors. This new rule has the effect of requiring basic phase identification at the service entrance and consistent conformity to that identification and sequence throughout the whole system (Fig. 408-6).

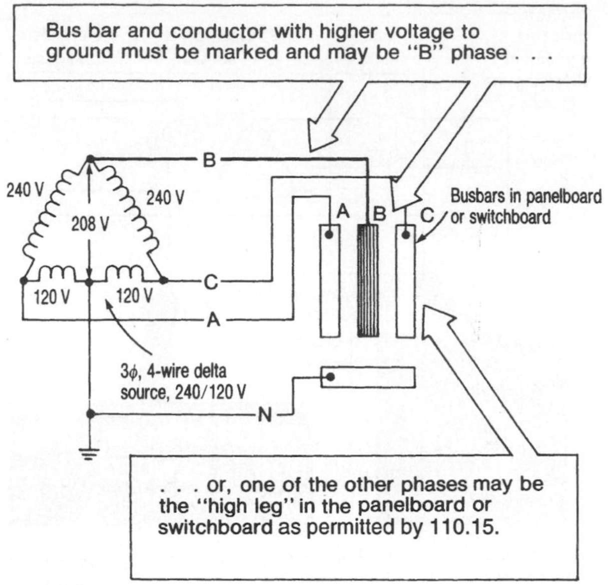

Bus bar and conductor with higher voltage to ground must be marked and may be "B" phase . . .
B
240 V
240 V
208 V
120 V
120 V
C
A
Busbars in panelboard or switchboard
A
B
C
3ϕ, 4-wire delta source, 240/120 V
N
. . . or, one of the other phases may be the "high leg" in the panelboard or switchboard as permitted by 110.15.

Fig. 408-5. Safety requires "high leg" identification on 4-wire delta systems. (Sec. 408.3, informational note.)

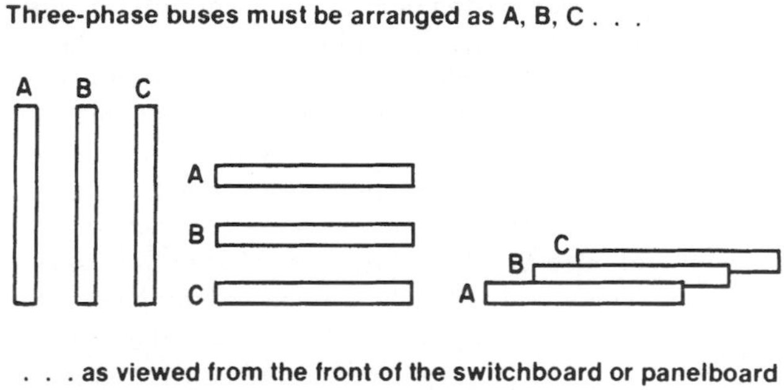

Three-phase buses must be arranged as A, B, C . . .
A
B
C
A
B
C
A
B
C
. . . as viewed from the front of the switchboard or panelboard

Fig. 408-6. Phase sequence in panelboards and switchboards must be fixed. (Sec. 408.3.)

Difficulty has been encountered with the rule of part **(E)(1)** requiring the high leg to be the B phase, because utility company rules may call for the high leg to be the C phase and the right-hand terminal in a meter socket—rather than the middle terminal. As shown in Fig. 408-7, the utility phase rotation can be converted to a Code phase rotation by applying the concept that phase rotation is relative, not absolute. If the utility C phase is designated as the NEC B phase, then the other phase legs are identified for NEC purposes as shown. The phase rotation C-A-B is the same as A-B-C, with voltage alternations such that wave B follows wave A by 120°, wave C follows wave B by 120°, wave A follows wave C by 120°, and so on. With the phase legs identified as at the bottom of the drawing, each is carried to the appropriately designated phase lug (A-B-C, left to right) at the panelboard shown in Fig. 408-5.

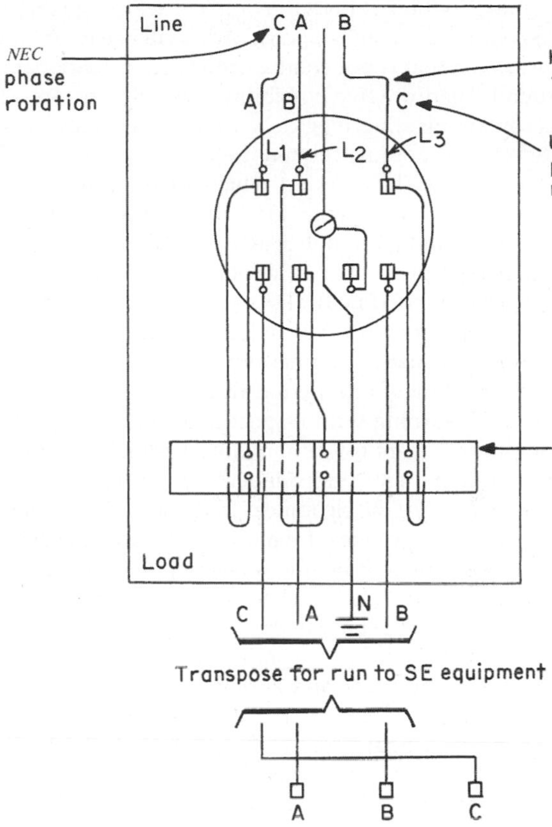
Line
C A B
NEC
phase
rotation
A B
C
L1 L2 L3
High leg, 208 V
to ground
Utility
phase
rotation
Manual circuit
closing device
This is
no longer
needed.
Load
C A N B
Transpose for run to SE equipment
A B C

Fig. 408-7. Utility C phase becomes NEC B phase for high-leg identification. [Sec. 408.3(E), Exception.] In the past, the "B" phase was required to be the "high leg." Now such transposition is not needed.

The exception to part **(E)(1)** will permit the high leg (the one with 208 V to ground) to be other than the B phase (such as the C phase) where the meter is within the same enclosure as the switchboard or panelboard and the phase configuration of the utility supply system requires other than the B phase for the high leg at the meter.

The concept behind this exception is to permit the same phase identification (such as C phase at 208 V) for the metering equipment and the busbars *within the switchboard or panelboard.* And it is the intent of the Code panel that the different service phase identification (such as C phase as the high leg) will apply to the entire switchboard or panelboard and that no transposition of phases B and C is needed within the switchboard or panelboard. However, beyond the service switchboard or panelboard, the basic rule of 408.3(E) must be observed to have the B phase (the middle busbar) as the high (208-V) leg, requiring phase transposition on the load side of the service switchboard or panelboard. Part **(A)(2)** is new in the 2014 **NEC** and covers dc bus arrangements for the first time. They are permitted in any order, but must be field marked as to polarity, grounding system, and nominal voltage classification.

Part **(F)**, greatly expanded in the 2014 **NEC**, requires a warning label in accordance with 110.21(B) to be posted on any switchboard, switchgear, or panelboard if it is operating on any system that is not solidly grounded or that operates with unequal voltages to ground. There are five possibilities. The first covers all switchboards or panelboards containing a high-leg delta system, and they must have a warning label to that effect, and not just rely on the color code for the phase leg with the higher voltage to ground. The marking, which is to be applied in the field, must read: "Caution: ___ Phase Has ___ Volts to Ground." The second paragraph covers ungrounded ac systems and gives notice as to the phase-to-phase voltage.

The third through fifth paragraphs were new as of the 2014 **NEC**, beginning with a notice for high-impedance grounded neutral systems. This notice must provide the phase-to-phase voltage together with the advice that the system may operate a specified voltage to ground, usually the phase-to-phase voltage, and that this operation may continue for an indefinite period during a fault. The fifth paragraph provides for a parallel warning with respect to a resistively grounded dc system. The fourth paragraph covers ungrounded dc systems and requires the same type of label as required in (2) for ungrounded ac systems.

Part **(G)** refers to the need for specific clearances in top and side gutters in both panelboards and switchboards and makes it mandatory that wire bending space at terminals and gutter spaces must afford the room required in 312.6. This is a repeated requirement throughout the Code and is aimed at ensuring safe termination of conductors as well as adequate space in the side gutters of panelboards and switchboards for installing the line and load conductors in such equipment. This concern for adequate wire bending space and gutter space is particularly important because of the very large size cables and conductors so commonly used today in panelboards and switchboards. Sharp turns to provide connection to terminal lugs do present possible damage to the conductor and do create strain and twisting force on the terminals themselves. Both of those objections can be eliminated by providing adequate wiring space.

408.4. Field Identification Required.

(A) Circuit directory or circuit identification This section specifically requires full and legible marking of a panelboard's *circuit directory* to show the loads supplied by each circuit originating in the panel. This is similar to the rule of 110.22(A), where the Code requires thorough identification of the loads fed by all circuit breakers and switches that serve a Code-required disconnecting means. Note that the wording also calls for "clear and legible marking" of circuit modifications. This requires that any load(s) added to or deleted from the panelboard in question must be shown in an updated circuit directory. A point commonly missed in the field is that this rule also applies to "circuit modifications." This means that any downstream rearrangement, additions, or subtractions from circuit coverage must result in an update to the circuit directory, even if the work was done without ever opening the circuit protection in the panel.

The prevalence of modern word processing software (and even circuit directory software) makes compliance with these rules far simpler than in the past (Fig. 408-8). All circuit directories, for both switchboards (labels required at each switch) and panelboards, must meet the following requirements:

- Every circuit must have a directory entry, which must be legible.
- Every directory entry must describe a "clear, evident, specific purpose or use."

WXYZ APARTMENT PANEL
(installed September 2003)
(revised August, 2004)
FED FROM WXYZ METER MAIN AT SERVICE

1. Peninsula counter receptacle, west	16. Overhead lights, hall receptacle, bathroom fan
2. Peninsula counter receptacle, east	17. Refrigerator; dining alcove receptacles (except NE)
3. East counter receps, left half of quad	18. Spare
4. East counter receps, right half of quad dining area, NE receptacle	19. Spare
5. Garbage disposer	20. Spare
6. Dishwasher	21. ______
7. Livingroom receps, left half of quad; south outdoor receptacle	22. ______
8. Livingroom receps, right half of quad; entry luminaires; north outdoor recept.	23. ______
9. Bedroom receps, right half of quad,	24. ______
10. Bedroom receps, left half of quad single duplex in south bedroom	25. ______
11. & Range	26. ______
12.	27. ______
13. Bathroom receptacle	28. ______
14. Microwave oven receptacle	29. ______
15. Oil burner for heat, hot water	30. ______

Fig. 408-8. The particular circuit that supplies every single outlet in the occupancy can be accurately identified using this circuit directory. There are no exceptions, and every description will remain valid after a change in tenancy. Because this panel is in a multifamily building, the feeder source is entered in the circuit directory header as well. (Sec. 408.4.)

- Every directory entry must include a sufficient degree of detail to distinguish it from all others. This could include outlet faceplate markings correlated with unique directory entries, such as "receptacle outlets in Room A identified as "Ckt ###."
- No directory entry can use a description that depends on transient conditions of occupancy. For example, terminology such as "Ed's bedroom" or "Annie's Hair Salon" must be rejected in favor of handing (left or right), or points of the compass, or the use of a map with room numbers, specific receptacle locations, etc.

- Spare overcurrent devices must be identified as such on the circuit directory.
- Circuit directories must be posted on the face or inside of the panel door (panelboards) and at each switch on a switchboard.

(B) Source of supply This part requires, for all but one- and two-family housing, that the source of supply for all nonservice panelboards and switchboards be labeled as to the device or equipment where the power originates. The wording "each device" was chosen to make it clear that a description of an electrical room or switchboard was insufficient because in the first case an electrical room is not a device, and in the second case a stated switchboard, while a device within the meaning of the Code, is not the device where the feeder conductors are connected; it merely houses the relevant device. The description must uniquely describe the actual source. The word "equipment" was added because some feeders originate at generators or other sources. The 2017 **NEC** added requirements for durability and that at least this part of the label not be handwritten. Note that this is a requirement in (B) and not the general directory requirements in (A).

408.5. Clearance for Conductors Entering Bus Enclosures. Figure 408-9 shows the rules of this section, which are aimed at eliminating high conduit stubups under equipment containing busbars to prevent contact or dangerous proximity between conduit stubups and the busbars. On this matter, UL says that "the acceptability of conduit stubs serving unit sections with respect to wiring space and spacing from live parts can be determined only by the local inspection authorities at the final installation."

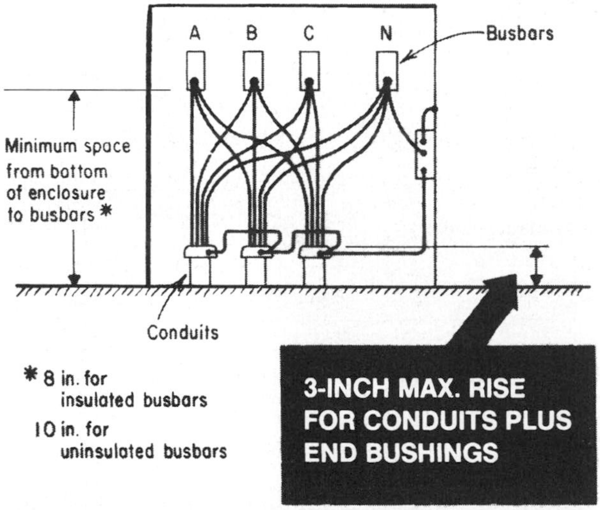
A
B
C
N
Busbars
Minimum space from bottom of enclosure to busbars *
Conduits
* 8 in. for insulated busbars
10 in. for uninsulated busbars
3-INCH MAX. RISE FOR CONDUITS PLUS END BUSHINGS

Fig. 408-9. Conduit stub-ups must have safe clearance from busbars. (Sec. 408.5.)

408.7. Unused Openings. The rule here requires that all unused openings in panelboards and switchboards—regardless of whether the opening is inside the cover or through the enclosure—must be closed to prevent persons from coming in contact with the enclosed energized components. The means utilized to satisfy this rule must be such that the enclosure opening is effectively returned to its original integrity. The general requirement is covered in 110.12(A); the rule here is aimed at missing circuit breaker openings in a deadfront.

408.16. Switchboards and Switchgear in Damp and Wet Locations. The rules in 312.2 apply to switchboards in these locations.

408.17. Location Relative to Easily Ignitible Material. A combustible floor under a switchboard must be protected against fire hazard, usually with sheet metal.

408.18. Clearances. Although it has long been a Code rule that a clearance of at least 3 ft be provided from the top of a switchboard to a "combustible" ceiling above, the opening phrase of part **(A)** excludes totally enclosed switchboards from this rule. The original rule requiring a 3-ft clearance was based on open-type switchboards and did not envision totally enclosed switchboards. The sheet-metal top of such switchboards provides sufficient protection against heat transfer to nonfireproof ceilings. As a result of this rule, now there is no minimum clearance required above totally enclosed switchboards.

As covered in part **(B)**, accessibility and working space are very necessary to avoid possible shock hazards and to provide easy access for maintenance, repair, operation, and housekeeping—as required by 110.26. It is preferable to increase the minimum space behind a switchboard where space is available or attainable.

408.20. Location of Switchboards and Switchgear. This section addresses the permitted environment for old-style open switchboards (Fig. 408-10).

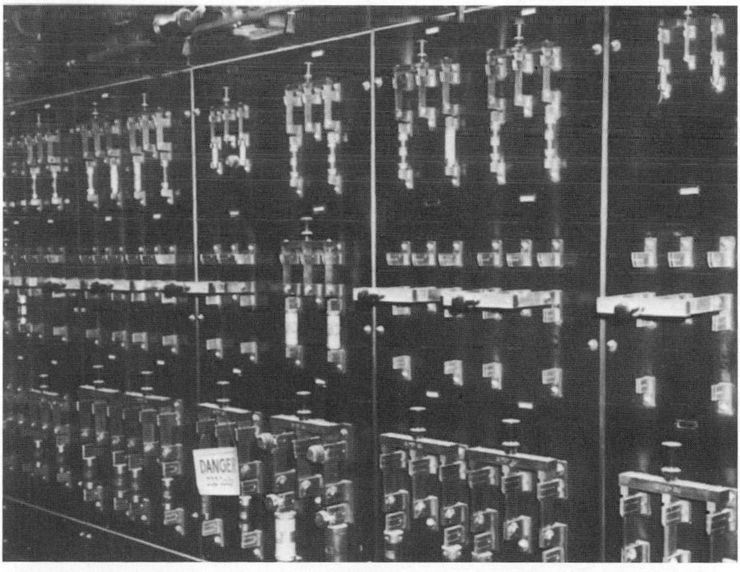
DANGER

Fig. 408-10. A switchboard with "*any*" exposed live parts is limited to use in "permanently dry" locations, accessible only to qualified persons. (Sec. 408.20.)

408.30. General. The first sentence here establishes the minimum acceptable rating of any panelboard.

All panelboards—lighting and power—are required by this section to have a rating (the ampere capacity of the busbars) not less than the NEC minimum feeder conductor capacity for the entire load served by the panel. That is, the panel busbars must have a nameplate ampere rating at least equal to the required ampere capacity of the conductors that feed the panel (Fig. 408-11). A panel may have a busbar current rating greater than the current rating of its feeder but must never have a current rating lower than that required for its feeder. [215.2(A) notes that a feeder for a continuous load must be rated at least 125 percent of the load current, which seems to mean that the panel busbars would have to be rated for 125 percent of the continuous load current.]

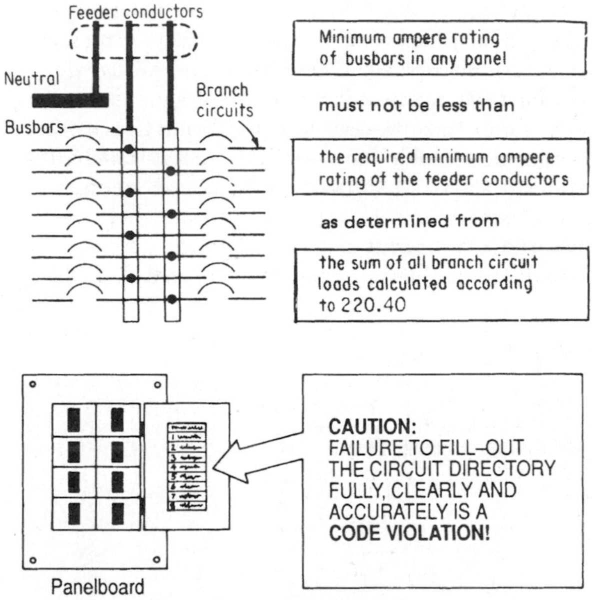
Feeder conductors
Neutral
Busbars
Branch circuits
Minimum ampere rating of busbars in any panel
must not be less than
the required minimum ampere rating of the feeder conductors
as determined from
the sum of all branch circuit loads calculated according to 220.40
CAUTION:
FAILURE TO FILL-OUT THE CIRCUIT DIRECTORY FULLY, CLEARLY AND ACCURATELY IS A CODE VIOLATION!
Panelboard

Fig. 408-11. Rating of panelboard bus must at least match required feeder ampacity and circuits must be identified. (Sec. 408.30.)

Although selection of a panelboard is based first on the number of circuits that it must serve, it must be ensured that the busbars in a panelboard for any application have at least the Code-minimum circuits.

With respect to panelboards, marking may appear on the individual terminals, but terminals can often be changed in the field, and wiring space and the means of mounting the terminals may not be suitable. Therefore, panelboards should be marked independently of the marking on the terminals to identify the terminals and switch or CB units which may be used with aluminum wire. If all terminals are suitable for use with aluminum conductors as well as with copper conductors, the panelboard will be marked "use copper or aluminum wire." A panelboard marked "use copper wire only" indicates that wiring space or other factors make the panelboard unsuitable for any aluminum conductors.

408.34. Classification of Panelboards (Deleted). For some 70 years, the NEC maintained special rules for a classification of panelboards described as *lighting and appliance branch-circuit panelboards*; however, effective with the 2008 NEC, this classification has been abolished. For example, these panelboards required individual protection in accordance with their busbar ratings, whereas power panels did not require individual protection; now all panelboards require individual protection, but the former NEC limit on the numbers of circuit positions (42) has been deleted. The distinctions were a response to a fire in a New York hotel in the 1930s involving large numbers of rubber-insulated wires in a panel, and no longer had technical validity with today's wire insulation and panel construction methods.

408.36. Overcurrent Protection. The trade-off for the elimination of rules setting a ceiling on the number of overcurrent devices permitted in a panel was changing the individual protection requirement that formerly only applied to lighting-and-appliance panelboards and some power panelboards so it now applied to all panelboards of any description. These panels are now in manufacture, including 56-circuit versions routinely shipped to Canada that never had this restriction, and there are many others, with even greater numbers of circuit positions. There are three exceptions to this general rule, as follows:

1. A panelboard with up to six circuit breakers may be installed without individual protection provided it is used as service equipment (see Fig. 408-12). In this case, with more than two main breakers installed in a new panelboard,

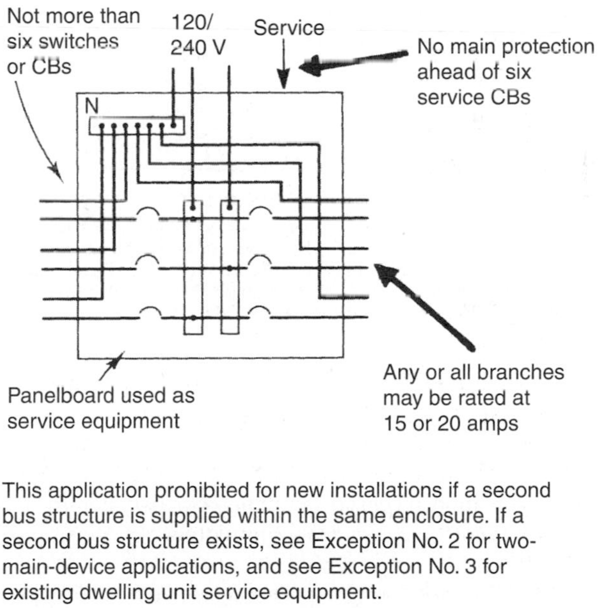
Not more than
six switches
or CBs
120/
240 V
Service
No main protection
ahead of six
service CBs
N
Panelboard used as
service equipment
Any or all branches
may be rated at
15 or 20 amps
This application prohibited for new installations if a second
bus structure is supplied within the same enclosure. If a
second bus structure exists, see Exception No. 2 for two-
main-device applications, and see Exception No. 3 for
existing dwelling unit service equipment.

Fig. 408-12. This application is permitted only where it comprises the service equipment. (Sec. 408.36, Exception No. 1.)

the use of a second bus structure within the panelboard as shown in Fig. 408-14 is now expressly prohibited. This type of arrangement is common in many parts of the country, with an outdoor service panel breaking out major loads and an indoor subpanel supplying the various lighting, small appliance, and receptacle circuits. A panelboard with two main breakers is covered in the next topic.

2. A panelboard with two main breakers, as shown in Fig. 408-13, is still permitted, provided it qualifies under the old 42-circuit limitation on numbers of circuits for lighting and appliance branch-circuit panelboards. The combined main breaker ratings must not exceed the panel rating.

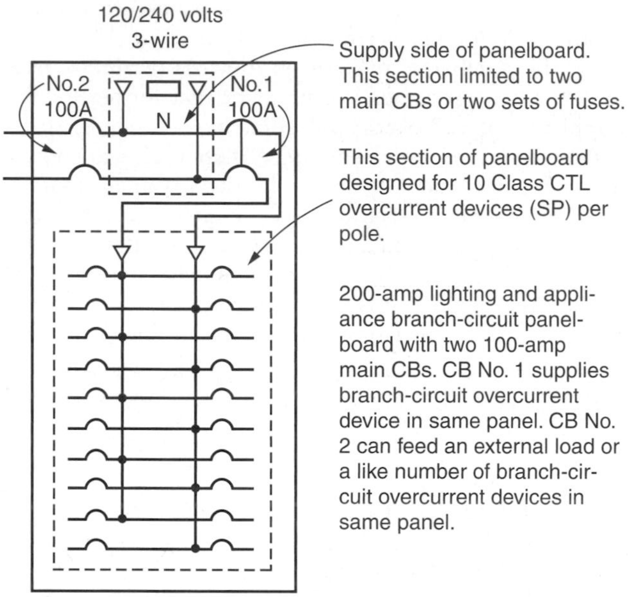
120/240 volts
3-wire
No.2
100A
N
No.1
100A
Supply side of panelboard. This section limited to two main CBs or two sets of fuses.
This section of panelboard designed for 10 Class CTL overcurrent devices (SP) per pole.
200-amp lighting and appliance branch-circuit panelboard with two 100-amp main CBs. CB No. 1 supplies branch-circuit overcurrent device in same panel. CB No. 2 can feed an external load or a like number of branch-circuit overcurrent devices in same panel.

Fig. 408-13. This application is permitted, subject to the old 42-circuit limitation, which, on the literal text, includes the mains. Note that the second main circuit breaker, at the upper left, is permitted to feed a second bus structure within the common enclosure under this exception, and some designs do use this arrangement. (Sec. 408.36, Exception No. 2.)

3. An existing panelboard with up to six circuit breakers may continue to be used without individual protection, provided it is used as service equipment for an individual dwelling unit. Figure 408-14 is an example of one of these old "split-bus" panels.

The individual protection required for panelboards generally may be placed at any location, provided the buses have the required protection. Figure 408-15 shows the possibilities.

The rule of part **(A)** of this section is covered in Fig. 408-16. Any panel, a lighting panel or a power panel, which contains snap switches (and CBs are not snap switches) rated 30 A or less, must have overcurrent protection and not in excess of 200 A.

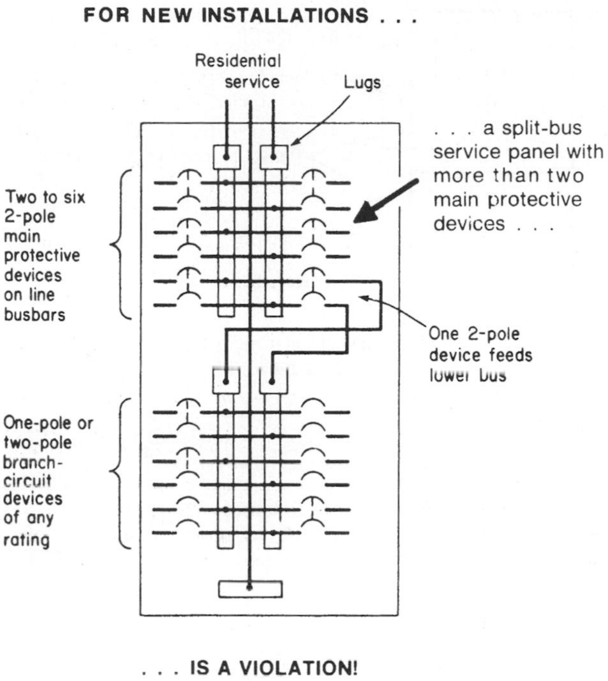
FOR NEW INSTALLATIONS . . .
Residential service
Lugs
. . . a split-bus service panel with more than two main protective devices . . .
Two to six 2-pole main protective devices on line busbars
One 2-pole device feeds lower bus
One-pole or two-pole branch-circuit devices of any rating
. . . IS A VIOLATION!

Fig. 408-14. Use of a split-bus loadcenter with more than two main overcurrent devices is permitted for residential service equipment *only* in "existing installations." (Sec. 408.36, Exception No. 3.)

Part **(B)** of this section applies to a panelboard fed from a transformer. The rule requires that overcurrent protection for such a panel, as required in **(A)** and **(B)** of the same section, must be located on the secondary side of the transformer. An exception is made for a panel fed by a 2-wire, single-phase transformer secondary and for "delta-delta" transformers. Such a panel may be protected by a primary-side OC device, as covered in 240.21(C)(1).

This concept of generally prohibiting use of panel protection located on the primary side of a transformer feeding the panel is consistent with the rules of 240.4, which cover conductor protection. Generally, the Code requires properly rated OC protection on the secondary side of the transformer for the secondary conductors because the primary-side device will not afford adequate protection. Similarly, the OC device on the primary side cannot adequately protect the panelboard on the secondary side. But where a delta-delta transformer is used, the per-unit current value on the secondary will be equal to or less than the per-unit current on the primary side. Therefore, with the delta-delta configured transformers, the secondary conductors and panelboard can be safely protected by the primary overcurrent protective device.

Part **(C)** is a rule prohibiting the installation of any 3-phase disconnect or 3-phase overcurrent device in a single-phase panelboard. It is now required that

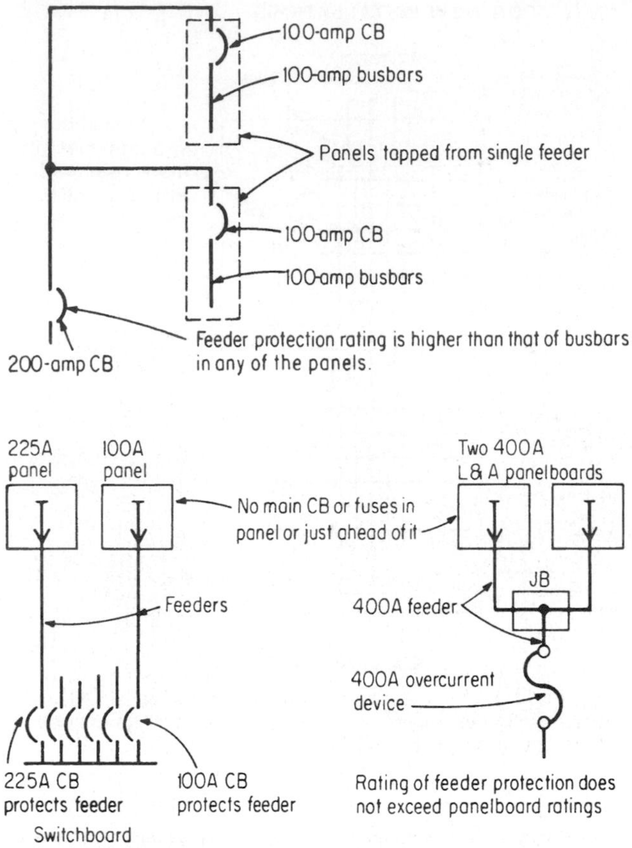

100-amp CB
100-amp busbars
Panels tapped from single feeder
100-amp CB
100-amp busbars
Feeder protection rating is higher than that of busbars in any of the panels.
200-amp CB
225A panel
100A panel
No main CB or fuses in panel or just ahead of it
Two 400A L& A panelboards
JB
Feeders
400A feeder
400A overcurrent device
225A CB protects feeder
100A CB protects feeder
Switchboard
Rating of feeder protection does not exceed panelboard ratings

Fig. 408-15. Panel protection may be provided in a variety of ways. (Sec. 408.36.)

any 3-pole disconnect or 3-phase protective device supplied by the bus within a panelboard may be used only in a 3-phase panelboard. The effect of this rule is to outlaw the so-called delta breaker, which was a special 3-pole CB with terminal layouts designed to be used in a single-phase panel fed by a 3-phase, 4-wire, 120/240-V delta supply where the loads served by the panel were predominantly single-phase, but where a single 3-phase motor or 3-phase feeder was needed and could readily be supplied from this type of delta breaker. The delta CB plugged into the space of three single-pole breakers, with the high leg of delta feeding directly through one pole of the common-trip assembly. The unit was used to protect the motor branch circuit or feeder to a 3-phase panel, rated up to 100 A.

Use of delta breakers has been found hazardous. When a delta breaker is used in a single-phase panel and the main disconnect for the single-phase panel is opened, there is still the high hot leg supplying the delta breaker. These breakers were likely to backfeed voltage from their load connections into the panelboard buswork even when the panel main was off, because the main did not interrupt the third phase.

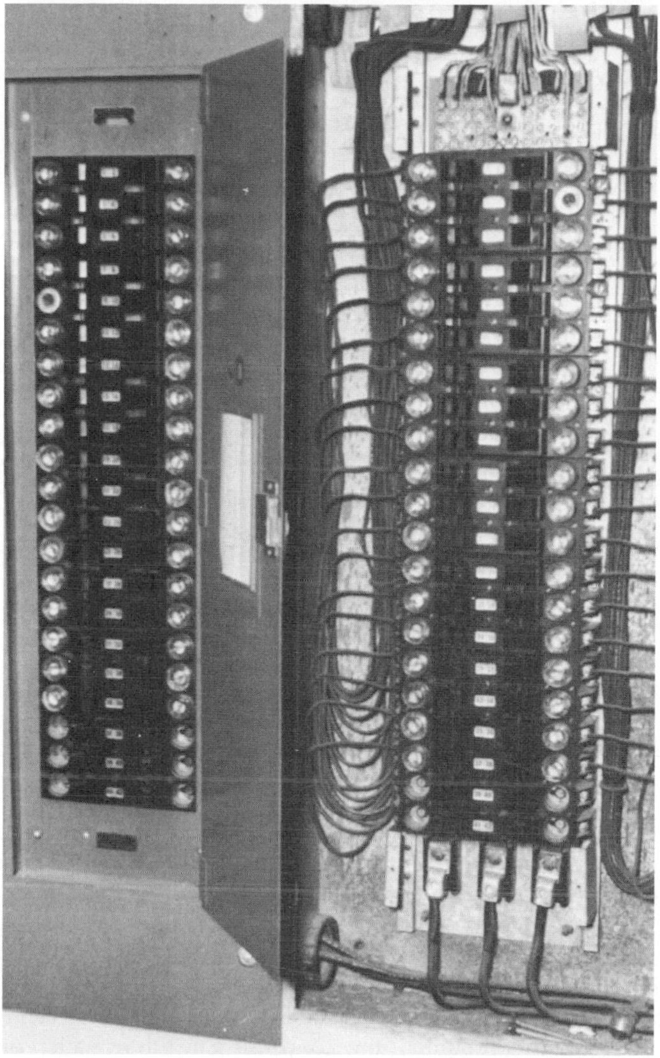

Fig. 408-16. Any panelboard containing snap switches rated 30 A or less must have main or feeder protection rated not over 200 A. [Sec. 408.36(A).]

As noted in part **(D),** a plug-in circuit breaker that is connected for "backfeed" (with the plug-in stabs being the load side of the CB) must be mechanically secured in its installed position. This rule requires all plug-in-type protective devices (CB or fusible) and/or main lug assemblies, in panelboards, to have some mechanical means that secure them in position, requiring more than just a pull to remove the devices. This is intended to eliminate the hazard of exposed, energized plug-in stabs of a device that is readily dislodged from its plug-in or connected position (Fig. 408-17).

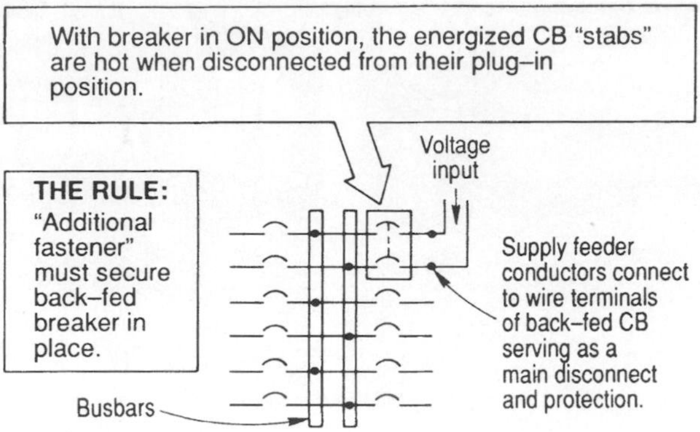
With breaker in ON position, the energized CB "stabs" are hot when disconnected from their plug–in position.
Voltage input
THE RULE:
"Additional fastener" must secure back–fed breaker in place.
Supply feeder conductors connect to wire terminals of back–fed CB serving as a main disconnect and protection.
Busbars

Fig. 408-17. This is an important safety rule for plug-in breakers. [Sec. 408.36(D).]

408.37. Panelboards in Damp or Wet Locations. UL data supplement the Code rules:

Enclosed panelboards marked "Raintight" will not permit entry of water when exposed to a beating rain. Enclosed panelboards marked "Rainproof" will not permit a beating rain to interfere with successful operation of the apparatus within the enclosure but may permit entry of water.

But note this carefully: In 408.37, the Code references 312.2. That rule requires that panelboard enclosures in "*damp or wet*" locations must be placed or equipped to "prevent moisture or water from *entering and accumulating* within" the enclosure, and there must be at least a 6-mm (¼-in.) air space between the enclosure and the wall or surface on which a metallic enclosure (or nonmetallic enclosure on an absorbent surface) is mounted. When they are installed exposed outdoors or in other wet locations, the NEC requires that panelboard enclosures must be weatherproof. The NEC definition of *weatherproof* is similar to the NEC and UL definitions of *rainproof.* Yet, NEC 312.2 requires exclusion of water entry—which clearly demands a "raintight" enclosure for outdoor, exposed panelboards (and not "rainproof"). These same considerations apply to other cabinets or enclosures used outdoors.

408.38. Enclosure. Because a panelboard (refer to Art. 100 definitions) is technically the busbar assembly and not the cabinet or cutout box in which the busbar assembly is customarily mounted, the UL Guide Card information on this category allows for field installation of panelboard bus assemblies in enclosures. This information goes on to explain that if that is how the panel was shipped, UL cannot assess gutter space within the enclosure. Therefore it would be judged by the electrical inspector in the field according to its compliance with Part IV of Art. 408 and other applicable requirements. Panelboards shipped in the usual way with enclosures bear the marking "Enclosed Panelboards."

408.39. Relative Arrangement of Switches and Fuses. For service equipment, switches are permitted on either the supply side or the load side of the fuses. In all other cases, if the panelboards are accessible to other than qualified persons, 240.40 requires that the switches shall be on the supply side so that when replacing fuses, all danger of shock or short circuit can be eliminated by opening the switch

408.40. Grounding of Panelboards. The effect of this rule is to *require* a panelboard to be equipped with a terminal bar for connecting all equipment grounding conductors run with the circuits connected in the panel. Such a bar must be one made by the manufacturer of the panel and must be installed in the panel in the position and in the manner specified by the panel manufacturer—to ensure its compliance with UL rules, as well as the NEC. The terminal bar for connecting equipment grounding conductors may be an inherent part of a panelboard, or terminal bar kits may be obtained for simple installation in any panelboard. Homemade or improvised grounding terminal bars are contrary to the intent of this Code section.

The terminal bar that is provided for connection of equipment grounding conductors must be bonded to the cabinet *and* frame of a metal panelboard enclosure. If such a panel enclosure is nonmetallic, the equipment grounding terminal bar must be connected to the equipment grounding conductor of the feeder supplying the panel.

Equipment grounding conductors must not be connected to terminals of a neutral bar—unless the neutral bar is identified for that purpose and is in a panel where Art. 250 requires or permits bonding and grounding of the system neutral (or grounded) conductor. These applications include a service panel or a panel fed from another building as part of an existing premises wiring system (250.24 or 250.32), or a panel fed from a separately derived system where the system disconnect is at the panel, and the bonding connection to the grounded circuit conductor is done at the system disconnect and not at the system source (250.30).

Figure 408-18 shows some details of grounding at panelboards. There have been many field problems relative to terminating grounding conductors in panelboards where nonmetallic wiring methods have been involved. The rule here requires an "equipment grounding terminal bar" in such panels so that these grounding conductors can be properly terminated and bonded to the panel.

In other than service equipment, the grounding conductor terminal bar must not be connected to the neutral bar (i.e., the neutral bar must not be bonded to the panel enclosure). Refer to 250.32, 250.30, 250.24, and 250.142. In a service panel, with the neutral bonded to the enclosure, equipment grounding conductors may, as previously noted, be connected to the bonded neutral terminal bar (or block). Note that manufacturers are only obliged to supply enough neutral and equipment grounding terminals to supply the minimum number of connections for an installation to meet some of the NEC wiring methods. Beyond that, the number of terminals supplied reflects the manufacturer's marketing goals. In other words, because a panel could be used to supply a house wired with EMT (and they do exist) which would require zero equipment grounding connections for separate wires, manufacturers are free to supply (but probably would not supply) a neutral bus that only supports white wire terminations. The field electrician must verify that the connections are sufficient and order a separate equipment grounding terminal bar of sufficient size if they are not. See also the discussion at 408.41 for more information on this topic.

The exception to this section allows an isolated ground conductor run with the circuit conductors to pass through the panelboard without being connected to the panelboard grounding terminal bar, in order to provide for the reduction of electrical noise (electromagnetic interference) on the grounding circuit as provided in 250.146(D).

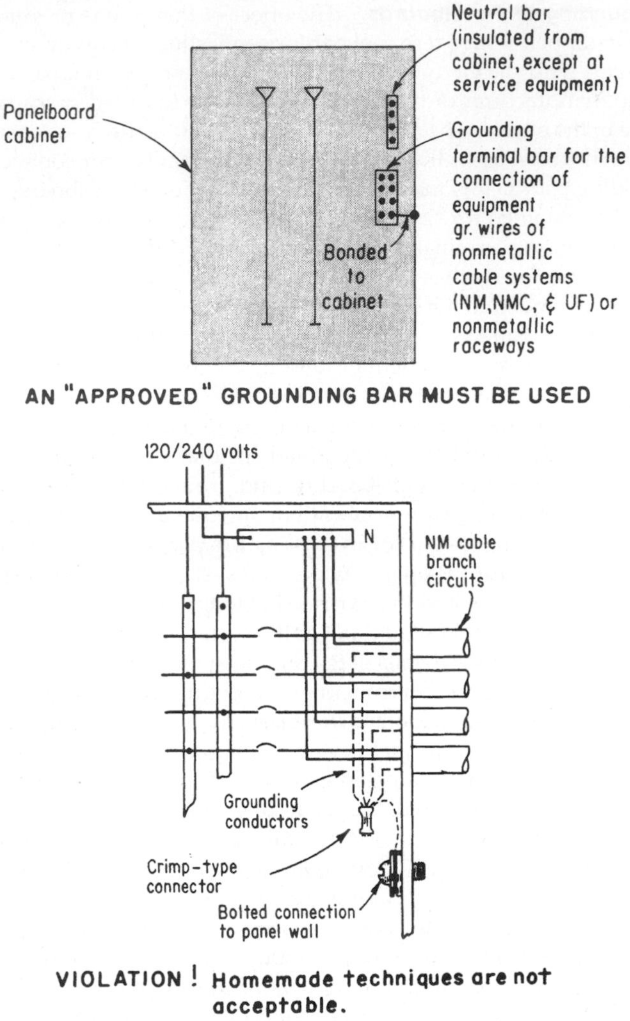
Neutral bar (insulated from cabinet, except at service equipment)
Panelboard cabinet
Grounding terminal bar for the connection of equipment gr. wires of nonmetallic cable systems (NM, NMC, & UF) or nonmetallic raceways
Bonded to cabinet
AN "APPROVED" GROUNDING BAR MUST BE USED
120/240 volts
N
NM cable branch circuits
Grounding conductors
Crimp-type connector
Bolted connection to panel wall
VIOLATION! Homemade techniques are not acceptable.

Fig. 408-18. Grounding in panelboards must use "identified" components. (Sec. 408.40.)

In order to maintain the isolation of the grounding wire necessary for a low-noise ground, the grounding wire must be connected directly to the grounding terminal bar in the service-entrance equipment if the service equipment is within the same building. To do this it may be necessary for the grounding wires to pass through one or more panelboards, but the grounding conductor *must not* leave the building in which the isolated ground is installed. Of course, such isolated grounding conductors may be spliced together by use of a terminal block installed in the panel but insulated from conductive contact with the metal enclosure of the panel. A "quiet ground" keeps grounding conductors apart from and independent of the metal raceways and enclosures (Fig. 408-19).

Fig. 408-19. "Quiet ground" terminal block for equipment grounding conductors provides for carrying isolated grounding conductors from circuits back to service bonded neutral, with single grounding conductor connecting terminal bar back to service. Terminal block is insulated from metal panel enclosure. Check with manufacturer to assure that isolated bar is approved. (Sec. 408.40.)

408.41. Grounded Conductor Terminations. Although in a service panelboard grounded conductors (white or gray) and equipment grounding conductors (green or bare) often arrive on the same terminal bar and with identical terminating provisions, actual permitted field practice considers them quite differently because only the grounded conductors are routinely current-carrying. The terminations for these conductors must not be opened inadvertently through service on another circuit. In addition, a terminal for a current-carrying conductor undergoes a heat-cycling test that is difficult to pass when multiple conductors are connected at the same point. For these reasons, all panelboard labels issued over the last 40 years or more have limited these current-carrying connections (but usually not small equipment grounding conductor connections) to one wire per termination. Many people never read those instructions, and so, just as 110.14(C) put termination temperature limitations that existed for decades into the NEC where they will be seen and enforced, 408.41 has taken similar requirements and brought them into the NEC where they will be seen and enforced. The exception refers to large, specially-shaped terminating openings for parallel conductors (therefore, 1/0 AWG and larger) that are specifically evaluated for such applications.

This limitation does not apply to equipment grounding terminations, because they are not current-carrying. This means, for applications where grounded and grounding conductors are permitted to be connected, that the same busbar might have one terminal with one white 14 AWG wire (and never more), and next to it an identical terminal with one, two, or three (depending on the instructions that come with the equipment; two is usual and three is common for small conductors) bare 14 AWG wires from NM cable. This is not a mistake, and the directions serve a purpose. If, after all the grounded conductor positions are used up (commonly the case) an additional equipment grounding terminal bar is necessary to comply with these rules; make sure that only equipment grounding conductors are terminated on it, even if it is in a service enclosure. Terminating a grounded conductor on a separate grounding bar forces its current to use the enclosure as a current-carrying conductor between the terminal bar and the service or system neutral. This is exactly what 200.2(B) is designed to prevent.

408.54. Maximum Number of Overcurrent Devices. Although manufacturers are now free to design panels with as many number of overcurrent device positions as they think appropriate, no manufacturer is to be responsible for field

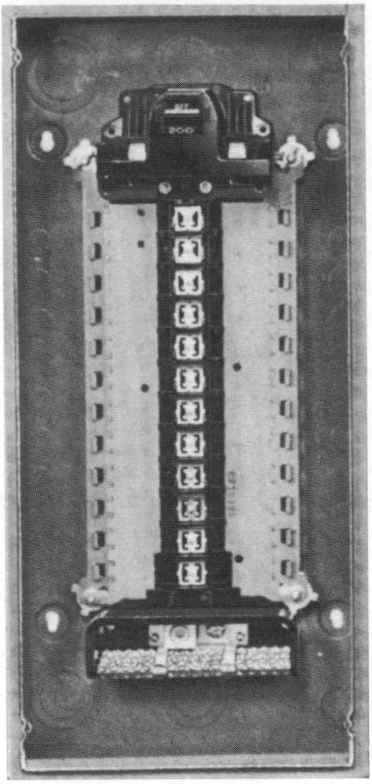

Fig. 408-20. Slots for push-in CB units have different configurations to limit the total number of poles to no more than 42. This is a CTL panelboard (or loadcenter). (Sec. 408.54.)

installations of more overcurrent devices than the panel was designed to hold. Figure 408-20 illustrates a panelboard with a 200-A main that provides for the insertion of class CTL overcurrent devices. The top stab receivers are of an F-slot configuration. Each F slot will receive only one breaker pole. The remainder of the slots are of an E configuration, which will receive two breaker poles per slot. Thus there is provision for installing not more than 42 overcurrent devices, which does not include the main CB in this case. If tandem breakers were put in slots intended for only full-size breakers, the manufacturer's design limitations would be defeated by an untested and very likely unsafe circuit fill. This is the reason for the physical rejection rule.

Class CTL is the Underwriters Laboratories, Inc., designation for the Code requirement for circuit limitation within a lighting and appliance branch-circuit panelboard. It means "circuit-limiting." The circuit-limiting (Class CTL) requirement first entered the **NEC** in its 1965 edition. In order to provide replacements for breakers already in use in panels that were installed previous to this time, UL continues to list the old designs, but they must be prominently marked to indicate that they are not CTL rated and are for replacement purposes only in non-CTL assemblies. Figure 408-21 shows two tandem breakers (two breakers designed to fit in one normal circuit position) from the same manufacturer. One is CTL-rated for current panels, and has a rejection feature that only allows its use where the manufacturer intends it to be applied. The other has no rejection feature, being

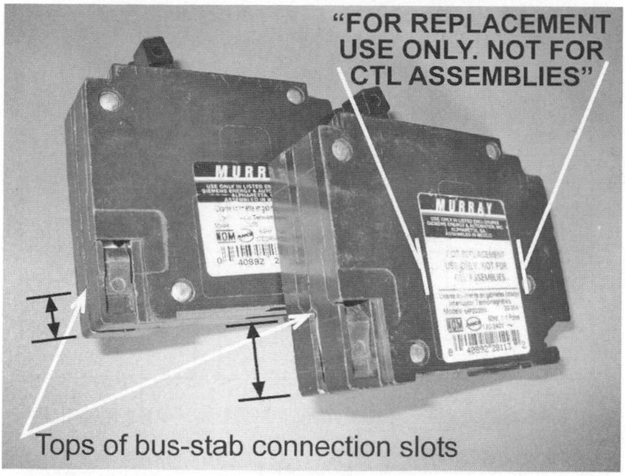
"FOR REPLACEMENT
USE ONLY. NOT FOR
CTL ASSEMBLIES"
MURRAY
Tops of bus-stab connection slots

Fig. 408-21. A 20-A Class CTL tandem (left) and a non-Class CTL tandem (right) circuit breaker shown for comparison. The CTL breaker set has a restricted-height bus connection slot (compare the dimension markings) that will only mount where a bus stab has been slotted by the panelboard manufacturer. (Sec. 408.54.)

thereby capable of occupying any position or any number of positions within a panel, easily defeating the intent of the limitation. They should never be used except on very old panels and only to the extent absolutely necessary.

408.55. Wire Bending Space Within an Enclosure Containing a Panelboard. This section of the Code correlates wire bending space at terminals in panelboards to the basic concepts of 312.6, as follows:

The basic rule requires the wire bending space at the top *and* bottom of a panel to satisfy the distances called for in Table 312.6(B), regardless of position of conduit entries. Exceptions are:

1. For a "42-circuit (maximum) panelboard rated 225 A or less," *either* the top *or* the bottom bending space may conform to Table 312.6(A)—but the other space at top or bottom (whichever is the terminal-lug space) must comply with Table 312.6(B).
2. For any panelboard, *either* the top *or* bottom bending space may conform to Table 312.6(A), provided that at least one of the side wiring terminal spaces satisfies Table 312.6(B), based on the largest conductor terminated in that space.
3. Depth of the wire bending space at the top and bottom of a panel enclosure may be as given in Table 312.6(A) rather than Table 312.6(B)—which is a deeper space requirement; but this may be done only where the panel is designed and constructed for a 90° bend (an L bend) of the conductors in the panel space and the panelboard wiring diagram is marked to show and describe the acceptable conditions of hookup.
4. Either the top or bottom space (but not both) can meet the more forgiving Table 312.6(A) where no conductors are terminated, in that space.

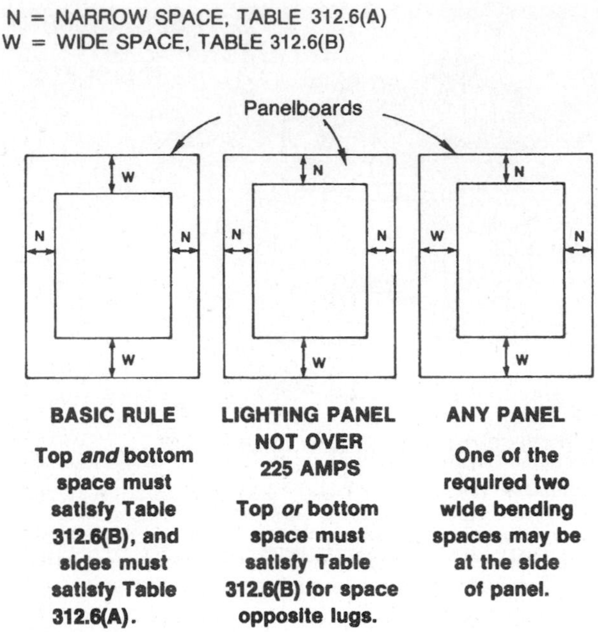
N = NARROW SPACE, TABLE 312.6(A)
W = WIDE SPACE, TABLE 312.6(B)
Panelboards
W
N
N
W
N
N
N
W
N
W
N
W
BASIC RULE
Top and bottom space must satisfy Table 312.6(B), and sides must satisfy Table 312.6(A).
LIGHTING PANEL NOT OVER 225 AMPS
Top or bottom space must satisfy Table 312.6(B) for space opposite lugs.
ANY PANEL
One of the required two wide bending spaces may be at the side of panel.

Fig. 408-22. This summarizes the requirements on wire-bending space at terminals where the feeder conductors supply a panelboard. (Sec. 408.55.) The reference to a "lighting" panel is one limited to 42 overcurrent devices.

Figure 408-22 shows the basic rule and Exceptions No. 1 and No. 2. Of course, the rules of 312.6(B) must be fully satisfied in the choice of location for the wire bending space. And the size of the largest conductor determines the minimum required space for all applications of the tables. Note that the title of this section reflects the fact that panelboards are actually contained within cabinets (refer to the Art. 100 definition) and therefore the wire bending space technically applies to the cabinet and not the panelboard.

The 2014 NEC has divided the rules for bending spaces in panelboards by location. The principal focus was and remains the top and bottom spaces as just covered, which are now (A). The single-side wire-bending space is now (**B**) and allows these gutters to have a width corresponding to the smaller dimensions of Table 312.6(A) based on the largest conductor terminated therein.

Part (**C**), new as of the 2014 NEC, is illustrated in Fig. 408-23. It covers the allowable entry points at the back of a panelboard enclosure, and two dimensions need to be considered. First, the rear entry is not permitted if the depth of the panel from the back wall to the removable cover fails to meet the minimum bend radius in Table 312.6(A) for one wire per terminal. This is the same rule as the one covering comparable work in 314.28(A)(2) Exception. Second, the distance between the center of the knockout and the nearest termination point of the entering conductor must comply with that given in Table 312.6(B). Note that in this case the word "one wire per terminal" are absent; if the raceway entry is comprised of a parallel makeup, then the single-wire column in Table 312.6(B) will not apply.

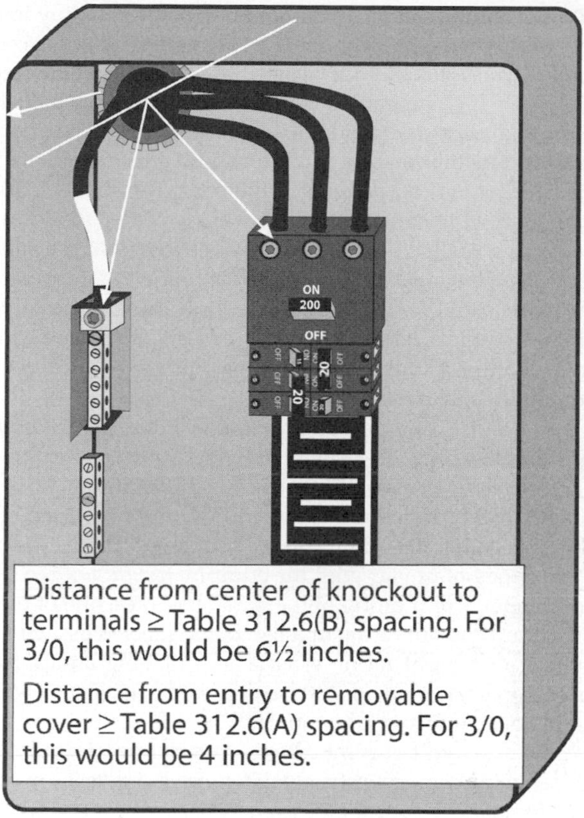
ON
200
OFF
Distance from center of knockout to terminals ≥ Table 312.6(B) spacing. For 3/0, this would be 6½ inches.
Distance from entry to removable cover ≥ Table 312.6(A) spacing. For 3/0, this would be 4 inches.

Fig. 408-23. The NEC now addresses the permitted location of rear entries in panelboard cabinets [408.55(C)]. The drawing uses dimensions for copper wire; the 312.6(B) spacing for 4/0 aluminum would be 5½ in., and the 312.6(A) spacing would be 3½ in.

\RTICLE 409. INDUSTRIAL CONTROL PANELS

09.1. Scope. This new article covers industrial control panels, and these rules ignificantly impact how these panels are assembled, and by whom. Refer to the efinition in Art. 100. Most of the article provisions duplicate other NEC rules and re covered elsewhere, primarily in Art. 430. The coverage here focuses on provi ons that are unique. The terminology is consistent with that used by testing labs, ıch as UL, and covered by NEMA/UL Standards 508 and 508A.

09.20. Conductor—Minimum Size and Ampacity. The wording here mimics ɪe wording of other Code rules that govern the minimum size and ampacity for eder conductors supplying heating and motor loads—either alone, in combina- ɔn with each other, or in combination with other loads.

9.21. Overcurrent Protection. Part B gives location requirements that are the me as those for panelboards and switchboards; note that they apply to each coming supply circuit and therefore encompass control panels with multiple ıpply circuits.

409.22. Short-Circuit Current Rating. Although this was certainly implied, no rule prior to the 2011 **NEC** specifically prohibited applying an industrial control panel in an environment for which it could not handle the available fault current. This, after all, was the reason 409.110(4) required the current rating to be calculated and marked. The 2017 **NEC** initiated a documentation requirement on the calculation as well.

409.110. Marking. The marking requirements will affect anyone who installs or significantly modifies these components in the field. The short-circuit current rating developed as part of this process will govern whether the industrial control panel is allowed to be installed based on the fault current that is available at its line terminals. Specifically, the label must include the following information:

1. The name of the manufacturer, or a trademark, or other descriptive marking by which the organization responsible for the product can be identified.
2. The supply voltage, number of phases, frequency, and full-load current for each incoming supply circuit.
3. If there is more than one power source and more than one disconnecting means required to fully disconnect the power, the control panel must be marked accordingly. The 2017 **NEC** clarified the requirement as pertaining to circuits of 50 volts or more, and also required that the location of these disconnects be "documented and available" so they could be operated quickly, even by someone unfamiliar with the equipment.
4. The short-circuit current rating of the panel based on one of the following:
   a. Short-circuit current rating of a listed and labeled assembly.
   b. Short-circuit current rating established utilizing an approved method. (An explanatory note follows, calling out *UL 508A, Suppl. SB* as an example of an approved method.)

   Note: A "short-circuit current rating" is a defined term in Art. 100. It is *"the prospective symmetrical fault current at a nominal voltage to which an apparatus or system is able to be connected without sustaining damage exceeding defined acceptance criteria."* This is a newer term for the older withstand values, and reflects an ability to continue to function after the fault.

   Exception: The short-circuit current rating marking *is not required* on an industrial control panel that contains only control circuit components and no power circuit components. This exception exists for the simple reason that *short-circuit current ratings are not assigned to components that are not in the power circuit*, and therefore if a panel is assembled with no power circuits entering, there would be no basis for determining such a rating. Such control panel would still be an industrial control panel, and must otherwise meet all the marking requirements, but the short-circuit current rating would not be included in those markings.
5. If the panel will be used as service equipment, it must be evaluated and marked accordingly.
6. The panel must include an electrical wiring diagram or the identification number of a separate such diagram or a designation referenced in a separate wiring diagram.
7. The enclosure type number (Table 110.20) must be marked on the enclosure

Most of these requirements are easily met, but the short-circuit current rating provisions are not. The wording here mimics the wording of other Code rules that govern the minimum size and ampacity for feeder conductors supplying heating and motor loads—either alone, in combination with each other, or in combination with other loads. If you are making these in the field, and industrial control electricians have been doing this every day, you will need to get a copy of *UL 508A, Suppl. SB* along with the control manufacturer's literature in order to meet these requirements. This now crucial publication for this work exempts some components such as voltmeters and reactors from short-circuit ratings, and describes how to apply established ratings for other components. The object is to assure that the components are not subjected to fault currents that will result in catastrophic failures. There are several pages of flow sheets that must be carefully followed in order to correctly arrive at a short-circuit current rating for one of these panels. Our sister book, the *American Electricians' Handbook*, contains copies of the three principal flowcharts, rearranged and reformatted to fit in this size book, but unchanged in terms of technical content.

ARTICLE 410. LUMINAIRES, LAMPHOLDERS, AND LAMPS

410.2. Definitions. This section includes definitions of "closet space" (applied in 410.16) and "lighting track" (applied in Part XV). These definitions are not new, but they were relocated because the **NEC** is moving toward standardized locations for certain material. Because it is more clear to discuss these terms in context, they will not be further discussed here. Note that the parenthetical expressions "(fixture)" or "(fixtures)" or "(lighting fixture)" have been dropped in this edition of the **NEC**, leaving "luminaire" as the correct terminology.

410.6. Listing Required. All luminaires must now be listed. This may turn out to be a major impediment to many historical reproductions of period lighting by small firms unable to afford the listing process, but it may assist overall quality. Time will tell whether this is a workable restriction, or overkill. The 2014 **NEC** has added retrofit kits to the listing requirement. Because these may involve extensive rearrangements of internal components, the directions that come with the listing must be followed and enforced.

410.8. Inspection. This was relocated from a subsection in the support section (now Sec. 36) because it applies to all luminaires, and therefore belongs to the general part of the article. The connections between the branch-circuit wiring and the luminaire supply conductors must be open for inspection without "requiring the disconnection of any part of the wiring" unless the luminaire is cord-and-plug-connected.

410.10. Luminaires in Specific Locations. Part **(A)** covers the kind of installations shown in Fig. 410-1. At left, the luminaire on the covered vehicle-loading dock is in a damp location and must be marked "SUITABLE FOR DAMP LOCATIONS" or "SUITABLE FOR WET LOCATIONS." At right, the luminaires at a vehicle-washing area are in a wet location and must be marked "SUITABLE FOR WET LOCATIONS"—unless the luminaires are so high mounted or otherwise protected so that there is no chance of water being sprayed on them.

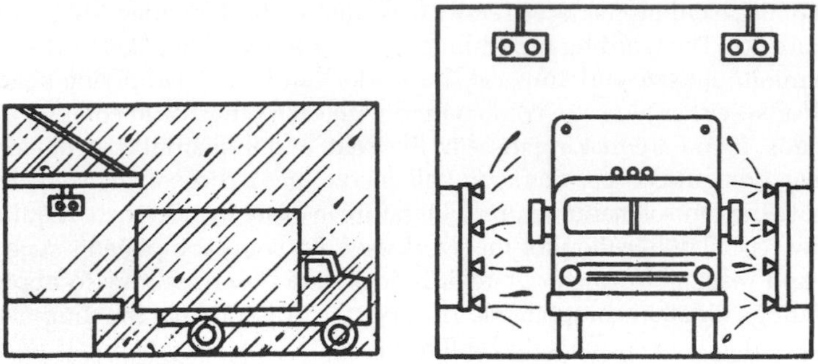

Damp location Wet location

Fig. 410-1. Luminaires must be marked as suitable for their place of application. (Sec. 410.10.)

An enclosed and gasketed luminaire would fulfill the requirement that water shall be prevented from entering the luminaire, although under some conditions water vapor might enter and a small amount of water might accumulate in the bottom of the globe.

Luminaires in the form of post lanterns, luminaires for use on service-station islands, and luminaires that are marked to indicate that they are intended for outdoor use have been investigated for outdoor installation.

An example of luminaires in "damp" locations would be those installed under canopies of stores in shopping centers, where they would be protected against exposure to rain but would be subject to outside temperature variation and corresponding high humidity and condensation. Thus, the internal parts of the luminaire need to be of nonhygroscopic materials, which will not absorb moisture and which will function under conditions of high humidity.

Part **(C)** recognizes use of luminaires in commercial and industrial ducts and hoods for removing smoke or grease-laden vapors from ranges and other cooking devices. The rule spells out the conditions for using luminaires and their associated wiring in all types of nonresidential cooking hoods. The requirement that such a luminaire be "identified for use" may be taken as listed by UL for such use, in view of 410.6. In fact, this is probably a misuse of the term "identified" and probably the intent is to require a marking on the product by its manufacturer indicating suitability for this use. Note that in addition to the requirement for special luminaires, the wiring method must not be exposed within the hood. Hoods wired properly have flush openings the size of outlet boxes so the boxes and the raceways with the high-temperature branch-circuit conductors are completely out of the air stream within the hood, and only the special luminaires remain exposed in the hood. See also the discussion at 402.11 on this point.

Part **(D)** covers the use of chandeliers, swag lamps, and pendants over bathtubs which could pose a potential hazard. Although there is no Code prohibition on use of luminaires over tubs and although a bathroom generally is not technically a damp or wet location, there is considerable concern over exposing persons in water to possible contact with energized parts. Where installed over a tub, a hanging luminaire or pendant must be at least 8 ft (2.5 m) above "the top of the tub." In addition, hanging luminaires must be excluded from a zone 3 ft (900 mm) horizontally and 8 ft (2.5 m) vertically from the top of a bathtub rim. This defines the volume of space from which a chandelier-type luminaire is excluded above and around a bathtub. This excludes the entire luminaire and its cord or chain suspension. Such application is difficult, if not impossible, in most interiors.

Part **(E).** Metal halide or mercury vapor luminaires subject to damage and used in sports venues or mixed-use facilities, whether over the court or over spectator seating area, must be designed to protect the lamp with a glass or plastic lens. Additional protection in terms of a wire or other guard is also permitted. This rule protects spectators and players from shards of broken glass and direct UV exposure that is radiated from the inner arc containment when the outer bulb is broken if the lamp continues to operate.

Part (E) is new as of the 2014 NEC and requires luminaires mounted under roof decking to be spaced at least 1½ in. below the low point on the decking to the top of the luminaire. This rule, which mirrors the one in 300.4(E), will help prevent damage from the widespread indiscriminate usage of long decking screws during reroofing operations.

410.11. Luminaires Near Combustible Material. This section provides the basis for an inspector to review a luminaire placement in cases where adjacent combustibles might become overheated. Note that this section does not literally apply to a lampholder. For that reason, the 2011 NEC inserted a new section (410.97) covering lampholders and imposing this requirement. The burn radius of a 60-W incandescent lamp is about 100 mm (4 in.).

410.12. Luminaires Over Combustible Material. This refers to pendants and fixed lighting equipment, not to portable lamps. Where the lamp cannot be located out of reach, the requirement can be met by equipping the lamp with a guard.

410.16. Luminaires in Clothes Closets. The intent is to prevent lamps from coming in contact with cartons or boxes stored on shelves and clothing hung in the closet, which would, of course, constitute a fire hazard. Use of luminaires in clothes closets is covered by this section, with specific rules based on the definition of "storage space" in a closet as given in 410.2 and the associated isometric drawing that illustrates the locations and dimensions of regulated space within the closet. Figure 410-2 shows the default storage space dimensions in one dimension. There are some aspects to the definition that are not widely appreciated and require emphasis:

1. The lower storage space (600 mm [2 ft] deep) does not simply run from right to left along the back closet wall, even if that is the only location of a closet pole. This space also covers both side walls of the closet, extending all the way to the front. It also covers from the floor to the height of the highest closet pole if one exists and to a height of 1.8 m (6 ft) if there is no pole.
2. The upper storage space (300 mm [2 ft] deep) sits above all applicable lower storage spaces, running from the rear and side walls out 300 mm (1 ft), and all the way up to the ceiling however high that may be. It begins 1.8 m (6 ft) up from the floor whether or not a shelf is provided. If shelving is provided and is deeper than 300 mm (1 ft), then the upper storage space deepens to the same extent.

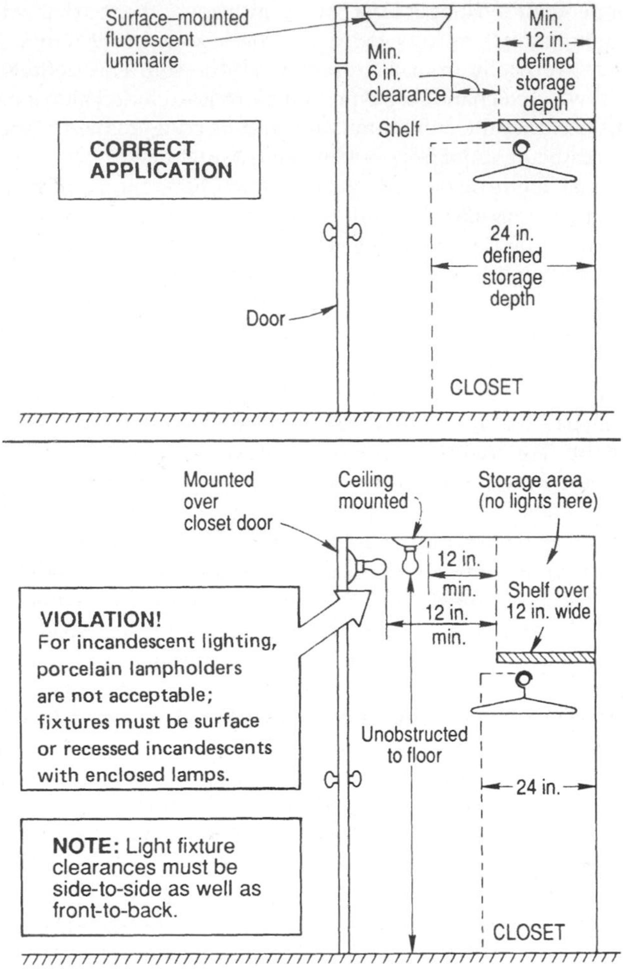
Surface-mounted fluorescent luminaire
Min. 6 in. clearance
Min. 12 in. defined storage depth
Shelf
CORRECT APPLICATION
24 in. defined storage depth
Door
CLOSET
Mounted over closet door
Ceiling mounted
Storage area (no lights here)
12 in. min.
12 in. min.
Shelf over 12 in. wide
VIOLATION!
For incandescent lighting, porcelain lampholders are not acceptable; fixtures must be surface or recessed incandescents with enclosed lamps.
Unobstructed to floor
24 in.
NOTE: Light fixture clearances must be side-to-side as well as front-to-back.
CLOSET

Fig. 410-2. These clearances apply to luminaires in closets. (Sec. 410.16.)

3. If the closet has doors on both sides so there is no rear wall, the lower spac is 300 mm (1 ft) on both sides (i.e., front and back) of the closet pole, with al other dimensions being the same.

This section presents rules that can be divided into two categories as follows:

1. Parts **(A)** and **(B)** describe the kind of luminaires that may be used and th kinds that are prohibited from use in closets. Incandescent or LED lumi naires with lamps that are not completely enclosed, or any pendant lumi naire or lampholder, are excluded regardless of the size of the closet and th position of the light source. Although not quite properly worded, it appear the same is intended to apply to an incandescent lampholder that is not pa of a pendant, and such lampholders do not appear in the list of permitte

locations in **(C)** following. On the other hand, both surface-mounted and recessed incandescent luminaires with completely enclosed lamps are permitted, along with fluorescent luminaires, either recessed or surface mounted, and whether or not the lamps are exposed. In addition, there is a new allowance for fluorescent and for LED luminaires to be accepted even within a storage area, but only if so "identified" (probably intended to mean "marked" because all luminaires must now be listed and any listed luminaire for some purpose automatically meets the definition of "identified" in Art. 100).

2. Part **(C)** defines four permitted locations for luminaires that depend on spacing from the storage area and on the type of luminaire. A fifth location, new in 2008, is in the storage area and will be discussed last. Because the defined storage areas extend to the closet ceiling on both side walls and the rear wall, the only possible locations left for luminaire positions are over the door (or just to either side if spacing to the side wall is available) or on the ceiling. The possibilities, expressed as clearances between the outer margins of the luminaires and the storage areas, are as follows:
   a. For surface-mounted incandescent (or LED) luminaires with a completely enclosed light source, 300 mm (12 in.)
   b. For surface-mounted fluorescent luminaires, 150 mm (6 in.)
   c. For recessed incandescent (or LED) luminaires with a completely enclosed light source, 150 mm (6 in.)
   d. For recessed fluorescent luminaires, 150 mm (6 in.)

The last possibility is the LED or fluorescent luminaire specifically rated to actually be installed within a storage space; if such a luminaire is developed, this language provides the authorization for testing laboratories to proceed with a listing.

For small clothes closets, proper lighting may be achieved by locating luminaires on the outside ceiling in front of the closet door—especially in hallways where such luminaires can serve a dual function. Flush recessed luminaires with a solid lens are considered outside of the closet (but still subject to the spacing rules) because the lamp is recessed behind the wall or ceiling line.

410.18. Space for Cove Lighting. Adequate space also improves ventilation, which is equally important for such equipment. This rule has been pushed beyond all reason by some installations. There are well-documented reports of fluorescent luminaires in cove lighting applications for which special tools and mirrors are required to service the ballasts, clearly amounting to a violation of this rule.

410.21. Temperature Limit of Conductors in Outlet Boxes. Luminaires equipped with incandescent lamps may cause the temperature in the outlet boxes to become excessively high. The remedy is to use luminaires of improved design, or in some special cases to use circuit conductors having insulation that will withstand the high temperature.

The first sentence of this rule is related to the rule of 410.11. Figure 410-3 shows how a luminaire may be "installed so" that the branch-circuit wires are not subjected to excessive temperature. That hookup relates to 410.117(C) for recessed luminaires, which requires a 1½- to 6-ft length of flex with high-temperature wire (say, Type FEP) to connect the hot luminaire junction point (150°C) to the

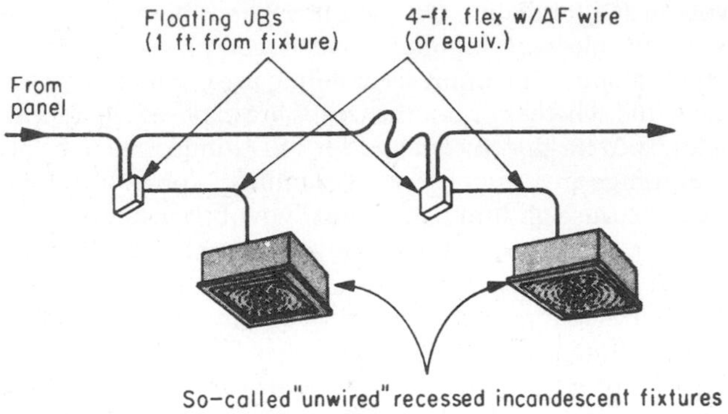
Floating JBs
(1 ft. from fixture)
4-ft. flex w/AF wire
(or equiv.)
From
panel
So-called "unwired" recessed incandescent fixtures

Fig. 410-3. Flex "whip" may be used to keep 60°C or 75°C wire away from 150°C terminal space in luminaires. (Sec. 410.21.)

lower-rated branch-circuit wires. These hookups are almost obsolete. Refer to the discussion at 410.117(C). However, the rule is very much alive in other ways. Refer to the discussion at 402.11 and 410.10(C) for one good example of this rule at work. Another frequent issue is the relative rarity today of luminaires that are not marked for 90°C supply connections. This is a real problem on rehab work where the NM cable was installed before the product standard revision in the 1980s and all the branch-circuit wiring is only rated 60°C. Any luminaire with a required supply conductor temperature rating above 60°C must be marked with the required temperature rating, in accordance with 410.74(A).

The second paragraph of this section applies to *prewired* recessed incandescent luminaires, which have been designed to permit 60°C (or 75°C, depending on the marking) supply conductors to be run into an outlet box attached to the luminaire. Such luminaires have been listed by UL on the basis of the heat contribution by the supply conductors at *not more than* the *maximum* permitted lamp load of the luminaire. Some luminaires have been investigated and listed by UL for *feed-through* circuit wiring. This does not mean what it may appear to mean, and a recent **NEC** change clarified how the luminaire industry, and this Code rule, use the terminology "through-wiring." It is frankly counterintuitive. "Through wiring" suitability is *not required* to daisy chain multiple luminaires together on the same branch circuit, or even on the same multiwire branch circuit. As long as the volume of the wiring compartment and the conductor fill markings accommodate the number of conductors required to go from one luminaire to the next on the circuits as described, no special marking is required. Only when yet an additional circuit, such as an unrelated lighting circuit or even a receptacle circuit passes through the luminaire wiring box is the "identified for through-wiring" provision activated (Fig. 410-4).

The current UL Luminaire Marking Guide includes the following information under the heading "Branch Conductors in Box":

THROUGH CONDUCTORS IN A WIRING COMPARTMENT—A luminair that is suitable for use with through branch conductors is marked "MAXIMUM O ___NO.___AWG THROUGH BRANCH CIRCUIT CONDUCTORS SUITABLE FO ___°C PERMITTED IN BOX.

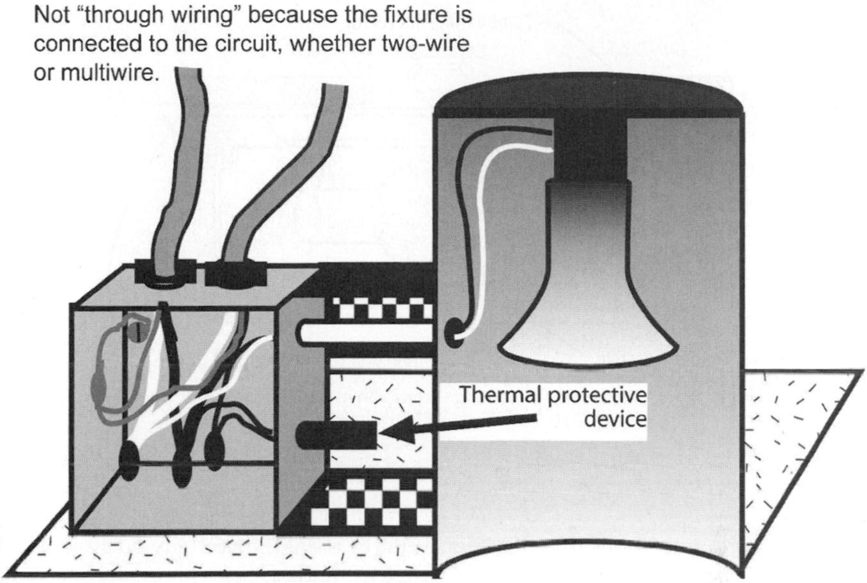
Not "through wiring" because the fixture is connected to the circuit, whether two-wire or multiwire.
Thermal protective device

Fig. 410-4. Care must be exercised in hooking up *prewired* types of luminaires. (Sec. 410.21.)

Note the word "THROUGH." This label applies to the number of "through" conductors (if any) and not the supply conductors. The branch-circuit conductors supplying the luminaire, including the associated conductors of a multiwire circuit, are not "through" conductors as defined in 410.21. The current **NEC** wording amended the 1996 **NEC** wording that said "Branch-circuit wiring shall not be passed through an outlet box that is an integral part of an incandescent fixture unless the fixture is identified for through-wiring." The substantiation for the proposal, which came from a panel member and manufacturer of these luminaires, stated in part:

> [This section] is commonly misinterpreted as applying to both branch circuits passing through a junction box and branch circuits supplying the fixture. This is a problem because the number of permitted through branch-circuit conductors is reduced by the number of wires in the circuit supplying the fixture. This negates the usefulness of the through-branch-circuit rating.

The proposal also broadened the applicability of the rule to other styles of luminaire beyond just incandescent luminaires because other types were being manufactured with the same provisions.

410.22. Outlet Boxes to Be Covered. This rule is similar to that of 314.25. The canopy may serve as the box cover, but if the ceiling or wall finish is of combustible material, the canopy and box must form a complete enclosure. The chief purpose of this is to require that no open space be left between the canopy and the box edge if the finish is wood or fibrous or any similar material.

410.23. Covering of Combustible Material at Outlet Boxes. If a luminaire canopy encloses over 180 sq. in. of a combustible surface, then that surface must be

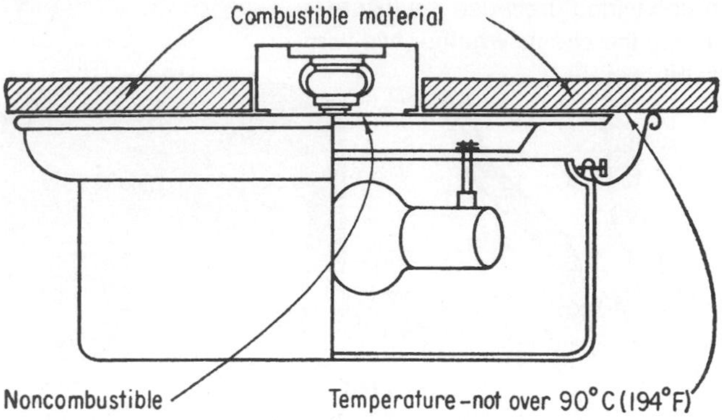
Combustible material
Noncombustible
Temperature-not over 90° C (194° F)

Fig. 410-5. A combustible mounting surface above a luminaire canopy must be protected if over 180 sq. in. is exposed to the inside of the canopy (410.23).

covered with noncombustible material. Prior to the 2014 **NEC** any such material had to be covered and this was a real problem. Because 180 sq. in. corresponds to a canopy (after allowing for about 10 sq. in. for the box at the center) of about 15-½ in. diameter, this problem has effectively disappeared. If the canopy is larger than this, then there aren't many good options short of visiting a sheet metal shop. Some manufacturers have round decorative backers (often for vinyl siding) or ceiling medallions that, depending on the canopy diameter, may fit acceptably.

410.24. Connection of Electric-Discharge and LED Luminaires. As 410.24 is worded, the rules presented apply to *both* indoor and outdoor applications of electric-discharge luminaires. The rules cover general lighting in commercial and industrial interiors as well as all kinds of outdoor floodlighting and area lighting. Part **(A)** of this section covers *only* connection of electric-discharge luminaires "... supported independently of the outlet box." Chain-hung luminaires, luminaires mounted on columns, poles, structures, or buildings, and any other luminaire that is not supported by the outlet box that provides the branch-circuit conductors to feed the luminaire are covered by part **(A).**

The basic rule requires a fixed or permanent wiring method to be used for supply to all "electric-discharge luminaires," which includes all luminaires containing mercury-vapor, fluorescent, metal-halide, high-pressure sodium, or low-pressure sodium lamps. BUT *incandescent* luminaires are *not* covered by 410.24. As a result, incandescent luminaires using cord connection are regulated only by 400.10(A)(2), 400.12(1), and 410.62(B).

Part **(B)** applies to surface-mounted fluorescent (and LED) luminaires where the ballast channel is not mounted *on* the box, and not designed to be supported solely by the box, but rather the luminaire is mounted *over* a concealed outlet box. Full access to the box must be afforded by opening the luminaire ballast channel. Usually this involves field punching the luminaire at the appropriate location with a metric designator 53 or 63 (trade size 2 or 2½) punch, depending on the box opening size. It does not mean relying on a metric designator 16 (trade size ½) knockout with a federal bushing to let the branch-circuit wiring into the ballast channel. In such a case the only access to the box is through completely removing the luminaire from the ceiling, in violation of this requirement.

410.30. Supports. Figure 410-6 shows a 3-kg (7-lb) luminaire shade that is 432 mm (17 in.) in diameter and supported by the screw-shell of the lamp and holder, on both counts violating the rule of part **(A)**.

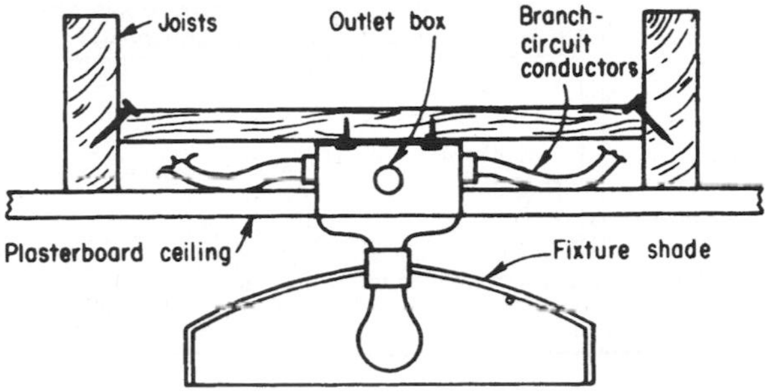
Joists
Outlet box
Branch-
circuit
conductors
Plasterboard ceiling
Fixture shade

Fig. 410-6. Luminaire shade assembly may be supported from a screw-shell lampholder if it is not too heavy or too big in diameter. [Sec. 410.30(A).]

In part **(B)** the rules cover use of metal poles for supporting luminaires. A metal pole supporting a luminaire must have a readily accessible handhole (minimum 2 by 4 in. [50 by 100 mm]) to provide access to the wiring within the pole or its base. A grounding terminal for grounding the metal pole must be provided and be accessible through the handhole. Any metal raceway supplying the pole from underground must be bonded to the pole with an equipment grounding conductor. And conductors run up within metal poles used as raceway must have vertical supports as required by 300.19 (Fig. 410-7).

As noted in Exception No. 1 to **(B)(1)**, a metal pole supporting a luminaire does not have to have an access handhole at its base if the pole is not over 8 ft (2.5 m) in height (such as a common post light) and the enclosed wiring is accessible at the luminaire end. This rule excludes the typical short (not over 8 ft [2.5 m] high) post light from the need for a wiring access handhole at its base. That handhole is important for higher poles used on commercial and industrial properties and does add safety. But it is unnecessary for a post light, and this exception allows omission of the handhole where the wiring runs "without splice or pull point" to a luminaire mounted on a metal pole not over 8 ft (2.5 m) high and where splices of the luminaire wires to the branch circuit supply conductors are accessible by removal of the luminaire. And, as noted in Exception No. 2, where the pole is provided with a hinged base and is *not* over 20 ft (6.0 m) tall, then the handhole may also be omitted. Such poles have the equipment grounding terminal in the base.

The effect of these rules is to recognize the pole as a raceway. Where such poles are used to also support power-limited circuits such as for the video signal from

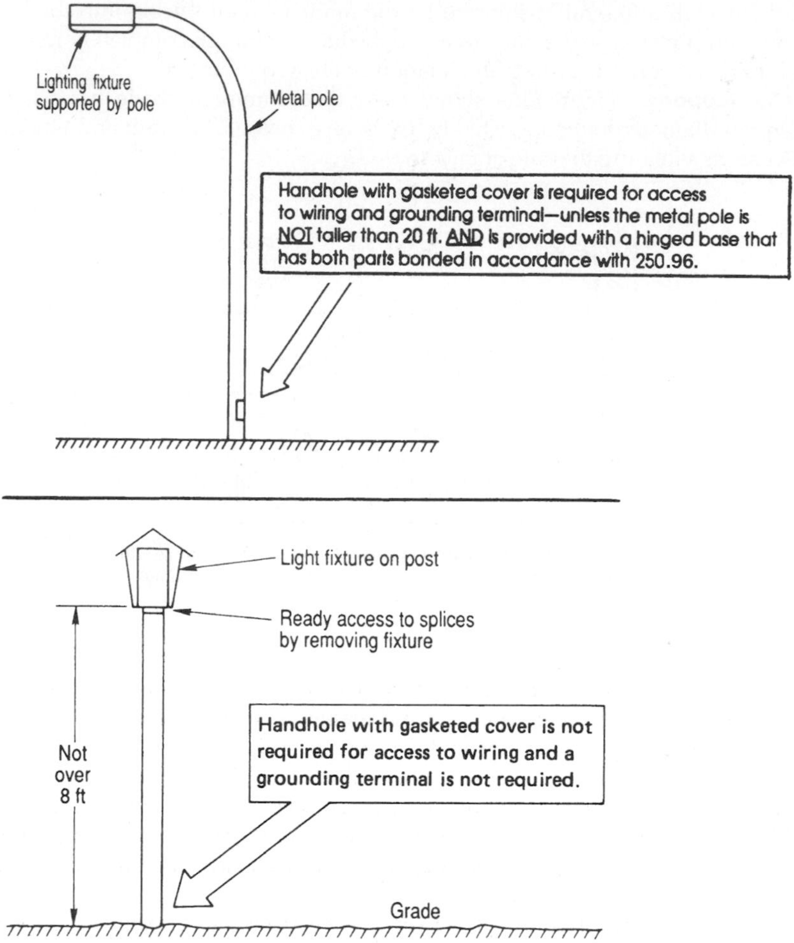
Lighting fixture supported by pole
Metal pole
Handhole with gasketed cover is required for access to wiring and grounding terminal—unless the metal pole is NOT taller than 20 ft. AND is provided with a hinged base that has both parts bonded in accordance with 250.96.
Light fixture on post
Ready access to splices by removing fixture
Not over 8 ft
Handhole with gasketed cover is not required for access to wiring and a grounding terminal is not required.
Grade

410.30(B)(1) Exception No. 1. This type of fixture–on–post does not need an access handhole at base of pole.

Fig. 410-7. Metal pole must provide access handhole and internal terminal for connecting grounding wire to metal of pole. [Sec. 410.30(B).]

a camera, and where the power circuit is run as individual conductors and not a Chap. 3 wiring method, full system separation must be maintained between the systems as required in 725.136 and other comparable requirements. Frequently this is addressed through the use of a flexible nonmetallic conduit to sleeve the power-limited wiring, because system separation in such cases must never depend on wire insulation alone.

410.36. Means of Support. A luminaire may be supported by attachment to an outlet box that is securely mounted in position (see 314.23), or a luminaire may be rigidly and securely attached or fastened to the surface on which it is mounted

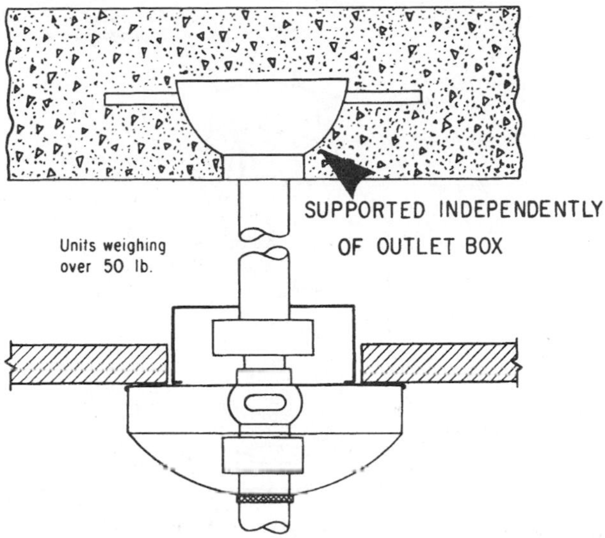
SUPPORTED INDEPENDENTLY
OF OUTLET BOX
Units weighing
over 50 lb.

Fig. 410-8. Any luminaire over 50 lb (22.7 kg) must be supported from the structure or some other means than the outlet box. (Sec. 410.36.)

or it may be supported by embedment in concrete or masonry. As shown in Fig. 410-8, heavy luminaires must have better support than the outlet box.

Various techniques are used for mounting luminaires independently of the outlet box, depending somewhat on the total weight of the individual luminaires. In general, pipe or rods are usually used to attach the luminaires to the building structure, and the electrical circuit is made by using flex between the luminaire and the outlet box concealed in the ceiling cavity. If provision is made for lowering the luminaire, by means of winch or otherwise, provision must also be made for disconnecting the electrical circuit.

The most common method of supporting luminaires is by means of fixture bars or straps bolted to the outlet boxes, as shown in Fig. 410-9. A luminaire weighing over 50 lb (23 kg) can be supported on a hanger such as is shown for boxes under 314.27 for a tile-arch ceiling. Care should be taken to see that the pipe used in the construction of the hanger is of such size that the threads will have ample strength to support the weight.

Any luminaire may be attached to an outlet box where the box will provide adequate support, but, as noted in 410.30(A), units that weigh more than 6 lb (3 kg) or exceed 16 in. (400 mm) in any dimension "shall not be supported by the screw-shell of a lampholder."

A normal method of securing an outlet box in place is to use strap iron attached to the back of the outlet box and fastened to studs, lathing channels, steel beams, and the like nearby. Lightweight units are sometimes attached to outlet box by means of screws passing through luminaire canopy, that thread into outlet box ears, or flanges, tapped for this purpose. For heavier luminaires, fixture studs, hickeys, tripods, or crowfeet are normally used.

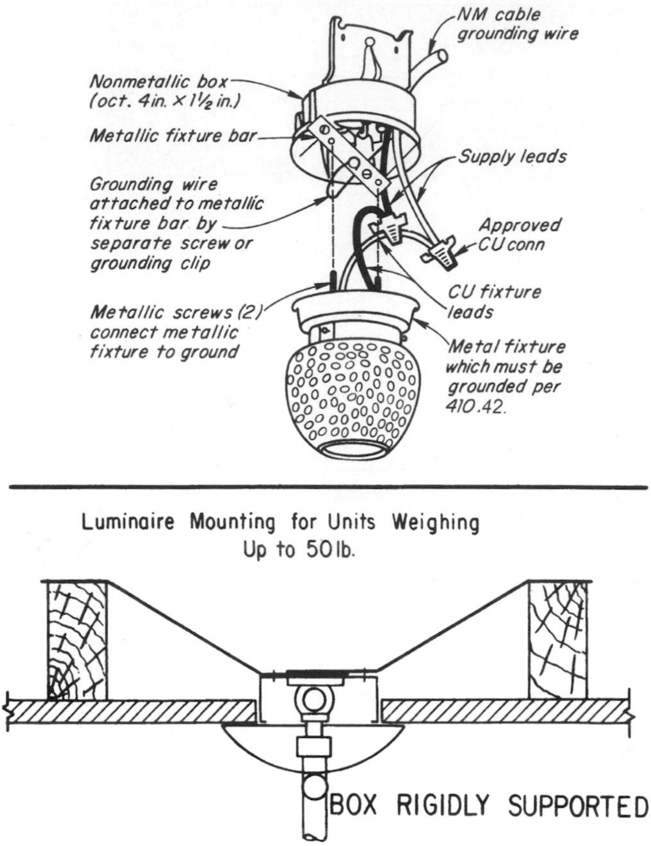
NM cable grounding wire
Nonmetallic box (oct. 4 in. × 1½ in.)
Metallic fixture bar
Supply leads
Grounding wire attached to metallic fixture bar by separate screw or grounding clip
Approved CU conn
CU fixture leads
Metallic screws (2) connect metallic fixture to ground
Metal fixture which must be grounded per 410.42.
Luminaire Mounting for Units Weighing Up to 50 lb.
BOX RIGIDLY SUPPORTED

Fig. 410-9. Luminaires must be supported by approved methods. (Sec. 410.36.)

Part **(B)** covers the support of luminaires installed in suspended ceilings. The Code rule wording was based on the following:

SUBSTANTIATION: The Uniform Building Code requires that suspended ceilings be adequately supported. This is usually in the form of an iron wire support attached to the structural ceiling members and the other end of the wire attached to the suspended ceiling frame members. The luminaires are then laid in the openings and secured only by light metal clips. There have been numerous accidents occur when these metal clips have been dislodged causing luminaires to fall to the floor. There have been several instances, where luminaires are installed in end-to-end rows, when one luminaire becomes dislodged from construction vibration causing the entire row to also fall to the floor.

There is also the danger of luminaires being shaken loose by seismic disturbances—Los Angeles, Oroville and Santa Rosa areas, to mention a few locations.

Having these luminaires attached to the framing members also becomes a severe problem to firemen. When the ceiling area becomes involved in a fire or enough heat generated from the fire, the framing members distort and cause the luminaires to fall through the openings.

The last sentence of part **(B)** does clearly recognize support of luminaires by means of "clips" that are listed for use with a particular framing member and type of luminaire. However, many jurisdictions across the country do *not* recognize the use of suspended-ceiling-supported luminaires—even where a ceiling system is specifically listed to provide such support. Check with your local inspector to determine the acceptability of ceiling-supported luminaires. Further, note that the rule requires the luminaire to be attached to the framing for the suspended ceiling, even if the luminaire is entirely supported to the building structure. This rule makes certain that in the event of a seismic disturbance, the ceiling and the luminaires move together. Otherwise luminaire movement out of step with the ceiling could easily result in the luminaires battering the ceiling to the point of causing it to fail. However, it must also be noted that UL, with respect to the T-bar clips mentioned here, makes the following statement: "The ability of these clips to withstand seismic disturbances has not been evaluated."

Part **(G)** correlates to 225.26, which prohibits trees from being used to support conductor spans, but, this rule specifically permits outdoor luminaires and their boxes and support means to be mounted on trees. Needless to say, branch-circuit wiring supplying such luminaires may also be supported by the tree. Note that 300.5(D) requires protection where such wiring emerges from grade and up the initial 2.5 m (8 ft).

410.42. Luminaires with Exposed Conductive Parts. Part **(A)** says all exposed metallic components of the luminaire—other than trim pieces, and so forth, which are exempted by the next sentence—must be grounded. This can be accomplished by connection of the luminaire to an equipment grounding means—metal raceway, metal cable armor, or a ground wire in NM cable.

410.44. Methods of Grounding. Luminaires must be mechanically connected to the equipment grounding system, with three exceptions. The first covers luminaires made of insulating material and with no exposed conductive parts, which are suitable for mounting to a wiring method such as concealed knob-and-tube wiring, that does not provide ready means for an equipment grounding connection. The other two cover replacements.

The first (Exception No. 2) recognizes the installation of what amounts to an "external" equipment ground, as described in 250.130(C), to provide the desired safety grounding of a noninsulating material luminaire where no equipment grounding means exists within the outlet box. The second (Exception No. 3) permits installation of metallic luminaires, provided the circuit supplying the luminaire is GFCI protected.

410.46. Equipment Grounding Conductor Attachment. This rule requires luminaires to have some terminal or other connection for an equipment grounding conductor when such luminaires have exposed metal parts and are supplied by NM cable or nonmetallic raceway, which must carry an equipment grounding conductor for grounding metal parts of the luminaire (Fig. 410-10). For luminaires supplied by metal raceway, proper connection of the raceway to the metal of the luminaire provides an equipment ground return path through the raceway, in accordance with 250.118.

410.50. Polarization of Luminaires. This method of wiring luminaires is required in order to ensure that the screw shells of sockets will be connected to the grounded

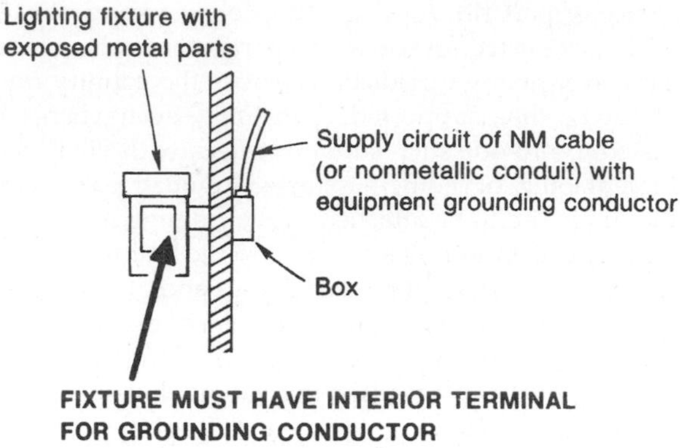
Lighting fixture with exposed metal parts
Supply circuit of NM cable (or nonmetallic conduit) with equipment grounding conductor
Box
FIXTURE MUST HAVE INTERIOR TERMINAL FOR GROUNDING CONDUCTOR

Fig. 410-10. Exposed metal parts of luminaires must incorporate some suitable means for connecting the equipment grounding conductor of a nonmetallic-enclosed supply circuit. (Sec. 410.46.)

circuit wire. This is extremely important because luminaires with multiple lamps are often relamped hot in order to simplify choosing the correct lamp to replace. If the screw shell is hot, the skirt of the burned-out lamp will be hot as it is unscrewed until it is entirely separated from its socket. And the skirt of the new lamp will be hot as soon as it touches the screw shell going in. At either stage, it is very easy to have a finger touch the screw shell of the lamp. If the person's other hand is braced on any grounded object, or perhaps bracing a chain-supported chandelier, the full voltage to ground will be present in a path running directly across the heart.

410.56. Protection of Conductors and Insulation. As noted in part **(E)**, a luminaire fed by a conduit stem suspended from a threaded swivel-type conduit body must be supplied by stranded, not solid, wires run through the conduit stem—because the swivel fitting permits movement of the conductors.

410.59. Cord-Connected Showcases. Figure 410-11 shows an arrangement of cord-supplied illuminated showcases in a store. The details of this layout are lettered, with dimensions as given in the answers and involve the following rules:

a. The first showcase is supplied by flexible cord plugged into grounding-type receptacle rated 20 A. And that is permitted by part **(B)**.

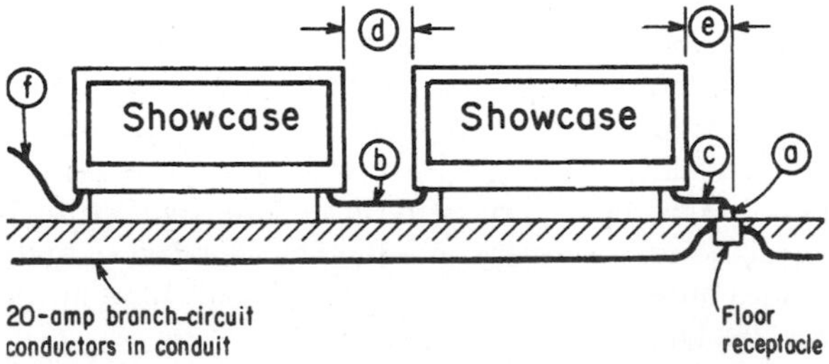
d
e
f
Showcase
Showcase
b
c
a
20-amp branch-circuit conductors in conduit
Floor receptacle

Fig. 410-11. Hookup of lighted showcases must satisfy a number of rules. (Sec. 410.59.)

b. Flexible cord feeds the second showcase; the cord is spliced in JBs in each showcase. That is a clear violation of the requirement that in such connections, separable locking-type (twist-type) cord connectors must be used, and spliced cord connections would be a violation of the wording used in the first paragraph of 410.59.
c. Cord is 14 AWG, hard-service type. That is a violation because the cord conductors must be 12 AWG, the size of the branch-circuit conductors for the 20-A circuit, as required in part **(A)**.
d. Showcases are separated by 2 in. (50 mm). That is okay, but it is the maximum permitted separation, as noted in part **(C)**.
e. The first case is 14 in. (350 mm) from the supply receptacle. No good! The maximum permitted distance is 12 in. (300 mm), as given in part **(C)**.
f. The second showcase feeds a spotlight. Violation! Part **(D)** says no other equipment may be connected to showcases.

410.62. Cord-Connected Lampholders and Luminaires. Part **(B)** recognizes the use of a fixed-cord connection for energy supply to luminaires that require aiming or adjustment after installation (Fig. 410-12). Use of a cord supply to luminaires has been a recurring controversial issue, although 400.10 has permitted cord supply to luminaires for a long time. 410.30(C) has required that electric-discharge luminaires, if suitable for supply by cord, must make use of plug-and-receptacle connection of the luminaire to the supply circuit. The rule here in part **(B)** permits floodlights—such as those used for outdoor and indoor areas for sporting events, for traffic control, or for area lighting—to have a fixed-cord connection from a bushed-hole cover of the branch-circuit outlet box to the wiring connection compartment in the luminaire itself. This rule gives adequate recognition to the type of cord connection that has long been used on floodlights, spotlights, and other luminaires used for area lighting applications.

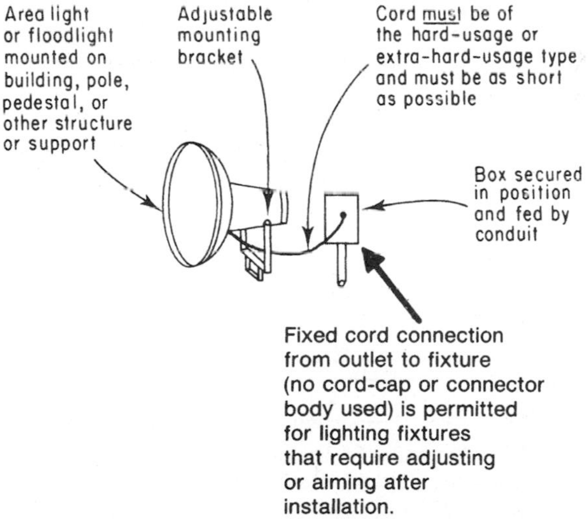
Area light or floodlight mounted on building, pole, pedestal, or other structure or support
Adjustable mounting bracket
Cord must be of the hard-usage or extra-hard-usage type and must be as short as possible
Box secured in position and fed by conduit
Fixed cord connection from outlet to fixture (no cord-cap or connector body used) is permitted for lighting fixtures that require adjusting or aiming after installation.

Fig. 410-12. Cord connection—either fixed cord or cord with plug cap—is permitted for adjustable luminaires. [Sec. 410.62(B).]

Prior to the 2017 **NEC**, Part **(C)(1)** was one of the most convoluted, poorly worded, and difficult to follow provisions in the entire **NEC**, but it has been completely rewritten and is now greatly improved in understandability. First, there are three global conditions that must be met for any of the permitted applications. First, the electric discharge luminaire (and not, therefore, an incandescent luminaire) must be directly below the outlet or busway. This last word correlates with 368.17(C) Exception No. 2 that recognizes busway connections to cord-and-plug-connected luminaires with fuses in the cord cap. "Directly below" clearly does not allow for cord to be run below (at a lower level) but swagged way off to the side. The second global condition is that the cord be visible over its entire length, and the third is that it not be subject to strain or damage.

If these three conditions are met, then there are three possibilities expressed in three separated paragraphs that are far more comprehensible than the 62-word run-on sentence that preceded the 2017 edition. The first is the oldest use, that of a cord-and-plug connection for the luminaire with the receptacle directly over the luminaire (Fig. 410-13). The original concept here is that maintenance personnel

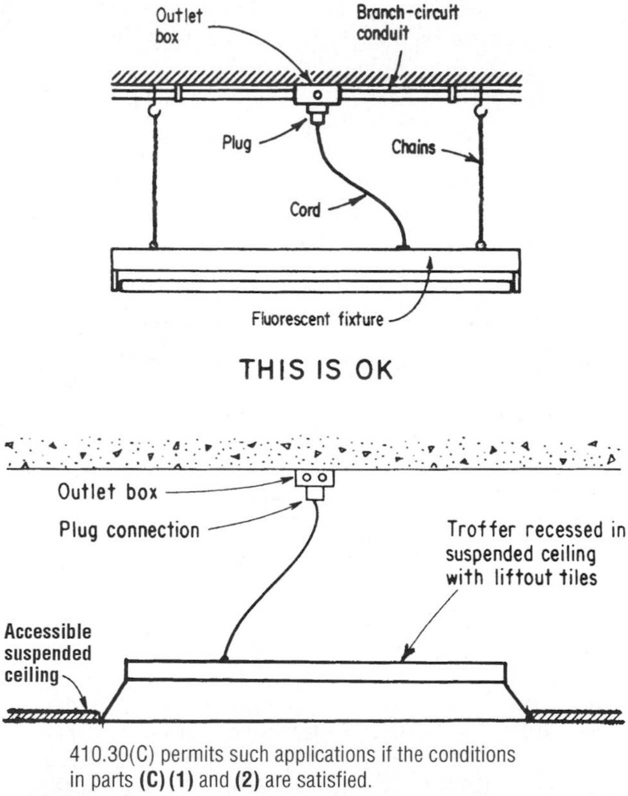
Outlet box
Branch-circuit conduit
Plug
Chains
Cord
Fluorescent fixture
THIS IS OK
Outlet box
Plug connection
Troffer recessed in suspended ceiling with liftout tiles
Accessible suspended ceiling
410.30(C) permits such applications if the conditions in parts (C)(1) and (2) are satisfied.

Fig. 410-13. Cord-and-plug luminaire supply is okay only if cord is "continuously visible." The suspended ceiling application would normally violate (C)(2)a and not be acceptable. (Sec. 410.62.)

would unplug the luminaire, unhook some jack chain, and take the luminaire to the bench for work or cleaning.

The next possibility is extensively used in upscale lighting designs with rows or other patterns of fluorescent luminaires suspended from aircraft cable capable of height adjustment so the luminaires can be precisely positioned. At a convenient point in the lineup, the canopy supplied with the luminaires is connected to the branch circuit using flexible cord between the canopy and the luminaire (Fig. 410-14). Some box manufacturers make outlet boxes with T-bar knockouts on the sides at right angles so the box can straddle the supports on a hung ceiling and mount flush with the ceiling tiles, and then this canopy can be mounted without further ado. However, a 2008 **NEC** amendment also recognizes a riser nipple from the canopy not over 150 mm (6 in.) long, allowing the cord to pass through the ceiling plane and connect to an outlet box just above the ceiling. The last possibility correlates with 604.100(A)(3) (requiring 12 AWG hard usage cord minimum) and uses the proprietary receptacle configurations covered in 604.100(C) that are uniquely polarized and locking for this purpose as part of a manufactured wiring system.

Part **(C)(2)** of this section permits electric-discharge lighting with mogul-base screw-shell lampholders (such as mercury-vapor or metal-halide units) to be supplied by branch circuits up to 50 A with the use of receptacles and caps of lesser ampere rating if such devices are rated not less than 125 percent of the luminaire full-load current. Note that there are no limitations placed on the length of the cord or the position of the receptacle relative to the luminaire under this provision, short of the generic wording in 400.12(1) prohibiting the use of cord as a substitute for fixed wiring.

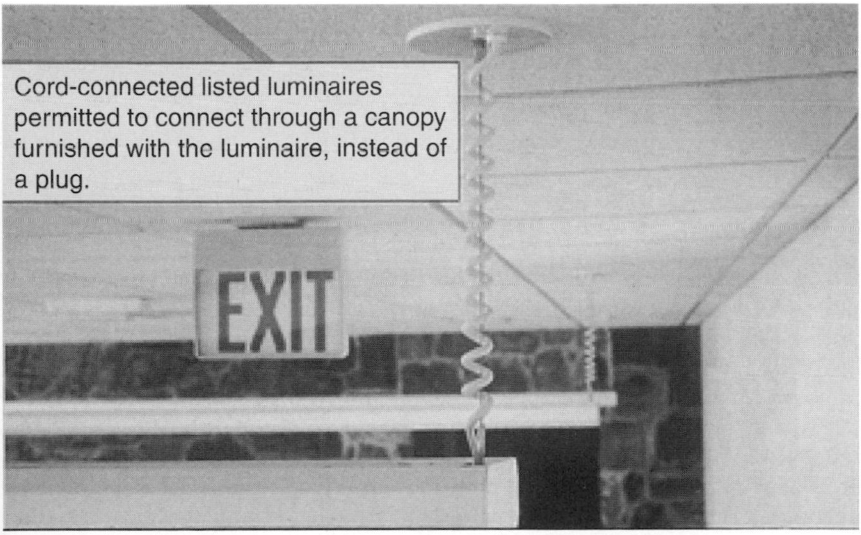
Cord-connected listed luminaires permitted to connect through a canopy furnished with the luminaire, instead of a plug.
EXIT

Fig. 410-14. This listed cord-connected luminaire lineup is arranged for precise vertical positioning. [Sec. 410.62(C)(1).]

Unlike part **(C)(2)**, paragraph **(3)** applies to fluorescent luminaires as well as HID applications. This rule essentially reverses the cord direction as covered in 410.62(C)(1)(c) (first clause) by using a flanged inlet in the back of the luminaire connected to a receptacle cord body fed by a drop from an unspecified source (Fig. 410-15). Here again there are no limitations placed on the length of the cord or the position of the cord connection to the branch circuit relative to the luminaire short of the generic wording in 400.12(1) prohibiting the use of cord as a substitute for fixed wiring. The use of a connector body and flanged inlet supply affords greater ease in maintenance of the luminaire, because maintenance people can disconnect the luminaire at the lower end of the cord to remove it for cleaning or repair.

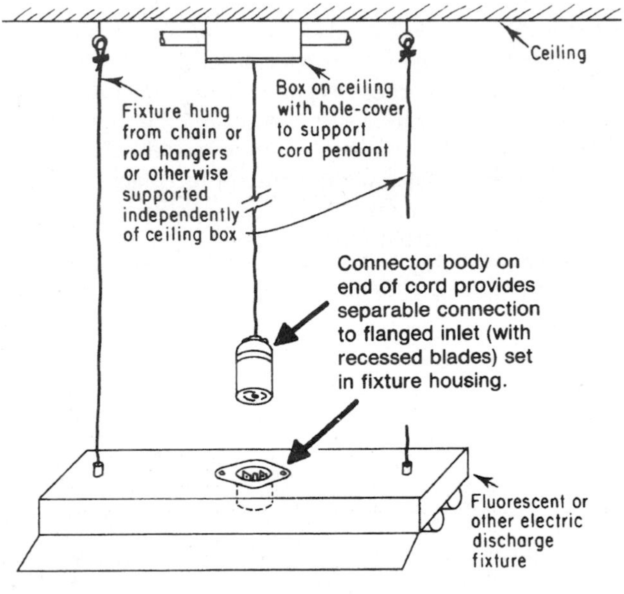
Ceiling
Box on ceiling with hole-cover to support cord pendant
Fixture hung from chain or rod hangers or otherwise supported independently of ceiling box
Connector body on end of cord provides separable connection to flanged inlet (with recessed blades) set in fixture housing.
Fluorescent or other electric discharge fixture

Fig. 410-15. This is another method recognized for cord supply to luminaires. [Sec. 410.62(C)(3).]

410.64. Luminaires as Raceways. This Code rule has long stated basically that luminaires shall not be used as a raceway for circuit conductors (Fig. 410-16) unless "listed and marked for use as a raceway." Note the correct use of the term "marked" in this rule. When a luminaire is specifically approved as a raceway, any number of branch-circuit conductors may be installed within the capacity of the raceway. When housings are approved as raceways, luminaires carry an Underwriters Laboratories label that states "Suitable for Use as Raceway." Any type of circuit may be run through the luminaire in such a case, provided listing limitations are met. UL rules require that luminaires that are suitable for use as raceway—that is, for carrying circuit wires other than the wires supplying the luminaires—must be so marked and must show the number, size, and type of conductors permitted.

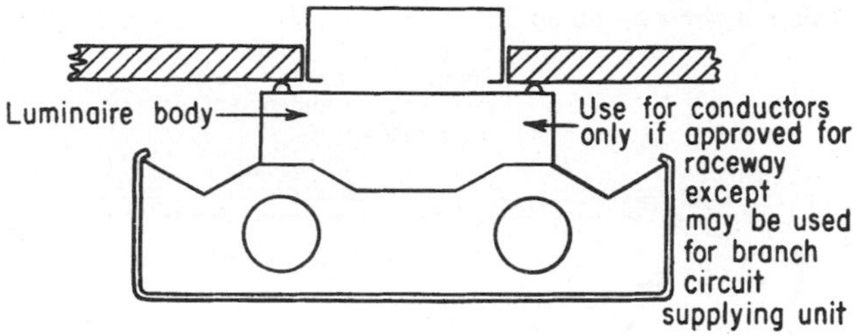
Luminaire body
Use for conductors only if approved for raceway except may be used for branch circuit supplying unit

Fig. 410-16. Use of wiring through luminaires is clearly limited. (Sec. 410.64.)

(C) Luminaires connected together This paragraph permits limited use of luminaire wiring compartments, even where such luminaires are *not* listed and marked for use as raceways, to carry through the circuit that supplies the luminaires. This allowance applies if the luminaires are designed for end-to-end assembly to form a continuous raceway or the luminaires are connected together by recognized wiring methods (such as rigid conduit and EMT). Most self-contained fluorescent luminaire units now available are designed for end-to-end assembly. Each luminaire contains a metal body, or housing, which serves as the structural member of the luminaire and provides a housing for the ballast, wiring, and so forth, and which is of sufficient size to permit running the branch-circuit wiring through the unit. Each luminaire is then tied to the branch circuit by means of a single tap (Fig. 410-17).

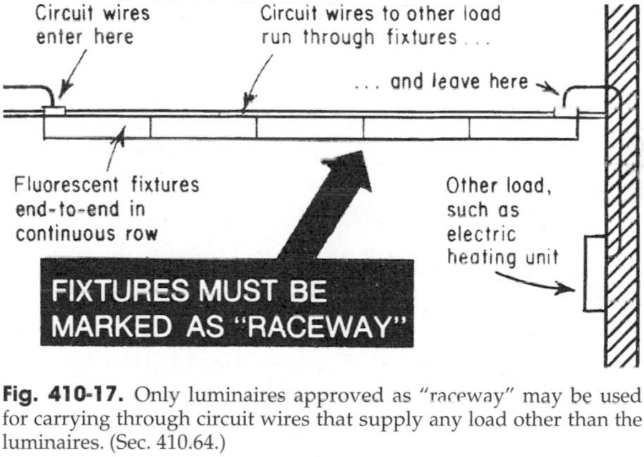
Circuit wires enter here
Circuit wires to other load run through fixtures . . .
. . . and leave here
Fluorescent fixtures end-to-end in continuous row
Other load, such as electric heating unit
FIXTURES MUST BE MARKED AS "RACEWAY"

Fig. 410-17. Only luminaires approved as "raceway" may be used for carrying through circuit wires that supply any load other than the luminaires. (Sec. 410.64.)

It should be noted that in 410.64(C) the permitted luminaire layouts may carry only conductors of either a 2-wire or a multiwire branch circuit where the wires of the branch circuit supply only the luminaires through which the circuit conductors are run. Thus, it is permissible to use a 3-phase, 4-wire branch circuit through luminaires so connected, with the total number of luminaires connected from all the phase legs to the neutral; that is, with the luminaire load divided among the

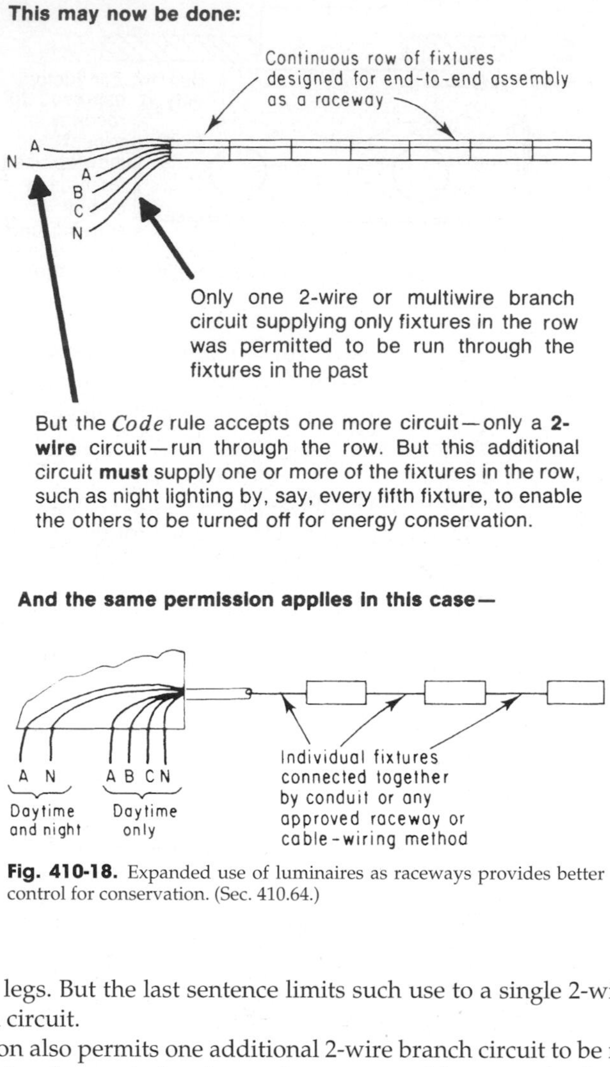
This may now be done:
Continuous row of fixtures designed for end-to-end assembly as a raceway
A
N
A
B
C
N
Only one 2-wire or multiwire branch circuit supplying only fixtures in the row was permitted to be run through the fixtures in the past
But the *Code* rule accepts one more circuit—only a **2-wire** circuit—run through the row. But this additional circuit **must** supply one or more of the fixtures in the row, such as night lighting by, say, every fifth fixture, to enable the others to be turned off for energy conservation.
And the same permission applies in this case—
A N
A B C N
Daytime and night
Daytime only
Individual fixtures connected together by conduit or any approved raceway or cable-wiring method

Fig. 410-18. Expanded use of luminaires as raceways provides better control for conservation. (Sec. 410.64.)

three phase legs. But the last sentence limits such use to a single 2-wire or multiwire branch circuit.

This section also permits one additional 2-wire branch circuit to be run through such luminaires (connected end-to-end or connected by recognized wiring methods) in addition to the 2-wire or multiwire branch circuit previously recognized. This additional 2-wire branch circuit may be used only to supply one or more of the connected luminaires throughout the total luminaire run supplied by the other branch circuit run through the raceway (Fig. 410-18). This was added to permit separate control of some of the luminaires fed by the additional branch circuit, providing the opportunity to turn off some of the luminaires for energy conservation during the night or other times when they are not needed.

410.68. Feeder and Branch-Circuit Conductors and Ballasts. The sentence of this section regulates use of wires run through or within luminaires where the wire would be exposed to possible contact with the ballast, which has a hot-spot surface temperature of 90°C. Thus, such wires must be rated at least 90°C—which is the temperature at which the wire will operate when carrying its rated current in an ambient not over 30°C (which is 86°F). Note that LED drivers, and associated power supplies and transformers, are potent heat sources and they are now grouped with ballasts as requiring equivalent separation from lower-temperature wiring. The use of conductors with lower-rated insulation is permissible only where the luminaire is specifically listed and marked for use with the lower-temperature insulation within 3 in. (76 mm) of the ballasts, or the specific wiring arrangement in the ballast channel precludes the lower-rated conductors from getting near the ballast. For example, if the outlet is 300 mm (12 in.) from the ballast, and only 200 mm (8 in.) of Type TW branch-circuit wire enters the channel, obviously this section is not in play because the branch-circuit wiring cannot get close enough to the ballast.

The question often arises, "May 150°C fixture wire be used for circuiting through end-to-end connected continuous-row fluorescent luminaires?" 402.10 permits "fixture wires" to be installed "in luminaires" where they will not be subject to bending or flexing in normal use. That would seem to approve fixture wire through the luminaires connected in a row. But, 402.11 prohibits fixture wires used as branch-circuit conductors. The conductors installed in the luminaire "ballast compartment" are referred to as "conductors of a . . . branch circuit" in 410.64, because they feed directly from the branch circuit and are tapped at each luminaire to feed each luminaire. The branch circuit extends from the point where it is protected by a CB or fuse to the last point where it feeds to the final outlet, device, apparatus, equipment, or luminaires. Under these conditions, it seems clear that the wiring must be that approved for branch circuits. Any fixture wires are not approved for branch-circuit wiring.

To satisfy 410.68, any of the 90°C-rated conductor insulations, such as Type RHH, Type THHN, or even 75°C-rated Type THW under the special 90°C allowance, must be installed. No fixture wire is permitted to be installed as branch-circuit conductors.

410.74. Luminaire Rating. This section specifically requires that any luminaire be suitably marked to indicate the need for supply wires rated higher than 60°C to withstand the heat generated in the luminaire. Such marking must be prominently made on the luminaire itself and also on the shipping carton in which the luminaire is enclosed (Fig. 410-19). This makes the information available to the consumer at the time the luminaire selection decision is being made. The label must include a statement that a qualified electrician should be consulted to determine the suitability of the branch-circuit wiring, or at least a warning label will require consultation with a qualified person. Most residential branch-circuit wiring installed prior to late 1983 is rated 60°C at best.

410.82. Portable Luminaires. Part **(A)** requires portable lamps (table lamps and floor lamps) to be wired with flexible cord approved for the purpose and to be equipped with polarized- or grounding-type attachment plugs. Two-prong nonpolarized attachment plugs are no longer permitted. Polarized-type plug caps permit a single orientation of the plug for insertion in the receptacle outlet. Such polarizing of the plug will provide for connecting the grounded conductor of the circuit to the screw shell of the lampholder in the lamp.

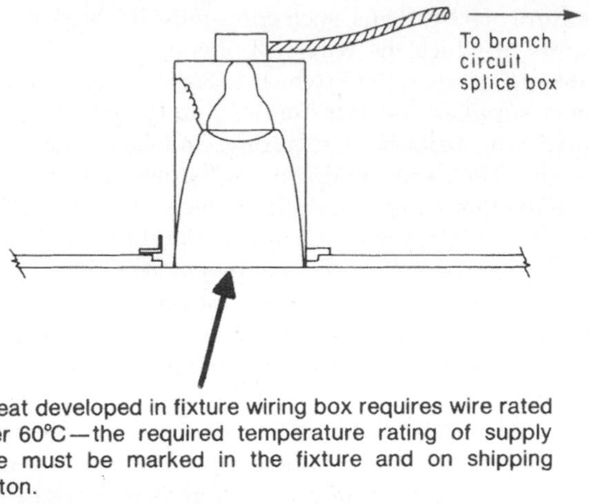
To branch circuit splice box
If heat developed in fixture wiring box requires wire rated over 60°C—the required temperature rating of supply wire must be marked in the fixture and on shipping carton.

Fig. 410-19. Where high-temperature wire is needed, the luminaire must be marked. [Sec. 410.74(A).]

In part **(B),** five specific rules are given on the use of portable handlamps, as shown in Fig. 410-20. All the requirements of part **(A),** including the need for polarized- or grounding-type attachment plugs, also are made applicable to portable handlamps.

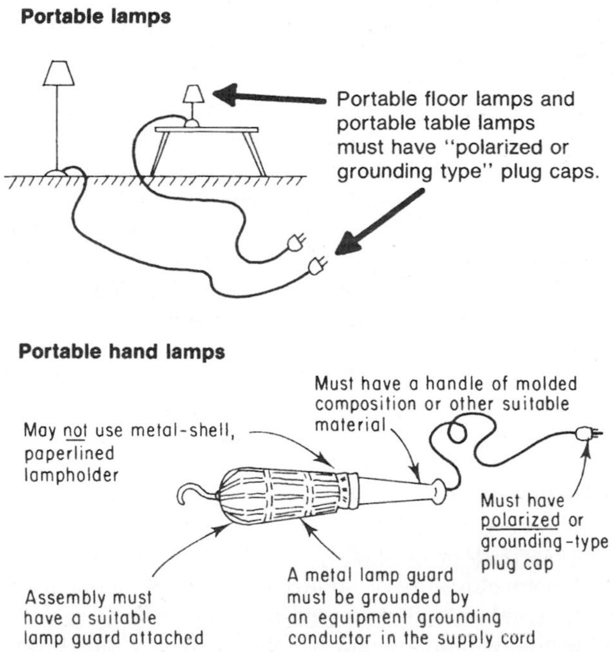
Portable lamps
Portable floor lamps and portable table lamps must have "polarized or grounding type" plug caps.
Portable hand lamps
Must have a handle of molded composition or other suitable material
May not use metal-shell, paperlined lampholder
Must have polarized or grounding-type plug cap
A metal lamp guard must be grounded by an equipment grounding conductor in the supply cord
Assembly must have a suitable lamp guard attached

Fig. 410-20. NEC rules aim at greater safety in use of portable lamps and portable handlamps. (Sec. 410.82.)

410.90. Screw-Shell Type. This warns against the previously common practice of installing screw-shell lampholders with screw-plug adapters in baseboards and walls for the connection of cord-connected appliances and lighting equipment, and thereby exposing live parts to contact by persons when the adapters were moved from place to place. See also 406.2(B).

410.93. Double-Pole Switched Lampholders. On a circuit having one wire grounded, the grounded wire must always be connected to the screw shell of the socket, and sockets having a single-pole switching mechanism may be used. (See 410.102.) On a 2-wire circuit tapped from the outside (ungrounded) wires of a 3-wire or 4-wire system, if sockets having switching mechanisms are used, these must be double-pole so that they will disconnect both of the ungrounded wires.

410.97. Lampholders Near Combustible Material. This rule prohibits, using identical language as in 410.11 for luminaires, the placement of a lampholder (unless provided with construction features or shades or guards) where adjacent combustible material will be heated above 90°C. A bare incandescent lamp in a plain vanilla lampholder can be one of the most potent heat sources available, and on the literal text of prior Codes there were no placement restrictions, because they are not luminaires.

410.104. Electric-Discharge Lamp Auxiliary Equipment. Part **(B)** requires that a switch controlling the supply to electric-discharge-lamp auxiliary equipment must simultaneously disconnect both hot conductors of a 2-wire ungrounded circuit. This is required to prevent an energized screw shell where only one circuit wire is disconnected—which would be a hazard during relamping. Note that this may require a relay in the event a conventional single-pole photo-control is used for this application.

410.110. General. Underwriters Laboratories rules comment on use of luminaires installed in hung ceilings, as follows:

All recessed luminaires, except those marked for use in concrete only, are suitable for use in suspended ceilings and may be marked "SUITABLE FOR SUSPENDED CEILING." Note that this is not a mandatory marking for this use. However, luminaires mounted in such locations are subject to the requirements of Parts X and XI of Art. 410, that is, from this section through Sec. 410.122. There was a very long-standing formal interpretation on this point. When a formal interpretation issues, NFPA regulations require that the committee responsible for the text that prompted the request for interpretation amend the text so as to avoid the lack of clarity in the future. Incorporating the words "including suspended ceilings" finally addresses that lack of clarity, and now that old formal interpretation can be officially "retired."

Air-handling luminaires must be used fully in accordance with the conditions marked on the luminaires. When used in fire-rated partitions, such luminaires must be specifically covered in the applicable listed design in the UL *Fire Resistance Directory*.

410.115. Temperature. Heat is a major problem in lighting system design, and with the trend to the use of recessed luminaires and equipment, the problem is increased. In the case of luminaires using incandescent lamps, the problem is primarily the prevention of concentrated spots of heat coming into contact with the building structure. In the case of fluorescent luminaires, the major heat problem is related to the ballast, which can build up severely high temperatures when not

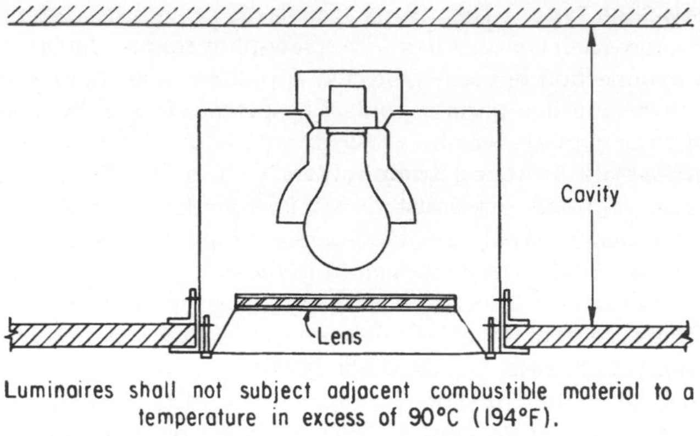
Cavity
Lens
Luminaires shall not subject adjacent combustible material to a temperature in excess of 90°C (194°F).

Fig. 410-21. Recessed luminaires must not threaten combustion of building materials. (Sec. 410.115.)

properly ventilated, or designed for cooler operation through adequate radiation and convection. These are problems that must be solved (1) through proper luminaire design, and (2) through proper installation methods and techniques (Fig. 410-21).

The rule of part **(C)** requiring thermal protection of recessed incandescent luminaires is intended to prevent fires that have been caused by overlamping or misuse of insulating materials. (See Fig. 410-22.) Exception No. 1 covers an installation with the luminaire embedded in poured concrete; no thermal protection is required for this application. As noted in Exception No. 2, a nonthermal-protected recessed incandescent luminaire may be used in direct contact with thermal

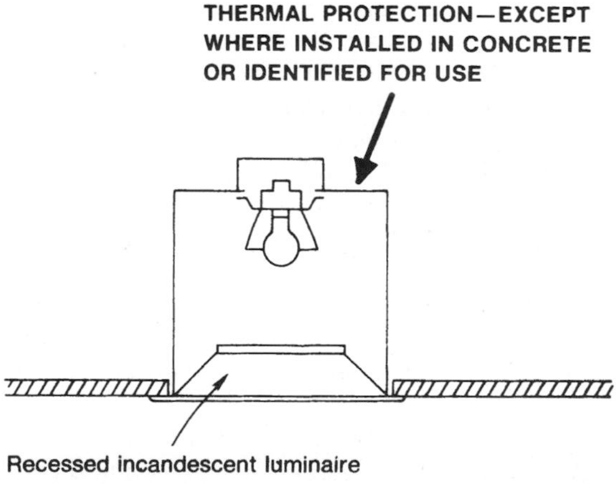
THERMAL PROTECTION—EXCEPT WHERE INSTALLED IN CONCRETE OR IDENTIFIED FOR USE
Recessed incandescent luminaire

Fig. 410-22. Recessed incandescent luminaires that incorporate thermal protection must be "identified as thermally protected," but a luminaire without thermal protection may be used if listed and identified for use as "inherently protected." [Sec. 410.115(C).]

insulation if it is listed as suitable by design for performance equal to thermally protected luminaires and is so identified. There are presently available recessed incandescent luminaires of such design as to prevent overheating even when installed in contact with thermal insulation. Such luminaires cannot be overlamped or mislamped to cause excessive temperature and are listed and marked for such application.

410.116. Clearance and Installation. As covered in part **(A)(1),** where "Non-Type IC" recessed luminaires are used with thermal insulation in the recessed space, thermal insulation must have a clearance of at least 3 in. (75 mm) on the side of the luminaire and at least 3 in. (75 mm) at the top of luminaire and shall be so arranged that heat is not trapped in this space. Free circulation of air must be provided with this 3-in. (75-mm) spacing. If, however, the luminaire is approved for installation with thermal insulation on closer spacing, it may be so used (Fig. 410-23). In parts **(A)(2)** and **(B),** the Code recognizes recessed incandescent luminaires in contact with thermal insulation or combustible material. Recessed incandescent luminaires must be listed and identified for use in contact with thermal insulation.

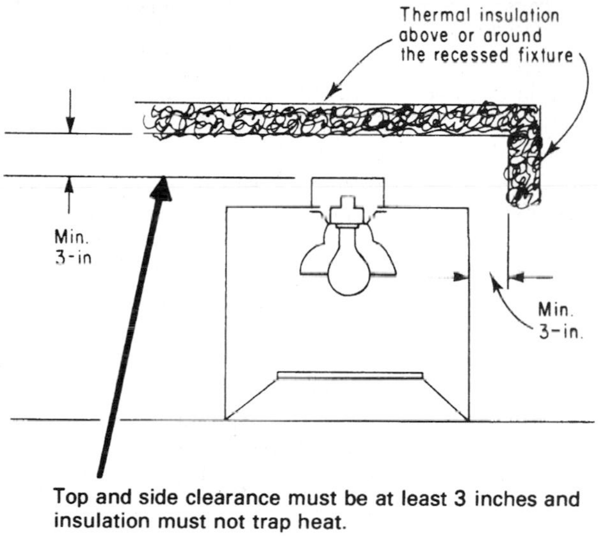
Thermal insulation above or around the recessed fixture
Min. 3-in
Min. 3-in.
Top and side clearance must be at least 3 inches and insulation must not trap heat.

Fig. 410-23. Clearance of recessed luminaire from thermal insulation is required unless luminaire is UL-listed for use in direct contact with insulation. (Sec. 410.116.)

In the past, thermal insulation has been installed in direct contact with recessed luminaires not approved for that use and has caused overheating in luminaires with resulting failures and fires. Obviously, the installer of thermal insulation will have to be educated on this subject, because electrical installers have little control over how the insulation will be applied.

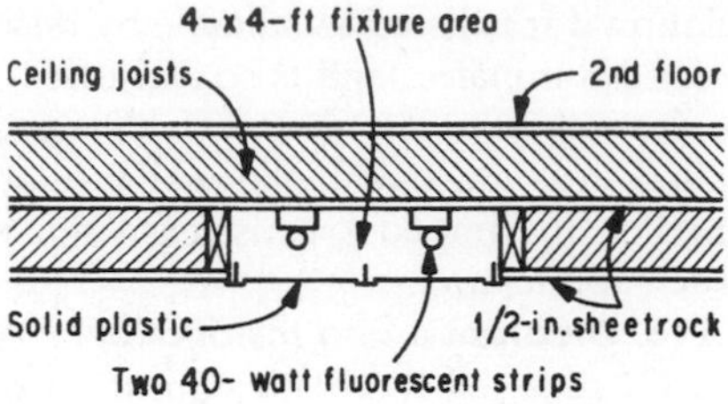
4-x 4-ft fixture area
Ceiling joists
2nd floor
Solid plastic
1/2-in. sheetrock
Two 40- watt fluorescent strips

Fig. 410-24. Clearances on custom recessed lighting applications must be evaluated carefully. (Sec. 410.116.)

Figure 410-24 shows an application that involves the ½-in. clearance covered in the first sentence of this section. It shows two 40-W fluorescent strips installed in a residential kitchen ceiling. The ceiling has been furred down on all sides of the 4- by 4-ft (1.22- by 1.22-m) luminaire area as shown. The question arises: Is this a "recessed" installation according to the Code? How small or large must such an enclosed space be to be considered a recess? The Code does not mention "recessed installations" but refers to "recessed luminaires." The installation as shown is basically a field-fabricated recessed luminaire. With only two 40-W lamps in this space, it is not likely there will be much of a heat problem. But sufficient information on the total construction of the cavity would be needed for an inspector to make an evaluation. The temperature limitations of 410.11 and 410.115 must be observed, and wiring must be in accordance with 410.117.

The next question is: Must a ½-in. (13-mm) clearance be maintained between the luminaires and the Sheetrock? Sheetrock is fire-rated by UL but is not fireproof. It is unlikely that inspectors would require the ½-in. (13-mm) spacing between the luminaires and the Sheetrock because the paperboard surfaces of the Sheetrock are still part of a rated material and 314.20 and many other Code rules treat it comparably to other noncombustible surfaces. Although the paper will char, the water of hydration in the substrate, when heated, will prevent transmission of fire, hence the rating. The local inspector would, of course, have the final say on this.

410.117. Wiring. Supply wiring to recessed luminaires may be the branch-circuit wires, if their 60°C, 75°C, or 90°C temperature rating at least matches the temperature that will exist in the luminaire splice compartment under operating conditions. Typically, fluorescent luminaires do not have very high operating temperatures where the branch-circuit wires splice to the luminaire wires, and branch-circuit conductors may be run right into the luminaire. But incandescent luminaires develop much higher localized heat because all the wattage is concentrated in a much smaller lamp, thereby requiring higher-temperature wire where the branch circuit splices to the luminaire leads.

The rules of part **(B)** and part **(C)** of this section are related to the details discussed under 410.21. Figure 410-25 shows branch-circuit wires rated at 75°C coming out of ceiling boxes (left) and then connecting directly to fluorescent strip units mounted over the boxes and attached to the ceiling (right). The 75°C THW branch-circuit wires are generally recognized as rated at 90°C for use within the fluorescent units [Table 310.104(A), THW properties row]. Therefore, this installation fully complies with 410.117(B).

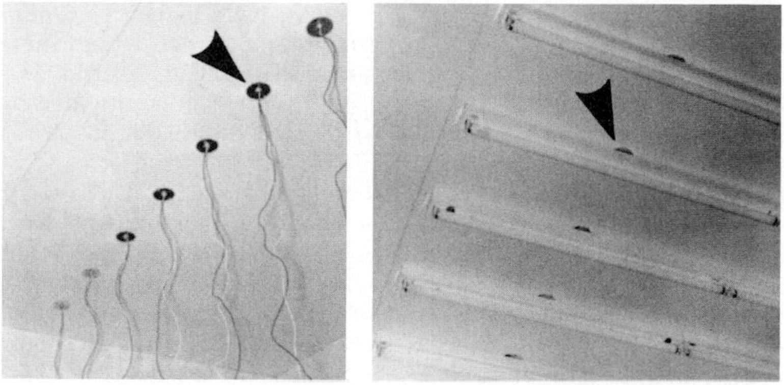

Fig. 410-25. Branch-circuit wires rated at 75°C are brought out of ceiling boxes (left) and then connected to luminaire leads within the relatively cool wiring compartment of the fluorescent units. (Sec. 410.117.)

Figure 410-26 is an old photo taken prior to the advent of 90°C Type NM-B cable, and so it shows 60°C branch-circuit wiring run directly to integral junction boxes of prewired incandescent luminaires. The box protects the branch-circuit wires from the heat generated within the luminaires. Therefore,

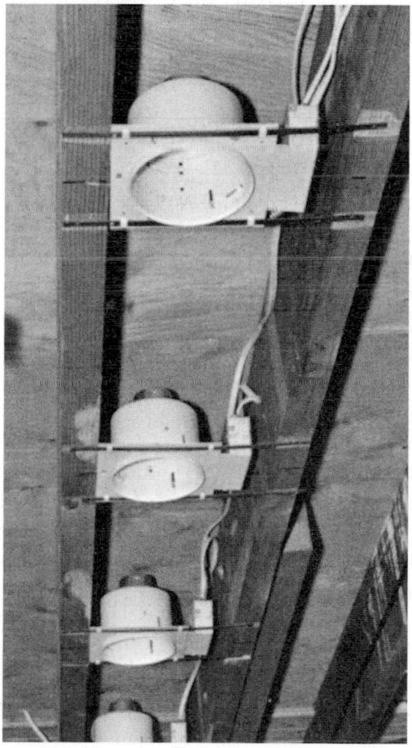

Fig. 410-26. Prewired recessed incandescent luminaires may have 60°C branch-circuit wiring run directly into their junction boxes. (Sec. 410.117.)

this installation fully complies with 410.117(A). Note that these luminaires are used for supply and feed-through of the branch-circuit wires. There are no tap conductors wired in the field, the junction box is not "placed" in position in the field, and the raceway, all 75 mm (3 in.) of it, is provided by the manufacturer and is certainly not "at least 450 mm (18 in.) but not more than 1.8 m (6 ft) long."

Figure 410-27 is another old photo that predates the requirement in 410.130(F)(1) for thermal protection on recessed HID luminaires, and for this reason required the field-wired cold lead as shown. This, in its day, was a completely legitimate application of 410.117(C), where a luminaire wiring compartment operates so hot that the temperature exceeds that of the branch-circuit-rated value, and high-temperature fixture wires must be run to the unit. The circuit outlet box supplying the luminaire must be "placed" in the field not less than 1 ft (300 mm) away. The flex whip may be metric designator 12 (trade size 3/8) flex for the number and type of luminaire wires as specified in Table 348.22. If the branch circuit supplying the luminaire is protected at 20 or 15 A, 18 AWG fixture wires may be used for luminaire loads up to 6 A, or 16 AWG fixture wires may be used for loads up to 8 A (402.5), and the metal flex may serve as the equipment grounding conductor (250.118). The flex may not be less than 18 in. (450 mm) and not more than 6 ft (1.8 m) long. This length assures enough of a loop to minimize convection of heat into the branch-circuit connections. The fixture wires could be Type PF for conditions requiring a 200°C rating or could be another type of adequate temperature rating for the luminaire's marked temperature, selected from Table 402.3. For flex, refer also to 348.12.

The 1½- to 6-ft (0.45- to 1.8-m) luminaire whip may be Type AC or Type MC cable, if ever such cables are manufactured with conductors rated higher than 90°C.

Having fully covered the requirements in 410.117(C), this subsection needs to be put in perspective. The rule is about wiring cold leads, and nothing else. If you don't need a cold lead, you don't need this rule. This rule has absolutely nothing to do with prewired luminaires like those shown in Fig. 410-26. Luminaires like the one shown in Fig. 410-27 have not been manufactured for many years. There is only one application where a cold lead might be necessary and that is for luminaires that will be embedded in poured concrete, and therefore exempt from thermal protection rules by 410.115(B) or 410.130(F)(3). Although there are references to 410.117(C) all over the **NEC**, they are usually erroneous. They generally describe taps that are not taps (whips above hung ceilings made up with branch-circuit wiring methods using the same size conductors are not taps) run to luminaires that are prewired and suitable for ordinary branch-circuit connections.

410.130. General. Paragraph **(E)** pertains only to fluorescent lamp ballasts used indoors. The protection called for must be a part of the ballast. Underwriters Laboratories, Inc., has made an extensive investigation of various types of protective devices for use within such ballasts, and ballasts found to meet UL requirements for these applications are listed and marked as "Class P." The protective devices are thermal trip devices or thermal fuses, which are responsive

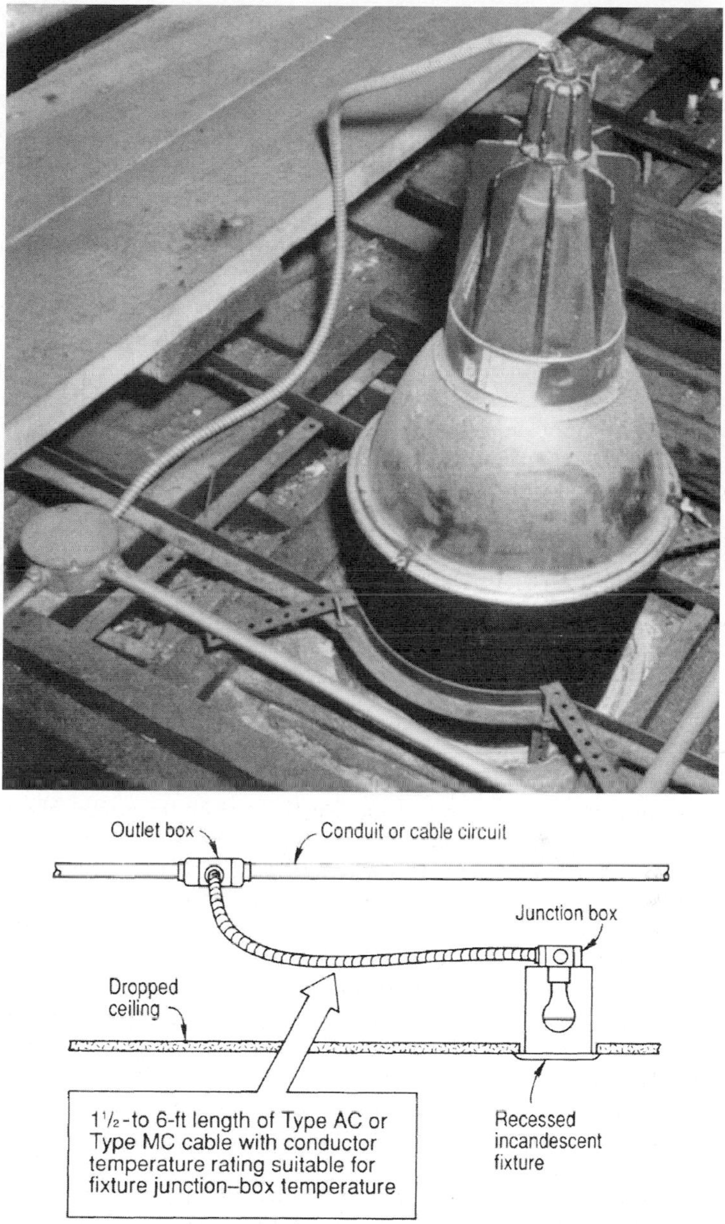
Outlet box
Conduit or cable circuit
Junction box
Dropped ceiling
1½-to 6-ft length of Type AC or Type MC cable with conductor temperature rating suitable for fixture junction–box temperature
Recessed incandescent fixture

This is an alternative to tap conductors run in metal flex or liquidtight metal flex.

Fig. 410-27. For luminaires with circuit-connection compartments operating very hot, high-temperature wires must be run in flex (other metal raceway or Type AC or MC metal armored cable) between 1½ and 6 ft (0.45 and 1.8 m) from an outlet box at least 12 in. (300 mm) away. [Sec. 410.117(C).]

to abnormal heat developed within the ballast because of a fault in components such as autotransformers, capacitors, reactors, and the like.

Simple reactance-type ballasts are used with preheat-type fluorescent lamp circuits for lamps rated less than 30 W. Also, a manual (momentary-contact) or automatic-type starter is used to start the lamp. The simple reactor-type ballast supplies one lamp only, has no autotransformer or capacitor, and is exempted from the protection rule of part **(E)**.

The thermal protection required for ballasts of fluorescent luminaires installed indoors must be within the ballast. Previous wording permitted the interpretation that the supplementary protection for the ballast could be in the luminaire and not necessarily within the ballast.

Note that part **(4)** of this rule prohibits the use of thermal protectors for emergency luminaires where such luminaires are energized only during an emergency. This rule presents a requirement similar to the concept of prohibiting overload protection on fire-pump motors. The prohibition of thermal and overload protection for emergency egress luminaires and fire-pump motors, respectively, is intended to ensure maximum operating continuity of the equipment during an emergency. It makes little sense to automatically de-energize such equipment under overload conditions during an emergency. Why save the luminaires or fire-pump motors from destruction at the risk of leaving people in the dark or cutting off the fire main's pressure? During an emergency, saving people and the building outweigh any concerns for the emergency luminaires or fire-pump motors. They are to be allowed to continue operating while overloaded, right to the point of failure.

Part **(F)** requires that recessed HID luminaires used *either* indoors or outdoors must have thermal protection. The luminaire—and not just the ballast—must be thermally protected and so identified. All *indoor, recessed* luminaires using mercury-vapor, metal-halide, or sodium lamps must be thermally protected. If remote ballasts are used with such luminaires, the ballasts must also be thermally protected. Paragraph (5) requires luminaires using metal halide lamps to provide secondary containment around the lamp to protect against what the industry euphemistically describes as "nonpassive end-of-life failures" where the inner arc tube ruptures with enough force to break the outer bulb. Several fires have been reported from the hot lamp fragments falling on combustible materials. The other permitted approach is to use a specially engineered lamp with an internal shroud around the arc tube, designated "Type O" (for open). Luminaires that use this approach have a screw shell that is deeper than conventional screw shells. The "O" lamps have an extended Edison base that will hit the center contact when screwed all the way in; other lamps will run out of thread before being energized.

In part **(G)** the Code requires, for nonresidential occupancies, a disconnecting means for fluorescent luminaires that are equipped with double-ended lamps and feature in-place replacement of ballasts. The location of this disconnecting means may be inside or outside of each such luminaire, and such a disconnect must simultaneously open all conductors to the ballast—including the grounded (neutral) conductor, if it is supplied by a multiwire branch circuit. The disconnect's line-side terminals must be guarded, and it must be accessible either in the ballast channel or mounted on or within sight of the luminaire while maintenance is being performed. In general these are usually small, pull-apart two-pole units that fit easily in the ballast area. They avoid the possibility of shock and/or open multiwire neutral failures when working a luminaire hot, which is the usual approach because the light from other luminaires avoids the necessity of additional portable illumination. **Watch Out!** Effective with the 2011 NEC, when a ballast that would be covered by this rule if it were being wired new is replaced, the disconnect required by this rule must be retrofitted if it is not already present.

Exceptions apply for hazardous (classified) locations, for two-wire circuiting with adjacent luminaires arranged to stay lit, and for emergency lighting. A cord-and-plug connection is recognized as well for obvious reasons. However, because manufacturers do not really know where their luminaires will end up, and because the little two-pole devices cost only pennies, the trend is to simply include these arrangements in most of the conventional luminaires sold today. The 2014 NEC has removed the prior exception for qualified industrial occupancies.

410.135. Open-Circuit Voltages Exceeding 300 Volts. This rule, limiting luminaires in dwelling occupancies with open circuit voltages above 300 unless special construction features are in place, may be dated. The only dwelling-unit location limitation markings in the current UL Luminaire Marking Guide are those that prohibit open circuit voltages over 1000 [and thereby enforce 410.140(B)] and those that prohibit luminaires marked for supply wiring rated over 90°C. The section is being retained, however, because it provides a basis for requirements in the relevant product standards, particularly as applicable to cold-cathode electric discharge lighting systems being listed for use in dwelling units.

410.136. Luminaire Mounting. Paragraph **(B)** of this section has been essentially unchanged for over 50 years, but remains valid although there are no luminaires now listed that qualify for such a marking. Until 2000, the UL Guide Card data on this product category read as follows: "Fluorescent luminaires suitable for mounting on combustible low-density cellulose fiberboard ceilings which have been evaluated for use with thermal insulation above the ceiling and which have been investigated for mounting directly on combustible low-density cellulose fiberboard ceilings are "marked 'Suitable for Surface Mounting on Combustible Low-Density Cellulose Fiberboard'" (Fig. 410-28). With no listings in sight, UL has dropped this explanation from its Guide Card information. This type of ceiling is not common, but if it is encountered, the spacing rule given here must be applied because no listed luminaire will likely be available to mount against this material.

410.137. Equipment Not Integral with Luminaire. Part **(C)** covers interconnection of "paired" luminaires, with a ballast supplying a lamp or lamps in both, using up to a 25-ft (7.5-m) length of metric designator 12 (trade size $^{3}/_{8}$) flexible metal conduit to enclose the circuit conductors. Lighting manufacturers have been making UL-listed pairs of luminaires for use with each other. Because some inspection agencies have questioned use of this equipment, this section describes Code-conforming use of such equipment to thereby resolve field controversy. Two-lamp ballasts are more efficient than one- or three-lamp versions, and the

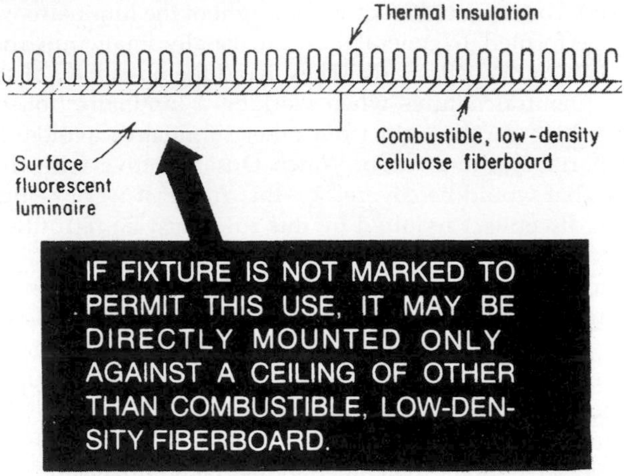
Thermal insulation
Combustible, low-density cellulose fiberboard
Surface fluorescent luminaire
IF FIXTURE IS NOT MARKED TO PERMIT THIS USE, IT MAY BE DIRECTLY MOUNTED ONLY AGAINST A CEILING OF OTHER THAN COMBUSTIBLE, LOW-DENSITY FIBERBOARD.

Fig. 410-28. There are no luminaires currently listed for this use. [Sec. 410.136(B).]

paired luminaire concept allows a two-lamp ballast in one luminaire to feed the odd lamp in its paired section.

410.138. Autotransformers. This rule ties in with the rules of 210.6 on voltage of branch circuits to luminaires. On neutral-grounded wye systems (such as 120/208 or 277/480) incandescent, fluorescent, mercury-vapor, metal-halide, high-pressure sodium, and low-pressure sodium equipment can be connected from phase to neutral on the circuits. If fluorescent or mercury-vapor luminaires are to be connected phase to phase, some Code authorities contend that autotransformer-type ballasts cannot be used when they raise the voltage to more than 300 V, because, they contend, the reference to "a grounded system" in this rule of 410.138 calls for connection to a circuit made up of a grounded wire and a hot wire (Fig. 410-29). On phase-to-phase connection they would require use of 2-winding (electrically isolating) ballast transformers. The wording of 410.138 does, however, lend itself to interpretation that it is only necessary for the supply *system* to the ballast to be *grounded*—thus permitting the two hot legs of a 208- or 480-V circuit to supply an autotransformer because the hot legs are derived from a neutral-grounded "system." Further, the rule in 210.9 restricts the use of autotransformers as a supply point for branch circuits. Because a luminaire ballast has no field wiring connected to its load side, it is difficult to see how this is a significant problem.

410.140. General. These sections apply to interior neon-tube lighting for general illumination, lighting with long fluorescent tubes requiring more than 1000 V, and cold cathode fluorescent-lamp installation arranged to operate with several tubes in series. Note that as soon as the neon tubing "outline[s] or call[s] attention to certain features such as the shape of a building or the decoration of a window" or if it conveys information in the form of a sign, then it is covered as outline lighting or as a sign, respectively, under the applicable provisions of Art. 600. As noted in part **(B)**, electric-discharge lighting equipment with open-circuit voltage exceeding

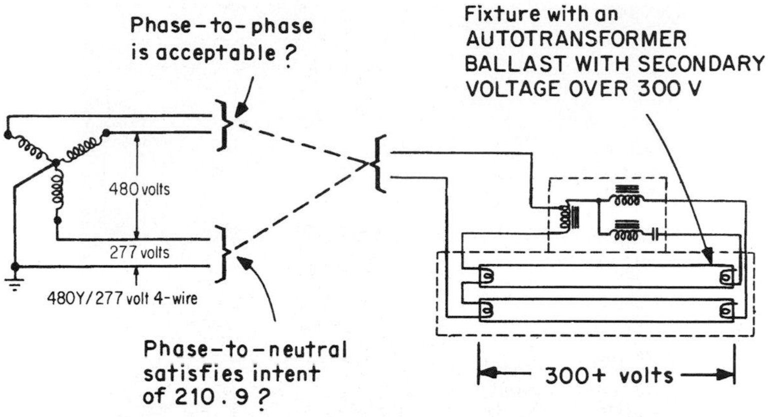
Phase-to-phase
is acceptable ?
Fixture with an
AUTOTRANSFORMER
BALLAST WITH SECONDARY
VOLTAGE OVER 300 V
480 volts
277 volts
480Y/277 volt 4-wire
Phase-to-neutral
satisfies intent
of 210.9 ?
300+ volts

Fig. 410-29. Intent of 210.9 raises questions about connection of autotransformer ballasts. (Sec. 410.138.)

1000 V shall not be installed in dwelling occupancies. This is correlated with the UL marking requirements, as discussed at 410.135. Note that **(C)** specifically requires that electric discharge lamp terminals are to be considered live parts, by which it is presumably intended to indicate "uninsulated live parts."

410.141. Control. When any part of the equipment is being serviced, the primary circuit should be opened and the servicers should have assurance that the disconnecting means will not be closed without their knowledge.

410.143. Transformers. See comments following 600.23.

410.144. Transformer Locations. In part **(B)** the Code addresses secondary conductors used with neon lighting. This type of cable is not included in the table in Chap. 3 listing various types of insulated conductors, but Underwriters Laboratories, Inc., have standards for such cables. It is now called out in 600.32(B) as Type GTO cable and is discussed at that location.

410.151. Installation. This covers the very popular and widely used track lighting (defined in Art. 100) made by a number of manufacturers and used so commonly today in residential and commercial interiors. In 410.151 through 410.155, extensive rules cover applications of track lighting mounted on ceilings or walls, while the rules for conductor sizing and overcurrent protection are given in 220.43(B). Specifically, these rules cover the installation and support of lighting track used to support and supply power to luminaires designed to be attached to the track at any point along the track length (Fig. 410-30). Unless otherwise specifically designed, lighting track must be supported at least every 1.2 m (4 ft) and it must be mounted at least 1.5 m (5 ft) above the elevation of the finished floor, although that distance can be lowered if there is appropriate protection from damage, or on low-voltage applications.

410.160. Decorative Lighting. Decorative lighting, such as holiday lighting in bushes, trees, and on architectural elements, must be listed even though it will only be used for temporary purposes pursuant to the 90-day limitation in 590.3(B).

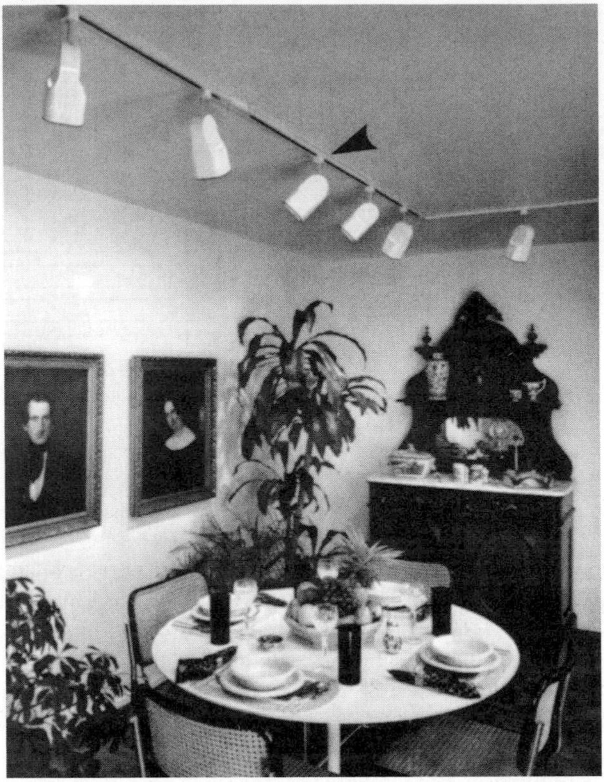

Fig. 410-30. Part XV of Art. 410 contains rules on installation, application, fastening, and circuit loading for standard and heavy-duty track lighting. However, rules for calculating required feeder circuit capacity are covered in 220.43(B). The rules in 220.43(B) are for feeders and do not limit the length of track connected to a single-branch circuit or the number of luminaires on a track. (Secs. 410.151 through 410.155.)

ARTICLE 411. LOW-VOLTAGE LIGHTING

411.1. Scope. This article, which was new in the 1996 NEC, covers the installation of so-called low-voltage lighting. This lighting, although power limited, is not necessarily power limited to the same degree as Class 2 or Class 3 circuits in Art. 725. For example, the upper limit on voltage is 30, but the upper limit on current is 25 A, making the potential available power roughly 6 times that of a Class 2 signaling circuit. Recent changes, largely driven by LED technology, have resulted in extensive systems covered by this article that operate entirely within Class 2 limits, and the article title now simply refers to low voltage. Generally the specified limits are 30 volts ac and 60 volts dc, but if wet contact is likely, then the voltage limits drop by half. A new informational note points to Art. 680 as the source of rules applicable to submersion.

411.3 Low-Voltage Lighting Systems. This sets the minimum components for these systems. They must have an isolating power supply, they must employ low-voltage luminaires, and any associated equipment must be identified for the application. The power supply output circuits must not exceed 25 amps under any load condition.

411.4. Listing Required. Effective with the 2008 **NEC**, there are two options regarding listings on these systems. The first option, is 411.4(A). Under this procedure the entire setup with all the components has to be listed as part of a complete system. The second option allows for a field assembly of listed parts, including the cord, cable (or other wiring method), luminaires and fittings, and the power supply. However, if the setup involves exposed bare wire as the system arrangement, then the luminaires and their associated fittings, along with the power supply, must all be listed for use as part of the same lighting system. An important clarification is that "fixed wiring method" (2014 **NEC**) is now "fixed Chap. 3 wiring method" with respect to the secondary circuit.

411.5. Specific Location Requirements. These systems must not encroach on the 3.0-m (10-ft) boundary around a swimming pool or similar location unless installed in accordance with a specific allowance in Art. 680. In other locations the conductors must not be concealed or run through partitions unless one of two approaches is taken. The first is to convert to a Chap. 3 wiring method. The second is to live within a Class 2 power limitation, in which case the low-voltage conductors can be concealed. Now that LED technology is steadily advancing, this is often feasible.

411.6. Secondary Circuits. These systems are a rare example in the **NEC** of a system that is forbidden to be grounded; neither of the two conductors on the load side of the transformer may be connected to the ground.

ARTICLE 422. APPLIANCES

422.1. Scope. See the definition of *appliance,* Art. 100. For purposes of the Code, the definition of an appliance indicates that it is utilization equipment other than industrial, which generally means small equipment such as may be used in a dwelling or office (clothes washer, clothes dryer, air conditioner, food mixer, coffee maker, etc.). See also the definition of *utilization equipment* in Art. 100.

422.5. Ground-Fault Circuit Interrupter (GFCI) Protection for Personnel. This new section as of the 2014 **NEC** requires that GFCI devices used to meet requirements in this article be readily accessible, correlating with the rule in 210.8. This is of particular importance to GFCI receptacles that supply older vending machines per 422.51(A), many of which are very difficult to move out of the way and provide access to the test buttons on the device. A GFCI circuit breaker may be the only practical option. This section now (2017 **NEC**) centralizes in one location appliance GFCI requirements in the article. As a result, former Sec. 23 on tire inflation and also automotive vacuum machines for public disappears, along with former Sec. 49 on spray washers, Sec. 51 on vending machines, and Sec. 52 on drinking water coolers (here renamed to match the product standard) all disappeared. The parent language in **(A)** sets consistent coverage limits for everything, namely, 250 volts or less, 60 amps. or less, and both single and three-phase. Part **(B)** on location, focusing on the actual equipment, allows the protection at the branch-circuit OCPD, or downstream in the supply circuit, or as an integral part of an attachment plug.

422.6. Listing. As of the 2017 NEC, if an appliance operates at 50 volts or more, it must be listed. This includes vending machines, for which the former definition was deleted. Instead, the NEC is, in effect, deferring to the product standard to define the term. There appear to be three UL Guide Card designations that apply to this equipment, and that may serve as a starting point to get a handle on what is included. The category that generally applies to vending machines is YWXV. Additionally, there are food and beverage applications covered under the TSYA designation, and if such dispensing is refrigerated, then the SQMX designation applies.

422.10. Branch-Circuit Rating. Part **(A)** states that the amp rating of an individual branch circuit to a single appliance must not be less than the marked ampere rating of the appliance.

422.11. Overcurrent Protection. The second sentence of part **(A)** presents a rule based on the fact that some appliances are marked to indicate the maximum permitted rating of a protective device (fuse or CB) for the branch circuit supplying that appliance.

The rule in part **(G)** requires application of the overload protection provisions of 430.32 and 430.33 to motors of motor-operated appliances. And overload protection for sealed hermetic compressor motors must satisfy 440.52 through 440.55. But motors that are not continuous-duty motors—and most appliances have intermittent, short-time, or varying-duty types of motor loads—do not require running overload protection. The branch-circuit protective device may perform that function for such motors. See 430.33.

Part **(E)** sets limits on the maximum permitted branch-circuit protection for an individual branch circuit supplying a single nonmotor-operated appliance. This rule is aimed at providing overcurrent protection for appliances that would not be adequately protected if too large a branch-circuit protective device were used ahead of them on their supply branch circuits.

The basic rule says that if a maximum rating or branch-circuit protective device is marked on the appliance, that value must be observed in selecting the branch-circuit fuse or CB. If, however, a particular appliance is *not* marked to show a maximum rating of branch-circuit protection, the values specified in this section must be used as follows:

For an appliance drawing more than 13.3 A—the branch-circuit fuse or CB mus not be rated more than 150 percent (1½ times) the full-load current rating of the appliance. For an appliance drawing up to 13.3 A—the branch-circuit fuse or CI must not be rated over 20 A.

Those values were chosen to limit the maximum rating of protection to not ove 150 percent of the appliance rating to afford greater protection to the appliance as well as to provide for continuous-load appliances (that operate for 3 h or mor continuously). In the latter case, 210.20(A) requires that a continuous load no exceed 80 percent of the branch-circuit fuse or CB; that is, the protective device i rated not less than 125 percent of the continuous-load current.

An appliance drawing 13.3 A must have protection of not over 150 percent c that value (1.5 × 13.3 = 20 A), which sets 20 A as the maximum protection. And 8 percent of a 20 A rating is 16 A—so that the 13.3-A load does not exceed that valu for a continuous load. For appliances drawing more than 13.3 A, the exception say that if 150 percent of the appliance current rating does not correspond to a standard rating of protective device (from 240.6), the next standard rating of protective device above that value may be used even though it is rated more than 150 percent of the appliance current rating.

Electric water heaters are typical nonmotor-operated appliances rated over 13.3 A, and 422.13 (see Fig. 422-1) says such water heaters of 120-gal (450-L) capacity or less must be classified as continuous loads, which means branch-circuit overcurrent device will be rated not less than 125 percent of the unit's current rating. And 422.11(E)(3) says the same branch-circuit device must not be rated more than 150 percent of the unit's current rating. So, for a 4500-W hot water tank on a 240 V system, the current draw will be 4500 VA ÷ 240 V = 19 A, 125 percent of that is 24 A and 150 percent of that is 28 A with the next higher standard size overcurrent device permitted. Therefore, in standard sizes the smallest size permitted is 25 A and the largest size is 30 A.

Fig. 422-1. Any fixed storage water heater with capacity of 120 gal (450 L) or less must be treated as a "continuous duty load" that does not load the circuit to more than 80 percent of its capacity. (Sec. 422.13.)

Part **(F)** of this section covers electric heating appliances using resistance-type heating elements. It requires that, where the elements are rated more than 48 A, the heating elements must be subdivided. Each subdivided load shall not exceed 48 A and shall be protected at not more than 60 A.

The rules of this section are generally similar to the rules contained in 424.22 for fixed electric space heating using duct heaters as part of heating, ventilating, and air-conditioning systems above suspended ceilings. But 422.11(F)(2) applies to commercial kitchen and cooking appliances using sheath-type heating elements. This section permits such heating elements to be subdivided into circuits not exceeding 120 A and protected at not more than 150 A under the conditions specified.

422.11(F)(3) of this same section permits a similar subdivision into 120-A loads protected at not more than 150 A for elements of water heaters and steam boilers employing resistance-type immersion electric heating elements contained in an ASME rated and stamped vessel. The 2014 **NEC** has broadened the scope of this allowance to include listed instantaneous water heaters and also water heaters incapable of building steam pressure. These include vented industrial process tanks ("low-pressure water heater tanks") or tanks directly connected to vented containers with no valves or restrictions ("open-outlet water heater vessels") separating the water from the atmosphere.

422.12. Central Heating Equipment. This rule requires a dedicated (individual) branch circuit to supply the electrical needs of "Central Heating Equipment" other than fixed electric space heating. This requires a separate circuit for the electrical ignition, control, fan(s), and circulating pump(s) of gas- and oil-fired central heating plants. The exception notes that auxiliary equipment "such as a pump, valve, humidifier, or electrostatic air cleaner directly associated with the heating equipment" *may* be connected to the same branch circuit (or another branch circuit). Using the noncoincident load concept of 220.60, permanently connected air-conditioning equipment can also be connected to the same circuit if desired.

The purpose of this rule is to prevent loss of heating when its circuit is opened due to a fault in a lamp or other appliance that is connected on the same circuit with the heater, as was permitted in the 1987 **NEC** and previous editions. All heating equipment should be supplied by one or more individual branch circuits that supply nothing but the heater and its auxiliary equipment.

422.13. Storage-Type Water Heaters. Water heaters with a capacity equal to or less than 450 L (120 gal) must be considered to be continuous loads. The branch-circuit overcurrent protective device (unless rated for full continuous loading) must be rated for not less than 125 percent of the water heater rating, and due to the operation of 210.19(A)(1) the sizing of the branch-circuit conductors (prior to the application of ampacity adjustment factors) follows the same procedure.

422.14. Infrared Lamp Industrial Heating Appliances (Deleted). This section now resides in Sec. 425.14 as part of the new article on industrial process heating equipment.

422.15. Central Vacuum Outlet Assemblies. This rule allows a local lighting circuit to supply the beater bar attachment or other accessories to a central vacuum system, in accordance with 210.23(A). Note that that reference carries with it the mandatory exception that disallows other connections to the dedicated appliance circuits. For example, if there is a central vacuum system outlet in a dining room, the local receptacle circuit is off-limits for this connection.

422.16. Flexible Cords. This section sets the rules for connecting certain appliances to their outlets through the use of a cord-and-plug connection. This is often a desirable design objective because when an appliance needs service, it can be disconnected without crossing trades. There are four specific headings in (B) that follow the general boilerplate rules in (A) that reiterate 400.10 and 12. Note that this language with respect to large appliances fastened in place specifically requires appropriate mechanical connections. This was aimed at disallowing cord-and-plug-connected hot water tanks.

Part **(B)(1)** covers "in-sink" waste disposers, as shown in Fig. 422-2. The designation shift from "kitchen" to "in-sink" reflects the fact that these appliances should follow the rules in this part whether or not they are located in a kitchen. There are four requirements for the arrangement, including receptacle accessibility

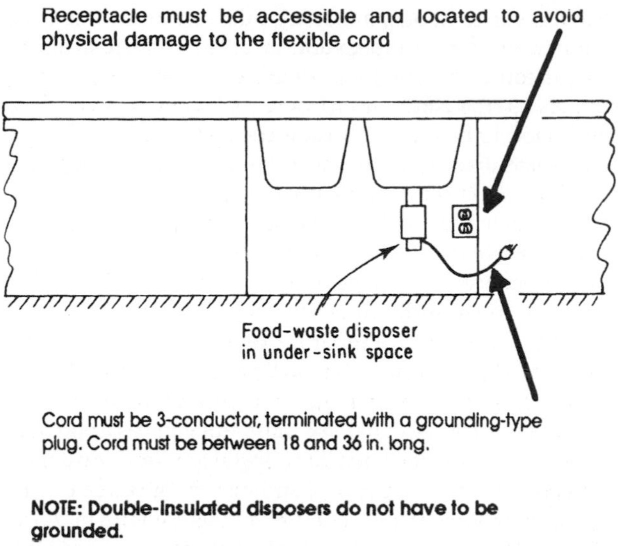
Receptacle must be accessible and located to avoid physical damage to the flexible cord
Food-waste disposer in under-sink space
Cord must be 3-conductor, terminated with a grounding-type plug. Cord must be between 18 and 36 in. long.
NOTE: Double-insulated disposers do not have to be grounded.

Fig. 422-2. Code rules aim at proper cord connections for in-sink garbage disposers. [Sec. 422.16(B)(1).]

location to avoid cord damage, permitted cord length [450 to 900 mm (18 in. to 3 ft)] and grounding or double insulation provisions. The parent language also requires that the cord set be identified in the installation instructions, so this is not an item that can be made in the field.

Part **(B)(2)** covers built-in dishwashers and trash compactors. The same conditions apply here, with two differences. The receptacle for a compactor must be in the same space, or it can be adjacent thereto, which is why these receptacles often end up in the sink base. This is often preferable in terms of design, because a receptacle under the sink is usually easier to get at quickly, and as of the 2017 NEC this is now mandatory. Make sure that the hole in the sink base to accommodate the cord is big enough, even after plumbing connections are in place, to pass the cord cap without having to unwire it. If the cord has to be unwired, it certainly takes on the flavor of permanent wiring, contrary to 400.12(1). The second difference is the cord can be longer, to better accommodate the adjacent receptacle. These cords can be 0.9 to 1.2 m (3 to 4 ft) long in the case of a compactor, and 0.9 to 2.0 m (3 to 6½ ft) for a dishwasher, and actually longer than that because the dimension is measured from the back plane of the appliance. Some modern kitchen designs place the dishwasher remote from the sink, which will make these rules challenging. One option always available is to hard-wire the appliance. Beyond that, ingenuity must prevail. The author once placed a receptacle behind a nearby filler panel spacing installed to accommodate a non-standard wall spacing dimension. With hidden hinges and a hidden latch the receptacle was easily accessible by a pull from the toe space. As always, communication with other trades is essential.

Part **(B)(3)** covers wall-mounted ovens and counter mounted cooking units. They are permitted to be cord-and-plug-connected, using either a receptacle or a mating cord body, but the assembly must be approved for the temperature of its location.

Part **(B)(4)** covers range hoods, including built-in microwave ovens with range hoods incorporated underneath them and installed as a single appliance. Historically range hoods could only be hard-wired; however, with the advent of these combined appliances, the NEC now allows for either a hood or a combination microwave and hood to be cord-and-plug-connected. The conditions are similar to the ones for waste disposers (the cord can be a foot longer), with one major additional requirement. The receptacle must be connected as an individual branch circuit, even if the original connection is to a 3-A range hood only. The intent is to prepare for a microwave oven combination later. And as discussed at the definition of "individual branch circuit" in Art. 100, a circuit supplying both halves of a duplex receptacle is not an individual branch circuit in most cases, because each half of the duplex is classified as a separate device.

422.18. Support of Ceiling Fans. This section has been rewritten to simply refer to 314.27(C) because this Code panel does not have the relevant technical jurisdiction. These rules have very little to do with paddle fans as appliances, and everything to do with how a box designed for a static load will respond to a dynamic load. Permission to use a locking support and mounting receptacle (see Fig. 100-28) is now recognized here for use with paddle fans, as covered in 314.27(E).

422.19. Space for Conductors. This rule correlates this article's coverage of paddle fans with 314.16(B)(1), which contains within it and its exception all the relevant requirements.

422.20. Outlet Boxes to be Covered. This rule correlates paddle fan requirements with 314.25.

422.21. Covering of Combustible Material at Outlet Boxes. This rule, new for the 2014 edition, was supposed to correlate the paddle fan rules with equivalent exposures for luminaires in Sec. 410.23. However, (refer to the discussion and drawing at that section) the luminaire rule no longer has any relevance to a paddle fan canopy given the new sizing threshold. The 2017 NEC now correlates the requirement here with 410.23, which as a practical matter removes the requirement from the Code entirely. No paddle fan canopy will be that large.

422.30. General. Appliances must be wired on the load side of a disconnecting means in accordance with Part **III** of the article, and if multiple disconnecting means are required for a given appliance, then those disconnecting means must be grouped and "identified" as such, with each disconnecting means simultaneously disconnecting the conductors it controls. The word "identified" here is incorrect and should say "marked."

422.31. Disconnection of Permanently Connected Appliances. This section divides the universe of permanently connected appliances into three parts. In part **(A)** lower-rated appliances (not over 300 VA or 1/8 hp) may use the "branch-circuit overcurrent device" as their disconnect means and such a device could be a plug-fuse or a CB. However, as of the 2017 NEC even such a trivial appliance must either be in sight of the overcurrent device, or the device must have a lock-open capability. In other words, there is no longer any operational difference between (A) and (B). This is somewhat at odds with customary NEC requirements. For example, 430.109(B) maintains the recognition of branch-circuit protective devices generally at 1/8 hp and below. In addition, 422.31(C) Exception grants relief to permanently connected appliances over 1/8 hp with unit switches; no such relief applies to smaller appliances. In part **(B)** higher-rated nonmotor-operated appliances does not permit use of a plug-fuse as the disconnect but requires a definite switch-action device—a switch or CB. A permanently connected nonmotor-operated appliance rated over 300 VA may use its branch-circuit switch or CB as the required disconnecting means *if* the switch or CB is within sight from the appliance or it can be locked in the open position. For 2008 in this and in countless other locations, additional language has been inserted clarifying that the lockout mechanism must be a permanent feature that stays with the switch or circuit breaker even when the lock is removed. Note that this section applies *only* to permanently connected appliances—that is, those with fixed-wiring connection (so-called hard wired) and not cord-and-plug connection.

In part **(A)**, the overcurrent device for the circuit to the appliance is not required to be "within sight"—that is, visible and not more than 50 ft (15.0 m) away. But the switch or CB in part **(B)** must be within sight or must be capable of being locked open.

Part **(C)** covers permanently connected appliances having motor loads rated over $^1/_8$ hp. The branch-circuit switch or circuit breaker is permitted to be used as the disconnect if it is in sight of the appliance, or equipped with a lock-open capability or it can also be out of sight if the appliance has a qualifying unit switch as covered in 422.34 (Fig. 422-3).

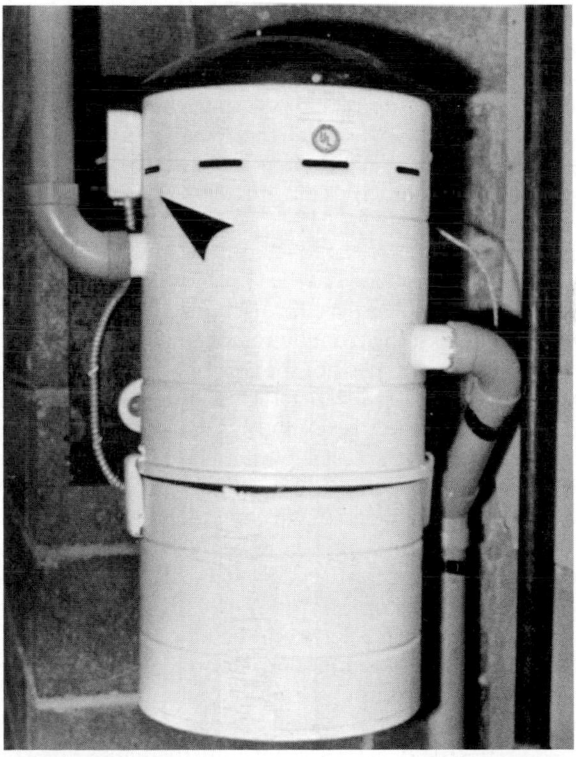

Fig. 422-3. Toggle switch in outlet box on this central vacuum cleaner serves as the disconnecting means within sight from the motor controller installed in the top of the unit. [Sec. 422.31(C).]

According to the exception in this rule, the branch-circuit switch or CB serving as the other disconnect required by 422.34(A), (B), (C), or (D) is permitted to be out of sight from the motor controller of an appliance that is equipped with a unit switch that has a marked OFF position and opens all ungrounded supply conductors to the appliance, as shown in Fig. 422-4.

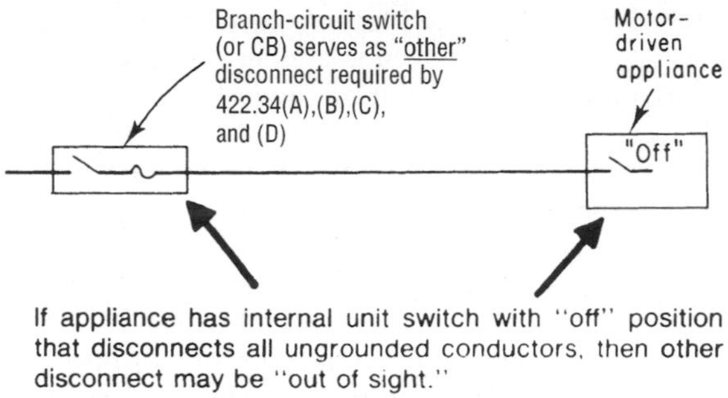
Branch-circuit switch (or CB) serves as "other" disconnect required by 422.34(A),(B),(C), and (D)
Motor-driven appliance
"Off"
If appliance has internal unit switch with "off" position that disconnects all ungrounded conductors, then other disconnect may be "out of sight."

Fig. 422-4. Disconnect for motor-driven appliance may be "out of sight from the motor controller." [Sec. 422.31(C) Exception; Sec. 422.34.]

422.33. Disconnection of Cord-and Plug-Connected Appliances. Examples of the application of this section to disconnecting means for appliances are found in the installation of household electric ranges and clothes dryers. The purpose of these requirements is to provide that for every such appliance there will be some means for opening the circuit to the appliance when it is to be serviced or repaired or when it is to be removed. This section also comprehensively recognizes the receptacle and locking attachment fittings shown in Fig. 100-28 and covered in 314.27(E).

In part **(B)**, household electric ranges may be supplied by cord-and-plug connection to a range receptacle located at the rear base of the range. The rule permits such a plug and receptacle to serve as the disconnecting means for the range if the connection is accessible from the front by removal of a drawer.

This rule refers to electric ranges but the concept also applies to gas ranges. For instance, there have been 115-V receptacle outlets installed behind gas ranges in mobile homes. Such a receptacle is used only as an outlet for the oven light and clock on the range and is not accessible after the range is installed. In order to disconnect the attachment plug or plugs from the receptacle, the range gas supply pipe has to be disconnected, the frame of the range has to be disconnected from its floor fastening, and then the range has to be moved in order to reach the receptacle outlet where the cords are plugged in (Fig. 422-5). Is such an installation in conformity with the intent of the Code? The answer seems surely to be No. The inaccessibility of the receptacle would be objectionable.

422.34. Unit Switch(es) as Disconnecting Means. As shown in Fig. 422-6, the ON-OFF switch on an appliance, such as a cooking unit in a commercial establishment, is permitted by part **(D)** to serve as the required disconnecting means if the user of the appliance has ready access to the branch-circuit switch or CB. Note that the wording does not recognize simply "the branch-circuit overcurrent device,"

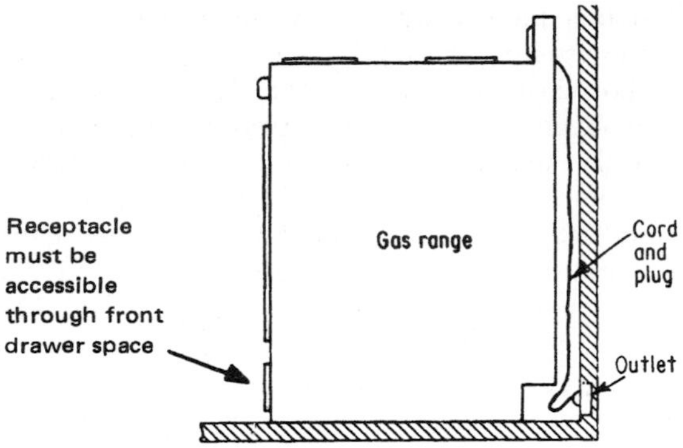
Receptacle must be accessible through front drawer space
Gas range
Cord and plug
Outlet

Fig. 422-5. Rule on receptacle behind electric ranges could be applied to gas ranges. (Sec. 422.33.)

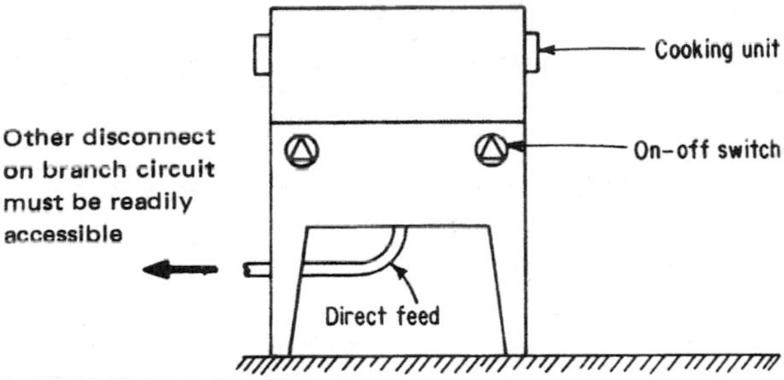
Cooking unit
On-off switch
Other disconnect on branch circuit must be readily accessible
Direct feed

Fig. 422-6. An ON-OFF switch on an appliance must be supplemented by an additional disconnect means. (Sec. 422.34.)

and a plug-fuse in the circuit to the appliance would not, therefore, be acceptable as the additional means for disconnection.

For residential occupancies the rules vary by the number of units. A single-family home may simply rely on its service disconnect; in a two-family home the disconnect can be within or without and may consist of a feeder disconnect that disconnects other loads. It can be a service disconnect if the two-family dwelling is equipped with multiple service disconnects such that opening the disconnect amounts to the opening of an "individual switch or circuit breaker for the dwelling unit." For multifamily applications an individual dwelling feeder disconnect is also permitted, but must be within the dwelling or on the same floor.

422.41. Cord-and-Plug-Connected Appliances Subject to Immersion. This is the rule that mandates the in-cord protection for hair dryers and portable hydromassage units.

422.42. Signals for Heated Appliances. The standard form of signal is a red light so connected that the lamp remains lighted as long as the appliance is connected to the circuit. No signal lamp is required if the appliance is equipped with a thermostatic switch that automatically opens the circuit after the appliance has been heated to a certain temperature. This rule does not apply in dwelling occupancies.

422.49. High-Pressure-Spray Washers. As required by this rule, a portable high-pressure-spray washing machine is required to have a "factory-installed" GFCI. *And* this factory-installed device *must* be "an integral part" of the plug cap or in the cord itself no more than 12 in. (300 mm) from the plug cap. As of the 2014 NEC, the rule now also applies to 3-phase equipment rated 208Y/120V and 60 A (or less), but it does not apply to washers operating over 250 V.

422.51. Cord-and-Plug-Connected Vending Machines. This rule requires an integral GFCI protector for all cord-and-plug-connected vending machines "manufactured or re-manufactured on or after January 1, 2005." This GFCI protection must be installed by the manufacturer within 12 in. of the plug cap and be identified for portable use, which means it will have open neutral protection. For nonintegral-GFCI-protected vending machines, a GFCI-protected receptacle must be used. The term "vending machine" now has a formal definition in 422.2. As of the 2014 NEC, circuit-level GFCI protection for hard-wired vending machines is now a requirement.

422.52. Electric Drinking Fountains. GFCI protection is now mandated for these units, either as part of the appliance, or at or on the line side of the outlet. In other words, if it is not built into the drinking fountain, then there must be a GFCI receptacle or GFCI circuit breaker on the line side. Note that this rule does not apply to bottled water coolers.

ARTICLE 424. FIXED ELECTRIC SPACE-HEATING EQUIPMENT

424.3. Branch Circuits. The basic rule of part **(A)** limits fixed electric space-heating equipment to use on 15-, 20-, 25-, or 30-A circuits, *if the circuit has more than one outlet.* The second sentence of the paragraph applies only to fixed infrared equipment on industrial and commercial premises, permitting use of 40- and 50-A circuits for multioutlet circuits to *fixed* space heaters.

The continuous load requirement in paragraph **(B)** means that branch circuits for electric space heating equipment cannot be loaded to more than 80 percent of the branch-circuit rating. Even though electric heating is thermostatically controlled and is a cycling load, it must be taken as a continuous load for sizing branch circuits.

The three parts of this rule are as follows:

1. The branch-circuit wires must have an ampacity not less than 1.25 times the total amp load of the equipment (heater current plus motor current).
2. The overcurrent protective device must have an amp rating not less than that calculated for the branch-circuit wires.
3. If necessary, the rating or setting of the branch-circuit overcurrent device may be sized according to part **(B)** of 240.4. That is, if the ampacity of the selected branch-circuit wires does not correspond to a standard rating or setting of protective device (240.6), the next higher standard rating or setting of protective device may be used. This accommodates those applications of large unit heaters, where often there is not agreement in ampacity of wires and ratings of standard fuses or CBs.

Figure 424-1 shows an example of branch-circuit sizing for an electric heat unit. Many line thermostats and contactors are approved for 100 percent load, and derating of such devices is not required.

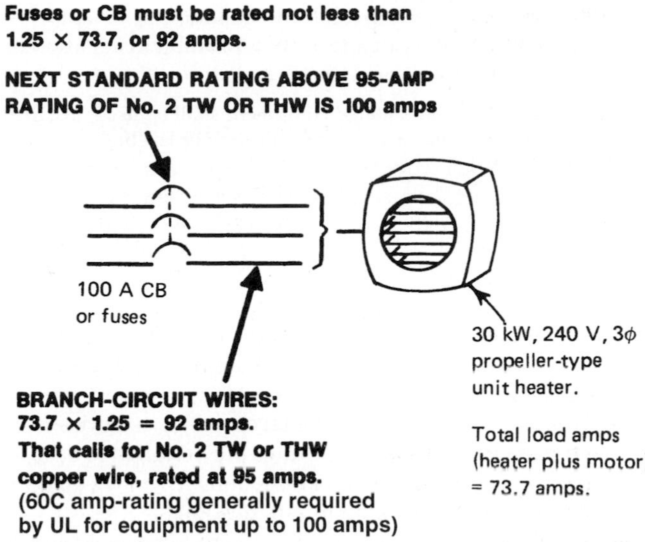
Fuses or CB must be rated not less than
1.25 × 73.7, or 92 amps.
NEXT STANDARD RATING ABOVE 95-AMP
RATING OF No. 2 TW OR THW IS 100 amps
100 A CB
or fuses
30 kW, 240 V, 3ϕ
propeller-type
unit heater.
BRANCH-CIRCUIT WIRES:
73.7 × 1.25 = 92 amps.
That calls for No. 2 TW or THW
copper wire, rated at 95 amps.
(60C amp-rating generally required
by UL for equipment up to 100 amps)
Total load amps
(heater plus motor)
= 73.7 amps.

Fig. 424-1. Branch-circuit conductors and overcurrent protection must be rated not less than 125 percent of heater unit nameplate current. (Sec. 424.3.)

Section 220.51 covers sizing of feeders for electric space-heating loads. The computed load of a feeder supplying such equipment shall be the total connected load on all branch circuits, with an exception left up to the authority enforcing the Code, which allows permission for feeder conductors to be of a capacity less than 100 percent, provided the conductors are of sufficient capacity for the load serving units operating on duty-cycle, intermittently, or from all units not operating at one time.

424.6. Listed Equipment. Baseboard heat, duct heaters, radiant heating systems, etc. must now be listed.

424.9. General. For instance, heating cable designed for use in ceilings may not be used in concrete floors and vice versa.

This rule ties into that of 210.52(A). The note that follows says that UL-listed baseboard heaters must be installed in accordance with their instructions, which *may* prohibit installation below outlets. This applies to all occupancies—residential, commercial, institutional, and industrial.

In buildings warmed by baseboard heaters, the question arises: How can wall receptacles be provided to satisfy the requirement of NEC 210.52(A) that no point along the floor line be more than 6 ft (1.8 m) from an outlet? Should the receptacles be installed in the wall above the heaters? The answer is that receptacles should not be placed above the heaters unless they have been designed to anticipate such usage, and such heaters probably don't exist. Fires have been attributed to cords being draped across heaters. Continued exposure to heat causes the cord insulation to become brittle, leading to possible short circuits or ground faults. Note that this limitation applies to electric baseboard heaters, and not to conventional central heating radiators configured as baseboard heating units using circulating hot water supplied by a furnace.

Article 210 does not specifically prohibit installation of receptacles over baseboard heaters, and Art. 424 does not specifically prohibit installation of baseboard heaters under receptacles. However, 110.3(B) says:

Installation and Use. Listed or labeled equipment shall be used, installed, or both, in accordance with any instructions included in the listing or labeling.

Underwriters Laboratories requires a warning that a heater is not to be located below an electrical convenience receptacle. A similar statement is included in the UL *Guide Card information* [see discussion at 110.3(B) for details]:

To reduce the likelihood of cords contacting the heater, the heater should not be located beneath electrical receptacles.

This instruction and the provisions of 110.3(B) make it very clear that installation of electric baseboard heaters under receptacles may be a Code violation.

Receptacles can be provided in electrically heated (baseboard-type) occupancies as required by 210.52 by making use of the receptacle accessories made available by baseboard heater manufacturers. These units are designed to be mounted at the end of a baseboard section or between two sections. Or, one could opt to utilize floor-mounted receptacles as described in 210.52(A)(3).

424.19. Disconnecting Means. The basic rule requires disconnecting means with all poles functioning simultaneously and rated for 125 percent of the total load for the heater and motor controller(s), plus supplementary overcurrent protective devices for all fixed electric space-heating equipment. If multiple circuits supply the equipment, then there will be multiple disconnects, and they must be grouped. As in other locations, the disconnects must be lockable and the locking provisions must not be removable when the lock itself is removed, as required in 110.25.

Part **(A)** of this section applies to heating equipment provided with supplementary overcurrent protection (such as fuses or CBs) to protect the subdivided resistance heaters used in duct heating, as required by 424.22. The basic rules are shown in Fig. 424-2. The disconnect in that drawing must comply with the following:

The disconnect must be "within sight from" the supplementary overcurrent panel. If circuit breakers are used as the supplementary overcurrent protective devices for the subdivided electric heating loads [424.22(B)], the circuit breakers may constitute the disconnects required by this section if they have the required locking provisions. This rule has to be correlated with the parent language and other rules based on whether there is a motor over $^1/_8$ hp. If there isn't (or the motor is smaller), then the disconnects in (A) are acceptable as the heater disconnects. If there is, then the disconnects qualify if one of four options is in place. However, because of the lock-open amendments in the 2008 **NEC** in the parent language of 424.19, the disconnects for (or constituting in the case of circuit breakers) the supplementary heat and other loads are guaranteed to meet (2) that covers, among other options, a lock-open disconnecting means, and it is pointless to review the other options in detail. Note that the options in 424.19(A)(2) now line up with 422.31(C), which is sensible given the nature of the loads. However, the disconnect in 424.19 appears to make these requirements superfluous.

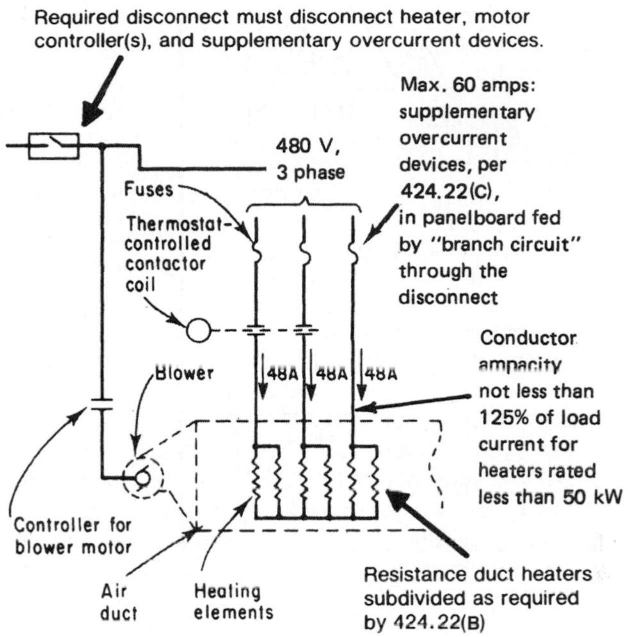
Required disconnect must disconnect heater, motor controller(s), and supplementary overcurrent devices.
Max. 60 amps: supplementary overcurrent devices, per 424.22(C), in panelboard fed by "branch circuit" through the disconnect
480 V, 3 phase
Fuses
Thermostat-controlled contactor coil
Blower
48A
48A
48A
Conductor ampacity not less than 125% of load current for heaters rated less than 50 kW
Controller for blower motor
Air duct
Heating elements
Resistance duct heaters subdivided as required by 424.22(B)

Fig. 424-2. Rules on disconnects for heating equipment demand careful study for HVAC systems with duct heaters and supplementary overcurrent protective devices. (Sec. 424.19.)

Part **(B)** applies to heating equipment *without* supplementary overcurrent protection.

Care must be taken to evaluate each of the specific requirements in this Code section to actual job details involved with electric heating installations. For room thermostats, see the discussion at 424.20.

Figure 424-3 shows two conditions that relate to the rules of part **(B)** of this section. For the two-family house, the service panel is located in a rear areaway and is accessible to both occupants. Circuit breakers in the panel constitute suitable means of disconnect for heaters in both apartments—*but* only if they are lockable. If the house uses baseboard heaters without motors in them, lockable breakers would be acceptable as the disconnects in accordance with the rule of **(B)(1).** For the four-family house (a "multifamily dwelling"), the service panels accessible to all tenants are grouped in the hallway under the first-floor stairway in this four-family occupancy. If lockable, the switches in the panels may constitute suitable means of disconnect for heaters in all apartments, under the rule of **(B)(1).** It should be noted that if the CBs are *not* capable of being locked open, then another disconnect that *is* lockable *or* within sight of the heater must be provided. And in either case, the locking provision must meet the requirements in the parent language in 424.19. But note that plug-fuses (without switches) on branch circuits to the heaters would not satisfy **(B)(1).**

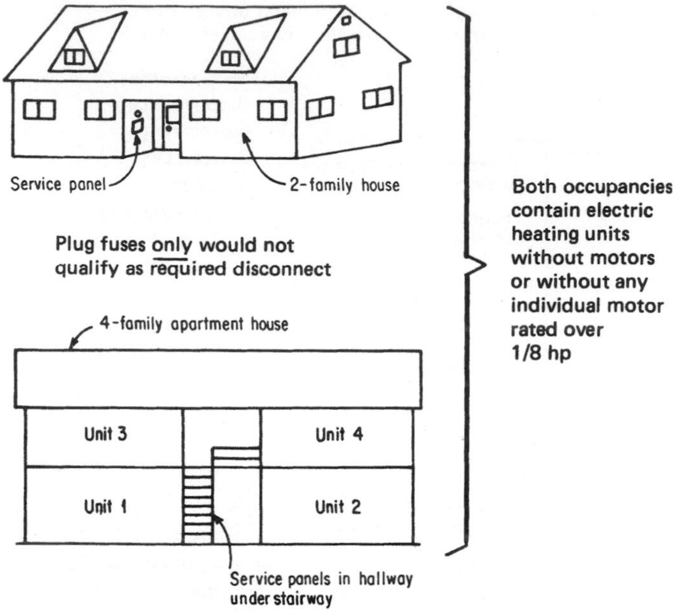
Service panel
2-family house
Plug fuses only would not qualify as required disconnect
4-family apartment house
Unit 3
Unit 4
Unit 1
Unit 2
Service panels in hallway under stairway
Both occupancies contain electric heating units without motors or without any individual motor rated over 1/8 hp

Fig. 424-3. For nonmotored electric heating units, a branch circuit switch or CB that is "within sight" from the heater or lockable may serve as disconnect. (Sec. 424.19.)

Note that **(B)(2)** requires a disconnect "within sight from" a motor controller serving a motor-operated space heater with a motor over $^1/_8$ hp, while permission is also given to use the rule of 424.19(A)(2).

Part **(C)** permits the ON-OFF switch on the heating unit to be used as the disconnecting means where "other means for disconnection" are provided, depending on the occupancy.

424.20. Thermostatically Controlled Switching Devices. Figure 424-4 shows a hookup of duct heaters that are controlled by a magnetic contactor that responds to a thermostatic switch in its coil circuit. The subdivided heater load of 48 A per leg satisfies the rule of 424.22(B), *but* the contactor does not constitute a combination controller-and-disconnect means because it does not open all the ungrounded conductors of the circuit. And because the heater contains the 60-A supplementary overcurrent protection, a disconnect ahead of the fuses would be required by 424.19(A) even if the contactor did open all three ungrounded conductors of the circuit. Note that the usual across-the-line room heating thermostats with a marked OFF position that open all ungrounded conductors in that position qualify as disconnects for baseboard heaters within sight of the thermostat, and thereby meet 424.19(B)(1) without further action to provide lockout mechanisms to circuit breakers, etc., elsewhere in the circuit.

424.22. Overcurrent Protection. Heating equipment employing resistance type heating elements rated more than 48 A must have the heating elements

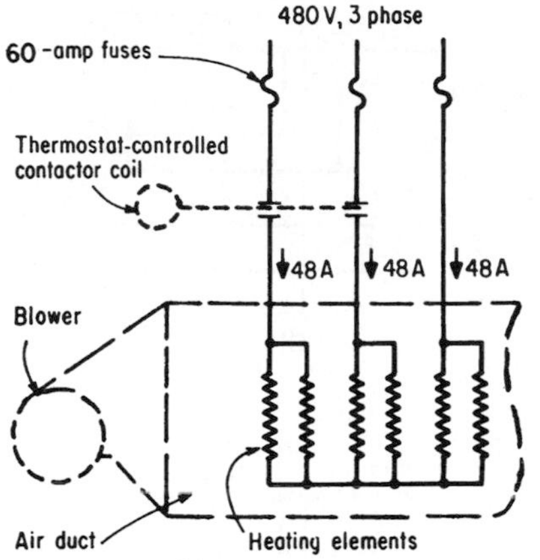
480 V, 3 phase
60-amp fuses
Thermostat-controlled contactor coil
48 A
48 A
48 A
Blower
Air duct
Heating elements

Fig. 424-4. This contactor may not serve as both controller and disconnect. (Sec. 424.20.)

subdivided, with each subdivided circuit loaded to not more than 48 A and protected at not more than 60 A. And each subdivided load must not exceed 80 percent of the rating of the protective device, to satisfy 424.3(B). Such a 60-A circuit could be classed as an individual branch circuit supplying a "single" outlet that actually consists of all the heater elements interconnected. By considering it as an "individual branch circuit" there is no conflict with 424.3(A), which sets a maximum rating of 30 A for a multioutlet circuit. The resistance-type heating elements on the market are not single heating elements in the 48-A size. They are made up of smaller wattage units in a single piece of equipment. The Code rule states that this single piece of equipment made up of smaller units must not draw more than 48 A and must be protected at not more than 60 A. Thus a heater of this type is limited to 48 A for each subdivided circuit. The subdivision is usually made by the manufacturer in the heater enclosure or housing.

Part **(B)** points to 424.3(B), whose continuous load requirement effectively requires that a resistance load less than 48 A have protection rated not less than 125 percent of heater current. The sentence at the end of this paragraph covers overcurrent protection sizing for subdivided resistance-heating-element loads that are less than 48 A.

Figure 424-5 shows an example of subdivision of heater elements in a heat pump with three 5-kW strip heaters in it. At 240 V, each 5-kW strip is a load of about 21 A. Two of them in parallel would be 42 A and that is not in excess of the 48-A maximum set by part **(B)** of this section. The three heaters would be a load of 63 A in parallel. There are a number of ways the total load might be supplied, but the Code rules limit the actual permitted types of hookup:

Section 424.3(A) states that an individual branch circuit may supply any load, but that permission is qualified by part **(B)** of this section, which requires resistance heating loads of more than 48 A to be subdivided so that no subdivided heater load will exceed 48 A, protected at not more than 60 A. As shown at *A*, one possible way to hook up the heaters is to use two strips on one circuit for a connected load of 2 × 21, or 42 A, and the other on one circuit with a connected

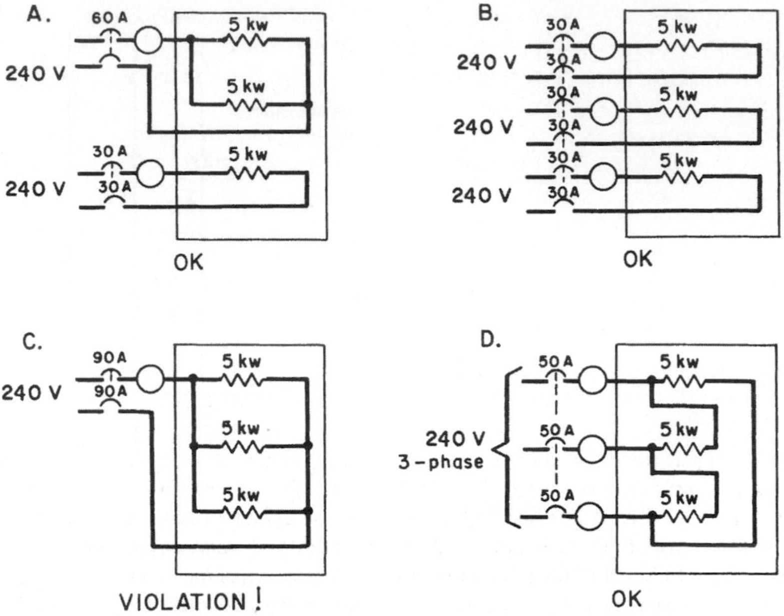
A.
60 A
5 kw
240 V
5 kw
30 A
5 kw
240 V
30 A
OK
B.
30 A
5 kw
240 V
30 A
30 A
5 kw
240 V
30 A
30 A
5 kw
240 V
30 A
OK
C.
90 A
5 kw
240 V
90 A
5 kw
5 kw
VIOLATION!
D.
50 A
5 kw
50 A
5 kw
240 V
3-phase
50 A
5 kw
OK

Fig. 424-5. Heater units must be limited to 48-A load with protection not over 60 A. (Sec. 424.22.)

load of 21 A. The two heaters would require a minimum ampacity for overcurrent protection of 42 × 1.25, or 53 A, calling for a 60-A overcurrent device. For the other circuit, 21 × 1.25 equals 26 A, requiring a 30-A fuse or breaker. Both circuits would thus be within the limits of a 48-A connected load and 60-A protection. As shown at *B*, it would also be acceptable to use a 30-A, 2-wire, 240-V circuit to each heater. But use of a single circuit sized at 1.25 × 63 A (79 A) and protected by 80-A fuses or breaker would clearly violate the Code rule of 60-A maximum protection, as at *C*.

Another possibility that would give a better balance to the connected load would be to feed the load with a 3-phase, 3-wire, 240-V circuit, if available. For such a circuit, the loading would be:

$$\frac{15{,}000}{1.732 \times 240} = 36\text{ A}$$

The minimum rating would be 36 A × 1.25, or 45 A, which calls for 45-A overcurrent protection. Since 50 A protective devices are more common, and since such a rating is still below the 60 A maximum for resistance elements, the 50 A rating in *D* is OK. Note that if the same furnace were connected to a 208 V 3-phase circuit, the lower voltage across the same resistance would lower the current to 31 A. Don't make the mistake of simply plugging 208 V into the above formula and expecting the current draw to be 42 A.

It should be noted that the rule of part **(B)** in this section applies to *any type* of space-heating equipment that utilizes resistance-type heating elements. The rule applies to duct heaters (as in Fig. 424-4), to the strip heaters in Fig. 424-5, and to heating elements in furnaces.

The purpose of paragraph **(C)** of this section is to require the heating manufacturer to furnish the necessary overcurrent protective devices where subdivided loads are required.

Per Part **(D)**, main conductors supplying overcurrent protective devices for subdivided loads are considered as branch circuits to avoid controversies about applying the 125 percent requirement in 424.3(B) to branch circuits *only*. It is not the intent, however, to deny the use of the *feeder tap* rules in 240.21 for these *main* conductors.

Paragraph **(E)** requires that the conductors used for the subdivided electric resistance-heat circuits specified in 424.22(C) must have an amp rating not less than 125 percent of the rating or setting of the overcurrent protective device protecting the subdivided circuit(s) (Fig. 424-6). The last paragraph provides what amounts to an exception to the basic rule, for heaters rated 50 kW or more where under the conditions specified it is permissible for the conductors to have an ampacity not less than the load of the respective subdivided circuits, rather than 100 percent of the rating of the protective devices protecting the subdivided circuits.

The wording of part **(E)** clarifies the need for field-wired conductors rated not less than 125 percent of the load-current rating, which must not exceed 48 A. The rating of supplementary overcurrent protection must protect these circuit wires at their ampacity, although the next higher standard rating of protection may be used where the ampacity of the circuit wires does not correspond to a standard protective device rating.

For instance, if the subdivided resistance heating load is 43 A per phase leg, the branch-circuit conductors must have an ampacity at least equal to 1.25 × 43, or 53.8 A. That would call for No. 6 TW with an ampacity of 55 A. The next standard rating of protection is 60 A, and that size protection may be used, although 60 A is the maximum size permitted for the subdivided load. The rule of the last paragraph is shown in the bottom part of Fig. 424-6.

424.35. Marking of Heating Cables. As of the 2017 NEC, the former color coding for voltage ratings table has been deleted; however, the marking for volts and either amperes or watts still applies.

424.36. Clearances of Wiring in Ceilings. Figure 424-7 shows the details of this rule. The wire at *a* is okay because it is not less than 2 in. (50 mm) above the ceiling, but it must be treated as operating at a 50°C ambient. The same is true of the wire at *c*, because it is within the insulation. The correction factors in Table 310.15(B)(2)(a) show that TW wire (60°C-rated wire) must be derated to 58 percent (0.58) of its normal table ampacity when operating in an ambient of 41°C to 50°C.

424.38. Area Restrictions. Figure 424-8 shows installations of heating cables and their relation to Code rules. The blanket prohibition against heating cables leaving the room from which they originate, formerly in (A), has been removed in the 2017 NEC in favor of more targeted limitations in (B) that address the real issues.

Heating cables shall not be installed under or through walls. They must not pass over the top of walls where the wall "intersects" the ceiling, but the single

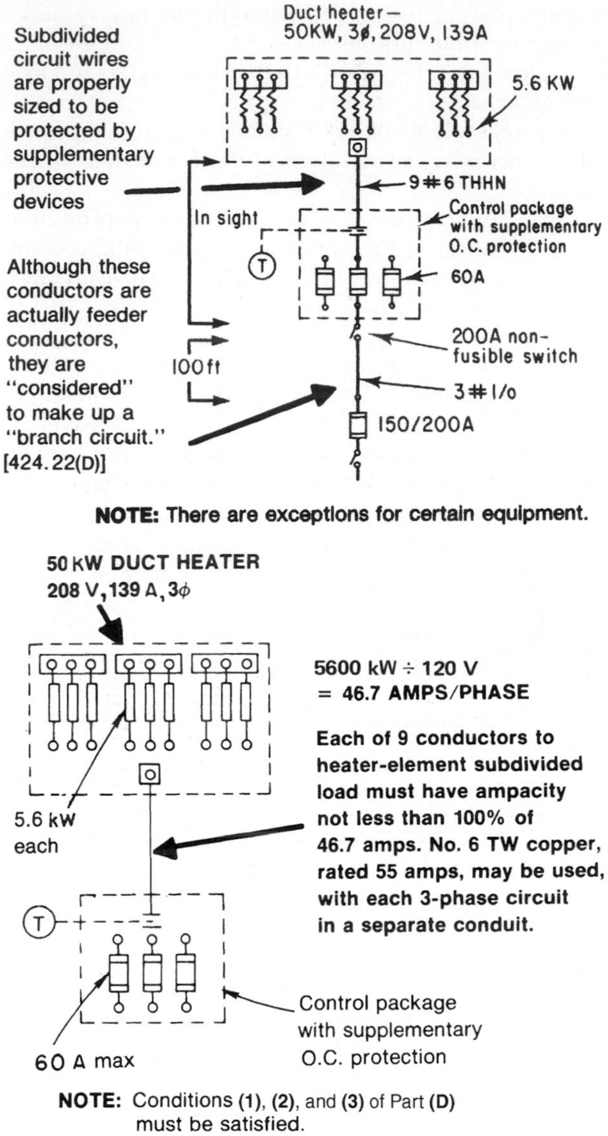

Duct heater—
50KW, 3ø, 208V, 139A
Subdivided circuit wires are properly sized to be protected by supplementary protective devices
5.6 KW
9 #6 THHN
In sight
Control package with supplementary O.C. protection
60A
Although these conductors are actually feeder conductors, they are "considered" to make up a "branch circuit." [424.22(D)]
100 ft
200A non-fusible switch
3 # 1/0
150/200A
NOTE: There are exceptions for certain equipment.
50 kW DUCT HEATER
208 V, 139 A, 3ϕ
5600 kW ÷ 120 V
= 46.7 AMPS/PHASE
Each of 9 conductors to heater-element subdivided load must have ampacity not less than 100% of 46.7 amps. No. 6 TW copper, rated 55 amps, may be used, with each 3-phase circuit in a separate conduit.
5.6 kW each
Control package with supplementary O.C. protection
60 A max
NOTE: Conditions (1), (2), and (3) of Part (D) must be satisfied.

Fig. 424-6. Conductors for subdivided heater circuits must fully match overcurrent device rating. (Sec. 424.22.)

isolated run of embedded cable is still permitted to pass over a "partition" that extends to the ceiling." The intent here is to avoid repeated crossings of cable over (or under) partitions, because radiation from these sections would be restricted or the cable would be unnecessarily exposed to possible physical damage. While the Code specifically speaks of partitions, the same reasoning would apply to arches

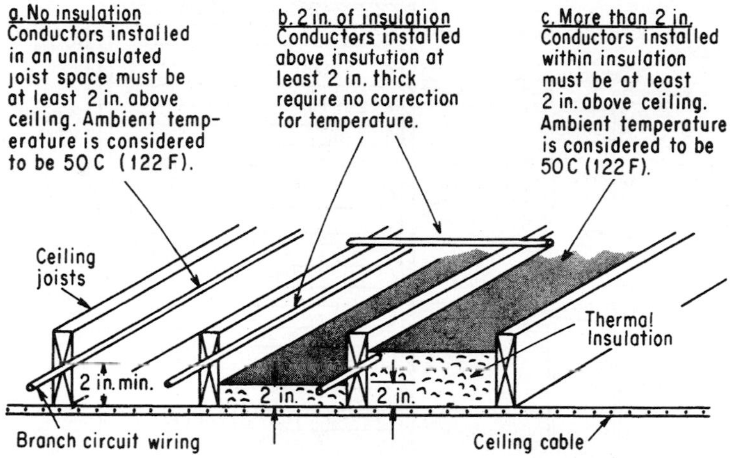
a. No insulation
Conductors installed in an uninsulated joist space must be at least 2 in. above ceiling. Ambient temperature is considered to be 50 C (122 F).
b. 2 in. of insulation
Conductors installed above insulation at least 2 in. thick require no correction for temperature.
c. More than 2 in.
Conductors installed within insulation must be at least 2 in. above ceiling. Ambient temperature is considered to be 50 C (122 F).
Ceiling joists
Thermal Insulation
2 in. min.
2 in.
2 in.
Branch circuit wiring
Ceiling cable

Fig. 424-7. Wiring above a heated ceiling may require derating because of heat accumulation. (Sec. 424.36.)

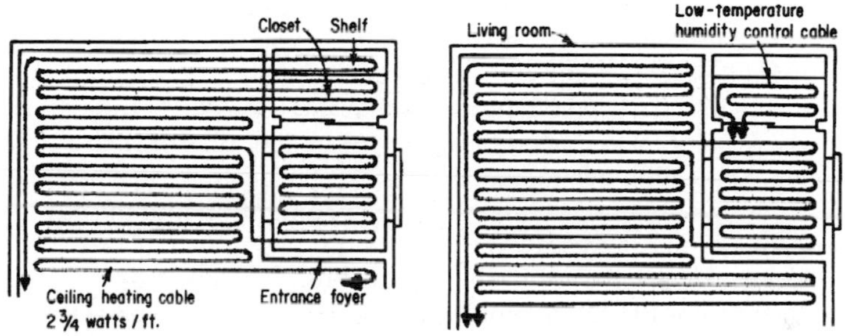
Closet
Shelf
Ceiling heating cable
2 3/4 watts / ft.
Entrance foyer
Living room
Low-temperature humidity control cable

VIOLATION — cable extends beyond room and is installed in the closet. Cable in foyer is OK

OK — this cable is permitted in closet. Single cable runs that are embedded may cross partition.

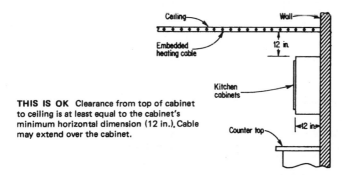

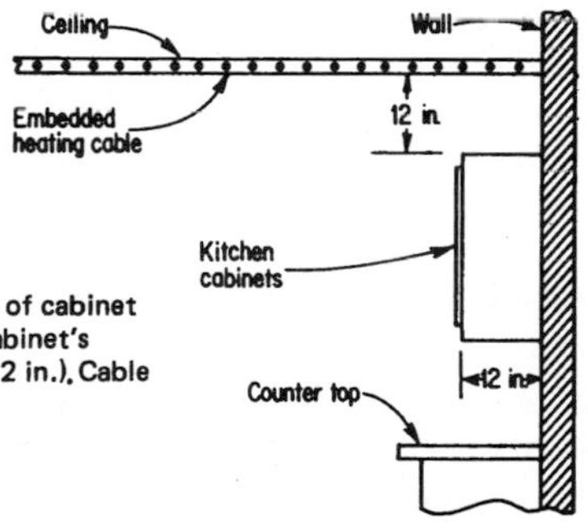
Ceiling
Wall
Embedded heating cable
12 in.
Kitchen cabinets
12 in.
Counter top

THIS IS OK Clearance from top of cabinet to ceiling is at least equal to the cabinet's minimum horizontal dimension (12 in.). Cable may extend over the cabinet.

Fig. 424-8. Layouts of heating cables must generally be confined to individual rooms or areas. (Sec. 424.38.)

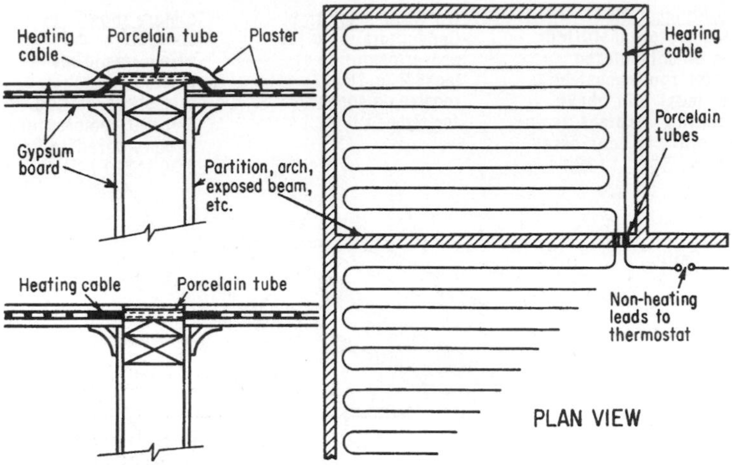
Heating cable
Porcelain tube
Plaster
Gypsum board
Partition, arch, exposed beam, etc.
Heating cable
Porcelain tube
Heating cable
Porcelain tubes
Non-heating leads to thermostat
PLAN VIEW

Fig. 424-9. Part **(B)(3)** of rule permits single runs across partitions. (Sec. 424.38.)

exposed ceiling beams, and so forth. The prohibitions now also include tub and shower walls, and under cabinets, etc. with no floor clearance.

However, there are times when a small ceiling area (such as over a dressing room or entryway) is separated from a larger room by such an arch or beam, yet it is impractical to install a separate heating cable and control. The rule of part **(3)** was intended as a solution to this problem. A typical floor plan of such a situation is shown in Fig. 424-9, with two methods of getting the heating cable past the partition or beam. In the upper drawing the cable is brought up into the attic space, through a porcelain tube, and back down through the gypsum board. Plaster is then forced into the tube and puddled over the exposed cable and tube in the attic. This should be the same plaster or joint cement that is used between the two layers of gypsum board.

In the lower drawing a hole is drilled through the top plate of the partition (or beam) and a porcelain tube is pressed into the hole. Plaster is packed into the tube after the cable has been passed through. In both cases, the plaster serves to conduct heat away from the cable, avoiding hot spots and possible burnouts.

424.39. Clearance from Other Objects and Openings. Figure 424-10 shows application of the specified clearance distances for different conditions. Although frequently within the scope of other trades and being therefore unenforceable, no surface-mounted equipment, whether electrical or not, is permitted to cover ceiling cables.

424.41. Installation of Heating Cables on Dry Board, in Plaster, and on Concrete Ceilings. All heating cables must observe these application methods. Figure 424-11 shows the rules of paragraph **(B)** and paragraph **(F)**.

Figure 424-12 shows the rule of paragraph **(D)** and ties into the rule of 424.43(C) at the outlet box.

Heating cable installed in plaster or between two layers of gypsum board must be kept clear of ceiling luminaires and side walls. In drywall construction, cable must be

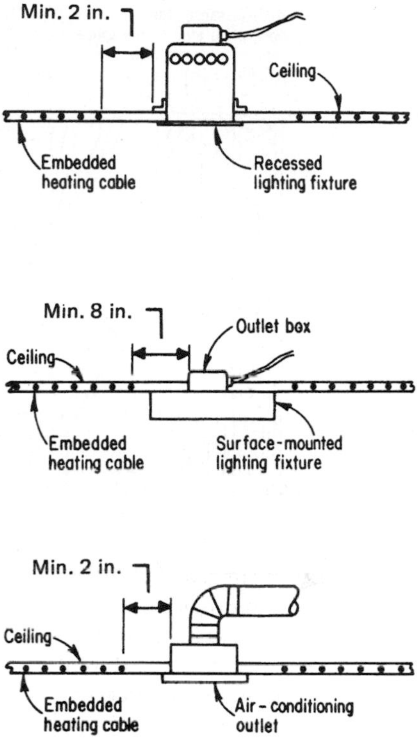
Min. 2 in.
Ceiling
Embedded heating cable
Recessed lighting fixture
Min. 8 in.
Outlet box
Ceiling
Embedded heating cable
Surface-mounted lighting fixture
Min. 2 in.
Ceiling
Embedded heating cable
Air-conditioning outlet

Fig. 424-10. Heating cables installed in ceilings must be kept clear of equipment. (Sec. 424.39.)

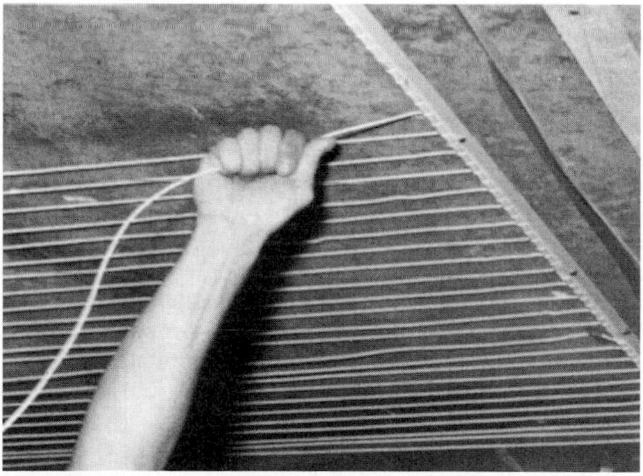

Fig. 424-11. The prescriptive rules on routing adjacent runs of cable (≤ 2¾ watts/ft to be ≥1½ in. on centers) have been removed in the 2017 **NEC** in favor of a simple reference to the manufacturer's instructions. [Sec. 424.41(B).]

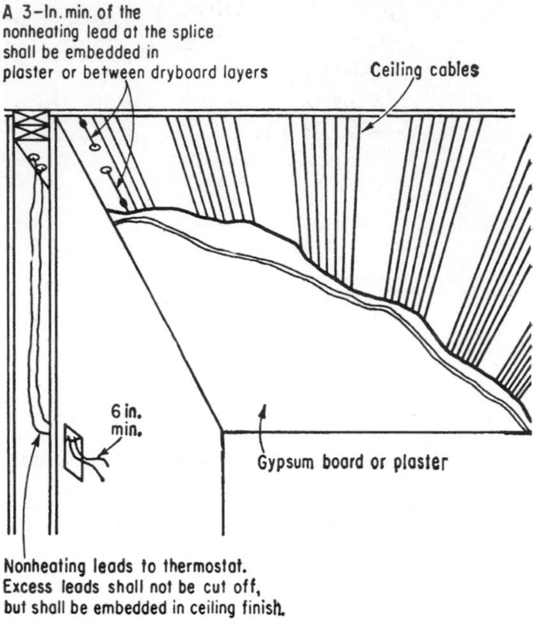
A 3-In. min. of the nonheating lead at the splice shall be embedded in plaster or between dryboard layers
Ceiling cables
6 in. min.
Gypsum board or plaster
Nonheating leads to thermostat. Excess leads shall not be cut off, but shall be embedded in ceiling finish.

Fig. 424-12. Ends of nonheating leads must be embedded in ceiling material. (Sec. 424.41.)

embedded in mastic or plaster. Without it, dead air space between the cable runs acts as a heat reservoir, increasing the possibility of cable burnouts (Fig. 424-13). Cable movement caused by the expansion and contraction accompanying temperature changes is also prevented. Laboratory tests on cable used without mastic have found properly spaced adjacent cable runs actually making contact with each other, producing a hot spot and subsequent burnout. The plaster and sand mixture normally used as a mastic is a good conductor of heat and thus accelerates the dissipation of heat from the entire circumference of the cable. In addition, it improves the conductance from the cable to the gypsum board where the cable does not make direct contact. Even in the most careful installations small irregularities in construction and material prevent perfect contact between the cable and both layers of gypsum board throughout the entire cable length. In no case should an insulating plaster be used. Thickness of the plaster coat should be just sufficient to cover the cable. Installations have been made, unfortunately, with plaster thickness as great as ¾ in. (19 mm). Nails are not capable of supporting the resulting excessive weight, and such ceilings have collapsed. Figure 424-14 shows rules that apply to installation of heating cables in drywall ceilings. Similar to in the case of 424.41(B), the 2017 **NEC** also recognizes alternatives as covered in the manufacturer's instructions, particularly when coping with metal lath or other conductive surfaces.

Where the cable is to be embedded between two layers of gypsum board (dry-wall construction), after the cable is stapled to the layer of gypsum lath, i

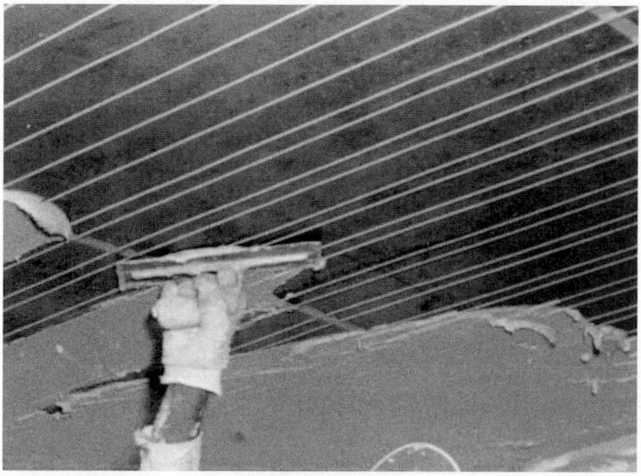

Fig. 424-13. In a drywall ceiling, the heating cable must be covered with a thermally conductive mastic before second course of gypsum board is applied to the ceiling, over the heating cable. (Sec. 424.41.)

is covered with noninsulating plaster or gypsum cement, and a finishing layer of gypsum board (Sheetrock) is screwed in place covering the cable and plaster. To make sure that screws driven to secure this gypsum board to the ceiling joists do not penetrate the cable, a clear space at least 2½ in. (65 mm) wide must be left between adjacent cable runs immediately beneath each joist. That is, while adjacent cable runs must in general be at least 1½ in. (38 mm) apart, the spacing beneath joists must be increased to at least 2½ in. (65 mm). This means, of course, that the cable must be run parallel to the joists, as in part **(I)**.

Part **(J)** requires that heating cables in ceilings must cross joists only at the ends of the room, except where necessary to satisfy manufacturer's instructions. If manufacturer's instructions advise the installer to keep the cable away from ceiling penetrations and light luminaires, the exception here will permit crossing joists at other than the ends of the room.

424.43. Installation of Nonheating Leads of Cables. Part **(C)** of this rule prohibits cutting off any of the length of heating leads that are provided by the manufacturers on the ends of heating cable. Any excess length of such leads must be secured to the ceiling and embedded in plaster or other approved material.

424.44. Installation of Cables in Concrete or Poured Masonry Floors. Details of these rules are shown in Fig. 424-15.

Paragraph **(B)** requires cables to be secured in place by nonmetallic frames or spreaders or other approved means. Metallic supports such as those commercially available for use in roadways or sidewalks are not to be used in floor space-heating installations. Lumber is often used, although a more common method is to staple the cable directly to the base concrete after it has set about 4 h. It was not the intention that this Code paragraph prohibit the use of metal staples. The object was to

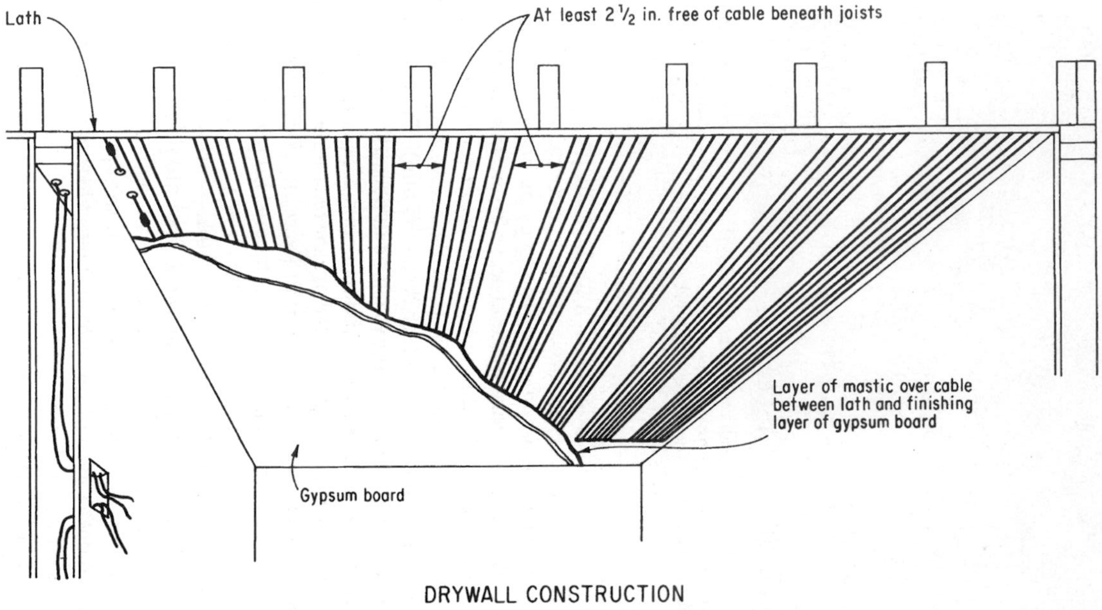
Lath
At least 2 1/2 in. free of cable beneath joists
Layer of mastic over cable between lath and finishing layer of gypsum board
Gypsum board
DRYWALL CONSTRUCTION

Fig. 424-14. Additional rules apply only to drywall ceiling construction. (Sec. 424.41.)

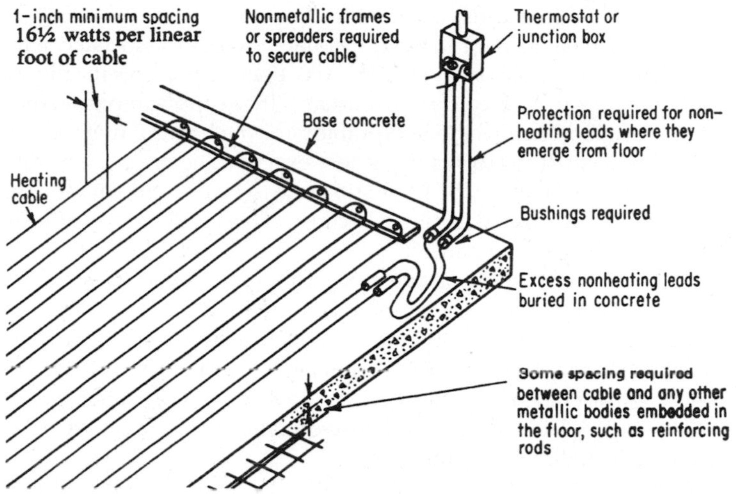
1-inch minimum spacing
16½ watts per linear foot of cable
Nonmetallic frames or spreaders required to secure cable
Thermostat or junction box
Base concrete
Protection required for non-heating leads where they emerge from floor
Heating cable
Bushings required
Excess nonheating leads buried in concrete
Some spacing required between cable and any other metallic bodies embedded in the floor, such as reinforcing rods

Fig. 424-15. Specific rules apply to heating cable in concrete floors. The drawing shows traditional heat density and spacing values; the 2017 NEC has removed these prescriptive requirements in favor of reliance on the manufacturer instructions. (Sec. 424.44.)

reduce the possibility of short circuits because of continuous metallic conducting materials spanning several adjacent cable runs.

The former requirement for spacing between the cables and metal embedded in the floor has been deleted in the 2017 NEC. Current product standards require a grounded braid or sheath over the cable as part of the assembly; this was not the case when the requirement first entered the Code.

Paragraph **(C)** requires leads to be protected where they leave the floor, and paragraph **(D)** adds that bushings shall be used where the leads emerge in the floor slab. These provisions refer to the nonheating leads that connect the branch circuit homerun to the heating cable. The splices connecting the nonheating leads to the cable are always buried in the concrete. About 6 in. (152 mm) of leads is left available in the junction box; any remaining length of nonheating leads is buried in the concrete.

The last part, part **(E),** presents a general rule covering in-the-floor heating leads in kitchens (new as of the 2011 edition), bathrooms and hydromassage bathtub "locations." Here the Code mandates GFCI protection for the supply circuits to the heating elements in such equipment. With electrically heated bathroom floors, the GFCI required will be in addition to any others required. Note that for pool areas, 680.27(C)(3), and as extended for spa and hot tub areas by 680.43, radiant cable heating is not permitted in floor decks in either location.

424.45. Installation of Cables Under Floor Coverings. This section, new in the 2017 NEC, recognizes the use of heating cables for under floor coverings, such as ceramic tile (popular in bathroom areas, and also under laminate flooring or even carpeting). The installation requirements, as reflected in installation instructions, clearly parallel heating panels, as have been recognized in Sec. 424.99 since the

1987 NEC. The installations must be done in accordance with the product listing and instructions, which account for and evaluate the risks associated with these installations. However, because Part V of Art. 424 did not specifically mention under floor coverings for cables, it was unclear if these applications were permissible. The new section clarifies the acceptability of the heating method. The specific requirements match up to 424.99 almost exactly. The only minor difference is the requirement in 424.99 for the floor substrate to be flat and smooth; this does not occur in 424.45, which makes sense because cables will be more forgiving of minor floor irregularities than panels.

424.47. Label Provided by Manufacturer. This requirement, which correlates with 424.92(D) for heating panels, augments the circuit directory rules in 408.4(A) and requires labels to be provided by the manufacturer that will adhere to branch-circuit panelboards and other circuit sources and provide prominent notice as to the existence of space heating cables and the identities of the branch circuits supplying those cables. A label need not be posted in the event the cables are visually evident after the installation is complete.

424.57. General. The rules of Part **VI** of Art. 424 apply to heater units that are mounted in air-duct systems, as shown in Fig. 424-16.

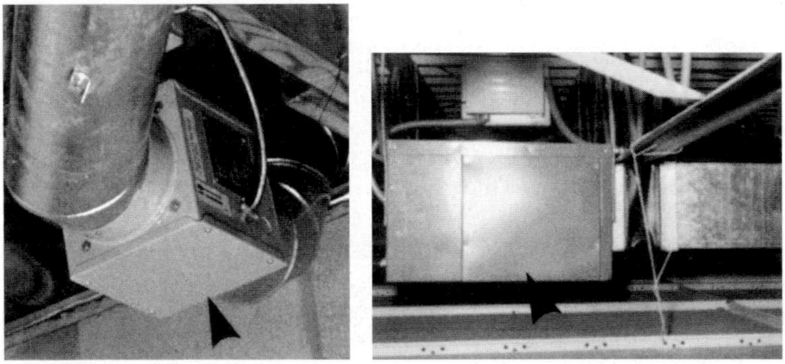

Fig. 424-16. Electric heaters designed and installed to heat air flowing through the ducts of forced-air systems are covered by Secs. 424.57 through 424.66. (Sec. 424.57.)

424.70. Scope. In applying Code rules, care must be taken to distinguish between "resistance-type" boilers and "electrode-type" boilers (Fig. 424-17).

424.72. Overcurrent Protection. Heating elements of resistance-type electric boilers must be arranged into load groups not exceeding the values specified in paragraph **(A)** or **(B)**. Figure 424-18 shows a 360-kW electric boiler used to heat a large school.

Part **(E)** requires that the ampacity of conductors used between the heater and the supplementary overcurrent protective devices for the subdivided heating circuits within such boilers must not be less than 125 percent of the load served. Again, however, the last paragraph permits conductors rated at 100 percent of the load for heaters rated 50 kW or more under certain given conditions.

Fig. 424-17. Electric boilers with resistance-type heating elements are regulated by Secs. 424.70 through 424.74. (Sec. 424.70.)

Fig. 424-18. Electric boiler contains subdivided heating-element circuits, totalling 360 kW. Fuses protect the subdivided circuit loads. (Sec. 424.72.)

24.80. Scope. This part of the article covers electrode-type boilers. The requireıents are the same as for resistance-type boilers with the major exception that the ꞉quirement to subdivide load over 48 A is not present.

24.90. Scope. The rules of 424.91 through 424.98 apply to radiant heating pans and heating panel sets. Rules on electric radiant heating panels are separated from the rules on heating cables. The NEC at one time did cover both heating cables and heating panels within the same set of rules. Panels are generically different from cables and require separate specific rules on installation and layout details essential to safety.

424.95. Location of Branch-Circuit and Feeder Wiring in Walls. When electric heating panels are mounted on interior walls of buildings, any wiring within the walls behind the heating panels is considered to be operating in an ambient of 40°C rather than the normal 30°C for which conductors are rated. Because of this, the ampacity of such conductors in wall spaces behind electric heating panels must be reduced in accordance with the correction factors given as part of Table 310.15(B)(2)(a) [424.95(B)] (Fig. 424-19). Exterior walls have no such prescriptive rule, only a reference to 310.15(A)(3). If the walls are uninsulated, the factor would be based on an analysis of climate extremes. The data discussed at 310.15(B)(2)(c) would be a good place to start. If the walls are insulated, then the quality of the insulation has a dramatic effect on how potential-derating factors might be assigned. Refer to the discussion at 310.15(A)(3) for an actual example and a general discussion of the impact of insulation on wiring.

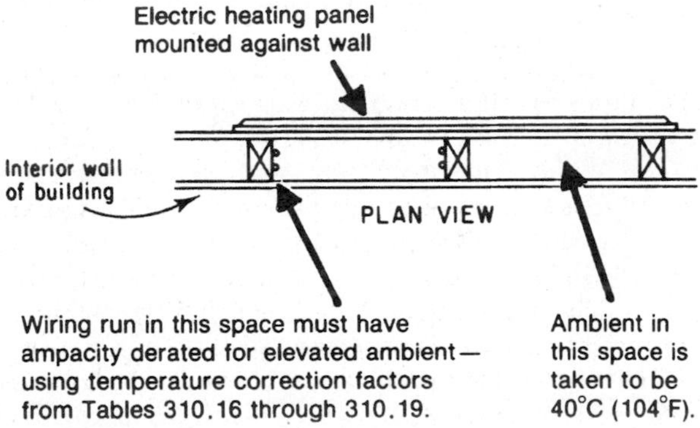
Electric heating panel mounted against wall
Interior wall of building
PLAN VIEW
Wiring run in this space must have ampacity derated for elevated ambient— using temperature correction factors from Tables 310.16 through 310.19.
Ambient in this space is taken to be 40°C (104°F).

Fig. 424-19. Wiring in walls behind heating panels must have ampacity corrected for more than 30°C. (Sec. 424.95.)

424.97. Nonheating Leads. Excess length of the nonheating leads of heating panels may be cut to the particular length needed to facilitate connection to the branch-circuit wires (Fig. 424-20).

424.100. Scope. This introduces a new Part X in Art. 424 in the 2017 NEC to cover "Low-Voltage Fixed Electric Space-Heating Equipment." The panel has made the assumption that these systems will operate in dry locations. This equipment must include an isolating power supply that will result in an ungrounded system in accordance with 424.103(B). The power supply must also be limited to maximum current of 25 amps at 30 volts ac or 60 volts dc, parameters that exactly coincide with those for low-voltage lighting in Art. 411. The entire system (424.102

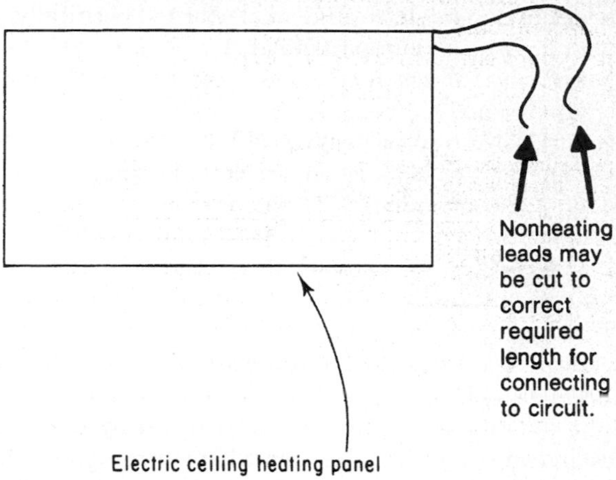
Nonheating leads may be cut to correct required length for connecting to circuit.
Electric ceiling heating panel

Fig. 424-20. The rule permits cutting of nonheating leads for "panels." (Sec. 424.97.)

including the heaters must be listed as a complete system and not, in contrast to Art. 411, as an assembly of listed parts.

One of the more intriguing aspects of Part X is 424.101(A), which clearly anticipates such applications as off-grid powering directly from wind or solar sources. Any power conversion equipment must meet the "applicable section" of the NEC for the source used; the source being limited to the general system limits covered above. Although this is extremely vague, the universal listing requirement will undoubtedly bring with it a plethora of installation specifications by the equipment manufacturer. The part omits any specific requirements for overload protection for the wiring. The substantiation indicated that the listing process would generate those parameters, which should then be followed. It also noted that the currents that could be generated may be enough to compromise the wiring, but the substantiation then continued to rely on the listing. This is not really a problem for a Chap. 4 (equipment) article; other rules in Chaps. 1 through 3 will still apply. f one of these systems is truly operating in a stand-alone mode, it will also be subject to the requirements of the new Art. 710, and 710.15(B) would require the vires to be sized based on the output rating, and that would avoid the need for overload protection.

ARTICLE 425. FIXED RESISTANCE AND LECTRODE INDUSTRIAL PROCESS HEATING QUIPMENT

25.1 Scope. This article, new in the 2017 NEC covers fixed electrically powered ndustrial process heating using resistance or electrode technology. The article oes not cover, for obvious reasons, technology for personnel comfort, including heating processes only designed for residential spaces. Specifically, the following Art. 424 provisions have no counterpart in Art. 425:

424.19(C)(1 through 4) on residential unit switches
424.20 on environmental air thermostats
424.34 through 424.47 on space heating cables
424.61 and 424.62 on duct heaters with air conditioners
424.90 through 424.99 on radiant heating panels
424.100 through 424.104 on low-voltage electric space hearing

After the remainder of Art. 424 is essentially cut and pasted as Art. 425 and carefully analyzed, the results are disturbing. The article is approximately 68 column inches long. Of that text, a careful comparison, after correcting what will almost certainly be acknowledged as formatting errors that failed to record carriage returns and in one case took normal text and converted it into an exception, shows that only 6 of the roughly 68 column inches occupied by the new article actually creates substantively different rules than what is already required in Art. 424, about 9 percent. The remainder of this analysis will be confined to offering background information on the areas that differ. Any section not included in the two mentioned here is substantively covered in Art. 424, and addressed in this book to the extent appropriate to its scope in the coverage of that article.

425.8 General. Part (A) is the duct heater installation language in 424.66 made broadly applicable to equipment covered in the article. See Fig. 424.16 for an example of cramped spaces often encountered about this equipment. The working space rules in (B) reiterate 110.26 and 110.34 based on whether or not the voltage exceeds 1000 volts. Part (C) sets practical rules for common industrial circumstances requiring working from ladders.

425.9. Approval. This topic is taken from Art. 424 and won't be analyzed here, but it does provide a basis to make an interesting observation about the new article. Art. 424 includes a wide-ranging listing requirement in 424.6. Article 425 does not have such a section, although some of its provisions use the term. This reflects the reality that a great deal of electrical equipment in the industrial environment is uniquely assembled for a complex task, and listing is impracticable. What is here, in this section (as in 424.5) is a simple requirement that covered equipment be installed in an approved manner, acceptable to the inspector.

ARTICLE 426. FIXED OUTDOOR ELECTRIC DEICING AND SNOW-MELTING EQUIPMENT

426.1. Scope. In addition to covering the longtime standard methods and equipment for electric deicing and snow melting, this article gives detailed coverage of *skin-effect heating*—a system for utilizing the alternating-current phenomenon of skin effect, which derives its name from the tendency of ac current to flow on the outside (the skin) of a conductor. This action is produced by electromagnetic induction, which has an increasing opposition to current flow from the outside to the core of the cross section of a conductor.

In the layout of the article, separate coverage is given to "Resistance Heating," "Impedance Heating," and "Skin-Effect Heating."

426.4. Continuous Load. This basic rule requires that where electric deicing or snow-melting equipment is connected to a branch circuit, the rating of the branch-circuit overcurrent protection and the branch-circuit conductors must be not less than 125 percent of the total load of the heaters (Fig. 426-1).

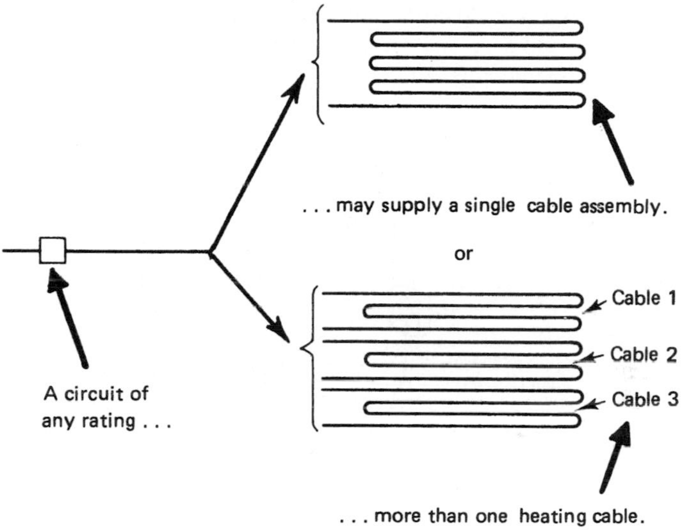
. . . may supply a single cable assembly.
or
Cable 1
Cable 2
Cable 3
A circuit of any rating . . .
. . . more than one heating cable.

Fig. 426-1. Electric deicing or snow-melting equipment may be fed by a circuit of any rating, provided it is sized for a "continuous" load. (Sec. 426.4.)

Both the circuit conductors and the overcurrent protection must be rated so the load is not over 80 percent of their amp rating. When the branch-circuit conductors are selected to have ampacity of at least 125 percent of the amp value of connected snow-melting and/or deicing load equipment, it is permissible to go to the next higher amp value of overcurrent protection where the ampacity of the conductors does not correspond to a standard rating of protective device from 240.6.

426.11. Use. Whether used in concrete, blacktop, or other building material, any deicing or snow-melting cable, panel, mat, or other assembly must be properly recognized (as by UL) for installation in the particular material. One key installation detail concerns concrete embedment in terms of whether a single or a double pour is required. If the installation instructions are silent on this topic then either method is acceptable, but if a double pour is intended, then the instructions will so state.

426.13. Identification. This rule requires the presence of a deicing or snow-melting installation to be effectively indicated to ensure safety and to prevent disruption.

After a snow-melting installation is made, some type of caution sign or other marking must be posted "where clearly visible" to make it evident to anyone that electric snow-melting equipment is present.

426.23. Installation of Nonheating Leads for Exposed Equipment. Part (A) of this rule permits cutting (shortening) of the nonheating leads provided the required marking on the leads (catalog numbers, volts, watts) is retained. That is not permitted for space-heating cables, which may not be cut [424.43(C)]. The marking on the nonheating leads of snow-melting cable must be within 3 in. (75 mm) of *each* end of the lead. The wording appears to permit cutting the nonheating leads back to the marking closest to the connection to the heating cable—which would mean a length of 3 in. (75 mm) plus needed for the marking is all that would be required. Or a length could be cut out between the two markings, provided that any splicing that would necessitate is made in boxes, as specified in 426.24(B). (The rule of 426.22 is shown in Figs. 426-2 and 426-3.)

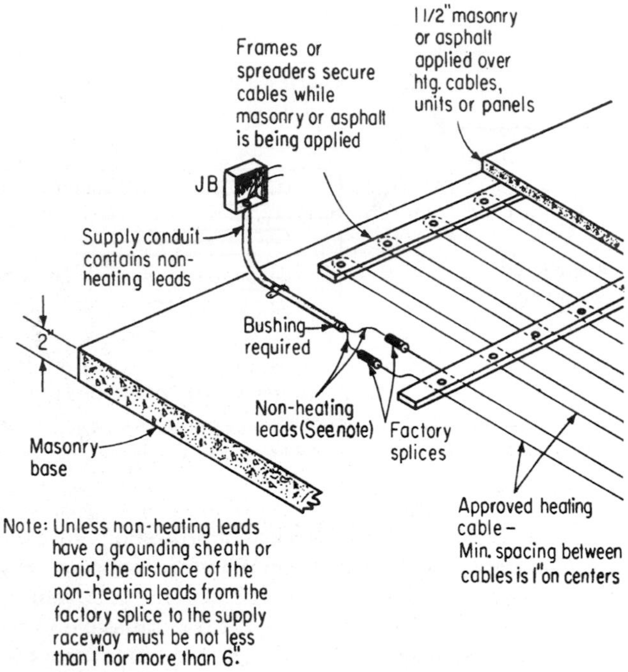
1 1/2" masonry or asphalt applied over htg. cables, units or panels
Frames or spreaders secure cables while masonry or asphalt is being applied
JB
Supply conduit contains non-heating leads
2
Bushing required
Non-heating leads (See note)
Factory splices
Masonry base
Approved heating cable - Min. spacing between cables is 1" on centers
Note: Unless non-heating leads have a grounding sheath or braid, the distance of the non-heating leads from the factory splice to the supply raceway must be not less than 1" nor more than 6".

Fig. 426-2. Detailed rules cover installation of heating and nonheating conductors. (Secs. 426.20, 426.21, and 426.22.)

426.28. Ground-Fault Protection of Equipment. Resistance snow-melting cables for rooftops typically involve long lengths of two-wire cable in a semiconductive filler that, when energized, results in a diffuse passage of current along the cable length. The actual wires embedded in the cable construction are quite small, and if the cable is damaged, any ground fault will result in a very low current even due to the overall resistance. These faults are far too low to open an overcurrent device; however, the sputtering arc can easily create fires because the damaged cable can restrike the arc due to the semiconductive nature of the surrounded cable filler. The solution is a residual current device similar to a GFCI, but with a 30-mA trip setting (manufacturers vary somewhat). These devices are a form of low-level ground-fault protection of equipment (GFPE), configured as molded-case

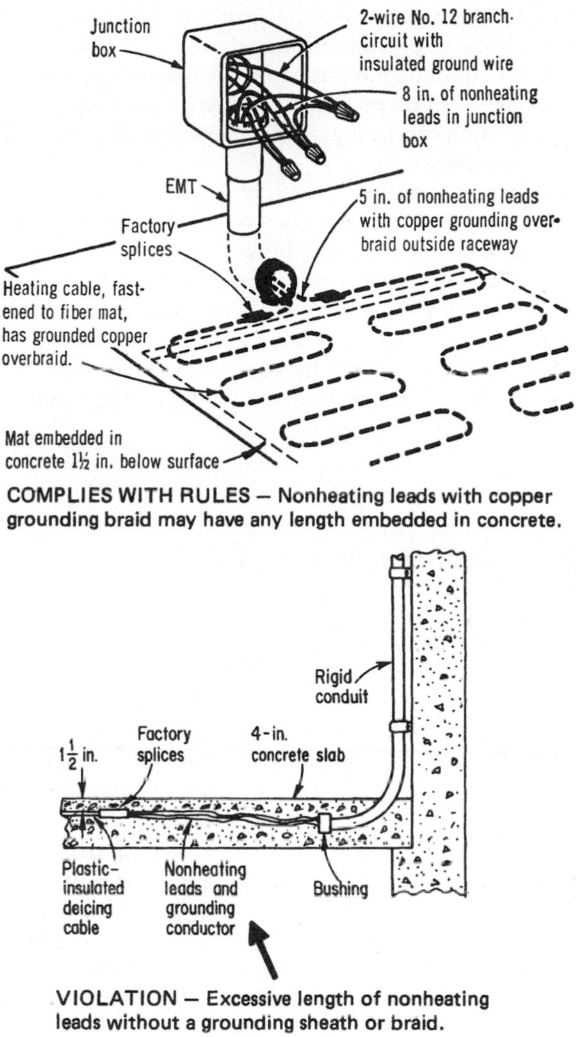
Junction box
2-wire No. 12 branch-circuit with insulated ground wire
8 in. of nonheating leads in junction box
EMT
5 in. of nonheating leads with copper grounding overbraid outside raceway
Factory splices
Heating cable, fastened to fiber mat, has grounded copper overbraid.
Mat embedded in concrete 1½ in. below surface
COMPLIES WITH RULES — Nonheating leads with copper grounding braid may have any length embedded in concrete.
Rigid conduit
1½ in.
Factory splices
4-in. concrete slab
Plastic-insulated deicing cable
Nonheating leads and grounding conductor
Bushing
VIOLATION — Excessive length of nonheating leads without a grounding sheath or braid.

Fig. 426-3. Installation of nonheating leads must observe all the rules. (Sec. 426.22.)

circuit breakers. They are not GFCIs, but look and act very similarly. In fact a GFCI would provide the required protection, but they are set in the 4- to 6-mA range, and would likely nuisance trip on an extensive layout. However, for a short run they have been known to work without incident. Some inspectors still fail them, because they don't want to sign off on a likely source of nuisance trips. Note that Type MI cable is now (as of the 2011 edition) subject to this requirement. MI cable can be made with several different sheaths, which vary in their ability to return fault current.

426.32. Voltage Limitations. The allowance for the voltage to be from 30 to 80 volts in systems using impedance heating if protected by GFCI has been deleted in the 2017 NEC. This equipment has too much leakage current and is not compatible with conventional GFCI protection. Note that the grounding rule in Sec. 426.34 still requires a grounding connection for such circuits, although the permission to use them no longer exists.

426.50. Disconnecting Means. For outdoor deicing and snow-melting equipment, the CB or fusible switch for the branch circuit to the equipment is adequate disconnecting means as long as it is readily accessible to the user. It must simultaneously open all ungrounded conductors. The cord-plug of plug-connected equipment rated up to 20 A and 150 V may serve as the disconnect device; this is the upper limit for UL on the cord-and-plug ratings in this area. The disconnecting means does not have to be in sight of the equipment, but it must be indicating and include a "positive lockout" which is understood to mean that it can be locked in the open position.

ARTICLE 427. FIXED ELECTRIC HEATING EQUIPMENT FOR PIPELINES AND VESSELS

427.22. Ground-Fault Protection of Equipment. Electric heating equipment applied to pipelines or vessels must be supplied by a branch circuit with ground fault protection. This is extensively covered for similar equipment in the discussion about deicing and snow-melting equipment; see 426.28 for the analysis.

427.23. Grounded Conductive Covering. These rules require an overall grounded metal jacket on any heating cables intended to be installed on pipelines or vessels. In addition, for heating panels, only the side that is *not* in contact with the pipe or vessel must have a grounded metal covering. The outer metal covering serves as an equipment ground-return path for fault current in the event of failure of the insulation on the heating conductors. Proper grounding will trip open the faulted circuit due to unbalanced current flowing through the grounded path and prevent the type of fires that have been reported—as well as provide greater personnel safety. Here again, the reason for the rule is protection against fire resulting from sputtering, uncleared arcing faults that require a residual current detector such as this to detect.

Although this article is generally thought of as an industrial article relevant to chemical process industries and oil refineries, it also applies to everyday frost protection applied to even residential water lines. A 6-m (20-ft) run of heat tracing on a length of water pipe at a single-family house lands you right here. Make sure to get cable with the grounded braid on the outside, and then make sure that the overcurrent device for the branch circuit has the low-level GFPE (~30 mA) built in. Also, don't forget the disconnection rules in 427.55, which are the same as for 426.50 and discussed at length there.

427.27. Voltage Limitations. The default secondary voltage limit is 30 volts. The 2017 NEC reevaluated the utility and practicality of conventional Class A GFCI for personnel protection on this equipment, and concluded that a more practic approach was to ask for GFPE instead, together with rules for qualified access. The former second paragraph on 30-80 volts is now the first exception, and the former exception for industrial applications up to 132 volts is now the second exception, with a guarding specification added.

ARTICLE 430. MOTORS, MOTOR CIRCUITS, AND CONTROLLERS

430.1. Scope. Two articles in the National Electrical Code are directed specifically to motor applications:

1. Article 430 of the NEC covers application and installation of motor circuits and motor control hookups—including conductors, short-circuit and ground-fault protection, starters, disconnects, and running overload protection. Article 430 also covers adjustable speed drives in addition to motors, motor circuits, and controllers. Specific Code rules are given throughout this article.
2. Article 440, covering "Air-Conditioning and Refrigerating Equipment," contains provisions for such motor-driven equipment and for branch circuits and controllers for the equipment, taking into account the special considerations involved with sealed (hermetic-type) motor compressors, in which the motor operates under the cooling effect of the refrigerant.

Diagram 430-1 in the NEC shows how various parts of Art. 430 cover the particular equipment categories that are involved in motor circuits. That Code diagram can be restructured, as shown in Fig. 430-1 in this handbook, to present the six basic elements that the Code requires the designer to account for in any motor circuit. Although these elements are shown separately here, there are certain cases where the Code will permit a single device to serve more than one function. For instance, in some cases, one switch can serve as both disconnecting means and controller.

In other cases, short-circuit protection and overload protection can be combined in a single CB or set of fuses.

Throughout this article, all references to "running overcurrent devices" or simply "overcurrent devices" have been changed to "overload devices." And references to motor "running overcurrent protection" have been changed to "overload protection." This has been done to correlate with the definition of *overload* given in Art. 100, which refers to "Operation of equipment in excess of normal, full-load rating. . . ." A fault, such as a short circuit or ground fault, is not an overload. *Overload* for motors means current due to overload, up to and including locked-rotor current—or failure to start, which is the same level of current.

430.4. Part-Winding Motors. A part-winding starter is an automatic type of starter for use with squirrel-cage motors which (as defined in 430.2) have two separate, parallel windings on the stator. It can be used with the commonly used 220/400-V (dual-voltage) motors when they are used at the lower voltage, with the two windings operating in parallel. Single-voltage motors and 440- and 550-V

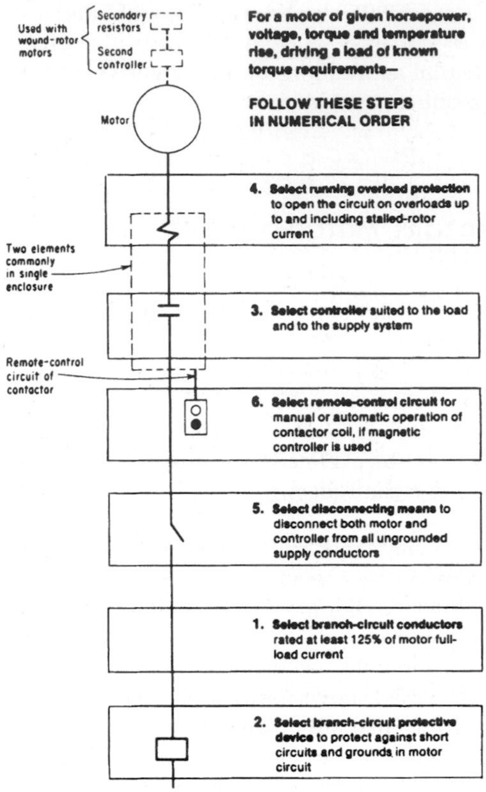
Used with wound-rotor motors
Secondary resistors
Second controller
Motor
For a motor of given horsepower, voltage, torque and temperature rise, driving a load of known torque requirements—
FOLLOW THESE STEPS IN NUMERICAL ORDER
4. Select running overload protection to open the circuit on overloads up to and including stalled-rotor current
Two elements commonly in single enclosure
3. Select controller suited to the load and to the supply system
Remote-control circuit of contactor
6. Select remote-control circuit for manual or automatic operation of contactor coil, if magnetic controller is used
5. Select disconnecting means to disconnect both motor and controller from all ungrounded supply conductors
1. Select branch-circuit conductors rated at least 125% of motor full-load current
2. Select branch-circuit protective device to protect against short circuits and grounds in motor circuit

Fig. 430-1. Code rules cover these considerations. (Sec. 430.1.)

motors must be ordered as specials if part-winding starting is to be used. The starter contains two magnetic contactors, each of which is rated for half the motor horsepower and is used to supply one winding.

430.6. Ampacity and Motor Rating Determination. For general motor applications (excluding applications of torque motors and sealed hermetic-type refrigeration compressor motors), whenever the current rating of a motor is used to determine the current-carrying capacity of conductors, switches, fuses, or CBs, the values given in Tables 430.247, 430.248, 430.249, and 430.250 must be used instead of the actual motor nameplate current rating. However, selection of separate motor-running overload protection *must* be based on the actual motor nameplate current rating.

There is a very common situation where a horsepower rating needs to be ignored, however, and that is when it is merely advertising hyperbole. As it turns out, when UL looks at nameplate information on appliances, including motor-operated appliances, it only polices the full-load current information. Manufacturers are free to tout bogus horsepower numbers such as "maximum peak horsepower," etc., that have no relationship to the horsepower/current tables at the end of the article whatsoever. This can make a very significant difference in wiring design. For example, suppose you were intending to put a 6-A garage door opener on the lighting circuit for the garage. This easily falls under the 210.23(A) limit of 50 percent of fixed load on such circuits and is very commonly done. However, if the unit were marked "maximum horsepower output 1 hp" or something equivalent, and you ended up in Table 430.248, all of a sudden the mandatory current number for that appliance would be 16 A and you would need a dedicated 20-A circuit for a 6-A load. Exception No. 3 to 430.6(A)(1) is what allows the use of the current rating from the nameplate in these cases.

As noted in part **(B)**, for any torque motor, the rated nameplate current is the locked-rotor current of the motor, and that value must be used in calculating the motor branch-circuit short-circuit and ground-fault protection in accordance with 430.52(B). The branch circuit for a torque motor must have its conductors and equipment protected at the motor nameplate rating (locked-rotor current) by selecting a fuse or CB in accordance with 240.4(G). The rule also requires that conductor ampacity and the setting or rating of the overload protective device be based on this value.

For shaded-pole motors, permanent-split-capacitor motors, and ac adjustable-voltage motors, the other rules apply. Part **(D)** covers valve-actuator motor assemblies, which use nameplate ratings, and are actually a special case of the motors referred to in the last sentence of 430.6(A)(1). Refer to the discussion at 430.102(A) Exception No. 3 for more information about this equipment.

430.7. Marking on Motors and Multimotor Equipment. This section covers markings that manufacturers are required to put on the equipment. There are two categories by which the **NEC** organizes this information. The first is general information, covered in (A), that lists information about the motor in 15 different respects as applicable, from the name of the manufacturer to the horsepower, and so forth, all the way down to the rating of condensation prevention heaters. The second involves item (8) in the above list, which addresses locked-rotor current. A manufacturer is free to specify an actual current value, but most use the Code letter instead, which has been standardized since the 1940 **NEC**. Code Table 430.7(B) lists the values associated with the various letters, and the five numbered paragraphs give specialized rules for implementing the Code letter designations. There are a number of applications where it is necessary to calculate the locked-rotor current of a motor, where that value of current is related to selection of overload protection or short-circuit protection. A typical example would be selection of an appropriately rated disconnecting means for a group motor installation, where 430.110(C)(1) requires as part of the evaluation, a summation of all currents being disconnected under locked rotor conditions, and then the summation is used to figuratively create one single larger motor that has a horsepower that corresponds to the larger locked rotor condition under similar voltage connections.

430.8. Marking on Controllers. Controllers now need to be marked with a short-circuit current rating, unless they are incorporated into larger equipment with that rating, or in limited instances of very small applications. This information may be required to establish the rating of an industrial control panel built in the field. See the discussion at 409.110 for more information on this concept.

430.9. Terminals. As required by part **(B)**, unless marked otherwise, copper conductors must be used with motor controllers, and screw-type terminals of control-circuit devices must be torqued. Part **(B)** is intended to ensure that only copper wires are used for wiring of motor starters, other controllers, and control-circuit devices because such equipment is designed, tested, and listed for use with copper only. If, however, equipment is tested and listed for aluminum wires, that must be indicated on the equipment, and use of aluminum wires is acceptable.

Part **(C)** requires that a torque screwdriver be used to tighten screw terminals of control-circuit devices used with No. 14 or smaller copper wires. Such terminals must be tightened to a value of 0.8 N-m (7 lb-in.)—unless marked for a different torque value. It should be noted that, although the rule literally mandates a "minimum" torque value, torque values are exact values. That is, equipment terminals should be torqued to the designated value and should *not* be overtorqued. (See top of Fig. 430-2.) In fact, the larger problem in the industry today is not under tightening, but over tightening that causes terminations to fail through metal fatigue and damaged conductor stranding.

430.10. Wiring Space in Enclosures. As noted in part **(A)**, standard types of enclosures for motor controllers provide space that is sufficient only for the branch-circuit conductors entering and leaving the enclosure and any control-circuit conductors that may be required. No additional conductors should be brought into the enclosure. 430.12 provides a comprehensive set of rules and tables for motor terminal housings to solve the complaint by installers that motor terminal housings are too small to make satisfactory connections.

Part **(B)** on "Wire Bending Space in Enclosures" is based on Table 430.10(B), which shows the minimum distance from the end of the lug or connector to the wall of the enclosure or to the barrier opposite the lug, for each size of conductor and for one or two wires per terminal lug (Fig. 430-2). The distances are greater than Table 312.6(A) for angle terminations, but run slightly smaller than Table 312.6(B) distances for straight-in terminations. If more than two conductors will be terminated per phase in a parallel makeup, then the distances revert to the requirements in 312.6 without modification.

A further rule applies to use of terminal lugs other than those supplied in the controller by its manufacturer. Such substitute terminal lugs or connectors must be of a type identified by the manufacturer for use with the controller, and use of such devices must *not* reduce the minimum wire bending space. That rule would apply, say, where mechanical set-screw lugs are replaced with crimp-on lugs (as for better connection of aluminum conductors). As with switches, CBs, and other equipment, the controller should be marked on its label to indicate acceptability of field changing of the lugs and to specify what type and catalog number of replacement terminal may be used, along with designation of the correct crimping tool and compression die.

430.11. Protection Against Liquids. Excessive moisture, steam, dripping oil, and the like on the exposed current-carrying parts of a motor may cause an insulation breakdown, which in turn may be the cause of a fire. To prevent such an occurence, here the Code calls for "protection"—presumably in the form of a fabricated structure above the motor, where the motor is exposed to dripping

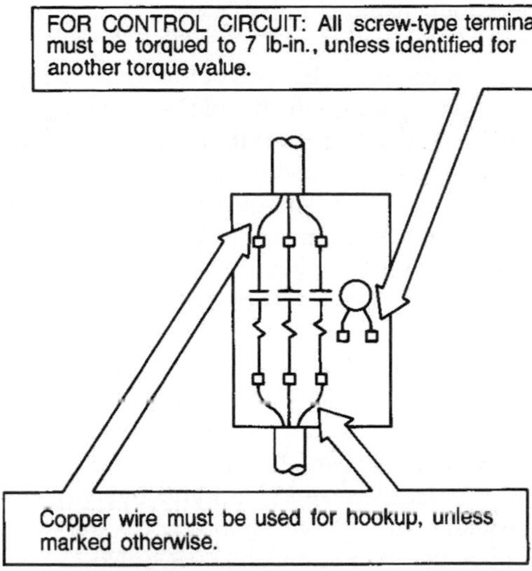
FOR CONTROL CIRCUIT: All screw-type terminals must be torqued to 7 lb-in., unless identified for another torque value.
Copper wire must be used for hookup, unless marked otherwise.

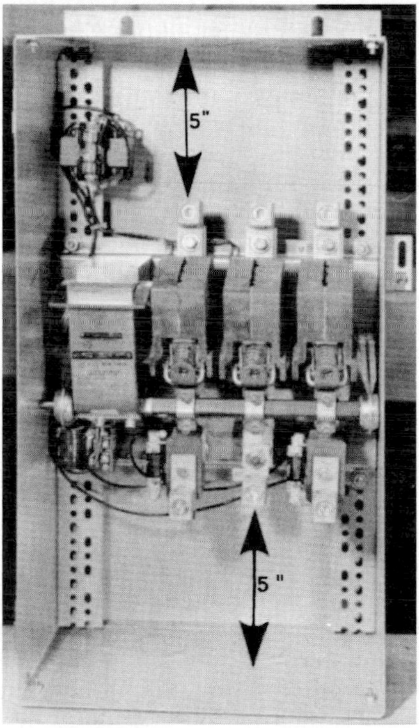
5"
5"

Fig. 430-2. For this size 5 motor starter, use of No. 1/0 THW for the line and load conductors would require a minimum gutter height of 5 in. (125 mm) from Table 430.10(B), to provide Code-acceptable wire bending space. (Sec. 430.10.)

liquid—or the motor may be listed as suitable for the environment to which it is exposed.

430.12. Motor Terminal Housings. The sizes of these enclosures are standardized as provided in the associated tables, which had their sizes significantly increased by about 40 percent in the 1996 NEC. In part **(E)** the Code rule requires some provision for connecting an equipment grounding conductor at the terminal box where the branch circuit supplies a motor. The grounding connection may be either a "wire-to-wire" connection or a "fixed terminal" connection, and the ground terminal provision may be either inside the junction box—for connection of an equipment grounding conductor run with the circuit wires within the supply raceway—or outside the junction box—for connection of an "equipment bonding jumper" on the outside of a length of flexible metallic conduit or liquidtight flex, either of which is so commonly used. As required by 348.60 and 350.60, a bonding jumper is required for even short lengths of flex (up to 6 ft [1.8 m]) when the wires within the flex are protected at their origin by fuses or CBs rated over 20 A; and liquidtight flex over metric designator 35 (trade size 1¼ in.) and overcurrent protection over 60 A (see 250.118 for details) must have a bonding jumper for the typical length (up to 6 ft [1.8 m]) used with motor connections. See also 430.245 and Fig. 430-70.

This rule permits either an inside or an outside connection of the bonding jumper to correlate with 250.102(E), which permits the bonding jumper (up to 6 ft [1.8 m] long) to be run either inside or outside the flex.

The exception to this rule eliminates the need for providing "a separate means for motor grounding" at the junction box where a motor is part of "factory-wired equipment" in which the grounding of the motor is already provided by some other conductive connection that is an element of the overall assembly.

430.13. Bushing. Refer also to 300.4(G).

430.16. Exposure to Dust Accumulations. The conditions described in this section could make the location a Class II, Division 2 location: the types of motors required are specified in Art. 502.

430.17. Highest Rated or Smallest Rated Motor. Note that the current rating, not the horsepower rating, determines the "highest rated" motor where Code rules refer to such. See 430.62.

430.22. Single Motor. The basic rule says that the conductors supplying a single-speed motor used for continuous duty must have a current-carrying capacity of not less than 125 percent of the motor full-load current rating, so that under full-load conditions the motor must not load the conductors to more than 80 percent of their ampacity. Be aware that this 125 percent factor has absolutely nothing to do with the 125 percent factor involved in calculating conductor sizing for overcurrent device terminations when those wires are subject to continuous loads. The fact that the two numbers are identical is a coincidence. The motor circuit is based on likely overload protective device settings, and addresses the fact that the 125 percent condition, or something close to it, could continue for a substantial amount of time, but probably not 3 h. Do not make the mistake of looking at a continuous duty motor, adding 25 percent as required for this section, and then adding another 25 percent of that (total factor = 156 percent) for the motor circuit conductor size. That amounts to double derating and the NEC does not require it to be done.

Part **(A)** covers dc motors, and requires that the conductors that supply the rectifier ahead of the motor have an ampacity of not less than 125 percent of the rated rectifier input current. This information is new with the 2011 edition of the NEC. If the motor is fed from a rectified single-phase supply, the field-wiring output conductors from the rectifier to the motor must be 150 percent of motor FLA for full-wave rectification, and 190 percent for half-wave rectification. This material is not new; it was reformatted from an exception format as of the 2011 edition.

For a multispeed motor, part **(B)** provides that the selection of branch-circuit conductors on the supply side of the controller must be based on the highest full-load current rating shown on the motor nameplate.

For a wye-start, delta-run motor with all six leads (twelve for dual-voltage motors) brought out, the starting voltage is $1/\sqrt{3}$ of the phase-to-phase voltage, and the starting current is $(1/\sqrt{3})^2$ or one-third of the full-load current. Because all the leads are brought out, during the normal running condition each phase connection has two wires connected to it, one to each of the two adjacent positions of the delta connection. Each of those two wires carries a current that does not add arithmetically in the final connection, because they connect to different phase windings in the motor. The current in each wire, when added together using three-phase procedures (i.e., multiplied by $\sqrt{3}$) will equal the full-load current taken from the line at the disconnect. This means that each of those wires carries $1/\sqrt{3}$ of the full-load current, or 58 percent of FLC. Part **(C)** formerly required that these same wires, which connect the controller and the motor, be "based" on this number for this reason. This was only the first step. For a customary continuous duty cycle application, one multiplied this number by 125 percent to get the actual wire size on the load side of the controller, or 72 percent of FLC. The NEC has, in effect, done this calculation out, specifying 72 percent, and adding an informative note explaining where the percentage comes from. If the duty cycle is other than continuous, one would expect to use the multiplier from Table 430.22(E) to get the final factor, but since 125 percent has now been applied, it will be necessary to use the underlying percentage (58%) to begin the calculation. In effect, by oversimplifying the requirement an implicit conflict has been created with (E).

Part **(D)** requires the comparable conductors that connect a part-winding connected motor to its controller be based on 50 percent of full-load current because each phase winding operates in parallel with its sister. The 2011 NEC restated this ampacity rule by doing out the 125 percent calculation to the base 50 percent (to equal 62.5%) and added a corresponding informational note to explain where the number came from. Just as in the case of the wye-delta motor, the result is a conflict with (E) if duty cycles are in play.

Figure 430-3 shows the sizing of branch-circuit conductors to four different motors fed from a panel. (Sizing is also shown for branch-circuit protection and running overload protection, as discussed in 430.34 and 430.52. Refer to Table 430.250 for motor full-load currents and Table 430.52 for maximum ratings of fuses.) Figure 430-3 is based on the following:

1. Full-load current for each motor is taken from Table 430.250.
2. Running overload protection is sized on the basis that nameplate values of motor full-load currents are the same as values from Table 430.250. If nameplate and table values are not the same, OL (overload) protection is sized according to nameplate.

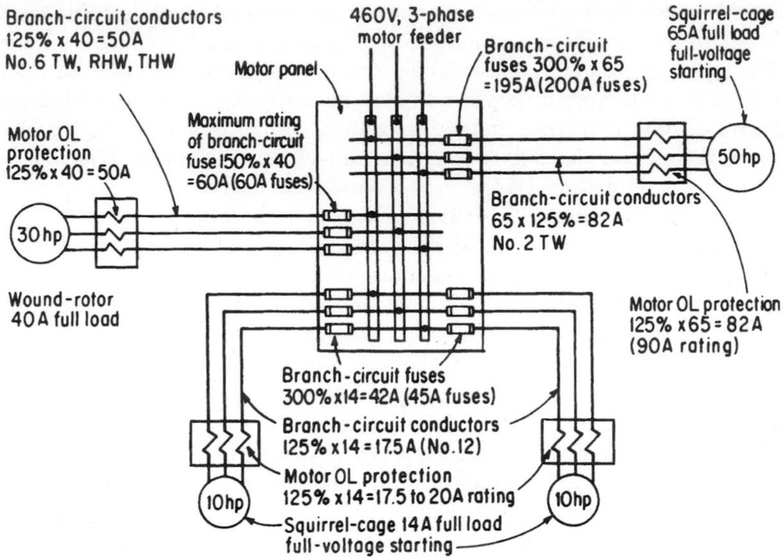
Branch-circuit conductors
125% x 40 = 50A
No. 6 TW, RHW, THW
460V, 3-phase
motor feeder
Squirrel-cage
65A full load
full-voltage
starting
Motor panel
Branch-circuit
fuses 300% x 65
= 195A (200A fuses)
Motor OL
protection
125% x 40 = 50A
Maximum rating
of branch-circuit
fuse 150% x 40
= 60A (60A fuses)
50hp
Branch-circuit conductors
65 x 125% = 82A
No. 2 TW
30hp
Wound-rotor
40A full load
Motor OL protection
125% x 65 = 82A
(90A rating)
Branch-circuit fuses
300% x 14 = 42A (45A fuses)
Branch-circuit conductors
125% x 14 = 17.5A (No. 12)
Motor OL protection
125% x 14 = 17.5 to 20A rating
10hp
10hp
Squirrel-cage 14A full load
full-voltage starting

Fig. 430-3. Circuit conductors are sized at 1.25 times motor current. (Sec. 430.22.) The fuses are assumed to be nontime delay.

3. Conductor sizes shown are for copper. Use the amp values given and Table 310.15(B)(16) to select correct size of aluminum conductors.

It is important to note that this rule establishes minimum conductor ratings based on temperature rise only and does not take into account voltage drop or power loss in the conductors. Such considerations frequently require increasing the size of branch-circuit conductors.

Part **(E)** includes requirements for sizing individual branch-circuit wires serving motors used for short-time, intermittent, periodic, or other varying duty. In such cases, frequency of starting and duration of operating cycles impose varying heat loads on conductors. Conductor sizing, therefore, varies with the application. But it should be noted that the note at the bottom of Table 430.22(E) says any motor is considered to be for continuous duty unless the nature of the apparatus that it drives is such that the motor cannot operate continuously with load under any condition of use.

When a motor is used for one of the classes of service listed in Table 430.22(E), the necessary ampacity of the branch-circuit conductors depends on the class of service and on the rating of the motor. A motor having a 5-min rating is designed to deliver its rated horsepower during periods of approximately 5 min each, with cooling intervals between the operating periods. The branch-circuit conductors have the advantage of the same cooling intervals and hence can safely be smaller than for a motor of the same horsepower but having a 60-min rating.

In the case of elevator motors, the many considerations involved in determining the smallest permissible size of the branch-circuit conductors make this a complex problem, and it is always the safest plan to be guided by the recommendations of the manufacturer of the equipment. This applies also to feeders supplying two or more elevator motors and to circuits supplying noncontinuous-duty motors used for driving some other machines. The duty cycle will essentially be intermittent because buildings are not of infinite height. Part **(G)** covers small motors capable of running on branch-circuit wiring smaller than 14 AWG. The motors must be physically protected by being located in a cabinet or other enclosure to use these allowances. The rule for short-circuit and ground-fault protection is complex, because in these cases, the type of protective device must be one of the three options given in 240.4(D)(1)(2) or (2)(2), but the size of those devices follows 430.52, and now this gets even more complicated. The type limitations in 240.4(D) call for certain specified classes of fast-acting fuses or other fuses or circuit breakers listed to protect the wire sizes being installed.

In order to use Table 430.52, however, you have to know which column to use in selecting a protective device. For a circuit breaker listed for protecting small wires, use the "inverse-time breaker" column, for "Class CC" fuses use the "nontime delay fuse" column [because of Note 1 to the table; also see 430.53(C)(1) Exception No. 2(a) if that condition applies], for a "Class T" fuse use the "nontime delay fuse" column, and for a "Class J" fuse use the applicable fuse column based on the specific fuse at hand because that classification fuse is available in either construction. At the present time no fuses other than those specified in 240.4(D) have been listed for use with these smaller conductors.

The limits on motor sizing are then based on the "full-load current rating" of the motor. Remember that by Table 400.5(A)(1) the 18 and 16 AWG three-phase cord ampacities are 7 and 10 A, which also line up with the limits in 240.4(D)

Moving forward on this basis, the current ranges will be based on the class of overload relay; if unfamiliar the reader should review this topic at 430.32(C). This in turn means that the term "overcurrent" is used incorrectly, instead of short-circuit and ground-fault protection. However, in order to make sense out of this we can assume that is what was intended. We are left with four possibilities for an upper limit on full-load current loading.

For Class 10 (fast) overload protection, 18 AWG wiring will support a full-load motor draw of 5 A and 16 AWG will support 8 A. For Class 20 (normal) overload protection 18 AWG wiring will support a full-load motor draw of 3.5 A and 16 AWG will support 5.5 A.

These provisions match up with the NFPA Standard for Industrial Machinery (NFPA 79) and are designed to facilitate competition with European enterprises using IEC conductor sizing that is smaller than 14 AWG. The smaller conductors are recognized as single conductors within enclosures, and the 2017 NEC extended this recognition to runs outside an enclosure, provided they are part of a jacketed cable assembly or in flexible cord.

430.23. Wound-Rotor Secondary. The full-load secondary current of a wound-rotor or slip-ring motor must be obtained from the motor nameplate or from the manufacturer. The starting, or starting and speed-regulating, portion of

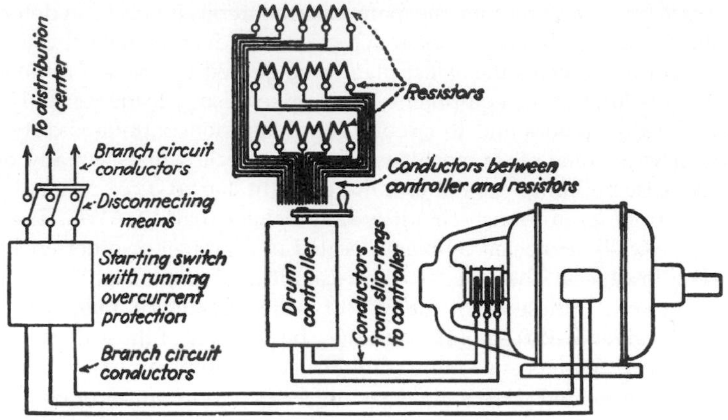
To distribution center
Resistors
Branch circuit conductors
Conductors between controller and resistors
Disconnecting means
Starting switch with running overcurrent protection
Drum controller
Conductors from slip-rings to controller
Branch circuit conductors

Fig. 430-4. Wound-rotor motor may be used with rotary drum switch for speed control. (Sec. 430.23.)

the controller for a wound-rotor motor usually consists of two parts—a dial-type or drum controller and a resistor bank. These two parts must, in many cases, be assembled and connected by the installer, as in Fig. 430-4.

The conductors from the slip rings on the motor to the controller are in circuit continuously while the motor is running and hence, for a continuous-duty motor, must be large enough to carry the secondary current of the motor continuously.

If the controller is used for starting only and is not used for regulating the speed of the motor, the conductors between the dial or drum and the resistors are in use only during the starting period and are cut out of the circuit as soon as the motor has come up to full speed. These conductors may therefore be of a smaller size than would be needed for continuous duty.

If the controller is to be used for speed regulation of the motor, some part of the resistance may be left in circuit continuously and the conductors between the dial or drum and the resistors must be large enough to carry the continuous load without overheating. In Table 430.23(C) the term *continuous duty* applies to this condition.

Conductors connecting the secondary of a wound-rotor induction motor to the controller must have a carrying capacity at least equal to 125 percent of the motor's full-load secondary current if the motor is used for continuous duty. If the motor is used for less than continuous duty, the conductors must have capacity not less than the percentage of full-load secondary nameplate current given in Table 430.22(E). Conductors from the controller of a wound-rotor induction motor to its starting resistors must have an ampacity in accordance with Table 430.23(C), as shown in Fig. 430-5 for a magnetic starter used for reduced inrush on starting but not for speed control.

430.24. Several Motors or a Motor(s) and Other Load(s). Conductors supplying two or more motors (such as feeder conductors to a motor control center, to a panel supplying a number of motors, or to a gutter with several branch circuits

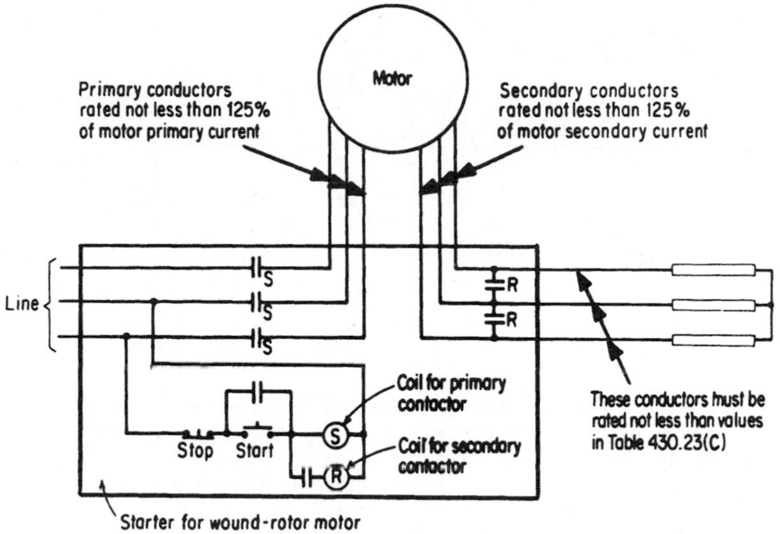
Motor
Primary conductors rated not less than 125% of motor primary current
Secondary conductors rated not less than 125% of motor secondary current
Line
S
S
S
R
R
Coil for primary contactor
Coil for secondary contactor
S
R
Stop
Start
These conductors must be rated not less than values in Table 430.23(C)
Starter for wound-rotor motor

Fig. 430-5. Rules cover conductor sizing for wound-rotor motors without speed control. (Sec. 430.23.)

tapped off) must have a current rating not less than 125 percent of the full-load current rating of the largest motor supplied, plus the sum of the full-load current ratings of the other motors supplied, plus capacity for the nonmotor loads taken at 100 percent for the noncontinuous and 125 percent of the continuous portion(s) of such other load.

Figure 430-6 shows an example of sizing feeder conductors for a load of four motors, selecting the conductors on the basis of ampacities given in Table 310.15(B)(16) and using conductors with a 60°C or 75°C insulating rating—or using 90°C-rated conductors at the ampacities of 75°C. UL rules generally prohibit use of 90°C conductors at the 90°C ampacities shown in Code Table 310.15(B)(16). [Refer to 110.14(C)(1)(a)(4).]

For the overcurrent protection of feeder conductors of the minimum size permitted by this section, the highest permissible rating or setting of the protective device is specified in 430.62. Where a feeder protective device of higher rating or setting is used because two or more motors must be started simultaneously, the size of the feeder conductors shall be increased correspondingly. Note that, as covered in Exception No. 1, if one or more motors is not classified as continuous duty, there is a two-step process involved in determining the highest motor rating. First determine the adjusted current requirements for all such motors by using 430.22(E). Then compare those results with the continuous duty motors taken at the customary 125 percent. The largest result is the number that goes forward as the largest motor for the feeder calculations.

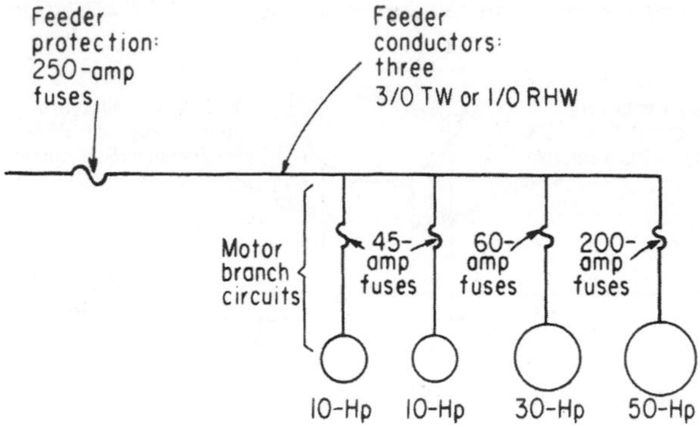
Feeder
protection:
250-amp
fuses
Feeder
conductors:
three
3/0 TW or 1/0 RHW
Motor
branch
circuits
45-
amp
fuses
60-
amp
fuses
200-
amp
fuses
10-Hp
10-Hp
30-Hp
50-Hp

The four motors supplied by the 3-phase, 440-volt, 60-cycle feeder, which are not marked with a Code letter (see Table 430.52), are as follows:

1 50-hp squirrel-cage induction motor (full-voltage starting)
1 30-hp wound-rotor induction motor
2 10-hp squirrel-cage induction motors (full-voltage starting).

Step 1. Branch-circuit loads

From Table 430-250, the motors have full-load current ratings as follows:

50-hp motor—65 amps
30-hp motor—40 amps
10-hp motor—14 amps

Step 2. Conductors

The feeder conductors must have a carrying capacity as follows (see Section 430-24):

1.25 × 65 = 81 amps
81 + 40 + (2 × 14) = 149 amps

The feeder conductors must be at least No. 3/0 TW, 1/0 THW or 1/0 RHH or THHN (copper).

Fig. 430-6. Feeder conductors are sized for the total motor load. (Sec. 430.24.)

These requirements and those of 430.62 for the overcurrent protection of powe feeders are based on the principle that a power feeder should be of such size tha it will have an ampacity equal to that required for the starting current of the larg est motor supplied by the feeder, plus the full-load running currents of all othe motors supplied by the feeder. Except under the unusual condition where two o more motors may be started simultaneously, the heaviest load that a power feeder will ever be required to carry is the load under the condition where the largest motor is started at a time when all the other motors supplied by the feeder are running and delivering their full-rated horsepower.

Where other loads are also supplied, conductor sizing is determined as follows:

1. The current-carrying capacity of feeder conductors supplying a single motor plus other loads must include capacity at least equal to 125 percent of the full-load current of the motor.
2. The current-carrying capacity of feeder conductors supplying a motor load and a lighting and/or appliance load must be sufficient to handle the lighting and/or appliance load as determined from the procedure for calculating size of lighting feeders, plus the motor load as determined from the previous paragraphs.

The Code permits inspectors to authorize use of demand factors for motor feeders—based on reduced heating of conductors supplying motors operating intermittently or on duty-cycle or motors not operating together. Where necessary this should be checked to make sure that the authority enforcing the Code deems the conditions and operating characteristics suitable for reduced-capacity feeders, as noted in 430.26.

For computing the minimum allowable conductor size for a combination lighting and power feeder, the required ampacity for the lighting load is to be determined according to the rules for feeders carrying lighting (or lighting and appliance) loads only. Where the motor load consists of one motor only, the required ampacity for this load is the capacity for the motor branch circuit, or 125 percent of the full-load motor current, as specified in 430.22. Where the motor load consists of two or more motors, or a motor(s) and other loads, the required ampacity for the motor load is the capacity computed according to 430.24.

Figure 430-7 shows a typical installation for which calculation of required feeder ampacity is as follows:

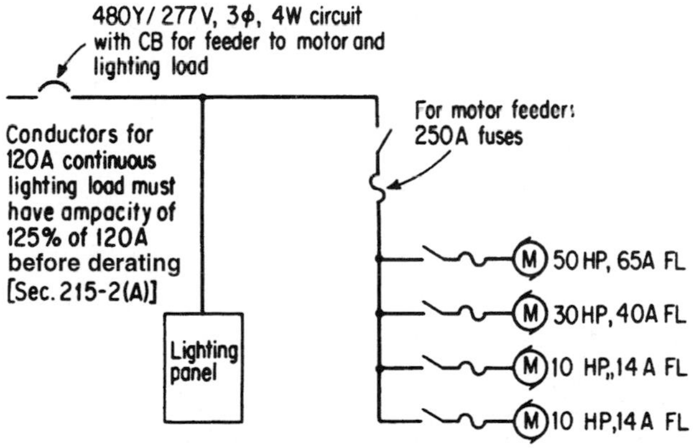
480Y/277V, 3ϕ, 4W circuit with CB for feeder to motor and lighting load
For motor feeder: 250A fuses
Conductors for 120A continuous lighting load must have ampacity of 125% of 120A before derating [Sec. 215-2(A)]
Lighting panel
M
50 HP, 65A FL
M
30 HP, 40A FL
M
10 HP, 14 A FL
M
10 HP, 14 A FL

Fig. 430-7. Other load must be properly combined with motor load. (Sec. 430.24.)

Step 1. Total Load

430.24 says that conductors supplying a lighting load and a motor load must have capacity for both loads, as follows:

$$\text{Motor load} = 65\text{ A} + 40\text{ A} + 14\text{ A} + 14\text{ A} + (0.25 \times 65\text{ A}) = 149\text{ A per phase}$$
$$\text{Lighting load} = 120\text{ A per phase} \times 1.25 = 150\text{ A}$$
$$\text{Total load} = 149 + 150 = 299\text{ A per phase leg}$$

Step 2. Conductors

Table 310.15(B)(16) shows that a load of 299 A can be served by the following copper conductors:

500-kcmil TW
350-kcmil THW, RHH, XHHW, or THHN

Table 310.15(B)(16) shows that this same load can be served by the following aluminum or copper-clad aluminum conductors:

700-kcmil TW
500-kcmil THW, RHH, XHHW, or THHN

Note that the 90°C rating for the THHN was not considered because of termination limitations. If there were adverse environmental conditions such as high temperature ambient conditions or mutual conductor heating, any required derating would have started at the 90°C rating.

430.26. Feeder Demand Factor. A demand factor of less than 100 percent may be applied in the case of some industrial plants where the nature of the work is such that there is never a time when all the motors are operating at one time. But the inspector must be satisfied with (and grant special permission for, including the provision of written notice) any application of a demand factor. Review also the coverage in 220.50 in this book on the topic of motor demand factors.

Sizing of motor feeders (and mains supplying combination power and lighting loads) may be done on the basis of maximum demand current, calculated as follows:

$$\text{Running current} = (1.25 \times I_f) + (DF \times I_t)$$

where I_f = full-load current of largest motor
DF = demand factor as permitted by 430.26
I_t = sum of full-load currents of all motors except largest

But modern design dictates use of the maximum-demand starting current in sizing conductors for improved voltage stability on the feeder. This current is calculated as follows:

$$\text{Starting current} = I_s + (DF \times I_t)$$

where I_s = average starting current of largest motor. (Use the percent of motor full-load current given for fuses in Table 430.52.)

430.28. Feeder Taps. This Code rule is an adaptation of 240.21(B)(1) and (B)(2), covering use of 10- and 25-ft (3.0- and 7.5-m) feeder taps with no overcurrent protection at the point where the smaller conductors connect to the higher-ampacity feeder conductors. The adaptation establishes that the tap conductors must have an ampacity as required by 430.22, 430.24, or 430.25.

In applying condition **(1)**, the conductor may have an ampacity less than one-tenth that of the feeder conductors but must be limited to not more than 10 ft (3.05 m) in length and be enclosed within a controller or raceway, and rated not less than 10 percent of the rating of the overcurrent device protecting the feeder.

If conductors equal in size to the conductors of a feeder are connected to the feeder, as in condition **(3)**, no fuses or other overcurrent protection are needed at the point where the tap is made, since the tap conductors will be protected by the fuses or CB protecting the feeder.

The more important circuit arrangement permitted by the preceding rule is shown in Fig. 430-8. Instead of placing the fuses or other branch-circuit protective device at the point where the connections are made to the feeder, conductors having at least one-third the ampacity of the feeder are tapped solidly to the feeder and may be run a distance not exceeding 25 ft (7.5 m) to the branch-circuit protective device. From this point on to the motor-running protective device and thence to the motor, conductors are run having the standard ampacity, that is, 125 percent of the full-load motor current, as specified in 430.22. If the tap conductors shown did not have an ampacity at least equal to one-third of that of the feeder conductors, then the tap conductors must not be over 10 ft (3.0 m) long.

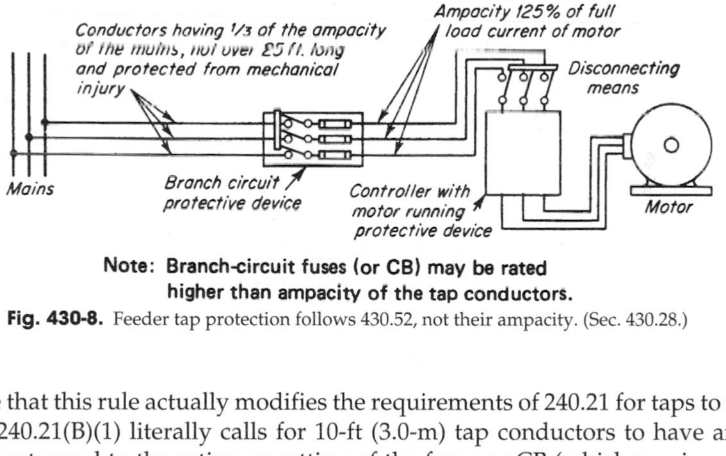
Ampacity 125% of full load current of motor
Conductors having 1/3 of the ampacity of the mains, not over 25 ft. long and protected from mechanical injury
Disconnecting means
Mains
Branch circuit protective device
Controller with motor running protective device
Motor

Note: Branch-circuit fuses (or CB) may be rated higher than ampacity of the tap conductors.

Fig. 430-8. Feeder tap protection follows 430.52, not their ampacity. (Sec. 430.28.)

Note that this rule actually modifies the requirements of 240.21 for taps to motor loads. 240.21(B)(1) literally calls for 10-ft (3.0-m) tap conductors to have ampacity at least equal to the rating or setting of the fuses or CB (whichever is used) at the load end of the tap. And such protection may be rated up to 4 times motor full-load current. But condition **(1)** of this Code section requires sizing of the 10-ft (3.0-m) tap conductors to be at least one-tenth the rating of the overcurrent device, protecting the feeder from which the tap conductors are supplied. And condition **(2)** does *not* require a 25-ft (7.5-m) tap to terminate in a protective device rated to protect the conductors at their ampacity (Fig. 430-9).

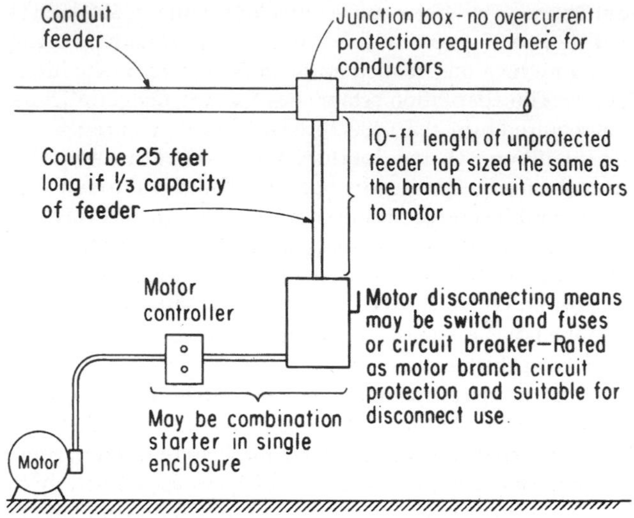
Conduit
feeder
Junction box - no overcurrent
protection required here for
conductors
Could be 25 feet
long if 1/3 capacity
of feeder
10-ft length of unprotected
feeder tap sized the same as
the branch circuit conductors
to motor
Motor
controller
Motor disconnecting means
may be switch and fuses
or circuit breaker—Rated
as motor branch circuit
protection and suitable for
disconnect use.
May be combination
starter in single
enclosure
Motor

Fig. 430-9. Tap conductors may terminate in protective device rated above their ampacity. (Sec. 430.28.)

example A 15-hp 230-V 3-phase motor with autotransformer starter is to be supplied by a tap made to a 250-kcmil feeder. All conductors are to be Type THW.

The feeder has an ampacity of 255 A; one-third of 255 A equals 85 A. Therefore the tap cannot be smaller than No. 4, which has an ampacity of 85 A for 75°C ratings.

The full-load current of the motor is 40 A and, according to part **(IV)** of Art. 430, assuming that the motor is not marked with a Code letter, the branch-circuit fuses should be rated at not more than 300 percent of 40 A, or 120 A, which calls for 125-A fuses (430.52) or less. With the motor-running protection set at 50 A (125 percent × 40 A), the tap conductors are well protected from overload.

The conductors tapped solidly to the feeder must never be smaller than the size of branch-circuit conductors required by 430.22.

The exception in this rule notes that a branch-circuit or subfeeder tap up to 100 ft (30.0 m) long may be made from a feeder to supply motor loads. The specific conditions are given for making a tap that is over 25 ft (7.5 m) long and up to 100 ft (30.0 m) long—where no protection is provided at the point of tap from the feeder conductors. This is a motor-circuit adaptation of the 100-ft (30.0-m) tap permission, which is fully described under 240.21(B)(4).

430.29. Constant-Voltage DC Motors—Power Resistors. These rules cover sizing of conductors from a dc motor controller to separate resistors for power accelerating and dynamic braking. This section, with its table of conductor ampacity percentages, ensures proper application of dc constant-potential motor controls and power resistors.

430.31. General. Detailed requirements for the installation of fire pumps are included in the National Electrical Code in Art. 695. Although electrical concerns are now covered in the NEC, more in-depth coverage of other concerns is given in NFPA 20.

As intended by 430.52, the motor branch-circuit protective device provides short-circuit protection for the circuit conductors. In order to carry the starting current of the motor, this device must commonly have a rating or setting so high that it cannot protect the motor against overload.

For a squirrel-cage induction motor, overload protection must be of the inverse time type with a setting of not over 20 s at 600 percent of the motor full-load current. It is the intent that the fire-pump motor be permitted to run under any condition of loading, even to complete failure, and not be automatically disconnected by an overload protection device. However, should a ground fault or short circuit develop in its conductors, it is also the intent to clear those faults before they result in another task for the fire pump to address. Refer to the extensive coverage of Art. 695 in this book for complete information.

Except where time-delay fuses provide both running overload protection and short-circuit protection as described in 430.55, in practically all cases where motor-running overload protection is provided the motor controller consists of two parts: (1) a switch or contactor to control the circuit to the motor, and (2) the motor-running protective device. Most of the protective devices make use of a heater coil, usually consisting of a few turns of high-resistance metal, though the heater may be of other form.

430.32. Continuous-Duty Motors. The Code makes specific requirements on motor running overload protection intended to protect the elements of the branch circuit—the motor itself, the motor control apparatus, and the branch-circuit conductors—against excessive heating due to motor overloads. Overload protection may be provided by fuses, CBs, or specific overload devices such as OL relays.

Overload is considered to be operating overload up to and including stalled-rotor current. When overload persists for a sufficient length of time, it will cause damage or dangerous overheating of the apparatus. Overload does not include fault current due to shorts or grounds.

Typical overload devices include:

1. Heaters in series with line conductors acting upon thermal bimetallic overload relays
2. Overload devices using resistance or induction heaters and operating on the solder-ratchet principle (Fig. 430-10)
3. Magnetic relays with adjustable instantaneous setting or adjustable time-delay setting
4. Microprocessor support in an adjustable speed drive or in other controllers capable of electronic settings

Of course, the provisions for overload protection are integrated in the enclosure of the controller.

Overload protective devices of the straight thermal type are available with varying tripping and time-delay characteristics. In such devices, the heater coils are made in many sizes and are interchangeable to permit use of the required heater sizes to provide running protection for different motor full-load current ratings. In some units, the heater coil can be adjusted to exact current values. Individual covers are used on the heating elements in some starters to isolate the relay from possible effects on its operation because of the temperature of surrounding air.

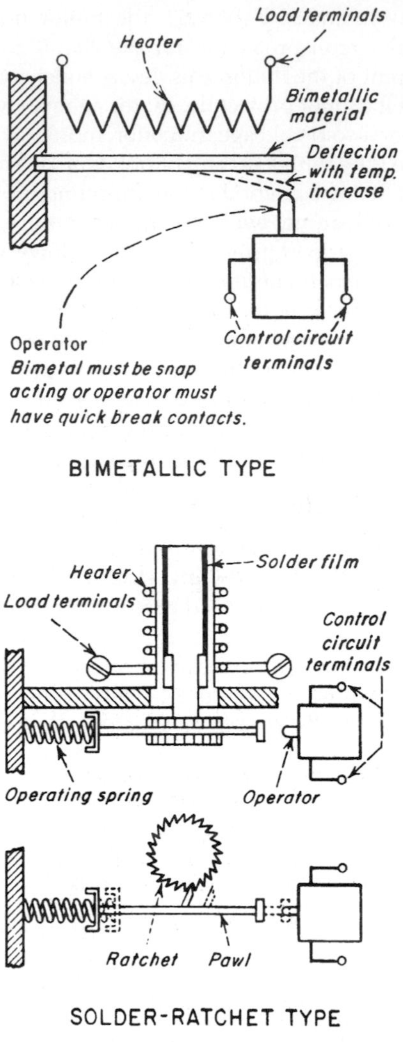
Load terminals
Heater
Bimetallic material
Deflection with temp. increase
Operator
Bimetal must be snap acting or operator must have quick break contacts.
Control circuit terminals
BIMETALLIC TYPE
Solder film
Heater
Load terminals
Control circuit terminals
Operating spring
Operator
Ratchet
Pawl
SOLDER-RATCHET TYPE

Fig. 430-10. Overload relay devices are made in various operating types. (Sec. 430.32.)

NOTE: For a manual starter, the contacts shown are the main load-current contacts of the switch—connected in series with the heater coil.

In general, it is required that every motor shall be provided with a running pro- tective device that will open the circuit on any current exceeding prescribed per- centages of the full-load motor current, the percentage depending on the type of motor. The running protective device is intended primarily to protect the windings of the motor; but by providing that the circuit conductors shall have an ampacit not less than 125 percent of the full-load motor current, it is obvious that these conductors are reasonably protected by the running protective device against any overcurrent caused by an overload on the motor.

Part **(A)** covers application for motors of more than 1 hp. If such a motor is used for continuous duty, running overload protection must be provided. This may be an external overcurrent device actuated by the motor running current and set to open at not more than 125 percent of the motor full-load current for motors marked with a service factor of not less than 1.15 and for motors with a temperature rise not over 40°C. See examples in Fig. 430-3. Sealed (hermetic-type) refrigeration compressor motors must be protected against overload and failure to start, as specified in 440.52. The overload device must be rated or set to trip at not more than 115 percent of the motor full-load current for all other motors, such as motors with a 1.0 service factor or a 55°C rise (Fig. 430-11).

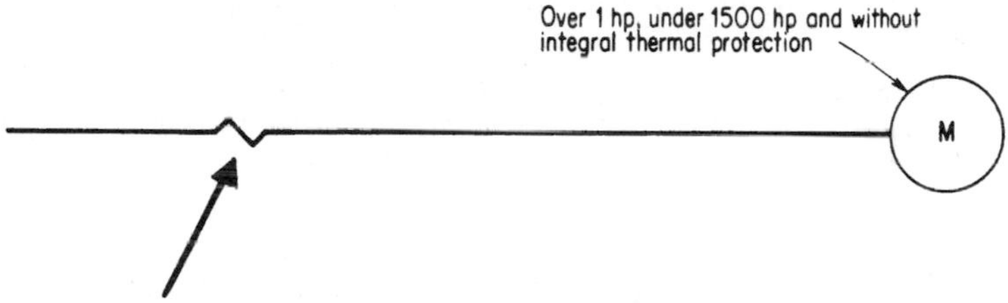
Over 1 hp, under 1500 hp and without integral thermal protection
M

Separate overcurrent device, responsive to motor current, rated or selected to trip at no more than the following percentage of the motor full-load current rating:

Motors with marked service factor not less than 1.15 .... 125%
Motors with marked temperature rise not over 40C 125%
All other motors 115%

Each winding of a multispeed motor must be considered separately. This value may be modified as permitted by Section 430.32(C).

Fig. 430-11. Specific rules apply to continuous-duty motors rated over 1 hp. (Sec. 430.32.)

Be careful when looking at overload element tables provided with controllers. s a general rule, those tables are intended to be directly read out. That is, the 25 percent has already been factored into the selection table. Look up the name-late FLA, and read out the relay element. If you take the nameplate FLA, multi-y by 125 percent, and then go to the table you will be selecting an overload based n 156 percent of the motor nameplate FLA (125 percent of 125 percent). The result ill protect neither the motor nor the conductors from sustained overload should e motor malfunction.

The term *rating,* or *setting,* as used here means the current at which the device ill open the circuit if this current continues for a considerable length of time.

Note: Refer to 460.9, which discusses the need to correct the sizing of running erload protection when power-factor capacitors are installed on the load side of e controller.

A motor having a temperature rise of 40°C when operated continuously at full load can carry a 25 percent overload for some time without injury to the motor. Other types of motors, such as enclosed types, do not have so high an overload capacity, and the running protective device should therefore open the circuit on a prolonged overload, which causes the motor to draw 115 percent of its rated full-load current.

Basic Code requirements are concerned with the rating or setting of overcurrent devices separate from motors. However, the Code permits the use of integral protection. Paragraph **(2)** of part **(A)** covers use of running overload protective devices within the motor assembly rather than in the motor starter. A protective device integral with the motor as used for the protection of motors is shown in Fig. 430-12. This device is placed inside the motor frame and is connected in series with the motor winding. It contains a bimetallic disk carrying two contacts, through which the circuit is normally closed. If the motor is overloaded and its temperature is raised to a certain limiting value, the disk snaps to the open position and opens the circuit. The device also includes a heating coil in series with the motor windings, which causes the disk to become heated more rapidly in case of a sudden heavy overload. For large motors, these devices may have a pair of wires brought out, allowing the sensor to operate within the control circuit and make the contactor or other controller stop the motor.

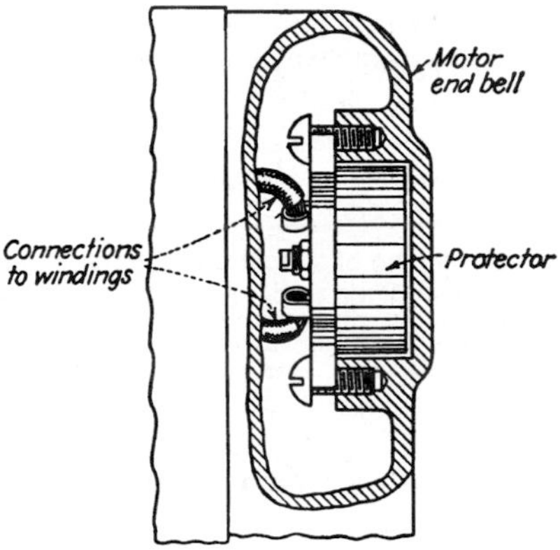
Motor end bell
Connections to windings
Protector

Fig. 430-12. Running overload protection may be built into the motor. (Sec. 430.32.)

Where the circuit-interrupting device is separate from the motor and is actuated by a device integral with the motor, the two devices must be so designed and connected that any accidental opening of the control circuit will stop the motor; otherwise, the motor would be left operating without any overcurrent protection.

There is special need for running protection on an automatically started motor because, if the motor is stalled when the starter operates, the motor will probably burn out if it has no running protection.

Part **(B)** of this section applies to smaller motors (1 hp or less) that are automatically started. Automatically started motors of 1 hp or less must be protected against running overload in the same way as motors rated over 1 hp—as noted in part **(B)**. That is, a separate or integral overload device must be used.

There are alternatives to the specific overload protection rules of parts **(A)** and **(B)**. Under certain conditions, no specific running overload protection need be used: The motor is considered to be properly protected if it is part of an approved assembly, which does not normally subject the motor to overloads and which has controls to protect against a stalled rotor. Or if the impedance of the motor windings is sufficient to prevent overheating due to failure to start, the branch-circuit protection is considered adequate.

In part **(C)**, the Code covers the procedure for "Selection of Overload Device." This rule sets the absolute maximum permitted rating of an overload relay where values are higher than the 125 or 115 percent trip ratings of 430.32(A)(1) and (B). Motors with a marked service factor not less than 1.15 and 40°C-rise motors may, if necessary to enable the motor to start or carry its load, be protected by overload relays with trip settings up to 140 percent of motor full-load current. Motors with a 1.0 service factor and motors with a temperature rise over 40°C (such as 55°C-rise motors) must have their relay trip setting at not over 130 percent of motor full-load current.

BUT WATCH OUT! The maximum settings of 140 or 130 percent apply only to OL relays, such as used in motor starters. Use of this option is discouraged, and an informational note advises that instead of raising the trip setting, a better option might be to change the class of the overload relay element. The usual element, a Class 20, will carry 6 times its rated current for 20 s before opening. Class 10 elements will carry the same current for 10 s, and Class 30 elements for 30 s. Changing from a Class 20 to a Class 30 increases the hold in time by 50 percent without varying the basic trip current, and this should be enough to solve legitimate problems of failure to start. The 2014 **NEC** has added Class 10A relays to the informational note. These relays clear locked rotor current in the same 10 s, and are therefore grouped with Class 10 devices, but undergo additional testing. They must open the circuit in less than 2 h at the rated overload conditions of 430.32, and after having reached thermal equilibrium they must open in less than 2 minutes when subjected to current equal to 150 percent of their setting.

Fuses or CBs may be used for running overload protection but may not be rated or set up to the 140 or 130 percent values. Fuses and breakers must have a maximum rating as shown in 430.32(A) and (B). If the value determined as indicated here does not correspond to a standard rating of fuse or CB, the next smaller size must be used. A rating of 125 percent of full-load current is the absolute maximum for fuses or breakers.

Part **(D)** covers motors of 1 hp or less that are manually started. They are considered to be protected against overload by the branch-circuit protection if the motor is within sight from the starter and the motor is not permanently installed (Fig. 430-13). Running overload devices are not required in such cases. A distance of over 50 ft (15.0 m) is considered out of sight. If the motor is out of sight of its controller, or is permanently installed, then the usual rules in 430.32(B) will apply.

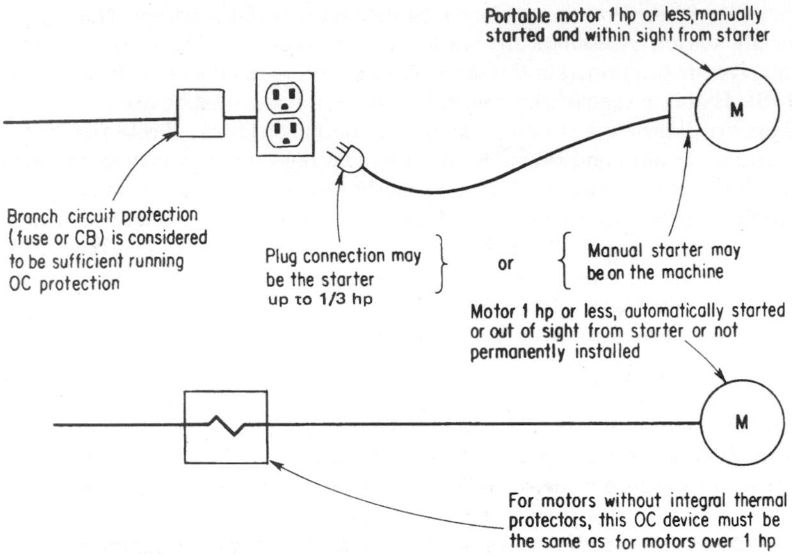
Portable motor 1 hp or less, manually started and within sight from starter
M
Branch circuit protection (fuse or CB) is considered to be sufficient running OC protection
Plug connection may be the starter up to 1/3 hp
or
Manual starter may be on the machine
Motor 1 hp or less, automatically started or out of sight from starter or not permanently installed
M
For motors without integral thermal protectors, this OC device must be the same as for motors over 1 hp

Fig. 430-13. The rules for automatic-start motors are different. (Sec. 430.32.)

430.33. Intermittent and Similar Duty. A motor used for a condition of service which is inherently short-time, intermittent, periodic, or varying duty does not require protection by overload relays, fuses, or other devices required by Sec. 430.32, but, instead, is considered as protected against overcurrent by the branch-circuit overcurrent device (CB or fuses rated in accordance with 430.52). Motors are considered to be for continuous duty unless the motor is completely incapable of operating continuously with load under any condition of use. One classic example is an elevator motor in a building of finite height.

430.35. Shunting During Starting Period. As covered in part **(A)** for motors that are *not* automatically started, where fuses are used as the motor-running protection, they may be cut out of the circuit during the starting period. This leaves the motor protected only by the branch-circuit fuses, but the rating of these fuses will always be well within the 400 percent limit specified in the rule. If the branch-circuit fuses are omitted, as allowed by the rule in 430.53(D), it is not permitted to use a starter that cuts out the motor fuses during the starting period unless the protection of the feeder is within the limits set by this rule. As shown in Fig. 430-14,

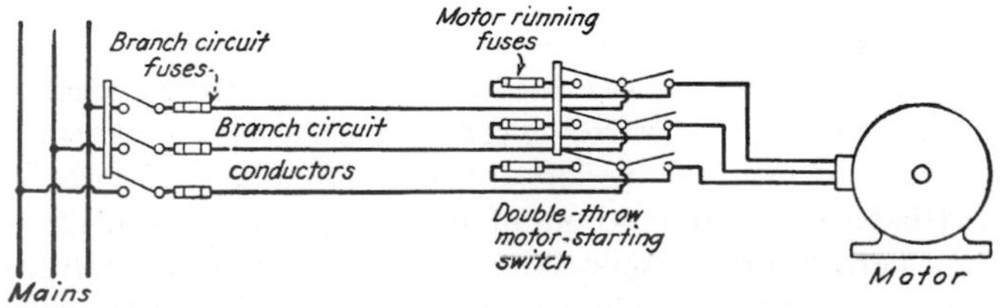
Branch circuit fuses
Motor running fuses
Branch circuit
conductors
Double-throw motor-starting switch
Mains
Motor

Fig. 430-14. Motor OL fuses may be shunted out for starting. (Sec. 430.35.)

a double-throw switch is arranged for across-the-line starting. The switch is thrown to the right to start the motor, thus cutting the running fuses out of the circuit. The switch must be so made that it cannot be left in the starting position.

In the exception to part **(B)**, conditions are given for shunting out overload protection of a motor that is automatically started. In previous Code editions, any motor that was automatically started was not permitted to have its overload protection shunted or cut out during the starting period. This exception now accommodates those motor-and-load applications that have a long accelerating time and would otherwise require an overload device with such a long trip time that the motor would not be protected if it stalled while running.

430.36. Fuses—In Which Conductor. This rule is listed in 240.22 as subpart **(2)** to the rule that prohibits use of an overcurrent device in an intentionally grounded conductor. When fuses are used for protection of service, feeder, or branch-circuit conductors, a fuse must never be used in a grounded conductor, such as the grounded leg of a 3-phase, 3-wire corner-grounded delta system. But, if fuses are used for OL protection for a 3-phase motor connected on such a system, a fuse must be used in all three phase legs—EVEN THE GROUNDED LEG. Figure 430-15 shows two conditions of such fuse application for OL protection for a motor.

430.37. Devices Other than Fuses—In Which Conductor. Complete data on the number and location of overcurrent devices are given in Code Table 430.37.

Table 430.37 requires three running overload devices (trip coils, relays, thermal cutouts, etc.) for all 3-phase motors unless protected by other approved means, such as specifically designed embedded detectors with or without supplementary external protective devices.

Figure 430-16 points out this requirement.

If fuses are used as the running protective device, 430.36 requires a fuse in each ungrounded conductor. If the protective device consists of an automatically operated contactor or CB, the device must open a sufficient number of conductors to stop the current flow to the motor and must be equipped with the number of overload units specified in Table 430.37.

430.42. Motors on General-Purpose Branch Circuits. Refer to Fig. 430-17, Type 3.

Branch circuits supplying lamps are usually 120-V single-phase circuits, and on such circuits the effect of subparagraphs **(A)** and **(B)** is that any motor larger than 6 A must be provided with a starter that is approved for group operation.

It is provided in 210.24 that receptacles on a 20-A branch circuit may have a rating of 20 A, and in such case subparagraph **(C)** requires that any motor or motor-driven appliance connected through a plug and receptacle must have running overcurrent protection. If the motor rating exceeds 1 hp or 6 A, the protective device must be permanently attached to the motor and subparagraph **(B)** must be complied with.

The requirements of 430.32 for the running overcurrent protection of motors must be complied with in all cases, regardless of the type of branch circuit by which the motor is supplied and regardless of the number of motors connected to the circuit.

430.43. Automatic Restarting. As noted in the comments to 430.32, an integral motor-running protective device may be of the type that will automatically restart, or it may be so constructed that after tripping out it must be closed by means of a

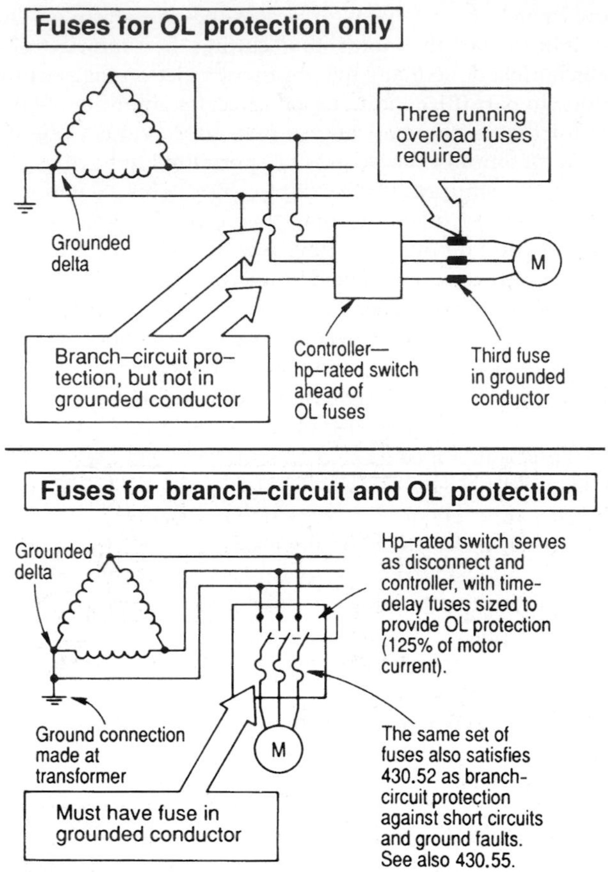
Fuses for OL protection only
Three running overload fuses required
Grounded delta
Branch–circuit protection, but not in grounded conductor
Controller—hp–rated switch ahead of OL fuses
Third fuse in grounded conductor
M
Fuses for branch–circuit and OL protection
Grounded delta
Hp–rated switch serves as disconnect and controller, with time-delay fuses sized to provide OL protection (125% of motor current).
Ground connection made at transformer
M
Must have fuse in grounded conductor
The same set of fuses also satisfies 430.52 as branch-circuit protection against short circuits and ground faults. See also 430.55.

Fig. 430-15. A fuse for OL protection must be used in each phase leg of circuit. (Sec. 430.36.)

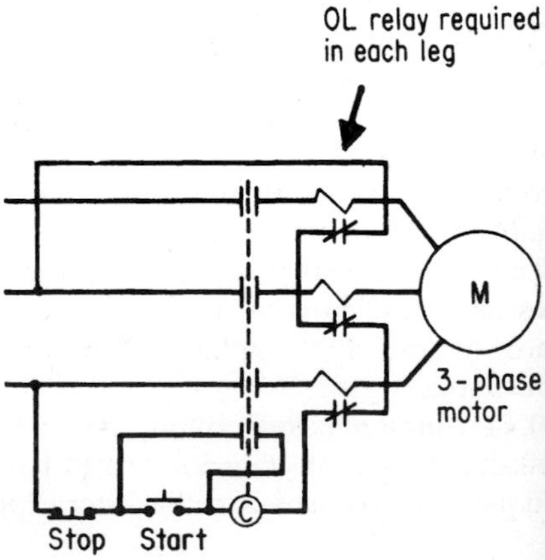
OL relay required in each leg
M
3-phase motor
C
Stop
Start

Fig. 430-16. Three OL units are required for 3-phase motors. (Sec. 430.37.)

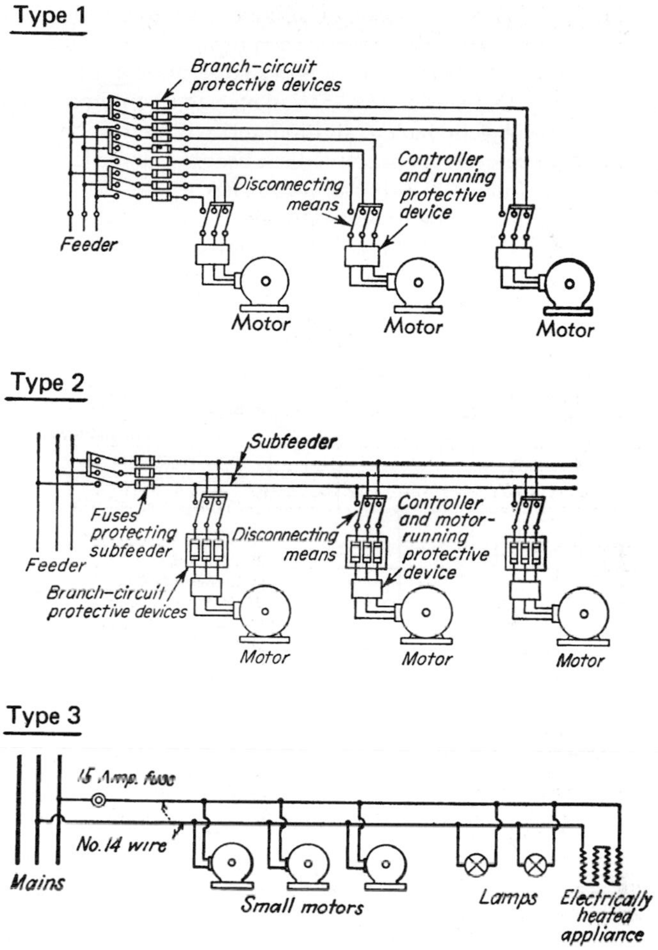

Type 1
Branch-circuit protective devices
Controller and running protective device
Disconnecting means
Feeder
Motor
Motor
Motor
Type 2
Subfeeder
Fuses protecting subfeeder
Feeder
Disconnecting means
Controller and motor-running protective device
Branch-circuit protective devices
Motor
Motor
Motor
Type 3
15 Amp. fuse
No. 14 wire
Mains
Small motors
Lamps
Electrically heated appliance

Fig. 430-17. Motor branch-circuit protection is used in various types of layouts. (Sec. 430.51.)

›set button. However, an automatic restart type of protective device must not be sed if an automatic restart could injure someone.

30.44. Orderly Shutdown. Although the NEC has all those requirements on use running overload protection of motors, this section recognizes that there are ıses when automatic opening of a motor circuit due to overload may be objec- ›nable from a safety standpoint. In recognition of the needs of many industrial ›plications the rule here permits alternatives to automatic opening of a circuit in e event of overload. This permission for elimination of overload protection is nilar to the permission given in 240.4(A) to eliminate overload protection when tomatic opening of the circuit on an overload would constitute a more serious hazard than the overload itself. However, it is necessary that the circuit be provided with a motor overload sensing device conforming to the Code requirement on overload protection to indicate by means of a supervised alarm the presence of the overload (Fig. 430-18). Overload indication instead of automatic opening will alert personnel to the objectionable condition and will permit corrective action, either immediately or at some more convenient time, for an orderly shutdown to resolve the difficulty. But, as is required in 240.4(A), short-circuit protection on the motor branch circuit must be provided to take care of those high-level ground faults and short circuits that would be more serious in their hazardous implications than simple overload.

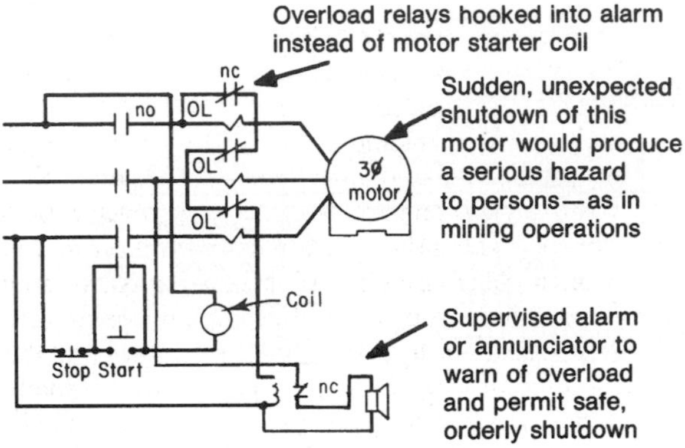
Overload relays hooked into alarm instead of motor starter coil
nc
no
OL
OL
OL
Sudden, unexpected shutdown of this motor would produce a serious hazard to persons—as in mining operations
3ø motor
Coil
Stop Start
nc
Supervised alarm or annunciator to warn of overload and permit safe, orderly shutdown

Fig. 430-18. This type of hookup may be used to warn of, but not open, an overload. (Sec. 430.44.)

Note: 445.12 also has an exception that permits this same use of an alarm instead of overcurrent protection where it is better to have a generator fail than stop operating.

430.51. General. This section indicates the coverage of Part **IV,** which requires "Motor Branch-Circuit Short-Circuit and Ground-Fault Protection." Although the phrase "ground-fault protection" is used in several of the sections of Part **IV,** it should be noted that it refers to the protection against ground fault that is provided by the set of fuses or CB that is used to provide short-circuit protection. The single CB or set of fuses is referred to as a "short-circuit and ground-fault protective device." The rule is *not* intended to require the type of ground-fault protective hookup required by 230.95 on service disconnects (such as a zero-sequence transformer and relay hookup).

Motor branch circuits are commonly laid out in a number of ways. With respect to branch-circuit protection location and type, the layouts shown in Fig. 430-17 are as follows:

Type 1

An individual branch circuit leads to each motor from a distribution center. This type of layout can be used under any conditions and is the one most commonly used.

Type 2

A feeder or subfeeder supports branch circuits tapped on at convenient points. This is the same as Type 1 except that the branch-circuit overcurrent protective devices are mounted individually at the points where taps are made to the subfeeder, instead of being assembled at one location in the form of a branch-circuit distribution center. Under certain conditions, the branch-circuit protective devices may be located at any point not more than 25 ft (7.5 m) distant from the point where the branch circuit is tapped to the feeder.

Type 3

Small motors, lamps, and appliances may be supplied by a 15- or 20-A circuit as described in Art. 210. Motors connected to these circuits must be provided with running overcurrent protective devices in most cases. See 430.42.

Figure 430-19 shows the typical elements of a motor branch circuit in their relation to branch-circuit protection, so that the protection is effective for the circuit conductors, the control and disconnect means, and the motor. Motor controllers provide protection for the motors they control against all ordinary overloads but are not intended to open short circuits. Fuses, CBs, or motor short-circuit protectors used as the branch-circuit protective device will open short circuits and therefore provide short-circuit protection for both the motor and the running protective device. Where a motor is supplied by an individual branch circuit having branch-circuit protection, the circuit protective devices may be either fuses or a CB, and the rating or setting of these devices must not exceed the values specified in 430.52. In Fig. 430-19, the fuses or CB at the panelboard must carry the starting current of the motor, and in order to carry this current the fuse rating or CB setting may be rated up to 300 or 400 percent of the running current of the motor, depending on the size and type of motor. It is evident that to install motor circuit conductors having an ampacity up to that percent of the motor full-load current would be unnecessary.

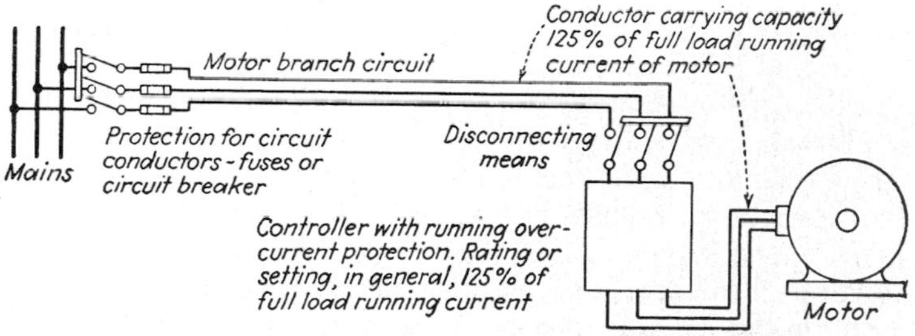
Conductor carrying capacity 125% of full load running current of motor
Motor branch circuit
Protection for circuit conductors - fuses or circuit breaker
Mains
Disconnecting means
Controller with running overcurrent protection. Rating or setting, in general, 125% of full load running current
Motor

Fig. 430-19. Branch-circuit protection is on the line side of other components. (Sec. 430.51.)

There are three possible causes of excess current in the conductors between the panelboard and the motor controller—a short circuit between two of these conductors, a ground fault on one conductor that forms a short circuit, and an overload on the motor. A short circuit would draw so heavy a current that the fuses or breaker at the panelboard would immediately open the circuit, even though the rating or setting is in excess of the conductor ampacity. Any excess current due to an overload on the motor must pass through the protective device at the motor controller, causing this device to open the circuit. Therefore, with circuit conductors having an ampacity equal to 125 percent of the motor-running current and with the motor-protective device set to operate at near the same current, the conductors are reasonably protected.

430.52. Rating or Setting for Individual Motor Circuit. The Code requires that branch-circuit protection for motor circuits must protect the circuit conductors, the control apparatus, and the motor itself against overcurrent due to short circuits or ground (430.51 through 430.58).

The first, and obviously necessary, rule is that the branch-circuit protective device for an individual branch circuit to a motor must be capable of carrying the starting current of the motor without opening the circuit. Then the Code proceeds to place maximum values on the ratings or settings of such overcurrent devices. It says that such devices must not be rated in excess of the values given in Table 430.52.

In case the values for branch-circuit protective devices determined by Table 430.52 do not correspond to the standard sizes or ratings of fuses, nonadjustable CBs, or thermal devices, or possible settings of adjustable CBs adequate to carry the load, the next higher size, rating, or setting may be used.

Under exceptionally severe starting conditions where the nature of the load is such that an unusually long time is required for the motor to accelerate to full speed, the fuse or CB rating or setting recommended in Table 430.52 may not be high enough to allow the motor to start. It is desirable to keep the branch-circuit protection at as low a rating as possible, but in unusual cases, it is permissible to use a higher rating or setting. Where absolutely necessary in order to permit motor starting, the device may be rated at other maximum values, as follows:

1. The rating of a fuse that is *not* a dual-element time-delay fuse (or a time-delay Class CC fuse) and is rated not over 600 A may be increased above the Code table value but must never exceed 400 percent of the full-load current.
2. The rating of a time-delay (dual-element) fuse may be increased but must never exceed 225 percent of full-load current.
3. The setting of an instantaneous trip CB (also known as a "motor circuit protector") that is part of a *listed* combination starter (which contains a magneti short-circuit trip element, without time delay, and independent overload device) may be increased but never over 1300 percent of the motor full-load current, unless supplying a Design B energy-efficient motor, in which th setting may be increased to not more than 1700 percent of motor full-load current. In this category, it is not necessary to actually experiment with lower-rated breaker; the need can be established by engineering evalua tion. And it is never necessary to decrease the protection below 15 A. Moto short-circuit protectors (also where part of a listed combination controlle and listed self-protected combination controllers are permitted to use th same parameters for protection.

4. The rating of an inverse time CB (a typical thermal-magnetic CB with a time-delay and instantaneous trip characteristic) may be increased but must not exceed 400 percent for full-load currents of 100 A or less and must not exceed 300 percent for currents over 100 A.
5. A fuse rated 601 to 6000 A may be increased but must not exceed 300 percent of full-load current.
6. Torque motors must be protected at the motor nameplate current rating, and if a standard overcurrent device is not made in that rating, the next higher standard rating of protective device may be used.
7. Multispeed motors may have a single short-circuit and ground-fault protective device if it meets the multipliers established in Table 430.52. As an alternative, the protection can be set based on the needs of the highest current winding, provided there is running overload protection for both speed windings, and both the controllers and the circuit conductors for every speed winding are sized in accordance with the comparable components of the highest current winding.
8. There are two global provisions to be considered before leaving this topic. The first is the issue of "Power Electronic Devices" that are special "semiconductor" fuses that can be substituted for any of the short-circuit and ground-fault protective devices considered so far. Their use is limited to solid-state motor controller systems, along with some associated electromechanical devices such as bypass and isolation contactors, and there must be a marking for replacement fuses provided adjacent to the fuseholders.
9. The second is 430.52(C)(2), which is a very important and widely overlooked item. This provision says that if the manufacturer of the overload relay components chooses to specify a maximum setting of short-circuit and ground-fault protective devices that can be used ahead of its equipment, and posts it as part of its relay table (or otherwise on the equipment), that number absolutely trumps any calculations made in accordance with multipliers in Table 430.52. That requirement is also specified in UL regulations that regulate the exposure of motor controllers to short-circuit currents to protect internal components, such as overload relays and contacts, from damage or destruction. Those rules state:

> Motor controllers incorporating thermal cutouts, thermal overload relays, or other devices for motor-running overcurrent protection are considered to be suitably protected against overcurrent due to short circuits or grounds by motor branch circuit, short circuit, and ground-fault protective devices selected in accordance with the National Electrical Code and any additional information marked on the product. Motor controllers may specify that protection is to be provided by fuses or by an inverse time circuit breaker. If there is no marking of protective device type, controllers are considered suitably protected by either type of device. Motor controllers may specify a maximum rating of protective device. If not marked with a rating, the controllers are considered suitably protected by a protective device of the maximum rating permitted by the National Electrical Code.

The rules of this section establish maximum values for branch-circuit protection, setting the limit of safe applications. However, use of smaller sizes of branch-circuit protective devices is obviously permitted by the Code and does offer opportunities for substantial economies in selection of CBs, fuses and the switches used with them, panelboards, and so forth. In any application, it is only necessary that the branch-circuit device which is smaller than the maximum permitted rating must have sufficient time delay in its operation to permit the motor starting current to flow without opening the circuit.

But a CB for branch-circuit protection must have a continuous current rating of not less than 115 percent of the motor full-load current, as required by 430.58.

Unless otherwise marked, motor controllers incorporating thermal cutouts or overload relays are considered suitable for use on circuits having available fault currents not greater than (refer to Fig. 430-20):

Horsepower rating	RMS symmetrical amperes
1 or less	1,000
1½ to 50	5,000
51 to 200	10,000
201 to 400	18,000
401 to 600	30,000
601 to 900	42,000
901 to 1600	85,000

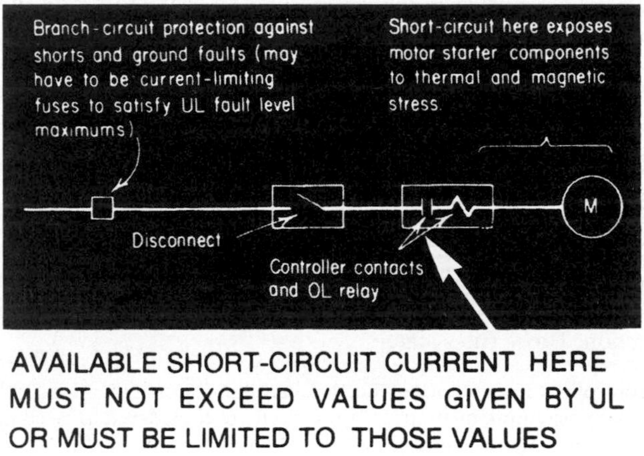
Branch-circuit protection against shorts and ground faults (may have to be current-limiting fuses to satisfy UL fault level maximums)
Short-circuit here exposes motor starter components to thermal and magnetic stress.
M
Disconnect
Controller contacts and OL relay
AVAILABLE SHORT-CIRCUIT CURRENT HERE MUST NOT EXCEED VALUES GIVEN BY UL OR MUST BE LIMITED TO THOSE VALUES

Fig. 430-20. UL specifies maximum short-circuit withstand ratings for controllers. (Sec. 430.52.)

Typical application of the basic rule of 430.52 on short-circuit protection fc motor circuits is shown in Fig. 430-3. Overcurrent (branch-circuit) protectio (from Table 430.52 and 430.52) using nontime-delay fuses is calculated as follow

1. The 50-hp squirrel-cage motor must be protected at not more than 200 (65 A × 300 percent, next higher standard size device allowed).
2. The 30-hp wound-rotor motor must be protected at not more than 60 (40 A × 150 percent).

3. Each 10-hp motor must be protected at not more than 45 A (14 A × 300 percent, next higher standard size device allowed).

As shown in Code Table 430.52, if thermal-magnetic CBs were used, instead of the fuses, for branch-circuit protection, the maximum ratings that are permitted by the basic rule are:

1. For the 50-hp motor—65 A × 250 percent, or 162.5 A, with the next higher standard CB rating of 175 A permitted.
2. For the 30-hp wound-rotor motor—40 A × 150 percent, or 60 A, calling for a 60-A CB.
3. For each 10-hp motor—14 A × 250 percent, or 35 A, calling for a 35-A CB.

Instantaneous Trip CBs

The **NEC** recognizes the use of an instantaneous trip CB (without time delay) for short-circuit protection of motor circuits. Such breakers—also called *magnetic-only* breakers—may be used only if they are adjustable and if combined with motor starters in combination assemblies. An instantaneous-trip CB or a motor short-circuit protector (MSCP) may be used *only* as part of a *listed* (such as by UL) combination motor controller. A combination motor starter using an instantaneous trip breaker must have running overload protection in each conductor (Fig. 430-21). Such a combination starter offers use of a smaller CB than would be possible if a standard thermal-magnetic CB were used. And the smaller CB offers faster operation for greater protection against grounds and short circuits—in addition to offering greater economy. Note that because these devices are only permitted as components in listed combination controllers, they are recognized components

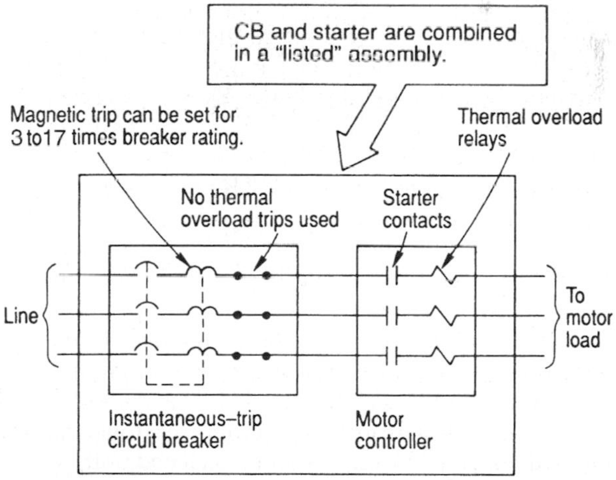
CB and starter are combined in a "listed" assembly.
Magnetic trip can be set for 3 to17 times breaker rating.
Thermal overload relays
No thermal overload trips used
Starter contacts
Line
To motor load
Instantaneous–trip circuit breaker
Motor controller

Fig. 430-21. Section 430.52 accepts use of a magnetic-only circuit breaker if it is part of a "listed" assembly of a combination starter. (Sec. 430.52.)

of listed equipment, but *they are not and can never be listed*. They will have a mark similar to the one shown in Fig. 100-29. Refer to the discussion of the new concept in receptacles as explained in Art. 100 for more information on the component recognition principle.

A combination motor starter, as shown in Fig. 430-21, is based on the characteristics of the instantaneous-trip CB, which is covered by the third percent column from the left in Code Table 430.52. Molded-case CBs with only magnetic instantaneous-trip elements in them are available in almost all sizes. Use of such a device requires careful accounting for the absence of overload protection in the CB, up to the short-circuit trip setting. Such a CB is designed for use as shown in Fig. 430-21. The circuit conductors are sized for at least 125 percent of motor current. The thermal overload relays in the starter protect the entire circuit and all equipment against operating overloads up to and including stalled rotor current. They are commonly set at 125 percent of motor current. In such a circuit, a CB with an adjustable magnetic trip element can be set to take over the interrupting task at currents above stalled rotor and up to the short-circuit duty of the supply system at that point of installation. The magnetic trip in a typical unit might be adjustable from 3 to 17 times the breaker current rating; that is, a 100-A trip can be adjusted to trip anywhere between 300 and 1700 A. Thus the CB serves as motor circuit disconnect and short-circuit protection.

Selection of such a listed assembly with an instantaneous-only CB is based on choosing a nominal CB size with a current rating at least equal to 115 percent of the motor full-load current to carry the motor current and to qualify under 430.58 and 430.110(A) as a disconnect means. Then the adjustable magnetic trip is set to provide the short-circuit protection—the value of current at which instantaneous circuit opening takes place, which should be just above the starting current of the motor involved—using a multiplier of something like 1.5 on locked-rotor current to account for asymmetry in starting current. Asymmetry can occur when the circuit to the motor is closed at that point on the alternating voltage wave where the inrush starting current is going through the negative maximum value of its alternating wave. That is the same concept as asymmetry in the initiation of a short-circuit current. Where supplying Design B energy-efficient motors, a greate inrush can be anticipated on start-up. As a result, higher initial and maximum set tings are recognized.

Listed equipment using an instantaneous CB type is available with very simpl instructions by the manufacturer to make proper selection and adjustment of th instantaneous-trip CB combination starter a quick, easy matter. The followin describes the concept behind the application of listed combination starters wit instantaneous-only CBs.

Given: A 30-hp, 230-V, 3-phase, squirrel-cage motor marked with the code le ter M, indicating that the motor has a locked-rotor current of 10 to 11.19 kVA p horsepower, from Code Table 430.7(B). A full-voltage controller is combined wi the CB, with running overload protection in the controller to protect the mot within its heating damage curve on overload in a listed unit.

Required: Select the maximum setting and minimum rating for the CB which w provide short-circuit protection and will qualify as the motor circuit disconne means.

Solution: The motor has a full-load current of 80 A (Code Table 430.250). A CB suitable for use as disconnect must have a current rating at least 115 percent of 80 A. As covered in 430.52(C)(3), for instantaneous-trip CBs, the initial setting from Table 430.52 would be limited to 800 percent of the 80-A full-load current. The maximum setting—for other than the high-efficiency Design B energy-efficient motors—is 1300 percent. For Design B energy-efficient motors, the initial setting may be 1100 percent of motor full-load current with a maximum setting of 1700 percent of the motor full-load current.

It should be noted that settings above 800 or 1100 percent of the motor's full-load current are permitted only if nuisance tripping occurs on starting *or* if evaluation of the motor's starting characteristics and the time-current trip curve of the breaker indicates that a greater setting is needed. Although not completely clear, the trip value established through the engineering evaluation should be considered as the maximum setting.

Because the use of a magnetic-only CB does not protect against low-level grounds and shorts in the circuit conductors on the line side of the starter running overload relays, the **NEC** rule permits such application only where the CB and starter are part of a *listed* combination starter in a single enclosure.

MSCPs

A motor short-circuit protector (MSCP), as referred to in part (7) of 430.52, is a fuselike device designed for use only in its own type of fusible-switch combination motor starter. As clarified in an informational note, it is not an instantaneous-trip circuit breaker. The combination offers short-circuit protection, running overload protection, disconnect means, and motor control—all with assured coordination between the short-circuit interrupter (the motor short circuit protector) and the running OL devices. It involves the simplest method of selection of the correct MSCP for a given motor circuit. This packaged assembly is a third type of combination motor starter—added to the conventional fusible-switch and CB types.

The **NEC** recognizes motor short-circuit protectors in 430.40 and 430.52, provided the combination is a *listed* assembly. This means a combination starter equipped with motor short-circuit protectors and listed by Underwriters Laboratories, Inc., or another nationally recognized testing lab, as a package called an *MSCP starter*. And here again, the motor short-circuit protector is a *recognized component* of the listed assembly, but it is ineligible for a listing because it cannot be used as a standalone device.

430.53. Several Motors or Loads on One Branch Circuit. A single branch circuit, which must be protected by fuses or inverse-time circuit breakers because instantaneous devices cannot be adequately coordinated, may be used to supply two or more motors as follows:

Part **(A):** Two or more motors, each rated not more than 1 hp and each drawing not over 6 A full-load current, may be used on a branch circuit protected at not more than 20 A at 125 V or less, or 15 A at 600 V or less. And the rating of the branch-circuit protective device marked on any of the controllers must not be exceeded. That is also a UL requirement.

Individual running overload protection is necessary in such circuits, unless: The motor is not permanently installed, is manually started, and is within sight from the controller location; or the motor has sufficient winding impedance to prevent overheating due to stalled rotor current; or the motor is part of an approved assembly that does not subject the motor to overloads and that incorporates protection for the motor against stalled rotor; or the motor cannot operate continuously under load.

Part **(B):** Two or more motors of any rating, each having individual running overload protection, may be connected to a branch circuit that is protected by a short-circuit protective device selected in accordance with the maximum rating or setting of a device that could protect an individual circuit to the motor of the smallest rating. This may be done only where it can be determined that the branch-circuit device so selected will not open under the most severe normal conditions of service that might be encountered.

This permission of part **(B)** offers wide application of more than one motor on a single circuit, particularly in the use of small integral-horsepower motors installed on 460-V, 3-phase systems. This application primarily concerns use of small integral-horsepower 3-phase motors as used in 208-, 220-, 460-, and 575-V industrial and commercial systems. Only such 3-phase motors have full-load operating currents low enough to permit more than one motor on circuits fed from 15-A protective devices.

There are a number of ways of connecting several motors on a single branch circuit, as follows:

In case I, Fig. 430-22, using a three-pole CB for branch-circuit protective device, application is made in accordance with part **(B)** as follows:

1. The full-load current for each motor is taken from NEC Table 430.250 [as required by 430.6(A)].

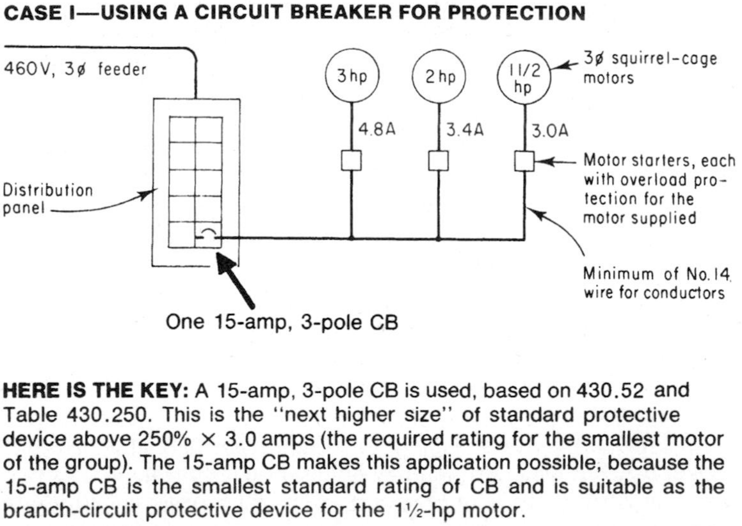
CASE I—USING A CIRCUIT BREAKER FOR PROTECTION
460V, 3ø feeder
3 hp
2 hp
1 1/2 hp
3ø squirrel-cage motors
4.8A
3.4A
3.0A
Motor starters, each with overload protection for the motor supplied
Distribution panel
Minimum of No. 14 wire for conductors
One 15-amp, 3-pole CB
HERE IS THE KEY: A 15-amp, 3-pole CB is used, based on 430.52 and Table 430.250. This is the "next higher size" of standard protective device above 250% × 3.0 amps (the required rating for the smallest motor of the group). The 15-amp CB makes this application possible, because the 15-amp CB is the smallest standard rating of CB and is suitable as the branch-circuit protective device for the 1½-hp motor.

Fig. 430-22. Three integral-horsepower motors may be supplied by this circuit makeup. (Sec. 430.53.)

2. Choosing to use a CB instead of fuses for branch-circuit protection, the rating of the branch-circuit protective device, 15 A, does not exceed the maximum value of short-circuit protection required by 430.52 and Table 430.52 for the smallest motor of the group—which is the 1½-hp motor. Although 15 A is greater than the maximum value of 250 percent times motor full-load current (2.5 × 3.0 A = 7.5 A) set by Table 430.52 (under the column "Inverse Time Breaker" opposite "polyphase squirrel-cage" motors), the 15-A breaker is the "next higher size, rating, or setting" for a standard CB—as permitted in 430.52. A 15-A CB is the smallest standard rating recognized by 240.6.
3. The total load of motor currents is:

$$4.8 \text{ A} + 3.4 \text{ A} + 3.0 \text{ A} = 11.2 \text{ A}$$

   This is well within the 15-A CB rating, which has sufficient time delay in its operation to permit starting of any one of these motors with the other two already operating. Torque characteristics of the loads on starting are not high. It was therefore determined that the CB will not open under the most severe normal service.
4. Each motor has individual running overload protection in its starter.
5. The branch-circuit conductors are sized in accordance with 430.24:

$$4.8 \text{ A} + 3.4 \text{ A} + 2.6 \text{ A} + (25 \text{ percent of } 4.8 \text{ A}) = 12.4 \text{ A}$$

Conductors must have an ampacity at least equal to 12 A. No. 14 THW, TW, RHW, RHH, THHN, or XHHW conductors will fully satisfy this application.

In case II, Fig. 430-23, a similar hookup is used to supply three motors—also with a CB for branch-circuit protection.

1. Section 430.53(B) requires branch-circuit protection to be not higher than the maximum amps set by 430.52 for the lowest rated of the motors.
2. From 430.52 and Table 430.52, that maximum protection rating for a CB is 250 percent × 1.1 A (the lowest rated motor), or 2.5 A. But, 2.8 A is not a "standard rating" of CB from 240.6; and the first exception to 430.52(C)(1) permits use of the "next higher size, rating, or setting" of standard protective device.
3. Because 15 A is the lowest standard rating of CB, it is the "next higher" device rating above 2.5 A and satisfies Code rules on the rating of the branch-circuit protection.

The applications shown in cases I and II permit use of several motors up to circuit capacity, based on 430.24 and 430.53(B) and on starting torque characteristics, operating duty cycles of the motors and their loads, and the time delay of the CB. Such applications greatly reduce the number of CB poles, the number of panels, and the amount of wire used in the total system. One limitation, however, is placed on this practice in 430.52(C)(2), as noted previously. Where more than one fractional- or small-integral-horsepower motor is used on a single branch circuit of 15-A rating in accordance with **NEC** 430.53(A) or (B), care must be taken to observe all markings on controllers that indicate a maximum rating of short-circuit protection ahead of the controller (Fig. 430-24).

CASE II—USING A CIRCUIT BREAKER FOR PROTECTION

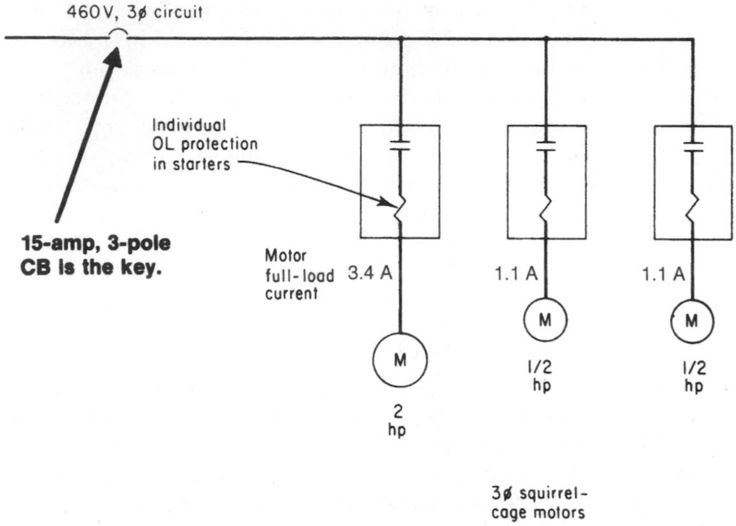
460V, 3ø circuit
Individual
OL protection
in starters
15-amp, 3-pole
CB is the key.
Motor
full-load
current
3.4 A
1.1 A
1.1 A
M
M
M
2
hp
1/2
hp
1/2
hp
3ø squirrel-
cage motors

Fig. 430-23. Fractional-horsepower and integral-horsepower motors may be supplied by the same circuit. (Sec. 430.53.)

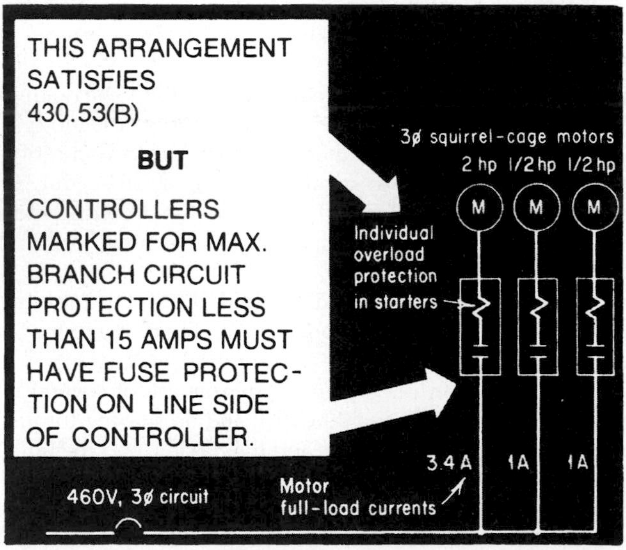
THIS ARRANGEMENT
SATISFIES
430.53(B)
BUT
CONTROLLERS
MARKED FOR MAX.
BRANCH CIRCUIT
PROTECTION LESS
THAN 15 AMPS MUST
HAVE FUSE PROTEC-
TION ON LINE SIDE
OF CONTROLLER.
3ø squirrel-cage motors
2 hp 1/2 hp 1/2 hp
M
M
M
Individual
overload
protection
in starters
3.4 A
1A
1A
460V, 3ø circuit
Motor
full-load currents

Fig. 430-24. Branch-circuit protection must not exceed marked maximum value. (Sec. 430.53.)

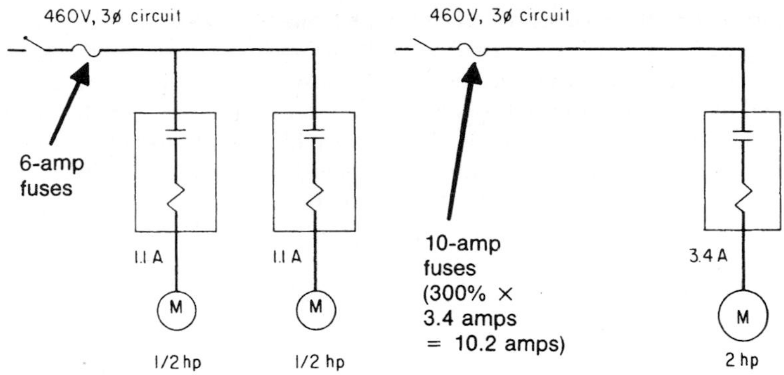
460V, 3ø circuit
6-amp fuses
1.1 A
1.1 A
M
M
1/2 hp
1/2 hp
460V, 3ø circuit
10-amp fuses (300% × 3.4 amps = 10.2 amps)
3.4 A
M
2 hp

Fig. 430-25. Fuse protection may require different circuiting for several motors. (Sec. 430.53.)

In case III, Fig. 430-25, the same three motors shown in case II would be subject to different hookup to comply with the rules of 430.53(B) when fuses, instead of a CB, are used for branch-circuit protection, as follows:

1. To comply with 430.53(B), fuses used as branch-circuit protection must have a rating not in excess of the value permitted by 430.52 and Table 430.52 for the smallest motor of the group—one of the ½-hp motors.
2. Table 430.52 shows that the maximum permitted rating of nontime-delay type fuses is 300 percent of full-load current for 3-phase squirrel-cage motors. Applying that to one of the ½-hp motors gives a maximum fuse rating of

$$300 \text{ percent} \times 1.1 \text{ A} = 3.3 \text{ A}$$

3. The fuse protection will need to be set at 6 A BECAUSE 6 A IS A "STANDARD" RATING OF FUSE (but not a standard rating of CB). 240.6 considers fuses rated at 1, 3, 6, and 10 A to be "standard" ratings.
4. The maximum branch-circuit fuse permitted by 430.53(B) for a ½-hp motor is 6 A (next higher standard size).
5. The two ½-hp motors may be fed from a single branch circuit with three 6-A fuses in a three-pole switch.
6. Following the same Code rules, the 2-hp motor would require fuse protection rated not over 10 A (300 percent × 3.4 A – 10.2 A).

Note: Because the standard fuse ratings below 15 A place fuses in a different elationship to the applicable Code rules, it may require interpretation of the Code ules to resolve the question of acceptable application in case II versus case III. ome jurisdictions may attempt to exclude CBs as circuit protection in these cases here use of fuses, in accordance with the precise wording of the Code, provides ower-rated protection than CBs—when applying the rule of the third paragraph f 430.52. However, there is no support in the current Code text for an enforceent position that would compel a change of overcurrent design from CBs to ises in this case, with one exception that tends to make the general point by virie of its clearly limited effect. Review the discussion and UL data in 430.52(C)(2) this book. If the manufacturer elects to specify the type and size of protection r its overloads, then and only then would a design change become enforceable.

Figure 430-26 shows one way of combining cases II and III to satisfy 430.53(B), 430.52, and 240.6; but the 15-A CB would then technically be feeder protection, because the fuses would be serving as the "branch-circuit protective devices" as required by 430.53(B). Those fuses might be acceptable in each starter, without a disconnect switch, in accordance with 240.40—which allows use of cartridge fuses at any voltage without an individual disconnect for each set of fuses, provided only qualified persons have access to the fuses. But 430.112 would have to be satisfied to use the single CB as a disconnect for the group of motors. And part **(B)** of that exception recognizes one common disconnect in accordance with 430.53(A) but not 430.53(B). Certainly, the use of a fusible-switch-type combination starter for each motor would fully satisfy all rules.

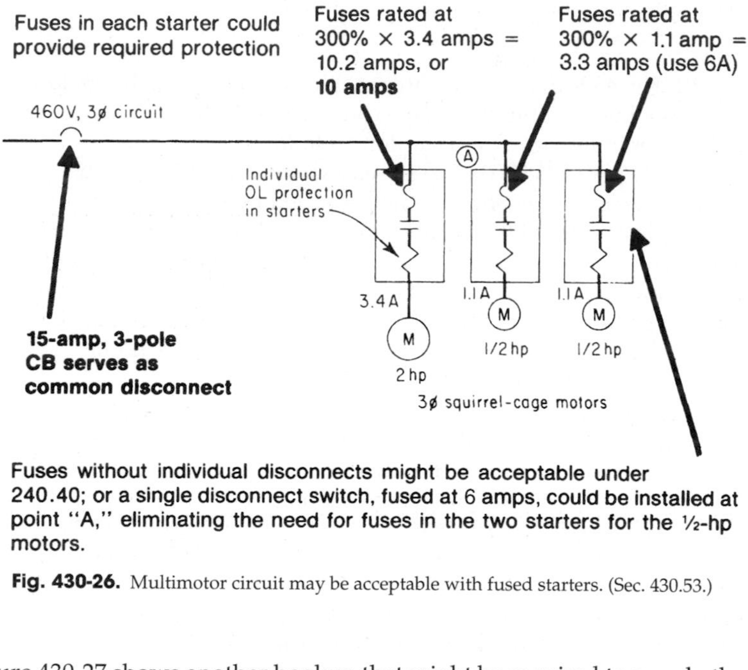
Fuses in each starter could provide required protection
Fuses rated at 300% × 3.4 amps = 10.2 amps, or **10 amps**
Fuses rated at 300% × 1.1 amp = 3.3 amps (use 6A)
460V, 3∅ circuit
Individual OL protection in starters
A
3.4 A
1.1 A
1.1 A
M
M
M
2 hp
1/2 hp
1/2 hp
3∅ squirrel-cage motors
15-amp, 3-pole CB serves as common disconnect
Fuses without individual disconnects might be acceptable under 240.40; or a single disconnect switch, fused at 6 amps, could be installed at point "A," eliminating the need for fuses in the two starters for the ½-hp motors.

Fig. 430-26. Multimotor circuit may be acceptable with fused starters. (Sec. 430.53.)

Figure 430-27 shows another hookup that might be required to supply the thre motors of Fig. 430-22.

Figure 430-28 shows another hookup of several motors on one branc circuit—an actual job installation that was based on application of 430.53(B). Tl installation was studied as follows:

Problem: A factory has 100 1½-hp, 3-phase motors, with individual motor star ers incorporating overcurrent protection, rated for 460 V. Provide circuits.

Solution: Prior to 1965, the NEC would not permit several integral-horsepow motors on one branch circuit fed from a three-pole CB in a panel. Each of the 1(motors would have had to have its own individual 3-phase circuit fed from a 15- 3-pole CB in a panel. As a result, a total of 300 CB poles would have been require calling for seven panels of 42 circuits each plus a smaller panel (or special pane of greater numbers than 42 poles per panel).

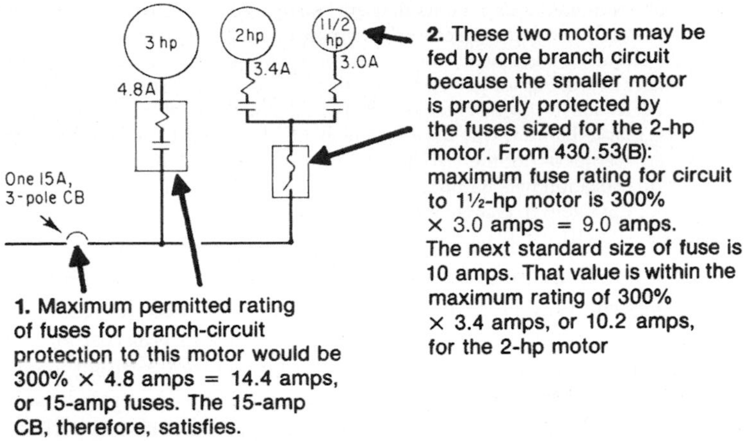
3 hp
2hp
11/2 hp
3.4A
3.0A
4.8A
One 15A, 3-pole CB
2. These two motors may be fed by one branch circuit because the smaller motor is properly protected by the fuses sized for the 2-hp motor. From 430.53(B): maximum fuse rating for circuit to 1½-hp motor is 300% × 3.0 amps = 9.0 amps. The next standard size of fuse is 10 amps. That value is within the maximum rating of 300% × 3.4 amps, or 10.2 amps, for the 2-hp motor
1. Maximum permitted rating of fuses for branch-circuit protection to this motor would be 300% × 4.8 amps = 14.4 amps, or 15-amp fuses. The 15-amp CB, therefore, satisfies.

Fig. 430-27. This hookup might be required to satisfy literal Code wording. (Sec. 430.53.)

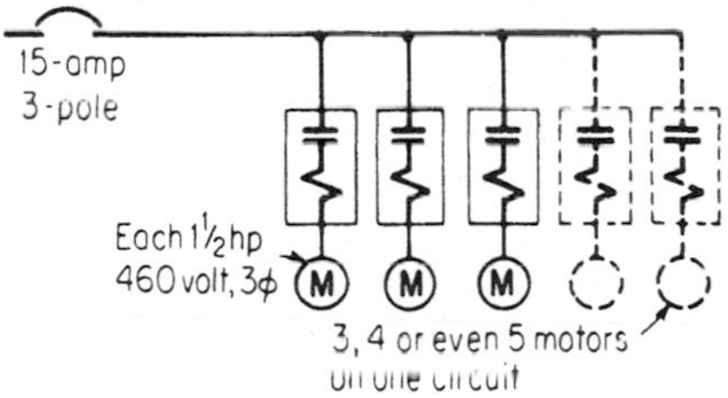
15-amp
3-pole
Each 1½hp
460 volt, 3ϕ
M
M
M
3, 4 or even 5 motors
on one circuit

Fig. 430-28. Multimotor circuits offer economical supply to small integral-horsepower motors. (Sec. 430.53.)

Under the present Code, depending on the starting torque characteristics and perating duty of the motors and their loads, with each motor rated for 3.0 A, hree or four motors could be connected on each 3-phase, 15-A circuit—greatly educing the number of panelboards and overcurrent devices and the amount of vire involved in the total system. Time delay of CB influences number of motors n each circuit.

BUT, an extremely important point that must be strictly observed is the requirenent that the rating of branch-circuit protection must not exceed any maximum alue that might be marked on the starters used with the motors.

Part **(C):** In selecting the wording for part **(C)**, it was the intent of the ode-making panel to clarify the intent that several motors should not be conected to one branch circuit unless careful engineering is exercised by qualified ersons to determine that all components of the branch circuit are selected and pecified to meet the present requirements and to function together. The intent is allow:

a. Completely factory-assembled equipment, or
b. A factory-assembled unit with a separate branch-circuit short-circuit and ground-fault protective device of a type and rating specified, or

c. Separately mounted components that are listed for use together and are specified for such use together by manufacturer's instructions and/or nameplate markings. Note that the circuit breaker need only be a listed inverse-time breaker, and need not be specifically evaluated for group installations. This was the procedure that generated the "HACR" label; it was established that all circuit breakers of the relevant sizes could be marked in that way because the product standard required the same testing anyway, whether or not the HACR label was applied, and therefore the group installations qualification was withdrawn.

It is not the intent to change requirements for supplemental overcurrent protection such as in 422.11(F) or 424.22(C).

Two or more motors of any rating may be connected to one branch circuit if each motor has running overload protection, if the overload devices and controllers are listed for group installation, and if the branch-circuit fuse or time-delay CB rating is in accordance with 430.52 for the largest motor plus the sum of the full-load current ratings of the other motors (Fig. 430-29). The branch-circuit fuses or CB must not be larger than the rating or setting of short-circuit protection permitted by 430.52 for the smallest motor of the group, unless the thermal device is approved for group installation with a given maximum size of fuse or time-delay CB for short-circuit protective device. (See 430.40.) Effective with the 2011 **NEC**, if the short-circuit and ground-fault protection protects each motor overload device and its associated motor in accordance with the parameters that would apply if the installation were not part of a group, then the overload devices in question need not be listed for group installations. The same permission is now effective for the controllers. In such cases, where in effect the line-up to each motor could be used by right as an individual motor circuit, nothing is added in terms of safety to require the group listing on either the overloads or the controllers.

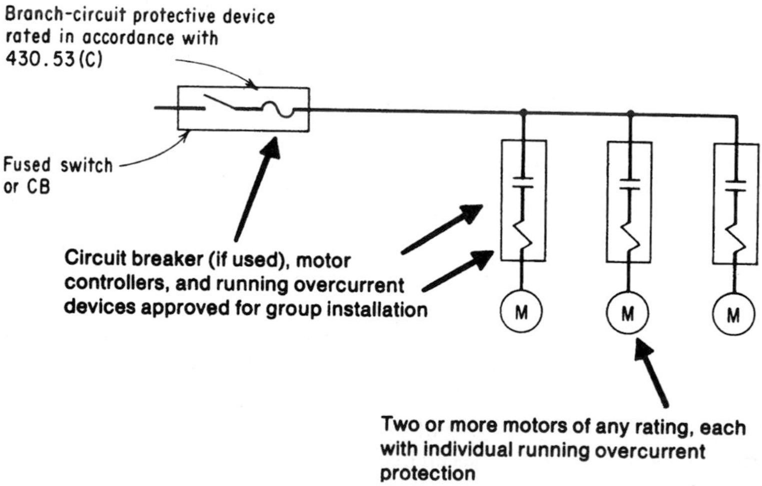
Branch-circuit protective device
rated in accordance with
430.53(C)
Fused switch
or CB
Circuit breaker (if used), motor
controllers, and running overcurrent
devices approved for group installation
M
M
M
Two or more motors of any rating, each
with individual running overcurrent
protection

Fig. 430-29. Motors of any horsepower rating require circuit equipment for group installation. (Sec. 430.53.) This is no longer a requirement for the circuit breaker, which now need only be listed and of the inverse-time type.

Part **(D):** For installations of groups of motors as covered in part **(C),** tap conductors run from the branch-circuit conductors to supply individual motors must be sized properly. Such tap conductors would, of course, be acceptable where they are the same size as the branch-circuit conductors themselves. However, tap conductors to a single motor may be smaller than the main branch-circuit conductors provided they have an ampacity of at least one-third of the branch-circuit conductors; their ampacity is not less than 125 percent of the motor full-load current; they are not over 25 ft (7.5 m) long; and they are in a raceway or are otherwise protected from physical damage (Fig. 430-30).

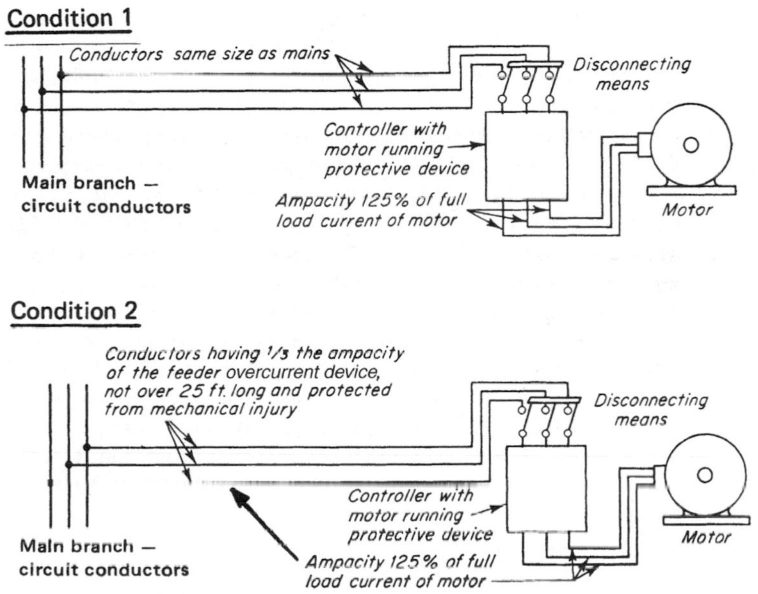
Condition 1
Conductors same size as mains
Disconnecting means
Controller with motor running protective device
Main branch – circuit conductors
Ampacity 125% of full load current of motor
Motor
Condition 2
Conductors having 1/3 the ampacity of the feeder overcurrent device, not over 25 ft. long and protected from mechanical injury
Disconnecting means
Controller with motor running protective device
Main branch – circuit conductors
Ampacity 125% of full load current of motor
Motor

Fig. 430-30. Overcurrent protection not required for taps to single motors of a group. (Sec. 430.53.)

An additional option is to use even smaller tap conductors based on not less than 10 percent of the branch-circuit short-circuit and ground-fault protective evice [but still in accordance with 430.22(A)] that run in a raceway or other suitbly sheltered location and not over 3 m (10 ft) in length, ending at a listed manual notor controller that is marked "Suitable for Tap Conductor Protection in Group nstallations." Effective with the 2011 **NEC**, the conductors may also land on a 'branch-circuit protective device." If the tap conductors are fully sized to be equal o the branch-circuit conductors, then the length and enhanced protection requirenents don't apply. This takes the principles of 430.28 and translates them here, vith the terminating branch circuit device in 430.28 replaced by a special issue nanual motor controller that has an instantaneous-trip component in it, according o the original substantiation.

The principle applied here is that, since the conductors are short and protected rom physical damage, it is unlikely that trouble will occur in the run between the mains and the motor protection which will cause the conductors to be overloaded, except some accident resulting in an actual short circuit. A short circuit will blow the fuses or trip the CB protecting the mains. An overload on the conductors caused by overloading the motor or trouble in the motor itself will cause the motor protective device to operate and so protect the conductors.

The 2017 NEC extended this concept of a tap to a similar manual motor controller to an arrangement that parallels the 25-ft tap rule. As you would expect, these tap conductors are at least one-third of the branch circuit conductors, up from one-tenth, and the permitted distance goes from 10 ft to 25 ft.

430.55. Combined Overcurrent Protection. A CB or set of fuses may provide both short-circuit protection and running overload protection for a motor circuit. For instance, a CB or dual-element time-delay fuse sized at not over 125 percent of motor full-load current (430.32) for a 40°C-rise continuous-duty motor would be acceptable protection for the branch circuit and the motor against shorts, ground faults, and operating overloads on the motor. See the bottom of Fig. 430-15 for a typical fuse application.

Figure 430-31 shows a CB used to fulfill four Code requirements simultaneously. For the continuous-duty, 40°C-rise motor shown, the CB may provide running overload protection if it is rated not over 125 percent of the motor's full-load running current. Therefore, 28 A × 1.25 = 35 A, which satisfies 430.32(A). Because the rating of the thermal-magnetic CB is not over 250 percent times the full-load current (from Table 430.52), the 35-A CB satisfies 430.52 and 430.58 as short-circuit and ground-fault protection. The CB may serve both those functions, as noted in 430.55. The CB may serve as the motor controller, as permitted by 430.83(A)(2). The CB also satisfies as the required disconnect means in accordance with 430.111 and has the rating "of at least 115 percent of the full-load current rating of the motor," as required by 430.110(A). And because it satisfies 430.110(A) on the disconnect minimum rating, it therefore satisfies 430.58, which sets the same minimum rating for a CB used as branch-circuit protection.

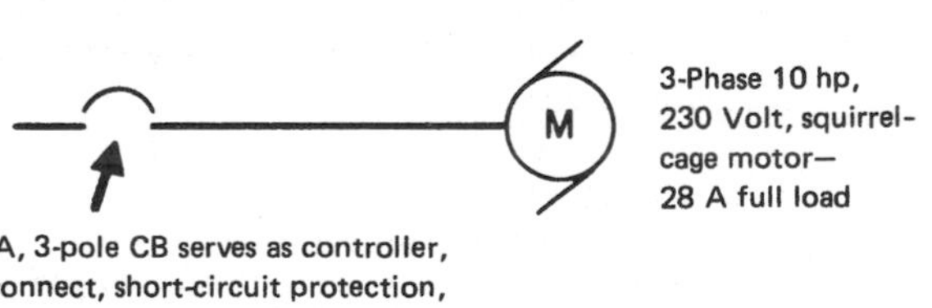
M
3-Phase 10 hp,
230 Volt, squirrel-
cage motor—
28 A full load
35-A, 3-pole CB serves as controller,
disconnect, short-circuit protection,
and running overload protection

Fig. 430-31. Overcurrent functions may be combined in a single CB or set of fuses. (Sec. 430.55.)

430.56. Branch-Circuit Protective Devices—In Which Conductor. Motor branch circuits are to be protected in the same way as other circuits with regard to the number of fuses and the number of poles and overcurrent units of CBs. If fuses are used, a fuse is required in each ungrounded conductor. If a CB is used, there must be an overcurrent unit in each ungrounded conductor (see 240.15).

430.57. Size of Fuseholder. The basic rule of this section covers sizing of fuse-holders for standard nontime-delay fuses used as motor branch-circuit protection. The exception recognizes that time-delay fuses permit use of smaller switches and lower-rated fuseholders.

A fusible switch can take either standard **NEC** fuses or time-delay fuses—up to the rating of the switch. Because a given size of time-delay fuse can hold on the starting current of a motor larger than that which could be used with a standard fuse of the same rating, fusible switches are given two horsepower ratings—one for use with standard fuses, the other for use with time-delay fuses. For example, a 3-pole, 30-A, 240-V fused switch has a rating of 3 hp for a 3-phase motor if standard fuses without time-delay characteristics are used. If time-delay fuses are used, the rating is raised to 7½ hp.

Consider a 7½-hp, 230-V, 3-phase motor (full-voltage starting, without code letters, or with code letters F to V), with a full-load current of 22 A. **NEC** Table 430.52 shows that such a motor may be protected by nontime-delay fuses with a maximum rating equal to 300 percent of the full-load current (66 A), or time-delay fuses with a maximum rating equal to 175 percent of the full-load current (38.5 A).

If standard, nontime-delay fuses were used, the maximum size permitted would be 70 A (the next standard size larger than 66 A). From the table, this would require a 100-A, 15-hp switch, which would have fuseholders that could accommodate the fuses, as required by the basic rule. Or, a 60-A, 7½-hp switch might be used with standard fuses rated 60 A maximum. But such a switch would be required by the basic rule to have fuseholders that could accommodate 70-A fuses. Because such a fuse has knife-blade terminals instead of end ferrules and is larger than a 60-A fuse, fuseholders in the 60-A switch could be held in conflict with the Code rule, even though the level of protection would be better with 60-A fuses in the 60-A switch. This is the reason for the exception that now clearly establishes that a switch designed for the smaller time-delay fuses can be used.

430.58. Rating of Circuit Breaker. This rule sets a maximum and minimum rating for a CB as branch-circuit protection. Refer to 430.55.

In the case of a CB having an adjustable trip point, this rule refers to the capacity of the CB to carry current without overheating and has nothing to do with the setting of the breaker. The breaker most commonly used as a motor branch-circuit protective device is the nonadjustable CB (see 240.6), and any breaker of this type having a rating in conformity with the requirements of 430.52 will have an ampacity considerably in excess of 115 percent of the full-load motor current.

430.62. Rating or Setting—Motor Load. Overcurrent protection for a feeder to several motors must have a rating or setting not greater than the largest rating or setting of the branch-circuit protective device for any motor of the group plus the sum of the full-load currents of the other motors supplied by the feeder.

The second paragraph notes that there are cases where two or more motors fed by a feeder will have the same rating as the branch-circuit device. And that can happen where the motors are of the same or different horsepower ratings. It is possible for motors of different horsepower ratings to have the same rating as the branch-circuit protective device, depending on the type of motor and the type of protective device. If two or more motors in the group are of different horsepower rating but the rating or setting of the branch-circuit protective device is the same for both motors, then one of the protective devices should be considered as the largest for the calculation of feeder overcurrent protection.

And because Table 430.52 recognizes many different ratings of branch-circuit protective devices (based on use of fuses or CBs and depending on the particular type of motor), it is possible for two motors of equal horsepower rating to have widely different ratings of branch-circuit protection. If, for instance, a 25-hp motor was protected by nontime-delay fuses, Table 430.52 gives 300 percent of the full-load motor current as the maximum rating or setting of the branch-circuit device. Thus, 250-A fuses would be used for a motor that had a 78-A full-load rating. But another motor of the same horsepower and even of the same type, if protected by time-delay fuses, must use fuses rated at only 175 percent of 78 A, which would be 150-A fuses, as shown in Fig. 430-32. If the two 25-hp motors were of different types, one being a wound-rotor motor, it would still be necessary to base selection of the feeder protection on the largest rating or setting of a branch-circuit protective device, regardless of the horsepower rating of the motor.

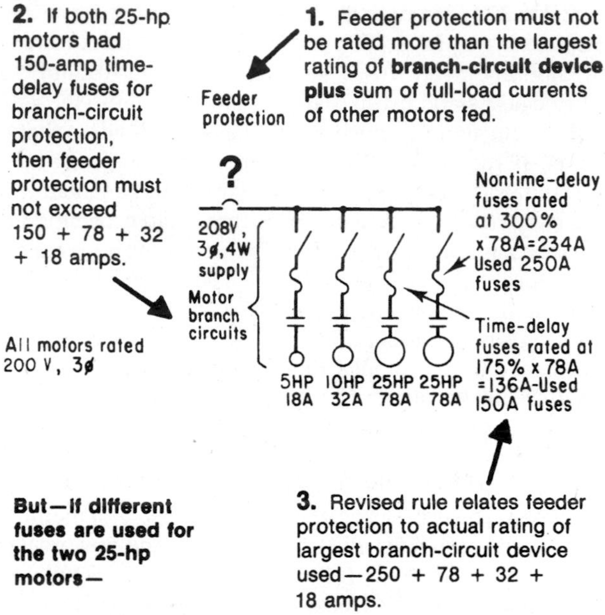
2. If both 25-hp motors had 150-amp time-delay fuses for branch-circuit protection, then feeder protection must not exceed 150 + 78 + 32 + 18 amps.
1. Feeder protection must not be rated more than the largest rating of branch-circuit device plus sum of full-load currents of other motors fed.
Feeder protection
?
208V, 3ø, 4W supply
Motor branch circuits
Nontime-delay fuses rated at 300% x 78A=234A Used 250A fuses
Time-delay fuses rated at 175% x 78A =136A-Used 150A fuses
All motors rated 200 V, 3ø
5HP 18A
10HP 32A
25HP 78A
25HP 78A
But—if different fuses are used for the two 25-hp motors—
3. Revised rule relates feeder protection to actual rating of largest branch-circuit device used—250 + 78 + 32 + 18 amps.

Fig. 430-32. Feeder protection is based on largest branch-circuit protection, not on motor horsepower ratings. (Sec. 430.62.)

Figure 430-33 shows a typical motor feeder calculation, as follows:

The four motors supplied by the 3-phase, 460-V, 60-cycle feeder, which are n marked with a code letter (see Table 430.52), are as follows:

- One 50-hp squirrel-cage induction motor (full-voltage starting)
- One 30-hp wound-rotor induction motor
- Two 10-hp squirrel-cage induction motors (full-voltage starting)

Step 1. Branch-Circuit Loads

From Table 430.250, the motors have full-load current ratings as follows:

50-hp motor—65 A
30-hp motor—40 A
10-hp motor—14 A

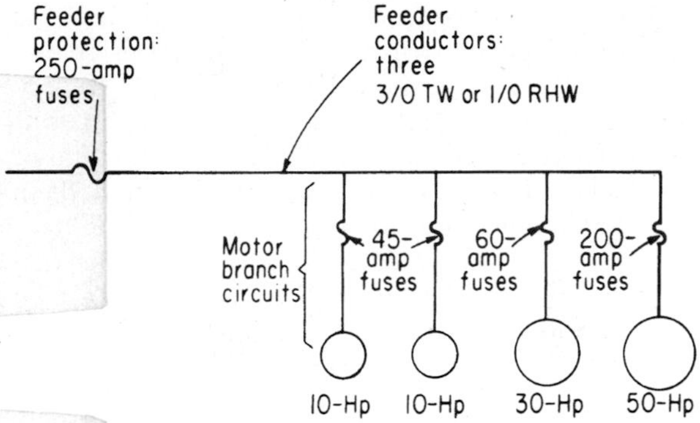
Feeder protection: 250-amp fuses
Feeder conductors: three 3/0 TW or 1/0 RHW
Motor branch circuits
45-amp fuses
60-amp fuses
200-amp fuses
10-Hp
10-Hp
30-Hp
50-Hp

Fig. 430-33. Rating of feeder protection is based on branch protection and motor currents. (Sec. 430.62.)

Step 2. Conductors

The feeder conductors must have a carrying capacity as follows (see 430.24):

$$1.25 \times 65 = 81 \text{ A}$$
$$81 + 40 + (2 \times 14) = 149 \text{ A}$$

The feeder conductors must be at least No. 3/0 TW, 1/0 THW, or 1/0 RHH or THHN (copper).

Step 3. Branch-Circuit Protection

Overcurrent (branch-circuit) protection (from Table 430.52 and 430.52) using nontime-delay fuses is as follows:

1. The 50-hp motor must be protected at not more than 200 A (65 A × 300 percent).
2. The 30-hp motor must be protected at not more than 60 A (40 A × 150 percent).
3. Each 10-hp motor must be protected at not more than 45 A (14 A × 300 percent).

Step 4. Feeder Protection

As covered in 430.62, the maximum rating or setting for the overcurrent device protecting such a feeder must not be greater than the largest rating or setting of branch-circuit protective device for one of the motors of the group plus the sum of the full-load currents of the other motors. From the preceding, then, the maximum allowable size of feeder fuses is 200 + 40 + 14 + 14 = 268 A.

This calls for a maximum standard rating of 250 A for the motor feeder fuses, which is the nearest standard fuse rating that does not exceed the maximum permitted value of 268 A.

Exception No. 1 to part **(A)** addresses those installations where instantaneous-trip CBs and/or MSCPs are used as short-circuit and ground-fault protection for the largest motor supplied by the feeder to be protected. Under certain conditions those devices may be set or rated to trip at 13 times or, for Design B energy-efficient motors, even as high as 17 times the motor full-load current. To prevent the feeder conductors from being underprotected, this exception requires that the rating or setting of the feeder protective device be based on the type of device used. That is, if nontime-delay fuses are used for feeder protection, the rating of those fuses must be based on the rating of fuse that would be permitted to protect the motor branch circuit if a fuse were used as branch-circuit protection instead of an instantaneous-trip CB or MSCP. For example, consider a 460-V, 3-phase, 100-hp motor, which draws approximately 124 A. If an instantaneous-trip breaker were used for branch-circuit short-circuit and ground-fault protection, it could be rated at over 1600 A (13 × 124 A).

When establishing the rating of protection for the feeder supplying a motor branch-circuit so protected, if nontime-delay fuses are used, the value of current that is summed with the ratings of the other type branch-circuit protective devices must be no more than that which would be permitted if nontime-delay fuses were used as branch-circuit protection instead of an instantaneous-trip CB or MSCP. In this example, Table 430.52 would permit a nontime-delay fuse to be 300 percent of the motor full-load current. If the branch circuit in question is protected by nontime-delay fuses, they would be rated at 350 A (3 × 125 A; rounded down because the exception gives no allowance to modify the Table 430.52 results). And 350 A, *not* 1300 A, would be used to calculate the maximum rating of nontime-delay fuses that are permitted to protect the feeder conductors.

Note: There is no provision in 430.62 whatsoever that would permit the use of "the next higher size, rating, or setting" of the protective device, for a motor feeder when the calculated maximum rating does not correspond to a standard size of device. The feeder calculations are rounded down, not up.

According to part **(B)** of this section, in large-capacity installations where extra feeder capacity is provided for load growth or future changes, the feeder overcurrent protection may be calculated on the basis of the rated current-carrying capacity of the feeder conductors. In some cases, such as where two or more motors on a feeder may be started simultaneously, feeder conductors may have to be larger than usually required for feeders to several motors.

In selecting the size of a feeder overcurrent protective device, the **NEC** calculation is concerned with establishing a maximum value for the fuse or CB. If a lower value of protection is suitable, it may be used. The motor example in Annex D (Example D8) is extremely informative and helpful, and was significantly improved further in the 2017 **NEC** with respect to the feeder calculations covered here. There are now two separate calculations based first on nontime-delay fuse protection (the traditional one), and a new one based on inverse-time circuit breakers, which produces a differently sized feeder protective device.

430.63. Rating or Setting—Motor Load and Other Load(s). Protection for a feeder to both motor loads and a lighting and/or appliance or other load must be rated on the basis of both of these loads. The rating or setting of the overcurrent device must be sufficient to carry the lighting and/or appliance load plus the rating or setting of the motor branch-circuit protective device if only one motor is supplied, or plus the highest rating or setting of branch-circuit protective device for any one motor plus the sum of the full-load currents of the other motors, if more than one motor is supplied.

Figure 430-34 presents basic NEC calculations for arriving at minimum requirements on wire sizes and overcurrent protection for a combination power and lighting load as follows:

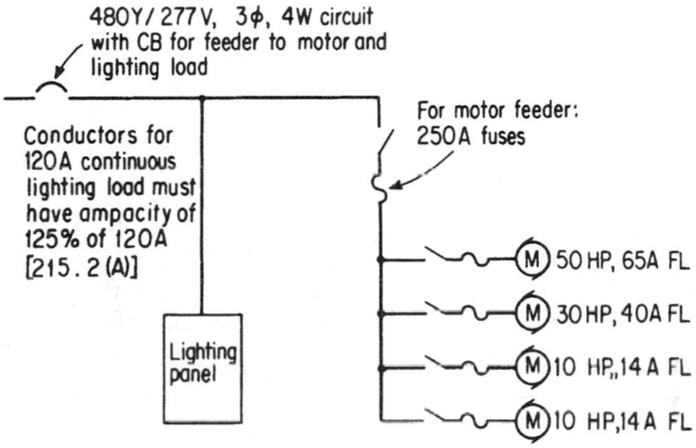
480Y/277V, 3φ, 4W circuit with CB for feeder to motor and lighting load
For motor feeder: 250A fuses
Conductors for 120A continuous lighting load must have ampacity of 125% of 120A [215.2 (A)]
Lighting panel
M 50 HP, 65A FL
M 30 HP, 40A FL
M 10 HP,,14 A FL
M 10 HP,14 A FL

Fig. 430-34. Feeder protection for combination load must properly add both loads. (Sec. 430.63.)

Step 1. Total Load

430.25(A) says that conductors supplying a lighting load and a motor must have capacity for both loads, as follows:

Motor load = 65 A + 40 A + 14 A + 14 A + (0.25 × 65 A) = 149 A per phase
Lighting load = 120 A per phase × 1.25 = 150 A
Total load = 149 + 150 = 299 A per phase leg

Step 2. Conductors

Table 310.15(B)(16) shows that a load of 299 A can be served by the following copper conductors:

500-kcmil TW
350-kcmil THW

Table 310.15(B)(16) shows that this same load can be served by the following aluminum or copper-clad aluminum conductors:

700-kcmil TW
500-kcmil THW, RHH, or THHN

Step 3. Protective Devices

430.63 says, in effect, that the protective device for a feeder supplying a combined motor load and lighting load may have a rating not greater than the sum of the maximum rating of the motor feeder protective device and the lighting load, as follows:

1. Motor feeder protective device = rating or setting of the largest branch-circuit device for any motor of the group being served plus the sum of the full-load currents of the other motors:

$$200 \text{ A (50-hp motor)} + 40 + 14 + 14 = 268 \text{ A maximum}$$

   This calls for a maximum standard rating of 250 A for the motor feeder fuses, which is the nearest standard fuse rating that does not exceed the maximum permitted value of 268 A.

2. Lighting load = 120 A × 1.25 = 150 A

$$\text{Rating of CB for combined load} = 268 + 150 = 418 \text{ A maximum}$$

   This calls for a 400-A CB, the nearest standard rating that does not exceed the 418-A maximum.

Again: There is no provision in 430.63 that permits the use of "the next higher size, rating, or setting" of the protective device for a motor feeder when the calculated maximum rating does not correspond to a standard size of device.

Such considerations as voltage drop, I^2R loss, spare capacity, lamp dimming on motor starting, and so forth would have to be made to arrive at actual sizes to use for the job. But, the circuiting as shown would be safe—although maybe not efficient or effective for the particular job requirements.

430.71. General. Figure 430-35 shows the *motor control circuit* part of a motor branch circuit, as defined in the second paragraph of this section. A control circuit, as discussed here, is any circuit that has as its load device the operating coil of a magnetic motor starter, a magnetic contactor, or a relay. Strictly speaking, it is a circuit that exercises control over one or more other circuits. And these other circuits controlled by the control circuit may themselves be control circuits or they may be *load* circuits—carrying utilization current to a lighting, heating, power, or signal device.

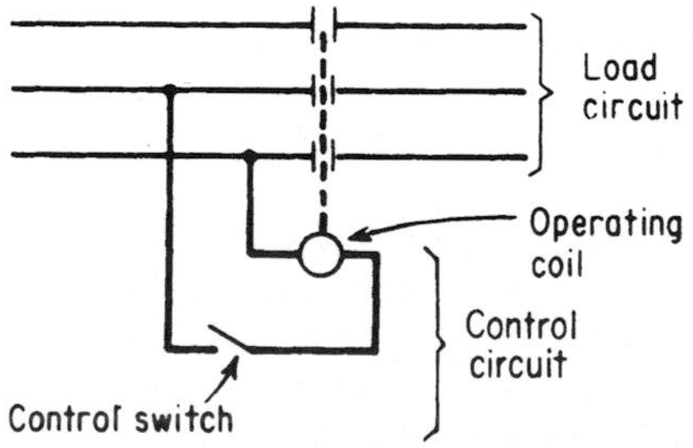
Load circuit
Operating coil
Control circuit
Control switch

Fig. 430-35. A control circuit governs the operating coil that switches the load circuit. (Sec. 430.71.)

The elements of a control circuit include all the equipment and devices concerned with the function of the circuit: conductors, raceway, contactor operating coil, source of energy supply to the circuit, overcurrent protective devices, and all switching devices that govern energization of the operating coil.

The **NEC** covers application of control circuits in Art. 725 and in 240.4 and 430.71 through 430.74. Design and installation of control circuits are basically divided into three classes (in Art. 725) according to the energy available in the circuit. Class 2 and 3 control circuits have low energy-handling capabilities; and any circuit, to qualify as a class 2 or 3 control circuit, must have its open-circuit voltage and overcurrent protection limited to conditions given in 725.121.

Most control circuits for magnetic starters and contactors could not qualify as class 2 or 3 circuits because of the relatively high energy required for operating coils. And any control circuit rated over 150 V (such as 230- or 460-V coil circuits) can never qualify, regardless of energy.

Class 1 control circuits include all operating coil circuits for magnetic starters that do not meet the requirements for class 2 or 3 circuits. Class 1 circuits must be wired in accordance with 725.41 to 725.52.

430.72. Overcurrent Protection. Part **(A)** tells the basic idea behind protection of the operating coil circuit of a magnetic motor starter, as distinguished from a manual (mechanically operated) starter:

1. 430.72 covers motor control circuits that are derived within a motor starter from the power circuit that connects to the line terminals of the starter. The rule here refers to such a control circuit as one "tapped from the load side" of the fuses or circuit breaker that provides branch-circuit protection for the conductors that supply the starter. See the top of Fig. 430-36.
2. The control circuit that is tapped from the line terminals within a starter is *not* a branch circuit itself.
3. Depending on other conditions set in 430.72, the conductors of the control circuit will be considered as protected by *either* the branch-circuit protective device ahead of the starter or the supplementary protection (usually fuses) installed in the starter enclosure.
4. Any motor control circuit that is not tapped from the line terminals within a starter must be protected against overcurrent in accordance with 725.43. Such control circuits would be those that are derived from a panelboard or a control transformer—as where, say, 120-V circuits are derived external to the starters and are typically run to provide lower-voltage control for 230-, 460-, or 575-V motors. See the bottom of Fig. 430-36.

Part **(B)** applies to overcurrent protection of conductors used to make up the control circuits of magnetic motor starters. Such overcurrent protection must be sized in accordance with the amp values shown in Table 430.72(B). And where that table makes reference to amp values specified in 310.15, as applicable, it does *not* specify that 310.15(B)(3) must be observed by derating conductor ampacity where more than three current-carrying conductors are used in a conduit. Previously, the rule in part **(B)** of this section specifically recognized the use of control-circuit wires in raceway "without derating factors." 725.51(A), however, does require class 1 remote-control wires to have their

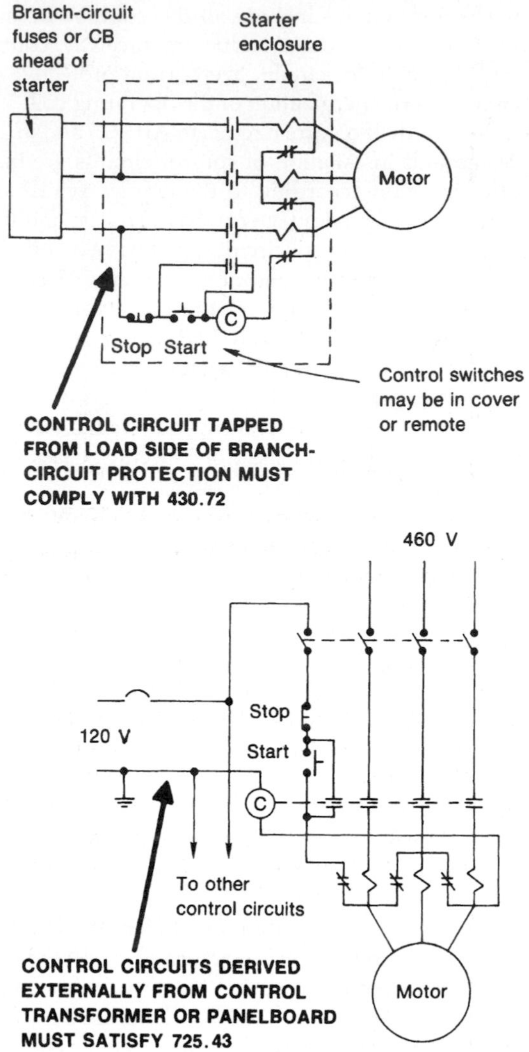
Branch-circuit
fuses or CB
ahead of
starter
Starter
enclosure
Motor
C
Stop Start
Control switches
may be in cover
or remote
CONTROL CIRCUIT TAPPED
FROM LOAD SIDE OF BRANCH-
CIRCUIT PROTECTION MUST
COMPLY WITH 430.72
460 V
Stop
120 V
Start
C
To other
control circuits
CONTROL CIRCUITS DERIVED
EXTERNALLY FROM CONTROL
TRANSFORMER OR PANELBOARD
MUST SATISFY 725.43
Motor

Fig. 430-36. Source of power supply to the control circuit determines which Code section applies to the coil circuit. (Sec. 430.72.)

ampacity derated in accordance with 310.15(B)(3), based on the number of conductors, when the conductors "carry continuous loads" in excess of 10 percent of each conductor's ampacity.

That makes the application in Fig. 430-37 compliant (almost, but not completely certain). The reason for the rule in 725.51(A) is that heat given off from a wire is a function of I^2R losses. A wire carrying 10 percent of its ampacity is putting out 1 percent (0.1^2) of the heat generated when fully loaded. That is a negligible

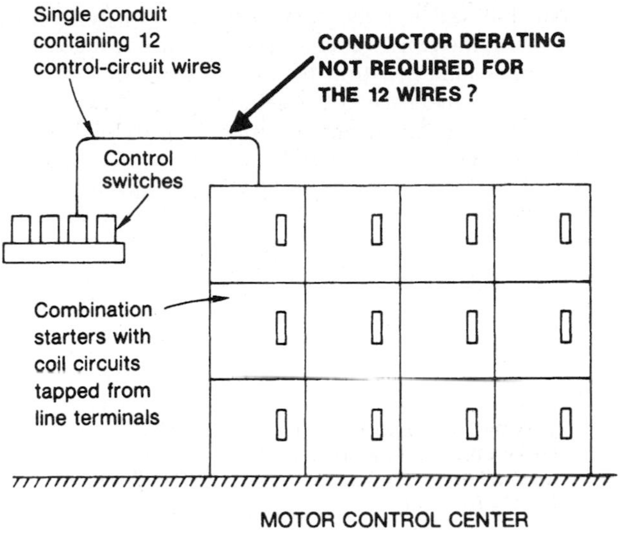
Single conduit
containing 12
control-circuit wires
CONDUCTOR DERATING
NOT REQUIRED FOR
THE 12 WIRES ?
Control
switches
Combination
starters with
coil circuits
tapped from
line terminals
MOTOR CONTROL CENTER

Fig. 430-37. Derating of control-wire ampacity is seldom necessary when more than three conductors are run within the same raceway. (Sec. 430.72.)

amount. Further, the realities of most control circuits are that they seldom carry much in the way of actual current. Most contactor coils and other such devices take a fraction of an ampere over time, and in order to trip over 310.15(B)(2)(a) the 10 percent limit not only has to be exceeded, it must be exceeded on a continuous basis, not just for the fraction of a second that a coil is actually pulling in. So, unless the control circuits are very unusual, the wires are as a practical matter not subject to derating for mutual conductor heating.

The basic rule of part **(B)(1)** requires coil-circuit conductors to have overcurrent protection rated in accordance with the maximum values given in Column A of Table 430.72(B). That table shows 7 A as the maximum rating of protection for 18 AWG copper wire and 10 A for 16 AWG wire and refers to Table 310.15(B)(16) for larger wires—20 A for 14 AWG copper, 25 A for No. 12, and so forth. These values are taken from the 60°C column to be consistent with other calculations in this table, but purposely not using the 240.4(D) special ampacities because the work here is governed by 240.4(G). The basic rules in part **(B)(2)** cover conditions under which other ratings of protection may be used, as follows:

The first sentence in part **(B)(2)** covers protection of control wires for magnetic starters that have their START-STOP buttons in the cover of the starter enclosure.

In part **(B)(2)**, the value of branch-circuit protection must be compared to the ampacity of the control-circuit wires that are factory-installed in the starter and connected to the START-STOP buttons in the cover. If the rating of the branch-circuit fuse or CB does not exceed the value of the current shown in

Column B of Table 430.72(B) for the particular size of either copper or aluminum wire used to wire the coil circuit within the starter, then other protection is not required to be installed within the starter (Fig. 430-38). If the rating of branch-circuit protection *does* exceed the value shown in Column B for the size of coil-circuit wire, then separate protection must be provided within the starter, and it must be rated not greater than the value shown for that size of wire in Column A of Table 430.72(B). For instance, if the internal coil circuit of a starter is wired with No. 16 copper wire and the branch-circuit device supplying the starter is rated over the 40-A value shown for 16 AWG copper wire in Column B of Table 430.72(B), then protection must be provided in the starter for the 16 AWG wire and the protective device(s) must be rated not over the 10-A value shown for 16 AWG copper wire in Column A of Table 430.72(B).

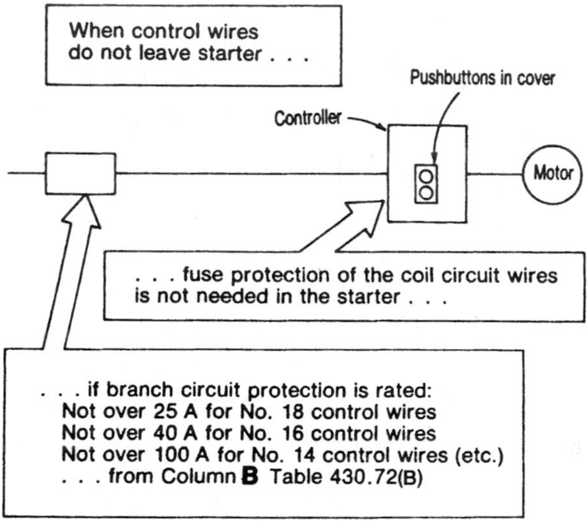
When control wires
do not leave starter . . .
Pushbuttons in cover
Controller
Motor
. . . fuse protection of the coil circuit wires
is not needed in the starter . . .
. . . if branch circuit protection is rated:
Not over 25 A for No. 18 control wires
Not over 40 A for No. 16 control wires
Not over 100 A for No. 14 control wires (etc.)
. . . from Column B Table 430.72(B)

Fig. 430-38. This is the rule of part **(B)(1)** of Sec. 430.72.

Because most starters are the smaller ones using 18 and 16 AWG wires for their coil circuits, part **(B)(2)** and its reference to Column B are particularly applicable to those wire sizes. For 16 AWG control wires, branch-circuit protection rated up to 40 A would eliminate any need for a separate control-circuit fuse in the starter. And for 18 AWG control wires, separate coil-circuit protection is not needed for a starter with branch-circuit protection rated not over 25 A. For 14, 12, and 10 AWG copper control wires, maximum protective-device ratings are given in Column B as 100, 120, and 160 A, respectively. For conductors larger than No. 10, the protection may be rated up to 400 percent of (or 4 times) the free-air ampacity of the size of conductor from Table 310.17.

The third sentence in part **(B)(2)** covers protection of control wires that run from a starter to a remote-control device (pushbutton station, float switch, limit switch, etc.). Such control wires may be protected by the branch-circuit protective device—without need for separate protection within the starter—if the branch-circuit device has a rating not over the value shown for the particular size of copper or aluminum control wire in Column C of Table 430.72(B) (Fig. 430-39). Note that the maximum ratings of 7 A for 18 AWG and 10 A for 16 AWG require that *fuse* protection at those ratings must always be used to protect those sizes of control-circuit wires connected to motor starters supplied by CB branch-circuit protection, because 15 A is the lowest available standard rating of CB. But branch-circuit fuses of 7- or 10-A rating could eliminate the need for protection in the starter where 18 AWG or 16 AWG control wires are used. Figure 430-40 shows an application that was permitted for many years under previous wording of the Code rule but is now contrary to the letter and intent of the rule.

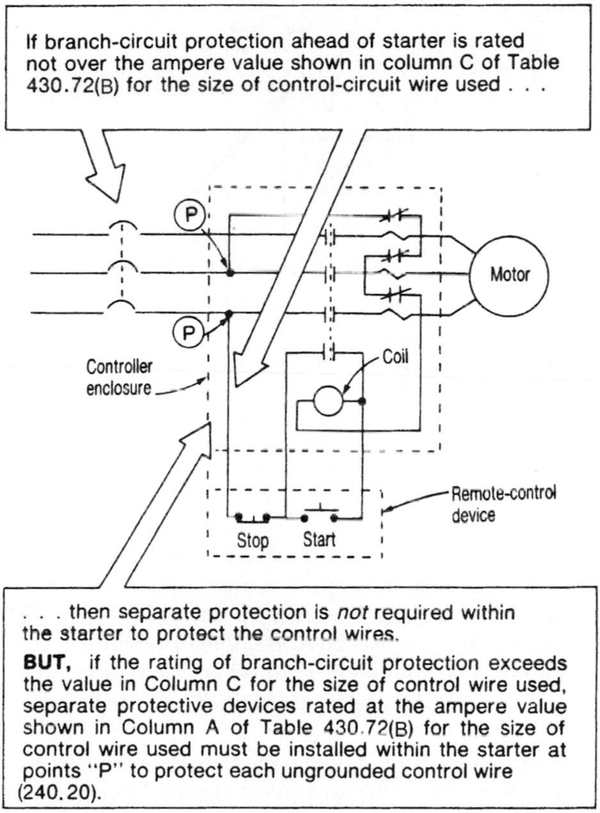
If branch-circuit protection ahead of starter is rated not over the ampere value shown in column C of Table 430.72(B) for the size of control-circuit wire used . . .
P
P
Motor
Coil
Controller enclosure
Remote-control device
Stop
Start
. . . then separate protection is *not* required within the starter to protect the control wires.
BUT, if the rating of branch-circuit protection exceeds the value in Column C for the size of control wire used, separate protective devices rated at the ampere value shown in Column A of Table 430.72(B) for the size of control wire used must be installed within the starter at points "P" to protect each ungrounded control wire (240.20).

Fig. 430-39. This is covered by part **(B)(2)** of Sec. 430.72.

For any size of control wire, if the branch-circuit protection ahead of the starter has a rating greater than the value shown in Column C of Table 430.72(B), then the control wire must be protected by a device(s) rated not over the amp value shown for that size of wire in Column A of Table 430.72(B). For instance, if 14 AWG

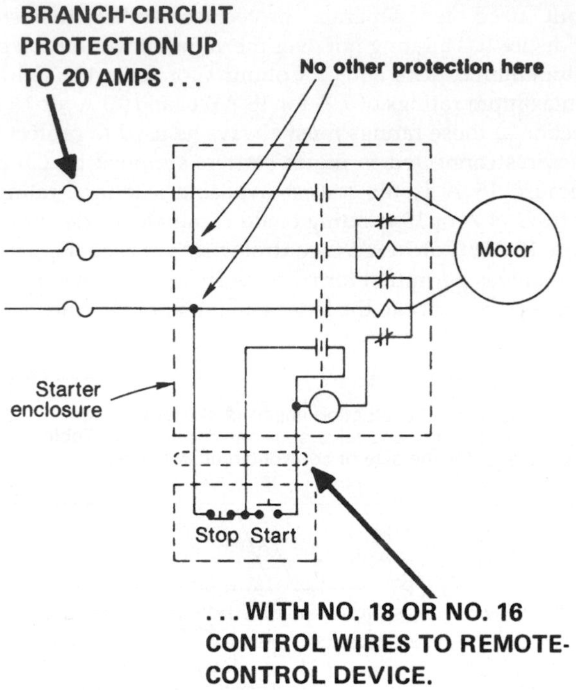
BRANCH-CIRCUIT PROTECTION UP TO 20 AMPS . . .
No other protection here
Motor
Starter enclosure
Stop Start
. . . WITH NO. 18 OR NO. 16 CONTROL WIRES TO REMOTE-CONTROL DEVICE.

Fig. 430-40. This was permitted by previous NEC editions but is now a violation of part **(B)(2).** (Sec. 430.72.)

copper wire is used for the control circuit from a starter to a remote pushbutton station and the branch-circuit protection ahead of the starter is rated at 40 A, then the branch-circuit device is not over the value of 45 A shown in Column C, and separate control protection is not required within the starter. But if the branch-circuit protection were, say, 100 A, then 14 AWG control wire would have to be protected at 15 A because Column A shows that 14 AWG must have maximum protection rating from Note 1—which refers to Table 310.15(B)(16) where 14 AWG wire in conduit is shown as 20 A. As noted previously, values are taken from the 60°C column to be consistent with other calculations in this table, but purposely not using the 240.4(D) special ampacities because the work here is governed by 240.4(G).

It should be noted that Column A gives the values to be used for overcurrent protection placed within the starter to protect control-circuit wires in any case where the rating of branch-circuit protection exceeds the value shown in either Column B (for starters with no external control wires) or Column C (for control wires run from a starter to a remote pilot control device).

Part **(C)** permits protection on the primary side of a control transformer to protect the transformer in accordance with 450.3 and the secondary conductors in accordance with the amp value shown in Table 430.72(B) for the particular size of the control wires fed by the secondary. This use is limited to transformers with

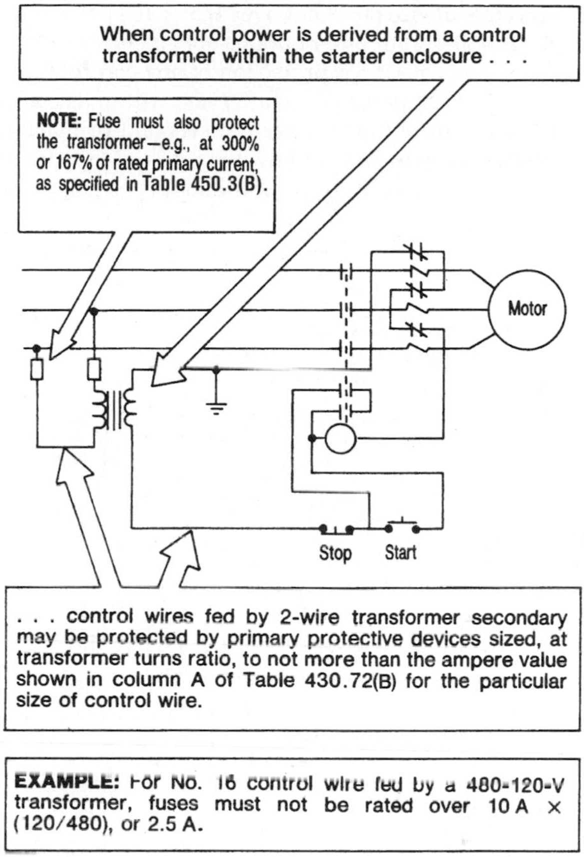
When control power is derived from a control transformer within the starter enclosure . . .
NOTE: Fuse must also protect the transformer—e.g., at 300% or 167% of rated primary current, as specified in Table 450.3(B).
Motor
Stop
Start
. . . control wires fed by 2-wire transformer secondary may be protected by primary protective devices sized, at transformer turns ratio, to not more than the ampere value shown in column A of Table 430.72(B) for the particular size of control wire.
EXAMPLE: For No. 16 control wire fed by a 480-120-V transformer, fuses must not be rated over 10 A × (120/480), or 2.5 A.

Fig. 430-41. Exception No. 2 to part **(B)** of 430.72 permits the secondary wires of the coil circuit to be protected by primary-side overcurrent protection. (Sec. 430.72.)

2-wire secondaries (Fig. 430-41). Because 430.72(A) notes that the rules of 430.72 apply to control circuits tapped from the motor branch circuit, the rule of part **(C)** must be taken as applying to a control transformer installed within the starter enclosure—although the general application may be used for any transformer because it conforms to 240.4(F), and to 450.3.

The exception eliminates any need for control-circuit protection where opening of the circuit would be objectionable, as for a fire-pump motor or other essential or safety-related operation.

Part **(C)** covers the use of control transformers and requires protection on the primary side. And, again, it must be taken to apply specifically to such transformers used in motor control equipment enclosures. The basic rule in part **(C)(2)** calls for each control transformer to be protected in accordance with 450.3 (usually by a primary-side protective device rated not over 125 or 167 percent of primary current), as shown in Fig. 430-41. But other options are given.

Part **(C)(3)** eliminates any need for protection of any control transformer rated less than 50 VA, provided it is part of the starter and within its enclosure.

Part **(C)(4)** permits a control transformer with a rated primary current of *less* than 2 A to be protected at up to 500 percent of rated primary current by a protective device in each ungrounded conductor of the supply circuit to the transformer primary, as shown in Fig. 430-42.

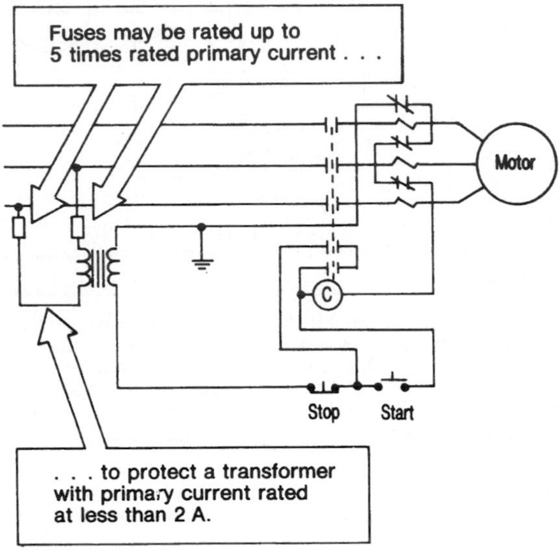
Fuses may be rated up to
5 times rated primary current . . .
Motor
C
Stop
Start
. . . to protect a transformer
with primary current rated
at less than 2 A.

Fig. 430-42. A control circuit fed by a transformer within the starter enclosure may have overcurrent protection in the primary rated up to 500 percent of the rated primary current of a small transformer. (Sec. 430.72.)

In the majority of magnetic motor controllers and contactors, the voltage of the operating coil is the voltage provided between two of the conductors supplying the load, or one conductor and the neutral. Conventional starters are factory wired with coils of the same voltage rating as the phase voltage to the motor. However, there are many cases in which it is desirable or necessary to use control circuits and devices of lower voltage rating than the motor. Such could be the case with high-voltage (over 600 V) controllers, for instance, in which it is necessary to provide a source of low voltage for practical operation of magnetic coils. And even in many cases of motor controllers and contactors for use under 600 V, safety requirements dictate the use of control circuits of lower voltage than the load circuit.

Although contactor coils and pilot devices are available and effectively used for motor controllers with up to 600-V control circuits, such practice has been prohibited in applications where atmospheric and other working conditions make it dangerous for operating personnel to use control circuits of such voltage. And certain OSHA regulations require 120- or 240-V coil circuits for the 460-V motors. In such cases, control transformers are used to step the voltage down to permit the use of lower-voltage coil circuits.

430.73. Protection of Conductors from Physical Damage. The condition under which physical protection of the control circuit conductor becomes necessary is where damage to the conductors would constitute either a fire or an accident hazard. Damage to the control circuit conductors resulting in short-circuiting two or more of the conductors or breaking one of the conductors would either cause the device to operate or render it inoperative, and in some cases either condition would constitute a hazard either to persons or to property; hence, in such cases the conductors should be installed in rigid or other metal conduit. On the other hand, damage to the conductors of the low-voltage control circuit of a domestic oil burner or automatic stoker does not constitute a hazard, because the boiler or furnace is equipped with an automatic safety control.

430.74. Electrical Arrangement of Control Circuits. This section focuses on the hazard of accidental starting of a motor. Figure 430-43 shows an example of a control circuit installation that should be carefully designed and is required to be observed for any control circuit that has one leg grounded. Whenever the coil is fed from a circuit made up of a hot conductor and a grounded conductor (as when the coil is fed from a panelboard or separate control transformer, instead of from the supply conductors to the motor), care must be taken to place the pushbutton station or other switching control device in the hot leg to the coil and not in the grounded leg to the coil. By switching in the hot leg, the starting of the motor by accidental ground fault can be effectively eliminated.

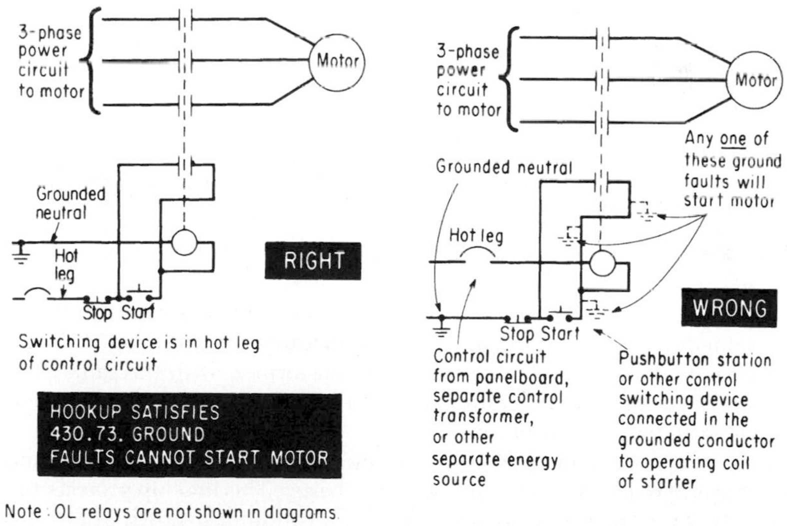
3-phase power circuit to motor
Motor
Grounded neutral
Hot leg
Stop Start
RIGHT
Switching device is in hot leg of control circuit
HOOKUP SATISFIES 430.73. GROUND FAULTS CANNOT START MOTOR
3-phase power circuit to motor
Motor
Any one of these ground faults will start motor
Grounded neutral
Hot leg
WRONG
Stop Start
Control circuit from panelboard, separate control transformer, or other separate energy source
Pushbutton station or other control switching device connected in the grounded conductor to operating coil of starter
Note: OL relays are not shown in diagrams

Fig. 430-43. Control hookup must prevent accidental starting. (Sec. 430.73.)

Note the words "remote from the motor controller." This means the rule does not require that the running overload relay contacts be wired on the ungrounded side of the control circuit. Very few are done this way and not much can go wrong to short leads that never leave their enclosure, but it is certainly permissible, if seldom done, to place them on the ungrounded side.

Combinations of ground faults can develop to short the pilot starting device—pushbutton, limit switch, pressure switch, and so forth—accidentally starting the motor even though the pilot device is in the OFF position. And because many remote-control circuits are long, possible faults have many points at which they might occur. Insulation breakdowns, contact shorts due to accumulation of foreign matter or moisture, and grounds to conduit are common fault conditions responsible for accidental operation of motor controllers.

Although not specifically covered by Code rules, there are many types of ground-fault conditions that affect motor starting and should be avoided.

As shown in Fig. 430-44, any magnetic motor controller used on a 3-phase, 3-wire ungrounded system always presents the possibility of accidental starting of the motor. If, for instance, an undetected ground fault exists on one phase of the 3-phase system—even if this system ground fault is a long distance from the controller—a second ground fault in the remote-control circuit for the operating coil of the starter can start the motor.

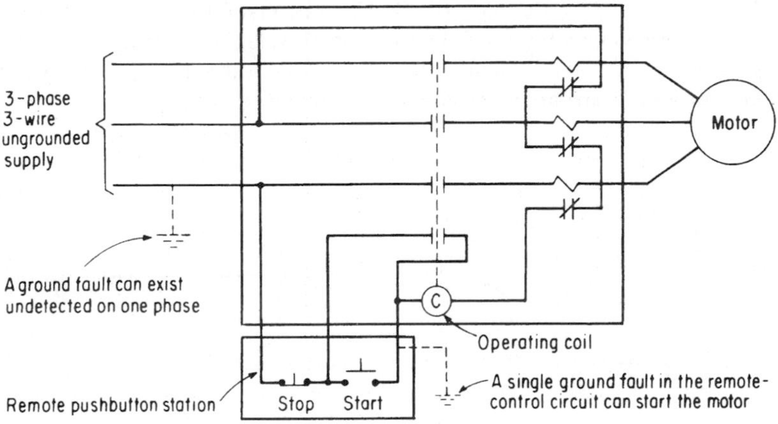
3-phase
3-wire
ungrounded
supply
Motor
A ground fault can exist
undetected on one phase
C
Operating coil
A single ground fault in the remote-
control circuit can start the motor
Remote pushbutton station
Stop
Start

Fig. 430-44. Accidental motor starting can be hazardous and contrary to Code rule. (Sec. 430.73.)

Figure 430-45 shows the use of a control transformer to isolate the control circuit from responding to the combination of ground faults shown in Fig. 430-46. This transformer may be a one-to-one isolating transformer, with the same primary and secondary voltage, or the transformer can step the motor circuit voltage down to a lower level for the control circuit.

In the hookup shown in Fig. 430-46, a 2-pole START button is used in conjunction with two sets of holding contacts in the motor starter. This hookup protects against accidental starting of the motor under the fault conditions shown in Fig. 430-44. The hookup also protects against accidental starting due to two ground faults in

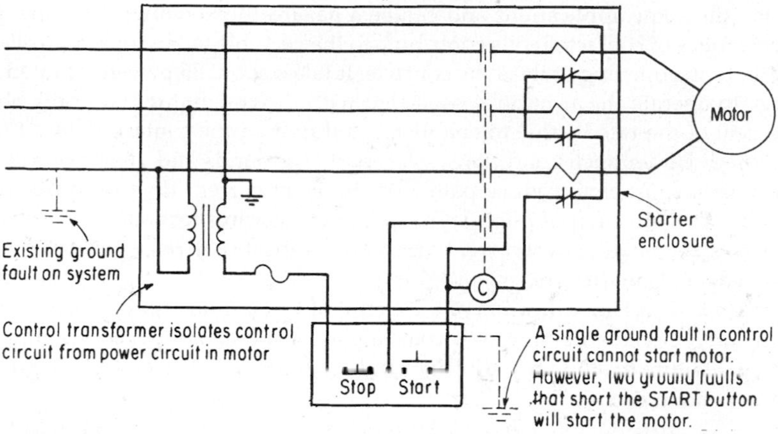

Motor
Starter enclosure
Existing ground fault on system
Control transformer isolates control circuit from power circuit in motor
Stop
Start
C
A single ground fault in control circuit cannot start motor. However, two ground faults that short the START button will start the motor.

Fig. 430-45. Control transformer can isolate control circuit from accidental starting. (Sec. 430.73.)

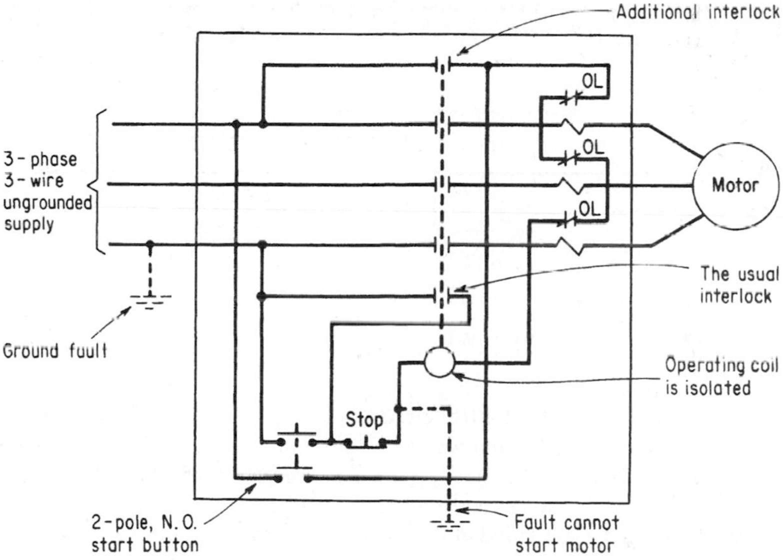

Additional interlock
OL
OL
OL
3-phase 3-wire ungrounded supply
Motor
The usual interlock
Ground fault
Operating coil is isolated
Stop
2-pole, N.O. start button
Fault cannot start motor

Fig. 430-46. Use of 2-pole start button can prevent accidental starting. (Sec. 430.73.)

the control circuit simply shorting out the START button and energizing the operating coil. This could happen in the circuit of Fig. 430-44 or the circuit of Fig. 430-45.

Another type of motor control circuit fault can produce a current path through the coil of a closed contactor to hold it closed regardless of the operation of the pilot device for opening the coil circuit. Again this can be done by a combination of ground faults that short the STOP device. Failure to open can do serious damage to motors in some applications and can be a hazard to personnel. The operating characteristics of contactor coils contribute to the possible failure of a controller to respond to the opening of the STOP contacts. It takes about 85 percent of rated coil voltage to operate the armature associated with the coil; but it takes only about 50 percent of the rated value to enable the coil to hold the contactor closed once it is closed. Under such conditions, even partial grounds and shorts on control contact assemblies can produce paths for sufficient current flow to cause shorting of the STOP position of pilot devices. And faults can short out running overload relays, eliminating overcurrent protection of the motor, its associated control equipment, and conductors.

Figure 430-47 is a modification of the circuit of Fig. 430-46, using a 2-pole START button and a 2-pole STOP button—protecting against both accidental starting and accidental failure to stop when the STOP button is pressed. Both effects of ground faults are eliminated.

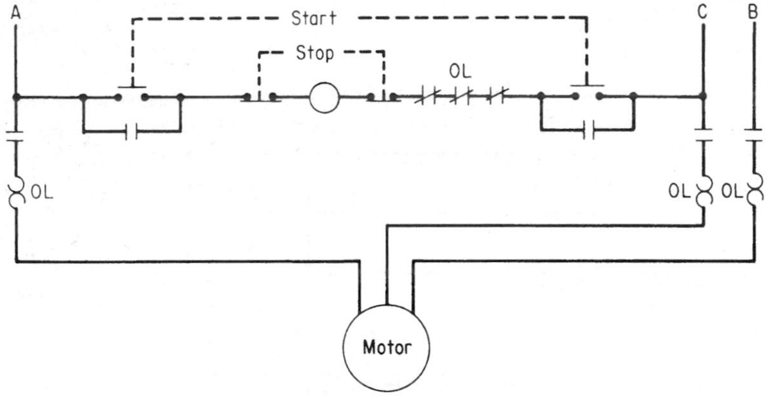
A
Start
Stop
OL
C
B
OL
OL
OL
Motor

Fig. 430-47. This circuit prevents accidental starting and assures stopping. (Sec. 430.73.)

430.75. Disconnection. The control circuit of a remote-control motor controller shall always be so connected that it will be cut off when the disconnecting means is opened, unless a separate disconnecting means is provided for the control circuit.

When the control circuit of a motor starter is tapped from the line terminals of the starter—in which case it is fed at line-to-line voltage of the circuit to the motor itself—opening of the required disconnect means ahead of the starter de-energizes the control circuit from its source of supply, as shown in Fig. 430-48. But, where voltage supply to the coil circuit is derived from outside the starter enclosure (as from a panelboard or from a separate control transformer), provision must be made to ensure that the control circuit is capable of being de-energized to permit safe maintenance of the starter. In such cases, the required power-circuit disconnect ahead of the starter can open the power circuit to the starter's line terminals; but, unless some provision is made to open the externally derived control circuit voltage supply, a maintenance worker could be exposed to the unexpected shock hazard of the energized control circuit within the starter.

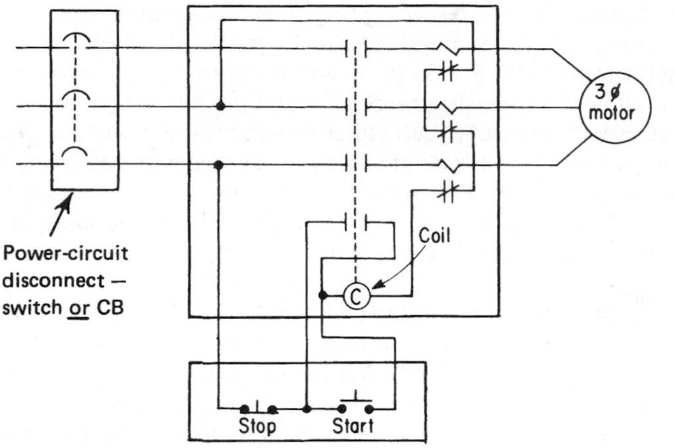
3 ø
motor
Coil
Power-circuit
disconnect –
switch or CB
Stop
Start

Fig. 430-48. Disconnect ahead of starter opens supply to line-voltage coil circuit. (Sec. 430.74.)

The disconnect for control voltage supply could be an extra pole or auxiliary contact in the switch or CB used as the main power disconnect ahead of the starter, as shown in Fig. 430-49. Or the control disconnect could be a separate switch (like a toggle switch), provided this separate switch is installed "immediately adjacent"

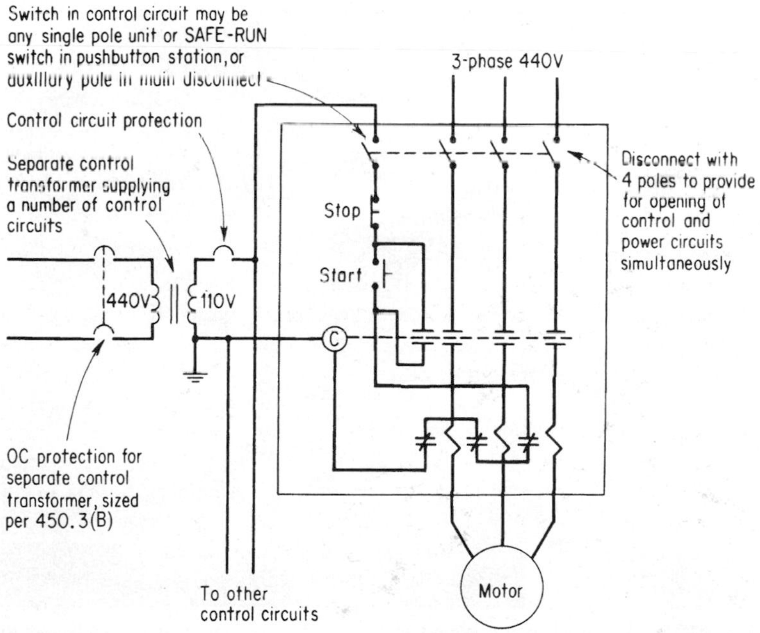
Switch in control circuit may be any single pole unit or SAFE-RUN switch in pushbutton station, or auxiliary pole in main disconnect
Control circuit protection
Separate control transformer supplying a number of control circuits
3-phase 440V
Disconnect with 4 poles to provide for opening of control and power circuits simultaneously
Stop
Start
440V
110V
OC protection for separate control transformer, sized per 450.3(B)
To other control circuits
Motor

Fig. 430-49. Control disconnect means must supplement power-circuit disconnect. (Sec. 430.74.)

to the power disconnect—so it is clear to maintenance people that *both* disconnects must be opened to kill *all* energized circuits within the starter. Control circuits operating contactor coils, and so forth, within controllers present a shock hazard if they are allowed to remain energized when the disconnect is in the OFF position. Therefore, the control circuit either must be designed in such a way that it is disconnected from the source of supply by the controller disconnecting means or must be equipped with a separate disconnect immediately adjacent to the controller disconnect for opening of both disconnects. [For grounding of the control transformer secondary in Fig. 430-49, refer to 250.20(B).]

Exception No. 1 of part **(A)** is aimed at industrial-type motor control hookups, which involve extensive interlocking of control circuits for multimotor process operations or machine sequences. In recognition of the unusual and complex control conditions that exist in many industrial applications—particularly process industries and manufacturing facilities—Exception No. 1 alters the basic rule that disconnecting means for control circuits must be located "immediately adjacent one to each other" (Fig. 430-50). When a piece of motor control equipment has more than 12 motor control conductors associated with it, remote locating of the disconnect means is permitted under the conditions given in Exception No. 1. As shown in Fig. 430-51, this permission is applicable only where qualified persons have access to the live parts and sufficient warning signs are used on the equipment to locate and identify the various disconnects associated with the control circuit conductors.

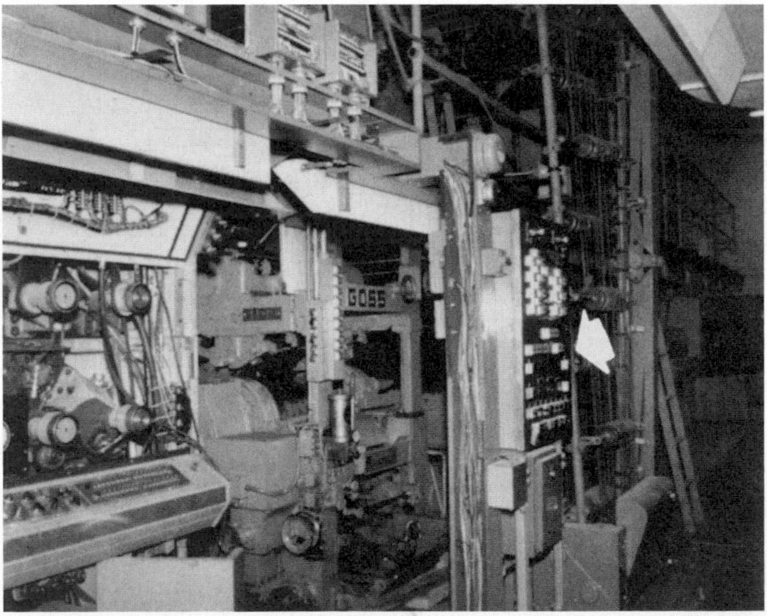
GOSS

Fig. 430-50. Industrial control layouts with more than 12 control circuit conductors for interlocking of controllers and operating stations (arrow) do not require control disconnects to be "immediately adjacent" to power disconnects. (Sec. 430.74.)

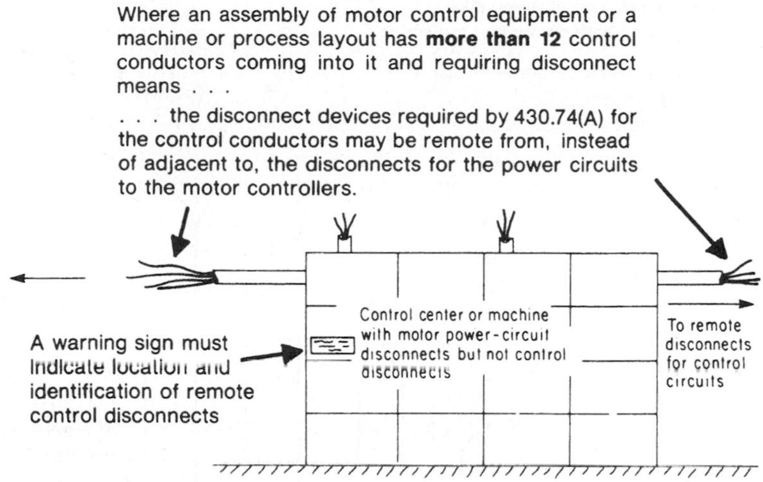
Where an assembly of motor control equipment or a machine or process layout has **more than 12** control conductors coming into it and requiring disconnect means . . .
. . . the disconnect devices required by 430.74(A) for the control conductors may be remote from, instead of adjacent to, the disconnects for the power circuits to the motor controllers.
A warning sign must indicate location and identification of remote control disconnects
Control center or machine with motor power-circuit disconnects but not control disconnects
To remote disconnects for control circuits

Fig. 430-51. For extensively interlocked control circuits, control disconnects do not have to be adjacent to power disconnects. (Sec. 430.74.)

Exception No. 2 presents another instance in which control circuit disconnects may be mounted other than immediately adjacent to each other. It notes that where the opening of one or more motor control circuit disconnects might result in hazard to personnel or property, remote mounting may be used where the conditions specified in Exception No. 1 exist, that is, that access is limited to qualified persons and that a warning sign is located on the outside of the equipment to indicate the location and the identification of each remote control circuit disconnect.

The requirement of part **(B)** of this section is shown in Fig. 430-52. When a control transformer is in the starter enclosure, the power disconnect means is on the line side and can de-energize the transformer control circuit. Grounding of the control circuit is not always necessary, as noted in 250.21(C), although there must be a showing of a need for power continuity and ground detectors must be installed when this option is used. Overcurrent protection must be provided for the control circuit when a control circuit transformer is used, as covered in 430.72(B). Such protection may be on the primary or secondary side of the transformer, as described. In 450.1, Exception No. 2 notes that the rules of Art. 450 do not apply to "dry-type transformers that constitute a component part of other apparatus. . . ." A control transformer supplied as a factory-installed component in a starter would therefore be exempt from the rules of 450.3(B), covering overcurrent protection for transformers, but would have to comply with 430.72(C).

430.81. General. As used in Art. 430, the term *controller* includes any switch or device normally used to start and stop a motor, in addition to motor starters and controllers as such. As noted, the branch-circuit fuse or CBs are considered an acceptable control device for stationary motors not over $^1/_8$ hp where the motor has sufficient winding impedance to prevent damage to the motor with its rotor continuously at standstill. And a plug and receptacle connection may serve as the controller for portable motors up to $^1/_3$ hp.

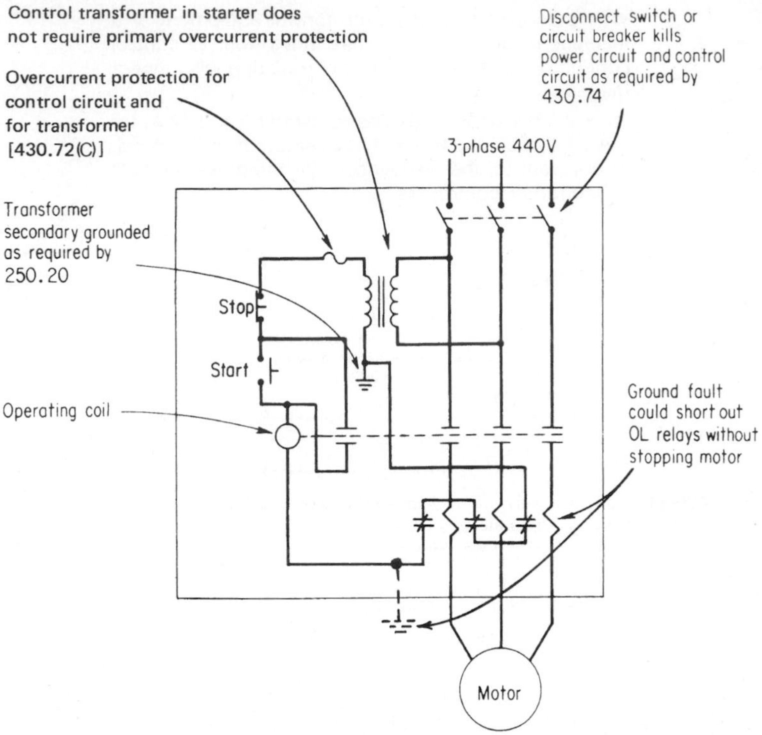
Control transformer in starter does not require primary overcurrent protection
Overcurrent protection for control circuit and for transformer [430.72(C)]
Disconnect switch or circuit breaker kills power circuit and control circuit as required by 430.74
3-phase 440V
Transformer secondary grounded as required by 250.20
Stop
Start
Operating coil
Ground fault could short out OL relays without stopping motor
Motor

Fig. 430-52. Control transformer in starter must be on load side of disconnect. (Sec. 430.74.)

As described in the definition in 430.2, a *controller* is a device that starts and stops a motor by "making and breaking the motor circuit current"—that is, the power current flow to the motor windings. A pushbutton station, a limit switch, a float switch, or any other pilot control device that "carries the electric signals directing the performance of the controller" (see the definition of *Motor Control Circuit* in 430.71) is not the controller where such a device is used to carry only the current to the operating coil of a magnetic motor controller. For purposes of Code application, the contactor mechanism is the motor "controller."

430.82. Controller Design. Every controller must be capable of starting and stopping the motor that it controls, must be able to interrupt the stalled-rotor current of the motor, and must have a horsepower rating not lower than the rating of the motor, except as permitted by 430.83.

430.83. Ratings. Figure 430-53 shows the basic requirements from part **(A)** for the required rating of a controller. Although the basic rule calls for a horsepower-rated switch or a horsepower-rated motor starter, there are acceptable alternative methods for specific applications as noted in 430.81 and as follows:

- The wording of 430.83(A)(1) calls for controllers—other than inverse time CBs and molded-case switches—to have a horsepower rating at least equal to that of the motor at the application voltage.

THESE ARE THE GENERAL RULES

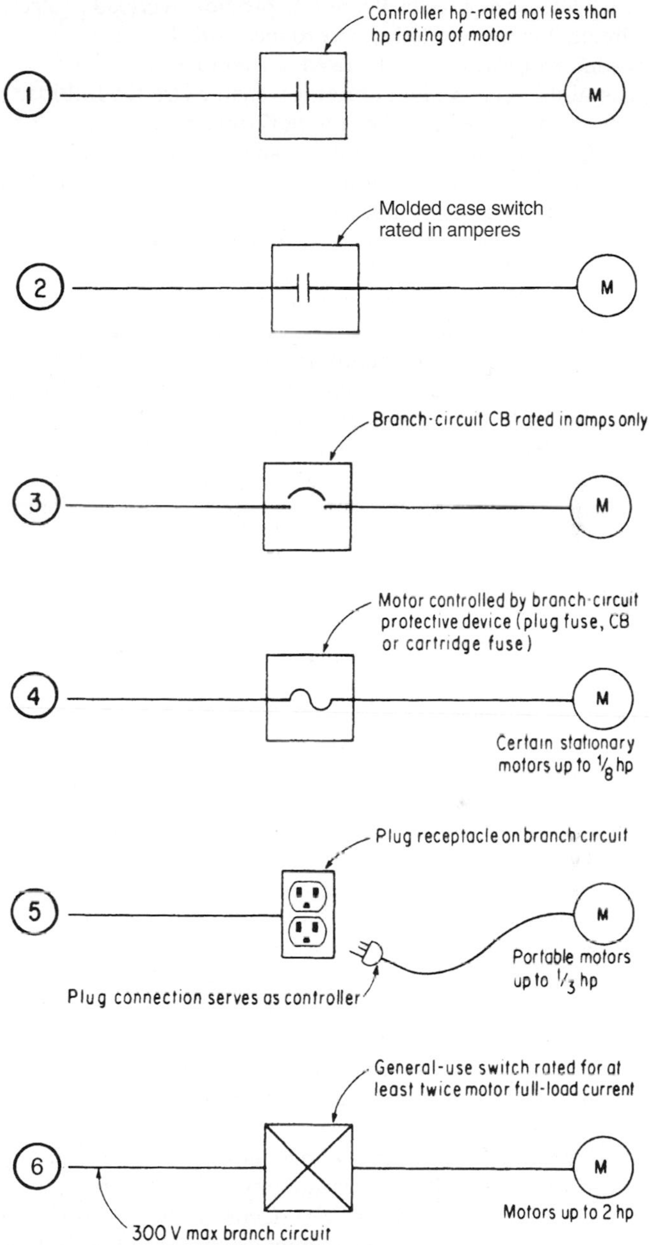
Controller hp-rated not less than hp rating of motor
1
M
Molded case switch rated in amperes
2
M
Branch-circuit CB rated in amps only
3
M
Motor controlled by branch-circuit protective device (plug fuse, CB or cartridge fuse)
4
M
Certain stationary motors up to 1/8 hp
Plug receptacle on branch circuit
5
M
Portable motors up to 1/3 hp
Plug connection serves as controller
General-use switch rated for at least twice motor full-load current
6
M
Motors up to 2 hp
300 V max branch circuit

Fig. 430-53. Controller must be a horsepower-rated switch or CB—but other devices may satisfy. (Sec. 430.83.)

- A branch-circuit CB rated in amperes only, may be used as a controller. If the same CB is used as a controller and to provide overload protection for the motor circuit, it must be rated in accordance with 430.32.
- A molded-case switch may be used as a controller.
- A general-use switch rated at not less than twice the full-load motor current may be used as the controller for stationary motors up to 2 hp, rated 300 V or less. On ac circuits, a general-use snap switch suitable only for use on ac may be used to control a motor having a full-load current rating not over 80 percent of the ampere rating of the switch.

In the UL Guide Card information there is essential additional information on the use of switches in motor circuits, as follows:

1. Enclosed switches with horsepower ratings in addition to current ratings may be used for motor circuits as well as for general-purpose circuits. Enclosed switches with ampere-only ratings are intended for general use but may also be used for motor circuits (as controllers and/or disconnects) as permitted by NEC 430.83(A)(1), 430.109, and 430.111.
2. A switch that is marked "MOTOR CIRCUIT SWITCH" is intended for use *only* in motor circuits.
3. For switches with dual-horsepower ratings, the higher horsepower rating is based on the use of time-delay fuses in the switch fuseholders to hold in on the inrush current of the higher-horsepower-rated motor.
4. Although 430.83 permits use of horsepower-rated switches as controllers and UL lists horsepower-rated switches up to 500 hp, UL does state in its Guide Card information [see discussion at 110.3(B) for details] that "enclosed switches rated higher than 100 hp are restricted to use as motor disconnect means and are not for use as motor controllers." But a horsepower-rated switch up to 100 hp may be used as both a controller and disconnect if it breaks all ungrounded legs to the motor, as covered in 430.111.

Figure 430-54 covers two of those points.

For selection of a controller for a sealed (hermetic-type) refrigeration compressor motor, refer to 440.41.

430.84. Need Not Open All Conductors. It is interesting to note that the NEC says that a controller need not open all conductors to a motor, except when the controller serves also as the required disconnecting means. For instance, a 2-pole starter of correct horsepower rating could be used for a 3-phase motor if running overload protection is provided in all three circuit legs by devices separate from the starter, such as by dual-element, time-delay fuses which are sized to provide running overload protection as well as short-circuit protection for the motor branch circuit. The controller must interrupt only enough conductors to be able to start and stop the motor.

However, when the controller is a manual (nonmagnetic) starter or is a manually operated switch or CB (as permitted by the Code), the controller itself also may serve as the disconnect means if it opens all ungrounded conductors to the motor, as covered in 430.111. This eliminates the need for another switch or CB to serve as the disconnecting means. But, it should be noted that only a manually operated switch or CB may serve such a dual function. A magnetic starter cannot

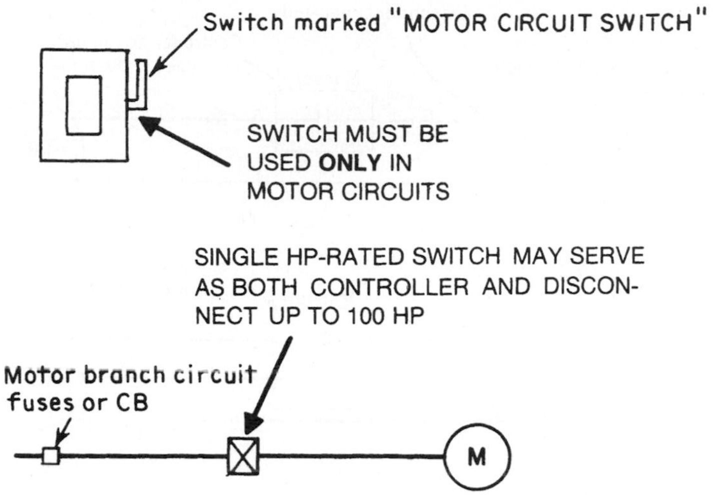
Switch marked "MOTOR CIRCUIT SWITCH"
SWITCH MUST BE USED ONLY IN MOTOR CIRCUITS
SINGLE HP-RATED SWITCH MAY SERVE AS BOTH CONTROLLER AND DISCONNECT UP TO 100 HP
Motor branch circuit fuses or CB
M

Fig. 430-54. UL rules limit Code applications. (Sec. 430.83.)

also serve as the disconnecting means, even if it does open all ungrounded conductors to the motor.

Figure 430-55 shows typical applications in which the controller does not have to open all conductors but a separate disconnect switch or CB is required ahead of the controller. In the drawing, the word *ungrounded* refers to the condition that none of the circuit conductors is grounded. These may be the ungrounded conductors of grounded systems.

Generally, one conductor of a 115-V circuit is grounded, and on such a circuit a single-pole controller may be used connected in the ungrounded conductor, or a 2-pole controller is permitted if both poles are opened together. In a 230-V circuit there is usually no grounded conductor, but if one conductor is grounded, 430.85 permits a 2-pole controller.

430.85. In Grounded Conductors. This rule permits a 3-pole switch, CB, or motor starter to be used in a 3-phase motor circuit derived from a 3-phase, 3-wire, corner-grounded delta system—with the grounded phase leg switched along with the hot legs, as in 430.36.

430.87. Number of Motors Served by Each Controller. Generally, an individual motor controller is required for each motor. However, for motors rated not over 600 V, a single controller rated at not less than the sum of the horsepower ratings of all the motors of the group may be used with a group of motors if any one of the conditions specified is met. Where a single controller is used for more than one motor connected on a single branch circuit as permitted under condition *b*, it should be noted that the reference is to part **(A)** of 430.53. That use of a single controller applies only to cases involving motors of 1 hp or less and does not apply for several motors used on a single branch circuit in accordance with parts **(B)** and **(C)** of 430.53—unless the several motors satisfy conditions *a* or *c* of this section.

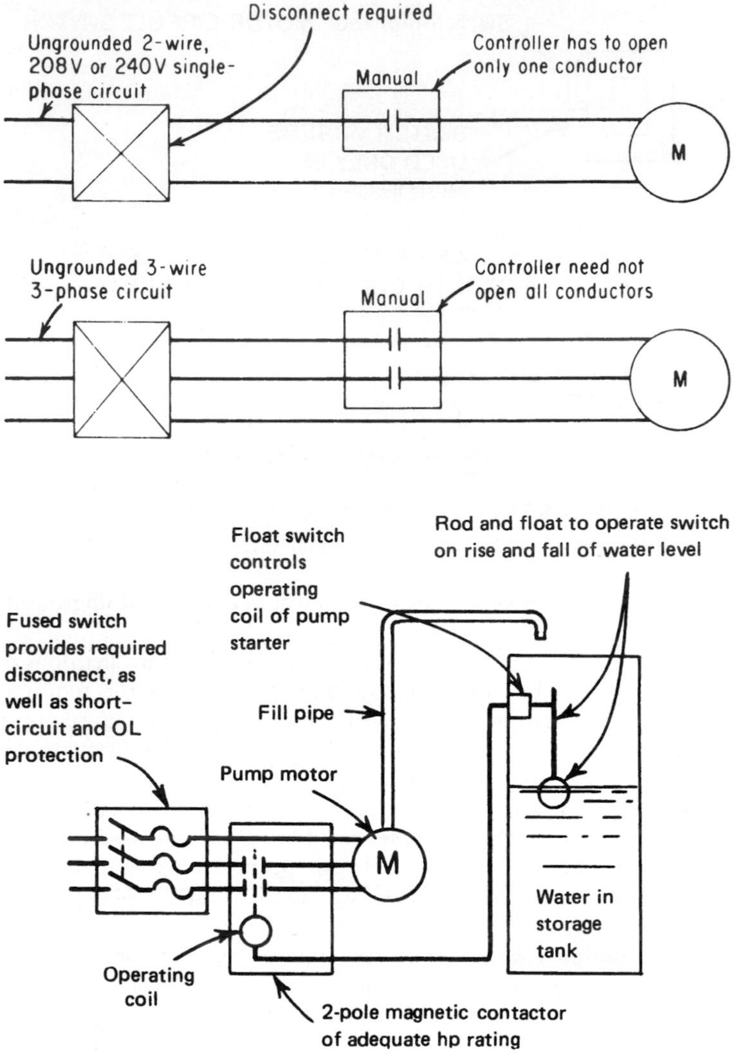
Disconnect required
Ungrounded 2-wire, 208V or 240V single-phase circuit
Controller has to open only one conductor
Manual
M
Ungrounded 3-wire 3-phase circuit
Controller need not open all conductors
Manual
M
Float switch controls operating coil of pump starter
Rod and float to operate switch on rise and fall of water level
Fused switch provides required disconnect, as well as short-circuit and OL protection
Fill pipe
Pump motor
M
Water in storage tank
Operating coil
2-pole magnetic contactor of adequate hp rating

Fig. 430-55. Controller does not have to break *all* legs of motor supply circuit. (Sec. 430.84.)

See 430.112, where the same conditions are set for a single disconnect means to serve a group of motors. Note that the reference to 430.110(C)(1) requiring an equivalent horsepower calculation can add unexpected sizing. For example, suppose you want to group control a 10-hp, a 15-hp, and a 20-hp, 460-V motor. The total horsepower by simple summation would be 45 hp, but that is not the correct result. The ampere ratings from Table 430.250 and the locked-rotor current numbers from Table 430.251(B) have to be summed as well: 14 A + 21 A + 27 A = 62 A;

and 81 A + 116 A + 145 A = 342 A. Making a reverse lookup in Tables 430.250 and 251(B) using these summations results in a requirement for a 50-hp-capable controller.

430.88. Adjustable-Speed Motors. Field weakening is quite commonly used as a method of controlling the speed of dc motors. If such a motor were started under a weakened field, the counter emf is reduced commensurately and the starting current would be excessive unless the motor is specially designed for starting in this manner.

430.89. Speed Limitation. A common example of a separately excited dc motor is found in a typical speed control system that is widely used for electric elevators, hoists, and other applications where smooth control of speed from standstill to full speed is necessary. In Fig. 430-56, G_1 and G_2 are two generators having their armatures mounted on a shaft which is driven by a motor, not shown in the diagram. M is a motor driving the elevator drum or other machine. The fields of generator G_1 and motor M are excited by G_1. By adjusting the rheostat R, the voltage generated by G_2 is varied, and this in turn varies the speed of motor M. It is evident that if the field circuit of motor M should be accidentally opened while the motor is lightly loaded, the motor would reach an excessive speed. In many applications of this system the motor is always loaded and no speed-limiting device is required.

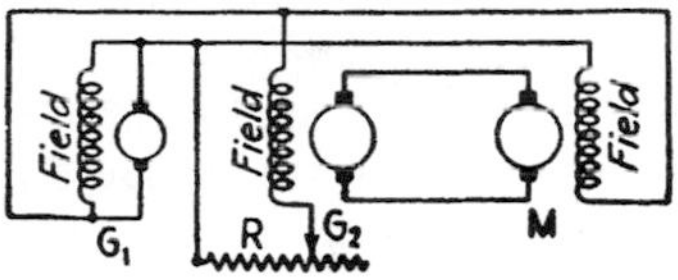
Field
G1
Field
R
G2
M
Field

Fig. 430-56. Typical speed control hookup involving the rule of 430.89.(Sec. 430.89.)

The speed of a series motor depends on its load and will become excessive at no load or very light loads. Traction motors are commonly series motors, but such a motor is geared to the drive wheels of the car or locomotive and hence is always loaded.

Where a motor generator, consisting of a motor driving a compound-wound dc generator, is operated in parallel with a similar machine or is used to charge a storage battery, if the motor circuit is accidentally opened while the generator is still connected to the dc buses or battery, the generator will be driven as a motor and its speed may become dangerously high. A synchronous converter operating under similar conditions may also reach an excessive speed if the ac supply is accidentally cut off.

A safeguard against overspeed is provided by a centrifugal device on the shaft of the machine, arranged to close (or open) a contact at a predetermined speed, thus tripping a CB, which cuts the machine off from the current supply.

430.90. Combination Fuseholder and Switch as Controller. The use of a fusible switch as a motor controller with fuses as motor-running protective devices is practicable when time-delay types of fuses are used. The rating of the fuses must not exceed 125 percent, or in some cases 115 percent, of the full-load motor current, and nontime-delay fuses of this rating would, in most cases, be blown by the starting current drawn by the motor, particularly where the motor turns on and off frequently. (See 430.35.)

It may be found that a switch having the required horsepower rating is not provided with fuse terminals of the size required to accommodate the branch-circuit fuses. For example, assume a 7½-hp, 230-V, 3-phase motor started at full-line voltage. A switch used as the disconnecting means for this motor must be rated at not less than 7½ hp, but this would probably be a 60-A switch and therefore, if fusible, would be equipped with terminals to receive 35- to 60-A fuses. 430.90 provides that fuse terminals must be installed that will receive fuses of 70-A rating. In such case a switch of the next higher rating must be provided, unless time-delay fuses are used.

430.92. General. This is the motor control center part of the article. Motor control centers are essentially switchboards adapted for this function, and many of the construction requirements came from Art. 408. It should be remembered that the clearances in 110.26 apply to this equipment generally, and 110.26(F), addressing what can go where in the dedicated wiring space above and below such equipment, specifically refers to motor control centers as a topic to which it applies. The fault current available at a motor control center must be determined and, together with the calculation date, be documented and made available to interested parties (430.99).

430.94. Overcurrent Protection. Motor control centers require individual overcurrent protection, either in the unit or ahead of it, rated or set not to exceed the rating of the common power bus. This is a significant departure from switchboards that do not have this limitation. Note that this limitation may limit the application of 430.62 or 430.63 because feeder protection determined by those sections, particularly with a very large motor in the load group, may significantly exceed the ampacity of the feeder conductors. The common power bus of a motor control center is a feeder, and this rule means that a motor control center supplying the same large motor will require a common power bus that is larger than the equivalent conductors installed as a conventional feeder. This is the reason for 430.62 Exception No. 2, which correlates these two provisions.

430.102. Location. Along with 430.101, this section specifically requires that a disconnecting means—basically, a motor-circuit switch rated in horsepower, or a CB—be provided in each motor circuit. Figure 430-57 shows the basic rule on "in-sight" location of the disconnect means. This applies always for all motor circuits rated up to 600 V—even if an "out-of-sight" disconnect can be locked in the open position.

Because the basic rule here requires a disconnecting means to be within sight from the "controller location," the question arises, "Is the magnetic contactor the controller or is the pushbutton station the controller?" The NEC makes clear that the contactor of a magnetic motor starter *is* the controller for the motor, *not* the pushbuttons that actuate the coil of the contactor. The NEC establishes that identification by the definition of *controller* in Art. 100 and by the definition of a *motor control circuit* in 430.2, as follows:

> *Controller:* A device or group of devices that serves to govern, in some predetermined manner, the electric power delivered to the apparatus to which it is connected. Note that this definition is modified "for the purposes of this article" in 430.2 by a definition that goes on to state that it is "any switch or device that is normally used to start and stop a motor by making and breaking the motor circuit current."
>
> *Motor control circuit:* The circuit of a control apparatus or system that carries the electric signals directing the performance of the controller, but does not carry the main power current.

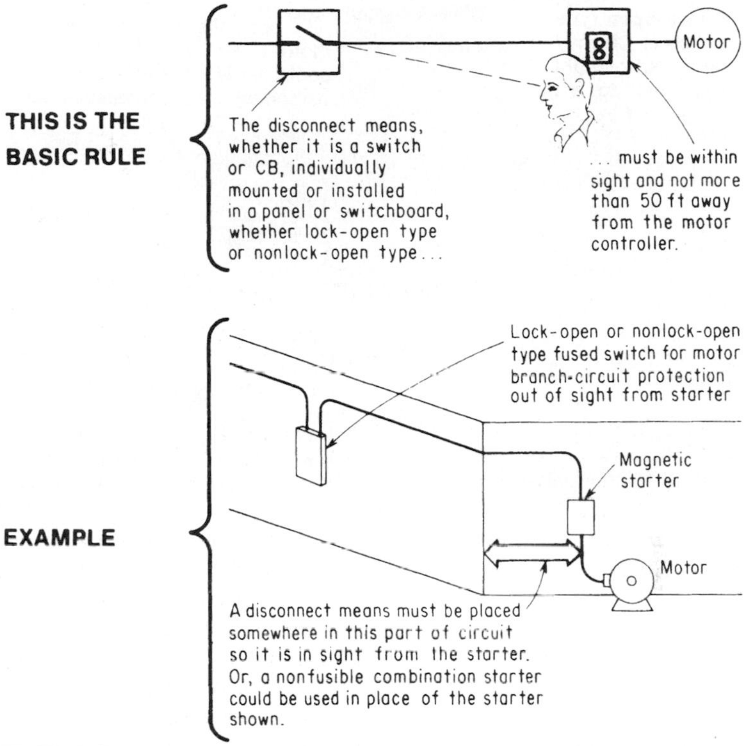
THIS IS THE BASIC RULE
Motor
The disconnect means, whether it is a switch or CB, individually mounted or installed in a panel or switchboard, whether lock-open type or nonlock-open type . . .
. . . must be within sight and not more than 50 ft away from the motor controller.
EXAMPLE
Lock-open or nonlock-open type fused switch for motor branch-circuit protection out of sight from starter
Magnetic starter
Motor
A disconnect means must be placed somewhere in this part of circuit so it is in sight from the starter. Or, a nonfusible combination starter could be used in place of the starter shown.

Fig. 430-57. The required disconnect must be visible from the controller. (Sec. 430.102.)

In a magnetic motor starter hookup, it is the contactor that actually governs the electric power delivered to the motor to which it is connected. The motor connects to the contactor and *not* to the pushbuttons, which are in the control circuit that carries the electric signals directing the performance of the *controller* (i.e., the contactor). The pushbuttons do *not* carry "the main power current," which is "delivered" to the motor by the contactor and which is, therefore, "the controller." It is well established that the intent of the Code rule, as well as the letter of the rule, is to designate the *contactor* and *not* the pushbutton station as the controller, and the disconnect must be within sight from it and not from a pushbutton station or some other remotely located pilot control device that connects into the contactor.

There are three exceptions to this basic Code rule, requiring a disconnect switch or CB to be located in sight from the controller:

Exception No. 1 permits the disconnect for a medium-voltage (over 600 V) motor to be out of sight from the controller location, as shown in Fig. 430-58. But, such use of a lock-open type switch as an out-of-sight disconnect for a motor circuit rated 600 V or less is a clear Code violation.

Exception No. 2 is aimed at permitting practical, realistic disconnecting means for industrial applications of large and complex machinery utilizing a number of

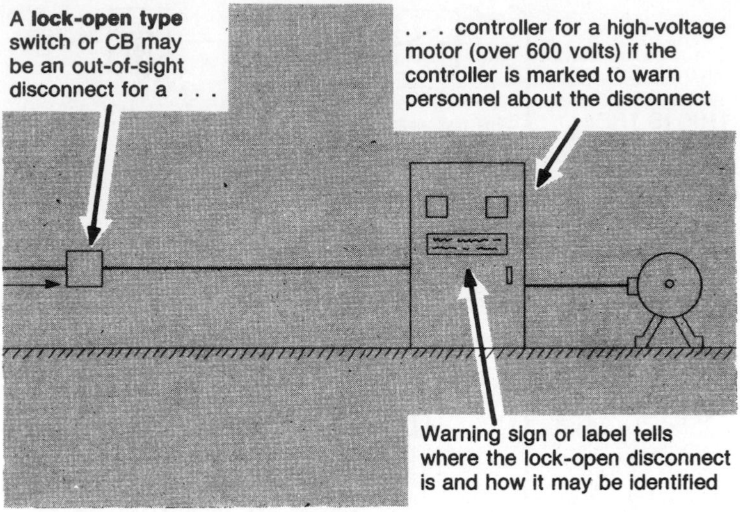
A lock-open type switch or CB may be an out-of-sight disconnect for a . . .
. . . controller for a high-voltage motor (over 600 volts) if the controller is marked to warn personnel about the disconnect
Warning sign or label tells where the lock-open disconnect is and how it may be identified

Fig. 430-58. An out-of-sight disconnect may be used for a high-voltage motor. (Sec. 430.102.)

motors to power the various interrelated parts of the machine. The exception recognizes that a single common disconnect for a number of controllers (as permitted by part **(A)** of the Exception of 430.112) is often impossible to install "within sight" of all the controllers, even though the controllers are "adjacent one to each other." On much industrial process equipment, the components of the overall structure obstruct the view of many controllers. Exception No. 2 permits the single disconnect to be technically out of sight from some or even all the controllers if the disconnect is simply *adjacent* to them—that is, nearby on the equipment structure, as shown in Fig. 430-59.

Exception No. 3 was new as of the 2008 edition of the NEC and addresses instances where a local disconnect for "valve actuator motors" would introduce additional hazards, and there is a label applied giving the location of a lock-open disconnect.

These motors, defined in 430.2, are common in industrial facilities, where they control process fluids. The actual motors are short-time duty with high torque. They generally have a controller built into the assembly, and the motor incorporates a self-resetting running overload protection. They are reversible for obvious reasons, and available in a range of voltages and either single-or three-phase, and up to eight poles. The gearing differs from unit to unit, and therefore the typical rating is not in horsepower but in torque. The short-circuit and ground-fault protective device ratings for these motors are determined by Table 430.52 using the nameplate current and not current taken from the tables at the end of the article. This procedure is not new; as covered in the last sentence of 430.6(A)(1), "motors built for low speeds (less than 1200 rpm) or high torques may have higher full-load currents, . . . in which case the nameplate current ratings shall be used."

None of these three exceptions allows for an upstream, out-of-sight lock-open disconnect ahead of a controller for conventional motors operating at 600 volts

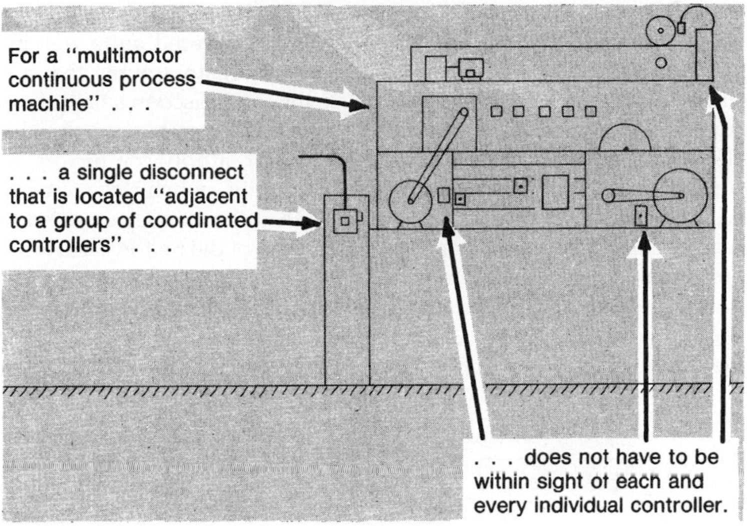
For a "multimotor
continuous process
machine" . . .
. . . a single disconnect
that is located "adjacent
to a group of coordinated
controllers" . . .
. . . does not have to be
within sight of each and
every individual controller.

Fig. 430-59. For multimotor machines, the disconnect may be "adjacent" to controller. (Sec. 430.102.)

and below. Do not confuse this controller disconnect with the motor disconnect in **(B)**. For more information on how this requirement and the disconnect provisions in Part **X** for variable speed drives may intersect, review the commentary at 430.128.

Part **(B)** basically requires a disconnect means (switch or CB) to be within sight and not more than 50 ft away from "the motor location and the driven machinery location." But the exception to that basic requirement says that a disconnect does not have to be within sight from the motor and its load *if* the required disconnect ahead of the motor controller is capable of being locked in the open position. This exception has been severely limited in recent Code cycles. The disconnect in sight of the controller must be "individually capable of being locked in the open position" which eliminates the possibility of a locked panel door qualifying. In addition, as in most instances throughout Chap. 4, the locking provisions must remain with the circuit breaker or switch even with the lock removed.

The above limitations apply in many comparable situations, but the following limitations are unique to the application of this exception. It only applies to (1) industrial occupancies with written safety procedures that ensure that only qualified persons will service the motor or (2) other installations if the additional disconnect would introduce additional hazards or would be impracticable.

For example, it would be plainly impracticable to place a disconnect 50 ft down a well shaft to be "in sight" (not over 50 ft distant) from a submersible pump motor 100 ft down the same shaft. Variable frequency drives should not have the motor disconnected unless the drive itself is disconnected, and multimotor equipment may cause hazards unless a coordinated stop is arranged. Large motors (over 100 hp) only need isolation switches anyway, so their disconnects can be remote, and additional disconnects in hazardous (classified) locations only exacerbate the explosion hazards. These issues are covered quite well in a note that follows the first part of the exception. Never again assume that a locking disconnect in the motor control center solves the local disconnect requirement.

According to the basic rule of part **(B)**, a manually operable switch, which will provide disconnection of the motor from its power supply conductors, must be placed within sight from the motor location. And this switch *may not* be a switch in the control circuit of a magnetic starter. (The **NEC** at one time permitted a switch in the coil circuit of the starter installed within sight of the motor. Such a condition is *not* acceptable to the present Code.)

These requirements are shown in Fig. 430-60. Specific layouts of the two conditions are shown in Fig. 430-61. (*Note:* The portions of these drawings covering the exception only apply when and if the restrictive conditions just discussed have

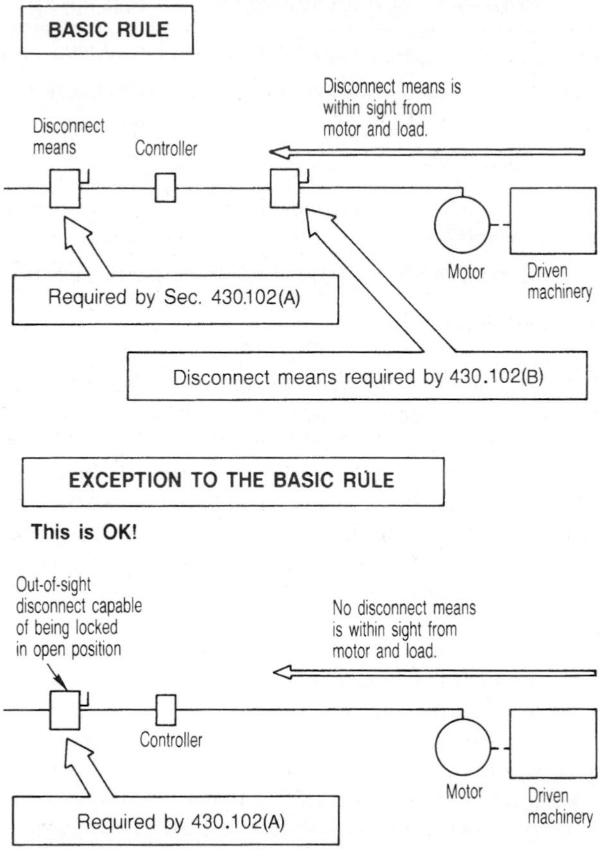
BASIC RULE
Disconnect means is within sight from motor and load.
Disconnect means
Controller
Motor
Driven machinery
Required by Sec. 430.102(A)
Disconnect means required by 430.102(B)
EXCEPTION TO THE BASIC RULE
This is OK!
Out-of-sight disconnect capable of being locked in open position
No disconnect means is within sight from motor and load.
Controller
Motor
Driven machinery
Required by 430.102(A)

Fig. 430-60. Disconnect means must be within sight from the motor and its driven load, unless out-of-sight disconnect can be locked open. [Sec. 430.102(B).] This exception is now severely limited (see text).

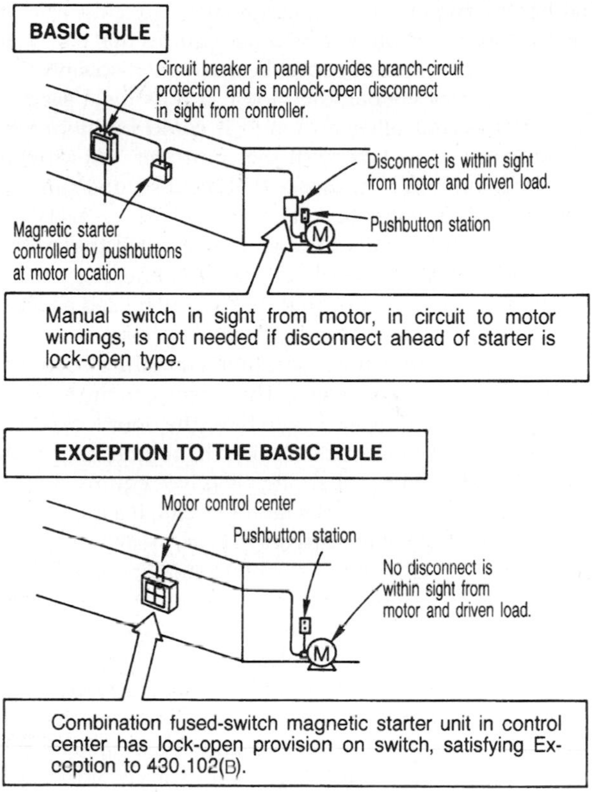
BASIC RULE
Circuit breaker in panel provides branch-circuit protection and is nonlock-open disconnect in sight from controller.
Disconnect is within sight from motor and driven load.
Pushbutton station
M
Magnetic starter controlled by pushbuttons at motor location
Manual switch in sight from motor, in circuit to motor windings, is not needed if disconnect ahead of starter is lock-open type.
EXCEPTION TO THE BASIC RULE
Motor control center
Pushbutton station
No disconnect is within sight from motor and driven load.
M
Combination fused-switch magnetic starter unit in control center has lock-open provision on switch, satisfying Exception to 430.102(B).

Fig. 430-61. Here's an example of the rules, showing physical layout. [Sec. 430.102(B).] As previously noted, this exception is now severely limited. As illustrated here, it would probably only apply in an industrial occupancy under written safety procedures and with qualified supervision and staff.

been met.) The intent of the exception to part **(B)** is to permit maintenance workers to lock the disconnecting means ahead of the controller in the open position and keep the key in their possession so that the circuit cannot be energized while they are working on it.

The pushbutton station in Fig. 430-61 operates only the holding coil in the magnetic starter. The magnetic starter *controls* the current to the motor; for example, the control wires to a pushbutton station could become shorted after the motor is in operation, and pushing the STOP button would not release the holding coil in the magnetic starter and the motor would continue to run. This is the reason that a disconnecting means is required to be installed within sight from the motor and its load or a lock-open switch installed ahead of the controller. In this case, operating the disconnecting means will open the supply to the controller and shut off the motor.

430.103. Operation. This rule actually defines the meaning of *disconnecting means*.

In order that necessary periodic inspection and servicing of motors and their controllers may be done with safety, the Code requires that a switch, CB, or other device shall be provided for this purpose. Because the disconnecting means must disconnect the controller as well as the motor, it must be a separate device and cannot be a part of the controller, although it could be mounted on the same panel or enclosed in the same box with the controller. The disconnect must be installed ahead of the controller. And note that the disconnect need open only the *ungrounded* conductors of a motor circuit.

In case the motor controller fails to open the circuit if the motor is stalled, or under other conditions of heavy overload, the disconnecting means can be used to open the circuit. It is therefore required that a switch used as the disconnecting means shall be capable of interrupting very heavy current.

A 2008 NEC amendment added the requirement that the disconnecting means be designed so it cannot close automatically. This addresses time-clock switches that can be placed in an open position, and even have the door locked shut, but unless the trippers are physically removed the clock will, in time, reclose the circuit.

430.105. Grounded Conductors. Although 430.103 requires a disconnect means only for the ungrounded conductors of a motor circuit, if a motor circuit includes a grounded conductor, one pole of the disconnect *may* switch the grounded conductor, provided all poles of the disconnect operate together—as in a multipole switch or CB. For instance, a 120-V, 2-wire circuit with one of its conductors grounded only requires a single-pole disconnect switch, but a 2-pole switch *could* be used, with one pole simultaneously switching the grounded leg.

430.107. Readily Accessible. Although a motor circuit may be provided with more than one disconnect means in series ahead of the controller—such as one at the panel where the motor circuit originates and one at the controller location—*only one* of the disconnects is required to be "readily accessible," as follows:

> *Readily accessible:* Capable of being reached quickly for operation, renewal, or inspection, without requiring those to whom ready access is requisite to climb over or remove obstacles or to resort to portable ladders, chairs, etc. (See "Accessible.")

The disconnecting means must be reached without climbing over anything, without removing crates or equipment or other obstacles, and without requiring the use of portable ladders.

Note carefully: A disconnect that has to be "readily accessible" must be so only for "those to whom ready access is requisite"—which clearly and intentionally allows for making equipment *not* readily accessible to other than authorized persons, such as by providing a lock on the door, with the key possessed by or available to those who require ready access.

Because the definition of *readily accessible* contains a last phrase that says "See 'Accessible,'" logic dictates that the installation must also satisfy the definition of *accessible.* And the wording of the definition clearly establishes that there is no Code violation in putting the disconnect means in a room or area under lock and key to make it accessible only to authorized persons.

The definition reads:

> *Accessible:* (As applied to Equipment.) Admitting close approach because not guarded by locked doors, elevation, or other effective means. (See "Readily Accessible.")

Again note carefully: That definition does not say that a door to an electrical room is prohibited from being locked. In fact, the wording of the definition, by referring to "locked doors," actually presumes the existence and, therefore, the acceptability of locked doors in electrical systems. The only requirement implied by the wording is that locked doors, where used, must not *guard* against access—that is, disposition of the key to the lock must be such that those requiring access to the room are not positively excluded. The rule is satisfied if the key is available to provide access to authorized persons.

In reference to the definition of *accessible,* the critical word is *guarded.* The definition is *not* intended to mean that equipment *cannot* be behind locked doors or that equipment *cannot* be mounted up high where it *can* be reached with a portable ladder. To make equipment "not accessible," a door lock or high mounting must be such that it positively *guards* against access. Equipment behind a locked door for which a key is not possessed by or available to persons who require access to the equipment is *not* "accessible." A common example of that latter condition occurs in multitenant buildings where a disconnect for the tenant of one occupancy unit is located behind the locked door of another tenant's occupancy unit from which the first tenant is effectively and legally excluded. And even that application *is* Code-acceptable if the disconnect is *not required* by the **NEC** to be "readily accessible."

Equipment may be fully "accessible" even though installed behind a locked door or at an elevated height. Equipment that is high-mounted but can be reached with a ladder that is fixed in place or a portable ladder *is* "accessible" (although the equipment would not be "readily accessible" if a portable ladder had to be used to reach it). Similarly, equipment behind a locked door *is* "accessible" to anyone who possesses a key to the lock or to a person who is authorized to obtain and use the key to open the locked door. In such cases, conditions do *not* guard against access.

Refer to the definitions of *accessible* and *readily accessible* in Art. 100 of this book.

430.109. Type. In a motor branch circuit, every switch or CB in the circuit, from where the circuit is tapped from the feeder to the motor itself, must satisfy the requirements on type and rating of disconnect means. A CB switching device with no automatic trip operation, a so-called molded-case switch, may be used as a motor disconnect instead of a conventional CB or a horsepower-rated switch. Such a device either must be rated for the horsepower of the motor it is used with or must have an amp rating at least equal to 115 percent of that of the motor with which it is used. Figure 430-62 covers the basic rules on types of disconnect means. An instantaneous-trip circuit breaker, as part of a listed combination motor controller, has the status of a circuit breaker in this classification and is an acceptable disconnecting means. The following items also qualify, with the limitations as described.

Self-protected combination controller—This combination device controls the motor's performance, and qualifies as a disconnecting means. The device must be listed to qualify. It is also on the list of short-circuit and ground-fault protective devices, as covered in 430.52(C)(6).

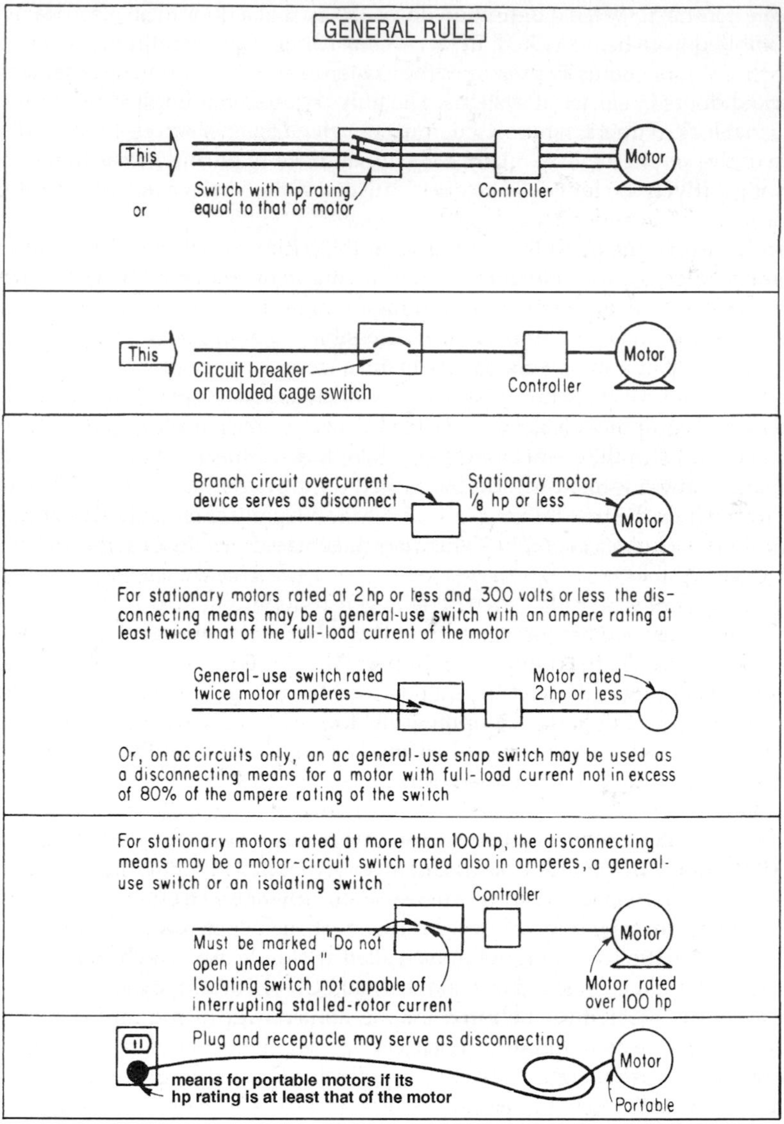
GENERAL RULE
This
Switch with hp rating equal to that of motor
Controller
Motor
or
This
Circuit breaker or molded cage switch
Controller
Motor
Branch circuit overcurrent device serves as disconnect
Stationary motor ⅛ hp or less
Motor
For stationary motors rated at 2 hp or less and 300 volts or less the disconnecting means may be a general-use switch with an ampere rating at least twice that of the full-load current of the motor
General-use switch rated twice motor amperes
Motor rated 2 hp or less
Or, on ac circuits only, an ac general-use snap switch may be used as a disconnecting means for a motor with full-load current not in excess of 80% of the ampere rating of the switch
For stationary motors rated at more than 100 hp, the disconnecting means may be a motor-circuit switch rated also in amperes, a general-use switch or an isolating switch
Controller
Must be marked "Do not open under load"
Isolating switch not capable of interrupting stalled-rotor current
Motor
Motor rated over 100 hp
Plug and receptacle may serve as disconnecting means for portable motors if its hp rating is at least that of the motor
Motor
Portable

Fig. 430-62. One of these disconnects must be used for a motor branch circuit. (Sec. 430.109.) There are others; see text.

Manual motor controller additionally marked SUITABLE AS MOTOR DISCONNECT—These devices carry principal listings as controllers, intended to turn a motor on and off manually. Even though it acts manually, and says OFF and ON, it does not qualify as a disconnecting means without meeting additional qualifications. Motor controllers, being designed as the manual equivalents of automatic controllers generally wired on the load side of a conventional disconnect, don't have as robust internal spacings as full-fledged disconnect switches. The NEC allows these devices, if so listed, to be used as formal disconnects in two circumstances. The first, covering small motors, allows them to be used as disconnects for motors of 2 hp or less, just as snap switches (described later in this list). The second, covering larger motors, allows them to be used as disconnects provided they are on the load side of the final branch-circuit short-circuit and ground-fault protective device. In either case, their horsepower rating must not be less than the motor.

General-use switch—A switch intended for use in general distribution and branch circuits. It is rated in amperes, and is capable of interrupting its rated current at its rated voltage. Its ampere rating must be not less than twice the full-load current rating of the motor. It generally cannot be used for a motor larger than 2 hp, unless it additionally qualifies as a motor-circuit switch, as described earlier. It also qualifies if the installation (2 hp up to 100 hp) involves an autotransformer-type controller and the motor is driving a generator with overload protection, provided the controller has (1) no-voltage release (shuts off when power is discontinued, with manual restart only), (2) running overload protection limited to 125 percent of motor full-load current, and (3) the capability to interrupt locked-rotor current. In addition, the branch circuit must include separate fuses or an inverse-time circuit breaker not over 150 percent of the motor full-load current.

General-use snap switch—A form of general-use switch constructed so that it can be installed in flush device boxes or on outlet box covers, or otherwise used in conjunction with wiring systems recognized by the NEC. They are for ac motors only. To qualify, the switch must be rated ac-only (general-use ac-dc snap switches are not acceptable) and the motor full-load current must not exceed 80 percent of the ampere rating of the switch.

Isolating switch—A switch intended for isolating an electric circuit from the source of power. It has no interrupting rating, and it is intended to be operated only after the circuit has been opened by some other means. This is permitted only for dc motors over 40 hp and for ac motors over 100 hp. These switches are available in larger ratings, as covered at the end of this list.

Plug and receptacle—A cord-and-plug-connected motor need not have an additional disconnecting means if the plug and receptacle have a suitable horsepower rating. Refer to the UL Guide Card information under the heading "Receptacles for Attachment Plugs and Plugs" (Guide Card Designator RTRT) for the complete list, and manufacturer's literature. The separate horsepower rating is not required for appliances, room air conditioners, or portable motors rated $^1/_3$ hp or less. Flanged inlets and cord connectors with equivalent motor circuit ratings, that follow the same UL Guide Card information, can equally be used for these purposes. See Fig. 406-3 for an example.

System isolation equipment—New in the 2005 NEC (and defined in 430.2), this concept uses a contactor on the load side of a motor circuit switch, circuit breaker, or molded case switch as a disconnecting means. Because contactors can be closed inadvertently, these devices must be listed and include means to redundantly monitor the contactor position. This equipment is designed for extremely large industrial applications, typically involving multiple personnel entry points to a piece of equipment and therefore making remote lockout provisions in a control circuit desirable.

Section 430.109(E) sets the maximum horsepower rating required for motor-circuit switches at 100 hp. Higher-rated switches are now available and will provide additional safety. The first sentence of this section makes a basic requirement that the disconnecting means for a motor and its controller be a motor-circuit switch rated in horsepower. For motors rated up to 500 hp, this is readily complied with, inasmuch as the UL lists motor-circuit switches up to 500 hp and the manufacturers mark switches to conform. But for motors rated over 100 hp, the Code does not require that the disconnect have a horsepower rating. It makes an exception to the basic rule and permits the use of an ampere-rated switch or isolation switch, provided the switch has a carrying capacity of at least 115 percent of the nameplate current rating of the motor [430.110(A)]. And UL notes that horsepower-rated switches over 100 hp *must not* be used as motor controllers. And part **(E)** notes that isolation switches for motors over 100 hp must be plainly marked "Do not operate under load," if the switch is not rated for safely interrupting the locked-rotor current of the motor. Figure 430-63 shows an example of disconnect switch application for a motor rated over 100 hp.

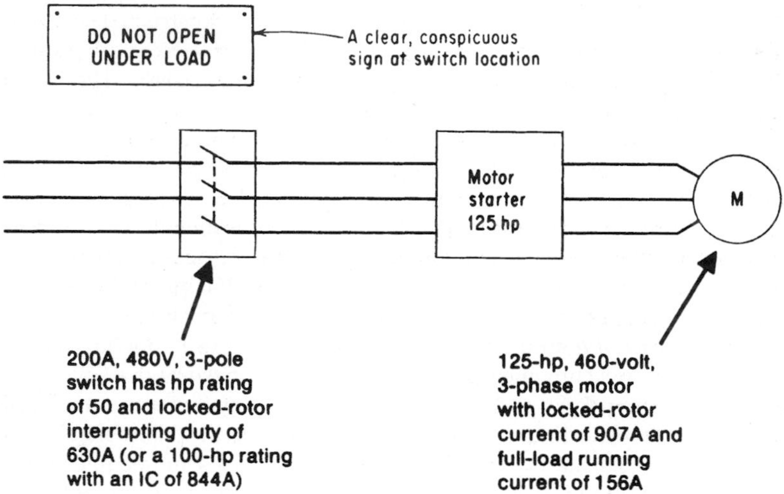
DO NOT OPEN
UNDER LOAD
A clear, conspicuous
sign at switch location
Motor
starter
125 hp
M
200A, 480V, 3-pole
switch has hp rating
of 50 and locked-rotor
interrupting duty of
630A (or a 100-hp rating
with an IC of 844A)
125-hp, 460-volt,
3-phase motor
with locked-rotor
current of 907A and
full-load running
current of 156A

Fig. 430-63. Above 100 hp, a switch does not have to be horsepower-rated. (Sec. 430.109.)

example Provide a disconnect for a 125-hp, 3-phase, 460-V motor. Use a nonfusible switch, inasmuch as short-circuit protection is provided at the supply end of the branch circuit.

The full-load running current of the motor is 156 A, from NEC Table 430.250. A suitable disconnect must have a continuous carrying capacity of 156 × 1.15, or 179 A, as required by 430.110(A).

This calls for a 200-A, 3-pole switch rated for 480 V. The switch may be a general-use switch, a current- and horsepower-marked motor-circuit switch, or an isolation switch. A 200-A, 3-pole, 480-V motor-circuit switch would be marked with a rating of 50 hp, but the horsepower rating is of no concern in this application because the switch does not have to be horsepower-rated for motors larger than 100 hp.

If the 50-hp switch were of the heavy-duty type, it would have an interrupting rating of 10 × 65 A (the full-load current of a 460-V, 50-hp motor), or 650 A. But the locked-rotor current of the 125-hp motor might run over 900 A. In such a case, the switch is required by part **(E)** to be marked "Do not operate under load."

If a fusible switch had to be provided for the example motor to provide both disconnect and short-circuit protection, the size of the switch would be determined by the size and type of fuses used. Using a fuse rating of 250 percent of motor current (which does not exceed the 300 percent maximum in Table 430.152) for standard fuses, the application would call for 400-A fuses in a 400-A switch. This switch would certainly qualify as the motor disconnect. However, if time-delay fuses are used, a 200-A switch would be large enough to take the time-delay fuses and could be used as the disconnect (because it is rated at 115 percent of motor current).

In the foregoing, the 400 A switch might have an interrupting rating high enough to handle the locked-rotor current of the motor. Or the 200-A switch might be of the CB-mechanism type or some other heavy current construction that has an interrupting rating up to 12 times the rated load current of the switch itself. In either of these cases, there would be no need for marking "Do not operate under load."

Up to 100 hp, a switch that satisfies the Code on rating for use as a motor controller may also provide the required disconnect means—the two functions being performed by the one switch—provided it opens all ungrounded conductors to the motor, is protected by an overcurrent device (which may be the branch-circuit protection or may be fuses in the switch itself), and is a manually operated air-break switch or an oil switch not rated over 600 V or 100 A—as permitted by 430.111.

430.110. Ampere Rating and Interrupting Capacity. An ampere-rated switch or a CB must be rated at least equal to 115 percent of a motor's full-load current if the switch or CB is the disconnect means for the motor.

When two or more motors are served by a single disconnect means, as permitted by 430.112, or where one or more motors plus a nonmotor load (such as electric heater load) make use of a single common disconnect, part **(C)** must be used in sizing the disconnect. Refer to the discussion at 430.87 for a worked-out example of the front half of this process, that of coming up with the equivalent horsepower rating. Now the second half of the calculation must be done, that is, determining the required ampere rating of the disconnect. This is the sum of the current ratings of the components, all multiplied by 1.15. In this case 62 A × 1.15 = 71 A. The combined disconnect must have a horsepower rating of 50 hp and the current rating must not be less than 71 A at the stated voltage of 460 V. If the disconnect is not rated in horsepower, such as a molded case switch, this ampere rating is sufficient.

430.111. Switch or Circuit Breaker as Both Controller and Disconnecting Means. As described under 430.84, a manual switch—capable of starting and stopping a given motor, capable of interrupting the stalled-rotor current of the motor, and having the same horsepower rating as the motor—may serve the functions of controller and disconnecting means in many motor circuits, if the switch opens all ungrounded conductors to the motor. That is also true of a manual motor starter. A single manually operated CB may also serve as controller and disconnect (Figs. 430-64 and 430-65). However, in the case of an autotransformer type of controller, the controller itself, even if manual, may not also serve as the disconnecting means. Such controllers must be provided with a separate means for disconnecting controller and motor.

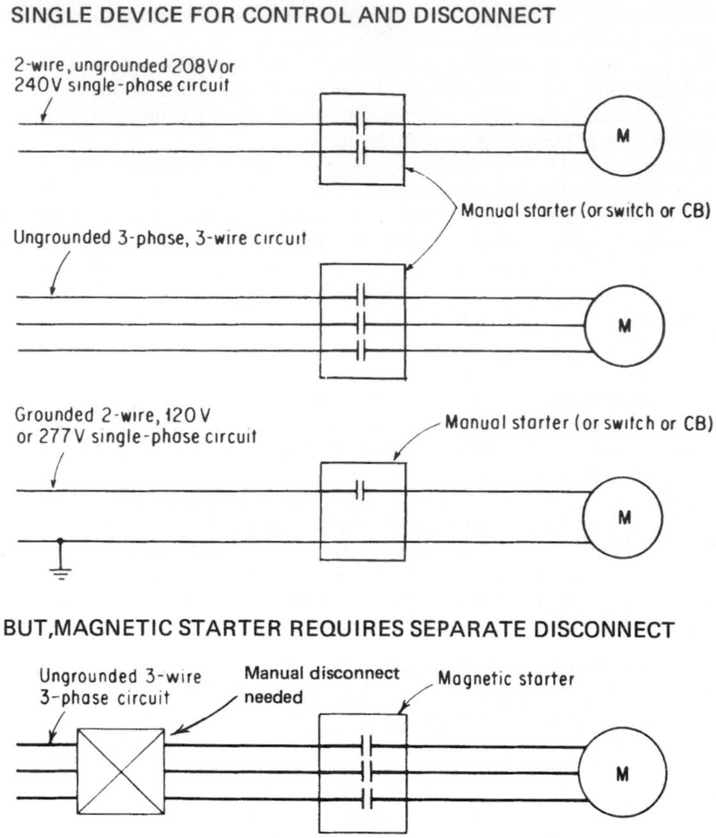
SINGLE DEVICE FOR CONTROL AND DISCONNECT
2-wire, ungrounded 208V or 240V single-phase circuit
M
Manual starter (or switch or CB)
Ungrounded 3-phase, 3-wire circuit
M
Grounded 2-wire, 120 V or 277 V single-phase circuit
Manual starter (or switch or CB)
M
BUT, MAGNETIC STARTER REQUIRES SEPARATE DISCONNECT
Ungrounded 3-wire 3-phase circuit
Manual disconnect needed
Magnetic starter
M

Fig. 430-64. A manual switch or CB may serve as both controller and disconnect means. (Sec. 430.111.)

Although this Code section permits a single horsepower-rated switch to be used as both the controller and the disconnect means of a motor circuit, UL rules note that "enclosed switches rated higher than 100 hp are restricted to use as motor disconnecting means and are not for use as motor controllers."

The acceptability of a single switch for both the controller and the disconnecting means is based on the single switch satisfying the Code requirements for a controller and for a disconnect. It finds application where general-use switches or horsepower-rated switches are used, as permitted by the Code, in conjunction with time-delay fuses that are rated low enough to provide both running overload protection and branch-circuit (short-circuit) protection. In such cases, a single fused switch may serve a total of four functions: (1) controller, (2) disconnect, (3) branch-circuit protection, and (4) running overload protection. And it is possible for a single CB to also serve these four functions.

For sealed refrigeration compressors, 440.12 gives the procedure for determining the disconnect rating, based on nameplate rated-load current or branch-circuit

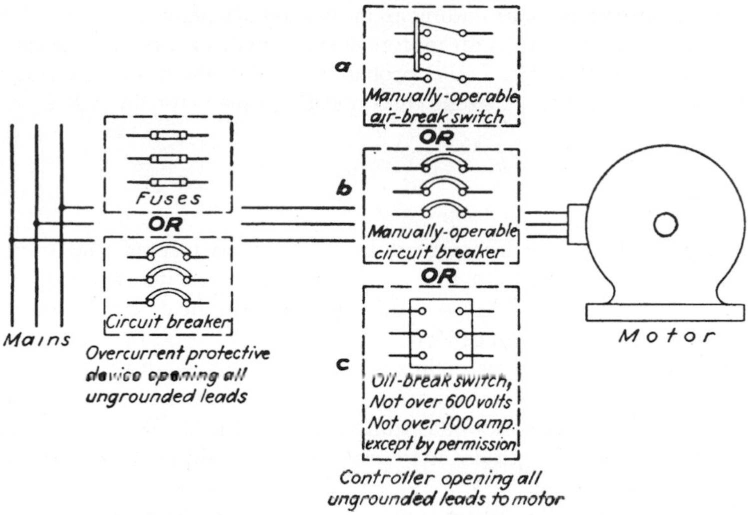
a
Manually-operable air-break switch
OR
b
Manually-operable circuit breaker
OR
c
Not over 600 volts
Not over 100 amp.
except by permission
Controller opening all ungrounded leads to motor
Fuses
OR
Circuit breaker
Mains
Overcurrent protective
ungrounded leads
Motor

Fig. 430-65. Use of a single controller disconnect is limited. (Sec. 430.111.)

selection current, whichever is greater, and locked-rotor current of the motor-compressor.

430.112. Motors Served by Single Disconnecting Means. In general, each individual motor must be provided with a separate disconnecting means. However, a single disconnect sometimes may serve a group of motors under the conditions specified, which are the same as in 430.87. Such a disconnect must have a rating sufficient to handle a single load equal to the sum of the horsepower ratings or current ratings.

Exception A

In 610.32 it is required that the main collector wires of a traveling crane shall be controlled by a switch located within view of the wires and readily operable from the floor or ground. This switch would serve as the disconnecting means for the motors on the crane. When repair or maintenance work is to be done on the electrical equipment of the crane, it is safer to cut off the current from all this equipment by opening one switch, rather than to use a separate switch for each motor. Also, in the case of a machine tool driven by two or more motors, a single disconnecting means for the group of motors is more serviceable than an individual switch for each motor, because repair and maintenance work can be done with greater safety when the entire electrical equipment is "dead."

Exception B

Such groups may consist of motors having full-load currents not exceeding 6 A each, with circuit fuses not exceeding 20 A at 125 V or less, or 15 A at 600 V or less. Because the expense of providing an individual disconnecting means for each motor is not always warranted for motors of such small size, and also because the entire group of small motors could probably be shut down for servicing without causing inconvenience, a single disconnecting means for the entire group is permitted.

Exception C

"Within sight" should be interpreted as meaning so located that there will always be an unobstructed view of the disconnecting switch from the motor, and Sec. 430.102 limits the distance in this case between the disconnecting means and any motor to a maximum of 50 ft.

These conditions are the same as those under which the use of a single controller is permitted for a group of motors. (See 430.87.) The use of a single disconnecting means for two or more motors is quite common, but in the majority of cases the most practicable arrangement is to provide an individual controller for each motor.

If a switch is used as the disconnecting means, it must be of the type and rating required by 430.109 for a single motor having a horsepower rating equal to the sum of the horsepower ratings of all the motors it controls. Thus, for six 5-hp motors the disconnecting means should be a motor-circuit switch rated at not less than 30 hp. If the total of the horsepower ratings is over 2 hp, a horsepower-rated switch must be used.

430.113. Energy from More than One Source. The basic rule of this section, which is similar to that of 430.74, requires a disconnecting means to be provided from each source of electrical energy input to equipment with more than one circuit supplying power to it, such as the hookup shown in Fig. 430-66, where two

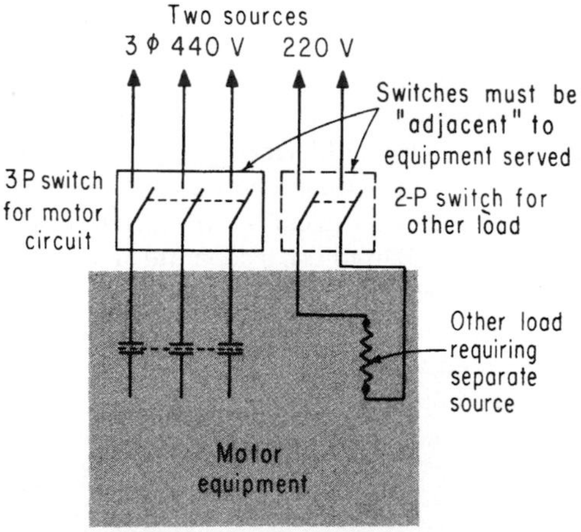
Two sources
3 Φ 440 V
220 V
Switches must be "adjacent" to equipment served
3P switch for motor circuit
2-P switch for other load
Other load requiring separate source
Motor equipment

Fig. 430-66. A disconnect must be used for each power input to motorized equipment. (Sec. 430.113.)

switches or a single 5-pole switch could be used. And each source is permitted to have a separate disconnecting means. This Code rule is aimed at the need for adequate disconnects for safety in complex industrial layouts. But an exception to the Code rule states that where a motor receives electrical energy from more than one source (such as a synchronous motor receiving both alternating current and direct current energy input), the disconnecting means for the main power supply to the motor shall *not* be required to be immediately adjacent to the motor—provided that the controller disconnecting means, which is the disconnect ahead of the motor starter in the main power circuit, is capable of being locked in the open position. If, for instance, the motor control disconnect can be locked in the open position, it may be remote; but the disconnect for the other energy input circuit would have to be adjacent to the machine itself, as indicated in Fig. 430-67.

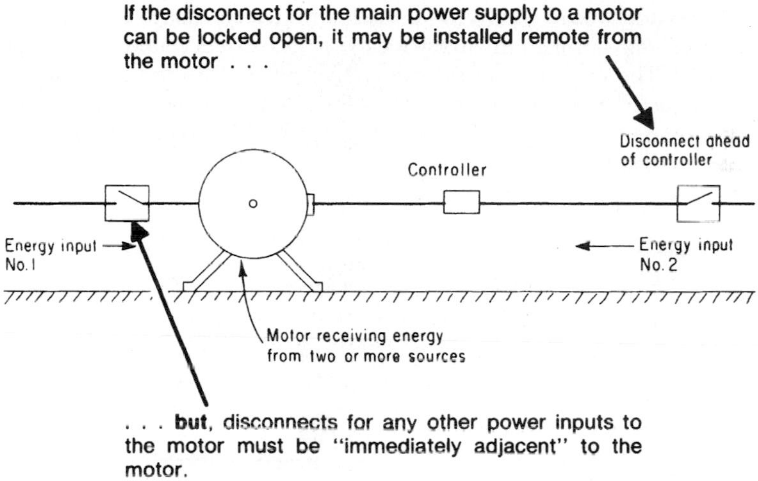
If the disconnect for the main power supply to a motor can be locked open, it may be installed remote from the motor . . .
Disconnect ahead of controller
Controller
Energy input No. 1
Energy input No. 2
Motor receiving energy from two or more sources
. . . but, disconnects for any other power inputs to the motor must be "immediately adjacent" to the motor.

Fig. 430-67. An exception is made for disconnects for multiple power sources. (Sec. 430.113.)

430.122. Conductors—Minimum Size and Ampacity. The majority of motors going into industrial and commercial occupancies are going in with adjustable speed drives for both customization of output and energy efficiency. The incoming branch circuit or feeder to power conversion equipment included as part of an adjustable-speed drive system must be based on the rated input to the power conversion equipment, and is to be taken at 125 percent of that value. If the drive has a bypass mechanism that allows the motor to run directly across the line, the conductors must meet the larger of two calculations, one being 125 percent of the input rating to the drive, and the other being 125 percent of the 430.6 determined rating (usually, therefore, the Table 430.250 rating), as applicable.

430.126. Motor Overtemperature Protection. Adjustable speed drive systems must be capable of detecting overheated motors that are in that condition not because of excessive load resulting in additional current draw, but for other reasons.

Traditional methods that monitor current will protect the circuit conductors from overload, but not the motor if it is drawing its nameplate current but operating below its rated speed because of factors involved with the application, because a low speed may not produce enough ventilation or coolant circulation, etc. A thermal protector (see 430.32) may be required in addition to the drive programming. This section is followed by an extensive informational note calling attention to the problem and offering suggestions as to causes and remedies.

430.128. Disconnecting Means. This rule (Fig. 430-68) merely permits a disconnect in the incoming line, but makes no distance or in-sight requirement. Frequently adjustable speed drives meet the definition of a controller in 430.2; they routinely start and stop the motor by making and breaking the power circuit, as well as changing its frequency for speed regulation purposes. The question then arises as to whether 430.102(A) applies to such a drive, requiring an in-sight disconnect. Many smaller drives are located in areas where a local disconnect might not be desired and would qualify for omission under 430.102(B) Exception. However, that exception applies only to the local motor disconnect and not to a controller disconnect. The vagueness of the disconnect rule in this part of the article has been used to justify the omission of local drive disconnects, with the understanding that a locking disconnect could be arranged at some point upstream, just not in sight of the remote drive.

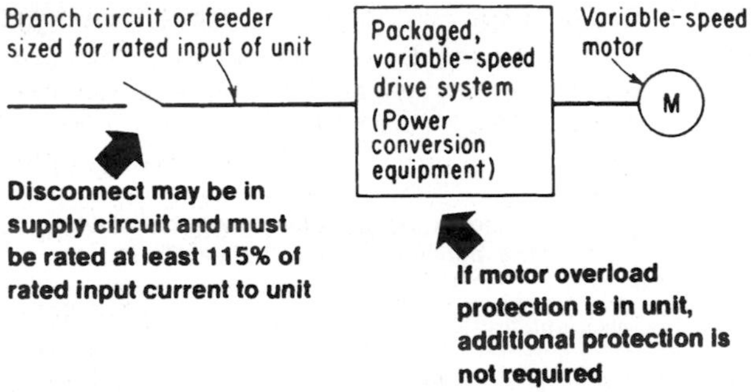
Branch circuit or feeder sized for rated input of unit
Packaged, variable-speed drive system (Power conversion equipment)
Variable-speed motor
M
Disconnect may be in supply circuit and must be rated at least 115% of rated input current to unit
If motor overload protection is in unit, additional protection is not required

Fig. 430-68. Circuit to packaged drive systems is sized for rating of unit, essentially considering the drive to be as a motor. (Art. 430 Part X.)

The conservative response is to apply 430.102(A) to the remote drive without variance, since this rule is vague and not violated in any way by such an interpretation. This approach also agrees with the general statement in 430.120 that Parts I through IX of the article are applicable unless modified, and no modifications to disconnect locations have been made in this part of the article. Further, an attempt to relax this rule failed in the 2014 NEC cycle.

430.130. Branch-Circuit Short-Circuit and Ground-Fault Protection for Single Motor Circuits Containing Power Conversion Equipment. This section is new as of the 2014 NEC and establishes unambiguous requirements for sizing this equipment. The first step is to bring the actual full-load current rating of the motor "as determined by 430.6" (which means the usual table approach by horsepower and voltage) into either 430.52(C)(1) (i.e., using Table 430.52) or (C)(3) (i.e., instantaneous trip circuit breakers) or (C)(5) (i.e., semiconductor fuses). Although (C)(6) (i.e., self-protected combination controller) is also listed, note that this is only permitted where it has been specifically identified by the power conversion equipment manufacturer. The final step, similar to that involved with 430.52(C)(2), is to review the manufacturers' instructions or equipment markings, because those cannot be exceeded even if the default calculations permitted in 430.52 would otherwise indicate a higher setting would be acceptable.

If the power conversion equipment includes a bypass contactor or similar circuiting, the bypass device must also be protected against faults based on the normally applicable rules in 430.52 for the type of control. However, in the frequent case where the fault protection protects both the bypass and the normal power conversion equipment, then that protection must be selected and arranged so both circuits are protected in accordance with relevant Code provisions.

The 2017 **NEC** added another provision intended to further curb the practice of field selection of certain overcurrent devices that are intended to be solely within the province of the manufacturer and testing laboratories. 430.52(C)(3) on instantaneous-trip circuit breakers and 430.52(C)(5) on semiconductor fuses have very strict limitations that preclude field selection decisions. These rules should apply here anyway, and now there is a fourth condition spelling this out. These devices are to be provided as an integral part of a single listed assembly. Review also Fig. 430-21 and the associated coverage for more information.

430.131. Several Motors or Loads on One Branch Circuit Including Power Conversion Equipment. If power conversion equipment is connected ahead of multiple motors, the rules in 430.53 continue to apply based on the full-load current rating of the connected motor load. Any applicable limits on the style of circuit protection permitted for the power conversion equipment will apply as well, and the final selection must be made accordingly. In this case the power conversion equipment is formally declared to be a motor controller, and the rules for the in-sight disconnecting means in 430.102(A) will apply.

430.224. Size of Conductors. For motors rated over 1000 V, the circuit conductors to the motor are selected to have a current rating equal to or greater than the trip setting of the running overload protective device for the motor.

430.225. Motor Circuit Overcurrent Protection. Overload protection must protect the motor and other circuit components against overload currents up to and including locked-rotor current of the motor. A CB or fuses must be used for protection against ground faults or short circuits in the motor circuit. Because there are no prescriptive settings in the Code for these motors, these protective device settings must be evaluated for each motor circuit under engineering supervision.

430.242. Stationary Motors. Usually, stationary motors are supplied by wiring in a metal raceway or metal-clad cable. The motor frames of such motors must be grounded, the raceway or cable armor being attached to the frame and serving as the grounding conductor. (See 250.118.)

Any motor in a wet location constitutes a serious hazard to persons and should be grounded unless it is so located or guarded that it is out of reach. *All* water pump motors, including those in the submersible-type pump, must be grounded, regardless of location, to comply with 250.112(L).

430.245. Method of Grounding. Good practice requires in nearly all cases that the wiring to motors that are not portable shall, at the motor, be installed in rigid or flexible metal conduit, electrical metallic tubing, or metal-clad cable and that such motors should be equipped with terminal housings. The method of connecting the conduit to the motor where some flexibility is necessary is shown in Fig. 430-69. The motor circuit is installed in rigid conduit and a short length of liquidtight flexible metal conduit is provided between the end of the rigid conduit and the terminal housing on the motor. But because the size of flex is over metric designator 35 (trade size 1¼), a separate equipment grounding conductor (or bonding jumper) must be used within or outside the flex as noted in 350.60 and 250.118. Refer to 430.12(E), which requires provision of a suitable termination for an equipment grounding conductor at every motor terminal housing, as shown in Fig. 430-70.

Fig. 430-69. Liquidtight flex provides flexible connection from rigid conduit supply to motor terminals but does require a separate equipment grounding conductor run within the flex with the circuit conductors or a separate external bonding jumper from the rigid metal conduit to the metal terminal box for each of the two runs. (Sec. 430.245.)

This section permits the use of fixed motors without terminal housings. If a motor has no terminal housing, the branch-circuit conductors must be brought to a junction box not over 6 ft (1.8 m) from the motor. Between the junction box and the motor, the specified provisions apply.

According to 300.16, the conduit, tubing, or metal-clad cable must terminate close to the motor in a fitting having a separable bushed hole for each wire. The method of making the connection to the motor is not specified; presumably, it is the intention that the wire brought out from the terminal fitting shall be connected

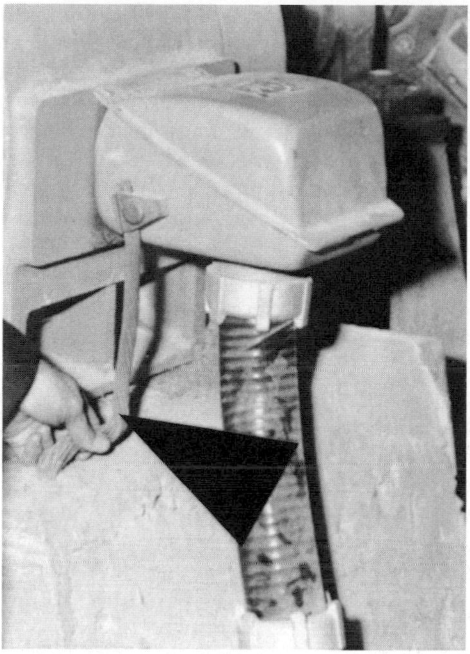

Fig. 430-70. Motor terminal housings must include some lug or terminal for connecting an equipment grounding conductor that may be run inside the raceway with the circuit wires or may be run as a bonding jumper around a length of flex or liquidtight flex, as commonly used for vibration-free motor connections. The terminal box here must have internal provision for connecting the equipment grounding conductor, required for this short length of liquidtight flex, that is larger than metric designator 35 (trade size 1¼) in size. The static grounding connection shown here (arrow) on the box does not satisfy 350.60 and 250.102(E) as a bonding jumper for the flex, and it does not satisfy 250.134 as an equipment ground for an ac motor. (Sec. 430.245.)

to binding posts on the motor or spliced to the motor leads. The conduit, tubing, or cable must be rigidly secured to the frame of the motor.

Table 430-251(B). This is the locked-rotor table that is necessary to determine equivalent horsepower ratings and other activities. It does not apply to Design A motors, which are not covered in this table and not limited to its provision; the manufacturer would need to be consulted in such cases.

ARTICLE 440. AIR-CONDITIONING AND REFRIGERATING EQUIPMENT

440.2. Definitions. The first definition for "Branch-Circuit Selection Current" is extremely important because of the differences between it and the last definition, covering "rated-load current." The latter corresponds to full-load current in many other places. The former includes the effects of the degree of sustained overload that the unit is capable of under defined test conditions. It always at least equals, and usually exceeds the rated-load current. The definition is critical because, where given, it becomes the number upon which the other size or ratings of elements of the circuit are determined, including conductors, overcurrent protective devices, and disconnects.

The third definition, "Leakage Current Detection and Interruption," describes a device capable of detecting and interrupting current flow "from the cord conductor," such as where a ground fault has occurred, and "between" the "cord conductors," such as where the device is unintentionally submerged. This definition ties into the rule of 440.65, which mandates such protective devices for all single-phase (generally, 120- or 220-V) cord-and-plug-connected room air conditioners. Although this is a manufacturer's concern, designers and installers must ensure that any such cord-and-plug-connected room air conditioners are factory equipped with a leakage current detector and interrupter (LCDI) or an arc-fault circuit interrupter (AFCI).

440.3. Other Articles. Article 440 is patterned after Art. 430, and many of its rules, such as on disconnecting means, controllers, conductor sizes, and group installations, are identical or quite similar to those in Art. 430. This article contains provisions for such motor-driven equipment and for branch circuits and controllers for the equipment, taking into account the special considerations involved with sealed (hermetic-type) motor-compressors, in which the motor operates under the cooling effect of the refrigeration.

It must be noted that the rules of Art. 440 are *in addition to* or are *amendments to* the rules given in Art. 430 for motors in general. The basic rules of Art. 430 also apply to A/C (air-conditioning) and refrigerating equipment unless exceptions are indicated in Art. 440.

Article 440 further clarifies the application of NEC rules to air-conditioning equipment and refrigeration equipment as follows:

1. A/C and refrigerating equipment that does not incorporate a sealed (hermetic-type) motor-compressor must satisfy the rules of Art. 422 (Appliances), Art. 424 (Space Heating Equipment), or Art. 430 (Conventional Motors)—whichever apply. For instance, where refrigeration compressors are driven by conventional motors, the motors and controls are subject to Art. 430, not Art. 440. Furnaces with air-conditioning evaporator coils installed must satisfy Art. 424. Other equipment in which the motor is not a sealed compressor and which must be covered by Arts. 422, 424, or 430 includes fan-coil units, remote forced-air-cooled condensers, remote commercial refrigerators, and similar equipment.
2. Room air conditioners are covered in Part **VII** of Art. 440 (440.60 through 440.64), but must also comply with the rules of Art. 422.
3. Household refrigerators and freezers, drinking-water coolers, and beverage dispensers are considered by the Code to be appliances, and their application must comply with Art. 422 and must also satisfy Art. 440, because such devices contain sealed motor-compressors.

Air-conditioning equipment (other than small room units and large custom installations) is manufactured in the form of packaged units having all necessary components mounted in one or more enclosures designed for floor mounting, for recessing into walls, for mounting in attics or ceiling plenums, for locating

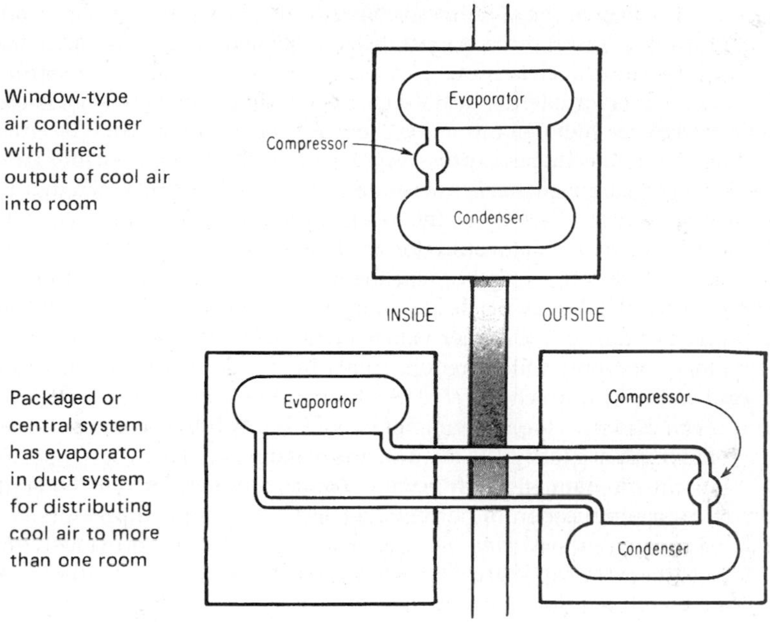
Window-type air conditioner with direct output of cool air into room
Evaporator
Compressor
Condenser
INSIDE
OUTSIDE
Packaged or central system has evaporator in duct system for distributing cool air to more than one room
Evaporator
Compressor
Condenser

Fig. 440-1. Code rules differentiate between unit room conditioners and central systems. (Sec. 440.3.)

outdoors, and so forth. Figure 440-1 shows the difference between room A/C units (such as window units) and the larger so-called packaged units or central air conditioners. Room units consist of a complete refrigeration system in a unit enclosure intended for mounting in windows or in the wall of the building, with ratings up to 250 V, single phase. Unitary assemblies may be console type for individual room use, rated up to 250 V, single phase, or central cooling units rated up to 600 V for commercial or domestic applications. This type may consist of one or more factory-made sections. If it is made up of two or more sections, each section is designed for field interconnection with one or more matched sections to make the complete assembly. Dual section systems consist of separate packaged sections installed remote from each other and interconnected by refrigerant tubing, with the compressor either within the outdoor section or within the indoor section.

Electrical wiring in and to units varies with the manufacturer, and the extent to which the electrical contractor need be concerned with fuse and CB calculations depends on the manner in which the units' motors are fed and the type of distribution system to which they are to be connected. A packaged unit is treated as a group of motors. This is different from the approach used with a plug-in room air conditioner, which is treated as an individual single-motor load of amp rating as marked on the nameplate.

A variation on this topic frequently arises with respect to whether the disconnect rule in 440.14 applies to a fan-forced heat exchange coil that is powered by the branch circuit serving the compressor, from the load side of a fully compliant, lock-open disconnect at the compressor. Specifically, since the coil fan is associated with the compressor, some inspectors have insisted on a full-fledged disconnect at the inside unit. Such a ruling ignores the content of this part [440.3(B)]. The issue of disconnects for internal heat exchange equipment that does not contain a hermetic refrigerant motor-compressor is entirely beyond the scope of Art. 440. In addition, 440.3(B) contains express language that sends the user into Art. 430 for this coverage, language that is unchanged since the 1971 **NEC** when this article first appeared. Article 440 contains the necessary special rules for motors that are deliberately operated in what would normally be an overloaded state, but that can function closer to the edge only because the windings are submerged in refrigerant. For example, this is why the default short-circuit protection in 440.22(A) is set at 175 percent instead of the higher values in 430.52(C) for conventional motors. Other motors need only follow requirements in Art. 430 and elsewhere in the Code. And 430.109(B), which governs this topic, allows the branch-circuit device to serve as the disconnect for motors not over $^{1}/_{8}$ hp. A full disconnect cluttering the wall of living space for a little fan motor is usually excessive.

440.4. Marking on Hermetic Refrigerant Motor-Compressors and Equipment. Important in the application of hermetic refrigerant motor-compressors are the terms *rated-load current* and *branch-circuit selection current*. As previously noted, definitions of these terms are given in 440.2. When the equipment is marked with the branch-circuit selection current, this greatly simplifies the sizing of motor branch-circuit conductors, disconnecting means, controllers, and overcurrent devices for circuit conductors and motors.

For some A/C equipment that is not required to have a branch-circuit selection current, the value of rated-load current will appear on the equipment nameplate; and that same value of current will also appear in the nameplate space reserved for branch-circuit selection current. In such cases, the branch-circuit selection current is to be equal to the rated-load current. The short-circuit current rating of either the motor controllers or the industrial control panel in which they are components must also be marked, so the ability of these components to function within the context of the fault current that is available at the point of application can be determined.

440.5. Marking on Controllers. Note that a controller may be marked with "full-load and locked-rotor current (or horsepower) rating." That possibility of two methods of marking requires careful application of the rules in 440.41 on selecting the correct rating of controller for motor-compressors.

440.6. Ampacity and Rating. Selection of the rating of branch-circuit conductors, controller, disconnect means, short-circuit (and ground-fault) protection, and running overload protection is *not* made the same for hermetic motor-compressors as for general-purpose motors. In sizing those components, the rated-load current marked on the equipment and/or the compressor must be used in the calculations covered in other rules of this article. That value of current must always be used, instead of full-load currents from Code Tables 430.248 to 430.250, which are used for sizing circuit elements for nonhermetic motors. And if a branch-circuit selection current is marked on equipment, that value must be used instead of rated-load current.

440.8. Single Machine. Even for split systems that have the compressor located remote from the associated fans to distribute conditioned air, the system is to be classified as a single machine, and therefore permitted to have a single controller under the provisions of part (a) of 430.87 Exception No. 1. The components are also permitted to operate under a single disconnect in accordance with part (a) of 430.112 Exception. For calculating required conductor ampacity of multimotor equipment, the provisions of 440.6(B) cover fan or blower components, and 440.7 covers the determination of which motor is to be taken as the largest. However, the mandatory exception to 440.7 again reverts the calculation to the branch-circuit selection current in all cases where one has been established and marked.

440.9. Grounding and Bonding. The 2017 NEC decided to disqualify metal raceways foe Art. 440 wiring on a rooftop as an equipment grounding return conductor, unless they had been assembled with threaded couplings and connectors. Primarily aimed at EMT, the panel responded to substantial documentation that conventional set-screw and compression fittings were allowing the raceway joints to open over time. EMT can still be used, but a wire-type equipment grounding conductor must be pulled into it so the steel does not constitute the sole ground return path.

440.10. Short-Circuit Rating. This new 2017 NEC requires that an available short-circuit current analysis be done to establish the fault current available at multimotor and combination load equipment locations as part of the installation process. The calculation requirement applies in instances when the equipment is required to be marked with that rating. This means that it will not be required in one- and two-family dwellings, and to comparatively small circuits (not over 60 amps.) as covered in 440.4(B) Exception No. 3. In other instances it will be required, and part (B) of the new rule requires that the calculated result be documented, including the date of the calculation, and made available to the inspector. Note that this calculation must be made, documented, and provided to the inspector, but not provided on a label on the equipment. This fundamentally differs from similar calculations made under 110.16(B). The existence of a label may be misconstrued by future workers having present-day accuracy. Periodic utility activities can easily affect any such calculations, and current conditions must always be re-evaluated prior to work.

440.12. Rating and Interrupting Capacity. Note that the rules here are qualifications that apply to the rules of 430.109 and 430.110 on disconnects for general-purpose motors.

A disconnecting means for a hermetic motor, as covered in part **(A)(2),** must be a motor-circuit switch rated in horsepower or a CB—as required by 430.109.

If a CB is used, it must have an amp rating not less than 115 percent of the nameplate rated-load current or the branch-circuit selection current—whichever is greater.

But, if a horsepower-rated switch is to be selected, the process is slightly involved for hermetic motors marked with locked-rotor current and rated-load current or rated-load current plus branch-circuit selection current—but *not* marked with horsepower. In such a case, determination of the equivalent horsepower rating of the hermetic motor must be made using the locked-rotor current and either the rated-load current or the branch-circuit selection current—whichever is greater—based on Code Tables 430.248, 430.249, or 430.250 for rated-load current or branch-circuit selection current and Table 430.251(B) for locked-rotor current, as follows:

For example, a 3-phase, 460-V hermetic motor rated at 11-A branch-circuit selection and 60-A locked-rotor is to be supplied with a disconnect switch rated in horsepower. The first step in determining the equivalent horsepower rating of that motor is to refer to Code Table 430.250. This table lists 7½ hp as the required size for a 460-V, 11-A motor. To ensure adequate interrupting capacity, Code Table 430.251(B) is used. For a 60-A locked-rotor current, this table also shows 7½ hp as the equivalent horsepower rating for any locked-rotor current over 45 to 66 A for a 400-V motor. Use of both tables in this manner thus establishes a 7½ hp disconnect as adequate for the given motor in both respects. Had the two ratings as obtained from the two tables been different, the higher rating would have been chosen.

Figure 440-2 shows an example of disconnect sizing for a horsepower-rated switch when a hermetic motor is used, in accordance with 430.53(C) and (D), along with fan motors on a single circuit, as covered in part **(B)** and in 440.34. Fan motors are usually wired to start slightly ahead of the compressor motor through use of interlock contacts or a time-delay relay. In some units, however, all motors start simultaneously, and that is covered by part **(B)** of this section in sizing the horsepower-rated disconnect switch. Where this is the case, the starting load will be treated like a single motor to the disconnect switch, and the sum of the locked-rotor currents of all motors should be used with Code Table 430.251(B) to determine the horsepower rating of the disconnect. The disconnect normally must handle the sum of the rated-load or branch-circuit selection currents; hence the rating as checked against Code Table 430.250 will be on the basis of the sum of the higher of those currents for all the motors. Code Table 430.250, using the full-load

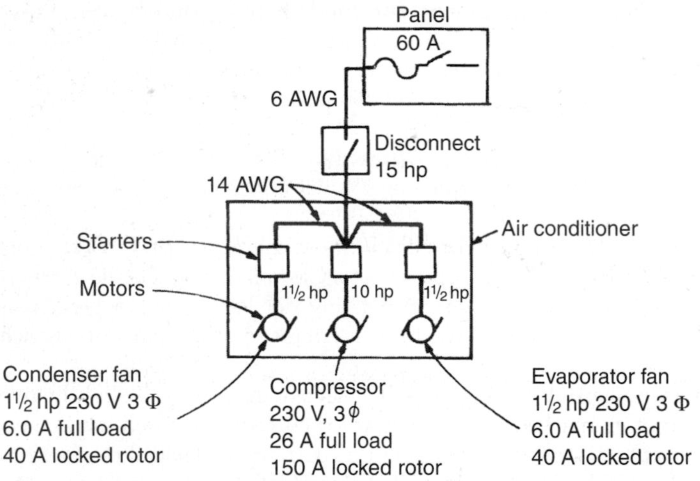
Panel
60 A
6 AWG
Disconnect
15 hp
14 AWG
Air conditioner
Starters
Motors
1½ hp
10 hp
1½ hp
Condenser fan
1½ hp 230 V 3 Φ
6.0 A full load
40 A locked rotor
Compressor
230 V, 3ϕ
26 A full load
150 A locked rotor
Evaporator fan
1½ hp 230 V 3 Φ
6.0 A full load
40 A locked rotor

Branch-circuit conductors (125% × 26) + 6.0 + 6.0 = 44.5 amps (6 AWG)
Compressor conductors (125% × 26) = 32.8 amps (8 AWG)
Fan conductors (125% × 6.0) = 7.5 amps but 1/3 × 50 = 16.7 A (14 AWG)
Fuses: (175% × 26) + 6.0 + 6.0 = 57.5 amps (60 amp fuses) but subject to group fusing restrictions of starters. Wire sizes based on 110.14(C)(1)(a) rules (60°C).

Fig. 440-2. Disconnect for multiple motors is sized from rated-load or branch-circuit selection currents and locked-rotor currents. (Sec. 440.12.)

total of 38 A (6.0 A + 26 A + 6.0 A) in this example, indicates a 15-hp disconnect. Code Table 430.251(B), assuming simultaneous starting of all three motors and using the total locked-rotor current of 230 A (40 A + 150 A + 40 A), also shows 15 hp as the required size.

If motors do not start simultaneously, the compressor locked-rotor current (150 A) used with Code Table 430.251(B) gives a 10-hp rating. However, the higher of the two horsepower ratings must be used; hence, the running currents impose the more severe requirements and dictate use of a 15-hp switch. See data under 440.22.

As required by part **(D)** of this section, all disconnects in a branch circuit to a refrigerant motor-compressor must have the required amp or horsepower rating and interrupting rating. This provides for motor-compressor circuits the same conditions that 430.108 requires for other motor branch circuits.

440.14. Location. 440.13 recognizes use of a cord-plug and receptacle as the disconnect for such cord-connected equipment as a room or window air conditioner. But this section (440.14) applies to fixed-wired equipment—such as central systems or units with fixed circuit connection. For conditioners with fixed wiring connection to their supply circuits, the rule poses a problem. If the branch-circuit breaker or switch which is to provide disconnect means is located in a panel that is out of sight (or more than 50 ft [15.0 m] away) from the unit conditioner, another breaker or switch must be provided at the equipment. If the panel breaker or switch does not satisfy the rule here, a separate disconnect means would have to be added in sight from the conditioner, as shown in Fig. 440-3. This is also true if the service switch is installed as shown in Fig. 440-4.

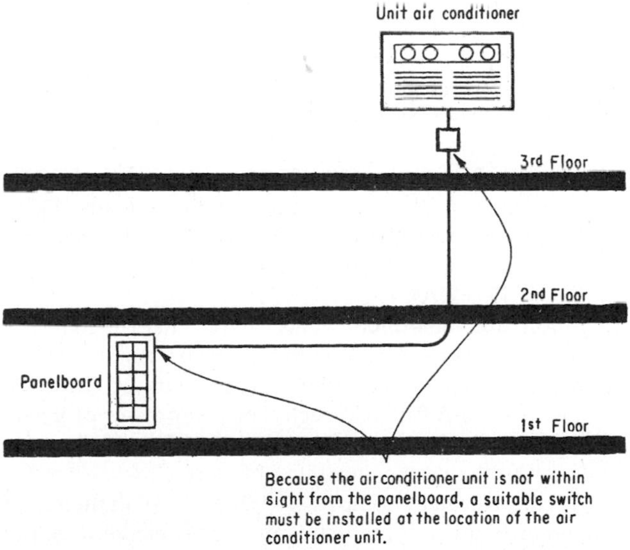
Unit air conditioner
3rd Floor
2nd Floor
Panelboard
1st Floor
Because the air conditioner unit is not within sight from the panelboard, a suitable switch must be installed at the location of the air conditioner unit.

Fig. 440-3. For any fixed-wire A/C equipment, disconnect must be "within sight." (Sec. 440.14.)

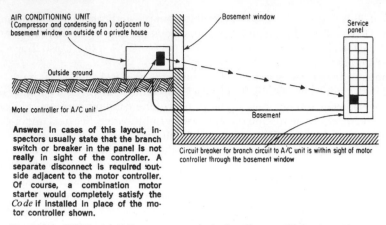

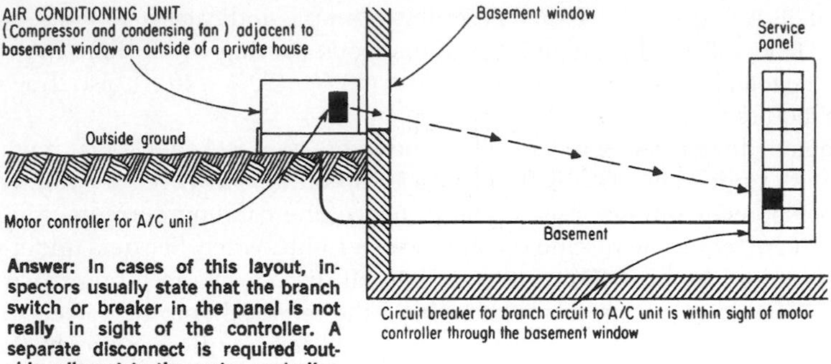
AIR CONDITIONING UNIT
(Compressor and condensing fan) adjacent to basement window on outside of a private house
Basement window
Service panel
Outside ground
Motor controller for A/C unit
Basement
Circuit breaker for branch circuit to A/C unit is within sight of motor controller through the basement window

Answer: In cases of this layout, inspectors usually state that the branch switch or breaker in the panel is not really in sight of the controller. A separate disconnect is required outside adjacent to the motor controller. Of course, a combination motor starter would completely satisfy the *Code* if installed in place of the motor controller shown.

Fig. 440-4. "Within sight" disconnect must also be "readily accessible" at the equipment. (Sec. 440.14.)

As stated in the second sentence, the required disconnect means for air-conditioning or refrigeration equipment may be installed on or within the equipment enclosure; however, it must not be mounted in a way that obstructs the removal of service panels or the visibility of nameplates. If suitably located on the unit, the location is recognized as an equivalent of the basic rule that the disconnect must be readily accessible and within sight (visible and not over 50 ft [15.0 m] away) from the A/C or refrigeration equipment. Such equipment is being manufactured now with the disconnect incorporated as part of the assembly.

440.21. General. Part III of this article covers details of branch-circuit makeup for A/C and refrigeration equipment; 440.3(A) says that the provisions of Art. 430 apply to A/C and refrigeration equipment for any considerations that are not covered in Art. 440. Thus, because Art. 440 does *not* cover feeder sizing and feeder overcurrent protection for A/C and refrigeration equipment, it is necessary to use applicable sections from Art. 430. 430.24 covers sizing of feeder conductors for standard motor loads and for A/C and refrigeration loads. 430.62 and 430.63 cover rating of overcurrent protection for feeders to both standard motors and A/C and refrigeration equipment. That fact is noted in the language at the end of the first paragraph of 430.62(A).

440.22. Application and Selection. Part **(A)** of this section is illustrated in Fig. 440-5, where a separate circuit is run to the compressor and to each fan motor of a packaged assembly, containing a compressor with 26-A rated-load current and fan motors rated at 6.0-A full load each. The compressor protection is sized at 1.75 × 26 A (175 percent of rated-load current), or 45.5 A—calling for 45-A fuses, maximum. Note carefully that the wording of this rule makes it a "not to exceed" or "round down" rule. Hermetic compressor windings sit in their own refrigerant, and in effect operate overloaded, which is only possible because of the refrigerant. This makes them more sensitive to problems and the overcurrent protection boundaries are more tightly drawn accordingly. In this case, if and only if the basic

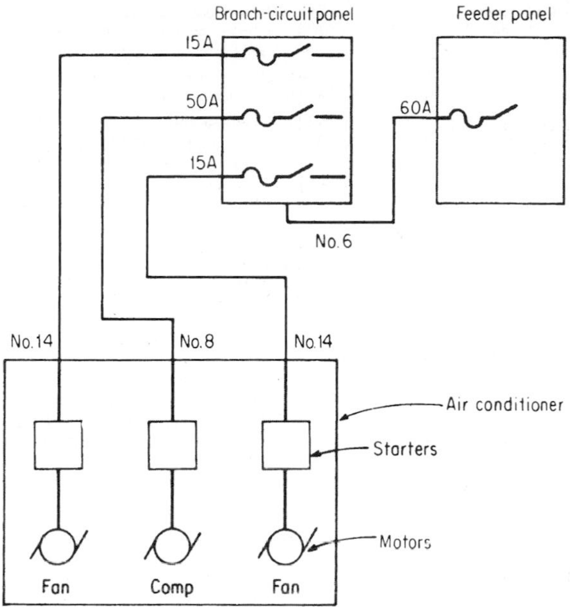
Branch-circuit panel
Feeder panel
15 A
50A
60A
15A
No. 6
No. 14
No. 8
No. 14
Air conditioner
Starters
Motors
Fan
Comp
Fan

Fig. 440-5. A separate circuit may be run to each motor of A/C assembly. (Sec. 440.22.)

rule doesn't work, then a modest increase to 225 percent is allowable, and that too is a not to exceed number. However, it is never necessary to decrease the protection below 15 A.

Sizing of branch-circuit protection for a single branch circuit to the same three motors is permitted by 430.53(C) as well as by 440.22(B) and is shown in Fig. 440-6. That layout is a specific example of the general rules covered in 440.22(B)(1), which ties the rules of 430.53(C) and (D) into the rules of 440.22(B), as shown in Fig. 440-7. Such application is based on certain factors, as covered in the UL *Electrical Appliance and Utilization Directory*, listed under "Heating and Cooling Equipment," as follows:

> In permanently connected units employing two or more motors or a motor(s) and other loads operating from a single supply circuit, the motor running overcurrent protective devices (including thermal protectors for motors) and other factory-installed motor circuit components and wiring are investigated on the basis of compliance with the motor-branch-circuit short circuit and ground fault protection requirements of 430.53(C) as referenced in Sec. 440.22 of the National Electrical Code. Such multimotor and combination load equipment is intended to be connected only to a circuit protected by fuses or a circuit breaker with a rating which does not exceed the value marked on the data plate. This marked protective device rating is the maximum for which the equipment has been investigated and found acceptable. Where the marking specifies fuses, or "HACR Type" circuit breakers, the circuit is intended to be protected by the type of protective device specified.

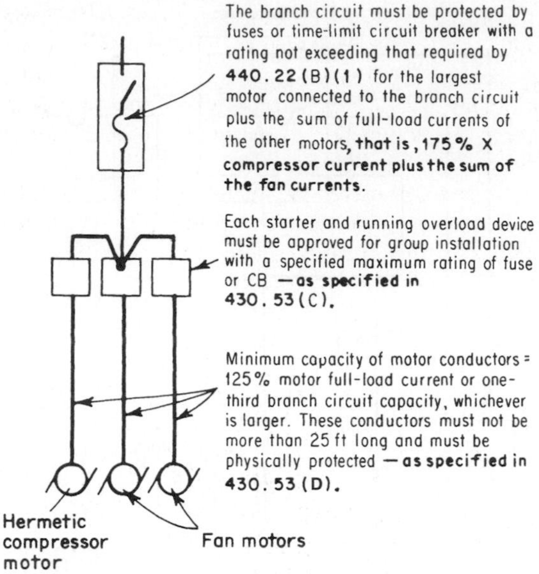
The branch circuit must be protected by fuses or time-limit circuit breaker with a rating not exceeding that required by 440.22(B)(1) for the largest motor connected to the branch circuit plus the sum of full-load currents of the other motors, that is, 175% X compressor current plus the sum of the fan currents.
Each starter and running overload device must be approved for group installation with a specified maximum rating of fuse or CB —as specified in 430.53(C).
Minimum capacity of motor conductors = 125% motor full-load current or one-third branch circuit capacity, whichever is larger. These conductors must not be more than 25 ft long and must be physically protected —as specified in 430.53(D).
Hermetic compressor motor
Fan motors

Fig. 440-6. Single multimotor branch circuit must conform to several rules. (Sec. 440.22.)

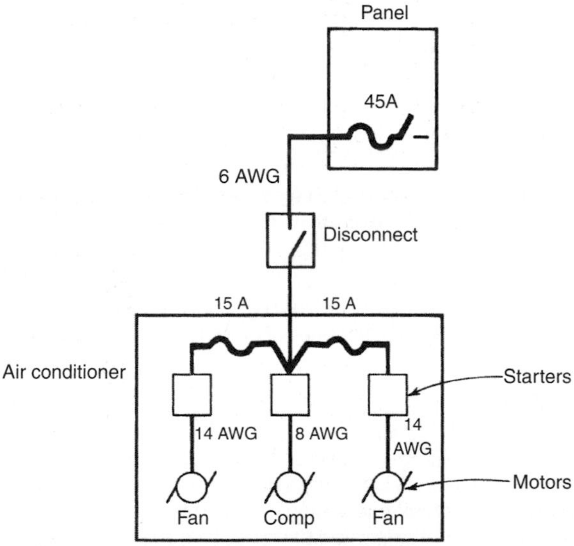
Panel
45A
6 AWG
Disconnect
15 A
15 A
Air conditioner
Starters
14 AWG
8 AWG
14 AWG
Motors
Fan
Comp
Fan

Main fuses: (175% × 26) = 45.5 amps (45-amp fuses)
Fan fuses: (300% × 6.0) =18 amps (15-amp fuses unless exceptions used)
Fan conductors: (125% × 6.0) = 7.5 amps (14 AWG)

Fig. 440-7. Fan circuits are sometimes fused in multimotor assemblies. (Sec. 440.22.) For the internal compressor conductors, refer to the discussion at 440.33.

The electrical contractor and inspector charged with wiring and approving such an installation can be sure that Code requirements have been met, provided that the branch-circuit protection as specified on the unit is not exceeded and the wiring and equipment is as indicated on the wiring diagram. Provision is made in such a unit for direct connection to the branch-circuit conductors; motors are wired internally by the manufacturer.

Units are sometimes encountered in which the manufacturer has wired separate fuse cutouts for the fan motors inside the enclosure to avoid meeting the requirements of 430.53(C) for group fusing, as shown in Fig. 440-8. The cutouts are normally fed from the line terminals of the compressor starter.

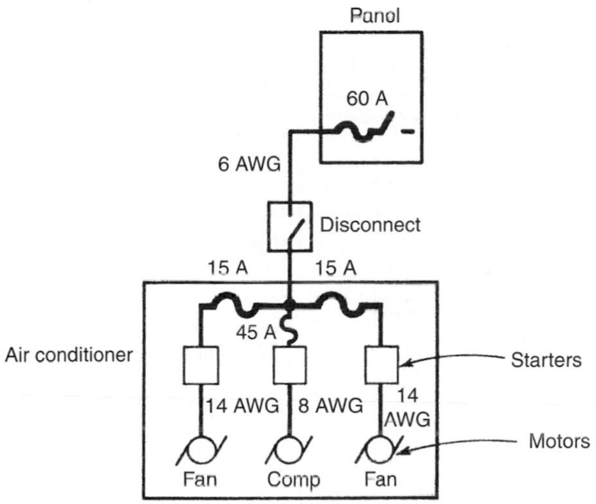
Panel
60 A
6 AWG
Disconnect
15 A
15 A
45 A
Air conditioner
Starters
14 AWG
8 AWG
14 AWG
Motors
Fan
Comp
Fan

Main fuses: [(Comp. fuses ≤ 175% · 26) = 45 A] + 6.0 + 6.0 = 57 amps (60-amp fuses)
Compressor fuses: (175% · 26) = 45.5 amps (45-amp fuses)

Fig. 440-8. Each motor may have individual short-circuit protection. (Sec. 440.22.)

Starter and disconnect sizes are the same as in Fig. 440-2, but starters and their overcurrent protection no longer need be approved for group fusing, and wiring inside the unit need not conform to 430.53(C). Fan motors may now be wired with 14 AWG wire and protected with 15-A fuses with no $^{1}/_{3}$ ratio calculation involved. The supply circuit, feeding the same motors, will again be 6 AWG.

Since the fan motors are not subject to group fusing requirements, they will not restrict the maximum value of the main fuses. However, these fuses provide the only short-circuit protection for the compressor starter and conductors. Unless the compressor starter is approved for group fusing at a higher fuse rating, the fuses must not exceed 175 percent of the compressor full-load rating, or 45.5 A, calling for 45-A fuses. The wire from the panel now gets sized more as a conventional motor feeder needing to accommodate the 45-A main compressor protective device plus the other full-load current supplying the fans, or 8.8 A, for a total of 54 A. The 6 AWG wire on a 60-A fuse is appropriate for this hookup.

Figure 440-9 shows an arrangement which includes, in addition to the branch-circuit panel, a feeder panel for distribution to other units. The breakers in the branch-circuit panel serve as branch-circuit protection as well as the disconnecting means, and their ratings are computed from 430.52 and 440.22. Code Tables 430.250 and 430.251 would not be involved, because breakers are not rated in horsepower. Ratings of CBs in the branch-circuit panel are computed at 175 percent of motor current for the hermetic motor and at 250 percent of full-load current for the fan motors, to satisfy Table 430.52. Breakers in subfeeder panel are rated using 430.62 and the conductors by 440.33. Because 440.22(A) does not distinguish between types of short-circuit and ground-fault protective devices in setting the upper limits on short-circuit protection, most of these numbers agree with those developed in Fig. 440-8 for fuses.

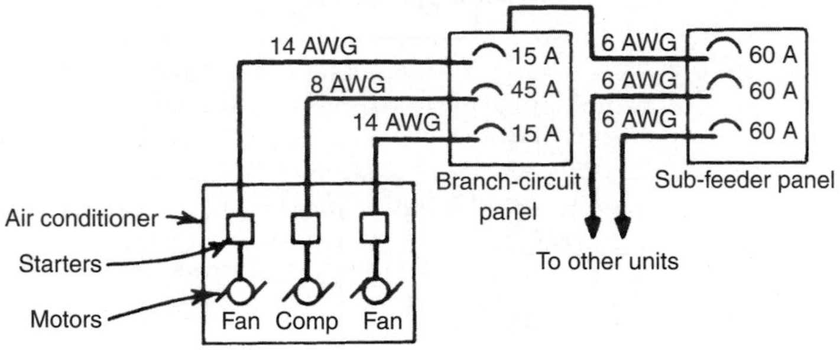
14 AWG
8 AWG
14 AWG
15 A
45 A
15 A
6 AWG
6 AWG
6 AWG
60 A
60 A
60 A
Branch-circuit panel
Sub-feeder panel
Air conditioner
Starters
Motors
Fan Comp Fan
To other units

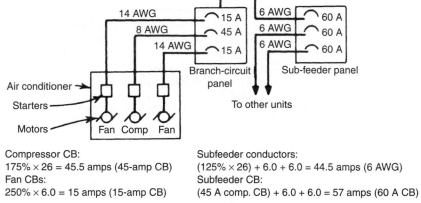

Compressor CB:
175% × 26 = 45.5 amps (45-amp CB)
Fan CBs:
250% × 6.0 = 15 amps (15-amp CB)

Subfeeder conductors:
(125% × 26) + 6.0 + 6.0 = 44.5 amps (6 AWG)
Subfeeder CB:
(45 A comp. CB) + 6.0 + 6.0 = 57 amps (60 A CB)

Fig. 440-9. Circuit breakers may be used for multimotor A/C assemblies. (Sec. 440.22.)

Part **(C)** points out that data on a manufacturer's heater table take precedence over the maximum ratings set by 440.22(A) or (B). This is the correlating language in this article to 430.52(C)(2) on the same subject.

440.32. Single Motor-Compressor. Branch-circuit conductors supplying a motor in a packaged unit are not sized in the same manner as other motor loads (430.22). Instead of using the full-load current from Code Tables 430.248 to 430.250, the *marked* rated-load current or the *marked* branch-circuit selection current must be used in determining minimum required conductor ampacity. Note that branch-circuit selection current must be used where it is given.

Examples are shown in the typical circuits shown in Figs. 440-2 and 440-6.

For a wye-start, delta-run compressor, the conductors between the controller and the unit are sized, just exactly as in the case of the same wires in 430.22(C), based on 58 percent of the full-load current of the compressor. When you apply the mandatory 125 percent factor to that base number, as covered in the first paragraph, the result is a 72 percent ampacity requirement. This number does not need to be increased further.

440.33. Motor-Compressor(s) With or Without Additional Motor Loads. Where more than one motor is connected to the same feeder or branch circuit, calculation of conductor sizes must provide ampere capacity at least equal to the sum of the nameplate rated-load currents or branch-circuit selection currents (using the higher of those values in all cases) plus 25 percent of the current (either rated-load or branch-circuit selection current for a hermetic motor or NEC table current for standard motor) of the largest motor of the group. Examples are shown in Figs. 440-2 and 440-7.

In Fig. 440-7, the question arises as to whether the No. 6 conductors feeding the unit may be decreased to No. 8 inside the unit to feed the compressor motor in the absence of fuses for this motor at the point of reduction. The status of the main feed to the unit—whether it should be considered a feeder or a branch circuit—is in doubt, since it is a branch circuit as far as the compressor motor is concerned and a feeder in that it also supplied the two fused fan circuits.

Considered solely as a branch circuit to the compressor, these conductors normally would be No. 8 to handle the 26-A compressor motor full-load current, protected at not more than 175 percent or 50-A fuses. Therefore, since 45-A fuses (or less) will actually be used for the main feed, they constitute proper protection for No. 8 conductors and their use should be permitted. The existence of No. 6 conductors over part of the circuit adds to its capacity and safety rather than detracting from it.

It is particularly important to keep in mind when selecting conductor sizes that the nameplate current ratings of air-conditioning motors are not constant maximum values during operation. Ratings are established and tested under standard conditions of temperature and humidity. Operation under weather conditions more severe than those at which the ratings are established will result in a greater running current, which can approach the maximum value permitted by the overcurrent device. Operating voltage less than the limits specified on the motor nameplate also contributes to higher full-load current values, even under standard conditions. Conductor capacity should be sufficient to handle these higher currents. Motor branch-circuit conductors are sized according to 440.33. Because overload protection may permit motors to run continuously overloaded (up to 140 percent full load), feeders must be sized to handle such overload. By basing calculations on the largest motor of the group, the extra capacity thus provided will normally be enough to handle any unforeseen overload on the smaller motors involved with enough diversity existing in any normal group of motors to make consistent overloads on all motors at one time unlikely. However, a group of air-conditioning compressor motors all of the same size on a single feeder has a common function—reducing the ambient temperature. Except for slight possible variations, weather conditions affect each conditioner to the same degree and at the same time. Therefore, if one unit is operating at an overload, it is likely that the rest are also. Here again, the branch-circuit selection current ratings are evaluated through the listing process to accommodate adverse climatic conditions, and where they have been established and duly marked on the equipment, they are to be used to make conductor-selection decisions.

440.35. Multimotor and Combination-Load Equipment. This rule ties into the data required by UL to be marked on such equipment. Refer to the UL data quoted in 440.22.

440.41. Rating. The basic rule calls for a compressor controller to have a full-load current rating and a locked-rotor current rating not less than the compressor nameplate rated-load current or branch-circuit selection current (whichever is greater) and locked-rotor current. But, as noted for the disconnect under 440.12, for sealed (hermetic-type) refrigeration compressor motors, selection of the size of controller is slightly more involved than it is for standard applications. Because of their low-temperature operating conditions, hermetic motors can handle heavier loads than general-purpose motors of equivalent size and rotor-stator construction. And because the capabilities of such motors cannot be accurately defined in terms of horsepower, they are rated in terms of full-load current and locked-rotor current for polyphase motors and larger single-phase motors. Accordingly, selection of controller size is different from the case of a general-purpose motor where horsepower ratings must be matched, because controllers marked in horsepower only must be carefully related to hermetic motors that are *not* marked in horsepower.

For controllers rated in horsepower, selection of the size required for a particular hermetic motor can be made after the nameplate rated-load current, or branch-circuit selection current, whichever is greater, and locked-rotor current of the motor have been converted to an equivalent horsepower rating. To get this equivalent horsepower rating, which is the required size of controller, the tables in Art. 430 must be used. First, the nameplate full-load current at the operating voltage of the motor is located in Code Tables 430.248, 430.249, or 430.250 and the horsepower rating which corresponds to it is noted. Then the nameplate locked-rotor current of the motor is found in Code Table 430.251(B), and again the corresponding horsepower is noted. In all tables, if the exact value of current is not listed, the next higher value should be used to obtain an equivalent horsepower, by reading horizontally to the horsepower column at the left side of those tables. If the two horsepower ratings obtained in this way are not the same, the larger value is taken as the required size of controller.

A typical example follows:

Given: A 230-V, 3-phase, squirrel-cage induction motor in a compressor has a nameplate rated-load current of 25.8 A and a nameplate locked-rotor current of 90 A.

Procedure: From Code Table 430.250, 28 A is the next higher current to the nameplate current of 25.8 under the column for 230-V motors and the corresponding horsepower rating for such a motor is 10 hp.

From Code Table 430.251(B), Art. 430, a locked-rotor current rating of 90 A for a 230-V, 3-phase motor requires a controller rated at 5 hp. The two values of horsepower obtained are not the same, so the higher rating is selected as the acceptable unit for the conditions. A 10-hp motor controller must be used.

Some controllers may be rated not in horsepower but in full-load current and locked-rotor current. For use with a hermetic motor, such a controller must simply have current ratings equal to or greater than the nameplate rated-load current and locked-rotor current of the motor.

440.52. Application and Selection. The basic rule of part **(A)(1)** calls for a running overload relay set to trip at not more than 140 percent of the rated-load current of a motor-compressor. As shown in the other subpart **(3)**, if a fuse or inverse time CB is used to provide overload protection, it must be rated not over 125 percent of the compressor rated-load current. Note that those are absolute maximum values of overload protection and no permission is given to go to "the next higher standard rating" of protection where 1.4 or 1.25 times motor current does not yield an amp value that exactly corresponds to a standard rating of a relay or of a fuse or CB.

Running overload protective devices for a motor are necessary to protect the motor, its associated controls, and the branch-circuit conductors against heat damage due to excessive motor currents. High currents may be caused by the motor being overloaded for a considerable period of time, by consistently low or unbalanced line voltage, by single-phasing of a polyphase motor, or by the motor stalling or failing to start.

Damage may occur more quickly to a hermetic motor which stalls or fails to start than to a conventional open-type motor. Due to the presence of the cool refrigerant atmosphere under normal conditions, a hermetic motor is permitted to operate at a rated current that is closer to the locked-rotor current than is the same rated current of an open-type motor of the same nominal horsepower rating. The curves of Fig. 440-10 show the typical relation between locked-rotor and full-loaded currents of small open-type and hermetic motors. Because a hermetic motor operates within the refrigerant atmosphere, it is constantly cooled by that atmosphere. As a result, a given size of motor may be operated at a higher current than it could be if it were used as an open, general-purpose motor without the refrigerant cycle to remove heat from the windings. In effect, a hermetic motor is operated overloaded because the cooling cycle prevents overheating. For instance, a 5-hp open motor can be loaded as if it were a 7½-hp motor when it is cooled by the refrigerant. The full-load operating current of such a motor is higher than the normal current drawn by a 5-hp load and is, therefore, closer to the value of locked-rotor current, which is the same no matter how the motor is used.

When the rotor of a hermetic motor is slowed down because of overload or is at a standstill, there is not sufficient circulation of the refrigerant to carry away the heat; and heat builds up in the windings. Special quick-acting thermal and hydraulic-magnetic devices have been developed to reduce the time required to disconnect the hermetic motor from the line before damage occurs when an overload condition develops.

Room air conditioners and packaged unit compressors are normally required to incorporate running overload protection, which will restrict the heat rise to definite maximum safe temperatures in case of locked-rotor conditions. Room conditioners normally use inherent protectors built into the compressor housing, which respond to the temperature of the housing. Larger units also often use inherent protection in addition to quick-acting overload heaters installed in the motor starter, which respond only to current. These protective methods are covered in paragraphs **(2)** and **(4)** of 440.52(A).

The electrical installer will normally be concerned with the running overcurrent protection of a hermetic motor only when it becomes necessary to replace

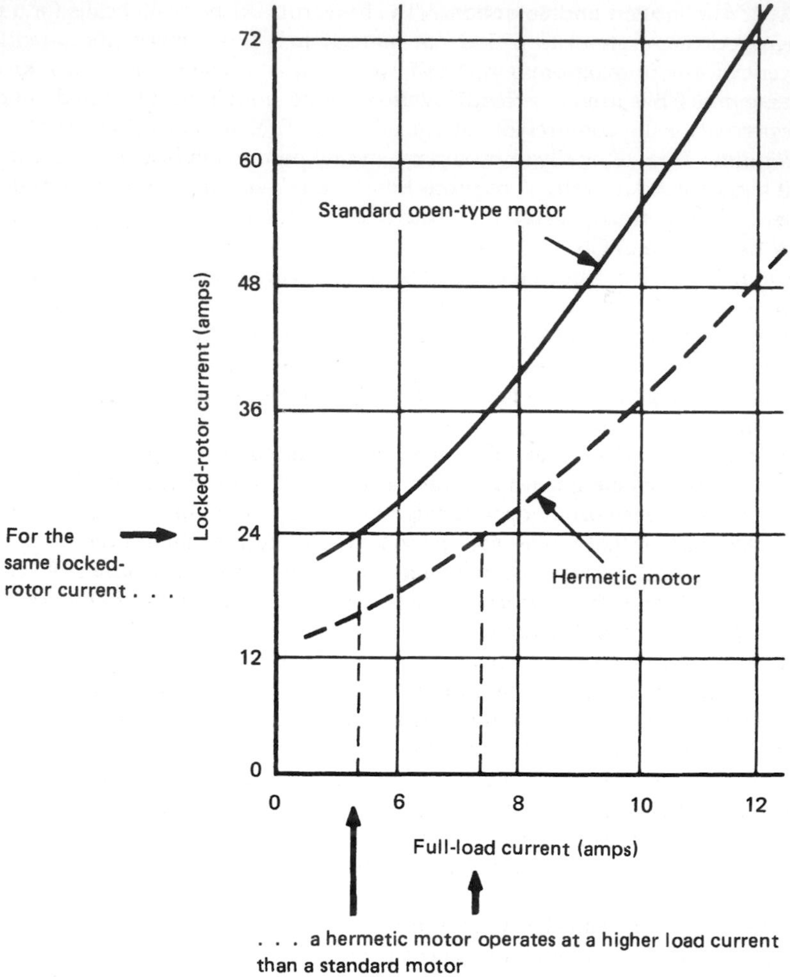
72
60
48
36
24
12
0
Locked-rotor current (amps)
Standard open-type motor
Hermetic motor
For the same locked-rotor current . . .
0
6
8
10
12
Full-load current (amps)
. . . a hermetic motor operates at a higher load current than a standard motor

Fig. 440-10. Hermetic motors operate at full-load currents closer to locked-rotor currents. (Sec. 440.52.)

the existing devices supplied with the equipment. For this purpose, compressor manufacturers' warranties explicitly specify catalog numbers of replacements that are to be used to ensure proper operation of the equipment.

440.60. General. These rules on room air-conditioning units recognize that such units are basically appliances, are low-capacity electrical loads, and may be supplied either by an individual branch circuit to a unit conditioner or by connection to a branch circuit that also supplies lighting and/or other appliances. For all Code discussion purposes, an air-conditioning unit of the window, console, or through-the-wall type is classified as a *fixed appliance*—which is described in Art. 100 as "fastened or otherwise secured at a specific location." Such an appliance may be cord-connected or it may be fixed-wired (so-called permanently connected). Any such units rated over 250 V or 3-phase must be directly connected and the provisions of this part of Art. 440 do not apply. There have been numerous attempts over the years to bring 277 V units within this part of the article, but they never succeeded.

210.23 in Art. 210 on "Branch Circuits" must also be applied in cases where a unit room air conditioner is connected to a branch circuit supplying lighting or other appliance load.

When a unit air conditioner is connected to a circuit supplying lighting and/or one or more appliances that are not motor loads, the rules of Art. 210 (and 406) must be observed:

1. For plug connection of the A/C unit, 406.4(A) says that receptacles installed on 15- and 20-A branch circuits must be of the grounding type and must have their grounding terminals effectively connected to a grounding conductor or grounded raceway or metal cable armor.
2. On 15- and 20-A branch circuits, the total rating of a unit air conditioner ("utilization equipment fastened in place") must not exceed 50 percent of the branch-circuit rating when lighting units or portable appliances are also supplied [210.23(A)]. It was on the basis of that rule that the 7½-A air conditioner was developed. Being 50 percent of a 15-A branch circuit, such units are acceptable for connection to a receptacle on a 15- or 20-A circuit that supplies lighting and receptacle outlets.
3. A branch circuit larger than 20 A may *not* be used to supply a unit conditioner plus a lighting load. Circuits rated 25, 30, 40, or 50 A may be used to supply fixed lighting or appliances—but not both types of loads.

440.61. Grounding. Air-conditioner units that are connected by permanent wiring must be grounded in accordance with the basic rules of 250.110 covering equipment that is "fastened in place or connected by permanent wiring." 250.114 covers grounding of cord-and-plug-connected air conditioners by means of an equipment grounding conductor run within the supply cord for each such unit.

440.62. Branch-Circuit Requirements. Even though a room air conditioner contains more than one motor (usually the hermetic compressor motor and the fan motor), this rule notes that for a cord-and-plug-connected air conditioner the entire unit assembly may be treated as a single-motor load under the conditions given. The nameplate marking of a room air conditioner shall be used in determining the branch-circuit requirements, and each unit shall be considered as a single motor unless the nameplate is otherwise marked. If the nameplate is marked to indicate two or more motors, 430.53 and 440.22(B)(1) must be satisfied, covering the use of several motors on one branch circuit.

Examples of the rule of part **(B)** are shown in Fig. 440-11. The total marked rating of any cord-and-plug-connected air-conditioning unit must *not* exceed 80 percent of the rating of a branch circuit that does not supply lighting units or other appliances, for units rated up to 40 A, 250 V, single phase.

As noted under 440.60, 210.2 seems to say that only Art. 440 and not Art. 210 applies when the circuit supplies only a motor-operated load. But, since Arts. 430 and 440 do not *rate* branch circuits—either on the basis of the size of the short-circuit

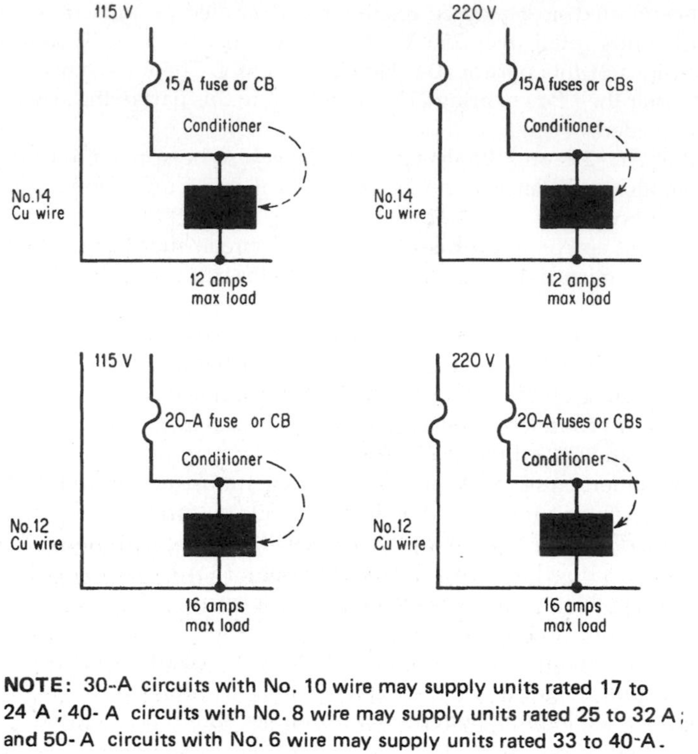
115 V
15 A fuse or CB
Conditioner
No.14
Cu wire
12 amps
max load
220 V
15A fuses or CBs
Conditioner
No.14
Cu wire
12 amps
max load
115 V
20-A fuse or CB
Conditioner
No.12
Cu wire
16 amps
max load
220 V
20-A fuses or CBs
Conditioner
No.12
Cu wire
16 amps
max load
NOTE: 30-A circuits with No. 10 wire may supply units rated 17 to 24 A; 40-A circuits with No. 8 wire may supply units rated 25 to 32 A; and 50-A circuits with No. 6 wire may supply units rated 33 to 40-A.

Fig. 440-11. Room air conditioners must not load an individual branch circuit over 80 percent of rating. (Sec. 440.62.)

protective device or the size of the conductors—a question arises about the meaning of the phrase "80 percent of the rating of a branch circuit." Does that mean 80 percent of the rating of the fuse or CB? The answer is: It means 80 percent of the rating of the protective device, which rating is not more than the amp rating of the circuit wire. The circuit as described here is taken to be a circuit with "rating" as given in Art. 210 and covered by part **(4)** of 440.62(A).

As part **(C)** of this section notes, the total marked rating of air-conditioning equipment must *not* exceed 50 percent of the rating of a branch circuit that *also* supplies lighting or other appliances. (See Fig. 440-12.) From the rule, we can see that the Code permits air-conditioning units to be plugged into existing circuits that supply lighting loads or other appliances. By the provisions of this section, such a conditioner must not draw more than 7½-A full load (nameplate rating) when connected to a 15-A circuit and not more than 10 A when connected to a 20-A circuit. This rule effectively makes the 50 percent limitation for fixed loads on general purpose branch circuits in 210.23(A) applicable to a load that is cord-and-plug-connected.

A problem exists in connecting two or more conditioners to the same circuit. Compressor and fan motors and their controls, when installed in the same

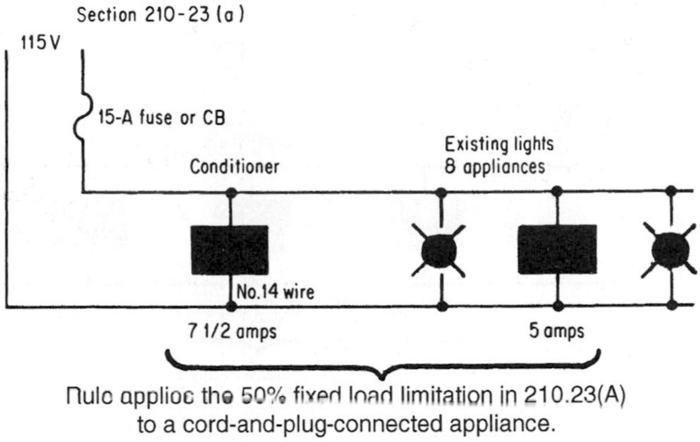
Section 210-23 (a)
115 V
15-A fuse or CB
Conditioner
Existing lights
8 appliances
No.14 wire
7 1/2 amps
5 amps
Rule applies the 50% fixed load limitation in 210.23(A)
to a cord-and-plug-connected appliance.

Fig. 440-12. Room air conditioner must not exceed 50 percent of circuit rating if other loads are supplied. (Sec. 440.62.)

enclosure and fed by one circuit, are approved by UL for group installation when tested as a unit appliance. However, an air conditioner's component parts carry no general group-fusing approval, which would permit the several separate conditioners to operate on the same circuit in accordance with 430.53(C). To connect more than one conditioner to the same branch circuit, the provisions of either 430.53(A) or 430.53(B) must be fulfilled, treating each cord-connected conditioner as a single-motor load. Note that 440.62(A) does classify a room air conditioner as a single motor load (that permission is conditional, but the terms are met in most cases).

According to 430.53(A), which applies only to motors rated not over 6 A, two 115-V, 6-A conditioners could be used on a 15-A circuit; three 5-A conditioners could be used on a 20-A circuit which supplies no other load; and two 220-V, 6-A units could be operated on a 15-A circuit, as shown in Fig. 440-13. But it could be argued that the maximum load in any such application may be calculated at 125 percent times the current of the largest air conditioner plus the sum of load currents of the additional air conditioners, with that total current being permitted right up to the rating of the circuit.

However, most conditioners sold today exceed 6-A full-load current. As a result, the application of two or more units as permitted by 430.53(A) is limited. But 430.53(B) does offer considerable opportunity for using more than one air conditioner on a single circuit. Figure 440-14 shows two examples of such application, which can be used if the branch-circuit protective device will not open under the most severe normal conditions that might be encountered. Although that usage is a complex connection among several Code rules and requires clearance with inspection authorities, it can provide very substantial economies.

Many local codes avoid the complications of connecting conditioners to existing circuits and connecting more than one conditioner to the same circuit by requiring a separate branch circuit for each conditioner. Multiple installations involving many room conditioners such as are frequently encountered in hotels, offices, and

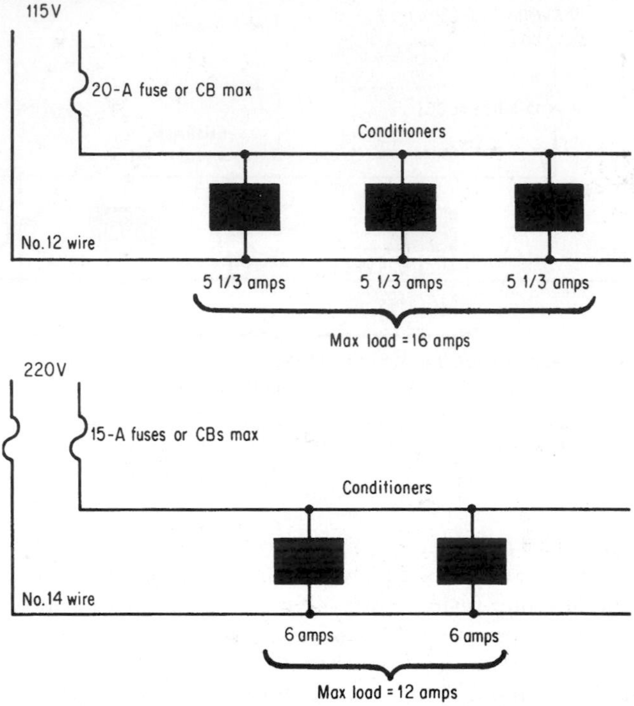
115V
20-A fuse or CB max
Conditioners
No.12 wire
5 1/3 amps
5 1/3 amps
5 1/3 amps
Max load = 16 amps
220V
15-A fuses or CBs max
Conditioners
No.14 wire
6 amps
6 amps
Max load = 12 amps

Fig. 440-13. Rules limit use of two or more room conditioners on single circuit. (Sec. 440.62.)

so forth, require careful planning to meet Code requirements and yet minimize expensive branch-circuit lengths.

Watch Out!

The **NEC** refers to motor-operated appliances and/or to room air conditioners in Arts. 210, 422, and 440. Great care must be exercised in correlating the various Code rules in these different articles to ensure effective compliance with the letter and spirit of Code meaning. There is much crossover in terminology and references, making it difficult to tell whether a room air conditioner should be treated as an appliance circuit load or a motor load. However, a step-by-step approach to the problem, which keeps in mind the intent of these provisions, can resolve confusing points. Since the manufacturer is required to supply the motor-running overcurrent protection, no problems should arise concerning these devices. For larger units connected permanently to the distribution system, these can be treated directly as hermetic motor loads, using the provisions of Art. 440.

It may be assumed that a window or through-the-wall unit will operate satisfactorily on a standard fuse of the same rating as its attachment plug cap if there is no marking to the contrary on the unit. In any event, a time-delay fuse of the same or smaller rating could be substituted. If CBs are used for branch-circuit protection,

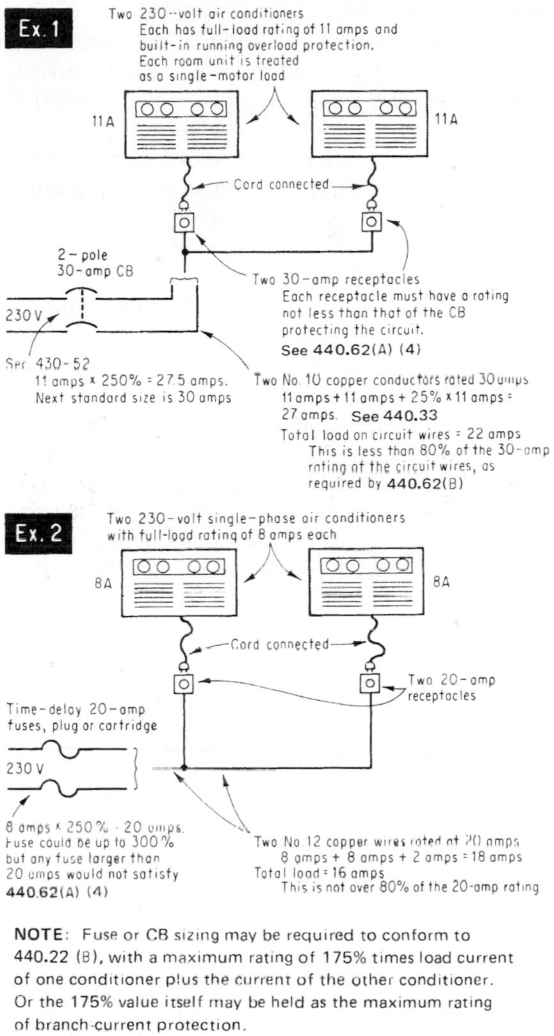
Ex. 1
Two 230-volt air conditioners
Each has full-load rating of 11 amps and built-in running overload protection.
Each room unit is treated as a single-motor load
11A
11A
Cord connected
2-pole 30-amp CB
230 V
Two 30-amp receptacles
Each receptacle must have a rating not less than that of the CB protecting the circuit.
See 440.62(A) (4)
Sec. 430-52
11 amps × 250% = 27.5 amps.
Next standard size is 30 amps
Two No. 10 copper conductors rated 30 amps
11 amps + 11 amps + 25% × 11 amps = 27 amps. See 440.33
Total load on circuit wires = 22 amps
This is less than 80% of the 30-amp rating of the circuit wires, as required by 440.62(B)
Ex. 2
Two 230-volt single-phase air conditioners with full-load rating of 8 amps each
8A
8A
Cord connected
Two 20-amp receptacles
Time-delay 20-amp fuses, plug or cartridge
230 V
8 amps × 250% = 20 amps.
Fuse could be up to 300% but any fuse larger than 20 amps would not satisfy 440.62(A) (4)
Two No. 12 copper wires rated at 20 amps
8 amps + 8 amps + 2 amps = 18 amps
Total load = 16 amps
This is not over 80% of the 20-amp rating
NOTE: Fuse or CB sizing may be required to conform to 440.22 (B), with a maximum rating of 175% times load current of one conditioner plus the current of the other conditioner. Or the 175% value itself may be held as the maximum rating of branch-current protection.

Fig. 440-14. These hookups have been accepted as conforming to rules of Arts. 440 and 430. (Sec. 440.62.) Generally speaking, if the branch-circuit and receptacle ratings match up to the rating of the plug on the cord as supplied by the manufacturer, the units can be expected to function safely. That would likely be the case in Example 2, but questionable in the first example.

a 15-A breaker will normally hold the starting current if a standard 15-A fuse will, because such breakers have inherent time delay. If the unit is marked to require a 15-A time-delay fuse and a 15-A breaker will not hold the starting current, few inspectors will object to the use of a 20-A breaker, because Art. 430 permits such a procedure for motor loads.

Normally, starting problems are not severe with these units, because the low inertia of present-day motor-compressor combinations permits them to reach full speed within a few cycles. Such a rapid drop in starting current is usually well within the time permitted by the trip or rupture characteristics of the breaker or fuse.

Similarly, the question of wire size may be resolved by application of Arts. 210, 422, or 440. Rarely do room conditioners even as large as 2 tons take more than 13-A running current; hence No. 12 copper or No. 10 aluminum conductors are more than sufficient. In addition, many local codes prohibit use of conductors smaller than No. 12. In localities where No. 14 wire may be used, provisions of 440.62(B), restricting the loading to 80 percent of the circuit rating, must determine the wire size, where "rating" is interpreted as referring to the conductor carrying capacity. If Art. 440 is used to determine the wire size, the 125 percent requirement of 440.32 gives the same result.

Figure 440-15 shows one feeder of an installation involving many room conditioners, which practically eliminates branch-circuit wiring and will serve to illustrate the complications of circuit calculations for a multiple-unit installation. Total running current of each unit is 12 A as shown; hence, No. 14 copper wire could be used for branch-circuit conductors, protected by a 15-A fuse—either standard or time-delay. However, if the appropriate conductors of a 4-wire, 3-phase feeder

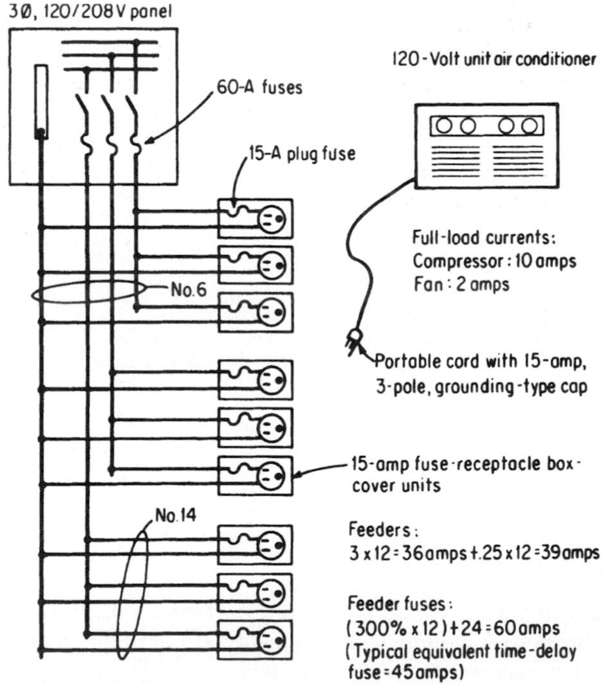
3Ø, 120/208V panel
60-A fuses
15-A plug fuse
120-Volt unit air conditioner
Full-load currents:
Compressor: 10 amps
Fan: 2 amps
No.6
Portable cord with 15-amp, 3-pole, grounding-type cap
15-amp fuse-receptacle box-cover units
No.14
Feeders:
3 x 12 = 36 amps + .25 x 12 = 39 amps
Feeder fuses:
(300% x 12) + 24 = 60 amps
(Typical equivalent time-delay fuse = 45 amps)

Fig. 440-15. This type of circuiting was used for air conditioners in a hotel modernization project. (Sec. 440.62.)

were routed to the location of each conditioner and a combination fuseholder and receptacle was installed as shown, the only existing branch-circuit conductors would be the jumpers between the feeder, the fuseholder, and the receptacles. These jumpers, then, could be No. 14 wire. Assuming that the fuse-receptacle unit is mounted directly on or in close proximity with the junction box in which the tap to the feeder is made, the No. 14 wire is justified from the fuse to the feeder since it is not over 10 ft (3.0 m) long and is sufficient for the load supplied [240.21(B)(1)]. Since all three motors might start simultaneously, to be conservative the total unit current is used to compute feeder conductor size and protection: 125 percent times 12 plus 24 is 39 A, permitting No. 8 conductors. However, this is practically the limit of the circuit's capacity; there is no provision for overload, and voltage drop is very likely to be a factor at the end of the feeder. Therefore No. 6 conductors should be used.

Feeder protection is calculated on the basis of 300 percent times 12 plus 24, or 60-A fuses. Substitution of time-delay fuses for this 39-A feeder load would likely permit 45-A fuses.

Important: The rules of 440.62 apply only to cord-and-plug-connected room air conditioners. A unit room air conditioner that has a *fixed* (not cord-and-plug) connection to its supply must be treated as a group of several individual motors and protected in accordance with 430.53 and 440.22(B), covering several motors on one branch. If the fixed units were wired to individual branch-circuit protection similar to that presented in Fig. 440-15, then the wires originating in the panel are feeders and the calculation follows 430.62.

440.63. Disconnecting Means. A disconnect is required for every unit air conditioner. An attachment plug and receptacle or a separable connector may serve as the disconnecting means (Fig. 440-16). However, if the manual controls are not both readily accessible and within 1.8 m (6 ft) of the floor, then an additional disconnecting means must be provided in the circuit that is readily accessible and within sight of the air conditioner.

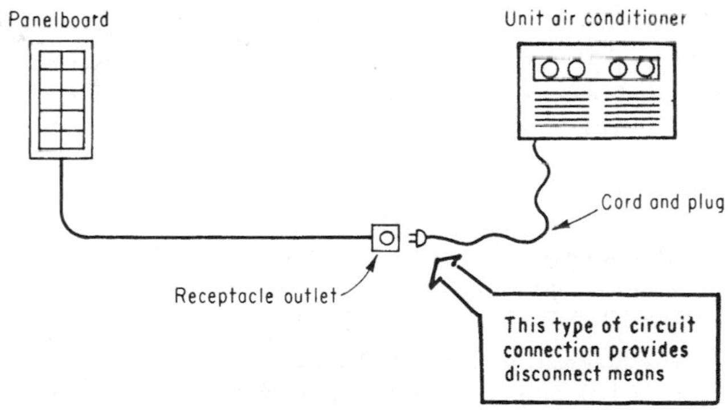
Panelboard
Unit air conditioner
Cord and plug
Receptacle outlet
This type of circuit connection provides disconnect means

Fig. 440-16. Plug-and-receptacle serves as required disconnect means. (Sec. 440.63.)

If a fixed connection is made to an A/C unit from the branch-circuit wiring system (i.e., not a plug-in connection to a receptacle), consideration must be given to a means of disconnect, as required in 422.33 and 422.34 for appliances:

- For unit air conditioners in any type of occupancy, the branch-circuit switch or CB may, where readily accessible to the user of the appliance, serve as the disconnecting means. Figure 440-17 shows this, but the switch or CB is permitted to be out of sight by the exception to 422.31(C) when the A/C unit has an internal OFF switch—which all units do have.

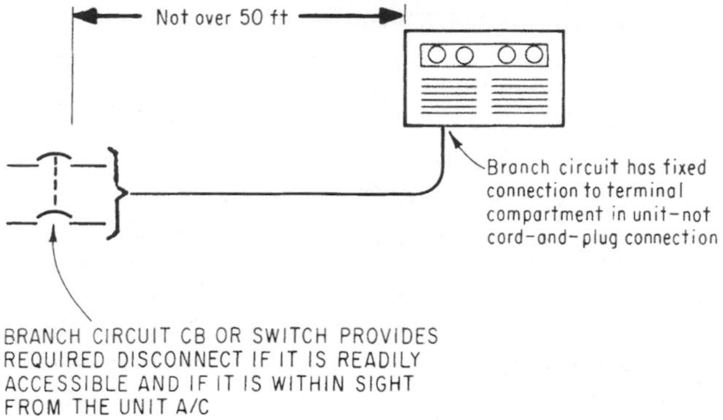
Not over 50 ft
Branch circuit has fixed connection to terminal compartment in unit–not cord-and-plug connection
BRANCH CIRCUIT CB OR SWITCH PROVIDES REQUIRED DISCONNECT IF IT IS READILY ACCESSIBLE AND IF IT IS WITHIN SIGHT FROM THE UNIT A/C

Fig. 440-17. Branch-circuit CB or switch may serve as disconnect. (Sec. 440.63.)

Because air conditioners have unit switches within them, the disconnect provisions of 422.34 may be applied. The internal unit switch with a marked OFF position that opens all ungrounded conductors may serve as the disconnect and is considered within sight as required by 422.34 in any case where there is another disconnect means as follows:

- In multifamily (more than two) dwellings, the other disconnect means must be within the apartment where the conditioner is installed or on the same floor as the apartment.
- In two-family dwellings, the other disconnect may be outside the apartment in which the appliance is installed. It may be the feeder disconnect, or the service disconnect if the two-family dwelling is equipped with multiple service disconnects such that opening the disconnect amounts to the opening of an "individual switch or circuit breaker for the dwelling unit."
- In single-family dwellings, the service disconnect may serve as the other disconnect means—whether the branch circuit to the conditioner is fed from plug fuses or from a breaker or switch (Fig. 440-18).

440.65. Protection Devices. This rule requires a factory installed LCDI or AFCI protective device in the cord and within 300 mm (12 in.) of the plug cap for all single-phase, cord-and-plug-connected room A/Cs. The 2017 **NEC** recognized a "heat detecting circuit interrupter" (HDCI) that is designed to function as a leakage current detector (LCDI), plus detect if the compressor is overheating. This is not mandatory, but it is a permitted option and joins the other two devices as a third protection option for this equipment.

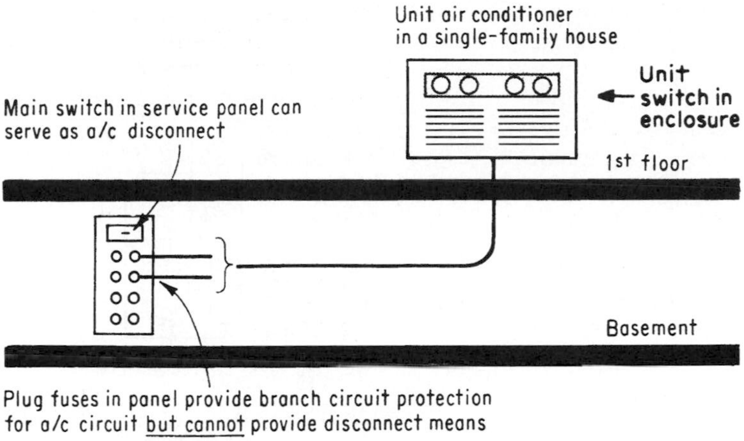
Unit air conditioner
in a single-family house
Unit
switch in
enclosure
Main switch in service panel can
serve as a/c disconnect
1st floor
Basement
Plug fuses in panel provide branch circuit protection
for a/c circuit but cannot provide disconnect means

Fig. 440-18. Service disconnect may be the "other disconnect" for A/C unit in a single-family dwelling. (Sec. 440.63.)

ARTICLE 445. GENERATORS

445.11. Marking. Generators must be marked with the manufacturer's name, the frequency and number of phases for ac units, the output rating in kilowatts or kilovolt-amps, the corresponding normal volts and amperes, the rated RPM, and the rated ambient temperature or temperature rise, along with the power factor. In addition, generators over 15 kW must have the reactances (transient, synchronous, and subtransient, along with zero sequence), the power rating category, and the insulation class marked as well. Inverter-based generators need the maximum short-circuit current provided, in place of all the reactances except the zero sequence number. The nameplate must also stipulate how the generator is protected against overload, whether by fuse or circuit breaker, or by a protective relay, or by its inherent design.

The 2014 **NEC** has added a critical marking requirement relative to whether or not the generator, when connected to the premises wiring system, must be considered to be the source of a separately derived system. The marking must state whether the neutral is bonded to the frame. And if that bonding is modified, then the resulting bonding status must be reflected in an additional corrective marking.

445.12. Overcurrent Protection. Alternating-current generators can be so designed that on excessive overload the voltage falls off sufficiently to limit the current and power output to values that will not damage the generator during a short period of time. Whether automatic overcurrent protection of a generator should be omitted in any particular case is a question that can best be answered by the manufacturer of the generator. It is a common practice to operate an exciter without overcurrent protection, rather than risk the shutdown of the main generator due to accidental opening of the exciter fuse or CB. In the case of constant-voltage generators, the **NEC** leaves this to the manufacturer, providing various options including "inherent design" as well as conventional overcurrent devices.

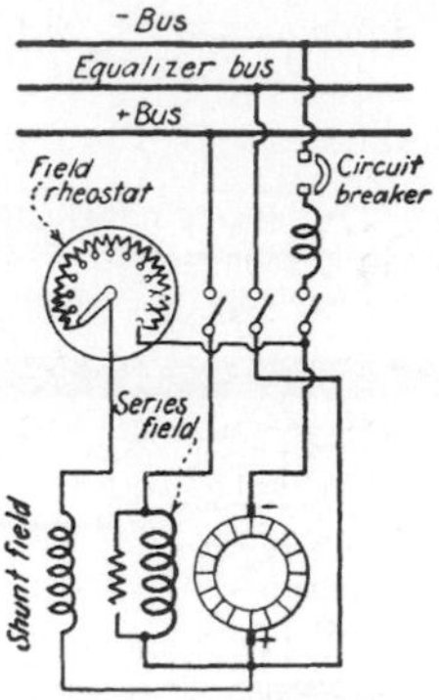
- Bus
Equalizer bus
+ Bus
Field rheostat
Circuit breaker
Series field
Shunt field

Fig. 445-1. With this connection, a single-pole CB can protect a 2-wire dc generator. (Sec. 445.12.)

Figure 445-1 shows the connections of a 2-wire dc generator with a single-pole protective device. If the machine is operated in multiple with one or more other generators, and so has an equalizer lead connected to the positive terminal, the current may divide at the positive terminal, part passing through the series field and positive lead and part passing through the equalizer lead. The entire current generated passes through the negative lead; therefore the fuse or CB, or at least the operating coil of a CB, must be placed in the negative lead. The protective device should not open the shunt-field circuit, because if this circuit were opened with the field at full strength, a very high voltage would be induced, which might break down the insulation of the field winding.

Paragraph **(C)** is intended to apply particularly to generators used in electrolytic work. Where such a generator forms part of a motor-generator set, no fuse or CB is necessary in the generator leads if the motor-running protective device will open when the generator delivers 150 percent of its rated full-load current.

In paragraph **(D),** use of a balancer set to obtain a 3-wire system from a 2-wire main generator is covered, as shown in Fig. 445-2. Each of the two generators used

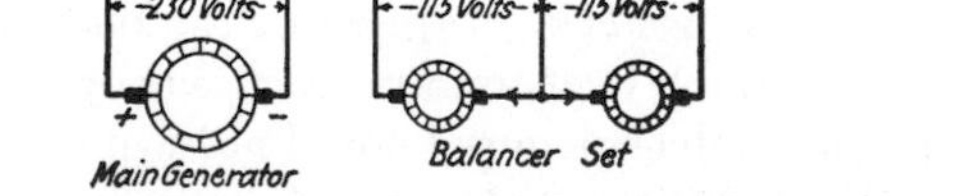
75 Amp.
+ Bus
100 Amp.
Neutral Bus
50 Amp.
Load
75 Amp.
- Bus
50 Amp.
Circuit-Breaker
A
B
A
25 Amp.
50 Amp.
25 Amp.
230 Volts
115 Volts
115 Volts
Main Generator
Balancer Set

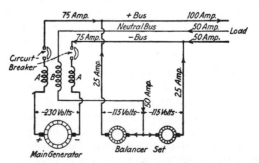

Fig. 445-2. A balancer set supplies the unbalanced neutral current of a 3-wire system, with each generator carrying 25 of the 50-A unbalance. (Sec. 445.12.)

as a balancer set carries approximately one-half the unbalanced load; hence these two machines are always much smaller than the main generator. In case of an excessive unbalance of the load, the balancer set might be overloaded while there is no overload on the main generator. This condition may be guarded against by installing a double-pole CB, with one pole connected in each lead of the main generator and with the operating coil properly designed to be connected in the neutral of the 3-wire system. In Fig. 445-2, the CB is arranged so as to be operated by either one of the coils *A* in the leads from the main generator or by coil *B* in the neutral lead from the balancer set.

445.13. Ampacity of Conductors. The first three sentences clarify sizing of circuit conductors connecting a generator to the control and protective device(s) it serves:

1. The ampacity of the conductors between the generator terminals and the first panelboard, switchboard, or other comparable equipment that contains overcurrent protective devices must not be less than 115 percent of the generator nameplate current rating. Note that there is no requirement that the conductors arrive at the terminals of a single overcurrent device of a particular rating, or that the sum of the ratings of the overcurrent devices at the terminating location not exceed some quantity. The tap rules in 240.21(G) recognize whatever protection is provided under 445.12, and allow the generator output conductors indefinite length without additional protection as long as they meet the 115 percent ampacity criterion.
2. The neutral of the generator feeder may have its size reduced from the minimum capacity required for the phase legs. As with any feeder or service circuit, the neutral has to have only enough ampacity for the unbalanced load it will handle—as covered by 220.61.
3. Where a generator is operating as a standalone power source on a system employing a grounded neutral, and that neutral is not connected to ground and the equipment grounding system at the generator housing, the generator neutral must be sized to carry ground-fault currents in addition to its normal function as feeder conductor sized in accordance with 220.61. Depending on the amount of line-to-neutral load, the fault-return capability requirements in 250.24(C) for nonseparately derived systems as shown in Fig. 445-3, or those for separately derived systems as covered in 250.30(A)(3), may require a larger neutral conductor. The third sentence of 445.13 attempts to convey this concept, but the Art. 250 reference is in part erroneous because it only points to separately derived system rules within Art. 250. For nonseparately derived systems the relevant reference is 250.24(C) as noted. Fortunately the sizing rules in both locations are essentially identical. Part (B) allows the load terminals of a generator overcurrent protective device, or a current transformer operating with an overcurrent relay, to serve as a feeder for code purposes. With this established, the installer therefore has a choice of feeder tap rules under 240.21(B) to apply when wiring the load conductors from the generator. This procedure is not permitted on smaller generators rated 15 kW or less with unavailable field wiring connections.

445.14. Protection of Live Parts. As a general rule, no generator should be "accessible to unqualified persons." If necessary to place a generator operating at over 50 V to ground in a location where it is so exposed, the commutator or

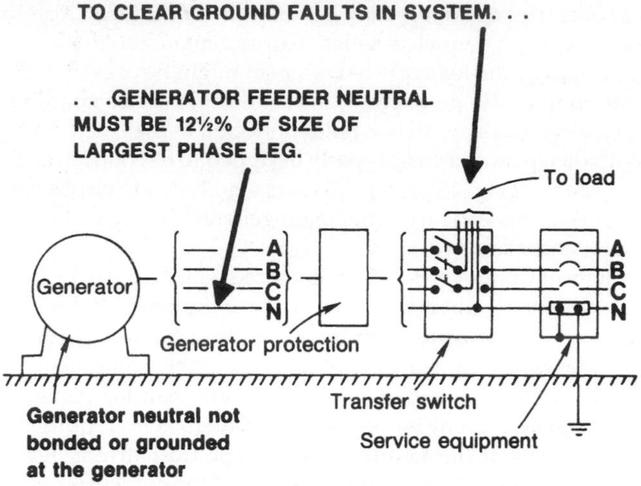
TO CLEAR GROUND FAULTS IN SYSTEM. . .
. . .GENERATOR FEEDER NEUTRAL MUST BE 12½% OF SIZE OF LARGEST PHASE LEG.
To load
Generator
A
B
C
N
Generator protection
Transfer switch
Service equipment
Generator neutral not bonded or grounded at the generator

Fig. 445-3. Neutral must have adequate capacity for generator that is not a "separately derived" system source (i.e., does not have its neutral bonded and grounded). (Sec. 445.13.)

collector rings, brushes, and any exposed terminals should be provided with guards that will prevent any accidental contact with these live parts.

445.18. Disconnecting Means and Shutdown of Prime Movers. Lockable disconnect(s) are required for generators other than cord-and-plug-connected portable generators, unless the generator's driving machine can be readily shut down and it is not running in parallel with another source of voltage, such as another generator. The phrasing "one or more" reflects the useful idea that an excellent design option is to establish multiple disconnects right at the generator so multiple systems can originate at the generator (such as emergency, legally required standby, and optional standby) without issues of system separation at the occupancy served.

Part (B) addresses the prime mover, which must have provisions to be shut down. The applicable means must be capable of two things, first, disabling all starting control circuits so as to make the prime mover incapable of starting. It must also initiate a shutdown, one that requires mechanical intervention to be reset. A prime mover shutdown mechanism that satisfies these rules is permitted to constitute the generator disconnecting means if it can be locked in the position that so disables the prime mover and prevents its restart. Generators over 15 kW must have an additional "requirement" (an obvious mistake, which presumably means a redundant mechanism) to disable the prime mover, located outside of the generator room or outside the generator enclosure, that also meets the full shutdown requirements of this part.

Part (C) addresses generators operating in parallel. The disconnecting requirements in part (A) must be capable of separating the generator output terminals from the paralleling equipment, and it need not be located on the generator.

445.20 Ground-Fault Circuit-Interrupter Protection for Receptacles on 15-kW or Smaller Portable Generators. This rule addresses the GFCI status of 125-volt 15- and 20-ampere convenience receptacles furnished as part of smaller portable generators covered here. That status depends in part on whether the generator's neutral connection is bonded to the generator frame, and that can be established by test, or preferably, by the marking, if present, that is required by 445.11. If the neutral is floating (unbonded), then two things are true. First, the 125-volt circuits will be functionally ungrounded and incapable of meeting the construction site requirements in 590.6(A)(3) because the GFCI receptacles will not trip reliably. Secondly, such a generator will be appreciated in terms of providing backup power for a house or other building because it will qualify to be wired as a non-separately derived system with a far less expensive transfer switch (need not transfer the neutral) and much simpler wiring connections.

If the neutral is bonded, then there will be a ground-fault return path over the equipment grounding conductor within any cord plugged into a generator receptacle and the GFCI functionality will be completely operational. The generator will fully qualify for use at construction sites. The unbonded (floating) option comes with an exception that offers some important flexibility. It recognizes a method of interlocking the receptacles on the generator so that when the generator is providing standby power to a house, for example, through the 125.250-volt 30- or 50-ampere 3-pole 4-wire grounding receptacle, the 125-volt GFCI receptacles are unpowered. This prevents those receptacles from being used under a circumstance where they would not be actually functional. There are also small generators manufactured with neutral bonding connections that are easily made or unmade in the field, together with directions that fully discuss the reasons and appropriate times for either connection.

This new rule must be understood in its historical context. UL released a new product standard for "Engine Generators for Portable Use" (UL2201, Guide Card designation FTCN) in March 2009. The Guide Card information includes a condition that "[The generators] are provided only with receptacle outlets for the ac output circuits." The standard also requires that the generator neutral be bonded to its frame. This has the effect of requiring portable generators used to power grounded systems, such as dwellings and small commercial enterprises, to be wired as separately derived systems through transfer equipment with a pole in the neutral conductor along with the implicit requirement to separate equipment grounding conductors and neutral conductors in what had been a service panel. The portable generator industry has generally opposed the attempted abolition of floating neutrals and has voted with its feet. In the entire history of UL 2201 there is yet, as of this writing, to be a single listee in the category.

The generator industry would prefer to be able to make unbonded generators generally on the basis that a functionally ungrounded 125-volt circuit is not a shock hazard. There is some technical substantiation for that point of view, but at this point it is very unlikely to persuade the panel, which voted overwhelmingly (only two negative votes) for the compromise language. Given the contentiousness of this issue, that was an impressive result. On the other hand, the fact that the entire generator industry has refused to participate in the UL listing process for 7 years is also impressive and shows real discipline. Time will tell whether the present wording is a workable compromise.

ARTICLE 450. TRANSFORMERS AND TRANSFORMER VAULTS (INCLUDING SECONDARY TIES)

450.1. Scope. The exceptions indicate those transformer applications that are not subject to the rules of Art. 450. The exceptions shown in Fig. 450-1 are as follows:

Exception No. 2 excludes any dry-type transformer that is a component part of manufactured equipment, provided that the transformer complies with the

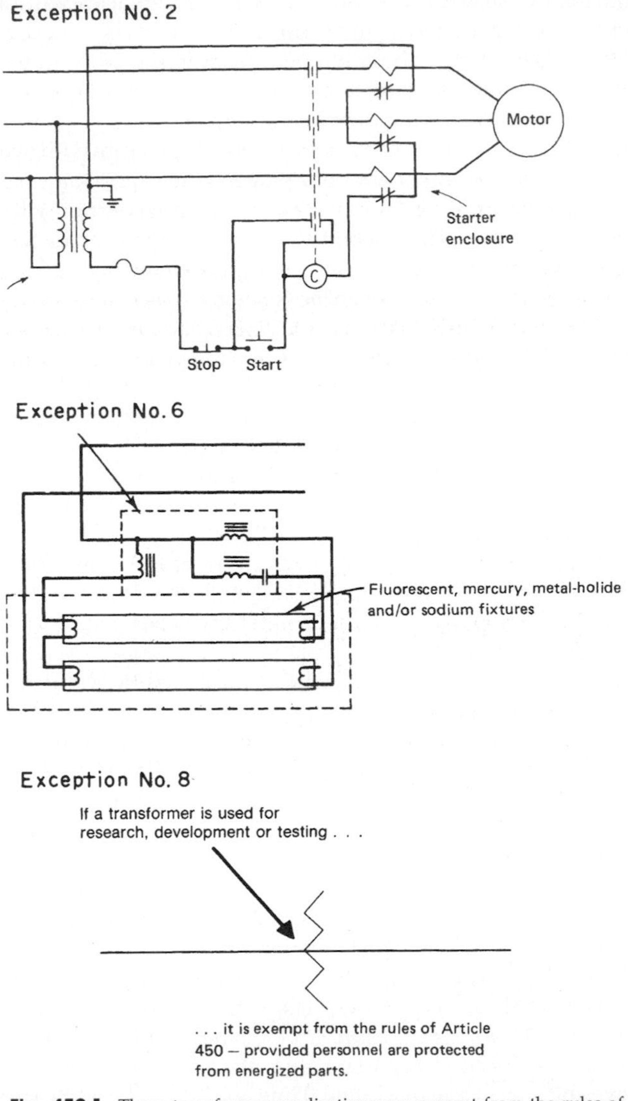
Exception No. 2
Motor
Starter enclosure
C
Stop
Start
Exception No. 6
Fluorescent, mercury, metal-holide and/or sodium fixtures
Exception No. 8
If a transformer is used for research, development or testing . . .
. . . it is exempt from the rules of Article 450 – provided personnel are protected from energized parts.

Fig. 450-1. These transformer applications are exempt from the rules of Art. 450. (Sec. 450.1.)

requirements for such equipment. Those requirements include UL standards on the construction of the particular equipment. This exclusion applies, for instance, to control transformers within a motor starter or within a motor control center. However, although such transformers do not have to be protected in accordance with 450.3(B), such control transformer circuits must have their control conductors protected as described under 430.72(B). But a separate control transformer—one that is external to other equipment and is not an integral part of any other piece of equipment—must conform to the protection rules of 450.3 and other rules in Art. 450.

Exception No. 6 points out that ballasts for electric-discharge lighting (although they *are* transformers—either autotransformers or separate-winding, magnetically coupled types) are treated as lighting accessories rather than transformers.

Exception No. 8 notes that liquid-filled or dry-type transformers used for research, development, or testing are exempt from the requirements of Art. 450, provided that effective arrangements are made to safeguard any persons from contacting energized terminals or conductors. Again, in the interest of the unusual conditions that frequently prevail in industrial occupancy, this rule recognizes that transformers used for research, development, or testing are commonly under the sole control of entirely competent individuals and exempts such special applications from the normal rules that apply to general-purpose transformers used for distribution within buildings and for energy supply to utilization equipment, controls, signals, communications, and the like. (See Fig. 450-2.)

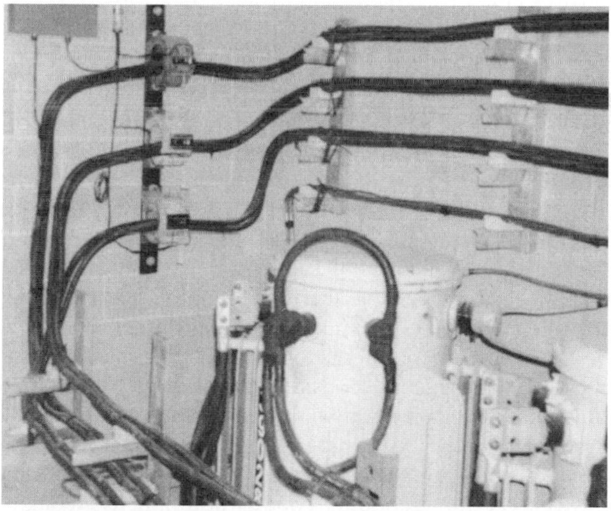

Fig. 450-2. Transformers that are set up in a laboratory to derive power for purposes of testing other equipment or powering an experiment are exempt from the rules of Art. 450, provided care is taken to protect personnel from any hazards due to exposed energized parts. (Sec. 450.1.)

UL listing The UL Guide Card information that covers this equipment carries the heading: "Transformers—Power and General Purpose, Dry Type." To satisfy NEC and OSHA regulations, as well as local code rules on acceptability of equipment, any transformers of the types and sizes covered by UL listing must be so listed. Use of an unlisted transformer of a type and size covered by UL listing would certainly be considered a violation of the spirit of NEC 110.3. UL listing covers *air-cooled* types rated up to 500 kVA for single-phase transformers and up to 1500 kVA for 3-phase units (all up to 1000-V rating).

450.2. Definitions. A *transformer* is an individual transformer, single or polyphase, identified by a single nameplate, unless otherwise indicated in this article. Three single-phase transformers connected for a 3-phase transformation must be taken as three transformers, not one. This definition helps to clarify the contents of some of the rules of Art. 450.

450.3. Overcurrent Protection. This section covers overcurrent protection in great detail, and other Code rules (240.21, 240.40, and 408.16 in particular) usually get involved in transformer applications. Both protection tables [450.3(A) and 450.3(B)] require primary protection in some form. This protection will now need to be considered in the context of 450.14, which as of the 2011 NEC imposes a disconnection requirement on most transformers. Both tables also cover secondary protection, which is required in many but not all cases. Where it is required, there is an allowance for multiple overcurrent devices to serve as secondary protection, up to a maximum of six devices, which must be grouped. Refer to Notes 2 to both tables for details.

It should be understood that the overcurrent protection required by this section is for transformers *only*. Such overcurrent protection will not necessarily protect the primary or secondary conductors or equipment connected on the secondary side of the transformer. Using overcurrent protection to the maximum values permitted by these rules would require much larger conductors than the full-load current rating of the transformer (other than permitted in the 25-ft [7.5-m] tap rule in 240.21). Accordingly, to avoid using oversized conductors, overcurrent devices should be selected at about 110 to 125 percent of the transformer full-load current rating. And when using such smaller overcurrent protection, devices should be of the time-delay type (on the primary side) to compensate for inrush currents, which reach 8 to 10 times the full-load primary current of the transformer for about 0.1 s when energized initially.

In approaching a transformer installation it is best to use a one-line diagram, such as shown in the accompanying sketches. Then by applying the tap rules in 240.21, proper protection of the conductors and equipment, which are part of the system, will be achieved. See comments following 240.21.

240.4(F) is the only Code rule that considers properly sized primary overcurrent devices to protect the secondary conductors without secondary protection and no limit to the length of secondary conductors.

On 3- and 4-wire transformer secondaries, it is possible that an unbalanced load may greatly exceed the secondary conductor ampacity, which was selected assuming balanced conditions. Because of this, the NEC does not permit the protection of secondary conductors by overcurrent devices operating through a transformer from the primary of a transformer having a 3-wire or 4-wire secondary.

For other than 2-wire to 2-wire or 3-wire to 3-wire delta-delta transformers, protection of secondary conductors has to be provided completely separately from any primary-side protection. In designing transformer circuits, the rules of 450.3 can be coordinated with 240.21(C), which provides special rules for tap conductors used with transformers. This general procedure allows for simultaneous protection of both the transformer and the conductors connected on both sides of it. Refer to the extensive discussion in 240.21 in Chap. 2 of this book. Those rules need not be covered again here.

Part **(A)** of this section mandates compliance with Table 450.3(A). That table sets rules for overcurrent protection of any transformer (dry-type or liquid-filled) rated over 1000 V, and, as covered in Note 4 to Table 450.3(A), electronically actuated fuses that are adjustable to a "specific current" may also be rated at 300 percent. Protection may be provided either by a protective device of specified rating on the transformer primary or by a combination of protective devices of specified ratings on both the primary and secondary. Figure 450-3 shows the basic rules of such overcurrent protection. The fact that E-rated fuses used for high-voltage circuits are given melting times at 200 percent of their continuous-current rating explains why this Code rule used to set 150 percent of primary current as the maximum fuse rating but permits CBs up to 300 percent. In effect, the 150 percent for fuses times 2 becomes 300 percent—the maximum value allowed for a CB. Now such fuses may be rated up to 250 percent (instead of 150 percent) for transformers with 6 percent or less impedance. The 2014 **NEC** clarified that the next higher rating of medium-voltage circuit protection, which is not covered in 240.6(A), is taken as the next higher "commercially available rating."

- The basic requirement for "Any Location" in Table 450.3(A) says that *any* high-voltage transformer must have *both* primary and secondary protection based on Table 450.3(A) for maximum ratings of the primary and secondary fuses or circuit breakers.
- The rules under "supervised locations only" give two alternative ways of protecting high-voltage transformers where "conditions of maintenance and supervision assure that only qualified persons will monitor and service the transformer." The two alternatives are as follows:
    1. Primary protection only may be used, with fuses set at not over 250 percent of the primary current or circuit breakers set at not over 300 percent of primary current. And if that calculation results in a fuse or CB rating that does not correspond to a standard rating or setting, the next higher standard rating or setting may be used.
    2. Primary and secondary protection based on Table 450.3(A) is the alternative.

These rules resulted from concerted industry action to provide better transformer protection. As stated in the substantiation for 450.3(A), "It is felt that this approach will aid in reducing the number of transformer failures due to overload, as well as maintaining the flexibility of design and operation by industry and the more-complex commercial establishments."

Part **(B)** of this section covers all transformers—oil-filled, high-fire-point liquid-insulated, and dry-type—rated up to 1000 V. The step-by-step approach to such protection is as follows:

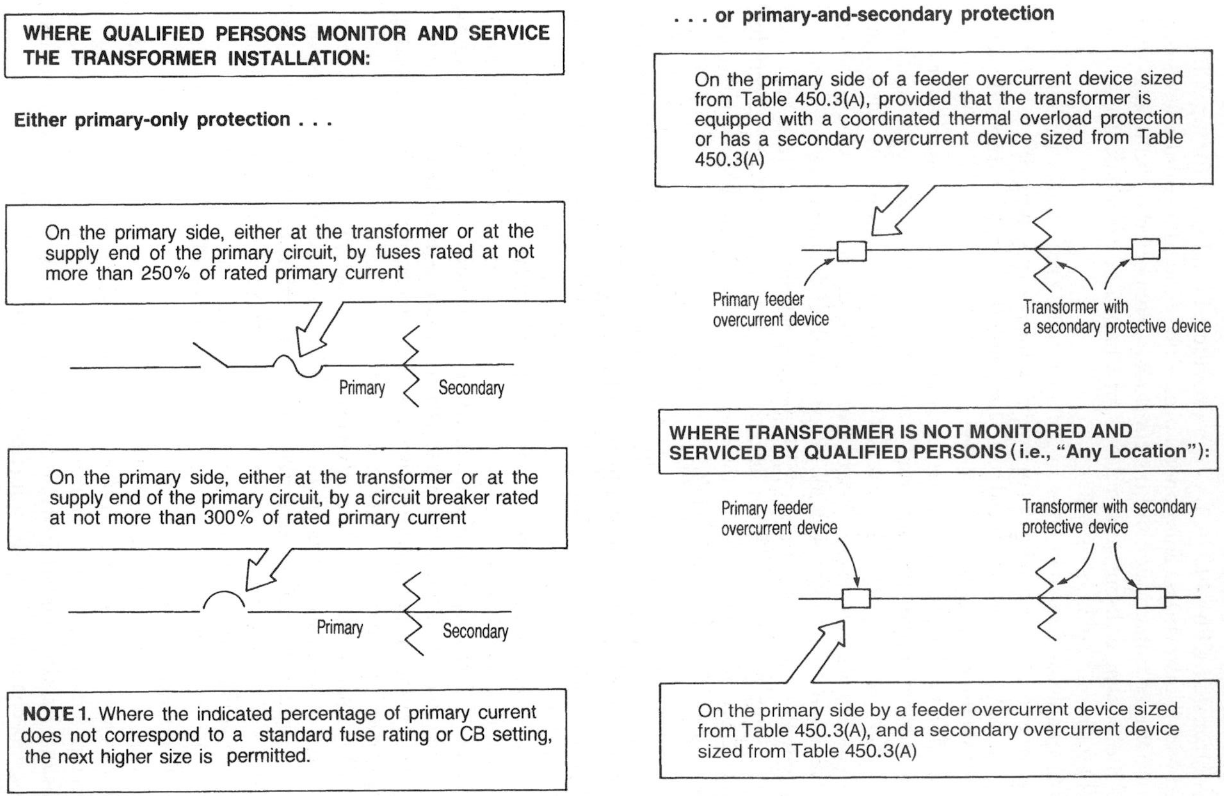
WHERE QUALIFIED PERSONS MONITOR AND SERVICE THE TRANSFORMER INSTALLATION:
Either primary-only protection . . .
On the primary side, either at the transformer or at the supply end of the primary circuit, by fuses rated at not more than 250% of rated primary current
Primary
Secondary
On the primary side, either at the transformer or at the supply end of the primary circuit, by a circuit breaker rated at not more than 300% of rated primary current
Primary
Secondary
NOTE 1. Where the indicated percentage of primary current does not correspond to a standard fuse rating or CB setting, the next higher size is permitted.
. . . or primary-and-secondary protection
On the primary side of a feeder overcurrent device sized from Table 450.3(A), provided that the transformer is equipped with a coordinated thermal overload protection or has a secondary overcurrent device sized from Table 450.3(A)
Primary feeder overcurrent device
Transformer with a secondary protective device
WHERE TRANSFORMER IS NOT MONITORED AND SERVICED BY QUALIFIED PERSONS (i.e., "Any Location"):
Primary feeder overcurrent device
Transformer with secondary protective device
On the primary side by a feeder overcurrent device sized from Table 450.3(A), and a secondary overcurrent device sized from Table 450.3(A)

Fig. 450-3. High-voltage transformers (rated over 1000 V, dry or fluid-filled) with any impedance must be protected in one of these ways. However, the value of OC protection for transformers with known internal impedances as indicated in Table 450.3(A) may be based on the particular impedance. [Table 450.3(A).]

Transformers operating 1000 V and less, with primary side protection only

For any transformer rated 1000 V or less (i.e., the rating of neither the primary nor the secondary winding is over 1000 V), the basic overcurrent protection may be provided just on the primary side [Table 450.3(B) (top row)] or may be a combination of protection on *both* the primary and secondary sides [Table 450.3(B) (bottom row)].

If a transformer is to be protected by means of a CB or set of fuses only on the primary side of the transformer, the basic arrangement is as shown in Fig. 450-4.

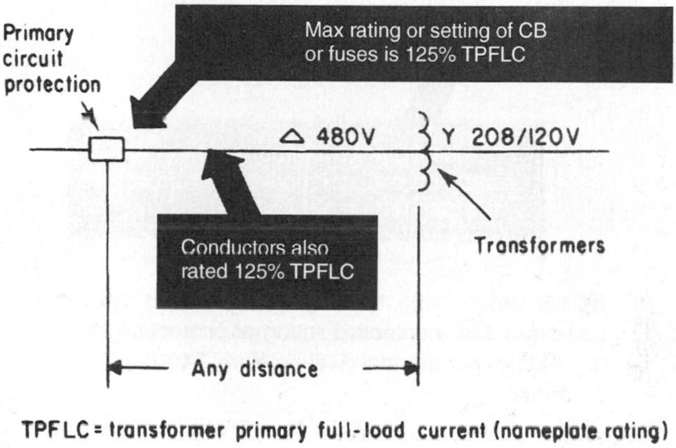
Primary
circuit
protection
Max rating or setting of CB
or fuses is 125% TPFLC
△ 480V
Y 208/120V
Conductors also
rated 125% TPFLC
Transformers
Any distance
TPFLC = transformer primary full-load current (nameplate rating)

Fig. 450-4. This is the basic rule on primary-side protection for transformers with primary current over 9 A. (Table 450.3.)

In that layout, a CB or a set of fuses rated not over 125 percent of the transformer rated primary full-load current provides all the overcurrent protection required by the NEC for the transformer. This overcurrent protection is in the feeder circuit to the transformer and is logically placed at the supply end of the feeder, so the same overcurrent device may also provide the overcurrent protection required for the primary feeder conductors. There is no limit on the distance between primary protection and the transformer. When the correct maximum rating for transformer protection is selected and installed at any point on the supply side of the transformer (either near or far from the transformer), then feeder circuit conductors must be sized so that the CB or fuses selected will provide the proper protection as required for the conductors. The ampacity of the feeder conductors must be at least equal to the amp rating of the CB or fuses unless 240.4(B) is satisfied. That is, when the rating of the overcurrent protection selected is not more than 125 percent of rated primary current, the primary feeder conductor may have an ampacity such that the overcurrent device is the next higher standard rating.

The rules set down for protection of a 1000-V transformer by a CB or set of fuses in its primary circuit are given in Fig. 450-5 for transformers with rated primary current of 9 A or more. Note 1 to Table 450.3(B) says that "the next higher standard" rating of protection may be used, if needed. Figure 450-6 shows the *absolute* maximum values of protection for smaller transformers. When using the 1.67 or

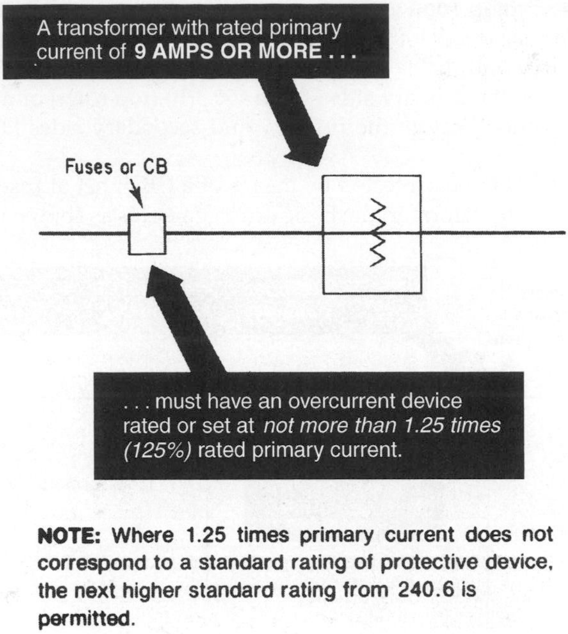
A transformer with rated primary current of 9 AMPS OR MORE . . .
Fuses or CB
. . . must have an overcurrent device rated or set at not more than 1.25 times (125%) rated primary current.
NOTE: Where 1.25 times primary current does not correspond to a standard rating of protective device, the next higher standard rating from 240.6 is permitted.

Fig. 450-5. Protection sizing for larger transformers is 125 percent of primary current. (Sec. 450.3.)

3 times factor, if the resultant current value is not exactly equal to a standard rating of fuse or CB, then the next *lower* standard rated fuse or CB must be used.

When the rules of Table 450.3(B) are observed, the transformer itself is properly protected and the primary feeder conductors, if sized to correspond, may be provided with the protection required by 240.4. But all considerations on the secondary side of the transformer then have to be separately and independently evaluated. When a transformer is provided with primary-side overcurrent protection, a whole range of design and installation possibilities are available for secondary arrangement that satisfies the Code. The basic approach is to provide required overcurrent protection for the secondary circuit conductors right at the transformer—such as by a fused switch or CB attached to the transformer enclosure, as shown in Fig. 450-7. Or 10- or 25-ft (3.0- or 7.5-m) taps may be made, as covered in 240.21.

Transformers operating 1000 V and less, with primary side protection only There is another acceptable way to protect a 1000-V transformer, described in Table 450.3(B). In this method, the transformer primary may be fed from a circuit which has overcurrent protection (and circuit conductors) rated up to 250 percent (next higher standard size permitted) of rated primary current (instead of 125 percent, as previously)—*but,* in such cases, there must be a protective device on the secondary side of the transformer, and that device must be rated or set at not more than 125 percent of the transformer's rated secondary current (Fig. 450-8). This secondary protective device must be located right at the transformer secondary

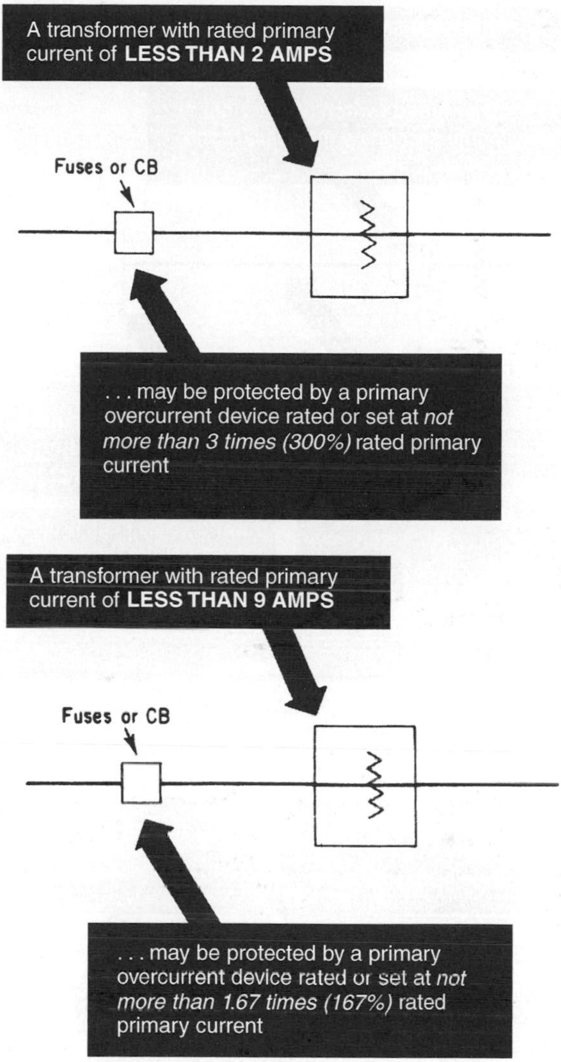
A transformer with rated primary current of LESS THAN 2 AMPS
Fuses or CB
. . . may be protected by a primary overcurrent device rated or set at not more than 3 times (300%) rated primary current
A transformer with rated primary current of LESS THAN 9 AMPS
Fuses or CB
. . . may be protected by a primary overcurrent device rated or set at not more than 1.67 times (167%) rated primary current

Fig. 450-6. Higher-percent protection is permitted for smaller transformers. (Table 450.3.)

terminals or not more than the length of a 10- or 25-ft (3.0- or 7.5-m) tap away from the transformer, and the rules of 240.21 on tap conductors must be fully satisfied.

The secondary protective device covered by Table 450.3(B) may readily be incorporated as part of other required provisions on the secondary side of the transformer, such as protection for a secondary feeder from the transformer to a panel or switchboard or motor control center fed from the switchboard. And a single secondary protective device rated not over 125 percent of

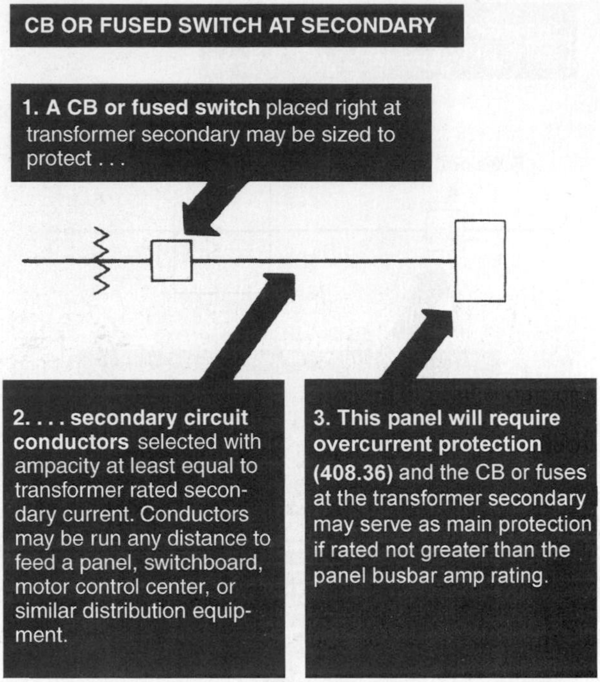
CB OR FUSED SWITCH AT SECONDARY
1. A CB or fused switch placed right at transformer secondary may be sized to protect . . .
2. . . . secondary circuit conductors selected with ampacity at least equal to transformer rated secondary current. Conductors may be run any distance to feed a panel, switchboard, motor control center, or similar distribution equipment.
3. This panel will require overcurrent protection (408.36) and the CB or fuses at the transformer secondary may serve as main protection if rated not greater than the panel busbar amp rating.

Fig. 450-7. Protection of secondary circuit must be independent of primary-side transformer protection. (Sec. 450.3.)

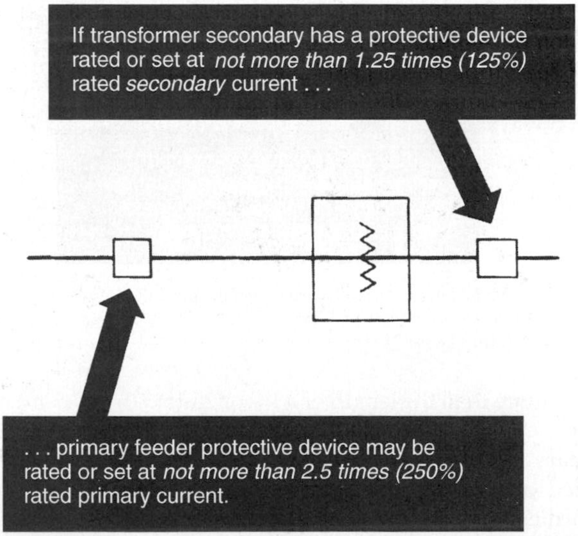
If transformer secondary has a protective device rated or set at not more than 1.25 times (125%) rated secondary current . . .
. . . primary feeder protective device may be rated or set at not more than 2.5 times (250%) rated primary current.

Fig. 450-8. Secondary protection permits higher-rated primary protective device. (Table 450.3.)

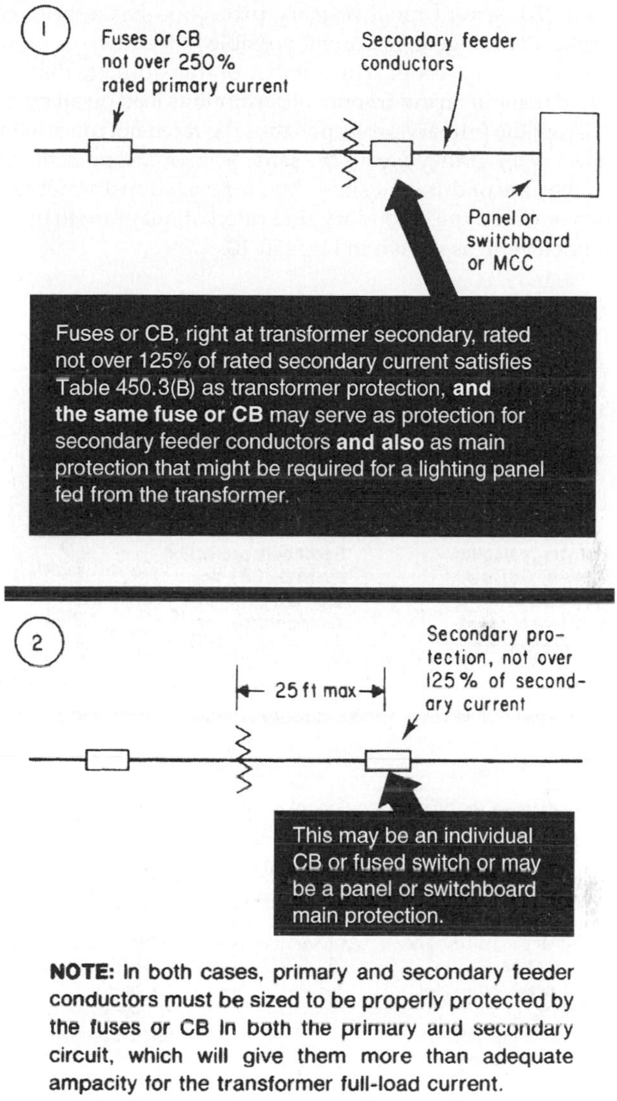
1
Fuses or CB not over 250% rated primary current
Secondary feeder conductors
Panel or switchboard or MCC
Fuses or CB, right at transformer secondary, rated not over 125% of rated secondary current satisfies Table 450.3(B) as transformer protection, **and the same fuse or CB** may serve as protection for secondary feeder conductors **and also** as main protection that might be required for a lighting panel fed from the transformer.
2
25 ft max
Secondary protection, not over 125% of secondary current
This may be an individual CB or fused switch or may be a panel or switchboard main protection.
NOTE: In both cases, primary and secondary feeder conductors must be sized to be properly protected by the fuses or CB in both the primary and secondary circuit, which will give them more than adequate ampacity for the transformer full-load current.

Fig. 450-9. With 250 percent primary protection, secondary protection may be located like this. (Table 450.3.)

secondary current may serve as a required panelboard main as well as the required transformer secondary protection, as shown at the bottom of Fig. 450-9.

The use of a transformer circuit with primary protection rated up to 250 percent of rated primary current offers an opportunity to avoid situations where a particular set of primary fuses or CB rated at only 125 percent would cause nuisance tripping or opening of the circuit on transformer inrush current.

But the use of a 250 percent rated primary protection has a more common and widely applicable advantage in making it possible to feed two or more transformers from the same primary feeder. The number of transformers that might be used in any case would depend on the amount of continuous load on all the transformers. But in all such cases, the primary protection must be rated not more than 250 percent of any one transformer, if they are all the same size, or 250 percent of the smallest transformer, if they are of different sizes. And for each transformer fed, there must be a set of fuses or CB on the secondary side rated at not more than 125 percent of rated secondary current, as shown in Fig. 450-10.

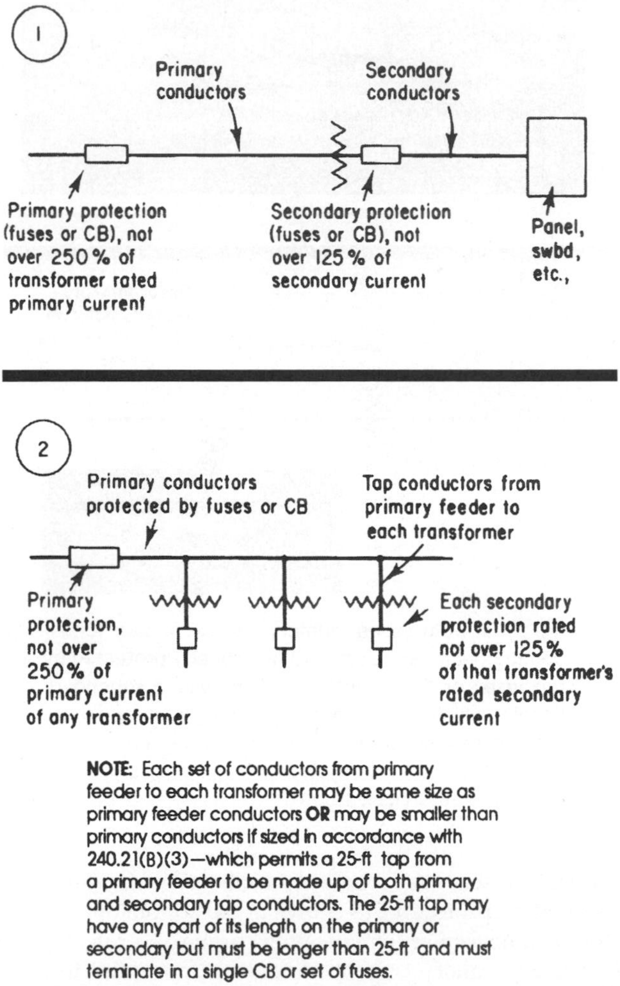
1
Primary conductors
Secondary conductors
Primary protection (fuses or CB), not over 250% of transformer rated primary current
Secondary protection (fuses or CB), not over 125% of secondary current
Panel, swbd, etc.,
2
Primary conductors protected by fuses or CB
Tap conductors from primary feeder to each transformer
Primary protection, not over 250% of primary current of any transformer
Each secondary protection rated not over 125% of that transformer's rated secondary current
NOTE: Each set of conductors from primary feeder to each transformer may be same size as primary feeder conductors OR may be smaller than primary conductors if sized in accordance with 240.21(B)(3)—which permits a 25-ft tap from a primary feeder to be made up of both primary and secondary tap conductors. The 25-ft tap may have any part of its length on the primary or secondary but must be longer than 25-ft and must terminate in a single CB or set of fuses.

Fig. 450-10. With primary 250 percent protection, primary circuit may vary. (Sec. 450.3.)

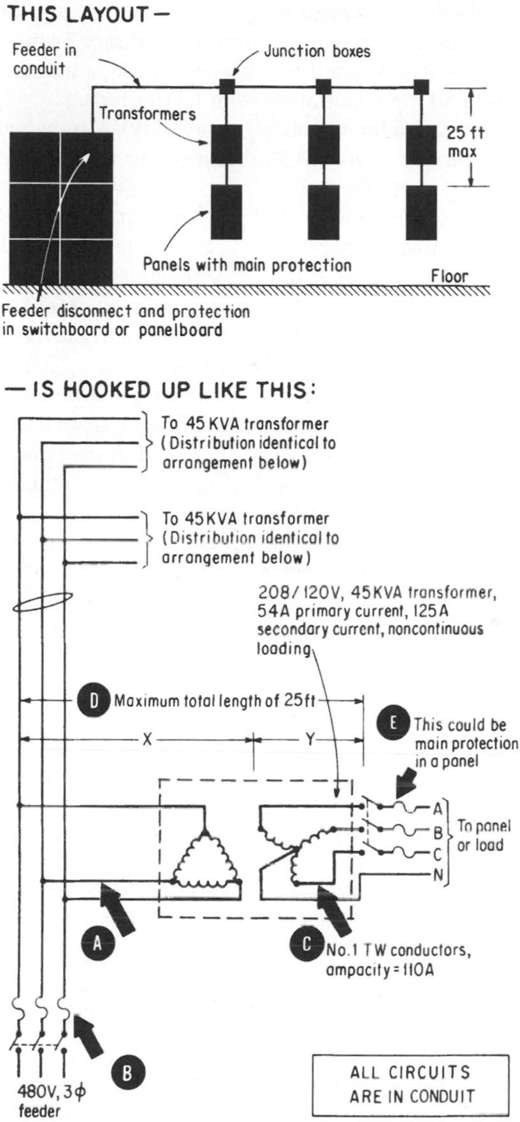
THIS LAYOUT—
Feeder in conduit
Junction boxes
Transformers
25 ft max
Panels with main protection
Floor
Feeder disconnect and protection in switchboard or panelboard
—IS HOOKED UP LIKE THIS:
To 45 KVA transformer (Distribution identical to arrangement below)
To 45KVA transformer (Distribution identical to arrangement below)
208/120V, 45KVA transformer, 54A primary current, 125A secondary current, noncontinuous loading
D Maximum total length of 25ft
X
Y
E This could be main protection in a panel
A
B
C
N
To panel or load
A
C No.1 TW conductors, ampacity=110A
B
480V, 3ϕ feeder
ALL CIRCUITS ARE IN CONDUIT

Fig. 450-11. Code rules must be tied together. (Sec. 450.3.)

Figure 450-11 shows an example of application of 250 percent primary protection to a feeder supplying three transformers (such as at the bottom of Fig. 450-10). The example shows how the rules of Table 450.3(B) must be carefully related to 240.21 and other Code rules:

Part **(B)(3)** of 240.21 of the NEC is a rule that covers use of a 25-ft (7.5-m) *unprotected* tap from feeder conductors, with a transformer inserted in the 25-ft

(7.5-m) tap. This rule does not eliminate the need for secondary protection—it makes a special condition for placement of the secondary protective device. It is a restatement of part **(B)(3)** as applied to a tap containing a transformer and applies to both single-phase and 3-phase transformer feeder taps.

Figure 450-11 shows a feeder supplying three 45-kVA transformers, each transformer being fed as part of a 25-ft (7.5-m) feeder tap that conforms to part (B)(3) of 240.21.

Although each transformer has a rated primary current of 54 A at full load, the demand load on each transformer primary was calculated to be 41 A, based on secondary loading. No. 1 THW copper feeder conductors were considered adequate for the total noncontinuous demand load of 3 × 41 A, or 123 A. A step-by-step analysis of this system follows. Refer to circled letters in the figure:

A. The primary circuit conductors are No. 6 TW rated at 55 A, which gives them "an ampacity at least $^1/_3$ that of the *overcurrent protection* from which they are tapped . . .," because these conductors are tapped from the feeder conductors protected at 125 A. No. 6 TW is okay for the 41-A primary current.

B. The 125-A fuses in the feeder switch properly protect the No. 1 THW feeder conductors, which are rated at 130 A.

C. The conductors supplied by the transformer secondary must have "an ampacity that, when multiplied by the ratio of the secondary-to-primary voltage, is at least $^1/_3$ the ampacity of the conductors *or* overcurrent protection from which the primary conductors are tapped . . ." The ratio of secondary-to-primary voltage of the transformer is

$$\frac{208\text{ V}}{480\text{ V}} = 0.433$$

Note that phase-to-phase voltage must be used to determine this ratio.
Then, for the secondary conductors, 240.21(B)(3)b says that

$$\text{Minimum conductor ampacity} \times 0.433 = \frac{1}{3} \times 125\text{ A}$$

$$\frac{1}{3} \times 125 = 41.67$$

Then, minimum conductor ampacity equals

$$\frac{41.67}{0.433} = 96\text{ A}$$

The No. 1 TW secondary conductors, rated at 110 A, are above the 96-A minimum and are, therefore, satisfactory.

D. The total length of the unprotected 25-ft (7.5-m) tap—that is, the primary conductor length *plus* the secondary conductor length ($x + y$) for any circuit leg—must not be greater than 25 ft (7.5 m).

E. The secondary tap conductors from the transformer must terminate in a single CB or set of fuses that will limit the load on those conductors to their rated ampacity from Table 310.15(B)(16). Note that there is no exception given to that requirement and the "next higher standard device rating" may not be used if the conductor ampacity does not correspond to the rating of a standard device.

The overcurrent protection required at E, at the load end of the 25-ft (7.5-m) tap conductors, must not be rated more than the ampacity of the No. 1 TW conductors.

Max. rating of fuses or CB at E = 110 A

But a 100-A main would satisfy the 96-A secondary load. *(Note: The overcurrent protective device required at E could be the main protective device required for a panel fed from the transformer.)*

Watch Out for This Trap!

Although the foregoing calculation shows how unprotected taps may be made from feeder conductors by satisfying the rules of 240.21(B)(3), the rules of 240.21 are all concerned with PROTECTION OF CONDUCTORS ONLY. Consideration must now be made of *transformer* protection, as follows:

1. Note 240.21(B)(3) makes no reference to *transformer* protection. [240.21(C) is followed by an informational note that references 450.3.] But 450.3 calls for protection of transformers, and there is no exception made for the conditions of part **(B)(3)** to 240.21.
2. It is clear from 450.3(B) and Table 450.3(B) that the transformer shown in Fig. 450-11 is *not* protected by a primary-side overcurrent device rated not more than 125 percent of primary current (54 A), because 1.25 × 54 A = 68 A, maximum.
3. But 450.3(B)(2) does offer a way to provide required protection. The 110-A protection at E *is* secondary protection *rated not over* 125 percent of rated secondary current (1.25 × 125 A secondary current = 156 A). With that secondary protection, a primary feeder overcurrent device rated not more than 250 percent of rated primary current will satisfy 450.3(B)(2). That would call for fuses in the feeder switch (or a CB) (at B in the diagram) rated not over

   250 percent × 54 A primary current = 135 A

   But, the fuses in the feeder switch are rated at 125 A—which are not in excess of 250 percent of transformer primary current and, therefore, satisfy Table 450.3(B).

In addition to the two basic methods for protecting transformers previously described, Note 3 to Table 450.3(B) also provides for protection with a built-in thermal overload protection, as shown in Fig. 450-12.

450.4. Autotransformers 1000 V, Nominal, or Less. This section sets the rules for the permitted connections to and the required overcurrent protection for autotransformers. It is important to place these rules in the context of provisions in 210.9 and 215.11 that limit their use in some circumstances. In general, a branch

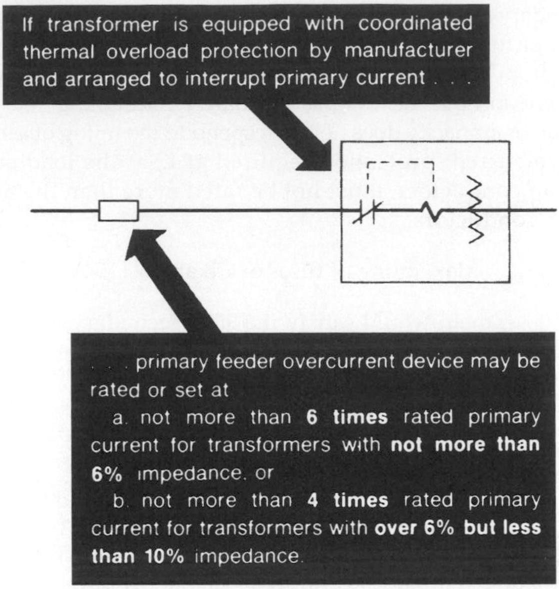
If transformer is equipped with coordinated thermal overload protection by manufacturer and arranged to interrupt primary current . . .
. . . primary feeder overcurrent device may be rated or set at
a. not more than **6 times** rated primary current for transformers with **not more than 6%** impedance, or
b. not more than **4 times** rated primary current for transformers with **over 6% but less than 10%** impedance.

Fig. 450-12. Built-in protection is another technique for transformers. [Note 3 to Table 450.3(B).]

circuit or a feeder may be supplied through an autotransformer only if the system supplied has a grounded conductor that is electrically connected to a grounded conductor of the system supplying the autotransformer, with the following exceptions:

1. An autotransformer can be used to extend or add a branch circuit for an equipment load without the connection to a similar grounded conductor when transforming from a nominal 208-V to a nominal 240-V supply or similarly from 240 to 208 V.
2. In industrial occupancies, where conditions of maintenance and supervision ensure that only qualified persons will service the installation, autotransformers can be used to supply nominal 600-V loads from nominal 480-V systems, and 480-V loads from nominal 600-V systems, without the connection to a similar grounded conductor.

This transformation is commonly employed to connect 480-V equipment in a factory with a 600-V distribution, or vice versa. A 480 to 120-V, two-winding transformer is connected with its windings in series so it functions as an autotransformer, as shown in Fig. 450-13. There is an easy way to determine how to size the two-winding transformer that will be connected in this way. If C = transformation capacity as an autotransformer; X = rated transformation capacity as a two-winding transformer; and R = the voltage ratio of the two windings, then

$$X = \frac{C}{(R+1)}$$

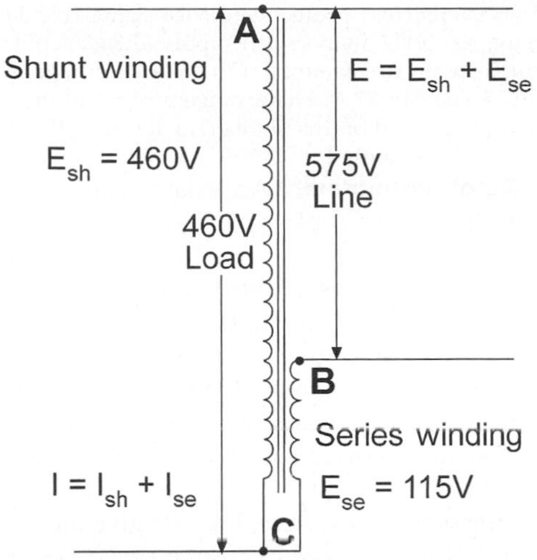
A
Shunt winding
E = E_sh + E_se
E_sh = 460V
575V
Line
460V
Load
B
Series winding
I = I_sh + I_se
E_se = 115V
C

Fig. 450-13. A 480- to 120-V transformer connected as an autotransformer to run 460-V equipment on a 575-V system. (Sec. 450.4.)

example Suppose a 50-hp and a 3-hp motor, both rated 460 V, will be connected to a 575-V, 3-phase system. If two 480 V/120 V two-winding transformers will be connected in open delta to service the loads, what sized transformers should be purchased?

The full-load ratings of the motors must be evaluated using the **NEC** Table 430.250. The 50-hp motor draws 65 A and the 3-hp motor 4.8 A. Converting to kilovolt-amperes, 65 A × 460√3/(1000) = 52 kVA and 4.8 A × 460√3/(1000) = 4 kVA.

Add 13 kVA (25 percent of the largest motor) to get 69 kVA total.

In the formula, *R* will be 480 V ÷ 120 V = 4, and *C* will be 69 kVA.

Then the capacity required will be 69 kVA ÷ 5 or 14 kVA.

Remember that the capacity of an open delta transformer bank is not the sum of the two transformers; instead, the capacity is that of the two transformers multiplied by half the square root of three. Two 10-kVA two-winding transformers would be a good choice in this case. The final capacity would be (10 kVA + 10 kVA)(½√3) = 17 kVA.

Note that this 480-V transformer (120-V secondary) will be connected to a 600-V supply voltage. The **NEC** requires that field-connected autotransformers be identified for use at the elevated voltage because the dielectric withstand test is often done at a lower level for windings rated 250 V or less, and in this case the 120-V winding must be tested as though it were a 600-V winding. Transformer manufacturers are very familiar with this application and appropriate products are readily available.

The **NEC** requires autotransformers rated 1000 V, nominal, or less be protected by an individual overcurrent device in series with each ungrounded conductor, rated or set at not more than 125 percent of the rated full-load input current of the autotransformer. If this calculation does not correspond to a standard rating and the rated input current is 9 A or more, the next higher standard device is permitted. No overcurrent device is permitted to be placed in series with the shunt winding.

Referring to Fig. 450-13, no overcurrent device should be placed between points "A" and "C." If the overcurrent device opened and the load impedance were low, the input 575 V would return to the autotransformer at some voltage approaching 575 V across the series winding (between "B" and "C" in Fig. 450-13). Given the 1:4 winding ratio, that could result in a voltage on the 460 V side of the autotransformer exceeding 2000 V.

As an example of an overcurrent protective device sizing calculation, the two 10-kVA transformers used in the Sec. 5.117 analysis can supply 17.3 kVA of load transformation. In the formula, rearrange the terms to solve for C: $C = X(R + 1)$, or in this case 17.3 kVA (4 + 1) = 87 kVA; 87,000 VA ÷ (575 V × √3) = 87 A. This arrangement of autotransformers must be protected at not more than 125 percent of this number, or 108 A, with the next higher standard size protective device (110 A) permitted.

450.5. Grounding Autotransformers. An existing, ungrounded, 480-V system derived from a delta transformer hookup can be converted to grounded operation in two basic ways:

First, one of the three phase legs of the 480-V delta can be intentionally connected to a grounding electrode conductor that is then run to a suitable grounding electrode. Such grounding would give the two ungrounded phases (A and B) a voltage of 480 V to ground. The system would then operate as a grounded system, so that a ground fault (phase-to-conduit or other enclosure) on the secondary can cause fault-current flow that opens a circuit protective device to clear the faulted circuit.

But corner grounding of a delta system does not give the lowest possible phase-to-ground voltage. In fact, the voltage to ground of a corner-grounded delta system is the same as it is for an ungrounded delta system because voltage to ground for ungrounded circuits is defined as the greatest voltage between the given conductor and any other conductor of the circuit. Thus, the voltage to ground for an ungrounded delta system is the maximum voltage between any two conductors, on the assumption that an accidental ground on any one phase puts the other two phases at full line-to-line voltage aboveground.

In recognition of increasing emphasis on the safety of grounded systems over ungrounded systems, 450.5 covers the use of zig-zag grounding autotransformers to convert 3-phase, 3-wire, ungrounded delta systems to grounded wye systems. Such grounding of a 480-V delta system, therefore, lowers the voltage to ground from 480 V (when ungrounded) to 277 V (the phase-to-grounded-neutral voltage) when converted to a wye system (Fig. 450-14).

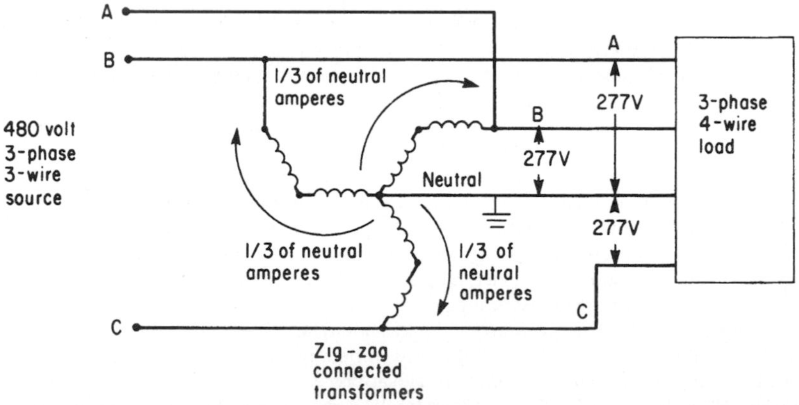
A
B
C
480 volt
3-phase
3-wire
source
1/3 of neutral
amperes
1/3 of neutral
amperes
1/3 of
neutral
amperes
Neutral
A
B
C
277V
277V
277V
3-phase
4-wire
load
Zig-zag
connected
transformers

Fig. 450-14. Zig-zag transformer changes voltage to ground from 480 to 277 V. (Sec. 450.5.)

A zig-zag grounding autotransformer gets its name from the angular phase differences among the six windings that are divided among the three legs of the transformer's laminated magnetic core assembly. The actual hookup of the six windings is an interconnection of two wye configurations, with specific polarities and locations for each winding. Just as a wye or delta transformer hookup has a graphic representation that looks like the letter Y or the Greek letter Δ, so a zig-zag grounding autotransformer is represented as two wye hookups with pairs of windings in series but phase-displaced, as in Fig. 450-15.

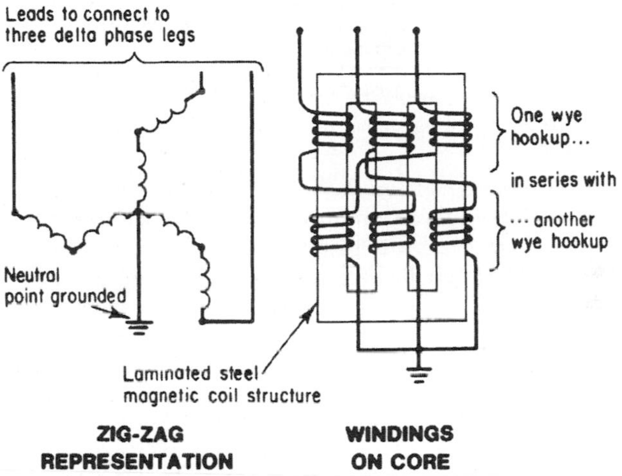
Leads to connect to three delta phase legs
Neutral point grounded
Laminated steel magnetic coil structure
One wye hookup...
in series with
... another wye hookup
ZIG-ZAG REPRESENTATION
WINDINGS ON CORE

Fig. 450-15. Windings of zig-zag transformer provide for flow of fault or neutral current. (Sec. 450.5.)

With no ground fault on any leg of the 3-phase system, current flow in the transformer windings is balanced, because equal impedances are connected across each pair of phase legs. The net impedance of the transformer under balanced conditions is very high, so that only a low level of magnetizing current flows through the windings. But when a ground fault develops on one leg of the 3-phase system, the transformer windings become a very low impedance in the fault path, permitting a large value of fault current to flow and operate the circuit protective device—just as it would on a conventional grounded-neutral wye system, as shown in Fig. 450-16.

Because the kilovolt-ampere rating of a grounding autotransformer is based on short-time fault current, selection of such transformers is much different from sizing a conventional 2-winding transformer for supplying a load. Careful consultation with a manufacturer's sales engineer should precede any decisions about the use of these transformers.

Such transformers shall have a continuous per-phase current rating and a continuous neutral current rating. Zig-zag connected transformers must never be connected on the load side of system grounding connections, including those made in accordance with 250.24(B), 250.30(A)(1), or 250.32(B) Exception No. 1.

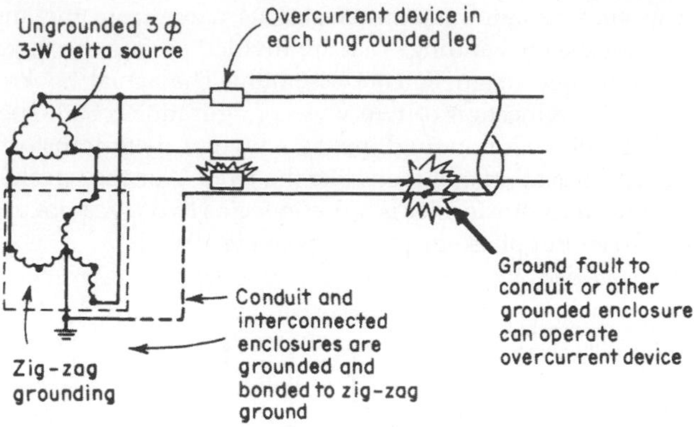
Ungrounded 3 ϕ
3-W delta source
Overcurrent device in
each ungrounded leg
Ground fault to
conduit or other
grounded enclosure
can operate
overcurrent device
Conduit and
interconnected
enclosures are
grounded and
bonded to zig-zag
ground
Zig-zag
grounding

Fig. 450-16. Zig-zag transformer converts ungrounded system to grounded operation. (Sec. 450.5.)

Part **(A)** of the section covers the use of these transformers when an actual 480Y/277V distribution system is being created. Zig-zag grounding autotransformers are comprised of two three-phase wye connections in series. Figure 450-17 shows most of the requirements as well as the angular displacements that give the transformation its name. The inner connections are a straightforward wye, and close inspection of the three outer segments of the pinwheel shows that they are also oriented in a wye configuration. In fact, at least one major transformer

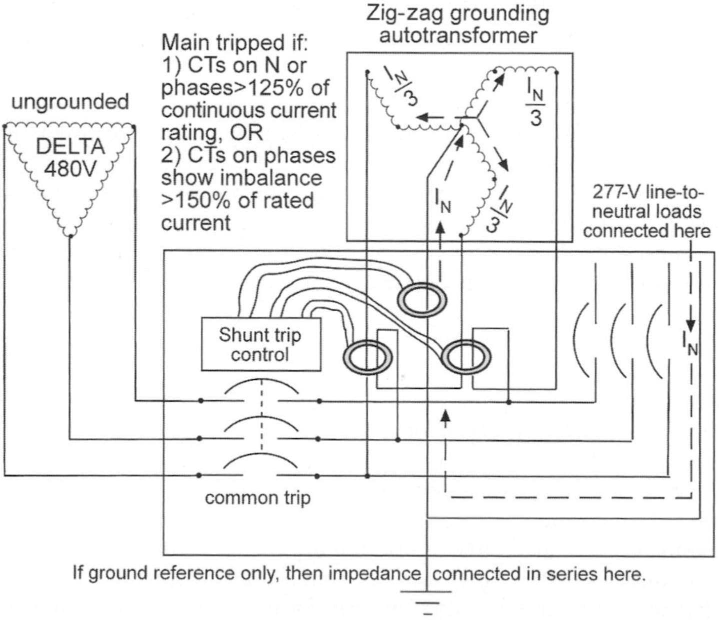
Zig-zag grounding
autotransformer
Main tripped if:
1) CTs on N or
phases>125% of
continuous current
rating, OR
2) CTs on phases
show imbalance
>150% of rated
current
ungrounded
DELTA
480V
277-V line-to-
neutral loads
connected here
Shunt trip
control
common trip
If ground reference only, then impedance connected in series here.

Fig. 450-17. Zig-zag connections must meet numerous NEC requirements. (Sec. 450.5.)

manufacturer now sells zig-zag customers three 2-winding transformers (thus, six windings) along with instructions as to how to connect them to achieve this result. (See also the right side of Fig. 450-15.) The simultaneous disconnection rule is crucial because if a single-pole overcurrent device were to open, the grounding transformer will no longer function as intended, including a strong likelihood of dangerous transient overvoltages on its load side. A grounding autotransformer used to create a 3-phase, 4-wire distribution system must comply with the following four requirements:

1. The transformer must be directly connected to the ungrounded phase conductors and never switched or provided with overcurrent protection that is independent of the main switch and common-trip overcurrent protection for the 3-phase, 4-wire system. A transformer used to limit transitory overvoltages must be of suitable rating and must be connected in accordance with this rule.
2. An overcurrent sensing device must be provided that will cause the main switch or common-trip overcurrent protection previously discussed to open if the load on the autotransformer reaches or exceeds 125 percent of the continuous current per-phase or neutral rating. Delayed tripping for temporary overcurrents sensed at the autotransformer overcurrent device is permitted for the purpose of allowing proper operation of branch or feeder protective devices on the 4-wire system.
3. A fault-sensing system that causes the opening of a main switch or common trip overcurrent device for the 3-phase, 4-wire system must be provided to guard against single phasing or internal faults. An explanatory note advises that this can be accomplished by the use of two subtractive-connected donut-type current transformers installed to sense and signal when an unbalance occurs in the line current to the autotransformer of 50 percent or more of rated current.
4. The autotransformer must have a continuous neutral-current rating that is sufficient to handle the maximum possible neutral unbalanced load current of the 4-wire system.

Note that there is now an express prohibition in the parent language regarding installing a zig-zag on the load side of any system grounding connection. These transformer hookups are never a substitute for running a neutral to a load, regardless of how remote. If the supply system is grounded at any point, then a parallel neutral return path will exist and this will desensitize the protective controls for the transformer.

This prohibition also follows from the fact that if a zig-zag grounding transformer is connected to a grounded system, and a ground fault occurs upstream, that fault current will share a ground-fault return path through both the supply (or system) main bonding jumper and also the derived neutral connection in the grounding transformer. If the upstream fault is large, such as a failure on a large capacity feeder, the current run through the grounding transformer can be enough to destroy it. In the past, zig-zag transformer hookups have been erroneously applied in instances where a remote location was served from a feeder that did not include a neutral, but it was later decided to change the local load profile to one that included line-to-neutral loads.

If a neutral is required on a remote point on a grounded system it must be brought to that point. If running the system neutral is truly onerous, another possibility is to install a fully-rated delta-wye transformer to create a separately derived 480Y/277V system at the desired location.

Part (**B**) covers grounding autotransformers that do not supply utilization loads, but instead merely provide a ground reference and a specified magnitude of ground-fault current for the operation of ground-responsive protective devices, typically so an ungrounded system can be grounded through an impedance. Refer to Fig. 450-18, but insert the grounding impedance as indicated and eliminate all load connections to the neutral. There are three requirements:

1. The autotransformer must have a continuous neutral current rating sufficient for the specified ground-fault current.
2. An overcurrent device having an interrupting rating suitable for the fault current available at its supply terminals, and that will open simultaneously all ungrounded conductors when it operates, must be connected in the grounding autotransformer branch circuit.
3. This overcurrent protection must be rated or set at a current not exceeding 125 percent of the autotransformer continuous per-phase current rating or 42 percent of the continuous-current rating of any series-connected devices in the autotransformer neutral connection. Delayed tripping for temporary overcurrents to permit the proper operation of ground-responsive tripping devices on the main system is permitted but must not exceed values that would be more than the short-time current rating of the grounding autotransformer or any series connected devices in the neutral connection.

The 42 percent rule reflects the fact that the current in each phase is 33 percent of the neutral return current, and 125 percent of this is about 42 percent. An exception follows for high-impedance grounded systems, for which the 125 percent rule is impracticable because the neutral current through the impedance is so low that 125 percent of this current does not come close to a conventional overcurrent device setting or rating.

For high-impedance grounded systems covered in 250.36, where the maximum ground-fault current is designed to be not more than 10 A, and where the grounding autotransformer and the grounding impedance are rated for continuous duty, an overcurrent device rated not more than 20 A that will simultaneously open all ungrounded conductors can be installed on the line side of the zig-zag.

450.6. Secondary Ties. In industrial plants having very heavy power loads it is usually economical to install a number of large transformers at various locations within each building, the transformers being supplied by primary feeders operating at voltages up to 13,800 V. One of the secondary systems that may be used in such cases is the network system.

The term *network system* as commonly used is applied to any secondary distribution system in which the secondaries of two or more transformers at different locations are connected together by secondary ties. Two network layouts of unit subs are shown in Fig. 450-18. In the spot network, two or three transformers in one location or "spot" are connected to a common secondary bus and divide the load. Upon primary or transformer fault, the secondary is isolated from the

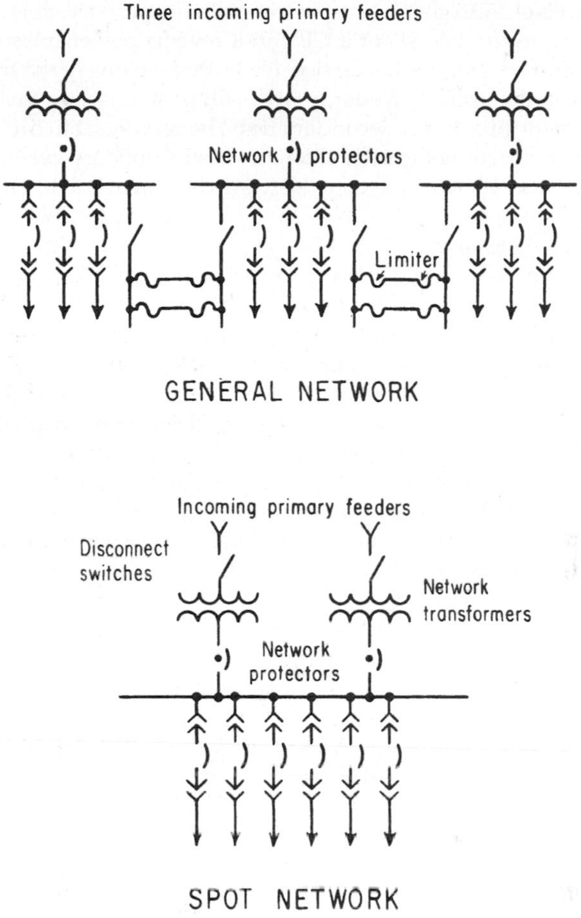
Three incoming primary feeders
Network protectors
Limiter
GENERAL NETWORK
Incoming primary feeders
Disconnect switches
Network transformers
Network protectors
SPOT NETWORK

Fig. 450-18. These are the two basic types of network systems. (Sec. 450.6.)

faulted section by automatic operation of the network protector, providing a high order of supply continuity in the event of faults. The general form of the network system is similar, except that widely separated individual substations are used with associated network protectors, and tie circuits run between the secondary bus sections. The system provides for interchange of power to accommodate unequal loading on the transformers. Limiters protect the ties. The purpose of the system is to equalize the loading of the transformers, to reduce voltage drop, and to ensure continuity of service. The use of this system introduces certain complications, and, to ensure successful operation, the system must be designed by an experienced electrical engineer.

The provisions of 450.3 govern the protection in the primary. Refer to Fig. 450-18. The network protector consists of a CB and a reverse-power relay. The protector is necessary because without this device, if a fault develops in the transformer, or, in some cases, in the primary feeder, power will be fed back to the fault from the other transformers through the secondary ties. The relay is set to trip the breaker on a reverse-power current not greater than the rated secondary current of the transformer. This breaker is not arranged to be tripped by an overload on the secondary of the transformer.

450.6(A)(3) provides that:

1. Where two or more conductors are installed in parallel, an individual protective device is provided at each end of each conductor.
2. The protective device (fusible link or CB) does not provide overload protection, but provides short-circuit protection only.

In the case of a short circuit, the protective device must open the circuit before the conductor reaches a temperature that would injure its insulation. The principles involved are that the entire system is so designed that the tie conductors will never be continuously overloaded in normal operation—hence protection against overloads of less severity than short circuits is not necessary—and that the protective devices should not open the circuit and thus cause an interruption of service on load peaks of such short duration that the conductors do not become overheated.

A limiter is a special type of fuse having a very high interrupting capacity. Figure 450-19 is a cross-sectional view of one type of limiter. The cable lug, the fusible section, and the extension for connection to the bus are all made in one piece from a length of copper tubing, and the enclosing case is also copper. A typical device of this type is rated to interrupt a current of 50,000 A without perceptible noise and without the escape of flame or gases from the case.

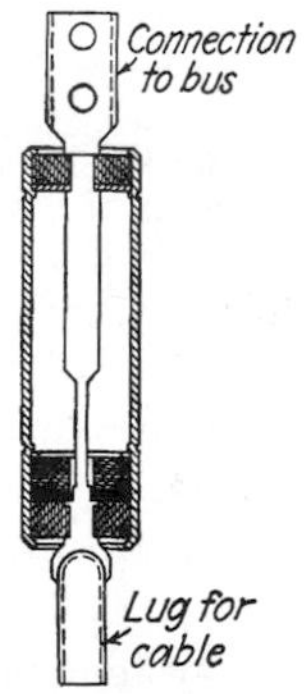
Connection to bus
Lug for cable

Fig. 450-19. A *limiter* is a cable connection device containing a fusible element. (Sec. 450.6.)

Figure 450-20 is a single-line diagram of a simple 3-phase industrial-plant network system. The primary feeders may operate at any standard voltage up to 13,800 V, and the secondary voltage would commonly be 480 V. The rating of the transformers used in such a system would usually be within the range

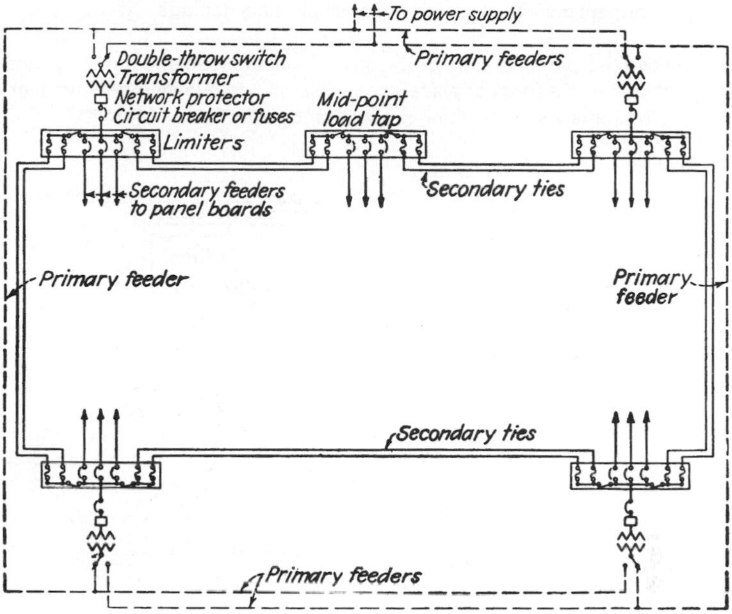

To power supply
Double-throw switch
Transformer
Network protector
Circuit breaker or fuses
Primary feeders
Mid-point
load tap
Limiters
Secondary feeders
to panel boards
Secondary ties
Primary feeder
Primary
feeder
Secondary ties
Primary feeders

Fig. 450-20. A typical industrial-plant network distribution system. (Sec. 450.6.)

of 300 to 1000 kVA. The diagram shows two primary feeders, both of which are carried to each transformer, so that by means of a double-throw switch each transformer can be connected to either feeder. Each feeder would be large enough to carry the entire load. It is assumed that the feeders are protected in accordance with 450.3(A) so that no primary overcurrent devices are required at the transformers. The secondary ties consist of two conductors in multiple per phase and it will be noted that these conductors form a closed loop. Switches are provided so that any section of the loop, including the limiters protecting that section, can be isolated in case repairs or replacements should be necessary.

450.7. Parallel Operation. To operate satisfactorily in parallel, transformers should have the same percentage impedance and the same ratio of reactance to resistance. Information on these characteristics should be obtained from the manufacturer of the transformers.

450.8. Guarding. Figure 450-21 summarizes these rules. Refer to 110.27 on guarding of live parts. Safety to personnel is always important, particularly where a transformer is to operate with live parts. To protect against accidental contact with such components, isolate the unit or units in a room or place accessible only to qualified personnel and guard live parts, such as with a railing. When elevation is used for safeguarding live parts, consult 110.34(E) and 110.27.

As noted in part **(C)**, switches and other equipment operating at up to 1000 V and serving only circuits within a transformer enclosure may be installed within the enclosure *if* it is accessible to qualified persons only. This is intended to be

1. Transformers must be protected against physical damage.

2. Exposed live parts must be protected against accidental contact by putting the transformer in a room or place accessible only to qualified personnel **or** by keeping live parts above the floor in accordance with Table 110.34(E).

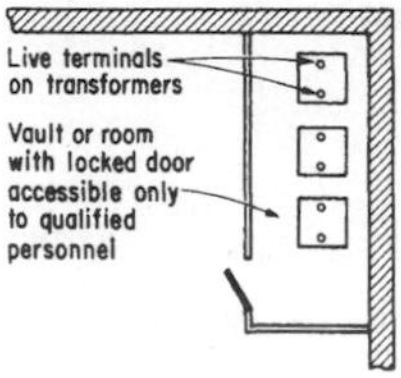
Live terminals on transformers
Vault or room with locked door accessible only to qualified personnel
ISOLATION

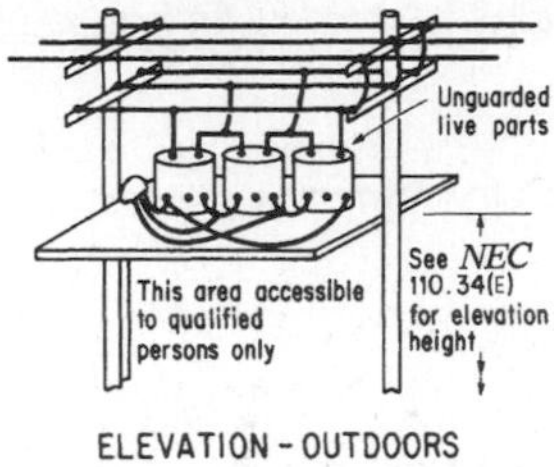
Unguarded live parts
See NEC 110.34(E) for elevation height
This area accessible to qualified persons only
ELEVATION - OUTDOORS

3. Signs or other visible markings must be used on equipment or structure to indicate the operating voltage of exposed live parts

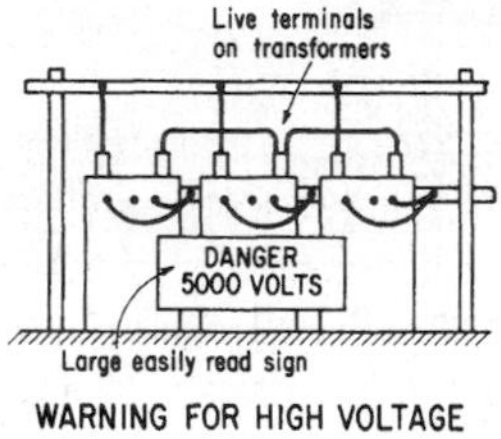
Live terminals on transformers
DANGER
5000 VOLTS
Large easily read sign
WARNING FOR HIGH VOLTAGE

Fig. 450-21. Transformer installations must be effectively guarded. (Sec. 450.8.)

part of the requirement that exposed, energized parts must be properly guarded. Part **(D)** requires that the operating voltage of exposed live parts be posted on the equipment or support structures.

450.9. Ventilation. As noted in the last paragraph, ventilation grilles or slots in the sides or back of transformer enclosures must have adequate clearance from walls and objects to ensure free and substantial airflow through the case. Further, the required clearance must be "clearly marked" on the transformer. Because of the tight quarters in today's electrical rooms and the tendency to give as little space as possible to electrical equipment, this is a critically important installation requirement to ensure safety and trouble-free operation of enclosed transformers. The substantiation for this proposal stated the following:

> Today's widespread use of dry type power transformers indoors has resulted in the common practice of their being installed directly up against walls completely blocking the rear vents. As inspectors, we frequently wind up trying to find out if the transformer installation instructions are anywhere around so we can see what clearances the manufacturer has specified.
>
> Clearly, this is not the best system. Furnaces, for example, commonly have required clearances marked on the nameplate. This proposal will let people know that clearances are required. Secondly, it will let us know what these clearances are.

450.10. Grounding. The 2014 NEC has completely reworked this section to address field problems involved with making equipment grounding connections in dry transformer connection enclosures. If the transformer makes its circuit connections using terminals and wired equipment grounding conductor or supply-side bonding conductors or both are used, then a terminal bar must be installed to receive the grounding connections. The paint must be scraped away to ensure a sound connection, and the bar must not land on or over any vented part of the enclosure. If the transformer uses leads instead of terminals, then any connection method compliant with 250.8 and bonded to the enclosure (if metallic) is sufficient.

450.11. Marking. This section appears in a list format as of the 2014 NEC, with a separate part (B) covering wiring the normal secondary side of a transformer as a primary; for example, taking a normal step-down transformer and wiring it to step-up the voltage. This is only permitted if the manufacturer has recognized the reverse transformation in the installation instructions. Such transformers are frequently marked "bidirectional." Note that higher than usual inrush currents may result from these connections, and some modifications may be necessary with respect to the time-delay characteristics of the primary side overcurrent protective devices. (See Fig. 450-22, top.)

450.13. Accessibility. Accessibility is an important location feature of transformer installation. The NEC generally requires a transformer (whether liquid-filled or dry-type) to be readily accessible to qualified personnel for inspection and maintenance (Fig. 450-22). That is, it must be capable of being reached quickly for operation, repair, or inspection without requiring use of a portable ladder to get at it, and it must not be necessary to climb over or remove obstacles to reach it. A transformer may be mounted on a platform or balcony, but there must generally be fixed stairs or a fixed ladder for access to the transformer (Fig. 450-23). However, there are some alternatives in subparts **(A)** and **(B)**.

Part **(A)** permits dry-type transformers rated 1000 V or less to be located "in the *open* on walls, columns, or structures"—without the need to be readily accessible. This allowance does not apply to a liquid-cooled transformer. A transformer suspended from the ceiling or hung on a wall—in which cases a ladder would be required to reach them because they are over 6½ ft (1.98 m) above the floor—would be okay, as shown in Fig. 450-24. There is no kilovolt-ampere limit on this provision, as opposed to the next paragraph. Part **(B)** permits dry-type transformers up to 1000 V, 50 kVA, to be installed in fire-resistant hollow spaces of buildings not permanently closed in by structure, provided the transformer is designed to have adequate ventilation for such installation. (See Fig. 450-25.)

Note in part **(A)** that a transformer *may* be *not* readily accessible only if it is located in the *open*. The words "in the open" do not readily and surely relate to such words as "concealed" or "exposed." But it is reasonable to conclude that "in the open" would be difficult to equate with "above a suspended ceiling." The latter location is in a generally smaller, confined space in which the transformers must be readily accessible; part **(A)** must be taken as a condition in which a transformer does *not* have to be readily accessible—that is, that it may be mounted up high where a portable ladder would be needed to get at it. And, the wording of part (B) specifically recognizes installation of transformers rated 50 kVA or less to be installed in "hollow spaces" such as hung ceilings (Fig. 450-26).

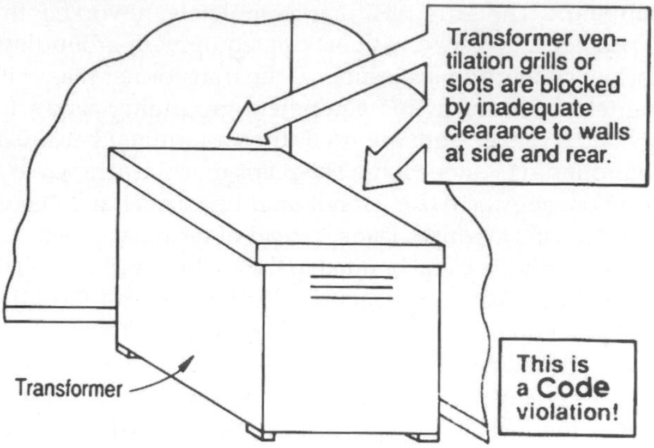
Transformer ventilation grills or slots are blocked by inadequate clearance to walls at side and rear.
Transformer
This is a Code violation!

DANGER

Fig. 450-22. Basic rule calls for every transformer to have adequate ventilation and ready, easy, direct access for inspection or maintenance. (Secs. 450.9 and 450.13.)

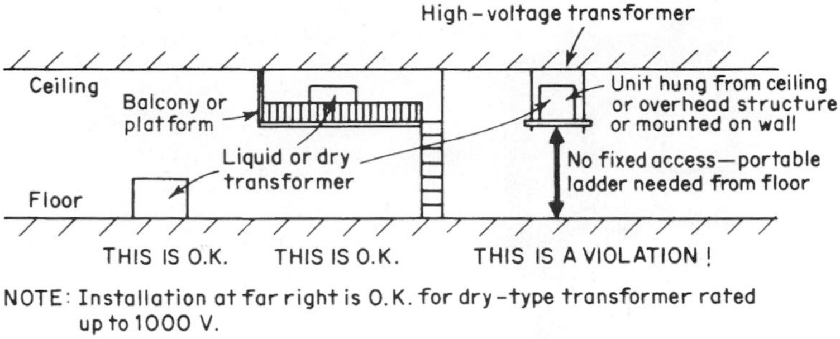
High-voltage transformer
Ceiling
Balcony or platform
Unit hung from ceiling or overhead structure or mounted on wall
Liquid or dry transformer
No fixed access—portable ladder needed from floor
Floor
THIS IS O.K.
THIS IS O.K.
THIS IS A VIOLATION !
NOTE: Installation at far right is O.K. for dry-type transformer rated up to 1000 V.

Fig. 450-23. Transformers must be "readily accessible" without need for portable ladder to reach them. (Sec. 450.13.)

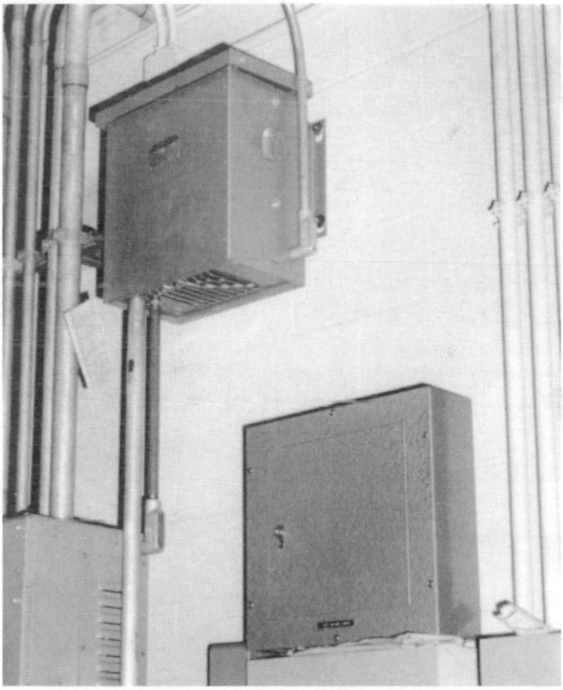

Fig. 450-24. Transformer mounted on wall (or suspended from ceiling) would be considered not readily accessible because a ladder would be needed to reach it. But, because it is "in the open," such use conforms to 450.13(A). (Sec. 450.13.)

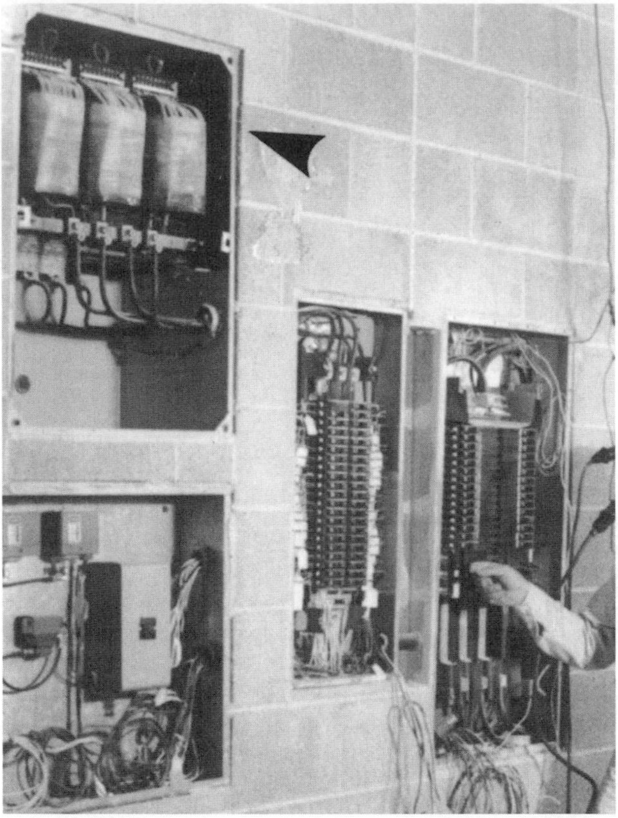

Fig. 450-25. Recessed mounting of dry-type transformers is permitted within hollow spaces of buildings, generally including spaces above suspended ceilings but not limited to those areas as this figure illustrates. (Sec. 450.13.)

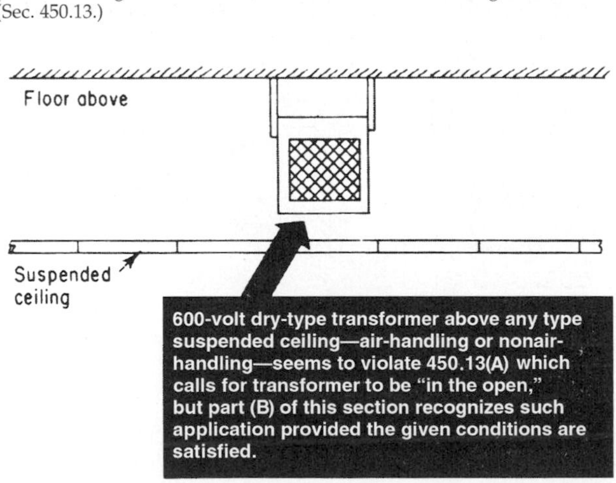
Floor above
Suspended ceiling
600-volt dry-type transformer above any type suspended ceiling—air-handling or nonair-handling—seems to violate 450.13(A) which calls for transformer to be "in the open," but part (B) of this section recognizes such application provided the given conditions are satisfied.

Fig. 450-26. Watch out for transformers above suspended ceilings. (Sec. 450.13.)

450.14. Disconnecting Means. Except for Class 2 or Class 3 control transformers, there is now (as of the 2011 NEC) a requirement for a disconnecting means for transformers for the first time. The disconnect need not be in sight of the transformer, but if it is remote, then it must be capable of being locked open, and the transformer must be field-marked to indicate the location of the remote disconnect.

450.21. Dry-Type Transformers Installed Indoors. This rule differentiates between dry-type transformers based on kilovolt-ampere rating. All dry-type transformers rated at 112½ kVA or less, at up to 35 kV, must be installed so that a minimum clearance of 12 in. (305 mm) is provided between the transformer and any combustible material, or a fire barrier must be provided.

The last sentence in 450.21(A) recognizes the use of fire-resistant heat-insulating barriers instead of space separation for transformers rated not over 112½ kVA. *But* be aware that clearances required to ensure proper ventilation of the transformer *must* be provided to satisfy 450.9. That is, the minimum clearance called for by the manufacturer to ensure proper airflow for cooling must be provided between the insulating barrier and the transformer's ventilation openings.

The exception permits those transformers rated 1000 V or less that are completely enclosed to be installed closer than the 12-in. (300-mm) minimum, but consideration must always be given to the requirements of 450.9 for those enclosed transformers with ventilation openings, as indicated in the previous paragraph.

For units rated over 112½ kVA, part **(B)** of the rule basically calls for such transformers to be installed in vaults, with two exceptions for high-temperature insulation systems provided. Figure 450-27 shows the rules of this section. Related application recommendations are as follows:

- Select a place that has the driest and cleanest air possible for installation of open-ventilated units. Avoid exposure to dripping or splashing water or other wet conditions. Outdoor application requires a suitable housing. Try to find locations where transformers will not be damaged by floodwater in case of a storm, a plugged drain, or a backed-up sewer.
- Temperature in the installation area must be normal, or the transformer may have to be derated. Modern standard, ventilated, dry-type transformers are designed to provide rated kilovolt-ampere output at rated voltage when the maximum ambient temperature of the cooling air is 40°C and the average ambient temperature of the cooling air over any 24 h period does not exceed 30°C. At higher or lower ambients, transformer loading can be adjusted by the following relationships:
    1. For each degree Celsius that average ambient temperature exceeds 30°C, the maximum load on the transformer must be reduced by 1 percent of rated kilovolt-amperes.
    2. For each degree Celsius that average ambient temperature is less than 30°C, the maximum load on the transformer may be increased by 0.67 percent of rated kilovolt-amperes.

Depending on the type of insulation used, transformer insulation life will be cut approximately in half for every 10°C that the ambient temperature exceeds the normal rated value—or doubled for every 10°C below rated levels. Estimates assume continuous operation at full load. With modern insulations this rule is

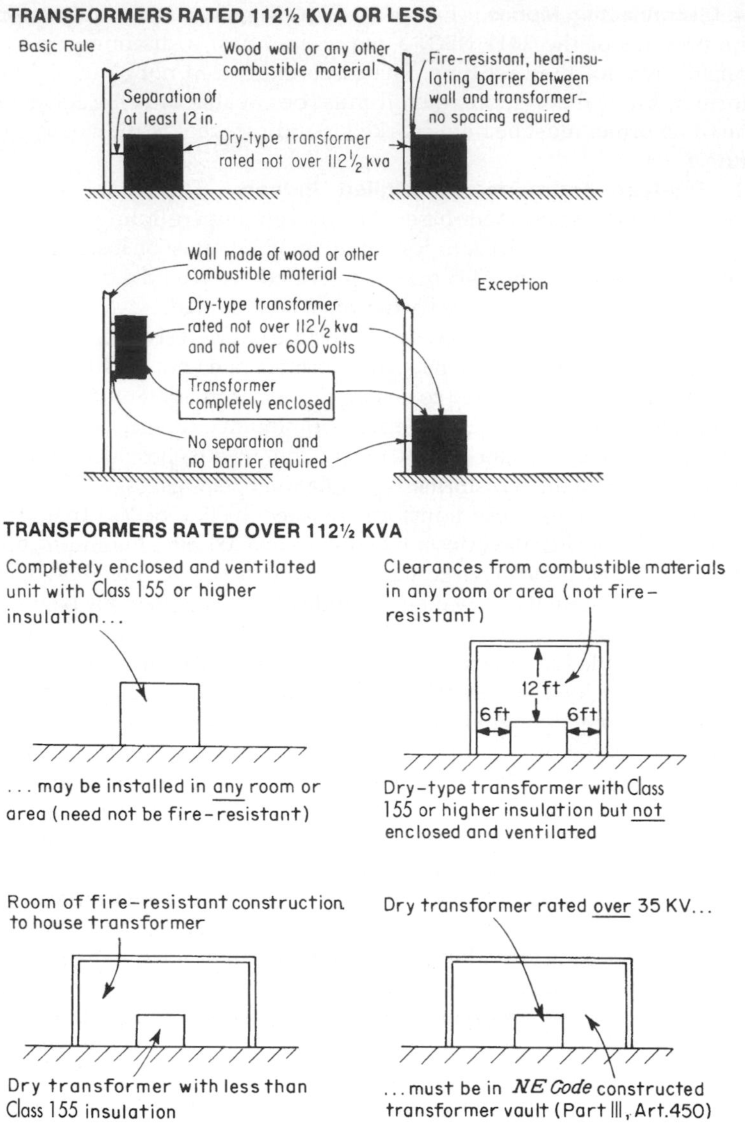
TRANSFORMERS RATED 112½ KVA OR LESS
Basic Rule
Wood wall or any other combustible material
Fire-resistant, heat-insulating barrier between wall and transformer–no spacing required
Separation of at least 12 in.
Dry-type transformer rated not over 112½ kva
Wall made of wood or other combustible material
Exception
Dry-type transformer rated not over 112½ kva and not over 600 volts
Transformer completely enclosed
No separation and no barrier required
TRANSFORMERS RATED OVER 112½ KVA
Completely enclosed and ventilated unit with Class 155 or higher insulation...
... may be installed in any room or area (need not be fire-resistant)
Clearances from combustible materials in any room or area (not fire-resistant)
12 ft
6 ft
6 ft
Dry-type transformer with Class 155 or higher insulation but not enclosed and ventilated
Room of fire-resistant construction to house transformer
Dry transformer with less than Class 155 insulation
Dry transformer rated over 35 KV...
... must be in NE Code constructed transformer vault (Part III, Art.450)

Fig. 450-27. Construction of dry-type transformer affects indoor installation rules. (Sec. 450.21.)

actually conservative for ambient temperature below normal operating temperatures and optimistic above it.

For proper cooling, dry-type transformers depend on circulation of clean air—free from dust, dirt, or corrosive elements. Filtered air is preferable and may be mandatory in some cases of extreme air pollution. In any case, it can reduce maintenance.

In restricted spaces—small basement mechanical rooms and the like—ventilation must be carefully checked to ensure proper transformer operating temperature. The usual requirement is for 100 cfm of air movement for each kilowatt of transformer loss. Areas of inlet and outlet vent openings should be at least 1 net sq ft per 100 kVA of rated transformer capacity.

Height of vault, location of openings, and transformer loading affect ventilation. One manufacturer calls for the areas of the inlet and outlet openings to be not less than 60 sq ft per 1000 kVA when the transformer is operating under full load and is located in a restricted space. And a distance of 1 ft should be provided on all sides of dry-type transformers as well as between adjacent units.

Freestanding, floor-mounted units with metal grilles at the bottom must be set up off the floor a sufficient distance to provide the intended ventilation draft up through their housings.

The installation location must not expose the transformer housing to damage by normal movement of persons, trucks, or equipment. Ventilation openings should not be exposed to vandalism or accidental or mischievous poking of rubbish, sticks, or rods into the windings. Adequate protection must be provided against possible entry of small birds or animals.

450.22. Dry-Type Transformers Installed Outdoors. A transformer that sustains an internal fault that causes arcing and/or fire presents the same hazard to adjacent combustible material whether it is installed indoors or outdoors. For that reason, a clearance of at least 12 in. (300 mm) is required between any dry-type transformer rated over 112½ kVA and combustible materials of buildings where installed outdoors.

As an alternative, the clearance of 12 in. (300 mm) from combustible building materials is *not* required for outdoor dry-type transformers that have a Class 155 insulation system and are completely enclosed, except for ventilation openings. The same consideration given for an 80°C-rise (Class 155) transformer is made outdoors as it is indoors, in Exception No. 2 in 450.21(B).

450.23. Less-Flammable Liquid-Insulated Transformers. 450.23 covers the liquid-filled transformers that have essentially replaced askarel-insulated transformers. Because oil-filled transformers used indoors require a transformer vault, the *less-flammable* (also called *high-fire-point*) insulated transformer offers an alternative to the oil-filled transformers, without the need for a vault. This Code section permits installation of these high-fire-point liquid-insulated transformers indoors or outdoors. Over 35 kV indoors, such a transformer must be in a vault.

The rules of this section recognize that these various high-fire-point liquid-insulated dielectrics are less flammable than the mineral oil used in oil-filled transformers but not as fire-resistant as askarel. Because these askarel substitutes will burn to some degree, Code rules are aimed at minimizing any fire hazards:

1. Less-flammable liquid dielectrics used in transformers must, first of all, be listed—that is, tested and certified by a testing laboratory or organization and shown in a published listing as suitable for application. *Less-flammable* liquids for transformer insulation are defined as having "a fire point of not less than 300°C."

2. Transformers containing the high-fire-point dielectrics may be used *without a vault* but only within buildings of noncombustible construction or limited combustibility (brick, concrete, etc.) and then only in rooms or areas that do not contain combustible materials. A Type I or Type II building is a building of noncombustible construction or limited combustibility, as described in the informational note and also in the Informative Annex E at the end of the Code book, and there must be no combustible materials stored in the area where the transformer is installed.
3. The entire installation must satisfy all conditions of use, as described in the listing of the liquid. There are two major testing laboratories with listing programs for these transformers (FM and UL) and there are very substantial differences between their procedures. Make certain to carefully explore the requirements that are imposed by the particular listing agency. The 2017 NEC added identical informational notes on this topic, one following 450.23(A)(1)d and one following 450.23(B)(1), which provide context allowing some advance understanding of the sort of conditions that one might expect as part of listing restrictions on the liquid. These could include the maximum tank pressure, whether pressure-relief valves must be provided, appropriate fuse characteristics, and how overcurrent protection is sized, among other possibilities.

   Be careful: The wording in 450.23(A) has been incorrectly formatted due major software issues involved in the processing of the 2017 edition, and it has become completely unusable as a result. NFPA has refused to classify this as an erratum, so it will stand until it is corrected with a TIA. The overview is that the mandatory provisions of the rule are supposed to be unchanged from the 2014 edition. Indoor installations are supposed to be permissible provided you comply with one of three approaches, (1), (2), or (3). The first approach should have four conditions, set up as a list of four lettered items (a., b., c., and d.). Then follows, correctly and new for the 2017 edition, is the informational note described in the prior paragraph, and directly addressing item d (hence its location.) Following this are supposed to come the other two approaches, which should have been numbered (2) (on sprinklers and containment) and (3) (on compliance with flammable liquid rules). Instead, the software caused these provisions to print as items e and f, making them additional factors in the first option, and removing them as options in their own right. Until this is corrected, you will need to work with the inspector on this.
4. A liquid-confinement area must be provided around such transformers that are not in a vault, because tests indicate these liquids are not completely nonpropagating—that is, if they are ignited, the flame will be propagated along the liquid. A propagating liquid must be confined to a given area to confine the flame of its burning (Fig. 450-28). The liquid-confinement area (a curb or dike around the transformer) must be of sufficient dimensions to contain the entire volume of liquid in the transformer.

A less-flammable liquid-insulated transformer installed in such a way that all of conditions 2 and 3 are not satisfied must be *either*

1. Provided with an automatic fire extinguishing system and a liquid-confinement area, or
2. Installed in a Code-specified transformer vault (Part III of Art. 450), without need for a liquid-confinement area

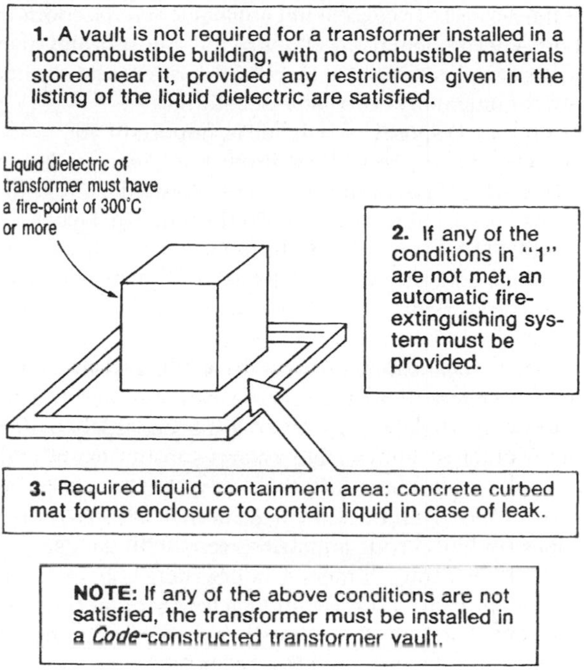
1. A vault is not required for a transformer installed in a noncombustible building, with no combustible materials stored near it, provided any restrictions given in the listing of the liquid dielectric are satisfied.
Liquid dielectric of transformer must have a fire-point of 300°C or more
2. If any of the conditions in "1" are not met, an automatic fire-extinguishing system must be provided.
3. Required liquid containment area: concrete curbed mat forms enclosure to contain liquid in case of leak.
NOTE: If any of the above conditions are not satisfied, the transformer must be installed in a Code-constructed transformer vault.

Fig. 450-28. Transformers containing askarel-substitute liquids must satisfy specific installation requirements. (Sec. 450.23.) This drawing shows the way the NEC is intended to read by the Code-making panel, and not the way it has been printed. See text.

Less-flammable liquid-insulated transformers rated over 35 kV and installed indoors must be enclosed in a Code-constructed transformer vault. All less-flammable liquid-insulated transformers installed outdoors may be attached to or adjacent to or on the roof of Type I or Type II buildings. Such installation at other than Type I or Type II buildings, where adjacent to combustible material, fire escapes, or door or window openings, must be guarded by fire barriers, space separation, and compliance with instructions for using the particular liquid.

Because these rules are general in nature and lend themselves to a variety of interpretations, application of these requirements may depend heavily on consultation with inspection authorities.

Although askarel-filled transformers up to 35 kV were used for many years for indoor applications because they do not require a transformer vault, there has been a sharp, abrupt discontinuance of their use over recent years. Growth in the ratings, characteristics, and availability of dry-type high-voltage transformers has accounted in major part for the reduction of askarel units. But another factor that has led to the rejection of askarel transformers in recent years is the environmental objections to the askarel liquid itself.

A major component of any askarel fluid is one or more polychlorinated biphenyls (PCBs), a family of chemical compounds designated as harmful environmental pollutants because they are nonbiodegradable and cannot be readily disposed of.

Thus, although the askarels are excellent coolants where freedom from flammability is important, environmental objections to the sale, use, and disposal of PCBs have eliminated new applications of askarel transformers and stimulated a search for a nontoxic, environmentally acceptable substitute.

Proper handling and disposal of askarel is important for units still in use. A regulation of the EPA (Environmental Protection Agency), No. 311, required that all PCB spills of 1 lb (454 g) or more must be reported. Failure to report a spill is a criminal offense punishable by a $10,000 fine and/or 1-year imprisonment. Both the EPA and OSHA have objected to the use of askarels. A manufacturer of askarel has established a program for disposal of spent or contaminated PCB fluid using an incinerator that completely destroys the fluid by burning it at over 2000°F.

Non-PCB dielectric coolant fluids for use in small- and medium-sized power transformers as a safe alternate to askarels are available, and transformers using these new high-fire-point dielectric coolants have been widely used.

Extensive data from tests on available askarel substitutes show that they provide a high degree of safety. Such fluids do have NEC and OSHA recognition. Responsibility for proper clearances with insurance underwriters, government regulating agencies, and local code authorities rests with the user or purchaser of the fluid in new or refilled transformers. Underwriters Laboratories does not test or list liquid-filled equipment. Both UL and Factory Mutual Research Laboratory have been involved in providing a classification service of flammability. The EPA has commented favorably on such high-fire-point fluids.

Few physical changes to transformers are necessary when using the new fluid dielectrics. However, load ratings on existing units may be reduced about 10 percent because of the difference in fluid viscosity and heat conductivity compared with askarel. The high-fire-point fluids cost about twice as much as askarel. Purchase price of a new transformer filled with the fluid (such as a typical 1000-kVA loadcenter unit) is about 10 to 15 percent more than an askarel-filled unit. But the economics vary for different fluids and must be carefully evaluated.

High-fire-point liquid-insulated transformers require no special maintenance procedures. The liquids exhibit good dielectric properties over a wide range of temperatures and voltage stress levels, and they have acceptable arc-quenching capabilities. They have a high degree of thermal stability and a high resistance to thermal oxidation that enables them to maintain their insulating and other functional properties for extended periods of time at high temperatures.

Because silicone liquids will ignite at 750°F (350°C), they are not classed as fire-resistant. However, if the heat source is removed or fluid temperature drops below 750°F, burning will stop. The silicone fluids are thus self-extinguishing.

450.24. Nonflammable Fluid-Insulated Transformers. This section permits indoor and outdoor use of transformers that utilize a noncombustible fluid dielectric, which is one that does not have a flash point or fire point and is not flammable in air (Fig. 450-29). As an alternative to askarel-insulated transformers, these transformers offer high BIL ratings and other features of operation similar to high-fire-point dielectric-insulated transformers—without concern for flammability. Such transformers do not, therefore, have the restrictions that are set down in 450.23 for the high-fire-point-liquid transformers.

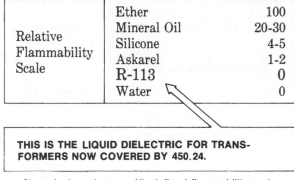

Relative Flammability Scale		
	Ether	100
	Mineral Oil	20-30
	Silicone	4-5
	Askarel	1-2
	R-113	0
	Water	0

THIS IS THE LIQUID DIELECTRIC FOR TRANSFORMERS NOW COVERED BY 450.24.

Chart is based on a UL-defined flammability scale as tested by ASTM.

Fig. 450-29. Comparison of the relative flammabilities of liquid dielectrics compared with ether. (Sec. 450.24.)

Nonflammable fluid-insulated transformers installed indoors must have a liquid-confinement area and a pressure-relief vent. In addition, such transformers must be equipped to absorb gases generated by arcing inside the tank, or the pressure-relief vent must be connected to a flue or duct to carry the gases to "an environmentally safe area." Units rated over 35 kV must be installed in a vault when used indoors.

450.25. Askarel-Insulated Transformers Installed Indoors. Although askarel transformers are being phased out, the Code rule says such transformers installed indoors must conform to the following:

1. Units rated over 25 kVA must be equipped with a pressure-relief vent.
2. Where installed in a poorly ventilated place, they must be furnished with a means for absorbing any gases generated by arcing inside the case, or the pressure-relief vent must be connected to a chimney or flue, which will carry such gases outside the building (Fig. 450-30).
3. Units rated over 35,000 V must be installed in a vault.

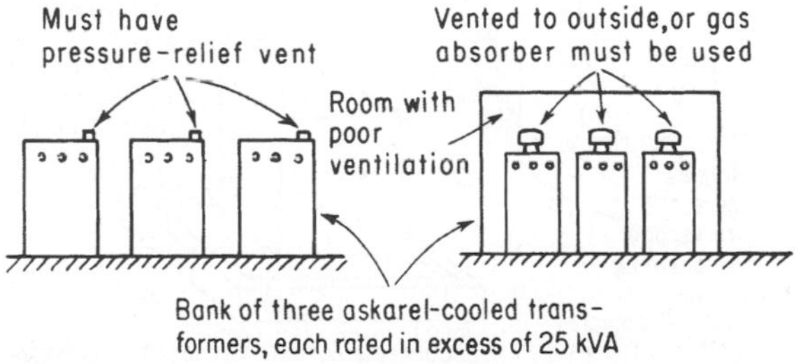
Must have
pressure-relief vent
Vented to outside,or gas
absorber must be used
Room with
poor
ventilation
Bank of three askarel-cooled transformers, each rated in excess of 25 kVA

UNITS RATED OVER 35,000 VOLTS MUST BE USED IN A VAULT

Fig. 450-30. Code rules still cover askarel transformers. (Sec. 450.25.)

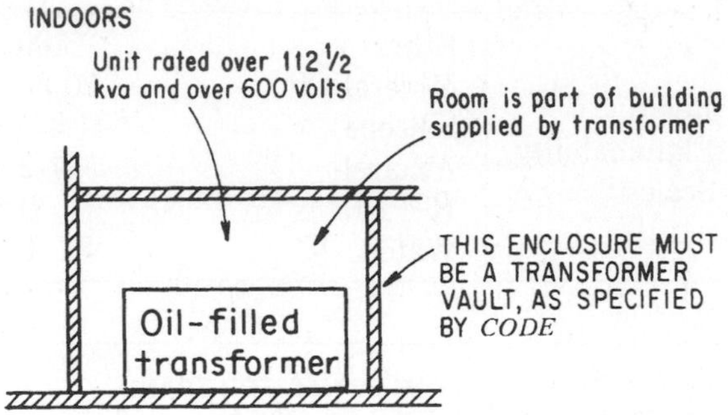
INDOORS
Unit rated over 112 1/2 kva and over 600 volts
Room is part of building supplied by transformer
THIS ENCLOSURE MUST BE A TRANSFORMER VAULT, AS SPECIFIED BY *CODE*
Oil-filled transformer

Fig. 450-31. Oil-filled transformers generally require installation in a *vault*. (Sec. 450.26.)

450.26. Oil-Insulated Transformers Installed Indoors. The basic rule is illustrated in Fig. 450-31. Oil-insulated transformers installed indoors must be installed in a vault constructed according to Code specs, but the exceptions note general and specific conditions under which a vault is not necessary. The most commonly applied exceptions are as follows:

1. A hookup of one or more units rated not over 112½ kVA may be used in a vault constructed of reinforced concrete not less than 4 in. thick.
2. Units installed in detached buildings used only for providing electric service do not require a Code-constructed vault if no fire hazard is created and the interior is accessible only to qualified persons.

450.27. Oil-Insulated Transformers Installed Outdoors. Figure 450-32 shows how physical locations of building openings must be evaluated with respect to potential fire hazards from leaking transformer oil.

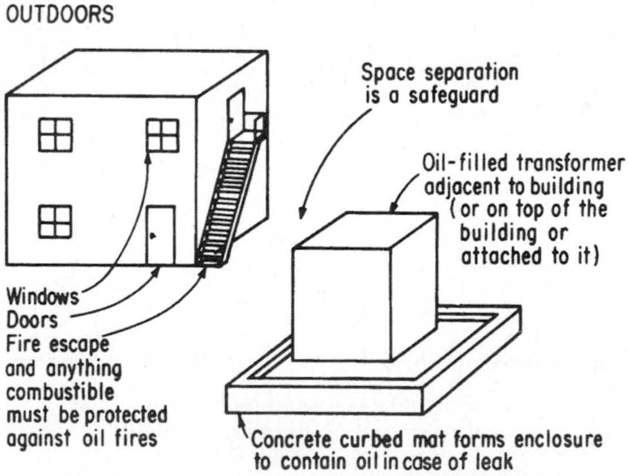
OUTDOORS
Space separation is a safeguard
Oil-filled transformer adjacent to building (or on top of the building or attached to it)
Windows
Doors
Fire escape
and anything combustible must be protected against oil fires
Concrete curbed mat forms enclosure to contain oil in case of leak

Fig. 450-32. Precautions must be taken for outdoor oil transformers. (Sec. 450.27.)

The outdoor locations need to be considered in light of the access requirement in 450.13. Some locations that comply with this section may be judged insufficiently accessible. The informational note following 450.27 points out that additional information can be found in the *National Electrical Safety Code* (*not* the **NEC**). That code covers use of such transformers as shown in Fig. 450-33. But, because this note is only informational and presents no rule, the requirements of 450.13 must be observed.

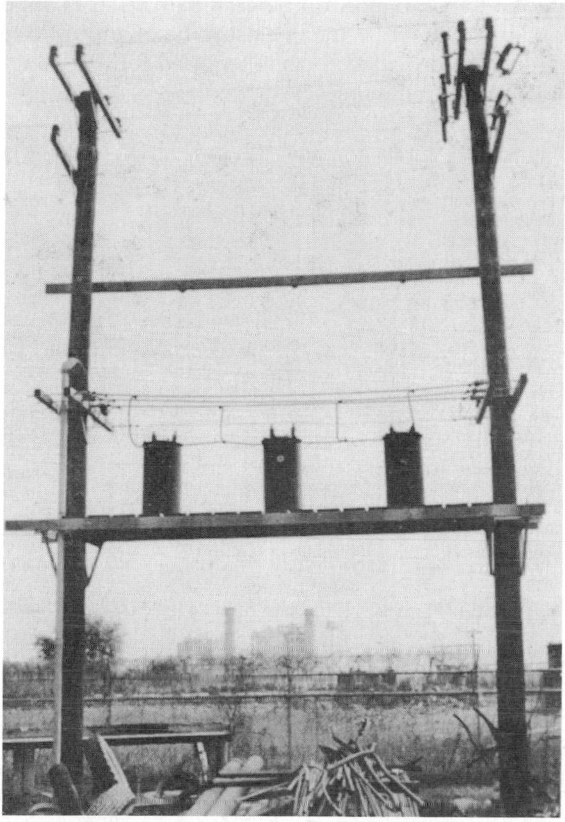

Fig. 450-33. High mounting of oil-filled transformers would require use of a portable ladder for access to the units. But 450.27 covers such units installed on poles or structures. (Sec. 450.27.)

450.28. Modification of Transformers. Askarel transformers that are drained and refilled with another liquid dielectric must be identified as such and must satisfy all rules of their retrofilled status. This rule is intended to maintain safety in all cases where askarel transformers are modified to eliminate PCB hazards. Marking must show the new condition of the unit and must not create Code violations. For instance, an indoor askarel transformer that is drained and refilled with oil may require construction of a vault, which is required for oil-filled transformers as specified in 450.26.

450.41. Location. Ideally, a transformer vault should have direct ventilating openings (grilles or louvers through the walls) to outdoor space. Use of ducts or flues for ventilating is not necessarily a Code violation, but should be avoided wherever possible.

450.42. Walls, Roof, and Floor. Basic mandatory construction details are established for an NEC-type transformer vault, as required for oil-filled transformers and for all transformers operating at over 35,000 V. The purpose of a transformer vault is to isolate the transformers and other apparatus. It is important that the door as well as the remainder of the enclosure be of proper construction and that a substantial lock be provided. Details required for any vault are shown in Fig. 450-34 and include the following:

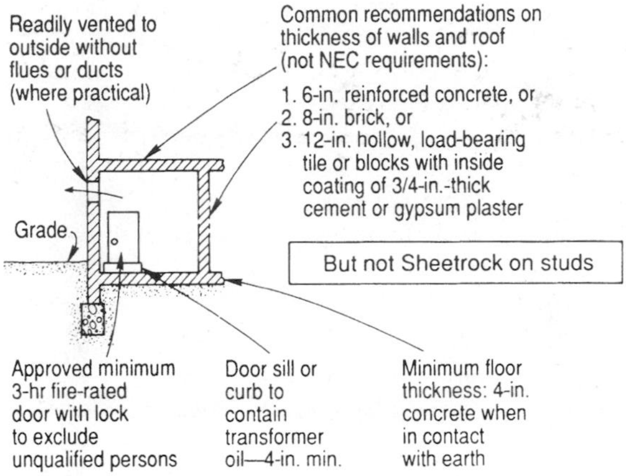
Readily vented to outside without flues or ducts (where practical)
Common recommendations on thickness of walls and roof (not NEC requirements):
1. 6-in. reinforced concrete, or
2. 8-in. brick, or
3. 12-in. hollow, load-bearing tile or blocks with inside coating of 3/4-in.-thick cement or gypsum plaster
But not Sheetrock on studs
Grade
Approved minimum 3-hr fire-rated door with lock to exclude unqualified persons
Door sill or curb to contain transformer oil—4-in. min.
Minimum floor thickness: 4-in. concrete when in contact with earth

Fig. 450-34. Transformer vault must assure containment of possible fire. (Sec. 450.42.)

1. Walls and roofs of vaults shall be constructed of reinforced concrete, brick, load-bearing tile, concrete block, or other fire-resistive constructions with adequate strength and a fire resistance of 3 h according to ASTM Standard E119-99. "Stud and wall board construction" may not be used for walls, roof, or other surfaces of a transformer vault. Although the rule here does *not* flatly mandate concrete or masonry construction, that is essentially the objective of the wording. The substantiation for this rule said the following:

> The only guidance provided in Sec. 450.42 as to the type of material to be used in vault construction is in an informational note which states that "six-in. (152-mm) thick reinforced concrete is a typical 3-h construction." This, of course, is only advisory. Sheetrock can be so installed as to have a 3-h fire rating, however, it should not be considered as being suitable for this type of installation. It would not have adequate structural strength in case of oil fire. It is not too difficult to break through such a wall, either intentionally or unintentionally. This is why the NEC specifically disallows it.
>
> Your attention is also called to Sec. 230.6(3), which states that conductors installed in a transformer vault shall be considered outside of a building. The major thrust of 230.6 has been that the conductors are to be masonry-encased.

2. A vault must have a concrete floor not less than 4 in. (100 mm) thick when in contact with the earth. When the vault is constructed with space below it, the floor must have adequate structural strength and a minimum fire resistance of 3 h. Six in. (150 mm) thick reinforced concrete is a typical 3-h-rated construction.
3. Building walls and floors that meet the preceding requirements may serve for the floor, roof, and/or walls of the vault.

An exception to the basic regulations establishing the construction standards for transformer fireproof vaults notes that the transformer-vault fire rating may be reduced where the transformers are protected with automatic sprinkler, water spray, or carbon dioxide. The usual construction standards for transformer vaults (such as 6 in. [150 mm] thick reinforced concrete) provide a minimum fire-resistance rating of 3 h. Where automatic sprinkler, water spray, or carbon dioxide is used, a construction rating of only 1 h will be permitted.

450.43. Doorways. Each doorway must be of 3-h fire rating as defined in the *Standard for the Installation of Fire Doors and Windows* (NFPA No. 80-1999). The Code-enforcing authority may also require such a door for doorways leading from the vault to the outdoors, in addition to any doorways into adjoining space in the building.

As required in part **(C)**, vault doors must swing *out* and must be equipped with "listed panic hardware."

This is intended to provide the greater safety of a push-open rather than a rotating-knob type of door release. As noted in the substantiation:

> Conventional rotating door knob hardware is used on transformer vault doors due to lack of specific wording in the paragraph as presently written. The *National Electrical Safety Code* is believed to be very specific, or has been formally interpreted to be, requiring "panic type" door hardware. In an electrical flash or arc an electrical worker may lose the use of hands for twisting a conventional door knob.

In accordance with the concept of providing greater protection for personnel within the vault, other Code rules have also adopted similar wording. See 110.26(C)(3).

450.45. Ventilation Openings. This rule sets the size and arrangement of vent openings in a vault where such ventilation is required by ANSI C57.12.00-1993—"General Requirements for Liquid-Immersed Distribution, Power, and Regulating Transformers," as noted in 450.9. Figure 450-35 shows the openings as regulated

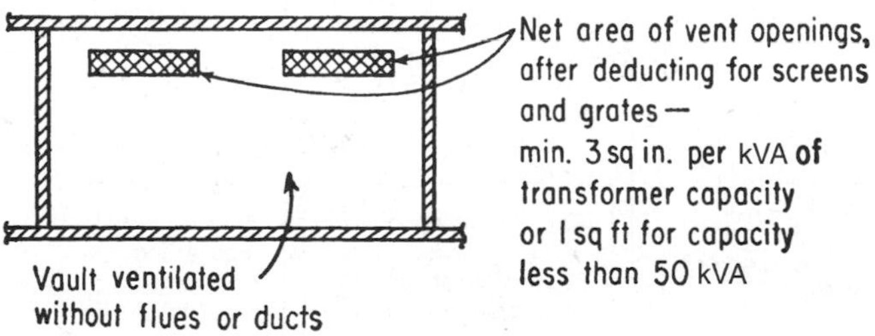
Net area of vent openings, after deducting for screens and grates —
min. 3 sq in. per kVA of transformer capacity or 1 sq ft for capacity less than 50 kVA
Vault ventilated without flues or ducts

Fig. 450-35. Vault vent opening(s) may be in or near the roof—or near floor level also, if one is at or near roof level. (Sec. 450.45.)

by part (C) of this section. One or more openings may be used, but if a single vent opening is used, it must be in or near the roof of the vault—and not near the floor.

450.47. Water Pipes and Accessories. This section addresses a crucial topic in electrical work. Sprinklers belong in transformer vaults and they are absolutely not to be considered as foreign to the electrical installation. For many electricians, the idea of providing water-based sprinkler protection for electrical equipment is anathema. "Water and electricity don't mix" goes the argument. However, it has been conclusively demonstrated that the loss experience associated with an electrical fire is far lower when that fire is controlled with an automatic sprinkler.

The express permission to use water-based sprinkler protection has been in this part of the NEC for about 65 years. A major challenge to this concept occurred during the 1987 NEC cycle, when the panel advanced a proposal requiring outdoor rated transformers in vaults that had sprinkler protection. Everyone understood that the cost differential in providing this rating would be a serious impediment to providing the protection. The initial panel action drew a large number of extremely negative comments, and when the panel reconvened to act on the comments it unanimously rejected the concept.

ARTICLE 455. PHASE CONVERTERS

Refer to Fig. 455-1, which shows details of the rules given in 455.2, 455.4, 455.6, 455.7, and 455.8.

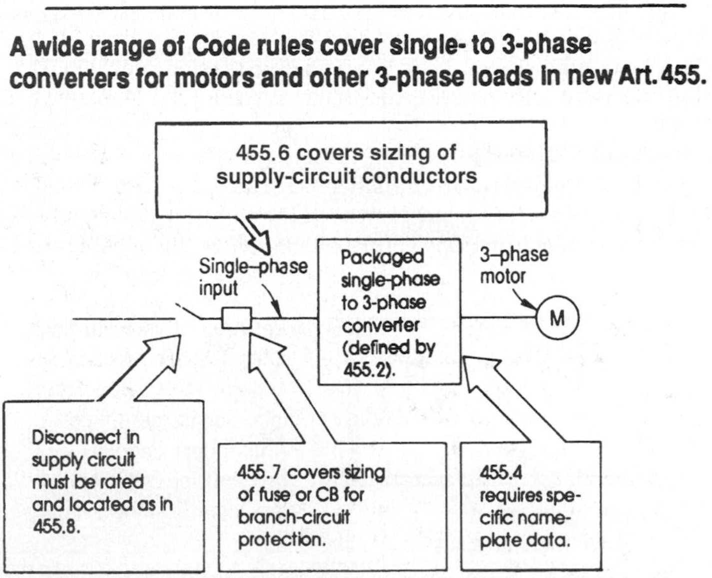
A wide range of Code rules cover single- to 3-phase converters for motors and other 3-phase loads in new Art. 455.
455.6 covers sizing of supply-circuit conductors
Single-phase input
Packaged single-phase to 3-phase converter (defined by 455.2).
3-phase motor
M
Disconnect in supply circuit must be rated and located as in 455.8.
455.7 covers sizing of fuse or CB for branch-circuit protection.
455.4 requires specific nameplate data.

Fig. 455-1. Code rules on phase converters cover these considerations. (Art. 455.)

455.2. Definitions. A phase converter is a static or a rotary device that converts single-phase power to 3-phase power. The informational note calls attention to the fact that phase converters, which do not produce perfect 3-phase power, alter the starting torque and locked-rotor current of motors connected to them.

455.6. Conductors. Part **(A)(1)** addresses the general case and requires that conductors supplying a phase converter must have an ampacity at least equal to 125 percent of the "phase converter nameplate single-phase input full-load amperes." Part **(A)(2)** covers instances where a specific load will be supplied, in which case the supply conductor ampacity must be at least 250 percent of the connected load. Currents are multiplied by the output-to-input voltage ratio as applicable if there is a voltage transformation involved. An informational note suggests 3 percent as a voltage drop allowance not to be exceeded in the single-phase portion of the circuit in order to promote proper starting and operation of the three-phase load.

Part **(B)** requires the manufactured phase conductor, which is the one originated by the converter and not connected to the line, to have a distinctive marking at all accessible locations and in a consistent manner throughout the system. The method of identification is not specified, but the color yellow appears to be the generally preferable choice. Section 455.9 prohibits connecting single-phase load to the manufactured phase, so it is essential to know which wire to avoid when making a single-phase load connection.

455.7. Overcurrent Protection. The rating of the overcurrent protective device protecting the phase converter supply conductors follows the same percentage rules (except stating the result as a maximum) as given for determining conductor ampacity, with an allowance for the next higher standard sized overcurrent device. The overcurrent protection cannot exceed 125 percent of the rated full-load input current.

There are no rules in this article for overcurrent protection on the load side of the converter. Presumably these conductors are a form of a tap and are subject to the rules in Secs. 240.4 and 240.21. Applying the 10-ft tap rule in 240.21(B) would clearly comply with Code rules.

455.8. Disconnecting Means. A readily accessible disconnecting means that opens the single-phase ungrounded supply conductors simultaneously must be located in sight of the phase converter. The disconnecting means must be a circuit breaker or molded-case switch, or a switch rated in horsepower. Nonmotor loads may use a switch rated in amperes. For general purposes, the disconnecting means must have a rating equal to 115 percent of the "phase converter nameplate single-phase input full-load amperes." If specific loads are supplied then the required rating depends on the type of disconnecting means used: Note that, in 455.20, the single-phase disconnecting means for a static phase converter supplying a single load may also serve as the disconnecting means for that load, provided it is in sight.

A circuit breaker or molded-case switch is to be based on at least 250 percent of the 3-phase full-load amperes, adjusted for voltage as required. The procedure for these disconnects parallels that for sizing overcurrent protection.

A horsepower-rated switch is to be based on 200 percent of the equivalent horsepower of the 3-phase load. The horsepower-rated switch rule uses a summation of up to three numbers to obtain a lock-rotor ampere rating: the nonmotor loads; the 3-phase locked rotor current of the largest motor, as given in Table 430.251(B); and the full-load currents of all other motors operating at the same time. The sum of these numbers is then multiplied by the output-to-input voltage ratio if necessary, and then taken back into Table 430.151(B) for reconversion into horsepower.

455.21. Start-Up. A rotary converter must be started before the 3-phase load is energized. If a rotary phase converter is started under load, the output phase angles will not usually allow the 3-phase motor to start properly, risking burnout of components and fire. The next section requires that rotary converter installations be arranged so that if power to the converter fails, the loads cannot be reenergized (until the converter has successfully restarted, per Sec. 455.21). An informational note advises that magnetic contactors, etc., with manual or time-delay restarting will comply with this rule.

455.23. Capacitors. Capacitors installed for a motor load (and not as an integral part of the phase conversion system) must be installed on the line side of the motor running overload protection. This avoids the necessity to apply 460.9 to correct for improved power factor in sizing running overload protection if capacitors are on the load side of that protection. As implied in the informational note following the phase converter definition in 455.2, that power factor may be difficult to accurately predict.

ARTICLE 460. CAPACITORS

460.1. Scope. The sections in this article apply chiefly to capacitors used for the power-factor correction of electric-power installations in industrial plants and for correcting the power factors of individual motors (Fig. 460-1). These provisions

Fig. 460-1. A typical power-factor correction capacitor bank is this 300-kvar bank of twelve 25-kvar, 480-V capacitor units installed in a steel enclosure in an outdoor industrial substation. (Sec. 460.1.)

apply only to capacitors used for surge protection where such capacitors are not component parts of other apparatus.

In an industrial plant using induction motors, the power factor may be considerably less than 100 percent, particularly when all or part of the motors operate most of the time at much less than their full load. The lagging current can be counteracted and the power factor improved by installing capacitors across the line. By raising the power factor, for the same actual power delivered the current is decreased in the generator, transformers, and lines, up to the point where the capacitor is connected.

Figure 460-2 shows a capacitor assembly connection to the main power circuit of a small industrial plant, consisting of capacitors connected in a 3-phase hookup and rated at 90 kVA for a 460-V system. An externally operable switch mounted on the wall is used as the disconnecting means, and the discharge device required by 460.6 consists of two high-impedance coils inside the switch enclosure, which consume only a small amount of power but, having a comparatively low dc resistance, permit the charge to drain off rapidly after the capacitor assembly has been disconnected from the line.

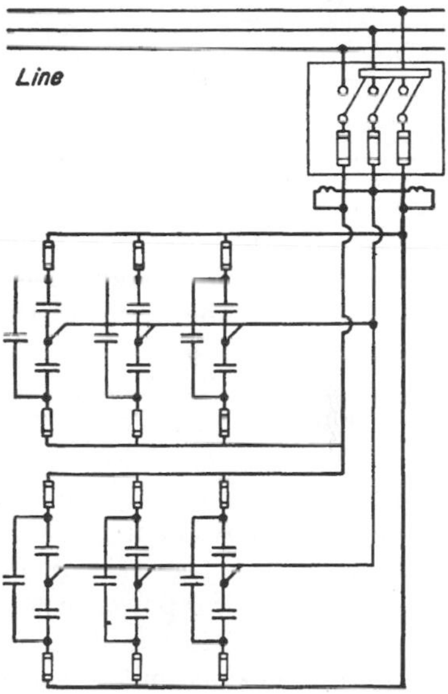
Line

Fig. 460-2. Six internally delta-connected capacitors form a 3-phase capacitor bank. (Sec. 460.1.)

460.6. Discharge of Stored Energy. If no means were provided for draining off the charge stored in a capacitor after it is disconnected from the line, a severe shock might be received by a person servicing the equipment, or the equipment might be damaged by a short circuit. If a capacitor is permanently connected to the windings of a motor, as in Fig. 460-3, the stored charge will drain off rapidly through the windings when the circuit is opened. Reactors or resistors used as discharge devices either must be permanently connected across the terminals

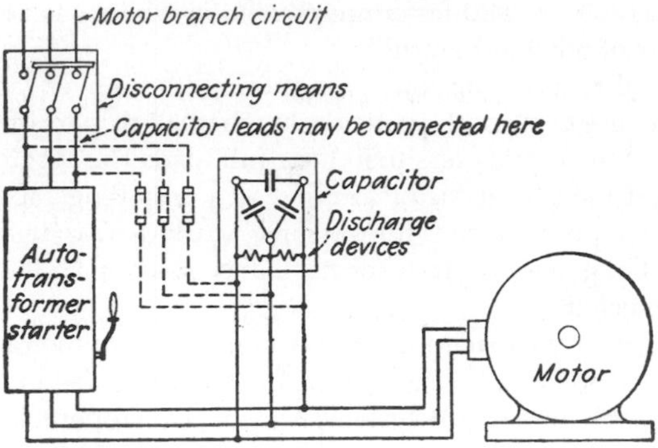
Motor branch circuit
Disconnecting means
Capacitor leads may be connected here
Capacitor
Discharge devices
Auto-transformer starter
Motor

Fig. 460-3. Capacitor voltage must be discharged when circuit is opened. (Sec. 460.6.)

of the capacitor (such as within the capacitor housing) or a device must be provided that will automatically connect the discharge devices when the capacitor is disconnected from the source of supply. Most available types of capacitors have discharge resistors built into their cases. When capacitors are not equipped with discharge resistors, a discharge circuit must be provided.

Figure 460-3 shows a capacitor used to correct the power factor of a single motor. The capacitor may be connected to the motor circuit between the starter and the motor or may be connected between the disconnecting means and the starter, as indicated by the dotted lines in the diagram. If connected as shown by the dotted lines, an overcurrent device must be provided in these leads, as required by 460.8(B). The capacitor is shown as having discharge devices consisting of resistors.

Power capacitors, in most applications, are installed to raise the system power factor, which results in increased circuit or system current-carrying capacity, reduced power losses, and lower reactive power charges (most utility companies include a power-factor penalty clause in their industrial billing). Also, additional benefits derived as a result of a power capacitor installation are reduced voltage drop and increased voltage stability. Figure 460-4 presents basic data on calculating size of capacitors for power-factor correction. However, manufacturers provide tables and graphs to help select the capacitor for a given motor load.

In the past, the Code limited power-factor correction to unity (100 percent, or 1.0) when there is no load on the motor. That will result in a power factor of 95 percent or better when the motor is fully loaded. The old rule recognized the use of capacitors sized *either* for the value that will produce 100 percent power factor of the circuit when the motor is running at no load *or* for a value equal to 50 percent of the kilovolt-ampere rating of the motor input for motors up to 50 hp, 1000 V.

That provision was removed as of the 1984 **NEC** based on an important and detailed proposal from NEMA. The proposal pointed out that the prescriptive limits then 460.7 were design considerations that were of chiefly economic impact, and did not address hazards to personnel. The proposal also described the impact

Power-factor capacitors can be connected across electric lines to neutralize the effect of lagging power-factor loads, thereby reducing the current drawn for a given kilowatt load. In a distribution system, small capacitor units may be connected at the individual loads or the total capacitor kilovolt-amperes may be grouped at one point and connected to the main. Although the total kvar of capacitors is the same, the use of small capacitors at the individual loads reduces current all the way from the loads back to the source and thereby has greater PF corrective effect than the one big unit on the main, which reduces current only from its point of installation back to the source.

Calculating Size of Capacitor:

Assume it is desired to improve the power factor a given amount by the addition of capacitors to the circuit.

Then $kvar_R = kw \times (\tan \theta_1 - \tan \theta_2)$

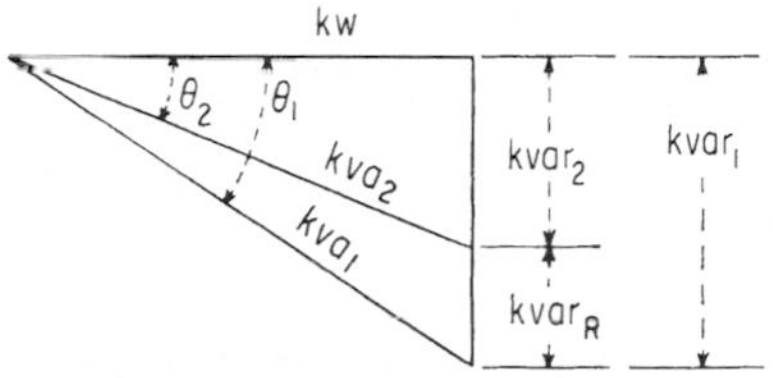
kw
θ2
θ1
kva2
kva1
kvar2
kvar1
kvarR

where $kvar_R$ = rating of required capacitor
$kvar_1$ = reactive kilovolt-amperes at original PF
$kvar_2$ = reactive kilovolt-amperes at improved PF
θ_1 = original phase angle
θ_2 = phase angle at improved PF
kw = load at which original PF was determined.

NOTE: The phase angles θ_1 and θ_2 can be determined from a table of trigonometric functions using the following relationships:
θ_1 = The angle which has its cosine equal to the decimal value of the original power factor (e.g., 0.70 for 70% PF; 0.65 for 65%; etc.)
θ_2 = The angle which has its cosine equal to the decimal value of the improved power factor.

Fig. 460-4. Capacitors reduce circuit current by supplying the magnetizing current to motors. (Sec. 460.6.)

of increasing motor efficiencies that would likely result in inadvertent "premature motor failure as the result of overvoltage and potential transient torques and currents as compared to previous designs." The substantiation also pointed out that the 50-hp parameter was no magic number. A 50-hp or a 1000-hp motor was subject to the same issues because the electromagnetic principle for both were identical. However, losing a 50-hp motor was far less costly than a larger motor, so in essence the Code was capping the risk of over-excitation at a point where the economic impact was judged acceptable. Taken as a whole the issue needed to come out of the electrical safety standard and remain with the owners and their engineers. The panel unanimously agreed.

Nevertheless, the entire commentary as well as three drawings illustrating its application have been retained in all subsequent editions of this book. After over 25 years, motor efficiencies have further increased and variable-speed drive technology has advanced and proliferated, which also impacts power factor in ways never considered in the 1970s. In addition, increased harmonic loading that comes in part from these developments creates new challenges to installing capacitors, because they can place an electrical system into resonance, creating overvoltages and overcurrents. These issues require careful engineering study and in some cases the installation of line reactors to remove the resonant condition.

The previously retained material in this location did not explain any topic now subject to NEC regulation, nor did it address these more contemporary topics. We have decided to remove this material as dated and likely counterproductive in some cases.

460.8. Conductors. Part **(A)** of this section covers sizing of circuit conductors. The current corresponding to the kilovolt-ampere rating of a capacitor is computed in the same manner as for a motor or other load having the same rating in kilovolt-amperes. If a capacitor assembly used at 460 V has a rating of 90 kVA, the current rating is 90,000 VA/(480 V × √3) = 90,000 VA/(277 V × √3 × √3) = 90,000 VA/831 V = 108 A. The minimum required ampacity of the conductors would be 1.35 × 108 A, or 146 A.

The manufacturing standards for capacitors for power-factor correction call for a rating tolerance of "0, + 15 percent," meaning that the actual rating in kilovolt-amperes is never below the nominal rating and may be as much as 15 percent higher. Thus, a capacitor having a nameplate rating of 100 kVA might actually draw a current corresponding to 115 kVA. The current drawn by a capacitor varies directly with the line voltage, so that, if the line voltage is higher than the rated voltage, the current will be correspondingly increased. Also, any variation of the line voltage from a pure sine wave form will cause a capacitor to draw an increased current. It is for these reasons that the conductors leading to a capacitor are required to have an ampacity not less than 135 percent of the rated current of the capacitor.

example Given the kvar rating of capacitors to be installed for a motor, determining the correct capacitor conductor size is relatively simple. The rule here requires that the ampacity of the capacitor conductors be not less than one-third the ampacity of the motor circuit conductors and not less than 135 percent of the capacitor rated current. The capacitor nameplate will give rated kvar, voltage, and current. It is then a simple matter of multiplying rated current by 1.35 to obtain the ampacity value of the conductor to be installed and selecting the size of conductor required to carry that value of current, from Table 310.15(B)(16). Then check that the ampacity is not less than one-third the ampacity of the motor circuit conductors.

For a motor rated 100 hp, 460 V, 121 A full-load current, a 25-kvar capacitor would correct power factor to between 0.95 and 0.98 at full load. The nameplate on the capacitor indicates that the capacitor is rated 460 V, 31 A. Then 31 × 1.35 = 42 A. From Code Table 310.15(B)(16), a No. 6 TW conductor rated to carry 55 A might appear to do the job. (No. 8 THW rated at 50 A would most likely be considered not acceptable because UL and the NEC generally call for use of 60°C wires in circuits up to 100 A.) The motor circuit conductors are found to be 2/0 THW [based on Table 430.250 (124A) increased by 25 percent to 155A], with an ampacity of 175 A. Because $^1/_3$ × 175 = 58 A, the No. 6 THW, with an ampacity of 65 A, should be used.

If these conductors are connected to the load terminals of the motor controller, the overload protection heaters may have to be changed (or if the OL is adjustable, its setting may have to be reduced), because the capacitor will cause a reduction in line current and adjustment of relay setting is required by 460.9.

Although part **(B)** of the rule requires overcurrent protection (fuses or a CB) in each ungrounded conductor connecting a capacitor assembly to a circuit, the exception considers the motor-running overload relay in a starter to be adequate protection for the conductors when they are connected to the motor circuit on the load side of the starter. Where separate overcurrent protection is provided, as required for line-side connection, the device must simply be rated "as low as practicable." When a capacitor is thrown on the line, it may momentarily draw an excess current. A rating or setting of 250 percent of the capacitor current rating will provide short-circuit protection. Being a fixed load, a capacitor does not need overload protection such as is necessary for a motor.

Most power capacitors are factory equipped with fuses that provide protection in case of an internal short circuit. These fuses are usually rated from 165 to 250 percent of the rated kilovar current to allow for maximum operating conditions and momentary current surges. When installed on the load side of a motor starter, as noted previously, capacitors do not require additional fusing. However, for bank installations, separate fuses are required.

Part **(C)** of the rule requires a disconnecting means for all the ungrounded conductors connecting a capacitor assembly to the circuit—but a disconnect is not needed when the capacitor is connected on the load side of a starter with overload protection. The disconnect must be rated at least equal to 1.35 times the rated current of the capacitor.

Note that part **(C)(1)** requires a multipole switching device for the disconnect. This rule was originally adopted because of the inherent danger of single-pole switching of low-voltage capacitors. Normal switching or closing on faults may cause arcs or splattering of molten metal.

Two accepted methods of wiring capacitors are illustrated in Fig. 460-5. Diagram *A* shows the method of connection at a central location, such as at a power center or on a busway feeder. In such an installation, the Code rule requires an overcurrent device in each ungrounded conductor, a separate disconnecting means, and a discharge resistor (usually furnished with capacitors). The current rating of both the capacitor disconnect switch and the conductors supplying the capacitor must be not less than 135 percent of the rated current of the capacitor. In *B*, the capacitor is connected directly to motor terminals. Installation on the load side of the motor starter eliminates the need for separate overcurrent protection and separate disconnecting means. However, motor-running overcurrent protection must take into account the lower running current of the motor, as required by 460.6.

460.9. Rating or Setting of Motor Overload Device. When a power-factor capacitor is connected to a motor circuit at the motor—that is, on the load or motor side of the motor controller—the reactive current drawn by the motor is provided by the capacitor, and, as a result, the total current flowing in the motor circuit up to the capacitor is reduced to a value below the normal full-load current of the motor. With that hookup, the total motor full-load current flows only over the conductors from the capacitor connection to the motor, and the entire motor circuit up to that connection carries only the so-called working current or resistive current. That is shown in the top part of Fig. 460-6.

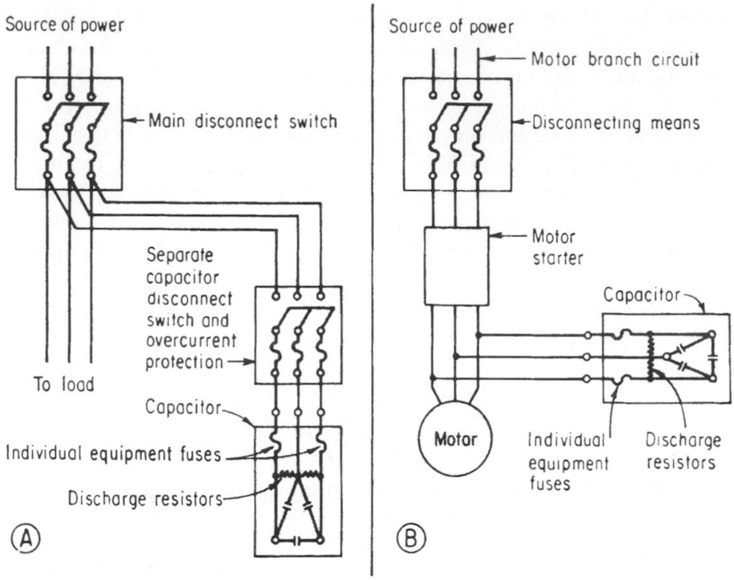
Source of power
Main disconnect switch
Separate capacitor disconnect switch and overcurrent protection
To load
Capacitor
Individual equipment fuses
Discharge resistors
A
Source of power
Motor branch circuit
Disconnecting means
Motor starter
Capacitor
Motor
Individual equipment fuses
Discharge resistors
B

Fig. 460-5. PF capacitor assembly may be connected on the line or load side of a starter. (Sec. 460.8.) Note that the discharge resistors are not required in drawing "B" because the charge will dissipate through the motor windings.

Under the conditions shown, it is obvious that setting the overload relay in the starter for 125 percent of the motor nameplate full-load current (as required by 430.32) would actually be an excessive setting for real protection of the motor, because considerably less than full-load current is flowing through the starter. The rule of this section clearly requires that the rating of a motor overload protective device connected on the line side of a power-factor correction capacitor must be based on 125 percent (or other percentage from 430.32) times the circuit current produced by the improved power factor—rather than the motor full-load current (see Fig. 460-6).

The 25-kvar capacitor used on the 100-hp, 460-V motor in the example in 460.8 will reduce the motor line current by about 9 percent. Section 430.32(A) (also 430.34 and 460.9) requires that the running overload protection be sized not more than 125 percent of motor full-load current produced with the capacitor. If the OL protection heaters were originally sized at 125 percent of the motor full-load current (1.25×121), they would have been sized at 151 A. With the motor current reduced by 9 percent (0.09×121, or 11 A), the motor full-load current with the capacitor installed would be $121 - 11$, or 110 A. Because 125 percent of 110 A is 137.5 A, the heaters must be changed to a size not larger than 137.5 A.

If the capacitor conductors could be connected on the line side of the heaters, the heaters would not have to be reduced in size, since the reduction of line current occurs only from the source back to the point of the capacitor connection. Conductor connections at this location are extremely difficult to make because of the lack of space and the large size of the connecting lugs. Controller load terminals are furnished with connectors that will accept an additional conductor, or they can be easily modified to permit a dependable connection.

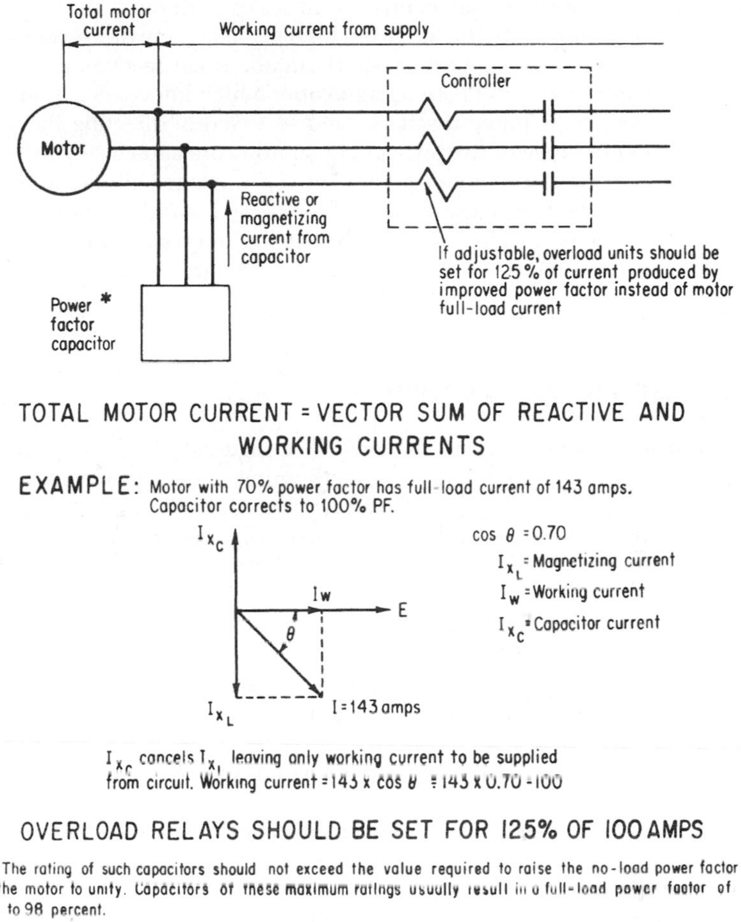
Total motor current
Working current from supply
Controller
Motor
Reactive or magnetizing current from capacitor
If adjustable, overload units should be set for 125% of current produced by improved power factor instead of motor full-load current
Power * factor capacitor
TOTAL MOTOR CURRENT = VECTOR SUM OF REACTIVE AND WORKING CURRENTS
EXAMPLE: Motor with 70% power factor has full-load current of 143 amps. Capacitor corrects to 100% PF.
cos θ = 0.70
I_{X_L} = Magnetizing current
I_W = Working current
I_{X_C} = Capacitor current
I_{X_C}
I_{X_L}
Iw
E
θ
I = 143 amps
I_{X_C} cancels I_{X_L} leaving only working current to be supplied from circuit. Working current = 143 x cos θ = 143 x 0.70 = 100
OVERLOAD RELAYS SHOULD BE SET FOR 125% OF 100 AMPS
* The rating of such capacitors should not exceed the value required to raise the no-load power factor of the motor to unity. Capacitors of these maximum ratings usually result in a full-load power factor of 95 to 98 percent.

Fig. 460-6. Motor overload protection must be sized for the current at improved PF. (Sec. 460.9.)

460.10. Grounding. The metal case of a capacitor is suitably grounded by locknut and bushing connections of grounded metal nipples or raceways carrying the conductors connecting the capacitor into a motor circuit or feeder.

ARTICLE 470. RESISTORS AND REACTORS

470.1. Scope. Except when installed in connection with switchboards or control panels that are so located that they are suitably guarded from physical damage and accidental contact with live parts, resistors should always be completely enclosed in properly ventilated metal boxes.

Large reactors are commonly connected in series with the main leads of large generators or the supply conductors from high-capacity network systems to assist in limiting the current delivered on short circuit. Small reactors are used with lightning arresters, the object here being to offer a high impedance to the passage of a high-frequency lightning discharge and so to aid in directing the discharge to ground. Another type of reactor, having an iron core and closely resembling a transformer, is used as a remote-control dimmer for stage lighting. Reactors, like resistors, are sources of heat and should therefore be mounted in the same manner as resistors. Reactors are also commonly installed in conjunction with variable-speed drives to improve power quality and performance.

ARTICLE 480. STORAGE BATTERIES

480.1. Scope. "Stationary installations of storage batteries" provide an independent source of power for emergency lighting, switchgear control, engine-generator set starting, signal and communications systems, laboratory power, and similar applications. They are an essential component of UPS systems. This Code article does not cover batteries used to supply the motive power for electric vehicles.

480.2. Definitions. **Nominal voltage (battery or cell)** The value assigned to a cell or battery for the purpose of having a convenient designation; the actual operating voltage may vary on either side of this number. It is now a generic definition, that applies to any storage battery technology. This definition replaced a definition that specified particular voltages.

Sealed cell or battery These have no provision for routine addition of water or electrolyte, or for external measurements of electrolyte specific gravity. They may contain a pressure relief venting arrangement. These batteries can have water or electrolyte added if necessary with a syringe-type tool, although this would not be routine maintenance.

Storage battery A battery with rechargeable cells.

Storage cells are of two general types: the so-called lead-acid type, in which the positive plates consist of lead grids having openings filled with a semisolid component, commonly lead peroxide, and the negative plates are covered with sponge lead, the plates being immersed in dilute sulfuric acid; and the alkali type, in which the active materials are nickel peroxide for the positive plate and iron oxide for the negative plate, and the electrolyte is chiefly potassium hydroxide (Fig. 480-1).

The most commonly used battery is the lead-acid type—either lead-antimony or lead-calcium. Nickel-cadmium batteries offer a variety of special features that, in many instances, offset their higher initial cost. Other types include silver-zinc, silver-cadmium, and mercury batteries.

The *lead-antimony battery* is readily available at a moderate price, has a high efficiency (85 to 90 percent), is comparatively small, and has a relatively long life if operated and maintained properly under normal conditions. Voltage output is about 2 V per cell; ratings range to about 1000 A-h (based on an 8-h discharge rate).

Fig. 480-1. Article 480 applies only to "stationary installations of storage batteries"—whether they are used for supply to lighting, generator cranking, switchgear control, or in UPS (uninterruptible power supply) systems. (Sec. 480.1.)

Lead-calcium batteries offer features similar to the lead-antimony type, and they require less maintenance. They do not require an "equalizing" charge (application of an overvoltage for a period of time to ensure that all cells in a battery bank will produce the same voltage). For this reason, they are often selected for use in UPS systems.

This type of cell can usually be operated for a year or more without needing water, depending on the frequency and degree of discharge. Sealed or maintenance-free batteries of this type never need water. Voltage output is 2 V per cell, with ratings up to about 200 A-h (8-h rate).

Nickel-cadmium batteries are particularly useful for application in temperature extremes. They are reputed to have been successfully operated at temperatures from –40°F to 163°F (–40°C to 73°C). They have a very high short-time current capability and are well suited to such applications as engine starting and UPS operation. Initial cost is higher than lead-acid types; however, they offer long life (25 to 30 years), reliability, and small size per unit. Voltage is about 1.2 V per cell.

480.3. Equipment. Other than lead-acid batteries, storage batteries and the equipment to manage their state of charge, etc., must be listed.

480.4. Battery and Cell Terminations. This section is new as of the 2014 NEC. Part (**A**) requires the use of an antioxidant compound where dissimilar metals are in contact at a battery terminal, provided the battery manufacturer recommends it. Part (**B**) requires that the conductors joining cell to cell and also tiers and shelves of a given battery rack ("intercell" and "intertier" connectors) to have a cross-sectional area sufficient to ensure that even during maximum load the conductors will not heat to the point of threatening the integrity of their insulation or supports. An informational note observes that customary voltage drop allowances based, in effect, on 215.2(A)(1) Informational Note No. 2 (here 3 percent maximum load; 5 percent to the most remote connection point) may not be appropriate for these systems and cites an IEEE publication for guidance. Part (**C**) requires that wiring connections to the battery terminals be arranged so as to prevent mechanical strain on the terminals. It also requires the use of terminal plates if practicable. An informational note suggests preforming conductor bends, or using finely stranded cables, or both, and consulting the manufacturer's literature.

480.5. Wiring and Equipment Supplied from Batteries. As indicated in Fig. 480-2, whatever kinds of circuits and loads a battery bank serves, all rules of the NEC covering operation at that voltage must be applied to the wiring and equipment.

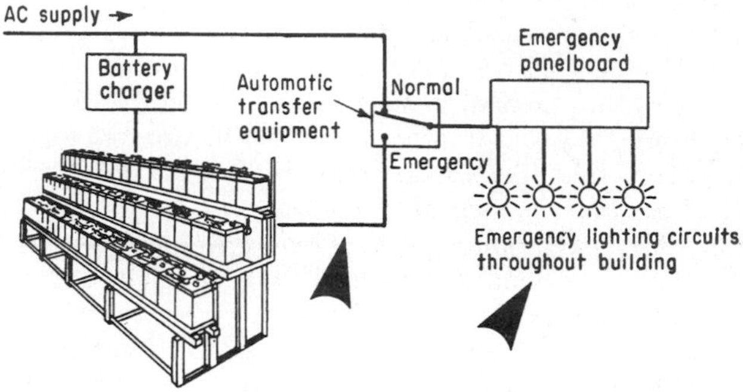
AC supply
Battery charger
Automatic transfer equipment
Normal
Emergency
Emergency panelboard
Emergency lighting circuits throughout building

Fig. 480-2. Applicable Code rules must be observed for load circuits fed by batteries. (Sec. 480.4.)

480.7. DC Disconnect Methods. Part (**A**) requires a disconnecting means to be provided for any ungrounded conductor supplied by a stationary battery bank operating over 60 V. It must be readily accessible and located within sight of the battery system. This rule correlates with the new 240.21(H) that specifies a comparable location for overcurrent protection for such conductors.

Part (**B**), new in the 2014 NEC, requires some background information in terms of context from its substantiating submittal to be comprehensible.

> DC busway is common in large UPS installations in which there are multiple strings of batteries. Each cell in a string is connected in series to create the necessary dc voltage and each string has a disconnecting means and/or overcurrent protective device. The strings are connected in parallel to a common dc bus which may also have a disconnecting means. The individual string disconnects allow manual disconnecting so that maintenance can be performed on a redundant battery string while the remaining battery strings support the load. It also functions as an OCPD to prevent the energy from other strings from feeding into a faulted cell in one string. The disconnect on a dc busway system can allow for a single point of shutdown for the entire dc supply.

The disconnecting means for an individual or a group of batteries in a large system may be managed remotely, and in such cases it must be capable of being locked open and it must bear a label describing the location of the remote controls.

Part (C) allows a busway switch to control the dc bus it supplies and thereby qualify as the system disconnect. Part (D) includes a label reminiscent of the one required for services in 110.24 that includes the nominal voltage and the maximum short-circuit current available from the battery system, along with a calculation date.

480.9. Battery Support Systems. This section now includes, in the second paragraph, a readily accessible requirement for battery terminals to allow for cleaning, inspection, and test readings. If a battery has a transparent container, then at least one side must be readily accessible for inspection of internal components.

480.10. Battery Locations. Although specific "battery rooms" or enclosures are no longer required for installation of any batteries, part (A) does require ventilation at battery locations. A specific battery room was previously required for open-tank or open-jar batteries, but such units are no longer made or in use.

This section has also received major attention in the 2014 NEC. Part (C) now includes rules for spacing battery racks such that a minimum of 1 in. separates a cell container and any wall or structure on a side not requiring maintenance access. The workspace rule (110.26 clearances) remains in effect and applies from the edge of a battery cabinet, rack, or tray. A battery stand can contact an adjacent structure, including for support, provided not over 10 percent of its length is not obstructed from air circulation. An informational note advises that greater space is often required to accommodate hoisting equipment and for other tasks. Part (D) requires that top-terminal batteries have open vertical spacings above them for maintenance in accordance with the manufacturer's recommendations. Part (E) applies the panic hardware rule in 110.26(C)(3) to egress doors from dedicated battery rooms. Part (F) prohibits gas piping in dedicated battery rooms. Part (G) incorporates the lighting requirement of 110.26(D), with the additional requirement that the luminaires be located so as to not expose those maintaining them to energized battery components, or where a failure in the luminaire would create a problem for the batteries.

ARTICLE 490. EQUIPMENT OVER 1000 VOLTS, NOMINAL

490.1. Scope. The requirements previously given in Art. 710 for equipment rated over 1000 V nominal are now covered in Art. 490. As defined in 490.2, the term *high voltage* applies to *any* equipment rated at more than 1000 V nominal. The systems customarily utilized as premises wiring and therefore subject to the NEC are usually not over 69 kV and therefore the technically correct terminology is "medium voltage," which is the usual term used in this book.

490.21. Circuit-Interrupting Devices. Medium-voltage power CBs provide load switching, short-circuit protection, electrical operation, adjustable time delays of trip characteristics for selectively coordinated protection schemes, quick reclosing after tripping, and various protective hookups such as differential relay protection of transformers. There are oil-type, oilless (or air-magnetic), and vacuum-break CBs. The air-magnetic CB is the common type for indoor applications in systems up to 15 kV and higher. Oil CBs are sometimes used for indoor and outdoor medium-voltage service equipment, where they provide economical disconnect and protection on the primary of a transformer.

Modern medium-voltage CB equipment meets all the needs of control and protection for electrical systems, from the simplest to the most complex and sophisticated. In particular, its use for selectively coordinated protection of services, feeders, and branch circuits is unique. In current ratings up to 3000 A, CB gear has the very high interrupting ratings required for today's high-capacity systems. Available in *metal-clad* assemblies, all live parts are completely enclosed within grounded metal enclosures for maximum safety. For applications exposed to lightning strikes or other transient overcurrents, CB equipment offers quick reclosing after operation. Drawout construction of the CB units provides ease of maintenance and ready testing of breakers. CB gear offers unlimited arrangements of source and load circuits and is suited to a variety of ac or dc control power sources. Accessory devices are available for special functions.

Figure 490-1 covers the basic rules of this section on use of CBs. Figure 490-2 shows an oil CB, with the line-side isolating switch required by 230.204(A).

Part **(B)** covers power fuses, which are available in current-limiting and noncurrent-limiting types. (See Fig. 490-3 for an example of the uses to which power fuses may be put.) The current-limiting types offer reduction of thermal and magnetic stresses on fault by reducing the energy let-through. They are constructed with a silver-sand internal element, similar to 600-V current-limiting fuses. Such fuses generally have higher interrupting ratings at some voltages, but their continuous current-carrying ratings are limited.

Noncurrent-limiting types of power fuses are made in two types of operating characteristics: expulsion type and nonexpulsion type. The expulsion fuse gets its name from the fact that it expels hot gases when it operates. Such fuses should not be used indoors without a "snuffer" or other protector to contain the exhaust, because there is a hazard presented by the expelled gases. At the end of part **(B)(1),** the rule says that vented expulsion-type power fuses used indoors, underground, or in metal enclosures must be "identified for the use." Vented power fuses are not safe for operation in confined space—unless specifically tested and *identified* for such use. Part **(B)(5)** of this section requires that fuses expelling flame in operation must be designed or arranged to prevent hazard to persons or property. The boric-acid fuse with a condenser or other protection against arcing and gas expulsion is a typical nonexpulsion, noncurrent-limiting fuse (Fig. 490-4).

Parts **(B)(6)** and **(B)(7)** cover very important safeguards for the use of fuses and fuseholders. The 2014 **NEC** relocated a substation requirement formerly at 225.70(A)(3) to (B)(6) requiring a warning label advising the need to disconnect the circuit before replacing a fuse. There are other considerations regarding the use of fuses:

In coordinating power fuses, care must be taken to account for ambient temperature adjustment factors, because time-current curves are based on an ambient of 25°C. Adjustment also must be made for preheating of fuses due to load current to ensure effective coordination of fuses with each other and/or with CBs. Manufacturer's curves of adjustment factors for ambient temperature and fuse preloading are available.

A. Indoor installations of circuit breakers must consist of metal-enclosed units or fire-resistant, cell-mounted units, except that open-mounted CBs may be used in places accessible to qualified persons only.

B. All CBs must be rated for short-circuit duty at point of application.

C. Circuit breakers controlling oil-filled transformers *must* either be located outside the transformer vault *or* be capable of being operated from outside the vault.

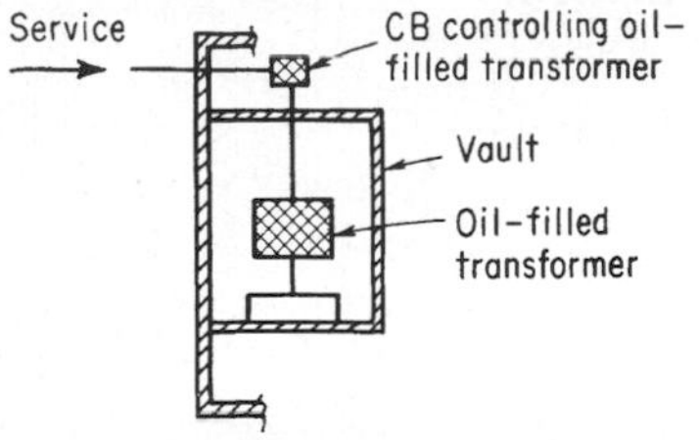
Service
CB controlling oil-filled transformer
Vault
Oil-filled transformer

D. Oil CBs must be arranged or located so that adjacent readily combustible structures or materials are safeguarded in an approved manner. Adequate space separation, fire-resistant barriers or enclosures, trenches containing sufficient coarse crushed stone, and properly drained oil enclosures such as dikes or basins are recognized as suitable safeguards.

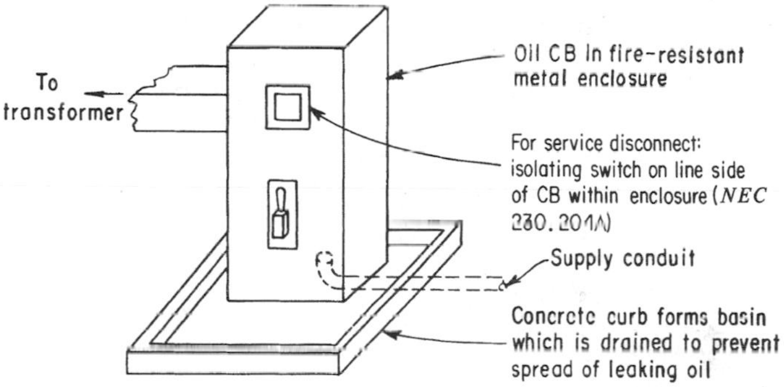
To transformer
Oil CB in fire-resistant metal enclosure
For service disconnect: isolating switch on line side of CB within enclosure (*NEC* 230.204(A))
Supply conduit
Concrete curb forms basin which is drained to prevent spread of leaking oil

Fig. 490-1. Detailed rules regulate use of medium-voltage circuit breakers. (Sec. 490.21.)

Figure 490-5 shows the time-current characteristic for an R-rated, current-limiting (silver-sand) fuse designed for 2400- and 4800-V motor applications. Such fuses must be selected to coordinate with the motor controller overload protection, with the controller clearing overloads up to 10 times motor current and the fuse taking over for faster opening of higher currents up to the interrupting rating of the fuse. The amp rating of R-rated fuses is given in values such as 2R or 12R or 24R. If the number preceding the R is multiplied by 100, the value obtained is the ampere level at which the fuse will blow in 20 s. Thus, the rating designation is *not* continuous current but is based on the operating characteristics of the R-rated fuse. Continuous current rating of such fuses is given by the manufacturer at some value of ambient temperature.

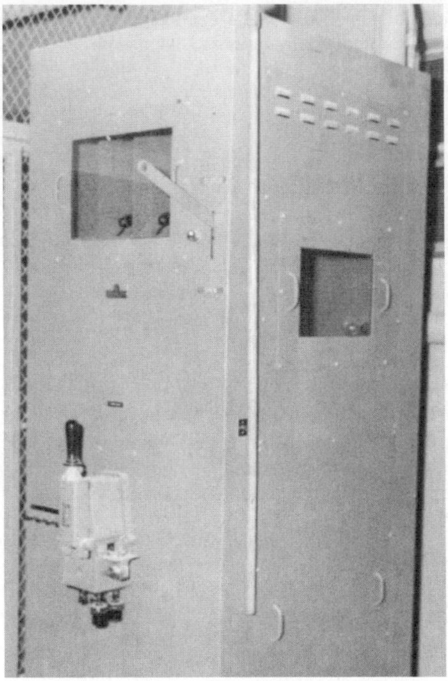

Fig. 490-2. Oil circuit breaker for medium-voltage application has a disconnecting switch on its supply side to isolate the line terminals of the breaker. (Sec. 490.21.)

Fig. 490-3. Typical power fuses are used in load-interrupter switchgear, which is an alternative to circuit-breaker gear for control and protection in indoor medium-voltage systems. (Sec. 490.21.)

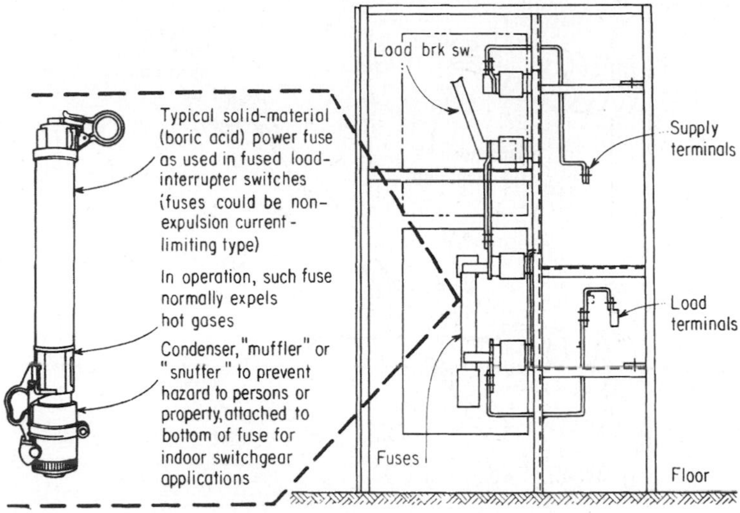
Load brk sw.
Typical solid-material (boric acid) power fuse as used in fused load-interrupter switches (fuses could be non-expulsion current-limiting type)
In operation, such fuse normally expels hot gases
Condenser, "muffler" or "snuffer" to prevent hazard to persons or property, attached to bottom of fuse for indoor switchgear applications
Supply terminals
Load terminals
Fuses
Floor

Fig. 490-4. Boric-acid fuse uses a device to protect against flame expulsion. (Sec. 490.21.)

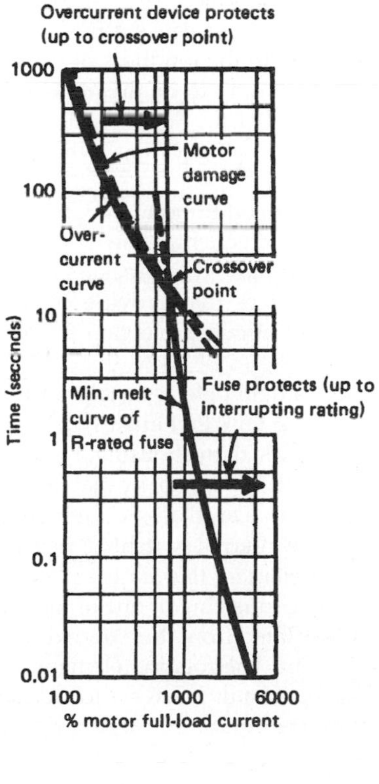
Overcurrent device protects (up to crossover point)
1000
100
10
1
0.1
0.01
Time (seconds)
Motor damage curve
Over-current curve
Crossover point
Fuse protects (up to interrupting rating)
Min. melt curve of R-rated fuse
100
1000
6000
% motor full-load current

Fig. 490-5. R-rated fuses are used in motor starters for 2400- and 4800-V motors. (Sec. 490.21.)

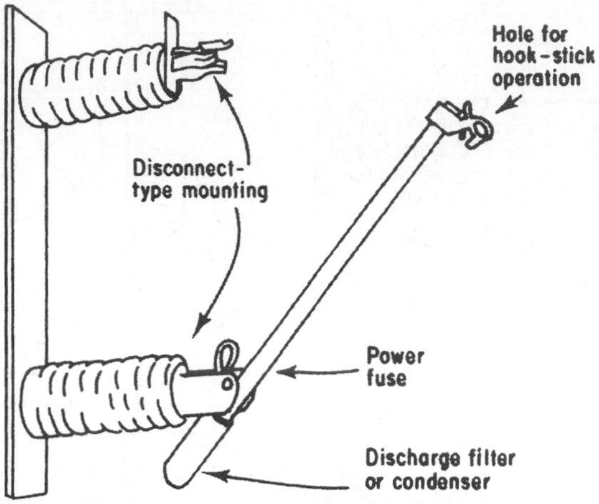
Hole for
hook-stick
operation
Disconnect-
type mounting
Power
fuse
Discharge filter
or condenser

Fig. 490-6. Expanded rules cover use of distribution cutouts. (Sec. 490.21.)

Fused cutouts for medium-voltage circuits, as shown in Fig. 490-6, are available for outdoor applications only, as regulated by part **(C)** of this section in general, and **(C)(1)** in particular, which flatly prohibits this equipment from being used indoors. Pull-type fuse cutouts are used outdoors on pole-line crossarms where the ionized gas expelled during operation is unlikely to cause a flashover in a confined space, or damage someone's hearing due to the loudness of the report. The use indoors in electric rooms where accessible only to qualified persons, as shown in Fig. 490-7, appears to be a violation of this rule. Such fused cutouts, if "designed for the purpose," are acceptable for use as an isolating switch, as permitted by 490.22.

Part **(D)** of this section covers oil-filled cutouts. In addition to air CBs, oil CBs, and fused load-interrupter switches, another device frequently used for control of medium-voltage circuits is the oil-filled cutout. Compared to breakers and fused switchgear, oil-filled cutouts are inexpensive devices that provide economical switching and, where desired, overload and short-circuit protection for primary voltage circuits.

The oil-filled cutout is a completely enclosed, single-pole assembly with a fusible or nonfusible element immersed in the oil-filled tank that makes up the major part of the unit, and with two terminals on the outside of the housing. Figure 490-8 shows the basic construction of a typical cutout with a listing of available entrance fittings for the terminals to suit them to various cable and job requirements. The circuit is broken or closed safely and rapidly by the internal switching mechanism. The switch mechanism is made up of a rotating element that, in the closed position, bridges two internal contacts—each contact connecting to one of the outside terminals. The rotating element is completely insulated from the external case and from the external handle that operates the element. The rotating element may be simply a shorting blade when the cutout is used as an unfused switch. When the cutout is to be used as a fused switching unit, the rotating bridging element is

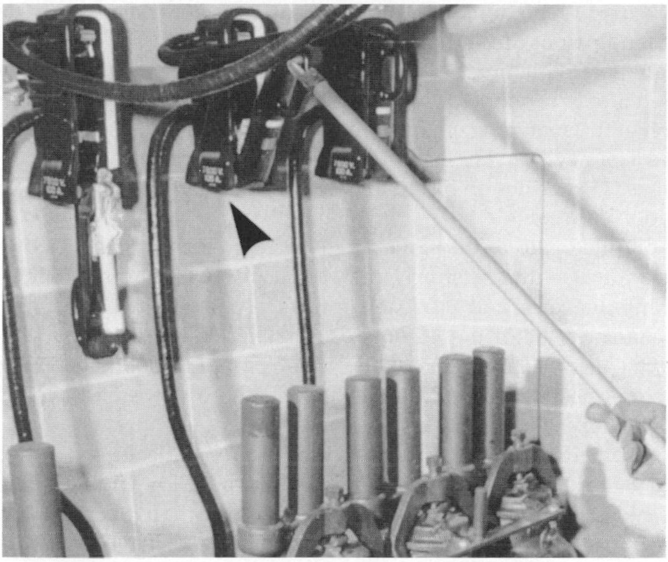

Fig. 490-7. Distribution cutouts are single-pole, fused, protective, and disconnect devices that are hook-stick-operable. Note voltage and current rating on case of each cutout (arrow), as required by part **(C)(5)** of 490.21. (Sec. 490.21.) The location appears to be indoors, and if so, would be controversial in light of the prohibition in 490.21(C)(1).

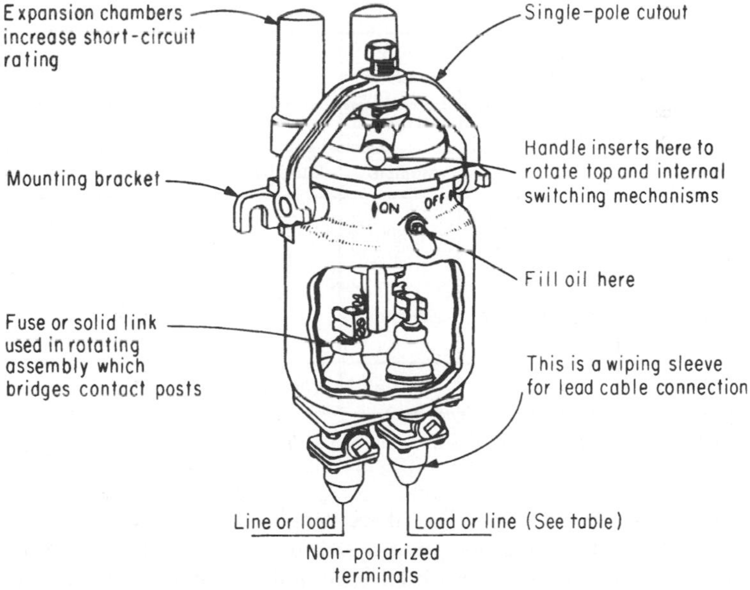
Expansion chambers increase short-circuit rating
Single-pole cutout
Handle inserts here to rotate top and internal switching mechanisms
Mounting bracket
ON
OFF
Fill oil here
Fuse or solid link used in rotating assembly which bridges contact posts
This is a wiping sleeve for lead cable connection
Line or load
Load or line (See table)
Non-polarized terminals

Fig. 490-8. Oil-filled cutout is a fused or unfused, single-pole disconnect device. (Sec. 490.21.)

fitted with a fuse. Operation of an oil-filled cutout is controlled at the top end of the shaft extending out through the top of the housing.

As a single-pole switching device, the oil-filled cutout is not polarized—that is, either terminal may be a line or load terminal. This is a result of the symmetrical construction of the switching element and suits the device to use in circuit sectionalizing or as a tie device in layouts involving two or more primary supply circuits. Note that these Code rules on oil-filled cutouts are different from those in part **(C)** on distribution cutouts.

With an oil-filled cutout, the switching of load current or the breaking of fault current is confined within a sturdy metal housing. Operation is made safe and quiet by confining arcs and current rupture forces within the enclosure. This operating characteristic of the cutouts especially suits them to use where there are explosive gases or flammable dusts, where complete submersion is possible, where severe atmospheric conditions exist, or where exposure of live electrical parts might be hazardous.

Oil-filled (sometimes called *oil-fuse*) cutouts are made in three sizes based on continuous current—100, 200, and 300 A, up to 15 kV. In one line there are three basic types. *Pole-type cutouts* are equipped with rubber-covered leads from the terminals for use in open wiring. *Pothead-type cutouts* have a cable lead from one terminal for open wiring and a sleeve on the other for connecting a lead or rubber-covered cable from an underground circuit. *Subway-type cutouts* are for underground vaults and manholes, particularly where submersion might occur, and are equipped with a sleeve on each terminal for rubber- or lead-sheathed cable. Figure 490-9 lists the various types of terminal connections that are available on oil-filled cutouts.

Application	Cable cover	No. of conduits	Entrance type	Method of sealing
Indoor	Rubber or neoprene	Single	Porcelain	Tape Always tape this connection!
		Multiple	Stud bushing	
Indoor or outdoor	Rubber, lead or neoprene	Single	Stuffing box	Compression fittings
	Polyethylene	Single		Compression fittings and tube seal
	Lead	Single	Wiping sleeve	Solder wipe

Fig. 490-9. Terminals on oil-filled cutouts must be matched to application and cable type. (Sec. 490.21.)

For multiphase circuits, two or three single-phase cutout units can be group-mounted with a gang-operating mechanism for simultaneous operation. Figure 490-10 shows three-gang assemblies. For pole mounting, linkage and a long handle are available for operating cutouts from the ground. Or cutouts can be

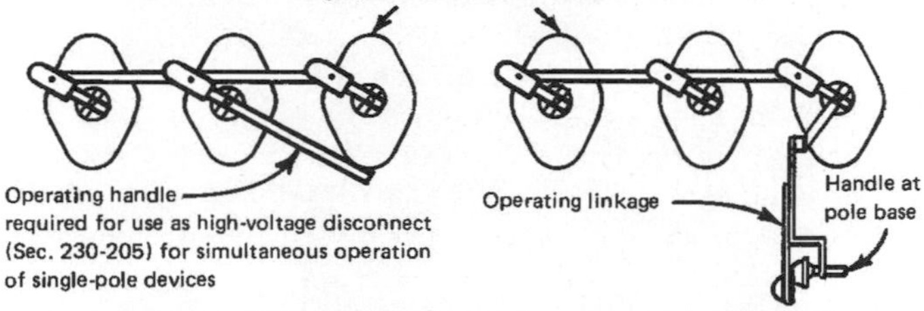
Single-phase cutouts (top view)
Operating handle
required for use as high-voltage disconnect
(Sec. 230-205) for simultaneous operation
of single-pole devices
Operating linkage
Handle at
pole base

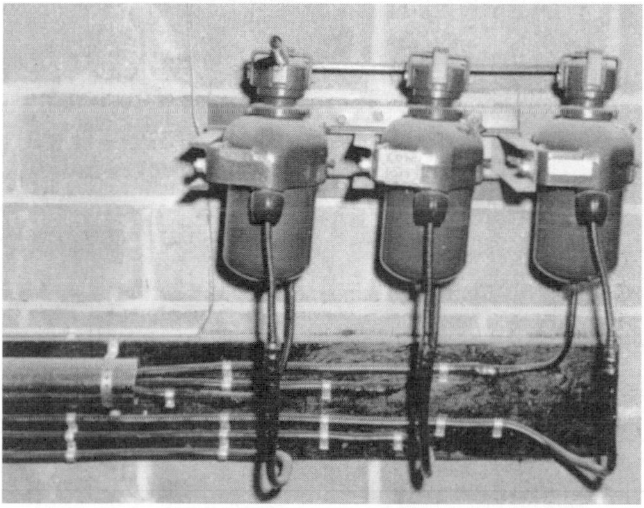

Fig. 490-10. Oil-filled cutouts can be assembled as a 3-pole device for 3-phase circuits. (Sec. 490.21.)

flange-mounted on a terminal box, as shown in Fig. 490-11, where the three-gang assembly was added to a medium-voltage switchgear on a modernization job.

Because oil-filled cutouts provide load-break capability and overcurrent protection, they may be used for industrial and commercial service equipment, for switching outdoor lighting of sports fields and shopping centers, for transformer loadcenters, for primary-voltage motor circuits, or for use in vaults and manholes of underground systems.

In 100- and 200-A ratings, oil-filled cutouts are available in combination with current-limiting power fuses in double-compartment indoor or outdoor enclosures. These fused oil interrupter switches provide moderate load-break and high fault-current interrupting capability in an economical package.

Part **(E)** of this section recognizes the use of so-called load-interrupter switches used in medium-voltage systems. In parts **(1)** to **(6)**, a wealth of specific data provides guidance to design engineers, electrical installers, and electrical inspectors on the proper installation, operation, and maintenance of medium-voltage interrupter switches—with particular emphasis on safety to operators and maintenance personnel.

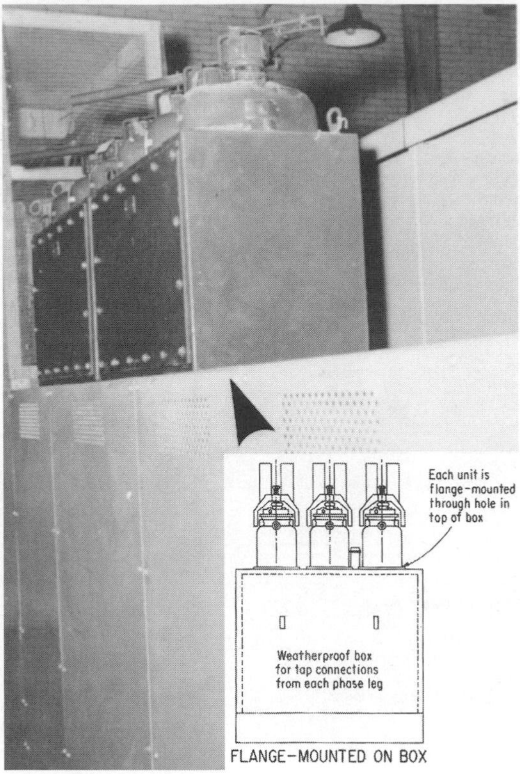
Each unit is flange-mounted through hole in top of box
Weatherproof box for tap connections from each phase leg
FLANGE-MOUNTED ON BOX

Fig. 490-11. Gang-operated 3-pole assembly of oil-filled cutouts provided addition of a new medium-voltage circuit on a modernization project, but location of the units was questioned because part **(D)(7)** imposes a 5-ft (1.5 m) maximum mounting height. (Sec. 490.21.)

Switching for modern medium-voltage electrical systems can be provided by a number of different equipment installations. For any particular case, the best arrangement depends on several factors: the point of application—either for outside or inside distribution or as service equipment; the voltage; the type of distribution system—radial, loop, selective, network; conditions—accessibility, type of actual layout of the equipment; job atmosphere; use; future system expansion; and economic considerations.

Types of switches used in medium-voltage applications include:

1. Enclosed air-break load-interrupter switchgear with or without power fuses
2. Oil-filled cutouts (fused or unfused)
3. Oil-immersed-type disconnect switches

Modern load-interrupter switchgear in metal safety enclosures finds wide application in medium-voltage distribution systems, in combination with modern power fuses (Fig. 490-12). 230.208 of the **NEC** covers use of air load-interrupter

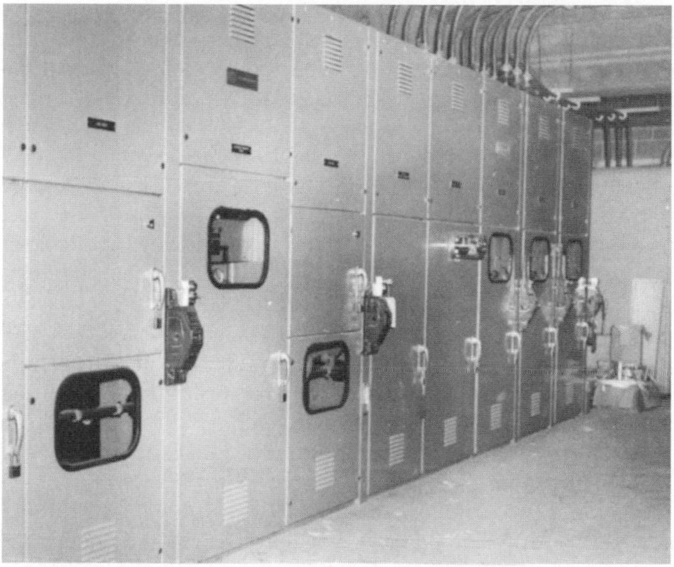

Fig. 490-12. Load-interrupter switchgear is generally used with fuses to provide protection as well as load-break switching for medium-voltage circuits. The fuses must be rated to provide complete protection for the load interrupter on closing, carrying, or interrupting current—up to the assigned maximum short-circuit rating. (Sec. 490.21.)

switches, with fuses, for disconnect and overcurrent protection of medium-voltage service-entrance conductors, and refers the reader to Part **II** of Art. 490. Part **(E)** of 490.21 also covers use of fused air load-interrupter switches for medium-voltage feeder in distribution systems, as well as at service equipment.

Metal-enclosed fused load interrupters offer a fully effective alternative to use of power CBs, with substantial economies, in 5- and 15-kV distribution systems for commercial, institutional, and industrial buildings. Typical applications for such switchgear parallel to those of power CBs include the following:

1. *In switching centers*—Switchgear is set up for control and protection of individual primary feeders to transformer loadcenters.
2. *In substation primaries*—Load-interrupter switchgear is used for transformer switching and protection in the primary sides of substations.
3. *In substation secondaries*—Here the switchgear is used as a switching center closely coupled to a medium-voltage transformer secondary.
4. *In service entrances*—This is a single-unit application of a switchgear bay for service-entrance disconnect and protection in a primary supply line.

Fused load-interrupter switchgear, typically rated up to 1200 A, can match the ratings and required performance capabilities of power CBs for a large percentage of applications in which either might be used.

Fuse-interrupter switches for medium-voltage circuits are available with manual or power operation—including types with spring-powered, over-center mechanisms for manual operation or motor-driven, stored-energy operators. Available in indoor and outdoor housings, assemblies can be equipped with a variety of accessory devices, including key-interlocks for coordinating switch operation with remote devices such as transformer secondary breakers.

Vacuum switchgear, with its contacts operating in a vacuum "bottle" that is enclosed in a compact cylindrical assembly, has gained wide acceptance as load interrupters for medium-voltage switching and sectionalizing. Available in 200- and 600-A ratings for use at 15.5, 27, and 38 kV, this switching equipment is suited to full-load interruption and is rated for 15,000- or 20,000-A short-circuit current under momentary and make-and-latch operations. BIL ratings are 95, 125, or 150 kV.

Vacuum switch assemblies, with a variety of accessories, including stored-energy operators and electric motor operators for remote control, are suited to all indoor and outdoor switching operations—including submersible operation for underground systems. The units offer fireproof and explosionproof operation, with virtually maintenance-free life for their rated 5000 load interruptions. Units are available in standard 2-, 3-, and 4-way configurations, along with automatic transfer options. Accessory CTs and relays can be used with stored-energy operators to apply vacuum switches for fault-interrupting duty.

490.22. Isolating Means. Air-break or oil-immersed switches of any type may be used to provide the isolating functions described in this section. Distribution cutouts or oil-filled cutouts are also used as isolating switches.

Oil-immersed disconnect switches are used for load control and for sectionalizing of primary-voltage underground-distribution systems for large commercial and industrial layouts (Fig. 490-13). Designed for high-power handling—such as 400 A up to 34 kV—this type of switch can be located at transformer loadcenter primaries or at other strategic points in medium-voltage circuits to provide a wide variety of sectionalizing arrangements to provide alternate feeds for essential load circuits.

Oil-immersed disconnect switches are available for as many as five switch positions and ground positions to ground the feeder or test-ground positions for grounding or testing. Ground positions are used in such switches to connect circuits to ground while they are being worked on to assure safety to personnel.

Oil switches for load-break applications up to 15 kV are available for either manual or electrically powered switching for all types of circuits. When electrical operation is used, the switch functions as a medium-voltage magnetic contactor.

Switches intended only for isolating duty must be interlocked with other devices to prevent opening of the isolating switch under load, or the isolating switch must be provided with an obvious sign warning against opening the switch under load.

490.24. Minimum Space Separation. For field fabricated installations, these are the distances that have to be met in terms of clearances between uninsulated live parts and from such parts to ground. If equipment has been formally manufactured and evaluated in accordance with recognized standards, then the dimensions in this table no longer apply.

490.25. Backfeed. This section was relocated from one of the substation provisions formerly in 225.70(A)(4) and also by borrowing appropriate prescriptive

Fig. 490-13. Oil switches are commonly used for isolating equipment and circuits for sectionalizing and for transfer from preferred to emergency supply. (Sec. 490.22.)

language regarding the required warning from 404.6(C) Exception. In this location it applies generally. A single-line diagram is required that identifies the connections to high voltage.

III. EQUIPMENT—METAL-ENCLOSED POWER SWITCHGEAR AND INDUSTRIAL CONTROL ASSEMBLIES

490.30. General. Where the previous sections presented regulations on the individual switching and protective devices, this section covers enclosure and interconnection of such unit devices into overall assemblies. Basically, the rules of this section are aimed at the manufacturers and assemblers of such equipment.

Use of all medium-voltage switching and control equipment must be carefully checked against information given with certification of the equipment by a test laboratory—such as data given by UL in its Guide Card information [see discussion at 110.3(B) for details]. Typical data are as follows:

Unit substations listed by UL have the secondary neutral bonded to the enclosure and have provision on the neutral for connection of a grounding conductor. A terminal is also provided on the enclosure near the line terminals for use with an equipment grounding conductor run from the enclosure of primary equipment feeding the unit sub to the enclosure of the unit sub. Connection of such an equipment grounding conductor provides proper bonding together of equipment enclosures where the primary feed to the unit sub is directly buried underground or is run in nonmetallic conduit without a metal conduit connection in the primary feed (Fig. 490-14).

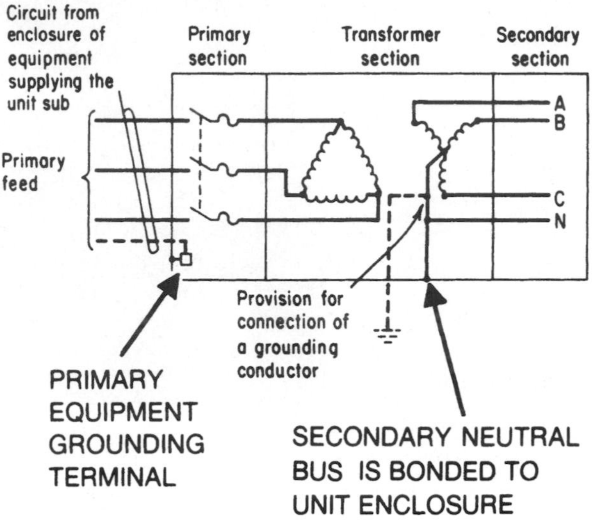
Circuit from enclosure of equipment supplying the unit sub
Primary section
Transformer section
Secondary section
A
B
C
N
Primary feed
Provision for connection of a grounding conductor
PRIMARY EQUIPMENT GROUNDING TERMINAL
SECONDARY NEUTRAL BUS IS BONDED TO UNIT ENCLOSURE

Fig. 490-14. NEC rules on equipment construction are supported by UL data. (Sec. 490.24.)

The rule of 490.44 is particularly aimed at the designers and installers of equipment, rather than at manufacturers. Part **(B)** emphasizes that careful layout and application of switching components of all types are important. Figure 490-15 shows the kind of condition that can be extremely hazardous in medium-voltage layouts where there is the chance of a secondary to primary feedback—such as the intentional one shown to provide emergency power to essential circuits in Building 2. Under emergency conditions, the main fused interrupter is opened and the secondary CB for the generator is closed, feeding power to the 480-V switchboard in Building 1 and then feeding through two transformers to supply power to the 480-V circuits in Building 2. This hookup makes the load side of the main interrupter switch live, presenting the hazard of electrocution to any personnel who might go into the switch thinking that it is dead because it is open. A second switch can eliminate this difficulty, if applied with interlocks.

490.35. Accessibility of Energized Parts. Equipment operating at 1000 V and below must be excluded from medium-voltage enclosures with exposed live parts [part **(A)**] unless access under hazardous conditions is eliminated due to interlocks or other procedures. However, space heaters, and also medium-voltage instrument or control transformers [part **(C)**] are permitted in such spaces without any

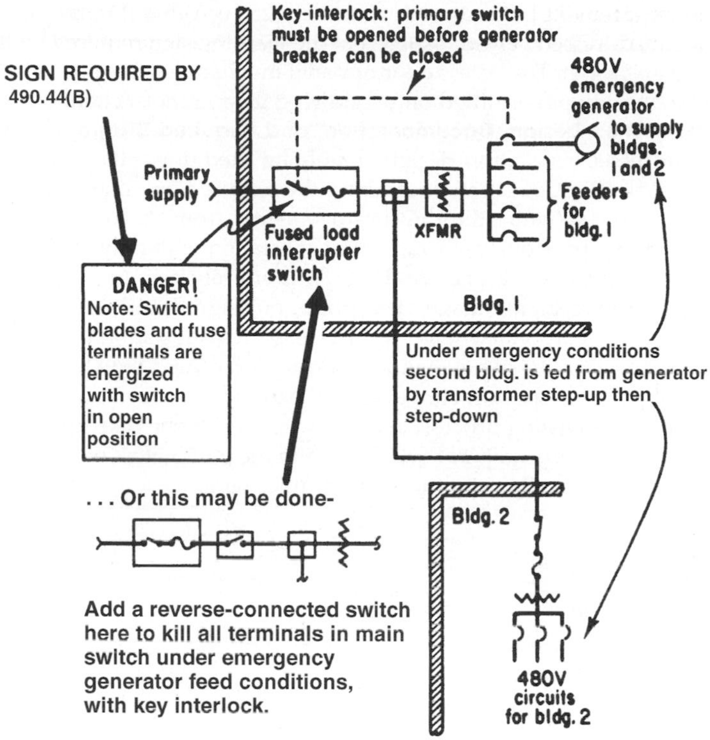
Key-interlock: primary switch must be opened before generator breaker can be closed
SIGN REQUIRED BY 490.44(B)
480V emergency generator
to supply bldgs. I and 2
Primary supply
Feeders for bldg. I
Fused load interrupter switch
XFMR
DANGER!
Note: Switch blades and fuse terminals are energized with switch in open position
Bldg. I
Under emergency conditions second bldg. is fed from generator by transformer step-up then step-down
. . . Or this may be done-
Bldg. 2
Add a reverse-connected switch here to kill all terminals in main switch under emergency generator feed conditions, with key interlock.
480V circuits for bldg. 2

Fig. 490-15. Feedback in medium-voltage hookups can be hazardous. (Sec. 490.21.)

access limitations beyond the locked doors required in part **(A)**. The 2014 NEC relocated a warning sign requirement from its former home [225.70(A)(5)(b)] addressing substations and made it generally applicable at these voltages by improving its editorial content and placing at the end of this section.

490.41. Location of Industrial Control Equipment. Operating handles for control and instrument transfer switches or pushbuttons must be readily accessible, and not over 2.0 m (78 in.) high. Operating handles requiring more than 23 kg (50 lb) of force to move must not be over 1.7 m (66 in.) high. Part **(B)**, however, waives this requirement for "infrequently operated devices" and sets out a list by way of explaining by example the intended application of the word "infrequently." The section was reorganized in the 2011 cycle to clarify its terms.

490.46. Circuit Breaker Locking. This is the medium-voltage version of the locking-mechanism-to-remain-in-place rule, addressing drawout gear, etc.

490.47. Metal-Enclosed and Metal-Clad Service Equipment. This rule requires medium-voltage service equipment manufacturers to provide a grounding busbar that extends into the compartment where the service conductors will terminate. The busbar will provide a termination point for cable shielding connections to earth, as well as an appropriate point to attach safety grounds for the protection of personnel who may need to work on the equipment. The 2014 NEC has added a marking requirement [relocated from former 225.70(A)(5)(c)] to the connection compartment addressed here and subject to the warning sign required by the new language in 490.35(A). The access door or panel must advise that access is limited to the serving utility or permitted only following their authorization.

490.48. Substation Design, Documentation, and Required Diagram. The 2014 NEC has addressed substation designs involving medium-voltage sections in a major way. Part **(B)** of this section covering the required one-line diagram is not new, having moved (with editorial improvements) from its former location in 225.70(A)(5)(a). As previously, a simple unit substation with only one set of high-voltage switching devices is exempt. Part **(A)** is entirely new to the NEC, having been suggested by a high-voltage task group, but significantly modified in the process of being incorporated into Art. 490. It places substation design into the hands of a licensed professional engineer and requires the design documentation to be available to the inspector. However, to the extent the substation has been the subject of an overall listing process, such as is frequently the case for unit substations, the underlying documentation falls within the parameters of 90.7 and is not required to be made available to fulfill this requirement.

Chapter Five

ARTICLE 500. HAZARDOUS (CLASSIFIED) LOCATIONS, CLASSES I, II, AND III, DIVISIONS 1 AND 2

The informational note provided before the main body of text informs the reader that certain rules within the scope of Arts. 500 through 504 contain references to rules in other NFPA standards, with slight editorial modifications. For more information regarding those rules, consult the specified portion and related rules within that standard.

500.1. Scope—Articles 500 Through 504. In the heading of this article, the word *classified* makes clear that hazardous locations are those which have been classified as hazardous by the inspection authority. Hazardous locations in plants and other industrial complexes are involved with a wide variety of flammable gases and vapors and ignitable dusts—all of which have widely different flash points, ignition temperatures, and flammable limits. These explosive or flammable substances are processed and handled under a wide range of operating conditions. In such places, fire or explosion could result in loss of lives, facilities, and/or production.

Effective with the 2017 NEC, the definitions formerly covered in this location have been moved to Art. 100. Refer to Table 100-1 at the end of Article 100 coverage in this book for a listing of definitions related to hazardous (classified) locations that are now in that location, including all the definitions formerly within Art. 500. This book will continue to comment on the requirements in these articles, and do so entirely within the coverage of these articles. There will be no commentary on hazardous (classified) location terminology within the coverage this book otherwise provides in Article 100.

500.4. General. This basic requirement of part **(A)** calls for documentation of all hazardous (classified) locations. The nature of the document may vary from jurisdiction to jurisdiction, but some paperwork must be produced to identify those areas that are hazardous (classified) locations within any facility. This paperwork should properly and clearly indicate those portions of the facility that are designated as classified locations and the specific classification affixed to that area, including the class/division or zone designation. Such documentation must be made available to design and maintenance personnel.

Part **(B)** contains a number of informational notes that identify other NFPA standards that bear on specifics related to various types of combustible materials. These documents provide standards and guidelines for the safe handling, processing, storage, and so on, of the combustible material in question. The 2017 NEC adds a sixth note here regarding the use of self-contained power supplies within hazardous (classified) locations.

500.5. Classifications of Locations. Part **(A)** addresses the classification of various explosive environments. Classification of hazardous areas must be approached very carefully, based on experience and a detailed understanding of electrical usage in the various kinds of locations. After study and analysis—and consultation with inspection authorities or other experts in such work—hazardous areas may be identified and delineated diagrammatically by defining the limits and degree of the hazards involved. In all cases, classification must be carefully based on the type of gas involved, whether the vapors are heavier or lighter than air, and similar factors peculiar to the particular hazardous substance.

Locations used for pyrophoric materials, which ignite spontaneously on contact with air, are exempted from designation as Class I hazardous locations. The status of adjacent electrical components has no bearing on whether these materials will ignite, and therefore any enhanced wiring precautions, including those in this chapter of the NEC, would be irrelevant to the likelihood of fire or explosion associated with their use. Therefore electrical equipment approved for classified locations is not needed for places where pyrophoric materials are handled. As an example, sodium metal is pyrophoric, and is usually stored under oil to prevent contact with air for this reason.

The NEC uses three basic classes to indicate the type of hazard involved, and those classes (I, II, and III) are in two cases (I and II) further subdivided into lettered groupings to specify the exact type of hazard. Additionally, each of the three classes is divided into two divisions to indicate the prevalence of hazard. Area classifications are seldom performed by an electrical inspector; they are done by qualified engineers subject to review. The reviewer may be the chief of the fire department or some other official.

Classification takes into account that all sources of hazards—gas, vapor, dust, fibers—have different ignition temperatures and produce different pressures when exploding. Electrical equipment must, therefore, be constructed and installed in such a way as to be safe when used in the presence of particular explosive mixtures. The source of hazard must be evaluated in terms of those characteristics that are involved with explosion or fire, and the following discussion regarding the classification of diesel fuel is a good example.

Diesel Oil and Heating Oil

Questions often arise about the need for hazardous location wiring for electrical equipment installed in areas containing diesel fuel oil or heating oil. The National Fire Code (NFC) classifies diesel fuel oil as a Class II liquid having a flash point at or above 100°F (37.8°C) and below 140°F (60°C). The same NFC standard on bulk plants states in part that in areas where Class II or Class III liquids are stored or handled, the electrical equipment *may* be installed in accordance with the provisions of the NEC for ordinary (i.e., nonhazardous) locations. Diesel fuel oil is classified as a Class II liquid and does not come under requirements for hazardous (classified) locations. With this type of liquid, explosionproof wiring methods *are not* required and the wiring methods listed in Chaps. 1 through 4 of the NEC may be used. The NFC does, however, caution that, if any Class II flammable liquid is heated, it may be necessary that Class I Group D wiring methods be used. In some geographic areas with hot climates, local regulations do require diesel fuel areas to be treated as hazardous areas because of high ambient temperatures. Temperatures in the Southwest often exceed 115°F (46°C), especially in closed, nonventilated areas.

The *flash point* of a liquid is the minimum temperature at which the liquid will give off sufficient vapor to form an ignitible mixture with air near the surface of the liquid or within the vessel used. (This characteristic is not applicable to gases.)

The *autoignition temperature* of a substance is the lowest temperature that will initiate explosion or cause self-sustained combustion of the substance.

Explosive limits: When flammable gases or vapors mix with air or oxygen, there is a minimum concentration of the gas or vapor below which propagation of flame does not occur upon contact with a source of ignition. There is also a maximum concentration above which propagation does not occur. These boundary-line mixtures are known as the lower and upper explosive (or flammable) limits and usually are expressed in terms of the percentage of gas or vapor in air, by volume.

Vapor density is the weight of a volume of pure vapor or gas (with no air present) compared to the weight of an equal volume of dry air at the same temperature and pressure.

A note in 500.5(A) indicates that maximum effort should be made to keep as much electrical equipment as possible out of the hazardous areas—particularly minimizing installation of arcing, sparking, and high-temperature devices in hazardous locations. It is generally economically and operationally better to keep certain electrical equipment out of hazardous areas. Figure 500-1 shows an example in which the drive shaft of the motor is extended through a packing gland in one of the enclosing walls. To prevent the accumulation of flammable vapors or gas within the motor room, it should be ventilated effectively by clean air or kept under a slight positive air pressure. A gas detector giving a visual and/or audible alarm would be an additional desirable safety feature.

The separate paragraph at this point covering rooms used for ammonia refrigeration systems has been revised for the 2017 NEC, and now provides specific, threshold

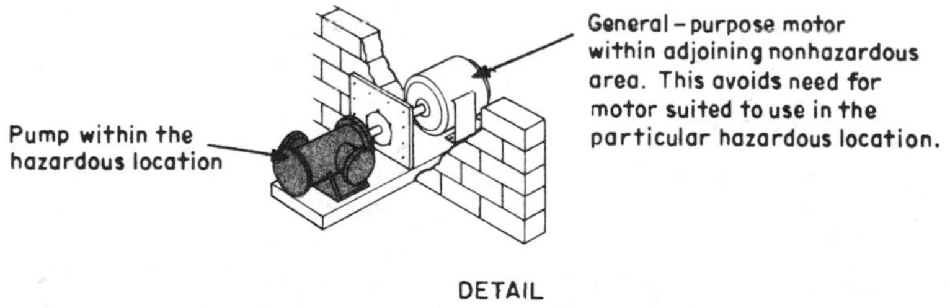
General-purpose motor within adjoining nonhazardous area. This avoids need for motor suited to use in the particular hazardous location.
Pump within the hazardous location
DETAIL

Fig. 500-1. Keeping electrical equipment in nonhazardous area eliminates costly hazardous types. (Sec. 500.5.)

parameters relative to acceptable ammonia concentrations. Specifically, if they are ventilated to prevent an ammonia concentration above 150 ppm, they can be unclassified. There are many vendors that sell gas detection systems capable of monitoring the area and initiating ventilation as required. Ammonia is an inexpensive refrigerant with wide industrial usage, and is often used for commercial activities as well, such as large commercial refrigeration activities including skating rinks. Note that there are other large-scale applications for ammonia beyond refrigeration. For just one, ammonia is used in many municipal water plants to help purify the water, and they face comparable area classification issues. It may be advisable in a future Code cycle to make this paragraph more generally worded.

Positive-pressure ventilation is also cited as a means of reducing the level of hazard in areas where explosive or flammable substances are or might be present. Air-pressurized building interiors can provide safe operation without explosionproof equipment. When explosionproof equipment is not justified financially, pressurized building interiors can alter the need. For example, if a motor control center is to be located in a building in a Class I hazardous location, the entire room may be pressurized. But construction must comply with certain specific requirements:

1. The building area or room must be kept as airtight as possible.
2. The interior space must be kept under slight overpressure by adequate positive-pressure ventilation from a source of clean air.
3. The pressurizing fans should be connected to an emergency supply circuit.
4. Ventilating louvers must be located near ground level to achieve effective airflow within the pressurized room.
5. Safeguards against ventilation failure must be provided.

The first part of this section concludes with important information regarding the use of ammonia for refrigeration. Ammonia normally presents a Class I explosion hazard, however, the lower explosive limit is relatively high, and far beyond the limits of toleration for human exposure. To correlate with an ASHRAE document governing refrigeration machinery, if gas detection equipment is in place that will initiate ventilation in the event the ammonia concentration reaches or exceeds 150 ppm, then the area can be considered unclassified. The current **NEC** limits this permission to refrigeration areas. Ammonia has many other uses, including water purification, that could, and probably should, benefit from inclusion in this allowance going forward. It might be a fruitful topic of a 90.4 (equivalent hazard) conversation with the inspector where it applies.

Part **(B)** of 500.5 describes Class I locations. In each of the three classes of hazardous locations discussed in parts **(B)**, **(C)**, and **(D)** of 500.5, the Code recognizes varying degrees of hazard; hence, under each class, two divisions are defined. In the installation rules that follow, the requirements for Division 1 of each class are more rigid than the requirements for Division 2.

The hazards in the three classes of locations are due to the following causes, and then divided into divisions (1 or 2) based on the prevalence of the combustible agent, as follows:

Class I locations, as covered in **(B)**, are those "in which flammable gases, flammable liquid-produced vapors, or combustible liquid-produced vapors are present or may be present in the air in quantities sufficient to produce explosive or ignitable mixtures." Class I, Division 1 locations are those in which hazardous concentrations of gas (1) exist continuously or intermittently under normal operations, (2) exist frequently because of maintenance or leakage, or (3) might exist because of breakdowns or faulty operation that might also result in simultaneous failure of electric equipment in such a way as to cause the electrical equipment to become an ignition source. Examples are paint-spraying areas, systems that process or transfer hazardous gases or liquids, portions of some cleaning and dyeing plants, and hospital operating rooms in countries that still use ethers and related flammable anesthetic agents. These anesthetizing materials are no longer used in the United States, but they are still used in some countries.

Class I, Division 2 locations are locations (1) in which flammable gases or volatile flammable liquids are normally confined within containers or closed systems from which they can escape only in case of breakdown, rupture, or abnormal operation; (2) in which the hazardous concentrations are prevented by mechanical ventilation from entering but which might become hazardous upon failure of the ventilation; or (3) that are adjacent to a Class I, Division 1 location and might occasionally become hazardous unless prevented by positive-pressure ventilation from a source of clean air by a ventilation system that has effective safeguards against failure. Examples are storage in sealed containers and piping without valves.

Class II locations, as covered in **(C)**, are those having combustible dust.

Class II, Division 1 locations are those in which combustible dust is or may be in suspension in the air in sufficient quantity to produce an explosive or ignitable mixture (1) continuously or intermittently during normal operation, (2) as a result of failure or abnormal operation that might also provide a source of ignition by simultaneous failure of electric equipment, or (3) in which combustible dust that is electrically conductive may be present. Examples are grain elevators, grain-processing plants, powdered-milk plants, coal-pulverizing plants, and magnesium-dust-producing areas.

Class II, Division 2 locations are those in which combustible dust is not normally in suspension in the air and is not likely to be under normal operations in sufficient quantity to produce an explosive or ignitable mixture but in which accumulations of dust (1) might prevent safe dissipations of heat from electric equipment, or (2) might be ignited by arcs, sparks, or burning material escaping from electric equipment. Examples are closed bins and systems using closed conveyors and spouts.

Class III locations are those in which there are easily ignitable fibers or flyings, such as lint, but in which the fibers or flyings are not in suspension in the air in sufficient quantity to produce an ignitable mixture. Class III, Division 1 locations are those in which easily ignitable fibers or materials that produce combustible flyings are manufactured, used, or otherwise handled, such as textile mills, cotton gins, and woodworking plants. Class III, Division 2 locations are those in which such fibers are stored, such as a warehouse for baled cotton, yarn, and so forth.

A frequent question arises as to whether a Class III location should be actually classified as Class II or the reverse. NFPA 499 includes a useful, objective definition on this point. A combustible finely divided dust that is comprised of particles no larger than 420 μm (0.017 in.) in diameter, thereby capable of passing through a U.S. No. 40 standard sieve, is a Class II exposure. This definition, extracted from NFPA 499, now appears in 500.2, so it is no longer necessary to search the source document for it. Larger fibers and flyings, such as in textile operations, are Class III.

Most sawdust is neither, but some woodworking operations with very fine powder from some sanding operations do fall into a Class II classification.

The classifications are easily understood, and, if a given location is to be classed as hazardous, it should not be difficult to determine in which of the three classes it belongs. Each of the classifications in this section includes extensive informational notes that are very well written and of great assistance in clarifying the conditions that produce a given area classification. However, it is obviously impossible to make rules that will in every case determine positively whether the location is or is not hazardous. Considerable common sense and good judgment must be exercised in determining whether the location under consideration should be considered as hazardous or likely to become hazardous because of a change in the processes carried on, and if so, what portion of the premises should be classed as coming under Division 1 and what part may safely be considered as being in Division 2.

500.6. Material Groups. Informational note No. 2 in part **(A)** explains the basic reason for the subdivisions. Equipment for use in Class I locations is divided into Groups A, B, C, and D; for Class II into Groups E, F, and G. Each group is for a specific hazardous material, which you can determine by consulting your copy of the **NEC**. All equipment for hazardous locations must be marked to show the class and the group for which it is approved; some equipment is suitable for more than one group and is so marked. Note that for Group E, which encompasses the electrically conductive metallic dusts, there are no Division 2 locations. These dusts are so hazardous that if they are present in any quantities, no matter how infrequently, the location must be wired to Class II, Division 1 standards, using equipment evaluated for Group E exposure.

The exception to 500.6, correlated with language in 500.8(C)(3) on required markings, allows an evaluation based on a specific substance (gas, vapor, dust, or fiber/flying) in lieu of a group classification. This is probably not practical for fiber/flying applications, but occasionally is used in other cases.

500.7. Protection Techniques. This section indicates those methods that provide protection against accidental ignition of a combustible atmosphere.

Purging of electrical raceways and enclosures is a strategy that can be used to reduce the degree of hazard (Fig. 500-2). But that requires both the manufacturer and the user of purged equipment to ensure the integrity of the system. Prepackaged purge controls for both Division 1 and Division 2 locations are in the market.

This technique is called "purged and pressurized" in the **NEC**. This equipment must be connected to a source of clean air under constant pressure. It must be arranged with automatic means to disconnect power if the air supply fails, and it must be equipped with a time-delay function so that its interior can purge prior to the restoration of power. There are three levels of purged and pressurized enclosures. Type X purging reduces a Division 1 environment to unclassified, Type Y purging reduces from Division 1 to Division 2, and Type Z purging goes from Division 2 to unclassified.

The purging medium—for example, inert gas, such as nitrogen, or clean air—must be essentially free from dust and liquids. The normal ambient air of an industrial interior is usually not satisfactory. And because the purge supply can contain only trace amounts of flammable vapors or gases, the compressor intake must be in a nonhazardous area. The compressor intake line should not pass through a hazardous atmosphere. If it does, it must be made of a noncombustible material, must be protected from damage and corrosion, and must prevent hazardous vapors from being drawn into the compressor.

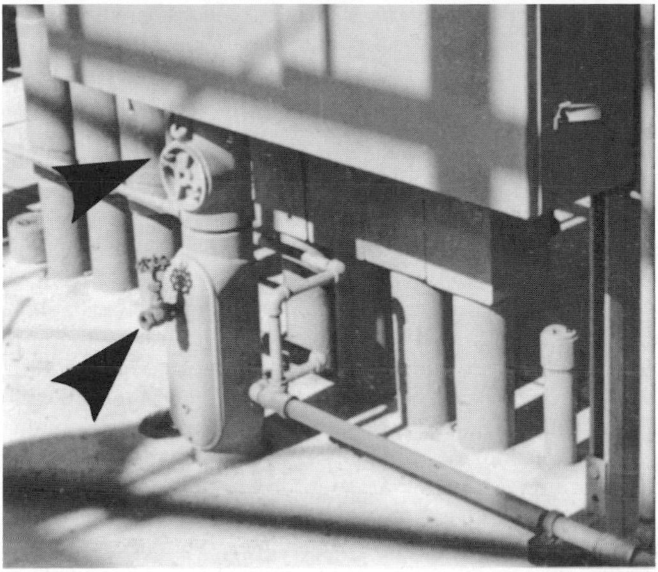

Fig. 500-2. A seal fitting (upper arrow) is used for equipment enclosure. But nitrogen purging of the conduit system was also applied at this installation at a space-rocket launch-pad. Equipment was specified to be listed for Class I, Division 1, Group B where exposed to hydrogen and for Group D where the equipment was in a rocket-fuel atmosphere. Where motors, panelboards, enclosures, etc., were not available in the proper group rating, nitrogen purging and pressurization were used in addition to sealing, to add another measure of safety. The valve (lower arrow) was used in the cover of each conduit body to provide continuous bleedoff of the nitrogen to maintain a pressure (2 in. of water) within the conduit, enabling the steady flow of nitrogen to keep the conduit free of any explosive mixture. (Sec. 500.7.)

Another widely used method involves the use of "intrinsically safe" equipment in hazardous locations and exempts such equipment from the rules of Arts. 500 through 517 where such areas of the various occupancies and locations covered by these articles are identified as "hazardous (classified) locations." Intrinsic safety is obtained by restricting the energy available in an electrical system to much less than that required for the ignition of specified flammable atmospheres such as gases and vapors that may exist in a particular processing area. If the energy level can be assured to be sufficiently low, any failure in the wiring system won't be ignition capable. These energy levels are very low. For example, an energy level of just ¼ milliwatt-second will ignite methane, and levels one-tenth of that will ignite unsaturated hydrocarbons such as ethylene. These circuits use Zener diodes in their supplies, arranged so that even a direct short circuit with its associated potential arcing and heating would not be ignition capable. These circuits can be used with ordinary wiring methods in hazardous (classified) locations for which they are rated. But such equipment must be "approved," which requires careful attention by the AHJ and the installer to any applicable UL listing and application data from the Hazardous Location Equipment Directory of UL.

Any applications of "intrinsically safe apparatus" in Class I, II, or III, or Class I, Zones 0, 1, and 2, or Zone 20, 21, and 22 locations must satisfy the rules of Art. 504 covering such apparatus. In this section, the reference to Art. 504 notes that the "provisions of

Arts. 501 through 503 and 510 through 516 shall not be considered applicable" to intrinsically safe apparatus and wiring "except as required by Art. 504."

Intrinsically safe circuits and equipment for use in Division 1 locations must be carefully applied. It is up to the designer and/or the installer to be sure that the energy level available in such equipment is below the level that could ignite the particular hazardous atmosphere. This must be ensured for both normal and abnormal conditions of the equipment. Testing of an intrinsically safe system by UL is based on a maximum distance of 5000 ft (1525 m) between the equipment installed in the nonhazardous, or Division 2, location and the equipment installed in the Division 1 location.

Nonincendive wiring is closely related to intrinsically safe wiring. Under all normal operating conditions it can't produce ignition-capable energy, but under very unusual (but foreseeable) operating conditions, might produce such an effect. These circuits are allowed in ordinary wiring in Division 2 locations, but not Division 1. The principle here is that given two low-likelihood probabilities, the probability of them occurring at the same moment is infinitesimal—the two low probabilities in this case being a Division 2 vapor release and a nonincendive circuit failing in a way that is ignition capable.

Wiring of intrinsically safe circuits must be run in separate raceways or otherwise separated from circuits for all other equipment to prevent imposing excessive current or voltage on the intrinsically safe circuits because of fault contact with the other circuits. In fact, no "fault" need occur; the adjacent circuits could cause a sufficient problem simply through induced voltages on the system believed to be intrinsically safe.

500.8. Equipment. Because of the inherently higher level of danger, design and installation of electrical circuits and systems in hazardous locations must be done in particularly strict compliance with the instructions given in product standards. Although NEC 110.3(B) requires all product applications to conform to the conditions and limitations specified in the directories issued by third-party testing labs (UL, Factory Mutual, ETL, etc.), the correlation of hazardous location electrical equipment is much more thoroughly dictated than that in nonhazardous areas.

In addition to the rules given in Arts. 500 through 504, an alternative method of classification is explained and permitted by Arts. 505 and 506. Application of the classifications indicated by Art. 505—that is, zone 0, 1, or 2, or 20, 21, and 22 (Art. 506)—is also permitted.

1. 500.8 requires that construction and installation of equipment in all hazardous areas "will ensure safe performance under conditions of proper use and maintenance." A note urges designers, installers, inspectors, and maintenance personnel to "exercise more than ordinary care" for hazardous location work. Parts **(A)** and **(B)** identify what type of equipment may be used in the various groups of hazardous locations and stipulate the markings required for equipment used in each such location. Explanatory material describes the nature of various hazardous atmospheres.
2. Paragraph **(A)(1)** of this section requires that all equipment in hazardous locations be approved not only for the class of location (such as Class I, II, or III) but also for the particular type of hazardous atmosphere (such as Group A, B, C, or D for locations involving gases or vapors, or Group E, F, or G if the atmosphere involves combustible or flammable dusts). The Code section describes the specific atmospheres that correspond to these letter designations.

Part **(A)(2)** indicates that Division 1 equipment may be used in Division 2 locations provided such equipment is suitable for the same "class, group, and temperature class" as would be required for the Division 2 equipment. And installation must be as described in subparts **(A)** and **(B)**, of this section.

3. An important regulation is given in part **(B)(3)**, which permits use of "general-purpose equipment" or "equipment in general-purpose enclosures" in Division 2 conditions of Class I, II, or III locations. That rule permits equipment that is *not* listed for hazardous locations but is listed for general use—*but* such use is acceptable only where a Code rule specifically mentions such application. For instance, 501.10(B)4 does say that boxes and fittings in Class I, Division 2 locations do *not* have to be of the explosionproof type; in other words, sheet metal boxes could be used. But controversy arises over that third paragraph of 500.8(A)(3) because general-use enclosures are permitted only where the equipment does not pose a threat of ignition "under *normal* operating conditions." Division 2 locations, however, *are* those where the hazardous atmosphere is not present under normal operating conditions.

   This Code text is simply an editorial choice to restate that general principle. This is another example of the larger principle discussed previously at nonincendive wiring, namely, that given two low-likelihood probabilities, the probability of them occurring at the same moment is infinitesimal—the two low probabilities in this case being a Division 2 vapor release and a wiring fault in the box. And this, in turn, is an application of the phrase in 90.1(B) that NEC adherence gives you an installation that is "essentially free from hazard"—not absolutely risk free, but essentially free from hazard.

4. In addition to being "approved" for the class and group of the hazardous area where it is installed, equipment is required by paragraph **(B)** of 500.8 to be "marked" with that data, along with its operating temperature at a 40°C ambient temperature, or at a higher ambient temperature if so rated and marked. Table 500.8(C) gives identification numbers that are used on heat-producing equipment nameplates to show the operating temperature for which the equipment is approved.

   Non-heat-producing equipment does not require a temperature marking, nor does equipment that will not exceed 100°C. More than one temperature marking may appear based on differing operating environments. Part **(D)** requires, in the case of gases and vapors, that the applied temperature marking not exceed that of the autoignition temperature of the hazardous material. In the case of combustible dusts, the marked temperature must be less than the ignition temperature of the applicable dust. For organic dusts that may dehydrate or carbonize, the temperature rating must not exceed 165°C even if the rated ignition temperature is higher. As of the 2014 NEC, due to changes in other standards, the term "ignition temperature" is limited to combustible dusts, and the comparable temperature for gases and vapors is "autoignition temperature."

Part **(E)** covers threads for conduit and other joints. In general all threaded joints must have at least five threads engaged, and also the joint must be made up wrenchtight whether or not a bonding jumper is installed. In this case the point is not simply to ensure equipment grounding continuity; it is also to ensure that the hot gases from an explosion inside a raceway only escape gradually and after they have cooled through the serpentine passage through the joint. There is an allowance for female threaded entries of listed equipment to only engage 4½ threads. The threads must use NPT specifications for taper of 1 in 16 (¾ in. taper per foot).

Metric-threaded equipment can be used, but the entries must be identified (probably means marked) as metric, or listed metric to NPT adapters must be provided. Note that there are cable fittings that use metric threads and that will mate with metric-threaded enclosures directly. Metric-threaded entries additionally require the engagement of eight threads for Group A (acetylene) and Group B (hydrogen) atmospheres. Metric entries also specify a minimum class of fit (6g/6H) which corresponds to a Class 2 (general use) fit in the NPT system. Not surprisingly unused openings must be plugged using metal plugs with comparable thread specifications.

Position of OSHA

With respect to equipment approval, great care must be taken to understand clearly the rules on hazardous locations equipment as covered in the electrical standards of the Occupational Safety and Health Administration (OSHA) of the U.S. Department of Labor. Those rules constitute federal law on this matter.

In the OSHA standard, Sec. 1910.307, Hazardous (Classified) Locations, is listed as one of the totally retroactive sections that apply to all electrical systems—both new ones and old ones—no matter when they were installed. On the matter of acceptability or approval of equipment used in hazardous locations, Sec. 1910.307(b) says:

> Equipment, wiring methods, and installations of equipment in classified locations shall be intrinsically safe, approved for the hazardous (classified) location, or safe for the hazardous (classified) location. Requirements for each of these options are as follows:
>
> **(1) Intrinsically safe.** Equipment and associated wiring approved as intrinsically safe shall be permitted in any hazardous (classified) location for which it is approved.
>
> **(2) Approved for the hazardous (classified) location.**
>
> [i] Equipment shall be approved not only for the class of location but also for the ignitible or combustible properties of the specific gas, vapor, dust, or fiber that will be present.
>
> **Note:** NFPA 70®, the National Electrical Code, lists or defines hazardous gases, vapors, and dusts by "Groups" characterized by their ignitible or combustible properties.
>
> [ii] Equipment shall be marked to show the class, group, and operating temperature or temperature range, based on operation in 40°C ambient, for which it is approved. The temperature marking shall not exceed the ignition temperature of the specific gas or vapor to be encountered. However, the following provisions modify this marking requirement for specific equipment:
>
> (A) Equipment of the non-heat-producing type, such as junction boxes, conduit, and fittings, and equipment of the heat-producing type having a maximum temperature not more than 100°C (212°F), need not have a marked operating temperature or temperature range.
>
> (B) Fixed lighting fixtures marked for use in Class I, Division 2 locations only need not be marked to indicate the group.
>
> (C) Fixed general-purpose equipment in Class I locations, other than lighting fixtures, which is acceptable for use in Class I, Division 2 locations need not be marked with the class, group, division, or operating temperatures.
>
> (D) Fixed dusttight equipment, other than lighting fixtures, which is acceptable for use in Class II, Division 2 and Class III locations need not be marked with the class, group, division, or operating temperature.

(3) **Safe for the hazardous (classified) location.** Equipment which is safe for the location shall be of a type and design which the employer demonstrates will provide protection from the hazards arising from the combustibility and flammability of vapors, liquids, gases, dusts, or fibers.

Those regulations apply to *all* electrical installations in hazardous locations—both new and existing systems. Note that there are actually *three* alternative ways for equipment to be acceptable for use in hazardous locations. A piece of equipment is acceptable if it is "intrinsically safe" *or* "approved for the hazardous (classified) location" *or* "safe for the hazardous (classified) location." A piece of equipment would have to satisfy only *one* of the three conditions described in the rule.

The three alternative conditions can be verified as follows:

Intrinsically safe equipment must be "approved" as such by UL, Factory Mutual Corporation, or some nationally recognized testing laboratory. In the UL *Hazardous Location Equipment Directory,* listings of various product categories indicate which equipment is intrinsically safe and the class and group of hazardous locations for which they are approved. Such equipment is evaluated and listed in accordance with Standard UL913, Intrinsically Safe Apparatus and Associated Apparatus for Use in Class I, II, and III, Division 1, Hazardous Locations. Another source of information on the subject is Installation of Intrinsically Safe Instrument Systems in Class I Hazardous Locations (ANSI/ISA RP12.6-1987).

Approved for the hazardous (classified) location equipment is basically "listed" and "labeled" equipment certified by a nationally recognized testing laboratory—such as equipment covered in the UL *Hazardous Location Equipment Directory.*

Safe for the hazardous (classified) location equipment is a category that becomes much more difficult to identify by firm, specific criteria. Just how does an employer demonstrate that equipment is safe for a hazardous location application? Although the OSHA rules themselves do not address that question anywhere, a note following Sec. 1910.307(b)(ii)(3) in the OSHA standard as published in the *Federal Register* does state that guidelines for making such a judgment are contained in Chap. 5 of the current **National Electrical Code (NFPA 70)**. However, these guidelines are not the only means of complying with the standard. Any equipment or installation shown by the employer to provide protection from the hazards involved will be acceptable. This performance-oriented approach will allow the employer maximum flexibility in providing safety for employees. Obviously, that is very general and vague explanation that ultimately leaves determination of acceptability completely up to the OSHA compliance officer.

A clear effect of OSHA regulations is to require "listed," "labeled," "accepted," and/or "certified" equipment to be used whenever available. If any electrical system component is of a kind that *any* nationally recognized testing lab "accepts, certifies, lists, labels, or determines to be safe," then that component *must* be so designated in order to be acceptable for use under OSHA regulations. Every electrical designer and installer must exercise great care in evaluating any and all equipment and products used in hazardous electrical work to ensure compliance with OSHA rules requiring certification by a nationally recognized testing lab.

In UL's Guide Card information [see discussion at 110.3(B) for details], limitations and application conditions are first set down for all equipment in general, as follows:

1. When equipment is listed and marked to show that it has been tested and is recognized for use in one or more of the Code-designated groups of hazardous locations, such marking indicates that such equipment is suitable for use in *either* a Division 1 or Division 2 location of the particular class of hazardous location, even though no reference is made to the division. Such equipment is, of course, also acceptable if used in a nonhazardous location. Figure 500-3 shows a typical nameplate used on such equipment as required by parts **(B)** and **(C)** of 500.8, showing suitability as "explosionproof" (Class I) and "dusttight" (Class II).

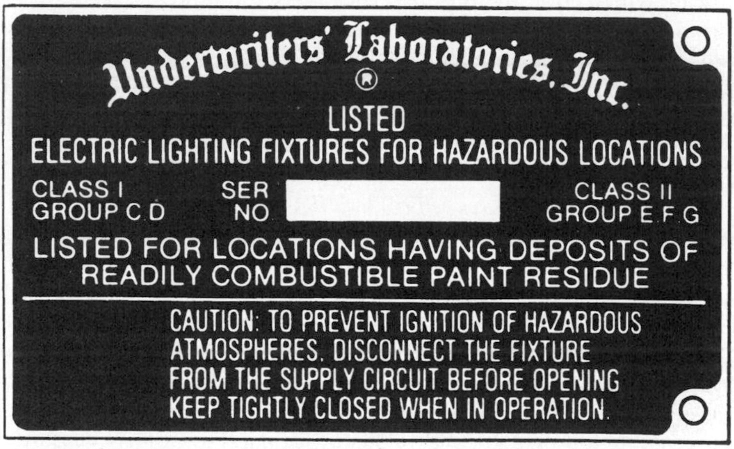
Underwriters' Laboratories, Inc.
®
LISTED
ELECTRIC LIGHTING FIXTURES FOR HAZARDOUS LOCATIONS
CLASS I
GROUP C D
SER
NO
CLASS II
GROUP E F G
LISTED FOR LOCATIONS HAVING DEPOSITS OF
READILY COMBUSTIBLE PAINT RESIDUE
CAUTION: TO PREVENT IGNITION OF HAZARDOUS
ATMOSPHERES. DISCONNECT THE FIXTURE
FROM THE SUPPLY CIRCUIT BEFORE OPENING
KEEP TIGHTLY CLOSED WHEN IN OPERATION.

Fig. 500-3. Typical UL part of equipment nameplate describes "Approval for Class and Properties." (Sec. 500.5.)

2. *But* equipment that is marked "Division 2" or "Div. 2" is suitable for use in only such a division and may *not* be used in a Division 1 location. However, a piece of equipment may have other marking to indicate its acceptability for other specific uses. Figure 500-4 shows a nameplate that is an addition to the nameplate in Fig. 500-3, noting that the same fixture is "Suitable for Wet Locations." 410.10(A) of the **NEC** requires such marking on "all fixtures installed in wet locations . . ." Care must be taken to distinguish between different parts of nameplates to precisely determine what constitutes third-party certification.
3. Equipment that is listed and marked for "Class I" locations (explosionproof) may be used for "Class II" locations if it is dusttight to exclude combustible dusts and if its external operating temperature is not at or above the

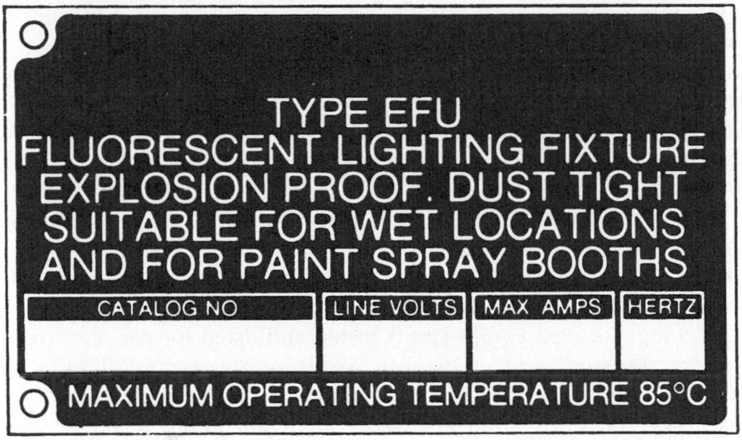
TYPE EFU
FLUORESCENT LIGHTING FIXTURE
EXPLOSION PROOF. DUST TIGHT
SUITABLE FOR WET LOCATIONS
AND FOR PAINT SPRAY BOOTHS
CATALOG NO
LINE VOLTS
MAX AMPS
HERTZ
MAXIMUM OPERATING TEMPERATURE 85°C

Fig. 500-4. Additional data on a nameplate may show other acceptability—as for "wet locations." (Sec. 500.5.)

ignition level of the particular dust that might accumulate on it. Obviously, these characteristics must be carefully established before Class I equipment is used in a Class II location.

It is commonly, and completely erroneously believed that a Class I suitability involves more robust qualifications than for Class II locations and, therefore, anything that will work for Class I will work for Class II. This is incorrect. Combustible dusts settle on equipment and, through an inherent thermal blanketing effect, raise its internal temperature if heat is being generated inside it. Combustible metal dusts add electrical conductivity to the equation. Flammable gases through the natural process of convection do not have these problems, and frequently the autoignition temperature of those gases will be higher than that of the very dusts that are impacting the surface temperature of the same equipment. On the other hand Class I chemicals can more easily reach the inside of an enclosure than Class II dusts. Class I and Class II threats are apples and oranges, and what works for one will not work for the other unless, by independent tests, it can be determined that the equipment under test is suitable for both locations.

4. Equipment listed for Class II, Group G (for flour, starch, or grain dust)—as used in a grain elevator—is also generally suitable for use in Class III locations, where combustible lint or flyings are present. The exception noted is for fan-cooled type motors, which might have their air passages choked or clogged by large amounts of the lint or flyings.
5. Because hazardous location equipment is critically dependent on proper operating temperature, a UL note warns that the ampere or wattage marking on power-consuming equipment is based on the equipment being supplied with voltage exactly equal to the rated voltage value. Voltage higher or lower than the rated value will produce other than rated amps or watts, with the possibility that the heating effect of the current within

the equipment will be greater than normal. Higher-than-normal current will be produced by overvoltage to resistive loads and by undervoltage to induction motors. Because of this, the actual circuit voltage, rather than the nameplate value, must be used when calculating the required ampacity of branch-circuit conductors, rating or setting of overcurrent protection, rating of disconnect, and so on—all to ensure adequate sizing and avoid overheating.

Low-temperature ambient conditions pose challenges in this area as well. Although in some cases liquids will be below their flash points, low-temperature conditions can increase the force of explosions and reduce the ability of metals and seals to retain adequate strength.

6. Hazardous location equipment is tested and listed for use at normal atmospheric pressure in an ambient temperature not over 104°F (40°C), unless indicated otherwise. Use of equipment under higher-than-normal pressure, in oxygen-enriched atmospheres, or at higher ambient temperatures can be dangerous. Such abnormal conditions may increase the chance of igniting the hazardous atmosphere and may increase the pressure of explosion within equipment.
7. Openings or modifications must not be made in explosionproof or dust-ignitionproof equipment, because any such field alterations would void the integrity and tested safety of the equipment. Field alteration of listed products for nonhazardous application is also generally prohibited.
8. All bolts as well as all threaded parts of enclosures must be tightly made up.
9. Indoor hazardous location equipment that is exposed to severe corrosive conditions must be listed as suitable for those conditions as well as for the hazardous conditions.

The requirements of 500.8 and information in Code Table 500.8(C) provide the means of properly identifying and classifying equipment for use in hazardous locations. The identification numbers in Code Table 500.8(C) pertain to temperature-range classifications as used by Underwriters Laboratories in UL *Hazardous Location Standards.*

While the Code rules for Class I locations do not differ for different kinds of gas or vapor contained in the atmosphere, note that it is necessary to select equipment designed for use in the particular atmospheric group to be encountered. This is necessary because explosive mixtures of the different groups have different flash points and explosion pressures. It is also necessary because the autoignition temperatures vary with the groups of explosive mixtures.

Underwriters Laboratories lists fittings and equipment as suitable for use in all groups of Class I, although the listings for Groups A and B are not as complete as those for Groups C and D. In addition, there is equipment that has its suitability for a particular group conditioned on seals being placed within a specified distance of the enclosure; if not, then the enclosure reverts to a different group with more challenging requirements.

In Class II locations, the Code, in a few cases, differentiates between the different kinds of dust—particularly dusts that are electrically conductive and those that are not conductive. Here again, as in Class I locations, care must be taken to determine that the equipment selected is suitable for use where a particular kind of dust is present.

In addition to taking more than ordinary care in selecting equipment for use in hazardous locations, special attention should be given to installation and maintenance details in order that the installations will be permanently free from electrical hazards. In making subsequent additions or changes, the high standards that were applied during the original installation must always be maintained.

For a more thorough knowledge of specific hazardous areas and equipment selection and location, it is essential to obtain copies of the various NFPA and ANSI standards referenced in Arts. 500 through 517.

ARTICLE 501. CLASS I LOCATIONS

501.1. Scope. In any Class I location, an explosive mixture of air and flammable gas or vapor may be present, which can be caused to explode by an arc or spark. To avoid the danger of explosions, all electrical apparatus that may create arcs or sparks should, if possible, be kept out of the rooms where the hazardous atmosphere exists, or, if this is not possible, such apparatus must be "of types approved for use in explosive atmospheres."

All equipment such as switches, CBs, or motors must have some movable operating part projecting through the enclosing case, and any such part—for example, the operating lever of a switch or the shaft of a motor—must have sufficient clearance so that it will work freely; hence, the equipment cannot be hermetically sealed. Also, the necessity for subsequent opening of the enclosures for servicing makes hermetic sealing impracticable. Furthermore, the enclosure of the equipment must be entered by a run of conduit, and it is practically impossible to make conduit joints absolutely air- and gastight. Due to slight changes in temperature, the conduit system and the apparatus enclosures "breathe"; that is, any flammable gas in the room may gradually find its way inside the conduit and enclosures and form an explosive mixture with air. Under this condition, when an arc occurs inside the enclosure an explosion may take place.

It is crucial to recognize that Class I Division 1 wiring methods and equipment are not and never have been designed or intended to have their suitability for the environment based on the exclusion of the flammable agent from their internal parts. On the contrary, such methods, especially the explosionproof ones, assume that the flammable agent will enter the enclosure and eventually reach an explosive concentration and be ignited. The wiring and equipment is tested by UL to ensure that when the internal explosion occurs, and it is assumed to be inevitable, the explosion will be contained safely, as described in greater detail below, and without igniting the surrounding atmosphere.

Nevertheless, some misguided installers look at threaded joints and think about plumbing, even in some instances to applying Teflon tape or compounds to the conduit and enclosure threads. Not only does this impair the equipment grounding continuity, it can also result in over-tightened conduit due to the lubricating effects of the pipe dope. There have been hydrostatic testing failures of these joints when they were made up in this way.

When the gas-and-air mixture explodes inside the enclosing case, the burning mixture must be confined entirely within the enclosure, so as to prevent the ignition of flammable gases in the room. In the first place, it is necessary that the enclosing case be so constructed that it will have sufficient strength to withstand the high pressure generated by an internal explosion. The pressure in pounds per square inch produced by the explosion of a given gas-and-air mixture has been quite definitely determined, and the enclosure can be designed accordingly.

Since the enclosures for apparatus cannot be made absolutely tight, when an internal explosion occurs, some of the burning gas will be forced out through any openings that exist. It has been found that the flame will not be carried out through an opening that is quite long in proportion to its width. This principle is applied in the design of so-called explosionproof enclosures for apparatus by providing a wide flange at the joint between the body and the cover of the enclosure and grinding these flanges to a definitely determined fit. In this case, the flanges are so ground that when the cover is in place the clearance between the two surfaces will at no point exceed 0.0015 in. (38.1 μm). Thus, if an explosion occurs within the enclosure, in order to escape from the enclosure the burning gas must travel a considerable distance through an opening not more than 0.0015 in. (38.1 μm) wide.

The basic construction characteristics of equipment for Class I hazardous locations are detailed in various sections of this article and in standards of testing laboratories. Application of the products hinges on understanding these details.

An explosionproof enclosure for Class I locations is capable of withstanding an explosion of a specified gas or vapor, which may occur within it, and of preventing the ignition of the specified gas or vapor surrounding the enclosure by sparks, flashes, or explosions of the gas or vapor within. Explosionproof equipment must provide three things: (1) strength, (2) joints which will not permit flame or hot gases to escape, and (3) cool operation, to prevent ignition of surrounding atmosphere.

UL requires that explosionproof enclosures must withstand a hydrostatic test of four times the maximum explosion pressure developed inside the enclosure. Explosionproof enclosures are not vapor- or gastight, and it is simply assumed that any hazardous gases in the ambient atmosphere will enter them either through normal breathing or when maintenance is performed on the enclosed equipment.

When an explosion occurs inside a rectangular explosionproof enclosure, the resulting force exerts pressure in all directions. The enclosure must be designed with sufficient strength to withstand these forces and avoid rupture (Fig. 501-1).

The energy generated by an explosion within an enclosure must be permitted to dissipate through the joints of the enclosure under controlled conditions. There are two generally recognized joint designs intended to provide this control—threaded and flat:

1. Threaded construction of covers and other removable parts that have five full threads engaged (4½ threads if part of a listed product assembly)

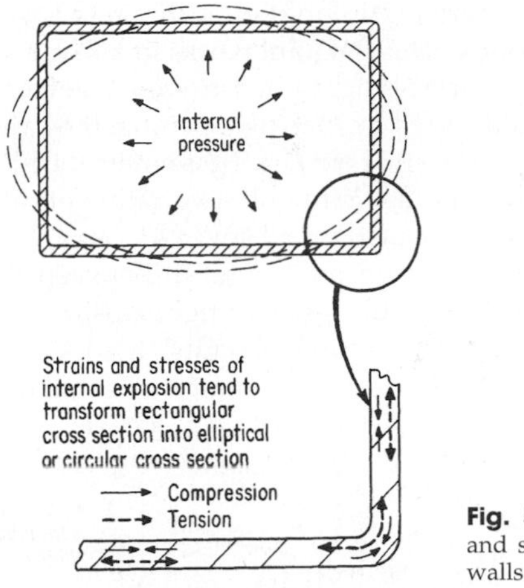
Internal pressure
Strains and stresses of internal explosion tend to transform rectangular cross section into elliptical or circular cross section
Compression
Tension

Fig. 501-1. An explosion creates strains and stresses in cross section of enclosure walls. (Sec. 501.1.)

produces a safe, flame-arresting, pressure-relieving joint. When an explosion occurs within a threaded enclosure, the flame and hot gases create an internal pressure against the cover, thus locking the threads and forcing the gases out through the path between the threaded surfaces. When the gases reach the outside hazardous atmosphere, they have been cooled by the heat-sink effect of the mass of metal, to a point below the autoignition temperature of the outside atmosphere, as shown in Fig. 501-2.

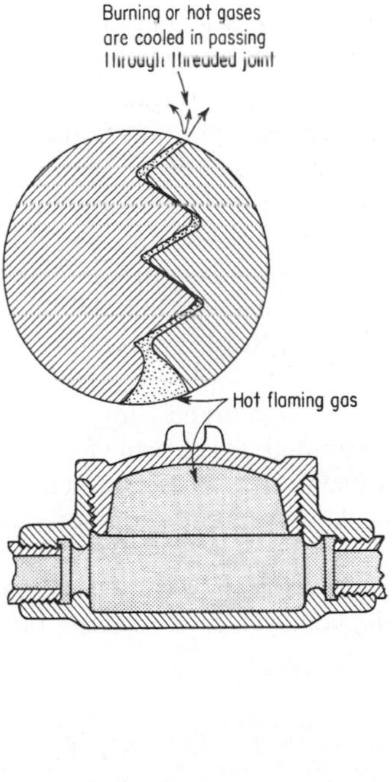
Burning or hot gases are cooled in passing through threaded joint
Hot flaming gas

Fig. 501-2. Threaded joints cool the heated gas as it escapes from enclosure under pressure. (Sec. 501.1.)

2. A flat joint is constructed by accurate grinding or machining of the mating surfaces of the cover and the body. This flat joint works in a manner similar to the threaded joint. The two surfaces are bolted closely together, and as flame and hot gases are forced through the narrow opening, they are cooled by the mass of the metal enclosure, so that only cool gas enters the hazardous atmosphere. Figure 501-3 shows the flat joint and a variation on it called a *rabbet* joint. Care must be taken to ensure that all cover screws are tight and that no particles of dirt or other foreign matter get in between the cover and the body. Even a small particle could prevent tight closing and might allow the joint to pass flame. Similarly, a careless scratch across the surface, such as a key might make, will destroy the explosionproof suitability of the enclosure.

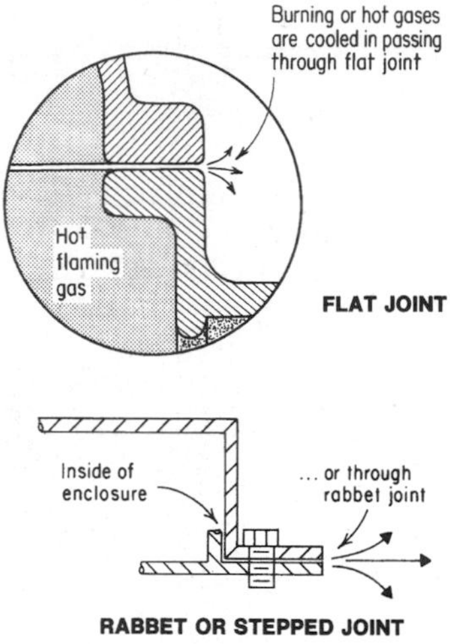
Burning or hot gases are cooled in passing through flat joint
Hot flaming gas
FLAT JOINT
Inside of enclosure
... or through rabbet joint
RABBET OR STEPPED JOINT

Fig. 501-3. Ground surfaces of flat or rabbet joint provide release and cooling of internal gases. (Sec. 501.1.)

UL standards on explosionproof enclosures contain rules on "Grease for Joint Surfaces":

Paint or a sealing material shall not be applied to the contacting surfaces of a joint. A suitable corrosion inhibitor (grease) such as petrolatum, soap-thickened mineral oils, or nondrying slushing compound may be applied to the metal joint surfaces before assembly. The grease shall be of a type that does not harden because of aging, does not contain an evaporating solvent, and does not cause corrosion of the joint surfaces.

501.5. Zone Equipment. Equipment "listed and marked" for installation in "Zone 0, 1, and 2 Locations" may be used, provided it is suitable for the location as determined in accordance with 505.9(C)(2). Specifically, Zone 0 equipment is suitable for Class I Division 1 or 2, and Zone 0, 1, and 2 equipment is suitable for Class I Division 2 locations. The more common Class I locations are those where some process is carried on involving the use of a highly volatile and flammable liquid, such as gasoline, petroleum naphtha, benzene, diethyl ether, or acetone, or flammable gases.

501.10. Wiring Methods. There are only three wiring methods generally permitted in Class I Division 1 locations, two raceways and one cable. The raceways are rigid metal conduit (RMC) and intermediate metal conduit (IMC). Type MI cable is the only cable assembly that is generally permitted in Class I, Division 1 locations (Fig. 501-4).

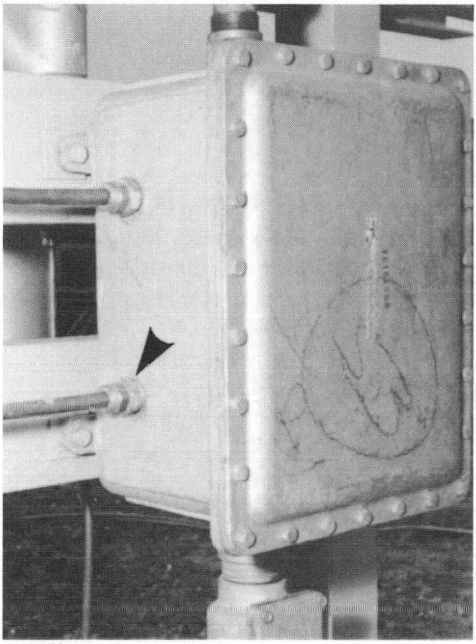

Fig. 501-4. Type MI cable is recognized for use in Class I, Division 1 locations, provided that the termination fittings (arrow) are listed as suitable for hazardous location use. (Sec. 501.10.)

There are limited exceptions that broaden the wiring method allowances under certain conditions. Polyvinyl chloride conduit (Type PVC), reinforced thermosetting resin conduit (Type RTRC), and along with high-density polyethylene conduit (Type HDPE) are now permitted to be used (with the customary grounding conductors) in Class I Division 1 locations that exist below grade, provided the conduit is embedded in a concrete pour not less than 50 mm (2 in.) thick around the conduit, and the last 600 mm (2 ft) of the run prior to emergence from grade must be threaded heavy-wall conduit that will accept the required seal fittings and other fittings as required at the point of emergence.

The other exceptions (formatted as positive text, but they still are variances from the normal rule) only apply to "industrial establishments with restricted public access" and with qualified maintenance and supervision in place. Although stated separately, the exceptions cover the same topic, that being special forms of metal-clad cable of particularly robust construction designed for hazardous locations and that feature a gas-tight corrugated armor with an outer nonmetallic jacket. The power version of the cable includes a separate equipment grounding conductor and is designated "Type MC-HL." The control version of the cable is designated "Type ITC-HL" and is a form of instrumentation tray cable that operates under severe energy limitations, as covered in Art. 727. Termination fittings listed for the application are required in either case.

Optical fiber cable has been recognized in 770.3(A) as acceptable in hazardous locations, and this use is now correlated in 501.10(A)(1)(e). Note that it must be installed within a 501.10(A) raceway, which circles back to threaded RMC or IMC, and then sealed in accordance with the usual requirements in 501.15. The same correlating provision was added as 501.10(B)(7) for Division 2 locations.

In general, approvals are to be based on the performance of a fitting or piece of equipment when subjected to a specific atmosphere. As applied to rigid metal conduit, to be explosionproof, threaded joints must be used at couplings, and for connection to fittings, the threads must be cleanly cut, five full threads must be engaged, and each joint must be made up tight. Conduit elbows and short-radius capped elbows provide for 90° bends in conduit but only where wires may be guided when being pulled into the conduit, to prevent damaging the conductors by pulling them around the sharp turn in the elbow. Figure 501-5 shows two types of fittings used in hazardous locations to facilitate pull-in of conductors that have stiff or heavy-wall insulation. The capped elbow is especially suited for use in tight quarters.

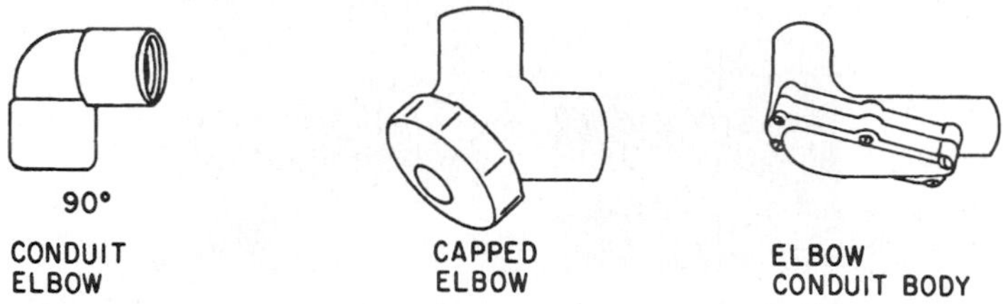
90°
CONDUIT
ELBOW
CAPPED
ELBOW
ELBOW
CONDUIT BODY

Fig. 501-5. Conduit elbows and similar fittings may be used where wires may be guided into conduit. (Sec. 501.10.) The conduit elbow at the left must only be used where the conductors may be guided through it during installation, and therefore it cannot be used in the middle of a run.

All fittings, such as outlet, junction, and switch boxes, and all enclosures for apparatus, should have threaded hubs to receive the conduit and must be explosionproof. Explosionproof junction boxes are available in a wide variety of types (Fig. 501-6). Box covers may have threaded connections with the boxes, or the cover may be attached with machine screws, in which case a carefully ground flanged joint is required.

A flexible, explosionproof fitting, [as covered in **(A)(2)**] suitable for use in Class I hazardous locations, is shown in Fig. 501-7. The flexible portion consists of a tube of bronze having deeply corrugated walls and reinforced by a braid of fine bronze wires. A heavy threaded fitting is securely joined to each end of the flexible tube, and a fibrous tubular lining, similar to "circular loom," is provided in order to prevent abrasion of the enclosed conductors that might result from long-continued vibration. The complete assembly is obtainable in various lengths up to a maximum of 914 mm (3 ft).

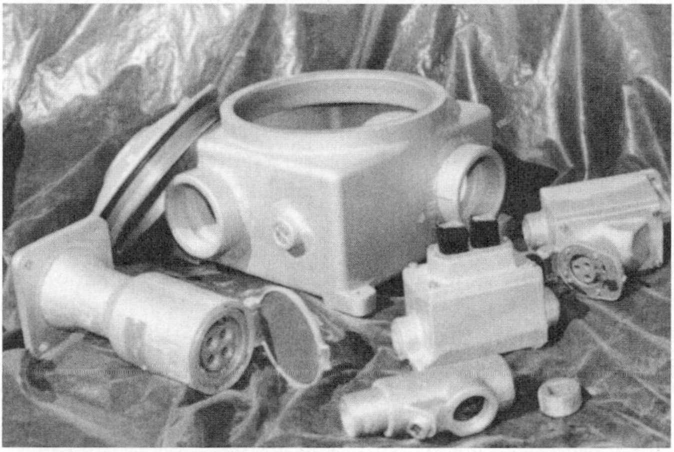

Fig. 501-6. A wide assortment of boxes, conduit bodies, fittings, and other enclosures are made in explosionproof designs listed for use in Class I locations. (Sec. 501.10.)

Part (A)(2)
to 501.10. Conductivity of flexible fiting is evaluated by voltage-drop test

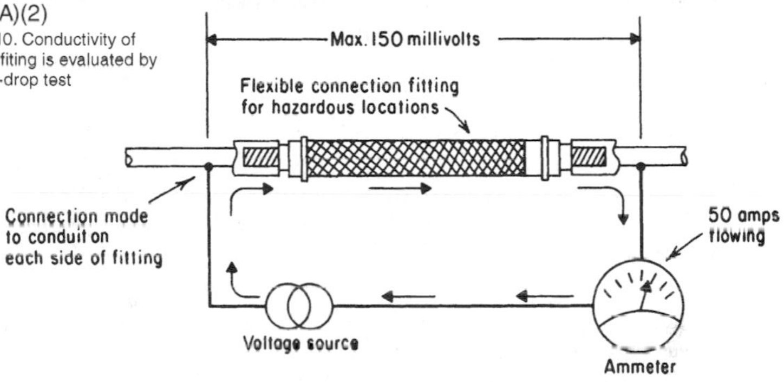
Max. 150 millivolts
Flexible connection fitting for hazardous locations
Connection made to conduit on each side of fitting
50 amps flowing
Voltage source
Ammeter

Voltage drop across fitting must not exceed 150 millivolts, measured between points on conduit 1/16 in. from each end of the fitting.

Fig. 501-7. A conductivity testing protocol for flexible fittings used in hazardous (classified) locations. (Sec. 501.10.)

Flexible connection fittings that are recognized by NEC 501.10(A)(2) for use in Class I, Division 1 locations are intended by UL and the NEC to be used where it is necessary to provide flexible connections in threaded rigid conduit systems—as at motor terminals. Use of such flexible fittings must observe the minimum inside radius of bend for which the fitting has been tested. Those data are provided with the fitting.

Note: UL warns that acceptability of the use of flexible connection fittings must be cleared with local inspection authorities. In general, use of such flexible fittings should be avoided wherever possible and should be limited to situations where use of threaded rigid conduit is completely ruled out by the needs or conditions of the application.

Where flexible connection fittings *are* used, the corrugated metal inner wall and the metal braid construction of the fitting provide equipment grounding continuity between the end connectors and the fitting. The UL test for conductivity through a flexible fitting is shown in Fig. 501-7. Although 501.30(B) requires either an internal or an external bonding jumper to be used with standard flexible metal conduit in Class I, Division 2 locations, that rule does not apply to listed flexible fittings.

Part **(A)(2)** indicates that 501.140 permits flexible cord in Class I, Division 1 locations for portable lighting equipment and portable utilization equipment. This rule eliminates any conflict between the clear and direct rules of 501.10(A) on wiring methods and the limited use of portable cord as an alternative to the wiring methods used for fixed wiring. The 2014 NEC has added the hazardous locations version of tray cable evaluated for exposed runs (TC-ER-HL) for applications in qualified industrial occupancies with restricted public access and physical protection by location or guarding.

In **Class I, Division 2** locations, explosionproof outlet boxes are not required at lighting outlets or at junction boxes containing no arcing device. However, where conduit is used, it should enter the box through threaded openings, as shown in Fig. 501-6, or if locknut-bushing attachment is used, a bonding jumper and/or fittings (such as bonding locknuts on knockouts without any concentric or eccentric rings left in the wall of the enclosure) must be provided between the box and conduits, as required in 501.30(B).

Part (B) actually leads off with a proforma recognition that wiring methods permitted in Division 1 are also permitted in Division 2. However, the 2017 NEC has added heavy walled steel conduits to the list a second time. As Division 1 wiring methods they must be threaded; in their second appearance in Division 2, they can be unthreaded with "listed threadless fittings," which presumably includes set-screw connectors and couplings. Interestingly, an attempt was made to include EMT at the same time and it failed. The proposed addition of cablebus, however, did succeed. That will be a significant assist for high-capacity feeders because it allows for free-air ampacities.

As noted in part **(B)** of this section, flexible connections permitted in Class I, Division 2 locations may consist of flexible conduit (Greenfield) and also liquidtight flexible metal conduit with approved fittings, and such fittings are not required to be specifically approved for Class I locations. Note that a separate grounding conductor is usually necessary to bond across such flexible connections (there is a control circuit exception for LFMC only), as required in 501.30(B).

501.10(B) also recognizes the use of power-limited tray cable (Type PLTC) on the list of wiring methods permitted in Class I, Division 2 locations, in accordance with the provisions of Art. 725 covering remote-control, signaling, and low-energy circuits. Similarly, instrumentation tray cable (Type ITC) is also permitted in Class I, Division 2 locations, as covered in Art. 727. And 501.10(B)(5) makes clear that medium-voltage circuits (i.e., circuits over 600 V) may employ the wiring methods covered in the first part of 501.10(B) and, where protected from physical damage, may be made up using medium-voltage cable (shielding governed only by rules in Chap. 3, a 2008 NEC change) in cable trays when installed in accordance with Art. 392. And Art. 328 dictates that such cable must be Type MV cable. Metal-clad cable,

Type MC and Power and Control Tray Cable, Type TC round out the former list; note that Type AC cable is not on the list. Two new approaches joined the list for the first time in 2008. The heaviest wall versions of Type PVC (i.e., "Schedule 80") and Type RTRC (i.e., suffix "-XW") conduits can now be used in Class I Division 2 locations exposed, without concrete encasement. This use is heavily restricted, being limited to industrial occupancies with limited public access and qualified maintenance and supervision, and in addition, these methods are only allowed where "metallic conduit does not provide sufficient corrosion resistance." In cases where boundary seals are necessary, the seal must be located in the Division 2 area, with the Division 1 wiring method continuing into the Division 2 area until it reaches the seal.

Figure 501-8 shows some applications of wiring methods that are covered by the rules of 501.10. At the top, use of standard flex (Greenfield) in a Division 1

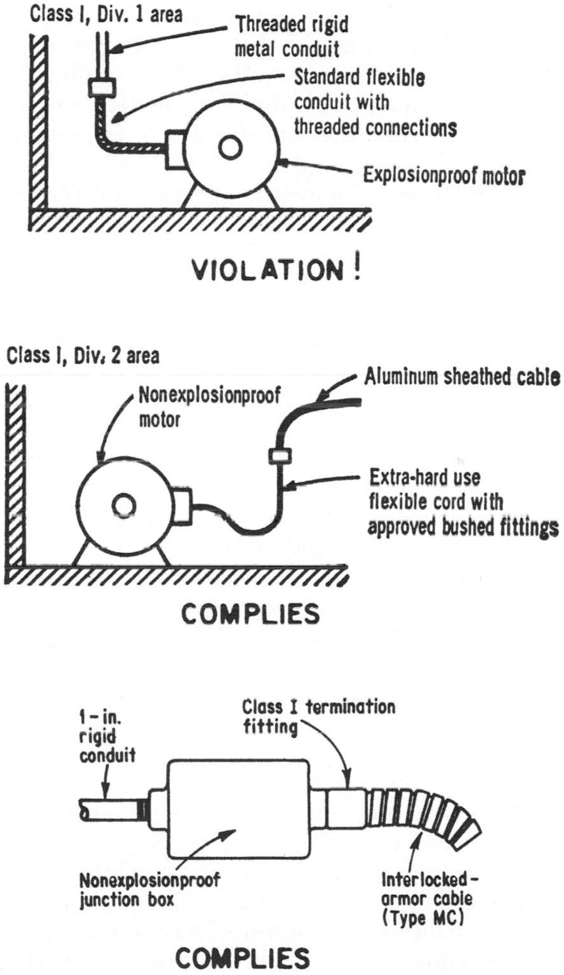
Class I, Div. 1 area
Threaded rigid metal conduit
Standard flexible conduit with threaded connections
Explosionproof motor
VIOLATION!
Class I, Div. 2 area
Nonexplosionproof motor
Aluminum sheathed cable
Extra-hard use flexible cord with approved bushed fittings
COMPLIES
1-in. rigid conduit
Class I termination fitting
Nonexplosionproof junction box
Interlocked-armor cable (Type MC)
COMPLIES

Fig. 501-8. Wiring methods in Class I locations are clearly regulated. (Sec. 501.10.) The bottom drawing applies to Division 2 locations only.

area violates part **(A)** of this section. At the center, use of aluminum-sheathed cable (Type ALS) is okay in a Division 2 area. Even though Type ALS is no longer mentioned in part **(B),** that type of cable is now covered by Art. 330 and is considered as one form of Type MC cable, which is mentioned in part **(B)** as acceptable in Division 2 areas. At the bottom, use of Type MC is okay in a Division 2 location.

For applications where flexibility is required in Division 2 locations, the acceptable wiring methods are itemized in **(B)(2)**. The 2014 NEC has added an interlocking armor Type MC cable to the list, along with certain elevator cables, which have been carried in Table 400.4 as acceptable for this use but never correlated with these rules.

Paragraph **(3)** covers nonincendive field wiring. Review the discussion at 500.7 for a better understanding of the principles governing this protection technique. To ensure that the nonincendive wiring functions as intended, it must conform to the control drawing and any components must not interconnect this wiring with other circuits. The NEC terminology at this point borrows from intrinsically safe provisions in Art. 504, which are discussed at greater length in this book under that article heading. Nonincendive wiring circuits must be run in separate cables unless they are shielded or employ insulation of the specified thickness.

501.15. Sealing and Drainage. The proper sealing of conduits in Class I locations is an important matter. Whether used in an enclosure or in conduit, seals are necessary to prevent gases, vapors, or flames from being propagated into an enclosure or conduit run and to confine an explosion that might occur within an enclosure.

The first note after the first paragraph points out that seal fittings properly installed are not normally capable of preventing the passage of liquids, gases, or vapors if there is a continuous pressure differential across the seal. However, as indicated, seals may be specifically designed and tested for preventing such passage. This explanation, along with the wording in such rules as 501.15(A)(4), makes clear that seals will only "minimize," not "prevent," passage of gases or vapors through the seal.

When an explosion takes place within an enclosure because of arc ignition of gas or vapor that has entered the enclosure, flames and hot gases could travel rapidly through unsealed conduits, and the resultant buildup of pressure could exceed the strength of conduit, wireways, or enclosures, causing explosive rupture. *Pressure piling* is the name given to the action that takes place when an explosion occurs inside an enclosure because of flammable gas within the enclosure being ignited by a spark or overheated wiring. When this happens, and there are no seals in the conduits connecting to the enclosure, exploding gas will compress the entire atmosphere within the conduit system and flames or heat will ignite compressed gas complete with the additional fuel that is present along the conduit run, causing another, more powerful explosion or making the initial explosion that much more powerful. Think, for a moment of the vastly improved power provided in a piston of an internal combustion engine when the fuel-air charge is compressed. The pressure builds as the flame front moves through the system of raceways and enclosures, increasing in intensity until a critical element in the system ruptures. To prevent such occurrences, it is mandatory that seal-off fittings be used in certain enclosures or conduit runs to block and confine potentially hazardous vapors.

The necessary sealing may be accomplished by inserting in the conduit runs special sealing fittings, as shown in Fig. 501-9, or provision may be made for sealing in the enclosure for the apparatus. An explosionproof motor is made with the leads sealed where they pass from the terminal housing to the interior of the motor, and no other seal is needed where a conduit terminates at the motor, except that if the conduit is metric designator 53 (trade size 2) or larger, a seal must be provided not more than 18 in. (450 mm) from the motor terminal housing.

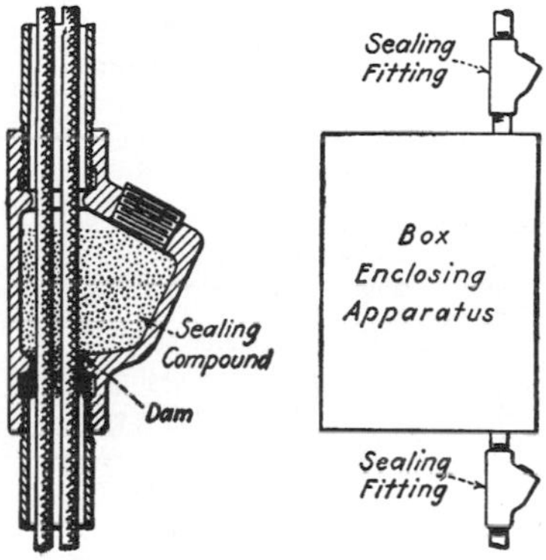
Sealing Compound
Dam
Sealing Fitting
Box Enclosing Apparatus
Sealing Fitting

Fig. 501-9. Seal fitting is filled with compound to prevent passage of flame or vapor through the conduit. (Sec. 501.15.)

Class I, Division 1

Part (A) of this section covers mandatory use of seals in Class I, Division 1 locations:

1. A seal is required in each and every conduit (regardless of the size of the conduit) entering (or leaving) an enclosure that contains one or more switches, CBs, fuses, relays, resistors, or any other device that is capable of producing an arc or spark that could cause ignition of gas or vapor within the enclosure or any device that might operate hot enough to cause ignition. The 2014 **NEC** quantifies this concept as exceeding 80 percent of the autoignition temperature of the specified gas or vapor. In each such conduit, a seal fitting must be placed never more than 18 in. (450 mm) from such enclosure. The Code rule has eliminated the phrase that said seals had to be installed "as close as practicable" to the enclosure—leaving the remainder of the requirement, that the seal must not be more than 18 in. (450 mm) from the enclosure, intact. As shown in Fig. 501-10, a conduit seal fitting is installed in the top conduit and one of the bottom conduits—close to the enclosure of the arcing device. But a seal is not used in the conduit to the pushbutton because that is a factory-sealed device and that seal is not over 18 in. (450 mm) from the starter. This complies with the intent of the Code rule, as well as the rule of 501.15(C), which recognizes "approved integral means for sealing," as in the pushbutton. Figure 501-11 shows a seal fitting as close as possible to a box housing a receptacle.

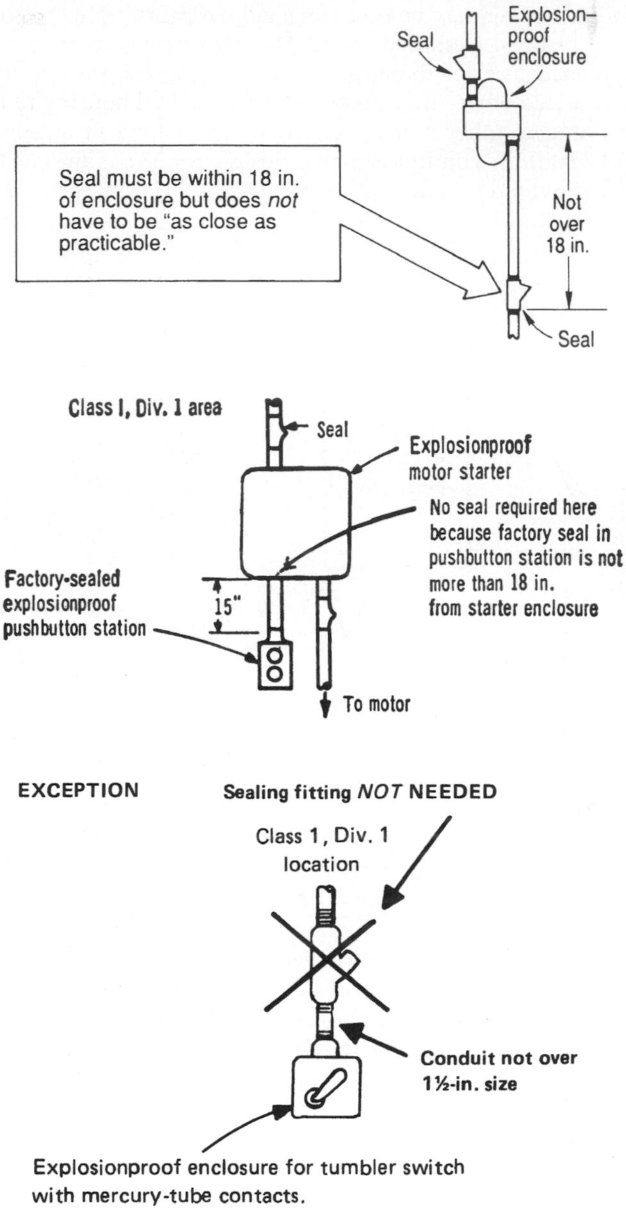
Seal
Explosion-proof enclosure
Seal must be within 18 in. of enclosure but does *not* have to be "as close as practicable."
Not over 18 in.
Seal
Class I, Div. 1 area
Seal
Explosionproof motor starter
No seal required here because factory seal in pushbutton station is not more than 18 in. from starter enclosure
Factory-sealed explosionproof pushbutton station
15"
To motor
EXCEPTION
Sealing fitting *NOT* NEEDED
Class 1, Div. 1 location
Conduit not over 1½-in. size
Explosionproof enclosure for tumbler switch with mercury-tube contacts.

Fig. 501-10. Seal should be in each conduit within 18 in. (450 mm) of the sealed enclosure, center, but may not be needed at all, bottom. (Sec. 501.15.)

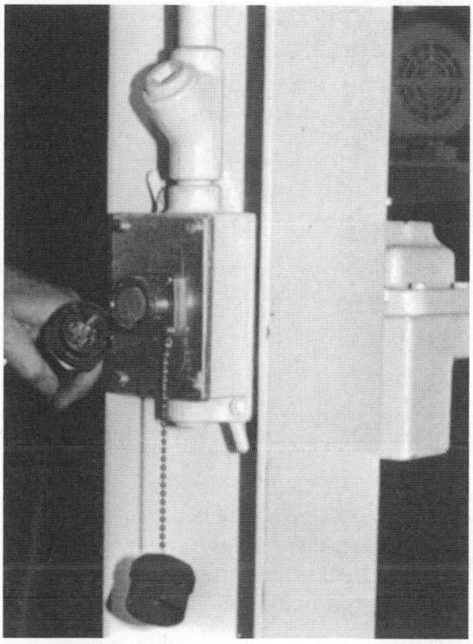

Fig. 501-11. Seal fitting is as close as possible to enclosure, providing maximum effectiveness. (Sec. 501.15.)

The exception to part **(A)(1)** notes that a seal is not required in a conduit entering an enclosure for a switching device in which the arcing or sparking contacts are internally sealed against the entrance of ignitible gases or vapors. Such conduit is applied to a condition similar to a conduit connection to an explosionproof junction box—that is, any gas that enters the enclosure will contact only wiring terminals and is not exposed to arcs. But because a conduit seal is always required for any conduit of metric designator 53 (trade size 2) size or larger that enters a junction box or terminal housing, this exception permits elimination of the conduit seal *only* for conduits of metric designator 41 (trade size 1½) size or smaller that enter an enclosure for switching devices with sealed or inaccessible contacts—such as mercury-tube switches, as shown at the bottom of Fig. 501-10. Any switch, CB, or contactor with its contacts in a hermetically sealed chamber or immersed in oil might be applied under this exception. The stand-alone paragraph following (2) on large entries applies generally and makes an important point in this regard. A factory-sealed enclosure is only recognized as sealing itself. However close it may be to another enclosure requiring a seal, it does not qualify as providing that seal for the adjacent enclosure.

Recognition of sealed-contact devices without a separate conduit seal is similar to recognition of "an integral means for sealing" [501.15(C)(1)], which is a seal provision that is manufactured directly into some enclosures for Class I equipment.

But questions have always risen about the acceptability of boxes or fittings between the sealing fittings and the enclosure being sealed. This section identifies the devices that may be used between the seal and the enclosure. Explosionproof unions, couplings, reducers, elbows, and capped elbows are the only enclosures or fittings permitted between the sealing fittings and the enclosure. A reducing bushing (a *reducer*) may be connected at a conduit entry to an explosionproof enclosure so that the bushing is connected between the seal and the enclosure. Because a reducer is commonly used to provide a conduit bushing in the wall of an enclosure and does not pose a threat to the integrity of the seal function, reducers have been added to the list of devices permitted to be used between a seal and the enclosure it supplies. The rule clearly rules out the use of a box or any similar large-volume enclosure between the seal fitting and the enclosure being sealed, as shown in Fig. 501-12. Undoing accepted practice since the 1978 **NEC**, the placement of a conduit body not larger than the raceway trade size between the seal and the enclosure is, as of the 2017 **NEC**, prohibited "due to the increased volume in the raceway system." No further substantiation was offered for this dramatic change.

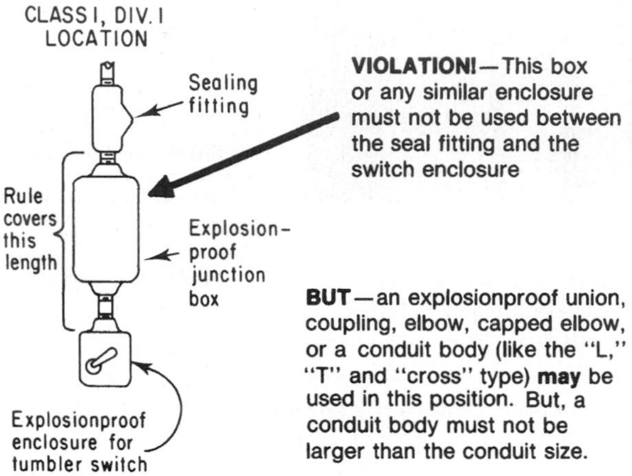
CLASS I, DIV. I LOCATION
Sealing fitting
VIOLATION!—This box or any similar enclosure must not be used between the seal fitting and the switch enclosure
Rule covers this length
Explosion-proof junction box
BUT—an explosionproof union, coupling, elbow, capped elbow, or a conduit body (like the "L," "T" and "cross" type) may be used in this position. But, a conduit body must not be larger than the conduit size.
Explosionproof enclosure for tumbler switch

Fig. 501-12. Any type of junction *box* may not be used between seal and enclosure. (Sec. 501.15.)

The fittings listed as acceptable for use between the seal and the enclosure were selected on the basis that their internal volume was sufficiently small as to prevent the accumulation of any dangerous volume of gas or vapor. Acceptability was based on limiting the volume of gas or vapor that may accumulate between the seal and the enclosure being sealed. It was on this basis also that conduit bodies are prohibited from being of a larger size than the conduit with which they are used. If they were permitted, they would present the opportunity for accumulation of a larger volume of gas or vapor, which is considered objectionable.

Figure 501-13 shows an interesting variation on this concern for use of a box between a seal and an enclosure. *No splices are permitted within seal-off fittings,* according to the UL *Hazardous Location Equipment Directory.* The illustration shows a round-box type of seal fitting that is used for pulling power and control wires. Such a fitting takes a large, round, threaded cover equipped with a pouring spout. The cover, shown removed, is readily unscrewed to provide maximum unobstructed access to the fitting interior, which facilitates damming either one or both conduit hubs. When the cover is replaced, it can be rotated so the spout points up to permit compound fill. This fitting can be used to seal conduit regardless of its direction or run.

Fig. 501-13. Seal fitting of the round box type is used to seal the conduit run into the bottom of an explosionproof starter enclosure, with flexible fitting connection to the Class I, Division 1, Group D motor below and a watertight Class I cord connector control cable. (Sec. 501.15.)

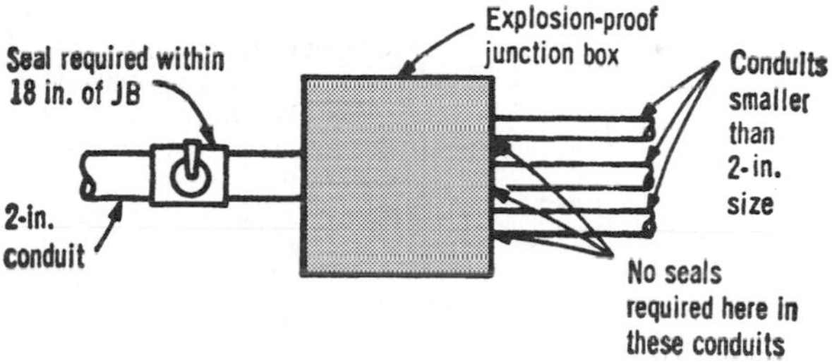
Explosion-proof junction box
Seal required within 18 in. of JB
Conduits smaller than 2-in. size
2-in. conduit
No seals required here in these conduits

Fig. 501-14. Seals for conduits to junction boxes or fittings are required only for conduits of 2-in. (53-mm) size and larger. (Sec. 501.15.)

2. A seal is required in any conduit run of metric designator 53 (trade size 2) size or larger, where such a conduit enters an "enclosure or fitting" that is required to be explosionproof and houses terminals, splices, or taps, as shown in Fig. 501-14. Note in such cases, however, that the rule does not call for the seal to be "as close as practicable" to the enclosure or fitting, as required for a housing of an arcing or sparking device. Here, it simply requires that the seal be not over 18 in. (450 mm) away from the enclosure.

Another example of seal application in accordance with parts **(1)** and **(2)** of 501.15(A) is shown in Fig. 501-15.

Part **(3)** of 501.15(A) covers use of a single seal to provide the required seal for a conduit connecting two enclosures. Where two such pieces of apparatus are connected by a run of conduit not over 3 ft (900 mm) long, a single seal in this run is considered satisfactory if located at the center of the run. Figure 501-16 shows this rule. Although the wording is not detailed, the reference to "within 18 in. (450 mm) of such enclosure" must be understood to be 18 in. (457 mm) measured along the conduit, to avoid the misapplication of the rule shown in Fig. 501-17. The single seal

Fig. 501-15. Seal fittings are very close to points where conduits enter the motor starter enclosures. Seals are not required for the conduits entering the junction box (arrow) because they are not metric designator 53 (trade size 2) or larger size—although seals may be used there. (Sec. 501.15.)

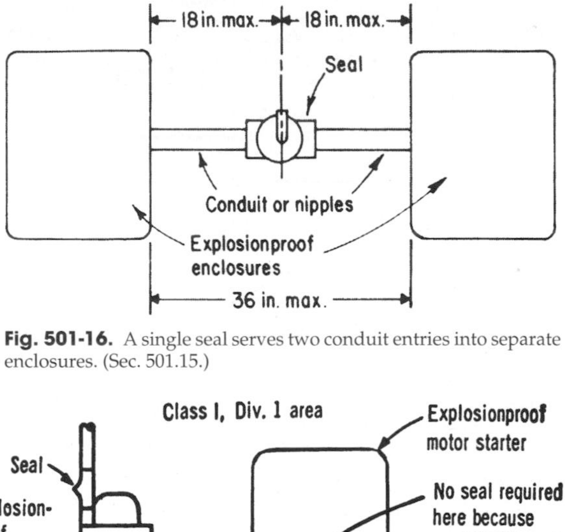

18 in. max.
18 in. max.
Seal
Conduit or nipples
Explosionproof enclosures
36 in. max.

Fig. 501-16. A single seal serves two conduit entries into separate enclosures. (Sec. 501.15.)

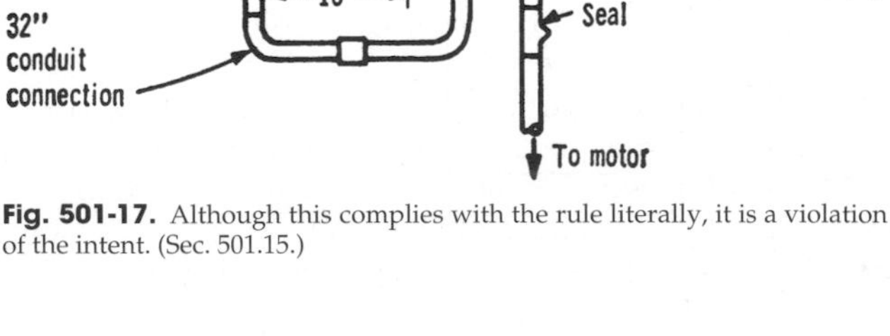

Class I, Div. 1 area
Explosionproof motor starter
Seal
Explosion-proof circuit breaker
Seal "A"
10"
No seal required here because seal "A" is not more than 18 in. from either enclosure
32" conduit connection
Seal
To motor

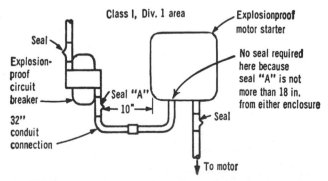

Fig. 501-17. Although this complies with the rule literally, it is a violation of the intent. (Sec. 501.15.)

at A is not over 18 in. (450 mm) from the CB enclosure and is literally not over 18 in. (450 mm) from the starter (it is 10 in. from the starter). But that use of a single seal for the two enclosures violates the Code intent that the 18 in. (450 mm) in each case *must* be measured along the conduit.

Part **(4)** requires a seal in each and every conduit that leaves the Class I, Division 1 location—whether it passes into a Division 2 location or into a nonhazardous location. This required seal may be installed on either side of the boundary *and* within 10 ft (3.0 m) of the boundary (Fig. 501-18). There must be no union, coupling, box, or fitting between the sealing fitting and the point where the conduit leaves the hazardous location. The rule does specify a maximum distance that must be observed between the sealing fitting and the boundary, and that distance is 10 ft (3.0 m). The purpose of this sealing is twofold: (1) The conduit usually terminates in some enclosure in the Division 2 or nonhazardous area containing an arc-producing device, such as a switch or fuse. If not sealed, the conduit and apparatus enclosure are likely to become filled with an explosive mixture and the ignition of this mixture may cause local damage in the Division 2 or nonhazardous location. (2) An explosion or ignition of the mixture in the conduit in the nonhazardous area would probably travel back through the conduit to the hazardous area and might cause an explosion there if, because of some defective fitting or poor workmanship, the installation is not completely explosionproof. If the conduit is unbroken (no union, coupling, etc.) between an enclosure seal and the point where the conduit leaves the hazardous area, an additional seal is not required at the boundary. Figure 501-19 shows two violations where conduit leaves the hazardous area.

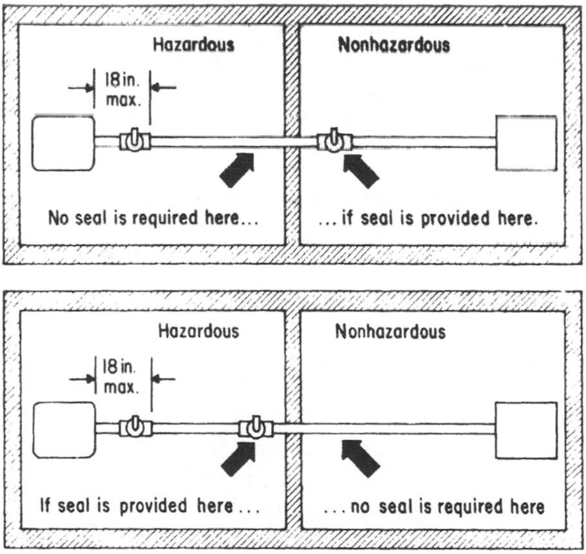

Hazardous
Nonhazardous
18 in. max.
No seal is required here...
... if seal is provided here.
Hazardous
Nonhazardous
18 in. max.
If seal is provided here ...
... no seal is required here

Fig. 501-18. Conduit must be provided with seal fitting where it crosses boundary. (Sec. 501.15.)

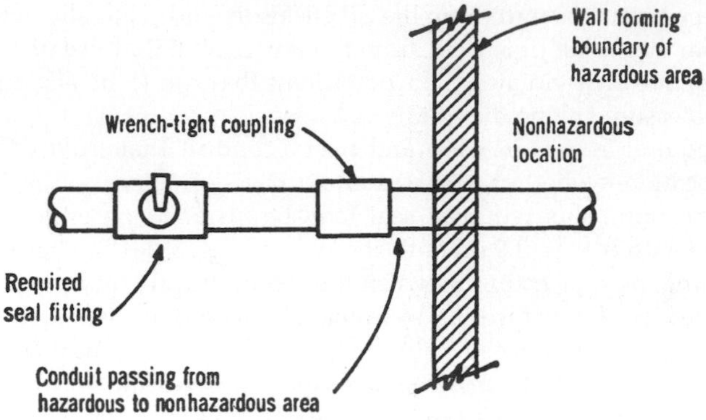
Wall forming
boundary of
hazardous area
Wrench-tight coupling
Nonhazardous
location
Required
seal fitting
Conduit passing from
hazardous to nonhazardous area

VIOLATION ! — Coupling not permitted between seal and the boundary

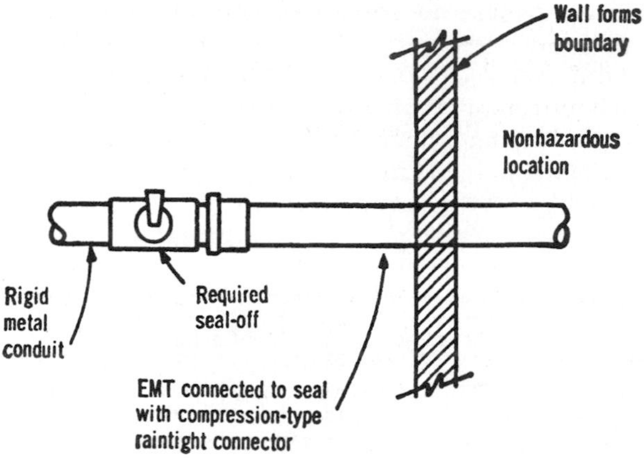
Wall forms
boundary
Nonhazardous
location
Rigid
metal
conduit
Required
seal-off
EMT connected to seal
with compression-type
raintight connector

VIOLATION! — EMT not permitted in Class I location

Fig. 501-19. These violate rules on seal fittings where conduit crosses boundary. (Sec. 501.15.)

Exception No. 1 to 501.15(A)(4) covers the case where a metal conduit system passes from a nonhazardous area, runs through a Class I, Division 1 hazardous area, and then returns to a nonhazardous area. Such a run is permitted to pass through the hazardous area without the need for a seal fitting at either of the boundaries where it enters and leaves the hazardous area. But the wording of this exception requires that such conduits, in order to be acceptable, must contain no union, coupling, box, or fitting in any part of the conduit run extending 300 mm (12 in.) into each of the nonhazardous areas involved (Fig. 501-20).

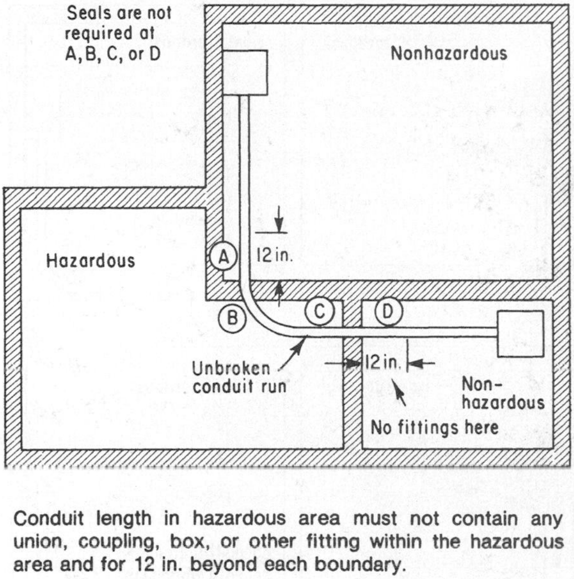
Seals are not required at A, B, C, or D
Nonhazardous
Hazardous
A
12 in.
B
C
D
Unbroken conduit run
12 in.
Non-hazardous
No fittings here
Conduit length in hazardous area must not contain any union, coupling, box, or other fitting within the hazardous area and for 12 in. beyond each boundary.

Fig. 501-20. Seals may be eliminated for conduit passing "completely through" hazardous area. (Sec. 501.15.)

For some installations of conduits crossing boundaries, straightforward application of NEC rules on conduit sealing is difficult. Most of these cases involve determination as to what constitutes the boundaries of a hazardous area; the Code provides no definition. The inspection authority should be consulted in cases not specifically covered by the Code. Because there are no provisions in the Code that *prohibit* the use of seals, "if in doubt, seal" would be a safe practice to follow.

At the top of Fig. 501-21, the conduit run is sealed within 18 in. (450 mm) of an explosionproof enclosure, as shown, and extends into a concrete floor slab, emerging in a nonhazardous area. It is not clear what constitutes the boundary of the Class I, Division 1 area. Must a seal be placed at A or might it instead be placed at B in the nonhazardous area? An NEC Official (now "Formal") Interpretation, pertaining to a hospital operating room, ruled that the entire concrete slab through which the conduit traveled constituted the boundary of the hazardous area, and that the seal could be placed either at A, where the conduit leaves the hazardous area, or at B, where it enters the nonhazardous area. But some authorities may require seals at A and B. With a seal at A and not at B, a heavier-than-air gas or liquid (such as gasoline) might penetrate a crack in the floor, enter the conduit through a coupling, and pass into the enclosure in the nonhazardous area. Or a seal at B but not at A might not prevent vapor in the conduit from entering the nonhazardous area through a coupling in the concrete and then through a crack in that floor. That kind of gas passage has occurred.

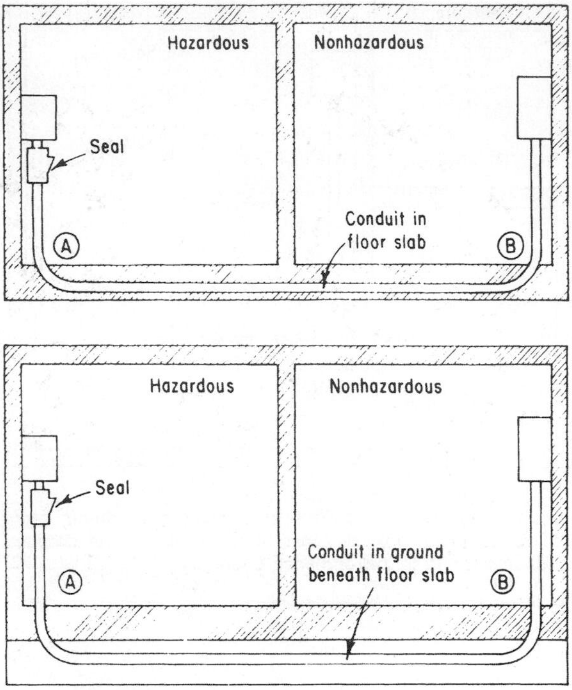
Hazardous
Nonhazardous
Seal
Conduit in
floor slab
A
B
Hazardous
Nonhazardous
Seal
Conduit in ground
beneath floor slab
A
B

Fig. 501-21. Conduit in floor-slab boundary may require seals at both A and B. (Sec. 501.15.)

In some instances, area classifications are already spelled out in the NEC, and it is interesting to see how this problem is addressed, because the panel has been inconsistent. For example, at an aircraft hangar (Art. 513) the hazardous location specifically extends below the hangar floor, as required in 513.8(B). On the other hand, in a comparable location at a vehicle service station, the area below the floor is unclassified as covered in 511.3. Over the last few Code cycles the Code-making panel has been moving away from classifying the below-grade and the below-floor areas unless there is an actual opening such as a pit. The belief has been that the availability of oxygen in such areas is too low (fuel-air mixture too rich) to support combustion. Although that is certainly true in the earth or concrete itself, it may very well not be true if a conduit, with its included air and couplings, traverses such a location. In areas where a classification is made in the NEC, it must be respected, but in general it raises concerns for other comparable areas where area classifications are done by engineers as refereed by local inspection authorities.

At the bottom of Fig. 501-21, the conduit run is not in the floor slab, but in the ground below the slab. Now what constitutes the boundary? Can the seal still be placed either at point A or at point B? Code rules applying to gasoline stations and aircraft hangars may be used as a guide. The real question is whether the ground beneath the slab is a hazardous or nonhazardous location. Section 514.8 used to define dispensing and service-station wiring and equipment, any portion of which is below the surface of a hazardous area, as a Class I, Division 1 location. Effective with the 2005 **NEC**, however, the classification stops at the grade boundary, as just noted. A similar change occurred at the same time in Art. 511, when the rule of 511.4(A)(1) in the 2002 **NEC** was deleted. However, as just noted, the below-grade classification continues for aircraft hangars.

Figure 501-22 is a wiring layout for a Class I, Division 1 location. The wiring is all rigid metal conduit with threaded joints. All fittings and equipment are explosionproof; this includes the motors, the motor controller for motor no. 1 (lower part of Fig. 501-22), the pushbutton control station for motor no. 2 (upper part of Fig. 501-22), and all outlet and junction boxes. The panelboard and controller for motor no. 2 are placed outside the hazardous area and hence need not be explosionproof.

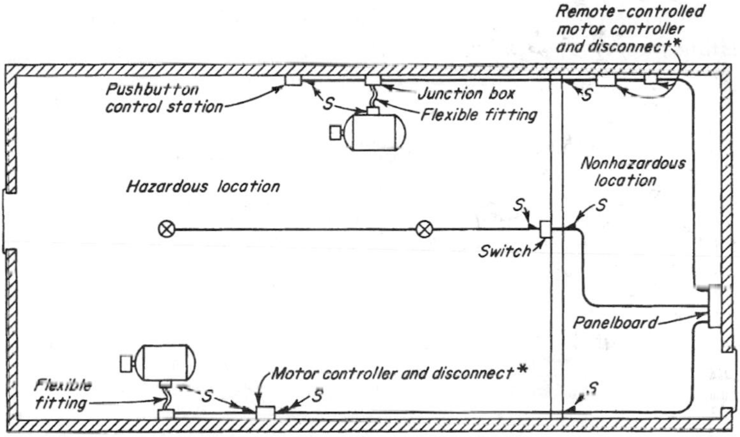
Remote-controlled motor controller and disconnect*
Pushbutton control station
Junction box
Flexible fitting
S
Hazardous location
Nonhazardous location
Switch
Panelboard
Flexible fitting
Motor controller and disconnect*

*Disconnecting means may be in same enclosure. See Sec. 430-103.

Fig. 501-22. Required seals are shown in points marked S. (Sec. 501.15.)

Each of the three runs of conduit from the panelboard is sealed just outside the hazardous area. A sealing fitting is provided in the conduit on each side of the controller for motor no. 1 (lower part of Fig. 501-22). The leads are sealed where they pass through the frame of the motor into the terminal housing, and no other seal is needed at this point, provided that the conduit and flexible fitting enclosing the leads to the motor are smaller than 2 in. (50.8 mm). The pushbutton control station for motor no. 2 (upper part of Fig. 501-22) is considered an arc-producing device assuming it is part of a conventional Class 1 control circuit (referring now to the energy level and not the area classification, that is with an Arabic and not a Roman numeral) and must comply with 501.115(A). Therefore, the conduit is sealed where it terminates at this device.

A seal is provided on each side of the switch controlling the lighting fixture. One of these seals is in the nonhazardous room and that single seal serves as both the seal for the arcing device and the seal for conduit crossing the boundary. The lighting fixtures are hung on rigid conduit stems threaded to the covers of explosionproof boxes on the ceiling.

About seal fittings In using seal fittings in conduits in hazardous locations, application data of UL must be observed, as follows:

- Conduit seal-off fittings to comply with **NEC** 501.15 must be used *only* with the sealing compound that is supplied with the fitting and specified by the fitting manufacturer in the instructions furnished with the fitting.
- Seal-off fittings are listed for sealing listed conductors in conduit, where the conductors are thermoplastic insulated, rubber covered, or lead covered.
- Any instructions supplied with a seal-off fitting must be carefully observed with respect to limitation on the mounting position (e.g., vertical only) or location (e.g., elbow seal). Figure 501-23 shows a variety of available seal fittings. Sealing fittings are designed for vertical orientation only, for optional vertical or horizontal positioning, or as combination elbow seals. Others are compatible with conduits installed at any angle, because covers can be rotated until sealing spouts point upward.

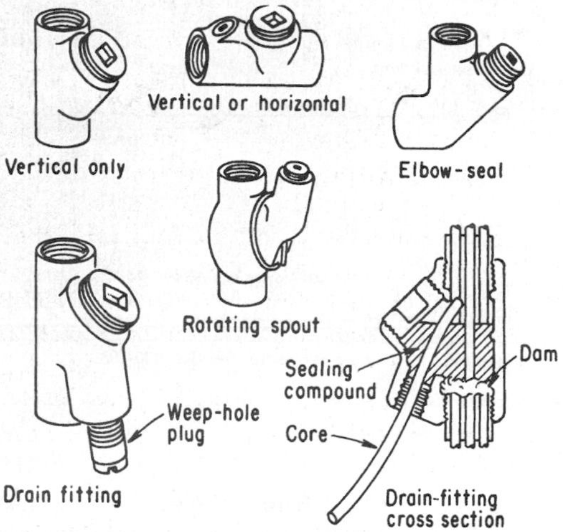
Vertical or horizontal
Vertical only
Elbow-seal
Rotating spout
Dam
Sealing
compound
Weep-hole
plug
Core
Drain fitting
Drain-fitting
cross section

Fig. 501-23. A variety of seal fittings are suited to different applications. (Sec. 501.15.)

Because conduits are installed vertically, horizontally, and at angles and require ells, tees, and offsets, the fittings used for sealing differ in construction features, orientation, and method of sealing.

Sealing fittings intended solely for vertical orientation have threaded, upward-slanting ports slightly larger than conduit hub openings to permit asbestos-fiber dams to be tamped into fitting bases. The dam prevents the fluid-sealing compound from running down into the conduit before the seal has solidified.

A second type of fitting is designed for either vertical or horizontal positioning. These units are identified by two seal-chamber plugs that can be removed to facilitate tamping dam fibers into both conduit hubs when the device is aligned horizontally. The compound is poured into the chamber through the larger of the two ports. The ports are then replugged, and the plugs tightened flush with their collars. When these fittings are oriented vertically, however, only lower conduit hubs need be dammed.

A third type of seal, which can be oriented in any position, is shown in the center of Fig. 501-23 and is described in Fig. 501-13. This same fitting may be used as a drain-type seal when its spout is turned down.

Elbow seals (as at upper right of Fig. 501-23) are double-duty devices that are practical either when horizontal conduits must elbow down to connect with an enclosure's top (as indicated) or when vertical conduits must turn to enter explosionproof enclosures horizontally. In either case, sealant application openings must slant upward.

Another fitting, designed for drainage purposes, is installed only in vertical runs of conduits. Where conduit is run overhead and is brought down vertically to an enclosure for apparatus, any condensation of moisture in the vertical run would be trapped by the seal above the apparatus enclosure. The lower part of Fig. 501-23 shows a sealing fitting designed to provide drainage for a vertical conduit run. Any water coming down from above runs over the surface of the sealing compound and down to an explosionproof drain, through which the water is automatically drained off. These fittings permit passage of condensation while also blocking the passage of explosive pressures or flames. They are equipped with plugs containing minute weep holes that can either be opened and closed periodically as need develops or allow continuous drainage.

Drain-type seal fittings must be oriented so that compound-application ports remain above the lower downward-slanting drainage plugs. To install the seal and drain, both ports are unplugged and the lower conduit hub is dammed. The drainage plug hole is then closed temporarily by a washer through which a rubber core is inserted. This core protrudes into the upper part of the sealing chamber, although it must be guided so as not to remain in contact with any of the conductors. Sealing compound is then poured into the chamber through the upper access port, which is replugged and screwed tight.

After the compound has initially set (but has not yet had time to permanently harden), the washer is removed and the rubber core is pulled down and out. This creates a clear drainage canal, which extends from above the seal down into the drainage weep hole. A drainage plug is then screwed into the threaded hole, with not less than five full threads engaged to fulfill the requirement for an explosionproof joint.

Class I, Division 2

Because Division 2 locations are of a lower degree of hazard than those in Division 1, the requirements for sealing as given in 501.5(B) are somewhat less demanding, as follows:

1. Where the rules of other sections in Art. 501 require an explosionproof enclosure for equipment in a Division 2 location, all conduits connecting to any

such enclosure must be sealed exactly the same as if it were in a Division 1 enclosure. And the conduit, the nipple, or any fitting between the seal and the enclosure being sealed must be approved for use in Class I, Division 1 locations—as specified in 501.15(A).

In Class I, Division 2 locations, each piece of apparatus that produces arcs or sparks, such as a motor controller, switch, or receptacle, must be in an explosionproof enclosure, and the seals required here serve to complete that explosionproof enclosure. Although the frequency of hazardous gas or vapor releases in Division 2 locations is severely curtailed by definition, the occurrence of arcs in equipment in switches, etc., is routine, and may be expected to occur in the presence of any gas or vapor release, however infrequent. The only way to respond to this set of conditions is to make certain that the arcing equipment is contained, and the **NEC** does exactly that.

As shown in Fig. 501-24, where a nonexplosionproof enclosure is permitted by other sections to be used in a Division 2 location a seal is *not* required in any size of conduit. Note that in Division 2 locations there is *not* always the need to seal conduits metric designator 53 (trade size 2) and larger, as required in Division 1 locations, by 501.15(A)(2). The reason for the permission to use conventional boxes and to avoid seals around them in Division 2 locations requires a review of the reason for such seals in Division 1 locations. The issue in Division 1 was the need to segment the conduit system so it would not fail due to pressure piling from an internal explosion. The only way to have such an explosion is to have large amounts of fuel-air mixtures migrating into the conduit system over time. That is possible in a Division 1 location where the hazardous chemicals are expected to be routine, and it is impossible in a Division 2 location where any vapor or gas release will be very infrequent.

2. Any and every conduit passing from a Class I, Division 2 area into a nonhazardous area must be sealed in the same manner as previously described for conduit passing from a Division 1 area into a Division 2 or nonhazardous

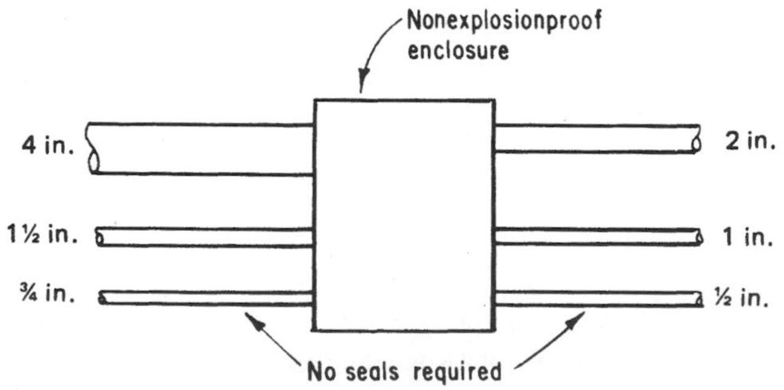
Nonexplosionproof enclosure
4 in.
2 in.
1½ in.
1 in.
¾ in.
½ in.
No seals required

Fig. 501-24. Seals are not required for conduits connected to nonexplosionproof enclosures that are permitted in Division 2 locations. (Sec. 501.15.)

area (Fig. 501-25). Rigid metal conduit or IMC must be used between the seal and the point where the conduit passes through the boundary. There is language to the effect that these boundary seals need not be explosionproof, provided they are "identified for the purpose of minimizing passage of gases under normal operating conditions." However, as of this writing no such seal has been developed and marketed, so the only practical option is the explosionproof variety.

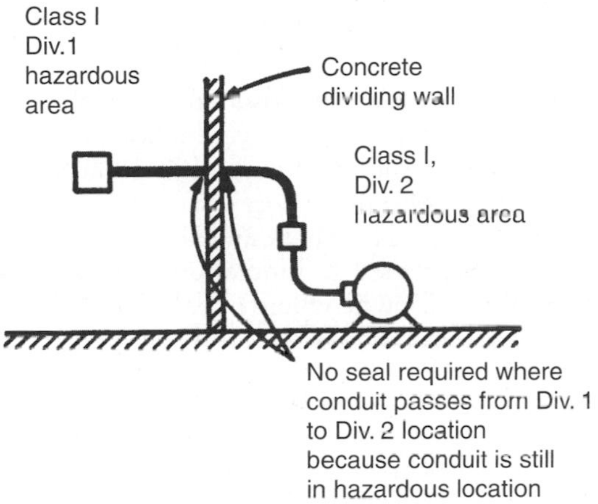
Class I
Div.1
hazardous
area
Concrete
dividing wall
Class I,
Div. 2
hazardous area
No seal required where
conduit passes from Div. 1
to Div. 2 location
because conduit is still
in hazardous location

Fig. 501-25. This is a violation; a seal *is* required. (Sec. 501.15.)

Exception No. 1 to part **(B)(2)** is worded the same as Exception No. 1 in part **(A)(4)** for Class I, Division 1 locations. Exception No. 1 to 501.15(B)(2) covers the case where a metal conduit system passes from a nonhazardous area, runs through a Class I, Division 2 hazardous area, and then returns to a nonhazardous area. Such a run is permitted to pass through the hazardous area without the need for a seal fitting at either of the boundaries where it enters and leaves the hazardous area, provided that the conduit in the hazardous area does not contain unions, couplings, boxes, or fittings. In a Class I, Division 2 location, the same prohibition against unions, couplings, and so on, is applicable, and the method in Fig. 501-26 is not acceptable if seals are omitted at the boundary crossings. A seal would not be needed at A, B, C, or D, if the conduit passed through the Class I, Division 2 location without any coupling or other fittings in the conduit.

Exception No. 2 to part **(B)(2)** addresses installations where the conduit passing from the Class I, Division 2 location into an unclassified location does not enter any enclosures that produce an arc or spark. Such an installation does *not* require a conduit seal, provided it is also installed outdoors, or if the conduit system is installed in a single room, it may be installed indoors. But in no case should the

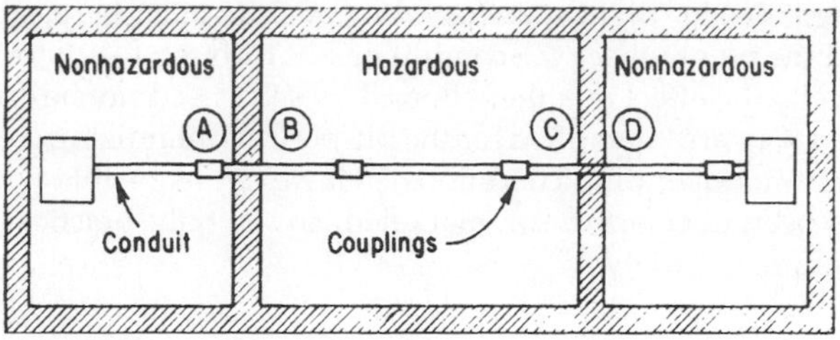
Nonhazardous
Hazardous
Nonhazardous
A
B
C
D
Conduit
Couplings

Fig. 501-26. Couplings in conduit through Division 2 location would require seals at the boundaries. (Sec. 501.15.)

unsealed conduit be connected to any enclosure that contains a source of ignition. The outdoor locations work by abundant ventilation and the indoor locations because if all the wiring is in the same room, it is all at the same atmospheric pressure and will not transmit fumes.

Exception No. 3 recognizes that where an enclosure or room is unclassified because it is pressurized to prevent accumulation of an explosive concentration, there is no need to seal the raceway where it leaves the Division 2 location and enters the pressurized spaces because the press will prevent gas from entering that unclassified area.

Exception No. 4 eliminates sealing for those portions of the conduit system that satisfy *all* the conditions given in parts **(1)** to **(5).** The conditions, taken together, describe a conduit layout for which it would be impossible for fumes to enter the raceway to begin with.

Part **(C)** of 501.15 sets regulations about the kind of seals that must be used where seals are required by foregoing rules. Part **(1)** calls for an integral seal within the enclosure itself or use of a separate seal fitting in each conduit connecting to the enclosure, as previously described. The use of factory-sealed devices eliminates the need for field sealing and generally is less expensive to install. In fittings of this type, the arcing device is enclosed in a chamber, with the leads or connections brought out to a splicing chamber. No external sealing fitting is required (Fig. 501-27). If separate seals are used, they must be accessible. Do not bury a seal fitting, or place it behind or within fixed partitions, unless an access door is provided.

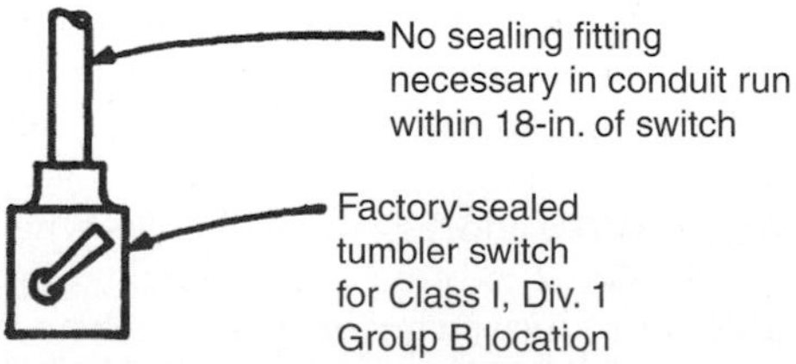
No sealing fitting
necessary in conduit run
within 18-in. of switch
Factory-sealed
tumbler switch
for Class I, Div. 1
Group B location

Fig. 501-27. Seal in conduit is not needed where enclosure has built-in seal. (Sec. 501.15.)

Where a seal fitting is used in the conduit, it must be *explosionproof.* The sealing compound must develop enough mechanical strength as it hardens to withstand the forces of explosions. Seals used only to prevent condensation accumulation do not have to be explosionproof; a vaportight seal is sufficient for that purpose.

A seal must be *vaportight* to minimize the passage of gases and vapors. To do that, the sealing compound must adhere to the fitting and to the conductors. It must expand as it hardens to close all voids without producing objectionable mechanical stresses in the fitting.

Liquid or condensed vapor may present a problem in Class I locations. Where such is the case, joints and conduit systems must be arranged to minimize entrance of liquid. Periodic draining may be necessary, which necessitates the inclusion of a means for draining in the original design of the motor or of the conduit system generally; explosionproof drain fittings are available for this purpose.

Installation instructions furnished by the manufacturer must be carefully followed. The seal fitting must be carefully packed with fibrous damming material, which packs more tightly and effectively around conductors when it is dampened, and then filled with the compound supplied with the fitting to a depth at least equal to the inside diameter of the conduit and never less than 5/8 in. deep, as required in part **(3)** and in the UL standard on seal fittings. New designs are now on the market that use self-expanding epoxy compounds that are explosionproof without the need of fiber damming to separate the conductors and keep the compound in place, and significantly reducing the labor time in the process.

Part **(2)** of 501.15(C) covers the compound used in seals (Fig. 501-28). The sealing compound used must be one that has a melting point of not less than 200°F (93°C) and is not affected by the liquid or gas that causes the location to be hazardous. Most of the insulating compounds commonly used in cable splices and potheads

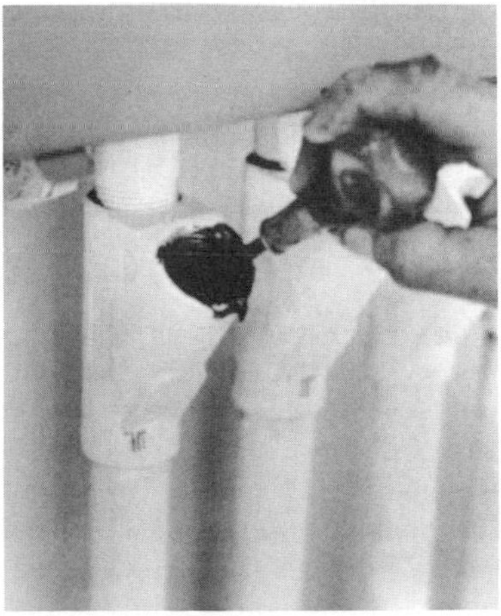

Fig. 501-28. Conduit seal fitting must be carefully packed with fibrous damming material, which packs more tightly and effectively around conductors when it is dampened and then filled with the compound supplied with the fitting to a depth at least equal to the inside diameter of the conduit and never less than ⅝-in. deep (UL standard). This type of seal is for vertical mounting only. (Sec. 501.15.)

are soluble in gasoline and lacquer solvents and hence are unsuitable for sealing conduits in locations where these liquids are used. A mixture of litharge and glycerin is insoluble in nearly all liquids and gases found in Class I locations and meets all other requirements, though this mixture is open to the objection that it becomes very hard and is difficult to remove if the wires must be pulled out. No sealing compounds are listed by UL as suitable for this use except in connection with the explosionproof fittings of specific manufacturers.

Part **(4)** prohibits splices or taps in seal fittings.

Part **(5)** recognizes use of listed Class I assemblies that have a built-in seal between a compartment housing devices that may cause arcs or sparks and a separate compartment for splicing or taps. Conduit connection to the splice or tap compartment requires a seal fitting only in conduit of metric designator 53 (trade size 2) or larger, as specified for junction boxes in 501.15(A)(1)(2).

Part **(6)** directs conduit fill to be not more than 25 percent of the raceway's cross-sectional area. This will require the use of Tables 4 and 5 in Chap. 9, simply because the conductor fill values in the tables of App. C are based on 40 percent fill. Ensure that the total conductor cross-sectional area—determined by adding together all the conductor cross-sectional area dimension(s) given in Table 5—is not more than 25 percent of the "Total Area" column for the particular raceway and size shown in Table 4.

A fitting is to be evaluated on the basis of a conduit with the same trade size, and conductor fill in that fitting must be no greater than 25 percent of the cross section, unless the fitting is specifically marked for greater fill. This is because the original work on current seal designs and test work was done to the 1968 **NEC**, which set a maximum conduit fill of 25 percent on new work and 40 percent on rework. Since a seal cannot be opened and repoured after it has been sealed, it cannot be reworked and it made sense to do the tests based on a 25 percent fill. Of course the **NEC** now allows a 40 percent fill by right even on initial installations. The industry response was to write the 25 percent limit into the **NEC**, along with an option for a higher fill if the manufacturer was willing to submit to the required testing.

Some manufacturers are meeting the higher fill by taking a seal fitting already designed and tested (at 25 percent) for a certain size of conduit and then drilling and tapping the female conduit entries one standard size smaller, and adjusting the installation directions accordingly. A similar approach can be done in the field using threaded reducers and then comparing the 40 percent fill on the smaller raceway with the permitted 25 percent fill on the larger seal fitting. One of the reasons for inserting the various mentions of "reducers" as being permitted on both sides of seal fittings was to accommodate this procedure.

The way the **NEC** rule is written makes it imperative to track the installation instructions, because the instructions will always trump the percentage called out in the rule. If the instructions allow a greater fill than 25 percent, then whatever the instructions allow is permitted by 501.15(C)(6), and if the instructions insist on a smaller fill, then that becomes the limit by 110.3(B).

Part **(D)** covers seals for cables in Class I Division 1 locations, whether run in conduit or as open cables as now permitted for Types MI and MC-HL cables—the only power cables permitted by 501.10(A)(1)(b) and 501.10(A)(1)(c) to be used in Class I, Division 1 locations. Type MI cable is inherently sealed with respect to its interior, and the listed fitting described in 501.10(A)(1)(b) is all that is needed. Type MC-HL cable is covered in (1) of this rule. The cable must be sealed at any terminations in the hazardous area with listed fittings that include provisions for removing the nonmetallic jacket, the cable armor, and incorporating sealant that envelops each conductor individually and makes up to the components of the armor such that the entry of gases and vapors into the cable assembly is minimized. Type MC-HL cable, like all metal-clad cable, is eligible to be produced by 330.104 in control sizes as small as 18 AWG.

If produced as a shielded twisted pair, it could be covered by the exception following, which allows such cable to be treated as a single wire at the cable connector/seal-off, provided there is an approved means in place at the actual termination area to prevent the propagation of flame into the cable core and to minimize the entry of gas and vapors. There are epoxy mastics that have been evaluated for this purpose and so this is a practical alternative; however, careful attention to installation instructions is crucial. With respect to data applications using twisted-pair cabling, each pair can be split out and sealed without untwisting the pair.

However, if the cable is shielded the proper procedure is more complicated because the shielding may be applied to each pair individually, or to the entire group of twisted pairs (usually four pairs) taken collectively. The product listing for this mastic and the installation instructions that come with it only cover shielding on the individual pairs. Sources within the Code panel indicate that this was the only form the panel considered in writing this part of the exception. If the shielding is applied just under the cable jacket and over the entire group, then as of this writing the product cannot be used based on the literal text of the product directions. Since the product acceptance criterion in the **NEC** is "approved" the decision comes down to the inspector. It would certainly not be unreasonable to approve the product conditioned on the shielding being formed into a wire at the point where the cable jacket ends, and then the mastic being applied to the conductor pairs, the shield made into a wire, and the cable jacket end all as a whole.

The 2017 **NEC** added a degree of flexibility regarding seal placement for cables that directly correlates with the rules for raceways in 501.15(A)(1). The seal for the cable assembly can be up to 18 in. from the enclosure, with the same general constraints regarding allowable fittings not larger than the entry trade size permitted between the seal and the enclosure.

The other two options for cables in Division 1 both involve cable run inside of conduit. Taken in reverse order, (3) covers cable that cannot transmit gases or vapors through its core. One example of this cable is the old "flooded" telephone cables that have been used for generations for below-grade voice telephone circuits. Such cables are available for Ethernet applications, although the content of the gel filling needs to be monitored because it varies and may not be chemically compatible with one or more chemicals on-site. The sealing process in this case works by sealing to the outside of an intact cable assembly.

The other possibility, in (2), involves a cable in conduit where the cable assembly can transmit gases and vapors. The default approach is to open up the cable assembly and seal to each of the enclosed conductors. And here again the cable can be treated as a single conductor run through a normal conduit seal, provided the same sort of approved means as in (D)(1) Exception is employed at the jacket termination within the enclosure to prevent substantial gas or vapor migration through the cable core.

Part **(E)** covers sealing of any of the cables permitted by 501.10(B) in Division 2 locations. The first topic concerns where such cables enter a box that encloses arcing or sparking equipment that must be explosionproof. In such cases the same rules and the same exception (No. 2 in this case) apply as for Division 1 terminations in the same locations, together with the same lack of clarity around shielding applied to multipair cables. In addition, an enclosure that is pressurized (Type Z in this case, Division 2 to nonhazardous) does not require a boundary seal. This directly correlates with 501.15(B)(2) Exception No. 3 for the equivalent situation involving conduit.

The second topic covers cable constructions that don't pass vapors or gases through their cores, in this case with a defined upper limit on the rate of passage being, logically, that allowed for seal fittings generally in the product standard. Cables that are not gas-blocked must have their terminations sealed to make them so, as described above. Cables that do not have an impervious sheath, such as conventional interlocking armor metal-clad cable, must simply have a boundary seal such that the act of leaving the classified area does not spread the classification to a previously unclassified location.

Part **(F)** covers drainage. Paragraphs (1) and (2) cover drainage provisions for control equipment; means must be provided for its periodic removal. The simplest way to accomplish this is to use a seal with a drain fitting in a conduit entry below the control enclosure. For motors, since they are factory sealed, the drainage provisions must be made at the factory.

501.17. Process Sealing. This section, new as of the 2011 NEC, is a comprehensive rewrite and expansion of what used to be a third numbered paragraph under the "Drainage" subsection in prior editions. The subject is electrical equipment that is directly connected to a process containment vessel or piping system involving chemicals that are or could become fuel for fire or explosions. Examples cited in the text include canned pumps, submersible pumps, and instrumentation that measure temperature, flow, chemical composition, or pressure of such process fluids. Liquefied natural gas systems are but one of many examples of these applications. If the process/electrical system interface relies on a single-sealing element to separate the process fluid from the electrical system, such as a diaphragm, compression seal, or tube or other mechanism, there must be an additional step taken to guard against the consequences of the single-element failure. Process-connected electrical equipment that does not rely on a single seal, or that is listed and marked "single seal" or "dual seal" can be used without additional steps. The NEC provides four options if additional steps need to be applied, but the list is not limiting and others can be submitted for approval. The four options available are:

- A barrier suitable for the temperature and pressures involved with the process, and arranged so the failure of the initial process seal will become obvious through visible leakage, an audible whistle, or other monitoring methods.
- Electrical connections through a listed Type MI cable assembly, rated to withstand not less than 125 percent of both the process pressure and maximum process temperature (in degrees Celsius).

- A drain or vent between the single-process seal and the conduit or cable seal such that the electrical seal is not pressurized above a 6-in. water column (1493 Pa) and arranged so the initial sealing failure is obvious using the same options as for the barrier method.
- An add-on secondary seal, which must be rated for the conditions in the event of a failure in the process seal. This reflects current technological developments in this area.

501.20. Conductor Insulation, Class I, Divisions 1 and 2. For economic reasons and greater ease in handling, nylon-jacketed Type THHN-THWN wire, suitable for use where exposed to gasoline, has in most cases replaced lead-covered conductors.

An excerpt from the UL Guide Card information that applies to this subject states as follows:

> Wires, Thermoplastic.
>
> Gasoline Resistant TW—Indicates a TW conductor with a jacket of extruded nylon suitable for use in wet locations, and for exposure to mineral oil, and to liquid gasoline and gasoline vapors at ordinary ambient temperature. It is identified by tag marking and by printing on the insulation or nylon jacket with the designation "Type TW Gasoline and Oil Resistant I."
>
> Also listed for such use is "Gasoline Resistant THWN" with the designation "Type THWN Gasoline and Oil Resistant II."

Note that other thermoplastic wires may be suitable for exposure to mineral oil, but, with the exception of those marked "Gasoline and Oil Resistant," reference to mineral oil does not include gasoline or similar light-petroleum solvents.

The conductor itself must bear the marking legend designating its use as suitable for gasoline exposure; such designation on the tag alone is not sufficient.

501.25. Uninsulated Exposed Parts, Class I, Divisions 1 and 2. Uninsulated exposed parts operating not over 30 V (15 V in wet locations) can be used, but only where intrinsically safe (Div. 1 or 2) or nonincendive (Div. 2) protection applies.

501.30. Grounding and Bonding, Class I, Divisions 1 and 2. The parent text of this section now (2014 NEC) includes express language that these provisions apply regardless of voltage. This requirement is also in 250.100, and as a result the entire technical content of this rule resides here in Chap. 5, and the former informational note pointing to 250.100 for additional requirements has been deleted. Special care in the grounding of all equipment is necessary in order to prevent the possibility of arcs or sparks when any grounded metal comes in contact with the frame or case of the equipment. All connections of a conduit to boxes, cabinets, enclosures for apparatus, and motor frames must be so made as to secure permanent and effective electrical connections. To be effective, this form of construction is not only necessary in the spaces that are classed as hazardous, but should also be carried out back to the point where the connection for grounding the conduit is made to the grounding electrode system serving the premises. Outside the space where the hazardous conditions exist, threaded connections should be used for the conduit, unless bonding techniques are used for connections to knockouts in sheet metal enclosures. Any conduit emerging from a Division 1 or Division 2 location must have a bonded path of equipment grounding from the hazardous location back to the bonded service equipment, to the bonded secondary of a transformer that supplies the circuit into the hazardous location, or, as covered in the exception, under certain conditions, to the bonded main building disconnect.

The last sentence in part **(A)** states that the service-type bonding required to the "point of grounding" means the special bonding requirements must be brought back to the point in the system where the grounded circuit conductor is connected to the grounding electrode. For disconnects grounded per 250.32(B)(1) Exception No. 1, that is the main building disconnect. But for hazardous locations in outbuildings where the outbuilding disconnect is within the main building, the service-type bonding must be carried back to the main building. That is, because the outbuilding disconnect is grounded as required by 250.32(D)—which prohibits connecting the neutral to the grounding electrode—the connection between the neutral and the grounding electrode conductor will be in the main building.

If the circuits to a hazardous location are supplied from a separately derived system (as from a transformer or from a generator on the load side of the service), bonded conduit connections are required only back to the transformer or generator and not back to the service ground. It is only necessary to bond back to the point where the neutral and equipment grounding conductor of the supply system are bonded together. In that case, the low-impedance ground-fault current return path to the transformer or generator neutral will ensure quick fault-clearing action of the circuit protective devices.

The net effect of the wording in part **(A)** of this section is to require the use of "service bonding" techniques [250.92(B)] throughout the length of a continuous path from the raceway and equipment in a hazardous location all the way back to the first point at which the system neutral is bonded to the system equipment grounding terminal or bus and both are connected to a grounding electrode—either at the transformer in the system or at a generator or at the service equipment if there is no voltage change in the system (Fig. 501-29). This means that every raceway termination in the ground return path to the service or transformer secondary must be a threaded metal conduit connection to a threaded hub or boss on a fitting or enclosure or any connection to a sheet metal KO must use one of the following methods:

1. A bonding bushing with a bonding jumper to a grounding terminal within the enclosure, on a KO that is clean or has rings left in the enclosure wall.
2. A bonding bushing that does not require a jumper when used on a clean KO. The various techniques that must be used instead of locknut and bushing or double-locknut-and-bushing are described in the figures accompanying Sec. 250.92.

In addition, although 250.118 allows flexible wiring methods in certain sizes and with certain overcurrent protection ahead of them to function as equipment grounding paths, these wiring methods are disallowed from using those provisions in hazardous locations. The wiring methods can be used, but only with a properly sized bonding conductor installed, regardless of the raceway size or the overcurrent protective ratings.

At the top, Fig. 501-30 shows a violation of part **(A).** Although 501.10(B) permits a sheet metal junction box (nonexplosionproof) in a Division 2 location,

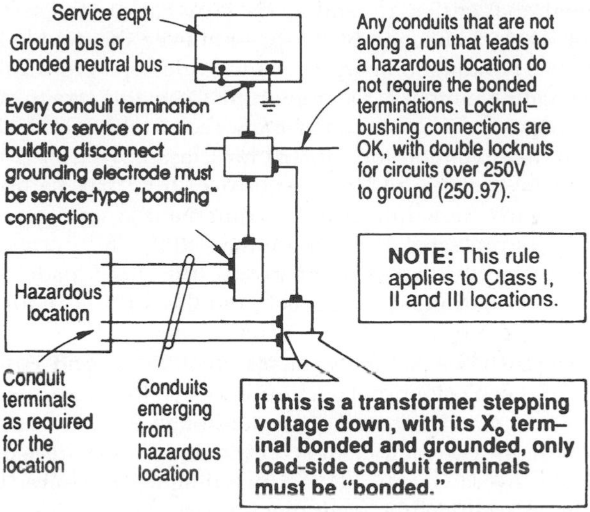
Service eqpt
Ground bus or bonded neutral bus
Every conduit termination back to service or main building disconnect grounding electrode must be service-type "bonding" connection
Any conduits that are not along a run that leads to a hazardous location do not require the bonded terminations. Locknut-bushing connections are OK, with double locknuts for circuits over 250V to ground (250.97).
NOTE: This rule applies to Class I, II and III locations.
Hazardous location
Conduit terminals as required for the location
Conduits emerging from hazardous location
If this is a transformer stepping voltage down, with its X_0 term-inal bonded and grounded, only load-side conduit terminals must be "bonded."

Fig. 501-29. Bonding of raceways and equipment must be made back to service ground. (Sec. 501.30.)

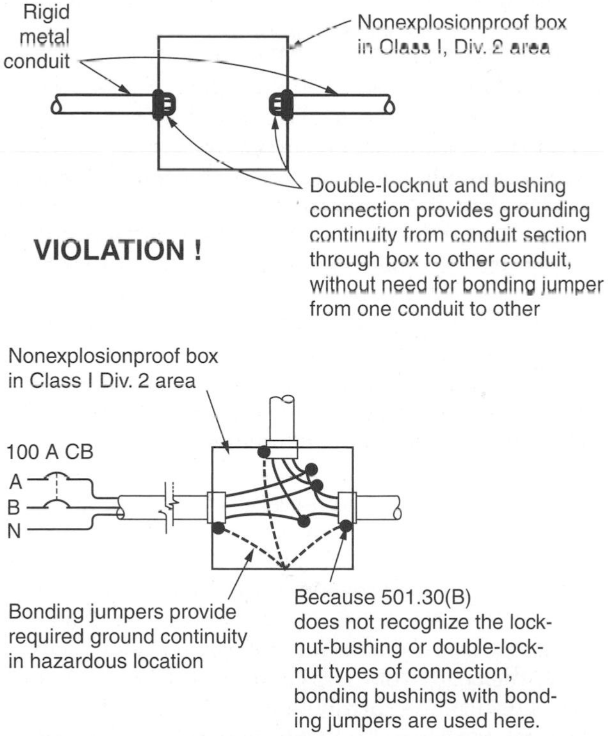
Rigid metal conduit
Nonexplosionproof box in Class I, Div. 2 area
Double-locknut and bushing connection provides grounding continuity from conduit section through box to other conduit, without need for bonding jumper from one conduit to other
VIOLATION !
Nonexplosionproof box in Class I Div. 2 area
100 A CB
A
B
N
Bonding jumpers provide required ground continuity in hazardous location
Because 501.30(B) does not recognize the lock-nut-bushing or double-lock-nut types of connection, bonding bushings with bond-ing jumpers are used here.

Fig. 501-30. Bonding bushings with jumpers comply with grounding rule. (Sec. 501.30.)

bonding through that enclosure (and all the way back to the service) may not be done simply by locknuts and bushings—not even the double-locknut type. At the bottom, the bonding jumpers satisfy the rule here. From Table 250.122, a No. 8 copper bonding jumper is the minimum acceptable size to be used with a 100-A-rated protective device, for connecting each bushing to the box.

According to the exception for part **(B)** for Class I, Division 2 areas, liquidtight flexible metal conduit, in lengths not over 6 ft, may be used without a bonding jumper (an internal or external equipment grounding conductor) to enclose conductors protected at not more than 10 A. This permission is given only for circuits to a load that is "not a power utilization load." The same use of liquidtight flex is permitted for Class II and Class III locations, in 502.30(B) and 503.30(B).

501.115. Switches, Circuit Breakers, Motor Controllers, and Fuses. Part **(A)** effectively requires explosionproof enclosures (listed for Class I locations, which means suitable for use in Division 1 areas), although purged enclosures are not ruled out. Explosionproof equipment is required in Class I areas only. And any of the equipment covered by 501.115(A), including its enclosure, must be listed as an assembly. In addition, the equipment must operate at a temperature low enough that it will not ignite the atmosphere around it.

Division 1 Electrical equipment described must be explosionproof and specifically designed for the respective class and group.

Division 2 Equipment selected must be explosionproof only in certain cases. Purged and pressurized enclosures are recognized as an alternative to explosionproof types, as noted in 500.7(D). General-purpose enclosures may be used only if all arcing parts are immersed under oil or enclosed within a chamber that is hermetically sealed against the entrance of gases or vapors. Oil-immersed contacts are permitted in general-purpose enclosures where the power circuit interruption takes place under at least 50 mm (2 in.) of oil and under at least 25 mm (1 in.) for control contacts. Another alternative, very popular today, is to leave the seal to the factory and mount a device with a factory-sealed explosionproof housing in a general-purpose enclosure. Solid-state switching, with no actual contacts involved, is permitted provided the surface temperature of the apparatus in the general-purpose enclosure does not exceed 80 percent of the autoignition temperature of the hazardous substances involved. If one of these conditions is not satisfied, the enclosure must be explosionproof.

Figure 501-31 shows two explosionproof panelboards. Each panelboard consists of an assembly of branch-circuit CBs, each pair of CBs being enclosed in a cast-metal explosionproof housing. Access to the CBs and to the wiring compartment is through handholes with threaded covers, and threaded hubs are provided for the conduits. Individual CBs and motor starters are also shown. Panelboards for light and power are limited in the UL Guide Card information [see discussion at 110.3(B) for details]. Listed panelboards for Class I and Class II hazardous locations are for "lighting and *low-capacity* power distribution." *High-capacity* panelboards (such as 1200-A floor-standing panels) and switchboards must be kept out of hazardous locations wherever possible. Enclosure requirements and details are generally the

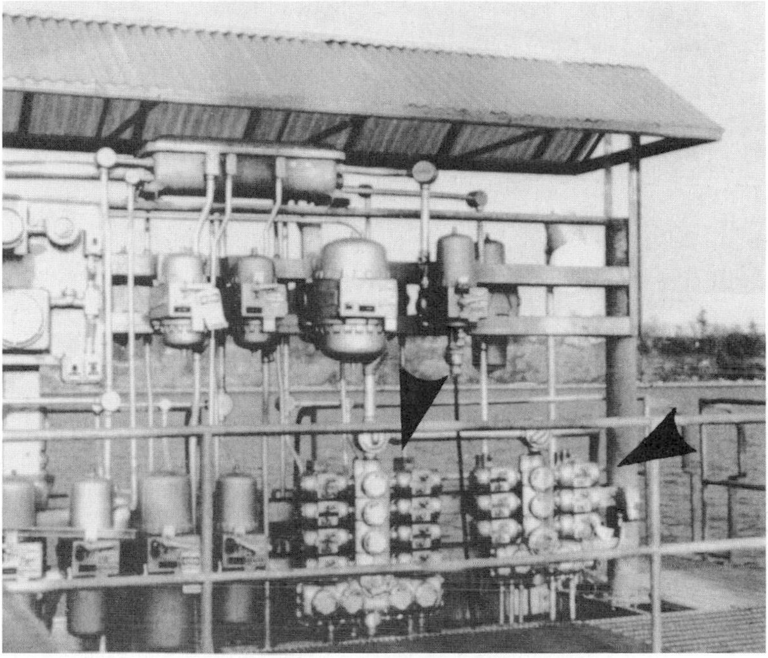

Fig. 501-31. Explosionproof panelboards (arrows) are assemblies made up of circuit-breaker housings coupled to wiring enclosures. Large explosionproof CB enclosure at center feeds the panelboards. Explosionproof motor controllers are at lower left. (Sec. 501.115.)

same as those described under "General Rules" and for CBs and boxes. Typical hazardous-location panelboards are shown at the right in Fig. 501-32.

"Industrial control equipment" is a broad category in the UL Guide Card information [see discussion at 110.3(B) for details], covering "control panels and assemblies" and "motor controllers." Control panels and assemblies include both enclosures and the components within them, such as motor controllers, pushbuttons, pilot lights, and receptacles (Fig. 501-33). Either a single enclosure or a group of interconnected enclosures may be used for mounting the components. Where a number of interconnected enclosures are included in an assembly, it is called a *modular assembly* and may be assembled either at the factory or in the field. An example of that is shown in Fig. 501-32.

Components are provided with the enclosures, to be installed either at the factory or in the plant. Wiring between components of modular assemblies is to be field-installed. Conduit seal-off fittings must be used in accordance with 501.15.

As noted in part **(1)** of part **(B)(1),** in a Class I, Division 2 location, a general-purpose enclosure may be used for a circuit breaker, a motor controller, or switch *if* "interruption of current occurs within a factory sealed explosionproof chamber." This is a condition under which a general-purpose enclosure may be used for a current-interrupting device—instead of requiring a Class I, Division 1 enclosure.

Fig. 501-32. Explosionproof panelboards (at right) are combined with separate enclosures for motor starters on a rack, to make up a "modular assembly," which is listed by UL under "Industrial control equipment"—which may be assembled either at the factory or in the field. (Sec. 501.115.)

Fig. 501-33. Combination motor starter is typical explosionproof control unit for Class I locations. Note the drain-type seal fittings that provide for draining the conduits of any accumulated condensation or other water. (Sec. 501.115.)

A snap switch in an explosionproof enclosure is shown in Fig. 501-34. If a snap switch has an internal factory seal between the switch contacts and its supply wiring connection in its enclosure, it will be so identified by a marking on it. Such switches (not the one in the illustration) do not require a seal fitting in a conduit entering the enclosure. The integral seal satisfies 501.15(A)(1) Exception item 3 and 501.15(C)(5).

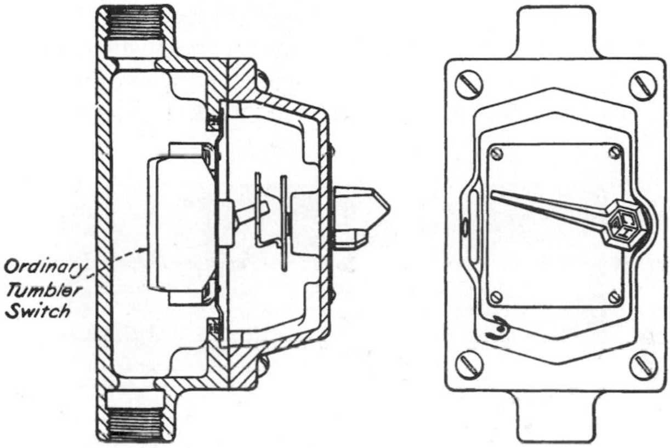
Ordinary
Tumbler
Switch

Fig. 501-34. Explosionproof enclosure suits snap switch to use in hazardous location. (Sec. 501.115.)

As noted in part **(B)(2)**, isolating switches in Class I, Division 2 locations may be used in general-purpose enclosures either with or without fuses in the enclosure. In previous Code editions, this rule recognized a general-purpose enclosure only for an isolating switch that did *not* contain fuses. But because the fuse in such a switch is for short-circuit protection and neither the switch nor the fuse operates as a "normal" interrupting device, fuses are permitted in isolating switches in general-purpose enclosures.

With reference to subparagraph **(B)(3)**, it is assumed that fuses will very seldom blow, or CBs will very seldom open, when used to protect feeders or branch circuits that supply only lamps in fixed positions. In Division 2 locations the conditions are not normally hazardous but may sometimes become so. Fuses are permitted in general-purpose enclosures if their operating element is oil-immersed, or hermetically sealed, or of a nonindicating and filled current-limiting type. This would exclude plug fuses, renewable cartridge fuses, or other common fuses, which would need to be in an enclosure suited for the location.

Part **(B)(4)** provides that listed fuses in luminaires used for supplementary overcurrent protection are permitted in Class I, Division 2 locations.

501.120. Control Transformers and Resistors. The term *control transformer* is commonly applied to a small dry-type transformer used to supply the control circuits of one or more motors, stepping down the voltage of a 480-V power circuit to 120 V.

Part **(A)** requires either explosionproof or purged and pressurized enclosures for Division 1 locations—the same as required for meters and instruments in 501.105(A).

501.125. Motors and Generators. Four different types of motor applications are recognized for use in a Class I, Division 1 location. The first is a motor approved for Class I, Division 1 locations, such as an explosionproof motor. The totally enclosed, fan-cooled motor (referred to as a TEFC motor) is available in a form listed by UL for use in explosive atmospheres. A motor of a type approved for use in explosive atmospheres of the TEFC type is shown in Fig. 501-35. The main frame and end housings are made with sufficient strength to withstand internal pressures due to ignition of a combustible mixture inside the motor. Wide metal-to-metal joints are provided between the frame and housings. Circulation of the air is maintained inside the inner enclosure by fan blades

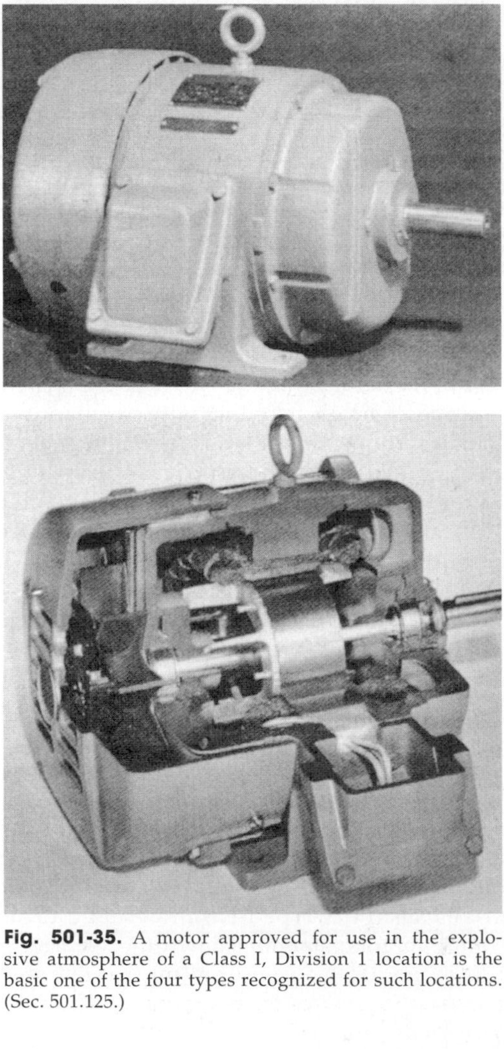

Fig. 501-35. A motor approved for use in the explosive atmosphere of a Class I, Division 1 location is the basic one of the four types recognized for such locations. (Sec. 501.125.)

on each end of the rotor. At the left side of the sectional view, a fan is shown in the space between the inner and outer housings. This fan draws in air through a screen and drives it across the surface of the stator punchings and out through openings at the drive end of the motor. The motors described in **(2)** and **(3)** of part **(A)** of this section are the only ones available for Class I, Group A and B locations and for medium-voltage, high-horsepower applications. The cost of ducts and ventilating systems limits their application in other areas.

A fourth option, available since the 1975 NEC, originated with the petroleum industry and consists of a motor operating in liquid and with controls that assure that it cannot actually start until entirely purged and comes off line in the event that its medium of submergence fails. The 2017 NEC modified this rule by conditioning it on use by industrial establishments with restricted public access and qualified maintenance and supervision.

The UL Guide Card information [see discussion at 110.3(B) for details] lists motors for Class I, Group C and D locations. To date, UL lists no motors for Groups A and B; therefore, where such conditions are encountered, motors must be located outside the hazardous area or must conform to the alternate arrangements and conditions of 501.125(A). Air or inert-gas purging are recognized as alternate methods. Motors suitable for Group C and D, Class I locations are designated as explosionproof.

In part **(B)**, the rule relaxes the requirements for Division 2 areas somewhat. In Class I, Division 2 locations, open or nonexplosionproof enclosed motors may be used if they have no brushes, switching mechanisms, or integral resistance devices. This includes conventional squirrel-cage three-phase motors. However, motors with any sparking or high-temperature devices must be approved for Class I, Division 1 areas, as previously described. A note (No. 4) addresses the potential issues arising in these environments from piston engines, such as typical power generators. The note references an ISA standard covering the applicable safety requirements in such cases.

As of the 2017 NEC, the issue of shaft bonding devices is addressed. These devices, installed either inside or outside the motor housing, have sliding contacts running on the shaft and keep the shaft at ground potential in order to reduce bearing failures from arcing that have been reported on motors fed from inverter based variable speed drives. The contact devices are permitted provided it can be determined that any energy discharged during operation is insufficient to be an ignition threat in the specific hazardous environment. A new informational note (No. 5) points to the applicable provisions of the UL standard that cover the process to determine incendivity.

"Motors and Generators, Rebuilt" May Be Listed

A number of years ago, a procedure was established to provide third-party certification of rewound or rebuilt motors in hazardous locations. Refer to the UL Guide Card information [see discussion at 110.3(B) for details] on hazardous-location equipment for motor-repair centers authorized to provide certified repairs.

501.130. Luminaires. In these locations, part **(A)** requires that each luminaire be *identified as a complete assembly* for the Class I, Division 1 location and marked to show the maximum wattage permitted for the lamps in the luminaire (Fig. 501-36). Reference to the listings in the UL Guide Card information shows many manufacturers listed for Class I luminaires for use in various groups of atmospheres.

Fig. 501-36. Fluorescent luminaire, fed by Type MI cable, is listed and marked as an explosionproof unit for use in a Class I, Division 1 location. (Sec. 501.130.)

Listings range from "Class I, Group C" to "Class I, Groups A, B, C, and D." For application of a luminaire in a particular group, it is simply a matter of ensuring that a manufacturer's luminaire is listed for Class I and the group. The designation "Class I" indicates that the luminaire is suitable for Division 1, except where the listing contains the phrase "Division 2 only" following the "Class I" reference or following the "Class I" plus group references.

Part **(A)(3)** permits support of a suspended luminaire on rigid metal conduit, IMC, or an explosionproof flexible connection fitting. Such application is acceptable provided the luminaire stem is no longer than 12 in. (300 mm), unless lateral bracing is provided. Where bracing *is* needed, it must be provided not more than 12 in. (300 mm) from the luminaire end of the stem. As indicated, where flexibility is needed, a flexible fitting, no longer than 12 in. (300 mm) is permitted.

Class I, Division 2 locations In these locations, the selection of a suitable luminaire becomes a little more involved and has caused problems in the field. Correlation between the requirements of 501.130 and the application data and listings of UL must be carefully established.

Watch Out! Controversy! 501.130(B)(1) does *not* say that a luminaire in a Division 2 location must "be listed for the Class I, Division 2 location." Instead, 501.130(B)(1) gives a description of the type of luminaire that would be acceptable, citing a number of requirements:

1. If the lamps under normal conditions reach a surface temperature exceeding 80 percent of the autoignition temperature (in degrees Celsius) of the surrounding gas or vapor, the luminaire must have been tested in order to determine the operating temperature, or "T code" [see NEC Table 500.8(C)]. The other possibility is to use a Division 1 luminaire rated for the location.

2. The luminaire must be protected from physical damage by suitable guards or "by location"—which can be taken to mean that mounting it high or otherwise out of the way of any object that might strike or hit it eliminates the need for a guard. If falling sparks or hot metal from the luminaire could possibly ignite local accumulation of the hazardous atmosphere, then an enclosure or other protective means must be used to eliminate that hazard.
3. If the luminaire is pendant mounted, it must meet a slightly relaxed version of 501.130(A)(3) for Division 1 equivalents. The only difference is that the set-screws to prevent the threaded joints from loosening in Division 1 are not required here.
4. Luminaires for portable lighting must generally meet Division 1 requirements; however, if they are mounted on movable stands such that they are protected from physical damage in any position, and connected by flexible cord in accordance with 501.140, they need only comply with (B)(1). The rule recognizes the common need for temporary lighting for maintenance work in Class I, Division 2 locations, where handlamps would not be adequate.
5. Any switches on the luminaire must have their contacts hermetically sealed.
6. Any ballasts or other control equipment must meet 501.120(B). Since ballasts would come under (B)(1) ("coils and windings") they would qualify for ordinary location enclosures.

Fig. 501-37. In a Class I, Division 2 location, a lighting fixture listed for Class I, Division 1 would satisfy—such as this factory-sealed mercury-vapor luminaire in a Class I, Group D location. Otherwise, a fixture listed for Class I, Division 2 must be used. (Sec. 501.130.)

Figure 501-37 shows a Class I, Division 1 luminaire that could be used in a Division 2 location. But it is not necessary to use a Division 1 luminaire and have questions develop.

For many years in Class I, Division 2 locations, simply an enclosed- and gasketed-type luminaire was the usual choice. The luminaire does not need to be explosionproof but must have a gasketed globe. The primary requirement is that any surface, including the lamp, must operate at not over 80 percent of the autoignition temperature of the gas or vapor that may be present. The effect of 501.130(B)(1) is to recognize the use of general-purpose luminaires if the conditions of luminaire-operating temperature and atmosphere-autoignition temperature are correlated as required or if the luminaire has been tested to verify its safety. At best, the described task of determining the suitability of a general-purpose luminaire for use in the hazardous location by relating its lamp-operating temperature to 80 percent of the atmosphere-ignition temperature could be difficult for any electrical designer and/or installer. It also seems highly unlikely that any of them would have the facilities or the experience to perform the testing described in the last part of the Code rule and make a sound judgment on the suitability, even though the manufacturer provides the necessary temperature data on the luminaire. It was the intention of the authors of that last part of the Code rule that the testing mentioned be done by a "qualified testing agency" (such as UL, Factory Mutual, ETL). As a result of such testing, luminaires for Class I, Division 2 locations would be approved and listed on the same basis that luminaires are certified for Class I, Division 1 locations.

Class I, Division 2 luminaires are listed in two ways in the Guide Card information [see discussion at 110.3(B) for details]. Some are listed as "Class I, Division 2 only," without reference to group or groups. Others are listed with an indication of the groups for which they are listed—for instance, "Class I, Groups A, B, C, and D, Division 2 only." Great care must be used in evaluating the detail of these listing designations. If such a luminaire is not marked otherwise, the temperature of the luminaire is lower than the autoignition temperature of any of the atmospheres for which it is listed. Where a Class I group designation is not mentioned, the luminaire must not be used where its marked operating temperature is greater than the autoignition temperature of the hazardous atmosphere. Class I, Division 2 luminaires with *internal* parts operating over 212°F (100°C) will be marked to show the actual operating temperature of internal parts.

Based on the foregoing, precise enforcement of the NEC and OSHA insistence on the maximum use of third-party-certified products would seem to suggest the following approach:

1. In Class I, Division 1 locations, only luminaires listed by a nationally recognized test lab may be used.
2. Because luminaires are listed for Class I, Division 2 locations (in the UL Guide Card information), any luminaire in a Class I, Division 2 location *must be listed* for that application (or, of course, a Class I, Division 1 luminaire could be used). Consistent with OSHA's rationale on the matter of listing, if a third-party-certified product is available (that is, Class I, Division 2 luminaires), then use of a nonlisted luminaire in a Class I, Division 2 location could draw an objection, but only from OSHA and not from someone enforcing the NEC.

Recessed Luminaires

Recessed luminaires of both incandescent and electric-discharge lamp types are listed in the UL Guide Card information and are suitable only for dry locations, unless marked "Suitable for Damp Locations" or "Suitable for Wet Locations." Other rules are as follows:

- Each luminaire is marked to show the minimum temperature rating of conductors used to supply the luminaire. Care must be taken to observe all such markings on these luminaires with respect to the number, size, and temperature rating of wires permitted in junction boxes or splice compartments that are part of such luminaires. Generally, no allowance is made in such boxes or compartments for heat produced by current to other loads that may be fed by taps or splices in the luminaire supply wires within the JB or splice compartment. Allowance is made only for the I^2R heat input of the current to the luminaire itself. If the luminaire is recognized for carrying through other conductors to other loads, the luminaire will be marked to cover the permitted conditions of wiring. This is consistent with the principle covered in NEC 410.21 (second paragraph).

- Luminaires are listed to ensure safe application in both Class I, Division 1 and Class I, Division 2 locations.
- In every Class I, Division 1 luminaire, the wiring compartment for supply circuit connections is internally sealed from the lamp chambers.
- Luminaires that may be used as raceways for carrying through circuit conductors other than those supplying the luminaires are marked "Suitable for Use as Raceway" and show the number, size, and type of wires permitted.
- Luminaires are marked to show suitability for installation in concrete and some may be used *only* in concrete.
- Luminaires are marked when they may be used *only* with fire-resistive building construction.
- Some luminaires are marked to show acceptable use *only* in Class I, Division 2 locations.

501.135. Utilization Equipment. This section covers devices that utilize electrical energy—other than luminaires and motors or motor-operated equipment. Electric heaters in Class I, Division 1 locations must be listed for such application. The UL Guide Card information [see discussion at 110.3(B) for details] lists convection-type heaters under "Heaters," for Groups C and D. Industrial and laboratory heaters—heat tracing systems, hot plates, paint heaters, and steam-heated ovens—are also listed.

According to 501.135(B)(1)(1), Exception No. 2, in a Class I, Division 2 location, electrically heated utilization equipment may be used *if* some current-limiting means is provided to prevent heater temperature from exceeding 80 percent of the autoignition temperature of the gas or vapor. Such current limitation may be part of the control equipment for the heater to ensure safe operation by preventing dangerously high temperatures.

501.140. Flexible Cords, Class I, Divisions 1 and 2. Although 501.10(A) does not mention flexible cord as an approved method of general wiring in Division 1 locations, it is mentioned for flexible connections in 501.10(A)(2). The allowances in Part **(A)** are not inconsistent and do permit such cord for connection of portable equipment. The concept is also applied in 501.105(B)(6), which permits cord-and-plug connection of process control instruments in Division 2 locations, to facilitate replacement of such units, which are not portable equipment but are subject to interchange. 501.10(B)(2)(5) also covers use of cord in Division 2 locations.

An explosionproof handlamp, listed for use in Class I locations, is an example of portable equipment covered by this rule, which requires that a three-conductor cord be used and that the device be provided with a terminal for the third, or grounding, conductor, which serves to ground the exposed metal parts.

Such handlamps are listed under "Portable Luminaires for Use in Hazardous Locations" in the UL Guide Card information [see discussion at 110.3(B) for details], which notes that flexible cords should be used only where absolutely necessary as an alternative to threaded rigid conduit hookups. Cords, plugs, and receptacles must be protected from moisture, dirt, and foreign materials. Frequent inspection and maintenance are critically important. Consultation with inspection authorities is always recommended where plug and receptacle applications are considered.

Flexible cord is recognized in industrial occupancies with qualified maintenance and supervision to supply equipment that requires a range or speed of motion that necessitates the use of flexible cord with its inherently flexible, very finely stranded conductors. The use of cord in these applications requires protection for the cord, either by location or through the use of guards. Flexible cord is also recognized with submersible pumps in wet-pits. For these applications the cord is permitted to pass through a conduit run unspliced back the industrial control panel or other power source supplying it. This is a rare example of a specific allowance for cord in conduit as covered in 400.12(6). Cord is also permitted for mixers that must travel in and out of open tanks. The 2011 **NEC** added another application, that of supplying power to maintenance carts for refineries and other industrial plants that contain equipment including receptacles to supply portable utilization equipment. In the event the cord enters or is supplied from an enclosure or fitting required to be explosionproof, the cord must terminate with a connector or attachment plug listed for the applicable protection scheme used to maintain the suitability of the enclosure for the surrounding environment. In a Division 2 location that does not require explosionproof equipment, the termination can be through a conventional listed plug, or a listed cord connector. The cord must be "of continuous length," but if the application is in conjunction with a maintenance cart as just described, one cord can feed the cart and another can feed the portable equipment fed from the cart.

When flexible cord is permitted in accordance with (A), Part **(B)** covers how to install it. It must be extra-hard-usage cord, such as Type SO, STO, or SOOW, containing an equipment ground, and arranged so there will not be any tension transmitted to the conductor connections at either end. Wherever connected to an enclosure or fitting required to be explosionproof, the connection must be through a cord connector (or attachment plug) that has been listed for that application and location. The cord may also be connected through a listed cord connector used in conjunction with a seal listed for the hazardous location. In Division 2 locations for applications where the terminating equipment is not required to be explosionproof, it is sufficient to apply a listed attachment plug or cord connector. The cord must be of continuous length. To connect the maintenance cart applications allowed in (A)(5), the continuous cord rule is applied twice, once from the main power supply to the temporary portable assembly, and once again for each connection from the assembly to each portable utilization equipment.

501.145. Receptacles and Attachment Plugs, Class I, Divisions 1 and 2. The basic rule calls for receptacles and plug caps to be approved for Class I locations, which suits them for use in either Division 1 or Division 2 locations. They must be part of the premises wiring system unless provided in accordance with 501.140(A). One of the motivations for this clarification was to disqualify receptacles becoming energized through energized cords, in violation of 406.7(B). The exception notes that cord connection of process control instruments in Class I, Division 2 locations does not require devices approved for Class I locations. General-purpose receptacles may be used as outlined in 501.105(B)(6).

Figure 501-38 shows a 3-pole 30-A receptacle and the attachment plug, which is so designed as to seal the arc when the circuit is broken, and therefore is suit-

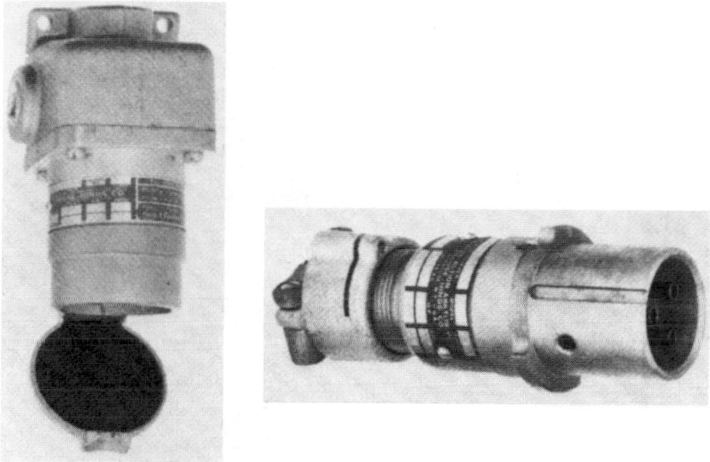

Fig. 501-38. Receptacle and plug must generally be explosionproof type for Divisions 1 and 2. (Sec. 501.145.)

able for use without a switch. The circuit conductors are brought into the base or body through rigid conduit screwed into a tapped opening and are spliced to pigtail leads from the receptacle. The receptacle housing is then attached to the base, the joint being made at wide flanges ground to a suitable fit. All necessary sealing is provided in the device itself, and no additional sealing is required when it is installed. The plug is designed to receive a three-conductor cord for a 2-wire circuit or a four-conductor cord for a 3-wire, 3-phase circuit, and is provided with a clamping device to relieve the terminals of any strain. The extra conductor is used to ground the equipment supplied.

UL data are as follows:

- Class I receptacles for Division 1 or Division 2 locations are equipped with boxes for threaded metal conduit connection, and a factory seal is provided between the receptacle and its box.
- Receptacles for Class I, Division 2 only may be used with general-purpose enclosures for supply connections, with factory sealing of conductors in the receptacle. The plugs for such receptacles are suitable for Class I, Division 1 locations.
- Frequent inspection is recommended for flexible cords, receptacles, and plugs, with replacement whenever necessary.

- For Class I, interlocked CBs and plugs are made for receptacles so that the plug cannot be removed from the receptacle when the CB is closed and the CB cannot be closed when the plug is not in the receptacle (Fig. 501-39).

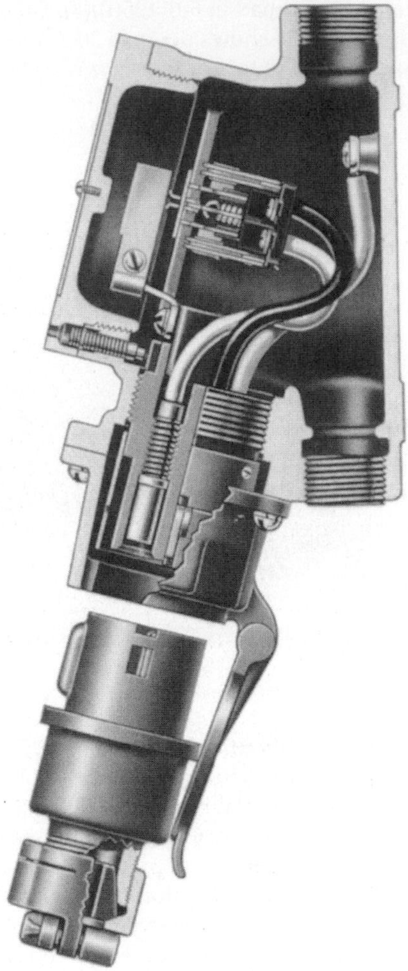

Fig. 501-39. A receptacle with a plug interlocked with a circuit breaker is an explosionproof assembly with operating safety features. (Sec. 501.145.)

Mechanical-interlock construction requires that the plug be fully inserted into the receptacle and rotated to operate an enclosed switch or CB that energizes the receptacle. The plug cannot be withdrawn until the switch or breaker has first deenergized the circuit.

The delayed-action type of plug and receptacle has a mechanism within the receptacle that prevents complete withdrawal of the plug until after electrical connection has been broken, permitting any arcs or sparks to be quenched inside the arcing chamber. Insertion of the plug seals the arcing chamber before electrical connection is made. Threaded conduit connection to the CB compartment is provided. The plug is for Type S flexible cord with an equipment grounding conductor.

501.150. Signaling, Alarm, Remote-Control, and Communication Systems. Nearly all signaling, remote-control, and communication equipment involves make-or-break contacts; hence, in Division 1 locations all devices must be explosionproof, and the wiring must comply with the requirements for light and power wiring in such locations, including seals.

Figure 501-40 shows a telephone having the operating mechanism mounted in an explosionproof housing. Similar equipment may be obtained for operating horns or sirens. Figure 501-41 shows fire-alarm hookups at a distillery.

Fig. 501-40. Explosionproof telephones are made and listed for Class I, Groups B, C, and D, and must be connected with the necessary seal fittings required in conduits to enclosures housing arcing or sparking devices. [Sec. 501.150(4).]

Referring to subparagraph **(B)**, covering Division 2 locations, it would usually be the more simple method to use explosionproof devices, rather than devices having contacts immersed in oil or devices in hermetically sealed enclosures, though mercury switches, which are hermetically sealed, may be used for some purposes. Of course, reference to 501.15(A) recognizes the use of purged enclosures as an alternative method.

The UL Guide Card information [see discussion at 110.3(B) for details] lists "Telephones" as follows:

- Telephones, sound-powered telephones, and communications equipment and systems are listed for Class I and Class II use in Division 1 locations and are explosionproof equipment. Such equipment complies with 501.14(A) and (B).
- Intrinsically safe sound-powered telephones are also listed for Class I, Division 1, Group D and may be used in both Division 1 and Division 2 locations in accordance with 501.14(A) and (B).

The Guide Card information also lists "Thermostats," "Signal Appliances" (which include fire alarms and fire detectors), "Solenoids," and "Sound Recording and Reproducing Equipment."

GRINNELL
PROTOMATIC
RELEASE

Fig. 501-41. Fire-alarm and control equipment at outdoor tank car delivery and pumping station of a distillery is housed and connected to comply with conduit and seal rules for a Class I, Division 1, Group D location. (Sec. 501.150.)

ARTICLE 502. CLASS II LOCATIONS

502.1. Scope. Referring to 500.5(C), the hazards in Class II locations are due to the presence of combustible dust. These locations are subdivided into three groups, as follows, and not all equipment is suitable for all groups:

- Group E, atmospheres containing metal dust
- Group F, atmospheres containing carbon black, coal dust, or coke dust
- Group G, atmospheres containing grain dust, such as in grain elevators

Any one of four hazards, or a combination of two or more, may exist in a Class II location: (1) an explosive mixture of air and dust, (2) the collection of conductive dust on and around live parts, (3) overheating of equipment because deposits of dust interfere with the normal radiation of heat, and (4) the possible ignition of deposits of dust by arcs or sparks. Combustible dust is now formally and objectively defined in Art. 500; refer to the discussion at 500.5 for more information. The 2014 NEC has revised upward the maximum size threshold for combustible dusts from passing a US #40 standard sieve or smaller (420 microns or 0.017 in.) to passing a US #35 standard sieve or smaller (500 microns or 0.020 in.).

A number of processes that may produce combustible dusts are listed in 500.6. Most of the equipment listed as suitable for Class I locations is also dusttight, but in accordance with 502.5 it should not be taken for granted that all explosionproof equipment is suitable for use in Class II locations. Some explosionproof equipment may reach too high a temperature if blanketed by a heavy deposit of dust. Grain dust will ignite at a temperature below that of many of the flammable vapors.

Location of service equipment, switchboards, and panelboards in a separate room away from the dusty atmosphere is always preferable.

In Class II locations, with the presence of combustible dust, UL standards call for a type of construction designed to preclude dust and to operate at specified limited temperatures. Dust-ignitionproof equipment is generally more economical to use in Class II areas; however, explosionproof devices are often used if such devices are approved for Class II areas and for the particular group involved.

Dust-ignitionproof equipment is enclosed in a manner that excludes dusts and does not permit arcs, sparks, or heat generated or liberated inside the enclosure to cause ignition of exterior accumulations or atmospheric suspensions of a specified dust on or in the vicinity of the enclosure. Any assemblies that generate heat, such as luminaires and motors, are tested with a dust blanket arranged to equal in thickness "the maximum amount of dust that can accumulate" to simulate the operation of the device in a Class II location.

502.6. Zone Equipment. Equipment "listed and marked" per 506.9(C)(2) for installation in Zone 20 locations may be used in Class II, Division 1, provided it is suitable for the same dust environment and a compatible temperature classification. Also, Zone 20, 21, and 22 location equipment is suitable for comparable dust exposures in Class II Division 2 locations.

502.10. Wiring Methods. Part **(A)** covers Division 1 locations. The wiring methods are essentially the same for Class II, Division 1 as they are for Class I, Division 1. Conduit connections to fittings and boxes must be made to threaded bosses, as covered in (A)(3). For fittings and boxes, only those used for taps, splices, or terminals in locations of electrically conductive dusts (metal, carbon, etc.) must be identified for Class II locations.

Optical fiber cable has been recognized in 770.3(A) as acceptable in hazardous locations, and this use is now correlated in 502.10(A)(1)(4). Note that it must be installed within a 502.10(A) raceway, which circles back to threaded RMC or IMC, and then sealed in accordance with the usual requirements in 502.15. The same correlating provision was added as 502.10(B)(8) for Division 2 locations. The 2017 **NEC** added recognition here [in (5)] of Type ITC-HL cable. This is an exact import from 501.10(A)(1)(d), and this book covers it in that location.

As covered in part **(A)(2),** where a flexible connection is necessary, it would usually be preferable to use a dusttight flexible fitting. Standard flexible conduit may not be used, but liquidtight flexible metal conduit or liquidtight flexible nonmetallic conduit may be used, with a bonding jumper as required by 502.30(B). The use of an extra-hard-service cord that has one conductor serving as a grounding conductor with listed dusttight fittings is permitted. In addition, the use of interlocking-armor Type MC cable with a nonmetallic jacket over the armor is permitted for flexible connections provided the termination fittings are listed for Class II Division 1 exposures.

In Division 2 locations, in order to provide adequate bonding, threaded fittings should be used with IMC or rigid metal conduit, but EMT may also be used

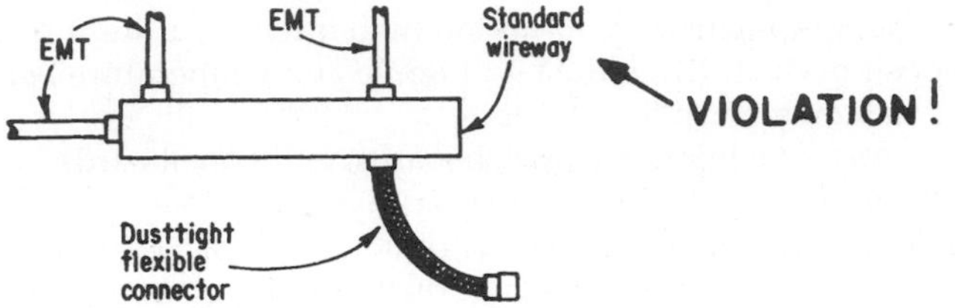
CLASS II, DIV. 2 LOCATION
EMT
EMT
Standard wireway
VIOLATION!
Dusttight flexible connector

Fig. 502-1. EMT is okay, but the wireway *must* be dusttight. (Sec. 502.10.)

(Fig. 502-1). The requirement for close-fitting covers could best be taken care of by using dusttight equipment. The standard type of pressed steel box cannot be used in any case where the box contains taps or splices. Where no taps, splices, or terminals are used in a box, a sheet metal box may be used, but all conduit connections to boxes that do not have threaded hubs or bosses must be of the bonded type, as required by 502.30(B) and described with illustrations under 501.30(B).

The list of permitted cables is also greatly increased. Type MC cable in any form is allowed, with listed termination fittings. It is also allowed in cable tray (see below). Type PLTC and Type ITC cables are allowed in tray as well for control work. Type MC cable (and also Type MI and TC) is acceptable cable for use in ladder, ventilated channel-type, and ventilated trough cable tray; however, there must be a cable diameter of open space separating each run of cable, or, only in cases where the Type MC is listed for use in Class II, Division 1 locations, it can be run without the spacings.

Two new approaches joined the list for the first time in 2011. The heaviest wall versions of Type PVC (i.e., "Schedule 80") and Type RTRC (i.e., suffix "-XW") conduits can now be used in Class II Division 2 locations exposed, without concrete encasement. This use is heavily restricted, being limited to industrial occupancies with limited public access and qualified maintenance and supervision, and in addition, these methods are only allowed where "metallic conduit does not provide sufficient corrosion resistance." Optical fiber cable joined the list in the 2014 **NEC** on the same basis as it arrived in Art. 501 and is now located at 501.10(B)(1)(8). The 2017 **NEC** added cablebus as an additional option.

Nonincendive field wiring is also permitted for these locations, under the same provisions as apply to Class I Division 2. Refer to the discussion at 501.10(B)(3) for more information on this topic.

All boxes and fittings must be dusttight, defined in Art. 100 as "constructed so that dust will not enter under specified test conditions." Flexible connections in Division 2 locations must observe the rules given in part **(A)(2)**.

502.15. Sealing, Class II, Divisions 1 and 2. Note that sealing or other isolation is required only for conduits entering a dust-ignitionproof enclosure that connects to an enclosure that is not dust-ignitionproof. Seals are not generally required in Class II areas; however, where a raceway connects an enclosure *required to be* dust-ignitionproof and one that is not, means must be provided to prevent dust from entering the dust-ignitionproof enclosure through the raceway. A seal in the conduit is one acceptable way of doing this, with the seal any

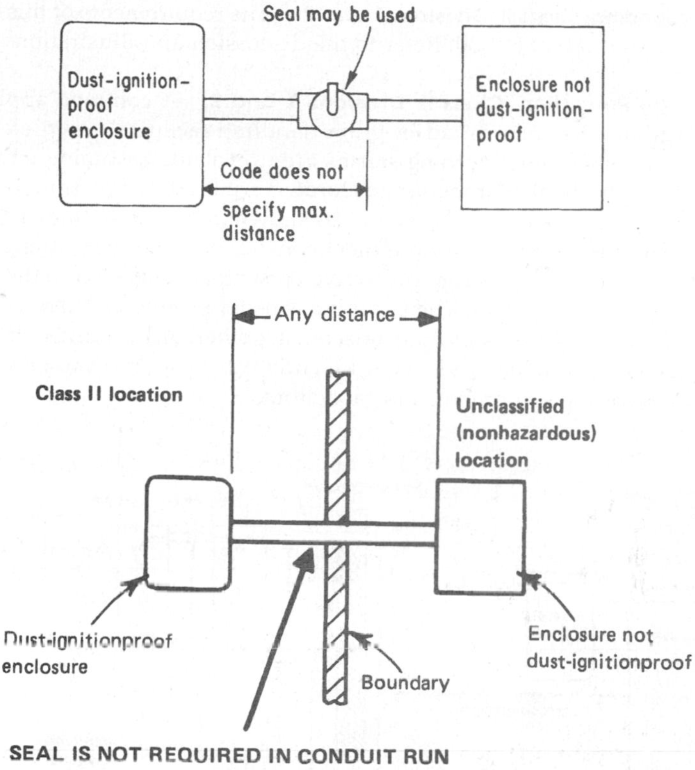
Seal may be used
Dust-ignition-proof enclosure
Enclosure not dust-ignition-proof
Code does not specify max. distance
Any distance
Class II location
Unclassified (nonhazardous) location
Dust-ignitionproof enclosure
Boundary
Enclosure not dust-ignitionproof
SEAL IS NOT REQUIRED IN CONDUIT RUN

Fig. 502-2. A seal fitting must be used in short [less than 10-ft (3.0 m)] connections, as shown at top, but is not required, as shown at bottom.

distance from the enclosure (Fig. 502-2). Where used, a seal must be accessible. This seal does not have to be explosionproof, and electrical sealing putty is sufficient. However, if the connecting raceway is horizontal and at least 10 ft (3.0 m) long, or if it is vertical, extending down from the dust-ignitionproof enclosure and at least 5 ft (1.5 m) long, or any combination of horizontal and downward length that provides an equivalent resistance to dust transmission, no sealing is necessary anywhere in the conduit. For example, a conduit that went 8 ft (2.5 m) horizontally and turned 2 ft (600 mm) down would certainly be an equivalent distance. The distance between and orientation of the enclosures are considered adequate protection against dust passage. And that applies in Division 1 and Division 2 locations wherever the "dust-ignitionproof" requirement is specified by one of the sections of this article.

The next-to-last sentence of this section permits use of conduit *without* a seal between a dust-ignitionproof enclosure that is required in a Class II location and an enclosure in an unclassified location (Fig. 502-2, bottom). This added permission is made because, unlike gases or vapors, dust does not travel very far within conduit.

502.30. Grounding, Class II, Divisions 1 and 2. The requirements of this section are the same as those of 501.30. Refer to the discussion and illustrations in that section.

502.35. Surge Protection, Class II, Divisions 1 and 2. A common application of this surge protection is found in grain-handling facilities (grain elevators) in localities where severe lightning storms are prevalent. Assuming a building supplied through a bank of transformers located a short distance from the building, the recommendations are, in general, as shown in the single-line diagram in Fig. 502-3. The surge-protective equipment consists of primary lightning arresters at the transformers and surge-protective capacitors connected to the supply side of the service equipment. The lightning arrester ground and the secondary system ground should be solidly connected together. All grounds should be bonded together and to the service conduit and to all boxes enclosing the service equipment, metering equipment, and capacitors.

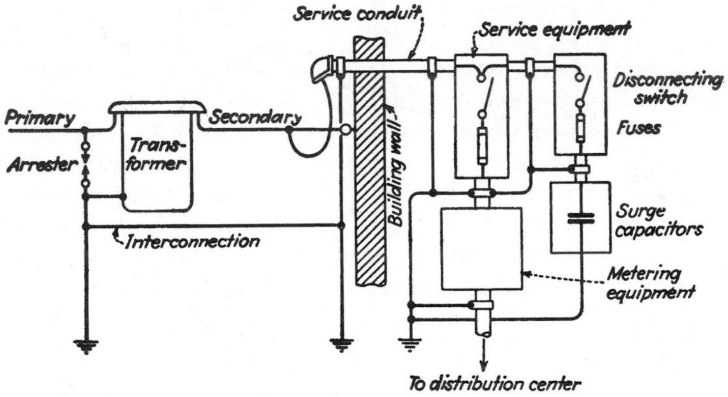
Service conduit
Service equipment
Disconnecting switch
Fuses
Primary
Secondary
Trans-former
Arrester
Building wall
Surge capacitors
Interconnection
Metering equipment
To distribution center

Fig. 502-3. Surge protection is often used to protect Class II systems against lightning. (Sec. 502.35.)

502.100. Transformers and Capacitors. Part **(A)** requires the use of a Code-constructed transformer vault for a transformer or capacitor that contains oil or other liquid dielectric that will burn, when used in a Class II, Division 1 location.

Dry-type transformers must either be approved as a complete assembly for Class II, Division 1 use, and UL now has product categories open in this area, both for medium-voltage transformers and for units 600 V and below, or a conventional transformer can be placed in a vault. A vault used for these purposes must be as covered in Art. 450 but with self-closing fire doors on both sides of the wall as well as weather-stripping or equal on those doors if they communicate with a Division 1 location. But transformers can be kept out of the hazardous areas, and part **(A)(3)** prohibits any use of a transformer or capacitor in a Class II, Division 1, Group E (metal dust) location. UL's category for capacitors does not appear to include hazardous locations, so they would have to be thoroughly separated from the Class II area, either by using a vault or some other method.

Part **(B)** requires a vault for any oil-filled or high-fire-point liquid-insulated transformer or capacitor in a Division 2 area. A dry-type unit in a Division 2 area may be used either in a vault or without a vault if the unit is enclosed within a tight metal housing without any openings and only if it operates at 600 V or less. There are no special requirements for capacitors in Division 2 locations, except that they must not contain oil or any other "liquid that will burn" if they are not within a vault.

502.115. Switches, Circuit Breakers, Motor Controllers, and Fuses. Note that part **(A)(1)** calls for "dust-ignitionproof" enclosures, which might involve the sealing called for in 502.15.

Most of the enclosed switches and circuit breakers approved for Class I, Division 1 locations are also approved for use in Class II locations.

Switches conforming to the definition of the term *isolating switch* would seldom be used in any hazardous location. Such switches are permitted for use as the disconnecting means for motors larger than 100 hp.

Enclosures for this equipment in a Division 2 application must be dusttight.

502.120. Control Transformers and Resistors. This part of Art. 502, like many others, used to use the expression "tight metal housings without ventilating openings" for Division 2 applications. That terminology was changed to "dusttight," which involves an actual testing protocol. The same change occurred in 502.150(B)(1).

502.125. Motors and Generators. In Class II, Division 1 locations, motors must be dust-ignitionproof (approved for Class II, Division 1) or totally enclosed with positive-pressure ventilation, and operating under the temperature limitations for Class II equipment that are set forth in 500.8(D)(2).

Part **(A)(2)** refers to a totally enclosed pipe-ventilated motor. A motor of this type is cooled by clean air forced through a pipe by a fan or blower. Such a motor has an intake opening, where air is delivered to the motor through the pipe from the blower. The exhaust opening is on the opposite side, and this should be connected to a pipe terminating outside the building, so that dust will not collect inside the motor while it is not running.

For Class II, Division 2 locations, part **(B)** permits use of totally enclosed, nonventilated motors and totally enclosed, fan-cooled (TEFC) motors in addition to totally enclosed, pipe-ventilated and dust-ignitionproof motors. In addition, a totally enclosed water-air-cooled motor would be permitted. This rule eliminates any interpretation that only a labeled motor (dust-ignitionproof) is acceptable for Class II, Division 2 locations. Experience has shown the other motors as covered in the exception to be entirely safe and effective for use in such locations. There are other limitations on the exception, however. The motor must be easily reached for routine maintenance, and the dust must be nonabrasive and nonconductive in character, and the expected accumulations of dust must be "moderate." Since these are Division 2 locations, that is not an unreasonable hurdle. The 2017 **NEC** added recognition within the exception to "sealed bearings, bearing isolators, and seals." This responds to current developments in motor designs for these environments.

Motors of the common totally enclosed type without special provision for cooling may be used in Division 2 locations, but to deliver the same horsepower, a plain totally enclosed motor must be considerably larger and heavier than a motor of the open type or an enclosed fan-cooled or pipe-ventilated motor.

The UL Guide Card information [see discussion at 110.3(B) for details] lists motors for Class II, Division 1 and 2, Group E, F, and G locations.

502.128. Ventilating Piping. In locations where dust or flying material will collect on or in motors to such an extent as to interfere with their ventilation or cooling, enclosed motors that will not overheat under the prevailing conditions must be used. It may be necessary to require the use of an enclosed pipe-ventilated motor or to locate the motor in a separate dusttight room, properly ventilated with clean air (430.16).

The reference to ventilation is clarified in this section. Vent pipes for rotating electrical machinery must be of metal not lighter than No. 24 MSG gauge, or equally substantial. They must lead to a source of clean air outside of buildings, be screened to prevent entry of small animals or birds, and be protected against damage and corrosion.

In Class II, Division 1 locations, vent pipes must be dusttight. In Division 2 locations, they must be tight to prevent entrance of appreciable quantities of dust and to prevent escape of sparks, flame, or burning material.

Typical conditions to which these requirements may apply include processing machinery or enclosed conveyors where dust may escape only under abnormal conditions, or storage areas where handling of bags or sacks may result in small quantities of dust in the air.

502.130. Luminaires. Luminaires used in Class II, Division 1 areas must prevent the entry of the hazardous dust and should prevent the accumulation of dust on the luminaire body. In Division 1 locations, all luminaires must be approved for that location, and display the lamp type and power. Where metal dusts are present, luminaires must be specifically identified for use in Group E atmospheres. Luminaires are listed by UL as suitable for use in all three of the locations classed as Groups E, F, and G, for Divisions 1 and 2.

The purpose of the latter part of subparagraph **(A)(3)** is to specify the type of cord to be used for wiring a chain-suspended luminaire. It is not the intention to permit a luminaire to be suspended by means of a cord pendant or drop cord. Except for the allowance for the use of cord to a chain-hung luminaire, the requirements are essentially the same for pendants here and in Art. 501. The cord termination at the luminaire must employ a listed dusttight fitting, in accordance with 502.10(A)(2)(5).

The Division 2 requirements for fixed lighting have been tightened, and now require dusttight enclosures together with a specified lamp type and maximum lamp wattage such that the worst-case exposed surface temperature will not exceed the limits in 500.8(D)(2). These requirements will be difficult to verify compliance to without the services of a testing laboratory. For Class II, Group G, Division 2 areas, luminaires normally are the enclosed and gasketed type. However, in addition, such luminaires must not have an exposed surface temperature exceeding 329°F (165°C).

There are portable handlamps approved for use in any Class II, Group G location—that is, where the hazards are due to grain dust.

The UL Guide Card information [see discussion at 110.3(B) for details] notes that a Class II luminaire for Divisions 1 and 2 is tested for dusttightness and safe operation in the dust atmosphere for which it is listed. A note points out the importance of effective maintenance—regular cleaning—to prevent buildup of combustible dust on such equipment.

502.135. Utilization Equipment. As noted in part **(B)(1),** Exception, dusttight metal-enclosed radiant heating panels may be used in Class II, Division 2 locations even though they are not approved for Class II locations. The exception to this section will permit such electric-heat panels provided that the surface temperature limitations from 500.8(D)(2) are satisfied. Heaters of this type are available with low surface temperature.

502.140. Flexible Cords—Class II, Divisions 1 and 2. Figure 502-4 shows flexible cord used in a Division 2 area of a grain elevator (Class II, Group G). As of the 2014 NEC, the provisions for the use of flexible cord in Class II locations almost exactly mirror those for Class I locations, with the continuous length rule of 501.140(B)(5) now repeated in this location. In this way the new provisions in 502.140(A)(5) for temporary portable assemblies such as maintenance carts line up with the requirement for continuous cords.

Fig. 502-4. Flexible cords and listed connectors are used in a grain elevator that handles combustible grain dust. (Sec. 502.140.)

502.10 permits its use in Division 1 and Division 2 areas of Class II locations. Listed cord connectors are recognized for use in Class II, Group G locations, using Types S, SO, ST, or STO multiconductor, extra-hard-usage cord *with* a grounding conductor. Cord connectors for connecting extra-hard-service type of flexible cord to devices in hazardous locations must be carefully applied, with cord connectors listed for the dust location or a listed cord connector entering a listed seal for the location; either option being permitted for Division 1. However, a Division 2 exposure simply requires a listed dusttight cord connector.

The UL Guide Card information [see discussion at 110.3(B) for details] notes, under "Receptacles with Plugs," that Type S flexible cord should be frequently inspected and replaced when necessary.

502.145. Receptacles and Attachment Plugs. Class II receptacles listed as approved for Division 1 locations are equipped with boxes for threaded metal conduit connection, and a factory seal is provided between the receptacle and its box. Only receptacles and plugs listed for Class II locations are permitted in Division 1 areas. The section has been restructured in the 2014 NEC, with the Class II suitability requirement moved to the parent text where the same requirement applies equally for both Division 1 and Division 2 exposures. The requirement for receptacles being part of the premises wiring reflects the same concerns about enforcement of 406.7(B) as motivated the same requirement in 501.145.

Frequent inspection is recommended for flexible cords, receptacles, and plugs, with replacement whenever necessary.

As shown in Fig. 501-39 for Class I locations, for Class II locations interlocked CBs and plugs are also made for receptacles so that the plug cannot be removed from the receptacle when the CB is closed and the CB cannot be closed when the plug is not in the receptacle.

ARTICLE 503. CLASS III LOCATIONS

503.5. General. The small fibers of cotton that are carried everywhere by air currents in some parts of cotton mills and the wood shavings and sawdust that collect around planers in woodworking plants are common examples of the combustible flyings or fibers that cause the hazards in Class III, Division 1 locations. Wood flour, as produced in some finishing operations, is classified as combustible dust if it meets the 500-μm particulate size limit of the definition (see 500.2 and the commentary in this book at 502.1). A cotton warehouse is a common example of a Class III, Division 2 location.

503.6. Zone Equipment. Equipment "listed and marked" for installation in combustible dust environments under the zone classifications system per 506.9(C)(2) as suitable for Zone 20 can be used in Class III Division 1 locations provided it does not operate at high temperature. Specifically it must not have a temperature classification over T120°C for equipment that is subject to overload and T165°C for other equipment. Similar to the case of combustible dust, the test protocol requires an accumulation of the subject fibers/flyings to be arranged so that they equal in thickness the maximum amount that can accumulate. Comparable equipment rated for Zone 20, 21, or 22 can be used for Class III Division 2 locations.

503.10. Wiring Methods. Type PVC and RTRC conduit, EMT, and Type MC (metal-clad cable) are permitted for Class III, Division 1 locations, in addition to rigid metal conduit, IMC, Type MI cable, and dusttight wireways. Type MC (Art. 330) includes interlocked armor cable, corrugated metal armor, smooth aluminum-sheathed cable (known until the 1978 NEC as Type ALS), and smooth copper-sheathed cable (also known until the 1978 NEC as Type CS).

Cable trays with ladder, ventilated trough, or ventilated channel construction are permitted in these locations, provided the allowable wiring methods are in a single layer with a maintained spacing not less than the diameter of the larger adjacent cable. The permitted cables are Type TC, MI, or MC; MC cable is also available in a form listed for use in comparable Class II locations, and these cables do not require spacing between adjacent cables. The 2017 NEC added cablebus to the list as well, without limitation.

Nonincendive field wiring is also permitted for these Division 1 locations, under the same provisions as apply to Class I and Class II Division 2. Refer to the discussion at 501.10(B)(3) for more information on this topic.

Fittings and boxes must be dusttight, whether or not they contain taps, joints, or terminal connections. As part **(A)(2)** notes about flexible connections, it is necessary to use dusttight flexible connectors, liquidtight flexible metal or nonmetallic conduit, or extra-hard-usage flexible cord that complies with 503.140. Interlocking-armor Type MC cable with an outer plastic jacket is also recognized for this purpose, provided the terminating fittings are dusttight.

Part **(B)** requires the same wiring methods for Division 2 as for Division 1. As indicated in Fig. 503-1, there are no seal requirements in Class III locations. By exception only in areas used only for storage and without machinery, open wiring on insulators is still permitted.

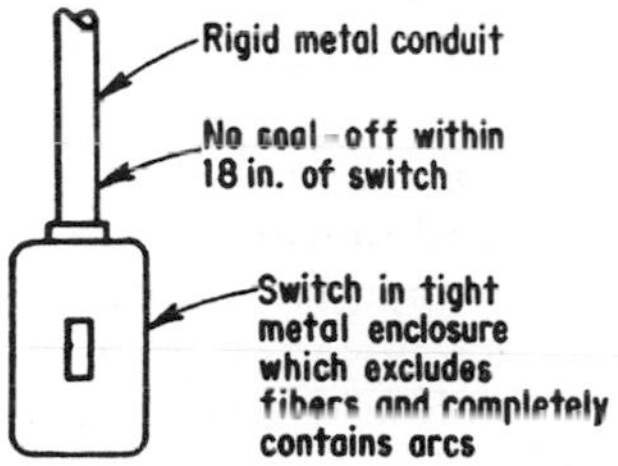
Rigid metal conduit
No seal-off within 18 in. of switch
Switch in tight metal enclosure which excludes fibers and completely contains arcs

Fig. 503-1. Seals are not required in Class III enclosures or conduit, but all boxes and fittings must be dusttight. (Sec. 503.10.)

503.30. Grounding and Bonding. These requirements are identical to Class I and II.

503.115. Switches, Circuit Breakers, Motor Controllers, and Fuses, Class III, Divisions 1 and 2. Enclosures for this equipment in a Class III location must be dusttight—as determined under defined test conditions.

503.120. Control Transformers and Resistors. These enclosures must be dusttight. In addition, equipment suitable for Class III locations must function at full rating without developing surface temperatures high enough to cause excessive dehydration or gradual carbonization of accumulated fibers or flyings. These devices have the same surface temperature limitations as Class II equipment, and the construction is similar.

503.125. Motors and Generators, Class III, Divisions 1 and 2. UL lists no Class III motors as such; however, totally enclosed pipe-ventilated motors or totally enclosed nonventilated motors, depending on the needs of the application, and the so-called lint-free or self-cleaning textile squirrel-cage motors are commonly used. The latter may be acceptable to the local inspection authority if only moderate amounts of flyings are likely to accumulate on or near the motor, which must be readily accessible for routine cleaning and maintenance. Or the motor may be a squirrel-cage motor, or a standard open-type machine having any arcing or heating devices enclosed within a tight metal housing without ventilating or other openings.

503.130. Luminaires—Class III, Divisions 1 and 2. In Class III, Division 1 and 2 areas, luminaires must minimize the entrance of fibers and flyings and prevent the escape of sparks or hot metal. And again, the surface temperature of the unit must be limited to 329°F (165°C). Available luminaires are third-party-certified (by a national test lab) for Class III locations, Divisions 1 and 2. In the past, enclosed and gasketed types of luminaires, of the type that was used in Class I, Division 2 areas, have been acceptable as suitable for use in this application. But because there are listed Class III luminaires available, inspection agencies and OSHA might insist on use of *only* listed luminaires in such applications, to be consistent with the trend to third-party certification. Pendants follow similar rules as in Class II, including the allowance for cord where it does not actually support the luminaire, and the requirements to avoid breakage of conduit stems at their threads.

503.155. Electric Cranes, Hoists, and Similar Equipment, Class III, Divisions 1 and 2. A crane operating in a Class III location and having rolling or sliding collectors making contact with bare conductors introduces two hazards:

1. Any arcing between a collector and a conductor rail or wire may ignite flyings of combustible fibers that have collected on or near the bare conductor. This danger may be guarded against by proper alignment of the bare conductor and by using a collector of such form that contact is always maintained, and by the use of guards or barriers that will confine the hot particles of metal that may be thrown off when an arc is formed.
2. Dust and flyings collecting on the insulating supports of the bare conductors may form a conducting path between the conductors or from one conductor to ground and permit enough current to flow to ignite the fibers. This condition is much more likely to exist if moisture is present. Operation on a system having no grounded conductor makes it somewhat less likely that a fire will be started by a current flowing to ground. A recording ground detector will show when the insulation resistance is being lowered by an accumulation of dust and flyings on the insulators, and a relay actuated by excessively low insulation resistance and arranged to trip a CB provides automatic disconnection of the bare conductors when the conditions become dangerous.

ARTICLE 504. INTRINSICALLY SAFE SYSTEMS

504.1. Scope. This NEC article covers design, layout, and installation of electrical equipment and systems that are not capable of releasing sufficient electrical or thermal energy to ignite flammable or combustible atmospheres.

504.2. Definitions. *Associated apparatus* is equipment that is not necessarily intrinsically safe, but that affects the energy levels in intrinsically safe circuits and that are therefore relied upon to maintain the intrinsic safety of those circuits. An example is an intrinsic safety barrier, such as the one shown schematically in

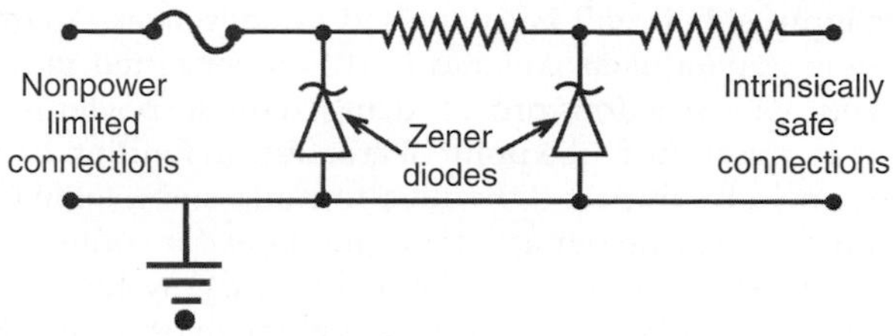
Nonpower
limited
connections
Zener
diodes
Intrinsically
safe
connections

Fig. 504-1. Zener diodes have a very high resistance until their breakdown voltage is reached, at which point they begin to conduct with very little resistance. They are set a few volts higher than the circuit design voltage. When they conduct, the circuit voltage cannot go higher than that for which the manufacturer selected the diodes, and the resistor (sized by Ohm's law to pass only the designed magnitude of current at that breakdown voltage) keeps the current within limits as well. This will be true even on a large overvoltage condition, which is usually defined for test purposes as 250 V [designated "U" in 504.10(A) informational note No. 2, an international symbol for potential difference]. There are two Zener diodes in these units so if one fails the protection is still available. This is only one of several approaches, but it is widely employed. If necessary to achieve an even higher level of redundancy, a third diode could be supplied so the unit will work as intended even after two internal faults.

Fig. 504-1, which limits the current and voltage in the protected circuit. This type of equipment must be listed (see 504.4). Note that the two diodes in this barrier both must fail in order for the system to lose its intrinsic safety. This is an example of a cardinal principle of these systems, that no single component failure regardless of the source of fault can introduce ignition-capable energy into an intrinsically safe circuit.

A *control drawing* is defined in 500.2. It is produced by the manufacturer of the system or apparatus and it details the allowable interconnections between the intrinsically safe circuit components and any associated apparatus or other such components.

Simple apparatus is a component with well-defined electrical characteristics that has minimal impact on the energy levels of the system. The definition includes useful parameters that describe the scope of simple apparatus, and the informational note that follows it provides clear examples and guidance. This terminology creates a class of components that do not require listing if they are on the control drawing (see 504.4 Exception) and that can be used to customize an intrinsically safe installation properly and without the necessity of approaching the manufacturer for a revised control drawing.

504.10. Equipment Installation. The basic rule requires adherence to the terms of the control drawing, as identified on the apparatus. General-purpose enclosures are permitted unless otherwise specified in the documentation provided by the manufacturer of apparatus considered for installation within them, as covered in **(C)**. Simple apparatus does not need to appear on the control drawing provided it is not used to interconnect two different intrinsically safe circuits. The energy level resulting from two such circuits in combination would require careful review.

Some associated apparatus is designed to work in the hazardous environment as part of the intrinsically safe circuit, in which case it can be in the classified environment in ordinary location enclosures. Other such apparatus must be in nonhazardous areas or otherwise protected by enclosures suitable for nonpower-limited exposures. The Zener barrier in Fig. 504-1 is a good example, because on its input side it will be connected to nonpower-limited circuits by definition. Simple apparatus as covered in **(B)**, is permitted in the hazardous environment without restrictions provided, based on the power levels available to it, that it cannot overheat to the point of creating an ignition hazard.

This section concludes in part **(D)** with a formula and a table to assist making the assessment as to whether apparatus qualifies for connection to a given intrinsically safe system without a reevaluation of the system and the control drawing. The table is based on a T4 (135°C) classification because that is below the autoignition temperatures of almost all likely agents. One notable exception is carbon disulfide (CS_2) with an autoignition temperature of 90°C needing a T6 (85°C) classification; however, that is a very unusual chemical and the T4 threshold is a reasonable one. The process also addresses energy generating devices, particularly thermocouples. Care must be taken when adding multiple such items to a circuit, because there may quickly be sufficient cumulative effects to make the system no longer intrinsically safe. The bottom line, as reflected in the parent language of the subsection, is that the surface temperature of the simple apparatus must not exceed the ignition temperature of the hazardous agent(s).

504.30. Separation of Intrinsically Safe Conductors. This section is perhaps the most important for electricians because IS systems can use conventional wiring methods including Class 2 cabling per Art. 725 (see 504.20), but the separation rules are absolutely critical in maintaining the safety of these systems. Routing these circuits near power circuits can induce ignition capable energy into wiring that is entirely unprotected. Even running these circuits next to each other can have unpredictable effects for the same reason. This section splits the separation rules accordingly.

Part **(A)** covers separation from nonintrinsically safe wiring, for which there are three possibilities which this analysis covers in the following order: separation between wiring methods (3), separation within wiring methods (1), and separation within enclosures (2). Note that these separation requirements are significantly more severe than even the requirements for Class 2 control circuits in 725.136.

Intrinsically safe wiring in the form of stand-alone wiring methods, including cables or raceways clipped to a building surface, has its own separation requirements. As covered in (A)(3), this wiring must run at least 50 mm (2 in.) from all other wiring that is not intrinsically safe, even limited-energy control circuits. This rule does not apply, however, if the intrinsically safe wiring is in Type MC or Type MI cable, and it does not apply if the nonintrinsically safe wiring is in a raceway, or in copper (not stainless steel) Type MI cable, or in one of the types of MC cable where the metal armor is a qualified equipment grounding conductor as provided in 250.118. This includes most instances of smooth or corrugated Type MC cable, and it also would include the new type of interlocking-armor Type MC cable with the full-sized aluminum bonding conductor run in direct contact with the cable armor.

Intrinsically safe wiring must also, in accordance with (A)(1), be excluded from any raceway, cable tray, or cable assembly containing nonintrinsically safe wiring. There are four exceptions to this general rule, two of which involve nonincendive wiring. This wiring operates under comparable power constraints as intrinsically safe wiring, but it lacks some of the safeguards with respect to maintaining its protective qualities under some fault conditions. Review the discussion at the end of 500.7 for more information. Because the power levels are comparable, the two types of systems are allowed to run together provided they meet the separation rules in part (B).

The other two exceptions address intrinsically safe wiring routed with nonintrinsically safe wiring of any description, including Class 2 control circuits that are not qualified as intrinsically safe. Exception No. 2 allows these systems to run together provided all of the conductors in at least one of the systems is in a metal-armored cable assembly that meets the equipment grounding qualifications in 250.118 as described above for systems run together in the open. Exception No. 1 covers primarily cable tray and some surface raceways and allows a common location if the intrinsically safe wiring is at least 50 mm (2 in.) away from other systems and secured in that location, such as by cable ties to the rungs of a ladder-type cable tray. The other option requires a physical barrier that compartmentalizes the cable tray or surface raceway; said barrier to be grounded, or approved if nonmetallic.

Part (A)(2) takes on the most challenging task, as in the case of Class 2 control work, of landing on common equipment or otherwise running in the same enclosure along with nonintrinsically safe wiring without compromising safety. The conductors must be tied down so that if one comes loose from a terminal it will be unlikely to come in contact with any other terminal. There is a list of four items at this point. Item (1) calls for the customary 50 mm (2 in.) spacing and items (2) and (3) call for barriers, either 20-gauge sheet metal or an approved insulating partition, which as of the 2014 **NEC** must extend to within 1/16 in. of the enclosure walls. In this context the third note at the end of the section suggests that the usual control panel wiring ducts are acceptable provided there is at least a 19 mm (¾ in.) separation maintained between systems. Item (4) recognizes the same metal-armored cable assemblies recognized in (A)(3) and previously discussed.

Part **(B)** covers separation between different intrinsically safe circuits. These circuits must be segregated through the use of a grounded metal shield, or through the use of 10 mil insulation unless another size is identified for this purpose. An additional basic requirement that applies to these connections generally states that clearances between terminations must not be less than 6 mm (¼ in.) unless the clearance is modified in the control drawing.

The 2014 **NEC** has added part **(C)**, which sets a minimum spacing between the uninsulated portions of field wiring where it has been stripped and connected to a terminal, and "grounded metal or other conducting parts." That distance must not be less than 1/8 in.

504.50. Grounding. The basic rule requires connection to the supply system equipment grounding conductor, but the note following reminds users that some control drawings specify an earth connection in addition to the equipment grounding conductor connection. This is particularly true with Zener diode barriers as shown in Fig. 504-1, which often want their ground plane to have a local earth reference. Associated apparatus and cable shielding must be grounded as required in the control drawing. If an earth connection is specified, it must be made as close as possible to the system, and to the nearest building steel or water pipe as set forth in 250.30(A)(4). The next choice of electrode would be a concrete-encased electrode or ground ring, and only if none of those were available could another one on the 250.52 list, such as rod, pipe, or plate electrodes, or other underground metal structures be used.

504.60. Bonding. Intrinsically safe apparatus must be bonded within the hazardous location in accordance with the relevant section of the applicable hazardous location article. If the circuits are wired using metal raceways, then the metal raceways must be bonded in accordance with the same rules. Somehow, this latter requirement applies even in nonhazardous locations, which begs the question of what exactly would be the "applicable" hazardous location article in such cases. However, Sec. 30 of the usual articles and Sec. 25 of the zone articles read pretty much the same, disallowing locknut bonding and asking for bonding jumpers back to the applicable main or system bonding jumper.

504.70. Sealing. This is a major change effective with the 2008 NEC. This section has always said that where seals were otherwise required for conventional wiring, they would be required for intrinsically safe circuit wiring to minimize the passage of gas or vapor or dust. The next sentence went on to say that the seal did not need to be explosionproof, and ended at that. This was usually taken to mean sealing putty. Now, using the identical wording as in 501.15(B)(2) for Class I Division 2 boundary seals, the seal must be "identified for the purpose of minimizing the passage of gases permitted under normal operating conditions and shall be accessible." As previously discussed in 501.15(B), there are no such seals on the market as of this writing, other than the usual explosionproof seal. If enforced as written, this means explosionproof seals.

This will have a major impact on wiring practice in many facilities, because it is customary to rewire these control circuits from time to time. Sealing putty stays flexible and can be removed easily. Explosionproof sealing compound cannot be removed; the entire fitting must be cut out of the pipe run. Then the conduits need to be rethreaded, which is at best inconvenient and time consuming, and at worst borders on the impossible and requires significant repiping. Take care to anticipate future needs as much as possible when wiring these systems the first time.

504.80. Identification. Intrinsically safe wiring must be prominently and permanently identified as such throughout the premises wiring systems wherever it is run. All terminals and junction points must be so identified. All wiring methods used, whether cable or raceway, for intrinsically safe wiring must be labeled as such in the field, at intervals not exceeding 7.5 m (25 ft) and at least once between all partitions, and so arranged that the IS circuits can be easily traced throughout the facility. Underground circuits get identified where they come out of the ground and are accessible. If the color light blue is not in use for other purposes, it is allowed to be an intrinsic safety color code, valid for raceways and cable trays and boxes, in addition to individual conductors.

ARTICLE 505. ZONE 0, 1, AND 2 LOCATIONS

This Code article is intended as an alternate method for classifying Class I hazardous locations. Under the traditional NEC rules, hazardous (classified) locations are divided into divisions depending on how frequent the hazardous atmosphere or dusts are expected to be present. If the hazardous material is merely in secure storage, that storage area gets a less severe classification than an area where people routinely work with it. Under the traditional rules the former location is Division 2 and the latter is Division 1. This article is now designed as a stand-alone article.

All requirements regarding seals and other provisions that at one time were incorporated by reference from Art. 501 have been duplicated here, so the article stands on its own with regard to technical requirements for the hazardous environment. Of course the Chap. 3 rules on conduit support, etc., continue to apply.

The zone concept goes one step further and divides the traditional Division 1 locations into Zone 1 and Zone 0, the latter classification reserved for areas such as inside a vented tank where the hazardous agent exists on a routine basis. They then assign more stringent requirements to the Zone 0 locations. Areas like Division 2 in the traditional approach have a Zone 2 classification. Places throughout the world that use this approach consistently also tend to have a work environment that is much more intensively engineered and supervised than areas using the Class and Division system. Having made these distinctions, and thereby having excluded the worst case (Zone 0) from Zone 1, and generally operating in a context of more comprehensive engineering support, the **NEC** allows products into Zone 1 locations that traditional Code practices would exclude from Division 1, including nonexplosionproof lighting.

At the time this came into the **NEC** (1996) it was bitterly contested. Many thought it would rapidly supersede the Class and Division approach. This has not happened, in part because the economic benefits of this system chiefly involve a cost comparison between Division 1 and Zone 1, with the Zone system the clear winner. However, Division 1 applications have steadily declined in extent, in comparison with Division 2 that has expanded to the same degree. The reason, in large part, is environmental regulation. Where large facilities would once classify (justifiably so based on the chemical releases) huge spaces as Division 1, they now have redesigned the equipment so all those flammable organic compounds are tightly contained, and the Division 1 area now is an infinitesimal proportion of what was there before. They are doing this to save on the costs of chemicals (much of which are derived from oil), and because if they ever continued to release the enormous quantities of chemicals of yesteryear, the government would fine them out of business. Since there is really no difference between Division 2 and Zone 2, the economic reason to abandon the traditional system largely disappeared. There are some major industrial players using the zone system, but the traditional system continues to do well. Another factor is that Zone 0 does not permit power connections, only control wiring of extremely limited energy, but the Division system does permit power connections in those areas.

505.2. Definitions. As in the case of 500.2, most of the defined terms here are the protection techniques that can be applied, as covered elsewhere, to address the hazardous environment. A good example is "Increased Safety 'e'" This is the technique frequently employed with luminaires in which they meet a standard of performance appropriate to Zone 1, and quite inappropriate to Division 1, which includes environments where the flammable agents are present on a routine basis.

505.5. Classification of Locations. This is where the differences between Zone 0, 1, and 2 are formally sorted out. Because the amount of time that a flammable agent is present has a major bearing on whether a location is Zone 0 or 1, there has been extensive discussion on this point in other publications. The current informal consensus on a rule of thumb is that if the hazardous gas or vapor is present for more than 1000 h per year (just over 10 percent of the time), it is Zone 0; and Zone 1 applies from there down to 10 h per year, with Zone 2 any time less than 10 h per year.

505.6. Material Groups. Article 505 uses different designations and groupings for the various hazardous gases than does Art. 500. 505.6 gives the group designation and spells out which gases are included in that group.

The familiar method given in Art. 500 has four gas groups, A, B, C, and D. But there are only three designations for gas groups in Art. 505: IIC, IIB, and IIA. Basically stated, Art. 500 groups can be equated to Art. 505 groups as follows:

Art. 500	Art. 505
Groups A and B	IIC
Group C	IIB
Group D	IIA

505.7. Special Precaution. This section covers three very different topics. Part **(A)** covers the implementation of the Zone system, and requires qualified persons to do the work, including the installation and field inspections involved. The former requirement for a licensed professional engineer was not continued in the 2008 NEC.

Part **(B)** covers dual classifications. In general the Zone system cannot be used on an opportunistic basis, with the zone concept applied to one machine as convenient, and the class and division method to an adjacent wet-well. Fig. 505-1 shows the principles at work. The two systems cannot ever overlap, and only Zone 2 may abut, but only Division 2 and never Division 1. As shown in Fig. 505-2, part **(C)** covers reclassifications, which are permitted, provided all of any given space classified because of a given gas or vapor becomes reclassified as part of this process.

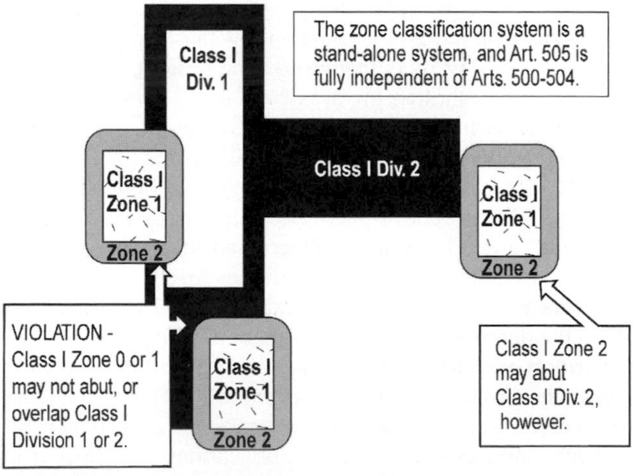
Class I
Div. 1
The zone classification system is a stand-alone system, and Art. 505 is fully independent of Arts. 500-504.
Class I Div. 2
Class I
Zone 1
Zone 2
Class I
Zone 1
Zone 2
VIOLATION - Class I Zone 0 or 1 may not abut, or overlap Class I Division 1 or 2.
Class I
Zone 1
Zone 2
Class I Zone 2 may abut Class I Div. 2, however.

Fig. 505-1. The Zone classification system is not a mix and match option with an existing class and division system. [Sec. 505.7(B).]

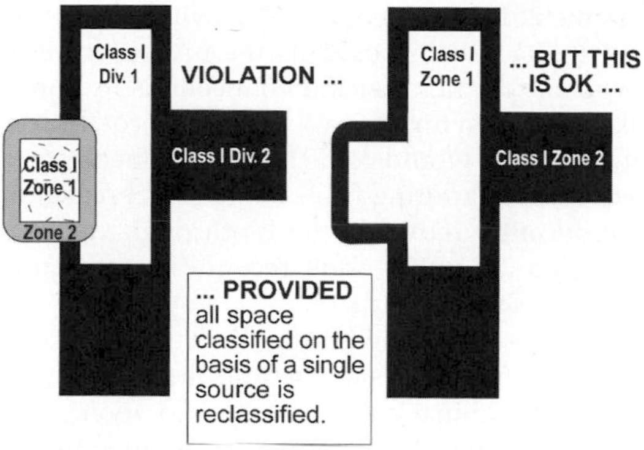
Class I
Div. 1
VIOLATION ...
Class I
Zone 1
Zone 2
Class I Div. 2
... PROVIDED
all space
classified on the
basis of a single
source is
reclassified.
Class I
Zone 1
... BUT THIS
IS OK ...
Class I Zone 2

Fig. 505-2. An existing facility can be reclassified under the Zone system. [Sec. 505.7(C).]

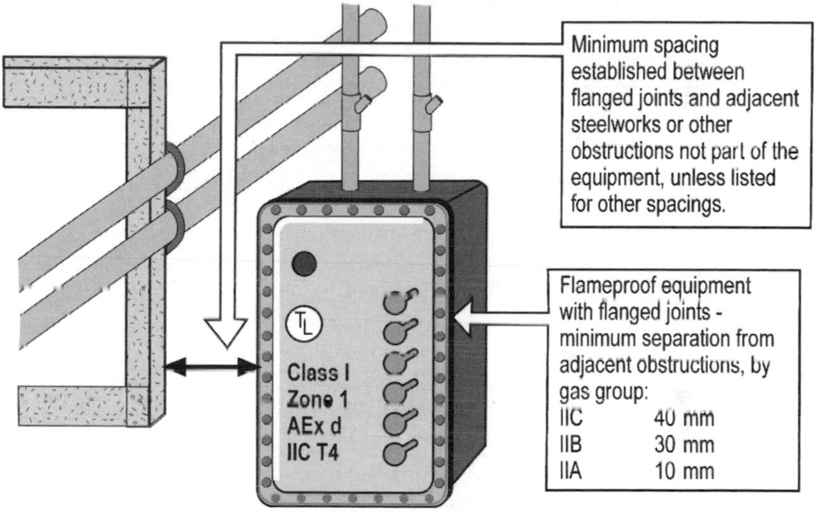
Minimum spacing
established between
flanged joints and adjacent
steelworks or other
obstructions not part of the
equipment, unless listed
for other spacings.
Class I
Zone 1
AEx d
IIC T4
Flameproof equipment
with flanged joints -
minimum separation from
adjacent obstructions, by
gas group:
IIC 40 mm
IIB 30 mm
IIA 10 mm

Fig. 505-3. Flameproof enclosures must observe clearance limits between the flanged joints and adjacent obstructions. [Sec. 505.7(D).]

Part **(D)** covers a rule that has no counterpart in the class and division coverage in the NEC. As shown in Fig. 505-3, flameproof equipment (the zone counterpart to explosionproof) has a minimum spacing between the flange openings and adjacent solid obstacles not a part of the wire equipment, including steel columns and strut, adjacent walls, etc. The NEC spacing applies unless the equipment is listed for a different interval.

Part **(E)** covers the simultaneous presence of both dust and gas or vapor. Both factors must enter the selection process for equipment and choice of wiring method. The operating temperature is very important because it is often lower for dusts (or fibers) than it is for gases and vapors.

Part **(F)**, new in the 2014 **NEC**, requires that where the commonly employed "increased safety" ("e") is being used as the protection technique, the short-circuit current available for field-wiring connections in Zone 1 locations must not exceed 10,000 A (RMS symmetrical). This limit correlates with the default limit given in UL 508A that would come into play when constructing an industrial control panel. Current limiting fuses and circuit breakers are available that are capable of significantly reducing the let-through currents to downstream connection points. The 2017 **NEC** adds recognition of equipment listed and marked for higher available fault current.

505.9. Equipment. As is the case with the division classification system given in Art. 500, the zone classification system permits the use of equipment suited for a more hazardous location within a lesser hazardous location. Any equipment listed for Zone 0 may be used in Zone 1 or 2 locations provided the equipment is listed for the same gas group. And any equipment listed for Zone 1 may be used in Zone 2 locations of the same gas group.

Part **(C)(2)** covers the extensive marking designations for zone-classified equipment in a series of five informational notes and a protection designation table. This material includes the area classification, the American standard designation (AEx), the protection technique, the gas or vapor classification involved from 505.6, and the temperature classification from (D) below. Since some associated apparatus to intrinsically safe systems is only appropriate where applied outside the hazardous area, but still requires a rating, it is shown in brackets, with an example that coordinates with the table. Some equipment includes a barrier that allows it to connect to Zone 0 located equipment but be installed in a lower-hazard Zone 1 environment; the marking in this case would be 0/1 with the slash mark indicating the barrier. Note 4 explains "equipment protection level" (or "EPL") markings. These are somewhat confusing, since no **NEC** rule takes note of them. However, the note does allow users to sort out the required **NEC** markings from these performance notations.

Table 505.9(D)(1) in the Code gives the temperatures that correspond to the "T" designation assigned to a piece of equipment. Note that because of a new 505.9(C)(2) Exception No. 3, as of the 2014 **NEC**, cable termination fittings do not require a marked operating temperature or temperature class. The major difference between this table and Table 500.8(C), which applies to the division system of classification, is the number of designations. The table for the zone system has only six different temperature ratings: T1 through T6, which range from 85°C (185°F) to 450°C (778°F). The division system temperature table has 14, which also range from 85°C (185°F) to 450°C (778°F), but with eight other intermediate values between T1 and T6.

505.15. Wiring Methods. This section identifies those wiring methods that may be used in the different zones. As covered in part **(A)**, for Zone 0, only intrinsically safe circuits may be installed in Zone 0 locations. The intrinsically safe circuits are as covered in Art. 504, and since the wiring is intrinsically safe, no specification as to wiring method need be made here.

Additionally, fiber-optic cable of the nonconductive type, which, as defined in 770.2, contains no metallic components, may be installed in Zone 0 locations. This system excludes any other equipment or wiring from being installed within the Zone 0 classified area, including power circuits of any type. Only the

Class and Division system accommodates power wiring in these environments, as part of Division 1.

The rule of part **(B)** indicates the acceptable wiring methods for Zone 1 locations. Intrinsically safe wiring would be permitted for obvious reasons. Also, any of those wiring methods permitted in Class I, Division 1 locations—501.10(A)—would also be suitable for use within a Zone 1 location.

The 2014 **NEC** has added Type TC-ER-HL cable to the list, but with a physical size restriction to cable diameters of 1 in. or less, and including the limitations of 336.10(7) which presumably focus on the mechanical protections envisioned in that article. This wiring method is permitted in 501.10(A), but with interesting modifications. Under the class and division system it appears without a physical size restriction and without the express reference to 336.10(7). However, it is not included for general wiring; it is only included in 501.10(A)(2) as a permitted wiring method for use at flexible connections. The zone system, having included this as general wiring, only mentions flexible cord and the usual flexible fittings for both Division 1 and Zone 1 in its version of these requirements, at 505.15(B)(2).

Part **(C)** covers Zone 2 wiring. In this case the rules correlate with 501.10(B) for Division 2 almost exactly. The zone rule for MV cable includes an additional requirement for shielding or armor.

It should be noted that all sealing requirements spelled out in **505.16** must be satisfied where the zone system of classification is applied. Those requirements line up almost exactly with the rules in Art. 501 but not in all cases. Figure 505-4 shows one interesting difference where an increased safety enclosure

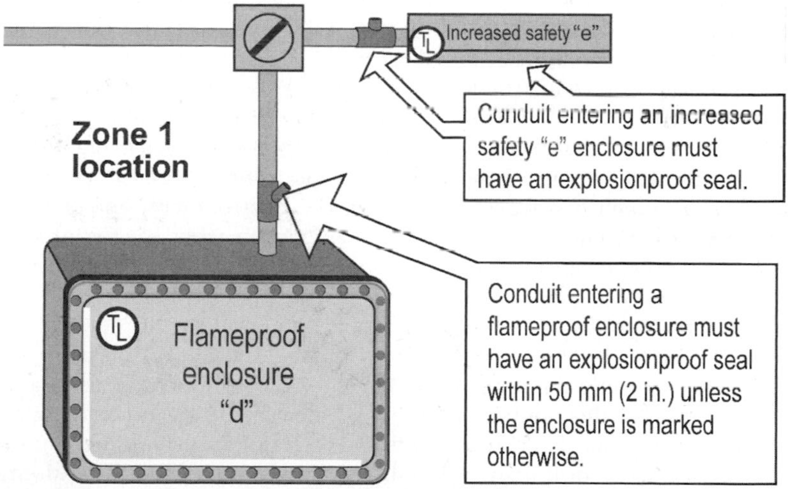
Increased safety "e"
Conduit entering an increased safety "e" enclosure must have an explosionproof seal.
Zone 1 location
Flameproof enclosure "d"
Conduit entering a flameproof enclosure must have an explosionproof seal within 50 mm (2 in.) unless the enclosure is marked otherwise.

Fig. 505-4. The seal at the flameproof enclosure completes the enclosure and must be within 50 mm (2 in.). The seal at the increased safety enclosure is only to protect that enclosure from the conduit system, and if only conventional, tapered-thread, heavy-wall conduit joints intervene, with both male and female threads tapered, need not be within 50 mm (2 in.). [505.16(B)(1).]

in a Zone 1 application gets a seal not because it is an ignition source, but because an ignition elsewhere must be arrested before it reaches that enclosure. Part of that consideration is what type of threads is used between the seal and the enclosure. The waiver from the 50 mm (2 in.) spacing limit is conditioned on only NPT male to NPT female, fully tapered joints on both sides. A typical conduit connection of NPT threads on the pipe to the typical straight threads in a coupling does not comply with the exception and reverts to the 50 mm (2 in.) limit. The reason for this limitation is that some increased safety equipment is very fussy about environmental contaminants getting into it, and it is believed a plumbing style limit on threaded joints is preferable.

505.17. Flexible Cables, Cords and Connections. This section lines up well with 501.140, combining both parts (A) and (B) in one grouping. As of the 2014 NEC it now contains a second part (B) that imports the content of 501.105(B)(6) for process instrumentation connections and makes it applicable to Zone 2 locations.

505.20. Equipment Requirements. The equipment permitted within any of the hazardous locations classified according to the zone system described in Art. 505 must comply with 505.20.

As required by 505.20(A), in Zone 0 locations, equipment specifically *listed* for use in Zone 0 locations, may be used. Equipment listed as intrinsically safe in accordance with Art. 504 with the evaluation having been done with respect to the same gas and suitable temperature class is permitted.

According to the basic rule in 505.20(B), any equipment that is listed for use in Zone 1 locations, or, as covered in the exception, any equipment approved for use in Division 1 or Zone 0 locations within the same gas group and with "suitable temperature class" may be used. Notice that the temperature marking does not have to correspond directly to the maximum permitted. It simply must be "suitable." This wording recognizes that the temperature designations in 500.8(C) are more finely graduated than those in 505.9(D)(1) and therefore there will not be 1:1 correspondence between markings.

Part **(D)** is new in the 2014 NEC although the concept is not. In addition to the normal hierarchy of groups, with IIC being more severe than IIB, which is in turn more severe than IIA, the rule allows for a specific mention of an outlying gas. The classic example, used in the informational note following, is hydrogen, which is a IIC gas. This is a IIC environment, and the IIC category includes extremely hazardous gases including acetylene for which there is very little to choose from in terms of equipment. On the other hand, due to the increased prevalence of hydrogen in connection with fuel cells, batteries, and other applications, there is a definite market for IIB equipment that although part of the next level down, it has been additionally adapted and tested to the specific demands of a hydrogen environment. This evaluation process allows for this, and thus, "IIB $+H_2$." Be careful: the underlying hierarchy does not necessarily correspond to the traditional system; there are many examples of Class I equipment that will not comply with Class II requirements, for example.

505.22. Increased Safety "e" Motors and Generators. This section covers the connections to motors in Zone 1 locations. The provisions here make a number of modifications of allowances permitted in Art. 430 (Fig. 505-5).

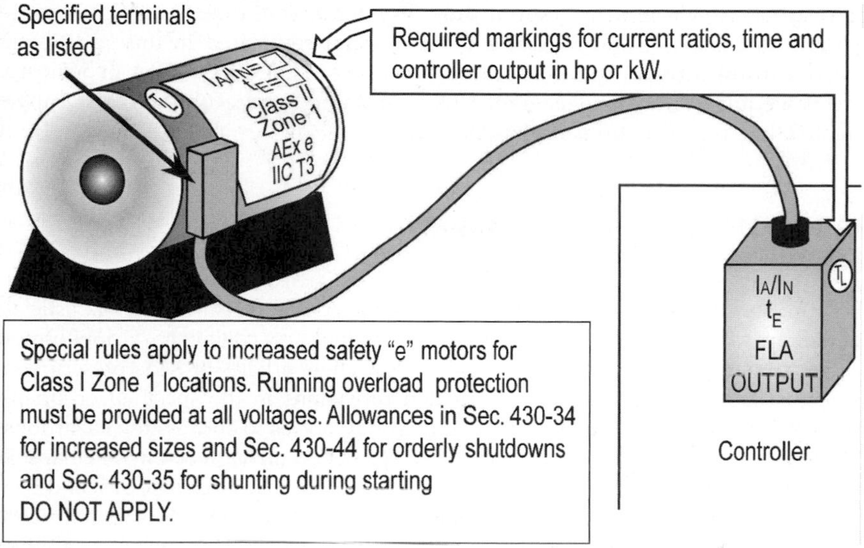
Specified terminals as listed
Required markings for current ratios, time and controller output in hp or kW.
Class II
Zone 1
AEx e
IIC T3
IA/IN
tE
FLA
OUTPUT
Controller
Special rules apply to increased safety "e" motors for Class I Zone 1 locations. Running overload protection must be provided at all voltages. Allowances in Sec. 430-34 for increased sizes and Sec. 430-44 for orderly shutdowns and Sec. 430-35 for shunting during starting DO NOT APPLY.

Fig. 505-5. Increased safety motors require disallowance of some permissions normally granted in motor circuits. (Sec. 505.22.)

505.26. Process Sealing. This repeats the new comprehensive treatment of this subject in 501.17. It must be repeated here because Art. 505 is a stand-alone article. Refer to the discussion at 501.17 for complete information on this topic.

ARTICLE 506. ZONE 20, 21, 22 LOCATIONS FOR COMBUSTIBLE DUSTS OR IGNITIBLE FIBERS/FLYINGS

506.1. Scope. This article extends the zone classification system to traditional Art. 502 coverage locations, with Art. 503 locations tagged on. The requirements line up with Art. 502 in a logical way, except that here again the Division 1 areas are divided into a zone (Zone 20) where the combustible agent is continuous for long periods, and a zone where the threat is routine (Zone 21) but not continuous. And as in Zone 0, the only wiring permitted in Zone 20 is intrinsically safe wiring, essentially for instrumentation. Zone 22 rounds out the area classifications, and lines up with Division 2. This article is in very limited usage in areas subject to **NEC** jurisdiction, if at all, and is at present primarily an academic exercise.

The 2014 **NEC** has added crucial information to round out the coverage of this article in a new Sec. 506.6. This presents the major material groupings for dusts, which agree with those in Arts. 502 and 503, but in a different order. Combustible metal dusts come first, in 506.6(A) as Group IIIC, and correspond to Group E in the traditional system. The next grouping includes both Group G dusts, such as flour, and Group F dusts (carbonaceous dusts that present an explosion hazard) in 506.6(B) as Group IIIB. The last is 506.6(C), which covers solid particles larger than the new cut point of 500 μm for dust. This category, Group IIIA, corresponds to Class III in the traditional system. Note that 506.20(D) is comparable to 505.20(D) and creates a hierarchy not replicated in the traditional system. In this article, IIIC equipment can be used for IIIB or IIIA exposures, and IIIB equipment can be used for IIIA exposures. Be careful not to generalize this to Class II applications under Art. 502.

ARTICLE 510. HAZARDOUS (CLASSIFIED) LOCATIONS—SPECIFIC

This unusual two-section article sets the organizational context for the six articles that follow, which all cover areas and occupancies that may contain hazardous atmospheric quantities of flammable gases or vapors. These articles provide prescriptive area classification requirements and related provisions in the specified occupancies; the procedures for actually wiring those occupancies to address the hazardous (classified) areas as described are included in Arts. 500 through 504, with Chaps. 1 through 4 as the default requirements as always. Note that Arts. 513, 515, and 516 have been harmonized with the zone system, and the reference to the earlier articles may need to be updated to include Arts. 505 and 506. The article concludes with the recognition that the AHJ may need to referee the application of some rules under unusual conditions in the specified occupancies.

ARTICLE 511. COMMERCIAL GARAGES, REPAIR AND STORAGE

511.1. Scope. This article covers occupancies engaged in service and repair operations for self-propelled vehicles fueled by flammable liquids or gases. Note that motorboats are self-propelled vehicles meeting this description, and 555.22 requires the service and repair facilities for such motor craft comply with this article.

511.2 Definitions. This article relies on extracted material from NFPA 30A for many of its requirements. Because that other standard organizes its requirements in this area around the distinction between a major and a minor repair garage, it made sense to extract those definitions and then organize the area classifications in the same way. One of the key distinctions between the two definitions is whether or not vehicle fuel tanks would be removed as part of anticipated repair operations, and that provides a useful and quick point of reference to decide which is which. The minor repair operations, typified by the quick-lube facilities, will not be doing that kind of work.

NFPA 30A includes a variety of important rules that are not within the scope of the NEC. For example, operations involving open flame or electric arcs, including fusion, gas, and electric welding, should be restricted to areas specifically provided for such purposes. Approved suspended unit heaters may be used, provided they are located not less than 8 ft (2.5 m) above the floor and are installed in accordance with the conditions of their approval.

The question often arises, "Does diesel fuel come within the classification of volatile flammable liquids, thereby requiring application of Art. 511 to places used exclusively for repair of diesel-powered vehicles?" 514.3(A) reads: "Where the authority having jurisdiction can satisfactorily determine that flammable liquids having a flash point below 38°C (100°F), such as gasoline, will not be handled, such location shall not be required to be classified."

The NFPA Inspection Manual, under identification of flammable liquids, says, "Minimum flash points for fuel oils of various grades are: No. 1 and No. 2, 100°F; No. 4, 110; No. 5, over 130; No. 6, 150 or higher. Actual flash points are commonly higher and are required to be higher by some state laws. No. 1 fuel is often sold as kerosene, range oil or coal oil."

Diesel fuel is a Class III flammable liquid, having flash points above 70°F (21°C). One listing of flash points of flammable liquids showed no diesel fuel below 120°F (49°C). Therefore, a diesel fuel installation may be classified as a nonhazardous area and wired as such, unless it can be firmly established that the particular fuel has a flash point under 100°F (38°C). But, of course, the authority enforcing the Code is the one responsible for classifying such areas as nonhazardous.

511.3. Classification of Locations. In part **(A)** the NEC clearly establishes that parking garages used for parking or storage need not be classified. And in part **(B)** the NEC brings Art. 514 into the picture in addition to Art. 511 when fuel dispensing takes place in a service bay. The remainder of the section is a very easily applied set of two tables that implement a three-part decision tree that, as text prior to the 2017 NEC, specifies the applicable area classifications. There are three questions to consider about any Art. 511 location: (1) Is it a major or minor repair garage; (2) Is it part of the floor, or the ceiling, or a pit in the floor; and (3) is it ventilated or not in terms of minimum NEC requirements. Figure 511-1 has the entire section diagrammed in an easy-to-apply chart.

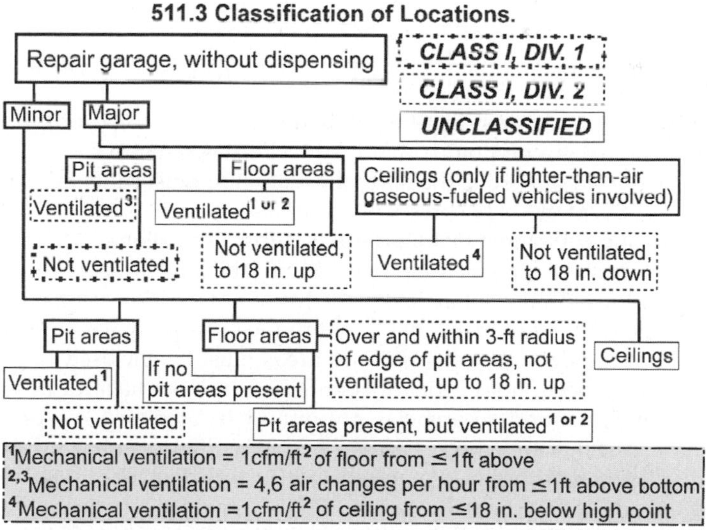
511.3 Classification of Locations.
Repair garage, without dispensing
CLASS I, DIV. 1
CLASS I, DIV. 2
UNCLASSIFIED
Minor
Major
Pit areas
Floor areas
Ceilings (only if lighter-than-air gaseous-fueled vehicles involved)
Ventilated3
Ventilated1 or 2
Not ventilated
Not ventilated, to 18 in. up
Ventilated4
Not ventilated, to 18 in. down
Pit areas
Floor areas
Over and within 3-ft radius of edge of pit areas, not ventilated, up to 18 in. up
Ceilings
Ventilated1
If no pit areas present
Not ventilated
Pit areas present, but ventilated1 or 2
1Mechanical ventilation = 1cfm/ft2 of floor from ≤ 1ft above
2,3Mechanical ventilation = 4,6 air changes per hour from ≤1ft above bottom
4Mechanical ventilation = 1cfm/ft2 of ceiling from ≤18 in. below high point

Fig. 511-1. Area classifications now follow a table, which implements the three-fork decision tree previously used here. Fig. 511-1 is a graphic representation of the same requirements. (Sec. 511.3.)

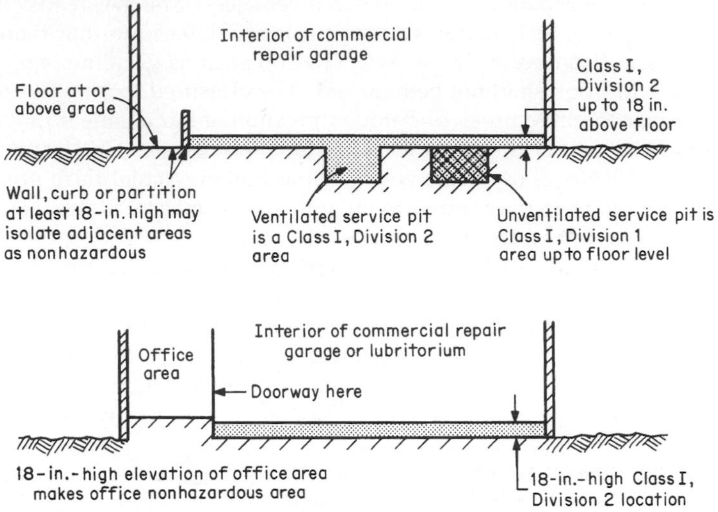
Interior of commercial repair garage
Floor at or above grade
Class I, Division 2 up to 18 in. above floor
Wall, curb or partition at least 18-in. high may isolate adjacent areas as nonhazardous
Ventilated service pit is a Class I, Division 2 area
Unventilated service pit is Class I, Division 1 area up to floor level
Office area
Interior of commercial repair garage or lubritorium
Doorway here
18-in.-high elevation of office area makes office nonhazardous area
18-in.-high Class I, Division 2 location

Fig. 511-2. Hazardous areas must be carefully established. (Sec. 511.3.)

Note that the areas below grade or in the floor slab that are below classified areas are not classified, and therefore an unbroken conduit running straight through the 450 mm (18-in.) zone and into the slab need not be sealed, either at its entry point or at its point of emergence if the conduit there is also unbroken. At the end of the section there are answers to two potential modifications to the area classifications. The first is that alcohol-based windshield washer fluid does not by itself generate an area classification. The second is that adjacent areas that are positively pressurized or independently ventilated or cut off by building construction do not need to be classified. This is covered in Fig. 511-2.

511.4 Wiring and Equipment in Class I Locations. Note (B)(2) of this section, which requires that if a luminaire on a cord reel can reach into Class I locations (usually 450 mm (18 in.) above the floor, the luminaire must be identified as suitable for Class I Division 1 exposures.

511.7. Wiring and Equipment Installed Above Class I Locations. The rules here apply to the lubritorium areas in service stations and *any* other space above the defined hazardous locations. Part **(A)** is the list of acceptable wiring methods.

Part **(B)** notes that equipment that may produce arcs or sparks and is within 12 ft (3.7 m) of the floor above hazardous areas must be enclosed or provided with guards to prevent hot particles from falling into the hazardous area, but lamps, lampholders, and receptacles are excluded. Standard receptacles are okay. Luminaires that are within 12 ft (3.7 m) of the floor over hazardous areas, over traffic lanes, or otherwise exposed to physical damage, must be totally enclosed, as required in part **(B)**.

511.8 Underground Wiring. This section brings over from Art. 514 the wiring method rules for underground wiring, including the allowance for nonmetallic raceways except over the last 2 feet that has to be steel, arranged with equipment grounding return continuity. It does not bring over the area classification rules. It does not, therefore, classify the underground area and require seals at a point of emergence, unless of course, it runs from a classified area within the service area and therefore would require a boundary seal.

511.10. Special Equipment. In part **(B)**, the requirements for battery-charging cables and connectors are similar to the requirements for outlets for the connection of portable appliances, except that when hanging free the battery-charging cables and connectors may hang within 6 in. (150 mm) from the floor. The common form is a plug that is inserted into a receptacle on the vehicle, and, because the prongs are "alive," they must be covered by a protecting hood.

511.12. Ground-Fault Circuit-Interrupter Protection for Personnel. This rule is shown in Fig. 511-3.

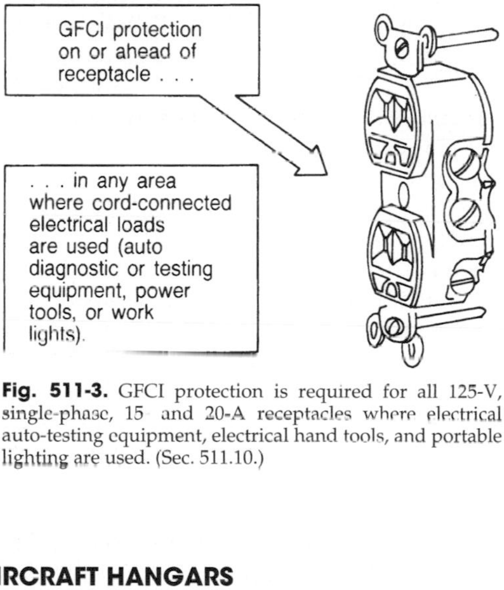
GFCI protection
on or ahead of
receptacle . . .
. . . in any area
where cord-connected
electrical loads
are used (auto
diagnostic or testing
equipment, power
tools, or work
lights).

Fig. 511-3. GFCI protection is required for all 125-V, single-phase, 15- and 20-A receptacles where electrical auto-testing equipment, electrical hand tools, and portable lighting are used. (Sec. 511.10.)

ARTICLE 513. AIRCRAFT HANGARS

513.3. Classification of Locations. Figure 513-1 shows the details of hangar classifications. The entire floor area up to 18 in. (450 mm) above the floor, and adjacent areas not suitably cut off from the hazardous area or not elevated at least 18 in. (450 mm) above it, are classified as Class I, Division 2 locations. Pits below the hangar floor are classified as Class I, Division 1. Within 5 ft (1.5 m) horizontally from aircraft power plants, fuel tanks, or structures containing fuel, the Class I, Division 2 location extends to a level that is 5 ft (1.5 m) above the upper surface of wings and engine enclosures.

There is a new part of this section as of the 2008 NEC to address the needs of hangars where aircraft will be painted. Note that the classified envelope in for these hangars is much larger than the conventional hangar, because the dimensions are both larger and they apply to the entire plane and not just the outer

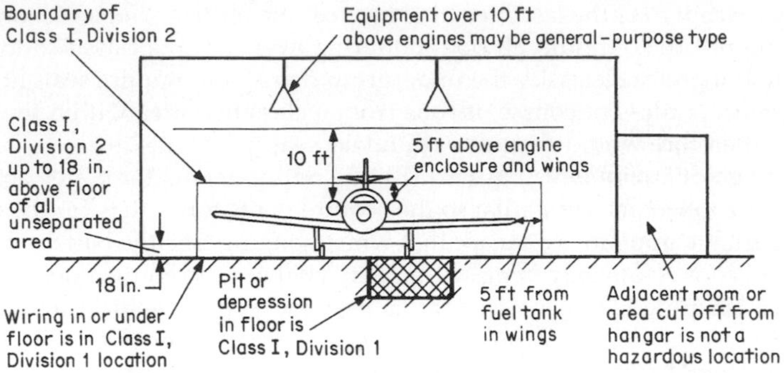
Boundary of
Class I, Division 2
Equipment over 10 ft
above engines may be general-purpose type
Class I,
Division 2
up to 18 in.
above floor
of all
unseparated
area
10 ft
5 ft above engine
enclosure and wings
18 in.
Wiring in or under
floor is in Class I,
Division 1 location
Pit or
depression
in floor is
Class I, Division 1
5 ft from
fuel tank
in wings
Adjacent room or
area cut off from
hangar is not a
hazardous location

Fig. 513-1. Boundaries of hazardous areas are clearly defined. (Sec. 513.3.)

margin of the engines, and in addition there is a very large part of the overall envelope that is Class I Division 1. In effect, this area classification melds the traditional aircraft rule with the open paint spray area rules in Art. 516, and this makes sense because that is exactly what these hangars are.

513.7. Wiring and Equipment Not Installed in Class I Locations. Luminaires and other equipment that produce arcs or sparks may not be general-use types but are required to be totally enclosed or constructed to prevent escape of sparks or hot metal particles if less than 10 ft (3.0 m) above aircraft wings and engine enclosures, as indicated in Fig. 513-1.

513.8 Underground Wiring. This section is different for the comparable ones in these articles in that a raceway running below a classified location is considered to be in that location, and therefore seals would be required upon emergence in an unclassified area.

ARTICLE 514. MOTOR FUEL DISPENSING FACILITIES

514.2. Definition. As noted under 511.2, there is a question about application of the rules of Arts. 511 and 514 to areas used for service of vehicles using diesel fuel and to dispensing pumps and areas for diesel fuel. Fuel with a flash point above 100°F (38°C) may be ruled to be *not* "a volatile flammable liquid," to which the regulations of Arts. 511 and 514 are addressed.

Note that vehicle repair rooms or areas and lubritoriums at gas stations must comply with Art. 511.

514.3. Classification of Locations. As noted, Table 514.3(B)(1) delineates and classifies the various areas at dispensing pumps and service stations. This table, which now includes zone classifications, brings the Code rules into agreement with data given in NFPA 30A, *Code for Motor Fuel Dispensing Facilities and Repair Garages*.

Figure 514-3 gives a visual representation of the dimensions of Class I locations, both Division 1 and Division 2, as described in the text of Table 514.3. In the wording of Table 514.3, the space within the dispensing-pump enclosure is a Class I, Division 1 location as described in ANSI/UL 87, "Power Operated Dispensing Devices for Petroleum Products."

Around the outside of a dispenser pump housing, the Division 2 location extends 18 in. (450 mm) horizontally in all directions, from grade up to the height of the pump enclosure or up to the height of "that portion of the dispenser enclosure containing liquid-handling components" (Fig. 514-1).

In the table, there is no direct statement that the ground under the Division 1 and Division 2 locations at a dispenser island is a Division 1 location. In fact, these areas are no longer classified because there isn't enough air to support combustion. The pits under the dispensers etc., are classified, however, and the underground raceway still gets sealed at its point of emergence, as covered in 514.8.

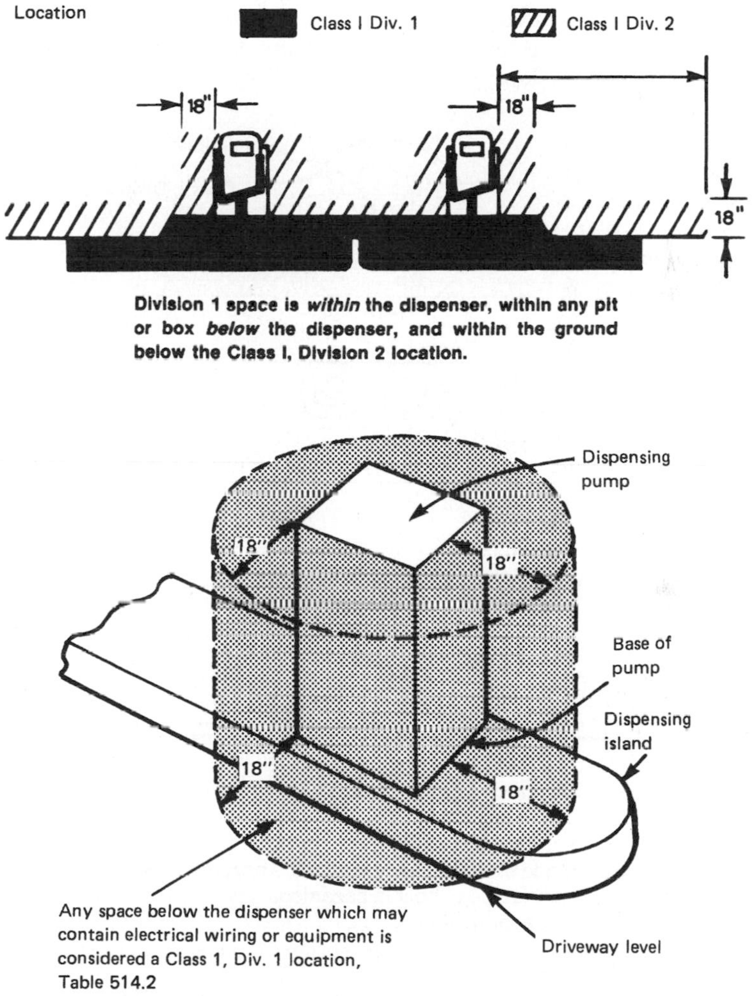
Location
Class I Div. 1
Class I Div. 2
18"
18"
18"
Division 1 space is *within* the dispenser, within any pit or box *below* the dispenser, and within the ground below the Class I, Division 2 location.
Dispensing pump
18"
18"
Base of pump
Dispensing island
18"
18"
Any space below the dispenser which may contain electrical wiring or equipment is considered a Class 1, Div. 1 location, Table 514.2
Driveway level

Fig. 514-1. The shaded space around the outside of the pump enclosures is a Cass I, Division 2 hazardous location. (Sec. 514.3.) The island structure and the ground beneath it is no longer classified, but the "entire space within and under dispenser pit or containment" is Class I Division 1 [Table 514.3(B)(1)].

Outdoor areas within 20 ft (6.0 m) of a pump are considered a Class I, Division 2 location and must be wired accordingly. Such a hazardous area extends 18 in. (450 mm) above grade. If a building with a below-grade basement is within this hazardous area, gasoline fumes could enter the building if there were any windows within the 18-in. high (450-mm high) classified space, thereby making the basement a Class I, Division 2 area. This condition could be eliminated by "suitably cutting off the building from the hazardous area" by installing an 18-in high (450-mm high) concrete curb between the service station and the residential property or enclosing the window openings up to that height.

Table 514.3(B)(1) sets a 20-ft diameter, 18-in. high (6.0-m diameter, 450-mm high), Division 2 area around each fill-pipe (this is the loose-fill dimension; the equivalent tight-fill connection is half of that) for the underground gasoline tanks at a gas station, as shown in Fig. 514-2.

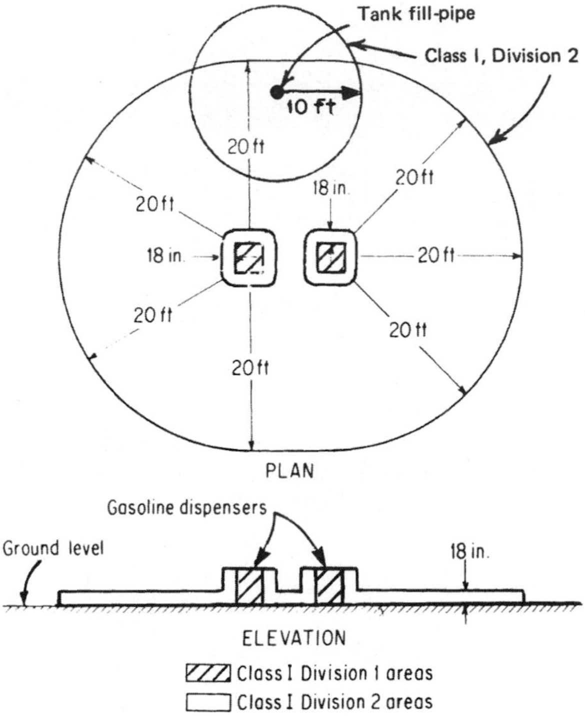
Tank fill-pipe
Class I, Division 2
10 ft
20 ft
18 in.
20 ft
20 ft
18 in.
20 ft
20 ft
20 ft
20 ft
20 ft
PLAN
Gasoline dispensers
Ground level
18 in.
ELEVATION
Class I Division 1 areas
Class I Division 2 areas

Note: Below-grade electrical wiring under the hazardous areas shown is not considered to be in a hazardous (classified) area, but sealing rules still apply at emergence from grade.

Fig. 514-2. Class I, Division 2 area extends around pumps and tank fill-pipes. (Sec. 514.3.)

As previously noted, any wiring or equipment that is installed beneath any part of a Class I, Division 1 or Division 2 location is no longer classified as being within a Class I, Division 1 location to the point where the wiring method is brought up out of the ground. However, there is still a sealing requirement that will apply in such cases, including if the conduit emerges under the panelboard, or even runs up to a lighting standard or a sign, as covered in Sec. 514.9.

As covered in Table 514.3(B)(1), under "Dispensing Device, Overhead Type," where the dispensing unit and/or its hose and nozzle are suspended from overhead, the space within the dispenser enclosure and "all electrical equipment integral with the dispensing hose or nozzle" are classified as Class I, Division 1 areas. The space extending 18 in. (450 mm) horizontally in all directions from the enclosure and extending down to grade level is classified as a Class I, Division 2 area. The horizontal area, for 18 in. (450 mm) above grade and extending 20 ft (6.0 m) measured from those points vertically below the outer edges of an overhead dispenser enclosure, is also classified as a Class I, Division 2 area. All equipment integral with an overhead dispensing hose or nozzle must be suitable for a Class I, Division 1 hazardous area.

Table 514.3(B)(1) notes that the hazardous space around any vent pipe for an underground tank at a gas station is simply a 3-ft radius (900-mm radius) *sphere* (a ball-like volume; Class I, Division 1) and a 5-ft radius (1.5-m radius) *sphere* (Class I, Division 2) around the top opening of a pipe that discharges upward. The space beyond the 5-ft (1.5-m) radius from tank vents that discharge upward and spaces beyond unpierced walls and areas below grade that lie beneath tank vents are not classified as hazardous. Although the area classifications are unchanged, note that the 2014 **NEC** has added a new drawing, Fig. 514-3(B), to illustrate the classification rules regarding aboveground fuel storage tanks and their dispensers.

Table 514-3(B)(1) designates a pit or depression below grade in a lubritorium as a Class I, Division 2 location if it is within an unventilated space, but does allow for the possibility of classification as unclassified, for a pit in a repair garage where ventilation exists according to the defined parameters. Refer to Sec. 511.3(C) and (D) on classification of this type of facility.

Sec. 514.3(B)(2) classifies dispensing operations for gaseous fuels. If a dispenser for lighter-than-air fuel is under a canopy, it must be designed to not entrap the gas, or else accept Division 2 classification. LPG operations are similar to gasoline dispensing, but such dispensers must be at least 1.5 m (5 ft) or 3.0 m (10 ft) away, depending on equipment design constraints in 514.3(B)(2). Natural gas classifications only focus on the dispenser.

The 2017 **NEC**, in 514.3(B)(3), added significant new rules for aboveground fuel storage tanks containing gaseous fuels, specifically CNG or LNG, hydrogen, or LP-gas. The first three limit the proximity of those tanks to any property line that "is or can be built upon, any public way, and the nearest important building on the same property." The specific dimensions are not given, but informational notes cite specific references within the applicable NFPA standards. The fourth rule creates separations (20 ft) between storage tanks containing fuels of different chemical composition, and also from dispensers using different fuels. An exception waives the separations mandated by the fourth rule if both the gaseous fuel storage and dispensers are at least 50 ft from all aboveground motor fuel storage and dispensing. The fifth and sixth rules needlessly duplicate the rules previously described in 514.3(B)(2) and need no further comment.

Parts **(C)**, **(D)**, and **(E)** represent the relocation in the 2014 **NEC** of all substantive content formerly within Secs. 555.21(A), (B)(1), and (B)(2) respectively to Art. 514, without substantive changes.

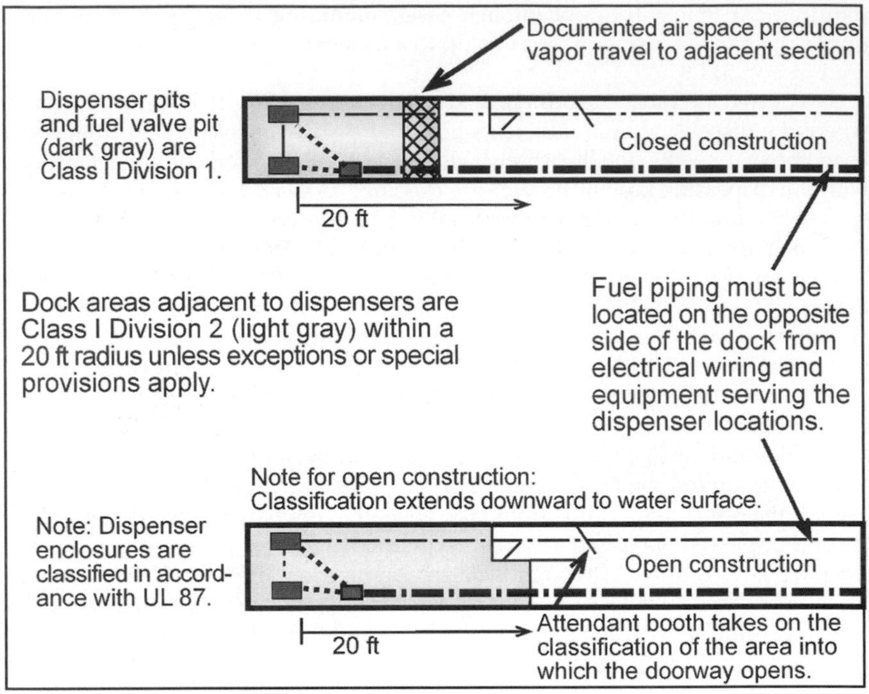
Documented air space precludes vapor travel to adjacent section
Dispenser pits and fuel valve pit (dark gray) are Class I Division 1.
Closed construction
20 ft
Dock areas adjacent to dispensers are Class I Division 2 (light gray) within a 20 ft radius unless exceptions or special provisions apply.
Fuel piping must be located on the opposite side of the dock from electrical wiring and equipment serving the dispenser locations.
Note for open construction:
Classification extends downward to water surface.
Note: Dispenser enclosures are classified in accordance with UL 87.
Open construction
20 ft
Attendant booth takes on the classification of the area into which the doorway opens.

Fig. 514-3. Gasoline pumping areas at a marina must utilize Class I wiring and equipment within the specific classified boundaries. (Sec. 514.3.)

Figure 514-3 shows the rules that define the need for hazardous location wiring at a marina. A fuel-dispensing area at the end of the two piers consists of gasoline dispensers at the pier edge, with a shack for service personnel mid pier. A panel installed in the shack supplies lighting and receptacles in the shack, as well as power for the fuel dispensers. Electrical connections from the dispenser pumps tie into the panel. Each shack takes advantage of provisions that exclude it from classification; the one on top uses an engineered air gap that precludes vapor travel (Sec. 514.3(D) Ex. 2) and the one at the bottom only has openings outside the classified location (Table 514.3(B)(1), "Sales, Storage, Rest Rooms"). Note the differences between open construction and solid construction that extends down to or below the water line.

514.8. Underground Wiring. Note that Exception No. 2 of this rule, which normally requires heavy-wall threaded steel conduits, permits Type PVC and Type RTRC nonmetallic conduit for circuits buried at least 2 ft (600 mm) deep in the earth as shown at the bottom of Fig. 514-4. The 2017 NEC added Type HDPE conduit to the list under the same rules as for other nonmetallic conduit materials.

Where these nonmetallic conduit types are buried at least 2 ft (600 mm) in the ground, as permitted for underground wiring at a gas station, a length of *threaded* rigid metal conduit or *threaded* IMC, at least 2 ft (600 mm) long, must be used at the end of the PVC or RTRC conduits where they turn up from the 2-ft (600-mm) burial depth. This clarifies that the entire length of these conduits must be down at least 2 ft (600 mm). See Fig. 514-5.

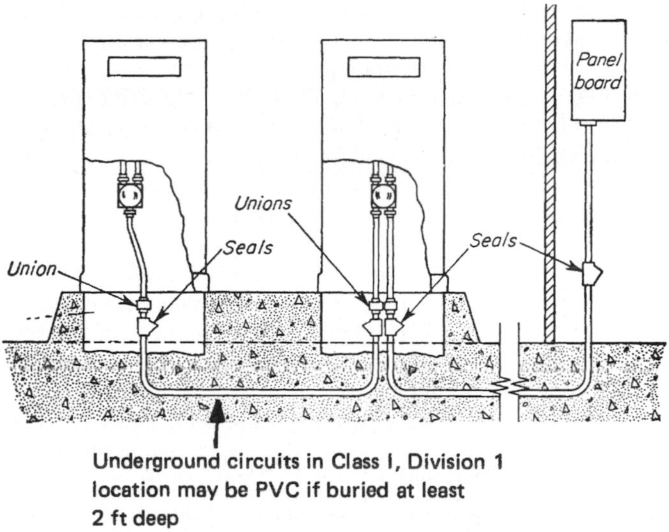
Panel board
Unions
Seals
Seals
Union
Underground circuits in Class I, Division 1 location may be PVC if buried at least 2 ft deep

Fig. 514-4. Conduit from pump island must be sealed at panelboard location. (Sec. 514.9.)

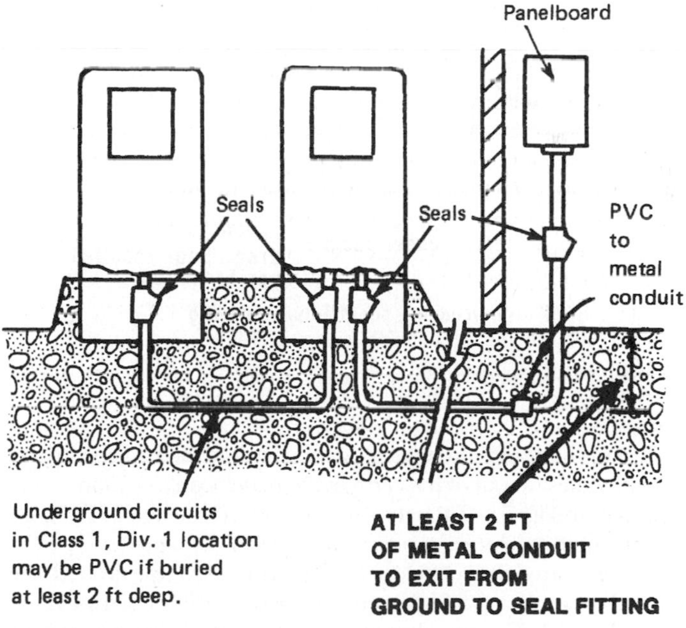
Panelboard
Seals
Seals
PVC to metal conduit
Underground circuits in Class 1, Div. 1 location may be PVC if buried at least 2 ft deep.
AT LEAST 2 FT OF METAL CONDUIT TO EXIT FROM GROUND TO SEAL FITTING

Fig. 514-5. The Type PVC or RTRC conduit permitted in the underground Class I, Division 2 location must never come above the required 2-ft burial depth. (Sec. 514.8.)

The phrase in this rule also requires the 2-ft (600-mm) length of metal conduit to be used on the end of the nonmetallic conduit where the conduit run does *not* turn up, but passes horizontally into the nonhazardous area of a basement. In that case or where the conduit turns up, a length of metal conduit is needed to provide for installation of a seal-off fitting because the conduit is, in effect, emerging from a Class I, Division 1 location below ground, even if we can't describe it that way officially; and a seal is required at the crossing of the boundary between the classified location and the nonhazardous location in, say, the office or other general area of a gas station.

514.9. Sealing. Every conduit connecting to a dispenser pump must have a seal in it, as shown in Fig. 514-6. Conduits connecting to gas pumps are commonly connected through an explosionproof junction box that is set in the pump island, as shown in Fig. 514-7. This box is approved as raintight and provided with integral sealing wells. All the conduits connecting to the box are sealed without need for separate individual sealing fittings. Additional individual seals are required where the conduits enter the pump cavity as shown. And, of course, a seal must be used in each conduit that leaves the hazardous area—such as in the conduit that feeds each lighting standard, with no fitting or coupling between the seal in the base of each standard and the boundary at the 18-in. (450-mm) height where the circuit crosses into nonhazardous areas.

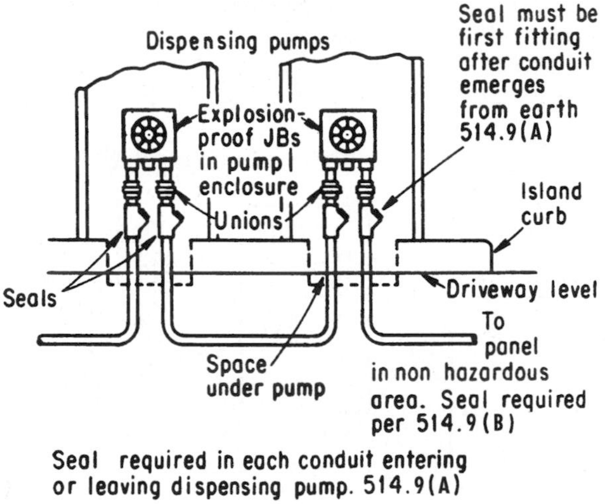
Dispensing pumps
Seal must be first fitting after conduit emerges from earth 514.9(A)
Explosion-proof JBs in pump enclosure
Unions
Island curb
Seals
Driveway level
To panel in non hazardous area. Seal required per 514.9(B)
Space under pump
Seal required in each conduit entering or leaving dispensing pump. 514.9(A)

Fig. 514-6. Seal fitting must be used for every conduit at dispenser. (Sec. 514.9.)

And the conduit at bottom right, which extends back to the panelboard, must also be sealed where it comes up out of the earth at the panelboard location.

The luminaires in Fig. 514-7 must satisfy 511.7(B) (by way of 514.7), which refers to "fixed lighting" that may be exposed to physical damage—such as impact by a vehicle. If the luminaires are not at least 12 ft (3.7 m) above the ground, they must be totally enclosed or constructed to prevent escape of sparks or hot metal.

In Fig. 514-4, four seals are shown. Normally, panelboards are located in a nonhazardous location so that a seal is shown where the conduit is emerging from underground, which is where Sec. 514.8 requires it to be.

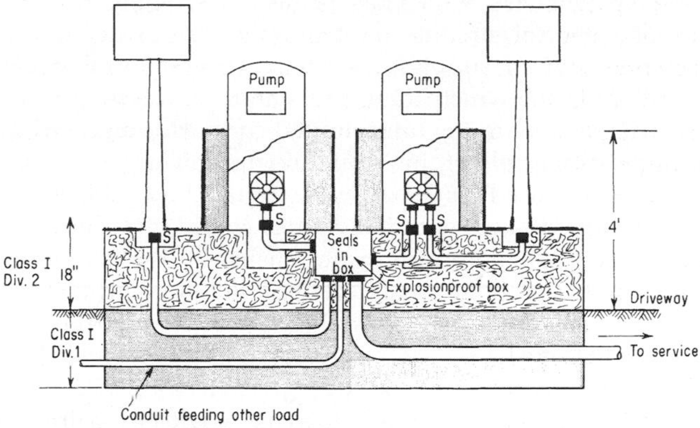

Pump
Pump
S
S
S
S
S
Seals in box
Explosionproof box
4'
Class I Div. 2
18"
Class I Div.1
Driveway
To service
Conduit feeding other load

Fig. 514-7. Seals are required in conduits to pumps and to lighting fixtures or signs at points marked *S*. (Sec. 514.9.)

514.11. Circuit Disconnects. When the electrical equipment of a pump is being serviced or repaired, it is very important that there be no hot wire or wires inside the pump. Because it is always possible that the polarity of the circuit wires may have been accidentally reversed at the panelboard, control switches or CBs must open all conductors, including the neutral.

To satisfy this Code rule, a special panel application is commonly used in gas stations. Figure 514-8 shows how the hookup is accomplished using a

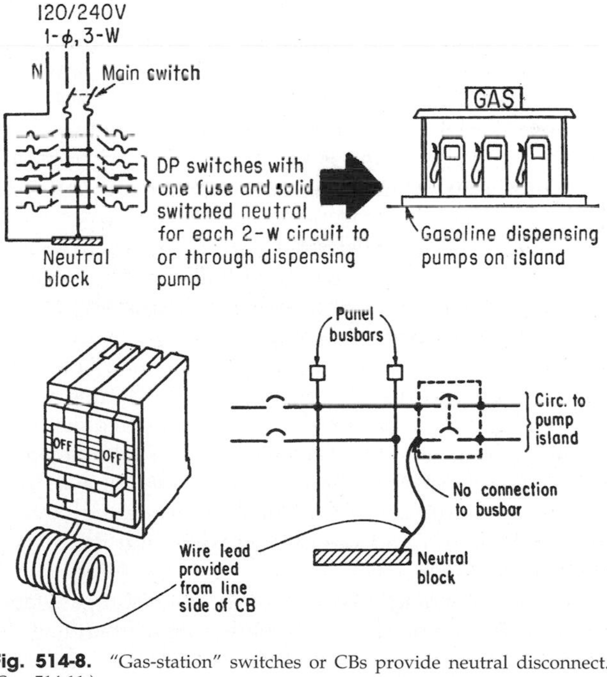

120/240V
1-ϕ, 3-W
N
Main switch
DP switches with one fuse and solid switched neutral for each 2-W circuit to or through dispensing pump
Neutral block
GAS
Gasoline dispensing pumps on island
Panel busbars
Circ. to pump island
No connection to busbar
Wire lead provided from line side of CB
Neutral block
OFF
OFF

Fig. 514-8. "Gas-station" switches or CBs provide neutral disconnect. (Sec. 514.11.)

gas-station-type panelboard, which has its bussing arranged to permit hookup of standard solid-neutral circuits in addition to the switch-neutral circuits required. Another way of supplying such switched-neutral circuits is with CB-type panelboards for which there are standard accessory breaker units, which have a trip element in the ungrounded conductor and only a switching mechanism in the other pole of the common-trip breaker, as shown. Either 2- or 3-pole units may be used for 2- or 3-wire circuits, rated 15, 20, or 30 A. Use of single-pole circuit breakers with handle ties would be a Code violation. No electrical connection is made to the panel busbar by the plug-in grip on the neutral breaker unit. A wire lead connects line side of neutral breaker to neutral block in panel, or two clamp terminals are used for neutral.

In addition, the **NEC** now has special rules to address self-service refueling stations. There must be "emergency shutoff devices or electrical disconnects" located as directed by the AHJ, which probably in this case would be the fire chief, because the purpose of this rule is to enable a bystander or firefighter to disconnect a pump island in the event of a conflagration. This must be located no less than 6 m (20 ft) or no more than 30 m (100 ft) from the fuel-dispensing equipment they serve. More than one such control is permitted, but they must be interconnected so the effect will be to shut off all power to all dispensers, not just at the island in question. The control must be clearly identified and readily accessible, and for an unattended station, it must be resettable only in a manner approved by the AHJ. As a practical matter, the only way to kill all the power to a given pump island like this is presumably through multiple relays that can, in the control circuit logic, be prevented from resetting the circuit until additional steps are taken as specified by the AHJ, which must include manual intervention. Whatever control method is used, it must open all conductors to the pump island, including any grounded circuit conductors if any, together with all associated power, communications, data, and video circuits. If the facility is attended, then a control point must be readily accessible to the attendant. For unattended facilities, the controls must be readily accessible to patrons, and at least one additional control must be readily accessible to each group of dispensers on an island. Note that the term "attended" requires interpretation, but often means someone responsible is constantly watching, not simply looking up from a cash register now and then. Where so interpreted the term would therefore seem to apply to only a small minority of such stations.

514.13. Provisions for Maintenance and Service of Dispensing Equipment. This rule mandates a lockable disconnecting means for the dispensing unit(s) to allow for safe maintenance or repair of the dispensing equipment. This disconnect must open all sources of supply to the dispensing unit, "including feedback" which would include data and communications sources that are frequently involved with dispenser controls. And the required disconnect may be located remote from, and not within sight of, the dispensing equipment and be lockable in the OPEN or OFF position. Note that in today's wiring layouts, frequently the motors are underground in enclosures set on or adjacent to the tanks, and the dispensers are for control and measurement only.

514.16. Grounding and Bonding. Because of the danger at gas stations, grounding is very important and the rule here calls for thorough grounding, "regardless of voltage."

ARTICLE 515. BULK STORAGE PLANTS

515.2. Definition. A flammable liquid is said to become volatile when the ambient temperature is equal to, or greater than, its flash point. Typical flash points are gasoline, –45°F (–43°C); kerosene, 100°F (38°C); diesel oil, 100°F (38°C). Thus, the status of gasoline is definitely established as a volatile flammable liquid regardless of geographical location of the storage facility. Other liquids may change from one state to another depending on the relation of the ambient temperature to their respective flash points. (See Fire Hazard Properties of Flammable Liquids, Gases and Volatile Solids, NFPA No. 325M.)

515.3. Class I Locations. This section sets its rules in a table indicating the extent of Division 1 and Division 2 locations at various equipment locations. As with the table in 514.3, this table brings the Code rules into agreement with the relevant NFPA standard, in this case, NFPA 30, *Flammable and Combustible Liquids Code*. Figure 515-1 shows rules from Table 515.3, covered under "Pumps, Bleeders, Withdrawal Fittings, Meters and Similar Devices, Indoors."

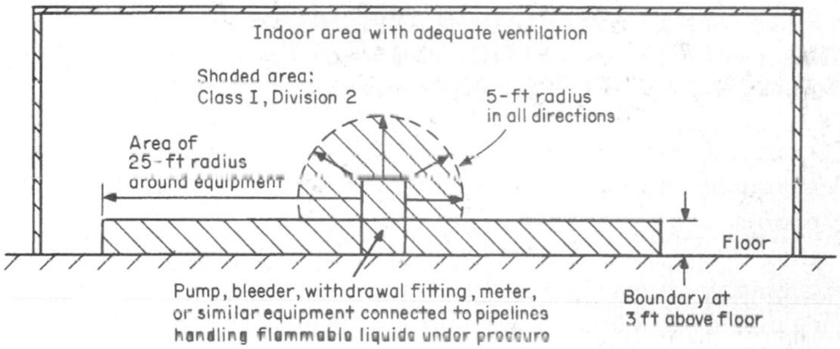
Indoor area with adequate ventilation
Shaded area:
Class I, Division 2
5-ft radius
in all directions
Area of
25-ft radius
around equipment
Floor
Pump, bleeder, withdrawal fitting, meter,
or similar equipment connected to pipelines
handling flammable liquids under pressure
Boundary at
3 ft above floor

Fig. 515-1. Space around indoor equipment is a Class I, Division 2 location if adequately ventilated; or a Class I, Division 1 location, if not. (Sec. 515.3.)

For outdoor use of the same kinds of equipment, the 5-ft (1.5-m) radius is reduced to 3 ft (900 mm), the 25-ft (7.5-m) radius is reduced to 10 ft (3.0 m), and the 3-ft high (900-mm high) level is reduced to 18 in. (450 mm). And where *transfer* of gasoline or similar liquid is done outdoors or in a ventilated indoor place, the space around the vent or fill opening becomes a Class I, Division 1 location for 3 ft (900 mm) in all directions from the opening and a Division 2 location out to 15 ft (4.5 m) from the opening.

The hazardous area around a volatile flammable liquid outdoor storage tank ("Tank-Aboveground" in Table 515.3) extends 10 ft (3.0 m) horizontally beyond the periphery of the tank (Fig. 515-2). Code designation is Class I, Division 2, and wiring installations within this range must conform to Code rules for this category. Space around the vent is a Division 1 location.

515.8. Underground Wiring. This rule is somewhat similar to that of 514.8. However, Sec. 515.9 requires that wiring run below a defined Class I location be considered to be in a Class I Division 1 environment, with seals placed accordingly. Although the semantics are different, the practical result should be the same. Note that 515.8 requires the customary 600-mm (2-ft) heavy wall threaded conduit tail

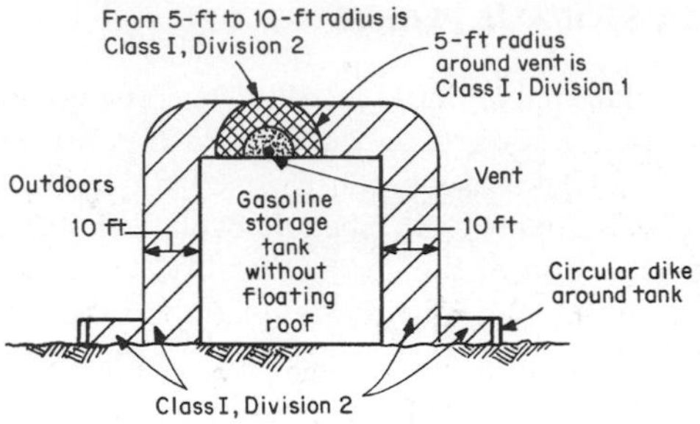
From 5-ft to 10-ft radius is Class I, Division 2
5-ft radius around vent is Class I, Division 1
Outdoors
Vent
10 ft
Gasoline storage tank without floating roof
10 ft
Circular dike around tank
Class I, Division 2

Fig. 515-2. Sec. 515.3.

to the point of emergence, and because this will normally be on the other side of a classified location boundary, an accessible seal will also be required at this point.

ARTICLE 516. SPRAY APPLICATION, DIPPING, COATING, AND PRINTING PROCESSES USING FLAMMABLE OR COMBUSTIBLE MATERIALS

516.1. Scope. Note that this article applies to "locations" used for finishing processes—which means open spraying areas as well as enclosed or semienclosed "booths."

The safety of life and property from fire or explosion in the spray application of flammable paints and finishes and combustible powders depends on the extent, arrangement, maintenance, and operation of the process.

An analysis of actual experience in industry demonstrates that the largest fire losses and greatest fire frequency have occurred where good practice standards were not observed.

This article consists largely of extracted text from two NFPA standards (No. 33 and 34). Because of exhaustive modernization within those standards, as of the 2014 NEC the front end of this article (Secs. 516.2 through 516.4) cover many different spray configurations and methods, with extensive both new and revised artwork. It is important to recognize that in this article the artwork is not just illustrative; in many cases it carries within it mandatory dimensional provisions. There is complete coverage in a separate subsection of zone classifications. Printing processes have been integrated as appropriate.

516.2. Definitions. The definitions largely carry over from the 2014 complete rewrite, including "Limited Finishing Workstation," "Unenclosed Spray Area," "Spray Area," and "Spray Booth." The 2017 NEC adds "Membrane Enclosure" and "Outdoor Spray Area." The content of these definitions will be discussed as necessary in the context of the revisions to the actual classification and wiring rules.

516.4. Area Classification. This section, the only one in Part II, covers open containers, spray gun cleaners, solvent recovery units, etc. The NEC illustrates these rules in Figure 516-4.

516.5. Area Classifications. This section opens Part III with a discussion focused on melding the two classification systems and the setting the general application considerations for these areas. Two different types of hazardous conditions are present in a paint-spraying operation: the spray and its vapor, which create explosive mixtures in the air, and combustible mists, dusts, or deposits. Each must be treated separately.

In part **(C),** the interior of every spray booth **(1)** and some area spraying **(3)** are Class I or Class II, Division 1 locations depending on whether the atmosphere contains vapors or dusts. When spray operations are not contained within a booth, there is greatly reduced control of flammable atmosphere, and the area of hazard is increased considerably. This is shown in Fig. 516-1, which is a typical specific application of the concept shown in Code Fig. 516-5(D)(1). A Class I or Class II, Division 1

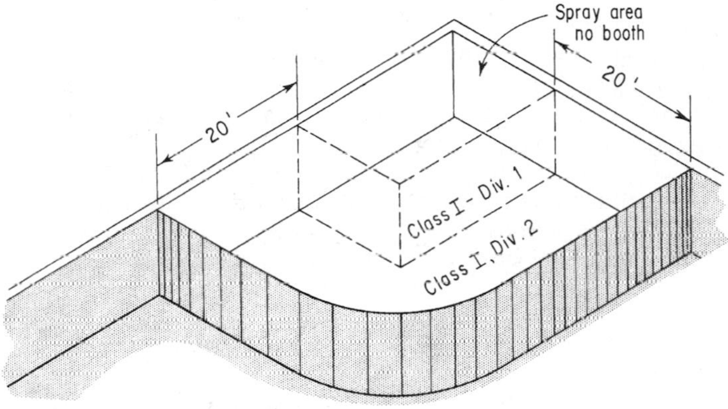
Spray area
no booth
20'
20'
Class I - Div. 1
Class I, Div. 2

Fig. 516-1. Open spraying involves Division 1 and Division 2 locations, for Class I or Class II conditions. (Sec. 516.3.)

area exists at the actual spraying operation plus a Division 2 area extending 20 ft (6.0 m) horizontally in any direction from the actual spraying area and for 10 ft (3.0 m) up. Although only one corner of the room is used for spraying, the entire room area inside the 20-ft (6.0-m) line around the spraying is classified as Class I or Class II, Division 2 and must be wired accordingly. And the Division 2 area extends 10 ft (3.0 m) above the spray operation.

The **NFPA** "Standard for Spray Application Using Flammable and Combustible Materials" (No. 33) notes that the inspection department having jurisdiction may, for any specific installation, determine the extent of the hazardous "spraying area."

The interior of any enclosed coating or dipping process must be considered a Class I or Class II, Division 1 location. A Division 2 location exists within 3 ft (900 mm) in all directions from any opening in such an enclosure, as shown in Fig. 516-5(D)(4). Note that today the majority of this type of work is carried out in dedicated spray booths and there are Class I Division 2 boundaries associated with the doorways into such spaces. A similar dimension applies to the openings to open-face and open-front spray rooms, as illustrated in Fig. 516-5(D)(2). Remember that actual spray booths and rooms, in accordance with their revised definitions, are power ventilated. The **NEC** prescribes area classifications for the entry and exit surroundings and the interiors of the associated duct work depending on whether or not the exhaust air is recirculated.

Part (5) provides extensive detail on both the fabrication of and the area classifications that apply to "limited finishing workstations." This equipment will confine the by-products of a spraying operation but that does not meet the strict requirements for spray booths or rooms; often the sides will consist of curtains, but still with mechanical exhaust and makeup air systems as illustrated in Fig. 516-5(D)(5).

516.6. Wiring and Equipment in Class I Locations. In part **(A)**, spray operations that constitute a hazardous location solely on the basis of the presence of flammable vapors (and no paint or finish residues) contain wiring and equipment for Class I locations, as specified in Art. 501.

Part **(B)** places tighter restrictions on spray booths or areas where readily ignitible deposits are present in addition to flammable vapors. In general, electrical equipment is not permitted inside any spray booth, in the exhaust duct from a spray booth, in the entrained air of an exhaust system from a spraying operation, or in the direct path of spray, unless such equipment is specifically approved for both readily ignitible deposits and flammable vapor.

However, for that part of the hazardous area where the luminaires or equipment may not be subject to readily ignitible deposits or residues, luminaires and equipment approved for Class I, Division 1 locations may be installed. The authority having jurisdiction may decide that because of adequate positive-pressure ventilation the possibility of the hazard referred to in paragraph **(B)** has been eliminated.

Sufficient lighting for operations, booth cleaning, and repair should be provided at the time of equipment installation in order to avoid the unjustified use of "temporary" or "emergency" electric lamps connected to ordinary extension cords. A satisfactory and practical method of lighting is the use of ¼-in thick (6.35-mm thick) wired or tempered glass panel in the top or sides of spray booths with electrical luminaires outside the booth, not in the direct path of the spray, as covered in (C)(4). Parts **(C)(1)** and **(2)** cover luminaires that illuminate the spray operation through "windows" in the top or walls of a spray booth, as illustrated in Fig. 516-6(C)(a). Any such luminaire used on the outside of the booth must be approved for a Class I, Division 2 location when used in any part of the top or sides of a booth that is within the Division 2 locations as shown in the various drawings included in 516.5(D) of the Code.

The other alternative is to work with luminaires mounted and serviced from within the spray booth. These are covered in 516.6(C)(3) and the associated artwork. They will be listed for the Division 2 exposure, and more importantly, the listing investigation will review their performance under conditions of paint residue on the glass, resulting in partial entrapment of infrared energy within the luminaire. Overheated paint residues can catch fire at relatively low temperatures, as can be inferred from the field exposure limits given in 516.6(C)(4). For comparable reasons, part **(D)** places severe limits on the use of portable lighting and other equipment in a spray area during operations.

Part **(F)** addresses the ignition hazards arising from static discharges associated with spraying and coating operations, and requires that both the operator and all electrically conductive surfaces be grounded. This includes all piping that conveys fluids used for flammable or combustible fluids and all metal parts of the process equipment involved in the coating processes together with all exhaust ductwork and material containers.

Although the 2014 NEC was the first time an express reference has been made to the problems of static electricity in Sec. 516.6 covering Class I locations, it has been an ongoing concern in Sec. 516.10 governing electrostatic equipment (both fixed and hand operated) and powder coating for a long time. Informational notes to this effect occur in 516.10(A)(6), in 516.10(B)(4) where in addition to the note the mandatory provisions now include a prescriptive reference to a resistance to ground not to exceed 1 megohm, and in 516.10(C)(4)(b).

516.7. Wiring and Equipment Not Within Classified Locations. This section, very similar (even the same section number) to content in Art. 511, focuses on wiring method and equipment specifications that prevent accidental failures from communicating into the nearby hazardous (classified) location.

Automobile undercoating spray operations in garages, conducted in areas having adequate natural or mechanical ventilation, may be exempt from the requirements pertaining to spray-finishing operations, when using undercoating materials not more hazardous than kerosene (as listed by Underwriters Laboratories in respect to fire hazard rating 30–40) or undercoating materials using only solvents listed as having a flash point in excess of 100°F (38°C). There should be no open flames or other sources of ignition within 20 ft (6.0 m) while such operations are conducted.

516.10. Special Equipment. The requirements in 516.10 cover electrostatic coating operations, often referred to as powder coating. This section is now included with other spray applications in Part III, and the integration is not yet perfected. For example, 516.10(A) requires automatic systems to comply with 516.6, but 516.6(A) requires adherence to provisions in Art. 501. The powders used in 516.10 fall within the scope of Art. 502, not 501.

516.18. Area Classification for Temporary Membrane Enclosure. This section, one of two in Part IV on this subject, is new as of the 2017 NEC. As soon becomes evident in Fig. 516-18 and the informational notes, it covers what can be very large contained areas, such as would be required to cover the large boat in the artwork.

516.29. Classification of Locations. This section, which opens Part V on printing, dipping, and coating processes, covers area classifications. Although comparable chemicals are used, these processes involve different risks than when discharged in the form of a spray and covers adjacent areas to open dipping and coating. They are visually presented in one old drawing showing the worst environmental case of no vapor containment or ventilation [516.29(a)] and three drawings covering operations with varying stages of vapor confinement. The first [516.29(b)] illustrates full peripheral vapor confinement. The second [516.29(c)] has partial vapor containment and ventilation, with the vapors not confined to the process equipment. The fourth drawing [516.29(d)] shows a printing process, a good example of the degree to which the NEC has broadened the focus of this article.

The same section includes four generic rules for the accessory areas around one of these operations. The first covers sumps, pits, and trenches; the second covers within 5 ft of a vapor source or the inside of an ink or a dip tank or the like (Division 1 for both). The third covers a 3-ft bubble around Division 1 locations (Division 2); and the fourth the area 3 ft above the floor and extending out 20 ft horizontally from the Division 1 location described in the third rule (also Division 2).

For small operations, 516.29(5) provides some relief on the extent of the Division 2 area. If the entire vapor source area does not exceed 0.5 m^2 (5 ft^2) and the total volume of contents in the dip tank does not exceed 19 L (5 gal), and the vapor concentration does not exceed 25 percent of the lower flammable limit, the area is permitted to be unclassified.

516.35. Areas Adjacent to Enclosed Dipping and Coating Processes. This section, illustrated by Fig. 516-35, provides a three-category area classification for surrounding areas. The interior is Division 1 and the exterior is unclassified, except near openings to the inside, which have the usual 3-ft Division 2 bubble around them.

516.36. Equipment and Containers in Ventilated Areas. This section covers open containers used in conjunction with spraying operations, and points to the comprehensive treatment of the topic in 516.4.

516.37. Luminaires. This section brings in the applicable material from 516.6(C).

516.38. Wiring and Equipment Not Within Classified Locations. This section essentially duplicates 516.7.

516.40. Static Electric Discharges. This section opens by duplicating the requirements in 516.6(F), but also specifically addresses printing operations and requires that provisions be made to remove static charges from nonconductive substrates.

ARTICLE 517. HEALTH CARE FACILITIES

517.1. Scope. Although the NEC is *not* intended as a design specification, the requirements for "electrical construction and installation" given in parts **II** and **III** of Art. 517, essentially present design requirements. For example, the rule in 517.31 mandating segregation of emergency and normal circuits necessitates an additional raceway system with all attendant components. Similarly, the rules for equipment ground-fault protection on the "next level of feeder disconnecting means down stream" as given in 517.17 present a design requirement. Therefore, in reality this article covers the *design and installation* of electric circuits and equipment in hospitals, nursing homes, residential custodial care facilities, mobile health care units, and doctors' and dentists' offices. *But* this article does not cover "performance, maintenance, and testing" of electrical equipment in such facilities. Such considerations are covered in other industry standards, such as NFPA 99.

Any specific type of health care location—such as a doctor's office or a dental office—must comply with Code rules whether the location is a sole occupancy itself or is part of a larger facility (such as a hospital containing other types of health care locations) or is within a school, office building, or the like.

To this end, the term "medical office (dental office)" as of the 2017 NEC has a formal definition. These offices are not for overnight stays or 24-hour operation. They are for examination, and for minor procedures that may involve local anesthesia or sedation under which the patient does not lose the capacity of self preservation in an emergency. The 2017 NEC also expanded the definition of a health care facility to include "mobile enclosures," reflecting the emergence of relocatable clinic facilities.

Veterinary facilities are not subject to the requirements of Art. 517.

517.2. Definitions.

Patient bed location The location of a sleeping bed in a health care facility. The term also includes the location of procedure tables in critical care areas, which would include an operating table.

Patient care space Any portion of a health care facility where people are examined or treated. An informational note makes it clear the business offices, corridors, lounges, etc. are not patient care areas. The **NEC** divides these areas into five defined classifications; basic (Category 3), general (Category 2), critical (Category 1), support (Category 4), and wet procedure location, which is no longer defined in this category, but given an entirely separate definition because a wet procedure may be carried out in more than one type of location within a patient care space. It is important to remember that electrical officials do not make these classifications. They are the exclusive province of the governing body of the facility based on the types of patient care they anticipate delivering.

Patient care vicinity This term provides a definite value for limiting the area, horizontally and vertically, in which special grounding requirements are to be observed in patient care areas.

Psychiatric hospital This is a facility used around the clock to provide only psychiatric care for not less than four resident patients.

Selected receptacles This phrase designates specific receptacles that will provide power to appliances used for patient care emergencies. A dissenting vote noted that the wording would allow task receptacles of any kind ("ordinarily required for local tasks") to be supplied by the emergency system, even receptacles as unimportant as those for floor cleaners—which is contrary to the basic concept that the essential system is intended to supply only extremely limited loads.

517.10. Applicability. All the Code rules on "Wiring and Protection" apply to the entire wiring system in hospitals and to "patient care spaces" of clinics, medical and dental offices, outpatient facilities, and doctor examining rooms or treatment rooms in nursing homes and residential care facilities. The basic rules of Part **II** apply to *all* health care facilities except those areas covered by 517.10(B)(1) and (2).

Part **(B)(1)** exempts those areas of a health care facility that are *not* intended for examining or treating patients. Areas that are dedicated to other purposes—business offices, corridors, waiting rooms, rest rooms, etc.—need not be wired as indicated in Part **II**.

The wording used in part **(B)(2)** is intended to exclude those health care facilities where patient rooms are used "exclusively" as sleeping quarters. The last sentence further reinforces the idea that such rooms are permitted to be exempt only where there is no intention to ever use the rooms as a treatment area.

517.13. Grounding of Receptacles and Fixed Electrical Equipment in Patient Care Spaces. In patient care spaces of *all* health care facilities, nonmetallic wiring methods are excluded, because they do not provide an equipment grounding return path over their outer margin. Refer to the discussion at 517.16 for the reasons behind this requirement. [See also part **(B)** of 517.10.]

Part **(B)** requires the use of a separate, green (no stripes of any color permitted) insulated equipment grounding conductor run with the branch-circuit conductors in a metal raceway or limited styles of metal-clad cable from a panelboard

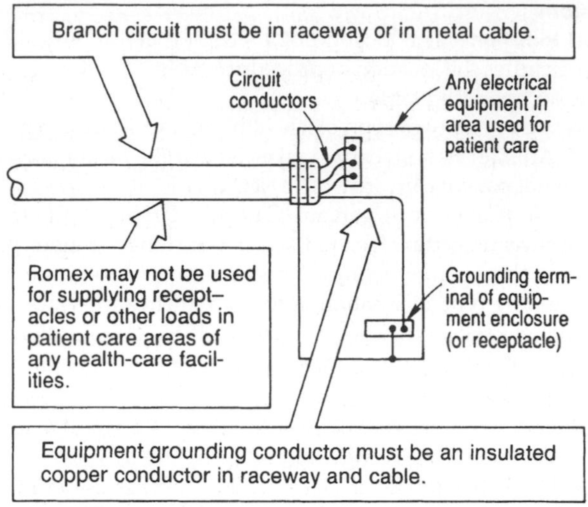
Branch circuit must be in raceway or in metal cable.
Circuit conductors
Any electrical equipment in area used for patient care
Romex may not be used for supplying recept– acles or other loads in patient care areas of any health-care facil- ities.
Grounding term- inal of equip- ment enclosure (or receptacle)
Equipment grounding conductor must be an insulated copper conductor in raceway and cable.

Fig. 517-1. Grounding conductor run with branch circuit in metal raceway or cable must ground receptacles and equipment. (Sec. 517.13.)

to any receptacle or metal surface of fixed electrical equipment operating over 100 V in all health care facilities. But a separate grounding conductor is not required in a feeder conduit to such a panel. For feeders, the metal conduit is a satisfactory grounding conductor, as recognized generally in 250.118. But for all branch circuits to "receptacles and all . . . fixed equipment . . ." in "areas used for patient care," neither metal conduit, jumpers with box clips (G-clips), nor a receptacle with self-grounding screw terminals [250.146(D)] may be used alone without the grounding wire run with the branch-circuit wires (Fig. 517-1), which must be in a metal raceway or in some styles of Type MC, Type MI, or Type AC cable (so-called BX). As required in **(A),** those metal-clad cables must have "cable armor or sheath assembly" that "itself qualifies as an equipment grounding return path." Type AC cable, Type MI cable, and Type MC cable with a smooth or corrugated continuous metal sheath all satisfy that grounding requirement. Type MC with a spiral-wrap metal sheath with two grounding conductors, one of which is insulated copper, has been a source of controversy. But the wording used in this section now makes it clear that the second ground return path must be through the metal raceway or cable armor. Use of two equipment grounding conductors in the spiral-wound Type MC cable does not satisfy the wording used here. However, the new style of interlocking-armor MC cable with a fully sized aluminum bonding conductor run outside the Mylar sheath and in contact with the cable armor would qualify for this purpose, as long as it was supplied with an additional green wire as part of the cable assembly.

The ground terminal of receptacles must be grounded to an equipment grounding conductor run in a *metal raceway or metal-covered cables.* Either metal raceway or metal cable must be used for circuits in patient care areas of hospitals, clinics, medical and dental offices, outpatient facilities, nursing homes, and residential custodial care facilities—*always* with an insulated *copper* equipment grounding conductor included in the raceway or cable. A new exception does clarify that the direct grounding connection language, with respect to connecting equipment to the equipment grounding conductor run in the raceway or cable, can be applied to a bonding pigtail from the receptacle, as long as the wire on the receptacle and the equipment grounding conductor running with the supply circuit directly connect to each other without relying on a bonding path through a metal box. This makes sense in the context that boxes and other enclosures containing receptacles are now expressly included in the patient care area enhanced grounding requirements.

This rule applies to "patient care spaces"—which, in hospitals, covers patient bedrooms and any other rooms, corridors, or areas where patients are treated, such as therapy areas or ECG areas. But for facilities other than hospitals, part **(B)(1)** of 517.10 excludes waiting rooms, admitting rooms, solariums, and recreation areas, as well as business offices and other places used solely by medical personnel or where a patient might be present but would not be treated.

The first exception is somewhat trivial and addresses a possible interpretation that there might be a problem with the branch-circuit equipment grounding conductor landing on the box, and then a bonding jumper running to the receptacle grounding terminal, because that would be an indirect connection. The critical return path performance is considered as upstream from the box and receptacle in combination, just as the wiring method is required to function with the redundant return path over its outer margin along with the wire next to the branch circuit conductors. This concept does not apply to isolated ground receptacles. Here again, refer to the comprehensive discussion at 517.16.

Exception No. 2 to this rule clarifies the use of metal faceplates on wall switches or receptacles without actually connecting an "insulated copper conductor" to each faceplate. They are acceptable as grounded simply by screw connection to a grounded box or grounded mounting strap of a grounded wiring device.

Exception No. 3 excludes luminaires from the rule for grounding by an insulated copper conductor, provided the luminaire is mounted "more than 7½ ft (2.30 m) above the floor." Although luminaires so located do *not* need an insulated grounding conductor, they must be fed by a conduit or cable that satisfies part **(A)** of this section. In other words, having gone to the bother of running a 250.118-compliant raceway, you can avoid running the redundant green wire in the conduit, as if that would be a significant concession.

Part **(A)** of this section emphasizes that a redundant metallic grounding path is required in patient care areas. Part **(A)** requires that *all* branch circuits supplying patient care areas must be run in a metal-enclosed wiring method—rigid metal conduit, IMC, EMT, or MI, MC, or AC cable—to provide a redundant metallic grounding path in parallel with the insulated copper ground wire required by part **(B)** in the wiring method. This rule emphasizes the need for high reliability of the ground-fault current return path as major protection against electrical shock.

517.14. Panelboard Bonding. Normal and essential electrical system panelboards serving either the same general care or critical care patient location must have their equipment grounding terminal bars bonded together with an insulated, continuous copper bonding jumper not smaller than No. 10 AWG.

Where more than one such panelboard supply a common patient care area, but are fed from separate transfer switches, the ground bus in each must be bonded together. Although required to be "continuous," the wording of the last sentence in this rule recognizes terminating this conductor at ground buses and terminals as satisfying the requirement for a "continuous" conductor.

517.16. Use of Isolated Ground Receptacles. After being prohibited for the 2011 NEC, the 2014 NEC resurrected them, but not for use in patient care vicinities. The 2017 NEC clarified the requirements. If such a receptacle is installed anywhere within a patient care space (not patient vicinity), its equipment grounding conductor must be distinguishable from other equipment grounding conductors of the wire type, in accordance with 517.16(B). It must be a green wire with yellow stripes, which will land on the ground terminal of the receptacle. This contrasts with the equipment grounding conductor of a wire type, colored solid green that runs with branch circuit conductors, and which in this case will also arrive at the receptacle location, but only terminate on the wall of the box.

Prior to the 2011 NEC, these receptacles were permitted, and a note followed cautioning that for these receptacles, "the grounding impedance is controlled only by the equipment grounding conductors and does not benefit functionally from any parallel grounding paths." The current prohibition within a patient care vicinity is a logical outcome of the concern expressed in the former note. This consideration also provides a valuable point of entry into the reasons that this article requires branch-circuit wiring methods that qualify, over their outer margins, as equipment grounding conductors. Testing was done at an older hospital wired generations earlier with rigid metal conduit in order to study how equipment grounding paths actually functioned on these systems. The test results were reported to the NFPA Electrical Section at the 1986 Annual Meeting at which the 1987 NEC was being voted on. The test consisted (in part) of removing the conductors from some branch-circuit raceways and breaking open the walls and floors at various points on the conduits' path back to the panelboard. When this work was completed, 50 A of fault current was deliberately injected into the box at the end of the branch circuit, and then the current flowing over the conduit was measured at the various access points on the way back to the panel. The conduit did show 50 A just below the outlet point, but nowhere else. As the readings progressed from outlet to panel with a steady 50 A going into the grounding return system, the current observed in the conduit dropped steadily until, at the panel, it was down to not more than a fifth of what was going into the wall.

The reason for this is that there were hundreds of ways back to that panel and electricity being electricity, every available path was used. Every piece of reinforcing steel that the conduit happened to touch, every water pipe, and so on, got into the act. Even concrete is somewhat conductive depending on moisture content. Now none of these alternate paths was very good by itself and we would never want to qualify any of them in Art. 250. But they exist, and by the principle of impedances in parallel, where the reciprocal of the total impedance of the circuit is the sum of the reciprocals of all the component impedances, we know that a large number of parallel impedances, even if all quite high individually, produce a total impedance that is very low. In fact the final impedance is much lower than the conduit, or that of a copper wire, or even that of both together.

517.17. Ground-Fault Protection. At least one additional level of ground-fault protection is required for health care facilities where ground-fault protection is used on service equipment (see 230.95). Where the installation of ground-fault protection is made on the normal service disconnecting means, each feeder must be provided with similar protective means. This requirement is intended to prevent a catastrophic outage. By applying appropriate selectivity at each level, the ground fault can be limited to a single feeder, and thereby service may be maintained to the balance of the health care facility.

As shown in Fig. 517-2, with a GFPE (ground-fault protection of equipment) hookup on the service, a GFPE hookup must be put on each feeder derived from the service. Note that the second-level requirement applies to health care facilities with critical care areas or utilizing life-support equipment. Questions have been raised about health care facilities taking up tenancy in a multiple-occupancy building with GFPE on its service. There is no question that if such a facility had this sort of critical power need, then a second level of GFPE would be required, presumably on each tenant main that was one step below the service, even though only one of the second-level GFPE devices had anything to do with a health care occupancy. However, many of these outlying medical offices do not have this degree of sophisticated care, and therefore do not trigger 517.17.

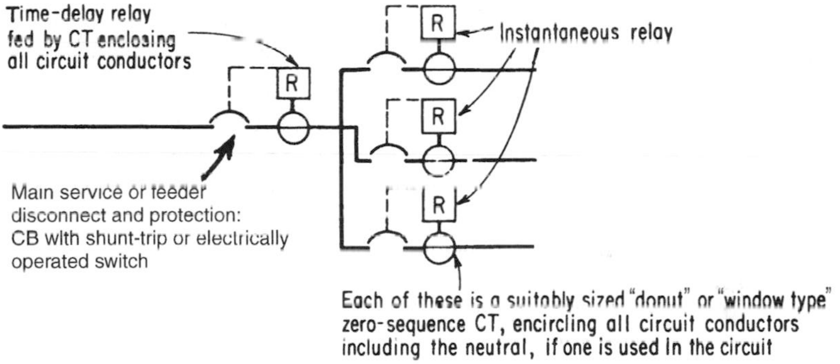
Time-delay relay
fed by CT enclosing
all circuit conductors
R
R
Instantaneous relay
R
R
Main service or feeder
disconnect and protection:
CB with shunt-trip or electrically
operated switch
Each of these is a suitably sized "donut" or "window type"
zero-sequence CT, encircling all circuit conductors
including the neutral, if one is used in the circuit

Fig. 517-2. GFPE on the service requires GFPE on main feeders also. (Sec. 517.17.)

The second paragraph is aimed at ensuring that essential systems are *not* isolated when the additional level of feeder GFPE is actuated. This requirement previously appeared as an FPN but now is included in the rule itself. Consequently, it is no longer advisory, but rather mandatory. Installation of GFPE on the load side of an emergency transfer switch is expressly prohibited, because if possible the fault should be cleared on the line side of the transfer switch. Closing a transfer switch into a fault may damage the switch and decrease the overall reliability of the system. However, the former prohibition against second levels of GFPE between the transfer switch and a standby generator, or on a voltage distribution not presently required to use GFPE, have been lifted. This should not be interpreted as a requirement to apply such protection, which is not an NEC requirement, but no harm results from lifting the prohibition. And part **(C)** requires that selection of the tripping time of the main GFPE be such that each feeder GFPE will operate to open a ground fault on the feeder, without opening the service GFPE. Instead of the former requirement for a six-cycle minimum separation, the **NEC** now creates a performance mandate for separation of the GFPE time-current characteristics in accordance with the recommendations of the manufacturer, consideration of required tolerances and operational timing so as to achieve complete selectivity. As shown, if the feeder GFPE relays are set for instantaneous operation, the relay on the service GFPE must have a delay such that it will only open in the rare event that the ground fault is between it and the downstream protective devices. A zone-selective GFPE system with a feedback lockout signal to an instantaneous relay on the service could satisfy the rule for selectivity.

Part **(D)** of this section calls for a test of the required GFPE when it is first installed. Such testing must demonstrate that the GFPE satisfies the rule of part **(C)** and provides "100 percent selectivity."

517.18. General Care (Category 2) Spaces. Two circuits must supply each bed used for inpatient care. A branch circuit supplying a patient bed location must not be part of a multiwire branch circuit. Given the common disconnect requirement in 210.4, a fault on one side of such a multiwire circuit would take out one or two other runs, depending on the distribution, reducing overall reliability. At least one of the circuits must be from the critical branch and one from the normal system, unless two circuits are available from the critical branch supplied through different transfer switches. If that arrangement is chosen, make certain it is correlated with the separation requirements in 517.31(C)(1). The receptacles, or the cover plates, at these bed locations that are supplied from the critical branch must have a "distinctive color or marking," and they must also indicate the identity of the panelboard and show the circuit number supporting them. But two branch circuits are not required for each patient bed in nursing homes, outpatient facilities, clinics, medical offices, limited care facilities, and the like. "Psychiatric, substance abuse, and rehabilitation hospitals" are also exempted from the branch-circuit and receptacle requirements for general care patient bed locations, including the requirement for panel and circuit identifying information.

As noted in part **(B)**, receptacles at patient bed locations in "General Care Areas" must be "listed hospital grade" and "so identified." The minimum of eight required receptacles at each such bed location may be single or duplex types, or a combination of the two. (Two duplex receptacles constitute a total of four receptacles.) *All* receptacles at *all* patient bed locations, general care and critical care, must be "listed hospital grade" devices. There are now quadruplex receptacle devices that are rated hospital grade, and are an additional option in these locations. Psychiatric facilities as covered in exceptions to (B) are not included in the minimum requirements for receptacle placements.

As noted in part **(C)**, only tamper-resistant receptacles are permitted in pediatric locations. This rule requires that all 15- or 20-A, 125-V receptacles in pediatric locations be tamper-resistant. Tamper-resistant receptacles make it extremely difficult, if not impossible, to insert a pin, paper clip, or similar small metal object into a slot on the receptacle and make contact with an energized part. Obviously, the concern here is to protect infants or children from shock hazard as a result of playful or inadvertent tampering with the receptacle. Another option, seldom used, is a listed tamper-resistant cover.

These receptacles are similar to those required by 406.12 for dwelling units, but for these locations they must additionally carry a hospital grade rating. As in the case of devices in dwelling units, the UL information on these devices indicates that they are marked "TR" on the yoke where the mark will be visible with the faceplate removed.

517.19. Critical Care (Category 1) Spaces. Patient bed locations in general care areas (517.18) must be supplied by eight single or two duplex receptacles. However, critical care area patient beds must be provided with at least 14 receptacles (single or duplex or quadruplex devices totaling 14 points for connecting a cord plug cap). The two or more branch circuits to each critical care area patient bed location must include one or more from the critical branch and one or more from the normal system (Fig. 517-3). In both cases, at least two branch circuits must supply these receptacles. As in the case of 517.18, a branch circuit supplying a patient bed location must not be part of a multiwire branch circuit. Given the common disconnect requirement in 210.4, a fault on one side of such a multiwire circuit would take out one or two other runs, depending on the distribution, reducing overall reliability. In the case of general care areas, additional receptacles serving other patient locations may be served by these branch circuits, but in the case of critical care areas, at least one of these branch circuits is required to be an individual branch having no other receptacles on it except those of a single bed location. Here again, if the critical branch is supplying two (or more) circuits through multiple transfer switches, the normal system circuit is no longer required. However if that arrangement is relied upon, make certain it is correlated with the separation requirements in 517.31(C)(1). In addition, the receptacles connected to the emergency source must be identifiable as such and have their circuit number and panelboard ID visible on the faceplate or equivalent.

The 2017 **NEC** added a requirement that in addition to the circuit numbering and identification rule, any receptacle connected to either the critical branch or the life safety branch must have a distinguishing color to make it more quickly identifiable. Either the receptacle face or the faceplate (or both, presumably) must use this color. The **NEC** does not specify any particular color, but presumably whichever color is chosen it must be applied consistently throughout the facility. This requirement, for these receptacles, is essentially a duplication of 517.31(E).

As covered in part **(B)**, "hospital-grade" receptacles must be used at patient bed locations in critical care areas. Fourteen single or seven duplex receptacles (or any combination totaling 14 receptacles, including the possibility of one or more quadruplex receptacle(s) taking care of some of the six receptacles required) that are UL-listed as "hospital-grade" devices must be used at each patient bed location and must be so identified at each patient bed location in critical care patient areas. The best point to call a reference grounding point is the grounding bus in the distribution panel, which is the transition connection point between the branch-circuit grounding wires and the feeder grounding system. This is also, in effect, how the term is defined in 517.2.

As of the 2014 **NEC**, operating rooms now have their own unique receptacle requirements in (C) that parallel those of critical care bed locations generally,

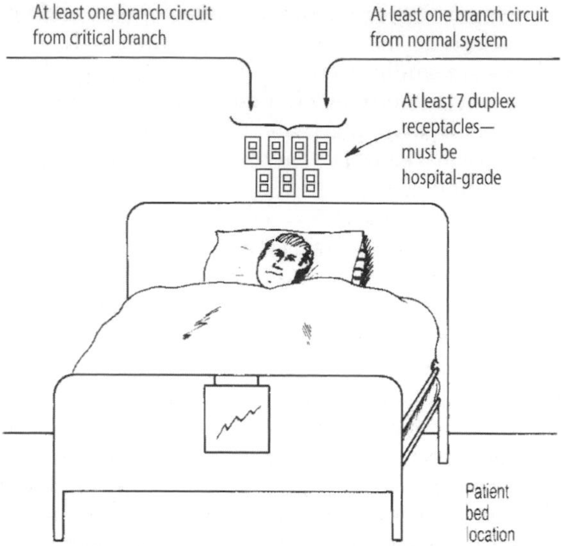
At least one branch circuit
from critical branch
At least one branch circuit
from normal system
At least 7 duplex
receptacles—
must be
hospital-grade
Patient
bed
location

517.19. Both emergency system and normal system must supply branch circuits to critical-care bed location.

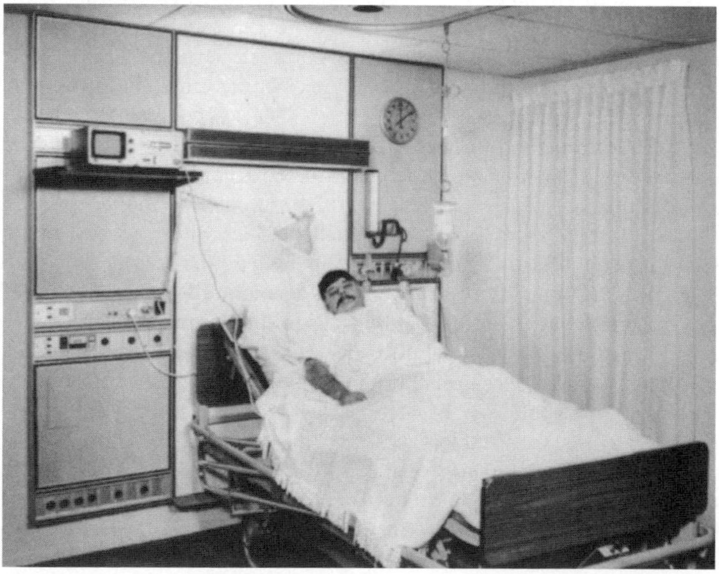

Fig. 517-3. Multipurpose patient care modules may incorporate a variety of the circuit, receptacle, and grounding requirements for patient care areas. In addition to a patient equipment grounding point and room bonding point, such a preassembled unit might include facilities for communication, patient monitoring, lighting, and lines for air, water, and medical gas. (Sec. 517.19.)

but that have significantly higher numbers of required hospital-grade receptacles. There must be at least 36 in the room divided between not less than two branch circuits, either laid out using single, duplex, or quadruplex receptacles, and at least 12, but no more than 24, need to be on the normal system or fed from a different transfer switch of the critical branch.

This set of requirements is for a room and not for a bed location. And a patient bed location includes a "procedure table of a critical care area," which would include an operating table. Therefore, the literal text puts at least 14 receptacles at the operating table, with the remainder elsewhere in the room as specified by design.

Part **(D)** means that a "patient equipment grounding point" (defined as a "point for redundant grounding of electric appliances") is *not* mandatory. Such a grounding point may be used, if desired, and connected as described in the rule. In its original application, years ago, the "patient equipment grounding point" was a special jack used to make a grounding connection to the metal enclosures of electrical medical equipment because, at the time, many power cords did not contain a grounding conductor. In addition, this was a carryover from procedures for operating rooms, where all metal surfaces had to be grounded to minimize static charge buildup. With today's universal use of good 3-wire (grounding type) power cords and plugs or double-insulated equipment, there is no real justification for requiring this patient equipment grounding jack. Each three-contact receptacle, in effect, becomes a patient equipment grounding point.

When a patient equipment grounding point is used in a patient vicinity, it must be grounded to the ground terminal of *all* grounding-type receptacles in the patient vicinity by means of a minimum No. 10 copper conductor looped to all of the receptacles or by individual No. 10 conductors run from the patient grounding point to each receptacle.

Regardless of what additional methods are employed, in order to keep potential differences within the required limits, equipotential grounding is essential to the electrical safety of critical care areas. Some of the earliest equipotential grounding installations consisted of copper busbars run around the walls of patient rooms to which furniture and equipment were attached by means of grounding jumpers. Based on experience obtained through these early installations as well as the refinements produced by the NFPA Committee on Hospitals, the National Electrical Code now contains the requirements that correlate with the pertinent NFPA standards on the subject. At the same time, these requirements also permit the achievement of the desired end with a minimum of expenditure in labor and materials.

Objection has been raised to the concept of grounding every piece of exposed metal in sight. Doing this may actually increase the hazard. Because a shock occurs when a person touches two surfaces with a voltage difference between the surfaces, the fewer surfaces that are deliberately grounded, the better. Thus, a door frame or window frame that is not likely to become energized is not required to be grounded. It was not good safety engineering to propose that metal furniture in a patient's room be deliberately grounded. Figure 517-4 shows the type of grounding and building points that past editions of the Code regulated, along with typical grounding hookups in older hospitals.

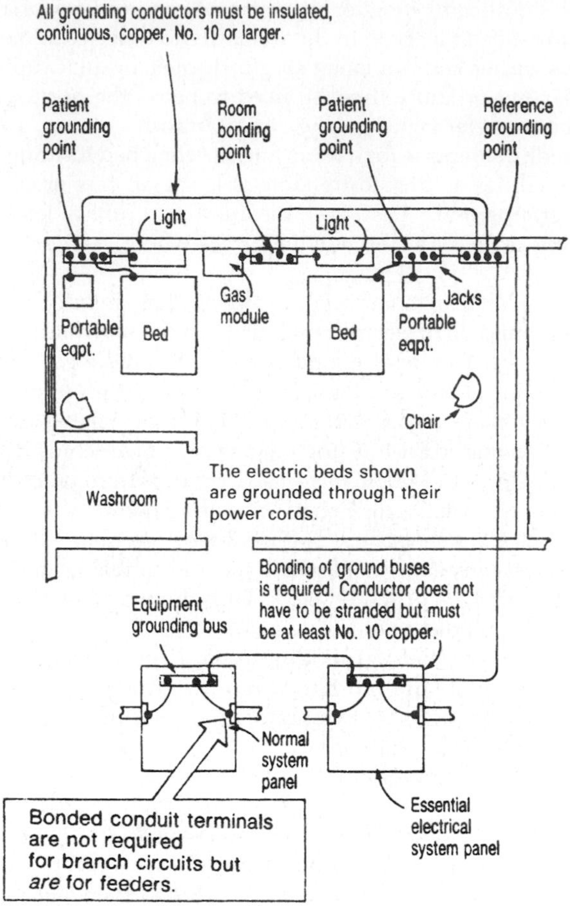
All grounding conductors must be insulated, continuous, copper, No. 10 or larger.
Patient grounding point
Room bonding point
Patient grounding point
Reference grounding point
Light
Light
Gas module
Jacks
Portable eqpt.
Bed
Bed
Portable eqpt.
Chair
Washroom
The electric beds shown are grounded through their power cords.
Bonding of ground buses is required. Conductor does not have to be stranded but must be at least No. 10 copper.
Equipment grounding bus
Normal system panel
Essential electrical system panel
Bonded conduit terminals are not required for branch circuits but *are* for feeders.

Fig. 517-4. The grounding techniques shown here were regulated by past editions of the NEC but are now optional at the patient bed locations. [Sec. 517.19(C).]

As required by part **(E)** of this section, a bonding-type connection is required for *feeder* metal conduit or metal cable (Type MC or MI) terminations, using a bonding bushing plus a *copper* bonding jumper from a lug on the bushing to the ground bus in the panelboard fed by the conduit (see illustrations in 250.92). Bonding locknuts or bonding bushings may be used on clean knockouts (all punched, concentric, or eccentric rings removed), and bonding may be provided by threaded connection to hubs or bosses on panel or switchboard enclosures. This is required for feeder conduits but not for branch-circuit conduits. And it seems clear that bonded terminals are required at both ends of each and every feeder to a panelboard that serves the critical care area (Fig. 517-5).

Part **(F)** makes use of an isolated power system for critical care areas a completely optional technique, simply noting that such systems are "permitted" to

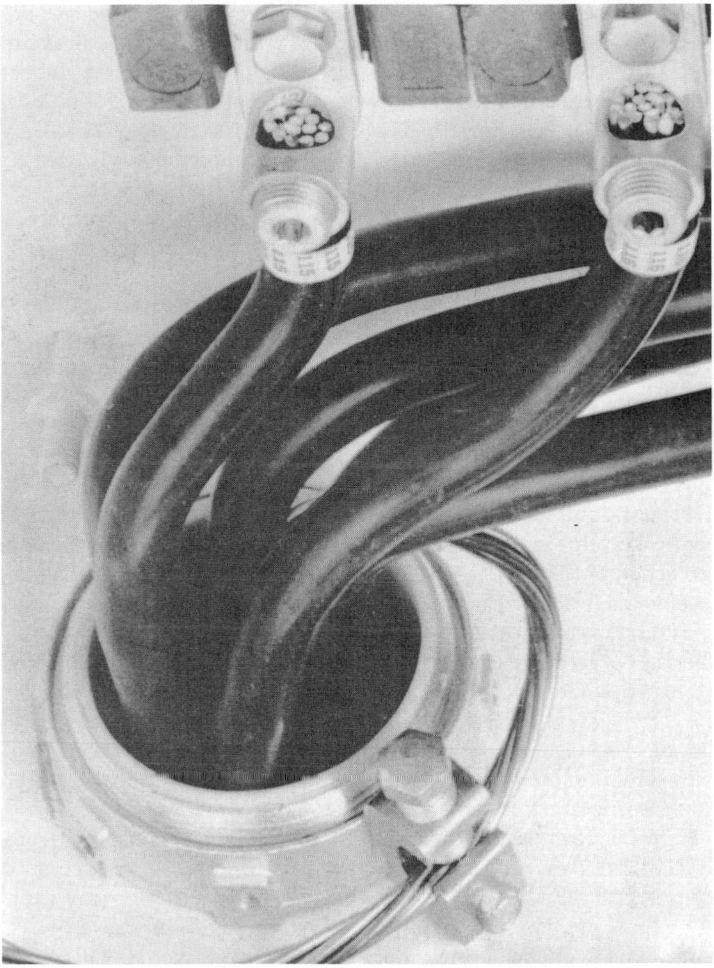

Fig. 517-5. Feeders to branch-circuit panelboards for critical care areas must be bonded. Figure 517-4 covers rules on grounding and bonding for the patient vicinity of a critical care area, with feeder conduit bonding as in this photo. [Sec. 517.19(D).]

be used if the design engineer or the hospital-client wants them. This approach ties in with the deletion of the maximum potential difference of 100 mV "under conditions of line-to-ground fault" in a critical care area—which until the 1978 **NEC** tended to make the isolated power system mandatory. Grounding of this optional power system must satisfy part **(G)**.

Part **(H)** covers receptacles that are intended for use by specific pieces of equipment. Those receptacles intended for special purposes—for example, mobile x-ray equipment—must have their equipment grounding conductor "extended" to the reference point within the branch-circuit panelboard that supplies the patient area.

517.20. Wet Procedure Locations. These locations, designated as such by the governing body of the hospital, are locations where certain medical procedures will be performed that involve copious amounts of fluids during the procedure, even though they are under cover and would generally meet the definition of a dry location. The reference to "wet" here is a medical procedure reference, and not an environmental description. Some surgeries, for example, involve extensive irrigation of the wound. The previous terminology "wet locations" was misleading and was revised for the 2008 NEC. Locations that are intended for protection would include hydrotherapy, dialysis facilities, selected wet laboratories, and special-purpose rooms where wet conditions prevail.

All receptacles (of any rating) *and fixed electrical equipment* at a wet location must "be provided with special protection against electric shock." What follows are the two acceptable responses to achieve this special protection status as in prior Codes, but using different language, as extracted from NFPA 99. The first describes an isolated power system, and the second GFCI. The only exception eliminates the need for GFCI protection or supply from isolated power for "listed, fixed" therapeutic and diagnostic equipment.

517.21. Ground-Fault Circuit Interrupter Protection for Personnel. This rule correlates with the rules in 210.8(B)(1) and 406.3. These two Code sections deal with GFCI protection of 15- and 20-A, 125-V receptacles installed in new bathrooms and replacements made in existing bathrooms in occupancies other than dwellings. The intent of 517.21 is to exempt receptacles installed in critical care areas—whether new installations or where replacement is made in an existing installation—from being GFCI-protected where a toilet and basin are within the same room. Literally, the rules of 210.8(B)(1) and 406.3 would mandate GFCI protection for any 15- and 20-A, 125-V receptacles installed in such a location. *But* for any "critical care" patient bed location where the bathroom is separate from the care area, critical or general, GFCI protection *is* required for 15- and 20-A, 125-V bathroom receptacles.

517.25. Scope. Essential electrical systems are covered for hospitals, clinics, medical and dental offices, outpatient facilities, nursing homes, residential custodial care facilities, and other health care facilities for patient care. In any facility, there must be enough standby lighting and power to allow for an orderly cessation of procedures. This might be nothing more than battery-powered wall units for some illumination, but the concept must be addressed in all health care facilities.

517.26. Application of Other Articles. The definition of "life safety branch" in 517.2 no longer directly references Art. 700. That allowed this section to authorize variances from Chap. 7 without contravening 90.3; with the restructuring of 90.3 that effort is no longer necessary, but does not change the intent. There are two principal areas of conflict between this article and Art. 700 that are now firmly resolved by this wording. Specifically, the conflict is between 700.5(D) that says an emergency system transfer switch must serve only emergency system loads, and 517.31(B) that expressly permits a single transfer switch at a small hospital (defined as not over 150 kVA of load on the essential electrical system). The second area involves the selective coordination requirements in 700.32, which are effectively set aside by 517.31(G).

517.29. Essential Electrical Systems for Hospitals and Other Health Care Facilities. This section, along with 517.29(B), makes clear that essential electrical systems that are to be installed in hospitals, and other health care facilities as specified elsewhere that are using essential electrical systems to serve patients on electrical life support, must observe the rules of Part **III** of Art. 517.

517.30. Sources of Power. Essential electrical systems must have no fewer than two sources of power, each capable of supporting the system. The normal configuration is a normal utility source and a standby source of sufficient size, but other arrangements are recognized, including on-site generation as the normal source and recourse to a utility (or additional on-site generation) as the backup. Fuel cell systems, comprised of no fewer than N+1 units (N being sufficient for the demand) are permitted as a source if they are listed for emergency system use and meet the other requirements in 517.30(B)(2). The power components must be positioned to minimize risk of failure from disasters and other factors that could result in the simultaneous interruption of plural sources of supply.

517.31. Requirements for the Essential Electrical System. This section begins by identifying the three separate branches comprising the system (life- safety, critical, and equipment). It should be noted that the critical branch in hospitals comprises different equipment and connections than does the critical system in nursing homes. One subject of endless debate over past decades has been whether critical branch loads were to be covered by Art. 700 requirements, largely driven by the former grouping of the critical branch and the life-safety branch together in a subgrouping of the essential system called the "emergency system." Essential electrical systems have three components, only one of which is directly comparable to Art. 700 (the life-safety branch). The other half of the former "emergency system" is the critical branch, which never did refer to Art. 700 within its definition, and was probably closer to Art. 701 (legally required standby) than Art. 700. Although it does have the same reconnection time of 10 s, it is not allowed to enter a common raceway with circuits on the life safety branch, by 517.31(C)(1) This left very much in question whether other rules in Art. 700, notably including 700.32 per the inclusive language in 517.26, ever could have applied to the critical branch because 700.9 (B) expressly allows all Art. 700 circuits in common enclosures. However, the "equipment system" is plainly not Art. 700 or even Art. 701, with numerous permissions for delayed automatic or even manual reconnection to power. The 2014 NEC firmly ended these debates. The term "emergency system" is no more. Each load grouping ("Life-Safety, "Critical," and "Equipment") is now followed by the same word in its title: "Branch." The NEC drawings in Figs. 517-31(a) and (b) show three coordinate branches, each with its own wiring rules. To add further clarification, 517.31(A) requires that the separation between branches occurs, for anything but a small facility using a single transfer switch, at the transfer switches.

Both the critical branch and the life-safety branch must be divorced from each other and all other wiring and equipment. These circuits must not enter raceways, boxes, or cabinets that are common to other systems. In addition, it is common in large hospitals to subdivide the critical system among multiple transfer switches to increase reliability. Both 517.18(a) and 517.19(A) allow branch circuits fed through different transfer switches to create the diversity of supply required for receptacles under their control. This section now requires that when this is done, the branch circuits from different transfer switches must be "kept independent of each other." This means separate raceways and boxes just as surely as it did, and still does, when circuits from both the normal system and the critical branch arrive at those bed locations. To correlate with this change, item (4) in 517.31(C)(1) now limits the double circuit allowance to applications originating from the same transfer switch.

In 517.31(C)(3), although wiring of the life-safety and the critical branches is required to be installed in a "nonflexible metal raceway" or MI cable, (C)(3)(1) permits Schedule 80 PVC or RTRC-XW for such circuits that do not supply patient care areas. Only the Schedule 80 version of rigid PVC conduit may be used without concrete encasement. (C)(3)(2) says Schedule 40 PVC conduit and electrical nonmetallic tubing are permitted for other than branch circuits in patient care areas, where encased in no less than 2 in. (50 mm) of concrete. *But* note that, although such a raceway may be used for life-safety and critical-branch circuits, it is *not permitted* for such circuits that are "branch circuits serving patient care areas."

Part **(C)(3)(3)** allows limited use of cable and flexible metal raceways to supply branch circuits in patient care areas, but don't forget that 517.13 still applies to branch circuits, so the metal cable armor would have to be recognized for redundant grounding, and the flexible metal raceways would be limited to 1.8 m (6 ft) because otherwise they would lose their status as qualified equipment grounding conductors under 250.118. And this rule also formally recognizes flexible metal conduit and cable assemblies in prefabbed "headwalls" and where flexibility is needed, as well as fished application in existing facilities, and luminaires in rigid ceilings that preclude access to the wiring after the initial installation. Parts **(C)(3)(4)** and **(C)(3)(5)** cover cord-and-plug connections for utilization equipment, and recognize the use of conventional limited energy cabling for nurse-call and other signaling systems operating under Class 2 and Class 3 power limitations, and also data, communications, and other power limited systems within the reach of Part **VI** of the article, whether or not run in a raceway.

Part **(D)** addresses the sizing requirements for the overall essential electrical system notably including its standby source and requires it to be capable of picking up the connected load based on any combination of historical data, Art. 220 results, the actual load ratings of connected equipment, and cautiously applied demand factors. Unlike 700.4 and 701.4, this section does not recognize automatic load shedding as a strategy to bring load demand into line with the capacity of the standby source, and the concluding sentence expressly disallows the applicability of those sections.

Part **(E)** applies to receptacles throughout the hospital and not just in the patient bed locations described in 517.18 and 19. It requires a distinctive color on the receptacle faces or their faceplates at every outlet that makes it immediately apparent that the receptacle is supplied by the essential electrical system. Note that when the 2014 **NEC** eliminated the former terminology used here, "emergency system," it broadened the applicability of this identification requirement to include receptacles on the equipment branch, including some that may be reenergized only following a delayed automatic or even a manual reconnection after a normal system outage.

Part **(F)**, new to the 2014 **NEC**, covers a single feeder supplied from a generator as a standby supply source for the essential electrical system. The transfer switch does not need to be at the alternate source, and the single feeder permission extends from the alternate source to the point where the sources for the life-safety branch, critical branch, and equipment branches divide.

Pert **(G)**, also new as of the 2014 NEC, covers system coordination. This terminology intentionally differs from comparable terminology throughout the NEC that applies requirements for "selective coordination." The latter term is now defined in Art. 100 and covers all overcurrent events capable of occurrence following the instant of energization. The requirement here in Art. 517 requires coordination for fault events for the period of time that extends beyond a tenth of a second. In this time period, six cycles, it is comparatively easy to accomplish coordination of feeder- and branch-circuit levels in a distribution without resorting to fuses, or to using circuit breakers with undesirable frame size differentials.

Strict selective coordination accomplishes protection for an electrical event that has never been recorded in an occupied building in the history of the use of electricity in buildings. Refer to the in-depth discussion of this topic as part of the coverage of 700.32 in Chap. 7 of this book as to why that is necessarily true. Unfortunately this sensible relaxation of coordination never would have come into the NEC voluntarily. The hospital industry, heavily impacted by the costs of compliance and well aware of the lack of observable benefit, put the amended rule in the *Health Care Facilities Code*, NFPA 99. The NFPA Standards Council, the ultimate arbiter in such things, decreed that this requirement related more to the performance of hospital electrical systems and not the essential installation criteria for those systems. As such, it can be extracted into the NEC but not substantively altered. This rule change was one of the single most important developments in the 2014 NEC cycle.

517.32. Branches Requiring Automatic Connection. Code diagrams 517-30 Nos. 1 and 2 in the Code book clarify interconnections and transfer switches required. Handbook Table 517.1 summarizes the loads supplied by the hospital life-safety and critical branches, which must restore electrical supply to the loads within 10 s of loss of normal supply. A minor clarification in the 2017 NEC, partially in the interest of comfort at night, resulted in task illumination on the critical branch now being capable of switch control [517.34(B)].

517.35. Equipment Branch Connection to Alternate Power Source. This is where the equipment system comes in, comprised primarily of three-phase power loads. Ventilation systems for operating and delivery rooms were added under delayed automatic connection for the 2008 NEC. The 2014 NEC has added HVAC systems for telephone and data equipment rooms and closets.

517.40. Type 2 Essential Electrical Systems for Nursing Homes and Limited Care Facilities. An extensive informational note in 517.40 describes the different levels of care available in nursing homes, and the level (now referred to as a category) of care offered, and how that potentially affects essential electrical system design.

517.42. Essential Electrical Systems. This section covers nursing homes and limited care facilities. In part **(B)**, a single transfer switch may be used for the entire essential electrical system instead of using a separate transfer switch for each branch. One transfer switch may supply one or more branches of the essential electrical system in a nursing home or residential custodial care facility

Table 517.1 A Hospital Emergency System Must Serve These Loads (Sec. 517.32)

Life-Safety Branch	Critical Branch
Lighting and receptacles and equipment for: —Means of egress illumination —Exit and directional signs —Alarm and alerting systems • Fire alarms • Nonflammable medical gas piping • Mechanical controls for life safety —Communications for emergency use —Generator—set battery charger, lighting, together with accessory loads for a specific generator that are essential to keep it in operation —Transfer switch lighting, selected receptacles —Elevator cab lighting and communications —Automatic doors used for building egress	—Isolating transformers in special locations —Critical care areas using anesthetizing gas, selected receptacles, equipment —Task illumination and selected receptacles in patient care areas in: • Nurseries • Medication preparation • Pharmacy • Acute nursing • Psychiatric beds (no receptacles) • Nurses stations • Ward treatment rooms —Additional specialized care —Nurse call systems —Bone, blood, and tissue banks —Telephone equipment areas —Task illumination, selected receptacles, and power for the following: • General care beds (one duplex per) • Angiographic labs • Cardiac catheter labs • Coronary care • Dialysis • Emergency rooms • Human physiology labs • Intensive care • Postoperative recovery rooms —Additional task illumination, receptacles, and selected power circuits as needed for effective operation; 1φ fractional hp motors permitted

where the essential electrical system has a maximum demand of 150 kVA. Separate transfer switches are required only if dictated by load considerations. For small facilities, the essential electrical system generally consists of the life-safety branch and the critical branch. For larger systems, the critical branch is divided into three separate branches for patients, heating, and sump pumps and alarms. Code diagrams 517.42(a) and 42(b) illustrate typical installations. Part **(E)** requires nonlocking 125-V, 15- and 20-A receptacles to incorporate either an illuminated face or an indicator light on a receptacle connected to the essential electrical system so it will be obvious whether or not there is power at the receptacle location. This requirement was also supposed to apply in hospitals, but the amendment to then 517.30 was rejected by the NFPA Standards Council on a complaint from the NFPA 99 Committee that this was a performance requirement at variance from the applicable provisions in the other standard. The same committee may have failed to notice this amendment, because no complaint was filed.

517.43. Automatic Connection to Life-Safety Branch. Part **(A)** describes the switching arrangements for night transfer of corridor lighting. The rule is intended to ensure that some lighting will always be provided in the corridor regardless of the mode of operation.

517.44. Connection to Equipment Branch. This section details the loads requiring transfer from normal source to the alternate power source. In part **(B)(2),** elevator operation must not trap passengers between floors. This is now mandatory.

517.45. Essential Electrical Systems for Other Health Care Facilities. This section is increasingly important as health care decentralizes. If these facilities have critical care areas or patients on life support, then an essential electrical system as covered in 517.30 through 517.35 will be applied, or a battery system is permitted.

517.60. Anesthetizing Location Classification. The rules for flammable anesthetizing locations in the NEC are virtually obsolete. No hospital in the United States is using or has used ether and other such agents for many years. However, the rules have not been removed from the NEC because the NEC is used in other countries, and some of them are still using flammable anesthetics. Figure 517-6 shows the classified hazardous locations of part **(A).**

Hazardous locations rules are separated from those for other-than-hazardous locations. A third category is also designated: "above-hazardous locations." The Code covers wiring and equipment in three relations to anesthetizing locations: 517.61(A), within-hazardous; 517.61(B), above-hazardous; and 517.61(C), other-than-hazardous.

517.61. Wiring and Equipment. Part **(A)** calls for explosionproof wiring methods, in general, for such locations.

Part **(A)(4)** notes an extension of the hazardous boundary. 517.60(A)(1) defines the area of a flammable anesthetizing location as a Class I, Division 1 location from the floor to a point 5 ft (1.5 m) above the floor. The question then arises, "Is the seal required in the upper conduit entering the switch box on the wall of a hospital operating room as shown in Fig. 517-7?" The box is partly below and partly above the 5-ft (1.5-m) level.

Part **(A)(4)** states that if a box or fitting is partially, but not entirely, beneath the 5-ft (1.5-m) level, the boundary of the Class I, Division 1 area is considered

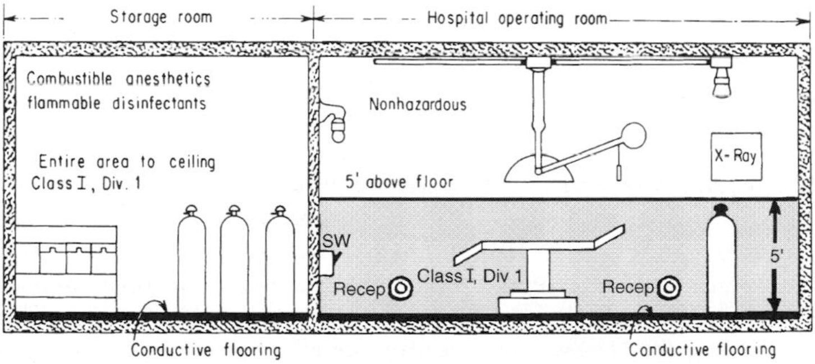
Storage room
Hospital operating room
Combustible anesthetics
flammable disinfectants
Entire area to ceiling
Class I, Div. 1
Nonhazardous
5' above floor
X-Ray
SW
Recep
Class I, Div 1
Recep
5'
Conductive flooring
Conductive flooring

Fig. 517-6. Two types of hazardous locations must be identified. (Sec. 517.60.)

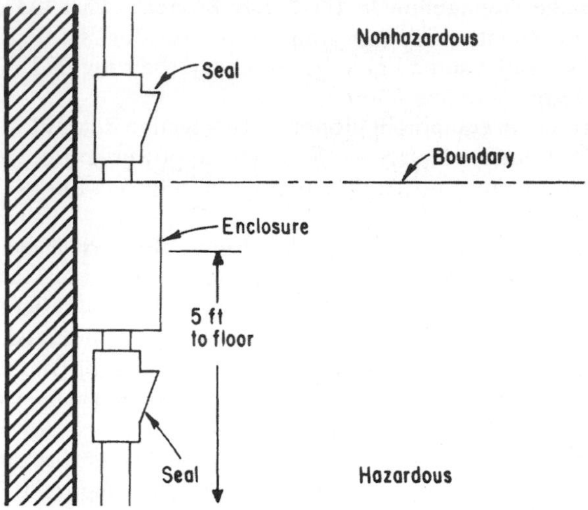
Nonhazardous
Seal
Boundary
Enclosure
5 ft
to floor
Seal
Hazardous

Fig. 517-7. Boundary of Class I, Division 1 location may be extended. (Sec. 517.61.)

to extend to the *top* of the box or fitting. Therefore, the box or fitting is entirely within the hazardous area, and a seal is required in conduit entering the enclosure from either above or below, as shown in Fig. 517-7.

If the box or fitting is entirely *above* the 5-ft (1.5-m) level, a seal would not be required at the box or fitting, but conduit running to the box from the hazardous area would have to be sealed at the boundary, on the hazardous-location side of the box. If the box shown were recessed in the wall instead of surface-mounted, some means would have to be provided to make the seals accessible [501.15(C)(1)], such as removable blank covers at the locations of the seals.

Part **(A)(5)** calls for explosionproof receptacles and plugs within hazardous locations described in 517.60(A).

In part **(B)(1)** of this section, rigid metal conduit, electrical metallic tubing (EMT), and intermediate metal conduit (IMC) or Type MI or MC cable that has a continuous "gas/vaportight" sheath are permitted in "above-hazardous anesthetizing locations." In "other-than-hazardous anesthetizing locations" [517.61**(C)(1)**], wiring may be in a metal raceway or cable assembly, but the cable armor must qualify as a ground return path per 250.118.

Hospital-Grade Receptacles

The **NEC** generally requires that "hospital-grade" receptacles (the UL-listed green-dot wiring devices) be used at patient bed locations in health care facilities. However, note that 517.61(B)(5) and 517.61(C)(2) of the Code also require use of "receptacles and attachment plugs" that are "listed for hospital use" *above* "hazardous" anesthetizing locations and *in* "other-than-hazardous" anesthetizing locations. As described, these rules require that all 2-pole, 3-wire grounding-type receptacles and plugs for single-phase 120-, 208-, or 240-V ac service must be

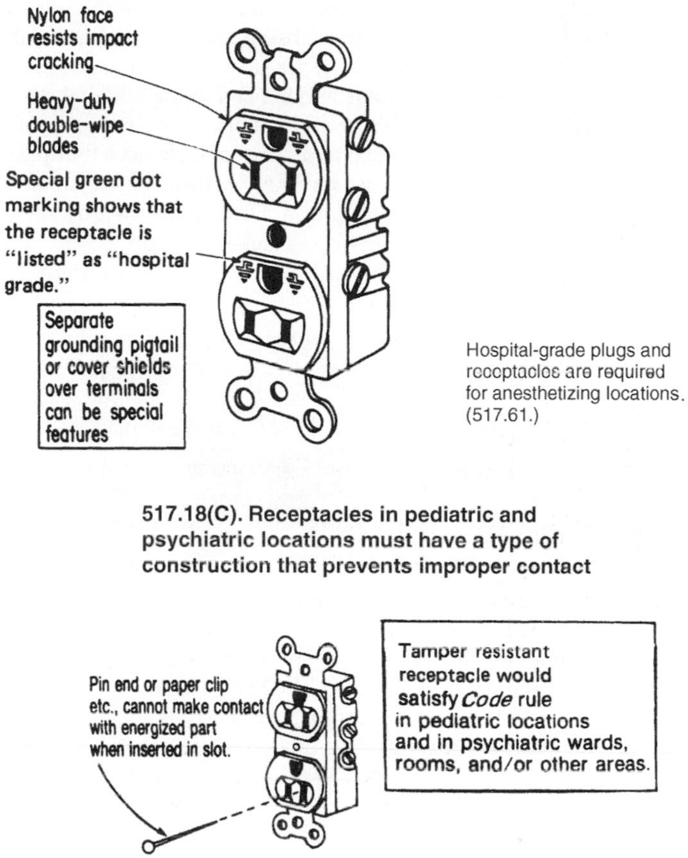
Nylon face resists impact cracking
Heavy-duty double-wipe blades
Special green dot marking shows that the receptacle is "listed" as "hospital grade."
Separate grounding pigtail or cover shields over terminals can be special features
Hospital-grade plugs and receptacles are required for anesthetizing locations. (517.61.)
517.18(C). Receptacles in pediatric and psychiatric locations must have a type of construction that prevents improper contact
Pin end or paper clip etc., cannot make contact with energized part when inserted in slot.
Tamper resistant receptacle would satisfy *Code* rule in pediatric locations and in psychiatric wards, rooms, and/or other areas.

Fig. 517-8. Hospital-grade plugs and receptacles are required for anesthetizing locations. (Sec. 517.61.)

marked "Hospital Only" or "Hospital Grade" and have a green dot on the face of each receptacle (Fig. 517-8). The relation between the phrase "listed for hospital use" and the phrase "Hospital Grade" is explained in the UL Guide Card information [see discussion at 110.3(B) for details], under the heading "Attachment Plug Receptacles and Plugs." It says:

> Receptacles for hospital use in other than hazardous (classified) locations in accordance with Art. 517 of the NEC are identified (1) by the marking "Hospital Only" (used to identify a specific grounding locking configuration rated 20 A, 125 V used for the connection of mobile x-ray and similar equipment) or (2) by the marking "Hospital Grade" and a green dot on the face of the receptacle. The identification is visible during installation on the wiring system or, in the case of the appliance outlet, after installation on the utilization equipment.

Of course, in the defined hazardous areas of flammable anesthetizing locations [517.60 and 517.61(A)(5)], receptacles must be explosionproof type, listed for Class I, Division 1 areas.

Underwriters Laboratories devised a special series of tests for wiring devices intended for hospital use. These tests are substantially more abusive than those performed on general-purpose devices and are designed to ensure the reliability of the grounding connection in particular, when used in the hospital environment. Hospital-grade receptacles have stability and construction in excess of standard specifications and can stand up to abuse and hard use. Devices that pass this test are listed as "Hospital Grade" and are identified with these words and a green dot, both of which are visible after installation. UL listings include 15- and 20-A, 125-V grounding, nonlocking-type plugs, receptacles, and connectors. This class of device is acceptable for use in any nonhazardous anesthetizing location (Fig. 517-9).

Fig. 517-9. Although the **NEC** requires use of hospital-grade receptacles at patient bed locations and in "anesthetizing" locations, the ruggedness and high degree of connection reliability strongly recommend their use for such critical applications as plug-connection of respiratory or life-sustaining equipment.

As noted in the exception to 517.61B)(2), receptacles above a hazardous location do not have to be totally enclosed and may be standard, available hospital-grade receptacles of the type that would be used to satisfy the rule of 517.61(B)(5).

As covered in part **(C)(1),** in anesthetizing locations that are not hazardous (no flammable agents used), Type AC cable is recognized along with Types MI and MC cable and rigid metal conduit, IMC, and EMT—but any such cable or raceway must contain an insulated-copper equipment grounding conductor and its outer jacket must be an approved grounding conductor. Note that, although the rule here does not specify a copper conductor, because an anesthetizing location is a patient care area, the copper wire is required by 517.13.

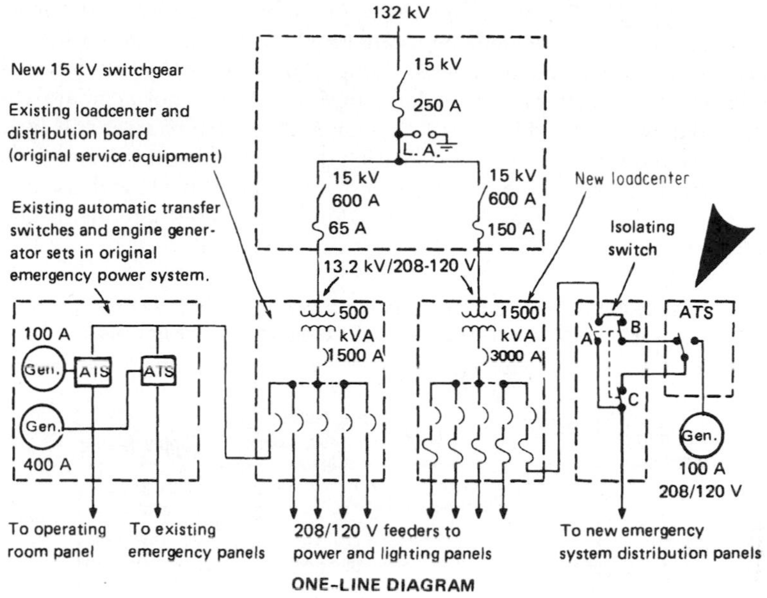
132 kV
New 15 kV switchgear
15 kV
250 A
Existing loadcenter and distribution board (original service equipment)
L. A.
15 kV
600 A
65 A
15 kV
600 A
150 A
New loadcenter
Existing automatic transfer switches and engine generator sets in original emergency power system.
Isolating switch
13.2 kV/208-120 V
500 kVA
1500 A
1500 kVA
3000 A
ATS
A
B
C
100 A
Gen.
ATS
ATS
Gen.
400 A
Gen.
100 A
208/120 V
To operating room panel
To existing emergency panels
208/120 V feeders to power and lighting panels
To new emergency system distribution panels
ONE-LINE DIAGRAM

Fig. 517-10. General-purpose lighting circuits may be fed from the normal service, or from a separate alternate source. [Sec. 517.63(C).]

517.63. Grounded Power Systems in Anesthetizing Locations. Part **(A)** used to call for a general-purpose lighting circuit, fed from the normal grounded service, to be installed in each operating room. And an exception used to recognize feed from an emergency generator or other emergency service that is separate from the source of the hospital's "Emergency System," as defined in 517.2. Figure 517-10 shows a layout where such an emergency supply (at lower right) is now the required source of supply to the lighting circuit in the operating room. Now, emergency illumination is required to be supplemented by at least one battery-powered emergency luminaire. This unit equipment (now formally defined in 517.2 as "Battery-Powered Lighting Units") is now permitted to be connected to the critical branch lighting circuit in the area, ahead of local switching. In the event of an extended normal power outage such units will run until their battery is discharged, at least 1½ h later; however, the lighting in the area will have been restored within 10 s by the emergency generator. Allowing these units to connect to the critical branch lighting circuit will avoid the pointless deep cycle discharge and long recharge time that would otherwise occur in the event of an extended normal system outage.

In part **(F)**, an isolated power system (ungrounded operation with a line isolation monitor) is required only in an anesthetizing location with flammable anesthetics. Long ago, isolated power systems were required for locations with both flammable and nonflammable anesthetics. But in 1984, a revision was made in the former 517.104(A)(1) to change the phrase "anesthetizing" to "flammable anesthetizing" in this section. As a result, anesthetizing locations that do not use flammable anesthetics (and as noted none do) don't require use of an isolated power system. 517.63(F) now makes the requirement for an isolated power system applicable only for anesthetizing locations where "flammable" anesthetics are used. For those cases where the anesthetizing location is used solely for nonflammable anesthetizing materials, an isolated power system is not mandatory in the NEC [although NFPA 99 and 517.20(B) for some wet procedure areas still require it under some circumstances]. It is also used by design in many facilities.

Figure 517-11 shows the application of a completely packaged transformer loadcenter to provide power for the ungrounded, isolated circuits in hospital operating suites.

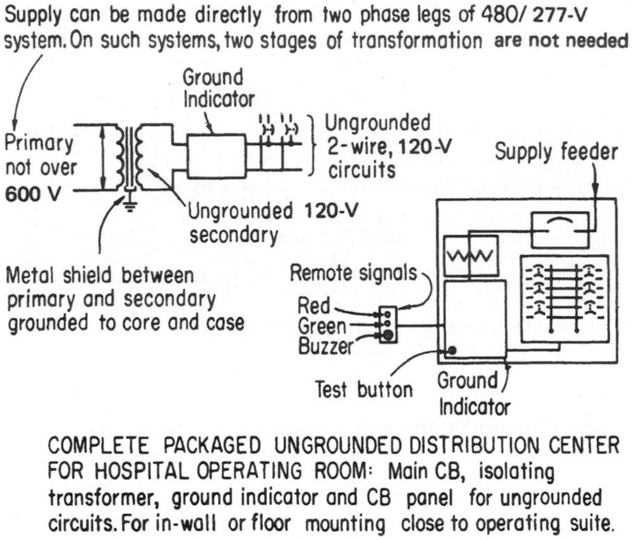
Supply can be made directly from two phase legs of 480/ 277-V system. On such systems, two stages of transformation are not needed
Ground Indicator
Primary not over 600 V
Ungrounded 2-wire, 120-V circuits
Supply feeder
Ungrounded 120-V secondary
Metal shield between primary and secondary grounded to core and case
Remote signals
Red
Green
Buzzer
Test button
Ground Indicator
COMPLETE PACKAGED UNGROUNDED DISTRIBUTION CENTER FOR HOSPITAL OPERATING ROOM: Main CB, isolating transformer, ground indicator and CB panel for ungrounded circuits. For in-wall or floor mounting close to operating suite.

Fig. 517-11. Isolated power supply is required for circuits only in flammable anesthetizing locations. (Sec. 517.63.)

517.64. Low-Voltage Equipment and Instruments. Specific details are given for use of low-voltage equipment in an anesthetizing location. Figure 517-12 shows some of the rules. 517.160(A)(2) limits isolating transformers to operation with primary at not over 600 V.

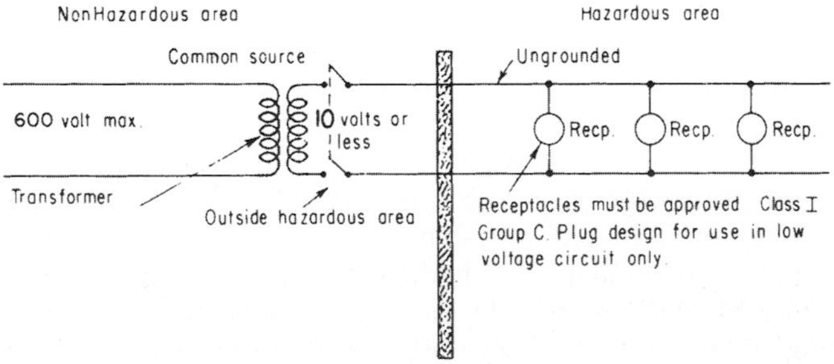
NonHazardous area
Hazardous area
Common source
Ungrounded
600 volt max.
10 volts or less
Recp.
Recp.
Recp.
Transformer
Outside hazardous area
Receptacles must be approved Class I Group C. Plug design for use in low voltage circuit only.

Fig. 517-12. Low-voltage circuits in anesthetizing locations must operate at 10 V or less or be otherwise approved. (Sec. 517.64.)

517.80. Patient Care Spaces. This section requires "equivalent insulation and isolation" to that required for patient care area power circuits, which has been interpreted as a requirement for equivalent mechanical protection, that is, raceways. For some time the raceway requirement has been debunked, but a new second paragraph here should decisively end all arguments. No special grounding requirements, no special mechanical protection requirements, and no special raceway requirement (unless required by Chap. 7 or 8 of the NEC) apply. See also the discussion in this book at 517.31(C)(1).

517.160. Isolated Power Systems. Each isolated power circuit must be controlled by a switch that has a disconnecting pole in each isolated circuit conductor to simultaneously disconnect all power. That is in part **(A)(1).** In addition, these circuits must not run within any raceway (or cable assembly or box or any other enclosure) that contains conductors that are not part of the isolated power system supplying those circuits. In fact, the literal text would even preclude conductors from two separate isolated power systems from running in a common raceway.

As covered in part **(A)(2),** any transformer used to obtain the ungrounded circuits must have its primary rated for not more than 600 V between conductors and must have proper overcurrent protection. This Code rule used to limit the transformer primary to 300 V between conductors, which often required two stages of voltage transformation to comply with the rule, as shown in Fig. 517-13. This diagram shows the circuit makeup used in a hospital to derive the 120-V ungrounded circuits, with transformation down from 480 to 240 V and then to 120 V. The ungrounded secondary system must be equipped with an approved ground contact indicator, to give a visual and audible warning if a ground fault develops in the ungrounded system.

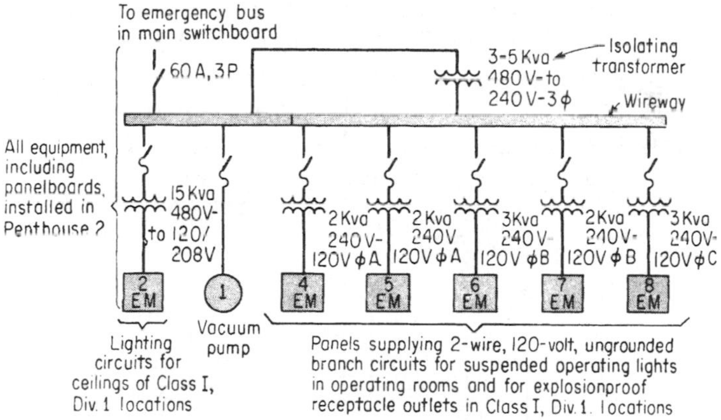
To emergency bus in main switchboard
60 A, 3P
3-5 Kva 480 V-to 240 V-3 ϕ
Isolating transformer
Wireway
All equipment, including panelboards, installed in Penthouse 2
15 Kva 480V-to 120/208V
2 Kva 240 V-120V ϕA
2 Kva 240V 120V ϕA
3Kva 240 V-120V ϕB
2Kva 240V-120V ϕB
3 Kva 240V-120V ϕC
2 EM
1
4 EM
5 EM
6 EM
7 EM
8 EM
Lighting circuits for ceilings of Class I, Div. 1 locations
Vacuum pump
Panels supplying 2-wire, 120-volt, ungrounded branch circuits for suspended operating lights in operating rooms and for explosionproof receptacle outlets in Class I, Div. 1 locations

Fig. 517-13. Two-stage transformation was often needed for isolated circuits but is no longer necessary. (Sec. 517.160.)

Isolating transformers must be installed out of the hazardous area [part **(A)(3)**]. The ground indicator and its signals must also be installed out of the hazardous area. In an anesthetizing location, the hazardous area extends to a height of 5 ft (1.5 m) above the floor.

Fixed lighting fixtures above the hazardous area in an anesthetizing location, other than the surgical luminaire, and certain x-ray equipment may be supplied by conventional grounded branch circuits [517.63(B) and 517.63(C)].

Part **(A)(5)** requires isolated circuit conductors to be identified by brown and orange colors. Each wire must be striped with some color other than white, gray, or green so it will not be confused, for example, with the high leg on a center-tapped delta system, and that striping must run for the entire length of the conductor. Although these systems are ungrounded and therefore by definition have no grounded circuit conductors, there is still a standard protocol as to which color to use when connecting to the "white" side of a conventional 125 V receptacle. Use the striped orange wire for this side of the receptacle.

Part **(B)(1)** details a line isolation monitor and clarifies line isolation monitor alarm values, specifying 5.0 mA as the lower limit of alarm for total hazard current. Figure 517-14 shows the basic concept behind detection and signal of a ground fault. The diagram shows major circuit components of a typical ground detector/alarm system. Partial ground energizes current-relay A, opening contact A2 (energizing red light and warning buzzer). Pressing the momentary-contact silencer switch energizes coil C, opening contact C1 (disconnecting buzzer), and closing holding contact C2. When ground is cleared, contacts resume the position shown in Fig. 517-14. This equipment must be listed.

The purpose of such a ground indicator is to provide warning of the danger of shock hazard and the possibility of a fault in the system due to accidental grounding of more than one conductor. If one conductor of an isolated system becomes grounded at one point, normal protective devices (fuses or CBs) will

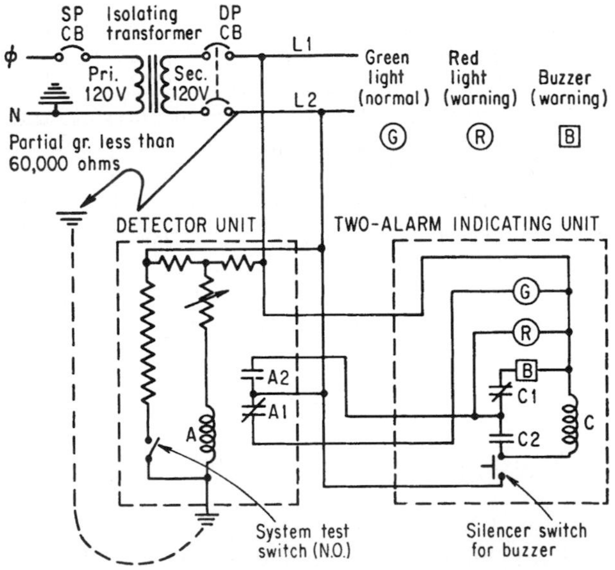
SP CB
Isolating transformer
DP CB
ϕ
N
Pri. 120V
Sec. 120V
L1
L2
Green light (normal)
Red light (warning)
Buzzer (warning)
G
R
B
Partial gr. less than 60,000 ohms
DETECTOR UNIT
TWO-ALARM INDICATING UNIT
A2
A1
A
C1
C2
C
System test switch (N.O.)
Silencer switch for buzzer

Fig. 517-14. Detection and alarm on ground fault is required for isolated power systems. (Sec. 517.160.)

not operate because there is no return path and, therefore, no flow of short-circuit fault current. However, if an accidental ground subsequently develops on the other conductor, a short circuit will occur, with possible disastrous consequences, such as ignition of ether vapors by arc or a lethal shock to personnel.

ARTICLE 518. ASSEMBLY OCCUPANCIES

518.1. Scope. This article covers places of "assembly," but does not apply to theaters, which are regulated by Art. 520. It covers any single indoor space (a whole building or part of a building) designed or intended for use by 100 or more persons for assembly purposes. That includes dining rooms, meeting rooms, entertainment areas (other than with a stage or platform or projection booth), lecture halls, bowling alleys, places of worship, dance halls, exhibition halls, museums, gymnasiums, armories, group rooms, funeral parlor chapels, skating rinks, pool rooms, transportation terminals, court rooms, sports arenas, and stadiums. A school classroom for less than 100 persons is not subject to this article. See 518.2. A supermarket with a rated occupancy load over 100 persons is not subject to this article because there is no assembly purpose and no likelihood of self-reinforcing panic during an emergency, as contrasted with the same number of people in an auditorium.

The clear differentiation given in this section points out that any such building or structure or part of a building that contains a projection booth or stage platform or even just an area that may, on occasion, be used for presenting theatrical or musical productions—whether the stage or platform is fixed or portable—must comply with the rules of Art. 520, as if it were a theater, and not Art. 518. A restaurant, say, that has a piano player for entertainment on Saturday night, could readily be classed as a theater and subject to Art. 520. This is covered in 518.2(C).

Article 518 directs attention "to a building or portion of a building" that would be used for the purposes outlined; therefore, you would have to determine how the occupancy is used.

A supermarket generally would not have a public assembly area. However, a department store could incorporate a community room for shows and similar audience functions. This room would be subject to Art. 518. The main areas of supermarkets and department stores, unlike theaters and assembly halls, have many aisles and exits that could be used in case of emergency evacuation of the building. It is these characteristics that permit conventional wiring methods to be accepted.

A proposal was once made to include supermarkets and department stores as "places of assembly," but it was rejected.

Note that places of assembly covered by this article must be for 100 or more people. As a practical matter, questions of occupancy load belong with the local building official.

518.3. Other Articles. Part **(B)** permits exhibition hall wiring for trade shows and the like to use the temporary wiring in accordance with Art. 590, and also recognizes the use of dedicated cable trays for hard- and extra-hard-usage cords run in such trays. It also waives the GFCI protection rules for the temporary wiring that is necessarily widespread at exhibitions. However, all other GFCI rules do apply. For example, if an exhibit at a home-improvement trade show featured a wet bar, 210.8(B)(5) requires receptacles within 1.8 m (6 ft) of the sink to have GFCI protection. This section makes it clear that that requirement continues in force.

In addition, to provide that requirement at a trade show, it is very likely that the protection would be established at the end of an extension cord. When this happens the real possibility exists for a loss of neutral continuity. When this happens a conventional GFCI will fail on, because the internal electronics will have lost the voltage required to function. This is why 590.6(A)(2) requires GFCI "identified for portable use." These devices have open neutral protection and disconnect in the event the neutral is open. This type of GFCI protective device is required to provide power in these cases. This is the practical solution and such devices are very much available; it is very unlikely that a field-assembled substitute, although allowed in this case, would be satisfactory.

518.4. Wiring Methods. The basic rule says that fixed wiring must be in a metal raceway, flex metal conduit, and nonmetallic raceways encased in *not less than 2 in. (50 mm) of concrete*, Type MI, MC or AC cable. Part **(B)** says that nonmetallic-sheathed cable, BX (Type AC cable), electrical nonmetallic tubing (ENT), and rigid nonmetallic conduit may be used in building areas that are *not* required by the local building code to be of fire-rated construction. Note that use of these methods no longer relates to the number of persons that the place holds, which was once in this rule. Another exception permits the use of other wiring methods for sound systems, communication circuits, Class 2 and 3 remote-control and signal circuits, and fire-alarm circuits.

Another allowance gives limited permission to use ENT and rigid PVC (and RTRC) conduits in smaller conference and meeting rooms in restaurants and in a variety of similar occupancies, provided the nonmetallic wiring is behind a thermal barrier providing at least 15 min of fire separation, such as conventional ½-in. drywall. It is also permitted above a suspended ceiling assembled in compliance with a recognized fire rating protocol. Of course, if the suspended ceiling is air-handling, 300.22(C) would not allow such wiring methods within the plenum cavity ceiling.

ARTICLE 520. THEATERS, AUDIENCE AREAS OF MOTION PICTURE AND TELEVISION STUDIOS, PERFORMANCE AREAS, AND SIMILAR LOCATIONS

520.1. Scope. Where only a part of a building is used as a theater or similar location, these special requirements apply only to that part and do not necessarily apply to the entire building. A common example is a school building in which there is an auditorium used for dramatic or other performances. All special requirements of this chapter would apply to the auditorium, stage, dressing rooms, and main corridors leading to the auditorium but not to other parts of the building that do not pertain to the use of the auditorium for performances or entertainment.

520.2. Definitions. These are the specialized terms used in this article. Of note for the 2008 NEC is the new term (and technology) "Solid-State Sine Wave Dimmer."

This equipment produces current that varies with voltage over time, so both the current and voltage traces are in step. These dimmers do not draw harmonic currents from the neutral and the feeders and branch circuits they feed do not need to consider their neutrals to be loaded due to harmonic, nonlinear currents. They are also recognized in 518.5 for assembly areas.

520.5. Wiring Methods. Building laws usually require theaters and motion picture houses to be of fireproof construction; hence, practical considerations limit the types of concealed wiring for light and power chiefly to metal raceways. Only Type MI or MC cables may be used, or Type AC with an insulated equipment grounding conductor. Cables were long ago found unsuitable for circuits in theaters because they do not readily offer increase in the size of conductors for load growth. Many instances of overfusing dictated the value of raceways, which do permit replacement of larger conductors for safely handling load growth. Other wiring methods are permitted in areas for which the local building code does not require fire rating.

Much of the stage lighting in a modern theater is provided by floodlights and projectors mounted in the ceiling or on the balcony front. In order that the projectors may be adjustable in position, they may be connected by plugs and short cords to suitable receptacles or "pockets."

There is an exception to the basic rule. Recording, communications, remote-control signaling, and fire alarm circuits may be installed using other wiring methods. And those installations described in parts **(B)** and **(C)** may also use other wiring methods.

520.6. Number of Conductors in Raceway. This rule is important because it removes the 30 conductor ceiling for waiving the mutual conductor heating limitations of 310.15(B)(3)(a). Therefore larger numbers of conductors can run within wireways and auxiliary gutters without a derating penalty. This arrangement is very advantageous for the large numbers of conductors, supplied by many different dimmers, that supply stage lighting that goes on and off as the lighting needs of a production change.

520.9. Branch Circuits. This section includes other important Chap. 5 modifications of Chaps. 1 through 4 rules (see 90.3), this one specifically waiving the 80 percent current limitation for receptacle loading on circuits with multiple receptacles or outlets that normally applies. For example, each receptacle in a 20 A duplex receptacle or comprising multiple single 20-A receptacles on a 20-A circuit must not be loaded over 16 A under normal circumstances, but for theatrical applications every receptacle could be loaded up to the full 20 A as long as the receptacle rating was fully equal to the size of the overcurrent device. These receptacles typically operate for short loading periods that would not cause nuisance tripping of the overcurrent protection. It should be noted that the usual plugs and receptacles for this purpose are not of the usual NEMA configuration, but rather a linear arrangement of three unevenly spaced pins and sleeves for which the spacing polarizes the device.

The express permission in the first sentence "of any size" allows a receptacle circuit to supply stage lighting equipment of large current ratings even if it exceeds the 40- and 50-A limitations for lighting connections in 210.23. Normally, however, the loading will be subdivided through branch-circuit overcurrent devices that are lower than those sizes. Note also that 520.41(B) does not waive the enhanced lampholder requirements in 210.23 that apply to circuits rated above 20 A. This section grants permission to operate stage set lighting at 100 percent of a receptacle configuration, and also, for outdoor stage sets, to do without GFCI protection for such receptacles that would normally apply.

520.10. Portable Equipment Used Outdoors. Portable stage equipment not rated for use outdoors is permitted outdoors on a temporary basis, provided it is appropriately supervised and the general public is excluded from access.

520.21. General. The 2014 NEC has greatly modernized the coverage of stage switchboards. This section is completely rewritten, including a listing requirement and requiring all circuits supplied by the switchboard to have their overcurrent devices contained within it. This correlates with the definition, which includes the potential for dimmers and relays, and the result is that if the switchboard controls a circuit, it also includes the overcurrent device for that circuit. This work also correlates with the new part (D) of Sec. 520.26 that allows for a "constant power" type of stage switchboard that has no dimmers or relays. The former Secs. 520.22, 23, and 24 covered features for former switchboard designs that haven't been necessary for over 40 years, and they have been deleted for correlation.

Major stage switchboards were commonly of the remote-control type. Pilot switches mounted on the stage board controlled the operation of contactors installed in any convenient location where space is available, usually below the stage. The contactors in turn controlled the lighting circuits.

The stage switchboard of many years ago was usually built into a recess in the proscenium wall, as shown in the plan view, Fig. 520-1. After passing

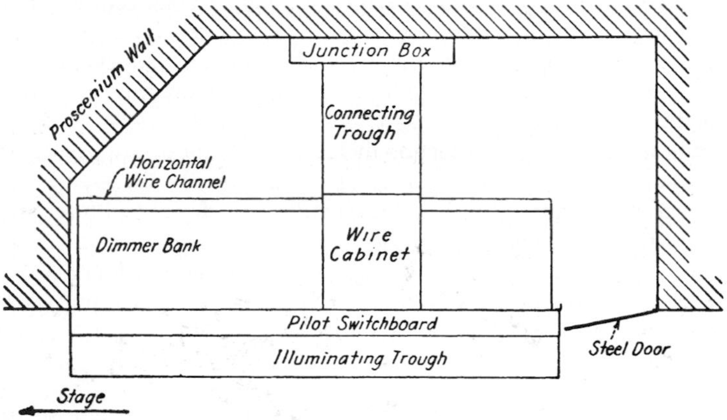
Proscenium Wall
Junction Box
Connecting Trough
Horizontal Wire Channel
Dimmer Bank
Wire Cabinet
Pilot Switchboard
Illuminating Trough
Steel Door
Stage

Fig. 520-1. A plan view for an old-fashioned stage switchboard, some of which are still in use. Modern switchboards are not required to be on stage and may be set up in multiple locations. (Sec. 520.21.)

through the switches and dimmers, many of the main circuits were subdivided into branch circuits so that no branch circuit will be loaded to more than 20 A. Where the board was of the remote-control type, the branch-circuit fuses were often mounted on the same panels as the contactors. Where a direct-control type of board was used, and sometimes where the board was remotely controlled,

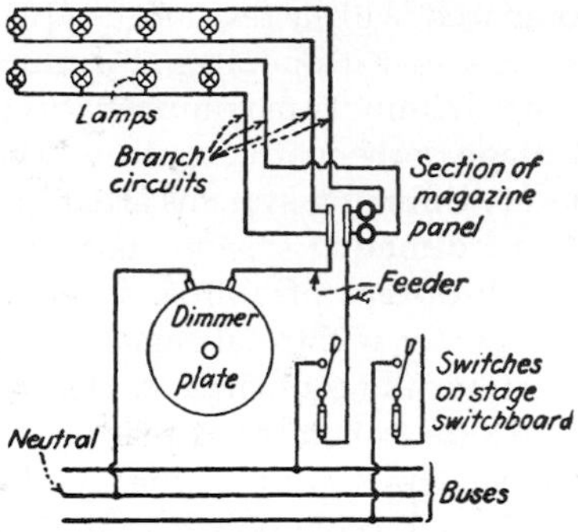
Lamps
Branch circuits
Section of magazine panel
Feeder
Dimmer plate
Switches on stage switchboard
Neutral
Buses

Fig. 520-2. Branch circuits of lighting may be controlled by single dimmer in grounded or ungrounded conductor. (Sec. 520.25.)

the branch-circuit fuses were mounted on special panelboards known as *magazine panels*, which were installed in the space in back of the switchboard, usually in the location of the junction box shown in Fig. 520-1.

520.25. Dimmers. Figure 520-2 shows typical connections of two branch circuits arranged for control by one switch and one dimmer plate or section, dating back many decades. These few lines of explanation are for those who may become involved in an antique system. The single-pole switch on the stage switchboard was connected to one of the outside buses, and from this switch a wire ran to a short bus on the magazine panel. The magazine panel was similar to an ordinary panelboard, except that it contained no switches and the circuits were divided into many sections, each section having its own separate buses. One terminal of the dimmer plate, or variable resistor, was connected to the neutral bus at the switchboard, and from the other terminal of the dimmer a wire ran to the neutral bus on the magazine panel. This neutral bus had to be well-insulated from ground and separated from other neutral buses on the panel; otherwise, the dimmer would have been shunted and would have failed to control the brightness of the lamps.

While the dimmer was permanently connected to the neutral of the wiring system, this neutral was presumed to be thoroughly grounded and, hence, the dimmer was dead. A dimmer in the grounded neutral does not require overcurrent protection, as noted in part **(A)**.

Figure 520-3 shows an autotransformer used as a dimmer. By changing the position of the movable contact, any desired voltage may be supplied to the lamps, from full-line voltage to a voltage so low that the lamps are "black-out."

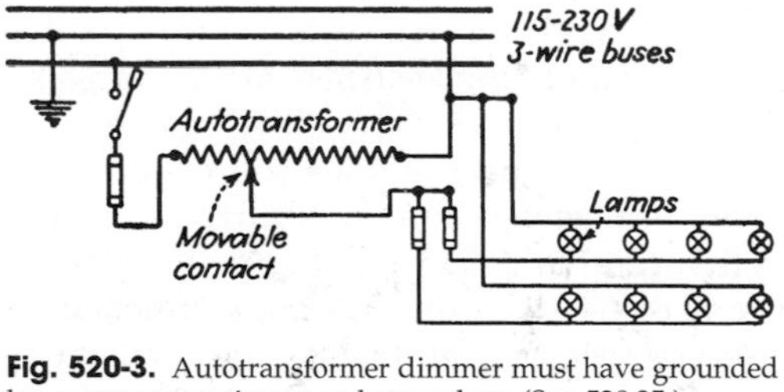
115-230 V
3-wire buses
Autotransformer
Movable contact
Lamps

Fig. 520-3. Autotransformer dimmer must have grounded leg common to primary and secondary. (Sec. 520.25.)

As compared with a resistance-type dimmer, a dimmer of this type has the advantages that it operates at a much higher efficiency, generates very little heat, and, within its maximum rating, the dimming effect is not dependent on the wattage of the load it controls. These dimmers are seldom used today; contemporary dimming systems are digitally controlled.

Modern dimming systems use digital controls to control multiple dimmer racks with dozens of dimmers per rack and where necessary, multiple racks bused together with total capacities measured in hundreds of kilowatts. Each dimmer typically controls a single 20 A circuit.

520.27. Stage Switchboard Feeders. Part **(A)(2)** of this section covers the supply of patch panels, formally referred to in the NEC as intermediate stage switchboards. They receive power from dimmer banks operating at various voltage levels and distribute it out over the various available branch circuits as selected, allowing sets of luminaires to operate at one of the available dimmer settings in the main stage switchboard. For this reason there will be a number of feeders, one for each dimmer setting. In addition, there will be a common neutral return sized to the maximum unbalanced load, but it need not be larger than the primary stage switchboard neutral. If run in the form of parallel conductors, the usual rules for parallel conductors apply, including the obligation to represent the neutral in all parallel raceways, to use wires no smaller than 1/0 AWG, and to divide the load on the associated ungrounded conductors as equally as possible. The common neutral means that the originating switchboard must be fed with a single feeder.

As digital controls take over this industry, patch panels are becoming obsolete. Instead, the power is increasingly distributed from the primary stage switchboard, avoiding the intermediate layer in the distribution altogether. Because literally hundreds of dimmers can be located at a convenient location, and because they can be addressed electronically with multiprocessor support and controlled individually with ease, this is the direction the theater industry is moving. These dimmer racks (switchboards) can be fed with a single feeder as in **(A)(1)**, or if of large size, fed with multiple feeders as in **(A)(3)**. In this case the switchboard must be prominently labeled with the disconnect location(s) and the number at each applicable location. Further, if the disconnects are from multiple switchboards, then the primary switchboards must include barriers to subdivide it in a manner that corresponds to the number of ungrouped remote disconnects that comprise its source of supply.

Part **(B)** used to require that all neutrals supplying these switchboards be counted as current-carrying conductors due to the harmonic loading produced by contemporary dimming systems. However, the advent of the new "solid-state sine wave 3-phase 4-wire dimming systems" made it possible for the 2008 NEC to waive this requirement, but only where the dimming is exclusively provided from such sources. If conventional dimming sources are used, in whole or even in part, then the neutrals must be counted as current-carrying.

Part **(C)**, which first entered the 1993 NEC, clarifies that switchboard feeders are permitted to be sized based on the maximum anticipated load, and not necessarily on the basis of every last watt that the dimmers can put out. There were many documented examples of inspection authorities making that insistence, which was completely unrealistic. To use this allowance the feeder must be protected at its ampacity (next higher standard size not allowed) and any egress or emergency lighting must be divorced from this feeder so it will be unaffected in the event it opens.

520.43. Footlights. A footlight of the disappearing type might produce so high a temperature as to be a serious fire hazard if the lamps should be left burning after the footlight is closed. Part **(C)** calls for automatic disconnect when the lights disappear.

There is no restriction on the number of lamps that may be supplied by one branch circuit. The lamp wattage supplied by one circuit should be such that the current will be slightly less than 20 A.

Individual outlets as described in part **(B)** are seldom used for footlights, as such construction would be much more expensive than the standard trough type.

A modern type of footlight is shown in Fig. 520-4. The wiring is carried in a sheet-iron wire channel in the face of which lamp receptacles are mounted. Each lamp is provided with an individual reflector and glass color screen, or *roundel.* The circuit wires are usually brought to the wire channel in rigid conduit. In the other type of footlight, still used to some extent, the lamps are placed vertically or nearly so, and an extension of one side of the wire channel is shaped so as to form a reflector to direct the light toward the stage.

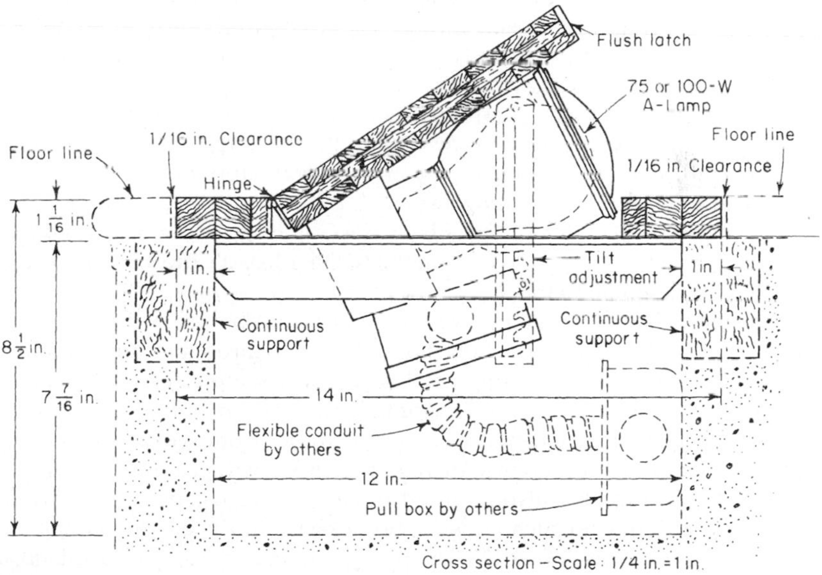
Flush latch
75 or 100-W A-Lamp
Floor line
1/16 in. Clearance
Floor line
1/16 in. Clearance
Hinge
1 1/16 in.
1 in.
Tilt adjustment
1 in.
Continuous support
Continuous support
8 1/2 in.
7 7/16 in.
14 in.
Flexible conduit by others
12 in.
Pull box by others
Cross section - Scale: 1/4 in. = 1 in.

Fig. 520-4. Footlights must be automatically deenergized when the flush latch is closed down. (Sec. 520.43.)

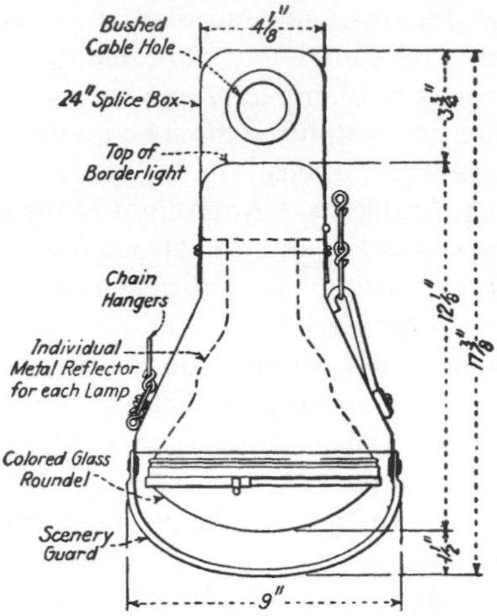
Bushed
Cable Hole
$4\frac{1}{8}''$
24" Splice Box
Top of
Borderlight
$3\frac{1}{4}''$
Chain
Hangers
Individual
Metal Reflector
for each Lamp
$12\frac{1}{8}''$
$17\frac{3}{8}''$
Colored Glass
Roundel
Scenery
Guard
$1\frac{1}{2}''$
9"

Fig. 520-5. Border lights must comply with NEC construction rules. (Sec. 520.44.)

520.44. Borders, Proscenium Sidelights, Drop Boxes, and Connector Strips. Figure 520-5 is a cross section showing the construction of a border light over the stage. This particular type is intended for the use of 200-W lamps. An individual reflector is provided for each lamp so as to secure the highest possible efficiency of light utilization. A glass roundel is fitted to each reflector; these may be obtained in any desired color, commonly white, red, and blue for three-color equipment and white, red, blue, and amber for four-color equipment. A splice box is provided on top of the housing for enclosing the connections between the border-light cable and the wiring of the border. From this splice box, the wires are carried to the lamp sockets in a trough extending the entire length of the border.

Border lights are usually hung on steel cables so that their height may be adjusted and so that they may be lowered to the stage for cleaning and replacing lamps and color screens; hence, the circuit conductors supplying the lamps must be carried to the border through a flexible cable. The individual conductors of the cable may be of 14 AWG, though 12 AWG is more commonly used. Note that there is a special ampacity table in the NEC for listed extra-hard-usage cord connected to these luminaires.

This section also covers connector strips and drop boxes. A drop box has a number of receptacles mounted on it and connects through a multiconductor cable. A connector strip is a longmetal wireway assembly (they are available in many lengths) with receptacles spaced along the strip, either directly on the strip or hanging as stage receptacle connector bodies from short-cord pendants. This equipment must be suitably stayed and supported, just as for border and proscenium lighting. It is also required to be listed as "stage and studio wiring devices." UL has no such classification exactly in the current Guide Card information [see discussion at 110.3(B) for details], but does show a "Stage and Studio Luminaires, Accessories and Connector Strips (IFDZ)" category whose products are being listed to the "Stage and Studio Luminaires and Connector Strips, UL1573" product standard.

The table governs the ampacity of extra-hard-usage multiconductor cords used for border lights, where they are not in direct contact with equipment containing heat-producing elements. The ampacities in this new table are identical to those in Table 400.5B in Column F (3 current-carrying conductors) for similar cords; however, where Table 400.5B begins at 8 AWG and goes up to 500 kcmil, this table begins at 14 AWG and goes up to 2 AWG. The ampacities below 8 AWG are proportionately reduced, but far higher than those in Table 400.5A which would otherwise apply to the smaller cords.

The table also sets the maximum overcurrent device ratings for each cord, which are identical to the values in Column A of Table 400.5A, that is, the traditionally determined cord ampacities for cord with three current-carrying conductors. These values are far lower than the allowable cord ampacities. For example, 10 AWG 75°C rated cord has an ampacity of 41A and maximum overcurrent device rating of 25A. These cords often run in areas with comparatively high temperatures, and these restrictions add a safety factor for this as well.

The value of the higher ampacities comes where large numbers of these cables travel together in limited space, resulting in derating penalties. This frequently occurs in these applications. The new table helps here as well, resurrecting the old mutual conductor heating table break points with the 70 percent bracket applicable for 7-24 conductors, the 60 percent bracket for 25-42 conductors, and 50 percent for more. This table now appears in Annex B as Table B.310.15(B)(3)(11).

These more liberal factors are conditioned on at least 50 percent load diversity. This is normally the case for the same reasons that allowed the change in calculations for stage switchboard feeders in 520.27(C).

A new informational note provides a definition of diversity for the use of this table. It is the percentage of total current of all simultaneously energized conductors within the cable being calculated, divided by the sum of the ampacity ratings of all "circuits" in the cable. There are some problems with this wording, beginning with the fact that circuits do not have ampacity. For now this should be applied as the number of conductors that can potentially carry current. What is important is that this factor, should it calculate below 50 percent, is a significant constraint on how the table can be applied. Many stage applications rely on assemblies with large numbers of conductors, and the prospect of mandatory applications of large derating factors is a significant constraint. Note that this table was titled Table 520.44 in prior Codes, and is now being titled Table 520.44(C)(3) even though that particular reference has nothing to do with ampacity calculations and the table should be retitled 520.44(C)(2).

If the current-carrying conductors exceed 50 percent of the total, then the derating factors in Table 400.5(A)(3) would have to be applied in order to avoid overheating the cables and violating the ultimate insulation temperature restriction at the bottom of the table.

Under typical stage conditions, it should almost always be possible to stay within this table for calculations using extra-hard-usage cords in general. The allowances are limited to extra-hard-usage cord, which includes the stage cables (SC, SCE, and SCT). Part **(C)(1)** generally requires extra-hard-usage cord for this equipment. Note that the new stage lighting hoists listed as complete assemblies with integral cable handling systems, as covered in 520.2 and 520.40

as of the 2014 NEC, are exempt. These hoists use specially configured cables, often flat, for which this requirement is impracticable.

Part **(C)(3)** provides further practical assistance for wiring this equipment, beyond the allowable ampacity modifications. For drop boxes and connector strips, the usual multiconductor cord configurations do not contain a white wire for every colored wire, and stage personnel would prefer to use these cables and reidentify some of the colored conductors. Access to this equipment is generally limited to qualified personnel. Therefore, 200.6(E) Exception No. 1 and 250.119(B)(B) already allow, under these conditions, both grounded and grounding conductors within multiconductor cables to be established through color coding methods applied in the field. However, allowance from the inspection community has been uneven and so this subsection provides unequivocal field permission to alter the color code on these wires at their points of termination.

520.49. Smoke Ventilator Control. A normally closed-circuit device has the inherent safety feature that, in case the control circuit is accidentally opened by the blowing of a fuse or in any other way, the device immediately operates to open the flue dampers.

520.50. Road Show Connection Panel (A Type of Patch Panel). This section effectively describes the interface between fixed stage wiring and portable stage switchboards as defined in 520.2 and covered in Sec. 520.53. Note that these provisions cover the construction of the connection panel and do not apply to wiring connected thereto. Part **(A)** requires the point where a circuit is connected to consist of an inlet that matches, by current and voltage, that of the receptacle supplied at the other end of the wiring outlet controlled by the panel. Part **(B)** requires that when a panel transfers power from the fixed switchboard in the building to that supplied by the portable switchboard, the transfer must include all circuit conductors including the grounded circuit conductor, and the transfer must be simultaneous with respect to all circuit conductors. Part **(C)** requires supplemental overcurrent protection for the supplied connections. Although this is normally provided in the supply switchboards, there are instances in old theaters where this cannot be relied upon, and the supplementary protection required here is important because as their name implies, these panels do not have a permanent address.

520.51. Supply. This section addresses the actual connection point for portable stage switchboards. These are often referred to as a "bull switch"; see the definition of this term in 530.2 for more information. The literal text calls for "power outlets" with the appropriate voltage and current rating. Be careful here because a "power outlet" in this context however, is not necessarily the equipment covered under this heading in the UL Guide Card information (guide card designation QPYV). Although the UL-covered equipment (typically enclosed assemblies incorporating receptacles and fused switches or circuit breakers) is certainly suitable for this application, it seems intended that this section also includes other connection methods including sequential interlocking connections of single conductors as covered in 520.53(K) and even direct connections to busbars, as covered in 520.53(H)(1). Although these latter references expressly cover only the load end of the feeder, the same feeder must have a supply connection. If the only possible connection were to an appropriately rated power outlet, then most of the motivation for special precautions and warning labels in 520.53(P) would disappear.

520.52. Overcurrent Protection for Branch Circuits. The portable switchboard is responsible for the overcurrent protection, but the usual limitations of 210.23 such as apply to receptacle loading at 80 percent do not apply to these circuits.

520.53. Construction. This section and the next cover the requirements for supplying portable switchboards (dimmer racks) used on stage, with this section focusing on the equipment. This equipment is covered by UL under the category "Power Distribution Equipment, Portable (Guide Card designator QPRW)." Part **(A)** requires a pilot light arranged to show the presence of power even if a main switch on the unit is in the open position.

Part **(B)** covers special rules for feeder neutral terminals. The neutral busbars and connecting supply terminal on a 3-phase 4-wire portable switchboard must "have an ampacity" not less than twice the "ampacity" of the largest ungrounded supply terminal. This oversizing allows these portable switchboards to be connected to single-phase systems where two phases will be connected to one phase line, potentially doubling the neutral current. If the portable switchboard has been specifically designed for field rearrangement from single- to three-phase and back again with the single-phase loading being left in balance when used, then the double size requirement is waived.

Part **(C)** covers single-pole separable connectors, which include three forms. This rule specifically waives the normal strain relief rule in 400.14, and thereby meets the exception thereto. It also waives the unique configuration and marking rules in 406.7 and 406.8 because such devices are rated in amperes but can be used on many circuit configurations including parallel applications where there will be any of a number of Phase "A" and "B" and "C" and "N" conductors, etc. For parallel applications, there must be a prominent label advising of the internal parallel connections. A compliant installation where these are used must meet at least one of three design criteria. The first is that the connections are interlocked with the source such that the feeder is deenergized until the connections are complete. The second option uses listed sequential interlocking mechanisms to assure appropriate sequencing of connections, with equipment grounding first, neutral second, and phase conductors last, with the reverse order for disconnection. The final option is single connectors with no interlocking mechanism, in which case a label at the line connections must detail the correct connection sequence. Note that any such application must only be serviced by qualified individuals, as covered in 520.54(K).

Part **(D)** allows an outlet for a feed-through feeder connection to avoid local overcurrent protection if it is of the same rating as its supply counterpart. This is entirely unnecessary; if the supply side is properly protected then the load side will be as well, but it is also harmless. Unlike the other parts of these two sections, this alone is new for 2017.

In Part **(E)**, specs cover conductors within portable stage dimmer switchboards, which must be stranded. Prior to the 2017 NEC, there were detailed rules covering the temperature ratings of conductors permitted in dimmer boards, based on the type of dimmer used. The rules recognized the difference in temperature of dimmers and permitted lower-rated conductors for solid-state dimmers. Those rules have been removed in favor of a comprehensive listing requirement in the parent language due to the sophistication involved in modern equipment.

520.54. Supply Conductors. Part **(A)** covers the general specifications for the feeder conductors to these units, which often are on casters to allow for easy repositioning. As such, the feeders must utilize a robust and flexible method of connection, and therefore require listed extra-hard-usage cord or cable. As in cases covered by 520.27(C), the supply ampacity must equal the load connected to the switchboard, and need not be sized in terms of its maximum capacity.

Part **(B)** requires that unless the switchboard is only using pure sine-wave dimmers, the neutral must be assumed as harmonically loaded and therefore required to be counted as current-carrying in terms of mutual conductor heating calculations.

Part **(C)** covers the use of single-conductor cable sets, provided the conductors are not smaller than 2 AWG. Single conductors do not experience mutual conductor heating and therefore have much higher ampacities than multiconductor make-ups. This makes them useful for these applications. For example, the ampacity of 2 AWG 75°C stage cable [from Table 400.5(A)(2)] is 170 A; if 75°C multiconductor cable were used, 520.54(B) would require counting the neutral on conventional dimmers, and therefore (170 A ÷ 0.8 = 213 A) a 3/0 AWG cable would be required. Such a cable weighs in at about 4.5 kg/m (3 lb/ft) and would not be practical for this use.

Parts **(D)** and **(E)** address tap rules for portable switchboards. They allow for smaller switchboards to be connected to larger sized power outlets covered in 520.51 without additional overcurrent protection. The feeders must not penetrate structural elements of the building, or run through doors or other traffic areas and must terminate at a single overcurrent device in the portable switchboard. This device must protect the feeder from overload by its size; no next higher standard size allowance applies here. If run as single conductors, the feeder conductors must not be bundled, and the terminations must be approved. Specifically, **(D)** covers 3.0 m (10 ft) taps, and in addition to the foregoing the feeder conductor ampacity must be not less than one-quarter of the rating of the supply-side protection, and the feeders must be kept off the floor in an approved way. Part **(E)** covers 6.0 m (20 ft) taps, which must adhere to the same rules except the protection ratio of one-half is more conservative, and these taps must be elevated not less than 2.1 m (7 ft) above the floor, except at terminations.

Part **(F)** allows feeders to pass through holes in a wall if specifically designed for this purpose, and provided the feeder is not using one of the tap allowances in the prior two paragraphs. The hole must be fire-stopped if a fire rating applies to it. This allows for special shows to run off an outdoor generator that would otherwise overtax the existing stage distribution system in the building.

Part **(G)** requires protection against physical damage to these conductors. For example, there are rubber mats designed for this purpose if the feeder runs on a floor, and if mutual conductor heating is not problematic, raceway protection is another option.

Part **(H)** limits mated conductor pairings (as in plugs and connector bodies) for feeders to three per run up to 30 m (100 ft). In the common instance where there is such a connection at each end, the rule would set a minimum cord length of 15 m (50 ft) for at least one set of cords. If the run is longer, one additional interconnection is permitted for every 30 m (100 ft).

Part (I), by specifically recognizing paralleled applications, also modifies 310.10(H)(1) to lower the threshold from 1/0 AWG to 2 AWG for these applications, although the equal length and the equal sizing rules still apply. In support of the principle that mutual conductor heating should not be a factor, the wording specifically prohibits bundling the single conductors, although they are to be grouped. The grounded conductors must be identified with at least 150 mm (6 in.) of marking at each end with white or gray, and the equipment grounding conductor (permitted to be not smaller than 6 AWG and otherwise sized by 250.122) the same way except the color is to be green. The ungrounded conductors must be identified by system if multiple voltage systems are present in the premises wiring.

Part (J)(1) duplicates the neutral marking rule in (C). The grounded conductors must be identified with at least 150 mm (6 in.) of marking at each end with white or gray. Part (J)(2) requires that the ampacity of a neutral on a multiphase system feeder to a portable stage switchboard must have an ampacity of 130 percent of the ungrounded conductors in cases where the feeder is run as single conductor cables. This rule only applies to single conductor uses, where the customary 125 percent increase in size for four conductors in a raceway or cable will not apply. This rule is due to the electrical characteristics of typical stage dimmer circuits and is not a response to the more general problem of harmonic currents. The U.S. Institute of Theatre Technology presented an analysis that the worst-case loading would occur when one phase was off, one phase was 100 percent loaded, and the third phase loaded to 55 percent; this would produce a 126 percent load on the neutral. For multiconductor cables, the harmonic content means the neutral is counted, resulting in a 125 percent factor already and this is judged sufficient to handle this loading. However, for single conductor cables, no such rule applies, so this section imposes a 130 percent factor. As in other comparable applications, the extra loading requirement is waived if the entire dimming system uses the new solid-state sine wave dimmers.

Part (K) requires portable stage switchboards to be connected by qualified personnel, and they must be conspicuously marked to indicate this requirement. An exception applies to smaller feeders 150 A and below using multipole receptacle outlets. The qualified personnel requirement applies to routing conductors, as well as to making and breaking connections and to energizing and deenergizing the switchboards. The exception only applies where a multipole, permanently installed receptacle outlet has overcurrent protection not above its rating and where it is not accessible to the general public. Extra-hard-usage cord with an ampacity equal to the receptacle rating must be used. The concept is that if the only task is to plug a multiconductor cord into an appropriately located power outlet, the work can be left to an amateur.

520.61. Arc Lamps. This section requires that all arc lamp fixtures must be listed, including any associated ballasts and interconnecting cables and cord sets. The term "arc lamp fixture" is intended to include all fixtures that rely on an arc lamp as the source of light, whether carbon-arc or HID. The motivation for the current wording came from concerns about extremely high-energy lighting sources that are being used in stage settings. Some of these sources use 24,000 W lamps with 90,000 V ignition circuits.

520.65. Festoons. "Lanterns or similar devices" are very likely to be made of paper or other flammable material, and the lamps should be prevented from coming in contact with such material. Staggering the joints lessens the likelihood of a line-to-line arcing fault that could start a fire in such lamps. The second sentence incorporates the requirement for stranded wire to be used with insulation-piercing lampholders as has long appeared in 225.24 for comparable outdoor uses.

520.68. Conductors for Portables. The basic rule is for listed extra-hard-usage flexible cord. Stand lamps are permitted with hard-usage cord. As of the 2014 **NEC** hard-usage cord in lengths not over a meter are allowed to connect to a listed luminaire or to an inlet on the luminaire with a "luminaire-specific listed connector." Luminaires operating at high temperatures are permitted with a 1.0 m (3.3 ft) connection typically consisting of some form of high-temperature fixture wire enclosed in a glass braid of some sort, and terminating in a pin-type stage plug. The braid is clamped by the plug housing and the luminaire connection, which provides a measure of strain relief. Paragraph 4 covers "breakout assemblies" and allows them to be made of hard-usage cord. They allow for one or multiple branch circuits to be taken out of a multiconductor cable that may have as many as 37 conductors within it. The use of hard-usage cord allows this to be done on a practical basis, and is appropriate given the conditions placed on their use. Limited to 20 A circuits, the branch cable cannot exceed 6.0 m (20 ft) and must be routed over its length attached to structure or other rigid supports. Note that this rule is not restricted to the supply of cord bodies, and that breakout plug assemblies are also available, in accordance with the same rules.

Part **(B)** allows cord ampacities and maximum currents to follow Table 520.44 unless they are in direct contact with "heat producing elements," in which case the usual 400.5 provisions apply. The fixture wires used in accordance with 520.68(A)(3) use conventional Code ampacity provisions, which in this case would be Table 402.5. This code rule now makes an express reference to the 50 percent diversity parameter in Table 520.44(C)(3). A low diversity factor applied to the cords commonly used under these rules may be a significant design constraint. Review the commentary provided at the table location in this book for additional information.

520.69. Adapters. The "two-fer" (a male plug made up with a molded splice generally configured in a wye arrangement to feed two female receptacles) is a good example of the use of this section. The current and voltage ratings of the plug and the receptacles supplied must be identical, with no reduction in current rating to either branch. See the definition in 520.2 and also Sec. 520.67 for additional requirements. The conductors must be listed, extra-hard-usage cord. Alternatively, hard-usage cord with its overall length limited to 2.0 m (6.6 ft) can be used.

520.72. Lamp Guards. Lamps in dressing rooms should be provided with guards that cannot easily be removed to prevent them from coming in contact with flammable material.

520.73. Switches Required; 520.74. Pilot Lights Required. The luminaires and receptacles located adjacent to the mirrors and above a dressing table in a dressing room must be controlled by one or more wall switches located in the dressing room but equipped with pilot circuits brought out to a point near the dressing room door. This allows supervisory personnel to quickly determine if likely loads in the dressing areas are still connected. As of the 2017 NEC, the pilot lights needs to be equipped with LEDs or extended-life lamps, recessed or guarded, and carry a marking to show what its switch controls.

ARTICLE 522. CONTROL SYSTEMS FOR PERMANENT AMUSEMENT ATTRACTIONS

This article was new as of the 2008 NEC. It covers the unique wiring requirements of permanent theme parks and comparable attractions, such as Disney World. There are almost 500 of these parks now operating in the United States, and many others in different countries that also use the NEC for their electrical installation standard. It is a very unusual article in that it does not cover power circuits, only the controls for equipment connected to power circuits. As such it recognizes and provides an ampacity table for conductors as small as 30 AWG.

ARTICLE 525. CARNIVALS, CIRCUSES, FAIRS, AND SIMILAR EVENTS

525.1. Scope. The rules given here apply to the wiring of temporary power used to supply power to rides and amusements at circuses, and so on. However, it should be remembered that, unless otherwise modified by Art. 525, the requirements given in Chaps. 1 through 4 also apply.

525.3. Other Articles. This article doesn't apply within permanent structures; Art. 518 and 520 apply instead. Art. 640 applies to audio signal and related wiring and equipment. Art. 680 (presumably focusing on Part I on general rules and Part III on storable applications, but the reference, to the entire article in violation of the whole-article reference prohibition in the NEC Style Manual, is extremely unclear) applies to attractions using contained volumes of water.

525.5. Overhead Conductor Clearances. Portable structures must not be placed under or within 4.5 m (15 ft) horizontally of overhead medium-voltage conductors, including utility primaries arranged on adjacent poles. This 4.5 m (15 ft) exclusion is to be applied from any point directly under the overhead conductors to a point on grade 4.5 m (15 ft) away, on both sides, all the way up to the level of the overhead conductor span. An example would be a standard 13.8 kV 3-phase utility primary run on cross-arms with a total span of 1.8 m (6 ft) between the outer phase wires [i.e., 900 mm (3 ft)] to either side of the actual pole line. No portion of any Ferris wheel, merry-go-round, portable ticket booth, or other "portable structure" shall be placed within 5.5 m (18 ft) of that pole line. The 4.5 m (15 ft) clearance also applies in any direction from overhead wiring at 600 V and below, except for the supply conductors to the structure.

525.10. Services. If in locations accessible to unqualified persons, service equipment must be lockable. It must be securely fastened to solid backing and protected from weather (unless weatherproof).

525.11. Multiple Sources of Supply. If multiple power sources, such as two different generator wagons, supply adjacent portable structures, the equipment grounding conductors of both sources must be bonded together if the structures are separated by less than 3.7 m (12 ft). The size of the bonding conductor must not be less than the required size of an equipment grounding conductor that would be used (per Table 250.122) based on the largest overcurrent device supplying load in either of the two structures, but in any case not smaller than 6 AWG.

525.20. Wiring Methods. The principal wiring method for traveling carnivals is flexible cord. As presented in **(A)**, for general use it should be listed for extra-hard service and for outside venues, listed for sunlight resistance and wet locations as well. Single-conductor cable is not permitted below 2 AWG, and splices are not permitted. Cord connectors must not be laid in traffic paths or within areas accessible to the public unless guarded. They must not be laid on the ground unless booted or otherwise made suitable for wet locations. They must be arranged so they do not create a tripping hazard, and they should be covered with low-profile matting that does not increase the tripping hazard. They can be pushed into a slit in the ground and then taken up because 525.20(G) waives 300.5 for these events. The conductors must only be spliced in boxes or in a comparable enclosure.

525.21. Rides, Tents, and Concessions. Each "structure," which would include rides such as Ferris wheels, must have a disconnecting means within 1.8 m (6 ft) of the operator's station at a readily accessible point while the ride is operating. A shunt trip device controlled from the operator's console is acceptable for this purpose. Wiring in tents and concession stands must be securely supported and provided with mechanical protection if subject to damage. Lamps must be either mounted within a luminaire or equipped with guards.

525.22. Portable Distribution or Termination Boxes. If used outdoors, these boxes must be arranged to sit at least 150 mm (6 in.) above the ground, and of weatherproof construction. The overcurrent protection ahead of the box must not exceed the current rating of any included busbars. Where the box includes receptacles, overcurrent protection must be included within the box, rated not higher than the receptacles.

525.23. Ground-Fault Circuit Interrupter (GFCI) Protection. As covered in part **(A),** all 15- and 20-A, 125-V rated non-locking-type receptacles and equipment accessible to the general public must be provided with GFCI protection for personnel protection. There is no waiver for equipment incompatible with a GFCI, as in the recent past. The idea is that if a GFCI is nuisance tripping, the appliance should be repaired or replaced. There are many old popcorn popping machines and comparable equipment with an electrical resistance inserted into refractory material that is somewhat conductive until it heats up and drives the water out of the refractory insulation. Until that happens, a GFCI will nuisance trip. However, it is also true that modern equipment designs do not have this shortcoming. In part **(B)** the Code states that there is no need for locking-type receptacles to be GFCI protected. This rule, for "quick disconnecting and reconnecting" is aimed at the multiple connections of lighting strips and other connections on large rides. Receptacles qualifying under this waiver must not be accessible from grade.

And (C) states that circuits supplying egress lighting are prohibited from employing GFCI protection.

The 2017 **NEC** added an additional part (**D**) to address widespread misunderstanding and improper use of conventional GFCI receptacles mounted in a box and fed using flexible cord, especially at venues governed by this article. That practice is unsafe because if the neutral connection opens, the GFCI will fail closed, removing all shock protection from the energized side of the receptacle. There have been documented fatalities from this problem, which is why 590.6(A)(1) only allows GFCIs identified for portable use in such cases on construction sites. A GFCI for portable use has a drop-out relay built into it that opens both sides of the circuit if circuit voltage fails.

This new provision is carefully written to address branch circuits. Portable distribution equipment supplied by feeders (see Fig. 590-5 for an example) and equipped with multiple GFCI devices supported with circuit breakers so as to originate multiple protected branch circuits fully comply with the new rule.

525.32. Grounding Conductor Continuity Assurance. Each time portable electrical equipment is deployed, the continuity of the grounding system must be verified. This would presumably take place during the final stages of set-up at each venue. The electrical inspection community may need to stipulate timing and access so as to be able to judge compliance with this requirement.

ARTICLE 530. MOTION PICTURE AND TELEVISION STUDIOS AND SIMILAR LOCATIONS

530.1. Scope. Article 520 covers theaters used for TV, motion picture, or live presentations where the building or part of a building includes an assembly area for the audience. Article 530, however, applies to TV or motion picture studios where film or TV cameras are used to record programs and to the other areas of similar application, but where the facility does not include an audience area.

The term *motion picture studio* is commonly used as meaning a large space, sometimes 100 acres or more in extent, enclosed by walls or fences within which are several "stages," a number of spaces for outdoor setups, warehouses, storage sheds, separate buildings used as dressing rooms, a large substation, a restaurant, and other necessary buildings. The so-called stages are large buildings containing numerous temporary and semipermanent setups for both indoor and outdoor views.

The Code rules for motion picture studios are intended to apply only to those locations where special hazards exist. Such special hazards are confined to the buildings in which films are handled or stored, the stages, and the outdoor spaces where flammable temporary structures and equipment are used. Some of these special hazards were due to the presence of a considerable quantity of highly flammable film, but this is rarely the case today since nitrocellulose film was discontinued in the early 1950s; otherwise, the conditions are much the same as on a theater stage and, in general, the same rules should be observed as in the case of theater stages. The rules in this article also amend provisions in Chap. 1 through 4 in ways that reflect some unusual load profiles and time factors that make possible special allowances.

530.11. Permanent Wiring. At one time, this section used the word *metal* between "approved" and "raceways," in the first sentence. Because the word *metal* no longer appears in the rule, rigid nonmetallic conduit is, therefore, acceptable for use in motion picture and TV studios. In addition to raceways, Type MI and Type MC cables and Type AC cable with an insulated grounding conductor are also permitted.

In the exception, Class 2 and 3 remote-control or signaling circuits and power-limited fire-protective signaling circuits are exempted from the basic rule requiring permanent wiring to be in raceways or certain cables. These circuits, along with communications and sound recording and reproducing circuits, are exempt from the basic rule.

530.12. Portable Wiring. Wiring that would be subject to physical damage needs to be extra-hard-usage cord; otherwise hard usage is permitted. Splices are permitted provided the load doesn't exceed the cable ampacity. Stage effect wiring can be either single or multiple conductor listed cord, provided it is protected from damage and secured to the scenery with cable ties or insulated staples. Splices are permitted only with listed devices on circuits not over 20A. For other equipment, cords that aren't suitable for extra hard usage are permitted if shipped as part of a listed assembly and protected at not over 20 A.

530.13. Stage Lighting and Effects Control. This section covers the use of "location boards" as defined in 530.2. They are a form of switchboard with fused switches or circuit breakers typically in the 200 A range protecting 3-wire circuits, so they can be used to source 200 A of line-to-neutral load on both ungrounded supply legs (total of 400 A). This equipment is usually mounted with casters and also often has a pulling eye in the upper surface so it can be hoisted to an elevated walkway. The common "deuce" board has two such switches, allowing it to supply double that amount of power. Usually contactors are arranged within the boards at a point where the board can be turned off or on all at once, or busbar by busbar. A local switch must be placed within 1.8 m (6 ft) of the board for each contactor, or a single switch that drops out all the contactors at once is also permitted. Today, it is common for these boards to be controlled at the stage level by incorporating suitable electronic controls that govern the switching activity within the board.

530.14. Plugging Boxes. At one time these were used on both ac and dc circuits; however they are unpolarized, which led to hazards when ac equipment designed with single-pole switching was connected. They were generally limited to dc applications, and effective with the 1987 NEC, ac connections are not permitted.

530.17. Portable Arc Lamps. At one time the only practical source of intensely white light in amounts large enough to illuminate studio sets came from drawing a dc arc between two carbon electrodes. The carbon electrodes would gradually burn back in the process, requiring them to be constantly advanced so the gap would remain constant. Early designs relied on manual supervision; later designs used mechanized methods to do the same thing. The current required for large-scale illumination was impressive, running on the order of 150A. The development of xenon filled high-intensity discharge lighting (introduced in 1954) made carbon arc light sources essentially obsolete by the 1970s, but many facilities still have the ac infrastructure in place. The xenon sources including their ballasts must be listed. These lamps operate with a color temperature of about 6200°K (about the same as the sun) and a color rendering index (CRI) only slightly below 100 percent.

530.18. Overcurrent Protection—General. Part **(A)** allows stage cables to be protected at 400 percent of their ampacity, and this percentage carries through in **(B)**, **(D)**, and **(E)**. It is a short-time rating and is based on the time that the conductors are energized during production, generally on the order of 10 min and not over 20 min. Part **(C)** covers cable protection and roughly parallels 520.53(H)(5), covered extensively in this book, in order to address similar concerns. Note that the final sentence of the parent text at the beginning of the section specifically requires that conductors be sized in accordance with the load to be supplied. This sentence clarifies that although the overcurrent protection can be rated up to 400 percent of ampacity, it is never permitted to impose loads on conductors greater than normal Code rules permit. The 400 percent rule allows for flexibility of arranging circuits, and does not justify overloading conductors. Studio equipment undergoes testing based on the higher overcurrent protection values permitted in this article.

530.19. Sizing of Feeder Conductors for Television Studio Sets. This section provides a table of demand factors that allow reductions in total feeder capacity for permanent studio and stage set lighting feeder loads. Portable feeders enjoy a blanket 50 percent demand factor.

530.21. Plugs and Receptacles. Part **(A)** covers ratings, which must not be less than the circuit voltage, and for ac circuits, the ampere ratings not less than the feeder or branch-circuit overcurrent device rating. Table 210-21(b)(2) does not apply. This is a similar allowance as in 520.9 for theatrical work. It allows fully rated and potentially fully loaded 20 A cord-and-plug-connected loads to plug into multioutlet 20 A branch circuits, which would otherwise be limited to 16 A by Table 210-21(b)(2). Some stage set lighting loads, such as 10,000-W 120-V luminaires, exceed 80 A.

Part **(B)** covers plugs and receptacles used in portable professional motion picture and television equipment, which can be interchangeable for ac or dc use on the same premises provided they are listed for ac/dc use and marked in a suitable manner to identify the system to which they are connected. Both types of power are commonly in use on the same premises in today's studios.

530.22. Single-Pole Separable Connectors. This section parallels 520.53(K), for which this book has complete coverage. Part **(B)** adds the ac/dc interchangeability principle in 530.21(B) to this location as well.

530.41. Lamps at Tables. This section, essentially unchanged for over 70 years, requires luminaire construction at film editing tables to be constructed in ways that make ignition of film scrap unlikely. This rule is rooted in the days of nitrocellulose film (a compound also known as guncotton), which is extremely flammable; however, it is unsuitable for color photography and has not been produced for almost 60 years.

530.51. Lamps in Cellulose Nitrate Film Storage Vaults. This section and the one following are another relic of another time, although some old prints are still archived. The goal is to keep all potential ignition sources away from the film. This film is so easy to ignite it has been known to spontaneously burst into flames and some spectacular (and horribly destructive) fires resulted as flames spread from reel to reel. These fires are extremely difficult to control because the nitrate component is a powerful oxidizing agent capable of sustaining combustion even under water. In addition, unless kept cool the film is chemically unstable because the nitrate component is also capable of forming nitric acid, which turns the film into a gelatinous mass. There have been many concerted efforts, largely successful, to convert the remaining old stock into modern film bases that do not have these problems.

530.61. Substations. This part of the article reminds us of the scale of motion picture sets and associated activities. The use of medium-voltage feeders to remote locations to support the large power demands during filming is often a necessary fact of life, with special allowances for work space reductions. The use of dc, still recognized here, is not as essential as it was during the time of carbon arc lighting. Here again, however, many facilities still operate dc distributions.

ARTICLE 540. MOTION PICTURE PROJECTION ROOMS

540.1. Scope. According to the definition of hazardous locations in Art. 500, a motion picture booth is not classed as a hazardous location, even though even the old nitrocellulose film is highly flammable. The film is not volatile at ordinary temperatures, hence, no flammable gases are present, and the wiring installation need not be explosionproof but should be made with special care to guard against fire hazards.

540.2. Definitions. Figure 540-1 shows a *professional projector*, which is subject to lengthier and stricter requirements than those of nonprofessional movie projectors. The one shown, with ventilating duct work over the top of it, is a carbon-arc projector. A special form of high-intensity discharge lamp using a xenon gas fill has supplanted these older projectors, although some still exist in art houses. Refer to the discussion in the previous chapter (at 530.17) for more information on carbon-arc light sources.

540.10. Motion Picture Projection Room Required. Professional projectors must be installed in a projector booth, which as noted does not have to be treated like a hazardous (classified) location.

Figure 540-2 shows the arrangement of the apparatus and wiring in the projection room of a large but antique motion picture theater. This room, or booth, contains three motion picture projectors P, one stereopticon or "effect machine" L, and two spot machines S. This method of projection has been functionally obsolete for over 40 years having been supplanted by xenon arc discharge lamps, just as in the case of motion picture studio set lighting as noted in the commentary in the previous chapter at 530.17. However, arc-lamp projection

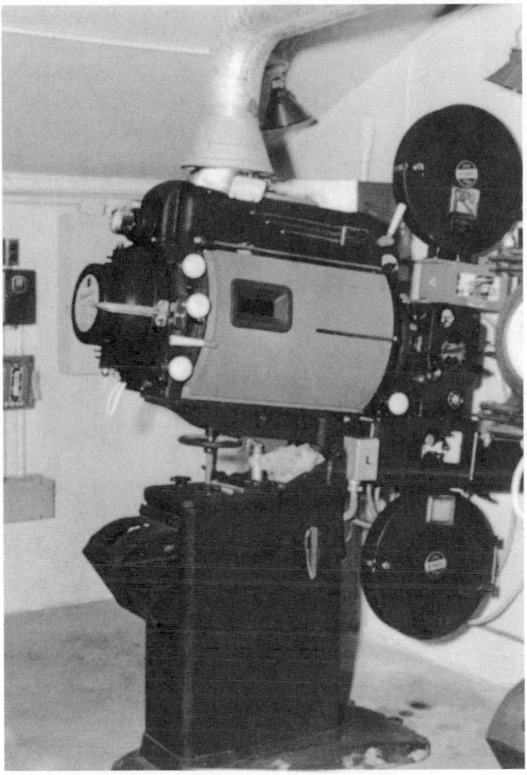

Fig. 540-1. Professional projector. But note that Art. 540 applies to *both* professional and nonprofessional movie projectors. The article is divided into Part **III** on professional equipment and Part **IV** on nonprofessional units. (Sec. 540.10.)

is still used in some art houses and since this **NEC** article is largely still written around this technology, coverage in this book has been retained.

The light source in each of the six machines is an arc lamp operated on dc. The dc supply is obtained from two motor-generator sets, which are installed in the basement in order to avoid any possible interference with the sound-reproducing apparatus. The two motor generators are remotely controlled from the generator panel in the projection room. From each generator, a feeder consisting of two 500-kcmil cables is carried to the dc panelboard in the projection room.

From the dc panelboard to each picture machine and to each of the two spot machines, a branch circuit is provided consisting of two 2/0 AWG cables. One of these conductors leads directly to the machine; the other side of the circuit is led through the auxiliary gutter to the bank of resistors in the rheostat room and from its rheostat to the machine. The resistors are provided with short-circuiting switches so that the total resistance in series with each arc may be preadjusted to any desired value.

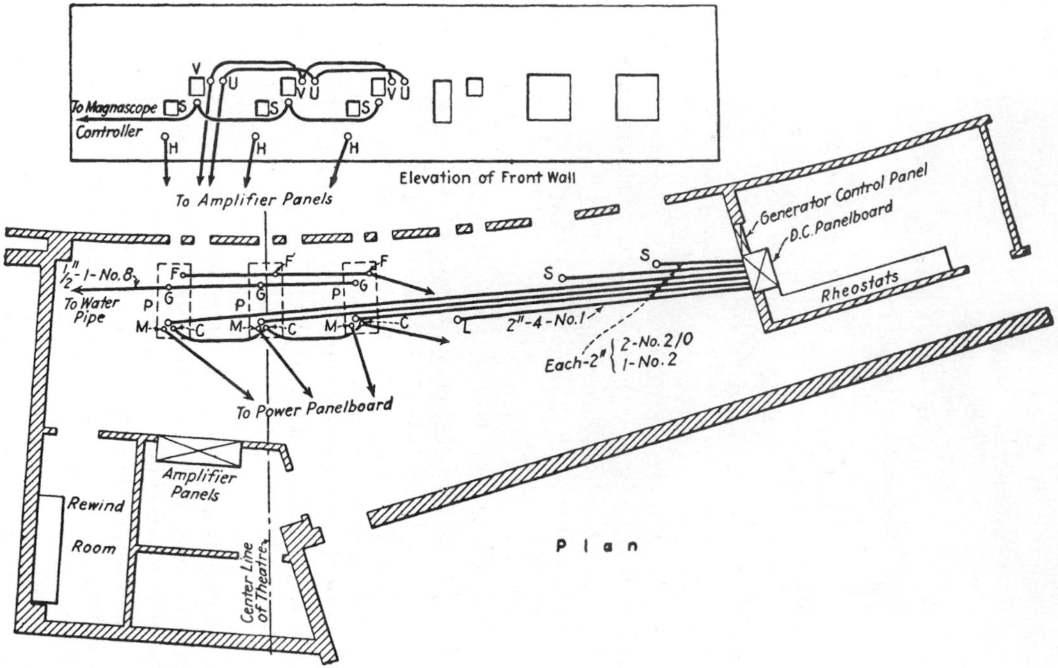

To Magnascope
Controller
Elevation of Front Wall
To Amplifier Panels
Generator Control Panel
D.C. Panelboard
Rheostats
1/2"-1-No.8
To Water Pipe
2"-4-No.1
Each-2" { 2-No.2/0 1-No.2
To Power Panelboard
Amplifier Panels
Rewind Room
Center Line of Theatre
Plan

Fig. 540-2. Code rules cover many electrical details in a motion-picture projection room (or "booth"). (Sec. 540.10.)

Two circuits consisting of 1 AWG conductors are carried to the stereopticon or "effect machine," since this machine contains two arc lamps.

The conduit leading to each machine is brought up through the floor.

540.13 specifies that the wires to the projector outlet should not be smaller than 8 AWG, but in every case the maximum current drawn by the lamp should be ascertained and conductors should be installed of sufficient size to carry this current. In this case, when suitably adjusted for the large pictures, the arc in each projector takes a current of nearly 150 A.

In addition to the main outlet for supplying the arc, four other outlets are installed at each projector machine location for auxiliary circuits. Outlets F are for foot switches that control the shutters in front of the lenses for changeover from one projector to another. Outlets G are for an 8 AWG grounding conductor, which is connected to the frame of each projector and to the water-piping system. From outlets C, a circuit is brought up to each machine for a small incandescent lamp inside the lamp house and a lamp to illuminate the turntable. Outlets M are for power circuits to the motors used to operate the projector machines.

Ventilation is provided by two exhaust fans and two duct systems, one exhausting from the ceiling of the projection room and one connected to the arc-lamp housing of each machine. (See Fig. 540-1.)

A separate room is provided for rewinding films, but since this room opens only into the projection room, it may be considered that the rewinding is performed in the projection room.

Modern projectors have integral rectifiers that eliminate the need for remote dc generators, and also different light sources than the old dc arc lamps, as previously noted in this chapter and in the discussion at 530.17. A typical branch circuit for a modern xenon professional projector would run between 20 and 60 A on a 208Y/120V 4-wire branch circuit (5000 W being a common size). The projector sizing is based on the film and screen size, which in turn governs the size of the projector and the related requirements for the lamp house. This then eliminates the motor generator, making both building and electrical construction far simpler. Note, however, that even under the usual conditions where xenon arc lamps are used as the light source, the requirements in this section for all projection ports, etc. to be closed with glass or other approved material still apply.

540.11. Location of Associated Electrical Equipment. All necessary equipment *may* be located in a projector booth, but equipment which is not necessary in the normal operation of the motion picture projectors, stage-lighting projectors, and control of the auditorium lighting and stage curtain *must* be located elsewhere. Equipment such as service equipment and panelboards for the control and projection of circuits for signs, outside lighting, and lighting in the lobby and box office must not be located in the booth, although remote controls for auditorium lights and curtain and screen adjustments are allowed for obvious reasons.

Part **A** of this section includes enhanced precautions to be used around nitrocellulose film, which parallel and build on the ones in 530.52. As noted in the discussion at 530.51, this film is long obsolete, but if an old print is screened, the precautions are appropriate. Part **(B)** includes an exception essentially waiving the equipment segregation rules if only safety film (i.e., other than nitrocellulose) will be used and provided the projection booth is provided with signage both within its confines and also on the door, clearly stating that only safety film is allowed in the room. Part **(C)** correlates these rules with Art. 700, including 700.21 that prohibits the placement of the only switch that can energize emergency lighting within a projection booth.

540.12. Work Space. The customary 750 mm (30 in.) of work space width about equipment that must be worked while energized also applies on each side of projection equipment, although the spaces to the sides may overlap where adjacent projection equipment is in use.

540.14. Conductors on Lamps and Hot Equipment. If equipment will be operating above 50°C (122°F) its supply conductors must be rated not less than 200°C (392°F). This could be a difficult requirement to meet if flexible cord is required for the application.

540.31. Motion Picture Projection Room Not Required. Assuming safety film is used, nonprofessional projectors do not require a projection booth. Such projectors must, however, be listed.

ARTICLE 545. MANUFACTURED BUILDINGS

545.1. Scope. This article covers prefabricated buildings delivered to a permanent foundation on flatbed trucks and put in place with a crane. They have no chassis framing and no running gear and therefore no capability to be towed. In this way they differ fundamentally from "Manufactured Homes" as covered in Art. 550.

545.4. Wiring Methods. Any Chap. 3 wiring method is permitted if otherwise permitted for the location. Where run in closed construction, cabled wiring is the equivalent of fished, and therefore a comparable allowance for securement only at the connectors. The cable size cannot exceed 10 AWG for this to be allowed.

545.10. Receptacle or Switch with Integral Enclosure. These are wiring devices without boxes, as permitted in 300.15(E).

545.13. Component Interconnections. When sections of a manufactured building must be mated in the field, the cabled interconnections (usually Type NM cable), need to be joined and there are listed devices that make up cable to cable and wire by wire within each cable, resulting in a field splice that is the equal of the cable as a whole. This section amends the usual requirements for a box at this point, and allows these devices to be used. Their track record is very good.

ARTICLE 547. AGRICULTURAL BUILDINGS

547.1. Scope. Any agricultural building without the environments covered in parts **(A)** and **(B)** must be wired in accordance with all other Code rules that apply to general building interiors. Article 547 covers *only* agricultural buildings with the dust, water, and/or corrosive conditions described in parts **(A)** and **(B).**

547.5. Wiring Methods. The wording leaves much of the determination of acceptability up to the inspection authority. But Type NMC cable (nonmetallic, corrosion-resistant—so-called barn wiring cable), UF, jacketed Type MC, or copper SE are specifically recognized for these buildings. PVC conduit and other nonmetallic or protected products including liquidtight flexible nonmetallic conduit would be suitable for the wet and corrosive conditions that prevail. The rule also accepts wiring for Class II hazardous locations in locations covered by 547.1(A) in the scope (excessive dust).

Note that boxes and fittings must be both dust- and watertight in instances where corrosive and wet conditions warrant the attention. Flexible connections must use dusttight flex or liquidtight flex or cord. Also note that *nonmetallic* boxes, fittings, and so on, are exempt from the provisions of 300.6(D). If such components and cables are made from a *metallic* material, then the ¼-in. (6.35-mm) clearance called for in 300.6(D) would apply.

Part **(F)** of this section requires that in all locations within the scope of Art. 547, any equipment grounding conductor that runs underground, whether or not directly buried, must be insulated. The preference for copper grounding conductors in this section has been removed in the 2014 **NEC** based on an argument from an aluminum wire manufacturer that aluminum has equal or superior resistance to the corrosive effects of gases common to agricultural environments, such as ammonia. This section also covers GFCI requirements for receptacles, at 547.5(G). Refer to the discussion of equipotential planes at 547.10 for more information on this topic.

547.6. Switches, Receptacles, Circuit Breakers, Controllers, and Fuses. In this part, the description of the type of enclosure required corresponds to the following NEMA designations on enclosures:

Type 4: Watertight and dusttight. For use indoors and outdoors. Protect against splashing water, seepage of water, falling or hose-directed water, and severe external condensation. They are sleet-resistant but not sleet- (ice-) proof.

Type 4X: Watertight, dusttight, and corrosion-resistant. These have the same provisions as Type 4 enclosures, but in addition are corrosion-resistant.

The rule of this section seems to clearly call for NEMA 4X enclosures (Fig. 547-1).

Fig. 547-1. The Code rule seems to make use of this type of enclosure mandatory in agricultural buildings. (Sec. 547.5.)

NEMA Type 4X

Stainless steel NEMA Type 4X enclosures are used in areas that may be regularly hosed down or are otherwise very wet, and where serious corrosion problems exist. Typical enclosures are made from 14-gauge stainless steel, with an oil-resistant neoprene door gasket.

Epoxy powdered resin–coated NEMA Type 4X enclosures are designed to house electrical controls, terminals, and instruments in areas that may be regularly hosed down or are otherwise very wet. These enclosures are also designed for use in areas where serious corrosion problems exist. They are suitable for use outdoors, or in dairies, packing plants, and similar installations. These enclosures are made from 14-gauge steel. All seams are continuously welded with no holes or knockouts. A rolled lip is provided around all sides of the enclosure opening. This lip increases strength and keeps dirt and liquids from dropping into the enclosure while the door is open.

547.9. Electrical Supply to Building(s) or Structure(s) from a Distribution Point. Figure 547-2 shows a typical overhead distribution at a farm. The central pole is defined in 547.2 as the *distribution point* for the farm. Because this pole top is both the metering point and the service point for the utility, there will be a switch here, as shown in Fig. 547-3. The disconnecting means may be provided by the utility or by the owner, depending on local practice. The switch has no overload protective devices within it, thereby varying from the normal rule for services in NEC 230.91. The NEC classifies this device in 547.2 as a *site isolating device* to distinguish it from a service disconnect. Even if supplied and maintained by the utility, and therefore beyond the scope of the NEC, the NEC avoids needless duplication by recognizing it as a disconnecting means provided it meets the requirements in NEC 547.9(A).

Since it is not an actual service disconnect, it follows that the wiring that leaves this device still has the status of service conductors and must meet the wiring method and clearance requirements in Art. 230. Although nothing

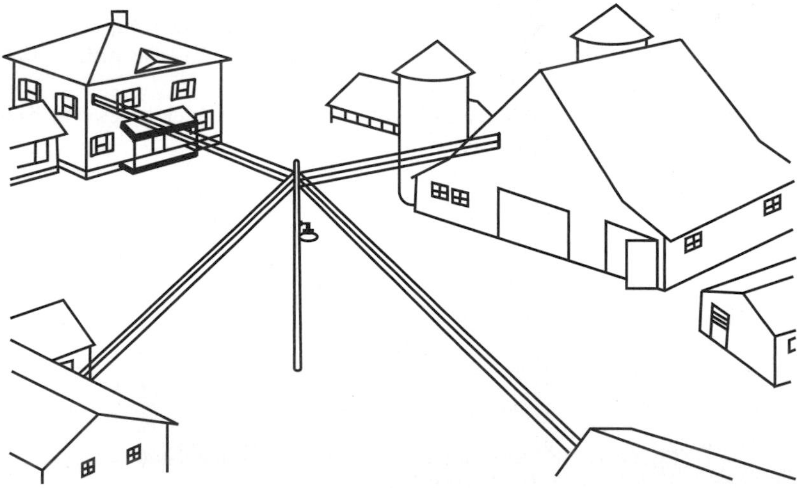

Fig. 547-2. Overhead distributions on farms usually involve one or more distribution points. (Sec. 547.9.)

technically prevents a farm from establishing a conventional service at the distribution pole, and then routing conventional overcurrent-protected feeders to each building, the arrangement shown here is widely used for overall cost effectiveness. Note that although the switch is at the top of the pole, it can be operated from a readily accessible point through the permanently installed linkage shown in Fig. 547-3. In addition, a grounding electrode conductor must

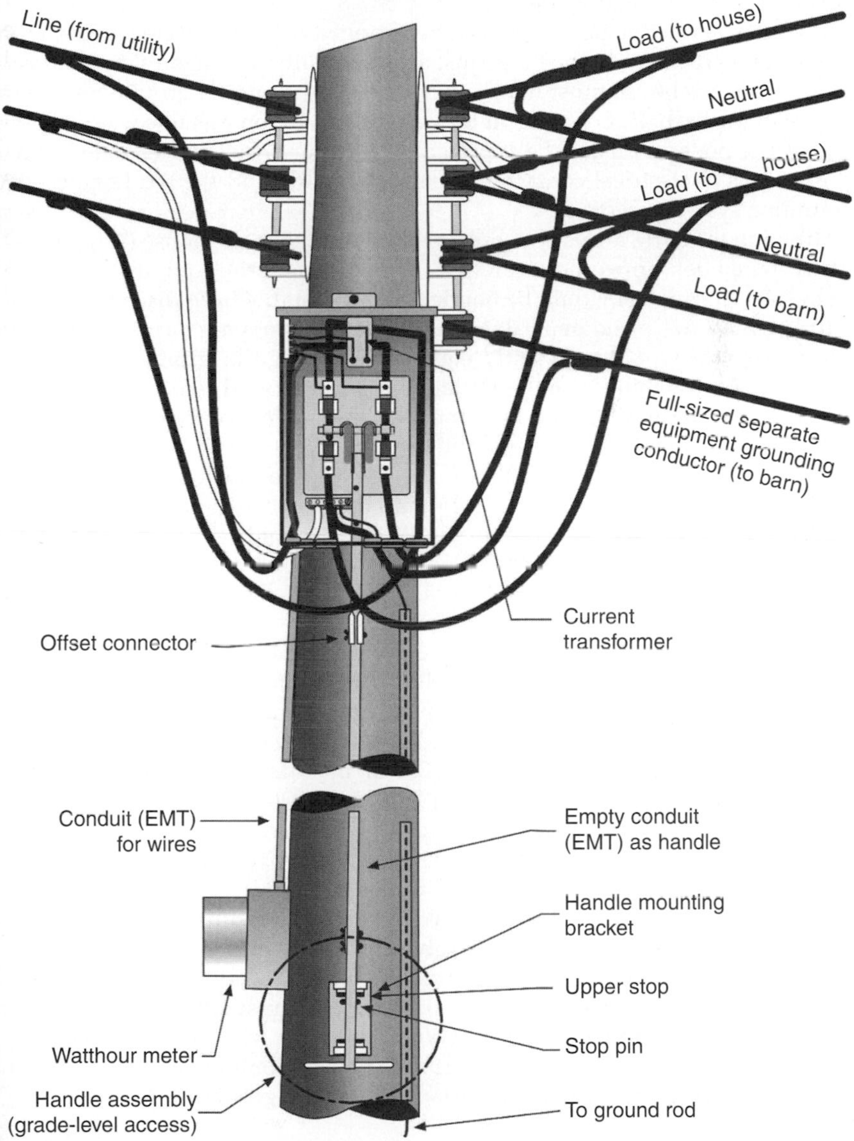
Line (from utility)
Load (to house)
Neutral
Load (to house)
Neutral
Load (to barn)
Full-sized separate equipment grounding conductor (to barn)
Offset connector
Current transformer
Conduit (EMT) for wires
Empty conduit (EMT) as handle
Handle mounting bracket
Upper stop
Stop pin
Watthour meter
Handle assembly (grade-level access)
To ground rod

Fig. 547-3. This is a site isolating device for a farm, showing the drops to the house and a barn. (Sec. 547.9.) (Adapted from *Practical Electrical Wiring*, 22nd ed., © Park Publishing, 2014)

be installed at this point and run from the neutral block of the switch to a suitable electrode at the pole base.

Part **(D)** requires that if there are multiple distribution points, each location must have reciprocal labeling setting out the location of the other point(s) and the buildings or structures served by each.

Part **(B)** governs the wiring of the service conductors from the site isolating device to the various buildings that will be served directly, as shown in Fig. 547-2, and as detailed at the site isolating device by Fig. 547-3.

As a general rule the farmhouse can be supplied by a three-wire service, with its neutral regrounded at the house just as if the utility had made a direct termination. The farmhouse must not, however, share a common grounding return path with the barn. If it does, as in the case of a common metallic water piping system, the house (1) has to be supplied with a four-wire service, and (2) have all instances of electrical contact between the neutral and the local equipment grounding system removed.

Although the barn, arguably, could also be wired like the house (three-wire), a three-wire hook-up would mean that the neutral and the equipment grounding system in the barn would be bonded together at the barn disconnect. That in turn would mean the neutral, in the process of carrying current across its own resistance, would constantly elevate the voltage to ground of all barn equipment by some finite amount relative to local ground, especially from the perspective of farm animals where they stand. The feet of livestock, being in close contact with moisture, urine, and other farm chemicals, are conductively rather well-coupled to local earth. Most livestock are much more sensitive to voltage gradients than are people. A potential difference in the range of a fraction of a volt can take a cow out of milk production, which no farmer can afford.

The NEC addresses this in two ways. First, it establishes the unique rules on farm distributions being covered here. Second, it establishes an equipotential plane for these environments, covered in 547.10. The service to the barn is normally wired four-wire, and that is (1) customary because of the reasoning discussed in the previous paragraph, and (2) mandatory unless there are no parallel grounding return paths over water systems, etc., a necessary condition to comply with NEC 250.32(B)(1) Exception No. 1. There is an additional condition [see 547.9(B)(3)] attached to the four-wire scheme that is unique to agricultural buildings. The separate equipment grounding conductor must be fully sized. That is, if the run to the barn is 3/0 AWG copper for a 200-A disconnect, and the neutral is 1/0 AWG copper (both sized on the basis of load), the equipment grounding conductor is not 6 AWG as normally required by NEC Table 250.122; nor is it 4 AWG, the size for a grounding conductor on the supply side of a service using 3/0 AWG wires; nor is it 1/0 AWG, the size of the neutral. It must not be smaller than the largest ungrounded line conductor, or 3/0 AWG. When this wire arrives at the barn, it must arrive at a local distribution with the neutral completely divorced from any local electrodes or equipment surfaces requiring grounding.

547.10. Equipotential Planes and Bonding of Equipotential Planes. Due to the sensitivity of livestock to very small "tingle" voltages, the NEC now requires an equipotential plane in livestock (does not include poultry) confinement areas, both indoors and out, if they are concrete floored and contain metallic equipment accessible to animals and likely to become energized. These areas must include wire mesh or other conductive elements embedded in (or placed under) the concrete floor, and those elements must be bonded to metal structures and fixed electrical equipment that might become energized, as well as to the grounding electrode system in the building. The bonding conductor must be solid copper, not smaller than 8 AWG. In the case of dirt confinement areas, the equipotential plane may be omitted. For outdoor areas the plane must encompass the area in which the livestock will be standing while accessing equipment likely to become energized.

Remember that the grounding system to which equipotential planes should be connected is usually (refer to the discussion on 547.9) electrically separated from neutral return currents. The idea is to minimize voltage gradients. Due to the well-grounded environment, the NEC also requires [see 547.5(G)] all general-purpose 15- and 20-A, 125-V receptacles in the area of an equipotential plane to have GFCI protection. This GFCI protection requirement also applies to similar receptacles in all damp or wet locations, including outdoors, and for dirt confinement areas whether indoors or out. Prior to the 2017 NEC, the wording of 547.5(G) effectively excluded receptacles that were other than "general purpose" from the GFCI requirement, which allowed single receptacles for such functions as electric fencing power controllers to avoid the coverage. That wording has been changed and the requirement now applies to all such receptacles.

ARTICLE 550. MOBILE HOMES, MANUFACTURED HOMES, AND MOBILE HOME PARKS

550.1. Scope. The provisions of this article cover the electric conductors and equipment installed within or on mobile homes and manufactured homes, and also the conductors that connect mobile homes and manufactured homes to a supply of electricity. But the service equipment that is located "adjacent" to the mobile home is covered in Art. 550 and all applicable Code rules on such service equipment—as in Arts. 230 and 250—must be observed.

550.2. Definitions. The difference between a mobile home and a manufactured home is that although both buildings have a chassis and are designed to move on running gear, the mobile home is not intended to be placed on a permanent foundation. Most buildings in this category today are manufactured homes, produced with the expectation that they will ultimately rest on a permanent foundation. A manufactured home is not a manufactured building covered in Art. 545.

They also enjoy a protected regulatory status, because the federal Department of Housing and Urban Development (HUD) has determined that they are a solution to the lack of housing. Therefore, acting under the supremacy clause in the U.S. Constitution with authority delegated by a congressional enactment, HUD has decreed that what passes muster under its regulations regarding the manufacture of these homes will be accepted by local enforcement agencies. This is why the version of the NEC that applies to any particular manufactured home is usually not synchronized with local adoption of the NEC.

The definition for "manufactured home" clarifies when the service equipment may be hung on the dwelling. The service may *never* be installed on, or in, a mobile home. But where a manufactured home meets the requirements of 550.32(B), which include provisions for a permanent foundation (and almost all do), it may be supplied by a service that is installed adjacent to or even on or in the manufactured home. Notice that the last sentence in the definition makes all rules for "mobile homes" applicable to "manufactured homes" as well.

550.4. General Requirements. Part (A) covers mobile homes that are not for residential purposes, such as construction trailers and the like. Such buildings need not comply with the rules regarding the number and capacity of circuits, but they must comply with the other applicable rules, including the requirement in 550.32(A) that the service not be mounted in or on the trailer (mobile home).

550.10. Power Supply. Part **(A)** requires that the power supply to the mobile home be a feeder circuit consisting of not more than *one* 50-A rated approved

Fig. 550-1. Service equipment for a mobile home lot consists of disconnect, overcurrent protection, and receptacle for connecting *one* 50-A (or 40-A) power-supply cord from a mobile home parked adjacent to the service equipment. (Sec. 550.10.) Note that the single-conduit support for the receptacle enclosure violates 314.23(F).

mobile home supply cord, or that feeder circuit could be a permanently installed circuit of fixed wiring (Fig. 550-1). An exception correlates with allowances for these dwellings to be supplied directly with service wiring if they meet the definition of a "manufactured home."

Part **(B)** covers use of a cord instead of permanent wiring. The power-supply cord to a mobile home is actually a feeder, and must be treated as such in applying Code rules. The service equipment must be located adjacent to the mobile home and could be either a fused or a breaker type in an appropriate enclosure or enclosures, with not more than 50-A overcurrent protection for the supply cord (or 40 A, as in the exception). The equipment must be approved service-entrance equipment with an appropriate receptacle for the supply cord, installed to meet Code rules the same as any installation of service equipment. The panel or panels in the home are feeder panels and are *never* to be used as service-entrance equipment, according to 550.10 and 550.11(A). This means that the neutral is isolated from the enclosure and the equipment grounding goes to a separate bus for that purpose only. As a result, there must be an equipment grounding conductor run from the service-entrance equipment to the panel or panels in the home. This is true whether there is cord connection or permanent wiring.

In some areas, mobile homes are permanently connected as permitted in paragraph **(I)**. Accordingly, local requirements must be checked in regard to the approved method of installing feeder assemblies where a mobile home has a calculated load over 50 A. In many such cases, a raceway is stubbed to the underside of a mobile home from the panelboard. It is optional whether the feeder conductors are installed in the raceway by the mobile home manufacturer or by field installers. When installed, four continuous, insulated, color-coded conductors, as indicated, are required. The feeder conductors may be spliced in a suitable junction box, but in no case within the raceway proper.

550.11. Disconnecting Means and Branch-Circuit Protective Equipment. As shown in Fig. 550-2, the required disconnect for a mobile home may be the main in the panelboard supplying the branch circuits for the unit. Details of this section must be observed by the mobile home builder.

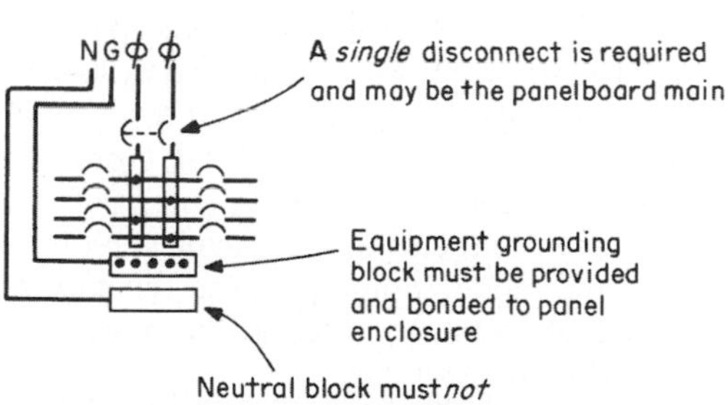
N G ϕ ϕ
A *single* disconnect is required and may be the panelboard main
Equipment grounding block must be provided and bonded to panel enclosure
Neutral block must *not* be bonded to enclosure

Fig. 550-2. A "panelboard," not "service panelboard" may be used *in* mobile home. (Sec. 550.11.)

550.12. Branch Circuits. The manufacturer of the mobile home must ensure this minimum circuiting.

550.13. Receptacle Outlets. These rules generally line up with comparable rules in Art. 210, but one cycle behind. As changes occur in Chap. 2, this article tends to be updated in the next cycle. For example, the disallowance of a snap-switch controlled receptacle to meet a placement rule in 210.52, a 2008 **NEC** change, is still not correlated with this article. On the other hand, the requirements for GFCI protection [see **(B)**] now correlate properly with comparable requirements in 210.8(A), including all sinks coverage and dishwashers as well. Part (F)(1) correctly correlates with 406.9(C). Part **(E)** for the heat tape receptacle outlet, however, is unique to this article. The heat tape outlet ("pipe heating cable") must now have GFCI protection, and it must be arranged so tripping the GFCI protection would be obvious to the occupant. Accordingly the outlet must go on an interior circuit arranged with all other outlets on that circuit on the load side of the GFCI protective device. A bathroom, but not a small appliance, branch circuit could be chosen for this purpose. Note that the GFCI protection, provided by a residual current device, also effectively complies with 427.22. Small-appliance branch circuits can only serve their intended outlets as covered in 550.12(B), so this correlates with Chap. 2. Note also that 550.12(B) makes no amendments to Chap. 2 in this regard. This section makes no provisions for tamper-resistant receptacles; however, the expansion of coverage in 406.12 to mobile homes may cure that if the jurisdictional question is resolved. Refer to the discussion at 406.12(1).

550.15. Wiring Methods and Materials. Part **(H)** governs under-chassis wiring that is exposed to weather. Previous Code editions required such wiring, including a hard-wired supply from an underground distribution, to be heavy wall steel conduit for the portion that was exposed to weather. There was an exception for nonmetallic conduit, but only where it closely followed the frame. Now Schedule 80 PVC and RTRC-XW (the forms designed for exposure to damage) can extend from direct burial at least 450 mm (18 in.) below grade (Table 300.5 requirement for such circuits) to a factory-intended termination point. These methods and also EMT and MI cable can also be used where protected by framing. The actual text was simplified, and now simply requires a raceway approved for wet locations and for resistance to physical damage if located where susceptible.

In the exception to part **(I)**, the smaller-dimensional box mentioned would usually be a box designed for a special switch or receptacle, or a combination box and wiring device. This application is recognized in 300.15(E) and 334.30(C). Such combinations can be properly evaluated and tested with a limited number of conductors and connections and a specific lay of conductors to ensure adequate wiring space in the spirit of 314.16.

550.16. Grounding. The white (neutral) conductor is required to be run from the insulated busbar in the mobile home panel to the service-entrance equipment, where it is connected to the terminal at the point of connection to the grounding electrode conductor.

The green-colored conductor is required to be run from the panel grounding bus in the mobile home to the service-entrance equipment, where it is connected to the neutral conductor at the point of connection to the grounding electrode conductor.

The requirements provide that the grounded (white) conductor and the grounding (green) conductor be kept separate within the mobile home structure in order to secure the maximum protection against electric-shock hazard if the supply neutral conductor should become open.

A common point of discussion among electrical authorities and electricians is whether the green-colored grounding conductor in the supply cord should be connected to the grounded circuit conductor (neutral) outside the mobile home—say, at the location of the service equipment. The grounding conductor in the supply cord or the grounding conductor in the power supply to a mobile home is always required to be connected to the grounded circuit conductor (neutral) outside the mobile home on the supply side of the service disconnecting means, but *not* in a junction box under the mobile home or at any other point on the *load side* of the service equipment (pedestal).

550.18. Calculations. This section provides the rules for making load calculations for mobile homes. These provisions generally track the provisions for comparable loads in other dwelling occupancies. For example a 12 kW range is an 8 kW load. However, for a home on a 40-A cord and no air-conditioner, 15 A per leg must be allowed for a future air-conditioning load. Part **(C)** allows the optional calculations in 220.82 to be used without amendment, but interestingly does not recognize the somewhat more forgiving provisions of 220.83 to be used for existing units.

550.25. Arc-Fault Circuit-Interrupter Protection. This section now requires adherence to 210.12, so it is now fully in line with the rest of the Code. However, even this version of AFCI, may or may not reflect current HUD 3280 rulemaking and therefore what is being shipped in interstate commerce. Refer to the discussion at 550.2 for more information on this point.

550.30. Distribution System. The mobile home park supply is limited to nominal 120/240-V, single-phase, 3-wire to accommodate appliances rated at nominal 240 V or a combination nominal voltage of 120/240 V. Accordingly, a 3-wire 120/208-V supply, derived from a 4-wire 208Y/120-V supply, would not be acceptable.

550.31. Allowable Demand Factors. While the demand factor for a single mobile home lot is computed at 16,000 VA (unless the 550.18 result for the typical home provided results in a larger number), it should be noted that 550.32(B) requires the feeder circuit conductors extending to each mobile home lot to be not less than 100 A. Table 550.31 provides demand factors for a mobile home lot with multiple units connected to a common service or feeder.

550.32. Service Equipment. The service equipment disconnect means for a mobile home must be mounted with the bottom of its enclosure at least 600 mm (2 ft) above the ground, because some very low mounted disconnects are subject to flooding and are difficult to operate. But the disconnect must not be higher than 2.0 m (6 ft 7 in.) above the ground or platform (Fig. 550-3). For mobile homes that are actually mobile and not mounted on a permanent foundation as covered in part **(B)** of this section, the service equipment must not be placed on the home, and the final connection must be through a feeder as covered in 550.33. Part **(B)**

covers what has become the more usual condition, where the home is located on a permanent foundation with no intention of relocation. Note the requirement in **(7)** for a red warning label advising site electricians not to make service connections until a grounding electrode is installed and connected properly.

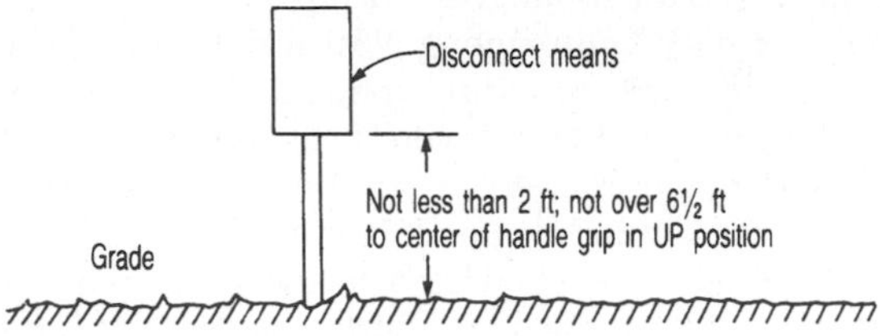
Disconnect means
Not less than 2 ft; not over 6½ ft
to center of handle grip in UP position
Grade

Fig. 550-3. Mobile home service disconnect must comply with minimum mounting height rule. (Sec. 550.32.)

ARTICLE 551. RECREATIONAL VEHICLES AND RECREATIONAL VEHICLE PARKS

551.1. Scope. Some states have laws that require factory inspection of recreational vehicles by state inspectors. Such laws closely follow NFPA 1192-2008, *Standard for Recreational Vehicles*. This standard contains electrical requirements in accordance with part **(I)** of Art. 551. It also contains requirements for plumbing and heating systems.

551.20. Combination Electrical Systems. As explained in the last exception of part **(B)**, "momentarily" operated electric appliances do not affect converter sizing. This exception excludes from calculation of the required converter rating any appliance that operates only momentarily (by a momentary contact switch) and cannot have its switch left in the closed position. Such appliances draw current for only momentary periods and do not have to be counted as load in sizing the converter rating.

551.30. Generator Installations. Part **(B)** requires a transfer function for safety, and adds the requirement that an automatic transfer switch be listed for either emergency systems or optional standby systems. Note that if the alternate source of power is some other system, such as an inverter, the same requirement applies, as covered in 551.33. It was noted that the previously acceptable methods involving relays were susceptible to a welded-contact type of failure where the connection to the utility side could have been maintained. Part **(E)** places an enclosed transfer switch on the same footing as a panelboard relative to the allowable length of supply conductors (18 in.) into the vehicle.

551.41. Receptacle Outlets Required. This section is primarily of interest to RV manufacturers. In addition to requirements that enhance placement requirements for countertops for obvious reasons in these cramped spaces, the 2014 NEC added a requirement for a receptacle meeting 406.9(B) for wet locations, and within the outer perimeter of a rooftop deck in cases where that deck is accessible from inside the RV.

551.42. Branch Circuits Required. This rule coordinates the rules on branch circuits to those of 551.45 on the panelboard. Note that if the branch circuits exceed five, the minimum supply becomes 50 A from a 120/240-V source (or 208 V, reflecting the allowance for three-phase distributions in RV parks). A sixth circuit is allowable if all it does is supply a power converter and the total load profile still agrees with the load as originally designed for the five circuits. Power converters are commonly available and allow for powering interior lights and some other loads at 12 V dc.

551.45. Panelboard. Note that Part **(B)** amends the normal workspace distances, reducing the width from 752 mm (30 in.) to 600 mm (24 in.) and the depth from 914 mm (3 ft) to 750 mm (30 in.). The metric differences between the two 30-in. spacings constitute another amendment, because unlike 110.26, this section uses hard conversions throughout (see 90.9 for more information on this topic). If the panel faces an aisle, the workspace in one of the required dimensions can be reduced to 22 in. if the panel front is not recessed more than 2 in. from the finished surface (or over 1 in. of the backside of a flush door enclosing it). In addition, this section does not require the dimensions to be applied before the RV is set up. This allows for panels in "side-out" RVs to be obstructed during travel, provided the proper clearances are provided when the vehicle is fully expanded on site.

551.46. Means for Connecting to Power Supply. Note that the configuration of the 125-V 30-A, 2-pole 3-wire grounding plug and receptacle is unique to the RV industry, and is incompatible with the standard NEMA 5-30 configuration for this voltage, amperage, and number of poles. Part **(E)** correlates with 551.77, assuring that properly configured RVs can park at properly configured RV parks with a minimum of cord travel.

551.47. Wiring Methods. RV wiring is chiefly of interest to their manufacturers and will not be considered at length here. For just one of many examples in a long section, part **(P)** covers the rules for wiring side-out provisions (previously mentioned here at 551.45).

551.71. Type Receptacles Provided. RV parks must meet certain minimum receptacle allotments, which in turn have a bearing on the number of sites that will go into the load calculations to determine the overall service size for the park. To begin with, every RV site with an electrical supply must have a 20-A 125-V 2-pole 3-wire grounding receptacle outlet. As covered in part **(A)**, this must be installed in recreational vehicle site supply equipment. This terminology is defined in 551.2 and includes overcurrent protection and disconnecting provisions for the site. Because this equipment is not for use in dwelling and has an overall NEMA 3R rating, these receptacles, which must be GFCI protected in accordance with part **(F)**, are exempted from tamper- and weather-resistance as would otherwise be required by 406.9 and 406.12.

Part **(B)** covers what is still the majority of RV hook-ups, the special 30-A 125-V receptacle shown in Fig. 551-46(C)(1). At least 70 percent of the sites with power must accommodate these supply ratings. Part (C) covers the NEMA 14-50 configuration (50-A 125/250-V 3-pole 4-wire grounding) receptacles which are in ever increasing demand, now running at 30 percent of all new units shipped, according to RV industry sources. The requirement for powered existing sites is that 20 percent of them have this configuration, and for any new sites developed the number rises to 40 percent. The informational note at the end of the section reminds designers that sites intended for "seasonal" use may approach 100 percent demand for 50-amp. connections by customers. A final requirement is that any site with one of these receptacles must also have a 30-A 125-V receptacle with it. The panel is trying to avoid an incentive to use bogus adapters that allow a 30-A. 125-V RV to plug in to a 50-A. receptacle due to what would be an oversized overcurrent device ahead of it.

Part (**D**) covers tent sites with 125-volt 15- or 20-amp. receptacles. These locations can be excluded when making the percentage calculations for the 30- and the 50-amp receptacles covered in (B) and (C). Part **(E)** covers additional receptacles for outdoor activities apart from the vehicle. These receptacles are permitted, but not mandatory.

551.72. Distribution System. As these parks increase in both area and power density, the incentive from the perspective of both the utility and the owner to employ three-phase distributions increases as well. For the first time, this section explicitly recognizes 208-volt circuits (with neutral, inaccurately described as "1-phase") from a wye system. Informational notes point to the issues of voltage drop (and calculation considerations) on long RV park feeders; ironically, the line-to-neutral voltage drop on a three-wire supply fed from two legs of a wye system is much higher than the same load fed from a single phase distribution, for which the neutral disappears for voltage drop purposes on equally loaded hot legs. Of course, this can be counteracted by feeding sites in groups of three.

551.73. Calculated Load. The load on a site is based on the highest rated receptacle. As of the 2017 NEC, for 50-A sites, use 12,000 VA; for 30-A sites, use 3600 VA; for 20-A sites, use 2400 VA; and for the dedicated tent sites with a 20-A 125-V receptacle use 600 VA. Add the results together and then apply the demand factor from Table 551.73 that matches the total number of powered sites. If one powered location can serve two vehicles, use the two highest-rated receptacles in the summation.

551.75. Grounding. Part (B) squarely addresses a frequent topic of discussion. Largely informed by the revision to the definition of "structure," the RV pedestal does not require a grounding electrode unless it is service equipment. An auxiliary electrode may be installed but is not required.

551.77. Recreational Vehicle Site Supply Equipment. This is where to find how to place the equipment, which differs, based on whether or not the site is a pull-through site.

ARTICLE 552. PARK TRAILERS

552.2 Definition. A park trailer is a unit that is built on a single chassis mounted on wheels and has a gross trailer area not exceeding 37 m^2 (400 ft^2) in the set-up mode. The article has no independent provisions or demand factors for the placement of these units in a group, although 552.47 does have a procedure for determining the rated load of any given unit. These calculations, as in the case of many other provisions in this part of the article, parallel comparable requirements for both mobile homes and for recreational vehicles. Here again, there are no requirements in this article that interface the loads in these units with other loads in the location where they are connected. For example, if a park trailer is set up at an RV park, how does its load relate to Table 551.73? Absent any demand factor allowances, and there are none, it must be taken at 100 percent. In addition, there are no rules for the feeder supply to the trailers, leaving open the question of whether direct connections are required or whether cord connections are permitted.

Gradually other rules are periodically correlated; as of the 2011 cycle, for example, 552.59(A) requires outdoor receptacle outlets to meet the weather-resistance rules and in-use cover rules in 406.9, and outdoor switches must comply with the weatherproof cover or enclosure rules in 404.4, depending on the installation.

ARTICLE 553. FLOATING BUILDINGS

553.1. Scope. This article covers the electrical system in a building—either residential (dwelling unit) or nonresidential—that floats on water, is moored in a permanent location, and has its electrical system supplied from a supply system on land. The rules apply to any floating building and are not limited only to floating "dwelling units."

553.4. Location of Service Equipment. The service-disconnect means and protection for a floating building must not be mounted on the unit. This ensures the ability to disconnect the supply conductors to the floating building in an emergency, such as in a storm, in the event that it is necessary to move the unit quickly (Fig. 553-1).

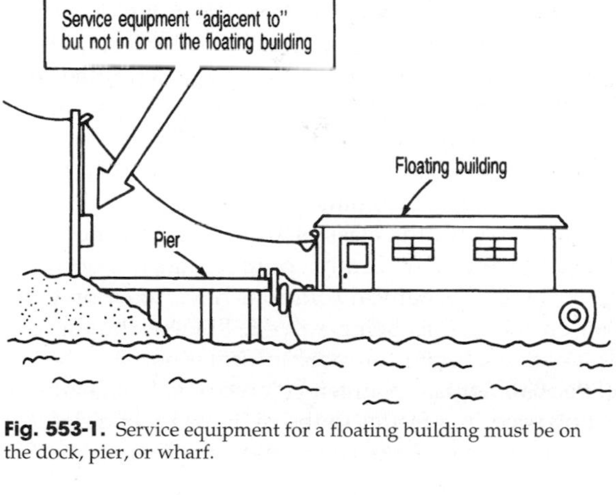
Service equipment "adjacent to" but not in or on the floating building
Floating building
Pier

Fig. 553-1. Service equipment for a floating building must be on the dock, pier, or wharf.

A major 2011 NEC change requires low-level GFPE/GFCI (the actual wording only says ground-fault protection) as a part of the main overcurrent protective device feeding the floating building, with a trip setting not above 100 mA. As an alternative, "ground-fault protection" can be provided on each and every branch circuit and feeder circuit. The change was motivated by documentation of 50 deaths from leakage currents "on and around marinas." Note that this issue is far more important in fresh water exposures. Fresh water has a much higher resistance than salt water, and as a consequence the voltage gradient in the vicinity of a fault is much steeper, to the point of being many volts across the width of a human body. A very few volts, especially with water in the ears, can bring on enough disorientation to cause drowning. Salt water has such low resistance that such a steep gradient across the width of a body is almost unheard of.

There are obvious major issues with this requirement. The proposal was offered by a manufacturer who just happens to have such a device. What does it actually do? It would appear to be a residual current device set to trip at 100 mA. This is higher than the usual setting for low-level GFPE used to protect heat tape and snow-melting cables, and far higher than usual GFCI settings, so it would be ineffectual at actually preventing an electrocution. It would function more as a "maintenance required" annunciator, hopefully tripping before someone was in the wrong place at the wrong time and got injured or worse. If it tripped, one supposes it would motivate major maintenance to be performed in order to reduce the leakage below the trip setting.

The 100 mA set point is obviously far too high to prevent an electrocution, but it is also far too low for its intended applications. There are many large floating buildings with services running over 1000 A and occupying considerable ocean front in major harbors, how could this be applied in those locations? Could every feeder and branch circuit be wired with the GFCI alternative? The same rule applies to marinas, there are likewise large marinas with service in the mid four figures, and undoubtedly in some cases with service taken at medium voltage.

Careful discussion with the local authority having jurisdiction is obviously called for in this instance, because a rule-making procedure may be required in order to modify or delete this requirement.

553.7. Installation of Services and Feeders. For obvious reasons the wiring between land and the floating building must be flexible, and this section makes that clear in **(A)**. Although flexible raceways are permitted, extra-hard-usage flexible portable power cable with its fine conductor stranding is the most resilient. Such cord must be marked "sunlight-resistant" and for "wet locations." If cord is used, however, the stranding will not terminate correctly in conventional mechanical lugs. According to UL data, a termination for that type of stranding must use a lug marked with the class of stranding and the number of strands, and generally only a few hydraulically crimped lugs has been available for this purpose. Recently some mechanical solutions have come on the market, so this problem is well on its way to being solved. Review the discussion regarding 110.14 and Fig. 110-8 in Chap. 1 for more information.

553.8. General Requirements. A green-colored, insulated equipment grounding conductor must be used in a feeder to the main panel of a floating building. For conductors larger than 6 AWG, the color can be applied afterward in accordance with the usual rules in 250.119. This equipment grounding conductor must be run to the panel from an equipment grounding terminal (or bonded neutral bus) in the building's service equipment on land.

ARTICLE 555. MARINAS, BOATYARDS, AND COMMERCIAL AND NONCOMMERCIAL DOCKING FACILITIES

555.1. Scope. This article covers both fixed and floating piers, wharfs, and docks—as in boat basins or marinas. As of the 2017 **NEC**, it covers these facilities of any size and in any location or occupancy, from a large ocean facility to a dock for a small boat at a single-family dwelling, and the article is retitled accordingly. In Fig. 555-1,

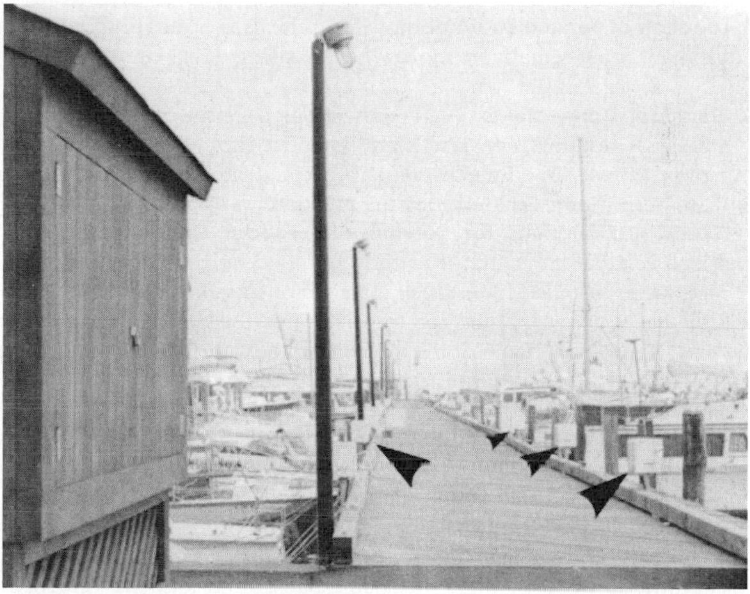

Fig. 555-1. This is one part of a 406-boat marina where shore power is supplied to moored boats from receptacle power pedestals (arrows) supplied by cables run under the pier from a panelboard in the shed at the left. (Sec. 555.1.)

branch circuits and feeder cables run from panelboards in the electrical shed at the left, down, and underground into a fabricated cable space running the length of the pier shown at the right, supplying shore-power receptacle pedestals (arrows) along both sides of the pier.

555.2. Definitions. The "electrical datum plane" is a horizontal plane that serves as the reference level for any rules governing height above water level. It is absolutely essential that this be determined before proceeding with other requirements. It is 606 mm (2 ft) above normal high water, except that on a floating pier, it is 762 mm (30 in.) above the water and also 305 mm (12 in.) above the floating pier.

555.3. Ground-Fault Protection. This is a major change as of the 2011 NEC and raises serious questions as to technical validity and workability. The same requirement applies to floating buildings. Review the discussion at 553.4. The 2017 NEC has doubled down on the issue and now requires protection at the 30 mA level (in marinas, not yet for floating buildings). This would be the traditional GFPE level so these devices are readily available. The requirement extends to the overcurrent devices that "supply the marina, boatyards, and commercial and noncommercial docking facilities." On the literal text, this rule applies to any circuit, whether or not it ventures anywhere near equipment that might generate a voltage gradient in the water, and it applies whether or not the water is fresh water. Salt water is sufficiently conductive that a hazardous voltage gradient is a virtual impossibility. Presumably the most effective solution is to place it on each shore power circuit. Because defective equipment on visiting boats is the likely source of most loss experience to date, in that way only the miscreant boat is inconvenienced.

555.7. Location of Service Equipment. As in the case of floating buildings, service equipment must remain on shore where floating docks or other facilities are supplied.

555.9. Electrical Connections. All connections must be at least 305 mm (1 ft) above a fixed or floating pier, and simultaneously above the datum plane. For floating piers, however, connections using sealed methods identified for submersion are permitted provided they are enclosed in approved junction boxes.

555.12. Load Calculations for Service and Feeder Conductors. This is an unusual load calculation in that it is entirely based on the current ratings of the power receptacles installed for each slip. If two (or more) receptacles are provided at a slip, the one whose configuration translates into the largest kVA load profile is the one used to enter the table. If the slips have kilowatt-hour meters, a reduction in electrical demand of 10 percent is assumed (through the use of a 0.9 multiplier). Note that on the literal text, the meters need not be read or used to bill the boat owners in order to achieve this presumed reduction in load. In addition, the neutrals must not be further reduced under the terms of 220.61(B); this demand factor table is the only permitted calculation for all circuit conductors.

The 2011 NEC clarified the appropriate application of the receptacle count in the left hand column of the demand factor table as being the number of shore power receptacles and no others. Apparently some had added local convenience receptacles into the count, artificially increasing the presumed load diversity and thereby reducing the demand factor to be applied.

555.13. Wiring Methods and Installation. The rules here present various options that are available for the circuiting to the loads at marinas and boatyards. This section recognizes any wiring method "identified" for use in wet locations. Examples of wiring methods that are recognized by the NEC for use in wet locations are as follows:

1. Type PVC or Type RTRC conduit.
2. Type MI cable.
3. Type UF cable.
4. Corrosion-resistant rigid metal conduit—which is taken to mean either rigid aluminum conduit or *galvanized* rigid steel conduit. The use of the word

corrosion-resistant is not intended to require a plastic jacket on galvanized rigid steel conduit, although such a jacket does provide significantly better resistance to natural corrosion, such as rusting.

5. Galvanized IMC.
6. Type MC (metal-clad) cable.
7. Extra-hard-usage portable power cable, such as Type W, where listed for both wet locations and sunlight resistance, has tremendous ability to flex because of its fine stranding. Take care to observe the UL termination limitations on fine-stranding at terminations, as covered in Chap. 1 at 110.14. If these cables are used, a corrosion-resistant junction box must be provided with permanently installed terminal blocks to facilitate the termination of the flexible cord and the feeder extensions. Alternatively, a listed power outlet box with suitable terminal bars or blocks built in can be used.

In the design and construction of a marina it is usually necessary to compare the material and labor costs involved in each of those methods. Emphasis is generally placed on long, reliable life of the wiring system—with high resistance to corrosion as well as high mechanical strength to withstand impact and to accommodate some flexing in the circuit runs. The need for great flexibility in running the circuits under the pier and coming up to receptacle pedestals and lighting poles is often extremely important in routing the circuits over, around, and below the many obstructions commonly built into pier construction. And that concern for flexibility in routing can weigh heavily as a labor cost if a rigid conduit system is used.

Any of the recognized types of cable offer the material-labor advantage of a preassembled, highly flexible "raceway and conductor" makeup that is pretested and especially suited to the bends, offsets, and saddles in the circuit routing at piers, as shown in Fig. 555-2. Cable with a metal armor can offer a completely sealed sheath over the conductors, impervious to fluids and water. For added protection for the metal jacket against oils and other corrosive agents, the cable assembly can have an overall PVC jacket.

555.15. Grounding. The purpose is to require an insulated equipment grounding wire that will ensure a grounding circuit of high integrity. Because of the corrosive influences around marinas and boatyards, metal raceways and boxes are not permitted to serve as equipment grounding conductors. The preference for copper grounding conductors in this section has been removed in the 2014 NEC based on an argument from an aluminum wire manufacturer that aluminum has equal or superior resistance to the corrosive effects of gases common to marine environments.

555.19. Receptacles. Figure 555-3 shows typical configurations of locking- and grounding-type receptacles and attachment plugs used in marinas and boatyards. A complete chart of these devices can be obtained from the National Electrical Manufacturers Association or wiring-device manufacturers. Locking-type receptacles and caps are required to provide proper contact and assurance that attachment plugs will not fall out easily and disconnect onboard equipment such as bilge pumps or refrigerators. Shore-power receptacles for boats must be rated at least 30 A.

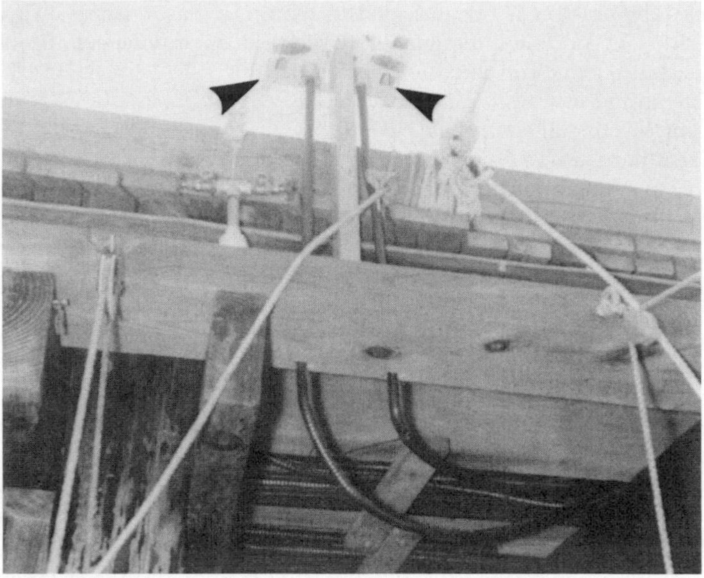

Fig. 555-2. PVC-jacketed metal-clad cable—with a continuous, corrugated aluminum armor that is completely impervious to any moisture and water and resistant to corrosive agents—is used for branch circuits and feeders under this marina pier. (Sec. 555.13.)

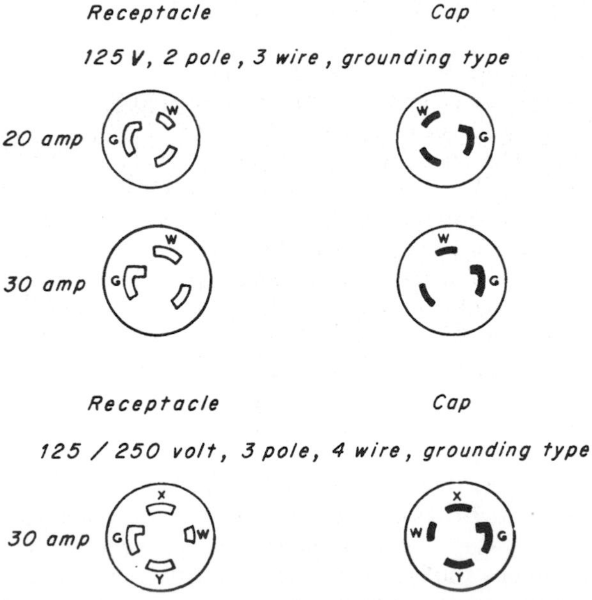
Receptacle
Cap
125 V, 2 pole, 3 wire, grounding type
20 amp
30 amp
G
W
Receptacle
Cap
125 / 250 volt, 3 pole, 4 wire, grounding type
30 amp
X
Y
W
G

Fig. 555-3. These types of connections provide shore power for boats. (Sec. 555.19.)

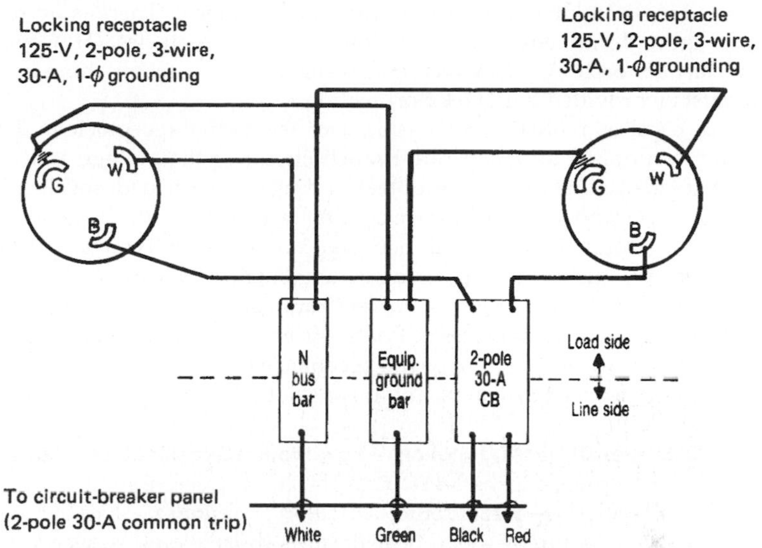
Locking receptacle
125-V, 2-pole, 3-wire,
30-A, 1-ϕ grounding
Locking receptacle
125-V, 2-pole, 3-wire,
30-A, 1-ϕ grounding
G
W
B
G
W
B
N
bus
bar
Equip.
ground
bar
2-pole
30-A
CB
Load side
Line side
To circuit-breaker panel
(2-pole 30-A common trip)
White
Green
Black
Red

Fig. 555-4. Receptacle providing shore power to each boat is contained in a "power pedestal" and is a locking and grounding type. (Sec. 555.19.)

As covered in part **(A)(3),** each single receptacle must be installed on an individual or multiwire branch circuit, with only the one receptacle on the circuit. As shown in Fig. 555-4, a receptacle pedestal unit (two mounted back to back at each location) contains receptacles providing plug-in power to boats at their berths along the pier, with CB protection and control in each housing. As required by NEC 555.19, each receptacle must be rated not less than 30 A and must be a single locking- and grounding-type receptacle. There is no requirement for ground-fault circuit interruption on these receptacles. (However, at a marina, any 15- or 20-A, 120-V receptacles that are *not* used for shore power to boats must be provided with GFCI protection.)

Also, as required by NEC 555.19(A)(3), each individual receptacle in the pedestal unit is supplied by a separate branch circuit of the voltage and current rating that corresponds to the receptacle rating. At each pedestal location, a separate bare, stranded No. 6 copper conductor (arrow) is available as a static grounding conductor bonded to all pedestals and lighting fixtures. The inset in Fig. 555-4 shows one receptacle wiring arrangement. Hookup details at pedestal units vary with voltage ratings, current ratings, and phase configuration of power required by different sizes of boats—from small motorboats up to 100-ft yachts. Note also that each shore power receptacle must be wired in conjunction with a properly marked disconnecting means not over 762 mm (30 in.) away, as covered in 555.17.

555.21. Motor Fuel Dispensing Stations—Hazardous (Classified) Locations. This material has been relocated in the 2014 NEC to 514.3(C), (D), and (E).

555.22. Repair Facilities—Hazardous (Classified Locations). This is the key to bringing Art. 511 requirements to bear on motorboats being serviced at a dry dock or comparable facility.

555.24. Signage. Permanent signs must be posted "at all approaches" to a marina facility to the effect that electrical currents might be in the water that present a shock hazard (see NEC for exact text). Refer to the discussion at 555.3 and 553.4 for more information. Here again, note that this is a nonexistent hazard at a salt water facility.

ARTICLE 590. TEMPORARY INSTALLATIONS

590.1. Scope. Although a temporary electrical system does not have to be made up with the detail and relative permanence that characterizes a so-called permanent wiring system, the specific rules of this article cover the only permissible ways in which a temporary wiring system may differ from a permanent system. Aside from the given permissions for variation from rules on permanent wiring, all temporary systems are required to comply in all other respects with Code rules covering permanent wiring (Fig. 590-1).

590.3. Time Constraints. In part **(A),** the words *maintenance* and *repair* indicate that the less rigorous methods of temporary wiring may be used and that all rules on temporary wiring must be observed wherever maintenance or repair work is in progress. This expands the applicability of temporary wiring beyond new construction, remodeling work, or demolition.

Part **(B)** recognizes use of temporary wiring for seasonal or holiday displays and decorations, as shown in Fig. 590-2.

Part **(C)** of this section permits temporary wiring to be used for other than simple construction work. Such wiring, as covered in this article, may also be used during emergency conditions or for testing, experiments, or development activities. As the proposal for this Code rule noted:

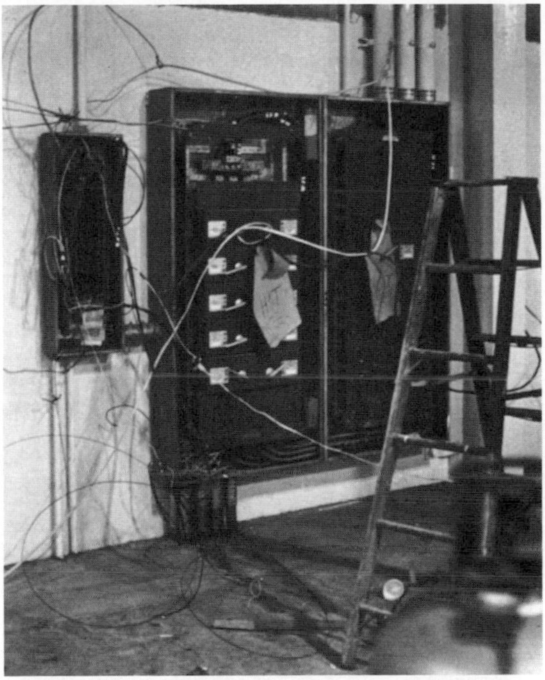

Fig. 590-1. Temporary wiring is not an "anything goes" condition and must comply with standard Code rules to prevent a rat-nest condition, which can pose hazard to life and property. (Sec. 590.1.)

Were it not permissible to use temporary wiring methods for testing purposes, it would be impossible to check, before placing in service, many electrical installations. Likewise, emergency conditions would remain without electric power and lighting until permanent installation could be made.

However, part **(D)** of this section is aimed at ensuring that the equipment and circuits installed under this article are really temporary and not a backdoor to low-quality permanent wiring systems.

590.4. General. Although part **(A)** requires a temporary service to satisfy all the rules of Art. 230, part **(B)** recognizes the use of temporary feeders that are conductor "cable assemblies" used as open wiring (Fig. 590-3), multiconductor cable assemblies (Type NM, UF, etc.), or multiconductor cord or cable of the type covered by Art. 400 for hard-usage or extra-hard-usage flexible cords and cables, which are not acceptable for use as feeder or branch-circuit conductors of permanent wiring systems. 400.12 specifically prohibits the use of such cords and cables "as a substitute for the fixed wiring of a structure." As shown in Fig. 590-4, prewired portable cables with plug and socket assemblies are available for power risers in conjunction with GFCI-protected branch-circuit centers, or cables can be run horizontally on a single floor to suit needs. GFCI breakers may be used in temporary panelboards interconnected with cable and feeding standard receptacles in portable boxes, as shown in Fig. 590-5. The 2017 **NEC**

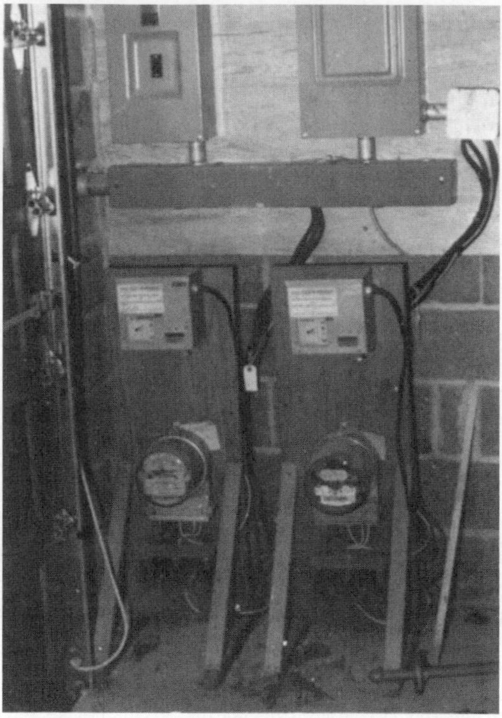

Fig. 590-2. Temporary wiring techniques are permitted for 90 days for such "experimental" work as energy demand analysis. (Sec. 590.3.)

added Type SE cable to both (B) and (C), feeders and branch circuits as permitted, including running in a raceway underground.

590.4(C) requires temporary branch circuits to consist of multiconductor cable assemblies (Types NM and UF) or cords or cables covered in Table 400-4, provided that they originate in a panelboard, switchboard, motor control center, fused switch, or "an approved power outlet," which is one of the manufactured assemblies made for job site temporary wiring. As shown in Fig. 590-6, the temporary branch circuits for receptacle outlets may be part of a manufactured temporary system, which consists of cable harnesses and power centers (or outlets). Several variations of protection may be provided by such portable receptacle boxes, as shown in Fig. 590-6. Box 1 may have GFCI protection for its own receptacles without providing downstream protection. Box 2 may have the same protection as box 1 and in addition have GFCI protection for its 50-A outlet, thus providing protection for box 3. With this arrangement, box 1 will sense the ground fault from the worker at upper left and will trip, allowing boxes 2 and 3 to continue to provide power. Or all three boxes could receive GFCI protection from a permanently mounted loadcenter feeding the 50-A receptacle outlet at upper left. In this case, the ground fault shown would interrupt the power to all boxes. It should be noted that use of a GFCI breaker may be viewed as a violation for temporary power applications

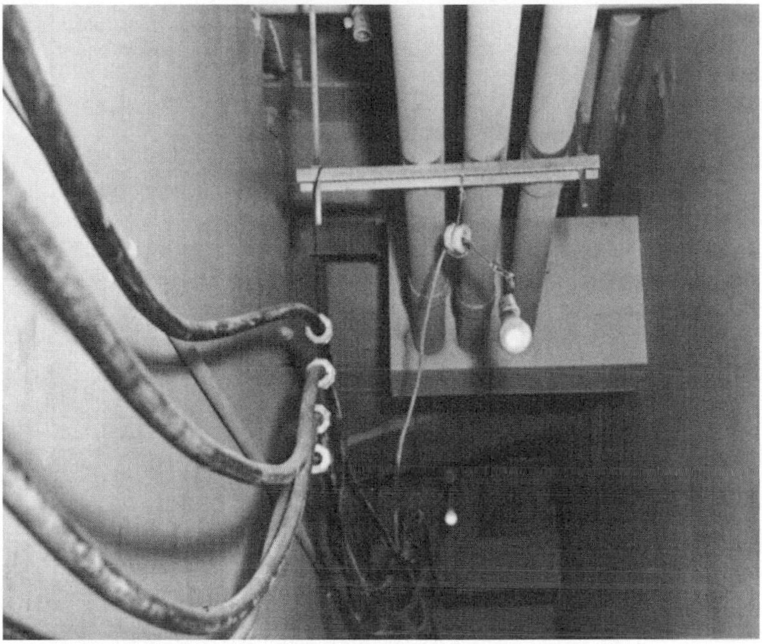

Fig. 590-3. The exception to 527.4(B) permits temporary feeders to be run as open individual conductors supported by insulators spaced not over 10 ft (3.0 m) apart where used during emergencies or testing. (Sec. 590.4.)

because it does not have the same characteristics—that is, "open neutral" and "reverse phasing" protection—as do listed temporary power GFCI devices.

590.6 makes it clear that only receptacles used under temporary job conditions require GFCI protection. The implication is that the nonmetallic-sheathed cable runs and pigtail connections traditionally associated with temporary power on the job site would not win awards for neatness and safety, but that once the permanent feeders and panelboards are in place and energized, the shock hazard is considerably reduced.

However, as long as portable tools are being used in damp locations in close proximity with grounded building steel and other conductive surfaces, the possibility of shock exists from faulty equipment whether it is energized from temporary or from permanent circuits.

Standard panelboards used for temporary power fitted with GFCI circuit breakers for the protection of entire circuits satisfies the rules of 590.6, but may be a violation of 110.3(B) because breaker type GFCI devices listed for permanent installation would not satisfy the UL requirements for temporary power use. However, the many varieties of portable power distribution centers and modules have been developed with integral GFCI breakers protecting single-phase, 15- and 20-A, 120-V circuits. Other circuits (higher amperage, higher voltage, and 3-phase) are also required by the **NEC** to have GFCI protection, or the Assured Equipment Grounding Conductor Program, described in 590.6(B), may

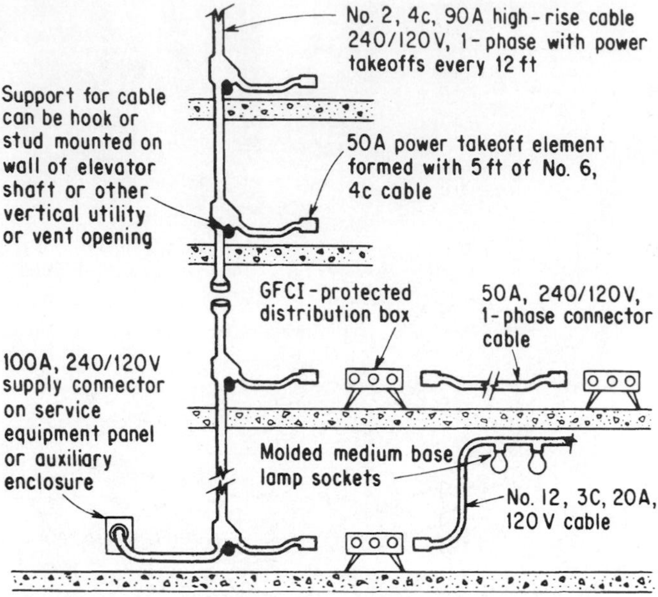
No. 2, 4c, 90A high-rise cable 240/120V, 1-phase with power takeoffs every 12 ft
Support for cable can be hook or stud mounted on wall of elevator shaft or other vertical utility or vent opening
50A power takeoff element formed with 5 ft of No. 6, 4c cable
GFCI-protected distribution box
50A, 240/120V, 1-phase connector cable
100A, 240/120V supply connector on service equipment panel or auxiliary enclosure
Molded medium base lamp sockets
No. 12, 3C, 20A, 120V cable

Fig. 590-4. Temporary feeders may be cord assemblies made especially for such use. (Sec. 590.4.)

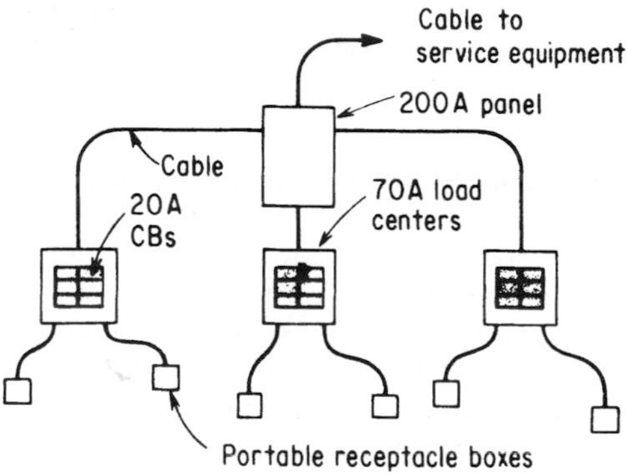
Cable to service equipment
200A panel
Cable
20A CBs
70A load centers
Portable receptacle boxes

Fig. 590-5. Distribution for temporary power may utilize cable or raceway feeders. (Sec. 590.4.)

be used. A variety of cord sets are also available for use with GFCI-protected plug-in units to supply temporary lighting and receptacle outlets.

While a manufactured system of cable harnesses and power-outlet centers costs more than nonmetallic-sheathed cable runs and pigtail sockets, it is completely recoverable, and its cost can be written off over several jobs. From then on, with the exception of costs for setup and removal, storage, and transportation, much of the temporary power charges included in bids could be profit.

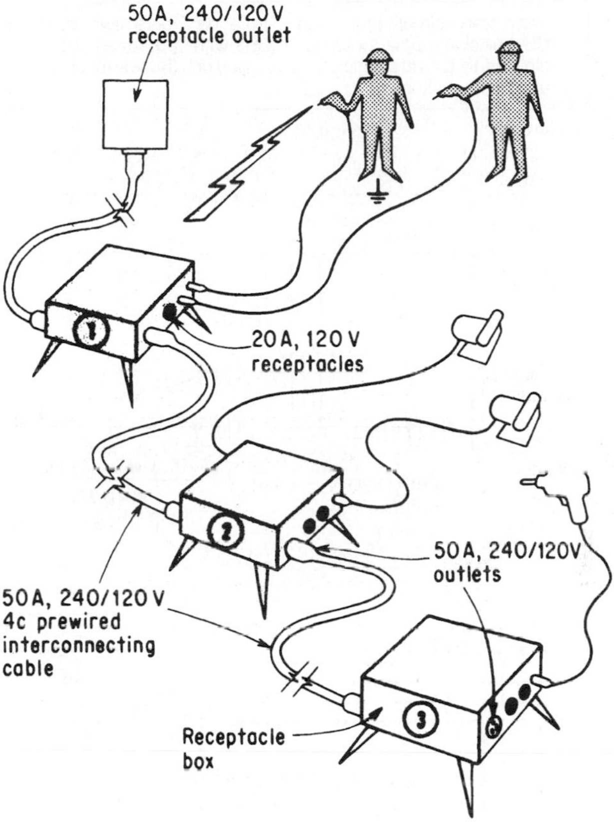

50A, 240/120V
receptacle outlet
20A, 120V
receptacles
50A, 240/120V
outlets
50A, 240/120V
4c prewired
interconnecting
cable
Receptacle
box

Fig. 590-6. Temporary branch circuits may be part of a manufactured system. (Sec. 590.4.)

In previous Code editions, part **(C)** of this section required temporary wiring circuits to be "fastened at ceiling height every 10 ft (3.0 m)." But now, if such circuits operate at not over 150 V to ground and are not subject to physical damage, the fourth sentence in this paragraph permits open-wiring temporary branch circuits to be run at any height "supported on insulators at intervals of not more than 10 ft (3.0 m)" where used for other than "during the period of construction" as covered in part **(A)** of 590.3. Open wiring must not be laid on the floor or ground.

In the interest of greater safety, part **(D)** prohibits the use of both receptacles and lighting on the same temporary branch circuit on construction sites. The purpose is to provide complete separation of the lighting so that operation of an overcurrent device or a GFCI due to a fault or overload of cord-connected tools will not simultaneously disconnect lighting (Fig. 590-7). In addition, all receptacles used for temporary power and located in wet locations must comply with the weather-resistance and the in-use cover requirements of 406.9(B)(1).

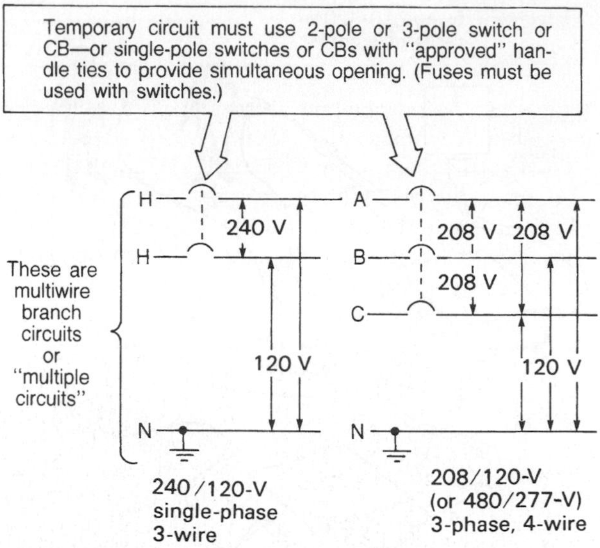
Temporary circuit must use 2-pole or 3-pole switch or CB—or single-pole switches or CBs with "approved" handle ties to provide simultaneous opening. (Fuses must be used with switches.)
H
H
N
240 V
120 V
A
B
C
N
208 V
208 V
208 V
120 V
These are multiwire branch circuits or "multiple circuits"
240/120-V single-phase 3-wire
208/120-V (or 480/277-V) 3-phase, 4-wire

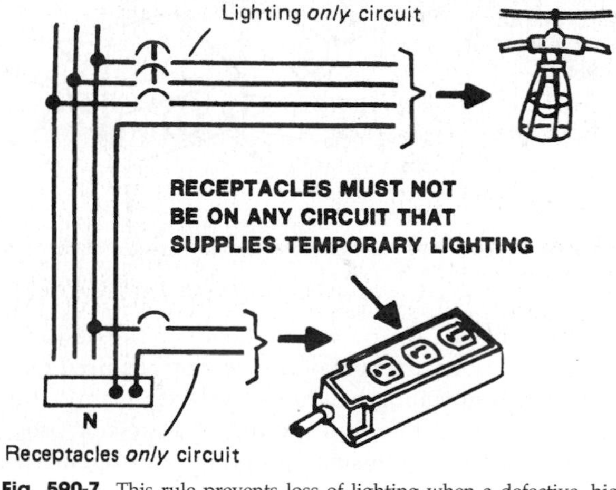
Lighting only circuit
RECEPTACLES MUST NOT BE ON ANY CIRCUIT THAT SUPPLIES TEMPORARY LIGHTING
N
Receptacles only circuit

Fig. 590-7. This rule prevents loss of lighting when a defective, high-leakage, or overloaded Code-connected tool or appliance opens the branch-circuit protection of a circuit supplying one or more receptacles. (Sec. 590.4.)

According to part **(E)**, every multiwire branch circuit must have a disconnect means that simultaneously opens all ungrounded wires of the temporary circuit. At the power outlet or panelboard supplying any temporary multiwire branch circuit (two hot legs and neutral, or three hot legs and neutral), a multipole disconnect means must be used. Either a 2-pole or a 3-pole switch or circuit breaker would satisfy the rule, or single-pole switches of single-pole CBs may be used with approved handle ties to permit the single-pole devices to operate together (simultaneously) for each multiwire circuit, as shown for the multiwire lighting circuit in Fig. 590-7.

Part **(F)** requires lamps for general lighting on temporary wiring systems to be "protected from accidental contact or breakage." Protection must be provided by a suitable fixture or lampholder with a guard (Fig. 590-8). OSHA rules also require the use of a suitable metal or plastic guard on each lamp. As shown in Fig. 590-9, commercial lighting strings provide illumination where required. Splice enclosure is equipped with integral support means, and a variety of lamp-guard styles provide protection for lamp bulbs.

Fig. 590-8. A lampholder with a guard is proper protection for a lamp at any height in a temporary wiring system (above). Unguarded lamps at any height constitute a Code violation (right). (Sec. 590.4.)

Part **(F)** requires grounding of metal lamp sockets. The high level of exposure to shock hazard on construction sites makes use of ungrounded metal-shell sockets extremely hazardous. When they are used, the shell must be grounded by a conductor run with the temporary circuit.

In part **(G)**, splices or tap-offs are permitted to be made in temporary wiring circuits of cord or cable without the use of a junction box or other enclosure at the point of splice or tap (Fig. 590-10). But this new permission applies only to nonmetallic cords and cables. A box, conduit body, or terminal fitting must be used when a change is made from a cord or cable circuit to a raceway system or to a metal-clad or metal-sheathed cable. The enclosure can generally be omitted

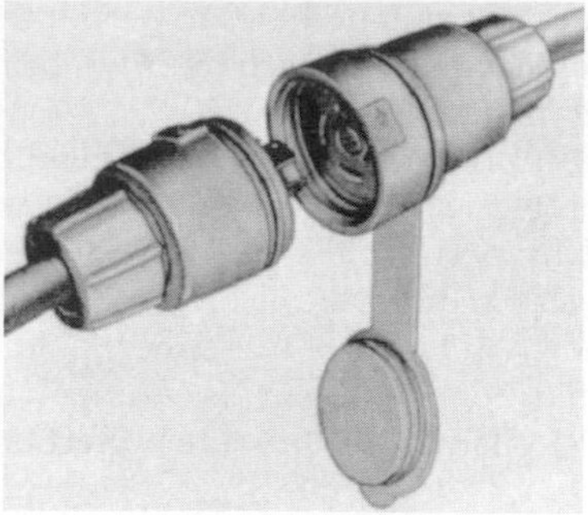

Special watertight plugs and connectors provide insurance against nuisance tripping caused by weather conditions on construction sites.

Fig. 590-9. Temporary lighting strings of cable and sockets are available from manufacturers. Note the multiconductor cable makeup; festoon lighting strings are not permitted. (Sec. 590.4.)

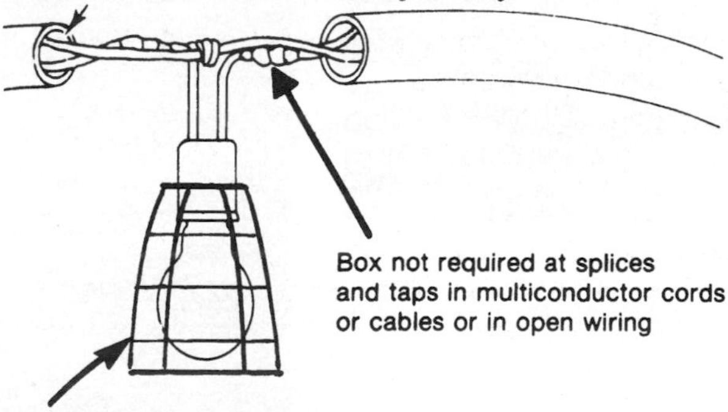
Temporary wiring for new construction or modernization or repair of existing buildings
Box not required at splices and taps in multiconductor cords or cables or in open wiring
NEC requires a guard on the lamp.

Fig. 590-10. Splices may be used without boxes for cord and cable runs on construction sites. (Sec. 590.4.)

with a major proviso: the equipment grounding continuity must not be compromised. This can be accomplished with nonmetallic wiring methods easily, and even with metallic wiring methods provided listed fittings are used to secure the cable sheath and maintain continuity.

Regulations in part **(H)** require protection of flexible cords and cables from damage due to pinching, abrasion, cutting, or other abuse.

Part **(I)** calls for the use of proper fittings to secure cables that enter enclosures containing receptacles and/or switches.

Part **(J)** requires cable assemblies and cords to be supported at intervals that assure protection from physical damage. You can use staples, cable ties, straps, or other fittings that don't damage the cable. The idea here is to allow a greater support interval than would be covered in the applicable article, because this is just for temporary applications.

Be careful! The 2014 NEC inserted a far-reaching prohibition against running any cable assemblies or flexible cords "installed as branch circuits or feeders" on the floor or on the ground. Although this language goes on to specifically exempt extension cords from its reach, it will definitely require the common applications of large SER cable assemblies to be supported so as to not run across a floor. This will have a definite impact on job costs. In addition, the boundary between an "extension cord" and a feeder or branch circuit is not always clear-cut on construction sites. For example, 50-A flexible cord assemblies connecting the distribution boxes in Fig. 590-6, although they connect and disconnect like extension cords, may very well be interpreted as feeders within the meaning of this rule. The supporting substantiation for the rule clearly intended the extension cord wording to only apply to the load side of GFCI protection, although the NEC text is silent on that point. OSHA regulations also do not allow feeders to be laid on the ground. Consultation with the inspector will be crucial to avoiding severe financial consequences during construction.

Vegetation, however, is prohibited from being used as temporary support of overhead runs of branch-circuit or feeder conductors. The only exception is holiday decorative lighting, which is permitted to run from tree to tree provided there is some tension take-up mechanism or other approved arrangement that will prevent tree movement from damaging the temporary lighting strings.

590.5. Listing of Decorative Lighting. Here the Code makes it absolutely clear that any manufactured decorative lighting used for temporary installation *must* be listed *and* so labeled on the product. And this would include a proper listing for the application. That is, where used outdoors such lighting must also be listed for outdoor use or wet locations, etc. Note that 410.160 imposes the identical requirement.

590.6. Ground-Fault Protection for Personnel. This section covers the rules that concern GFCI protection for all receptacles supplied from temporary wiring systems.

Part **(A)** is a construction site GFCI requirement that applies to temporary power provided for receptacle outlets, whether from a utility or on-site generator source, that is used by personnel to construct, demolish, remodel, maintain, or repair buildings or structures, or perform similar work. This analysis will use the term "construction activities" as a shorthand reference to the preceding list of qualified activities. The requirements for GFCI protection apply to 125-V, single-phase, 15-, 20-, and 30-A receptacle outlets as specifically described in three-numbered paragraphs itemized below.

Industrial occupancies with qualified maintenance and supervision are permitted to use the assured equipment grounding conductor program. However, that permission is strictly limited to those specific receptacle outlets that supply equipment having a design that is incompatible with GFCI protective devices, or that supply power to equipment that, if suddenly disconnected, would create a greater hazard than that addressed by the shock protection inherent to the GFCI device. The substantiation for this exception (originated in the 2002 NEC) pointed to small 120-V MIG welders as examples of the first problem, and magnetic-based drills and vacuum-held coring drills during operation as examples of the second problem.

Paragraph **(1)** applies to receptacle outlets that are not part of the permanent wiring on site (Fig. 590-11) and are used for construction activities, such as receptacles in a spider box. Any such receptacles must "have" GFCI protection for personnel at all times when they are being used. Reversing the content of the 2011 NEC, the 2014 NEC now expressly permits listed cord sets or devices incorporating GFCI protection identified for portable use to supply temporary loads fed from a receptacle not considered part of the permanent wiring. (Fig. 590-11). However, the permission to use the portable use devices does not excuse compliance with the basic requirement that temporary receptacles in use as part of the construction project must have GFCI protection.

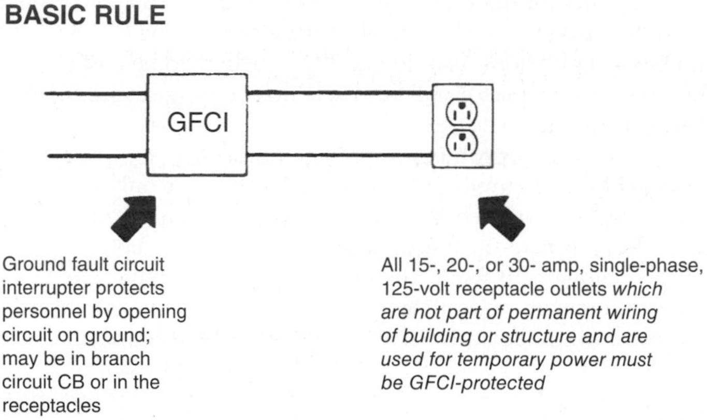
BASIC RULE
GFCI
Ground fault circuit interrupter protects personnel by opening circuit on ground; may be in branch circuit CB or in the receptacles
All 15-, 20-, or 30- amp, single-phase, 125-volt receptacle outlets which are not part of permanent wiring of building or structure and are used for temporary power must be GFCI-protected

Fig. 590-11. GFCI protection on construction sites for receptacles in use that are not part of the permanent wiring of the building or structure. Listed GFCI cord sets or devices identified for portable use are now, as of the 2014 NEC, permitted to be used. [Sec. 590.6(A)(1).]

Paragraph **(2)** applies to receptacle outlets that are part of the permanent wiring for the building, but are being used for construction activities. GFCI protection must be provided, but a portable GFCI device or cord set that is "identified for portable use" is permitted so as to avoid changing out receptacles or circuit breakers (Fig. 590-12).

Both paragraphs only directly apply to receptacles that: [Paragraph **(1)** "are in use by personnel"] or are [Paragraph **(2)** "used for temporary electric power"]. For either paragraph the Code wording clearly limits required GFCI protection to receptacles that are actually being used at any particular time. Receptacles not in use do not have to be GFCI-protected. This means that portable GFCI protectors may be used at only those permanently installed outlets being used (Fig. 590-13). There is no need to use GFCI breakers in the panel to protect all receptacles or to use all GFCI-type receptacles. This might seem to seriously confuse the task of electrical inspection. If all cord-connected tools and appliances are unplugged from receptacles when the inspector comes on the job, then none of the receptacles

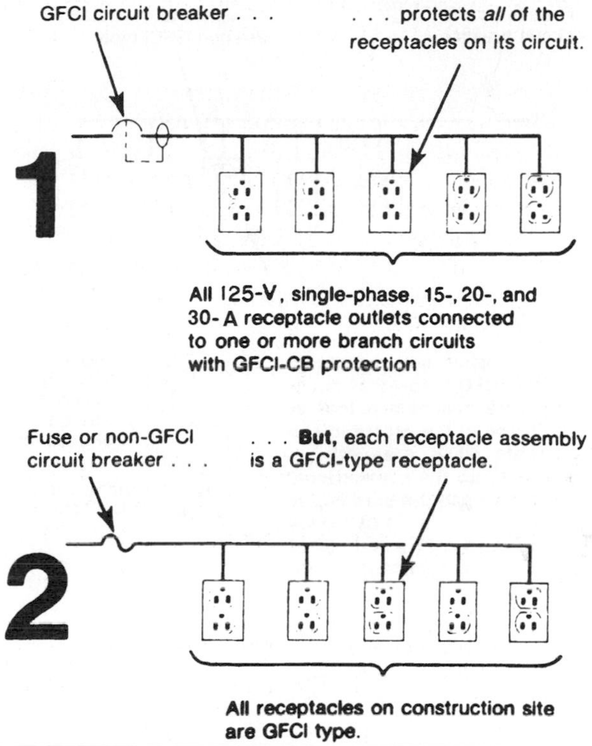
GFCI circuit breaker . . .
. . . protects *all* of the receptacles on its circuit.
1
All 125-V, single-phase, 15-, 20-, and 30-A receptacle outlets connected to one or more branch circuits with GFCI-CB protection
Fuse or non-GFCI circuit breaker . . .
. . . But, each receptacle assembly is a GFCI-type receptacle.
2
All receptacles on construction site are GFCI type.

Fig. 590-12. Two ways to satisfy the basic rule on personnel shock protection at *temporary* receptacles on construction sites that are part of the permanent wiring of the building or structure subject to construction activities. [Sec. 590.6(A)(2).] Listed GFCI cord sets or devices identified for portable use can also be used. That means they will have internal circuitry that opens all poles of the device if the grounded circuit conductor is open for any reason. Conventional GFCI devices for permanent connection in a panel or at an outlet do not have this feature, and will fail closed (energized) if the grounded conductor opens because their circuitry will not have the 120 V they need to work. This is why making a short extension cord with an ordinary GFCI receptacle in a box does not comply with this requirement.

is "in use" and none of them has to have GFCI protection, and there is no Code violation. Most inspectors can see through such foolishness.

The 2008 **NEC** added the requirement that GFCI protection must apply "to power derived from an electric utility company or from an on-site-generated power source." Paragraph **(3)** now carries the generator portion of that requirement in the **NEC**. Now all receptacles for construction activities that are a part of a 15-kW or smaller portable generator must have listed GFCI protection for personnel. Such portable generators must also provide listed GFCI protection for 125/250-V receptacles if they provide them. Note also that for portable generators that will operate in damp or wet locations, 15- and 20-A 125- and 250-V single-phase receptacles that are "part" of the generator must have the weather-resistance and the in-use cover features required in 406.9.

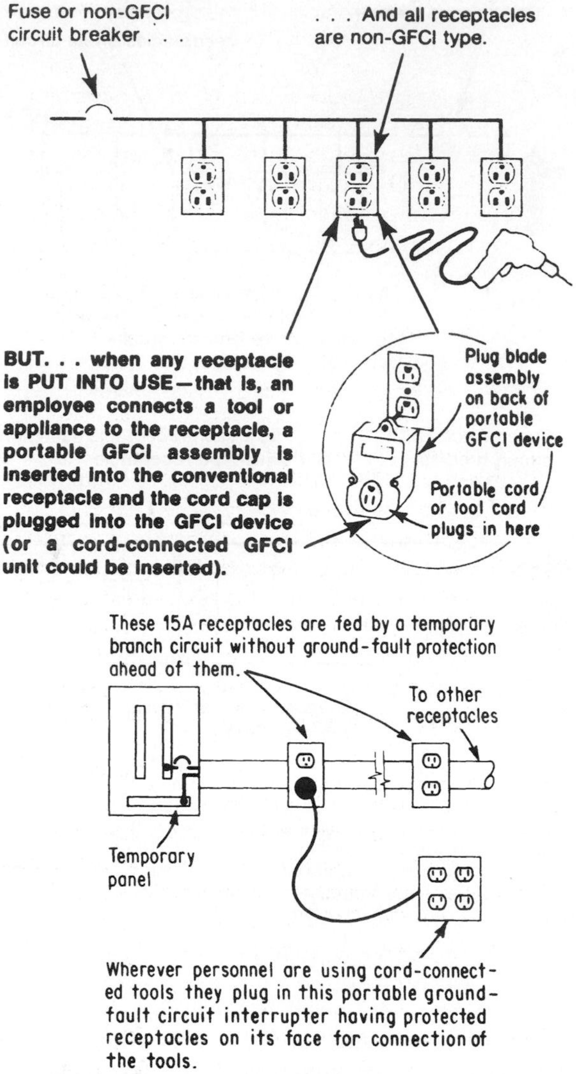
Fuse or non-GFCI circuit breaker . . .
. . . And all receptacles are non-GFCI type.
BUT. . . when any receptacle is PUT INTO USE—that is, an employee connects a tool or appliance to the receptacle, a portable GFCI assembly is inserted into the conventional receptacle and the cord cap is plugged into the GFCI device (or a cord-connected GFCI unit could be inserted).
Plug blade assembly on back of portable GFCI device
Portable cord or tool cord plugs in here
These 15A receptacles are fed by a temporary branch circuit without ground-fault protection ahead of them.
To other receptacles
Temporary panel
Wherever personnel are using cord-connected tools they plug in this portable ground-fault circuit interrupter having protected receptacles on its face for connection of the tools.

Fig. 590-13. Portable GFCI devices may be used to satisfy GFCI rule, but only when plugged into receptacles that are permanently installed. The drawing at the top, presumed to show receptacles that are permanently installed, shows a portable GFCI applied correctly. The violation for lack of GFCI protection at the temporary receptacles in the bottom drawing cannot be cured with portable GFCI devices. (Sec. 590.6.)

GFCI receptacles don't provide meaningful protection on ungrounded circuits, and if the generator neutral is not bonded to the frame, then the generator output circuits are functionally ungrounded. For this reason, the recently released UL Standard on Portable Engine Generator Assemblies (UL 2201) requires all 15-kW or smaller portable generators to have bonded neutral connections. This makes them suitable to supply GFCI receptacles that will actually trip on a ground fault, and therefore agrees with the new rules in this paragraph. This change is extremely controversial in the generator community. Refer to the discussion in this book at 250.34 on portable generators for more information on this point. Because of the large numbers of generators without bonded neutrals, Paragraph **(3)** also allows portable-use GFCI protective devices to be used on all generators manufactured or rebuilt prior to Jan. 1, 2011. There are further developments in this area, covered in some depth in this book in the commentary regarding the new Sec. 445.20.

Part **(B)** extends the requirements to all receptacles other than those for construction activities [as defined in the discussion at **(A)** for these purposes]. That includes 3-phase and phase-to-phase receptacles of any current value temporarily installed. Alternately, the Assured Equipment Grounding Conductor Program explained in the remainder of this section may be used.

As of 2013, some manufacturers are making new classes of GFCIs available. They use a 20 mA trip threshold, which will not cause ventricular fibrillation but can be enough to cause the muscles to lock on to the faulted equipment. They work on the principle that if there is a suitable equipment grounding return path in parallel with the person being shocked, then the fault will be cleared before permanent damage is done. In effect the ground trip is the let-go protection. UL anticipated these by creating the relevant categories in the 2009 edition of the standard. For example, a Class C device can be used on circuits not over 300V to ground (e.g., 480Y/277V). These circuits must use an EGC at least equal in size to the circuit conductors and the device must trip if the ground opens. There are also Class D devices for 600V systems.

The 2017 **NEC** recognized this option by adding, as a new (B)(2) "Special purpose ground-fault circuit-interrupter protection for personnel" or "SPGFCI." This is the first time this technology has made it into the **NEC**, and undoubtedly it won't stop here. They are covered in UL Standard 943C.

Still another option for avoiding use of GFCI protection on temporary wiring systems is given in part **(B)** of this section for other than 15-, 20-, or 30-A, 125-V receptacles. GFCI protection at the higher ampere and voltage ratings of receptacles may be omitted totally if a "written procedure" is established to assure testing and maintenance of "equipment grounding conductors" for receptacles, cord sets, and cord-and-plug-connected tools and appliances used on the temporary wiring systems (Fig. 590-14). In effect, the **NEC** accepts such an equipment grounding conductor program as a measure that provides safety that is equivalent to the safety afforded by GFCI protection. GFCI protection is not required for other than the 15-, 20-, and 30-A, 125-V receptacles if all the following conditions are satisfied:

1. The inspection authority having jurisdiction over a construction site must approve a written procedure for an equipment grounding program.
2. The program must be enforced by a single designated person at the construction site.
3. "Electrical continuity" tests must be conducted on all equipment grounding conductors and their connections. The requirements on making such tests are vague, but they do call for:
   a. Testing of fixed receptacles where there is any evidence of damage.
   b. Testing of extension cords before they are first used and again where there is evidence of damage or after repairs have been made on such cords.

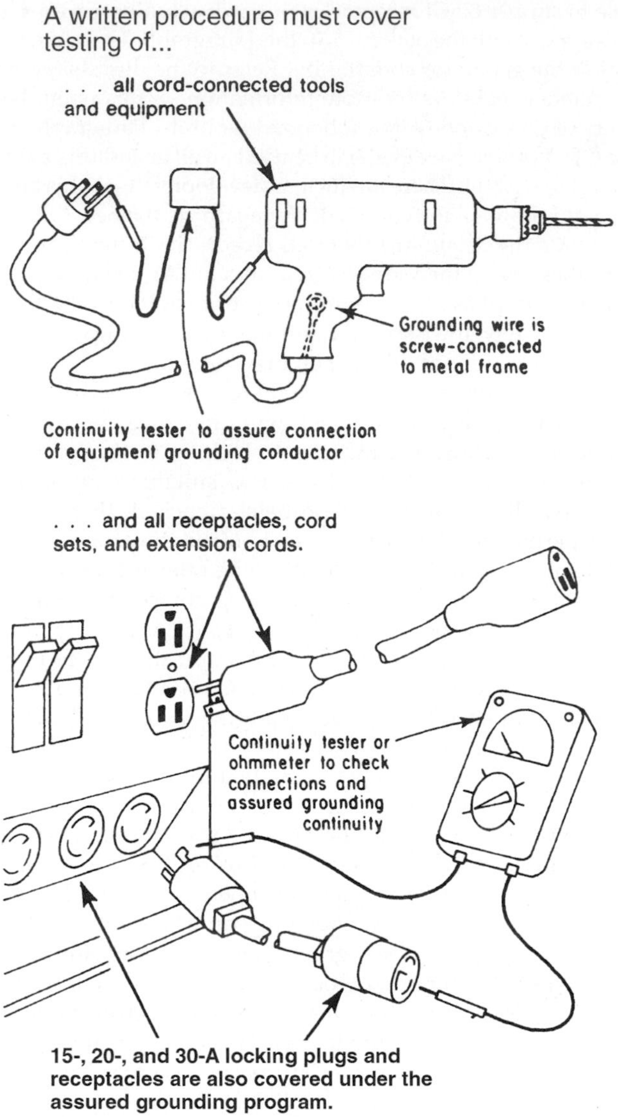
A written procedure must cover testing of...
. . . all cord-connected tools and equipment
Grounding wire is screw-connected to metal frame
Continuity tester to assure connection of equipment grounding conductor
. . . and all receptacles, cord sets, and extension cords.
Continuity tester or ohmmeter to check connections and assured grounding continuity
15-, 20-, and 30-A locking plugs and receptacles are also covered under the assured grounding program.

Fig. 590-14. Assured grounding program eliminates the need for GFCI. Note: The assured equipment grounding program may *only* be used on 15-, 20-, and 30-A, 125-V receptacles in industrial establishments. (Sec. 590.6.)

c. Testing of all tools, appliances, and other equipment that connect by cord and plug before they are first used on a construction site, again any time there is any evidence of damage, after any repair, and at least every 3 months.

Obviously, these rules are very general and could be satisfied in either a rigorous, detailed manner or a fast, simple way that barely meets the qualitative criteria. The electrical contractor who has responsibility for the temporary wiring on any job site is the one to develop, write, and supervise the Assured Equipment Grounding Program, where that option is chosen as an alternative to use of GFCI protection. This whole **NEC** approach to use of either GFCI or an Assured Equipment Grounding Program directly parallels the OSHA approach to the matter of receptacle protection on construction sites.

It should be noted that the Assured Equipment Grounding Program described in part **(B)** is one of the most frequently cited violations during OSHA inspections. Implementation is a bureaucratic nightmare and is rarely successfully executed.

Chapter Six

ARTICLE 600. ELECTRIC SIGNS AND OUTLINE LIGHTING

600.1. Scope. In the case of signs that are constructed at a shop or factory and sent out complete and ready for erection, the inspection department must require listing and installation in conformance with the listing. In the case of outline lighting and signs that are constructed at the location where they are installed—the so-called skeleton tubing covered by Part II—the inspection department must make a detailed inspection to make sure that all requirements of this article are complied with.

Rules governing the installation of electric signs vary widely from jurisdiction to jurisdiction. In some cities, inspection departments inspect signs in local shops as well as performing installation site inspections. Likewise, in some cities, the electrical inspector conducts the sign inspection alone. In others, a sign inspector will review plans for proposed signs for compliance with local ordinances, and the electrical inspector will then do the field inspection in conjunction with the sign inspector. Contact the building department in the municipality having jurisdiction to establish the exact procedure that must be followed.

600.2. Definitions. This section contains a number of definitions that are critical to the proper application of the Code rules for electric signs and outline lighting.

600.3. Listing. All electric signs, section signs, and outline lighting, whether fixed, mobile or portable, and all retrofit kits applied for similar purposes must be listed unless exempted by special permission. However, field-installed skeleton tubing is not required to be listed if installed per NEC rules, and outline lighting need not be listed as a system if it consists of listed luminaires wired using Chap. 3 methods. Required markings and listing labels must be visible to the installer and servicer, but need not be visible on the outside after the installation is completed. The 2014 NEC notably includes a requirement, directed at manufacturers, that essentially all products within the scope of Art. 600 must be marked to say that "field-wiring and installation instructions are required." The intended purpose is to put installers firmly on notice to abide by 110.3(B) and follow the directions. The only exception to this rule is for portable, cord-connected signs.

600.4. Markings. As of the 2017 edition, this section addresses signs retrofitted from fluorescent lamps to tubular-style LED strips. After the retrofit procedure, the sign may have characteristics that will cause a conventional fluorescent lamp to fail violently if mounted in the lamp sockets now wired for the LED replacements. Sign retrofits done in this matter must have a marking applied that alerts service personnel to the retrofit, identifies the equipment manufacturer and the installer, and provides a suitable warning label in accordance with 110.21(B) advising not to attempt the substitution of fluorescent lamps for the LED strips.

600.5. Branch Circuits. In part **(A),** the Code mandates that any commercial occupancy that is "accessible to pedestrians" must be provided with one outlet for the purpose of supplying an electric sign, which must be accessible. Additionally, this section requires that the branch circuit supplying this outlet be a dedicated circuit, with no other loads supplied. The wording of this section requires that a sign outlet be installed for every ground-level store—even if an outdoor electric sign is not actually installed or planned. Note that no limit is placed on the number of outlets that may be connected on one circuit for a sign or for outline lighting, except that the total load should not exceed 16 A where incandescent or fluorescent lighting loads are to be supplied. A 30-A maximum is established for branch circuits supplying neon tubing. However, the minimum size for this circuit is 20 A. Note that for a mall-type environment, the wording is for "each entrance" so the outlet could be on the central hallway.

Where the loads to be supplied are "continuous loads," that is, where in normal operation, the load will continue for 3 h or more, the load should not exceed 80 percent of the branch-circuit rating. Given that commercial lighting is generally considered to be a "continuous load," circuits and OC devices used to supply and protect such loads must be sized on the basis described for continuous loading, and 600.5(B) removes all doubt by classifying this load. That is, the conductors must have current-carrying capacity that is at least 125 percent of the continuous load, before derating. And the OC device must also have a long-time trip rating of at least 125 percent of the rated lighting load. This effectively translates into a 16-A maximum lighting load and No. 12 copper conductors for a 20-A circuit, and 24 A using No. 10 copper on a 30-A circuit. See 210.19(A), 210.20(A), and 240.4(D).

Part **(B)** indicates the minimum ratings for the required sign circuit. The 20-A branch circuit for sign and/or outline lighting for commercial occupancies with ground-floor pedestrian entry—required by part **(A)**—may supply one or more outlets for the purpose, but not any other loads. The intent is that the required, dedicated 20-A circuit supply one or more outlets intended for electric signs. However, if the intended sign is to be made of neon tubing, the rating could be 30 A (and no more).

As noted in part **(C),** the wiring method used to supply signs must conform with the requirements of parts **(1), (2),** and **(3).**

In part **(C)(1),** the Code requires that the wiring method used to supply the sign—which may be any of the wiring methods recognized in Chap. 3 suitable for the type of location in which the sign is installed—must be terminated either in the sign, in a box provided with the sign, or in a typical junction/outlet box or conduit body. In part **(C)(2),** the Code recognizes the use of signs and electric-sign transformer enclosures as raceways to supply adjacent signs and associated equipment. And part **(C)(3)** mandates that metal poles used to support electric signs comply with the rules for poles used to support lighting, as covered in 410.30(B). (See Fig. 600-1.)

Every store with ground-floor access to customers must have at least one sign outlet fed by a separate 20-A circuit—but may have two or more outlets on the circuit, which must not supply loads other than sign outlets.
STORE
Show window

Fig. 600-1. Commercial buildings must have outdoor sign outlet. (Sec. 600.5.)

600.6. Disconnects. This section is somewhat difficult to understand and expect consistent interpretation of because the parent language at the subsection level provides an unclear context for the three parts that follow. The major point of disagreement turns on whether it is sufficient to comply with any one of the three location rules. Each of the three carries mandatory language.

Taking the last one (3) first, it is clear that there must be a disconnect in sight of a sign controller remote from the sign, that the disconnect must be ahead of the sign as well as the controller, and that disconnect must have a lock-open capability. The next case is (2), which requires a sign ahead of the sign, and if the sign is configured in such a way that part of it is out of the line of sight from the disconnect, then that disconnect must also have a lock-open capability, such as where a multisection sign extends around the corner of a building.

The NEC has applied these concepts in a very consistent way since sign disconnects first appeared as Rule 3807 of the 1928 edition. For almost 90 years, signs have been required to have disconnects, either in sight or remote with a locking feature. The parent text of the section, which provides the general requirements for disconnection, dictates that they must be externally operable, control no other load, and open all legs of a multiwire circuit simultaneously. Every instance of the term "disconnect" is handled grammatically as singular in number.

This was the context into which the 2014 NEC placed 600.6(A)(1), renumbering the previous options as (2) and (3). It either required or covered, interpretation being required, a disconnect at the physical location of the sign itself, at the point where the electrical supply circuit for the sign enters the sign enclosure, or its supporting pole. And by the term "enclosure," the intent is a reference to the outer margin of the sign itself, and not any electrical enclosure within the sign body. The rule does address very large signs, some larger than houses, that are supplied by multiple feeders running in Chap. 3 raceway wiring methods supporting different parts of what is technically a single sign.

Following discussions with panel members during the 2014 NEC cycle and reviewing the supporting substantiation, it was apparent at the time that the panel intended the first requirement to supersede the existing options and to apply to all signs. However, it was also the case that the second and third options were not removed, nor was any prioritizing language inserted to follow the subsection title, "Location." All three numbered items, based on the literal text, had to be regarded as compliant, and there was no basis to require multiple disconnects.

Three years later (2017 NEC), the literal text is still unclear, but the context has improved considerably. There is now parent language ahead of the three options stating "shall be permitted to be located" in reference to the three options. The sentence, however, does use the conjunction "and" instead of "or," potentially suggesting all of the options need to be installed. The substantiation supporting the 2017 NEC language is very helpful here, because it comes from the the major trade organization of professional sign installers, and the intent is very clear:

> The current text of the rule does not specify that the subsections are choices for the disconnect location. Texts added to identify the subsections that follow are separate, alternate rules for the location of the sign's disconnecting means. "Shall be permitted" effectively conveys the rule is allowed as an option/choice for the location of the disconnecting means as intended by the additional text in (A).

In addition to this substantiation, there is also a highly significant group of text changes within all three of the options. For options (2) and (3), when they are out of sight of the sign, a warning label meeting the requirements of 110.21(B) must be inserted where it will be visible during service work, and it must spell out the location of the disconnect. This is even true of the first option (right at the sign), which carries that language in its exception addressing very large signs with internal panelboards located remotely. If items (2) and (3) were superseded by (1), that language would not have been necessary.

Option (1) is particularly directed at very large signs, and to that end now carries an exception directed at extremely large signs with a panelboard within them. The panel can be remote from the entry point, but the feeder wiring must be in a raceway or metal jacketed cable with a warning label applied that advises that the enclosed wiring may be energized.

It is now clear that enforcing authorities should not attempt to prioritize (A)(1) and disallow any other location for the required disconnect. The technical merit for such an interpretation is compelling. A close examination of instances cited in the supporting substantiation for the 2014 **NEC** change demonstrates that the conclusion reached was fallacious and, on the contrary, that strict adherence to the provisions of (A)(2) would have been fully sufficient to address the problems cited. If a worker does not choose to avail himself or herself of reasonable provisions for disconnecting a sign and keeping that disconnect locked open as appropriate, the responsibility for the consequences of a failure does not rest with the **NEC**. However, as always, the final decision rests with the authority enforcing the Code. Figure 600-2 depicts the disconnecting means that should be within sight of the sign, outline lighting, or remote controller, as covered in part **(A)(2).** The phrase "within sight" is clearly defined in Art. 100, and it is well understood that it means the same thing as the phrase "in sight from," which specifies that it shall be visible and not more than 50 ft (15.0 m) distant from the other. Some signs are comprised of sections as part of a listed unit (see the "Section Sign" definition in 600.2) and those sections may in some cases be placed where they continue around the corner of a building. In such cases, part of the sign will be "out of the line of sight" from other sections, and the disconnect must be capable of being locked in the open position. Note that section signs must be marked to indicate that field wiring and installation instructions are required.

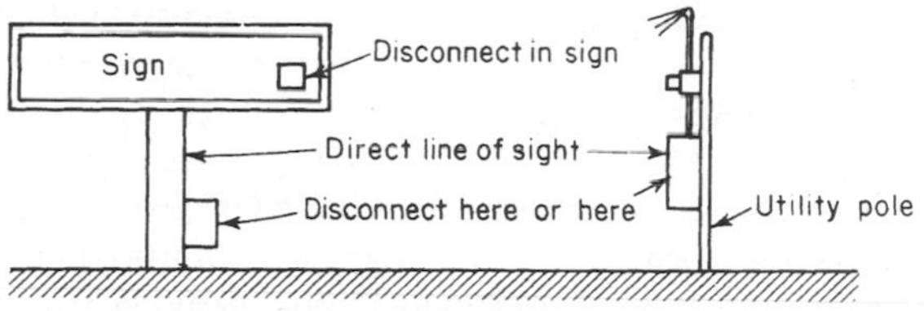
Sign
Disconnect in sign
Direct line of sight
Disconnect here or here
Utility pole

Fig. 600-2. An "in sight" disconnect may be *in* the sign or visible from the sign. (Sec. 600.6.)

Figure 600-3 illustrates the conditions recognized by part **(A)(3),** which allows the disconnecting means to be located within sight of the controller where the signs are operated by electronic or electromechanical controllers located external to the sign. Note that any sign disconnect located within sight of the controller must be capable of being locked in the OFF or OPEN position, as specified in part **(A)(2)(3).**

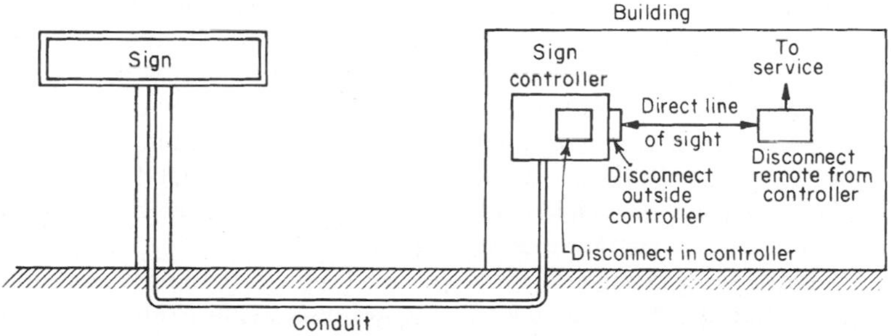
Building
Sign
Sign controller
To service
Direct line of sight
Disconnect remote from controller
Disconnect outside controller
Disconnect in controller
Conduit

Fig. 600-3. Controller disconnect location may vary, but disconnect must be lock-open type. (Sec. 600.6.)

With respect to part **(B)**, any switching device controlling the primary of a transformer that supplies a luminous gas tube operates under unusually severe conditions. In order to avoid rapid deterioration of the switch or flasher due to arcing at the contacts, the device must be rated for inductive loads or have a current-rating of at least twice the current rating of the transformer it controls. General-use ac snap switches, as covered in 404.14, are rated for inductive loads and will handle motor loads up to 80 percent of their rating, making them suitable for this application provided a permanent provision has been made for locking them open. There are handy-box and square-box raised covers available with factory mounted escutcheons around the toggle slot that will accept padlocks.

The question of what constitutes "operated by electronic or electromagnetic controllers" frequently arises in the context of modern energy management systems that turn signs on and off along with other building lighting. The purpose of the disconnect rule is to provide maintenance personnel with a secure means to ensure that the equipment they are working on is disconnected and that it will stay that way until they are ready to reenergize it. By long usage and custom, a maintenance disconnect is unique to its equipment. No one would seriously suggest that because a service disconnect could be locked in the OPEN position, all **NEC** requirements for disconnects of specific loads were met. Any maintenance worker on any downstream equipment would feel compelled to work the downstream equipment hot rather than inconvenience the enterprise to that extent. Therefore, if the energy management system operates a contactor for a sign among other control devices for other loads, this section requires that contactor to be in sight of a disconnect, that disconnect will be capable of being locked open, and a second disconnect would not be required at the sign location. On the other hand, if the contactor operated a lighting panel for which the sign was one load among many, then a local disconnect for the sign must be installed, and apply the warning signs now required within the sign enclosure that will clearly indicate where the disconnect can be found. This principle has also been reinforced and the requirement clarified through the added wording "and controls no other load" to the parent wording of 600.6.

600.7. Grounding and Bonding. The general principles in this section agree with the grounding principles in Art. 250, but there are specific provisions that are unique to this article, particularly with respect to neon signs that operate at high voltage and very low current. Part **(A)** addresses equipment grounding connections to sign parts that may be exposed to line voltage and follow the usual rules for equipment grounding connections. Specific language forbids the use of metal building parts as an equipment grounding return path, reiterating 250.136(A). Small metal parts, such as the metal feet on tubing support clips, do not require bonding connections, and remote metal parts in a section sign powered by a Class 2 power supply need not be bonded either. Although skeleton tubing does not have metal parts over the length of the tubing, if it is enclosed in a metal housing or has a transformer enclosure or such items as metal tubing receptacles, then those items also fall within the equipment grounding provisions.

Part **(B)** addresses bonding and addresses neon signage with provisions that are unique throughout the **NEC**. Flexible metal conduit is permitted in total accumulated lengths of up to 30 m (100 ft) as a bonding conductor, although the run from the transformer to the first termination will not exceed 6 m (20 ft) per 680.32(J)(1).

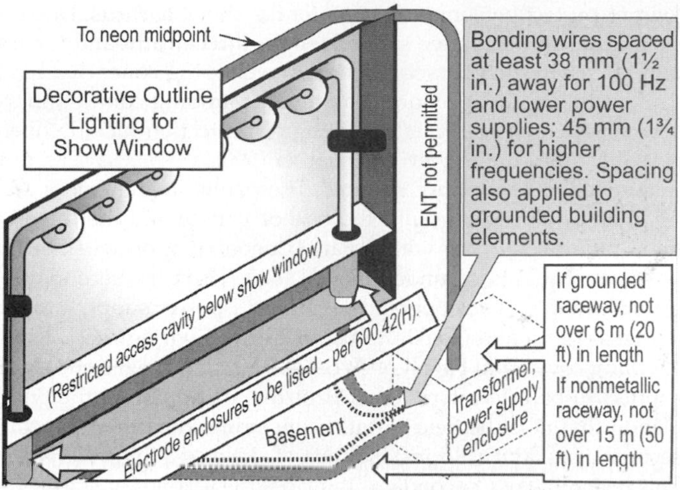
To neon midpoint
Decorative Outline Lighting for Show Window
ENT not permitted
Bonding wires spaced at least 38 mm (1½ in.) away for 100 Hz and lower power supplies; 45 mm (1¾ in.) for higher frequencies. Spacing also applied to grounded building elements.
If grounded raceway, not over 6 m (20 ft) in length
If nonmetallic raceway, not over 15 m (50 ft) in length
(Restricted access cavity below show window)
Electrode enclosures to be listed – per 600.42(H).
Basement
Transformer, power supply enclosure

Fig. 600-4. Many unique bonding requirements apply to neon installations. (Sec. 600.7.)

The currents are very small, and the performance of this wiring method in this context was investigated by a testing laboratory. If the wiring to the sign is in a nonmetallic wiring method such as PVC conduit, the bonding conductor must run outside the conduit, with the spacing governed by the frequency of the power supply (Fig. 600-4). For transformers with no change in frequency, the spacing from conductor to conduit must be at least 38 mm (1½ in.), and for electronic power supplies with frequencies over 100 Hz, the spacing must be at least 45 mm (1¾ in.). This reduces the voltage gradient between the GTO cable and the bonding conductor, and in doing so, it reduces the likelihood of damaging corona discharge. Metal raceways distribute the ground plane equally around the cable, and are also shorter, as noted in Fig. 600-4. Research showed that most of the damage occurs in the initial cable run from the power supply.

The **NEC** does address, courtesy (originally) of Las Vegas, the wiring of signs in fountains. Most of the requirements are in 680.57, covered later in this chapter. However the bonding requirements are here, in that the metal piping system, required to be bonded by 680.53, is permitted as the bonding connection required here. If the piping is nonmetallic, then a connection must be arranged to the equipment grounding conductor. This section also addresses the increasing use of LED and other sign components that can function within Class 2 power limitations. Here 600.7(B)(1) Exception waives the bonding requirements for "remote" metal parts of a section sign or outline lighting system from any need to be connected to an equipment grounding conductor. Presumably, "remote" means remote from the power supply and not from the first display component of the sign itself, but this is not completely clear.

600.10. Portable or Mobile Signs. This rule applies to outdoor portable or mobile signs that are plug-connected. The rule calls for manufacturer provided GFCI protection in or within 12 in. (300 mm) of the plug cap at the end of the supply cord from the sign to protect personnel from potential shock hazards. Documentation for the need for this new rule cited six accidents—three deaths and three shocks—due to ground faults in such outdoor signs that were plug-connected but in which there was no grounding connection or a failed grounding connection. Note that the product standard requires that this integral protection include open-neutral protection, as discussed in the previous chapter (590.6).

600.12. Field-Installed Secondary Wiring. The parent language has added skeleton tubing, thereby correlating with the title of Part II, and the three categories in this section point to sections within Part II depending on whether the wiring is over (600.32) or equal to or under 1000 V (600.31) on the secondary side. For LED sign illumination systems operating on Class 2 power supplies recognized in 600.24, the Part II destination is 600.33 using Chap. 3 wiring methods or Class 2 compliant cabling covered in Part II of Art. 725. In general, signs and outline lighting are required to be listed items. Part II of the article covers the portions of a sign installation that must be field installed, including the interconnecting wiring between section signs, and field installations of skeleton tubing signs and outline lighting. This terminology ("secondary") implies the secondary side of transformers, as are usually involved somewhere in the voltage transformations, either up or down. However, both 600.23 and 600.24 apply the term "secondary" equally to both "transformers and electronic power supplies" for elevated voltage and to "transformers, power supplies, and power sources" in the case of Class 2 sources.

600.21. Ballasts, Transformers, Electronic Power Supplies, and Class 2 Power Sources. The transformers used to supply luminous gas tubes are, in general, constant-current devices and, up to a certain limit, the voltage delivered by the transformer increases as the impedance of the load increases. The impedance of the tube increases as the length increases and is higher for a tube of small diameter than for one of larger diameter. Hence, a transformer should be selected that is designed to deliver the proper current and voltage for the tube. If the tube is too long or of too small a diameter, the voltage of the transformer may rise to too high a value. The maximum voltage permitted is center-grounded 15 kV (max. voltage to ground on each leg 7500 V) with 300 mA as the maximum current allowed, as covered in 600.23(C&D). The various power sources must be either self-contained or enclosed in a listed sign body or in a listed enclosure.

This section includes numerous rules on location of and access to this equipment. The first consideration is to minimize the length of the secondary conductors, which places the power supply as close as possible to the sign or outline lighting. Locations above suspended ceilings are acceptable as long as the branch-circuit connections use Chap. 3 wiring methods and not flexible cord. Wherever located, the equipment must have work space for servicing, defined as a cubical volume 1 yd (900 mm) on a side, and if located in an attic, there must be access to this space through an access door no thinner than the spacing between standard framing on 24-in. (610-mm) centers, and at least 900 mm (3 ft) high. There must be a permanent walkway to the location at least 300 mm (1 ft) wide and the work space must be lit with the lighting controlled either by a snap switch or a pull chain, as long as the control can be activated at the point of entry. Note that because Class 2 power sources have been added throughout 600.21, this special access provision even applies to a Class 2 transformer, which may seem excessive.

When challenged on this highly controversial point, the Code panel defended it as justified based on the line-voltage, primary-side connections.

600.23. Transformers and Electronic Power Supplies. This equipment must be listed, and must incorporate, with limited exceptions, secondary-circuit ground-fault protection. Arcing failures on the secondary side of these transformers involve very small but ignition-capable currents. This major design innovation responds to numerous documented fires from this source. The power supply must be marked accordingly.

600.24. Class 2 Power Sources. Thanks to the evolution of LED light sources, some signs and outline lighting can actually operate under Class 2 power limitations, and those are covered here.

600.31. Neon Secondary-Circuit Wiring, 1000 Volts or Less, Nominal. This section covers small neon applications that operate at not over 1000 V and can use conventional Chap. 3 wiring methods. It is included to completely fill out the wiring requirements for the full range of sign applications covered in 600.12.

600.32. Neon Secondary-Circuit Wiring, over 1000 Volts, Nominal. These are the wiring rules governing the secondary side of a neon-lighting power supply. Most Chap. 3 raceways are permitted, but ENT is not, based on concerns it may sag over time in a way that would not have been apparent during the original installation or inspection, and drop the contained single-conductor GTO cable too close to a grounded surface. As noted in Fig. 600-4, these surfaces must also meet the distances required of bonding conductors, and for the same reasons. In running nonmetallic raceways, nonmetallic methods must be used to secure the raceways away from grounded building elements to the extent required. This rule does not, however, prevent the use of grounded metal raceways, because the ground plane evenly surrounds the cable. The final choice of wiring method involves consideration of permitted length of cable run (advantage: nonmetallic) versus spacing requirements (advantage: metallic) along with any other design factors. These may include recommendations of the power supply manufacturer, which may include limitations based on the amount of capacitive reactance its power supply can cope with. Note that the maximum distances specified here are not accumulative distances as in 600.7, but only to the first tubing connection. Where the GTO cable emerges from a metallic raceway, the insulated cable must extend not less than 65 mm (2½ in.) beyond the raceway end.

The 2014 **NEC** deleted PVC and RTRC conduits from the acceptable list of wiring methods for neon secondary-circuit wiring. The change was very poorly substantiated based on the 105°C rating of the enclosed GTO cabling exceeding that of the PVC and RTRC conduits. However, just because a wire has a temperature rating it does not follow that it will operate at this temperature; the rating may be directed to the behavior of the expected terminations. This is certainly the case of wiring terminating at an arc discharge point on neon tubing, which is far different from heating produced by current draw over the length of these conductors. In addition, the UL Guide Card information shows RTRC as inherently capable of 110°C operation, and the temperature limits for LFNC, the last nonmetallic wiring method remaining, are generally similar to PVC conduit. This is not really unexpected given that both wiring methods start with the same chemical conduit (polyvinyl chloride) and then plasticizer is added to the flexible product. In addition,

LFNC is inherently more flexible than ENT, and ENT was omitted from the list because of its flexibility compared to PVC.

600.33. Class 2 Sign Illumination Systems, Secondary Wiring. This section centralizes the requirements that apply to LED circuits used in signs, incorporating a reference Part III of Art. 725 and the sign manufacturer's instructions together with a general reference allowing any Chap. 3 wiring methods as would apply to the installation conditions to be used on the supply side of the sign, as covered in 600.12. Class 2 wiring must be listed and no smaller than 22 AWG or the amount required by the load served if larger. Note that the **NEC** does have an ampacity table that applies to small conductors such as this (Table 522.22) in the event the installation instructions are not clear. Applications in wet locations must use cable identified for that usage or a moisture-impervious outer sheath is required. For all other locations, any cable permitted in Table 725.154(G) is permitted. The 2017 **NEC** has imported the relevant information from Chap. 7 regarding cable types and allowable substitutions into this section. Nothing really changed except the code book got a little longer so sign installers will not have to pay attention to the rest of the book. This is an unfortunate trend.

Part **(B)** incorporates the contents of 725.24 (Mechanical Execution of Work), applying them to "support wiring." The 2017 **NEC** added a maximum support interval of 6 ft. Connections are to be made with insulating devices and only where accessible after installation. Connections in a wall must be made in a listed box. The remaining material requires wiring protection in accordance with 300.4 in locations subject to physical damage, and incorporates the grounding and bonding rules in 600.7. Note that 600.7(B)(1) Exception waives the bonding requirements for "remote" metal parts of a section sign or outline lighting system from any need to be connected to an equipment grounding conductor.

600.34. Photovoltaic (PV) Powered Sign. This section adds little to the requirements of Art. 690, to which this section promptly refers. The components must be listed for PV applications. The wiring must be as short as possible, routed close to the sign body, and protected from damage. If used, flexible cord must be extra hard usage, and suitable for outdoor use. A disconnecting means must be provided as covered in Art. 690. Note that these self-contained signs, if operating at 30 or less, are not required to be grounded by the UL product standard. Part **(F)** does require that the design be such that a tool must be used to open the battery compartment.

600.41. Neon Tubing. Part **(D)** requires protection in the form of guards or other approved means if the skeleton tubing is accessible to other than qualified persons. Also note that 600.32(I) forbids the use of this type of installation in (or on) a dwelling occupancy.

600.42. Electrode Connections. These requirements have been toughened over the years, and now require listed components for most of this part of the installation.

ARTICLE 604. MANUFACTURED WIRING SYSTEMS

604.1. Scope. This article covers modular prefab wiring systems for ceiling spaces, raised floors, and increasingly walls as well. Both the system wiring and the associated components must be listed (604.6).

These manufactured wiring systems were logically dictated by a variety of needs in electrical systems for commercial-institutional occupancies. In the interest of giving the public a better way at a better price, a number of manufacturers developed basic wiring systems to provide plug-and-receptacle interconnection of branch-circuit wires to luminaires in suspended-ceiling spaces. Such systems afford ready connection between the hard-wired circuit homerun and cables and/or ducts that form a grid- or tree-like layout of circuiting to supply incandescent, fluorescent, or HID luminaires in the ceiling.

Acknowledged advantages of modular wiring systems are numerous and significant:

- Factory-prewired raceways and cables provide highly flexible and accessible plug-in connection to multicircuit runs of 120- and/or 227-V conductors.
- Drastic reductions can be made in conventional pipe-and-wire hookups of individual circuits, which are costly and inflexible.
- Plug receptacles afford a multiplicity of connection points for luminaires to satisfy needs for specific types and locations of lighting units to serve any initial layout of desks or other workstations while still offering unlimited, easy, and extremely economical changes or additions of luminaires for any future rearrangements of office landscaping or activities.
- Systems may also supply switches and/or convenience receptacles in walls or partitions, with readily altered switching provisions to provide energy conservation through effective ON-OFF control of any revised lighting layout.
- Work on the systems has been covered by agreement between the IBEW and associated trades.
- Such systems have a potential for a tax advantage of accelerated depreciation as office equipment rather than real estate.

604.10. Uses Permitted. Modular systems may be used in air-handling ceilings. Equipment may be used in the specific applications and environments for which it is listed by UL. Note that one end of such a system can be concealed, such as when it is fished down a wall that is open above a suspended ceiling, and used to connect either a switch or an outlet.

604.100. Construction. Prewired plug-in connections may be AC or MC cable or metal flex. The permitted conductors must be copper, and limited to a minimum of 12 AWG and running up to 8 AWG, and including a fully sized copper equipment grounding conductor (insulated or bare) is always required in each cable or flex length—even though flex itself is otherwise permitted to be used by the **NEC** without an equipment grounding conductor in lengths not over 6 ft (1.8 m), provided the wires within it are protected at not over 20 A, and AC cable in other uses is recognized by the Code and by UL for equipment grounding through its armor and enclosed aluminum bonding wire (Fig. 604-1). Type MC cable of the smooth or corrugated variety with a qualified armor and equipment grounding conductor combination is also permitted, provided it is equivalent to the ungrounded conductor sizing. The purpose for the 8 AWG is the reduction of voltage drop and not high ampere loading, and therefore the equivalent sizing is appropriate in these branch circuit sizes to meet 250.122(B).

The same sizing applies to conductors in flexible metal raceways. Here, however, a whip to a luminaire is permitted at not over 1.8 m (6 ft) using not less than 18 AWG taps. In addition, 12 AWG or larger flexible cord suitable for hard usage,

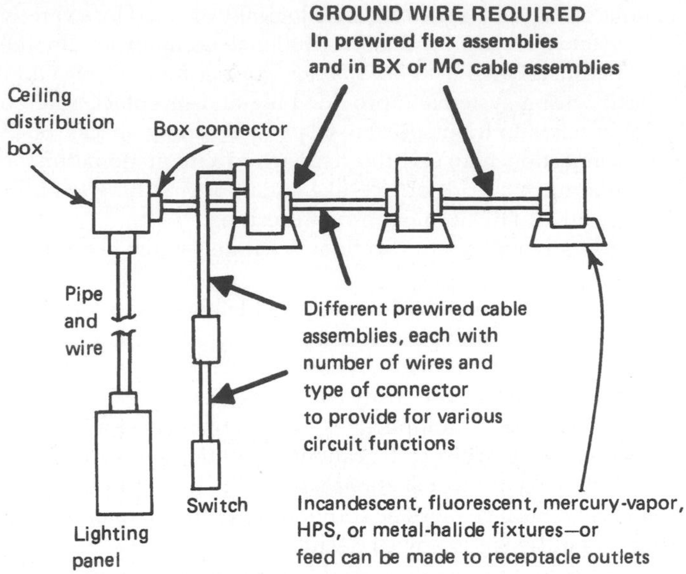
GROUND WIRE REQUIRED
In prewired flex assemblies
and in BX or MC cable assemblies
Ceiling distribution box
Box connector
Pipe and wire
Different prewired cable assemblies, each with number of wires and type of connector to provide for various circuit functions
Switch
Lighting panel
Incandescent, fluorescent, mercury-vapor, HPS, or metal-halide fixtures—or feed can be made to receptacle outlets

Fig. 604-1. Modular wiring systems are fully recognized by the Code. (Sec. 604.100.)

not over 1.8 m (6 ft) long and visible over its entire length, is permitted as a supply to equipment not fixed in place, such as some display cases. This wiring system is also permitted to contain signaling and communications wiring, within the limits set in Art. 725 and Chap. 8.

Listed manufactured wiring systems may also incorporate unlisted flex of smaller than NEC recognized trade sizes, and/or noncircular cross-section, provided the systems are supplied with fittings and conductors when manufactured. Flexible cord suitable for hard usage is also permitted in sizes not smaller than 12 AWG to provide for wiring transitions to utilization equipment that is not permanently installed and secured to the building. It must be visible over its entire length and provided with appropriate strain relief.

These systems may incorporate busways, and also prewired surface metal raceways that meet the requirements of 386.100 as part of their design, and they must be installed in accordance with Arts. 386.12, 386.30, and 386.60. Note that as worded, 386.100 appears as an installation criterion, which is obviously a mistake because an installer has no control over the extent to which a surface metal raceway complies with the manufacturing criteria in that section.

ARTICLE 605. OFFICE FURNISHINGS

605.1. Scope. This Code article covers electrical equipment that is part of manufactured partitions used for subdividing office space, as shown in Fig. 605-1. The 2014 NEC has revised the article throughout to replace the terms "relocatable wired partitions" and "partition" and related terminology with "office furnishings" to

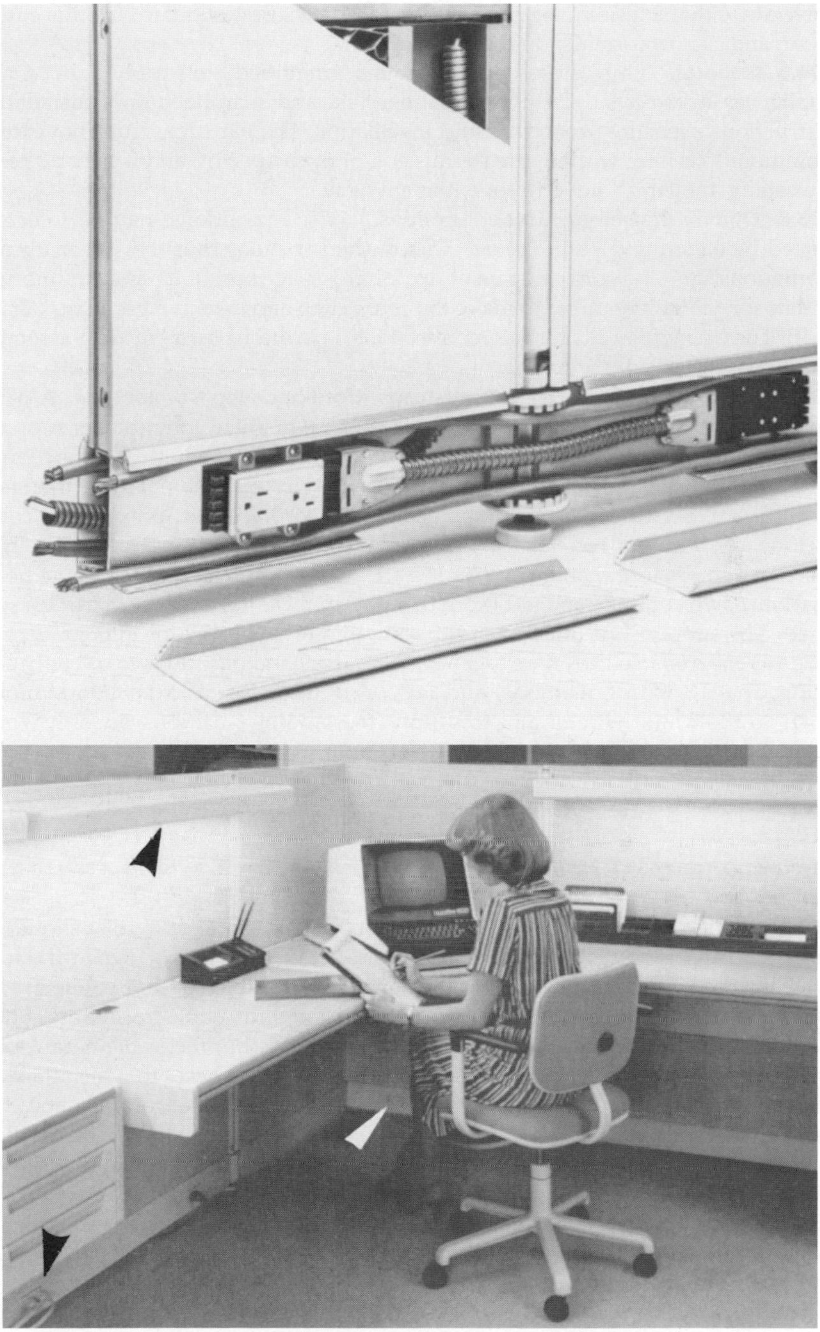

Fig. 605-1. This article covers electrical wiring and electrical components within or attached to manufactured partitions, desks, cabinets, and other equipment that constitute "office furnishings." Photo at top shows interior wiring in base of partitions, to supply luminaires and receptacle outlets—as shown at arrows in bottom photo of a typical electrified office work station.

correlate with the terminology used in the product safety standard for this equipment, and as now formally defined in Sec. 605.2.

605.3. General. Only those wiring systems "identified" to supply lighting and appliances may do so. Check the listing data and manufacturer's installation instructions to ensure proper use and installation. The partitions must not extend from floor to ceiling without the permission of the inspector, and if they do reach the ceiling, they must not go above it in any way.

605.5. Office Furnishing Interconnections. Wired partitions may be interconnected by a cord and plug. If cord is used, the partitions must be "mechanically contiguous" and the cord must be of extra-hard-usage and of 12 AWG minimum, and no longer than required to make the connection and never longer than 600 mm (2 ft). The basic rule calls for interconnection of partitions by a "flexible assembly identified for use with wired partitions."

605.9. Freestanding-Type Office Furnishings, Cord-and-Plug-Connected. A partition or group of connected partitions [not over 9.0 m (30 ft) long] that is supplied by cord-and-plug connection to the building electrical system must not be wired with multiwire circuits (all wiring must be 2-wire circuits) and not more than thirteen 15-A, 125-V receptacles may be used. The receptacles included in the count include either duplex receptacles (therefore counted as one) or two adjacent single receptacles not more than 0.3 m (1 ft) apart. The supply receptacle(s) must be on its (their) own circuit(s) and not more than 300 mm (12 in.) from the partition supplied. The supply cord must meet the same requirements as for interconnecting cords as covered in 605.5. Note that 605.7 requires fixed partitions to be connected using Chap. 3 wiring methods, and 605.7 & 8 [see also 605.9(D) informational note] do allow multiwire circuits to supply freestanding partitions, but only where Chap. 3 wiring methods bring them (must comply with 210.4) to the partition.

ARTICLE 610. CRANES AND HOISTS

610.11. Wiring Method. In general, the wiring on a crane or a hoist should be raceways, or Type AC with a grounding conductor, Type MC or Type MI. However, for practical considerations, short lengths of flexible conduit or metal-clad cable and even open conductors may be used for connections to motors, brake magnets, or other devices where a rigid connection is impracticable because the devices are subject to some movement with respect to the bases to which they are attached. In outdoor or wet locations, liquidtight flexible metal conduit should be used for flexible connections. Part **(E)** also recognizes listed festoon cable for flexible uses. This is defined in 610.2 as either single or multiconductor, and it is available with temperature ratings from 60°C up through 105°C and it carries a 600-V insulation rating as well. Do not confuse this wiring method, which has been listed expressly for Art. 610 context, with "festoon lighting" covered in 225.6(B), 225.24, and 520.65. It is also recognized in 610.13(C) for flexible applications, and it can be rolled into a take-up device or reel.

610.14. Rating and Size of Conductors. Crane conductors operate for very limited time intervals, and the NEC includes a special ampacity table for this purpose. Note that there is exactly one place in the NEC where 5 AWG wire has formal recognition and it is here in this table. Part **(E)** covers motor calculations such as where multiple motors could operate at one time. As covered here using nameplate data, take the largest motor or group of motors for any particular crane motion, and add 50 percent of the next largest motor or group of motors, in all cases using the column of Table 610.14(A) that applies to the longest time-rated motor.

610.21. Installation of Contact Conductors. Part **(F)** permits use of the track as one of the circuit conductors. In some cases, particularly where a monorail crane or conveyor is used for handling light loads, for the sake of convenience and simplicity it may be desirable to use the track as one conductor of a 3-phase system. Where this arrangement is used, the power must be supplied through a transformer or bank of transformers so that there will be no electrical connection between the primary power supply and the crane circuit, as in Fig. 610-1. The secondary voltage would usually be 220 V, and the primary of the transformer would usually be connected to the power distribution system of the building or plant. The leg connected to the track must be grounded at the transformer only, except as permitted in 610.21(F)(4).

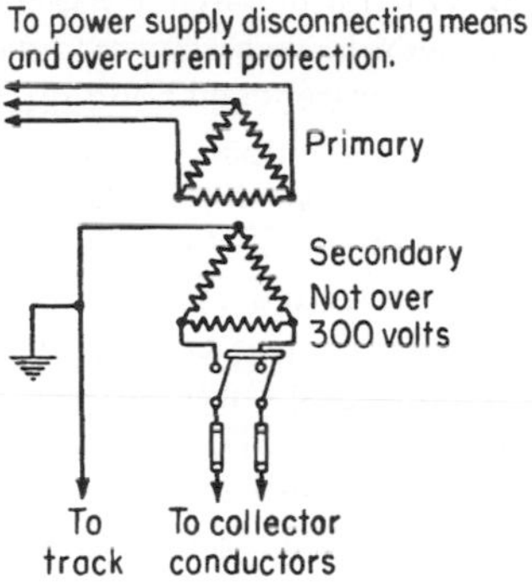
To power supply disconnecting means
and overcurrent protection.
Primary
Secondary
Not over
300 volts
To
track
To collector
conductors

Fig. 610-1. Isolating transformer is used to power track of crane or conveyor. (Sec. 610.21.)

610.32. Disconnecting Means for Cranes and Monorail Hoists. This disconnect is an emergency device provided for use in case trouble develops in any of the electrical equipment on the crane or monorail hoist, or to permit maintenance work to be done safely. A motor-circuit switch, molded-case switch, or circuit breaker (or other 430.109 compliant method) must be provided in the leads from the runway contact conductors or other power supply on all cranes and monorail hoists. The disconnecting means must be capable of being locked in the open position. As in other places where the **NEC** discusses required disconnecting means, the provision for locking or adding a lock to the disconnecting means must be installed on or at the switch or circuit breaker used as the disconnecting means and must remain in place with or without the lock installed. Portable means for adding a lock to the switch or circuit breaker is not permitted.

Where a monorail hoist or hand-propelled crane bridge installation meets all of the following, the disconnecting means need not be installed:

1. The unit is controlled from the ground or floor level.
2. The unit is within view of the power supply disconnecting means.
3. No fixed work platform has been provided for servicing the unit.

Where the disconnecting means is not readily accessible from the crane or monorail hoist operating station, means must be provided at the operating station to open the power circuit to all motors of the crane or monorail hoist.

610.33. Rating of Disconnecting Means. It is possible that all the motors on a crane might be in operation at one time, but this condition would continue for only a very short while. A switch or CB having a current rating not less than 50 percent of the sum of full-load current rating of all the motors will have ample capacity. The continuous ampere rating of the switch or circuit breaker specified above shall be not less than 50 percent of the combined short-time ampere ratings of the motors nor less than 75 percent of the short-time ampere rating of the motors required for any single motion.

Note that Art. 610 uses the wording "within view" in several places instead of "in sight" consciously in order to avoid the 15 m (50 ft) limitation built into the NEC definition of "in sight." Many large industrial cranes are too big to make the 15 m (50 ft) limit workable.

610.61. Grounding. All exposed non-current-carrying metal parts of cranes and hoists must be bonded together, usually by the usual mechanical connections, to make an effective ground-fault current path. Moving parts may have their equipment grounding continuity established through metal-to-metal contact on bearing surfaces, making it unnecessary to run very long, strain-relieved bonding conductors that would need to accommodate major frame movements of perhaps hundreds of feet at large industrial facilities. However, as of the 2005 NEC, the contact between the wheels of a trolley (the part that makes the load go up and down) and its associated bridge girder can no longer be depended upon for grounding continuity, and now a bonding conductor must be installed between these two parts. The wording change only covers the trolley frame, and not the bridge girder as its wheels turn on the runway, even though the contact surfaces seem identical. There was no substantiation to distinguish one from the other, nor was there any loss experience presented to suggest that the prior allowance, unchanged since the 1962 NEC, was deficient.

ARTICLE 620. ELEVATORS, DUMBWAITERS, ESCALATORS, MOVING WALKS, PLATFORM LIFTS, AND STAIRWAY CHAIR LIFTS

620.1. Scope. These provisions may also be considered as applying to console lifts, equipment for raising and lowering or rotating portions of theater stages, and all similar equipment.

620.2. Definitions. Informational note Fig. 620.2 presents an overview of the components of a modern elevator control system, and the definitions in 620.2 create a context that correlates provisions of the principal elevator safety document (ASME A17.1, *Safety Code for Elevators and Escalators*) with provisions in Art. 620.

620.11. Insulation of Conductors. The major concern in this section is with the integrity of electrical systems in a hoistway that is a natural chimney in the event of fire. The hoistway door interlock wiring must be suitable for 200°C for this reason, and the traveling cables must be one of the kinds listed in Table 400.4 for this purpose. Other insulation must be flame retardant. Alternatively, an approved conductor assembly can be used if suitable for the same temperature exposure.

620.12. Minimum Size of Conductors. Code Tables 310.15(B)(16) to 310.15(B)(19) do not include the ampacity for 20 AWG copper conductors. However, it is generally considered that 20 AWG conductors up to two conductors in cable or cord may safely carry 3 A (and Table 522.22 allows 5 A). This section amends 310.10(H) and permits those smaller conductors to be paralleled to equal the capacity of a 14 AWG wire in this case. This type of allowance makes for more flexible traveling cable make-ups, which is essential in today's very tall buildings.

Because of wider use of advanced semiconductor computer equipment, use of wire smaller than 20 AWG is permitted for other than lighting circuits by part **(B)**, with 24 AWG as the minimum, and even smaller if listed.

The development of elevator control equipment, which has been taking place for many years, has resulted in the design and use of equipment, including electronic unit contactors, requiring very much smaller currents (milliamperes) for their operation.

620.13. Feeder and Branch Circuit Conductors. This section and the next track the requirements in Art. 430 quite closely, but include a demand factor table that is unique to multicar elevator groups, in Sec. 620.14. There are two examples in Annex D (Examples D9 and D10) that demonstrate multiple elevator loading calculations as covered in these Code sections very well.

620.14. Feeder Demand Factor. When multiple elevators are connected to the same feeder, not all of them will be in use at the same time and this section through its demand table allows for demand factors to be applied to these loads.

620.21. Wiring Methods. The parent language here provides the default wiring method choices for this wiring, including heavy-wall steel and "nonmetallic" conduit (an error, should be changed to PVC or RTRC conduit), EMT, wireways, or MC, MI, or AC cable. Hoistways add flexible methods for door operating controls, etc., as well as instrumentation on cars, and the length in many cases is not limited to 1.8 m (6 ft). For obvious reasons the supply cords of listed cord-and-plug-connected equipment need not run in a raceway.

620.22. Branch Circuits for Car Lighting, Receptacle(s), Ventilation, Heating, and Air Conditioning. This section establishes that each elevator car will be served by two dedicated circuits, one for heating and air conditioning, and one for lighting, the receptacle on the cab top or similar for service work, and other accessory loads. The wiring sequence on this circuit must be such that the service receptacle, which will be a GFCI receptacle per 620.85, must not disconnect the cab light if it trips. The overcurrent protective devices for these circuits must be located in the machine room or other control space for that car. A similar requirement governs lighting and receptacles in machine rooms and similar control spaces, as given in the next section (620.23), and another similar requirement governs hoistway pit lighting and its required receptacle outlet as covered in the section after that (620.24). As of the 2017 NEC, the lighting must not share the branch circuit with the receptacles in either the machine room or the pit.

620.32. Metal Wireways and Nonmetallic Wireways. This section allows the wire fill in wireways to more than double, from the 20 percent allowed in

376.22(A) to 50 percent. Note, however, that no mention is made of the derating factors in 310.15(B)(3)(a) that apply to the entire fill as soon as the 30 current-carrying conductors threshold is reached, and to all conductors in a nonmetallic wireway. Because no exception to this Chap. 3 rule is taken in this Chap. 6 article, the derating factors will apply, and they could easily hit 0.35 (example: for 41 wires, 12 AWG THHN, 30 A × 0.35 = 10.5 A) if the wireway were actually stuffed to 50 percent fill.

620.36. Different Systems in One Raceway or Traveling Cable. It would be difficult, if not practically impossible, to keep the wires of each system completely isolated from the wires of every other system in the case of elevator control and signal circuits. Hence, such wires may be run in the same conduits and cables if all wires are insulated for the highest voltage used and if all live parts of apparatus are insulated from ground for the highest voltage, provided that the signal system is an integral part of the elevator wiring system. These are very sophisticated, listed cables, and if multiple cables were required, they could easily tangle in the long lengths that are required in today's high-rise applications.

620.37. Wiring in Hoistways, Machine Rooms, Control Rooms, Machinery Spaces, and Control Spaces. Although hoistways are very tempting chases to run from floor to floor, they are not permitted for this purpose. The only wiring permitted in an elevator hoistway is wiring for the elevator functions. Neither are hoistways permitted to contain, nor the vertical elevator rails permitted to be, down conductors for a lightning protection system. However, if a hoistway happens to run close enough to a down conductor for a lightning protection system such that NFPA 780 requires that rails in the hoistway be bonded to the down conductor to eliminate the possibility of side flash, then that bonding may proceed.

620.51. Disconnecting Means. A disconnecting means must be provided for every elevator (or any other form of lift covered in the article), and it must open all loads connected with that elevator except three, which are forbidden to be disconnected as a result of this disconnect being in the open position. The three exceptions are the cab lighting and accessories circuit, the cab heating and air conditioning circuit, and the hoistway and machine room lighting and service receptacle circuit(s). These must stay on so cab occupants will stay comfortable and not panic in the event of a malfunction, and so maintenance can proceed on the failed elevator under emergency conditions.

This main disconnecting means can be a switch or a circuit breaker. It can open automatically but must only close manually. It cannot be arranged to open from "any other part of the premises." If there is an elevator control panel, it must have a field marking showing the available fault current along with the date the calculation was made. If there are modifications that affect this number, then it must be revisited, recalculated, and redated. Of course, the most likely impact on this number is utility activity, which will not be known to a facility electrician absent a formal request. Refer to the commentary at 110.24 for additional information on this type of requirement. The adjacent hoistway is not considered an "other part of the premises." Further, if sprinklers are installed in the hoistway, this disconnect must open prior to waterflow. The usual protocol to bring this together involves a smoke detector in the hoistway that will detect smoke considerably prior to when the sprinkler head fuses. Upon detecting smoke, the smoke detector, supervised as part of a fire alarm system, initiates a control sequence that parks the elevator on the recall floor and discharges passengers. Adjacent to the sprinkler head, a heat detector, also supervised by the fire alarm system, activates the shunt trip on the disconnect if the temperature continues to rise, doing so before the sprinkler head fuses to start the flow of water. This prevents waterflow onto an active elevator, which would be an extreme hazard as the brakes could slip and the controls fail. Effective with the 2017 NEC, if one or more of the elevators has been designated an emergency system load, surge protection must be provided. A survey of facility managers found that nearly a quarter of them had experienced surge damage to elevators, and this change responds to that finding with a new 620.51(E).

620.53. Car Light, Receptacle(s), and Ventilation Disconnecting Means. There must be a disconnecting means for this circuit, described in 620.22, located in the machine room or control space for the associated elevator. The disconnecting means must be permanently capable of being locked in the open position. A factory-designed locking hasp for the circuit breaker originating this dedicated circuit would allow the circuit breaker to fulfill both requirements, provided it is in the machine room or control space for the associated car. It must be plainly labeled as to its function and the identity of the elevator for which it provides this function. As in the case of the required branch circuit for heating and air conditioning, the next section (620.54) provides parallel requirements for the heating and air conditioning circuit. An exception permits this disconnecting means to be a manual motor controller, or a general-use snap switch, or an ac snap switch, all as covered in 430.109(C) for this purpose.

620.61. Overcurrent Protection. These requirements directly track other sections in the NEC.

620.62. Selective Coordination. This was the first mandatory selective coordination rule in the history of the NEC. It was substantiated on the basis that if an elevator feeder opened because of a fault, and if the upstream overcurrent protection opened due to a lack of coordination, people would be trapped in all the elevators stranded by the upstream protection opening, and service personnel might not be able to figure out on a timely basis how to find and reclose the upstream protection. Of course a sign on the elevator disconnect could solve that problem, but the panel insisted on solving a problem in a way that was both unprecedented and poorly substantiated. Depending on the relative size of the two levels of overcurrent protection and the trip curves for each this may be a simple or an intractable problem. Refer to the discussion at 700.32 for more information on a very controversial topic. The 2014 NEC has added more expense by making it mandatory to have a selectivity evaluation by a licensed professional engineer.

620.85. Ground-Fault Circuit-Interrupter Protection for Personnel. All 125-V 15- and 20-A receptacle outlets in pits, hoistways, "on the cars of elevators and dumbwaiters associated with wind turbine elevators," on the platforms or in the runways and machinery spaces of platform lifts and stairway chairlifts, and in escalator and moving-walk wellways must be provided with GFCI protection, and this protection must be at the point of use, that is, through the use of a GFCI receptacle. A GFCI circuit breaker ahead of the receptacle is not permitted. The issue is the difficulty in resetting a tripped GFCI protective device that is nowhere near the worker who may need to reset the device immediately. On the other hand, for the receptacle outlets in the machine room or in the machinery space a worker can more easily reach the originating panelboard and reset the device, so either approach is permitted in those areas.

The material in quotation marks in the preceding paragraph is a verbatim restatement of new 2017 NEC text, and it is astonishing. Since most elevators are not associated with wind tower elevators, the comma placements in the literal text effectively remove the requirement from every conventional elevator. The public input that put this change in motion, from an elevator professional, used the following wording: "installed on the cars of elevators and dumbwaiters, in the wellways or the landings of escalators or moving walks, in the travel path or on the cars of wind turbine tower elevators, or in the runways and machinery spaces of platform lifts and stairway chairlifts shall be of the ground-fault circuit-interrupter type." This is substantively what the sentence is actually supposed to say, based on technical merit. How and when this gets corrected is unknown for now.

ARTICLE 625. ELECTRICAL VEHICLE CHARGING SYSTEMS

625.1. This article regulates the sizing and installation of equipment and conductors used to supply electrical vehicle charging systems and the electrical vehicle. This article does not apply to battery charging systems that are used for fork-lifts, etc., but rather automobile-type electrical vehicles, including plug-in hybrid electric vehicles (PHEVs).

625.10. Electric Vehicle Coupler. This is the interface between the premises wiring system and the vehicle electrical system for propulsion. The couplers must be configured to avoid inadvertent contact by untrained persons with uninsulated live parts. The coupling from connector to vehicle inlet must be positive so as to prevent unintentional disconnection. Unless listed otherwise, it must have a grounding pole that is first make/last break. Presumably the inductive paddles qualify as listed otherwise (see 625.16).

625.15. Markings. In addition to an electric vehicle usage label, there will also be a marking with respect to whether or not ventilation is required. The vehicle owner should know which type of battery he or she has. Some are sealed or of a chemistry that does not emit hydrogen, and others do emit hydrogen. As provided in 625.52(A) if a vehicle requiring ventilation to remove hydrogen gas accumulations plugs into a charger marked ventilation not required, the charging station will listen for a coded signal indicating the appropriate battery is present, and not receiving it, will not charge the battery.

625.16. Means of Coupling. Again, the coupling can be inductive or conductive, or even wireless, and the connecting devices must be listed for the purpose.

625.17. Cords and Cables. This section covers both the power supply *cord* that runs from the premises outlet to the vehicle charging equipment, and the output *cable* that runs from the charging equipment to an electric vehicle connector, defined as the device that either inductively or through conduction transfers power to and exchanges information with an electric vehicle. Part (**A**) covers the supply cord, and the permitted length depends greatly on the location of the protective system required by 625.22. That section requires protection in the cord cap or within 12 in.; this section, to not be inconsistent, sets the cord length limit at 12 in. if the protection is in the charging equipment and 15 ft if it is in the cord cap. If the shorter length applies, clearly the location of the premises wiring outlet must be exact.

Part (**B**) covers the output cable. Special cords have been developed for these connections, easily identifiable by the "EV" (electric vehicle) letters in their type designation. They can be either hard usage or extra-hard-usage. They must not exceed 25 ft as the overall total length of the flexible connection on both sides of the charging equipment (unless equipped with a cable management system), and they can use the higher power cable ampacities in Table 400-5(A)(2) (unless they are No. 10 or smaller). The rule also allows for hybrid cables that include signaling circuits (or optical fiber cables). The measurement begins at the charging equipment where it is fastened in place and from the end of the supply attachment cord cap if the charging equipment is free to move. In either case, the measurement ends at the face of the vehicle connector.

625.18. Interlock. The charging system must be designed with an interlock that deenergizes the cable and connector whenever it is not connected to the vehicle. This is not required on the 15- and 20-A 125-V charging systems. This interlocking system will also prevent charging a vehicle that uses batteries that outgas hydrogen unless the installation is arranged with ventilation as provided in 625.52(B). It is also not required on a dc supply of less than 60 V. However, 625.52(D)(4) does require a switching mechanism for both these dc systems and the 15- and 20-A ac systems that will disable or disconnect the charger unless the required ventilation is operating "during the entire electric vehicle charging cycle." There should be better correlation between the wording here and that in 625.52(D)(4).

625.19. Automatic De-energization of Cable. The electric vehicle supply cord and system must include some method of deenergization in the event of excessive strain on the cable, such as by driving away while plugged in. This is not required on the 15- and 20-A 125-V charging systems, or on portable systems covered in 625.44(A).

625.22. Personnel Protection System. The electric vehicle supply system must incorporate shock protection that may differ somewhat in trip thresholds from conventional GFCI levels, but that has the same effect. The test labs and manufacturers are being given some needed design flexibility here given that the output current may be dc or at different voltages. If the charging equipment is cord-and-plug-connected, then this function must be built into the attachment plug or into the supply cord within a foot of the plug. This system is not required for power supplies operating at less than 60 volts dc.

625.40. Electric Vehicle Branch Circuit. This section (relocated from 210.17) requires the use of individual branch circuits for this equipment. In general, this is relatively high capacity equipment; a typical rating is 32 A. with an intended connection to a 40-A. 240-V circuit. However, there are some portable chargers that are far smaller and capable of being connected very comfortably to a 15- or a 20-A 125-V circuit. This section may require relief in the future for such equipment.

625.41. Overcurrent Protection. As noted in the discussion under Sec. 625.42, these chargers are defined as a continuous load, and therefore the conductor ampacity and the overcurrent device must be increased by an additional

25 percent. As covered in 625.40, these vehicle recharging circuits must be wired as individual branch circuits.

625.42. Rating. This must be high enough for the load to be served and must be considered as continuous. Although some highway quick-charge protocols assume a 10- or 15-minute recharge at very high ampere values, and with component specifications established accordingly, the rule is for a continuous classification on any load. The more usual anticipated charging protocol involves 32-A charging current from a 208-V wye or delta (or even single-phase) connection on a 40-A branch circuit. In addition, there are many vehicle charging stations manufactured for and connected to ordinary 15- or 20-A, 120-V circuits. The rule also accommodates automatic load management systems, and where in use, the loading to be entered in a feeder or service calculation is that presented by the maximum draw of the installed management system.

625.43. Disconnecting Means. High-capacity charging equipment (over 60 A or over 150 V to ground) must have a disconnecting means in a readily accessible location. It must be able to be locked in the open position. This is a disconnecting means for the equipment, and therefore a unit switch in the equipment would not comply, even if it opened all ungrounded conductors. Maintenance personnel must be free to service the entire unit without hazard.

Note that this equipment would meet the definition of an appliance (other than industrial, produced in standard sizes, etc.) and therefore must comply with 422.31(B). That rule also requires a local disconnect, which can only be the branch-circuit protective device if it is within sight or can be locked open. These provisions can be enforced on any capacity charging system.

625.44. Electric Vehicle Supply Equipment (EVSE) Connection. This section is extremely poorly written and governs all supply connections for vehicle supply equipment made to a premises wiring outlet, which as of the 2014 NEC must be in one of three categories. Part **(A)** allows mating 15- and 20-A 125-V plugs and receptacles (via flexible cord described in Sec. 625.17) by right. This category also includes dc supplies not over 60 V maximum, at 15 or 20 amperes.

Part (**B**) covers stationary equipment, defined as "fastened in place in such a way as to permit ready removal for interchange, facilitation of maintenance or repair, or repositioning." This equipment will be cord-and-plug-connected to one of three system outlets, as follows:

1. Single-phase ac: A nonlocking 2-pole 3-wire grounding receptacle rated from 15 to 50 amperes and either 125 or 250 volts.
2. Three-phase ac: A nonlocking 3-pole 4-wire grounding receptacle rated from 15 to 50 amperes, 250 volts.
3. DC: 2-pole, 3-wire grounding receptacle rated from 15 to 20 amperes, not over 60 volts.

Note that the requirement uses the phrase "fastened in place." This terminology is specifically defined for use in this article in 625.2 as "specifically designed to permit periodic removal for relocation, interchangeability, maintenance, or repair without the use of a tool." It presumably would have been better to simply use the defined term, which will still apply anyway.

Part (**C**) covers all other EVSE. This would include equipment rated for higher voltages, or current, or both. The equipment must be fastened in place and hard wired, and it must have no exposed live parts.

625.46. Loss of Primary Source. The charging equipment must be arranged so the energy stored in the car batteries cannot backfeed into the supply wiring if the supply power fails. The vehicle cannot be allowed to serve, even if so desired, as a standby power source unless (as covered in 625.48) listed for this purpose, in which case the provisions in Art. 702 must be met for standby power and Art. 705 for interactive power, as applicable, must be met. An informational note identifies numerous applicable standards that would be involved in these connections.

625.47. Multiple Feeder or Branch Circuits. If identified for such connections, multiple supply circuits are permitted.

625.50. Location. The charging equipment must be located so the charging cable can connect directly to the vehicle. Unless listed and marked differently, the vehicle coupling means must be stored or located at a height of not less than 18 in. above the floor (indoor locations) and 24 in. above grade (outdoor locations). Note that these requirements do not apply to portable equipment covered in 625.44(A). This term is specifically defined in Art. 625 as designed to be transported in the vehicle when not being used because it is intended to be carried from one charging location to another.

625.52. Ventilation. This section covers indoor enclosed spaces. Part **(A)** covers instances where ventilation is not required because of the nature of the battery system in the vehicle, as noted in 625.15(B); mechanical ventilation is not required in such cases. A great deal of effort has gone into the production of batteries for these vehicles that do not out-gas significant quantities of hydrogen, and thereby allow this provision to be utilized. The majority of batteries now being produced do qualify as not requiring enhanced ventilation. The ventilation required otherwise is a substantial burden on the acceptance of this technology.

Mechanical ventilation as covered in part **(B)** must be provided with systems that are suitable for charging electric vehicles that outgas hydrogen and that are identified accordingly in 625.15(C). It must be permanently installed and it must include both supply and exhaust equipment arranged to take air from and exhaust air directly out to the outdoors. This ventilation must be interlocked with the charging system and it must operate during the entire charging cycle. The required volume of air to be exchanged is provided in a table based on the ampere rating and voltage of the branch circuit supplying the charging equipment. For example, a 20-A 120-V supply requires 49 cfm, and a 200-A 480-V 3-phase supply requires 3416 cfm. These numbers apply for each space equipped to charge an electric vehicle. If there are two spaces, then the required ventilation doubles. The 2014 NEC has also inserted ventilation information for dc supplies operating not over 50 V.

A large proportion of battery charging involves the release of hydrogen gas. This is a Class I Group B gas, extremely dangerous, and its lower explosive limit is only 4 percent. That means that a hydrogen-air mixture over 4 percent hydrogen by volume can explode. Although hydrogen is much less dense than air and dissipates rapidly, charging operations can generate enormous quantities. Actual testing showed ignitable concentrations of hydrogen near the ceiling even on 15-A branch circuits in residential garages with the door open! The mechanical ventilation requirements in this section need to be taken seriously.

Part IV. Wireless Power Transfer Equipment. Although the energy transfer process may be wireless, the infrastructure to support it is anything but. 625.102 covers the transmitting (primary) pad and how it is to be positioned, together with the mounting requirements for the charger power converter, and protection requirements for the output cable. 625.101 covers the composition of and the grounding requirements for the primary pad base plate, which must be of non-ferrous metal. This part involves unfamiliar technology and will require careful attention to the definitions in 625.2 and the installation directions supplied with this equipment.

ARTICLE 626. ELECTRIFIED TRUCK PARKING SPACES

626.1. Scope. This article, new in the 2008 NEC, covers what is defined in 626.2 as a "truck parking space that has been provided with an electrical system that allows truck operators to connect their vehicles while stopped and use off-board power sources in order to operate on-board systems such as air conditioning, heating, and appliances, without any engine idling."

Environmental concerns about diesel exhaust together with skyrocketing costs for diesel fuel are creating a very strong market for this type of service. This new article creates the necessary standardization of approach because a truck moves from jurisdiction to jurisdiction and needs to be able to connect to this infrastructure in any state in order for this to work.

626.10. Branch Circuits. Each stand must be supplied from a 208Y/120V system or a 480Y/277V system, with an exception for existing 120-V facilities.

626.11. Feeder and Service Load Calculations. Each parking space must be calculated on the basis of not less than 11 kVA each, although part **(B)** applies a demand factor to this load based on expected heating and air-conditioning burden. This is related to the "USDA Plant Hardiness Zone Map" and decreases from a high of 70 percent (Fairbanks, Alaska) to a low of 20 percent (Houston, Texas) with some small increases to 24 percent for the highest zones (Miami, Florida and Honolulu, Hawaii) where an increased air-conditioning load would take over from decreased heating load. The selection of a plant-hardiness map based on worst-case winter temperature is probably appropriate in terms of predicting worst-case winter loading, but may prove inaccurate in terms of predicting summer air-conditioning load. It would seem to have been questionable in terms of predicting maximum demand, but data submitted with the proposal that tabulated actual measured demand from existing facilities showed close agreement with table predictions. These facilities are typically 100 percent occupied at night and therefore all available spaces go into the calculation prior to applying the demand factor.

Note that the 11 kVA is figured on the basis of the maximum power capability of the receptacles required by 626.24(B), that is, two 20-A receptacles on 120-V circuits, and one 30-A circuit on a 208-V circuit, as follows:

$$2(120\ \text{V} \times 20\ \text{A}) + (208\ \text{V} \times 30\ \text{A}) = 11{,}000\ \text{VA}$$

Note that some gantry operations with umbilical assemblies drop from overhead and supply recirculated air from the truck cab after heating or cooling it as required. In practice such heating and air-conditioning units in the gantry are not centralized, but remain on a site-by-site basis, because each truck is charged on the basis of specific services provided by the site operator. In these cases the heating and air conditioning is still part of the electrical load for each site, although it will not appear in the truck cab as a receptacle. As long as the feeder to a group of sites includes within its load profile the heating and air conditioning load for each of those sites, this calculation and the demand factors that follow will be correct.

626.22. Wiring Methods and Materials. If the supply is from a pedestal or raised concrete pad, the mounting height must be at least 600 mm (2 ft) above grade or above the prevailing high-water mark in areas subject to flooding. Supplies that drop from a gantry are obviously not subject to this limitation. The supply equipment must be accessible through an entryway not smaller than 600 mm (2 ft) wide and 2 m (6½ ft high) There must be a remote, permanently lock-open capable disconnecting means, readily accessible, that will open the supply to one or more spaces.

626.24. Electrified Truck Parking Space Supply Equipment Connection Means. This section begins with a requirement for extra-hard-usage service cords to each connection, run together as a "single separable power supply cable assembly." The receptacle requirements are unusual. For example, consider the requirement in (B)(1) for three receptacles (minimum), each two-pole three-wire grounding and two of the three connected to at least two different branch circuits. The intention is for a conventional duplex GFCI receptacle in one case, and a single receptacle on another gang strap because some loads expected to use these devices have plugs too large to connect to adjacent halves of a duplex receptacle. Note that (D) requires that any receptacle outlets installed in accordance with this section have GFCI protection. No single GFCI receptacles are now in production, so this would mandate protection in the form of a faceless GFCI or a GFCI circuit breaker at a considerable distance away, wherever this distribution originates.

There is a widely circulated set of full-color photographs of one of these umbilical-supplied assemblies that drop from an overhead gantry and are designed to mount in the passenger-side window. It is equipped with a conditioned air supply, a touch-screen computer terminal with a USB port and a card stripe reader, Ethernet port, etc. It also has two GFCI duplex receptacles, at least of the kind the manufacturer apparently thought were two receptacles, because actually there are two GFCI duplex receptacles, for a total of four receptacles in violation of (1). But wait. If you look at the outside photo carefully, there is a double flap wet location cover for both halves of a duplex receptacle, not GFCI, with no provisions for it to be weatherproof while in use. This is where you plug in the block heater, as it turns out. It is apparent that we have quite a way to go before the equipment on the ground meets the **NEC**, and frankly, to where the **NEC** should be in terms of fine tuning this article.

This section also recognizes a single receptacle, "3-pole 4-wire grounding-type, single-phase rated either 208Y/120V or 125/250V." Because there is no NEMA configuration for a 208Y/120V single-phase receptacle at any amperage, this is confusing at best. There is an informational note following that describes a standard for pin-and-sleeve devices. However, that is not a requirement, and 30 ampere 125/250V plugs and receptacles are used by the million on identical distributions as these. The majority of multifamily housing is supplied through 208Y/120V three-phase services, and almost without exception, the feeder to each apartment consists of two-phase conductors and a neutral. Every conventional dryer receptacle outlet will have one of these devices providing the same sort of connectivity on the identical distribution system. It appears the reference to 208 V may have been an attempt by a proprietary interest to game the process and should be ignored. As of the inception of the 2008 NEC, there was very little in the way of an installed user base and so there was no consensus as to what these receptacles should have been. No clarification succeeded in the 2011 cycle; however, in the 2014 cycle in a panel statement on a related action, the panel effectively stated that it now intended the provision of three receptacles exactly as defined in Art. 100, which would mean three single receptacles or a single and a duplex receptacle.

This receptacle is intended to supply heating and air-conditioning equipment in the truck cab. If the conditioned air will be supplied by the truck stop through an umbilical connection instead, then this receptacle need not be provided. This is covered in 626.24(B) Exception, which requires careful study. The exception applies to the entirety of 626.24(B), and therefore waives the 125/250V 30-A receptacle in (B)(2), along with one of the three 20-A, 125-V receptacles in (B)(1), leaving only two in place as the minimum standard. Just as the exception covers all of 626.24(B), note also that the literal text of 626.24(D) reaches not only the 125-V receptacles, but the 125/250V 30-A receptacle as well.

626.25. Separable Power-Supply Cable Assembly. These are the rules, not inconsistent with the receptacle configuration rules in 626.24, that govern the power supply cords run from the site outlets to the truck. 15-A assemblies are permitted to operate an engine block heater on existing vehicles. The overall length of the cord is to be 7.5 m (25 ft) unless, if longer, a listed cable management system is employed.

626.26. Loss of Primary Power. This section and the next directly correspond with 625.46 and 625.48. It is not clear why these sections, which cover battery-powered vehicles for which there could be a use for the energy stored in the battery to power dwelling unit loads either on a stand-alone basis (625.46) or possibly in parallel with the utility (625.48), are being duplicated at a diesel truck stop. However, they do no harm.

626.30. Transport Refrigerated Units. These are the refrigerated trailers that are one of the principal reasons for diesel trucks to sit idle, and they represent a very significant load. This load is not included in the site load calculation of 626.11. There are two conventional voltage options given here, either 30-A on a 480-V 3-phase system or 60-A on a 208-V 3-phase system, so these are significant power loads. The cord connections will be through extra-hard-usage assemblies with a 90°C conductor rating along with an outer jacket evaluated for sunlight resistance and wet locations, and additional ratings for cold weather, oil and gasoline, ozone, acids, other chemicals, and abrasion. Worthy of note is that effective with the 2017 NEC, this section gives full recognition to a 20-ampere, 1000-volt, 3-pole, 4-wire grounding receptacle, configured as a pin and sleeve type.

ARTICLE 630. ELECTRIC WELDERS

630.1. Scope. There are two general types of electric welding: arc welding and resistance welding. In arc welding, an arc is drawn between the metal parts to be joined together and a metal electrode (a wire or rod), and metal from the electrode are deposited on the joint. In resistance welding, the metal parts to be joined are pressed tightly together between the two electrodes, and a heavy current is passed through the electrodes and the plates or other parts to be welded. The electrodes make contact on a small area—thus the current passes through a small cross section of metal having a high resistance—and sufficient heat is generated to raise the parts to be welded to a welding temperature. Either type of welding and cutting power equipment used with the scope of Art. 630 must be listed (630.6).

In arc welding with ac, an individual transformer is used for each operator; in other words, a transformer supplies current for one arc only. When dc is used, there is usually an individual generator for each operator, though there are also multioperator arc-welding generators. Note that the scope also includes plasma cutting operations where the electrical equipment is involved in creating and maintaining the arc that creates the ionized gas that does the cutting.

630.11. Ampacity of Supply Conductors. The term *transformer arc welder* is commonly used in the trade and, hence, is used in the Code, though the equipment might more properly be described as an *arc-welding transformer.* Refer to the informational note following 630.31, where the term *duty cycle* is explained.

It is evident that the load on each transformer is intermittent. Where several transformers are supplied by one feeder, the intermittent loading will cause much less heating of the feeder conductors than would result from a continuous load equal to the sum of the full-load current ratings of all the transformers. The ampacity of the feeder conductors may therefore be reduced if the feeder supplies three or more transformers. Note that if the value "I_{1eff}" is provided on the welder name plate, then this value must be used instead of the value determined from the table. This value can also be calculated using the formula in the informational note under 630.12(B). This is the effective current for the welder; it contrasts with "I_{1max}" which is essentially the rated primary current, as further described in the same note.

630.12. Overcurrent Protection. Arc-welding transformers are so designed that as the secondary current increases, the secondary voltage decreases. This characteristic of the transformer greatly reduces the fluctuation of the load on the transformer as the length of the arc, and consequently the secondary current, is varied by the operator.

The rating or setting of the overcurrent devices specified in this section provides short-circuit protection. It has been stated that with the electrode "frozen" to the work, the primary current will in most cases rise to about 170 percent of the current rating of the transformer. This condition represents the heaviest overload that can occur, and, of course, this condition would never be allowed to continue for more than a very short time. However, rating of the OC protection can be based on 200 percent of the maximum value of supply current or primary current of the welder, at the maximum rated output (next higher standard size permitted). The OC device may be located at the welder or at the line-end of the supply circuit.

630.31. Ampacity of Supply Conductors. The explanatory information that formerly resided in the parent language about the relationship between sizing for acceptable voltage drop compared with sizing for a potential overload condition has been, quite properly, relocated as an informational note. Subparagraph **(A)(1)** applies where a resistance welder is intended for a variety of different operations, such as for welding plates of different thicknesses or for welding different metals.

In this case, the branch-circuit conductors must have an ampacity sufficient for the heaviest demand that may be made on them. Because the loading is intermittent, the ampacity need not be as high as the rated primary current. A value of 70 percent is specified for any type of welding machine that is fed automatically. For a manually operated welder, the duty cycle will always be lower and a conductor ampacity of 50 percent of the rated primary current is considered sufficient.

example 1 A spot welder supplied by a 60-Hz system makes 400 welds per hour, and in making each weld, current flows during 15 cycles.

The number of cycles per hour is $60 \times 60 \times 60 = 216{,}000$ cycles.

During 1 h, the time during which the welder is loaded, measured in cycles, is $400 \times 15 = 6000$ cycles.

The duty cycle is therefore $(6000/216{,}000) \times 100 = 2.8$ percent.

example 2 A seam welder operates 2 cycles "on" and 2 cycles "off," or in every 4 cycles the welder is loaded during 2 cycles.

The duty cycle is therefore $2/4 \times 100 = 50$ percent.

Transformers for resistance welders are commonly provided with taps by means of which the secondary voltage, and consequently the secondary current, can be adjusted. The rated primary current is the current in the primary when the taps are adjusted for maximum secondary current.

When a resistance welder is set up for a specific operation, the transformer taps are adjusted to provide the exact heat desired for the weld; then in order to apply subparagraph **(A)(2),** the actual primary current must be measured. A special type of ammeter is required for this measurement because the current impulses are of very short duration, often a small fraction of a second. The duty cycle is controlled by the adjustment of the controller for the welder.

The procedure in determining conductor sizes for an installation consisting of a feeder and two or more branch circuits to supply resistance welders is first to compute the required ampacity for each branch circuit. Then the required feeder ampacity is 100 percent of the highest ampacity required for any one of the branch circuits, plus 60 percent of the sum of the ampacities of all the other branch circuits.

Some resistance welders are rated as high as 1000 kVA and may momentarily draw loads of 2000 kVA or even more. Voltage drop must be held within rather close limits to ensure satisfactory operation.

630.32. Overcurrent Protection. In this case, as in the case of the overcurrent protection of arc-welding transformers (630.12), the conductors are protected against short circuits. The conductors of motor branch circuits are protected against short circuits by the branch-circuit overcurrent devices and depend on the motor-running protective devices for overload protection. Although the resistance welder is not equipped with any device similar to the motor-running protective device, satisfactory operation of the welder is a safeguard against overloading of the conductors. Overheating of the circuit could result only from operating the welder that either the welds would be imperfect or parts of the control equipment would be damaged, or both.

630.42. Installation. This section governs the placement of welding cables in cable tray, which must then be labeled accordingly. This cable is *not* fine-stranded building wire that happens to be easy to bend and install in motor control centers, etc. The UL data on this cable clearly limits its use to the "secondary circuit of electric welders" and is as follows:

> This category covers welding cable, which is a single-conductor cable intended for use in the secondary circuit of electric welders in accordance with Art. 630, Part IV of ANSI/NFPA 70, "National Electrical Code." The conductors are flexible-stranded copper, 8 AWG through 250 kcmil, the individual strands of which are 34 through 30 AWG.
>
> Welding cable is rated 60, 75 or 90°C and 100 or 600 V.
>
> The voltage and temperature ratings, if higher than 100 V and 60°C, respectively, are identified by printing on the surface of the insulation.

ARTICLE 640. AUDIO SIGNAL PROCESSING, AMPLIFICATION, AND REPRODUCTION EQUIPMENT

640.1. Scope. Centralized distribution systems consist of one or more disc or tape recorders and/or radio receivers, the audio-frequency output of which is distributed to a number of reproducers or loudspeakers.

A public-address system includes one or more microphones, an amplifier, and any desired number of reproducers or speakers. A common use of such a system is to render the voice of a speaker clearly audible in all parts of a large assembly room. The article does not cover fire and burglar alarm signaling devices.

640.2. Definitions. "Abandoned Audio Distribution Cable" is previously installed cable that is not terminated at equipment and not identified for future use with a tag.

"Audio Signal Processing Equipment" essentially covers the range of equipment covered by this article, and the note that follows gives a full picture of the coverage. Note that a sound signal processor such as a computer creating sound signals from a MIDI (musical instrument digital interface) file is included.

"Technical Power System" covers systems using isolated grounding as covered in 250.146(D). However, the terminology is intriguing because it appears in another location in the NEC, namely, 647.6(B) and 647.7(A)(2). The provisions of Art. 647 are predominantly the creation of an audio engineer who was looking for a way to eliminate audio hum in recording studios. He succeeded so well that the grounding system in that article (center-tapped 120 V operating with both circuit wires running 60 V to ground), which started out in Art. 530, is now a stand-alone article with provisions to balance loads across a three-phase distribution. It can certainly be used to power Art. 640 applications, and with far more effective results than simply relying on 250.146(D), whose effectiveness is now widely questioned.

640.3. Locations and Other Articles. In general, the power-supply wiring from the building light or power service to the special equipment named in 640.1, and between any parts of this equipment, should be installed as required for light and power systems of the same voltage. Certain variations from the standard requirements are permitted by the following sections. For radio and television receiving equipment, the requirements of Art. 810 apply except as otherwise permitted here.

640.9 covers wiring to loudspeakers and microphones and signal wires between equipment components—tape recorder or record player to amplifier, and so on.

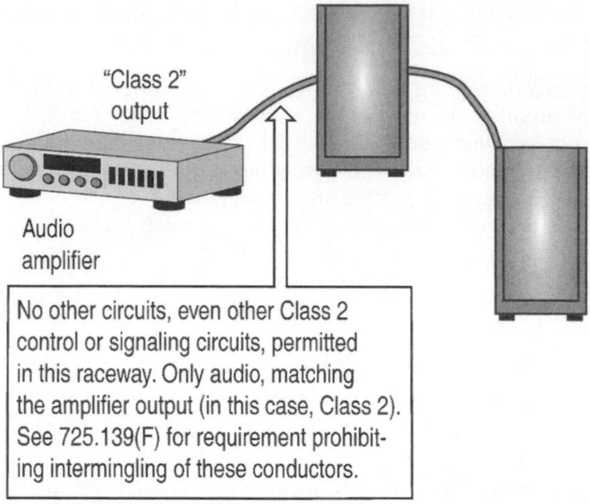
"Class 2"
output
Audio
amplifier
No other circuits, even other Class 2 control or signaling circuits, permitted in this raceway. Only audio, matching the amplifier output (in this case, Class 2). See 725.139(F) for requirement prohibiting intermingling of these conductors.

Fig. 640-1. Sound-system speaker wiring may be either Class 2 or 3 signal system, but must not intermingle with those systems. (Sec. 640.9.)

As shown in Fig. 640-1, amplifier output wiring to loudspeakers handles energy limited by the power (wattage) of the amplifier and must conform to the rules of Art. 725. As shown in 725.41(A), the voltage and current rating of a signal circuit will establish it as either Class 2 or Class 3 signal circuit. Amplifier output circuits rated not over 70 V, with open-circuit voltage not over 100 V, may use Class 3 wiring as set forth in 725.41(A).

Article 725 of the Code covers, among other things, signal circuits. A signal circuit is defined as any electrical circuit that supplies energy to a device—such as a loudspeaker or an amplifier—that gives a recognizable signal.

640.6. Mechanical Execution of Work. Part **(A)** requires cables that are exposed (including above a hung ceiling) to be supported such that normal building use will not damage them. Above the hung ceiling, the cabling must meet 300.11(A), which means staying off ceiling support wires that are an integral support of the ceiling system, but additional wires are allowed; refer to that discussion in Chap. 3 for more information. Abandoned cables (not identified for future use with a tag) must be removed. Tagged cables left in place must have tags that will hold up in their environment and the tags must give the date of identification, the date of intended use, and information relative to the intended use.

640.7. Grounding. Part **(A)** has the usual grounding requirements for wireways and auxiliary gutters. Parts **(B)** and **(C)** address implementation of Art. 647. See the comments offered here in the definition of "Technical Power System" for more information.

640.8. Grouping of Conductors. In this class of work, the wires of different systems are in many cases closely associated in the apparatus itself; therefore, little could be gained by separating them elsewhere.

640.9. Wiring Methods. **Part (A)(1)** requires wiring connected to the premises wiring system to comply with Chaps. 1 through 4 except as modified.

Part **(A)(2)** requires separately derived systems to comply with Code rules generally. This subsection also recognizes the procedures in Art. 647, which effectively allows the full 60/120-V system procedures to be used in the context of this article.

Part **(A)(3)** requires all other wiring follow the rules in Art. 725.

Part **(B)** requires auxiliary power supply wiring to equipment with a separate input therefore must follow Art. 725 rules; batteries must be installed in accordance with Art. 480. Auxiliary inputs mean that normal premises wiring supplies are also intended as a supply. An UPS is not an auxiliary input unless it is a direct part of the equipment and providing a dc supply. True auxiliary inputs may include fire alarm or other paging inputs typically involving a 10- to 50-V supply.

Part **(C)** allows amplifier output wiring to follow the normal rules in Art. 725 for the Class of circuit generated, whether Class 1, 2, or 3, (and the amplifier must be listed and marked accordingly) except that common raceways or enclosures must contain only audio circuits of the same class (see Fig. 640-1). Note that there are subtle differences in the capacities of audio power supplies and output circuits that, while allowing them to use limited energy wiring protocols, make them unsuitable for intermixing in common raceways. Therefore even in cases where you have a Class 2 audio output, that circuit may not be run in the same raceway as a normal Class 2 signaling circuit. This is addressed directly in 725.139(F).

Part **(D)** requires audio transformers and autotransformers to be used so as not to exceed the product limitations. System grounding is not required for circuits on the load side of this equipment, in contrast with the usual requirement for power circuits in 210.9. These transformers are generally used to match impedances between an amplifier and a speaker, and should not be grounded unless so designed by the system manufacturer.

640.10. Audio Systems near Bodies of Water. This section covers audio systems near bodies of water. An exception waives the requirements for watercraft, even if supplied by shore power. Wiring on these vessels would be exempt anyway, because Sec. 90.2(b)(1) provides that such wiring is beyond the scope of the Code. The rules in this section turn on whether the equipment is supplied by "branch-circuit power." If so, it has to be kept at least 5 ft away from the water; otherwise, if rated as Class 2, then it is limited only by the manufacturer's instructions.

640.21. Use of Flexible Cords and Cables. Cords used for connection to branch circuits are permitted according to the normal rules. Cords running between amplifiers and speakers, or speaker to speaker, or equipment to equipment, follow applicable Art. 725 requirements, with an allowance for other cabling. In addition, for equipment to equipment applications, other cabling can be used if specified by the manufacturer. Battery and other power sources need to follow the applicable code rules. The note points out that some of these additional sources will end up being the only source, but that they may be supplied in turn by intermittent or continuous power from a branch circuit.

Part **(E)** permits cords to be used to connect permanently installed equipment racks to facilitate equipment access or for "isolating the technical power system of the rack from the premises ground." This relates to the definition of "Technical Power System" and therefore appears intended to refer to isolation in the sense of connecting to an isolated-ground receptacle, as covered in 250.146(D). That wiring protocol still requires an equipment grounding conductor running with the circuit conductors. Therefore this would refer to a grounded cord plugged into an isolated ground receptacle. That receptacle might be supplied, however, by a "Technical Power System." Refer to the discussion at 640.22, below.

640.22. Wiring of Equipment Racks and Enclosures. This covers the wiring of equipment racks, which if made of metal (the usual condition) must be grounded. The racks have to be neat and workmanlike and there needs to be "reasonable access" to equipment power switches and overcurrent protective devices. The supply cords have to terminate within the equipment rack enclosure in an "identified connector assembly." They have to be able to carry the load and have overcurrent protection, which could be on the branch circuit.

Bonding is not required if the rack is supplied by a "technical power ground" (Fig. 640-2). Sec. 640.2 defines a "Technical Power System" as one in which insulated grounding receptacles are used in accordance with 250.146(D). However, 640.7(B) specifically refers to 647.6, which mandates the use of the term "Technical Equipment Ground" in reference to separately derived 120-V systems operating with a midpoint ground (and therefore 60/120 V). Since the original proposal to add this system to the NEC came from the audio engineering community, probably the reference here to a "technical power ground" is indeed to a system covered by Art. 647.

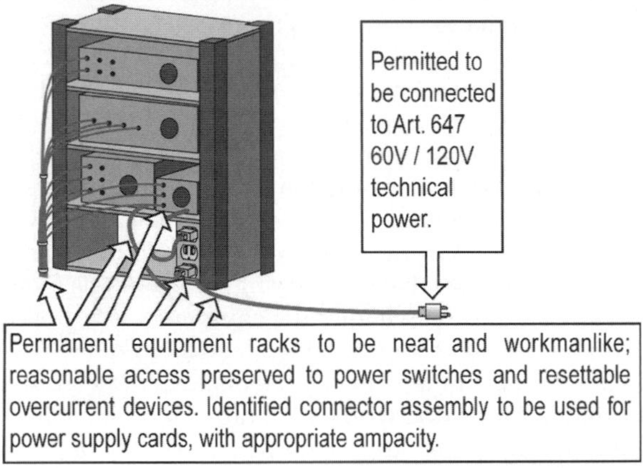
Permitted to be connected to Art. 647 60V / 120V technical power.
Permanent equipment racks to be neat and workmanlike; reasonable access preserved to power switches and resettable overcurrent devices. Identified connector assembly to be used for power supply cards, with appropriate ampacity.

Fig. 640-2. Grounding in racks may correlate with Art. 647 provisions. (Sec. 640.22.)

640.24. Wireways and Auxiliary Gutters. Wireways and auxiliary gutters follow the normal fill requirements; the 75 percent allowance in all versions of this article prior to the 1999 rewrite was not reinstated.

640.25. Loudspeaker Installation in Fire Resistance-Rated Partitions, Walls, and Ceilings. Speakers placed in fire-resistant partitions must be listed for the purpose or else go into enclosures (or a recess in the wall) that will maintain the fire-resistance rating.

640.41. Multipole Branch-Circuit Cable Connectors. Multipole branch-circuit cable connectors must be polarized and must not transmit strain to terminations, and the female half must be attached to the load end of the power supply cord.

Differently rated devices must be uniquely configured so they cannot intermate, and they must not intermate with nonlocking devices used for speaker connections, or with connectors rated 250 V whether locking or not. This suggests 125-V locking configured devices could be acceptable for speaker connections.

Signal cabling not intended for such speaker connections must not be intermateable with multipole branch-circuit cable connectors of "any accepted configuration." Note that this first part of this section substantially duplicates the language found in 520.67.

640.42. Use of Flexible Cords and Cables. This section covers the use of flexible cord for portable and temporary audio system applications, and parts **(A)** through **(D)** are essentially the same as Sec. 640.21 for permanent installations. Part **(E)** requires that flexible cord run to supply a portable equipment rack must use listed extra-hard-usage cord. For outdoor use, the cord must also be suitable for wet locations and sunlight resistant. If the racks also supply lighting or power equipment, Arts. 520 and 525 apply as appropriate. The use of any cable extensions, adapters, and breakout assemblies must meet the applicable provisions of those articles as well, because of the references in 640.3(F) and (G). Note that the relevant requirements principally appear in 520.68(A)(4) and 520.69.

640.43. Wiring of Equipment Racks. This section covers wiring of portable or temporary equipment racks, and it is similar to Sec. 640.22, which covers the permanent variety. There are significant differences, however. If the rack is nonmetallic and equipped with a cover, removal of that cover must not allow access to Class 1, Class 3, or "primary circuit power" without removal of terminal covers or the use of tools. Wiring that leaves such equipment to other equipment or a power supply must have strain relief or other arrangements so pull on the cord will not increase the risk of damage.

640.44. Environmental Protection of Equipment. This section requires protection of portable or temporary equipment used outdoors from adverse weather conditions. If the equipment is expected to remain operational, arrangements need to be made including ventilating heat-producing equipment.

640.45. Protection of Wiring. If there is public accessibility, cord cannot be laid on the ground or floor without being covered with approved nonconductive mats and there must not be a tripping hazard. This is essentially the same as 525.20(G), and as of the 2011 NEC, there is the same waiver of 300.5 allowing for temporary shallow burial.

640.46. Equipment Access. Equipment likely to be hazardous to the public needs to be protected with barriers or supervised by qualified personnel to prevent public access.

ARTICLE 645. INFORMATION TECHNOLOGY EQUIPMENT

645.1. Scope. This article applies only to an information technology equipment room. Because this article covers all of the designated "equipment," "wiring," and "grounding" that is contained in a room, there is no question about the mandatory application of these rules to such "rooms." But the specific nature of that word *room* in the 645.1 statement of "Scope" leaves some uncertainty about electronic computer/data processing equipment and systems that are *not* installed in a dedicated room. In such cases, one must follow all the rules given in Chaps. 1 through 4 and ignore the rules and permissions given in Art. 645. The point being: If you do not have a computer room, as defined by 645.42, then Art. 645 does *not* cover the installation, because, as given in the first sentence of 645.1, only equipment and the like installed in "an information technology equipment room" *is* covered.

Furthermore, this article is unusual in that it is, in effect, voluntary. If you do not want to comply with its provisions, simply make sure that one (or more) of the conditions in 645.4 is not met, and the article does not and never will apply. The article offers certain wiring advantages that may be beneficial, but there is no obligation to implement those advantages. To assist in the orderly application of this article, information technology equipment now has a formal definition in 645.2. Note that it carefully excludes the processing of communications circuits. These are the exclusive province of Art. 800, and the power-limited cabling connected to this equipment is generally covered by Art. 725 (and also Art. 770 if applicable) with additional conditions and provisions found here. However, due to provisions in 645.3, any power-limited cabling, as covered in Arts. 725, 760, 770, 800, etc. must adhere to the listing protocols established within those articles.

645.4. Special Requirements for Information Technology Equipment Room. Six conditions are described that must be complied with in order for Art. 645 to be applied to data processing equipment rooms:

1. Disconnects in accordance with 645.10 are provided.
2. A dedicated HVAC system must be used for the room, or strictly limited use involving automatic actuation of fire/smoke dampers may be made of an HVAC system that "serves other occupancies."
3. Only "listed" (such as by UL) information technology equipment may be installed.
4. The computer room must be "occupied by, and accessible to, only those personnel needed for operating and maintaining the computer/data processing equipment."
5. The room must have *complete* fire-rated separation from "other occupancies."
6. Electrical equipment not associated with the information technology room is installed within that room.

It is very clear from the wording that if any of these conditions is *not* met, the entire computer/data processing installation is *not* subject to the rules of Art. 645. But such an installation would be subject to *all* other applicable rules of the NEC.

645.5. Supply Circuits and Interconnecting Cables. Part **(A)** limits every branch circuit supplying data processing units to a maximum load of not over 80 percent of the conductor ampacity (which is an ampacity of 1.25 times the total connected load).

Part **(B)** covers use of computer or data processing cables and flexible cords, which can be up to 4.5 m (15 ft) long, and must be protected from damage if run across the floor. As shown in Fig. 645-1 (under a raised floor), part **(C)** permits data processing units to be "interconnected" by flexible connections that are "of a type permitted for information technology equipment."

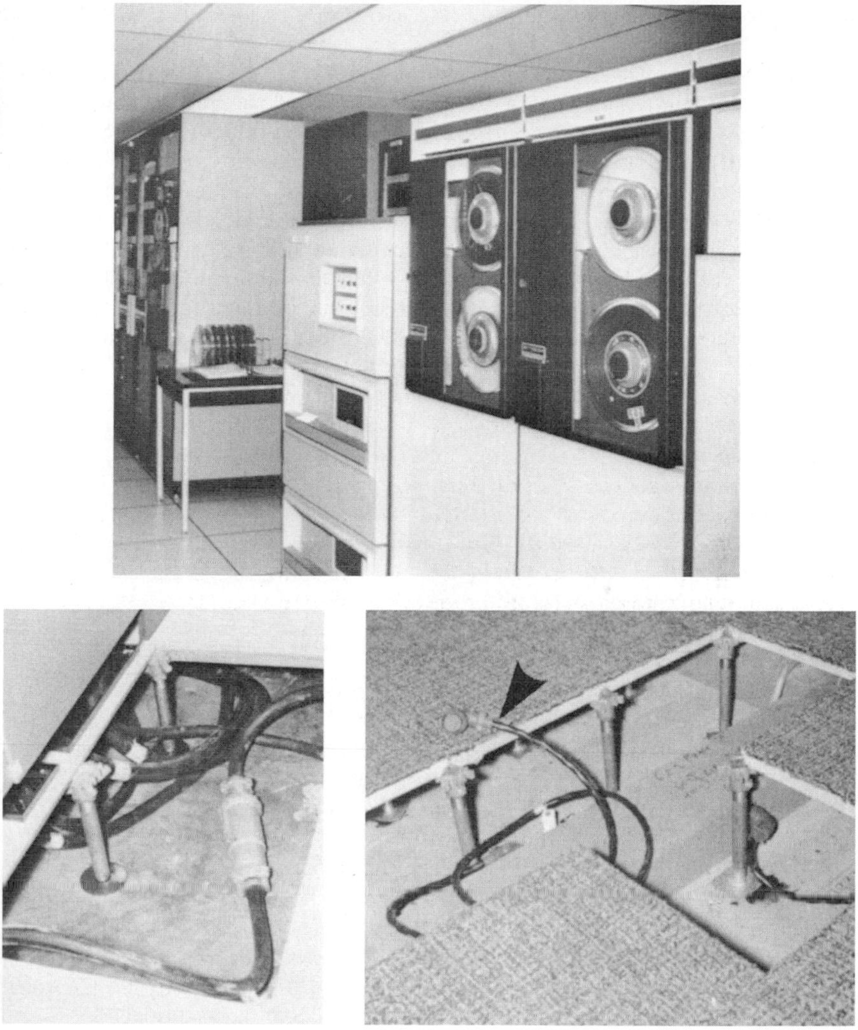

Fig. 645-1. Connection of data-processing units to their supply circuits and interconnection between units (power supply, memory storage, etc.) may be made only with cables or cord-sets specifically listed [see (B)(2)] as parts of the data processing system. (Sec. 645.5.)

Part **(C)** covers interconnecting cables, which are not subject to the 4.5 m (15 ft) limitation. Part **(D)** requires supply circuits and interconnecting cables to be protected if they are exposed to damage, such as running across a traveled walkway. Presumably this rule can also be applied to insist floor openings have protection against abrasion of cords and cables passing through them. This used to be explicitly covered in the same sentence that required minimizing debris entry beneath the floor. The latter concept now lives in the parent language in 645.5(E), but the other half of that sentence covering cable abrasion was lost when 645.5(E) was reorganized in the 2017 **NEC**.

Part **(E)** permits a variety of wiring methods under an accessible and approved raised floor serving a data processing system, provided the openings are configured so they will minimize the entry of debris into the floor. This part now, for editorial clarity, divides the universe of possible wiring types under raised floors into three categories, for which the first covers the wiring methods for branch-circuit supply conductors. All branch-circuit wiring covered here must be secured as covered in 300.11. The permitted list of wiring methods is extensive because it is a list of 17 items, stated as being in addition to the "wiring methods of 300.22(C)." This would correspond to 300.22(C)(1), and after disregarding the six methods that appear in both places, the result is a total of 20 wiring methods (see Fig. 645-2) that

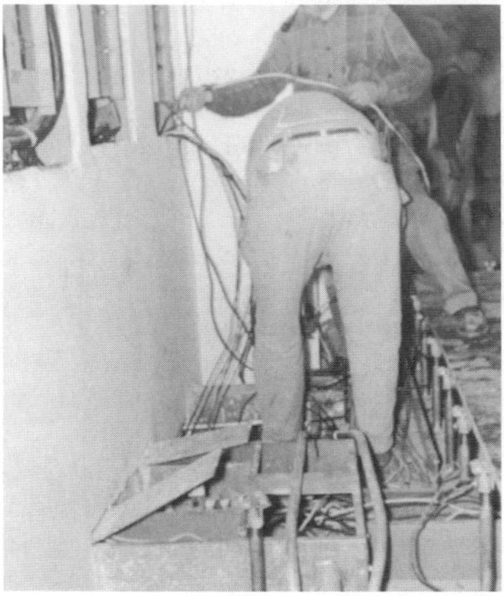

Fig. 645-2. Branch circuits from a panelboard to data processing receptacle outlets can be in a wide variety of Chap. 3 wiring methods. (Sec. 645.5.)

end up as permissible for branch-circuit wiring under a computer floor. Adding the three additional methods looks like this:

(18) Totally enclosed nonventilated busway, with no plug-in connections

(19) Other factory-assembled multiconductor control or power cable specifically listed for air-handling spaces

(20) Listed prefabricated manufactured wiring systems without nonmetallic sheath [Art. 604]

In addition, in accordance with 300.22(C)(b):

(21) Solid side and bottom metal cable tray systems with solid metal covers, which brings in all the wiring methods in 392.10(A)

The second category of underfloor wiring consists of "electrical supply cords, data cables, interconnecting cables, and grounding conductors." This includes a five-part list, as follows:

(1) Listed information technology equipment supply cords, as covered in 645.5(B)

(2) Interconnecting cables enclosed in a raceway

(3) Equipment grounding conductors

(4) Plenum-grade power limited cabling in 725.136(C), plus higher ranked cables in the 725.154(A) substitution hierarchy

(5) Listed Type DP cable

Note that as presented in 645.5(E)(2), item (4) above contains potential conflicts unless it is carefully applied. Only the plenum grade cable (CMP) can be used directly; the others listed can be used, but only if completely enclosed in metal raceways or solid metal cable tray as covered in 300.22(C).

The third category of underfloor wiring covers plenum grade optical fiber cable. The reference correctly points to 770.113(C), but adds OFNR, OFCR, OFN, and OFC. Although these additional cable can be used, as in the case of 645.5(E)(1)(4), these non-plenum graded cables must run in metal raceways or solid metal cable trays, as specifically addressed in 770.113(C)(5) and (6).

It appears that the technology of cable construction has advanced to the point that high-speed data transfers can occur over cables that easily meet plenum standards. This was not always the case. Years ago, cables that met plenum standards often had unacceptable performance characteristics in computer rooms, and this part of Art. 645 allowed other than plenum graded cabling under the raised floors for that reason. However, the model building codes had rules for acceptable materials in plenum cavities that, when it came to electrical cables, strictly required plenum ratings. Wiring methods could be changed, but this meant that the equipment would not even function properly. This direct conflict between construction disciplines set the stage for an extended stand-off that continued for years. The solution was arrived at in the 1990s through extended discussion between electrical code experts and leaders of the model building code organizations. The building officials decided that they would not need to classify the underfloor space as a plenum cavity if it would not function in that way during a fire condition. The electrical experts said, in effect, done! There was already a drop-out relay under these floors to shut down the air circulation, as required by 645.10. All that was required was to add a smoke detector under the floor and have it drop out the ventilation if products of combustion were detected. The fan would stop, the air movement would cease, and the plenum cavity designation need not apply. This was the case for four code cycles. Unfortunately, with absolutely zero substantiation that took these past discussions into account, the smoke detector rule was deleted in the 2014 **NEC**. Fortunately, the science of cable construction has, for now, made the conflict academic.

Part **(F)** adds important information by stating clearly that any cable, boxes, connectors, receptacles, or other components that are "listed as part of, or for, information technology equipment" are *not* required to be secured in place, *but* any cable or equipment that is *not* "listed as part of" the computer equipment *must* be secured in accordance with all Code rules covering it.

Part (G) requires that the accessible portions of abandoned supply circuits and interconnecting cables, defined as unterminated and unidentified for future use, be removed. The next part **(H)** builds on this by requiring that in order to qualify as identified for future use, the cables needed to be tagged in a manner with sufficient durability for the location and stating the date of identification, the expected date of use in the future, and some information as to what that use is expected to be.

645.10. Disconnecting Means. Part (A) of this section covers conventional ITE rooms and its provisions are similar to those that have applied since the article began in the 1968 NEC. Over the intervening decades, information processing has become increasingly critical, and a survey of those who are responsible for these facilities revealed that the overwhelming majority of activations of these EPO switches were either accidental, or acts of sabotage. It was further determined that the average outage was about 4 hours, with a resultant loss of revenue in the seven-figure range. Relocating these control switches is a first step in responding to the changing nature of the importance of these rooms. As shown in Fig. 645-3, a master means of disconnect (which could be one or more switches or breakers) must provide disconnect for all computer equipment, ventilation, and air-conditioning (A/C) in the data processing (DP) room.

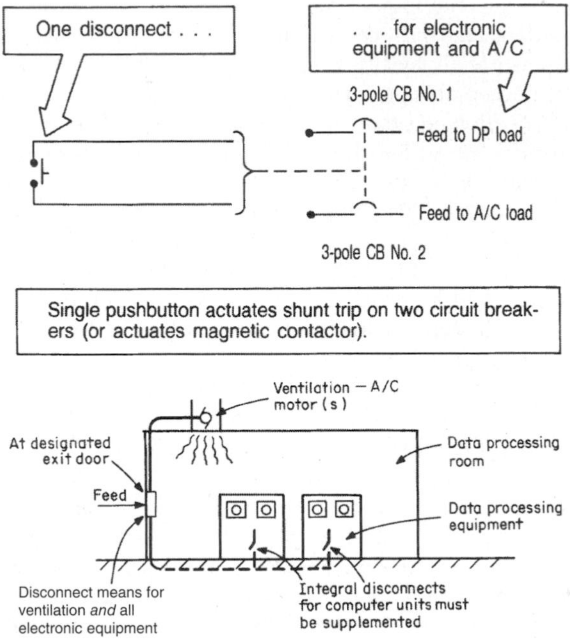
One disconnect . . .
. . . for electronic equipment and A/C
3-pole CB No. 1
Feed to DP load
Feed to A/C load
3-pole CB No. 2
Single pushbutton actuates shunt trip on two circuit breakers (or actuates magnetic contactor).
Ventilation – A/C motor (s)
At designated exit door
Feed
Data processing room
Data processing equipment
Disconnect means for ventilation *and* all electronic equipment
Integral disconnects for computer units must be supplemented

Fig. 645-3. Data processing room must have arrangements like those shown here. (Sec. 645.10.) The disconnect control is no longer mandated to be at the principal exit door.

The disconnects called for in this rule are required to shut down the DP system and its dedicated HVAC and to close all required fire/smoke dampers under emergency conditions, such as fire in the equipment or in the room. For that reason, the rule further requires that the disconnect for the electronic equipment and "a similar" disconnect for A/C (which could be the same control switch or a separate one) must be grouped and identified and must be "controlled" from locations that are readily accessible to the computer operator(s) or DP manager. The location of these controls is no longer required to be at a principal exit door;

the requirement now is for a location that is "readily accessible to authorized personnel and emergency responders." Large rooms requiring multiple protection zones are also recognized [in (A)(3)] provided an approved means exists to confine a fire to its zone of origin. Figure 645-4 shows two control switches—one in the control circuit of the A/C system and the other a shunt-trip pushbutton in the CB of the feeder to the DP branch circuits—with a collar guard to prevent unintentional operation.

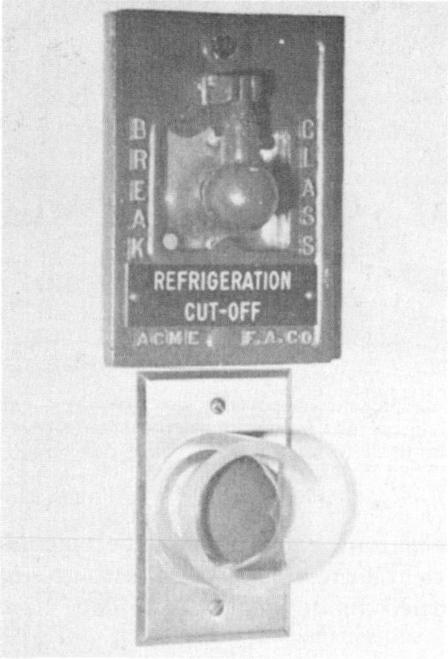
BREAK
GLASS
REFRIGERATION
CUT-OFF
ACME F.A.CO

Fig. 645-4. Adjacent to the door of a DP room, a break-glass station (at top) provides emergency cutoff of the A/C system in the DP room; and a mushroom-head pushbutton—with an extended collar guard that requires definite, intentional pushing action—energizes a shunt trip coil in the feeder circuit breaker supplying branch circuits for the electronic DP equipment. (Sec. 645.10.)

Although the present wording of this rule readily accepts the use of a single disconnect device (pushbutton) that will actuate one or more magnetic contactors that switch the feeder or feeders supplying the branch circuits for the computer equipment and the circuits to the A/C equipment, the wording also recognizes the use of separate disconnect control switches for electronic equipment and A/C. Control of the branch circuits to electronic equipment may be provided by a contactor in the feeder to the transformer primary of a computer power center, as shown in Fig. 645-5.

A single means used to control the disconnecting means for *both* the electronic equipment and the air-conditioning system offers maximum safety. In the event of a fire emergency, having two separate disconnecting means (or their remote operators) at the principal exit doors will require the operator to act twice and thus increase the hazard that only the electronic equipment or the air-conditioning system will be shut down. If only the electronic equipment is disconnected, a smoldering fire will become intensified by the air-conditioning system force-ventilating the origin of combustion. Similarly, if only the air-conditioning system

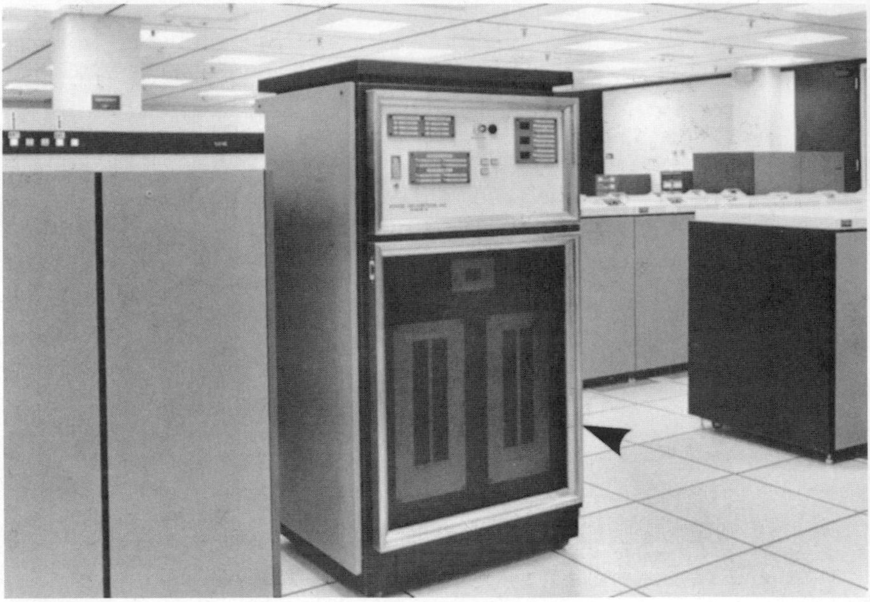

Fig. 645-5. Rapidly accelerating application of DP equipment in special DP rooms with wiring under a raised floor of structural tiles places great emphasis on Art. 645. "Computer power centers" (arrow) are complete assemblies for the supply of branch circuits to DP equipment, with control, monitoring, and alarm functions. (Sec. 645.10.)

is disconnected, either a fire within the electronic equipment will become intensified (since the electric energy source is still present) or the electronic equipment could become dangerously overheated due to the lack of air conditioning in this area.

Wording of this rule requires means to disconnect the "dedicated" A/C "system serving the room." If the DP room has A/C from the ceiling for personnel comfort and A/C through the raised floor space for cooling of the DP equipment, both A/C systems would have to be disconnected. There have been rulings that *only* the A/C serving the raised floor space must be shut down to minimize fire spread within the DP equipment and that the general room A/C, which is tied into the whole building A/C system, does not have to be interrupted. In other cases, it has been ruled that the general room A/C must be shut down, while the floor space A/C may be left operating to facilitate the dispersion of fire suppressant and extinguishing materials within the enclosures of DP equipment that is on fire. Because of the possibility of various specific interpretations of very general rules, this whole matter has become extremely controversial. Review, as well, the history behind the former wording on 645.5(E)(4), which bears on this question.

The exception to this rule waives the need for disconnect means in any "integrated electrical systems" (Art. 685), where orderly shutdown is necessary to ensure safety to personnel and property. In such cases, the entire matter of type of disconnects, their layout, and their operation is left to the designer of the specific installation.

Part **(B)** covers "Critical Operations Data Systems." As defined in 645.2, it includes three occupancies (public safety, emergency management, and national security) that are completely logical given its name, and one at the end (business continuity) that is loose enough to include virtually every ITE room in existence, or so their IT staff would undoubtedly insist. These rooms do not require an EPO. Five conditions must be met in order to trade off the lack of the EPO control:

1. AHJ—In this case probably a building or fire inspector and not an electrical inspector, has agreed to a protocol under which power is removed and air movement halted within the room, or within a designated zone, defined as a physically identifiable area in terms of distance or actual barriers and having a discrete power and cooling system.
2. Qualified operators are available 24/7 to advise first responders as to disconnecting procedures.
3. Smoke detectors [already required by 645.5(E)(4) for underfloor spaces] are in place or other "sensing system."
4. There is an automatic fire suppression system in place and the AHJ, again, probably not the electrical inspector, has approved it.
5. The limited energy wiring under a raised floor is listed for plenum applications and the power circuits meet 645.5(E)(2), or for supply cords, 645.5(E)(3). Note that the reference to 725.135(B) is probably an error and should point to 725.135(C) to correlate with the referenced 800.113(C).

These are actually reasonable requirements and presumably well within the construction budget of any enterprise looking at a multimillion dollar loss for an outage of a few hours.

645.11. Uninterruptible Power Supplies (UPSs). This rule requires disconnects for "supply and output circuits" of any UPS rated over 750 VA "within" the computer room. The UPS disconnecting means *must* satisfy 645.10, and it *must* "disconnect the battery from its load." The wording of this rule leaves questions about a disconnecting means for a UPS installed *outside* the computer room (Fig. 645-6). The UPS output circuit—which presumably feeds the computer room—would

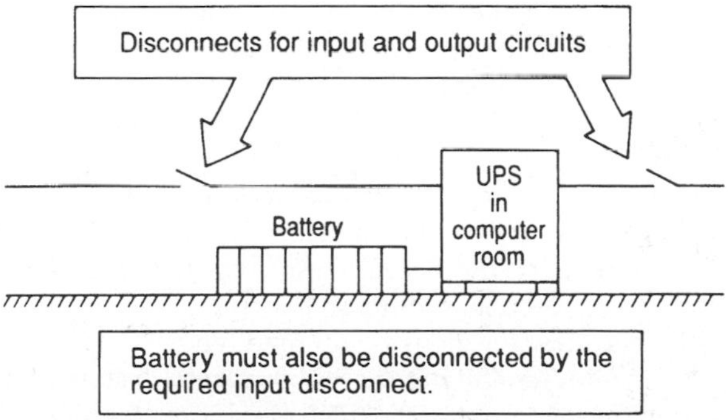
Disconnects for input and output circuits
UPS
in
computer
room
Battery
Battery must also be disconnected by the required input disconnect.

Fig. 645-6. Considerable interpretation latitude is inherent in the rule that requires "grouped and identified" switches or circuit breakers "at principal exit doors" to control an uninterruptible power supply. (Sec. 645.11.)

have to be disconnected. But it seems as if the supply to the UPS and the battery charge/supply circuit need not be automatically disconnected when the UPS is located outside the room. On the other hand, the wording clearly does require the UPS output circuit to meet 645.10 even if the battery were not required to be disconnected from its load in such cases.

645.14. System Grounding. The default status of ITE power systems is that they meet the conventional rules in Art. 250. However, immediately following this statement is an allowance that is so important it frequently drives the decision about whether to qualify for Art. 645 coverage or not. Power systems derived within listed IT equipment are not required to comply with 250.30, which sets in motion a chain of events that includes the permission to not bother installing a grounding electrode or making a connection to an existing electrode. The note following explains that listing requirements adequately address the safety issues involved. The principal issue is whether or not the power system includes a system bonding jumper between the neutral and the equipment grounding system between the output terminals of the transformer and the secondary disconnect. If this jumper is properly installed, then any electrical fault will have an effective fault current path to return to the system source. The product standards do ensure that this jumper will be installed.

645.15. Equipment Grounding and Bonding. Data processing equipment must *either* be grounded in full compliance with Art. 250 or be "double insulated." The last sentence of 680.15 points out the need for bonding of "signal reference structures" where such grids are installed within the information technology room. Presumably this bonding should be provided between the equipment ground bus at the power systems source and the reference grid. As for sizing this bonding jumper, in the absence of a specific rule on this matter, it seems reasonable to size the bonding jumper based on the size of the source's main or system bonding jumper. Where the room is supplied by a feeder, a bonding jumper not less than the size of the equipment grounding conductor required by 250.122 for the feeder would also seem to satisfy this rule. Such sizing should be viewed as acceptable especially since there is no wording provided to establish a method for the sizing of this required bonding conductor. The 2014 NEC added a requirement requiring any supplemental grounding electrodes to comply with 250.54, including the fact that the earth is not to be relied upon as a fault-current return path.

645.17. Power Distribution Units. This section recognizes the use of power distribution units (PDUs) that are equipped with multiple panelboards. That is, PDUs may have more than one panelboard within a single enclosure. The former limitation to 42 circuits per panel was removed following changes in Art. 408 that removed that limitation generally. Note that because a panelboard is actually the bus structure and not the enclosure, which is an enclosing cabinet, the allowance for multiple panels in a common enclosure was always unnecessary to begin with. Listing for these locations is still required.

645.18. Surge Protection for Critical Operations Data Systems. As of the 2017 NEC, the protection must be provided, which also correlates with 708.20(D).

645.25. Engineering Supervision. This section permits an engineered load calculation to be made in lieu of the usual procedures in Art. 220 with respect to the loads in one of these information technology equipment rooms. A great deal of attention is now being paid to reducing energy budgets in these rooms, both directly in terms of processing power, and indirectly in terms of the removal of heat. As developments proceed, there will be many solutions, all headed in the direction of reduced energy expenditure, and this section will allow those designs to receive the dividends they should in terms of reduced capital expenditures.

645.27. Selective Coordination. This is the latest application of a popular buzz term. Refer to the discussion at 700.32 for in-depth coverage of the merits, and expense, of protecting against a failure mode that has never occurred in an occupied building.

ARTICLE 646. MODULAR DATA CENTERS

646.1. Scope. This article, new in the 2014 NEC, can be thought of as taking the equipment and electrical wiring in Art. 645 and compressing it into prefabricated whole that approximates the size of a shipping container. The informational note attached to the definition in 646.2 is very helpful in rounding out the picture. They can be set up for either indoor or outdoor applications, and they can also be configured with support equipment in a separate enclosure.

646.3. Other Articles. This section, over a full page in length, is harmless but far longer than it needs to be, and clearly is intended as one-stop shopping for designers, installers, and inspectors of this equipment. All the cross-referenced requirements would necessarily apply in accordance with 90.3 with rare exceptions. For example, **(L)(1)** requires all wiring to be listed and labeled. Although Chap. 3 requirements typically include that for raceways, they often do not for cables. Another provision that will be extremely consequential is **(L)(8)** that brings in the alternate wiring allowances of Art. 645, but on the same proviso governing that article, namely, that the separation requirements in 645.4 are met. Part **(N)** requires compliance with 645.10 for the electrical system disconnect and then appears to require "a similar" HVAC disconnect. This will raise the question as to whether the two can be combined if desired. Section 645.10 allows them to be combined, so presumably that permission would carry over.

646.4. Applicable Requirements. The content and format of this section is without precedent in the NEC, and it is therefore very interesting. It establishes the line between MDCs that are submitted for a listing, and those that although a prefabricated whole, are field wired. If the prefab unit (and it will be a prefab unit per 646.2) is wired off-site without the benefit of a listing examination, then the inspector will be looking at field compliance with every provision. If it does go through a listing examination, 646.5 through 646.9 (the remainder of Part I of the article) would have been evaluated by the test lab and the results obvious and available for the AHJ, who will then focus the inspection on the other sections. It follows that compliance to those other sections should always be scrutinized in the field by all parties whether or not a listing is evident.

646.5. Nameplate Data. This section governs the content of a required nameplate, focusing on how the unit needs to be supplied with power. It must take into account any special loading profiles, including some that might not be obvious for a data processing facility including motor loads complete with 430.26 duty cycling allowances and a reference to 430.26. Note that the reference to duty cycles [and 430.22(E)] is almost certainly incorrect. A duty cycle is unique to a very particular type of motor load and extremely unlikely to be a factor in the load profile of an Art. 646 application. The motor loads in such cases would probably be covered in Art. 440 and enter feeder load evaluations through 430.24. On the other hand, the reference to 430.26 in the informational note is essentially correct, in that 430.26 applies to demand factors and they should absolutely be considered while evaluating these loads. The only quibble would be that as written the reference is in the form of a note and not a Chap. 6 amendment to Chap. 4. If this evaluation is being performed by the inspector under 646.4(2), then such factors can be applied. If the evaluation is performed by a testing lab under 646.4(1), the status of 430.26 as being discretionary on the part of the inspector makes this problematic because a testing lab and an inspector are very different entities. That said, if a testing lab determines through the listing process that a certain rated current is correct it should always be respected.

646.6. Supply Conductors and Overcurrent Protection. This is the expected 125 percent ampacity rule for the supply conductors, presumably based on the likely continuous operating nature of these loads, and the fact that the wires will need to land on a device that will have a correlating requirement for the wires connected to it. The note in part **(A)** does repeat the erroneous reference to motor duty cycles and could be omitted entirely because the required evaluations would have been part of the nameplate creation process. The provisions of 646.6(B)(2) allow for the orderly implementation of 240.21 in instances where the MDC occurs at the end of a tap.

646.7. Short-Circuit Current Rating. These provisions mirror those for industrial control panels in 409.110 as they would be evaluated for use under 409.22. As in those cases, UL 508A, Supplement SB is essential to making these evaluations.

646.9. Flexible Power Cords and Cables for Connecting Equipment Enclosures of an MDC System. Extra-hard-usage flexible cord, evaluated for outdoor use if necessary, is permitted for one function, and only one function, namely, to interconnect enclosures of an MDC system where the routing is judged to not make it subject to physical damage. As with 520.53(H)(2) single-conductor cord is permitted, but not smaller than 2 AWG.

646.11. Distribution Transformers. This section covers the allowable transformer placements within or even in the vicinity of an MDC. Part **(A)** is a unique provision in the NEC and it forbids a utility-owned transformer within any MDC. The next two parts cover non-utility transformers, the first (part **B**) being "premises transformers." This term is also unique, as well as undefined, but the context indicates that this is referring to a transformer on the load side of the service point, and therefore a part of the premises wiring system, that serves the MDC as well as additional loads at the occupancy that are unrelated to the MDC, and that are located near, but not in the MDC. These transformers simply follow Art. 450, but they must be either dry (450.21 or 450.22) or use a nonflammable insulating medium (450.24). Since such transformers are necessarily a type of power transformer also covered in part **(C)**, this interpretation relies on the obvious necessity to distinguish this provision from part **(C)**. Part **(C)** allows power transformers that supply power only to the MDC to be located within it, provided they are dry and located within "the MDC equipment enclosure." Although undefined, there is some useful descriptive text on the topic in the informational note within 646.9(A). Note, however, that such an enclosure need not be separate and could be a component within the overall MDC.

646.12. Receptacles. Every work area within an MDC must include a service receptacle.

646.14. Installation and Use. More one-stop shopping, with an obvious restatement of 110.3(B).

646 Part III. Lighting. This part contains interesting requirements some of which generally belong to building codes, but that are judged necessary because an MDC may arrive as a stand-alone piece of equipment, albeit large enough for employees to work within. The rules meld workspace illumination rules in 110.26(D) with typical building code rules for exit discharge illumination together with typical emergency standby rules for such lighting. Aside from the one-stop shopping considerations, the rules are necessary here because the lighting requirements are coming from the NEC and not from a building code, and thereby arguably sidestep the literal application of Art. 700. Clearly 646.17 recapitulates the relevant provisions of that article.

646.19. Entrance to and Egress from Working Space. Due to the incorporation by reference of 110.26 in 646.18, the workspace depth and height dimensions of that section are completely enforceable. The double-end access rule in this section closely parallels all of 110.26(C)(2) but has important differences. First, the rule applies not just to a wide line-up, but to one equally deep as well. In addition, it applies regardless of the amperage rating. Since the amperage parameter is removed from the equation, the panic hardware rule is then applied to all exit doors from an MDC that contains an equipment line up that trips the 6-ft workspace threshold.

646.20. Working Space for ITE. Workspace for voltage-limited circuits (≤30 V RMS; 42 V peak; or 60 V dc) does not need to meet the prescriptive requirements in 646.19; circuits of greater voltage will, however, require that workspace.

646.21. Work Areas and Working Space Around Batteries. In agreement with 480.9(C), workspace around a battery system must comply with 110.26 measured from the edge of the rack.

646.22. Workspace for Routine Service and Maintenance. Work such as air filter cleaning and replacements must have workspace to accomplish those tasks, but there is no prescriptive dimension provided.

ARTICLE 647. SENSITIVE ELECTRONIC EQUIPMENT

647.1. Scope. This article covers a new 60/120-V distribution system designed to eliminate the effects of electronic noise, originally on audio and video production when this material was a Part (G) in Art. 530, and now in a more broadly applicable format. This system, as shown in Fig. 647-1, uses two ungrounded circuit

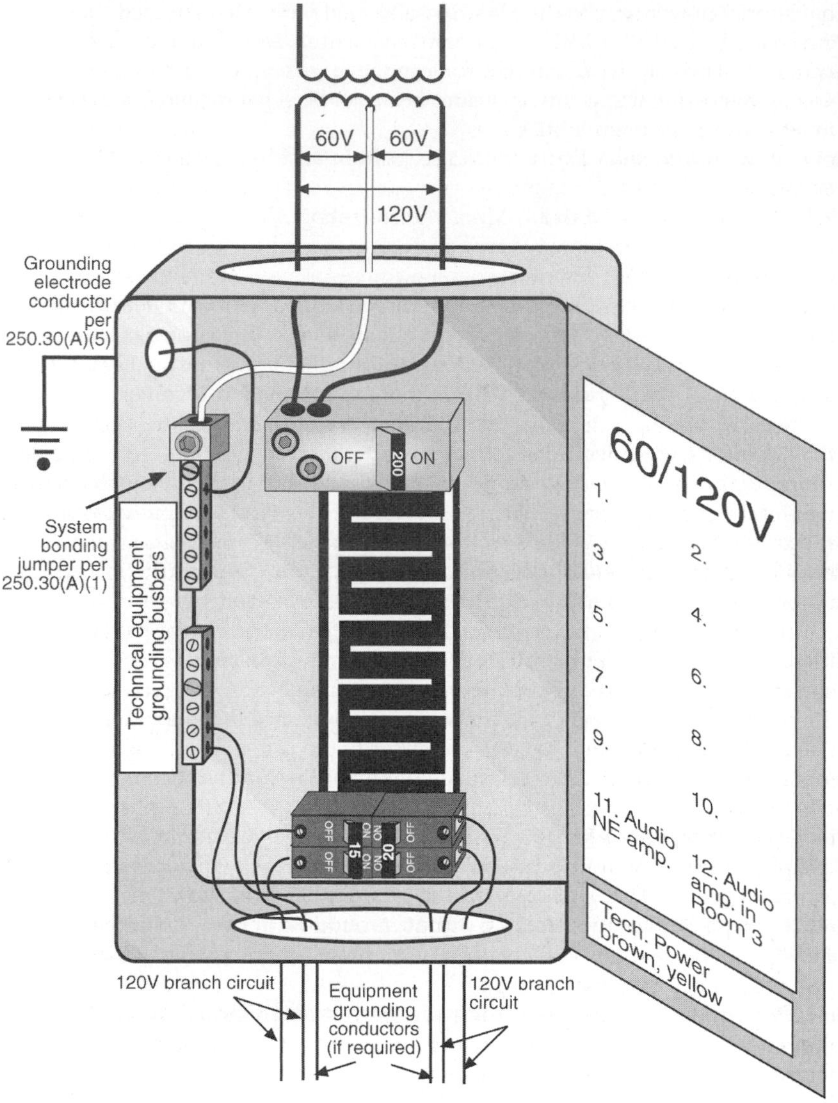
60V
60V
120V
Grounding electrode conductor per 250.30(A)(5)
System bonding jumper per 250.30(A)(1)
Technical equipment grounding busbars
OFF
200
ON
60/120V
1.
2.
3.
4.
5.
6.
7.
8.
9.
10.
11. Audio NE amp.
12. Audio amp. in Room 3
Tech. Power brown, yellow
15
20
120V branch circuit
Equipment grounding conductors (if required)
120V branch circuit

Fig. 647-1. A panelboard properly wired to distribute power using circuit conductors with 120 V between them and with each wire running 60 V to ground. The system must not be designed to be used for line to neutral load. (Sec. 647.1.)

conductors at a potential of 120 V, and the system secondary is midpoint tapped and grounded to hold the system to 60 V to ground.

These systems have proven extremely effective at reducing audio hum and video interference, which is why this system started its days of NEC recognition in Art. 530. The noise from the filtering circuits tends to cancel on this type of balanced system and it routinely outperformed the most elaborate grounding arrangements on traditional distributions. All harmonic frequencies cancel. The transformers with the center-tapped 120-V secondaries are increasingly available as well. The location here as a standalone article means that other applications can use its benefits.

647.3. General. These systems can be used to reduce electronic noise in "sensitive electronic equipment locations" provided they are installed only in industrial or commercial occupancies, and under close supervision of qualified personnel.

647.4. Wiring Methods. Part **(A)** provides that standard panelboards, such as 120/240 single-phase panelboards are permitted, but the voltage system must be clearly marked on the panel face or door. Common-trip two-pole circuit breakers must be used for all circuits (or a two-pole fused switch), and they must be identified for use at the system voltage. In general, circuit breakers work OK as long their voltage rating is not exceeded, and that will not be a problem on this system. As shown in the drawings, all loads supplied by these systems will use two ungrounded circuit conductors.

Part **(B)** covers junction box covers, which must be clearly marked to indicate the distribution panelboard and the system voltage.

Part **(C)** covers the identification rules for feeder and branch-circuit conductors, which must be identified as belonging to this kind of system at all splices and terminations by color, tagging, or equally effective means. The means chosen must be marked on all panelboards, and on the building disconnect. This is intended to reduce the possibility of confusion during future alterations.

Part **(D)** is one of very few instances in the NEC of a mandatory voltage drop requirement. The voltage drop on any branch circuit must not exceed 1.5 percent, and the total drop including the feeder contribution must not exceed 2.5 percent. These circuits are operating with only one-half the voltage to ground, which means that in any ground fault, only one-half the fault current would flow across the same fault. The intent of the rule is to maximize the available voltage to ground, which is already starting out at only one-half the normal amount, so overcurrent devices will trip open as quickly as possible in the event of a ground fault. For receptacles the limits are even tougher, with 1 percent on the branch circuit and 2 percent total from source to outlet if there is an intervening feeder.

647.5. Three-Phase Systems. To equalize the load on a three-phase distribution, these systems can be arranged with three transformers connected on their primary side to a premises conventional wye distribution with one primary transformer circuit (see Fig. 647-1) supplied by each of the three phases (Fig. 647-2). A maximum of two disconnects per winding are permitted, for a total of six. Note that with two disconnects per winding, the primary side of the transformers would be limited to 125 percent protection per 450.3(B). The more usual case would be three secondary disconnects, one per phase winding. Six-phase transformers operating 60 V to the star point are commercially available to support these wiring systems, with the secondary windings identified as in Fig. 647-2.

647.6. Grounding. These are separately derived systems, and as shown in the drawing, they must meet the requirements in 250.30. The neutral connection at the center tap of the transformer has no circuit load function; its sole function is to stabilize the voltage to ground and to provide an equipment grounding return path. It supplies, on the downstream side of the main bonding jumper for the system, equipment grounding terminal bars. They must be marked "Technical Equipment Ground."

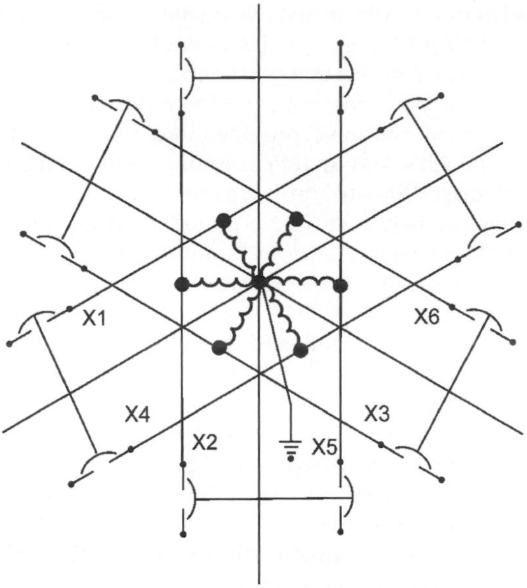
X1
X6
X4
X3
X2
X5

Fig. 647-2. Three systems derived from three transformers connected to different phase pairs result in three different technical power systems, each one displaced 120 electrical degrees from the next and capable of supplying two disconnects each. (Sec. 647.5.)

If there were a succession of panelboards and the system designer decided to isolate the return path from the enclosing raceways and panelboard enclosures for further noise reduction, this section allows the use of insulated grounding conductors and having the Technical Equipment Grounding busbars isolated from the enclosures. This is the same procedure allowed in 250.146(D) and 408.40. Note, however, that such raceways and enclosures must still be equipment grounding system. This can happen using any of the recognized equipment grounding conductors in 250.118.

Conversely, the last sentence allows for a raceway grounding return path without any additional grounding conductors, provided that the impedance of the grounding return path over the raceway does not exceed the impedance of a separate grounding conductor installed to meet these minimum requirements. This is why the equipment grounding conductors in Fig. 647-1 have the notation "(if required)." Remember, if the circuit conductors have been increased to meet the voltage-drop requirement, then 250.122 requires a corresponding increase in the size of grounding conductors on the same circuit. In this case the raceway impedance must be compared with the lower impedance of the larger equipment grounding conductor that would otherwise have been substituted. The informational notes that follow reinforce this concept.

647.7. Receptacles. Where receptacles are used to connect equipment, all of the following conditions need to be met. The reason for these requirements is to make as certain as possible that the receptacles are not used for cord-and-plug-connected loads designed for use on systems with only one ungrounded conductor. For example, if a floor lamp were connected, the screw shell would remain alive at 60 V to ground, even with the switch off, assuming a single-pole lampholder:

(1) All 15- and 20-A receptacles shall be GFCI protected. This item then adds an additional level of safety by requiring GFCI protection at or ahead of all receptacle outlets. Be careful if you are considering using a 2-pole GFCI for this purpose. Its electronic circuitry may not work on a system voltage of only 60 V to ground. The best choice is the so-called "master" GFCI units that are, in effect, feed through GFCI receptacles without the contact slots. They will work because they would see only their design voltage of 120 V. Remember, using an actual GFCI receptacle would violate the configuration rule in (4).
(2) All outlet strips, adapters, receptacle covers, and faceplates shall be marked: "WARNING—TECHNICAL POWER"; "Do not connect to lighting equipment"; "For electronic equipment use only"; "60/120 V 1Φac"; and "GFCI protected." These markings are self-explanatory.
(3) A conventional 125-V 15-A or 20-A receptacle must be located within 6 ft of all permanently-installed 15-A or 20-A 60/120-V technical power system receptacles. This rule relieves the temptation to plug the proverbial floor lamp into the wrong receptacle, because an appropriate receptacle must be nearby.
(4) All 125-V receptacles used for 60/120-V technical power must be uniquely configured and identified for use with this class of system. These configurations have yet to be developed. When this system first entered the **NEC** (1996 edition), an exception followed, allowing a conventionally configured receptacle in "machine rooms, control rooms, equipment rooms, equipment racks, and similar locations that are restricted to use by qualified personnel." That exception format has been restored, greatly clarifying how this rule should be applied. In the long term, it is to be hoped that a unique configuration is developed, allowing the main rule to take effect.

647.8. Lighting Equipment. Luminaires and associated equipment operated on these systems must have a disconnect that opens both circuit conductors. They must be permanently installed and listed for operation on these systems. No screw shell for such equipment is permitted to be exposed. Because fluorescent ballasts are a renowned source of electronic noise on wiring systems, placing them on one of these systems and thereby reducing that noise component to zero would benefit the overall system. However, the listing requirement may be excessive and unrealistic in terms of persuading conventional luminaire manufacturers to have their products reexamined.

ARTICLE 650. PIPE ORGANS

650.3. Other Articles. Some pipe organs now utilize sound equipment of the type covered in Art. 640, and those aspects are to be treated in accordance with that article.

650.4. Source of Energy. As now specified here, electronic organs must be supplied from a listed dc power supply with a maximum output of 30 VDC.

650.5. Grounding or Double Insulation of the DC Power Supply. The power supply must be either double insulated, or its metallic case must be bonded to the supply-side equipment grounding conductor.

650.6. Conductors. In part **(A)** 28 AWG wiring for electronic signal circuits and 26 AWG for electromagnetic valve supply are recognized minimums. These instruments operate with common return conductors that may see currents from multiple sources at the same time, resulting in a 14 AWG minimum in the return from the electromagnetic supply. In part **(C),** the wires of the cable are normally all of the same polarity; hence, they need not be heavily insulated from one another. The full voltage of the control system exists between the wires in the cable and the common return wire; therefore, the common wire must be reasonably well-insulated from the cable wires.

650.7. Installation of Conductors. A 30-V system involves very little fire hazard, and the cable may be run in any manner desired; but for protection against injury and convenience in making repairs, the cable should preferably be installed in a metal raceway. Control equipment and bus work involving the common return conductor may be secured directly to the organ structure without additional insulation, and splices need not be made in enclosures. Unterminated and abandoned cables must be identified as such, but need not be removed. The panel determined that in many cases it was not practical to do so in this context.

650.8. Overcurrent Protection. Circuits must be arranged so the 26 and 28 AWG wiring is protected at not over 6 A, with other size wiring protected at its ampacity. The common return wires do not require protection.

650.9. Protection from Accidental Contact. The sounding apparatus, which has exposed wiring, must not be exposed to unqualified contact. However, part of what can prevent unqualified access in such cases is not a sound-obstructing enclosure, but other parts of the organ, such as the pipework or other decorative millwork.

ARTICLE 660. X-RAY EQUIPMENT

660.1. Scope. This covers industrial or other nonmedical applications exclusively; health care provisions are covered in Art. 517, Part V. An x-ray tube of the hot-cathode type, as now commonly used, is a two-element vacuum tube in which a tungsten filament serves as the cathode. Current is supplied to the filament at low voltage. In most cases, unidirectional pulsating voltage is applied between the cathode and the anode. The applied voltage is measured or described in terms of the peak voltage, which may be anywhere within the range of 10,000 to 1 million V, or even more. The current flowing in the high-voltage circuit may be as low as 5 mA or may be as much as 1 A, depending on the desired intensity of radiation. The high voltage is obtained by means of a transformer, usually operating at 230-V primary, and usually is made unidirectional by means of two-element rectifying vacuum tubes, though in some cases an alternating current is applied to the x-ray tube. The x-rays are radiations of an extremely high frequency (or short wavelength), which are the strongest in a plane at right angles to the electron stream passing between the cathode and the anode in the tube.

660.4. Connection to Supply Circuit. Fixed and stationary equipment must be connected through a Chap. 3 wiring method, with an allowance for 30 A and lesser ratings to use hard-service cord and a plug. Portable and relocatable equipment can use cord-and-plug connections up to a 60 A rating.

660.5. Disconnecting Means. The minimum rating is 50 percent of the short-time current rating or 100 percent of the long-time rating, whichever is higher, with an allowance for cord-and-plug connections at 30 A or less and on a 120 V branch circuit. It must be within sight of the control, and readily accessible. An exception waives the requirement entirely with qualified maintenance and supervision, or on a finding of additional hazard or impracticability.

660.6. Rating of Supply Conductors and Overcurrent Protection. Branch-circuit wiring is sized the same as the disconnecting means. Feeders supplying multiple units are sized on the basis of 100 percent of the largest two units (determined by making the comparison required in the branch-circuit analysis), plus 20 percent of all others.

660.20. Fixed and Stationary Equipment. In radiography it is important that the exposure be accurately timed, and for this purpose a switch is used that can be set to open the circuit automatically in any desired time after the circuit has been closed.

660.24. Independent Control. If multiple x-ray units or other equipment are operated from a single circuit operating over 600 V, each such piece of equipment must have its own switch or equivalent. This device shall be arranged and installed such that its live parts will not be contacted by persons using the equipment or others.

660.35. General. A power transformer supplying electrical systems is usually supplied at a high primary voltage; hence, in case of a breakdown of the insulation on the primary winding, a large amount of energy can be delivered to the transformer. X-ray units are supplied at lesser voltages and can do without the protections normally required for transformers in Art. 450.

660.47. General. This section definitely requires that all new x-ray equipment be so constructed that all high-voltage parts, except leads to the x-ray tube, are in grounded metal enclosures. Conductors leading to the x-ray tube are shielded and heavily insulated.

ARTICLE 665. INDUCTION AND DIELECTRIC HEATING EQUIPMENT

665.1. Scope. Induction and dielectric heating are systems, wherein a workpiece is heated by means of a rapidly alternating magnetic or electric field.

665.2. Definitions

Dielectric heating In contrast, dielectric heating is used to heat materials that are nonconductors, such as wood, plastic, textiles, and rubber, for such purposes as drying, gluing, curing, and baking. It uses frequencies from 1 to 200 MHz, especially those from 1 to 50 MHz. Vacuum-tube generators are used exclusively to supply dielectric heating power, with outputs ranging from a few hundred watts to several hundred kilowatts.

Whereas induction heating uses a varying magnetic field, dielectric heating employs a varying electric field. This is done by placing the material to be heated between a pair of metal plates, called *electrodes,* in the output circuit of the generator. When high-frequency voltage is applied to the electrodes, a rapidly alternating electric field is set up between them, passing through the material to be heated. Because of the electrical charges within the molecules of this material, the field causes the molecules to vibrate in proportion to its frequency. This internal molecular action generates the heat used for dielectric heating.

Induction heating Induction heating is used to heat materials that are good electrical conductors, for such purposes as soldering, brazing, hardening, and annealing. Induction heating, in general, involves frequencies ranging from 3 to about 500 kHz, and power outputs from a few hundred watts to several thousand kilowatts. In general, motor-generator sets are used for frequencies up to about 30 kHz; spark-gap converters, from 20 to 400 kHz; and vacuum-tube generators, from 100 to 500 kHz. Isolated special jobs may use frequencies as high as 60 to 80 MHz. Motor-generator sets normally supply power for heating large masses for melting, forging, deep hardening, and the joining of heavy pieces, whereas spark-gap and vacuum-tube generators find their best applications in the joining of smaller pieces and shallow case hardening, with vacuum-tube generators also being used where special high heat concentrations are required.

To heat a workpiece by induction heating, it is placed in a work coil consisting of one or more turns, which is the output circuit of the generator (Fig. 665-1).

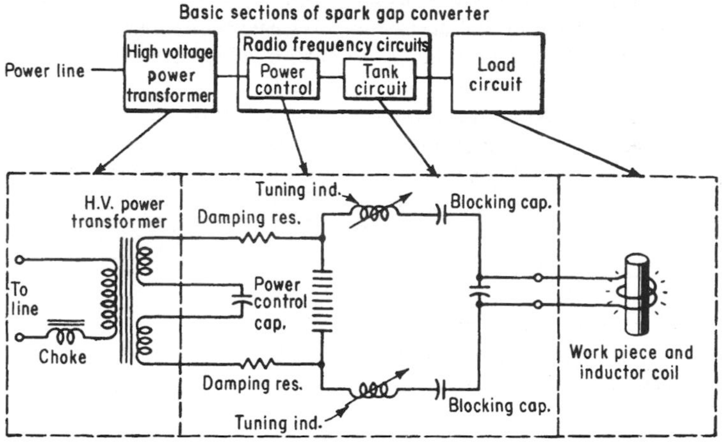
Basic sections of spark gap converter
Power line
High voltage power transformer
Radio frequency circuits
Power control
Tank circuit
Load circuit
H.V. power transformer
To line
Choke
Tuning ind.
Damping res.
Power control cap.
Damping res.
Tuning ind.
Blocking cap.
Blocking cap.
Work piece and inductor coil

Fig. 665-1. A *generator* circuit supplies the *work coil* of an induction heater. (Sec. 665.2.)

The high-frequency current that flows through this coil sets up a rapidly alternating magnetic field within it. By inducing a voltage in the workpiece, this field causes a current flow in the piece to be heated. As the current flows through the resistance of the workpiece, it generates heat (I^2R loss) in the piece itself. It is this heat that is utilized in induction heating.

665.5. Output Circuit. The induction circuit can be isolated from ground, as is often done with induction furnaces that employ water cooling. This mitigates the energy available in a ground fault until the furnace can be shut down. Otherwise, the ground fault could compromise the water jacket. Water hitting molten metal will result in a steam explosion ejecting molten metal from the furnace and endangering anyone nearby. If not isolated, the system must be arranged so there is no shock hazard of over 50 V to ground on any accessible surface, even if the coil and heated object touch.

665.7. Remote Control. In part (A), if interlocking were not provided, there would be a definite danger to an operator at the remote-control station. It might then be possible, if the operator had turned off the power and was doing some work in contact with a work coil, for someone else to apply power from another point, seriously injuring the operator.

665.10. Ampacity of Supply Conductors. Quite often where several pieces of equipment are operated in a single plant, it is possible to conserve on power-line requirements by taking into account the load or use factor of each piece of equipment. The time cycles of operation on various machines may be staggered to allow a minimum of current to be taken from the line. In such cases the Code requires sufficient capacity to carry all full-load currents from those machines that will operate simultaneously, plus the standby requirements of all other units.

665.12. Disconnecting Means. An in-sight or lock-open disconnect must be provided for each heating equipment, sized to the equipment nameplate.

665.22. Access to Internal Equipment. This section allows the manufacturer the option of using interlocked doors or detachable panels. Where panels are used and are not intended as normal access points, they should be fastened with bolts or screws of sufficient number to discourage removal. They should not be held in place with any type of speed fastener.

665.25. Dielectric Heating Applicator Shielding. This section is intended primarily to apply to dielectric heating installations where it is absolutely essential that the electrodes and associated tuning or matching devices are properly shielded.

RF Lines

When it is necessary to transmit the high-frequency output of a generator any distance to the work applicator, a radio-frequency line is generally used. This usually consists of a conductor totally enclosed in a grounded metal housing. This central conductor is commonly supported by insulators, mounted in the grounded housing, and periodically spaced along its length. Such a line, rectangular in cross section, may even be used to connect two induction generators to the load.

While contact with high-voltage radio frequencies may cause severe burns, contact with high-voltage dc could be fatal. Therefore, it is imperative that generator output (directly, capacitively, or inductively coupled) be effectively grounded with respect to dc so that, should generator failure place high-voltage dc in the tank oscillating circuit, there will still be no danger to the operator. This grounding is generally internal in vacuum-tube generators. In all types of induction generators, one side of the work coil should usually be externally grounded.

In general, all high-voltage connections to the primary of a current transformer should be enclosed. The primary concern is the operator's safety. Examples would be interlocked cages around small dielectric electrodes and interlocking safety doors.

On induction heating jobs, it is not always practical to completely house the work coil and obtain efficient production operation. In these cases, precautions should be taken to minimize the chance of operator contact with the coil.

665.26. Grounding and Bonding

Bonding

At radio frequencies, and especially at dielectric-heating frequencies (1 to 200 MHz), it is very possible for differences in radio-frequency potential to exist between the equipment proper and other surrounding metal objects or other units of the complete installation. These potentials exist because of stray currents flowing between units of the equipment or to ground. Bonding is therefore essential, and such bonding must take the form of very wide copper or aluminum straps between units and to other surrounding metal objects such as conveyors and presses. The most satisfactory bond is provided by placing all units of the equipment on a flooring or base consisting of copper or aluminum sheet, thoroughly joined where necessary by soldering, welding, or adequate bolting. Such bonding reduces the radio-frequency resistance and reactance between units to a minimum, and any stray circulating currents flowing through this bonding will not cause sufficient voltage drop to become dangerous.

Shielding

Shielding at dielectric-heating frequencies is a necessity to provide operator protection from the high radio-frequency potentials involved, and also to prevent possible interference with radio communication systems. Shielding is accomplished by totally enclosing all work circuit components with copper sheet, copper screening, or aluminum sheet.

ARTICLE 668. ELECTROLYTIC CELLS

668.1. Scope. This article provides effective coverage of basic electrical safety in electrolytic cell rooms.

The presentation of these requirements was accompanied by a commentary from the technical subcommittee that developed them. Significant background information from that commentary is as follows:

> In the operation and maintenance of electrolytic cell lines, however, workmen may be involved in situations requiring safeguards not provided by existing articles of the NEC. For example, it is sometimes found that in the matter of exposed conductors or surfaces it is the man or his workplace which has to be insulated rather than the conductor. Work practices and rules such as are included in IEEE Trial Use Standard pertinent to such specific situations have been developed which offer the same degree of safety provided by the traditional philosophy of the NEC.

As a corollary to this concept, overheating of conductors, overloading of motors, leakage currents and the like may be required in cell lines to maintain process safety and continuity.

Proposed Art. 668 introduces such concepts as these as have been proven in practice for electrolytic cell operation.

668.2. Definitions. The subcommittee noted:

> An electrolytic cell line and its dc process power supply circuit, both within a cell line working zone, comprise a single functional unit and as such can be treated in an analogous fashion to any other individual machine supplied from a single source. Although such an installation may cover acres of floor space, may have a load current in excess of 400,000 amperes dc or a circuit voltage in excess of 1000 volts dc, it is operated as a single unit. At this point, the traditional NEC concepts of branch circuits, feeders, services, overload, grounding, disconnecting means are meaningless, even as such terms lose their identity on the load side of a large motor terminal fitting or on the load side of the terminals of a commercial refrigerator.

It is important to understand that the cell line process current passes through each cell in a series connection and that the load current in each cell is not capable of being subdivided in the same fashion as is required, for example, in the heating circuit of a resistance-type electric furnace by Sec. 424.72(A).

668.3. Other Articles. Electrical equipment and applications that are not within the space envelope of the "cell line working zone," as dimensioned in 668.10, must comply with all the other regulations of the NEC covering such work.

668.11. Direct-Current Cell Line Process Power Supply. These conductors are not required to be grounded. If that power supply is operating over 50 V between conductors, then the metal enclosures must be grounded either through protective relaying or by using a copper bonding conductor not less than 2/0 AWG in size.

668.12. Cell Line Conductors. These can be of any suitable conductive material, joined by bolting, welding, clamping, or compression, and of such size that under maximum load and ambient temperature their insulating supports will not be unsafely heated.

668.13. Disconnecting Means. As shown in Fig. 668-1, each dc power supply to a single cell line must be capable of being disconnected. And the disconnecting means may be a removable link in the busbars of the cell line.

668.14. Shunting Means. Similar means as for disconnecting can be used to make a shunted current path around one or more cells.

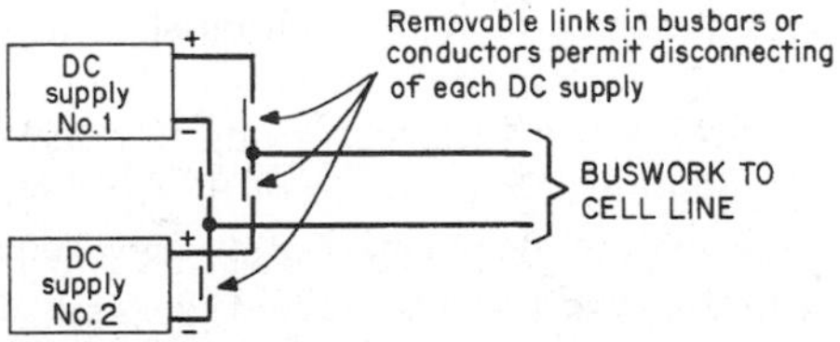
Removable links in busbars or conductors permit disconnecting of each DC supply
DC supply No.1
DC supply No.2
BUSWORK TO CELL LINE
+
−
+
−

Fig. 668-1. Removal of busbar sections may provide disconnect of each supply. (Sec. 668.13.)

668.15. Grounding. The requirements of Art. 250 apply to any equipment or structure or other apparatus required to be grounded; however, an otherwise qualified water pipe electrode is not required to be included in the grounding electrode system.

668.20. Portable Electrical Equipment. This section and the rules of 668.21, 668.30, 668.31, and 668.32 cover installation and operating requirements for cells with exposed live conductors or surfaces. These rules are necessary for the conditions as noted by the subcommittee:

In some electrolytic cell systems, the terminal voltage of the cell line process power supply can be appreciable. The voltage to ground of exposed live parts from one end of a cell line to the other is variable between the limits of the terminal voltage. Hence, operating and maintenance personnel and their tools are required to be insulated from ground. If the cell-line voltage does not exceed 200 V dc, grounded enclosures are permitted but not required, and such enclosures are permitted to be grounded where guarded. To this end, receptacle circuits that supply this equipment as used within the cell-line working zone must only use ungrounded conductors in their supply created by the use of isolating transformers with ungrounded secondaries. Such tools and equipment must be marked to this effect, and the plugs and receptacles must be so configured that they will not be connectable to conventional receptacles, nor inadvertently interchanged with other equipment designed for conventional connections.

668.21. Power-Supply Circuits and Receptacles for Portable Electrical Equipment. Part **(A)** reiterates the requirement to use isolating power supplies for the ungrounded receptacles in the cell line working zone, and adds the requirement that the transformer primary cannot operate over 600 V with appropriate overcurrent protection. The ungrounded secondary conductors must also have overcurrent protection, and not be operating over 300 V.

668.30. Fixed and Portable Electrical Equipment. AC systems in the cell line working zone, and all exposed conductive surfaces, are not required to be grounded. The wiring to such equipment must be by flexible cord or nonmetallic raceway or cable assemblies. If metal raceways of cables are used, insulating breaks must be installed to avoid a hazardous condition. Fixed electrical equipment operating in the cell line zone is permitted to be bonded to the energized conductive surfaces of the cell line, and required to be so if it is attached to such a surface. Control and instrumentation circuits operating within the cell line working zone do not require overcurrent protection.

668.32. Cranes and Hoists. Conductive surfaces that enter the cell line working zone are not required to be grounded, and any part that contacts an energized cell or equipment attached to such a cell must be insulated from ground. Remote crane or hoist controls must employ (1) isolated circuits; or (2) pendants with nonconductive supports and surfaces, or ungrounded surfaces; or (3) radio controls; or (4) a rope operator. Note that as of the 2014 NEC, 610.31 Exception provides a conditional waiver to the normal within view requirement for runway disconnecting means for this equipment.

668.40. Enclosures. General-purpose enclosures are permitted where natural drafts prevent accumulations of gas from the process.

ARTICLE 669. ELECTROPLATING

669.1. Scope. This article covers all electroplating activity including anodizing. These operations often involve very high current at relatively low voltage, to the point that conventional conductors are not practicable.
669.5. Branch-Circuit Conductors. This wiring must be rated for 125 percent of the connected load; busbar ampacity is that of busbars in auxiliary gutters in 366.23.
669.6. Wiring Methods. For systems not over 60 V dc, open wires are permitted without insulated support if they are protected from damage, and bare conductors are permitted where supported on insulators. For systems running at a higher voltage, open insulated conductors can be used where run on insulated supports and protected from damage. Bare conductors are permitted where run on insulated supports and as guarded against accidental contact up to the point of termination using methods in 110.27.
669.7. Warning Signs. Warning signs must be posted indicating the presence of bare conductors.
669.8. Disconnecting Means. A disconnecting means must be provided from each dc power supply, which can consist of removable links or conductors.
669.9. Overcurrent Protection. DC conductors must have overcurrent protection in the form of fuses or circuit breakers, or a current sensing means that operates a disconnecting means, or some other approved means.

ARTICLE 670. INDUSTRIAL MACHINERY

670.2. Definitions

Industrial machinery (machine) This definition is sufficiently broadly written that it should not be necessary to routinely revisit the definition as technology changes. The provisions do not apply to any machine or tool that is not normally used in a fixed location and can be carried from place to place by hand. The concept in this article is that the rules in this article and Art. 430 apply to the point of connection to the main terminals of the machine. On the load side of those terminals, including control circuit protocols and workspace rules, NFPA 79, *Electrical Standard for Electrical Machinery* takes over. This principle applies to machinery prepared as a unit by its manufacturer, but then disassembled for shipment and reassembled on-site. However, normal NEC provisions apply when various components are provided from different vendors on-site and then field assembled into a process. For example, motor controllers applied to an extensive conveyor system may or may not come directly under NEC provisions depending on whether they were furnished by the OEM as part of its engineered system.
670.3. Machine Nameplate Data. This information has been correlated with 409.110 with respect to the short-circuit current rating requirements. If overcurrent protection is provided as covered in 670.4(B), the machine must be marked accordingly.
670.4. Supply Conductors and Overcurrent Protection. Part **(A)** requires the supply wiring to have an ampacity not less than 125 percent of resistance loads,

plus 125 percent of the largest motor plus 100 percent of all other motors and apparatus, based on their duty cycle and likelihood of simultaneous operation. Part **(B)** defines the machine as a single entity for the purposes of disconnecting rules, and therefore must be provided with a disconnecting means. Although not stated here, NFPA 79 is correlated with 404.8(A) and uses the same 2.0 m (6 ft 7 in.) height limit, measured in the same way (the center of the grip in its uppermost position), and arranged to be lockable in the open position. When so locked, the machine cannot be energized by any local or remote action. Part **(C)** reiterates the rules of 430.62 for motor feeder overcurrent protection limits, because that is, after all, the function that such an overcurrent device on one of these units performs.

670.5. Short-Circuit Current Rating. If the available fault current (see 110.9) exceeds the short-circuit current rating on the equipment nameplate [see 670.3(A)(4)], the equipment is not to be installed and connected. This equipment requires this current to be part of a field marking, which is to be made after an analysis of local overcurrent protection and other factors, and the date of the determination must be included on the label.

670.6. Surge Protection. Safety interlock circuits must be provided with surge protection.

ARTICLE 675. ELECTRICALLY DRIVEN OR CONTROLLED IRRIGATION MACHINES

675.1. Scope. This article covers electrically driven irrigation machines that are not portable by hand, and used primarily for agricultural purposes. It does not cover water pumps that bring water to the machines.

675.3. Irrigation Cable. This is the cable used to interconnect enclosures on an irrigation machine. It has a number of construction specifications predicated on its expected use in a wet location and where subject to flexing, and requiring metallic armoring in its inner construction. It must be supported at least every 1.2 m (4 ft). The cable is permitted to be a composite with control and grounding conductors included, and must be terminated in fittings designed for the cable.

675.7. Equivalent Current Ratings. For continuous duty the normal rules in Art. 430 apply. Where the duty cycle is inherently intermittent, part (**A**) determines a continuous current rating by taking 125 percent of the largest motor (using the nameplate rating), adding the sum of all other motor nameplate ratings, and then multiplying the final summation by the maximum percent duty cycle for which they can operate. Part (**B**) determines an equivalent locked-rotor current by adding the locked-rotor currents of the two largest motors to the full-load current ratings of the remaining motors.

675.8. Disconnecting Means. Part **(A)** requires the main controller governing the complete machine to at least match the equivalent continuous current rating determined as above, or per 675.22(A) for center pivot machines, and a horsepower rating taken out of the locked-rotor tables at the end of Art. 430 based on the equivalent locked rotor current determined above or by 675.22(B) for center pivot machines. If a listed molded case switch is used for the disconnecting means, it must have the ampere rating as described but it need not have a horsepower rating.

Part **(B)** requires a main disconnect, which must include overcurrent protection, and can be at the main power connection to the machine or not over 15 m (50 ft) away, provided it is within view and readily accessible with a permanent lock-open feature. Its ratings must not be less than as required for the main controller, although it is recognized that circuit breakers do not carry horsepower ratings.

Part **(C)** requires individual disconnects for each motor and controller, as covered in Part IX of Art. 430. These disconnects need not be readily accessible.

675.9. Branch-Circuit Conductors. The branch-circuit conductor minimum ampacity follows the continuous current ratings determined by 675.7(A) or 675.22(A).

675.10. Several Motors on One Branch Circuit. This section varies the normal limits in 430.53, and allows several motors, each not over 2 hp and protected not over 30 A at not over 600 V, to operate off of taps made up with copper wire not smaller than 14 AWG and not longer than 7.5 m (25 ft). Each motor so connected must have normal running overload protection and its full-load current must be limited to 6 A. If these conditions apply, individual short-circuit and ground-fault protective devices need not be installed.

675.11. Collector Rings. Collector rings must carry their motor loads, as determined in one of three ways: 1) the result from 675.7(A); or 2) the result from 675.22(A); or 3) 125 percent of the largest controller (NEC says "device"; a motor is not a device but a controller is) plus the full load currents of all other "devices" served. The grounding ring must be fully sized to the same rating. Control and signaling rings must carry their load taken as 125 percent of the largest device plus 100 percent of all other devices. The rings must be in a suitable enclosure so they will withstand the environment they will operate within.

675.12. Grounding. All electrical equipment and enclosures on the machine must be grounded, although a machine that is electrically controlled but not electrically driven is exempt from grounding if the voltage is 30 V or less and the control circuits are power limited in accordance with Chap. 9, Tables 11(A) and 11(B). This last provision means that the control circuits operate within Class 2 or Class 3 parameters, but need not meet the source requirements for these circuits in 725.121(A) that would actually qualify them for this designation.

675.13. Methods of Grounding. This rule requires on-machine grounding conductors to be fully sized to the ungrounded conductors, and not as permitted in Table 250.122. However, feeder circuits are permitted to use the normal sizing rules.

675.15. Lightning Protection. Irrigation machines with a stationary point must be connected to a grounding electrode system as covered in Part III of Art. 250. Note that this connection alone will not meet the minimum requirements of NFPA 780, *Standard for the Installation of Lightning Protection Systems*, but it is the NEC minimum.

675.22. Equivalent Current Ratings. This section covers the calculations for center pivot machines that differ from other types of irrigation machinery. Part **(A)** determines a continuous current rating by taking 125 percent of the largest motor (using the nameplate rating) and adding 60 percent of the sum of all other motor nameplate ratings. Part **(B)** determines an equivalent locked-rotor current by adding double the locked-rotor currents of the largest motor to 80 percent of the full-load current ratings of the remaining motors.

ARTICLE 680. SWIMMING POOLS, FOUNTAINS, AND SIMILAR INSTALLATIONS

680.1. Scope. Electrification of swimming, wading, therapeutic, and decorative pools, along with fountains, hot tubs, spas, and hydromassage bathtubs, has been the subject of extensive design and Code development over recent years. Details on circuit design and equipment layout are covered in NEC Art. 680. Careful reference to this article should be made in connection with any design work on pools, fountains, and so forth. As of the 2017 NEC, all equipment and products either in the water or in the structure or decks of pools and other contained bodies of water covered by Art. 680 must be listed (680.4).

Research work conducted by Underwriters Laboratories and others indicated that an electric shock could be received in two different ways. One of these involved the existence in the water of an electrical potential with respect to ground, and the other involved the existence of a potential gradient in the water itself.

A person standing in the pool and touching the energized enclosure of faulty equipment located at poolside would be subject to a severe electrical shock because of the good ground which his or her body would establish through the water and pool to earth. Accordingly, the provisions of this article specify construction and installation that can minimize hazards in and adjacent to pools and fountains.

The potential gradient in the water presents primarily a drowning hazard and not an electrocution hazard. It was determined by actual human volunteers immersed in the water, particularly with any water in their ears, that a gradient of as little as 4 V was enough to cause disorientation to the point that the individual may not be able to leave the water. This is why the bonding requirements for a hydromassage bathtub are less severe than for a spa or hot tub; the bathtub does not present the drowning hazard that a spa or hot tub, or larger pool presents.

Very important: Therapeutic pools in a health care facility are not exempt from this article. Therapeutic pools in hospitals are subject to all applicable rules in Art. 680, Part VI.

As noted at the end of the first sentence, the rules here also govern the installation of "metallic auxiliary equipment, such as pumps, filters, and similar equipment." That wording has the effect of requiring that any such "metallic auxiliary equipment" satisfy the requirements of Art. 680. Where a particular installation detail addresses say, a circulating pump, the pump must be installed as described. This is very clearly stated by the first sentence.

One last note applies to the phrase *body of water*. Where a rule refers to a "body of water," the rule applies to all types of pools, and the like, that are covered by Art. 680. Where individual types of pools, and the like, are identified, then the rule given there applies only to the specified types of pools, and the like.

680.2. Definitions. These definitions are important to correct, effective application of Code rules of Art. 680. Figure 680-1 shows a typical dry-niche swimming pool luminaire, which are now permitted in the pool floor as well as the wall. Figure 680-2 shows a forming shell for a wet-niche luminaire.

The definition of *cord-and-plug-connected lighting assembly* covers a luminaire of all-plastic construction for use in the wall of a spa, hot tub, or storable pool. This type of luminaire operates from a cord-and-plug-connected transformer, and it does not require a metal niche around the luminaire.

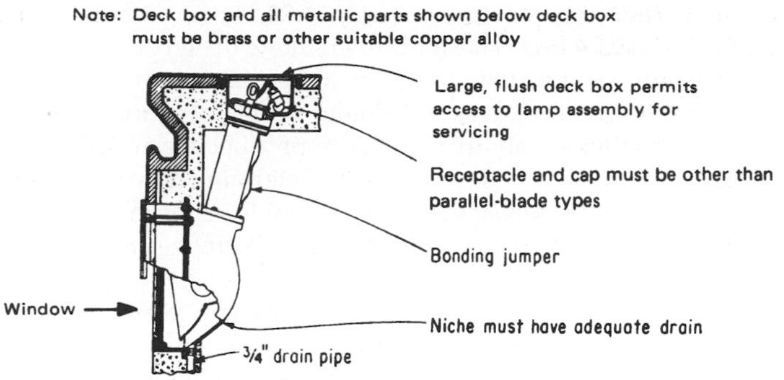
Note: Deck box and all metallic parts shown below deck box must be brass or other suitable copper alloy
Large, flush deck box permits access to lamp assembly for servicing
Receptacle and cap must be other than parallel-blade types
Bonding jumper
Window
Niche must have adequate drain
3/4" drain pipe

Fig. 680-1. Dry-niche luminaire lights underwater area through glass "window." (Sec. 680.2.)

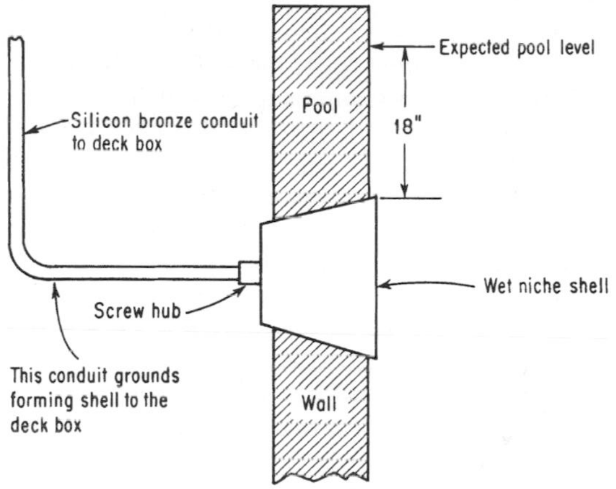
Expected pool level
Pool
18"
Silicon bronze conduit to deck box
Wet niche shell
Screw hub
This conduit grounds forming shell to the deck box
Wall

Fig. 680-2. Forming shell is a support for the lamp assembly of a wet-niche luminaire. (Sec. 680.2.)

A *hydromassage bathtub* is a "whirlpool" or "Jacuzzi" bath for an individual bather, which is smaller than a hydromassage pool (spa or hot tub), but is covered in 680.70 through 680.74. It is designed to discharge its water after use.

The definition of *maximum water level* is, essentially, the deck level, because that is the level where the water can "spill out." It is not, as generally previously supposed, the maximum fill level of the skimmer trough, because the water does not actually spill out at that point. This definition has serious repercussions on the placement of swimming-pool junction boxes, as covered later.

The *low-voltage contact limit* is the upper limit of voltage that is considered safe for human exposure under wet-contact conditions, and therefore generally applicable to a pool environment. It is taken from the wet-contact limits for power-limited circuits as reflected in Notes 2 and 4 of Tables 11(A) and 11(B) in Chap. 9. These values have a very long pedigree, extending back to the 1975 NEC. Under current usage, Tables 11(A) and 11(B) are not for field applications; they remain in the NEC only to set policy parameters for qualified testing laboratories to apply in making listing evaluations. These numbers have been brought forward into Art. 680 in order for some installation rules to use as a point of departure for a change in installation protocol, but no installer should attempt actual field measurements to interpret the suitability of a certain product design. Look for a listing, and then review the installation instructions to determine how to proceed.

The definition of *no-niche luminaire* covers a luminaire for installation above or below the water without any niche. This definition correlates to 680.23(D), which provides installation criteria.

Because there are differences in the requirements for "permanently installed" pools and "storable" pools, there has been some confusion in the past as to just what a "storable" pool is. A *storable swimming or wading pool* must not hold a greater depth than 1.0 m (3½) ft of water, or have plastic or inflatable walls of any dimension, and be "constructed on or above the ground." The plastic/inflatable wall part of this definition identifies specific types of pool construction that are considered to be storable regardless of water depth. The storable definition has been broadened in the 2014 NEC to clarify that it also applies to storable/portable spas and hot tubs.

This definition, as well as two others (*pool* and *permanently installed swimming, wading, immersion, and therapeutic pools*), were changed to add the word "immersion" into the definition (pool), or into the title of the definition. The new title for this definition is "*storable swimming, wading, immersion, and therapeutic pools*." The changes were made to clarify that a baptistery, whether permanently installed and maintained, or as part of a storable assembly, was covered by the applicable parts of Art. 680.

Figure 680-3 shows a "wet-niche luminaire."

Fig. 680-3. Wet-niche luminaire consists of forming shell set in pool wall with cord-connected lamp-and-lens assembly that attaches to the forming shell, with cord coiled within the shell housing. (Sec. 680.2.)

680.5. Ground-Fault Circuit Interrupters. This section describes ground-fault circuit interrupters (GFCIs) that are required to be used by other rules of this article. Additional protection may be accomplished, even where not required, by the use of a GFCI. Since the GFCI operates on the principle of line-to-ground leaks or breakdowns, it senses, at low levels of magnitude and duration, any fault currents to ground caused by accidental contact with energized parts of electrical equipment. Because the ground-fault interrupter operates at a fraction of the current required to trip a 15-A CB, its presence is mandatory under certain Code rules and is generally very desirable. As indicated by the wording here, either receptacle, CB, or "other listed type" of GFCIs may be used to satisfy the rules of Art. 680 where a GFCI is required.

680.7. Grounding and Bonding Terminals. These conductor termination points obviously play a crucial role in how the electrical systems will perform over time in a wet and corrosive environment. UL 467, the grounding and bonding product standard, recognizes copper, brass, bronze, and stainless steel as generally suitable for direct burial exposures, and this rule makes it clear that a listing accordingly is required.

680.8. Cord-and-Plug-Connected Equipment. The 3-ft (900-mm) cord limitation mentioned in this rule would not apply to swimming pool filter pumps used with storable pools under Part **III** of Art. 680, because these pumps are considered as portable instead of *fixed or stationary*. See the comments following 680.30. The UL Guide Card information [see discussion at 110.3(B) for details] echoes this requirement for motors that will be connected by a cord and plug. Of course, the manufacturer may also omit the attached cord and simply provide a hub or knockout to accommodate a permanent wiring connection, and the decision to add a cord may also be made in the field providing all code rules are followed including the provisions in 110.14 for connections involving fine-stranded conductors. Note that the plugs may be conventionally configured regardless of location now that the locking requirement has been withdrawn from 680.22(A). The equipment grounding conductor in the cord, which as a practical matter usually drives the overall cord size, must not be smaller than 12 AWG, or larger if required by 250.122.

680.9. Overhead Conductor Clearances. The general rule states that service drops and open overhead wiring must not be installed above a swimming pool or surrounding area extending 10 ft (3.0 m) horizontally from the pool edge, or diving structure, observation stands, towers, or platforms. Note that although these rules could limit the placement of a pool, the **NEC** has no jurisdiction over how utility supply lines might be subsequently installed.

Item C in Code Table 680.9 and the diagram clarify the horizontal dimensions around the pool to which the clearances of the table apply for overhead wiring over a pool area (Fig. 680-4, top). The dimension C, measured horizontally around a pool and its diving structure, establishes the area above which such wiring is permitted, provided the clearance dimensions of A or B in the table are observed. The dimensions A and B do *not* extend to the ground as radii, and the dimension C is the sole ruling factor on the "horizontal limit" of the area above which the clearances of A and B apply.

As the basic rule is worded—and the table and diagram specify—the clearances of A and B must be observed for utility lines (and overhead service

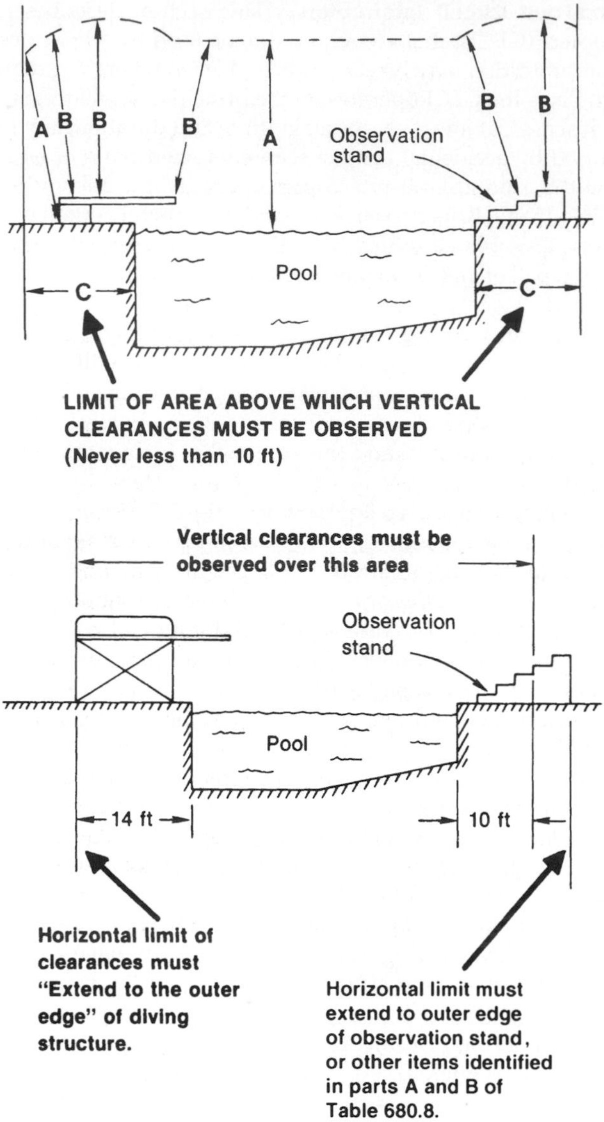
A
B
B
B
A
Observation stand
B
B
Pool
C
C
LIMIT OF AREA ABOVE WHICH VERTICAL CLEARANCES MUST BE OBSERVED (Never less than 10 ft)
Vertical clearances must be observed over this area
Observation stand
Pool
14 ft
10 ft
Horizontal limit of clearances must "Extend to the outer edge" of diving structure.
Horizontal limit must extend to outer edge of observation stand, or other items identified in parts A and B of Table 680.8.

Fig. 680-4. Clearances from Table 680.9 apply as indicated in these diagrams. (Sec. 680.9.)

conductors) above the water and above that area at least 10 ft (3.0 m) back from the edge of the pool, all around the pool. But the horizontal distance would have to be greater if any part of the diving structure extended back farther than 10 ft (3.0 m) from the pool's edge. If, say, the diving structure extended back 14.5 ft (4.4 m) from the edge, then the overhead line clearances of A and B would be required above the area that extends 14.5 ft (4.4 m) back from the pool edge, not just 10 ft (3.0 m) back (Fig. 680-4, bottom).

Part (C) of the table says that the horizontal limit of the area over which the required vertical clearances apply extends to the "outer edge of the structures listed in A and B." That wording clearly includes observation stands in the need to extend the horizontal limit over 10 ft (3.0 m), as shown at the bottom of Fig. 680-4.

The next paragraph in this section (B) provides guidance on use of telephone company overhead lines and community antenna system cables above swimming pools. Although the first sentence of 680.9 generally prohibits "service-drop or other open overhead wiring" above pools, it was never the intent that the rules of this section apply to telephone lines. The general concept of this wording is to specifically permit such lines above pools *provided that* such conductors and their supporting messengers have a clearance of *not less* than 10 ft (3.0 m) above the pool and above diving structures and observation stands, towers, or platforms. However, network-powered broadband communications drops [covered in (C)] can operate at significantly higher voltages and must meet the normal clearances for the 0-750 V column in Table 680.9.

680.10. Electric Pool Water Heaters. A swimming pool heater requires branch-circuit conductor ampacity and rating of the CB or fuses at least equal to 125 percent of the nameplate load current. An electrically powered swimming pool heater is considered to be a continuous load and is therefore made subject to the same requirements given in 422.13 for hot water tanks. In addition, large units must meet the usual load subdivision rules, as for example in 422.11(F)(1).

680.11. Underground Wiring Location. This section is aimed at eliminating the hazard that underground wiring can present under fault conditions that create high potential fields in the earth and in the deck adjacent to a pool. Aside from the electric circuits associated with pool equipment, underground wiring must not be run under the pool. The 5-ft zone around the pool reserved for special coverage in this section was removed in the 2017 **NEC** cycle in favor of a simple prohibition against running extraneous circuits under the pool. The depth table was also withdrawn in favor of a simple reference to Table 300.5. The complete raceway system rule was also discontinued. Instead, there is a list of permitted underground wiring methods consisting of RMC, IMC, PVC, and RTRC conduits, and Type MC cable suitable for direct burial at that location. There is no horizontal dimension provided to indicate the reach of the rule at this point. One reasonable interpretation would be to apply the permitted raceway list to the horizontal outer limit of pool decks and other comparable structures within the scope of Art. 680.

680.12. Equipment Rooms and Pits. If these areas house electrical equipment, they must not be subject to water accumulations, either from weather, or pool spillage, or filter maintenance, etc. A new (2017) informational note points to the effects of chlorinating chemicals if stored near electrical equipment, and a published standard covering ventilation of such areas and the effectiveness of ventilation in terms of reduction of corrosive vapors.

680.13. Maintenance Disconnecting Means. Here the Code mandates that a disconnect—one or more, as needed—must be provided to disconnect all ungrounded conductors for all pool-associated equipment other than lighting. This disconnecting means can be a unit switch where the switch disconnects all ungrounded conductors, or it can be a CB in a control panel, or other Code-recognized disconnecting means. If no such disconnect is available, then additional disconnecting means are required to be installed. The required maintenance disconnect must be "within sight" of the utilization equipment, which means not more than 50 ft (15 m) from the equipment, visible, and not guarded by a door that prevents direct access. It must be at least 1.5 m (5 ft) from the pool measured horizontally from the water's edge, although it can be nearer if behind a permanent barrier that provides a reach path of at least that extent. Because this rule is in Part I of the article, it also applies to spas and hot tubs, as well as fountains and therapeutic tubs.

680.14. Corrosive Environment. As of the 2014 NEC, this new section describes the likely corrosive agents in swimming pool/hot tub/fountain/etc. locations, and provides prescriptive information as to appropriate wiring methods. Essentially, the corrosive agents are described in terms of the presence of acids, with or without one or both of two halogens (chlorine and bromine) that are used for water purification and sanitation. There are many acids in the world and no specification is offered in this section. However, generally in this case, the acids are derived from the presence of the halogens, which in free form react with water to form an acidic solution that is very corrosive. The Code focuses on storage areas for pool chemicals, along with circulators, chlorinators (or brominators), filters, and under deck areas abutting the pool as being of particular concern. Part **(B)** cites four raceways as being resistant to the corrosive effects. The two nonmetallic conduits (PVC and RTRC) are obvious; the two steel conduits would probably benefit from supplemental coatings. Zinc, the galvanizing material used with steel conduits, is not particularly resistant to attack from either hydrochloric or hydrobromic acid, and neither is aluminum [hence, 680.25(B)]. However, red brass heavy wall conduit is quite resistive to this sort of chemical attack. Currently no manufacturers are making it in its listed 10-ft electrical put up, but essentially the same heavy-wall product, often sold in 12-ft lengths, is available from some plumbing supply houses.

680.20. General. This rule mandates compliance with the provisions of parts **I** and **II** in this article where installing electrical and pool-associated equipment at a permanently installed swimming pool.

680.21. Motors. Part **(A)** covers wiring methods, and in an obvious error, the paragraph references additional to **(A)(1)** point to the four numbered paragraphs that existed in the 2014 edition, only two of which still exist in the 2017 edition. These general requirements are dramatically different as of the 2017 NEC, and now focus entirely on whether or not the wiring under discussion is in a corrosive location as defined in 680.14. Wiring within such areas is generally limited to the four rigid raceways included in 680.14(B), along with permission to use type MC cable "listed for that location."

This is presumably intended as a reference to typically PVC jacketed MC cables that are suitable for sunlight exposure and wet locations. Although these conditions may not apply indoors, the construction features of such cables make them more stable in even an indoor environment. Although product literature periodically cites the utility of these constructions for swimming pool applications, current UL Guide Card information does not formally describe any category of MC cable as such. This will need to be sorted out with the inspector on a case-by-case basis. Figure 680-5 illustrates some of these rules.

As soon as the circuit leaves the corrosive environment, the second paragraph of (A)(1) clearly removes all wiring method limitations at this point beyond normal Chap. 3 provisions. In addition, there are no particular limitations on the size and

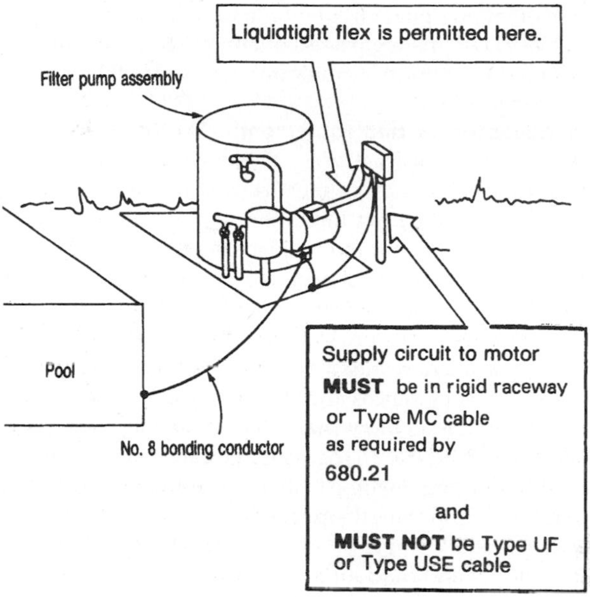
Liquidtight flex is permitted here.
Filter pump assembly
Pool
No. 8 bonding conductor
Supply circuit to motor
MUST be in rigid raceway
or Type MC cable
as required by
680.21
and
MUST NOT be Type UF
or Type USE cable

Fig. 680-5. The wiring method requirements at the motor generally follow 680.14(B). As of the 2017 NEC, no special requirements apply to any portions of the circuit ahead of the corrosive environmental boundary. (Sec. 680.21.)

character of the equipment grounding conductor, or even whether it must be of a wire type at all in the home run.

Flexible connections are permitted, as covered in part **(A)(2)**, only as necessary at or adjacent to the motor, and are limited to liquidtight flexible metal and nonmetallic conduits terminating in listed fittings. Although unmentioned in this paragraph, the copper, insulated, 12 AWG minimum equipment grounding conductor required in **(A)(1)** is a legacy this paragraph inherits, and must be used with such flexible connections.

The last part, **(A)(3),** of this section permits use of flexible cord for cord-and-plug-connected pump motors, under the same conditions as described for general cord-and-plug connection in 680.7. The wording of the requirement here for sizing the equipment grounding conductor within the flexible cord had often been interpreted to permit the use of a 14 AWG copper, but that was never the case. The general wording in 680.21(A)(1) requires a 12 AWG minimum equipment ground, and the rules there apply to all motors unless modified in (2) and (3). Because this provision was not expressed as a modification, no such modification was ever allowed and the 12 AWG sizing minimum prevailed. However, in the interest of clarity and consistent application of the rules, the size requirement is now repeated in this paragraph.

The wording of 680.21(B) excludes cord-and-plug-connected double-insulated pumps from the need for bonding. Where such a pump is used, all other components of the pool and its associated equipment must be bonded together, but are required to connect to the pump's equipment grounding conductor. Presumably, this can be done at the receptacle outlet supplying the pump.

Part **(C)** requires that any pool circulation motor on single-phase 120- through 240-V (includes 208-V) circuits regardless of amperage (as of the 2014 **NEC**) must incorporate GFCI protection for the motor outlet whether the motor is hard-wired or plugged into a receptacle.

680.22. Lighting, Receptacles, and Equipment. Part **(A)(1)** requires, for all occupancies as of the 2014 **NEC**, at least one 125-V receptacle to be "located" within the 6- to 20-ft (1.8- to 6.0-m) band around the pool for *any* permanent pool wherever constructed. The long-standing limitation on this rule to dwellings only was removed. The word *located* was put in to replace the word *installed*, because there is no need to install such a receptacle if there is already one located within that area around the pool. This rule ensures that a receptacle will be available at the pool location to provide for the use of cord-connected equipment. It was found that the absence of such a requirement resulted in excessive use of long extension cords to make power available for appliances and devices used at pool areas. This required receptacle may not be mounted more than 2.0 m (6½ ft) above the pool and should be located so that it is visible after all the pool-associated equipment is "hidden" by shrubs or other natural or man-made partitions. The basic rules in Part **(A)** prohibit receptacles within 6 ft (1.8 m) from the pool edge, and part **(A)(4)** calls for GFCIs to protect all 15- and 20-A 125-V single-phase receptacles located within 20 ft (6.0 m) of the inside walls of indoor and outdoor pools. But part **(A)(2)** inexplicably required the installation of a receptacle for a swimming pool or fountain recirculating pump, "or other loads directly related to circulation and sanitation," to be single and locking if less than 10 ft (3.0 m) but not closer than 6 ft (1.83 m) from the inside wall of the pool. This provision was a relic of former provisions that excluded such receptacles from a zone within 3 m (10 ft) at dwellings, and the allowance for a reduced distance of 1.8 m (6 ft) came with additional conditions. Now that such receptacles are allowed generally by right at the shorter distance, these conditions no longer make any sense. The 2014 **NEC** has partially addressed this by withdrawing the locking configuration rule for these receptacles, but leaving the 6-ft spacing rule in place, along with the GFCI and grounding rules, all of which taken together duplicate other requirements. The 2017 **NEC** completed the process by removing the single receptacle rule. Now the circulation and sanitation equipment receptacle rules still have their own numbered paragraph, but the requirements in that paragraph are identical to those that apply to receptacles generally.

Part **(A)(3)** allows other receptacles, of other configurations, voltages, etc., to be no closer than 1.83 m (6 ft) from the pool.

Part **(A)(4)** requires that all receptacles within 20 ft (6.0 m) of the inside wall of the pool must be protected by a GFCI. Of course, under current requirements in 210.8, virtually, *all* outdoor receptacles must have GFCI protection—at any distance from the pool.

Part **(A)(5)** says measurement of the prescribed distances of a receptacle from a pool is made over an unobstructed route from the receptacle to the pool, with hinged or sliding doors, windows and walls, floors, and ceilings considered to be "effective permanent barriers." If a receptacle is physically only, say, 3 ft (900 mm) from the edge of the pool but a hinged or sliding door is between the pool edge and the receptacle, then the distance from the receptacle to the pool is considered to be infinite, and the receptacle is thus more than 20 ft (6.0 m) from the inside wall of the pool and does not require GFCI protection (Fig. 680-6, bottom).

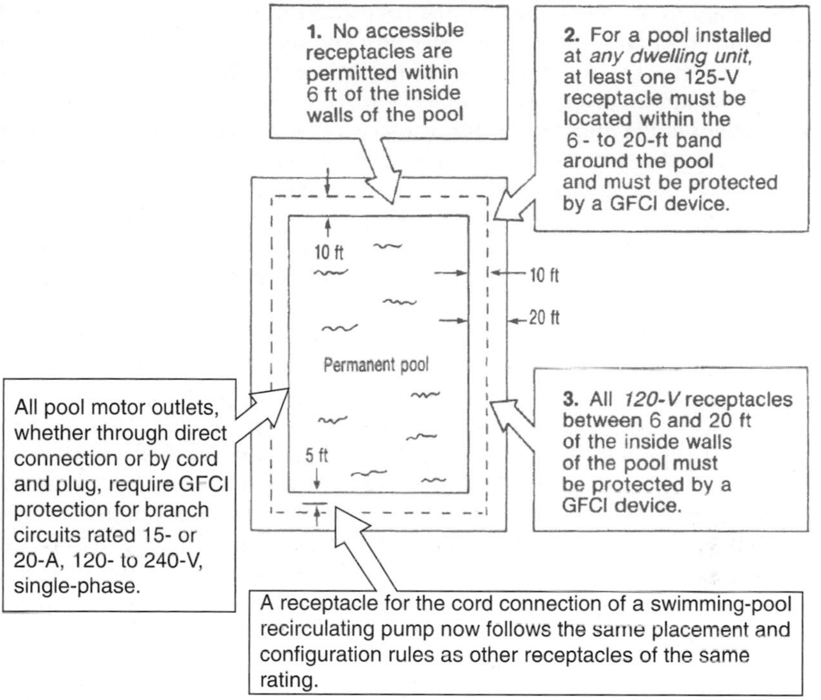
1. No accessible receptacles are permitted within 6 ft of the inside walls of the pool
2. For a pool installed at *any dwelling unit*, at least one 125-V receptacle must be located within the 6- to 20-ft band around the pool and must be protected by a GFCI device.
10 ft
10 ft
20 ft
Permanent pool
5 ft
All pool motor outlets, whether through direct connection or by cord and plug, require GFCI protection for branch circuits rated 15- or 20-A, 120- to 240-V, single-phase.
3. All *120-V* receptacles between 6 and 20 ft of the inside walls of the pool must be protected by a GFCI device.
A receptacle for the cord connection of a swimming-pool recirculating pump now follows the same placement and configuration rules as other receptacles of the same rating.

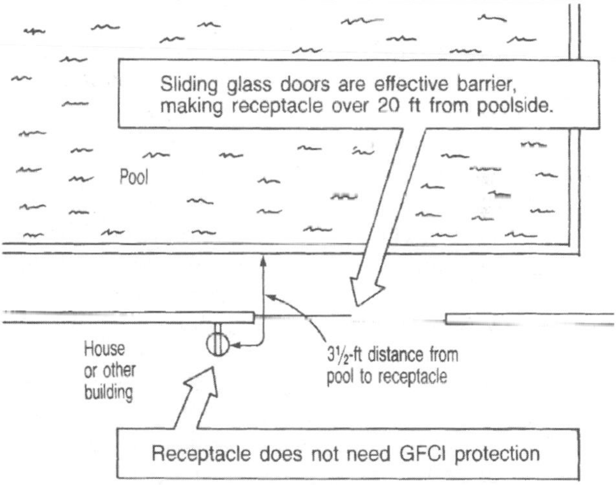
Sliding glass doors are effective barrier, making receptacle over 20 ft from poolside.
Pool
House
or other
building
3½-ft distance from
pool to receptacle
Receptacle does not need GFCI protection

680.22(A)(5) Doors or windows block a receptacle from connection of appliances that might be hazardous at poolside.

Fig. 680-6. Rules cover all receptacles within 20 ft (6.0 m) of the pool's edge. (Sec. 680.22.)

The reference to "existing installations" in 680.22**(B)(3)** must be understood to refer to luminaires that are already in place on a building or structure or pole at the time construction of the pool begins. Where a pool is installed close to, say, a home or country club building, luminaires attached to the already existing structure may fall within the shaded area for a band of space 5 ft (1.5 m) wide, extending from 5 ft (1.5 m) above the water level to 12 ft (3.7 m) above water level all around the perimeter of the pool, as described in part **(B)(1).** GFCI protection is required for these existing luminaire outlets, as shown in Fig. 680-8. However, new luminaires may not be installed in that space band around the pool.

Under the conditions given in part **(B)(2)** for indoor pools, luminaires may be installed less than 12 ft (3.7 m) above the water of indoor pools. Luminaires that are totally enclosed and supplied by a circuit with GFCI protection may be installed where there is at least 7½ ft (2.3 m) of clearance between the maximum water level and the lowest part of the luminaire. Paddle fans are also permitted in this zone, provided they have been identified as described in this paragraph. They are subjected to a water-spray test and will be marked as acceptable for such applications.

Figure 680-7 applies the foregoing rules to lighting at an enclosed pool, including part **(B)(4)** on GFCI protection for certain outlets in the area adjacent to that immediately above and around the pool itself. Part **(B)(5)** applies the general short cord and 12 AWG minimum size rules in 680.7 to all cord-and-plug-connected luminaires within 16 ft of any point on the water surface, measured in a radius. Part **(B)(6)**, new as of the 2014 **NEC**, correlates with 411.5(B) by providing a specific

Fig. 680-7. The luminaires over the pool and for 5 ft (1.5 m) back from the edge must be at least 12 ft (3.7 m) above the maximum water level if their supply circuit is without GFCI protection. If GFCI protection is provided by a GFCI-type circuit breaker in the supply circuit(s) to these totally enclosed luminaires, their mounting height may be reduced to a minimum of only 7½ ft (2.3 m) clearance above water level for an indoor (but not an outdoor) pool. The luminaires at right side do not require GFCI protection because they are over 5 ft (1.5 m) above the water level and rigidly attached to the structure. Refer to Fig. 680-8. [Sec. 680.22(B)(2) and (B)(4).]

instance of an acceptable low-voltage lighting system that is acceptable immediately adjacent to a pool.

Part **(B)(7)**, new as of the 2017 **NEC**, covers listed low-voltage gas-fire luminaires, decorative fireplaces, fire pits, and similar equipment using low-voltage ignitors without the need for grounding, and using transformers or power supplies that could qualify for use with underwater luminaires. This equipment can be located inside the usual 5-ft exclusion zone around the pool. Gas piping, if any, requires bonding, both in accordance with 250.104(B) as well as the 680.26(B)(7) rules for the bonding grid.

Part **(C)** requires that switching devices be at least 5 ft (1.5 m) from the pool's edge or be guarded [Fig. 680-8(c)]. To eliminate possible shock hazard to persons in the water of a pool, all switching devices—toggle switches, CBs, safety switches, time switches, contactors, relays, and so on—must be at least 5 ft (1.5 m) back from the edge of the pool, or they must be behind a wall or barrier that will prevent a person in the pool from contacting them. However, where the switch is specifically listed for installation within the 5-ft (1.5-m) horizontal exclusion area, then the switch may be located closer than 5 ft (1.5 m), where listed for such installation, as indicated by the last sentence.

Part **(D)** requires "other outlets" to observe a 3-m (10 ft) spacing minimum. An informational note describes such "outlets" and such items as fire alarm and communications circuits. Whether this article has jurisdiction in this area is unproven, and the substantiation for requiring a greater spacing for an RJ-11 telephone jack than for a 125-V receptacle outlet is not evident. On the other hand, such circuits

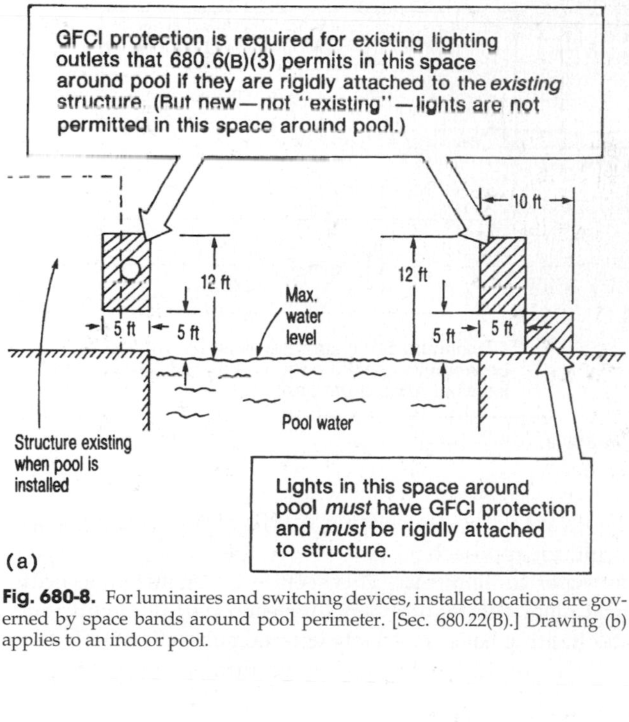
GFCI protection is required for existing lighting outlets that 680.6(B)(3) permits in this space around pool if they are rigidly attached to the *existing* structure. (But new—not "existing"—lights are not permitted in this space around pool.)
10 ft
12 ft
12 ft
Max. water level
5 ft
5 ft
5 ft
5 ft
Pool water
Structure existing when pool is installed
Lights in this space around pool *must* have GFCI protection and *must* be rigidly attached to structure.
(a)

Fig. 680-8. For luminaires and switching devices, installed locations are governed by space bands around pool perimeter. [Sec. 680.22(B).] Drawing (b) applies to an indoor pool.

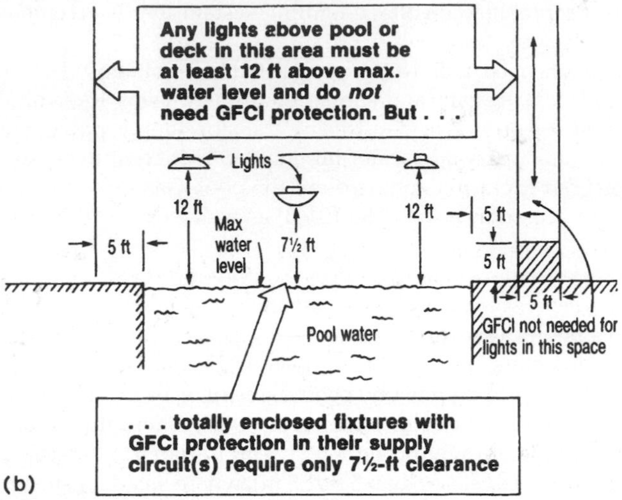
Any lights above pool or deck in this area must be at least 12 ft above max. water level and do *not* need GFCI protection. But . . .
Lights
12 ft
Max water level
7½ ft
12 ft
5 ft
5 ft
5 ft
5 ft
Pool water
GFCI not needed for lights in this space
. . . totally enclosed fixtures with GFCI protection in their supply circuit(s) require only 7½-ft clearance

(b)

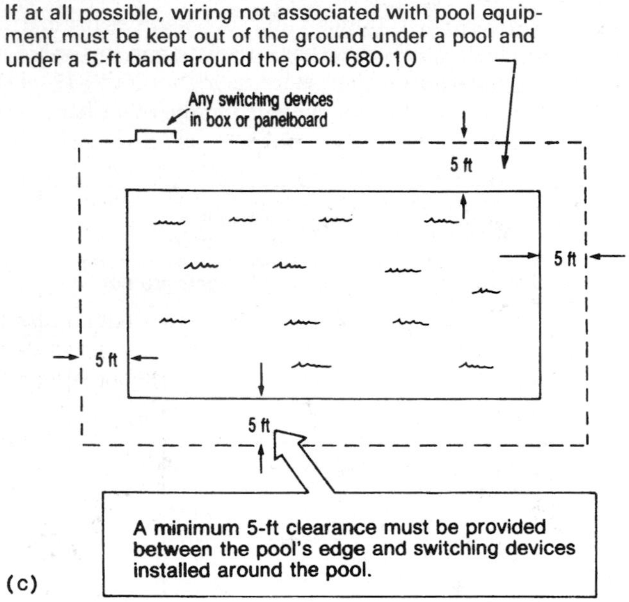
If at all possible, wiring not associated with pool equipment must be kept out of the ground under a pool and under a 5-ft band around the pool. 680.10
Any switching devices in box or panelboard
5 ft
5 ft
5 ft
5 ft
A minimum 5-ft clearance must be provided between the pool's edge and switching devices installed around the pool.

(c)

Fig. 680-8. *(Continued)*

do present significant voltage exposures. A typical ring voltage on a conventional telephone circuit can approach 90 V.

680.23. Underwater Luminaires. This section is organized according to a definite and logical plan. The requirements for each part of circuits and equipment for underwater lighting have separately lettered subsection designations. Part **(A)** begins with the actual luminaire and its equipment protection requirements.

For example, **(A)(1)** covers the luminaire operation, and **(A)(2)** covers the transformers and power supplies (because the **NEC** now includes LED luminaires in these rules). Parts **(B)**, **(C)**, **(D)**, and **(E)** cover the various types of underwater luminaires that are recognized, with the numbered paragraphs under each lettered subhead covering special wiring requirements that apply to the underwater installation. Part **(F)** covers the branch-circuit wiring requirements for underwater luminaires on the line side of the underwater environment. To complete this overall picture of Part II, Sec. 680.24 covers the swimming pool junction box (and equivalent enclosure) requirements, Sec. 680.25 covers feeder circuits, Sec. 680.26 covers bonding, and Sec. 680.27 covers specialized equipment. In part **(A)** of this section, the wording must be followed carefully to avoid confusion about the intent.

Part **(1)** starts by requiring that any underwater luminaire must be of such design as to ensure freedom from electric shock hazard when it is in use and must provide that protection without a GFCI. Part **(A)(2)** recognizes voltage limitation as a protective scheme and, with the advent of modern technology, also recognizes the low-voltage dc power supplies used to run LED light sources. *But* in part **(A)(3),** a GFCI *is* required for all line-voltage luminaires (any operating over the low-voltage contact limit, such as a 120-V luminaire) to provide protection against shock hazard during relamping. A GFCI is not required for these typically 12-V low-voltage swimming pool lights.

The UL Guide Card information presents certain essential data on use of GFCI devices, which must be factored into application of such devices, as follows:

> A GFCI is a device whose function is to interrupt the electric circuit to the load when a fault current to ground exceeds some predetermined value that is less than that required to operate the overcurrent protective device of the circuit.
>
> A GFCI is intended to be used only in circuits where one of the conductors is solidly grounded.
>
> Class A GFCIs trip when the current to ground has a value in the range of 4 through 6 mA. Class A GFCIs are suitable for use in branch and feeder circuits, including swimming pool circuits. However, swimming pool circuits installed before local adoption of the 1965 **NEC** may include sufficient leakage current to cause a Class A GFCI to trip.
>
> Class B GFCIs trip when the current to ground exceeds 20 mA. These devices are suitable for use with underwater swimming pool luminaires installed before the local adoption of the 1965 **NEC**.
>
> GFCIs of the enclosed type that have not been found suitable for use where they will be exposed to rain are so marked.

It should be noted that *only* a "listed" luminaire should be used—which means only a luminaire listed by UL or other test lab for use at a permanently installed pool, as called for by part **(A)(8).** In fact, the requirements given in parts **(A)(5)** through **(A)(7)** are very similar to UL data regarding the installation of listed underwater luminaires. UL data on listed luminaires must be carefully observed.

Part **(4)** sets 150 V as the maximum permitted for a pool luminaire, which means that the usual 120-V listed luminaires are acceptable. Note that the actual wording limits "supply circuits," which allows for a HID luminaire with a higher starting voltage, provided the line connections remained at 120 V.

Part **(5)** repeats the UL limitation on mounting distance of a luminaire below water level. When installed, the top edge of the luminaire must be at least 18 in. (450-mm) below the *normal* level of the pool water (Fig. 680-9). This 18-in. (450-mm) rule was adopted to keep the luminaire away from a person's "chest area," because this is the vital area of the body concerning electric shocks in swimming pools. Keeping the top of the luminaire 18 in. (450 mm) below the normal water level avoids a swimmer's chest area when he or she is hanging onto the edge of the pool while in the water. *But,* as the last sentence notes, an underwater luminaire may be used at less than 18 in. (450 mm) below the water

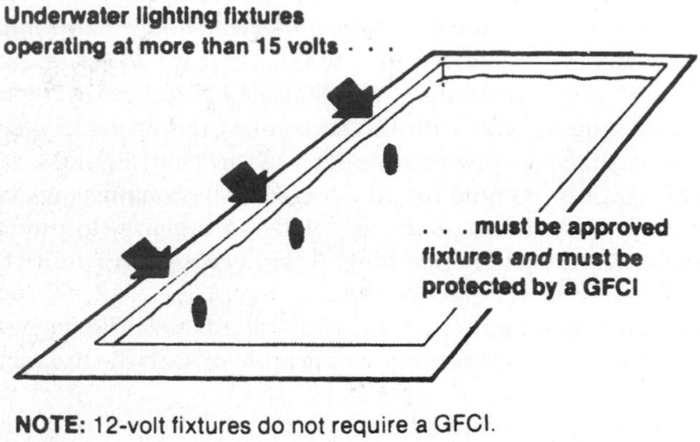
Underwater lighting fixtures
operating at more than 15 volts . . .
. . . must be approved
fixtures and must be
protected by a GFCI
NOTE: 12-volt fixtures do not require a GFCI.

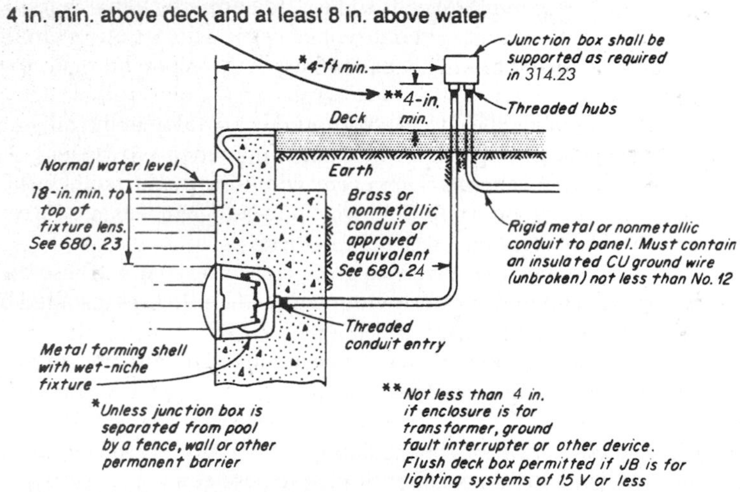
4 in. min. above deck and at least 8 in. above water
*4-ft min.
**4-in. min.
Deck
Junction box shall be supported as required in 314.23
Threaded hubs
Normal water level
Earth
18-in. min. to top of fixture lens. See 680.23
Brass or nonmetallic conduit or approved equivalent See 680.24
Rigid metal or nonmetallic conduit to panel. Must contain an insulated CU ground wire (unbroken) not less than No. 12
Threaded conduit entry
Metal forming shell with wet-niche fixture
*Unless junction box is separated from pool by a fence, wall or other permanent barrier
**Not less than 4 in. if enclosure is for transformer, ground fault interrupter or other device. Flush deck box permitted if JB is for lighting systems of 15 V or less

Fig. 680-9. Mounting of luminaire and circuit components must observe all Code rules and their specific dimensions. (Sec. 680.23.) Note that with the definition of maximum water level now at the point where the water spills out, the water height and the deck height are one and the same in most cases, and therefore the swimming pool junction box will be 200 mm (8 in.) above the deck. Also note that only rigid (or intermediate) metal conduit can support a box directly, without additional support direct to structure, per 314.23(E). Therefore, unless both conduits are metal [two is the minimum for box support in 314.23(E)] additional direct support is required for this swimming pool junction box.

surface if it is a unit that is identified for use at a depth of not less than 4 in. (100 mm). Part **(6)** requires upward-facing luminaires to have their lens guarded, or be listed for use without a guard.

Part **(7)** presents an interesting requirement on the use of wet-niche luminaires. The rule here requires that some type of cutoff or other inherent means be provided to protect against overheating of wet-niche luminaires that are not submerged but are types that depend on submersion in water for their safe operation. Note that the UL rules quoted here require some luminaires to be marked "Submerse Before Lighting." Manufacturers of such luminaires should incorporate this protection—such as in the form of a bimetal switch similar to those used in motor end-bells for motor overload protection.

Part **(B)** details the use of wet-niche luminaires. A wet-niche underwater lighting assembly consists of two parts: a forming shell, which is a metal structure designed to support a wet-niche luminaire in the pool wall, and a luminaire, which usually consists of a lamp within a housing furnished with a waterproof flexible cord and a sealed lens that is removable for relamping.

Part **(B)(2)** requires that the conduit between the forming shell and the junction box or transformer enclosure must be *approved* (1) rigid metal conduit or IMC and made of brass or other approved corrosion-resistant metal, or (2) liquidtight flexible nonmetallic or rigid nonmetallic conduit with a No. 8 *insulated* copper conductor installed in the conduit and connected to the junction box or transformer enclosure and to the forming shell enclosures. The No. 8 insulated copper wire may be stranded or solid. Each enclosure—the forming shell as well as the box—must contain approved grounding terminals.

Note the term "approved" with respect to metal conduit. This is a deliberate Chap. 6 amendment of a Chap. 3 rule, in this case 344.6, that requires all rigid metal conduit to be listed. Listed red brass conduit has not been generally available for decades. However, as heavy wall brass water pipe, it is available from some plumbing supply houses. It threads very well with the usual NPT threading dies, and is an extremely robust product. Without very heavy foot pressure on the bender shoe, an attempt to bend this product by pulling on the bender handle will bend the steel handle and not the pipe. The inside of this pipe is smoother than listed steel heavy wall conduits. The wording in this section at least makes this an option, subject to inspectional approval.

Figure 680-10 shows a typical connection from a forming shell to a transformer enclosure supplying the 12-V lamp in the luminaire. If a 120-V luminaire is used, the conduit from the shell terminates in a junction box, as shown in Fig. 680-9. In the drawing of Fig. 680-10, from the forming shell, a length of metric designator 27 (trade size 1) PVC conduit extends directly to a 120/12-V transformer mounted on the back wall of a planter adjoining the pool [observing the 4-ft (1.2-m) back and 8-in. (200-mm) high provisions of 680.24(B)(2)]. Where the nonmetallic conduit stubs up out of the planter soil, an LB (could be nonmetallic) connects the conduit to the transformer. The required 8 AWG conductor in the PVC conduit is terminated at the grounding bar in the transformer enclosure and on the *inside* terminal of an inside/outside grounding/bonding terminal on the forming shell. The external bonding lug provides for connecting the forming shell to the common bonding grid, as required by 680.26(A) and (B). The 8 AWG

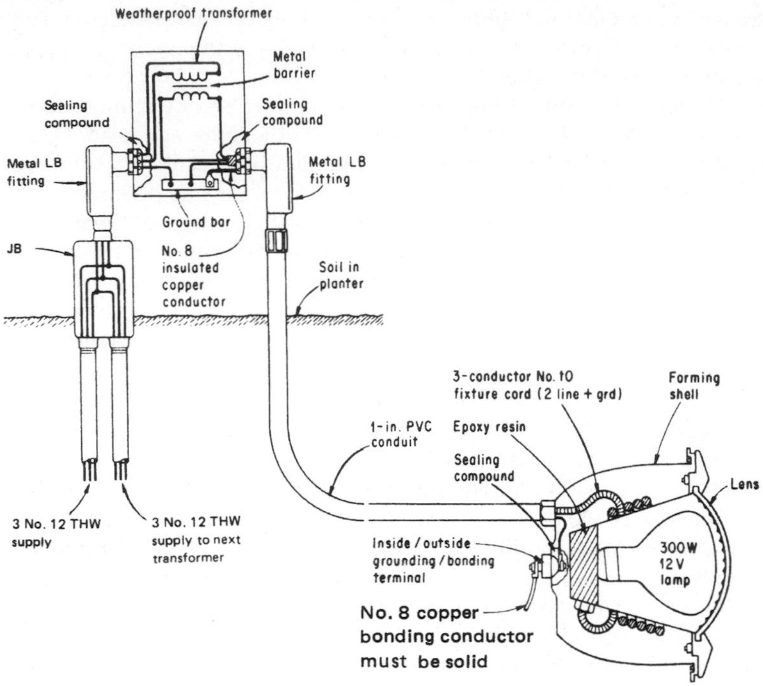
Weatherproof transformer
Metal barrier
Sealing compound
Sealing compound
Metal LB fitting
Metal LB fitting
Ground bar
JB
No. 8 insulated copper conductor
Soil in planter
3-conductor No. 10 fixture cord (2 line + grd)
Forming shell
1-in. PVC conduit
Epoxy resin
Sealing compound
Lens
3 No. 12 THW supply
3 No. 12 THW supply to next transformer
Inside / outside grounding / bonding terminal
300 W 12 V lamp
No. 8 copper bonding conductor must be solid

Fig. 680-10. Grounding and bonding is required in a typical hookup of low-voltage wet-niche luminaire with PVC conduit. (Sec. 680.23.) Note that in order to make the direct connection between the transformer and the luminaire as shown here, the transformer must be listed for such a direct connection. Further, it is to be presumed that such enclosures are not suitable for this task, unless they are specifically marked otherwise.

in the PVC conduit bonds the forming shell up to the transformer enclosure. Note that this 8 AWG conductor is not needed if metal conduit connects the shell to the transformer enclosure. One of the 12 AWG conductors in the supply circuit is an equipment grounding conductor that runs back to the panelboard grounding block and thereby grounds the 8 AWG and the metal fittings and transformer enclosure.

Part **(B)(2)(b)** requires that the inside forming shell termination of the No. 8 be covered with, or encapsulated in, a UL-listed potting compound. Experience has shown that corrosion occurs when connections are exposed to pool water. Listed epoxies are available to achieve this protection; however, some inspection agencies do accept a waterproof, permanently pliable silicone caulk compound.

Note that the illustrated assembly includes three noncurrent-carrying conductors: (1) an 8 AWG *bonding* conductor connecting the forming shell to the bonding grid; (2) an 8 AWG insulated conductor in PVC conduit between the forming shell and the transformer enclosure; and (3) a *grounding* conductor in the luminaire flexible supply cord. This is why 680.24(D) requires these enclosures to have at least one more grounding terminal than the number of conduit entries; with nonmetallic wiring, the run to the luminaire usually has two grounding conductors; however, for a listed low-voltage lighting system, the equipment grounding conductor in the cord may be omitted.

Part **(B)(4)** requires sealing of the luminaire cord end and terminals within the wet-niche to prevent water from entering the luminaire. And grounding terminations must also be protected by potting compounds.

Part **(B)(5)** states that an underwater luminaire must be secured and grounded to the forming shell by a positive locking device which will ensure a low-resistance contact and require a tool to remove the luminaire from the forming shell. This provides added assurance that luminaires will remain grounded because, in the case of wet-niche luminaires, the metal forming shell provides a bond between the raceway (or No. 8 conductor in PVC) connected to the forming shell and the noncurrent-carrying metal parts of the luminaire.

Part **(B)(6)** calls for the luminaires to be so located that they can be maintained or relamped without the need to get wet. That is, consideration must be given to ensure that such "servicing" activities can be performed from a "dry location." Further, the relamping location must be accessible without going into the water. In some instances, including luminaires in the bottom of a pool, these rules may require extensive lengths of cord, perhaps longer than what could be accommodated in the wet niche, and this will require careful design planning and consultation with the owners.

Part **(C)** permits use of an approved dry-niche luminaire that may be installed outside the walls or in the bottom of the pool in closed recesses that are adequately drained and accessible for maintenance. For a dry-niche luminaire, a "deck box," set in the concrete deck around the pool, may be used and fed by metal (rigid or IMC) or nonmetallic conduit from the service equipment or from a panelboard. Where the circuit conductors to the luminaire are run on or within a building, the rule [680.23(F)(1)] permits the conductors to be enclosed in EMT—but rigid metal or IMC or rigid nonmetallic conduit must be used outdoors when not on a building. And such a deck box does not have to be 4 in. (100 mm) up and 4 ft (1.2 m) back from the pool edge, as required for a junction box for a wet-niche luminaire, because 680.24(A) and (B) only apply to wet-niche and no-niche luminaire wiring. (See Fig. 680-1.)

Some approved dry-niche luminaires are provided with an integral flush deck box used to change lamps. Such luminaires have a drain connection at the bottom of the luminaire to prevent accumulation of water or moisture. The Code also recognizes listed low-voltage luminaires that do not require grounding.

Part **(D)** covers no-niche luminaires. They follow the same rules as wet-niche luminaires, except where a niche is mentioned, the no-niche bracket replaces it. Part **(E)** covers a through-wall lighting assembly, which is designed for above-grade use. Here again the rules in 680.23 apply, except in this case where the connecting point to the wiring system is specified, the conduit hub on the luminaire replaces the niche or the bracket. Review the UL data on this luminaire, quoted in this book at 680.33 below. These luminaires are used for both storable and permanent pools, and frequently wired improperly.

Part **(F)** covers branch-circuit wiring that runs on the supply side of transformer enclosures and junction boxes as shown in Figs. 680-9 and 680-10, and on the supply side of dry-niche wiring compartments. The first paragraph covers permitted wiring methods, which include both metallic and nonmetallic heavy wall conduits (RMC, IMC, PVC, RTRC) along with liquidtight flexible nonmetallic conduit, which in the earth must be listed for burial as discussed in Art. 356. An exception follows that permits liquidtight flexible metal conduit to make connections to pool-lighting transformers in lengths up to 1.8 m (6 ft) for any single use and up to 3 m (10 ft) for the total run. Here again, the wiring method limits beyond normal Chap. 3 considerations is, as of the 2017 **NEC**, the limit of the corrosive environment. However the rules for equipment grounding in (2) survive the transition to a noncorrosive environment and will deservedly constrain the choices of wiring methods. An in-pool light is potentially one of the most dangerous of electrical applications, and the requirements for raceways and unbroken equipment grounding return conductors have been in the **NEC** for about 50 years.

In all instances including the cabled methods an insulated 12 AWG (or larger if so required by 250.122) equipment grounding conductor must be installed, unless the luminaire is a listed low-voltage lighting assembly not requiring grounding. The exception that follows permits an equipment grounding conductor to remain sized in accordance with the branch-circuit protection where run between a transformer enclosure and a swimming pool junction box. Often the wire sizes are increased on transformer secondaries because of the higher currents that follow from the transformation down to 12 V.

The second paragraph adds joint and termination rules for grounding conductors. The basic rule requires the grounding conductor to be installed unbroken at any point. There are two locations, and only two, where this is not so. The first is where the same branch circuit daisy chains through swimming pool junction boxes, transformer enclosures, or dry-niche wiring compartments. In every instance, *the connections must be on "grounding terminals."* Do not use twist-on wire connectors to make these joints. The second location for joints in the grounding conductor is at the one or more points where the luminaire(s) are controlled by various means including a simple snap switch. Here again, *the connections must be on "grounding terminals."* This applies even in the case of snap switches, which means field installing a short segment of equipment grounding bus in the device box or some comparable method.

680.24. Junction Boxes and Enclosures for Transformers or Ground-Fault Circuit Interrupters. Part **(A)** covers junction boxes that *connect to a conduit that extends directly to a pool-lighting forming shell* or no-niche luminaire mounting bracket, such as shown in Fig. 680-9. The junction box must be of corrosion-resistant material provided with threaded hubs for the connections of conduit, or nonmetallic hubs for nonmetallic wiring methods. Part **(A)(3)** requires these boxes to ensure continuity between metal conduit entries, even if the box is plastic, through the use of an "integral" means. This would mean a box constructed to meet 314.3 Exception No. 2, (refer to the discussion at that point) and it is unlikely such a box will be encountered. Plan on using metal-based swimming pool junction boxes if you are using metal conduits for wiring.

For line-voltage (120-V) pool luminaires, the so-called deck box (set in the concrete deck around the pool) is no longer permissible (except where approved dry-niche luminaires include flush boxes as part of an approved assembly), because deck boxes, which were installed flush in the concrete adjacent to the pool, were the major source of failure of branch-circuit, grounding, and luminaire conductors due to water accumulation within them. The rule of part **(A)(2)**, covering low-voltage and line-voltage lighting, states that these junction boxes must be located not less than 4 in. (100 mm) above the ground level or above the pool deck, and not less than 8 in. (200 mm) above the maximum pool water level (whichever provides the greatest elevation), and not less than 4 ft (1.2 m) back from the pool perimeter.

Watch out for this placement issue on junction boxes. Now that the "maximum water level" has been defined as where the water can spill out, these two dimensions will use the same starting point (the deck) in most cases, thereby adding 100 mm (4 in.) to the elevation of most deck boxes. Swimming pool junction boxes require a domed cover to lift up and off the enclosed wiring, so a 200 mm (8 in.) spacing to the bottom translates to about 350 mm (14 in.) of minimum clearance from the deck to the underside of a diving board or other structure the box may be placed under. This will often complicate the location of these boxes, which are a tripping hazard if left in the open. This was the reason for the 100 mm (4-in.) height reduction when it first went into the **NEC**; with the new definition, it no longer has much practical effect.

The wording of part **(A)(2)** does make clear that the elevated junction box could be less than 4 ft (1.2 m) from the pool's edge if a fence or wall were constructed around the pool, with the box on the side of the wall away from the pool, isolating the box from contact by a person in the pool. Or the box could be within 4 ft (1.2 m) of the edge if the box were on the other side of a permanent nonconductive barrier.

Important: The last part of the rule in 680.24(A)(2)(c) still permits flush deck boxes where underwater lighting systems are at the low-voltage contact limit or less if approved potting compound is used in the deck boxes and the deck boxes are located 4 ft (1.2 m) from the edge of the pool. In Fig. 680-11, a deck box for a 12-V luminaire could be used in the deck but the use of the box less than 4 ft (1.2 m) from the pool's edge might be considered a violation, which does not recognize the fence along the pool in the same way as the first part in part **(A)(2)(b)**. That is, the fence is not mentioned in the first part as sufficient isolation of the box from the pool—although the installation certainly does comply with the basic concept in part **(A)(2)(b)**.

Part **(B)** covers installation of enclosures for 12-V lighting transformers and for GFCIs that are required for line-voltage luminaires. Such enclosures may be installed indoors or at the pool location. If a ground-fault interrupter is utilized at a pool, its enclosure must be located not less than 4 ft (1.2 m) from the perimeter of the pool, unless separated by a permanent means, and must be elevated not less than 8 in. (200 mm), measured from the inside bottom of the box down to the pool deck or maximum water level, whichever provides higher mounting. These rules cover installation of transformer or GFCI enclosures that connect to a conduit that "extends directly" to a forming shell.

Part **(B)(1)** specifically and clearly mandates that "other enclosures" be listed *and* labeled for the purpose. Use of any enclosures for GFCIs or transformers permanently installed in swimming pools that are not listed specifically for use at swimming pools is a violation. Enclosures that are so listed will be engineered to meet the following requirements.

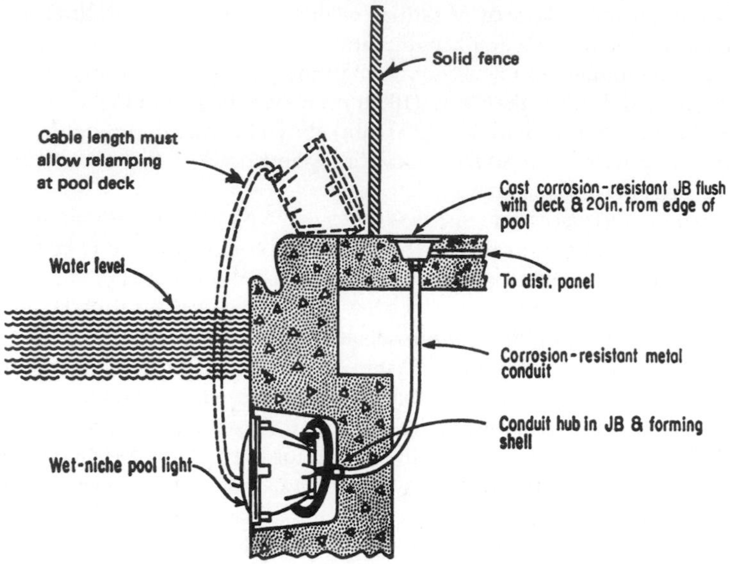
Solid fence
Cable length must allow relamping at pool deck
Cast corrosion-resistant JB flush with deck & 20in. from edge of pool
Water level
To dist. panel
Corrosion-resistant metal conduit
Conduit hub in JB & forming shell
Wet-niche pool light

Fig. 680-11. The fence here permits the box to be closer than 4 ft (1.2 m) from pool's edge. (Sec. 680.24.)

Part **(B)(1)(1)** requires any such enclosure connected to a conduit that extends directly to an underwater pool-light forming shell to have threaded hubs or bosses or a nonmetallic hub. An enclosure of cast construction with raised, threaded hubs or with threaded openings in the enclosure wall would satisfy that rule. But because approved swimming pool transformers are usually available *only* in sheet metal enclosures with knockouts, this is usually not practical. Such a transformer connection will not meet other requirements in this section and violate the UL rule that "unless marked otherwise, these transformers are not suitable for connection to a conduit that extends directly to a wet-niche or no-niche luminaire." In general, plan on running a raceway between the transformer enclosure and a listed swimming pool junction box that will receive the conduit running to the luminaire.

Part **(B)(1)(2)** requires corrosion-resistant construction, using brass, copper, stainless steel, plastic, comparable material. Often a combination of such materials is used.

Part **(B)(1)(3)** also requires that transformer or GFCI enclosures be provided with an approval seal (such as duct seal) at conduit connections to prevent circulation of air between the conduit and the enclosure; that they must have electrical continuity between every connected metal conduit and the grounding terminals by means of copper, brass, or other approved corrosion-resistant metal that is integral with the enclosures; that they must be located not less than 4 ft (1.2 m) from the inside walls of the pool (unless separated by a solid fence, wall, or other permanent barrier); and that they must be located not less than 8 in. (200 mm) from the ground level, pool deck, or maximum pool water level, whichever provides the greatest elevation. This distance is measured from the inside bottom of the enclosure. (See Fig. 680-10.)

Note that part **(B)(1)(4)** intends to ensure a grounding path *from the enclosure and its grounding terminals* to any metal conduit. The section specifically states "metal conduit." Where PVC conduit is used, the provision is not applicable, and the No. 8 ground wire in the PVC bonds to the forming shell. However, the section requires electrical continuity between an enclosure and "*every* connected metal conduit." The conduit feeding the transformer primary does not seem to be involved with that rule because the concern is with the grounding path between the transformer or GFCI enclosure and the forming shell and because the No. 12 equipment grounding conductor in the primary supply will carry any current from a fault originating within the transformer enclosure. Local Code authorities should be consulted on the point.

The phrase "integral with the enclosures" is meant to cover a situation where the enclosure is nonmetallic. In this case, electrical continuity between the metal conduits and the grounding terminals must be provided by one of the metals specified, and this "jumper" must be permanently attached to the nonmetallic box so that it is "integral." Refer to the discussion on the same topic at (A)(3) for swimming pool junction boxes; this is not a practical alternative.

In Fig. 680-10, the transformer enclosure is being used as a junction box to an underwater light, with the equipment grounding conductors terminated at the grounding bar and carried through. The figure is predicated on the unlikely (but conceivable) assumption that the transformer is listed for a direct connection to a forming shell, and therefore meets all the requirements for such enclosures. However, the primary purpose of this enclosure is to house the transformer. Parts **(B)**, **(C)**, **(D)**, **(E)**, and **(F)** of 680.24 still apply. 680.24(A) would apply to boxes connected directly to underwater lights and is intended to cover situations where splices, terminations, or pulling of conductors might be required. Again, plan on providing a listed swimming pool junction box, which will be designed to meet all these rules, at every conduit termination that runs to a forming shell.

Part **(C)** of this section warns against creating a tripping hazard or exposing enclosures to damage where they are elevated as required. It is also important to remember that these junction boxes must be afforded additional protection against damage if located on the walkway around the pool. For protection against impact, they may be installed under a diving board or adjacent to a permanent structure such as a lamppost or service pole. As noted in the discussion at (A)(2), panel action on the maximum water level definition has greatly complicated the placement of these boxes under many diving boards.

Part **(D)** can be satisfied simply by using listed equipment; this is the rule requiring one more terminal than conduit entries, as discussed at 680.23(B)(2)(b). Part **(E)** calls for strain relief to be added to the flexible cord of a wet-niche lighting luminaire at the termination of the cord within a junction box, a transformer enclosure, or a GFCI. This mechanism will be furnished with or as an integral part of any listed swimming pool junction box.

680.25. Feeders. The permitted wiring methods permitted for feeders to panels in cabanas or other locations that supply branch circuits covered by Art. 680 consist of heavy-wall metal and nonmetallic conduits (RMC except aluminum in the pool area, IMC, PVC, RTRC) and liquidtight flexible nonmetallic conduit. Here again, the special wiring method rules only apply to the corrosive environment, and normal Chap. 3 techniques can be used elsewhere. And in the case of feeders, there is no enhanced equipment grounding conductor rule that applies to the home run. Aluminum heavy-wall conduit is not permitted in pool areas where subject to corrosion.

680.26. Equipotential Bonding. The 8 AWG conductor and bonding system required in this section are not part of the effective fault current path that the equipment grounding system establishes, although they are ultimately connected to it. When all the required bonding connections are made at a pool, the entire interconnected hookup will be grounded by the "equipment grounding conductor" that is required to be run to the filter pump and to the lighting junction boxes and is connected to the 8 AWG bonding conductor at the pump and in the boxes. In a pool without underwater lighting, the equipment grounding conductor run with a pump-motor circuit will be the sole grounding connection for the bonded parts—and that is all that is required [see, however, 680.26(B)(6)]. The 8 AWG bonding conductor does not have to be run to a panelboard, service equipment ground block, or grounding electrode.

Its sole function is to reduce voltage gradients in the pool vicinity. A pool that is 750 V to ground because of a voltage transient is only a danger to a swimmer if one side is 752 V and the other side is 748 V, resulting in a 4 V gradient that could result in drowning, as noted in the discussion at the beginning of Art. 680. A bonding conductor extended to a remote panelboard in a basement could definitely save the life of an alien being with indefinitely long and stretchable arms who decided to reach through a cellar bulkhead and touch the panel while swimming in the pool. For the rest of us such a connection is useless. Likewise, the bonding connections are irrelevant to whether the pool is connected to one or more grounding electrodes. The frequently imposed, and bogus, requirement to drive ground rods around a pool perimeter may assist users in avoiding shocks from the adjacent azalea bushes, but nothing else.

The 8 AWG conductor mandated in this section is a bonding conductor, not an equipment grounding conductor. It must be solid in order to better withstand chemical attack from pool chemicals. Certain controversies regarding this wire in terms of solid versus stranded, insulated or not, color coded or not are resolvable if its status as a bonding conductor is kept in mind.

For example, no rule requires it to be insulated. If it is insulated, inspectors might require green color coding at any permanently exposed termination, if a rigid interpretation is put on 250.119 and if the "bonding" conductor is considered to be an "equipment grounding" conductor. That view is incorrect because 250.102 on "bonding jumpers" does not specify color of insulation or covering any more than there is a color code for a grounding electrode conductor.

Conflict has also arisen in the past over use of this 8 AWG bonding conductor because 310.106(C) requires 8 AWG and larger conductors to be *stranded* where installed in raceways. That rule had the effect of limiting use of *solid* No. 8 conductors, and their manufacture seemed to be curtailed. 310.106(C) has an exception that exempts conductors from being stranded in raceways where other Code rules permit such application. However, while the "8 AWG equipment grounding conductor" covered by 680.23(B)(2)(b) is permitted to be either stranded or solid, the 8 AWG that serves as the common bonding grid is not so exempt and must always be solid.

Part **(A)** gives a general statement aimed at defining the nature and performance of the bonding connection. This connection is not intended to act as an equipment grounding conductor, which, of course, carries fault current to ensure operation of the circuit OC device in the case of a faulted conductor. The bonding connection required here is only intended to ensure that all noncurrent-carrying parts of the pool and its electrical system are at the same potential with respect to ground at the pool location, thereby reducing voltage gradients in and around the pool (Fig. 680-12). Note that the bonding requirements for components required as part of the bonding grid are not necessarily dependent on any particular proximity to the pool, such as 1.5 m (5 ft). For example, a recirculating pump motor for a pool and the associated piping functionally extends the pool water zone, and requires bonding however far it is from the pool.

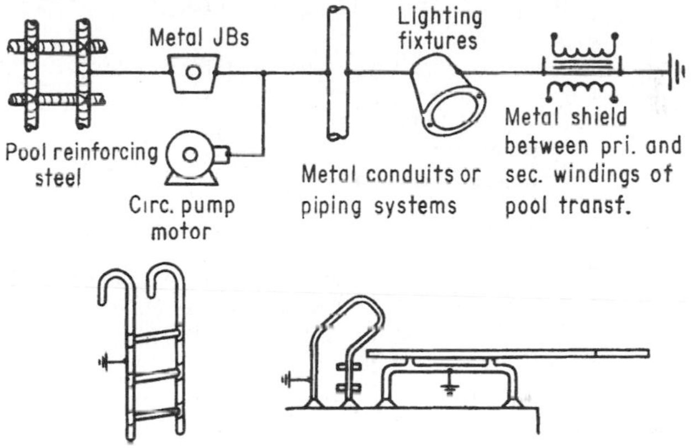
Metal JBs
Lighting fixtures
Pool reinforcing steel
Metal shield between pri. and sec. windings of pool transf.
Metal conduits or piping systems
Circ. pump motor

Fig. 680-12. All of these metallic, noncurrent-carrying parts of a pool installation must be "bonded together" through the "equipotential bonding grid." (Sec. 680.26.)

Part **(B)** reiterates the concept that bonding conductors need not run to remote panels because that would be irrelevant to their intended function. It then spells out in detail the pool components that must be bonded together, using 8 AWG solid copper wire. It also mentions brass conduit as a bonding conductor, which is incomprehensible in terms of functioning as a wire, but apparently intended to address a brass conduit as a collection point for wired bonding connections. Figure 680-13 shows what could be an example. This would be a reasonable application of the wording. There are seven categories of components that must be included in the bonding connections (Fig. 680-14).

Paragraph **(1)** covers conductive pool shells. These include pools made of concrete, including concrete block, whether or not it is supported by steel reinforcing. Bonding must be accomplished to the reinforcing steel using bonding connections that are listed for direct burial and for rebar connections; refer to the discussion

Fig. 680-13. 8 AWG insulated bonding conductors from several bonded items connect to the brass conduit, thereby using it as a bonding conductor. 680.26(B)(7) requires bonding of all "metal conduit" within 5 ft (1.5 m) of pool edge. (Sec. 680.26.)

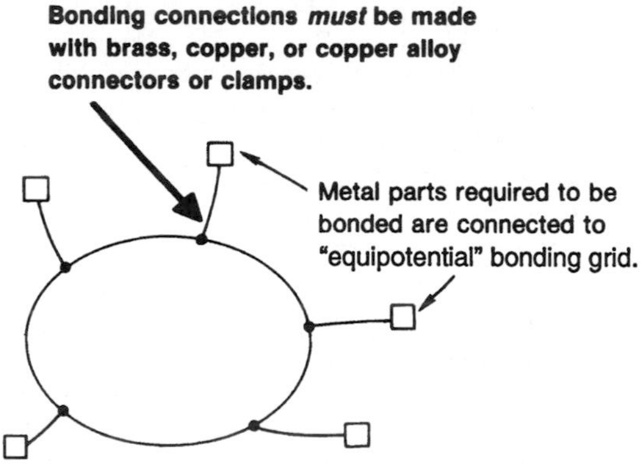
Bonding connections *must* be made with brass, copper, or copper alloy connectors or clamps.
Metal parts required to be bonded are connected to "equipotential" bonding grid.

Fig. 680-14. All designated parts must be connected to an equipotential bonding grid. (Sec. 680.26.)

in this book at 250.52(A)(3) (concrete-encased electrodes) for more information on this point. The reinforcing steel used for this purpose must be the usual, unencapsulated product with crossing points tied by steel tie wires or equivalent methods. If the reinforcing steel is the epoxy coated (typically green) product commonly used for highway bridge construction, then it is ignored and an alternative method is substituted.

The Code does not require each individual reinforcing bar to be bonded. It recognizes that the steel tie wires used to secure the rebars together where they cross each other provide the required bonding of the individual rods. Tests conducted over a period of several years by the NEC Technical Subcommittee on Swimming Pools have shown the resistance of the path from one end to the other through the structural steel to remain at less than 0.001 Ω. The use of conventional steel reinforcing steel members as a permitted basis of a collective bonding grid has been recognized in the **NEC** since Art. 680 first appeared in the 1962 edition.

If conventional steel reinforcing is not available on a pool defined as having a "conductive pool shell," then a mandatory alternative method applies. Warning: This method is so expensive that the cost of the required bonding grid alone is likely to be extremely costly, especially after considering required changes to normal trade scheduling in order to accommodate the additional work. Try to make customers aware, as early in the design process as possible, that the lack of conventional reinforcing steel may have this effect. The alternative requires a copper bonding grid to cocoon the pool structure, below as well as on the sides, and arranged to approximate the coverage of normal steel reinforcing. Specifically, 8 AWG bare solid wires must envelop the pool at a distance of no more than 150 mm (6 in.) using a 300 mm by 300 mm (1 ft by 1 ft) pattern (with a tolerance of 100 mm or 4 in.), and with all points of crossing connected in a manner suitable for direct burial. The only practical way to cope with this is through the use of substation grounding mats. These are available in 6-ft wide rolls with a range of bare copper wires and welded grid patterns to choose from, including the 8 AWG 300 mm (1 ft) pattern mandated here.

Paragraph **(2)** covers the pool deck and any other areas that immediately surround the pool perimeter. If there is steel reinforcing in the deck, then it gets the same bonding treatment as for the walls of a pool that use steel reinforcing. The deck reinforcing must be bonded to the method used for the pool walls at no fewer than four points evenly distributed around the edge of the pool. If there is no steel reinforcing in the deck, or if it is epoxy coated, then an alternate method is provided. Unlike the alternate method for the pool walls, the alternate for the deck reinforcing is simple, inexpensive, practical, and well-rooted in historical pool bonding practice. An 8 AWG bare solid copper wire is run around the pool at a distance of 450 to 600 mm (18 in. to 24 in.) from the inside of the pool wall. It must be buried 100 mm to 150 mm (4 in. to 6 in.) "below the subgrade," which presumably means the distance below the final surface treatment such as coping stones. Any splices must be listed for the application, which means listed and marked "DB" or "direct burial" and comprised of all copper, brass, or bronze components, or of stainless steel. Note that if portions of the perimeter area are separated from the pool by a permanent wall at least 1.5 m (5 ft) high, perimeter bonding is not required for that portion.

Part **(3)** covers metallic components of the pool structure not already covered This would include the walls of metal-walled pools (Fig. 680-15). As before, epoxy-coated reinforcing steel is ignored; thereby avoiding a conflict with 680.26(B)(1)(a). The rule does not require individual sections of such pools to be bonded. However, the overlapping ends of each section to be bolted must not be painted. If they are, the paint must be removed completely to restore conductivity. In addition, resistance tests should be made across each bolted section after assembly to ensure low resistance. These sections normally are fastened together by corrosion-resistant bolts at least ⅜ in. (9.5 mm) in diameter, and such an installation satisfies the bonding objectives. But electrical parts of such a pool must be tied into that common bonding grid by 8 AWG bonding conductors.

Part **(4)** covers the back side of metal forming shells and no-niche luminaire brackets. An exception exempts listed luminaire forming shells that are part of a low-voltage lighting system that does not require bonding.

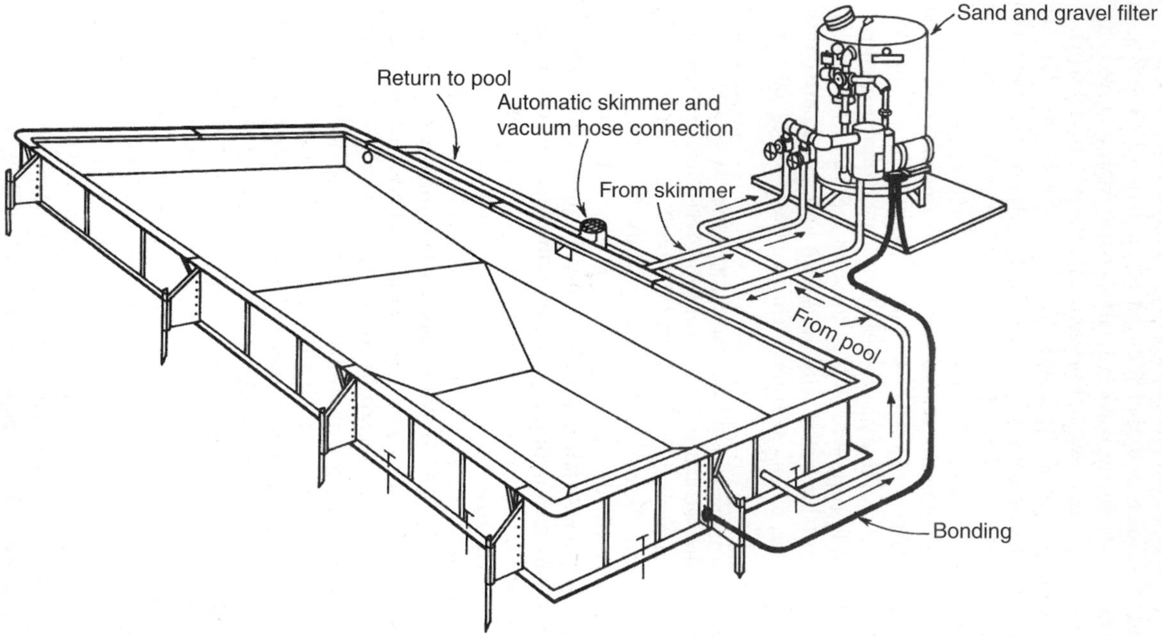
Sand and gravel filter
Return to pool
Automatic skimmer and
vacuum hose connection
From skimmer
From pool
Bonding

Fig. 680-15. Walls of bolted or welded metal pool are metallic parts of the pool structure and require bonding. (Sec. 680.26.)

Fig. 680-16. Fittings for pool ladders have bonding strips attached to them for connection of the No. 8 bonding conductor. When the supports are set in the concrete deck, the bonding connections tie them all together. The 8 AWG bonding connections are shown (arrow) and a protective coating is painted on the connectors to protect against corrosion. (Sec. 680.26.)

Part **(5)** covers all metal fittings in or attached to the pool structure (Fig. 680-16), with the usual exemption (built into the wording of the rule) for small hand grips, etc., that are not over 100 mm (4 in.) in any dimension and that do not extend over 25 mm (1 in.) into the pool wall.

Part **(6)** covers electrical equipment associated with the pool, including pump motors and the associated circulation system (Fig. 680-17) and conductive surfaces that are part of pool covers and their motors. Double insulated equipment does not, however, require incorporation into the bonding grid. This includes double-insulated pump motors; however, for those applications a run of 8 AWG bare copper must be extended from the bonding grid to the pump location, and left coiled there long enough so that if the double-insulated motor is replaced with a conventional motor, a bonding connection will be readily accessible.

In addition, in cases where no component in the bonding grid has a connection to the equipment grounding system for the premises, such a connection will be made at this point using this bonding conductor. This could be achieved by connecting it at the outlet passing through a wet-location extension box, or for hard-wired applications entering the raceway and extending to an enclosure through a tee conduit body using a suitable watertight gland connector, or any other workmanlike solution appropriate for the location.

This part also covers pool water heaters that generally require bonding. However, for heaters rated above 50 A and with specialized instructions, only the specifically designated parts are to be bonded, or grounded, or both.

Part **(7)** includes metal raceways, metal-clad cable assemblies, metal plumbing, metal building components including door bucks, awnings, window frames and

Fig. 680-17. Water-fill pipe and metal housings associated with the water-circulation system are "bonded" with an 8 AWG solid copper conductor, as required. (Sec. 680.26.)

all other "fixed metal parts." This also includes swimming pool junction boxes (Fig. 680-18) and swimming pool transformer enclosures, as well as metal fences, etc., if not separated by a barrier "that prevents contact by a person" or more than 1.5 m (5 ft) away, measured horizontally from the pool walls. In addition, parts more than 3.7 m (12 ft) above the pool or observation stands, diving platforms, etc., are not required to be bonded. Figure 680-19 shows an example of the entire system put together.

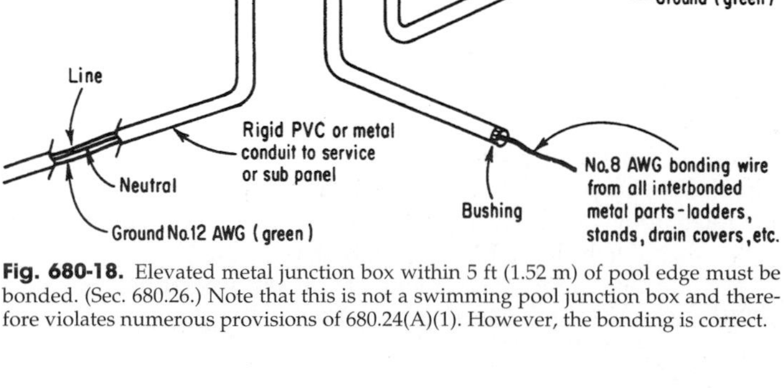
Screw terminals to terminate No.8 bonding conductor and No.12 grounding conductor
Underwater light circuit-cord assembly
Line
Neutral
Ground (green)
Line
Rigid PVC or metal conduit to service or sub panel
Neutral
Ground No.12 AWG (green)
Bushing
No.8 AWG bonding wire from all interbonded metal parts-ladders, stands, drain covers, etc.

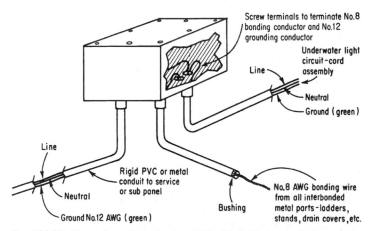

Fig. 680-18. Elevated metal junction box within 5 ft (1.52 m) of pool edge must be bonded. (Sec. 680.26.) Note that this is not a swimming pool junction box and therefore violates numerous provisions of 680.24(A)(1). However, the bonding is correct.

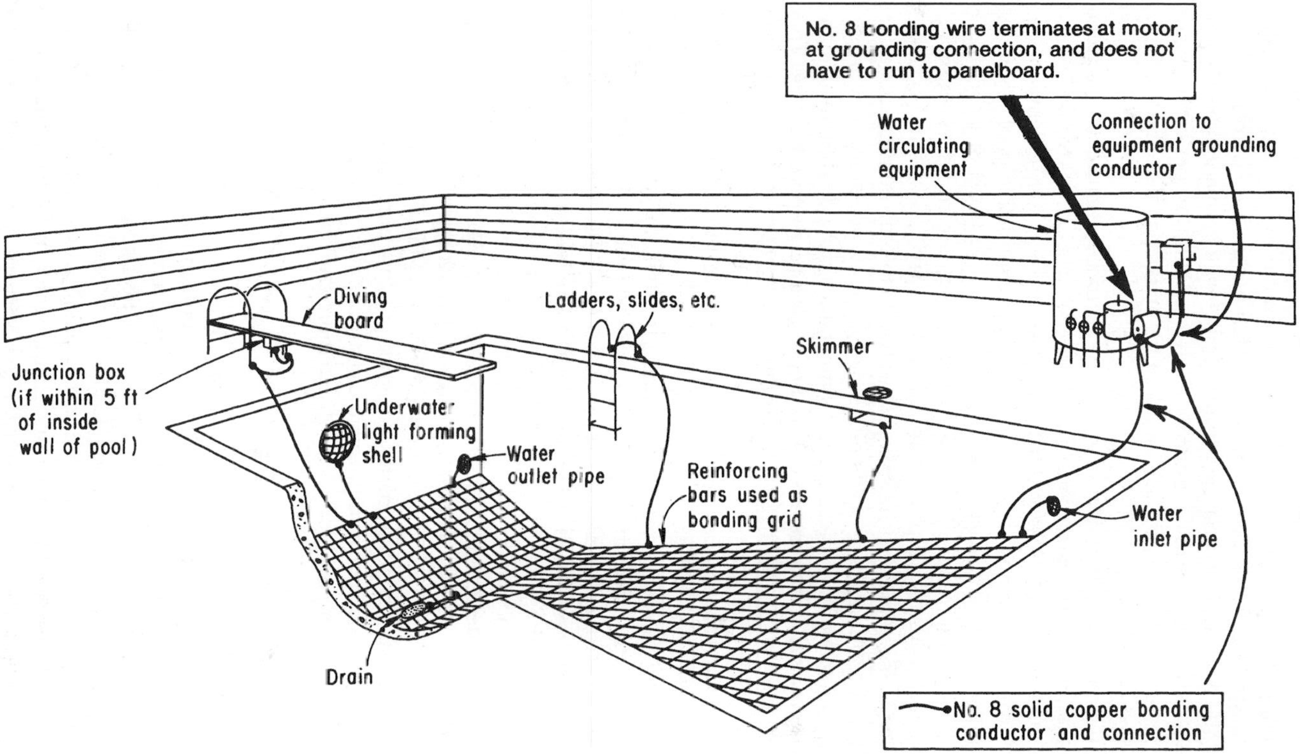

No. 8 bonding wire terminates at motor, at grounding connection, and does not have to run to panelboard.
Water circulating equipment
Connection to equipment grounding conductor
Diving board
Ladders, slides, etc.
Skimmer
Junction box (if within 5 ft of inside wall of pool)
Underwater light forming shell
Water outlet pipe
Reinforcing bars used as bonding grid
Water inlet pipe
Drain
No. 8 solid copper bonding conductor and connection

Fig. 680-19. An 8 AWG bonding jumper ties each of the indicated parts to the rebar grid, completing the bonding. (Sec. 680.26.)

In that drawing, the steel reinforcing rods, tied together with steel tie wires at intersections, are used as a common equipotential bonding grid to bond together pool equipment. Equipment shown here is required to be bonded. In addition, any metal parts (lighting standards, pipes, etc.) within 5 ft (1.5 m) of the inside walls of the pool and not separated from the pool by a permanent barrier must be bonded. All connections made must be completed with proper connectors, lugs, and so on, as required in 250.8 and as called for by the parent language in part **(B)** of this section.

Part **(C)** of this section is unfortunately worded, because the terminology "bonded water" is an oxymoron, like "married bachelor." Nevertheless reasonable substantiation was presented that there was a potential problem if there were no electrical connection whatsoever between the conductive pool water and the surrounding bonding grid [which will be required in (B)(2) at least, and probably (B)(6) and (B)(7) as well], because there are significant issues with potential water-to-deck voltages. In past years, this would have been unheard of because something in the pool structure was always tied to the surrounding bonding grid. Today this is no longer an academic concern. Many pools are now being constructed with all nonmetallic framing, nonmetallic liners, double-insulated pump motors, nonmetallic ladders, no diving board and no in-pool lighting. This rule ensures that in such cases, there will be a functional contact between the water and the surrounding grid.

Nevertheless, there were issues with the requirement, even beyond the terminology about bonding to that which cannot be bonded. The rule was missing any provisions about corrosion resistance of the connecting medium, whether it would likely be disturbed during swimming, etc. There are also practical issues of not wanting an 8 AWG conductor draped over the corner of a pool. The 2014 NEC has addressed these issues and provided the tools so inspectors can ensure enforcement.

One way to address this in a very simple, workmanlike way is to put a current collector in the drain line to the pool, as close to the pool drain as possible. If the drain is a metric designator 53 (trade size 2) pipe, a short brass or stainless steel nipple (check with the pool contractor for an assessment/preference as to reactivity with his or her chemicals) placed in the drain line will suffice. There are several patented and listed bonding plates designed for this purpose as well. The area exposed on the inside of this nipple is the circumference multiplied by the length. A nipple just 50 mm (2 in.) long will expose more than enough area to the water.

$$A = \pi Dh = 3.14 \times 2 \text{ in.} \times 2 \text{ in.} = 12.5 \text{ in.}^2$$

Attach an 8 AWG solid bonding wire to this current collector using a pipe clamp listed for this pipe size, and further manufactured using all copper, brass, or bronze, or stainless steel components, enabling a further listing for direct burial and with the appropriate markings. This is critical because replacing this item would involve extensive excavation. Connect the other end of the bonding conductor to the bonding grid, using a direct burial rated splicing device at the other end. Note that this will not only connect the water to the bonding grid, but to the equipment grounding system for the premises as well. This is because, if the installation is fully compliant, the connection mandated in 680.26(B)(6)(a) will have been made as well. Note that a bonded pool ladder frame may also work.

680.27. Specialized Pool Equipment. Part **(A)** of this section treats connection of loudspeakers for underwater audio output in the same way as a wet-niche pool luminaire. Wording and rules are almost identical to those in 680.23(B)(5). Connection from the speaker forming shell is made to a junction box installed as set forth in 680.21(A) for a luminaire. The wiring methods correlate with 680.23(B)(2).

Part **(B)** covers pool covers, and the control location must be such that the operator can see the entire pool. The motor(s) must be located to provide at least 1.5 m (5 ft) of space between them and the pool wall unless separated by a "wall, cover, or other permanent barrier," and they must have GFCI protection. There are now motors that are part of listed systems and that operate within the low-voltage contact limit, and that have power supplies comparable to those for wet-niche luminaires. Exceptions in the 2017 NEC permit these motors to be installed within the normal 5-ft exclusion zone around the pool, and to be not subject to GFCI protection requirements.

Part **(C)** covers deck area heating. These rules cover safe application of unit and radiant heaters. Such units must be securely installed, must be kept at least 5 ft (1.5 m) back from the edge of the pool, and must not be installed over a pool. Permanently wired radiant heaters have the same horizontal clearance, but add a 3.7-m (12-ft) vertical clearance as well. Radiant heating cables are prohibited from use embedded in the concrete deck.

680.28. Gas-Fired Water Heater. If these heaters operate above the low-voltage contact limit, they must have GFCI protection.

680.31. Pumps. There are portable filter pumps listed by Underwriters Laboratories, and they comply with 680.31. They are both double insulated and equipped with an equipment grounding conductor that picks up the inaccessible dead metal parts, and they must have GFCI protection built in to their power supply cord. Note that the 2014 NEC has added storable spas and storable hot tubs throughout Part III so the same rules as for storable pools also apply to this equipment.

680.32. Ground-Fault Circuit Interrupters Required. All 125-V receptacles rated 15 or 20 A within 6 m (20 ft) of the storable pool must have GFCI protection; the distances follow the same rules for measurement as in 680.22(A)(5) and do not pass through a sliding glass doorway, for example.

680.33. Luminaires. Underwater luminaires are available for storable pools in both 12-V versions ("within the low-voltage contact limit") with a transformer and in 120-V ("over the low-voltage contact limit") versions with long cords matching those that come with the motor. Usually these will not be subject to the default 450-mm (18-in.) submersion rule in 680.23(A)(5). Review the "through-wall lighting assembly" definition in 680.2 and the wiring rules for permanent pools in 680.23(E), and then factor in the following information from UL on products that are frequently installed improperly:

> Underwater Luminaires for Aboveground Storable Swimming Pools—These luminaires are a type of through-wall lighting assembly as described in Art. 680 of the NEC. They have been investigated for use with an aboveground storable pool (a pool that is constructed on or above the ground and is capable of holding water to a maximum depth of 1.0 m (42 in.), or a pool with nonmetallic, molded polymeric walls regardless of dimension). They include all three of the following factory-provided parts:

1. Lamp assembly for temporary installation on or through the wall of an aboveground pool
2. Transformer or ground-fault circuit interrupter assembly provided with a 0.9–1.8 m (3–6 ft) power supply cord for connection to a source of supply and for temporary mounting away from the pool (the remote assembly)
3. Jacketed flexible cord of not less than 7.6 m (25 ft) in length connecting the lamp assembly and the remote assembly

These luminaires have been investigated for installation with the top of the lens not less than 200 mm (8 in.) below the top of the pool. A hole through the pool wall may be required for luminaire installation. Unless otherwise indicated in the luminaire's installation instructions, the luminaire design has been investigated for the lower edge of any hole that a luminaire installer must cut in the pool wall to be no more than 360 mm (14 in.) below the top of the pool wall. The pool wall manufacturer may provide, at a greater depth, a properly sized hole or a reinforced wall section designed for field-cutting a properly sized hole for a luminaire or plumbing fitting. Unless otherwise marked for a maximum installation depth, these luminaires have been investigated for installation in such a hole at a greater depth where the pool installation instructions provide for the hole placement and usage.

Underwater Luminaires for Aboveground Nonstorable Swimming Pools—These luminaires are a type of through-wall lighting assembly as described in Art. 680 of the NEC. They have been investigated for permanent installation through or on the wall of an aboveground nonstorable pool. The information provided above for underwater luminaires for aboveground storable swimming pools regarding installation depth and using an existing hole or cutting a new hole for installation also applies to underwater luminaires for aboveground nonstorable swimming pools.

Convertible Underwater Luminaires for Aboveground Swimming Pools—These luminaires are initially configured as an underwater luminaire for aboveground storable swimming pool for use as described above. They include provisions for the one-time field conversion of the luminaire to an underwater luminaire for aboveground nonstorable swimming pool for use as described above. Once converted, these luminaires are not suitable for being modified back to their original configuration.

680.34. Receptacles. Receptacle outlets must be at least 1.8 m (6 ft) from the pool. How this can be enforced with what is in effect an appliance that can be set anywhere the ground is flat and for which no electrical permit will issue is anyone's guess.

680.41. Emergency Switch for Spas and Hot Tubs. This switch is not a 680.13 maintenance disconnect. It is an emergency stop button for the spa motor, and may or (usually) may not operate across the line. It must be clearly labeled, adjacent to the unit and within sight, but at least 1.5 m (5 ft) away. The rule does not apply to single-family dwellings.

Its sole function is to allow one or more people using a spa, or even a passerby, usually in a commercial environment, to shut down the motor in the event someone becomes entrapped by suction from the intake of the pump. There have been a number of tragedies from this problem, with documented instances of children being eviscerated after sitting over an intake and the event in New Jersey that directly led to this requirement, where a woman was trapped and drowned as a number of football players tried and failed to pry her away from a broken intake. The real solution to this problem is nonelectrical and has been implemented in the product standards by now, in the form of multiple intakes widely separated so the complete obstruction of one only sends the suction elsewhere, changes to the intake port designs, and other safety improvements.

680.42. Outdoor Installations. Outdoor units must meet the rules in Part II governing permanently installed pools, except as they are varied in this section. There are four such changes. Part **(A)(1)** allows flexible connections in any length; as of the 2014 **NEC** the former 6-ft limitation has been removed. This procedure is often used to bring a feeder up under a skirt of a spa or hot tub to a panel (makes the wiring a feeder) or to a "control panel" (the wiring lateral may be a branch circuit depending on conditions). Note that this permission only applies to listed units, both the packaged equipment assembly versions and the self-contained spa or hot tub versions that include the tub vessel.

Part **(A)(2)** allows flexible cord up to 4.6 m (15 ft), thereby extending the usual 900 mm (3 ft) limit in 680.7. GFCI protection is required. Remember that the circuit ratings on this equipment typically run in the 40 to 60 A range. Since this is an outdoor installation, 406.9(B)(2)(a) will ask for an in-use cover for this cord. This may necessitate a mobile-home type receptacle with an angle cord cap on the cord with its grounding pin set opposite to the cord exit.

Part **(B)** allows the omission of bonding connections to the metal bands or hoops that hold the wooden staves in place at a hot tub. All other requirements in 680.26 regarding equipotential bonding must be met. As of the 2014 **NEC**, the bonding requirements in 680.26(B)(2) no longer apply to listed *self-contained* spas and hot tubs that are evaluated for aboveground use outdoors and that are located on or above grade in accordance with all listing requirements. To qualify for this waiver, mentally extend a plane 30 in. horizontally around the perimeter of the tub or hot tub from the height of the rim. Bonding of adjacent conductive objects can be omitted if all permanently installed surrounding equipment or other surfaces are not less than 28 in. below that plane. Nonmetallic steps are ignored in making the measurements.

Part **(C)** allows cabled wiring through the interior of a dwelling, whether single-family or part of a two-family or multifamily building, or an accessory building thereto, such as a detached garage. Any Chap. 3 method is allowed and the requirement for a 12 AWG minimum copper equipment grounding conductor as part of the circuit was discontinued in the 2017 **NEC**. This rule is intended to apply to a branch circuit, which is why there is an informational note pointing elsewhere (680.25) for feeders. This allows Type NM cable to run indoors to an outlet on the side of the house or outbuilding, or at that point a transition can also be made to other wiring methods that are recognized in 680.25(A) if the spa or hot tub location is some distance away.

Underwater luminaires, however, must follow the rules for underwater luminaires generally, which would involve a more restrictive list of wiring methods not including Type NM cable. Therefore they would need to be split from the feeder because the last sentence requires compliance with 680.23. This is why the sentence was placed in this section to begin with; there was never any intent to make a variance for something as potentially hazardous as an underwater luminaire. The interior circuit would require a 680.23(F)(1) wiring method and, perhaps more importantly, a 680.23(F)(2) unbroken equipment grounding conductor return path that extends back to the point where the branch circuit originates. Another option given is a luminaire for a storable pool; this could be brought out on a conventional lighting circuit and the requisite luminaire, or luminaire and transformer, could be plugged in.

680.43. Indoor Installations. Indoor applications require Chap. 3 wiring with an allowance for a cord-and-plug connection for a listed unit rated 20 A or less. The first exception is legitimate because the rule is for a Chap. 3 connection, and the little plug-in models used indoors by definition do not rely on a Chap. 3 connection. The second exception is another matter entirely. It is placed as an exception to the parent language of 680.43, covering indoor installations. The text under exception requires compliance with Parts I and II (which do, of course include 680.26) "*except as modified by this section.*" In fact, 680.43(D), which is incontrovertibly part of Sec. 43, does not require a perimeter bonding grid, only bonding of the specified items on the five-part list. The exception is an exception to nothing, was never required, and should be deleted. The third exception, new in the 2014 NEC, extends the principle of 680.42(C) to indoor applications.

Part **(A)** requires that at least *one* 15- or 20-A, 125-V convenience receptacle must be installed at a spa or hot tub—not closer than 1.83 m (6 ft) from the inside wall of the unit and not more than 10 ft (3.0 m) away from it. This is intended to prevent the hazards of extension cords that might otherwise be used to operate radios, TVs, and so on (Fig. 680-20).

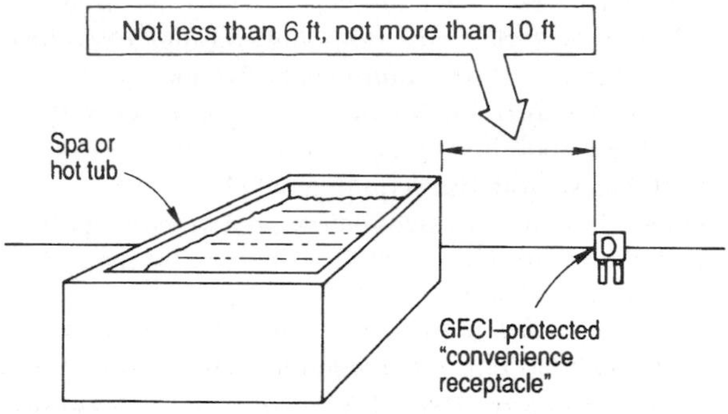
Not less than 6 ft, not more than 10 ft
Spa or hot tub
D
GFCI–protected "convenience receptacle"

Fig. 680-20. At least one 15- or 20-A general-purpose receptacle must be installed at a spa or hot tub. (Sec. 680.43.)

As required by part **(2)** of this section, this receptacle and any others within 10 ft (3.05 m) of the spa or tub must be GFCI protected, including the one supplying the power to the unit. The distance measurements do not pass through doorways, etc., similar to other comparable provisions in the article.

Part **(B)(1)** classifies luminaires over a spa or hot tub by height, with the limitations becoming more severe as the luminaire descends. First, if the luminaire is at or over 3.7 m (12 ft) above (measured to maximum water level, which in this case would be the tub rim as per the definition), there is no further limitation. Second, if the lighting outlet is GFCI protected, the luminaire can be mounted as close as 2.3 m (7 ft 6 in.). Finally, if the luminaire is GFCI protected and recessed with either an electrically isolated metal trim or a nonmetallic trim, it can be even closer to the spa or hot tub. The same condition applies to surface-mounted luminaires with a glass or plastic globe and a nonmetallic body or a metallic body isolated from contact. In both of these circumstances the luminaire must be listed for a damp location.

Part **(B)(2)** governs underwater luminaires. Refer to the commentary on 680.42(C) at the end; the last sentence of that rule is the same as this wording and will be applied in the same way.

Part **(C)** requires the same 1.5-m (5-ft) exclusion zone around hot tubs and spas as 680.22(C) does for pools. There is no provision in this location that would recognize a barrier, so the distance cannot be decreased.

Part **(D)** requires bonding of the usual metal parts with 1.5 m (5 ft) of the spa or hot tub, including the pump motors. And there is the usual exception for incidental metal parts such as air and water jets, isolated plumbing fittings, towel bars, etc. Control devices within the 1.5 m (5 ft) zone must be bonded; if further away bonding is optional. A major change for the 2008 **NEC** was that by virtue of what is now an exclusion in (2) a listed self-contained spa or hot tub need not have its pump motors and the associated equipment meet the normal bonding rules.

Part **(E)** allows metal-to-metal contact on a frame or base and threaded piping interconnections for bonding methods, along with the usual 8 AWG solid copper. Part **(F)** requires grounding connections to the spa or hot tub equipment along with other electrical equipment within the 1.5 m (5 ft) zone, which would include such items as a thermostat for electric heat. Part **(G)** requires compliance with all rules in 680.27(A) for underwater audio equipment.

680.44. Protection. The default rule for all spas and hot tubs given here is that they will have GFCI protection in place for the outlet that supplies them. There are two circumstances where this will not be the case. The first concerns listed units, both self-contained units with the tub and equipment assemblies without, that have a system of integral GFCI protection for all components. Such units must be marked accordingly. The second exemption concerns field-assembled units that exceed in rating what is available in GFCI protective devices, specifically, over 250 V, or 3-phase, or over 50 A in heater load.

680.50. General. Part V of the article covers fountains, with a few exclusions. A fountain that is part of a body of water in common with a pool is covered by the requirements for pools generally because they are more severe. Portable fountains are appliances and covered in Art. 422.

680.51. Luminaires, Submersible Pumps, and Other Submersible Equipment. All of this equipment must be GFCI protected unless listed for operation at the low-voltage contact limit or less and fed with a transformer that meets 680.23(A)(2) for equivalent pool equipment. Luminaires are limited in operating voltage to 150 V for the "supply circuits" and for submersible equipment including pumps to 300 V. The "supply circuit" wording allows for an HID luminaire with a higher starting voltage provided its line connection remained at 120 V. Note that the reach of this paragraph, that used to apply to all fountain equipment, now only applies to submersible equipment of the types specified. For example, there are fountains in outdoor northern areas that have 480-V heaters keeping them from freezing. This is not submersible equipment and beyond the reach of this paragraph.

Part **(C)** requires that luminaires must be installed with the lens below the normal water line, unless listed for use above water. If the luminaire faces up, the lens must be guarded against contact by the public, or it can be specifically listed for use without a guard. 680.23(A)(6) for regular swimming pools has a similar requirement, and UL requires distinctive installation instructions for those luminaires that can be used in the floor. Luminaires that are listed for use in fountains carry the term "submersible" in their description. For example, you might see a "wet-niche submersible luminaire," or a "special use submersible luminaire." These are not listed to the same requirements as luminaires listed for swimming pools, and they must not be confused with the other luminaires.

Part **(D)** states that luminaires and other equipment that must be submerged in order to avoid overheating must have a low-water cutoff or equivalent protection. In the case of luminaires, they will be marked "Submerse Before Lighting" and this will be visible after installation. This rule is similar to 680.23(A)(7) for lights in conventional swimming pools that must be submerged before lighting, which must be "inherently protected against the hazards of overheating."

Part **(E)** limits equipment wiring entries such that they must either be threaded hubs or have suitable flexible cord, and any metal parts in contact with the water must be brass or other corrosion-resistant material. Where flexible cord is used, it must not extend more than 3.0 m (10 ft) in the open within the fountain. If the cord extends beyond the fountain, it must run in an "approved wiring enclosure." This could be a conduit, provided the conduit material was suitable for its location.

Part **(F)** requires equipment to be removable from the water for normal maintenance or relamping. Luminaires must not be embedded in the walls of the fountain such that the fountain must be drained in order to reach them for maintenance. Part **(G)** requires equipment in the fountain to be stable so it will not be likely to tip over, or it must be secured.

680.52. Junction Boxes and Other Enclosures. Junction boxes that are not located in the water must meet the same construction requirements as for enclosures connected to conduits leading to forming shells as covered in 680.24.

Part **(B)** covers underwater enclosures. These junction boxes and other submersible enclosures must:

- Be made with threaded conduit entries, or else be provided with appropriate gland or other sealing connectors for flexible cord.
- Be made of copper, brass, or other suitable corrosion-resistant material such as some types of stainless steel.
- Be potted with an approved compound. Wax is often used for this purpose.
- Be firmly attached to the fountain surface or supports, and bonded as required.

The remainder of the section applies 314.23(E) and (F). Nonmetallic conduit cannot serve as the sole support of a box, and if it is used, the enclosure must have additional supports that are suitably corrosion-resistant so as to survive in the fountain water. If metal conduit is the support, two entries would be required, and it must have the requisite corrosion resistance.

680.53. Bonding. This rule requires piping to be bonded to the branch-circuit equipment grounding conductor. An external bonding conductor would not be required, since the box must be connected to that conductor in its role as an equipment grounding conductor anyway, per 314.4.

The rule for bonding fountains is a very simple one, but it contrasts dramatically with the allowance in 680.26(A)(6) Exception that allows DI motors to be left out of the bonding grid. Metallic piping systems associated with a fountain must be bonded to the equipment grounding conductor of the branch circuit supplying the fountain. Note that this part of the article on fountains does not specify how the bonding is to be achieved, or any minimum size bonding conductor, etc. It only states that certain piping is to be bonded.

680.54. Grounding. All electrical equipment within 1.5 m (5 ft) of the inside walls of the fountain, and all electric equipment associated with the recirculation system must be grounded. Listed low-voltage luminaires not requiring grounding are exempt from this requirement. In addition, item (3) requires panelboards that supply a fountain to be grounded. Note that 408.40 imposes a grounding rule on all panelboards anyway.

680.55. Methods of Grounding. Essentially the rules that govern an underwater luminaire apply here. Specifically, the following rules for conventional pools, as listed in Part **(A)** come forward:

- 680.21(A), pool pump motor wiring
- 680.23(B)(3), equipment grounding provisions in cords to underwater luminaires
- 680.23(F)(1), wiring method rules for branch-circuit wiring for underwater luminaires
- 680.23(F)(2), equipment grounding rules in the above branch circuits
- 680.24(F), equipment grounding provisions for supply circuits
- 680.25, feeder wiring methods and grounding

Part **(B)** has its own rule for grounding equipment supplied by cord and plug. The equipment grounding conductor must be an integral part of the cord, and it must connect the equipment grounding terminal at its supply end with all exposed noncurrent-carrying metal parts of the equipment. This is comparable to rules in 680.7, but it does not include the 900 mm (3 ft) limitation in 680.7(A). By setting the rule up this way, the cord can exceed 900 mm (3 ft) in length. This, in turn, correlates with the allowance for up to 3 m (10 ft) of flexible cord to run within the fountain, as covered in 680.51(E).

680.56. Cord-and-Plug-Connected Equipment. All electrical equipment under this heading, including the power-supply cords, must have GFCI protection ahead of it. Although this only applies to cord-and-plug-connected equipment, 680.51(A) essentially requires GFCI protection on almost everything else.

Part **(B)** requires that flexible cord that is immersed in or subject to being splashed by the water must be rated for hard-service and also listed with a "W" suffix. This qualifies the cord for continuous submersion.

The ends of the flexible cord jacket, the conductor terminations, and the grounding terminations must all be set in a suitable potting compound so if water somehow gets into the cord, it cannot then get into the equipment. Now that listed potting compound is available for wet-niche grounding terminations [680.23(B)(2)(b)], this might be a good choice for these terminations as well.

Connections made with flexible cord for underwater equipment must be permanent. Although this paragraph allows for attachment plugs and receptacles, this is only for maintenance or storage of equipment that is not in an area of the fountain that contains water. In other words, submersible equipment supplied by a flexible cord cannot have that cord run out of the fixture to a receptacle in some remote, dry location.

680.57. Signs. This section covers signs in fountains, and also signs adjacent to fountains out to a distance of 3 m (10 ft) from the fountain. Part **(B)** requires GFCI protection on all circuits supplying a sign covered in this section, whether installed on the feeder level or on the branch circuit(s). Part **(C)** prohibits fixed or stationary signs in fountains from being within 1.5 m (5 ft) from the inside walls of a fountain. Portable signs must not be placed within a fountain or outside a fountain within 1.5 m (5 ft) of the inside walls of the fountain. Part **(D)** requires local disconnects for the sign in accordance with both 680.13 and 600.6; both requirements can be met with a single device. Part **(E)** requires bonding connections in accordance with 600.7. Note that 600.7(B)(8) permits bonding in accordance with 680.53, which in turn enables a bonding connection to local metal piping.

680.58. GFCI Protection for Adjacent Receptacle Outlets. All 15- and 20-A 125-V through 250-V receptacle outlets within 6 m (25 ft) of a fountain must have GFCI protection. Note that this applies not just outdoors, but also within shopping malls and other indoor fountain locations. There is no allowance here for a waiver based on a locking configuration. This may prove problematic when running some maintenance equipment if the receptacle near the fountain is the only one for a great distance.

680.60. General. This is the beginning of Part VI on therapeutic tubs and pools. These are used for therapy in health care facilities, athletic training areas and similar areas. Portable therapeutic appliances need only comply with Art. 422 and do not come under these requirements.

680.61. Permanently Installed Therapeutic Pools. If a therapeutic pool is installed in or on the ground, or within a building, such that it cannot be readily disassembled, then it must comply with all the rules for normal permanently installed pools, as covered in Parts I and II of the article. The only exception is for luminaires; they need not observe either the placement or the GFCI restrictions in 680.22(B)(1 through 4) if they are totally enclosed. The special rules in Part VI of the article discussed here only apply to therapeutic tubs that are essentially stationary. That is, they can be disassembled but are normally left in one place, per 680.62 (below).

680.62. Therapeutic Tanks (Hydrotherapeutic Tanks). These rules roughly correspond to 680.44 for (A), and 680.43 for (B) through (F) on indoor spas and hot tubs.

Part **(A)** requires that any of the subspecies of therapeutic tanks (See 680.2 for the following definitions: Packaged Therapeutic Tub or Hydrotherapeutic Tank Equipment Assembly; Self-Contained Therapeutic Tubs or Hydrotherapeutic Tanks, and then consider the undefined term used here, Field-Assembled Therapeutic Tub or Hydrotherapeutic Tank) must normally have GFCI protection. Listed self-contained units or listed packaged equipment assemblies that incorporate integral GFCI protection for electrical components need not be wired with additional GFCI protection. Field-assembled units with higher heating loads than 50 A, or that are three-phase or rated over 250 V do not require GFCI protection.

Part **(B)** covers bonding. Just as swimming pools need a bonded environment for safety, tubs for therapeutic uses do as well. For stationary therapeutic tubs such as those at athletic training rooms and health care facilities (whether indoors or not), the following items need to be bonded together:

Any metallic fittings within or attached to the tub structure.

Metal parts of the circulation system, including the motor. Note that there is no allowance, at least as yet, for an unbonded DI pump motor.

Metal conduit, metal sheathed cables, and metal piping within 1.5 m (5 ft) of the inside walls of the tub, except items isolated by a permanent barrier. Note that this wording is not included in the bonding rules for conventional tubs in 680.43(D)(3). However, because 300.12 requires cable armors to be continuous between enclosures, and because the equipment must be bonded, presumably any metallic cable armor connected to a conventional spa or hot tub will be effectively bonded anyway.

All metal surfaces within 1.5 m (5 ft) and not separated by a permanent barrier from the tub area. A new small parts exception waives the requirement for many small, incidental parts such as metallic rims on water jets and drain fittings (provided they are not connected to metallic piping), towel bars, mirror frames, etc. However, unlike the comparable exception for indoor hot tubs, this exception does not exempt small parts if they are connected to metal framing members, such as metal studs. For this reason, 680.43(D)(4) Exception No. 1 covering conventional spas and hot tubs makes generally the same waiver, although as noted, that exception has no mention of metal framing members.

Electric devices and controls not associated with the tubs must be at least 5 ft away, unless they are bonded to the tub. This means that such items as a thermostat for electric heat must be bonded if they are within the 5 ft radius.

Note on the 1.5 m (5 ft) dimension. The Code is very unclear as to how this should be measured, in a radius or horizontally. The wording of 680.43(B)(1) implies a horizontal measurement by default, because luminaires are excluded from a 1.5-m (5-ft) zone to a height of 2.3 m (7½ ft) above the maximum water level. If we were to measure the 1.5-m (5-ft) distance in a radius, then a luminaire outside the tub would have to be 2.3 m (7½ ft) above the water outside the tub, but could be just over 1.5 m (5 ft) above the water level within the tub enclosure, which makes no sense. Nevertheless, the word "horizontal" is absent from this subsection, as well as subsection (D) on bonding, and the local authority must decide whether it should be inferred from the context. There is not any height limitation expressed either. The 3.7 m (12-ft) GFCI limitation in 680.43(B)(1)(a), and the same bonding limit in 680.26(B)(7) Exception No. 3 suggest that dimension as a reasonable limit. In addition, 680.43(A)(1) uses the phrase "measured horizontally" in the rule on receptacle placements.

Part **(C)** covers the methods of bonding. The bonding methods are similar for conventional and therapeutic tubs, but not identical. Both types of spas and tubs recognize the continuity of threaded metal piping entering a mated fitting. Both types also allow for metal-to-metal mounting on a common base or frame, just as for outdoor spas and hot tubs. Only therapeutic tubs, however, recognize "suitable metal clamps" for bonding. This may be a distinction without any practical difference, since metal-to-metal connections to a frame are recognized. Both types also require solid 8 AWG copper as the minimum for a separate bonding conductor. The solid conductor has far greater inherent resistance to chemical attack.

Part **(D)** covers grounding, and requires that all electrical equipment within 1.5 m (5 ft) of the inside walls of the facility, and that all electrical equipment associated with the recirculation system, must be connected to the equipment grounding conductor. Note that Chap. 6 rules cannot supersede rules in Chap. 5. If a therapeutic tub is used for patient care in a health care facility, then the enhanced grounding requirements in 517.13 will apply. An insulated copper equipment grounding conductor must be run to the unit, and the enclosure or armor of the wiring method must independently qualify as an equipment grounding return path. This includes metal-clad, armored, or mineral insulated cables with the outer armor fully qualified as a grounding path, most rigid metallic raceways, and some flexible metallic raceways up to 1.8 m (6 ft) in length.

Portable equipment must comply with 250.114, which requires equipment grounding connections for some equipment, and also the possibility of double-insulated appliances that will not have the grounding connection.

Part **(E)** requires receptacles within a 1.83 m (6 ft) radius to be GFCI protected. Part **(F)** requires luminaires to be totally enclosed, but does not impose any GFCI requirements on the lighting.

680.70. General. Unlike other general statements for parts within Art. 680, this one covering hydromassage bathtubs does not bring in other parts of the article. On the contrary, it positions this part as a stand-alone part, and no other part of Art. 680 applies to this equipment.

680.71. Protection. Any hydromassage bathtub and its associated electrical equipment must be supplied by an individual branch circuit and the outlet must be protected by a GFCI that must be readily accessible. All receptacles within 1.5 m (5 ft) of the tub and rated 125 V and not over 30 A must have GFCI protection as well. Note that most such tubs go in bathrooms where all the receptacles will be GFCI protected anyway, but these tubs do show up in other areas, and additional receptacles in the vicinity of the tub will require GFCI protection.

It is quite common to place the GFCI protection for one of these tubs as a GFCI receptacle adjacent to the pump motor, which is accessible (assuming 680.73 is met) but rarely readily accessible. The readily accessible requirement, new in the 2008 NEC, will change this practice. In theory, it should be possible in some locations to provide an access panel that is so well made, with obvious handles and unlatching mechanisms, set to open into a completely open area, and so uncluttered behind it, that an untrained person using no tools can quickly remove the panel and immediately find and put hands on the receptacle. That is possible, subject to the inspector's judgment, but not likely. In reality, too many of these receptacle connections require a double-jointed midget with a mirror and a flashlight to find and reset. There are several good solutions to this. One is to use a GFCI circuit breaker, since the panelboard devices must be readily accessible. Another is to use a master-trip GFCI device, the so-called faceless GFCI that takes up one device gang and presents test and reset buttons but no receptacle slots. Remember that a remote GFCI feed through receptacle is not an option because, by virtue of its installation, the circuit it protects is no longer an individual branch circuit as this section requires.

680.72. Other Electric Equipment. Hydromassage bathtubs are not subject to the requirements for spas and hot tubs. Receptacles do not have to be at least 5 ft (1.5 m) from the tub's inside wall, and luminaires do not have to be mounted at least 7½ ft (2.3 m) above the tub's water level.

680.73. Accessibility. This rule reiterates the requirement of 110.26, but does so specifically directed to this equipment. Many hydromassage tub motors have been set behind tiled partitions for which there was no access whatsoever. Effective communication with the general contractor on this as well as many other issues is absolutely essential to the progress of work. In addition, if the hydromassage bathtub is cord-and-plug-connected to a supply receptacle accessible through a service opening, the receptacle must be installed so it is within 300 mm (1 ft) of the opening, and with its face in direct view. The substantiation pointed to the same issues discussed at 680.71 above. An electric appliance that is difficult to disconnect will likely be an appliance worked hot.

680.74. Bonding. Both the metal piping system and the metal parts in contact with the circulation system must be bonded together with an 8 AWG solid copper bonding conductor, connected to the bonding terminal on the motor. Metal wiring methods, exposed metal surfaces, and electrical devices and controls within a 5-ft limit (and not separated by a permanent barrier) whether or not part of the tub assembly join the list. There is a small parts exception, but an adjacent metal faceplate would be included. The 5-ft dimension applies "from the inside walls" and in this case the literal text applies by mentally taking a string that long and include anything it touches, up and out, as it is rotated around the perimeter of the tub. The result, significantly increased in the 2017 NEC and clearly based on 680.26(B)(7) for swimming pools, is plainly excessive. For a bathtub a better model might be the self-contained spa provisions in 680.42(B) that approximate a reaching exposure for someone in the tub, or even back to the 2014 edition, which put metal piping systems together with conductive surfaces in contact with the circulating water and called it a day. The principal reason for an equipotential grid is to mitigate a drowning hazard cause by disorientation from voltage gradients. That is a valid concern for a hot tub, but much less so here. Double-insulated motors are exempt from this requirement. However, the bonding conductor must be run to the pump location, long enough to bond a conventional, non-DI motor if one is purchased to replace the original. This is the same principle as that used in 680.26(A)(6)(a) for permanently installed pools. Note that the requirement is to bond piping systems. With today's increasing use of nonmetallic water piping systems there is frequently nothing to bond under the skirt anyway. A metal escutcheon around the faucet with no metal piping behind it is not a metal piping system. It is very possible that even with a motor with a bonding lug, there will be no opportunity to run a bonding conductor. A bond wire must bond at least two things, and increasingly, there is no second item requiring bonding. Neither is it required to run a bonding conductor to the panel or anywhere else. Refer to the bonding discussion at 680.26 for the reasons why this is so.

Part VIII. Electrically Powered Pool Lifts. This part, new with the 2017 NEC, addresses the increasing use of powered lifts, now often required by law for public pools, to allow disabled individuals to enjoy the water. They must be listed for the application in a swimming pool environment unless there are no issues in terms of exceeding the low-voltage contact limit. Lifts run on power circuits must have GFCI protection, and they must be connected to the bonding grid regardless of power source. Switches operating above the low-voltage contact limit must be separated 5 ft, or listed for a closer approach. Battery power lifts must indicate a specific reference and voltage rating, along with the wattage rating.

ARTICLE 682. NATURAL AND ARTIFICIALLY MADE BODIES OF WATER

682.1. Scope. This article, new in the 2005 NEC, provides the first systematic coverage of electrical installations in or adjacent to bodies of water that do not qualify under Art. 680, such as artificial water features as part of landscaping, aeration ponds and storm retention basins, and fish farms in artificial ponds. As defined, the water depths can vary seasonally or be controllable. The article also addresses wiring in natural bodies of water, such as submersible pump in a natural pond for irrigation or other reasons.

682.2. Definitions. Aside from the descriptions of the bodies of water, three definitions are essential to apply this article. The most important is the *electrical datum plane*, which is (oversimplified) basically the horizontal plane that is 600 mm (2 ft) above the highest normal water level, such as the highest high tide. It is not the height of some catastrophic flooding event, such as the level of the 200-year flood as may have been established for other regulatory purposes such as flood insurance boundaries. The actual definition is in four parts and is much more complicated, but this will usually work.

An *equipotential plane*, for the purposes of this article is an area where wire mesh or other conductive elements are on, within, or under a walk surface not more than 75 mm (3 in.) below the top surface, and these conductive elements are bonded to all metal structures and fixed nonelectrical equipment that may become energized, and connected to the electrical grounding system to prevent a difference in voltage from developing within the plane.

The *shoreline* is the farthest extent of standing water under the applicable conditions that determine the electrical datum plane for the specified body of water.

682.10. Electrical Equipment and Transformers. This is a good example of how the electrical datum plane is used in this article. Here electrical equipment and transformers including their enclosures must be specifically approved for the intended location. Only enclosures that are identified by their manufacturer's published information as being suitable not just for a wet location, but rather for actual operation while submerged, can be used below the electrical datum plane, which is 2 ft above the normal high water mark.

682.11. Location of Service Equipment. This is a good example of where the shoreline definition is used. The service equipment for floating structures and submersible electrical equipment must be on land not closer than 1.5 m (5 ft) horizontally from the shoreline. The live parts must be elevated a minimum of 300 mm (12 in.) above the electrical datum plane, and there must be a shunt trip or equivalent arranged so the service disconnect will open if the water level reaches the height of the datum plane.

682.12. Electrical Connections. All electrical connections not intended for operation while submerged must be located at least 300 mm (12 in.) above the deck of a floating or fixed structure, but also, wherever located, not below the datum plane.

682.13. Wiring Methods and Installation. Liquidtight flexible metal conduit or liquidtight flexible nonmetallic conduit with approved fittings can be used for feeders and where flexible connections are required for services. Extra-hard-usage portable power cable listed for both wet locations and sunlight resistance is also permitted for a feeder or a branch circuit where flexibility is required. Other wiring methods, suitable for the location can be installed where flexibility is not required. Temporary wiring that meets the provisions in 590.4 can also be used.

682.14. Submersible or Floating Equipment Power Connections. Submersible and floating equipment must be cord-and-plug-connected, using extra-hard-usage cord, and with a "W" suffix. Note that this suffix, traditionally denoting water resistance, also guarantees, in accordance with Note 15 to Table 400.4, that the cord is sunlight-resistant. The plug and receptacle combination must be suitable for the location while in use. For the usual outdoor application, this would mean compliance with 406.9(B). In addition, the receptacle outlet must be on the load side of a disconnecting means such that the load can be deenergized without requiring the plug to be removed from the receptacle.

An exception applies to equipment anchored in place and incapable of routine movement by water currents or wind. This equipment can be connected (but does not need to be) through one of the Chap. 3 wiring methods cited in 682.13, such as liquidtight flexible nonmetallic conduit.

As covered in **(A)**, the receptacle disconnecting means (for cord-and-plug connections) and the equipment disconnecting means (for hard-wired equipment) must be a circuit breaker, switch, or both, or a molded case switch, and it must be specifically labeled to show exactly which receptacle or outlet it controls, reiterating 110.22(A).

Part **(B)** covers the disconnect locations. It must be readily accessible, on land, and within sight of but at least 1.5 m (5 ft) back from the shoreline, and not more than 750 mm (30 in.) from the receptacle it disconnects if applicable. It must be elevated at least 300 mm (12 in.) above the datum plane.

682.15. Ground-Fault Circuit Interrupter (GFCI) Protection. This covers 15-, 20-A 125- to 250-V receptacles that are outdoors, or if located in or on a floating building or a structure used for maintenance, storage, or repair. The rule applies to receptacles falling within the datum plane. GFCI protection must be provided for these outlets, with the protective device located at least 300 mm (1 ft) above the datum plane. As defined in 680.2, the datum plane is 600 mm (24 in.) above water level generally, and in the case of floating structures the plane is 750 mm (30 in.) above the water level and a minimum of 300 mm (12 in.) above the deck. The literal text of this section now imposes a GFCI requirement on all 15- and 20-amp 125- and 250-V receptacles installed outdoors and within the datum plane, which includes floating buildings. Article 553 has no such requirement, and in the absence of any mention, this rule will apply to floating buildings generally.

682.30. Grounding. This section brings over rules from Arts. 553 (floating buildings) and 555 (marinas and boatyards), specifically 555.15. The 555.15 reference includes unremarkable, well known grounding rules, with one major exception, found in 555.15(B). Due to the corrosive effects of a marine environment, metallic raceways are not permitted to be relied on as equipment grounding conductors, and must, where used, be backed up by separate equipment grounding conductors, sized per Table 250.122 and [per 555.15(C)] not smaller than 12 AWG. The rules in Part III of Art. 553 contribute a requirement (from 553.11) that all metal parts in contact with water and any metal piping, as well as the usual metal parts that may become energized, must be connected to the equipment grounding system. No cross connections between neutral conductors and equipment grounding conductors are permitted, courtesy of 553.9.

682.31. Equipment Grounding Conductors. These rules are obvious and are taken from 555.15(C), 555.15(D), 555.15(E), and 553.10(B), as already incorporated by reference in the prior section.

682.32. Bonding of Noncurrent-Carrying Metal Parts. This is taken from 553.11 (see above).

682.33. Equipotential Planes and Bonding of Equipotential Planes. This, start to finish, is critical information that breaks new ground. Much of this, although used in the NESC for utility work, is without precedent in the **NEC**. Review the definition in 682.2 before beginning. The section begins with the basic requirement to install such a plane as set forth in this section to mitigate "step and touch voltages" at electrical equipment. These terms are not defined in the **NEC**, but they are in the IEEE dictionary, as follows:

> The *step voltage* is the potential difference between two points on the earth's surface separated by a distance of one pace (assumed to be one meter) in the direction of maximum potential gradient. This potential difference could be dangerous when current flows through the earth or material upon which a worker is standing, particularly under fault conditions.
>
> The *touch voltage* is the potential difference between a grounded metallic structure and a point on the earth's surface separated by a distance equal to the normal maximum reach, approximately one meter. This potential difference could be dangerous and could result from induction or fault conditions, or both.

Part **(A)** requires an equipotential plane to be installed at the service equipment if outdoors, and (assuming access to personnel is provided) at metallic outdoor enclosures for disconnecting means that are likely to become energized and that control equipment in or on the water. The equipotential plane must begin below the equipment and extend outwards not less than 900 mm (3 ft) in all directions from which one could come in contact with the equipment.

Part **(B)** covers where the planes are not required, specifically, at the equipment supplied and disconnected by the items specified in Part **(A)**. However, all circuits rated 120 V through 250 V, single phase, equal to or less than 60 A, must have GFCI protection. Note that this is a circuit requirement, so point-of-use GFCI devices, such as GFCI receptacles, will not comply with this requirement.

Part **(C)** requires the equipotential planes to be bonded to the local grounding system, as also covered in the definition. The bonding conductor(s) must be solid copper, no smaller than 8 AWG, and the connections must be either exothermically welded or made up using copper, brass, or bronze, or stainless steel parts listed for the condition and application. Such connectors will be marked "DB" or "direct burial."

ARTICLE 685. INTEGRATED ELECTRICAL SYSTEMS

685.1. Scope. This covers a unitized segment of an industrial process and the attendant electrical system that is under highly qualified supervision and maintenance, and for which an orderly shutdown is essential to minimize hazards to personnel and damage to equipment. These issues arise at nuclear facilities and in the chemical process industries. As itemized in **Table 685.3**, many other Code articles have orderly shutdown allowances built into their provisions, and these are incorporated to the extent relevant for any particular application.

685.10. Location of Overcurrent Devices in or on Premises. This rule varies the customary readily accessible requirements and permits overcurrent devices to be out of reach, in elevated locations where they will not by casually operated by unqualified personnel.

The final two sections permit the use of ungrounded circuits, both two-wire dc circuits and 150 V or lesser control circuits from a separately derived system, as required for operational continuity. Note that the rule in 250.21(B) is not varied here, so ground detectors would be required on the ungrounded ac circuits.

ARTICLE 690. SOLAR PHOTOVOLTAIC (PV) SYSTEMS

690.1. Scope. This complete, detailed article covers this developing technology for direct conversion of the sun's light into electric power. Large-scale PV systems are now (2017 NEC) covered by Art. 691. For that reason most of the medium voltage requirements for PV systems are no longer included in this article.

690.2. Definitions. Solar photovoltaic sources require the collection of solar energy in equipment that can convert it to electricity. A group of *cells*, the base unit of solar production is put together by their manufacturer into *modules*, which are defined as a complete, environmentally protected unit capable of producing power when exposed to sunlight. As a practical matter, modules are usually assembled into *panels*, which are grouped sets of modules that are wired together to make a field installable entity. Multiple panels can be ganged in the field on framing to comprise the *array* that becomes the dc power production source for the occupancy. Figure 690-1 shows four modules in each of six panels to make up an array that is making power that is being used to offset energy that would otherwise have been purchased from a utility.

The wiring between the modules and the common dc connection point attached at the left of the array in Fig. 690-1 is *photovoltaic source circuits*, and the dc wiring from that point to the dc utilization equipment (if any) and to the inverter (if used, true in this case) is the *photovoltaic output circuit*.

A very important definition, describing a new term as of the 2017 NEC, is a *functional grounded PV system*. This is a system with an electrical reference to ground, but that is not solidly grounded. The informational note following explains that the functional connection will be through a variety of devices (Fig. 690-2 shows a fused connection) that result in the designated side of the system being at ground potential during normal operation, but likely at some voltage above ground during a fault condition.

This change, rather ingenious in its wording, is easily the most important in the entire article and one of the most important in the 2017 NEC cycle. It enables all PV systems to share a common topology. The article no longer refers to grounded and ungrounded systems. Most are all wired the same way. In fact, the overwhelming majority of NEC systems have been wired under this protocol for some time.

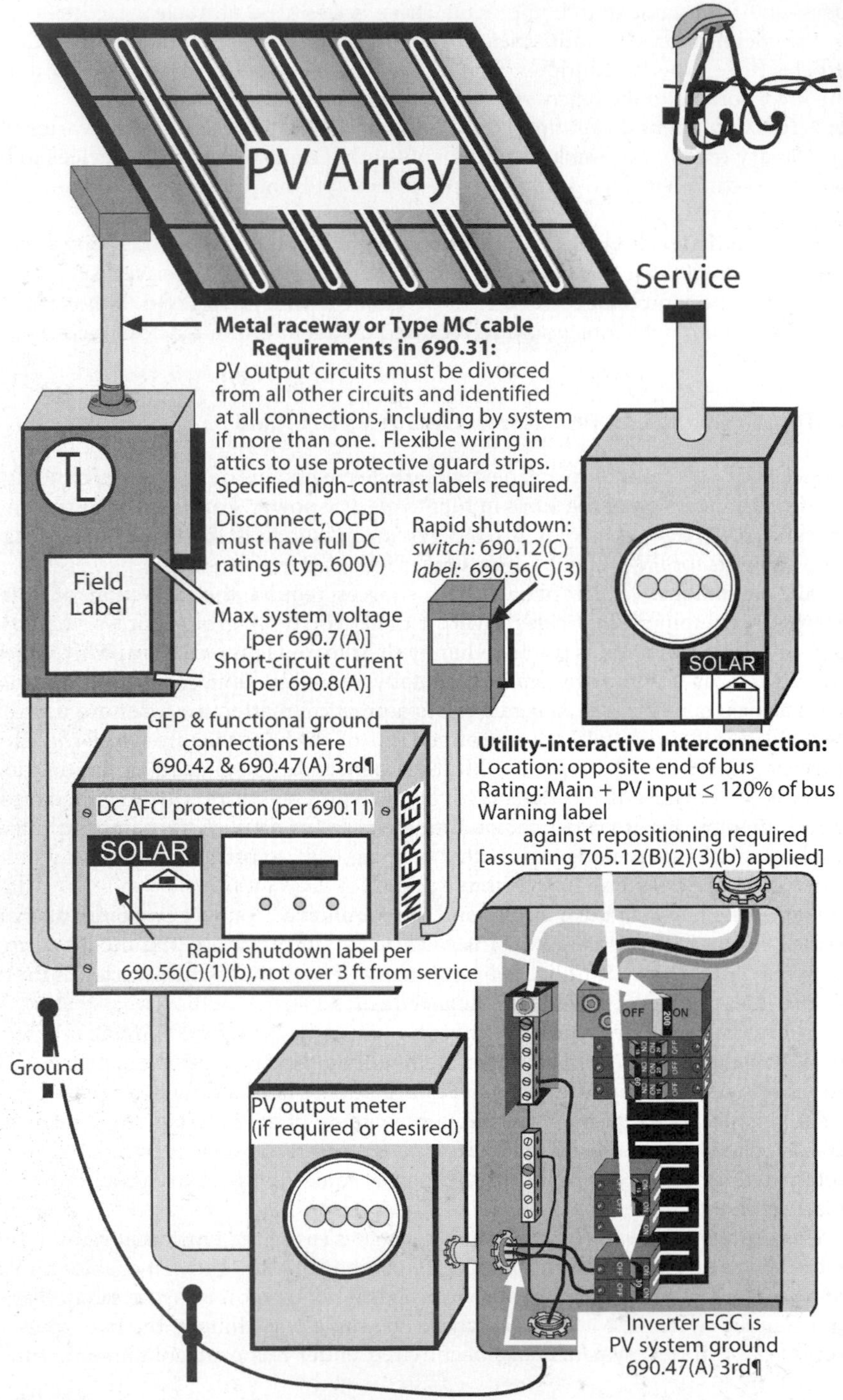
PV Array
Service
Metal raceway or Type MC cable
Requirements in 690.31:
PV output circuits must be divorced from all other circuits and identified at all connections, including by system if more than one. Flexible wiring in attics to use protective guard strips. Specified high-contrast labels required.
Disconnect, OCPD must have full DC ratings (typ. 600V)
Rapid shutdown: switch: 690.12(C) label: 690.56(C)(3)
Field Label
Max. system voltage [per 690.7(A)]
Short-circuit current [per 690.8(A)]
SOLAR
GFP & functional ground connections here 690.42 & 690.47(A) 3rd¶
Utility-interactive Interconnection:
Location: opposite end of bus
Rating: Main + PV input ≤ 120% of bus
Warning label against repositioning required
[assuming 705.12(B)(2)(3)(b) applied]
DC AFCI protection (per 690.11)
INVERTER
SOLAR
Rapid shutdown label per 690.56(C)(1)(b), not over 3 ft from service
OFF ON
Ground
PV output meter (if required or desired)
Inverter EGC is PV system ground 690.47(A) 3rd¶

Fig. 690-1. An interactive system involves many NEC rules (see text). (Adapted from *Practical Electrical Wiring*, 22nd ed., © Park Publishing, 2014, all rights reserved.)

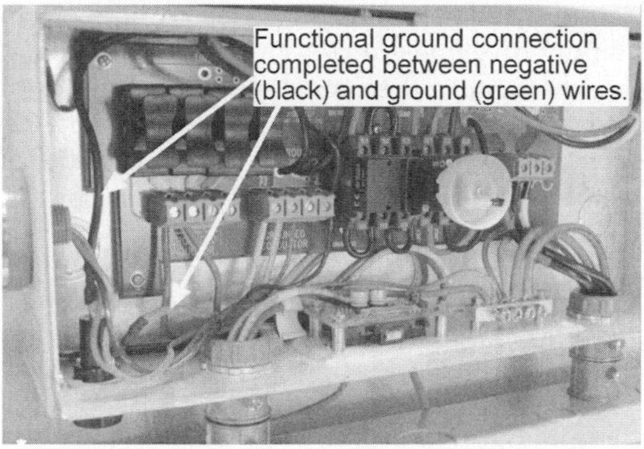
Functional ground connection
completed between negative
(black) and ground (green) wires.

Fig. 690-2. The black cylinder at the lower left is a fuseholder containing a small fuse (likely 0.5 or 1.0 amp.) inserted between the negative vertical (black) wire and the system grounding connection wire (green) leaving the fuseholder skirt at a 30 deg. angle upwards to the right. Note that some of the negative array conductors entering from below use white insulation. This wire color was permitted at the time the photo was taken, but now violates 690.31(B)(1). Courtesy, Solar Design Associates, Inc.

For just one example, any system wired in accordance with the ground fault provisions of the former 690.5 [now moved into 690.41(B)] was, and still is, one of these systems because one of its poles is connected to ground through a small overcurrent device. European standards have been arranged in this manner for some time.

690.4. General Requirements. Part **(B)** requires the equipment doing the heavy lifting in PV work, including the modules and panels, the combiner equipment, and inverters, be listed or field labeled for the application. Note that PV has many applications beyond **NEC** work, and some listings may not be relevant to premises wiring work governed by its provisions.

Part **(C)** requires that PV systems be installed by qualified persons. This work is becoming increasingly specialized, and just as in the case of medium-voltage splicing, specialized training beyond generic trade licensure is a necessary part of the work.

Part **(D)** permits the installation of multiple PV systems. If they are remote from each other, the reciprocal signage required by 705.10 must be provided at their disconnecting means. This will augment the labeling already required by 230.2(E), that provides information regarding other inverter locations.

Part **(E)** prohibits PV system equipment and disconnecting means in bathrooms.

690.6. Alternating-Current (ac) Modules. These modules have inverters built in, and their output is classified in Part **(B)** as an inverter output circuit. The former provision for ac ground-fault detection has been removed as of the 2014 **NEC** because no products are available to meet the requirement and conventional GFCI devices were being misapplied for this purpose.

690.7. Maximum Voltage. This section begins by defining the maximum voltage of PV systems in much the same way as in Art. 100 for conventional ac systems, but without the need for rms allowances. It is the highest potential difference (expressed as voltage) between two circuit conductors, or between any one conductor and ground. The section also limits limits the PV output circuit voltage in one- and two-family dwellings to 600 V dc. Other occupancies are limited to 1000 V dc. The parent language then concludes with an interesting sentence that does not actually address maximum voltage but does make a Chap. 6 amendment of Chap. 4 provisions, by excluding the rules in Art. 490 addressing the majority of medium voltage equipment, switchgear, industrial control assemblies, and substations (all of Parts II and III) from applicability to the listed dc equipment operating at 1500 V or less. This is perfectly appropriate given both the listing requirement and the comparatively limited range of voltage excluded.

Part **(A)** addresses the rating of PV source and output circuits. Modules are rated under standardized laboratory conditions at moderate temperatures. The photochemistry of the crystalline forms of silicon is such that under conditions of constant irradiance, the voltage increases as the temperature drops. For **NEC** application purposes, at a temperature of about –40°C or °F (the temperature scales cross at this point) is generally considered the worst case, and at this temperature the open circuit voltage put out by a photocell is about 125 percent of its rated voltage. For areas with other design temperature minimums, the table factors can be used accordingly. However, the 125 percent factor is the one usually applied. If the module comes with directions to use different adjustments, then those directions trump the numbers in Table 690.7. Looking at Fig. 690-1, the open-circuit rated voltage on a module is given as 64 V. The panel with four modules wired in series will generate 256 V under test conditions. However, applying the 125 percent factor gives a final result of 320 V, and that becomes the voltage rating of the PV output circuit derived from this system. An informational note points to an ASHRAE publication that is useful in determining the extreme cold weather temperature that would be applicable in any locality. Large scale PV systems (≥100 kW) qualify for an evaluation of voltage by a licensed PE, presumably using the standard covered in the informational note. The maximum voltage is what will be carried forward to determine the minimum voltage ratings of system components. The process as described in published literature anticipates the use of computer modeling and significantly increased accuracy.

Part **(B)** covers dc-to-dc converters, capable of changing voltages and thus doing for dc what transformers have done for ac since the days of Tesla. For a single converter the maximum voltage is simply its maximum rated output. However, two of these devices can be series connected in concert with any directions to this end that have been made part of the listing process, and those directions will establish the total maximum voltage. In the event such documentation is unavailable, then the maximum voltage is the arithmetic sum of the two maximum voltages at hand.

Part **(C)** covers bipolar source and output circuits. The maximum voltage is taken as the voltage between the two wires that run from each monopole subarray and not the combined voltage that results from the connections made in the inverter. See 690.31(I) for more information.

690.8. Circuit Sizing and Current. Just as voltage is determined under standard conditions that do not always reflect field conditions, the same is true for current ratings on modules. In this case maximum conditions of solar irradiance for the period running about 2 hours on either side of solar noon can exceed rated current output by, and the recurrence of this number yet again is an extraordinary coincidence, 125 percent. Therefore, this factor must always be applied to rated current outputs on modules. Referring to Fig. 690-1, the short-circuit current rating on the modules is given as 5.4 A. The panels are wired in series, so no change in current there. However the array has six panels wired in parallel [see Part **(A)(2)**], for a total of 32.4 A under test conditions. However, by this rule, this number must be increased by a 125 percent factor, with a resulting current for **NEC** purposes of 41A. Similar to the maximum voltage calculations in 690.7(A)(3), large scale PV systems (≥100 kW) qualify for an evaluation of current by a licensed PE, presumably using the standard covered in the informational note. The maximum current is what will be carried forward to determine the amperage ratings of system components. Note, however, that the engineered result must not be taken as less than 70 percent of the traditional calculation.

Part **(A)(3)** requires that the inverter continuous output current rating be taken as the maximum rated output of the inverter, or as in Fig. 690-3 inverters, regardless of whether the array size is capable of delivering that much energy to the inverter. In this case a 5 kW inverter operating on a 240 V circuit will be calculated as 5000 ÷ 240 = 21 A in output current. Part **(A)(5)** makes a similar requirement for dc-to-dc converters, and Part **(A)(6)** sums those results when they are connected in parallel. Part **(A)(4)** reflects the fact that stand-alone units are constant power devices that increase current when battery voltage declines.

Part **(B)** reflects the fact that under worst (best?) case conditions sunlight is a continuous phenomenon, and that therefore the circuit rating calculated in (A)(1) must be viewed as a continuous load. For conventional overcurrent devices, the

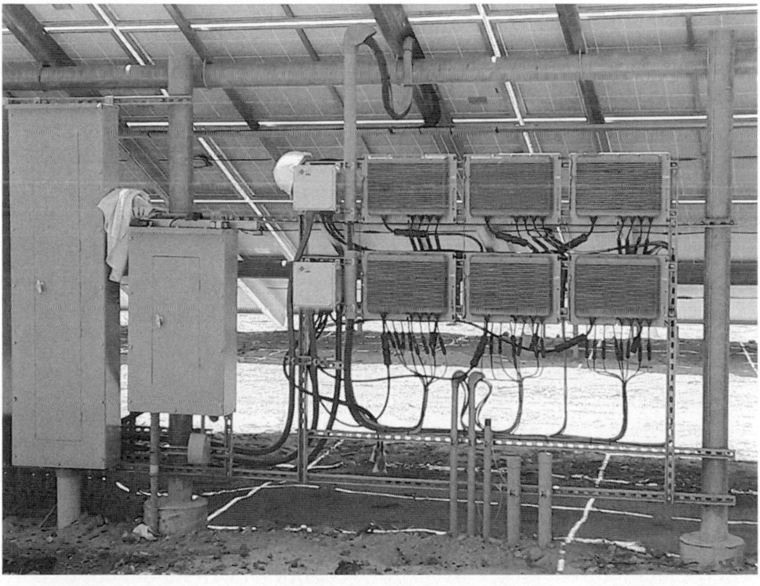

Fig. 690-3. Six string inverters with their ac outputs combined at junction boxes into two feeders, sized in accordance with the sum of the continuous output current ratings comprising their supply, and increased by 25 percent for continuous duty as required by 690.9(B)(1) and 215.2(A)(1)(a). Courtesy, Solar Design Associates, Inc.

result is another 125 percent multiplier, and for the array in Fig. 690-1, that is 41 A × 1.25 = 51 A. The other two paragraphs remind users of the special sizing rules regarding the next higher standard sizing and small conductors, along with UL guide card information about operating circuit breakers in high-temperature ambients. The second numbered topic in (B) also governs conductor ampacity. Once again start with a 125 percent factor for temperature, increase it again for continuous loading because that extra size will be needed as a heat sink when it arrives at a device, but compare that with the size developed from the conditions of use. Some design temperature allowances for array wiring anticipate a 70°C ambient, and after that is factored in, the wire may be larger than it is for continuous loading. And at the end of the process, make sure the overcurrent device protects the wire after all has been said and done.

Part **(3)** recognizes adjustable electronic trip circuit breakers with settings that can match the steps in 240.6, and that are rated for direct current. They do not require the 125 percent factor to be added, and therefore it need not be factored in to a conductor ampacity calculation where these devices are in use.

690.9. Overcurrent Protection. PV circuits are normally required to have overcurrent protection per Art. 240, with an allowance for conditions where the conductor could not be overloaded even if short-circuited due to the inherent lack of capacity in the system. These devices must now be listed for use in PV systems. Therefore the former separate rule requiring dc ratings has been discontinued because it is inherent to the PV listing. They need to be accessible but not readily accessible. These rules accommodate placements at the array in the form of fuses or circuit breakers, and this location will be required by the terms of the final sentence of 690.9(A) in the majority of cases. They are available in a wide variety of sizes for these applications. In Fig. 690-1, the current determined by 8(A)(1)—5.4 A × 1.25 = 6.75 A is then increased by 125 percent per (9)(B): 6.75 A × 1.25 = 8.4 A. The likely next higher available size is a 9-A device, so that is what would be used. Note that (9)(B)(3) correlates with (8)(B)(3) to recognize the adjustable electronic trip devices that, if used, could moderate this calculation.

Note that this fuse will require an acceptable fault clearing rating on a dc circuit. In the case of Fig. 690-1, the fault current is minimal because there are no batteries in the system. However, if there were standby batteries with thousands of amperes available to a fault, a calculation would need to be done as to the available current after the length of the circuit between battery and fuse is considered. Current-limiting fuses can also be installed on the battery connections that have a let-through low enough to coordinate with the fuses at the array. DC ratings on overcurrent devices significantly differ from their ac ratings because a dc circuit with no natural current zero is more difficult to interrupt, and this equipment must be listed for the dc duty that will be required.

Part **(C)** takes this one step further and requires a specific listing for PV use with respect to overcurrent devices provided for PV source and output circuits. A single device is an option to protect multiple module conductors. However this protection is configured, either all positive conductors, or all negative conductors of a PV system must be protected. Either side can be protected, and the protected side must always be the same polarity throughout the entire system. Disconnecting means, however, must open both sides (690.15). These devices are typically placed near the array, and as such they are subject to wide variations in current draw, ambient temperature, high open-circuit voltages, etc. Product standards are now in place to enable such listings. This provision, rewritten in the 2014 **NEC**, no longer permits the use of supplemental overcurrent devices in 1-A increments for this purpose. Choose from among the large numbers of fully PV-rated overcurrent devices now available to provide this protection.

690.11. Arc-Fault Circuit Protection (Direct Current). PV systems with dc source circuits or output circuits or a combination of both operating (maximum voltage) at or above 80 V must have listed dc arc-fault protection or equivalent. As of the 2014 **NEC** this requirement applies generally and not just to wiring on or penetrating a building. The protective device(s) must respond to any continuity failure and interrupt arcing faults. If it operates, that fact must display at an annunciator. Both the annunciator and the system device must not reset automatically. The product standard is now published and products are available for this function. The fact that this article now uses a common functional grounding topology will continue to assist the development of this technology, which generally relies on a device placed between one of the circuit conductors and ground. The listed combiner box shown in Fig. 690-5 includes this functionality. This is a good example of equipment that can be installed in multiple, thereby mitigating the amount of power reduction in the event of a failure.

An exception, new as of the 2017 **NEC**, does waive the requirement in instances where there would be little practical advantage in applying it. Specifically, PV systems not located on buildings, and that in addition have no system components that enter buildings, are exempt. In addition there are restrictive wiring method limitations. The exception only applies to output circuits run in metal raceways, or enclosed metal cable trays, or "directly buried." This raises the important question as to whether the exception is voided if the wiring is in PVC conduit run underground. On the one hand, such wiring is not, strictly speaking, "directly buried." On the other hand, the PVC conduit in this instance would be directly buried, and little seems to be gained by disallowing it. This will need to be resolved in the next code cycle. The exception also clarifies that wiring entering a utility building with no function other than to house PV equipment does not void the exception.

690.12. Rapid Shutdown of PV Systems on Buildings. This is the "rapid shutdown" section. In effect, PV system circuits on or within buildings must be capable of quickly and dramatically limiting their voltage, regardless of how bright the sun is shining, upon the command of a first-responder seeking access to the vicinity of an array for fire-fighting activities. This is proving to be one of the most technically demanding tasks of electrical standards writing. Everyone agrees on the objective, but the fact that the continuing source of power (sun) is beyond the reach of human intervention complicates the task. Note that the location aspect of the rule applies to "system circuits" and not just to the arrays. This means that a ground-mounted array that extends circuits to a building [see also **(A)** and the exception] will be usually be subject to the requirements.

Remember that the sun is still shining on the installed array, constantly generating voltage at all times during the day. To completely shutdown every element of a PV array is extremely challenging. To make the project manageable based on the current state of technology, Part **(B)** divides the implementation of rapid shutdown requirements into two separate zones based on a new concept the **NEC** refers to as the "array boundary." Wiring is inside this boundary if it is closer than one foot from the array in any direction, including downwards into, or even through the roof into the space below depending on construction and array separation distances. The definition clearly includes the supports, so the array boundary will almost always penetrate the roof to some degree. Conductors outside the array boundary, and more than 3 ft from an entry point into a building must respond to a rapid shutdown command sequence by falling to 30 V or less (line-to-line or line-to ground) within 30 seconds of the control activation. In fact the 3-ft limitation is to prevent someone from hugging the roof sheathing on the attic side and claiming to have not left the 1-ft halo. This technology is available and no implementation delay will be needed.

The actual PV arrays and associated wiring harnesses, etc. fall on the inside of the array boundary, and controlling what happens here, with the sun out, requires significant additional engineering and product standards work. For this reason a two-year delay, until Jan. 1, 2019 applies to this portion of the rapid shutdown provisions. There are three options, all of which have the two-year window before becoming mandatory. The first is to use a PV array that is either listed or field evaluated and labeled as a "rapid shutdown PV array." These would consist of arrays equipped with control features that allow them to respond, at the array, to a shutdown control command. An elaborate informational note at this point provides useful information regarding the anticipated design features that may be involved as the technology evolves. The second option is to install control equipment that can be placed at the array location and that will result in the voltage decreasing to 80 volts within 30 seconds of a shutdown being initiated. The third option involves the creation of what might be described as double-insulated PV arrays, that is arrays that are nonmetallic with no exposed wiring methods or other conductive parts, and positioned over 8 ft from any exposed conductive parts, or the earth. Any arrays meeting these requirements would be exempt from the entire rapid shutdown requirements, because they would not present any issues to first responders.

Part **(C)** covers the initiation device(s) for the procedure, which must be indicating. An "OFF" indication denotes a command procedure is in effect. The initiating device must consist of at least one of three options (more than one is permitted), beginning with the service disconnect. Also included is the PV system disconnecting means (690.13), and also a readily accessible switch intended for this function and so marked. An informational note correctly explains that this would be appropriate if an Article 702 standby system was in place that would keep premises loads energized regardless of the status of utility power. Whatever option is used, the switch(es) that actually invoke a rapid shutdown must comply with the marking requirements in 690.56(C). If there are multiple PV systems, then the shutdown devices must not number more than six, be grouped, and if auxiliary devices come into play, they must also control all PV systems involved with that service and having rapid shutdown provisions.

Part **(D)** requires equipment that performs the shutdown function, other than the simple initiation sequence, be listed for that purpose. An informational note advises that inverters not designed for this function often keep the energized input

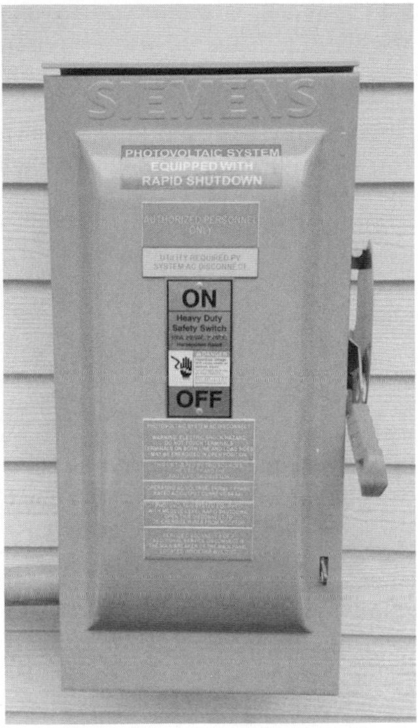
SIEMENS
PHOTOVOLTAIC SYSTEM
EQUIPPED WITH
RAPID SHUTDOWN
ON
Heavy Duty
Safety Switch
OFF

Fig. 690-4. The switch (shown in the open position) is the PV system disconnecting means. The utility (in this jurisdiction) required an open blade, outside mounted switch, connected ahead of the service equipment [permitted by 230.82(6)]. The switch is also wired as the rapid shutdown initiation device. Although installed under the 2014 NEC, the labeling largely meets the intent of 690.56(C)(3), although it misses the ⅜-in. height rule. The labeling does compliy with the 2017 NEC with respect to its system disconnecting function [690.13(B)]. Courtesy, Solar Design Associates, Inc.

conductors energized for up to 5 minutes, far longer than the shutdown provisions in this section permit.

690.13. Figure 690-4 shows the system disconnect for a system equipped with module-level microinverters. This system, far more expensive than a conventional system, is capable of discontinuing power at the module when the switch is OPEN. Remember a conventional inverter will shut down when it loses access to utility power, and these do the same. This means that this particular installation effectively meets the performance objectives of 690.12(B)(2) already. It offers improved function in terms of shadow tolerance; the effect of a shadow of a branch will have far less impact on the array output as a whole than would occur on a conventional string. It also offers benefits in terms of avoiding problems of module mismatch and unbalanced numbers of module connections in a given area.

Photovoltaic System Disconnecting Means. The PV system, at a readily accessible location **(A)**, must be fully disconnectable from all other premises wiring components, including energy storage systems, standby power sources, utility power, etc. This switch is intended to provide for safe servicing procedures on the PV system. It is not intended to limit the exposure of the building to energized conductors, and therefore there is no longer any proximity provision relative to the entry point of the conductors attached to this requirement. As explained in the informational note, if there is an emergency, there is now a rapid shutdown procedure to address it. Part **(B)** adds an advisory marking label to the switch

(Fig. 690-4) that voltage may be present on all terminals when the switch is in the open position. If the switch is placed on the line side of the service, as permitted in 230.82(6), Part **(C)** requires that it must be suitable for use as service equipment. *BE CAREFUL: This rule addresses certain construction features of the switch, and does not make it service equipment.* The reference to 230.82 necessarily places it on the line side of the service equipment, and 230.40 Exception No. 5 recognizes these connections as well. It does not need to be grouped with the premises service equipment. The requirements in 705.31 govern the distance and protection requirements for this wiring where it is connected in this way.

As covered in Part **(D)**, each PV system requires a disconnect, which can consist of up to six switches, etc. Do not confuse this with a limitation on how many PV set-ups can be installed, which is unlimited. The informational note makes clear that multiple inverter outputs can be combined into a single PV system disconnecting means and input to the electrical system. Any such disconnect **(E)** must have a rating that corresponds to its application, and for dc applications **(F)** it must have operating poles both conductors, marked for PV use, or suitable for backfeed.

690.15. Disconnection of Photovoltaic Equipment. Isolating devices must be provided for disconnecting inverters, charge controllers, etc., from all conductors that are not solidly grounded, and for a functionally grounded PV system, neither conductor is solidly grounded. For dc combiner outputs, or for charge controller or inverter inputs, a disconnect is mandatory if the current will exceed 30 amp. If energy comes from more than one source, then those disconnects must be grouped and identified. In Fig. 690-1, this applies to the inverter, which is why the dc disconnects to the left of the inverter and the circuit breaker to the right must be adjacent to the inverter, or additional disconnect(s) would need to be provided.

Part **(A)** locates equipment disconnects either within the equipment or in sight and not over 10 ft away, with an allowance for a locally mounted remote trip provision for a more distant disconnect. Figure 690-5 shows a dc combiner box with an integrated ac-to-dc power supply at the upper left. The door handle mechanism includes on-off-test switch function and interfaces with the contact block at the center right. The AFCI detection contactor, connected to the main positive busbar of the combiner, opens if the switch is open or if an arc fault is detected. Each source circuit is wired through a current sensor, monitored by the AFCI detection system.

Part **(B)** assures the appropriate interrupting rating, however, a device that will not be used to actually open the circuit under load does not require that rating.

Part **(C)** covers the various isolating devices for system components. If not designed to open under load, it must be marked accordingly.

Part **(D)** includes the essential equipment characteristics for this equipment, including simultaneous operation of poles, indication, externally operable, and lockable. Other than connectors (690.33), if both sides are energized in the OPEN position, the warning in 690.13(B) must be provided.

690.31. Methods Permitted. In general, Chap. 3 wiring methods are permitted. In addition, wiring that is nonstandard but part of a listed system is permitted. In the event that PV source or output circuits in readily accessible locations are operating above 30 V, they must be guarded or installed in Type MC cable or a raceway. This would include, on the literal text, MC cable run in accordance with 690.31(G), which probably results from an unintentional failure to correlate this

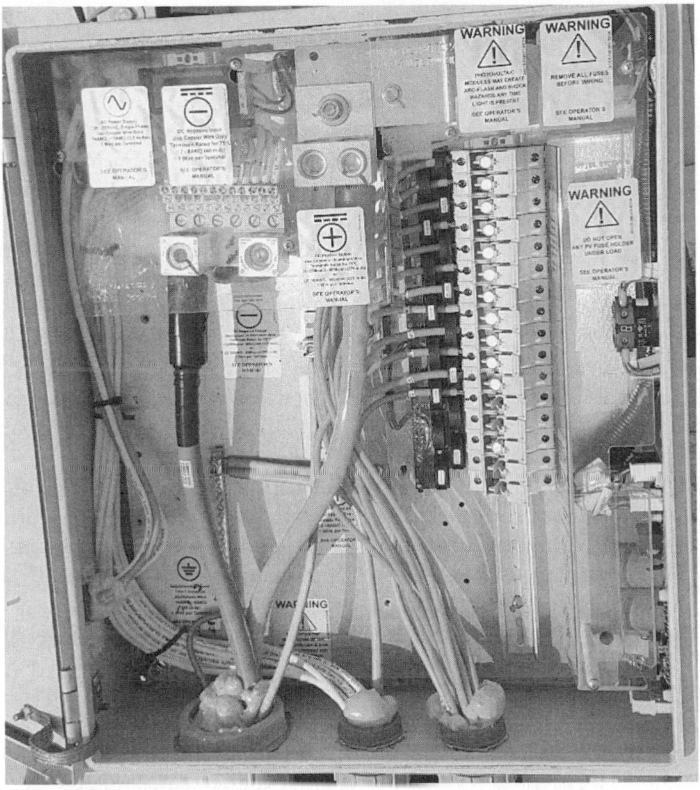
WARNING
WARNING
WARNING

Fig. 690-5. PV equipment disconnection as a part of a dc AFCI combiner box wired under the 2014 **NEC**. This integral disconnection function meets 690.15(A) on the positive side, but not on the negative side. Because the negative input conductors are not to be considered solidly grounded, this equipment may require redesign for the 2017 edition, and may be a candidate for acceptance by early adopting jurisdictions under 90.4 (3rd ¶). Courtesy, Solar Design Associates, Inc.

wording and the express permission in the other section. In the general context of Part **(A)**, there are some general wiring issues to explore in this context, one being substantially elevated ambient temperatures, which is a major concern. All wiring terminating on or running with a PV array should be assumed to be operating in a 70°C ambient temperature or higher due to solar heating. This means that the wiring (and any nonmetallic raceway if used) must have a minimum temperature rating of 90°C as a practical matter, and that the effects of temperature on final conductor ampacity will be pronounced. For example, 12 AWG THWN-2, a 90°C conductor with a Table 310.15(B)(16) ampacity of 30 A would end up with (using a 0.58 derating factor) a final ampacity of 17.4 A or even 12.3 A if a higher temperature were assumed. Looking at Fig. 690-1, the wiring between the PV source and the inverter is a feeder for a separately derived system and follows normal rules for sizing, although, depending on how it is routed, temperature may have an important impact. These applications often involve high temperature ambients, for which ampacities require often substantial correction from basic table values. The table provided matches the one in 310.15(B)(2)(a) except a 105°C column is added.

As we have seen, each of the six strings generates 6.75 A, resulting in a total array load of 41 A. Since these wires will be connected to an overcurrent device and switch terminals, and transmit continuous loads, we need to add 25 percent of this number to determine conductor sizes. The minimum ampacity of the feeder conductors would be 125 percent of 41 or 51 A. This corresponds to 6 AWG copper conductors, evaluated under the 75°C column of Table 310.15(B)(16). If these feeder conductors were routed along the array the evaluation would change, because ambient temperature would become a major factor. In such a case (assuming 70°C conditions) the derating factor to be applied to THWN-2 conductors is 0.58. The actual value of maximum continuous current is used here, or 41 A. 41 A ÷ 0.58 = 71 A. Since this is less than the ampacity (in the 90°C column) of these conductors, the 6 AWG conductors can still be used, but the result is far closer to the table limits. Refer to Chap. 9, Annex D, Example D3(a) for a comprehensive explanation of these considerations.

Part **(B)** requires that photovoltaic source and output circuits, including inverter output circuits, must be divorced from other systems, unless partitioned off, or unless connected together, such as at a common piece of equipment. It also requires PV system conductors to be identified and grouped in several ways. First, PV system circuits must be identified at connections and splices. Because functionally grounded conductors are not solidly grounded, this section [in (B)(1)] now explicitly prohibits them from using white insulation or otherwise using an identification method reserved for truly grounded circuit conductors, as covered in 200.6 Next, if there are multiple PV systems on the premises, and the conductors that are part of more than one such system enter a common enclosure or raceway, the wiring must be identifiable by system at all connections and splices unless the identification is self-evident because of unique entry and exit points, arrangement within the enclosure, or other factors. These identification requirements can be met through tags, marking tape, or other approved means.

Finally, if multiple PV systems occupy a common wireway or box or other wiring method with a removable cover, the wiring must be discernible by both systems and, if both current types are present, by type of current (ac or dc). This grouping must be by wire ties or equal and take place at least once and then at intervals not to exceed 1.8 m (6 ft), unless the entry is from a unique cable or raceway that makes this grouping obvious. Note that any wireway of a length long enough to invoke the 1.8-m (6-ft) marking requirement is probably one for which a unique entry point would be lost in the middle and therefore no longer obvious.

Part **(C)** addresses single-conductor cable. Effective as of the 2008 **NEC**, unless conventional multiconductor wiring methods are used, module and panel interconnections must use either single-conductor Type USE-2 cable or the new Type PV (photovoltaic) single conductor wire in exposed outdoor locations for PV source circuits and interconnections; the previous allowance for Types UF, SE, and USE was discontinued. Do not use welding cable. (See commentary at the end of Art. 630 for the UL information on this point.) Note that USE-2 wire must not be used inside a building unless it is dual rated as building wire, as shown by an additional marking of XHHW-2 or RHW-2 or comparable. Note also that the

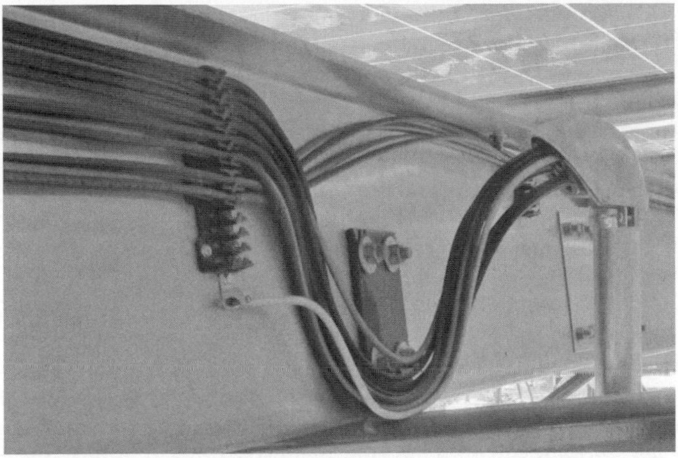

Fig. 690-6. Single conductors run on structures must comply with Type NM cable support distances [690.31(C)]. The installation pictured is only accessible to qualified persons. Courtesy, Solar Design Associates, Inc.

new Type PV conductors have a thicker than usual insulation, and as such require special raceway fill calculations if pulled into a raceway.

Type PV or USE-2 single conductors routed on structures (see Fig. 690-6) must comply, as of the 2017 **NEC**, with the support distances for NM cable (334.30), certainly met in this case. However, this rule now also references 338.10(B)(4)(b), which includes 334.30 anyway, and which would automatically apply to USE-2 in any event. This does additionally bring in all of Art. 225 Part I, including the requirement of 225.10 to use one of the specified Chap. 3 wiring methods. There is no comparable wiring method on the list; open wiring on insulators is the closest match but it is limited to industrial and agricultural occupancies. Even there, it is a poor match for conventional practice on solar arrays because, historically, it is an antiquated wiring method designed for timber frame buildings. This rule will require careful discussion with the inspector. A sound approach would be to argue that the wording "permitted in exposed outdoor locations ... within the array" constitutes a Chap. 6 amendment of Art. 225.

These wires have 90°C temperature ratings and tolerate sunlight and wet-location exposure, and they can be connected to the positive and negative leads coming from the modules. However, if the array is readily accessible to passersby, the conductors must usually run within a raceway, as covered in **(A)**. Single-conductor entries into a pull box must have strain relief provided, such as through listed gland-type connectors that maintain the wet-location integrity of the box. Leave enough slack in the module leads so the module can be removed for servicing or replacement. PV source and output circuits are also permitted in outdoor locations to run in cable tray whether or not they carry the CT mark for that purpose. They must be secured not less than every 4½ ft and for ladder tray, the rung spacing must not exceed 12 in. This provision will probably find significant application in industrial occupancies with large solar arrays.

Part **(D)** allows any jacketed multiconductor cable assembly "listed and identified" for the location to be used outdoors to carry PV inverter output circuits from utility-interactive inverters located where they are not readily accessible. It must be secured every 6 ft and include an equipment grounding conductor.

Part **(E)** covers flexible cord for the moving parts of tracking PV equipment, provided the cord meets the ampacity limitations for the high ambient temperatures usually applicable to these applications. A special temperature correction table is available for this purpose, with a 105°C column for cords so rated.

Part (G) is a very extensively used provision that provides a method of routing PV output circuits indoors from a rooftop without providing an additional disconnecting means. The entry point can be remote from its disconnect, thereby allowing the inverter disconnect to serve both functions, provided the wiring between the point of entry and the disconnect is in a metal raceway, such as EMT, or Type MC cable, as of the 2011 **NEC**. This is now less important than in the past, because the shift to functional grounding and the implementation of rapid shutdown provisions have led to the result that the rule requiring protection near the point of conductor entry into the building envelope was no longer valid, and has been removed. See 690.13(A) and the clarifying informational note. However, that note to the contrary not withstanding, 690.31(A) still asks for raceway or MC cable, which is consistent with this requirement.

Part **(1)** covers PV source or output circuits actually running within a built-up roof surface. Their location, if not "covered by PV modules and associated equipment," must be evident to a firefighter attempting to ventilate an attic during a fire. This presents a real challenge as far as durability in the presence of UV radiation and precipitation; since there is no current listing category the only possible acceptance criterion at this time is approval by the AHJ.

Additional requirements in paragraph **(2)** apply to Type MC cable and greenfield, and perhaps liquidtight flexible metal conduit although that is not mentioned, If the greenfield is not over metric designator 16 (trade size ½) and if the MC cable is smaller than 25 mm (1 in.) in diameter, then it must be protected by guard strips not less than the cable size where it crosses joists. Note that the size limits in the **NEC** text are not to be equalled dimensions, as opposed to the more usual not be exceeded specifications. Other than that they must follow framing or building surfaces; there is no permission for drilling the centers of framing, other than the alternative for an approved protection scheme, so that could be an area for discussion with the inspector as well.

The PV wiring method, if exposed, must be labeled as such, every 3 m (10 ft) for continuous runs and at least once in shorter runs between vertical or horizontal partitions. In addition, box and other enclosure covers, and even the covers of conduit bodies with an unused hub get the same marking. Parts **(3)** and **(4)** are new for the 2014 **NEC** that is extremely prescriptive labeling information for the required labels. They must be reflective, in all capital letters at least ⅜ in. high, and done with white printing on a red background. The label must read "WARNING: PHOTOVOLTAIC POWER SOURCE." It will be interesting to see just how long such labels will stick to MC cable with an interlocking armor.

Part **(H)**, new as of the 2008 **NEC**, addresses fine-stranding. Since the topic has been generically addressed in 110.14, this paragraph now simply refers to that location. Refer to the commentary at 110.14 in Chap. 1 for more information on this point.

Part **(I)** covers bipolar PV systems. If the total open-circuit voltage of the monopole subarrays (Fig. 690-7) exceeds system parameters (typically 600 V), then the two monopole outputs must be physically separated into separate raceways that go all the way to the inverter, with separate disconnects at the inverter. If the inverter is listed for the total open-circuit voltage and has an internal barrier, then the two raceways may land at the inverter. If the positive and negative conductors from the two subarrays come into contact, the full open-circuit voltage will appear across circuit components that cannot handle that exposure. Actual fires have resulted from these occurrences.

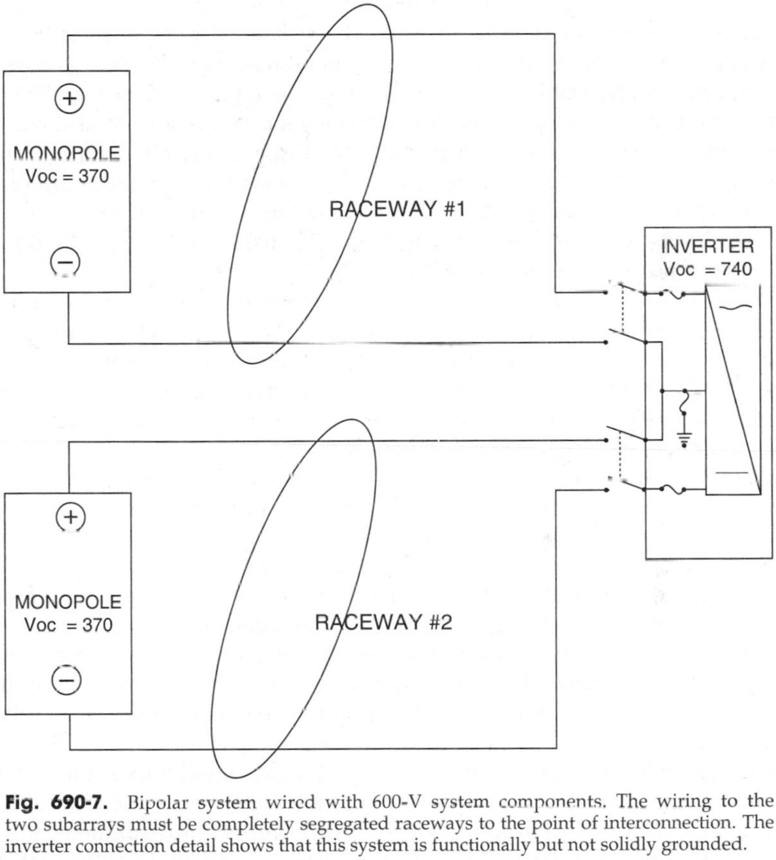

MONOPOLE
Voc = 370
RACEWAY #1
INVERTER
Voc = 740
MONOPOLE
Voc = 370
RACEWAY #2

Fig. 690-7. Bipolar system wired with 600-V system components. The wiring to the two subarrays must be completely segregated raceways to the point of interconnection. The inverter connection detail shows that this system is functionally but not solidly grounded.

A warning label must be applied advising as to the consequences of disconnecting the grounded conductor. For the 2017 edition, this warning only applies to solidly grounded systems. Some legacy systems with an actual grounded conductor will need to take this advice seriously. Going forward, the "solidly grounded" parameter will not affect modern systems that are functionally grounded.

An exception follows that allows listed equipment to be used that is rated for the maximum open circuit voltage. This equipment creates a functional grounded center point and it can be positioned near the array. With one of the two wires from each monopole terminated within it, the home run conductors running to the inverter are reduced in number by half. Note, however, that the home run will now be operating well over 600 V, although never over 1000 V to ground. In the case of Fig. 690-2, that voltage would be 740 volts. Type PV wire is rated 600, 1000, or 2000 volts (refer to the UL Guide Card information, code ZKLA) and would be the logical choice, 1000V in this case. Note also that the exception refers to "switchgear." This equipment does not meet the terms of that definition; better wording would probably be "control equipment." However, the intent is clear and this equipment, if appropriately listed, should be interpreted as being in compliance with these terms.

690.33. Connectors. These are different from the ones covered in 690.32 that are concealable, and are used to interconnect panels, etc. They must be polarized and not interchangeable with other receptacles used in different electrical systems. They must be latching or locking, and if used on systems operating over 30 V must require the use of a tool to open if in locations that are readily accessible. They must be rated for interrupting current, or, they must be labeled to not open under load and require a tool to open. If opened improperly these devices will expose anyone fooling with them to full PV circuit voltages.

690.34. Access to Boxes. This is the box access rule in the realm of PV arrays. If not accessible directly, a box is accessible if the module or panel over it is secured using removable fasteners and wired using flexible wiring methods.

690.41. System Grounding. This section now consists of its 2014 **NEC** predecessor melded with the GFP rule formerly in 690.5 (2014 **NEC** in both cases), and then heavily influenced by the new definition of functional grounding. Part **(A)** lists the system allowable grounding configurations, which are roughly arranged by predicted frequency of implementation going forward.

Option (1) is the 2-wire system with one conductor functionally grounded. This is by far the predominant system and is likely to retain that distinction.

Option (2) is a bipolar array with a functional grounding reference at the center tap.

Option (3) is a non-isolated system that uses the ground on the inverter output circuit as the reference. This system has isolation between the ac and dc sides, and no inserted resistance in the ground connection on the dc side for operation. The inverter internal circuitry looks for unequal dc flowing on each polarity; those currents become unequal in the event of a fault. The result is reliable detection of ground faults.

Option (4) is for truly ungrounded systems, and primarily used on very large systems (see Art. 691). These systems typically operate in the context of an ungrounded ac system more common in IEC (European) distributions and may interface with an ac isolation (1:1) transformer. Similar to Option (3) systems, the floating array operates with one or more continuously operating residual current monitor(s) that will disconnect the system in the event of substantial current leakage from either the positive or negative poles.

Option (5) correlates with 690.41(B) Exception and allows for up to two source circuits that are solidly grounded in an application that is only permitted outdoors and unattached to any buildings. This system was formerly 690.5 Exception, and it is used for such applications as PV powered school crossing signals.

Option (6) is a final catch-all for other methods meeting 250.4(A) with listed equipment, and is unchanged from 2014.

The remainder, Part **(B)** is the GFP mandate from former 690.5(B) applied in a world of functional grounding. Every PV system must have GFP in some form, detect ground faults in both ungrounded and functionally grounded conductors and components, and be listed accordingly. This is done by either disconnecting the current carrying conductors of the faulted circuit, or having the inverter (or charge controller) open output circuits. For a functionally grounded dc system, it will isolate the system from the ground reference. Fig. 690-8 shows equipment that senses such faults and isolates the system by opening the functional grounding connection. Note that the system grounding conductor makes a second turn through the current transformer to increase the sensitivity.

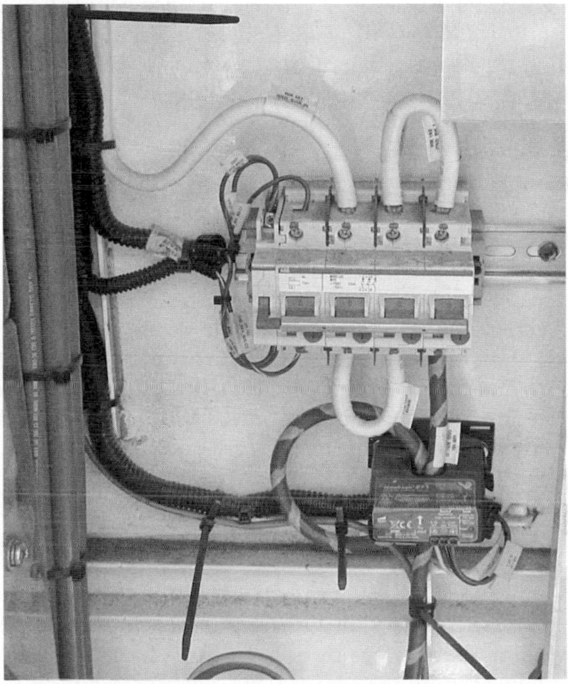

Fig. 690-8. A GFDI (ground-fault detector interrupter) breaker. The far left hand pole opens at (in this case) 0.5 amps. and is mechanically interlocked with the power poles so they will open simultaneously. The negative circuit conductor looping through the multiple poles of the breaker accommodates what is necessary for the breaker to achieve a 600 V dc rating. Since dc circuits have no zero crossing points, elevated voltages may (as in this case) require the interrupting action to be distributed over multiple contact sets. The wording of the warning label and the use of white wire reflects 2014 **NEC** requirements locally applicable to this installation. The 2017 **NEC** [at 690.31(B)(1)] changes the permitted color for this wire, and discontinues this specific warning in favor of the similar warning label required at the PV system disconnecting means at 690.13(B). Courtesy, Solar Design Associates, Inc.

690.42. Point of System Grounding Connection. Systems with GFP will have the connection made where the protective device calls for it because the protective device will be responsible for the insertion of the resistance and/or overcurrent device etc. and for how it functions, all as covered in its listing. In the case of systems with an inverter, this will normally be inside the inverter in accordance with the installation directions. For solidly grounded dc systems, the dc "circuit grounding connection" must occur at only one point, and the location of where that point is, is a design choice. The terminology "circuit grounding connection" refers to the connection of the dc grounding electrode conductor. The language here correlates this mandated connection with the operation of the GFP in 690.41(B), as discussed extensively in this book at that point. Left unsaid is the connection between the grounding electrode conductor, the grounded dc conductor, and the equipment grounding system. However, this is a dc system, and where Art. 690 leaves a vacuum, Art. 250 rushes in. This connection is covered in 250.168, where an unspliced bonding jumper must be installed at this point, sized in the same way as the grounding electrode conductor, which per 250.166(B) will be the size of the PV conductors with the usual allowances for mechanical permanence and electrode quality. That said, these systems are now quite unusual, as functional grounding predominates.

690.43. Equipment Grounding. Equipment grounding conductors in PV systems follow well understood installation requirements, even being required to stay with the system conductors as they leave the roof although part of a dc system. The frames of electrical equipment and arrays are connected as one would expect, with refinements for listed bonding procedures that may be idiosyncratic to the actual metal framework encountered. Take care in terminating these conductors at an array. Listed modules will provide for grounding connections and they must be used, unless otherwise provided as part of a racking system listed for bonding. Other modules, typically aluminum bodied, will require field terminations. Look for a copper-bodied lug that has a tin coating and is listed for direct burial. These lugs are compatible with an aluminum surface and their stainless steel hardware will survive outdoor conditions. To attach these to the module, drill a small pilot hole, then seat the lug with a 10–32 thread-forming sheet metal screw treated with an antioxidant compound rated for aluminum connections. This is not a "tek" screw with sheet-metal threads; it is an actual machine screw, and the compound will ensure that no oxygen reaches the raw aluminum surface during the thread-making process. This prevents the formation of aluminum oxide and ensures a low-impedance, long-lasting connection. Fig. 690-9 shows an example of this principle as applied to specialized hardware.

690.45. Size of Equipment Grounding Conductors. Equipment grounding conductors must follow Table 250.122, based on the PV maximum circuit current if there is no overcurrent protective device, which is 125 percent of the short circuit or other applicable value as determined per 690.8(A). In the case of Fig. 690-1 the maximum circuit current is 41 A. A 10 AWG conductor would be the correct size.

690.46. Array Equipment Grounding Conductors. Equipment grounding conductors run to PV modules must comply with 250.120(C). This rule was originally

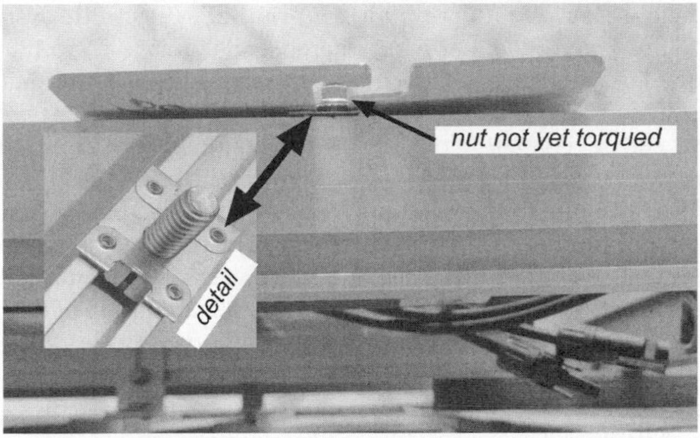
nut not yet torqued
detail

Fig. 690-9. Specialized listed bonding hardware can be used in lieu of wired bonding jumpers for components of an array [690.43(A)]. Note that the spacing between the upper frame and the rail means that the nut has not yet been torqued. This hardware is usually stainless steel. When driven down its teeth embed themselves into the structural aluminum with an air-tight bond, preventing aluminum oxide from forming, which would make the connection resistive. Depending on specific equipment hardware design and applicable listing limitations, this equipment is frequently not reusable, or only reusable a very limited number of times. The lack of torque in this photo is not an oversight; the photo was taken before the final position had been determined. Courtesy, Solar Design Associates, Inc.

designed to address equipment grounding conductors run to create 250.130(C) grounding connections, which is why it discusses hollow spaces within walls, etc. It is true that dc grounding conductors need not run with their circuit conductors, as covered in 250.134(B) Exception No. 2. However, this will, as a practical matter, generally mandate raceway protection for smaller equipment grounding conductors, certainly out at the array where there are no hollow wall cavities. Further, 690.43 requires equipment grounding conductors to run with the circuit conductors after leaving the array, which strengthens the argument that all this is about is providing mechanical protection at the array. Note that 690.46 covers the array-to-array routing of the grounding conductors with a reference to 250.120(C). This allows for dc conductors smaller than 6 AWG to run apart from the associated circuit conductors but only where protected from damage or if run within raceways, etc. to provide this protection. Some rooftop designs specify 6 AWG grounding conductors to avoid the issue entirely.

690.47. Grounding Electrode System. After numerous code making cycles that attempted to mandate additional grounding electrodes as lightning mitigation measures for PV arrays on buildings, a subject entirely outside the scope of the **NEC** (it belongs to NFPA 780: Standard for the Installation of Lightning Protection Systems), sanity has largely returned and this section, at least in so far as its mandatory requirements, is completely defensible. First, a building or structure supporting an array must have a grounding electrode system that meets Part III of Art. 250. Well, since the PV system is generally supporting an **NEC** recognized electrical system, there will be a grounding electrode system. Second, the equipment grounding conductors must be bonded to the building grounding electrode system as covered in Part VII of Art. 250. The relevant reference in this case is 250.130(A), calling ultimately for a connection to the grounded service conductor and the grounding electrode conductor. This means that the equipment grounding conductors, which have already been aggregated and connected at the functional grounding connection point in accordance with 690.42, would then seem to be required to be extended into the service equipment, but read on. The next paragraph allows the inverter output equipment grounding conductor, which inevitably moves upstream to the service equipment, to provide the required connection. The informational note takes due note of the historical context and very well explains all of this.

Note that for ground-mounted arrays, which will constitute structures under the Article 100 definition, the array mounting structure will generally require, at some convenient point, a local grounding electrode. Fig 690-10 shows an example of such a connection.

Fig. 690-10. This wire is properly connected to at least 20 ft of reinforcing steel within the pour and that therefore qualifies as a concrete-encased electrode as covered in 250.52(A)(3). This becomes the grounding electrode system for the outdoor structure supporting the array [690.47(A)]. Courtesy, Solar Design Associates, Inc.

Part **(B)** then allows what has always been allowed but only sometimes not required, namely, for auxiliary grounding electrode connections directly from an array frame to a grounding electrode. Sec. 250.54 has permitted additional "auxiliary grounding electrodes" right along, without limit. Work as hard as you want, put as many as you want in place, and connect them to the array frames which are part of the equipment grounding system. These are entirely optional connections. However, no one should believe that this work is going to remotely substitute for a qualified NFPA 780 Lightning Protection System. In fact, some studies have demonstrated that auxiliary electrodes may exacerbate property risks to connected equipment by providing an unbalanced entry point for voltage wave fronts from nearby lightning strikes. NFPA 780 rules will protect PV arrays, but only if fully implemented. **NEC** 690.47(B) is a fool's paradise, but at least this work is no longer mandatory.

690.50. Equipment Bonding Jumpers. If used, they must follow 250.120(C). See comment on 690.46.

690.53. Direct Current Photovoltaic Power Source. This is a field marking requirement, illustrated in Fig. 690-1. If there were battery storage in the illustrated installation, then item (3) would also go on the label specifying the rated output current of the charge controller. If there are multiple outputs, then the label is applied to each.

690.54. Interactive System Point of Interconnection. The point of interconnection must be labeled as a power source at the disconnecting means, with the rated ac output current and nominal voltage. In the case of Fig. 690-1, that label would go at the location of the 30-A circuit breaker in the panel, perhaps incorporated in the circuit directory.

690.55. Photovoltaic Systems Connected to Energy Storage Systems. Such systems must have a marking showing the polarity of the grounded circuit conductor the maximum operating voltage, and the equalizing voltage if any.

690.56. Identification of Power Sources. For stand-alone systems there must be a plaque declaring that fact and giving the location of the system disconnecting means. The location is to be outside the building at a readily accessible location suitable to the AHJ. For buildings with both a service and a PV system there must be a plaque giving the location of the PV and service disconnects if they are not in the same location. This last item is already required, in better form because its requirement is reciprocal, in 230.2(E). It is also covered, as cited, in 705.10.

Part **(C)** covers the field labeling requirements for the rapid shutdown systems required by 690.12. There are two labels required at minimum. One alerts the fire service that the building supports a PV system, and that a rapid shutdown system is in place. This label will be located either on or not more than 3 ft from the building service. There are two versions of this sign, with a graphic provided on the Code pages. For both of the signs, the largest lettering at the top reads the same, "SOLAR PV SYSTEM EQUIPPED WITH RAPID SHUTDOWN." The first version covers shutdown systems covered in 690.12(B)(2)(1), where the array itself shuts down as well as the wiring outside the array boundary. This version has that lettering using black lettering on a yellow background, and the sign on the roof also in yellow. This color scheme is recognized as appropriate for a "caution" level of risk in the standards referenced in 110.21(B). The second version covers shutdown systems covered in 690.12(B)(2)(2), where the array does not shut down and the wiring inside the array boundary can remain at as much as 80 volts. This version has the same basic lettering and graphics, but uses white printing on a red background. In addition, the interior graphic that shows a panel on the roof shows color scheme with a red dashed line around the panel as if to express the idea of voltage, and there is also a red exclamation point on a triangle on a red square. This color scheme is recognized as appropriate for a "danger" level of risk.

The second label serves to identify the switch that is supposed to be activated in order to initiate the shutdown sequence. It must be applied to the switch, or be posted within 3 ft, and carry the legend: "RAPID SHUTDOWN SWITCH FOR SOLAR SYSTEM." This label must be reflective and use white lettering on a red background, and with the same font size as the principal lettering in the larger labels. In the event the building has multiple systems with more than one style of shutdown system, or some with and some without shutdown, then a more complicated label must be created that shows a plan view of the roof delineating the extent of each system, and using the dashed line to encircle the system(s) that will remain energized during the day.

690.59. Connection to Other Sources. The requirements for utility-interactive inverters and other aspects of such interconnections are now completely within the scope of Article 705.

690.72. Self-Regulated PV Charge Control. Energy storage system requirements are generally the province of Art. 706. The general requirement in 706.23(A) is that provisions must be made to control the charging process of the energy storage system. This section provides a path whereby the PV system may be able, in effect, to claim inherent compliance with the control requirements. If successful, this would be a major design plus for the PV system. Specifically, the charge control would be deemed to appropriately "control the charging process" if the PV source circuit is matched in terms of required voltage and charging current to the battery line, and the maximum charging current times 1 hour is less than 3 percent of the total rated capacity of the battery line. If the battery manufacturer has a different parameter, then that would be substituted.

ARTICLE 691. LARGE-SCALE PHOTOVOLTAIC (PV) ELECTRIC POWER PRODUCTION FACILITY

691.1. Scope. This article covers extremely large-scale PV farms, at least 5 MW in capacity, and not under the exclusive control of an electric utility. The article covers a power production operation organized entirely to supply electrical energy that an electrical utility will transfer and resell through its distribution system. The "sole purpose" wording in the first informational note makes this very clear. The need to bring operations of this nature within a regulatory framework is genuine, and the size of the problem based on their ever increasing numbers, especially in rural states with favorable geography and weather is equally clear. It is also very clear that the sophisticated engineering required to make one of these solar farms function on a scale that involves medium voltage at the least and some recent installations have involved 69 kV interconnections, which is a transmission voltage, the scale of these facilities is difficult to even imagine. Renewable energy mandates from government policymaking have the attention of the utilities, and the market economies of scale are driving the creation of these systems in ever increasing sizes. The article submitters are quite correct when they argue that the NEC was never and never will be in a position to write prescriptive rules for something like this. The technical merit that underlies the structure of the entirety of this article, open ended provisions that essentially say, "go get competent engineering, qualified personnel, and make sure you are transparent to the local authorities in terms of your design process and documentation" is frankly compelling.

The larger question is whether this article, which effectively regulates a utility generation component, will fit under the usual regulatory structures governing the implementation of **NEC** provisions. It might fit better as regulatory framework for a department of public utilities. This is a question with at least 50 different answers in the United States alone. As this article was moving through the process, it very clearly involved projects outside the scope of the **NEC**. It was only at the very last **NEC** Committee stage, after final panel balloting, that the Correlating Committee stepped in and wisely inserted the words "and not under exclusive utility control" into the scope. This theoretically mitigated the conflict with 90.2(B)(5), which is also flagged in the second informational note. However, this may turn out to be a distinction without a difference because the local utility is necessarily operating as a natural monopsony; no other entity will ever be able to buy its product, and therefore it exercises virtual exclusive control. In fact, one of the substantiating arguments presented in favor of the article stated, "They are under the direct control of a utility via transfer trip relays controlled by the utility." Further, the purchase agreements involved will necessarily be the subject of review and approval of the regulatory authority governing utility operations.

The analysis that follows of the specific requirements in the article largely quotes from the arguments made by the panel. These arguments largely support the above analysis, but perhaps not its conclusion. One substantially underlies most of the others: "The **NEC** includes a number of specific requirements intended to demonstrate a conservative approach to safety in environments where unqualified persons are present. These provisions can severely constrain the design of large-scale PV generating stations; therefore, if a less conservative approach is allowed, unqualified individuals must be excluded from accessing the station."

691.4. Special Requirements for Large-Scale PV Electric Supply Stations. These facilities are expressly prohibited from having any public access. The panel argument for this accurately sets the context for this article: "Unlike smaller scale PV systems, large scale PV electric supply stations are designed and operated similarly to traditional utility power generating assets. Unqualified individuals must not access the system for their own safety and for protection of the system which is crucial to grid stability." The realities of the power levels involved dictate a utility interconnection at medium or higher voltages and using switching facilities capable of performing the task safely. Article 690 does not address these issues, and neither does Art. 490, which supports the safe use of electricity for premises wiring. Any electrical loads connected at an Art. 691 location are entirely purposed for converting power received from the sun to electricity and making it available to the grid. These systems must never be installed on buildings. The panel decided that such locations could never be adequately secured against unauthorized access.

691.5. Equipment Approval. The usual process of seeking listed products, either off the shelf or as a result of a field evaluation will continue, but the third option of engineering review to ascertain conformance to recognized standards will be the more frequent approach.

691.6; 691.7. Engineered Design; Conformance of Construction to Engineered Design. These systems are relatively uncommon. These sections are controversial in that they allow the enforcing jurisdiction to impose the costs of an external engineering review of both the design and the conformity assessment on the permit holder. This is understandable given the financial realities underlying the usual processes involved in field enforcement of NEC provisions, and in these cases such costs could be substantial, especially because multiple engineering disciplines are involved in these large-scale systems. In addition, these systems often create local environmental impacts that dwarf the electrical complexities. There are a number of even small-scale projects (smaller than 5 MV) currently embroiled in disputes such as the mandated height of the fencing vs. local zoning ordinances prescribing a lesser height. The ultimate engineering design and conformity will need to account for all of these variables to the extent the electrical design is impacted.

691.8. Direct Current Operating Voltage. Here again, the panel decided against a reference to Art. 490, and left it to the engineering process. This is appropriate in this case, because we are not discussing a premises wiring issue.

691.9. Disconnection of Photovoltaic Equipment. This is an excellent demonstration of the extent to which this article varies from Art. 690. Typical installations of this sort involve inverter shacks spread out over a large geographic area, buildings constructed to protect the inverters from the weather but never for continuous occupancy, and rapid shutdown provisions are not possible or appropriate. The numbers alone result in separations from disconnecting equipment. On the ac side, the ac disconnect is usually at the substation switchgear, whose distance from the inverters might be measured in miles. On the dc side, some designs have dc disconnects located at large combiner boxes. These might number in the dozens in order to supply an inverter that might be rated in the MW range. Again quoting the panel: "Utilization of one-line diagrams by qualified electricians is the most effective method to ensuring proper isolation of equipment in shelters."

691.10. Arc-Fault Mitigation. The site review process typically includes fire breaks and vegetation management agreements that are subject to review by fire departments. Attempting conventional arc-fault control on this scale is impossible. And, the fact that these systems are ground and not building mounted reduces the potential for a conflagration, the effects of which would be felt most keenly by the system owner and not the public.

691.11. Fence Grounding. This is a complicated issue, because there might be more than 10 miles of fencing at one of these facilities. Given the realities of utility transmission systems, the presence of overhead high voltage lines often results in NESC mandates not just for general bonding but also invoking specific NESC requirements relative to step and touch potentials. Here again, this must be part of the engineering design; no prescriptive requirements can be written to adequately address it.

ARTICLE 692. FUEL CELL SYSTEMS

692.1 Scope. This article covers fuel cell systems, which have much in common with solar photovoltaic systems, and as a result many of the article provisions directly correlate. Fuel cells, like PV sources, are inherently dc sources with a finite current output. They also are used in conjunction with energy storage systems and with utility interaction.

692.4. Installation. Part **(B)** reiterates 230.2(E), but not as well, since it lacks a requirement for reciprocity. The systems are required to be installed by qualified persons only.

692.6. Listing. Fuel cell systems must be listed or field labeled for the intended application prior to installation. Apparently no field listings are acceptable in this category.

692.8. Circuit Sizing and Current. The rated current of a fuel cell is that which is marked on it, with no adjustments for temperature or other complications. The required ampacity of conductors supplied by fuel cells is the greater of (1) the marked rating(s) or (2) the rating of the fuel cell system(s) overcurrent protective device ahead of the conductor. Part **(C)** of this section matches Sec. 705.95(A); refer to the commentary at that location for details.

692.9. Overcurrent Protection. If conductors are protected by the fuel cell overcurrent device, no further protection is necessary. Overcurrent devices must be readily accessible. If more than one source is contributing power, overcurrent protection must be provided.

692.10. Stand-Alone System. These rules used to be the same as in 690.10; however, most of those moved to the new Art. 710, and that is where this book now addresses the topic. This section needs to be revisited for this reason in the next cycle.

692.13. All Conductors. This rule provides for complete disconnection of all current-carrying conductors of a fuel cell system from all other conductors on-site.

692.17. Switch or Circuit Breaker. This is the same basic requirement as in 690.17; refer to the coverage at that location for more information. The issues here are less complex than with PV systems.

692.31. All Chap. 3 wiring methods are permitted. The function of the last sentence is anyone's guess. It appears to line up with 334.30(C), but what that might have to do with a fuel cell system is unknown. A review of the original proposal submittal shows no explanation or substantiation to support this wording, however; the recognition of all Chap. 3 wiring methods is clear.

692.41. System Grounding. These rules simply apply separately derived system rules to a fuel cell system, which is a separately derived system. Part **(C)** is a much simpler version of 690.47(C). Refer to the commentary at that location for more information.

692.47. Grounding Electrode System. Some manufacturers are still enthralled at what they think a ground rod here or there will do for them. This section correctly describes such foolishness as an "auxiliary" grounding electrode, which is covered in 250.54, and correctly directs that it be connected in a manner that will do no harm, that is, to the equipment grounding conductor.

692.53. Fuel Cell Power Source. This requires a plaque to be posted at the fuel cell power source (or other accessible location "on the site") with the fuel cell system and its principal electrical characteristics such as voltage and power rating.

692.54. Fuel Shut-Off. This plaque is to be posted at the "primary disconnecting means" of the building or circuits supplied, giving the location of the manual fuel shut-off valve.

692.56. Stored Energy. If the fuel cell system stores electric energy, then a plaque as specified in the NEC must be posted warning of "energy storage devices" at the service disconnect for the premises. The fact that this says service and the previous sign said primary strongly suggests that the difference is intended. The proposal for the article used "primary" in both locations; the panel reworded this one to say "service" but failed to document why it made the change. The overall concept is that a fuel cell system will continue to produce energy for a period of time after its fuel supply is shut off, and the original submittal did not want someone to be surprised by the continued generation of power. On the other hand, no one should assume that a circuit is dead, especially when its disconnect is still closed, without testing.

692.59. Transfer Switch. If the fuel cell is used in a noninteractive system that also has a service connection as a backup supply, the fuel cell system is to be connected to the premises system through one side of a transfer switch that keeps the two supply sources separated. This simply requires what 702.5 already requires anyway. Where service conductors are connected to one side of a transfer switch (and the fuel cell on the other side), the switch must meet Part V of Art. 230, and that in effect means that the switch must be listed as "suitable for use as service equipment."

692.60. Identified Interactive Equipment. Only fuel cells listed and marked as interactive are permitted to be used in an interactive system.

692.62. Loss of Interactive System Power. This rule is the same as 690.61 with some editorial changes in the first paragraph. Refer to the discussion at that location for more information.

692.80. General. This is the medium-voltage part of the article and it contains no useful information, only the requirement to do what the Code requires should be done at these voltages anyway.

ARTICLE 694. WIND ELECTRIC SYSTEMS

694.1. Scope. This article addresses wind generation together with the associated generators (dc), alternators (ac), inverters, and controllers. These systems can be interconnected with a utility supply or function as an island on a stand-alone basis, and for that reason may have provisions for energy storage, such as through the use of batteries.

To formulate the original proposal, Art. 690 was largely cut, duplicated, and pasted as this article. There are important distinctions between those two systems (PV vs. wind) that make this more of a difficult project than seems immediately obvious.

694.2. Definitions. The definition of "Charge Controller" has been relocated to Art. 100. The definitions of a **Diversion Charge Controller,** and an **Inverter Output Circuit** are essentially the same as in Art. 690. Four definitions apply to the windmill structure: A **Diversion Load Controller** is the same as a diversion charge controller, except that the load is not limited to energy storage equipment such as batteries and this equipment is the means of regulating the output of a wind generator. The **Tower** (where needed) supports the **Nacelle,** which houses the **Wind Turbine** that does the work of converting the mechanical energy of wind into usable electrical energy, and the tower may be reinforced by **Guy** wires as required for the height and prevailing conditions. The **Wind Turbine Output Circuit** carries the electrical output running between the internal components of the turbine and other equipment. Three definitions provide the basis for important ratings that are used elsewhere in the article. The **Maximum Voltage** is the maximum voltage the turbine will produce in operation, whether open circuited or not. The **Maximum Output Power** is the highest one-minute average power a wind turbine will produce in operation while functioning in normal steady-state conditions, and the **Rated Power** is the output at its rated wind speed. When all components are put together as required, the result is a wind turbine system, which was defined in the 2011 NEC and has been deleted in the 2014 NEC because it is not used in the article and therefore need not be defined. However, in lower case wording, that is what is created when everything goes together. It could also be a "wind electric system," which is used in the article at 694.7(B), although not formally defined.

694.7. Installation. Part **(B)**, which used to require that just the inverters be listed, now as of the 2014 NEC requires the entire wind electric system to be listed and labeled. Part **(C)** covers diversion load controllers used as the principal method of regulating the speed of a wind turbine rotor. Unlike solar panels that can be open circuited without damage, wind turbines will frequently overspeed in such cases, causing excessive noise and very possibly self-destructing. This rule requires that wind systems provide some "additional independent reliable means" to prevent excessive rotor speed, and an interconnected utility power grid is not to be assumed to be adequate as the alternate diversion load for such purposes. There is precedent for this sort of power dump in 620.91 governing regenerative power from a descending elevator. Some form of electrical resistance load is usually configured into the system, and to maximize the energy benefit, electrical hot-water heating is often factored into the mix.

Part **(D)** breaks new ground, in that it mandates surge protection ahead of all loads served by the premises wiring system that are supplied by the wind turbine system, and as such it is one of very few surge protective devices (Art. 285) mandated in the NEC. Part **(E)** permits the installation of a receptacle, presumably a 125-V 15- or 20-A single-phase style although that is not specified, to the output circuit for data or maintenance applications. This receptacle must be GFCI protected. Part **(F)** allows a supporting pole for a wind turbine to be treated as a raceway, generally to be addressed as part of the now required listing on the turbine. Part **(G)** allows a reduction in work space clearances for activity within the equipment by qualified personnel.

694.10. Maximum Voltage. For one- and two-family applications, the maximum voltage on a wind-turbine output circuit is limited to 600 V, and live parts (this is presumably an incorrect application of the term and probably the phrase "uninsulated live parts" is intended) operating above 150 V to ground must not be accessible to other than qualified persons while energized. The default upper voltage limit for other occupancies is 1000 V. DC utilization circuits, used in some off-grid applications, are limited as described in 210.6.

694.12. Circuit Sizing and Current. Part **(B)** classifies system currents as continuous and generally requires conductors and overcurrent devices to be sized on the basis of 125 percent of the maximum current, with the next-higher-standard-size device allowable up to 800 A.

694.15. Overcurrent Protection. Part **(A)** requires overcurrent protection for all circuit components, with particular attention to the possibility of backfeed. An informational note calls attention to some inverters that may operate in reverse for turbine startup or speed control. Similarly, Part **(B)** requires that transformers must be evaluated in both directions in terms of whether their overcurrent protection meets 450.3. Part **(C)** requires that overcurrent devices applied to the dc portion of the wiring must have the appropriate ratings. Just as in the case of PV circuit conductors that cannot be damaged by the available current [see 690.9(A) Exception], conductors that cannot be exposed to overcurrent due to the available system capacity need not have overcurrent protection.

694.20. All Conductors. Wiring from a wind turbine must be disconnectable from other premises wiring, and no pole is permitted in a grounded circuit conductor if such a conductor would become or remain energized with the disconnect open. For cases where connected loads on the turbine output circuit are used for speed regulation, a disconnect need not be placed in the output circuit.

694.22. Additional Provisions. The disconnect must be suitable for use as service equipment and readily accessible. It must be externally operable without exposing the operator to live parts (here again, probably "uninsulated live parts" is the intent). It must plainly indicate its open or closed status and it must be appropriately load-break rated. It must be plainly marked with a warning that both line and load terminals may be energized when it is open. Part **(B)** permits rectifiers, controllers, output circuit isolating and shorting switches, and overcurrent devices (that list pointedly not being exclusive) on the line side of the disconnecting means and part **(D)** allows rectifiers, controllers, and inverters to be up in the nacelle or otherwise not readily accessible.

Part **(C)** places the system disconnecting means at a readily accessible location either on or adjacent to the tower, or on the outside of the supplied building or structure, or inside nearest the point of conductor entry. If the installation uses dc turbine output circuits in a building, they must be run in a metal raceway or other metal enclosure (including Type MC cable) to the first readily accessible disconnecting means in accordance with 694.30(C), and that disconnecting means need not be close to the entry point for the wiring. In general, a disconnect is not required at the tower, and it is not permitted in a bathroom.

694.23. Turbine Shutdown. Wind turbines must have a manual means to shut them down. It must also require the shutdown procedure be documented and posted both at the disconnect or turbine controller and at the stop control if it is in a different place. The result must be either the turbine stops, or it retains limited movement combined with a means to deenergize the rotor circuit. An exception waives the requirement for smaller turbines with a swept area of less than 50 m^2 (538 sq ft)—Note that this metric conversion is a classic example of an illegal soft conversion as described in 90.9.

694.24. Disconnection of Small Wind Turbine Electric System Equipment. Components of the wind system must be disconnectable, including items such as inverters, batteries, and charge controllers, from all sources of power and if multiple sources potentially feed such equipment, multiple disconnects are required and they must be grouped and identified. For components that are part of a system that uses the turbine output circuit to govern rotor speed, a shorting switch or plug is permitted to substitute for a disconnecting means. Equipment in a nacelle does not require a disconnecting means.

694.26. Fuses. If a fuse is energized from both directions and accessible to unqualified persons, switches or other disconnects rated for the application must be installed on both sides.

694.28. Installation and Service of a Wind Turbine. Open circuiting, short circuiting, or mechanical brakes (at least one of those three possibilities) must be in place to allow a turbine to be disabled for both the original installation and also for subsequent maintenance.

694.30. Permitted Methods. Chapter 3 wiring methods are generally permitted, with the restriction that turbine output circuits operating at voltages higher than 30 V in readily accessible locations must be installed in a raceway. Flexible cord is permitted for the logical uses, including the connection of moving parts and to provide ready removal of components for repair, all in accordance with Art. 400. The rule specifies hard service cord suitable for extra-hard-usage or portable power cable, listed for outdoor use, water- and sunlight-resistant. This would include the Type S series and Type G cords, with a "W" in the type designation. In accordance with Footnote 9 to Table 400.4, the wet location designation now necessarily includes a sunlight-resistance evaluation. Part **(C)** on dc output circuits is covered at 694.22(C)(1) in this analysis.

694.40. Equipment Grounding. This is the point of departure where a cut-and-paste approach from Art. 690 into Art. 694 breaks down, and serious issues remain unresolved in terms of conflicts with established nomenclature and requirements in Art. 250. PV arrays are typically large in area and not thought of in the sense of a point source such as a generator, although everyone will agree that they are separately derived systems. In addition, there is usually some form of overcurrent protection at the arrays. Therefore, Art. 690 is generally organized around the terminology of an equipment grounding conductor. A wind turbine is a point source, and so this portion of the article addresses system bonding jumpers and not equipment grounding conductors. In short, it is a permanently installed generator. Where a generator is outside a building and the first disconnect is at the building, Art. 250 has highly evolved terminology for the grounding connections between the two locations, and "equipment grounding conductors" are not used, because they are on the line side of any overcurrent protection. Instead, the appropriate terminology is a "supply-side bonding jumper" and the sizing should follow 250.30(A)(2) and the overall installation should follow 250.30(C). Refer to the commentary in Chap. 2 of this book for more information. The 2017 **NEC** made incremental progress by describing the required connection in terms of bonding "to the premises grounding and bonding system." For systems with dc turbine output circuits, the applicable language is in Part VIII of Art. 250. The reference to 250.168 in this case would be particularly robust, since it would bring in parts of 250.28 that in turn would bring in 250.8. There has been some progress, but we still have quite a ways to go. The references in 694.40(A) omit Part II of Art. 250 referencing system grounding. On the other hand, the same references now include for the first time Part V, which includes supply-side bonding jumper sizing. Section 694.40(B)(2), which references both the incorrect term and the correct one in the same sentence, is a perfect example of this ongoing work.

Part **(B)** covers tower grounding and bonding. The requirement for a grounding electrode at the tower is technically justified, but needs to reference the electrode options comprehensively and the connecting conductor needs to be described correctly so it will be sized and installed correctly. The guy wire provision is much improved because it is no longer presumed to be a grounding electrode.

694.56. Instructions for Disabling Turbine. Instructions must be posted at the "turbine location" (one hopes this will be interpreted in the field as the base of a tower and not up at the nacelle) that explain how to disable the turbine.

694.62. Installation. Small wind systems must meet the cogeneration requirements in Art. 705 when other sources are connected.

694.66. Operating Voltage Range. Where small wind systems connect to the end of a dedicated feeder or branch circuit, this section grants permission for the nominal voltage limits for such circuits established in other parts of the NEC to be exceeded at the point of connection, provided the voltage at the panel where those circuits interconnect with other circuits in the occupancy remains within normal limits. An informational note explains that wind turbines may use the utility grid to dump energy from a short-term wind gust. Longer-term exposures would presumably require a more comprehensive solution. Review the discussion at 694.7(C) for more information.

ARTICLE 695. FIRE PUMPS

695.1. Scope. This article covers a part of the electrical system that is almost unique in that it is designed to self-sacrifice and run to complete failure. The requirements in this article are organized around the principle that when a fire pump is called into service, it must run until told not to, or die trying. On the other hand, the fire pump and its circuits cannot become a source of ignition that might call the fire pump into service. For this reason, for example, fire pump circuits have no running overload protections but they do have short-circuit protection.

The 2014 NEC has made some clarifications in part **(B)** as to what is not covered in the article. The installation of pressure-maintenance (jockey) pumps (as distinguished from how their existence may need to be factored into the fire-pump electrical installation should they be present) is not covered in Art. 695. A new informational note directs the reader to Art. 430 for the installation of such motors. Another change establishes that transfer equipment installed upstream from any fire-pump transfer switch(es) is not covered within the scope of Art. 695.

695.2. Definitions. A *fault-tolerant external control circuit* is one that however open or shorted after leaving the controller, the fire pump will still be able to start from other means, including the possibility that it may start anyway.

An *on-site power production facility* is a nonutlity on-site source such as a power plant that is the normal source of electric supply for the facility.

An *on-site standby generator* only runs when it is called to run for a specific reason, and as a backup supply of electric power. It does not constantly produce power.

695.3. Power Source(s) for Electric Motor-Driven Fire Pumps. The power source(s) must be reliable. The definition of the term "reliable" is not within the province of the NEC, because the performance requirements for fire pumps belong to the NFPA 20 Committee. In the current edition of NFPA 20-2010, *Standard for the Installation of Stationary Pumps for Fire Protection*, there is extensive information on this concept in the annex material on its Chap. 9, specifically A.9.3.2. What follows is an extremely abbreviated summary of material that should be read in its entirety. The qualities of a reliable source include four general characteristics.

First, the source of power should not have shown any outages longer than 4 continuous hours in the year prior to plan submittal. This is the time threshold in NFPA 25 for which fire watches and other special undertakings become necessary. Second, no outages were experienced in the area in the utility grid, unless such outages were due to a natural disaster. The annex goes on to say that the normal source is not required to be infallible. The east coast blackout of 1965, for example, would not condemn a utility distribution to forever be deemed unreliable. On the other hand, if the power grid is known to have general reliability issues caused by faulty switching procedures or animals shorting transformer bushings, and those issues were not resolved promptly, then the utility supply could reasonably be described as unreliable.

The third characteristic describes a utility supply fed through overhead drops to the protected premises as inherently unreliable due to fire department operations that result in these supplies being deenergized routinely in order to clear the fireground for ladders and other aerial apparatus. The fourth characteristic concerns local switching arrangements that result in a likelihood that a fire pump will be inadvertently disconnected on-site through erroneous opening of a disconnect. Great care has been taken to coordinate the electrical disconnection rules between the NEC and NFPA 20, so it would be fair to say that if the local distribution system complies fully with the current NEC, then the facility must be presumed to be compliant with the fourth characteristic, and only the first three need to be further evaluated in a reliability determination.

It should be further noted, that a backup generator does not, in and of itself, make an unreliable supply reliable. Backup generators have a long record, and that record is replete with issues around failure to start and other reliability problems. However, under the circumstances defined in this section, it can add enough to the mix to make the overall picture one of reliability.

Part **(A)(1)** is the first preference, and it is for an individual, reliable source. For example, a separate utility service connection from a reliable utility grid through an underground supply would presumably pass muster as "reliable" and be acceptable. Note that such a connection must be ahead of the service and if tapped from the utility lateral, the tap cannot be in the same cabinet, enclosure or vertical switchboard section as the service disconnecting means. The actual service disconnect for the fire pump must comply with 230.72(B), which makes it remote from the normal service, and usually right at the fire pump so someone opening service disconnects will not inadvertently disconnect it. Part **(A)(2)** allows for a reliable on-site power plant with a comparable record of reliability, especially in terms of the first characteristic of reliability covered above. In such cases, and with a comparable disconnect location [230.72(B) does not apply because this is not a service supply, but 695.4(A) does], and a direct connection from such a source would be viewed as reliable.

Part **(A)(3)** covers a dedicated feeder originating at a service connection. Such a feeder must meet the wiring requirements in 695.4(B)(1), which are covered in detail at that location below. Part **(B)** covers the case where the sources in (A) are available, but judged unreliable. The first option is two of the sources in (A) wired as alternate sources of supply and landing at a listed fire pump transfer switch. The second option is an approved combination of one (or more) part **(A)** source(s) and an on-site standby generator that qualified under 695.3(D) (see below). Another option, covered extensively in NFPA 20, is where a backup fire pump is in place driven by a nonelectric prime mover, either engine-driven or steam turbine-driven.

Part **(C)** is where the process gets more complicated. This is where the requirements begin to accommodate large industrial or institutional campus distributions, where a service connection is out of the question. Consider, for example, the matter of a fire pump in a college academic building thousands of meters (feet, does not matter) from the service for the campus. This service is 13.8 kV because the college owns its distribution. A literal service connection is inconceivable, but a connection to gear fed from the medium-voltage loop is quite feasible. Similar conditions apply at countless industrial plants throughout the country.

There are two options, and in both cases it must be demonstrated that the overcurrent devices in the chain of supply to the fire pump selectively coordinate. This is a performance requirement from NFPA 20 and because this material is extracted from that document, the NEC can not change it. Refer to the extensive presentation at 700.32 in this book for an extensive discussion of this concept and the practical difficulties in implementing it. This application is particularly bizarre, because it only applies to the fire pump circuit. Essentially, it says this: In the event of a short-circuit in the fire pump branch circuit, the branch-circuit protection will clear without opening the feeder protection above it and causing a wider outage. And the same process applies for the chain of protective devices above that. In effect, this rule protects the rest of the facility from the fire pump, instead of protecting the fire pump from the rest of the facility, which would have at least a faint basis in logic. Possibly what is intended is a requirement for all overcurrent protective devices on the supply side of the fire pump to be selectively coordinated with all overcurrent protective devices supplied by them that serve non-fire-pump loads. However, that is most certainly not what this rule now says. And if it were so worded, the consequences of effectively imposing a selective coordination requirement on an entire industrial distribution just to provide a fire pump supply would be so severe as to discourage the installation of such protection.

The first option is to arrive at the fire pump controller and transfer switch with two feeders that are derived from separate utility services (Fig. 695-1). The chain of supply to each feeder must meet the requirements in 695.4(B). The second option is a single feeder arriving at the fire pump transfer switch with the alternate source being a standby generator (Fig. 695-2) or other alternate, completely independent source of power.

Part **(D)** covers the usual case where the alternate source of power for a fire pump load is a standby generator, and the requirements are quite reasonable. The

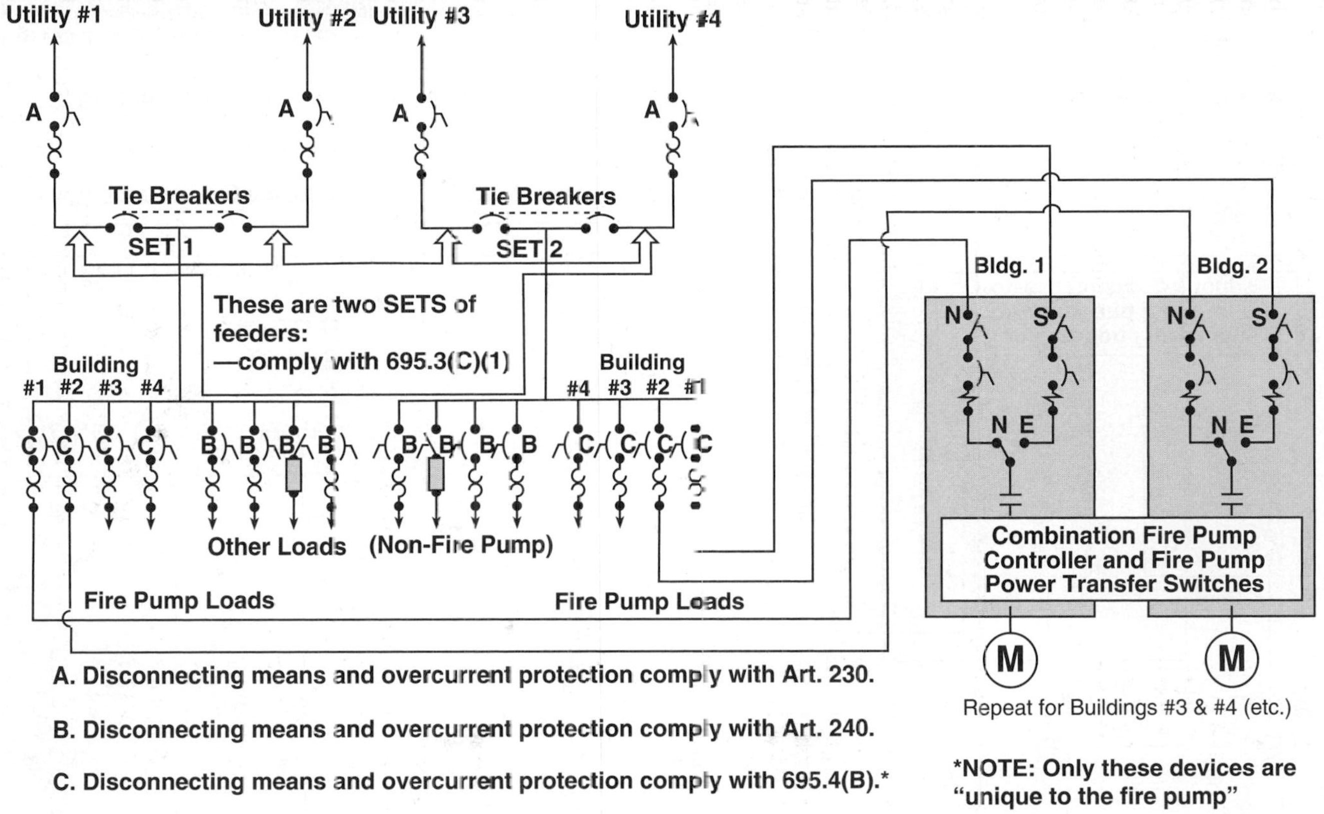
Utility #1
Utility #2
Utility #3
Utility #4
A
A
A
A
Tie Breakers
SET 1
Tie Breakers
SET 2
These are two SETS of feeders:
—comply with 695.3(C)(1)
Building
#1 #2 #3 #4
Building
#4 #3 #2 #1
C C C C
B B B B
B B B B
C C C C
Other Loads (Non-Fire Pump)
Fire Pump Loads
Fire Pump Loads
Bldg. 1
Bldg. 2
N S
N E
N S
N E
Combination Fire Pump Controller and Fire Pump Power Transfer Switches
M
M
Repeat for Buildings #3 & #4 (etc.)
A. Disconnecting means and overcurrent protection comply with Art. 230.
B. Disconnecting means and overcurrent protection comply with Art. 240.
C. Disconnecting means and overcurrent protection comply with 695.4(B).*
*NOTE: Only these devices are "unique to the fire pump"

Fig. 695-1. Two or more feeders comprise a reliable source if fed from different utility services. The NEC no longer requires multiple sets of feeders, so the double-ended service arrangements shown here are not required, but still permitted. The minimum requirement is one service for each feeder.

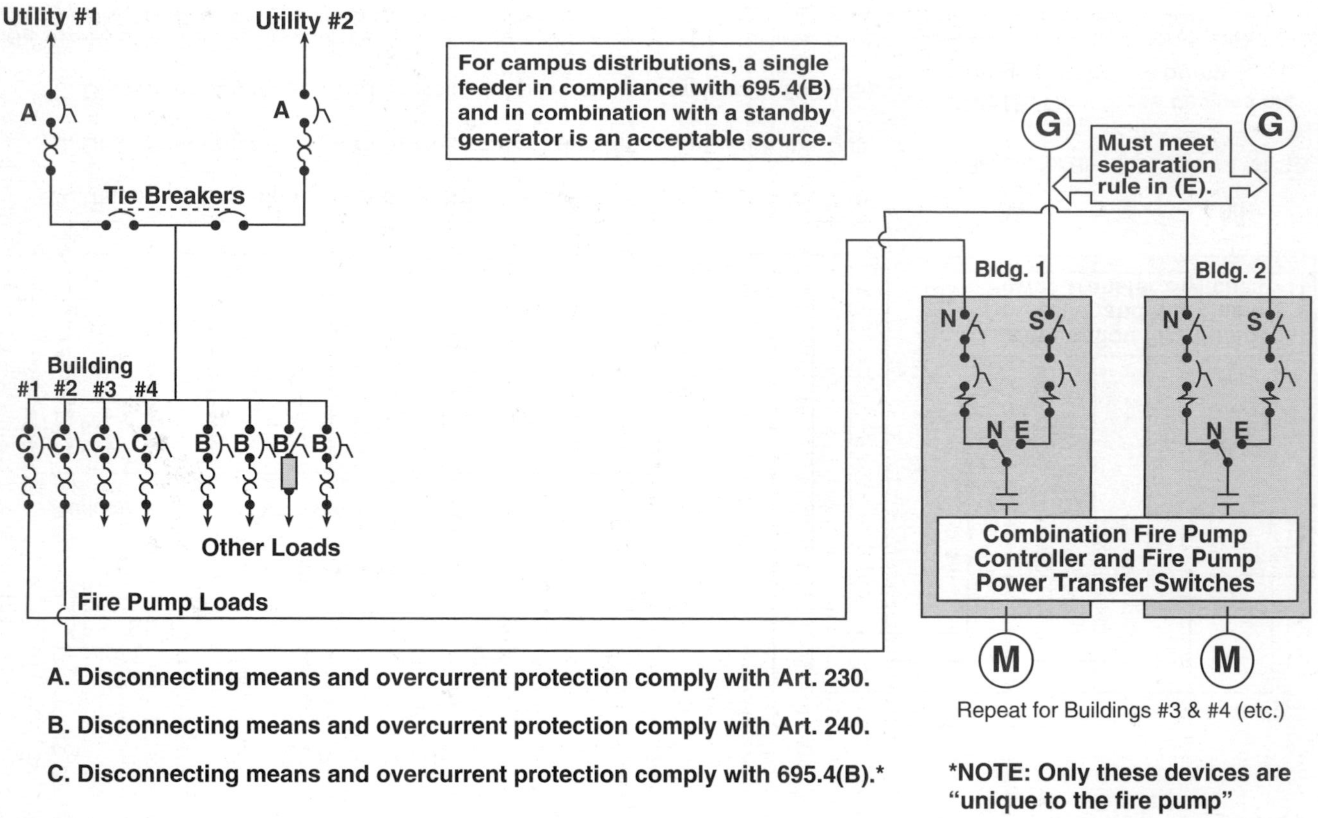
Utility #1
Utility #2
A
A
Tie Breakers
For campus distributions, a single feeder in compliance with 695.4(B) and in combination with a standby generator is an acceptable source.
G
G
Must meet separation rule in (E).
Bldg. 1
Bldg. 2
N
S
N E
Building
#1 #2 #3 #4
C
B
Other Loads
Fire Pump Loads
Combination Fire Pump Controller and Fire Pump Power Transfer Switches
M
M
A. Disconnecting means and overcurrent protection comply with Art. 230.
B. Disconnecting means and overcurrent protection comply with Art. 240.
C. Disconnecting means and overcurrent protection comply with 695.4(B).*
Repeat for Buildings #3 & #4 (etc.)
*NOTE: Only these devices are "unique to the fire pump"

Fig. 695-2. A single feeder can comprise a reliable source if backed up with an alternate source such as the standby generator shown here. The NEC no longer requires multiple sets of feeders, so the double-ended service arrangement shown here is not required, but still permitted. The minimum requirement is one service for the feeder, provided an alternate power source is provided.

generator must be able to pick up the entire fire pump load including associated jockey pumps and controls that have to operate at the same time. It does not have to be sized to carry the locked-rotor current of the fire pump. The generator can supply additional loads in the facility as long as it has the required capacity. Or, it can be fitted with automatic load-shedding controls, similar to what is allowable for emergency generators in 700.4(B), that ensure the fire pump start will not be encumbered by other loads on a generator with insufficient capacity. The connection for the feeder circuit from the generator does not need to be made ahead of the disconnecting means for the generator. The requirement in 430.113 for plural disconnects near a motor fed from two sources, one for each, does not apply to fire pump supplies. In this case the feed from both sources to the fire pump will be through a transfer switch so it is questionable whether 430.113 is in play, although waiving it does no harm.

Part **(E)** requires that power sources to a fire pump be located and arranged so they are protected against fire damage and "exposing hazards." This is another performance requirement extracted from NFPA 20, so it means what the NFPA 20 committee says it means and the NEC cannot vary it. In this case, there is a note in the Annex to NFPA 20 that indicates this requirement is to be applied to an "on-site power production facility" as well as the fire pump wiring and components. In addition, the multiple sources must be arranged so a fire at one location does not disrupt the alternate source. Part **(F)** requires that a transfer of power between the "individual source" and one alternate source takes place within the pump room. In this context the "individual source" is one of the three covered in 695.3(A). The selection of sources must be performed by a transfer switch that is specifically listed for fire pump applications. And that transfer switch, if it is configured as a listed assembly to include the short-circuit protection normally required for fire pump circuits, is permitted to use instantaneous circuit breakers as that overcurrent protection. Part **(G)** prohibits phase converters from sourcing a fire pump load because of the effects of voltage imbalances that can occur, particularly in the manufactured phase.

695.4. Continuity of Power. This section tries to make sure that human error will not inadvertently compromise a fire pump connection. Part **(A)** asks for a direct connection to the power supply system, and essentially works for 695.3(A) sources [or (B) sources through a transfer switch] that are therefore reliable. In addition, by virtue of being directly available to the fire pump controls at the point of entry to the building, they are capable of being terminated directly on the fire pump controller (or listed fire-pump power transfer switch) with no intervening overcurrent protection and disconnecting means. However, in more complex systems, it is not always possible to bring those supply conductors to the pump controller without some intervening disconnect and associated overcurrent protection. In such cases, covered in part **(B)**, the disconnect must be installed and conductors protected, but only in the context of additional requirements that ensure that the disconnect will not be opened by human error and that the overcurrent device will not open except under the most extreme conditions. There are three types of installation to address under this topic.

The first is the relatively uncomplicated condition where the fire pump is remote from its supply conductor entry but is otherwise served by a reliable source. In such cases a single disconnecting means, as covered in **(B)(1)**(a), may be installed. The associated overcurrent device must be sized so that it will only open in the event of a short-circuit or ground fault, but not a simple pump overload condition. As covered in **(B)(2)**(a), this device must carry the locked-rotor current of the largest fire pump motor, plus the locked rotor current of the pressure maintenance (often referred to as a jockey pump) motor, plus the full load current of all other associated accessory equipment for an indefinite time period without opening. This rule only covers the overcurrent device; it is not necessary to cable the motors based on this calculation or upsize any other devices. The 2014 **NEC** has added another option here. It is now possible to specify "an assembly listed for fire-pump service." The four performance criteria that follow would be part of the listing evaluation and not judged in the field.

Having assured that the overcurrent device will not open under any condition but the most extreme circumstances, where the supply conductors would actually be creating a fire condition, the next step is to remove human error from the operation of the disconnecting means, as is covered in **(B)(3)**(a), (c), (d), and (e). Specifically, the disconnect must be SUSE rated, lockable in the closed position, not located in any equipment, such as a common panelboard or switchboard, that feeds any other loads than the fire pump, and physically located sufficiently remote from such other equipment that inadvertent operation at the same time would be unlikely. It must also be plainly marked with its function (c), the controller must be marked with its location and the location of the key if it is locked closed (d), and its status must be supervised by using one (or more) of the strategies itemized in (e). Note that the 2014 **NEC** added criteria to the lock-out mechanism that has been standardized throughout the **NEC** by references to 110.25. For some reason that reference is still missing here and will need to await action in the 2020 cycle.

The next possibility to address is when the service (or on-site power plant) is not judged reliable, and an on-site generator is required to establish the required reliability. This case also allows for an additional overcurrent device and disconnecting means, as covered in **(B)(1)**(c). This overcurrent device need only be sized to carry the full fire pump equipment load, as covered in (2)(b). Its disconnecting means, addressed in **(B)(3)**(b), must comply with 700.10(B)(5) as though it were an emergency circuit, which means it must not be located within the same vertical switchboard section, although that vertical section may share a common bus with other sections. It must not be located in a common panelboard or other common enclosure as other loads supplied by the generator as well. However, the generator may serve as the originating point for other feeders originating at multiple overcurrent devices on the generator, as covered in 445.18. The overall concept is to recognize that many large facilities use highly reliable central generator facilities, and enhanced separation rules only apply at the point where the physical separation of functional feeder conductors, in this case for the fire pump backup supply function, originate and separate from other generator loads. The generator disconnecting means must, however, comply with the supervisory requirements in (c), (d), and (e).

The final possibility is the existence of a large multibuilding campus type distribution, as covered in **(B)(1)**(b) to address 695.3(C) applications. In these cases there will be additional disconnects and overcurrent protective devices in order to comply with various Code rules regarding feeder protection and building disconnects, etc. However, any disconnecting means in the chain of supply "that are unique to the fire pump loads" must comply with the enhanced features and supervisory requirements in **(B)(3)**(a) and (c), (d), and (e).

To reiterate and provide a calculation example for the basic requirement in (B)(2), the overcurrent devices must carry indefinitely the locked-rotor currents of the fire pump, the jockey pump (if used) and all associated accessory equipment when connected. The next higher standard sized protective device must be used. Note that overcurrent devices ahead of fire pump controls, etc., are not part of the locked rotor rule except insofar as their currents add arithmetically to the motor currents in determining the required rating of the next higher sized device.

Example: A 75 hp 575 V fire pump motor and a 5 hp (575 V) jockey pump are protecting a facility along with 15 A of accessory load as supplied on the primary side of a control transformer that supplies 120 V accessories. Calculate the setting of the short-circuit and ground-fault protective device for this feeder.

Solution: From Table 430.251(B), the locked rotor current for the motors is 434 A + 37 A = 471 A. Add the accessory load of 15 A for a total of 486 A, so a 500 A short-circuit and ground-fault protective device is required for this load.

695.5. Transformers. Transformers are necessary where the service or feeder voltage differs from that of the fire pump. This rule has a decision tree built into it. If the installation is on a campus distribution as covered in 695.3(B)(2), then the transformer requirements are taken from part **(C)**. If the fire pump supply is from any other source, then the transformer is dedicated to the fire pump load, and the transformer size and the rating of the upstream overcurrent device is calculated here.

Part **(A)** requires the size of the transformer be rated in terms of 125 percent of the motor load, and 100 percent of accessory equipment. Using the previous example, the motor load from Table 430.250 is 72 A + 6 A = 78 A; take this at 125 percent = 104 A, and add the accessory load of 15 A for a total of 119 A

$$\text{kVA transformer rating} = \frac{119\text{A} \times 600\text{V} \times \sqrt{3}}{1000} = 124\text{ kVA}$$

The next higher commercially available transformer size is 150 kVA.

Part **(B)** determines the overcurrent protection based on the locked rotor current determined previously, or 471 A reflected through the winding ratio of the transformer. The primary voltage in this case is 2400 V. 600 V ÷ 2400 V = 0.25, so the minimum primary side overcurrent protection is 0.25 × 471 A = 118 A. A 125-A "E" rated fuse for this voltage will provide the appropriate protection.

Part **(C)** covers campus-style feeder distributions. In these cases, and only these cases, the transformers are permitted to supply other load. However, their minimum size is still based on the same calculation as above, and their primary side protection must still carry the locked rotor currents indefinitely. However, the conductor and the transformer overcurrent protection will be in accordance with normal procedures in 215.3 and 450.3. In other words, for these applications, the locked rotor currents of the fire pumps are carried as loads, with the transformer sizing and wire sizing figured accordingly. Remember that the usual applications for these rules are large institutional and industrial occupancies where the typical transformers are measured in MVA, and even locked rotor current on a fire pump among all the other motors is essentially lost. This is a practical approach, used without incident for almost 10 years now. In terms of reliability, these transformers are the least likely to be out of service, given the likely perceived criticality of other simultaneously connected loads. (See Fig. 695-3.)

695.6. Power Wiring. These are the wiring methods and related requirements. The first set of cross references, contrary to the wording of the sentence, does not involve any requirements because the items are instances of where fire pump wiring receives special dispensation in other Code sections.

Part **(A)(1)** covers fire pump wiring that is normally connected to a utility supply and therefore consists of service conductors, but it also applies to a separately derived source originating from an on-site power production facility. As is true for service conductors generally, this wiring must be routed outside a building either in fact, or for Code purposes as covered in 230.6, but even there, only where encased in concrete or run below grade. The allowance in 230.6 for installation in a vault does not apply here, since a transformer vault may be the very location of the fire the fire pump will be trying to extinguish.

Part **(A)(2)** applies to feeder wiring, including wiring from an on-site standby generator on the load side of the final disconnecting means allowed in 695.4(B). This wiring must be divorced from all other wiring, and supply only fire pump and related loads. This wiring must be arranged to avoid damage by fire, structural failures, and operational accidents. Where routed through the building, they must employ construction methods that provide a 2-h fire separation. Options here include 50 mm (2 in.) concrete encasement or in a listed electrical circuit protective assembly such as MI cable installed in accordance with a special protocol, or construction methods that provide an equivalent separation. However, once the wiring arrives in the pump room, the additional protection can be discontinued unless required by 700.10(D). Note that these rules protect the wiring from the building; the rules in (A), in part, protect the building from unprotected wiring.

Part **(B)** should remove all doubt that the locked-rotor overcurrent protection rules earlier do not imply oversized motor circuit conductors. The sizing here reflects time-honored standard practice with respect to motor circuit conductor sizing. In the example worked previously, the equipment feeder covered in (A) would use the 119 A load (1 AWG wire), and the motor circuit would use 125 percent of 77 A = 96 A, for a 3 AWG wire.

Part **(C)** waives the overload requirements, for the fire pump only. The only protection will be the short-circuit protection. This is why a tap off of such a feeder is treated as a service conductor for the purpose of defending the building. However, having now protected the tap in this way, the trade-off is that the distance limitations in 240.21 do not apply. In addition, the vault allowance is restored in this citation of 230.6, so the wiring could run through a vault without additional protection. This correlates this section with the rules for engineered systems. In addition, the last sentence facilitates the placement of a fire pump disconnecting means remote from other disconnecting means in cases where it will be fed from feeder circuits instead of being comprised of service conductors. This makes inadvertent

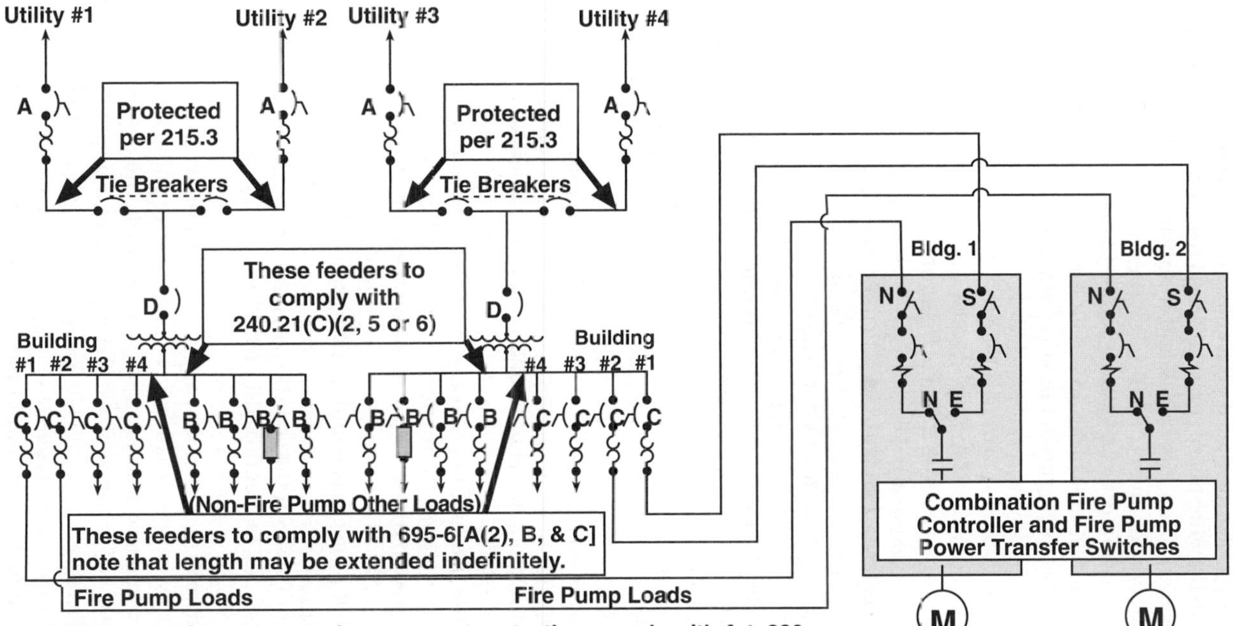

Utility #1
Utility #2
Utility #3
Utility #4
A
A
A
A
Protected
per 215.3
Protected
per 215.3
Tie Breakers
Tie Breakers
These feeders to
comply with
240.21(C)(2, 5 or 6)
D
D
Building
#1 #2 #3 #4
Building
#4 #3 #2 #1
C C C C B B B B
B B B B C C C C
(Non-Fire Pump Other Loads)
These feeders to comply with 695-6[A(2), B, & C]
note that length may be extended indefinitely.
Fire Pump Loads
Fire Pump Loads
Bldg. 1
Bldg. 2
N
S
N
S
N E
N E
Combination Fire Pump
Controller and Fire Pump
Power Transfer Switches
M
M

A. Disconnecting means and overcurrent protection comply with Art. 230.

B. Disconnecting means and overcurrent protection comply with Art. 240.

C. Disconnecting means and overcurrent protection comply with 695.4(B).*

D. Overcurrent protection complies with 450.3 AND 695.5(C)(2).
Secondary OCPDs may be limited to six.

Repeat for Buildings #3 & #4 (etc.)

*NOTE: Only these devices are "unique to the fire pump"

Fig. 695-3. Large transformers and feeders per 695.5(C)(2) with conventional overcurrent protection may supply diverse loads.

contemporaneous operation by the fire service, as covered in 695.4(B)(3)(4), far less of a possibility. (See also Fig. 695-3.)

An example of the problem is that if conventional tap rules are applied to a 4000 A multiple-ended switchboard, even a 3-m (10-ft) tap would need to be 600 kcmil (at least 400 A). Even if the 600 kcmil conductors could be terminated in the fire pump disconnect, the 3-m (10-ft) length limitation would probably leave the fire pump disconnect in the same room. Remember also, 230.6 protects the building from the least protected of all electrical conductors, namely, service conductors. If the tap is run in the same way, surely the building will be equally protected.

An exception applies to battery wiring for cranking engines. In addition, if the standby generator rated current capacity exceeds 225 percent of the fire pump full load current, the **Code** assumes a short-circuit or ground-fault hazard exists on the order of a service exposure. Therefore, in such cases the conductors running between the transfer switch and the generator must comply with 695.6(B) limitations.

Part **(D)** covers pump wiring methods, which include RMC, IMC, EMT, LFMC, LFNC-B for raceways, and Types MI and MC cables, the latter with an impervious covering. As of the 2014 **NEC**, twist-on wire connectors, along with insulation-piercing connectors and soldered connections are expressly prohibited from the terminal boxes of fire pump motors. Pick another method, and make sure it is listed. Note that 110.14 will come into play given the fine-stranding leads common with most motors. Refer to the extensive coverage in Chap. 1 of this book on that topic.

Part **(E)** prohibits a fire pump transfer switch and/or a fire pump controller from being used for any other purpose than operating the fire pump equipment.

Part **(F)** requires wiring from batteries and engine controllers to be protected from damage and be installed in accordance with directions.

Part **(G)** forbids the use of GFPE "in any fire pump power circuit." This is absolutely to be distinguished from any upstream feeder, including upstream feeders in campus distributions. If the upstream wiring is subject to GFPE requirements such as 215.10, and the fire pump power circuit happens to fall underneath them, that protection is unimpeded by this rule, which is specific to the fire pump circuit.

Part **(H)** incorporates some of the specific requirements carried in the electrical circuit protective systems, as mentioned in 695.6(A)(2)(d)(3), and as itemized in **NFPA** 20 at 9.8.2 and 9.8.3.

Part **(I)** includes six requirements that apply to a junction or pull box through which fire pump wiring passes, whether from or to the fire pump controller. The pull box must, at a minimum (if in the room, and at least equal to the enclosure type for the controller), be "drip-proof" which presumably corresponds to a **NEMA** 2 or better enclosure as shown in Table 110.28. The box must be securely mounted and in a way that does not compromise the integrity of the enclosure for the fire pump controller. Terminals, "junction blocks" [presumably power distribution blocks, as covered in 314.28(E)], and wire connectors must be listed. Item (6) on the list directly recalls 695.6(E) and prohibits the same other load wiring from even passing through it.

Part (J) covers raceway terminations, and requires the use of hubs with ratings at least equal to that of the controller enclosure for all raceway entries. All installation instructions of the controller manufacturer must be observed and any other alterations of the enclosure beyond the conduit entries as contemplated by the manufacturer are subject to specific approvals by the inspector.

695.7. Voltage Drop. Fire pump voltage as measured at the controller line terminals must not drop more than 15 percent below the normal, that is, the controller rated voltage, during motor starting conditions. In addition, the voltage at the motor controller load terminals while running at 115 percent of full-load current must not fall below 95 percent of the voltage rating of the motor. However, this limitation does not apply for emergency run mechanical starting. The issue in this section is not the fire pump itself; it is the likelihood that under reduced voltage a magnetic contactor would chatter and drop out, preventing the pump from functioning. This is why a mechanical start is exempted from the rule.

695.10. Listed Equipment. Diesel engine fire pump controllers, electric fire pump controllers, electric motors, fire pump power transfer switches, foam pump controllers, and limited-service controllers are to be listed for fire pump duty.

695.12. Equipment Location. Fire pump transfer switches and controllers must be as close as practical and within sight of the fire pump motors they control, and engine drive controllers must be as close as practical to the engines they control. Storage batteries must be up off the floor, secured, and located where they will not be damaged.

All energized equipment must be not less than 300 mm (12 in.) above the floor level. Fire pump control enclosures must be securely mounted on noncombustible surfaces. Fire pump motors often leak water at their main bearings by design, and their control equipment must be mounted or protected to avoid damage from escaping water.

695.14. Control Wiring. External control circuits that leave the fire pump area (the NEC says "room" but a room is not an actual Code mandate under the terms of this article) must be fault-tolerant as defined in 695.2, and the loss of power in a control circuit may cause the pump to run continuously after starting but the controller will still be able to initiate a start. Any control conductors in the room must be protected. The motor must always try to start, and therefore no phase-loss relay or other sensor is permitted to prevent actuation of the contactor. No remote device can be installed that would inhibit the operation of the transfer switch. Engine drive wiring must be stranded and capable of continuously carrying the charging or control loads, and protected from damage.

For electric fire pump control wiring, only RMC, IMC, LFMC, EMT, LFNC, MI cable, and MC cable with an impervious covering are permitted. Note that this list correlates with the list for wiring to the pump itself in 695.6(D). For generator control wiring connecting the transfer switch and the generator, this wiring must be divorced from all other wiring, and protected in a manner to resist damage from fire or collapse. If run through the building, they must be encased in 50 mm (2 in.) of concrete, or other 2-h fire rated assembly dedicated to the pump circuits, or in an electrical circuit protective system with a 2-h rating, and wired in accordance with the rules for the applicable assembly.

The integrity of this wiring must be continuously monitored, and if compromised, visual and audible alarms must activate and the generator must start. An informational note points to the relevant provision in NFPA 20 regarding "fault-tolerant external control circuits." The same requirement appears in 700.10(D)(3) for generator control wiring on emergency systems.

695.15. Surge Protection. As of the 2017 NEC, surge protection must be arranged "in or on" the fire pump controller.

Chapter Seven

ARTICLE 700. EMERGENCY SYSTEMS

700.1. Scope. Note that all the regulations of this article apply to the designated emergency systems. As defined in Sec. 700.2, these are the systems "legally required and classed as emergency by municipal, state, federal, or other codes, or by any governmental agency having jurisdiction." *The* NEC, *itself, does* not *require emergency light, power, or exit signs.*

The effect of this definition is to exclude from all those rules any emergency circuits, systems, or equipment that is installed on a premises but is not legally mandated for the premises. Of course, any emergency provisions that are provided at the option of the designer (or the client) must necessarily conform to all other NEC regulations that apply to the work.

The placement or location of exit lights is not a function of the National Electrical Code but is covered in the Life Safety Code, NFPA No. 101-2012. But where exit lights are required by law, the NEC considers them to be parts of the emergency system. The NEC indicates how the installation will be made, not where the emergency lighting is required, except as specified in Part III of Art. 517 for essential electrical systems in health care facilities.

An emergency lighting system in a theater or other place of public assembly includes exit signs, the chief purpose of which is to indicate the location of the exits, and lighting equipment commonly called "emergency lights," the purpose of which is to provide sufficient illumination in the auditorium, corridors, lobbies, passageways, stairways, and fire escapes to enable persons to leave the building safely (Fig. 700-1).

These details, as well as the various classes of buildings in which emergency lighting is required, are left to be determined by state or municipal codes, and where such codes are in effect, the following provisions apply.

EXIT

Fig. 700-1. Exit lights and wall-hanging battery-pack emergency lighting units are covered by Art. 700 whenever such equipment or provisions for emergency application are legally required by governmental authorities. (Sec. 700.1.)

As indicated in the informational note following the definition, an emergency system to supply loads classified as emergency in nature by virtue of such a building code or related provision is required to function at all times. Emergency loads do not lose their status as emergency loads just because the generator or other backup source is not running. If a fire is in progress at one end of the building, and the normal wiring system has not yet been compromised, the exit signs still need to work just as well as when the generator is the source of supply. Therefore, all the rules in this article apply to those loads regardless of the status of backup power equipment. This discussion will come up again in 700.32.

700.3. Tests and Maintenance. Part **(A)** of 700.4 calls for inspection of an operational test at the time of system installation and commissioning, generally referred to as an acceptance test *and* operational testing at periodic intervals thereafter. The frequency of these witnessed tests is to be determined by the local inspection authority [part **(B)**]. Part **(D)** requires documentation of such tests. Part **(C)**, broadened in the 2017 **NEC** from a focus on batteries to all emergency system equipment, requires maintenance, to be enforced by the AHJ, of all emergency system equipment, including the batteries that start prime movers and those contained within unit equipment. Emergency systems must be tested "during maximum anticipated load conditions," as called for in part **(E)**. Testing under less than full-load conditions can be misleading and is not a true test. Part **(D)** requires written documentation of testing and maintenance.

There has been a requirement, formerly residing in 700.4(**B**) as a third paragraph, that an alternate source must be available when the emergency generator is out of service for major maintenance or repair. The 2017 **NEC** no longer has that paragraph, and instead includes almost a full column and a half [as Part (**F**)] of prescriptive text and a graphic to cover the same ground. The basic requirement applies to any location where the alternate source of power is a single generator, and in such cases the site wiring must include a permanently wired switching arrangement and a method of connection of the temporary alternate source. One example would be a permanently wired inlet, with either a conventional or pin-and-sleeve configuration of appropriate rating. The inlet (or other connection point) must be marked to provide notice of the phase rotation and system bonding requirements. The switching can be done manually, but in the event of emergency conditions, the allowable pick-up time for the temporary alternate source must still meet the 10-s parameter in 700.12. The switching method must also preclude an inadvertent interconnection of sources. To add flexibility and ongoing utility for the hard-wired temporary standby connection point, there is express permission to utilize this point for a load bank connection.

These provisions come with a one-line diagram showing a typical compliant arrangement, and also with a detailed exception that provides four options to avoid the new requirement. The first is to demonstrate that for any process or load that relies on the emergency source being available, that function is capable of being disabled during the major maintenance or repair. The second is for an unoccupied building, with the additional condition that any sprinkler (or equal) system will remain operational. For example, a dry system reliant on compressed air would need to qualify under a different provision. Other temporary (unspecified) means are permitted, obviously with the inspector's approval. Finally, and almost not needing to be said, where there is a permanent additional source the need for the connections goes away. In this case, the system would presumably not fall within the triggering provisions ("relies on a single alternate source of power") and qualify for exemption on that basis.

700.4. Capacity. It is extremely important that the supply source be of adequate capacity. There are two main reasons for adequate capacity:

1. It is important that power be available for the necessary supply to exit lights and emergency and egress lighting, as well as to operate such equipment as required for elevators and other equipment connected to the emergency system.
2. In such occupancies as hospitals, there may be a need for an emergency supply for lighting in hospital operating rooms, and also for such equipment as inhalators, iron lungs, and incubators.

It is also essential for safety and effective, long-time, successful operation of emergency system equipment that it be rated to sustain the maximum available short circuit that the supply circuit could deliver at the terminals of the equipment.

Part (**B**) of this section permits one generator to be used as a single power source to supply emergency loads, essential (legally required) standby loads, and optional standby loads when control arrangements for *selective* load pickup and load shedding are provided to ensure that adequate power is available first for emergency loads, then legally required standby loads, and finally optional standby loads. [See Fig. 700-2 (top) and Arts. 701 and 702.]

There is an important difference between how this article and Art. 517 handle the question of automatic load shedding and mandated provisions for selective load pickup. As covered in 517.31(B)(1), if there will be non-Art. 517 loads supplied by a generator, there must be one or more separate transfer switches for those loads [agrees with 700.5(D)] and there must be automatic load control regardless of the generator capacity. That last part of the rule is more severe than the one here, which waives the automatic load control provisions on a showing of adequate capacity, which could be determined by an Art. 220 load calculation.

An on-site generator may be used for peak load shaving, in addition to its use for supplying emergency legally required standby and optional standby loads. The title of part **(B)** refers to "Peak Load Shaving," which is recognized as a function of the generator in the last sentence of the paragraph. For peak shaving use, however, the generator must be equipped with the selective load-control equipment described in the paragraph to ensure the order of priority for various purposes. The second paragraph essentially permits the peak load shaving operation to satisfy the testing and maintenance requirements of 700.3.

Generators used for peak load reductions often have the virtue of being routinely exercised under load for significant periods. This increases their overall reliability because any maintenance deficiencies tend to become apparent prior to an emergency.

700.5. Transfer Equipment. Any switch or other control device that transfers emergency loads from the normal power source of a system to the emergency power supply *must* operate automatically on loss of the normal supply. Transfer equipment must also be automatic for legally required standby systems, as covered in 701.5, but a manual transfer switch may be used for switching loads from the normal to an optional standby power source, as in 702.5. The transfer switch must also be listed for emergency use. Listed switches are now available for use on medium-voltage systems, and therefore the voltage limitation on the listing requirement has been removed as of the 2017 NEC.

As described in part **(B)**, a bypass switch is recognized on an automatic transfer switch (ATS) to provide for repair or maintenance of the ATS (Fig. 700-2, bottom). Because hospitals and many industrial systems contain transfer switches that cannot be shut down, some means is required to isolate a transfer switch for routine maintenance or for some specific task like contact replacement. When a bypass switch is used, inadvertent parallel operation must not be possible.

Part **(C)** requires that transfer switches be electrically operated and mechanically held. This ensures that the contacts stay closed in the event of a coil failure, or a temporary loss of voltage in the control circuit directing the emergency connection resulted in a generator running full speed and not picking up the load. This way the switch stays held where it belongs until an affirmative unlatch signal comes in conjunction with a normal system latch command.

Part **(D)** requires transfer switches that are used in emergency systems to be used exclusively for this purpose and supply no other load. This is the case even in instances where multiple systems are supplied from a single generator. Note that this is a direct conflict with the permission granted in 517.31(B)(2) in small hospital facilities. See the discussion under that heading, and also the discussion at 700.4(B) regarding a related issue where the rule in 517.31(B)(3) is more severe than the requirements here.

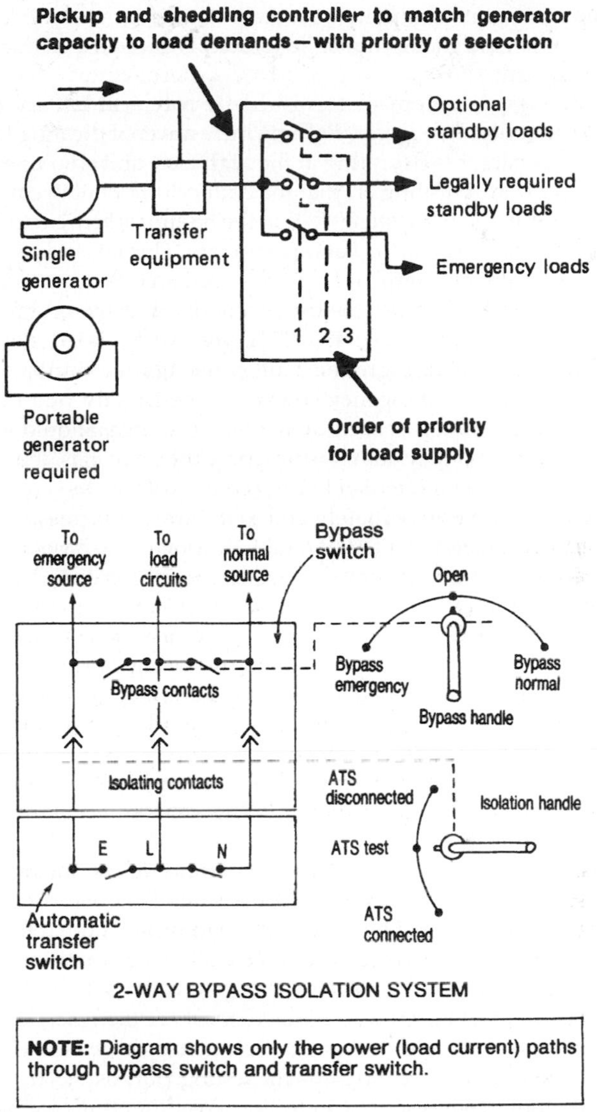
Pickup and shedding controller to match generator capacity to load demands—with priority of selection
Optional standby loads
Legally required standby loads
Emergency loads
Transfer equipment
Single generator
Portable generator required
1 2 3
Order of priority for load supply
To emergency source
To load circuits
To normal source
Bypass switch
Open
Bypass emergency
Bypass normal
Bypass contacts
Bypass handle
Isolating contacts
ATS disconnected
Isolation handle
ATS test
E
L
N
ATS connected
Automatic transfer switch
2-WAY BYPASS ISOLATION SYSTEM
NOTE: Diagram shows only the power (load current) paths through bypass switch and transfer switch.

Fig. 700-2. Special load-control must be used where a single generator supplies emergency and standby loads (top). And a bypass switch may be used to isolate an automatic transfer switch. (Secs. 700.4 and 700.5.)

Part (E) requires transfer equipment to be field marked on its exterior with its short-circuit current rating. The field marking is required because the characteristics of the applied upstream overcurrent protection will affect the rating.

700.6. Signals. In order to be effective, the signal devices should be located in some room where an attendant is on duty. Lamps may readily be used as signals to indicate the position of an automatic switching device. An audible signal in any place of public assembly should not be so located or of such a character that it will cause a general alarm.

The standard signal equipment furnished by a typical battery manufacturer with a 60-cell battery for emergency lighting includes an indicating lamp, which is lighted when the charger is operating at the high rate, and a voltmeter marked in three colored sections indicating (1) that the battery is not being charged or is discharging into the emergency system, (2) that the battery is being trickle charged, or (3) that the battery is being charged at the high rate. This last indication duplicates the indication given by the lamp.

Part **(D)** of this section requires the use of a hookup for ground-fault indication on the output of a grounded-wye 480/277V generator. 700.31 states that emergency systems do not require ground-fault protection of equipment. Yet when the load is served by the emergency source, the possibility of a ground fault is no less than when the load is served by the utility source. Should a ground fault occur within the emergency system during a power failure, it is debatable whether essential loads should be immediately disconnected from the emergency power supply. However, for reasons of safety and to minimize the possibility of fire and equipment damage, at least an alarm should indicate if a hazardous ground-fault condition exists, so appropriate corrective steps will be initiated. When ground-fault indication is used on the generator output, the fault sensor must be located within or on the line side of the generator main disconnect. For a large facility with multiple generators connected with a paralleling bus, the neutral-ground bond cannot be located at or ahead of the main. The rule also makes it mandatory to have specific instructions for dealing with ground-fault conditions. (Refer to 700.31 and Fig. 700-17.)

700.7. Signs. A sign at the service entrance is required to designate the type and location of emergency power source(s). This sign, however, need not display information regarding individual unit equipment as shown in Fig. 700-1 for obvious reasons. Part **(B)** requires that if the grounded circuit conductor from an emergency source is connected to a grounding electrode conductor anywhere except at that source, then a sign must be posted at the location of the remote grounding connection identifying all emergency and normal sources connected at that location. Figure 700-3 explains the issue.

This requirement resulted from an actual case where the emergency source was supplying power, and during that period maintenance personnel disconnected the normal source grounded conductor for testing purposes. The personnel did not realize that they were also disconnecting the grounding connection for the emergency source at the same time, since the grounded system conductor was only connected to the grounding electrode conductor in the main switchboard. The NEC now specifies the wording for this sign.

700.8. Surge Protection. As of the 2014 NEC, a listed SPD is required in every Art. 700 switchboard and panelboard. This wording was not correlated with the addition of the term "switchgear" throughout the NEC, but the oversight remains to be corrected and it should be applied in any event.

700.10. Wiring, Emergency System. Part **(A)** requires that all boxes and enclosures for emergency circuits must be readily identified as parts of the emergency

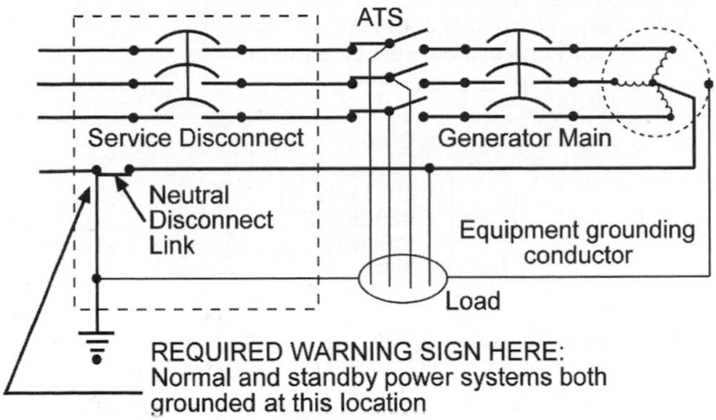
ATS
Service Disconnect
Generator Main
Neutral
Disconnect
Link
Equipment grounding
conductor
Load
REQUIRED WARNING SIGN HERE:
Normal and standby power systems both
grounded at this location

Fig. 700-3. This emergency system is not separately derived and with the generator supplying the load, an open neutral disconnect link will unground the system. [Sec. 700.7(B).]

system. Labels, signs, or some other *permanent* marking must be used on all enclosures containing emergency circuits to "readily" identify them as components of an emergency system. The "boxes and enclosures" that must be marked include enclosures for transfer switches, generators, and power panels. *All* boxes and enclosures for emergency circuits must be painted red, marked with red labels saying "Emergency Circuits," or marked in some other manner clearly identifiable to electricians or maintenance personnel (Fig. 700-4, top). The 2017 **NEC** dramatically expanded this rule by requiring any receptacles connected to an emergency system have an equivalent marking, using a distinctive color, either on the device or on the faceplate. In addition, all exposed wiring, whether cable or raceway, must also be identifiable by a periodic marking, at intervals not to exceed 25 ft. At least one manufacturer already has red (and other colors) EMT on the market.

The bottom of Fig. 700-4 shows a clear violation of the rule of part **(B)** because the wiring to the emergency light (or to an exit light, which is classed as an emergency light) is run in the same raceway and boxes as the wiring to the decorative floodlight.

This section requires that the wiring for emergency systems be kept entirely independent of the regular wiring used for lighting, and it thus needs to be in separate raceways, cables, and boxes. This requirement is to ensure that where faults may occur on the regular wiring, they will not affect the emergency system wiring, as it will be in a separate enclosure.

Part **(B)(1)** for transfer switches is intended to permit normal supply conductors to be brought into the transfer-switch enclosure and to specify that these conductors would be the only ones within the transfer-switch enclosure that were not part of the emergency system. Parts **(B)(2)** and **(3)** permit two sources supplying emergency or exit lighting to enter the luminaire, or a load control relay as covered in 700.26, and its common junction box. Part **(B)(4)** also allows a "common junction box" on unit equipment to contain both emergency and nonemergency circuits.

Part **(B)(5)** allows wiring from a common source location to supply both emergency and nonemergency load distributions. This provision is very poorly

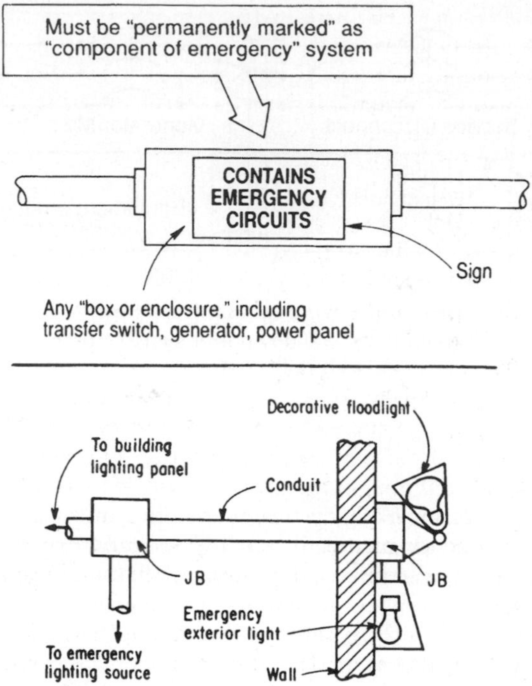
Must be "permanently marked" as "component of emergency" system
CONTAINS EMERGENCY CIRCUITS
Sign
Any "box or enclosure," including transfer switch, generator, power panel
Decorative floodlight
To building lighting panel
Conduit
JB
JB
Emergency exterior light
To emergency lighting source
Wall

Fig. 700-4. Emergency wiring may not be run in enclosures with wiring for nonemergency circuits, and emergency enclosures must be marked. (Sec. 700.9.)

written in that there is no parent language to lay out how the four provisions go together, but based on its history it appears that the four provisions are intended to be read as one rule with differing particulars. In effect, the Art. 700 function is permitted to coexist in the same conductor set with Art. 701 and Art. 702 functions, and the system separation rules do not take effect until that point in the wiring system where the load on a particular set of conductors becomes exclusively Art. 700 in character. Sec. 700.10(B)(5) provides great flexibility as to where that is permitted to occur. This recognizes that many enterprises maintain fully functioning standby power generation capable of serving the entire standby needs of the facility, and any attempt to prematurely divorce emergency circuits from one of these distributions could actually decrease reliability by reducing supply diversity. This follows from the parent wording "to supply emergency and other (nonemergency) loads." The parenthetical word "nonemergency," added in the 2017 **NEC**, proves that the separation rules in 700.10 do not apply ahead of the point where a particular set of loads becomes uniquely emergency in character.

Item "a." recognizes separate vertical switchboard (or switchgear) sections, with or without a common power bus, or in a sequence of disconnects mounted in separate enclosures. In these cases, the separation rules begin at the one or more vertical section(s) of a switchboard that is/are to supply Art. 700 loads, or in the one or more disconnect enclosure(s) that serve(s) Art. 700 loads. The common power bus of the switchboard and the wiring that supplies it are permitted to transmit power for any mixture of power needs, as would the common feeder supplying a line-up of individual disconnects through a wireway with taps extending to each.

Item "b." says that those switchboard (or switchgear) sections and enclosures, whether singular or plural, can be supplied by single or multiple feeders based on the tap rules in 240.21(G) that do not require overcurrent protection at the line side of a generator tap, provided they are sized at 115 percent of the generator capacity. This is now covered in a separate paragraph with its own graphic (1) to illustrate this point. The second paragraph shows how to achieve the same result with an overcurrent device ahead of the bus, and a graphic (2) to illustrate this arrangement. In this event, the only additional requirement is that the one or more nonemergency overcurrent device(s) selectively coordinate with the upstream device. With that having been established, any fault below the level of the common bus will be isolated from tripping the common OCPD and taking down the emergency system.

Item "c." clarifies that although the feeders addressed in these provisions can carry a mixture of loads, a vertical switchboard section or a separate enclosure is prohibited from supplying a mixture of loads. At this point of separation, Art. 700 circuits must be fully divorced from other loads, except where otherwise permitted in 700.10(B)(1-4).

Item "d." is new as of the 2011 **NEC** and carries the basic concepts developed here to another level. In general "distribution equipment" can intervene between an emergency source and where the separation is made. It serves to recognize in particular multibuilding installations with centralized generating facilities, including medium-voltage generating facilities, which have intervening distribution equipment, including transformation, placed well before any subdivision that produces unique Art. 700 load profiles.

Part **(C)** is motherhood and apple pie wording advising care in the system design to minimize their likelihood of damage. Part **(D)**, however is an entire set of rules regarding the physical separation of emergency system components in order to increase the reliability of the system in higher-hazard exposures. These locations are defined in the parent text as assembly occupancies with an occupancy load of not less than 1000 persons, and also high-rise buildings over 23 m (75 ft) high. As of the 2014 **NEC** the former list of occupancy classifications (assembly, educational, etc.) has been removed, so the enhanced separation rules apply in any building of the designated height. As of the 2017 **NEC**, two other classifications join the list, educational facilities with an occupancy load over 300 people, and health care occupancies serving patients incapable of self-preservation. The enhanced separation rules divide into two categories, the first for feeder circuits [never branch circuits except as in (D)(3)] and the second for distribution equipment fed by the emergency system feeders. In evaluating the various options do not be confused by beam spray products commonly applied to steel framing. These products are intended to prevent the deflection of those members at high temperatures, and as such keep the steel below temperatures in the range of 540°C (1000°F). At such temperatures an electrical circuit in conventional wiring will be lucky to survive 1 or 2 min.

Part **(D)(1)** on feeders provides five options for protection as follows (somewhat simplified): (1) Run them within an area fully protected by sprinklers. This means the area and not the feeder; an emergency system feeder run above a suspended ceiling does not need the sprinklers to spray on the feeder, only that the area involved meet the requirements in NFPA 13 or similar as a fully protected space. (2) Run them as a listed electrical circuit protective system with a minimum 2-h rating. These systems require great attention to detail and involve some surprising aspects. Refer to more detailed coverage in this book at 728.4.

Other possibilities include: (3) Install them in a listed fire-resistive cable system. These systems may be evaluated for a 2-h fire exposure, but not necessarily so. The previous wording (2014 NEC) mandated a 2-h classification; that criterion was omitted, apparently by mistake, from the 2017 NEC. The parent language was under discussion for a global 2-hour requirement, but that did not proceed because the last item (concrete encasement) may not always be, strictly speaking, a 2-h fire barrier. So, the 2-h rule didn't make it into the parent language, and it appears the panel forgot to revisit this line item and include it. It was always intended to be a 2-h condition. Until this is fixed, there is a way to get this back to 2-h. The second informational note in 728.4 specifically identifies fire-resistive cable systems as part of an electrical circuit protective system, and as previously noted here these systems, in turn, are subject to a 2-h minimum requirement. (4) Install them with a listed fire-rated assembly defined in the applicable UL literature as having the required time parameter, such as multiple layers of drywall in accordance with detailed installation criteria. (5) Embed them within 50 mm (2 in.) of concrete.

Part **(D)(2)** on distribution equipment requires panelboard, transformers, transfer switches, and comparable equipment to either be in a space with full sprinkler protection, or in a space with a 2-h fire-resistive rating. Part **(D)(3)** requires generator control wiring running between a transfer switch and its generator to be kept segregated from all other wiring, and to meet one of the feeder protective schemes covered in (D)(1). As of the 2017 NEC, the integrity of this wiring must be continuously monitored, and if compromised, visual and audible alarms must activate and the generator must start. The same requirement appears in 695.14 for generator control wiring for fire pumps.

700.12. General Requirements. This section lists and describes the types of emergency supply systems that are acceptable—with one or more of such systems required where emergency supply is mandated by law. It specifies that the normal-to-emergency transfer must not exceed 10 s. It also repeats the generic language of 700.10(C) regarding avoidance of hazards in terms of siting this equipment.

The last paragraph presents important information on protection of emergency power sources for certain occupancies. Inexplicably, and apparently through oversight, this paragraph has never been correlated with the changes in 700.10(D) that struck out the list of occupancy classes within the high-rise building classification. So, on the literal text, this requirement only applies in the designated occupancy classes. Such protection may be provided by (1) automatic sprinkler or (2) enclosing the equipment in a room with 1-h fire-rated walls. It quite logically makes the generator subject to the same rules as apply to the transformers, transfer switches, and panelboards supplied by it in accordance with 700.10(D)(2), except that the protection parameter is reduced by an hour [1-h fire resistance here, vs. 2-h fire resistance for feeders and equipment covered by 700.10(D)]; in both locations, however, a fully sprinklered building results in a full waiver.

Part **(A)** recognizes storage batteries for an emergency source. A storage battery for emergency power must maintain *voltage to the load* at not less than 87½ percent of the normal rated value. This was changed from "87½ percent of system voltage" (i.e., battery voltage) because the concern is to keep the voltage to the lamps at 87½ percent. The "electronics" between the battery and the lamps maintain the required "load voltage," and the battery voltage is not in itself the major concern. Automotive batteries are not permitted.

Part **(B)** covers use of engine-generator sets for emergency supply as an alternate to utility supply. Engine-driven generators (diesel, gasoline, or gas) are commonly used to provide an alternate source of emergency or standby power when normal utility power fails. Gas-turbine generators are also used. Note that the system may delay a retransfer to normal power for up to 15 min to avoid damaging short-time operation, especially of internal combustion prime movers.

The first step in selecting an on-site generator is to consider applicable requirements of the **National Electrical Code**, which differ depending on whether the generating set is to function as an emergency system, as a standby power system, or as a power source in a health care facility such as a hospital.

For example, an *internal-combustion*-type engine-generator set selected for use under Art. 700 must be provided with *automatic starting and, if required by 700.4(B), automatic load transfer,* with enough on-site fuel to power the full demand load for at least 2 h (Fig. 700-5). If a standby power system selected under the regulations of

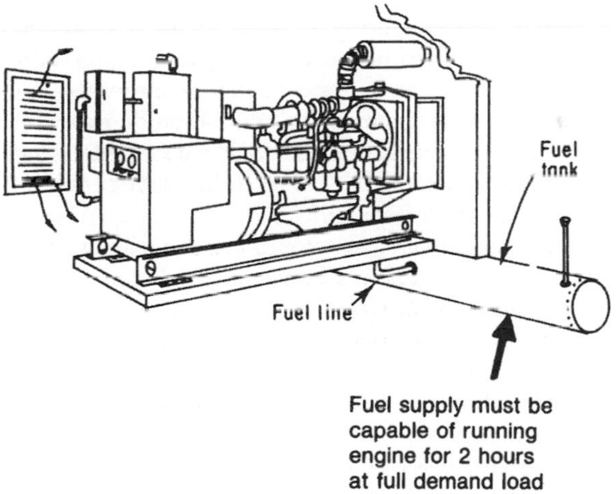
Fuel tank
Fuel line
Fuel supply must be capable of running engine for 2 hours at full demand load

Fig. 700-5. Generator must have automatic start and adequate fuel supply. (Sec. 700.12.)

Art. 701 is *legally* required, it must be provided with enough on-site fuel to power the full demand operation of the load for *not less than 2 h.* Note that 700.12(B)(2) specifically requires that fuel transfer pumps that run fuel to the generator day tank be connected to the emergency system for the obvious reason that if that tank runs out of fuel, the generator stops. This rule had been overlooked for legally required standby systems in 701.12(B)(2), but this was corrected in the 2011 NEC.

In part **(B)(3),** the engine driving an emergency generator must not be dependent on a public water supply for its cooling. That means a roof tank or other on-site water supply must be used and its pumps connected to the emergency source (Fig. 700-6).

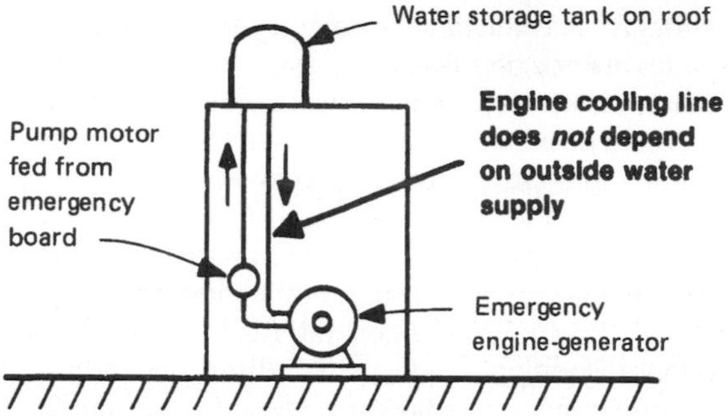
Water storage tank on roof
Engine cooling line does *not* depend on outside water supply
Pump motor fed from emergency board
Emergency engine-generator

Fig. 700-6. Engine cooling for an emergency generator set must be ensured for continuous operation of the generator. (Sec. 700.12.)

An exception in part **(B)(3)** permits use of a utility gas supply to the engine of an emergency generator—at the discretion of the local inspector—where simultaneous outage of both electric power and gas supply is highly unlikely.

As required by part **(B)(4),** a battery used with a generator set must have adequate capacity whether it cranks the generator for starting or is simply used for control and signal power for another means (such as compressed air) for starting the generator. A separate charging system must be provided, and this system, together with the circuit (if used) powering ventilation dampers for the generator set must be connected to the emergency system. In this way the generator will keep the circuits functioning that are essential to its continued operation.

Part **(B)(5)** of this section requires another power supply to pick up an emergency load in not over 10 s where the main generator cannot come up to power output in 10 s.

Part **(B)(6)** amends the rules in 225.32 that require a disconnecting means for conductors supplying a building or structure to be located at a readily accessible location nearest the point where the conductors enter the building. For emergency generators, the disconnect(s) on the generator, if suitable in accordance with 445.18, can serve as the disconnect at the building, provided the generator (and its disconnect) is within sight (and therefore not over 15 m [50 ft] distant). Remember, 445.18 now allows for a generator shutdown switch as a disconnecting method provided it renders the generator incapable of restarting and it can be locked in the OFF position. A new (2011) exception makes the existing 225.32 Exception No. 1 covering qualified supervision and safe switching procedures apply to emergency generator disconnects. This will provide some flexibility and relief, particularly in crowded real estate markets where the 50-ft limit is difficult. At the point in the building where the generator conductors eventually terminate, they must do so at a switch that is compliant with 225.36. Note that 225.36 has evolved and no longer requires an SUSE rating.

Part **(C)** allows an uninterruptible power supply to supply emergency power, provided battery capacity satisfies part **(A)**, and if a generator is also used—which is generally the case—it must satisfy part **(B)**.

As recognized by part **(D)**, two separate services brought to different locations in the building are always preferable, and these services should at least receive their supply from separate transformers where this is practicable. In some localities, municipal ordinances require either two services from independent sources of supply, auxiliary supply for emergency lighting from a storage battery, or a generator driven by a steam turbine, internal-combustion engine, or other prime mover. Figure 700-7 shows two different forms of the separate-service type of emergency

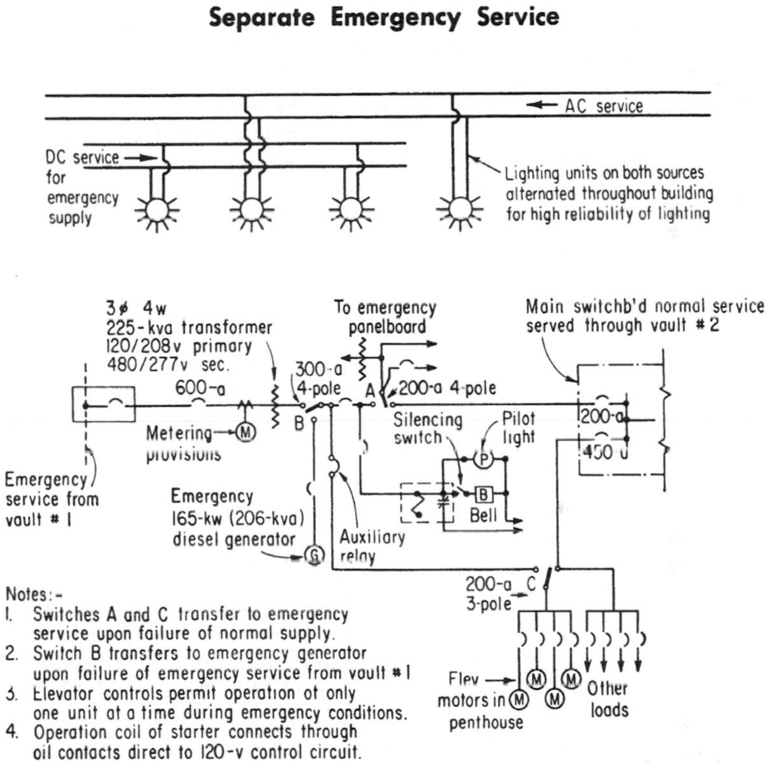
Separate Emergency Service
AC service
DC service for emergency supply
Lighting units on both sources alternated throughout building for high reliability of lighting
3ϕ 4w
225-kva transformer
120/208v primary
480/277v sec.
600-a
Metering
To emergency panelboard
300-a 4-pole
A
B
200-a 4-pole
Main switchb'd normal service served through vault # 2
200-a
450 a
Silencing switch
Pilot light
P
B
Bell
Emergency service from vault # 1
Emergency 165-kw (206-kva) diesel generator
G
Auxiliary relay
200-a C 3-pole
Elev motors in penthouse
M
Other loads
Notes:-
1. Switches A and C transfer to emergency service upon failure of normal supply.
2. Switch B transfers to emergency generator upon failure of emergency service from vault # 1
3. Elevator controls permit operation of only one unit at a time during emergency conditions.
4. Operation coil of starter connects through oil contacts direct to 120-v control circuit.

Fig. 700-7. Dual-service emergency provisions can take many different forms. (Sec. 700.12.)

supply. The method at bottom makes use of two sources of emergency input. This method requires the authorization of the AHJ. In many jurisdictions this approval is routinely denied based on common experience that utility outages usually involve significant areas, and even two services from different transformers would seldom make a difference.

The method shown in Fig. 700-8 was recognized in previous Codes as a means of providing emergency power. However, this permission in the NEC was at odds

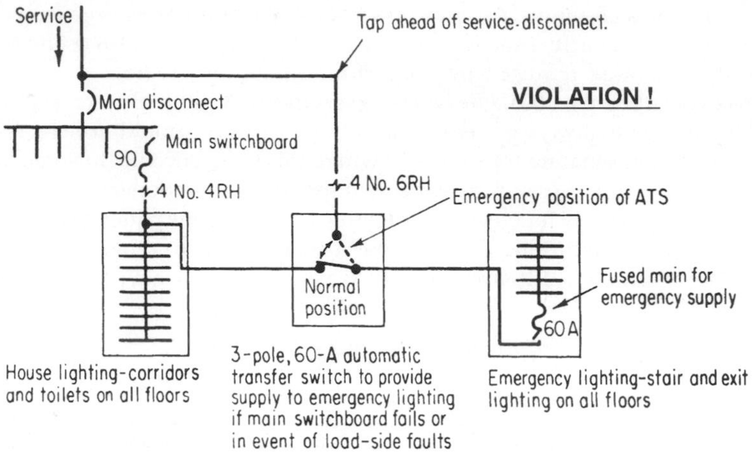
Service
Tap ahead of service-disconnect.
Main disconnect
VIOLATION !
Main switchboard
90
4 No. 4RH
4 No. 6RH
Emergency position of ATS
Normal position
Fused main for emergency supply
60 A
House lighting-corridors and toilets on all floors
3-pole, 60-A automatic transfer switch to provide supply to emergency lighting if main switchboard fails or in event of load-side faults
Emergency lighting-stair and exit lighting on all floors

Fig. 700-8. A tap ahead of the service main protects only against internal failures and is no longer recognized. (Sec. 700.12.)

with the NFPA's *Life Safety Code*. In an effort to harmonize the two standards, this permission was deleted in the 1996 NEC. Under that arrangement, in Fig. 700-8, the tap ahead of the main could have supplied the emergency panel directly without need for the transfer switch, but this is no longer permitted.

Note that fire pumps use connections ahead of the main routinely in instances where service conductors are accessible to the protected building. Authorities disagree as to whether a fire pump is an Art. 700 load. Fire pumps are generally classified as part of a system for property protection, allowing sprinkler systems to work properly in order to limit the degree of damage from a fire. Sprinklers do not help people find their way out of a burning building. On the other hand, statistics show that buildings protected with sprinklers have better loss experience in terms of fire related injuries and fatalities. Some local regulations do classify fire pump loads as emergency loads, which would require an on-site generator regardless of any finding as to reliability of the local service (review the discussion at 695.3 on this point). Review the local requirements in order to make this decision.

Part **(E)** recognizes the use of a fuel cell power system to supply the emergency loads provided it has adequate capacity to supply the entire load for not less than 2 h. The second paragraph reiterates the need for compliance with Art. 692, Parts II through IV, which covers the installation of fuel cell systems. The last paragraph of 700.12(E) puts forth a prohibition against using a single fuel cell system as both a normal and emergency source. In such a case, some other means of supply must be provided to power the emergency loads. The source may be another fuel cell system, a generator, a separate service, etc.

Part **(F)** covers typical wall-hanging battery-pack emergency lighting units, as shown in Fig. 700-9. The 1971 NEC accepted only connection of emergency light units by means of fixed wiring. Part **(F)** now recognizes permanent wiring connection *or* cord-and-plug connection to a receptacle.

Even though the unit equipment is allowed to be hooked up with flexible cord-and-plug connection, it is still necessary that the unit equipment be permanently fixed in place.

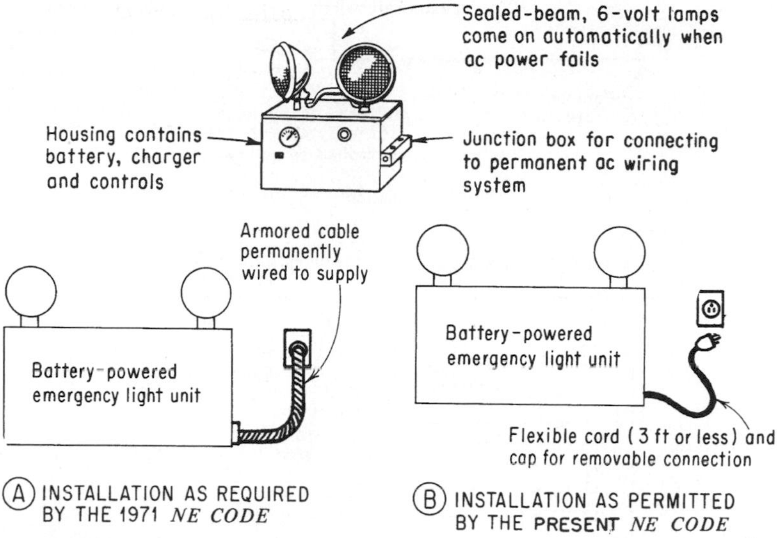

Sealed-beam, 6-volt lamps come on automatically when ac power fails
Housing contains battery, charger and controls
Junction box for connecting to permanent ac wiring system
Armored cable permanently wired to supply
Battery-powered emergency light unit
Battery-powered emergency light unit
Flexible cord (3 ft or less) and cap for removable connection
(A) INSTALLATION AS REQUIRED BY THE 1971 NE CODE
(B) INSTALLATION AS PERMITTED BY THE PRESENT NE CODE

Fig. 700-9. Unit emergency lights may serve as required source of emergency supply. (Sec. 700.12.)

Individual unit equipment provides emergency illumination only for the area in which it is installed; therefore, it is not necessary to carry a circuit back to the service equipment to feed the unit. This section clearly indicates that the branch circuit feeding the normal lighting in the area to be served is the same circuit that should supply the unit equipment.

In part **(F)** the intent of the 87½ percent value is to ensure proper *lighting output* from lamps supplied by unit equipment. It is generally considered acceptable to design equipment that will produce acceptable lighting levels for the required 1½ h, even though the 87½ percent rating of the battery would not be maintained during this period. The objective is adequate light output to permit safe egress from buildings in emergencies. The UL Guide Card information for this equipment puts it this way: "Emergency power equipment with batteries provides 90 minutes (more if so marked) of rated operating power for emergency lighting equipment (integral or remote) sufficient to meet the illuminance performance requirements of ANSI/NFPA 101 and the IBC, when installed as part of a facility's emergency lighting system."

As shown at the top of Fig. 700-10, a battery-pack emergency unit must not be connected on the load side of a local wall switch that controls the supply to the unit or the receptacle into which the emergency unit is plugged. Such an arrangement exposes the emergency unit to accidental energization of the lamps and draining of the battery supply. But, as permitted by the exception, if a panelboard supplies at least three normal lighting circuits for a given area, and they are not part of a multiwire branch circuit, emergency battery-pack lighting units for that area may be connected to a separate branch circuit from the panel, with lock-on provision for that circuit.

The multiwire prohibition originated in the 2014 NEC and responds to current requirements for multiwire circuits to have a common disconnecting means.

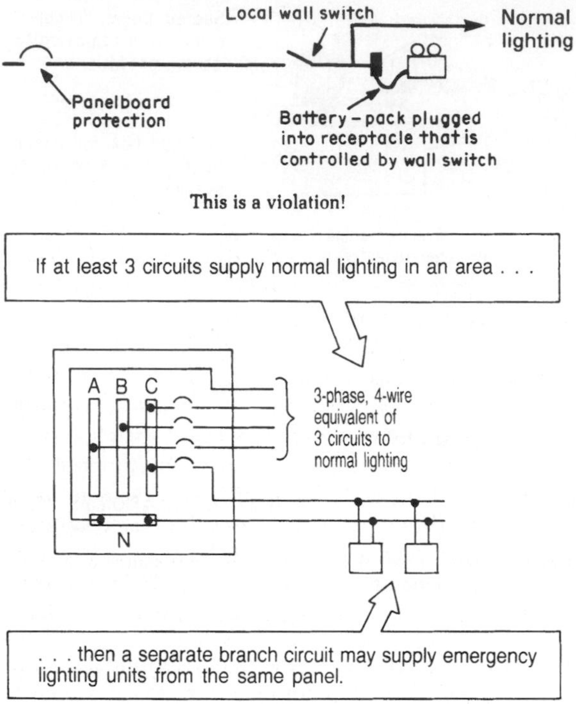
Local wall switch
Normal lighting
Panelboard protection
Battery – pack plugged into receptacle that is controlled by wall switch
This is a violation!
If at least 3 circuits supply normal lighting in an area . . .
A B C
N
3-phase, 4-wire equivalent of 3 circuits to normal lighting
. . . then a separate branch circuit may supply emergency lighting units from the same panel.

Fig. 700-10. Circuiting of battery-pack emergency lighting may be on normal lighting circuit or separate circuit. The minimum of three circuits must not be multiwired. (Sec. 700.12.)

When three circuits require three different faults to put an area into darkness, the odds are pretty good that the only time the area is dark is when the panel itself has lost power and the emergency lighting has activated. A common disconnect means that only a single fault would trip the same three circuits, thereby defeating the objective of the requirement.

Note that whatever source is used to supply this equipment, the circuit directory covered in 408.4 must clearly identify the source circuit. The three-circuit allowance is an exception to the rule of the previous paragraph, which says that unit equipment must be connected on a branch circuit supplying normal lighting in the area. By allowing unit equipment on a separate branch circuit from the same panel, the unit equipment will sense loss of power to the panel and activate. The advantage of such application from a design standpoint is that there is no need to observe the rule that the unit equipment must be "connected ahead of any local switches," which applies when unit equipment is connected on normal lighting circuits. The final paragraph recognizes that a remote head fed from unit equipment is permitted to light the other side (and therefore the egress pathway) of an exit door from a protected space.

700.15. Loads on Emergency Branch Circuits. Figure 700-11 shows a clear violation of this rule, because appliances are excluded from emergency lighting circuits.

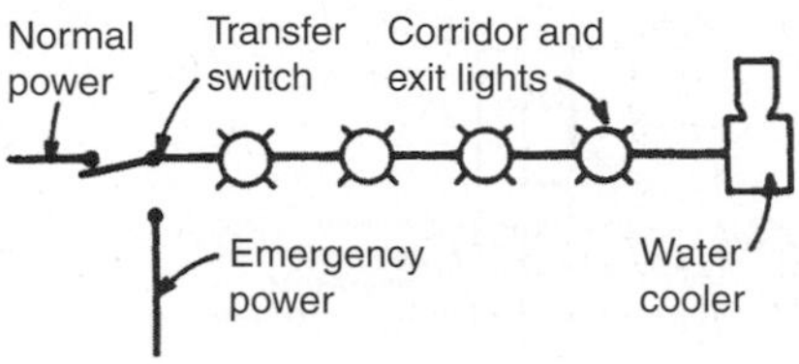
Normal
power
Transfer
switch
Corridor and
exit lights
Emergency
power
Water
cooler

Fig. 700-11. The water cooler may not be on an emergency circuit. (Sec. 700.15.)

700.16. Emergency Illumination. Note that all exit lights are designated as part of the emergency lighting, and, as such, their circuiting must conform to 700.17. As of the 2014 NEC, emergency illumination at indoor service and building disconnecting means locations is required if emergency illumination is provided for the interior of the building or structure. Referring to the definition in 700.2, the literal text of this requirement applies even to a building whose sole emergency lighting system is entirely comprised of unit equipment, even a single instance.

Where HID lighting is the *sole* source of *normal* illumination, the emergency light system must continue to operate for a sufficient time after return of normal power to enable the HID lighting to come up to brightness. This rule in the third paragraph is intended to prevent the condition that return of normal power and the disconnect of the emergency lighting leave the building in darkness because of the inherent, normal time delay in the light output upon energizing HID lamps. The exception after that rule permits "alternative means" to keep emergency lighting on. This often involves special HID luminaires that incorporate auxiliary tungsten-halogen incandescent lamps within them that come on immediately and shut off when the HID lamp comes up to brightness.

700.17. Branch Circuits for Emergency Lighting. Figure 700-12 shows the basic rule on transfer of emergency lighting from the normal source to the emergency source. If a single emergency system is installed, a transfer switch should be provided, which, in case of failure of the source of supply on which the system is operating, will automatically transfer the emergency system to the other source. Where the two sources of supply are two services, the single

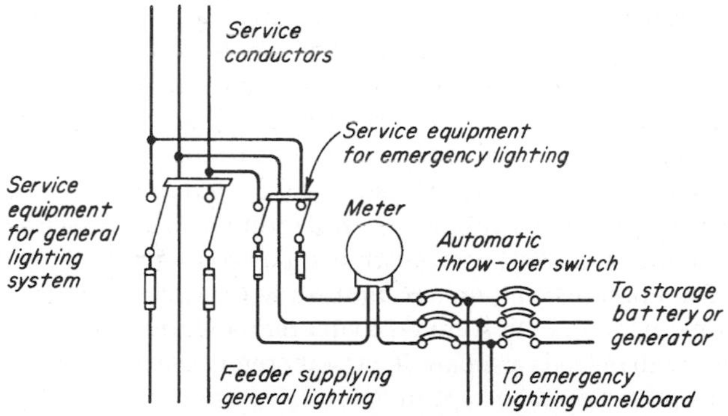
Service
conductors
Service equipment
for emergency lighting
Service
equipment
for general
lighting
system
Meter
Automatic
throw-over switch
To storage
battery or
generator
Feeder supplying
general lighting
To emergency
lighting panelboard

Fig. 700-12. Emergency lighting is automatically switched from normal service to the battery or generator. (Sec. 700.17.)

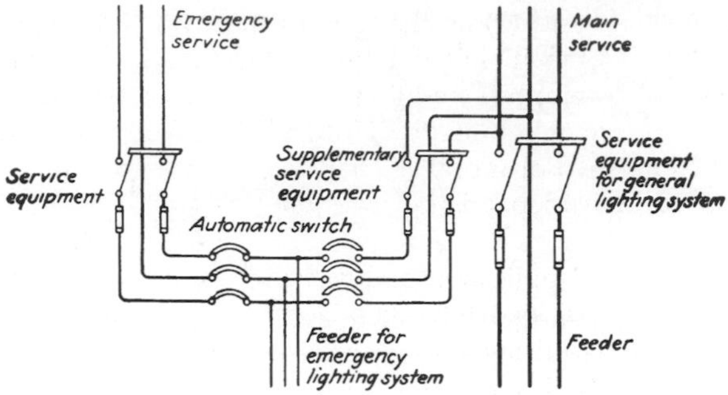
Emergency service
Main service
Service equipment
Supplementary service equipment
Service equipment for general lighting system
Automatic switch
Feeder for emergency lighting system
Feeder

Fig. 700-13. Emergency lighting may be supplied from an emergency service. (Sec. 700.17.)

emergency system may normally operate on either source, as in Fig. 700-13. Where the two sources of supply are one service and a storage battery, or one service and a generator set, as in Fig. 700-12, the single emergency system would, as a general rule, be operated normally on the service, using the battery or generator only as a reserve in case of failure of the service. Figure 700-14 shows an emergency hookup that has two separate supplies tied into the emergency lighting system.

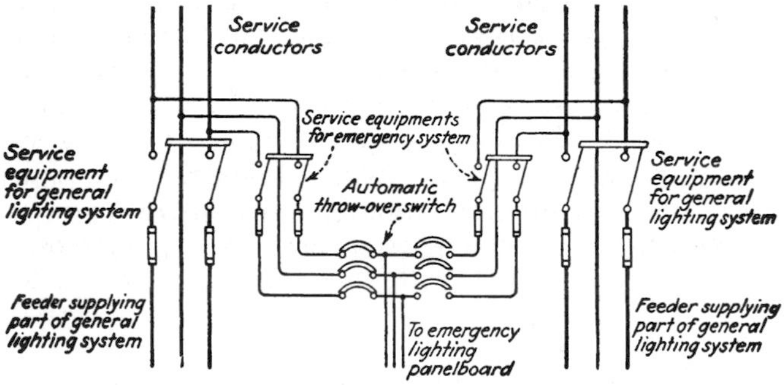
Service conductors
Service conductors
Service equipments for emergency system
Service equipment for general lighting system
Automatic throw-over switch
Service equipment for general lighting system
Feeder supplying part of general lighting system
To emergency lighting panelboard
Feeder supplying part of general lighting system

Fig. 700-14. A single emergency lighting system may be fed from two services. (Sec. 700.17.)

Part **(2)** of this section provides rules for use of two or more "branch circuits supplied from separate and complete" systems with independent power sources, of which one is from the emergency system, and one is permitted to be part of the normal distribution. If two emergency lighting systems are installed, each system should operate on a separate source of supply, as where each emergency disconnect in Fig. 700-13 feeds a separate, independent emergency lighting system. Either both systems should be kept in operation or switches should be provided that would automatically place either system in operation upon failure of the other system.

Watch Out! Both (1) and (2) as changed in the 2011 NEC, together with the addition of the word "Branch" in the section title, make it clear that this rule is a branch-circuit rule. Paragraph (1) requires a transfer and alternate lighting source pickup whenever a normal lighting *branch circuit* fails. More extensive changes in Paragraph (2) require two or more "*branch circuits* supplied from" separate and complete systems with independent power, at least one of which is part of the emergency system.

Consider the common application of this section to lighting outlets in building stairwells, a typical Art. 700 load. It has been commonly believed that Paragraph (1), with its prior wording requiring a transfer in the event of a failure in the "general lighting system supply," only required a transfer of the feeder to the emergency panel, thereby transferring the "system supply." Frequently local lighting panels were wired with phase-loss relays that activated the transfer switch upon sensing a power failure in the lighting panel. Now that transfer must operate with a single branch-circuit failure.

The installation of unit equipment wired to the local branch circuit is one way to address these changes; the familiar emergency backup ballasts in fluorescent luminaires in the stairwell are another.

The other option would be to use Paragraph (2) with two supplies to each stairwell luminaire outlet, the circuits being fed from a panel on the normal distribution and a panel on the emergency distribution. This arrangement would clearly meet the "separate and complete systems" rule. Automatic load control relays (see discussion at 700.26) could be installed on each stairwell emergency lighting branch circuit, arranged to sense a failure in the normal branch circuit and energize the luminaire from the emergency panel. If the failure resulted from a building outage, then the transfer switch would have operated and the luminaire would be powered by the generator. If the failure was idiosyncratic to a particular branch circuit, such as a breaker being turned off, the light would still be returned to service through the transfer to the other panel, which would be on normal power.

If the branch-circuit breaker tripped during a wiring fault that was uncleared before the breaker tripped, then the corresponding breaker in the other panel may trip as well because it might close into the same fault. But this does offer a level of redundancy that is a distinct improvement over the way these systems have often been wired previously.

Another possibility, if the remaining illumination meets the minimum brightness required by the applicable building code, would be to wire double-ballasted luminaires to opposite branch circuits, or to alternate branch-circuit coverage by floor level. In any event, it is now clear that emergency branch-circuit lighting coverage now requires fully redundant branch-circuit wiring. Note also that the second paragraph of (2) regarding system transfers, is correctly formatted as of the 2011 NEC printings as a second paragraph of (2) and not, as occurred in prior editions, as a second paragraph to the entire section.

700.19. Multiwire Branch Circuits. Again, because of the problems arising from the common disconnect rule for multiwire circuits, they cannot be used to supply emergency circuits.

700.20. Switch Requirements. Figure 700-15 shows a violation of the last sentence of this rule, which follows the exceptions, where use of three- and four-way

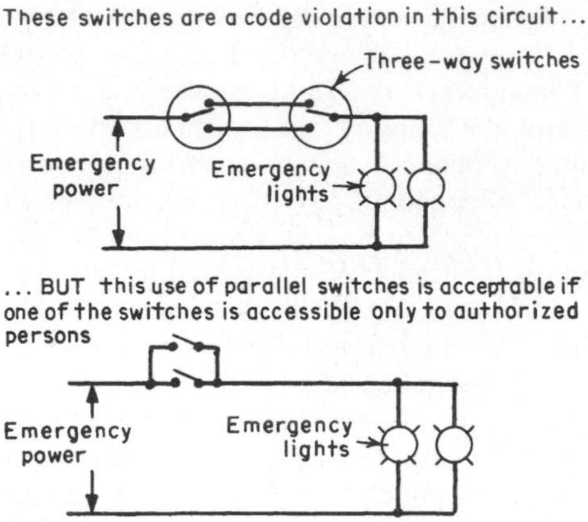
These switches are a code violation in this circuit...
Three-way switches
Emergency power
Emergency lights
... BUT this use of parallel switches is acceptable if one of the switches is accessible only to authorized persons
Emergency power
Emergency lights

Fig. 700-15. Watch out for switches in emergency lighting circuits. (Sec. 700.20.)

switches is prohibited. One good example of Exception No. 2 involves a single-pole service switch in parallel with a photocell that turns on luminaires that light secondary egress pathways from multifamily housing. Only the rising sun can turn the lights off after darkness, but the janitor can turn them on during the day to see which lamps need replacement.

700.21. Switch Location. The sole switch for emergency lighting control in a theater may not be placed in a projection booth. Figure 700-16 shows the exception, where an emergency lighting switch may be used in a projection booth or on a stage if it can energize such lighting but cannot deenergize the lighting if another switch located elsewhere is in the closed (ON) position.

700.23. Dimmer and Relay Systems. This 2008 addition to the NEC recognizes dimmer systems comprised of multiple dimmers and listed as an emergency lighting control mechanism. On failure of the normal power, the system energizes selected lighting outlets such that the required degree of illumination is provided. The wiring to the designated outlets must comply with other rules in Art. 700. The 2014 NEC has added relays to this concept, because they are commonly employed and provide the same functionality.

700.24. Directly Controlled Luminaires. This new (2014 NEC) section is intended to recognize LED luminaires with on-board dimming capabilities, and that can also respond to a change of state in an emergency system transfer switch and turn on to full illumination during the emergency event. The electronics to make this possible are complicated, and for this reason such luminaires together with their external controls must be listed for use in emergency systems.

700.25. Branch Circuit Emergency Lighting Transfer Switch. This is a new term, first defined in the 2017 NEC. These devices take automatic load control relays (LCRs) (700.26) to the next level. LCRs are not transfer equipment and for that reason they are not being evaluated in terms of capability for transferring between asynchronous power sources. A special task group was assembled to develop a testing protocol that would allow similar equipment to be evaluated as full-fledged

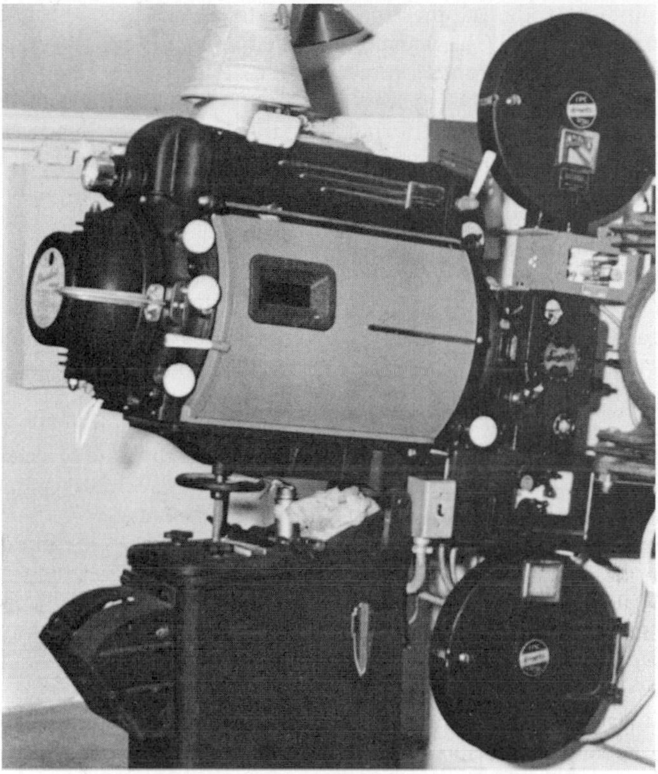

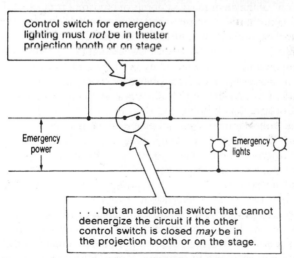

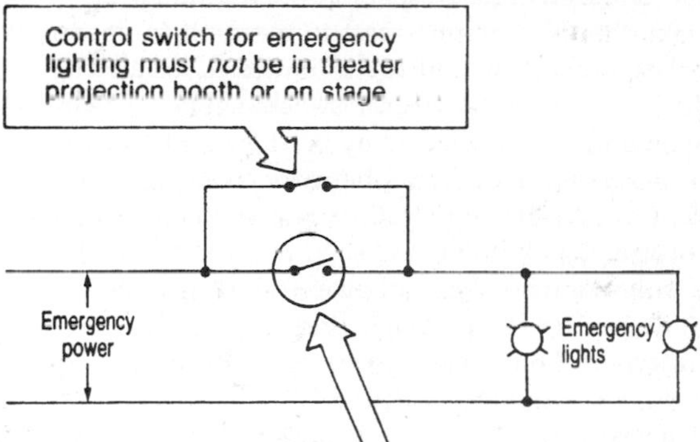
Control switch for emergency lighting must *not* be in theater projection booth or on stage . . .
Emergency power
Emergency lights
. . . but an additional switch that cannot deenergize the circuit if the other control switch is closed *may* be in the projection booth or on the stage.

Fig. 700-16. The sole switch that controls emergency lighting in a theater may *not* be installed in a projection booth or on the stage, with the exception shown. (Sec. 700.21.)

transfer equipment, and this is the result, limited to 20-A circuits. The mechanically held requirement in 700.5(C) is waived for this equipment.

700.26. Automatic Load Control Relay. These devices are defined in 700.2 and provide a relatively inexpensive way to bypass a local switch component and place full voltage on a luminaire during a condition when normal power has been lost. They have a lead or contact point that is connected to the normal source at the relevant upstream transfer switch, and if power is lost, they change state and close the ungrounded conductor across the line. When normal power is restored, they change state again, opening the bypassing ungrounded conductor. As a result, the only power to the luminaire at that point, however, was being controlled before the outage. For example, if the luminaire was on a dimmer and set to 50 percent brightness, a loss of power would force the luminaire to full brightness until normal power was restored.

These devices do not switch the grounded conductor; they do not qualify as and cannot be used for transfer equipment functions. The definition in 700.2 has been revised to emphasize this point. They are used to change the state of a lighting load after a transfer has occurred, including energizing loads within their ratings that had been shut off and needed to be turned on during the outage.

700.27. Ground-Fault Protection of Equipment. Emergency generator disconnect does not require ground-fault protection. This rule clarifies the relationship between 230.95, calling for GFP on service disconnects, and the disconnect means for emergency generators. To afford the highest reliability and continuity for an emergency power supply, GFP is not *required* on any 1000-A or larger disconnect for an emergency generator, although it is permissible to use it if desired (Fig. 700-17). However, a ground-fault condition must be annunciated as required in 700.6(D) in the event automatic GFPE is not provided.

700.32. Selective Coordination. The rule put forth in this section mandates "selective coordination" of the emergency system's overcurrent protective devices, which essentially requires the OC device closest to the fault to operate before line-side devices. This will ensure that any outage is limited to the affected circuit only, and thereby minimize the effect on the overall system supply. The exception addresses circumstances where the loads affected by an upstream outage are identical to an outage from a downstream device, and therefore coordination creates no benefit.

This principle, prior to the 2005 **NEC**, was presented as a desirable condition in a fine print note at this location. As such it could not be literally enforced in all installations. Following a successful proposal from a fuse company employee, however, it is now mandatory and the shortcomings are becoming increasingly apparent, particularly if the choice of overcurrent protective devices on a system operating where exposed to high available fault conditions, was, at least initially, to use circuit breakers.

Circuit breakers of the usual thermal-magnetic type are inverse-time devices until approximately 10 times their nominal ratings. Current somewhat over their nominal ratings takes a long time to clear; and much larger current clears very quickly. Current at or above the 10-times parameter unlatches the breaker in a fraction of a cycle in response to magnetic forces. Further, the unlatch sequences use the same amount of time for all values of fault current within their magnetic trip ranges. This means that a branch circuit experiencing such a fault may trip not

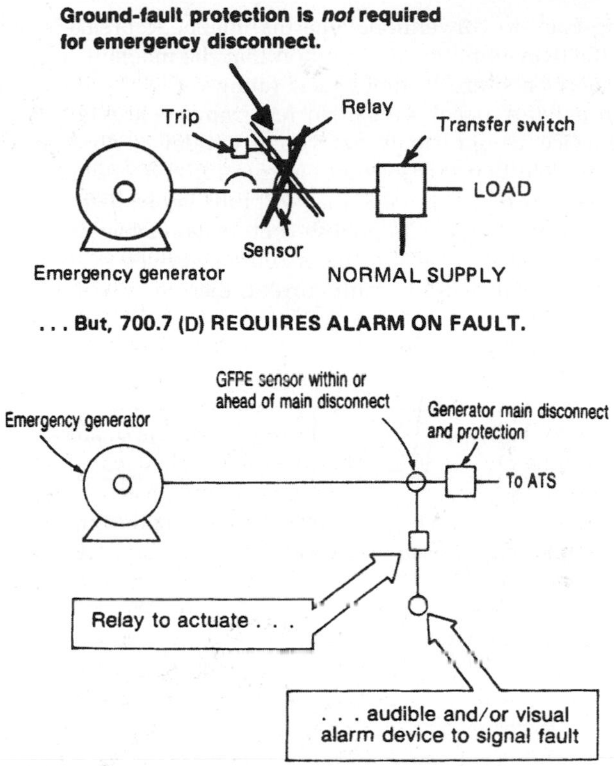
Ground-fault protection is *not* required
for emergency disconnect.
Trip
Relay
Transfer switch
LOAD
Sensor
Emergency generator
NORMAL SUPPLY
. . . But, 700.7 (D) REQUIRES ALARM ON FAULT.
GFPE sensor within or
ahead of main disconnect
Emergency generator
Generator main disconnect
and protection
To ATS
Relay to actuate . . .
. . . audible and/or visual
alarm device to signal fault

Fig. 700-17. Ground-fault protection is optional for any 480/277V grounded-wye generator disconnect rated 1000 A or more, but ground-fault "indication" is required. (Sec. 700.26.)

just the branch-circuit protection, but upstream feeder protection as well as far up into the system as necessary until and unless the current falls into the range of the inverse time tripping range of a much larger breaker. How far upstream is very difficult to predict, however, because the smaller breaker parts have less inertia and their contacts will begin to open first, introducing the dynamic impedance caused by the internal arcing. This is another wrinkle on the dynamic impedance problem discussed at length in this book at 240.86(B). Note that the 10-times parameter is an oversimplification and actual values will differ between makes and sizes, but the principle holds.

One of the reasons that this is an oversimplification is that it only considers thermal-magnetic breakers. Some electronic-trip breakers include a second instantaneous trip function, in some cases referred to as an instantaneous override, that operates more quickly than the familiar instantaneous settings that are field adjustable. In addition, very high levels of fault current will begin to open breaker contacts even before they actually unlatch. In fact, some breakers rely on this feature and incorporate designs that keep the contacts separated until the full unlatching sequence is completed.

Getting back to conventional thermal-magnetic breakers, and for example, making the oversimplified assumption that the magnetic pick-up points on the devices mentioned are 10 times their ratings, a 5000-A fault on a 20-A branch circuit in a 100-A panel with main fed from a 400-A feeder supplied from a 2500-A service, would trip every overcurrent device downstream of the service protection. Note that this example is true for conventional breaker selections, but there are breakers available that will fully coordinate in this particular fault range. Breaker manufacturers publish coordination tables to assist in this analysis. Fuses, on the other hand, are inverse-time devices over their entire clearing range. For every increase in fault current, there will be some shortening to the

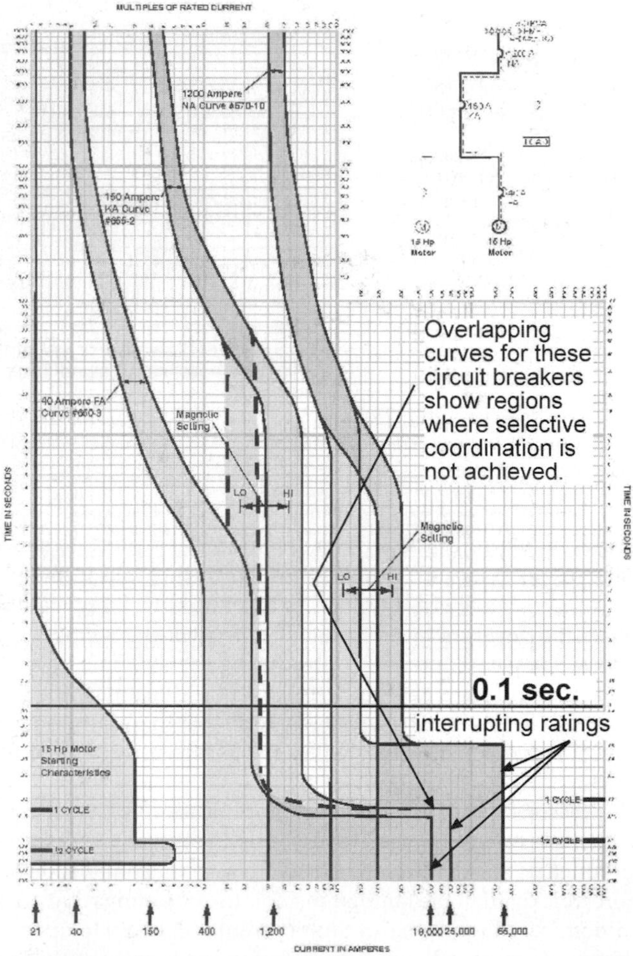
MULTIPLES OF RATED CURRENT
1200 Ampere
150 Ampere
40 Ampere
Overlapping curves for these circuit breakers show regions where selective coordination is not achieved.
Magnetic Setting
LO
HI
Magnetic Setting
LO
HI
0.1 sec.
interrupting ratings
15 Hp Motor Starting Characteristics
1 CYCLE
½ CYCLE
1 CYCLE
½ CYCLE
15 Hp Motor
15 Hp Motor
TIME IN SECONDS
TIME IN SECONDS
21
40
150
400
1,200
10,000
25,000
65,000
CURRENT IN AMPERES

Fig. 700-18A. Overlapping trip curves showing three levels of a distribution protected with thermal-magnetic circuit breakers. The right-hand vertical boundaries of the "feet" are the interrupting ratings.

clearing time. This is a simple matter of physics; fuses need to see an amount of energy equal to their latent heat of fusion in order to clear. The available energy is a function of the of current squared multiplied by time (I^2t). An increase in current exponentially decreases the amount of time to release the required energy into the link. And, when such functions are plotted logarithmically, the result is a sloping straight line. Further, if the amount of comparable eutectic alloy changes, the sloping line will move to the side, never intersecting one for a different size fuse. Refer to Fig. 700-18(B) for an example. This is what makes the selective coordination of fuses comparatively simple. Fuse manufacturers publish ratio tables to assist in establishing permitted fuse combinations for this purpose and because

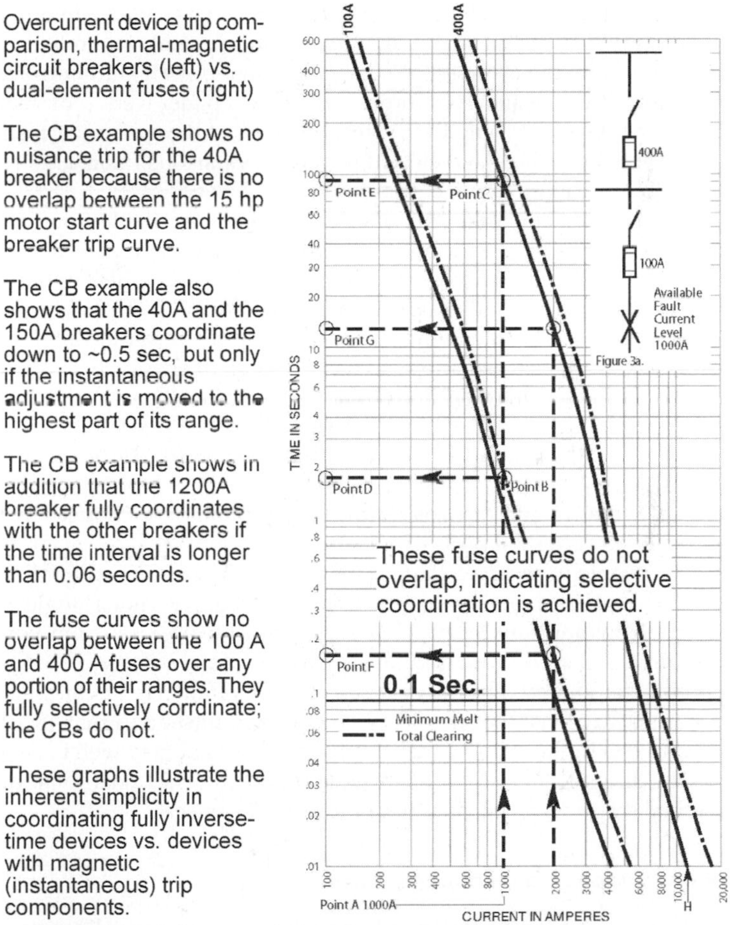
Overcurrent device trip comparison, thermal-magnetic circuit breakers (left) vs. dual-element fuses (right)
The CB example shows no nuisance trip for the 40A breaker because there is no overlap between the 15 hp motor start curve and the breaker trip curve.
The CB example also shows that the 40A and the 150A breakers coordinate down to ~0.5 sec, but only if the instantaneous adjustment is moved to the highest part of its range.
The CB example shows in addition that the 1200A breaker fully coordinates with the other breakers if the time interval is longer than 0.06 seconds.
The fuse curves show no overlap between the 100 A and 400 A fuses over any portion of their ranges. They fully selectively corrdinate; the CBs do not.
These graphs illustrate the inherent simplicity in coordinating fully inverse-time devices vs. devices with magnetic (instantaneous) trip components.
100A
400A
Point E
Point C
Point G
Point D
Point B
Point F
These fuse curves do not overlap, indicating selective coordination is achieved.
0.1 Sec.
Minimum Melt
Total Clearing
400A
100A
Available Fault Current Level 1000A
Figure 3a.
TIME IN SECONDS
CURRENT IN AMPERES
Point A 1000A
H

Fig. 700-18B. Trip curves for two dual-element fuses. As inverse-time devices, there is little chance of overlap. For example, a 2000-A fault clears the 100-A fuse in ~0.17 sec. and the 400-A fuse in ~13 sec.

conventional trip curve data is not published below 0.01 s. Figures 700-18A and 18B show examples of overcurrent device trip curves in terms of coordination.

The only way to definitively coordinate the circuit breaker sequences mentioned is through the use of circuit breakers equipped with electronic controls that allow for zone selective interlocking, allowing a lower level in the protection hierarchy to restrain the tripping of a higher level until the lower level clears. This is incredibly expensive and not even available in all frame sizes, and it is not practicable under all instantaneous tripping scenarios. It must be noted here that even zone selectivity does not always guarantee coordination without careful attention being paid to the settings on every protective element in the system. It does, however, offer an important reduction in the amount of energy that can enter the electrical system at feeder levels, because a fault at that level can be cleared quickly since no downstream device will restrain the feeder protection in such cases. This is a very important issue when considering the amount of energy available in an arc flash event, for example. The other way, also not always practical, is to make sure that the next level up is in the inverse-time part of its curve for all fault-currents available. For example, in the above example if the 100-A panel were replaced with a 400-A panel, thereby decreasing the protection levels to three (service, feeder, and branch), and the 400-A breaker selected so its instantaneous (magnetic) response were at 15 times its rating, coordination could be achieved.

This does, however, increase the amount of energy available in an arc blast, so this is not a purely positive design change. In fact the number of distributions with intermediate protective levels being designed upward to deal with phantom coordination issues is becoming so common that we now are having to write special rules regarding breaker controls that turn on the instantaneous trip function during servicing operations so as to reduce incident energy levels, levels that otherwise would be far more difficult because the instantaneous trips were shut off to allow for coordination. See the discussion at 240.87 for more information on this point. If you think this is beginning to look like Rube Goldberg, this author would not disagree.

Economics is not, in and of itself, sufficient motivation to change the NEC. If there were numerous field reports of coordination failures that adversely affected the ability of people to escape a dangerous condition, then the adverse affect on circuit breaker market share would be warranted. However, not a single example of such a coordination failure exists in the record of proposals and comments submitted on this change. What is known, however, is the adverse effects of introducing additional energy into the electrical systems under fault conditions. The difference in the arc blast that would be caused by a mishap in that 400-A panel as opposed to one in the 100-A panel originally proposed may well be, and often has been, the difference between life and death for the electrician who dropped the proverbial screwdriver across the proverbial line buses.

Why it can't happen in an occupied building?

There is a good reason for the lack of loss experience around coordination failures. Coordination studies necessarily involve the worst-case electrical fault condition, which is a bolted short circuit, such as can exist on a cross-wired parallel installation.

These failures are the sort that are rare and that would be expected to be confined to system startups prior to building occupancy. Buildings do not gradually descend into bolted short-circuit conditions over the years and as a consequence of lack of maintenance. However, ground faults do occur under these conditions, and not infrequently. For this reason, most electrical failures do not involve short-circuits at all, but arcing ground faults instead.

A ground fault frequently involves just enough impedance at the point of arcing contact that the overcurrent device next upstream from the branch-circuit protective device will still be acting within its inverse-time characteristics. Because this is the predominant failure mode on electrical systems, coordination is maintained. Further, it can be shown that even systems perfectly coordinated in terms of short-circuits will fail to coordinate on some ground-fault conditions, even with fuses. For example take a large, heavily loaded feeder, say, 1200 A loaded to 1000 A, with four 400-A subfeeders and a 600-A subfeeder originating at a distribution panel. If the 600-A subfeeder with all loads off sees a 800-A ground fault, it will likely take out the 1200-A main because the other loads that are making it heavily loaded will continue, and the ground fault will push it over the edge. All things being equal, a 600-A fuse loaded to 800 A (+133 percent) during the ground fault event will take much longer to clear than a 1200-A fuse loaded to 1800 A (+150 percent).

As written, this rule conflicts with NFPA 110, *Standard for Emergency and Standby Power Systems*, which only requires that coordination be "optimized." Many jurisdictions are limiting the reach of this rule in order to achieve practical results. It bears repeating that not a single instance of loss experience supported its inclusion in the NEC. You will need to sort this out at the local level with inspectional authorities to see what will be permitted on particular jobs. There is no problem with keeping the selective coordination concept in mind and using it as one consideration in an overall system design, but mandatory application at every level in every distribution is extremely problematic.

The NEC promises an electrical system that is "essentially free from hazard" and not one that is absolutely free from hazard. Enforcing absolute selectivity is comparable to relentless enforcement of a 20-mph speed limit on an interstate highway. The costs to the industry do not remotely justify this "benefit," particularly because it addresses a nonexistent loss experience history.

Were other factors involved in the selective coordinations decision?

The proposal was submitted in the later half of 2002 by a fuse company employee. Prior to that time, fusible panelboards were unusual in occurrence in comparison with circuit breaker panelboards. They were widely used though the middle third of the last century in residential applications, equipped with plug fuses for most of the circuits, and cartridge fuses for the service protection and larger circuits. They were gradually eclipsed by panelboards with circuit breakers. It takes a number of years to develop, test, submit to testing laboratories, and come to market with a new product line. However, a full product line came to market just as the 2005 NEC was coming into full effect. The new product line was the subject of a full-color article in *IAEI News*, the March-April issue of 2005. The article featured

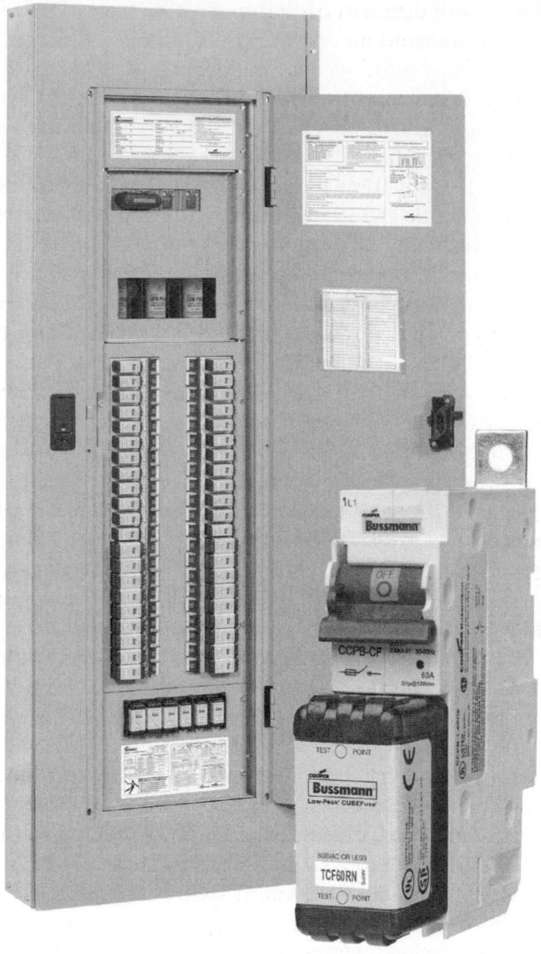
1L1
Bussmann
OFF
CCPB-CF
60A
TEST POINT
Bussmann
TCF60RN
TEST POINT

Fig. 700-19 A modern fusible panelboard, suitable for a full range of occupancies and applications (700.32). (Adapted from Practical Electrical Wiring, 22nd ed., © Park Publishing, 2014, all rights reserved.)

fusible panelboards that were anything but residential in design (See Fig. 700-19) and very specifically presented as a solution to selective coordination issues occasioned by the NEC change.

If there were any doubt as to the motivation and product timing in relationship to the NEC, this article, focused on selective coordination, largely removed it. After presenting extensive instruction as to how to selectively coordinate fuses, the article then commenced a discussion of circuit breakers. In the end, the article opined "It is generally difficult to achieve selective coordination with circuit breakers that incorporate instantaneous trip settings." There is a term in formal economics for this sort of action, called "rent seeking." This is a process whereby one generally avails themselves of government action to obtain a greater rent (i.e., return) than would have been possible on the free market.

Transcendent code issues

As of the 2014 NEC, the engineering costs involved in providing these systems received an upward click of the ratchet because now a licensed PE or other similarly qualified person must design the system. In fairness, these systems are so complicated that qualified engineering support was almost inevitable.

Certain Code controversies over time become what the author refers to as transcendent issues. These are issues that take on a public policy aspect that attracts the attention of those outside the electrical industry. A classic example was the three-story limit on NM cable, which was in the NEC from the 1975 NEC edition until it was removed in the 2002 edition. The author recalls being contacted by a representative of the Roman Catholic Archdiocese of Boston regarding eliminating this provision in the Massachusetts Electrical Code because of its adverse effects on the costs of providing affordable housing. The specific issue was moot because Massachusetts never did accept the NEC prohibition, but the larger issue was that the Archdiocese of Boston was intently interested in a technical requirement of the electrical code.

The adoption of an electrical code is the exercise of a police power implicitly reserved to the states under the 10th amendment. At any time a legislature can exercise that authority and enact a law that changes the electrical code that is in effect. The author has no problem with a competent group of electrical professionals in a state reaching a different conclusion about what a code provision should say and enacting the applicable code in a transparent manner that supports those conclusions. That is the sort of federalism enshrined in our frame of government. However, once a state legislature gets involved on a purely political level, it is another story. And after the first time such an amendment passes, it is only easier for it to happen a second time. This makes it critical to avoid the first time, and that means watching transcendent issues very carefully. The NM cable limitation provides historical precedent for these concerns.

At the end of the twentieth century, the National Multi-Housing Council realized what the three-story NM cable restriction was costing them in new construction electrical expense and challenged the provision. The electrical industry adamantly refused to support the change, and it failed in the Code-making panel and also at the NFPA Annual Meeting. The NFPA Standards Council, after looking at the entirety of the technical record, agreed that there was no credible technical substantiation to support the prohibition. The opposition of the electrical industry was motivated by a mixture of romance and self-interest; metallic wiring methods required greater expertise to install and brought the manufacturers more profit. In the author's opinion, the Standards Council action on the 2002 NEC, where the entire prohibition was thrown out and the section rewritten at the Council level, was an act of great courage and integrity, and it saved our industry from itself. No longer would that provision attract the attention of the lay community.

Selective coordination is well on its way to becoming the next transcendent issue. The engineering and design costs to protect against a loss that has never occurred and will never occur during the occupancy of a building using electrical power are in the uncounted millions of dollars and rising. The health care industry was one area particularly hard hit, and it reacted. It has its own standard, NFPA 99, and the industry ratcheted down the coordination to events longer than 0.1 sec. That time period is a full six cycles, and coordination is simple across almost all ranges of overcurrent devices with that parameter in place (see Fig. 700-18A.)

The history of the NFPA 99 intervention is important. The final vote on the proposed action in the NFPA 99 Electrical Systems Committee passed 18-1, with 3 installer/maintainers, 5 special experts, 4 users, 3 manufacturers, 1 consumer, 1 research/testing, and 1 enforcer voting yes and only 1 negative vote (labor). The vote on the action included the following substantiation:

> Selective coordination is only one of several competing factors that must be considered in the selection of appropriate overcurrent protective devices (OCPDs) in health care facilities. Other factors that must be considered in the selection of overcurrent protective devices include: arc flash risk hazard, equipment damage, and reduced risk of extended outages; all of which have direct effects on both staff and patient safety. Mandating selective coordination below 0.1 s as the sole determining factor in OCPD selection will result in diminished reliability of the essential electrical system. The method of application of selective coordination directly affects the performance of the essential electrical system in a health care facility. Establishment and management of this type of performance criterion traditionally belongs under the purview of this committee.

The battle went on, however, on the floor of the NFPA Annual Meeting. On a vote of 87 yes and 119 no, the attempt, by (who else) the National Electric Fuse Association to overturn the committee action failed. The same association then pursued an appeal to the NFPA Standards Council, the highest authority in the NFPA Standards process, and lost with all electrical members having recused themselves.

The hospital industry has escaped the selective coordination noose, but once again, only over the entrenched opposition of the electrical industry. Fusible branch-circuit panelboards are once again readily available, and the circuit breaker industry has responded with complicated correction schemes of its own with enormous value added. The vote on the rule in NFPA 99 did not, however, necessarily bind the NEC to follow suit. Now there was a conflict between NFPA 99 and NFPA 70, the NEC. In this case the health care industry went to the Standards Council with the argument that coordination was a performance issue belonging to NFPA 99. At the bitter end, the fuse industry was in front of the Council with a final appeal against the NFPA 99 action. And once again, the Standards Council saved the electrical industry from itself. It supported NFPA 99 and effectively forced what is now 517.31(G) into the 2014 NEC. However, the general issue is still very present, still transcendent, and still very dangerous to the autonomy of the Code-making process.

How does this fit with the organizing principles of the NEC?

This entire issue recalls first principles of Code creation and application. The first NEC edition to have a written introduction (now Art. 90) was the 1937 major rewrite, which included the following opening sentence of an unnumbered second paragraph: "The requirements of this Code constitute a minimum standard. Compliance therewith and proper maintenance will result in an installation *reasonably free from hazard* but not necessarily efficient or convenient." [emphasis supplied]

This concept has survived essentially intact of the intervening 80 years; today that provision reads as follows:

(B) Adequacy. This *Code* contains provisions that are considered necessary for safety. Compliance therewith and proper maintenance result in an installation that is essentially free from hazard but not necessarily efficient, convenient, or adequate for good service or future expansion of electrical use.

The NEC does not now and never has promised an electrical installation that is without any risk, only one essentially free from hazard. Mandatory selective coordination is an attempt at removing all hazard, and with rent seeking as one of the motivations. The hospital industry has sidestepped the issue, but other users do not have another NFPA standard to shield them, and the battle will likely continue above the level of the NEC Committee, as this terminology slowly metastasizes across various Code provisions. This entire issue is a wonderful validation of the wisdom of the old French philosopher Voltaire, who famously observed in 1768 that "the perfect is the enemy of the good."

ARTICLE 701. LEGALLY REQUIRED STANDBY SYSTEMS

701.1. Scope. This article covers *standby* power systems that are required by law. Legally required standby power systems are those systems required and so classed as legally required standby by municipal, state, federal, or other codes or by any government agency having jurisdiction. These systems are intended to supply power automatically to selected loads (other than those classed as emergency systems) in the event of failure of the normal source.

Legally required standby power systems are typically installed to serve such loads as heating and refrigeration systems, communication systems, ventilation and smoke-removal systems, sewage disposal, lighting, and industrial processes that, when stopped during any power outage, could create hazards or hamper rescue or firefighting operations.

This article covers the circuits and equipment for such systems that are permanently installed in their entirety, including power source.

701.3. Tests and Maintenance. These tests are in some ways similar to those for emergency systems. The differences are as follows: Part **(A)** does not require full system tests to be done after the acceptance test after the initial installation. Part **(C)** does not address unit equipment, which is odd given that the article does encompass this equipment in 701.12(G). Part **(E)** requires periodic load testing, but there is no express requirement to test the system under the maximum anticipated load conditions.

701.4. Capacity and Rating. This is also comparable to the standby source for an emergency system. There is no express permission to use the standby source for peak load shaving, nor is there a requirement for an alternate source to be available when the generator for the standby system is out for major maintenance or repairs.

701.5. Transfer Equipment. These requirements are the same as for Art. 700 systems except the product acceptance rules are in terms of standby instead of emergency use. However, there is no part (D) and therefore no limitation that a legally required standby system transfer switch will only supply legally required standby system load. This is logical because Art. 701 circuits need not be divorced from normal circuits, as covered in 701.10.

701.6. Signals. This rule is identical to 700.6, including a comparable requirement for ground-fault notification. The same additional paragraph as in 700.6(D) regarding paralleling bus applications entered this rule. Refer to the explanatory information at that location for more information.

701.7. Signs. This rule is identical to the one in 700.7. Refer to the discussion and drawing at that location.

701.10. Wiring Legally Required Standby Systems. This is one the few very major differences between this wiring and emergency system wiring, in that there is no separation requirement from ordinary wiring.

701.12. General Requirements. The parent language contains another major difference between these circuits and emergency circuits, that being these systems must transfer to the standby source within 1 minute, as opposed to 10 s for emergency loads, in the event of an outage. Part **(A)** on storage batteries as a source of supply is identical to comparable requirements in emergency systems. Part **(B)** omits some requirements that appear for emergency systems, including the requirement that a battery charger needed to recharge the battery and that power to cooling, power-operated dampers connected to the standby system are omitted as well. There is also no requirement for an auxiliary power supply, because the 60 s permitted time delay should be plenty of time to get a generator running. The outdoor generator set rule, at (5), is the same as 700.12(B)(6) but lacks a comparable exception for safe switching procedures under the control of qualified persons. It does, however, carry the same reference to 445.18 and the same reference to 225.36 as covered in 700.12(B)(6). Refer to the comparable material in Art. 700 of this book for a complete analysis. Parts **(C)** and **(D)** are the same as their counterparts in 700.12. Part **(E)** covers connections ahead of the service disconnect, which are permitted here but were removed from 700.12 in the 1996 NEC. This provision requires permission of the AHJ; it is discussed at 700.12(D), which gives some of the details and limitations for this type of connection. Part **(F)** is the same as its Art. 700 counterpart. Part **(G)** only differs by omitting the circuit identification rule; however, the present language in 408.4 is strong enough to mandate the same result. It also lacks the additional exception for including an exit door light on an interior system.

701.25. Accessibility. Branch-circuit protective devices in legally required standby systems must not be accessible to unauthorized persons.

701.26. Ground-Fault Protection of Equipment. Here again GFPE may be omitted; however, in this, article there is no requirement to sense and annunciate the fact that a fault is in progress, although it is certainly permitted. As in the case of emergency systems, ground-fault indication is required and it must be annunciated in accordance with 701.6(D) in all instances where automatic GFPE shutdown is not provided.

701.27. Coordination. This is the same rule and the same comments apply to it as for the rule in 700.32. The same requirement for professional engineering support applies to these systems; refer to the analysis at 700.32 for extensive commentary on this subject.

ARTICLE 702. OPTIONAL STANDBY SYSTEMS

702.1. Scope. Life safety is *not* the purpose of optional standby systems. Optional standby systems are intended to protect private business or property where life safety does not depend on the performance of the system. Optional standby systems are intended to supply on-site generated power to selected loads, either automatically or manually.

Optional standby systems are typically installed to provide an alternate source of electric power for such facilities as industrial and commercial buildings, farms, and residences to serve such loads as heating and refrigeration systems, data processing and communications systems, and industrial processes that, when stopped during any power outage, could cause discomfort, serious interruption of the process, or damage to the product or process. To be covered in this article, a system must be permanently installed in its entirety, including any prime movers.

Because of the constant expansion in electrical applications in all kinds of buildings, the use of standby power sources is growing at a constantly accelerating rate. Continuity of service has become increasingly important with the widespread development of computers and intricate, automatic production processes. More thought is being given to and more money is being spent on the provision of on-site power sources to back up or supplement purchased utility power to ensure the needed continuity as well as provide for public safety in the event of utility failure.

At this point a distinction should be clear: Contrary to both common public opinion and frequent trade slang among electricians, the majority of standby power systems are not emergency systems. Just because a system will back up some or all loads on an electrical system, it is not necessarily an emergency backup system. In fact, statistically it is probably an optional standby system. Only a standby system that will supply emergency loads and transfer within 10 s, etc., is an emergency system. This article covers optional standby systems, even those of immense size that may protect millions of dollars worth of data, or extensive plant equipment that would be damaged in the event of a disorderly shutdown, or any other load that the owner may deem a dire emergency should it be disconnected. Until and unless such backup systems also supply life-safety equipment as defined in local laws and regulations [and sometimes they do under the allowances in 700.4(B)], these systems are as presented in Art. 702 and they are not emergency systems. Further, even where one generator, as noted, is the standby source for both optional and emergency systems, only the wiring to the designated emergency loads is part of the emergency system under Art. 700. The rest of the standby wiring, even if it dwarfs the life-safety equipment wiring, is nothing more than an optional standby wiring system, covered here.

The most basic application of standby power is the portable alternator used for residential standby power where electric utility supply is not sufficiently reliable or is subject to frequent outages (Fig. 702-1).

702.4. Capacity and Rating. Any standby power source and system must be capable of fully serving its demand load. This can generally be satisfied relatively easily for generator loads. But the task can be complex for uninterruptible power supply (UPS) systems. The UPS is an all-solid-state power conversion system designed to protect computers and other critical loads from blackouts, brownouts,

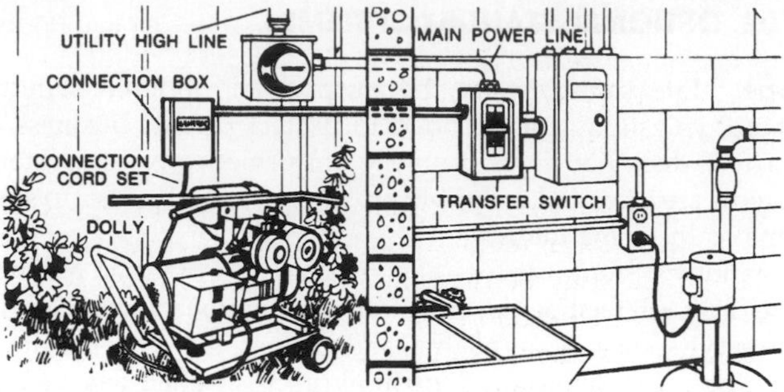
UTILITY HIGH LINE
CONNECTION BOX
CONNECTION CORD SET
DOLLY
MAIN POWER LINE
TRANSFER SWITCH

Fig. 702-1. Standby generator with manual transfer is common residential application. (Sec. 702.1.)

and transients. It is usually connected in the feeder supplying the load, with bypass provisions to permit the load to be fed directly. Figure 702-2 shows a typical basic layout for a UPS system. Such a system utilizes a variety of power sources to ensure continuous power. Circuits are shown for normal power operation. If normal power fails, the static switch transfers the load to the inverter within ¼ cycle. The battery is the power source while the engine-generator is being started. When the generator is running properly, it is brought onto the line through its transfer switch, and the static switch transfers the load to the generator supply. The dashed line indicates the synchronizing signal, which maintains phase and frequency of inverter output.

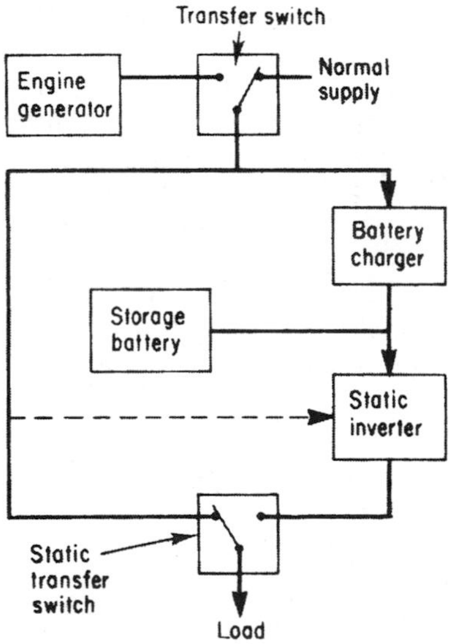
Transfer switch
Engine generator
Normal supply
Battery charger
Storage battery
Static inverter
Static transfer switch
Load

Fig. 702-2. UPS system is a common standby power system. (Sec. 702.2.)

Calculation of the required capacity of a UPS system must be carefully made. Data processing installations normally require medium-to-large 3-phase UPS systems ranging from 37.5 to over 2000 kVA. Some of the typical ratings available from UPS manufacturers are 37.5 kVA/30 kW, 75 kVA/67.5 kW, 125 kVA/112.5 kW, 200 kVA/180 kW, 300 kVA/270 kW, 400 kVA/360 kW, and 500 kVA/450 kW. Larger systems are configured by paralleling two or more of these standard-size single modules.

The necessary rating is chosen based on the size of the critical load. If a power profile itemizing the power requirements is not available from the computer manufacturer, the load may be measured using a kilowattmeter and a power-factor meter. Because UPS modules are both kVA- (apparent power) and kW- (real power) limited, a system should be specified with both a kVA and a kW rating. The required kVA rating is obtained by dividing the actual load kW by the actual load power factor. For example, an actual 170-kW load with an actual 0.85 power factor would require a 200-kVA-rated UPS. The system should be specified as 200 kVA/170 kW, and the standard 200 kVA/180 kW UPS module should be selected for the application.

The full load requirement applies to the connections as noted in Fig. 702-2 and the preceding discussion because an automatic transfer switch was used and illustrated. Part **(B)(2)** requires this capacity in the standby source if loads are transferred automatically, unless there is an energy management system in place, in which case the system capacity must match the maximum load that the management system will allow to be connected.

On the other hand, if a manual transfer switch is used, as in Fig. 702-1, the only rule, in **(B)(1)**, is that the generator supply the load intended to be operated at one time. This gives a homeowner, for example the ability to turn on a breaker supplying electric heat for a while, and then turn that off and energize a specific appliance for some purpose, and then restore the electric heat, etc. Whether this is advisable is another question. Murphy's law generally results in the most unqualified person being the one present during an outage, and an overloaded generator will not continue in service. At the very least, a qualified person should post instructions advising what combinations of circuits will achieve the desired results without overloading the generator.

702.5. Transfer Equipment. This equipment is mandated for all optional standby systems for which an electric utility is one of the electric supply options. This ensures that a standby source does not backfeed power into a utility system, which can be a source of extreme hazard for utility personnel attempting to cope with an outage. However, for facilities with qualified personnel, a portable generator can be connected without a transfer switch provided the normal supply can be locked out, or where the normal supply conductors are actually disconnected from the distribution equipment. Transfer equipment for single-circuit applications and connected on the load side of the branch-circuit overcurrent protection is also recognized and is permitted to contain supplementary overcurrent devices in recognition of the low available fault currents that are possible from generators of such sizes.

702.7. Signs. These are the same requirements as in 700.7; refer to the discussion and drawing at that location. Large optional standby sources are very important, in size frequently the most important, applications of these requirements. Part **(C)** is new as of the 2014 **NEC** and mandates a sign on a power inlet designed to receive a female cord body attached to a cord-and-plug-connected generator that will be used to power its connected load on a temporary basis. The warning label (the prescriptive language is in the NEC) alerts the user whether the transfer switch is arranged for a separately derived source (bonded neutral) or nonseparately derived source (floating neutral). Of course, it remains to be seen if those connecting the cord body from the generator will understand the meaning of the sign. Refer also to the discussion at the new 445.20 for additional background information on this issue.

702.10. Wiring Optional Standby Systems. As in the case of Art. 701 systems, there is no requirement to segregate this wiring.

702.11. Portable Generator Grounding. These rules effectively turn on whether the portable generator will or will not be connected as a separately derived system, and therefore whether it will or will not need a grounding electrode connection in its capacity as a separately derived system.

702.12. Outdoor Generator Sets. This is the same rule as in 700.12(B)(6) although it lacks a comparable exception for safe switching procedures under the control of qualified persons; refer to the discussion at that point. The reference to 445.18 is there as well, allowing the shut-off switch on the generator to provide this function.

Part **(B)** allows a flanged inlet and cord body connection to serve as the disconnecting means for portable generators 15 kW or less. Part **(C)** is new as of the 2017 NEC and requires a listed, interlocked disconnecting means for power inlets that are rated 100 amps and up and intended for use with portable generators. If the inlet is listed as a disconnecting means, or if the application is far a supervised industrial installation with permanent space set off for the portable generator(s) that is within line of sight of the power inlets, then these requirements do not apply. The rule is aimed at unqualified persons disconnecting a portable generator connected to an inlet under load.

ARTICLE 705. INTERCONNECTED ELECTRIC POWER PRODUCTION SOURCES

705.1. Scope. This NEC article covers interconnected electric power sources, particularly utility supplies, and on-site power sources including solar photovoltaic systems, fuel cell sources, and other sources intended to operate in parallel with an active utility connection. Another common name for this type of connection is "cogeneration."

The overall intent is to have this article cover issues of utility-interactive, interconnected parallel production sources that apply to all such connections, with technology specific variances addressed in the Chap. 6 articles as required.

705.2. Definitions. There are two new definitions as of the 2017 NEC that interface with the equally new Art. 712, which pertain to premises wiring systems. A *microgrid system* is such a system that has electrical generation, storage, and one or more loads or any combination that includes the ability to both disconnect from and operate in parallel with the prevailing primary (presumably usually the utility) source. And the interconnecting device that allows that interconnection to function is the *microgrid interconnect device* (*MID*). The term "microgrid" has superseded the prior terminology that referred to intentionally islanded systems. Power production equipment is the term chosen to describe the power sources that provide electricity from other than utility-supplied services, including PV, fuel cells, and wind.

705.8. System Installation. Installation of cogenerating production is to be done by qualified persons, using (Sec. 706.6) listed equipment evaluated for this function.

705.10. Directory. A permanent plaque or directory must be posted at the service equipment showing the location of all local parallel production source disconnecting means capable of interconnection, and reciprocal information must be posted at any remote disconnecting means, if any, for all other power production sources capable of interconnection.

705.12. Point of Connection. Utilities have different policies about where an interactive connection is to be located, either ahead of the service disconnect or after it, and this section covers both. Part **(A)** covers supply side connections, which are permitted in accordance with 230.82(6). The sum of the ratings of all overcurrent devices connected to power production sources must never exceed the service rating. No further requirement applies to this connection in this section, although other requirements will apply, the most important being the disconnection rule in 705.22.

Part **(B)** covers load side connections. If the interconnection is through a panelboard or other comparable equipment that is simultaneously supplied by a primary source, and where this equipment also supplies (or is capable of supplying) other circuits, whether feeders or branch circuits or both, then specialized and relatively complicated rules apply to exactly how the supply connections are sized and located. The reasons for this are based in two overarching considerations. First, the busbars of the interconnecting equipment must be protected from overload by excessive loads being taken from those busbars due to the multiple supplies. The second concern is related to the first, but less obvious. The amount of power distributed from the distribution equipment must not exceed the power dispersion that was being taken out of it at the time it was being investigated for its listing. The **NEC** generally allows a 20 percent buffer before drawing the line in this latter area to make the requirements practicable. However, consider a 200-A panel fed with 200-A breakers at both ends, The 400 A that could be taken from intermediate positions in this panel would divide towards both ends and not overload the busbars in terms of amperes. However, the power distributed from this panel would double the amount it underwent during heat rise testing based on the normal configuration of a main OCPD limiting the busbars to the nominal rating.

Interactive inverters and other power sources are permitted to connect to the line side of a service disconnect, as covered in 230.82(6) (and correlated in 230.40 Exception No. 5). Connections on the load side of a service disconnect, as illustrated in Fig. 690-1 for a utility interactive inverter, must comply with a number of requirements, including the requirement at 705.32 to make the connection on the line side of any GFPE. If made on the load side, the inverter output may have unanticipated effects on the GFPE sensor, including damaging the equipment or interfering with the residual current readings. The connections are permitted only if there is GFPE from all sources and the load side equipment must be listed as suitable for backfeeding conditions.

The interconnecting breaker must be suitable for backfeeding, as at (4). Most breakers are tested in both directions. The ones that are not are marked "line" and "load" and if so marked they can only be used in the direction indicated. In Fig. 690-1, a 5-kW inverter operating on a continuous basis on a 240-V system would calculate as a (5000 kW/240 V) × 1.25 = 26-A load for termination purposes, and a 30-A circuit breaker would be used. Note that this would be OK for a 150-A or larger panel, but too large for a smaller panel. The hold-down feature in 408.36(D) designed to prevent breakers with energized bus stabs from getting into trouble while not parked on a busbar is waived in (5) on these installations. If the breaker were lifted from the bus, it would be the same as the ac power going off, and the inverter or other source would immediately shut down anyway, per 705.40. When a panel has multiple sources providing power to the bus, it must be marked to show those sources, as covered in (3). A note following (4) indicates that fused switches are also suitable for backfeed, unless marked otherwise.

705.12(B)(2) was completely rewritten for the 2014 **NEC** in a comprehensive manner to cover every conceivable permutation of busbar connections from interconnected electric power sources. The presentation is so complex that it has to be taken apart a bit at a time in order to be fully understood. At the outset, there is a new requirement in the parent language of (2) to use 125 percent of the inverter output circuit current in calculations to determine the minimum ampacity for conductors and busbars. With this in mind, take the inverter output and land it somewhere. There are three basic possibilities, fully illustrated in the eight figures that follow. Each of the illustrations shows two interconnected power sources, a service (indicated by the breaker fed from a revenue meter [dollar sign]), and also an interconnected electric power source (abbreviated "IEPS"). The numbers used in this presentation are those in the **NEC**.

(1) Covers landing on a feeder. There are two possibilities in this case, the first being a connection being made at any point along the length of a feeder *that is not the opposite end of the feeder from its point of supply*. There are two logical solutions. The first (Fig. 705-1) is at (a) and requires the ampacity of the feeder now supplied from a normal source and from the inverter to have an ampacity equal to the sum of the normal source OCPD and 125 percent of the inverter-rated output. The second (Fig. 705-2) is at (b) and requires an OCPD on the load side of the inverter connection based on the feeder ampacity for the remainder of the feeder circuit.

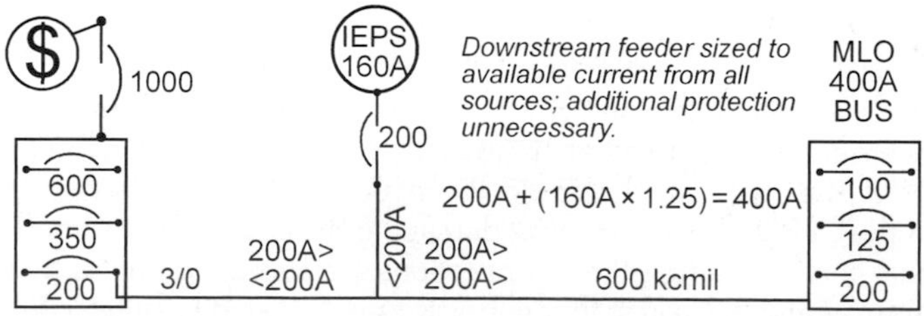
$
1000
600
350
200
3/0
200A>
<200A
IEPS
160A
200
<200A
200A>
200A>
Downstream feeder sized to available current from all sources; additional protection unnecessary.
200A + (160A × 1.25) = 400A
600 kcmil
MLO
400A
BUS
100
125
200

Fig. 705-1. Illustration of 705.12(B)(2)(1)a.

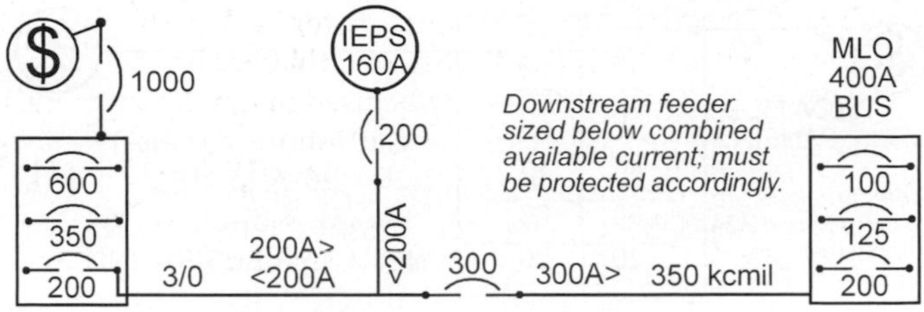

$
1000
IEPS
160A
200
<200A
Downstream feeder
sized below combined
available current; must
be protected accordingly.
MLO
400A
BUS
600
350
200
3/0
200A>
<200A
300
300A>
350 kcmil
100
125
200

Fig. 705-2. Illustration of 705.12(B)(2)(1)b.

(2) Covers a connection made to a feeder and a tap is extended from such a connection (see Fig. 705-3); the allowable tap size and length are to be based on 125 percent of the inverter output and the current rating of the supply side OCPD with calculations being based (see also Fig. 705-4) on the sum of both taken into the relevant part of 240.21(B).

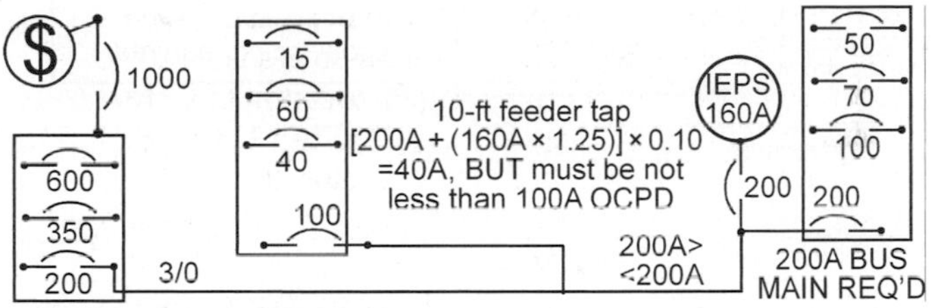

$
1000
600
350
200
3/0
15
60
40
100
10-ft feeder tap
[200A + (160A × 1.25)] × 0.10
=40A, BUT must be not
less than 100A OCPD
IEPS
160A
200
200A>
<200A
50
70
100
200
200A BUS
MAIN REQ'D

Fig. 705-3. Illustration of 705.12(B)(2)(2) as used with the 10-ft tap rule.

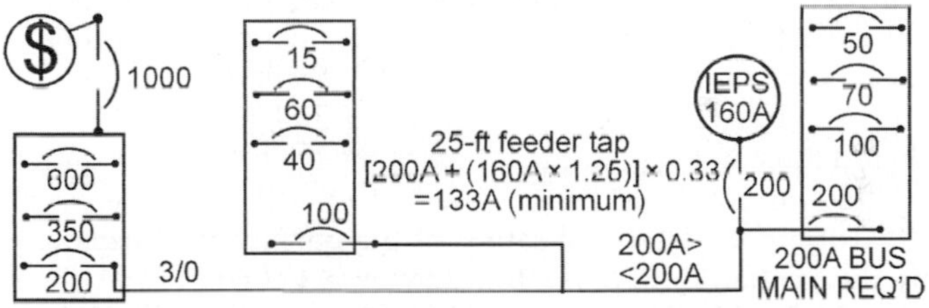

$
1000
600
350
200
3/0
15
60
40
100
25-ft feeder tap
[200A + (160A × 1.25)] × 0.33
=133A (minimum)
IEPS
160A
200
200A>
<200A
50
70
100
200
200A BUS
MAIN REQ'D

Fig. 705-4. Illustration of 705.12(B)(2)(2) applied to the 25-ft tap rule.

(3) Covers a connection to a busbar, including a busbar in a panelboard. Such a connection itself presents four connection types, as covered in (3)(a), (3)(b), (3)(c), and (3)(d) below:

For the first type of connection, (3)(a), (Fig. 705-5) 125 percent of the inverter output current plus the rating of the normal source overcurrent device must not exceed the rating of the busbar. As explained in a new informational note, this scenario assumes no limitation on the numbers of loads or sources (and logically no limitation on the positioning of any of the entry or exit points) as applied to the buswork.

The second type of connection, (3)(b) (Fig. 705-6), is a slightly modified presentation of the requirement prior to the 2014 **NEC**: The inverter output connection is

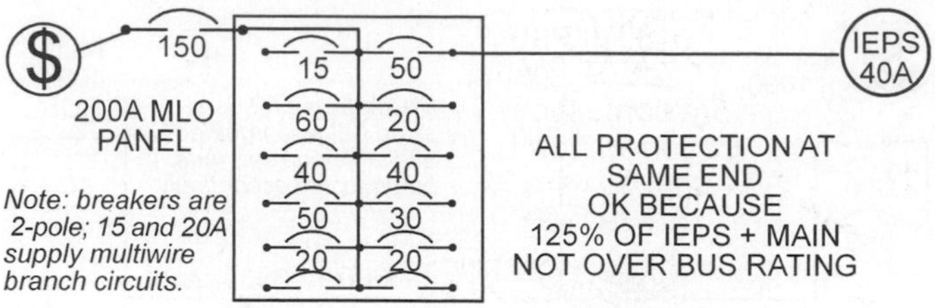
150
15
50
60
20
40
40
50
30
20
20
200A MLO
PANEL
Note: breakers are
2-pole; 15 and 20A
supply multiwire
branch circuits.
IEPS
40A
ALL PROTECTION AT
SAME END
OK BECAUSE
125% OF IEPS + MAIN
NOT OVER BUS RATING

Fig. 705-5. Illustration of 705.12(B)(2)(3)(a).

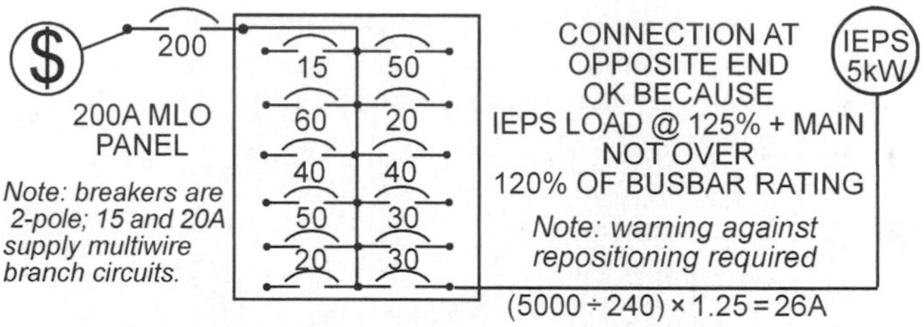
200
15
50
60
20
40
40
50
30
20
30
200A MLO
PANEL
Note: breakers are
2-pole; 15 and 20A
supply multiwire
branch circuits.
CONNECTION AT
OPPOSITE END
OK BECAUSE
IEPS LOAD @ 125% + MAIN
NOT OVER
120% OF BUSBAR RATING
Note: warning against
repositioning required
IEPS
5kW
(5000 ÷ 240) × 1.25 = 26A

Fig. 705-6. Illustration of 705.12(B)(2)(3)(b), and the connection shown in Fig. 690-1.

made at the opposite end of the busbar from the normal supply protection, and that value, when added to 125 percent of the inverter output current, is not to exceed 120 percent of the busbar rating, and a warning sign is to be placed in compliance with 110.21(B) prohibiting relocation of the inverter connection.

The third type of connection, (3)(c), (Fig. 705-7) requires that the maximum rating of all overcurrent protective devices (other than the main OCPD) installed in a panelboard must not exceed the rating of the busbar. This requirement also specifies that a warning label be provided to indicate that the combined ratings of OCPDs cannot exceed the rating of the panelboard busbars. Fig. 705-8 shows why this section now refers to interconnected sources in general, since no inverter is involved in this application, but the connection requirements raise the same issues.

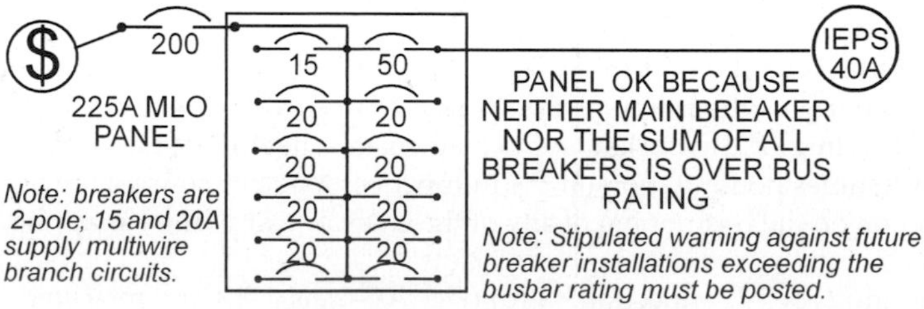
200
15
50
20
20
20
20
20
20
20
20
225A MLO
PANEL
Note: breakers are
2-pole; 15 and 20A
supply multiwire
branch circuits.
IEPS
40A
PANEL OK BECAUSE
NEITHER MAIN BREAKER
NOR THE SUM OF ALL
BREAKERS IS OVER BUS
RATING
Note: Stipulated warning against future
breaker installations exceeding the
busbar rating must be posted.

Fig. 705-7. Illustration of 705.12(B)(2)(3)(c).

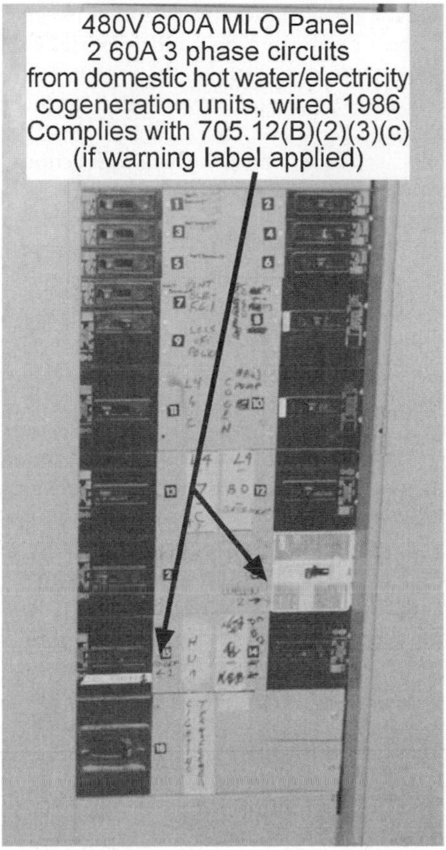
480V 600A MLO Panel
2 60A 3 phase circuits
from domestic hot water/electricity
cogeneration units, wired 1986
Complies with 705.12(B)(2)(3)(c)
(if warning label applied)

Fig. 705-8. A cogeneration installation at a college dormitory in continuous operation for 30 years, it would comply with this section if the required warning label were applied.

The fourth type of connection, (3)(d), (Fig. 705-9) covers center-fed panels that are common in some regions of the country. They often ship as combination equipment with a meter socket to the side. They benefit from the same procedure as at (b), but only one end of the panel is eligible for this treatment. This arrangement

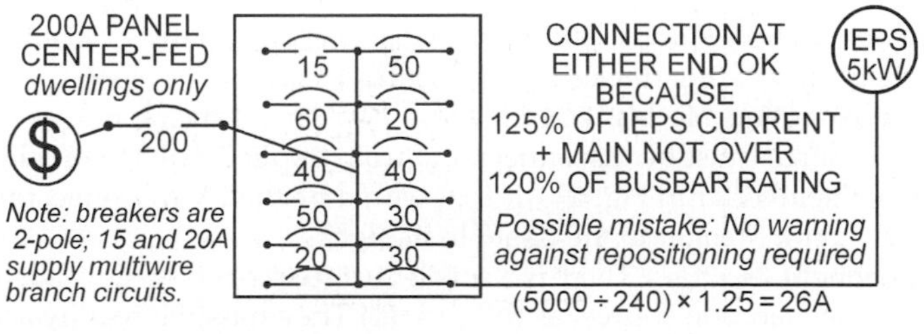
200A PANEL
CENTER-FED
dwellings only
$
200
15
50
60
20
40
40
50
30
20
30
Note: breakers are 2-pole; 15 and 20A supply multiwire branch circuits.
CONNECTION AT EITHER END OK BECAUSE 125% OF IEPS CURRENT + MAIN NOT OVER 120% OF BUSBAR RATING
IEPS 5kW
Possible mistake: No warning against repositioning required
(5000 ÷ 240) × 1.25 = 26A

Fig. 705-9. Illustration of 705.12(B)(2)(3)(d).

has a greater possible load concentration on half the bus, and is limited to dwellings where the likelihood of intensive demand is much less. For no observable reason, and this may very well be a mistake, the requirement for a warning label against repositioning the interconnected power source connection breaker was not repeated in this provision.

The remainder of this discussion focuses on the connections made in accordance with 705.12(B)(2)(3)(b), which has become the usual mode of connection, and for that reason it deserves a bit more attention by way of explaining the concepts. Most panelboards have busbars with ratings equal to the rating of the main overcurrent protective device ahead of them in the panel. Refer to Fig. 690-1, which shows a common 200-A panelboard. The NEC now requires, at (3)(b), that the input breaker from the inverter be at the opposite end of the bus from the normal main. In this way, no configuration of branch circuit or feeder loads could draw more than the current the busbars were designed to carry. In addition, the panel must be marked with a warning label advising against repositioning this breaker. There is, however, still a 20 percent limitation on the size of the interconnecting breaker, also at (3)(b). This is because, although the busbars could not be overloaded, the panel would still be delivering more power than the conditions under which it was tested. The 20 percent is a reasonable limit until more testing is carried out.

Note that what this refers to as a 20 percent rule is actually stated in the NEC as a requirement that all input breakers must not have a combined rating not over 120 percent of the busbars—which makes the effective limit on the branch to be 20 percent of the bus after taking out 100 percent of the main. Note also that main lug (MLO) panels in higher busbar ampere ratings are available. For example, in reference to Fig. 690-1, it is possible to obtain a 200-A MLO panel and then install a comparable 150-A breaker (with hold-downs) as a side-feeding device at the upper end. Now an inverter requiring a 40-A interconnecting circuit breaker could be connected.

705.14. Output Characteristics. The interconnecting source must be controlled so as to be compatible with the applicable utility system. A note indicates that it is not necessary to match the wave shape exactly and still have a compatible system for interconnection.

705.16. Interrupting and Short-Circuit Current Rating. Cogenerating sources will, to varying degrees, elevate the available fault current on the system being connected to; as such sources are added this section requires a reevaluation of compliance of existing circuit components in terms of their ability to break currents at fault levels (110.9) and in terms of the required short-circuit current ratings of various types of equipment.

705.20. Disconnecting Means, Sources. Cogenerating sources must be disconnectable using a qualifying device under 705.22.

705.21. Disconnecting Means, Equipment. Equipment in the power train of a cogeneration source must be disconnectable from applied voltage on either side. In the case of the inverter in Fig. 690-1, that would be the PV disconnect to its left, and the 30-A circuit breaker in the panel to its right.

705.22. Disconnect Device. There is a seven-part rule covering the system disconnects from a cogeneration source, as follows: (1) They must be readily accessible; (2) They must be externally operable by hand even if their power supply fails;

(3) They must be indicating; (4) They must be fully rated in terms of load and available fault current; (5) They must equipped with a warning sign that all contacts can be energized if that is the case (and it usually is on these systems); (6) They require simultaneous action across all poles when the disconnect opens; and (7) They require a lock-open capability.

705.23. Interactive System Disconnecting Means. A readily accessible means must be installed to separate the interactive system from all other wiring systems on the premises. The point is to differentiate between the equipment disconnecting means that have their own requirements, and the system disconnecting means.

705.30. Overcurrent Protection. Conductors must have protection at both ends to cope with the multiple supplies that are inherent to this article. Part **(A)** (PV systems) and part **(C)** (Fuel cells) defer to those articles, as discussed at length in Chap. 6 of this book. Transformers, covered in part **(B)** are considered from both sides as both primary and secondary in terms of overcurrent protection provided in accordance with 450.3. Interactive (and not just utility-interactive) inverters, per part **(D)**, must comply with 705.65 and generators [part **(E)**] must comply with 705.130.

705.31. Location of Overcurrent Protection. This rule is new as of the 2014 NEC. Where premises power production sources interconnect on the supply side of the service equipment as covered in 705.12(A) and 230.82(6), overcurrent protection must be arranged within 10 ft of the connection. If that dimension is not met, then cable limiters or current-limiting circuit breakers must be located at the connection point. The likely difficulty and expense of applying that exception means that probably 10 ft will be the limit for these connections. These conductors have been unlimited by any express Code rule because they are not, strictly speaking, service conductors, but they are subject to the full utility available fault current should they fail. This change addresses that hazard.

705.32. Ground-Fault Protection. The preference is to make cogeneration connections on the line side of GFPE sensors; however, if there is GFPE arranged so as to offer protection from all current sources, then the cogeneration output may enter on the load side of the sensors. This may prove problematic on large cogeneration systems connected to remote panels.

705.40. Loss of Primary Source. When the primary source is interrupted, the cogenerating source must cease power production. If this were not to happen for any reason, the cogenerating source would be exporting power out into utility lines that are subject to service by utility line crews, an extreme hazard. Remember that these systems normally do not operate through transfer switches. There is an exception that allows for the development of listed interactive inverters that are capable of continuing to produce ac power, but without any power reaching the service conductors and beyond. In this way the cogeneration facility becomes, in effect, an Art. 702 optional standby system.

705.42. Loss of 3-Phase Primary Source. If the primary source is a 3-phase system, and a phase is dropped, the entire interactive system must disconnect unless it is a listed interactive inverter as in 705.40 Exception that will not export power back into the utility system, but will function instead as an island, at least until all utility phases are restored. If the parallel production source is supplying an Art. 700 or Art. 701 load, however, the requirement to shut down on a utility phase drop does not apply.

705.50. Grounding. Interconnected power production sources must follow the usual system and equipment grounding provisions that normally apply from Art. 250. However, for interactive dc systems connected through an inverter to a grounded ac system, equivalent methods using equipment listed to provide comparable system protection are permitted.

705.60. Circuit Sizing and Current. These systems are presumed to be continuous loads and the calculations must be done with the usual 125 percent factor.

705.65. Overcurrent Protection. Overcurrent protective devices are sized according to usual procedures, unless the source current is inherently so limited that the conductor ampacity will not be exceeded, and there are no parallel source connections, such as from batteries or inverter backfeed. Transformers also follow the usual overcurrent protection considerations, but the protection chosen must work as if either side could be primary or secondary, depending on conditions.

Because this topic now involves power sources not so inherently limited in capacity, **(C)** now creates a more open-ended calculation procedure that will work for any interactive source. The approach here marries the result in 705.60(B) with the feeder tap rules in 240.21(B).

705.70. Interactive Inverters Mounted in Not Readily Accessible Locations. This section covers utility-interactive inverters mounted in difficult locations such as rooftops. In such cases dc and ac disconnects must be mounted in sight of, and on the appropriate sides of the inverter. Then an additional ac disconnect must be mounted in accordance with the usual rules in 705.22 for a readily accessible location, and reciprocal labeling as required in 705.10 must be applied to this accessible disconnect and at the service equipment.

705.80. Utility-Interactive Power Systems Employing Energy Storage. Such systems must have a marking showing the polarity of the grounded circuit conductor, and the maximum operating voltage including the equalizing voltage if any.

705.82. Hybrid Systems. These systems are defined in 705.2 and involve multiple power production sources from PV to micro-hydro generators. They may include energy storage, but if so, the storage equipment is not a power source within this definition. This section allows hybrid systems to be connected to an interactive inverter.

705.95. Ampacity of Neutral Conductor. If a single-phase 2-wire inverter output is connected phase or line to neutral in a 3-wire or a 3-phase 4-wire system, the maximum connected load between the neutral and any one ungrounded conductor of that system, plus the inverter contribution, must not exceed the ampacity of the system neutral. This prevents such an inverter connection from overloading a neutral. For example, if a 100-A panel is wired with 3 AWG copper and has 80 A of line-to-neutral load and a 30-A line-to-neutral inverter input, 110 A of current can be taken out of the panel on line-to-neutral loads, overloading the busbar if the inverter connection is made at the same end as the service neutral connection.

A neutral used strictly for instrumentation and sized equal to or larger than the equipment grounding conductor is permitted to be sized smaller than its associated ungrounded conductors.

705.100. Unbalanced Interconnections. Single-phase hybrid inverters and ac modules in interactive systems must not be connected to a 3-phase system unless the interconnected system is designed to avoid significant unbalanced voltages. Three-phase interactive inverters must automatically deactivate if there is a loss of or unbalanced voltages in one or more phases, unless the system is designed to avoid significant unbalanced voltages. As of the 2014 NEC, when single-phase inverters are connected to a 3-phase system, it must be done in such a way as to limit the unbalanced voltages to not over 3 percent. An informational note references an ANSI document that provides guidance on this point. The prior Code language was routinely interpreted as effectively prohibiting a practice that utilities routinely apply on their systems.

705.130. Overcurrent Protection. This is a reiteration of 705.30, with a sentence tacked on at the end requiring generators to be protected in accordance with 445.12. Because the rules in Chaps. 1 through 4 apply unless modified here, this rule technically adds absolutely nothing to the NEC. The material here was copied from 705.30 in order to begin the population of a new Part III of the article so rotating sources would be grouped together.

705.143. Synchronous Generators. This requires controls on synchronous generators to maintain their synchronicity while in operation. This is not new material; it was relocated from its former location to group rules for rotating sources in a common location.

Part IV. Microgrid Systems.

705.150. System Operation. This is the first recognition of systems that can intentionally separate themselves from the normal power environment and then reconnect as needed. They are receiving increasing attention as a way to increase resiliency in premises wiring systems.

705.160. Primary Power Source Connection. These connections must comply with 705.12.

705.165. Reconnection to Primary Power Source. Intentionally islanded systems, as these were formerly known, that then intentionally reconnect must do so in a synchronous manner that does not disrupt the surrounding connections.

705.170. Microgrid Interconnect Devices (MID). These devices must be listed or field labeled for this use and comply with requirements that reflect 705.30 regarding overcurrent protection from all available directions. The key here will be the listings. There is an informational note at this point advising that these functions may be found in multimode inverters and other systems involved with interactive operation.

ARTICLE 706. ENERGY STORAGE SYSTEMS

706.1 Scope. This article, new in the 2017 NEC, applies to energy storage systems operating over 50 volts ac or 60 volts dc whether acting on a stand-alone basis or interactive with other electric power production sources. There is an extensive list of definitions, but key to the scope is the definition of energy storage system itself. It is:

1) comprised of one or more components
2) capable of storing energy
3) operational using electrical equipment (batteries, capacitors, etc.) or kinetic energy storage (flywheels, compressed air, etc.)
4) may have either ac or dc output
5) may include inverters or other components for interconnection

Energy storage systems are classified into one of three general forms:

1) Self-contained, pre-assembled into a single container or unit (ESS, self-contained)
2) Pre-engineered using matched components, intended to be field assembled (ESS, pre-engineered of matched components)
3) Other systems, that would be completely field assembled as a system from constituent parts (ESS, other)

These systems form points of mediation between variable sources of power, such as day vs. night for PV, windy vs. calm for wind, and variable demands for power. This will make them an important part of the future of energy distribution, on both sides of the service point.

706.5 Equipment. The components, with the exception of lead-acid batteries, must be listed, or a self-contained ESS can qualify for an overall listing.

706.7 Disconnecting Means. This section is inspired by 480.7 for batteries. Refer to the commentary in this book at that location for details. An exception to (D) waives some of the prescriptive information for the labeling if an arc label of the sort specified in NFAP 70E is applied. The informational note describes the requested content as falling entirely within the parameters of such labeling. In addition, the primary safety objective of the label is the ability to properly anticipate and mitigate the hazards of arc flashes. Large battery banks are a classic source of extremely high available fault current exposures. Part (E) is unique to this article and covers instances where the ESS is on the other side of a partition or at a distance and otherwise out of sight. Disconnects then become mandatory at both ends of the circuit; a lock-open alternative is not offered. Reciprocal placards are required advising of the location of the relevant disconnects.

706.8. Connection to Other Energy Sources. A load disconnect with multiple sources of power supplying it must disconnect all such sources from its load when it is in the OFF position. Inverters and ac modules that will be used in an interactive system must be listed for this duty. The remainder of this section points to material in Art. 705 and is discussed at that location in this book.

706.10. Energy Storage System Locations. This section is completely based on 480.9, but with some additional text that makes necessary allowances for pre-engineered and self-contained ESS applications.

706.11. Directory. Part (A) requires the usual power source reciprocal labeling at the service and at all locations where power production sources are capable of interconnection. This largely duplicates 705.10. If a facility has a stand-alone system, then a directory must be provided on the outside of the building as approved by the inspector. It must advise that there is a stand-alone electrical power system is on site, and specify the location for its disconnecting means. It copies 690.56(A), and it is somewhat surprising that this requirement is not in Art. 710.

706.20. Circuit Sizing and Current. These requirements are what might be expected, and were taken from other locations and adapted for this equipment. For example, 706.20(A)(1) comes from 692.8(A) and simply points to the nameplate; the inverter parameters all come from 690.8(A) and 705.60(A). The conductor ampacity rules are copied from 692.8(B) and (C).

706.21. Overcurrent Protection. This section, in part (A), effectively makes all tap rules off limits for ESS circuit conductors, because they must "be protected at the source." Part **(E)** includes an unusual, but understandable requirement for fuses in this context. If a fuse is accessible to unqualified personnel, then it must be replaceable without exposing the operator to live parts, even where energized in both directions. The paragraph goes on to mention pullouts and other suitably rated devices. The location rule in **(F)** (not over 5 ft and within view) matches the disconnect location rule in 706.7(E).

706.23. Charge Control. Charge controllers are required and their controls must be accessible only to qualified persons.

Part **(B)** covers diversion charge controllers. If they are the sole means of controlling the charge, then a second, independent method must be in place to prevent overcharging. If there is also a diversion load, that load must have a higher voltage rating than that of the batteries, and it must not exceed the current rating of the diversion load charge controller. The power rating of the diversion load must be at least half again the power rating of the PV array. The conductor ampacity and overcurrent device ratings for the charging circuit must be at least half again the maximum current rating of the diversion charge controller.

If an ESS is utility interactive, and is using interactive inverters to control battery charge by diverting excess power into the utility system, then the charge regulation circuits do not have to comply with these rules. However, these systems must also have a second, independent means of preventing battery overcharging if the utility is off line, or if the primary charge controller fails for any reason.

Part **(C)** covers dc charge controllers used for battery charging or comparable converters. For example, a PV system might produce 5 A of current at 250 V, and the equipment could pass that on to the storage batteries as 24 V, 50 A. The output conductor ampacity must be based on the continuous rated output current for the selected voltage range, and the output voltage rating must be based on the maximum voltage of the charge controller or converter-selected output voltage setting.

706.30. Installation of Batteries. Any parts of an ESS for a dwelling unit that are accessible during routine maintenance must not exceed 100 V (double the voltage of prior codes). Any series battery string with a combined voltage exceeding 240 V, either between conductors or to ground, must have provisions to subdivide the string so the voltage limitation can be met. The disconnecting means do not require a load break rating.

Storage systems exceeding 100 V must have a similar switch, accessible only to qualified persons, that will open both grounded and ungrounded conductors for maintenance. The ESS can operate on an ungrounded system, provided there is a ground-fault detector in place that will detect a ground fault.

706.31. Battery and Cell Terminations. As recommended by the manufacturer, antioxidant compound must be used on terminals, and the wires landing on them must be arranged so as to prevent mechanical stress on the terminals. The wire size selected for terminal jumpers and rack take-offs must have been evaluated for temperature rise during full load conditions to avoid overheating of insulation or conductor supports.

706.32. Battery Interconnections. Single-conductor flexible cord is permitted, in sizes not smaller than 2/0 AWG, to be used as interconnecting cables on batteries, or to run from battery terminals to nearby enclosures to connect to normal building wire. The cord must be listed for hard service and identified as moisture resistant. The cord terminations must be made using components that are listed and marked for the use. Refer to the discussion at 110.14 for more information on this topic.

706.34. Battery Locations. This material comes straight from 480.10, for which there is additional information in this book.

706.40. General. This section and the ones that follow address flow batteries. These are defined in 706.2 and consist of an enclosed reactor surface that separates two liquids that create a potential difference on either side of the membrane as they flow past and react. This means they can be instantly recharged by renewing the chemicals as they flow, while removing the spent chemicals for renewal (Fig. 706-1). The process has similarities to fuel cells, and Art. 692 provisions are incorporated into these rules by reference. The reagent tanks must be clearly identified by chemical compound, and posted at any point where it can be inserted into or removed from the system. The installation must also provide a method for electrolyte containment in the event of a spill, and the system must be alarmed in the event of such a failure. In addition, in the event of a blockage, the system must

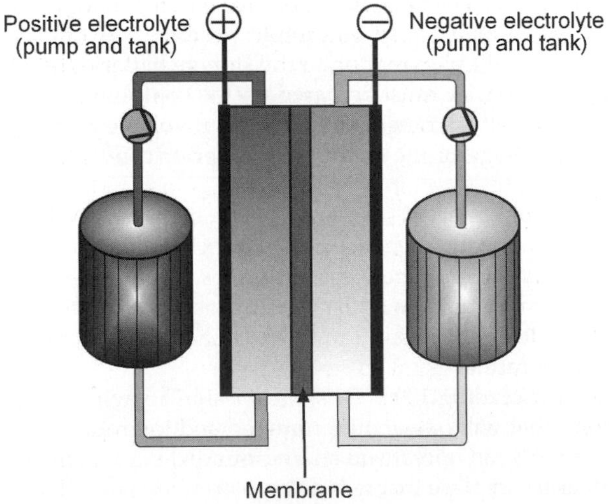
Positive electrolyte
(pump and tank)
Negative electrolyte
(pump and tank)
Membrane

Fig. 706-1. Flow battery components. Flow batteries store/release charge in/from the electrolyte solutions, instead of the electrodes.

be arranged to shut down. The equipment must be evaluated for the chemical exposure, and all wiring must be routed to avoid contact with the chemicals.

706.50. General. Other ESS technology must comply with the NEC in general, and specifically Part III of Art. 705.

ARTICLE 708. CRITICAL OPERATIONS POWER SYSTEMS (COPS)

708.1. Scope. This new 2008 NEC article covers wiring in facilities that must be kept in operation in a "designated critical operations area (DCOA)" in the event normal operations are disrupted. This is a response to studies of infrastructure that followed the 9-11 attacks. These systems must be so classified by a governmental authority, or even by a facility engineering document establishing the need. For example, it would certainly be plausible for an oil refinery, for example, to document and provide such coverage that would be expected to survive a major conflagration. The scope of this article includes both power systems and also HVAC, communications, security, and alarm systems in such locations.

708.2. Definitions. Often seen in its acronym, *supervisory control and data acquisition (SCADA)* is an electronic system that monitors and controls the operation of the critical operations power system. This includes start and stop functions in the power system, annunciation and communications to personnel, both within the facility and for remote operators. Annex G provides additional information on this topic.

708.4. Risk Assessment. Risk assessment must be documented with respect to identifying and quantifying the likelihood of various hazards to the electrical system, and assessing the vulnerability of the electrical system to those hazards. Then, based on the foregoing, a strategy must be implemented to mitigate hazards deemed insufficiently addressed by the applicable requirements of the NEC, both within this article and elsewhere. A note points to Chap. 5 of NFPA 1600-2007, *Standard on Disaster/Emergency Management and Business Continuity Programs*, for guidance concerning risk assessment and hazard analysis.

708.5. Physical Security. Based on the risk assessment, a strategy for physically securing COPS must be developed, documented, and implemented. Electrical circuits and equipment for these systems must be made accessible exclusively to qualified persons.

708.6. Testing and Maintenance. This rule is the same as 700.4 except part (C) on maintenance applies to the entire COPS and not just the batteries, and a note points to NFPA 70B-2006, *Recommended Practice for Electrical Equipment Maintenance* for guidance on testing and maintenance procedures.

708.8. Commissioning. This is a defined term in 708.2. It includes the acceptance testing, operational tune-up work, and start-up testing in order to create a baseline of testing results that verify the proper operation and sequencing of electrical equipment operation, as well as creating a set of data such that future trends analysis can be applied to identify deteriorating equipment. This section requires a commissioning plan to be developed and documented, and a note here also points to NFPA 70B, this time for guidance on developing a commissioning program. The remainder of this section requires component and system tests to ensure proper functioning, and the documentation of all baseline test results. The section concludes with a requirement to create, document, and carry out a functional performance test program upon completion of the initial installation to provide the definitive baseline for future performance criteria. Annex F provides additional information on creating and carrying out a functional performance testing program.

The annex material includes formulas that use failure statistics, such as mean time between failures (*MTBF*). The IEEE has been accumulating this sort of data for several decades, as published in its *Gold Book*. In addition, the U.S. Army Corps of Engineers funded an exhaustive reliability survey in order to support its own mission. These are potential sources of information that can be used to perform the mandated statistical analysis. This article will require an extensive learning curve because it breaks new ground in the NEC by the extent it relies on performance-based rules, rather than the usual prescriptive requirements elsewhere. This in turn means that a heavy emphasis will be placed on the judgments of qualified engineering staff, because it will not be possible to compare their analysis with black-letter code. The AHJ will be in the position of umpiring analytical techniques few in that position have been traditionally familiar with.

708.10. Feeder and Branch-Circuit Wiring. All boxes and enclosures must be marked with permanent methods so they are readily identifiable as components of the COPS, but only in the event that a particular building has both COPS and normal power systems present. If the building is classified as entirely a COPS facility, then specific identification of boxes and enclosures is pointless. Receptacles must also be so identified using a distinctive color or marking; however, in a stand-alone facility where all the receptacles are necessarily part of the system, no special identification scheme is required. As of the 2017 NEC, nonlocking 15- and 20-A. receptacles (no voltage parameter provided, therefore, literally applies to any voltage) must have either a lit face or an indicator light that shows the receptacle is powered.

COPS wiring must be completely divorced from any other wiring; however, if there are two distinct COPS in a particular location, then wiring from those two systems may share the same location. COPS wiring and normal wiring may enter the same transfer switch for obvious reasons.

The wiring methods are robust, consisting of heavy-wall steel conduit, or Type MI cable. PVC, RTRC, EMT, and LFNC or LFMC are permitted, but only where embedded in not less than 50 mm (2 in.) of concrete. A jacketed metallic cable assembly is also permitted in concrete if so listed for concrete embedment. If flexibility is required at a termination, flexible metal fittings, flexible metal, and liquidtight flexible metal conduits are permitted.

Feeders must have a fire separation generally using one of the same strategies enumerated in the list of techniques in 700.10(D)(1); specifically items (2), (3), (4), or (5). There is no waiver if the facility is fully sprinklered (1). If the facility is below the level of the 100-year floodplain, the conductors and wiring methods must be suitable for wet locations.

The wiring methods and fire separation rules for feeders also apply to branch circuits that are part of the system, if they are outside the DCOA. If they are inside the DCOA, then any methods in Chap. 3 are permissible.

708.11. Branch-Circuit and Feeder Distribution Equipment. Branch circuit distributions must be within the same DCOA as is supplied by the circuit. Feeder circuits, including transfer equipment, panels and transformers, must be in spaces with a 2-h fire resistance rating and located above the 100-year floodplain.

708.12. Feeders and Branch Circuits Supplied by COPS. No equipment shall be supplied by either unless it is a required element of the COPS.

708.14. Wiring of HVAC, Fire Alarm, Security, Emergency Communications, and Signaling Systems. All such circuits must use rigid or intermediate metal conduit, or Type MI cable unless embedded in concrete. The former requirement for exclusively shielded cables has been relaxed, as of the 2014 NEC, in view of the reality that they are inconsistent with some data transmission protocols. The enclosing raceway methods are not changing, and it is now generally left to the performance requirements of the system and the published instructions of the manufacturer to determine the characteristics of the preferred cable assemblies, with fiber required between buildings on a property under single management. Listed primary protectors must be applied to all communications circuits. Listed secondary protectors are required on communications cable terminations. All control circuits operating over 50 V must use 600-V wiring. Relay contact ratings must exceed the control voltage and currents in the controlled circuit. Riser emergency communications cables must be Type CMR-CI or be run in a 2-h listed electrical circuit protective assembly. All fire alarm, security, and signaling systems must be riser rated and utilize a listed 2-h rated electrical circuit protective system.

HVAC system control and monitoring, as well as the power circuits to this equipment, must be in a qualifying 2-h electrical circuit protective assembly. Note that even concrete encasement, or routing within a listed assembly of building materials with a 2-h rating, does not comply with this rule.

708.20. Sources of Power. The permitted sources of power in the event of failure in the normal supply largely line up with emergency systems in 700.12. Part **(A)** is the first paragraph of 700.12 minus unit equipment. A major difference, however, is the absence of an arbitrary time-for-restoration parameter. In this location, the rule is the system must function in the "time required for the application." Part **(B)** is the enhanced system for distribution and generation equipment in high-hazard occupancies as covered in the fourth paragraph of 700.12.

Part **(C)** on grounding is unique to Art. 708, and requires a COPS to be wired as a separately derived system with its own grounding and bonding equipment within it. However, if the service equipment containing the main bonding jumper meets the enhanced reliability provisions of 708.10(C) and 708.11(B), then an additional step of converting to a separately derived system is not required. Note that this exception also points to the location of a system bonding jumper. Since a system bonding jumper only pertains to a separately derived system, it is difficult to see how a system bonding jumper for a separately derived system belongs as an exception to a rule that requires a separately derived system. The panel failed to substantiate its insertion of this wording, so the importance of this particular reference remains unclear.

Part **(D)** is also unique to this article, and requires surge protection devices at all distribution levels within a COPS.

Part **(E)** covers storage batteries, using the language in 700.12(A). The first paragraph of the Art. 700 materials addresses the 1½-h rule, which is not appropriate for facilities that may need to be running for days or more.

Part **(F)** covers on-site generator sets, largely the same as in Art. 700. The following seven differences apply: (1) The 2-h fuel source rule is not brought over because these systems may need to run far longer. (2) There is no exception for an off-site fuel delivery system. (3) There is no provision for an auxiliary power supply for cases where a generator will not start in 10 s, which is appropriate given the provisions of the first paragraph. (4) There is no requirement for the generator disconnect in an outdoor set to be rated as suitable for use as service equipment. (5) There is an additional requirement that if a single generator is the alternate supply for a COPS, then permanent provisions must be made to connect a portable or vehicle-mounted generator. (6) There is an allowance for a portable generator to power the facility through a flanged inlet; this is new as of the 2014 NEC. (7) If internal combustion engines constitute the prime mover, the fuel must be stored on site, secured and protected as covered in the risk assessment documentation.

Part **(G)** covers UPS systems, and where they are the sole supply of a COPS, they must comply with the applicable rules of (E) and (F) just discussed. Part **(H)** recognizes fuel cell systems installed per Art. 692, but with no express performance requirements as appear in 700.12(E), which is somewhat unexpected. The original proposal contained this language, which the panel deleted without documenting its action.

708.21. Ventilation. The standby power source must be equipped with sufficient ventilation such that it will be able to operate continuously on even the hottest day.

708.22. Capacity of Power Sources. Part **(C)** drives much of the differences between these rules and their counterparts in 700.5, because these systems must be able to run for at least 72 h at the full load of the DCOA while holding voltage to within 10 percent, plus or minus, of the nominal utilization voltage. This number was recommended by the U.S. Army Corps of Engineers based on what can be expected in terms of the replenishment of fuel, and the system may well be expected to function for longer periods. With that in mind, part **(A)** requires full load capacity for an indefinite period, and a redundant source available whenever the normal standby source is out of service for maintenance or repair. Note that this requirement is here instead of (B) as is its counterpart in 700.4, and also that it applies at any time the standby source is down, not just for occasions of "major" maintenance or repair. Part **(B)** is the same as 700.4(B) except that the COPS wiring is co-equal with the emergency system in the pecking order, and they are both to be energized first.

708.24. Transfer Equipment. These rules are the same as 700.6, although the requirements for listing and for electrical operation/mechanical holding are waived if the sources are inherently synchronized. This can occur in large facilities with multiple switchboards supplied from multiple power sources with other methods in operation to assure that the sources are synchronized. Part **(E)** requires the short-circuit current rating applicable to the transfer equipment under the extant types and settings of overcurrent protection to be field-marked on its exterior.

708.30. Branch Circuits Supplied by COPS. This, with respect to branch circuits, is a duplication of 708.12.

708.50. Accessibility. Only qualified persons are permitted access to overcurrent devices, a concept already essentially covered in 708.5(B).

708.52. Ground-Fault Protection of Equipment. This is a requirement for a second level of GFPE when a first level is installed to meet 230.95 or 215.10. It is similar to 517.17.

708.54. Selective Coordination. This correlates with 700.32 and 701.27. Refer to the discussion at 700.32 for more information on this controversial requirement. The same requirement for professional engineering support applies to these systems.

708.64. Emergency Operations Plan. A facility with a COPS must have an emergency operations plan, properly documented. NFPA 1600-2007 provides the guidance for developing and implementing emergency plans.

ARTICLE 710. STAND-ALONE SYSTEMS

710.1. Scope. This is the new home for stand-alone systems, which have inherently common features even though they have come up in several articles over the years, most notably former 690.10. Not all systems are utility interactive like the one in Fig. 690-1. Many such systems serve off-grid occupancies, and the NEC had to decide what to tell people wiring homes that were off-grid, but that might later be on-grid. The decision was to have all the branch-circuit wiring installed as if the occupancy were supplied with a normal service, including the panelboard. However, the supply arrangements at the panel would reflect the realities of a PV system. The feeder size from the off-grid source and the main overcurrent protection (to be located at the inverter) would be based in accordance with the size of the source output. The only limitation on the local source is that it be able to handle at least the highest single connected utilization equipment. The other general rule (710.6) is for all equipment to be listed or field labeled.

This raises the question of what to do with a 120/240-V panelboard, because inverters are usually single-voltage equipment, and in this case would usually be 120-V rated. Feeding only one of the ungrounded buses would waste half the capacity of the board. The decision here was to allow both buses to be connected to the same line. This is contingent on the rating of the inverter overcurrent device being less than the current rating of the neutral bus, there being no 240-V outlets connected, and there being no multiwire branch circuits connected. There must also be a warning sign against such circuit connections, because if they were made, the grounded conductor would carry up to double its expected load. Although PV systems are well known for overriding the hold-down rule for back-fed panel mains in 408.36(D) [See 705.12(B)(5)], this does not happen for stand-alone systems, as clarified in 710.15(E).

ARTICLE 712. DIRECT CURRENT MICROGRIDS

712.1. Scope. This article, new as of the 2017 NEC, and in reference to the definition, covers dc power distribution systems that are powered by at least two interconnected dc power sources and that actively supply dc loads (which can include dc-to-ac inverters that supply ac loads). These systems do not generally have a connection to an ac primary power supply such as a utility service entrance. However, some are equipped with dc-ac bidirectional converters or dc-ac inverters. These systems offer inherent energy efficiency, because many renewable energy sources are inherently, or easily configured as, dc sources, including wind turbines, fuel cells, and photovoltaic systems. Every time dc is changed to ac (or vice-versa) there are inherent losses in the conversion that are avoided with direct dc load utilization. LED lighting is a classic example of the potential use of dc distributions, especially since part of the process of making an LED luminaire turn on is the rectification of ac to dc, and a dc circuit eliminates that step.

These systems are ever increasing in size and number, driven by steady advances in energy storage systems (Art. 706), reliable methods for large-scale dc power transformation, and decreasing costs of PV as a dc power source. The substantiation cited such seemingly unlikely examples as an auto assembly plant in Michigan and a jail in California that is currently integrating power from four dc sources (PV, fuel cells, wind, and diesel generators).

712.2. Definitions. Most topics on this list are described within the context where they are discussed here. However, one definition will be key to understanding a good deal of what applies generally, "Reference-Grounded dc System." This is exactly the same as the "Functional Grounded PV System" that is discussed at great length in Art. 690, except it is not a photovoltaic system. However, many of these systems will be powered by exactly that sort of system. Understand that the two concepts are identical and completely compatible. Undoubtedly a common term will be agreed on by the end of the next Code cycle.

712.3. Other Articles. If a dc microgrid has the capacity to operate in parallel with a conventional utility source, it must do so in accordance with the provisions of Art. 705. Key to this type of arrangement will be some sort of, presumably listed, Island Interconnection Device (IID) that will serve as the gatekeeper. See also 705.170, which covers the ac counterpart, the MID.

712.10. Directory. This is a reciprocal labeling requirement, with all potentially primary dc sources included.

712.25. Identification of Circuit Conductors. The usual rules for dc conductors apply. However, there is a semantic conflict between Art. 690 and this article. This section refers to "ungrounded conductors" as following 210.5(C)(2). In many cases, one of the conductors in the circuit will be "reference grounded." It is not an ungrounded conductor subject to the red and black default color code in Art. 210. So now, suppose the negative (and therefore black) wire on a PV supplied microgrid is functionally grounded, and in accordance with 690.31(B)(1) it is, indeed, not colored white. When it arrives at the microgrid it is powering, now it might become a white wire. This will obviously need to be sorted out with the inspector. It should, in fact, remain a black wire. This is why 712.35 correctly insists that this wire be disconnected along with its red brother at any disconnecting means. It is not a grounded circuit conductor for these purposes. Note also that reidentification is permitted for 6 AWG and smaller wires; this is to assist retrofitting this technology into existing wiring.

Part III. Disconnecting Means. One of the most interesting and important sections here is 712.37, and its informational note. This section actually reaches beyond simple disconnect switches and covers control equipment as well. It turns out that there is a great deal of dc-rated equipment that is extremely polarity sensitive, and must be installed accordingly.

712.52. System Grounding. Part **(A)** requires grounding in accordance with 250.162. That is not particularly enlightening, because, the only system grounding requirement conferred by that reference is one for 300 volts and below, which must be grounded. However, it is far from clear whether a reference ground satisfies this rule, and that may rule out reference ground connections at lower voltages. Note that reference (or functional) grounding is essential for the ground fault detection requirement in 712.55 to proceed. The PV ground fault requirement in 690.41(B) functions through a ground reference that the literal text here compromises at lower voltages. Above 300 volts, there are no issues with 250.162 and reference/functional grounding is mandatory.

712.55. Ground-Fault Detection Equipment. This section builds on the long-standing success that has come from PV systems covered in 690.41(B). Any system operating over 60 volts must have this equipment. The actual grounding system must be identified and marked "in accordance with 250.167(C)." There may be a problem with that reference. 250.167(C) comes with an informational note pointing to the list of dc grounding systems in NFPA 70E. That list is reproduced in this book, and it does not include reference (or functional) grounding. This may be another topic of conversation with the inspector.

712.57. Arc-Fault Protection. This protection is required "where required elsewhere in this Code." The type of protection the input submitters had in mind is clear; it is the type required for PV systems by 690.11. It is unclear whether any other section in the NEC requires dc arc-fault protection. This rule is presumably a placeholder to drive market interest in an expansion of the applicability of the technology supporting 690.11.

Part V. Marking. The maximum available dc short-circuit current on the microgrid must be determined and field-marked at the dc source equipment, together with the date the calculation was made. The calculation is to be revisited if changes that would affect the result take place.

Part VI. Protection. For multiple sources of power (and remember, the definition insists that there be more than just one) there must be overcurrent protection from all of them. The contributions of all sources must be considered in determining the applicable available fault current and the required short-circuit current ratings of all equipment for which that is relevant, principally equipment intended to break current at fault levels.

ARTICLE 720. CIRCUITS AND EQUIPMENT OPERATING AT LESS THAN 50 VOLTS

720.1. Scope. This article covers low-voltage applications that are not power-limited circuits as defined in 725.2 and are not remote-control or signal circuits. Determination that Art. 720 applies to any circuit or equipment must be carefully made on the basis of the particular load being supplied and circuit conditions.

This article covers circuits operating at more than 30 V but not over 50 V—such as 32-V circuiting. Be aware that any low-voltage lighting system that specifically is "listed" as a "low-voltage lighting system" must comply with the requirements given in Art. 411. If, however, the low-voltage lighting in question is not listed as a complete system, then the requirements given here must be observed. Other installations that are exempted from the requirements given here are those covered by the articles indicated by 720.2.

This article looks very much the same today as it did in the 1930s, when it was introduced to deal with rural farms wired on 32-V battery systems charged with windmills. Since the advent of the Rural Electrification Administration (REA) in the later part of that decade, this article has become increasingly irrelevant.

720.4. Conductors. The minimum No. 12 conductor size, rather than No. 14 as permitted for standard power and light wiring, is aimed at the higher current required for a given wattage load at low voltage. For instance, at 32 V, the current corresponding to a given wattage is 3.6 times the current for the same wattage at 115 V. It should also be noted that for a given load in watts and a given size of wire and circuit length, the voltage drop in percentage is about 13 times as great at 32 V as at 115 V.

720.5. Lampholders. Where medium-base sockets are used, there is no good reason for using any but those having a 660-W rating. The ampere ratings of candelabra and intermediate-base sockets would permit the use of 25-W lamps at 32 V, but it is not considered safe to allow the installation of these low-wattage sockets on circuits operating at 50 V or less.

720.7. Receptacles Required. Receptacles must be rated 20 A where used in kitchens, laundries, and other locations where portable appliances are likely to be used. This is a good example of a section in this article that is unchanged since at least as far back as the 1937 NEC, and that largely dates the article.

720.11. Mechanical Execution of Work. This section essentially reiterates the requirement given in 110.12. Although, according to 90.3, all requirements given in Chaps. 1 through 4 generally apply to installations and equipment covered by Chaps. 5 through 7, the inclusion of this requirement here, clarifies that applicability of the general requirement given in 110.12 to the work covered by Art. 720. Cables must be supported in a way that ensures they will not be damaged by normal building use.

ARTICLE 725. CLASS 1, CLASS 2, AND CLASS 3 REMOTE-CONTROL, SIGNALING, AND POWER-LIMITED CIRCUITS

725.1. Scope. Article 725 of the Code covers power-limited circuits and remote-control and signal circuits. A signal circuit is defined as any electrical circuit that supplies energy to an appliance or device that gives a visual and/or audible signal. Such circuits include those for doorbells, buzzers, code-calling systems, signal lights, annunciators, fire or smoke detection, fire or burglar alarm, and other detection indication or alarm devices.

A "remote-control" circuit is any circuit that has as its load device the operating coil of a magnetic motor starter, a magnetic contactor, or a relay. Strictly speaking, it is a circuit that exercises control over one or more other circuits. And these other circuits controlled by the control circuit may themselves be control circuits or they may be "load" circuits—carrying utilization current to a lighting, heating, power, or signal device. Figure 725-1 clarifies the distinction between control circuits and load circuits.

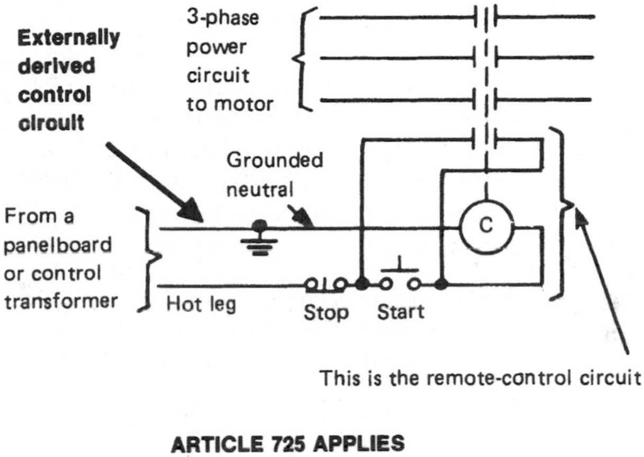

Externally derived control circuit
3-phase power circuit to motor
Grounded neutral
From a panelboard or control transformer
Hot leg
Stop
Start
C
This is the remote-control circuit
ARTICLE 725 APPLIES

Fig. 725-1. A control circuit governs the operating coil or some other element, to switch the load circuit. (Sec. 725.1.)

The elements of a control circuit include all the equipment and devices concerned with the function of the circuit: conductors, raceway, contactor operating coil, source of energy supply to the circuit, overcurrent protective devices, and all switching devices that govern energization of the operating coil. Typical control circuits include the operating-coil circuit of magnetic motor starters (NEC 430.71 and 430.72), magnetic contactors (as used for switching lighting, heating, and power loads), and relays. Control circuits include wiring between solid-state control devices as well as between magnetically actuated components. Low-voltage relay switching of lighting and power loads is also classified as remote-control wiring (Fig. 725-2).

Power-limited circuits are circuits used for functions other than signaling or remote control, but in which the source of the energy supply is limited in its power (volts times amps) to specified maximum levels. Low-voltage lighting, using 12-V lamps in fixtures fed from 120/12-V transformers, is a typical "power-limited circuit" application.

725.2. Definitions. The provisions of this section divide all signaling and remote-control systems into three classes.

Class 1 includes all signaling and remote-control systems that do not have the special current limitations of Class 2 and Class 3 systems.

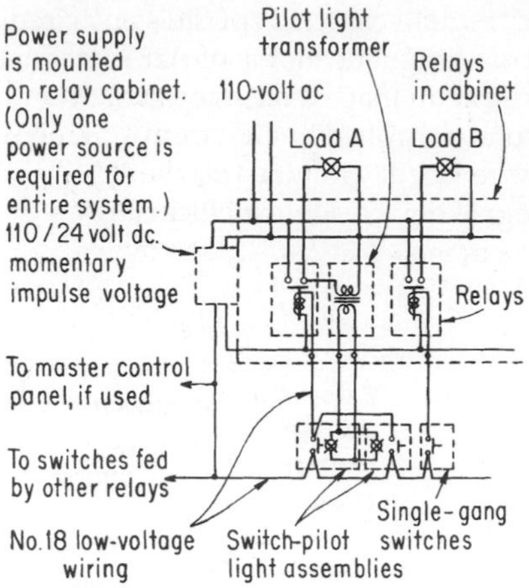
Power supply
is mounted
on relay cabinet.
(Only one
power source is
required for
entire system.)
110/24 volt dc.
momentary
impulse voltage
Pilot light
transformer
110-volt ac
Relays
in cabinet
Load A
Load B
Relays
To master control
panel, if used
To switches fed
by other relays
No.18 low-voltage
wiring
Switch-pilot
light assemblies
Single-gang
switches

Fig. 725-2. Low-voltage switching involves typical "remote-control circuits." (Sec. 725.1.)

Class 2 and Class 3 systems are those systems in which the current is limited to certain specified low values by fuses or CBs, and by supply through transformers that will deliver only very small currents on short circuit, or by other means that are considered satisfactory. All Class 2 and 3 circuits must have a power source with power-limiting characteristics as described in parts **(A)(1)** through **(4)** in 725.121. Class 2 circuits consider power limitation in terms of both shock and fire hazard; Class 3 circuits have higher power ratings and consider power levels largely in terms of fire initiation.

725.3. Other Articles. All the applications under this article must observe the specified sections that also rule on use of general power and light wiring.

Part **(B)** prohibits any installation of remote-control, signaling, or power-limited wiring in such a way that there is an appreciable reduction in the fire rating of floors, walls, or ceilings.

Part **(C)** requires that Class 1 circuits covered by this article must be run in metal raceway or metal cable assembly when used in an air-handling ceiling, as required by 300.22. Class 2 and/or 3 circuits (as defined in 725.2) are permitted to be used without a metal raceway or metal cable cover in ducts, plenums, or ceiling spaces used for environmental air *provided that the conductors are "listed" as Type CL2P or CL3P, as required in 725.154.* Because of the definition of the word "listed" (see Art. 100, "Definitions"), this rule would require that any such nonmetallic assembly of conductors for these circuits must be specifically described in the relevant UL product documentation (or with similar third-party certification) as having the specified characteristics for use without a metal raceway or covering in ducts, plenums, and air-handling ceilings.

Note: For any such application, check that the conductors of the circuit are definitely listed by UL or others.

Part **(F)** notes that Art. 725 does not apply to control circuits tapped from line terminals in motor starters. As described under 430.72, the control circuit for the operating coil in a magnetic motor starter where the coil voltage supply is tapped from the line terminals of the starter is regulated by the rules of 430.72 and not by the remote-control rules of Art. 725. Where a control circuit for the operating coil of one or more magnetic motor starters is derived from a separate control transformer, one that is not fed from a motor branch circuit, the control circuit(s) and all components are covered by the rules of Art. 725. In the same way, a control circuit that is taken from a panelboard for power supply to the operating coil of one or more motor starters is also covered by Art. 725 and not by the rules of 430.72. (See Fig. 725-1.)

Part **(G)** correlates control circuits wired under the provisions of Art. 727, which is a wiring method used for certain industrial control circuits, with this article.

Part **(H)** incorporates the rules in 300.7(A) for sealing raceway interiors that are subject to convective air migration, particularly where condensation is a likely result.

Part **(I)** brings in the vertical support rules for cables run in raceway, but only for the circuit-integrity (CI) cables. This makes sense because in a fire condition, these cables become extremely brittle and vertical supports can easily make a significant difference in how long they survive. These cables must be used in close agreement with the directions that come with them from their manufacturer in terms of installation methods, and these may be more restrictive than the limits in 300.19.

Part **(J)** requires a bushing in accordance with 300.15(C) where cables emerge from being sleeved in a raceway riser for physical protection. Note that the actual 300.15(C) rule requires a fitting to protect the cable from abrasion, which could be some sort of change-over fitting, and does not mention a bushing.

Part **(K)** brings in the requirements on 300.8 to divorce electrical raceways from any conveyance of nonelectrical systems. Part **(L)** brings in corrosion-resistance provisions in Chaps. 1 and 3 and applies them to Class 2 and Class 3 circuits.

Parts **(M)** and **(N)** bring in cable routing assemblies and communications raceways as suitable for use with Art. 725 cabling.

725.21. Access to Electrical Equipment Behind Panels Designed to Allow Access. It is contrary to the NEC to drape endless quantities of these cables across suspended ceiling panels in a way that limits the intended access to the ceiling cavity above them.

725.24. Mechanical Execution of Work. In addition to the usual neat and workmanlike manner requirement, this section also requires exposed cabling (which includes the area above a suspended ceiling) to be supported to structure such that normal building operations will not damage the cable. In addition, if the wiring is run parallel to framing, it must meet the 32 mm (1¼ in.) spacing requirement from the leading edge of the framing.

725.25. Abandoned Cables. The last sentence here has the effect of requiring that any "abandoned" conductors—as defined in 725.2 as not being terminated or identified for future use with a tag—must be removed. It is no longer permissible to leave abandoned Class 2 and Class 3 conductors in place when they are no longer in use, unless they have been "identified" for future use. See 725.2.

725.30. Class 1, Class 2, and Class 3 Circuit Identification. Circuits covered by this article must be identifiable at terminations so as to prevent inadvertent interference with other systems during service and testing.

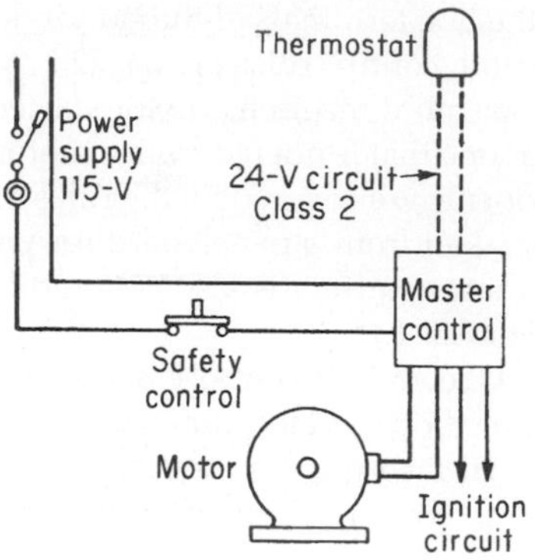
Thermostat
Power
supply
115-V
24-V circuit
Class 2
Master
control
Safety
control
Motor
Ignition
circuit

Fig. 725-3. Remote-control circuit *must* be Class 1 if failure would create a hazard. (Sec. 725.31.)

725.31. Safety-Control Equipment. The application of this rule is illustrated by Fig. 725-3, which is a simplified diagram of a common type of automatic control for a domestic oil burner. Assuming a steam boiler, the safety control is a switch that opens automatically when the steam pressure reaches a predetermined value and, preferably, also opens if the water level is allowed to fall too low. The master control includes a transformer of the current-limiting type that supplies the thermostat circuit at a voltage of 24 V. When the thermostat contacts close, a relay closes the circuits to the motor and to the ignition transformer.

Failure of the safety control or ignition to operate would introduce a direct hazard; hence, the circuits to this equipment are Class 1. Part (B) gives the permitted wiring methods to be used if the safety-control rule applies and the control circuit is reclassified as Class 1.

The thermostat circuit fulfills all requirements of a Class 2 circuit and can be short-circuited or broken without introducing any hazard. The wiring of this circuit can therefore be done with any type of wire or cable that is sufficiently protected from physical damage to ensure serviceability.

725.41. Class 1 Circuit Classifications and Power Source Requirements. Class 1 systems may operate at any voltage not exceeding 600 V. They are, in many cases, merely extensions of light and power systems, and, with a few exceptions, are subject to all the installation rules for light and power systems.

Part **(A)** requires that Class 1 power-limited circuits must have energy limitation on the power source that supplies them. And such circuits may be supplied from either a transformer or another type of power supply—such as a generator, batteries, or manufactured power supply. Note that a Class 1 power-limited circuit must be supplied at *not over* 30 V, 1000 VA. There are other sources for these circuits, as presented in the second numbered topic. These rules set the power limitations for these alternate sources.

Part **(B),** however, permits Class 1 remote-control or signaling circuits to operate at up to 600 V, and no limitation is placed on the power rating of the source to such circuits (Fig. 725-4).

The most common example of a Class 1 remote-control system is the circuit wiring and devices used for the operation of a magnetically operated motor controller. The term *remote-control switch* is used in various Code references to

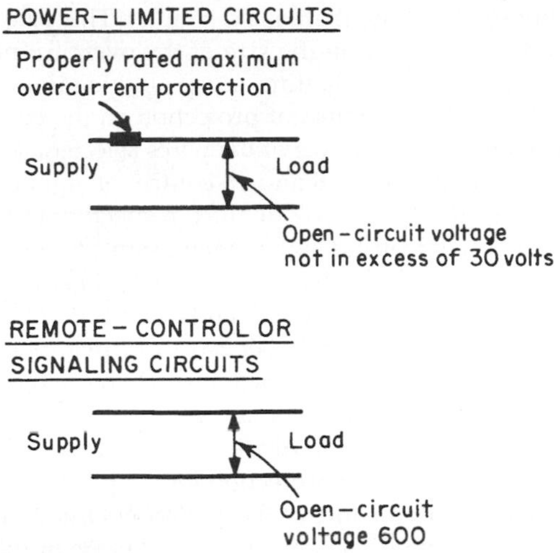
POWER-LIMITED CIRCUITS
Properly rated maximum overcurrent protection
Supply
Load
Open-circuit voltage not in excess of 30 volts
REMOTE-CONTROL OR SIGNALING CIRCUITS
Supply
Load
Open-circuit voltage 600

Fig. 725-4. Class 1 circuits are divided into two maximum voltages. (Sec. 725.41.)

designate a switch or contactor used for the remote control of a feeder or branch circuit, with the operating-coil circuit as a Class 1 remote-control circuit.

The signaling systems that are included in Class 1 operate at 120 V with 20-A overcurrent protection, although they are not necessarily limited to this voltage and current. Some of the signaling systems that may be so operated include electric clocks, bank alarm systems, and factory call systems. An example of a lower-voltage Class 1 signaling system is a nurses' call system, as used in hospitals. Such systems commonly operate at not over 25 V.

Most control circuits for magnetic starters and contactors could not qualify as Class 2 or Class 3 circuits because of the relatively high energy required for operating coils. And any control circuit rated over 150 V (such as 220- or 440-V coil circuits) can never qualify, regardless of energy.

Class 1 control circuits include all operating-coil circuits for magnetic starters or contactors that do not meet the requirements for Class 2 or 3 circuits. Class 1 circuits must be wired in accordance with the requirements of Part II of Art. 725.

725.43 and 725.45. Class 1 Circuit Overcurrent Protection and Location. In general, conductors for any Class 1 remote-control, signaling, or power-limited circuit must be protected against overcurrent. No. 14 AWG and larger wires must generally be protected at their ampacities from **NEC** Table 310.15(B)(16). In the second clause of this rule, an important statement indicates that it is *not* necessary to take any ampere derating—for either elevated ambient temperature or for more than three wires in a conduit or cable. The wires may simply be used and protected at the ampacity values given in the table.

Important: The clause "without applying the derating factors of 310.15 to the ampacity calculations" is a strange permission, which can be disregarded without violating this or any Code rule. And there seems to be a conflict between this clause and the requirement in 725.51 that Class I conductors must be derated under some conditions. Actually, there is no conflict. This rule says derating factors need not be applied in figuring the size of the overcurrent protection; however, they must be applied in evaluating acceptable loading on those wires, as covered in 725.51. Figure the overcurrent protection on the table numbers, but be sure to factor any derating into the size of the wires selected.

It is important to note that Nos. 18 and 16 control or signal-circuit conductors must always be protected at not over 7 or 10 A, respectively. These smaller sizes of wire may be used for control and signal circuits supplying coils, relays, or signal devices. Note that these are OCPD limits; the actual load must not exceed the limits in Table 402.5 (6 A and 8 A, respectively) as provided in 725.49(A).

The rule of 725.45(D) is the same as that of 240.3(F), but applies to the case where the 2-wire transformer secondary supplies a control circuit to one or more operating coils in motor starters or magnetic contactors. A properly sized circuit breaker or fuses may be used at the supply end of the circuit that feeds the transformer primary and may provide overcurrent protection for the primary conductors, for the transformer itself, and for conductors of the control circuit, which is run from the transformer secondary to supply power to motor starters or other control equipment, as follows:

1. The primary-side protection must not be rated greater than that required by 450.3(B)(1) for transformers rated up to 600 V. For a transformer rated 9 A or more, the rating of the primary CB or fuses must not be greater than 125 percent of (1.25 times) the rated transformer primary current. And if 1.25 times rated primary current does not yield a value exactly the same as a standard rating of fuse or CB, the next-higher-rated standard protective device may be used. Where the transformer-rated primary current is less than 9 A—as it would be for all the usual control transformers rated 5000 VA and stepping 480 V down to 120 V—the maximum permitted rating of primary protection must not exceed 167 percent of (1.67 times) the rated primary current. For a transformer with a primary rated less than 2 A, the primary protection must never exceed 300 percent of (3 times) the rated primary current. With most control transformers, with primary ratings well below 10 A, fuse protection will be required on the primary because the smallest standard CB rating is 15 A, and that will generally exceed the maximum values of primary protection permitted by 450.3(B)(1). (See Fig. 725-5.)
2. Primary protection must not exceed the amp rating of the primary circuit conductors. And when protection is sized for the transformer, as previously described, 14 AWG copper primary conductors will be protected well within their 15-A rating.
3. Secondary conductors for the control circuit can then be selected to have an ampacity at least equal to the rating of primary protection times the primary-to-secondary transformer voltage ratio. Of course, larger conductors may be used if needed to keep voltage drop within limits.

In Fig. 725-6, covering use of a magnetic contactor, 14 AWG and larger remote-control conductors may be properly protected by the feeder or branch-circuit overcurrent devices if the devices are rated or set at not more than 300 percent of (3 times) the ampacity of the control conductors. If the branch-circuit overcurrent devices were rated or set at more than 300 percent of the rating of the

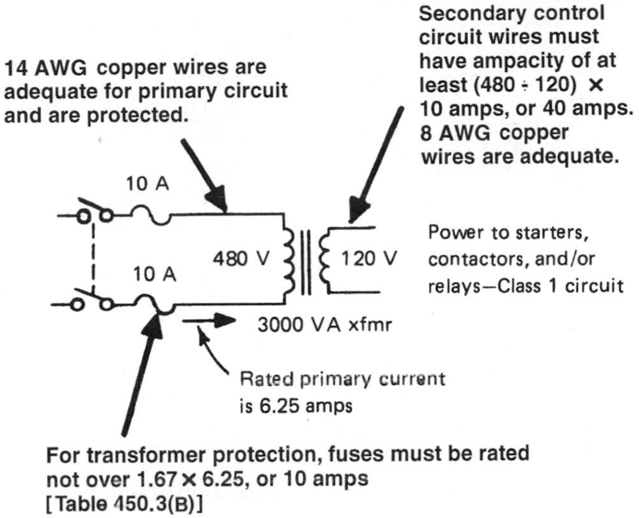

14 AWG copper wires are adequate for primary circuit and are protected.
Secondary control circuit wires must have ampacity of at least (480 ÷ 120) × 10 amps, or 40 amps. 8 AWG copper wires are adequate.
10 A
10 A
480 V
120 V
Power to starters, contactors, and/or relays—Class 1 circuit
3000 VA xfmr
Rated primary current is 6.25 amps
For transformer protection, fuses must be rated not over 1.67 × 6.25, or 10 amps [Table 450.3(B)]

Fig. 725-5. A separate control transformer supplying a number of coil circuits for motor starters or magnetic contactors must have primary protection that protects the secondary control conductors as well as the transformer. (Sec. 725.43.)

control conductors, the control conductors would have to be protected by separate protective devices located within the contactor enclosure at the point where the conductor to be protected receives its supply.

This is covered by 725.45(C), which applies to the remote-control circuit that energizes the operating coil of a magnetic contactor, as distinguished from a magnetic motor starter. Although it is true that a magnetic starter is a magnetic contactor with the addition of running overload relays, part **(C)** covers only the coil circuit of a magnetic contactor. That applies to control wires for magnetic contactors used for control of lighting or heating loads, but not motor loads. 430.72 covers that requirement for motor-control circuits.

In Fig. 725-6, for instance, 45-A fuses at A in the feeder or branch circuit ahead of the contactor would be adequate protection if 14 AWG wire, with its ampacity of 15 A [reduced from 20 A in Table 310.15(B)(16) for 75°C and 90°C conductors by 240.4(D)(3)], were used for the remote-control circuit, because 45 A is not more than 300 percent of (3 times) the 15-A ampacity. Larger fuses ahead of the contactor would require overcurrent protection in the hot leg of the control circuit, at B, rated not over 15 A. [See 240.4(D).] Similar calculations apply to electronic sources, as covered in **(E)**.

Note that the overcurrent protection is required for the control conductors and not for the operating coil. Because of this, the size of control conductors can be selected to allow application without separate overcurrent protection. When overcurrent protection is added in the enclosure, its rating must be such that it conforms to the first paragraph of this rule.

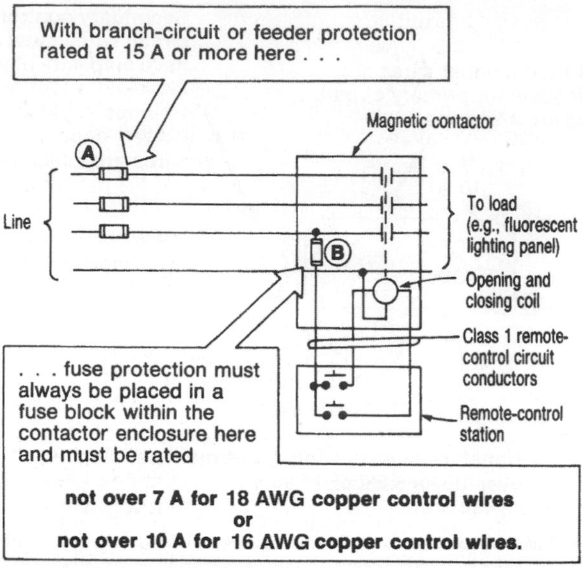
With branch-circuit or feeder protection rated at 15 A or more here . . .
Magnetic contactor
A
Line
B
To load (e.g., fluorescent lighting panel)
Opening and closing coil
Class 1 remote-control circuit conductors
Remote-control station
. . . fuse protection must always be placed in a fuse block within the contactor enclosure here and must be rated
not over 7 A for 18 AWG copper control wires
or
not over 10 A for 16 AWG copper control wires.

NOTE: If 14 AWG or larger control-circuit wires are used, 725.45(C) permits omission of separate protection in the control circuit when the rating of the branch-circuit or feeder protection does not exceed three times the ampacity of the particular size of control wire from Table 310.15(B)(16).

EXAMPLE: 30-A fuses at "A" would be adequate protection if 14 AWG wire, rated at 15 A, is used for the remote-control circuit, because 30 A is *less than* 3 × 15 A. If fuses at "A" were rated over 45 A, 15-A protection would be required at "B" for 14 AWG wire.

Fig. 725-6. Protection of coil circuit of a magnetic contactor is similar to that of a starter. (Sec. 725.43 and 725.45.)

725.46. Class 1 Circuit Wiring Methods. In general, wiring of Class 1 signal systems must be the same as power and light wiring, using any of the cable or raceway wiring methods that are Code-recognized for general-purpose wiring. The two exceptions refer to the details of wiring permitted by 725.48, 725.49, and 725.51.

725.48. Conductors of Different Circuits in Same Cable, Cable Tray, Enclosure, or Raceway. Any number and any type of Class 1 circuit conductors—for remote control, for signaling, and/or for power-limited circuits—may be installed in the same conduit, raceway, box, or other enclosure *if* all conductors are insulated for the maximum voltage at which any of the conductors operates.

Class 1 circuit wires (starter coil-circuit wires, signal wires, and power-limited circuits) may be run in raceways by themselves in accordance with the first sentence of this section. A given conduit, for instance, may carry one or several sets of Class 1 circuit wires. And 725.46 says use of Class 1 wires must conform to the same basic rules from **NEC** Chap. 3 that apply to standard power and light wiring.

But note that two specific sections of the **NEC** cover the use of Class 1 circuit conductors in the same raceway, cable, or enclosure containing circuit wires carrying power to a lighting load, a heating load, or a motor load (Fig. 725-7). 300.3 covers

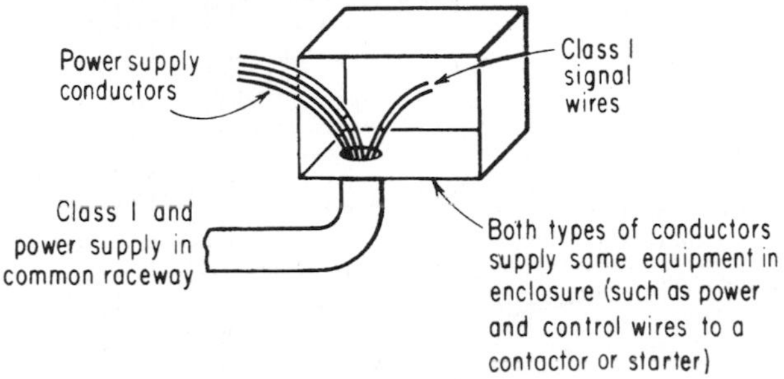
Power supply
conductors
Class I
signal
wires
Class I and
power supply in
common raceway
Both types of conductors
supply same equipment in
enclosure (such as power
and control wires to a
contactor or starter)

Fig. 725-7. This is permitted by 300.3(C) and 725.48. (Sec. 725.48.)

the general use of "conductors of different systems" in raceways as well as in cable assemblies and in equipment wiring enclosures (i.e., cabinets, housings, starter enclosures, junction boxes). But Class 1 circuit wires are also regulated by this section, which strictly limits use of Class 1 wires in the same box and/or raceway with power wires. Figure 725-8 shows a clear violation, if the annunciator has no relationship to the motor load.

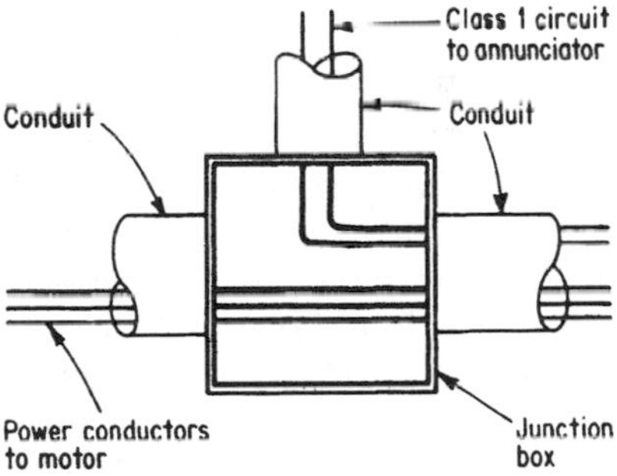
Class 1 circuit
to annunciator
Conduit
Conduit
Power conductors
to motor
Junction
box

Fig. 725-8. Class 1 wires must generally *not* be used in raceways with "unrelated" wires. (Sec. 725.48.)

Note that 725.48(B)(1) permits Class 1 circuit wires to be installed in the same raceway or enclosure as "power supply" conductors *only* if the Class 1 wires and the power wires are "functionally associated" with each other. That would be the case where the power conductors to a motor are run in the same conduit along with the Class 1 circuit wires of the magnetic motor starter used to control or to start or stop the motor. Refer to the commentary in 300.3.

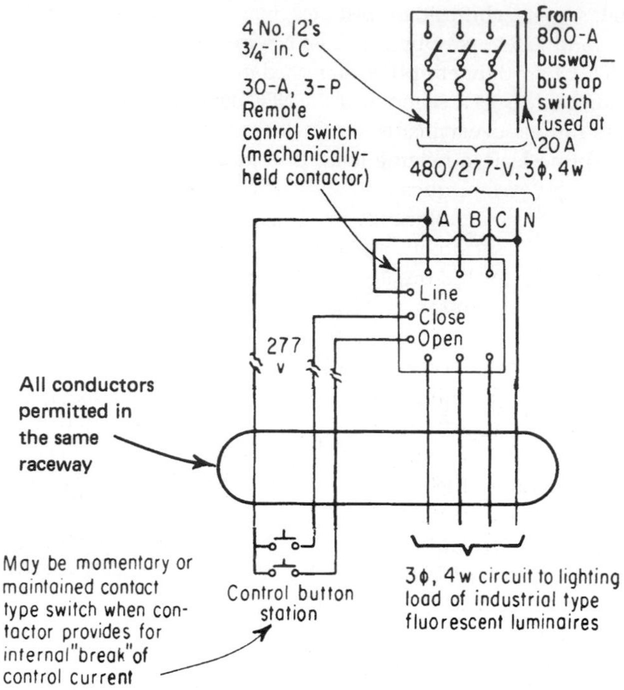
4 No. 12's
3/4- in. C
30-A, 3-P
Remote
control switch
(mechanically-
held contactor)
From
800-A
busway—
bus tap
switch
fused at
20 A
480/277-V, 3ϕ, 4w
A
B
C
N
Line
Close
Open
277
V
All conductors
permitted in
the same
raceway
May be momentary or
maintained contact
type switch when con-
tactor provides for
internal "break" of
control current
Control button
station
3ϕ, 4 w circuit to lighting
load of industrial type
fluorescent luminaires

Fig. 725-9. Class 1 wires and power wires may be used in same raceway for "functionally associated" equipment. (Sec. 725.48.)

The same permission would apply to the hookup of a magnetic contactor controlling a lighting or heating load, as shown in Fig. 725-9. There the circuit wires for the Class 1 remote-control run to the pushbutton station may be run in the same conduit carrying the wires supplying the lighting fixtures. A typical application would have the magnetic contactor adjacent to a panelboard, with the control and power wires run in the same raceway to a box at some point where it is convenient to bring the control wires down to the control switch and carry the power wires to the lighting fixtures being controlled. The contactor can be located at the approximate center of its lighting load to keep circuit wiring as short as possible for minimum voltage drop, and the control wires are then carried to one or more control points. In such a layout, the control and power wires are definitely "functionally associated" because the control wires provide the ON-OFF function for the lighting. *But* other control or power wires are prohibited from being in the same conduit, boxes, or enclosures with the single set of associated Class 1 and power wires.

There is another issue, for which the NEC offers no clear answer. First, some background; if you have two unrelated motors at the same general location, it is beyond question that the two sets of branch-circuit conductors to the two motors can run in a common raceway. Now, add a set of control conductors for each motor to the same conduit. Are these control conductors functionally associated with the power conductors? Obviously. However, is every control conductor functionally associated with every power conductor? Obviously not. Is this, therefore, a violation of 725.48(B)(1)? The words of the NEC say "functionally associated" and leave it at that, so every inspector gets to make a call here. The answer cannot be determined based on the literal text of the NEC.

This is a very important question, so it is worth exploring the technical merits of this question. The usual objections to this practice run to the undesirability of exposing unrelated control conductors to a fault and thereby disabling multiple motor functions in unrelated processes. Although that is a reasonable design argument, there are serious limitations to this argument as a matter of NEC minimum standards. The power and control wiring that goes to those two motors can originate in the same vertical motor control center section, as covered in 725.48(B)(2). There are no limits on running the two power circuits together to the motors, as already covered. And a fault in one of those motor circuits will certainly disable the other motor, yet that is clearly allowed. Why then object to the control conductors? There is no supportable argument that multiple functions in a single raceway (other than very rare exceptions as with fire pump and emergency circuits) rise to the level of a fire or electrocution hazard, which, as covered in 90.1, is and ought to be the controlling principle. Nevertheless, some inspectors will turn it down, so be sure to ask before committing yourself to a particular circuit design.

Note: Part **(B)(2)** in 725.48 permits power and control wires for more than one motor in a common raceway. Factory- or field-assembled control centers may group power and Class 1 control conductors that are not functionally associated. This rule recognizes the use of listed motor control centers that have power and control wiring in the same wireway or gutter space supplying motors that are *not* "functionally associated." The basic rule generally prohibits that condition when hooking up motor circuits.

Part **(B)(3)** of 725.48 says that Class 1 circuit conductors and unassociated power-supply conductors are permitted in a manhole if either of them is in metal-enclosed cable or Type UF cable *or* if effective separation is provided between the Class 1 conductors and the power conductors. This rule covers the conditions under which Class 1 conductors and unrelated power conductors may be used in the same enclosure (a manhole).

Part **(B)(4)** covers the same issues in cable tray. Item (1) is a functional association rule that raises the same issue as covered in (B)(1). Item (2), which applies in the event there is no functional association as illustrated in Fig. 725-8, avoids a barrier if all the wiring is in separate multiconductor cables, including Type TC. This is an improvement over prior versions of this rule that asked for, exclusively, cables with metal armor.

725.49. Class 1 Circuit Conductors. Figure 725-10 shows this basic rule of part **(A)**, which accepts use of building wire to a minimum 14 AWG size. But 16 or 18 AWG fixture wires of the types specified in part **(B)** *may* be used for running starter coil circuits, signal circuits, and any other Class 1 circuits. Of course, use of 16 or 18 AWG fixture wire for Class 1 circuits depends on such conductors having sufficient ampacity for the current drawn by the contactor or relay operating coil or by whatever control device is involved. Wires larger than 16 AWG must be building

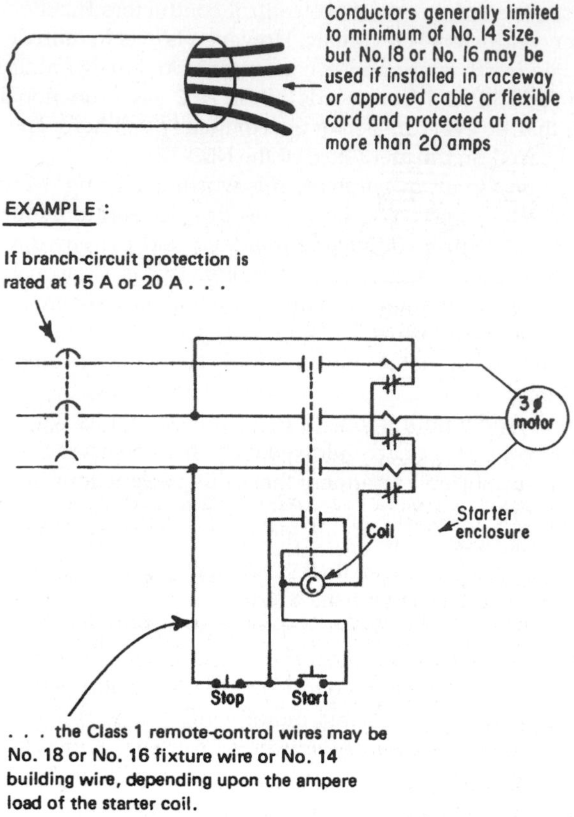
Conductors generally limited to minimum of No. 14 size, but No. 18 or No. 16 may be used if installed in raceway or approved cable or flexible cord and protected at not more than 20 amps
EXAMPLE :
If branch-circuit protection is rated at 15 A or 20 A . . .
3ø motor
Starter enclosure
Coil
C
Stop
Start
. . . the Class 1 remote-control wires may be No. 18 or No. 16 fixture wire or No. 14 building wire, depending upon the ampere load of the starter coil.

Fig. 725-10. 16 or 18 AWG fixture wire may be used for Class 1 circuits. (Sec. 725.49.)

types (TW, THW, THHN, etc.). Class 1 circuits may not use fixture wires larger than 16 AWG. And ampacity of any Class 1 circuit wires larger than 16 AWG must have that value shown in Table 310.15(B)(16).

Note: 402.5 shows that the ampacity of any 18 AWG fixture wire is 6 A and the ampacity of any 16 AWG fixture wire is 8 A. Any 18 or 16 AWG fixture wire or 14 AWG building wire used for a Class 1 circuit is considered adequately protected by a fuse or CB rated not over 20 A. See 725.43 and 240.4.

725.51. Number of Conductors in Cable Trays and Raceway, and Derating. The number of Class 1 remote-control, signal, and/or power-limited circuit conductors in a conduit must be determined from Tables 1 through 5 in Chap. 9 of the NEC, or Annex C, where all conductors are of the same size.

When more than three Class 1 circuit conductors are used in a raceway, ampacity derating of 310.15(B)(3)(a) applies only if the conductors carry continuous loads in excess of 10 percent of the ampacity of each conductor (Fig. 725-11). This rule is aimed at relieving the need to derate conductors that are usually carrying

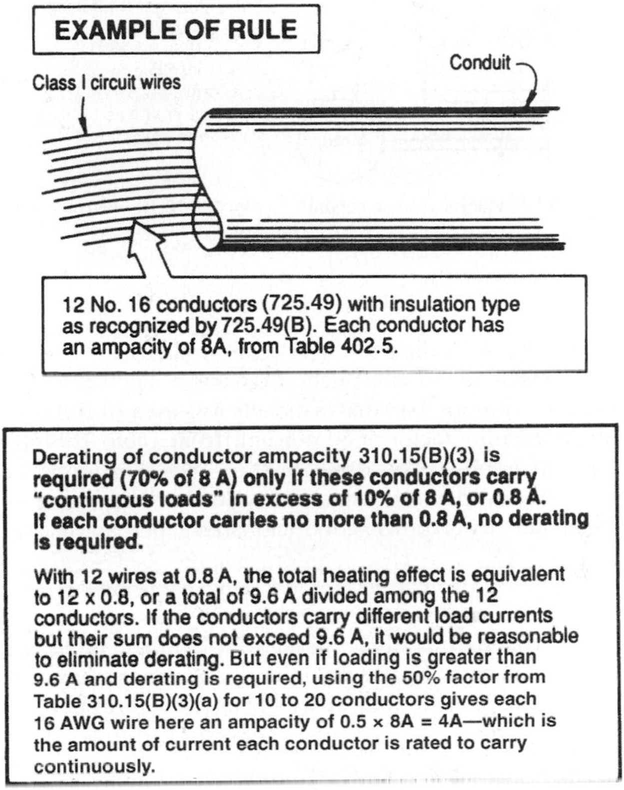
EXAMPLE OF RULE
Conduit
Class I circuit wires
12 No. 16 conductors (725.49) with insulation type as recognized by 725.49(B). Each conductor has an ampacity of 8A, from Table 402.5.
Derating of conductor ampacity 310.15(B)(3) is required (70% of 8 A) only if these conductors carry "continuous loads" in excess of 10% of 8 A, or 0.8 A. If each conductor carries no more than 0.8 A, no derating is required.
With 12 wires at 0.8 A, the total heating effect is equivalent to 12 x 0.8, or a total of 9.6 A divided among the 12 conductors. If the conductors carry different load currents but their sum does not exceed 9.6 A, it would be reasonable to eliminate derating. But even if loading is greater than 9.6 A and derating is required, using the 50% factor from Table 310.15(B)(3)(a) for 10 to 20 conductors gives each 16 AWG wire here an ampacity of 0.5 × 8A = 4A—which is the amount of current each conductor is rated to carry continuously.

Fig. 725-11. Determining conductor ampacity for more than three Class 1 circuit wires in a raceway requires careful evaluation of conditions and the Code rule. (Sec. 725.51.)

very low values of current (such as up to 2 A for most coil circuits of contactors and motor starters). The wording does not spell out whether all the conductors are carrying continuous current or if only some of them are.

The same concept of "10 percent of the ampacity of each conductor" has been applied in part **(B)(1)** and **(2)** of 725.51. This number was chosen because based on the I^2R relationship, a wire carrying 10 percent of its ampacity is putting out 1 percent of the heat it would be putting out at full load, a number low enough to comfortably ignore.

When power conductors and Class 1 circuit conductors are used in a single conduit or EMT run (as permitted by 725.48), the derating factors of Table 310.15(B)(3)(a) must be applied as follows:

1. Table 310.15(B)(3)(a) must be applied to all conductors in the conduit when the remote-control conductors carry continuous loads in excess of 10 percent of each conductor's ampacity and the total number of conductors (remote-control and power wires) is more than three. For example, in Fig. 725-12, the conduit size

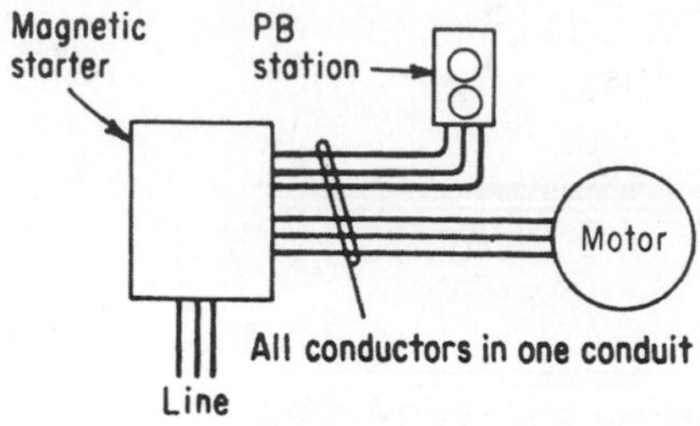
Magnetic starter
PB station
Motor
All conductors in one conduit
Line

Fig. 725-12. Derating of conductor ampacity is usually not required for this circuit makeup. (Sec. 725.51.)

must be selected according to the number and sizes of the wires. Because two of the control wires to the pushbutton and the power wires to the motor will carry a continuous load that is usually less than 10 percent of conductor ampacity, a derating factor of 80 percent [from Table 310.15(B)(3)(a)] does not have to be applied. This drawing shows the most important impact of the 10 percent rule, namely, where the control circuit wiring would impose a significant penalty on power conductor sizing, perhaps changing wire and conduit sizes.

2. Table 310.15(B)(3)(a) must be applied only to the power wires when the remote-control wires do not carry continuous load and when the number of power wires is more than three. In Fig. 725-13, no derating at all is applied because the control wires do not carry continuous current (only for the instant of switching operation), and there are only three power wires.

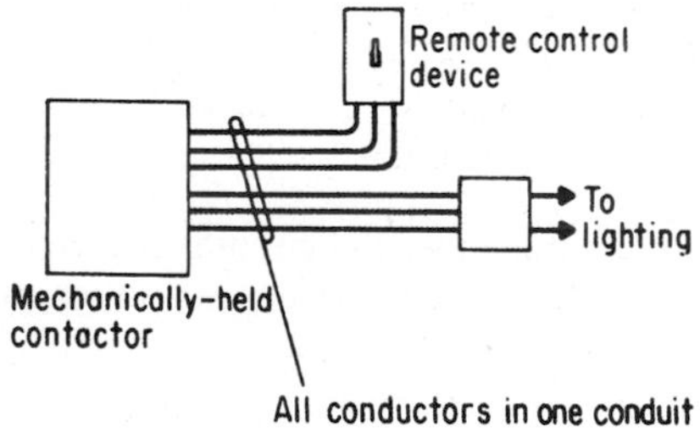
Remote control device
To lighting
Mechanically-held contactor
All conductors in one conduit

Fig. 725-13. Derating is not required here if Class 1 conductors do not carry continuous load. (Sec. 725.51.)

These rules of part **(B)** have created controversy. It usually starts with this question: If a conduit from a starter carries the three power wires of a motor circuit and also contains three control wires run from the starter to a pushbutton station, is it necessary to derate any of the conductor ampacities?

Answer: 725.51(B) covers this. (Read that rule several times.) If the starter is the usual magnetically held type of contactor, the two control wires to the STOP button at the pushbutton station will carry the holding current to the coil as long as the starter is closed. 725.51(B)(1) says that all conductors in the raceway must be derated in ampacity if the total number of conductors (power wires plus control wires) is more than three—*but only if the Class 1 circuit conductors carry continuous loads* of more than 10 percent of conductor ampacities. The continuous control current in this case is the sealed current for the contactor, likely a fraction of an ampere. As noted in the explanation of where the 10 percent rule came from, the effect on the adjacent conductors is negligible.

725.52. Circuits Extending Beyond One Building. Class 1 circuits that extend aerially are subject to the requirements in Art. 225.

725.121. Power Sources for Class 2 and Class 3 Circuits. Specific data regarding maximum current and voltage ratings for Class 2 and 3 circuits are given in Tables 11(A) and 11(B) in Chap. 9. These tables were removed from this part of the Code and sent to Chap. 9 in order to avoid the frequent misinterpretation that Class 2 and Class 3 power supplies can be manufactured in the field, given appropriate fuses and transformers, etc. This is not and has not been true for many decades, if ever. These circuits normally require a power supply that has been investigated by a qualified testing laboratory. As far as installation is concerned, the marking on any listed piece of equipment will be the determining factor as to how a power source is classified. And the installation must satisfy all rules related to that classification. Note that Class 2 and Class 3 power supplies must not have their outputs paralleled or interconnected, unless listed for this application. To do so would likely exceed the allowable power limitations for the circuits, creating a hazard.

There are some exceptions to the requirement for listed power sources. Thermocouples are a Class 2 source without a specific marking, as are limited power circuits from listed equipment with energy levels as in Tables 11(A) and 11(B); this is explained in an extensive note that gives examples of circuit cards in listed assemblies, and listed computer equipment circuits such as the Ethernet cable leaving desktop #1 and talking to desktop #2. Also, No. 6 carbon-zinc dry cells, 1½ V each, can be wired in series up to 30 V and still be considered a Class 2 source.

With regard to the computer wiring allowance, as of the 2017 NEC, this category has been expanded by inserting "audio/video" ahead of "information technology (computer)," and also adding "communications, and industrial" after it, so it now reads "Listed audio/video information technology (computer), communications, and industrial equipment limited power circuits." Unfortunately a comma was omitted after the word "video," which literally makes "audio/video" a modifier of the phrase "information technology (computer)." The intent was to address the current title of the applicable UL standard (referenced in the panel statement), which does have a comma after audio/video. The clear intention was to cover audio/video equipment as an expansion of the reach of this provision, and not a reduction of the types of computers that (4) covers.

725.124. Circuit Marking. Equipment supplying these circuits must be plainly marked as to the classification of the circuits supplied.

725.127. Wiring Methods on Supply Side of the Class 2 or Class 3 Power Source. Conductors and equipment on the line side of devices supplying Class 2 systems must conform to rules for general power and light wiring. Transformers that are Class 2 or Class 3 sources must be protected by supply side overcurrent protection not over 20 A; if fed with pigtails, however, the supply leads can be as small as 18 AWG if fully insulated and not over 305 mm (12 in.) long.

725.130. Wiring Methods and Materials on Load Side of the Class 2 or Class 3 Power Source. There is a decision point here, namely, whether or not to use limited energy wiring methods. Part **(A)** covers the option to use Class 1 wiring methods and materials, as covered in 725.46. Two exceptions immediately bear on this practice. The first waives all mutual conductor heating derating penalties. The second, although written as a permissive exception is actually a mandatory part of the reclassification decision, because the reclassification is effectively conditioned on the obliteration of the Class 2 and/or Class 3 markings as covered in 725.124. Note also that the exception correctly mandates Class 1 "wiring methods and materials." The wiring method is only the first issue; equal attention must be paid to the equipment on which the wiring is terminated. A Class 2 thermostat circuit run in EMT will satisfy the wiring method rule, but when it arrives at the thermostat there will be a problem unless it is rated for a Class 1 wiring connection. Further, once this is done, the circuit is forever Class 1 even though the source will have been listed as a Class 2 (or 3) source.

This procedure is widely used in industrial wiring, where all the wiring is Class 1 to begin with, and the reclassification allows these control circuits, usually run with 14 or 12 AWG 600 V THHN, to run with other control wiring with which it is functionally associated. The normal circuit separation rules for power-limited wiring in 725.136 no longer apply. The overcurrent protection rules in 725.43 admittedly do apply. Nevertheless, the majority of industrial control panels with wiring duct between relays and other control components employ large numbers of Class 1 circuits, both native and manufactured under this rule and all mixed together and terminating on the same control equipment; it would be nearly impossible to do this kind of wiring otherwise.

Some equipment manufacturers deliberately avoid having their components evaluated as Class 2 or 3 for similar reasons. For example, a major producer of automated gasoline station dispensing equipment has its circuits routinely evaluated with Class 1 spacings just so its equipment can be retrofitted in older stations that only have a single conduit going out to the dispenser.

Part **(B)** requires that conductors be installed in accordance with 725.133 and 725.154, using insulation recognized in 725.179. There are three exceptions. The first (unclear because not written as a complete sentence as required in the Style Manual) apparently removes all limitations on this wiring when it is run in the various elevator wiring methods detailed in 620.21. The second exception is apparently intended to recognize equivalent wiring that complies with 725.3, including Chap. 3 methods where the installer does not wish to reclassify the circuit, but only to extend or replace something existing. The third exception recognizes small bare wires as part of a listed intrusion-protection system.

725.135. Installation of Class 2, Class 3, and PLTC Cables. When power-limited cabling began to be subject to fire and smoke performance criteria in the 1987 NEC, it was extremely controversial, because the cables were never the source of a fire or an electric shock. It raised serious questions about whether this was appropriate in terms of the scope and history of the NEC. In the end, actual loss experience persuaded the NEC Committee to add these criteria, and with them, logically, coverage of optical fiber cables, many of which have no metallic parts whatsoever. The loss experience that provoked this change in policy was a severe fire in a telephone company facility where the major issue in the severity of the fire was the fuel load represented in the enormous quantity of cables present. In effect, the NEC Committee decided that the mere existence of certain types of cable jackets and conductor insulation was a fire hazard, thereby meeting the implicit test in 90.1. This same thought process also supports the recent requirements to remove abandoned cables.

A hierarchy of cable insulation was established, with essentially four levels. The low-smoke, low- flame type suitable for plenum cavity ceilings ("P" suffix designation) was at the top. The next step down was a cable that would not carry fire from floor to floor, designated a riser ("R" suffix designation) cable. The next step down was a cable suitable for general commercial use (no suffix), and the bottom of the totem pole went to a cable, constructed much like cables before the rule changes went forward and suitable for exposed areas in individual dwelling units ("X" suffix designation). Each level in the hierarchy included in the testing all lower-level tests, so any cable listed for an upper level could substitute for one in a lower level, and within any level, a cable tested for a higher voltage could substitute for a lower-voltage application. This general approach still holds today, although there are many complications.

This section is a new method of presenting this information, with the various classifications of installation circumstance listed in lettered subsections, roughly in order of decreasing levels of resistance to flame and smoke. So this list begins with actual ductwork (no suffix, only special rules) and then plenum installations (suffix "P"), then riser installations (suffix "R"), then a catch-all of intermediate industrial/commercial installations, then general building installations (no suffix), and finally residential applications (suffix "X" mixed, and then all "X"). In theory this approach supposedly separates installation requirements from application requirements, which remain in 725.154. However, that appears to be a distinction without a difference, because the same information largely appears in both locations, only formatted differently. For example, riser cables in vertical runs are now covered in 725.135(D), and at that location (1) cover the eligible naked cables, and (2) cover allowable uses in various raceways. Moving over to Table 725.154 in the horizontal block "In risers" one reads "In vertical runs" and/or then the sub-row for the raceway of interest to read off where what cable can be used. Whether this has been a useful exercise is far from proven.

725.136. Separation from Electric Light, Power, Class 1, Non-Power-Limited Fire Alarm Circuit Conductors, and Medium-Power Network-Powered Broadband Communications Cables. Part **(A)** of 725.136 says that Class 2 or 3 conductors must not be used in any raceway, compartment, outlet box, or similar fitting with light and power conductors or with Class 1 signal or control conductors, unless the conductors of the different systems are separated by a partition. This section sets out the separation rules that cope with one of the most fundamental requirements for installing power-limited cables, namely, that it is not permitted to establish the required system separations between power-limited wiring and non-power-limited wiring through the use of insulation alone. Some additional affirmative step must be taken, and this section provides different ways to meet this requirement.

Part **(B)** allows for a common enclosure where the conductors of different systems are separated through the use of a barrier.

Part **(C)** allows a raceway to run within an enclosure, where the raceway wall is, in effect, a barrier. This approach is often used when a power circuit must enter at an inconvenient point and it is necessary to route the power conductors to a set of terminals on the other side; a run of flexible metal conduit run from the entry point around to the power terminals will maintain system separation.

Part **(D)** allows a common enclosure where the power-limited and non-power-limited wiring must terminate on the same equipment, as in the case of a small power contactor whose 24-V coil is being controlled by a Class 2 circuit. In this case there is a general solution and a limited alternative. The general solution is to restrain the wiring with duct, tie-downs, or other methods such that a 6-mm (¼-in.) air separation is maintained between systems. The limited solution only applies if the non-power-limited wiring is operating not over 150 V to ground, and the limited energy cabling uses Class 3 rated cabling under the further condition that the individual control conductors, where they extend outside the cable jacket, maintain additional separation by maintaining not less than a 6-mm (¼-in.) air separation, or they are separated by a nonconductive sleeve or barrier from other conductors. Note that this provision is for dead-end wiring only, not daisy chains ("solely to connect" does not mean to connect and then to feed another piece of equipment).

This rule also presents the option of installing these wires as Class 1 conductors. This provision has effectively been superseded by the requirements of 725.130(A), and should probably be deleted from this rule. It came into the NEC before 725.130(A), and apparently has been overlooked.

Part **(E)** covers the case where an enclosure containing equipment requiring multiple system connections only has a single conduit entry point, so the connection will take place through a tee with the different systems arriving through opposite sides. In this case (another instance of dead-end wiring only) the power-limited wiring must be enclosed in a "firmly fixed nonconductor" such as loom of some sort or other flexible tubing.

Part **(F)** covers manholes, and multiple systems are permitted provided one of three options is applied. The first is that the non-power-limited wiring is running in a metal clad cable assembly or in Type UF cable. The second is that the power-limited wiring runs in loom of some sort or other firmly fixed non-conductive tubing in addition to the cable jackets. The third is that the power-limited wiring is secured to racks or insulators in the manhole such that effective separation is maintained.

Part **(G)** covers cable tray and requires separation through the use of a divider within the tray, unless the power-limited wiring is run within Type MC cable.

Part **(H)** covers hoistways, and conductors must be installed in rigid conduit, IMC, PVC, RTRC, LFNC, or EMT, except as provided for elevators in 620.21.

In part **(I)**, Class 2 or 3 circuit conductors must be separated not less than 2 in. (50 mm) from conductors for light, power, Class 1, or non-power-limited fire-protective signaling circuits. The present wording of 725.136(I) clarifies many past misunderstandings about the intent of the rule. Previous Code editions only mentioned raceways, porcelain tubes, or "loom" as a means of separating Class 2 circuits, such as "bell wiring," from conductors of light and power systems where such systems were closer together than 2 in. (50 mm). And on that basis, some inspectors required such bell wiring or similar Class 2 wires to have a 2-in. (50-mm) clearance from any type of cable (NM, UF, ac, etc.) that contained conductors for power or lighting circuits. The old rule was also commonly applied to prohibit bell wires and NM cables in the same bored holes through studs, and so on. With the present wording, the 2-in. (50-mm) clearance from Class 2 wiring applies only to "open" light, power, and Class 1 circuit conductors. Power and light circuits or Class 1 circuits that are in raceway or cable do not require 2-in. (50-mm) separation from Class 2 and/or 3 circuits (Fig. 725-14).

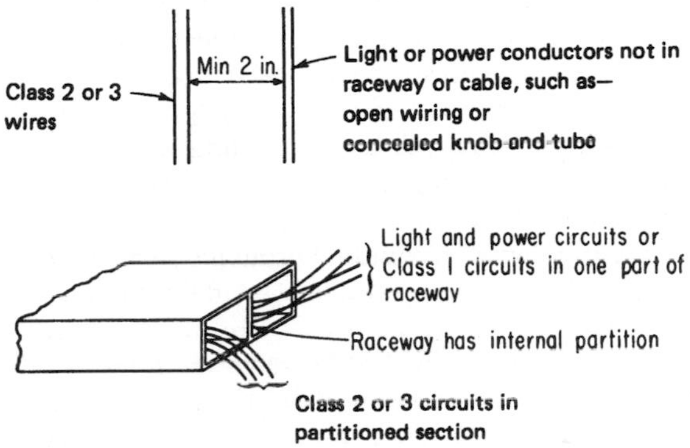
Class 2 or 3
wires
Min 2 in.
Light or power conductors not in
raceway or cable, such as—
open wiring or
concealed knob and tube
Light and power circuits or
Class I circuits in one part of
raceway
Raceway has internal partition
Class 2 or 3 circuits in
partitioned section

Fig. 725-14. Class 2 or 3 wiring must be separated from open wiring for power and light. (Sec. 725.136.)

Note, however, that there is no express permission to include Class 2 or Class 3 conductors within a common cable assembly with power conductors. Paragraph (2) here comes the closest, because it recognizes a "continuous and firmly fixed nonconductor." This is crucial to the production of hybrid cables, where additional separation beyond the conductor insulation is applied to the power-limited conductors in accordance with the spirit of these principles. For example, 334.116(C) expressly recognizes this type of construction for Type NMS cable, and UL has been listing such constructions for many years.

There is another close approximation, with several available cable assemblies based on Type MC cable and Type TC cable that include a twisted pair of control wires (could be shielded) intended as Class 2 and run within an overall plastic jacket. This jacket is being evaluated for conformity to the chemistry and thickness (30 mil) of Type NM cable of equivalent diameter. In effect, this construction complies with the literal text of Paragraph (1) that allows (among others) "nonmetallic-sheathed ... cables" for the power-limited wiring. There are obvious applications including 0-10V dimming controls for luminaires. Once out of the cable assembly at both ends, the Class 2 wiring needs to be run and terminated in accordance with other applicable rules in Sec. 725.136.

725.139. Installation of Conductors of Different Circuits in the Same Cable, Enclosure, Cable Tray, Raceway, or Cable Routing Assembly. Class 2 circuits can coexist in a common location, as can Class 3 circuits. The rule of 725.139**(C)** requires separation of Class 2 and 3 circuits, *unless* the Class 2 wires have insulation that is at least equivalent to that required for Class 3 wires. Part **(D)** covers instances such as a cable made up with eight-conductor Ethernet data wires cabled with a four-conductor, Art. 800 telephone connection; in such cases the combination is allowed but the entire assembly becomes reclassified as a communications cable; and the Class 2 or Class 3 circuit is now an Art. 800 circuit. On the other hand, there are composite cables comprised of jacketed signaling cables along side of jacketed communications cables, all within a common overall jacket, and in this case the signaling circuit is permitted, but not required, to be reclassified as a communications circuit.

Part **(E)** covers the common enclosure rules between power-limited wiring under this article and wiring under related power-limited articles. The following systems can share an enclosure with jacketed Class 2 and Class 3 cables: (1) Power-limited fire alarm wiring—Art. 760; (2) Optical fiber cabling both conductive and nonconductive—Art. 770; (3) Communications circuits—Art. 800; (4) CATV systems—Art. 820; and (5) Low-power network-powered broadband communications—Art. 830.

Part **(F)** forbids co-locating power-limited cabling under this article and comparably marked audio circuits as detailed in 640.9(C); the testing protocols showed they should not run together.

725.141. Installation of Circuit Conductors Extending Beyond One Building. Aerial extensions (the type vulnerable to contact with light and power wiring, or exposed to lightning) must meet comparable requirements for Art. 800 wiring, as covered in a list, or Art. 820 wiring, as covered in a list for comparable coaxial wiring exposures.

725.143. Support of Conductors. Class 2 and Class 3 cables must not be tie-wrapped to or otherwise supported by electrical raceways, as also disallowed in 300.11(B). However, the allowance for a Class 2 cable to run on the outside of a power raceway where the control cable is functionally related to the power conductors within the raceway remains valid.

Low-voltage relay switching is a common application of Class 2 remote control circuits regulated by Art. 725. Low-voltage relay switching is commonly used where remote control or control from a number of spread-out points is required for each of a number of small 120- or 277-V lighting or heating loads. In this type of control, contacts operated by low-voltage relay coils are used to open and close the hot conductor supplying the one or more luminaires or load devices controlled by the relay. The relay is generally a 3-wire, mechanically held, ON-OFF type, energized from a step-down control transformer.

In some cases, all the relays may be mounted in an enclosure near the panelboard supplying the branch circuits, which the relays switch with a single transformer mounted there to supply the low voltage. Where a single panelboard serves a large number of lighting branch circuits over a very large area—such as large office areas in commercial buildings—a number of relays associated with each section of the overall area may be group-mounted in an enclosure in that area.

Figure 725-15 shows 24-V control of 277-V fixtures, with constantly illuminated switch plates alongside doorways to define interior exit routes and practical hollow-partition clip-in switch boxes in interior labs of a medical research center. Control relays are in compact boxes atop luminaires. Relays are connected to switches and to 50-VA continuous-duty 120/24V transformers by Class 2 remote-control circuits routed through overhead non-air-handling ceiling plenums and

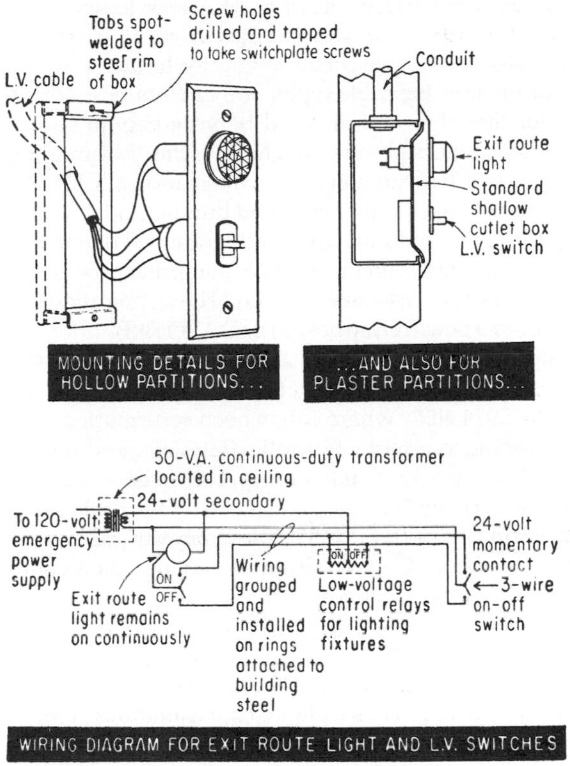
Screw holes
drilled and tapped
to take switchplate screws
Tabs spot-
welded to
steel rim
of box
L.V. cable
Conduit
Exit route
light
Standard
shallow
outlet box
L.V. switch
MOUNTING DETAILS FOR
HOLLOW PARTITIONS...
...AND ALSO FOR
PLASTER PARTITIONS...
50-V.A. continuous-duty transformer
located in ceiling
24-volt secondary
To 120-volt
emergency
power
supply
ON
OFF
Exit route
light remains
on continuously
Wiring
grouped
and
installed
on rings
attached to
building
steel
ON OFF
Low-voltage
control relays
for lighting
fixtures
24-volt
momentary
contact
3-wire
on-off
switch
WIRING DIAGRAM FOR EXIT ROUTE LIGHT AND L.V. SWITCHES

Fig. 725-15. Low-voltage relay switching is a typical application of Class 2 conductors. (Sec. 725.154.)

supported by insulator rings attached to fixture hangers by spring clips. Wiring of the Class 2 circuits may be done with multiwire low-voltage cable, such as 3- or 4-conductor thermostat cable. Interior-area route indicating lights and general lighting switches are mounted together in thin boxes set in partitions.

725.144. Transmission of Power and Data. There is now a comprehensive set of requirements for what is widely known as "power over Ethernet." There are many electronic devices for which this is applicable. Although the word "power" sounds incompatible with the limited power circuits covered in this location, in fact even the most restrictive voltage and current limits for the Class 2 envelope are large enough to allow for small electronic equipment to be powered. This section sets the ground rules for how all this is to be applied. The current must not exceed the ratings of the connectors, with presumably points to the normal RJ45 jacks used for terminations. An informational note suggests a default ampere rating for these jacks would be 1.3 amperes. For conventional cables, the ampacities of typical Ethernet cables are displayed in the table, with the four usual conductor sizes at the left, and number of cables in the bundle running across the top. The notes at the bottom address the frequent circumstance where only half the wires

(4 out of 8) in a cable are current carrying. At higher temperature ambients, the rules invoke the temperature adjustment rules in 310.15(B)(2).

Part **(B)** recognizes a new cable construction (for limited power, denoted by the marking suffix of LP after the cable type). For small bundles this wire would not be helpful, but for large bundles, it could be very helpful because its ampacity never falls below the numerical value that follows the "LP" marking. For example, as covered in the second informational note, the marking "CL2-LP(0.5A)" means a Class 2 cable evaluated for a highly populated bundle, and rated to carry 0.5A. The other rules, including the cable substitution hierarchy, are unaffected by this and still apply. The listing requirements for PoE equipment are in 725.170, and that new section correlates with this new section. The listing requirements for these cables now appear in a new (re-purposed) Sec. 725.179(I).

725.154. Applications of Listed Class 2, Class 3, and PLTC Cables. Almost all the narrative subsection material that was located here has been relocated to Sec. 725.135 in the 2014 NEC, where it has been reformatted as a series of lists. What remained behind is a table of applications of questionable utility. Three important subsections did not make the trip, however.

Part **(A)** and its associated table presents visually the cable hierarchy discussed at length in the commentary on 725.135. This part authorizes a substitution table that covers that discussion. Communications cabling has a comparable voltage capability to CL3 cabling, and therefore can be used in lieu of CL3 where otherwise ranked on an equal level, for example, CMR can be used in a CL3R application, but CMR cannot substitute for CL3P.

Part **(B)** covers circuit integrity (CI) cabling. This cable has a ceramifiable insulation that when heated turns into a glassy compound that, although brittle, will maintain the insulation values between the enclosed conductors for the specified fire exposure period. Some must be installed in raceways, and if so, that fact will be included in the instructions. These cables can be substituted for others without this designation but with equivalent evaluations in the hierarchy. Rounding out the list, cables that are part of a listed electrical circuit protective system, such as Type MI cable with enhanced installation specifications, can also be used.

Part **(C)** recognizes a style of Type PLTC that uses thermocouple extension wire for Class 2 circuits. As implicitly recognized in 725.121(A)(3) Exception No. 1, thermocouples are sufficiently power-limited to automatically qualify as a Class 2 source.

That general principle survives, and shows clearly in the cable substitution hierarchy table, Fig. 725.154(A). However, the specific rules that once resided here have been transferred to 725.135, and all that remains is a table that summarizes the uses permitted.

725.179. Listing and Marking of Class 2, Class 3, and Type PLTC Cables; Communications Raceways; and Cable Routing Assemblies. This section covers the listing requirements for the cables used in the applications just covered, as follows:

Part **(A)** covers Types CL2P and CL3P for plenum cavity applications.
Part **(B)** covers Type CL2R and CL3R for riser applications.
Part **(C)** covers Type CL2 and CL3 for general-purpose applications.

Part **(D)** covers Type CL2X and CL3X for limited-use residential applications.

Part **(E)** covers Type PLTC for cable tray and some additional applications.

Part **(F)** covers cables additionally rated for circuit integrity (CI). This part is now split in order to make a useful distinction between CI cable itself and CI cable as a component of an electrical circuit protective system. Any given CI cable that relies on a raceway to protect it must be specifically listed and marked as part of such a system. Refer to the coverage of this topic in this book at 728.4 for more details on this protective technique.

Part **(G)** covers the required voltage ratings on Class 2 and Class 3 cables, that being not less than 150 V and 300 V respectively.

Part **(H)** Class 3 single conductors must be no smaller than 18 AWG and Type CL3. Fixture wire covered in 725.49(B) can also be listed in this category.

Part **(I)** covers the new limited power cable constructions for power over Ethernet as covered in 725.144 and 725.170.

Part **(J)** summarizes the marking rules, and the associated table reinforces the earlier comments about the cable hierarchy. This rule also disallows marking voltage ratings on power-limited cables because they may lead unqualified users to confuse them with the ratings on actual power cables covered in Chap. 3. Some cables have multiple listings and if such a listing requires a voltage marking, then that marking can appear.

ARTICLE 727. INSTRUMENTATION TRAY CABLE: TYPE ITC

727.1. Scope. This is a specialized cable used for control circuit applications, particularly in industrial occupancies. It is limited to 150 V and 5 A. It uses power levels that are in a range between Class 3 and Class 1, and was originally conceived as a fourth class of wiring within Art. 725. That proved too complicated, and in the 1996 NEC it entered as a stand-alone article for what is in effect a power-limited Class 1 control cable with power limitations even more severe than the ones in 725.41(A). It originated as a work-around for control circuiting limitations for power-limited circuits in Art. 725 that were unacceptable, particularly in industrial occupancies. It also has a far less severe system separation requirement than other power-limited wiring. When it went into the NEC, it was only after a pitched battle that resulted in a vote at an NFPA Annual Meeting that effectively overruled a decision of the Technical Correlating Committee.

Review the commentary at 725.130(A) on this point, because the ability to reclassify control circuits as Class 1 now addresses some of the concerns with respect to system separation that led to the creation of this article. From a true Code purist point of view, this article should never have been necessary and the necessary accommodations should have been made in Art. 725. However, with this now in place, the major industrial players adamantly refuse to give it up, and several serious attempts to eliminate it have failed. So now Art. 725 gives it a grudging, half-hearted recognition, in 725.3(G). Every circuit running on Art. 727 cabling is, in fact, classifiable under the provisions of Art. 725. That is a fact, and because that is so, this article is an ongoing contradiction in terms given a very thin veneer of consistency through the 725.3(G) reference.

727.4. Uses Permitted. This cable is permitted to be used roughly where Type PLTC cabling is permitted to be used, as covered in 725.135. It even has an –ER variant that matches up to 725.135(J), along with the usual 6-ft out-of-tray exception.

727.5. Uses Not Permitted. The cable can't be used above 150 V or over 5 A. It can run with other cabling unless the other cabling rules say it can't, and none of them do. This cabling must be divorced from power or non-power-limited Class 1 circuits unless it is of the metal-armored variety, and it is also permitted to enter common enclosures with power circuits when mutual terminations must be made; in such cases a separation must be maintained. Note that it is permitted to be used with Class 1 power-limited circuits limited only by the 30-V 1000-VA parameters in 725.41(A) (but see 727.7 below).

727.6. Construction. The size range for this cable is 22 AWG up to 12 AWG, copper (or thermocouple alloy) with 300 V insulation, with optional shielding, and with recognition of a metal-clad type of cable armor.

727.7. Marking. The markings are the same as for comparable cables, as set forth in 310.120(A)(2 through 5). As in 725.179(L), voltage ratings must not be marked on the cable.

727.8. Allowable Ampacity. This cable is not permitted to exceed 5 A down to 20 AWG conductors, and 3 A on the 22 AWG size conductors.

727.9. Overcurrent Protection. The overcurrent protection permitted matches the allowable ampacity.

727.10. Bends. Bends must not damage the cable.

ARTICLE 728. FIRE-RESISTIVE CABLE SYSTEMS

728.1. Scope. This new article (2014 NEC) is designed to centralize the requirements for fire-resistive cable systems, which are defined as a cable and components employed to ensure survivability of critical circuits for the prescribed period of time during a fire.

728.4. General. One of the most important concepts in this area is that every single component of a designated system must work with all other components in a tested and predictable way. These systems require great attention to detail in order to install properly, particularly when one considers that the only test after installation will likely be their performance under the one and only fire event they will be subjected to. This section includes two informational notes. The second identifies the cables addressed in this article as components of an "electrical circuit protective system." This is defined in Art. 100 as "a system consisting of components and materials intended for installation as protection for specific electrical wiring systems with respect to the disruption of electrical circuit integrity upon exterior fire exposure." These systems are evaluated by UL and other laboratories under UL Subject 1724, "Outline of Investigation for Fire Tests for Electrical Circuit Protective Systems." Additional commentary on practical issues that arise during design and installation is included in the next section.

728.5. Installations. A careful examination of the eight subsections supports the above assertion. The fire testing is very robust, and also considers likely firefighting activity by incorporating hose stream testing after the full fire exposure testing is completed. To make certain the installation in the field performs as tested in the laboratory, the installation instructions are exacting and require extraordinary attention to detail. By way of example, consider Type MI cable, which is the basis for one widely used design. The **NEC** support distance for MI cable (332.30) is a 6 ft maximum spacing. However, when used for this purpose, the maximum spacing drops to 1 m (39 in.), and the cable must be routed along a concrete or masonry wall. In addition, the supports must use steel clamps of a prescribed configuration, and be secured to the wall with a steel concrete screw of a stipulated length and diameter. In addition, the testing applies only to continuous cable runs; if splicing is required, it must occur not less than 12 in. into a protected space on the cold side of a rated partition.

Other designs use conductors with ceramifiable insulation installed in tubular steel raceways, with some recognizing EMT and others not. These systems use special conductors, such as RHH but with nonstandard diameters, so the manufacturer's information needs to be consulted for wire fill. The raceway fill tables in Annex C have an end note warning not to use their values for these applications for this reason. This insulation, upon exposure to extreme heat, breaks down but the residue is akin to colored glass, not capable of significant movement but capable of maintaining the integrity of the circuit for a rated period of time. The designs have interesting additional requirements for running the circuits.

For example, "The raceway should be connected together using the coupling type referenced in the system, such as steel set-screw type for EMT or threaded types of couplings for IMC and RMC. No other coupling should be used unless noted in the specific system." This is because under the hose stream testing compression couplings usually fail, and even set-screw couplings must have their screws driven very hard with a significant dimple or the raceway will pull apart and fail the test.

These designs also address equipment grounding conductors installed in the same raceway, and the special requirements are anything but intuitive. "The bare or insulated ground wire may be of special manufacture to be compatible with the system. The system will specify the manufacturer of an allowable ground wire. If not specified, the ground should be the same as the fire-rated wire described in the system." This is because a separate equipment grounding conductor of random insulation type, under fire conditions, may be incompatible with the tested cables it is in intimate contact with; unless tested, this cannot be predicted. This is the reason that some designs have disallowed the use of a THHN equipment grounding conductor, because its outgassing products from heating proved incompatible with its RHH companions. Wire pulling compounds, if used, must also be included in the design for the same reasons.

728.60. Grounding. For systems using raceways and separate equipment grounding conductors, the separate grounding conductor must be recognized in the system design so as to assure that an insulation breakdown during a fire does not react with the special insulation required for the ungrounded conductors. The system must specify the conductor to be used, and any alternative must be marked with the system number. See also the discussion at 250.120(A).

728.120. Marking. In addition to standard markings, cables and conductors used for these systems must be surface marked with fire rating expressed as hours, a suffix "FRR" (for fire-resistive rating) and the system identification number.

ARTICLE 750. ENERGY MANAGEMENT SYSTEMS

750.1. Scope. This article, new in the 2014 NEC, covers these systems, which have a very expansive definition. The simple fact of a monitor creates a system based on the literal text of the definition. Management systems presumably do more than watch; they also control, as effectively defined as turning power of or on or varying its magnitude.

750.20. Alternate Power Systems. This section places the availability of alternate power sources for five designated loads as off limits for energy management interference.

750.30. Load Management. Article 700 allows for emergency generators to support nonemergency loads; however, when the emergency loads call for generator support, nonemergency loads must be shed automatically to the extent necessary to allow for full support of the emergency loads. 750.30(A)(2) prohibits the management system from interfering in that function, as well as comparable actions for three other areas.

750.50. Field Markings. If a management system is capable of controlling remote loads, a directory must be posted that displays the controlled circuits or devices on the "enclosure of the controller, disconnect, or branch-circuit overcurrent device." The use of the conjunction "or" suggests that any of the locations is acceptable. Note that if a receptacle is to be controlled, it must display the special mark required by 406.3(E).

ARTICLE 760. FIRE ALARM SYSTEMS

760.1. Scope. As indicated by informational note No. 1, the various elements associated with the fire alarm system are now all covered in a single NFPA Standard: NFPA 72. The overall article here has been revised to consolidate the various requirements to facilitate comprehension as well as to correlate with changes made in Art. 725. This article covers fire alarm systems, including the initiating and notification appliances and the fire alarm control panel. It does not cover a run of reciprocally alarming smoke detectors connected to a conventional branch circuit in a residential occupancy.

Most of the provisions in this article line up with equivalent provisions in Art. 725, with Non-Power-Limited Fire Alarm Circuits corresponding to Class 1 signaling circuits, and Power-Limited Fire Alarm Circuits corresponding to Class 3 signaling circuits. The commentary that follows highlights the differences, so for any section not commented on here, the comments on the identical section number in Art. 725 apply.

760.2. Definitions. A fire alarm circuit is the wiring system between the power supply or overcurrent device and the farthest connected equipment that is part of that system.

760.3. Other Articles. Parts **(A)** and **(B)** incorporate the usual limitations on usage in ducts and plenum cavities, including correlation with the new 300.22(B) Exception. There is no reference to cable tray, motor control, or instrumentation tray cable as at 725.3(E, F, and G), which is appropriate. On the other hand, this section does reference Art. 770 because optical fiber cables are used in fire alarm systems, as is Art. 725 in the context of building controls that frequently must respond following a fire alarm activation. In addition, 300.8 is now specifically referenced in order to keep foreign systems out of fire alarm raceways.

Part **(H)** incorporates the rules in 300.7(A) for sealing raceway interiors that are subject to convective air migration, particularly where condensation is a likely result.

Part **(I)** brings in the vertical support rules for cables run in raceway, but only for the circuit-integrity (CI) cables. This makes sense because in a fire condition, these cables become extremely brittle and vertical supports can easily make a significant difference in how long they survive. These cables must be used in close agreement with the directions that come with them from their manufacturer in terms of installation methods, and these may be more restrictive than the limits in 300.19.

Part **(J)** incorporates the general prohibition against overfilling raceways in 300.17, without making an explicit mention of raceway fill limitations in Chap. 9. Note that although 300.17 is mentioned in 760.51 on non-power-limited (NPLFA) systems, it does not appear for the power-limited ones (PLFA).

Part **(K)** requires a bushing in accordance with 300.15(C) where cables emerge from being sleeved in a raceway riser for physical protection. Note that the actual 300.15(C) rule requires a fitting to protect the cable from abrasion, which could be some sort of change-over fitting, and does not mention a bushing.

760.24. Mechanical Execution of Work. In addition to the usual content of this section, the 2014 NEC has added part (B) to address the support of circuit integrity (CI) cables. They must be supported every 24 in. and every 18 in. in places where they run within 7 ft of the floor, using steel supports. The wording is supposed to put some of the requirements for the fire-resistive system into the NEC, but the total requirements are always far more complicated. Refer to the commentary at 728.5 on this point. In addition, the original proposal included wording referring to running on a surface, but the final wording left that out and the result does not correlate with the actual fire-resistance specification.

760.32. Fire Alarm Circuits Extending Beyond One Building. This treatment differs from its counterparts at 725.52 and 725.141 in that the Part I location makes it apply to all power levels with only a single entry. Power-limited fire alarm circuits running outdoors are either installed as communications circuits or as conventional circuits according to the rules of Part I of Art. 300. Non-power-limited fire alarm circuits must meet those requirements and also the applicable provision in Part I of Art. 225.

To correlate with developments in NFPA 72, this section now has a new (2014 NEC) informational note pointing to UL 497B, Protectors for Data Communications.

The intent is not obviously based simply on the information presented on the printed page; however, the UL standard points in turn to the Guide Card category QVGQ. This covers the category "Isolated Loop Circuit Protectors for Communications Circuits." This category includes the following general description:

> The purpose of the isolated loop circuit protector is to suppress abnormal voltages caused by hazards such as lightning and other EMI transients. An isolated loop circuit protector is intended for use on data or communication lines that are not exposed to accidental contact with electric light or power conductors operating at over 300 V to ground.

Although there is no direct requirement here, the intent is to mandate that transient surge protection based not on the usual Art. 285 approach, but instead on the protectors used for communications circuits that are to be provided. In particular the protection should be one of these devices, particularly because these are not line-voltage systems. The QVGQ category is one of five in a broader category of protectors that specifically include those protectors covered in Part II of Art. 800, which closes the loop because 760.32 does make that reference.

760.41. NPLFA Circuit Power Source Requirements. Part **(A)** lines up with 725.41(B), with unlimited power and voltage not to exceed 600 V. It includes express permission to secure a fire alarm circuit disconnect in the closed position. Part **(B)** forbids the use of either arc-fault circuit interrupter or ground-fault circuit interrupter protective devices for its branch circuit, which must be an individual branch circuit. It also includes other requirements not previously mandated by the NEC. The location of the branch-circuit overcurrent device supplying a fire alarm control panel shall be marked on the control panel. The circuit disconnecting means must be marked "FIRE ALARM CIRCUIT" and must be accessible only to qualified personnel, which raises obvious questions in many occupancies. It must be marked in red, but not so as to damage or obscure any manufacturer's markings.

760.48. Conductors of Different Circuits in Same Cable, Enclosure, or Raceway. Part **(A)** of this is functionally the same as 725.48; however, part **(B)** is much more limiting. A non-power-limited fire alarm circuit conductor can run in the same raceway as a power supply conductor only if they are connected to the same equipment.

760.49. NPLFA Circuit Conductors. Part **(A)** differs from 725.49 only in that it does not recognize flexible cord. Part **(B)** is essentially the same except Type FFH-2 is not recognized, and Types PAF and PTF are only recognized for high-temperature operation. Part **(C)** simply requires copper conductors; the old "special stranding" rules that required solid or bunch-tinned copper up to 7-strand 16 and 18 AWG, and up to 19-strand 14 AWG, were deleted in the 1996 NEC.

760.53. Multiconductor NPLFA Cables. This section has no counterpart in Art. 725. Part **(A)** covers the wiring methods for these cables. They can be fished through concealed spaces and run on exposed building surfaces. All cable terminations must be in listed equipment, including fire alarm devices, boxes, or utilization equipment. Where exposed, the cable layout must be such that the wiring is protected as much as possible, such as by routing next to door framing. If less than 2.1 m (7 ft) from the floor, the cable must be secured at 450-mm (18-in.) intervals or less. A kick pipe must be installed where passing through a floor such that the cable is protected to the 2.1-m (7-ft) height using a metal raceway or Type PVC or RTRC conduit. Only RMC, PVC, RTRC, IMC LFNC, or EMT are permitted in hoistways, but the cables can also enter elevator raceways as included in 620.21.

Part **(B)** creates the same type of cable hierarchy for these cables as in Art. 725, with a plenum grade on top (NPLFP), a riser grade (NPLFR), and a general-purpose grade (NPLF) at the bottom. Given the limited need, there is no limited grade ("X" suffix) in this category. Note that the riser grade now applies to cabling extending vertically through "one or more" instead of the wording "more than one" floor, thereby increasing the demand for riser-rated cabling.

760.121. Power Sources for PLFA Circuits. The permitted power sources here are much simpler than in Art. 725. In addition, as with 760.41, AFCI and GFCI are off-limits for protecting the fire alarm control panel. This part also includes other requirements not previously mandated by the **NEC**. The location of the branch-circuit overcurrent device supplying a fire alarm control panel shall be marked on the control panel. The circuit disconnecting means must be marked "FIRE ALARM CIRCUIT" and must be accessible only to qualified personnel, which raises obvious questions in many occupancies. It must be marked in red, but not so as to damage or obscure any manufacturer's markings.

760.130. Wiring Methods and Materials on Load Side of the PLFA Power Source. This wiring can use NPLFA wiring methods and materials, including the ones covered in 760.49 and 760.53, with no mutual conductor heating derating penalties applied. As with their Class 2 and Class 3 counterparts, if the PLFA markings per 760.124 are removed, a PLFA circuit can be reclassified as an NPLFA circuit.

Part **(B)** requires devices to be installed in accordance with their directions, and also 300.11(A) with respect to staying off the designed supports of a suspended ceiling, and in addition 300.15, which effectively requires a box at termination points. The cabling rules are identical to the ones for NPLFA runs as covered in 760.53(A).

760.135. Installation of PLFA Cables in Buildings. This article uses the same hierarchy of cable insulation covered at length at 725.135. As in that case, this section is a new method of presenting this information, with the various classifications of installation circumstance listed in lettered subsections, roughly in order of decreasing levels of resistance to flame and smoke. So this list begins with actual ductwork (no suffix, only special rules) and then plenum installations (suffix "P"), then riser installations (suffix "R"), then a catch-all of intermediate industrial/commercial installations, and then general building installations (no suffix). The section ends with coverage of nonconcealed spaces and portable fire alarm systems. Item (I) does not line up with 725.135(L&M) because instead of an "X" category, there is an allowance for any cable that meets the copper content and 26 AWG size minimum (18 AWG if single conductor cable) requirements to be run for exposed uses in these areas up to 3 m (10 ft) in length. Item (J) recognizes portable fire alarm systems for a stage or set protection, wired per 530.12.

760.136. Separation from Electric Light, Power, Class 1, NPLFA, and Medium-Power Network-Powered Broadband Communications Circuit Conductors. These rules line up almost exactly with 725.136 except that manholes and cable trays are not covered.

760.139. Installation of Conductors of Different PLFA Circuits, Class 2, Class 3, and Communications Circuits in the Same Cable, Enclosure, Cable Tray, Raceway, or Cable Routing Assembly. PLFA circuits are treated in the same category as Class 3 circuits in Art. 725, and most of the requirements in this section follow logically from that principle. Part **(A)** allows two or more PLFA circuits in the same location, together with Class 3 circuits and communications circuits. Part **(B)** lines up exactly with 725.139(C) on Class 2 and Class 3 circuits in common locations for the same reason. Part **(C)** covers intermingled network-powered broadband circuits (low-power type) and permits the PLFA wiring in those locations. Part **(D)** lines up with 725.139(F) and forbids installing PLFA wiring with audio amplifier output wiring as covered in 640.9(C).

760.142. Conductor Size. This section recognizes conductors as small as 26 AWG if part of a multiconductor cable. The equipment terminations must be listed as suitable for such terminations, or the 26 AWG must be joined to a larger conductor using a step-up splice listed as suitable for the conversion. These cables are increasingly used with multiplexed, addressable fire alarm systems.

760.145. Current-Carrying Continuous Line-Type Fire Detectors. These include listed, continuous line-type copper pneumatic tubing that also carries signaling circuits. They can be used with PLFA circuiting.

760.154. Applications of Listed PLFA Cables. Almost all the narrative subsection material that was located here has been relocated to Sec. 760.135 in the 2014 NEC, where it has been reformatted as a series of lists. What remained behind is a table of applications of questionable utility. One important subsection did not make the trip, however. Part **(A)** and its associated table presents visually the cable hierarchy discussed at length in the commentary on 725.135. This part authorizes a substitution table that covers that discussion. Communications cabling has a comparable voltage capability to FPL cabling, and therefore can be used in lieu of FPL where otherwise ranked on an equal level; for example, CMR can be used in an FPLR application, but CMR cannot substitute for FPLP. Circuit integrity cables are also recognized in all three levels (FPLP-CI, FPLR-CI, and FPL-CI).

Figure 760.154(A) is the customary cable substitution hierarchy presented in graphic form, with communications cables shown as the next level up.

760.176. Listing and Marking of NPLFA Cables. This section covers the listing requirements for the cables used in Part II of the article, as follows:

Part **(A)** covers the requirement for 18 AWG and larger copper wire, solid or stranded.

Part **(B)** covers the insulation levels required, that is, 600 V.

Part **(C)** covers Type NPLFP for plenum cavity applications.

Part **(D)** covers Type NPLFR for riser applications.

Part **(E)** covers Type NPLF for general-purpose applications.

Part **(F)** covers cables additionally rated for circuit integrity (CI). This part is now split in order to make a useful distinction between CI cable itself and CI cable as a component of an electrical circuit protective system. Any given CI cable that relies on a raceway to protect it must be specifically listed and marked as part of such a system. The best approach is to follow Art. 728 and the listing specifications exactly. Refer to the commentary in that article for more information on this point.

Part **(G)** covers voltage markings on NPLFA cables, that being for a maximum of 150 V, with CI cables listed for circuit integrity required to show this with a "CI" designation. Temperature ratings and markings for these cables follow those in 760.179(I); refer to that location for more details.

760.179. Listing and Marking of PLFA Cables and Insulated Continuous Line-Type Fire Detectors. In addition to the specific requirements below, cable used in a wet location, including underground, must be listed for use in such locations or use a moisture-impervious metal sheath.

Part **(A)** covers the requirement for copper conductors, solid or stranded.

Part **(B)** covers the requirement for 18 AWG and larger wire for single conductors, and 26 AWG or larger for multiconductor cable.

Part **(C)** covers the required voltage rating of 300 V.

Part **(D)** covers Type FPLP for plenum cavity applications.

Part **(E)** covers Type FPLR for riser applications.

Part **(F)** covers Type FPL for general-purpose applications.

Part **(G)** covers cables additionally rated for circuit integrity (CI). This category is split in the same way and for the same reasons as the comparable part of 760.176.

Part **(H)** covers coaxial cables, which will have the customary hierarchy listing. The rule also recognizes that the center conductor can be permitted to use 30 percent conductivity copper covered steel.

Part **(I)** has the indicated cable types in a table of required markings, along with the CI designation if appropriate, and the rule that voltage ratings not be marked together with the same exception as in 725.179(L) for cables with multiple ratings, some of which may require a voltage marking as part of a listing. PLFA cables, if rated higher than 60°C, must have the temperature marked on the cable, but the conductor size must always be marked.

Part **(J)** covers insulated continuous line-type fire detectors equipped with 300-V insulation per 760.179(C), which must be highly resistant to abrasion. The insulation must also be resistant to the spread of fire and smoke, as covered in 760.179(D, E, or F as applicable).

ARTICLE 770. OPTICAL FIBER CABLES AND RACEWAYS

770.1. Scope. This article is interesting in that it covers "wiring" that in some cases does not bring with it even a single conductive element, although the cabling will always be used in conjunction with electrical circuits. Figure 770-1 shows operation of a fiber-optic (FO) link in the transmission of a telephone signal.

770.2. Definitions. There are several that need to be kept in mind:

Abandoned Optical Fiber Cable is not terminated at any equipment other than a connector, and is not identified for future use with a tag.

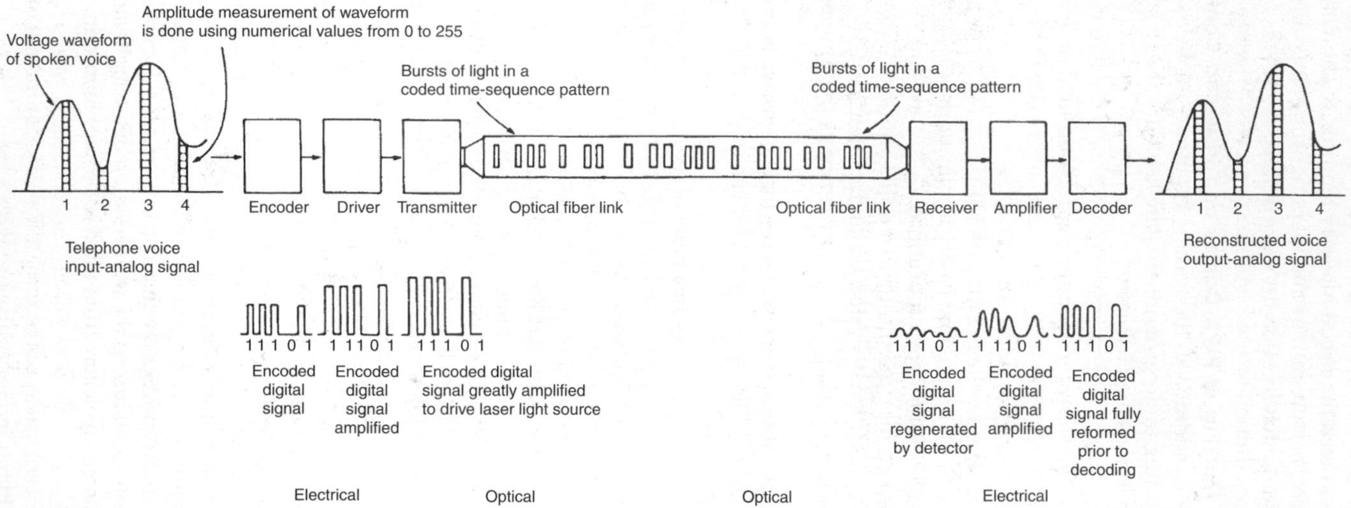
Amplitude measurement of waveform
is done using numerical values from 0 to 255
Voltage waveform
of spoken voice
Bursts of light in a
coded time-sequence pattern
Bursts of light in a
coded time-sequence pattern
1 2 3 4
Encoder
Driver
Transmitter
Optical fiber link
Optical fiber link
Receiver
Amplifier
Decoder
1 2 3 4
Telephone voice
input-analog signal
Reconstructed voice
output-analog signal
1 1 1 0 1
1 1 1 0 1
1 1 1 0 1
1 1 1 0 1
1 1 1 0 1
1 1 1 0 1
Encoded digital signal
Encoded digital signal amplified
Encoded digital signal greatly amplified to drive laser light source
Encoded digital signal regenerated by detector
Encoded digital signal amplified
Encoded digital signal fully reformed prior to decoding
Electrical
Optical
Optical
Electrical

Fig. 770-1. The transmission of a signal (left to right) along an optical fiber link (a cable) at center. At the left, an electrical signal is converted to light pulses that are sent through the fiber cable by a laser diode (a light signal generator), and then, at the right, the light pulses are received and reconstructed into the original electrical signal from the left. (Sec. 770.2.)

Cable Routing Assembly is a single channel or connected multiple channels that form a structural system to support high densities of optical fiber cables and other limited energy cables including Class 2 and Class 3 cables within the scope of Art. 725 and often associated with information technology and communications equipment. This definition has been moved to Art. 100 because it applies to multiple articles.

Composite Optical Fiber Cable contains optical fibers together with current-carrying electrical conductors.

Conductive Optical Fiber Cable contains metallic strength members, or a metallic vapor barrier, or a metallic sheath or armor, or any combination of these conductive elements. Figure 770-2 shows an example.

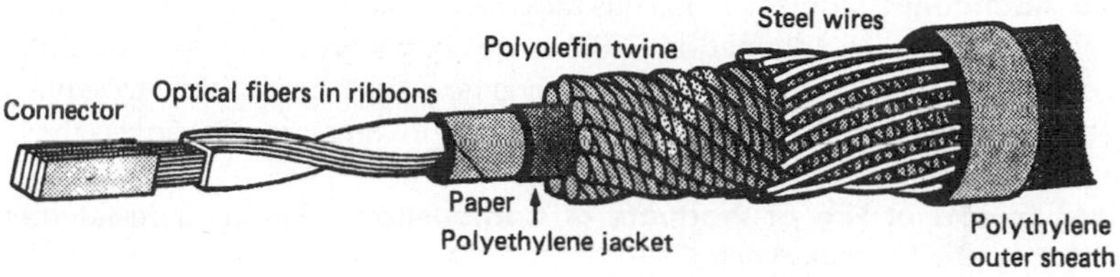
Steel wires
Polyolefin twine
Optical fibers in ribbons
Connector
Paper
Polyethylene jacket
Polythylene
outer sheath

Fig. 770-2. This fiber-optic cable, which is used in a mile-long telephone communication line, is classified as a *conductive* type of optical fiber cable because of the steel wires used to provide an outer mechanical sheath over the fiber assembly. (Sec. 770.5.)

Exposed (to Accidental Contact) applies to a conductive optical fiber cable that is in such a position that it would be likely to become energized in the event that a support or insulation or both failed.

Nonconductive Optical Fiber Cable contains no electrically conductive members in any form.

Optical Fiber Cable. A factory or field assembly of one or more optical fibers having an overall covering. This definition has been changed in the 2014 NEC and now includes the concept of a field-assembled cable for which the jacket is installed first, and then the optical fibers are drawn in.

Point of Entrance is the point within a building at which cable emerges from an external wall, or a floor slab, or from RMC or IMC. In such cases the use of heavy-wall steel conduit artificially extends the point of entrance into a building, doing for this wiring what 230.6 does for service entrance conductors. This is an important concept, and it holds here and for all of Chap. 8. See also Sec. 770.49, which now contains the bonding requirement for the conduit.

770.12. Innerduct for Optical Fiber Cables. Optical fiber cables are sensitive to pulling tension, and communications raceways, which are a corrugated nonmetallic product closely related to electrical nonmetallic tubing, are frequently pulled into large ducts, particularly on underground applications where there may be debris in the duct. This provides a uniform pulling environment with a very low coefficient of friction.

770.21. Access to Electrical Equipment Behind Panels Designed to Allow Access. It is contrary to the NEC to drape endless quantities of these cables across suspended ceiling panels in a way that limits the intended access to the ceiling cavity above them.

770.24. Mechanical Execution of Work. In addition to the usual neat and workmanlike manner requirement, this section also requires exposed cabling (which includes the area above a suspended ceiling) to be supported to structure such that normal building operations will not damage the cable. In addition, if the wiring is run parallel to framing, it must meet the 32-mm (1¼-in.) spacing requirement from the leading edge of the framing. Finally, the wiring must stay off the tie wires designed to support a suspended ceiling, in accordance with 300.11. Where cable ties and other nonmetallic components are used in 300.22(C) spaces, they must be listed for that environment.

770.25. Abandoned Cables. This has the effect of requiring that any "abandoned" optical fiber cables—as defined in 770.2 as not being terminated or identified for future use with a tag—must be removed. It is no longer permissible to leave abandoned optical fiber cables in place when they are no longer in use, unless they have been "identified" for future use.

770.26. Spread of Fire or Products of Combustion. This is a restatement of 300.21, as applied to this article.

770.44. Overhead (Aerial) Optical Fiber Cables. This addresses a construction that contains a metallic noncurrent-carrying strength member and is designed for overhead work. It corresponds to 800.44, where this book provides more installation details.

770.47. Underground Optical Fiber Cables Entering Buildings. This section title should add the word "Conductive" ahead of "Optical Fiber" because the requirements only address conductive cables and exactly parallel other comparable provisions such as 820.47. Where such cables enter a building from a manhole or other enclosure they must be separated from non-power-limited circuits and power circuits, and the enclosures must be partitioned. For direct burial applications they must observe a 1-ft separation unless the power wiring (divided into two exceptions, one for services and one for everything else but the requirements are essentially the same) is in a raceway or using a cable method that includes metal cable armor.

770.48. Unlisted Cables and Raceways Entering Buildings. Part (A) allows unlisted outside plant cables, either conductive or nonconductive, to enter a building and move through general-purpose spaces (not risers or plenums) up to 15 m (50 ft) from the point of entrance. This allows a transition length along which a convenient point can be provided for a transition to the listed cables required for most interior work. If such a point is unavailable, the point of entrance can be artificially extended through the use of grounded heavy-wall steel conduit.

Part (B) allows unlisted nonconductive outside plant cables to enter buildings and continue without any length restriction, if they are run within IMC, RMC, PVC, or EMT. Note that for risers and plenum cavities EMT and PVC are excluded.

770.49. Metallic Entrance Conduit Grounding. Heavy-wall metal conduit containing optical fiber entrance cables must be bonded to a grounding electrode as covered in 770.100(B). This, in combination with the point of entrance definition, creates the remote entry point for this work that is the parallel to 230.6 for power service work. In fact, it used to be part of the definition, but was separated because definitions cannot contain requirements. Similar sections occur throughout Chap. 8.

770.93. Grounding or Interruption of Noncurrent-Carrying Metallic Members of Optical Fiber Cables. This rule is split into a termination outside or inside a building, but the action to be taken is the same in either event. Make a grounding connection as covered in 770.100, or place an insulating joint as near as practicable to the point of entry.

770.100. Entrance Cable Bonding and Grounding. If grounded (and not merely interrupted as permitted in 770.93) the grounding must comply with the rules of this section.

Part **(A)** requires the grounding conductor to be insulated, made of copper (or other corrosion-resistant material), and stranded or solid. The size must be no smaller than 14 AWG (or larger if needed to equal the current-carrying capacity of the grounded elements in the cable assembly) and having an ampacity roughly equal (or larger) than the conductive elements of the cable assembly, but it need not be larger than 6 AWG. It must run in as straight a line as practical, and be guarded from physical damage. If run in a metal raceway, both ends must be bonded even though 250.64(E) imposes that requirement only on ferrous raceways.

Bonding conductors must be as short as practicable, and in one- or two-family dwellings they must not exceed 20 ft in length. If that distance is too short, then a separate electrode must be provided within that distance in accordance with 770.100(B)(3)(2), with that electrode then bonded to the local power system electrode as per 770.100(D). An informational note advises that other buildings will benefit from observing this length limitation, but there is no mandatory rule.

Part **(B)** covers electrode terminations. If there is an intersystem bonding termination, then the termination is to that point. If the intersystem bonding point is lacking but there is one or more grounding electrodes, the connection must be made to that grounding system through one of the optional methods listed. If the premises served have no intersystem bonding termination and no grounding electrode system, then the bonding connection is to the nearest traditional building electrode (water pipe, building steel, concrete-encased electrode, ground ring), and if all those are missing then to a 1.5-m (5-ft) ground rod, driven into permanently damp soil to the extent practicable, using connections that comply with 250.70.

Part **(D)** requires that this electrode, although required to be at least 1.8 m (6 ft) distant from other electrodes, must be bonded to the power system grounding electrode system with a conductor no smaller than 6 AWG copper. Steam and hot water pipes are off limits for grounding conductors, as are lightning protection conductors. The power system bonding requirement is presumably waived at mobile homes for which the connection would be impractical. This is the apparent result of the wording of an exception that, in violation of the Style Manual, is not in the form of a complete sentence and therefore unclear as to effect. Refer to the discussion at 770.106 for more information.

770.106. Grounding and Bonding of Entrance Cables at Mobile Homes. This is very difficult to follow, because there is no straightforward explanation of what happens when the mobile home has an electrical supply that complies with the NEC. Specifically, mobile homes are supposed to have a bonding terminal on their chassis at an accessible location, as required in 550.16(C)(1). They are also supposed to have either service equipment or a feeder disconnect located within 9.0 m (30 ft), per 550.32(A). If they have actual service equipment at that location, or if they have a feeder disconnect with a regrounded neutral at that point, there will be a grounding electrode system connected at that disconnect. In fact, even if the disconnect does not include a regrounded neutral, 250.32(A) still requires a grounding electrode connection at this point. Assuming the optical fiber entry comes from this location, a bonding connection between the conductive elements in the optical fiber cable and the grounding system at the disconnect will meet all requirements and no further action needs to be taken at the mobile home. If, and only if, the disconnect is located at an excessive distance or grounded incorrectly does the necessity of creating a local electrode at the mobile home kick in.

This brings up the separate question of bonding. If the mobile home is permanently connected to a Code-compliant disconnect, then the permanent equipment grounding system connecting the mobile home and the disconnect is sufficient for these purposes. However, if the mobile home is cord-and-plug-connected, removing the plug would disrupt this continuity. In this case there is a conundrum because this wording was brought over from 800.106. Article 800 (telephone) systems have primary protectors that will not exist on these systems. That leaves only one option, a grounding electrode, and therefore, through this back door, a local grounding electrode becomes required in the event of a cord-and-plug connection. The grounding electrode must be connected to the conductive elements of the optical fiber cable, and also to the mobile home bonding terminal as previously mentioned.

This requirement was copied to here from Art. 800. The reason it is difficult to follow is that the essential context was not brought with it, and should have been. See 800.90(B) second paragraph for this information, substituting "grounding electrode" for "primary protector."

770.110. Raceways and Cable Routing Assemblies for Optical Fiber Cables. Raceways can be Chap. 3 methods, as shown in Fig. 770-3, or communications raceways as covered later. Communications raceways, as mentioned in the discussion about 770.12, are very close to ENT in construction and this section even mandates the use of the installation provisions in Art. 362 having to do with bends, supports, joints, and perhaps most intriguingly, 362.56 which invokes the box-at-termination rules in 300.15. In addition, if current-carrying conductors are in a raceway with optical fiber cables, the wire fill rules apply to the combination. On the other hand, if the raceway only contains optical fiber cables, the fill limitations do not apply. Figure 770-4 shows a termination enclosure for optical fiber cables.

Note that as of the 2014 NEC, the term of art for these raceways is "communications raceway" because the product employed is common throughout this article and Chap. 8. The definition has been relocated to Art. 100 because it does now apply across numerous articles.

Part **(B)** covers raceway fill, and limits NEC fill limitations to applications where nonconductive optical fiber cables are run in the same raceway as "electric light or

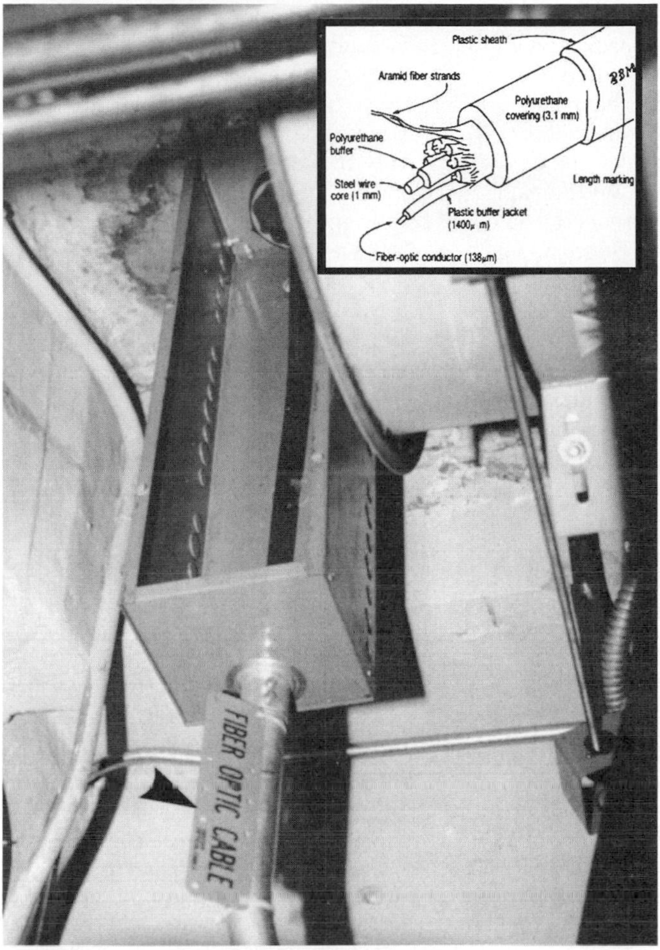
Plastic sheath
Aramid fiber strands
Polyurethane covering (3.1 mm)
Polyurethane buffer
Steel wire core (1 mm)
Plastic buffer jacket (1400μ m)
Fiber-optic conductor (138μm)
Length marking
FIBER OPTIC CABLE

Fig. 770-3. A pullbox in a metric designator 38 (trade size 1½) EMT run carrying the 10-conductor cables, with a bright orange tag identifying the FO conduit (inset shows FO cable). (Sec. 770.5.)

power conductors." This appears to be a sort of shorthand notation for distinctions made in 770.133. Specifically, fill limits should apparently be applied to applications covered by 770.133(A) and they need not be applied to applications covered by 770.133(B) and 770.133(C).

Part **(C)**, new as of the 2014 NEC, covers cable routing assemblies. These are usually open on top and serve as continuous support for cables. They are customized by application, as in general-purpose, riser, or plenum, with a comparable substitution hierarchy. The default support intervals are 3 ft for horizontal runs and 4 ft for vertical runs, with an allowance for longer spans if part of the listing, but never over 10 ft.

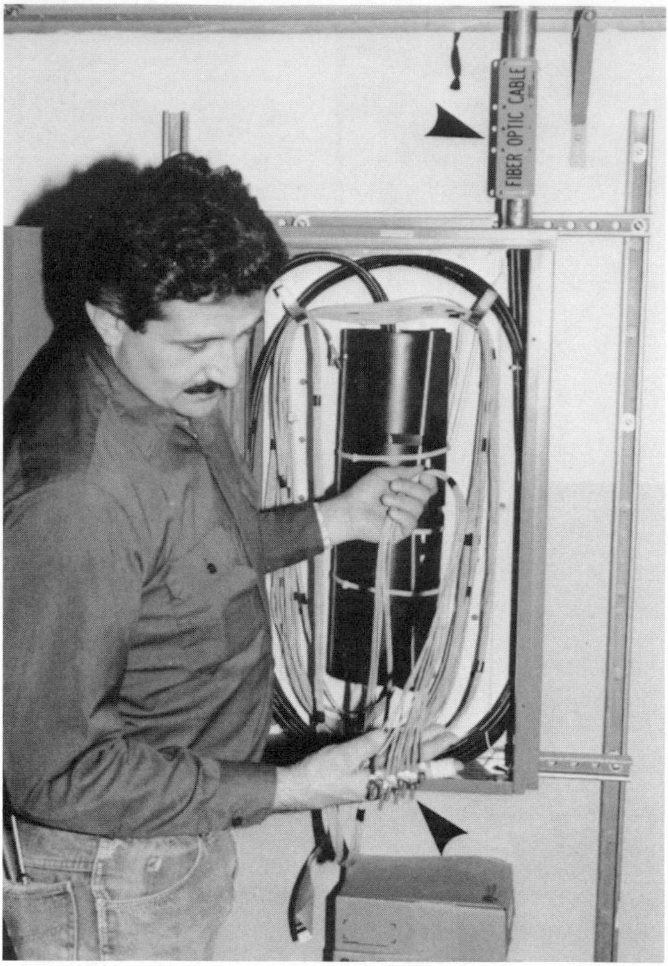
FIBER OPTIC CABLE

Fig. 770-4. A NEMA 3R splice box in the telecommunications room of one of the buildings, with two 10-conductor cables entering at top (arrow) and the individual FO conductors terminated in special connectors (bottom arrow). FO conductors are used as data communication links between five different buildings in which a banking firm has branch locations in a large city. The overall circuiting includes the runs within the buildings (connecting to computers, video equipment, telephones, and telecommunications equipment) and over 5 mi of underground 30-conductor FO cable with a metal armor sheath run under the city in ducts.

770.113. Installation of Optical Fiber Cables and Raceways, and Cable Routing Assemblies. This section totaled four lines in the 2008 edition, and in subsequent editions runs to almost two full pages. There are some general principles at work that make this less complicated than it appears. The entire two pages boils down to the following:

1. The hierarchy of cable compositional categories, and by logical extension the hierarchy of compositional categories for the nonmetallic raceways and supporting assemblies that contain those cables, continues without change. Refer to the discussion at 725.135 regarding the history and applications of these categories, which have not changed since the advent of this work in the 1987 NEC. Specifically, there are plenum exposures, which require the most severe testing for both smoke generation and vertical flammability, and result in a "P" suffix on cable assemblies. There are riser exposures that are the next most severely tested, focusing on vertical flammability, and result in a "R" suffix on cable assemblies. Then there are general-purpose applications that receive general fire load testing and result in either a "G" suffix or no suffix. This is a hierarchy, so riser suitability implies general purpose suitability as well, and the plenum grade can be used in all locations. At this time there are no low-grade residential optical fiber cables (suffix "X") recognized.
2. If a cable is pulled into a Chap. 3 raceway, its suitability follows that of the enclosing Chap. 3 raceway, just as surely as the THHN or RHW or XHHW-2 that could otherwise be pulled into the same raceway without a separate evaluation of the fire resistance of its insulation.
3. Fireproof shafts with firestopping at every floor cancel the need for special riser ratings, and one- and two-family dwelling exposures, even vertical ones, do not merit a riser rating either.

770.114. Grounding. Conductive optical fiber cables must have their conductive members grounded, either to a grounded equipment rack, or as specified in 770.100(B)(2). The reason for a reference that only applies to a building with a grounding means, according to the proposal substantiation, is that this section is only supposed to apply to cables within buildings, and cables outside of buildings use 770.100. That substantiation is both true and beside the point; however, the section says what it says.

770.133. Installation of Optical Fibers and Electrical Conductors. There are three types of optical cables, each with their own system separation rules, as follows:

a. *Composite optical fiber cables* containing non-power-limited control or power circuits are only permitted to be used where the optical fiber and the non-power-limited components are functionally associated. However, once a composite cable qualifies for use under this criterion, it may be used in the same raceway or cable tray as non-power-limited wiring conductor, and these cables may enter other wiring enclosures for termination or other purposes. In industrial occupancies only, with qualified maintenance and supervision, composite cables are even permitted with the power conductor operating at medium voltage (over 600 V).
b. *Conductive optical fiber cables* are not permitted in the same raceway or cable tray as non-power-limited wiring of any kind.
c. *Nonconductive optical fiber cables* are permitted within a common raceway or the same cable tray as non-power-limited wiring of any kind operating at 600 V or less. If the nonconductive fiber cables are functionally associated with the non-power-limited wiring, the cables are permitted in non-power-limited wiring enclosures. Nonconductive optical fiber cables are also permitted in

common wiring enclosures where installed in a factory- or field-assembled control center. In industrial occupancies only, with qualified maintenance and supervision, nonconductive optical fiber cables are also permitted with the power conductors operating at medium voltage (over 600 V).

d. *With power-limited circuits* optical fiber cables are permitted without limitation as to being either conductive or nonconductive, and are permitted to be part of a composite cable with, and to run in the same wiring raceways, cable trays, and other enclosures with copper wiring that is part of a power-limited system. Specifically, optical fiber cabling is permitted with a Class 2 or Class 3 system per Art. 725, or a power-limited fire alarm system per Art. 760, or a communications system as covered in Art. 800, or a CATV system covered in Art. 820, or a network-powered broadband communications system having a low-power source as defined by the applicable parameters in Table 830.15.

Where these requirements specify a separation between optical fiber cabling of any type and electrical power or control circuits of any type, those separation requirements can be met by providing a permanent or listed barrier between the systems. This can be accomplished in cable tray or surface raceway systems by using barriers compatible with the tray or surface raceway, and in an enclosure by either arranging a fixed and grounded permanent barrier or by enclosing the optical fiber cables in a grounded raceway as required. For example, flexible metal conduit is often used within a non-power-limited enclosure to extend either the non-power-limited source wiring or the power-limited or optical fiber cabling from the entry point to a point where system separation need no longer be maintained. Refer to 800.133(B) for detailed coverage of how 770.133(B) should be applied.

Part **(D)** echoes 300.11(B) and disallows the use of a electrical raceways as a support for optical fiber cables tied to its outside. However, there is an exception for optical fibers to run up a mast on its outside, if "intended for the attachment and support of such cables." This can never include a service mast, for which 230.28 absolutely disallows any connections thereto except the service drop conductors, and 225.17 places the same limits for other power circuits.

770.154. Application of Listed Optical Fiber Cables and Raceways. The installation requirements for these cables now reside at 770.110 and 770.113, and the applications have been converted to a full-page table that summarizes the results.

770.179. Optical Fiber Cables. This section covers the listing requirements for the cables used in the applications just covered. It also, as of the 2014 **NEC**, contains essential requirements for field-assembled optical fiber cables. The overall covering must be marked on the surface to indicate the specific "optical fiber conductors" with which it is listed, and the "optical fiber conductors" must bring with them a marking tape that indicates the outer cover with which they are listed. In that way the inspector can verify that the proper components are used with each other. The provision goes on to require that overall covering of a field-assembled optical fiber cable meets the listing requirements for "optical fiber raceways." Of course, this raises several questions, such as how an optical fiber can be a "conductor" and why an "optical fiber raceway" is not actually a "communications raceway" when the former term was supposedly obsolete. The following subsections address the listing requirements for the possible applications:

Part **(A)** covers Types OFNP and OFCP for plenum cavity applications.

Part **(B)** covers Type OFNR and OFCR for riser applications.

Part **(C)** covers Type OFNG and OFCG for general-purpose applications.

Part **(D)** covers Type OFN and OFC for general-purpose applications. The presence or absence of a "G" suffix depends on the test protocol used. A "no suffix" cable may not have been tested to a certain Canadian standard and could be restricted in certain applications within the Canadian market.

Part **(E)** covers optical fiber circuit integrity (CI) cables, which will retain their communications function for a rated time period during a fire exposure through a self-sacrificing ceramifiable outer jacket. This classification follows others throughout the limited energy cabling articles. Technically, CI cables must be installed in free air to retain the validity of that designation. A separate paragraph (2) covers fire-resistive cables. They will have the protective system number on them or on the smallest unit container (note, however, that 728.100 requires the marking on the cable). In either case they must be installed under all the terms of the listing. Refer to the commentary on the new Art. 728 for more information.

Part **(F)** provides specific marking and listing information for field-assembled optical fiber cables that fully address the general information presented in the parent language of the section. Note that in this case the reference is not to optical fiber raceways, but to communications raceways instead.

770.180. Grounding Devices. Equipment employed to bond metallic components of cable assemblies covered in Art. 770 to grounding and bonding conductors must be listed, or be part of a larger component that is listed.

Chapter Eight

ARTICLE 800. COMMUNICATIONS CIRCUITS

800.1. Scope. This article covers communications circuits and equipment. The definition of a "communications circuit" (discussed below) is essential to understanding this scope statement.

Considerable attention has been paid to modernizing the usage of grounding and bonding terms in the communications article to correctly line up with usage now in effect throughout the rest of the **NEC** and within Art. 250. New informational drawings explain the usage to those who were accustomed to old practice.

The paragraph titled "Code Arrangement" of the "Introduction to the Code" (90.3) states that Chap. 8, which includes Art. 800, "Communication Circuits," is independent of the preceding chapters except as they are specifically referred to.

800.2. Definitions. A *communications circuit* extends voice, audio, video, data, interactive services, telegraph (except radio), outside wiring for fire alarm and burglar alarm from the communications utility to the customer's communications equipment up to and including terminal equipment such as a telephone or facsimile machine.

The connection to the facilities of a serving utility is an essential component of this definition. Ethernet connections between computers or between computers and a router are not communications circuits and are covered as signaling circuits in Art. 725.

A *communications raceway* is an enclosed nonmetallic channel for communications cabling.

Exposed (to Accidental Contact) applies to a communications cable that is in such a position that it would be likely to become energized in the event that a support or insulation or both failed.

800.3. Other Articles. This section correlates with Art. 770 to bring in its provisions if optical fiber is used as part of the communications circuit, along with the terminology "cable routing assembly" as defined in Art. 100, together with its installation rules. It also brings in the rules of 300.22 because due to this Chap. 8 location, only Code rules from Chaps. 1 through 7 specifically referenced apply to this work.

800.12. Innerduct. Listed communications raceways can be used by right where pulled into larger Chap. 3 raceways; such applications are now formally defined as innerduct. The number pulled in to the larger raceway is not specified, which is appropriate because any applicable wire/cable fill will be calculated in reference to the specific innerduct being used. The specific style must agree with the application (plenum, riser, etc.).

800.18. Installation of Equipment. With the exception of test equipment used temporarily, all equipment electrically connected to a telecommunications network must be listed as suitable for that purpose.

800.21. Access to Electrical Equipment Behind Panels Designed to Allow Access. It is contrary to the NEC to drape endless quantities of these cables across suspended ceiling panels in a way that limits the intended access to the ceiling cavity above them.

800.24. Mechanical Execution of Work. In addition to the usual neat and workmanlike manner requirement, this section also requires exposed cabling (which includes the area above a suspended ceiling) to be supported to structure such that normal building operations will not damage the cable. In addition, if the wiring is run parallel to framing, it must meet the 32-mm (1¼-in.) spacing requirement from the leading edge of the framing. Finally, the wiring must stay off the tie wires designed to support a suspended ceiling, in accordance with 300.11. Where cable ties and other nonmetallic components are used in 300.22(C) spaces, they must be listed for that environment. An informational note (2) provides a cross-reference to the controlling rule in NFPA 90A.

800.25. Abandoned Cables. This has the effect of requiring that the accessible portions of any "abandoned" conductors—as defined in 800.2 as not being terminated or identified for future use with a tag—must be removed. It is no longer permissible to leave abandoned communications cables in place when they are no longer in use, unless they have been "identified" for future use.

800.26. Spread of Fire or Products of Combustion. This is a restatement of 300.21, as applied to this article.

800.44. Overhead Communications Wires and Cables. This section divides the topic into pole wiring and over rooftops. Part **(A)** covers pole work and requires communications wires to run below power wiring on the same pole if practicable, and never attached to the same cross arm. The climbing space is to be as required in 225.14(D), and there must be a minimum of a 300-mm (12-in.) separation from power wiring at any point in a span, including at their point of attachment at the load end. This reduced spacing only applies if the spacing at the pole meets the required 1-m (40-in.) spacing on the pole. Part **(B)** covers rooftops, and the default clearance is 2.5 m (8 ft), with exceptions for auxiliary buildings such as garages, and also for mast applications and steep roof applications that exactly duplicate the comparable exceptions for service drops operating at not over 300 V.

800.47. Underground Circuits Entering Buildings. Where in manholes with power circuits, barriers must be arranged to create a separate section for the communications circuits. Note that although 800.110 requires installers to run Chap. 3 raceways in accordance with the applicable Chap. 3 rules, which include rules for wet locations, that requirement was never intended to extend to mandating a wet-location criterion for the enclosed cabling. UL 444, the applicable product standard, does not recognize such a rating. This rule specifically waives the normal wet-location rule in Chap. 3. Although this is not required because Chap. 8 stands alone, the cabling manufacturers requested an explicit reference due to repeated misapplications by the inspection community. Note, however, that gel-filled cable constructions ("flooded" cables, so-called) for this work have been around for generations, and have a well-proven track record of endurance in wet and underground applications.

800.48. Unlisted Cables and Raceways Entering Buildings. Unlisted outside plant cables may enter a building and move through general-purpose spaces (not risers or plenums) up to 15 m (50 ft) from the point of entrance. This allows a transition length along which a convenient point can be provided for a transition to the listed cables required for most interior work. If such a point is unavailable, the point of entrance can be artificially extended through the use of grounded heavy-wall steel conduit. The last sentence addresses the artificial extension into a building of a point of entrance, much as 230.6 does for service entrance conductors with concrete encasement, etc. For communications circuits, this is done with heavy-wall conduit. This is an important concept, and it holds for all of Chap. 8. See also the next section, which contains the bonding requirement for the conduit.

800.49. Metallic Entrance Conduit Grounding. Heavy-wall metal conduit containing optical fiber entrance cables must be bonded to a grounding electrode as covered in 800.100(B). This, in combination with the point of entrance definition, creates the remote entry point for this work that is parallel to 230.6 for power service work. In fact, it used to be part of the definition, but was separated because definitions cannot contain requirements. Similar sections occur throughout Chap. 8.

800.50. Circuits Requiring Primary Protectors. The requirement that creates the need for this section is 800.90, discussed below. Part **(A)** covers the drop wire from the last outdoor support to the protector and it must be listed for this purpose, as covered in 800.173. Part **(B)** covers the required 100-mm (4-in.) system separation from open power wiring (not in raceways or cable assemblies) on the outside of buildings. Note that communications wiring outdoors that is exposed (see definition above) to contact with power wiring over 300 V to ground, as on most utility poles, must be separated from building woodwork by wiring on porcelain knobs, etc., unless fuseless protectors are used or where the exposed wiring has a grounded metallic sheath.

Part **(C)** covers circuits entering a building ahead of the primary protector. The building must be protected from the unprotected wiring through a bushing or metal raceway, unless the entry is through masonry, or uses metal sheathed cable, or qualifies for a fuseless primary protector (see 800.173 for a cable assembly that inherently meets this requirement) or is listed drop cable extending from circuits running with a metallic sheath. Raceways must slope up in the out-to-in direction or else drip loops must be provided outside. Raceways must also have a service head, and if ahead of the protector, must be grounded.

800.90. Protective Devices. Part **(A)** covers where primary protectors are required, which can be summarized with respect to the entering circuits as wherever there is a lightning exposure or exposure to contact by power wiring over 300 V to ground. The second note provides good information as to where a lightning exposure might be disregarded, such as in high-rise construction areas where lightning will be intercepted, or for short interbuilding cable runs not over 42 m (140 ft) apart and running as direct burial or in underground conduit with either a metallic cable shield or the conduit being metallic. Areas on the west coast and in other locations with five or fewer thunderstorm days a year are also candidates to avoid this requirement. (A)(1) gives five instances where a fuseless protector is permitted; (A)(2) covers the fused protectors that are required in the event that local conditions do not permit qualification under one of those five allowances.

Part **(B)** covers the location of primary protectors, which is in, on, or immediately adjacent to the building served, as near as practicable to the point of entrance. And note the definition of that term; it is possible to extend that point through the use of grounded heavy-wall conduit. The second paragraph covers mobile homes, and accepts a protector at the remote disconnecting means as long as the distance is within NEC requirements in 550.32(A).

800.93. Grounding or Interruption of Metallic Sheath Members of Communications Cables. This rule is split into a termination outside or inside a building, but the action to be taken is the same in either event: Make a grounding connection as covered in 800.100, or place an insulating joint as near as practicable to the point of entry. Here again that point of entry can be extended through the use of conduit.

800.100. Cable and Primary Protector Bonding and Grounding. Both the protector and any metallic members of a cable sheath must be grounded in accordance with the rules of this section.

Part **(A)** requires the grounding conductor to be listed and insulated, made of copper (or other corrosion-resistant material), and stranded or solid. The size must be no smaller than 14 AWG, or larger if needed to equal the current carrying capacity of the grounded metallic elements of the sheath, or of the protected conductors in the cable. It must run in as straight a line as practical, and be guarded from physical damage. If run in a metal raceway, both ends must be bonded even though 250.64(E) only imposes that requirement on ferrous raceways. It must be as short as practicable, and not over 6 m (20 ft) in one- and two-family dwellings. If the 6-m (20-ft) limitation can't be met, place an additional electrode as covered in 800.100 and then bond this electrode to the power system electrode with a 6 AWG or larger copper wire.

Part **(B)** covers electrode terminations. If there is an intersystem bonding termination, then the termination is to that point. If the intersystem bonding point is lacking but there is one or more grounding electrodes, the connection must be made to that grounding system through one of the optional methods listed. If the premises served have no intersystem bonding termination and no grounding electrode system, then the bonding connection is to the nearest traditional building electrode (water pipe, building steel, concrete-encased electrode, ground ring), and if all those are missing then to a 1.5-m (5-ft) ground rod, driven into permanently damp soil to the extent practicable, using connections that comply with 250.70.

Part **(D)** requires that this electrode, although required to be at least 1.8 m (6 ft) distant from other electrodes, must be bonded to the power system grounding electrode system with a conductor no smaller than 6 AWG copper. The power system bonding requirement is presumably waived at mobile homes for which the connection would be impractical. This is the apparent result of the wording of an exception that, in violation of the Style Manual, is not in the form of a complete sentence and therefore unclear as to effect. Refer to the discussion at 800.106 for more information.

800.106. Grounding of Entrance Cables at Mobile Homes. This is very difficult to follow, because there is no straightforward explanation of what happens when the mobile home has an electrical supply that complies with the NEC. Specifically, mobile homes are supposed to have a bonding terminal on their chassis at an accessible location, as required in 550.16(C)(1). They are also supposed to have either service equipment or a feeder disconnect located within 9.0 m (30 ft), per 550.32(A). If they have actual service equipment at that location, or if they have a feeder disconnect with a regrounded neutral at that point, there will be a grounding electrode system connected at that disconnect. In fact, even if the disconnect does not include a regrounded neutral, 250.32(A) still requires a grounding electrode connection at this point. Assuming the communications entry comes from this location, a bonding connection between the conductive elements in the communications cable jacket and the grounding system at the disconnect will meet all requirements and no further action needs to be taken at the mobile home. If, and only if, the disconnect is located at an excessive distance or grounded incorrectly does the necessity of creating a local electrode at the mobile home kick in. The context for this rule is at 800.90(B), second paragraph.

This brings up the separate question of bonding. If the mobile home is permanently connected to a Code-compliant disconnect, then the permanent equipment grounding system connecting the mobile home and the disconnect is sufficient for these purposes. However, if the mobile home is cord-and-plug-connected, removing the plug would disrupt this continuity. Article 800 systems have primary protectors that will not exist on most other systems. That leaves only one option, a grounding electrode, and therefore, through this back door, a local grounding electrode becomes required in the event of a cord-and-plug connection. The grounding electrode must be connected to the conductive elements of the communications cable, and also to the mobile home bonding terminal as previously mentioned.

800.110. Raceways for Communications Wires and Cables. Raceways can be Chap. 3 methods or communications raceways as covered later. These raceways are very close to ENT in construction and this section even mandates the use of the installation provisions in Art. 362 having to do with bends, supports, joints, and perhaps most intriguingly, 362.56 which invokes the box-at-termination rules in 300.15. However, no raceway fill limitations apply.

Part **(C)**, new as of the 2014 NEC, covers cable routing assemblies. These are usually open on top and serve as continuous support for cables. They are customized by application, as in general-purpose, riser, or plenum, with a comparable substitution hierarchy. The default support intervals are 3 ft for horizontal runs and 4 ft for vertical runs, with an allowance for longer spans if part of the listing, but never over 10 ft.

800.113. Installation of Communications Wires, Cables and Raceways. There are some general principles at work that make this less complicated than it appears. The entire two pages boils down to the following:

1. The hierarchy of cable compositional categories, and by logical extension the hierarchy of compositional categories for the nonmetallic raceways and supporting assemblies that contain those cables, continue without change. Refer to the discussion at 725.135 regarding the history and applications of these categories, which have not changed since the advent of this work in the 1987 NEC. Specifically, there are plenum exposures, which require the most severe testing for both smoke generation and vertical flammability, and result in a "P" suffix on cable assemblies. There are riser exposures that are the next most severely tested, focusing on vertical flammability, and resulting in an "R" suffix on cable assemblies. Then there are general-purpose applications that receive general fire load testing and result in no suffix. Residential grades (suffix "X") apply to one- and two-family dwellings less than 6 mm (¼ in.) in diameter, and for multifamily dwellings in nonconcealed spaces. This is a hierarchy, so general-purpose acceptability includes residential, riser suitability implies general-purpose suitability as well, and the plenum grade can be used in all locations. Under carpet cable (suffix "UC") is suitable for general-purpose applications as described.
2. If a cable is pulled into a Chap. 3 raceway, its suitability follows that of the enclosing Chap. 3 raceway, just as surely as the THHN or RHW or XHHW-2 that could otherwise be pulled into the same raceway without a separate evaluation of the fire resistance of its insulation.
3. Fireproof shafts with firestopping at every floor cancel the need for special riser ratings, and one- and two-family dwelling exposures, even vertical ones, do not merit a riser rating either.

800.133. Installation of Communications Wires, Cables, and Equipment. Part **(A)** sets forth the system separation rules, which are comparable to Art. 725 provisions. Part **(A)(1)** covers raceways, cable trays, boxes, and cable assemblies, in three headings, the first being uses with "other circuits." This raises the editorial problem of beginning a discussion of three alternatives with "other," as in other than what? So, this analysis begins with the last item (c) covering non-power limited circuits. The rule is no common locations allowed, but there are three exceptions. The first recognizes a permanent barrier or listed divider. The second exception allows a common enclosure where the power-limited and non-power-limited wiring must terminate on the same equipment, as in the case of some powered equipment with a communications interface. In this case there is a general solution and a limited alternative. The general solution is to restrain the wiring with duct, tie-downs, or other methods such that a 6-mm (¼-in.) air separation is maintained between systems. Note that this provision is for dead-end wiring only, not daisy chains ("solely for power supply to" does not mean to connect and then to feed another piece of equipment). The third exception correlates with 620.36 with respect to the traveling cable.

The second, "non-other" item is (b), covering communications circuits sharing a cable assembly with Class 2 and Class 3 circuits. In such cases the combination is allowed but the entire assembly becomes reclassified as a communications cable; and the Class 2 or Class 3 circuit is now an Art. 800 circuit and must meet the listing requirements of this article. On the other hand, there are composite cables comprised of jacketed signaling cables alongside of jacketed communications cables, all within a common overall jacket, and in this case the signaling circuit is permitted, but not required, to be reclassified as a communications circuit. Note that even though this rule does not expressly cover the topic, Type NMS cable relies on this concept, and UL has listed this sort of hybrid cable for many years, as expressly covered in 334.116(C). Such cable assemblies do not depend on conductor insulation alone to establish the required system separation, squarely in accordance with the spirit of these rules; however, the literal text of this wording is in direct conflict with the Chap. 3 permission.

What remains therefore, in (a), is that which is therefore "other." And this would cover the use of communications cables (as distinguished from conductors sharing a cable assembly) with other power-limited circuits in raceways and enclosures. The context for these rules is that communications circuits have a voltage exposure that is comparable to power-limited fire alarm or Class 3 signaling circuits. Therefore, the following systems can share an enclosure with communications cables: (1) Class 2 and Class 3 cabling—Art. 725 (or in Art. 645); (2) Power-limited fire alarm wiring—Art. 760; (3) Optical fiber cabling both conductive and nonconductive—Art. 770; (4) CATV systems—Art. 820; and (5) Low-power network-powered broadband communications—Art. 830.

Part **(A)(2)** requires a 50-mm (2-in.) separation from open power wiring and other open, non-power-limited circuit conductors. Two exceptions apply, the first where a raceway on either system provides separation, or where the non-power-limited wiring uses Chap. 3 cabled wiring methods. The second applies where fixed barriers such as porcelain tubes or flexible tubing are installed.

Part **(B)** covers conductor support and states that cables must not be tie-wrapped to or otherwise supported by electrical raceways, as also disallowed in 300.11(B). However, there is an exception for communications cables to run up a mast on its outside if it is "intended for the attachment and support of such cables." This can never include a service mast, for which 230.28 absolutely disallows any connections thereto except the service drop conductors. Part **(C)** brings in the prohibition against ductwork locations subject to loose stock and vapor removal in 300.22(A). This may seem odd and commonsensical, unworthy of mention here, but Chap. 8 articles carry no baggage from the first seven chapters, except only that which they go back to get. So, this is necessary.

800.154. Applications of Listed Communications Wires and Cables and Communications Raceways and Listed Cable Routing Assemblies. The installation requirements for these cables now reside at 800.110 and 800.113, and the applications have been converted to several tables that summarize the result, beginning with Table 154a covering the wires and cables, and then Table 154b covering communications raceways, and then Table 154c covering cable routing assemblies. The cable substitution hierarchy table and its associated figure remain here, as Table 800.154d (literally appearing in Fig. 800.154).

800.156. Dwelling Unit Communications Outlet. In new construction, at least one communications outlet must be provided within all dwelling units and cabled to the service demarcation point of the communications utility. This is true regardless of plans to rely on cellular phone service.

800.170. Equipment. Reiterating 800.18, all communications equipment, even a lowly RJ-11 jack that will be field installed, must be listed as suitable for connection to a telecommunications network. As of the 2014 NEC, cable ties used in plenum cavities must be listed for that application as well.

800.173. Drop Wire and Cable. Communications wires without a metallic shield that run from the last outdoor support to the primary protector must be listed as being suitable for the purpose, and shall have the current-carrying capacity as covered in 800.90(A)(1)(b) or (c). The last condition means that they will "safely fuse on all currents greater than the current-carrying capacity of the primary protector" and thereby qualify the installation as not requiring a fused protector.

800.179. Communications Wires and Cables. This section includes general rating provisions that apply to all iterations of this cabling. Communications wires and the cables (other than the outer shield on coaxial cable) must be insulated for 300 volts, but the actual voltage rating must not be marked on the cable, unless required by other listings. The concern is the likelihood unqualified persons may interpret such a marking as indicative as suitable for power applications. This wiring must be suitable for 60°C or higher, and if higher than that default value, the temperature rating must be marked accordingly. The listing requirements for these cables follow:

Part **(A)** covers Types CMP for plenum cavity applications.

Part **(B)** covers Type CMR for riser applications.

Part **(C)** covers Type CMG for general-purpose applications.

Part **(D)** covers Type CM for general-purpose applications. The presence or absence of a "G" suffix depends on the test protocol used. A "no suffix" cable may not have been tested to a certain Canadian standard and could be restricted in certain applications within the Canadian market.

Part **(E)** covers Type CMX for limited use in residential applications.

Part **(F)** covers Type CMUC undercarpet wires and cables.

Part **(G)** covers communications circuit integrity (CI) cables. These are additionally marked with the designation "CI." This part is now split in order to make a useful distinction between CI cable itself and CI cable as a component of an electrical circuit protective system. Any given CI cable that relies on a raceway to protect it must be specifically listed and marked as part of such a system. The best approach is to follow Art. 728 and the listing specifications exactly. Refer to the commentary in that article for more information on this point.

Part **(H)** covers communications wires, as opposed to cables, as used in cross-connect arrays and distributing frames.

Part **(I)** covers hybrid power and communications wire, typically cross-listed as a composite of both Type NM and Type CM.

800.180. Grounding Devices. Equipment employed to bond metallic components of cable assemblies covered in Art. 800 to grounding and bonding conductors must be listed, or be part of a larger component that is listed.

800.182. Communications Raceways and Cable Routing Assemblies. Parts **(A)**, **(B)**, and **(C)** cover communications raceways, one for each basic level of the hierarchy. As previously noted, the use of one of these raceways does not relieve the installer of the responsibility to pull in communications cables with equivalent fire resistance into the raceway. For example, a plenum communications raceway in an air-handling plenum cavity per 300.22(C) must have CMP cabling pulled into it. This section also applies to cable routing assemblies.

ARTICLE 810. RADIO AND TELEVISION EQUIPMENT

810.1. Scope. This article covers antenna systems for radio and television reception, and amateur radio transmission, as well as citizen band equipment. The article does not cover equipment that couples signaling impulses to power line conductors. The 2014 NEC has clarified that parabolic and flat antennas are included in the scope of this article.

810.4. Community Television Antenna. The antenna of such a system is covered here. Any coaxial distribution is covered under Art. 820.

810.5. Radio Noise Suppressors. This equipment must be listed and in a protected location.

810.6. Antenna Lead-In Protectors. This section, new as of the 2014 NEC, provides the listing requirements for these devices, along with the field grounding requirement to install per 810.21(F).

810.7. Grounding Devices. Equipment employed to bond metallic components of cable assemblies, or metal parts of equipment or antennas to grounding and bonding conductors must be listed, or be part of a larger component that is listed.

810.11. Material. Except as used for a short lead-in (under 11 m or 35 ft), antenna and lead-in conductors must be hard drawn copper or other high-strength, corrosion-resistant material.

810.12. Supports. Outdoor antennas must be securely supported and not attached to a service mast, or any pole or similar structure carrying power circuits over 250 V line-to-line. Insulators must be strong enough to secure the antenna and the lead-in must be securely attached to the antenna.

810.13. Avoidance of Contacts with Conductors of Other Systems. For service drops and conductors on the exteriors of buildings, the requirements for insulating covering and methods of installation depend on the likelihood of crosses occurring between signal conductors and light or power conductors. Antenna wiring must be kept where it will not come into contact with power wiring, and a 600-mm (2-ft) minimum spacing is required.

810.14. Splices. The antenna may unavoidably be so located that in case of a break in the wire, it may come in contact with electric light or power wires. For this reason, the wire should be of sufficient size to have considerable mechanical strength. Splices must be made so the conductors are not appreciably weakened.

810.15. Grounding. Masts and metal structures supporting antennas must be grounded as covered in 810.21. This includes antennas for satellite TV reception.

The 2017 NEC, for the first time, is incorporating a specific lightning protection concept from NFPA 780, *Standard for the Installation of Lightning Protection Systems* into a mandatory provision. In this case a satellite television provider successfully argued for relief from the necessity to ground all antennas.

Specifically, an antenna within the scope of Art. 810 need not be grounded if it is within a zone of protection extending from an adjacent structure. The concept, known as the rolling sphere, is that a lightning discharge occurs after a downward leader from a cloud, moving in jagged steps based on the voltage, contacts an upward leader from an object on the ground. The downward progression is not smooth, advancing in a series of jumps in varying directions. These jumps typically don't exceed 150 ft. Therefore, if you roll a sphere of that radius around adjacent objects, and your (in this case) antenna never pierces the underside of the sphere as it rolls by, then any downward leader will first intercept the adjacent structure. To some extent the protection is a function of probability. A leader with enormous charge behind it will move with longer branching jumps, suggesting the radius of the sphere be decreased to decrease the probability of being struck. However, the radius chosen in NFPA 780 (based on a charge that will result in a 10 kA peak discharge) will intercept 91 percent of all lightning events.

810.16. Size of Wire-Strung Antenna—Receiving Stations. For wire-strung antennas, Table 810.16(A) gives the minimum AWG sizes based on span. Part (B) covers self-supporting antennas, including the ones for satellite TV reception. Here again they must be located away from overhead power drops.

810.17. Size of Lead-In—Receiving Station. The tensile strength of a lead-in must at least equal that of the antenna in 810.16.

810.18. Clearances—Receiving Station. Lead-ins must not swing closer than 600 mm (2 ft) to power circuits up to 250 V wire-to-wire, and not within 3 m (10 ft) of higher voltage wiring. If the wiring is tied down on insulators or otherwise secured and not over 150 V between conductors, the spacing can come down to 100 mm (4 in.). A lead-in must never be closer than 1.8 m (6 ft) to a lightning down conductor. Underground lead-ins must be at least 300 mm (12 in.) from power, lighting, or Class 1 circuits, unless either the power side or the lead-in side is run in a raceway or metal cable armor.

Part **(B)** covers the lead-ins where they run indoors. In this case the separation is 50 mm (2 in.), also allowed closer if the systems are permanently separated through the use of fixed nonconductors or the power side is using metal raceways or cable armor. The wiring can enter the same box as long as a permanently installed barrier keeps the systems apart.

810.19. Electrical Supply Circuits Used in Lieu of Antenna—Receiving Stations. The coupling device between the supply circuit wiring and the antenna leads must be listed.

810.20. Antenna Discharge Units—Receiving Stations. Every lead-in conductor from an outdoor antenna must be provided with a listed antenna discharge unit, unless it is enclosed in a continuous metallic shield that is grounded per 810.21, or protected with its own antenna discharge unit. Antenna discharge units must be located as near as practicable to the point of entrance into the building they protect, and ahead of any radio set. They must be grounded in accordance with 810.21.

810.21. Bonding Conductors and Grounding Electrode Conductors—Receiving Stations. Grounding conductors must comply with (A) through (K).

The NEC requires the grounding conductor to be made of copper (or other corrosion-resistant material), and stranded or solid. Insulation is not required and it can run inside or outside the building. The size must be no smaller than 10 AWG (or 8 AWG aluminum, or 17 AWG bronze or copper-clad steel). It must run in as straight a line as practical, securely fastened in place, and be guarded from physical damage. If run in a metal raceway, both ends must be bonded even though 250.64(E) only imposes that requirement on ferrous raceways. A grounding electrode conductor or bonding conductor can also be used for operational purposes at a ground station.

Part **(F)** covers electrode terminations. If there is an intersystem bonding termination, then the termination is to that point. If the intersystem bonding point is lacking but there are one or more grounding electrodes, the connection must be made to that grounding system through one of the optional methods listed. If the premises served have no intersystem bonding termination and no grounding electrode system, then the bonding connection is to the nearest traditional building electrode (water pipe, building steel, concrete-encased electrode, ground ring), and if all those are missing then to a 1.5 m (5 ft) ground rod, driven into permanently damp soil to the extent practicable, using connections that comply with 250.70.

810.52. Size of Antenna. The minimum size of antenna conductors for citizen band and for ham radio transmission/reception operations is given in Table 810.52, which is also the size of the lead-in as specified in 810.53.

810.54. Clearance on Building. Antenna conductors must be firmly mounted at least 75 mm (3 in.) clear of the surface wired over, and the lead-in conductors must observe the same clearance. If, however, coaxial cable ("continuous metal shield") is used for the lead-in, the clearance rule disappears provided the coaxial cable shield is grounded to the station grounding conductor as covered in 810.58.

810.55. Entrance to Building. If a coaxial cable with a 810.58 grounding connection is not used for the lead-in wire, the lead-in conductor must enter through an insulating nonabsorbent tube or bushing, or through an opening arranged so the entrance wires can be firmly secured with a 50-mm (2-in.) clearance, or through a drilled hole in a pane of glass.

810.56. Protection Against Accidental Contact. Antenna lead-ins must be positioned so inadvertent contact with them would be difficult.

810.57. Antenna Discharge Units—Transmitting Stations. Every lead-in must have an antenna discharge unit, or other suitable method to drain off static charge. This could be as simple as a connection to the station ground when the station is not operating.

810.58. Bonding Conductors and Grounding Electrode Conductors—Amateur and Citizen Band Transmitting and Receiving Stations. The grounding conductors must comply with the minimum provisions of 810.21, with the size no smaller than as in 810.21(H), but also no smaller than the lead-in. The operating grounding conductor must not be smaller than 14 AWG.

810.70. Clearance from Other Conductors. All station conductor wires must be at least 100 mm (4 in.) from power or signaling wiring, unless separated from them with a securely fastened insulator such as a porcelain tube or loom. This section also has an exception, not in the form of a complete sentence and therefore of uncertain meaning, pointing to Art. 640. This exception is probably obsolete. It was in the NEC over 50 years ago when Art. 640 was firmly rooted in the vacuum tube era, and it is unlikely to have any meaning with today's technology.

810.71. General. Transmitters must be enclosed in a grounded metal enclosure, and any external controls must have their operating handles connected to an equipment grounding conductor as well, if the controls are accessible to operating personnel, assuming the transmitter is powered from the premises wiring system, or grounded using the 810.21 system if otherwise. The transmitter enclosure doors must have interlocks that will shut down any voltages above 350 between conductors when any access door opens.

ARTICLE 820. COMMUNITY ANTENNA TELEVISION AND RADIO DISTRIBUTION SYSTEMS

820.2. Definitions. A *coaxial cable* is a cylindrical assembly comprising a central signaling conductor surrounded by a dielectric and then set within a metallic tube or shield, and usually covered by an insulating jacket.

Exposed (to Accidental Contact) applies to a CATV cable that is in such a position that it would be likely to become energized in the event that a support or insulation or both failed.

820.3. Other Articles. This section correlates with Art. 770 to bring in the terminology "cable routing assembly" as defined in 770.2, together with its installation rules. It also brings in the rules of 300.22 because due to this Chap. 8 location, only Code rules from Chaps. 1 through 7 specifically referenced apply to this work.

820.15. Power Limitations. Coaxial cable on systems within the scope of this article is permitted to provide up to 60 V in the process of providing limited power to some equipment directly associated with the system.

820.21. Access to Electrical Equipment Behind Panels Designed to Allow Access. It is contrary to the NEC to drape endless quantities of these cables across suspended ceiling panels in a way that limits the intended access to the ceiling cavity above them.

820.24. Mechanical Execution of Work. In addition to the usual neat and workmanlike manner requirement, this section also requires exposed cabling (which includes the area above a suspended ceiling) to be supported to structure such that normal building operations will not damage the cable. In addition, if the wiring is run parallel to framing, it must meet the 32-mm (1¼-in.) spacing requirement from the leading edge of the framing. Finally, the wiring must stay off the tie wires designed to support a suspended ceiling, in accordance with

300.11. Where cable ties and other nonmetallic components are used in 300.22(C) spaces, they must be listed for that environment. An informational note (2) provides a cross-reference to the controlling rule in NFPA 90A.

820.25. Abandoned Cables. This has the effect of requiring that any "abandoned" conductors—as defined in 820.2 as not being terminated or identified for future use with a tag—must be removed. It is no longer permissible to leave abandoned CATV cables in place when they are no longer in use, unless they have been "identified" for future use.

820.26. Spread of Fire or Products of Combustion. This is a restatement of 300.21, as applied to this article.

820.44. Overhead (Aerial) Coaxial Cables. Part **(A)** covers pole and drop work and requires CATV wires to run below power wiring on the same pole if practicable, and never attached to the same cross arm. The climbing space is to be as required in 225.14(D), and there must be a minimum of 300 mm (12 in.) separation from power wiring at any point in a span, including at their point of attachment at the load end. This reduced spacing only applies if the spacing at the pole meets the required 1 m (40 in.) spacing on the pole.

Part **(B)** covers roof clearances, and the default clearance is 2.5 m (8 ft), with exceptions for auxiliary buildings such as garages, and also for mast applications and steep roof applications that exactly duplicate the comparable exceptions for service drops operating at not over 300 V.

Part **(C)** covers coaxial cables to run up a mast on its outside if it is "intended for the attachment and support of such cables." This can never include a service mast, for which 230.28 absolutely disallows any connections thereto except the service drop conductors. This prohibition is accurately conveyed in the statement of the rule.

Part **(D)** covers interbuilding runs, and requires secure supports at the points of attachment. The cable must be identified for this purpose, which has been clarified in the equivalent passage in Art. 830 [830.44(F)] as being "suitable for outdoor aerial applications." The exception following provides for a messenger in the event the span is too much for the CATV on its own.

Part **(E)** covers wiring on the outside of buildings. The CATV cabling must be not less than 100 mm (4 in.) from non-power-limited circuits, or else separated with some fixed nonconductive barrier. In addition, the CATV wiring must be routed to avoid interference with adjacent communications wiring. Finally, cables on buildings must avoid separations of less than 1.8 m (6 ft) from lightning protection system conductors wherever practicable.

820.47. Underground Circuits Entering Buildings. Where in manholes with power circuits, barriers must be arranged to create a separate section for the communications circuits. Direct-buried cables must be separated not less than 300 mm (12 in.) from direct-buried non-power-limited circuits, such as service laterals run as single-conductor Type USE. If the power circuits are in a recognized wiring method, either raceway or recognized cable assembly including multiconductor Type USE or jacketed Type MC listed for direct burial, then the spacing rule does not apply.

820.48. Unlisted Cables and Raceways Entering Buildings. Unlisted outside plant cables may enter a building and move through general-purpose spaces (not risers or plenums) up to 15 m (50 ft) from the point of entrance. This allows a transition length along which a convenient point can be provided for a transition to the listed cables required for most interior work. If such a point is unavailable, the point of entrance can be artificially extended through the use of grounded heavy-wall steel conduit. The last sentence addresses the artificial extension into a building of a point of entrance, much as 230.6 does for service entrance conductors with concrete encasement, etc. For communications circuits, this is done with heavy-wall conduit. This is an important concept, and it holds for all of Chap. 8. See also the next section, which contains the bonding requirement for the conduit.

820.49. Metallic Entrance Conduit Grounding. Heavy-wall metal conduit containing optical fiber entrance cables must be bonded to a grounding electrode as covered in 820.100(B). This, in combination with the point of entrance definition, creates the remote entry point for this work that is parallel to 230.6 for power service work. In fact, it used to be part of the definition, but was separated because definitions cannot contain requirements. Similar sections occur throughout Chap 8.

820.93. Grounding of the Outer Conductive Shield of Coaxial Cables. This rule is split into a termination outside or inside a building, but the action to be taken is the same in either event: Make a grounding connection as covered in 820.100 as near as practicable to the point of entry. Here again that point of entry can be extended through the use of conduit. Part **(C)** covers a primary protector, should one be installed (not required as of this Code). If used, it must be listed and applied at the point of entrance. And, per **(D)**, it must not be located in a hazardous (classified) location.

820.100. Cable Bonding and Grounding. Both the protector and any metallic members of a cable sheath must be grounded in accordance with the rules of this section. A new exception effectively correlates this grounding requirement with allowances in 840.101(A). Refer to the discussion at that point. The grounding procedures that follow address coaxial cables that originate off-site and that are subject to lightning exposures. This exception has been extended as of the 2017 NEC through the incorporation of a concept taken from NFPA 780. Refer to a more detailed discussion in 810.15.

Part **(A)** requires the grounding conductor to be listed and insulated, made of copper (or other corrosion-resistant material), and stranded or solid. The size must be no smaller than 14 AWG. It must run in as straight a line as practical, and be guarded from physical damage. If run in a metal raceway, both ends must be bonded even though 250.64(E) only imposes that requirement on ferrous raceways. It must be as short as practicable and not over 6 m (20 ft) in one- and two-family dwellings. If the 6-m (20-ft) limitation can't be met, place an additional electrode as covered in 820.100 and then bond this electrode to the power system electrode with a 6 AWG or larger copper wire.

Part **(B)** covers electrode terminations. If there is an intersystem bonding termination, then the termination is to that point. If the intersystem bonding point is lacking but there is one or more grounding electrodes, the connection must be made to that grounding system through one of the optional methods listed. If the premises served have no intersystem bonding termination and no grounding electrode system, then the bonding connection is to the nearest traditional building electrode (water pipe, building steel, concrete-encased electrode, ground ring), and if all those are missing then to a 1.5-m (5-ft) ground rod, driven into permanently damp soil to the extent practicable, using connections that comply with 250.70.

New language added in the 2014 **NEC** clarifies that steam or hot water piping, or down conductors from air terminals on a lightning protection system, must never be employed for the required bonding and grounding functions. This latter provision correlates with 250.60 for conventional grounding systems. Note, however, that 250.106 does require the lightning protection grounding system terminals to be bonded to the grounding electrode system in the building. And the system set up in this section will either use that system as well or be bonded to it.

Part **(D)** requires that this electrode, although required to be at least 1.8 m (6 ft) distant from other electrodes, must be bonded to the power system grounding electrode system with a conductor no smaller than 6 AWG copper. The power system bonding requirement is presumably waived at mobile homes for which the connection would be impractical. This is the apparent result of the wording of an exception that, in violation of the Style Manual, is not in the form of a complete sentence and therefore unclear as to effect. Refer to the discussion at 820.106 for more information.

Part **(E)** correlates with 820.93(C) and opens the door to protective devices that are being developed to shunt excessive shield currents so the shields do not incinerate in response to current imposed by a failure in a power system neutral with the coaxial shield running in parallel and fully accessible through the electrode bonding rules. These devices divert 60 Hz current while blocking high-frequency current, thereby keeping the principal function of the cable shielding in operation.

820.103. Equipment Grounding. This section specifically recognizes that the coaxial shield is an equipment grounding conductor when connected to the equipment being supplied by it.

820.106. Grounding of Entrance Cables at Mobile Homes. This is very difficult to follow, unless you factor in the wording in the first paragraph of 820.93. Specifically, mobile homes are supposed to have a bonding terminal on their chassis at an accessible location, as required in 550.16(C)(1). They are also supposed to have either service equipment or a feeder disconnect located within 9.0 m (30 ft), per 550.32(A). If they have actual service equipment at that location, or if they have a feeder disconnect with a regrounded neutral at that point, there will be a grounding electrode system connected at that disconnect. In fact, even if the disconnect does not include a regrounded neutral, 250.32(A) still requires a grounding electrode connection at this point. Assuming the communications cable entry comes from this location, a bonding connection between the conductive elements in the communications cable and the grounding system at the disconnect will meet all requirements and no further action needs to be taken at the mobile home. If, and only if, the disconnect is located at an excessive distance or grounded incorrectly does the necessity of creating a local electrode at the mobile home kick in. The philosophical context for this rule is at 800.90(B), second paragraph.

This brings up the separate question of bonding. If the mobile home is permanently connected to a Code-compliant disconnect, then the permanent equipment grounding system connecting the mobile home and the disconnect is sufficient for these purposes. However, if the mobile home is cord-and-plug-connected, removing the plug would disrupt this continuity. In this case there is a conundrum because this wording was brought over from 800.106. Article 800 (telephone) systems have primary protectors that will not exist on these systems. That leaves only one option, a grounding electrode, and therefore, through this back door, a local grounding electrode becomes required in the event of a cord-and-plug connection. The grounding electrode must be connected to the conductive elements of the CATV cable, and also to the mobile home bonding terminal as previously mentioned.

820.110. Raceways and Cable Routing Assemblies for Coaxial Cables. Raceways can be Chap. 3 methods or CATV raceways as covered later. These raceways are very close to ENT in construction and this section even mandates the use of the installation provisions in Art. 362 having to do with bends, supports, joints, and perhaps most intriguingly, 362.56 which invokes the box-at-termination rules in 300.15. On the other hand, conduit fill limitations do not apply. Innerducts are also suitable for routing these cables.

Part **(C)**, new as of the 2014 **NEC**, covers cable routing assemblies. These are usually open on top and serve as continuous support for cables. They are customized by application, as in general-purpose, riser, or plenum, with a comparable substitution hierarchy. The default support intervals are 3 ft for horizontal runs and 4 ft for vertical runs, with an allowance for longer spans if part of the listing, but never over 10 ft.

820.113. Installation of Coaxial Cables. There are some general principles at work that make this less complicated than it appears. The entire two pages of Code text boils down to the following:

1. The hierarchy of cable compositional categories, and by logical extension the hierarchy of compositional categories for the nonmetallic raceways and supporting assemblies that contain those cables, continues without change. Refer to the discussion at 725.135 regarding the history and applications of these categories, which have not changed since the advent of this work in the 1987 **NEC**. Specifically, there are plenum exposures, which require the most severe testing for both smoke generation and vertical flammability, and result in a "P" suffix on cable assemblies. There are riser exposures that are the next most severely tested, focusing on vertical flammability, and result in a "R" suffix on cable assemblies. Then there are general-purpose applications that receive general fire load testing and result in either a "G" suffix or no suffix. Residential grades (suffix "X") apply to cables less than 10 mm (3⁄8 in.) in diameter, and for general usage up 3 m (10 ft) in nonconcealed spaces. This is a hierarchy, so general-purpose acceptability includes residential, riser suitability implies general-purpose suitability as well, and the plenum grade can be used in all locations.
2. If a cable is pulled into a Chap. 3 raceway, its suitability follows that of the enclosing Chap. 3 raceway, just as surely as the THHN or RHW or XHHW-2 that could otherwise be pulled into the same raceway without a separate evaluation of the fire resistance of its insulation.

3. Fireproof shafts with firestopping at every floor cancel the need for special riser ratings, and one- and two-family dwelling exposures, even vertical ones, do not merit a riser rating either.

820.133. Installation of Communications Wires, Cables, and Equipment. Part **(A)** sets forth the system separation rules, which are comparable to Art. 725 provisions. Part **(A)(1)** covers raceways, cable trays, boxes, and cable assemblies, in three headings, the first being uses with "other circuits." This raises the editorial problem of beginning a discussion of two alternatives with "other," as in other than what? So, this analysis begins with the last item (b) covering non-power-limited circuits. The rule is no common locations allowed, but there are two exceptions. The first recognizes a permanent barrier or listed divider. The second exception allows a common enclosure where the power-limited and non-power-limited wiring must terminate on the same equipment, as in the case of some powered equipment with a coaxial cable interface. In this case the solution is to restrain the wiring with duct, tie-downs, or other methods such that a 6-mm (¼-in.) air separation is maintained between systems. Note that this provision is for dead-end wiring only, not daisy chains ("solely for power supply to" does not mean to connect and then to feed another piece of equipment).

What remains therefore, in (a), is that which is therefore "other." And this would cover the use of coaxial cables with other power-limited circuits in raceways and enclosures. The context for these rules is that communications circuits have a voltage exposure that is comparable to power-limited fire alarm or Class 3 signaling circuits. Therefore, the following systems can share an enclosure with communications cables: (1) Class 2 and Class 3 cabling—Art. 725 (or in Art. 645); (2) Power-limited fire alarm wiring—Art. 760; (3) Optical fiber cabling both conductive and nonconductive—Art. 770; (4) communications systems—Art. 800; and (5) Low-power network-powered broadband communications—Art. 830.

Part **(A)(2)** requires a 50-mm (2-in.) separation from open power wiring and other open, non-power-limited circuit conductors. Two exceptions apply, the first where a raceway on either system provides separation, or where the non-power-limited wiring uses Chap. 3 cabled wiring methods. The second applies where fixed barriers such as porcelain tubes or flexible tubing are installed.

Note that even though this rule does not expressly cover the topic, Type NMS cable relies on this concept, and UL has listed this sort of hybrid cable for many years, as expressly covered in 334.116(C). Such cable assemblies do not depend on conductor insulation alone to establish the required system separation, squarely in accordance with the spirit of these rules; however, the literal text of this wording is in direct conflict with the Chap. 3 permission.

Part **(B)** covers conductor support and states that cables must not be tie-wrapped to or otherwise supported by electrical raceways, as also disallowed in 300.11(B). However, there is an exception for coaxial cables to run up a mast on its outside if it is "intended for the attachment and support of such cables." This can never include a service mast, for which 230.28 absolutely disallows any connections thereto except the service drop conductors. The correct statement of this rule is found at 820.44(C).

820.154. Applications of Listed CATV Cables. The installation requirements for these cables now reside at 820.113, and the applications have been converted to a full-page table that summarizes the results. The cable substitution hierarchy table and its associated figure remain here.

820.179. Coaxial Cables. This section includes general rating provisions that apply to all iterations of this cabling. The voltage rating must not be marked on the cable, unless required by other listings. The concern is the likelihood unqualified persons may interpret such a marking as indicative as suitable for power applications. This wiring must be suitable for 60°C or higher, and if higher than that default value, the temperature rating must be marked accordingly. The listing requirements for these cables follow:

Part **(A)** covers Types CATVP for plenum cavity applications.
Part **(B)** covers Type CATVR for riser applications.
Part **(C)** covers Type CATV for general-purpose applications.
Part **(D)** covers Type CATVX for limited use in residential applications.

820.180. Grounding Devices. Equipment employed to bond metallic components of cable assemblies covered in Art. 800 to grounding and bonding conductors must be listed, or be part of a larger component that is listed.

ARTICLE 830. NETWORK-POWERED BROADBAND COMMUNICATIONS SYSTEMS

830.1. Scope. These systems are being designed and implemented to provide a combination of voice, audio, video, data, and interactive services. Powering, grounding, bonding, and electrical protection issues for these new systems had not been adequately addressed in the NEC articles that heretofore covered communications systems. As explained in the informational note, the systems typically involve a cable that supplies some power from the network as well as signal to a network interface unit, which converts the broadband signal to their components, such as voice, data, and video. The second informational note calls attention to the exclusions in 90.2(B)(4) that apply to some facilities of communications utilities.

830.2. Definitions

Exposed to Accidental Contact with Electrical Light or Power Conductors. This is a similar definition to that in Art. 800, but there are differences. The defined term is much more specific than the general term "Exposed" in Art. 830, and the possibility of contact could occur through a failure of a support or of insulation; Art. 830 requires both elements to fail.

Fault Protection Device. This is an electronic device intended for personnel protection. It is used with low power network-powered broadband communications circuits, and in the event of a fault such as a cable short or open condition, it limits the current and/or voltage and provides acceptable protection from shock.

Network Interface Unit (NIU). This is the device that breaks the broadband signal into its component parts, such as voice, data, and video. It also isolates the network power from the premises signal circuits. It may, but does not necessarily, contain primary and secondary protectors. See Fig. 830-1 for application.

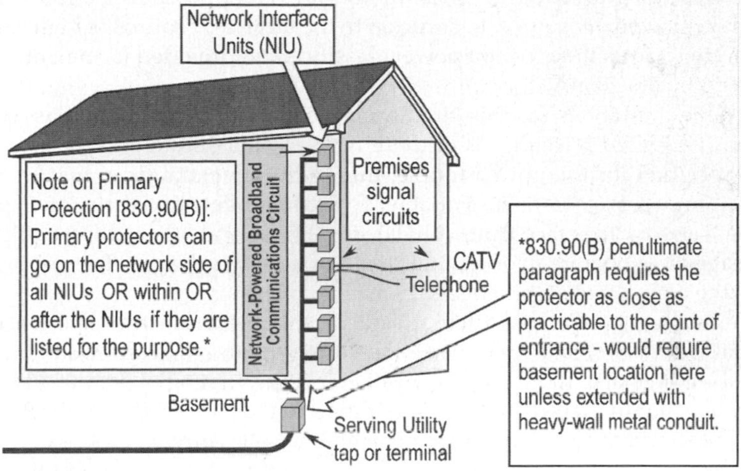
Network Interface Units (NIU)
Note on Primary Protection [830.90(B)]: Primary protectors can go on the network side of all NIUs OR within OR after the NIUs if they are listed for the purpose.*
Network-Powered Broadband Communications Circuit
Premises signal circuits
CATV
Telephone
*830.90(B) penultimate paragraph requires the protector as close as practicable to the point of entrance - would require basement location here unless extended with heavy-wall metal conduit.
Basement
Serving Utility tap or terminal

Fig. 830-1. Network Interface Units are where the component signals are separated. (Sec. 830.2.)

Network-Powered Broadband Communications Circuit. The circuit extending from the communications utility's serving terminal or tap up to and including the NIU. A power system analogy might be a service drop or lateral, plus the associated service entrance conductors and ending at the service disconnect. There are important differences, however, in terms of how far and under what conditions the respective circuits can enter a building, based on the relative degree of risk posed by the different systems. A note describes how a typical single-family circuit would fit into the requirements.

Point of Entrance is the point within a building at which cable emerges from an external wall, or a floor slab, or from RMC or IMC. In such cases the use of heavy-wall steel conduit artificially extends the point of entrance into a building, doing for this wiring what 230.6 does for service entrance conductors. This is an important concept, and it holds for all of Chap. 8. See also Sec. 830.49, which now contains the bonding requirement for the conduit.

830.3. Other Articles. This section covers the application of other articles, and incorporates a generic hazardous (classified) location reference, Sec. 300.22, and specific references to other material as appropriate for specific circuits derived from the NIU. The two items in Chap. 8 include Art. 800 for communications, and Art. 820 for CATV and for radio distributions, unless protection has been arranged in accordance with Sec. 830.90(B)(3) in the output side of the NIU.

Three items apply in Chap. 7, including Art. 770 for optical fiber cables, Art. 760 for power-limited fire alarm circuits, and Art. 725 for Class 2 and Class 3 signaling circuits. The requirements to properly apply listing requirements [110.3(B)] and to protect against cable damage (300.4) are included as well.

830.15. Power Limitations. This sets the allowable power limitations for low and medium power circuits. Both systems share a 100 VA maximum power rating, but the medium power source can be up to 150 V, as opposed to the 100 V limitation on the low power source. In addition to the table, a dc source as high as 200 V is permitted within the medium power classification, provided its maximum current to ground is 10 mA.

The table limitations for this circuit are identical to those for a Class 3 "Not Inherently Limited" circuit except there isn't any maximum current nameplate rating specified for the power source, due to the potential drop that normally occurs along a service transmission line. To accommodate this variation in potential, the Network Interface Units (NIUs), amplifiers, and other utility equipment are designed to operate over a wide range of supply voltages. The supply current to the equipment varies inversely as the voltage.

For example, an NIU connected at a point on the transmission line where the potential is 50 V would require twice the current of an NIU connected at a point where the potential is 100 V, but the input volt ampere rating of both would be the same and is not to exceed 100 VA. In addition to the 100 VA rating, the required 100 ÷ V overcurrent protection (or equivalent) limits the current to 100 ÷ V.

The low power source is figured as the likely source for a single-family dwelling, perhaps for an existing CATV subscriber, and the medium power source for similar occupancies with greater functionality or for multiple NIUs or for greater distances from the transmission lines. As proposed, there was originally a high power source option with a 150-V limit (same as medium power) but a 2250-VA power limit. High power cable is a coaxial construction with a grounded shield and 600-V insulation on the center conductor, which provide a high level of mechanical protection and electrical isolation of the center conductor. During the comment period there were serious technical issues raised regarding some of the details and the panel elected to proceed with the new article (1999 NEC) but only after stripping out the high power provisions.

830.21. Access to Electrical Equipment Behind Panels Designed to Allow Access. It is contrary to the NEC to drape endless quantities of these cables across suspended ceiling panels in a way that limits the intended access to the ceiling cavity above them.

830.24. Mechanical Execution of Work. In addition to the usual neat and workmanlike manner requirement, this section also requires exposed cabling (which includes the area above a suspended ceiling) to be supported to structure such that normal building operations will not damage the cable. In addition, if the wiring is run parallel to framing, it must meet the 32-mm (1¼-in.) spacing requirement from the leading edge of the framing. Finally, the wiring must stay off the tie wires designed to support a suspended ceiling, in accordance with 300.11. Where cable ties and other nonmetallic components are used in 300.22(C) spaces, they must be listed for that environment. An informational note (2) provides a cross-reference to the controlling rule in NFPA 90A.

830.25. Abandoned Cables. This has the effect of requiring that any "abandoned" conductors—as defined in 830.2 as not being terminated or identified for future use with a tag—must be removed. It is no longer permissible to leave abandoned network-powered broadband communications cables in place when they are no longer in use, unless they have been "identified" for future use.

830.26. Spread of Fire or Products of Combustion. This is a restatement of 300.21, as applied to this article.

830.40. Entrance Cables. Both the equipment and the cables used on these systems need to be listed for the purpose, although there is the usual cable substitution table (Table 830.154) that allows a better cable to substitute for a lesser one. The first exception, however, grandfathers all CATV system coaxial cables installed per Art. 820 before Jan. 1, 2000, for low power usage.

Although the subsections following essentially cover the usual four tier hierarchy for this type of cabling (-P, -R, - , -X), there are some interesting gaps. There isn't a BMP cable, for example, so medium-power cabling installed in plenum cavities would go back to Sec. 300-22(c) wiring methods. In addition, there isn't a BLR cable, so running low power systems in a riser would require the better grade BMR cable normally identified with medium power applications.

830.44. Overhead (Aerial) Cables. Part **(A)** covers running cable on and from poles and the wording follows 820.44(A). Where practicable, these cables should run below any non-power-limited circuits, and they are forbidden to run on a common crossarm.

Part **(B)** rule and its exception correspond to 820.44(B).

Part **(C)** provides a set of grade clearances based on NESC dimensions. Neither Art. 800 nor Art. 820 have any required clearances from grade. They run somewhat lower than similar dimensions in 225.18, which is appropriate because they are supposed to run below power conductors where practicable.

Part **(D)** points to Table 680.9(A) for its pool clearances.

Part **(E)** covers the final span requirement, which came directly from the 1996 version of 230.9. It needs to be correlated with important changes to that section subsequently that added what is now 230.9(B).

Part **(F)** covers runs between buildings; the wording comes partially from 820.44(D).

Part **(G)** covers cabling routed on buildings. The first three paragraphs come directly from 820.44(E), and the fourth paragraph was added. It requires protecting network-powered broadband communications cables from damage, to a height of 8 ft from grade, as shown in Fig. 830-2. The exception (another incomplete sentence) recognizes that the use of a listed fault protection device to the circuit provides an acceptable level of protection from electric shock. With this level of protection on a low power network-powered broadband communications circuit, the mechanical protection requirements can be relaxed without sacrificing safety.

830.47. Underground Network-Powered Broadband Communications Cables Entering Buildings. Due to somewhat greater power levels on these systems, this section includes an additional set of requirements based on the 1996 NEC version (uncorrelated with 1999 and subsequent changes) of 300.5 for physical protection of underground power circuits, including a requirement to protect

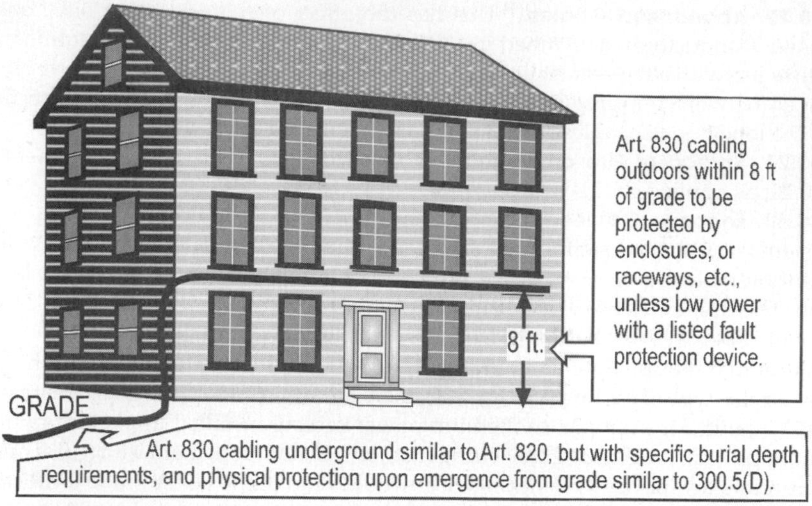
Art. 830 cabling
outdoors within 8 ft
of grade to be
protected by
enclosures, or
raceways, etc.,
unless low power
with a listed fault
protection device.
8 ft.
GRADE
Art. 830 cabling underground similar to Art. 820, but with specific burial depth
requirements, and physical protection upon emergence from grade similar to 300.5(D).

Fig. 830-2. Higher power levels lead to enhanced protection rules [Sec. 830.44(I)(4); 830.47].

direct-buried cables that make a transition to above grade by using raceway or equal to the point of entrance (or at least to a height of 8 ft).

The section comes with a burial depth table similar to Table 300.5 but greatly simplified and with many burial depths reduced. There is an exception here, like the one in 830.44(G)(4), that waives the mechanical protection requirements if there is a fault protection device on a low-power circuit. The last subsection, part **(D)**, picks up what is now the second sentence of 680.11 for underground clearances to swimming pools. Because the rule points to the entire section in Art. 680, the entire rule in 680.11 applies, including the 2008 modification requiring the raceway system to be complete and not just a sleeve.

830.49. Metallic Entrance Conduit Grounding. Heavy-wall metal conduit containing optical fiber entrance cables must be bonded to a grounding electrode as covered in 820.100(B). This, in combination with the point of entrance definition, creates the remote entry point for this work that is parallel to 230.6 for power service work. In fact, it used to be part of the definition, but was separated because definitions cannot contain requirements. Similar sections occur throughout Chap. 8.

830.90. Primary Electrical Protection. Part **(A)** includes the requirements for installing primary protectors for broadband circuits run as unprotected aerial circuits. This is a simplified version of 800.90(A), although without the requirement for protection at each end of an interbuilding circuit with a lightning exposure. This presumably reflects an expectation that these circuits principally involve utility-to-customer connections under the control of utilities, given the "network powered" nature of these systems, and not premises-wired PBX equivalents. However, the section does go on to capture both informational notes (unenforceable, of course) from 800.90(A), the second of which is appropriately modified by deleting the provision covering the likelihood of interbuilding exposures.

Part **(B)** is based on 800.90(B). The text has been revised to reflect the presence of the NIU and the options for providing primary protection, on the network side of the NIU, incorporated within the NIU, or on the derived circuits (output side of the NIU). The options help to resolve technical problems, such as signal attenuation by a protector placed on the network-side broadband cable, while providing identical electrical protection and safety. Mobile homes are protected just as they are in Sec. 800.90(B).

Part **(C)** works within the context of hazardous (classified) locations and limits the placement of either a primary protector or other equipment serving that function, in the same way as 800.90(C).

830.93. Grounding or Interruption of Metallic Members of Network-Powered Broadband Communications Cables. This is based on 820.93. The text has been revised to accommodate additional metallic cable members "not used for communications or powering." These members must be either grounded or interrupted by an insulating joint or equivalent device. A coaxial cable shield is part of the signal path and cannot be interrupted. Therefore, the rule requires grounding for coaxial cable shields, and permits all other noncommunications or nonpower conductors to be either grounded or interrupted. This approach is also applied in 800.93. A paragraph addresses mobile homes to clarify the treatment of metallic members in those locations.

830.100. Cable, Network Interface Unit, and Primary Protector Bonding and Grounding. These grounding rules were directly modeled on the equivalent language in what is now Sec. 800.100 and Sec. 800.106.

The only difference is that instead of just referring to primary protectors, the text also refers to NIUs with protectors and NIUs with metallic enclosures as requiring grounding according to the rules in this part. In addition, 830.100(A)(3), in addition to setting 14 AWG as the minimum size (also true in Art. 800), additionally requires that the size be such that its current-carrying capacity be approximately equal to that of the grounded metallic member(s) and protected conductor(s) of the cable, but it need not be larger than 6 AWG. Correlating language now applies in both Art. 800 and Art. 820.

830.106. Grounding and Bonding at Mobile Homes. This is very difficult to follow, unless you factor in the wording in the second paragraph of 830.93. Specifically, mobile homes are supposed to have a bonding terminal on their chassis at an accessible location, as required in 550.16(C)(1). They are also supposed to have either service equipment or a feeder disconnect located within 9.0 m (30 ft), per 550.32(A). If they have actual service equipment at that location, or if they have a feeder disconnect with a regrounded neutral at that point, there will be a grounding electrode system connected at that disconnect. In fact, even if the disconnect does not include a regrounded neutral, 250.32(A) still requires a grounding electrode connection at this point. Assuming the network-powered broadband communications cable entry comes from this location, a bonding connection between the conductive elements in the cable and the grounding system at the disconnect will meet all requirements and no further action needs to be taken at the mobile home. If, and only if, the disconnect is located at an excessive distance or grounded incorrectly does the necessity of creating a local electrode at the mobile home kick in. The context for this rule is at 830.90(B), second full paragraph just above (C).

This brings up the separate question of bonding. If the mobile home is permanently connected to a Code-compliant disconnect, then the permanent equipment grounding system connecting the mobile home and the disconnect is sufficient for these purposes. However, if the mobile home is cord-and-plug-connected, removing the plug would disrupt this continuity. In this case there is a conundrum because this wording was brought over from 820.106. Article 800 (telephone) systems have primary protectors that will not exist on these systems. That leaves only one option, a grounding electrode, and therefore, through this back door, a local grounding electrode becomes required in the event of a cord-and-plug connection. The grounding electrode must be connected to the conductive elements of the network-powered broadband communications cable, and also to the mobile home bonding terminal as previously mentioned.

830.110. Raceways and Cable Routing Assemblies for Network-Powered Broadband Communications Cables. Low-power network-powered broadband communications cables can run in Chap. 3 raceways installed in accordance with the applicable article, but fill restrictions will not apply. Medium-power network-powered broadband cables in a raceway must use a Chap. 3 type, installed according to its requirements in Chap. 3, and conduit fill limitations will apply to these cables.

Part **(C)**, new as of the 2014 **NEC**, covers cable routing assemblies. These are usually open on top and serve as continuous support for cables. They are customized by application, as in general-purpose, riser, or plenum, with a comparable substitution hierarchy. The default support intervals are 3 ft for horizontal runs and 4 ft for vertical runs, with an allowance for longer spans if part of the listing, but never over 10 ft.

830.113. Installation of Network-Powered Broadband Communications Cables. There are some general principles at work that make this less complicated than it appears. The entire two pages of Code text boils down to the following:

1. The hierarchy of cable compositional categories, and by logical extension the hierarchy of compositional categories for the nonmetallic raceways and supporting assemblies that contain those cables, continues without change. Refer to the discussion at 725.135 regarding the history and applications of these categories, which have not changed since the advent of this work in the 1987 **NEC**. Specifically, there are plenum exposures, which require the most severe testing for both smoke generation and vertical flammability, and result in a "P" suffix on cable assemblies. There are riser exposures that are the next most severely tested, focusing on vertical flammability, and result in an "R" suffix on cable assemblies. Then there are general-purpose applications that receive general fire load testing and result in no suffix. Residential grades (suffix "X") apply to cables less than 10 mm (3⁄8 in.) in diameter. This is a hierarchy, so general-purpose acceptability includes residential, riser suitability implies general-purpose suitability as well, and the plenum grade can be used in all locations. These cables also come in an underground configuration, with a "U" suffix.
2. If a cable is pulled into a Chap. 3 raceway, its suitability follows that of the enclosing Chap. 3 raceway, just as surely as the THHN or RHW or XHHW-2 that could otherwise be pulled into the same raceway without a separate evaluation of the fire resistance of its insulation.

3. Fireproof shafts with firestopping at every floor cancel the need for special riser ratings, and one- and two-family dwelling exposures, even vertical ones, do not merit a riser rating either.

830.133. Installation of Network-Powered Broadband Communications Cables and Equipment. Part **(A)(1)** includes the rules regarding separation of broadband cables from other wiring, as illustrated in Fig. 830-3.

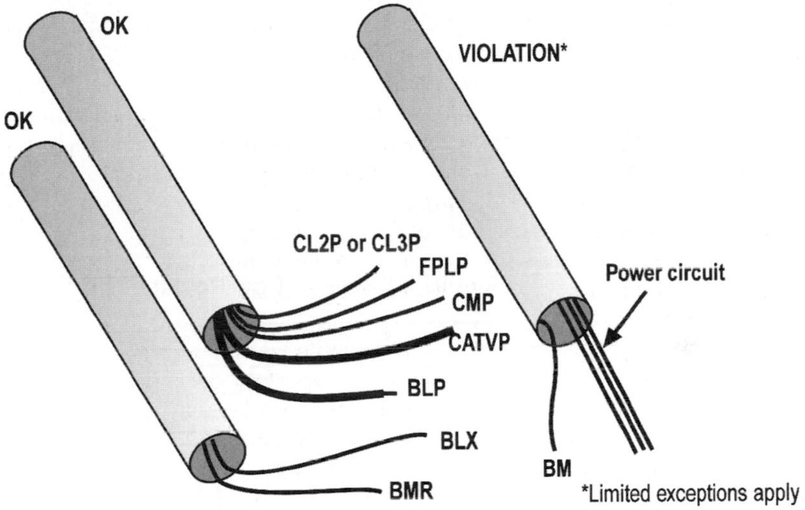
OK
OK
VIOLATION*
CL2P or CL3P
FPLP
CMP
CATVP
BLP
BLX
BMR
Power circuit
BM
*Limited exceptions apply

Fig. 830-3. System separation rules are always important. [Sec. 830.133(A)(1).]

Low power network-powered broadband communications circuits are essentially Class 3 circuits. Medium-power network-powered broadband communications circuits are not Class 3 circuits (similar in some ways but not the same; Class 3 status was denied), but they do have a 600-V jacket. With those principles in mind, the following separation rules make sense:

First, low and medium power cables (not circuit wires) are permitted in the same raceway.

Second, low power cables are permitted in raceways with power-limited fire alarm, CATV, Class 2 and 3, Art. 800 communications, and both conductive and nonconductive optical fiber.

Third, [this paragraph combines (c) and (d) of Code text] medium power circuits cannot be installed in a common raceway or enclosure with any of those same systems, with the only exception being nonconductive optical fiber.

Fourth, both low power and medium power broadband circuits have to be divorced from non-power-limited wiring entirely, but with the customary exceptions for barriered separations and for entry into an enclosure housing equipment connected to both systems, with a maintained 6-mm (¼-in.) air separation.

Part **(A)(2)** on system separation for other applications follows the same rules as for similar CATV applications in 820.133(A)(2), with a normal 2-in. separation and allowances for routing the other circuits in independent Chap. 3 wiring methods or raceways, or for the use of raceways or for porcelain or flexible tubing in addition to the insulation on the broadband cabling.

Part **(B)** is the wording to keep broadband conductors off interior raceways. This wording is based on similar wording used in 820.133(B).

830.154. Applications of Network-Powered Broadband Communications System Cables. The installation requirements for these cables now reside at 820.113, and the applications have been converted to a full-page table that summarizes the results. The cable substitution hierarchy table remains here.

830.160. Bends. This section, a first in the communications articles, requires that any bends in the cable be made so as not to damage it. This is in Part V, so it only applies indoors.

830.179. Network-Powered Broadband Communications Equipment and Cables. Cables outside and entering buildings must be listed accordingly. Low power circuits get the usual permitted substitutions (see Fig. 830-4) plus the grandfathering in 830.179 Exception No. 1.

	Medium Power	Low Power
Plenum Cavities	NONE-use 300.22(C) wiring methods	BLP
Riser	BMR	BLR
General purpose	BM	BL
Dwelling Unit	NONE-use BM	BLX
Outdoor underground	BMU	BLU

Note: 830.179 Exception No. 1 grandfathers all Art. 820 cabling installed prior to 1/1/2000 per Art. 820 for low-power uses.

Fig. 830-4. Cable substitutions are incomplete for the medium power level. (Sec. 830.179.)

There are just two listing sequences in this section, and they leave gaps on the medium power side. The medium power varieties include BMR for risers, BM for general-purpose applications and BMU for underground use, leaving the limited residential use and the plenum cavity use without a cable type. On the other hand, the low-power list is fully populated, from the plenum cavity BLP, to the riser BLR, to general-purpose BL, to residential limited application BLX, and the underground BLU.

830.180. Grounding Devices. Equipment employed to bond metallic components of cable assemblies covered in Art. 800 to grounding and bonding conductors must be listed, or be part of a larger component that is listed.

ARTICLE 840. PREMISES-POWERED BROADBAND COMMUNICATIONS SYSTEMS

840.1. Scope. As explained in an informational note that follows, these systems extend from an unpowered optical fiber cable, either conductive or nonconductive, that carries a broadband signal through to an optical network terminal that has a battery backup and is powered from the premises wiring system. The network terminal divides the incoming signal into conventional telephone, high-speed internet, video, and interactive services.

840.2. Definitions. In addition to other applicable definitions from Arts. 100, 770, 800, and 820, there are three definitions unique to Art. 840:

Network Terminal is a device that converts network-provided signals (optical, electrical, or wireless) into component communications elements, and is considered a network device on the premises that is connected to a communications service provider and is powered at the premises.

Premises Communications Circuit is a circuit that extends communication functions from the network terminal to premises communications equipment such as telephones, facsimile machines, and answering machines, and also extends outside wiring if present, as for burglar and fire alarm systems.

Premises Community Antenna Television (CATV) Circuit is the circuit that extends CATV signals for audio, video, data, and interactive systems to on-premises equipment and appliances as necessary.

840.3. Other Articles. Part **(E)** incorporates Arts. 725, 760, 770, 800, and 820 as they would be expected to apply to premises wiring supplied from an Art. 840 system. Part **(F)** brings in other Chap. 8 articles, and part **(G)** requires reclassification of limited power conductors running with communication wires as such.

840.44. Overhead Optical Fiber Cables. This section is identical to 800.44; refer to the coverage in this book for those applications. Note that, for some reason, this section applies to aerial conductive optical cables, but is silent regarding nonconductive cables, although its provisions, such as climbing space, would seemingly apply generally. The likely explanation is that a metallic strength member is considered essential to any optical fiber cable run overhead. The next two sections (840.45 and 840.46) bring in by reference the appropriate requirements for other cables run overhead.

840.47. Underground Optical Fiber Cables Entering Buildings. Part **(A)** requires a barrier between the conductive versions of these cables and non-power-limited wiring in manholes, handholes, and raceways. Part **(B)** covers direct burial applications. Conductive cable styles require a 300 mm (12 in.) separation from non-power-limited wiring unless that wiring is in a raceway or is cabled with a metallic armor. No separation is required from Type UF cable, or Type USE cables provided they are not part of service wiring. Part **(C)** requires underground entrances to have at least 150 mm (6 in.) of cover. Parts **(B)** and **(C)** bring in by reference the appropriate requirements for communications and coaxial cables run underground, including completing the process of establishing a remote entry point for a cable entrance, comparable to what 230.6 accomplishes for service entrance work.

840.48. Unlisted Wires and Cables Entering Buildings. This section incorporates the appropriate requirements for communications, coaxial, and optical fiber cables entering buildings.

840.49. Metallic Entrance Conduit Grounding. By the cross references made, the customary bonding requirements are applied.

840.90. Protective Devices. This incorporates the panoply of protective devices for communications circuits generally into this article.

840.93. Grounding or Interruption. This section simply incorporates the requirements of 770.93, 800.93, and 820.93 as they would apply to the cable at hand.

840.101. Premises Circuits Not Leaving the Building. These provisions apply where the supply optical cable is nonconductive or conductive and the conductive member has been interrupted by an insulating joint or equivalent, and where all building circuits that are supplied by the network terminal do not leave the building. Part **(A)** covers within-building runs of CATV cable (Art. 820), and its shield must be grounded using one of the procedures covered in 820.100 (or 820.106 for mobile homes), or a "fixed" connection to an equipment grounding conductor covered in 250.118 (apparently, any EGC that is convenient), or a grounding terminal on the network terminal, provided the network terminal is itself grounded using an 820.100 (or 820.106 for a mobile home) method. If this is done, the network terminal grounding connection must be made to an equipment grounding conductor (here again, apparently, any EGC that is convenient), using a "listed grounding device" that maintains continuity even if the network terminal is unplugged. Communications (Art. 800) circuits need not be grounded, and the network terminal need not be grounded unless otherwise required by its listing. Assuming the coaxial cable in use has its shield grounded in a manner that does not depend on the network terminal, a cord-and-plug connection is permitted for the network.

840.106. Grounding and Bonding at Mobile Homes. This section shares the problem in 770.106 regarding requirements for mobile homes with supply connections that do comply with current NEC provisions. The cross-references to 800.106(A)(1), 820.106(A)(1), 800.106(A)(2), and 820.106(A)(2) share this problem as well. Refer to the discussion at 770.106 and 800.106 for more information.

840.113. Installation on the Customer Premises Side of the Network Terminal. The usual installation procedures in 800.113 and 800.133, and 820.113 and 820.133 for premises communications wiring and premises CATV wiring, respectively, apply.

840.160. Powering Circuits. Communications cables are authorized to provide up to 60 watts of power for powering communications equipment. In the event that a higher power capability is entertained, the limits designed in 725.144 for power over Ethernet are permitted to be used.

840.179. Equipment and Cables. The network terminal must be listed for applications covered by this article. The associated cables must be listed in accordance with 770.179, 800.179, and 820.170 as applicable.

840.180. Grounding Devices. Equipment employed to bond metallic components of cable assemblies covered in Art. 800 to grounding and bonding conductors must be listed, or be part of a larger component that is listed.

Chapter Nine

TABLES

The basic requirements governing the maximum permitted fill for raceways and tubing recognized by the NEC are covered in Table 1 of Chap. 9. That table shows 53 percent as the maximum cross-sectional area of a raceway or tube that may be occupied by a conductor, where a single conductor is run within a raceway or tube. Where there are two conductors within the raceway or tube, then the maximum permitted *fill*—the area occupied by the conductors—is reduced to no more than 31 percent of the raceway or tube's cross-sectional area, which is shown in the applicable part of Table 4. And the most commonly encountered installation—three or more conductors in the raceway or tube—permits a maximum fill of no more than 40 percent of the cross-sectional area of the conduit.

Table 1 does not apply where conduit sleeves are used to protect various types of cables from physical damage. As indicated in Note 2, it applies only to "complete systems." And Note 1 directs the reader to the new Annex C for conduit fill where all conductors are the same size and insulation type.

While Note 3 mentions bare (as well as insulated) equipment grounding or bonding conductors, Note 8 to Code Table 1 applies to all forms of *bare* conductors (equipment grounding conductors and neutral or grounded conductors). Where any bare conductors are used in conduit or tubing, the dimensions given in Table 8 may be used. Because *all* wires utilize space in raceways, they must be counted in calculating raceway sizes whether the conductors are insulated or bare.

In regard to Note 5, there are conductors (particularly high-voltage types) that do not have dimensions listed in Chap. 9, and this note also applies to optical fiber cables. Conduit sizes for such conductors may be determined by computing the cross-sectional area of each conductor as follows:

$$D^2 \times 0.7854 = \text{cross-sectional area}$$

where D = outside diameter of conductor, including insulation. Then the proper conduit size can be determined by applying Tables 1 and 4 for the appropriate number of conductors.

example: Three single-conductor, 5-kV cables are to be installed in conduit. The outside diameter D of each conductor is 0.750 in. Then $0.750^2 \times 0.7854 \times 3 = 1.3253$ in^2. From Tables 1 and 4 (40 percent fill) a metric designator 53 (trade size 2) rigid metal conduit would be permitted. Code Tables in Annex C are based on Table 1 allowable percentage fills, and have been provided for the sake of convenience. Although revisions have been made to eliminate conflict, in any calculation, Table 1 is the table to be used where any conflict may occur.

Table 1 is also used for computing conduit sizes where various sizes of conductors or conductor types are to be used in the same conduit. It is also used, and has been retitled accordingly, to calculate allowable fill where cable assemblies are drawn into raceway systems that are complete. To this end, Note 9 now expressly includes optical fiber cables as a potential application.

An example of Note 7 would be to determine how many 14 AWG Type TW conductors would be permitted in a ½-in. rigid metal conduit. For three or more such conductors, Table 1 permits a 40 percent fill. From Table 4, 40 percent of the internal cross-sectional area of a ½-in. rigid metal conduit is 0.125 sq in. From Table 5, the cross-sectional area of a 14 AWG Type TW conductor is 0.0139 sq in. Thus, 0.125/0.0139 = 8.9, or 9 such conductors would be permitted in a ½-in. conduit. Where the decimal is less than 0.8 (such as 0.7), the decimal would be dropped and the whole number would be the maximum number of equally sized conductors permitted; for example, 8.7 would be 8 conductors (see Note 7). This same procedure can apply to a single conductor evaluated for a raceway fill calculation.

The following is an example for computing a conduit size for various conductor sizes:

Number	Wire Size and Type	Table 5 Cross-sectional Area (ea.) (in^2)	Subtotal Cross-sectional Area (in^2)
3	10 AWG TW	0.0243	0.0729
3	12 AWG TW	0.0181	0.0543
3	6 AWG TW	0.0726	0.2178
		Total cross-sectional area	0.345

Table 1 permits a 40 percent fill for three or more conductors. Following the 40 percent column in Table 4, 1-in. rigid metal conduit would be required for these nine conductors, which have a combined cross-sectional area of 0.345 sq in.

Table 5A gives the maximum number of compact conductors permitted in trade sizes of conduit or tubing. Conductors with "compact-strand" construction have the cross-section areas of their strands shaped as trapezoids to provide tight "nesting" of the strands when they are twisted together. Such construction eliminates the air voids that occur when individual strands of circular cross section are twisted together and results in a smaller overall diameter of the total bundle of strands. Thus, a 600-kcmil compact-strand assembly has an overall cross-sectional area of a conventional 500 kcmil with circular strands.

Note 6 of Table 1 in Chap. 9 recognizes the fact that compact-strand conductors have increased conduit fill because of their overall smaller area. Compact-strand conductors are covered by Table 5A in Chap. 9 of the NEC.

Note 9, intended to address fill issues of multiconductor cables, added a new sentence in the 2017 NEC, as follows:

"Assemblies of single insulated conductors without an overall covering shall not be considered a cable when determining conduit or tubing fill area. The conduit or tubing fill for the assemblies shall be calculated based upon the individual conductors." This revision clarifies that assembled conductors should not be considered to be a cable, but instead treated as individual conductors for wire fill calculation.

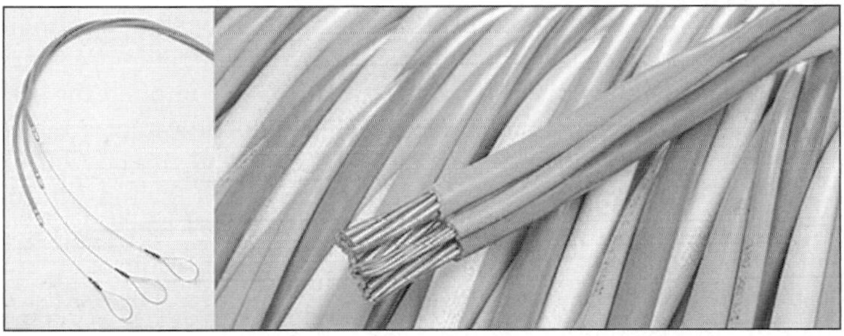

Fig. 9-1. This photo combination showed up as an advertisement from several leading trade publications. It included the following announcement: "Pulls Like a Single Cable—With 53 percent conduit fill (per NEC Chap. 9)."

The new sentence in the NEC completely invalidates the assertion regarding raceway fill in the advertisement. The correct fill limit is (and was) the customary 40 percent. Figure 9-1 shows the incorrect advertising assertion.

Note 10, new in the 2014 NEC, clarifies a frequent topic of interest, namely, are the wire dimensions in Table 5 for the smaller sizes based on stranded wire or solid wire? It turns out that they are worst case sizes based on stranded configurations. In coordination with the insertion of the words "actual dimensions or" into Note 6, it is now clear that if solid data from the manufacturer are available relative to the actual wiring being installed, that data can be used.

As indicated in the previous example, **Table 4** in Chap. 9 provides a variety of dimensional data for the various Code-recognized raceways and tubing. As of the 2014 NEC, this table is now arranged with the 40 percent fill column closest to the raceway size, significantly improving user-friendliness. It correlates the metric dimensions (millimeters) with imperial system units (inches) and gives specific values for various fills. These values represent the maximum useable area for the percent fill indicated at the top of the table. The total cross-sectional area of the contained conductors or fixture wires must not exceed the value indicated for the percent fill permitted by Table 1, Chap. 9.

Table 5 in Chap. 9 provides dimensional data—diameter and cross-sectional area—for all Code-recognized conductor and fixture wire sizes and insulation types. Simply locate the size and insulation type of the conductors in question and use the value shown for the specific conductor size and insulation type to determine the total cross-sectional area of the conductors. And this table has been rearranged for user-friendliness as well, with the cross-sectional area columns closest to the wire size columns. Then, Table 4 can be used to determine the minimum permitted size of raceway or tubing that is needed to satisfy the Code-prescribed fill limitations given in Table 1 and its accompanying notes.

Table 2 contains the Code-prescribed minimum bending radii for all conduits and tubings recognized in Chap. 3 of this Code. These data, which were located in 344.24 of the 2002 **NEC**, indicate the minimum radius permitted, as measured to the centerline of the raceway. Notice that the one-shot and full shoe benders are permitted to provide a smaller radius than "other bends" typically made with a hickey in the smaller sizes.

Table 9, which is not directly referred to in any **NEC** location and therefore has no mandatory application, is nevertheless an invaluable reference source for determining accurate voltage drop on ac circuits. It was compiled for the 1987 **NEC** using Neher-McGrath calculations, as also underlie the Annex B tables and the 310.15(C) formula. The effective impedances that read directly out of the table use an 85 percent power factor, which is very good for most industrial work, particularly where conductors are loaded near 75°C, which is the usual upper limit for terminations. For conductors operating at room temperature, frequently more realistic, the numbers need to be adjusted. The example and analysis that follows is complicated because it does not use standard values. If you carefully digest this analysis, you should always be able to work any calculations that come your way, especially those that allow for direct applications of the table under simpler conditions.

For this example, suppose a 480V 3-phase feeder comprised of a double run of 250 kcmil copper in steel raceway to a motor control center is loaded to 370 A, and the wires are operating at 25°C. If the motors are lightly loaded such that a 75 percent power factor is measured, determine the voltage drop for a 200 ft (60 m) run. Note that the impedance for the double run is the same as that for a single run with the total current halved, so this analysis can proceed on the basis of 185 A flowing over a set of 250 kcmil conductors.

To adjust the resistance values in Table 9, use the footnote to Table 8. In this example, to reflect the 50°C lower operating temperature the formula works like this:

$$R_2 = 0.054[1 + 0.00323(25 - 75)] = 0.054(0.8385) = 0.045 \text{ ohms/kFT}$$

The next step is to read out the reactance, which is 0.052 ohms/kFT. The power factor angle of the circuit must be determined next. To do this, take the power factor as a decimal (0.75) into a table of cosines, or use a scientific calculator (including the ones built into most computer operating systems;

just enter the decimal, set the "INV" check box, and click on the cosine button) to get this angle, which is about 41.4 degrees. This is also the arccosine of the power factor. Now use the formula at the bottom of Note 2 to determine the effective impedance for the circuit in ohms/kFT to neutral under the revised conditions (Z_c), as follows:

$$Z_c = (R \times 0.75) + [X_L \times \sin(41.4°)] = 0.054 \times 0.75 + 0.052 \times 0.66 = 0.068$$

Using Ohm's Law (E = IZ), find the line-to neutral voltage drop under the specified loading:

$$E_D = 185\text{ A} \times 0.068 = 12.6\text{ volts/kFT} = 0.0126\text{ volts/FT}$$

Because the circuit is 200 ft long, the drop (to neutral) is 200 FT × 0.0126 volts/FT = 2.5 V. The final step is to convert the line-to-neutral drop to more useful line-to-line values in a 3-phase circuit. To do this, simply apply the usual ÷3 multiplier:

$$E_{D\text{ (line to line)}} = \sqrt{3}E_{D\text{ (line to neutral)}} = 1.73 \times 2.5 = 4.3\text{ volts}$$

Referring to 215.2(A) informational note No. 2, this number, which is less than 1 percent of the 480-V circuit voltage, is certainly acceptable.

Table 10 on conductor stranding gives the standard stranding configurations for Class B and Class C copper stranding, along with Class B aluminum stranding. With reference to the fine-stranded conductor termination limitations in 110.14 (second paragraph), any conductors more finely stranded than the configurations presented in this table are required to use terminations identified for the class of stranding employed by the conductor at hand.

Tables 11(A) and (B). These tables are not for general field use. Refer to the discussion at 725.121 for more information.

INFORMATIVE ANNEX A. PRODUCT SAFETY STANDARDS

This annex provides a listing of the various standards that are used to test products by third-party testing facilities, such as Underwriters Laboratories. These tests are designed to establish the suitability of a particular manufacturer's product for use in a specific installation or application covered by the NEC. A certain amount of information regarding proper use and installation can be gleaned from the actual testing and the pass/fail criteria given in these publications, which will serve to enhance one's knowledge and prevent misapplication or misuse, as required by 110.3(B).

INFORMATIVE ANNEX B. APPLICATION INFORMATION FOR AMPACITY CALCULATION

The data given here are for use where conductor ampacity is calculated "under engineering supervision," as covered by 310.15(C). It should be noted that many jurisdictions do not recognize the use of 310.15(C)—especially if the conductor ampacity determined by that method is greater than the ampacity that would be permitted by the more familiar procedure given in part **(B)** of 310.15. Always check with the local authority having jurisdiction before determining conductor ampacity in accordance with 310.15(C). And, where such calculation is permitted, it must be conducted under "engineering supervision," which would preclude application by a good percentage of the electrical design and installation community.

Table B.310.15(B)(3)(11). Adjustment Factors for More Than Three Current-Carrying Conductors in a Raceway or Cable with Load Diversity.

[Adapted with permission from the *Engineers's Edition, Illustrated Changes in the 1999* National Electrical Code, © 1998, Penton Business Media. All rights reserved.]

The current edition of the NEC generally makes it prohibitive to install more than nine conductors in a raceway due to the prohibitive derating factors for mutual conductor heating [see coverage of Example D3(a) later in this chapter] that take effect at that point. For example, just 10 current-carrying conductors invoke a 50 percent derating factor, and the derating penalties get worse from there. However, there is an alternative, at this location. As just noted, access to Annex B is through engineering supervision, but if that is achievable, the old ampacity derating factors are revivable.

This table is almost the same table as the one the industry used under the old Note 8 for about 30 years, from the late 1950s to the late 1980s. Originally, this table was based on an assumption that when 10 or more conductors were in a raceway, half of them were control circuits energized in the 1 to 2 A range. If all the conductors were, however, energized and supplying power loads, then the allowable currents had to be further decreased. To that effect, the 1987 NEC added a cautionary asterisk note stating that table factors were based on 50 percent diversity. Eventually the table was exiled to Annex B where it resides today, and Table 310.15(B)(3)(a) (in the main body of the Code) reflects the worst possible case of zero diversity on any of the wires. As a practical matter, that is not a realistic assumption.

The table differs from the one in Art. 310 in its three break points over the nine-conductor fill level. Specifically, the next cut point is 10 to 24 conductors and the derating factor is 70 percent, followed by 25 to 42 conductors taken at 60 percent, and 43 to 85 conductors taken at 50 percent. These three levels are qualified by a note that indicates that they assume a "load diversity" of 50 percent. The purpose of this analysis is to make that statement usable so the table itself can be used easily.

Verification of table validity The table comes with an informational note that includes a formula that underlies the table and that can be used to deal with more complicated "real-world" diversity questions, such as all the conductors being energized, but carrying varying loads.

$$A_2 = \sqrt{\frac{0.5N}{E}} \times (A_1) \text{ or } A_1\text{, or } A_1\text{, whichever is less}$$

where:

A_1 = ampacity from Table 310.15(B)(16), etc., multiplied by the appropriate factor from Table B.310.15(B)(3)(11)

N = total number of conductors used to obtain the multiplying factor from Table B.310.15(B)(3)(11)

E = desired number of conductors in the raceway or cable

A_2 = ampacity limit for the current-carrying conductors in the raceway or cable

For example, if twenty 12 AWG THHN conductors are involved having an ampacity (from Table 310-16) of 30 A, a diversity of 50 percent could mean only 10 are carrying current at a particular time. From Table B.310.15(B)(3)(11), the derating factor (under engineering supervision) to be applied is 70 percent.

Allowable ampacity = 30 A times 0.70 = 21 A

Figuring the heating effect of the 10 conductors fully loaded to 21 A produces the following:

I^2R = 10 times 21^2 times R = 4410 R (R being the resistance per unit length of 12 AWG copper, which will be taken as a constant for these calculations).

Expressed in English, the formula says that the desired ampacity (A_2) is the square root of one half the ratio of the total number of current-carrying conductors (N) to the number of energized conductors (E), all times the "old-fashioned" ampacity as calculated from App. B.310.15(B)(3)(11) (A_1).

Therefore, working from this formula, first calculate the "old-fashioned" ampacity [Table 310.15(B)(16) ampacity times B.310.15(B)(3)(11) factor]. In this case, that would be 30 A times 0.7 = 21 A. This is the number that goes into the formula as "A_1."

Then, the final ampacity is:

$$A_2 = \sqrt{\left[0.5N\sqrt{E}\right]} \text{ times } A_1$$

$$A_2 = \sqrt{\left[0.5 \text{ times } \left(20\sqrt{20}\right)\right]} \text{ times 21 A}$$

$$= 0.71 \text{ times 21 A} \approx 15 \text{ A}$$

The ratio between the developed ampacity here (15 A) and the Table 310.15(B)(16) ampacity (30 A) is 50 percent, which is identical to that shown in Table 310.15(B)(3)(a) for a similar conductor fill, and the heating effect is:

$$I^2R = 20 \times 15^2 \times R = 4500\ R$$

The close agreement between the two results (4410 R vs. 4500 R) validates this approach and holds on other numbers of conductors in a raceway. This is no accident, because both Sec. 310.15(B)(3)(a) and the formula in Annex B are based on the same laws of physics. The principle can be broadened to allow a qualified engineer to develop corrected ampacities under many other load conditions and numbers of conductors in a raceway or cable, as covered in the next section.

Applying the formula to general cases. The table footnote actually refers to a "load diversity" of 50 percent. This is not a defined term in the Code, and can refer to circuit configurations other than half fully on/half off. The overall approach is to first figure the limiting amount of heat $W = N(I^2R)$ that the conductors in the raceway to be evaluated will generate assuming a 50 percent diversity and using the adjustment factor in Table B.310.15(B)(3)(11). This sets the limit on the amount of heat that is allowable. If there are different size conductors involved, then begin with the 50 percent that will give the largest I^2R heating when loaded as allowed in the table.

The next step is to work backward and calculate the ampacity of the conductors under the actual conditions. If all the conductors are the same size, plug the amount of heat just calculated ("W") into the rearranged formula $I = \sqrt{(W \div NR}$; this gives the equivalent allowable heating with "N" conductors equally loaded. Calculations involving combinations of conductor sizes are more complicated but the principle is the same. Remember to factor in the resistances as ratios (based on NEC Chap. 9 Table 8) to the conductor resistance used in the initial determination of allowable heating. This way the resistance per unit length ("R") is a constant and need not be actually calculated. Figure 9-2 shows a worked-out example of 12 and 14 AWG conductors mixed in the same raceway, along with other loading versus heating examples of interest.

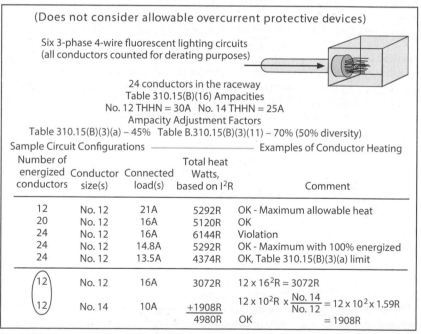

(Does not consider allowable overcurrent protective devices)

Six 3-phase 4-wire fluorescent lighting circuits
(all conductors counted for derating purposes)

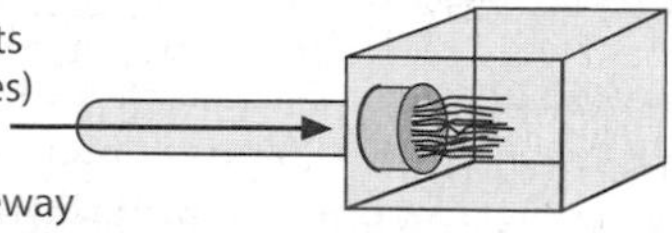

24 conductors in the raceway
Table 310.15(B)(16) Ampacities
No. 12 THHN = 30A No. 14 THHN = 25A
Ampacity Adjustment Factors
Table 310.15(B)(3)(a) – 45% Table B.310.15(B)(3)(11) – 70% (50% diversity)

Sample Circuit Configurations ——— Examples of Conductor Heating

Number of energized conductors	Conductor size(s)	Connected load(s)	Total heat Watts, based on I^2R	Comment
12	No. 12	21A	5292R	OK - Maximum allowable heat
20	No. 12	16A	5120R	OK
24	No. 12	16A	6144R	Violation
24	No. 12	14.8A	5292R	OK - Maximum with 100% energized
24	No. 12	13.5A	4374R	OK, Table 310.15(B)(3)(a) limit
12	No. 12	16A	3072R	$12 \times 16^2R = 3072R$
12	No. 14	10A	+1908R	$12 \times 10^2R \times \frac{\text{No. 14}}{\text{No. 12}} = 12 \times 10^2 \times 1.59R$
			4980R	OK $= 1908R$

Fig. 9-2. Examples of multiple-conductor heating. The bottom example (below the line) shows mixed size conductors in a common raceway. (Chap. 9.)

As shown in Fig. 9-2, many different combinations of numbers of conductors energized versus reduced loads can amount to "a load diversity of 50 percent." As currents fall away from their theoretical maximums, the heating falls by the square of those decreases. This is the reason that under actual loading conditions there was very little documentation of any problems with the traditional approach. The example also shows how a possible design load at zero diversity (all 24 wires carrying 16 A each) could overheat the conductors. This is why Table 310.15(B)(3)(a) was changed a few cycles back.

INFORMATIVE ANNEX C. CONDUIT AND TUBING FILL TABLES FOR CONDUCTORS AND FIXTURE WIRES OF THE SAME SIZE

The tables given in this annex are based on the rule of Table 1, Chap. 9; the dimensions given for raceways and tubing in Table 4, Chap. 9; and the dimensional data provided for conductors and fixture wires in Table 5, Chap. 9. Where conductors of the same size and insulation type are run in a common raceway or tube, the maximum number permitted within any sized raceway or tube can be readily determined by consulting the table that covers the type of raceway or tubing that is to be used. Because all types of liquidtight flexible nonmetallic conduit are allowed without a 6-ft limitation, LFMC-C values were added to this annex as to the 2017 NEC.

As of the 2005 NEC, notes were added to most of the tables calling attention to the fact that 2-h fire rated RHH conductors have much larger cross-sectional areas than conventional RHH conductors, and that the cable manufacturer's published data must be consulted to determine allowable raceway fill. This cable has ceramifiable insulation that turns to a brittle, glass-like consistency after a fire exposure, avoiding a failure for a limited period of time in order to facilitate a building evacuation or other necessary functions during the time period specified in its fire rating.

INFORMATIVE ANNEX D. EXAMPLES

Example D3(a) Industrial Feeders in a Common Raceway

Perhaps the most vital task confronting electrical personnel on the design side is correctly choosing the size of a wire to use in any particular application. Unfortunately, that is also one of the most complex tasks as well, because so many different factors influence the result, from termination limits to insulation style to continuous loading issues. There have always been two parts to this puzzle, beginning with knowing how to calculate the load that the wire would have to serve, and then selecting the wire. The majority of examples in this annex are quite good, but aimed at the first half of the problem, determining loads. Now there is an example that tackles the other side of the problem. Example D3(a) essentially stipulates the load at the outset, and goes through figuring out which conductor to install. The text that follows uses the exact approach given in Example D3(a), and expands on every step in the process, so you can really see how it is done. In general, Example D3(a) keeps the load data in the form of volt-amperes until the end, whereas this analysis converts to amperes sooner because those numbers are more intuitively familiar to users. In addition, a result in amperes is immediately recognizable in terms of whether a change in wire size is required. However, the procedures are identical. It is not simple, but there are a few key concepts that will keep you headed in the right direction.

The 2011 **NEC** inadvertently added another level of difficulty because of changes to a handful of ampacity ratings in Table 310.15(B)(16), one of which (1 AWG @ 90°C, 150A reduced to 145 A) changes a number of calculations within the scope of Example D3(a). The loads that make up Example D3(a) were carefully selected for their value as instructional material. Specifically, any procedural misstep would result in an actual change of wire size and not simply a difference of volt-amperes. The **NEC** example was still correct as written, but it no longer had the same instructional value because its load profile no longer generated wire size changes at certain critical points. The solution, implemented in the 2014 **NEC**, reduced the noncontinuous load group by 4 A, which restored those wire size break points at intermediate stages of the calculations. Looking at the load profile in D3(a), this was done by lowering the air compressor size from 7.5 hp to 5 hp. This lowered the current from 11 A to 7.6 A (from Table 430.250), and since this is still the largest motor, the 25 percent adder line (from 430.24) decreased as well, providing the necessary 4 A reduction. Example D3(a) (as volt-amperes and not amperes) now uses 68 A continuous and 47 A noncontinuous in its feeder calculations for ungrounded conductors. All feeder calculations and procedures presented here line up exactly with Example D3(a).

NEC rules for the ends of a wire differ from those for the middle (Adapted from *Practical Electrical Wiring,* 22nd edition, © Park Publishing, 2014, all rights reserved). The key to applying these rules, and the new **NEC** Example D3(a) in Informative Annex D on this topic is to remember that the end of a wire is different from its middle. Special rules apply to calculating wire sizes based on how the terminations are expected to function. Entirely different rules aim at ensuring that wires, over their length, don't overheat under prevailing loading and conditions of use. These two sets of rules have nothing to do with each other—they are based on entirely different thermodynamic considerations. Some of the calculations use, purely by coincidence, identical multiplying factors. Sometimes it is the termination requirements that produce the largest wire, and sometimes it is the requirements to prevent conductor overheating. You can't tell until you complete all the calculations and then make a comparison. Until you are accustomed to doing these calculations, do them on separate pieces of paper.

Current is always related to heat Every conductor has some resistance and as you increase the current, you increase the amount of heat, all other things being equal. In fact, you increase the heat by the square of the current. The ampacity tables in the NEC reflect heating in another way. As you can see in NEC Table 310.15(B)(16), the tables tell you how much current you can safely (meaning without overheating the insulation) *and continuously* draw through a conductor under the prevailing conditions—which is essentially the definition of ampacity in NEC Art. 100: *The maximum current, in amperes that a conductor can carry continuously under the conditions of use without exceeding its temperature rating.*

Ampacity tables show how conductors respond to heat The ampacity tables do much more than what is described in the previous paragraph. *They show, by implication, a current value below which a wire will run at or below a certain temperature limit.* Remember, conductor heating comes from current flowing through metal arranged in a specified geometry (generally, a long flexible cylinder of specified diameter and metallic content). In other words, for the purposes of thinking about how hot a wire is going to be running, you can ignore the different insulation styles. As a learning tool, let's make this into a "rule" and then see how the NEC makes use of it:

A conductor, regardless of its insulation type, runs at or below the temperature limit indicated in an ampacity column when, after adjustment for the conditions of use, it is carrying equal or less current than the ampacity limit in that column.

For example, a 90°C THHN 10 AWG conductor has an ampacity of 40 A. Our "rule" tells us that when 10 AWG copper conductors carry 40 A under normal-use conditions, they will reach a worst-case, steady-state temperature of 90°C just below the insulation. Meanwhile, the ampacity definition tells us that no matter how long this temperature continues, it won't damage the wire. That's often not true of the device, however. If a wire on a wiring device gets too hot for too long, it could lead to loss of temper of the metal parts inside, cause instability of non-metallic parts, and result in unreliable performance of overcurrent devices due to calibration shift.

Termination restrictions protect devices Because of the risk to devices from overheating, manufacturers set temperature limits for the conductors you put on their terminals. Consider that a metal-to-metal connection that is sound in the electrical sense probably conducts heat as efficiently as it conducts current. If you terminate a 90°C conductor on a circuit breaker, and the conductor reaches 90°C (almost the boiling point of water), the inside of the breaker won't be much below that temperature. Expecting that breaker to perform reliably with even a 75°C heat source bolted to it is expecting a lot.

Testing laboratories take into account the vulnerability of devices to overheating, and there have been listing restrictions for many, many years to prevent use of wires that would cause device overheating. These restrictions now appear as 110.14(C) in the NEC. Smaller devices (generally 100 A and lower, or with termination provisions for 1 AWG or smaller wire) historically weren't assumed to operate with wires rated over 60°C such as TW. Higher-rated equipment assumed 75°C conductors but generally no higher for 600-V equipment and below. This is still true for the larger equipment. (Note that medium-voltage equipment, over 600 V, has larger internal spacings and the usual allowance is for 90°C per 110.40, but that equipment will not be further considered at this point.) Smaller equipment increasingly has a "60/75°C" rating, which means it will function properly even where the conductors are sized based on the 75°C column of Table 310.15(B)(16).

Figure 9-3 shows a "60/75°C" marking on a 20-A circuit breaker, which means it can be used with 75°C conductors, or with 90°C conductors used under the 75°C ampacity column. In the case of a circuit breaker, using the allowance for 75°C terminations requires that both the CB as well as its enclosing panelboard carry this designation. Always remember, though, that wires have two ends. Successfully using the smaller (higher ampacity) wires requires equivalent markings on both the circuit breaker and panelboard (or fused switch) and the device at the other end. Refer to Fig. 9-4 for an example of this principle at work.

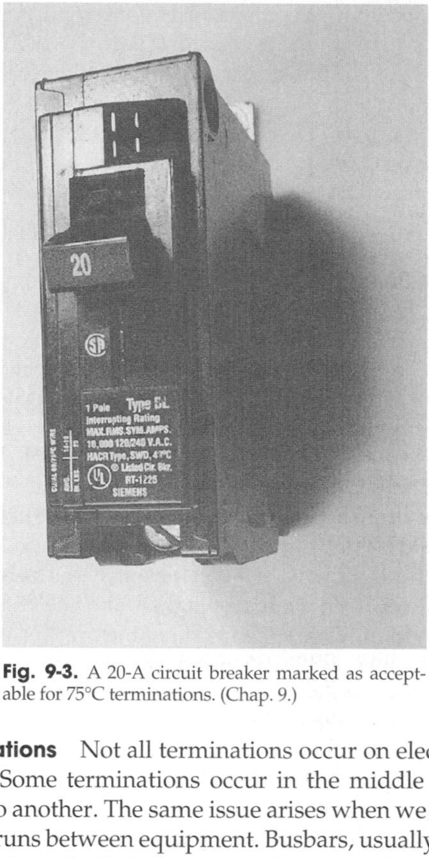
20

Fig. 9-3. A 20-A circuit breaker marked as acceptable for 75°C terminations. (Chap. 9.)

Splices as terminations Not all terminations occur on electrical devices or utilization equipment. Some terminations occur in the middle of a run where one conductor is joined to another. The same issue arises when we make a field connection to a busbar that runs between equipment. Busbars, usually rectangular in cross section, are often used to substitute for conventional wire in applications involving very heavy current demands. When you make a connection to one of these busbars (as distinct from a busbar within a panel), or from one wire to another, you only have to be concerned about the temperature rating of the compression connectors or other splicing means involved. Watch for a mark such as "AL9CU" on the lug. This one means you can use it on either aluminum or copper conductors, at up to 90°C, but only where the lug is "separately installed" (**NEC** text).

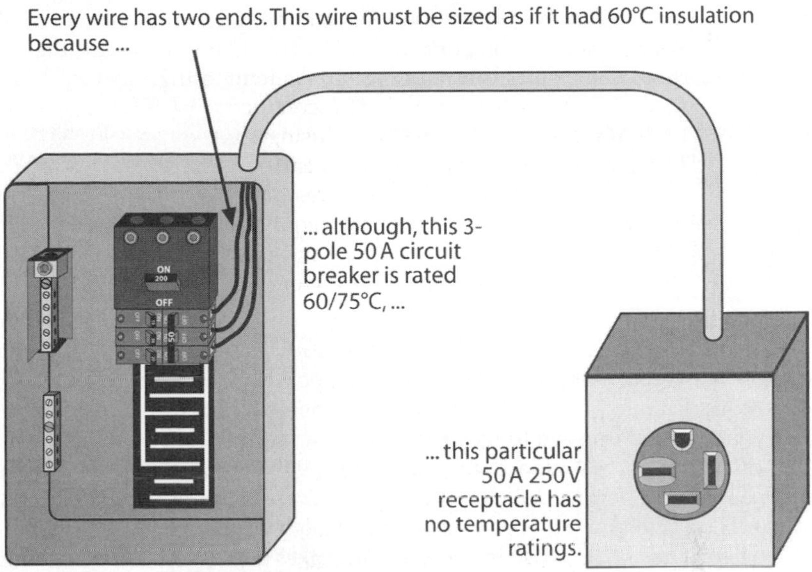
Every wire has two ends. This wire must be sized as if it had 60°C insulation because ...
... although, this 3-pole 50 A circuit breaker is rated 60/75°C, ...
... this particular 50 A 250 V receptacle has no temperature ratings.
ON
OFF

Fig. 9-4. Always consider both ends of a wire when making termination temperature evaluations. (Chap. 9.)

Lug temperature markings usually mean less than they appear to mean. Many contactors, panelboards, etc., have termination lugs marked to indicate a 90°C acceptance. Ignore those markings because the lugs aren't "separately installed." Apply the normal termination rules for this kind of equipment. What's happening here is that the equipment manufacturer is buying lugs from another manufacturer who doesn't want to run two production lines for the same product. The lug you field install on a busbar, and use safely at 90°C, also works when furnished by your contactor's OEM. But on a contactor you don't want the lug running that hot. The lug won't be damaged at 90°C, but the equipment it is bolted to won't work properly.

Protecting devices under continuous loads The NEC defines a continuous load as one that continues for 3 h or longer. Most residential loads aren't continuous, but many commercial and industrial loads are. Consider, for example, the banks of fluorescent lighting in a store. Not many stores always stay open less than 3 h at a time. Although continuous loading doesn't affect the ampacity of a wire (defined, as we've seen, as a continuous current-carrying capacity), it has a major impact on electrical devices. Just as a device will be affected mechanically by a heat source bolted to it, it also is affected mechanically when current near its load rating passes through it continuously. To prevent unremitting thermal stress on a device from affecting its operating characteristics, the NEC restricts the connected load to not more than 80 percent of the circuit rating. The reciprocal of 80 percent is 125 percent, and you'll see the restriction stated both ways. Restricting the continuous portion of a load to 80 percent of the device rating means the same thing as saying the device has to be rated 125 percent of the continuous portion of the load. If you have both continuous and noncontinuous load on the same circuit, take the continuous portion at 125 percent, and then add the noncontinuous portion. The result must not exceed the circuit rating.

example Suppose, for example, a load consists of 47 A of noncontinuous load and 68 A of continuous load (115 A total) as shown in Fig. 9-5. Calculate the minimum capacity we need to allow for our connected equipment as follows:

Step 1:	47 A × 1.00 = 47 A
Step 2:	68 A × 1.25 = 85 A
Step 3:	Minimum = 132 A

A device such as a circuit breaker that will carry this load profile must be rated no less than 132 A, even though only 115 A actually passes through the device. In the case of overcurrent protective devices, the next higher standard size would be 150 A. In general, for overcurrent protective devices not over 800 A, the **NEC** allows you to round upward to the next higher standard overcurrent device size. In this case, suppose a standard wire size had a 90°C ampacity of 140 A (there doesn't happen to be one, but this is only for discussion). The **NEC** normally allows a 150-A fuse or circuit breaker to protect such a conductor.

Having gone this far, it's easy to make two mistakes at this point. First, although you can round up in terms of the overcurrent device rating, you can't round up in terms of conductor loading, not even 1 A. Number 1 AWG conductors in the 75°C column can carry 130 A. If your actual load runs at 131 A, you have to use a larger wire. Second, when continuous loads are a factor, you have to build in additional headroom on the conductor sizes to ensure that the connected devices perform properly. This last point continually results in confusion because it may seem to contradict what was said about conductor ampacity tending to be the factor that determines minimum wire size.

In our work we handle conductors, and we worry about conductors getting overheated. Device manufacturers don't worry about conductors in this sense; they worry about their devices getting overheated and not performing properly. Continuous loads pose real challenges in terms of heat dissipation from the inside of mechanical equipment. Remember that when you bolt a conductor to a device, the two become one in the mechanical as well as the electrical sense. Device manufacturers rely on those conductors as a heat sink, particularly under continuous loading. The **NEC** allows for this by requiring conductors carrying continuous loads to be oversized at those terminations according to the same formula that applies to the device, namely an additional 25 percent of the continuous portion of the load.

example Our 10 AWG THHN conductor, for example, will carry 40 A for a month at a time without damage to itself. But under those conditions the conductor would represent a continuous 90°C heat source. Now watch what happens when we (1) size the conductor *for termination purposes* at 125 percent of the continuous portion of the load, and (2) use the 75°C column for the analysis. This calculation assumes the termination is rated for 75°C instead of the default value of 60°C:

Step 1:	1.25 × 40 A = 50 A
Step 2:	Table 310.15(B)(16) at 75°C = 8 AWG

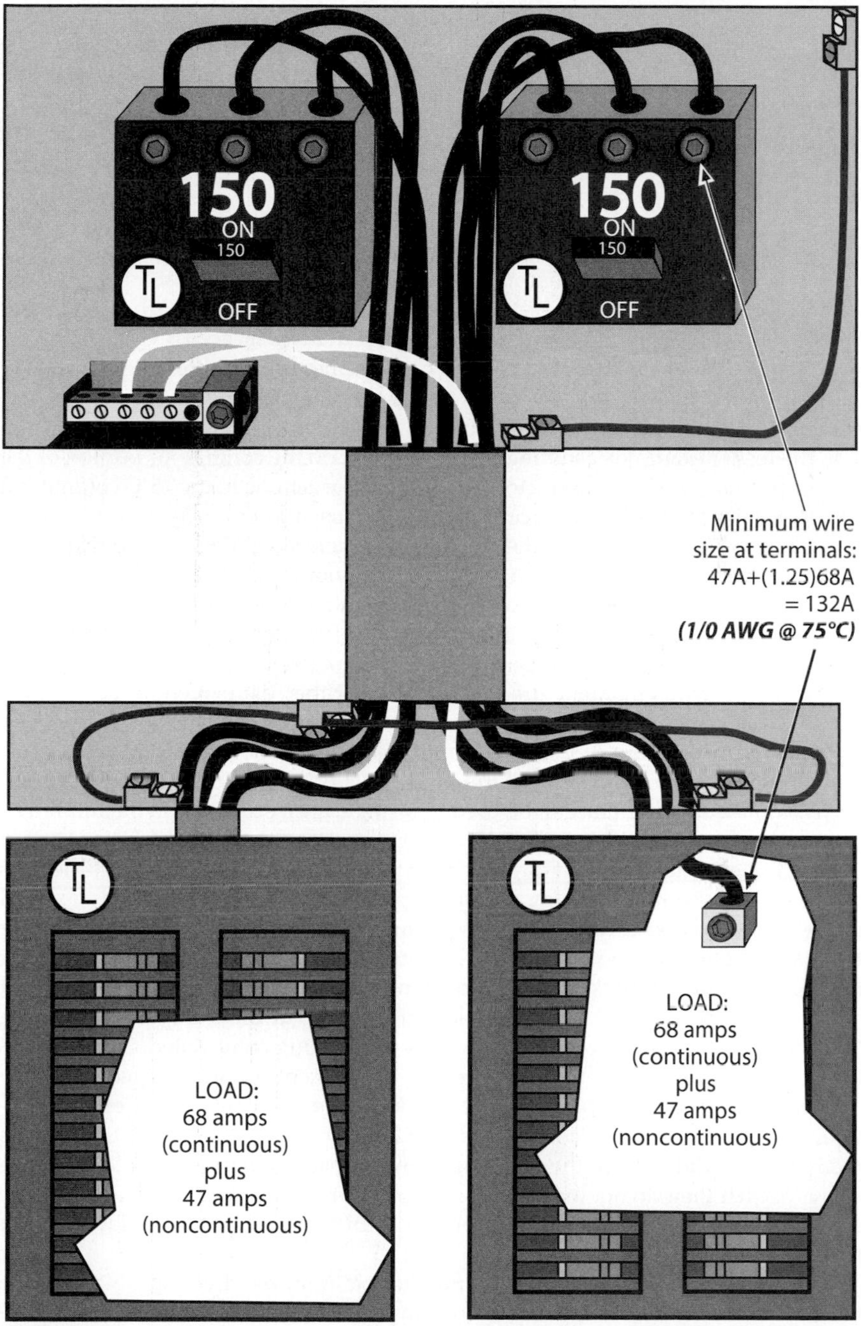
150
ON
150
OFF
150
ON
150
OFF
Minimum wire
size at terminals:
47A+(1.25)68A
= 132A
(1/0 AWG @ 75°C)
LOAD:
68 amps
(continuous)
plus
47 amps
(noncontinuous)
LOAD:
68 amps
(continuous)
plus
47 amps
(noncontinuous)

Fig. 9-5. Continuous loads require increased wire sizes, so terminating devices will not overheat. This is true even if the size based on the ampacity required over the run seemingly allows a smaller wire.

We go from a 10 AWG conductor to an 8 AWG conductor (6 AWG if the equipment doesn't have the allowance for 75°C terminations). That is just one customary wire size, but look at it from the device manufacturer's perspective. Number 10 AWG carrying 40 A continuously is a continuous 90°C heating load. What about the 8 AWG? Use the ampacity table in reverse, according to our "rule." Forty amperes happens to be the ampacity of an 8 AWG, 60°C conductor. Therefore, any 8 AWG wire (THHN or otherwise) won't exceed 60°C when its load doesn't exceed 40 A. By going up just one wire size, the termination temperature dropped from 90°C to 60°C. The NEC allows manufacturers to count on this headroom.

To recap, if you have a 40-amp continuous load, the circuit breaker must be sized at least at 125 percent of this value, or 50 A. In addition, the conductor must be sized to carry this same value of current *based on the 75°C ampacity column (or 60°C if not evaluated for 75°C).* The manufacturer and testing laboratory count on a relatively cool conductor to function as a heat sink for heat generated within the device under these continuous operating conditions.

In the feeder example, including the 125 percent on the continuous portion of the load brings us to a 132-A conductor, and the next larger one in the 75°C column is a 1/0. Remember to use the 75°C column here because the 150-A device exceeds the 100-A threshold (below which the rating is assumed to be 60°C). Remember, only 115 A (68 A + 47 A) of current actually flows through these devices. The extra 17 A (the difference between 115 A and 132 A) is phantom load. You include it only so your final conductor selection is certain to run cool enough to allow it to operate in accordance with the assumptions made in the various device product standards.

There are devices manufactured and listed to carry 100 percent of their rating continuously, and the NEC recognizes their use in exceptions. Typically these applications involve very large circuit breaker frame sizes in the 600-A range (although the trip units can be smaller). Additional restrictions accompany these products, such as on the number that can be used in a single enclosure and on the minimum temperature rating requirements for conductors connected to them. Learn how to install conventional devices first, and then apply these 100 percent-rated devices if you run across them, making sure to apply all installation restrictions covered in the directions that come with this equipment. The warning about wires having two ends applies here with special urgency; be aware that one of these devices at one end of a circuit doesn't imply anything about the suitability of equipment at the other end. Review the discussion in Chap. 2, at 215.2 under the heading "A Comparison: 100 Percent-Rated versus Non-100 Percent-Rated OC Devices" for an extensive analysis of where these devices may prove advantageous.

Grounded conductors have a special allowance The NEC includes an exception for feeder grounded conductors [215.2(A)(1) Exception No. 2] that allows them to be excluded from the upsizing requirement for continuously loaded wiring, provided they do not terminate on an overcurrent device. This is appropriate for feeder conductors that run from busbar to busbar and have no relevance as a heat sink for a connected device.

The middle of a wire—preventing conductors from overheating None of the preceding discussion has anything to do with preventing a conductor from overheating. That's right. All we've done is to be sure the device works as the manufacturer and the test lab anticipate in terms of the terminations. Now we have to be sure the conductor doesn't overheat. Again, ampacity is by definition a continuous capability. *The heating characteristics of a device at the end of the run don't have any bearing on what happens in the middle of a raceway or cable assembly.*

To reiterate, you have to compartmentalize your thinking at this point. We just covered the end of the wire; now we'll get to the middle of the wire. Remember being asked to do these on separate pieces of paper? Lock the first one up, and forget everything you just calculated. It has absolutely no bearing on what comes next. Only after you've made the next series of calculations should you retrieve the first sheet of paper. And only then should you go back and see which result represents the worst case and therefore governs your conductor choice.

If you have trouble making this distinction, and many do, apply an imaginary pull box at each end of the run (Fig. 9-6). Actually, with the cost of copper near historic highs, the imaginary pull boxes are becoming real. One large contractor, faced with installing a 300-ft triple run of Type MC cable for a 1200 A feeder, installed three sets of 500 kcmil cable made up with 90°C THHN (ampacity = 1290 A) between pull boxes at each end. From each pull box he extended 600-kcmil conductors into the device terminations (75°C ampacity = 1260 A). The savings in copper more than paid for the labor and materials expense of providing the pull boxes. The **NEC** now plainly permits this procedure to take place by right, with a new Exception No. 2 in 215.2(A)(1)(a). In Fig. 9-6, make the pull boxes real. A set of feeder conductors run between the boxes can be sized for the conditions that truly affect ampacity, in this case mutual conductor heating and ambient temperature. Larger wires, based

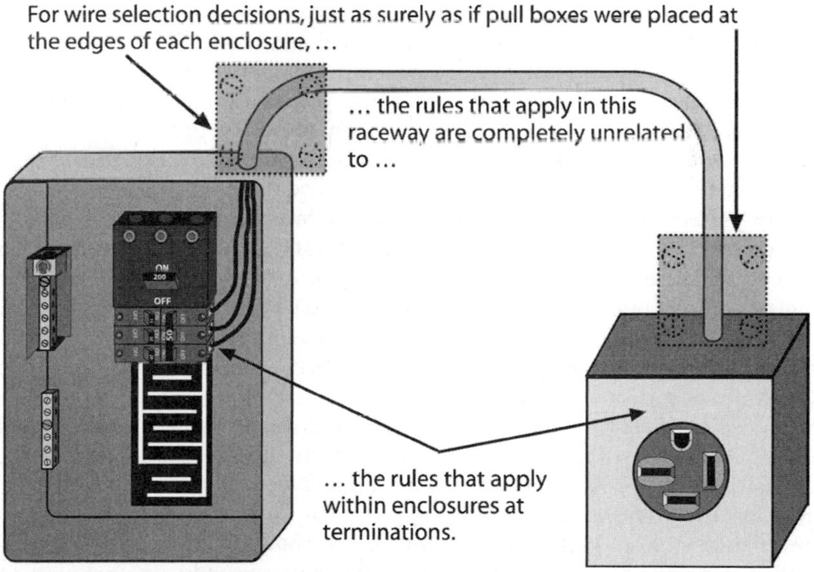
For wire selection decisions, just as surely as if pull boxes were placed at the edges of each enclosure, ...
... the rules that apply in this raceway are completely unrelated to ...
... the rules that apply within enclosures at terminations.
ON
200
OFF

Fig. 9-6. These imaginary pull boxes at each end of the run illustrate how to separate raceway/cable heating calculations from termination calculations. The **NEC** now allows the pull boxes to become real. Because the allowance is in Art. 215, it only applies to feeders and not branch circuits. (Chap. 9.)

on the termination rules, land on the line and load side equipment terminals, and exit the terminating enclosures (this is both appropriate and mandatory under the terms of the exception), arriving in the pull boxes. The conductors are then spliced with connectors rated purely in terms of allowable temperature encountered, as covered in 110.14(C)(2). Another very convenient option would be to use power distribution blocks, as covered in 314.28(e). And there are many other examples that use this procedure. Cablebus operates safely using free-air ampacity numbers (see 370.80) that allow for much smaller wires than would ever be considered for terminations, and this allows for an orderly transition to the equipment. A similar procedure applies to MI cable terminations where necessary (see the discussion at 332.80).

To begin this part of the analysis, review the ampacity definition. Conductor ampacity is its current-carrying capacity *under the conditions of use.* For NEC purposes, two field conditions routinely affect ampacity, and they are mutual heating and ambient temperature. Either or both may apply to any electrical installation. Both of these factors reduce ampacities from the table numbers.

Mutual conductor heating A conductor under load dissipates its heat through its surface into the surrounding air; if something slows or prevents the rate of heat dissipation, the temperature of the wire increases, possibly to the point of damage. The more conductors there are in the same raceway or cable assembly, the lower the efficiency with which they can dissipate their heat. To cover this mutual heating effect, the NEC imposes derating penalties on table ampacity values. Penalties increase with the number of conductors in a raceway or cable assembly. NEC Table 310.15(B)(3)(a) limits the permissible load by giving adjustment factors that apply to table ampacities. For example, if the number of wires exceeds 3 but is fewer than 7, the ampacity is only 80 percent of the table value; if the number exceeds 6 but is fewer than 10, 70 percent; more than 9 but fewer than 21, 50 percent, and so on. However, if the raceway is not over 24 in. long (classified as a nipple), the NEC assumes heat will escape from the ends of the raceway and the enclosed conductors need not have their ampacity derated.

Apply the footnote to Table 310.15(B)(3)(a) and count only conductors that are connected to electrical equipment and that will actually carry current at one time, plus any spares left in the raceway. Conductors that will be subject to noncoincident energization are not counted. For example, only count one of the two travelers in a three-way switch loop. Another example would be a variable speed drive used to control two identical motors through interlocking contactors, such that only one of the two motors could operate at once, but all six wires are running in the same raceway; only count three of the six wires.

Grounding conductors are never counted. A neutral wire that carries only the unbalanced current of a circuit (such as the neutral wire of a three-wire, single-phase circuit, or of a four-wire, three-phase circuit) is not counted for derating purposes in some cases. But remember: Grounded wires are not always neutrals. Suffice it to say for now that in most modern systems one conductor is deliberately connected to ground, and neutrals are almost always grounded. The grounded ("white") wire of a two-wire circuit carries the same current as the hot wire and, although a neutral under the NEC definition if it connects to a neutral point, is not a neutral in the sense of carrying only unbalanced current. If you install two such two-wire circuits in a conduit, they must be counted as four wires.

The grounded ("white") wire of a three-wire corner-grounded delta power circuit is a phase conductor carrying the same current as the two ungrounded phase conductors and is *not* a neutral. It must be counted for derating purposes.

How (and when) to count neutrals Although neutral conductors are counted for derating purposes only if they are actually current-carrying, it is increasingly common in commercial distribution systems derived from three-phase, four-wire, wye-connected transformers to find very heavily loaded neutrals. If the circuit supplies mostly electric-discharge lighting (fluorescent, mercury, and similar types), you must always count the neutral. In such circuits, the neutral carries a "third harmonic" or 180-Hz current produced by such luminaires. Be aware that the neutrals of such circuits or feeders must be counted because those harmonic currents add together in the neutral instead of canceling out, and therefore require derating. The remainder of the feeder examples in this analysis of Example D3(a) (see Figs. 9-6, 9-7, and 9-9) generally assume (due to typical industrial load profiles) that the neutrals must be counted on the same basis as the ungrounded conductors.

In addition, harmonic currents from other nonlinear loads often add instead of canceling in three-phase, four-wire, wye-connected neutrals as well. In fact, there are conditions where such loads cause the neutral to carry *more load* than the ungrounded conductors. There are now Type MC cable assemblies manufactured with oversized neutral conductors to address this problem. Finally, any time you run just two of the three-phase conductors of a three-phase, four-wire system together with the system neutral, that neutral always carries approximately the same load as the ungrounded conductors and must always be counted. This arrangement is very common on large apartment buildings where the feeder to each apartment consists of two-phase wires along with a neutral, but the overall service is three-phase, four-wire.

However, the neutral of a true single-phase, three-wire system need not be counted, because harmonic currents fully cancel in these systems. The overwhelming majority of single-family and small multifamily dwellings and most smaller farms have these distributions.

How (and when) to count control wires Many types of control wires, such as motor control wires and similar wires that carry only intermittent or insignificant amounts of current need not be counted, provided the continuous control current does not exceed 10 percent of the conductor ampacity, as covered in 725.51(B). For example, a motor starter with a 120-V coil might require 100 VA to operate, or less than 1 A, and even less to hold in. Since that is an intermittent load, and in any event it is less than 10 percent of the ampacity of 14 AWG, you could use 14 AWG wires for this control circuit and run them in the same raceway as the power wires without counting them for derating purposes. Because derating applies to every wire in the conduit unless it is a qualified neutral or grounding conductor, for a typical three-wire control circuit this special allowance has a major impact on the ampacity of those power wires. For a normal three-phase motor circuit running in the same conduit, it allows the power wires to be counted at 100 percent of their table value instead of 80 percent.

Conductor ampacity adjustment Now that you know how to count the number of wires in a conduit for derating purposes, it's time to learn how to apply NEC rules to the result. Using the NEC directly means going from the ampacity table

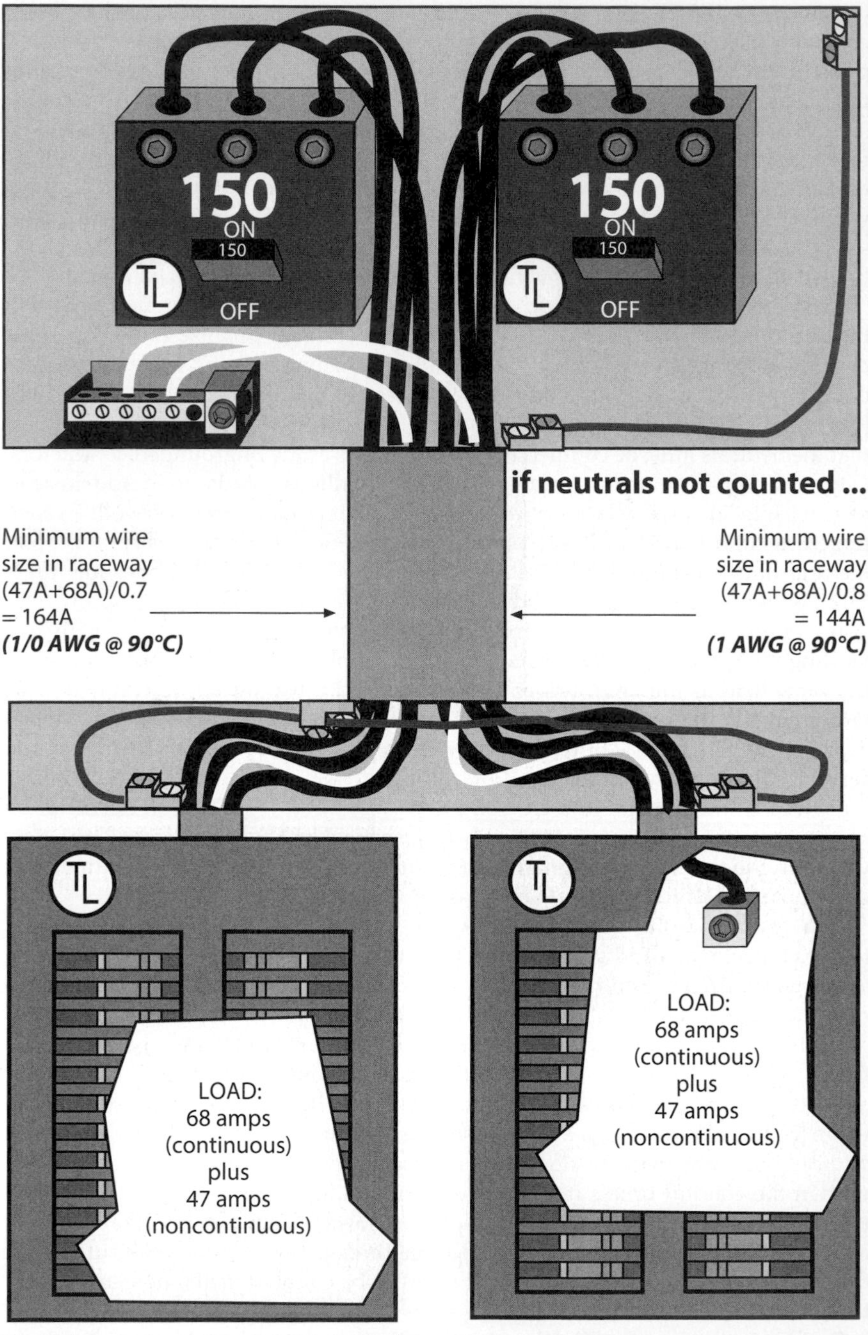
150
ON
150
OFF
150
ON
150
OFF
if neutrals not counted ...
Minimum wire
size in raceway
(47A+68A)/0.7
= 164A
(1/0 AWG @ 90°C)
Minimum wire
size in raceway
(47A+68A)/0.8
= 144A
(1 AWG @ 90°C)
LOAD:
68 amps
(continuous)
plus
47 amps
(noncontinuous)
LOAD:
68 amps
(continuous)
plus
47 amps
(noncontinuous)

Fig. 9-7. Examples of more than three countable conductors in a common raceway. Assume the raceway is longer than 24 in. and that on the left the neutrals must be counted due to harmonic load content.

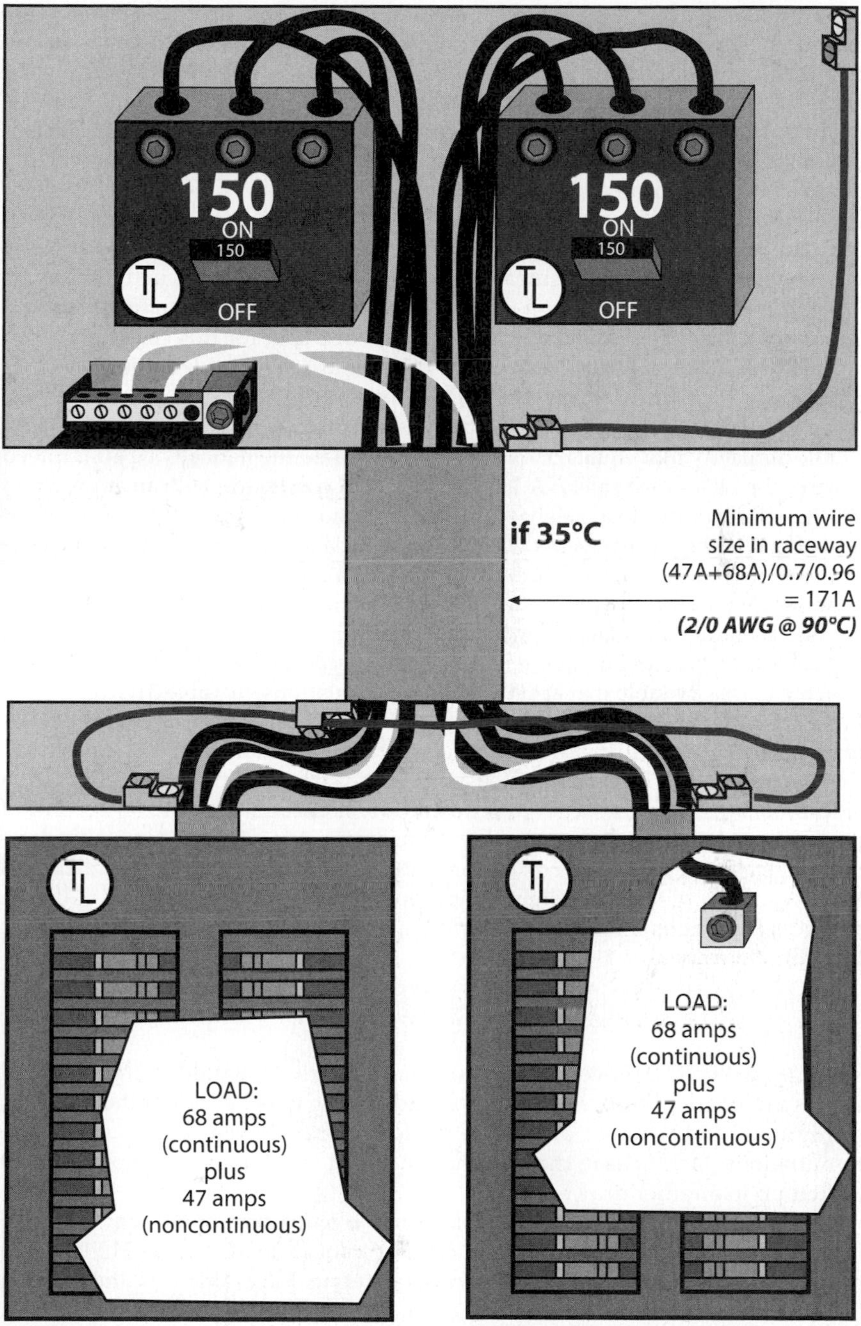
150
ON
150
OFF
150
ON
150
OFF
if 35°C
Minimum wire size in raceway (47A+68A)/0.7/0.96 = 171A
(2/0 AWG @ 90°C)
LOAD:
68 amps
(continuous)
plus
47 amps
(noncontinuous)
LOAD:
68 amps
(continuous)
plus
47 amps
(noncontinuous)

Fig. 9-8. The two feeders in Fig. 9-7 as they would be affected by adding an elevated ambient temperature.

to the adjustment factor (by which you multiply), and comparing the result to the load. That's great for the inspector who checks your work, but it doesn't help you pick the right conductor in the first place. You want to go the other way: Knowing the load, you want to select the proper conductor. Consider the Example D3(a) intermediate results as previously mentioned, using a feeder with 47 A of noncontinuous load and 68 A of continuous load. Figure 9-6 shows the impact of whether or not neutrals are counted for adjustment purposes.

Refer to Fig. 9-7. Suppose you have two of those feeders supplying identical load profiles and run in the same conduit. That would be (as previously noted we are counting the neutrals) eight countable conductors in the raceway. *For this part of the analysis, ignore continuous loading and termination problems.* Remember, you should be using a fresh sheet of paper for this calculation.

Start with 115 A of actual load (47 A + 68 A), and *divide* (you're going the other direction, so you use the opposite of multiplication) by 0.7 [the factor in NEC Table 310.15(B)(3)(a)], to get 164 A in this case. In other words, any conductor with a table ampacity that equals or exceeds 164 A will mathematically be guaranteed to carry the 115-A load safely. A 1/0 AWG THHN conductor, with an ampacity of 170 A, will carry this load safely under the conditions of use, and it might appear to work. Whether it represents your final choice depends on the outcome of the analysis explained in the later topic on choosing a conductor.

Ambient temperature problems High ambient temperatures hinder the dissipation of conductor heat just as in the case of mutual heating. To prevent overheating, the NEC provides ambient temperature correction factors for Table 310.15(B)(16), the ampacity table that applies to these calculations, in Table 310.15(B)(2)(a). In addition, the NEC has a Table [310.15(B)(3)(c)] of mandatory temperature adders for rooftop wiring (with a specific and glaring exception for XHHW-2 insulation). Getting back to the feeder example, suppose the circuit conductors go through a 35°C ambient. Their ampacity goes down (for 90°C conductors) to 96 percent of the base number in the ampacity table. Here again, start with 115 A, and *divide* by 0.96 to get 120 A. Any 90°C conductor with an ampacity equal to or higher than 120 A would carry this load safely. What happens if, as shown in Fig. 9-8, you have *both* a high ambient temperature *and* mutual heating? Divide twice, once by each factor. In this case:

$$115 \text{ A} \div 0.7 \div 0.96 = 171 \text{ A}$$

A 2/0 AWG THHN conductor (ampacity = 195 A) would carry this load without damaging itself. Again, this would be true whether or not the load was continuous, and whether or not the devices were allowed for use with 90°C terminations. Don't cheat; the termination calculations are still supposed to be locked up in another drawer.

When diminished ampacity applies to only a small part of the run You will run into installations where most of a circuit conforms to NEC Table 310.15(B)(16) but a small portion requires very significant derating. For example as illustrated in Fig. 9-9, your circuit may be 207 ft long, with 200 ft in normal environments and 7 ft of it passing through a corner of a boiler room with a very high ambient temperature. The NEC generally observes the weak-link-in-the-chain principle and requires

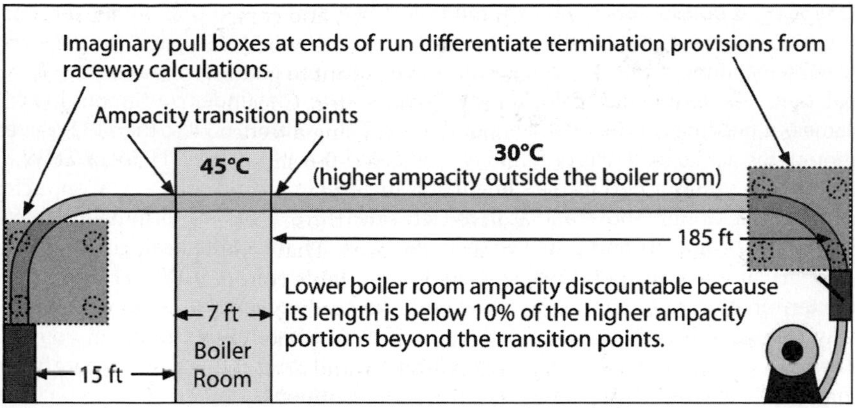
Imaginary pull boxes at ends of run differentiate termination provisions from raceway calculations.
Ampacity transition points
45°C
30°C
(higher ampacity outside the boiler room)
185 ft
Lower boiler room ampacity discountable because its length is below 10% of the higher ampacity portions beyond the transition points.
7 ft
Boiler Room
15 ft

Fig. 9-9. There is a limited exception to the weak link in the chain principle. [310.15(A)(2) Exception.]

the lowest ampacity anywhere over a run to be the allowed ampacity. However, for very short runs where the remainder of the circuit can function as a heat sink, the **NEC** allows the higher ampacity to be used.

Specifically, any time ampacity changes over a run, determine all points of transition. On one side of each point the ampacity will be higher than on the other side. Now measure the length of the wire having the higher ampacity (in this example, the portions *not in* the boiler room) and the length having the lower ampacity (in this example, *in* the boiler room). Compare the two lengths. **NEC** 310.15(A)(2) Exception allows you to use the higher ampacity value beyond the transition point for a length equal to 10 ft or 10 percent of the circuit length having the higher ampacity, whichever is less.

The 2017 has extended the reach of this exception to non-contiguous segments of the run. This preposterous change, which will be aggressively challenged in the 2020 cycle, means that the key feature justifying the exception, namely, the ability of an adjacent part of the run to be a heat sink, no longer applies. The drawing and discussion presented on this topic are still correct, but for the duration of the 2017 cycle there will be an additional, unjustified allowance to apply the allowed ampacity increase to a segment of the circuit it should never apply to. The exception had been in the **NEC** without change for ten Code editions (since 1990) without change. An installation made under the revised terms might be resisted under the general principles in 310.15(A)(3).

In this case of the 200-ft run beyond the 7 ft in the boiler room, 10 percent of the circuit length having the higher ampacity would be 20 ft, but you can't apply the rule to anything over 10 ft. Since 7 ft is less than or equal to 10 ft (and less than the 10 percent limit of 20 ft), the exception applies and you can ignore the ambient temperature in the boiler room in determining the allowable ampacity of the wires passing through it. In the words of the exception, the "higher ampacity" (which applies to the run outside the boiler room) can be used beyond the transition point (the boiler room wall) for "a distance equal to 3.0 m (10 ft) or 10 percent of the circuit length figured at the higher ampacity, whichever is less."

Choosing a conductor Now you can unlock the drawer and pull out the termination calculations. Put both sheets of paper in front of you, and design for the worst case by installing the largest conductor that results from those two independent calculations. The termination calculation ("Protecting devices under continuous loads") came out needing conductors sized, under the 75°C column, no less than 132 A even though the actual load was only 115 A. You could use 1/0, either THHN or THW.

Suppose you put two feeders (eight conductors) in a conduit, as in "Conductor ampacity derating," above and as in the left side of Fig. 9-10. The termination calculation comes out 1/0, and as we've seen, the raceway adjustment calculation under ordinary ambient temperature conditions also comes out 1/0 AWG THHN. Here, the termination rules happen to agree with the raceway analysis. If the same raceway also goes through the area with high ambient temperature, however, you will need 2/0 THHN. Now the raceway conditions are limiting and you size accordingly. Figure 9-10 shows the calculations under both conditions.

Finally, suppose for a moment that the neutrals do not carry heavy harmonic loading and therefore need not be counted, and suppose the ambient temperature does not exceed 30°C, as shown in the right side of Fig. 9-7. With only six countable conductors, the adjustment factor drops to 0.8, so the required ampacity becomes 115 A/0.8 = 144 A, seemingly allowing 1 AWG THHN (90°C) conductors. Now the termination rules (as shown in the right side of Fig. 9-3 and mandating a 75°C temperature evaluation) would be the worst case and would govern the conductor selection process.

Remember, the terminations and the conductor ampacity are two entirely separate issues. Just because you need to use the 75°C column for the terminations doesn't mean you start in that column to determine the overall conductor ampacity. Go ahead and make full use of the 90°C temperature limits on THHN, and its resulting ampacity, to solve mutual conductor heating problems over the middle of a run.

Don't get confused by the fact that the derating factor for continuous loads (0.8) is the same as the adjustment factor for four to six conductors in a raceway (0.8). This is only a coincidence. One applies to terminal heating at the end of a wire, and the other applies over the interior of the raceway. They never apply at the same point in a circuit because the technical basis for each is entirely different. The middle is never the end, so never apply the rules for one to the other. The calculations in this discussion illustrate this through different instances of similar connected loading. Review again Figs. 9-5, 9-7, 9-8, and 9-10.

In one case the termination requirements force an increase in conductor size. In another case the termination requirements agree with the results of the ampacity calculations; and in yet another case the ampacity calculation forces you to use a larger conductor than what the termination rules would predict.

The conductor must always be protected Don't lose sight of the fact that the overcurrent device must always protect the conductor. For 800-A and smaller circuits, NEC 240.4(B) allows the next higher standard size overcurrent device to protect conductors. Above that point, NEC 240.4(C) requires the conductor ampacity to be no less than the rating of the overcurrent device. As a final check, be sure the size of the overcurrent device selected to accommodate continuous loads protects the conductors in accordance with these rules; if it doesn't you will need

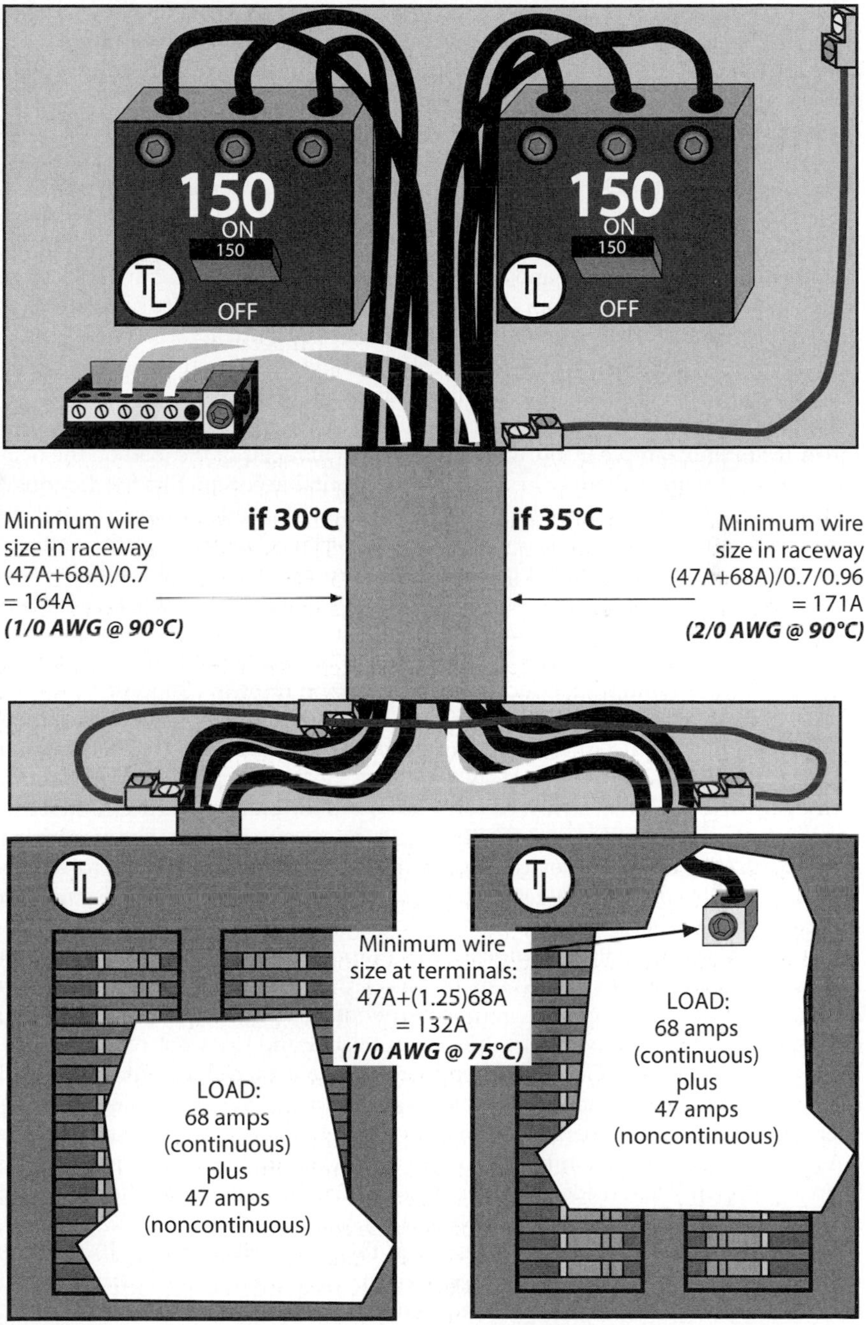
150
ON
150
OFF
150
ON
150
OFF
Minimum wire
size in raceway
(47A+68A)/0.7
= 164A
(1/0 AWG @ 90°C)
if 30°C
if 35°C
Minimum wire
size in raceway
(47A+68A)/0.7/0.96
= 171A
(2/0 AWG @ 90°C)
Minimum wire
size at terminals:
47A+(1.25)68A
= 132A
(1/0 AWG @ 75°C)
LOAD:
68 amps
(continuous)
plus
47 amps
(noncontinuous)
LOAD:
68 amps
(continuous)
plus
47 amps
(noncontinuous)

Fig. 9-10. Calculations for termination limitations must be divorced from run calculations. In some cases the termination restrictions produce the worst case, and in others the run calculations will be the limiting factor. As previously, these calculations assume a raceway length greater than 24 in.

to increase the conductor size accordingly. Refer to the discussion of noncontinuous loads, in "Noncontinuous loads," below, for an example of where, even after doing both the termination and the ampacity calculations, this consideration forces you to change the result.

Small conductors Small conductors (14, 12, and 10 AWG) present an additional wrinkle. The NEC imposes special limitations on overcurrent protection beyond the values in the ampacity tables for these small wires. Generally, the overcurrent protective device for 14 AWG wire can't exceed 15 A; for 12 AWG wire, 20 A; for 10 AWG wire, 30 A. The higher ampacities of these conductors remain what the table says they are, however, and there are cases, notably including motor circuits, where this restriction doesn't apply. In general, perform all ampacity calculations as described previously, based on the ampacity table limits. But at the very end, make sure your overcurrent device doesn't exceed these specific ampere limits unless you fall into one of the exceptions specifically tabulated in NEC 240.4(G).

Noncontinuous loads Suppose none of the 115 A load is continuous on our 150-A feeder in Example D3(a), and suppose the character of the load on the neutrals does not require them to be classified as countable conductors for derating/adjustment purposes. Looking at the two feeders in "Conductor ampacity adjustment," above (and illustrated on the right side of Fig. 9-7), suppose the ambient temperature doesn't exceed 30°C. The termination need not include any phantom load allowance, but it still needs to assume 75°C termination restrictions. A 1 AWG conductor (or even a 2 AWG conductor) will carry the actual 115-A load without exceeding 75°C, and therefore would seem to be usable until you consider the mutual effects of multiple conductors in the common raceway. Suppose you went to a 1 AWG THHN conductor, ampacity 145. Will it carry the 115-A load safely? Yes, because 145 A × 0.8 (the adjustment factor for six conductors) = 116 A. Will it overheat the breaker terminations? No, because 1 AWG copper is 1 AWG copper no matter what insulation style it has around it, and it won't rise to 75°C until it carries 130 A. But its final derated ampacity within the conduit is 116 A. The next higher standard-sized overcurrent device is 125 A. The 150-A breaker does not protect this wire under these conditions of use, and has to be reduced to 125 A, or else you need to increase the wire size to 1/0 AWG.

Neutral sizing As noted previously this example provided the basis for what is now Example D3(a) in NEC Annex D. The NEC example also calculates neutral sizing, based on a worst-case-scenario continuous neutral load of 42 A. The D3(a) example notes that this wire runs busbar to busbar and does not require upsizing by 25 percent (see "Grounded conductors have a special allowance," stated earlier) leading to a preliminary size of 8 AWG under the 75°C column. However, this size would be inadequate to clear a line-to-neutral short circuit because 8 AWG is smaller than the minimum equipment grounding conductor that could be installed on a 150-A feeder as covered in NEC 250.122. Therefore, the minimum size becomes 6 AWG, as required in NEC 215.2(A)(2).

There is another issue to be addressed at this point. NEC 250.122(B) requires that if ungrounded conductors are increased in size, a proportional increase must be made in the size of the equipment grounding conductor, and as we have just seen, in the case of the Example D3(a) load profile, this size will control the size of the neutral. The size of the ungrounded conductors determined in this example (in some cases as large as 2/0 AWG copper) are the smallest size allowed by the NEC under the specified conditions. No increase for voltage drop or other design consideration has been made, and Example D3(a) does not show a further increase either. On the other hand, such sizes are larger than the sizes specified in NEC Table 310.15(B)(16). The historical context for the rule addressed instances where increases to lower voltage drop implied a similar benefit for a long equipment grounding conductor. Fortunately, the 2014 *NEC* changed 250.122(B) to clarify that the place to start is the minimum size required to meet *NEC* requirements and not necessarily a simple number printed in the ampacity tables. No further increase in size is required in this context.

INFORMATIVE ANNEX E. TYPES OF CONSTRUCTION

The information given in Table E.1 is taken from another NFPA standard, NFPA 5000, and can serve to help identify the "type" of building construction employed on a given project. The primary use for this table is determining whether or not Type NM cable (Romex) may be used. As given in 334.10, Type NM cable may be used in other than one- and two-family dwellings provided the "construction type" of the building in question is *permitted to be* either a Type III, IV, or V. Any building that can be so constructed under the prevailing building code, whether or not it actually was constructed that way, meets the threshold in 334.10 and may use Type NM cable, with certain restrictions.

As an alternative to deciphering Table E.1 and researching all applicable data from NFPA 5000, one could consult the local building department as its inspectors are completely familiar with such matters and will be able to clearly indicate the type of construction its building code mandates for any building or structure within its jurisdiction. If the building department identifies a particular structure as qualifying for Type III, IV, or V construction, then the permission given in 334.10 may be exercised and Romex may be used.

INFORMATIVE ANNEX F. AVAILABILITY AND RELIABILITY FOR OPERATIONS POWER SYSTEMS; AND DEVELOPMENT AND IMPLEMENTATION OF FUNCTIONAL PERFORMANCE TESTS (FPTS) FOR CRITICAL OPERATIONS POWER SYSTEMS

This new annex provides guidance in setting up functional performance tests that are required for Critical Operations Power Systems, and as referenced in informational notes in 708.1 (No. 8) and in 708.8(D).

INFORMATIVE ANNEX G. SUPERVISORY CONTROL AND DATA ACQUISITION (SCADA)

This new annex supports the provisions of new Art. 708, as referenced in 708.1, informational note No. 9.

INFORMATIVE ANNEX H. ADMINISTRATION AND ENFORCEMENT

80.1. Scope. Annex H provides a standardized approach for the makeup of the inspecting authorities including organization, operating procedures, means of redress, individual qualifications for board members and inspectors, as well as other administrative necessities related to the enforcement of the National Electrical Code (NEC), also referred to here as the Code.

Coverage of this Annex is limited to those portions that may be useful to installers and designers. Sections not covered are either self-explanatory or don't apply to designers and installers. The following sections express ideas and concepts that can serve to provide guidance for designers and installers on basic application and enforcement concerns.

80.5. Adoption. This part makes clear that Annex H must be specifically and separately adopted to be considered part of the Code. That is, if the legislative body responsible for adopting the NEC does not specifically and clearly call for the adoption of the wording in Annex H, then this annex is not a binding or mandatory part of the Code within that jurisdiction. Where the legislative body responsible for promulgation of electrical safety standards does adopt Annex H, all existing enforcement and administrative elements employed by the local inspection agency must be reviewed by the inspecting entity to determine whether their procedures, chain-of-command, personnel, and so forth satisfy the various rules given herein.

80.9. Application. The wording here indicates exactly what parts of a given installation are subject to inspection and what standard applies. This is a question that has been raised many times in the past, but the concepts expressed by this wording were nowhere to be found within the NEC itself. Because these sections are intended to be used as guidance for enforcement where Annex H is accepted, even if it is not adopted, it seems reasonable to assume that the guidance provided here can be used by designers, installers, and inspectors to identify those installations, or parts of installations, that are intended to be subject to inspection and what edition of the Code must be satisfied.

As given in 80.9(A), all new construction is subject to inspection according to the Code edition (2002, 2005, 2008, etc.) that is in effect at the time the construction permit was dated. That is, if, say, the next edition of the Code was adopted on, say, 1 June, and a building permit was issued on 10 June, then the electrical installation within that building must conform to the new Code. Any permits issued before the adoption of the new Code—in the preceding example before 1 June—would be subject to the preceding edition of the NEC and must, in all ways, satisfy that edition of the Code. It's worth noting that such interpretation has been in place in the vast majority of jurisdictions across the United States. Inspection authorities have long applied this simple logic—especially at that point in time when a new Code is adopted—to establish which installations are subject to which Code.

Part **(B)** of 80.9 applies to "Existing Installations." Here the wording gives a general statement regarding the circumstances under which an existing installation may continue to operate even though it is not compliant with the prevailing edition of the Code.

Retroactive compliance has not been mandated by any edition of the NEC. That is, if as the result of a Code change, a certain application is no longer acceptable, it is not generally necessary to change out equipment or otherwise rearrange system connections to conform to the new Code. This wording essentially states that the inspector may require corrective action for an existing installation only where the inspector determines that continued operation of the system, in its present configuration, presents a safety hazard. In such cases, the inspector is empowered to mandate corrective action, but must allow a reasonable amount of time to achieve compliance.

80.9(C) covers "Additions, Alterations, or Repairs." This portion of Annex H states that all renovations, modernizations, or expansions of existing electrical systems are subject to the Code that is in effect at the time the work is performed. Additionally, any such projects must not "adversely affect" the existing system, as determined by the inspector. This essentially means that the inspector can require, say, a service upgrade if the existing service is incapable of supplying the additional load, or a higher-rated feeder and feeder protective device, or an up-sized panelboard or switchboard, or so forth to ensure that the existing system will not be "adversely affected" by the addition, alteration, or repair.

80.29. Liability for Damages. This section presents an extremely important point. Although not very clearly stated, what is said here is that "any party owning, operating, controlling, or installing any electrical equipment" must have an inspection of that completed installation by an authorized electrical inspector. And, such inspection does not relieve the owner, controller, operator, or installer from any legal liability.

This is a very important concept and one that prevails throughout the litigation of lawsuits filed as the result of electrical accidents or incidents. The electrical installer/designer is the party typically named as defendant in such suits—even though the completed installation was inspected and satisfactorily "passed" the inspection. The point is: Regardless of whether Annex H is adopted within any given jurisdiction, the installer and designer are responsible for the integrity and safety of the work they perform. The inspector is never held legally liable, provided an inspection is performed. Therefore, be aware that if the local inspector accepts an eccentric or unconventional interpretation and application of a given Code rule, the designer and/or installer is still legally liable should an accident or incident occur. And the designer and/or installer may be asked, at some point, to explain why he or she chose the unconventional method. It may be hard to justify such action, and will be nearly impossible to do so to a jury should the accident or incident be attributed to some electrical safety deficiency that resulted from your interpretation and application. Don't be fooled into thinking that the "green" sticker and issuance of a certificate of occupancy provides legal protection. Your best protection is strict adherence to the "letter" of Code rules where clearly identifiable, or the "spirit" or "intent" where the application is not clearly or specifically regulated by the Code rule in question.

INFORMATIVE ANNEX I. RECOMMENDED TIGHTENING TORQUE TABLES FROM UL STANDARD 486A-B

In general, always use the torque specifications that come with the equipment being installed. However, these tables are very useful for older equipment or for new equipment that, for whatever reasons, does not come with specifications.

This annex also notes that "because it is normal for some relaxation to occur in service, checking torque values sometime after installation is not a reliable means of determining the values of torque applied at installation." This phenomenon is illustrated in Fig. 9-11, where the resistance decreases on initial tightening of the connector, but as the connection relaxes, the resistance remains low until the connector becomes very loose. Connector torque specifications anticipate this phenomenon.

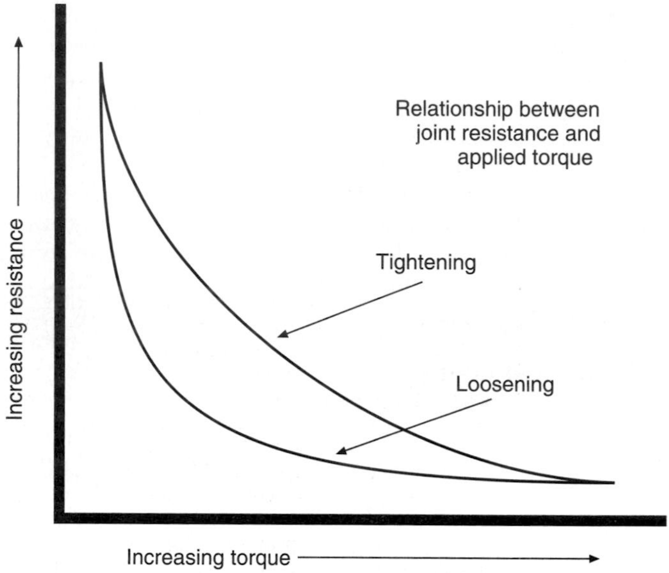
Relationship between
joint resistance and
applied torque
Tightening
Loosening
Increasing resistance
Increasing torque

Fig. 9-11. Some relaxation of tightness in a connection is anticipated and does not degrade the quality of the connection. (Informative Annex I.)

The prior version of the torque table in the 2011 NEC had two columns, A and B; the values in Column A were for connector certification testing and were not relevant to field conditions. That column has been removed as of the 2014 NEC, and only the former Column B values go forward.

INFORMATIVE ANNEX J. ADA STANDARDS FOR ACCESSIBLE DESIGN

This annex, new with the 2014 NEC, presents ADA design criteria. They are not part of the minimum safety standards in the NEC, although they clearly inform some requirements. For example, 210.52(C)(5) Exception No. 1 (1) takes them into account relative to required receptacle placements on dwelling unit kitchen counters in "construction for the physically impaired." They are presented as a service to users, but they do not constitute the full regulation set.

Index